Recommended Dietary Allowances (RDA) and Adequate Intakes (AI) for Vitamins

Age (yr)	Thiamin RDA (mg/day)	Riboflavin RDA (mg/day)	Niacin RDA (mg/day)[a]	Biotin AI (µg/day)	Pantothenic acid AI (mg/day)	Vitamin B6 RDA (mg/day)	Folate RDA (µg/day)[b]	Vitamin B12 RDA (µg/day)	Choline AI (mg/day)	Vitamin C RDA (mg/day)	Vitamin A RDA (µg/day)[c]	Vitamin D AI (µg/day)[d]	Vitamin E RDA (mg/day)[e]	Vitamin K AI (µg/day)
Infants														
0–0.5	0.2	0.3	2	5	1.7	0.1	65	0.4	125	40	400	5	4	2.0
0.5–1	0.3	0.4	4	6	1.8	0.3	80	0.5	150	50	500	5	5	2.5
Children														
1–3	0.5	0.5	6	8	2	0.5	150	0.9	200	15	300	5	6	30
4–8	0.6	0.6	8	12	3	0.6	200	1.2	250	25	400	5	7	55
Males														
9–13	0.9	0.9	12	20	4	1.0	300	1.8	375	45	600	5	11	60
14–18	1.2	1.3	16	25	5	1.3	400	2.4	550	75	900	5	15	75
19–30	1.2	1.3	16	30	5	1.3	400	2.4	550	90	900	5	15	120
31–50	1.2	1.3	16	30	5	1.3	400	2.4	550	90	900	5	15	120
51–70	1.2	1.3	16	30	5	1.7	400	2.4	550	90	900	10	15	120
>70	1.2	1.3	16	30	5	1.7	400	2.4	550	90	900	15	15	120
Females														
9–13	0.9	0.9	12	20	4	1.0	300	1.8	375	45	600	5	11	60
14–18	1.0	1.0	14	25	5	1.2	400	2.4	400	65	700	5	15	75
19–30	1.1	1.1	14	30	5	1.3	400	2.4	425	75	700	5	15	90
31–50	1.1	1.1	14	30	5	1.3	400	2.4	425	75	700	5	15	90
51–70	1.1	1.1	14	30	5	1.5	400	2.4	425	75	700	10	15	90
>70	1.1	1.1	14	30	5	1.5	400	2.4	425	75	700	15	15	90
Pregnancy														
≤18	1.4	1.4	18	30	6	1.9	600	2.6	450	80	750	5	15	75
19–30	1.4	1.4	18	30	6	1.9	600	2.6	450	85	770	5	15	90
31–50	1.4	1.4	18	30	6	1.9	600	2.6	450	85	770	5	15	90
Lactation														
≤18	1.4	1.6	17	35	7	2.0	500	2.8	550	115	1200	5	19	75
19–30	1.4	1.6	17	35	7	2.0	500	2.8	550	120	1300	5	19	90
31–50	1.4	1.6	17	35	7	2.0	500	2.8	550	120	1300	5	19	90

NOTE: For all nutrients, values for infants are AI. The glossary on the inside back cover defines units of nutrient measure.

[a] Niacin recommendations are expressed as niacin equivalents (NE), except for recommendations for infants younger than 6 months, which are expressed as preformed niacin.

[b] Folate recommendations are expressed as dietary folate equivalents (DFE).

[c] Vitamin A recommendations are expressed as retinol activity equivalents (RAE).

[d] Vitamin D recommendations are expressed as cholecalciferol and assume an absence of adequate exposure to sunlight.

[e] Vitamin E recommendations are expressed as α-tocopherol.

Recommended Dietary Allowances (RDA) and Adequate Intakes (AI) for Minerals

Age (yr)	Sodium AI (mg/day)	Chloride AI (mg/day)	Potassium AI (mg/day)	Calcium AI (mg/day)	Phosphorus RDA (mg/day)	Magnesium RDA (mg/day)	Iron RDA (mg/day)	Zinc RDA (mg/day)	Iodine RDA (µg/day)	Selenium RDA (µg/day)	Copper RDA (µg/day)	Manganese AI (mg/day)	Fluoride AI (mg/day)	Chromium AI (µg/day)	Molybdenum RDA (µg/day)
Infants															
0–0.5	120	180	400	210	100	30	0.27	2	110	15	200	0.003	0.01	0.2	2
0.5–1	370	570	700	270	275	75	11	3	130	20	220	0.6	0.5	5.5	3
Children															
1–3	1000	1500	3000	500	460	80	7	3	90	20	340	1.2	0.7	11	17
4–8	1200	1900	3800	800	500	130	10	5	90	30	440	1.5	1.0	15	22
Males															
9–13	1500	2300	4500	1300	1250	240	8	8	120	40	700	1.9	2	25	34
14–18	1500	2300	4700	1300	1250	410	11	11	150	55	890	2.2	3	35	43
19–30	1500	2300	4700	1000	700	400	8	11	150	55	900	2.3	4	35	45
31–50	1500	2300	4700	1000	700	420	8	11	150	55	900	2.3	4	35	45
51–70	1300	2000	4700	1200	700	420	8	11	150	55	900	2.3	4	30	45
>70	1200	1800	4700	1200	700	420	8	11	150	55	900	2.3	4	30	45
Females															
9–13	1500	2300	4500	1300	1250	240	8	8	120	40	700	1.6	2	21	34
14–18	1500	2300	4700	1300	1250	360	15	9	150	55	890	1.6	3	24	43
19–30	1500	2300	4700	1000	700	310	18	8	150	55	900	1.8	3	25	45
31–50	1500	2300	4700	1000	700	320	18	8	150	55	900	1.8	3	25	45
51–70	1300	2000	4700	1200	700	320	8	8	150	55	900	1.8	3	20	45
>70	1200	1800	4700	1200	700	320	8	8	150	55	900	1.8	3	20	45
Pregnancy															
≤18	1500	2300	4700	1300	1250	400	27	12	220	60	1000	2.0	3	29	50
19–30	1500	2300	4700	1000	700	350	27	11	220	60	1000	2.0	3	30	50
31–50	1500	2300	4700	1000	700	360	27	11	220	60	1000	2.0	3	30	50
Lactation															
≤18	1500	2300	5100	1300	1250	360	10	14	290	70	1300	2.6	3	44	50
19–30	1500	2300	5100	1000	700	310	9	12	290	70	1300	2.6	3	45	50
31–50	1500	2300	5100	1000	700	320	9	12	290	70	1300	2.6	3	45	50

Tolerable Upper Intake Levels (UL) for Vitamins

Age (yr)	Niacin (mg/day)[a]	Vitamin B_6 (mg/day)	Folate (μg/day)[a]	Choline (mg/day)	Vitamin C (mg/day)	Vitamin A (μg/day)[b]	Vitamin D (μg/day)	Vitamin E (mg/day)[c]
Infants								
0–0.5	—	—	—	—	—	600	25	—
0.5–1	—	—	—	—	—	600	25	—
Children								
1–3	10	30	300	1000	400	600	50	200
4–8	15	40	400	1000	650	900	50	300
9–13	20	60	600	2000	1200	1700	50	600
Adolescents								
14–18	30	80	800	3000	1800	2800	50	800
Adults								
19–70	35	100	1000	3500	2000	3000	50	1000
>70	35	100	1000	3500	2000	3000	50	1000
Pregnancy								
≤18	30	80	800	3000	1800	2800	50	800
19–50	35	100	1000	3500	2000	3000	50	1000
Lactation								
≤18	30	80	800	3000	1800	2800	50	800
19–50	35	100	1000	3500	2000	3000	50	1000

[a] The UL for niacin and folate apply to synthetic forms obtained from supplements, fortified foods, or a combination of the two.

[b] The UL for vitamin A applies to the preformed vitamin only.
[c] The UL for vitamin E applies to any form of supplemental α-tocopherol, fortified foods, or a combination of the two.

Tolerable Upper Intake Levels (UL) for Minerals

Age (yr)	Sodium (mg/day)	Chloride (mg/day)	Calcium (mg/day)	Phosphorus (mg/day)	Magnesium (mg/day)[d]	Iron (mg/day)[b]
Infants						
0–0.5	—[e]	—[e]	—	—	—	40
0.5–1	—[e]	—[e]	—	—	—	40
Children						
1–3	1500	2300	2500	3000	65	40
4–8	1900	2900	2500	3000	110	40
9–13	2200	3400	2500	4000	350	40
Adolescents						
14–18	2300	3600	2500	4000	350	45
Adults						
19–70	2300	3600	2500	4000	350	45
>70	2300	3600	2500	3000	350	45
Pregnancy						
≤18	2300	3600	2500	3500	350	45
19–50	2300	3600	2500	3500	350	45
Lactation						
≤18	2300	3600	2500	4000	350	45
19–50	2300	3600	2500	4000	350	45

[d] The UL for magnesium applies to synthetic forms obtained from supplements or drugs only.
[e] Source of intake should be from human milk (or formula) and food only.

NOTE: An Upper Limit was not established for vitamins and minerals not listed and for those age groups listed with a dash (—) because of a lack of data, not because these nutrients are safe to consume at any level of intake. All nutrients can have adverse effects when intakes are excessive.

www.wadsworth.com

www.wadsworth.com is the World Wide Web site for Thomson Wadsworth and is your direct source to dozens of online resources.

At www.*wadsworth.com* you can find out about supplements, demonstration software, and student resources. You can also send email to many of our authors and preview new publications and exciting new technologies.

www.wadsworth.com
Changing the way the world learns®

NUTRITION
Concepts and Controversies

10TH EDITION

Frances Sienkiewicz Sizer

Ellie Whitney

THOMSON
WADSWORTH

Australia • Canada • Mexico • Singapore • Spain
United Kingdom • United States

THOMSON

WADSWORTH

Nutrition: Concepts and Controversies, Tenth Edition
Frances Sienkiewicz Sizer and Ellie Whitney

PUBLISHER Peter Marshall

DEVELOPMENT EDITOR Elizabeth Howe

ASSISTANT EDITOR Elesha Feldman

EDITORIAL ASSISTANT Lauren Vogelbaum

TECHNOLOGY PROJECT MANAGER Travis Metz

MARKETING MANAGER Jennifer Somerville

MARKETING ASSISTANT Michele Colella

MARKETING COMMUNICATIONS MANAGER Shemika Britt

PROJECT MANAGER, EDITORIAL PRODUCTION Sandra Craig

ART DIRECTOR Lee Friedman

PRINT BUYER Barbara Britton

PERMISSIONS EDITOR Sarah Harkrader

PRODUCTION SERVICE The Book Company

TEXT DESIGNER Diane Beasley

ART EDITOR Carolyn Deacy

PHOTO RESEARCHER Myrna Engler

COPY EDITOR Patricia Lewis

ILLUSTRATIONS Imagineering

COVER DESIGNER Lee Friedman

COVER IMAGE Diego Rivera (1866–1957), *Watermelons* (Las sandias), 1957, © Banco de Mexico Trust. Schalkwijk/Art Resource, NY

COVER PRINTER Phoenix Color Corp

COMPOSITOR Parkwood Composition Service

PRINTER Courier Corporation/Kendallville

Library of Congress Control Number: 2004114404

Student Edition: ISBN 0-534-64506-2

Instructor's Edition: ISBN 0-534-40672-6

International Student Edition: ISBN 0-495-01183-5

THOMSON HIGHER EDUCATION
10 Davis Drive
Belmont, CA 94002-3098
USA

ASIA (INCLUDING INDIA)
Thomson Learning
5 Shenton Way
#01-01 UIC Building
Singapore 068808

AUSTRALIA/NEW ZEALAND
Thomson Learning Australia
102 Dodds Street
South Melbourne, Victoria 3006
Australia

CANADA
Thomson Nelson Learning
1120 Birchmount Road
Toronto, Ontario M1K 5G4
Canada

UK/EUROPE/MIDDLE EAST/AFRICA
Thomson Learning
High Holborn House
50–51 Bedford Row
London WC1R 4LR
United Kingdom

LATIN AMERICA
Thomson Learning
Seneca, 53
Colonia Polanco
11560 Mexico
D. F. Mexico

SPAIN (INCLUDING PORTUGAL)
Thomson Paraninfo
Calle Magallanes, 25
28015 Madrid, Spain

About the Authors

To Lauren Annabelle, that clever little cutie who owns my heart, and to her great grandmother Bernice Sienkiewicz, the pillar of our family—thanks, Mom.

Fran

FRANCES SIENKIEWICZ SIZER, M.S., R.D., F.A.D.A., attended Florida State University where, in 1980, she received her B.S., and in 1982, her M.S. in nutrition. She is certified as a charter Fellow of the American Dietetic Association. She is a founding member and vice president of Nutrition and Health Associates, an information and resource center in Tallahassee, Florida, that maintains an ongoing bibliographic database tracking research in more than 1,000 topic areas of nutrition. Her textbooks include *Life Choices: Health Concepts and Strategies; Making Life Choices; The Fitness Triad: Motivation, Training, and Nutrition;* and others. She is a primary author of *Nutrition Interactive*, an instructional college-level nutrition CD-ROM. In addition to writing, she lectures at universities and at national and regional conferences, and serves actively on the board of directors of ECHO, a local hunger and homelessness relief organization in her community.

To all of my grandchildren and stepgrandchildren: Max, Zoey, Emily, Rebecca, Sarah, Will, TootToot, and Jacob.

Ellie

ELEANOR NOSS WHITNEY, PH.D., received her B.A. in Biology from Radcliffe College in 1960 and her Ph.D. in Biology from Washington University, St. Louis, in 1970. Formerly on the faculty at Florida State University, and a dietitian registered with the American Dietetic Association, she now devotes full time to research, writing, and consulting in nutrition, health, and environmental issues. Her earlier publications include articles in *Science, Genetics,* and other journals. Her textbooks include *Understanding Nutrition, Understanding Normal and Clinical Nutrition, Nutrition and Diet Therapy,* and *Essential Life Choices* for college students and *Making Life Choices* for high-school students. Her most intense interests presently include energy conservation, solar energy uses, alternatively fueled vehicles, and ecosystem restoration.

Contents in Brief

Contents

Preface

 A BILLBOARD IN Louisiana reads, "Come as you are. Leave different," meaning that once you've seen, smelled, tasted, and listened to Louisiana, you'll never be the same. This book extends the same invitation to its readers: Come to nutrition science as you are, with all of the knowledge and enthusiasm you possess, with all of your unanswered questions and misconceptions, and with the habits and preferences that now dictate what you eat.

But leave different. Take a more complete understanding of nutrition with you. Take a greater ability to discern between nutrition truth and fiction, to ask sophisticated questions, and to find the answers. Finally, take with you a better sense of how to feed yourself in ways that not only please you and soothe your spirit but nourish your body as well.

For over a quarter of a century, *Nutrition Concepts and Controversies* has been a cornerstone in nutrition classes across North America, serving the needs of students and professors in building a healthier future. In keeping with our tradition, in this, our 10th edition, we continue to explore the ever-changing frontier of nutrition science while maintaining our sense of personal connection with students and instructors alike. We address the learner in a clear, engaging writing style that has become our trademark, but with a fresh crispness that we hope you will enjoy.

Pedagogical Features

Throughout these chapters, features tickle the reader's interest and inform. For both verbal and visual learners, our logical presentation and our redesigned figures keep interest high and understanding at a peak. New figures in many chapters demonstrate current concepts, and improvements to classic figures enhance their usefulness to the learner. New photos also adorn many of our pages, adding pleasure to reading. Each chapter begins with "Do You Ever . . ." questions to pique interest and set a personal tone for the information that follows. Little reminders, called *Think Fitness*, appear from time to time to alert readers to ways in which physical activity links with nutrition to support health. For example, the *Think Fitness* feature of Chapter 1 points out the health benefits that one can expect from regular physical activity. In the remaining chapters, these sections emphasize the roles of the nutrients in supporting physical activity and spell out the advantages of an active lifestyle throughout life.

Many tried-and-true features return in this edition: The practical *Food Features* invite students to apply chapter con-

tents in their everyday encounters with food. The popular *Self Check* features provide review questions at the end of each chapter, with answers in Appendix G to provide immediate feedback to the learner.

By popular demand, we have retained our *Snapshots* of vitamins and minerals, but in an improved format to ease studying. These concentrated capsules of information depict food sources of vitamins and minerals, present the DRI recommended intakes and UL, and offer the chief functions of each nutrient along with deficiency and toxicity symptoms.

The *Food Feature* sections that appear in most chapters act as bridges between theory and practice; they are practical applications of the chapter concepts that help readers to choose foods according to nutrition principles. *Consumer Corners* present information on whole-grain breads, fat replacers, amino acid supplements, vitamin C and the common cold, bottled water, irradiation of foods, and other nutrition-related marketplace issues to empower students to make informed decisions.

New or major terms in chapters are defined in the margins of the pages where they are introduced and also in the Glossary at the end of the book. Terms in Controversy sections are grouped together and defined in tables within the sections and in the Glossary. The reader who wishes to locate any term can do so by consulting the index, which lists the page numbers of definitions in boldface type.

A new feature to notice is the *Media Menu* found at the end of each chapter. The *Media Menu* contains an annotated list of CD-ROM activities available for each chapter, as well as a description of helpful study tools and video clips available on the student website.

Chapter Contents

Chapter 1 begins the text with a personal challenge to students. It asks the question so many people ask of nutrition educators: "Why should people care about nutrition?" We answer with a lesson in the ways in which nutritious foods affect diseases, and present a continuum of diseases from those purely genetic in origin to those almost totally preventable by nutrition. After offering some beginning facts about the genes, nutrients, and foods, the chapter goes on to present the *Dietary Guidelines for Americans 2005* and the *Healthy People 2010* goals for the nation. It concludes with a discussion of scientific research in nutrition to lend a perspective on the context in which study results may be rightly viewed. Chapter 2 brings together the concepts of nutrient allowances, such as the *Dietary Reference Intakes* (DRI), and diet planning using the 2005 USDA Food Guide. Chapter 3 presents a thorough, but brief, introduction to the workings of the human body from the genes to the organs, with major emphasis on the digestive system. Chapters 4 through 6 are devoted to the energy-yielding nutrients—carbohydrates, lipids, and protein. Chapters 7 and 8 present the vitamins, minerals, and water, including the new DRI constructs for water. Chapter 9 relates energy balance to body composition, obesity, and underweight and provides guidance to lifelong weight maintenance. Chapter 10 presents the relationships between physical activity, athletic performance, and nutrition. Chapter 11 applies the essence of the first ten chapters to two broad and rapidly changing areas within nutrition: immunity and disease prevention. Chapter 12 delivers urgently important concepts of food safety. Chapters 13 and 14 emphasize the importance of nutrition throughout the life span. Chapter 15 touches on the vast problems of the global food supply and world and U.S. hunger and links each reader to the meaningful whole through the daily choices available to them.

Controversies

The *Controversies* of this book's title invite you to explore beyond the safe boundaries of established nutrition knowledge. These optional readings, which appear at the end of each chapter and are printed with colored borders, delve into current scientific topics and emerging controversies. All are up to date, and some are new to this edition. Of special current interest is Controversy 5, which presents the science behind the current dietary guidance concerning fats. Controversy 7 explores the issues of vitamin supplements, raising the question, "Who benefits?" Controversy 11 presents the current thinking about the influence of society on the obesity epidemic. Controversy 12 considers the pros and cons of both organic foods and the products of biotechnology. Controversy 13 discusses the pressing problems of childhood obesity and premature onset of chronic diseases. Controversy 15 explores ways in which agriculture can ensure a high-quality food supply throughout this century and beyond.

Appendixes

The appendixes are a collection of useful resources to enhance your understanding of food and nutrition science. Appendix A presents complete and accurate listings of the nutrient contents of more than 2,000 foods in units compatible with the DRI intake recommendations. Appendix B, *Canadiana*, supplies the RNI, the Guidelines, the Food Guide, Food Labels, and the Choice System for our Canadian readers. Appendix C demonstrates nutrition calculations, with special emphasis on finding the percentage of calories from fat in a diet and other percentages. Appendix D provides full coverage with applications of the

U.S. Exchange System. Appendix E offers an invaluable list of current addresses, telephone numbers, and Internet websites for those interested in additional information. To save space, we have collected all reference notes into Appendix F. (Older source notes have been removed but are easily available by consulting older editions of this book or by contacting the publisher.) Appendix H, "Physical Activity and Energy Requirements," has been added to help students accurately calculate energy expenditures according to the new DRI PAL (Physical Activity Level). Appendix I, also new to this edition, presents food patterns to meet the *Dietary Guidelines for Americans 2005*.

Nutrition Connections CD-ROM

The *Nutrition Connections* CD-ROM is found in every copy of the textbook and serves as a springboard to nutrition investigation in an electronic environment. This interactive CD-ROM contains 12 Do It! activities tied to chapter topics. For example, the Chapter 1 Do It! asks students to assess the credibility of nutrition news found in newspapers and magazines. Other Do It! activities ask students to analyze food labels, assess the fibers in food, and add up the protein in a typical day's meals. The CD-ROM also contains a hot glossary, student quizzes, and a library of web links, as well as 14 dynamic interactions. These activities demonstrate important nutrition concepts through animation; for example, the Chapter 3 interaction, *Digestion and Absorption,* follows the progress of food as it moves through the GI tract and illustrates the functions of the primary organs involved in digestion.

Ancillary Materials

A number of helpful teaching and learning resources are available. Students and instructors alike will appreciate the Do It! booklet containing the popular "Do It!" activities. Some of these include analyzing the vitamin content of restaurant menus, finding the fiber in foods, and developing a personal plan to employ the HACCP food safety model. Students will also find the Study Guide to be a helpful review of chapter material, reiterating chapter objectives and key concepts.

For instructors, the Nutrition: Concepts and Controversies Multimedia Manager makes it easy to assemble, edit, and present custom lectures—bringing together art, video clips, and animations from this CD-ROM, the web, and your own material. JoinIn on TurningPoint provides book-specific instant response system content, which reinforces key concepts using illustrations and tests students' comprehension with challenging questions. In addition a complete set of transparency acetates is available and includes 180 vibrant visual aids taken from the text. WebTutor for Blackboard and WebCT platforms offer robust preloaded quizzing and assignment content for instructors administering their course online. ExamView® Computerized Testing makes it possible to create custom tests and study guides (both print and online) in minutes. A printed test bank provides instructors with a complete and thorough test for each chapter of the text, and an instructor's manual contains annotated lecture outlines, 30 new handouts, and helpful classroom activities.

Closing Thoughts

As always, our purpose in writing this text is to enhance our readers' understanding of nutrition science and their motivation to apply it. We hope the information on this book's pages will reach beyond the classroom to our readers' lives. Take the information you find inside this book home with you. Use it in your life: nourish yourself, educate your loved ones, and nurture others to be healthy. Stay up with the news, too. For despite all the conflicting messages, inflated claims, and even quackery that abound in the marketplace, true nutrition knowledge progresses with a genuine scientific spirit, and important new truths are constantly unfolding.

Acknowledgments

OUR THANKS to Linda Kelly DeBruyne for the newly updated Chapter 10 and Chapter 13 and for her work on our Table of Food Composition. More thanks to Linda and also to Sharon Rady Rolfes for their timely efforts in helping to apply the new *Dietary Guidelines* and USDA Food Guide to this edition. Thank you, Gamble Gluesenkamp, for your competent dedication and positive energy in preparing and proofing our text. Thanks also to Lynn Earnest for help when time ran short.

Thanks to our associate Lori Turner for much of the Instructor's Manual; thanks to Margaret Hedley, University of Guelph, who prepared the Canadian material and reviewed the Canadian resources listed in the text. We thank Judy Kaufman, Monroe Community College, for developing the Self Check questions for this edition. Thanks to Pat Brewer and Pat Lewis for their creative and excellent editing help. Thank you, Carolyn Deacy, Myrna Engler, Matthew Farruggio, Imagineering, and associates for bringing our figures and photos to their full potential. Thanks to Diane Beasley for the fresh design of the text.

Special thanks to our publisher, Peter Marshall, and to our editors, Elizabeth Howe, Sandra Craig, and Dusty Friedman, for their unflagging efforts to ensure the highest quality of all facets of this book. Our gratitude goes to Jennifer Somerville for her excellent work in marketing this text and to Travis Metz for creating the media package. In addition, we thank Elesha Feldman for her excellent work on the Instructor's Edition and supplement package.

Thanks to Jana Kicklighter, Georgia State University, for preparing the Student Study Guide and Test Bank. We are grateful to First Data for the creation of Appendix A and for the computerized diet analysis program that accompanies this book. To our reviewers, heartfelt thanks for your many thoughtful ideas and suggestions.

Kwaku Addo	University of Kentucky
Raga M. Bakhit	Virginia Polytechnic Institute & State University
Cynthia L. Brown	The College of St. Scholastica
Thomas W. Castonguay	University of Maryland, College Park
Melissa A. Chabot	State University of New York at Buffalo
Mary Ellen Clark	Monroe Community College
Gina L. Craft	Oklahoma Baptist University
Margaret C. Craig-Schmidt	Auburn University
Diane Dembicki	Dutchess Community College
Bernard Frye	The University of Texas at Arlington
Juliet M. Getty	University of North Texas
Diane L. Golzynski	California State University, Fresno
Evette M. Hackman	Seattle Pacific University
Nancy Harris	East Cardina University
Catherine R. Heinlein	Mt. San Antonio College
Karen Heller	University of New Mexico
Candice F. Hines-Tinsley	Saddleback College, Mission Viejo, CA
David H. Holben	Ohio University
Danita Saxon Kelley	Western Kentucky University
Laura J. Kruskall	University of Nevada, Las Vegas
Melody K. Kyzer	University of North Carolina, Wilmington
Christina O. Lengyel	University of North Carolina at Greensboro
Mark S. Meskin	California State Polytechnic University, Pomona
Cherie L. Moore	Cuesta College, San Luis Obispo, CA
Michelle Neyman	California State University, Chico
Glen J. Peterson	Century College, White Bear Lake, MN
Leonard A. Piche	Brescia University
Robert D. Reynolds	University of Illinois at Chicago
Suzanne E. Rhodes	University of North Carolina at Greensboro
Jennifer Ricketts	University of Arizona

Antonio S. Santo — University at Buffalo

Marie L. Smith — Morrisville State College

Margaret K. Snooks — University of Houston, Clear Lake

Colleen Spees — Columbus State Community College

Norman J. Temple — Athabasca University, Alberta, Canada

Kathy Timmons — Murray State University

Amy A. Vaughan — Radford University

Eric Vlahov — The University of Tampa

Carolyn A. Weiglein — College of Charleston

Fred H. Wolfe — University of Arizona

Gloria Young — Virginia State University

Nancy Zwick — Northern Kentucky University

NUTRITION

Concepts and Controversies

Food Choices and Human Health

Georgia O'Keeffe (1887–1986), *Apple Family–2*, 1920. © 2005 The Georgia O'Keeffe Foundation/Artists Rights Society (ARS), New York. art Resource, NY

CONTENTS

Do You Ever . . .

Question whether your diet can make a real difference between getting sick or staying healthy?

Purchase supplements, believing that they are more powerful than food for ensuring good nutrition?

Wonder why you prefer the foods you do?

Become alarmed or confused by the news and media reports of nutrition science?

KEEP READING.

IF YOU CARE ABOUT your body, and if you have strong feelings about **food,** then you have much to gain from learning about **nutrition**—the study of how food nourishes the body. Nutrition is a fascinating, much talked-about subject. Each day, newspapers, radio, and television present stories of new findings on nutrition and heart health or nutrition and cancer prevention, and at the same time advertisements and commercials bombard us with multi-colored pictures of tempting foods—pizza, burgers, cakes, and chips. If you are like most people, when you eat you sometimes wonder, "Is this food good for me?" or you berate yourself, "I probably shouldn't be eating this."

When you study nutrition, you learn which foods serve you best, and you can work out ways of choosing foods, planning meals, and designing your **diet** wisely. Knowing the facts can enhance your health and your enjoyment of eating while relieving your feelings of guilt or worry that you aren't eating well.

This chapter addresses these "why, what, and how" questions about nutrition:

- *Why* care about nutrition? The **nutrients** interact with body tissues, adding a little or subtracting a little, day by day, and thus change the very foundations upon which the health of the body is built.
- *What* are the nutrients in foods, and what roles do they play in the body? Meet the nutrients, and discover their general roles in building body tissues and maintaining health.
- *What* constitutes a nutritious diet? Can you choose foods wisely, for nutrition's sake? And what motivates your choices?
- *How* do governments suggest that their citizens choose their diets to meet the national health objectives?
- And finally, *how* do we know what we know about nutrition? Scientific research reports provide an important foundation for understanding nutrition science.

The Controversy section demonstrates the differences between trustworthy sources of nutrition information and those that are less reliable.

When you choose foods with nutrition in mind, you can enhance your own well-being.

A Lifetime of Nourishment

If you live for 65 years or longer, you will have consumed more than 70,000 meals, and your remarkable body will have disposed of 50 tons of food. The foods you choose have cumulative effects on your body. As you age, you will see and feel those effects—if you know what to look for.

Your body renews its structures continuously, and each day it builds a little muscle, bone, skin, and blood, replacing old tissues with new. It may also add a little fat, if you consume excess food energy (calories), or subtract a little, if you consume less than you require. Some of the food you eat today becomes part of "you" tomorrow. The best food for you, then, is the kind that supports the growth and maintenance of strong muscles, sound bones, healthy skin, and sufficient blood to cleanse and nourish all parts of your body. This means you need food that provides not only energy but also sufficient nutrients, that is, enough water, carbohydrates, fats, protein, vitamins, and minerals. If the foods you eat provide too little or too much of any nutrient today, your health may suffer just a little today. If the foods you eat provide too little or too much of one or more nutrients every day for years, then, in later life, you may suffer severe disease effects.

A well-chosen array of foods supplies enough energy and enough of each nutrient to prevent **malnutrition.** Malnutrition includes deficiencies, imbalances, and excesses of nutrients, any of which can take a toll on health over time.

food medically, any substance that the body can take in and assimilate that will enable it to stay alive and to grow; the carrier of nourishment; socially, a more limited number of such substances defined as acceptable by each culture.

nutrition the study of the nutrients in foods and in the body; sometimes also the study of human behaviors related to food.

diet the foods (including beverages) a person usually eats and drinks.

nutrients components of food that are indispensable to the body's functioning. They provide energy, serve as building material, help maintain or repair body parts, and support growth. The nutrients include water, carbohydrate, fat, protein, vitamins, and minerals.

malnutrition any condition caused by excess or deficient food energy or nutrient intake or by an imbalance of nutrients. Nutrient or energy deficiencies are classed as forms of undernutrition; nutrient or energy excesses are classed as forms of overnutrition.

▶ **KEY POINT**   The nutrients in food support growth, maintenance, and repair of the body. Deficiencies, excesses, and imbalances of nutrients bring on the diseases of malnutrition.

How Powerful Is a Nutritious Diet in Preventing Diseases?

Your choice of diet profoundly influences your long-term health prospects.[*1] Only two common lifestyle habits are more influential: smoking and other tobacco use, and excessive drinking of alcohol. Of the leading causes of death listed in Table 1-1, four are related to nutrition, and another—motor vehicle and other accidents—is related to drinking alcohol.

Many older people suffer from debilitating conditions that could have been largely prevented had they known and applied the nutrition principles of today. The **chronic diseases**—heart disease, diabetes, some kinds of cancer, dental disease, and adult bone loss—all have a connection to poor diet. These diseases cannot be prevented by a good diet alone; they are to some extent determined by a person's genetic constitution, activities, and lifestyle. Within the range set by your genetic inheritance, however, the likelihood of developing these diseases is strongly influenced by your food choices.

▶ **KEY POINT** Nutrition profoundly affects health.

Genetics and Individuality

Consider the role of genetics. Genetics and nutrition affect different diseases to varying degrees (see Figure 1-1). The anemia caused by sickle-cell disease, for example, is purely hereditary and thus appears at the left of Figure 1-1 as a genetic disease unrelated to nutrition. Nothing a person eats affects the person's chances of developing this anemia, although nutrition therapy may help ease its course. At the other end of the spectrum in Figure 1-1, iron-deficiency anemia most often results from undernutrition. Diseases and conditions of poor health appear all along this continuum from almost entirely genetic to almost purely nutritional; the more nutrition-related a disease or health condition is, the more successfully sound nutrition can prevent it.

Furthermore, some diseases, such as heart disease and cancer, are not one disease but many. Two people may both have heart disease, but not the same form. One person's heart disease or cancer may be nutrition-related, but another's may not be. Individual people differ genetically from each other in thousands of subtle ways, so no simple statement can be made about the extent to which diet can help any one person avoid a disease or slow its progress.

[*]Reference notes are found in Appendix F.

TABLE 1·1 LEADING CAUSES OF DEATH, U.S., 2001

Blue shading indicates that a cause of death is related to nutrition; the light yellow indicates that it is related to alcohol. (The percentage of total deaths from each cause is in parentheses.)[a]

1. Heart disease	(29.0%)
2. Cancers	(22.9%)
3. Strokes	(6.8%)
4. Chronic lower respiratory diseases	(5.1%)
5. Motor vehicle and other accidents	(4.2%)
6. Diabetes mellitus	(3.0%)
7. Pneumonia and influenza	(2.6%)
8. Alzheimer's disease	(2.2%)
9. Kidney disease	(1.6%)
10. Infections of the blood	(1.3%)

SOURCE: National Center for Health Statistics, 2003.
[a]Hypertension (high blood pressure), a nutrition-related cause of death, ranks number 13.

■ **Anemia** is a blood condition in which red blood cells, the body's oxygen carriers, are inadequate or impaired, and so cannot meet the oxygen demands of the body. More about the anemia of sickle-cell disease in Chapter 6; iron-deficiency anemia is described in Chapter 8.

chronic diseases long-duration degenerative diseases characterized by deterioration of the body organs. Examples include heart disease, cancer, and diabetes.

FIGURE 1·1 NUTRITION AND DISEASE

Not all diseases are equally influenced by diet. Some are almost purely genetic, like the anemia of sickle-cell disease. Some may be inherited (or the tendency to develop them may be inherited in the genes) but may be influenced by diet, like some forms of diabetes. Some are purely dietary, like the vitamin and mineral deficiency diseases.

Less nutrition-related (genetic)			More nutrition-related
Down syndrome Hemophilia Sickle-cell anemia	Adult bone loss (osteoporosis) Cancer Infectious diseases	Diabetes Hypertension Heart disease	Iron-deficiency (anemia) Vitamin deficiencies Mineral deficiencies Toxicities Poor resistance to disease

■ The human genome is 99.9% the same in all people; all of the normal variations such as differences in hair color, as well as variations that result in diseases lie in the 0.1% that varies.

■ Only about 2% of the human genome contains genes. Scientists are asking, "What does the rest do?"

The recent completion of the human **genome** establishes the entire sequence of human **DNA** that holds the **genes,** bits of genetic material that hold the instructions for making a human being. This leap of knowledge promises greater control over health and disease as well as a greatly increased understanding of the interactions between a person's genetic makeup and the nutrients in the body.[2]

▶ KEY POINT   Choice of diet influences long-term health within the range set by genetic inheritance. Nutrition has little influence on some diseases but strongly affects others.

The Importance of Nutritional Genomics

The integration of nutrition, genomic science, and molecular biology has launched a new area of study, **nutritional genomics.** Scientists working in this area are describing how nutrients affect the activities of genes and how genes affect the activities of nutrients.[3] Soon, such revelations are expected to help pinpoint nutrient needs more precisely for growing children, healthy adults, and those fighting diseases, among others.[4] Food manufacturers look forward to providing food and supplement products tailored to an individual's specific nutrient needs.[5] Later chapters expand the story of nutritional genomics, beginning in Chapter 6 with a description of the link between the body's proteins and the activities of genes.

▶ KEY POINT   Nutritional genomics holds great promise for advances in nutrition science.

Other Lifestyle Choices

■ Alcohol use and abuse and their effects on body tissues are topics of Controversy 3.

Besides food choices, other lifestyle choices also affect people's health. Tobacco use and alcohol and other substance abuse can destroy health. Physical activity, sleep, stress levels, and conditions at home and at work, including the quality of the air and water and other aspects of the environment, can help prevent or reduce the severity of some diseases.

Putting together a diet that supports health depends partly on knowing which foods to choose. The next section provides an overview of the nutrients in foods, and later sections present dietary guidelines that can provide goals to shoot for.

▶ KEY POINT   Personal life choices, such as staying physically active or using tobacco or alcohol, also affect health for the better or worse.

genome (GEE-nome) the full complement of genetic material in the chromosomes of a cell. The study of genomes is *genomics.*

DNA an abbreviation for deoxyribonucleic (dee-OX-ee-RYE-bow-nu-CLAY-ick) acid, the molecule that encodes genetic information in its structure.

genes units of a cell's inheritance, made of the chemical DNA (deoxyribonucleic acid). Each gene directs the making of one or more proteins, which perform important tasks in the body.

nutritional genomics the science of how nutrients affect the activities of genes and how genes affect the activities of nutrients. Also called *molecular nutrition* or *nutrigenomics.*

energy the capacity to do work. The energy in food is chemical energy; it can be converted to mechanical, electrical, heat, or other forms of energy in the body. Food energy is measured in calories, defined on page 5.

organic carbon containing. Four of the six classes of nutrients are organic: carbohydrate, fat, protein, and vitamins. Strictly speaking, organic compounds include only those made by living things and do not include carbon dioxide and a few carbon salts.

The Human Body and Its Food

As your body moves and works each day, it must use **energy.** The energy that fuels the body's work comes indirectly from the sun by way of plants. Plants capture and store the sun's energy in their tissues as they grow. When you eat plant-derived foods such as fruits, grains, or vegetables, you obtain and use the solar energy they have stored. Plant-eating animals obtain their energy in the same way, so when you eat animal tissues, you are eating compounds containing energy that came originally from the sun.

The body requires six kinds of nutrients—families of molecules indispensable to its functioning—and foods deliver these. Table 1-2 lists the six classes of nutrients. Four of these six are **organic;** that is, the nutrients contain the element carbon derived from living things. The human body and foods are made of the same materials, arranged in different ways (see Figure 1-2).

TABLE 1·2 ELEMENTS IN THE SIX CLASSES OF NUTRIENTS

The nutrients that contain carbon are organic.

	Carbon	Oxygen	Hydrogen	Nitrogen	Minerals
Water		✔	✔		
Carbohydrate	✔	✔	✔		
Fat	✔	✔	✔		
Protein	✔	✔	✔	✔	b
Vitamins	✔	✔	✔	✔a	b
Minerals					✔

a All of the B vitamins contain nitrogen; *amine* means nitrogen.
b Protein and some vitamins contain the mineral sulfur; vitamin B_{12} contains the mineral cobalt.

The Nutrients in Foods

Foremost among the six classes of nutrients in foods is *water*, which is constantly lost from the body and must constantly be replaced. Of the four organic nutrients, three are **energy-yielding nutrients,** meaning that the body can use the energy they contain. The *carbohydrates* and *fats* (fats are properly called *lipids*) are especially important energy-yielding nutrients. As for *protein,* it does double duty: it can yield energy, but it also provides materials that form structures and working parts of body tissues. (Alcohol yields energy, too, but it is a toxin, not a nutrient—see the note to Table 1-3, p. 6).

The fifth and sixth classes of nutrients are the *vitamins* and the *minerals*. These provide no energy to the body. A few minerals serve as parts of body structures (calcium and phosphorus, for example, are major constituents of bone), but all vitamins and minerals act as regulators. As regulators, the vitamins and minerals assist in all body processes: digesting food; moving muscles; disposing of wastes; growing new tissues; healing wounds; obtaining energy from carbohydrate, fat, and protein; and participating in every other process necessary to maintain life. Later chapters are devoted to these six classes of nutrients.

When you eat food, then, you are providing your body with energy and nutrients. Furthermore, some of the nutrients are **essential nutrients,** meaning that if you do not ingest them, you will develop deficiencies; the body cannot make these nutrients for itself. Essential nutrients are found in all six classes of nutrients. Water is an essential nutrient; so is a form of carbohydrate; so are some lipids, some parts of protein, all of the vitamins, and the minerals important in human nutrition.

To support understanding of discussions throughout this book, two definitions and a set of numbers are useful. Food scientists measure food energy in **calories,** units of heat. Food and nutrient quantities are often measured in **grams,** units of weight. The most energy-rich of the nutrients is fat, which contains 9 calories in each gram. Carbohydrate and protein each contain only 4 calories in a gram (see Table 1-3, p. 6).

Scientists have worked out ways to measure the energy and nutrient contents of foods. They have also calculated the amounts of energy and nutrients various types of people need—by gender, age, life stage, and activity. Thus, after studying human nutrient requirements (in Chapter 2), you will be able to state with some accuracy just what your own body needs—this much water, that much carbohydrate and fat, so much protein, and so forth. So why not simply take pills or **dietary supplements** in place of food? Because, as it turns out, food offers more than just the six basic nutrients.

▶ **KEY POINT** Food supplies energy and nutrients. Foremost among the nutrients is water. The energy-yielding nutrients are carbohydrates, fats (lipids), and protein. The regulator nutrients are vitamins and minerals. Food energy is measured in calories; food and nutrient quantities are often measured in grams.

FIGURE 1·2 MATERIALS OF FOOD AND THE HUMAN BODY

Foods and the human body are made of the same materials.

Water
Carbohydrate
Fat
Protein
Minerals
Vitamins

energy-yielding nutrients the nutrients the body can use for energy. They may also supply building blocks for body structures.

essential nutrients the nutrients the body cannot make for itself (or cannot make fast enough) from other raw materials; nutrients that must be obtained from food to prevent deficiencies.

calories units of energy. Strictly speaking, the unit used to measure the energy in foods is a kilocalorie (*kcalorie* or *Calorie*): it is the amount of heat energy necessary to raise the temperature of a kilogram (a liter) of water 1 degree Celsius. This book follows the common practice of using the lowercase term *calorie* (abbreviated *cal*) to mean the same thing.

grams units of weight. A gram (g) is the weight of a cubic centimeter (cc) or milliliter (ml) of water under defined conditions of temperature and pressure. About 28 grams equal an ounce.

dietary supplements pills, liquids, or powders that contain purified nutrients or other ingredients (see Controversy 7).

TABLE 1·3 CALORIE VALUES OF ENERGY NUTRIENTS

The energy a person consumes in a day's meals comes from these three energy-yielding nutrients; alcohol, if consumed, also contributes energy.

Energy Nutrient	Energy
Carbohydrate	4 cal/g
Fat (lipid)	9 cal/g
Protein	4 cal/g

NOTE: Alcohol contributes 7 calories/gram that the human body can use for energy. Alcohol is not classed as a nutrient, however, because it interferes with growth, maintenance, and repair of body tissues.

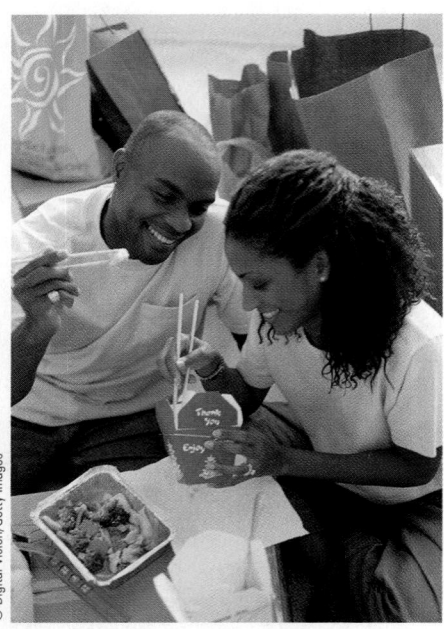

When you eat foods, you are receiving more than just nutrients.

© Digital Vision/Getty Images

elemental diets diets composed of purified ingredients of known chemical composition; intended to supply all essential nutrients to people who cannot eat foods.

nonnutrients a term used in this book to mean compounds other than the six nutrients that are present in foods and have biological activity in the body.

phytochemicals nonnutrient compounds in plant-derived foods that have biological activity in the body (*phyto* means "plant").

Can I Live on Just Supplements?

Nutrition science can state what nutrients human beings need to survive—at least for a time. Scientists are becoming skilled at making **elemental diets**—diets with a precise chemical composition that are life-saving for people in the hospital who cannot eat ordinary food. These formulas, administered to severely ill people for days or weeks, support not only continued life but also recovery from nutrient deficiencies, infections, and wounds.

Lately, marketers have taken these liquid formulas out of the medical setting and have advertised them heavily to healthy people of all ages as "meal replacers" or "insurance" against malnutrition. The truth is that such products are not superior to a sound diet of real foods. Formula diets are essential to help sick people to survive, but they do not enable people to thrive over long periods. Elemental diet formulas do not support optimal growth and health, and they often lead to medical complications.[6] Although these problems are rare and can be detected and corrected, they show that the composition of these diets is not yet perfect for all people in all settings. Healthy people who eat a healthful diet do not need such formulas and, in fact, most need no dietary supplements.[7]

Even if a person's basic nutrient needs are perfectly understood and met, concoctions of nutrients still lack something that foods provide. Hospitalized clients who are fed nutrient mixtures through a vein often improve dramatically when they can finally eat food. Something in real food is important to health—but what is it? What does food offer that cannot be provided through a needle or a tube? Science has some partial explanations, some physical and some psychological.

In the digestive tract, the stomach and intestine are dynamic, living organs, changing constantly in response to the foods they receive—even to just the sight, aroma, and taste of food. When a person is fed through a vein, the digestive organs, like unused muscles, weaken and grow smaller. Lack of digestive tract stimulation may even weaken the body's defenses against certain infections, such as infections of the respiratory tract.[8] Medical wisdom now dictates that a person should be fed through a vein for as short a time as possible and that real food taken by mouth should be reintroduced as early as possible. The digestive organs also release hormones in response to food, and these send messages to the brain that bring the eater a feeling of satisfaction: "There, that was good. Now I'm full." Eating offers both physical and emotional comfort.

Food does still more than maintain the intestine and convey messages of comfort to the brain. Foods are chemically complex. In addition to their nutrients, they contain **nonnutrients,** including the **phytochemicals.** These compounds confer color, taste, and other characteristics on foods, and many are believed to affect health by reducing disease risks (see Controversy 2). Even an ordinary baked potato contains hundreds of different compounds. In view of all this, it is not surprising that food gives us more than just nutrients. If it were otherwise, that would be surprising.

▶ **KEY POINT** In addition to nutrients, food conveys emotional satisfaction and hormonal stimuli that contribute to health. Foods also contain phytochemicals that give them their tastes, aromas, colors, and other characteristics. Some phytochemicals may play roles in reducing disease risks.

The Challenge of Choosing Foods

Well-planned meals convey pleasure and are nutritious, too, fitting your tastes, personality, family and cultural traditions, lifestyle, and budget. Given the astounding numbers and varieties available, consumers can lose track of what individual foods contain and how to put them together into health-promoting diets. A few guidelines can help.

The Abundance of Foods to Choose From

A list of the foods available a hundred years ago would be relatively short. It would consist of **basic foods**—foods that have been around for a long time such as vegetables, fruits, meats, milk, and grains. These foods have also been called unprocessed, natural, whole, or farm foods. An easy way to obtain a nutritious diet is to consume a variety of selections from among these foods each day. On a given day, however, almost half of our population consume no fruits or fruit juices. Also, although people generally consume a few servings of vegetables, the vegetable they most often choose is potatoes, usually prepared as french fries. Such dietary patterns increase the risk of chronic disease.[9]

The number of foods supplied by the food industry today is astounding. Thousands of foods now line the market shelves—many are processed mixtures of the basic ones, and some are even constructed mostly from artificial ingredients. Ironically, this abundance may make it more difficult, rather than easier, to plan a nutritious diet.

Table 1-4 presents a glossary of terms related to foods. The terms reveal that all types of food—including **fast foods** and **processed foods**—offer various constituents to the eater. You may also hear about **functional foods,** a term coined in an attempt to identify those foods that might lend protection against chronic diseases by way of the nutrients or nonnutrients they contain. The trouble is, scientists trying to single out the most health-promoting foods find that almost every naturally occurring food—even chocolate—is functional in some way with regard to human health. Controversy 2 provides more information about functional foods.

The extent to which foods support good health depends on the calories, nutrients, and nonnutrients they contain. In short, to select well among foods, you need to know more than their names; you need to know the foods' inner qualities.

Even more importantly, you need to know how to combine foods into nutritious diets. Foods are not nutritious by themselves; each is of value only in so far as it contributes to a nutritious diet. A key to wise diet planning is to make sure that the foods you eat daily, your **staple foods,** are especially nutritious.

Some foods offer beneficial nonnutrients called phytochemicals.

■ In 1900, Americans chose from among 500 or so different foods; today, they choose from more than 50,000.

TABLE 1·4   GLOSSARY OF FOOD TYPES

The purpose of this little glossary is to show that good-sounding food names don't necessarily signify that foods are nutritious. Read the comment at the end of each definition.

- **basic foods** milk and milk products; meats and similar foods such as fish and poultry; vegetables, including dried beans and peas; fruits; and grains. These foods are generally considered to form the basis of a nutritious diet. Also called *whole foods.*
- **enriched foods** and **fortified foods** foods to which nutrients have been added. If the starting material is a whole, basic food such as milk or whole grain, the result may be highly nutritious. If the starting material is a concentrated form of sugar or fat, the result may be less nutritious.
- **fast foods** restaurant foods that are available within minutes after customers order them—traditionally, hamburgers, french fries, and milkshakes; more recently, salads and other vegetable dishes as well. These foods may or may not meet people's nutrient needs, depending on the selections made and on the energy allowances and nutrient needs of the eaters.
- **functional foods** a term that reflects an attempt to define as a group the foods known to possess nutrients or nonnutrients that might lend protection against diseases. However, all nutritious foods can support health in some ways; Controversy 2 provides details.

- **natural foods** a term that has no legal definition, but is often used to imply wholesomeness.
- **nutraceutical** a term that has no legal or scientific meaning, but is sometimes used to refer to foods, nutrients, or dietary supplements believed to have medicinal effects (see Chapter 11). Often used to sell unnecessary or unproved supplements.
- **organic foods** understood to mean foods grown without synthetic pesticides or fertilizers. In chemistry, however, all foods are made mostly of organic (carbon-containing) compounds. (See Controversy 12 for details.)
- **partitioned foods** foods composed of parts of whole foods, such as butter (from milk), sugar (from beets or cane), or corn oil (from corn). Partitioned foods are generally overused and provide few nutrients with many calories.
- **processed foods** foods subjected to any process, such as milling, alteration of texture, addition of additives, cooking, or others. Depending on the starting material and the process, a processed food may or may not be nutritious.
- **staple foods** foods used frequently or daily, for example, rice (in East and Southeast Asia) or potatoes (in Ireland). If well chosen, these foods are nutritious.

All foods once looked like this . . .

. . . but now many foods often look like this.

> **KEY POINT** Foods come in a bewildering variety in the marketplace, but the foods that form the basis of a nutritious diet are basic foods, such as ordinary milk and milk products; meats, fish, and poultry; vegetables and dried peas and beans; fruits; and grains.

How, Exactly, Can I Recognize a Nutritious Diet?

A nutritious diet has five characteristics. First is **adequacy:** the foods provide enough of each essential nutrient, fiber, and energy. Second is **balance:** the choices do not overemphasize one nutrient or food type at the expense of another. Third is **calorie control:** the foods provide the amount of energy you need to maintain appropriate weight—not more, not less. Fourth is **moderation:** the foods do not provide excess fat, salt, sugar, or other unwanted constituents. Fifth is **variety:** the foods chosen differ from one day to the next. In addition, to maintain a steady supply of nutrients, meals should occur with regular timing throughout the day.

Adequacy Any nutrient could be used to demonstrate the importance of dietary *adequacy.* Iron provides a familiar example. It is an essential nutrient: you lose some every day, so you have to keep replacing it; and you can get it into your body only by eating foods that contain it.[*] If you eat too few of the iron-containing foods, you can develop iron-deficiency anemia: with anemia you may feel weak, tired, cold, sad, and unenthusiastic; you may have frequent headaches; and you can do very little muscular work without disabling fatigue. Some foods are rich in iron; others are notoriously poor. If you add iron-rich foods to your diet, you soon feel more energetic. Meat, fish, poultry, and **legumes** are in the iron-rich category, and an easy way to obtain the needed iron is to include these foods in your diet regularly.

Balance To appreciate the importance of dietary *balance,* consider a second essential nutrient, calcium. A diet lacking calcium causes poor bone development during the growing years and increases a person's susceptibility to disabling bone loss in adult life. Most foods that are rich in iron are poor in calcium. Calcium's richest food sources are milk and milk products, which happen to be extraordinarily poor iron sources. Clearly, to obtain enough of both iron and calcium, people have to balance their food

adequacy the dietary characteristic of providing all of the essential nutrients, fiber, and energy in amounts sufficient to maintain health and body weight.

balance the dietary characteristic of providing foods of a number of types in proportion to each other, such that foods rich in some nutrients do not crowd out of the diet foods that are rich in other nutrients. Also called *proportionality.*

calorie control control of energy intake; a feature of a sound diet plan.

moderation the dietary characteristic of providing constituents within set limits, not to excess.

variety the dietary characteristic of providing a wide selection of foods—the opposite of monotony.

legumes (leg-GOOMS, LEG-yooms) beans, peas, and lentils, valued as inexpensive sources of protein, vitamins, minerals, and fiber that contribute little fat to the diet. Also defined in Chapter 6.

---

[*] A person can also take supplements of iron, but as later discussions demonstrate, this is not as effective as eating iron-rich foods.

choices. Balancing the whole diet to provide enough but not too much of every one of the 40-odd nutrients the body needs for health requires considerable juggling. As you will see in Chapter 2, food group plans can help you achieve dietary adequacy and balance because they recommend specific amounts of foods of each type.

Calorie Control Energy intakes should not exceed energy needs. Nicknamed calorie control, this diet characteristic ensures that energy intakes from food balance energy expenditures in activity. Eating such a diet helps to control body fat content and weight. The many strategies that promote this goal appear in Chapter 9.

Moderation Intakes of certain food constituents such as fat, cholesterol, sugar, and salt should be limited for health's sake. A major guideline for healthy people is to keep fat intake below 35 percent of total calories.[10] Some people take this to mean that they must never indulge in a delicious beefsteak or hot-fudge sundae, but they are misinformed: *moderation,* not total abstinence, is the key. A steady diet of steak and ice cream might be harmful, but once a week as part of an otherwise moderate diet plan, these foods may have little impact; as once-a-month treats, these foods would have practically no effect at all. Moderation also means that limits are necessary, even for desirable food constituents. For example, a certain amount of fiber in foods contributes to the health of the digestive system, but too much fiber leads to nutrient losses.

Variety As for *variety,* nutrition scientists agree that people should not eat the same foods, even highly nutritious ones, day after day. One reason is that variety increases the likelihood that diet will be adequate in nutrients.[11] In addition, some nonnutrient food components are probably important to health; some foods may be better sources of these than others. Another reason is that a monotonous diet may deliver large amounts of toxins or contaminants. Each such undesirable item in a food is diluted by all the other foods eaten with it and is even further diluted if the food is not eaten again for several days. Last, variety adds interest—trying new foods can be a source of pleasure. Table 1-5 takes an honest look at obstacles to eating well.

A caution is in order. Any one of these dietary principles alone cannot ensure a healthful diet. For example, the most likely outcome of relying solely on variety could easily be a low-nutrient, high-calorie diet consisting of a variety of snack foods and nutrient-poor sweets.[12] If you establish the habit of using all of the principles just

■ A nutritious diet follows the A, B, C, M, V principles:
- Adequacy.
- Balance.
- Calorie control.
- Moderation.
- Variety.

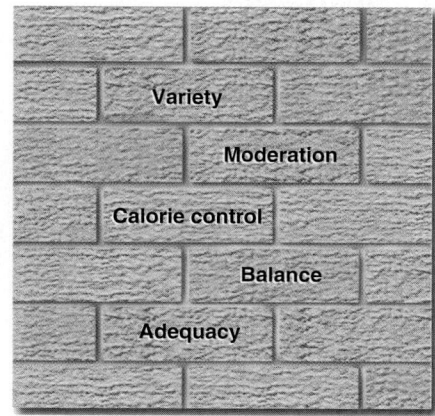

These factors are the building blocks of a nutritious diet.

TABLE 1·5	WHAT'S TODAY'S EXCUSE FOR NOT EATING WELL?

If you find yourself saying, "I know I should eat well, but I'm too busy" (or too fond of fast food, or have too little money, or a dozen other excuses), take note:
- *No time.* Everyone is busy. In truth, eating well takes little time. Convenience packages of frozen vegetables, jars of pasta sauce, and prepared meats and salads are abundant in markets today and take no longer to pick up than snack chips and colas. Priorities change drastically and instantly when illness strikes—better to spend a little time now nourishing your body's defenses than to spend time later treating illness.
- *Crave fast food.* Occasional fast-food meals can support health, if you choose wisely (see Chapter 5).
- *Too little money.* Eating right costs no more than eating poorly. Chips, colas, fast food, and premium ice cream are expensive. And serious illness costs more than a well person can imagine. By a 2005 USDA estimate, the needed fruits and vegetables can cost as little as 64 cents a day.
- *Like to eat large portions.* An occasional splurge, say, once a month, is a healthy part of moderation.
- *Take vitamins instead.* Vitamin pills cannot make up for consistently poor food choices. Food constituents such as fiber and phytochemicals are also important to good health.
- *Love sweets.* Sweets in moderation are an acceptable, and even desirable, part of a balanced diet.

described, you will find that choosing a healthful diet becomes as automatic as brushing your teeth or falling asleep.

▶ **KEY POINT**   A well-planned diet is adequate in nutrients, is balanced with regard to food types, offers food energy that matches energy expended in activity, is moderate in unwanted constituents, and offers a variety of nutritious foods.

Why People Choose Foods

Eating is an intentional act. Each day, people choose from the available foods, prepare the foods, decide where to eat, which customs to follow, and with whom to dine. Many factors influence food-related choices.

■ Figure 2-5 in Chapter 2 depicts some ethnic foods that have become an integral part of the "American diet."

Cultural and Social Meanings Attached to Food   Like wearing traditional clothing or speaking a native language, enjoying traditional **cuisines** and **foodways** can be a celebration of your own or a friend's heritage. Sharing **ethnic food** can be symbolic: people offering foods are expressing a willingness to share cherished values with others. People accepting those foods are symbolically accepting not only the person doing the offering but the person's culture.

Cultural traditions regarding food are not inflexible; they keep evolving as people move about, learn about new foods, and teach each other. Today, some people are ceasing to be **omnivores** and are becoming **vegetarians.** Vegetarians often choose this lifestyle because they honor the lives of animals or because they have discovered the health and other advantages associated with diets rich in beans, whole grains, fruits, nuts, and vegetables. The Controversy of Chapter 6 explores the pros and the cons of both the vegetarian's and the meat eater's diets.

Factors That Drive Food Choices   Consumers today value convenience so highly that they are willing to spend over half of their food budget on meals that require little or no preparation.[13] They frequently eat out, bring home ready-to-eat meals, or have food delivered. In their own kitchens, they want to prepare a meal in 15 to 20 minutes, using only four to six ingredients. Such convenience limits food choices but doesn't necessarily mean that nutrition is out the window. This chapter's Food Feature addresses the time and nutrition tradeoff.

Convenience is only one consideration.[14] Physical, psychological, social, and philosophical factors all influence how you choose the foods you generally eat:

Sharing ethnic food is a way of sharing culture.

- *Advertising.* The media have persuaded you to eat these foods.
- *Availability.* There are no others to choose from.
- *Economy.* They are within your means.
- *Emotional comfort.* They can make you feel better for a while.
- *Habit.* They are familiar; you always eat them.
- *Personal preference and genetic inheritance.* You like the way these foods taste, with some preferences possibly determined by the genes.[15]
- *Positive associations.* They are eaten by people you admire, or they indicate status, or they remind you of fun.
- *Region of the country.* They are foods favored in your area.
- *Social pressure.* They are offered; you feel you can't refuse them.
- *Values or beliefs.* They fit your religious tradition, square with your political views, or honor the environmental ethic.
- *Weight.* You think they will help to control body weight.
- *Nutritional value.* You think they are good for you.

cuisines styles of cooking.

foodways the sum of a culture's habits, customs, beliefs, and preferences concerning food.

ethnic foods foods associated with particular cultural subgroups within a population.

omnivores people who eat foods of both plant and animal origin, including animal flesh.

vegetarians people who exclude from their diets animal flesh and possibly other animal products such as milk, cheese, and eggs.

Only the last two of these reasons for choosing foods assign a high priority to nutritional health. Similarly, the choice of where, as well as what, to eat is often based more on social considerations than on nutrition judgments. College students often choose to eat at fast-food and other restaurants to socialize, to get out, to save time, or to date; they are not always conscious of the need to obtain healthful food.

If you wish to choose intentionally for health's sake, the final section of this chapter delivers some broad guidelines for choosing an adequate diet and some larger goals for the nutrition of the nation. Setting goals for our national nutritional health depends upon a firm base of scientific knowledge. The next section describes how such knowledge comes to light, and addresses the final "how" question of this chapter: How do we know what we know about nutrition?

▶ KEY POINT Cultural traditions and social values revolve around food. Some values are expressed through foodways. Many factors other than nutrition drive food choices.

The Science of Nutrition

Nutrition is a science—a field of knowledge composed of organized facts. Unlike sciences such as astronomy and physics, nutrition is a relatively young science. Most nutrition research has been conducted since 1900. The first vitamin was identified in 1897, and the first protein structure was not fully described until the mid-1940s. Because nutrition science is an active, changing, growing body of knowledge, scientific findings often seem to contradict one another or are subject to conflicting interpretations.

For this reason, people may despair as they try to decipher current reports to learn what is really going on: "When the scientists themselves can't agree on what is true, how am I supposed to know?" Yet many facts in nutrition are known with great certainty. To understand why apparent contradictions sometimes arise in nutrition science, we need to look first at what scientists do.

What Nutrition Scientists Do

Everyone stampedes for oat bran, red wine, or fish oil based on today's news that these products are good for health. Then tomorrow's news reports, "It isn't true after all," and everyone drops oat bran, red wine, or fish oil and takes up the next craze. Meanwhile, bewildered consumers complain in frustration, "Those scientists don't know anything."

In truth, though, it is a scientist's business not to know. Scientists obtain facts by systematically asking questions—that's their job. They use the scientific method outlined in Figure 1-3. To answer the questions, they design and conduct experiments to test for various possible answers (see Figure 1-4, p. 12 and Table 1-6, p. 13). When they have ruled out some possibilities and found evidence for others, they submit their findings, not to the news media, but to boards of reviewers composed of

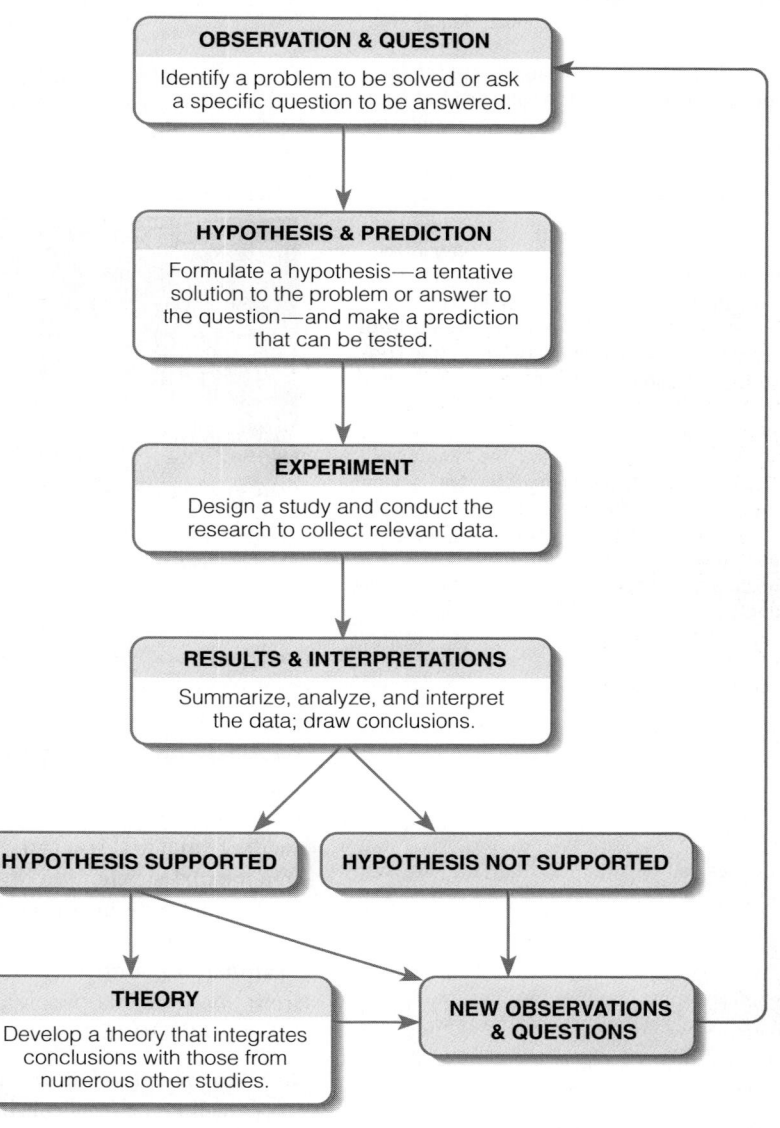

FIGURE 1·3 THE SCIENTIFIC METHOD

Research scientists follow the scientific method. Note that most research projects result in new questions, not final answers. Thus, research continues in a somewhat cyclical manner.

OBSERVATION & QUESTION
Identify a problem to be solved or ask a specific question to be answered.

HYPOTHESIS & PREDICTION
Formulate a hypothesis—a tentative solution to the problem or answer to the question—and make a prediction that can be tested.

EXPERIMENT
Design a study and conduct the research to collect relevant data.

RESULTS & INTERPRETATIONS
Summarize, analyze, and interpret the data; draw conclusions.

HYPOTHESIS SUPPORTED

HYPOTHESIS NOT SUPPORTED

THEORY
Develop a theory that integrates conclusions with those from numerous other studies.

NEW OBSERVATIONS & QUESTIONS

FIGURE 1·4 EXAMPLES OF RESEARCH DESIGN

The source of valid nutrition information is scientific research. Nutrition is a science, an organized body of knowledge composed of facts that we believe to be true because they have been supported, time and again, in experiments designed to rule out all other possibilities. Each fact has been established by many different kinds of experiments. For example, we know that eyesight depends partly on vitamin A because animals deprived of that vitamin and only that vitamin begin to go blind; when it is restored soon enough to their diet, they regain their sight. The same fact holds true in observations of human beings.

The type of study chosen for research depends upon what sort of information the researchers require. Studies of individuals (**case studies**) yield observations that may lead to possible avenues of research. A study of a man who ate gumdrops and became a famous dancer might suggest that an experiment be done to see if gumdrops contain dance-enhancing power.

Studies of whole populations (**epidemiological studies**) provide another sort of information. Such a study can reveal a **correlation.** For example, an epidemiological study might find no worldwide correlation of gumdrop eating with fancy footwork but, unexpectedly, might reveal a correlation with tooth decay.

Studies in which researchers actively intervene to alter people's eating habits (**intervention studies**) go a step further. In such a study, one set of subjects (the **experimental group**) receive a treatment, and another set (the **control group**) go untreated or receive a **placebo** or sham treatment. If the study is a **blind experiment,** the subjects do not know who among the members receives the treatment and who receives the sham. If the two groups experience different effects, then the treatment's effect can be pinpointed. For example, an intervention study might show that withholding gumdrops, together with other candies and confections, reduced the incidence of

tooth decay in an experimental population compared to that in a control population.

Finally, **laboratory studies** can pinpoint the mechanisms by which nutrition acts. What is it about gumdrops that contributes to tooth decay: their size, shape, temperature, color, ingredients? Feeding various forms of gumdrops to rats might yield the information that sugar, in a gummy carrier, promotes tooth

decay. In the laboratory, using animals or plants or cells, scientists can inoculate with diseases, induce deficiencies, and experiment with variations on treatments to obtain in-depth knowledge of the process under study. Intervention studies and laboratory experiments are among the most powerful tools in nutrition research because they show the effects of treatments.

Epidemiological Study

"This country's food supply contains more nutrient X, and these people suffer less illness Y."

Case Study

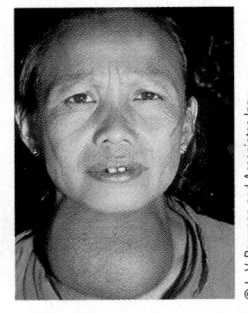

"This person eats too little of nutrient X and has illness Y."

Laboratory Study

"Now let's prove that a nutrient X deficiency causes illness Y by inducing a deficiency in these rats."

Intervention Study

"Let's add foods containing nutrient X to some people's food supply and compare their rates of illness Y with the rates of others who don't receive the nutrient."

other scientists who try to pick the findings apart. If these reviewers consider the conclusions to be well supported by the evidence, they endorse the work for publication in scientific journals where still more scientists can read it. Then the news media read it and you can read it, too; Table 1-7 explains what you can expect to find in a journal article.

As you study nutrition, you are likely to hear of findings based on two ongoing national scientific research projects. The first, the National Health and Nutrition Examination Surveys (NHANES), is a nationwide project that gathers information from about 50,000 people using diet histories, physical examinations and measurements, and laboratory tests.[16] Boiled down to its essence, NHANES involves:

- Asking people what they have eaten.
- Recording measures of their health status.

TABLE 1·6 RESEARCH DESIGN TERMS

- **blind experiment** an experiment in which the subjects do not know whether they are members of the experimental group or the control group. In a *double-blind experiment,* neither the subjects nor the researchers know to which group the members belong until the end of the experiment.
- **case studies** studies of individuals. In clinical settings, researchers can observe treatments and their apparent effects. To prove that a treatment has produced an effect requires simultaneous observation of an untreated similar subject (a *case control*).
- **control group** a group of individuals who are similar in all possible respects to the group being treated in an experiment but who receive a sham treatment instead of the real one. Also called *control subjects.* See also *experimental group* and *intervention studies.*
- **correlation** the simultaneous change of two factors, such as the increase of weight with increasing height (a *direct* or *positive* correlation) or the decrease of cancer incidence with increasing fiber intake (an *inverse* or *negative* correlation). A correlation between two factors suggests that one may cause the other, but does not rule out the possibility that both may be caused by chance or by a third factor.
- **epidemiological studies** studies of populations; often used in nutrition to search for correlations between dietary habits and disease incidence; a first step in seeking nutrition-related causes of diseases.
- **experimental group** the people or animals participating in an experiment who receive the treatment under investigation. Also called *experimental subjects.* See also *control group* and *intervention studies.*
- **intervention studies** studies of populations in which observation is accompanied by experimental manipulation of some population members—for example, a study in which half of the subjects (the *experimental subjects*) follow diet advice to reduce fat intakes while the other half (the *control subjects*) do not, and both groups' heart health is monitored.
- **laboratory studies** studies that are performed under tightly controlled conditions and are designed to pinpoint causes and effects. Such studies often use animals as subjects.
- **placebo** a sham treatment often used in scientific studies; an inert harmless medication. The *placebo effect* is the healing effect that the act of treatment, rather than the treatment itself, often has.

TABLE 1·7 THE ANATOMY OF A RESEARCH ARTICLE

Here's what you can expect to find inside a research article:

- *Abstract.* The abstract provides a brief overview of the article.
- *Introduction.* The introduction clearly states the purpose of the current study.
- *Review of literature.* A review of the literature reveals all that science has uncovered on the subject to date.
- *Methodology.* The methodology section defines key terms and describes the procedures used in the study.
- *Results.* The results report the findings and may include summary tables and figures.
- *Conclusions.* The conclusions drawn are those supported by the data and reflect the original purpose as stated in the introduction. Usually, they answer a few questions and raise several more.
- *References.* The references list relevant studies (including key studies several years old as well as current ones).

The second project is the Continuing Survey of Food Intakes by Individuals (CSFII); it involves:

- Recording what people have actually eaten for two days.
- Comparing the foods they have chosen with recommended food selections.

Nutrition monitoring makes it possible for researchers to assess the nutrient status, health indicators, and dietary intakes of the U.S. population. The agencies involved with these efforts are listed in the margin.

Scientific Challenge

Once a new finding is published, it is still only preliminary. One experiment does not "prove" or "disprove" anything. The next step is for other scientists to attempt to duplicate and support the work of the first researchers or to challenge the finding by designing experiments to refute it.

Only when a finding has stood up to rigorous, repeated testing in several kinds of experiments performed by several different researchers is it finally considered confirmed. Even then, strictly speaking, science consists not of facts that are set in stone but of *theories* that can always be challenged and revised. Some findings, though, like the theory that the earth revolves about the sun, are so well supported by observations and experimental findings that they are generally accepted as facts. What we "know" in nutrition is confirmed in the same way—through years of replicating study findings. This slow path of repeated studies stands in sharp contrast to the media's desire for today's latest news.

■ Agencies active in nutrition policy, research, and monitoring:

- Department of Health and Human Services (DHHS).
- United States Department of Agriculture (USDA).
- Centers for Disease Control and Prevention (CDC).

■ Ongoing national nutrition research projects:

- National Health and Nutrition Examination Surveys (NHANES).
- Continuing Survey of Food Intakes by Individuals (CSFII).

Can I Trust the Media to Deliver Nutrition News?

■ Some newspapers, magazines, talk shows, Internet websites, and other media strive for accuracy in reporting, but others specialize in sensationalism that borders on quackery—see this chapter's Controversy for details.

The news media are hungry for new findings, and reporters often latch onto ideas from scientific laboratories before they have been fully tested. Also, a reporter who lacks a strong understanding of science may misunderstand complex scientific principles. To tell the truth, sometimes scientists get excited about their findings, too, and leak them to the press before they have been through a rigorous review by the scientists' peers. As a result, the public is often exposed to late-breaking nutrition news stories before the findings are fully confirmed.[17] Then, when the hypothesis being tested fails to hold up to a later challenge, consumers feel betrayed by what is simply the normal course of science at work.

It also follows that people who take action based on single studies are almost always acting impulsively, not scientifically. The real scientists are trend watchers. They evaluate the methods used in each study, assess each study in light of the evidence gleaned from other studies, and, little by little, modify their picture of what is true. As evidence accumulates, the scientists become more and more confident about their ability to make recommendations that apply to people's health and lives. The Consumer Corner offers some tips for evaluating news stories about nutrition.

■ The links between lipids and heart disease are discussed in Chapter 5.

Sometimes media sensationalism overrates the importance of even true, replicated findings. For example, a few years ago the media eagerly reported that oat bran lowers blood cholesterol, a lipid indicative of heart disease risk. Although the reports were true, oat bran is only one of several hundred factors that affect blood cholesterol. News reports on oat bran often failed to mention that cutting intakes of certain fats is still the major step to take to lower blood cholesterol.

Also, new findings need refinements. Oat bran is truly a cholesterol reducer, but how much bran must a person eat to produce the desired effects? Do little oat bran pills or powders meet the need? Do oat bran cookies? If so, how many cookies must be eaten? As for oatmeal, it takes a bowl-and-a-half daily to affect blood lipids. A few cookies cannot provide nearly so much and certainly cannot undo all the damage from a high-fat meal.

Today, oat bran's cholesterol-lowering effect is considered to be established, and labels on food packages can proclaim that a diet high in oats may reduce the risk of heart disease. The whole process of discovery, challenge, and vindication took almost ten years of research. Some other lines of research have taken many years longer. In science, a single finding almost never makes a crucial difference to our knowledge as a whole, but like each individual frame in a movie, it contributes a little to the big picture. Many such frames are needed to tell the whole story.

▶ **KEY POINT** Scientists uncover nutrition facts by experimenting. Single studies must be replicated before their findings can be considered valid. New nutrition news is not always to be believed; established nutrition news has stood up to the test of time.

Dietary Guidelines and Nutrition Objectives

Many countries set forth dietary guidelines to help their citizens answer the question, "What should I eat to stay healthy?" Governments also set health objectives for the nation. The guidelines and objectives are related: if everyone fol-lowed the guidelines for individuals, many of the goals for the nation would fall into place.

Dietary Guidelines

The U.S. Department of Agriculture's *Dietary Guidelines for Americans 2005* (Figure 1-5, p. 16), offers science-based advice to promote health and to reduce risk for major chronic diseases through diet and physical activity.[18] People who balance their energy

CONSUMER CORNER

Reading Nutrition News with an Educated Eye

A NEWS READER, who had sworn off butter years ago for his heart's sake, bemoaned this headline: "Margarine Fat as Bad as Butter for Heart Health." "Do you mean to say that I could have been eating butter all these years? That's it. I quit. No more diet changes for me." His response is understandable—diet changes, after all, take effort to make and commitment to sustain. Those who do make changes may feel betrayed when, years later, science appears to have turned its advice upside down.

It bears repeating that the findings of a single study never prove or disprove anything. Study results may constitute strong supporting evidence for one view or another, but they rarely merit the sort of finality implied by journalistic phrases such as "Now we know" or "The answer has been found." Misinformed readers who look for simple answers to complex nutrition problems often take such phrases literally.

To read news stories with an educated eye, keep these points in mind:

- The study being described should be published in a peer-reviewed journal such as the *American Journal of Clinical Nutrition*. An unpublished study or one from a less credible source may or may not be valid; the reader has no way of knowing because the study has not been challenged or reviewed by other experts in the field.
- The news report should state the purpose of the study and describe the research methods used to obtain the data, although, in truth, few provide these details.[1] It should also note their limitations (in the Methodology section—look again at Table 1-7 on page 13). For example, it matters whether the study participants numbered 8 or 8,000, or whether the researchers personally observed the participants' behaviors or relied on self-reports collected over the telephone.
- The report should clearly define the subjects of the study—single cells, animals, or human beings. If the study subjects were human beings, the more you have in common with them (age and gender, for example), the more applicable the findings may be for you.
- Valid reports also describe previous research and put the current research in proper context. Some reporters regularly follow developments in a research area and thus acquire the background knowledge to report meaningfully in that area.
- Useful for their broad perspective on a single topic are review articles appearing in journals such as *Nutrition Reviews*. Such articles allow judgment about a single study within the context of many other studies on the same topic.

Finally, ask yourself if the study makes common sense. Even if it turns out that the fat of margarine is damaging to the heart, do you eat enough margarine to worry about its effects? Before making a decision, learn more about the effects of fats on the arteries in Chapters 5 and 11 and then ask the critical questions about yourself.

When a headline touts a shocking new "answer" to a nutrition question, read the story with a critical eye. It may indeed be a carefully researched report, but often it is a sensational story intended to catch the attention of newspaper and magazine buyers, not to offer useful nutrition information.

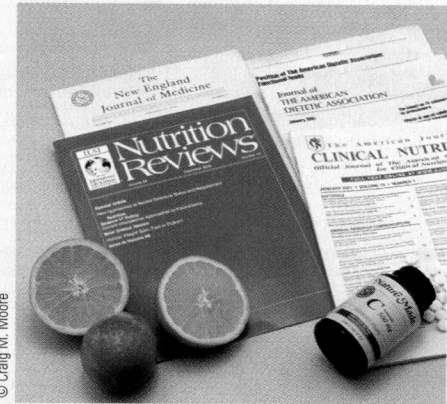

A person wanting the whole story on a nutrition topic is wise to seek articles from peer-reviewed journals such as these. A review journal examines all available evidence on major topics. Other journals report details of the methods, results, and conclusions of single studies.

(calorie) intakes with expenditures, consume diets that meet nutrient recommendations, and engage in regular physical activity most often enjoy optimum health. In fact, physical activity is so closely linked with nutrition in supporting health that most chapters of this book offer features called *"Think Fitness,"* such as the one on page 17. The *Dietary Guidelines* apply to most people age two years or older.

A major recommendation of the *Dietary Guidelines* is to choose a healthy diet based on the diet-planning guide, the *USDA Food Guide,* explained in Chapter 2. To meet its recommendations, most U.S. consumers need to limit calorie intakes and obtain more

FIGURE 1·5 *DIETARY GUIDELINES FOR AMERICANS, 2005*

ADEQUATE NUTRIENTS WITHIN ENERGY NEEDS

- Consume a variety of nutrient-dense foods and beverages within and among the basic food groups; limit intakes of saturated and *trans* fats, cholesterol, added sugars, salt, and alcohol.
- Meet recommended intakes within energy needs by adopting a balanced eating pattern, such as the USDA Food Guide (Chapter 2).

WEIGHT MANAGEMENT

- To maintain body weight in a healthy range, balance calories from foods and beverages with calories expended (Chapter 9).
- To prevent gradual weight gain over time, make small decreases in food and beverage calories and increase physical activity.

PHYSICAL ACTIVITY

- Engage in regular physical activity and reduce sedentary activities to promote health, psychological well-being, and a healthy body weight (Chapter 10).
- Achieve physical fitness by including cardiovascular conditioning, stretching exercises for flexibility, and resistance exercises or calisthenics for muscle strength and endurance.

FOOD GROUPS TO ENCOURAGE

- Consume a sufficient amount of fruits, vegetables, milk and milk products, and whole grains while staying within energy needs (Chapter 2).
- Select a variety of fruits each day. Include vegetables from all five subgroups (dark green, orange, legumes, starchy vegetables, and other vegetables) several times a week. Make at least half of the grain selections whole grains. Select fat-free or low-fat milk products.

FATS

- Keep saturated fat, *trans* fat, and cholesterol consumption low—less than 10 percent of calories from saturated and *trans* fats and less than 300 milligrams of cholesterol per day (Chapter 5).
- Keep total fat intake between 20 to 35 percent of calories, mostly from foods that provide unsaturated fats, such as fish, nuts, olives, and vegetable oils.
- Select and prepare foods that are lean, low fat, or fat-free.

CARBOHYDRATES

- Choose fiber-rich fruits, vegetables, and whole grains often (Chapter 4).
- Choose and prepare foods and beverages with little added sugars.
- Reduce the incidence of dental caries by practicing good oral hygiene and consuming sugar- and starch-containing foods and beverages less frequently.

SODIUM AND POTASSIUM

- Choose and prepare foods with little salt (less than 2300 milligrams sodium or approximately 1 teaspoon salt). At the same time, consume potassium-rich foods, such as fruits and vegetables (see Chapter 8).

ALCOHOLIC BEVERAGES

- Those who choose to drink alcoholic beverages should do so sensibly and in moderation.
- Some individuals should not consume alcoholic beverages (see Controversy 3).

FOOD SAFETY

- To avoid microbial foodborne illness, keep foods safe: clean hands, food contact surfaces, and fruits and vegetables; separate raw, cooked, and ready-to-eat foods; cook foods to a safe internal temperature; chill perishable food promptly; and defrost food properly (Chapter 12).

© Otto Greule Jr./Time Life Pictures/Getty Images

The *Dietary Guidelines* suggest that physical activity should be part of a healthy lifestyle.

and varied daily selections among fruits, vegetables, whole grains, and nonfat or low-fat milk or milk products (the reasons will become clear as you move through this book). A basic premise of both the *Dietary Guidelines* and of this book is that foods, not supplements, should provide the needed nutrients whenever possible.

Another focus of the *Dietary Guidelines* is on limiting potentially harmful dietary constituents. A healthful diet is carefully chosen to supply the kinds of carbohydrates that the body needs but little sugar and to offer the needed fats and oils while limiting saturated fat, *trans* fat, and cholesterol (Chapter 4 and 5 explain these distinctions). People are also asked to consume less salt and choose sensibly if they use alcohol. Finally, foods should be kept safe from spoilage or contamination (see Chapter 12).

Canadian readers can find *Canada's Guidelines for Healthy Eating* in Table 1-8. The *Nutrition Recommendations for Canadians* and other Canadian resources are located in Appendix B. Notice that the *Dietary Guidelines* do not require that you give up your favorite foods or eat strange, unappealing foods. With some planning and adjustments, almost anyone's diet and lifestyle can fit most of these recommendations.

If the experts who develop such guidelines were to ask us, we would add one more recommendation to their lists: take time to enjoy and savor your food. The joys of eating are physically beneficial

TABLE 1·8	CANADA'S GUIDELINES FOR HEALTHY EATING

- Enjoy a variety of foods.
- Emphasize cereals, breads, other grain products, vegetables, and fruits.
- Choose lower-fat dairy products, leaner meats, and foods prepared with little or no fat.
- Achieve and maintain a healthy body weight by enjoying regular physical activity and healthy eating.
- Limit salt, alcohol, and caffeine.

SOURCE: These guidelines derive from *Action Towards Healthy Eating: The Report of the Communications/Implementation Committee and Nutrition Recommendations. . . . A Call for Action: Summary Report of the Scientific Review Committee and the Communications/Implementation Committee*, which are available from Branch Publications Unit, Health Services and Promotion Branch, Department of Health and Welfare, 5th Floor, Jeanne Manice Building, Ottawa, Ontario K1A 1B4.

to the body because they trigger health-promoting changes in the nervous, hormonal, and immune systems. When the food is nutritious as well as enjoyable, then the eater obtains all the nutrients needed for the healthy skin, glossy hair, and natural good looks that accompany good health. Remember to enjoy your food.

▶ KEY POINT   The *Dietary Guidelines for Americans, Nutrition Recommendations for Canadians*, and other recommendations address the problems of overnutrition and undernutrition. To implement them requires exercising regularly; following the USDA Food Guide, seeking out whole grains, fruits, and vegetables, healthful fats, and limiting intakes of saturated fats, sugar, and salt, and moderating alcohol intake.

THINK FITNESS

Why Be Physically Active?

WHY SHOULD PEOPLE bother to be physically active? Though a person's daily food choices can powerfully affect health, the combination of nutrition and physical activity is more powerful still. People who are physically active can expect to receive at least some of the benefits listed in the margin. If even half of these benefits were yours for the asking, wouldn't you step up to claim them? In truth, they are yours to claim, at the price of including physical activity in your day. Chapter 10 comes back to the benefits of fitness and provides scientific evidence to support them.

Healthy People 2010: Nutrition Objectives for the Nation

The U.S. Department of Health and Human Services sets ten-year health objectives for the nation in its document *Healthy People*.[19] The objectives for the year 2010, listed in Table 1-9 (p. 18), provide a quick scan of the nutrition-related objectives set for this decade. The inclusion of nutrition and food-safety objectives shows that public health officials consider these areas to be top national priorities.[20]

By the close of the twentieth century, the U.S. population had either met or was approaching the targets of the previous *Healthy People 2000* objectives.[21] Today, positive trends toward reduced rates of certain food-borne infections and cancers are continuing. Deaths from heart disease and stroke are also declining, but on the negative side, heart disease is still the leading cause of death among adults and the numbers of overweight people and people diagnosed with diabetes are soaring.[22] To fully meet the current *Healthy People 2010* goals, our nation must take steps to reverse current trends toward increasingly overweight and diabetes.

▶ KEY POINT   The U.S. Department of Health and Human Services sets nutrition objectives for the nation each decade.

■ Potential benefits of physical activity include:
- Reduced risk of cardiovascular diseases.
- Increased cardiovascular endurance.
- Increased muscle strength and endurance.
- Increased flexibility.
- Reduced risk of some types of cancer (especially colon and breast).
- Improved mental outlook and lessened likelihood of depression.
- Improved mental functioning.
- Feeling of vigor.
- Feeling of belonging—the companionship of sports.
- Strong self-image and belief in one's abilities.
- Reduced body fatness, increased lean tissue.
- A more youthful appearance, healthy skin, and improved muscle tone.
- Greater bone density and lessened risk of adult bone loss in later life.
- Increased independence in the elderly.
- Sound, beneficial sleep.
- Faster wound healing.
- Lessening or elimination of menstrual pain.
- Improved resistance to infection.

TABLE 1·9 *HEALTHY PEOPLE 2010* NUTRITION-RELATED OBJECTIVES

- Increase *nutrition education* among consumers and in educational settings at all levels.
- Increase the proportion of children, adolescents, and adults who are at a *healthy weight.*
- Reduce *growth retardation* among low-income children under age 5 years.
- Increase the proportion of persons aged 2 years and older who consume at least two daily servings of *fruit.*
- Increase the proportion of persons aged 2 years and older who consume at least three daily servings of *vegetables,* with at least one-third being dark green or orange vegetables.
- Increase the proportion of persons aged 2 years and older who consume at least six daily servings of *grain products,* with at least three being whole grains.
- Increase the proportion of persons aged 2 years and older who consume less than 10% of calories from *saturated fat.*
- Increase the proportion of persons aged 2 years and older who consume no more than 30% of calories from *total fat.*
- Increase the proportion of persons aged 2 years and older who consume 2,400 milligrams or less of *sodium.*
- Increase the proportion of adults with *high blood pressure* who are taking action to control their blood pressure.
- Increase the proportion of persons aged 2 years and older who meet dietary recommendations for *calcium.*
- Reduce *iron deficiency* among young children, females of childbearing age, and pregnant females.
- Reduce *anemia* among low-income pregnant females in their third trimester.
- Reduce *key vitamin and mineral deficiencies* in pregnant women.
- Increase the proportion of children and adolescents aged 6 to 19 years whose intake of *meals and snacks at school* contributes to good overall dietary quality.
- Increase the proportion of worksites that offer *nutrition or weight management classes or counseling.*
- Increase the proportion of physician office visits made by patients with a diagnosis of cardiovascular disease, diabetes, or hyperlipidemia that include *counseling or education related to diet and nutrition.*
- Reduce deaths from anaphylaxis caused by *food allergies.*
- Increase the number of consumers and retail establishments who follow key *food-safety* practices and reduce key foodborne illnesses.
- Increase *food security* among U.S. households and in so doing reduce hunger.

SOURCE: Details about these and hundreds of other objectives are available from the U.S. Department of Health and Human Services, *Healthy People 2010,* 2nd ed. (Washington, D.C.: Government Printing Office, 2000), online at www.health.gov/healthypeople or call (800) 367-4725.

Conclusion

According to the experts, people in the United States are not very successful at meeting the nutrition goals set forth in this chapter. In particular, only a tiny percentage of adults manage to achieve both adequacy and moderation. People seem to behave as though they must choose between two alternatives—getting all their nutrients but overconsuming calories in the process, or keeping their calories in line but running short on nutrients.[23] That finding defines the challenge for the health-conscious eater: try to achieve adequacy and moderation at the same time. Because this challenge is the key to good nutrition, this chapter's Food Feature offers a tool to help make it easier—the concept of **nutrient density**. The next chapter takes up the challenge again, offering details about diet planning for adequacy and moderation.

nutrient density a measure of nutrients provided per calorie of food.

FOOD FEATURE

How Can I Get Enough Nutrients without Consuming Too Many Calories?

IN TRYING TO control calories while balancing the diet and making it adequate, certain foods are especially useful. These foods are rich in nutrients relative to their energy contents: that is, they are foods with high nutrient density. Figure 1-6 is a simple depiction of this concept. Consider calcium sources, for example. Ice cream and fat-free milk both supply calcium, but the milk is "denser" in calcium per calorie. A cup of rich ice cream contributes more than 350 calories, a cup of fat-free milk only 85—and with almost double the calcium. Most people cannot, for their health's sake, afford to choose foods without regard to their energy contents. Those who do very often exceed calorie allowances while leaving nutrient needs unmet.

Nutrient density is such a useful concept in diet planning that this book encourages you to think in those terms. Watch for the tables and figures in later chapters that show the best buys among foods, not necessarily in nutrients per dollar (although nutrient-dense foods are often among the least expensive), but in nutrients per calorie. This viewpoint can help you distinguish between more and less nutritious foods. For people who wish to eat larger meals, yet not exceed their energy bud-gets, the concept of nutrient density can help them to identify foods that provide bulk without a lot of calories.

The foods that offer the most nutrients per calorie are the vegetables, especially the nonstarchy vegetables such as broccoli,

FIGURE 1·6 A WAY TO JUDGE WHICH FOODS ARE MOST NUTRITIOUS

Some foods deliver more nutrients for the same number of calories than others do. These two breakfasts provide about 500 calories each, but they differ greatly in the nutrients they provide per calorie. Note that the sausage in the larger breakfast is lower-calorie turkey sausage, not the high-calorie pork variety. Making small choices like this at each meal can add up to large calorie savings, making room in the diet for more servings of nutritious foods and even some treats.

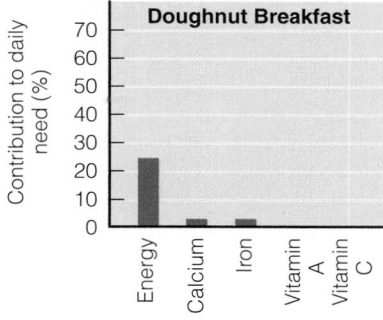

carrots, mushrooms, peppers, and tomatoes. These foods are also rich in phytochemicals thought to protect against diseases. These inexpensive foods take time to prepare, but time invested this way pays off in nutritional health. Twenty minutes spent peeling and slicing vegetables for a salad is a better investment in nutrition than 20 minutes spent fixing a fancy, high-fat, high-sugar dessert. Besides, the dessert ingredients cost more money and strain the calorie budget, too.

In today's households, although both men and women spend some 70 hours a week sleeping and taking care of personal needs, women still do most of the cooking and food shopping. Few households can afford a stay-at-home spouse, so families have very little time for food preparation. Busy chefs should seek out convenience foods that are nutrient dense, such as bags of ready-to-serve salads, refrigerated prepared meats, and frozen vegetables. To round out the meal, fat-free milk is both nutritious and convenient. Other selections, such as most potpies, many frozen pizzas, and "pocket" style sandwiches, are less nutritious overall because they contain too few vegetables and too much fat, making them high in calories and low in nutrient density.

All of this discussion leads to a principle that is central to achieving nutritional health: It is not the individual foods you choose, but the way you combine them into meals and the way you arrange meals to follow one another over days and weeks that determine how well you are nourishing yourself. Nutrition is a science, not an art, but it can be used artfully to create a pleasing, nourishing diet. The remainder of this book is dedicated to helping you make informed choices and combine them artfully to meet all the body's needs.

MEDIA MENU

@ 🏠 ◀ ▶

Nutrition Connections

The Nutrition Connections CD provides access to these additional resources.

Do It!
The Do It! exercise for this chapter, *Analyze Nutrition News,* asks you to assess nutrition articles that appear in newspapers and magazines for credibility.

Interaction
The animation, *Consumer Concerns: Quackery and Sensationalism,* examines the fraudulent and misleading claims used to market some nutrition products.

Chapter Quiz
Take the quiz for this chapter to assess your understanding of chapter concepts.

Nutrition on the Net
Annotated links to reputable nutrition-related websites.

Online Glossary
Test your knowledge of key vocabulary through this comprehensive, interactive glossary.

Internet Connections

A variety of helpful study tools for this chapter are available at our website
http://nutrition.wadsworth.com/ncc10

Practice Test
Take the practice test for this chapter and gauge your knowledge of key concepts; answers are provided.

Student Notebook
A lecture notebook is available to help organize note taking while reading or attending lecture.

ABC Video Clip
"Weight Loss Ads" (1:37) explores the marketing tactics used by promoters of diet products.

Study Tools
Study resources for students include Chapter Summaries, Crossword puzzles, Flashcards, Glossary, Internet Exercises, and Learning Objectives.

SELF CHECK

Answers to these Self Check questions are in Appendix G.

1. Energy-yielding nutrients include all of the following *except:*
   a. vitamins
   b. carbohydrates
   c. fat
   d. protein

2. Organic nutrients include all of the following *except:*
   a. minerals
   b. fat
   c. carbohydrates
   d. protein

3. One of the characteristics of a nutritious diet is that the diet provides no constituent in excess. This principle of diet planning is called:
   a. adequacy
   b. balance
   c. moderation
   d. variety

4. A slice of peach pie supplies 357 calories with 48 units of vitamin A; one large peach provides 42 calories and 53 units of vitamin A. This is an example of:
   a. calorie control
   b. nutrient density
   c. variety
   d. essential nutrients

5. Which of the following adjustments in one's diet would agree with the *Dietary Guidelines for Americans*?
   a. eating baked potatoes rather than french fries
   b. eating fruits rather than cakes and pies
   c. drinking fat-free rather than 2% milk
   d. all of the above

6. Studies of populations in which observation is accompanied by experimental manipulation of some population members are referred to as:
   a. case studies
   b. intervention studies
   c. laboratory studies
   d. epidemiological studies

7. Both heart disease and cancer are due to genetic causes, and diet cannot influence whether they occur.   T   F

8. Both carbohydrates and protein have 4 calories per gram.   T   F

9. If a food is labeled "natural," you can be confident that it is more nutritious than a product not carrying that label.   T   F

10. According to the Healthy People 2010 nutrition-related objectives, it is recommended that the proportion of persons aged 2 years and older who eat at least two daily servings of fruit be increased.   T   F

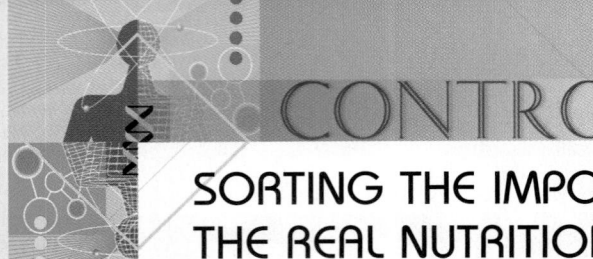

CONTROVERSY 1

SORTING THE IMPOSTORS FROM THE REAL NUTRITION EXPERTS

NUTRITION QUACKERY has plagued this nation from the era of the snake oil salesman hawking wares from the back of a wagon to today's global Internet marketplace. Despite decades of attempts at regulation and enforcement, nutrition consumers today must still learn to distinguish among dishonest scams, well-meaning but misinformed advice, and honest and helpful ideas and products.

Research shows that people rely on television to provide most of their nutrition information, with magazines a close second, and the Internet gaining quickly in popularity.[1*] Innumerable products are sold on late-night television **infomercials,** in magazine and newspaper **advertorials,** and on the Internet as **urban legends** promising astonishing results with minimal effort, and at bargain prices. When scam products are garden tools or stain removers, hoodwinked consumers may lose a few dollars and some pride. But when lapses in judgment lead to consumption of ineffective, untested, or even hazardous "dietary supplements," a person stands to lose much more. The sham products and procedures of quacks not only rob people of the very health they are seeking but also delay

their use of legitimate strategies that could be of real help.

People want to know what nutrition news they can believe and safely use. This heightened interest in nutrition translates into a deluge of dollars spent on services and products peddled by both legitimate and fraudulent businesses. Unfortunately, nutrition and other health **fraud** (or **quackery,** defined in Table C1-1) rings cash registers to the tune of $27 billion annually. Consumers with questions about fraud or suspicions about a product or individual can contact one of the consumer fraud organizations listed in Appendix E.

How can people learn to distinguish valid nutrition information from misinformation? Some quackery may be easy to identify—like the salesman in Figure C1-1—but most fraudulent nutrition claims are not so blatant.

Between the extremes of intentional quackery and scientific data lies an abundance of less easily recognized nutrition misinformation.[2] An instructor at a gym, a physician, a health-store clerk, an author of a book (and seller of juice machines) may all believe that the nutrition regimens they recommend are beneficial. What qualifies these people to give advice? Would following the advice be helpful or harmful? To sift the meaningful nutrition information from the rubble, you must first learn to recognize quackery wherever it presents itself.

Identifying Valid Nutrition Information

Nutrition derives information from scientific research, which has these characteristics:

- Scientists test their ideas by conducting properly designed scientific experiments. They report their methods and procedures in detail so that other scientists can verify the findings through replication.
- Scientists recognize the inadequacy of **anecdotal evidence** or testimonials.
- Scientists who use animals in their research do not apply their findings directly to human beings.
- Scientists may use specific segments of the population in their research. When they do, they are careful not to generalize the findings to all people.
- Scientists report their findings in respected scientific journals. Their work must survive a screening review by their peers before it is accepted for publication.

With each report from scientists, the field of nutrition changes a little—each finding contributes another piece to the whole body of knowledge. Table C1-2 (p. 24) lists some sources of credible nutrition information.

Nutrition on the Net

Hundreds of millions of websites await users of the Internet. Be forewarned: much of the nutrition "information" found on the Internet is pure fiction, aimed at separating consumers from

---

TABLE C1·1	MISINFORMATION TERMS

- **advertorials** lengthy advertisements in newspapers and magazines that read like feature articles but are written for the purpose of touting the virtues of products and may or may not be accurate.
- **anecdotal evidence** information based on interesting and entertaining, but not scientific, personal accounts of events.
- **fraud** or **quackery** the promotion, for financial gain, of devices, treatments, services, plans, or products (including diets and supplements) that alter or claim to alter a human condition without proof of safety or effectiveness. (The word *quackery* comes from the term *quacksalver,* meaning a person who quacks loudly about a miracle product—a lotion or a salve.)
- **infomercials** feature-length television commercials that follow the format of regular programs but are intended to convince viewers to buy products, and not to educate or entertain them. The statements made may or may not be accurate.
- **urban legends** stories, usually false, that may travel rapidly throughout the world via the Internet gaining strength of conviction solely on the basis of repetition.

Who is speaking on nutrition?

*Reference notes are found in Appendix F.

The more of these claims you hear about nutrition information, the less likely it is to be valid.

Too good to be true
The claim presents enticingly simple answers to complex problems. It says what most people want to hear. It sounds magical.

Suspicions about food supply
The person or institution pushing the product or service urges distrust of the current methods of medicine or suspicion of the regular food supply, with "alternatives" for sale (providing profit to the seller) under the guise that people should have freedom of choice.

Testimonials
The evidence presented to support the claim is in the form of praise by people who "felt healed," "were younger," "lost weight," and the like as a result of using the product or treatment.

Fake credentials
The person or institution making the claim is titled "doctor," "university," or the like, but has simply created or bought the title and is not legitimate.

Unpublished studies
Scientific studies are cited, but are not published anywhere and so cannot be critically examined.

A **SCIENTIFIC BREAKTHROUGH**! FEEL **STRONGER**, LOSE WEIGHT. **IMPROVE** YOUR MEMORY ALL WITH THE HELP OF **VITE-O-MITE**! OH SURE, YOU MAY HAVE HEARD THAT **VITE-O-MITE** IS NOT ALL THAT WE SAY IT IS, BUT THAT'S WHAT THE FDA WANTS YOU TO THINK! **OUR DOCTORS** AND SCIENTISTS SAY IT'S THE ULTIMATE VITAMIN SUPPLEMENT. SAY NO! TO THE WEAKENED VITAMINS IN TODAY'S FOODS. **VITE-O-MITE** INCLUDES **POTENT SECRET INGREDIENTS** THAT YOU CANNOT GET WITH ANY OTHER PRODUCT! ORDER RIGHT NOW AND WE'LL SEND YOU ANOTHER FOR FREE!

Logic without proof
The claim seems to be based on sound reasoning but hasn't been scientifically tested and shown to hold up.

Persecution claims
The person or institution pushing the product or service claims to be persecuted by the medical establishment or tries to convince you that physicians "want to keep you ill so that you will continue to pay for office visits."

Authority not cited
The studies cited sound valid, but are not referenced, so that it is impossible to check and see if they were conducted scientifically.

Motive: personal gain
The person or institution making the claim stands to make a profit if it is believed.

Advertisement
The claim is being made by an advertiser who is paid to make claims for the product or procedure. (Look for the word "Advertisement," probably in tiny print somewhere on the page.)

Unreliable publication
The studies cited are published, but in a newsletter, magazine, or journal that publishes misinformation.

their money. Sales of unproven and dangerous products over the Internet have reached huge proportions, partly because the Internet is not regulated. The Internet also offers access to high-quality information, however, so it pays to learn to use it wisely. Table C1-3 (p. 25) provides some clues to these reliable nutrition information websites. One of the most trustworthy sites, used by scientists and others, is the National Library of Medicine's PubMed website, which provides free access to over 10 million abstracts (short descriptions) of research papers published in scientific journals around the world. Many abstracts provide links to websites from which full articles can be purchased. Figure C1-2 (p. 25) provides a brief glimpse of this valuable Internet resource.

Hoaxes and scare stories abound on unsound websites and in e-mails. Be suspicious of the content of an Internet source when:

- The contents were written by someone other than the sender or some authority you know.
- Something like the phrase, "Forward this to everyone you know" appears anywhere in the piece.
- The piece states something like "This is not a hoax"; chances are the opposite is true.

- The information seems shocking or something that you've never heard from legitimate sources.
- The language is overly emphatic or sprinkled with capitalized words or exclamation marks.
- No references are offered or, if present, are of questionable validity.
- The message has been debunked on websites such as www.quackwatch.com or www.urbanlegends.com.[3]

Of course, these hints alone are insufficient for judging material on the Net. The user must also scrutinize "nutrition experts" who post materials, even when they possess legitimate degrees, as described next.

TABLE C1-2 CREDIBLE SOURCES OF NUTRITION INFORMATION

Professional health organizations, government health agencies, volunteer health agencies, and consumer groups provide consumers with reliable health and nutrition information. Credible sources of nutrition information include:

- Professional health organizations, especially the American Dietetic Association's National Center for Nutrition and Dietetics (NCND) www.eatright.org/ncnd.html also the Society for Nutrition Education www.sne.org and the American Medical Association www.ama-assn.org
- Government health agencies such as the Federal Trade Commission (FTC) www.ftc.gov the U.S. Department of Health and Human Services (DHHS) www.os.dhhs.gov, the Food and Drug Administration (FDA) www.fda.gov and the U.S. Department of Agriculture (USDA) www.usda.gov
- Volunteer health agencies such as the American Cancer Society www.cancer.org the American Diabetes Association www.diabetes.org, and the American Heart Association www.americanheart.org
- Reputable consumer groups such as the Better Business Bureau www.bbb.org the Consumers Union www.consumersunion .org the American Council on Science and Health www.acsh.org, and the National Council Against Health Fraud www.ncahf.org

Appendix E provides addresses and websites for these and other organizations.

TABLE C1-3 IS THIS SITE RELIABLE?

To judge whether an Internet site offers reliable nutrition information, answer the following questions.

- **Who is responsible for the site?** Clues can be found in the three-letter "tag" that follows the dot in the site's name. For example, "gov" and "edu" indicate government and university sites, usually reliable sources of information.
- **Do the names and credentials of information providers appear? Is an editorial board identified?** Many legitimate sources provide e-mail addresses or other ways to obtain more information about the site and the information providers behind it.
- **Are links with other reliable information sites provided?** Reputable organizations almost always provide links with other similar sites because they want you to know of other experts in their area of knowledge. Caution is needed when you evaluate a site by its links, however. Anyone, even a quack, can link a webpage to a reputable site without the organization's permission. Doing so may give the quack's site the appearance of legitimacy, just the effect the quack is hoping for.
- **Is the site updated regularly?** Nutrition information changes rapidly, and sites should be updated often.
- **Is the site selling a product or service?** Commercial sites may provide accurate information, but they also may not, and their profit motive increases the risk of bias.
- **Does the site charge a fee to gain access to it?** Many academic and government sites offer the best information, usually for free. Some legitimate sites do charge fees, but before paying up, check the free sites. Chances are good you'll find what you are looking for without paying.
- **Some credible websites include:**

National Council Against Health Fraud www.ncahf.org	Tufts University www.navigator.tufts.edu
Stephen Barrett's Quackwatch www.quackwatch.com	Federal Trade Commission's Operation Cure All www.ftc.gov/opa/2001/06/cureall.htm
Centers for Disease Control and Prevention's Current Health Related Hoaxes and Rumors www.cdc.gov/hoax_rumors.htm	

SOURCE: Adapted from M. Larkin, Health information on-line, *FDA Consumer,* June 1996. Available from http://vm.cfsan.fda.gov/list.html.

Who Are the True Nutrition Experts?

Most people turn to their physicians for dietary advice. Physicians are expected to know all about health-related matters. But only about a quarter of all medical schools in the United States require students to take even one nutrition course, and less than half provide an elective nutrition course.[4] Students attending these classes receive an average of 20 hours of nutrition instruction—an amount most graduates consider inadequate. More than a decade ago, Congress passed a law mandating that "students enrolled in United States medical schools and physicians practicing in the United States [must] have access to adequate training in the field of nutrition and its relationship to human health." So far, this goal has not been fully achieved.

Enlarging on this idea, the **American Dietetic Association (ADA),** the professional association of dietitians, asserts that nutrition education should be part of the curriculum for healthcare professionals: physician's assistants, dental hygienists, physical and occupational therapists, social workers, and all others who provide services directly to clients. This plan would bring access to reliable nutrition information to more people.

Physicians who specialized in clinical nutrition in medical school are highly qualified to advise on nutrition. Membership in the American Society for Clinical Nutrition, whose journal is cited many times throughout this text, is another sign of nutrition knowledge. Still, few physicians have the knowledge, time, or experience to develop diet plans and provide detailed diet instruction for clients, and they often refer their clients to nutrition specialists. Table C1-4 (p. 26) lists the best specialists to choose.

Fortunately, the credential that indicates a qualified nutrition expert is easy to spot—you can confidently call on a **registered dietitian (RD).** Additionally, some states require that **nutritionists,** as well as **dietitians,** obtain a **license to practice.** Meeting state-established criteria certifies that an expert is the genuine article.

Dietitians are easy to find in most communities because they perform a multitude of duties in a variety of settings. They work in foodservice operations, pharmaceutical companies, sports nutrition programs, corporate wellness programs, the food industry, home health agencies, long-term care institutions, private practice, community and public health settings, cooperative extension offices,† research centers, universities and other educational settings, and hospitals, health maintenance organizations (HMOs), and other health-care facilities.

Dietitians in hospitals have many subspecialties. Administrative dietitians manage the foodservice system; clinical dietitians provide client care (see Table C1-5, p. 26); and nutrition support team dietitians coordinate nutrition care with the efforts of other health-care professionals. In the food industry, dietitians conduct research, develop products, and market services. In government, **public health nutritionists** play key roles in delivering nutrition services to people in the community. A public health nutritionist may plan, coordinate, administer, and evaluate food assistance programs; act as a consultant to other agencies; manage finances; and much more.

In some facilities, a **dietetic technician** assists registered dietitians in both administrative and clinical responsibilities. A dietetic technician has been educated and trained to work under the guidance of a registered dietitian; upon passing a national examination, the technician earns the title **dietetic technician, registered (DTR)**.

Detecting Fake Credentials

In contrast to RDs, thousands of people possess fake nutrition degrees and claim to be nutrition counselors, nutritionists, or "dietists." These and other such titles may sound meaningful, but most of these people lack the established credentials of the ADA-sanctioned dietitian. If you look closely, you can see signs that their expertise is fake.

Take, for example, a nutrition expert's educational background. The minimum standards of education for a dietitian specify a bachelor of science (BS) degree in food science and human nutrition (or related fields) from an **accredited** college or university (Table C1-6, p. 26 defines this and related

---

†Cooperative extension agencies are associated with land grant colleges and universities and may be found in the phone book's government listings.

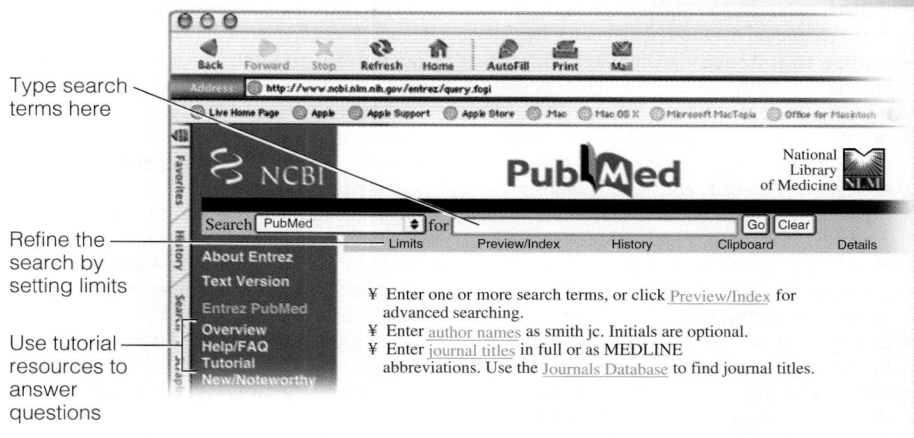

FIGURE C1·2 PUBMED (www.pubmed.org): INTERNET RESOURCE FOR SCIENTIFIC NUTRITION REFERENCES

The U.S. National Library of Medicine's PubMed web site offers tutorials to help teach the beginner to use the search system effectively. Often, simply visiting the site, typing a query in the "Search for" box, and clicking "GO" will yield satisfactory results.

For example, to find research concerning calcium and bone health, typing in "calcium bone" nets almost 3,000 results. To refine the search, try setting limits on dates, types of articles, languages, and other criteria to obtain a more manageable number of abstracts to peruse.

Type search terms here

Refine the search by setting limits

Use tutorial resources to answer questions

---

terms). Such a degree generally requires four to five years of study. In contrast, a fake nutrition expert may display a degree from a six-month correspondence course; such a degree is simply not the same. In some cases, schools posing as legitimate **correspondence schools** offer even less. They are actually **diploma mills**—fraudulent businesses that sell certificates of competency to anyone who pays the fees, from under a thousand dollars for a bachelor's degree to several thousand for a doctorate. Buyers ordering multiple degrees are given discounts. To obtain these "degrees," a candidate need not read any books or pass any examinations.

Lack of proper accreditation is the identifying sign of a fake educational institution. To guard educational quality, an accrediting agency recognized by the U.S. Department of Education certifies that certain schools meet the criteria defining a complete and accurate schooling, but in the case of nutrition, quack accrediting agencies cloud the picture. Fake nutrition degrees are available from schools "accredited" by

more than 30 phony accrediting agencies.‡

To dramatize the ease with which anyone can obtain a fake nutrition degree, one writer enrolled for $82 in a nutrition diploma mill that billed itself as a correspondence school. She made every attempt to fail, intentionally answering all the examination questions incorrectly. Even so, she received a "nutritionist" certificate at the end of the course, together with a letter from the "school" officials explaining that they were sure she must have misread the test.

---

‡To find out whether a correspondence school is accredited, write the Distance Education and Training Council, Accrediting Commission, 1601 Eighteenth Street, NW, Washington, DC 20009; call (202) 234-5100; or visit their website (www.detc.org).

To find out whether a school is properly accredited for a dietetics degree, write the American Dietetic Association, Division of Education and Research, 216 West Jackson Boulevard, Chicago, IL 60606; call (312) 899-4870; or visit their website (www.eatright.org/caade).

The American Council on Education publishes a directory of accredited institutions, professionally accredited programs, and candidates for accreditation in Accredited Institutions of Postsecondary Education Programs (available at many libraries). For additional information, write the American Council on Education, One Dupon Circle NW, Suite 800, Washington, DC 20036; call (202) 939-9382; or visit their website (www.acenet.edu).

TABLE C1·4 TERMS ASSOCIATED WITH NUTRITION ADVICE

- **American Dietetic Association (ADA)** the professional organization of dietitians in the United States. The Canadian equivalent is the Dietitians of Canada (DC), which operates similarly.
- **dietetic technician** a person who has completed a two-year acadmic degree from an accredited college or university and an approved dietetic technician program. A **dietetic technician, registered** (DTR) has also passed a national examination and maintains registration through continuing professional education.
- **dietitian** a person trained in nutrition, food science, and diet planning. See also *registered dietitian*.
- **license to practice** permission under state or federal law, granted on meeting specified criteria, to use a certain title (such as *dietitian*) and to offer certain services. Licensed dietitians may use the initials LD after their names.
- **medical nutrition therapy** nutrition services used in the treatment of injury, illness, or other conditions; includes assessment of nutrition status and dietary intake, and corrective applications of diet, counseling, and other nutrition services.
- **nutritionist** someone who engages in the study of nutrition. Some nutritionists are RDs, whereas others are self-described experts whose training is questionable and who are not qualified to give advice. In states with responsible legislation, the term applies only to people who have masters of science (MS) or doctor of philosophy (PhD) degrees from properly accredited institutions.
- **public health nutritionist** a dietitian or other person with an advanced degree in nutrition who specializes in public health nutrition.
- **registered dietitian (RD)** a dietitian who has graduated from a university or college after completing a program of dietetics. The program must be approved or accredited by the American Dietetic Association (or Dietitians of Canada). The dietitian must serve in an approved internship, coordinated program, or preprofessional practice program to practice the necessary skills; pass the five parts of the association's *registration* examination; and maintain competency through continuing education.[a] Many states also require licensing for practicing dietitians.
- **registration** listing with a professional organization that requires specific course work, experience, and passing of an examination.

[a]The five content areas of the registration examination for dietitians are food and nutrition: clinical and community nutrition; education and research; food and nutrition systems; and management. New emphasis is placed on genetics, complementary care, and reimbursement.

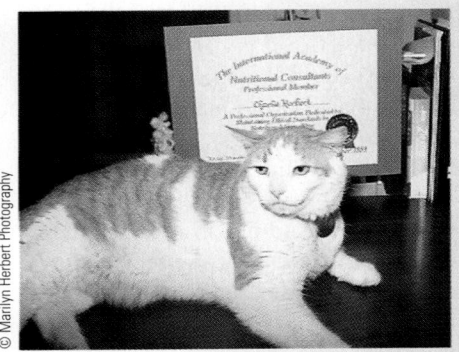

© Marilyn Herbert Photography

Charlie displays his professional credentials.

TABLE C1·5 RESPONSIBILITIES OF A CLINICAL DIETITIAN

The first six items on this list play essential roles in **medical nutrition therapy** as part of a medical treatment plan.

- Assesses clients' nutrition status.
- Determines clients' nutrient requirements.
- Monitors clients' nutrient intakes.
- Develops, implements, and evaluates clients' medical nutrition therapy.
- Counsels clients to cope with unique diet plans.
- Teaches clients and their families about nutrition and diet plans.
- Provides training for other dietitians, nurses, interns, and dietetics students.
- Serves as liaison between clients and the foodservice department.
- Communicates with physicians, nurses, pharmacists, and other health-care professionals about clients' progress, needs, and treatments.
- Participates in professional activities to enhance knowledge and skill.

TABLE C1·6 TERMS DESCRIBING INSTITUTIONS OF HIGHER LEARNING, LEGITIMATE AND FRAUDULENT

- **accredited** approved; in the case of medical centers or universities, certified by an agency recognized by the U.S. Department of Education.
- **correspondence school** a school that offers courses and degrees by mail. Some correspondence schools are accredited; others are *diploma mills*.
- **diploma mill** an organization that awards meaningless degrees without requiring its students to meet educational standards.

In a similar stunt, Ms. Sassafras Herbert was named a "professional member" of a nutrition association. For her efforts, Sassafras received a wallet card and is listed in a fake "Who's Who in Nutrition" that is distributed at health fairs and trade shows nationwide. Sassafras is a poodle. Her master, Victor Herbert, MD, paid $50 to prove that she could be awarded these honors merely by sending in her name. Mr. Charlie Herbert is also a professional member of such an organization; Charlie is a cat.

State laws do not necessarily help consumers distinguish experts from fakes; some states allow anyone to use the title *dietitian* or *nutritionist*. But other states have responded to the need by allowing only RDs or people with certain graduate degrees and state licenses to call themselves dietitians. Licensing provides a way to identify people who have met minimum standards of education and experience.

Summary

In summary, to stay one step ahead of the nutrition quacks, check a provider's qualifications. First look for the degrees and credentials listed after the person's name (such as MD, RD, MS, PhD, or LD). Next find out what you can about the reputations of the institutions that awarded the degrees. Then call your state's health-licensing agency and ask if dietitians are licensed in your state. If they are, find out whether the person giving you dietary advice has a license—and if not, find someone better qualified. Your health is your most precious asset, and protecting it is well worth the time and effort it takes to do so.

Nutrition Tools—Standards and Guidelines

Jacob Lawrence, *Street to M'bari* (1964). Gift of Mr. and Mrs. James T. Dyke. Image © 2005 Board of Trustees, National Gallery of Art, Washington, DC

CONTENTS

Do You Ever . . .

Wonder how scientists decide how much of each nutrient you need to consume each day?

Dismiss government dietary recommendations as too simplistic to help you plan your diet?

Consume the portions offered in restaurants and fast-food places, believing that food professionals must offer recommended amounts?

Wish that your foods could provide a boost to health from substances beyond the nutrients they provide?

KEEP READING.

27

EATING WELL is easy in theory—just choose a selection of foods that supplies appropriate amounts of the essential nutrients, fiber, phytochemicals, and energy without excess intakes of fat, sugar, and salt, and be sure to get enough exercise to balance the foods you eat. In practice, eating well proves harder than it appears. Many people are overweight, or undernourished, or suffer from nutrient excesses or deficiencies that impair their health—that is, they are malnourished. You may not think that this statement applies to you, but you may already have less-than-optimal nutrient intakes and activity without knowing it. Accumulated over years, the effects of your habits can seriously impair the quality of your life.

Putting it positively, you can enjoy the best possible vim, vigor, and vitality throughout your life if you learn now to nourish yourself optimally. To learn how, you first need answers to several basic questions. How much energy and how much of each nutrient do you need? How much physical activity do you need to balance the energy you take in from foods? Which types of foods supply which nutrients? How much of each type of food do you have to eat to get enough? And how can you eat all these foods without gaining weight? This chapter begins by identifying some ideals for nutrient intakes and ends by showing how to achieve them.

Nutrient Recommendations

Nutrient recommendations are sets of yardsticks used as standards for measuring healthy people's energy and nutrient intakes. Nutrition experts use the recommendations to assess intakes and to offer advice on amounts to consume. Individuals may use them to decide how much of a nutrient they need to consume and how much is too much.

The standards in use in the United States and Canada are the **Dietary Reference Intakes (DRI).** A committee of nutrition experts from the United States and Canada develops and publishes the DRI.[*] The DRI committee has set values for the vitamins and minerals, as well as for carbohydrates, fiber, lipids, protein, water, and energy. Values for other food constituents that may play roles in health maintenance are forthcoming.

Another set of nutrient standards is practical for the person striving to make wise choices among packaged foods. These are the **Daily Values,** familiar to anyone who has read a food label. (Read about the Daily Values and other nutrient standards in Table 2-1.) Nutrient standards—the DRI and Daily Values—are used and referred to so often that they are printed on the inside front covers of this book.

> ▶ KEY POINT The Dietary Reference Intakes are nutrient intake standards set for people living in the United States and Canada. The Daily Values are U.S. standards used on food labels.

Goals of the DRI Committee

For each nutrient, the DRI establish a number of values, each serving a different purpose. Most people need to focus on only two kinds of DRI values: those that set nutrient intake goals for individuals (RDA and AI, described next) and those that define an upper limit of safety for nutrient intakes (UL, addressed later). The following sections address the different DRI values, arranged by the goals of the DRI committee.

[*]This is a committee of the Food and Nutrition Board, Institute of Medicine of the National Academy of Sciences, working in association with Health Canada.

A directory of recommendations:
- DRI lists—inside front cover pages A, B, and C.
- Daily Values—inside back cover.

Dietary Reference Intakes (DRI) a set of four lists of values for measuring the nutrient intakes of healthy people in the United States and Canada. The four lists are Estimated Average Requirements (EAR), Recommended Dietary Allowances (RDA), Adequate Intakes (AI), and Tolerable Upper Intake Levels (UL). Descriptions of the DRI values are found in Table 2-1.

Daily Values nutrient standards that are printed on food labels. Based on nutrient and energy recommendations for a general 2,000-calorie diet, they allow consumers to compare the nutrient and energy contents of packaged foods.

TABLE 2·1   NUTRIENT STANDARDS

Standards from the DRI Committee

Dietary Reference Intakes (DRI) a set of four lists of nutrient intake values for healthy people in the United States and Canada. These values are used for planning and assessing diets:

1. **Recommended Dietary Allowances (RDA)** nutrient intake goals for individuals; the average daily nutrient intake level that meets the needs of nearly all (97 to 98 percent) of healthy people in a particular life stage and gender group.[a] Derived from the Estimated Average Requirements (see below).

2. **Adequate Intakes (AI)** nutrient intake goals for individuals; the recommended average daily nutrient intake level based on intakes of healthy people (observed or experimentally derived) in a particular life stage and gender group and assumed to be adequate.[a] Set whenever scientific data are insufficient to allow establishment of an RDA value.

3. **Tolerable Upper Intake Levels (UL)** the highest average daily nutrient intake level that is likely to pose no risk of toxicity to almost all healthy individuals of a particular life stage and gender group. Usual intake above this level may place an individual at risk of illness from nutrient toxicity.

4. **Estimated Average Requirements (EAR)** the average daily nutrient intake estimated to meet the requirement of half of the healthy individuals in a particular life stage and gender group; used in nutrition research and policymaking and is the basis upon which RDA values are set.

5. **Acceptable Macronutrient Distribution Ranges (AMDR)** values for carbohydrate, fat, and protein expressed as percentages of total daily caloric intake; ranges of intakes set for the energy-yielding nutrients that are sufficient to provide adequate total energy and nutrients while reducing the risk of chronic diseases.

Daily Values

Daily Values (DV) nutrient standards used on food labels, in grocery stores, and on some restaurant menus. The DV allow comparisons among foods with regard to their nutrient contents.

[a] For simplicity, this book combines the two sets of nutrient goals for individuals (AI and RDA) and refers to them as the *DRI recommended intakes*. The AI values are not the scientific equivalent of the RDA, however.

Goal #1. Setting Recommended Intake Values—RDA and AI   One of the great advantages of the DRI values lies in their applicability to the diets of individuals.[*1] The committee offers two sets of values specifying intake goals for individuals: **Recommended Dietary Allowances (RDA)** and **Adequate Intakes (AI).**

The RDA are the indisputable bedrock of the DRI recommended intakes, for they are based on solid experimental evidence and reliable observations. The AI values are also as scientifically based as possible, but setting them requires some educated guesswork. The committee establishes an AI value whenever scientific evidence is insufficient to generate an RDA. Although not scientifically equivalent, both the RDA and AI values are intended to be used as nutrient goals in planning nutritious diets for individuals so, for the consumer, there is no practical need to distinguish between them. This book refers to the RDA and AI values collectively as the DRI recommended intakes.

Goal #2. Facilitating Nutrition Research and Policy—EAR   Another set of values established by the DRI committee, the **Estimated Average Requirements (EAR),** establishes nutrient requirements for given life stages and gender groups that researchers and nutrition policymakers use in their work. Public health officials may also use them to assess nutrient intakes of populations and make recommendations. The EAR values form the scientific basis upon which the RDA values are set (a later section explains how).

Goal #3. Establishing Safety Guidelines—UL   Beyond a certain point, it is unwise to consume large amounts of any nutrient, so the DRI committee sets the **Tolerable**

■ The DRI table on the inside front cover distinguishes the RDA and AI values, but both kinds of values are intended as nutrient intake goals for individuals.

[*]Reference notes are found in Appendix F.

Consuming too much of a nutrient endangers health, just as consuming too little does. The DRI recommended intake values fall within a safety range with the UL marking the tolerable upper levels.

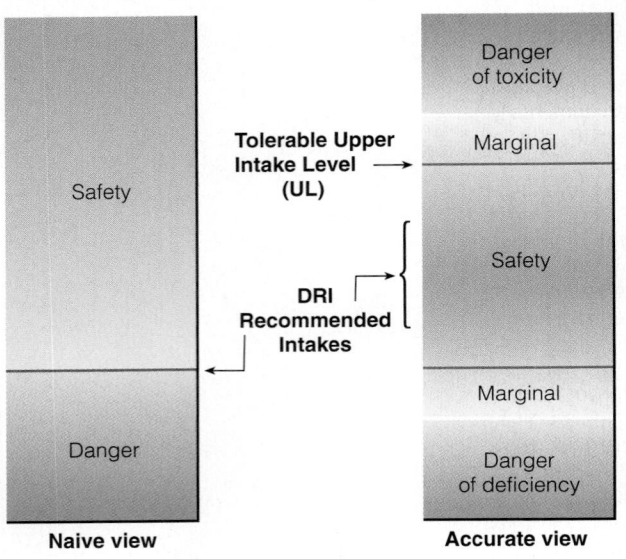

Naive view | Accurate view

■ Tolerable Upper Intake Levels (UL) are listed on page C, inside the front cover.

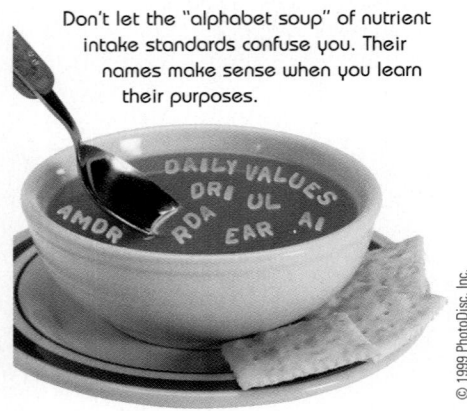

Don't let the "alphabet soup" of nutrient intake standards confuse you. Their names make sense when you learn their purposes.

© 1999 PhotoDisc, Inc.

Upper Intake Levels (UL) to identify potentially hazardous levels of nutrient intake (see Table 2-1, p. 29). The UL are indispensable to consumers who take supplements or consume foods and beverages to which vitamins or minerals have been added—a group that includes almost everyone. Public health officials also rely on UL values to set safe upper limits for nutrients added to our food and water supplies.

Nutrient needs fall within a range, and a danger zone exists both below and above that range. Figure 2-1 illustrates this point. People's tolerances for high doses of nutrients vary, so caution is in order when nutrient intakes approach the UL values.

Some nutrients do not have UL values. The absence of a UL for a nutrient does not imply that it is safe to consume it in any amount, however. It means only that insufficient data exist to establish a value.

Goal #4. Preventing Chronic Diseases The DRI committee also takes into account chronic disease prevention, wherever appropriate. In the last decade, abundant new research has linked nutrients in the diet with the promotion of health and the prevention of chronic diseases, and the DRI committee uses this research in setting intake recommendations. For example, the committee set lifelong intake goals for the mineral calcium at the levels believed to lessen the likelihood of osteoporosis-related fractures in the later years.

The DRI committee also set healthy ranges of intake for carbohydrate, fat, and protein known as **Acceptable Macronutrient Distribution Ranges (AMDR)**. Each of these three energy-yielding nutrients contributes to the day's total calorie intake, and their contributions can be expressed as a percentage of the total. According to the committee, a diet that provides adequate energy nutrients in the following proportions can provide adequate nutrients while reducing the risk of chronic diseases:

● 45 to 65 percent from carbohydrate.
● 20 to 35 percent from fat.
● 10 to 35 percent from protein.

All in all, the DRI values are designed to meet the diverse needs of individuals, the scientific and medical communities, and others. Table 2-1 sums up the names and purposes of the nutrient intake standards just introduced. A later section comes back to the Daily Values, also listed in the table.

▶ **KEY POINT** The DRI provide nutrient intake goals for individuals, provide a set of standards for researchers and makers of public policy, establish tolerable upper limits for nutrients that can be toxic in excess, and take into account evidence from research on disease prevention. The DRI are composed of the RDA, AI, UL, and EAR lists of values, along with the AMDR ranges for energy-yielding nutrients.

Understanding the DRI Intake Recommendations

Nutrient recommendations have been much misunderstood. One young woman posed this question: "Do you mean that some bureaucrat says that I need exactly the same amount of vitamin D as every other young woman in my group? Do they really think that 'one size fits all'?" The DRI committee acknowledges differences between individuals. It has made separate recommendations for specific sets of people: men, women, pregnant women, lactating women, infants, and children, and for specific age ranges. Children aged four to eight years, for example, have their own DRI recommended intakes. Each individual can look up the recommendations for his or her own age and gender group.

Within your own age and gender group, the committee advises adjusting nutrient intakes in special circumstances that may increase or decrease nutrient needs, such as illness, smoking, or vegetarianism. Later chapters provide details about which nutrients may need adjustment.

For almost all healthy people, a diet that consistently provides the RDA or AI amount for a nutrient is very likely to be adequate in that nutrient.[2] On average, you should try to get 100 percent of the DRI recommended intake for every nutrient to ensure an adequate intake over time. The following facts will help put the DRI recommended intakes into perspective:

- The values are based on available scientific research to the greatest extent possible and are updated periodically in light of new knowledge.
- The values are based on the concepts of probability and risk. The DRI recommended intakes are associated with a low probability of deficiency for people of a life stage and gender group, and they pose almost no risk of toxicity for that group.
- The values are recommendations for optimal intakes, not minimum requirements. They include a generous margin of safety and meet the needs of virtually all healthy people in a specific age and gender group.
- The values are set in reference to specific indicators of nutrient adequacy, such as blood nutrient concentrations, normal growth, and reduction of certain chronic diseases or other disorders when appropriate, rather than prevention of deficiency symptoms alone.
- The values reflect daily intakes to be achieved, on average, over time. They assume that intakes will vary from day to day, and they are set high enough to ensure that body nutrient stores will meet nutrient needs during periods of inadequate intakes lasting a day or two for some nutrients and up to a month or two for others.
- The recommendations apply to healthy persons only.

The DRI are designed for health maintenance and disease prevention in healthy people, not for the restoration of health or repletion of nutrients in those with deficiencies. Under the stress of serious illness or malnutrition, a person may require a much higher intake of certain nutrients or may not be able to handle even the DRI amount. Therapeutic diets take into account the increased nutrient needs imposed by certain medical conditions, such as recovery from surgery, burns, fractures, illnesses, malnutrition, or addictions.

▶ **KEY POINT** The DRI represent up-to-date, optimal, and safe nutrient intakes for healthy people in the United States and Canada.

How the Committee Establishes DRI Values—An RDA Example

A theoretical discussion will help to explain how the DRI committee goes about its work of setting the DRI values. Suppose we are the DRI committee members with the task of setting an RDA for nutrient X (an essential nutrient).[*] Ideally, our first step will be to find out how much of that nutrient various healthy individuals need. To do so, we review studies of deficiency states, nutrient stores and their depletion, and the factors influencing them. We then select the most valid data for use in our work. Of the DRI family of nutrient standards, the setting of an RDA value demands the most rigorous science and tolerates the least guesswork.

One experiment we would review or conduct is a **balance study.** In this type of study, scientists measure the body's intake and excretion of a nutrient to find out how much intake is required to balance excretion. For each individual subject, we can determine a **requirement** to achieve balance for nutrient X. With an intake below the requirement, a person will slip into negative balance or experience declining stores that could, over time, lead to deficiency of the nutrient.

balance study a laboratory study in which a person is fed a controlled diet and the intake and excretion of a nutrient are measured. Balance studies are valid only for nutrients like calcium (chemical elements) that do not change while they are in the body.

requirement the amount of a nutrient that will just prevent the development of specific deficiency signs; distinguished from the DRI recommended intake value, which is a generous allowance with a margin of safety.

[*]This discussion describes how an RDA value is set; to set an AI value, the committee would use some educated guesswork as well as scientific research results to determine an approximate amount of the nutrient most likely to support health.

FIGURE 2·2 INDIVIDUALITY OF NUTRIENT REQUIREMENTS

Each square represents a person. A, B, and C are Mr. A, Mr. B, and Mr. C. Each has a different requirement.

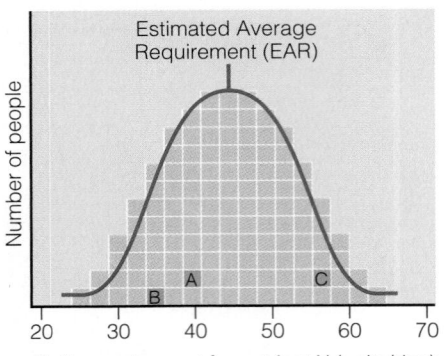

We find that different individuals, even of the same age and gender, have different requirements. Mr. A needs 40 units of the nutrient each day to maintain balance; Mr. B needs 35; Mr. C, 57. If we look at enough individuals, we find that their requirements are distributed as shown in Figure 2-2—with most requirements near the midpoint (here, 45), and only a few at the extremes.

To set the value, we have to decide what intake to recommend for everybody. Should we set it at the mean (45 units in Figure 2-2)? This is the Estimated Average Requirement (EAR) for nutrient X, mentioned earlier as valuable to scientists, but not appropriate as an individual's nutrient goal. The EAR value is probably close to everyone's minimum need, assuming the distribution shown in Figure 2-2. (Actually, the data for most nutrients indicate a distribution that is much less symmetrical.) But if people took us literally and consumed exactly this amount of nutrient X each day, half the population would begin to develop internal deficiencies and possibly even observable symptoms of deficiency diseases. Mr. C (at 57) would be one of those people.

Perhaps we should set the recommendation for nutrient X at or above the extreme, say, at 70 units a day, so that everyone will be covered. (Actually, we didn't study everyone, and some individual we didn't happen to test might have an even higher requirement.) This might be a good idea in theory, but what about a person like Mr. B, who requires only 35 units a day? The recommendation would be twice his requirement, and to follow it, he might spend money needlessly on foods containing nutrient X to the exclusion of foods containing other nutrients he needs.

The decision we finally make is to set the value high enough so that 97 to 98 percent of the population will be covered but not so high as to be excessive (the graph of Figure 2-3 illustrates such a value). In this example, a reasonable choice might be 63 units a day. Moving the DRI further toward the extreme would pick up a few additional people, but it would inflate the recommendation for most people, including Mr. A and Mr. B. The committee makes judgments of this kind when setting the DRI recommended intakes for many nutrients. Relatively few healthy people have requirements that are not covered by the DRI recommended intakes.

▶ KEY POINT The DRI are based on scientific data and are designed to cover the needs of virtually all healthy people in the United States and Canada.

FIGURE 2·3 NUTRIENT RECOMMENDED INTAKE: RDA EXAMPLE

Intake recommendations for most vitamins and minerals are set so that they will meet the requirements of nearly all people (boxes represent people).

Setting Energy Requirements

In contrast to the recommendations for nutrients, the value set for energy, the **Estimated Energy Requirement (EER)**, is not generous; instead it is set at a level predicted to maintain body weight for an individual of a particular age, gender, height, weight, and physical activity level consistent with good health. The energy DRI values reflect a balancing act: enough food energy is critical to support health and life, but too much energy causes unhealthy weight gain. Because even small amounts of excess energy consumed day after day cause weight gain and associated diseases, the DRI committee did not set a Tolerable Upper Intake Level for energy.

People don't eat energy directly. They derive energy from foods containing carbohydrate, fat, and protein, each in proportion to the others. The Acceptable Macronutrient Distribution Ranges, listed earlier, are designed to provide a healthy balance among these nutrients and to minimize a person's risks of chronic diseases. These ranges resurface in later chapters of this book wherever intakes of the energy-yielding nutrients are discussed with regard to chronic disease risks.

▶ **KEY POINT**   Estimated Energy Requirements are energy intake recommendations predicted to maintain body weight and to discourage unhealthy weight gain.

■ The DRI values for energy intakes are found on page C, inside the front cover.

Why Are Daily Values Used on Labels?

Most careful diet planners are already familiar with the Daily Values because they are used on U.S. food labels. After learning about the DRI, you may wonder why yet another set of standards is needed for food labels. One answer is that the DRI values vary from group to group, whereas on a label, one set of values must apply to everyone. The Daily Values reflect the needs of an "average" person—someone eating 2,000 to 2,500 calories a day. Soon, the Daily Values will be updated to reflect current DRI intake recommendations.[3]

The Daily Values are ideal for allowing comparisons among *foods*. This strength is also their limitation, however. Because the Daily Values apply to all people, from children of age four through aging adults, they are much less useful as nutrient intake goals for individuals. Details about how to use the Daily Values appropriately in making comparisons among foods are offered in this chapter's Consumer Corner.

▶ **KEY POINT**   The Daily Values are standards used only on food labels to enable consumers to compare the nutrient values among foods.

Other Nutrient Standards

Many nations and international groups have published sets of standards similar to the DRI. They differ from the DRI in some respects, partly because of different interpretations of the data from which they are derived and partly because people in different parts of the world have somewhat different food intakes and energy expenditures. In addition to setting nutrient standards, experts around the world recommend daily physical activity to help people stay healthy and live long (see the Think Fitness box on page 34).

▶ **KEY POINT**   Many nations and groups issue recommendations for nutrient and energy intakes appropriate for specific groups of people.

■ Appendix E, Nutrition Resources, provides addresses for the World Health Organization (WHO), the Food and Agriculture Organization (FAO), and other agencies that put forth nutrient recommendations for the world's people.

Estimated Energy Requirement (EER) the average dietary energy intake predicted to maintain energy balance in a healthy adult of a certain age, gender, weight, height, and level of physical activity consistent with good health.

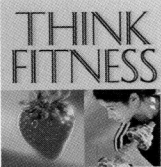
Diet Planning with the USDA Food Guide

Diet planning connects nutrition theory with the food on the table, and a few minutes invested in meal planning can pay off in better nutrition. To help people achieve the goals set forth by the *Dietary Guidelines for Americans 2005* (listed on page 16), the USDA provides a **food group plan**—the USDA Food Guide, updated in 2005.[4] Figure 2-4 displays this plan. By using it wisely and by learning about the energy nutrients and vitamins and minerals in various foods (as you will in coming chapters), you can achieve the goals of a nutritious diet first mentioned in Chapter 1: adequacy, balance, calorie control, moderation, and variety.

Canada issues its own food group plan, *Food Guide to Healthy Eating*, outlined briefly in Table 2-2 (p. 37) and presented in full in Appendix B. A different kind of planning tool, the **exchange system** (see Appendix D), focuses on controlling the carbohydrate, fat, protein, and energy (calories) in the diet.

How Can the USDA Food Guide Help Me to Eat Well?

As a nation, Americans eat too few foods that supply certain key nutrients (see the margin) and too many foods that are rich in calories and fats. For most people, then, meeting the diet ideals of the *Dietary Guidelines* requires choosing *more*:

- Vegetables (especially dark green vegetables, orange vegetables, and legumes).
- Fruits.
- Whole grains.
- Fat-free or low fat milk and milk products.

It also requires choosing *less* of these:

- Refined grains.
- Total fats (especially saturated fat, *trans* fat, and cholesterol).
- Added sugars.

In addition, many people should reduce total calorie intakes. The diet planner can achieve these ideals with the help of the USDA Food Guide.

Achieving Adequacy, Balance, and Variety: The Food Groups and Subgroups
The USDA Food Guide of Figure 2-4 defines the major food groups and their subgroups and dictates equivalent portions of foods from each group. This standardization is needed to ensure that a diet based on the plan will deliver a certain amount of a given nutrient.

▓ Another eating plan, the DASH eating plan of Appendix I at the back of the book, also meets the goals of the *Dietary Guidelines for Americans 2005*.

▓ The nutrients most often lacking in the U.S. diet: fiber, vitamin A, vitamin C, vitamin E, and the minerals calcium, magnesium, and potassium.

food group plan a diet-planning tool that sorts foods into groups based on their nutrient content and then specifies that people should eat certain minimum numbers of servings of foods from each group.

exchange system a diet-planning tool that organizes foods with respect to their nutrient contents and calorie amounts. Foods on any single exchange list can be used interchangeably. See the U.S. Exchange System, Appendix D, for details.

FIGURE 2·4 USDA FOOD GUIDE, 2005

Key:

● Foods generally high in nutrient density (choose most often)

▲ Foods lower in nutrient density (limit selections)

FRUITS

© Polara Studios, Inc.

Consume a variety of fruits and no more than one-third of the recommended intake as fruit juice.

These foods contribute vitamin A, vitamin C, potassium, and fiber.

> $\frac{1}{2}$ c fruit is equivalent to $\frac{1}{2}$ c fresh, frozen, or canned fruit; 1 medium fruit; $\frac{1}{4}$ c dried fruit; $\frac{1}{2}$ c fruit juice.

● Apples, apricots, avocados, bananas, blueberries, cantaloupe, grapefruit, grapes, guava, kiwi, mango, oranges, papaya, peaches, pears, pineapples, plums, raspberries, strawberries, watermelon; dried fruit; unsweetened juices.

Canned or frozen fruit in syrup; juices, punches, ades, and fruit drinks with added sugars; fried plantains.

VEGETABLES

© Polara Studios, Inc.

Choose a variety of vegetables from all five subgroups several times a week.

These foods contribute folate, vitamin A, vitamin C, magnesium, potassium, and fiber.

> $\frac{1}{2}$ c vegetables is equivalent to $\frac{1}{2}$ c cut-up raw or cooked vegetables; $\frac{1}{2}$ c cooked legumes; $\frac{1}{2}$ c vegetable juice; 1 c raw, leafy greens.

● Dark green vegetables: Broccoli and leafy greens such as arugula, beet greens, collard greens, kale, mustard greens, romaine lettuce, spinach, and turnip greens.

● Orange and deep yellow vegetables: Carrots, carrot juice, pumpkin, sweet potatoes, and winter squash.

● Legumes: Black beans, black-eyed peas, garbanzo beans (chickpeas), kidney beans, lentils, pinto beans, soybeans and soy products such as tofu, and split peas.

● Starchy vegetables: Cassava, corn, green peas, hominy, and potatoes.

● Other vegetables: Artichokes, asparagus, bamboo shoots, bean sprouts, beets, bok choy, brussels sprouts, cabbages, cactus, cauliflower, celery, cucumbers, eggplant, green beans, iceburg lettuce, mushrooms, okra, onions, peppers, seaweed, snow peas, tomatoes, vegetable juices, zucchini.

Baked beans, candied sweet potatoes, coleslaw, french fries, potato salad, refried beans, scalloped potatoes, tempura vegetables.

GRAINS

© Polara Studios, Inc.

Make at least half of the grain selections whole grains.

These foods contribute folate, niacin, riboflavin, thiamin, iron, magnesium, and fiber.

> 1 oz grains is equivalent to 1 slice bread: $\frac{1}{2}$ c cooked rice, pasta, or cereal; 1 oz dry pasta or rice; 1 c ready-to-eat cereal.

● Whole grains (barley, brown rice, bulgur, millet, oats, rye, wheat) and whole-grain, low-fat breads, cereals, crackers, and pastas.

● Enriched bagels, breads, cereals, pastas (couscous, macaroni, spaghetti), rice, rolls, tortillas.

▲ Biscuits, cakes, cookies, cornbread, crackers, croissants, doughnuts, french toast, fried rice, granola, muffins, pancakes, pastries, pies, presweetened cereals, taco shells, waffles.

continued

FIGURE 2·4 USDA FOOD GUIDE, 2005—continued

MEAT, POULTRY, FISH, DRIED PEAS AND BEANS, EGGS, AND NUTS

© Polara Studios, Inc.

Make lean or low-fat choices.

Meat, poultry, fish, and eggs contribute protein, niacin, thiamin, vitamin B_6, vitamin B_{12}, iron, magnesium, potassium, and zinc; legumes and nuts are notable for their protein, folate, thiamin, vitamin E, iron, magnesium, potassium, zinc, and fiber.

> 1 oz meat is equivalent to 1 oz cooked lean meat, poultry, or fish; 1 egg; ¼ c cooked legumes or tofu; 1 tbs peanut butter; ½ oz nuts or seeds.

● Poultry (no skin), fish, shellfish, legumes, eggs, lean meat (fat-trimmed beef, game, ham, lamb, pork); low-fat tofu, tempeh, peanut butter, nuts or seeds.

▲ Bacon; baked beans; fried meat, fish, poultry, eggs, or tofu; refried beans; ground beef; hot dogs; luncheon meats; marbled steaks; poultry with skin; sausages; spare ribs.

MILK, YOGURT, AND CHEESE

© Polara Studios, Inc.

Make fat-free or low-fat choices.

These foods contribute protein, riboflavin, vitamin B_{12}, calcium, magnesium, potassium, and, when fortified, vitamin A and vitamin D.

> 1 c milk is equivalent to 1 c fat-free milk or yogurt; 1½ oz fat-free natural cheese; 2 oz fat-free processed cheese.

● Fat-free milk and fat-free milk products such as buttermilk, cheeses, cottage cheese, yogurt; fat-free fortified soy milk.

▲ 1% low-fat milk, 2% reduced-fat milk, and whole milk; low-fat, reduced-fat, and whole-milk products such as cheeses, cottage cheese, and yogurt; milk products with added sugars such as chocolate milk, custard, ice cream, ice milk, milk shakes, pudding, sherbet; fortified soy milk.

OILS

Matthew Farruggio

Select the recommended amounts of oils from among these sources.

These foods contribute vitamin E and essential fatty acids (see Chapter 5), along with abundant calories.

> 1 tsp oil is equivalent to 1 tbs low-fat mayonnaise; 2 tbs light salad dressing; 1 tsp vegetable oil; 1 tsp soft margarine.

● Liquid vegetable oils such as canola, corn, flaxseed, nut, olive, peanut, safflower, sesame, soybean, and sunflower oils; mayonnaise, oil-based salad dressing, soft *trans*-free margarine.

● Unsaturated oils that occur naturally in foods such as avocados, fatty fish, nuts, olives, and shellfish.

SOLID FATS AND ADDED SUGARS

Matthew Farruggio

Limit intakes of food and beverages with solid fats and added sugars.

Solid fats deliver saturated fat and *trans* fat, and intake should be kept low. Solid fats and added sugars contribute abundant calories but few nutrients, and intakes should not exceed the discretionary calorie allowance—calories to meet energy needs after all nutrient needs have been met with nutrient-dense foods. Alcohol also contributes abundant calories but few nutrients, and its calories are counted among discretionary calories. See Table 2-3 for some discretionary calorie allowances; Table I-1 of Appendix I includes others.

Solid fats that occur in foods naturally such as milk fat and meat fat (see ● in previous lists).

Solid fats that are often added to foods such as butter, cream cheese, hard margarine, lard, sour cream, and shortening.

Added sugars such as brown sugar, candy, honey, jelly, molasses, soft drinks, sugar, and syrup.

Alcoholic beverages include beer, wine, and liquor.

This doesn't mean that you can never choose more than ½ cup of pasta, for example. Instead, it means that if you choose 1½ cups, you will have received the nutrients (and the calories) of three ounces of grains.

The USDA Food Guide also teaches people to recognize key nutrients provided by foods within the groups. These are listed in the figure. Note that the figure also sorts foods within each group by nutrient density (the key to Figure 2-4 explains).

The foods in each group are well-known contributors of the key nutrients listed in the figure, but you can count on these foods to supply many other nutrients as well. If you design your diet around this plan, it is assumed that you will obtain adequate amounts not only of the nutrients of greatest concern but also of the other two dozen or so essential nutrients, as well as beneficial phytochemicals, because all of these are distributed among the food groups.

Applying the USDA Food Guide requires looking at foods in a new way. Vegetables, for example, are sorted into subgroups according to their nutrient contents. All vegetables provide valuable fiber and the mineral potassium, but the vegetables of each subgroup reliably provide a target nutrient as well, such as vitamin A from the "orange and deep yellow vegetables," the vitamin folate from the "dark green vegetables," abundant carbohydrate energy from the "starchy vegetables," iron and protein from "legumes," and many of the same nutrients but few calories from "other vegetables." The diet planner should strive to vary vegetable choices among and within the vegetable subgroups to achieve adequacy.

Spices, herbs, coffee, tea, and diet soft drinks, excluded from the USDA Food Guide, provide few, if any, nutrients, but can add flavor and pleasure to meals. They can also provide some potentially beneficial phytochemicals, such as those in tea or certain spices—see this chapter's Controversy section.

Controlling Calories: The Discretionary Calorie Allowance To help people control calories and prevent unhealthy weight gain, the USDA developed the concept of the **discretionary calorie allowance** (illustrated in Figure 2-5). As the figure demonstrates, a person needing 2,000 calories a day to maintain weight may need only 1,700 calories or so of nutrient-dense foods to supply the day's required nutrients. The difference between the calories needed to maintain weight and those to supply nutrients from nutrient-dense foods (in this case, 267 calories) is the person's discretionary calorie allowance.

A person with a discretionary calorie allowance to spend may choose to consume the following within the limits of the allowance:

1. Extra servings of the same nutrient-dense foods that make up the base of the diet, for example, an extra piece of skinless chicken or a second ear of corn.
2. Fats from two sources (within the limits recommended for health):

TABLE 2·2 CANADA'S *FOOD GUIDE TO HEALTHY EATING*[a]	
Food Group	Servings/Day
Grain products	5–12
Vegetables and fruits	5–10
Milk products	
Children aged 4–9 years	2–3
Youth aged 10–16 years	3–4
Adults	2–4
Pregnant and breast-feeding women	3–4
Meat and alternatives	2–3

[a]Canada's *Food Guide* and other Canadian guidelines are presented in full in the Canadiana appendix (Appendix B).

■ Chapter 1 defined phytochemicals as nonnutrient compounds that exert biological effects on the body.

■ Legumes were also defined in Chapter 1 as dried beans, peas, and lentils.

FIGURE 2·5 DISCRETIONARY CALORIE ALLOWANCE

discretionary calorie allowance the balance of calories remaining in a person's energy allowance after accounting for the number of calories needed to meet recommended nutrient intakes through consumption of nutrient-dense foods.

■ The USDA offers suggestions on how people might allocate their discretionary calorie allowances between added sugars and fats (see Table I-2 of Appendix I).

- Foods higher in naturally occurring fats such as regular hamburger instead of lean hamburger or whole milk or reduced-fat milk instead of fat-free milk.
- Added fats, including solid fats such as butter, hard margarine, lard, and shortening, and oils consumed in excess of need.

3. Added sugars, such as jams, colas, and honey.
4. Alcohol, within limits (some people should *not* make this choice, read Controversy 3). Alternatively, a person wishing to lose weight might choose to:
5. Omit the discretionary calories from the diet. This is a safe strategy because discretionary calories are not essential for delivering needed nutrients to the diet.

Discretionary calories are distinguished from the calories of the nutrient-dense foods of which they may be a part. A fried chicken leg, for example, provides discretionary calories from two sources: the naturally occurring fat of the chicken skin and the added fat absorbed during frying. The calories of the skinless chicken underneath are not discretionary (unless consumed in excess of need)—they are necessary to provide the nutrients of chicken.

Physical activity affects an individual's discretionary calorie allowance. Physically active people burn more calories in a day than do sedentary people, and so can afford to consume more discretionary calories each day. People who need fewer calories to maintain their weight have fewer discretionary calories to spend.

■ Chapter 1 explained the concept of nutrient density.

Achieving Moderation: Nutrient Density   To control calories and prevent overweight or obesity, the USDA Food Guide instructs diet planners to choose the most nutrient-dense foods from each group. Unprocessed or lightly processed foods are generally best because some processes strip foods of beneficial nutrients and fiber, while others add many calories in the form of sugar or fat. Figure 2-4 (pp. 35–36) identifies a few of the most nutrient-dense food selections in each food group and some foods of lower nutrient density to give you an idea of which are which. Oil is a notable exception. Oil is pure fat and therefore rich in calories, but a small amount of oil from sources such as avocado, olives, nuts, fish, or vegetable oil provides vitamin E and other important nutrients that other foods lack.

In using the USDA Food Guide, it may help to think of the leanest meats as "meats" and to view fattier cuts as "meats with added fat." Likewise, fat-free milk is "milk," and whole milk and reduced-fat milk are "milk with added fat." Pudding made with whole milk provides discretionary calories from the naturally occurring fat of whole milk and from sugar added for sweetness. Even fruits, vegetables, and grains can carry discretionary calories into the diet. Examples include the sugary syrup of canned peaches, the added fat of buttered corn, and the shortening added to flour to make muffins. Now it becomes clear why the USDA encourages consumers to build a diet of mostly nutrient-dense foods: less nutrient-dense choices deliver many discretionary calories without increasing intakes of needed nutrients.

■ Chapter 9 will help you determine your energy needs. For a quick approximation, look up the DRI Estimated Energy Requirement on the inside front cover for your age and gender group.

How Much Food Do I Need Each Day?   The USDA Food Guide specifies the amounts from each food group needed to create a healthful diet at a number of calorie levels. Look at the top line of Table 2-3 and find yourself among the people described there (for other calorie levels, see Table I-1 of Appendix I). Then look at the column of numbers below for the approximate amounts to take from each food group to meet your calorie needs. Table 2-3 also specifies a discretionary calorie allowance for each calorie level. If you are physically active, you can afford to eat more calories and the more active you are, the higher the discretionary calorie allowance you "earn." For vegetables, intakes should be divided among all the vegetable subgroups over a week's time, as shown in Table 2-4—it isn't necessary to eat vegetables from every subgroup every day.

With judicious selections, the diet can supply all the necessary nutrients and provide some luxury items as well. A sample diet plan demonstrates how the theory of the USDA Food Guide translates to food on the plate. The USDA Food Guide ensures that a certain amount from each of the food groups and subgroups is represented in the

TABLE 2·3 HOW MUCH FOOD FROM EACH GROUP DAILY?

	Sedentary Women: 51+ yr	Sedentary Women: 31–50 yr	Sedentary Women: 19–30 yr / Active Women: 51+ yr / Sedentary Men: 51+ yr	Active Women: 31–50 yr / Sedentary Men: 31–50 yr	Active Women: 19–30 yr / Active Men: 51+ yr / Sedentary Men: 19–30 yr	Active Men: 31–50 yr	Active Men: 19–30 yr
Calories[a]	1,600	1,800	2,000	2,200	2,400	2,600	2,800
Fruits	1½ c	1½ c	2 c	2 c	2 c	2 c	2½ c
Vegetables	2 c	2½ c	2½ c	3 c	3 c	3½ c	3½ c
Grains	5 oz	6 oz	6 oz	7 oz	8 oz	9 oz	10 oz
Meats and legumes	5 oz	5 oz	5½ oz	6 oz	6½ oz	6½ oz	7 oz
Milk	3 c	3 c	3 c	3 c	3 c	3 c	3 c
Oils[b]	4½ tsp	5 tsp	5½ tsp	6 tsp	6 tsp	7 tsp	7 tsp
Discretionary calorie allowance	132 cal	195 cal	267 cal	290 cal	362 cal	410 cal	426 cal

NOTE: In addition to gender, age, and activity levels, energy needs vary with height and weight (see Chapter 9 and Appendix H).
[a] Assumes high nutrient density choices—lean, low fat, and fat-free with no added sugars.
[b] Approximate measures; the gram values are 22, 24, 27, 29, 31, 34, and 36, respectively.

diet. The diet planner begins by assigning each of the food groups to meals and snacks, as shown in Table 2-5 (p. 40). Then the plan can be filled out with real foods to create a menu. For example, the breakfast calls for 1 ounce grains, 1 cup milk, and ½ cup fruit. Here's one possibility for this meal:

1 cup ready-to-eat cereal = 1 ounce grains.
1 cup of fat-free milk = 1 cup milk.
1 medium banana = ½ cup fruit.

Then the planner moves on to complete the menu for lunch, supper, and snacks, as shown in Figure 2-6 (p. 40). This day's choices are explored further as "Monday's Meals" in the Food Feature at the end of the chapter.

▶ KEY POINT The USDA Food Guide specifies the amounts of foods from each group people need to consume to meet their nutrient requirements without exceeding their calorie allowances.

TABLE 2·4 WEEKLY AMOUNTS FROM VEGETABLE SUBGROUPS

Table 2-3 specifies the recommended amounts of total vegetables per *day*. This table shows those amounts dispersed among five vegetable subgroups per *week*.

Vegetable Subgroups	1,600 cal	1,800 cal	2,000 cal	2,200 cal	2,400 cal	2,600 cal	2,800 cal
Dark green	2 c	3 c	3 c	3 c	3 c	3 c	3 c
Orange and deep yellow	1½ c	2 c	2 c	2 c	2 c	2½ c	2½ c
Legumes	2½ c	3 c	3 c	3 c	3 c	3½ c	3½ c
Starchy	2½ c	3 c	3 c	6 c	6 c	7 c	7 c
Other	5½ c	6½ c	6½ c	7 c	7 c	8½ c	8½ c

TABLE 2·5 **SAMPLE DIET PLAN**

This diet plan is one of many possibilities for a day's meals. It follows the amounts suggested for a 2,000-calorie diet (with an extra ½ cup of vegetables).

Food Group	Recommended Amounts	Breakfast	Lunch	Snack	Dinner	Snack
Fruits	2 c	½ c		½ c	1 c	
Vegetables	2½ c		1 c		2 c	
Grains	6 oz	1 oz	2 oz	½ oz	2 oz	½ oz
Meat and legumes	5½ oz		2 oz		3½ oz	
Milk	3 c	1 c		1 c		1 c
Oils	5½ tsp		1½ tsp		4 tsp	
Discretionary calorie allowance	267 cal					

FIGURE 2·6 **A SAMPLE MENU**

This sample menu provides about 1,850 calories of the 2,000-calorie plan. About 150 discretionary calories remain available to spend on more nutrient-dense foods or luxuries such as added sugars and fats.

AMOUNTS	✳	SAMPLE MENU	✳	ENERGY (cal)
Breakfast				
1 oz whole grains		1 c whole-grain cereal		108
1 c milk		1 c fat-free milk		100
1/2 c fruit		1 medium banana (sliced)		105
Lunch				
2 oz meats, 2 oz whole grains		1 turkey sandwich on whole-wheat roll		272
11/2 tsp oils		11/2 tbs low-fat mayonnaise		71
1 c vegetables		1 c vegetable juice		50
Snack				
1/2 oz whole grains		4 whole-wheat reduced-fat crackers		86
1 c milk		11/2 oz low-fat cheddar cheese		74
1/2 c fruit		1 medium apple		72
Dinner				
1/2 c vegetables		1 c raw spinach leaves		8
1/4 c vegetables		1/4 c shredded carrots		11
1 oz meats		1/4 c garbanzo beans		71
2 tsp oils		2 tbs oil-based salad dressing and olives		76
3/4 c vegetables, 21/2 oz meat, 2 oz enriched grains		Spaghetti with meat and tomato sauce		425
1/2 c vegetables		1/2 c green beans		22
2 tsp oils		2 tsp soft margarine		67
1 c fruit		1 c strawberries		49
Snack				
1/2 oz whole grains		3 graham crackers		90
1 c milk		1 c fat-free milk		100

Note: This plan meets the recommendations to provide 45 to 65 percent of calories from carbohydrate, 20 to 35 percent from fat, and 10 to 35 percent from protein.

Conveying USDA Messages to Consumers

For consumers, the USDA makes applying its Food Guide easier through a graphic image that depicts the highlights of the Food Guide. As this book goes to press, the well-known Food Guide Pyramid (shown in Figure 2-7) is undergoing revision, but it is still useful for conveying the major messages of the USDA Food Guide in a fashion that consumers can easily assimilate. The pyramid shape dictates that nutritious foods, such as whole and enriched grains, vegetables, fruits, lean meats, and milk and milk products, should contribute the bulk of the day's calories. The Pyramid also demonstrates that fats and sugars are not typically nutrient-dense foods, and so their intakes should be limited. It also conveys the idea that even nutritious foods should be eaten in limited amounts to avoid overconsuming calories and prevent unneeded weight gain and overweight. By developing consumer messages such as these, the USDA aims to educate U.S. consumers about how to eat well and encourage the adoption of healthy habits.

Armed with this information, consumers can make appropriate food choices, obtain needed nutrients, and stay within calorie limits. If all consumers in the United States would begin to apply the USDA Food Guide today, their rewards in terms of greater quality of life, less heart disease, less cancer, and better overall health would be well worth the effort.

▶ **KEY POINT** The messages of the USDA Food Guide are conveyed to consumers to improve the health of the nation.

Flexibility of the USDA Food Guide

Although it may appear rigid, the USDA Food Guide can actually be very flexible once its intent is understood. For example, the user can substitute fat-free cheese for fat-free milk because both supply the key nutrients for the milk, yogurt, and cheese group. Legumes provide many of the nutrients of the meat group and also constitute a vegetable subgroup, so a meal of legumes can be counted among either one. Consumers can adapt the plan to mixed dishes such as casseroles and to national and cultural foods as well, as Figure 2-8 (pp. 42–43) demonstrates.

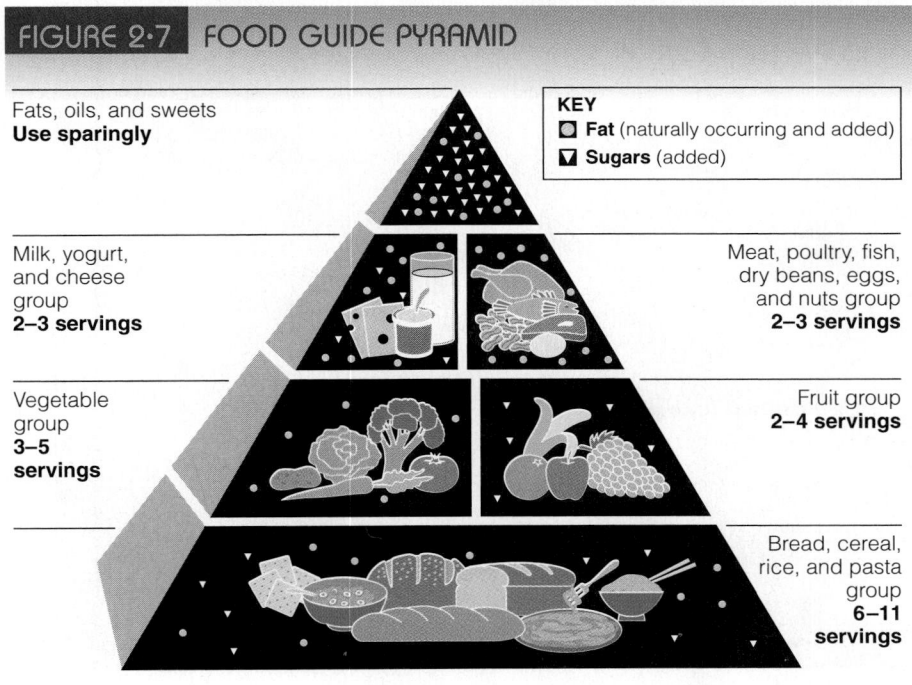

FIGURE 2·7 FOOD GUIDE PYRAMID

Fats, oils, and sweets
Use sparingly

KEY
◻ **Fat** (naturally occurring and added)
◥ **Sugars** (added)

Milk, yogurt, and cheese group
2–3 servings

Meat, poultry, fish, dry beans, eggs, and nuts group
2–3 servings

Vegetable group
3–5 servings

Fruit group
2–4 servings

Bread, cereal, rice, and pasta group
6–11 servings

FIGURE 2·8 ETHNIC AND REGIONAL FOODS IN THE FOOD GROUPS

Foods from every cuisine can fit into the USDA Food Guide. Many other countries have developed their own food guides to healthy diets, and an American research team developed the pyramid for Mediterranean diets shown below.

Key: Nutrient Density

● Foods generally high in nutrient density (choose most often)

▲ Foods lower in nutrient density (limit selections)

China

ASIAN[a]

Grains

● Barley; glass (mung bean) noodles; millet; rice dumplings; rice or wheat noodles; rice rolls (sushi); steamed buns.
▲ Fried rice; fried noodles.

Vegetables

● Baby corn; bamboo shoots; bean sprouts; bok choy; cabbages; dried fungus; lentils (northern Asia); lotus root; miso; scallions; seaweed; snow peas; soybeans; tempeh; water chestnuts; wild yam.
▲ Fried vegetables.

Fruits

● Oranges, pears, plums, and other fresh fruit.

Milk, Yogurt, and Cheese

● Soy milk.

Meat, Poultry, Fish, Dried Peas and Beans, Eggs, and Nuts

● Broiled or stir-fried beef, fish, pork, and seafood; egg whites; egg yolks; peanuts, pine nuts, cashews, other nuts; tofu.
▲ Deep-fried meats and seafood; egg foo yung.

Oils / Solid fats

● Vegetable oils.[b]
▲ Lard for deep-frying.

Seasonings and Sauces[c]

● Bean sauce; fish sauce;[d] garlic; ginger root; hoisin sauce;[d] oyster sauce;[d] plum sauce;[d] rice wine; scallions; soy sauce.[d]
▲ Sesame oil; other oils; oily gravies.

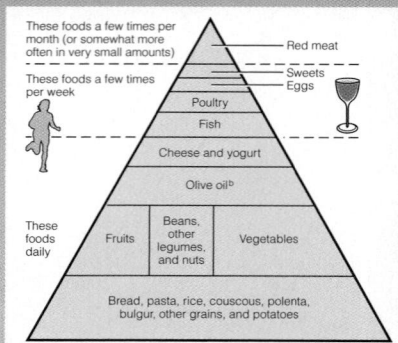

These foods a few times per month (or somewhat more often in very small amounts) — Red meat

These foods a few times per week — Sweets / Eggs

Poultry

Fish

Cheese and yogurt

Olive oil[b]

These foods daily — Fruits | Beans, other legumes, and nuts | Vegetables

Bread, pasta, rice, couscous, polenta, bulgur, other grains, and potatoes

MEDITERRANEAN

Grains

● Bulgur, couscous, focaccia, Italian bread, pasta, pita pocket bread, polenta, rice.
▲ Baklava (honey-soaked nut pastry); cakes.

Vegetables

● Cucumbers; eggplant; grape leaves; lentils and beans; onions; peppers; tomatoes.
▲ Olives.

Fruits

● Dates; figs; grapes; lemons; melons; raisins.

Milk, Yogurt, and Cheese

● Fat free or low-fat yogurt.
▲ Feta, goat, mozzerella, parmasan, provolone, and ricotta cheeses; yogurt.

Meat, Poultry, Fish, Dried Peas and Beans, Eggs, and Nuts

● Beef; eggs; fish; and seafood; lamb; lentils and beans; poultry; almonds; walnuts.
▲ Ground lamb; ground beef; gyros (spicy roasted meat and yogurt sauce, usually rolled in flat bread); sausages.

Oils

● Olive oil.[b]

Seasonings and Sauces

● Garlic; herbs; lemons; egg and lemon sauce.
▲ Olive oil.

[a]Traditional cuisines of China and of West African influence exclude fluid milk as a beverage for adults and use few or no milk products in cooking. Calcium and certain other nutrients of milk are supplied by other foods, such as small fish eaten with the bones or large servings of leafy green vegetables.

[b]Consumed in amounts recommended for caloric intakes.

[c]Many Chinese sauces are fat-free.

[d]May be high in sodium.

FIGURE 2·8   ETHNIC AND REGIONAL FOODS IN THE FOOD GROUPS—continued

Mexico

In Mexico, Great Britain, and most European countries, a circle depicts the food guide principles.

MEXICAN

Grains

● Cereal; corn or flour tortillas; graham crackers; macaroni and other pasta; masa (corn flour); posole (hominy); rice.
△ Churros (doughnuts); fried tortilla shells; pastries; tortilla chips.

Vegetables

● Cabbage; cactus; cassava; corn; iceburg lettuce; legumes; malanga; onions; potatoes; scallions; squash; tomatoes; yucca.
△ Olives.

Fruits

● Bananas; guava; mango; oranges; papaya; pineapple.
△ Avocados.

Milk, Yogurt, and Cheese

● Evaporated low fat milk; powdered fat-free milk.
△ Cheddar or jack cheese; flan (caramel custard); cocoa drink; leche quemada (burnt-milk candy); queso asadero and other Mexican cheeses.

Meat, Poultry, Fish, Dried Peas and Beans, Eggs, and Nuts

● Eggs; fish; lean beef, poultry, lamb, and pork; many bean varieties.
△ Bacon; fried fish; pork, or poultry; nuts; chorizo (sausages); refried beans.

Oils / Solid Fats

● Vegetable oil.[b]
△ Butter, cream cheese; hard margarine; lard; sour cream.

Seasonings and Sauces[d]

● Herbs; hot peppers; garlic; mole (seasoned chili and chocolate sauce); pico de gallo (finely chopped tomatoes, peppers, and onions with seasonings); salsas; spices.
△ Guacamole.

U.S. DEEP SOUTH (WEST AFRICAN INFLUENCE)[a]

Grains

● Grits; macaroni; rice
△ Biscuits; cornbread; hoe cakes; pastries.

Vegetables

● Beans; black-eyed peas; collards (other leafy greens); corn; hominy; okra; onions; pole beans; potatoes; snap beans; summer squash; sweet potatoes; tomatoes.
△ Fried green tomatoes; fried okra; fried potatoes.

Fruits

● Apples; bananas; berries; melons; peaches; pears.
△ Cakes; fried pies; fruit pastries.

Milk, Yogurt, and Cheese

● Fat-free buttermilk; low-fat cheeses; fat-free millk.
△ Full-fat American cheese; cheddar cheese.

Meat, Poultry, Fish, Dried Peas and Beans, Eggs and Nuts

● Braised or roasted meats (beef, lean ham, lean pork, poultry); beans and peas; boiled peanuts; grilled or smoked poultry and fish; peanut butter.
△ Bacon; chitterlings; fat back; fried chicken, fish, or pork; ham hocks; pork rinds; salted pork; sausages; spareribs.

Oils / Solid Fats

● Vegetable oil.[b]
△ Butter; hard margarine; lard; meat fats, gravies, shortening.

---

[d]Many Mexican sauces are fat-free.

■ Vegetarians will find information about meeting their nutrient needs in the chapters to come.

Portion Controller

An ice cream scoop is useful for portioning foods, including mashed potatoes, pasta, vegetables, rice, and cereal. The typical ice cream scoop filled to level holds ¼ cup. One-half cup (two scoops) is the USDA equivalent amount for many food items. Test the size of your scoop by filling it with water and pouring it into a measuring cup.

To estimate the sizes of food portions, remember these common objects:

- 3 ounces of meat = the size of the palm of a woman's hand or a deck of cards.
- 1 medium piece of fruit or potato = the size of a regular (60-watt) lightbulb.
- 1½ ounces cheese = the size of a nine-volt battery.
- 1 ounce lunch meat or cheese = 1 slice.
- 1 pat (1 teaspoon) butter or margarine = a slice from a quarter pound stick of butter about as thick as 280 pages of this book (pressed together).

Because the USDA Food Guide encourages consumption of fruits, vegetables, and whole grains and provides alternates to meats, milk, and other animal products, it can assist vegetarians in their food choices. The food group that includes the meats also includes legumes, nuts, seeds, and products made from soybeans. In the food group that includes milk, soy drinks—beverages made from soybeans—can fill the same nutrient needs, provided that they are fortified with calcium, riboflavin, vitamin A, vitamin D, and vitamin B_{12}. Thus, people who choose to eat no meats or products taken from animals can still use the USDA Food Guide to make their diets adequate. For any careful diet planner, then, the USDA Food Guide can provide a general road map for planning a healthful diet.

▶ KEY POINT   The USDA Food Guide can be used with flexibility by people with different eating styles.

Portion Control

To control calories the diet planner must learn to control food portions (USDA serving equivalents were listed in Figure 2-4). Restaurants often deliver colossal helpings to ensure repeat business; a server on a cafeteria line may be instructed to deliver "about a spoonful"; fast-food burgers range from a one-ounce child-sized burger to a three-quarter-pound triple deluxe. The trend in the United States has been toward consuming larger food portions, especially of foods rich in fat and sugar (see Figure 2-9). At the same time, body weights have been creeping upward, suggesting an increasing need to control portion sizes. In contrast to the random-sized helpings found elsewhere, the quantities recommended in the USDA Food Guide are specific, precise, and reliable for delivering certain amounts of key nutrients in foods. The margin offers some tips for estimating portion sizes.

FIGURE 2·9   LIVING LARGE: U.S. TREND TOWARD COLOSSAL CUISINE

Chapter 9 and Controversy 11 discuss the consequences of increasing portion sizes in terms of body fatness.

Food	Typical 1970s	Today's colossal
Cola	10 oz bottle, 120 cal	40–60 oz fountain, 580 cal
French fries	about 30, 475 cal	about 50, 790 cal
Hamburger	3–4 oz meat, 330 cal	6–12 oz meat, 1,000 cal
Bagel	2–3 oz, 230 cal	5–7 oz, 550 cal
Steak	8–12 oz, 690 cal	16–22 oz, 1,260 cal
Pasta	1 c, 200 cal	2–3 c, 600 cal
Baked potato	5–7 oz, 180 cal	1 lb, 420 cal
Candy bar	1½ oz, 220 cal	3–4 oz, 580 cal
Popcorn	1½ c, 80 cal	8–16 c tub, 880 cal

NOTE: Calories are rounded values for the largest portions in a given range.

1970s   Today

1970s   Today

1970s   Today

Among volumetric measures, 1 "cup" refers to an 8-ounce measuring cup (not a teacup or drinking glass), filled to level (not heaped up, or shaken, or pressed down). Tablespoons and teaspoons refer to measuring spoons (not flatware), filled to level (not rounded or heaping). Ounces signify weight, not volume. Two ounces of meat, for example, refers to ⅛ pound of cooked meat. One ounce (weight) of crispy rice cereal measures a full cup (volume), but take care: 1 ounce of granola cereal measures only ¼ cup. Also, some foods are specified as "medium," as in "one medium apple," but the word *medium* means different things to different people. When college students are asked to bring medium-sized foods to class, they bring bagels weighing anywhere from 2 to 5 ounces, muffins from about 2 to 8 ounces, baked potatoes from 4 to 9 ounces, and so forth. The USDA Food Guide specifies one grain portion as one ounce. This means that if you consume a 5-ounce bagel, you have consumed five of your grain ounces, not one. The Table of Food Composition, Appendix A, can help in determining serving sizes because it lists both weights and volumes of a wide variety of foods.

▶ **KEY POINT** People wishing to avoid overconsuming calories must pay attention to the sizes of their food servings.

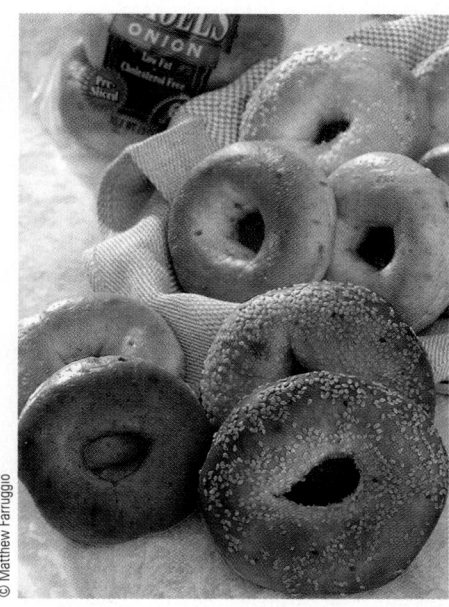

A serving of grains is 1 ounce, yet most bagels today weigh 4 ounces or more—meaning that a single bagel can easily supply more than half of the total grains that many people need in a day.

A Note about Exchange Systems

Exchange systems can be useful to careful diet planners, especially those wishing to control calories (weight watchers), those who must control carbohydrate intakes (people with diabetes), and those who should control their intakes of fat and saturated fat (almost everyone). An exchange system, as presented in Appendix D (Appendix B for Canada), lists the estimated carbohydrate, fat, saturated fat, and protein contents of food portions, as well as their calorie values. The values in the exchange lists differ from the exacting values given for individual foods in Appendix A because exchange lists estimate values for whole groups of foods. With these estimates, exchange system users can make an educated approximation of the nutrients and calories in almost any food they might encounter.

The exchange system also highlights a fact that the USDA Food Guide also points out: most foods provide more than just one energy nutrient. Meat, for example, is famous for protein, but meats like bacon and sausage deliver many more calories from fat than from protein. A slice of bread provides most of its calories as carbohydrate, but biscuits provide many of their calories as fat, and so on. This focus on energy nutrients leads to some unexpected food groupings in the exchange lists. The high-fat meats mentioned above and also many cheeses are listed together as "high-fat meats" because fat constitutes the predominant form of energy in these foods, followed by protein. Potatoes and other vegetables high in starch are listed with the breads because one serving of bread and one serving of a starchy vegetable contain about the same amount of carbohydrate. To explore the usefulness of this powerful aid to diet planning, spend some time studying Appendix D or B.

▶ **KEY POINT** Exchange lists facilitate calorie control by providing an understanding of how much carbohydrate, fat, and protein are in each food group.

The Healthy Eating Index for Assessing Diet Quality

To judge the quality of a diet, researchers use an assessment tool known as the **Healthy Eating Index (HEI),** developed by the USDA.[5] The HEI scores a person's diet according to its adherence to the principles of the USDA Food Guide and the *Dietary Guidelines for*

Healthy Eating Index (HEI) a dietary assessment tool that evaluates a diet's adherence to the principles of the USDA Food Guide and the *Dietary Guidelines for Americans,* as well as the variety of foods the diet contains.

■ To learn more about the Healthy Eating Index or to test your own diet with an online assessment tool, the *Interactive Healthy Eating Index,* visit the HEI website on the Internet: www.usda.gov/cnpp/.

Americans. Of ten HEI scoring components, the first five judge whether the diet provides the recommended amounts from each of the food groups. The next four scores reflect moderation in intakes of fats, saturated fat, cholesterol, and sodium. A final score is earned for variety of foods because variety ensures intakes of many beneficial constituents while minimizing intakes of those that could pose problems.

According to the HEI, the great majority (more than 70 percent) of Americans eat diets that "need improvement," and only 12 percent score as "good" or "excellent."[6] The remainder are just "fair." The greater people's awareness of nutrition, the higher their scores, underscoring the importance of nutrition education for improving the nation's diet.

Wise consumers also read labels on packaged foods to help them determine the foods' nutrient and energy contents and to decide how the foods may fit into their total eating plan. This chapter's Consumer Corner explains how to gain insight from the information on food labels.

▶ KEY POINT   The USDA Healthy Eating Index is a tool for assessing diet quality.

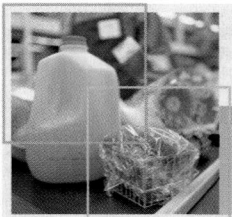

CONSUMER CORNER

Checking Out Food Labels

A POTATO IS A POTATO and needs no label to tell you so. But what can a package of potato chips tell you about its contents? By law, its label must list the chips' ingredients—potatoes, fat, and salt—and its **Nutrition Facts** panel must also reveal details about their nutrient composition (see Table 2-6). If the oil is high in saturated fat or *trans* fat, the label will tell you so (more about fats in Chapter 5). In addition to required information, labels may make optional statements about the food being delicious, or good for you in some way, or a great value. Some of these comments, especially some that are regulated by the Food and Drug Administration (FDA), are reliable. Many others are based on less convincing evidence.

This Consumer Corner introduces food labels and points out the accurate, tested, regulated, and therefore helpful information consumers need to make wise choices. It then turns the spotlight on claims whose purpose is to attract consumer dollars by treading beyond established nutrition science into the

realm of pure marketing. Consumers must hone their label-reading skills and acquire some tools for digging out the truth from among the rubble.

What Food Labels *Must* Include

The Nutrition Education and Labeling Act of 1990 set the requirements for certain label information to ensure that food labels truthfully inform consumers about the nutrients and ingredients in the package. This information remains reliable and true today. According to the law, every packaged food must state the following:

- The common or usual name of the product.
- The name and address of the manufacturer, packer, or distributor.
- The net contents in terms of weight, measure, or count.
- The nutrient contents of the product (Nutrition Facts panel).

Then, the label must list the following in ordinary language:

- The ingredients, in descending order of predominance by weight.

Not every package need display information about every vitamin and mineral. A large package, such as the box of cereal in Figure 2-10 (p. 48), must provide all of the information listed above. A smaller label, such as the label on a can of tuna, provides some of the information in abbreviated form. A label on a roll of candy rings provides only a phone number, which is allowed for the tiniest labels. The Canadian version of a food label can be found in Appendix B.

The Nutrition Facts Panel

Most food packages are required to display a Nutrition Facts panel, like the one shown in Figure 2-10. Grocers also voluntarily post placards or offer hand-

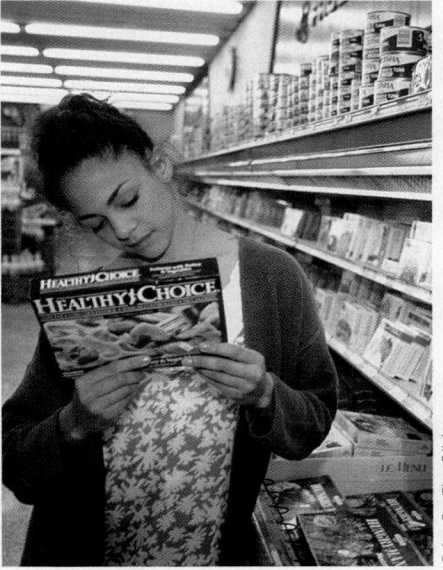

Food labels provide clues for nutrition sleuths.

outs in fresh-food departments to provide consumers with similar sorts of nutrition information for the most popular types of fresh fruits, vegetables, meats, poultry, and seafoods.

When you read a Nutrition Facts panel, be aware that only the top portion of the panel conveys information specific to the food inside the package. The bottom portion is identical on every label—it stands as a reminder of the Daily Values.

The highlighted items in this section correspond with those of Figure 2-10, which shows the location of the items listed below.

- Serving size. Common household and metric measures to allow comparison of foods within a food category. This amount of the food constitutes a single serving and the portion that contains the nutrient amounts listed. A serving of chips may be 10 chips, so if you eat 50 chips, you will have consumed five times the nutrient amounts listed on

---

TABLE 2·6   NUTRITION FACTS AND CLAIMS ON FOOD LABELS

- **health claims** claims linking food constituents with disease states; allowable on labels within the criteria established by the Food and Drug Administration.
- **nutrient claims** claims using approved wording to describe the nutrient values of foods, such as a claim that a food is "high" in a desirable constituent or "low" in an undesirable one.
- **Nutrition Facts** on a food label, the panel of nutrition information required to appear on almost every packaged food. Grocers may also provide the information for fresh produce, meats, poultry, and seafoods.
- **structure-function claim** a legal but unproven claim permitted on labels of dietary supplements and functional foods. See Controversy 7.

FIGURE 2·10 WHAT'S ON A FOOD LABEL?

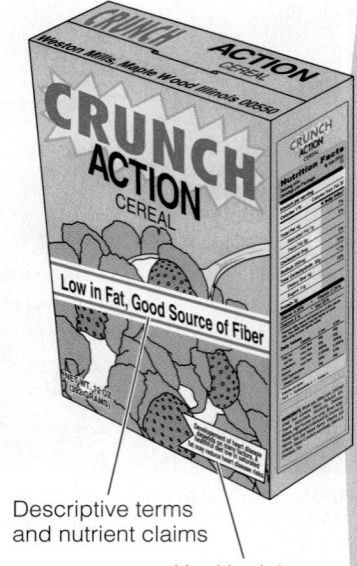

Descriptive terms and nutrient claims

Health claims

CRUNCH ACTION CEREAL

Nutrition Facts

Serving size   ¾ cup (55g)
Servings per Package   10

Amount per serving

Calories 176   Calories from Fat 36

	% Daily Value*
Total Fat 4g	7%
Saturated Fat 1g	5%
Trans Fat 1g	
Cholesterol 0mg	0%
Sodium 250mg	10%
Total Carbohydrate 32g	11%
Dietary fiber 4g	16%
Sugars 11g	
Protein 3g	

Vitamin A 25%	●	Vitamin C 25%
Calcium 2 %	●	Iron 25%

*Percent daily values are based on a 2,000 calorie diet. Your daily values may be higher or lower depending on your calorie needs.

Daily Values

		Calories	2,000	2,500
Total Fat	Less than		65g	80g
Sat Fat	Less than		20g	25g
Cholesterol	Less than		300mg	300mg
Sodium	Less than		2,400mg	2,400mg
Total Carbohydrate			300g	375g
Dietary Fiber			25g	30g

Calories per gram
Fat 9   ●   Carbohydrate 4   ●   Protein 4

INGREDIENTS, Whole oats, Milled corn, Enriched wheat flour (contains Niacin, Reduced iron, Thiamin mononitrate, Riboflavin), Dextrose, Maltose, High-Fructose corn syrup, Brown sugar, Partially hydrogenated cottonseed oil, Coconut oil, Walnuts, Salt, and Natural flavors. Vitamins and minerals: Vitamin C (sodium ascorbate), Vitamin A (Palmitate), Iron.

Serving size

Servings per container

Serving sizes in household and metric measures
1 teaspoon (tsp) = 5ml
1 tablespoon (tbs) = 15ml
1 cup (c) = 240ml
1 fluid ounce (fl. oz.) = 30ml
1 ounce (oz) = 28g

Calories and calories from fat

Nutrient amounts and percentages of Daily Values

Daily Values and calories-per-gram reminder

Ingredients

the label. As of 2005, serving sizes on labels are under review.

- Servings per container. Number of servings per box, can, package, or other unit.
- Calories/calories from fat. Total food energy per serving, and energy from fat per serving.
- Nutrient amounts and percentages of Daily Values. This section provides the core information concerning these nutrients:
  - *Total fat.* Grams of fat per serving with a breakdown showing grams of *saturated fat* and *trans* fat per serving.
  - *Cholesterol.* Milligrams of cholesterol per serving.
  - *Sodium.* Milligrams of sodium per serving.
  - *Total carbohydrate.* Grams of carbohydrate per serving, including starch, fiber, and sugars, with a breakdown showing grams of dietary *fiber* and *sugars.* The sugars include those that occur naturally in the food plus any added during processing. The terms "net carbs," "impact carbs," and related terms have not been defined scientifically but may appear on a label to imply that a food contains less digestible carbohydrate than similar foods (this may or may not be the case).
  - *Protein.* Grams of protein per serving.

In addition, the label must state the contents of these nutrients expressed as percentages of the Daily Values:

- *Vitamin A.*
- *Vitamin C.*
- *Calcium.*
- *Iron.*

Other nutrients present in significant amounts in the food may also be listed on the label. The percentages of the Daily Values (inside front cover, page C) are given in terms of a person requiring 2,000 calories each day.

- Daily Values and calories-per-gram reminder. This portion lists the Daily Values for a person needing 2,000 or 2,500 calories a day and provides a calories-per-gram reminder as a handy reference for label readers.

Ingredients List

An often neglected but highly valuable body of information is the list of:

- Ingredients. The product's ingredients must be listed in descending order of predominance by weight.

Knowing how to read an ingredients list puts you many steps ahead of the naive buyer. Consider the ingredients list on an orange drink powder whose first three ingredients are "sugar, citric acid, orange flavor." You can tell that sugar is the chief ingredient. Now consider a canned juice whose ingredients list begins with "water, orange juice concentrate, pineapple juice concentrate." This product is clearly made of *reconstituted* juice. Water is first on the label because it is the main constituent of juice. Sugar is nowhere to be found among the ingredients because sugar has not been added to the product. Sugar occurs naturally in juice, though, so the label does specify sugar grams; details are in Chapter 4.

Now consider a cereal whose entire list contains just one item: "100 percent shredded wheat." No question, this is a whole-grain food with nothing added. Finally, consider a cereal whose first three ingredients are "puffed milled corn, sweeteners (sugars: corn syrup, sucrose, honey, dextrose), salt." If you recognize that sugar, corn syrup, honey, and dextrose are all different versions of sugar (and you will, after Chapter 4), you might guess that this product contains close to half its weight as sugar.

More about Percentages of Daily Values

Some of the Daily Values are printed on each label in the Nutrition Facts panel. The entire list can be found on the inside back cover of this text. The calculations used to determine the "% Daily Value" figures for nutrient contributions from a serving of food are based on a 2,000-calorie diet. For example, if a food contributes 13 milligrams of vitamin C per serving, and the Daily Value is 60 milligrams, then a serving of that food provides about 22 percent of the Daily Value for vitamin C.

The Daily Values are of two types. Some, such as those for fiber, protein, vitamins, and most minerals, are akin to other nutrient intake recommendations. They suggest an intake goal to strive to reach; below that level, some people's needs may go unmet. Other Daily Values, such as those for cholesterol, total fat, saturated fat, and sodium, constitute healthy daily maximums.

Of course, though the Daily Values are based on a 2,000-calorie diet, people's actual calorie intakes vary widely; some people need fewer calories and some need many more. This makes the Daily Values most useful for comparing one food with another and less useful as nutrient intake targets for individuals. Still, by examining a food's general nutrient profile, you can determine whether the food contributes "a little" or "a lot" of a nutrient, whether it contributes "more" or "less" than another food, and how well it fits into your overall diet.

What Food Labels *May* Include

So far, this Consumer Corner has presented the accurate and reliable facts on nutrition labels. This section looks at reliable claims and also describes the unreliable but legal claims that can be made on food labels.

Nutrient Claims on Food Labels

If a food meets specified criteria, the label may display certain approved **nutrient claims** concerning the product's nutritive value. The Daily Values serve as the basis for claims that a food is "low" in cholesterol or a "good source" of vitamin A. Table 2-7 (p. 50) provides a list of these regulated, reliable label terms along with their definitions. By remembering the meanings of these terms, consumers can make informed choices among foods. For example, any food providing 10 percent or more of the Daily Value for a nutrient can boast that it is "a good source" of the nutrient; a food providing 20 percent is considered "high" in the nutrient.

For nutrients that can be harmful if consumed excessively, such as saturated fat or sodium, foods providing less than 5 percent are desirable. For hard-to-get nutrients such as iron or calcium, a reasonable goal might be to choose foods that are "good sources" of or "high" in those nutrients several times a day. (See the Snapshot features of Chapters 7 and 8 for foods qualifying as "good sources" or better for the vitamins and minerals.)

Health Claims: The FDA's "A" through "D" Lists

Until recently, the FDA held manufacturers to the highest standards of scientific evidence before allowing them to place **health claims** on food labels. When a label stated, "Diets low in sodium may reduce the risk of high blood pressure," for example, consumers could be sure that the FDA had examined much scientific evidence and found substantial scientific support for the claim. Such reliable health claims still appear on some food labels, and they have a high degree of scientific validity (see Table 2-8).

Recently, however, the FDA created a new ranking system that includes three new categories of allowable health claims that are supported by weaker evidence.[1] Somewhat reminiscent of a schoolchild's report card, the new system assigns each claim a letter grade reflecting the degree to which the claim is backed by science (see Figure 2-11, p. 52). The FDA can no longer demand that only health claims with the highest degree of scientific support appear on food labels.[2]

Under the new system, the reliable health claims shown in Table 2-8 receive an "A" grade. As for the "B" through "D" claims, the FDA is relying on consumers to notice the printed disclaimers required to accompany these "qualified" health claims. These claims are "qualified" in the sense that they must bear a statement explaining the degree of scientific evidence backing them up. Consumers will have to look for the fine print to identify which health claims are scientifically reliable and which are best ignored.

Conclusion

The Nutrition Facts panels and ingredients lists on food labels provide reliable information on which consumers can base their food choices. Regrettably, more and more of the health-related claims printed on labels are based on less-than-convincing scientific evidence. In the world of food marketing, label rulings put the consumer on notice, "let the buyer beware."

TABLE 2·7	DESCRIPTIVE TERMS USED ON FOOD LABELS

Energy Terms

- **low calorie** 40 calories or fewer per serving.
- **reduced calorie** at least 25% lower in calories than a "regular," or reference, food.
- **calorie free** fewer than 5 calories per serving.

Fat Terms (Meat and Poultry Products)

- **extra lean**
  less than 5 g of fat *and*
  less than 2 g of saturated fat and *trans* fat combined, *and*
  less than 95 mg of cholesterol per serving.
- **lean**[a]
  less than 10 g of fat *and*
  less than 4 g of saturated fat and *trans* fat combined, *and*
  less than 95 mg of cholesterol per serving.

Fat and Cholesterol Terms (All Products)

- **cholesterol free**
  less than 2 mg of cholesterol *and*
  2 g or less saturated fat and *trans* fat combined per serving.
- **fat free** less than 0.5 g of fat per serving.
- **less saturated fat** 25% or less saturated fat and *trans* fat combined than the comparison food.
- **low cholesterol**
  20 mg or less of cholesterol *and*
  2 g or less saturated fat per serving.
- **low fat** 3 g or less fat per serving.[a]
- **low saturated fat** 1 g or less saturated fat and less than 0.5 g of *trans* fat per serving.
- **percent fat free** may be used only if the product meets the definition of *low fat* or *fat free.* Requires disclosure of grams of fat per 100 g food.
- **reduced** or **less cholesterol**
  at least 25% less cholesterol than a reference food *and*
  2 g or less saturated fat per serving.
- **reduced saturated fat**
  at least 25% less saturated fat *and*
  reduced by more than 1 g saturated fat per serving compared with a reference food.
- **saturated fat free**
  less than 0.5 g of saturated fat *and*
  less than 0.5 g of *trans* fat.
- ***trans* fat free**
  less than 0.5 g of *trans* fat *and*
  less than 0.5 g of saturated fat per serving.

[a]Exceptions may be granted for foods with low levels of saturated and *trans* fats.

TABLE 2·7 DESCRIPTIVE TERMS USED ON FOOD LABELS (continued)

Fiber Terms

- **high fiber** 5 g or more per serving. (Foods making high-fiber claims must fit the definition of low fat, or the level of total fat must appear next to the high-fiber claim.)
- **good source of fiber** 2.5 g to 4.9 g per serving.
- **more** or **added fiber** at least 2.5 g more per serving than a reference food.

Other Terms

- **free, without, no, zero** none or a trivial amount. *Calorie free* means containing fewer than 5 calories per serving; *sugar free* or *fat free* means containing less than half a gram per serving.
- **fresh** raw, unprocessed, or minimally processed with no added preservatives.
- **good source** 10 to 19% of the Daily Value per serving.
- **healthy** low in fat, saturated fat, *trans* fat, cholesterol, and sodium and containing at least 10% of the Daily Value for vitamin A, vitamin C, iron, calcium, protein, or fiber.
- **high in** 20% or more of the Daily Value for a given nutrient per serving; synonyms include "rich in" or "excellent source."
- **less, fewer, reduced** containing at least 25% less of a nutrient or calories than a reference food. This may occur naturally or as a result of altering the food. For example, pretzels, which are usually low in fat, can claim to provide less fat than potato chips, a comparable food.
- **light** this descriptor has three meanings on labels:
  1. A serving provides one-third fewer calories or half the fat of the regular product.
  2. A serving of a low-calorie, low-fat food provides half the sodium normally present.
  3. The product is light in color and texture, so long as the label makes this intent clear, as in "light brown sugar."
- **more, extra** at least 10% more of the Daily Value than in a reference food. The nutrient may be added or may occur naturally.

Sodium Terms

- **low sodium** 140 mg or less sodium per serving.
- **reduced sodium** at least 25% lower in sodium than the regular product.
- **sodium free** less than 5 mg per serving.
- **very low sodium** 35 mg or less sodium per serving.

aThe word *lean* as part of the brand name (as in "Lean Supreme") indicates that the product contains fewer than 10 grams of fat per serving.

TABLE 2·8 HEALTH CLAIMS ON FOOD LABELS—THE "A" LIST

- Calcium and reduced risk of osteoporosis
- Sodium and reduced risk of hypertension
- Dietary saturated fat and cholesterol and reduced risk of coronary heart disease
- Dietary fat and reduced risk of cancer
- Fiber-containing grain products, fruits, and vegetables and reduced risk of cancer
- Fruits, vegetables, and grain products that contain fiber, particularly soluble fiber, and reduced risk of coronary heart disease
- Fruits and vegetables and reduced risk of cancer
- Folate and reduced risk of neural tube defects
- Sugar alcohols and reduced risk of tooth decay
- Soluble fiber from whole oats and from psyllium seed husk and reduced risk of heart disease
- Soy protein and reduced risk of heart disease
- Whole grains and reduced risk of heart disease and certain cancers
- Plant sterol and plant stanol esters and heart disease
- Potassium and reduced risk of hypertension and stroke

FIGURE 2·11 THE FDA'S HEALTH CLAIMS REPORT CARD

Grade	Level of Confidence in Health Claim	Label Disclaimers Required by the FDA
A	**High** Significant scientific agreement	These health claims do not require disclaimers; see Table 2-8 for examples.
B	**Moderate** Evidence is supportive, but not conclusive	"[Health claim.] Although there is scientific evidence supporting this claim, the evidence is not conclusive."
C	**Low** Evidence is limited and not conclusive	"Some scientific evidence suggests [health claim]. However, the FDA has determined that this evidence is limited and not conclusive."
D	**Extremely low** Little scientific evidence supporting this claim	"Very limited and preliminary scientific research suggests [health claim]. The FDA concludes that there is little scientific evidence supporting this claim."

Getting a Feel for the Nutrients in Foods

FIGURE 2-12 ILLUSTRATES a playful contrast between two days' meals. "Monday's Meals" were selected according to the recommendations of this chapter and follow the sample menu of Figure 2-6 shown earlier (p. 40). "Tuesday's Meals" were chosen more for convenience and familiarity than out of concern for nutrition.

How can a person compare the nutrients that these sets of meals provide? One way is to look up each food in a table of food composition, write down the food's nutrient values, and compare each one to a standard such as the DRI recommended intakes for nutrients as we've done in Figure 2-12. By this measure, Monday's meals are the clear winners in terms of meeting nutrient needs within a calorie budget. Tuesday's meals oversupply calories and saturated fat while undersupplying fiber and critical vitamins and minerals.

Another useful exercise is to reverse the process of diet planning that was demonstrated in an earlier section. This time, the diet planner starts with the completed day's meals, adding up the amounts of foods from each food group consumed during the day. The result of this process is demonstrated in Table 2-9 (p. 55). By comparing the totals with the recommendations of the USDA Food Guide, a meaningful assessment of the adequacy of the day's meals can be made. Other evaluations, such as identifying sources of whole grains, comparing vegetables with subgroups, and watching for the discretionary calories of solid fats and added sugars complete the assessment.

In this analysis, Monday's meals again prove best for nutrition, providing the necessary amounts from each food group along with a small amount of raw oil needed for health. The calorie intake falls well within the 2,000-calorie allowance. A closer look at Monday's foods reveals that the whole-grain cereal at breakfast, whole-grain sandwich roll at lunch, and whole-grain crackers at snack time meet the recommendation to obtain at least half of the day's grain servings from whole grains. For the vegetable subgroups, dark green vegetables, orange vegetables, and legumes are represented in the dinner salad and "other vegetables" are prominent in other meals. To repeat: it isn't necessary to choose vegetables from each subgroup each day, and the person eating this day's meals will need to include vegetables from other subgroups throughout the week. In addition, Monday's meals have room to spare for some discretionary calories, such as additional nutritious foods or added sugars or fats.

Tuesday's meals, though abundant in oils, meats, and enriched grains, completely lack fruit and whole grains and are too low in vegetables and milk to provide adequate nutrients. Tuesday's meals supply too much saturated fat and sugar, as well as excess amounts of meat and enriched grains, pushing the calorie total to well above the day's allowance. Although a single day of such fare poses little threat to the eater, a steady diet of "Tuesday meals" presents a high probability of nutrient deficiencies and weight gain, and greatly increased risks of chronic diseases in later life.

If you have access to a computer, it can be a time saver—it performs all of these calculations with lightning speed. This convenience may make working with paper, pencils, and erasers seem a bit old-fashioned, but computers are rarely available for diners in cafeterias or at fast-food counters where real-life decisions must be made. Those who work out such analyses for themselves can often remember their target amounts of foods to meet USDA Food Guide recommendations and learn to "see" the nutrients in foods (and you should develop this skill by the time you reach Chapter 10 of this text), allowing them to make informed choices at mealtimes. People who fail to develop such skills must wait until they visit their computers to find out how well they did in retrospect.

FIGURE 2·12 TWO DAYS' MEALS

Breakfast

© Polara Studios, Inc.

Lunch

© Polara Studios, Inc.

Afternoon snack

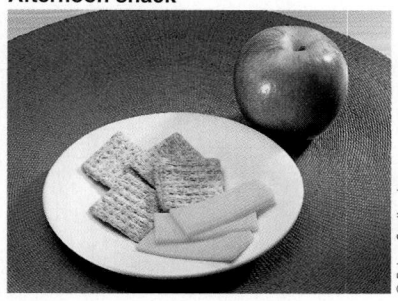

© Polara Studios, Inc.

Dinner

© Matthew Farruggio

Bedtime snack

© Quest

Monday's Meals

Monday's meals reflect nutrient-dense choices.

Foods	Energy (cal)	Saturated Fat (g)	Fiber (g)	Vitamin C (mg)	Calcium (mg)
Before heading off to class, a student eats breakfast:					
1 c whole grain cold cereal	108	–	3	14	95
1 c fat free milk	100	–	–	2	306
1 medium banana (sliced)	105	–	3	10	6
Then goes home for a quick lunch:					
1 roasted turkey sandwich on whole-grain roll with low-fat mayonnaise	343	4	2	–	89
1 c low-salt vegetable juice	50	–	1	60	27
While studying in the afternoon, the student eats a snack:					
4 whole-wheat reduced-fat crackers	86	1	2	–	–
1½ oz low-fat cheddar cheese	74	2	–	–	176
1 apple	72	–	3	6	8
That night, the student makes dinner:					
A salad:					
1 c raw spinach leaves, shredded carrots	19	–	2	18	61
¼ c garbanzo beans	71	–	3	2	19
5 lg olives and 2 tbs oil-based salad dressing	76	1	1	–	2
A main course:					
1 c spaghetti with meat sauce	425	3	5	15	56
½ c green beans	22	–	2	6	29
2 tsp soft margarine	67	1	–	–	–
And for dessert:					
1 c strawberries	49	–	3	89	24
Later that evening, the student enjoys a bedtime snack:					
3 graham crackers	90	–	–	–	–
1 c fat free milk	100	–	–	2	306
Totals:	1,857	12	30	224	1,204
DRI recommended intakes:[a]	2,000	<20[b]	25	75	1,000
Percentage of DRI recommended intakes:	93%	60%	120%	299%	120%

[a]DRI values for a sedentary woman, age 19–30. Other DRI values are listed on the inside front cover.

[b]The 20-gram value listed is the maximum allowable saturated fat for a 2,000-calorie diet. The DRI recommends consuming less than 10 percent of calories from saturated fat.

FIGURE 2·12 TWO DAYS' MEALS—continued

Breakfast

Lunch

Afternoon snack

Dinner

Bedtime snack

Tuesday's Meals

Tuesday's meals are lower on the nutrient density scale.

Foods	Energy (cal)	Saturated Fat (g)	Fiber (g)	Vitamin C (mg)	Calcium (mg)
Today, the student starts the day with a fast-food breakfast:					
1 c coffee	5	–	–	–	–
1 English muffin with egg, cheese, and bacon	436	9	2	–	266
Between classes, the student returns home for a quick lunch:					
1 peanut butter and jelly sandwich on white bread	426	4	3	–	93
1 c whole milk	156	6	–	4	290
While studying, the student has:					
12 oz diet cola	–	–	–	–	–
Bag of chips (14 chips)	105	2	–	4	–
That night for dinner, the student eats:					
A salad:					
1c lettuce					
1 tbs blue cheese dressing	84	2	1	2	23
A main course:					
6 oz steak	349	6	–	–	27
1/2 baked potato (large)	161	–	4	17	26
1 tbs butter	102	7	–	–	3
1 tbs sour cream	31	2	–	–	17
12 oz diet cola	–	–	–	–	–
And for dessert:					
4 sandwich-type cookies	158	2	1	–	–
Later on, a bedtime snack:					
2 cream-filled snack cakes	250	2	2	–	20
1 c herbal tea	–	–	–	–	–
Totals:	2,263	42	13	27	765
DRI recommended intakes:[a]	2,000	<20[b]	25	75	1,000
Percentage of DRI recommended intakes:	113%	210%	52%	36%	77%

[a]DRI values for a sedentary woman, age 19–30. Other DRI values are listed on the inside front cover.

[b]The 20-gram value listed is the maximum allowable saturated fat for a 2,000-calorie diet. The DRI recommends consuming less than 10 percent of calories from saturated fat.

TABLE 2·9 COMPARISON OF MONDAY'S AND TUESDAY'S MEALS WITH THE USDA FOOD GUIDE

Notice that Monday's meals meet the recommendations of the USDA Food Guide and include an extra half-cup of vegetables, with room to spare for some discretionary calories in the 2,000-calorie allowance. Tuesday's meals fall short of the Food Guide ideals while overshooting the 2,000-calorie allowance.

Monday's Meals

Food Group	Breakfast	Lunch	Snack	Dinner	Snack	Totals	Recommended Amounts
Fruits	½ c		½ c	1 c		2 c	2 c
Vegetables		1 c		2 c		3 c	2½ c
Grains	1 oz	2 oz	½ oz	2 oz	½ oz	6 oz	6 oz
Meat and legumes		2 oz		3½ oz		5½ oz	5½ oz
Milk	1 c		1 c		1 c	3 c	3 c
Oils		1½ tsp		4 tsp		5½ tsp	5½ tsp
Calorie allowance						1,857 cal	2,000 cal

Tuesday's Meals

Food Group	Breakfast	Lunch	Snack	Dinner	Snack	Totals	Recommended Amounts
Fruits						0 c	2 c
Vegetables			a	1 c		1 c	2½ c
Grains	2 oz	2 oz		1 oz	2 oz	7 oz	6 oz
Meat and legumes	2 oz	1 oz		6 oz		9 oz	5½ oz
Milk	1 c	1 c				2 c	3 c
Oils						7½ tsp[b]	5½ tsp
Calorie allowance						2,263 cal	2,000 cal

[a] The potato in 14 potato chips provides less than ½ cup of vegetable.
[b] Does not include solid fats.

Nutrition Connections

The Nutrition Connections CD provides access to these additional resources.

Do It!
The Do It! exercise for this chapter, *Evaluate the Nutrients Listed on Labels,* provides practice understanding the descriptive terms used on food labels.

Interaction
The animation, *Diet Planning Using the Healthy Eating Index,* provides practice in planning healthy meals using the Healthy Eating Index diet-planning tool.

Chapter Quiz
Take the quiz for this chapter to assess your understanding of chapter concepts.

Nutrition on the Net
Annotated links to reputable nutrition-related websites.

Online Glossary
Test your knowledge of key vocabulary through this comprehensive, interactive glossary.

Internet Connections

A variety of helpful study tools for this chapter are available at our website
http://nutrition.wadsworth.com/ncc10

Practice Test
Take the practice test for this chapter and gauge your knowledge of key concepts; answers are provided.

Student Notebook
A lecture notebook is available to help organize note taking while reading or attending lecture.

ABC Video Clip
"Restaurant Nutrition" (3:54) examines the portion sizes and nutritional value of meals served in restaurants.

Study Tools
Study resources for students include Chapter Summaries, Crossword puzzles, Flashcards, Glossary, Internet Exercises, and Learning Objectives.

SELF CHECK

Answers to these Self Check questions are in Appendix G.

1. The nutrient standards in use today include all of the following *except:*
   a. Adequate Intakes (AI)
   b. Daily Minimum Requirements (DMR)
   c. Daily Values (DV)
   d. (a) and (c)

2. The Dietary Reference Intakes were devised for which of the following purposes?
   a. to set nutrient goals for individuals
   b. to suggest upper limits of intakes, above which toxicity is likely
   c. to set average nutrient requirements for use in research
   d. all of the above

3. According to the USDA Food Guide, which of the following may be counted among either the meats or the vegetables?
   a. chicken
   b. avocados
   c. black beans
   d. potatoes

4. The USDA Food Guide recommends a small amount of daily oil from which of these sources?
   a. olives
   b. nuts
   c. vegetable oil
   d. all of the above

5. Which of the following values is found on food labels?
   a. Daily Values
   b. Dietary Reference Intakes
   c. Recommended Dietary Allowances
   d. Estimated Average Requirements

6. The energy intake recommendation is set at a level predicted to maintain body weight.   T   F

7. The Dietary Reference Intakes (DRI) are for all people, regardless of their medical history.   T   F

8. People who choose not to eat animals or their products need to find an alternative to the USDA Food Guide when planning their diets.   T   F

9. By law, food labels must state as a percentage of the Daily Values the amounts of vitamin C, vitamin A, niacin, and thiamin present in food.   T   F

10. To be labeled "low fat," a food must contain 3 grams of fat or less per serving.   T   F

CONTROVERSY 2

PHYTOCHEMICALS AND FUNCTIONAL FOODS: WHAT DO THEY PROMISE? WHAT DO THEY DELIVER?

CONSUMERS OFTEN READ exciting headlines reporting some remarkable new discovery about potential health benefits from eating foods that contain **phytochemicals**—biologically active chemical compounds found in plants. Should we be eating more soy, tomatoes, grapes, and other foods from plants for their phytochemicals? What about taking supplements of phytochemicals—does this strategy offer benefits? Store shelves are overflowing with such supplements alongside **functional foods** whose labels often make fantastic claims for their health-promoting powers.

Consumers need information about the potential benefits and hazards of these substances and products, and this Controversy invites you to consider the scientific evidence concerning their effects. To begin, Table C2-1 defines some terms, and Table C2-2 (p. 58) provides a sampling of foods with potential phytochemical effects. The next sections focus on a few food sources of the most interesting of the thousands of known phytochemicals and then discuss the concept of functional foods. The conclusion of this Controversy and of every other valid assessment of phytochemicals is that *foods*, not supplements, are the best and safest source of these potentially beneficial substances.

First, Some Cautions

Foods consist of thousands of different chemicals, and each one has the potential of being beneficial, neutral, or harmful to the body. Some may even be mixed: beneficial in some ways and

TABLE C2·1	PHYTOCHEMICAL AND FUNCTIONAL FOOD TERMS

- **antioxidants** (anti-OX-ih-dants) compounds that protect other compounds from damaging reactions involving oxygen by themselves reacting with oxygen (*anti* means "against"; *oxy* means "oxygen"). *Oxidation* is a potentially damaging effect of normal cell chemistry involving oxygen (more in Chapters 5 and 7).
- **broccoli sprouts** the sprouted seed of *Brassica italica,* or the common broccoli plant; believed to be a functional food by virtue of its high phytochemical content.
- **conjugated linoleic acid (CLA)** a type of fat in butter, milk, and other dairy products believed by some to have biological activity in the body. Not a phytochemical, but a biologically active chemical produced by animals.
- **drug** any substance that when taken into a living organism may modify one or more of its functions.
- **flavonoid** (FLAY-von-oyd) any member of a chemical family of yellow pigments in foods; phytochemicals that may exert physiological effects on the body. *Flavus* means "yellow."
- **flaxseed** small brown seed of the flax plant; used in baking, cereals, or other foods. Valued by industry as a source of linseed oil and fiber.
- **functional foods** no official U.S. definition exists, but used generally to describe foods with beneficial physical or psychological effects beyond providing nutrients. Health Canada defines functional foods as foods that appear similar to conventional foods, consumed as part of the usual diet, with demonstrated physiological benefits or with the ability to reduce chronic disease risks beyond basic nutrient functions. Also defined in Chapter 1.
- **genistein** (GEN-ih-steen) a phytoestrogen found primarily in soybeans that both mimics and blocks the action of estrogen in the body.
- **kefir** (KEE-fur) a liquid form of yogurt, based on milk, probiotic microorganisms, and flavorings.
- **lignans** phytochemicals present in flaxseed, but not in flax oil, that are converted to phytosterols by intestinal bacteria and are under study as possible anticancer agents.
- **lutein** (LOO-teen) a plant pigment of yellow hue; a phytochemical believed to play roles in eye functioning and health.
- **lycopene** (LYE-koh-peen) a pigment responsible for the red color of tomatoes and other red-hued vegetables; a phytochemical that may act as an antioxidant in the body.
- **miso** fermented soybean paste used in Japanese cooking. Soy products are considered to be functional foods.
- **organosulfur compounds** a large group of phytochemicals containing the mineral sulfur. Organosulfur phytochemicals are responsible for the pungent flavors and aromas of foods belonging to the onion, leek, chive, shallot, and garlic family and are thought to stimulate cancer defenses in the body.
- **phytochemicals** (FIGH-toe-CHEM-ih-cals) biologically active compounds of plants believed to confer resistance to diseases on the eater, also defined in Chapter 1. *Phyto* means "plant."
- **phytoestrogens** (FIGH-toe-ESS-troh-gens) phytochemicals structurally similar to mammalian hormones, such as the female sex hormone estrogen. Phytoestrogens weakly mimic hormone activity in the human body.
- **probiotics** consumable products containing live microorganisms in sufficient numbers to alter the bacterial colonies of the body in ways believed to benefit health. A *prebiotic* product is a substance that may not be digestible by the host, such as fiber, but serves as food for probiotic bacteria and thus promotes their growth.
- **soy drink** a milk-like beverage made from soybeans, claimed to be a functional food. Soy drink should be fortified with vitamin A, vitamin D, riboflavin, and calcium to approach the nutritional equivalency of milk. Also called *soy milk.*
- **sterol esters** compounds derived from vegetable oils that lower blood cholesterol in human beings by competing with cholesterol for absorption from the digestive tract. The term *sterol esters* often refers to both stanol esters and sterol esters.
- **tofu** a white curd made of soybeans, popular in Asian cuisines, and considered to be a functional food.

TABLE C2·2 A SAMPLING OF PHYTOCHEMICALS—POSSIBLE EFFECTS AND FOOD SOURCES

Name	Possible Effects	Food Sources
Capsaicin	May modulate blood clotting, may reduce the risk of fatal clots in heart and artery disease.	Hot peppers
Carotenoids (including beta-carotene, lutein, lycopene, and hundreds of related compounds)[a]	Act as antioxidants; possibly reduce risks of heart disease, age-related eye disease,[b] cancer, and other diseases.	Deeply pigmented fruits and vegetables (apricots, broccoli, cantaloupe, carrots, pumpkin, spinach, sweet potatoes, tomatoes)
Curcumin	May inhibit enzymes that activate carcinogens.	Turmeric, a yellow-colored spice
Flavonoids (including flavones, flavonols, isoflavones, catechins, and others)[c,d]	Act as antioxidants; may scavenge carcinogens; bind to nitrates in the stomach, preventing conversion to nitrosamines; inhibit cell proliferation; flavonoids of blueberries may improve memory.	Berries, black tea, celery, chocolate, citrus fruits, green tea, olives, onions, oregano, purple grapes, purple grape juice, soybeans and soy products, vegetables, whole wheat, wine
Indoles	May trigger production of enzymes that block DNA damage from carcinogens; may inhibit estrogen action.	Broccoli and other cruciferous vegetables (brussels sprouts, cabbage, cauliflower), horseradish, mustard greens
Isothiocyanates (including sulforaphane)	May inhibit enzymes that activate carcinogens; trigger production of enzymes that detoxify carcinogens.	Broccoli and other cruciferous vegetables (brussels sprouts, cabbage, cauliflower), horseradish, mustard greens
Monoterpenes (including limonene)	May trigger enzyme production to detoxify carcinogens; may inhibit cancer promotion and cell proliferation.	Citrus fruit peels and oils
Organosulfur compounds (including allicin)	May speed production of carcinogen-destroying enzymes or slow production of carcinogen-activating enzymes.	Chives, garlic, leeks, onions
Phenolic acids[d] (including ellagic acid)	May trigger enzyme production to make carcinogens water soluble, facilitating excretion.	Coffee beans, fruits (apples, blueberries, cherries, grapes, oranges, pears, prunes, strawberries), oats, potatoes, soybeans
Phytic acid	Binds to minerals, preventing free-radical formation, possibly reducing cancer risk.	Whole grains
Phytoestrogens (members of the flavonoid family, genistein and diadzein)	May inhibit estrogen and produce these actions: inhibit cell replication in GI tract; reduce risk of breast, colon, ovarian, prostate, and other estrogen-sensitive cancers; reduce cancer cell survival; may reduce risk of osteoporosis. May also alter blood lipids favorably and reduce heart disease risk when consumed in soy foods.	Soybeans, soy flour, soy milk, tofu, textured vegetable protein, other legume products
Phytoestrogens (lignans)	Block estrogen activity in cells, possibly reducing the risk of cancer of the breast, colon, ovaries, and prostate.	Flaxseed, whole grains
Protease inhibitors	May suppress enzyme production in cancer cells, slowing tumor growth; inhibit hormone binding; inhibit malignant changes in cells.	Broccoli sprouts, potatoes, soybeans and other legumes, soy products
Resveratrol[e]	May offset artery-damaging effects of high-fat diets.	Red wine, peanuts
Saponins	May interfere with DNA replication, preventing cancer cells from multiplying; stimulate immune response.	Alfalfa sprouts, other sprouts, green vegetables, potatoes, tomatoes
Tannins[d]	May inhibit carcinogen activation and cancer promotion; act as antioxidants.	Black-eyed peas, grapes, lentils, red and white wine, tea

[a]Other carotenoids include alpha-carotene, beta-cryptoxanthin, and zeaxanthin.
[b]The age-related eye disease is macular degeneration.
[c]Other flavonoids of interest include ellagic acid and ferulic acid.
[d]A subset of the larger group *polyphenolic phytochemicals*.
[e]A member of the chemical group stilbene, which is a subset of the larger group *polyphenolic phytochemicals*.

harmful in others. To complicate matters further, some chemicals may exert different effects on different people or when taken in differing doses or at different life stages.[*1] All of these possibilities have yet to be clarified by science.

*Reference notes are found in Appendix F.

Research on phytochemicals is in its infancy, and what is current today will likely be challenged a year from now by further studies. In most cases, the health benefits observed with intakes of certain foods cannot be ascribed to individual phytochemicals, much less to purified supplements of them.

When considering concentrated supplements of phytochemicals, be aware that any normally beneficial substance, even water, can be toxic when taken in too high a dose (Chapter 8 explains). Also, though most naturally occurring substances are assuredly safe for most healthy people when consumed in foods, virtually no safety studies exist to support the taking of any *purified* phytochemical, nor have safe dosages been established.

How Scientists View Phytochemicals in Foods

At one time, scientists believed that phytochemicals in foods played limited roles in human health. Phytochemical functions were thought to include only their well-known sensory properties, such as taste, aroma, texture, and color. Thank phytochemicals for the burning sensation of hot peppers, the pungent flavor of onions and garlic, the bitter tang of chocolate, the aromatic qualities of herbs, and the deep red color of tomatoes. Today, researchers recognize that some phytochemicals have profound effects on the body through such actions as acting as **antioxidants,** mimicking hormones, and altering blood constituents in ways that may protect against some diseases. For example, cancer and heart disease are linked to processes involving oxygen (oxidation) in the body, and antioxidants are thought to oppose these actions (Chapters 5 and 7 provide details).

The scientific path to discovery often begins when scientists notice that people with a diet rich in foods that contain a particular phytochemical rarely suffer from "disease X." Researchers may then examine the diets of people with the condition to find out how much of the phytochemical they consume. If the diets are low in the substance, the researchers perform laboratory experiments in which they expose animals or cell cultures to the phytochemical and observe what happens to uncover plausible biological mechanisms for disease prevention. Science demands the answers to such questions as:

- What is the function of the phytochemical in the plant tissues that produce it, and might such a function also occur in the tissues of human beings? For example, if the substance acts as an antioxidant in plant tissue, it is likely that it may act the same way in human tissues.
- Can human beings absorb and metabolize enough of the phytochemical to have an effect? The substance may be present in the diet but not in a form that people can absorb or use.
- What safety issues pose concerns—is the phytochemical safe for consumption over many years, or in pregnancy, childhood, or old age?

Then, if the evidence warrants, scientists may perform clinical studies to evaluate effects of the substance on human beings. Currently, the evidence is insufficient to say with any degree of certainty whether any phytochemical is effective in fighting diseases or if it is safe to consume in concentrated doses. Research findings on a few phytochemical-rich foods that seem likely to benefit health are discussed next.

Whole Foods, Wine, and Tea

Historically, diets containing whole grains, fruits, vegetables, herbs, spices, teas, and red wine have been reputed to possess health-promoting qualities. Science has revealed that these foods and beverages all have something in common: phytochemicals of the flavonoid family. Figure C2-1 (p. 60) depicts some of the best-known phytochemical-rich foods. Epidemiological evidence spanning many countries indicates that deaths from cancer, heart disease, and heart attacks are less common where these foods are plentiful in the diet, where tea is a beverage, or where red wine is consumed in moderation.[2] Many **flavonoids** act as antioxidants and may protect against cancers and heart disease by this mechanism. Nevertheless, a recent review of the literature concludes that more evidence is needed before any claims can be made for flavonoids themselves as the protective factor in foods, particularly when they are extracted from foods or herbs and sold as supplements.[3]

Research suggests that one flavonoid of grapes and red wine in particular may have disease-fighting qualities;

however, the amount of this and other flavonoids in the wine may be too low to benefit human health.[*4] The same flavonoid has been credited with greatly extending the life of yeast cells, but no one knows if such an effect is possible in other species.[5] Because flavonoids often impart a bitter taste to foods, food producers may refine away natural flavonoids to please consumers who generally prefer milder flavors.[6] To produce white grape juice or white wine, makers remove the red, flavonoid-rich grape skins to lighten the flavor and the color of the product, while greatly reducing its flavonoid content.

Whether or not research confirms the cancer-fighting and heart-defending nature of the flavonoids, consumers should seek out a variety of whole fruits, vegetables, and other plant-derived foods with their flavonoids intact in place of their more refined counterparts.[7] Beyond any doubt, such diets are consistently associated with low rates of diseases. Flavonoid supplements have not been proved effective or safe. As for red wine, read Controversy 3 before making a decision. The potential health benefits may not be worth alcohol's immediate and substantial risks.

Chocolate

Imagine the delight of research subjects who were instructed to eat three ounces of dark (bittersweet) chocolate chips as part of an experiment. In a less appealing aspect of the study, researchers drew blood from the subjects for analysis. They discovered that a flavonoid antioxidant from chocolate had accumulated in the blood; at the same time, the level of certain harmful oxidizing compounds dropped substantially (40 percent).[8] The antioxidant effects of dark chocolate may turn out to be as powerful as those of tea or red wine.[9] In theory, chocolate may also "thin the blood" by reducing the tendency of blood to clot, and blood clots are a major cause of heart attacks and strokes (see Chapter 11).[10] No evidence exists, however, to indicate that people who eat chocolate actually suffer fewer heart attacks or strokes than people who do not.[11]

*The flavonoid is resveratrol.

FIGURE C2·1 AN ARRAY OF PHYTOCHEMICALS IN A VARIETY OF FRUITS AND VEGETABLES

Broccoli and broccoli sprouts contain an abundance of sulforaphane.

An apple a day—rich in flavonoids.

The phytoestrogens genistein and diadzein are found in soybeans and soy products.

Garlic, leeks, and onions, with their abundant organosulfur compounds, may benefit health.

The phytochemicals of grapes, red wine, and peanuts include resveratrol.

Strawberries are a source of flavonoids .

Citrus fruits provide limonene.

The flavonoids in black tea differ from those in green tea, with different potential effects on health.

Cooked tomatoes are the best source of lycopene, a carotenoid that may reduce cancer risk.

The flavonoids in cocoa, chocolate, fruits and vegetables, legumes, and tea, may benefit health.

Spinach and other green, orange, red, and yellow fruits and vegetables contain carotenoids.

Flaxseed is the richest source of lignans.

Blueberries are among the richest sources of flavonoids.

Garlic, Citrus fruits, © EyeWire, Inc.; Flaxseed, Courtesy of Flax Council of Canada; Broccoli, Courtesy of Brassica Protection Products. Apples, Strawberries, Tomatoes, Blueberries, Black Tea, Soybeans, Grapes, Digital Imagery © 2001 PhotoDisc, Inc.

SOURCE: Data from P. M. Kris-Etherton and coauthors, Bioactive compounds in nutrition and health—research methodologies for establishing biological function: The antioxidant and anti-inflammatory effects of flavonoids on atherosclerosis, *Annual Review of Nutrition* 24 (2004): 511–538.

© Courtesy of Brassica Protection Products. © 2001 PhotoDisc, © EyeWire, Inc., Courtesy of Flax Council of Canada, Photodisc/Getty Images. Matthew Farruggio

If consuming some daily chocolate sounds like a prudent and harmless idea, consider another centuries-old medicinal use of chocolate: promoting weight gain.[12] Three ounces of sweetened chocolate candy contain over 400 calories, a significant portion of most people's daily calorie allowance. At the same time, chocolate contributes few nutrients save two—fat and sugar. For most people,

antioxidant phytochemicals are best obtained from nutrient-dense low-calorie fruits and vegetables and calorie-free green or black tea—with chocolate enjoyed as an occasional treat.

Soybeans

Compared with people in the West, Asians living in Asia suffer less frequently from osteoporosis (adult bone loss); can-

cers, especially of the breast, colon, and prostate; and heart disease. Asian women also suffer less from symptoms related to menopause, the midlife decline in women's estrogen secretion accompanied by the cessation of menstruation.[13] When Asians immigrate to the United States and adopt Western diets and habits, however, they experience these diseases and problems at the same rate as native Westerners. Among many differences between the diets of the two regions, Asians consume far more soybeans (a legume) and soy products such as **miso, soy drink,** and **tofu** than do Westerners.

Soybeans contain phytochemicals known as **phytoestrogens.** Researchers suspect that the phytoestrogens of soy foods, their protein content, or a combination of these factors may be responsible for health effects in soy-eating peoples. However, research, though ongoing, is limited. So far, we know with certainty that phytoestrogens are plant-derived chemical relatives of the human hormone estrogen, that they weakly mimic or modulate the hormone's effects on some body tissues, and that they also act as antioxidants. We also know that breast cancer, colon cancer, and prostate cancer are estrogen-sensitive—that is, they grow when exposed to estrogen. Whether any of these or other actions of phytoestrogens may alter the course of estrogen-sensitive cancers remains unknown, but recent results from breast cancer studies do not support the idea.[14]

As for symptoms of menopause, phytoestrogens may alter a woman's monthly hormonal cycle in ways that may reduce her risk of adult bone loss and the sensation of elevated body temperature known as "hot flashes," common in menopause.[15] Although more research is needed, one review of the literature concluded that a *diet* high in soy foods may offer bone protection rivaling that of hormone replacement therapy (HRT, the administration of female hormone drugs to prevent bone loss and symptoms of menopause).[16] Because HRT involves some serious health risks, supplements of phytoestrogens are often sold to menopausal women as a "natural" alternative, but research does not support taking phytoestrogen *supplements* for bone mineral retention or hot flashes.[17]

Not only do phytoestrogen supplements seem ineffective for preserving health, but their use may also involve some risk. While studying one soy phytoestrogen, **genistein,** researchers found that, instead of suppressing cancer growth, high doses of genistein appeared to speed division of breast cancer cells in laboratory cultures and in mice.[18] Also,

the female offspring of mice treated with high doses of genistein during pregnancy seem prone to developing cancer of the uterus with an incidence even greater than that of a known cancer-causing drug.*[19] Pregnant women should never take chances with unproven supplements of any kind. Chapter 13 provides many reasons why.

The paired, opposing findings on the health effects of phytoestrogens should raise a red warning flag against taking supplements, especially in women whose close relatives have developed breast cancer.[20] In truth, little is known about the potential effects of doses of these compounds in human beings of any age.[21] Until more is known, a safer route to obtaining soy phytoestrogens is to include moderate amounts of soy-based foods in the diet, as generations of Asian people have safely done through the ages.[22]

Flaxseed

Historically, people have used **flaxseed** for relieving constipation or digestive distress. Currently, flaxseed and its oil are under study for potential health benefits.[23] Flaxseed contains **lignans,** compounds converted into biologically active phytoestrogens by bacteria that normally reside in the human intestine. A review of the literature offers this evidence:[24]

- Compared with rats fed an ordinary chow, rats fed chow high in flaxseed develop fewer cancerous changes and smaller tumors in mammary tissue after exposure to chemicals known to cause cancer.
- Cancerous tumors of the lung diminish in size, and new tumor development is significantly reduced in rats fed flaxseed chow.
- Studies of populations suggest that women who excrete more phytoestrogens in the urine (an indicator of phytoestrogen intake from flaxseed and other sources) have lower rates of breast cancer.

Studies of the direct effects of giving flaxseed to *people* are lacking, however,

*The drug is DES, or diethylstilbesterol, once given to pregnant women before the discovery of greatly increased risk of uterine and breast cancer among their daughters.

and some risks are possible with its use. Flaxseed contains compounds that may interfere with vitamin or mineral absorption, and thus high daily intakes could cause nutrient deficiency diseases. Large quantities of flaxseed can also cause digestive distress.

Although no clear role has been established for flaxseed in the prevention of human cancer, including a spoonful or two of flaxseed in the diet may not be a bad idea. Flaxseed richly supplies linolenic acid, a needed nutrient often lacking in the U.S. diet (see Chapter 5).

Tomatoes

People around the world who eat the most tomatoes, say, about five tomato-containing meals per week, are less likely to suffer from cancers of the esophagus, prostate, or stomach than those who avoid tomatoes.[25] Among the phytochemical candidates for promoting this effect is **lycopene,** a red pigment with antioxidant activity found in guava, papaya, pink grapefruit, tomatoes (especially cooked tomatoes and tomato products), and watermelon.

Lycopene may inhibit the reproduction of cancer cells.[26] Some research suggests that an increased risk of breast cancer may follow low blood concentrations of lycopene and related compounds.[27] Low blood lycopene also correlates with an elevated incidence of heart disease, heart attack, and stroke.[28] Lycopene may also protect against the damaging sun rays that cause skin cancer.

In one study, women who consumed a diet rich in fruits and vegetables had high blood lycopene concentrations and a greatly reduced concentration of an indicator of cervical cancer risk.[29] Do scientists conclude, therefore, that lycopene prevents cervical cancer and that women should take lycopene supplements? No, the authors of the study and another like it rightly suggest that a diet rich in fruits and vegetables with their host of nutrients and phytochemicals reduces women's risk of cervical cancer and many other diseases.[30]

A lesson about supplements can be learned from experience with lycopene's chemical cousin, the normally beneficial vitamin A relative beta-carotene. Diets high in fruits and vegetables that contain beta-carotene often correlate with low rates of lung cancer; when beta-carotene supplements were given to smokers in studies, however, lung cancer rates *increased.* For safety, forgo supplements of lycopene until research confirms their effectiveness and rules out hazards.

Garlic

For thousands of years, people have credited garlic with medicinal properties. Descriptions of its uses for headaches, heart disease, and tumors are recorded in early Egyptian medical writings.[31] In modern medical research, over 3,000 publications have investigated potential health benefits of garlic, and many have reported positive findings.[32] Among garlic's active constituents are **organosulfur compounds** that are reported to inhibit cancer development in laboratory animals.[33] These compounds may work by suppressing the formation of certain harmful oxidizing compounds that damage genetic material in animal cells and trigger cancerous changes. This evidence hints that eating garlic may be beneficial against some forms of cancer in human beings.[34] Investigations into potential roles for garlic against allergies, heart disease, and even the bacterial cause of ulcers are ongoing.[35]

Other constituents of garlic also seem promising in promoting heart health. Some may fight fungal infections, reduce the clotting of the blood, or improve levels of blood cholesterol in people whose cholesterol is too high for heart health.[36] Thus, if you like garlic in foods, you can consume them with confidence; history and some research are on your side.

Though these findings hold promise, studies of garlic *supplements,* such as powders and oil, have been disappointing, save one type.[37] Positive results seem to be associated with an aged preparation of garlic that lacks garlic's characteristic odor, but is rich in antioxidants and other compounds.[38] All in all, from the scientists' point of view, no certainty exists about whether large doses of concentrated chemicals from garlic may improve a person's health or injure it.[39]

Supporters and Detractors of Phytochemical Supplements

Users and sellers of supplements argue that the existing evidence is good enough to recommend that people take supplements of purified phytochemicals. Eager for the potential benefits, they seem to discount the potential for harm. They point out that people have been consuming foods containing phytochemicals for tens of thousands of years. The body is clearly accustomed to handling phytochemicals in foods, the reasoning goes, so it follows that supplements of those phytochemicals must be safe for the body as well.

Such thinking raises concerns among scientists. They point out that although the body is equipped to handle phytochemicals when diluted among all of the other constituents of natural foods, it is not adapted to receiving concentrated doses of phytochemicals in supplement form. Consider these facts about phytochemical supplements and health:

1. Phytochemicals can alter body functions, sometimes powerfully, in ways that are only partly understood.
2. Evidence for the safety of isolated phytochemicals in human beings is lacking.
3. No regulatory body oversees the safety of phytochemicals sold to consumers. No studies are required to prove that they are safe or effective before marketing them.
4. Phytochemical labels may make claims about contributing to the body's structure or functioning, but research to support such claims may be weak.

Researchers who study phytochemicals conclude that the best-known, most effective, and safest sources for phytochemicals are foods, not supplements.

If consuming whole *foods* that are rich sources of phytochemicals is wise, would it be wiser still to choose foods that have been enriched with extra doses of phytochemicals? The next section explores some issues surrounding functional foods.

The Concept of Functional Foods

Today, the American food supply is being transformed by a proliferation of functional foods—foods claimed to provide health benefits beyond those of the traditional nutrients.[40] In truth, all the foods mentioned so far in this Controversy, from grains to garlic, are functional—they stand out as having a potentially greater impact on the health of the body (see Table C2-3).[41] Manufacturers have taken this concept a step further by adding phytochemicals and other substances perceived to be beneficial to processed foods.

Not long ago, most of us could agree on what was a food and what was a **drug.** As these new foods come to market, this distinction is becoming less clear.

Cholesterol-Reducing Margarine, Juice, and Candies

Consider some options for lowering high blood cholesterol, an indicator of an increased risk of heart disease. Replacing butter with ordinary liquid margarine in the diet may gradually lower blood cholesterol by a few percentage points over several months. Cholesterol-lowering medication, on the other hand, takes just weeks to dramatically lower cholesterol by as much as 40 percent—clearly, a drug action. But regular daily use of foods such as margarine, juice, or candies enhanced with **sterol esters** or **stanol esters** may achieve a sizable reduction in blood cholesterol, say, 10 to 15 percent over a relatively short time.[42]

The sterol esters added to these foods act like a drug in the body. Also like drugs, sterol esters may have an undesirable side effect: reports link high blood levels of these compounds with early signs of heart disease in people who have an inherited disability to clear them from the blood; they may also lower the blood concentration of some beneficial carotenes, such as lycopene, mentioned earlier.[43]

Sterol esters may pose some degree of risk to certain people. No one yet knows how these substances may affect children and teenagers, yet children and teens may encounter them in the diet in the form of margarine, juice, and can-

TABLE C2-3 CATEGORIES OF FUNCTIONAL FOODS

These categories of functional foods have been identified by the American Dietetic Association:[a]

1. Foods with naturally occurring beneficial compounds, which qualify to bear on their labels an FDA "A" level health claim. These are listed in Table 2-8 of the Consumer Corner of Chapter 2.
2. Foods with naturally occurring beneficial compounds, such as fish that provide fish oils, which have significant evidence supporting their benefits but lack the degree of certainty necessary for an "A" health claim.
3. Foods with biological plausibility for an effect, such as chocolate, garlic, and other foods discussed earlier in this Controversy.
4. Foods fortified with nutrients or other constituents associated with the prevention or treatment of a disease, such as calcium-fortified orange juice marketed to reduce the risk of osteoporosis. Other examples include folate-enriched cereal to reduce birth defects and margarine enhanced with sterol esters to lower blood cholesterol.
5. Fortified foods marketed as "dietary supplements." Examples include beverages, candies, and bars enhanced with antioxidants, minerals, or herbs.

[a]Position of the American Dietetic Association: Functional foods, *Journal of the American Dietetic Association* 104 (2004): 814–826.

dies. As is so often true in nutrition, a cautious approach to these new, potentially beneficial products is warranted.

Yogurt

Yogurt is a special case among functional foods because it contains living *Lactobacillus* or other bacteria that ferment milk into yogurt or the liquid yogurt beverage called **kefir.** Such microorganisms, or **probiotics,** are believed to alter the native bacterial colonies or other conditions in the digestive tract in ways that may reduce diseases.[44] *Lactobacillus* organisms may be useful for improving the diarrhea that often occurs from the use of antibiotic drugs or from other causes, especially among children.[45] Some research also suggests benefits from yogurt against certain chronic intestinal diseases, colon cancer, ulcers, and other digestive problems.[46] More research is needed to verify suggestions that probiotic preparations may alleviate lactose intolerance and allergies, enhance immune function, protect against digestive tract cancers (particularly colon cancer) and ulcers, reduce urinary and vaginal infections in women, and lower blood cholesterol.[47]

Is Every Food Functional?

Many functional foods occur in nature. A serving of **broccoli sprouts,** for example, provides a concentrated source of a phytochemical associated with cancer prevention. Broccoli itself contains this type of phytochemical in less impressive but still healthy doses.[48] Drinking a half-cup of cranberry juice daily may reduce the incidence of urinary tract infections in women because cranberries contain a phytochemical that dislodges bacteria from the tract.[49] Cooked tomatoes, as mentioned, provide lycopene, along with **lutein** (an antioxidant associated with healthy eye function), vitamin C (an antioxidant vitamin), and many other healthful attributes.[50] Lutein also occurs in leafy greens, along with other phytochemicals and beneficial nutrients. In fact, all vegetables, fruits, and whole foods of every kind possess characteristic arrays of thousands of potentially healthful constituents.

Who can say whether even butter and cheese, foods known to contain fats damaging to the heart and arteries when eaten in excess, may qualify as "functional foods" by virtue of their content of **conjugated linoleic acid (CLA)?** In some animal studies, CLA inhibits cancer, lowers cholesterol, improves diabetes, improves immune function, and nudges body composition toward leanness.[51] In other studies, results are negative, however, and little research has been done on the effects of CLA on human beings.[52]

Virtually all foods, even a chocolate bar, have some special value in supporting health.[53] How can a consumer sort all this out and choose wisely? The final section of this Controversy provides some guidance, but first a look at some of the problems concerning functional foods is in order.

Functional Food and Phytochemical Concerns

Serious problems exist concerning manufactured functional foods. Large doses of purified phytochemicals added to foods may produce effects vastly different, even potentially toxic, from those of the phytochemicals in whole foods. For example, concentrated cranberry tablets may significantly increase the taker's risk of developing kidney stones instead of providing the bladder benefits of cranberry juice mentioned earlier.[54]

Aside from the safety issues, conclusions about effectiveness cannot be drawn until research uncertainties are remedied. Even for soy, a popular focus of research, serious problems exist.[55] For example, soy extracts commonly used in experiments come from only the soy germ and lack many of the phytoestrogens present in whole soybeans, rendering comparisons among studies employing various soy products all but meaningless. Also, the way animals metabolize soy phytochemicals differs substantially from most people, short-circuiting human application of some animal results as well. Cell studies, too, continuously expose cells to 10 to 1,000 times the levels of phytochemicals typically achieved by people who eat soy foods in the diet. Until such issues are resolved, the most likely way to benefit from soy is to eat small servings of regular soy foods, such as soy drinks, tofu, or soy burgers, not by taking a large dose from a supplement or highly fortified "functional" food.[56]

Foods sold as functional foods often contain untested medicinal herbs. Such herbs have in recent years caused serious damage to health and even some deaths among consumers (see Chapter 11). Research has not matured nearly enough to identify which isolated phytochemicals, medicinal herbs, or other

Functional foods currently on the market promise to "enhance mood," "promote relaxation and good karma," "increase alertness," and "improve memory," among other claims.

constituents may appropriately be added to foods, and consuming these substances can be risky in unimaginable ways.

One might also question the wisdom of dousing foods of low nutrient density, such as fried snack chips or candies, with phytochemicals and then labeling them "functional," implying that they will enhance health. Such products often seem to be aimed more at consumers' wallets than at improved consumer health.

Until research determines more about functional foods, consumers are on their own to make sure that the products they use are safe and effective. Start by finding answers to the following questions:[57]

- *Does it work?* Well-controlled, peer-reviewed research is generally lacking or inconclusive for manufactured functional foods (see Chapter and Controversy 1 for guidelines to help identify such research).
- *Is it safe?* Check the research for well-controlled safety studies. The active ingredients of functional foods may cause allergies, drug interactions (see Chapter 14's Controversy section), dizziness, and other side effects.
- *Has the Food and Drug Administration (FDA) issued warnings about any of the ingredients?* Check the FDA's MEDWATCH website (www.fda.gov/medwatch) or call the FDA (1-888-INFO-FDA) to find out.
- *How much of what does it contain?* Manufacturers are required to list the names of added herbs and phytochem-

icals on labels, but not the quantities added. Beware, especially, of combinations of "functional" ingredients.
- *Is the food in keeping with the* Dietary Guidelines for Americans? A candy bar or brownie or "smoothie" shake may be fortified with herbs and phytochemicals, but it is still made mostly of sugar and fat.

The Final Word

In light of all of the evidence for and against phytochemicals and functional foods, it seems clear that a moderate approach is warranted. People who eat the recommended amounts of a variety of fruits and vegetables may cut their risk of many diseases by as much as half. Replacing some meat with soy foods or other legumes may also lower heart disease and cancer risks. In the context of a healthy diet, foods are time-tested for safety, posing virtually no risk of toxic levels of nutrients or phytochemicals (although some contain natural toxins: see Chapter 12). Table C2-4 offers some tips for consuming the foods known to provide phytochemicals.

Beneficial constituents are widespread among foods. Don't try to single out one phytochemical for its magical health effect. Instead, take a no-nonsense approach where your health is concerned: Choose a wide variety of whole grains, legumes, fruits, and vegetables in the context of an adequate, balanced, and varied diet and receive all of the health benefits that these foods offer.

TABLE C2·4	TIPS FOR CONSUMING PHYTOCHEMICALS

- **Eat more fruit.** The average U.S. diet provides little more than ½ cup fruit a day. Remember to choose juices and raw, dried, or cooked fruits and vegetables at mealtimes as well as for snacks. Choose dried fruit in place of candy.
- **Increase vegetable portions.** Double the normal portion of cooked plain vegetables to 1 cup.
- **Use herbs and spices.** Cookbooks offer ways to include parsley, basil, garlic, hot peppers, oregano, and other beneficial seasonings.
- **Replace some meat.** Replace some of the meat in the diet with grains, legumes, and vegetables. Oatmeal, soy meat replacer, or grated carrots mixed with ground meat and seasonings make a luscious, nutritious meat loaf, for example.
- **Add grated vegetables.** Carrots in chili or meatballs, celery and squash in spaghetti sauce, etc. add phytochemicals without greatly changing the taste of the food.
- **Try new foods.** Try a new fruit, vegetable, or whole grain each week. Walk through vegetable aisles and visit farmers' markets. Read recipes. Try tofu, fortified soy drink, or soybeans in cooking.

The Remarkable Body

Paul Gauguin (1848–1903 French), *Spring of Miracles*, Hermitage Museum, St. Petersburg, Russia/SuperStock, Inc.

CONTENTS

Do You Ever . . .

Feel your heart beat and wonder where the blood goes?

Hear people say, "You are what you eat," and think it is just an old saying?

Wonder how your food becomes nourishment for your body?

Take antacids to relieve heartburn?

KEEP READING.

AT THE MOMENT of conception, you received genes, in the form of DNA, from your mother and father. Since that moment, your genes have been working behind the scenes, directing your body's development and basic functions. Many of your genes are ancient in origin and little changed from genes of thousands of centuries ago, but here you are—living with the food, the luxuries, the smog, the additives, and all the other pleasures and problems of the twenty-first century. There is no guarantee that a diet haphazardly chosen from today's foods will meet the needs of your "ancient" body. Unlike your ancestors, who nourished themselves from the wild plants and animals surrounding them, you must learn how your body works, what it needs, and how to select foods to meet its needs.

The Body's Cells

The human body is composed of trillions of **cells,** and none of them knows anything about food. *You* may get hungry for fruit, milk, or bread, but each cell of your body needs nutrients—the vital components of foods. The ways in which the body's cells cooperate to obtain and use nutrients are the subjects of this chapter.

Each of the body's cells is a self-contained, living entity (see Figure 3-1), but at the same time it depends on the rest of the body's cells to supply its needs. Among the cells' most basic needs are energy and the oxygen with which to burn it. Cells also need water to maintain the environment in which they live. They need building blocks and control systems. They especially need the nutrients they cannot make for themselves, the essential nutrients first described in Chapter 1, which must be supplied from food. The first principle of diet planning is that the foods we choose must provide energy and the essential nutrients, including water.

As living things, cells also die off, although at varying rates. Some skin cells and red blood cells must replenish themselves every 10 to 120 days. Cells lining the digestive

FIGURE 3·1 A CELL (SIMPLIFIED DIAGRAM)

A membrane encloses each cell's contents.

These fingerlike projections are typical of cells that absorb nutrients in the intestines.

A separate, inner membrane encloses the cell's nucleus.

Inside the nucleus is the hereditary material, which contains the genes. The genes control the inheritance of the cell's characteristics and its day-to-day workings. They are faithfully copied each time the cell duplicates itself.

On these membranes, instructions from the genes are translated into proteins that perform functions in the body.

Many other structures are present. This is a mitochondrion, a structure that takes in nutrients and releases energy from them.

cells the smallest units in which independent life can exist. All living things are single cells or organisms made of cells.

tract replace themselves every three days. Under ordinary conditions, many muscle cells reproduce themselves only once every few years. Liver cells have the ability to reproduce quickly and do so whenever repairs to the organ are needed. Certain brain cells do not reproduce at all; if damaged by injury or disease, they are lost forever.

The Workings of the Genes

In the human body, cells work in cooperation with each other to support the whole. A cell's genes determine the nature of that work. Each gene is a blueprint that directs the production of a piece of protein machinery, often an **enzyme,** that helps to do the cell's work. Genes also provide the instructions for all of the structural components cells need to survive (see Figure 3-2). Each cell contains a complete set of genes, but different ones are active in different types of cells. For example, in some intestinal cells, the genes for making digestive enzymes are active; in some of the body's **fat cells,** the genes for making enzymes that metabolize fat are active.

Genes affect the body's handling of nutrients. Certain genetic variations alter the absorption, use, or excretion of nutrients. Some variations cause lifelong maladies that require special diets to minimize damage to health. Nutrients also affect the genes. For

enzyme any of a great number of working proteins that speeds up a specific chemical reaction, such as breaking the bonds of a nutrient, without undergoing change itself. Enzymes and their actions are described in Chapter 6.

fat cells cells that specialize in the storage of fat and form the fat tissue. Fat cells also produce enzymes that metabolize fat and hormones involved in appetite and energy balance (see Chapter 9).

FIGURE 3·2 FROM DNA TO LIVING CELLS

If the human genome were a book of instructions on how to make a human being, then the 23 chromosomes of DNA would be chapters. Each gene would be a word, and the individual molecules that form the DNA would be letters of the alphabet.

1 Nucleus   Cell

DNA 2

3 Genes

Protein 4 Protein Protein

6 Living tissue

5 Living cell

1 Each cell's nucleus contains DNA — the material of heredity in all living things.

2 Long strands of human DNA coil into 23 pairs of chromosomes. If the strands of DNA in all the body's cells were uncoiled and laid end to end, they would stretch to the sun and back four hundred times. Yet DNA strands are so tiny that about 5 million of them could be threaded at once through the eye of a needle.

3 Genes contain instructions for making proteins. Genes are sections along the strands of DNA that serve as templates for the building of proteins. Some genes are involved in building just one protein; others are involved in building more than one.

4 Many other steps are required to make a protein. See Figure 6-6 of Chapter 6.

5 Proteins do the work of living cells. Cells employ proteins to perform essential functions and provide structures.

6 Communities of functioning cells make up the living tissue.

■ Interactions between vitamins and minerals and the genes are addressed in Chapters 7 and 8; other nutrient and gene interactions are addressed in the chapters on pregnancy and disease prevention.

example, the concentrations of certain vitamins and minerals influence the genes' production of nutrient metabolizing proteins, with sweeping effects on body systems that originate at the cellular level.

Cells, Tissues, Organs, Systems

Cells are organized into **tissues** that perform specialized tasks. For example, individual muscle cells are joined together to form muscle tissue, which can contract. Tissues, in turn, are grouped together to form whole **organs.** In the organ we call the heart, for example, muscle tissues, nerve tissues, connective tissues, and other types all work together to pump blood. Some body functions are performed by several related organs working together as part of a **body system.** For example, the heart, lungs, and blood vessels cooperate as parts of the cardiovascular system to deliver oxygen to all the body's cells. The next few sections present the body systems with special significance to nutrition.

▶ KEY POINT   The body's cells need energy, oxygen, and nutrients, including water, to remain healthy and do their work. Genes direct the making of each cell's machinery, including enzymes. Genes and nutrients interact in ways that affect health. Specialized cells are grouped together to form tissues and organs; organs work together in body systems.

tissues systems of cells working together to perform specialized tasks. Examples are muscles, nerves, blood, and bone.

organs discrete structural units made of tissues that perform specific jobs. Examples are the heart, liver, and brain.

body system a group of related organs that work together to perform a function. Examples are the circulatory system, respiratory system, and nervous system.

blood the fluid of the cardiovascular system; composed of water, red and white blood cells, other formed particles, nutrients, oxygen, and other constituents.

lymph (LIMF) the fluid that moves from the bloodstream into tissue spaces and then travels in its own vessels, which eventually drain back into the bloodstream.

arteries blood vessels that carry blood containing fresh oxygen supplies from the heart to the tissues (see Figure 3-3, p. 69).

veins blood vessels that carry blood, with the carbon dioxide it has collected, from the tissues back to the heart (see Figure 3-3, p. 69).

capillaries minute, weblike blood vessels that connect arteries to veins and permit transfer of materials between blood and tissues (see Figures 3-3 and 3-4, pp. 69 and 70).

plasma the cell-free fluid part of blood and lymph.

extracellular fluid fluid residing outside the cells that transports materials to and from the cells.

intracellular fluid fluid residing inside the cells that provides the medium for cellular reactions.

lungs the body's organs of gas exchange. Blood circulating through the lungs releases its carbon dioxide and picks up fresh oxygen to carry to the tissues.

intestine the body's long, tubular organ of digestion and the site of nutrient absorption.

liver a large, lobed organ that lies just under the ribs. It filters the blood, removes and processes nutrients, manufactures materials for export to other parts of the body, and destroys toxins or stores them to keep them out of the circulation.

The Body Fluids and the Cardiovascular System

Body fluids supply the tissues continuously with energy, oxygen, and nutrients, including water. The fluids constantly circulate to pick up fresh supplies and deliver wastes to points of disposal. Every cell continuously draws oxygen and nutrients from those fluids and releases carbon dioxide and other waste products into them.

The body's circulating fluids are the **blood** and the **lymph.** Blood travels within the **arteries, veins,** and **capillaries,** as well as within the heart's chambers (see Figure 3-3). Lymph travels in separate vessels of its own. Circulating around the cells are other fluids such as the **plasma** of the blood, which surrounds the white and red blood cells, and the fluid surrounding muscle cells (see Figure 3-4, page 70). The fluid surrounding cells (**extracellular fluid**) is derived from the blood in the capillaries; it squeezes out through the capillary walls and flows around the outsides of cells, permitting exchange of materials. Some of the extracellular fluid returns to the blood by reentering the capillaries. The fluid remaining outside the capillaries forms lymph, which travels around the body by way of lymph vessels. The lymph eventually returns to the bloodstream near the heart where large lymph and blood vessels join. In this way, all cells are served by the cardiovascular system.

The fluid inside cells (**intracellular fluid**) provides a medium in which all cell reactions take place. Its pressure also helps the cells to hold their shape. The intracellular fluid is drawn from the extracellular fluid that bathes the cells on the outside.

All the blood circulates to the **lungs** where it picks up oxygen and releases carbon dioxide wastes from the cells, as Figure 3-5 (page 71) shows. Then the blood returns to the heart, where the pumping heartbeats push this fresh oxygenated blood from the lungs out to all body tissues. As the blood travels through the rest of the cardiovascular system, it delivers materials cells need and picks up their wastes.

As it passes through the digestive system, the blood delivers oxygen to the cells there and picks up most nutrients other than fats and their relatives from the **intestine** for distribution elsewhere. Lymphatic vessels pick up most fats from the intestine and then transport them to the blood. All blood leaving the digestive system is routed directly to the **liver,** which has the special task of chemically altering the absorbed

FIGURE 3·3   BLOOD FLOW IN THE CARDIOVASCULAR SYSTEM

Lungs

Heart

Liver

Kidneys

Intestines

Head and **Arms**

Lungs
Oxygenate
  blood
Remove carbon dioxide
  from blood
Return blood to heart

Right   Left

Heart
Right side pumps
  blood to lungs
Left side pumps
  oxygenated
  blood to body

Liver
Filters toxins from blood
Stores, transforms, and
  mobilizes nutrients

Intestines
Absorb nutrients

Kidneys
Filter wastes from
  blood
Form urine

Pelvis and **Legs**

The blood is routed through the body as follows:
• Heart to tissues to heart to lungs to heart
  (repeat).
The portion of the blood that flows through the
blood vessels of the intestine travels from:
• Heart to intestine to liver to heart.

materials to make them better suited for use by other tissues. Later, in passing through
the **kidneys,** the blood is cleansed of wastes (look again at Figure 3-3). Note that the
blood carries nutrients from the intestine to the liver, which releases them to the heart,
which pumps them to the waiting body tissues.

To ensure efficient circulation of fluid to all your cells, you need an ample fluid
intake. This means drinking sufficient water to replace the water lost each day.

kidneys a pair of organs that filter wastes
from the blood, make urine, and release it to
the bladder for excretion from the body.

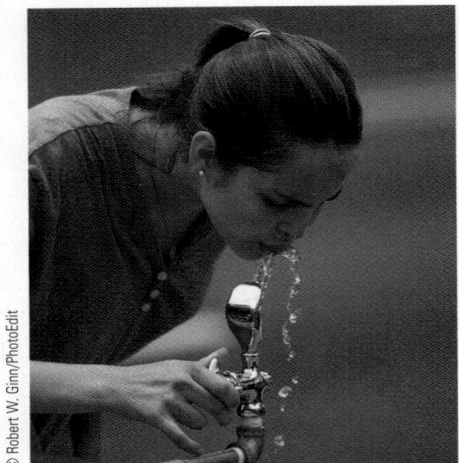

All the body's cells live in water.

FIGURE 3·4 HOW THE BODY FLUIDS CIRCULATE AROUND CELLS

The upper box shows a tiny portion of tissue with blood flowing through its network of capillaries (greatly enlarged). The lower box illustrates the movement of the extracellular fluid. Exchange of materials also takes place between cell fluid and extracellular fluid.

Lymph vessel.

Blood circulates among cells by way of capillaries.

Blood collects into veins for return to heart.

Blood enters tissues by way of artery.

Lymph vessel.

Inside capillary.

Capillary wall has spaces between its flat cells.

Cells of surrounding tissue.

Lymph vessel.

Fluid filters out of blood through the capillary whose walls are made of cells with small spaces between them.

Fluid may flow back into capillary or into lymph vessel. Lymph enters the bloodstream later through a large lymphatic vessel that empties into a large vein.

■ Chapter 8 offers guidelines for water intake.

Cardiovascular fitness is essential, too, and constitutes an ongoing project that requires attention to both nutrition and physical activity. Healthy red blood cells also play a role, for they carry oxygen to all the other cells, enabling them to use fuels for energy. Since red blood cells arise, live, and die within about four months, your body replaces them constantly, a manufacturing process that requires many essential nutrients from food. Consequently, the blood is very sensitive to malnutrition and often serves as an indicator of disorders caused by dietary deficiencies or imbalances of vitamins or minerals.

▶ KEY POINT   Blood and lymph deliver nutrients to all the body's cells and carry waste materials away from them. Blood also delivers oxygen to cells. The cardiovascular system ensures that these fluids circulate properly among all organs.

hormones chemicals that are secreted by glands into the blood in response to conditions in the body that require regulation. These chemicals serve as messengers, acting on other organs to maintain constant conditions.

The Hormonal and Nervous Systems

In addition to nutrients, oxygen, and wastes, the blood also carries chemical messengers, **hormones,** from one system of cells to another. Hormones communicate changing conditions that demand responses from the body organs.

FIGURE 3·5 OXYGEN–CARBON DIOXIDE EXCHANGE IN THE LUNGS

| body tissue | tissue capillary | lung capillary | air sac in lung |

In body tissues, red blood cells give up their oxygen (O_2) and absorb carbon dioxide (CO_2).

In the air sacs of the lungs, the red blood cells give up their load of carbon dioxide (CO_2) and absorb oxygen (O_2) from air to supply to body tissues.

What Do Hormones Have to Do with Nutrition?

Hormones are secreted and released directly into the blood by organs known as glands. Glands and hormones abound in the body. Each gland monitors a condition and produces one or more hormones to regulate it. Each hormone acts as a messenger that stimulates various organs to take appropriate actions.

For example, when the **pancreas** (a gland) detects a high concentration of the blood's sugar, glucose, it releases **insulin,** a hormone. Insulin stimulates muscle and other cells to remove glucose from the blood and to store it. The liver also stores glucose. When the blood glucose level falls, the pancreas secretes another hormone, **glucagon,** to which the liver responds by releasing into the blood some of the glucose it stored earlier. Thus, normal blood glucose levels are maintained.

Nutrition affects the hormonal system. Fasting, feeding, and exercise alter hormonal balances. People who become very thin have an altered hormonal balance that may make them unable to maintain their bones, for example.

Hormones also affect nutrition. Along with the nervous system, hormones regulate hunger and affect appetite. They carry messages to regulate the digestive system, telling the digestive organs what kinds of foods have been eaten and how much of each digestive juice to secrete in response. A hormone produced by the fat tissue informs the brain about the degree of body fatness and helps to regulate appetite. Hormones also regulate the menstrual cycle in women, and they affect the appetite changes many women

pancreas an organ with two main functions. One is an endocrine function—the making of hormones such as insulin, which it releases directly into the blood (*endo* means "into" the blood). The other is an exocrine function—the making of digestive enzymes, which it releases through a duct into the small intestine to assist in digestion (*exo* means "out" into a body cavity or onto the skin surface).

insulin a hormone from the pancreas that helps glucose enter cells from the blood (details in Chapter 4).

glucagon a hormone from the pancreas that stimulates the liver to release glucose into the bloodstream.

■ Details about hormones, menstruation, and the bones appear in Controversy 8 and Controversy 9.

experience during the cycle and in pregnancy. An altered hormonal state is thought to be at least partially responsible, too, for the loss of appetite that sick people experience. Hormones also regulate the body's reaction to stress, suppressing hunger and the digestion and absorption of nutrients. When there are questions about a person's nutrition or health, the state of that person's hormonal system is often part of the answer.

▶ KEY POINT   Glands secrete hormones that act as messengers to help regulate body processes.

How Does the Nervous System Interact with Nutrition?

The body's other major communication system is, of course, the nervous system. With the brain and spinal cord as central controllers, the nervous system receives and integrates information from sensory receptors all over the body—sight, hearing, touch, smell, taste, and others—which communicate to the brain the state of both the outer and inner worlds, including the availability of food and the need to eat. The nervous system also sends instructions to the muscles and glands, telling them what to do.

The nervous system's role in hunger regulation is coordinated by the brain. The sensations of hunger and appetite are perceived by the brain's **cortex,** the thinking, outer layer. Deep inside the brain, the **hypothalamus** (see Figure 3-6) monitors many body conditions, including the availability of nutrients and water. To signal hunger, the physiological need for food, the digestive tract sends messages to the hypothalamus by way of hormones and nerves. The signals also stimulate the stomach to intensify its contractions and secretions, causing hunger pangs (and gurgling sounds). When your brain's cortex perceives these hunger sensations, you want to eat. The conscious mind

FIGURE 3·6   CUTAWAY SIDE VIEW OF THE BRAIN SHOWING THE HYPOTHALAMUS AND CORTEX

The hypothalamus monitors the body's conditions and sends signals to the brain's thinking portion, the cortex, which decides on actions. The pituitary gland is called the body's master gland, referring to its roles in regulating the activities of other glands and organs of the body.

cortex the outermost layer of something. The brain's cortex is the part of the brain where conscious thought takes place.

hypothalamus (high-poh-THAL-uh-mus) a part of the brain that senses a variety of conditions in the blood, such as temperature, glucose content, salt content, and others. It signals other parts of the brain or body to adjust those conditions when necessary.

of the cortex, however, can override such signals, and a person can choose to delay eating despite hunger or to eat when hunger is absent.

In a marvelous adaptation of the human body, the hormonal and nervous systems work together to enable a person to respond to physical danger. Known as the **fight-or-flight reaction,** or the *stress response*, this adaptation is present with only minor variations in all animals, showing how universally important it is to survival. When danger is detected, nerves release **neurotransmitters,** and glands supply the compounds **epinephrine** and **norepinephrine.**[*] Every organ of the body responds and **metabolism** speeds up. The pupils of the eyes widen so that you can see better; the muscles tense up so that you can jump, run, or struggle with maximum strength; breathing quickens and deepens to provide more oxygen. The heart races to rush the oxygen to the muscles, and the blood pressure rises so that the fuel the muscles need for energy can be delivered efficiently. The liver pours forth glucose from its stores, and the fat cells release fat. The digestive system shuts down to permit all the body's systems to serve the muscles and nerves. With all action systems at peak efficiency, the body can respond with amazing speed and strength to whatever threatens it.

In ancient times, stress usually involved physical danger, and the response to it was violent physical exertion. In the modern world, stress is seldom physical, but the body reacts the same way. What stresses you today may be a checkbook out of control or a teacher who suddenly announces a pop quiz. Under these stresses, you are not supposed to fight or run as your ancient ancestor did. You smile at the "enemy" and suppress your fear. But your heart races, you feel it pounding, and hormones still flood your bloodstream with glucose and fat.

Your number-one enemy today is not a saber-toothed tiger prowling outside your cave, but a disease of modern civilization: heart disease. Years of fat and other constituents accumulating in the arteries and stresses that strain the heart often lead to heart attacks, especially when a body accustomed to chronic underexertion experiences sudden high blood pressure. Daily exercise as part of a healthy lifestyle releases pent-up stress and helps to protect the heart.

> ▶ **KEY POINT** The nervous system joins the hormonal system to regulate body processes through communication among all the organs. Together, the hormonal and nervous systems respond to the need for food, govern the act of eating, regulate digestion, and call for the stress response.

The Immune System

Many of the body's tissues cooperate to maintain defenses against infection. The skin presents a physical barrier, and the body's cavities (lungs, digestive tract, and others) are lined with membranes that resist penetration by invading **microbes** and other unwanted substances. These linings are highly sensitive to vitamin and other nutrient deficiencies, and health-care providers inspect both the skin and the inside of the mouth to detect signs of malnutrition. (Later chapters present details of the signs of deficiencies.) If an **antigen,** or foreign invader, penetrates the body's barriers, the **immune system** rushes in to defend the body against harm.

Of the 100 trillion cells that make up the human body, one in every hundred is a white blood cell. The actions of two types of white blood cells, the phagocytes and the **lymphocytes** known as T-cells and B-cells, are of interest:

- **Phagocytes.** These scavenger cells travel throughout the body and are the first to defend body tissues against invaders. When a phagocyte recognizes a foreign particle, such as a bacterium, the phagocyte forms a pocket in its own outer membrane,

fight-or-flight reaction the body's instinctive hormone- and nerve-mediated reaction to danger. Also known as the *stress response*.

neurotransmitters chemicals that are released at the end of a nerve cell when a nerve impulse arrives there. They diffuse across the gap to the next cell and alter the membrane of that second cell to either inhibit or excite it.

epinephrine (EP-ih-NEFF-rin) the major hormone that elicits the stress response.

norepinephrine (NOR-EP-ih-NEFF-rin) a compound related to epinephrine that helps to elicit the stress response.

metabolism the sum of all physical and chemical changes taking place in living cells; includes all reactions by which the body obtains and spends the energy from food.

microbes bacteria, viruses, or other organisms invisible to the naked eye, some of which cause diseases. Also called *microorganisms*.

antigen a microbe or substance that is foreign to the body.

immune system a system of tissues and organs that defend the body against antigens, foreign materials that have penetrated the skin or body linings.

lymphocytes (LIM-foh-sites) white blood cells that participate in the immune response; B-cells and T-cells.

---

[*]Strictly speaking, norephinephrine is a neurotransmitter.

■ More about oxidation in Chapter 5 and Chapter 7.

A killer T-cell (the smaller cell on the bottom) has recognized a cancer cell and is releasing toxic chemicals that punch holes in the cancer cell's surface, destroying it. A nutritious diet supports immune system functioning.

Boehringer Ingelheim International GmbH, photo Lennart Nilsson/ Bonnier Alba AB

engulfing the invader. Then, the phagocytes may attack the invader with oxidative chemicals in an "oxidative burst" or may otherwise digest or destroy them. Phagocytes also leave a chemical trail that helps other immune cells to join the defense against infection.

- **T-cells.** Killer T-cells recognize chemical messages from phagocytes and "read" and "remember" the identity of an invader from the messages. The killer T-cells then seek out and destroy all foreign particles having the same identity. T-cells defend against fungi, viruses, parasites, some bacteria, and some cancer cells (see the photo). They also pose a formidable obstacle to a successful organ transplant—the physician must prescribe immunosuppressive drugs following surgery to hold down the T-cells' attack against the "foreign" organ. Another group, helper T-cells, do not attack invaders but help other immune cells to do so. People suffering from the disease AIDS (acquired immunodeficiency syndrome) are rendered defenseless against other diseases because the human immunodeficiency virus (HIV) that causes AIDS selectively attacks and destroys their helper T-cells.
- **B-cells.** B-cells respond rapidly to infection by dividing and releasing invader-fighting proteins, **antibodies,** into the bloodstream. Antibodies travel to the site of the infection and stick to the surface of the foreign particles, killing or inactivating them. Like T-cells, the B-cells also retain a chemical memory of each invader, and if the encounter recurs, the response is swift. Immunizations work this way—a disabled or harmless form of a disease-causing organism is injected into the body so that the B-cells can learn to recognize it. Later, if the real, live infectious organism invades, the B-cells quickly release antibodies to destroy it.

Chapter 11 describes the roles of nutrition in supporting the body's defense system.

▶ **KEY POINT** The immune system enables the body to resist diseases.

The Digestive System

When your body needs food, your brain and hormones alert your conscious mind to the sensation of hunger. Then, when you eat, your taste buds guide you in judging whether foods are acceptable.

On the surfaces of the taste buds are structures that detect four basic chemical tastes: sweet, sour, bitter, and salty. A fifth taste is sometimes included on this list: the taste of monosodium glutamate, sometimes called *savory* or *umami* (ooh-MOM-ee), its Asian name. Aroma, texture, temperature, and other flavor elements can also affect a food's flavor. In fact, the human ability to detect a food's aroma is thousands of times more sensitive than our sense of taste. The nose can detect just a few molecules responsible for the aroma of frying bacon, for example, even when they are diluted in several rooms full of air.

Why Do People Like Sugar, Fat, and Salt?

phagocytes (FAG-oh-sites) white blood cells that can ingest and destroy antigens. The process by which phagocytes engulf materials is called *phagocytosis*. The Greek word *phagein* means "to eat."

T-cells lymphocytes that attack antigens. *T* stands for the thymus gland of the neck, where the T-cells are stored and matured.

B-cells lymphocytes that produce antibodies. *B* stands for bursa, an organ in the chicken where B-cells were first identified.

antibodies proteins, made by cells of the immune system, that are expressly designed to combine with and inactivate specific antigens.

Sweet, salty, and fatty foods seem to be universally desired, but most people have aversions to bitter and sour tastes in isolation (see Figure 3-7).[*1] The enjoyment of sugars and fat encourages people to consume ample energy, especially in the form of foods containing sugars, which provide the energy fuel for the brain. Likewise, foods containing fats provide energy and essential nutrients needed by all body tissues. The pleasure of a salty taste prompts eaters to consume sufficient amounts of two very

*Reference notes are found in Appendix F.

FIGURE 3·7 THE INNATE PREFERENCE FOR SWEET TASTE

This newborn baby is (a) resting, (b) tasting distilled water, (c) tasting sugar, (d) tasting something sour, and (e) tasting something bitter.

(a)   (b)   (c)

(d)   (e)

SOURCE: Taste-induced facial expressions of neonate infants from the classic studies of J. E. Steiner, in *Taste and Development: The Genesis of Sweet Preference,* ed. J. M. Weiffenbach, HHS publication no. NIH 77–1068 (Bethesda, Md.: U.S. Department of Health and Human Services, 1977), pp. 173–189, with permission of the author.

important minerals—sodium and chloride. The aversion to bitterness discourages consumption of foods containing bitter toxins and also affects people's food preferences. People born with great sensitivity to bitter tastes are apt to avoid foods with slightly bitter flavors, such as turnips and broccoli.

The instinctive liking for sugar, fat, and salt can lead to drastic overeating of these substances. Sugar has become widely available in pure form only in the last hundred years, so it is relatively new to the human diet. Although fat and salt are much older, today all three substances are being added liberally to foods by manufacturers to tempt us to eat their products.

▶ KEY POINT   The preference for sweet, salty, and fatty tastes seems to be inborn and can lead to overconsumption of foods that offer them.

The Digestive Tract

Once you have eaten, your brain and hormones direct the many organs of the **digestive system** to **digest** and **absorb** the complex mixture of chewed and swallowed food. A diagram showing the digestive tract and its associated organs appears in Figure 3-8 (page 76). The tract itself is a flexible, muscular tube extending from the mouth through the throat, esophagus, stomach, small intestine, large intestine, and rectum to the anus, for a total length of about 26 feet. The human body surrounds this digestive canal. When you swallow something, it still is not inside your body—it is only inside the inner bore of this tube. Only when a nutrient or other substance passes through the wall of the digestive tract does it actually enter the body's tissues. Many things pass into the digestive tract and out again, unabsorbed. A baby playing with beads may swallow one, but the bead will not really enter the body. It will emerge from the digestive tract within a day or two.

The digestive system's job is to digest food to its components and then to absorb the nutrients and some nonnutrients, leaving behind the substances, such as fiber, that are

digestive system the body system composed of organs that break down complex food particles into smaller, absorbable products. The *digestive tract* and *alimentary canal* are names for the tubular organs that extend from the mouth to the anus. The whole system, including the pancreas, liver, and gallbladder, is sometimes called the *gastrointestinal,* or *GI,* system.

digest to break molecules into smaller molecules; a main function of the digestive tract with respect to food.

absorb to take in, as nutrients are taken into the intestinal cells after digestion; the main function of the digestive tract with respect to nutrients.

FIGURE 3·8 THE DIGESTIVE SYSTEM

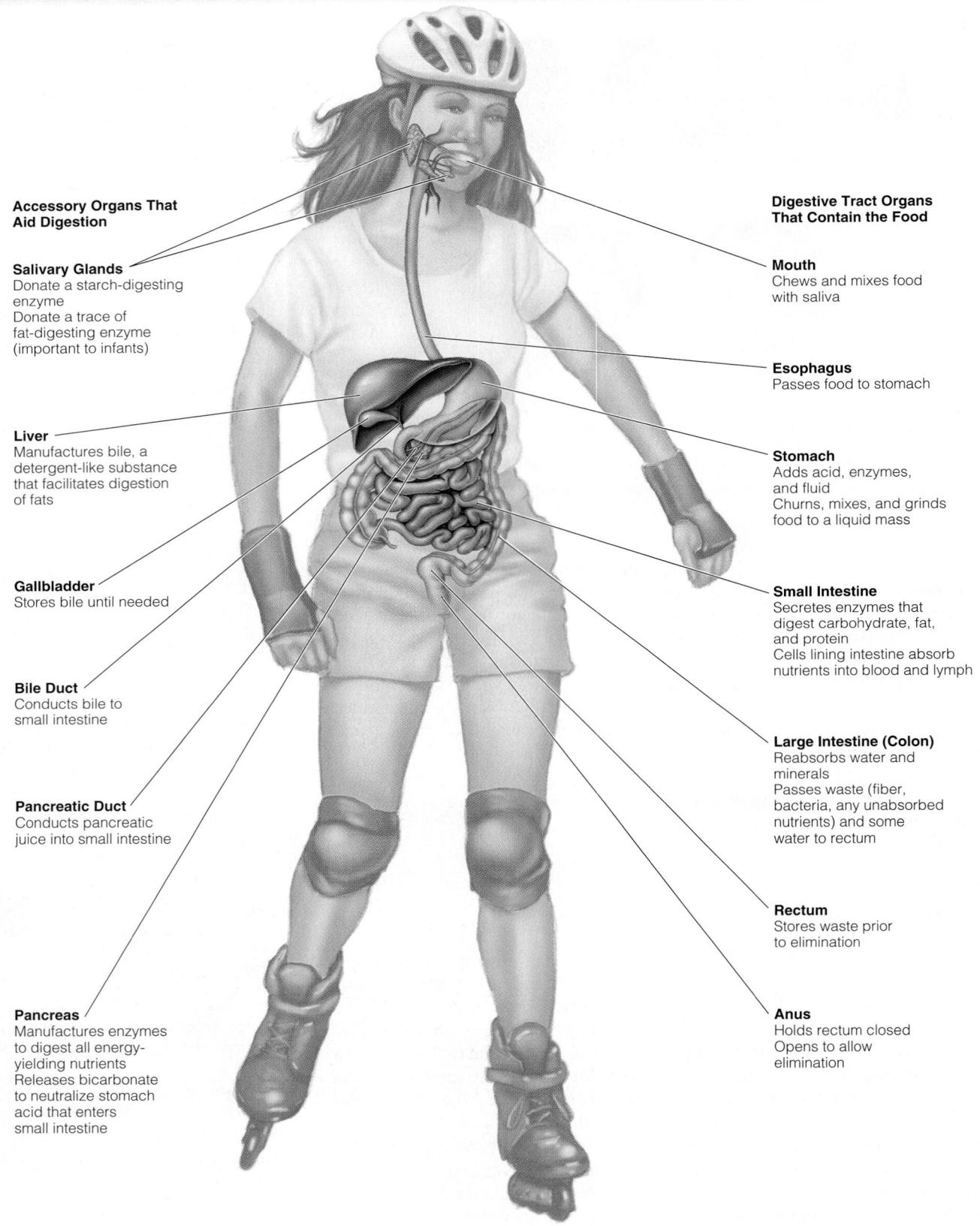

Accessory Organs That Aid Digestion

Salivary Glands
Donate a starch-digesting enzyme
Donate a trace of fat-digesting enzyme (important to infants)

Liver
Manufactures bile, a detergent-like substance that facilitates digestion of fats

Gallbladder
Stores bile until needed

Bile Duct
Conducts bile to small intestine

Pancreatic Duct
Conducts pancreatic juice into small intestine

Pancreas
Manufactures enzymes to digest all energy-yielding nutrients
Releases bicarbonate to neutralize stomach acid that enters small intestine

Digestive Tract Organs That Contain the Food

Mouth
Chews and mixes food with saliva

Esophagus
Passes food to stomach

Stomach
Adds acid, enzymes, and fluid
Churns, mixes, and grinds food to a liquid mass

Small Intestine
Secretes enzymes that digest carbohydrate, fat, and protein
Cells lining intestine absorb nutrients into blood and lymph

Large Intestine (Colon)
Reabsorbs water and minerals
Passes waste (fiber, bacteria, any unabsorbed nutrients) and some water to rectum

Rectum
Stores waste prior to elimination

Anus
Holds rectum closed
Opens to allow elimination

appropriate to excrete. To do this, the system works at two levels: one, mechanical; the other, chemical.

> **KEY POINT** The digestive tract is a flexible, muscular tube that digests food and absorbs its nutrients and some nonnutrients. Ancillary digestive organs aid digestion.

The Mechanical Aspect of Digestion

The job of mechanical digestion begins in the mouth, where large, solid food pieces such as bites of meat are torn into shreds that can be swallowed without choking. Chewing also adds water in the form of saliva to soften rough or sharp foods, such as fried tortilla chips, to prevent them from tearing the esophagus. Saliva also moistens and coats each bite of food, making it slippery so that it can pass easily down the esophagus.

Nutrients trapped inside indigestible skins, such as the hulls of seeds, must be liberated by breaking these skins before they can be digested. Chewing bursts open kernels of corn, for example, which would otherwise traverse the tract and exit undigested. Once food has been mashed and moistened for comfortable swallowing, longer chewing times provide no additional advantages to digestion. In fact, for digestion's sake, a relaxed, peaceful attitude during a meal aids digestion much more than chewing for an extended time.

The stomach and intestines then take up the task of liquefying foods through various mashing and squeezing actions. The best known of these actions is **peristalsis,** a series of squeezing waves that start with the tongue's movement during a swallow and pass all the way down the esophagus (see Figure 3-9). The stomach and the intestines also push food through the tract by waves of peristalsis. Besides these actions, the stomach holds swallowed food for a while and mashes it into a fine paste; the **stomach** and intestines also add water so that the paste becomes more fluid as it moves along.

Figure 3-10 (p. 78) shows the muscular stomach. Notice the circular **sphincter** muscle at the base of the esophagus. It squeezes the opening at the entrance to the stomach to narrow it and prevent the stomach's contents from creeping back up the esophagus as the stomach contracts. The stomach stores swallowed food in a lump in its upper portion and

peristalsis (perri-STALL-sis) the wavelike muscular squeezing of the esophagus, stomach, and small intestine that pushes their contents along.

stomach a muscular, elastic, pouchlike organ of the digestive tract that grinds and churns swallowed food and mixes it with acid and enzymes, forming chyme.

sphincter (SFINK-ter) a circular muscle surrounding, and able to close, a body opening.

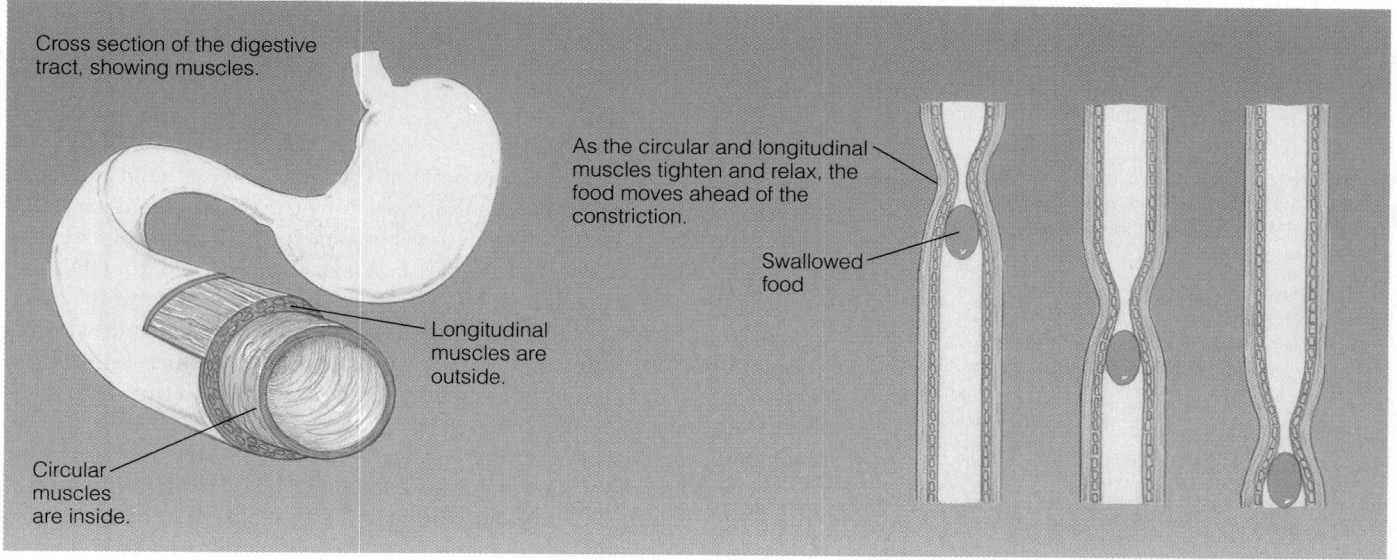

FIGURE 3·9 PERISTALTIC WAVE PASSING DOWN THE ESOPHAGUS AND BEYOND

Peristalsis moves the digestive tract contents.

Cross section of the digestive tract, showing muscles.

As the circular and longitudinal muscles tighten and relax, the food moves ahead of the constriction.

Swallowed food

Longitudinal muscles are outside.

Circular muscles are inside.

FIGURE 3·10 THE MUSCULAR STOMACH

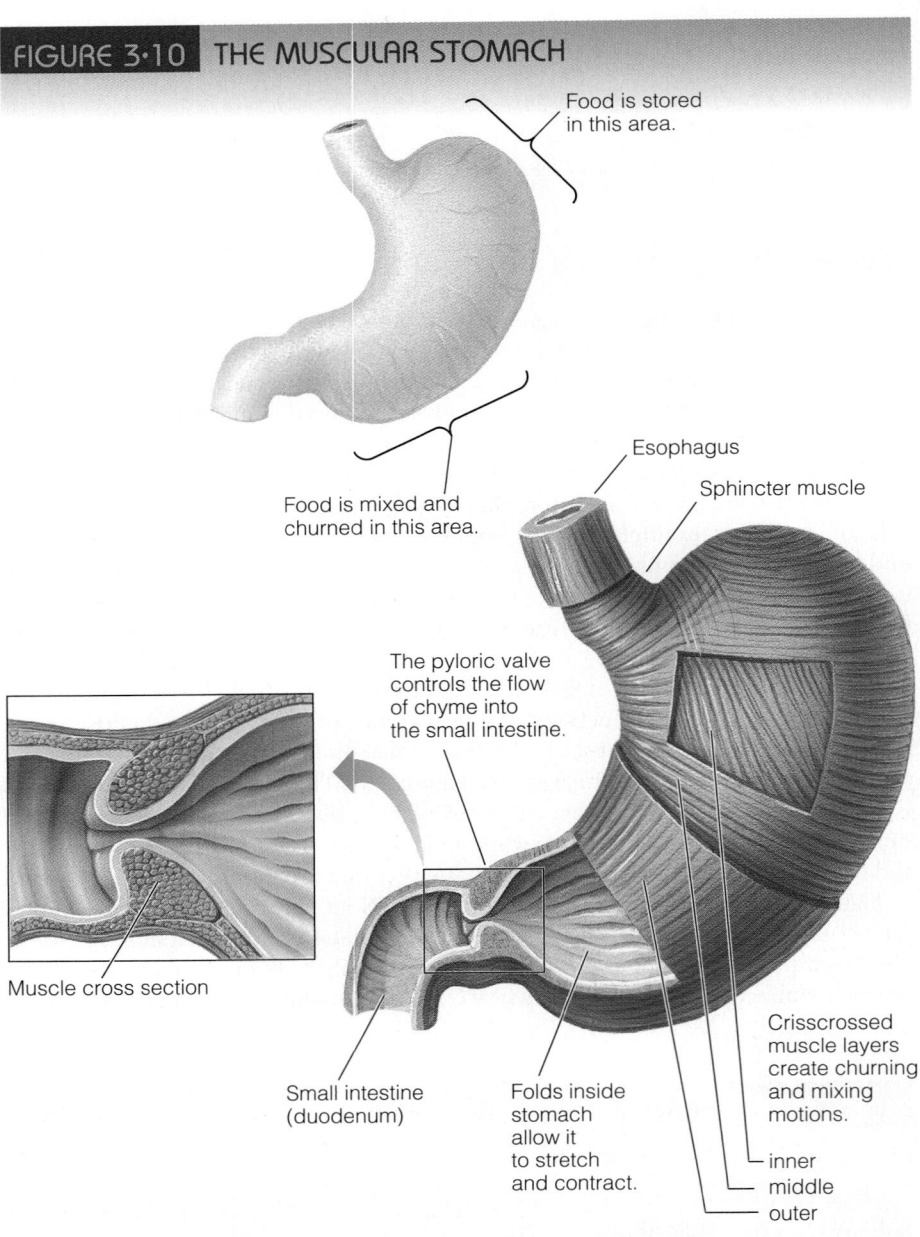

Food is stored in this area.

Food is mixed and churned in this area.

Esophagus

Sphincter muscle

The pyloric valve controls the flow of chyme into the small intestine.

Muscle cross section

Small intestine (duodenum)

Folds inside stomach allow it to stretch and contract.

Crisscrossed muscle layers create churning and mixing motions.

inner
middle
outer

chyme (KIME) the fluid resulting from the actions of the stomach upon a meal.

pyloric (pye-LORE-ick) **valve** the circular muscle of the lower stomach that regulates the flow of partly digested food into the small intestine. Also called *pyloric sphincter.*

small intestine the 20-foot length of small-diameter intestine, below the stomach and above the large intestine, that is the major site of digestion of food and absorption of nutrients.

large intestine the portion of the intestine that completes the absorption process.

colon the large intestine.

feces waste material remaining after digestion and absorption are complete; eventually discharged from the body.

squeezes the food little by little to its lower portion. There the food is ground and mixed thoroughly, ensuring that digestive chemicals mix with the entire thick liquid mass, now called **chyme.** Chyme bears no resemblance to the original food. The starches have been partly split, proteins have been uncoiled and clipped, and fat has separated from the mass.

The stomach also acts as a holding tank. The muscular **pyloric valve** at the stomach's lower end (look again at Figure 3-10) controls the exit of the chyme, allowing only a little at a time to be squirted forcefully into the **small intestine.** Within a few hours after a meal, the stomach empties itself by means of these powerful squirts. The small intestine contracts rhythmically to move the contents along its length.

By the time the intestinal contents have arrived in the **large intestine** (also called the **colon**), digestion and absorption are nearly complete. The colon's task is mostly to reabsorb the water donated earlier by digestive organs and to absorb minerals, leaving a paste of fiber and other undigested materials, the **feces,** suitable for excretion. The fiber provides bulk against which the muscles of the colon can work. The rectum stores this fecal material to be excreted at intervals. From mouth to rectum, the transit of a meal is accomplished in as short a time as a single day or as long as three days.

Some people wonder whether the digestive tract works best at certain hours in the day and whether the timing of meals can affect how a person feels. Timing of meals is important to feeling well, not because the digestive tract is unable to digest food at certain times, but because the body requires nutrients to be replenished every few hours. Digestion is virtually continuous, being limited only during sleep and exercise. For some people, eating late may interfere with normal sleep. As for exercise, it is best pursued a few hours after eating because digestion can inhibit physical work (see Chapter 10 for details).

▶ **KEY POINT**   The digestive tract moves food through its various processing chambers by mechanical means. The mechanical actions include chewing, mixing by the stomach, adding fluid, and moving the tract's contents by peristalsis. After digestion and absorption, wastes are excreted.

The Chemical Aspect of Digestion

Several organs of the digestive system secrete special digestive juices that perform the complex chemical processes of digestion. Digestive juices contain enzymes that break nutrients down into their component parts. The digestive organs that release digestive juices are the salivary glands, the stomach, the pancreas, the liver, and the small intestine. Their secretions were listed previously in Figure 3-8 (page 76).

How Do "Digestive Juices" Work?   Digestion begins in the mouth. An enzyme in saliva starts rapidly breaking down starch, and another enzyme initiates a little digestion of fat, especially the digestion of milk fat, important in infants. Saliva also helps maintain the health of the teeth in two ways: by washing away food particles that would otherwise foster decay and by neutralizing decay-promoting acids produced by bacteria in the mouth.

In the stomach, protein digestion begins. Cells in the stomach release **gastric juice,** a mixture of water, enzymes, and hydrochloric acid. This strong acid mixture is needed to activate a protein-digesting enzyme and to initiate digestion of protein—protein digestion is the stomach's main function. The strength of an acid solution is expressed as its **pH.** The lower the pH number, the more acidic the solution; solutions with higher pH numbers are more basic. As Figure 3-11 (p. 80) demonstrates, saliva is only weakly acidic, while the stomach's gastric juice is much more strongly acidic. Notice on the right side of Figure 3-11 that the range of tolerance for the blood's pH is exceedingly small.

Upon learning of the powerful digestive juices and enzymes within the digestive tract, students often wonder how the tract's own cellular lining escapes being digested along with the food. The answer: Specialized cells secrete a thick, viscous substance known as **mucus,** which coats and protects the digestive tract lining.

In the small intestine, the digestive process gets under way in earnest. The small intestine is *the* organ of digestion and absorption, and it finishes what the mouth and stomach have started. The small intestine works with the precision of a laboratory chemist. As the thoroughly liquefied and partially digested nutrient mixture arrives there, hormonal messengers signal the gallbladder to contract and to squirt the right amount of **bile,** an **emulsifier,** into the intestine. Other hormones notify the pancreas to release **pancreatic juice** containing the alkaline compound **bicarbonate** in amounts precisely adjusted to neutralize the stomach acid that has reached the small intestine. All these actions alter the intestinal environment to perfectly support the work of the digestive enzymes.

Meanwhile, as the pancreatic and intestinal enzymes act on the chemical bonds that hold the large nutrients together, smaller and smaller pieces are released into the intestinal fluids. The cells of the intestinal wall also hold some digestive enzymes on their

▨ Alcohol needs no assistance from digestive juices to ready it for absorption; its handling by the body is described in this chapter's Controversy section.

gastric juice the digestive secretion of the stomach.

pH a measure of acidity on a point scale. A solution with a pH of 1 is a strong acid; a solution with a pH of 7 is neutral; a solution with a pH of 14 is a strong base.

mucus (MYOO-cus) a slippery coating of the digestive tract lining (and other body linings) that protects the cells from exposure to digestive juices (and other destructive agents). The adjective form is *mucous* (same pronunciation). The digestive tract lining is a *mucous membrane.*

bile a cholesterol-containing digestive fluid made by the liver, stored in the gallbladder, and released into the small intestine when needed. It emulsifies fats and oils to ready them for enzymatic digestion (described in Chapter 5).

emulsifier (ee-MULL-sih-fire) a compound with both water-soluble and fat-soluble portions that can attract fats and oils into water, combining them.

pancreatic juice fluid secreted by the pancreas that contains both enzymes to digest carbohydrate, fat, and protein and sodium bicarbonate, a neutralizing agent.

bicarbonate a common alkaline chemical; a secretion of the pancreas; also, the active ingredient of baking soda.

FIGURE 3·11 pH VALUES OF DIGESTIVE JUICES AND OTHER COMMON FLUIDS

A substance's acidity or alkalinity is measured in pH units. Each step down the scale indicates a tenfold increase in concentration of hydrogen particles, which determine acidity. For example, a pH of 2 is 1,000 times stronger than a pH of 5.

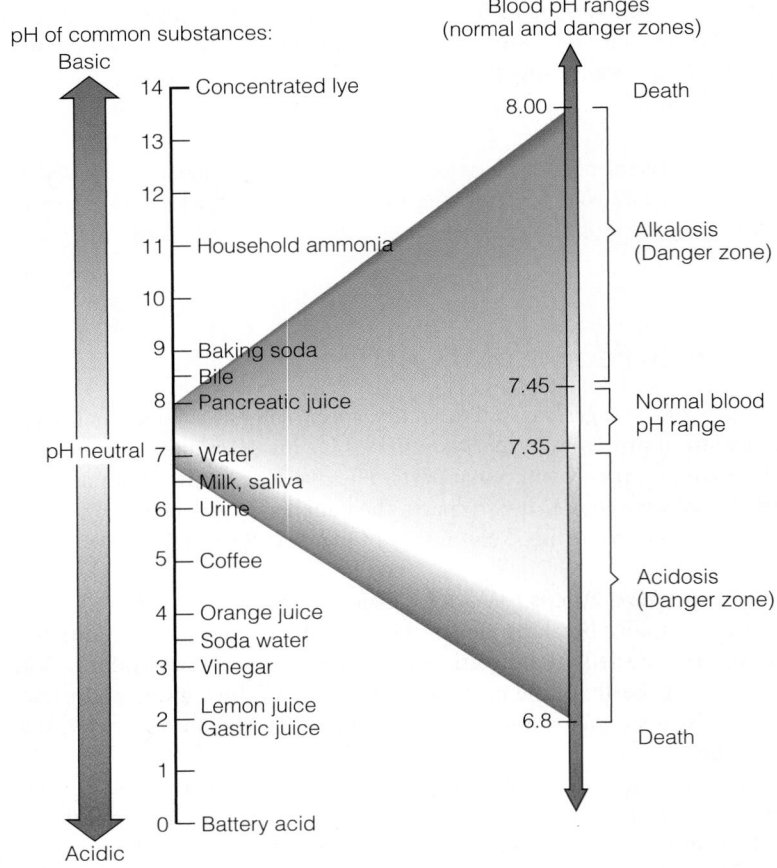

pH of common substances:

Basic

pH	Substance
14	Concentrated lye
13	
12	
11	Household ammonia
10	
9	Baking soda
	Bile
8	Pancreatic juice
7	Water
	Milk, saliva
6	Urine
5	Coffee
4	Orange juice
	Soda water
3	Vinegar
2	Lemon juice
	Gastric juice
1	
0	Battery acid

pH neutral 7

Acidic

Blood pH ranges (normal and danger zones)

8.00 — Death

Alkalosis (Danger zone)

7.45 — Normal blood pH range

7.35 —

Acidosis (Danger zone)

6.8 — Death

surfaces; these enzymes perform last-minute breakdown reactions required before nutrients can be absorbed. Finally, the digestive process releases pieces small enough for the cells to absorb and use. Digestion by human enzymes and absorption of carbohydrate, fat, and protein are essentially complete by the time the intestinal contents enter the colon. Water, fiber, and some minerals, however, remain in the tract.

Certain kinds of fiber, which cannot be digested by human enzymes, can often be broken down by the billions of living inhabitants of the human digestive tract, the resident bacteria. So active are these inhabitants in breaking down substances from food that they have been likened to "an organ of intense metabolic activity that is involved in nutrient salvage." The intestinal cells then absorb the small fat fragments released from the fiber to provide a tiny bit of energy. Table 3-1 provides a summary of all the processes involved.

▶ **KEY POINT** Chemical digestion begins in the mouth, where food is mixed with an enzyme in saliva that acts on carbohydrates. Digestion continues in the stomach, where stomach enzymes and acid break down protein. Digestion then continues in the small intestine; there the liver and gallbladder contribute bile that emulsifies fat, and the pancreas and small intestine donate enzymes that continue digestion so that absorption can occur. Bacteria in the colon break down certain fibers.

TABLE 3·1   SUMMARY OF CHEMICAL DIGESTION

	Mouth	Stomach	Small Intestine, Pancreas, Liver, and Gallbladder	Large Intestine (Colon)
Sugar and Starch	The salivary glands secrete saliva to moisten and lubricate food; chewing crushes and mixes it with a salivary enzyme that initiates starch digestion.	Digestion of starch continues while food remains in the upper storage area of the stomach. In the lower digesting area of the stomach, hydrochloric acid and an enzyme of the stomach's juices halt starch digestion.	The pancreas produces a starch-digesting enzyme and releases it into the small intestine. Cells in the intestinal lining possess enzymes on their surfaces that break sugars and starch fragments into simple sugars, which then are absorbed.	Undigested carbohydrates reach the colon and are partly broken down by intestinal bacteria.
Fiber	The teeth crush fiber and mix it with saliva to moisten it for swallowing.	No action.	Fiber binds cholesterol and some minerals.	Most fiber is excreted with the feces; some fiber is digested by bacteria in the colon.
Fat	Fat-rich foods are mixed with saliva. The tongue produces traces of a fat-digesting enzyme that accomplishes some breakdown, especially of milk fats. The enzyme is stable at low pH and is important to digestion in nursing infants.	Fat tends to rise from the watery stomach fluid and foods and float on top of the mixture. Only a small amount of fat is digested. Fat is last to leave the stomach.	The liver secretes bile; the gallbladder stores it and releases it into the small intestine. Bile emulsifies the fat and readies it for enzyme action. The pancreas produces fat-digesting enzymes and releases them into the small intestine to split fats into their component parts (primarily fatty acids), which then are absorbed.	Some fatty materials escape absorption and are carried out of the body with other wastes.
Protein	Chewing crushes and softens protein-rich foods and mixes them with saliva.	Stomach acid works to uncoil protein strands and to activate the stomach's protein-digesting enzyme. Then the enzyme breaks the protein strands into smaller fragments.	Enzymes of the small intestine and pancreas split protein fragments into smaller fragments or free amino acids. Enzymes on the cells of the intestinal lining break some protein fragments into free amino acids, which then are absorbed. Some protein fragments are also absorbed.	The large intestine carries undigested protein residue out of the body. Normally, almost all food protein is digested and absorbed.
Water	The mouth donates watery, enzyme-containing saliva.	The stomach donates acidic, watery, enzyme-containing gastric juice.	The liver donates a watery juice containing bile. The pancreas and small intestine add watery, enzyme-containing juices; pancreatic juice is also alkaline.	The large intestine reabsorbs water and some minerals.

Are Some Food Combinations More Easily Digested Than Others? People sometimes wonder if the digestive tract has trouble digesting certain foods in combination—for example, fruit and meat. Proponents of fad "food-combining" diets claim that the digestive tract cannot perform certain digestive tasks at the same time, but this is a gross underestimation of the tract's capabilities. The digestive system adjusts to whatever mixture of foods is presented to it. The truth is that all foods, regardless of identity, are broken down by enzymes into the basic molecules that make them up.

Scientists who study digestion suggest that the organs of the digestive tract analyze the diet's nutrient contents and deliver juice and enzymes appropriate for digesting those nutrients. The pancreas is especially sensitive in this regard and has been observed to adjust its output of enzymes to digest carbohydrate, fat, or protein to an amazing degree. The pancreas of a person who suddenly consumes a meal unusually high in carbohydrate, for example, would begin increasing its output of carbohydrate-digesting enzymes within 24 hours, while reducing outputs of other types. This sensitive mechanism ensures that foods of all types are used fully by the body. The next section reviews the major processes of digestion by showing how the nutrients in a mixture of foods are handled.

▶ **KEY POINT** The healthy digestive system is capable of adjusting to almost any diet and can handle any combination of foods with ease.

If "I Am What I Eat," Then How Does a Sandwich Become "Me"?

The process of rendering foods into nutrients and absorbing them into the body fluids is remarkably efficient. Within about 24 to 48 hours of eating, a healthy body digests and absorbs about 90 percent of the carbohydrate, fat, and protein in a meal. Here, we follow a peanut butter and banana sandwich on whole-wheat, sesame seed bread through the tract.

In the Mouth   In each bite, food components are crushed, mashed, and mixed with saliva by the teeth and the tongue. The sesame seeds are crushed and torn open by the teeth, which break through the indigestible fiber coating so that digestive enzymes can reach the nutrients inside the seeds. The peanut butter is the "extra crunchy" type, but the teeth grind the chunks to a paste before the bite is swallowed. The carbohydrate-digesting enzyme of saliva begins to break down the starches of the bread, banana, and peanut butter to sugars. Each swallow triggers a peristaltic wave that travels the length of the esophagus and carries one chewed bite of sandwich to the stomach.

In the Stomach   The stomach collects bite after swallowed bite in its upper storage area, where starch continues to be digested until the gastric juice mixes with the salivary enzymes and halts their action. Small portions of the mashed sandwich are pushed into the digesting area of the stomach where gastric juice mixes with the mass. Acid in gastric juice unwinds proteins from the bread, seeds, and peanut butter; then, an enzyme clips the protein strands into pieces. The sandwich has now become chyme. The watery carbohydrate- and protein-rich part of the chyme enters the small intestine first; a layer of fat follows closely behind.

In the Small Intestine   Some of the sweet sugars in the banana require so little digesting that they begin to cross the linings of the small intestine immediately on contact. Nearby, the liver donates bile through a duct into the small intestine. The bile blends the fat from the peanut butter and seeds with the watery enzyme-containing digestive fluids. The nearby pancreas squirts enzymes into the small intestine to break down the fat, protein, and starch in the chemical soup that just an hour ago was a sandwich. The cells of the small intestine itself produce enzymes to complete these processes. As the enzymes do their work, smaller and smaller chemical fragments are liberated from the chemical soup and are absorbed into the blood and lymph through the cells of the small intestine's wall. Vitamins and minerals are absorbed here, too. They all eventually enter the bloodstream to nourish the tissues.

In the Large Intestine (Colon)   Only fiber fragments, fluid, and some minerals are absorbed in the large intestine. The fibers from the seeds, whole-wheat bread, peanut butter, and banana are partly digested by the bacteria living in the colon, and some of the products are absorbed. Most fiber is not absorbed, however, and, along with some other components, it passes out of the colon, excreted as feces.

Time in mouth, less than a minute.

Time in stomach, about 1–2 hours.

Time in small intestine, about 7–8 hours.*

Time in colon, about 12–14 hours.*

*Based on a 24-hour transit time. Actual times vary widely.

Absorption and Transportation of Nutrients

Once the digestive system has broken food down to its nutrient components, the rest of the body awaits their delivery. First, though, every molecule of nutrient must traverse one of the cells of the intestinal lining. These cells absorb nutrients from the mixture within the intestine and deposit them in the blood and lymph. The cells are selective: they recognize some of the nutrients that may be in short supply in the diet. The mineral calcium is an example. The less calcium in the diet, the greater the percentage of calcium the intestinal cells absorb from foods. The cells are also extraordinarily efficient: they absorb enough nutrients to nourish all the body's other cells.

The cells of the intestinal tract lining are arranged in sheets that poke out into millions of finger-shaped projections (**villi**). Every cell on every villus has a brushlike covering of tiny hairlike projections (**microvilli**) that can trap the nutrient particles. Each villus (projection) has its own capillary network and a lymph vessel so that as nutrients move across the cells, they can immediately mingle with the body fluids. Figure 3-12 (p. 84) provides a close look at these details.

The small intestine's lining, villi and all, is wrinkled into thousands of folds, so its absorbing surface is enormous. If the folds, and the villi that poke out from them, were spread out flat, they would cover a third of a football field. The billions of cells of that surface weigh only 4 to 5 pounds, yet they absorb enough nutrients to nourish the other 150 or so pounds of body tissues.

After the nutrients pass through the cells of the villi, the blood and lymph start transporting the nutrients to their ultimate consumers, the body's cells. The lymphatic vessels initially transport most of the products of fat digestion and a few vitamins, later delivering them to the bloodstream. The blood vessels carry the products of carbohydrate and protein digestion, most vitamins, and the minerals from the digestive tract to the liver. Thanks to these two transportation systems, every nutrient soon arrives at the place where it is needed.

The digestive system's millions of specialized cells are themselves exquisitely sensitive to an undersupply of energy, nutrients, or dietary fiber. In cases of severe undernutrition of energy and nutrients, the absorptive surface of the small intestine shrinks. The surface may be reduced to a tenth of its normal area, preventing it from absorbing what few nutrients a limited food supply may provide. Without sufficient fiber to provide an undigested bulk for the tract's muscles to push against, the muscles become weak from lack of exercise. Malnutrition that impairs digestion is self-perpetuating because impaired digestion makes malnutrition worse. In fact, the digestive system's needs are few, but important. The body has much to say to the attentive listener, stated in a language of symptoms and feelings that you would be wise to study. The next section takes a lighthearted look at what your digestive tract might be trying to tell you.

A Letter from Your Digestive Tract

To My Owner,

You and I are so close; I hope that I can speak frankly without offending you. I know that sometimes I *do* offend with my gurgling noises and belching at quiet times and, oh yes, the gas. But, as you can read for yourself in Table 3-2, when you chew gum,

villi (VILL-ee, VILL-eye) fingerlike projections of the sheets of cells that line the intestinal tract. The villi make the surface area much greater than it would otherwise be (singular: *villus*).

microvilli (MY-croh-VILL-ee, MY-croh-VILL-eye) iny, hairlike projections on each cell of every villus that greatly expand the surface area available to trap nutrient particles and absorb them into the cells (singular: *microvillus*).

FIGURE 3·12   DETAILS OF THE SMALL INTESTINAL LINING

stomach

small intestine

folds with villi
on them

© Bill Crew/SuperStock

If you have ever watched a sea
anemone with its fingerlike
projections in constant motion,
you have a good picture of how
the intestinal villi move.

The wall of the small intestine is
wrinkled into thousands of
folds and is carpeted with villi.

a villus

capillaries

lymphatic vessel

muscle layers
beneath folds

Between the villi are tubular
glands that secrete enzyme-
containing intestinal juice.

artery

vein

lymphatic
vessel

© Don W. Fawcett

microvilli

This is a photograph of part of
an actual human intestinal cell
with microvilli.

Each villus, in turn is covered
with even smaller projections,
the microvilli.

TABLE 3•2 FOODS AND INTESTINAL GAS

Recent experiments have shed light on the causes and prevention of intestinal gas. Here are some recent findings.

- Milk intake causes gas in those who cannot digest the milk sugar lactose. Most people, however, can consume up to a cup of milk without producing excessive gas.
  Solution: Drink up to 4 ounces of fluid milk at a sitting, or substitute reduced-fat cheeses or yogurt without added milk solids. Use lactose-reduced products, or treat regular products with lactose-reducing enzyme products.

- Beans cause gas because some of their carbohydrates are indigestible by human enzymes, but are broken down by intestinal bacteria. The amount of gas may not be as much as most people fear, however.
  Solution: Use rinsed canned beans, or dried beans that are well-cooked, because cooked carbohydrates are more readily digestible. Try enzyme drops that can help break down the carbohydrate before it reaches the intestine.

- Air swallowed during eating or drinking can cause gas, as can the gas of carbonated beverages. Each swallow of a beverage can carry three times as much air as fluid, which some people belch up.
  Solution: Slow down during eating and drinking, and don't chew gum or suck on hard candies that may cause you to swallow air.

- Vegetables may or may not cause gas in some people, but research is lacking.
  Solution: If you feel certain vegetables cause gas, try eating small portions of the cooked product. Do try the vegetable again: the gas you experienced may have been a coincidence and had nothing to do with eating the vegetable.

SOURCE: Adapted from B. Liebman, Who ya gonna call? GasBusters, an interview with Michael Levitt, *Nutrition Action Healthletter,* May 2003, pp. 1, 3–5.

What is your digestive tract trying to tell you?

drink carbonated beverages, or eat hastily, you gulp air with each swallow. I can't help making some noise as I move the air along my length or release it upward in a noisy belch. And if you eat or drink too fast, I can't help getting **hiccups.** Please sit and relax while you dine. You will ease my task, and we'll both be happier.

Also, when someone offers you a new food, you gobble away, trusting me to do my job. I try. It would make my life easier, and yours less gassy, if you would start with small amounts of new foods, especially those high in fiber. The breakdown of fiber by bacteria produces gas, so introduce fiber-rich foods slowly. But please: if you do notice more gas than normal from a specific food, avoid it. If the gas becomes excessive, check with a physician. The problem could be something simple—or serious.

When you eat or drink too much, it just burns me up. Overeating causes **heartburn** because the acidic juice from my stomach backs up into my esophagus. Acid poses no problem to my healthy stomach, whose walls are coated with thick mucus to protect them. But when my too-full stomach squeezes some of its contents back up into the esophagus, the acid burns its unprotected surface. Also, those tight jeans you wear constrict my stomach, squeezing the contents upward into the esophagus. Just leaning over or lying down after a meal may allow the acid to escape up the esophagus because the muscular sphincter separating the two spaces is much looser than other sphincters. And if we need to lose a few pounds, let's get at it—excess body fat can also squeeze my stomach, causing acid to back up. When heartburn is a problem, do me a favor: try to eat smaller meals; drink liquids an hour before or after, but not during, meals; wear reasonably loose clothing; and relax after eating, but sit up (don't lie down).

Sometimes your food choices irritate me. Specifically, chemical irritants in foods, such as the "hot" component of chili peppers, chemicals in coffee, fat, chocolate, carbonated soft drinks, and alcohol, may worsen heartburn in some people. Avoid the ones that cause trouble. Above all, do not smoke. Smoking makes my heartburn worse—and you should hear your lungs bellyache about it.

By the way, I can tell you've been taking heartburn medicines again. You must have been watching those misleading TV commercials. You need to know that **antacids** are designed only to temporarily relieve pain caused by heartburn by neutralizing stomach acid for a while. But when the antacids reduce my normal stomach acidity, I respond by

hiccups spasms of both the vocal cords and the diaphragm, causing periodic, audible, short, inhaled coughs. Can be caused by irritation of the diaphragm, indigestion, or other causes. Hiccups usually resolve in a few minutes, but can have serious effects if prolonged. Breathing into a paper bag (inhaling carbon dioxide) or dissolving a teaspoon of sugar in the mouth may stop them.

heartburn a burning sensation in the chest (in the area of the heart) area caused by backflow of stomach acid into the esophagus.

antacids medications that react directly and immediately with the acid of the stomach, neutralizing it. Antacids are most suitable for treating occasional heartburn. More about antacids appears in Controversy 8.

producing *more* acid to restore the normal acid condition. Also, the ingredients in antacids can interfere with my ability to absorb nutrients. Please check with our doctor if heartburn occurs more than just occasionally and certainly before you decide that we need to take the heavily advertised **acid reducers;** these restrict my normal ability to produce acid so much that my job of digesting food becomes harder. They may also reduce our defense against certain infections.[2]

Given a chance, my powerful acid can help fight bacterial infections from contaminated food and other sources—some dangerous bacteria won't survive a bath in my caustic stomach juices.[3] Acid reducing drugs may allow more bacteria to pass through. And even worse, self-prescribed heartburn medicine can mask the symptoms of **ulcer, hernia,** or the severe destructive form of chronic heartburn known as **gastroesophageal reflux disease (GERD).**[4] This can be serious because, if not treated with antibiotic drugs, the bacterium that causes stomach ulcer may also cause stomach cancer.[5] A hernia can cause food to back up into the esophagus, so it can feel like heartburn, but many times hernias require corrective treatment by a physician. GERD can feel like heartburn, too, but requires the correct drug therapy to prevent respiratory problems, severe damage to tissues, or less likely, cancer.[6]

When you eat too quickly, I worry about choking (see Figure 3-13). Please take time to cut your food into small pieces, and chew it until it is crushed and moistened with saliva. Also, refrain from talking or laughing before swallowing, and never attempt to eat when you are breathing hard. Also, for our sake and the sake of others, learn the Heimlich maneuver shown in Figure 3-14.

When I'm suffering, you suffer, too. When **constipation** or **diarrhea** strikes, neither of us is having fun. Slow, hard, dry bowel movements can be painful, and failing to have a movement for too long brings on headaches and ill feelings. Worse, chronic constipation with fewer than three bowel movements each week has been associated with a more-than-doubled risk of colon cancer.[7]

Laxatives may help in the short term, but they often contain stimulants that can cause side effects.* Instead of relying on laxatives, listen carefully for my signal that it is time to defecate, and make time for it even if you are busy. The longer you ignore my signal, the more time the colon has to extract water from the feces, hardening them. Also, please choose foods that provide enough fiber (Chapter 4 lists some of these foods). Fiber

---
*One such stimulant, phenolphthalein, was banned because of an association with colon cancer.

■ Some bacteria survive stomach conditions to cause foodborne illnesses. See Chapter 12.

■ For more information concerning ulcers and medication, call the Centers for Disease Control and Prevention at 1-888-MY-ULCER. The call is toll-free.

acid reducers prescription and over-the-counter drugs that reduce the acid output of the stomach; effective for treating severe, persistent forms of heartburn but not for neutralizing acid already present. Side effects are frequent and include diarrhea, other gastrointestinal complaints, and reduction of the stomach's capacity to destroy alcohol, thereby producing higher-than-expected blood alcohol levels from each drink (see this chapter's Controversy section). Also called *acid controllers.*

ulcer an erosion in the topmost, and sometimes underlying, layers of cells that form a lining. Ulcers of the digestive tract commonly form in the esophagus, stomach, or upper small intestine.

hernia a protrusion of an organ or part of an organ through the wall of the body chamber that normally contains the organ. An example is a *hiatal* (high-AY-tal) *hernia,* in which part of the stomach protrudes up through the diaphragm into the chest cavity, which contains the esophagus, heart, and lungs.

gastroesophageal reflux disease (GERD) (GAS-tro-eh-SOFF-ah-jee-al) a severe and chronic splashing of stomach acid and enzymes into the esophagus, throat, mouth, or airway that causes inflammation and injury to those organs. Untreated GERD may increase the risk of esophageal cancer; treatment may require surgery or management with medication.

constipation infrequent, difficult bowel movements often caused by diet, inactivity, dehydration, or medication. (Also defined in Chapter 4.)

diarrhea frequent, watery bowel movements usually caused by diet, stress, or irritation of the colon. Severe, prolonged diarrhea robs the body of fluid and certain minerals, causing dehydration and imbalances that can be dangerous if left untreated.

FIGURE 3·13 SWALLOWING AND CHOKING

Tongue

Epiglottis closes over larynx

Trachea (to lungs)

Food

Esophagus (to stomach)

A normal swallow. The epiglottis acts as a flap to seal the entrance to the lungs (trachea) and direct food to the stomach via the esophagus.

Choking. A choking person cannot speak or gasp because food lodged in the trachea blocks the passage of air. The red arrow points to where the food should have gone to prevent choking.

attracts water, creating softer, bulkier stools that stimulate my muscles to contract, pushing the contents along. Fiber helps my muscles to stay fit, too, making elimination easier. Be sure to drink enough water because dehydration causes the colon to absorb all the water it can get from the feces. And please make time to be physically active; exercise strengthens not just the muscles of your arms, legs, and torso, but those of the colon, too.

When I have the opposite problem, diarrhea, my system will rob you of water and salts. In diarrhea my intestinal contents have moved too quickly, drawing water and minerals from your tissues into the contents. When this happens, please rest a while and drink fluids. To avoid diarrhea, try not to change my diet too drastically or quickly. I'm willing to work with you and learn to digest new foods, but if you suddenly change your diet, we're both in for it. I hate even to think of it, but one likely cause of diarrhea is dangerous food poisoning (*please* read, and use, the tips in Chapter 12 to keep us safe). Also, if diarrhea lasts longer than a day or two, or if it alternates with constipation, the cause could be **irritable bowel syndrome,** and you should go see a physician.

Thank you for listening. I know we'll both benefit from communicating like this because you and I are in this together for the long haul.

Affectionately,
Your Digestive Tract

▶ **KEY POINT** The digestive tract has many ways to communicate its needs. By taking the time to listen, you will obtain a complete understanding of the mechanics of the digestive tract and its signals.

The Excretory System

Cells generate a number of wastes, and all of them must be eliminated. Many of the body's organs play roles in removing wastes. Carbon dioxide waste from the cells travels in the blood to the lungs, where it is exchanged for oxygen. Other wastes are pulled out of the bloodstream by the liver. The liver processes these wastes and either tosses them out into the digestive tract with bile, to leave the body with the feces, or prepares them to be sent to the kidneys for disposal in the urine. Organ systems work together to dispose of the body's wastes, but the kidneys are waste- and water-removal specialists.

The kidneys straddle the cardiovascular system and filter the passing blood. Waste materials, dissolved in water, are collected by the kidneys' working units, the **nephrons.** These wastes become concentrated as urine, which travels through tubes to the urinary **bladder.** The bladder empties periodically, removing the wastes from the body. Thus, the blood is purified continuously throughout the day, and dissolved materials are excreted as necessary. One dissolved mineral, sodium, helps to regulate blood pressure, and its excretion or retention by the kidneys is a vital part of the body's blood pressure–controlling mechanism.

Though they account for just 0.5 percent of the body's total weight, the kidneys use up 10 percent of the body's oxygen supply, indicating intense metabolic activity.[8] The kidney's waste-excreting function rivals breathing in importance to life, but the kidneys act in other ways as well. By sorting among dissolved substances, retaining some while excreting others, the kidneys regulate the fluid volume and concentrations of substances in the blood and extracellular fluids with great precision. Through these mechanisms, the kidneys help to regulate blood pressure (see Chapter 11 for details). As you might expect, the kidneys' work is regulated by hormones secreted by glands that respond to conditions in the blood (such as the sodium concentration). The kidneys also release certain hormones.

Because the kidneys remove toxins that could otherwise damage body tissues, whatever supports the health of the kidneys supports the health of the whole body. A strong cardiovascular system and an abundant supply of water are important to keep blood flushing swiftly through the kidneys. In addition, the kidneys need sufficient energy to do their complex sifting and sorting job, and many vitamins and minerals serve as the cogs in their machinery. Exercise and nutrition are vital to healthy kidney function.

FIGURE 3·14 THE HEIMLICH MANEUVER

Rescuer positions fist directly against victim's abdomen as shown.

1. Rescuer stands behind victim and wraps her or his arms around victim's waist.
2. Rescuer makes a fist with one hand and places the thumb side of the fist against the victim's abdomen, slightly above the navel and below the rib cage.
3. Rescuer grasps fist with other hand and rapidly squeezes it inward and upward three or four times in rapid succession.
4. Rescuer repeats the process if necessary.

If the victim is alone, the victim positions himself or herself over edge of fixed horizontal object, such as a chair back, railing, or table edge, and presses abdomen into edge with quick movement.

■ The lungs also excrete some small percentage of ingested alcohol—the basis for the "breathalyzer" test given to drivers to determine if they've been drinking.

irritable bowel syndrome intermittent disturbance of bowel function, especially diarrhea or alternating diarrhea and constipation; associated with diet, lack of physical activity, or psychological stress.

nephrons (NEFF-rons) the working units in the kidneys, consisting of intermeshed blood vessels and tubules.

bladder the sac that holds urine until time for elimination.

Storage Systems

The human body is designed to eat at intervals of about four to six hours, but cells need nutrients around the clock. Providing the cells with a constant flow of the needed nutrients requires the cooperation of many body systems. These systems store and release nutrients to meet the cells' needs between meals. Among the major storage sites are the liver and muscles, which store carbohydrate, and the fat cells, which store fat and other fat-related substances.

When I Eat More Than My Body Needs, What Happens to the Extra Nutrients?

Nutrients collected from the digestive system sooner or later all move through a vast network of capillaries that weave among the liver cells. This arrangement ensures that liver cells have access to the newly arriving nutrients for processing. The body tissues store excess energy-containing nutrients in two forms (details in later chapters). The liver makes some into **glycogen** (a carbohydrate), and some is stored as fat. The liver stores the glycogen to meet the body's ongoing glucose needs. Liver glycogen can sustain cell activities when the intervals between meals become long. Without glucose absorbed from food, the cells (including the muscle cells) draw on liver glycogen. Should no food be available, the liver's glycogen supply dwindles; it can be effectively depleted within as few as three to six hours. Muscle cells make and store glycogen, too, but selfishly reserve it for their own use.

Whereas the liver stores glycogen, it ships out fat in packages (see Chapter 5) to be picked up by cells that need it. All body cells may withdraw the fat they need from these packages, and the fat cells of the **adipose tissue** pick up the remainder and store it to meet long-term energy needs. Unlike the liver, fat tissue has virtually infinite storage capacity. It can continue to supply the body's cells with fat for days, weeks, or possibly even months when no food is eaten.

These storage systems for glucose and fat ensure that the body's cells will not go without energy even if the body is hungry for food. Body stores also exist for many other nutrients, each with a characteristic capacity. For example, liver and fat cells store many vitamins, and bones provide reserves of calcium and other minerals. Stores of nutrients are available to keep the blood levels constant and to meet cellular demands.

Variations in Nutrient Stores

Some nutrients are stored in the body in much larger quantities than others. For example, certain vitamins are stored without limit, even if they reach toxic levels within the body. Other nutrients are stored in only small amounts, regardless of the amount taken in, and these can readily be depleted. As you learn how the body handles various nutrients, pay particular attention to their storage so that you can know your tolerance limits. For example, you needn't eat fat at every meal because fat is stored abundantly. On the other hand, you normally do need to have a source of carbohydrate at intervals throughout the day because the liver stores less than one day's supply of glycogen.

■ Later chapters provide details about the storage of energy nutrients.

glycogen a storage form of carbohydrate energy (glucose), described more fully in Chapter 4.

adipose tissue the body's fat tissue, consisting of masses of fat-storing cells and blood vessels to nourish them.

Conclusion

In addition to the systems just described, the body has many more: bones, muscles, reproductive organs, and others. All of these cooperate, enabling each cell to carry on its own life. For example, the skin and body linings defend other tissues against microbial invaders, while being nourished and cleansed by tissues specializing in these tasks. Each system needs a continuous supply of many specific nutrients to maintain itself and carry out its work. Calcium is particularly important for bones, for example; iron for muscles; glucose for the brain. But all systems need all nutrients, and every system is impaired by an undersupply or oversupply of them.

While external events clamor and vie for attention, the body quietly continues its life-sustaining work. Most of the body's work is directed automatically by the unconscious portions of the brain and nervous system, and this work is finely regulated to achieve a state of well-being. But you need to involve your brain's cortex, your conscious thinking brain, to cultivate an understanding and appreciation of your body's needs. In doing so, attend to nutrition first. The rewards are liberating—ample energy to tackle life's tasks, a robust attitude, and the glowing appearance that comes from the best of health. Read on, and learn to let nutrition principles guide your food choices.

▶ **KEY POINT** To achieve optimal function, the body's systems require nutrients from outside. These have to be supplied through a human being's conscious food choices.

MEDIA MENU

Nutrition Connections

The Nutrition Connections CD provides access to these additional resources.

Interaction
The animation, *Digestion and Absorption,* follows the progress of food as it moves through the G.I. tract. The functions of the primary organs involved in digestion are illustrated.

Chapter Quiz
Take the quiz for this chapter to assess your understanding of chapter concepts.

Nutrition on the Net
Annotated links to reputable nutrition-related websites.

Online Glossary
Test your knowledge of key vocabulary through this comprehensive, interactive glossary.

Internet Connections

A variety of helpful study tools for this chapter are available at our website
http://nutrition.wadsworth.com/ncc10

Practice Test
Take the practice test for this chapter and gauge your knowledge of key concepts; answers are provided.

Student Notebook
A lecture notebook is available to help organize note taking while reading or attending lecture.

Study Tools
Study resources for students include Chapter Summaries, Crossword puzzles, Flashcards, Glossary, Internet Exercises, and Learning Objectives.

SELF CHECK

Answers to these Self Check questions are in Appendix G.

1. All of the following are correct concerning ulcers *except:*
   a. they usually occur in the large intestine
   b. some are caused by a bacterium
   c. if not treated correctly, they can lead to stomach cancer
   d. their symptoms can be masked by using antacids regularly

2. Which of the following increases the production of intestinal gas?
   a. chewing gum
   b. drinking carbonated beverages
   c. eating certain vegetables
   d. all of the above

3. Chemical digestion of all nutrients mainly occurs in which organ?
   a. mouth
   b. stomach
   c. small intestine
   d. large intestine

4. Which chemical substance released by the pancreas neutralizes stomach acid that has reached the small intestine?
   a. mucus
   b. enzymes
   c. bicarbonate
   d. bile

5. Which nutrient passes through the large intestine mostly unabsorbed?
   a. starch
   b. vitamins
   c. minerals
   d. fiber

6. T cells are immune cells that ingest and destroy antigens in a process known as phagocytosis.   T   F

7. Bile starts the process of protein digestion in the stomach.   T   F

8. To digest food efficiently, people should not combine certain foods, such as meat and fruit, at the same meal.   T   F

9. The gallbladder stores bile until it is needed to emulsify fat.

10. Absorption of the majority of nutrients takes place across the mucus-coated lining of the stomach.   T   F

CONTROVERSY 3

ALCOHOL AND NUTRITION: DO THE BENEFITS OUTWEIGH THE RISKS?

ON AVERAGE, people in the United States consume from 6 to 10 percent of their total daily energy intake as alcohol. Drinking habits span a wide spectrum: many adults drink no alcohol whatsoever, some take a glass of wine only with meals, others drink on social occasions, and others take in large quantities of alcohol daily because of a life-shattering addiction. A third of U.S. college students drink alcohol in a pattern that identifies them as **binge drinkers,** although, when asked, more than 90 percent say that they are not binge drinkers.[*1] Whether or not they recognize the problem, many college students will pay a high price in terms of their health and safety as a result of episodes of heavy drinking. **Moderate drinkers** usually consume the calories of alcohol in addition to their normal food intake, so the alcohol contributes to their daily calorie totals.[2] (Alcohol-related terms are defined in Table C3-1.) Alcohol is not just an energy source, however; it is also a psychoactive drug and a toxin to the body.

Despite its toxicity, people want to know if there is an amount of alcohol they can drink safely or whether they may derive benefits, particularly for the health of the heart, by drinking. This Controversy first defines some terms and examines alcohol's actions within the body and its effects on the brain and other organs. It then summarizes the long-term effects of alcohol on the body and nutrition and concludes with research on moderate drinking.

Defining Drinks and Drinking

When people congregate to enjoy conversation and companionship, beverages are usually a part of that

*Reference notes are found in Appendix F.

TABLE C3·1   ALCOHOL AND DRINKING TERMS

- **acetaldehyde** (ass-et-AL-deh-hide) a substance to which ethanol is metabolized on its way to becoming harmless waste products that can be excreted.
- **alcohol dehydrogenase** (dee-high-DRAH-gen-ace) **(ADH)** an enzyme system that breaks down alcohol. The antidiuretic hormone listed below is also abbreviated ADH.
- **alcoholism** a dependency on alcohol marked by compulsive uncontrollable drinking with negative effects on physical health, family relationships, and social health.
- **antidiuretic** (AN-tee-dye-you-RET-ick) **hormone (ADH)** a hormone produced by the pituitary gland in response to dehydration (or a high sodium concentration in the blood). It stimulates the kidneys to reabsorb more water and so to excrete less. (This hormone should not be confused with the enzyme alcohol dehydrogenase, which is also abbreviated ADH.)
- **beer belly** central-body fatness associated with alcohol consumption.
- **binge drinkers** people who drink four or more drinks in a short period.
- **CAGE questions** a set of four questions often used internationally for initial screening for alcoholism.
- **cirrhosis** (seer-OH-sis) advanced liver disease, often associated with alcoholism, in which liver cells have died, hardened, turned an orange color, and permanently lost their function.
- **congeners** (CON-jen-ers) chemical substances other than alcohol that account for some of the physiological effects of alcoholic beverages, such as appetite, taste, and aftereffects.
- **drink** a dose of any alcoholic beverage that delivers ½ ounce of pure ethanol.
- **ethanol** the alcohol of alcoholic beverages, produced by the action of microorganisms on the carbohydrates of grape juice or other carbohydrate-containing fluids.
- **euphoria** (you-FOR-ee-uh) an inflated sense of well-being and pleasure brought on by a moderate dose of alcohol and some other drugs.
- **fatty liver** an early stage of liver deterioration seen in several diseases, including kwashiorkor and alcoholic liver disease, in which fat accumulates in the liver cells.
- **fibrosis** (fye-BROH-sis) an intermediate stage of alcoholic liver deterioration. Liver cells lose their function and assume the characteristics of connective tissue cells (fibers).
- **formaldehyde** a substance to which methanol is metabolized on the way to being converted to harmless waste products that can be excreted.
- **gout** (GOWT) a painful form of arthritis caused by the abnormal buildup of the waste product uric acid in the blood, with uric acid salt deposited as crystals in the joints.
- **MEOS** (microsomal ethanol oxidizing system) a system of enzymes in the liver that oxidize not only alcohol but also several classes of drugs.
- **methanol** an alcohol produced in the body continually by all cells.
- **moderate drinkers** people who do not drink excessively and do not behave inappropriately because of alcohol. A moderate drinker's health is not harmed by alcohol over the long term.
- **nonalcoholic** a term used on labels of beverages, such as wine or beer, indicating that the product contains less than 0.5% alcohol. The terms *dealcoholized* and *alcohol removed* mean the same thing. *Alcohol free* means that the product contains no detectable alcohol.
- **problem drinkers** or **alcohol abusers** people who suffer social, emotional, family, job-related, or other problems because of alcohol. A problem drinker is on the way to alcoholism.
- **proof** a statement of the percentage of alcohol in an alcoholic beverage. Liquor that is 100 proof is 50% alcohol, 90 proof is 45%, and so forth.
- **social drinkers** people who drink only on social occasions. Depending on how alcohol affects a social drinker's life, the person may be a moderate drinker or a problem drinker.
- **urethane** a carcinogenic compound that commonly forms in alcoholic beverages.
- **Wernicke-Korsakoff** (VER-nik-ee KOR-sah-koff) **syndrome** a cluster of symptoms involving nerve damage arising from a deficiency of the vitamin thiamin in alcoholism. Characterized by mental confusion, disorientation, memory loss, jerky eye movements, and staggering gait.

sociability. All beverages seem to ease conversation, whether they contain alcohol or not. Many people are **social drinkers** who choose alcohol over cola, juice, milk, tea, or coffee as a pleasant accompaniment to a meal, a drink of celebration, or a way to relax with friends. Taken in moderation, alcohol reduces inhibitions, encourages social interactions, and produces feelings of **euphoria,** a pleasant sensation that people seek. The term *moderation* is important because alcohol *worsens* social interactions at higher intakes. The **non-alcoholic** beers and wines now on the market also elevate mood and encourage social interaction, a testimony to the placebo effect at work.

In contrast to moderate social drinking, the effect of alcohol on **problem drinkers** or people with **alcoholism** is overwhelmingly negative. For these people, drinking alcohol brings irrational and often dangerous behavior, such as driving a car while intoxicated, and regrettable human interactions, such as arguments and violence. With continued drinking, such people face psychological depression, physical illness, severe malnutrition, and demoralizing erosion of self-esteem. To screen for alcoholism, clinicians often ask a set of four questions known as **CAGE,** listed in Table C3-2, as well as assessing other indicators.

Moderation

Moderation is not easily defined because tolerance to alcohol differs among individuals. In general, women cannot handle as much alcohol as men and should not try to match drinks with male companions. People of Asian or Native American descent may have lower than average tolerance to alcohol. Health authorities have set limits at not more than two drinks a day for the average-sized, healthy man and not more than one drink a day for the average-sized, healthy woman. These amounts are supposed to be enough to elevate mood without incurring long-term harm to health—note that these are not average amounts, but daily maximums. In other words, a person who drinks no alcohol during the week but then takes seven drinks on Saturday night

---

TABLE C3·2	CAGE QUESTIONS

Cut Down—Ever felt you ought to cut down on your drinking?

Annoyed—Have people annoyed you by criticizing your drinking?

Guilt—Ever felt bad or guilty about your drinking?

Eye-Opener—Ever had an eye-opener to steady nerves in the A.M.?

Interpretation: One "Yes" answer suggests a possible problem with alcohol. More than one "Yes" answer indicates a high likelihood of a problem with alcohol.

SOURCE: J. A. Ewing, Detecting alcoholism: The CAGE questionaire, *Journal of the American Medical Association* 252 (1984): 1905–1907.

---

is not a moderate drinker—instead, that alcohol intake pattern characterizes binge drinking.

Doubtless some people can safely consume slightly more than the alcohol dose called moderate; others, especially those prone to alcohol addiction, cannot handle nearly so much without significant risk. Table C3-3 lists those people advised by the *Dietary Guidelines for Americans 2005* not to drink at all. If you think your own drinking might not be moderate or normal or if alcohol has caused problems in your

life, you may want to seek a professional evaluation.[†] Table C3-4 contrasts some behaviors of moderate drinkers with those of problem drinkers.

Binge Drinking

Binge drinking is defined as consuming four or more drinks in a short time and poses serious threats. Compared with nondrinkers or moderate drinkers, binge drinkers are more likely to damage property, to assault other people, to cause fatal automobile accidents, and to engage in risky unprotected and unplanned sexual intercourse. Female binge drinkers are more likely to be victims of rape than others. A binge is most likely to occur at a party, a sporting event, or other social occasion.

Binge drinkers skew the statistics on alcohol use on college campuses. The median number of drinks consumed by all college students is 1.5 per week, but for binge drinkers, it is 14.5 per week. Nationally, only 20 percent of all students are frequent binge drinkers; yet they account for two-thirds of all the alcohol students report consuming and most of the alcohol-related problems.[3]

Binge drinkers on and off college campuses may find it difficult to recog-

---

[†]The U.S. center for facts on alcohol is the National Clearinghouse for Alcohol and Drug Information: (800) 729-6686.

TABLE C3·3	WHO SHOULD NOT DRINK ALCOHOL?

The *Dietary Guidelines for Americans 2005* suggest that these people not drink alcoholic beverages at all:

- *Children and adolescents.* The earlier in life drinking begins, the greater the risk of alcoholism later on.
- *People of any age who cannot restrict their drinking to moderate levels.* Especially, people recovering from alcoholism, problem drinkers, and people whose family members have alcohol problems.
- *Women who may become pregnant or who are pregnant or breastfeeding.* A safe level of alcohol intake has not been established for women during pregnancy (see Chapter 13), and alcohol may be especially hazardous during the first few weeks, before a woman knows she is pregnant.
- *People who plan to drive, operate machinery, or take part in other activities that require attention, skill, or coordination to remain safe.* Alcohol remains in the blood for several hours after taking even a single drink.
- *People taking medications that can interact with alcohol.* Alcohol alters the effectiveness or toxicity of many medications, and some drugs may increase blood alcohol levels.
- *People with medical conditions* worsened by alcohol, such as liver disease.

SOURCE: U.S. Department of Agriculture and U.S. Department of Health and Human Services, *2005 Dietary Guidelines for Americans,* 6th ed., (Washington, D.C.: 2005), available at www.usda.gov/cnpp or call (888) 878-3256.

| TABLE C3·4 | BEHAVIORS TYPICAL OF MODERATE DRINKERS AND PROBLEM DRINKERS |

Moderate Drinkers Typically	Problem Drinkers Typically
• Drink slowly, casually. • Eat food while drinking or beforehand. • Don't binge drink; know when to stop. • Respect nondrinkers. • Avoid drinking when solving problems or making decisions. • Do not admire or encourage drunkenness. • Remain peaceful, calm, and unchanged by drinking. • Cause no problems to others or themselves by drinking.	• Gulp or "chug" drinks. • Drink on an empty stomach. • Binge drink; drink to get drunk. • Pressure others to drink. • Turn to alcohol when facing problems or decisions. • Consider drunks to be funny or admirable. • Become loud, angry, violent, or silent when drinking. • Physically or emotionally harm themselves, family members, or others when drinking.

nize themselves as problem drinkers (refer to Table C3-4) until their drinking behavior causes a crisis, such as a car crash, or until they've binged long enough to have caused substantial damage to their health.

What Is Alcohol?

In chemistry, the term *alcohol* refers to a class of chemical compounds whose names end in "-ol." The glycerol molecule of a triglyceride is an example. Alcohols affect living things profoundly, partly because they act as lipid solvents. Alcohols can easily penetrate a cell's outer lipid membrane, and once inside, they denature the cell's protein structures and kill the cell. Because some alcohols kill microbial cells, they make useful disinfectants and antiseptics.

The alcohol of alcoholic beverages, **ethanol,** is somewhat less toxic than others. Sufficiently diluted and taken in small enough doses, its action in the brain produces euphoria. Used in this way, alcohol is a drug, and like many drugs, alcohol presents both benefits and hazards to the taker. Its effects depend on the quantity of alcohol consumed.

What Is a "Drink"?

Alcoholic beverages contain a great deal of water and some other substances, as well as the alcohol ethanol. In wine, beer, and wine coolers, alcohol contributes a relatively low percentage of the beverage's volume. In contrast, as much as 50 percent of the volume of whiskey, vodka, rum, or brandy may be alcohol. The percentage of alcohol is stated as **proof.** Proof equals twice the percentage of alcohol; for example, 100 proof liquor is 50 percent alcohol.

A serving of an alcoholic beverage, commonly called a **drink,** delivers ½ ounce of pure ethanol. Figure C3-1 depicts servings of alcoholic beverages that are considered to be one drink. These standard measures may have little in common with the drinks served by enthusiastic bartenders, however. Many wine glasses easily hold 6 to 8 ounces of wine; wine coolers may come packaged 12 ounces to a bottle; a large beer stein can hold 16, 20, or even more ounces; a strong liquor drink may contain 2 or 3 ounces of various liquors.

Alcohol Enters the Body

From the moment an alcoholic beverage is swallowed, the body pays special attention to it. Unlike food, which requires digestion before it can be absorbed, the tiny alcohol molecules can diffuse right through the stomach walls and reach the brain within a minute. Ethanol is a toxin, and a too-high dose of alcohol triggers one of the body's primary defenses against poison—vomiting. Many times, though, alcohol arrives gradually and in a beverage dilute enough that the vomiting reflex is delayed and the alcohol is absorbed.

A person can become intoxicated almost immediately when drinking, especially if the stomach is empty. When the stomach is full of food, molecules of alcohol have less chance of touching the stomach walls and diffusing through, so alcohol reaches the brain more gradually (see Figure C3-2). By the time the stomach contents are emptied into the small intestine, however, alcohol is absorbed rapidly whether food is present or not.

FIGURE C3·1 SERVINGS OF ALCOHOLIC BEVERAGES THAT EQUAL ONE DRINK

12 oz beer
10 oz wine cooler
5 oz wine (12% alcohol)
1½ oz hard liquor (80 proof whiskey, gin, brandy, rum, vodka)

© Polara Studios, Inc.

FIGURE C3·2 FOOD SLOWS ALCOHOL'S ABSORPTION

The alcohol in a stomach filled with food has a low probability of touching the walls and diffusing through. Food also holds alcohol in the stomach longer, slowing its entry into the highly absorptive small intestine.

food
alcohol

A person who wants to drink socially and not become intoxicated should eat the snacks provided by the host (avoid the salty ones; they make you thirstier). Other tips include adding ice or water to alcoholic drinks to dilute them and alternating alcoholic with nonalcoholic beverages to quench thirst.

Anyone who has had an alcoholic drink has experienced one of alcohol's physical effects: alcohol increases urine output (because alcohol depresses the brain's production of **antidiuretic hormone**). Loss of body water leads to thirst. The only fluid that relieves dehydration is water, so alternating alcoholic beverages with nonalcoholic ones will quench thirst. Otherwise, each drink may worsen the thirst.

The water lost due to hormone depression takes with it important minerals, such as magnesium, potassium, calcium, and zinc, depleting the body's reserves. These minerals are vital to fluid balance and to nerve and muscle coordination. When drinking incurs mineral losses, the losses must be made up in subsequent meals to avoid deficiencies.

If a person drinks slowly enough, the alcohol, after absorption, will be collected by the liver and processed without much effect on other parts of the body. If a person drinks more rapidly, however, some of the alcohol bypasses the liver and flows for a while through the rest of the body and the brain.

Alcohol Arrives in the Brain

Some people use alcohol as a kind of social anesthetic to help them relax or to relieve anxiety. One drink relieves inhibitions, which gives people the impression that alcohol is a stimulant. It gives this impression by sedating the *inhibitory* nerves, allowing excitatory nerves to take over. This effect is temporary, and ultimately, alcohol acts as a depressant and sedates all the nerve cells (see Figure C3-3).

It is lucky that the brain centers respond to rising blood alcohol in the order shown in the figure, because a person usually passes out before drinking a lethal dose. If a person drinks fast enough, though, the alcohol continues to be absorbed, and its effects continue to accelerate after the person has gone to

FIGURE C3·3 ALCOHOL'S EFFECTS ON THE BRAIN

When alcohol flows to the brain, it first sedates the frontal lobe of the cortex, the reasoning part. As the alcohol molecules diffuse into the cells of this lobe, they interfere with reasoning and judgment.

With continued drinking, the speech and vision centers of the brain become sedated, and the area that governs reasoning becomes more incapacitated.

Still more drinking affects the cells of the brain responsible for large-muscle control; at this point people under the influence stagger or weave when they try to walk.

Finally the conscious brain becomes completely subdued, and the person passes out. Now the person can drink no more. This is fortunate because a higher dose would anesthetize the deepest brain centers that control breathing and heartbeat, causing death.

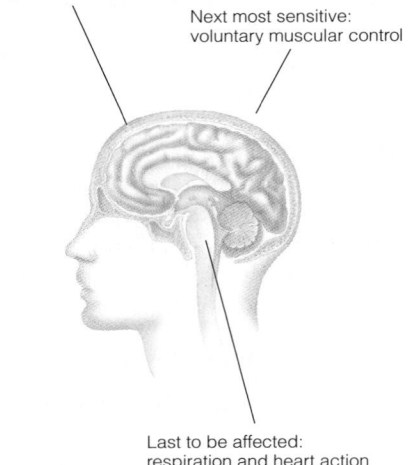

Most sensitive: judgment and reasoning

Next most sensitive: voluntary muscular control

Last to be affected: respiration and heart action

sleep. Every year, deaths attributed to this effect take place during drinking contests. Before passing out, the drinker drinks fast enough to receive a lethal dose. Table C3-5 shows blood alcohol levels that cor-

respond with progressively greater intoxication, and Table C3-6 shows brain and nervous system responses that occur at these levels.

Brain cells are particularly sensitive to excessive exposure to alcohol. The brain shrinks, even in people who drink only moderately. The extent of the shrinkage is proportional to the amount drunk. Abstinence, together with good nutrition, reverses some of the brain damage, and possibly all of it, if heavy drinking has not continued for more than a few years. Prolonged drinking beyond an individual's capacity to recover, however, can do severe and irreversible harm to vision, memory, learning ability, and other brain functions.

Alcohol Arrives in the Liver

The capillaries that surround the digestive tract merge into veins that carry the alcohol-laden blood to the liver. Here the veins branch and rebranch into capillaries that touch every liver cell. The liver cells make the largest share of the body's alcohol-processing machinery. The routing of blood through the liver allows the cells to go right to work detoxifying substances before they reach other body organs such as the heart and brain.

The Liver Metabolizes Alcohol

The liver makes and maintains two sets of equipment for metabolizing alcohol. One is an enzyme that removes hydrogens from alcohol to break it down; the name, **alcohol dehydrogenase (ADH),** almost says what it does.[‡] This enzyme handles about 80 percent or more of the

[‡]There are actually two ADH enzymes; each performs a specific task in alcohol breakdown.

TABLE C3·5 ALCOHOL DOSES AND AVERAGE BLOOD LEVELS

Number of Drinks[a]	Average Percent Blood Alcohol by Body Weight				
	100 lb	120 lb	150 lb	180 lb	200 lb
2	0.08	0.06	0.05	0.04	0.04
4	0.15	0.13	0.10	0.08	0.08
6	0.23	0.19	0.15	0.13	0.11
8	0.30	0.25	0.20	0.17	0.15
12	0.45	0.36	0.30	0.25	0.23
14	0.52	0.42	0.35	0.34	0.27

[a]Taken within an hour or so; each drink equal to ½ ounce pure ethanol.

TABLE C3·6 BLOOD ALCOHOL LEVELS AND BRAIN RESPONSES

Blood Alcohol Level (%)	Brain Response
0.05[a]	Judgment impaired
0.10	Emotional control impaired
0.15	Muscle coordination and reflexes impaired
0.20	Vision impaired
0.30	Drunk, lacking control
0.35	In a stupor
0.50–0.60	Loss of consciousness; death.

[a] A 0.08 percent level is the legal limit for intoxication according to most states' highway safety ordinances; however, driving ability may be impaired at blood alcohol levels lower than 0.08 percent.

alcohol in the body. The other set of alcohol-metabolizing equipment is a chain of enzymes, named **MEOS,** which is thought to handle about 10 percent of alcohol. The remaining 10 percent is excreted through the breath and in the urine. Because the alcohol in the breath is directly proportional to the alcohol in the blood, the breathalyzer test that law enforcement officers administer when someone may be driving under the influence of alcohol accurately reveals the person's degree of intoxication.

The amount of alcohol a person's body can process in a given time is limited by the amount of ADH enzymes residing in the liver. If more molecules of alcohol arrive at the liver cells than the enzymes can handle, the extra alcohol must wait. It circulates again and again through the brain, liver, and other organs until enzymes are available to degrade it.

Some ADH enzymes reside in the stomach and break down some alcohol before it enters the blood. Research shows that people with alcoholism make less stomach ADH than others and that women make less than men. Earlier, this Controversy warned that women should not try to keep up with male drinkers, and here is the reason why: women absorb about one-third more alcohol than men do, even when the women are the same size as the men and drink the same amount of alcohol.

The amount of ADH enzymes present is also affected by whether a person eats. Fasting for as little as a day causes degradation of body proteins, including the ADH enzymes, which can reduce the rate of alcohol metabolism by half.

The body takes about an hour and a half to metabolize one drink, depending on the person's size, previous drinking experience, how recently the person has eaten, and the person's current state of health. The liver is the only organ that can dispose of significant quantities of alcohol, and its maximum rate of alcohol clearance cannot be speeded up. This explains why only time restores sobriety. Walking will not because muscles cannot metabolize alcohol. Nor will drinking a cup of coffee help. Caffeine is a stimulant, but it won't speed up the metabolism of alcohol. The police say that a cup of coffee only makes a sleepy drunk into a wide-awake drunk. Table C3-7 presents other alcohol myths.

Alcohol Affects Body Functions

Upon exposure to alcohol, the liver speeds up its synthesis of fatty acids. Fat is known to accumulate in the livers of young men after a single night of heavy drinking and to remain there for more than a day. The first stage of liver deterioration seen in heavy drinkers is therefore known as **fatty liver;** it interferes with the distribution of nutrients and oxygen to the liver cells. If the condition lasts long enough, fibrous scar tissue invades the liver. This is the second stage of liver deterioration, called **fibrosis.** Fibrosis is reversible with good nutrition and abstinence from alcohol, but the next (last) stage, **cirrhosis,** is not. In cirrhosis, the liver cells harden, turn orange, and die, losing function forever (shown in the photo on page 96).

The presence of alcohol alters amino acid metabolism in the liver cells. Synthesis of some immune system proteins slows down, weakening the body's defenses against infection. Synthesis of blood lipids speeds up, increasing the concentration of triglycerides and high-density lipoproteins (see Chapter 5). In addition, excess alcohol adds to the body's acid burden and interferes with normal uric acid metabolism, causing symptoms like those of **gout.**

The reproductive system is also vulnerable to alcohol's effects. Heavy drinking in women may lead to infertility and spontaneous abortion. Alcohol may also suppress the male reproductive hormone testosterone, leading to decreases in muscle and bone tissue, altered immunity, abnormal prostate gland, and decreased reproductive ability. All of these effects demonstrate the importance of moderation in the use of alcohol.

TABLE C3·7 MYTHS AND TRUTHS CONCERNING ALCOHOL

Myth:	A shot of alcohol warms you up.
Truth:	Alcohol diverts blood flow to the skin making you *feel* warmer, but it actually cools the body.
Myth:	Wine and beer are mild; they do not lead to addiction.
Truth:	Wine and beer drinkers worldwide have high rates of death from alcohol-related illnesses. It's not what you drink, but how much, that makes the difference.
Myth:	Mixing drinks is what gives you a hangover.
Truth:	Too much alcohol in any form produces a hangover.
Myth:	Alcohol is a stimulant.
Truth:	Alcohol depresses the brain's activity.
Myth:	Alcohol is legal; therefore, it is not a drug.
Truth:	Alcohol is legal, but it alters body functions and is medically defined as a depressant drug.

© A. Glaubermann/Photo Researchers, Inc.

Left, normal liver; center, fatty liver; right, cirrhosis.

The Fattening Power of Alcohol

Alcohol should probably be counted as fat in the diet because metabolic interactions occur between fat and alcohol in the body. Presented with both fat and alcohol, the body stores the comparatively harmless fat and rids itself of the toxic alcohol by burning it off as fuel.[4] Alcohol may promote fat storage particularly in the central abdominal area—the **"beer belly"** effect seen in moderate drinkers. The risks of excess abdominal fat to the heart are described in Chapter 9.

Alcohol yields 7 calories of energy per gram to the body, so many alcoholic drinks can be much more fattening than their nonalcoholic counterparts. A general guideline states that each ounce of ethanol in a drink represents the same number of calories as about half an ounce of fat. An observant reader, knowing that, in the laboratory, a gram of fat and a gram of alcohol yield 9 and 7 calories, respectively, may wonder why alcohol in a drink is worth only half the calorie value of fat. The answer is that the body rids itself of a small but measurable amount of the alcohol by way of the breath and urine.

The Hangover

The hangover—the awful feeling of headache, pain, unpleasant sensations in the mouth, and nausea the morning after drinking too much—is a mild form of drug withdrawal. (The worst form is a delirium with severe tremors that presents a danger of death and demands medical management.) Hangovers are caused by several factors. One is the toxic effects of **congeners** that accompany the alcohol in alcoholic beverages. Mixing or switching drinks will not prevent hangover because congeners are only one factor. Dehydration of the brain is a second factor: alcohol reduces the water content of the brain cells. When they rehydrate the morning after and swell back to their normal size, nerve pain results.

Another contributor to the hangover is **formaldehyde,** the same chemical that laboratories use to preserve dead animals. Formaldehyde comes from **methanol,** an alcohol produced constantly by normal chemical processes in all the cells. Normally, a set of liver enzymes converts this methanol to formaldehyde, and then a second set immediately converts the formaldehyde to carbon dioxide and water, harmless waste products that can be excreted. But these same two sets of liver enzymes are the ones that process ethanol to its own intermediate (also highly toxic) waste product, **acetaldehyde,** and then to carbon dioxide and water.[5] The enzymes prefer ethanol 20 times over methanol. Both alcohols are metabolized without delay until the excess acetaldehyde monopolizes the second set enzymes, leaving formaldehyde to wait for later detoxification. At that point, formaldehyde starts accumulating and the hangover begins.

Time alone is the cure for a hangover. Vitamins, tranquilizers, aspirin, drinking more alcohol, breathing pure oxygen, exercising, eating, and drinking something awful are all useless. Fluid replacement can help to normalize the body's chemistry. The headache, unpleasantness in the mouth, and nausea of a hangover come simply from drinking too much.

Alcohol's Long-Term Effects

By far the longest-term effects of alcohol are those felt by the child of a woman who drinks during pregnancy. When a pregnant woman takes a drink, her fetus takes the same drink within minutes, and its body is defenseless against the effects. Pregnant women should not drink alcohol—this topic is so important that it has its own section in Chapter 13. For nonpregnant adults, however, what are the effects of alcohol over the long term?

A couple of drinks set in motion many destructive processes in the body. The next day's abstinence can reverse them only if the doses taken are moderate, the time between them is ample, and nutrition is adequate.

If the doses of alcohol are heavy, however, and the time between them is short, complete recovery cannot take place, and repeated onslaughts of alcohol take a toll on the body. For example, alcohol is directly toxic to skeletal and cardiac muscle, causing weakness and deterioration that is greater, the larger the dose. Alcoholism makes heart disease likely, probably because chronic alcohol use raises blood pressure. At autopsy, the heart of a person with alcoholism appears bloated and weighs twice as much as a normal heart.

Alcohol attacks brain cells directly, and heavy drinking can result in dementia. Cirrhosis also develops after 10 to 20 years from the cumulative effects of frequent episodes of heavy drinking.

Alcohol abuse also leads to cancers of the breast, mouth, throat, esophagus, rectum, and lungs.[6] Daily human exposure to ethanol ranks high among carcinogenic hazards. Once cancer has started, alcohol also seems to promote its development.

A convincing body of evidence implicates alcohol intake by women in the causation of cancer of the breast—women who drink less than one drink per day elevate their risk slightly, and those who drink more increase their risk accordingly.[7] Cancer of the rectum increases with intakes of more than 15 ounces of beer each day. In the case of beer, alcohol may be acting together with other compounds formed during brewing to promote the cancer. One compound, **urethane,** often found in alcoholic beverages, is known to cause cancer in animals, but the risk to human beings is unknown.

Other long-term effects of alcohol abuse include the following:

- Bladder, kidney, pancreas, and prostate damage.
- Bone deterioration and osteoporosis.
- Brain disease, central nervous system damage, and strokes.
- Deterioration of the testicles and adrenal glands.
- Diabetes (type 2 or noninsulin-dependent).
- Disease of the muscles of the heart.
- Feminization and sexual impotence in men.
- Impaired immune response.
- Impaired memory and balance.
- Increased risks of death from all causes.
- Malnutrition.
- Nonviral hepatitis.
- Severe psychological depression.
- Skin rashes and sores.
- Ulcers and inflammation of the stomach and intestines.

This list is by no means all-inclusive. Alcohol abuse exerts direct toxic effects on all body organs. Monetarily, alcoholism costs our society an estimated $166 *billion* every year in medical services, lost wages, criminal offenses, auto crashes, and other losses.

Alcohol's Effect on Nutrition

Alcohol abuse also does damage indirectly via malnutrition. The more alcohol a person drinks, the less likely that he or she will eat enough food to obtain adequate nutrients. Like pure sugar and pure fat, alcohol provides empty calories; it displaces nutrients. Table C3-8 shows the calorie amounts of typical alcoholic beverages.

Alcohol abuse also disrupts every tissue's metabolism of nutrients. Stomach cells oversecrete both acid and histamine, an agent of the immune system that produces inflammation. Intestinal cells fail to absorb thiamin, folate, vitamin B_6, and other vitamins. Liver cells lose efficiency in activating vitamin D and alter their production and excretion of bile. Rod cells in the retina, which normally process vitamin A alcohol (retinol) to the form needed in vision, find themselves processing drinking alcohol instead. Liver cells, too, suffer a

TABLE C3·8 CALORIES IN ALCOHOLIC BEVERAGES AND MIXERS

Beverage	Amount (oz)	Energy (cal)
Beer	12	150
Dessert wine	3½	140
Fruit-flavored soda, Tom Collins mix	8	115
Gin, rum, vodka, whiskey (86 proof)	1½	105
Cola, root beer	8	100
Light beer	12	100
Table wine	3½	85
Tonic, ginger ale	8	80
Club soda, plain seltzer, diet drinks	8	1

reduced capacity to process and use vitamin A.[8] The kidneys excrete magnesium, calcium, potassium, and zinc.

The inadequate food intake and impaired nutrient absorption that accompany chronic alcohol abuse frequently lead to a deficiency of the B vitamin thiamin. In fact, the cluster of thiamin-deficiency symptoms commonly seen in chronic alcoholism has its own name—the **Wernicke-Korsakoff syndrome.** This syndrome is characterized by paralysis of the eye muscles, poor muscle coordination, impaired memory, and damaged nerves; the syndrome and other alcohol-related memory problems may respond to treatment with thiamin supplements.[9]

Most dramatic is alcohol's effect on folate. When an excess of alcohol is present, the body actively expels folate from all of its sites of action and storage. The liver, which normally contains enough folate to meet all needs, leaks its folate into the blood. As blood folate rises, the kidneys are deceived into excreting it, as if it were in excess. The intestine normally releases and retrieves folate continuously, but it becomes so damaged by folate deficiency and alcohol toxicity that it fails to absorb folate. Alcohol also interferes with the action of what little folate is left, causing a buildup in the blood of a compound suspected of involvement with many diseases, including heart disease, stroke, and birth defects.§[10] This interference inhibits the production of new cells, especially the rapidly dividing cells of the intestine and the blood.

---

§The compound is homocysteine; see Chapter 7.

Nutrient deficiencies are thus an inevitable consequence of alcohol abuse, not only because alcohol displaces food but also because alcohol interferes directly with the body's use of nutrients. People treated for alcohol addiction also need nutrition therapy to reverse deficiencies and even deficiency diseases rarely seen in others: night blindness, beriberi, pellagra, scurvy, and protein-energy malnutrition.

Does Moderate Alcohol Use Benefit Health?

What are the potential risks and benefits to heart health of moderate drinking? Does age matter?

Age does matter. Young people do not benefit their health by drinking; rather, they increase their risk of dying from all causes. Young nondrinkers are found to have a lower risk of dying than even light drinkers (fewer than 15 drinks per month) of the same age. Alcohol is related to car crashes, homicides, and other violence that account for the great majority of deaths of young people each year. In fact, the sum of all alcohol-related U.S. deaths tops 75,000 annually, amounting to the third leading cause of mortality.[11] Young women in particular should not drink alcohol for the sake of their heart. For women before menopause, the risk of heart disease is low, but the risk of breast cancer is substantial. Even one drink a day, the amount providing heart benefits to older people, raises the risk of breast cancer in young women by about 10 percent.[12]

Alcohol and Heart Disease

One to two standard drinks of alcoholic beverages a day are credited with reducing the risk of death from heart disease in people over 60 years old who have an increased risk of heart disease. Increasing alcohol beyond this amount *increases* the risk of heart disease substantially.[13] Wine is often credited with aiding heart health, but research indicates that even beer may reduce heart attack risk in some populations.[14]

Though many studies support a beneficial effect of moderate alcohol intake on heart health, the matter is not yet settled. Researchers followed the alcohol intakes and health histories of almost 6,000 men for over 20 years.[15] The results showed no beneficial relationship between mortality from cardiovascular disease and *any* level of alcohol consumption. The study did show an increased risk of death from all causes with more than 22 drinks per week and that men drinking more than 35 drinks a week had double the mortality from stroke compared with nondrinkers. Strokes are associated with elevated blood pressure, and heavy alcohol intake is known to increase both.[16]

The Health Effects of Wine

Red wine has been credited with special health-supporting properties, and recently, white wine has won attention for an antioxidant effect.[17] The following two statements concerning wine and health have been approved to appear on U.S. wine labels:

- "The proud people who made this wine encourage you to consult your family doctor about the health effects of wine consumption."
- "To learn the health effects of wine consumption, send for the Federal Government's Dietary Guidelines for Americans, Center for Nutrition Pol-

icy and Promotion, USDA, 1120 20th Street, NW, Washington, DC 20036 or visit its website."

These statements seem to promise that good news about wine and health may await the information seeker, but the science on wine and health is mixed. For example, the high potassium content of grape juice may lower high blood pressure, and potassium persists when the grape juice is made into wine. Since alcohol in large amounts raises blood pressure, however, the grape juice may be more suitable than the wine for people with hypertension. Dealcoholized wine also facilitates the absorption of potassium, calcium, phosphorus, magnesium, and zinc. So does wine, but the alcohol in it promotes the quick *excretion* of these minerals, so the dealcoholized version is preferred.

In addition to alcohol, wine contains flavonoids that are under study as protectors against events that are thought to trigger heart disease. An antioxidant effect of flavonoids in wine has been offered as an explanation for why wine-drinking French and other Mediterranean peoples have a lower incidence of heart disease despite having many risk factors. (Controversy 5 comes back to the issue of Mediterranean diets and disease.) Compared with other food sources such as onions or other vegetables, though, wine may deliver only small amounts of antioxidant flavonoids to the body.[18] In addition, alcohol itself increases oxidation in the body.

Furthermore, recent evidence calls into question whether heart protection from antioxidants is at work. In a study of mice genetically prone to develop heart-damaging changes leading to heart disease, dealcoholized red wine effectively reduced the changes without reducing oxidation.[19] Dealcoholized wine, purple grape juice, and the grapes themselves contain phytochemicals sim-

ilar to those of wine but without the potential dangers of alcohol.

Alcohol and Appetite

Alcoholic beverages affect the appetite. Usually, they reduce it, making people unaware that they are hungry. But in people who are tense and unable to eat or in the elderly who have lost interest in food, a small dose of wine taken 20 minutes before meals may improve appetite. For undernourished people and for people with severely depressed appetites, wine may facilitate eating even when psychotherapy fails to do so.

Another example of the beneficial use of alcohol comes from research showing that moderate use of wine in later life improves morale, stimulates social interaction, and promotes restful sleep. In nursing homes, improved patient and staff relations have been attributed to greater self-esteem among elderly patients who drink moderate amounts of wine. Researchers hypothesize that chronic fatigue may be responsible for some behaviors associated with old age. The positive effects of wine on sleep may alleviate fatigue.

The Final Word

This discussion has explored some of the ways alcohol affects health and nutrition. In contrast to some possible benefits of moderate alcohol consumption, excessive alcohol consumption presents a great potential for harm. Alcohol is guilty of contributing not only to deaths from health problems but also to most of the other needless deaths of young people each year, including car crashes, falls, suicides, homicides, drownings, and other accidents. The surest way to escape the harmful effects of alcohol is, of course, to refuse alcohol altogether. If you choose to drink, do so with care and strictly in moderation.

The Carbohydrates: Sugar, Starch, Glycogen, and Fiber

CHAPTER
4

© Ilya Mashkov (1881–1944), *Bread*. Russian State Museum, St. Petersburg, Russia. Scala/Art Resource, NY

CONTENTS

Do You Ever . . .

Think of carbohydrates as providing nothing but calories to the body?

Wonder why nutrition authorities unanimously recommend foods high in fiber?

Have trouble choosing among breads at the grocery store?

Try to reduce your sugar intake by choosing foods sweetened with purified fruit concentrates?

KEEP READING.

FIGURE 4·1 CARBOHYDRATE —MAINLY GLUCOSE—IS MADE BY PHOTOSYNTHESIS

The sun's energy becomes part of the glucose molecule—its calories, in a sense. In the molecule of glucose on the leaf here, black dots represent the carbon atoms; bars represent the chemical bonds that contain energy.

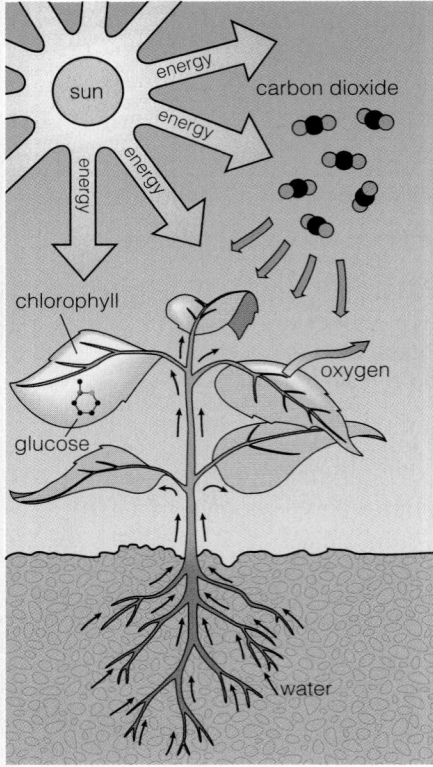

■ Single sugars are monosaccharides.

■ Pairs of sugars are disaccharides.

carbohydrates compounds composed of single or multiple sugars. The name means "carbon and water," and a chemical shorthand for carbohydrate is CHO, signifying carbon (C), hydrogen (H), and oxygen (O).

complex carbohydrates long chains of sugar units arranged to form starch or fiber; also called *polysaccharides*.

simple carbohydrates sugars, including both single sugar units and linked pairs of sugar units. The basic sugar unit is a molecule containing six carbon atoms, together with oxygen and hydrogen atoms.

photosynthesis the process by which green plants make carbohydrates from carbon dioxide and water using the green pigment chlorophyll to capture the sun's energy (*photo* means "light"; *synthesis* means "making").

CARBOHYDRATES ARE THE IDEAL nutrients to meet your body's energy needs, feed your brain and nervous system, keep your digestive system fit, and, within calorie limits, help keep your body lean. Digestible carbohydrates, together with fats and protein, add bulk to foods and provide energy and other benefits for the body. Indigestible carbohydrates, which include most of the fibers in foods, yield little or no energy but provide other important benefits.

All carbohydrates are not equal in terms of nutrition. This chapter invites you to learn to distinguish between foods containing the **complex carbohydrates** (starch and fiber) and those made of the **simple carbohydrates** (sugars), and to consider the effects of both on the body. The Controversy asks whether sugar harms health and whether alternative sweeteners are preferable.

This chapter on the carbohydrates is the first of three on the energy-yielding nutrients. Chapter 5 deals with the fats and Chapter 6 with protein. The Controversy of Chapter 3 addressed one other contributor of energy, alcohol.

A Close Look at Carbohydrates

Carbohydrates contain the sun's radiant energy, captured in a form that living things can use to drive the processes of life. Green plants make carbohydrate through **photosynthesis** in the presence of **chlorophyll** and sunlight. In this process, water (H_2O), absorbed by the plant's roots, donates hydrogen and oxygen. Carbon dioxide gas (CO_2), absorbed into its leaves, donates carbon and oxygen. Water and carbon dioxide combine to yield the most common of the **sugars,** the single sugar **glucose.** Scientists know the reaction in the minutest detail but have never been able to reproduce it— green plants are required to make it happen (see Figure 4-1).

Light energy from the sun drives the photosynthesis reaction. The light energy becomes the chemical energy of the bonds that hold six atoms of carbon together in the sugar glucose. Glucose provides energy for the work of all cells of the stem, roots, flowers, and fruits of the plant. For example, in the roots, far from the energy-giving rays of the sun, each cell draws upon some of the glucose made in the leaves, breaks it down (to carbon dioxide and water), and uses the energy thus released to fuel its own growth and water-gathering activities.

Plants do not use all of the energy stored in their sugars, so it remains available for use by the animal or human being that consumes the plant. Thus, carbohydrates form the first link in the food chain that supports all life on earth. Carbohydrate-rich foods come almost exclusively from plants; milk is the only animal-derived food that contains significant amounts of carbohydrate. The next few sections describe the forms assumed by carbohydrates: sugars, starch, glycogen, and fiber.

▶ **KEY POINT** Through photosynthesis, plants combine carbon dioxide, water, and the sun's energy to form glucose. Carbohydrates are made of carbon, hydrogen, and oxygen held together by energy-containing bonds: *carbo* means "carbon"; *hydrate* means "water."

Sugars

Six sugar molecules are important in nutrition. Three are single sugars, or **monosaccharides.** The other three are double sugars, or **disaccharides.** All of their chemical names end in *ose*, which means "sugar." Although they all sound alike at first, they exhibit distinct characteristics once you get to know them as individuals. Figure 4-2 shows the relationships among the sugars.

The three monosaccharides are glucose, **fructose,** and **galactose.** Fructose or fruit sugar, the intensely sweet sugar of fruit, is made by rearranging the atoms in glucose molecules. Fructose occurs mostly in fruits, in honey, and as part of table sugar. Other sources include soft drinks, ready-to-eat cereals, and other products sweetened with high-fructose corn syrup (defined on page 128). Glucose and fructose are the most common monosaccharides in nature.

The other monosaccharide, galactose, has the same number and kind of atoms as glucose and fructose but in another arrangement. Galactose is one of two single sugars that are bound together to make up the sugar of milk. It rarely occurs free in nature but is tied up in milk sugar until it is freed during digestion.

The three other sugars important in nutrition are disaccharides, which are linked pairs of single sugars. All three contain glucose. In **lactose,** the milk sugar just mentioned, glucose is linked to galactose.

Malt sugar, or **maltose,** has two glucose units. Maltose appears wherever starch is being broken down. It occurs in germinating seeds and arises during the digestion of starch in the human body.

The last of the six sugars, **sucrose,** is familiar table sugar, the product most people think of when they refer to *sugar*. In sucrose, fructose and glucose are bonded together. Table sugar is obtained by refining the juice from sugar beets or sugarcane, but sucrose also occurs naturally in many vegetables and fruits. It tastes sweet because it contains the sweetest of the monosaccharides, fructose.

When you eat a food containing single sugars, you can absorb them directly into your blood. When you eat disaccharides, though, you must digest them first. Enzymes in your intestinal cells must split the disaccharides into separate monosaccharides so that they can enter the bloodstream. The blood delivers all products of digestion first to the liver, which possesses enzymes to modify nutrients, making them useful to the body. Glucose is the most-used monosaccharide inside the body, so the liver quickly converts fructose or galactose to glucose or to smaller pieces that can serve as building blocks for glucose, fat, or other needed molecules.

Although it is true that the energy of fruits and many vegetables comes from sugars, this doesn't mean that eating them is the same as eating concentrated sweets such as candy or cola beverages. From the body's point of view, fruits are vastly different from purified sugars, as a later section makes clear, except that both provide glucose in abundance.

▶ **KEY POINT** Glucose is the most important monosaccharide in the human body. Most other monosaccharides and disaccharides become glucose in the body.

chlorophyll the green pigment of plants that captures energy from sunlight for use in photosynthesis.

sugars simple carbohydrates; that is, molecules of either single sugar units or pairs of those sugar units bonded together. By common usage, *sugar* most often refers to sucrose.

glucose (GLOO-cose) a single sugar used in both plant and animal tissues for energy; sometimes known as blood sugar or *dextrose*.

monosaccharides (mon-oh-SACK-ah-ride) single sugar units (*mono* means "one"; *saccharide* means "sugar unit").

disaccharides pairs of single sugars linked together (*di* means "two").

fructose (FROOK-tose) a monosaccharide; sometimes known as fruit sugar (*fruct* means "fruit"; *ose* means "sugar").

galactose (ga-LACK-tose) a monosaccharide; part of the disaccharide lactose (milk sugar).

lactose a disaccharide composed of glucose and galactose; sometimes known as milk sugar (*lact* means "milk"; *ose* means "sugar").

maltose a disaccharide composed of two glucose units; sometimes known as malt sugar.

sucrose (SOO-crose) a disaccharide composed of glucose and fructose; sometimes known as table, beet, or cane sugar and, often, simply *sugar*.

■ Digestive enzymes were introduced in Chapter 3.

FIGURE 4·2 HOW MONOSACCHARIDES JOIN TO FORM DISACCHARIDES

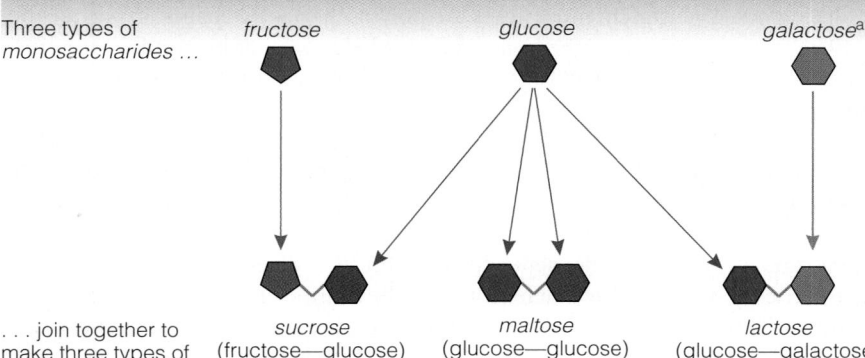

Three types of *monosaccharides* ...

fructose   glucose   galactose[a]

... join together to make three types of *disaccharides*.

sucrose (fructose—glucose)   maltose (glucose—glucose)   lactose (glucose—galactose)

A note on the glucose symbol:
The glucose molecule is really a ring of 5 carbons and 1 oxygen plus a carbon "flag."

carbons   oxygen

For convenience, glucose is symbolized as

or

[a]Galactose does not occur in foods singly but only as part of lactose.

Starch

In addition to occurring in sugars, the glucose in food also occurs in long strands of thousands of glucose units. These are the **polysaccharides** (see Figure 4-3). **Starch** is a polysaccharide as are glycogen and most of the fibers.

Starch is a plant's storage form of glucose. As a plant matures, it not only provides energy for its own needs but also stores energy in its seeds for the next generation. For example, after a corn plant reaches its full growth and has many leaves manufacturing glucose, it stores packed clusters of starch molecules in **granules** and packs the granules into its seeds. These giant starch clusters are packed side by side in the kernels of corn. For the plant, starch is useful because it is an insoluble substance that will stay with the seed in the ground and nourish it until it forms shoots with leaves that can catch the sun's rays. Glucose, in contrast, is soluble in water and would be washed away by the rain while the seed lay in the soil. The starch of corn and other plant foods is nutritive for people as well because they can digest the starch to glucose and extract the sun's energy stored in its chemical bonds. A later section describes starch digestion in greater detail.

▶ **KEY POINT** Starch is the storage form of glucose in plants and is also nutritive for human beings.

Glycogen

Just as plants store glucose in long chains of starch, animal bodies store glucose in long chains of **glycogen.** Glycogen resembles starch in that it consists of glucose molecules linked together to form chains, but its chains are longer and more highly branched (see Figure 4-3). Unlike starch, which is abundant in grains, potatoes, and other foods from plants, glycogen is nearly undetectable in meats because glycogen breaks down rapidly when the animal is slaughtered. A later section describes how the human body handles its own packages of stored glucose.

▶ **KEY POINT** Glycogen is the storage form of glucose in animals and human beings.

Fiber

Some of the **fibers** of a plant form the supporting structures of its leaves, stems, and seeds. Other fibers play other roles, for example, to retain water and thus protect seeds from drying out. Like starch, most fibers are polysaccharides—chains of sugars—but they differ from starch in that the sugar units are held together by bonds that human digestive enzymes cannot break. Most fibers therefore pass through the human body without providing energy for its use.[*]

Billions of bacteria residing within the human large intestine, however, do possess enzymes that can digest fibers to varying degrees by fermenting them. Through this process, the fibers are broken down to waste products, mainly small fatlike fragments that the large intestine (colon) absorbs.

Many animals, such as cattle, depend heavily on their digestive system's bacteria to make the energy of glucose available from the abundant cellulose, a form of fiber, in their fodder. Thus, when we eat beef, we indirectly receive some of the sun's energy that was originally stored in the fiber of the plants. Beef itself contains no fiber, however; no meats or dairy products contain fiber.

[*]Recently, the committee on Dietary Reference Intakes (DRI) proposed these fiber definitions: total fiber refers to the sum of fibers from two sources: naturally occurring fibers in intact foods (*dietary fibers*) and added fibers that have health benefits (functional fibers).

© Gala/SuperStock

foods such as these fruits provide carbohydrates and other nutrients to the body.

■ Strands of many sugar units are polysaccharides.

polysaccharides another term for complex carbohydrates; compounds composed of long strands of glucose units linked together (*poly* means "many"). Also called *complex carbohydrates*.

starch a plant polysaccharide composed of glucose. After cooking, starch is highly digestible by human beings; raw starch often resists digestion.

granules small grains. Starch granules are packages of starch molecules. Various plant species make starch granules of varying shapes.

glycogen (GLY-co-gen) a highly branched polysaccharide composed of glucose that is made and stored by liver and muscle tissues of human beings and animals as a storage form of glucose. Glycogen is not a significant food source of carbohydrate and is not counted as one of the complex carbohydrates in foods.

fibers the indigestible parts of plant foods, largely nonstarch polysaccharides that are not digested by human digestive enzymes, although some are digested by resident bacteria of the colon. Fibers include cellulose, hemicelluloses, pectins, gums, and mucilages and the non-polysaccharide lignin.

FIGURE 4·3 HOW GLUCOSE MOLECULES JOIN TO FORM POLYSACCHARIDES

Glucose

Starch (unbranched)   Starch (branched)   Glycogen   Cellulose

Starch Glucose units are linked in long, occasionally branched chains to make starch. Human digestive enzymes can digest these bonds, retrieving glucose. Real glucose units are so tiny that you can't see them, even with the highest-power light microscope.

Glycogen Glycogen resembles starch in that the bonds between its glucose units can be broken by human enzymes, but the chains of glycogen are more highly branched.

Cellulose (fiber) The bonds that link glucose units together in cellulose are different from the bonds in starch or glycogen. Human enzymes cannot digest them.

Fibers are often divided into two general groups by their chemical and physical properties. In the first group are fibers that dissolve in water (**soluble fibers**). These form gels (are **viscous**) and are easily digested by bacteria in the human colon (are easily fermented). Commonly found in barley, legumes, fruits, oats, and vegetables, these fibers are often associated with lower risks of chronic diseases (as discussed in a later section). In foods, soluble fibers add pleasing consistency, such as the pectin that puts the gel in jelly and the gums that are added to salad dressings and other foods to thicken them.

Other fibers do not dissolve in water (**insoluble fibers**), do not form gels (are not viscous), and are less readily fermented. Insoluble fibers, such as cellulose and hemicellulose, are found in the outer layers of whole grains (bran), the strings of celery, the hulls of seeds, and the skins of corn kernels. These fibers retain their structure and rough texture even after hours of cooking. In the body, they aid the digestive system by easing elimination.[*1]

■ Soluble, viscous, fermentable fibers are often gummy or add thickness to foods.

■ Insoluble, nonviscous, less fermentable fibers are often tough, stringy, or gritty in foods.

▶ **KEY POINT**   Human digestive enzymes cannot break the bonds in fiber, so most of it passes through the digestive tract unchanged. Some fiber, however, is susceptible to fermentation by bacteria in the colon.

In summary, plants combine carbon dioxide, water, and the sun's energy to form glucose, which may be stored in the polysaccharide starch. Then animals or people eat the

soluble fibers food components that readily dissolve in water and often impart gummy or gel-like characteristics to foods. An example is pectin from fruit, which is used to thicken jellies. Soluble fibers are indigestible by human enzymes but may be broken down to absorbable products by bacteria in the digestive tract.

viscous (VISS-cuss) having a sticky, gummy, or gel-like consistency that flows relatively slowly.

insoluble fibers the tough, fibrous structures of fruits, vegetables, and grains; indigestible food components that do not dissolve in water.

*Reference notes are found in Appendix F.

■ Chapter 15 revisits humankind's relationship with the earth's food chain.

plants and retrieve the glucose. In the body, the liver and muscles may store the glucose as the polysaccharide glycogen, but ultimately it becomes glucose again. The glucose delivers the sun's energy to fuel the body's activities. In the process, glucose breaks down to the waste products carbon dioxide and water, which are excreted. Later, these compounds are used again by plants as raw materials to make carbohydrate. Fibers are plant constituents that are not digested directly by human enzymes, but their presence in the diet contributes to the health of the body.

The Need for Carbohydrates

Glucose from carbohydrate is an important fuel for most body functions. Only two other nutrients provide energy to the body: protein and fats. Protein-rich foods are usually expensive, and when used to make fuel for the body, they provide no advantage over carbohydrates. Moreover, overuse of proteins has disadvantages, as explained in Chapter 6. Fats normally are not used as fuel by the brain and central nervous system. Thus, glucose is a critical energy source for nerve cells including those of the brain. And starchy foods, or complex carbohydrates, and especially the fiber-rich ones, are the preferred source of glucose in the diet.

■ Chapter 10 describes how the body uses carbohydrate and other fuels in muscular work.

Scientists have only recently uncovered other roles for sugars in living tissues.[2] For example, sugar molecules dangle from many of the body's fat and protein molecules. Once thought to be mere hitchhikers, these sugars can often have dramatic effects. They can change the shape of a protein, for instance, thereby altering its function. The protein of mucus, which protects body membranes and the digestive tract with a slippery coating, depends on a sugar molecule for its functional properties. Sugars also bind to cell membranes, affecting cellular interactions that may play roles in disease processes.

If I Want to Lose Weight and Stay Healthy, Should I Avoid Carbohydrates?

■ The DRI committee recommends that 45% to 65% of daily calories come from carbohydrate. Details about controlling body fatness are in Chapter 9.

Popular books and magazines often wrongly accuse complex carbohydrates of being the "fattening" ingredient of foods, thereby misleading millions of weight-conscious people into avoiding all kinds of carbohydrate-rich foods.[3] Despite intense marketing of diet books and low-carbohydrate foods, the truth remains that people who wish to lose fat and to maintain lean tissue and the health of the body can do no better than to attend closely to portion sizes, control total calories, and design their diets around foods that supply carbohydrates in balance with other energy nutrients.[4]

Gram for gram, carbohydrates donate fewer calories than do dietary fats, and converting glucose into fat for storage is metabolically costly. Hence, bite for bite, a moderate balanced diet based on foods high in complex carbohydrates is likely to be lower in total calories than a diet based on high-fat foods.

■ 1 g carbohydrate = 4 calories.
1 g fat = 9 calories.

Recommendations to select complex carbohydrates do not extend to refined sugars. Whereas complex carbohydrates contribute needed nutrients, pure sugars displace nutrient-dense foods from the diet. Purified, refined sugars (mostly sucrose or fructose) contain no other nutrients—protein, vitamins, minerals, or fiber—and thus qualify as foods of low nutrient density. A person choosing 400 calories of sugar in place of 400 calories of whole-grain bread loses the protein, vitamins, minerals, phytochemicals, and fiber of the bread. You can afford to do this only if you have already met all of your nutrient needs for the day and still have calories to spend.

Overuse of sugars may have other effects as well. Some evidence suggests that, for many obese people, a diet too high in added sugars may alter blood lipids in ways that may worsen their heart disease risk (see the Controversy for details).[5] For these people, weight loss on a calorie-controlled diet high in whole grains, fruits, and vegetables reduces the blood lipid response to sugars and lowers their heart disease risk.

TABLE 4·1 RECOMMENDATIONS CONCERNING INTAKES OF CARBOHYDRATES

1. Recommendations for total carbohydrates

   World Health Organization
   - 55 to 75% of total calories from carbohydrate.

   Dietary Guidelines for Americans 2005
   - Consume between 45% and 65% of calories from carbohydrate.

   Dietary Reference Intakes (DRI)
   - At a minimum, 130 grams per day for adults and children, to provide glucose to the brain.
   - For health, most people should consume between 45% and 65% of total calories from carbohydrate.

   USDA Food Guide
   - Grains, fruit, starchy vegetables, and milk contribute to the day's total carbohydrate intake.

2. Recommendations for added sugars

   World Health Organization
   - Less than 10% of total calories from sugars.[b]

   Dietary Guidelines for Americans 2005
   - Choose and prepare foods and beverages with little added surgars.

   Dietary Reference Intakes (DRI)
   - Insufficient evidence exists to set an upper limit for added sugars; however, the DRI committee suggests a high maximum of 25% or less of calories for people who otherwise meet their nutrient needs, maintain a healthy body weight, and need additional energy.[c]

   USDA Food Guide
   - Added sugars may provide discretionary calories within the energy recommendation after meeting all nutrient recommendations with nutritious foods.

3. Recommendations for fiber

   World Health Organization
   - More than 25 grams of fiber from whole grains, fruits, and vegetables.

   USDA Food Guide
   - Increase intakes of whole fruits and vegetables, make at least half the grain choices whole grains, and choose legumes several times per week.

   Dietary Reference Intakes (DRI)
   - 38 grams of total fiber per day for men through age 50; 30 grams for men 51 and older.
   - 25 grams of total fiber per day for women through age 50; 21 grams for women 51 and older.

   Canada's Food Guide to Healthy Eating is presented in Appendix B.

[a] Serving sizes were presented in Chapter 2.
[b] The World Health Organization uses the term *free sugars* to mean all monosaccharides and disaccharides added to foods by the manufacturer, cook, or consumer, plus sugars naturally present in honey, syrups, and fruit juices.
[c] An example might be an athlete in training whose high energy need allows greater amounts of added sugars from sports drinks without compromising nutrient intakes; for most sedentary people, maximums of 3 to 12 teaspoons per day are suggested.

For health's sake, then, most people should increase their intakes of fiber-rich sources of carbohydrates such as whole grains, fruits, and vegetables and use moderation in choosing refined foods such as white flour and sugars, and the kinds of fats associated with heart disease (see Chapter 5).[6] Table 4-1 presents carbohydrate recommendations and guidelines from several authorities. This chapter's Consumer Corner (pp. 110–112) describes various breads, and the Food Feature comes back to the sugars in foods.

▶ **KEY POINT** The body tissues use carbohydrates for energy and other functions; the brain and nerve tissues prefer carbohydrate as fuel. Nutrition authorities recommend a diet based on foods rich in complex carbohydrates and fiber.

Why Do Nutrition Experts Recommend Fiber-Rich Foods?

Fiber's best-known health benefits include:

1. Promotion of normal blood cholesterol concentrations (reduced risk of heart disease).
2. Modulation of blood glucose concentrations (reduced risk of diabetes).
3. Maintenance of healthy bowel function (reduced risk of bowel diseases).
4. Promotion of a healthy body weight.

■ Appendix A lists the fiber contents of over 2,000 foods.

■ The roles of saturated fat, *trans* fat, cholesterol, and other lipids in heart disease are discussed in Chapters 5 and 11. The role of vegetable proteins in heart disease is presented in Chapter 6. The benefits of phytochemicals in disease prevention are featured in Controversy 2.

As mentioned, carbohydrate-rich foods offer additional benefits if they are also rich in fiber. Foods such as whole grains, vegetables, legumes, and fruits supply valuable vitamins, minerals, and phytochemicals, along with a healthy dose of fiber, and with little or no fat. Apples, barley, carrots, legumes, and oats are rich in viscous fibers that have a significant cholesterol-lowering effect. Cooked legumes can also help regulate the blood glucose following a carbohydrate-rich meal. Wheat bran, composed mostly of insoluble nonviscous fiber, is one of the most effective stool-softening fiber-rich foods. These benefits of fiber are the best known: promotion of normal blood cholesterol concentrations, modulation of blood glucose concentrations, and maintenance of healthy bowel function.[7] Add one other benefit of a fiber-rich diet to the list—help in maintaining a healthy body weight—and the obvious choice for anyone placing a value on health is to obtain fibers from a variety of sources each day.[8]

Table 4-2 shows the diverse effects of different fibers, and Figure 4-4 provides a brief guide to finding these fibers in foods. Most unrefined plant foods contain a mix of fiber types. The following paragraphs describe health benefits associated with daily intakes of these foods.

▶ **KEY POINT** Fiber-rich diets benefit the body by helping to normalize blood cholesterol and blood glucose and by maintaining healthy bowel function. They are also associated with healthy body weight.

Lower Cholesterol and Heart Disease Risk Diets rich in legumes, vegetables, and whole grains and therefore rich in complex carbohydrates may protect against heart disease and stroke, although sorting out the reasons why has proved difficult.[9] Such diets are generally low in saturated fat, *trans* fat, and cholesterol and high in fibers, vegetable proteins, and phytochemicals—all factors associated with a lower risk of heart disease.

Relatively certain is that foods rich in viscous fibers lower blood cholesterol by binding with cholesterol-containing bile in the intestine and carrying it out with the feces (see Figure 4-5, page 108).[10] Bile is needed in digestion, so the liver responds to its loss by drawing on the body's cholesterol to synthesize more. Another mechanism

TABLE 4·2 CHARACTERISTICS, SOURCES, AND HEALTH EFFECTS OF FIBERS

Fiber Characteristics	Major Food Sources	Actions in the Body	Health Benefits
Viscous, soluble, more fermentable			
• Gums • Pectins • Psyllium • Some hemicelluloses	Barley, oats, oat bran, rye, fruits (apples, citrus), legumes (especially young green peas and black-eyed peas), seaweeds, seeds and husks, vegetables; fibers used as food additives.	• Lower blood cholesterol by binding bile. • Slow glucose absorption. • Slow transit of food through upper GI tract, lending satiety. • Hold moisture in stools, softening them. • Yield small fatlike molecules after fermentation that the colon can use for energy.	• Lower risk of heart disease. • Lower risk of diabetes.
Nonviscous, insoluble, less fermentable			
• Cellulose • Lignin • Resistant starch • Many hemicelluloses	Brown rice, fruits, legumes, seeds, vegetables (cabbage, carrots, brussels sprouts), wheat bran, whole grains; extracted fibers used as food additives.	• Increase fecal weight and speed fecal passage though colon. • Provide bulk and feelings of fullness.	• Alleviate constipation. • Lower risks of diverticulosis, hemorrhoids, and appendicitis. • May help with weight management.

FIGURE 4·4   FIBER COMPOSITION OF COMMON FOODS

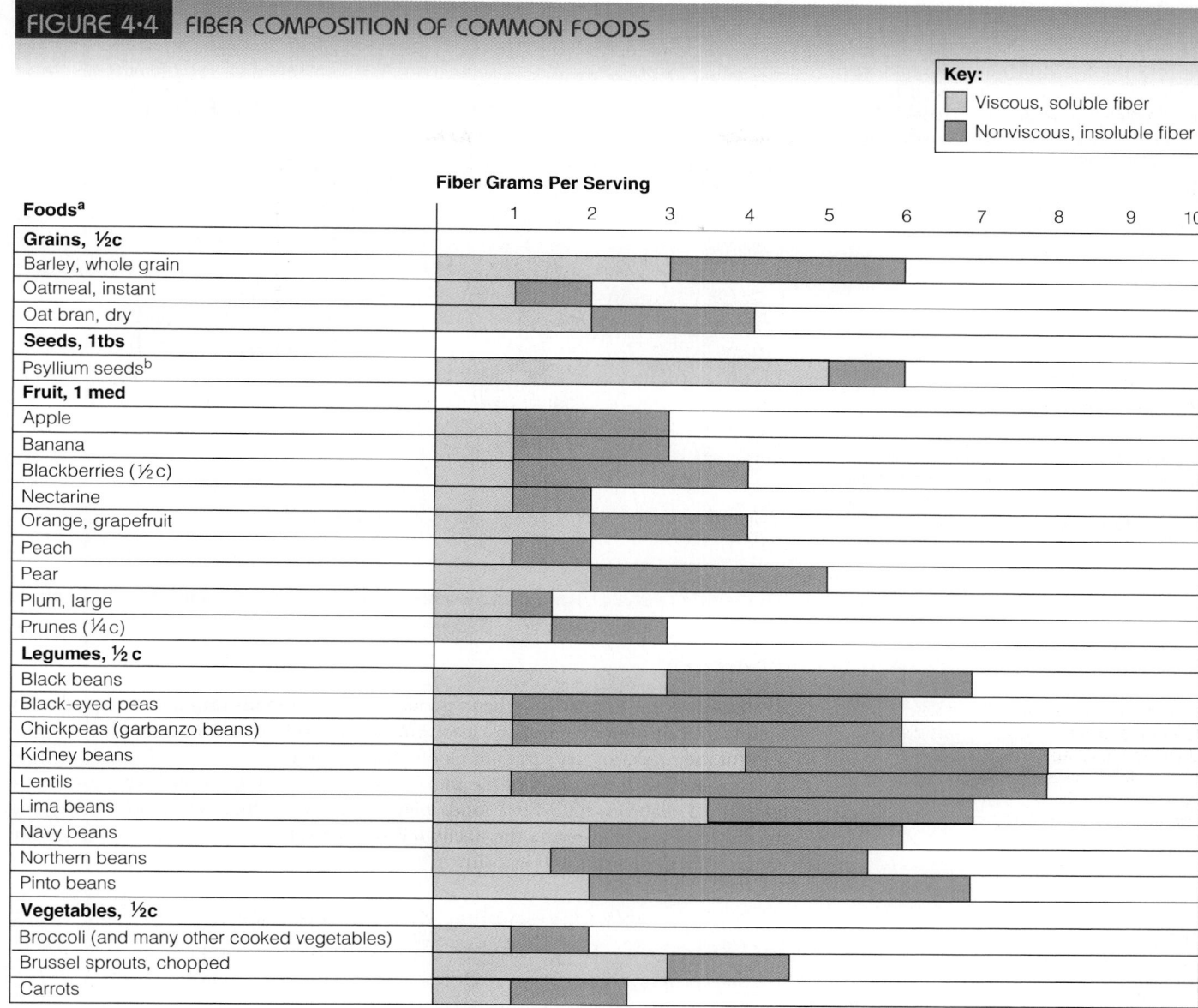

Key:
- Viscous, soluble fiber
- Nonviscous, insoluble fiber

Fiber Grams Per Serving

Foods[a]	1	2	3	4	5	6	7	8	9	10
Grains, ½c										
Barley, whole grain										
Oatmeal, instant										
Oat bran, dry										
Seeds, 1tbs										
Psyllium seeds[b]										
Fruit, 1 med										
Apple										
Banana										
Blackberries (½c)										
Nectarine										
Orange, grapefruit										
Peach										
Pear										
Plum, large										
Prunes (¼c)										
Legumes, ½ c										
Black beans										
Black-eyed peas										
Chickpeas (garbanzo beans)										
Kidney beans										
Lentils										
Lima beans										
Navy beans										
Northern beans										
Pinto beans										
Vegetables, ½c										
Broccoli (and many other cooked vegetables)										
Brussel sprouts, chopped										
Carrots										

[a] Values are for cooked or ready-to-serve foods unless specified.
[b] Psyllium is used as a fiber laxative and fiber-rich food additive.
SOURCE: Data from the National Heart, Lung and Blood Institute.Third Report of the National Cholesterol Education Proigram (NCEP) Expert Panel on Detection, Evaluation and Treatment of High Blood Cholesterol in Adults (Adult Treatment Panel 10, NIH publication no. 02-5215(2002); V-6; ESHA Research, 2004.

by which blood cholesterol may be reduced is through the actions of a small fatty acid released during bacterial fermentation of fiber. This fatty acid is absorbed and travels to the liver where it may help to reduce cholesterol synthesis.[11] By whatever mechanisms, the net result is lower blood cholesterol.[12]

▶ KEY POINT   Foods rich in soluble viscous fibers help control blood cholesterol.

FIGURE 4·5 ONE WAY FIBER IN FOOD MAY LOWER CHOLESTEROL IN THE BLOOD

In some ways, the liver is like a vacuum cleaner, sucking up cholesterol from the blood, converting the cholesterol to bile, and discharging the bile into its storage bag, the gallbladder. The gallbladder empties its bile into the intestine, where bile performs necessary digestive tasks. In the intestine, some of the bile associates with fiber and is carried out of the body in feces.

A. When the diet is rich in fiber, more cholesterol (as bile) is carried out of the body.
B. When the diet is low in fiber, most of the cholesterol is reabsorbed and returned to the bloodstream.

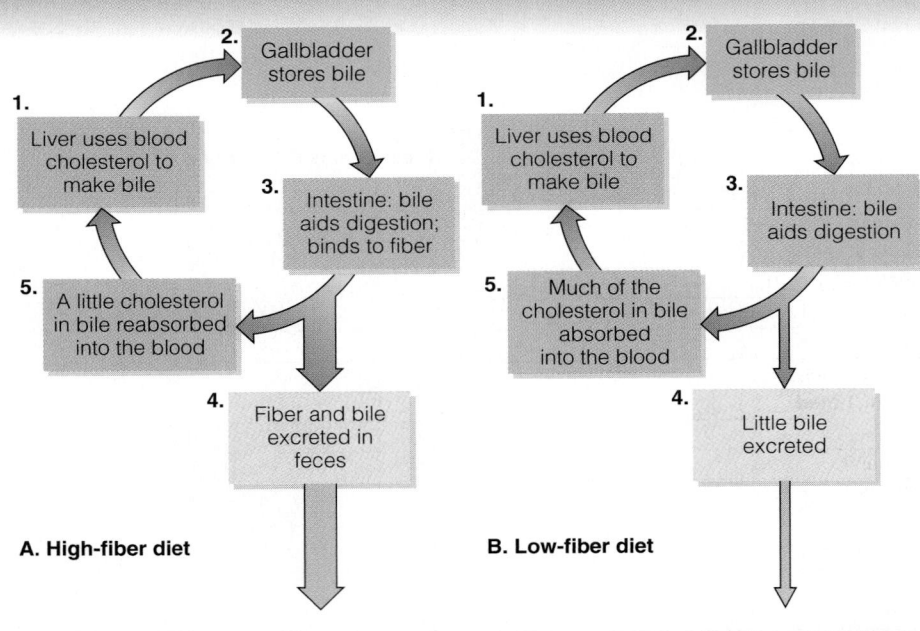

A. High-fiber diet

1. Liver uses blood cholesterol to make bile
2. Gallbladder stores bile
3. Intestine: bile aids digestion; binds to fiber
4. Fiber and bile excreted in feces
5. A little cholesterol in bile reabsorbed into the blood

B. Low-fiber diet

1. Liver uses blood cholesterol to make bile
2. Gallbladder stores bile
3. Intestine: bile aids digestion
4. Little bile excreted
5. Much of the cholesterol in bile absorbed into the blood

Blood Glucose Control When viscous fibers from foods trap nutrients and delay their transit through the digestive tract, glucose absorption is slowed, and this helps to prevent the roller-coaster surges in blood glucose and the hormone insulin thought to be associated with the onset of the most common form of diabetes.[13] In people with established diabetes, high-fiber foods play a role in modulating blood glucose and insulin levels, thus lowering the likelihood of medical complications. A later section comes back to diabetes and its control.

▶ KEY POINT   Foods rich in viscous fibers help to modulate blood glucose concentrations.

Maintenance of Digestive Tract Health All kinds of fibers, along with an ample fluid intake, probably play roles in maintaining proper colon function. Fibers such as cellulose (as in cereal brans, fruits, and vegetables) enlarge and soften the stools, easing their passage out of the body and speeding up their transit time through the intestine. Thus, foods rich in these fibers help to alleviate or prevent **constipation.**

Large, soft stools ease the task of elimination for the rectal muscles. Pressure is then reduced in the lower bowel (colon), making it less likely that rectal veins will swell (**hemorrhoids**). Fiber prevents compaction of the intestinal contents, which could obstruct the appendix and permit bacteria to invade and infect it (**appendicitis**). In addition, fiber stimulates the GI tract muscles so that they retain their strength and resist bulging out into pouches known as **diverticula** (illustrated in Figure 4-6 in the margin).[14]

Populations consuming high-fiber diets generally have lower rates of colon cancer than similar populations consuming low-fiber diets.[15] Most research also suggests that increasing dietary fiber protects against colon cancer.[16] On completing a study of almost 520,000 people, researchers concluded that doubling the naturally occurring fiber in diets of populations with low fiber intakes could reduce the risks of colon and rectal cancers by 40 percent.[17] They caution, though, that they studied only fiber in foods, not supplements, and so suggest food as a fiber source. Fiber-rich foods also

constipation difficult, incomplete, or infrequent bowel movements, associated with discomfort in passing dry, hardened feces from the body.

hemorrhoids (HEM-or-oids) swollen, hardened (varicose) veins in the rectum, usually caused by the pressure resulting from constipation.

appendicitis inflammation and/or infection of the appendix, a sac protruding from the intestine.

diverticula (dye-ver-TIC-you-la) sacs or pouches that balloon out of the intestinal wall, caused by weakening of the muscle layers that encase the intestine. The painful inflammation of one or more of these diverticula is known as *diverticulitis*.

contain important nutrients and phytochemicals that supplements lack, and these substances may reduce the risk of colon cancer.

How fiber may help prevent colon cancer is under investigation. One focus of research is fiber's ability to dilute and speed removal of potential cancer-causing agents from the colon. Other possibilities involve the intestine's resident bacteria. In fiber-rich intestinal contents, feasting bacteria reproduce rapidly, and in so doing, they bind nitrogen and carry it out of the body in the feces. Nitrogen is under study as a potential factor in cancer causation.

As mentioned earlier, soluble fibers yield small fatlike molecules when fermented by bacteria in the colon. The cells of the colon prefer one of these little fats, **butyrate,** to provide energy.[18] A colon well supplied with butyrate from a diet high in soluble fibers may resist chemical injury that could otherwise lead to cancer formation.[19] Another possible link: a well-fed colon frequently replaces its own lining, sloughing damaged cells before they can initiate the cancer process.

Other processes may also be at work. As research progresses, cancer experts recommend a high-fiber diet that includes five to nine half-cup servings of vegetables and fruits each day and generous portions of whole grains and legumes.[20]

▶ KEY POINT   Fibers in foods help to maintain digestive tract health.

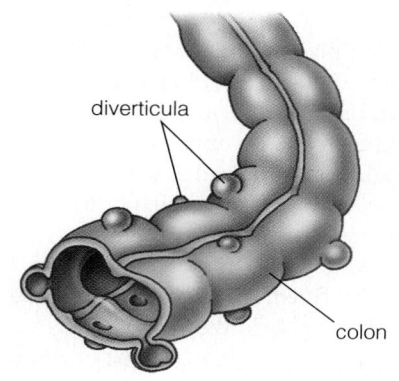

FIGURE 4·6 DIVERTICULA

Diverticula are abnormal bulging pockets formed in the colon wall. These pockets can entrap feces and become painfully infected and inflamed, requiring hospitalization, antibiotic therapy, or surgery.

diverticula

colon

Healthy Weight Management   It bears repeating that foods rich in complex carbohydrates tend to be low in fats and added sugars and can therefore promote weight loss by delivering less energy per bite. Fibers also create feelings of fullness and delay hunger because they swell as they absorb water from the digestive juices.[21]

Some weight-loss products on the market today contain bulk-inducing fibers, but buying pure fiber compounds like this is neither necessary nor advisable.* To use fiber in a weight-loss plan, select fresh fruits, vegetables, legumes, and whole-grain foods. High-fiber foods not only add bulk to the diet but are economical and nutritious, bringing many benefits to the eater.[22]

▶ KEY POINT   Diets that are adequate in fiber assist the eater in maintaining a healthy body weight.

Recommendations and Intakes

Few people in the United States or Canada eat a diet providing all of their needed fiber. Compare the average intake of about 14 to 15 grams of fiber with these recommendations:

- The DRI fiber recommendations are based on energy needs, and so vary widely among gender and age groups (see the margin).
- The American Dietetic Association suggests 20 to 35 grams of fiber daily, or about two times higher than the average intake.[23]

An effective way to add fiber while lowering fat is to substitute plant sources of proteins (legumes) for some of the animal sources of protein (meats and cheeses) in the diet. Another way is to focus on consuming the recommended amounts of fruits and vegetables each day. People choosing high-fiber foods are wise to seek out a variety of fiber sources and to drink extra fluids to help the fiber do its job.

Fiber recommendations are given in terms of total fiber without distinction between fiber types. This makes sense because most fiber-rich foods supply a mixture of fibers (recall Figure 4-4, page 107). The Consumer Corner near here provides detailed information about choosing wisely among grain foods.

■ DRI recommendations for fiber:

Men, age 19–50:	38 g/day.
Men, age 51 and up:	30 g/day.
Women, age 19–50:	25 g/day.
Women, age 51 and up:	21 g/day.

butyrate (BYOO-tier-ate) a small fat fragment produced by the fermenting action of bacteria on viscous, soluble fibers; the preferred energy source for the colon cells.

*An example is methylcellulose, a fiber that swells up and fills the stomach.

CONSUMER CORNER

Refined, Enriched, and Whole-Grain Bread

FOR MANY PEOPLE, bread supplies much of the carbohydrate, or at least most of the starch, in a day's meals. Any food used in such abundance in the diet should be scrutinized closely, and if it doesn't measure up to high nutrition standards, it should be replaced with a food that does. The meanings of the words **refined, enriched, fortified,** and **whole grain** hold the key to understanding the nutritional value of wheat breads (see Table 4-3).

The part of the wheat plant that is made into flour and then into bread and other baked goods is the seed or kernel. The wheat kernel (a whole grain) has four main parts: the **germ,** the **endosperm,** the **bran,** and the **husk,** as shown in Figure 4-7. The germ is the part that grows into a wheat plant and

therefore contains concentrated food to support the new life—it is especially rich in vitamins and minerals. The endosperm is the soft, white, inside portion of the kernel, containing starch and proteins that help nourish the seed as it sprouts. The kernel is encased in the bran, a protective coating that is similar in function to the shell of a nut; the bran is also rich in nutrients and fiber. The husk, commonly called chaff, is the dry outermost layer and is inedible for human beings but can be used in animal feed.

In earlier times, people milled wheat by grinding it between two stones, blowing or sifting out the chaff, and retaining the nutrient-rich bran and germ as well as the endosperm. Then milling machinery was "improved," and

it became possible to remove the dark, heavy germ and bran as well, leaving a whiter, smoother-textured flour with a higher starch content and far less fiber.[1] People looked on this flour as more desirable than the crunchy, dark brown, "old-fashioned" flour.

In turning to white bread, bread eaters suffered a tragic loss of needed nutrients. Many people developed deficiencies of iron, thiamin, riboflavin, and niacin—nutrients formerly received from whole-grain bread. Finally, the problem was recognized, and Congress

TABLE 4-3 TERMS THAT DESCRIBE GRAIN FOODS

- **bran** the protective fibrous coating around a grain; the chief fiber donator of a grain.
- **brown bread** bread containing ingredients such as molasses that lend a brown color; may be made with any kind of flour, including white flour.
- **endosperm** the bulk of the edible part of a grain, the starchy part.
- **enriched, fortified** refers to the addition of nutrients to a refined food product. As defined by U.S. law, these terms mean that specified levels of thiamin, riboflavin, niacin, folate, and iron have been added to refined grains and grain products. The terms *enriched* and *fortified* can refer to the addition of more nutrients than just these five; read the label.[a]
- **germ** the nutrient-rich inner part of a grain.
- **husk** the outer, inedible part of a grain.
- **refined** refers to the process by which the coarse parts of food products are removed. For example, the refining of wheat into flour involves removing three of the four parts of the kernel—the chaff, the bran, and the germ—leaving only the endosperm, composed mainly of starch and a little protein.
- **stone ground** refers to a milling process using limestone to grind any grain, including refined grains, into flour.
- **unbleached flour** a beige-colored endosperm flour with texture and nutritive qualities that approximate those of regular white flour.
- **wheat flour** any flour made from wheat, including white flour.
- **white flour** an endosperm flour that has been refined and bleached for maximum softness and whiteness.
- **whole grain** refers to a grain milled in its entirety (all but the husk), not refined.
- **whole-wheat flour** flour made from whole-wheat kernels; a whole-grain flour.

[a]Formerly, *enriched* and *fortified* carried distinct meanings with regard to the nutrient amounts added to foods, but a change in the law has made these terms virtually synonymous.

FIGURE 4-7 A WHEAT PLANT AND A SINGLE KERNEL OF WHEAT

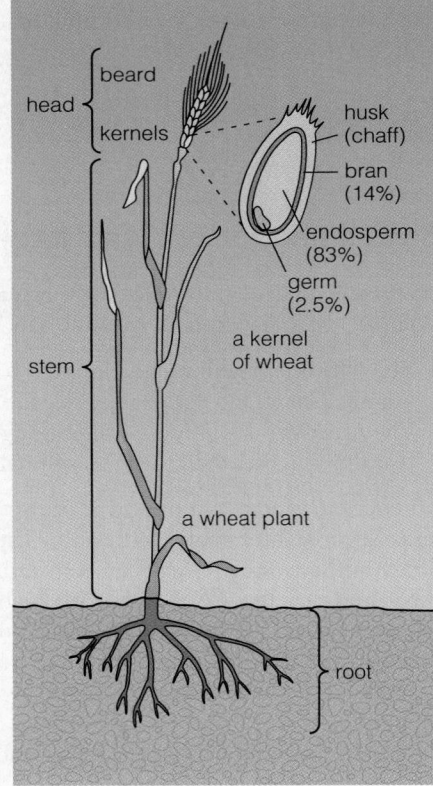

beard
head
kernels
stem
husk (chaff)
bran (14%)
endosperm (83%)
germ (2.5%)
a kernel of wheat
a wheat plant
root

Only *whole-grain* flour contains all the nutritive portions of the grain. Notice the distinctions between **wheat flour** and **whole-wheat flour** and **white flour** and **unbleached flour** among the terms that describe grain foods; also notice that the terms wheat bread, **brown bread,** and **stone ground** on a label do not guarantee that the bread has been made with whole-grain flour

In Western societies, bread is the staff of life.

passed the Enrichment Act requiring that iron, niacin, thiamin, and riboflavin be added to refined grain products before they were sold. The U.S. Enrichment Act of 1942 was amended in 1996 to include the vitamin folate (sometimes called folic acid on food labels), and is currently under review for an update.[2]

A single slice of refined bread is not "rich" in the enrichment nutrients, but people who eat several slices of bread a day obtain significantly more of the nutrients than they would from unenriched white bread, as Figure 4-8 shows. Today, breads, grain products such as rice, macaroni, and spaghetti, and most types of cereals have been enriched with at least these nutrients.

To a great extent, the enrichment of grain products eliminated known deficiency problems, but other deficiencies went undetected for many more years. The trouble with enriched flour is that it is comparable to whole grain only with respect to the added nutrients and not with respect to others. Enriched products still contain less magnesium, zinc, vitamin B_6, vitamin E, and chromium than whole-grain products do. When a grain is refined, fiber is lost, too (see Table 4-4).

TABLE 4·4 GRAMS OF FIBER IN ONE CUP OF FLOUR
Dark rye, 18 g
Whole wheat, 15 g
Light rye, 14 g
Buckwheat, 12 g
Whole-grain cornmeal, 9 g
Enriched white, 3 g

FIGURE 4·8 NUTRIENTS IN WHOLE-GRAIN, ENRICHED WHITE, AND UNENRICHED WHITE BREAD

Key:
- Whole-grain bread
- Enriched white bread
- Unenriched white bread

Nutrients in bread

Iron
- 100%
- 83%
- 21%

Thiamin
- 100%
- >100%
- 26%

Riboflavin
- 100%
- >100%
- 28%

Niacin
- 100%
- 98%
- 2%

Vitamin B_6
- 100%
- 18%
- 18%

Folate
- 100%
- >100%
- 64%

Fiber
- 100%
- 24%
- 24%

Magnesium
- 100%
- 23%
- 23%

Zinc
- 100%
- 36%
- 36%

Percentage of nutrients
(100% represents nutrient levels of whole-grain bread)

(see Figure 4-9). Bread sold for weight-reduction dieting may be fortified with pure cellulose, but adding cellulose alone is not enough; the bread still lacks other beneficial substances that occur in whole wheat. If bread is a staple food in your diet—that is, if you eat it every day—you are well advised to learn to like the hearty flavor of whole-grain bread.

FIGURE 4·9 BREAD LABELS COMPARED

Although breads may appear similar, their ingredients vary widely. Breads made mostly from whole-grain flours provide more benefits to the body than breads made of enriched refined wheat flour.

Some "high-fiber" breads may contain purified cellulose or more nutritious whole grains. "Low carbohydrate" breads may be regular white bread, thinly sliced to reduce carbohydrates per serving, or may contain lower starch soy flour, barley flour, or flaxseed.

A trick for estimating a bread's content of a nutritious ingredient, such as whole grain flour, is to read the ingredients list (ingredients are listed in order of predominance). Bread recipes generally include one teaspoon of salt per loaf. Therefore, when a bulky nutritious ingredient, such as whole grain, is listed *after* the salt, you'll know that less than a teaspoonful of the nutritious ingredient was added to the loaf—not enough to significantly change the nutrient value per slice.

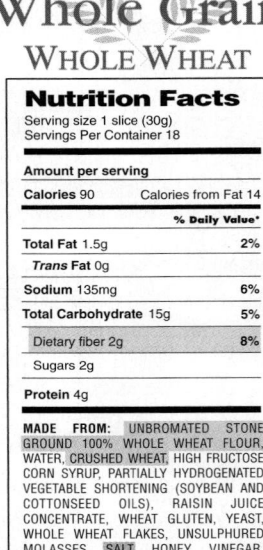

Whole Grain
WHOLE WHEAT

Nutrition Facts
Serving size 1 slice (30g)
Servings Per Container 18

Amount per serving

Calories 90	Calories from Fat 14

	% Daily Value*
Total Fat 1.5g	2%
Trans Fat 0g	
Sodium 135mg	6%
Total Carbohydrate 15g	5%
Dietary fiber 2g	8%
Sugars 2g	
Protein 4g	

MADE FROM: UNBROMATED STONE GROUND 100% WHOLE WHEAT FLOUR, WATER, CRUSHED WHEAT, HIGH FRUCTOSE CORN SYRUP, PARTIALLY HYDROGENATED VEGETABLE SHORTENING (SOYBEAN AND COTTONSEED OILS), RAISIN JUICE CONCENTRATE, WHEAT GLUTEN, YEAST, WHOLE WHEAT FLAKES, UNSULPHURED MOLASSES, SALT, HONEY, VINEGAR, ENZYME MODIFIED SOY LECITHIN, CULTURED WHEY, UNBLEACHED WHEAT FLOUR AND SOY LECITHIN.

Natural
Wheat Bread

Nutrition Facts
Serving size 1 slice (30g)
Servings Per Container 15

Amount per serving

Calories 90	Calories from Fat 14

	% Daily Value*
Total Fat 1.5g	2%
Trans Fat 0g	
Sodium 220mg	9%
Total Carbohydrate 15g	5%
Dietary fiber less than 1g	2%
Sugars 2g	
Protein 4g	

INGREDIENTS: UNBLEACHED ENRICHED WHEAT FLOUR [MALTED BARLEY FLOUR, NIACIN, REDUCED IRON, THIAMIN MONONITRATE (VITAMIN B1), RIBOFLAVIN (VITAMIN B2), FOLIC ACID], WATER, HIGH FRUCTOSE CORN SYRUP, MOLASSES, PARTIALLY HYDROGENATED SOYBEAN OIL, YEAST, CORN FLOUR, SALT, GROUND CARAWAY, WHEAT GLUTEN, CALCIUM PROPIONATE (PRESERVATIVE), MONOGLYCERIDES, SOY LECITHIN.

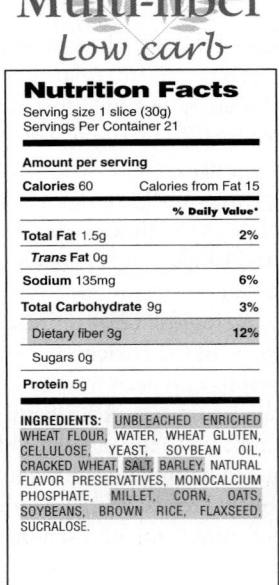

Multi-fiber
Low carb

Nutrition Facts
Serving size 1 slice (30g)
Servings Per Container 21

Amount per serving

Calories 60	Calories from Fat 15

	% Daily Value*
Total Fat 1.5g	2%
Trans Fat 0g	
Sodium 135mg	6%
Total Carbohydrate 9g	3%
Dietary fiber 3g	12%
Sugars 0g	
Protein 5g	

INGREDIENTS: UNBLEACHED ENRICHED WHEAT FLOUR, WATER, WHEAT GLUTEN, CELLULOSE, YEAST, SOYBEAN OIL, CRACKED WHEAT, SALT, BARLEY, NATURAL FLAVOR PRESERVATIVES, MONOCALCIUM PHOSPHATE, MILLET, CORN, OATS, SOYBEANS, BROWN RICE, FLAXSEED, SUCRALOSE.

Can My Diet Have Too Much Fiber? Adding purified fibers, such as oat or wheat bran, to foods can be taken to extremes. One enthusiastic eater of purified oat bran in muffins required emergency surgery for a blocked intestine; too much oat bran and too little fluid overwhelmed his digestive system. This doesn't mean that you should avoid bran-containing foods, of course, but that you should use bran with moderation and drink an extra beverage with it. Less extreme concerns are that purified fiber might displace nutrients from the diet or cause them to be lost from the digestive tract by binding the nutrients and speeding up transit. Purified fibers are like refined sugars in one way: the nutrients that originally accompanied the fibers have been lost. Also, a purified fiber may not affect the body the same way as the fiber in its original food product. Most experts agree that the health benefits attributed to a fiber may come from other constituents of fiber-containing foods, and not from the fiber alone.[24] You can make a quick approximation of a day's fiber intake by following the instructions in Table 4-5.

Binders in Fiber   Binders in some fibers act as **chelating agents.** This means that they link chemically with important nutrient minerals (iron, zinc, calcium, and others) and then carry them out of the body. The mineral iron is mostly absorbed at the beginning of the intestinal tract, and excess insoluble fibers may limit its absorption by speeding foods through the upper part of the digestive tract. Too much bulk in the diet can also limit the total amount of food consumed and cause deficiencies of both nutrients and energy. People with marginal intakes, such as the malnourished, the elderly, and children who consume no animal products, are particularly vulnerable to this chain of events. Fibers also carry water out of the body and can cause dehydration. Add an extra glass or two of water to go along with the fiber added to your diet.

The next section focuses on the handling of carbohydrates by the digestive system. Table 4-6 (p. 114) sums up the points made so far concerning the functions of carbohydrates in the body and in foods.

▶ **KEY POINT**   Most adults need between 21 and 38 grams of total fiber each day, but few consume this amount. Fiber needs are best met with whole foods. Purified fiber in large doses can have undesirable effects. Fluid intake should increase with fiber intake.

From Carbohydrates to Glucose

You may eat bread or a baked potato, but the body's cells cannot use foods or even whole molecules of lactose, sucrose, or starch for energy. They need the glucose in those molecules, and they need it continuously. The various body systems must make glucose available to the cells, not all at once when it is eaten, but at a steady rate all day.

Digestion and Absorption of Carbohydrate

To obtain glucose from newly eaten food, the digestive system must first render the starch and disaccharides from the food into monosaccharides that can be absorbed through the cells that line the small intestine. The largest of the digestible carbohydrate molecules, starch, requires the most extensive breakdown. Disaccharides, in contrast, need be split only once before they can be absorbed.

Starch   Digestion of most starch begins in the mouth, where an enzyme in saliva mixes with food and begins to split starch into maltose. While chewing a bite of bread, you may notice that a slightly sweet taste develops—maltose is being liberated from starch by the enzyme. The salivary enzyme continues to act on the starch in the bite of bread while it remains tucked in the stomach's upper storage area. As each chewed lump is pushed downward and mixed with the stomach's acid and other juices, the salivary enzyme (made of protein) is deactivated by the stomach's protein-digesting acid. Not all digestive enzymes are susceptible to digestion in the stomach. For example, a protein-digesting enzyme is designed to work best in the stomach—its structure protects it from the stomach's acid. With the breakdown of the salivary enzyme, starch digestion ceases in the stomach, but it resumes at full speed in the small intestine, where another starch-splitting enzyme is delivered by the pancreas. This enzyme breaks starch down into disaccharides and small polysaccharides. Other enzymes liberate monosaccharides for absorption.

Some forms of starch are easily digested. The starch in bread made of refined white flour, for example, breaks down rapidly to glucose that is absorbed high up in the small

---

TABLE 4·5   A QUICK METHOD FOR ESTIMATING FIBER INTAKE

To quickly estimate fiber in a day's meals:

1. Multiply serving equivalents[a] of fruits and vegetables (excluding juice) by 1.5 g.[b]
   Example: 5 servings of fruits and vegetables × 1.5 = 7.5 g fiber
2. Multiply serving equivalents of refined grains by 1.0 g.
   Example: 4 servings of refined grains × 1.0 = 4.0 g fiber
3. Multiply serving equivalents of whole grains by 2.5 g.
   Example: 3 servings of whole grains × 2.5 = 7.5 g fiber
4. Add fiber values for serving equivalents of legumes, nuts, seeds, and high-fiber cereals and breads; look these up in Appendix A.
   Example: ½ c navy beans = 6.0 g fiber
5. Add up the grams of fiber from the previous lines.
   Example: 7.5 + 4.0 + 7.5 + 6.0 = 25 g fiber

Day's total fiber = 25 g fiber

[a] Use standard serving equivalents presented in Figure 2-4 of Chapter 2.
[b] Most cooked and canned fruits and vegetables contain about this amount, while whole raw fruits and some vegetables contain more.

■ Chelating agents are often sold by supplement vendors to "remove poisons" from the body. Some valid medical uses such as treatment of lead poisoning exist, but most of the chelating agents sold over-the-counter are based on unproven claims.

chelating (KEE-late-ing) **agents** molecules that attract or bind with other molecules and are therefore useful in either preventing or promoting movement of substances from place to place.

TABLE 4·6 USEFULNESS OF CARBOHYDRATES

Carbohydrates in the Body	Carbohydrates in Foods
• *Energy source.* Sugars and starch from the diet provide energy for many body functions; they provide glucose, the preferred fuel for the brain and nerves. • *Glucose storage.* Muscle and liver glycogen store glucose. • *Raw material.* Sugars are converted into other compounds, such as amino acids (the building blocks of proteins), as needed. • *Structures and functions.* Sugars interact with fat and protein molecules, affecting their structures and functions. • *Digestive tract health.* Fibers help to maintain healthy bowel function (reduce risk of bowel diseases). • *Blood cholesterol.* Fibers promote normal blood cholesterol concentrations (reduce risk of heart disease). • *Blood glucose.* Fibers modulate blood glucose concentrations (help control diabetes). • *Satiety.* Fibers and sugars contribute to feelings of fullness. • *Body weight.* A fiber-rich diet may promote a healthy body weight.	• *Flavor.* Sugars provide sweetness. • *Browning.* When exposed to heat, sugars undergo browning reactions, lending appealing color, aroma, and taste to foods. • *Texture.* Sugars help make foods tender. Cooked starch lends a smooth, pleasing texture. • *Gel formation.* Starch molecules expand when heated and trap water molecules, forming gels. The fiber pectin forms the gel of jellies when cooked with sugar and acid from fruit. • *Bulk and viscosity (thickness).* Carbohydrates lend bulk and increased viscosity to foods. Soluble, viscous fibers lend thickness to foods such as salad dressings. • *Moisture.* Sugars attract water and keep foods moist. • *Preservative.* Sugar in high concentrations dehydrates bacteria and preserves the food. • *Fermentation.* Carbohydrates are fermented by yeast, a process that causes bread dough to rise and beer to brew, among other uses.

intestine.[25] Some starch, such as that of cooked beans, digests more slowly and releases its glucose later in the digestion process. Less digestible starch, called **resistant starch,** is technically a kind of fiber and may behave similarly in the body.[26] The starch of raw potatoes, for example, resists digestion; so does the resistant starch that forms during overheating of foods and the starch tucked inside the unbroken hulls of swallowed seeds.[27] Some resistant starch may be digested, but slowly, and most remains intact until the bacteria of the colon eventually break it down. The rate of starch digestion may affect the body's handling of its glucose, as a later section explains.

Sugars Sucrose and lactose from food, along with maltose and small polysaccharides freed from starch, undergo one more split to yield free monosaccharides before they are absorbed. This split is accomplished by enzymes attached to the cells of the lining of the small intestine. The conversion of a bite of bread to nutrients for the body is completed when monosaccharides cross these cells and are washed away in a rush of circulating blood that carries them to the waiting liver. Figure 4-10 presents a quick review of carbohydrate digestion.

The absorbed carbohydrates (glucose, galactose, and fructose) travel to the liver, which converts fructose and galactose to glucose or related products. The circulatory system transports the glucose and other products to the cells. Liver and muscle cells may store circulating glucose as glycogen; all cells may split glucose for energy.

Fiber Although molecules of many fibers are not changed by human digestive enzymes, many of them can be digested (fermented) by the bacterial inhabitants of the human digestive tract. A by-product of metabolism by the bacteria in the colon can be any of several odorous gases. Don't give up on high-fiber foods if they cause gas. Instead, start with small servings and gradually increase the serving size over several weeks; chew foods thoroughly to break up hard-to-digest lumps that can ferment in the intestine; and try a variety of fiber-rich foods until you find some that do not cause the problem. Some people also find relief from excessive gas by using commercial enzyme preparations sold for use with beans. Such products contain enzymes that help to break down some of the indigestible fibers in foods before they reach the colon. In other people, persistent painful gas may indicate that the digestive tract has undergone a change in its ability to digest the sugar in milk, a condition known as **lactose intolerance.**

resistant starch the fraction of starch in a food that is digested slowly, or not at all, by human enzymes.

lactose intolerance impaired ability to digest lactose due to reduced amounts of the enzyme lactase.

FIGURE 4·10   HOW CARBOHYDRATE IN FOOD BECOMES GLUCOSE IN THE BODY

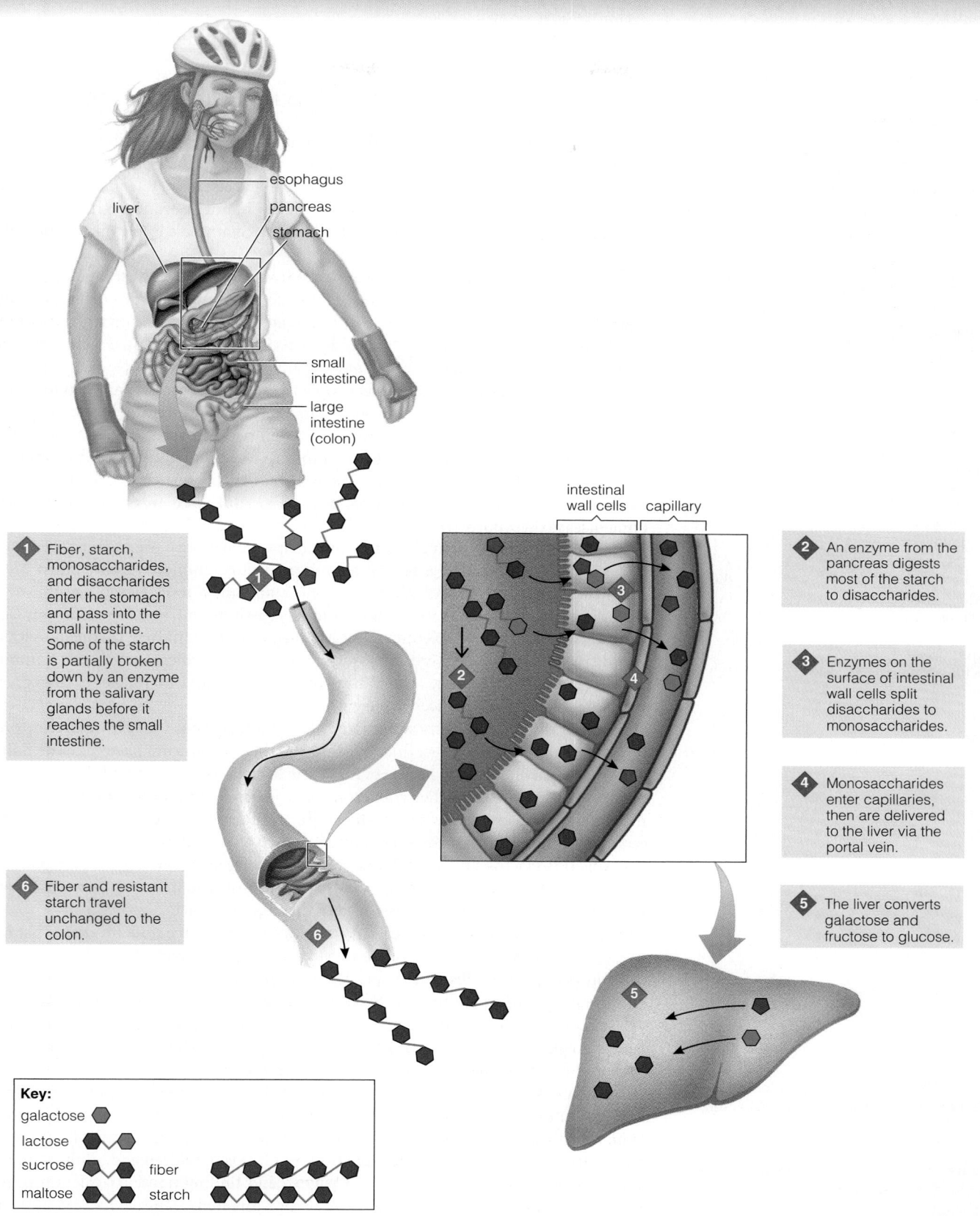

esophagus

liver

pancreas

stomach

small intestine

large intestine (colon)

intestinal wall cells

capillary

1 Fiber, starch, monosaccharides, and disaccharides enter the stomach and pass into the small intestine. Some of the starch is partially broken down by an enzyme from the salivary glands before it reaches the small intestine.

2 An enzyme from the pancreas digests most of the starch to disaccharides.

3 Enzymes on the surface of intestinal wall cells split disaccharides to monosaccharides.

4 Monosaccharides enter capillaries, then are delivered to the liver via the portal vein.

6 Fiber and resistant starch travel unchanged to the colon.

5 The liver converts galactose and fructose to glucose.

Key:

galactose

lactose

sucrose   fiber

maltose   starch

> ► KEY POINT   With respect to starch and sugars, the main task of the various body systems is to convert them to glucose to fuel the cells' work. Fermentable fibers may release gas as they are broken down by bacteria in the intestine.

Why Do Some People Have Trouble Digesting Milk?

Among adults, the ability to digest the carbohydrate of milk varies widely. As they age, upwards of 75 percent of the world's people lose much of their ability to produce the enzyme **lactase** to digest the milk sugar lactose.[28] Lactase, which is made by the small intestine, splits the disaccharide lactose into its component monosaccharides glucose and galactose, which are then absorbed. Almost all mammals lose some of their ability to produce lactase as they age.

People with lactose intolerance experience some amount of nausea, pain, diarrhea, and excessive gas on drinking milk or eating lactose-containing products. The undigested lactose remaining in the intestine demands dilution with fluid from surrounding tissue and the bloodstream. Intestinal bacteria use the undigested lactose for their own energy, a process that produces gas and intestinal irritants.

The failure to digest lactose affects people to differing degrees. Many can tolerate as much as a cup or two of milk a day; some can tolerate lactose-reduced milk; only a rare few cannot tolerate lactose in any amount. Often people overestimate the severity of their lactose intolerance, blaming it for symptoms most probably caused by something else—a mistake that could cost them the health of their bones (see Chapter 8 and its Controversy, which examine the topic of milk in adult diets in relation to the adult bone disease osteoporosis).[29]

Infants produce abundant lactase, which helps them absorb the sugar of breast milk and milk-based formulas; a few suffer inborn lactose intolerance and must be fed solely on lactose-free formulas. Because milk is an almost indispensable source of the calcium every child needs for growth, a milk substitute must be found for any child who becomes lactose intolerant. Disadvantaged young children of the developing world sustain the most severe consequences of lactose intolerance when it combines with disease, malnutrition, or parasites to produce a loss of nutrients that greatly reduces the children's chances of survival.

Women who fail to consume enough calcium during youth may later develop weak bones, so young women must find substitutes if they become unable to tolerate milk.[30] Yogurt or aged cheese may be acceptable—the bacteria or molds that help create these products digest lactose as they convert milk to a fermented product. Bacterial cultures in some yogurts may take up residence in the intestinal tract where they seem to reduce symptoms of lactose intolerance.[31] Yogurts that contain added milk solids also contain added lactose; milk solids and live cultures are listed among the ingredients on the label. Drinking milk with other foods at meals may also increase tolerance because the foods slow the transit of milk through the digestive tract.[32]

People with lactose intolerance can also choose lactose-free milk products that have undergone treatment with lactase, or they can treat the products themselves with over-the-counter enzyme pills and drops. The pills are taken with milk-containing meals, and the drops are added to milk-based foods; both products help to digest lactose by replacing the missing natural enzyme. In all cases, the trick is to find ways of splitting lactose to glucose and galactose so that the body can absorb the products, rather than leaving the lactose undigested to feed the bacteria of the colon.

Sometimes sensitivity to milk is due not to lactose intolerance but to an allergic reaction to the protein in milk. Milk allergy arises the same way other allergies do—from sensitization of the immune system to a substance. In this case, the immune system overreacts when it encounters the protein of milk. Children and adults with milk allergy often cannot tolerate cheese or yogurt, and finding nondairy calcium sources becomes imperative. Good choices are calcium-fortified orange juice, calcium- and vitamin-fortified soy drink, and canned sardines or salmon with the bones.

Approximate percentages of adults with lactose intolerance:

- 90% Asian Americans.
- 80% Native Americans.
- 80% African Americans.
- 70% Mediterranean peoples.
- 60% Inuits (most Alaskan natives).
- 50% Hispanics.
- 25% U.S. population.
- <15% Northern Europeans.

Lactose in selected foods:

Whole-wheat bread, 1 slice	0.5 g
Dinner roll, 1	0.5 g
Cheese, 1 oz	
Cheddar or American	0.5 g
Parmesan or cream	0.8 g
Doughnut (cake type), 1	1.2 g
Chocolate candy, 1 oz	2.3 g
Sherbet, 1 c	4.0 g
Cottage cheese (low-fat), 1 c	7.5 g
Ice cream, 1 c	9.0 g
Milk, 1 c	12.0 g
Yogurt (low-fat, 1 c	
with added milk solids)	15.0 g

■ Milk *allergy* is a serious problem unrelated to lactose intolerance; see Chapter 14.

lactase the intestinal enzyme that splits the disaccharide lactose to monosaccharides during digestion.

The Body's Use of Glucose

Glucose is the basic carbohydrate unit that each cell of the body uses for energy. The body handles its glucose judiciously—maintaining an internal supply for use in case of need and tightly controlling its blood glucose concentration to ensure that glucose remains available for ongoing use. As mentioned, carbohydrates serve structural roles in the body, too, such as forming part of the mucus that protects the body's linings and organs, but they are best known for their role in providing energy.

Splitting Glucose for Energy

Glucose fuels the work of most of the body's cells. When a cell splits glucose for energy, it performs an intricate sequence of maneuvers that are of great interest to the biochemist—and of no interest whatever to most people who eat bread and potatoes. What everybody needs to understand, though, is that there is no good substitute for carbohydrate. Carbohydrate is *essential*, as the following details illustrate.

The Point of No Return   At a certain point, glucose is forever lost to the body. Inside a cell, glucose is broken in half, releasing some energy. Two pathways are then open to these glucose halves. They can be put back together to make glucose again, or they can be broken into smaller fragments. If they are broken further, they cannot be reassembled to form glucose. The smaller fragments can yield still more energy and in the process break down completely to carbon dioxide and water; they can be formed into building blocks of protein, or they can be hitched together into units of body fat. Figure 4-11 shows how glucose is broken down to yield energy and carbon dioxide.

Below a Healthy Minimum   Although glucose can be converted into body fat, body fat cannot be converted into glucose to feed the brain adequately. When the body faces a severe carbohydrate deficit, it has two problems. Having no glucose, it must turn to protein to make some (the body has this ability), diverting protein from critical functions of its own such as maintaining the body's immune defenses. Protein's functions in the body are so indispensable that carbohydrate should be kept available precisely to prevent the use of protein for energy. This is called the **protein-sparing action** of carbohydrate.

The second problem arises because fat fragments normally combine with a compound derived from glucose before being used by the cells to supply energy. Without the help of this compound, fat fragments combine with each other instead, producing increased amounts of the normally scarce acidic products, **ketone bodies.** Ketone bodies can accumulate in the blood (**ketosis**) to reach levels high enough to disturb the normal acid-base balance.[33] Adults consuming a diet that produces chronic ketosis may also face deficiencies of vitamins and minerals, loss of bone minerals, altered blood lipids, increased risk of kidney stones, an impaired mood and sense of well-being, and glycogen stores that are too scanty to meet a metabolic emergency or to support maximal high-intensity muscular work.[34] Ketosis isn't all bad, however. A therapeutic ketosis-inducing diet has long been used along with medication to reduce the occurrence of seizures in children with severe epilepsy.[35]

The minimum amount of digestible carbohydrate set by the DRI committee to adequately feed the brain and reduce ketosis has been set at 130 grams a day for an

FIGURE 4·11   THE BREAK-DOWN OF GLUCOSE YIELDS ENERGY AND CARBON DIOXIDE

Cell enzymes split the bonds between the carbon atoms in glucose, liberating the energy stored there for the cell's use. ◆ The first split yields two 3-carbon fragments. The two-way arrows mean that these fragments can also be rejoined to make glucose again. ◆ Once they are broken down further into 2-carbon fragments, however, they cannot rejoin to make glucose. ◆ The carbon atoms liberated when the bonds split are combined with oxygen and released into the air, via the lungs, as carbon dioxide. Although not shown here, water is also produced at each split.

carbon atoms

bonds

glucose
(6-carbon compound)

3–carbon compound + 2–carbon + Energy ◆ 1

carbon dioxide + 2–carbon compound + Energy ◆ 2

2 molecules of carbon dioxide + Energy ◆ 3

protein-sparing action the action of carbohydrate and fat in providing energy that allows protein to be used for purposes it alone can serve.

ketone (kee-tone) **bodies** acidic, fat-related compounds that can arise from the incomplete breakdown of fat when carbohydrate is not available.

ketosis (kee-TOE-sis) an undesirable high concentration of ketone bodies, such as acetone, in the blood or urine.

average-sized person.[36] Several times this minimum is recommended to maintain health and glycogen stores (see next section). The amounts of vegetables, fruits, legumes, grains, and milk recommended in the USDA Food Guide of Chapter 2 (pages 35–36) deliver abundant carbohydrates.

▶ KEY POINT   Without glucose, the body is forced to alter its uses of protein and fats. To help supply the brain with glucose, the body breaks down protein to make glucose and converts its fats into ketone bodies, incurring ketosis.

Storing Glucose as Glycogen

After a meal, as blood glucose rises, the pancreas is the first organ to respond. It releases the hormone **insulin,** which signals the body's tissues to take up surplus glucose. Muscle and liver cells use some of this excess glucose to build the polysaccharide glycogen. The muscles hoard two-thirds of the body's total glycogen and use it just for themselves. The brain stores a tiny fraction of the total, thought to provide an emergency glucose reserve sufficient to fuel the brain for an hour or two in severe glucose deprivation.[37] The liver stores the remainder and is generous with its glycogen, making it available as blood glucose for the brain or other tissues when the supply runs low.

Unlike starch, which has long chains with only occasional branches that are cleaved linearly during digestion, glycogen is highly branched with hundreds of ends extending from each molecule's surface (review this structure in Figure 4-3 on p. 103). When the blood glucose concentration drops and cells need energy, a pancreatic hormone, **glucagon,** floods the bloodstream. Then, enzymes within the liver cells respond by attacking a multitude of glycogen ends simultaneously to release a surge of glucose into the blood for use by all the body's other cells. Thus, the branched structure of glycogen uniquely suits the purpose of releasing glucose on demand.

▶ KEY POINT   Glycogen is the body's storage form of glucose. The liver stores glycogen for use by the whole body. Muscles have their own glycogen stock for their exclusive use. The hormone glucagon acts to liberate stored glucose from liver glycogen.

Returning Glucose to the Blood

Should your glucose supplies ever fall too low, you would feel dizzy and weak. Should your blood glucose ever climb abnormally high, you might become confused or have difficulty breathing. The healthy body guards against both conditions.

Regulation of Blood Glucose   Maintaining normal blood glucose concentration depends on two safeguards: replenishment from liver glycogen stores and siphoning off of excess glucose into the liver (to be converted to glycogen or fat) and into the muscles (to be converted to glycogen).

When blood glucose starts to fall too low, the hormone glucagon triggers the breakdown of liver glycogen to free glucose. Hormones that promote the conversion of protein to glucose are also released, but only a little protein can be spared. When body protein is used, it is taken from blood, organ, or muscle proteins; no surplus of protein is stored specifically for emergencies. As for fat, it cannot regenerate enough glucose to feed the brain and prevent ketosis.

Another hormone, epinephrine, also breaks down liver glycogen as part of the body's defense mechanism in times of danger.* To a person living in the Stone Age, this

insulin a hormone secreted by the pancreas in response to a high blood glucose concentration. It assists cells in drawing glucose from the blood.

glucagon (GLOO-cah-gon) a hormone secreted by the pancreas that stimulates the liver to release glucose into the blood when blood glucose concentration dips.

*Epinephrine is also called adrenaline.

internal source of quick energy was indispensable. Life was filled with physical peril, and the person who stopped and ate before running from a saber-toothed tiger did not survive to produce our ancestors. The quick-energy response in a stress situation works to our advantage today as well. For example, it accounts for the energy you suddenly have to clean up your room when you learn that a special person is coming to visit. To meet such emergencies, we are advised to eat and store carbohydrate at regularly timed meals throughout the day because the liver's glycogen stores can be depleted within half a waking day.

You may be asking, "What kind of carbohydrate?" Candy, "energy bars," and sugary beverages supply sugar energy quickly but are not the best choices. Balanced meals, eaten on a regular schedule, help the body to maintain its blood glucose. Meals containing starch and fiber along with some protein and a little fat slow digestion so that glucose enters the blood gradually in an ongoing steady supply.

▶ KEY POINT   Blood glucose regulation depends mainly on the hormones insulin and glucagon. Most people have no problem regulating their blood glucose when they consume mixed meals at regular intervals.

The Glycemic Response   Some carbohydrate-rich foods elevate blood glucose and insulin concentrations higher relative to others. When this effect is measured, each food's score can be ranked on a scale known as the **glycemic index (GI).** Scores are then compared with that of a standard food, usually white bread or glucose (given a score of 100).[38] A food's ranking depends on a number of factors working together, and the effect is not always what one might expect. Ice cream, for example, is a high-sugar food, but it ranks lower than baked potatoes, a high-starch food. Mashed potatoes produce more of a glucose response than pure sugar (sucrose), perhaps because the monosaccharide fructose makes up half of each sucrose molecule, and fructose has little effect on blood glucose. The starch of the potatoes is all glucose. Figure 4-12 shows generally where some foods fall on the glycemic index scale, although individuals' responses often vary.[39]

The glycemic index, and its mathematical offshoot **glycemic load,** may be meaningful to people with diabetes who must take steps to regulate their blood glucose.[40] A theory states that the lower the glycemic load of the diet, the less glucose builds up in the blood and, therefore, the less insulin is needed to maintain normal blood glucose concentrations. Evidence also suggests that a low glycemic load may even help to control body weight by ensuring a steady flow of glucose into the bloodstream and extend feelings of fullness.[41] In contrast, a rapid rise in blood glucose seems to promote overeating in some overweight people and may lead to the increased presence of a marker of cardiovascular disease risk.[42] New research is focusing on differences in the rate of blood glucose clearance by the tissues rather than absorption as a key factor in glycemic load.[43]

The concept of glycemic load offers an accounting of both a food's glycemic index and its carbohydrate content. Take carrots, for example. In laboratory tests, 50 grams of carbohydrate from boiled carrots ranks in the middle of the glycemic index scale, indicating a moderate capacity for elevating blood glucose. However, the glycemic index fails to reveal that a typical half-cup serving of carrots presents the body with only 8 grams of absorbable carbohydrate—to receive a test portion of 50 grams requires eating more than 3 cups of boiled carrots at a sitting. Therefore, the *glycemic load* of a serving of carrots is low, a point often overlooked.

Today, the utility of the glycemic index or glycemic load for reducing weight or risks of diseases remains uncertain.[44] One problem is that many factors can affect glycemic index test results, including the time of day of the test, body size and weight, blood volume, and metabolic rate.[45] Another problem is that a food's glycemic effect may vary from person to person and also depends on how the food is prepared, its ripeness, and which *other* foods accompany it in a meal.[46] Even a cup of coffee can alter glucose absorption.[47] Research is evolving to address these issues, but meanwhile, dietary

FIGURE 4·12   GLYCEMIC INDEX OF SELECTED FOODS

HIGH

Glucose

Baked potato

Cornflakes

Sports drinks, jelly beans
Pumpkin, doughnut

Watermelon, popcorn, bagel
White bread

Couscous
Raisins, white rice
Ice cream

Cola, pineapple
Brown rice
Wheat bread, corn, pound cake
Banana
Rye bread, orange juice
Green peas, carrots, baked beans, pasta
Grapes
Chocolate pudding
Bran cereals, black-eyed peas, peaches
Apple juice
Tomato juice, navy beans, apples, pears
Yogurt

Butter beans

Milk, kidney beans, garbanzo beans
Barley

Cashews, cherries

Soybeans

Peanuts

LOW

glycemic index (GI) a ranking of foods according to their potential for raising blood glucose relative to a standard such as glucose or white bread.

glycemic load a mathematical expression of both the glycemic index and the carbohydrate content of a food, meal, or diet (glycemic index multiplied by grams carbohydrate).

guidelines already suggest choosing foods ranking fairly low on the glycemic index scale—whole grains, legumes, vegetables, fruits, and milk products.[48]

▶ **KEY POINT** The glycemic index is a measure of blood glucose response to foods relative to the response to a standard food. The glycemic load is the product of the glycemic index multiplied by the carbohydrate content of a food.

Handling Excess Glucose

■ Appendix D provides a valuable tool for estimating the carbohydrate and other energy nutrients in foods—the U.S. Food Exchange System.

Suppose you have eaten dinner and are now sitting on the couch, munching pretzels and drinking cola as you watch a ball game on television. Your digestive tract is delivering molecules of glucose to your bloodstream, and your blood is carrying these molecules to your liver and other body cells. The body cells use as much glucose as they can for their energy needs of the moment. Excess glucose is linked together and stored as glycogen until the muscles and liver are full to capacity with glycogen. Still, the glucose keeps coming. To handle the excess glucose, body tissues shift to burning more glucose for energy instead of fat. As a result, more fat is left to circulate in the bloodstream until it is picked up by the fatty tissues and stored there.

If these measures still do not accommodate all of the incoming glucose, the liver has no choice but to handle the excess. The liver breaks the extra glucose into small fragments and puts them together into its durable energy-storage compounds—fats.[49] These newly made fats are then released into the blood, carried to the fat tissues, and deposited. The fat cells may also take up glucose and convert it to fat directly.[50] Unlike the liver cells, which can store only about four to six hours' worth of glycogen, the fat cells can store practically unlimited quantities of fats.

Human beings possess enzymes to convert excess glucose to fat, but the process requires many enzymatic steps costing a great deal of energy. The body is thrifty by nature, so when presented with both glucose and fat from a mixed meal, it prefers to store the fat and use the glucose to meet immediate energy needs. In this way, the maximum available food energy is retained because the dietary fat slips easily into storage with few conversions—its energy is conserved.[51] Moral: you had better play the game if you are going to eat the food. (The Think Fitness feature offers tips to help you play.)

You had better play the game if you are going to eat the food.

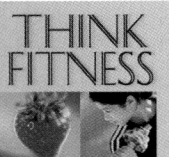

THINK FITNESS

What Can I Eat to Make Workouts Easier?

A WORKING BODY needs carbohydrate fuel to replenish glycogen, and when it runs low, physical activity can seem more difficult. If your workouts seem to drag and never get easier, take a look at your diet. Are your meals regularly timed? Do they provide abundant carbohydrate to fill up glycogen stores so they last through a workout?

Here's a trick: about two hours before you work out, eat a small snack of about 300 calories of foods rich in complex carbohydrates, and drink some extra fluid (see Chapter 10 for ideas). The snack provides glucose at a steady rate to spare glycogen, and the fluid helps to maintain hydration.

A balanced diet that is high in complex carbohydrates helps control body weight and maintain lean tissue. Chapter 5 presents a few more details, but the main point is that, bite for bite, carbohydrate-rich foods contribute less to the body's available energy than do fat-rich foods. Thus, if you want to stay healthy and remain lean, you should make every effort to choose a calorie-appropriate diet providing 45 to 65 percent of its calories from mostly unrefined sources of complex carbohydrates and less than 35 percent from the right kind of fats.[52] This chapter's Food Feature provides the first set of tools required for the job of designing such a diet. Once you have learned to identify the carbohydrates in foods, you must then learn which fats are which (Chapter 5) and how to obtain adequate protein without overdoing it (Chapter 6). Chapter 9 puts it all together with regard to a healthy body weight.

> ▶ KEY POINT   The liver has the ability to convert glucose into fat; under normal conditions, most excess glucose is stored as glycogen or used to meet the body's immediate needs for fuel.

Diabetes and Hypoglycemia

What happens if the body cannot handle carbohydrates normally? One result is **diabetes,** which is common in developed nations and can be detected by a blood test. Another is hypoglycemia, which is rare as a true disease condition, but many people believe they experience its symptoms at times.

What Is Diabetes?

Diabetes is a chronic disease characterized by elevated blood glucose concentrations that often leads to or contributes to a number of other serious diseases. Diabetes is one of the top ten killers of adults, and its prevalence is rising both here and worldwide (see Figure 4-13).[53] Diabetes causes more new cases of blindness in the United States than any other cause, and possible complications include amputations, complications in pregnancy, heart disease, kidney disease, and premature death. People with diabetes are three times more likely to die of flu or pneumonia than people without diabetes.[54] Each year, diabetes costs the United States nearly $132 billion in healthcare services.[55]

Fully half of those suffering from diabetes are unaware of their condition and so fail to take action to prevent its damage. The early stage of the most common form of diabetes often presents few or no warning signs typically associated with diabetes (see Table 4-7, p. 122). Therefore, the American and Canadian diabetes associations are calling for everyone over 45 years old (40 in Canada) and younger people with risk factors such as overweight to be tested regularly for diabetes.[56]

The common forms of diabetes are type 1 and type 2. Both are disorders of blood glucose regulation. Their characteristics are summarized in Table 4-8.

Type 1 Diabetes   **Type 1 diabetes** is less common overall (about 10 percent of cases) but seems to be on the rise and is currently the leading chronic disease among children and adolescents.[57] In type 1 diabetes, the person's own immune system attacks the cells of the pancreas that synthesize the hormone insulin. Soon the pancreas can no longer produce insulin, and after each meal, blood glucose remains elevated, even though body tissues are simultaneously starving for glucose. The person must receive insulin from an external source to assist the cells in taking up the needed glucose from the blood; therefore, an older name for this type of diabetes is *insulin-dependent diabetes mellitus (IDDM)*. Researchers concur that genetics, viral infection, other diseases, toxins, allergens, and a disordered immune system are all probable culprits in provoking an immune system attack on the pancreas.[58]

Insulin is a protein, and if it were taken orally, the digestive system would digest it. Insulin must therefore be injected, either by daily shots or by a small insulin pump, about the size of a personal pager; the pump is worn next to the abdomen and delivers insulin through an implanted needle or tube. New fast-acting and long-lasting

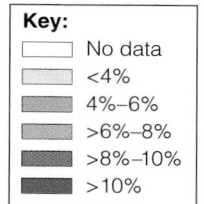

FIGURE 4·13   PREVALENCE OF DIABETES AMONG U.S. ADULTS, 1990 AND 2001

The maps below depict regional changes in U.S. diabetes incidence. In Canada, as of 2001, 4.5 percent of the population over age 18 (or about 1.1 million people) were reported to have diabetes.

Key:	
	No data
	<4%
	4%–6%
	>6%–8%
	>8%–10%
	>10%

1990

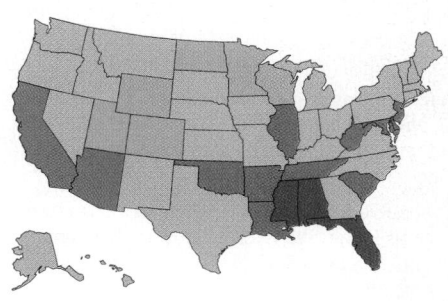

2001

SOURCES: Centers for Disease Control and Prevention, www.cdc.gov/needphp/ aag/aag_ddt.htm; (Canadian) W. J. Millar and T. K. Young, Tracking diabetes: Prevalence, incidence, and risk factors, *Health Reports* 14 (2003): 35–47.

diabetes (dye-uh-BEET-eez) a disease (technically termed *diabetes mellitus*) characterized by elevated blood glucose and inadequate or ineffective insulin, which impairs a person's ability to regulate blood glucose normally.

type 1 diabetes the type of diabetes in which the pancreas produces no or very little insulin; often diagnosed in childhood, although some cases arise in adulthood. Formerly called *juvenile-onset* or *insulin-dependent diabetes*.

TABLE 4·7   WARNING SIGNS
OF DIABETES

These signs appear reliably in type 1 diabetes and, often, in the later stages of type 2 diabetes.

- Excessive urination and thirst
- Glucose in the urine
- Weight loss with nausea, easy tiring, weakness, or irritability
- Cravings for food, especially for sweets
- Frequent infections of the skin, gums, vagina, or urinary tract
- Vision disturbances; blurred vision
- Pain in the legs, feet, or fingers
- Slow healing of cuts and bruises
- Itching
- Drowsiness
- Abnormally high glucose in the blood

■ The first use of genetic engineering was to alter the DNA of a bacterium to produce insulin for treatment of diabetes. More on genetic engineering in Controversy 12.

type 2 diabetes the type of diabetes in which the pancreas makes plenty of insulin, but the body's cells resist insulin's action; often diagnosed in adulthood. Formerly called *adult-onset* or *noninsulin-dependent diabetes.*

insulin resistance a condition in which a normal or high level of insulin produces a less-than-normal response by the tissues; thought to be a metabolic consequence of obesity.

impaired glucose tolerance blood glucose levels higher than normal but not high enough to be diagnosed as diabetes; sometimes called *prediabetes.*

forms of insulin and other drugs allow more flexibility in managing meals and treatments, but users must still plan ahead to balance blood insulin and glucose concentrations. Medical advances may one day eliminate the need for insulin shots—an insulin nasal spray and an inhaler that delivers insulin to the lungs are proving useful in clinical studies.[59] Surgery to transplant insulin-producing cells can eliminate the need for insulin and normalize blood indicators of the disease for more than three years, though the technique is rarely employed because of the scarcity of human insulin-producing cells.[60] A more abundant supply is under development in genetic laboratories.[61] Also on the horizon is a vaccine to prevent type 1 diabetes.

Type 2 Diabetes   The predominant type of diabetes mellitus, **type 2 diabetes,** has increased by over 60 percent since 1991 and may double by 2050 if current trends continue.[62] **Insulin resistance** of the body's cells is the telling characteristic of type 2 diabetes. Insulin may be present in varying amounts, but it is not effective in moving glucose from the bloodstream into the cells. Blood glucose rises too high, as in type 1, but in this case blood insulin also rises. Eventually, the pancreas may become less able to make insulin. At some point, as people with type 2 diabetes age, they may require insulin to supplement their own supply. If drugs are necessary, a preferred therapy is to take a drug that stimulates the person's own pancreas to secrete insulin sufficient to overwhelm the cells' resistance or one that improves the uptake of glucose by the tissues.

Type 2 Diabetes and Obesity   Type 2 diabetes tends to occur late in life, but overweight children and adolescents are now being diagnosed with the condition, and no one really knows why (see Controversy 13).[66] People with the disease often tend to be obese, and obesity often brings on insulin resistance. Compared with normal-weight people with diabetes, obese people require much more insulin to maintain normal blood glucose. The greater the accumulation of body fat, the more insulin resistant the cells become, and the higher the blood glucose rises.[67] Even moderate overweight often brings a foreshadowing of diabetes with the slightly elevated blood glucose known as **impaired glucose tolerance.**

One theory of how obesity and type 2 diabetes may worsen each other is depicted in Figure 4-14.[68] Many factors may contribute to obesity (see Chapter 9), but according to the theory, once obesity sets in, metabolic changes trigger the tissues to resist insulin. As insulin resistance develops, glucose builds up in the blood, while the tissues are deprived of glucose (type 2 diabetes). Meanwhile, blood lipid levels rise to meet the energy demands of the glucose-starved tissues, resulting in an overabundance of circulating fuels available to be stored as fat in the fat cells. As a result, fat mass increases, insulin resistance worsens, and obesity is perpetuated. Even moderate weight gain in adults has been observed to predict diabetes. Given this series of events, is it any wonder that obese people with type 2 diabetes have trouble losing weight?

TABLE 4·8   DIABETES TYPES 1 AND 2 COMPARED

	Type 1 Diabetes	Type 2 Diabetes
Age of onset	Childhood or midlife	Adulthood or, increasingly, childhood
Body cells	Responsive to insulin action	Resistant to insulin action
Body fatness	Generally low to average	Generally high
Insulin shots required	Yes	Possibly[a]
Insulin-stimulating drugs or other drugs may be effective	No	Yes
Natural insulin	Pancreas makes too little or none.	Pancreas may make enough or too much.
Pancreatic function	Insulin-producing cells impaired or nonfunctional	Insulin-producing cells may be normal
Severity of symptoms	Relatively severe; many are apparent on diagnosis.	Relatively mild; few or none may be present on diagnosis.

[a] People past age 40 who suffer from type 2 diabetes may lose pancreatic function and become dependent on insulin.

Type 2 Diabetes and the Genes   Type 2 diabetes tends to run in families. Particularly, people of African, Hispanic, Native American, Asian, or Pacific Island descent may inherit certain genetic variations that make them susceptible to developing type 2 diabetes.[63] Not every person with these genetic variations suffers the disease, however. Having the genes may increase the risk for diabetes by about 30 percent, but other factors, such as diet and exercise, strongly affect its development.[64]

In the great majority of cases, prevention is not only possible but is also likely to occur when individuals take action to control their lifestyle choices. Recent research findings indicate that U.S. men and women who maintain a healthy body weight and who also choose a diet high in vegetables, fruit, fish, poultry, and whole grains, exercise regularly, restrict alcohol, and abstain from smoking have a greatly reduced incidence of type 2 diabetes compared with those with less healthy lifestyles.[65]

Diagnosis and Perils of Diabetes   Diabetes can be diagnosed when two or more fasting blood glucose tests register positive. In this test, a clinician draws blood after a night of fasting and measures an indicator of blood glucose to determine whether it falls within the normal range.

The effects of diabetes can be severe, but, in general, the tighter the control over blood glucose, the milder those effects tend to be. In diabetes, blockage or destruction of capillaries that feed the body organs and tissues causes these tissues to die from lack of nourishment.[69] Problems include impaired circulation leading to disease of the feet and legs, often necessitating amputation; kidney disease, sometimes requiring a machine to assist the failing kidney in cleansing the blood (**dialysis**) or kidney transplant; impaired vision or blindness due to cataracts and damaged retinas; nerve damage; skin damage; disease of the gums and tooth loss; and strokes and heart attacks. Some of this damage may result from oxidation that accompanies elevated blood glucose.[70] A diet high in antioxidant-containing vegetables and fruits may be protective, but much more research is needed to confirm the oxidation theory. Meanwhile a person with diabetes is advised to control not only weight but also all possible lifestyle risk factors that might contribute to heart and blood vessel disease (atherosclerosis and hypertension, discussed in Chapter 11). In addition, aggressive medical treatment of risk factors that can lead to heart attacks and strokes may reduce the occurrence of these killers by as much as half.[71]

Managing Diabetes   Nutrition intervention plays a central role in managing diabetes, with a focus on weight control in type 2 diabetes.[72] Often, weight loss alone in overweight people with type 2 diabetes helps to control the disease. Constructed of a balanced pattern of foods, the same diet that best controls diabetes can also help to control body weight and support physical activity. This diet:

- Is adequate in nutrients (to avoid deficiencies).
- Is adequate in fiber (from whole grains, fruits, legumes, and vegetables).
- Is moderate in added sugars (must be counted among the day's carbohydrates).
- Is controlled in total carbohydrate (to regulate glucose concentration).[73]
- Is low in saturated fat (thought to worsen insulin response and cardiovascular disease) and should provide some unsaturated oils (to provide essential nutrients).[74]
- Is adequate but not too high in protein (protein may help modulate blood glucose, but too much may damage kidneys weakened by diabetes).[75]

Such a diet also has all the characteristics important to prevention of chronic diseases and meets most of the recommendations of the United States and Canada. The diet can vary, depending on personal tastes and on how much restriction is required to control an individual's blood glucose and lipid values. A person at risk for diabetes can do no better than to adopt such a diet long before symptoms appear.

The role of regular physical activity in preventing and controlling diabetes cannot be overstated.[76] Not only does exercise help to achieve and maintain a healthy body

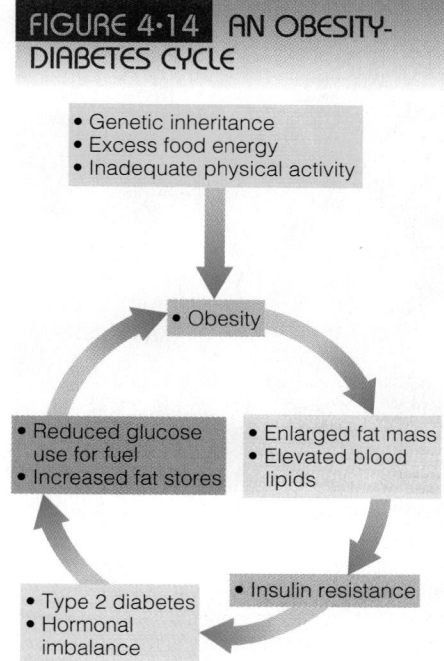

FIGURE 4·14   AN OBESITY-DIABETES CYCLE

- Genetic inheritance
- Excess food energy
- Inadequate physical activity

- Obesity

- Reduced glucose use for fuel
- Increased fat stores

- Enlarged fat mass
- Elevated blood lipids

- Type 2 diabetes
- Hormonal imbalance

- Insulin resistance

SOURCE: Ideas from J. P. Girod and D. J. Brotman, The metabolic syndrome as a vicious cycle: Does obesity beget obesity: *Medical Hypotheses* 60 (2003): 584–589.

■ The exchange system introduced in Chapter 2 and presented in full in Appendix D was developed to help people with diabetes control calorie, carbohydrate, sugar, and fat intakes.

■ Chapter 13 discusses a form of diabetes seen only in pregnancy—gestational diabetes.

dialysis (die-AL-ih-sis) in kidney disease, treatment of the blood to remove toxic substances or metabolic wastes; more properly, *hemodialysis*, meaning "dialysis of the blood."

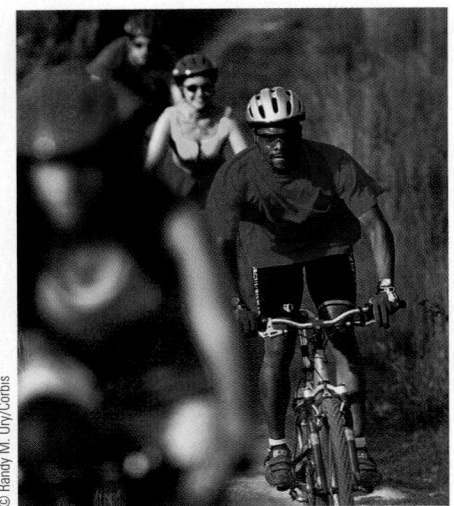

Physical activity: a key player in controlling diabetes.

weight, but it also heightens tissue sensitivity to insulin. With even modest weight loss, increasing physical activity in overweight people seems to delay diabetes onset, and in those with the disease, increased activity, even without weight loss, often helps to control it.[77] Like a juggler who keeps three balls in motion, the person with diabetes must constantly balance three lifestyle factors—diet, exercise, and medication—to control the blood glucose level.

▶ KEY POINT   Diabetes is an example of the body's abnormal handling of glucose. Inadequate or ineffective insulin leaves blood glucose high and cells undersupplied with glucose energy. Weight control and exercise may be effective in preventing the predominant form of diabetes (type 2) and the illnesses that accompany it. A person diagnosed with diabetes must establish patterns of eating, exercise, and medication to control blood glucose.

If I Feel Dizzy between Meals, Do I Have Hypoglycemia?

The term **hypoglycemia** refers to abnormally low blood glucose. People with the condition **postprandial hypoglycemia**, literally, "low blood glucose after a meal," may experience fatigue, weakness, dizziness, irritability, a rapid heartbeat, anxiety, sweating, trembling, hunger, or headaches. Mental symptoms, such as confusion and impairment of intellectual tasks, are also evident.[78] These symptoms are so general and common to many conditions that people can easily misdiagnose themselves as having postprandial hypoglycemia. The condition is rare, however, except among very lean people or those who have recently lost a great deal of weight, often due to an exaggerated sensitivity to insulin.[79] A true diagnosis requires a test to detect low blood glucose while the symptoms are present to confirm that both occur simultaneously.

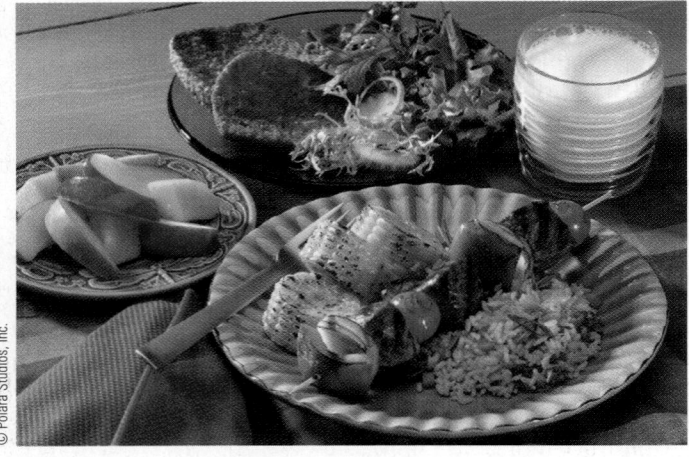

Regularly timed, balanced meals help to hold blood glucose steady.

A person who has symptoms while fasting (for example, overnight) has a different kind of hypoglycemia—**fasting hypoglycemia.** Its symptoms are headache, mental dullness, fatigue, confusion, amnesia, and even seizures and unconsciousness. Serious diseases and conditions, such as cancer, pancreatic damage, uncontrolled diabetes, infection of the liver with accompanying damage (hepatitis), and advanced alcohol-induced liver disease, can all produce true hypoglycemia.

To bring on even mild hypoglycemia with symptoms in normal, healthy people requires extreme measures—administering drugs that overwhelm the body's glucose-controlling hormones, insulin and glucagon. Without such intervention, those hormones hardly ever fail to keep blood glucose within normal limits. Symptoms that people ascribe to hypoglycemia rarely correlate with low blood glucose in blood tests, and such conditions may best be classified as *non*hypoglycemia. Still, people who feel such symptoms may benefit from eating regularly timed, balanced, protein-containing meals; they especially should avoid oscillating between low-carbohydrate dieting and sudden, large, refined carbohydrate doses.[80] It may also help to minimize alcoholic beverages because a high alcohol intake can injure an otherwise healthy pancreas.[81]

▶ KEY POINT   Postprandial hypoglycemia is an uncommon medical condition in which blood glucose falls too low. It can be a warning of organ damage or disease. Many people believe they experience symptoms of hypoglycemia, but their symptoms do not accompany below-normal blood glucose.

hypoglycemia (HIGH-poh-gly-SEE-mee-uh) a blood glucose concentration below normal, a symptom that may indicate any of several diseases, including impending diabetes.

postprandial hypoglycemia an unusual drop in blood glucose that follows a meal and is accompanied by symptoms such as anxiety, rapid heartbeat, and sweating; also called *reactive hypoglycemia.*

fasting hypoglycemia hypoglycemia that occurs after 8 to 14 hours of fasting.

Part of eating right is choosing wisely among the many foods available. The previous section explained that largely without your awareness, the body responds to the carbohydrates supplied by your diet. Now you take the controls by learning how to integrate carbohydrate-rich foods into a diet that meets your body's needs.

FOOD FEATURE

Finding the Carbohydrates in Foods

TO SUPPORT HEALTH, a diet must supply enough of the right kinds of carbohydrate-rich foods. According to the DRI committee, a health-promoting 2,000-calorie diet delivers between 225 and 325 grams of carbohydrate each day. The World Health Organization recommends even more: between 275 and 375 grams per 2,000 calories. People needing more or less energy need proportionately more or less carbohydrate. To find your own DRI range, use your DRI estimated energy requirement (inside front cover) and multiply first by 45 percent and then by 65 percent; then divide both answers by 4 calories per gram (see the example in the margin).

This Food Feature illustrates how you can obtain the carbohydrate-rich foods you need. Breads, cereals, starchy vegetables, fruits, and milk are all good contributors of starches and dilute sugars, both valuable energy-yielding carbohydrates. Many foods also provide fiber in varying amounts, as Figure 4-15 (page 126) demonstrates. Concentrated sweets contain sugars but little else, as the last section demonstrates.

Grains

Breads and other starchy foods are famous for their carbohydrate. Nutrition authorities encourage people to eat grains often, and recommend that half of the grain choices should be whole grains. A slice of bread, half an English muffin, a 6-inch tortilla, or a third-cup of rice or pasta, or a half-cup of cooked cereal—provides about 15 grams of carbohydrate, mostly as starch.

Do not mistake all high fiber foods for whole grains. One hundred percent bran cereal and bran muffins may be high-fiber foods but their added bran doesn't qualify them as whole grains. These foods provide beneficial fiber but they do not contribute to the day's need for whole grain foods. Conversely, puffed wheat cereal, a whole grain food, registers low in fiber per cup because the air that puffs up the grains takes up space in the measuring cup.

Most grain choices should be low in fat and sugar. When extra calories are required to meet energy needs, some selections higher in fat (unsaturated fat—see Chapter 5) and sugar can supply discretionary calories. These choices include biscuits, croissants, muffins, and snack crackers.

Vegetables

Starchy vegetables are major contributors of starch in the diet. Just one small white or sweet potato or a half-cup of cooked dry beans, corn, peas, plantain, or winter squash provides 15 grams of carbohydrate, as much as in a slice of bread, though as a mixture of sugars and starch. A half-cup of carrots, okra, onions, tomatoes, cooked greens, or most other nonstarchy vegetables or a cup of salad greens provides about 5 grams as a mixture of starch and sugars.

Fruits

A typical fruit serving—one-half cup of juice; a small banana, apple, or orange; a half-cup of most canned or fresh fruit; or a quarter-cup of dried fruit—contains an average of about 15 grams of carbohydrate, mostly as sugars, including the fruit sugar fructose. Fruits vary greatly in their water and fiber contents and in their sugar concentrations. No more than a third of a day's fruit should be from juice. With the exception of avocado, which is high in fat, fruits contain insignificant amounts of fat and protein.

Meat, Poultry, Fish, Dry Beans, Eggs, and Nuts

With two exceptions, foods of this group provide almost no carbohydrate to the diet. The exceptions are nuts, which provide a little starch and fiber along with their abundant fat, and dry beans, revered by diet watchers as low-fat sources of both starch and fiber. Just a half cup of beans provides 15 grams of carbohydrate, an amount equaling the

DRI example for a 2,700-calorie diet:

2,700 cal × .45 = 1,215 cal.
1,215 cal ÷ 4 cal/g = 304 g.
2,700 cal × .65 = 1,755 cal.
1,755 cal ÷ 4 cal/g = 439 g.

The DRI recommended carbohydrate intake for a 2,700-calorie diet is approximately 300 to 440 grams per day.

■ The U.S. Food Exchange System (Appendix D) lists carbohydrate values for a variety of foods. Gram values in this section are from the exchange system.

■ Fiber recommendations are listed in the margin on page 109.

■ Recall from Chapter 2 that discretionary calories are the balance of calories remaining in a person's energy allowance after consuming the nutrient-dense foods sufficient to meet the day's nutrient needs. See Appendix I.

FIGURE 4·15 FIBER IN THE FOOD GROUPS

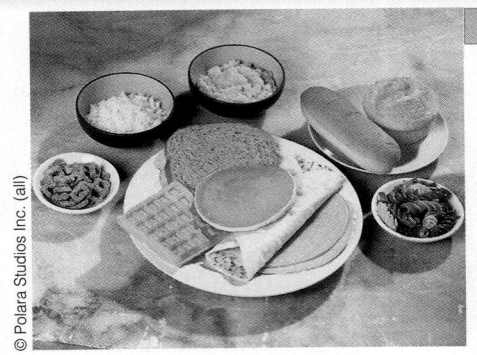

© Polara Studios Inc. (all)

Grains

Food	Fiber [a](g)	Food	Fiber (g)
100% bran cereal, 1 oz	10	Pumpernickel bread, 1 slice	2
Barley, pearled, 1/2 c	3	Shredded wheat, 1 large biscuit	2
Cheerios, 1 oz	3	Corn flakes, 1 oz	1
Whole-wheat bread, 1 slice	3	Muffin, blueberry, 1	1
Wheat flakes, 1 oz	3	Pasta,[b] 1/2 c	1
Brown rice, 1/2 c	2	Puffed wheat, 11/2 c	1
Light rye bread, 1 slice	2	Cream of wheat, 1/2 c	<1
Muffin, bran, 1 small	2	White bread, 1 slice	<1
Oatmeal, 1/2 c	2	White rice, 1/2 c	<1
Popcorn, 2 c	2		

Vegetables

Food	Fiber (g)	Food	Fiber (g)
Baked potato with skin, 1	4	Mashed potatoes, home recipe, 1/2 c	2
Broccoli, chopped, 1/2 c	3	Bell peppers, 1/2 c	1
Brussels sprouts, 1/2 c	3	Broccoli, raw, chopped, 1/2 c	1
Spinach, 1/2 c	3	Carrot juice, 1/2 c	1
Asparagus, 1/2 c	2	Celery, 1/2 c	1
Baked potato, no skin, 1	2	Dill pickle, 1 whole	1
Cabbage, red, 1/2 c	2	Eggplant, 1/2 c	1
Carrots, 1/2 c	2	Lettuce, romaine, 1 c	1
Cauliflower, 1/2 c	2	Onions, 1/2 c	1
Corn, 1/2 c	2	Tomato, raw, 1 medium	1
Green beans, 1/2 c	2	Tomato juice, canned, 3/4 c	1

Fruits

Food	Fiber (g)	Food	Fiber (g)
Pear, raw, 1 medium	5	Other berries, raw, 1/2 c	2
Blackberries/raspberries, raw, 1/2 c	4	Peach, raw, 1 medium	2
Prunes, cooked, 1/4 c	4	Strawberries, sliced, 1/2 c	2
Figs, dried, 3	3	Cantaloupe, raw, 1/2 c	1
Apple, 1 medium	3	Cherries, raw, 1/2 c	1
Apricots, raw, 4 each	3	Fruit cocktail, canned, 1/2 c	1
Banana, raw, 1	3	Peach half, canned	1
Orange, 1 medium	3	Raisins, dry, 1/4 c	1
		Orange juice, 3/4 c	<1

Meat, Poultry, Fish, Dry Peas and Beans, Eggs, and Nuts Group

Food	Fiber (g)	Food	Fiber (g)
Lentils, 1/2 c	8	Soybeans, 1/2 c	5
Kidney beans, 1/2 c	8	Almonds or mixed nuts, 1/4 c	4
Pinto beans, 1/2 c	8	Peanuts, 1/4 c	3
Black beans, 1/2 c	7	Peanut butter, 2 tbs	2
Black-eyed peas, 1/2 c	6	Cashew nuts, 1/4 c	1
Lima beans, 1/2 c	5	Meat, poultry, fish, and eggs	0

[a]All values are for ready-to-eat or cooked foods, unless otherwise noted. Fruit values include edible skins. All values are rounded values
[b]Pasta includes spaghetti noodles, lasagna, and other noodles made from enriched white flour. Whole-wheat pastas have significantly more fiber.

richest carbohydrate sources. Among sources of fiber, beans and other legumes are peerless, providing as much as 8 grams in a half-cup.

Milk, Cheese, and Yogurt

A cup of milk or yogurt is a generous contributor of carbohydrate, donating about 12 grams. Cottage cheese provides about 6 grams of carbohydrate per cup, but most other cheeses contain little if any carbohydrate. These foods also contribute high-quality protein (a point in their favor), as well as several important vitamins and minerals. Calcium-fortified soy beverages are options for providing calcium and 14 grams of carbohydrate. Milk products vary in fat content, an important consideration in choosing among them; Chapter 5 provides the details.

Butter and cream cheese, although dairy products, are not equivalent with milk because they contain little or no carbohydrate and insignificant amounts of the other nutrients important in milk. They are appropriately associated with the solid fats.

Oils, Solid Fats, and Added Sugars

Oils and fats are devoid of carbohydrate but added sugars supply carbohydrate. Most people enjoy sweets, so it is important to learn something of their nature. First, the definition of "sugar" comes into play. (Table 4-9 defines sugar terms.) A sugar molecule arising in a grape by way of photosynthesis, one of the **naturally occurring sugars,** is chemically indistinguishable from one added at the jam factory to sweeten grape jam. The term **added sugars** refers to all sugars added to foods, regardless of their origin. The combined total of naturally occurring and added sugars appears on food labels in the line reading "sugars." The body handles all the sugars in the same ways, whatever their source.

The *USDA Food Guide* distinguishes between naturally occurring and added sugars, designating the calories from added sugars as discretionary calories.[82] Concentrated sugars add calories but no other

nutrients, and therefore contribute discretionary calories to the diet. Because the law requires manufacturers to list only total sugars on food labels, consumers are hard pressed to determine how much added sugar, and therefore how many discretionary calories, their foods contain.

Whether they come from beets, corn, grapes, honey, or sugarcane, the carbohydrate sweeteners are all alike. All arise naturally and, through processing, are purified of most or all of the original plant material—bees process honey, and machines process the other types. Nutrition authorities recommend that we limit the use of added sugars (see Controversy 4).

The Nature of Sugar

Each teaspoonful of any sweet can be assumed to supply about 16 calories and 4 grams of carbohydrate. You may not think of candy or molasses in terms of *teaspoons,* but this helps to emphasize that all sugary items are like white sugar—in spite of many people's belief that some are different or "better." If you use ketchup liberally, remember that a tablespoon of it contains a teaspoon of sugar. And for the soft drink user, a 12-ounce can of sugar-sweetened cola contains about 8 or more teaspoons of sugar. Figure 4-16 shows that processed foods contain surprisingly large amounts of sugar. Figure 4-17 shows that strawberry spread claiming to be "100% fruit" can contain even more sugars per tablespoon than regular sucrose-sweetened jam.

What about the nutritional value of a product such as molasses, honey, or concentrated fruit juice sweetener compared to white sugar? Molasses contains 1 milligram of iron per tablespoon so, if used frequently, it can contribute some of this important nutrient. Molasses is less sweet than the other sweeteners, however, so more molasses is needed to provide the same sweetness as sugar. Also, the iron comes from the iron machinery in which the molasses is made and is in the form of an iron salt that is not easily absorbed by the body.

Honey is no better for health than sugars by virtue of being "natural"—honey is chemically almost indistinguishable from sucrose. Honey contains the two monosaccharides glucose and fructose in approximately equal amounts. Sucrose contains the same monosaccharides but joined together in the disaccharide form.

■ The Canadian Choice System is in Appendix B.

FIGURE 4·16 SUGAR IN PROCESSED FOODS

½ c canned corn = 3 tsp sugar[a]
12 oz cola = 8 tsp sugar
1 tbs ketchup = 1 tsp sugar
1 tbs creamer = 2 tsp sugar
8 oz sweetened yogurt = 7 tsp sugar
2 oz chocolate = 8 tsp sugar

[a]Values based on 1 tsp = 4 g.

naturally occurring sugars sugars that are not added to a food but are present as its original constituents, such as the sugars of fruit or milk.

added sugars sugars and syrups added to a food for any purpose, such as to add sweetness or bulk or to aid in browning (baked goods). Also called *carbohydrate sweeteners,* they include glucose, fructose, corn syrup, concentrated fruit juice, and other sweet carbohydrates.

| TABLE 4·9 | TERMS THAT DESCRIBE SUGAR |

Note: The term *sugars* here refers to all of the monosaccharides and disaccharides. On a label's ingredients list, the term *sugar* means sucrose. See Controversy 4 for terms related to *artificial sweeteners* and *sugar alcohols*.

- **brown sugar** white sugar with molasses added, 95% pure sucrose.
- **concentrated fruit juice sweetener** a concentrated sugar syrup made from dehydrated, deflavored fruit juice, commonly grape juice; used to sweeten products that can then claim to be "all fruit."
- **confectioner's sugar** finely powdered sucrose, 99.9% pure.
- **corn sweeteners** corn syrup and sugar solutions derived from corn.
- **corn syrup** a syrup, mostly glucose, partly maltose, produced by the action of enzymes on cornstarch. *High-fructose corn syrup (HFCS)* is a mixture of fructose, glucose (dextrose), and maltose.
- **dextrose** an older name for glucose.
- **evaporated cane juice** raw sugar from which impurities have been removed.
- **fructose, galactose, glucose** the monosaccharides.
- **granulated sugar** common table sugar, crystalline sucrose, 99.9% pure.
- **honey** a concentrated solution primarily composed of glucose and fructose, produced by enzymatic digestion of the sucrose in nectar by bees.
- **invert sugar** a mixture of glucose and fructose formed by the splitting of sucrose in an industrial process. Sold only in liquid

form and sweeter than sucrose, invert sugar forms during certain cooking procedures and works to prevent crystallization of sucrose in soft candies and sweets.
- **lactose, maltose, sucrose** the disaccharides.
- **levulose** an older name for fructose.
- **maple sugar** a concentrated solution of sucrose derived from the sap of the sugar maple tree, mostly sucrose. This sugar was once common but is now usually replaced by sucrose and artificial maple flavoring.
- **molasses** a syrup left over from the refining of sucrose from sugarcane; a thick, brown syrup. The major nutrient in molasses is iron, a contaminant from the machinery used in processing it.
- **raw sugar** the first crop of crystals harvested during sugar processing. Raw sugar cannot be sold in the United States because it contains too much filth (dirt, insect fragments, and the like). Sugar sold as "raw sugar" is actually evaporated cane juice.
- **turbinado** (ter-bih-NOD-oh) **sugar** raw sugar from which the filth has been washed; legal to sell in the United States.
- **white sugar** pure sucrose, produced by dissolving, concentrating, and recrystallizing raw sugar.

Spoon for spoon, however, sugar contains fewer calories than honey because the dry crystals of sugar take up more space than the sugars of honey dissolved in its water.

As for concentrated juice sweeteners, these are highly refined and have lost virtually all of the beneficial nutrients and nonnutrients of the original fruit. No form of sugar is "more healthy" than white sugar, as Table 4-10 shows.

It would be absurd to rely on any sugar for nutrient contributions. A tablespoon of

■ *Sugars* on the Nutrition Facts panel of a food label reflect both added and naturally occurring sugars in foods. Sugars listed among the ingredients are all added. Products listing sugars among the first few ingredients contain substantial amounts of added sugars.

■ Sugar alcohols, discussed in this chapter's Controversy, help protect against tooth decay.

| FIGURE 4·17 | JAM AND FRUIT SPREAD LABELS COMPARED |

Notice that a product claiming to contain "100% fruit" can contain concentrated purified fruit juice sweeteners that contribute sweet flavor but few nutrients, just as ordinary sugar does.

Strawberry Jam

Nutrition Facts

Serving size 1 Tbsp (20g)
Servings Per Container About 14

Amount per serving

Calories 40	Calories from Fat 0

	% Daily Value*
Total Fat 0g	0%
Sodium 1mg	1%
Total Carbohydrate 10g	4%
Sugars 7g	
Protein 0g	

*Percent Daily Values are based on a 2,000 calorie diet.

INGREDIENTS: Strawberries, Corn Syrup, Sugar, High Fructose Corn Syrup, Citric Acid, Fruit Pectin.

Strawberry 100% Fruit Spread

Nutrition Facts

Serving size 1 Tbsp (18g)
Servings Per Container About 16

Amount per serving

Calories 40	Calories from Fat 0

	% Daily Value*
Total Fat 0g	0%
Sodium 0mg	0%
Total Carbohydrate 10g	3%
Sugars 8g	
Protein 0g	

*Percent Daily Values are based on a 2,000 calorie diet.

INGREDIENTS: Clarified Grape Juice Concentrate, Strawberries, Clarified Pear Juice Concentrate, Pectin, Natural Flavor, Citric Acid.

TABLE 4·10   THE EMPTY CALORIES OF SUGAR

At first glance, honey, jelly, and brown sugar look more nutritious than plain sugar, but when compared with a person's nutrient needs, none contributes anything to speak of. The cola beverage is clearly an empty-calorie item, too.

Food	Energy (cal)	Protein (g)	Fiber (g)	Calcium (mg)	Iron (mg)	Magnesium (mg)	Potassium (mg)	Zinc (mg)	Vitamin A (µg)	Thiamin (mg)	Riboflavin (mg)	Niacin (mg)	Vitamin B6 (mg)	Folate (µg)	Vitamin C (mg)
Sugar (1 tbs)	46	0	0	0	0.0	0	0	0.0	0	0	0	0.0	0	0	0
Honey (1 tbs)	64	0	0	1	0.1	0	11	0.0	0	0	0	0.0	0	<1	0
Molasses (1 tbs)	55	0	0	42	1.0	50	300	0.1	0	0	0	0.2	0.1	0	0
Concentrated grape or fruit juice sweetener (1 tbs)	30	0	0	0	0	0	0	0	0	0	0	0	0	0	
Jelly (1 tbs)	49	0	0	1	0.0	1	12	0.0	0	0	0	0.0	0	0	<1
Brown sugar (1 tbs)	34	0	0	8	0.2	3	31	0.0	0	0	0	0.0	0	0	0
Cola beverage (12 fl oz)	153	0	0	11	0.1	4	4	0	0	0	0	0.0	0	0	0
Daily Values	2,000	56	25	1,000	18.0	400	3,500	15.0	1,000	1.5	1.7	20.0	2.0	400	60

honey (64 calories) does offer 0.1 milligram of iron, but it would take 180 tablespoons of honey—11,500 calories—to provide 100 percent of a young woman's DRI value of 18 milligrams of iron. The nutrients of honey just don't add up as fast as its calories. Thus, if you choose molasses, brown sugar, or honey, choose them not for their nutrient contributions but for the pleasure they give. These tricks can help magnify the sweetness of foods without boosting their calories:

- Serve sweet food warm (heat enhances sweet tastes).
- Add sweet spices such as cinnamon, nutmeg, allspice, or clove.
- Add a tiny pinch of salt; it will make food taste sweeter.
- Try reducing the sugar added to recipes by one-third.
- Select fresh fruits or fruit juice, or those prepared without added sugar.
- Use small amounts of sugar substitutes in place of sucrose.
- Read food labels for clues on sugar content.

Finally, enjoy whatever sugar you do eat. Sweetness is one of life's great sensations, so enjoy it in moderation.

■ The USDA Food Guide suggests that, within calorie limits, small amounts of added sugars can be enjoyed as part of the discretionary calories in a nutrient dense diet:

- 3 tsp for 1,600 cal
- 5 tsp for 1,800 cal
- 8 tsp for 2,000 cal
- 9 tsp for 2,200 cal
- 12 tsp for 2,400 cal

MEDIA MENU @ 🏠 ◀ ▶

Nutrition Connections

The Nutrition Connections CD provides access to these additional resources.

Do It!
The Do It! exercise for this chapter, *Find the Fiber in Two Lunches,* provides practice in estimating the amount of fiber in foods.

Interaction
The animation, *Carbohydrate Digestion,* follows the digestion of starch as it begins in the mouth, and is completed in the small intestine with the release of glucose.

Chapter Quiz
Take the quiz for this chapter to assess your understanding of chapter concepts.

Nutrition on the Net
Annotated links to reputable nutrition-related websites.

Online Glossary
Test your knowledge of key vocabulary through this comprehensive, interactive glossary.

Internet Connections

A variety of helpful study tools for this chapter are available at our website
http://nutrition.wadsworth.com/ncc10

Practice Test
Take the practice test for this chapter and gauge your knowledge of key concepts; answers are provided.

Student Notebook
A lecture notebook is available to help organize note taking while reading or attending lecture.

Study Tools
Study resources for students include Chapter Summaries, Crossword puzzles, Flashcards, Glossary, Internet Exercises, and Learning Objectives.

SELF CHECK

Answers to these Self Check questions are in Appendix G.

1. The dietary monosaccharides include:
   a. sucrose, glucose, and lactose
   b. fructose, glucose, and galactose
   c. galactose, maltose, and glucose
   d. glycogen, starch, and fiber

2. The polysaccharide that helps form the supporting structures of plants is:
   a. cellulose
   b. maltose
   c. glycogen
   d. sucrose

3. Digestible carbohydrates are absorbed as _____ through the small intestinal wall and are delivered to the liver where they are converted to _____ .
   a. disaccharides; sucrose
   b. glucose; glycogen
   c. monosaccharides; glucose
   d. galactose; cellulose

4. When blood glucose concentration rises, the pancreas secretes _____ , and when blood glucose levels fall, the pancreas secretes _____ .
   a. glycogen; insulin
   b. insulin; glucagon
   c. glucagon; glycogen
   d. insulin; fructose

5. When the body uses fat for fuel without the help of carbohydrate, this results in the production of _____ :
   a. ketone bodies
   b. glucose
   c. starch
   d. galactose

6. Foods rich in fiber lower blood cholesterol.
   T   F

7. Type 1 diabetes is most often controlled by successful weight-loss management.
   T   F

8. Around the world, most people are lactose intolerant.   T   F

9. By law, enriched white bread must equal whole-grain bread in nutrient content.
   T   F

10. The fiber-rich portion of the wheat kernel is the bran layer.   T   F

CONTROVERSY 4

SUGAR AND ALTERNATIVE SWEETENERS: ARE THEY BAD FOR YOU?

ALMOST EVERYONE finds sweet tastes pleasing—after all, the preference for sweets is inborn.[*1] To a child's taste, the sweeter the food, the better. In adults, the preference for sweets is somewhat diminished, but adult consumption of sugars and sweeteners is high and increasing in the United States.

Imagine pouring almost three-quarters of a measuring cup (31 teaspoons) of sugar onto your foods and into your beverages before consuming them each day.[†] This represents the average daily amount of added sugars in the U.S. food supply, enough on average to provide every man, woman, and child with more than 100 pounds per year. This number may be somewhat higher than actual intakes because it does not account for waste, such as the syrup drained from sweet pickles or jam that molds and is tossed out.[2] Still, such numbers are useful for approximating average U.S. sugars consumption and for tracking changes from year to year (see Figure C4-1).

The steady upward trend in U.S. sugars consumption parallels a dramatic increase in purchases of commercially prepared foods and beverages that contain added sugars. By one account, consumption of regular soft drinks and sugar-sweetened fruit drinks is responsible for 80 percent of this increase.[3] Desserts and jams and jellies make up most of the remainder. In contrast, people are adding less sugar to foods from the sugar bowl at home.

In addition to increasing their intake of sugars, people are also consuming more artificial sweeteners, such as aspartame and saccharine. These substances are intended to reduce sugar

intakes, but instead of substituting artificial sweeteners for sugar, people seem to be choosing more of both.

The committee on the *Dietary Guidelines for Americans 2005* offers clear advice on added sugars: treat them as discretionary calories.[4] In other words, fill the day's nutrient needs with foods presenting little or no added sugar. Then, if the diet still lacks energy, sugar-sweetened foods may provide some of the remaining calories. Most people can afford only a little added sugar in their diets if they are to meet nutrient needs within calorie limits. The USDA Food Guide suggests about 9 teaspoons of sugar, or one soft drink's worth, in a nutrient dense 2,200 calorie diet (page 129 listed other amounts).[5]

In response to a world-wide trend toward increasing sugar consumption, the World Health Organization also took a stand on sugar intake echoing the advice of more than 23 countries: consume no more than 10 percent of total calories from added sugar.[§6]

Does sugar harm people's health? And if it does, are sugar substitutes a better choice? This Controversy addresses these questions and, in the process, demonstrates how nutrition researchers pursue their answers, step by step, via scientific inquiry.

Evidence Concerning Sugar

Sugar has been accused of causing these nutrition problems: (1) promoting and maintaining obesity, (2) causing and aggravating diabetes, (3) increasing the risk of heart disease, (4) disrupting behavior in children and adults, and (5) causing dental decay and gum disease. Is sugar guilty or innocent of these charges?

Does Sugar Cause Obesity?

Over the past decade, obesity rates have risen sharply in the United States; during the same period, sugar consumption has reached an all-time high. Does this mean that the increasing sugar intakes are responsible for the increase in obesity?

[§]The World Health Organization uses the term *free sugars* to mean all monosaccharides and disaccharides added to foods by the manufacturer, cook, or consumer plus the naturally occurring sugars in honey, syrups, and fruit juices.

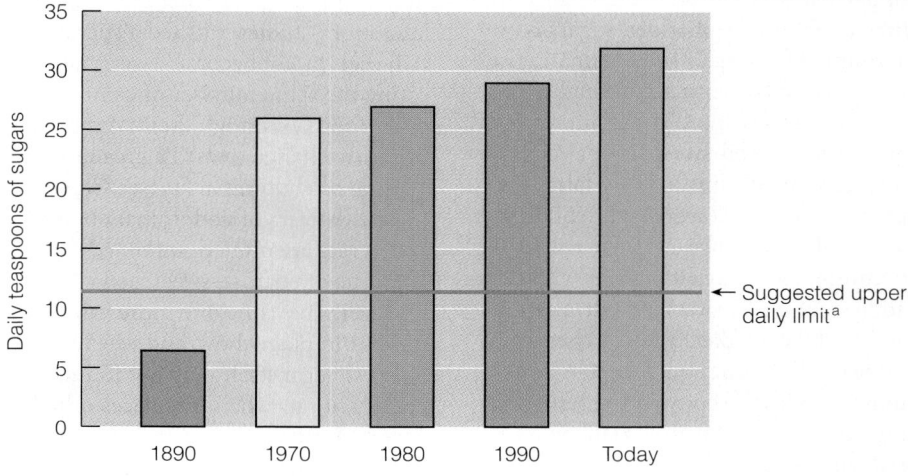

FIGURE C4·1 ADDED SUGARS: AVERAGE SUPPLY PER PERSON IN THE UNITED STATES, 1890–PRESENT

← Suggested upper daily limit[a]

[a]The World Health Organization recommends an upper limit of about 12 teaspoons in an adequate 2,200-calorie diet.
SOURCE: Data from J. Putnam and S. Haley, Estimating consumption of caloric sweeteners, *Amber Waves,* April 2003, available at www.ers.usda.gov/Amberwaves/April03/Indicators/BehindData.htm.

---

* Reference notes are found in Appendix F.
†This estimate from the USDA Economic Research Service includes all caloric sweeteners in the U.S. human food supply, including cane and beet sugars, corn sweeteners, honey, and syrups. Other estimates derived from self-reporting of sugars eaten and other databases may differ.

Sugary soft drinks are the leading source of added sugars in the United States; cakes, cookies, pies, and other baked goods come next; and sweetened fruit drinks and punches follow closely behind.

Human studies are in conflict on this question. When eaten in excess of need, calories from sugar contribute to body fat stores, just as calories from other sources do. From available research, however, it seems unlikely that moderate amounts of sugar would cause obesity when total calorie intakes are controlled.[7]

People with diets high in added sugars do often consume more calories each day than people with lower sugar intakes.[8] For example, adolescents who drink upwards of 26 ounces (about two cans) of sugar-sweetened soft drinks daily consume 400 more calories a day than teens who abstain from such drinks.[9] Soft drinks have been the focus of particular attention because the increase in intake of added sugars is accounted for largely by a dramatic rise in high-fructose corn syrup intake, mostly in soft drinks. The trend toward greater use of this sweetener parallels unprecedented gains in body fatness in the population.[10] Overweight children and adolescents increase their risk of becoming obese by 60 percent with each additional syrup-sweetened drink they add to the daily diet.[11] Researchers theorize that sugar in liquid form may go unnoticed by the body, in the sense that the body does not compensate by reducing food energy intakes later in the day.[12] Between 1977 and 2001, as people grew fatter, their intake of calories from fruit drinks and punches doubled while calories from soft drinks

nearly tripled.[13] Investigators are evaluating these and other possible links between fructose in the syrupy sweeteners of soft drinks and weight gain.[14]

Such findings appear to be a smoking gun implicating added sugars as the culprit in the rise of obesity, but before concluding that sugar is guilty, consider that some population studies have found a *negative* correlation between intakes of added sugar and degree of body fatness.[15] These reports have been questioned, however, because people are notoriously poor at recalling their actual food intakes, and overweight people often underestimate their intakes of sugary treats.[16]

Findings from populations of many developing nations seem to strengthen the case against sugar: the prevalence of obesity often increases as sugar consumption rises, but this evidence is still insufficient to name sugar as the cause. Often, a rise in a nation's sugar intake mirrors a rise in income and, with it, greater consumption of calories, processed foods, meats, and fats.[17] Another confounding factor: as people begin to gain weight, their physical activity often declines, making it impossible to tell whether the sugar, excess calories from other sources, or too little physical activity is at fault.

Research is easier with laboratory rats—they are fed controlled diets and need not remember what they ate. In laboratory studies, rats fed a sucrose-rich diet do not become fatter than rats eating the same number of calories of other foods, although their fat distribution may change toward a greater percentage of abdominal fat (see Chapter 9 for the dangers of abdominal obesity).[18] When rats are offered *unlimited* amounts of sucrose-sweetened chow, however, they consume more calories than from plain chow, and weight gain follows. Sugar itself may not induce obesity in rats when it replaces other energy sources calorie for calorie, but sugar stimulates overeating when food is unlimited, and the excess calories cause weight gain—at least in rats.

Sugar acts on areas of the brain responsible for pleasure and reward.

When stimulated, this area of the brain releases chemicals that produce a pleasurable sensation that, in animals, causes repetition of the behavior producing the sensation. When animals are given sugar-sweetened chow, they increase their intake compared with regular chow because the sweet taste rewards them with a pleasure sensation.[19] If the same holds true for people, this mechanism could explain why some people crave sweets, even though they rarely sit down to bowlfuls of plain sugar, as the photo on this page comically reminds us. Instead, they seek out sweet-flavored high-fat treats, such as cakes, candies, chocolate bars, cookies, doughnuts, ice cream, and muffins. Sugar added to foods high in fat may be what produces sufficient rewards to trigger overconsumption. More research is needed to determine connections between human brain chemistry, the sweetness of sugar, and obesity.

Does Sugar Cause Type 2 Diabetes?

Recall from the chapter that in diabetes, insulin secretion or tissue responsiveness to it becomes abnormal, affecting the body's ability to manage sugar. At one time, people thought that eating sugar caused diabetes by "overstraining the pancreas," but now we know that this is not the case. Body fatness is more closely related to diabetes than diet is; high rates of diabetes have not been reported in any society where obesity is rare.

Still, in some populations around the world, and especially among Native Americans, a profound increase in the prevalence of diabetes has occurred simultaneously with an increase in sugar consumption. Yet, in some other populations, no relationship has been found between sugar intake and diabetes.

Wherever starch and the fiber that accompanies it are the major carbohydrates in the diet, diabetes is rare.[20] But this does not prove that sugar causes diabetes or that starch prevents it. The apparent protective effect of starch might be due, for example, to the fiber or the mineral chromium that comes with it. Likewise, the apparent causative effect of sugar may reflect a sudden availability of calorie-rich foods in societies gaining in both disposable income and body fatness.

Recent popular books have claimed links among sugar intake, the glycemic index, and type 2 diabetes, but science in this area is inconclusive. A decade ago, a study tracking the dietary habits of over 65,000 women and almost 43,000 men seemed to indicate that consuming a diet high with a high glycemic load (based on mashed potatoes, white rice, highly refined cold breakfast cereals, and white bread) was associated with diabetes, while diet with a low glycemic load (more whole grains, legumes, and fruits and vegetables) was protective. In contrast, no effect on diabetes risk from glycemic load or glycemic index was reported in a subsequent study of almost 36,000 women in Iowa.[21] In this study, a diet high in whole grain foods and dietary fiber predicted a reduced risk of type 2 diabetes, but the mechanism is not known. As for sugar itself, remember from the chapter that sucrose elicits only a moderate glycemic effect, so the fairest conclusion is that sugar alone is not culpable in type 2 diabetes causation. If a high energy intake from added sugars cause gains of excess body fat, however, then sugar does indeed elevate the risk for type 2 diabetes.[22] And both obesity and diabetes increase the risk of heart disease.

Does Eating Sugar Increase the Risk of Heart Disease?

Several years ago, results from research on rats launched investigations into the relationship between sugar and heart disease. When researchers fed rats a diet with sucrose as the only carbohydrate source, the rats sustained microscopic damage to their arteries, and their blood tested high for saturated fat. Rats fed starch instead of sugar did not develop the damage or the highly altered blood lipids. When researchers applied similar study conditions to human beings, they confirmed that some people, too, respond to diets high in sugars, particularly sugars that contain fructose, by releasing saturated fat (probably made from the sugar) into the blood.[23] Studies have also found that with a high-sugar diet, the level of a *protective* blood lipid (HDL, the "good cholesterol," see the next chapter) falls

in some people.[24] The amount of sugar required to evoke these changes is about double the nation's *average* intake, so, for most people, moderate sugar intakes pose little risk of altering their blood lipids by much.[25] Throughout many years of research, no one has conclusively shown that an average intake of sugar adversely affects heart health in otherwise healthy human beings.[26]

Saturated fat, however, clearly remains the major *dietary* culprit in heart disease susceptibility, but there is a *genetic* culprit, too: some people seem genetically prone to increasing the production of saturated fat in the body in response to dietary sugar. This is especially true of obese people who secrete abnormal amounts of insulin in response to sugar intake, because one of insulin's actions is to increase the liver's synthesis of saturated fat, which then enters the bloodstream.[27] In the same people, insulin's other action on saturated fat, that is, to clear it from the blood into storage in the fat tissue, is diminished.

No one can say with certainty whether the blood lipid changes observed with high sugar intakes are benign or whether they indeed predict heart disease until more research reveals the needed evidence.[28]

What about Sugar and Behavior?

Many years ago, claims began appearing that eating sugary foods caused children to become unruly and adolescents and adults to exhibit antisocial and even criminal behavior. Science has put the "sugar-behavior" theory to rest, but many teachers, parents, grandparents, and others still believe that some children react behaviorally to sugar.[29] Sugar might influence behavior in many ways: by altering the levels of chemicals in the brain that affect mood, by inducing nutrient deficiencies, by stimulating the release of the series of hormones the body secretes after consuming sugar, or by providing pure energy (the Halloween effect). Children allowed to load up on colas and chocolate candies may become overstimulated by the action of the stimulant caffeine found in these treats (see Chapter and Controversy 14).

Research results do not suggest that sugar itself negatively affects behavior.[30]

Indeed, in several well-controlled studies, sugar administered to normal children calmed them down, a finding consistent with biochemical evidence. In adults, carbohydrate-rich foods seem to improve mood and memory.[31] In the end, the most rational conclusion seems to be that while occasional behavioral reactions to sugar may be possible, studies have failed to demonstrate any consistent effects of sucrose on behavior in either normal or hyperactive children.

Does Sugar Cause Dental Caries?

Dental caries are a serious public health problem afflicting the majority of people in the country, half by the age of two (see Figure C4-2). A very lucky few *never* get dental caries because they have an inherited resistance; others have sealant applied to teeth during childhood that stop caries before they can begin. Table C4-1, p. 134 defines some terms related to caries.

Another successful measure taken to reduce the incidence of dental decay is fluoridation of community water. But sugar has something to do with dental caries, too.

Caries develop as acids produced by bacterial growth in the mouth eat into tooth enamel. Bacteria form colonies

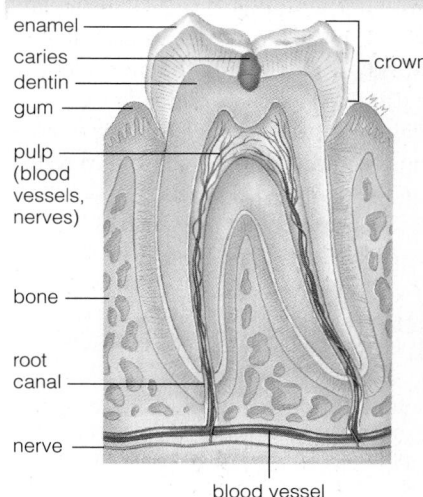

FIGURE C4·2 DENTAL CARIES

enamel
caries
dentin
gum
crown

pulp (blood vessels, nerves)

bone

root canal

nerve

blood vessel

Caries begin when acid dissolves the enamel that covers the tooth. If not repaired, the decay may penetrate the dentin and spread into the pulp of the tooth, causing inflammation and an abscess.

known as **plaque** whenever they can get established on tooth surfaces. Once established, they multiply and affix themselves more and more firmly unless they are brushed, flossed, or scraped away. Eventually, the acid of plaque creates pits that deepen into cavities. Below the gum line, plaque works its way down until the acid erodes the roots of teeth and the jawbone in which they are embedded, loosening the teeth and leading to infections of the gums. Gum disease severe enough to threaten tooth loss afflicts the majority of our population by their later years.

Bacteria thrive on carbohydrate. Carbohydrate as sugar has been named as the main causative factor in the formation of caries, but starch also supports bacterial growth if the bacteria are allowed sufficient time to work on it. Of prime importance is the length of time the food stays in your mouth, and this depends on the food's composition, how sticky it is, how often you eat it, and especially on whether you brush your teeth soon afterward.[32]

Bacteria produce acid for 20 to 30 minutes after exposure to sugar. Thus, when you eat three pieces of candy, one right after the other, your teeth are exposed to approximately 30 minutes of acid demineralization; when you eat the candy pieces at half-hour intervals, the acid exposure time is 90 minutes. Likewise, slowly sipping a sugary soft drink may be more harmful than drinking quickly and emptying the mouth of sugar. Foods such as milk and cheese may be particularly helpful in minimizing the effects of the acids and in restoring the damaged enamel.[33]

Some forms of candy, such as milk chocolate and caramels, may be less

TABLE C4·1 DENTAL TERMS

- **dental caries** decay of the teeth (*caries* means "rottenness").
- **plaque** (PLACK) a mass of microorganisms and their deposits on the crowns and roots of the teeth, a forerunner of dental caries and gum disease. (The term *plaque* is also used in another connection—arterial plaque in atherosclerosis. See Chapter 11.)

harmful because the sugar dissolves completely and is washed away in saliva. Breads, granola bars, sugary cereals, oatmeal cookies, raisins, salted crackers, and chips, however, may be worse because particles get stuck in the teeth and do not dissolve. These particles may remain in contact with tooth surfaces for hours, providing a feast for bacteria and greatly increasing the likelihood of caries. A table in Chapter 14 lists foods of both high and low caries potential. The best advice seems to be: Brush your teeth after eating. Regular brushing (twice a day, with a fluoride toothpaste) and flossing (once a day) may be more effective in preventing dental caries and gum disease than restricting sugary foods.

Total sugar intake does play a major role in the prevalence of dental caries. Populations whose diets provide no more than 10 percent of calories from sugar have a low prevalence of dental caries.[34] Worldwide, many governing agencies urge their citizens to consume no more than 10 percent of calories from sugar because of sugar's link with dental caries. Sugar is an energy source for the bacteria that cause tooth decay, and thus it is clearly implicated in this major public health problem.[35]

Personal Strategy for Using Sugar

Based on research, no guilty verdict can yet be issued for any of the first four accusations against sugar, but of the fifth, it is guilty as charged—sugar does cause dental caries. Should you avoid sugar or reduce your intakes? The *Dietary Guidelines* suggest that you limit the calories from added sugars, not avoid them altogether. As mentioned, authorities agree that added sugars should contribute no more than 10 percent of the total calorie intake.[36] Thus, a person who eats 2,000 calories of energy a day is allowed 200 calories from sugar. Those 200 calories of sugar, about 13 teaspoons (50 grams), sound like quite a lot. But when you add up all the teaspoons of sugars added to common foods, 200 calories may seem restrictive.

To meet these sugar intake recommendations, many people should reduce their intakes. One option is to replace

sugar's sweetness by choosing from two sets of alternative sweeteners: the sugar alcohols, which are energy-yielding sweeteners sometimes referred to as *nutritive sweeteners,* and the artificial sweeteners, which provide virtually no energy and are thus sometimes referred to as *nonnutritive sweeteners.* These options do provide sweetness without sucrose, but what about their safety?

Evidence Concerning Sugar Alcohols

The sugar alcohols are familiar to people who use special dietary products. Many new low-sugar food products depend on sugar alcohols for their bulking and sweetening powers. Contrary to the meaning implied by the name *sugar alcohols,* these sweeteners do not contain ethanol (the alcohol of alcoholic beverages) or any other intoxicant. They are members of the chemical family *alcohol* by virtue of their structures.

Many people are familiar with **isomalt, lactitol, maltitol, mannitol, sorbitol,** and **xylitol,** sugar alcohols that provide sweetness and bulk to cookies, sugarless gum and hard candies, and jams and jellies. Their sweetness relative to sugar is shown in Table C4-2. Sugar alcohols can be metabolized by human beings, and most provide a little less energy than sucrose.[37]

Sugar alcohols evoke a low glycemic response. The body absorbs sugar alcohols slowly; consequently, they are slower to enter the bloodstream than other sugars. Side effects such as gas, abdominal discomfort, and diarrhea arise from ingesting large quantities, however. For this reason, regulations require that if a food is likely to present more than 50 grams of sugar alcohols in a day, its label must state that "Excess consumption may have a laxative effect."

The real benefit of using sugar alcohols is that they do not contribute to dental caries. Bacteria in the mouth cannot metabolize sugar alcohols as rapidly as sugar. They are therefore valuable in chewing gums, breath mints, toothpaste, and other products that people keep in their mouths for a while.

If you want to reduce energy intake, remember that the sugar alcohols *do*

TABLE C4·2 SUGAR ALCHOLS

Sugar Alcohols	Relative Sweetness[a]	Energy (cal/g)	Approved Uses
Isomalt	0.5	2.0	Candies, chewing gum, ice cream, jams and jellies, frostings, beverages, baked goods
Lactitol	0.4	2.0	Candies, chewing gum, frozen dairy desserts, jams and jellies, frostings, baked goods
Maltitol	0.9	2.1	Particularly good for candy coating
Mannitol	0.7	1.6	Bulking agent, chewing gum
Sorbitol	0.5	2.6	Special dietary foods, candies, gums
Xylitol	1.0	2.4	Chewing gum, candies, pharmaceutical and oral health products

[a]The relative sweetness depends on the temperature, acidity, and other flavors of the foods in which the substance occurs. The sweetness of pure sucrose is the standard with which the approximate sweetness of sugar substitutes is compared.

provide energy. The body handles them differently from sugar, but unlike artificial sweeteners, they are not calorie-free.

Evidence Concerning Artificial Sweeteners

Like the sugar alcohols, artificial sweeteners make foods taste sweet without promoting dental decay. Unlike sugar alcohols, they are practically calorie-free, and the human taste buds perceive them as supersweet. But how do we know they are safe?

All substances are toxic if high enough doses are consumed. Artificial sweeteners, their components, and their metabolic by-products are not exceptions. The questions to ask are whether artificial sweeteners are harmful to human beings at the levels normally used, and how much is too much. The Food and Drug Administration (FDA) has proposed answers by setting **acceptable daily intake (ADI)** levels for some of the artificial sweeteners used in the United States. Table C4-3 defines some sugar substitute terms. The major synthetic sweeteners on the market today are **saccharin, aspartame, acesulfame-potassium, sucralose, neotame,** and **tagatose.**

Saccharin

Saccharin has had a rocky history of acceptance, although it is now consumed by millions of people in the United States, primarily in prepared foods and beverages and secondarily as a tabletop sweetener. Questions about its safety surfaced in the late 1970s, when experiments suggested that it caused bladder tumors in rats. As a result, the FDA proposed banning it. The public outcry in favor of retaining it was so loud, however, that Congress imposed a moratorium on any action, and the ban proposal was eventually withdrawn. For 25 years, saccharin was widely used but remained on the government's roster of "anticipated carcinogens." Products containing it carried a warning label about saccharin as a cancer hazard. In the year 2000, government officials reviewed the research and reversed their opinion, removing saccharin from the carcinogen list and freeing it from the labeling requirement.[38] Opponents to these changes, however, maintain that saccharin causes cancer in mice and rats and so should be avoided.

Does saccharin cause cancer? The evidence in animals is as follows. Rats fed diets high in saccharin from the time of weaning to adulthood were mated. The offspring of those rats were then fed high-saccharin diets throughout their lives and were found to have a higher incidence of bladder tumors than comparable animals not fed saccharin. In Canada, on the basis of these findings, saccharin was banned except for use as a tabletop sweetener to be sold in pharmacies with a warning label.

TABLE C4·3 SUGAR SUBSTITUTE TERMS

- **acceptable daily intake (ADI)** the estimated amount of sweetener that can be consumed daily over a person's lifetime without any adverse effects.
- **acesulfame** (AY-sul-fame) **potassium,** also called **acesulfame-K** a zero-calorie sweetener approved by the FDA and Health Canada.
- **alitame** a noncaloric sweetener formed from the amino acids L-aspartic acid and L-alanine. In the United States, the FDA is considering its approval.
- **aspartame** a compound of phenylalanine and aspartic acid that tastes like the sugar sucrose but is much sweeter. It is used in both the United States and Canada.
- **cyclamate** a zero-calorie sweetener under consideration for use in the United States and used with restrictions in Canada.
- **isomalt, lactitol, maltitol, mannitol, sorbitol, xylitol** sugar alcohols that can be derived from fruits or commercially produced from a sugar; absorbed more slowly and metabolized differently than other sugars in the human body and not readily used by ordinary mouth bacteria.
- **neotame** (NEE-oh-tame) an artificial sweetener composed of two amino acids (phenylalanine and aspartic acid) linked in such a way as to make them indigestible by human enzymes.
- **saccharin** a zero-calorie sweetener used freely in the United States but restricted in Canada.
- **stevia** (STEEV-ee-uh) the sweet-tasting leaves of a shrub sold as a dietary supplement, but lacking FDA approval as a sweetener.
- **sucralose** a noncaloric sweetener derived from a chlorinated form of sugar that travels through the digestive tract unabsorbed. Approved for use in the United States and Canada.
- **tagatose** an incompletely absorbed monosaccharide sweetener derived from lactose with a caloric value of 1.5 calories per gram. About 80% of the ingested tagatose travels to the large intestine where bacterial colonies ferment it. Tagatose is not readily used by mouth bacteria, and so does not promote dental caries.

Some years ago, the National Cancer Institute released results from a large-scale population study involving 9,000 people. The results seemed to show that women who drank two or more saccharin-sweetened diet sodas a day and men and women who both smoked heavily and used artificial sweeteners had a slightly elevated risk of some cancers.[39] Other studies involving more than 5,000 people showed no excess risk of bladder cancers.

A solid clue from the laboratory is based on some physiological differences between the urinary systems of rats and human beings. Proportionally, rats excrete far less water in their urine than people do. As a result, rats can make highly concentrated solutions of substances in just small amounts of water in their urine. Dissolved substances in such high concentrations are likely to crystallize.[40] In safety tests, saccharin overdoses caused crystals to form in the rats' bladders, and the crystals probably caused the tumors. Human beings cannot concentrate urinary substances to such a degree, so they would never form saccharin crystals, even if they consumed larger-than-normal doses of saccharin. They would, however, lose large amounts of water as the kidneys struggled to free the blood of the overload.

Overloading on huge saccharin doses is probably not safe, but consuming moderate amounts almost certainly does not cause bladder cancer in human beings. An ADI has been set for saccharin in the amount of 5 milligrams per kilogram of body weight (see Table C4-4). The amount of saccharin that can be commercially added to foods or drinks is limited to about 30 milligrams per serving.

Aspartame

Aspartame is one of the most thoroughly studied substances ever to be approved for use in foods.[41] Manufacturers use aspartame under the name *NutraSweet* to sweeten foods that require sweetness but are not exposed to cooking temperatures, as aspartame is not heat stable. A gram of aspartame provides 4 calories, as does a gram of protein, but because so little is needed, calories are negligible. Under various brand names, such as *Equal* or *NutraSweet*, aspartame is also available as a powder to use at home in place of sugar. In powdered form, it is mixed with lactose, so a 1-gram packet contains 4 calories.

The amazing popularity of aspartame is mostly due to its flavor, which is almost identical to that of sugar. Furthermore, aspartame is touted as safe for children, so families wishing to limit their children's sugar intakes are offering them NutraSweet products instead.

Aspartame is a simple chemical compound: two protein fragments (the amino acids phenylalanine and aspartic acid) joined together. In the digestive tract, the two fragments are split apart, absorbed, and metabolized just as they would be if they had come from protein in food. The flavors of the components give no clue to the combined effect; one of them tastes bitter, and the other is tasteless. Yet aspartame is 200 times sweeter than sucrose.

With its phenylalanine base, aspartame poses problems for people with an inherited metabolic disease known as phenylketonuria (PKU). People with PKU have the hereditary inability to dispose of phenylalanine eaten in excess of need. Unusual products made from phenylalanine build up and damage the tissues. PKU causes irreversible, progressive brain damage if left untreated in early life. Newborns are tested for PKU; if they have it, the treatment is to limit dietary intake of phenylalanine.

| TABLE C4·4 | U.S. APPROVED ARTIFICIAL SWEETENERS AND SUGAR-BASED SWEETENERS |

Artificial Sweeteners	Energy (cal/g)	Acceptable Daily Intake (ADI)	Average Amount to Replace 1 tsp Sucrose[a]	Approved Uses
Saccharin (SugarTwin, Sweet N' Low, others)	0	5 mg/kg body weight (341 mg for a 150 lb person)	12 mg	Tabletop sweeteners, wide range of foods, beverages, cosmetics, and pharmaceutical products
Aspartame (NutraSweet, Equal, others)	4	50 mg/kg body weight (3,409 mg for a 150 lb person)	18 mg	General-purpose sweetener in all foods and beverages. Warning to population with PKU
Acesulfame-potassium (Sunette, Sweet One)	0	15 mg/kg body weight (1,023 mg for a 150 lb person)	25 mg	Alcoholic beverages, baked goods, candies, chewing gum, desserts, gelatins, puddings, tabletop sweeteners
Sucralose (Splenda)	0	5 mg/kg body weight (341 mg for a 150 lb person)	6 mg	Baked goods, carbonated beverages, chewing gum, coffee and tea, dairy products, frozen desserts, fruit spreads, salad dressing, syrups, tabletop sweeteners
Neotame	0	18 mg/day	0.5 μg	Baked goods, beverages (nonalcoholic), candies, chewing gum, frostings, frozen desserts, gelatins, puddings, jams and jellies, syrups
Tagatose		7.5 g/day	1 tsp	Bakery products, beverages, cereals, chewing gum, confections, dairy products, dietary supplements, health bars, tabletop sweetener

[a] Rounded values.

Children with PKU should not get their phenylalanine from aspartame. Phenylalanine occurs in such protein-rich and nutrient-rich foods as milk and meat, and the PKU child is allowed only a limited amount of these foods. The child has difficulty obtaining the many essential nutrients, such as calcium, iron, and the B vitamins, found along with phenylalanine in these foods. To suggest that such a child squander any of the limited phenylalanine allowance on the purified phenylalanine of aspartame, with none of the associated nutrients to support normal growth, would be to invite nutritional disaster. Product labels carry special warnings for people with PKU.

Other concerns about aspartame's safety have had to do with compounds that arise briefly during its metabolism. These compounds (methyl alcohol, formaldehyde, and diketopiperazine) are not toxic at the levels generated from the ADI amount of aspartame, and concerns about them have been laid to rest.

Urban legends circulating on the Internet accuse aspartame of causing everything from Alzheimer's disease and brain cancer to nerve disorders and skin warts. Orderly scientific investigations into these issues find no relationship between aspartame intake and brain tumors, behavior, mood, or brain chemistry. Nor are any other physical symptoms evident from research. No experimental evidence has shown a connection between headaches and aspartame, but the Centers for Disease Control (CDC) has received many thousands of individual complaints. Every day, millions of people use aspartame. Every day, millions of people have headaches. Anyone who claims, on this basis, that aspartame causes headaches is using personal experience to jump to conclusions. Some of the headache sufferers might indeed be reacting to the artificial sweetener, but they might also be reacting to another substance such as caffeine in aspartame-sweetened beverages or to factors in their lives unrelated to foods. People who believe aspartame gives them headaches should use a different sweetener.

On approving aspartame for U.S. consumers, the FDA set the ADI at 50 milligrams per kilogram of body weight in a day. In Canada, the acceptable level is set at 40 milligrams per kilogram. These are reasonable numbers. For a 132-pound person, 50 milligrams equals 80 packets of aspartame sweetener or 15 soft drinks sweetened only with aspartame. It is possible to exceed the ADI amount, however: a child who drinks a quart of Kool-Aid on a hot day and who also has pudding, chewing gum, cereal, and other products sweetened with aspartame can pack in more than the daily ADI limit. Infants or toddlers under two years old should not be fed artificially sweetened foods and drinks.

Acesulfame-Potassium

During 15 years of testing and use, the artificial sweetener acesulfame-potassium (or acesulfame-K) has been used without reported health problems. An ADI of 15 milligrams per kilogram of body weight was set for acesulfame-potassium on its approval. Marketed under the trade names *Sunette* and *Sweet One,* this sweetener is about as sweet as aspartame and is used in chewing gum, beverages, instant coffee and tea, gelatins, and puddings, as well as for table use. Acesulfame-potassium holds up well during cooking.

Acesulfame-potassium is 200 times as sweet as sucrose, but to some, it leaves a slight aftertaste. Blending it with other sweeteners solves the problem. Acesulfame-potassium is not recognized by the body's metabolic equipment and therefore is excreted unchanged by the kidneys.

Sucralose

Approved for use as a sweetener in the United States and Canada, sucralose (trade name *Splenda*) is the only artificial sweetener made from sucrose. Three chlorine atoms substitute for three hydrogen and oxygen groups on the structure of sucrose, making a product that provides 600 times the sweetness of sugar. Many years of testing have deemed sucralose safe to use and, specifically, not a cause of cancer. Sucralose is not recognized by the body as sugar and therefore passes through unchanged. The ADI for sucralose has been set at 5 milligrams per kilogram of body weight per day for all ages, including pregnant and lactating women. Sucralose is heat stable and so is useful for cooking and baking; it is used in commercially prepared products and as a tabletop sweetener.

Neotame

In 2002, the FDA approved neotame on the strength of findings from over 110 safety studies conducted on both animals and human beings. Neotame is so intensely sweet—about 7,000 to 13,000 times sweeter than sugar—that very little is needed to sweeten foods and beverages. Currently, neotame is available to food manufacturers for sweetening processed foods but not to consumers for home use.

A chemical cousin of aspartame, neotame also contains the amino acids phenylalanine and aspartic acid. Unlike aspartame, however, neotame is indigestible. Whereas digestive enzymes can separate the amino acids in aspartame, neotame's chemistry is sufficiently different to block the enzyme's action. This slight chemical difference makes neotame a better choice for people with phenylketonuria because it provides no phenylalanine to the body.

Tagatose

The FDA has granted tagatose, a relative of fructose, the status of "generally recognized as safe," making it available as a lower-calorie sweetener for foods, beverages, confections, dietary supplements, and other uses.[42] Tagatose is derived from lactose, but unlike fructose or lactose, 80 percent of tagatose remains unabsorbed until it reaches the large intestine. There the normal bacterial colonies of the colon ferment tagatose, releasing gases and small products that are absorbed. At high doses, tagatose causes flatulence, rumbling, and loose stools. Otherwise, no adverse side effects have been noted by the maker.

Other Sweeteners

Two other artificial sweeteners are awaiting FDA approval—**cyclamate** and **alitame.** Cyclamate was once approved in the United States but later was banned when it was suspected, but never proved, to cause cancer in rats. In Canada, cyclamate is restricted to use as a tabletop sweetener on the advice of a physician and as a sweetening additive in medicines. Alitame resembles aspartame in being composed of two amino acids, but unlike aspartame, it remains stable when heated.

A naturally sweet herb called **stevia** is gaining in popularity as a sugar substitute, especially in beverages. Food additives must provide evidence of their safety and effectiveness before receiving FDA approval, and stevia lacks this approval for use as a sweetener because so little is known about its effects on human health, save that it can be absorbed by the human digestive tract.[43] But stevia can be sold as a dietary supplement because dietary supplements need not present any supporting evidence of safety or effectiveness, which highlights a problem in the FDA's regulation of supplements. Stevia may be harmful, safe, or even beneficial in some way, but no one can say for sure until science reveals more about this sweetener.[44]

Do Artificial Sweeteners Help with Weight Control?

Many people eat and drink products sweetened with artificial sweeteners in the belief that the products help control weight. Do they work? Ironically, some studies of rats report that intense sweeteners, such as saccharin, stimulate appetite and lead to weight *gain* instead of loss. Many studies on *people,* however, find either no change or a decline in feelings of hunger. Food energy intake has also been reported to be lower when artificial sweeteners replace sugar, resulting in greater weight losses when obese people eat or drink artificially sweetened products instead of their sugar-sweetened counterparts.[45]

In studying the effects of artificial sweeteners on food intake and body weight, researchers ask different questions and take different approaches. It matters, for example, whether the people used in a study are of a healthy weight and whether they are following a weight-loss diet. Motivations for using

sweeteners can affect total calorie intakes, too. For example, a person wishing to eat a high-calorie food might try to compensate for its calories by drinking an artificially sweetened beverage. This person's energy intake might stay the same or increase. In contrast, another person who wants to cut total calorie intake might choose the artificial sweetener in the context of a whole diet of low-calorie foods. This person's calorie intake might well be lower than it otherwise would have been.

Researchers must also distinguish between the effects of the experience of tasting something sweet and the physiological effects of a particular substance on the body. If a person experiences hunger or feels full shortly after eating an artificially sweetened snack, is that because tasting something sweet stimulates or depresses the appetite? Or is it because the artificial sweetener itself somehow affects the appetite through nervous, hormonal, or other means? Furthermore, if appetite is stimulated, does that actually lead to increased food intake?

A recent study reports sizable weight losses when artificial sweeteners replace sizable amounts of sugar in the diet.[46] Researchers supplemented the diets of overweight men and women with snacks, mostly beverages, containing either sucrose (snacks totaling about 600 calories) or artificial sweeteners (snacks totaling about 200 calories), and then allowed free choice to determine the rest of the diet. After 10 weeks, the sucrose group had gained about 3½ pounds, while the artificial sweetener group had lost about 2 pounds. As a side benefit to weight loss, blood pressure dropped in the artificial sweetener group, too. The researchers speculate that sugars in the form of fluids may be less satisfying than those in solid foods, making it easier to overconsume calories later in the day. To say with any certainty whether artificial sweeteners

© Matthew Farruggio

Can artificial sweeteners help people lose weight?

reduce energy intakes or body fatness requires more evidence, however, because many other studies report no weight differences between people fed sugar and those fed artificial sweeteners.[47]

In contrast to the findings for the overweight subjects just described, sugar consumed before a meal by normal-weight people reliably dampens the appetite.[48] The common belief that sugar "spoils the appetite" proves true for most people, and especially for children.

Personal Strategies for Using Artificial Sweeteners

Current evidence indicates that moderate intakes of artificial sweeteners pose no health risks. For those who choose to include artificial sweeteners in their diets, moderation is the key. Using artificial sweeteners does not automatically lower energy intake; controlling energy intake successfully requires informed diet and activity decisions throughout the day (as Chapter 9 explains). Although not magic bullets in fighting overweight, artificial sweeteners probably do not hinder weight-loss efforts either, and they are safer for the teeth than carbohydrate sweeteners.

The Lipids: Fats, Oils, Phospholipids, and Sterols

David Tindle, R. A. (b. 1932), *Plate with Avocado Pears*, Christie's Images, NY

CONTENTS

Do You Ever . . .

Think of fats as unhealthy food constituents that are best eliminated from the diet?

Wonder about the differences between "bad" and "good" cholesterol?

Choose fish for health's sake without fully knowing why?

Recognize the invisible fats in your foods?

KEEP READING.

YOUR BILL FROM a medical laboratory reads, "Blood **lipid** profile—$200." A health-care provider reports, "Your blood **cholesterol** is high." Your physician advises, "You must cut down on the saturated **fats** in your diet and replace them with **oils** to lower your risk of **cardiovascular disease (CVD).**" Blood lipids, cholesterol, saturated fats, and oils—what are they, and how do they relate to health?

No doubt you are expecting to hear that fat-related compounds have the potential to harm your health, but lipids are also valuable. In fact, lipids are absolutely necessary. The diet recommended for health is moderate in fats, but it is by no means a "no-fat" diet. Luckily, at least traces of fats and oils are present in almost all foods, so you needn't make an effort to eat any extra as long as the diet is balanced among nutritious foods.

Introducing the Lipids

The lipids in foods and in the human body fall into three classes. About 95 percent are **triglycerides.** The other classes of the lipid family are the **phospholipids** (of which **lecithin** is one) and the **sterols** (cholesterol is the best known of these). Some of these names may sound unfamiliar, but most people will recognize at least a few functions of lipids in the body and in food that are listed in Table 5-1. More details on each of the lipid classes follow later.

Usefulness of Fats in the Body

When people speak of fat, they are usually talking about triglycerides. The term *fat* is more familiar, though, and we will use it in this discussion. Fat is the body's chief storage form for the energy from food eaten in excess of need. The storage of fat is a valuable survival mechanism for people who live a feast-or-famine existence: stored during times of plenty, fat enables them to remain alive during times of famine. Fats also provide most of the energy needed to perform much of the body's work, especially muscular work.

Most body cells can store only limited fat, but some cells are specialized for fat storage. These fat cells seem able to expand almost indefinitely—the more fat they store, the larger they grow. An obese person's fat cells may be many times the size of a thin person's. Far from being a collection of inert sacks of fat, however, adipose (fat) tissue secretes hormones that help regulate the appetite and influence other body functions.[1] A fat cell is shown in Figure 5-1.

*Reference notes are found in Appendix F.

---

lipid (LIP-id) a family of organic (carbon-containing) compounds soluble in organic solvents but not in water. Lipids include triglycerides (fats and oils), phospholipids, and sterols.

cholesterol (koh-LESS-ter-all) a member of the group of lipids known as sterols; a soft, waxy substance made in the body for a variety of purposes and also found in animal-derived foods.

fats lipids that are solid at room temperature (70°F or 25°C).

oils lipids that are liquid at room temperature (70°F or 25°C).

cardiovascular disease (CVD) disease of the heart and blood vessels; disease of the arteries of the heart is called *coronary heart disease (CHD)*.

triglycerides (try-GLISS-er-ides) one of the three main classes of dietary lipids and the chief form of fat in foods and in the human body. A triglyceride is made up of three units of fatty acids and one unit of glycerol (fatty acids and glycerol are defined later). Triglycerides are also called *triacylglycerols*.

phospholipids (FOSS-foh-LIP-ids) one of the three main classes of dietary lipids. These lipids are similar to triglycerides, but each has a phosphorus-containing acid in place of one of the fatty acids. Phospholipids are present in all cell membranes.

lecithin (LESS-ih-thin) a phospholipid manufactured by the liver and also found in many foods; a major constituent of cell membranes.

sterols (STEER-alls) one of the three main classes of dietary lipids. Sterols have a structure similar to that of cholesterol.

TABLE 5·1 THE USEFULNESS OF FATS	
Fats in the Body	**Fats in Food**
• *Energy stores* Fats are the body's chief form of stored energy.	• *Nutrient* Fats provide essential fatty acids.
• *Muscle fuel* Fats provide most of the energy to fuel muscular work.	• *Energy* Fats provide a concentrated energy source in foods.
• *Emergency reserve* Fats serve as an emergency fuel supply in times of illness and diminished food intake.	• *Transport* Fats carry fat-soluble vitamins A, D, E, and K along with some phytochemicals, and assist in their absorption.
• *Padding* Fats protect the internal organs from shock through fat pads inside the body cavity.	• *Raw materials* Fats provide raw material for making needed products.
• *Insulation* Fats insulate against temperature extremes through a fat layer under the skin.	• *Sensory appeal* Fats contribute to the taste and smell of foods.
• *Cell membranes* Fats form the major material of cell membranes.	• *Appetite* Fats stimulate the appetite.
• *Raw materials* Fats are converted to other compounds, such as hormones, bile, and vitamin D, as needed.	• *Satiety* Fats contribute to feelings of fullness.
	• *Texture* Fats help make foods tender.

FIGURE 5·1 A FAT CELL

Within the fat cell, lipid is stored in a droplet. This droplet can greatly enlarge, and the fat cell membrane will grow to accommodate its swollen contents. More about fat tissue (also called *adipose tissue*) and body functions in Chapter 9.

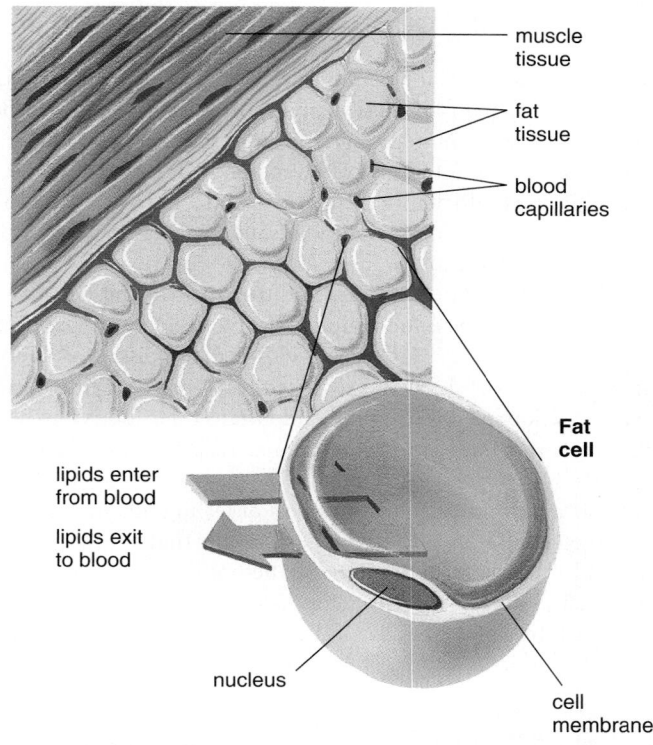

muscle tissue

fat tissue

blood capillaries

Fat cell

lipids enter from blood

lipids exit to blood

nucleus

cell membrane

essential fatty acids fatty acids that the body needs but cannot make in amounts sufficient to meet physiological needs.

You may be wondering why the carbohydrate glucose is not the body's major form of stored energy. As mentioned in Chapter 4, glucose is stored in the form of glycogen. Because glycogen holds a great deal of water, it is quite bulky and heavy, and the body cannot store enough to provide energy for very long. Fats, however, pack tightly together without water and can store much more energy in a small space. Gram for gram, fats provide more than twice the energy of carbohydrate, making fat an efficient storage form of energy. The body fat found on a normal-weight, healthy person contains more than sufficient energy to fuel an entire marathon run or to battle disease should the person become ill and stop eating for a while.

Fat serves many other purposes in the body, too. Pads of fat surrounding the vital internal organs serve as shock absorbers. Thanks to these fat pads, you can ride a horse or a motorcycle for many hours with no serious internal injuries. The fat blanket under the skin also insulates the body from extremes of temperature, thus assisting with internal climate control. Lipids are also important to all the body's cells as part of their surrounding envelopes, the cell membranes.

Some essential nutrients are soluble in fat. They are therefore found mainly in foods that contain fat and are absorbed most efficiently from them. These nutrients are the fat-soluble vitamins: A, D, E, and K. Other essential nutrients, the **essential fatty acids,** constitute parts of the fats themselves. As a later section explains, the essential fatty acids serve as raw materials from which the body makes molecules it requires. Fat also aids in the absorption of some phytochemicals, plant constituents believed to benefit health (see Controversy 2).

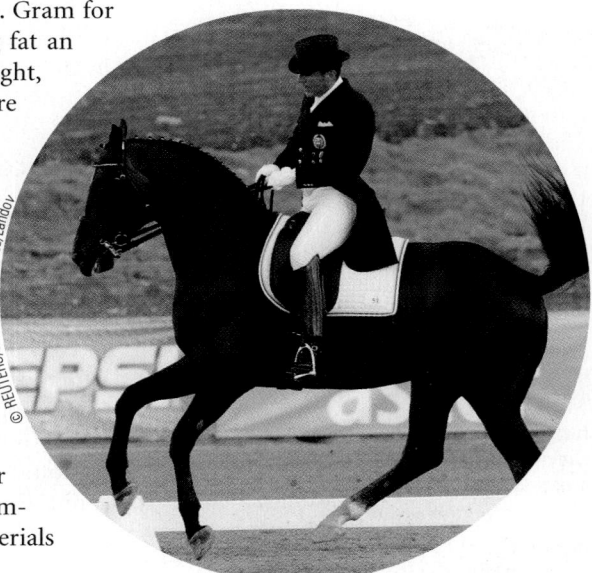

Thanks to internal fat pads, vital organs are cushioned from shock.

FIGURE 5·2 TWO LUNCHES

Both lunches contain the same number of calories, but the fat-rich lunch takes up less space and weighs less.

Carbohydrate-rich lunch
1 low-fat muffin
1 banana
2 oz carrot sticks
8 oz fruit yogurt

calories   = 550
weight (g) = 500

Fat-rich lunch
6 butter-style crackers
1¹/₂ oz American cheese
2 oz trail mix with candy

calories   = 550
weight (g) = 115

■ A reminder from Chapter 1:
1 g carbohydrate = 4 calories.
1 g fat = 9 calories.
1 g protein = 4 calories.

satiety (sat-EYE-uh-tee) the feeling of fullness or satisfaction that people experience after meals.

fatty acids organic acids composed of carbon chains of various lengths. Each fatty acid has an acid end and hydrogens attached to all of the carbon atoms of the chain.

glycerol (GLISS-er-all) an organic compound, three carbons long, of interest here because it serves as the backbone for triglycerides.

▶ KEY POINT   Lipids not only serve as energy reserves but also cushion the vital organs, protect the body from temperature extremes, carry the fat-soluble nutrients and phytochemicals, serve as raw materials, and provide the major component of which cell membranes are made.

Usefulness of Fats in Food

The energy density of fats makes foods rich in fat valuable in many situations. A gram of fat or oil delivers more than twice as many calories as a gram of carbohydrate or protein. A hunter or hiker needs to consume a large amount of food energy to travel long distances or to survive in intensely cold weather. As Figure 5-2 shows, such a person can carry energy most efficiently in fat-rich foods. But for a person who is not expending much energy in physical work, those same high-fat foods may deliver many unneeded calories in only a few bites.

People naturally like high-fat foods. Around the world, as fat becomes less expensive and more available in a given food supply, people seem to choose diets providing greatly increased amounts of fat. Fat carries with it many dissolved compounds that give foods enticing aromas and flavors, such as the aroma of frying bacon or french fries. In fact, when a sick person refuses food, dietitians offer foods flavored with some fat to tempt that person to eat again. Fat also lends tenderness to foods such as meats and baked goods.

Fat also contributes to **satiety,** the satisfaction of feeling full after a meal. The fat of swallowed food triggers a series of physiological events that slow down the emptying of the stomach and promote satiety. Even so, before the sensation of fullness stops them, people can easily overeat on fat-rich foods because the delicious taste of fat stimulates eating and each bite of a fat-rich food delivers many calories. Chapter 9 revisits the topic of appetite and its control.

▶ KEY POINT   Lipids provide more energy per gram than carbohydrate and protein, enhance the aromas and flavors of foods, and contribute to satiety, or a feeling of fullness, after a meal.

A Close Look at Lipids

Each class of lipids—triglycerides, phospholipids, and sterols—possesses unique characteristics. As mentioned, the term *fat* refers to triglycerides, the major form of lipid found in the body and in foods. Triglycerides, in turn, are made of fatty acids and glycerol.

Triglycerides: Fatty Acids and Glycerol

Very few **fatty acids** are found free in the body or in foods; most are incorporated into large, complex compounds: triglycerides. The name almost explains itself: three fatty acids *(tri)* are attached to a molecule of **glycerol** to form a triglyceride molecule (Figure 5-3). Tissues all over the body can easily assemble triglycerides or disassemble them as needed.

Fatty acids can differ from one another in two ways: in chain length and in degree of saturation (explained next). Triglycerides usually include mixtures of various fatty acids. Depending on which fatty acids are incorporated into a triglyceride, the resulting fat will be soft or hard. Triglycerides containing mostly the shorter-chain fatty acids or the more unsaturated ones are softer and melt more readily at lower temperatures. Each species of animal (including people) makes its own characteristic kinds of triglycerides, a function governed by genetics. Fats in the diet, though, can affect the types of triglycerides made because dietary fatty acids are often incorporated into triglycerides in the body.[2] For example, many animals raised for food can be fed diets containing softer or harder triglycerides to give the animals softer or harder fat, whichever consumers demand.

FIGURE 5·3 TRIGLYCERIDE FORMATION

Glycerol, a small, water-soluble carbohydrate derivative, plus three fatty acids, equals a triglyceride.

glycerol

3 fatty acids
of differing
lengths

A triglyceride
formed from
1 glycerol +
3 fatty acids

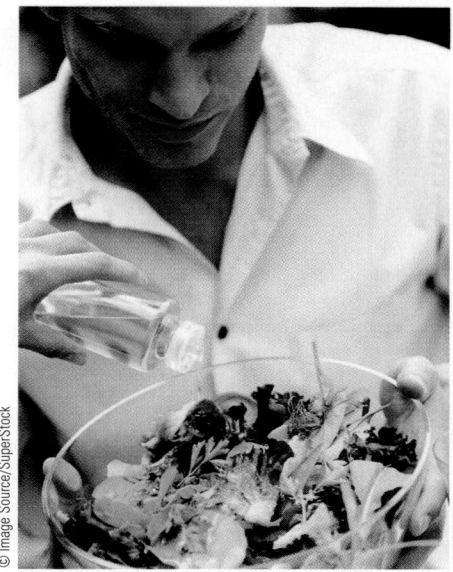

© Image Source/SuperStock

When people add fat or oil to foods, they are
adding triglycerides.

▶ KEY POINT The body combines three fatty acids with one glycerol to make a triglyceride, its storage form of fat. Fatty acids in food influence the composition of fats in the body.

■ Three types of fatty acids: saturated,
monounsaturated, and polyunsaturated.

Saturated versus Unsaturated Fatty Acids

Saturation refers to whether or not a fatty acid chain is holding all of the hydrogen
atoms it can hold. If every available bond from the carbons is holding a hydrogen, the
chain forms a **saturated fatty acid;** it is filled to capacity with hydrogen. The zigzag
structure on the left in Figure 5-4 (p. 144) represents a saturated fatty acid.

Sometimes, especially in the fatty acids of plants and fish, the chain has a place
where hydrogens are missing: an "empty spot" or **point of unsaturation.** A fatty acid
carbon chain that possesses one or more points of unsaturation is an **unsaturated
fatty acid.** With one point of unsaturation, the fatty acid is a **monounsaturated fatty
acid** (see the second structure in Figure 5-4). With two or more points of unsaturation, it is a **polyunsaturated fatty acid,** sometimes abbreviated PUFA (see the third
structure in Figure 5-4; other examples are given later in the chapter).

The degree of saturation of the fatty acids in a fat affects the temperature at which the
fat melts. Generally, the more unsaturated the fatty acids, the more liquid the fat is at room
temperature. In contrast, the more saturated the fatty acids, the firmer the fat. Thus, looking at three fats—beef tallow (a type of beef fat), chicken fat, and safflower oil—beef tallow is the most saturated and the hardest; chicken fat is less saturated and somewhat soft;
and safflower oil, which is the most unsaturated, is a liquid at room temperature. If a
health-care provider recommends limiting **saturated fats** or *trans* **fats** and using
monounsaturated fats or **polyunsaturated fats** instead, you can generally judge by the
hardness of the fats which ones to choose. Figure 5-5 (p. 145) compares the percentages of
saturated, monounsaturated, and polyunsaturated fatty acids in various fats and oils. To
determine whether an oil you use contains saturated fats, place the oil in a clear container
in the refrigerator and watch for cloudiness. The least saturated oils remain the clearest.

Most vegetable and fish oils are rich in polyunsaturates; some vegetable oils, olive oil
and canola oil in particular, are also rich in monounsaturates; animal fats are generally
the most saturated. But you have to know your oils—it is not enough to choose foods
with labels claiming plant oils over those containing animal fats. Some nondairy

saturated fatty acid a fatty acid carrying the
maximum possible number of hydrogen atoms
(having no points of unsaturation).

point of unsaturation a site in a molecule
where the bonding is such that additional
hydrogen atoms can easily be attached.

unsaturated fatty acid a fatty acid that lacks
some hydrogen atoms and has one or more
points of unsaturation. An unsaturated fat is a
triglyceride that contains one or more unsaturated fatty acids.

monounsaturated fatty acid a fatty acid
containing one point of unsaturation.

polyunsaturated fatty acid (PUFA) a fatty
acid with two or more points of unsaturation.

saturated fats triglycerides in which most of
the fatty acids are saturated.

***trans* fats** fats that contain unusual fatty
acids, *trans* fatty acids, formed during processing; a later section provides details.

monounsaturated fats triglycerides in which
most of the fatty acids have one point of
unsaturation (are monounsaturated).

polyunsaturated fats triglycerides in which
most of the fatty acids have two or more
points of unsaturation (are polyunsaturated).

© Quest Photographic, Inc.

Fats melt at different temperatures. The more unsaturated a fat, the more liquid it is at room temperature. The more saturated a fat, the higher the temperature at which it melts.

FIGURE 5·4 THREE TYPES OF FATTY ACIDS

The more carbon atoms in a fatty acid, the longer it is. The more hydrogen atoms attached to those carbons, the more saturated the fatty acid is.

Saturated   **Monounsaturated**   **Polyunsaturated**

Point of unsaturation

Points of unsaturation

whipped dessert toppings use coconut oil in place of cream (butterfat). Coconut oil does come from a plant, but it disobeys the rule that plant oils are less saturated than animal fats; the fatty acids of coconut oil are more saturated than those of cream and seem to add to heart disease risk.[3] Palm oil, a vegetable oil used in food processing, is also highly saturated and has been shown to elevate blood cholesterol.[4] Likewise, shortenings, stick margarine, and commercially fried or baked products may claim to be "all vegetable fat," but much of their fat may be of the harmful saturated or *trans* kind.

Researchers report a benefit to heart health when monounsaturated or polyunsaturated fats replace saturated fat and *trans* fat in the diet. When olive oil, rich in monounsaturated fatty acids, replaces other fats in the diet, it may even offer a degree of protection against heart disease, as evidence from Mediterranean regions suggests. In addition to abundant monounsaturated fatty acids, dark-colored olive oils deliver valuable phytochemicals as well.[5] Canola oil, rich in both monounsaturated and polyunsaturated fatty acids, also supports heart health when replacing saturated fats in the diet.

▶ **KEY POINT**   Fatty acids are energy-rich carbon chains that can be saturated (filled with hydrogens) or monounsaturated (with one point of unsaturation) or polyunsaturated (with more than one point of unsaturation). The degree of saturation of the fatty acids in a fat determines the fat's softness or hardness.

Phospholipids and Sterols

Thus far we have dealt with the largest of the three classes of lipids—the triglycerides and their component fatty acids. The other two classes—phospholipids and sterols— also play important roles in the body.

▥ *Trans* fats are the topic of a later section.

▥ The Controversy section explores the effects of fats in foods on the health of the heart.

FIGURE 5·5 COMPARISON OF DIETARY FATS

Most fats are a mixture of saturated, monounsaturated, and polyunsaturated fatty acids.

Key:

■ Saturated fatty acids	■ Polyunsaturated, omega-6 fatty acids[a]
■ Monounsaturated fatty acids	■ Polyunsaturated, omega-3 fatty acids[a]

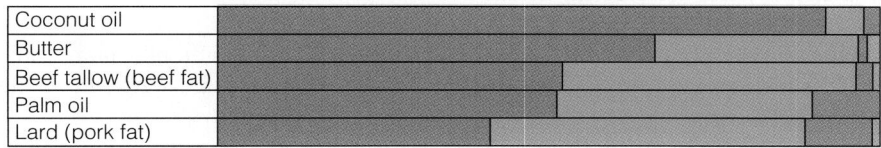

Animal fats and the tropical oils of coconut and palm contain mostly saturated fatty acids.

Coconut oil
Butter
Beef tallow (beef fat)
Palm oil
Lard (pork fat)

Some vegetable oils, such as olive and canola, are rich in monounsaturated fatty acids.

Olive oil
Canola oil
Peanut oil

Many vegetable oils are rich in polyunsaturated fatty acids.

Safflower oil
Sunflower oil
Corn oil
Soybean oil
Cottonseed oil

[a]These families of polyunsaturated fatty acids are explained in a later section.

Phospholipids A phospholipid, like a triglyceride, consists of a molecule of glycerol with fatty acids attached, but it contains two, rather than three, fatty acids. In place of the third is a molecule containing phosphorus, which makes the phospholipid soluble in water, while its fatty acids make it soluble in fat. This versatility permits any phospholipid to play a role in keeping fats dispersed in water—it can serve as an **emulsifier.**

Food processors blend fat with watery ingredients by way of **emulsification.** Some salad dressings separate to form two layers—vinegar on the bottom, oil on the top. Other dressings, such as mayonnaise, are also made from vinegar and oil but never separate. The difference lies in a special ingredient of mayonnaise, the emulsifier lecithin in egg yolks. Lecithin, a phospholipid, blends the vinegar with the oil to form the stable, spreadable mayonnaise.

Lecithin and other phospholipids also play key roles in the structure of cell membranes. Because phospholipids are emulsifiers, they have both water-loving and fat-loving characteristics, which enable them to help fats travel back and forth across the lipid-containing membranes of cells into the watery fluids on both sides. Health-promoting properties, such as the ability to lower blood cholesterol, are sometimes attributed to lecithin, but the people making the claims profit from selling supplements. Lecithin supplements have no special ability to promote health—the body can make all of the lecithin it needs.

Sterols Sterols such as cholesterol are large, complicated molecules consisting of interconnected *rings* of carbon atoms with side chains of carbon, hydrogen, and oxygen attached. Cholesterol serves as the raw material for making another emulsifier, **bile,** which is important to digestion (see the next section for details). Other sterols include vitamin D, which is made from cholesterol, and the familiar steroid hormones, including the sex hormones.

Cholesterol is important in the structure of cell membranes and so is a part of every cell and necessary to the body's functioning. Like lecithin, cholesterol can be made by the body,

emulsifier a substance that mixes with both fat and water and permanently disperses the fat in the water, forming an emulsion.

emulsification the process of mixing lipid with water by adding an emulsifier.

bile an emulsifier made by the liver from cholesterol and stored in the gallbladder. Bile does not digest fat as enzymes do but emulsifies it so that enzymes in the watery fluids may contact it and split the fatty acids from their glycerol for absorption.

so it is not an essential nutrient. Cholesterol also forms the major part of the plaques that narrow the arteries in atherosclerosis, the underlying cause of heart attacks and strokes.

> **KEY POINT** Phospholipids, including lecithin, play key roles in cell membranes; sterols play roles as part of bile, vitamin D, the sex hormones, and other important compounds.

Lipids in the Body

From the moment they enter the body, lipids affect the body's functioning and condition. They also demand special handling, because fat separates from water, and body fluids consist largely of water.

Digestion and Absorption of Fats

A bite of food in the mouth first encounters the enzymes of saliva. One enzyme, produced by the tongue, plays a major role in digesting milk fat in infants but is of little importance to digestion in adults.[6]

After being chewed and swallowed, the food travels to the stomach, where the fat separates from the watery components and floats as a layer on the top. Because fat does not mix with the stomach fluids, little fat digestion takes place in the stomach.

As the stomach contents empty into the small intestine, the digestive system faces a problem: how to thoroughly mix fats, which are now separated and floating, with its own watery fluids. The solution is bile. A bile molecule, made from cholesterol, works because one of its ends attracts and holds fat, while the other end is attracted to and held by water.

By the time fat enters the small intestine, the gallbladder, which stores the liver's output of bile, has contracted and squirted its bile into the intestine. Bile mixes fat particles with watery fluid by emulsifying them (see Figure 5-6), suspending them in the fluid until the fat-digesting enzymes contributed by the pancreas can split them into smaller particles for absorption. To review: first, the digestive system mixes fats with bile-containing digestive juices to emulsify the fats; then, fat-digesting enzymes can break the fats down.

People sometimes wonder how a person without a gallbladder can digest food. The gallbladder is just a storage organ. Without it, the liver still produces bile but delivers it

■ Chapter 3 first described the action of bile and gave details of the digestive system.

FIGURE 5·6   THE ACTION OF BILE IN FAT DIGESTION

Detergents are emulsifiers and work the same way, which is why they are effective in removing grease spots from clothes. Molecule by molecule, the grease is dissolved out of the spot and suspended in the water, where it can be rinsed away.

In the stomach, the fat and watery digestive juices tend to separate. Enzymes are in the water and can't get at the fat.

When fat enters the small intestine, the gallbladder secretes bile. Bile has an affinity for both fat and water, so it can bring the fat into the water.

After emulsification, more fat is exposed to the enzymes, and fat digestion proceeds efficiently.

---

continuously into the small intestine. People who have had their gallbladders removed must initially reduce their fat intakes because they can no longer store bile and release it at mealtimes. As a result, their systems can handle only a little fat at a time.

Once the intestine's contents are emulsified, fat-splitting enzymes act on triglycerides to split fatty acids from their glycerol backbones. Free fatty acids, glycerol, and **monoglycerides** cling together in balls surrounded by bile. At this point, the fats face another watery barrier, the watery layer of mucus that coats the absorptive lining of the digestive tract. Fats must traverse this layer to enter the cells of the digestive tract lining. The solution again depends on bile, this time in the balls of digested lipids. The bile shuttles the lipids across the watery mucus layer to the waiting absorptive cells of the intestinal villi. The cells then extract the lipids. The bile may be absorbed and reused by the body, or it may exit with the feces as was shown in Figure 4-4 of Chapter 4.

The digestive tract absorbs triglycerides from a meal with up to 98 percent efficiency. In other words, little fat is excreted by a healthy system. The process of fat digestion takes time, though, so the more fat taken in at a meal, the slower the digestive system action becomes. The efficient series of events just described is depicted in Figure 5-7 (p. 148).

▶ **KEY POINT**   In the stomach, fats separate from other food components. In the small intestine, bile emulsifies the fats, enzymes digest them, and the intestinal cells absorb them.

Transport of Fats

The smaller products of lipid digestion, glycerol and shorter-chain fatty acids, pass directly through the cells of the intestinal lining into the bloodstream where they travel unassisted to the liver. The larger lipids, however, present a problem for the body. As mentioned, fat floats in water. Without some mechanism to keep it dispersed, large lipid globules would separate out of the watery blood as it circulates around the body, disrupting the blood's normal functions. The solution to this problem lies in an ingenious use of proteins: many fats travel from place to place in the watery blood as passengers in **lipoproteins,** assembled packages of lipid and protein molecules.

The larger digested lipids, monoglycerides and long-chain fatty acids, must form lipoproteins before they can be released into the lymph that leads to the blood. Inside the intestinal cells, they are re-formed into triglycerides and clustered together with proteins and phospholipids to form **chylomicrons** that can safely travel in the watery blood. Chylomicrons form one type of lipoprotein (shown in Figure 5-7); other types receive attention later with regard to their profound effects on health.

▶ **KEY POINT**   Small lipids travel in the bloodstream unassisted. Large lipids are incorporated into chylomicrons for transport in the lymph and blood. Blood and other body fluids are watery, so fats need special transport vehicles—the lipoproteins—to carry them in these fluids.

How Can I Use My Stored Fat for Energy?

Many triglycerides eaten in foods are transported by the chylomicrons to the fat depots—muscles, breasts, the insulating fat layer under the skin, and others—where they are stored by the body's fat cells for later use. When a person's body starts to run out of available fuel from food, it begins to retrieve this stored fat to use for energy. (It also draws on its stored glycogen, as the last chapter described.) Fat cells respond to the call for energy by dismantling stored fat molecules and releasing fat components into the blood. Upon receiving these components, the energy-hungry cells break them down further into small fragments. Finally, each fat fragment is combined with a fragment derived from glucose, and the energy-releasing process continues, liberating energy, carbon dioxide, and water. The way to use more of the energy stored as body

monoglycerides (mon-oh-GLISS-er-ides) products of the digestion of lipids; consist of glycerol molecules with one fatty acid attached (*mono* means "one"; *glyceride* means "a compound of glycerol").

lipoproteins (LYE-poh-PRO-teens, LIH-poh-PRO-teens) clusters of lipids associated with protein, which serve as transport vehicles for lipids in blood and lymph. Major lipoprotein classes are the chylomicrons, the VLDL, the LDL, and the HDL.

chylomicrons (KYE-low-MY-krons) clusters formed when lipids from a meal are combined with carrier proteins in the cells of the intestinal lining. Chylomicrons transport food fats through the watery body fluids to the liver and other tissues.

FIGURE 5·7   THE PROCESS OF LIPID DIGESTION AND ABSORPTION

1 In the mouth and stomach:

Little fat digestion takes place.

2 In the small intestine:

Digestive enzymes accomplish most fat digestion in the small intestine. There, bile emulsifies fat, making it available for enzyme action. The enzymes cleave triglycerides into free fatty acids, glycerol, and monoglycerides.

3 At the intestinal lining:

The parts are absorbed by intestinal villi. Small lipid particles such as glycerol and short-chain fatty acids are small enough to enter directly into the bloodstream.

4 The cells of the intestinal lining convert large lipid fragments, such as monoglycerides and long-chain fatty acids back into triglycerides and combine them with protein, forming chylomicrons (a type of lipoprotein) that travel in the lymph vessels to the bloodstream.

5 In the large intestine:

A small amount of cholesterol trapped in fiber exits with the feces

Note: In this diagram, molecules of fatty acids are shown as large objects, but, in reality, molecules of fatty acids are too small to see even with a powerful microscope, while villi are visible to the naked eye.

fat, then, is to create a greater demand for it in the tissues by decreasing intake of food energy, by increasing the body's expenditure of energy, or both.

Whenever body fat is broken down to provide energy, carbohydrate must be available as well. Without carbohydrate, as described in the last chapter, products of incomplete fat breakdown (ketones) will build up in the blood and urine. Because this process and its consequences are so important in weight control, Chapter 9 describes them in greater detail.

The body can also store excess carbohydrate as fat, but this conversion is not energy efficient. Figure 5-8 illustrates a simplified series of steps from carbohydrate to fat. Before excess glucose can be stored as fat, it must first be broken into tiny fragments and then reassembled into fatty acids, steps that require energy to perform. Fat requires fewer chemical steps before storage. For weight-loss dieters, however, knowing these differences in metabolic costs of energy storage is less important than remembering what common sense tells us: successful weight loss depends on a low energy intake—not the proportion of energy nutrients.[7] Still, the proportions of certain lipids in the diet matter greatly, particularly to the health of the heart, as the next section makes clear.

FIGURE 5·8 GLUCOSE TO FAT

Glucose can be used for energy, or it can be changed into fat and stored.

fatty acid

Glucose is broken down into fragments.

The fragments can provide immediate energy for the tissues.

Energy

Or, if the tissues need no more energy, the fragments can be reassembled, not back to glucose but into fatty acid chains.

▶ **KEY POINT** When low on fuel, the body draws on its stored fat for energy. Carbohydrate is necessary for the complete breakdown of fat.

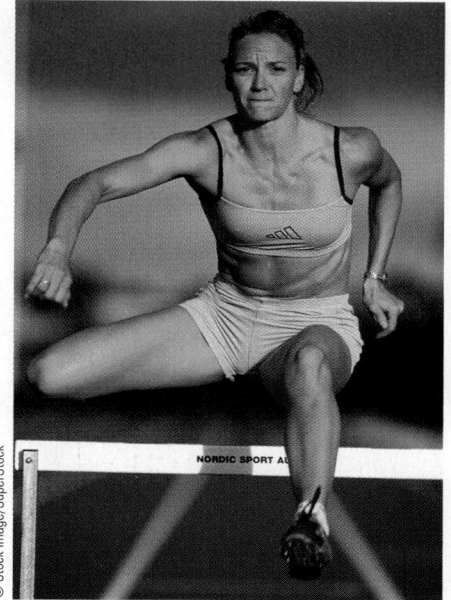

© Stock Image/SuperStock

Body fat supplies much of the fuel that muscles need to do their work.

Dietary Fat, Cholesterol, and Health

High intakes of certain dietary fats are associated with serious diseases. The person who chooses a diet too high in saturated fats or *trans* fats invites the risk of heart and artery disease (CVD), and heart disease is the number-one killer of adults in the United States and Canada. As for cancer, research is mixed, but some studies suggest that the person who eats a high-fat diet may also incur a greater-than-average risk of developing some forms of cancer.[8] Obesity carries serious risks to health, and the high energy density of fatty foods makes it easy for people to exceed their energy needs and so gain unneeded weight. Much research has focused on the links between diet and disease, and Chapter 11 is devoted to these connections. Some points about fats and heart health are presented here because they underlie dietary recommendations concerning fats (see Table 5-2, page 150). More specifically, concerns surround the lipoproteins, their functions, and their roles in heart health.

Lipoproteins and Heart Disease Risk

As previously mentioned, the monoglycerides and long-chain fatty acids liberated from digested food fat must travel in the bloodstream as chylomicrons. These protein and phospholipid clusters act as emulsifiers, attracting both water and fat, to enable their large lipid passengers to travel dispersed in the watery body fluids. The tissues of the body can extract whatever fat they need from chylomicrons passing by in the bloodstream. The remnants are then picked up by the liver, which dismantles them and reuses their parts.

Major Lipoproteins: VLDL, LDL, HDL In addition to the chylomicrons, the body uses three other types of lipoproteins to carry fats:

- **Very-low-density lipoproteins (VLDL),** which carry triglycerides and other lipids made in the liver to the body cells for their use.
- **Low-density lipoproteins (LDL),** which transport cholesterol and other lipids to the tissues. LDL are made from VLDL after they have donated many of their triglycerides to body cells.
- **High-density lipoproteins (HDL),** which are critical in the process of carrying cholesterol away from body cells to the liver for disposal.[9]

The last of two of these lipoproteins, LDL and HDL play major roles with regard to heart health and are the focus of most recommendations made for reducing the risk

■ Here's a trick:
Remember **HDL** is **H**ealthy.
LDL is **L**ess healthy.

very-low-density lipoproteins (VLDL) lipoproteins that transport triglycerides and other lipids from the liver to various tissues in the body.

low-density lipoproteins (LDL) lipoproteins that transport lipids from the liver to other tissues such as muscle and fat; contain a large proportion of cholesterol.

high-density lipoproteins (HDL) lipoproteins that return cholesterol from the tissues to the liver for dismantling and disposal; contain a large proportion of protein.

TABLE 5·2 RECOMMENDATIONS CONCERNING INTAKES OF FATS FOR HEALTHY PEOPLE

1. Total fat[a]

   American Heart Association
   • Limit fat to 30% or less of total energy.

   Dietary Guidelines for Americans 2005
   • Keep total fat intake between 20 to 35 percent of calories from mostly polyunsaturated and monounsaturated fat sources such as fish, nuts, and vegetable oils.
   •Select and prepare foods that are lean, low-fat, or fat-free.

   Dietary Reference Intakes
   • An acceptable range of fat intake is estimated at 20 to 35% of total calories.

   World Health Organization
   • 15 to 30% of total calories from fat.

2. Saturated fat

   American Heart Association
   • Limit saturated fat to less than 10% of total energy.[b]

   Dietary Reference Intakes;[c] Dietary Guidelines for Americans 2005
   • Keep saturated fat intake low, less than 10% of calories, within the context of an adequate diet.

   World Health Organization
   • Less than 10% of total calories from saturated fat.

3. *Trans* fat

   Dietary Guidelines for Americans 2005
   • Keep *trans* fat intake as low as possible.

   World Health Organization
   • Less than 1% of total calories from *trans*-fatty acids.

4. Polyunsaturated fatty acids

   Dietary Reference Intakes[c]
   • Linoleic acid (5 to 10% of total calories):
     17 grams per day for young men.
     12 grams per day for young women.
   • Linolenic acid (0.6 to 1.2% of total calories):
     1.6 grams per day for men.
     1.1 grams per day for women.

   World Health Organization
   • 6 to 10% of total calories from omega-6 polyunsaturated fatty acids.[d]
   • 1 to 2% of total calories from omega-3 polyunsaturated fatty acids.[e]

5. Cholesterol

   American Heart Association, Dietary Guidelines for Americans 2005, and World Health Organization
   • Limit cholesterol to less than 300 milligrams per day.

   Dietary Reference Intakes[c]
   • Minimize cholesterol intake within the context of a healthy diet.

   Canada's Food Guide to Healthy Living is presented in Appendix B.

[a]Includes monounsaturated fatty acids.
[b]The *Dietary Guidelines for Americans,* 2005 use the term *solid fats* to describe sources of saturated fatty acids. Solid fats include milk fat, fats of high-fat meats and cheeses, hard margarines, butter, lard, and shortening.
[c]For DRI values set for other life stages, see the inside front cover. Linoleic and linolenic acids are defined on p. 154.
[d]The fatty acid family that includes linoleic acid (p. 155).
[e]The fatty acid family that includes linolenic acid (p. 155).

of heart disease. Figure 5-9 depicts typical lipoproteins and demonstrates how a lipoprotein's density changes with its lipid and protein contents.

The LDL and HDL Difference   The separate functions of LDL and HDL are worth a moment's attention because they carry important implications for the health of the heart and blood vessels. Both LDL and HDL carry lipids in the blood, but LDL are larger, lighter, and richer in cholesterol; HDL are smaller, denser, and packaged with more protein. To repeat: LDL deliver triglycerides and cholesterol from the liver to the tissues; HDL scavenge excess cholesterol and phospholipids from the tissues for disposal.

Both LDL and HDL carry cholesterol, but elevated LDL concentrations in the blood are a sign of high risk of heart attack, whereas elevated HDL concentrations are associated with a low risk. Thus, some people refer to LDL as "bad" cholesterol and HDL as "good" cholesterol—yet they carry the same kind of cholesterol. The difference to health between LDL and HDL lies in the proportions of lipids they contain and the tasks they perform, not in the *type* of cholesterol they carry.

The Importance of LDL and HDL Cholesterol   The importance of blood cholesterol to heart health cannot be overstated.[*] The blood lipid profile, a medical test mentioned at the beginning of this chapter, tells much about a person's blood concen-

---

[*] *Blood, plasma,* and *serum* all refer to about the same thing; this book uses the term *blood* cholesterol. Plasma is blood with the cells removed; in serum, the clotting factors are also removed. The concentration of cholesterol is not much altered by these treatments.

FIGURE 5·9 LIPOPROTEINS

As the graph shows, the density of a lipoprotein is determined by its lipid-to-protein ratio. All lipoproteins contain protein, cholesterol, phospholipids, and triglycerides in varying amounts. An LDL has a high ratio of lipid to protein (about 80 percent lipid to 20 percent protein) and is especially high in cholesterol. An HDL has more protein relative to its lipid content (about equal parts lipid and protein).

A typical lipoprotein

trations of cholesterol and the lipoproteins that carry it.[10] Blood LDL and HDL cholesterol account for two major risk factors for CVD (listed in the margin). A high blood LDL cholesterol concentration is a predictor of the likelihood of suffering a fatal heart attack or stroke, and the higher the LDL, the earlier the episode is expected to occur. Conversely, high HDL cholesterol signifies a *lower* disease risk.

▶ KEY POINT   The chief lipoproteins are chylomicrons, VLDL, LDL, and HDL. Blood LDL and HDL concentrations are among the major risk factors for heart disease.

What Does *Food* Cholesterol Have to Do with *Blood* Cholesterol?

The answer may be, "Not as much as most people think." Most saturated food fats (and *trans* fats, see later section) raise blood cholesterol more than food *cholesterol* does. When told that cholesterol doesn't matter as much as saturated fat, people may then jump to the wrong conclusion—that blood cholesterol doesn't matter. It does matter. High *blood* cholesterol is an indicator of risk for CVD. The main dietary factors associated with elevated blood cholesterol are high saturated fat and *trans* fat intakes. LDL cholesterol indicates a risk of heart disease because the LDL are carrying cholesterol, made mostly from saturated fat in the diet, to the body tissues to be deposited there.

Dietary cholesterol makes a smaller but still significant contribution to elevated blood cholesterol.[11] The margin lists the top five contributors to dietary cholesterol in the U.S. diet. Genetic inheritance modifies everyone's ability to handle dietary cholesterol somewhat. Many people (about 60 percent) exhibit little increase in their blood cholesterol even with a high dietary intake.[12] A smaller percentage responds to the same diet with greatly increased blood cholesterol. A few individuals have inherited a total inability to clear from their blood the cholesterol they have eaten and absorbed.

People with a genetic tendency toward high blood cholesterol must strictly limit fats and refrain from eating foods rich in cholesterol. For most others, a limited amount of eggs, liver, and other cholesterol-containing foods poses no threat of incurring high

The more of these factors present in a person's life, the more urgent the need for changes in diet and other controllable factors to reduce heart disease risk:

- High blood LDL cholesterol.
- Low blood HDL cholesterol.
- High blood pressure (hypertension).
- Diabetes (insulin resistance).
- Obesity.
- Physical inactivity.
- Cigarette smoking.
- An "atherogenic" diet (high in saturated fats, including *trans* fats, and low in vegetables, legumes, fruit, and whole grains).

Family history, older age, and male gender are risk factors that cannot be changed.

These five foods contribute about 70% of the food cholesterol in the U.S. diet:

- Eggs, 30%.
- Beef, 16%.
- Poultry, 12%.
- Cheese, 6%.
- Milk, 5%.

SOURCE: P. A. Cotton and coauthors, Dietary sources of nutrients among U.S. adults, 1994–1996, *Journal of the American Dietetic Association* 104 (2004): 921–930.

■ Standards for blood lipids are found in Figure 11-6 of Chapter 11.

blood cholesterol because the body slows its cholesterol synthesis when the diet provides greater amounts. Moderation, not elimination, is key for most people as far as cholesterol-containing foods are concerned.

▶ KEY POINT   Elevated blood cholesterol is a risk factor for cardiovascular disease. Among major dietary factors that raise blood cholesterol, saturated fat and *trans* fat intakes are the most influential. Dietary cholesterol raises blood cholesterol to a lesser degree.

Lowering LDL Cholesterol

■ Maximum saturated fat intakes set by the *Dietary Guidelines for Americans*:

   1,600-calorie diet: 18 grams
   2,000-calorie diet: 20 grams
   2,200-calorie diet: 24 grams
   2,500-calorie diet: 25 grams
   2,800-calorie diet: 31 grams

To repeat, dietary saturated fat and *trans* fat are potent in triggering a rise in LDL cholesterol in the blood.[13] A dietary tactic often effective against high blood cholesterol is to trim the fat, and especially the saturated fat and *trans* fat, from foods. As could be expected, the success of this tactic is also affected by genetics; some people respond better than others to dietary means of controlling blood lipids.[14]

The photos of Figure 5-10 show that food trimmed of fat is also trimmed of much of its saturated fat and energy. A pork chop trimmed of its border of fat loses almost 70 percent of its saturated fat and 220 calories. A plain baked potato has no saturated fat and about 40 percent of the calories of one with butter and sour cream. Choosing fat-free milk over whole milk provides large savings of fat, saturated fat, and calories. The single most effective step you can take to reduce a food's potential for elevating blood cholesterol is to eliminate as much of its saturated fat and *trans* fat as possible. Figure 5-11 (p. 154) identifies top sources of saturated fat in the U.S. diet. *Trans* fat sources are identified in a later figure.

As for HDL cholesterol, dietary measures are generally ineffective at significantly raising its concentrations.[15] Instead, regular physical activity defends against heart disease partly because it effectively raises HDL, as the Think Fitness feature (p. 154) points out.

■ Antioxidant nutrients are topics of Chapters 7 and 8; phytochemicals are discussed in Controversy 2; Chapter 11 provides perspective on the role of diet in heart disease prevention.

An important detail about LDL concerns its susceptibility to damage by **oxidation.** Evidence points to oxidation of the lipid part of LDL as injurious to the arteries of the heart. Adequate intakes of **dietary antioxidants,** such as vitamin C, vitamin E, the mineral selenium, and antioxidant phytochemicals, may slow LDL oxidation.

▶ KEY POINT   Trimming fat from food trims calories and, often, saturated fat and *trans* fat as well.

■ Chapter 11 presents many details about lowering LDL cholesterol, and the Controversy of this chapter helps distinguish between fats that may be harmful and those compatible with good health.

Recommendations Applied

Some health authorities recommend that all adults take steps to reduce their LDL cholesterol; others say that only those medically identified as having an elevated risk for heart disease should do so. In any case, most people are wise to choose a diet that provides 20 to 35 percent of its calories from fat and that keeps saturated fat and *trans* fat as low as possible (less than 10 percent and 1 percent of calories, respectively).[16] Controversy 5 suggests substituting monounsaturated or polyunsaturated fats for artery-clogging saturated fats. No beneficial change in blood lipids occurs when monounsaturated or polyunsaturated fat is *added* to a diet rich in saturated or *trans* fat, however. The best diet for heart health is also rich in fruits, vegetables, nuts, and whole grains that offer many health advantages by supplying abundant nutrients and antioxidants along with beneficial fiber.

oxidation interaction of a compound with oxygen; in this case, a damaging effect by a chemically reactive form of oxygen (Chapter 7 provides details).

dietary antioxidant (anti-OX-ih-dant) a substance in food that significantly decreases the damaging effects of reactive compounds, such as reactive forms of oxygen and nitrogen on tissue functioning (*anti* means "against"; *oxy* means "oxygen").

What about cholesterol intake? The best course is to proceed with caution. Eggs, shellfish, liver, and other cholesterol-containing foods are nutritious. Cholesterol differs from salt and added fats and sugar in this respect: it cannot be omitted from the diet without omitting nutritious foods. As mentioned, cholesterol is not an essential

FIGURE 5·10 FOOD FAT, SATURATED FAT, AND CALORIES

You can find much of the saturated fat and calories in food by looking for the fat. The fats of meat, milk, and added fats are the main contributors of saturated fat to the U.S. diet. When you trim fat, you trim calories and often saturated fat.

Nutrition Facts
Amount Per Serving

Pork chop (5 ounces) with ½ inch of fat

Calories 450	Calories from Fat 315
	% Daily Value*
Total Fat 35g	**54%**
Saturated Fat 13g	**65%**

Potato (5 ounces) with 1 tablespoon butter and 1 tablespoon sour cream

Calories 400	Calories from Fat 250
	% Daily Value*
Total Fat 28g	**43%**
Saturated Fat 18g	**90%**

Whole milk (1 cup)

Calories 150	Calories from Fat 70
	% Daily Value*
Total Fat 8g	**12%**
Saturated Fat 5g	**25%**

Pork chop (4 ounces) with fat trimmed off

Calories 230	Calories from Fat 100
	% Daily Value*
Total Fat 11g	**17%**
Saturated Fat 4g	**20%**

Plain potato (5 ounces)

Calories 150	Calories from Fat 0
	% Daily Value*
Total Fat 0g	**0%**
Saturated Fat 0g	**0%**

Fat-free milk (1 cup)

Calories 90	Calories from Fat 0
	% Daily Value*
Total Fat 0g	**0%**
Saturated Fat 0g	**0%**

SOURCE: Data from ESHA Research, The Food Processor Nutrition and Fitness Software, version 8.3, 2004.

nutrient—the body makes plenty for itself—but the foods rich in cholesterol are often also rich in nutrients. A caution is in order, however: many high-cholesterol foods, such as cheeseburgers, are also high in saturated fat.

▶ KEY POINT   Dietary measures to lower LDL in the blood involve reducing saturated fat and *trans* fat and substituting monounsaturated and polyunsaturated fats. Cholesterol-containing foods are nutritious and are best used in moderation by most people.

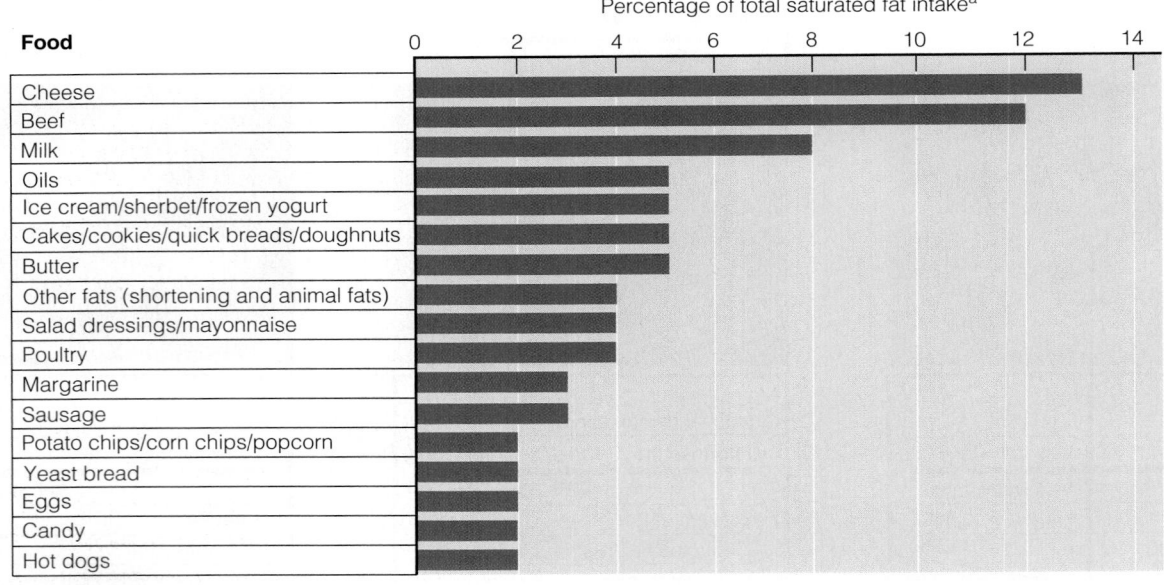

FIGURE 5·11 TOP CONTRIBUTORS OF SATURATED FATS TO THE U.S. DIET

These foods supply about 80 percent of the saturated fat in the U.S. diet. The remainder comes from foods supplying less than 2 percent each, such as cold cuts, pork, cream, bacon, ham, nuts and seeds, pies and cobblers, and nondairy creamers and toppings. Note that fruits, grains, and vegetables are insignificant sources, unless saturated fats are added during processing or preparation.

[a]Rounded values

SOURCE: Data from P. A. Cotton and coauthors, Dietary sources of nutrients among U.S. adults, 1994–1996, *Journal of the American Dietetic Association* 104 (2004): 921–930.

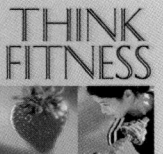

THINK FITNESS — Why Exercise the Body for the Health of the Heart?

EVERY LEADING AUTHORITY recommends physical activity to promote and maintain the health of the heart. The blood, arteries, heart, and other body tissues respond to exercise in these ways:

- Blood HDL concentration increases, shifting blood lipids in a healthy direction.
- The circulation improves, easing the delivery of blood to the heart.
- A larger volume of blood is pumped with each heartbeat, reducing the heart's workload.
- The body grows leaner, reducing overall risk of cardiovascular disease.

Essential Polyunsaturated Fatty Acids

The human body needs fatty acids, and it can use carbohydrate, fat, or protein to synthesize nearly all of them. Two are well-known exceptions: **linoleic acid** and **linolenic acid.** Body cells cannot make these two polyunsaturated fatty acids from scratch, nor can the cells convert one to the other. Linoleic and linolenic acids must be supplied by the diet and are therefore essential nutrients.

The essential fatty acids serve many functions in the body. They serve as raw materials from which the body makes substances known as **eicosanoids** that act somewhat like hormones, affecting a wide range of diverse body functions, such as muscle relaxation and muscle contraction, blood vessel dilation and blood vessel constriction,

linoleic (lin-oh-LAY-ic) **acid** and **linolenic** (lin-oh-LEN-ic) **acid** polyunsaturated fatty acids that are essential nutrients for human beings. The full name of linolenic acid is *alpha-linolenic acid.*

eicosanoids (eye-COSS-ah-noyds) biologically active compounds that regulate body functions.

blood clot formation, blood lipid regulation, and immune response to injury and infection including fever, inflammation, and pain.[17] A familiar drug, aspirin, relieves fever, inflammation, and pain by slowing the synthesis of these eicosanoids.

Table 5-3 summarizes the many roles of the essential polyunsaturated fatty acids, and new functions continue to emerge. So important are these two fatty acids that the DRI committee has set recommended intake levels for them (see the margin).

Deficiencies of Essential Fatty Acids

A deficiency of an essential fatty acid in the diet leads to observable changes in cells, some more subtle than others. When the diet is deficient in *all* of the polyunsaturated fatty acids, symptoms of reproductive failure, skin abnormalities, and kidney and liver disorders appear. In infants, growth is retarded and vision is impaired. The body stores some essential fatty acids, so extreme deficiency disorders are seldom seen except when intentionally induced in research or on rare occasions when inadequate diets have been provided to infants or hospital patients by mistake.

The DRI recommended intakes reflect average intakes of healthy people in the United States because deficiencies of essential polyunsaturated fatty acids severe enough to cause symptoms among healthy adults in this country are unknown. The story doesn't end there, however. Research has established that a higher intake of certain fatty acids, the omega-3 fatty acids, can often improve the health of the heart.

Omega-6 and Omega-3 Fatty Acid Families

Linoleic acid is the "parent" member of the **omega-6 fatty acid** family, so named for the chemical structure of these compounds. Given dietary linoleic acid, the body can produce other needed members of the omega-6 family. One of these is **arachidonic acid,** notable for its role as a starting material from which a number of eicosanoids are made.

Linolenic acid is the parent member of the **omega-3 fatty acid** family. Given dietary linolenic acid, the body can make other members of the omega-3 series. Two family members of great interest to researchers in the field of heart health are **EPA** and **DHA.**[18] The body makes only limited amounts of these omega-3 fatty acids, but they are found abundantly in the oils of certain fish.

Years ago, someone thought to ask why the native peoples of Greenland and Alaska, who eat a diet very high in fat, have such low rates of death from heart disease.[19] The trail led to the abundant fish and other marine life in their diets, then to the oils in fish, and finally to EPA and DHA in fish oils.[20] Furthermore, as younger generations of native peoples of the north abandon traditional marine-based diets for more modern foodways, their rates of high blood pressure, elevated blood lipids, diabetes, and obesity soar.[21]

Although not every study supports a benefit from fish and fish oil, results from many population studies and controlled clinical trials support a recommendation to eat fish.[22] It seems safe to say that a diet that includes two meals of fatty fish each week can reduce deaths and illness from heart disease, especially in people who have already suffered a heart attack. It also seems clear that fish is more beneficial than supplements of fish oil.[23]

Current research suggests that linolenic acid, the chemical parent of the omega-3 family, may benefit heart health as well.[24] Evidence is also mounting to suggest that omega-3 fatty acids may support immunity and inhibit development of certain cancers.[25] Potential mechanisms for beneficial effects of these remarkable lipids are listed in Table 5-4 (p. 156).

Recommendations for Omega-3 Fatty Acid Intake

For healthy people, a normal balanced diet that includes grains, seeds, nuts, leafy vegetables, oils, and fish supplies all the needed forms of fatty acids in abundance and prevents deficiencies. Including fatty fish in a meal two or three times a week, for a total of about 12 to 14 ounces of fish per week, can help to achieve a healthy balance

---

TABLE 5·3 FUNCTIONS OF THE ESSENTIAL FATTY ACIDS

These roles for the essential fatty acids are known, but others are under investigation.

Linoleic and linolenic acid:
- Provide raw material for eicosanoids.
- Serve as structural and functional parts of cell membranes.
- Contribute lipids to the brain and nerves.
- Promote normal growth and vision.
- Assist in gene regulation.
- Maintain outer structures of the skin, thus protecting against water loss.
- Help regulate genetic activities affecting metabolism.
- Support immune cell functions.

■ Dietary Reference Intakes:

Linoleic acid (5 to 10% of total calories):
- 17 g per day for young men.
- 12 g per day for young women.

Linolenic acid (0.6 to 1.2% of total calories):
- 1.6 g per day for men.
- 1.1 g per day for women.

For other life stages, see the inside front cover.

---

omega-6 fatty acid a polyunsaturated fatty acid with its endmost double bond six carbons from the end of the carbon chain. Linoleic acid is an example.

arachidonic (ah-RACK-ih-DON-ik) **acid** an omega-6 fatty acid derived from linoleic acid.

omega-3 fatty acid a polyunsaturated fatty acid with its endmost double bond three carbons from the end of the carbon chain. Linolenic acid is an example.

EPA, DHA eicosapentaenoic (EYE-cossa-PENTA-ee-NO-ick) acid, docosahexaenoic (DOE-cossa-HEXA-ee-NO-ick) acid; omega-3 fatty acids made from linolenic acid in the tissues of fish.

Fish is a good source of omega-3 fatty acids.

TABLE 5·4 POTENTIAL HEALTH BENEFITS OF FISH OILS

Research suggests these benefits from fish or fish oil as part of a healthy diet:

Against heart disease (supported by most studies):

- A shift toward omega-3 eicosanoids by reducing production of omega-6 eicosanoids. This shift may reduce abnormal blood clotting, help sustain more regular heartbeats, and reduce inflammation of many body tissues, including the arteries of the heart.[a]
- Reduced blood triglycerides (in some studies, fish oil supplements elevated blood LDL cholesterol, an opposing, detrimental outcome).[b]
- Retarded hardening of the arteries (atherosclerosis).[c]
- Relaxation of blood vessels, mildly reducing blood pressure.[d]

In infant growth and development (well researched and accepted):

- Normal brain development in infants. DHA concentrates in the brain's cortex, the conscious thinking part.[e]
- Normal vision development in infants. DHA helps to form the eye's retina, the seat of normal vision.

Against cancer (early research promising, but requires further investigation):

- Eicosanoid shift that may oppose cancer.[f]
- Altered genetic activities that affect cell metabolism, possibly inhibiting cancer development.
- Other potential effects that inhibit certain cancers.[g]

[a]J. N. Din, D. E. Newby, and A. D. Flapan, Omega 3 fatty acids and cardiovascular disease—Fishing for a natural treatment, *British Medical Journal* 328 (2004): 30–35; H. Tapiero and coauthors, Polyunsaturated acids (PUFA) and eicosanoids in human health and pathologies, *Biomedicine and Pharmacotherapy* 56 (2002): 215–222; I. Hiroyasu and coauthors, Intake of fish and omega-3 fatty acids and risk of stroke in women, *Journal of the American Medical Association* 285 (2001): 304–312; R. De Caterina, J. K. Liao, and P. Libby, Fatty acid modulation of endothelial activation, *American Journal of Clinical Nutrition* 71 (2002): 213S–223S.

[b]M. Laidlaw and F. J. Holub, Effects of supplementation with fish oil–derived n-3 fatty acids and gamma-linolenic acid on circulating plasma lipids and fatty acid profiles in women, *American Journal of Clinical Nutrition* 77 (2003): 37–42; T. A. Mori and L. J. Beilin, Long-chain omega 3 fatty acids, blood lipids and cardiovascular risk reduction, *Current Opinions in Lipidology* 12 (2001): 11–17.

[c]R. A. Christon, Mechanisms of action of dietary fatty acids in regulating the activation of vascular endothelial cells during atherogenesis, *Nutrition Reviews* 61 (2003): 272–279.

[d]M. M. Engler and coauthors, Effects of docosahexaenoic acid on vascular pathology and reactivity in hypertension, *Experimental Biology and Medicine* 228 (2003): 229–307; T. A. Mori and coauthors, Differential effects of eisosapentanoic acid and docosahexaenoic acid on vascular reactivity of the forearm microcirculation in hyperlipidemic, overweight men, *Circulation* 102 (2002): 1264–1269.

[e]R. Uauy and P. Mena, Lipids and neurodevelopment, *Nutrition Reviews* 59 (2001): S34–S48.

[f]S. C. Larsson and coauthors, Dietary long-chain n-3 fatty acids for the prevention of cancer: A review of potential mechanisms, *American Journal of Clinical Nutrition* 79 (2004): 935–945.

[g]M. F. Leitzman and coauthors, Dietary intake of n-3 and n-6 fatty acids and the risk of prostate cancer, *American Journal of Clinical Nutrition* 80 (2004): 204–216.

between omega-3 and omega-6 fatty acid intakes needed for health.[26] To improve the balance further, foods high in omega-6 fatty acids, such as most vegetable oils and other fats, should be consumed in moderation.

Greater heart benefits can be expected when fish is grilled, baked, or broiled, partly because the varieties prepared this way often contain more EPA and DHA than species used for fried fish in fast-food restaurants and frozen products. Additionally, benefits result from avoiding commercial frying fats, which are often laden with saturated fat and *trans* fat. Further benefits arise when fish displaces high-fat meats or other foods rich in saturated fats in several meals each week.[27] Table 5-5 lists sources of both omega-6 and omega-3 fatty acids.

Fish Oil Supplements

Taking fish oil supplements is not recommended although many claims are made for their power to cure diseases. The Food and Drug Administration (FDA) does not permit labels to claim that fish oil supplements can prevent or cure diseases, but it does allow the claim that research is suggestive but inconclusive regarding heart disease.

TABLE 5·5	FOOD SOURCES OF OMEGA-6 AND OMEGA-3 FATTY ACIDS
Omega-6	
Linoleic acid	Leafy vegetables, seeds, nuts, grains, vegetable oils (corn, cottonseed, safflower, sesame, soybean, sunflower), poultry fat
Omega-3	
Linolenic acid[a]	Oils (canola, flaxseed, soybean, walnut, wheat germ; liquid or soft margarine made from canola or soybean oil) Nuts and seeds (butternuts, flaxseeds, walnuts, soybeans) Vegetables (soybeans)
EPA and DHA	Human milk Fatty coldwater fish[b] (mackerel, salmon, bluefish, mullet, sablefish, menhaden, anchovy, herring, lake trout, sardines, tuna)

[a] Alpha-linolenic acid. Also found in the seed oil of the herb *evening primrose*.
[b] All of these fish except tuna provide at least 1 gram of omega-3 fatty acids in 100 grams of fish (3.5 ounces); the fish oil content of each species varies with the season and site of harvest. Tuna provides fewer omega-3 fatty acids, but because it is commonly consumed, its contribution can be significant.

Supplements of fish oil may raise LDL cholesterol and may not be the best choice for a number of reasons. Most important among them, high intakes of omega-3 polyunsaturated fatty acids may increase bleeding time, interfere with wound healing, and suppress immune function.[28] Excessive amounts of either omega-6 or omega-3 fatty acids can interfere with normal functions that depend on a proper balance between the two.[29] Supplements also lack the *other* beneficial nutrients that fish provides, such as the minerals iodine and selenium and fish protein.

Yet another drawback is that fish oil supplements are made from fish skins and livers, which may have accumulated toxic concentrations of pesticides, heavy metals such as mercury, and other industrial contaminants. Unless the oils are refined to eliminate them, such contaminants can become further concentrated in the pills.[30]

In addition to contamination, fish oil naturally contains high levels of the two most potentially toxic vitamins, A and D. Lastly, supplements of fish oil are expensive. Considering these known drawbacks and that so little is known about the long-term effects of fish oil supplements, only one reasonable conclusion can be drawn: taking them is risky.

Is Fish Safe to Eat?

On learning of the contamination of fish oils, consumers may wonder whether fish is safe to eat. The answer is a qualified yes. For most healthy people, most ocean fish is safe to consume several times a week; for freshwater species in some areas, consumption should be held to government limits. Also, choosing from a variety of fish is a good idea to minimize exposure to any single toxin that may accumulate in a favored species. With regard to mercury, the margin lists the most heavily contaminated species and those that are lower in mercury.

Some people face greater risks. Pregnant women and children are more sensitive to contaminants than others, but even they can benefit from safer fish varieties within recommended limits (see Chapter 13 for details). People with existing heart disease are a special concern because mercury may worsen heart disease.[31] While researchers work to establish precisely how much mercury poses a hazard, consumers with heart disease should choose the fish varieties lowest in mercury to receive fish benefits while minimizing risk to health.

■ Chapter 12 provides more information on the contaminants from human activities that end up in the earth's oceans and lakes.

- Species most heavily contaminated with mercury: shark, swordfish, king mackerel, fresh tuna steaks, and tilefish.
- Lower in mercury: shrimp, canned light tuna, salmon, pollock, and catfish. Canned albacore ("white") tuna, contains more mercury than light tuna.

▌▌ Government limits on freshwater fish intakes are in Chapter 12, page 458.

Most people can safely consume fish to receive its benefits, but no such claim can be made for fish oil supplements. Better to go to the source for fish oils: eat safer varieties of fish.

> ▶ KEY POINT   Two polyunsaturated fatty acids, linoleic acid (an omega-6 fatty acid) and linolenic acid (an omega-3 fatty acid), are essential nutrients used to make substances that perform many important functions. The omega-6 family includes linoleic acid and arachidonic acid. The omega-3 family includes linolenic acid, EPA, and DHA. The principal food source of EPA and DHA is fish, but some species have become contaminated with environmental pollutants.

The Effects of Processing on Unsaturated Fats

Vegetable oils make up most of the added fat in the U.S. diet because fast-food chains use them for frying, food manufacturers add them to processed foods, and consumers tend to choose margarine over butter.[32] Consumers of vegetable oils may feel safe in choosing them because they are generally less saturated than animal fats. If consumers choose a liquid oil, they may be justified in feeling secure. If the choice is a processed food, however, their security may be questionable, especially if the word *hydrogenated* appears on the label's ingredient list.

Baked goods often contain hydrogenated fats.

What Is "Hydrogenated Vegetable Oil," and What's It Doing in My Chocolate Chip Cookies?

When manufacturers process foods, they often alter the fatty acids in the fat (triglycerides) the foods contain through a process called **hydrogenation.** Hydrogenation of fats makes them stay fresher longer and also changes their physical properties.

Points of unsaturation in fatty acids are weak spots that are vulnerable to attack by oxygen. Oxidative damage is not confined to fats within body tissues but occurs anywhere oxygen mixes with fats. When the unsaturated points in the oils of food are oxidized, the oils become rancid and the food tastes "off." This is why cooking oils should be stored in tightly covered containers that exclude air. If stored for long periods, they need refrigeration to retard oxidation.

One way to prevent spoilage of unsaturated fats and also to make them harder and more stable when heated to high temperatures is to change their fatty acids chemically by hydrogenation, as shown on the left side of Figure 5-12. When food producers want to use a polyunsaturated oil such as corn oil to make a spreadable margarine, for example, they hydrogenate it by forcing hydrogen into the liquid oil. Some of the unsaturated fatty acids become more saturated as they accept the hydrogen, and the oil hardens. The resulting product is more saturated and more spreadable than the original oil. It is also more resistant to damage from oxidation or breakdown from high cooking temperatures. Hydrogenated oil has a high **smoking point,** so it is suitable for purposes such as frying.

Hydrogenated oils are thus easy to handle, easy to spread, and store well. Makers of peanut butter often replace a small quantity of the liquid oil from the ground peanuts with hydrogenated vegetable oils to create a creamy paste that does not separate into layers of oil and peanuts as the "old-fashioned" types do. Neither type of peanut butter is high in saturated fat, however.

Once fully hydrogenated, oils lose their unsaturated character and the health benefits that go with it. Hydrogenation may affect not only the fatty acids in oils but also vitamins, such as vitamin K, decreasing their activity in the body.[33] If you, the consumer, are looking for health benefits from polyunsaturated oils, hydrogenated oils such as those in shortening or stick margarine will not meet your need.

hydrogenation (high-dro-gen-AY-shun) the process of adding hydrogen to unsaturated fatty acids to make fat more solid and resistant to the chemical change of oxidation.

smoking point the temperature at which fat gives off an acrid blue gas.

FIGURE 5·12 HYDROGENATION YIELDS BOTH SATURATED AND *TRANS* FATTY ACIDS

Unsaturated fatty acid
Points of unsaturation are places on fatty acid chains where hydrogen is missing. The bonds that would normally be occupied by hydrogen in a saturated fatty acid are shared, reluctantly, as a double bond between two carbons that both carry a slightly negative charge.

Usual shape[a]

Point of unsaturation

Trans shape

Hydrogenated fatty acid (now fully saturated)
When a positively charged hydrogen is made available to an unsaturated bond, it readily accepts the hydrogen and, in the process, becomes saturated. The fatty acid no longer has a point of unsaturation.

Trans fatty acid
The hydrogenation process also produces some *trans* fatty acids. The *trans* fatty acid retains its double bond but takes a twist instead of becoming fully saturated. It resembles a saturated fatty acid both in shape and in its effects on health.

[a] The usual shape of the double bond structure is known as a cis (pronounced *sis*) formation.

An alternative to hydrogenation is to add a chemical preservative that will compete for oxygen and thus protect the oil. The additives are antioxidants, and they work by reacting with oxygen before it can do damage. Examples are the additives BHA and BHT[*] listed on snack food labels. Another alternative, already mentioned, is to keep the product refrigerated.

■ A famous antioxidant, Vitamin E, occurs naturally in foods; see Chapter 7.

▶ **KEY POINT** Vegetable oils become more saturated when they are hydrogenated. Hydrogenated fats resist rancidity better, are firmer textured, and have a higher smoking point than unsaturated oils, but they also lose the health benefits of unsaturated oils.

What Are *Trans* Fatty Acids, and Are They Harmful?

When polyunsaturated oils are hardened by hydrogenation, some of the unsaturated fatty acids end up changing their shapes instead of becoming saturated (look at the right side of Figure 5-12). This change in chemical structure creates unusual unsaturated fatty acids that are similar in shape to saturated fatty acids. They are not made by the body but occur naturally in tiny amounts mainly in dairy foods and beef. These changed fatty acids, or ***trans* fatty acids,** affect the body's health.

Consuming fats with *trans* fatty acids, often called *trans* fats, poses a risk to the health of the heart and arteries by raising LDL and lowering HDL cholesterol, and by producing inflammation.[34] When processing changes essential fatty acids into their *trans* counterparts, the consumer derives none of the original health benefits of the oil

***trans* fatty acids** fatty acids with unusual shapes that can arise when polyunsaturated oils are hydrogenated.

[*] BHA and BHT are butylated hydroxyanisole and butylated hydroxytoluene.

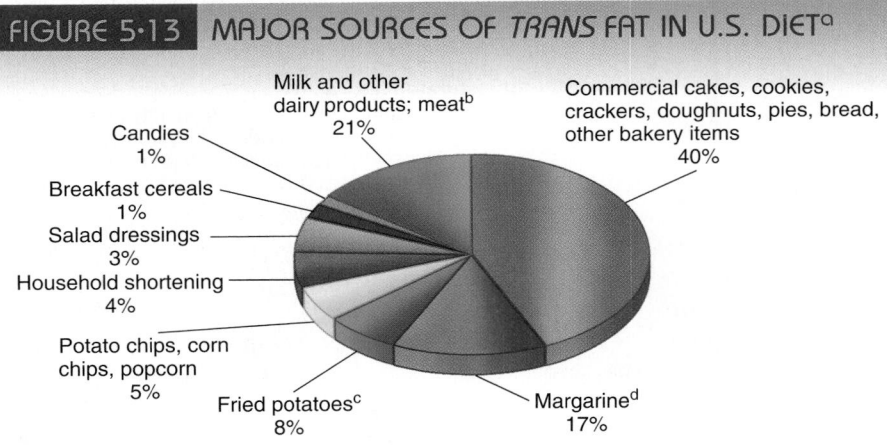

FIGURE 5·13 MAJOR SOURCES OF *TRANS* FAT IN U.S. DIET^a

Milk and other dairy products; meat^b
21%

Candies
1%

Breakfast cereals
1%

Salad dressings
3%

Household shortening
4%

Potato chips, corn chips, popcorn
5%

Fried potatoes^c
8%

Commercial cakes, cookies, crackers, doughnuts, pies, bread, other bakery items
40%

Margarine^d
17%

^aMany manufacturers are currently reformulating products to reduce *trans* fats. Other fats used to replace them may be saturated or unsaturated—read the labels.
^bMostly from commercial fried chicken or fried fish products.
^cMostly commercial french fries, hashbrowns, and "tots."
^dExcept liquids, sprays, and some soft margarines.
SOURCE: Revealing *trans* fats, *FDA Consumer,* September/October 2003, p. 24.

and runs risks similar to those posed by saturated fats. The DRI committee concludes that people should consume as little *trans* fat as possible.

A diet high in *total* fat and especially in saturated fat may be associated with cancer susceptibility, and *trans* fats contribute to the total fat intake.[35] To date no evidence suggests that *trans* fatty acids by themselves play a specific role in promoting or causing cancer.

Popular reports have alerted consumers to the *trans* fat in margarine. While some margarines, especially the solid stick varieties, can contain almost half their fat as *trans* fat, many other popular foods contribute a greater percentage of *trans* fat to the diet (see Figure 5-13). In fact, consumers are now eating more *trans* fats than ever before, amounting to about 3 percent of daily calories, and they are eating these *trans* fats in the form of processed foods.[36] Fast foods, chips, baked goods, and other commercially prepared foods are high in fats containing up to 50 percent *trans* fatty acids. Fast-food chains currently fry foods in hydrogenated vegetable oil that contains abundant *trans* fatty acids, although healthier commercial fats are under development.

As of 2006, all food labels must list grams of *trans* fat in foods to help consumers make informed choices.[37] Replacing both saturated and *trans* fats with monounsaturated and polyunsaturated fats may be the wisest strategy in preventing heart disease.[38] To this end, many manufacturers are reformulating foods to reduce their contents of harmful *trans* fats. At the same time, food scientists are perfecting fat replacers intended to eliminate added fats altogether (described in the Consumer Corner, p. 162).

▶ KEY POINT The process of hydrogenation also creates *trans* fatty acids. *Trans* fats act somewhat like saturated fats in the body.

Fat in the Diet

The remainder of this chapter and the Controversy show you how to choose fats wisely with the goals of providing optimal health and pleasure in eating. As you read, notice which foods offer unsaturated fat and which offer saturated fat and *trans* fat. Also, remember always that fat delivers many calories per bite of food, and limiting energy intake may be important to maintaining good health.[39] Your choices can make a difference in the unseen condition of your arteries.

▥ This chapter's Controversy takes up the issue of whether butter or margarine is best for heart health.

▥ The words *hydrogenated vegetable oil* or *shortening* in an ingredients list indicate *trans* fatty acids in the product.

Remember, too, that some fat is necessary for health. People who take fat recommendations to an extreme and try to eliminate all traces of fat from food do so at their peril. Most people need about 20 percent of their daily energy in the form of unsaturated fat.

In the U.S. diet, fat provides, on average, 34 percent of total daily calories, down from 45 percent in 1965. At first glance, this appears to be a healthy trend—until the actual grams of fat and carbohydrate are inspected. The total number of fat grams people take in has actually increased, rather than decreased, but the number of carbohydrate grams has increased even more. The net result is a higher total calorie intake and a misleading relative drop in the percentage of fat calories. Bottom line: People should learn to recognize the fats in foods and to distinguish harmful saturated fats and *trans* fats, which should be kept to a minimum, from more beneficial unsaturated fats that provide the needed essential fatty acids. Perhaps most important for many people is learning to control portion sizes, particularly of fatty foods that can pack hundreds of calories into just a few bites.

The *Dietary Guidelines for Americans 2005* urge that, beyond a healthy minimum, people should limit their fat intakes to limit saturated fat, *trans* fat, and calories. To do this requires consistently making nutrient dense choices such as fat-free milk, fat-free cheese, the leanest meats, and so forth and refraining from adding solid fats, such as butter, hard margarines, or shortening to foods during preparation or at the table. Higher fat foods may be included in the diet, but the fat calories they provide must fit within a person's discretionary calorie allowance. Exceptions are the fats of fatty fish, nuts, and vegetable oils: they provide beneficial EPA and DHA, linoleic acid and linolenic acid, and vitamin E needed for a healthful diet, and so, within calorie limits, the calories these fats provide are necessary, not discretionary.* Many people, especially sedentary people, have few or no discretionary calories to spend on high-fat foods.

Keep in mind that the fat of some foods, such as the rim of fat on a steak, is visible (and therefore removable); other fats are invisible, such as the fats in the marbling of meat, the fat ground into lunch meats and hamburger, the fats blended into sauces of mixed dishes, and the fats in avocados, biscuits, cheese, coconuts, other nuts, olives, and fried foods.* Invisible fats are on the rise in U.S. diets.[40]

▧ The Canadian Food Guide is found in Appendix B.

▧ Table 2-3 of Chapter 2 specifies how much oil is required to meet nutrient needs at several calorie levels.

▧ The concept of discretionary calories was addressed in Chapter 2.

Added Fats

A dollop of dessert topping, a spread of butter on bread, oil or shortening in a recipe, dressing on a salad—all of these are examples of *added* fats. All sorts of fats can be added to foods during commercial or home preparation or at the table. The following amounts of these fats contain about 5 grams of pure fat, providing 45 calories and negligible protein and carbohydrate:

- 1 teaspoon oil or shortening.
- 1½ teaspoon mayonnaise, butter, or margarine.
- 1 tablespoon regular salad dressing, cream cheese, or heavy cream.
- 1½ tablespoon sour cream.

The majority of added fats in the diet are invisible. They are the hidden fats of fried foods and baked goods, sauces and mixed dishes, and dips and spreads.

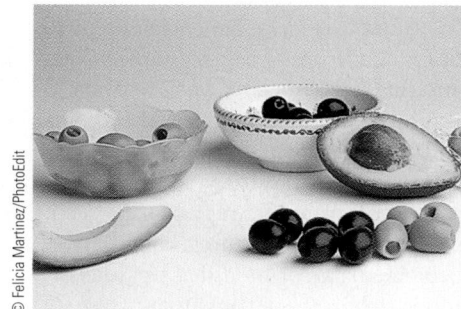

Ten small olives or a sixth of an avocado each provide about 5 grams of mostly monounsaturated fat.

▶ KEY POINT Fats added to foods during preparation or at the table are a major source of fat in the diet.

Meat, Poultry, Fish, Dried Peas and Beans, Eggs, and Nuts

Meats conceal a good deal of the fat—and much of the saturated fat—that people consume. To help "see" the fat in meats, it is useful to think of them in four categories according to their fat contents: very lean, lean, medium-fat, and high-fat meats, as the

---

*A new diglyceride-rich oil, brand name Enova, is similar in calories and nutrients to soybean oil, but makes health and weight loss claims that remain unproved.

CONSUMER CORNER

Fat Replacers

TODAY, CONSUMERS can choose from thousands of fat-reduced products. Many bakery goods, lunch meats, cheeses, spreads, frozen desserts, and other products made with **fat replacers** (see Table 5-6) offer less than half a gram of fat, saturated fat, and *trans* fat in a serving. Some of these products contain **artificial fats,** and others use conventional ingredients in unconventional ways to reduce fats and calories. Among the latter, manufacturers can:

- Add water or whip air into foods.
- Add fat-free milk to creamy foods.
- Use lean meats and soy protein to replace high-fat meats.
- Bake foods instead of frying them.

Common food ingredients such as fibers, sugars, or proteins can also take the place of fats in some foods. In particular, fat replacers made from oats or barley not only cut down on fats in foods but introduce beneficial viscous fibers, while imparting desirable tastes and textures associated with real fats.[1] Products made from sugars or proteins still provide calories, but far fewer calories than if they were made from fats.

Manufactured fat replacers consist of chemical derivatives of carbohydrate, protein, or fat, or modified versions of foods rich in those constituents (see Table 5-7). To gain the FDA's consent for the use of a new fat replacer, U.S. manufacturers must prove that their fat replacer contributes little food energy, is nontoxic, is not stored in body tissues, and does not rob the body of needed nutrients. Olestra serves as an example of an artificial fat that has won the FDA's approval.

An Artificial Fat: Olestra

Olestra, brand name Olean, is a member of the **sucrose polyester** chemical family. Chemically, olestra bears some resemblance to ordinary fat; it consists of a core molecule of the carbohydrate sucrose to which up to eight fatty acid molecules are bonded (see Figure 5-14). Human enzymes of the digestive tract do not recognize molecules of olestra and so cannot split its fatty acids from its sucrose. All of the olestra eaten in a food passes through the digestive tract and exits intact.

Olestra's properties are identical to those of fats and oils when used in frying, cooking, and baking. It can be heated to frying temperatures without breaking down. Aside from a slight aftertaste, to many people it tastes like fat.

More than two decades of research have revealed that olestra is safe in most regards, but when consumed in

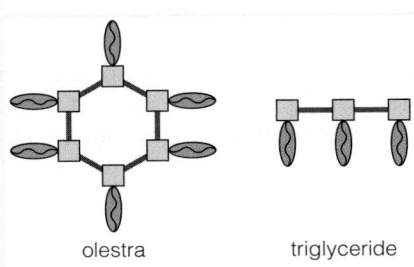

FIGURE 5·14 OLESTRA AND TRIGLYCERIDE COMPARED

Human digestive enzymes easily break apart triglycerides but cannot digest olestra.

olestra   triglyceride

large quantities, it can cause digestive distress, nutrient losses, and losses of phytochemicals. The presence of indigestible olestra in the large intestine can theoretically cause diarrhea, gas, cramping, and an urgent need for defecation. Oily olestra can creep through the feces and leak uncontrollably from the anus, producing smelly, dark yellow stains on underwear. No significant increase in digestive distress was detected in an experiment with some 3,000 volunteers who ate olestra-containing snacks.[2] In ordinary consumers, however, about 10 percent of olestra users calling a toll-free number reported digestive disturbances after consuming olestra-containing foods.[3] These effects are not considered serious or dangerous, so the FDA no longer requires olestra-containing foods to carry a warning on their labels.[4]

Olestra acts as a potent solvent for some of the fat-soluble substances in foods, such as the vitamins that dissolve in fat (vitamins A, D, E, and K). The absorption of these vitamins from a meal that includes olestra is reduced because olestra dissolves them and carries them out of the digestive tract

---

TABLE 5·6   TERMS RELATED TO FAT REPLACERS

- *Artificial fats* zero-energy fat replacers that are chemically synthesized to mimic the sensory and cooking qualities of naturally occurring fats, but are totally or partially resistant to digestion. Also called *fat analogues.*
- *Fat replacers* ingredients that replace some or all of the functions of fat and may or may not provide energy. Often used interchangeably with *fat substitutes,* but the latter technically applies only to ingredients that replace all of the functions of fat and provide no energy.
- *Olestra* a noncaloric artificial fat made from sucrose and fatty acids; formerly called *sucrose polyester.*
- *Sucrose polyester* any of a family of compounds in which fatty acids are bonded with sugars or sugar alcohols. Olestra is an example.

TABLE 5·7 A SAMPLING OF FAT REPLACERS

Fat replacers can dramatically reduce the saturated fat, *trans* fats, and calories in processed foods. Further, fiber-based replacers may provide some of the health benefits of viscous fibers (see Chapter 4). For comparison, remember that fat has 9 calories per gram.

Fat Replacers	Energy (cal/g)	Uses in Foods
Carbohydrate-Based Fat Replacers		
• *Fruits* purees and pastes of apples, bananas, cherries, plums, or prunes; add bulk and tenderness to baked goods.	1–4	*Foods:* Baked goods, candy, dairy products. *Uses:* Replace bulk of fat; add moisture and tenderness.
• *Maltodextrins* made from corn; powdered and flavored to resemble the taste of butter.	1–4	*Foods:* Butter-flavored "sprinkles" for melting on hot foods. *Uses:* Add "buttery" flavor.
Fiber-Based Fat Replacers		
• *Gels* derived from cellulose or starch to mimic the texture of fats in regular margarine and other products.	0–4[a]	*Foods:* Fat-free margarine, salad dressing, frozen desserts. *Uses:* Replace bulk; lend thickness.
• *Gums* extracted from beans, sea vegetables, or other sources.	0–4	*Foods:* Salad dressing, processed meats, desserts. *Uses:* replace bulk; thicken salad dressings.
• *Oatrim* derived from oat fiber; has the added advantage of providing satiety.	4	*Foods:* Dips, dressings, baked goods, cereal bars, beverages, meats. *Uses:* Lends creaminess; replaces bulk of fat; can be used in baking but not frying.
• *Z-trim* a modified form of insoluble fiber; is powdered and feels like fat in the mouth.	0	*Foods:* Cheese, meats, sauces, dips, baked goods, dressings, spreads, desserts, ice cream. *Uses:* Lends creaminess; replaces bulk of fat; can be used in baking but not frying.
Fat-Based Replacers		
• *Olestra*[b] a noncaloric artificial fat made from sucrose and fatty acids; formerly called *sucrose polyester*.	0	*Foods:* Potato chips, tortilla chips, snack crackers. *Uses:* Same properties as fats; heat stable in frying, cooking, and baking. Not available for home use.
• *Salatrim*[c] derived from fat and contains short- and long-chain fatty acids.	5	*Foods:* Confections, dairy products, bakery products. *Uses:* Same properties as fats; can be used in baking but not frying.
Protein-Based Fat Replacers		
• *Microparticulated protein*[d] processed from the proteins of milk or egg white into mistlike particles that feel and taste like fat when they roll over the tongue.	4	*Foods:* Ice cream, dairy products, mayonnaise, salad dressing, baked goods, spreads. *Uses:* Lends creaminess; heat stable in some cooking and baking but not frying.

[a] Energy made available by action of colonic bacteria.
[b] Trade name: Olean. Not available in Canada.
[c] Trade name: Benefat.
[d] Trade names: Simplesse and K-Blazer.

unabsorbed. To compensate for this effect, olestra is fortified with vitamins A, D, E, and K.

Olestra also causes the loss of health-promoting phytochemicals from foods. One study showed that just 3 grams of olestra a day strongly reduced (by about 40 percent) the blood concentrations of lycopene, a phytochemical believed to defend the body against some forms of cancer (see Controversy 2). To date, no studies exist to predict the effects, if any, of lifelong olestra exposure or the effects of olestra on growing children, important because children often favor the foods in which olestra is allowed.

Olestra's pros and cons are summed up in Figure 5-15 (p. 164).

Fat Replacers and Weight Control

People choosing reduced-fat foods, but not necessarily foods made with fat replacers, generally consume fewer

FIGURE 5·15 OLESTRA'S PROS AND CONS

FIGURE 5·15 OLESTRA'S PROS AND CONS

Pros of Olestra

- Zero calories
- Zero fat, saturated fat, and *trans* fat
- Zero cholesterol
- Withstands frying
- Withstands baking
- Tastes like fat

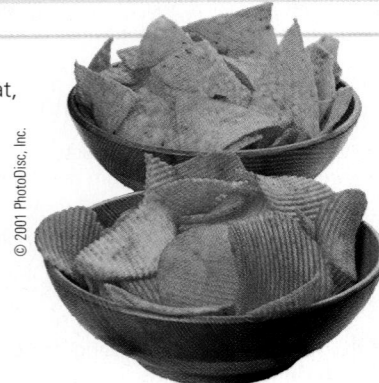

© 2001 PhotoDisc, Inc.

Cons of Olestra

- Vitamin losses
- Phytochemical losses
- Possible digestive upset
- Possible anal leakage
- Slight aftertaste
- Expensive
- No long-term studies in children

calories, less saturated fat, and more nutrients than nonusers of such foods.[5] Whether using fat replacers assists in weight control, however, seems to depend on whether people compensate for the resulting energy and fat deficit by eating more food later on. In two small studies, researchers secretly replaced one-third of young men and women's regular intake of daily fat with olestra.[6] The subjects seemed not to notice the energy shortfall as the study progressed—they didn't fully compensate for the shortfall and so lost weight.

As for blood lipids, people who used olestra heavily, consuming more than 2 grams each day for a year, significantly lowered their blood cholesterol. These subjects did not lose significant weight.[7]

A natural thought seems to be, "If my food has less fat, then I can eat more and remain lean." Labels reveal, however, that many reduced-fat foods deliver appreciable fat and about as many calories as a comparable regular product. If the U.S. experience with artificial sweeteners, described in Controversy 4, is any guide, consumers are likely to eat fat-replacer products *in addition* to other high-fat foods they prefer, negating the potential benefits of the fat replacers. Used wisely, though, fat replacers can help consumers achieve some of their dietary goals, especially when the user learns to cut calories and saturated fat from the diet in other ways as well (see this chapter's Food Feature).

exchange lists do in Appendix D. Meats in all four categories contain about equal amounts of protein, but their fat contents differ and their saturated fat and calorie amounts vary significantly. Figure 5-16 shows fat and calorie data for some ground meats. Table 2-10 on pages 000 and 000 in Chapter 2 provided some definitions concerning the fat contents of meats.

The 2005 USDA Food Guide suggests that most adults limit a day's intake of meats or equivalents to about 5 to 7 ounces. For comparison, the smallest fast-food hamburger weighs about 3 ounces. A steak served in a restaurant often runs 8, 12, or 16 ounces, more than a whole day's meat allowance. You may have to weigh a serving or two of meat to see how much you are eating.

People think of meat as protein food, but calculation of its nutrient contents reveals a surprising fact. A big (4-ounce), fast-food hamburger sandwich contains 23 grams of protein and 20 grams of fat. Because protein offers 4 calories per gram and fat offers 9, the sandwich provides 92 calories from protein and about twice that amount from fat. The calorie total, counting carbohydrates from the bun and condiments, is over 400 calories, with more than 50 percent of them from fat. Hot dogs, fried chicken sandwiches, and fried fish sandwiches are also high-fat choices. Because so much of the energy in a meat eater's diet is hidden from view, people can easily overeat on high-fat food, making weight control difficult.

When choosing beef or pork, look for lean cuts named *loin* or *round* from which the fat can be trimmed. Eat small portions, too. As for chicken and turkey, these meats are naturally lean, but commercial processing and frying add fats, especially in "patties," "nuggets," "fingers," and "wings." Chicken wings are mostly skin, and a chicken stores most of its fat just under its skin. The tastiest wing snacks have also been fried in cooking fat (often a hydrogenated saturated type with *trans* fatty acids), smothered with a buttery, spicy sauce, and then dipped in blue cheese dressing, making wings an

FIGURE 5·16 CALORIES, FAT, AND SATURATED FAT IN COOKED GROUND MEAT PATTIES

Only the ground round, at 10 percent fat by raw weight, qualifies to bear the word *lean* on its label. To be called "lean," products must contain fewer than 10 grams of fat, 4 grams of saturated fat, and 95 milligrams of cholesterol per 100 grams of food. The red labels on these packages list rules for safe meat handling, explained in Chapter 12.

Higher in fat   Lower in fat

Regular ground beef 23% fat	Ground chuck 16% fat	Commercial ground turkey[a] (with skin ground in) 13% fat	Ground round 10% fat
260 cal/3 oz[b] — 4½ tsp fat, 8 g saturated fat	220 cal/3 oz[b] — 3 tsp fat, 6 g saturated fat	200 cal/3 oz[b] — 2¼ tsp fat, 3 g saturated fat	180 cal/3 oz[b] — 1½ tsp fat, 4 g saturated fat

© Polara Studios Inc. (all)

[a]Values for 3 ounces of cooked turkey breast ground without skin are 108 calories, ½ teaspoon fat, and 1 gram saturated fat.
[b]Larger servings will, of course, provide more fat, saturated fat, and calories than the values listed here.
SOURCE: Data from ESHA Research, The Food Processor Nutrition and Fitness Software, version 8.3, 2004.

extraordinarily high-fat snack. If you snack on wings, plan on eating low-fat foods at several other meals to balance them out.

Watch out for ground turkey or chicken products. The skin is often ground in to add moistness when cooked, and these products can be much higher in fat than even lean beef—see Figure 5-16. Some people (even famous chefs) misinterpret Figure 5-5 on page 145, believing that because the fats of poultry and pork are less saturated than beef fat they are harmless to the heart. Nutrition authorities, however, emphatically state that all sources of saturated fat pose a risk to the heart and that even poultry with skin should be limited for heart health.

▶ **KEY POINT** Meats account for a large proportion of the hidden fat and saturated fat in many people's diets. Most people consume meat in larger amounts than recommended.

Milk, Yogurt, and Cheese

Some milk products contain fat, as Figure 5-17 (p. 166) shows. In homogenizing whole milk, milk processors blend in the cream, which otherwise would float and could be removed by skimming. A cup of whole milk contains the protein and carbohydrate of fat-free milk, but in addition it contains about 60 extra calories from fat. A cup of reduced-fat (2 percent fat) milk falls between whole and fat-free, with 45 calories of

FIGURE 5·17 LIPIDS IN MILK, YOGURT, AND CHEESE

Red boxes below indicate foods with higher lipid contents that warrant moderation in their use. Green indicates lower-fat choices.

Fat-free, skim, zero-fat, no-fat, or nonfat milk, 8 oz (<0.5% fat by weight)

Calories 80	Calories from Fat 0
	% Daily Value*
Total Fat 0g	**0%**
Saturated Fat 0g	**0%**
Cholesterol 5mg	**2%**

Low-fat milk, 8 oz (1% fat by weight)

Calories 105	Calories from Fat 20
	% Daily Value*
Total Fat 2g	**3%**
Saturated Fat 1.5g	**8%**
Cholesterol 10mg	**3%**

Low-fat cheddar cheese, 1.5 oz

Calories 70	Calories from Fat 30
	% Daily Value*
Total Fat 3g	**5%**
Saturated Fat 2g	**10%**
Cholesterol 10mg	**3%**

Nutrition Facts
Amount Per Serving

© Polara Studios, Inc.

Strawberry yogurt, 8 oz

Calories 250	Calories from Fat 45
	% Daily Value*
Total Fat 5g	**8%**
Saturated Fat 3g	**15%**
Cholesterol 15mg	**5%**

Whole milk, 8 oz (3.3% fat by weight)

Calories 150	Calories from Fat 70
	% Daily Value*
Total Fat 8g	**12%**
Saturated Fat 5g	**25%**
Cholesterol 24mg	**8%**

Reduced fat, less-fat milk, 8 oz (2% fat by weight)

Calories 120	Calories from Fat 45
	% Daily Value*
Total Fat 5g	**8%**
Saturated Fat 2g	**10%**
Cholesterol 20mg	**7%**

Cheddar cheese, 1.5 oz

Calories 165	Calories from Fat 130
	% Daily Value*
Total Fat 14g	**22%**
Saturated Fat 9g	**45%**
Cholesterol 40mg	**13%**

Low-fat strawberry yogurt, 8 oz

Calories 240	Calories from Fat 20
	% Daily Value*
Total Fat 2.5g	**4%**
Saturated Fat 2g	**10%**
Cholesterol 15mg	**5%**

Milk's names:
- Milk, whole milk (whole milk).
- Reduced-fat, less-fat milk (2% milk).
- Low-fat milk (1% milk).
- Fat-free, zero-fat, or no-fat milk (skim or nonfat milk).

fat. The fat of whole milk occupies only a teaspoon or two of the volume but nearly doubles the calories in the milk. Depending on its fat content, milk bears one of the names listed in the margin (older names are listed in parentheses).

Milk and yogurt appear in the milk group, but cream and butter do not. Milk and yogurt are rich in calcium and protein, but cream and butter are not. Cream and butter are fats, as are whipped cream, sour cream, and cream cheese, so they are grouped together with the solid fats. Cheeses are the single greatest contributor of saturated fat in the diet.[41] Among food fats, only the lipids of palm oil and coconut oil rank higher for saturation than the butterfat in fatty dairy products (review Figure 5-5 on page 145).

▶ **KEY POINT** The choice between whole and fat-free milk products can make a large difference to the fat and saturated fat content of a diet. Cheeses are a major contributor of saturated fat.

Grains

Grain foods in their natural state are very low in fat, but fat, including saturated fat and *trans* fat, may be added during manufacturing, processing, or cooking (see Figure 5-18). The fats in these foods can be particularly hard to detect, so diners must remember which foods stand out as being high in fat. Notable are granola and certain other ready-to-eat cereals, croissants, biscuits, cornbread, fried rice, pasta with creamy or oily sauces, quick breads, snack and party crackers, muffins, pancakes, and homemade waffles.

FIGURE 5·18 LIPIDS IN GRAINS

Red boxes below indicate foods with higher lipid contents that warrant moderation in their use. Green indicates lower-fat choices.

Low-fat granola, 1/2 c

Calories 195	Calories from Fat 35
	% Daily Value*
Total Fat 3g	5%
Saturated Fat 1g	5%
Cholesterol 0mg	0%

Crispy oat bran, 1/2 c

Calories 150	Calories from Fat 45
	% Daily Value*
Total Fat 5g	8%
Saturated Fat 1.5g	8%
Cholesterol 0mg	0%

Buttery crackers, 5 crackers

Calories 80	Calories from Fat 35
	% Daily Value*
Total Fat 4g	6%
Saturated Fat 1g	5%
Cholesterol 0mg	0%

Fried rice, 1/2 c[a]

Calories 140	Calories from Fat 65
	% Daily Value*
Total Fat 7g	11%
Saturated Fat 1g	5%
Cholesterol 20mg	7%

Nutrition Facts
Amount Per Serving

A home-made waffle

Calories 220	Calories from Fat 100
	% Daily Value*
Total Fat 11g	17%
Saturated Fat 2g	10%
Cholesterol 50mg	17%

A dinner roll

Calories 80	Calories from Fat 20
	% Daily Value*
Total Fat 2g	3%
Saturated Fat 0g	0%
Cholesterol 0mg	0%

Fettuccine alfredo, 1/2 c

Calories 250	Calories from Fat 130
	% Daily Value*
Total Fat 14g	22%
Saturated Fat 8g	40%
Cholesterol 60mg	20%

A breakfast bar

Calories 150	Calories from Fat 55
	% Daily Value*
Total Fat 6g	9%
Saturated Fat 2.5g	13%
Cholesterol 0mg	0%

A muffin

Calories 160	Calories from Fat 54
	% Daily Value*
Total Fat 6g	9%
Saturated Fat 1g	5%
Cholesterol 20mg	7%

A large biscuit

Calories 260	Calories from Fat 80
	% Daily Value*
Total Fat 11g	17%
Saturated Fat 2.5g	13%
Cholesterol 0mg	0%

A small croissant

Calories 230	Calories from Fat 108
	% Daily Value*
Total Fat 12g	18%
Saturated Fat 7g	35%
Cholesterol 38mg	13%

© Polara Studios, Inc.

[a]The calorie and fat contents of fried rice vary by preparation method.

Packaged breakfast bars often resemble vitamin-fortified candy bars in their fat and sugar contents.

▶ **KEY POINT** Fat in breads and cereals can be well hidden. Consumers must learn which foods of this group contain fats.

Now that you know where the fats in foods are found, how can you reduce or eliminate the harmful ones from your diet? The Food Feature provides some pointers.

FOOD FEATURE

Defensive Dining

Four reasons to keep intakes of fats low:

1. Diets lower in fat are generally lower in calories and thereby help to achieve and maintain a healthy body weight.
2. Diets low in saturated and *trans* fats reduce the risk of heart disease.
3. Diets lower in fat, and particularly saturated fat, may lower the risks of some cancers.
4. Diets with fewer calories from fat have more room for health-promoting foods such as fruits, fish, legumes, low-fat milk products, nuts, vegetables, and whole grains.

TO MEET TODAY'S dietary recommendations concerning fats requires not only identifying total fat in foods but learning to recognize and limit saturated and *trans* fats, too. People with diseases, such as heart disease or obesity, are advised to reduce their intakes of total fat to help them minimize illness. Even healthy people, if they are consuming diets that supply over 35 percent of calories as fat, should cut back to prevent health and nutrition problems.

The recommendation to limit daily intakes of saturated and *trans* fats for the health of the heart applies to all people. To repeat: No amount of these fats is needed for health, and to stay healthy, intakes should be minimized in a balanced, adequate diet. Though this advice is easy to dispense, following it may not be so easy because even the unsaturated vegetable oils, needed for their essential fatty acids, contain some saturated fat.[42] The first step in achieving low intakes of saturated and *trans* fats while obtaining necessary fatty acids is to learn which foods contain heavy doses of the heart-clogging fats. Then keep these foods to a minimum. This Food Feature can help.

The first arena is the grocery store. The right choices here can save you many grams of fat, saturated fat, and *trans* fats, while the wrong ones can undermine your efforts. Food labels can reveal much about a processed food's fat contents. With that knowledge, you can decide whether to use a food as a staple item in your diet or as an occasional treat. Choose foods lowest in harmful fats for everyday use; limit others to occasional use only. For example, choose frozen vegetables (a staple food) without butter or other high-fat sauces, which are often highly saturated and generally drive up the calorie content of the vegetables, as well as the price; add your own flavorings such as a touch of herbs, olive oil, garlic, or lemon pepper at home. If you choose precooked meats, avoid those that are coated and fried or prepared in fatty sauces. Try rotisserie chicken from the deli—rotisserie cooking lets much of the fat and saturated fat drain away. Look for new innovations aimed at reducing saturated fats, for new ones appear all the time.

Once at home, one of the most effective steps is to eliminate solid saturated fats used as seasonings. This means eating cooked vegetables without butter, bacon, or stick margarine; replacing shortening with oils such as olive or canola oil; omitting high-fat meat gravies and cheese or cream sauces; and leaving off most other last-minute fatty additions. As for calories, butter and regular margarine contain the same number of calories (about 35 per teaspoon); diet margarine contains fewer calories because water, air, or fillers have been added. Imitation butter-flavored sprinkles contain no fat and few calories.

For snacks, make it a habit to choose lower-fat microwave popcorn, and then sprinkle on butter or cheese flavoring, if you like it. Keep that flavoring on hand together with other substitutes such as diet, soft, or liquid margarine (generally low in saturated and *trans* fats), reduced-fat sauce mixes or recipes, and nonstick spray or olive oil for frying. Table 5-8 provides a list of possible substitutes for high-fat ingredients in recipes. These

TABLE 5·8 SUBSTITUTES FOR HIGH-FAT INGREDIENTS

Use	Instead of
Fat-free milk products	Whole-milk products
Evaporated fat-free ("skim") milk (canned)	Cream
Yogurt[a] or fat-free sour cream replacer	Sour cream
Reduced-calorie margarine; butter replacers	Butter
Wine, lemon juice, or broth	Butter
Fruit butters	Butter
Part-skim or fat-free ricotta; low-fat or fat-free cottage cheese[a]	Whole-milk ricotta
Part-skim or reduced-fat cheeses; "filled" cheeses in which vegetable oil has replaced saturated fat	Regular cheeses
1 tbs cornstarch (for thickening sauces)	1 egg yolk
Low-fat or fat-free mayonnaise	Regular mayonnaise
Low-fat or fat-free salad dressing (for salads and marinades)	Regular salad dressing
Water-packed canned fish and meats	Oil-packed fish and meats
Lean ground meat and grain mixture	Ground beef
Low-fat frozen yogurt or sherbet	Ice cream
Herbs, lemons, spices, fruits, liquid smoke flavoring, or ham-flavored bouillon cubes	Butter, bacon, bacon fat
Baked tortilla or potato chips; pretzels	Regular chips

[a]If the recipe calls for the food to be boiled, the yogurt or cottage cheese must be stabilized with a small amount of cornstarch or flour.

replacements will not change the taste or appearance of the finished product very much, but they will dramatically lower the calories and saturated fat.

When you add fats to foods, be sure that they are detectable and that you enjoy them. For example, if you use strongly flavored fat, a little goes a long way. Sesame oil, peanut butter, nut oils, and the fats of strong cheeses are equal in calories to others, but they are so strongly flavored that you can use much less. Try small amounts of grated sapsago, romano, or other hard cheeses to replace larger amounts of less flavorful cheeses. A new "filled" cheese product has been processed to remove saturated fat and cholesterol, which are then replaced by unsaturated vegetable oils to maintain a taste and texture close to the original cheese. These cheeses provide the same calories as regular cheese but help to reduce saturated fat intakes.

When choosing oils, trade off among different types to obtain the benefits different oils offer. Peanut and safflower oils are especially rich in vitamin E. Olive oil presents the heart with health benefits (see the Controversy section for details), and canola oil is rich with monounsaturates and the essential fatty acids. High temperatures, such as those used in fry-ing, destroy some omega-3 acids and other beneficial constituents, so treat your oils gently. Especially important: Take care to *substitute* oils for saturated fats in the diet; do not add oils to an already fat-rich diet. No benefits are expected unless oils replace other more saturated fats.

Here are some other tips to revise high-fat recipes that contribute excess calories and saturated fats:

- Grill, roast, broil, boil, bake, stir-fry, microwave, or poach foods. Don't fry in solid fats, such as shortening or butter. If you must fry, use a little liquid oil for pan frying.
- Choose large portions of salad greens and other vegetables, and dress lightly. Reduce or eliminate "add-ons" such as butter, creamy sauces, cheese, sour cream and bacon that drive up the calories and saturated fat.
- Cut recipe amounts of meat in half; use only lean meats. Fill in the lost bulk with soy meat replacers, shredded vegetables, legumes, pasta, grains, or other low-fat items.
- Replace a thick slice of ham with two or three wafer-thin slices. The serving will be smaller and thus provide less fat, but the taste will be the same because the surface area that imparts flavor to the taste buds is greater.[43]

USDA facts:

- 70% of teenage males eat meals away from home each day.
- 57% of all Americans, and 40% of those over 60 years old, do so.
- The foods chosen away from home are higher in fat, saturated fat, *trans* fat, and cholesterol, and lower in vitamins and minerals, than meals typically eaten at home.

- Refrigerate meat pan drippings and broth, and lift off the fat when it solidifies. Then add the defatted broth to a recipe.
- Make prepared mixes, such as rice or potato mixtures, without the fats called for on the label, or substitute liquid oils for solid fats in preparing them.

All of these suggestions work well when a person plans and prepares each meal at home. But in the real world, people fall behind schedule and don't have time to cook, so they eat fast food. Figure 5-19 compares some fast-food choices and offer tips to reduce the calories and saturated fat to make fast-food meals healthier.

Keep these facts about fast food in mind:

- Salads are a good choice, but beware of toppings such as fried noodles, bacon bits, greasy croutons, sour cream, or shredded cheese that can drive up the calories, saturated fat, and *trans* fat contents. To reduce calories, avoid mixed salad bar items, such as macaroni salad. Use only about a quarter of the dressing provided with fast-food salads or use low-fat dressing.
- If you are really hungry, order a small hamburger or "veggie burger" and a side salad. Hold the cheese; use mustard or ketchup as condiments.
- A small bowl of chili (hold the cheese and sour cream) poured over a plain baked potato can also satisfy a bigger appetite. Top with chopped raw onions or hot sauce for spice.
- Tacos and other Mexican treats can often be ordered with salsa instead of cheese and sour cream.
- Fast-food fried fish or chicken sandwiches provide at least as much fat as hamburgers—and more *trans* fat. Broiled sandwiches are far less fatty if you order them made without spreads, dressings, cheese, bacon, or mayonnaise.

Because fast foods are short on variety, let them be part of a lifestyle in which they complement the other parts. Eat differently, often, elsewhere.

By this time you may be wondering if you can realistically make all the changes recommended for your diet. In truth, even small changes yield big dividends in terms of reducing harmful fats in the diet (see Table 5-9). Be assured that most such changes can become habits after a few repetitions. You do not have to give up all high-fat treats, nor should you strive to eliminate all fats; you need only learn to exercise moderation. You decide what the treats should be and then choose them judiciously, just for pure pleasure. Meanwhile, make sure that your everyday, ordinary choices are those whole, nutrient-dense foods suggested throughout this book. That way you'll meet all your body's needs for nutrients and never feel deprived.

TABLE 5·9 CHOOSING UNSATURATED FATS INSTEAD OF SATURATED FATS

Unsaturated fats can easily substitute for saturated fats in everyday foods, for example, cooking with olive oil instead of butter or slicing avocado on a salad instead of cheese. The fat grams listed in this table are for 100-calorie portions.

Foods (100 cal portions)	Saturated Fat (g/100 cal)	Unsaturated Fat (g/100 cal)	Total Fat (g/100 cal)
Olive oil (1 tbs) vs. butter (1 tbs)	2 vs. 7	9 vs. 4	11 vs. 11
Sunflower seeds (2 tbs) vs. bacon (2 slices)	1 vs. 3	7 vs. 6	8 vs. 9
Mixed nuts (2 tbs) vs. potato chips (10 chips)	1 vs. 2	8 vs. 5	9 vs. 7
Avocado (6 small slices) vs. cheese (1 slice)	2 vs. 4	8 vs. 4	10 vs. 8
Salmon (2 oz) vs. steak (1½ oz)	1 vs. 2	3 vs. 3	4 vs. 5

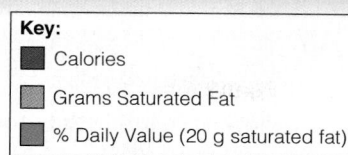

Key:
- Calories
- Grams Saturated Fat
- % Daily Value (20 g saturated fat)

When ordering Mexican-style fast food, you can reduce both calories and saturated fat by limiting cheese, meat, and sour cream.

Higher in saturated fat   **Lower in saturated fat**

Burrito choices

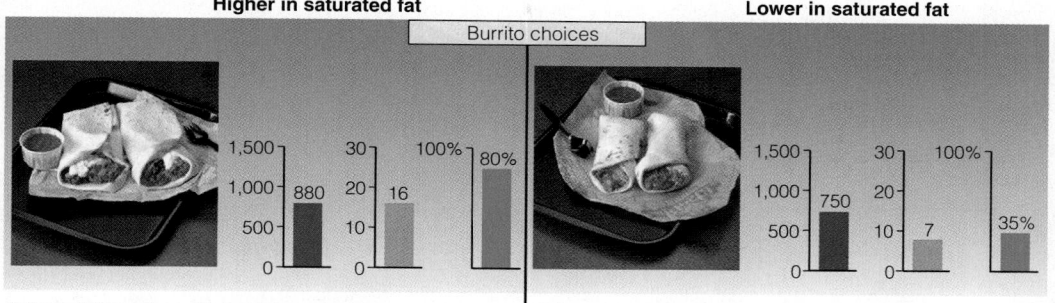

880   16   80%   750   7   35%

2 "grande" burritos with beef, beans, cheese, and sour cream; salsa

2 bean burritos; salsa

A broiled chicken breast sandwich with spicy mustard is just as tasty as a burger but delivers far less saturated fat and fewer calories. Beware of fried chicken sandwiches or "patties"—these can be as fatty as the hamburger choice.

Sandwich choices

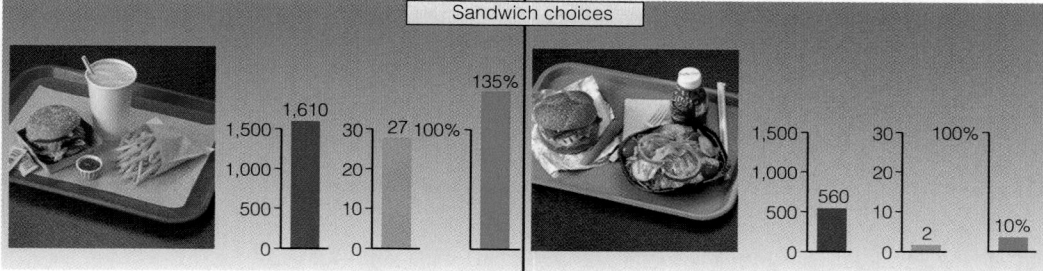

1,610   27   135%   560   2   10%

Big double bacon cheeseburger, large fries, regular milkshake

Big broiled chicken breast sandwich, pickle, side salad with low-calorie dressing, fat-free milk

Don't let add-ons, such as greasy croutons, chips, bacon bits, full-fat cheese, and sour cream pile the calories and saturated fat onto your otherwise healthy fast-food salad. To cut fats and calories, leave off most of the toppings and use just half the dressing.

Salad choices

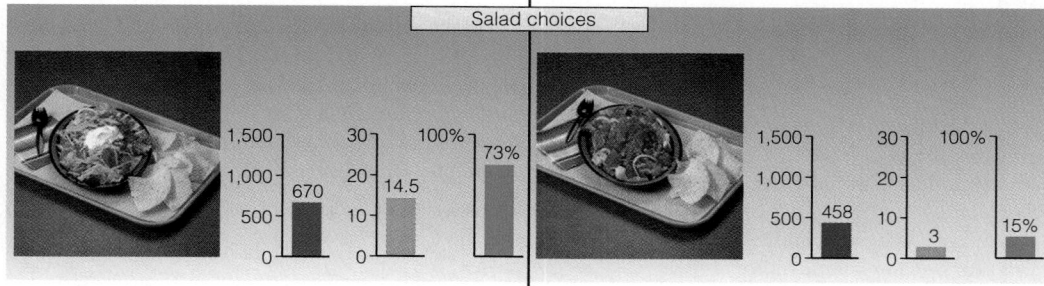

670   14.5   73%   458   3   15%

Taco salad with chili, cheese, sour cream, salsa, and taco chips

Taco salad with chili, salsa, and taco chips

Reduce calories and saturated fat even further: try ordering your veggie pizza with half the regular melted cheese and sprinkle it with parmesan cheese, herbs, or hot peppers for flavor.

Pizza choices

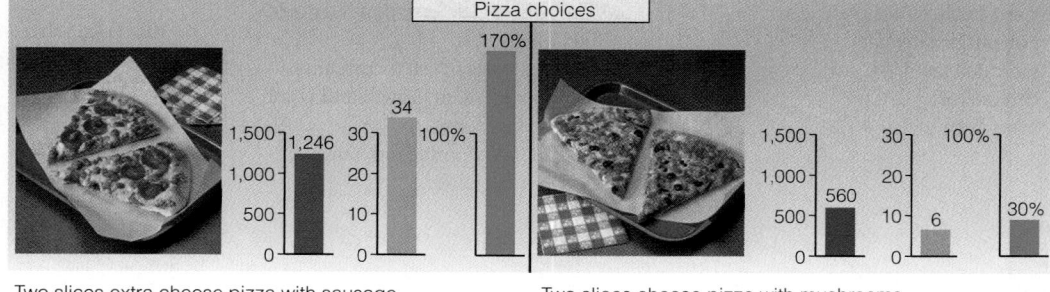

1,246   34   170%   560   6   30%

Two slices extra cheese pizza with sausage and pepperoni

Two slices cheese pizza with mushrooms, olives, onions, and peppers

© Matthew Farruggio (all photos)

Nutrition Connections

The Nutrition Connections CD provides access to these additional resources.

Do It!
The Do It! exercise for this chapter, *Evaluate Saturated and Trans Fats on Food Labels,* helps you calculate a personal set of guidelines for saturated fat, *trans* fat, and dietary cholesterol.

Interaction
The animation, *Lipoprotein Function in the Body,* shows the various types of lipoproteins as they leave their tissue of origin and deliver their cargo of lipids throughout the body.

Chapter Quiz
Take the quiz for this chapter to assess your understanding of chapter concepts.

Nutrition on the Net
Annotated links to reputable nutrition-related websites.

Online Glossary
Test your knowledge of key vocabulary through this comprehensive, interactive glossary,

Internet Connections

A variety of helpful study tools for this chapter are available at our website
http://nutrition.wadsworth.com/ncc10

Practice Test
Take the practice test for this chapter and gauge your knowledge of key concepts. Answers are provided.

Student Notebook
A lecture notebook is available to help organize note taking while reading or attending lecture.

Study Tools
Study resources for students include Chapter Summaries, Crossword puzzles, Flashcards, Glossary, Internet Exercises, and Learning Objectives.

SELF CHECK

Answers to these Self Check questions are in Appendix G.

1. Which of the following is *not* one of the ways fats are useful in foods?
   a. Fats contribute to the taste and smell of foods.
   b. Fats carry fat-soluble vitamins.
   c. Fats provide a low-calorie source of energy compared to carbohydrates.
   d. Fats provide essential fatty acids.

2. Generally speaking, vegetable and fish oils are rich in which of these?
   a. polyunsaturated fat
   b. saturated fat
   c. cholesterol
   d. *trans*-fatty acids

3. A benefit to health is seen when _____ is used in place of _____ in the diet.
   a. saturated fat/monounsaturated fat
   b. saturated fat/polyunsaturated fat
   c. monounsaturated fat/saturated fat
   d. polyunsaturated fat/cholesterol

4. Chylomicrons, a class of lipoproteins, are produced in the:
   a. gallbladder
   b. small intestinal cells
   c. large intestinal cells
   d. liver

5. The roles of the essential fatty acids include:
   a. form parts of cell membranes.
   b. support infant growth and vision development.
   c. support immune function.
   d. all of the above.

6. LDL deliver triglycerides and cholesterol from the liver to the body's tissues.   T   F

7. Taking supplements of fish oil is recommended for those who don't like fish.   T   F

8. Consuming large amounts of *trans*-fatty acids lowers LDL cholesterol and thus lowers the risk of heart disease and heart attack.   T   F

9. When the fat-replacer olestra is present in the digestive tract it enhances the absorption of vitamin E.   T   F

10. Fried fish from fast food restaurants and frozen fried fish products are often low in omega-3 and high in *trans*- and saturated, fatty acids.   T   F

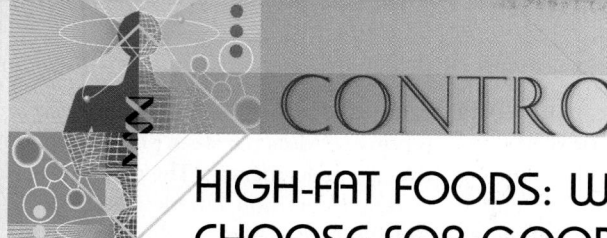

CONTROVERSY 5

HIGH-FAT FOODS: WHICH TO CHOOSE FOR GOOD HEALTH?

TO CONSUMERS, advice about dietary fats appears to change almost daily. "Eat less fat—buy more fatty fish." "Give up butter—use margarine instead." "Give up margarine—replace it with olive oil." "Avoid saturated fat—increase omega-3 fat." To researchers, however, the evolution of such advice reflects many decades of following leads and testing theories about the health effects of dietary fats. As scientific understanding has grown, dietary guidelines have become more specific and therefore more meaningful.

This Controversy begins with a look at the latest guidelines for lipid intakes from a scientific point of view. It also singles out the Mediterranean diet as an eating style famous for supporting the health of the heart while including foods high in fat. It concludes with strategies for choosing the right amounts of the right kinds of fats within the context of a heart-healthy diet and lifestyle.

The Changing Fat Guidelines

For years, consumers were urged to cut their fat intakes in everything from hot dogs to salad dressings. This advice was straightforward—cut the fat, improve your health. Saturated fat was a well-established culprit behind elevated blood cholesterol, but the guideline makers focused on total fat in the diet, limiting intake to 30 percent or less of calories because when total fat is reduced, saturated fat intake also falls. Did this strategy work to cut saturated fat intake? Yes, but only in people who applied the advice consistently. Many who tried failed, finding it impossible to maintain a low-fat diet over the long run.[1]

Low-fat diets present several other problems.[2] For example, low-fat diets are not necessarily low-calorie diets, and many people with heart disease are overweight and need to reduce body fatness. Also, diets high in carbohydrates but low in fiber, especially those high in refined

sugars, cause blood triglycerides to rise with unknown effects on heart health.[3] Finally, taken to an extreme, a low-fat diet may exclude nutritious foods, such as fatty fish, nuts, seeds, and vegetable oils, that provide the essential fatty acids along with many phytochemicals, vitamins, and minerals.

Low-Fat Diet Questions

Low-fat diets remain a centerpiece of treatment plans for people with elevated blood lipids or heart disease and are therefore important in nutrition.[4] But what about healthy people? Evidence from around the world has led researchers to question whether a low-fat diet is the only way, or even the best way, for healthy people to stay healthy. In a classic study of the effects of diet on the world's people, the Seven Countries Study, death rates from heart disease were strongly associated with diets high in saturated fats but only weakly linked with total fat.[5] In fact, the two countries with the highest fat intakes were Finland and the Greek island of Crete, yet Finland had the highest rate of death from heart disease in the study while Crete had the lowest.

In both countries, the people consumed 40 percent or more of their calories from fat. These findings indicated that total fat was clearly not to blame for a high rate of heart disease—something else was affecting the risk. When researchers examined the diets of these fat-loving peoples, they found that diets high in olive oil but low in saturated fat (less than 10 percent of calories) were consistently associated with relatively low disease risks. These results have since been supported by many other studies.

Today's Fat Guidelines

On reviewing the evidence, the DRI committee concluded that a diet containing a slightly greater percentage of fat—up to 35 percent of total calories—but reduced in saturated fat and *trans* fat

and controlled in energy (calories) is compatible with low rates of heart disease, diabetes, obesity, and cancer. The *Dietary Guidelines for Americans 2005* therefore suggest replacing the "bad" saturated and *trans* fats with "good" unsaturated fats and enjoying these fats within calorie limits.[6] The *Dietary Guidelines* make clear, however, that the human body has no need of dietary saturated fat or *trans* fat, so the less consumed the better, as long as the diet is adequate in nutrients, including the essential fatty acids.[7] Following this advice requires that consumers first learn which foods contain which fats and then make appropriate selections among them.

High-Fat Foods and Heart Health

Avocados, bacon, walnuts, potato chips, and mackerel are all high-fat foods, yet the fats of these foods differ markedly in their effects on health. The following evidence can help to clarify why some high-fat foods rightly belong in a heart-healthy diet and why others are best left on the shelf.

Olive Oil: The Potential Connection

The traditional diets of Greece and the Mediterranean region are exemplary in their use of "good" fats, especially olives and their oil. In population and laboratory studies, use of dark green olive oil instead of other cooking fats, especially butter, stick margarine, and meat fats, has been linked with numerous potential health benefits.[8] When olive oil replaces saturated fats, such as those of butter, coconut or palm oil, hydrogenated stick margarine, lard, or shortening, it may help protect against heart disease by some of these mechanisms:

- Lowering total and LDL cholesterol and not lowering HDL cholesterol or raising triglycerides.[9]
- Reducing LDL cholesterol's vulnerability to oxidation.[10]
- Lowering blood-clotting factors.[11]
- Providing phytochemicals that act as antioxidants (see Controversy 2).[12]
- Lowering blood pressure.[13]

Although such findings are promising, controlled human studies are too

CONTROVERSY 5

scarce to support population-wide recommendations to switch to a high-fat diet rich in olive oil.[14] Other vegetable oils, in their liquid unhydrogenated states, are also generally low in saturated fats and high in unsaturated fats, but they lack the phytochemicals that olive oil provides. When choosing olive oils, go for the darker "extra virgin" kind because it contains the highest levels of potentially beneficial phytochemicals.

People who hope that olive oil will be a magic potion against heart disease are bound to be disappointed. Drizzling olive oil on a food high in saturated fat, such as a cheese and sausage pizza, does not make such foods healthier. Also, like other fats, olive oil delivers 9 calories per gram, which can contribute to weight gain in people who fail to balance their energy intake with their energy output.

The Mediterranean Diet: Beyond Olive Oil

Factors other than olive oil probably deserve some of the credit for the low rates of heart disease among people consuming a traditional Mediterranean diet.[15] Such factors as amounts and types of meats, nuts, vegetables, fruit, and sweets differ widely between the traditional Mediterranean diets of the past century and modern North American diets.

Though each of the countries bordering the Mediterranean Sea has its own culture and dietary traditions, researchers have identified some common characteristics. Most traditional Mediterranean people focus their diets

Olives and their oil may benefit heart health.

on crusty breads, whole grains, nuts, potatoes, and pastas; a variety of vegetables (including wild greens) and legumes; feta and mozzarella cheeses and yogurt; nuts; and fruits (especially grapes and figs). They eat some fish, other seafood, poultry, a few eggs, and a little meat. Along with olives and olive oil, their principal sources of fat are nuts and fish; they rarely use butter or encounter hydrogenated fats. Consequently, traditional Mediterranean diets are:

- Low in saturated fat.
- Very low in *trans* fat.
- Rich in unsaturated fat.
- Rich in starch and fiber.
- Rich in nutrients and phytochemicals that support good health.

These characteristics are consistent with many indicators of heart health.[16]

In addition, because the animals graze in fields, their meat, dairy products, and eggs are richer in omega-3 fatty acids than those from animals kept in feedlots and fed grain, as is done elsewhere. Omega-3 fatty acids in the Mediterranean diet also derive from other typical foods, such as wild plants and snails, that are unavailable to U.S. consumers. Apparently, each of the foods in Mediterranean diets contributes some small benefit that harmonizes with others to produce a substantial cumulative or synergistic benefits.[17]

Nuts are extraordinarily popular in Mediterranean cuisines and show up in everything from sauces to desserts. Recent scientific findings have fostered a turnabout in the attitude toward these high-fat foods.

Nuts: More Than a High-Calorie Snack Food

Nuts and peanuts traditionally have no place in a low-fat diet, with good reason. Nuts provide up to 80 percent of their calories from fat, and a quarter cup (about an ounce) of mixed nuts provides over 200 calories.

Nevertheless, scientists are finding links between nuts and heart health. A recent review of the literature suggests that people who eat a one-ounce serving of nuts on five or more days a week have lower heart disease risks than those con-

suming no nuts.[18] Even as little as two servings of nuts per week provide a positive, if smaller, benefit. The nuts under study are common varieties: almonds, Brazil nuts, cashews, hazelnuts, macadamia nuts, pecans, pistachios, walnuts, and even peanuts. On average, such nut varieties contain mostly monounsaturated fat (59 percent), some polyunsaturated fat (27 percent), and just a small amount of saturated fat (14 percent).[19]

Walnuts and almonds in particular may prove beneficial. In study after study, walnuts, when substituted for other fats in the diet, produce favorable effects on blood lipids—even in people whose total and LDL cholesterol were elevated at the outset.[20] Results are similar for almonds. Researchers gave men and women one of three kinds of snacks, all of equal calories: whole-wheat muffins, almonds (about 2½ ounces), or half muffins and half almonds.[21] At the end of a month, people receiving the full almond snack had the greatest drop in blood LDL cholesterol; those eating the half almond snack had a lesser, but still significant, drop in blood lipids; and those eating the muffin-only snack demonstrated no change in blood lipids.

Studies on peanuts, macadamia nuts, pecans, and pistachios also indicate that consuming a variety of nuts may be a wise choice. Nuts may lower heart disease risk because they are:

- Low in saturated fats.
- High in fiber, vegetable protein, and other valuable nutrients, including the antioxidant vitamin E.
- High in phytochemicals that act as antioxidants.

Walnuts also supply the essential fatty acid linolenic acid, which may quell blood vessel inflammation associated with heart disease.[22]

In addition to their heart benefits, nuts may also benefit other body organs—people who frequently consume nuts and other healthy fats suffer fewer gallbladder problems.[23]

Before deciding to include nuts in your diet, remember that a snack of nuts delivers substantially more calories per bite than, say, a snack of whole-grain pretzels, and many more still than crunchy raw

Matthew Farruggio

174 • Chapter 5 The Lipids: Fats, Oils, Phospholipids, and Sterols

Stay mindful of calories when snacking on nuts.

Matthew Farruggio

vegetables. For people who struggle to maintain enough body weight for health, the calories of nuts are a welcome addition. For the majority, however, caution is in order. In studying the effects of nuts on heart disease, researchers carefully adjust the calories of experimental diets to make room for the nuts—that is, they use nuts instead of, not in addition to, other fat sources (such as meats, potato chips, oils, margarine, and butter) to keep calories constant. People who decide to start eating nuts are urged to do likewise to prevent unneeded weight gain, because weight gain elevates blood lipids and raises the risk of heart disease.

Butter or Margarine: Which to Choose?

When news of the possible effects of *trans*-fatty acids on heart health first emerged, some people stopped using margarine and switched back to butter, believing oversimplified reports that margarine provides no heart health advantage over butter. Admittedly, hardened margarines and virtually all shortenings are made largely from hydrogenated fats and are therefore saturated and contain substantial *trans*-fatty acids—up to 40 percent by weight. Some margarines, however, especially the soft or liquid varieties, are made from unhydrogenated oils, which are mostly unsaturated and so are less likely to elevate blood cholesterol than the saturated fats of butter.

When oils (but not hydrogenated oils) are the first ingredient listed on a margarine label, the margarine is, in all probability, low in saturated fat and *trans* fat and therefore a good choice for a healthy heart. On average, margarine

contributes less *trans* fat to the diet than do other contributors (see Figure 5-13 on page 160 of the chapter).

In addition to soft and liquid margarine choices, some stick margarines are now specially formulated to contain few or no *trans*-fatty acids. Other types contain an added phytochemical ingredient, sterol esters, which reduces blood cholesterol when consumed as part of a low-fat diet.* These compounds belong to the sterol family of lipids and so are chemical relatives of cholesterol. Unlike cholesterol, though, sterol esters come from plants and are not recognized by the intestine and not absorbed. They may also block the absorption of cholesterol itself.[24]

Simply adding sterol ester–containing margarine to a diet high in saturated fat is unlikely to bring health benefits. Sterol esters lower blood cholesterol only when people cut their fat intakes as well. Drawbacks to the sterol ester margarine include the price (three or four times higher than regular margarine), a high fat and calorie content (the full-fat kind equals the fat and calories in regular margarine), and an unproven record of safety for use by certain populations, such as growing children. Recently, sterol esters have also been added to some kinds of orange juice and candies, which may make these compounds more accessible to more people who may or may not need them.

Fish: Benefits and Cautions

The preceding chapter made clear that fish oils hold the potential to improve health, and particularly the health of the heart. Research studies have provided strong evidence that increasing omega-3 fatty acids in the diet supports heart health and lowers the risk of death from heart disease.[25] People who eat some fish each week lower their risks of heart attack and

stroke.[26] For this reason, the American Heart Association and other authorities recommend including two fatty fish servings totaling 12 ounces a week in a heart-healthy diet. Table 5-5 on page 157 of the chapter identified fish varieties that supply at least 1 gram of omega-3 fatty acids per serving.

Fish is the best source of EPA and DHA in the diet, but fish is also a major source of mercury and other potentially toxic environmental contaminants, as mentioned in the chapter. Most fish have at least trace amounts of mercury, but tilefish, swordfish, king mackerel, marlin, shark, and tuna have especially high levels of contamination. Freshwater fish may contain PCBs and other pollutants, so local advisories warn sport fishers of species that can pose problems. The chapter listed safer species of fish. To minimize risks while obtaining fish benefits, vary your choices among fatty fish species often.

© www.comstock.com

Fish is the only good food source of EPA and DHA.

Fats to Avoid: Saturated Fats and *Trans* Fats

The number-one dietary determinant of LDL cholesterol is saturated fat. Figure C5-1 (p. 176) shows that each 1 percent increase in energy from saturated fatty acids in the diet produces an estimated 2 percent jump in heart disease risk by way of elevating blood LDL cholesterol. Conversely, reducing saturated fat intake by 1 percent is estimated to

---

*The term *sterol esters* includes related *stanol esters*. Controversy 2 defined these terms.

FIGURE C5·1 IMPACT OF CHANGE IN SATURATED FATTY ACID INTAKE ON BLOOD LDL CHOLESTEROL AND HEART DISEASE RISK

aPercentage of change in total energy intake from saturated fatty acids.
bThe change in an individual's risk may compound when blood lipid changes are sustained over time.
SOURCE: Third Report of the National Cholesterol Education Program (NCEP) Expert Panel on Detection, Evaluation, and Treatment of High Blood Cholesterol in Adults (Adult Treatment Panel III), NIH publication no. 02-5215 (Bethesda, Md.: National Heart, Lung, and Blood Institute, 2002), pp. II-4 and V-8.

produce a 2 percent drop in heart disease risk by the same mechanism. Even a 2 percent drop in LDL represents a significant improvement for the health of the heart.[27] Similarly, *trans* fats also raise heart disease risk by elevating LDL cholesterol. A heart-healthy diet limits foods rich in these two types of fat.

To limit saturated fat intake, consumers must choose carefully among high-fat foods. The major sources of saturated fats in the U.S. diet are fatty meats, whole-milk products, coconut and palm oils, and products made from any of these foods. Over a third of the fat in most meats is saturated (review Figure 5-5 of the chapter). Over half of the fat in whole milk and other high-fat dairy products, such as cheese, butter, cream, half-and-half, cream cheese, sour cream, and ice cream, is saturated. The saturated fats of palm and coconut oils are rarely used by consumers in the kitchen, but their stability and other properties make them useful to food manufacturers, so commercially prepared foods provide these fats in abundance.

To restate the message of the chapter, when choosing meats, milk products, and commercially prepared foods, read labels and choose foods lowest in saturated fat and *trans* fat. Appendix A lists the saturated fat in foods that bear no labels.

Keeping total fat within the recommended level of 35 percent of total caloriers is essential for controlling saturated fat intake. Even with careful selections, a nutritionally adequate diet will necessarily provide some saturated fat because oils that provide the essential polyunsaturated fatty acids also supply a mixture of other fatty acids, including saturated fatty acids.[28] Designing an adequate diet with zero saturated fat is not possible even for experts.[29] Diets based on fruits, greens, legumes, nuts, soy products, vegetables, and whole grains can, and often do, deliver less saturated fat than diets that

TABLE C5·1 FOOD SOURCES OF FATTY ACIDS

Healthful Fatty Acids		
Monounsaturated	**Omega-6 Polyunsaturated**	**Omega-3 Polyunsaturated**
Avocado	Margarine (nonhydrogenated)	Fatty fish (herring, mackerel,
Nuts (almonds, cashews,	Mayonnaise	salmon, tuna)
filberts, hazelnuts,	Nuts (walnuts)	Flaxseed
macadamia nuts, peanuts,	Oils (corn, cottonseed,	Nuts
pecans, pistachios)	safflower, soybean)	
Oils (canola, olive, peanut,	Salad dressing	
sesame)	Seeds (pumpkin, sunflower)	
Olives		
Peanut butter (old-fashioned)		
Seeds (sesame)		

Harmful Fatty Acids	
Saturated	**Trans**
Bacon	Commercial baked goods, including cookies,
Butter	cakes, pies, or other goodies made with
Cheese	margarine or vegetable shortening
Chocolate	Fried foods, particularly restaurant and fast
Coconut	foods
Cream, half-and-half	Many fried or processed snack foods, includ-
Cream cheese	ing microwave popcorn, chips, and crackers
Lard	Margarine (hydrogenated or partially
Meat	hydrogenated)
Milk fat (whole milk products)	Nondairy creamers
Oils (coconut, palm, palm kernel)	Shortening
Shortening	
Sour cream	

NOTE: Keep in mind that foods contain a mixture of fatty acids; see Figure 5-5, p. 145.

depend heavily on animal-derived foods, however.

Finally, keeping *trans* fat intake low requires restricting foods known to contain it. Table C5-1 summarizes which foods provide which fats. Substituting unsaturated fats for saturated fats at each meal and snack can help protect against heart disease. Table 5-9 in Chapter 5's Food Feature showed how such substitutions can lower saturated fat even when total fat and calories remain unchanged.

Conclusion

Are some fats "good" and others "bad" from the body's point of view? Certainly, the saturated and *trans* fats seem mostly bad for the health of the heart. Aside from providing energy, which unsaturated fats can do equally well, saturated and *trans* fats bring no indispensable benefits to the body. Furthermore, no harm can come from consuming diets low in saturated fats and *trans* fats.

In contrast, unsaturated fats are mostly good for the health of the heart when consumed in moderation and within a sensible calorie total. To date, their one proven fault seems to be that they, like all fats, provide abundant energy and so may promote obesity if they drive calorie intakes higher than energy needs.[30] Obesity, in turn, often begets many body ills (see Chapter 9).

When judging foods by their fatty acids, keep in mind that food fats present the body with a mixture, providing both saturated and unsaturated fatty acids.

Even olive oil and other vegetable oils deliver some saturated fat. Consequently, even when a person chooses foods with mostly unsaturated fats, saturated fat can still add up if total fat is high.

Food manufacturers may soon come to the assistance of consumers wishing to avoid the health threats from saturated and *trans* fats. Most major snack manufacturers are reducing the saturated and *trans* fats in some of their products and offering snack foods in single-serving packages.

Adopting some of the Mediterranean eating habits may serve the needs of those who enjoy somewhat more fat in the diet. Figure C5-2 presents a Mediterranean food pyramid for guidance. Basing the diet on vegetables, fruits, whole grains, and legumes as part of a balanced daily diet is a good idea, as is *replacing* saturated fats such as butter, shortening, and meat fat with unsaturated fats like olive oil and the oils from nuts and fish. These foods also provide vitamins, minerals, and phytochemicals—all valuable in protecting the body's health.

The authors of this book would not stop there, however. They would urge you to reduce fats from convenience foods and fast foods; choose small portions of meats, fish, and poultry; and to select portion sizes that do not exceed your energy requirement. Also, Mediterranean peoples lead physically active lifestyles, and physical activity reduces disease risks. Therefore, if you want to include olive oil and generally eat like a Greek, you'd better walk, garden, bicycle, and swim like one, too.[31]

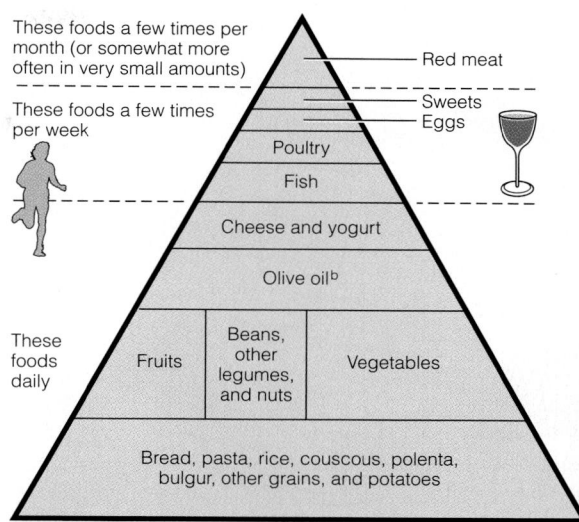

FIGURE C5-2 A MEDITERRANEAN DIET PYRAMID[a]

MEDITERRANEAN DIET PYRAMID[a]

These foods a few times per month (or somewhat more often in very small amounts) — Red meat

These foods a few times per week — Sweets, Eggs

Poultry

Fish

Cheese and yogurt

Olive oil[b]

These foods daily — Fruits | Beans, other legumes, and nuts | Vegetables

Bread, pasta, rice, couscous, polenta, bulgur, other grains, and potatoes

[a]The authors of this pyramid also recommend regular physical exercise and moderate consumption of wine.
[b]Other oils rich in monounsaturated fats, such as canola or peanut oil, can be substituted for olive oil. People who are watching their weight should limit their oil consumption.

SOURCE: 1994 Oldways Preservation & Exchange Trust, by permission.

The Proteins and Amino Acids

CHAPTER 6

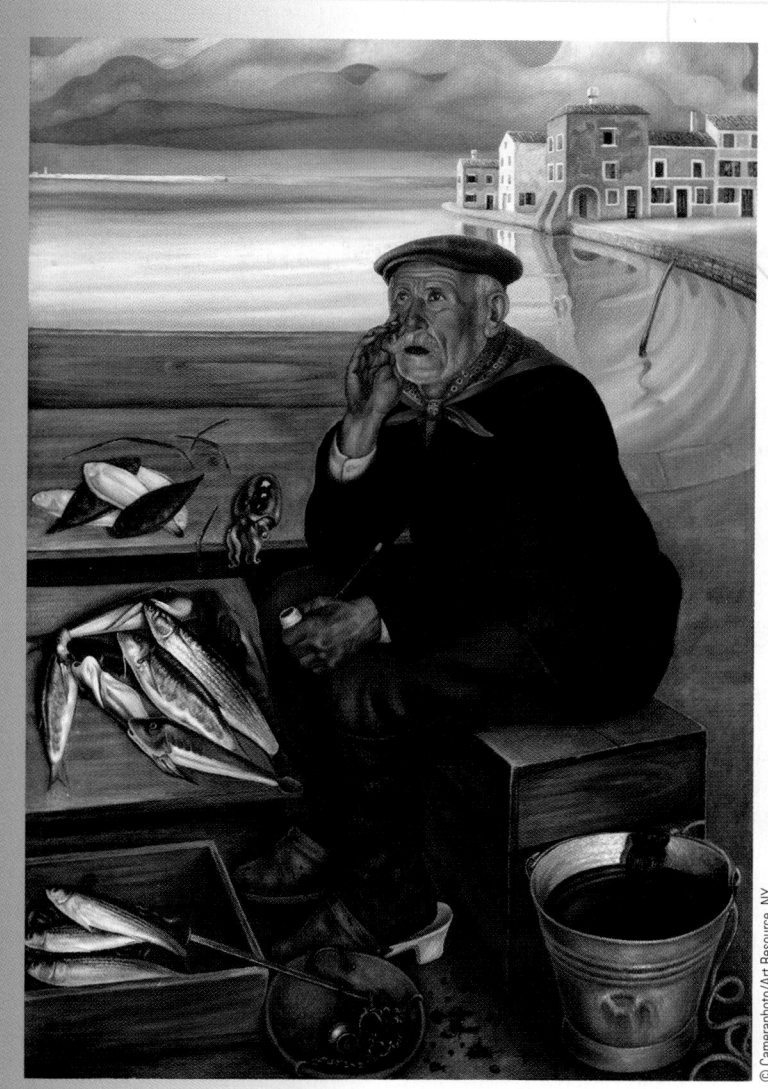

© Cameraphoto/Art Resource, NY

CONTENTS

Do You Ever . . .

Wonder why protein is important to the health of the body?

Find it curious that heating an egg changes it from a liquid to a solid?

Take protein or amino acid supplements to bulk up muscles or lose weight?

Fear that your diet will lack protein unless you eat at least some meat?

KEEP READING.

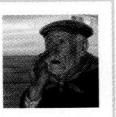

FIRST NAMED 150 years ago after the Greek word *proteios* ("of prime importance"), the proteins are versatile and vital cellular working molecules. Without them, life would not exist. Research on **proteins** has revealed countless secrets of the processes of life and has helped to answer many questions about nutrition. How do we grow? How do our bodies replace the materials they lose? How does blood clot? What gives us immunity? What makes one person different from another? Understanding the nature of the proteins in the body helps to solve these mysteries.

Some of the body's proteins are working proteins; others form structures. Working proteins include the body's enzymes, antibodies, transport vehicles, hormones, cellular "pumps," and oxygen carriers. Structural proteins include tendons and ligaments, scars, the fibers of muscles, the cores of bones and teeth, the filaments of hair, the materials of nails, and more. Yet, however different their functions, all protein molecules have much in common.

The Structure of Proteins

The structure of proteins enables them to perform many vital functions. One key difference from carbohydrates and fats is that proteins contain nitrogen atoms in addition to the carbon, hydrogen, and oxygen atoms that all three energy-yielding nutrients contain. These nitrogen atoms give the name *amino* ("nitrogen containing") to the **amino acids,** the building blocks of proteins. Another key difference is that in contrast to the carbohydrates, whose repeating units, glucose molecules, are identical, the amino acids in a strand of protein are different from one another. A strand of amino acids that makes up a protein may contain 20 *different* kinds of amino acids.

Amino Acids

All amino acids have the same simple chemical backbone consisting of a single carbon atom with both an **amine group** (the nitrogen-containing part) and an acid group attached to it. Each amino acid also has a distinctive chemical **side chain** attached to the center carbon of the backbone (see Figure 6-1). It is this side chain that gives identity and chemical nature to each amino acid. About 20 amino acids, each with its own different side chain, make up most of the proteins of living tissue.[*1] Other rare amino acids appear in a few proteins.

The side chains make the amino acids differ in size, shape, and electrical charge. Some are negative, some are positive, and some have no charge (are neutral). The first part of Figure 6-2 is a diagram of three amino acids, each with a different side chain attached to its backbone. The rest of the figure shows how amino acids link to form protein strands. Long strands of amino acids form large protein molecules, and the side chains of the amino acids ultimately help to determine the molecules' shapes and behaviors.

Essential Amino Acids   The body can make about half of the 20 amino acids for itself, given the needed parts: fragments derived from carbohydrate or fat to form the backbones and nitrogen from other sources to form the amine groups. The healthy adult body cannot make some amino acids or makes them too slowly to meet its needs. These are the **essential amino acids** (listed in the margin of the facing page). Without these essential nutrients, the body cannot make the proteins it needs to do its work. Because the essential amino acids can only be replenished from foods, a person must frequently eat the foods that provide them.

Under special circumstances, a nonessential amino acid can become essential. For example, the body normally makes tyrosine (a nonessential amino acid) from the essen-

---
*Reference notes are found in Appendix F.

FIGURE 6·1   AN AMINO ACID

The "backbone" is the same for all amino acids. The side chain differs from one amino acid to the next. The nitrogen is in the amine group.

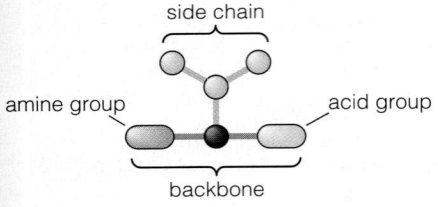

side chain

amine group   acid group

backbone

---

proteins compounds composed of carbon, hydrogen, oxygen, and nitrogen and arranged as strands of amino acids. Some amino acids also contain the element sulfur.

amino (a-MEEN-o) **acids** the building blocks of protein. Each has an amine group at one end, an acid group at the other, and a distinctive side chain.

amine (a-MEEN) **group** the nitrogen-containing portion of an amino acid.

side chain the unique chemical structure attached to the backbone of each amino acid that differentiates one amino acid from another.

essential amino acids amino acids that either cannot be synthesized at all by the body or cannot be synthesized in amounts sufficient to meet physiological need. Also called *indispensable amino acids.*

tial amino acid phenylalanine. If the diet fails to supply enough phenylalanine, or if the body cannot make the conversion for some reason (as happens in the inherited disease phenylketonuria, see Controversy 4), then
tyrosine becomes a **conditionally essential amino acid.**

Recycling Amino Acids   The body not only makes some amino acids
but also breaks protein molecules apart and reuses those amino acids. Both
food proteins, after digestion, and body proteins, when they have finished
their cellular work, are dismantled to liberate their component amino
acids. Amino acids from both sources provide the cells with raw materials
from which they can build the protein molecules they need. Cells can also
use the amino acids for energy and discard the nitrogen atoms as wastes.
By reusing amino acids to build proteins, however, the body recycles and
conserves a valuable commodity while easing its nitrogen disposal burden.

This recycling system also provides access to an emergency fund of
amino acids in times of fuel, glucose, or protein deprivation. At such
times, tissues can break down their own proteins, sacrificing working
molecules before the ends of their normal lifetimes, to supply energy and
amino acids to the body's cells. The body employs a priority system in
selecting the tissue proteins to dismantle—it uses the most dispensable
ones first, such as the small proteins of the blood and muscles.[2] It guards
the structural proteins of the heart and other organs until forced, by dire need, to
relinquish them.

FIGURE 6·2 DIFFERENT AMINO ACIDS JOIN TOGETHER

This is the basic process by which proteins are assembled.

valine   leucine   tyrosine
Single amino acids with different
side chains...

can bond to form...

a strand of amino acids, part of a protein.

▶ **KEY POINT**   Proteins are unique among the energy nutrients in that they possess nitrogen-containing
amine groups and are composed of 20 different amino acid units. Of the 20 amino acids, some are essential and some are essential only in special circumstances.

How Do Amino Acids Build Proteins?

In the first step of making a protein, each amino acid is hooked to the next (as shown in
Figure 6-2). A chemical bond, called a **peptide bond,** is formed between the amine group
end of one amino acid and the acid group end of the next. The side chains bristle out
from the backbone of the structure, giving the protein molecule its unique character.

The strand of protein does not remain a straight chain. Figure 6-2 shows only the first
step in making proteins—the linking of amino acid units with peptide bonds until the
strand contains from several dozen to as many as 300 amino acids. Amino acids at different places along the strand are chemically attracted to each other, and this attraction
causes some segments of the strand to coil, somewhat like a metal spring. Also, each spot
along the coiled strand is attracted to, or repelled from, other spots along its length.
These interactions cause the entire coil to fold this way and that, forming either a globular structure, as shown in Figure 6-3 (page 182), or a fibrous structure (not shown).

The amino acids whose side chains are electrically charged are attracted to water.
Therefore, in the body's watery fluids, they orient themselves on the outside of the protein structure. The amino acids whose side chains are neutral are repelled by water and
are attracted to one another; these tuck themselves into the center, away from the body
fluids. All these interactions among the amino acids and the surrounding fluids give
each protein a unique architecture, a form to suit its function.

One final detail may be needed for the protein to become functional. Several
strands may cluster together into a functioning unit, or a metal ion (mineral) or a vitamin may join to the unit and activate it.

▶ **KEY POINT**   Amino acids link into long strands that coil and fold to make a wide variety of different
proteins.

The essential amino acids:

- Histidine (HISS-tuh-deen).
- Isoleucine (eye-so-LOO-seen).
- Leucine (LOO-seen).
- Lysine (LYE-seen).
- Methionine (meh-THIGH-oh-neen).
- Phenylalanine (fen-il-AL-ah-neen).
- Threonine (THREE-oh-neen).
- Tryptophan (TRIP-toe-fan, TRIP-toe-fane).
- Valine (VAY-leen).

Other amino acids important in nutrition:

- Alanine (AL-ah-neen).
- Arginine (ARJ-ih-neen).
- Asparagine (ah-SPAR-ah-geen).
- Aspartic acid (ah-SPAR-tic acid).
- Cysteine (SIS-the-een).
- Glutamic acid (GLU-tam-ic acid).
- Glutamine (GLU-tah-meen).
- Glycine (GLY-seen).
- Proline (PRO-leen).
- Serine (SEER-een).
- Tyrosine (TIE-roe-seen).

NOTE: In special cases, some nonessential amino acids may
become conditionally essential (see the text).

conditionally essential amino acid   an
amino acid that is normally nonessential, but
must be supplied by the diet in special circumstances when the need for it exceeds the
body's ability to produce it.

peptide bond   a bond that connects one
amino acid with another, forming a link in a
protein chain.

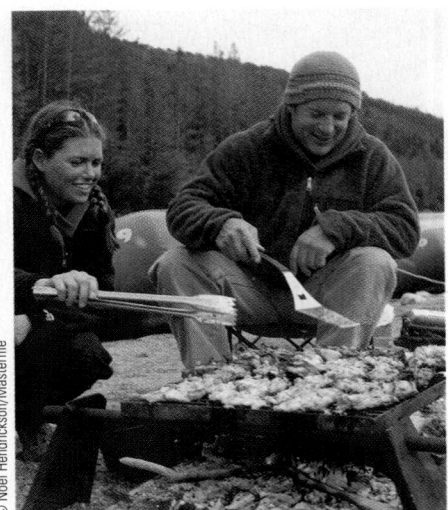

Hair, skin, eyesight, and the health of the whole body depend on protein from food.

FIGURE 6·3 THE COILING AND FOLDING OF A PROTEIN MOLECULE

A portion of a strand of amino acids.

The strand coils, as this "ribbon" demonstrates.

The completed protein.

Coiling the strand. The strand of amino acids takes on a spring-like shape as their side chains variously attract and repel each other.

Folding the coil. Once coiled and folded, the protein may be functional as is, or it may need to join with other proteins or add a vitamin or mineral to become active.

The Variety of Proteins

The particular shapes of proteins enable them to perform different tasks in the body. Those of globular shape, such as some proteins of blood, are water soluble. Some form hollow balls, which can carry and store materials in their interiors. Some proteins, such as those of tendons, are more than ten times as long as they are wide, forming stiff, rodlike structures that are somewhat insoluble in water and very strong. A form of the protein **collagen** acts somewhat like glue between cells. The hormone insulin, a protein, helps to regulate blood glucose. Among the most fascinating proteins are the **enzymes,** which act on other substances to change them chemically.

Some protein strands function alone, while others must associate with others to become functional. One molecule of **hemoglobin**—the large, globular protein molecule that, by the billions, packs the red blood cells and carries oxygen—is made of four associated protein strands, each holding the mineral iron (see Figure 6-4).

The great variety of proteins in the world is possible because an essentially infinite number of sequences of amino acids can be found. To understand how so many different proteins can be designed from only 20 or so amino acids, think of how many words are in an unabridged dictionary—all of them constructed from just 26 letters. If you had only the letter "G," all you could write would be a string of Gs: G–G–G–G–G–G–G–G. But with 26 different letters available, you can create poems, songs, or novels. Similarly, the 20 amino acids can be linked together in a huge variety of sequences—many more than are possible for letters in a word, which must alternate consonant and vowel sounds. Thus, the variety of possible sequences for amino acid strands is tremendous. A single human cell may contain as many as 10,000 different proteins, each one present in thousands of copies.

collagen (KAHL-ah-jen) a type of body protein from which connective tissues such as scars, tendons, ligaments, and the foundations of bones and teeth are made.

enzymes (EN-zimes) protein catalysts. A catalyst is a compound that facilitates a chemical reaction without itself being altered in the process.

hemoglobin the globular protein of red blood cells, whose iron atoms carry oxygen around the body via the bloodstream (more about hemoglobin in Chapter 8).

Inherited Amino Acid Sequences For each protein there exists a standard amino acid sequence, and that sequence is specified by heredity. Often, if a wrong amino acid is inserted, the result can be disastrous to health.

Sickle-cell disease, in which hemoglobin, the oxygen-carrying protein of the red blood cells, is abnormal, is an example of an inherited variation in the amino acid sequence. Normal hemoglobin contains two kinds of protein strands. In sickle-cell disease, one of the strands is an exact copy of that in normal hemoglobin, but in the other strand, the sixth amino acid is valine, rather than glutamic acid. This replacement of one amino acid so alters the protein that it is unable to carry and release oxygen. The red blood cells collapse from the normal disk shape into crescent shapes (see Figure 6-5). If too many crescent-shaped cells appear in the blood, the result is abnormal blood clotting, strokes, bouts of severe pain, susceptibility to infection, and early death.[3]

You are unique among human beings because of minute differences in your body proteins. These differences are determined by the amino acid sequences of your proteins, which are written into the genetic code you inherited from your parents and they from theirs. Your unique combination of genes directs the making of all your body's proteins, as shown in Figure 6-6 (page 185). Ultimately, the genes determine the sequence of amino acids in each finished protein, and some genes are involved in making more than one protein. As scientists completed the human genome, they recognized that a still greater task lies ahead: the identification of every protein made by the human body.[*]

Nutrients and Gene Expression When a cell makes a protein as shown in Figure 6-6, scientists say that the gene for that protein has been "expressed." Every cell nucleus contains the DNA for making every human protein, but cells do not make them all. Instead, cells specialize in making certain proteins typical of their cell types. For example, cells of the pancreas express the gene for the protein hormone insulin; in other cells, that gene is idle.

Nutrients influence genetic expression and the resulting protein synthesis, and scientists are working out the details of these fascinating interactions.[4] Living cells, including the cells of the body, continuously monitor nutrient concentrations in the fluids circulating around them. A nutrient lack or overabundance triggers a cascade of molecular events inside the cell that ultimately leads to expression or suppression of certain genes. The cell's genetic material speeds up or slows down its synthesis of certain proteins to cope with the nutrient imbalance.

Two examples may clarify this relationship. The genes of specialized cells of the pancreas, just mentioned, respond to molecular messages generated when blood glucose is overabundant. The cells' genetic material steps up synthesis of the protein insulin, whose job it is to bring the glucose level down. The mineral iron also plays roles in regulating protein synthesis. Normally, immature red blood cells pack their interiors with hemoglobin but when the body's iron stores run low the cells respond by suppressing hemoglobin synthesis.[†5] Conversely, iron abundance stimulates hemoglobin synthesis. Later chapters come back to topics of nutrient-gene interactions. The Think Fitness feature (p. 184) addresses a related concern of athletes: Can more dietary protein trigger the synthesis of muscle tissue?

---

*The identification of the entire collection of human proteins, the *human proteonome* (PRO-tee-oh-nome), is a work in progress.

†Mature red blood cells in the bloodstream have lost their nuclei and DNA and so lack the ability to make hemoglobin or other proteins.

FIGURE 6·4 THE STRUCTURE OF HEMOGLOBIN

Four highly folded protein strands form the globular hemoglobin protein.

Iron

A heme structure holds the iron.

The amino acid sequence causes the strands to coil and loop, forming the globular protein structure.

■ DNA was defined in Chapter 1.

FIGURE 6·5 NORMAL RED BLOOD CELLS AND SICKLE CELLS

Normal red blood cells are disk shaped. In sickle-cell disease, one amino acid in the protein strands of hemoglobin takes the place of another, causing the red blood cells to change shape and lose function.

Sickle-shaped blood cells   Normal red blood cells

© Dr. Stanley Flegler/Visuals Unlimited

What a difference one amino acid can make!

Amino acid sequence of normal hemoglobin:

Val — His — Leu — Thr — Pro — Glu — Glu

Amino acid sequence of sickle-cell hemoglobin:

Val — His — Leu — Thr — Pro — Val — Glu

Denaturation of Proteins

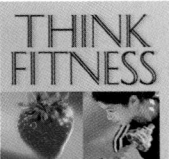

© Matthew Farruggio

Heat denatures egg protein, making it firm.

Proteins can be denatured (distorted in shape) by heat, alcohol, acids, bases, or the salts of heavy metals. The **denaturation** of a protein is the first step in its destruction; thus, these agents are dangerous because they damage the body's proteins. In digestion, however, denaturation is useful to the body. During the digestion of a food protein, the stomach acid opens up the protein's structure, permitting digestive enzymes to make contact with the peptide bonds and cleave them. Denaturation also occurs during the cooking of foods. Cooking an egg denatures the proteins of the egg and makes it firm as the margin photo demonstrates. More important for nutrition is that heat denatures two proteins in raw eggs: one binds the B vitamin biotin and the mineral iron, and the other slows protein digestion. Thus, cooking eggs liberates biotin and iron and aids digestion.

Many well-known poisons are salts of heavy metals like mercury and silver; these denature protein strands wherever they touch them. The common first-aid antidote for swallowing a heavy-metal poison is to drink milk. The poison then acts on the protein of the milk rather than on the protein tissues of the mouth, esophagus, and stomach. Later, vomiting can be induced to expel the poison that has combined with the milk.

■ Chapter 10 discusses the nutrient needs of athletes in detail.

THINK FITNESS

Can Eating Extra Protein Make Muscles Grow Larger?

CAN ATHLETES AND FITNESS seekers stimulate their muscles to grow larger simply by consuming more protein or amino acids? No. Hard work, not excess dietary protein, is the trigger for the genes to build more muscle tissue. Exercise generates cellular messages that stimulate DNA to begin the process of building up muscle fibers (muscle fibers are made of protein). Nevertheless, a snack rich in both protein and carbohydrate (half a turkey sandwich, for example) eaten directly *after* exercise may be of some help in this regard (more on this in Chapter 10). The path to bigger muscles is rigorous physical training with adequate energy and nutrients from balanced, well-timed meals, not consuming excess protein.

Digestion and Absorption of Protein

■ Chapter 3 discussed the use of medicines to control the stomach's acidity and also defined pH as a measure of acidity. See page 79.

denaturation the irreversible change in a protein's shape brought about by heat, acids, bases, alcohol, salts of heavy metals, or other agents.

Each protein performs a special task in a particular tissue of a specific kind of animal or plant. When a person eats food proteins, whether from cereals, vegetables, beef, fish, or cheese, the body must first alter them by breaking them down into amino acids; only then can it rearrange them into specific human body proteins.

The whole process of digestion is an ingenious solution to a complex problem. Proteins (enzymes), activated by acid, digest proteins from food, denatured by acid. The coating of mucus secreted by the stomach wall protects its proteins from attack by either acid or enzymes. The normal acid in the stomach is so strong (pH 1.5) that no food is acidic enough to make it stronger; for comparison, the pH of vinegar is about 3.

FIGURE 6·6 PROTEIN SYNTHESIS

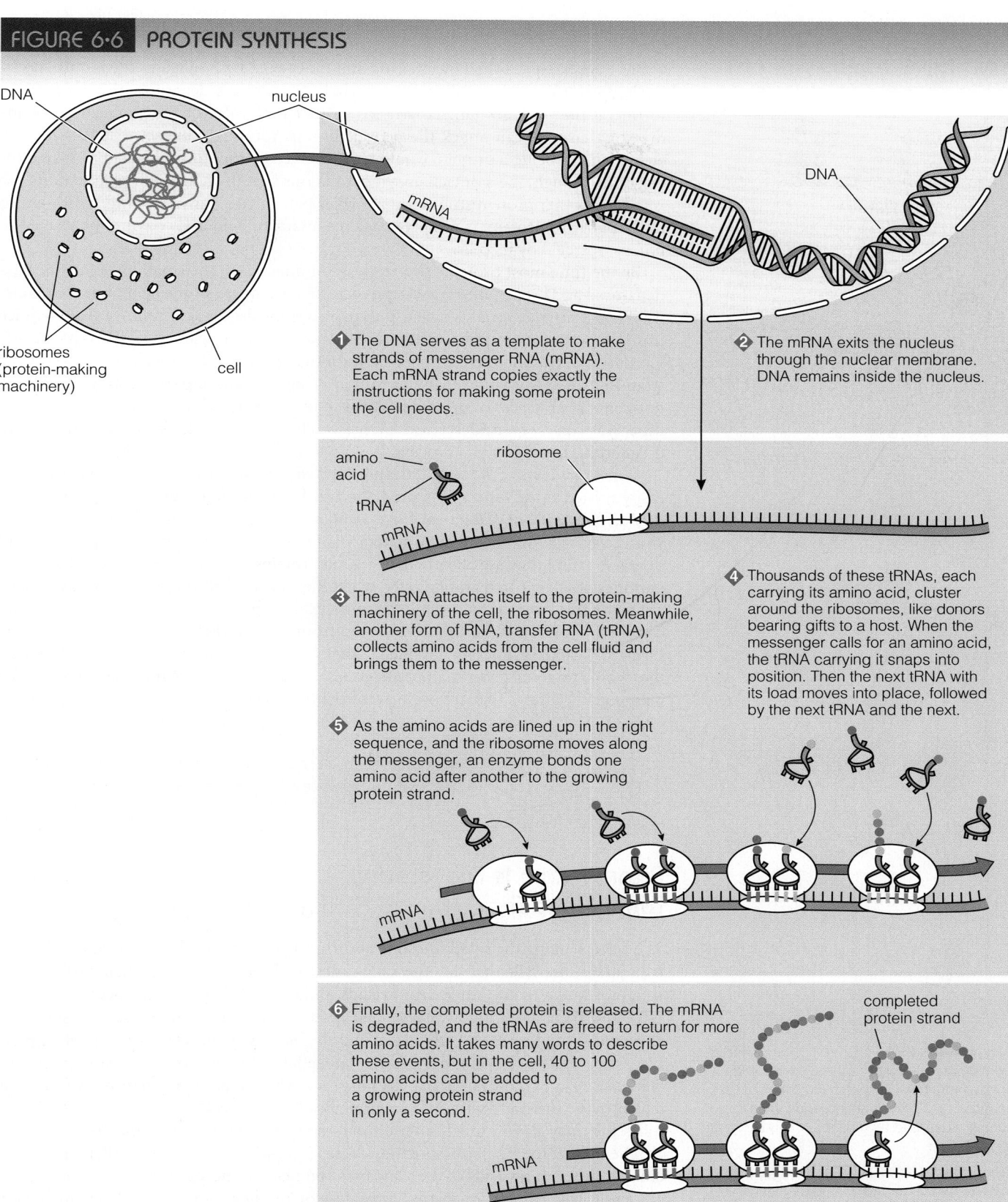

DNA

nucleus

ribosomes (protein-making machinery)

cell

mRNA

DNA

1 The DNA serves as a template to make strands of messenger RNA (mRNA). Each mRNA strand copies exactly the instructions for making some protein the cell needs.

2 The mRNA exits the nucleus through the nuclear membrane. DNA remains inside the nucleus.

amino acid

ribosome

tRNA

mRNA

3 The mRNA attaches itself to the protein-making machinery of the cell, the ribosomes. Meanwhile, another form of RNA, transfer RNA (tRNA), collects amino acids from the cell fluid and brings them to the messenger.

4 Thousands of these tRNAs, each carrying its amino acid, cluster around the ribosomes, like donors bearing gifts to a host. When the messenger calls for an amino acid, the tRNA carrying it snaps into position. Then the next tRNA with its load moves into place, followed by the next tRNA and the next.

5 As the amino acids are lined up in the right sequence, and the ribosome moves along the messenger, an enzyme bonds one amino acid after another to the growing protein strand.

mRNA

6 Finally, the completed protein is released. The mRNA is degraded, and the tRNAs are freed to return for more amino acids. It takes many words to describe these events, but in the cell, 40 to 100 amino acids can be added to a growing protein strand in only a second.

completed protein strand

mRNA

FIGURE 6·7 A DIPEPTIDE AND TRIPEPTIDE

dipeptide

tripeptide

Protein Digestion

Other than being crushed by chewing and moistened with saliva in the mouth, nothing happens to protein until the strong acid of the stomach denatures it. The acid helps to uncoil the protein's tangled strands so that molecules of the stomach's protein-digesting enzyme can attack the peptide bonds. You might expect that the stomach enzyme itself, being a protein, would be denatured by the stomach's acid. Unlike most enzymes, though, the stomach enzyme functions best in an acid environment. Its job is to break other protein strands into smaller pieces. The stomach lining, which is also made partly of protein, is protected against attack by acid and enzymes by its coat of mucus, secreted by its cells.

By the time most proteins slip from the stomach into the small intestine, they are denatured and broken into smaller pieces. A few are single amino acids, but the majority remain in large strands called **polypeptides.** In the small intestine, alkaline juice from the pancreas neutralizes the acid delivered by the stomach. The pH rises to about 7 (neutral), enabling the next enzyme team to accomplish the final breakdown of the strands. Protein-digesting enzymes from the pancreas and intestine continue working until almost all pieces of protein are broken into strands of two or three amino acids, **dipeptides** and **tripeptides,** respectively (see Figure 6-7), or single amino acids. Figure 6-8 summarizes the whole process.

Consumers who fail to understand the basic mechanism of protein digestion are easily misled by advertisers of products and books urging, "Take enzyme A to help digest your food" or "Don't eat foods containing enzyme C, which will digest cells in your body." The writers of such statements fail to realize that enzymes (proteins) are digested before they are absorbed, just as all proteins are. Even the stomach's digestive enzymes are denatured and digested when their jobs are through. Similar false claims are made that predigested proteins (amino acid supplements) are "easy to digest" and can therefore protect the digestive system from "overworking." Of course, the healthy digestive system is superbly designed to digest whole proteins with ease. In fact, it handles whole proteins *better* than predigested ones because it dismantles and absorbs the amino acids at rates that are optimal for the body's use.

▶ **KEY POINT** Digestion of protein involves denaturation by stomach acid, then enzymatic digestion in the stomach and small intestine to amino acids, dipeptides, and tripeptides.

After Protein Is Digested, What Happens to Amino Acids?

The cells all along the small intestine absorb single amino acids. As for dipeptides and tripeptides, the cells that line the small intestine have enzymes on their surfaces that split most of them into single amino acids, and the cells absorb them, too. Some dipeptides and tripeptides are also absorbed as is into the cells where they are split into amino acids and released with others into the bloodstream. A few larger peptide molecules can escape the digestive process altogether and enter the bloodstream intact.[6] Scientists believe these larger particles may act as hormones to regulate body functions and provide the body with information about the external environment. The larger molecules may also stimulate an immune response and thus play a role in food allergy.

The cells of the small intestine possess separate sites for absorbing different types of amino acids. Amino acids of the same type compete for the same absorption sites. Consequently, when a person ingests a large dose of any single amino acid, that amino acid may limit absorption of others of its general type. The Consumer Corner (page 193) cautions against taking single amino acids as supplements, partly for this reason.

Once amino acids are circulating in the bloodstream, they are carried to the liver where they may be used or released into the blood to be taken up by other cells of the

polypeptides (POL-ee-PEP-tides) protein fragment of many (more than ten) amino acids bonded togeteher (*poly* means "many"). A peptide is a strand of amino acids. A strand of between four and ten amino acids is called an *oligopeptide.*

dipeptides (dye-PEP-tides) protein fragments that are two amino acids long (*di* means "two").

tripeptides (try-PEP-tides) protein fragments that are three amino acids long (*tri* means "three").

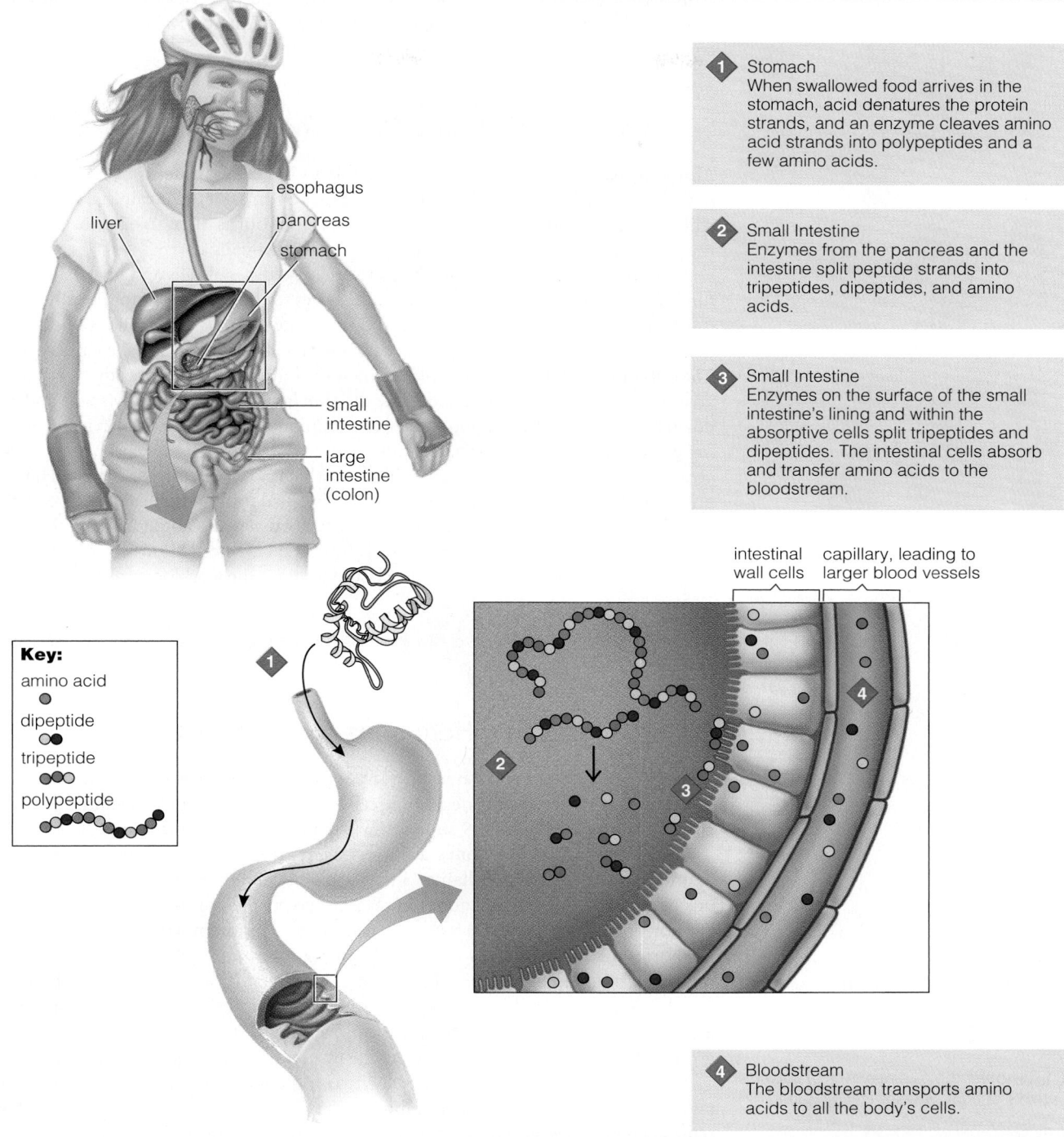

esophagus

liver

pancreas

stomach

small intestine

large intestine (colon)

1 Stomach
When swallowed food arrives in the stomach, acid denatures the protein strands, and an enzyme cleaves amino acid strands into polypeptides and a few amino acids.

2 Small Intestine
Enzymes from the pancreas and the intestine split peptide strands into tripeptides, dipeptides, and amino acids.

3 Small Intestine
Enzymes on the surface of the small intestine's lining and within the absorptive cells split tripeptides and dipeptides. The intestinal cells absorb and transfer amino acids to the bloodstream.

intestinal wall cells   capillary, leading to larger blood vessels

Key:

amino acid

dipeptide

tripeptide

polypeptide

4 Bloodstream
The bloodstream transports amino acids to all the body's cells.

body. The cells can then link the amino acids together to make proteins that they keep for their own use or liberate into lymph or blood for other uses. When necessary, the body's cells can also use amino acids for energy.

▶ **KEY POINT**   The cells of the small intestine complete digestion, absorb amino acids and some larger peptides, and release them into the bloodstream for use by the body's cells.

The Roles of Proteins in the Body

Only a sampling of the many roles proteins play can be described here, but these illustrate their versatility, uniqueness, and importance in the body. No wonder their discoverers called proteins the primary material of life.

Supporting Growth and Maintenance

Amino acids must be continuously available to build the proteins of new tissue. The new tissue may be in an embryo; in the muscles of an athlete in training; in a growing child; in new blood needed to replace blood lost in burns, hemorrhage, or surgery; in the scar tissue that heals wounds; or in new hair and nails.

Less obvious is the protein that helps to replace worn-out cells and internal cell structures. Each of your millions of red blood cells lives for only three or four months. Then it must be replaced by a new cell produced by the bone marrow. The millions of cells lining your intestinal tract live for only three days; they are constantly being shed and replaced.[7] The cells of your skin die and rub off, and new ones grow from underneath. Nearly all cells arise, live, and die in this way, and while they are living, they constantly make and break down their proteins. In addition, cells must continuously replace their own internal working proteins as old ones wear out. Amino acids conserved from these processes provide a great deal of the required raw material from which new structures are built. The entire process of breakdown, recovery, and synthesis is called **protein turnover.** Each day, about a quarter of available amino acids are irretrievably diverted to other uses, such as being used for fuel. For this reason, amino acids from food are needed each day to support all the new growth and maintenance of cells and to make the working parts within them.

> ▶ **KEY POINT**   The body needs dietary amino acids to grow new cells and to replace worn-out ones.

Building Enzymes, Hormones, and Other Compounds

Enzymes are among the most important of the proteins formed in living cells. Thousands of enzymes reside inside a single cell, each one a catalyst that facilitates a specific chemical reaction. Figure 6-9 shows how a hypothetical enzyme works.

The body's many **hormones** are messenger molecules, and some are made from amino acids. (Recall from Chapter 5 that some hormones are made from lipids.) Various body glands release hormones in response to changes in the internal environment; the hormones then elicit the responses necessary to restore normal conditions. Among the hormones made of amino acids is the thyroid hormone, which regulates the body's metabolism. An opposing pair of hormones, insulin and glucagon, maintain blood glucose levels, as described in Chapter 4. For interest, Figure 6-10 shows how many amino acids are linked in sequence to form human insulin. It also shows how certain side groups attract one another to complete the insulin molecule and make it functional.

In addition to serving as building blocks for proteins, amino acids also perform other tasks in the body. For example, the amino acid tyrosine forms parts of the chemical messengers epinephrine and norepinephrine, which relay messages throughout the nervous system. The body also uses tyrosine to make the brown pigment melanin, which is responsible for skin, hair, and eye color. Tyrosine is also converted into the hormone thyroxine, which helps to regulate the body's metabolic rate. The amino acid tryptophan serves as starting material for the neurotransmitter **serotonin** and the vitamin niacin.

> ▶ **KEY POINT**   The body makes enzymes, hormones, and chemical messengers of the nervous system from its amino acids.

▩ Neurotransmitters were introduced in Chapter 3; the vitamin niacin is discussed in Chapter 7.

protein turnover the continuous breakdown and synthesis of body proteins involving the recycling of amino acids.

hormones chemical messengers secreted by a number of body organs in response to conditions that require regulation. Each hormone affects a specific organ or tissue and elicits a specific response. Also defined in Chapter 3.

serotonin (SARE-oh-TONE-in) a compound related in structure to (and made from) the amino acid tryptophan. It serves as one of the brain's principal neurotransmitters.

FIGURE 6·9   ENZYME ACTION

Enzymes are catalysts: they speed up reactions that would happen anyway, but much more slowly. This enzyme works by positioning two compounds, A and B, so that the reaction between them will be especially likely to take place.

    Compounds A and B are attracted to the enzyme's active site and park there for a moment in the exact position that makes the reaction between them most likely to occur. They react by bonding together and leave the enzyme as the new compound, AB.

    A single enzyme can facilitate several hundred such synthetic reactions in a second. Other enzymes break compounds apart into two or more products or rearrange the atoms in one compound to make another one.

  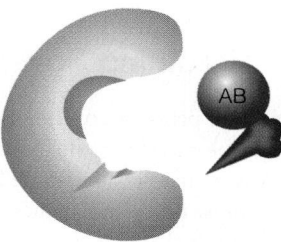

Enzyme plus
two compunds
A and B

Enzyme
complex with
A and B

Enzyme plus
new compound
AB

Building Antibodies

Of all the proteins in living organisms, the **antibodies** best demonstrate that proteins are specific to one organism. Antibodies recognize every protein that belongs in "their" body and leave it alone, but they attack foreign particles (usually proteins) that invade that body. The foreign protein may be part of a bacterium, a virus, or a toxin, or it may be present in a food that causes allergy. The body, upon recognizing that it has been invaded, manufactures antibodies specially designed to inactivate the foreign protein.

    Each antibody is designed to destroy one specific invader. An antibody active against one strain of influenza is of no help to a person ill with another strain. Once the body has learned how to make a particular antibody, it remembers. The next time the body encounters that same invader, it destroys the invader even more rapidly. In other words, the body develops **immunity** to the invader. This molecular memory underlies the principle of immunizations, injections of drugs made from destroyed and inactivated microbes or their products that activate the body's immune defenses. Some immunities are lifelong; others, such as that to tetanus, must be "boosted" at intervals.

▶ **KEY POINT**   Antibodies are proteins that defend against foreign proteins and other foreign substances within the body.

Maintaining Fluid and Electrolyte Balance

Proteins help to maintain the **fluid and electrolyte balance** by regulating the quantity of fluids in the compartments of the body. To remain alive, cells must contain a constant amount of fluid. Too much can cause them to rupture; too little makes them unable to function. Although water can diffuse freely into and out of cells, proteins cannot; and proteins attract water. By maintaining stores of internal proteins and also of some minerals, cells retain the fluid they need. By the same mechanism, fluid is kept inside the blood vessels by proteins too large to move freely across the capillary walls. The proteins attract water and hold it in within the vessels, preventing it from freely flowing into the spaces between the cells. Should any part of this system begin to fail, too much fluid will soon collect in the spaces between the cells of tissues, causing **edema.**

FIGURE 6·10   AMINO ACID SEQUENCE OF HUMAN INSULIN

This picture shows a refinement of protein structure not mentioned in the text. The amino acid cysteine (cys) has a sulfur-containing side group. The sulfur groups on two cysteine molecules can bond together, creating a bridge between two protein strands or two parts of the same strand. Insulin contains three such bridges.

■ The immune system is a topic of Chapter 3.

■ The control of water's location by electrolytes is discussed further in Chapter 8.

antibodies (AN-te-bod-ees) large proteins of the blood, produced by the immune system in response to an invasion of the body by foreign substances (antigens). Antibodies combine with and inactivate the antigens. Also defined in Chapter 3.

immunity protection from or resistance to a disease or infection by development of antibodies and by the actions of cells and tissues in response to a threat.

fluid and electrolyte balance the distribution of fluid and dissolved particles among body compartments (see also Chapter 8).

edema (eh-DEEM-uh) swelling of body tissue caused by leakage of fluid from the blood vessels; seen in protein deficiency (among other conditions).

Not only is the quantity of body fluids vital to life, but so also is their composition. Transport proteins in the membranes of cells maintain this composition by continuously transferring substances into and out of cells (see Figure 6-11). For example, sodium is concentrated outside the cells, and potassium is concentrated inside. A disturbance of this balance can impair the action of the heart, lungs, and brain, triggering a major medical emergency. Cell proteins avert such a disaster by holding fluids and electrolytes in their proper chambers.

▶ **KEY POINT**   Proteins help to regulate the body's electrolytes and fluids.

Maintaining Acid-Base Balance

Normal processes of the body continually produce **acids** and their opposite, **bases,** that must be carried by the blood to the organs of excretion. The blood must do this without allowing its own **acid-base balance** to be affected. This feat is another trick of the blood proteins, which act as **buffers** to maintain the blood's normal pH. The protein buffers pick up hydrogens (acid) when there are too many in the bloodstream and release them again when there are too few. The secret is that negatively charged side chains of amino acids can accommodate additional hydrogens, which are positively charged.

Blood pH is one of the most rigidly controlled conditions in the body. If blood pH changes too much, **acidosis** or the opposite basic condition, **alkalosis,** can cause coma or death. These conditions constitute medical emergencies because of their effect on proteins. When the proteins' buffering capacity is filled—that is, when they have taken on all the acid hydrogens they can accommodate—additional acid pulls them out of shape, denaturing them and disrupting many body processes. Table 6-1 sums up the functions of protein.

▶ **KEY POINT**   Proteins buffer the blood against excess acidity or alkalinity.

Clotting of Blood

To prevent dangerous blood loss, special blood proteins respond to an injury by clotting the blood. In an amazing series of chemical events, these proteins form a stringy net that traps blood cells to form a clot. The clot acts as a plug to stem blood flow from

■ Chapter 3 provides a discussion of pH.

acids compounds that release hydrogens in a watery solution.

bases compounds that accept hydrogens from solutions.

acid-base balance equilibrium between acid and base concentrations in the body fluids.

buffers compounds that help keep a solution's acidity or alkalinity constant.

acidosis (acid-DOH-sis) the condition of excess acid in the blood, indicated by a below-normal pH (*osis* means "too much in the blood").

alkalosis (al-kah-LOH-sis) the condition of excess base in the blood, indicated by an above-normal blood pH (alkalinity—*alka* means "base"; *osis* means "too much in the blood").

FIGURE 6·11   PROTEINS TRANSPORT SUBSTANCES INTO AND OUT OF CELLS

The transport protein within the cell membrane acts as a sort of two-door passageway—substances enter on one side and are released on the other, but the protein never leaves the membrane. The protein differs from a simple passageway in that it actively escorts the substances in and out of cells; therefore, this form of transport is often called active transport.

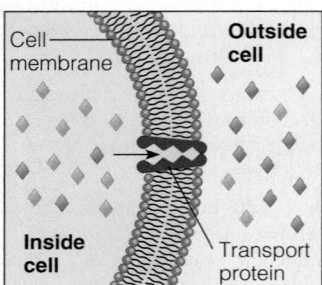

Molecule enters protein from inside cell.

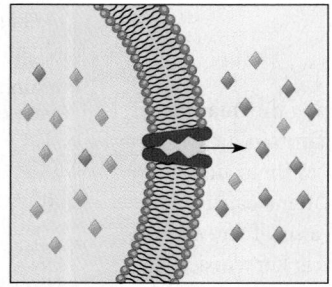

Protein changes shape; molecule exits protein outside the cell.

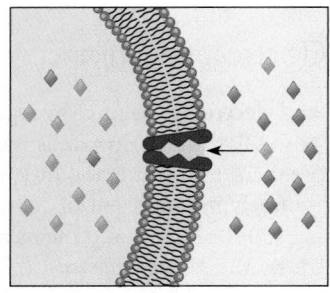

Molecule enters protein from outside cell.

Molecule exits protein; proper balance restored.

the wound. Later, as the wound heals, the protein collagen finishes the job by replacing the clot with scar tissue.

▶ **KEY POINT** Proteins that clot the blood prevent uncontrolled bleeding from injuries.

Providing Energy and Glucose

Only protein can perform all the functions just described, but protein will be surrendered to provide energy if need be. For most people eating a normal mixed diet, protein provides about 15 percent of the daily need for energy. Under conditions of inadequate energy or carbohydrate, protein use speeds up.[8] The body must have energy to live from moment to moment, so obtaining that energy is a top priority.

When amino acids are degraded for energy, their amine groups are stripped off and used elsewhere or are incorporated by the liver into **urea** and sent to the kidneys for excretion in the urine. The fragments that remain are composed of carbon, hydrogen, and oxygen, as are carbohydrate and fat, and can be used to build glucose or fatty acids or can be metabolized like them.

Not only can amino acids supply energy, but many of them can be converted to glucose, as fatty acids can never be. Thus, if the need arises, protein can help to maintain a steady blood glucose level and serve the glucose need of the brain.

The similarities and differences of the three energy-yielding nutrients should now be clear. Carbohydrate offers energy; fat offers concentrated energy; and protein can offer energy plus nitrogen (see Figure 6-12, p. 192).

Glucose is stored as glycogen and fat as triglycerides, but there is no specialized protein energy-storage compound. Body protein is available only as the active working molecular and structural components of the tissues. When protein-sparing energy from carbohydrate and fat is lacking and the need becomes urgent, as in starvation, prolonged fasting, or severe calorie restriction, the body must dismantle its tissue proteins to obtain amino acids for energy. As mentioned, each protein is taken in its own time: first, small proteins from the blood; then, proteins from the muscles, liver, and other organs. Thus, energy deficiency (starvation) always incurs wasting of lean body tissue as well as loss of fat.

If amino acids are oversupplied, the body cannot store them. It has no choice but to remove and excrete their amine groups and then either use the residues to meet ongoing energy needs, convert them to glucose for storage as glycogen, or convert them to fat for energy storage. The body readily converts amino acids to glucose. As for converting them to fat, the body possesses enzymes to do so and can produce fatty acids for storage as triglycerides in the fat tissue. Another important contribution of amino acids to fat stores is indirect—the body speeds up its use of amino acids for fuel, burning them instead of fat, which is then abundantly available for storage in the fat tissue.

▶ **KEY POINT** When insufficient carbohydrate and fat are consumed to meet the body's energy need, food protein and body protein are sacrificed to supply energy. The nitrogen part is removed from each amino acid, and the resulting fragment is oxidized for energy. No storage form of amino acids exists in the body.

The Fate of an Amino Acid

To review the body's handling of amino acids, let us follow the fate of an amino acid that was originally part of a protein-containing food. When the amino acid arrives

TABLE 6·1 SUMMARY OF FUNCTIONS OF PROTEINS

- *Acid-base balance.* Proteins help maintain the acid-base balance of various body fluids by acting as buffers.
- *Antibodies.* Proteins form the immune system molecules that fight diseases.
- *Blood clotting.* Proteins provide the netting on which blood clots are built.
- *Energy.* Proteins provide some fuel for the body's energy needs.
- *Enzymes.* Proteins facilitate needed chemical reactions.
- *Fluid and electrolyte balance.* Proteins help to maintain the water and mineral composition of various body fluids.
- *Growth and maintenance.* Proteins form integral parts of most body tissues and serve as building materials for growth and repair of body tissues, such as skin, connective tissues, muscles, organs, and bone.
- *Hormones.* Proteins regulate body processes. Some hormones are proteins or are made from amino acids.
- *Transportation.* Proteins help transport needed substances, such as lipids, minerals, and oxygen, around the body.

■ More about the effects of fasting in Chapter 9.

urea (yoo-REE-uh) the principal nitrogen-excretion product of protein metabolism; generated mostly by removal of amine groups from unneeded amino acids or from amino acids being sacrificed to a need for energy.

FIGURE 6·12 THREE DIFFERENT ENERGY SOURCES

Carbohydrate offers energy; fat offers concentrated energy; and protein, if necessary, can offer energy plus nitrogen. The compounds at the left yield the 2-carbon fragments shown at the right. These fragments oxidize quickly in the presence of oxygen to yield carbon dioxide, water, and energy.

carbohydrate — Energy + (4 calories per gram)

fat — Energy Energy + (9 calories per gram)

protein — nitrogen + Energy (4 calories per gram)

in a cell, it can be used in one of several ways, depending on the cell's needs at the time:

- The amino acid can be used as is to build part of a growing protein.
- The amino acid can be altered somewhat to make another needed compound, such as the vitamin niacin.
- The cell can dismantle the amino acid in order to use its amine group to build a different amino acid. The remainder can be used for fuel or, if fuel is abundant, converted to glucose or fat.

Amino acids in a cell can be:

- Used to build protein.
- Converted to other amino acids or small nitrogen-containing compounds.

Stripped of their nitrogen, amino acids can be:

- Burned as fuel.
- Converted to glucose or fat.

In a cell that is starved for energy but has no glucose or fatty acids, the cell strips the amino acid of its amine group (the nitrogen part) and uses the remainder of its structure for energy. The amine group is excreted from the cell and then from the body in the urine. When the body has a surplus of amino acids and energy, the cell takes the amino acid apart, excretes the amine group, and uses the rest for energy or converts it to glucose or fat for storage.[9]

When not used to build protein or make other nitrogen-containing compounds, amino acids are "wasted" in a sense. This wasting occurs under any of four conditions:

1. When the body does not have enough energy from other sources.
2. When the diet provides more protein than the body needs.
3. When the body has too much of any single amino acid, such as from a supplement.
4. When the diet supplies protein of low quality, with too few essential amino acids, as described in the next section.

Amino acids are wasted when:

- Energy is lacking.
- Protein is overabundant.
- An amino acid is oversupplied in supplement form.
- The quality of the diet's protein is too low (too few essential amino acids).

To prevent the wasting of dietary protein and permit the synthesis of needed body protein, the dietary protein must be adequate in quantity, it must supply all essential amino acids in the proper amounts, and it must be accompanied by enough energy-yielding carbohydrate and fat to permit the dietary protein to be used as such.

▶ **KEY POINT** Amino acids can be metabolized to protein, nitrogen plus energy, glucose, or fat. They will be metabolized to protein only if sufficient energy is present from other sources. The diet should supply all essential amino acids and a full measure of protein according to guidelines.

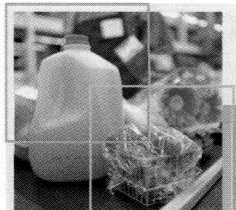

CONSUMER CORNER

Protein and Amino Acid Supplements

WHY DO PEOPLE take protein or amino acid supplements? Athletes often take them when trying to build muscle. Dieters may take them in hopes of speeding the process of weight loss. Some consumers believe the products will cure herpes virus infections, induce restful sleep, or relieve pain or depression. Do protein and amino acid supplements really do any of these things? Probably not. Are they safe? Not always.

In the skilled hands of clinical dietitians, formulas with supplemental amino acids may help to reverse malnutrition in some critically ill patients.[1] Not every patient is a candidate for such therapy, though, because supplemental amino acids may also stimulate inflammation and so worsen some illnesses.[2] When used by healthy people as a replacement for foods, especially in weight-loss diets, such formulas can be downright dangerous. A "liquid protein" diet, advocated some years ago for weight loss, caused the deaths of many users and heart problems in many others.

Enthusiastic popular reports have led to widespread use of certain amino acids. One is lysine, touted to prevent or relieve the infections that cause herpes sores on the mouth or genital organs. Lysine does not cure herpes infections. Whether it reduces outbreaks or even whether it is safe is unknown because scientific studies are lacking. Blends of the branched-chain amino acids leucine, isoleucine, and valine have become popular with athletes hoping for a competitive edge, but the value of these potions also remains unclear (see Controversy 10).

Tryptophan supplements are advertised to relieve pain, depression, and insomnia. Tryptophan plays a role as a precursor for the brain neurotransmitter serotonin, which makes it an important regulator of sleep, appetite, mood, and sensory perception.[3] In the case of tryptophan supplements, however, advertising and popular reports have long preceded careful scientific experiments. The DRI committee concludes that high doses of tryptophan may induce sleepiness, but they may also have side effects such as severe nausea and skin disorders.[4]

In the not-too-distant past, people who elected to take supplements of the amino acid tryptophan developed a blood disorder (EMS, short for *eosinophilia-myalgia syndrome*), and at least 15 of the supplement takers died. Contaminants in the supplement were blamed for the disease, and the Food and Drug Administration (FDA) recalled tryptophan supplements and formulas to which it was added. Impurities of the same sort are still detected in some tryptophan-containing products currently on the market.[5]

The body is designed to handle whole proteins best. It breaks them into manageable pieces (dipeptides and tripeptides), then splits these a few at a time, simultaneously releasing them into the blood. This slow bit-by-bit assimilation is ideal because groups of chemically similar amino acids compete for the carriers that absorb them into the blood. An excess of one amino acid can tie up a carrier and disturb amino acid absorption, at least temporarily. Digestive disturbances, excess water in the digestive tract, and an increased need for the vitamin thiamin also pose potential problems from amino acid supplements.[6]

The DRI committee reviewed the available research on amino acids but with next to no safety research in existence, the committee was unable to set Tolerable Upper Intake Levels for supplemental doses. Until research becomes available, no level of amino acid supplementation can be assumed safe for all people. Growth or altered metabolism makes these groups of people especially likely to suffer harm from self-prescribed amino acid supplements:

- All women of childbearing age.
- Pregnant or lactating women.
- Infants, children, and adolescents.
- Elderly people.
- People with inborn errors of metabolism that affect their bodies' handling of amino acids.
- Smokers.
- People on low-protein diets.
- People with chronic or acute mental or physical illnesses.

A careful review of the literature concluded that not enough research exists to support recommending long-term consumption of amino acid supplements by healthy people and that more research into adverse effects is needed.[7] Canada bans sales of single amino acids to consumers. The warning is this: much is still unknown, and the taker of amino acid supplements cannot be certain of their safety or effectiveness.

Many chapters of this book present evidence on purified nutrients added to foods or taken singly. The Consumer Corner in Chapter 4 showed that a nutritionally inferior food (refined bread) enriched with a few added nutrients is still inadequate in many others. The Consumer Corner in Chapter 5 showed that artificial fats can have side effects. The same is true of amino acids. Even with all that we know about science, it is hard to improve on nature.

Cooking with moist heat improves protein digestibility, whereas frying makes protein harder to digest.

Food Protein: Quality, Use, and Need

The body's response to protein depends on many factors: the body's state of health, the other nutrients and energy taken with the protein, and the protein's quality. To know whether, say, 60 grams of a particular protein is enough to meet a person's daily needs, one must consider the effects of these other factors on the body's use of the protein.

Regarding a person's state of health, malnutrition or infection may greatly increase the need for protein while making it hard to eat even normal amounts of food. In malnutrition, secretion of digestive enzymes slows as the tract's lining degenerates, impairing protein digestion and absorption. When infection is present, extra protein is needed for enhanced immune functions.

Regarding the other nutrients taken with protein, the need for ample energy, carbohydrate, and fat has already been emphasized. To be used efficiently by the cells, protein must also be accompanied by the full array of vitamins and minerals.

The remaining factor, protein quality, helps determine how well a diet supports the growth of children and the health of adults. Two factors influence protein quality: a protein's digestibility and its amino acid composition.

Which Kinds of Protein-Rich Foods Are Easiest to Digest?

The digestibility of protein varies from food to food and bears profoundly on protein quality. The protein of oats, for example, is less digestible than that of eggs. In general, amino acids from animal proteins are most easily digested and absorbed (over 90 percent). Those from **legumes** are next (about 80 to 90 percent). Those from grains and other plant foods vary (from 70 to 90 percent). Cooking with moist heat improves protein digestibility, whereas dry heat methods can impair it.

In measuring a protein's quality, digestibility is important. Simple measures of the total protein in foods are not useful by themselves—even animal hair and hooves would receive a top score by those measures alone. They are made of protein, but not in a form that people can use.

> ▶ **KEY POINT** The body's use of a protein depends in part on the user's health, the protein quality, and the other nutrients and energy taken with it. Digestibility of protein varies from food to food, and cooking can improve or impair it.

Amino Acid Composition

Put simply, **high-quality proteins** provide enough of all the essential amino acids needed by the body to create its own working proteins, whereas low-quality proteins don't. In making their required proteins, the cells need a full array of amino acids from food, from their own **amino acid pools,** or from both. If a nonessential amino acid (that is, one the cell can make) is unavailable from food, the cell synthesizes it and continues attaching amino acids to the protein strands being manufactured. If the diet fails to provide enough of an essential amino acid (one the cell cannot make), the cells begin to adjust their activities. Within a single day of restricted intake of an essential amino acid, the cells begin to conserve it by limiting the breakdown of their working proteins and by reducing their use of amino acids for fuel.

Limiting Amino Acids Can Limit Protein Synthesis The measures just described help the cells to channel the available **limiting amino acid** to its wisest use: making new proteins. Even so, the normally fast rate of protein synthesis slows to a

legumes (leg-GOOMS, LEG-yooms) plants of the bean, pea, and lentil family that have roots with nodules containing special bacteria. These bacteria can trap nitrogen from the air in the soil and make it into compounds that become part of the plant's seeds. The seeds are rich in protein compared with those of most other plant foods. Also defined in Chapter 1.

high-quality proteins dietary proteins containing all the essential amino acids in relatively the same amounts that human beings require. They may also contain nonessential amino acids.

amino acid pools amino acids dissolved in the body's fluids that provide cells with ready raw materials from which to build new proteins or other molecules.

limiting amino acid an essential amino acid that is present in dietary protein in an insufficient amount, thereby limiting the body's ability to build protein.

crawl as the cells make do with the proteins on hand. When the limiting amino acid once again becomes available in abundance, the cells resume their normal protein-related activities. If the shortage becomes chronic, however, the cells begin to break down their protein-making machinery. Consequently, when protein intakes become adequate again, protein synthesis lags behind until the needed machinery can be rebuilt. Meanwhile, the cells function less and less effectively as their proteins wear out and are only partially replaced.

Thus, a diet that is short in any of the essential amino acids limits protein synthesis. An earlier analogy likened amino acids to letters of the alphabet. To be meaningful, words must contain all the right letters. For example, a print shop that has no letter "N" cannot make personalized stationery for Jana Johnson. No matter how many Js, As, Os, Hs, and Ss are in the printer's possession, they cannot replace the missing Ns. Likewise, in building a protein molecule, no amino acid can fill another's spot. If a cell that is building a protein cannot find a needed amino acid, synthesis stops, and the partial protein is released.

Partially completed proteins are not held for completion at a later time when the diet may improve. Rather, they are dismantled, and the component amino acids are returned to the circulation to be made available to other cells. If they are not soon inserted into protein, their amine groups are removed and excreted, and the residues are used for other purposes. The need that prompted the call for that particular protein will not be met. Since the other amino acids are wasted, the amine groups are excreted, and the body cannot resynthesize the amino acids later.

Complementary Proteins It follows that if a person does not consume all the essential amino acids in proportion to the body's needs, the body's pools of essential amino acids will dwindle until body organs are compromised. Consuming the essential amino acids presents no problem to people who regularly eat proteins containing ample amounts of all of the essential amino acids such as those of meat, fish, poultry, cheese, eggs, milk, and most soybean products. An equally sound choice is to eat a combination of foods from plants so that amino acids that are low in some foods will be supplied by the others. The protein-rich foods are combined to yield **complementary proteins** (see Figure 6-13), or proteins containing all the essential amino acids in amounts sufficient to support health.[10] This concept, called **mutual supplementation,** is illustrated in Figure 6-14 (p. 196). The complementary proteins need not even be eaten together, so long as the day's meals supply them all, and the diet provides enough energy and total protein from a variety of sources.

Just as each letter of the alphabet is important in forming whole words, each amino acid must be available to build finished proteins.

FIGURE 6·13 COMPLEMENTARY PROTEIN COMBINATIONS

Healthful foods like these contribute substantial protein (42 grams total) to this day's meals without meat. Additional servings of nutritious foods, such as milk, bread, and eggs, can easily supply the remainder of the day's need for protein (14 additional grams for men and 4 for women).

¾ c oatmeal	=	5 g
Protein total		5 g

1 c rice	=	4 g
1 c beans	=	16 g
Protein total		20 g

1½ c pasta	=	11 g
1 c vegetables	=	2 g
2 tbs parmesan cheese	=	4 g
Protein total		17 g

complementary proteins two or more proteins whose amino acid assortments complement each other in such a way that the essential amino acids missing from one are supplied by the other.

mutual supplementation the strategy of combining two incomplete protein sources so that the amino acids in one food make up for those lacking in the other food. Such protein combinations are sometimes called *complementary proteins*.

FIGURE 6·14 PROTEINS THAT COMPLEMENT EACH OTHER WORK TOGETHER

In general, legumes provide plenty of the amino acids isoleucine (Ile) and lysine (Lys), but fall short in methionine (Met) and tryptophan (Trp). Grains have the opposite strengths and weaknesses, making them a perfect match for legumes.

	Ile	Lys	Met	Trp
Legumes				
Grains				
Together				

Concern about the quality of individual food proteins is of only theoretical interest in settings where food is abundant. Most people in the United States and Canada eat a variety of nutritious foods to meet their energy needs. Healthy adults would find it next to impossible *not* to meet their protein requirements, even if they were to eat no meat, fish, poultry, eggs, cheese, or soy products. They need not pay attention to mutual supplementation, so long as the diet is varied, nutritious, and adequate in energy and other nutrients—not made up just of, say, cookies, potato chips, and alcoholic beverages. Protein sufficiency follows effortlessly behind.

For people in areas where food sources are less reliable, protein quality can make the difference between health and disease. When food energy intake is limited (where malnutrition is widespread) or when the selection of foods available is severely limited (where a single food such as potatoes provides 90 percent of the calories), the primary food source of protein must be checked because its quality is crucial.

▶ **KEY POINT** A protein's amino acid assortment greatly influences its usefulness to the body. Proteins lacking essential amino acids can be used only if those amino acids are present from other sources.

Measuring Protein Quality

Researchers have developed many methods of evaluating the quality of food protein. The most important one for consumers is the **protein digestibility–corrected amino acid score,** or **PDCAAS,** which is used by the DRI committee to evaluate people's protein intakes.

On the PDCAAS scale of 100 to 0, with 100 representing protein sources that are most readily digested and most perfectly balanced for meeting human needs, egg white, ground beef, chicken products, fat-free milk, and tuna fish all score 100. Soybean protein isn't far behind at 94. Most legumes rank in the 60s and 50s. The wheat protein, gluten, formed during bread making, ranks 25. Something interesting happens when pea flour (67) and whole-wheat flour (40) are combined: the score for the resulting flour is 82. Why? They are complementary proteins.

In trying to choose between peanut butter and chili in the grocery store, you may think you have no use for the PDCAAS. When you read the "% Daily Value" for protein listed on the labels, however, you are using it indirectly. Manufacturers that list values for protein on food labels must use the PDCAAS to determine the protein quality of their products. Thus, protein values on labels reflect both the digestibility and the amino acid composition of the foods. Another measure relevant to the average well-fed North American is the DRI recommended intake for protein, discussed next.

▶ **KEY POINT** The quality of a protein is measured by its amino acids, by its digestibility, and by how well it meets human needs. The protein value listed on a food label reflects the PDCAAS of the protein in the food.

How Much Protein Do People Really Need?

The DRI recommendation for protein intake is designed to cover the need to replace protein-containing tissue that healthy adults lose and wear out every day. Therefore, it depends on body size: larger people have a higher protein need. For adults, the DRI recommended intake is set at 0.8 gram for each kilogram (or 2.2 pounds) of body weight (see inside front cover, page A). The minimum amount is set at 10 percent of total calories. Athletes may need slightly more according to authorities other than the DRI committee, but the increased need is well covered by a regular diet. For infants and growing children, the protein recommendation, like all nutrient recommendations, is higher per unit of body weight. The DRI committee set an upper limit for protein intake of no more than 35 percent of total calories, an amount significantly

protein digestibility–corrected amino acid score (PDCAAS) a measuring tool used to determine protein quality. The PDCAAS reflects a protein's digestibility as well as the proportions of amino acids that it provides.

TABLE 6·2 RECOMMENDATIONS CONCERNING INTAKES OF PROTEIN FOR ADULTS

DRI recommended intakes:[a]
- 0.8 gram protein per kilogram of body weight per day.
- Women: 46 grams per day; men: 56 grams per day.

2005 USDA Food Guide
- Every day most adults should eat 5 to 6½ ounce equivalents of lean meat, poultry without skin, fish, legumes, eggs, nuts, or seeds.
- Every day most adults need 3 cups of fat-free or low fat milk or yoguart, or the equivalent of fat free cheese or vitamin and mineral fortified soy beverage.
- Eat a variety of foods to provide small amounts of protein from other sources.

World Health Organization
- Lower limit: 10% of total calories from protein.
- Upper limit: 15% of total calories from protein.

Canada's Food Guide to Healthy Eating is presented in Appendix B.

[a]Protein recommendations for infants, children, and pregnant and lactating women are higher; see inside front cover, page B.

■ The DRI recommended intake for protein (adult) = 0.8 g/kg.

To figure your protein need:
1. Find your body weight in pounds.
2. Convert pounds to kilograms (by dividing pounds by 2.2).
3. Multiply kilograms by 0.8 to find total grams of protein recommended.

For example:
- Weight = 130 lb.
- 130 lb ÷ 2.2 = 59 kg.
- 59 kg × 0.8 = 47 g.

higher than average intakes. Table 6-2 reviews recommendations for protein intakes and the margin provides a method for determining your own protein need.

Recommendations for protein intake assume a normal mixed diet; that is, a diet that includes a combination of animal and plant protein. Because not all proteins are used with 100 percent efficiency, the recommendation is quite generous. Many healthy people can consume less than this amount and still meet their bodies' protein needs. What this means in terms of food selections is presented in this chapter's Food Feature.

Underlying the protein recommendation are **nitrogen balance** studies, which compare nitrogen lost by excretion with nitrogen eaten in food. In healthy adults, nitrogen-in (consumed) must equal nitrogen-out (excreted). Scientists measure the body's daily nitrogen losses in urine, feces, sweat, and skin under controlled conditions and then estimate the amount of protein needed to replace these losses.[*][11]

Under normal circumstances, healthy adults are in nitrogen equilibrium, or zero balance; that is, they have the same amount of total protein in their bodies at all times. When nitrogen-in exceeds nitrogen-out, people are said to be in positive nitrogen balance; somewhere in their bodies more proteins are being built than are being broken down and lost. When nitrogen-in is less than nitrogen-out, people are said to be in negative nitrogen balance; they are losing protein. Figure 6-15 (p. 198) illustrates these different states.

Growing children add new blood, bone, and muscle cells to their bodies every day, so children must have more protein, and therefore more nitrogen, in their bodies at the end of each day than they had at the beginning. A growing child is therefore in positive nitrogen balance. Similarly, when a woman is pregnant, she must be in positive nitrogen balance until after the birth when she once again reaches equilibrium.

Negative nitrogen balance occurs when muscle or other protein tissue is broken down and lost. Illness or injury triggers the release of powerful messengers that signal the body to break down some of the less vital proteins, such as those of the skin and even muscle.[†][12] This action floods the blood with amino acids and energy needed to fuel the body's defenses and fight the illness. The result is negative nitrogen balance. Astronauts, too, experience negative nitrogen balance. In the stress of space flight and

Growing children end each day with more bone, blood, muscle, and skin cells than they had at the beginning of the day.   © IT International eStock Photography/PictureQuest

nitrogen balance the amount of nitrogen consumed compared with the amount excreted in a given time period.

---

[*]The average protein is 16 percent nitrogen by weight; that is, each 100 grams of protein contain 16 grams of nitrogen. As a rule of thumb, multiply the nitrogen's weight by 6.25 to estimate the protein's weight.
[†]The messengers are cytokines.

FIGURE 6·15 NITROGEN BALANCE

Positive Nitrogen Balance
These people, a growing child, a person building muscle, and a pregnant woman, are all retaining more nitrogen than they are excreting.

Nitrogen Equilibrium
These people, a healthy college student and a young retiree, are in nitrogen equilibrium.

Negative Nitrogen Balance
These people, an astronaut and a surgery patient, are losing more nitrogen than they are taking in.

with no need to support the body's weight against gravity, the astronauts' muscles waste and weaken. To minimize the inevitable loss of muscle tissue, the astronauts must do special exercises in space.[13]

▶ KEY POINT   The amount of protein needed daily depends on size and stage of growth. The DRI recommended intake for adults is 0.8 gram of protein per kilogram of body weight. Protein recommendations are based on nitrogen balance studies, which compare nitrogen excreted from the body with nitrogen ingested in food.

Protein Deficiency and Excess

Protein deficiencies are well known because, together with energy deficiencies, they are the world's leading form of malnutrition. The health effects of too much protein are far less well known. Both deficiency and excess are of concern.

What Happens When People Consume Too Little Protein?

Protein deficiency and energy deficiency go hand in hand. This combination—**protein-energy malnutrition (PEM)**—is the most widespread form of malnutrition in the world today. Over 500 million children face imminent starvation and suffer the effects of severe malnutrition and **hunger.** Most of the 33,000 children who die each day are malnourished. PEM is prevalent in Africa, Central America, South America, the Middle East, and East and Southeast Asia, but developed countries, including those in North America, are not immune to it.

PEM strikes early in childhood, but it endangers many adults as well. Inadequate food intake leads to poor growth in children and to weight loss and wasting in adults. Stunted growth due to PEM is easy to overlook because a small child can look normal. The small stature of children in impoverished nations was once thought to be a normal adaptation to the limited availability of food; now it is known to be an avoidable failure of growth due to a lack of food during the growing years.

Scant supplies of donated food save some from starvation, but many others go hungry.

protein-energy malnutrition (PEM) the world's most widespread malnutrition problem, including both marasmus and kwashiorkor and states in which they overlap; also called *protein-calorie malnutrition (PCM).*

hunger the physiological craving for food; the progressive discomfort, illness, and pain resulting from the lack of food. (See also Chapters 9 and 15.)

The extreme loss of muscle and fat characteristic of marasmus is apparent in this child's "matchstick" arms.

The edema and enlarged liver characteristic of kwashiorkor are apparent in this child's swollen belly.

PEM takes two different forms, with some cases exhibiting a combination of the two. In one form, the person is shriveled and lean all over—this disease is called **marasmus.** In the second, a swollen belly and skin rash are present, and the disease is named **kwashiorkor.**[*][14] In the combination, some features of each type are present. Marasmus reflects a chronic inadequate food intake and therefore inadequate energy, vitamins, and minerals as well as too little protein. Kwashiorkor may result from severe acute malnutrition, with too little protein to support body functions.

Marasmus   Marasmus occurs most commonly in children from 6 to 18 months of age in overpopulated city slums. Children in impoverished nations subsist on a weak cereal drink with scant energy and protein of low quality; such food can barely sustain life, much less support growth. A starving child often looks like a wizened little old person—just skin and bones.

Without adequate nutrition, muscles, including the heart muscles, waste and weaken.[15] Brain development is stunted and learning is impaired. Metabolism is so slow that body temperature is subnormal. There is little or no fat under the skin to insulate against cold, and hospital workers have found that children with marasmus need to be wrapped up and kept warm. They also need love because they have often been deprived of parental attention as well as food.

The starving child faces this threat to life by engaging in as little activity as possible—not even crying for food. The body collects all its forces to meet the crisis and so cuts down on any expenditure of protein not needed for the heart, lungs, and brain to function. Growth ceases; the child is no larger at age four than at age two. The skin loses its elasticity and moisture, so it tends to crack; when sores develop, they fail to heal. Digestive enzymes are in short supply, the digestive tract lining deteriorates, and absorption fails. The child can't assimilate what little food is eaten.

Blood proteins, including hemoglobin, are no longer produced, so the child becomes anemic and weak. If a bone breaks, healing is delayed because the protein needed to heal it is lacking.[16] Antibodies to fight off invading bacteria are degraded to provide amino acids for other uses, leaving the child an easy target for infection. Then **dysentery,** an infection of the digestive tract, causes diarrhea, further depleting the

marasmus (ma-RAZ-mus) the calorie-deficiency disease; starvation.

kwashiorkor (kwash-ee-OR-core, kwashee-or-CORE) a disease related to protein malnutrition, with a set of recognizable symptoms, such as edema.

dysentery (DISS-en-terry) an infection of the digestive tract that causes diarrhea.

[*]A term gaining acceptance for use in place of kwashiorkor is *hypoalbuminemic-type PEM.*

body of nutrients, especially minerals. Measles, which might make a healthy child sick for a week or two, kills a child with PEM within two or three days. In the marasmic child, once infection sets in, kwashiorkor often follows and the immune response weakens further.[17] Infections that occur with malnutrition are responsible for two-thirds of the deaths of young children in developing countries.

Ultimately, marasmus progresses to the point of no return when the body's machinery for protein synthesis, itself made of protein, has been degraded. At this point, attempts to correct the situation by giving food or protein fail to prevent death. If caught before this time, however, the starvation of a child can be reversed by careful nutrition therapy.[18] The fluid balances are most critical. Diarrhea will have depleted the body's potassium and upset other electrolyte balances. The combination of electrolyte imbalances, anemia, fever, and infections often leads to heart failure and sudden death. Careful correction of fluid and electrolyte balances usually raises the blood pressure and strengthens the heartbeat within a few days. Later, fat-free milk, providing protein and carbohydrate, can safely be given; fat is introduced still later, when body protein is sufficient to provide carriers. Years after PEM is corrected, a child may experience deficits in thinking and school achievement compared with well-nourished peers.[19]

Kwashiorkor Kwashiorkor is the Ghanaian name for "the evil spirit that infects the first child when the second child is born." In countries where kwashiorkor is prevalent, each baby is weaned from breast milk as soon as the next one comes along. The older baby no longer receives breast milk, which contains high-quality protein designed perfectly to support growth, but is given a watery cereal with scant protein of low quality. Small wonder the just-weaned child sickens when the new baby arrives. Though rare in the United States, kwashiorkor is not entirely unknown, usually occurring when fad diets replace sound nutrition in children.

Some kwashiorkor symptoms resemble those of marasmus (see Table 6-3) but often without severe wasting of body fat. Proteins and hormones that previously maintained fluid balance are now diminished, so fluid leaks out of the blood and accumulates in the belly and legs, causing edema, a distinguishing feature of kwashiorkor. The kwashiorkor victim's belly often bulges with a fatty liver, caused by lack of the protein carriers that transport fat out of the liver. The fatty liver loses some of its ability to clear poisons from the body, prolonging their toxic effects. Without sufficient tyrosine to make melanin, the child's hair loses its normal color; inadequate protein synthesis leaves the skin patchy and scaly; sores fail to heal.

■ The term *electrolyte balance,* defined earlier, refers to the proper concentrations of salts (dissolved particles) within the body fluids (see Chapter 8 for details).

■ Melanin, a brown pigment of hair, skin, and eyes, was mentioned earlier as a product made from tyrosine.

TABLE 6·3 FEATURES OF MARASMUS AND KWASHIORKOR IN CHILDREN

Separating PEM into two classifications oversimplifies the condition, but at the extremes, marasmus and kwashiorkor exhibit marked differences. Marasmus-kwashiorkor mix presents symptoms common to both marasmus and kwashiorkor. In all cases, children are likely to develop diarrhea, infections, and multiple nutrient deficiencies.

Marasmus	Kwashiorkor
Infants and toddlers (less than 2 yr)	Older infants and young children (1 to 3 yr)
Severe deprivation or impaired absorption of protein, energy, vitamins, and minerals	Inadequate protein intake or, more commonly, infections
Develops slowly; chronic PEM	Rapid onset; acute PEM
Severe weight loss	Some weight loss
Severe muscle wasting with fat loss	Some muscle wasting, with retention of some body fat
Growth: <60% weight-for-age	Growth: 60 to 80% weight-for-age
No detectable edema	Edema
No fatty liver	Enlarged, fatty liver
Anxiety, apathy	Apathy, misery, irritability, sadness
Appetite may be normal or impaired	Loss of appetite
Hair is sparse, thin, and dry; easily pulled out	Hair is dry and brittle; easily pulled out; changes color; becomes straight
Skin is dry, thin, and wrinkled	Skin develops lesions

PEM at Home   PEM occurs among some groups in the United States and Canada: the poor living on U.S. Indian reservations, in inner cities, and in rural areas; many elderly people; hungry and homeless children; and those suffering from the eating disorder anorexia nervosa. Some well-meaning but misinformed parents have inflicted PEM and other deficiency diseases on their toddlers by replacing their milk with unenriched, protein-poor "health food" soy or rice drinks.[20] Also at risk for PEM are those with wasting diseases such as cancer or AIDS and those addicted to drugs and alcohol. In a downward spiral, PEM and serious illness worsen each other, so treating the PEM often reduces medical complications and suffering even when the underlying disease is untreatable.[21]

Today, in the United States, millions of people who work to support their children earn so little that they cannot afford nutritious food—well over 2.5 million children go hungry at least some of the time.[22] Hunger, especially in children, threatens everyone's future. Hungry children do not learn as well as fed children, nor are they as competitive. They are ill more often, they have higher absentee rates from school, and when they attend, they cannot concentrate for long. The forces driving poverty and hunger will require many great minds working together to find solutions. In fighting hunger, programs that tailor interventions to the local people and involve them in the process of identifying problems and devising solutions report the most success.[23]

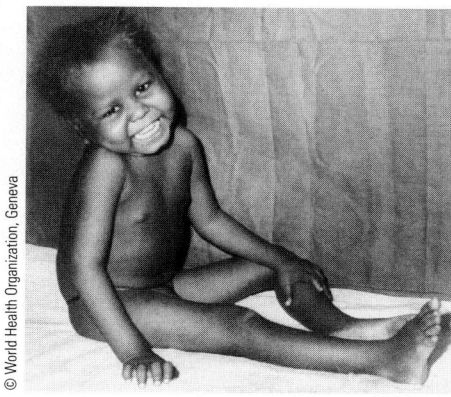

Given appropriate nutrition and care, this child has successfully recovered from kwashiorkor.

■ Turn to Chapter 15 for details concerning the causes of hunger at home and abroad.

▶ **KEY POINT**   Protein-deficiency symptoms are always observed when either protein or energy is deficient. Extreme food energy deficiency is marasmus; extreme protein deficiency is kwashiorkor. The two diseases overlap most of the time, and together are called PEM.

Is It Possible to Consume Too Much Protein?

The DRI committee recommends that the diet contain no more than 35 percent of calories from protein to decrease risks of chronic diseases. Overconsumption of protein offers no benefits and may pose health risks for the heart, for weakened kidneys, and for the bones. Diets high in protein-rich foods are often associated with obesity and its many accompanying health risks.[24] Animal protein sources in particular can be high in saturated fat, a known contributor to atherosclerosis and heart disease. The effect of animal protein itself on heart health is uncertain. Conversely, substituting plant protein for animal protein improves indicators of heart disease risk, as this chapter's Controversy explains.[25]

Animals fed experimentally on high-protein diets often develop enlarged kidneys or livers. In human beings, a high-protein diet worsens existing kidney problems and may accelerate a decline in only mildly impaired kidneys.[26] One of the most effective treatments for people with established kidney problems is to reduce protein intakes to slow down the progression of their disease.

Evidence is mixed about whether high intakes of protein from animal sources, especially when accompanied by low calcium and low fruit and vegetable intakes, can accelerate adult bone loss. No doubt exists about the effect of feeding purified protein to human subjects—purified protein causes calcium to be spilled from the urine. Also, eating diets high in protein, particularly animal protein, correlates with a greater incidence of hip fractures in some populations. In one population, however, the reverse may be true—in malnourished elderly individuals, protein deficiency and hip fractures often occur together, and restoring dietary protein and giving calcium and vitamin D supplements can improve bone status.*[27] In establishing protein recommendations, the DRI committee considered the effect of excess protein on the bones but found the evidence insufficient to set a Tolerable Upper Intake Level.

A balance between protein and at least one other energy nutrient seems necessary for survival. Native Alaskans survive for long periods on a diet constructed almost entirely

*The calcium supplement is citrate malate.

of meats; they choose fatty meats and obtain a significant percentage of their calories from fat to balance their high intake of protein from meat. No one yet knows exactly what percentage of dietary protein is hazardous, but the meat-eating Alaskans typically consume less than 50 percent of calories from protein. The average protein intake in U.S. adult men is about 16 percent. Health recommendations typically advise a protein intake between 10 and 35 percent of energy intake. Popular low-carbohydrate, high-protein weight-loss diets often suggest up to 65 percent of energy from protein, an amount that may border on being too high for health.[28]

■ More about high-protein diets for weight loss in Chapter 9.

▶ **KEY POINT** Health risks may follow the overconsumption of protein-rich foods.

FOOD FEATURE

Getting Enough, but Not Too Much Protein

■ This chapter's Controversy section describes the benefits and pitfalls of vegetarian and meat-containing diets.

PEOPLE IN DEVELOPED nations usually eat more than ample protein. The DRI recommendation for protein is generous and more than adequately covers the estimated needs of most people, even those with unusually high requirements. Most foods contribute at least some protein to the diet.

Protein-Rich Foods

Foods in the meat, poultry, fish, dry peas and beans, eggs, and nuts group and in the milk, yogurt, and cheese group contribute an abundance of high-quality protein. Two others, the vegetable group and the grains group, contribute smaller amounts of protein, but they can add up to significant quantities. What about the fruit group? Don't rely on fruit for protein—fruit contains only small amounts. Figure 6-16 shows that a wide variety of foods contribute protein to the diet. Figure 6-17 (page 204) shows the top protein contributors in the U.S. diet.

Protein is critical in nutrition, but too many protein-rich foods can displace other important foods from the diet. Foods richest in protein carry with them a characteristic array of vitamins and minerals, including vitamin B_{12} and iron, but they lack others—vitamin C and folate, for example. In addition, many protein-rich foods such as meat are high in calories, and to overconsume them is to invite obe-

sity. Because American consumption of protein is ample, you can plan meatless or reduced-meat meals with pleasure. Of the many interesting, protein-rich meat equivalents available, one has already been mentioned: the legumes.

The Advantages of Legumes

The protein of some legumes is of a quality almost comparable to that of meat, an unusual trait in a fiber-rich vegetable. For practical purposes, the quality of soy protein can be considered equivalent to that of meat. Figure 6-18 (page 205) shows a legume plant's special root system that enables it to make abundant protein. Legumes are also excellent sources of many B vitamins, iron, calcium, and other minerals, making them exceptionally nutritious. On average, a cup of cooked legumes contains about 30 percent of the Daily Values for both protein and iron.[*] Like meats, though, legumes do not offer every nutrient, and they do not make a complete meal by themselves. They contain no vitamin A, vitamin C, or vitamin B_{12}, and their balance of amino acids can be much improved by using grains and other vegetables with them.

Soybeans are versatile legumes, and many nutritious products are made from

[*]Data from the *Food Processor Plus*, ESHA research, version 7.11.

FIGURE 6•16 FINDING THE PROTEIN IN FOODS[a]

Grains Group

Food		Protein g	%DV[b]
Pancakes	2 sm	6	12
Bagel	½	4	8
Brown rice	½ c	3	6
Grain bread	1 sl	3	6
Noodles, pasta	½ c	3	6
Oatmeal	½ c	3	6
Barley	½ c	2	4
Cereal flakes	1 oz	2	4

Vegetable Group

Food		Protein g	%DV[b]
Corn	½ c	3	6
Broccoli	½ c	2	4
Collard greens	½ c	2	4
Sweet potato	½ c	2	4
Baked potato	½ c	1	2
Bean sprouts	½ c	1	2
Winter squash	½ c	1	2

Fruit Group

Food		Protein g	%DV[b]
Avocado	½ c	2	4
Cantaloupe	½ c	1	2
Orange sections	½ c	1	2
Strawberries	½ c	1	2

Meat, Poultry, Fish, Dry Peas and Beans, Eggs, and Nuts Group

Food		Protein g	%DV[b]
Roast beef	2 oz	19	33
Turkey leg	2 oz	16	32
Chicken breast	2 oz	15	30
Pork meat	2 oz	15	30
Tuna	2 oz	14	28
Lentils, beans, peas	½ c	9	18
Peanut butter	2 tbs	8	16
Almonds	¼ c	8	16
Hot dog	1 reg	7	14
Lunch meat	2 oz	6	12
Egg	1 lg	6	12
Cashew nuts	¼ c	5	10

Milk, Yogurt, and Cheese Group

Food		Protein g	%DV[b]
Cheese, processed	2 oz	13	26
Milk, yogurt	1 c	10	20
Pudding	1 c	5	10

Fats, Oils, Sweets

Not a significant source

[a] All foods are prepared and ready to eat.
[b] The Daily Value (DV) for protein is 50g, based on an energy intake of 2,000 calories per day.

© Polara Studios Inc. (all)

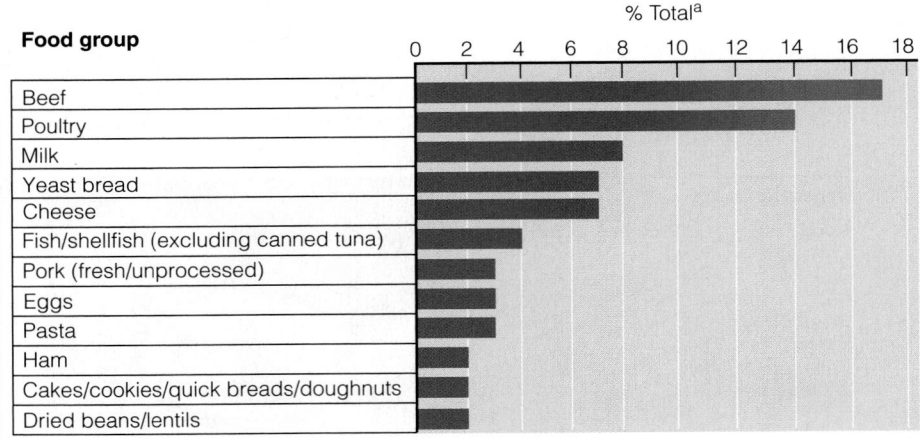

FIGURE 6·17 TOP CONTRIBUTORS OF PROTEIN TO THE U.S. DIET

These foods supply about 70 percent of the protein in the U.S. diet. The remainder comes from foods contributing less than 2 percent of the total such as cold cuts; ready-to-eat cereal; white potatoes; sausage; flour and baking ingredients; ice cream, sherbet, and frozen yogurt; nuts and seeds; cooked rice and other grains; and canned tuna.

[a]Rounded values.

SOURCE: Data from P. A. Cotton and coauthors, Dietary sources of nutrients among U.S. adults, 1994–1996, *Journal of the American Dietetic Association* 104 (2004): 921–930.

them. Heavy use of soy products in place of meat, however, inhibits iron absorption. The effect can be alleviated by using small amounts of meat and/or foods rich in vitamin C in the same meal with soy products. Vegetarians sometimes use convenience foods made from **textured vegetable protein** (soy protein) formulated to look and taste like hamburgers or breakfast sausages. Many of these are intended to match the known nutrient contents of animal protein foods, but they often fall short.[†] A wise vegetarian uses such foods sparingly and learns to use combinations of whole foods to supply the needed nutrients.

The nutrients of soybeans are also available as bean curd, or **tofu,** a staple

---

[†]In Canada, regulations govern the nutrient contents of such products.

used in many Asian dishes. Thanks to the use of calcium salts when some tofu is made, it can be high in calcium. Check the Nutrition Facts panel on the label.

The Food Features presented so far show that the recommendations for the three energy-yielding nutrients occur in balance with each other. The diets of most people, however, supply too little fiber, too much fat, too many calories, and abundant protein. To bring their diets into line with recommendations, then, requires changing the bulk of intake from calorie-rich fried foods, fatty meats, and sweet treats to lower calorie complex carbohydrates and fiber-rich choices, such as whole grains, legumes, and vegetables. With these changes protein totals automatically come into line with the requirements.

FIGURE 6·18   A LEGUME

The legumes include such plants as the kidney bean, soybean, garden pea, lentil, black-eyed pea, and lima bean. Bacteria in the root nodules can "fix" nitrogen from the air, contributing it to the beans. Ultimately, thanks to these bacteria, the plant accumulates more nitrogen than it can get from the soil and also leaves more nitrogen in the soil than it takes out. The legumes are so efficient at trapping nitrogen that farmers often grow them in rotation with other crops to fertilize fields. Legumes are included with the meat group in Figure 6-16.

© Photodisc Blue/Getty Images

Legumes: Protein-rich and exceptionally nutritious.

Seed pods (peas), where nitrogen is stored

Root nodules, which capture nitrogen

Nutrition Connections

The Nutrition Connections CD provides access to these additional resources.

Do It!
The Do It! exercise for this chapter, *Add Up the Protein in a Day's Meals,* provides practice in assessing the protein content of foods.

Interaction
The animation, *Protein Digestion,* follows the digestion of proteins in the stomach and small intestine.

Chapter Quiz
Take the quiz for this chapter to assess your understanding of chapter concepts.

Nutrition on the Net
Annotated links to reputable nutrition-related websites.

Online Glossary
Test your knowledge of key vocabulary through this comprehensive, interactive glossary.

Internet Connections

A variety of helpful study tools for this chapter are available at our website
http://nutrition.wadsworth.com/ncc10

Practice Test
Take the practice test for this chapter and gauge your knowledge of key concepts. Answers are provided.

Student Notebook
A lecture notebook is available to help organize note taking while reading or attending lecture.

Study Tools
Study resources for students include Chapter Summaries, Crossword puzzles, Flashcards, Glossary, Internet Exercises, and Learning Objectives.

SELF CHECK

Answers to these Self Check Questions are in Appendix G.

1. The basic building blocks for protein are:
   a. glucose units
   b. amino acids
   c. side chains
   d. saturated bonds

2. Protein digestion begins in the:
   a. mouth
   b. stomach
   c. small intestine
   d. large intestine

3. To prevent wasting of dietary protein, which of the following conditions must be met?
   a. Dietary protein must be adequate in quantity.
   b. Dietary protein must supply all essential amino acids in the proper amounts.
   c. The diet must supply enough calories from carbohydrate and fat.
   d. All of the above.

4. For healthy adults, the DRI recommended intake for protein has been set at:
   a. 0.8 gram per kilogram of body weight
   b. 2.2 pounds per kilogram of body weight
   c. 12 to 15 percent of total calories
   d. 100 grams per day

5. Which of the following statements is correct regarding protein and amino acid supplements?
   a. They help athletes build muscle.
   b. They help dieters lose weight quicker.
   c. They can assist in relieving depression.
   d. None of the above.

6. Under certain circumstances, protein can be converted to glucose and so serve the energy needs of the brain.   T   F

7. Too little protein in the diet can have severe consequences, but excess protein has not been proven to have adverse effects.   T   F

8. Although protein-energy malnutrition (PEM) is prevalent in underdeveloped nations, it is not seen in the United States.   T   F

9. Partially completed proteins are not held for completion at a later time when the diet may improve.   T   F

10. An example of a person in positive nitrogen balance is a pregnant woman.   T   F

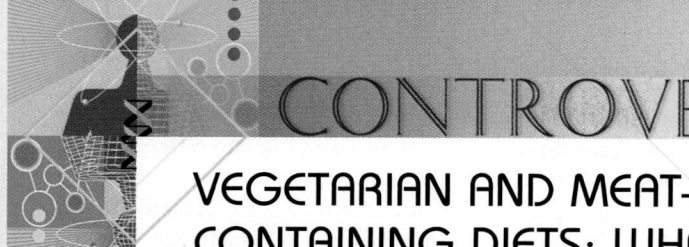

CONTROVERSY 6

VEGETARIAN AND MEAT-CONTAINING DIETS: WHAT ARE THE BENEFITS AND PITFALLS?

IN RECENT YEARS, obesity, heart disease, high blood pressure, and cancer have taken a great toll on the health of Americans and others living in developed nations. The scientific community has observed that, in affluent countries, people who eat well-planned **vegetarian** diets often have reduced rates of these killers and may even live longer lives.[1] Should people consider eating a vegetarian diet? If so, is it beneficial to simply stop eating meat, or is more demanded of the vegetarian diet planner? And what positive contributions do animal products make to the diet?

This Controversy looks first at the positive health aspects of vegetarian diets, then at the positive aspects of meat eaters' diets. It lends practical assistance to both vegetarians and meat eaters in making beneficial choices that minimize their risks.

Vegetarianism is often mistakenly associated with a particular culture or belief system, but individuals choose it for many different reasons. Some omit meat to reduce their intakes of calories or saturated fat. Some believe that we should not kill animals to eat their meat. Some believe that we should not partake of animal products such as milk, cheese, eggs, or honey or use items made from leather, wool, or silk. Many people object to inhumane treatment of livestock in pens, feedlots, and slaughterhouses. Some believe we should eat less meat for health reasons or for environmental reasons—production of meat protein requires a much greater input of resources than does an equal amount of vegetable protein. Some fear contracting diseases, such as food poisoning or "mad cow disease" from meats. (The effects of food choices on the earth's resources are topics of Chapter 15 and its Controversy, and Chapter 12 provides the whole story on the threat from foodborne illnesses.) In any case, vegetarian diets are not categorized by motivation but by the foods they exclude, and even the experts struggle to agree on the terms to describe them (see Table C6-1).[2]

People who eat meat also do so for a variety of reasons. Some find that a hamburger makes a convenient lunch while providing a concentrated source of energy and nutrients. Others enjoy the taste of roast chicken or beef stew. Others wouldn't know what to eat without meat; they are accustomed to seeing it on the plate. Still others mistakenly think that eating meats instead of grains, potatoes, and breads will help them lose weight (Chapter 9 takes up the issues of weight-loss dieting). Whatever your diet or reasons for choosing it, the foods most important to your health are the foods that you choose regularly.

Positive Health Aspects of Vegetarian Diets

Strong evidence linking vegetarian diets with reduced incidences of chronic diseases has prompted several nutrition authorities to state with confidence that a well-chosen vegetarian diet can meet nutrient needs while supporting health superbly.[3] Such evidence, though abundant, is not easily obtained. It would be easy if vegetarians differed from others only in not consuming meat, but they often have *increased* intakes of fruits and vegetables as well. These foods are the primary contributors of nutrients and phytochemicals believed to reduce disease risks. Vegetarian diets often contain more fiber, potassium, and several vitamins associated with reduced disease risks. Also, though there are exceptions, vegetarians typically use no tobacco, use alcohol in moderation if at all, and may be more physically active than other adults. Researchers must account for the effects of a total health-conscious lifestyle on disease development before they can see how diet correlates with health. Despite these limitations, much evidence weighs in favor of vegetarian diets as the next sections make clear.

Weight Control

In general, vegetarians maintain a healthier body weight than nonvegetarians.[4] Studies report that body weights for height are higher for people eating a mixed diet than for vegetarians and that weight increases as frequency of meat consumption increases.[5] Lower body weights also correlate with high intakes of fiber and low intakes of fat. Obesity impairs health in a number of ways, and vegetarians who maintain a healthy weight are at an advantage.

TABLE C6·1	TERMS USED TO DESCRIBE VEGETARIANS AND DIETS

Some of the terms below are in common usage, but others are useful only to researchers.

- **fruitarian** includes only raw or dried fruits, seeds, and nuts in the diet.
- **lacto-ovo vegetarian** includes dairy products, eggs, vegetables, grains, legumes, fruits, and nuts; excludes flesh and seafood.
- **lacto-vegetarian** includes dairy products, vegetables, grains, legumes, fruits, and nuts; excludes flesh, seafood, and eggs.
- **macrobiotic diet** a vegan diet composed mostly of whole grains, beans, and certain vegetables; taken to extremes, macrobiotic diets have resulted in malnutrition and even death.
- **ovo-vegetarian** includes eggs, vegetables, grains, legumes, fruits, and nuts; excludes flesh, seafood, and milk products.
- **partial vegetarian** a term sometimes used to mean an eating style that includes seafood, poultry, eggs, dairy products, vegetables, grains, legumes, fruits, and nuts; excludes or strictly limits certain meats, such as red meats.
- **pesco-vegetarian** same as partial vegetarian, but eliminates poultry.
- **vegan** includes only food from plant sources: vegetables, grains, legumes, fruits, seeds, and nuts; also called *strict vegetarian*.
- **vegetarian** includes plant-based foods and eliminates some or all animal-derived foods.

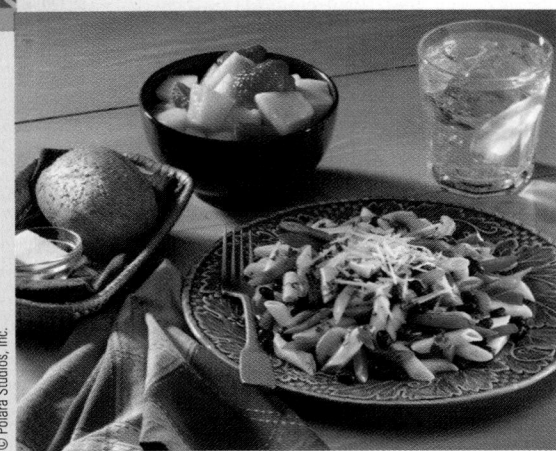

© Polara Studios, Inc.

A balanced meal need not include meat to be nutritious.

Heart Disease

People consuming plant-based diets die less often from heart disease and related illnesses than do meat-eating people.[6] In general, plant-based diets, such as vegetarian diets, are lower in saturated fat and cholesterol than are diets based on fatty meats, and dietary saturated fat intake and to a lesser extent cholesterol intake are directly linked with heart disease. Plant-based diets contain fats, but they are generally the unsaturated fats of soybean products, seeds, nuts, and vegetable oils that are associated with lower risk of heart disease.[7] Furthermore, such diets are generally higher in dietary fibers, a bonus to blood lipids and the condition of the heart. Vegetarians who eat fish and milk products have blood lipids between the low blood lipids of **vegans** and the higher blood lipids of people eating typical Western diets high in meats.[8]

Vegetables may also lower disease risks by virtue of their phytochemicals (see Controversy 2). For example, vegetarians often seek tofu (the curd of soybeans) for its protein value, but its heart health benefits may exceed protein alone.[9] The phytochemicals of soy products help lower cholesterol, and when tofu replaces red meat in the diet, blood cholesterol and triglyceride levels drop even when intakes of energy, protein, carbohydrate, total fat, saturated fat, unsaturated fat, alcohol, and fiber are held constant.[10] Tofu may have a reputation as a high-fat food, but Appendix A shows that tofu contains just slightly less fat than an equal por-

tion of skinless chicken thigh meat. Furthermore, the fat of tofu is far less saturated than the fat of the chicken.

Heart Benefits from Plant Proteins

Over 60 years of experimentation have revealed that when soy protein *replaces* animal protein in the diet, total blood cholesterol concentration is reduced along with LDL cholesterol, triglycerides, blood pressure, and other indicators of heart disease risk.[11] At the same time, blood HDL cholesterol rises or stays the same. The question remains, however, whether these potentially beneficial effects are attributable to soy protein itself or to the phytochemicals that accompany it.

Weighing in on this issue are results from a study on a breed of mice that develop the hardening of the arteries associated with heart disease. The mice were divided into three groups: one group was fed chow made of regular soy protein, another group was fed soy protein from which the active phytochemicals had been washed away, and a control group was fed soy-free chow. When the groups were compared at the end of the study, both soy-fed groups had lower blood lipids and had developed less hardening of the arteries than the controls. The mice fed regular soy chow developed significantly less artery damage than did the washed-soy group, suggesting an extra benefit from soy's phytochemicals.[12] Similar findings from similar experiments with other species confirm this idea, but here's a puzzle: when researchers extract the soy phytochemicals and feed them to animals, the benefits of whole soy foods do not occur.[13]

Human studies tip the scale in favor of replacing some of the animal protein in the diet with whole soy foods.[14] Many of the beneficial blood lipid changes mentioned earlier often occur among people who regularly consume soy foods.[15] A firm base of such findings prompted the Food and Drug Administration to approve the food label claim that "25 grams of soy

protein a day, as part of a diet low in saturated fat and cholesterol, may reduce the risk of heart disease."

The benefits of vegetable proteins may also reach beyond just soy foods. Even the protein gluten, formed in wheat bread, may reduce oxidation and other indicators of heart disease risk when it replaces animal protein in the diet.[16] Likewise, in people, fat-free milk *lowers* blood cholesterol when it replaces other animal proteins, and meals of fish consumed for its protein provide benefits to heart health in other ways, as Chapter 5 described.[17] These developments have led researchers to look beyond dietary protein for answers about the benefits to vegetarians with regard to heart disease risk.

Defense against Blood Pressure

Vegetarians tend to have lower blood pressure and lower rates of hypertension than nonvegetarians. Appropriate body weight helps to maintain a healthy blood pressure, as does a diet low in total fat and saturated fat and high in fiber, fruits, and vegetables. Lifestyle factors also influence blood pressure: smoking and alcohol intake raise blood pressure, and physical activity lowers it.

Defense against Cancer

Vegetarians have significantly lower rates of certain cancers than the general population. Again, though, low rates of cancer may be associated more with the abundance of fruits and vegetables in vegetarian diets than with the exclusion of all animal products. Colon cancer appears to correlate with moderate-to-high intakes of:

- Alcohol.[18]
- Total food energy.[19]
- Fatty red meats and processed meats (but not poultry or fish).[20] (A diet high in red meat and processed meat may be positively associated with stomach cancer as well.)[21]

Possible correlations with colon cancer have been found with:

- High intakes of refined grain products.
- Low intakes of whole grains.[22]
- Low intakes of vitamin D.[23]

Certain cancers appear to occur less frequently among people who eat mostly plant-based diets that are rich in vegetables, fruit, whole grains, and legumes (soybeans in particular). The protective effect may be related to the activities of phytochemicals in those foods, activities that occur whether or not the diet also contains some meat.[24]

Positive Health Aspects of the Meat Eater's Diet

Meat eaters generally consume other foods in addition to meat. The healthiest meat eaters base their diets on abundant fruits, vegetables, whole grains, and milk products with the addition of small servings of fish, meats, and poultry. True meat lovers and misguided weight-loss dieters who shun all green and yellow vegetables, fruits, and grains place themselves in immediate peril of malnutrition and chronic diseases.

Unlike vegetarians who can find suitable replacements for meats in a healthy diet, those who would exclude fruits and vegetables have no adequate substitutes for these foods (not even antioxidant supplement pills, as the next chapter points out). The following sections consider a *balanced, adequate* diet whether or not meats, eggs, and milk products play a part.

What Diet Offers the Best Support during Critical Times?

Both meat eaters and **lacto-ovo vegetarians** can generally rely on their diets during critical times of life. In contrast,

a vegan diet can pose challenges. The chapter made clear that protein from animal sources, such as meat, milk, and eggs, is the clear winner in tests of digestibility and availability to the body, with soy protein a close second. In addition to providing protein, meat provides abundant iron, zinc, and vitamin B_{12} needed by everyone, but in particular by pregnant women, infants, children, and adolescents.

In Pregnancy and Infancy

Unlike vegans, women who eat meat, eggs, and milk products can be sure of receiving enough vitamin B_{12}, vitamin D, calcium, iron, and zinc, as well as protein, to support pregnancy and breastfeeding. A woman following a well-planned lacto-ovo vegetarian diet can also relax in the knowledge that she is supplied with all necessary nutrients. If she also habitually consumes more folate in the form of vegetables and fruits, she can relax further, knowing that her developing fetus is well protected from birth defects associated with low folate intakes.

A too-low body weight is generally not a problem faced by meat-eating women. A vegan woman who doesn't meet her nutrient needs, however, may enter pregnancy too thin; have inadequate stores of iron, zinc, and vitamin B_{12}; and fail to consume the omega-3 fatty acids needed to support normal fetal development. She may then find it difficult to gain enough weight to support a normal pregnancy.

Obtaining enough vitamin B_{12}, abundant in foods of animal origin but absent from vegetables, poses a challenge to vegans of all ages, and deficiencies of this vitamin are on the rise.[25] The need for vitamin B_{12} is critical for pregnant and lactating women.[26] A severe disorder that is often irreversible and sometimes fatal is reported among breastfed infants of vegan mothers who fail to obtain sufficient vitamin B_{12}.[27] The infants exhibit a syndrome of body tremors and facial twitches involving the tongue and throat that com-

bine to make nursing or eating difficult. After some months, deprivation of vitamin B_{12} leads to infant growth failure, psychomotor retardation, and shrinkage of the brain. In some cases, the retardation from the deficiency lingers long after treatment with the missing vitamin begins.

In Childhood

Children who eat small servings of meat, poultry, and fish receive abundant protein, iron, vitamin B_{12}, and food energy, making these foods reliable, convenient sources of these key nutrients needed for growth. Likewise, children eating well-planned vegetarian diets that include eggs, milk, and milk products also receive adequate nutrients and grow as well as their meat-eating peers.[28]

Compared with meat-eating children, well-fed vegan children tend to be shorter and lighter in weight but not excessively so. Most studies find that their growth falls within normal ranges, as long as they consume ample energy from food.

Child-sized servings of vegan foods can fail to provide sufficient energy or several key nutrients to support normal growth. A child's small stomach can hold only so much food, and the vegan child may feel full before eating enough to meet his or her nutrient needs. Frequent meals of fortified breads, cereals, or pastas with legumes, nuts, nut butters, and sources of unsaturated fats can help to meet protein and energy needs in a smaller volume at each sitting.[29]

Because vegan children derive protein only from plant foods, their daily protein requirement may be somewhat higher than the DRI indicates for the general meat-eating population.[30] Specific recommendations to support the growth of vegan children have not been established.

Children reared on vegan diets and then switched to more liberal diets may have difficulty achieving an adequate vitamin B_{12} status, even with moderate consumption of animal products.[31] Other nutrients of concern for vegan children include vitamin D, calcium, iron, and zinc. Numerous fortified animal-free foods and supplements may help meet these nutrient needs.[32]

This 5-ounce steak provides almost all of the meat recommended for a day's intake in a 2,000 calorie diet.

In Adolescence

Like the diets of meat-eating adolescents, adolescent vegetarian diets range from healthy eating plans to those causing health-destroying deficiencies. On the positive side, some vegetarian adolescents choose diets with more fruits and vegetables and fewer sweets, fast foods, and salty snacks than is common among their meat-eating peers. As a result, these teens enjoy robust growth and health. In fact, such adolescents are the most likely among their age group to meet dietary objectives, such as those of *Healthy People 2010* (listed in Chapter 1).[33]

Other adolescent vegans, however, adopt poorly planned diets lacking not only vitamin B_{12} but calcium and vitamin D as well. Omissions of calcium and vitamin D lead to weak bone development at precisely the time when bones must develop strength to protect them through later life.[34] As Controversy 8 makes clear, bone density is a now-or-never proposition for adolescents—bone density rarely increases in adulthood.

Urban legends may convince vegan adolescents that lack of vitamin B_{12} poses no problems, but they are far off the mark. The well-known threat to a developing infant was already described, but vitamin B_{12} deficiency in older children and adolescents also causes problems such as impaired reasoning and memory.[35] In fact, even grown men are vulnerable, as evidenced by a recent report of a vegan man suffering permanent blindness apparently brought on by vitamin B_{12} deficiency.[36]

Some teens may adopt a vegetarian identity to serve as a sort of camouflage to hide an eating disorder.[37] Even college age women may disguise their eating disorders as vegetarianism.[38] By doing so, these individuals can greatly limit their food choices while distracting their families and peers from an eating disorder that threatens their health and, in the case of adolescents, their growth and bone strength.

In the Elderly and in Illness

For elderly people with diminished appetites or for people recovering from illnesses, soft or ground meats can provide a well-liked, well-tolerated concentrated source of nutrients. Variety is key to nutritional adequacy in any diet, of course. People battling life-threatening diseases may encounter testimonial stories of cures attributed to restrictive eating plans, such as **macrobiotic diets,** but these diets limit food selections to a few grains and vegetables and cannot possibly deliver the full array of nutrients needed for recovery.

A Few Nutrients of Interest

The quality of a vegetarian or meat-containing diet depends not on whether it includes meat, but on whether the other food choices are nutritionally sound. Both vegetarian and meat-containing diets, if not properly balanced, can lack nutrients. Poorly planned vegetarian diets typically lack iron, zinc, calcium, vitamin B_{12}, and vitamin D; without planning, the meat eater's diet may lack vitamin A, vitamin C, folate, and fiber, among others. Quite simply, the negative health aspects of any diet, including vegetarian diets, reflect poor diet planning—the diet omits foods that supply important nutrients.

Protein

The DRI recommended intake for protein is the same for vegetarians as for others, although some have suggested that it should be higher because of the lower digestibility of plant proteins.[39] Vegetarians who use animal-derived foods such as milk and eggs receive high-quality protein and are likely to meet their protein needs. Even those who adopt only plant-based diets are likely to meet protein needs provided that energy intakes are adequate and the protein sources varied.[40]

As mentioned in the preceding chapter, vegetarians sometimes use foods made from textured vegetable protein, which is formulated to look and taste like meat. Though fortified, these foods often fall short of the nutrients in meats and may also be high in salt, sugar, or other additives. A wise vegetarian learns to use a variety of whole, unrefined foods often and commercially prepared foods less frequently. Vegetarians may also use soybeans in the form of plain tofu (bean curd) to bolster protein intake without consuming unwanted salt, sugar, or other additives.

Iron

Getting enough iron can be a problem even for meat eaters, and vegetarians must be especially vigilant about obtaining iron. The iron in plant foods such as legumes, dark green leafy vegetables, iron-fortified cereals, and whole-grain breads and cereals is poorly absorbed.[41] Such foods contain inhibitors of iron absorption, so the DRI committee recommended that iron intake for vegetarians be adjusted upward (see Chapter 8 for more details). The committee suggests that vegetarians need 1.8 times the amount of iron recommended for the general population.

At some point, the body may adapt to a vegetarian diet by absorbing iron more efficiently.[42] Also, iron absorption is enhanced by vitamin C consumed with iron-rich foods, and vegetarians typically eat many vitamin C–rich fruits and vegetables.[43] Consequently, vegetarians suffer no more iron deficiency than other people do.[44]

Zinc

Zinc is similar to iron in that meat is its richest food source, and zinc from plant sources is not well absorbed.[45] In addition, soy, which is commonly used as a meat alternate in vegetarian meals, interferes with zinc absorption. Zinc can be a problem for growing children, but few vegetarian adults are zinc deficient. Perhaps the best advice to vegetarians regarding zinc is to eat a variety of nutrient-dense foods; include whole grains, nuts, and legumes such as black-eyed peas, pinto beans, and kidney beans; and maintain an adequate energy intake.

Calcium

The calcium intakes of vegetarians who use milk and milk products are similar to those of the general population, but those who do not use milk products risk deficiency. Careful planners select calcium-rich foods, such as calcium-fortified juices, soy milk, and breakfast cereals, in ample quantities regularly. This is especially important for children and adolescents. Soy formulas for infants are fortified with calcium and can be used in cooking, even for adults. Other good sources of absorbable calcium include figs, calcium-set tofu, some legumes, some green vegetables (broccoli, kale, and turnip greens, but not spinach—see Chapter 8), some nuts such as almonds, and certain seeds such as sesame seeds.* The choices should be varied because calcium absorption from some plant foods is limited.

Vitamin B₁₂

The requirement for vitamin B_{12} is small, but this vitamin is found only in animal-derived foods, such as meats, milk, and eggs. Those who consume these foods are rarely deficient in the vitamin. For vegans, fermented soy products may contain some vitamin B_{12} from the bacteria that did the fermenting, but unfortunately, much of the vitamin B_{12} found in these products may be an inactive form. Seaweeds such as nori and chlorella supply some vitamin B_{12}, but not much, and excessive intakes of these foods can lead to iodine toxicity. To defend against vitamin B_{12} deficiency, vegans must rely on vitamin B_{12}–fortified foods (such as soy milk or breakfast cereals) or pills that contain it. As described earlier, without vitamin B_{12}, the nerves suffer damage, leading to such health consequences as loss of vision.[46]

Vitamin D

People who do not use vitamin D–fortified foods and do not receive enough exposure to sunlight to synthesize adequate vitamin D may need supplements to defend against bone loss.[47] This is particularly important for infants, children, and older adults. In northern climates during winter months, young children on vegan diets can readily develop rickets, the vitamin D–deficiency disease.

Omega-3 Fatty Acids

Vegetarian diets typically provide enough omega-6 fatty acids but lack omega-3 fatty acids. This imbalance slows production of EPA and DHA in the body, and without fish, eggs, or sea vegetables in the diet, intake of EPA and DHA falls short as well.[48] A vegetarian's daily diet should include good sources of linolenic acid, such as canola oil and flaxseed, walnuts, soybeans, and their oils, because linolenic acid is an essential nutrient.[49]

Planning a Vegetarian Diet

Vegetarians are often advised to follow the Daily Food Guide presented in Chapter 2 with a few modifications. Those who include milk products and eggs can follow the regular plan, using legumes, nuts, and seeds and products made from them, such as peanut butter and tofu, in place of meat. Soy milk and tofu fortified with calcium, vitamin D, and vitamin B_{12} can substitute for cow's milk products. Dark green vegetables and legumes help meet iron and zinc needs.

Several food guides have been developed to ease the task of planning nutritious vegetarian diets.[50] Figure C6-1 presents one version to consider.[51] Notice that green leafy vegetables in the vegetable group are distinguished because some of these foods provide almost five times as much calcium per serving as other vegetables. Similarly, dried fruits receive special notice in the fruit group because some of them deliver six times as much iron as other fruits. A separate group for nuts and seeds was established to provide additional sources of protein, iron, zinc, and essential fatty acids. A group for oils at the tip encourages the use of vegetable oils rich in unsaturated fats and omega-3 fatty acids. The meat and meat alternate group is transformed into "beans and protein foods," and soy milk is found in milk and milk products.

A word of caution in planning a vegetarian diet: choose fresh, whole foods as often as possible and avoid heavy reliance on highly processed convenience foods that contain added sugars, saturated fats, and *trans* fats. An example clarifies this point: a snack of a medium sliced banana provides 100 calories and practically no fat. A snack of a quarter cup of banana chips, often sold as a "healthy" or "natural" alternative to other snack foods, provides 150 calories with 7 grams of saturated fat. (A big quarter pound hamburger delivers only one gram more of saturated fat—8 grams.) Banana chips are made from regular bananas but are sweetened with sugar and fried in coconut oil, one of

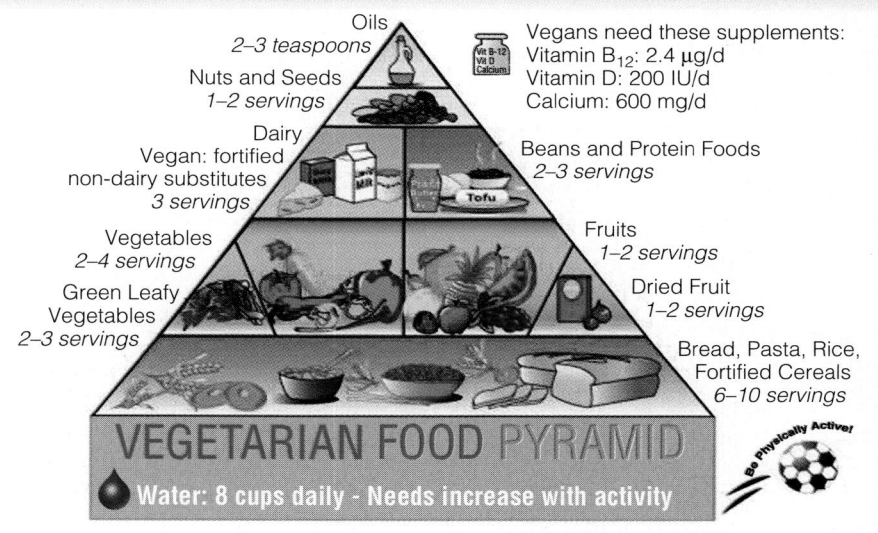

FIGURE C6·1 A FOOD GUIDE PYRAMID FOR VEGETARIANS

Oils
2–3 teaspoons

Nuts and Seeds
1–2 servings

Dairy
Vegan: fortified non-dairy substitutes
3 servings

Vegetables
2–4 servings

Green Leafy Vegetables
2–3 servings

Vegans need these supplements:
Vitamin B_{12}: 2.4 μg/d
Vitamin D: 200 IU/d
Calcium: 600 mg/d

Beans and Protein Foods
2–3 servings

Fruits
1–2 servings

Dried Fruit
1–2 servings

Bread, Pasta, Rice, Fortified Cereals
6–10 servings

VEGETARIAN FOOD PYRAMID

Water: 8 cups daily - Needs increase with activity

Be Physically Active!

SOURCE: C. A. Venti and C. S. Johnston, Modified food guide pyramid for lactovegetarians and vegans. *Journal of Nutrition* 132 (2002): 1050–1054.

---

*Calcium salts are often added during processing to coagulate the tofu.

the most saturated fats available. To avoid such tricks of processing, scrutinize the labels of convenience and prepared vegetarian foods just as you would those of ordinary foods, and choose foods that provide needed nutrients without a lot of unneeded calories from sugar or excess saturated fat or *trans* fat.

Finally, to ensure adequate intakes of vitamin B_{12}, vitamin D, and calcium, vegans need to select fortified foods or use supplements daily. The vegetarian food guide also includes two other lifestyle factors that contribute to good health: physical activity and water intake.

Conclusion

This comparison has shown that both a meat eater's diet and a vegetarian's diet are best approached scientifically. Some people make much of the distinctions between types of vegetarians; although these distinctions are useful academically, they do not represent uncrossable lines. Some people use meat as a condiment or seasoning for vegetable or grain dishes. Some people eat meat only once a week and use plant protein foods the rest of the time. Many people rely mostly on milk products to meet their protein needs but eat fish occasionally, and so forth. To force people into the categories of "vegetarians" and "meat eaters" leaves out all these in-between styles of eating that have much to recommend them.

If you are just beginning to study nutrition, consider adopting the attitude that the choice to make is not whether to be a meat eater or a vegetarian, but where along the spectrum to locate yourself. Your preferences should be honored with only these caveats: that you plan your own diet, and the diets of those in your care, to be adequate, balanced, and varied, and that you use moderation when choosing foods high in saturated fat or calories.

The Vitamins

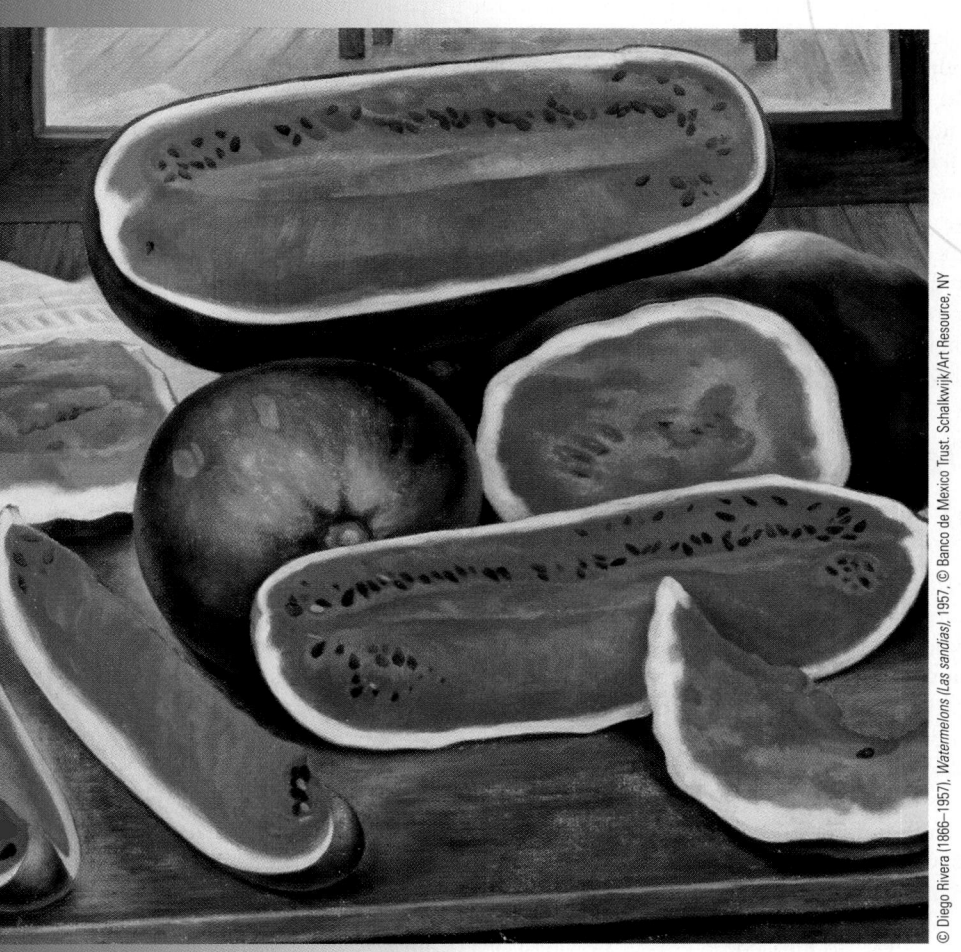

© Diego Rivera (1886–1957), *Watermelons (Las sandías)*, 1957. © Banco de Mexico Trust. Schalkwijk/Art Resource, NY

CONTENTS

Do You Ever . . .

Wonder how vitamins work in the body?

Associate sunshine with good health?

Take vitamin C tablets to ward off a cold?

Eat highly fortified foods or "meal replacers," regarding them as a harmless form of vitamin insurance?

KEEP READING.

AT THE BEGINNING OF the twentieth century, the thrill of the discovery of the first **vitamins** captured the world's imagination as seemingly miraculous cures took place. In the usual scenario, a whole group of people were unable to walk (or were going blind or bleeding profusely) until an alert scientist stumbled onto the substance missing from their diets. The scientist confirmed the discovery by feeding vitamin-deficient feed to laboratory animals, which responded by becoming unable to walk (or going blind or bleeding profusely). When the missing ingredient was restored to their diet, they soon recovered. People, too, were quickly cured when they received the vitamins they lacked.

In the following decades, advances in chemistry and biology allowed scientists to isolate the vitamins and define their chemical structures. More scientific advances brought an understanding of the biological roles that vitamins play in maintaining health and preventing deficiency diseases. Today, research hints that two of the major scourges of humankind, cardiovascular disease (CVD) and cancer, may be linked with low intakes of vitamins. Respected thinkers in genetics suspect that chronic deficiencies of vitamins and minerals may be major contributors to genetic damage that can lead to cancer (others suspect that some vitamin excesses may do the same thing).[*1] Can it be that foods rich in vitamins will protect us from life-threatening diseases? What about vitamin pills? For now, we can say only this with certainty: the only disease a vitamin will *cure* is the one caused by a deficiency of that vitamin. As for chronic disease *prevention,* the evidence is still emerging. The Controversy explores the questions of who may need supplements and which ones may be most beneficial.

■ The only disease a vitamin can cure is the one caused by a deficiency of that vitamin.

■ Vitamin intake recommendations are found on the inside front cover.

vitamins organic compounds that are vital to life and indispensable to body functions, but are needed only in minute amounts; noncaloric essential nutrients.

precursors, provitamins compounds that can be converted into active vitamins.

Definition and Classification of Vitamins

A child once defined a vitamin as "what, if you don't eat, you get sick." Although the grammar left something to be desired, the definition was accurate. Less imaginatively, a vitamin is defined as an essential, noncaloric, organic nutrient needed in tiny amounts in the diet. The role of many vitamins is to help make possible the processes by which other nutrients are digested, absorbed, and metabolized or built into body structures. Although small in size and quantity, the vitamins accomplish mighty tasks.

As they were discovered, the vitamins were named, and many were also given letters and numbers. This led to the confusing variety of vitamin names that still exists today. This chapter uses the names in Table 7-1; alternative names are given in Tables 7-3 and 7-4 (pp. 245, 247) at the end of the chapter.

Certain vitamins occur in foods in a form known as **precursors,** or **provitamins.** Once inside the body, these are transformed chemically to one or more active vitamin forms. Thus, to measure the amount of a vitamin found in food, we often must count not only the amount of the true vitamin but also the vitamin activity potentially available from its precursors. Tables 7-3 and 7-4 specify which vitamins have precursors.

The vitamins fall naturally into two classes: fat soluble and water soluble (listed in Table 7-1). Solubility confers on vitamins many of their characteristics. It determines how they are absorbed into and transported around by the bloodstream, whether they can be stored in the body, and how easily they are lost from the body. In general, like other lipids, fat-soluble vitamins are absorbed into the lymph, and they travel in the blood in association with protein carriers. Fat-soluble vitamins can be stored in the liver or with other lipids in fatty tissues, and some can build up to toxic concentrations. The water-soluble vitamins are absorbed directly into the bloodstream, where they travel freely. Most are not stored in tissues to any great extent; rather, excesses are

*Reference notes are found in Appendix F.

excreted in the urine. Thus, the risks of immediate toxicities are not as great as for fat-soluble vitamins. This chapter examines the fat-soluble vitamins first and then the water-soluble ones. The tables at the end of the chapter sum up the basic facts about all of them.

▶ **KEY POINT**   Vitamins are essential, noncaloric nutrients that are needed in tiny amounts in the diet and help to drive cell processes in the body. The fat-soluble vitamins are vitamins A, D, E, and K; the water-soluble vitamins are vitamin C and the B vitamins.

Vitamins fall into two classes—fat soluble and water soluble.

The Fat-Soluble Vitamins

The fat-soluble vitamins—A, D, E, and K—are found in the fats and oils of foods and require bile for absorption. Once absorbed, these vitamins are stored in the liver and fatty tissues until the body needs them. The body can survive weeks of consuming foods that lack these vitamins, as long as the diet as a whole provides *average* amounts that approximate the recommended intakes. This capacity to be stored also sets the stage for toxic buildup if you take in too much. Excesses of vitamins A and D from supplements can reach toxic levels especially easily.

Deficiencies of the fat-soluble vitamins are likely when the diet is consistently low in them. We also know that any disease that produces fat malabsorption (such as liver disease that prevents bile production) can cause the loss of vitamins dissolved in undigested fat and so bring about deficiencies. In the same way, a person who uses mineral oil (which the body can't absorb) as a laxative risks losing fat-soluble vitamins because they dissolve into the oil and are excreted. Deficiencies are also likely when people eat diets that are extraordinarily low in fat because such diets interfere with absorption of these vitamins.

Fat-soluble vitamins play diverse roles in the body. Vitamins A and D may act somewhat like hormones, directing cells to convert one substance to another, to store this, or to release that. Many of their effects are exerted at the level of the genes, influencing protein production.[2] Vitamin E flows throughout the body, preventing oxidative destruction of tissues. Vitamin K is necessary for blood to clot. Each is worth a book in itself.

Characteristics fat-soluble vitamins share:
- Dissolve in lipid.
- Require bile for absorption.
- Are stored in tissues.
- May be toxic in excess.

■ Look back at Figure 5-6 (p. 146) of Chapter 5 to see how bile acts in fat absorption.

Vitamin A

Vitamin A has the distinction of being the first fat-soluble vitamin to be recognized. Today, after a century of research, vitamin A and its plant-derived precursor, **beta-carotene,** are still very much a focus of research.

Three forms of vitamin A are active in the body; one of the active forms, **retinol,** is stored in the liver. The liver makes retinol available to the bloodstream and thereby to the body's cells. The cells convert retinol to its other two active forms, retinal and retinoic acid, as needed.

A Jack of All Trades—Vitamin A   Vitamin A is a versatile vitamin, with roles in gene expression, vision, maintenance of body linings and skin, immune defenses, growth of bones and of the body, and normal development of cells. It is of critical importance for reproduction.[3] In short, vitamin A is needed everywhere (its chief functions in the body are listed in the Snapshot on page 219 and in Table 7-3 on page 245).

■ Chapter 6 describes gene expression, protein synthesis, and the work of proteins within the body.

Regulation of Gene Expression   Vitamin A exerts considerable influence on body functions through its regulation of the activities of the genes.[4] Genes direct the synthesis of proteins, including enzymes, and enzymes perform the metabolic work of the tissues (see Chapter 6). Hence, factors that influence gene expression also

beta-carotene an orange pigment with antioxidant activity; a vitamin A precursor made by plants and stored in human fat tissue.

retinol one of the active forms of vitamin A made from beta-carotene in animal and human bodies; an antioxidant nutrient. Other active forms are *retinal* and *retinoic acid.*

An eye (sectioned).

affect the metabolic activities of the tissues, and the health of the body. Hundreds of genes have been suggested as regulatory targets of the retinoic acid form of vitamin A.[5]

Researchers have long known that simply possessing the genetic equipment needed to make a particular protein does not guarantee that the protein will be produced, any more than owning a car guarantees you a trip across town. To get the car rolling, you must also use the right key to trigger the events that start up its engine or turn it off at the appropriate time. Some dietary components, including the retinoic acid form of vitamin A, are now known to be such keys—they help to activate or deactivate certain genes and thus affect the production of specific proteins.[6]

Eyesight The most familiar function of vitamin A, however, is in eyesight. Vitamin A plays two indispensable roles: in the process of light perception at the **retina** and in the maintenance of a healthy, crystal-clear outer window, the **cornea** (see the margin drawing).

When light falls on the eye, it passes through the clear cornea and strikes the cells of the retina, bleaching many molecules of the pigment **rhodopsin** contained within those cells. Vitamin A is a part of the rhodopsin molecule. When bleaching occurs, the vitamin is broken off, initiating the signal that conveys the sensation of sight to the optic center in the brain. The vitamin then reunites with the pigment, but a little vitamin A is destroyed each time this reaction takes place, and fresh vitamin A must regenerate the supply. If the supply begins to run low, a lag occurs before the eye can see again after a flash of bright light at night (see Figure 7-1). This lag in the recovery of night vision, termed **night blindness,** may indicate a vitamin A deficiency. A bright flash of light can temporarily blind even normal, well-nourished eyes, but if you experience a long recovery period before vision returns, your health-care provider may want to check your vitamin A intake.

A more profound deficiency of vitamin A is exhibited when the protein **keratin** accumulates and clouds the eye's outer vitamin A–dependent part, the cornea. The condition is known as **keratinization,** and if the deficiency of vitamin A is not corrected, it can worsen to **xerosis** (drying) and then progress to thickening and permanent blindness, **xerophthalmia.** Tragically, a half million of the world's vitamin A–deprived children become blind each year from this often preventable condition. If the deficiency is discovered early, capsules providing 60,000 micrograms of vitamin A taken twice each year can reverse it. Better still, a child fed fruits and vegetables regularly is virtually assured protection.

Skin and Body Linings Vitamin A is needed by all **epithelial tissue** (external skin and internal linings), not just by the cornea.[7] The skin and all of the protective linings of

FIGURE 7·1 NIGHT BLINDNESS

This is one of the earliest signs of vitamin A deficiency.

© David Farr/Image Smythe

In dim light, you can make out the details in this room.

A flash of bright light momentarily blinds you as the pigment in the retina is bleached.

You quickly recover and can see the details again in a few seconds.

With inadequate vitamin A, you do not recover but remain blind for many seconds; this is night blindness.

the lungs, intestines, vagina, urinary tract, and bladder serve as barriers to infection or damage from other sources. An example of vitamin A's work behind the scenes at the genetic level is the process of **cell differentiation,** in which each type of cell develops to perform a specific function.[8] For example, when goblet cells (cells found in linings of body organs) mature, they specialize in synthesizing and releasing mucus to protect delicate tissues from toxic particles or bacteria and other microbial invaders.

If vitamin A is deficient, the cell differentiation and maturing process is impaired. Goblet cells, among others, fail to mature, then fail to make protective mucus, and eventually die off. Some of the cells in these areas are displaced by cells that secrete keratin, the protein mentioned earlier. Keratin is the same protein that provides toughness in hair and fingernails, but in the wrong place, like the skin, keratin makes the tissue surfaces dry, hard, and cracked. As dead cells accumulate on the surface, the tissue becomes vulnerable to infection (see Figure 7-2). In the cornea, keratinization leads to xerophthalmia; in the lungs, the displacement of mucus-producing cells makes respiratory infections likely; in the vagina, the same process leads to vaginal infections.

The process of cell differentiation also has links to cancer. Cancer researchers are investigating roles of retinoic acid in regulating genes that control cell differentiation because such genes may suppress or even reverse malignant cell changes leading to cancer.[9]

Immunity   Vitamin A has gained a reputation as an "anti-infective" vitamin because so many of the body's defenses against infection depend on an adequate supply. Much research supports the need for vitamin A in the regulation of the genes that produce proteins involved in immunity. Without sufficient vitamin A, these genetic interactions produce an altered response to infection that weakens the body's defenses against disease.

When the defenses are weak, especially in vitamin A–deficient children, an illness such as measles can become severe. A downward spiral of malnutrition and infection can set in. The child's body must devote its scanty store of vitamin A to the immune system's fight against the measles virus, but the infection causes vitamin A to be lost from the body. As vitamin A dwindles, the infection worsens. Even if the child survives the measles infection, blindness is likely. The corneas, already damaged by the chronic vitamin A shortage, degenerate rapidly as their meager supply of vitamin A is diverted to the immune system.

Growth   Vitamin A also assists in growth of bone (and teeth). Normal children's bones grow longer, and the children grow taller, by remodeling each old bone into a new, bigger version. To do so, the body dismantles the old bone structures and replaces them with new, larger bone parts. Growth cannot take place just by adding on to the original small bone; vitamin A is needed in the critical dismantling steps. In children, failure to grow is one of the first signs of poor vitamin A status. Restoring vitamin A to such children is imperative, but correcting dietary deficiencies may be more effective than giving vitamin A supplements alone because other nutrients from nutritious food are also needed for children to gain weight and grow taller.

Vitamin A Deficiency around the World   Vitamin A deficiency presents a vast problem worldwide, placing a heavy burden on society.[10] Between 3 and 10 million of the world's children suffer from signs of severe vitamin A deficiency—not only xerophthalmia and blindness but diarrhea and reduced food intake that rapidly worsen their condition.[11] A staggering 275 million more children suffer from milder deficiency that impairs immunity and promotes infections. In some areas of the world, vitamin A and other deficiencies seem to be the rule rather than the exception among new mothers and infants.[12]

In countries where children receive vitamin A supplements, childhood death rates have declined by half. Even in the United States, vitamin A supplements are recommended for certain groups of infants and for children with measles. Vitamin A supplementation also offers some protection against the complications of other life-threatening infections, including malaria, lung diseases, and HIV.[13] The World Health Organization (WHO) and UNICEF (United Nations International Children's Emergency Fund) are working to eliminate vitamin A deficiency; achieving this goal would improve child survival throughout the developing world.

FIGURE 7·2 THE SKIN IN VITAMIN A DEFICIENCY

The hard lumps on the skin of this person's arm reflect accumulations of keratin in the epithelial cells.

© H. Sanstead, U. of Texas/Galveston

■ Using new technologies, researchers are close to developing a vitamin A–rich rice to serve as a staple food for the world's children who lack vitamin A. See the Controversy section of Chapter 12 for details.

■ World hunger is a topic of Chapter 15.

epithelial (ep-ith-THEE-lee-ull) **tissue** the layers of the body that serve as selective barriers to environmental factors. Examples are the cornea, the skin, the respiratory tract lining, and the lining of the digestive tract.

cell differentiation (dih-fer-en-she-AY-shun) the process by which immature cells are stimulated to mature and gain the ability to perform functions characteristic of their cell type.

FIGURE 7·3   VITAMIN A DEFICIENCY AND TOXICITY

Danger lies both above and below a normal range of intakes of vitamin A.

Vitamin A intake, μg/day	Deficient 0–500		Normal 500–3,000		Toxic 3,000 and over	
	Effects on cells	**Health consequences**	**Effects on cells**	**Health consequences**	**Effects on cells**	**Health consequences**
	Decreased cell division and deficient development	Night blindness	Normal cell division and development	Normal body functioning	Overstimulated cell division	Skin rashes
		Keratinization				Hair loss
		Xerophthalmia				Hemorrhages
		Impaired immunity				Bone abnormalities
		Reproductive and growth abnormalities				Birth defects
		Exhaustion				Fractures
		Death				Liver failure
						Death

Vitamin A Toxicity   For people who take excess vitamin A in supplements or fortified foods, Figure 7-3 shows that toxicity compromises the tissues just as deficiency does and is equally dangerous. Vitamin A's toxicity's many symptoms include abdominal pain, hair loss, joint pain, stunted growth, bone and muscle soreness, cessation of menstruation, nausea, diarrhea, rashes, damage to the liver, and enlargement of the spleen. The earliest symptoms of overdoses are loss of appetite, blurred vision, growth failure in children, headache, itching of the skin, and irritability. Over the years, even relatively small excesses may weaken the bones and contribute to hip fractures.[14] Small daily excesses are easily supplied by today's heavily fortified food supply—even some fortified candy bars and bubble gum deliver substantial vitamin A (see Table 7-2). Some experts are asking whether this amount of fortification is doing more harm than good in the population, and others suggest that vitamin A supplements be reserved for treatment of true deficiencies caused by malabsorption or malnutrition.[15]

Pregnant women, especially, should be wary—chronic use of vitamin A supplements providing three to four times the amount recommended for pregnancy has caused malformations of the fetus. Even a single massive vitamin A dose (100 times the need) will do so. Children, who often mistake chewable vitamin pills for candy, are also likely to be hurt by vitamin A excesses because they need less and are more sensitive to overdoses. Adolescents may take massive vitamin A doses in the mistaken belief that vitamin A can correct acne. An effective acne medicine, Accutane, is *derived* from vitamin A, but it is chemically altered and given in carefully controlled dosages—vitamin A itself has no effect on acne.

Healthy people can eat foods naturally rich in vitamin A in large amounts without risking toxicity, with the possible exception of liver. When laboratory pigs eat chow made from salmon parts, including the livers, the animals stop growing and fall ill from vitamin A toxicity. Inuit people and Arctic explorers know that polar bear livers are a dangerous food source because the bears eat fish whole with the livers and concentrate large amounts of vitamin A. One ounce of ordinary liver offers three times the DRI recommended intake of preformed vitamin A along with abundant nutrients of many kinds, and eating it occasionally can boost nutrient status. Daily use invites vitamin A toxicity, however, especially in young children and pregnant women who eat other fortified foods or take supplements.[16]

◼ The effects of excessive vitamin A intakes during pregnancy are discussed in Chapter 13.

TABLE 7·2   FORTIFICATION SOURCES OF VITAMIN A

Vitamin A from highly fortified foods and other rich sources can add up. The UL for Vitamin A is 3,000 μg per day.

High-potency vitamin pill	3,000 μg
Calf's liver, 1 oz. cooked	2,300 μg
Regular multivitamin pill	1,500 μg
Vitamin gumball, 1	1,500 μg
Chicken liver, 1 oz cooked	1,400 μg
"Complete" liquid supplement drink, 1 serving	350–1,500 μg
Instant breakfast drink, 1 serving	600–700 μg
"Diet" low-carbohydrate drink, 1 serving	500–700 μg
Cereal breakfast bar, 1	350–400 μg
"Energy" candy bar, 1	350 μg
Milk, 1 c	150 μg
Vitamin-fortified cereal, 1 serving	150 μg
Margarine, 1 tsp	55 μg

Vitamin A Recommendations   The ability of vitamin A to be stored in the tissues means that although the vitamin A intake recommendation is given as a daily amount, you need not consume the vitamin every day. An average intake that meets the daily need over several months is sufficient. The amount of vitamin A you need is proportional to your body weight. According to the DRI committee, a man needs a daily average of about 900 micrograms; a woman needs about 700 micrograms. During lactation, her need is higher. Children need less. A regular balanced diet that includes the recommended servings of vegetables and fruits each day supplies more than adequate amounts.

Vitamin A recommendations are expressed in micrograms, but some food tables and supplement labels still express vitamin A contents using a different unit, the **IU (international unit)**. Be careful to notice whether food tables or supplement labels use micrograms or IU. See the Aids to Calculations (Appendix C) for help in converting many kinds of units. When comparing vitamin A in foods, make sure that the amounts are all expressed in the same units.

As for vitamin A supplements, the DRI committee and other nutrition agencies recommend that people avoid taking supplements that exceed the Tolerable Upper Intake Level of 3,000 micrograms (for adults over age 18). The best way to ensure a safe intake of vitamin A is to steer clear of supplements and obtain it instead from foods.

Food Sources of Vitamin A   Active vitamin A is provided in foods of animal origin. The richest sources are liver and fish oil, but milk and milk products and other fortified foods such as cereals can also be good sources. Even butter and eggs provide some vitamin A to the diet. Plants contain no active vitamin A, but many vegetables and fruits contain the vitamin A precursor, beta-carotene. Snapshot 7-1 is the first of a series of figures that show a sampling of foods that provide more than 10 percent of the Daily Value for a vitamin in a standard-size portion and therefore qualify to be called "good" or "rich" sources.

■ The U.S. and Canadian standard for vitamin A intake is the DRI recommended intake, listed on the inside front cover, page B.

■ 1 IU = 0.3 μg retinol.

IU (international unit) a measure of fat-soluble vitamin activity sometimes used on supplement labels.

SNAPSHOT 7·1

VITAMIN A AND BETA-CAROTENE

DRI RECOMMENDED INTAKES:
Men: 900 μg/day[a]
Women: 700 μg/day[a]

TOLERABLE UPPER INTAKE LEVEL:
Adults: 3,000 μg vitamin A/day

CHIEF FUNCTIONS: Vision; maintenance of cornea, epithelial cells, mucous membranes, skin; bone and tooth growth; regulation of gene expression; reproduction; immunity

DEFICIENCY: Night blindness, corneal drying (xerosis), and blindness (xerophthalmia); impaired bone growth and easily decayed teeth; keratin lumps on the skin; impaired immunity

TOXICITY:
Vitamin A: Increased activity of bone-dismantling cells causing reduced bone density and pain; liver abnormalities; birth defects
Beta-carotene: Harmless yellowing of skin

*These foods provide 10 percent or more of the vitamin A Daily Value in a serving. For a 2,000-calorie diet, the DV is 900 μg/day.

[a]Vitamin A recommendations are expressed in retinol activity equivalents (RAE).

[b]This food contains preformed vitamin A.

GOOD SOURCES*

FORTIFIED MILK[b]
1 c = 150 μg

CARROTS[c] (cooked)
½ c = 671 μg

SWEET POTATO[c] (baked)
½ c = 961 μg

SPINACH[c] (cooked)
½ c = 472 μg

BEEF LIVER[b] (cooked)
3 oz = 6,582 μg

BOK CHOY (cooked)
½ c = 180 μg

APRICOTS[c]
3 apricots = 100 μg

Colorful foods are often rich in vitamins.

The definitive fast-food meal—a hamburger, fries, and cola—lacks vitamin A. Many fast-food restaurants, however, now offer salads with cheese and carrots and other vitamin A–rich foods. These selections greatly improve the nutritional quality of a fast-food meal.

▶ KEY POINT Vitamin A is essential to vision, integrity of epithelial tissue, bone growth, reproduction, and more. Vitamin A deficiency causes blindness, sickness, and death and is a major problem worldwide. Overdoses are possible and cause many serious symptoms. Foods are preferable to supplements for supplying vitamin A.

■ Vitamin A conversion factors are provided in Appendix C.

Beta-Carotene In plants, vitamin A exists only in its precursor forms. Beta-carotene, the most abundant of these **carotenoid** precursors, has the highest vitamin A activity. The conversion of beta-carotene to retinol in the body entails losses, however, so vitamin A activity for vitamin A precursors is measured in **retinol activity equivalents (RAE)**.[17] It takes about 12 micrograms of beta-carotene from food to supply the equivalent of one microgram of retinol to the body.

Retinol in excess is toxic, but beta-carotene in food is not converted to retinol efficiently enough to cause vitamin A toxicity symptoms. A steady diet of abundant pumpkin, carrots, carrot juice, and the like has been known to turn people bright yellow because beta-carotene builds up in the fat just beneath the skin and imparts a harmless yellow cast; see Figure 7-4 in the margin.[18] Concentrated beta-carotene supplements, however, may have adverse effects, as another section points out.

FIGURE 7·4 EXCESS BETA-CAROTENE SYMPTOM—DISCOLORATION OF THE SKIN

The hand on the right shows skin discoloration from excess beta-carotene. Another person's normal hand (left) is shown for comparison.

Does Eating Carrots Really Promote Good Vision? Bright orange fruits and vegetables derive their color from beta-carotene and are so colorful that they decorate the plate. Carrots, sweet potatoes, pumpkins, mango, cantaloupe, and apricots are all rich sources of beta-carotene—and therefore contribute vitamin A to the eyes and to the rest of the body—so, yes, eating carrots does promote good vision. Another colorful group, *dark* green vegetables, such as spinach, other greens, and broccoli, owe their color to the green pigment chlorophyll and to beta-carotene. The green and orange pigments together give a deep dark green color to the vegetables.

Other colorful vegetables, such as red cabbage, beets, and yellow corn, can fool you into thinking they contain beta-carotene, but these foods derive their colors from other pigments and are poor sources of beta-carotene. As for "white" plant foods such as grains and potatoes, they have none. Some confusion exists concerning the term *yam*. A white-fleshed Mexican root vegetable called "yam" is devoid of beta-carotene, but the orange-fleshed sweet potato called "yam" in the United States is one of the richest beta-carotene sources known. Recommendations state that a person should eat *deep* orange or *dark* green vegetables and fruits regularly.

carotenoid (CARE-oh-ten-oyd) a member of a group of pigments in foods that range in color from light yellow to reddish orange and are chemical relatives of beta-carotene. Many with a degree of vitamin A activity in the body. Also defined in Controversy 2.

retinol activity equivalents (RAE) a new measure of the vitamin A activity of beta-carotene and other vitamin A precursors that reflects the amount of retinol that the body will derive from a food containing vitamin A precursor compounds.

macular degeneration a common, progressive loss of function of the part of the retina that is most crucial to focused vision (the macula is shown on page 216). This degeneration often leads to blindness.

Carotenoids and Diseases The important roles of active vitamin A in vision have been emphasized, but another link is worth mentioning. People whose diets lack foods rich in beta-carotene have a high incidence of the most common form of untreatable age-related blindness, **macular degeneration**.[19] The macula is a yellow spot located at the focal center of the retina (identified in the drawing on page 216). In macular degeneration, the macula loses integrity, causing impairment of the most important field of vision, the central focus. For a while, beta-carotene was credited with providing the macula's protective yellow pigment, but research soon determined that carotenoids other than beta-carotene deserve the credit.[20] These other carotenoids occur together with beta-carotene in colorful foods such as leafy green vegetables.*

Likewise, abundant beta-carotene from foods and elevated beta-carotene in the blood are associated with lower cancer incidence, but beta-carotene supplements are not. In

---
*The other carotenoid is zeaxanthin.

they appear to pose a cancer hazard to smokers. This seems to indicate that beta-carotene is not itself responsible for an anticancer effect but simply tags along as a marker for another unknown factor or combination of factors that occurs along with it. Beta-carotene is just one of the major **dietary antioxidants** present in foods—others include vitamin E, vitamin C, the mineral selenium, and many phytochemicals. Further, dietary antioxidants are just one class of a complex array of beneficial constituents in whole foods that seem to benefit health synergistically. In the case of beta-carotene, diets rich in this antioxidant and its chemical relatives correlate with low rates of eye diseases and cancer.

> ▶ KEY POINT   The vitamin A precursor in plants, beta-carotene, is an effective antioxidant in the body. Brightly colored plant foods are richest in beta-carotene, and diets containing these foods are associated with eye health.

■ Chapter 5 introduced antioxidants as substances that protect body compounds from damage by oxidation. Controversy 2 introduced the carotene family of phytochemicals and explores more of their disease-fighting potential. This chapter's Controversy comes back to supplements and disease prevention.

Vitamin D

Vitamin D is different from all the other nutrients in that the body can synthesize all it needs with the help of sunlight. Therefore, in a sense, vitamin D is not an essential nutrient. Given enough sun each day, most people need consume no vitamin D at all from foods. As simple as this may sound, recent surveys indicate that many people, particularly African Americans and Mexican Americans, may border on vitamin D insufficiency.[21]

Roles of Vitamin D   The best-known role of vitamin D is as a member of a large cast of nutrients and hormones that interact to regulate blood calcium and phosphorus levels and thereby maintain bone integrity. Many of these interactions take place at the genetic level in ways that are under investigation.[22]

Calcium is indispensable to the proper functioning of cells in all tissues of the body, including muscles, nerves, and glands, which draw calcium from the blood as they need it. When more calcium is needed to replenish the supply, vitamin D acts at three body locations to raise the blood calcium level. First, the skeleton serves as a vast warehouse of stored calcium that can be tapped when blood calcium begins to fall. Only two other organs can act to raise the level of blood calcium: the digestive tract, where food brings calcium in, and the kidneys, which can recycle calcium into the body from blood filtrate destined to become urine.

Vitamin D functions as a hormone, that is, a compound manufactured by one organ of the body that acts on other organs or tissues. In addition to its actions in the bones, intestines, and kidneys, vitamin D plays roles in the workings of the brain, heart, stomach, pancreas, skin, and reproductive organs. Like vitamin A, vitamin D stimulates maturation of cells, including cells of the immune system. Research is hinting (sometimes strongly) that to incur a deficit of vitamin D is to invite problems of many kinds, including high blood pressure, some common forms of cancer, type 1 diabetes, heart disease, rheumatoid arthritis, inflammatory bowel disease, and even multiple sclerosis.[23] The well-established problems, however, concern impaired calcium balance and the bones, especially during growth.

Too Little Vitamin D—A Danger to Bones   The most obvious sign of vitamin D deficiency is abnormality of the bones in the disease **rickets,** shown in Figure 7-5 (p. 222). Children with rickets develop bowed legs because they are unable to mineralize newly forming bone material, a rubbery protein matrix. As gravity pulls their body weight against these weak bones, the legs bow. Many such children also have a protruding belly because of lax abdominal muscles.

As early as the 1700s, rickets was known to be curable with cod-liver oil, which is rich in vitamin D. More than a hundred years later, a Polish physician linked sunlight exposure to prevention and cure of rickets. Today, the bowed legs, knock-knees, beaded ribs,

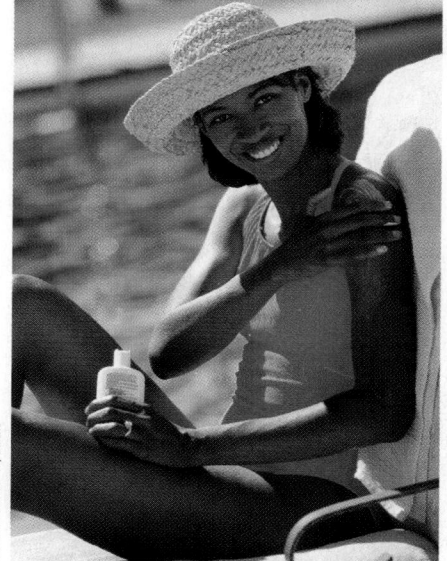

The sunshine vitamin: vitamin D.

© Michael Keller/Corbis

■ Chapter 8 and Controversy 8 present more about bone minerals and their regulation, and about osteoporosis, the bone-weakening disease.

dietary antioxidants compounds typically found in plant foods that significantly decrease the adverse effects of oxidation on living tissues. The major antioxidant vitamins are vitamin E, vitamin C, and beta-carotene.

rickets the vitamin D–deficiency disease in children; characterized by abnormal growth of bone and manifested in bowed legs or knock-knees, outward-bowed chest, and knobs on the ribs.

FIGURE 7·5 RICKETS

This child has the bowed legs of the vitamin D–deficiency disease rickets.

This child displays the beaded ribs common in rickets.

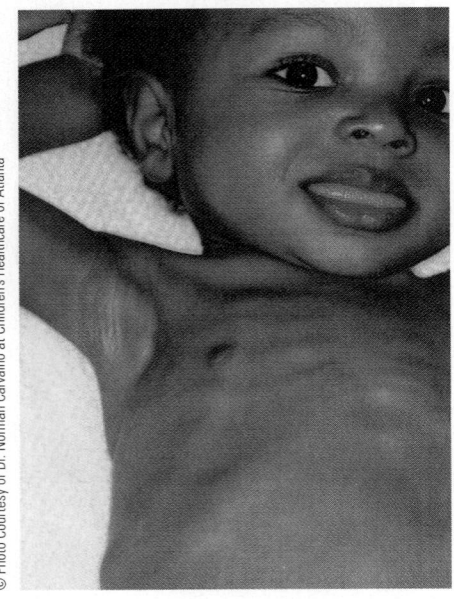

osteomalacia (OS-tee-o-mal-AY-shuh) the adult expression of vitamin D–deficiency disease, characterized by an overabundance of unmineralized bone protein (*osteo* means "bone"; *mal* means "bad"). Symptoms include bending of the spine and bowing of the legs.

and protruding (pigeon) chests of children with rickets are no longer common sights in the United States, although rickets occasionally occurs among breastfed black infants not supplemented with vitamin D, and infants and toddlers fed unfortified soy and rice beverages instead of formula or milk.[24] Many children worldwide suffer the ravages of rickets because of inadequate food combined with a lack of sunlight.

Adolescents, who often abandon vitamin D–fortified milk in favor of soft drinks, may also prefer indoor pastimes such as video games to outdoor activities during daylight hours. Such teens often lack vitamin D and so fail to develop the bone density needed to prevent bone loss in later life.[25]

Even older people can suffer painful joints and muscles if their vitamin D levels are low, although the condition is easily missed during examinations and may be mistaken for arthritis or other painful conditions.[26] The adult form of rickets, **osteomalacia,** occurs most often in women who have low calcium intakes and little exposure to sun and who go through repeated pregnancies and periods of lactation. Given this combination of risk factors, bone proteins fail to mineralize normally, and the leg bones may soften to such an extent that a young woman who was tall and straight at 20 may become bent and bowlegged before she turns 30.

Too Much Vitamin D—A Danger to Soft Tissues Vitamin D is the most potentially toxic of all vitamins. Chronic ingestion of excesses may be directly toxic to the bones, kidneys, brain, nerves, and the heart and arteries. Symptoms of toxicity include appetite loss, nausea, vomiting, and increased urination and thirst; a severe form of psychological depression can result from vitamin D's effects on the central nervous system. If overdoses continue, vitamin D raises the blood mineral level to dangerous extremes, forcing calcium to be deposited in soft tissues such as the heart, blood vessels, lungs, and kidneys. Even the soft pulp of the teeth hardens, while tooth enamel thins. Calcium deposited in critical organs may cause them to malfunction, with serious consequences to health and life.

The likeliest victims of vitamin D poisoning are infants whose well-intentioned but misguided parents think that if some is good, more is better. Also, older people who take vitamin D supplements to stem adult bone loss may also easily overdose. Intakes of only five times the recommended amount have been associated with signs of vitamin D toxicity in young children and adults. Two people died and others became ill after drinking milk from a dairy that had mistakenly overfortified the milk with up to 500 times the usual dosage of vitamin D. One infant survivor later developed dental problems because the vitamin D overdose disturbed the normal development of her permanent teeth. Such occurrences are rare, but the incident renewed awareness of the potential for harm from vitamin D and the need for close monitoring of those who fortify the nation's foods with vitamins. The DRI committee has set a Tolerable Upper Intake Level for vitamin D at 50 micrograms per day (2,000 IU on supplement labels).

How Can People Make a Vitamin from Sunlight? Most of the world's population relies on natural exposure to sunlight to maintain adequate vitamin D nutrition. When ultraviolet light from the sun shines on a cholesterol compound in human skin, the compound is transformed into a vitamin D precursor and is absorbed directly into the blood. Slowly, over the next day and a half, the liver and kidneys finish converting the precursor to the active form of vitamin D. Diseases that affect either the liver or the kidneys can impair the conversion of the inactive precursor to the active vitamin and therefore produce symptoms of vitamin D deficiency.

Unlike concentrated supplements, sunlight presents no risk of vitamin D toxicity; the sun itself begins breaking down excess vitamin D made in the skin. Sunbathers run *other* risks, of course, such as premature wrinkling of the skin and the increased risk of skin cancer. Sunscreens with sun protection factors (SPF) of 8 and above can reduce these risks, but they also prevent vitamin D synthesis. Production of vitamin D doesn't demand idle hours of sunbathing, however. In the warmer months in most locations, just being outdoors when the sun is overhead, even in lightweight clothing, is sufficient.

The pigments of dark skin provide protection from ultraviolet radiation but also reduce vitamin D synthesis. Dark-skinned people require longer exposure to direct sun (up to three hours, depending on the climate) for several days' worth of vitamin D, but light-skinned people need much less time (10 or 15 minutes). Thus, a person can wait until just enough time has elapsed to make some vitamin D and then apply sunscreen. Tanning booths may or may not promote vitamin D synthesis, but the Food and Drug Administration (FDA) has declared them risky because their unfiltered rays may promote skin cancer and damage the blood vessels and eyes.

The ultraviolet rays of the sun that promote vitamin D synthesis cannot penetrate clouds, smoke, smog, heavy clothing, window glass, or even window screens. In the United States and Canada, many cases of rickets show up in dark-skinned people who live in smoggy northern cities or who lack exposure to sunlight.[27] A surprisingly high number of otherwise healthy northern adults, even those drinking milk fortified with vitamin D, have recently been reported with low blood levels of vitamin D, mostly at the end of the winter season.[28] It may be that milk does not deliver enough vitamin D to prevent a drop in blood levels through the winter months. In addition, people who are housebound or institutionalized and those who work at night may incur (over years) a vitamin D deficiency, as do many elderly people, who have limited exposure to sunlight and lose efficiency in activating vitamin D as they age. For these people, dietary vitamin D is essential.[29]

Because of increased risk with age, the DRI committee set recommended intakes for vitamin D that increase over the years: 5 micrograms per day for adults 19 to 50 years, 10 micrograms for those 51 to 70 years, and 15 micrograms for those over 70.

Snapshot 7-2 shows the few significant food sources of vitamin D. Butter, cream, and fortified margarine contribute small amounts. In the United States and Canada, milk, whether fluid, dried, or evaporated, is fortified with vitamin D. Young adults who drink the recommended 3 cups a day receive half of their daily requirement; the other half comes from exposure to sunlight and other food sources. A daily quart (or liter)

Factors affecting sun exposure and vitamin D synthesis:

- *Air pollution.* Particles in the air screen out the sun's rays.
- *City living.* Tall buildings block sunlight.
- *Clothing.* Most clothing blocks sunlight.
- *Geography.* Lack of direct sunlight prevents vitamin D synthesis:
  - September through March at latitudes above 50 degrees (most of Canada).
  - November through February at latitudes between 35 and 50 degrees (most U.S. locations).
  - In locations south of 35 degrees (northern borders of Alabama and Georgia), direct sun exposure is sufficient for vitamin D synthesis year round.
- *Homebound.* Living indoors prevents sun exposure.
- *Season.* Warmer seasons of the year bring more direct sun rays.
- *Sunscreen.* Use reduces or prevents skin exposure to sun's rays.
- *Time of day.* Midday hours provide maximum direct sun exposure.

SNAPSHOT 7·2

VITAMIN D

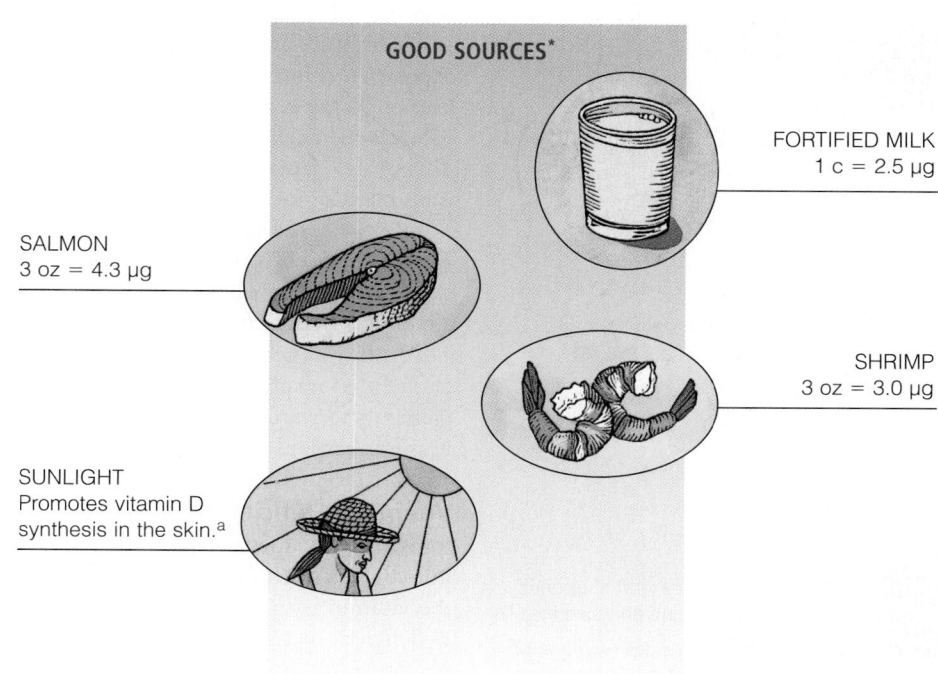

DRI RECOMMENDED INTAKES:
Adults: 5 μg/day (19–50 yr)
　　　　10 μg/day (51–70 yr)
　　　　15 μg/day (>70 yr)

TOLERABLE UPPER INTAKE LEVEL:
Adults: 50 μg/day

CHIEF FUNCTIONS: Mineralization of bones and teeth (raises blood calcium and phosphorus by increasing absorption from digestive tract, withdrawing calcium from bones, stimulating retention by kidneys)

DEFICIENCY: Abnormal bone growth resulting in rickets in children, osteomalacia in adults; malformed teeth, muscle spasms

TOXICITY: Elevated blood calcium; calcification of soft tissues (blood vessels, kidneys, heart, lungs, tissues of joints), excessive thirst, headache, nausea, weakness

GOOD SOURCES[*]

SALMON
3 oz = 4.3 μg

SUNLIGHT
Promotes vitamin D
synthesis in the skin.[a]

FORTIFIED MILK
1 c = 2.5 μg

SHRIMP
3 oz = 3.0 μg

[*]These foods provide 10 percent or more of the vitamin D Daily Value in a serving. For a 2,000-calorie diet, the DV is 10 μg/day.

[a]Avoid prolonged exposure to sun.

of milk will supply the entire recommended amount. Children who drink 2 cups or more of milk a day will have a head start toward meeting their vitamin D needs for growth. Yogurt and cheese products are often not fortified, so read the labels. Strict vegetarians and their children may have low vitamin D intakes because only two fortified plant sources exist: some margarines and, in the United States, certain fortified cereals.

> ▶ **KEY POINT** Vitamin D raises mineral levels in the blood, notably calcium and phosphorus, permitting bone formation and maintenance. A deficiency can cause rickets in childhood or osteomalacia in later life. Vitamin D is the most toxic of all the vitamins, and excesses are dangerous or deadly. People exposed to the sun make vitamin D from a cholesterol-like compound in their skin; fortified milk is an important food source.

Vitamin E

More than 80 years ago, researchers discovered a compound in vegetable oils necessary for reproduction in rats. This compound was named **tocopherol** from *tokos,* a Greek word meaning "offspring." A few years later, the compound was named vitamin E. Four tocopherol compounds have been identified, and each is designated by one of the first four letters of the Greek alphabet: alpha, beta, gamma, and delta. Of these, alpha-tocopherol is the gold standard for vitamin E activity. For this reason, DRI intake recommendations are expressed as alpha-tocopherol.

The Extraordinary Bodyguard Vitamin E is an antioxidant and thus serves as one of the body's main defenders against oxidative damage. Such damage occurs when highly unstable molecules known as **free radicals,** formed during normal cell metabolism, run amok and disrupt the structures of lipids in cell membranes, molecules of DNA, or the proteins that perform cellular work. A long-standing hypothesis states that such free-radical activity, if unchecked, may lead to cancer, heart disease, or other diseases. Vitamin E, by being oxidized itself, quenches free radicals, thus protecting vulnerable cell components and membranes from destruction. Figure 7-6 provides an overview of the activity of vitamin E and its potential role in disease prevention.

Vitamin E's antioxidant effect is especially crucial in the lungs, where the cells are exposed to high oxygen concentrations that can destroy molecules in their membranes. As the red blood cells carry oxygen from the lungs to other tissues, vitamin E protects their cell membranes, too. The white blood cells that defend the body against disease also depend on vitamin E, and it may play other roles in immunity. Vitamin E is also crucial for normal nerve development.

Over a decade ago, preliminary research aroused hope that supplements of vitamin E might protect against heart disease. Subsequent research, however, indicates that except for people whose blood measures indicate low vitamin E levels, people generally do not benefit from vitamin E supplements and may risk harm from high doses.[30] After a thorough investigation, a U.S. Preventive Services Task Force of scientists concluded that insufficient support exists to recommend taking vitamin E or other supplements to ward off chronic diseases.[31] Supplements of vitamin E may improve the immune response in healthy elderly people, however.

Vitamin E Deficiency A deficiency of vitamin E produces a wide variety of symptoms in laboratory animals. Most of these symptoms have not been reproduced in human beings, however, despite many attempts. Three reasons have been given: First, the vitamin is so widespread in food that it is almost impossible to create a vitamin E–deficient diet. Second, the body stores so much vitamin E in its fatty tissues that a person would find it difficult to eat a vitamin E–free diet for long enough to deplete these stores and produce a deficiency. Third, the cells recycle their working supply of vitamin E, using the same molecules over and over to ward off deficiency.

■ For perspectives on possible risks and benefits of vitamin E supplements, see the Controversy.

tocopherol (tuh-KOFF-er-all) a kind of alcohol. The active form of vitamin E is alpha-tocopherol.

free radicals atoms or molecules with one or more unpaired electrons that make the atom or molecule unstable and highly reactive.

FIGURE 7·6 FREE-RADICAL DAMAGE AND ANTIOXIDANT PROTECTION

Free-radical formation occurs during metabolic processes, and it accelerates when diseases or other stresses strike.

Free radicals cause chain reactions that damage cellular structures.

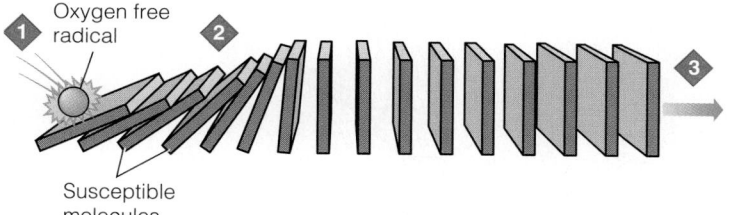

Oxygen free radical

Susceptible molecules

1 A chemically reactive oxygen free radical attacks fatty acid, DNA, protein, or cholesterol molecules, which form other free radicals in turn.

2 This initiates a rapid, destructive chain reaction.

3 The result is disabling injury to lipids of cell membranes and cellular proteins, damage to DNA, or oxidation of cholesterol. These changes may initiate steps leading to diseases such as heart disease, cancer, macular degeneration, and others.

Antioxidants quench free radicals and protect cellular structures.

Antioxidant

4 Antioxidants, such as vitamin E, stop the chain reaction by changing the nature of the free radical.

The classic vitamin E–deficiency symptom in human beings occurs in premature babies who are born before the transfer of the vitamin from the mother to the infant, which takes place in the last weeks of pregnancy. Without sufficient vitamin E, the infant's red blood cells rupture (**erythrocyte hemolysis**), and the infant becomes anemic. The few symptoms of vitamin E deficiency that have been observed in adults include loss of muscle coordination and reflexes with impaired movement, vision, and speech. All of these symptoms may be caused by oxidative damage; vitamin E treatment corrects them.

In adults, vitamin E deficiency is usually associated with diseases that cause malabsorption of fat, including disease of or injury to the liver (which makes bile, necessary for digestion of fat), the gallbladder (which delivers bile into the intestine), and the pancreas (which makes fat-digesting enzymes). In people without diseases, deficiencies of vitamin E are most likely when diets extremely low in fat are consumed for years. People who rely solely on fat replacers, such as diet margarines and fat-free salad dressings, to the exclusion of all oils may have low vitamin E intakes. Those consuming diets of highly processed or "convenience" foods may lack sufficient vitamin E because vitamin E is destroyed by extensive heating in the processing of these foods.

Chronically low intakes of vitamin E may play other roles in diseases. One line of research suggests that when body stores of vitamin E and the mineral selenium are low, viruses respond by becoming more virulent—even normally harmless viruses appear to undergo changes that make them more likely to cause diseases.[32] No one yet knows the details of these effects, but they may be related to "oxidative stress," caused when oxidative activities outstrip the capacity of the tissues' antioxidant defenses.

Extravagant claims are often made for vitamin E because its deficiency affects animals' muscles and reproductive systems. Research in human beings has discredited claims that vitamin E improves athletic endurance and skill, enhances sexual performance, or cures sexual dysfunction in males, although such claims are still being used to sell supplements containing vitamin E.

Vitamin E Requirements, Toxicity, and Sources The DRI recommended intake (inside front cover) for vitamin E is 15 milligrams a day for adults. This amount seems sufficient to maintain blood values for both vitamin E and indicators of oxidation

■ The USDA Food Guide recommends small daily intakes of oils to supply vitamin E.

■ More about oxidative stress in Chapter 11.

erythrocyte (eh-REETH-ro-sight) **hemolysis** (HE-moh-LIE-sis, he-MOLL-ih-sis) rupture of the red blood cells, caused by vitamin E deficiency (*erythro* means "red"; *cyte* means "cell"; *hemo* means "blood"; *lysis* means "breaking").

VITAMIN E

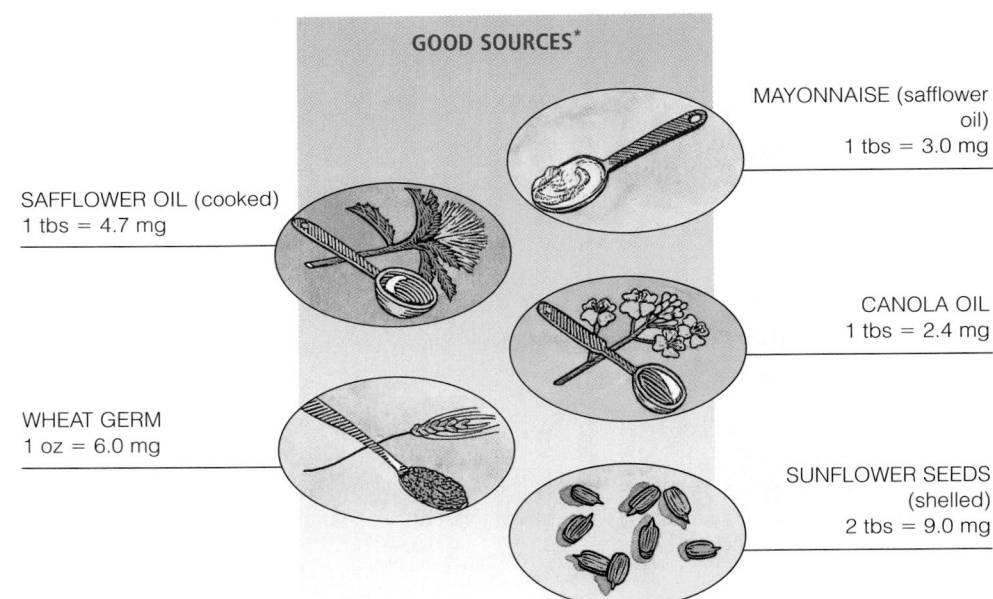

GOOD SOURCES*

DRI RECOMMENDED INTAKE:
Adults: 15 mg/day

TOLERABLE UPPER INTAKE LEVEL:
Adults: 1,000 mg/day

CHIEF FUNCTIONS: Antioxidant (protects cell membranes, regulates oxidation reactions, protects polyunsaturated fatty acids)

DEFICIENCY: Red blood cell breakage, nerve damage

TOXICITY: Augments the effects of anti-clotting medication

*These foods provide 10 percent or more of the vitamin E Daily Value in a serving. For a 2,000-calorie diet, the DV is 30 IU or 20 mg/day.

MAYONNAISE (safflower oil)
1 tbs = 3.0 mg

SAFFLOWER OIL (cooked)
1 tbs = 4.7 mg

CANOLA OIL
1 tbs = 2.4 mg

WHEAT GERM
1 oz = 6.0 mg

SUNFLOWER SEEDS (shelled)
2 tbs = 9.0 mg

■ Vitamin E conversion factors are in Appendix C.

FIGURE 7·7 VITAMIN E RECOMMENDATION AND U.S. INTAKES COMPARED

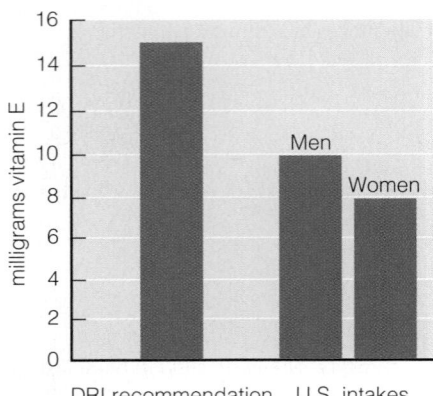

milligrams vitamin E

Men

Women

DRI recommendation   U.S. intakes

SOURCE: USDA Center for Nutrition Policy and Promotion, 2004, available at www.cnpp.usda.gov.

reactions within healthy normal limits. On average, U.S. intakes of vitamin E fall substantially below the recommendation (see Figure 7-7).

The need for vitamin E rises as people consume more polyunsaturated oil because the oil requires antioxidant protection by the vitamin. Most raw oils contain vitamin E, so people who eat raw oils also receive the vitamin. As mentioned, heat processing, such as frying, destroys vitamin E, as does oxidation, so most of the processed, fast, deep-fried, and convenience foods popular among U.S. consumers retain little intact vitamin E.

No adverse effects are known to arise from naturally-occurring vitamin E in foods. Also, studies of vitamin E supplements in doses up to the DRI recommended intake value generally indicate safety. Larger doses may be risky, however. Vitamin E supplements increase the effects of anticoagulant medication used to oppose unwanted blood clotting, so people taking such drugs risk uncontrollable bleeding if they also take large doses of vitamin E. A decade ago, researchers reported an increase in brain hemorrhages, a form of stroke, among smokers taking just 50 milligrams of vitamin E per day. Recently, the pooled results from 19 experiments involving almost 136,000 people revealed that those taking vitamin E in doses greater than 400 IU (about a third of the Tolerable Upper Intake Level) per day were at an increased risk of death from all causes compared with people taking smaller doses.[33] To err on the safe side until more is known about the safety of vitamin E supplements, intakes should probably be kept low, and they certainly should not exceed the Tolerable Upper Intake Level of 1,000 milligrams alpha-tocopherol per day.

Vitamin E is widespread in foods. About 20 percent of the vitamin E people consume comes from vegetable oils and products made from them, such as salad dressings and shortening (see Snapshot 7-3). Another 20 percent comes from fruits and vegetables although none of these is a good source by itself. Fortified cereals* and other grain products contribute about 15 percent of vitamin E in the diet, and meats, poultry, fish, eggs, milk products, nuts, and seeds contribute smaller percentages. Wheat germ is a good source of vitamin E; animal fats have almost none.

*Cereals fortified with vitamin E may not be available in Canada.

Vitamin K

Have you ever thought about how remarkable it is that blood can clot? The liquid turns solid in a life-saving series of reactions—if blood did not clot, wounds would just keep bleeding, draining the blood from the body. The main function of vitamin K is to help synthesize proteins that help clot the blood. Hospitals measure the clotting time of a person's blood before surgery and sometimes administer vitamin K before operations to reduce bleeding in surgery. Vitamin K may be of value at this time, but only if a vitamin K deficiency exists. Vitamin K does not improve clotting in those with other bleeding disorders, such as the inherited disease hemophilia.

Some people with heart problems need to *prevent* the formation of clots within their circulatory system—this is popularly referred to as "thinning" the blood. One of the best-known medicines for this purpose is warfarin, which interferes with the action of vitamin K in promoting clotting. Vitamin K therapy may be needed for people on warfarin if uncontrolled bleeding should occur.[34] People taking warfarin who self-prescribe vitamin K supplements risk interfering with the action of the drug.

Vitamin K is also necessary for the synthesis of key bone proteins. Vitamin K, together with the more famous bone nutrient, vitamin D, ensures the health of the bones. Without vitamin K, the bones produce an abnormal protein that cannot bind to the minerals that normally form bones, so bone density is low.[35] Adequate vitamin K may reduce the risk of hip fracture: people who consume abundant vitamin K, often in the form of green leafy vegetables, are reported to suffer fewer hip fractures than those with lower intakes.[36]

Like vitamin D, vitamin K can be obtained from a nonfood source—in this case, the intestinal bacteria. Billions of bacteria normally reside in the intestines, and some of them synthesize vitamin K. Newborn infants present a unique case with regard to vitamin K because they are born with a sterile intestinal tract, and the vitamin K–producing bacteria take weeks to establish themselves. To prevent hemorrhage, the newborn is given a single dose of vitamin K at birth.

As Snapshot 7-4 (page 228) shows, vitamin K's richest plant food sources include dark green leafy vegetables such as cooked spinach and collard greens, which provide an average of 300 micrograms per half-cup serving. Lettuces, broccoli, brussels sprouts, and other members of the cabbage family are also good sources. There is also one rich animal food source, liver. Canola and soybean oils (unhydrogenated liquid oils) provide smaller but still significant amounts, while fortified cereals can be rich sources of added vitamin K. One egg and a cup of milk contain about equal amounts: 25 micrograms each. Tables of food composition do not include the vitamin K contents of foods because they are not well enough known.

Few U.S. adults are likely to experience vitamin K deficiency, even if they seldom eat vitamin K–rich foods. Exceptions are newborn infants, as mentioned above, and people who have taken antibiotics that have killed both the beneficial and the harmful bacteria in their intestinal tracts. In certain medical conditions, bile production falters, making fats, including fat-soluble vitamins, unabsorbable. Supplements of the vitamin are needed in these cases because a vitamin K deficiency can be fatal.

Reports of vitamin K toxicity among healthy adults are rare, and the DRI committee has not set a Tolerable Upper Intake Level. For infants and pregnant women, however, vitamin K toxicity can result when supplements of a synthetic version of vitamin K are given too enthusiastically.* Toxicity induces breakage of the red blood cells and

■ *K* stands for the Danish word *koagulation* (clotting). Chapter 6 describes the roles of protein in blood clotting.

Soon after birth, newborn infants receive a dose of vitamin K.

---

*The version of vitamin K responsible for this effect is menadione.

VITAMIN K

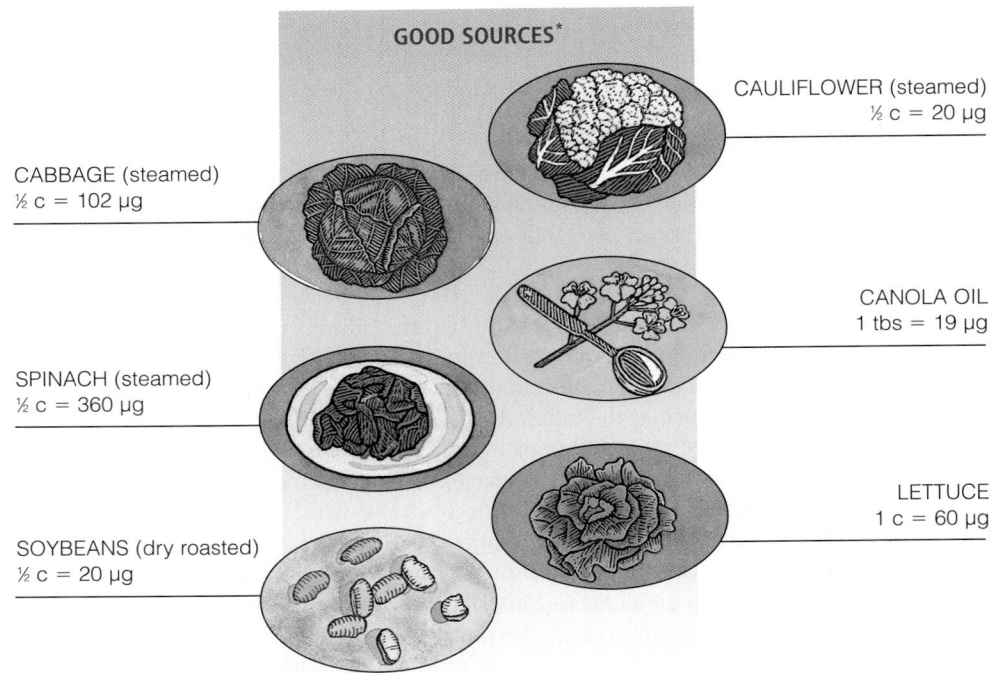

DRI RECOMMENDED INTAKES:
Men: 120 µg/day
Women: 90 µg/day

CHIEF FUNCTIONS: Synthesis of blood-clotting proteins and bone proteins

DEFICIENCY: Hemorrhage; abnormal bone formation

TOXICITY: Opposes the effects of anti-clotting medication

*These foods provide 10 percent or more of the vitamin K Daily Value in a serving. For a 2,000-calorie diet, the DV is 80 µg/day.

Data from Standing Committee on the Scientific Evaluation of Dietary Reference Intakes, Food and Nutrition Board, Institute of Medicine, *Dietary Reference Intakes for Vitamin A, Vitamin K, Arsenic, Boron, Chromium, Copper, Iodine, Iron, Manganese, Molybdenum, Nickel, Silicon, Vanadium, and Zinc* (Washington, D.C.: National Academy Press, 2001), pp. 5–18.

GOOD SOURCES*

CABBAGE (steamed)
½ c = 102 µg

SPINACH (steamed)
½ c = 360 µg

SOYBEANS (dry roasted)
½ c = 20 µg

CAULIFLOWER (steamed)
½ c = 20 µg

CANOLA OIL
1 tbs = 19 µg

LETTUCE
1 c = 60 µg

release of their pigment, which colors the skin yellow. A toxic dose of synthetic vitamin K causes the liver to release the blood cell pigment (bilirubin) into the blood (instead of excreting it into the bile) and leads to **jaundice.** When bilirubin invades the brain of an infant, the condition may lead to brain damage or death of the infant.

▸ **KEY POINT**   Vitamin K is necessary for blood to clot; deficiency causes uncontrolled bleeding. The bacterial inhabitants of the digestive tract produce vitamin K. Toxicity causes jaundice.

■ The water-soluble vitamins require special consideration in food preparation to avoid losing or destroying them. See the Food Feature of Chapter 12.

Characteristics water-soluble vitamins share:

• Dissolve in water.
• Are easily absorbed and excreted.
• Are not stored extensively in tissues.
• Seldom reach toxic levels.

jaundice (JAWN-dis) yellowing of the skin due to spillover of the bile pigment bilirubin (bill-ee-ROO-bin) from the liver into the general circulation.

The Water-Soluble Vitamins

Vitamin C and the B vitamins are water soluble. Cooking and washing with water can leach them out of foods. The body absorbs these vitamins easily and just as easily excretes them in the urine. Some of the water-soluble vitamins can remain in the lean tissues for a month or more, but these tissues are actively exchanging materials with the body fluids at all times. At any time, the vitamins may be picked up by the extracellular fluids, washed away by the blood, and excreted in the urine. Advice for meeting the need for these nutrients is straightforward: choose foods that are rich in water-soluble vitamins to achieve an average of the recommended intakes over three days' time. The snapshots in this section can help to guide your choices.

Foods never deliver toxic doses of the water-soluble vitamins, but the large doses concentrated in some vitamin supplements can reach toxic levels. Normally, though, the most likely hazard to the supplement taker is to the wallet: "If you take supplements of the water-soluble vitamins, you may have the most expensive urine in town." The Think Fitness feature discusses whether athletes need vitamin supplements.

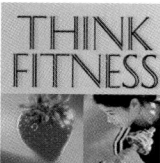

THINK FITNESS — Vitamins for Athletes

DO ATHLETES WHO strive for top performance need more vitamins than foods can supply? Competitive athletes who choose their diets with reasonable care almost never need nutrient supplements. The reason is elegantly simple. The need for energy to fuel exercise requires that people eat extra calories of food, and if that extra food is of the kind shown in this chapter's snapshots—fruits, vegetables, milk, eggs, whole or enriched grains, lean meats, and some oils—then the vitamins to support activity follow automatically. Chapter 10 comes back to roles of vitamins in physical activity.

Vitamin C

More than two hundred years ago, any man who joined the crew of a seagoing ship knew he had only half a chance of returning alive—not because he might be slain by pirates or die in a storm but because he might contract **scurvy,** a disease that might kill as many as two-thirds of a ship's crew on a long voyage. Ships that sailed on short voyages, especially around the Mediterranean Sea, were safe from this disease. The special hazard of long ocean voyages was that the ship's cook used up the fresh fruits and vegetables early and relied on cereals and live animals for the duration of the voyage.

The first nutrition experiment to be conducted on human beings was devised nearly 250 years ago to find a cure for scurvy. A physician divided some British sailors with scurvy into groups. Each group received a different test substance: vinegar, sulfuric acid, seawater, oranges, or lemons. Those receiving the citrus fruits were cured within a short time. Sadly, it took 50 years for the British navy to make use of the information and require all its vessels to provide lime juice to every sailor daily. British sailors were mocked with the term *limey* because of this requirement. The name later given to the vitamin that the fruit provided, **ascorbic acid,** literally means "no-scurvy acid." It is more commonly known today as vitamin C.

Long voyages without fresh fruits and vegetables spelled death by scurvy for the crew.

The Work of Vitamin C   Vitamin C performs a variety of functions in the body, but it is best known for its work in maintaining the connective tissues and as an antioxidant. As one of its tasks, vitamin C assists several enzymes in performing their jobs. In particular, the enzymes involved in the formation and maintenance of the tissue protein **collagen** depend on vitamin C for their activity. Collagen forms the base for all connective tissues: bones, teeth, skin, and tendons. Collagen forms the scar tissue that heals wounds, the reinforcing structure that mends fractures, and the supporting material of capillaries that prevents bruises. Vitamin C also acts as a cofactor in the production of carnitine, important for transporting fatty acids within the cells.

In addition to assisting enzymes, vitamin C also acts in a more general way as an antioxidant.[37] Vitamin C protects substances from oxidation by being oxidized itself. For example, cells of the immune system maintain high levels of vitamin C to protect themselves from free radicals generated during assaults on bacteria and other invaders. While some of the oxidized vitamin C is degraded irretrievably and must be replaced by the diet, much is not lost but recycled back to the active form for reuse.

In the intestines, vitamin C protects iron from oxidation and so promotes its absorption. In the blood, vitamin C protects sensitive blood constituents from oxidation and helps to protect and recycle vitamin E to its active form. The antioxidant roles of vitamin C are the focus of extensive study, especially in relation to disease prevention. Unfortunately, research has yielded only disappointing results: vitamin C supplements seem useless against heart disease, cancer, and other diseases, unless they are prescribed to treat a deficiency.[38] In test tubes, a high concentration of vitamin C has the opposite effect from antioxidants; that is, it acts as a **prooxidant** by activating oxidizing elements, such as iron and copper. One study revealed an increase in markers of oxidation in men

scurvy the vitamin C–deficiency disease.

ascorbic acid one of the active forms of vitamin C (the other is *dehydroascorbic* acid); an antioxidant nutrient.

collagen (COLL-a-jen) the chief protein of most connective tissues, including scars, ligaments, and tendons, and the underlying matrix on which bones and teeth are built.

prooxidant a compound that triggers reactions involving oxygen.

FIGURE 7·8 SCURVY SYMPTOMS—GUMS AND SKIN

Vitamin C deficiency causes the breakdown of collagen, which supports the teeth.

Small pinpoint hemorrhages (red spots) appear in the skin indicating that invisible internal bleeding may also be occurring.

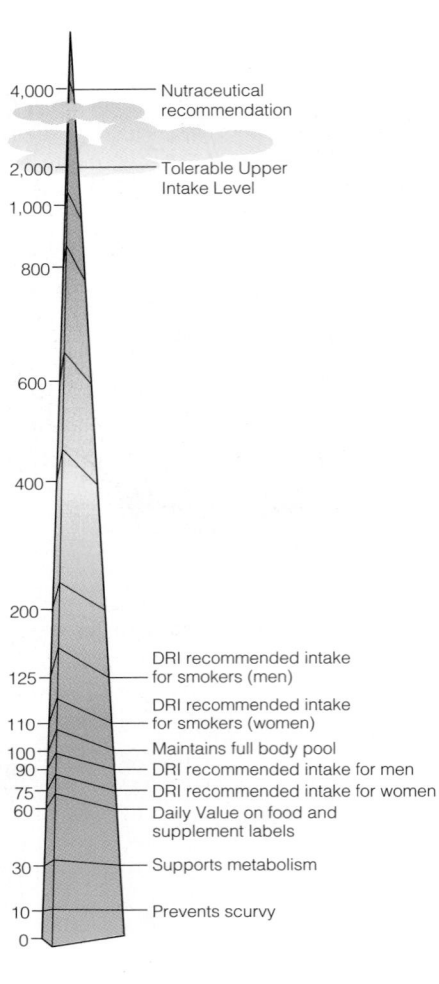

4,000 — Nutraceutical recommendation

2,000 — Tolerable Upper Intake Level
1,000 —

800 —

600 —

400 —

200 —

125 — DRI recommended intake for smokers (men)

110 — DRI recommended intake for smokers (women)

100 — Maintains full body pool
90 — DRI recommended intake for men
75 — DRI recommended intake for women
60 — Daily Value on food and supplement labels

30 — Supports metabolism

10 — Prevents scurvy

0 —

■ The term *nutraceutical* is used to market nutrients having pharmacological effects.

■ The DRI Tolerable Upper Intake Level for vitamin C is set at 2,000 mg (2 g) a day.

given 500 milligrams of vitamin C daily. The question of what, if anything, such findings may mean for human health remains unanswered.

Deficiency Symptoms Most of the symptoms of scurvy can be attributed to the breakdown of collagen in the absence of vitamin C: loss of appetite, growth cessation, tenderness to touch, weakness, bleeding gums (shown in Figure 7-8), loose teeth, swollen ankles and wrists, and tiny red spots in the skin where blood has leaked out of capillaries (also shown in the figure). One symptom, anemia, reflects an important role worth repeating—vitamin C helps the body to absorb and use iron. Table 7-4 at the end of the chapter summarizes deficiency symptoms and other information about vitamin C.

In the United States, scurvy is seldom seen today except in a few elderly people, people addicted to alcohol or other drugs, and a few infants who are fed only cow's milk.[39] Breast milk and infant formula supply enough vitamin C, but infants who are fed cow's milk and receive no vitamin C in formula, fruit juice, or other outside sources are at risk. Low intakes of fruits and vegetables and a poor appetite overall lead to low vitamin C intakes and are not uncommon among people aged 65 and older.

Vitamin C also supports immune system functions and so protects against infection. A long-claimed relationship between vitamin C and the common cold is the topic of this chapter's Consumer Corner.

Is Too Much Vitamin C Hazardous to Health? The easy availability of vitamin C in pill form and the publication of books recommending vitamin C as a "nutraceutical" treatment to prevent and cure colds and cancer have led thousands of people to take huge doses of vitamin C (see the margin drawing). These "volunteer" subjects enabled researchers to study potential adverse effects of large vitamin C doses. One effect observed with a 2-gram dose is alteration of the insulin response to carbohydrate in people with otherwise normal glucose tolerances. Other adverse effects include digestive upsets, such as nausea, abdominal cramps, excessive gas, and diarrhea. Massive doses of vitamin C may interfere with medications to prevent blood clotting. Vitamin C supplements in any dosage may be dangerous for people with an overload of iron in the body because vitamin C increases iron absorption from the intestine and releases iron from storage.

The published research on large doses of vitamin C reveals few instances in which consuming more than 100 to 300 milligrams a day is beneficial, although the range of safe vitamin C intakes seems to be broad, from the absolute minimum of 10 milligrams a day to the Tolerable Upper Intake Level of 2,000 milligrams (2 grams). People with kidney disorders and those with a condition of too much iron in their blood may be more

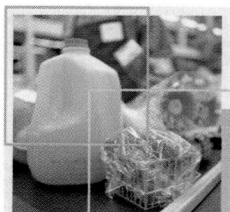

CONSUMER CORNER

Vitamin C and the Common Cold

WHY DO SO MANY PEOPLE take vitamin C supplements to relieve colds? Does any research support this practice? More than 30 years ago, Linus Pauling, a Nobel Prize winner, became a vocal supporter of vitamin C supplements.[1] The scientific community all but discounted his claims because research comparing daily intakes of vitamin C with numbers of colds fails to support Pauling's theories.[2] No study to date has shown conclusively that vitamin C can prevent colds or reduce their severity. One group of researchers, however, has reached two interesting conclusions.[3] First, they confirm that daily doses of vitamin C, taken regularly, do not prevent colds. Second, they find some small therapeutic benefit from vitamin C in high doses (1 gram) taken at the onset of a cold. In this case, cold sufferers may shorten the duration of colds by about half a day and reduce the severity of symptoms by about 40 percent. The effect may be greater in children than in adults; in adults, doses teetering on the edge of the Tolerable Upper Intake Level (2 grams a day) may be required to produce any effect at all.[4]

Vitamin C (2 grams taken daily for two weeks) seems to reduce blood histamine. Anyone who has ever had a cold knows the effects of histamine: sneezing, a runny or stuffy nose, and swollen sinuses. Antihistamine medications provide relief from just those symptoms. In druglike doses, vitamin C may work like a weak antihistamine by reducing histamine levels. Alternatively, vitamin C's antioxidant or other activities may boost the body's immunity or somehow improve its defenses.[5]

One other effect is hard at work with supplements of all kinds—the placebo effect.[6] Half of the experimental subjects in one study received a placebo but thought they were receiving vitamin C.

Can vitamin C ease the suffering of a person with a cold?

These subjects reported having fewer colds than the group who had received vitamin C but thought they were receiving the placebo. At work was the powerful healing effect of faith.

Much more research is required on vitamin C and the common cold before any recommendations are possible. One thing is certain, though—no drug is risk-free, and vitamin C in large doses qualifies as a drug that may have side effects.[7]

---

susceptible to adverse effects and should avoid vitamin C supplements altogether. Doses approaching 10 grams can be expected to be unsafe. Vitamin C from food is always safe.

The Need for Vitamin C The adult DRI intake recommendation for vitamin C is 90 milligrams for men and 75 milligrams for women. These amounts are far higher than the 10 or so milligrams per day needed to prevent the symptoms of scurvy. In fact, they are close to the amount at which the body's pool of vitamin C is full to overflowing: about 100 milligrams per day.

Tobacco use, among its many harmful effects, introduces oxidants that deplete the body's vitamin C.[40] Even "passive smokers," who live and work with smokers, and those who regularly chew tobacco need more vitamin C than others.[41] Intake recommendations for smokers are set high, at 125 milligrams for men and 110 milligrams for women, in order to maintain blood levels comparable to those of nonsmokers. Sufficient intake of vitamin C can normalize blood levels, but it cannot protect against the often serious damage caused by exposure to tobacco smoke.

VITAMIN C

DRI RECOMMENDED INTAKES:
Men: 90 mg/day
Women: 75 mg/day
Smokers: +35 mg/day

TOLERABLE UPPER INTAKE LEVEL:
Adults: 2,000 mg/day

CHIEF FUNCTIONS: Collagen synthesis (strengthens blood vessel walls, forms scar tissue, provides matrix for bone growth), antioxidant, restores vitamin E to active form, supports immune system, boosts iron absorption

DEFICIENCY: Scurvy, with pinpoint hemorrhages, fatigue, bleeding gums, bruises; bone fragility, joint pain; poor wound healing, frequent infections

TOXICITY: Nausea, abdominal cramps, diarrhea; rashes; interference with medical tests and drug therapies; in susceptible people, aggravation of gout or kidney stones.

*These foods provide 10 percent or more of the vitamin C Daily Value in a serving. For a 2,000-calorie diet, the DV is 60 mg/day.

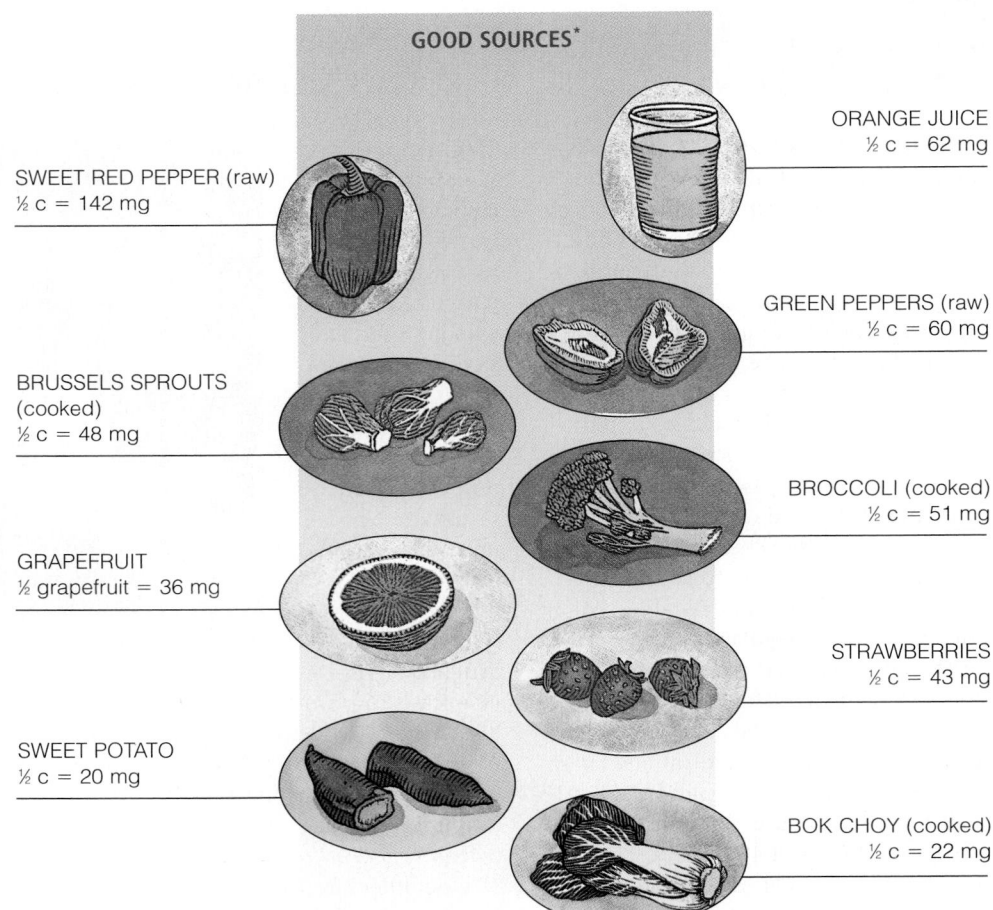

GOOD SOURCES*

ORANGE JUICE
½ c = 62 mg

SWEET RED PEPPER (raw)
½ c = 142 mg

GREEN PEPPERS (raw)
½ c = 60 mg

BRUSSELS SPROUTS (cooked)
½ c = 48 mg

BROCCOLI (cooked)
½ c = 51 mg

GRAPEFRUIT
½ grapefruit = 36 mg

STRAWBERRIES
½ c = 43 mg

SWEET POTATO
½ c = 20 mg

BOK CHOY (cooked)
½ c = 22 mg

Food Sources of Vitamin C Fruits and vegetables are the foods to remember for vitamin C, as Snapshot 7-5 shows. A cup of orange juice at breakfast, a salad for lunch, a stalk of broccoli and a potato for dinner easily provide 300 milligrams, making pills unnecessary. People commonly identify citrus fruits as sources of vitamin C, but they often overlook other rich sources (see the photos on the facing page). For maximum vitamin C, consumers should treat their fruits and vegetables gently, as vitamin C is vulnerable to heat and destroyed by oxygen. Fresh, raw, and quickly cooked fruits, vegetables, and juices retain the most vitamin C, and they should be stored properly and consumed promptly after purchase.[42] Chapter 12 gives tips for maximizing the vitamins in foods.

Because of their enormous popularity, white potatoes contribute significantly to vitamin C intakes, despite providing less than 10 milligrams per half-cup serving. The sweet potato, often ignored in favor of its paler cousin, is a gold mine of nutrients: a single half-cup serving provides about a third of many people's recommended intake for vitamin C, in addition to its lavish contribution of vitamin A.

▶ **KEY POINT** Vitamin C, an antioxidant, helps to maintain collagen, the protein of connective tissue; protects against infection; and helps in iron absorption. The theory that vitamin C prevents or cures colds or cancer is not well supported by research. Taking high vitamin C doses may be unwise. Ample vitamin C can be obtained from foods.

When dietitians say "vitamin C," people think "oranges . . ."

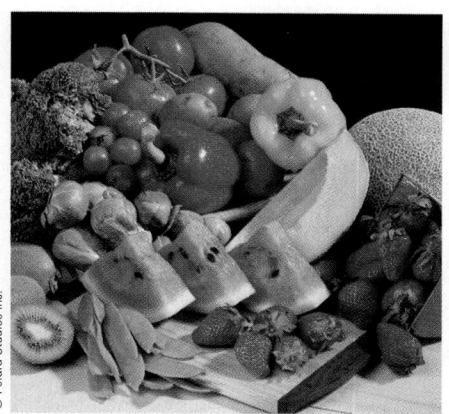

but these foods also supply vitamin C.

FIGURE 7·9 COENZYME ACTION

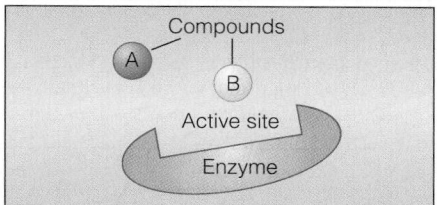

Without the coenzyme, compounds A and B don't respond to the enzyme.

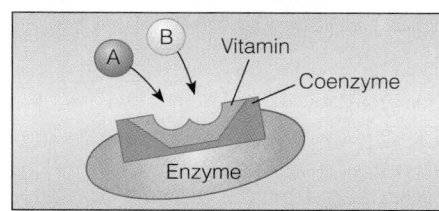

With the coenzyme in place, compounds A and B are attracted to the active site on the enzyme, and they react.

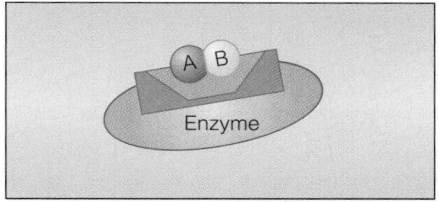

The reaction is completed with the formation of a new product. In this case the product is AB.

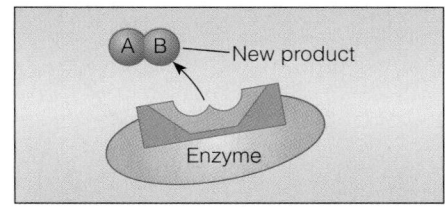

The product AB is released.

The B Vitamins in Unison

The B vitamins act as part of coenzymes. A **coenzyme** is a small molecule that combines with an enzyme and activates it. (Recall from Chapter 6 that enzymes are large proteins that do the body's building, dismantling, and other work.) Figure 7-9 shows how a coenzyme enables an enzyme to do its job. Sometimes the vitamin part of the enzyme is the active site, where the chemical reaction takes place. The substance to be worked on is attracted to the active site and snaps into place; the reaction proceeds instantaneously. The shape of each enzyme predestines it to accomplish just one kind of job. Without its coenzyme, however, the enzyme is as useless as a car without wheels.

Each of the B vitamins has its own special nature, and the amount of detail known about each one is overwhelming. To simplify things, this introduction describes the teamwork of the B vitamins and emphasizes the consequences of deficiencies. The sections that follow present more details about the vitamins as individuals.

B Vitamin Roles in Metabolism   Figure 7-10 (p. 234) shows some body organs and tissues in which the B vitamins help the body metabolize carbohydrates, lipids, and amino acids. The purpose of the figure is not to present a detailed account of metabolism but to give you an impression of where the B vitamins work together with enzymes in the metabolism of energy nutrients and in the making of new cells.

Many people mistakenly believe that B vitamins supply the body with energy. They do not, at least not directly. The B vitamins are "helpers." The energy-yielding nutrients—carbohydrate, fat, and protein—give the body fuel for energy; the B vitamins *help* the body use that fuel. More specifically, active forms of five of the B vitamins—thiamin, riboflavin, niacin, pantothenic acid, and biotin—participate in the release of energy from carbohydrate, fat, and protein. Vitamin B_6 helps the body use amino acids to make protein; the body then puts the protein to work in many ways—to build new tissues, to make hormones, to fight infections, or to serve as fuel for energy, to name only a few.

Folate and vitamin B_{12} help cells to multiply, which is especially important to cells with short life spans that must replace themselves rapidly. Such cells include both the red blood cells (which live for about 120 days) and the cells that line the digestive tract (which replace themselves every three days). These cells deliver energy to all the others. In short, each and every B vitamin is involved, directly or indirectly, in energy metabolism.

B Vitamin Deficiencies   As long as B vitamins are present, their presence is not felt. Only when they are missing does their absence manifest itself in a lack of energy

■ To memorize the names of the eight B vitamins, try remembering this sentence or make up one of your own:

Tender	(thiamin)
Romance	(riboflavin)
Never	(niacin)
Fails,	(folate)
with 6 or 12	(B_6 and B_{12})
Beautiful	(biotin)
Pearls.	(pantothenic acid)

coenzyme (co-EN-zime) a small molecule that works with an enzyme to promote the enzyme's activity. Many coenzymes have B vitamins as part of their structure (*co* means "with").

FIGURE 7·10 SOME ROLES OF THE B VITAMINS IN METABOLISM: EXAMPLES

This figure does not attempt to teach intricate biochemical pathways or names of B vitamin–containing enzymes. Its sole purpose is to show a few of the many tissue functions that depend on B vitamin–containing enzymes. The B vitamins work in every cell, and this figure displays less than a thousandth of what they actually do.

Every B vitamin is part of one or more coenzymes that make possible the body's chemical work. For example, the niacin, thiamin, and riboflavin coenzymes are important in the energy pathways. The folate and vitamin B_{12} coenzymes are necessary for making RNA and DNA and thus new cells. The vitamin B_6 coenzyme is necessary for processing amino acids and, therefore, protein. Many other relationships are also critical to metabolism.

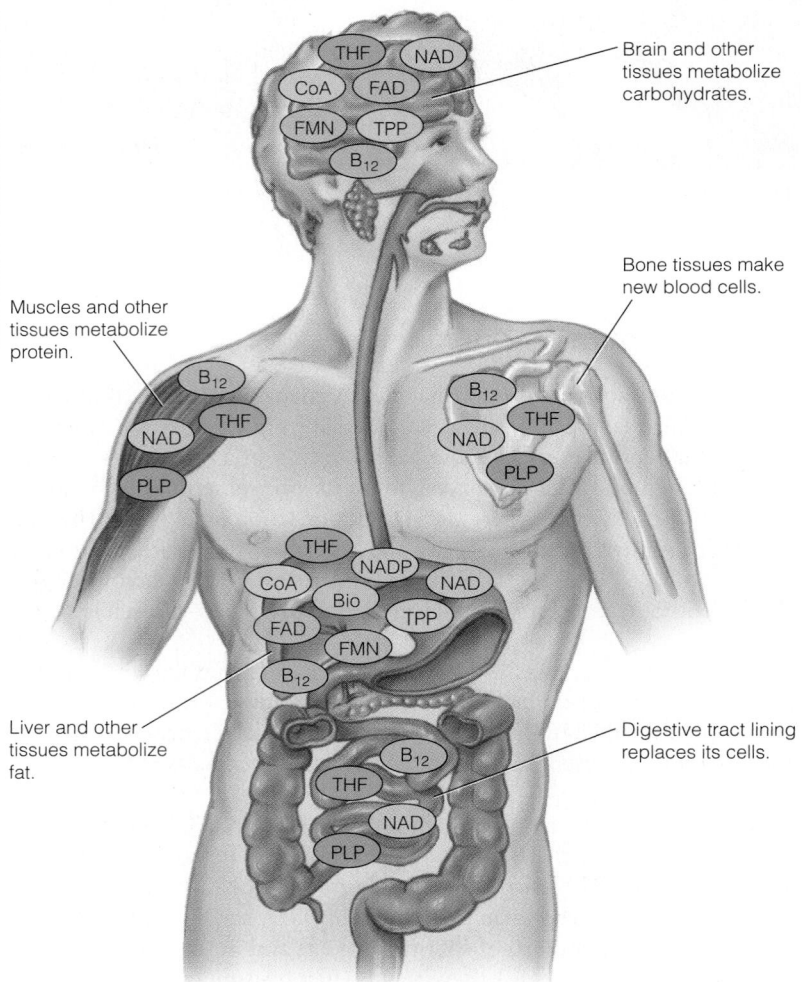

Key:

Coenzyme name		Vitamin name
TPP	=	thiamin
FAD FMN	=	riboflavin
NAD NADP	=	niacin
PLP	=	vitamin B_6
THF	=	folate
CoA	=	pantothenic acid
Bio	=	biotin
B_{12}	=	vitamin B_{12}

Brain and other tissues metabolize carbohydrates.

Bone tissues make new blood cells.

Muscles and other tissues metabolize protein.

Liver and other tissues metabolize fat.

Digestive tract lining replaces its cells.

and a multitude of other symptoms, as you can imagine after looking at Figure 7-10. The reactions by which B vitamins facilitate energy release take place in every cell, and no cell can do its work without energy. Thus, in a B vitamin deficiency, every cell is affected. Among the symptoms of B vitamin deficiencies are nausea, severe exhaustion, irritability, depression, forgetfulness, loss of appetite and weight, pain in muscles, impairment of the immune response, loss of control of the limbs, abnormal heart action, severe skin problems, swollen red tongue, and teary or bloodshot eyes. Because cell renewal depends on energy and protein, which in turn depend on the B vitamins, the digestive tract and the blood are invariably damaged. In children, full recovery may be impossible. In the case of a thiamin deficiency during growth, permanent brain damage can result.

In academic discussions of the vitamins, different sets of deficiency symptoms are given for each one. Such clear-cut sets of symptoms are found only in laboratory animals that have been fed fabricated diets that lack just one nutrient.[43] In real life, a deficiency of any one B vitamin seldom shows up by itself because people don't eat nutrients singly; they eat foods that contain mixtures of nutrients. A deficiency of one B vitamin may appear to be responsible for a cluster of symptoms, but subtler, undetected deficiencies may accompany it. If treatment involves giving wholesome food rather than a single supplement, the subtler deficiencies will be corrected along with the major one. The symptoms of B vitamin deficiencies and toxicities are listed in Table 7-4 at the end of the chapter.

▶ **KEY POINT**   As part of coenzymes, the B vitamins help enzymes do their jobs. The B vitamins facilitate the work of every cell. Some help generate energy; others help make protein and new cells. B vitamins work everywhere in the body tissue to metabolize carbohydrate, fat, and protein.

The B Vitamins as Individuals

Although the B vitamins all work as part of coenzymes and share other characteristics, each B vitamin has special qualities. The next sections provide details about each of the B vitamins.

Thiamin and Riboflavin   **Thiamin** plays a critical role in the energy metabolism of all cells. Thiamin also occupies a special site on nerve cell membranes. Consequently, nerve processes and their responding tissues, the muscles, depend heavily on thiamin.

The classic thiamin-deficiency disease, **beriberi,** was first observed in East Asia, where rice provided 80 to 90 percent of the total calories most people consumed and was therefore their principal source of thiamin. When the custom of polishing rice (removing its brown coat, which contained the thiamin) became widespread, beriberi swept through the population like an epidemic. Scientists wasted years of effort hunting for a microbial cause of beriberi before they realized that the cause was not something present in the environment but something absent from it. Figure 7-11 depicts beriberi and describes its two forms.

Just before 1900, an observant physician working in a prison in East Asia discovered that beriberi could be cured with proper diet. The physician noticed that the chickens at the prison had developed a stiffness and weakness similar to that of the prisoners who had beriberi. The chickens were being fed the rice left on prisoners' plates. When the rice bran, which had been discarded in the kitchen, was given to the chickens, their paralysis was cured. The physician met resistance when he tried to feed the rice bran, the "garbage," to the prisoners, but it worked—it produced a miracle cure like those described at the beginning of the chapter. Later, extracts of rice bran were used to prevent infantile beriberi; still later, thiamin was synthesized.

In developed countries today, alcohol abuse often leads to a severe form of thiamin deficiency, Wernicke-Korsakoff syndrome, defined in Controversy 3. Alcohol contributes energy but carries almost no nutrients with it and often displaces food. In addition, alcohol impairs absorption of thiamin from the digestive tract and hastens its excretion in the urine, tripling the risk of deficiency. The syndrome is characterized by symptoms almost indistinguishable from alcohol abuse itself: apathy, irritability, mental confusion, disorientation, loss of memory, jerky eye movements, and a staggering gait. Unlike alcohol toxicity, the syndrome responds quickly to an injection of thiamin, and some experts recommend a precautionary dose for any patients suspected of having the syndrome.

Thiamin occurs in small amounts in many nutritious foods. Ham and other pork products, sunflower seeds, enriched or whole-grain cereals, and legumes are especially rich in thiamin (see Snapshot 7-6, p. 236). If you keep empty-calorie foods to a minimum and focus your meals on nutritious foods each day, you will easily meet your thiamin needs. The DRI committee set the thiamin intake recommendation at 1.2 milligrams per day for men and at 1.1 milligrams per day for women. Pregnancy and lactation require somewhat more thiamin (see the DRI, inside front cover, page B).

FIGURE 7·11   BERIBERI

Beriberi takes two forms: wet beriberi, characterized by edema (fluid accumulation), and dry beriberi, without edema. This person's ankle retains the imprint of the physican's thumb, showing the edema of wet beriberi.

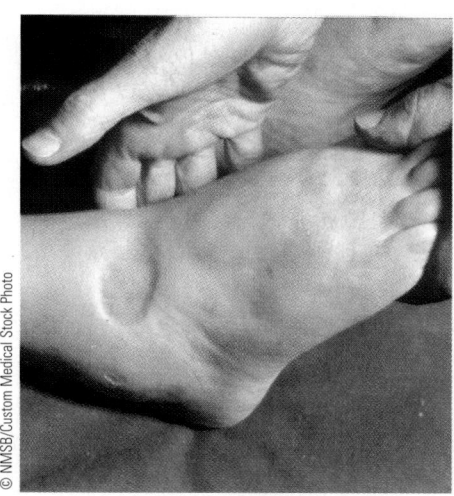

© NMSB/Custom Medical Stock Photo

thiamin (THIGH-uh-min) a B vitamin involved in the body's use of fuels.

beriberi (berry-berry) the thiamin-deficiency disease; characterized by loss of sensation in the hands and feet, muscular weakness, advancing paralysis, and abnormal heart action.

THIAMIN

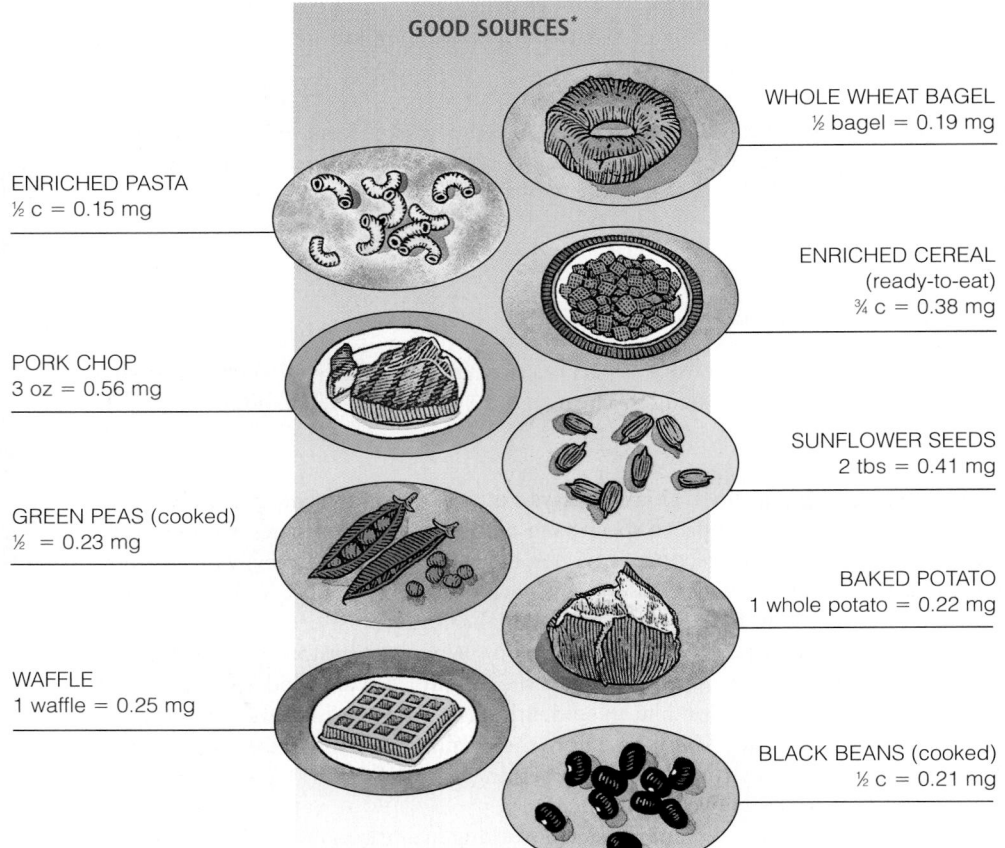

DRI RECOMMENDED INTAKES:
Men: 1.2 mg/day
Women: 1.1 mg/day

CHIEF FUNCTIONS: Part of coenzyme active in energy metabolism

DEFICIENCY: [a]Beriberi with possible edema or muscle wasting; enlarged heart, heart failure, muscular weakness, pain, apathy, poor short-term memory, confusion, irritability, difficulty walking, paralysis, anorexia, weight loss

TOXICITY: None reported

*These foods provide 10 percent or more of the thiamin Daily Value in a serving. For a 2,000-calorie diet, the DV is 1.5 mg/day.

[a]Severe thiamin deficiency is often related to heavy alcohol consumption.

GOOD SOURCES*

WHOLE WHEAT BAGEL
½ bagel = 0.19 mg

ENRICHED PASTA
½ c = 0.15 mg

ENRICHED CEREAL
(ready-to-eat)
¾ c = 0.38 mg

PORK CHOP
3 oz = 0.56 mg

SUNFLOWER SEEDS
2 tbs = 0.41 mg

GREEN PEAS (cooked)
½ = 0.23 mg

BAKED POTATO
1 whole potato = 0.22 mg

WAFFLE
1 waffle = 0.25 mg

BLACK BEANS (cooked)
½ c = 0.21 mg

 Table 7-4 on p. 247 lists the symptoms of riboflavin deficiency.

riboflavin (RIBE-o-flay-vin) a B vitamin active in the body's energy-releasing mechanisms.

niacin a B vitamin needed in energy metabolism. Niacin can be eaten preformed or can be made in the body from tryptophan, one of the amino acids. Other forms of niacin are *nicotinic acid, niacinamide,* and *nicotinamide.*

pellagra (pell-AY-gra) the niacin-deficiency disease (*pellis* means "skin"; *agra* means "rough"). Symptoms include the "4 Ds": diarrhea, dermatitis, dementia, and death.

Like thiamin, **riboflavin** plays a role in the energy metabolism of all cells. When thiamin is deficient, riboflavin may be lacking, too, but its deficiency symptoms may go undetected because those of thiamin deficiency are more severe. Worldwide, riboflavin deficiency has been documented among children whose diets lack milk products and meats, and researchers suspect that it occurs among some U.S. elderly as well.[44] A diet that remedies riboflavin deficiency invariably contains some thiamin, and so clears up both deficiencies. People obtain over a quarter of their riboflavin from enriched breads, cereals, pasta, and other grain products, while milk and milk products supply another 20 percent. Some vegetables, eggs, and meats contribute most of the rest (see Snapshot 7-7).[45]

Niacin The vitamin **niacin,** like thiamin and riboflavin, participates in the energy metabolism of every cell of the body. The niacin-deficiency disease, **pellagra,** appeared in Europe in the 1700s when corn from the New World became a staple food. In the early 1900s in the United States, pellagra was devastating lives throughout the South and Midwest. Hundreds of thousands of pellagra victims were thought to be suffering from a contagious disease until this dietary deficiency was identified. The disease still occurs among poorly nourished people living in urban slums and particularly in those with alcohol addiction. Pellagra is also still common in parts of Africa and Asia. Its symptoms are known as the four "Ds": diarrhea, dermatitis, dementia, and, ultimately, death.

Figure 7-12 shows the skin disorder associated with pellagra. For comparison, Figure 7-15 (on page 242) and Figure 7-2 (p. 217) show skin disorders associated with vitamin

RIBOFLAVIN

DRI RECOMMENDED INTAKES:
Men: 1.3 mg/day
Women: 1.1 mg/day

CHIEF FUNCTIONS: Part of coenzymes active in energy metabolism; supports vision and skin health

DEFICIENCY: Cracks and redness at corners of mouth; painful, smooth, purplish red tongue; sore throat; inflamed eyes and eyelids, sensitivity to light; skin rashes

TOXICITY: None reported

*These foods provide 10 percent or more of the riboflavin Daily Value in a serving. For a 2,000-calorie diet, the DV is 1.7 mg/day.

GOOD SOURCES*

BEEF LIVER (cooked)
3 oz = 2.9 mg

COTTAGE CHEESE
1 c = 0.38 mg

ENRICHED CEREAL
(ready-to-eat)
¾ = 0.43 mg

SPINACH (cooked)
½ c = 0.21 mg

MILK
1 c = 0.45 mg

YOGURT (plain)
1 c = 0.60 mg

PORK CHOP (lean only)
3 oz = 0.23 mg

MUSHROOMS (cooked)
½ c = 0.23 mg

B$_6$ and vitamin A deficiency, respectively, a reminder that any nutrient deficiency affects the skin and all other cells. The skin just happens to be the organ you can see. Table 7-4 at the end of the chapter lists many more symptoms of niacin deficiency.

The key nutrient that prevents pellagra is niacin, but any protein containing sufficient amounts of the amino acid tryptophan will serve in its place. Tryptophan, which is abundant in almost all proteins (but is limited in the protein of corn), is converted to niacin in the body, and it is possible to cure pellagra by administering tryptophan alone. Thus, a person eating adequate protein (as most people in developed nations do) will not be deficient in niacin. The amount of niacin in a diet is stated in terms of **niacin equivalents (NE),** a measure that takes available tryptophan into account.

Early workers seeking the cause of pellagra observed that well-fed people never got it. From there the researchers defined a diet that reliably produced the disease—one of cornmeal, salted pork fat, and molasses. Corn is not only low in protein but also lacks tryptophan. Salt pork is almost pure fat and contains too little protein to compensate; and molasses is virtually protein-free. Snapshot 7-8 (p. 238) shows some good food sources of niacin.

Physicians may administer large doses of a form of niacin as part of a treatment regimen to lower blood lipids associated with cardiovascular disease.* When used this way, niacin leaves the realm of nutrition to become a pharmacological agent—a drug. As with any drug, self-dosing with niacin is ill-advised; large doses can injure the liver,

*The form of niacin is nicotinic acid.

FIGURE 7·12 PELLAGRA

The typical dermatitis of pellagra develops on skin that is exposed to light.

© George L. Blackburn M.D., Ph.D., Harvard Medical School

niacin equivalents (NE) the amount of niacin present in food, including the niacin that can theoretically be made from its precursor tryptophan that is present in the food.

NIACIN

GOOD SOURCES*

DRI RECOMMENDED INTAKES:
Men: 16 mg/day[a]
Women: 14 mg/day

TOLERABLE UPPER INTAKE LEVEL:
Adults: 35 mg/day

CHIEF FUNCTIONS: Part of coenzymes needed in energy metabolism

DEFICIENCY: Pellagra, characterized by flaky skin rash (dermatitis) where exposed to sunlight; mental depression, apathy, fatigue, loss of memory, headache; diarrhea, abdominal pain, vomiting; swollen, smooth, bright red or black tongue

TOXICITY: Painful flush, hives, and rash ("niacin flush"); excessive sweating; blurred vision; liver damage, impaired glucose tolerance

[a]Niacin DRI Recommended Intakes are expressed in niacin equivalents (NE); the Tolerable Upper Intake Level refers to preformed niacin.

*These foods provide 10 percent or more of the niacin Daily Value in a serving. For a 2,000-calorie diet, the DV is 20 mg/day. The DV values are for preformed niacin, not niacin equivalents.

CHICKEN BREAST
3 oz = 8.9 mg

PORK CHOP
3 oz = 3.9 mg

BAKED POTATO
1 whole potato = 3.3 mg

TUNA (in water)
3 oz = 11.3 mg

ENRICHED CEREAL
(ready-to-eat)
¾ c = 5.0 mg

MUSHROOMS (cooked)
½ c = 3.5 mg

cause peptic ulcers, or cause vision loss.[46] Certain forms of niacin supplements in amounts two to three times the DRI intake recommendation cause "niacin flush," a dilation of the capillaries of the skin with perceptible tingling that, if intense, can be painful. For safety's sake, anyone taking large doses of niacin should do so only under the care of a physician.

Folate To make new cells, tissues must have the vitamin **folate.**[47] Each new cell must be equipped with new genetic material—copies of the parent cell's DNA—and folate helps to synthesize DNA. Folate is also critical to the normal metabolism of several amino acids. Because immature red and white blood cells and the cells of the digestive tract divide most rapidly, they are most vulnerable to deficiency.

Deficiencies of folate cause anemia, diminished immunity, and abnormal digestive function. In the United States, a significant number of cases of folate-deficiency anemia occur yearly. This anemia is related to the anemia of vitamin B$_{12}$ malabsorption because the two vitamins work as teammates in producing red blood cells—see Figure 7-14 later in the chapter. Research suggests that a diet deficient in folate may also elevate the risk of cardiovascular disease and cancer of the colon and increase a woman's risk for cervical cancer.[48] Folate deficiencies may result from an inadequate intake or from illnesses that impair folate's absorption, increase its excretion, require medication that interacts with folate, or otherwise increase the body's need.

Folate deficiency is also associated with a group of devastating birth defects known as **neural tube defects.**[49] Neural tube defects range from slight problems in the spine to mental retardation, severely diminished brain size, and death shortly after birth. These neural tube defects arise in the first days or weeks of pregnancy, long before most women suspect that they are pregnant, and most women eat too few fruits and vegetables from day to day to supply even half the folate needed to prevent them. In the late 1990s, the FDA ordered all enriched grain products such as breads, cereals, and

folate (FOH-late) a B vitamin that acts as part of a coenzyme important in the manufacture of new cells. The form added to foods and supplements is *folic acid*.

neural tube defects abnormalities of the brain and spinal cord apparent at birth and believed to be related to a woman's folate intake before and during pregnancy. Also defined in Chapter 13.

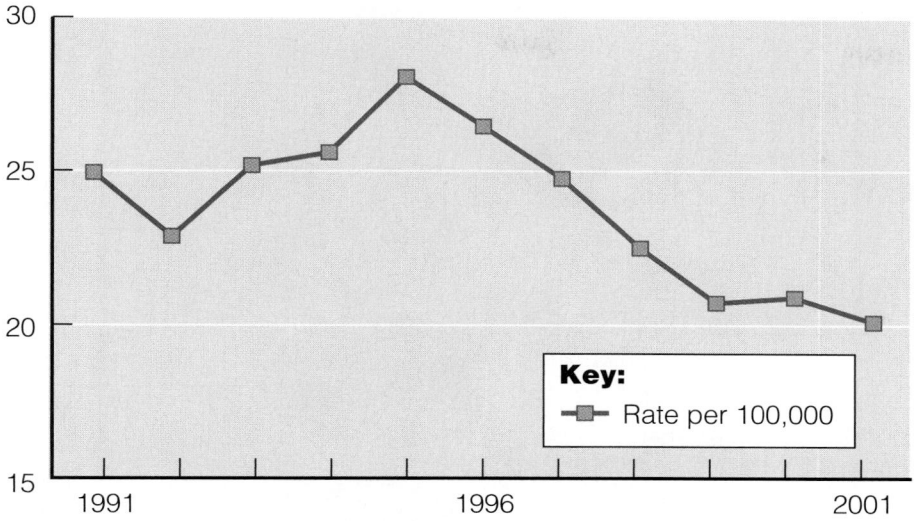

FIGURE 7·13 INCIDENCE OF A COMMON NEURAL TUBE DEFECT, SPINA BIFIDA, 1991–2001

Neural tube defects have declined since folate fortification began in 1996.

Key:
■— Rate per 100,000

SOURCE: National Vital Statistics System, Centers for Disease Control and Prevention.

pastas to be fortified with an especially absorbable synthetic form of folate, folic acid. Since this fortification began, typical folate intakes from fortified foods have increased dramatically—by more than double the predicted levels—and observers report an almost 25 percent drop in the national incidence of neural tube defects, even among women receiving late or no prenatal care (see Figure 7-13).[50] Researchers expect to see declines in some other birth defects and miscarriages as well.[51]

As for folate toxicity, a Tolerable Upper Intake Level for synthetic folate from supplements and enriched foods is set at 1,000 micrograms a day for adults. Of major importance are concerns about folate's ability to mask deficiencies of vitamin B_{12} (more about this effect later). The possibility also exists that, once in the blood, excess folate may negate actions of some anticancer drugs that work by blocking the activities of folate in rapidly dividing cancer cells. Time will tell whether the apparent benefits of folate enrichment outweigh the risks.

The difference in absorption between naturally occurring food folate and the synthetic folate that enriches foods and is added to supplements necessitates compensation when measuring folate. The unit of measure, **dietary folate equivalent,** or **DFE,** converts all forms of folate into micrograms that are equivalent to the folate in foods. Watch out for food and supplement labels that still express folate values in micrograms not converted to DFE; Appendix C offers a conversion factor.

Folate's name is derived from the word *foliage,* and sure enough, folate is naturally abundant in leafy green vegetables such as spinach and turnip greens (see Snapshot 7-9, p. 240). Fresh, uncooked vegetables and fruits are good natural sources because the heat of cooking and the oxidation that occurs during storage destroy much of the folate in foods. Eggs also contain some folate. Milk included in a meal may enhance the absorption of folate.

The DRI recommendation for folate intake for healthy adults is set at 400 micrograms per day. The DRI committee also advises all women of childbearing age to consume 400 micrograms of synthetic folate, or *folic acid,* from supplements or enriched foods each day in addition to the folate that occurs naturally in their foods. Of all the vitamins, folate is most likely to interact with medications. Ten major groups of drugs,

■ The B vitamins thiamin, riboflavin, niacin, and folate (as folic acid) are among the enrichment nutrients added to grain foods such as breads and cereals sold in the United States. Chapter 4 presented more details on enrichment of grain foods.

dietary folate equivalent (DFE) a unit of measure expressing the amount of folate available to the body from naturally occurring sources. The measure mathematically equalizes the difference in absorption between less absorbable food folate and highly absorbable synthetic folate added to enriched foods and found in supplements.

FOLATE

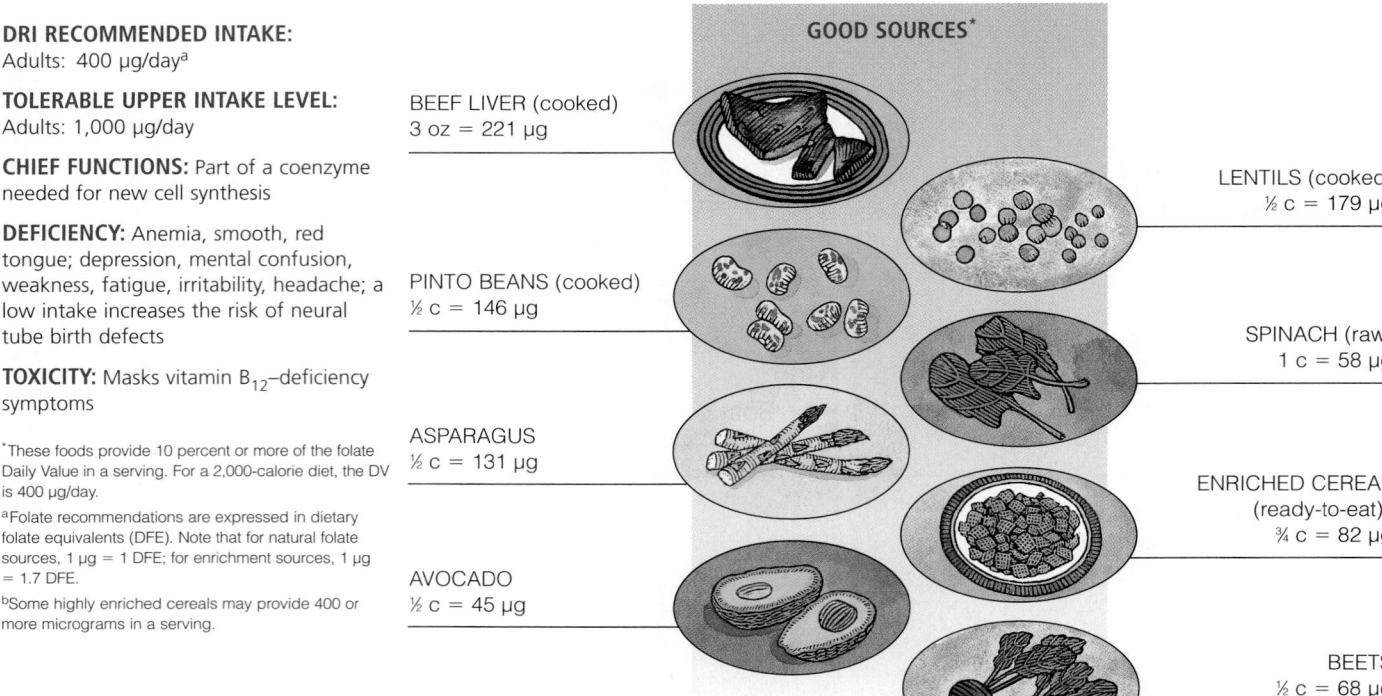

DRI RECOMMENDED INTAKE:
Adults: 400 µg/day[a]

TOLERABLE UPPER INTAKE LEVEL:
Adults: 1,000 µg/day

CHIEF FUNCTIONS: Part of a coenzyme needed for new cell synthesis

DEFICIENCY: Anemia, smooth, red tongue; depression, mental confusion, weakness, fatigue, irritability, headache; a low intake increases the risk of neural tube birth defects

TOXICITY: Masks vitamin B_{12}–deficiency symptoms

*These foods provide 10 percent or more of the folate Daily Value in a serving. For a 2,000-calorie diet, the DV is 400 µg/day.

[a]Folate recommendations are expressed in dietary folate equivalents (DFE). Note that for natural folate sources, 1 µg = 1 DFE; for enrichment sources, 1 µg = 1.7 DFE.

[b]Some highly enriched cereals may provide 400 or more micrograms in a serving.

GOOD SOURCES*

BEEF LIVER (cooked)
3 oz = 221 µg

PINTO BEANS (cooked)
½ c = 146 µg

ASPARAGUS
½ c = 131 µg

AVOCADO
½ c = 45 µg

LENTILS (cooked)
½ c = 179 µg

SPINACH (raw)
1 c = 58 µg

ENRICHED CEREAL
(ready-to-eat)[b]
¾ c = 82 µg

BEETS
½ c = 68 µg

---

including antacids and aspirin and its relatives, have been shown to interfere with the body's use of folate. Occasional use of these drugs to relieve headache or upset stomach presents no concern, but frequent users may need to pay attention to their folate intakes. These include people with chronic pain or ulcers who rely heavily on aspirin or antacids as well as those who smoke or take oral contraceptives or anticonvulsant medications.

■ Controversy 6 describes the devastating effects on infants born to women who were deficient in vitamin B_{12} during pregnancy.

Vitamin B_{12} **Vitamin B_{12}** and folate are closely related: each depends on the other for activation. By itself vitamin B_{12} also helps to maintain the sheaths that surround and protect nerve fibers. Without sufficient vitamin B_{12}, nerves become damaged and folate fails to do its blood-building work, so vitaminB_{12} deficiency causes an anemia identical to that caused by folate deficiency. The blood symptoms of a deficiency of either folate or vitamin B_{12} include the presence of large, immature red blood cells. Administering extra folate often clears up this blood condition but allows the deficiency of vitamin B_{12} to continue undetected.[52] Vitamin B_{12}'s other functions then become compromised, and the results can be devastating: damaged nerve sheaths, creeping paralysis, and general malfunctioning of nerves and muscles. Scientists are currently investigating whether the addition of folic acid to U.S. grain foods is masking vitamin B_{12} deficiencies, especially among vegetarians and the elderly, who generally depend heavily on grain foods and consume little or no meat or other animal products.[53]

vitamin B_{12} a B vitamin that helps to convert folate to its active form and also helps maintain the sheath around nerve cells. Vitamin B_{12}'s scientific name, not often used, is *cyanocobalamin*.

intrinsic factor a factor found inside a system. The intrinsic factor necessary to prevent pernicious anemia is now known to be a compound that helps in the absorption of vitamin B_{12}.

Absorption of vitamin B_{12} requires an **intrinsic factor,** a compound made by the stomach with instructions from the genes. With the help of the stomach's acid to liberate vitamin B_{12} from the food proteins that bind it, intrinsic factor attaches to the vitamin; the complex is then absorbed from the small intestine into the bloodstream. A few people have an inherited defect in the gene for intrinsic factor, which makes vitamin B_{12}

absorption abnormal, beginning in mid-adulthood. In later years, many others lose the ability to produce enough stomach acidity and intrinsic factor, leaving them open to both chronic stomach infections and reduced ability to absorb vitamin B_{12}.[*]

Without normal absorption of vitamin B_{12} from food, people develop deficiency symptoms. In these cases, vitamin B_{12} must be supplied by injection to bypass the defective absorptive system. The anemia of the vitamin B_{12} deficiency caused by lack of intrinsic factor is known as **pernicious anemia** (see Figure 7-14).

Diagnosing a vitamin B_{12} problem is difficult, and often the damage will proceed unchecked. In an effort to prevent excessive folate intakes that could mask symptoms of a vitamin B_{12} deficiency, the FDA specifies the exact amounts of folate that can be added to enriched foods.

As Snapshot 7-10 (p. 242) shows, vitamin B_{12} is present only in foods of animal origin, not in foods from plants. Worldwide, vitamin B_{12} deficiency among vegetarians is a growing problem.[54] The uninformed, strict vegetarian is at special risk and may not show signs of deficiency right away because the body stores up to six years' worth of vitamin B_{12}.[55] A pregnant or lactating vegetarian woman who eats no foods of animal origin should be aware that her infant can develop a vitamin B_{12} deficiency, even if the mother appears healthy. A deficiency of this vitamin can cause irreversible nervous system damage in the developing fetus, which can only be diagnosed after birth when the infant displays nerve and cognitive problems that can last into childhood.[56] All strict vegetarians, and especially pregnant women, must be sure to use vitamin B_{12}–fortified products, such as vitamin B_{12}–fortified soy drink, or to take the appropriate supplements.

The way folate masks the anemia of vitamin B_{12} deficiency underscores a point worth repeating. It takes a skilled professional to correctly diagnose a nutrient deficiency or imbalance, and you take a serious risk when you diagnose yourself or listen to self-proclaimed experts. A second point: Since vitamin B_{12} deficiency in the body may be caused by either a lack of the vitamin in the diet or a lack of the intrinsic factor necessary to absorb the vitamin, a change in diet alone may not correct the deficiency, another reason for seeking professional diagnosis when you have physical symptoms.

Vitamin B_6   **Vitamin B_6** participates in over 100 reactions in body tissues and is needed to help convert one kind of amino acid, which cells have in abundance, to other nonessential amino acids that the cells lack.[57] In addition, vitamin B_6 functions in these ways:

- Aids in the conversion of tryptophan to niacin.

- Plays important roles in the synthesis of hemoglobin and neurotransmitters, the communication molecules of the brain. For example, vitamin B_6 assists in the conversion of the amino acid tryptophan to the neurotransmitter **serotonin.**

- Assists in releasing stored glucose from glycogen and thus contributes to the regulation of blood glucose.

- Has roles in immune function and steroid hormone activity.

- Is critical to the developing brain and nervous system of a fetus; deficiency during this stage causes behavioral problems later.

Because of these diverse functions, vitamin B_6 deficiency is expressed in general symptoms, such as weakness, psychological depression, confusion, irritability, and insomnia. Other symptoms include anemia, the greasy dermatitis depicted in Figure 7-15 (p. 242), and, in advanced cases of deficiency, convulsions. A shortage of vitamin B_6 may also weaken the immune response. Some evidence suggests that low vitamin B_6 intakes may also be related to increased incidence of heart disease. Some people have taken

[*]This condition is atrophic gastritis (a-TRO-fik gas-TRY-tis), a chronic inflammation of the stomach accompanied by a diminished size and functioning of the stomach's mucous membrane and glands.

FIGURE 7·14   ANEMIC AND NORMAL BLOOD CELLS

The anemia of folate deficiency is indistinguishable from that of vitamin B_{12} deficiency.

Carolina Biological/Visuals Unlimited

Blood cells of pernicious anemia. The cells are larger than normal and irregular in shape.

© Martin M. Rotker

Normal blood cells. The size, shape, and color of these red blood cells show that they are normal.

pernicious (per-NISH-us) **anemia** a vitamin B_{12}–deficiency disease, caused by lack of intrinsic factor and characterized by large, immature red blood cells and damage to the nervous system (*pernicious* means "highly injurious or destructive").

vitamin B_6 a B vitamin needed in protein metabolism. Its three active forms are *pyridoxine, pyridoxal,* and *pyridoxamine.*

serotonin (SER-oh-tone-in) a neurotransmitter important in sleep regulation, appetite control, and mood regulation, among other roles. Serotonin is synthesized in the body from the amino acid tryptophan with the help of vitamin B_6.

VITAMIN B₁₂

DRI RECOMMENDED INTAKE:
Adults: 2.4 µg/day

CHIEF FUNCTIONS: Part of coenzymes needed in new cell synthesis; helps to maintain nerve cells

DEFICIENCY: Pernicious anemia;[a] anemia (large-cell type);[b] smooth tongue; tingling or numbness; fatigue, memory loss, disorientation, degeneration of nerves progressing to paralysis

TOXICITY: None reported

*These foods provide 10 percent or more of the vitamin B₁₂ Daily Value in a serving. For a 2,000-calorie diet, the DV is 6 µg/day.

[a]The name *pernicious anemia* refers to the vitamin B₁₂ deficiency caused by a lack of stomach intrinsic factor, but not to anemia from inadequate dietary intake.

[b]Large cell–type anemia is known as either *macrocytic* or *megaloblastic anemia.*

GOOD SOURCES*

CHICKEN LIVER
3 oz = 14.0 µg

SARDINES
3 oz = 7.5 µg

SIRLOIN STEAK
3 oz = 2.0 µg

TUNA (in water)
3 oz = 3.0 µg

COTTAGE CHEESE
1 c = 2.0 µg

SWISS CHEESE
1½ oz = 1.5 µg

PORK ROAST (lean)
3 oz = 1.0 µg

FIGURE 7·15 VITAMIN B₆ DEFICIENCY

In this dermatitis, the skin is greasy and flaky, unlike the skin affected by the dermatitis of pellagra.

© George L. Blackburn, M.D., Harvard Medical School

carpal tunnel syndrome a pinched nerve at the wrist, causing pain or numbness in the hand. It is often caused by repetitive motion of the wrist.

vitamin B₆ supplements to relieve **carpal tunnel syndrome** and sleep disorders even though such treatment seems to be ineffective.[58]

Moreover, large doses of vitamin B₆ from supplements can be dangerous. Years ago it was generally believed that, like most of the other water-soluble vitamins, vitamin B₆ could not reach toxic concentrations in the body. Then a report told of women who took more than 2 grams of vitamin B₆ daily (the Tolerable Upper Intake Level is set at 100 *milligrams*, or 0.1 gram) for two months or more in an attempt to cure the symptoms of premenstrual syndrome (PMS). The women developed numb feet, then lost sensation in their hands, and eventually became unable to walk or work. Since the first report of vitamin B₆ toxicity, researchers have seen toxicity symptoms in more than 100 women who took vitamin B₆ supplements for more than five years. The women recovered after they stopped taking the supplements.

The potential for a toxic dose from supplements is great. Consider that one small capsule can easily deliver 2 grams of vitamin B₆, but it would take almost 3,000 bananas, more than 1,600 servings of liver, or more than 3,800 chicken breasts to supply an equivalent amount. Moral: stick with food. Table 7-4 on page 247 lists common deficiency and toxicity symptoms of vitamin B₆ and other water-soluble vitamins.

Vitamin B₆ plays so many roles in protein metabolism that the body's requirement for vitamin B₆ is roughly proportional to protein intakes. The DRI committee set the vitamin B₆ intake recommendation high enough to cover most people's needs, regardless of differences in protein intakes (see the inside front cover). Meats, fish, and poultry (protein-rich foods), potatoes, leafy green vegetables, and some fruits are good sources of vitamin B₆ (that is, one serving provides at least 10 percent of the Daily Value; see Snapshot 7-11). Other foods such as legumes and peanut butter provide smaller amounts.

VITAMIN B$_6$

DRI RECOMMENDED INTAKE:
Adults (19–50 yr): 1.3 mg/day

TOLERABLE UPPER INTAKE LEVEL:
Adults: 100 mg/day

CHIEF FUNCTIONS: Part of a coenzyme needed in amino acid and fatty acid metabolism; helps to convert trypto-phan to niacin and to serotonin; helps to make hemoglobin for red blood cells

DEFICIENCY: Anemia, depression, con-fusion, abnormal brain wave pattern, convulsions; greasy, scaly dermatitis

TOXICITY: Depression, fatigue, impaired memory, irritability, headaches, nerve damage causing numbness and muscle weakness progressing to an inability to walk and convulsions; skin lesions

*These foods provide 10 percent or more of the vitamin B$_6$ Daily Value in a serving. For a 2,000-calorie diet, the DV is 2 mg/day.

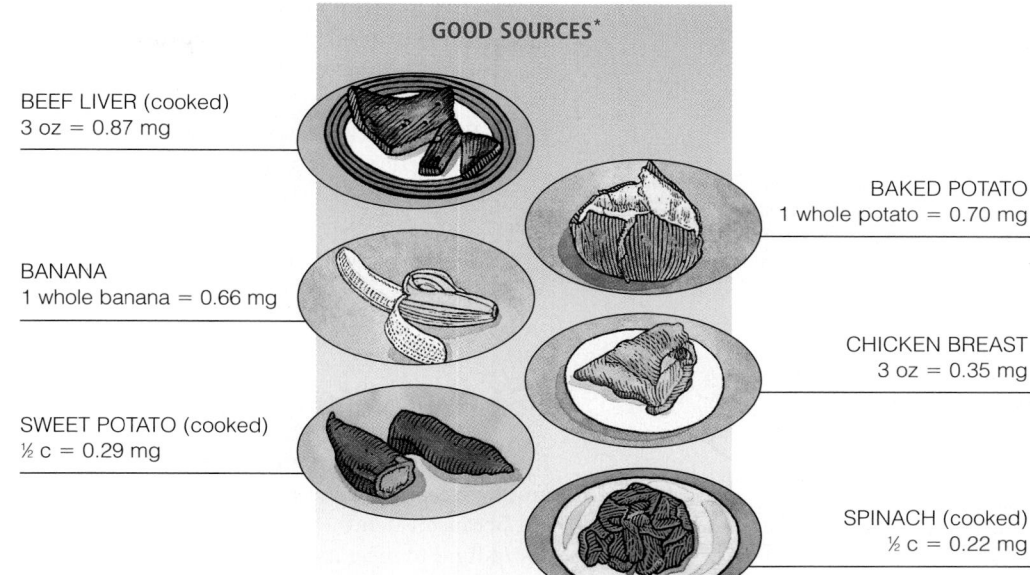

GOOD SOURCES*

BEEF LIVER (cooked)
3 oz = 0.87 mg

BANANA
1 whole banana = 0.66 mg

SWEET POTATO (cooked)
½ c = 0.29 mg

BAKED POTATO
1 whole potato = 0.70 mg

CHICKEN BREAST
3 oz = 0.35 mg

SPINACH (cooked)
½ c = 0.22 mg

How Are B Vitamins Related to Heart Disease?

People who inherit a rare dis-order that raises the level of a special amino acid, **homocysteine,** in the blood almost invariably suffer from a severe early form of cardiovascular disease (CVD).* Also, some other CVD sufferers without the inherited disorder accumulate homocysteine in the blood. Some researchers suspect that elevated homocysteine may increase the risk for CVD and that a lack of some B vitamins in the diet contributes to elevating it.[59]

A deficiency of folate, vitamin B$_{12}$, or vitamin B$_6$ causes excess homocysteine to build up in the blood. When healthy men are given supplements of B vitamins (folate, vitamin B$_6$, and vitamin B$_{12}$), their homocysteine values drop significantly. Some research sup-ports the idea that folate-fortified foods and folate supplements both raise blood folate and produce a drop in homocysteine. Whether these effects reduce a person's risk of developing CVD is unknown.[60] Higher dietary intakes of folate and vitamin B$_6$ also correlate with lower blood values for other substances associated with heart disease and with a lower incidence of heart disease itself.† The missing link is how, or even whether, B vitamin deficiencies or elevated homocysteine directly affects processes leading to heart disease.[61]

Emerging data indicate that U.S. blood values for folate have risen since the folate enrichment of food started in the 1990s. Meanwhile, homocysteine values are drop-ping.[62] The missing evidence is a concurrent drop in CVD. Until more firm evidence of benefit exists, caution is in order when considering taking B vitamin supplements to improve heart health.

Biotin and Pantothenic Acid

Two other B vitamins, **biotin** and **pantothenic acid,** are, like thiamin, riboflavin, and niacin, important in energy metabolism. Biotin is a cofactor for several enzymes in the metabolism of carbohydrate, fat, and protein. Pantothenic acid is a component of a key coenzyme that makes possible the release of

homocysteine (hoe-moe-SIS-teen) an amino acid produced as an intermediate compound during amino acid metabolism. A buildup of homocysteine in the blood is associated with deficiencies of folate and other B vitamins and may increase the risk of diseases.

biotin (BY-o-tin) a B vitamin; a coenzyme nec-essary for fat synthesis and other metabolic reactions.

pantothenic (PAN-to-THEN-ic) **acid** a B vitamin.

*Although chemically an amino acid, homocysteine is not incorporated into body proteins but is metabolized to other compounds.
†One such substance is C-reactive protein.

■ The DRI recommended intakes for biotin and pantothenic acid are listed on the inside front cover, page B.

energy from the energy nutrients. It also participates in more than 100 steps in the synthesis of lipids, neurotransmitters, steroid hormones, and hemoglobin.

Although rare diseases may precipitate deficiencies of biotin and pantothenic acid, both vitamins are widespread in foods. A steady diet of raw egg whites, which contain a protein that binds biotin, can produce biotin deficiency, but you would have to consume more than two dozen raw egg whites daily to produce the effect. Cooking eggs denatures the protein. Healthy people eating ordinary diets are not at risk for deficiencies.

▶ **KEY POINT**   Historically, famous B vitamin–deficiency diseases are beriberi (thiamin), pellagra (niacin), and pernicious anemia (vitamin B_{12}). Pellagra can be prevented by adequate protein because the amino acid tryptophan can be converted to niacin in the body. A high intake of folate can mask the blood symptom of vitamin B_{12} deficiency but will not prevent the associated nerve damage. Vitamin B_6 is important in amino acid metabolism and can be toxic in excess. Biotin and pantothenic acid are important to the body and are abundant in food.

Non-B Vitamins

In addition to the B vitamins just discussed, a few compounds that are topics of debate among researchers deserve mention. **Choline** could be considered an essential nutrient because when the diet is devoid of choline, the body cannot make enough of the compound to meet its need. Choline is common in foods, though, and deficiencies are practically unheard of outside the laboratory. DRI intake recommendations have been set for choline (see inside front cover).

The compounds **carnitine, inositol,** and **lipoic acid** might appropriately be called *nonvitamins* because they are not essential nutrients for human beings. Carnitine, sometimes called "vitamin B_T," is an important piece of cell machinery, but it is not a vitamin. Although deficiencies can be induced in laboratory animals for experimental purposes, these substances are abundant in ordinary foods. Even if these compounds were essential in human nutrition, supplements would be unnecessary for healthy people eating a balanced diet. Vitamin companies often include these substances to make their formulas appear more "complete," but there is no physiological reason to do so.

In addition to carnitine, inositol, and lipoic acid, other substances have been mistakenly thought to be essential in human nutrition because they are needed for growth by bacteria or other life-forms. These substances include PABA (para-aminobenzoic acid), bioflavonoids ("vitamin P" or hesperidin), and ubiquinone (coenzyme Q). Other names you may hear are "vitamin B_{15}" and pangamic acid (hoaxes) or "vitamin B_{17}" (laetrile or amygdalin, not a cancer cure as claimed and not a vitamin by any stretch of the imagination).*[63]

choline (KOH-leen) a nonessential nutrient used to make the phospholipid lecithin and other molecules.

carnitine a nonessential nutrient that functions in cellular activities.

inositol (in-OSS-ih-tall) a nonessential nutrient found in cell membranes.

lipoic (lip-OH-ic) **acid** a nonessential nutrient.

▶ **KEY POINT**   Choline is needed in the diet, but it is not a vitamin and deficiencies are unheard of outside the laboratory. Many other substances that people claim are B vitamins are not. Among these substances are carnitine, inositol, and lipoic acid.

This chapter has addressed all 13 of the vitamins. The basic facts about each one are summed up in Tables 7-3 and 7-4.

---

*Read the truth about these and many other claims at the website of the National Council Against Health Fraud: www.ncahf.org.

TABLE 7·3   THE FAT-SOLUBLE VITAMINS—FUNCTIONS, DEFICIENCIES, AND TOXICITIES

Vitamin A

Other Names
Retinol, retinal, retinoic acid; main precursor is beta-carotene

Chief Functions in the Body
Vision; health of cornea, epithelial cells, mucous membranes, skin; bone and tooth growth; regulation of gene expression; reproduction; immunity
Beta-carotene: antioxidant

Deficiency Disease Name
Hypovitaminosis A

Significant Sources
Retinol: fortified milk, cheese, cream, butter, fortified margarine, eggs, liver
Beta-carotene: spinach and other dark, leafy greens; broccoli; deep orange fruits (apricots, cantaloupe) and vegetables (winter squash, carrots, sweet potatoes, pumpkin)

Deficiency Symptoms

Anemia (small-cell type)[a]

Cessation of bone growth, painful joints; impaired enamel formation, cracks in teeth, tendency toward tooth decay

Diarrhea, changes in intestinal and other body linings

Depression; frequent respiratory, digestive, bladder, vaginal, and other infections

Night blindness (retinal)

Keratinization, corneal degeneration leading to blindness,[b] rashes

Kidney stones, impaired growth

Toxicity Symptoms

Blood/Circulatory System
Red blood cell breakage, cessation of menstruation, nosebleeds

Bones/Teeth
Bone pain; growth retardation; increased pressure inside skull; headaches; possible bone mineral loss

Digestive System
Abdominal pain, nausea, vomiting, diarrhea, weight loss

Immune System
Overreactivity

Nervous/Muscular Systems
Blurred vision, muscle weakness, fatigue, irritability, loss of appetite

Skin and Cornea
Dry skin, rashes, loss of hair; cracking and bleeding lips, brittle nails; hair loss

Other
Liver enlargement and liver damage; birth defects

Vitamin D

Other Names
Calciferol, cholecalciferol, dihy-droxy vitamin D; precursor is cholesterol

Chief Functions in the Body
Mineralization of bones (raises blood calcium and phosphorus via absorption from digestive tract and by withdrawing calcium from bones and stimulating retention by kidneys)

Deficiency Disease Name
Rickets, osteomalacia

Significant Sources
Self-synthesis with sunlight; fortified milk or margarine, liver, sardines, salmon, shrimp

Deficiency Symptoms

Abnormal growth, misshapen bones (bowing of legs), soft bones, joint pain, malformed teeth

Muscle spasms

Toxicity Symptoms

Blood/Circulatory System
Raised blood calcium; calcification of blood vessels and heart tissues

Bones/Teeth
Calcification of tooth soft tissues; thinning of tooth enamel

Nervous System
Excessive thirst, headaches, irritability, loss of appetite, weakness, nausea

Other
Kidney stones; calcification of soft tissues (kidneys, lungs, joints); mental and physical retardation of offspring

[a]Small-cell anemia is termed *microcytic anemia;* large-cell type is *macrocytic* or *megaloblastic anemia.*
[b]Corneal degeneration progresses from *keratinization* (hardening) to *xerosis* (drying) to *xerophthalmia* (thickening, opacity, and irreversible blindness).

(Continued on next page)

Vitamin E

Other Names
Alpha-tocopherol, tocopherol

Chief Functions in the Body
Antioxidant (quenching of free radicals), stabilization of cell membranes, support of immune function, protection of polyunsaturated fatty acids; normal nerve development

Deficiency Disease Name
(No name)

Significant Sources
Polyunsaturated plant oils (margarine, salad dressings, shortenings), green and leafy vegetables, wheat germ, whole-grain products, nuts, seeds

Deficiency Symptoms

Red blood cell breakage, anemia

Nerve degeneration, weakness, difficulty walking, leg cramps

Toxicity Symptoms

Blood/Circulatory System
Augments the effects of anticlotting medication

Digestive System
General discomfort, nausea

Eyes
Blurred vision

Nervous/Muscular Systems
Fatigue

Vitamin K

Other Names
Phylloquinone, naphthoquinone

Chief Functions in the Body
Synthesis of blood-clotting proteins and proteins important in bone mineralization

Deficiency Disease Name
(No name)

Significant Sources
Bacterial synthesis in the digestive tract; green leafy vegetables, cabbage-type vegetables, soybeans, vegetable oils

Deficiency Symptoms

Hemorrhage

Poor skeletal mineralization

Toxicity Symptoms

Blood/Circulatory System
Interference with anticlotting medication

Bones

Vitamin C

Other Names
Ascorbic acid

Chief Functions in the Body
Collagen synthesis
(strengthens blood vessel walls,
forms scar tissue, matrix for
bone growth), antioxidant, restores
vitamin E to active form, hormone
synthesis, supports immune cell
functions, helps in absorption of iron

Deficiency Disease Name
Scurvy

Significant Sources
Citrus fruits, cabbage-type
vegetables, dark green vegetables,
cantaloupe, strawberries,
peppers, lettuce, tomatoes,
potatoes, papayas, mangoes

Deficiency Symptoms

Anemia (small-cell type)[a]

Suppression, frequent infections

Bleeding gums, loosened teeth

Muscle degeneration and pain,
depression, disorientation

Bone fragility, joint pain

Pinpoint hemorrhages, rough skin,
blotchy bruises

Failure of wounds to heal

Toxicity Symptoms

Blood/Circulatory System

Digestive System
Nausea, abdominal cramps, diarrhea,
excessive urination

Immune System

Mouth, Gums, Tongue

Nervous/Muscular Systems
Headache, fatigue, insomnia

Skeletal System
Aggravation of gout

Skin
Rashes

Other
Interference with medical tests; kidney
stones in susceptible people

Thiamin

Other Names
Vitamin B$_1$

Chief Functions in the Body
Part of a coenzyme needed in
energy metabolism,
supports normal appetite
and nervous system function

Deficiency Disease Name
Beriberi (wet and dry)

Significant Sources
Occurs in all nutritious foods
in moderate amounts; pork,
ham, bacon, liver, whole and
enriched grains, legumes, seeds

Deficiency Symptoms

Edema, enlarged heart, abnormal heart
rhythms, heart failure

Degeneration, wasting, weakness, pain,
apathy, irritability, difficulty walking,
loss of reflexes, mental confusion,
paralysis

Anorexia; weight loss

Toxicity Symptoms

Blood/Circulatory System
(No symptoms reported)

Nervous/Muscular Systems
(No symptoms reported)

Other

Riboflavin

Other Names
Vitamin B$_2$

Chief Functions in the Body
Part of a coenzyme needed in
energy metabolism, supports
normal vision and skin health

Deficiency Disease Name
Ariboflavinosis

Significant Sources
Milk, yogurt, cottage cheese,
meat, liver, leafy green
vegetables, whole-grain or
enriched breads and cereals

Deficiency Symptoms

Cracks at corners of mouth,[b]

Hypersensitivity to light,
reddening of cornea

Skin rash

Toxicity Symptoms

Mouth, Gums, Tongue

(No symptoms reported)
smooth magenta tongue[c]; sore throat

Nervous System and Eyes
(No symptoms reported)

Skin
(No symptoms reported)

[a]Small-cell anemia is termed *microcytic anemia;* large-cell type is *macrocytic* or *megaloblastic anemia.*
[b]Cracks at the corners of the mouth are termed *cheilosis* (kee-LOH-sis).
[c]Smoothness of the tongue is caused by loss of its surface structures and is termed *glossitis* (gloss-EYE-tis).
[d]The name *pernicious anemia* refers to the vitamin B$_{12}$ deficiency caused by lack of intrinsic factor, but not to that caused by inadequate dietary intake.

(Continued on next page)

Niacin

Other Names
Nicotinic acid, nicotinamide, niacinamide, vitamin B_3; precursor is dietary tryptophan

Chief Functions in the Body
Part of coenzymes needed in energy metabolism

Deficiency Disease Name
Pellagra

Significant Sources
Synthesized from the amino acid tryptophan; milk, eggs, meat, poultry, fish, whole-grain and enriched breads and cereals, nuts, and all protein-containing foods

Deficiency Symptoms

Diarrhea; vomiting; abdominal pain

Black or bright red swollen smooth tongue[c]

Irritability, loss of appetite, weakness, headache, dizziness, mental confusion progressing to psychosis or delirium

Flaky skin rash on areas exposed to sun

Toxicity Symptoms

Digestive System
Nausea, vomiting

Mouth, Gums, Tongue

Nervous System

Skin
Painful flush and rash, sweating

Other
Liver damage; impaired glucose tolerance

Folate

Other Names
Folic acid, folacin, pteroylglutamic acid

Chief Functions in the Body
Part of a coenzyme needed for new cell synthesis

Deficiency Disease Name
(No name)

Significant Sources
Asparagus, avocado, leafy green vegetables, beets, legumes, seeds, liver, enriched breads, cereal, pasta, and grains

Deficiency Symptoms

Anemia (large-cell type),[a] elevated homocysteine

Heartburn, diarrhea, constipation

Suppression, frequent infections

Smooth red tongue[c]

Increased risk of neural tube birth defects

Depression, mental confusion, fatigue, irritability, headache

Toxicity Symptoms

Blood/Circulatory System
Masks vitamin B_{12} deficiency

Digestive System

Immune System

Mouth, Gums, Tongue

Nervous System

Vitamin B_{12}

Other Names
Cyanocobalamin

Chief Functions in the Body
Part of coenzymes needed in new cell synthesis, helps maintain nerve cells

Deficiency Disease Name
(No name)[d]

Significant Sources
Animal products (meat, fish, poultry, milk, cheese, eggs)

Deficiency Symptoms

Anemia (large-cell type)[a]

Smooth tongue[c]

Fatigue, nerve degeneration progressing to paralysis

Tingling or numbness

Toxicity Symptoms

Blood/Circulatory System
(No toxicity symptoms known)

Mouth, Gums, Tongue

Nervous System

Skin

[a]Small-cell anemia is termed *microcytic anemia;* large-cell type is *macrocytic* or *megaloblastic anemia.*
[b]Cracks at the corners of the mouth are termed *cheilosis* (kee-LOH-sis).
[c]Smoothness of the tongue is caused by loss of its surface structures and is termed *glossitis* (gloss-EYE-tis).
[d]The name *pernicious anemia* refers to the vitamin B_{12} deficiency caused by lack of intrinsic factor, but not to that caused by inadequate dietary intake.

Vitamin B$_6$

Other Names	Deficiency Symptoms	Toxicity Symptoms
Pyridoxine, pyridoxal, pyridoxamine		**Blood/Circulatory System**
	Anemia (small-cell type)[a]	Bloating
Chief Functions in the Body		**Nervous/Muscular Systems**
Part of a coenzyme needed in amino acid and fatty acid metabolism, helps convert tryptophan to niacin and to serotonin, helps make red blood cells	Depression, confusion, abnormal brain wave pattern, convulsions	Depression, fatigue, impaired memory, irritability, headaches, numbness, damage to nerves, difficulty walking, loss of reflexes, restlessness, convulsions
		Skin
Deficiency Disease Name	Rashes, greasy, scaly dermatitis	Lesions
(No name)		
Significant Sources		
Meats, fish, poultry, liver, legumes, fruits, potatoes, whole grains, soy products		

Pantothenic Acid

Other Names	Deficiency Symptoms	Toxicity Symptoms
(None)		**Digestive System**
	Vomiting, intestinal distress	
Chief Functions in the Body		**Nervous/Muscular Systems**
Part of a coenzyme needed in energy metabolism	Insomnia, fatigue	
		Other
Deficiency Disease Name		
(No name)	Hypoglycemia, increased sensitivity to insulin	Water retention (infrequent)
Significant Sources		
Widespread in foods		

Biotin

Other Names	Deficiency Symptoms	Toxicity Symptoms
(None)		**Blood/Circulatory System**
	Abnormal heart action	(No toxicity symptoms reported)
Chief Functions in the Body		**Digestive System**
A cofactor for several enzymes needed in energy metabolism, fat synthesis, amino acid metabolism, and glycogen synthesis	Loss of appetite, nausea	
		Nervous/Muscular Systems
	Depression, muscle pain, weakness, fatigue, numbness of extremities	
Deficiency Disease Name		**Skin**
(No name)	Dry around eyes, nose, and mouth	
Significant Sources		
Widespread in foods		

Choosing Foods Rich in Vitamins

ON LEARNING HOW important the vitamins are to their health, most people want to choose foods that are vitamin-rich whether they are eating at home or in restaurants. Look down the columns of vitamins and calories in Table 7-5 to identify some vitamin-rich foods in popular restaurant meals. If you are interested in folate, for instance, you can see that cornflakes are an especially good source (folic acid is added to cornflakes), as is orange juice (folate occurs naturally in this food).

Another way of looking at such data appears in Figure 7-16 (page 252)—the long bars show some foods that are rich sources of a particular vitamin, and the short or nonexistent bars indicate poor sources. The colors of the bars represent the various food groups.

Which Foods Should I Choose?

After looking at Figure 7-16, don't think that you must memorize the richest sources of each vitamin and include those foods daily. That false notion would lead you to limit your variety of foods while overemphasizing the components of a few foods. Although it is reassuring to know that your carrot-raisin salad at lunch provided more than your entire day's need for vitamin A, it is a mistake to think that you must then select equally rich sources of all the other vitamins. Such rich sources do not exist for many vitamins—rather, foods

TABLE 7·5 VITAMIN CONTENTS OF RESTAURANT MEALS

	Energy (cal)	Vitamin A (µg)	Thiamin (mg)	Vitamin C (mg)	Folate (µg)	Vitamin B$_{12}$ (µg)
Breakfast Foods						
Hotcakes with syrup and butter, scrambled egg and sausage patty	1,140	150	0.41	0	32	1.7
Egg, ham, and cheese muffin	290	100	0.49	0	30	0.7
Oatmeal, brown sugar	107	2	0.13	0	5	0
Cornflakes	110	225	0.87	15	100	0
Hash browned potatoes	130	0	0.08	3	8	0
2 small cinnamon sweet rolls	300	50	0.25	2	20	0.1
Large blueberry muffin	400	10	0.2	2	20	0.1
English muffin	130	0	0.25	0	20	0.1
Orange juice	80	40	0.17	90	60	0
Milk (low-fat)	120	140	0.1	2	10	0.8
Lunch Foods						
Homemade chili/crackers	350	150	0.26	25	40	0.5
Cold cut hoagie sandwich/chips	460	80	1.0	12	55	1.1
Peanut butter and jelly sandwich on whole wheat; fruit cocktail	450	30	0.3	3	62	0
Tuna sandwich on white; banana	470	40	0.32	11	49	0.6
Chef's salad with cheese, ham, turkey, and dressing/crackers	580	140	0.43	16	100	0.8
Fat-free milk	85	150	0.1	0	10	0.9
Apple juice	120	0	0.05	2	0	0
Supper Foods						
Baked ham dinner (ham, potatoes, brussels sprouts)	440	150	0.27	80	70	1.4
Vegetable plate	400	680	0.26	53	120	0.1
Spaghetti and meatballs; small salad	600	220	0.38	25	50	1.3
Fried fish, tartar sauce, corn, and macaroni salad	510	70	0.26	6	83	1
Rolls	100	0	0.15	0	0	0
Garlic bread	190	4	0.4	0	0	0
Corn muffins	300	0	0.12	0	0	0
Lemon pie	350	60	0.16	4	11	0.2
Chocolate cake	250	20	0.02	0	5	0.1

work in harmony to provide most nutrients. For example, a baked potato, not a star performer among vitamin C providers, contributes substantially to a day's need for this nutrient and contributes some thiamin, too. By the end of the day, assuming that your food choices were made with reasonable care, the bits of thiamin, vitamin B_6, and vitamin C from each serving of food have accumulated to make a more-than-adequate total diet.

A Variety of Foods Works Best

With a few exceptions, nutritious foods provide small quantities of thiamin, as shown in Figure 7-16. Members of the pork family are an exceptionally good thiamin source with one small pork chop (275 calories) providing over half of the Daily Value for thiamin—but again, this does not suggest that you eat pork every day. Legumes and grains are also good, low-fat sources, and they provide beneficial fiber and nutrients lacking from meats. Beans lack the vitamin B_{12} provided by meats,

however. Peanut butter is a good source of thiamin, and of most other B vitamins as well, but its high fat and calorie contents call for moderation in its use.

The vitamin B_6 data provide another insight to support the argument for variety. From just the few foods listed here, you can see that no one source can provide the whole day's requirement for vitamin B_6, but that many small servings of a variety of meats, fish, and poultry along with potatoes and a few other vegetables and fruits consumed throughout the day can work together to supply it.

The last two graphs of Figure 7-16 show sources of folate and vitamin C. These nutrients are both richly supplied by fruits and vegetables. The richest source of either may be only a moderate source of the other, but the recommended amounts of fruits and vegetables in the USDA Food Guide of Chapter 2 cover both needs amply. As for vitamin E, vegetable oils and some seeds and nuts are the richest sources, and some vegetables and fruits contribute a little, too.

FIGURE 7·16 FOOD SOURCES OF VITAMINS SELECTED TO SHOW A RANGE OF VALUES

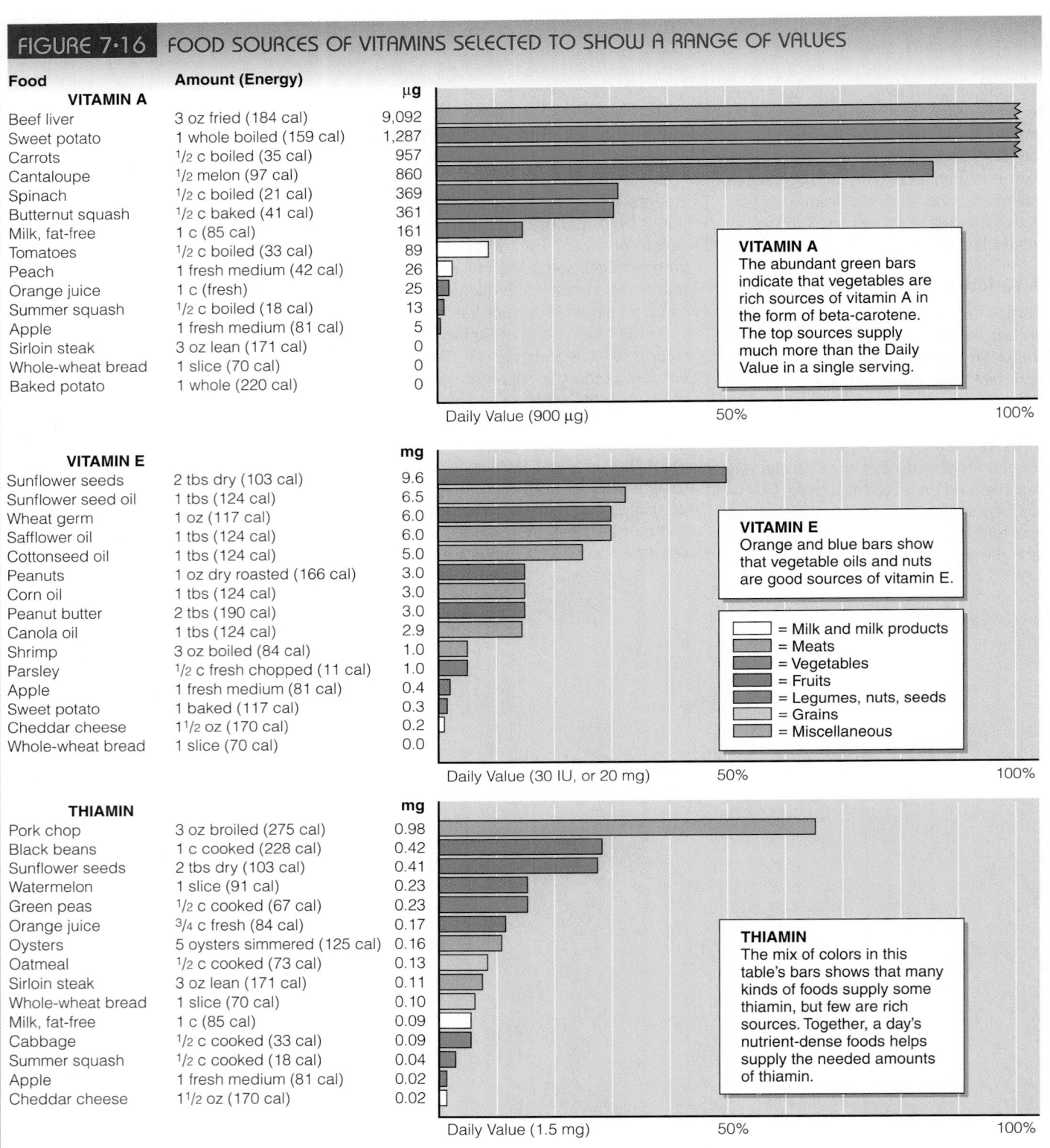

Food	Amount (Energy)	μg
VITAMIN A		
Beef liver	3 oz fried (184 cal)	9,092
Sweet potato	1 whole boiled (159 cal)	1,287
Carrots	1/2 c boiled (35 cal)	957
Cantaloupe	1/2 melon (97 cal)	860
Spinach	1/2 c boiled (21 cal)	369
Butternut squash	1/2 c baked (41 cal)	361
Milk, fat-free	1 c (85 cal)	161
Tomatoes	1/2 c boiled (33 cal)	89
Peach	1 fresh medium (42 cal)	26
Orange juice	1 c (fresh)	25
Summer squash	1/2 c boiled (18 cal)	13
Apple	1 fresh medium (81 cal)	5
Sirloin steak	3 oz lean (171 cal)	0
Whole-wheat bread	1 slice (70 cal)	0
Baked potato	1 whole (220 cal)	0

VITAMIN A
The abundant green bars indicate that vegetables are rich sources of vitamin A in the form of beta-carotene. The top sources supply much more than the Daily Value in a single serving.

Daily Value (900 μg)   50%   100%

Food	Amount (Energy)	mg
VITAMIN E		
Sunflower seeds	2 tbs dry (103 cal)	9.6
Sunflower seed oil	1 tbs (124 cal)	6.5
Wheat germ	1 oz (117 cal)	6.0
Safflower oil	1 tbs (124 cal)	6.0
Cottonseed oil	1 tbs (124 cal)	5.0
Peanuts	1 oz dry roasted (166 cal)	3.0
Corn oil	1 tbs (124 cal)	3.0
Peanut butter	2 tbs (190 cal)	3.0
Canola oil	1 tbs (124 cal)	2.9
Shrimp	3 oz boiled (84 cal)	1.0
Parsley	1/2 c fresh chopped (11 cal)	1.0
Apple	1 fresh medium (81 cal)	0.4
Sweet potato	1 baked (117 cal)	0.3
Cheddar cheese	1 1/2 oz (170 cal)	0.2
Whole-wheat bread	1 slice (70 cal)	0.0

VITAMIN E
Orange and blue bars show that vegetable oils and nuts are good sources of vitamin E.

☐ = Milk and milk products
▨ = Meats
▨ = Vegetables
▨ = Fruits
▨ = Legumes, nuts, seeds
☐ = Grains
▨ = Miscellaneous

Daily Value (30 IU, or 20 mg)   50%   100%

Food	Amount (Energy)	mg
THIAMIN		
Pork chop	3 oz broiled (275 cal)	0.98
Black beans	1 c cooked (228 cal)	0.42
Sunflower seeds	2 tbs dry (103 cal)	0.41
Watermelon	1 slice (91 cal)	0.23
Green peas	1/2 c cooked (67 cal)	0.23
Orange juice	3/4 c fresh (84 cal)	0.17
Oysters	5 oysters simmered (125 cal)	0.16
Oatmeal	1/2 c cooked (73 cal)	0.13
Sirloin steak	3 oz lean (171 cal)	0.11
Whole-wheat bread	1 slice (70 cal)	0.10
Milk, fat-free	1 c (85 cal)	0.09
Cabbage	1/2 c cooked (33 cal)	0.09
Summer squash	1/2 c cooked (18 cal)	0.04
Apple	1 fresh medium (81 cal)	0.02
Cheddar cheese	1 1/2 oz (170 cal)	0.02

THIAMIN
The mix of colors in this table's bars shows that many kinds of foods supply some thiamin, but few are rich sources. Together, a day's nutrient-dense foods helps supply the needed amounts of thiamin.

Daily Value (1.5 mg)   50%   100%

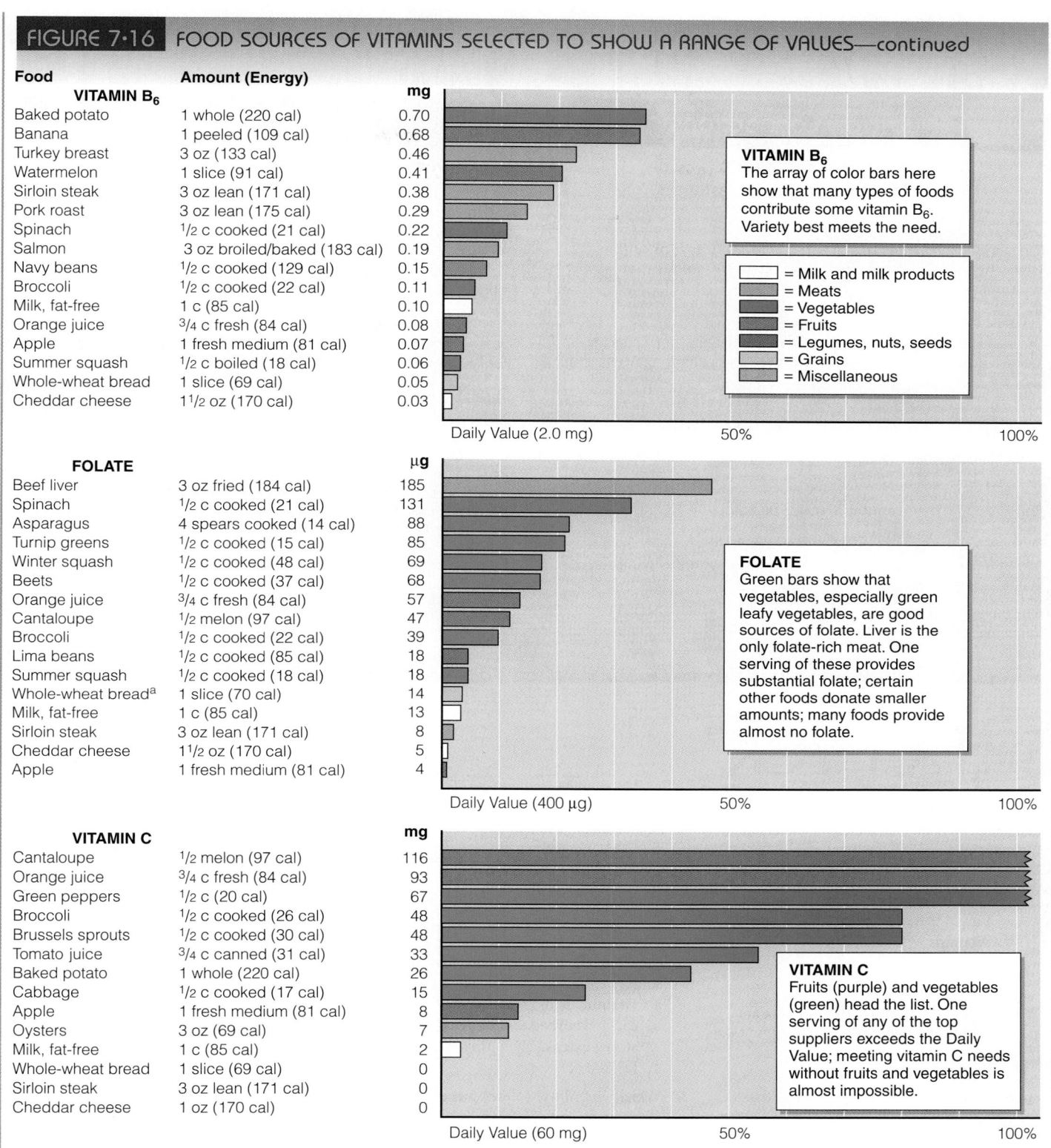

Food	Amount (Energy)	mg
VITAMIN B6		
Baked potato	1 whole (220 cal)	0.70
Banana	1 peeled (109 cal)	0.68
Turkey breast	3 oz (133 cal)	0.46
Watermelon	1 slice (91 cal)	0.41
Sirloin steak	3 oz lean (171 cal)	0.38
Pork roast	3 oz lean (175 cal)	0.29
Spinach	1/2 c cooked (21 cal)	0.22
Salmon	3 oz broiled/baked (183 cal)	0.19
Navy beans	1/2 c cooked (129 cal)	0.15
Broccoli	1/2 c cooked (22 cal)	0.11
Milk, fat-free	1 c (85 cal)	0.10
Orange juice	3/4 c fresh (84 cal)	0.08
Apple	1 fresh medium (81 cal)	0.07
Summer squash	1/2 c boiled (18 cal)	0.06
Whole-wheat bread	1 slice (69 cal)	0.05
Cheddar cheese	1 1/2 oz (170 cal)	0.03

VITAMIN B6
The array of color bars here show that many types of foods contribute some vitamin B6. Variety best meets the need.

☐ = Milk and milk products
= Meats
= Vegetables
= Fruits
= Legumes, nuts, seeds
= Grains
= Miscellaneous

Daily Value (2.0 mg)   50%   100%

FOLATE		μg
Beef liver	3 oz fried (184 cal)	185
Spinach	1/2 c cooked (21 cal)	131
Asparagus	4 spears cooked (14 cal)	88
Turnip greens	1/2 c cooked (15 cal)	85
Winter squash	1/2 c cooked (48 cal)	69
Beets	1/2 c cooked (37 cal)	68
Orange juice	3/4 c fresh (84 cal)	57
Cantaloupe	1/2 melon (97 cal)	47
Broccoli	1/2 c cooked (22 cal)	39
Lima beans	1/2 c cooked (85 cal)	18
Summer squash	1/2 c cooked (18 cal)	18
Whole-wheat bread[a]	1 slice (70 cal)	14
Milk, fat-free	1 c (85 cal)	13
Sirloin steak	3 oz lean (171 cal)	8
Cheddar cheese	1 1/2 oz (170 cal)	5
Apple	1 fresh medium (81 cal)	4

FOLATE
Green bars show that vegetables, especially green leafy vegetables, are good sources of folate. Liver is the only folate-rich meat. One serving of these provides substantial folate; certain other foods donate smaller amounts; many foods provide almost no folate.

Daily Value (400 μg)   50%   100%

VITAMIN C		mg
Cantaloupe	1/2 melon (97 cal)	116
Orange juice	3/4 c fresh (84 cal)	93
Green peppers	1/2 c (20 cal)	67
Broccoli	1/2 c cooked (26 cal)	48
Brussels sprouts	1/2 c cooked (30 cal)	48
Tomato juice	3/4 c canned (31 cal)	33
Baked potato	1 whole (220 cal)	26
Cabbage	1/2 c cooked (17 cal)	15
Apple	1 fresh medium (81 cal)	8
Oysters	3 oz (69 cal)	7
Milk, fat-free	1 c (85 cal)	2
Whole-wheat bread	1 slice (69 cal)	0
Sirloin steak	3 oz lean (171 cal)	0
Cheddar cheese	1 oz (170 cal)	0

VITAMIN C
Fruits (purple) and vegetables (green) head the list. One serving of any of the top suppliers exceeds the Daily Value; meeting vitamin C needs without fruits and vegetables is almost impossible.

Daily Value (60 mg)   50%   100%

MEDIA MENU

Nutrition Connections

The Nutrition Connections CD provides access to these additional resources.

Do It!
The Do It! exercise for this chapter, *Find the Vitamins on the Menu,* helps you identify foods on restaurant menus that are rich in vitamins.

Interaction
The animation, *Protecting the Vitamins in Foods,* demonstrates food-handling procedures that preserve the vitamin content of foods.

Chapter Quiz
Take the quiz for this chapter to assess your understanding of chapter concepts.

Nutrition on the Net
Annotated links to reputable nutrition-related websites.

Online Glossary
Test your knowledge of key vocabulary through this comprehensive, interactive glossary.

Internet Connections

A variety of helpful study tools for this chapter are available at our website
http://nutrition.wadsworth.com/ncc10

Practice Test
Take the practice test for this chapter and gauge your knowledge of key concepts. Answers are provided.

Student Notebook
A lecture notebook is available to help organize note taking while reading or attending a lecture.

ABC Video Clip
"Boost the Immune System" (2:59) examines whether it is possible to boost the immune system through diet.

Study Tools
Study resources for students include Chapter Summaries, Crossword puzzles, Flashcards, Glossary, Internet Exercises, and Learning Objectives.

SELF CHECK

Answers to these Self Check Questions are in Appendix G.

1. Which of the following vitamins are classified as fat soluble?
   a. vitamins B, D
   b. vitamins A, D, E, and K
   c. vitamins B, E, D, and C
   d. vitamins B and C

2. Night blindness and xerophthalmia are the result of a deficiency of which vitamin?
   a. niacin
   b. vitamin C
   c. vitamin A
   d. vitamin K

3. Which of the following foods is (are) rich in beta-carotene?
   a. sweet potatoes
   b. pumpkin
   c. spinach
   d. all of the above

4. A deficiency of niacin may result in which disease?
   a. pellagra
   b. beriberi
   c. scurvy
   d. rickets

5. Which of the following describes the fat soluble vitamins?
   a. vitamins B and C
   b. easily absorbed and excreted
   c. stored extensively in tissues
   d. (a) and (c)

6. Which vitamin(s) is (are) present only in foods of animal origin?
   a. the active form of vitamin A
   b. vitamin B_{12}
   c. riboflavin
   d. (a) and (b)

7. The theory that vitamin C prevents or cures colds is well supported by research.
   T   F

8. Xerophthalmia results from advanced vitamin A deficiency and can lead to permanent blindness.   T   F

9. A high blood level of homocysteine is associated with a deficiency of B vitamins and may increase the risk of heart disease.
   T   F

10. Vitamin A supplements can help treat acne.   T   F

CONTROVERSY 7

VITAMIN SUPPLEMENTS: WHO BENEFITS?

ALMOST 40 PERCENT of the U.S. population dose themselves with vitamin supplements, collectively spending billions of dollars a year on them. Some take a single pill containing a multitude of nutrients; others take huge doses of single nutrients in an attempt to ward off diseases. Do most people need these supplements? What about the safety of today's supplements? Can taking a daily vitamin pill prevent chronic diseases? And who is looking out for the consumer? This Controversy examines evidence surrounding these questions and concludes with some advice for those choosing to take a supplement.

What the Experts Are Saying

On reviewing the evidence concerning supplements, researchers come to different conclusions. Several have concluded that all healthy people might benefit from a multiple vitamin tablet or two every day as a sort of nutritional insurance.[*1] They judge that the risks of deficiencies outweigh those of any potential for overload from moderate daily doses.

Other experts vigorously disagree. They argue that the risks incurred by taking supplements are considerable and often unpredictable. Too little evidence of true benefit exists, they say, to support taking supplements except in cases where an established risk of deficiency exists.[2] Additionally, for those who may need supplements, these scientists recommend a thorough assessment of the individual's vitamin status to determine which nutrients need attention, and even then, they recommend dietary changes as well as supplementation to correct immediate deficiencies.[3]

Indisputably, certain people derive benefits from supplements, and these people are listed in Table C7-1. For them, nutrient supplements have been shown with certainty to prevent or

Which is the best source of vitamins to support good health: supplements or foods?

reverse illnesses. Consider a woman who loses a lot of blood and therefore a lot of iron and other blood-building nutrients in menstruation each month. This woman may be able to eat in such a way as to make up all nutrient losses except that of iron; we can say with certainty that for iron, she needs an assessment of her iron status and a supplement prescribed by a health-care provider. Similarly, few women of childbearing age obtain the folic acid they need for healthy pregnancies without enriched foods or supplements in addition to folate from food.

Experts also almost always agree on these points:

- People who eat diets that omit any major food group probably need certain supplements to ensure adequate intakes of vitamins and minerals. For example, vegetarians who eat no animal products most likely need vitamin B_{12}, calcium, and vitamin D.
- Beta-carotene supplements are probably hazardous, especially to those who smoke and consume alcoholic beverages, and are not recommended either alone or in combination with other nutrients.[4]

Many experts also agree that people who take multinutrient supplements should limit their intakes of preformed vitamin A for the sake of their bones—

TABLE C7·1 SOME VALID REASONS FOR TAKING SUPPLEMENTS

These People May Need Supplements:

- People with nutrient deficiencies.
- Women in their childbearing years (supplemental or enrichment sources of folic acid are recommended to reduce risk of neural tube defects in infants).
- Pregnant or lactating women (they may need iron and folate).
- Newborns (they are routinely given a vitamin K dose).
- Infants (they may need various supplements, see Chapter 13).
- Those who are lactose intolerant (they need calcium to forestall osteoporosis).
- Habitual dieters (they may eat insufficient food).
- Elderly people often benefit from some of the vitamins and minerals in a balanced supplement; (they may choose poor diets, have trouble chewing, or absorb or metabolize less efficiently; see Chapter 14).
- Victims of AIDS or other wasting illnesses (they lose nutrients faster than foods can supply them).
- Those addicted to drugs or alcohol (they absorb fewer and excrete more nutrients; nutrients cannot undo damage from drugs or alcohol).
- Those recovering from surgery, burns, injury, or illness (they need extra nutrients to help regenerate tissues).
- Strict vegetarians (they may need vitamin B_{12}, vitamin D, iron, and zinc).
- People taking medications that interfere with the body's use of nutrients.

*Reference notes are found in Appendix F.

elevated blood vitamin A predicts a significant increase in hip fractures.[5] Vitamin A also reliably and quickly produces liver injury at doses greater than 10,000 micrograms, but liver problems can appear with much lower doses taken regularly over many years.

The experts part company regarding who should *not* take a routine supplement. The remainder of this Controversy presents evidence on both sides of the argument, starting with an issue of paramount importance—supplement safety.

Supplements Must Be Safe, or the Government Would Not Allow Their Sales, Right?

Many consumers believe that government scientists, in particular those of the Food and Drug Administration (FDA), test supplements to ensure their safety and effectiveness. This belief is false. The FDA has no mandate to test consumer products; instead, under federal law, the FDA is charged with ensuring that supplements do not pose a "significant or unreasonable risk" of illness or injury to consumers. To do this, FDA needs scientific evidence of supplement safety from manufacturers and the records of adverse health effects reported by consumers but just a few ethical companies provide this information voluntarily. Consumers can report adverse reactions from supplements directly to FDA via their hotline or website but most people are unaware of these options (see the note on this page).*

Recent documented cases of serious illness and death from supplements, particularly from ephedrine sold for weight loss and as an "energy booster," have raised an alarm, and many scientists are calling for greater FDA regulation of the supplement industry. The FDA's task would be easier if makers of supplements were required to provide scientific evidence of safety, as is true for prescription drugs. Laws governing supplements, however, assume them to be safe until proved otherwise. In truth, many vitamin supplements, but not all, probably are safe when used as

---

*Consumers should report suspected harm from dietary supplements to their health providers or to the FDA's MedWatch program at (800) FDA-1088 or on the Internet at www.fda.gov/medwatch/.

directed.[6] The trouble is, consumers have no way of knowing which may be harmful before purchasing them.

Dietary Supplement Health and Education Act

In 1994, Congress passed the Dietary Supplement Health and Education Act (DSHEA, often pronounced *dee-SHAY*), which specifies the FDA's roles with regard to supplements. Under DSHEA, supplements:

- Must provide a Supplement Facts panel of nutrient information (see Figure C7-1).
- Must not be adulterated; that is, they should contain the ingredients stated on the labels (although few products conformed to this standard during recent testing).
- Must not make label claims for the diagnosis, prevention, treatment, or cure of illness.
- May bear "structure/function" claims (described in a following section).[7]

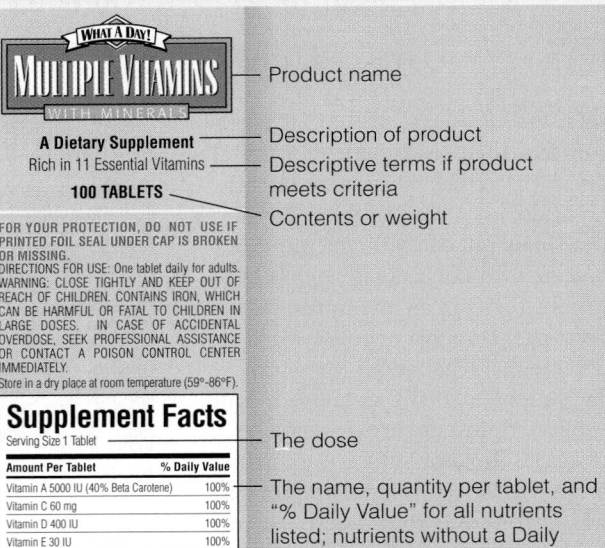

FIGURE C7·1 A SUPPLEMENT LABEL

- Product name
- Description of product
- Descriptive terms if product meets criteria
- Contents or weight
- The dose
- The name, quantity per tablet, and "% Daily Value" for all nutrients listed; nutrients without a Daily Value may be listed below.
- All ingredients must be listed on the label, but not necessarily in the ingredient list nor in descending order of predominance; ingredients named in the nutrition panel need not be repeated here.
- Name and address of manufacturer

DSHEA also defined the term **dietary supplement** (see Table C7-2), but the definition is too broad to be meaningful.

The stated intent of DSHEA was to enable consumers to make informed choices about nutrient supplements without too much government interference. Its effect, however, was to deregulate the supplement industry. In 1994, when DSHEA was passed, about 4,000 supplements, mostly vitamins and minerals, were on the market. Today, over 29,000 supplements are available for

sale.[8] More than 1,000 new products enter the market every year. Nutrients, herbs, amino acids, dried organ tissues of animals, microorganisms, and even hormones are sold for human consumption freely over-the-counter. (Consult the glossary at the back of the book for descriptions of many other supplement products.)

Unable to require credible scientific evidence to support supplement safety prior to marketing, the FDA has identified hazardous supplements only after

TABLE C7·2 DIETARY SUPPLEMENT TERMS

- **aristolochic acid** a Chinese herb ingredient known to attack the kidneys and to cause cancer; U.S. consumers have required kidney transplants and must take lifelong anti-rejection medication after use. Banned by the FDA but available in supplements sold on the Internet.
- **coenzyme Q-10** an enzyme made by cells and important for its role in energy metabolism. With diminished coenzyme Q-10 function, oxidative stress increases, as may occur in aging. Preliminary research suggests that it may be of value for treating certain conditions; toxicity in animals appears to be low. No safe intake levels for human beings have been established.
- **DHEA**[a] a hormone secretion of the adrenal gland whose level falls with advancing age. DHEA may protect antioxidant nutrients. Real DHEA is available only by prescription; the herbal DHEA imitator for sale in health-food stores is not active in the body. No safety information exists for either.
- **dietary supplement** a product, other than tobacco, that is added to the diet and contains one of the following ingredients: a vitamin, mineral, herb, botanical (plant extract), amino acid, metabolite, constituent, or extract, or a combination of any of these ingredients.
- **ephedrine** one of a group of compounds with dangerous amphetamine-like stimulant effects: extracted from the herb ma huang and recently banned by the FDA, but still available from Internet sources. The most severe reported side effects of ephedrine include heart attack, stroke, and sudden death.
- **garlic oil** an extract of garlic; may or may not contain the chemicals associated with garlic; claims for health benefits unproved.
- **green pills, fruit pills** pills containing dehydrated, crushed vegetable of fruit matter. An advertisement may claim that each pill equals a *pound* of fresh produce, but in reality a pill may equal one small forkful—minus nutrient losses incurred in processing.
- **kelp tablets** tablets made from dehydrated kelp, a kind of seaweed used by the Japanese as a foodstuff.
- **ma huang** an evergreen plant that supposedly boosts energy and helps with weight control. Ma huang, also called *ephedra,* contains ephedrine (see above), and is especially dangerous in combination with kola nut or other caffeine-containing substances.
- **melatonin** a hormone of the pineal gland believed to help regulate the body's daily rhythms, to reverse the effects of jet lag, and to promote sleep. Claims for life extension or enhancement of sexual prowess are without merit.
- **nutritional yeast** a preparation of yeast cells, often praised for its high nutrient content. Yeast is a source of B vitamins as are many other foods. Also called *brewer's yeast;* not the yeast used in baking.
- **organ and glandular extracts** dried or extracted material from brain, adrenal, pituitary, or other glands or tissues providing few nutrients but posing a theoretical risk of "mad cow disease." See Chapter 12.
- **SAM-e** an amino acid derivative that may have an antidepressant effect on the brain in some people, but is not recommended as a substitute for standard antidepressant therapy.
- thousands of others.

[a]Dehydroepiandrosterone.

NOTE: According to legal definitions, all of the substances listed qualify as dietary supplements, even though some appear to have the effects of drugs, not nutrients. Table 11-7 on page 415 describes many more medicinal herbs, including their effects and their hazards.

of mislabeled or adulterated products. These efforts address some problems in the supplement industry, but they do not eliminate all potential dangers to consumers. Additionally, consumers face a confusing array of label claims made for the health effects of supplements.

Structure/Function Claims

A consumer attempting to select a nutrient or other supplement is likely to encounter claims on labels resembling the FDA's most reliable, Grade A *health* claims (described in Chapter 2's Consumer Corner). The claims found on supplement labels, however, are likely to be the less regulated structure/function claims. Before printing a Grade A health claim on a label, a food manufacturer must submit scientific evidence and petition the FDA for permission in advance, at much effort and expense. If the same manufacturer elects to print a similar-appearing structure/function claim instead, it need not obtain prior approval and must comply with only a few regulations, most importantly the requirement to include a disclaimer stating that the FDA has not evaluated the claim.[10] Our consumer, however, might reasonably assume that the claims are identical:

TABLE C7·3 PROBLEMS IMPEDING THE FDA'S OVERSIGHT OF SUPPLEMENTS

The FDA currently faces four major problems with regard to supplement oversight strategies:

1. The volume of supplement sales and use is too high to ensure adequate post-marketing safety within the current legal structure.
2. The FDA's funding and staff are insufficient to allow identification of supplement hazards and implementation of appropriate actions against them.
3. The broad array of ingredients in the products sold impedes identification of hazards.
4. Failure to require manufacturers to report serious adverse effects associated with supplement use undermines FDA efforts to ensure supplement safety.

SOURCE: Committee on the Framework for Evaluating the Safety of Dietary Supplements, Institute of Medicine and National Research Council, *Dietary Supplements: A Framework for Evaluating Safety* (Washington, D.C.: National Academies Press, (2004), pp. ES1–ES14.

actual harm has come to consumers. In addition to ephedrine, the dangerous weight-loss supplement mentioned earlier, the FDA is currently taking action against usnic acid supplements now known to cause liver damage in dieters; kava, an herb sold to improve mood, is also under FDA scrutiny for causing liver failure; others are also being investigated.[9]

The FDA is currently striving to improve its ability to take action on a suspected supplement hazard before serious illnesses or deaths occur, but it still faces the problems listed in Table C7-3. A new plan for a science-based approach to protecting consumers from unsafe supplements is under development. In the first half of 2004, the FDA sent warning letters to over a hundred U.S. makers of dietary supplements, refused entry of over a thousand foreign supplement shipments, and supervised destruction of almost $20 million worth

- *"Lowers cholesterol."*
- *"Helps maintain normal cholesterol levels."*

Yet the first requires full FDA process and approval before printing—it is a Grade A health claim. The second is a structure/function claim, and so its accuracy is uncertain.

What, then, is a consumer to do? To make a rational choice concerning whether to take a supplement, the consumer must weigh potential benefits against the potential for harm. Doing so requires studying all available scientific evidence about the supplement ingredients, while ignoring marketing claims. For most supplements, unfortunately, no or few scientific studies exist. Others have been well researched, including the antioxidant vitamins often sold with promises of disease prevention. The next sections present the known evidence both for and against the theory that taking these nutrients may enhance the health of the body.

Arguments in Favor of Supplements for Chronic Disease Prevention

There is considerable scientific interest in identifying which nutrient deficiencies may be linked with diseases other than classic deficiency syndromes. The idea that **subclinical,** or **marginal, deficiencies** may compound risks for chronic diseases is currently fostering much research. In exploring these links, scientists are asking two questions important in nutrition:

- Do subclinical deficiencies of nutrients increase the risks for chronic diseases, and can supplements reverse these risks?
- Do well-nourished people gain extra protection against chronic diseases by taking supplements of nutrients?

Marginal Deficiencies and Oxidative Stress

Subclinical deficiencies present no or only subtle symptoms easily overlooked by physicians during regular examinations. For example, persons consuming a diet lacking in vitamin C silently incur an increase in **oxidative stress** (terms are in Table C7-4) in the tissues long before the symptoms of scurvy appear, but no symptoms warn the individual or alert health-care providers.[11] People lacking vitamin C in the diet often suffer greater rates of cancer and other diseases possibly due to increased oxidative stress, but perhaps due to other causes.[12] A diet low in vitamin C probably also lacks other nutrients, making it difficult to determine exactly which nutrient deficiency may be causing the problem.

The link between suboptimal intakes of **antioxidant nutrients** and chronic diseases follows from a theory linking chronic diseases, such as heart disease, with oxidative stress in the body. As body cells use oxygen to produce energy, they produce free radicals (highly unstable molecules of oxygen). Radiation, pollution, tobacco smoke, and other environmental factors can also cause free-radical formation. When

free-radical activity in the body exceeds antioxidant defenses, the theory goes, a destructive chain reaction of oxidation increases oxidative stress in the tissues, triggering the onset of diseases.

Today, the world's laboratories have put forth an extensive, but largely inconclusive, body of evidence on antioxidant nutrients, oxidative stress, and chronic diseases. Clearly, diets rich in fruits, vegetables, and whole grains with their abundant antioxidant nutrients, such as vitamin C, vitamin E, beta-carotene and other carotenoids, and the mineral selenium, provide benefits. Those eating such diets have low disease rates. Additionally, people whose blood tests below normal for these nutrients often have elevated rates of chronic diseases, and restoring these nutrients with foods or supplements reliably produces a drop in chronic disease incidence.[13] Such research results support an affirmative answer to the first question asked above: people whose blood tests reveal low concentrations of antioxidant nutrients benefit in terms of reduced chronic disease risks when the missing nutrients are restored through either food or supplements.

Marginal Deficiencies, Cancer, and Other Diseases

In agreement is a leading geneticist who observes that subclinical deficiencies of vitamin B_{12}, vitamin B_6, folate, vitamin C, vitamin E, or the minerals zinc or iron, common among low-income groups in the United States, appear to cause extensive breaks in DNA strands. The DNA damage from marginal deficiencies is said to mimic the well-known cancer-causing damage from radiation.[14] Not only do more breaks in DNA occur when nutrients are lacking, but they are more slowly repaired than damage arising in well-nourished cells. Consequently, metabolic damage to the cell builds up in ways that could cause disease.

Many forms of cancer, which can arise from breaks in DNA, occur more often among poorly nourished low-income populations. The hope is that by reversing slight nutrient deficits with balanced supplements in this group, the excess burden of cancer, among other

TABLE C7-4 ANTIOXIDANT TERMS

- **antioxidant nutrients** vitamins and minerals that oppose the effects of oxidants on human physical functions. The antioxidant vitamins are vitamin E, vitamin C, and beta-carotene. The mineral selenium also participates in antioxidant activities.
- **electrons** parts of an atom; negatively charged particles. Stable atoms (and molecules, which are made of atoms) have even numbers of electrons in pairs. An atom or molecule with an unpaired electron is an unstable *free radical.*
- **oxidants** compounds (such as oxygen itself) that oxidize other compounds. Compounds that prevent oxidation are called *anti*oxidants, whereas those that promote it are called *pro*oxidants (anti means "against"; pro means "for").
- **oxidative stress** damage inflicted on living systems by free radicals.
- **subclinical,** or **marginal, deficiency** a nutrient deficiency that has no outward clinical symptoms. The term is often used to market unneeded nutrient supplements to consumers.

diseases, could be partially relieved. Other groups, such as menstruating women, young children, and the elderly, may suffer from the effects of low vitamin and mineral status in many other ways. Presumably, resolving their slight deficiencies might relieve a variety of illnesses, from minor complaints to premature loss of cognition in aging.[15]

Arguments against Supplements for Chronic Disease Prevention

While no one doubts the benefit from nutrient supplements for preventing deficiencies in people with suboptimal intakes, the second question posed earlier with regard to potential benefits of supplements in well-nourished people arouses considerable controversy. The argument concerns some theoretical benefits of supplements with regard to the prevention of chronic diseases. The scientists recommending supplements believe strongly that vitamins and certain minerals in amounts greater than those available from food are very likely to protect against chronic diseases, particularly heart disease, in ways explained in the following sections.[16] If the risks are low, they seem to imply, why not take supplements now to be ahead of the game in case research one day reveals a benefit?

More conservative voices point out critical gaps in the current body of evidence. Studies of the health and nutrient intakes of populations support the theory that higher intakes of vitamins and minerals correlate with lower rates of diseases—people with high intakes of fruits and vegetables that supply the antioxidant nutrients generally enjoy better health than others. Also supporting the theory are results from cell and animal studies that demonstrate the physiology of the nutrients in disease prevention—they lend biological plausibility to the argument.

Mixed or negative results, however, emerge from the gold standard for evaluating the effectiveness of a medical treatment for human beings—the controlled clinical human trial.[17] The general finding of five years of research is that routinely providing supplements of

nutrients to well-nourished populations produces few, if any, observed benefits. Death rates from heart disease and other chronic diseases remain unchanged whether or not well-nourished people are given supplements.[18] One exception may involve the healthy elderly: two preliminary studies have revealed a slight reduction in the advance of common causes of vision loss in those taking a combined antioxidant supplement, and these findings warrant further investigation.[19] The following paragraphs illustrate the problems surrounding research on the potential effects of two antioxidant nutrients, beta-carotene and vitamin E, with regard to chronic disease.

Beta-Carotene and Cancer: A Disappointing Story

Low rates of many cancers are reported in people whose diets include abundant vegetables and fruits, the richest sources of antioxidants.[20] In particular, intakes of vegetables and fruits rich in beta-carotene are linked with low rates of cancers (see Figure C7-2).[21] For a time, all evidence seemed to point to beta-carotene as "the" anticancer substance in these foods, and consumers eagerly bought and took beta-carotene supplements. In a sudden reversal, support for

beta-carotene crumbled as studies of supplements of beta-carotene revealed no benefit and a possibility of serious harm.[22] Major clinical trials of beta-carotene supplements were immediately ceased upon finding a 28 percent *increase* in lung cancer among smokers taking beta-carotene compared with placebos.

Were the original findings wrong? No. Intake of fruits and vegetables is undoubtedly linked with lower cancer risks, but beta-carotene itself is clearly not the agent of this effect. Fruits and vegetables contain hundreds of potentially beneficial phytochemicals, such as the other carotenoids described in Controversy 2. Such foods are also rich in fibers, and they are generally low in calories and so may assist in weight control—diet characteristics linked with low cancer rates. In truth, perhaps the difference in cancer development should be attributed to the entire *diet* chosen by eaters of fruits and vegetables or even to their entire lifestyle. People choosing foods they perceive to be good for them probably also do not smoke, may exercise regularly, and generally take better care of themselves than other people do. In other words, the beta-carotene story provides an illustration of why population evidence alone

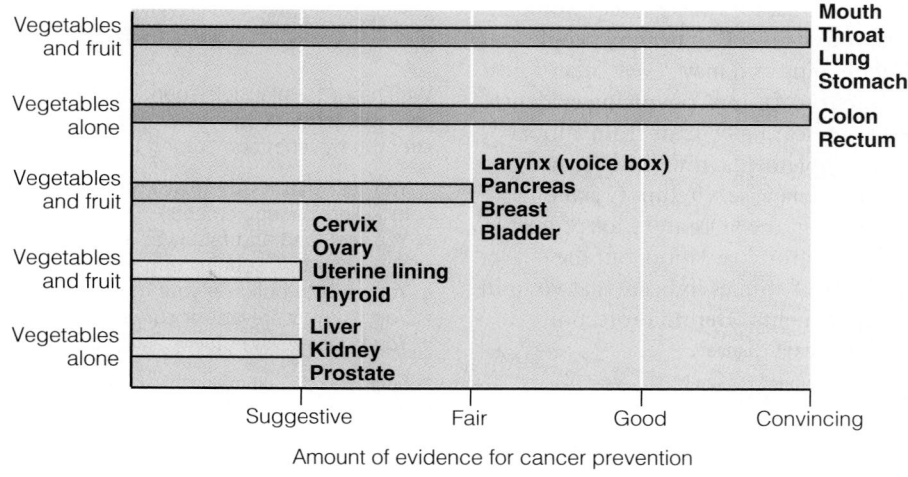

FIGURE C7·2 VEGETABLE AND FRUIT INTAKES AND CANCER IN POPULATION STUDIES

Groups of people with high fruit and vegetable intakes often have low rates of cancer. Well-controlled laboratory and clinical studies are needed to verify these findings.

is insufficient to establish causal relationships. The entire picture emerges only when controlled clinical trials confirm population findings. In the end, supplements of beta-carotene provide no benefits and may cause harm to certain people, especially smokers.

Vitamin E and Heart Disease: Unconvincing Evidence

High blood cholesterol carried in low-density lipoproteins (LDL) is a major risk factor for cardiovascular disease, but how does LDL exert its damage? One scenario is that free radicals within the blood vessel walls oxidize LDL, changing its structure and activities. The oxidized LDL then speeds the formation of artery-clogging plaques that impede the flow of blood to the heart muscle. Factors such as a diet high in saturated fat and toxins from tobacco smoke heighten this oxidative damage within the artery walls. In contrast, a diet high in fruits and vegetables, especially when fat is kept low, strengthens antioxidant defenses against LDL oxidation.[23] The question, then, is do antioxidant nutrients taken as supplements also slow the onset of heart disease?

Population studies demonstrate that people who eat foods rich in vitamin E have low rates of death from heart disease.[24] In the laboratory, vitamin E has been shown to defend against LDL oxidation, inflammation, arterial injuries, and blood clotting.[25] Yet, despite the many talents of vitamin E, supplements of the vitamin have generally proved useless against cardiovascular disease in people at risk for the disease.[26] Some trials of vitamin E in healthy people are still ongoing and may reveal other findings, but so far no convincing evidence exists indicating benefits from antioxidant supplements in well-nourished people.* Likewise, vitamin C exhibits potential in the laboratory for protecting lipids from oxidation, but the majority of studies indicate that vitamin C supplements offer no protection against heart disease.[27]

---

*Results from the Physicians Health Study are expected in 2007.

Should We Take Supplements To Ensure Adequate Nutrients?

A scientific approach clearly identifies foods, not supplements, as overwhelmingly the best source of nutrients. Opponents of supplements energetically oppose them on the grounds that side effects, which may not be immediately evident, might endanger supplement takers' health in these ways:

- High doses of vitamin E may interfere with vitamin K functions, delay blood clotting, increase the risk of brain hemorrhage (a form of stroke) over time, or even increase the risk of death from all causes.[28]
- Vitamin C supplements enhance iron absorption, making iron overload likely in some people.
- Vitamin C supplements in high doses *increase* markers of oxidation in the blood.
- High doses of vitamin C taken by women who have diabetes may increase their likelihood of dying of cardiovascular disease.[29]
- Biotin-supplemented cell cultures suffer DNA damage of a type related to cancer formation.[30]
- Daily supplements of beta-carotene may increase lung cancer in smokers.
- Supplements of vitamin D, vitamin A, and many minerals can be toxic in large doses.

Besides, while an orange and a pill may both contain vitamin C, the orange presents a balanced array of chemicals and fiber that modulate vitamin C's effects. The pill provides only vitamin C, a lone chemical.

Still, sometimes supplements are appropriate and necessary to health. For those who need supplements, the next section provides some guidance to making a selection.

Selection of a Multinutrient Supplement

If you fall into one of the categories listed earlier in Table C7-1 (p. 255) and if you absolutely cannot meet your nutrient needs from foods, a supplement containing *nutrients only* can prevent serious problems. In these cases, the benefits probably outweigh the risks.

Which supplement to choose? The first step is to remain aware that sales of vitamin supplements often approach the realm of quackery because the profits are high and the industry is largely free of oversight. To escape the clutches of the health hustlers, use your imagination and delete the picture on the label of sexy people on the beach and the meaningless, glittering generalities stating "Advanced formula," "Maximum power," and the like. Don't be misled into buying and taking unneeded supplements (Table C7-5 provides some invalid reasons for taking supplements).

If you see a USP symbol on the label, it means that a manufacturer has voluntarily paid an independent laboratory to

---

TABLE C7·5 SOME INVALID REASONS FOR TAKING SUPPLEMENTS

Watch out for plausible-sounding, but false reasons given by marketers trying to convince you, the consumer, that you need supplements. The invalid reasons listed below have gained strength by repetition among friends, on the Internet, and by the media:

- You fear that foods grown on today's soils lack nutrients (a common false statement made by sellers of supplements).
- You feel tired and falsely believe that supplements can provide energy.
- You hope that supplements can help you cope with stress.
- You wish to build up your muscles faster or without physical exercise.
- You want to prevent or cure self-diagnosed illnesses.
- You hope excess nutrients will produce unnamed mysterious beneficial reactions in your body.

People who should never take supplements without a physician's approval include those with kidney or liver ailments (they are susceptible to toxicities), those taking medications (nutrients can interfere with their actions), and smokers (who should avoid products with beta-carotene).

This symbol means that a supplement contains the nutrients stated and that it will dissolve in the system—the symbol does not guarantee safety or health advantages.

test the product and affirm that it contains the ingredients listed and that it will dissolve or disintegrate in the digestive tract to make the ingredients available for absorption. The symbol does not imply that the supplement has been tested for safety or effectiveness with regard to health, however.

Now all you have left is the list of ingredients, the form they are in, and the price—the plain facts. You have two basic questions to answer. The first question: What form do you want—chewable, liquid, or pills? If you'd rather drink your vitamins and minerals than chew them, fine. If you choose a fortified liquid meal replacer or "energy bar" (a candy bar to which vitamins and other nutrients are added), you must then proportionately reduce the calories you consume as food, or you may gain unwanted weight. If you choose chewable pills, be aware that vitamin C can erode tooth enamel. Swallow promptly and flush the teeth with a drink of water. Avoid vitamin-fortified bubble gum to protect both the teeth and gum-loving children who may chew a whole box and receive too large a dose for their small bodies.

The second question: Who are you? What vitamins and minerals do you need? The DRI nutrient intake recommendations listed in the tables on the inside front cover are the standards appropriate for all reasonably healthy people.

For most people, an appropriate supplement provides all the vitamins and

minerals in amounts smaller than, equal to, or very close to the intake recommendations. For those who require a higher dose, such as young women who need supplemental folate in the child-bearing years, choose a supplement with just the needed nutrient or in combination with a reasonable dose of others. Avoid any preparation that, in a daily dose, provides more than the DRI recommended intake of vitamin A, vitamin D, or any mineral or more than the Tolerable Upper Intake Level for any nutrient. Warning: Expect to reject about 80 percent of available preparations when you choose according to these criteria; be choosy where your health is concerned.

In addition, avoid these:

- High doses of iron (more than 10 milligrams per day) except for menstruating women. People who menstruate need more iron, but people who don't, don't.
- "For low-carb diets." Preparations containing extra biotin are claimed to better metabolize the excess protein in such diets, but no evidence supports these claims.
- "Organic" or "natural" preparations with added substances. They are no better than standard types, but they cost much more.
- "High-potency" or "therapeutic dose" supplements. More is not better.
- Items not needed in human nutrition, such as carnitine and inositol. These particular items won't harm you, but they reveal a marketing strategy that makes the whole mix suspect. The manufacturer wants you to believe that its pills contain the latest "new" nutrient that other brands omit, but, in fact, for every valid discovery of this kind, there are 999,999 frauds.
- "Time release." Medications such as some antibiotics or pain relievers often must be sustained at a steady concentration in the blood to be effective, but nutrients are incorporated into the tissues where they are needed whenever they arrive.
- "Stress formulas." Although the stress response depends on certain B vitamins and vitamin C, the recommended amount provides all that is

needed of these nutrients. If you are under stress (and who isn't?), generous servings of fruits and vegetables will more than cover your need.

- Pills containing extracts of parsley, alfalfa, and other vegetable components.
- Geriatric "tonics." They are generally poor in vitamins and minerals and yet may be so high in alcohol as to threaten inebriation.
- Any supplement sold with claims that today's foods lack the nutrients they once contained. Plants make vitamins for their own needs, not ours. A plant lacking a mineral or failing to make a needed vitamin dies before it can bear food for our consumption.

As for price, be aware that local or store brands may be just as good or better than nationally advertised brands.[31] If they are less expensive, it may be because the price does not have to cover the cost of national advertising. To get the most from a supplement of vitamins and minerals take it with food. A full stomach retains and dissolves the pill with its churning action. Also, if you take an iron supplement, choose foods that will assist in its absorption, such as meats, fish, poultry, or foods containing vitamin C.

Choosing Doses

When people choose supplements, they choose doses of nutrients that can vary fivefold or more, a fact revealed by a comparison of the back panels of the products. The higher the dose, the greater the risk of toxicity, of course, and studies bear out the logic that supplement users can teeter on the edge of toxicity for nutrients.[32] Although tolerance for high doses of nutrients varies, and no one can with certainty say what level may be safe for an individual person, some guidance is offered in the DRI Tolerable Upper Intake Levels (see inside front cover, page C).

Even more worrisome than the possibility of short-term, acute overdoses is the potential for chronic, low-level nutrient toxicity whose subtle effects develop slowly and go unrecognized. Because of the potential hazards that supplements may present, some authorities believe supplements and highly fortified foods, particularly those high in vitamin A or vitamin D, should be required to bear warning labels, but such labels have not been seriously considered.

One final problem with supplements is that they may lull their takers into a false sense of security. A person may eat irresponsibly, thinking, "My supplement will cover my needs." More often, supplements supply precisely the nutrients people need least—those they consume in food—while failing to provide those missing from the diet. Further, if a person's diet is inadequate in vitamins, it probably lacks minerals, phytochemicals, and fiber as well.

Conclusion

People in developed nations are far more likely to suffer from *overnutrition* and poor lifestyle choices than from nutrient deficiencies. People wish that swallowing vitamin pills would boost their health. The truth—that they need to improve their eating and exercise habits—is harder to swallow.

Don't waste time and money trying to single out a few nutrients to take as supplements. Invest energy in eating a wide variety of fruits and vegetables in generous quantities, along with the recommended daily amounts of whole grains, lean meats, and milk products every day, and take supplements only when they are truly needed.

Water and Minerals

CHAPTER 8

© Kuzma Petrov-Vodkin (1878–1930 Russian), *Still Life in Pink, Branch of an Apple Tree.* Tretyakov Gallery, Moscow, Russia/SuperStock, Inc.

CONTENTS

Do You Ever . . .

Buy bottled water because you think it is safer than tap water?

Blame "water weight" when you've gained a few pounds?

Skip milk products, believing that you no longer need the nutrients they supply?

Feel tired and wonder if you need an iron supplement?

KEEP READING.

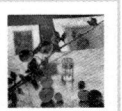

"ASHES TO ASHES and dust to dust"—it is true that when the life force leaves the body, what is left behind becomes nothing but a small pile of ashes. Carbohydrates, proteins, fats, vitamins, and water are present at first, but they soon disappear.

The carbon atoms in all the carbohydrates, fats, proteins, and vitamins combine with oxygen to produce carbon dioxide, which vanishes into the air; the hydrogens and oxygens of those compounds unite to form water; and this water, along with the water that was a large part of the body weight, evaporates. The ashes left behind are the **minerals,** a small pile that weighs only about 5 pounds. The pile is not impressive in size, but those minerals are critical to the functioning of living tissue.

Consider calcium and phosphorus. If you could separate these two minerals from the rest of the pile, you would take away about three-fourths of the total. Crystals made of these two minerals, plus a few others, form the structure of the bones and so provide the architecture of the skeleton.

Run a magnet through the pile that remains and you pick up the iron. It doesn't fill a teaspoon, but it consists of billions and billions of iron atoms. As part of hemoglobin, these iron atoms are able to attach to oxygen and make it available at the sites inside the cells where metabolic work is taking place.

If you then extract all the other minerals from the pile of ashes, leaving only copper and iodine, close the windows first. A slight breeze would blow these remaining bits of dust away. Yet the amount of copper in the dust is necessary for iron to hold and to release oxygen, and iodine is the critical mineral in the thyroid hormones. Figure 8-1 shows the amounts of the seven **major minerals** and a few of the **trace minerals** in the human body. Other minerals such as gold and aluminum are present in the body but are not known to be nutrients.

The distinction between major and trace minerals doesn't mean that one group is more important in the body than the other. A daily deficiency of a few micrograms of iodine is just as serious as a deficiency of several hundred milligrams of calcium. The major minerals are present in larger total quantities, however, and so they influence the body fluids, which in turn affect the whole body.

minerals naturally occurring, inorganic, homogeneous substances; chemical elements.

major minerals essential mineral nutrients found in the human body in amounts larger than 5 grams.

trace minerals essential mineral nutrients found in the human body in amounts less than 5 grams.

FIGURE 8·1   MINERALS IN A 60-KILOGRAM (132-POUND) PERSON (GRAMS)

The major minerals are those present in amounts larger than 5 grams (a teaspoon). The essential trace minerals number a dozen or more: only four are shown. A pound is about 454 grams; thus, only calcium and phosphorus appear in amounts larger than a pound.

[a]Sulfur is a nonmetallic, yellow element; sulfate, a compound of sulfur and oxygen, is required by the body for making sulfur-containing molecules.
[b]Chlorine appears in the body as the chloride ion.

This chapter begins with a discussion of water. Water is unique among the nutrients—standing alone as the most indispensable of all. The body needs more water each day than any other nutrient—50 times more water than protein and 5,000 times more water than vitamin C. You can survive a deficiency of any of the other nutrients for a long time, in some cases for months or years, but you can survive only a few days without water. In less than a day, a lack of water alters the body's chemistry and metabolism.

Our discussion begins with water's many functions. Next we examine how water and the major minerals mingle to form the body's fluids and how cells regulate the distribution of those fluids. Then we take up the specialized roles of each of the minerals.

Water is the most indispensable nutrient.

Water

You began as a single cell bathed in a nourishing fluid. As you became a beautifully organized, air-breathing body of trillions of cells, each of your cells had to remain next to water to stay alive.

Water makes up about 60 percent of an adult person's weight—that's almost 80 pounds of water in a 130-pound person. All this water in the body is not simply a river coursing through the arteries, capillaries, and veins. Some of the water is incorporated into the chemical structures of compounds that form the cells, tissues, and organs of the body. For example, proteins hold water molecules within them—water that is locked in and not readily available for any other use. Water also participates actively in many chemical reactions.

Why Is Water the Most Indispensable Nutrient?

Water brings to each cell the exact ingredients the cell requires and carries away the end products of the cell's life-sustaining reactions. The water of the body fluids is thus the transport vehicle for all the nutrients and wastes. Without water, cells quickly die.

Water is nearly a universal solvent: it dissolves amino acids, glucose, minerals, and many other substances needed by the cells. Fatty substances are specially packaged with water-soluble proteins so that they too can travel freely in the blood and lymph.

Water is also the body's cleansing agent. Small molecules, such as the nitrogen wastes generated during protein metabolism, dissolve in the watery blood and must be removed before they build up to toxic concentrations. The kidneys filter these wastes from the blood and excrete them, mixed with water, as urine. When the kidneys become diseased, as can happen in diabetes and other disorders, toxins can build to life-threatening levels. A machine must then take over the task of cleansing the blood by filtering wastes into water contained in the machine.*

Water molecules resist being crowded together. Thanks to this incompressibility, water can act as a lubricant and a cushion for the joints, and it can protect sensitive tissue such as the spinal cord from shock. The fluid that fills the eye serves in a similar way to keep optimal pressure on the retina and lens. From the start of human life, a fetus is cushioned against shock by the bag of amniotic fluid in the mother's uterus. Water also lubricates the digestive tract, the respiratory tract, and all tissues that are moistened with mucus.

Yet another of water's special features is its ability to help maintain body temperature. The water of sweat is the body's coolant. Heat is produced as a by-product of energy metabolism and can build up dangerously in the body. To rid itself of this excess heat, the body routes its blood supply through the capillaries just under the skin. At the same time, the skin secretes sweat and its water evaporates. Converting water to vapor takes energy; therefore, as sweat evaporates, heat energy dissipates, cooling the skin and the underlying blood. The cooled blood then flows back to cool the body's core. Sweat evaporates continuously from the skin, usually in slight amounts that go unnoticed; thus, the

*The machine that cleanses the blood is a kidney dialysis machine.

Human life begins in water.

skin is a major organ through which water is lost from the body. Lesser amounts are lost by way of exhaled breath and the feces.[1]

To sum up, water:

- Carries nutrients throughout the body.
- Serves as the solvent for minerals, vitamins, amino acids, glucose, and other small molecules.
- Cleanses the tissues and blood of wastes.
- Actively participates in many chemical reactions.
- Acts as a lubricant around joints.
- Serves as a shock absorber inside the eyes, spinal cord, joints, and amniotic sac surrounding a fetus in the womb.
- Aids in maintaining the body's temperature.

▶ KEY POINT   Water provides the medium for transportation, acts as a solvent, participates in chemi-cal reactions, provides lubrication and shock protection, and aids in temperature regulation in the human body.

The Body's Water Balance

Water is such an integral part of us that people seldom are conscious of water's impor-tance, unless they are deprived of it. Since the body loses some water every day, a per-son must consume at least the same amount to avoid life-threatening losses, that is, to maintain **water balance.**

The total amount of fluid in the body is kept balanced by delicate mechanisms. Imbalances such as **dehydration** and **water intoxication** can occur, but the balance is restored as promptly as the body can manage it. The body controls both intake and excretion to maintain water equilibrium.

The amount of the body's water varies by pounds at a time, especially in women who retain water during menstruation. Eating a meal high in salt can temporarily increase

*Reference notes are found in Appendix F.

water balance the balance between water intake and water excretion, which keeps the body's water content constant.

dehydration loss of water. The symptoms progress rapidly, from thirst to weakness to exhaustion and delirium, and end in death.

water intoxication a dangerous dilution of the body's fluids resulting from excessive inges-tion of plain water. Symptoms are headache, muscular weakness, lack of concentration, poor memory, and loss of appetite.

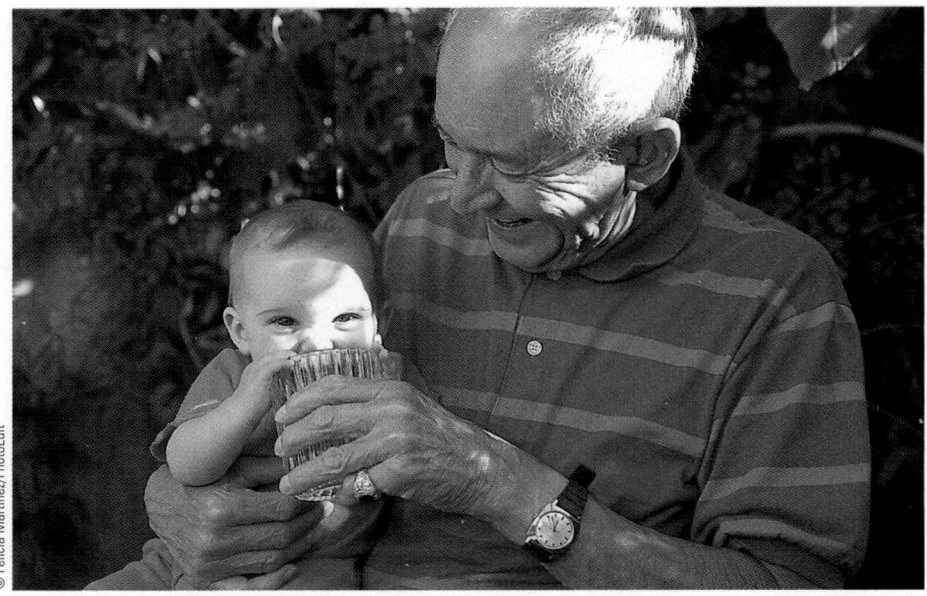

An extra drink of water benefits both young and old.

the body's water content; the body sheds the excess over the next day or so as the sodium is excreted. These temporary fluctuations in body water show up on the scale, but gaining or losing water weight does not reflect a change in body fat. Fat weight takes days or weeks to change noticeably, whereas water weight can change overnight.

▶ **KEY POINT** Water makes up about 60 percent of the body's weight. A change in the body's water content can bring about a temporary change in body weight.

Quenching Thirst and Balancing Losses

Thirst and satiety govern water intake.[2] When the blood is too concentrated (having lost water but not salt and other dissolved substances), the molecules and particles in the blood attract water out of the salivary glands, and the mouth becomes dry. The brain center known as the hypothalamus (described in Chapter 3) plays the major role in monitoring the concentration of the blood. When the blood is too concentrated, or when the blood volume or pressure is too low, the hypothalamus initiates nerve impulses to the brain that register as "thirst." The hypothalamus also signals the pituitary gland to release a hormone that directs the kidneys to shift water back into the bloodstream from the pool destined for excretion. The kidneys themselves respond to the sodium concentration in the blood passing through them and secrete regulatory substances of their own. The net result is that the more water the body needs, the less it excretes. Figure 8-2 shows how intake and excretion naturally balance out.

Thirst lags behind a lack of water. When too much water is lost from the body and is not replaced, dehydration can threaten survival. A first sign of dehydration is thirst, the signal that the body has already lost up to 2 cups of its total fluid and that the need to obtain fluid is urgent. But suppose a thirsty person is unable to obtain fluid or, as in many elderly people, fails to perceive the thirst message. With a loss of just 5 percent of body fluid, perceptible symptoms appear: headache, fatigue, confusion or forgetfulness, and an elevated heart rate. Instead of "wasting" precious water in sweat, the dehydrated body diverts most of its water into the blood vessels to maintain the life-supporting blood pressure. Meanwhile, body heat builds up because sweating has ceased, creating the possibility of serious consequences (see Table 8-1). A water deficiency that develops slowly can switch on drinking behavior in time to prevent serious dehydration, but one that develops quickly may not.

FIGURE 8·2 WATER BALANCE —A TYPICAL EXAMPLE

Each day, water enters the body in liquids and foods, and some water is created in the body as a by-product of metabolic processes. Water leaves the body through the evaporation of sweat, in the moisture of exhaled breath, in the urine, and in the feces.

Water input (Total = 1,450–2,800 ml)
foods (700–1,000 ml)
water created by metabolism (200–300 ml)
liquids (550–1,500 ml)

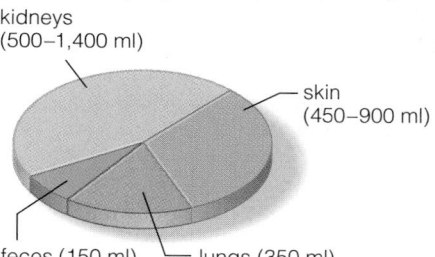

Water output (Total = 1,450–2,800 ml)
kidneys (500–1,400 ml)
skin (450–900 ml)
feces (150 ml)
lungs (350 ml)

■ A 150 lb person contains 90 lb of water; a 5% loss of body fluid for this person amounts to about 4½ lb of water.

TABLE 8·1 THREATS FROM MILD AND SEVERE DEHYDRATION AND CHRONIC LACK OF FLUID		
Mild Dehydration (Loss of <5% Body Weight)	**Severe Dehydration** (Loss of >5% Body Weight)	**Chronic Low Fluid Intake** May Increase the Likelihood of:
Thirst	Pale skin	Bladder, colon, and other cancers
Sudden weight loss	Bluish lips and fingertips	Cardiac arrest (heart attack) and other
Rough dry skin	Confusion; disorientation	heart problems
Dry mouth, throat, body linings	Rapid, shallow breathing	Gallstones
Rapid pulse	Weak, rapid, irregular pulse	Kidney stones
Low blood pressure	Thickening of blood	Urinary tract infections
Lack of energy; weakness	Shock; seizures	
Impaired kidney function	Coma; death	
Reduced quantity of urine; concentrated urine		
Decreased mental functioning		
Decreased muscular work and athletic performance		
Fever or increased internal temperature		
Fainting		

SOURCE: Standing Committee on the Scientific Evaluation of Dietary Reference Intakes, Food and Nutrition Board, Institute of Medicine, *Dietary Reference Intakes: Water, Potassium, Sodium, Chloride, and Sulfate* (Washington, D.C.: National Academies Press, 2004): pp. 4-31–4-48.

■ For more about water intoxication, see Chapter 10.

To ignore the thirst signal is to invite dehydration. People should stay attuned to thirst and drink whenever they feel thirsty to replace fluids lost throughout the day.[3] Older adults in whom thirst is blunted should drink regularly throughout the day, regardless of thirst.

At the other extreme from dehydration, water intoxication occurs when too much plain water floods the body's fluids and disturbs their normal composition. Most victims have consumed several gallons of plain water in a few hours' time.[4] Water intoxication is rare, but when it occurs, immediate action is needed to reverse dangerously diluted blood before death ensues.

▶ **KEY POINT** Water losses from the body necessitate intake equal to output to maintain balance. The brain regulates water intake; the brain and kidneys regulate water excretion. Dehydration and water intoxication can have serious consequences.

How Much Water Do I Need to Drink in a Day?

Water needs vary greatly depending on the foods a person eats, the environmental temperature and humidity, the altitude, the person's activity level, and other factors (see Table 8-2). Because individual needs vary, adequate hydration can be maintained over a wide range of fluid intakes. As a general guideline, the DRI committee recommends that, given a normal diet and moderate environmental conditions, men need about 13 cups of fluid from beverages and drinking water, and women need about 9 cups. This amount of fluid provides about 80 percent of the day's need for water. Most of the rest is provided by the water consumed in foods. Nearly all foods contain some water: water constitutes up to 95 percent of the volume of most fruits and vegetables, and at least 50 percent of many meats and cheeses (see the margin). A small percentage of the day's fluid is generated in the tissues as energy-yielding nutrients in foods release water as a product of chemical breakdown.

Sweating increases water needs. Especially when performing physical work outdoors in hot weather, people can lose two to four *gallons* of fluid in a day. An athlete training in the heat can sweat out more than a half gallon of fluid each hour.[5] The importance of maintaining hydration for athletes exercising in the heat cannot be overemphasized, and Chapter 10 provides detailed instructions on how, exactly, to hydrate the exercising body.

When people who normally abstain from caffeine drink a caffeine-containing beverage such as coffee, tea, or soda, their urine output increases somewhat more than it would for a similar amount of plain water. This is because caffeine acts as a **diuretic.** Research is mixed on whether any but the highest caffeine intakes, say, four or five cups of coffee, cause a net water deficit in the body; most people make up for small water losses by drinking additional fluid later. Also, people who habitually consume caffeine may adapt to its diuretic effects, losing no more fluid than when drinking other beverages.[6] Therefore, for most people, an occasional caffeinated beverage can contribute to the day's fluid requirement.[7] The Controversy section of Chapter 14 comes back to the effects of caffeine.

▶ **KEY POINT** Many factors influence a person's need for water. The water of beverages and foods meets nearly all of the need for water, and a little more is supplied by the water formed during cellular breakdown of energy nutrients.

Are Some Kinds of Water Better for My Health Than Others?

Water occurs as **hard water** or **soft water,** a distinction that affects your health with regard to three minerals. Hard water has high concentrations of calcium and magnesium. Soft water's principal mineral is sodium. In practical terms, soft water makes

Water content of various foods and beverages:

- 100% = water, diet soft drinks, seltzer (unflavored), plain tea.
- 95–99% = sugar-free gelatin dessert, clear broth, Chinese cabbage, celery, cucumber, lettuce, summer squash, decaffeinated black coffee.
- 90–94% = Gatorade, grapefruit, fresh strawberries, broccoli, tomato.
- 80–89% = sugar-sweetened soft drinks, milk, yogurt, egg white, fruit juices, low-fat cottage cheese, fresh apple, carrot.
- 60–79% = low-calorie mayonnaise, instant pudding, banana, shrimp, lean steak, pork chop, baked potato.
- 40–59% = diet margarine, sausage, chicken, macaroni and cheese.
- 20–39% = bread, cake, cheddar cheese, bagel, cooked oatmeal.
- 10–19% = butter, margarine, regular mayonnaise, cooked rice.
- 5–9% = peanut butter, popcorn.
- 1–4% = ready-to-eat cereals, pretzels.
- 0% = cooking oils, meat fats, shortening, white sugar.

The Table of Food Composition, Appendix A, lists the water contents of most other foods and beverages.

diuretic (dye-you-RET-ic) a compound, usually a medication, causing increased urinary water excretion; a "water pill."

hard water water with high calcium and magnesium concentrations.

soft water water with a high sodium concentration.

more bubbles with less soap; hard water leaves a ring on the tub, a jumble of rocklike crystals in the teakettle, and a gray residue in the wash.

Soft water may seem more desirable, and some homeowners purchase water softeners that remove magnesium and calcium and replace them with sodium. Some evidence suggests, however, that soft water, even when it bubbles naturally from the ground, may aggravate hypertension and heart disease.[8] Mineral-rich hard water may oppose these conditions by virtue of its calcium content.[9]

Soft water also more easily dissolves certain contaminant metals, such as cadmium and lead, from pipes. Cadmium can harm the body, affecting enzymes by displacing zinc from its normal sites of action. Cadmium is also suspected of promoting bone fractures, kidney problems, and hypertension.[10] Lead is another toxic metal, and the body seems to absorb it more readily from soft water than from hard water, possibly because the calcium in hard water protects against its absorption. Old plumbing may contain cadmium or lead. People who live in old buildings should run the cold water tap a minute to flush out harmful minerals before drawing water for the first use in the morning and whenever no water has been drawn during the previous six hours.

▶ KEY POINT Hard water is high in calcium and magnesium. Soft water is high in sodium, and it dissolves cadmium and lead from pipes.

TABLE 8-2 FACTORS THAT INCREASE FLUID NEEDS

These conditions increase a person's need for fluids:
- Alcohol consumption
- Cold weather
- Dietary fiber
- Diseases that disturb water balance, such as diabetes and kidney diseases
- Forced-air environments, such as airplanes and sealed buildings
- Heated environments
- High altitude
- Hot weather, high humidity
- Increased protein, salt, or sugar intakes
- Ketosis
- Medications (diuretics)
- Physical activity (see Chapter 10)
- Pregnancy and breastfeeding (see Chapter 13)
- Prolonged diarrhea, vomiting, or fever
- Surgery, blood loss, or burns
- Very young or old age

■ Lead poisoning is especially harmful to children (see Chapter 14).

Safety and Sources of Drinking Water

Remember that water is practically a universal solvent: it dissolves almost anything it encounters to some degree. Hundreds of contaminants—including disease-causing bacteria and viruses from human wastes, toxic pollutants from highway fuel runoff, spills and heavy metals from industry, organic chemicals such as pesticides from agriculture, and manure bacteria from farm animals—have been detected in public drinking water.

Public water systems remove some hazards; treatment includes the addition of a disinfectant (usually chlorine) to kill most microorganisms. Private well water is usually not chlorinated, so the 40 million Americans who drink water from private wells are likely to encounter microorganisms, mostly harmless, in their water.

All public drinking water must be tested regularly for contamination, and the Environmental Protection Agency (EPA) is responsible for ensuring that public water systems meet minimum standards for protection of public health. Public utilities are required to provide their customers with a yearly statement, written in plain language, that names the chemicals and bacteria found in local water. This document makes fascinating reading for those interested in the purity of their tap water.

The law also requires a utility to notify the public within 24 hours of discovering any dangerous contaminants in drinking water. The intent is to reduce the threat from such harmful contaminants as *Cryptosporidium*, a chlorine-resistant parasite common in lakes and rivers. Several years ago, *Cryptosporidium* invaded the public water supply of Milwaukee, Wisconsin, and caused 400,000 people to fall ill. Some even died. Since that time, awareness of the threat from *Cryptosporidium* has prevented all but a few outbreaks of illness from this organism, and no more lives have been lost.[11] Nevertheless, the incident stands as testimony that even our sophisticated water systems cannot always guarantee 100 percent safety, especially during floods and other natural disasters, excessive runoff, chemical spills, or intentional tampering.[*]

Some people fear that chlorine itself presents a danger to health. Large doses of byproducts of water chlorination have been found to cause cancer-related changes in

---

[*]Concerned consumers can call the Safe Drinking Water Hotline toll-free at 1-800-426-4791, ask experts water safety questions by e-mail at hotline-sdwa@epamail.epa.gov, or visit the EPA's Drinking Water Homepage at www.epa.gov.

If electric service goes out for any length of time in your community, the water supply may become contaminated with bacteria, making water purification at home necessary. The Centers for Disease Control and Prevention recommend using bottled water during these times. If bottled water is unavailable, or, if you are not sure that your water is safe, follow these directions:

- Boil the available water vigorously for one to three minutes.
- If you cannot boil it, add 8 drops of clean household bleach (regular strength) per gallon of water; stir it well and let it sit for 30 minutes before using it.
- You can also purchase purifying tablets from your pharmacy to keep with other emergency supplies.

More information is available at the EPA website: www.epa.gov/ogwdw000/faq/emerg.html.

TABLE 8·3 WATER SOURCES

- **aquifers** underground rock formations containing water that can be drawn to the surface for use.
- **ground water** water that comes from underground aquifers.
- **surface water** water that comes from lakes, rivers, and reservoirs.

bottled water drinking water sold in bottles.

human cells and cancer in laboratory animals.[12] People consuming large amounts of chlorinated tap water a day have been reported to be somewhat more likely to develop colon, brain, and other cancers.[13] Conversely, men who take in 8 cups of water from any source, chlorinated or not, have been found to be half as likely to develop bladder cancer as men who restrict water intake to less than a cupful.[14] Although most investigators acknowledge the possibility of a connection between consumption of chlorinated drinking water and cancer incidence, they also passionately defend chlorination as a benefit to public health. In areas of the world without chlorination, an estimated 25,000 people die *each day* from diseases caused by organisms carried by water and easily killed by chlorine. Substitutes for chlorine exist, but they are too expensive or too slow to be practical for treating a city's water, and some may create by-products of their own.

Meanwhile, what is a consumer to drink? One option is to drink tap water because municipal water is held to minimum standards for purity. Another option is to further purify tap water with home purifying equipment, which ranges in price from about $20 to $5,000. Some home systems do an adequate job of removing lead, chlorine, and other contaminants, but others only improve the water's taste. Many are not designed to remove microorganisms that are not affected by chlorine. Each system has advantages and drawbacks, and all require periodic maintenance or filter replacements that vary in price. Not all companies or representatives are legitimate—some perform water tests that yield dramatic-appearing but meaningless results to sell unneeded systems. Verify all claims of contamination by checking reports from local municipal water agencies or by independently testing well water before buying any purifying system.

A third option is to use **bottled water.** Many people turn to bottled water as an alternative to tap water. Read the Consumer Corner for more about bottled water. Whether water comes from the tap or is poured from a bottle, all water comes from the same sources—**surface water** and **ground water** (see Table 8-3).

Surface water flowing from lakes, rivers, and reservoirs fills about half of the nation's need for drinking water, mostly in major cities. Surface water is exposed to contamination by acid rain, petroleum products, pesticides, fertilizer, human and animal wastes, and industrial wastes that run directly from pavements, septic tanks, farmlands, and industrial areas into streams that feed surface water bodies. Surface water generally moves faster than ground water and stays aboveground where aeration and exposure to sunlight can cleanse it. The plants and microorganisms that live in surface water also filter it. These processes can remove some contaminants, but others stay in the water.

Ground water comes from protected **aquifers,** deep underground rock formations saturated with water. People in rural areas rely mostly on ground water pumped from private wells, and some cities tap this resource, too. Ground water can become contaminated from hazardous waste sites, dumps, oil and gas pipelines, and landfills, as well as downward seepage from surface water bodies. Ground water moves slowly and is not aerated or exposed to sunlight, so contaminants break down more slowly than in surface water. To mingle with water in the aquifer, surface water must first "percolate," or seep, through soil, sand, or rock, which filters out some contaminants.

Given water's importance in the body, the world's supply of clean wholesome water is a precious resource to be guarded. The remainder of this chapter addresses other important nutrients—the minerals.

▶ KEY POINT   Public drinking water is tested and treated for safety. All drinking water originates from surface water or ground water that is vulnerable to contamination from human activities.

CONSUMER CORNER

Bottled Water

ABOUT 1 IN 15 HOUSEHOLDS uses bottled water as its main drinking water source, believing it to be safer than tap water and therefore worth its substantial price—typically 250 to 10,000 times the cost of tap water.[1] A consumer group tested bottled water, however, and disproved the notion of superior safety. Of 1,000 bottles and 103 brands tested, about a third were contaminated with bacteria, arsenic, or synthetic organic chemicals.*[2] At least a quarter of bottled water is drawn directly from the tap.

Bottled water sold in interstate commerce is regulated by the Food and Drug Administration (FDA).[3] The FDA requires yearly tests of bottled water for purity and sanitation standards, but the standards are substantially less rigorous than those applied to U.S. tap water. For example, bottled water does not have to be filtered to remove disease-causing organisms, such as *Cryptosporidium*, or tested for the presence of asbestos contamination as tap water sources must be. Still, the great majority of people who buy bottled water say that it tastes better than the water from their taps. Most water-bottling plants disinfect their products with ozone, which, unlike chlorine, leaves no flavor or odor in the water.

As a consumer, what should you look for? Look for the trademark of the International Bottled Water Association (IBWA), a trade organization supporting the FDA's regulations and enforcement efforts. Next, look for the water's place of origin. Water bottled in another state might be the safest choice because only water sold across state lines must meet the FDA's sanitation and safety

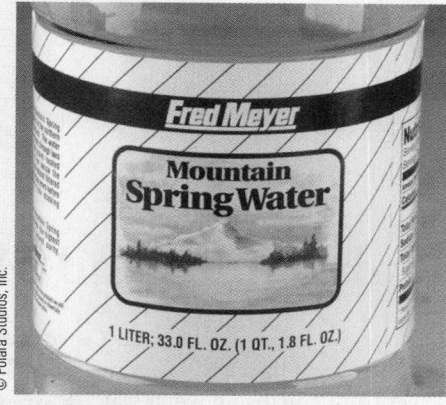

The label on a water bottle may imply purity, but the product inside often falls far short.

requirements. Then try to determine the water's source. If the water you buy is from a spring or a stream in your state, is the area agricultural, residential, industrial, or undeveloped? Agricultural, industrial, and residential activities can expose water sources to contamination. Finally, ask whether your state strictly enforces standards for purity and sanitation of bottled water. Many states have unenforced rules on the books.

Table 8-4 defines some terms that appear on labels. What you are unlikely to find on the label, however, is the

TABLE 8·4 WATER TERMS THAT MAY APPEAR ON LABELS

- **artesian water** water drawn from a well that taps a confined aquifer in which the water is under pressure.
- **baby water** ordinary bottled water treated with ozone to make it safe but not sterile.
- **caffeine water** bottled water with caffeine added.
- **carbonated water** water that contains carbon dioxide gas, either naturally occurring or added, that causes bubbles to form in it; also called *bubbling* or *sparkling water*. Seltzer, soda, and tonic waters are legally soft drinks and are not regulated as water.
- **distilled water** water that has been vaporized and recondensed, leaving it free of dissolved minerals.
- **filtered water** water treated by filtration, usually through *activated carbon filters* that reduce the lead in tap water, or by *reverse osmosis* units that force pressurized water across a membrane removing lead, arsenic, and some microorganisms from tap water.
- **fitness water** lightly flavored bottled water enhanced with vitamins, supposedly to enhance athletic performance (see Chapter 10).
- **mineral water** water from a spring or well that typically contains 250 to 500 parts per million (ppm) of minerals. Minerals give water a distinctive flavor. Many mineral waters are high in sodium.
- **natural water** water obtained from a spring or well that is certified to be safe and sanitary. The mineral content may not be changed, but the water may be treated in other ways such as with ozone or by filtration.
- **public water** water from a municipal or county water system that has been treated and disinfected.
- **purified water** water that has been treated by distillation or other physical or chemical processes that remove dissolved solids. Because purified water contains no minerals or contaminants, it is useful for medical and research purposes.
- **spring water** water originating from an underground spring or well. It may be bubbly (carbonated), or "flat" or "still," meaning not carbonated. Brand names such as "Spring Pure" do not necessarily mean that the water comes from a spring.
- **vitamin water** bottled water with a few vitamins added; does not replace vitamins from a balanced diet and may worsen overload in people receiving vitamins from enriched food, supplements, and other enriched products such as "energy" bars.
- **well water** water drawn from ground water by tapping into an aquifer.

*The group was the National Resources Defense Council, and its report, *Bottled Water: Pure Drink or Pure Hype?* is available at www.nrdc.org/water/drinking/nbw.asp.

water's mineral content. Some bottling companies will provide mineral information if a consumer requests it. For nutrition's sake the best choice is a water rich in calcium and magnesium but low in sodium.[4] The calcium is as available from the water as from milk; absorption of magnesium is lower but improves when the water is consumed with a meal.[5]

Bottled water is unpredictable in its content of fluoride, a mineral impor-tant to the health of teeth and bones, especially in growing children. As for vitamin-fortified bottled waters, these are simply liquid supplements. Read the Controversy section of Chapter 7 to determine if you need a supplement before paying for high-priced vitamin-fortified water.

If your water is dispensed from a water cooler, cleanse the cooler once a month by running half a gallon of white vinegar through it. Remove the vinegar residue by rinsing the cooler with 4 or 5 gallons of tap water. The microbial content of water coolers has been found to be considerably higher than that recom-mended by the government. Regular cleaning reduces bacterial and mold growths that can cause serious infection and disease in those who ingest water contaminated with them.

Body Fluids and Minerals

Most of the body's water weight is contained inside the cells, and some water bathes the outsides of the cells. The remainder fills the blood vessels. How do cells keep themselves from collapsing when water leaves them and from swelling up when too much water enters them? The cells cannot regulate the amount of water directly by pumping it in and out because water slips across membranes freely. The cells can, however, pump minerals across their membranes. The major minerals form **salts** that dissolve in the body fluids; the cells direct where the salts go; and this determines where the fluids flow because water follows salt.

When mineral (or other) salts dissolve in water, they separate into single, electri-cally charged particles known as **ions.** Unlike pure water, which conducts electricity poorly, ions dissolved in water carry electrical current; for this reason, these electrically charged ions are called **electrolytes.** As Figure 8-3 shows, when dissolved particles, such as electrolytes, are present in unequal concentrations on either side of a water-permeable membrane, water flows toward the more concentrated side to equalize the concentrations. Cells and their surrounding fluids work in the same way. Think of a cell as a sack made of a water-permeable membrane. The sack is filled with watery

Water follows salt. Notice the beads of "sweat," formed on the right-hand slices of eggplant, which were sprinkled with salt. Cellular water moves across each cell's mem-brane (water-permeable divider) toward the higher concentration of salt (dissolved particles) on the surface.

salts compounds composed of charged parti-cles (ions). An example is potassium chloride (K+Cl−).

ions (EYE-ons) electrically charged particles, such as sodium (positively charged) or chloride (negatively charged).

electrolytes compounds that partly dissociate in water to form ions, such as the potassium ion (K+) and the chloride ion (Cl−).

FIGURE 8·3   HOW ELECTROLYTES GOVERN WATER FLOW

Water flows in the direction of the more highly concentrated solution.

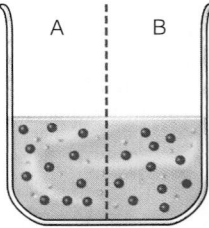

❶ With equal numbers of dissolved particles on both sides of a water-permeable divider, water levels remain equal.

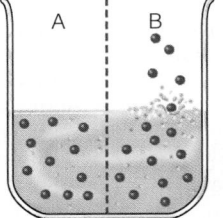

❷ Now additional particles are added to increase the concentration on side B. Particles cannot flow across the divider (in the case of fluid inside and outside a cell, the divider is a cell membrane).

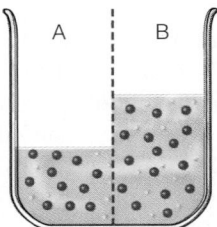

❸ Water can flow both ways across the divider, but tends to move from side A to side B, where there is a greater concentration of dissolved particles. The *volume* of water increases on side B, and the *concentrations* on sides A and B become equal.

fluid and suspended in a dilute solution of salts and other dissolved particles. Water flows freely between the fluids inside and outside the cell, but generally moves from the more dilute solution toward the more concentrated one (the photo of salted eggplant slices on the previous page shows this effect). To control the flow of water, the body must spend energy moving its electrolytes from one compartment to another (see Figure 8-4). Figure 6-11 (p. 190) introduced the proteins that form the pumps that move mineral ions across cell membranes. The result is **fluid and electrolyte balance,** the proper amount and kind of fluid in every body compartment.

If the fluid balance is disturbed, severe illness can develop quickly because fluid can shift rapidly from one compartment to another. For example, in vomiting or diarrhea, the loss of water from the digestive tract pulls fluid from between the cells in every part of the body. Fluid then leaves the cell interiors to restore balance. Meanwhile the kidneys detect the water loss and attempt to retrieve water from the pool destined for excretion. To do this, they raise the sodium concentration outside the cells, and this pulls still more water out of them. The result is **fluid and electrolyte imbalance,** a medical emergency. Water and minerals lost in vomiting or diarrhea ultimately come from every body cell. This loss disrupts the heartbeat and threatens life. It is a cause of death among those with eating disorders.

The minerals help manage still another balancing act, the **acid-base balance,** or pH, mentioned in Chapters 3 and 6. In pure water, a small percentage of water molecules (H_2O) exist as positive (H) and negative (OH) ions, but they exist in equilibrium—the positive charges exactly equal the negatives. When dissolved in watery body fluids, some of the major minerals give rise to acids (H, or hydrogen, ions), and others to bases (OH). Excess H ions in a solution make it an acid; they lower the pH. Excess OH ions in a solution make it a base; they raise the pH.

Maintenance of body fluids at a nearly constant pH is critical to life. Even slight changes in pH drastically change the structure and chemical functions of most biologically important molecules. The body's proteins and some of its mineral salts help prevent changes in the acid-base balance of its fluids by serving as **buffers**—molecules that gather up or release H ions as needed to maintain the correct pH. The kidneys help to control the pH balance by excreting more or less acid (H ions). The lungs also help by excreting more or less carbon dioxide. (Dissolved in the blood, carbon dioxide forms an acid, carbonic acid.) This tight control of the acid-base balance permits all other life processes to continue.

▶ **KEY POINT**   Mineral salts form electrolytes that help keep fluids in their proper compartments and buffer these fluids, permitting all life processes to take place.

The Major Minerals

Though all the major minerals help to maintain the fluid balance, each one also has some special duties of its own. Table 8-11 (pp. 297–298) summarizes the roles of the minerals discussed below.

Calcium

As Figure 8-1 (p. 264) showed, calcium is by far the most abundant mineral in the body. Nearly all (99 percent) of the body's calcium is stored in the bones and teeth, where it plays two important roles. First, it is an integral part of bone structure. Second, bone calcium serves as a bank that can release calcium to the body fluids if even the slightest drop in blood calcium concentration occurs. Many people have the idea that, once deposited in bone, calcium (together with the other minerals of bone) stays there forever—that once a bone is built, it is inert, like a rock. Not so. The minerals of bones are in constant flux (Figure 8-5, p. 274), with formation and dissolution taking place every minute of the day and night.

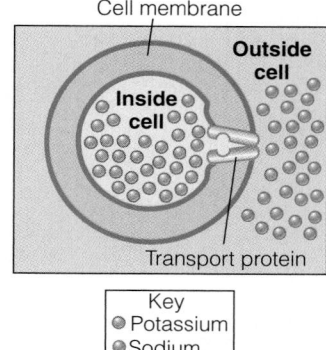

FIGURE 8·4   ELECTROLYTE BALANCE

Transport proteins in cell membranes maintain the proper balance of sodium (mostly outside the cells) and potassium (mostly inside the cells).

Cell membrane

Outside cell

Inside cell

Transport protein

Key
● Potassium
● Sodium

■ Controversy 9 describes the problems of eating disorders.

■ Figure 3-11 (p. 80) showed the pH of common substances; Figure 3-3 (p. 69) depicted fluid movement in and around cells.

Major minerals:
- Calcium.
- Chloride.
- Magnesium.
- Phosphorus.
- Potassium.
- Sodium.
- Sulfate.

fluid and electrolyte balance maintenance of the proper amounts and kinds of fluids and minerals in each compartment of the body.

fluid and electrolyte imbalance failure to maintain the proper amounts and kinds of fluids and minerals in every body compartment; a medical emergency.

acid-base balance maintenance of the proper degree of acidity in each of the body's fluids.

buffers molecules that can help to keep the pH of a solution from changing by gathering or releasing H ions.

FIGURE 8·5 A BONE

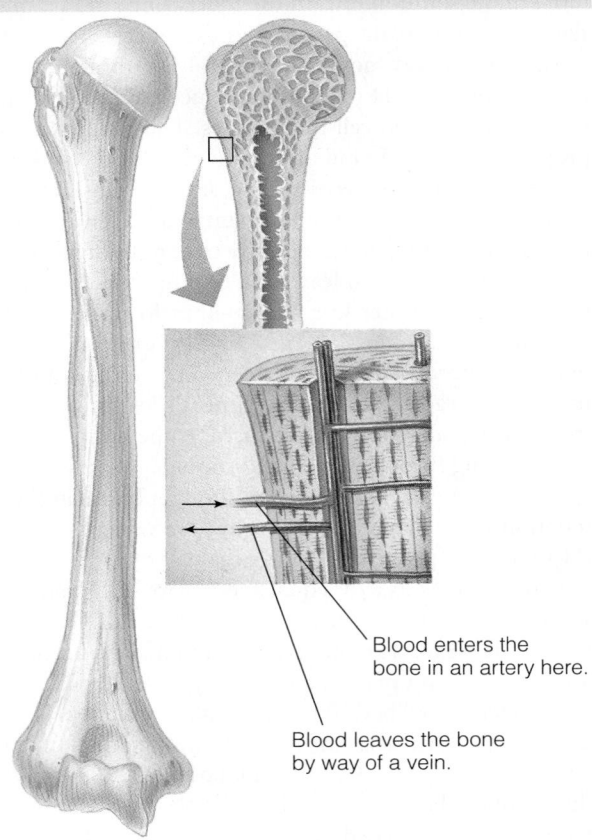

Bone is active, living tissue. Blood travels in capillaries throughout the bone, bringing nutrients to the cells that maintain the bone's structure and carrying away waste materials from those cells. It picks up and deposits minerals as instructed by hormones.

Bone derives its structural strength from the lacy network of crystals that lie along its lines of stress. If minerals are withdrawn to cover deficits elsewhere in the body, the bone will grow weak and ultimately will bend or crumble.

Blood enters the bone in an artery here.

Blood leaves the bone by way of a vein.

Calcium and phosphorus are both essential to bone formation: calcium phosphate salts crystallize on a foundation material composed of the protein collagen. The resulting **hydroxyapatite** crystals invade the collagen and gradually lend more and more rigidity to a youngster's maturing bones until they are able to support the weight they will have to carry. During and after the bone-strengthening processes, fluoride may displace the "hydroxy" parts of these crystals, making **fluorapatite.** Fluorapatite resists bone-dismantling forces to help maintain bone integrity.

Teeth are formed in a similar way: hydroxyapatite crystals form on a collagen matrix to create the dentin that gives strength to the teeth (see Figure 8-6). The turnover of minerals in teeth is not as rapid as in bone, but some withdrawal and redepositing do take place throughout life. As in bone, fluoride hardens and stabilizes the crystals of teeth and makes the enamel resistant to decay.

Other roles for calcium are emerging as well. People with low intakes of calcium may be somewhat more likely to develop cancer of the colon and rectum than those with higher intakes.[15] Calcium from low-fat milk and milk products may also play a role in maintaining a healthy body weight.[16] Large, well-designed clinical studies are needed to confirm the effects of dietary calcium on cancer and body fatness.

Calcium in Body Fluids Only about 1 percent of the body's calcium is in the fluids that bathe and fill the cells, but this tiny amount plays these major roles:

- Regulates the transport of ions across cell membranes and is particularly important in nerve transmission.
- Helps maintain normal blood pressure (see Chapter 11).
- Plays an essential role in the clotting of blood.

hydroxyapatite (hi-DROX-ee-APP-uh-tight) the chief crystal of bone, formed from calcium and phosphorus.

fluorapatite (floor-APP-uh-tight) a crystal of bones and teeth, formed when fluoride displaces the "hydroxy" portion of hydroxyapatite. Fluorapatite resists being dissolved back into body fluid.

- Is essential for muscle contraction and therefore for the heartbeat.
- Allows secretion of hormones, digestive enzymes, and neurotransmitters.
- Activates cellular enzymes that regulate many processes.

Because of its importance, blood calcium is tightly controlled.

Calcium and the Bones   The key to bone health lies in the body's calcium balance. Cells need continuous access to calcium, so the body maintains a constant calcium concentration in the blood. The skeleton serves as a bank from which the blood can borrow and return calcium as needed. Blood calcium is regulated, not by a person's daily calcium intake or bone density, but by hormones sensitive to blood calcium concentrations.* Thus, a person whose calcium intake is inadequate maintains normal blood calcium but at the expense of the bones.

The body is sensitive to an increased need for calcium, although it sends no signals to the conscious brain indicating calcium need. Instead, the body quietly increases the absorption of calcium from the intestine and prevents its loss from the kidneys. For example, more calcium is needed for growth, so infants and children absorb about 60 percent of ingested calcium and pregnant women absorb about 50 percent, compared with 25 percent in other healthy adults. The body of an adolescent hungers for calcium, absorbing and retaining more calcium from each meal than does the body of an adult.[17] The body also absorbs a higher percentage of calcium when less total calcium is provided in the diet.[18] Deprived of calcium for months or years, an adult may double the calcium absorbed; when supplied for years with abundant calcium, the same person may absorb only about one-third the normal amount. These adjustments take time, though, and increased absorption cannot fully compensate for a reduced calcium intake. A person who suddenly cuts back on calcium is likely to lose calcium from the bones.

Despite the body's adjustments, some bone loss is an inevitable consequence of aging. By the late 20s, or 10 years after adult height is achieved, the skeleton no longer adds significantly to bone density.[19] After about 40 years of age, regardless of calcium intake bones begin to lose density, but the loss can be slowed somewhat by a diet high in calcium along with sufficient physical activity.

To protect against bone loss, high calcium intakes early in life are recommended. A calcium-poor diet during the growing years may prevent a person from achieving **peak bone mass.**[20] Furthermore, a person whose calcium savings account is not sufficient is more likely to develop the fragile bones of **osteoporosis,** or **adult bone loss.** Too little calcium packed into the skeleton during childhood and young adulthood strongly predicts susceptibility to osteoporosis in adulthood. Osteoporosis constitutes a major health problem for many older people—its possible causes and prevention are the topics of this chapter's Controversy.

How Much Calcium Do I Need?   Setting recommended intakes for calcium is difficult because absorption varies not only with age but also with a person's vitamin D status and the calcium content of the diet. The DRI committee took such variations into account and set recommendations for calcium at levels that produce maximum calcium retention. At lower intakes, the body does not store calcium to capacity; at greater intakes, the excess calcium is excreted and thus is wasted.

Recommended intakes are high for children and adolescents because people develop their peak bone mass during their growing years. Obtaining enough calcium at that time helps to ensure that the skeleton starts adulthood with a high bone density. Despite the importance of consuming adequate calcium during the growing years, average intakes among today's youth are too low to meet recommendations. Snapshot 8-1 (p. 276) provides a look at some foods that are good or excellent sources of calcium, and the Food Feature at the end of the chapter focuses on foods that can help to meet calcium needs.

---

*Calcitonin, made in the thyroid gland, is secreted whenever the calcium concentration in the blood rises too high. It acts to stop withdrawal from bone and to slow absorption somewhat from the intestine. Parathormone, from the parathyroid glands, has the opposite effect.

FIGURE 8·6   A TOOTH

The inner layer of dentin is bonelike material that forms on a protein (collagen) matrix. The outer layer of enamel is harder than bone. Both dentin and enamel contain hydroxyapatite crystals (made of calcium and phosphorus). The crystals of enamel may become even harder when exposed to the trace mineral fluoride.

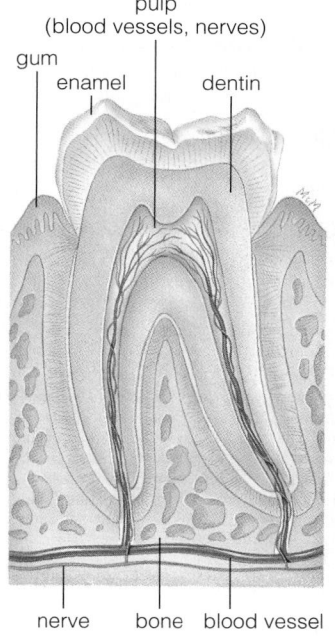

pulp
(blood vessels, nerves)
gum
enamel   dentin

nerve   bone   blood vessel

In osteoporosis:

- Bones of older adults become brittle and fragile.

■ The importance of vitamin D in calcium absorption was described in Chapter 7.

peak bone mass the highest attainable bone density for an individual; developed during the first three decades of life.

osteoporosis (OSS-tee-oh-pore-OH-sis) a reduction of the bone mass of older persons in which the bones become porous and fragile (*osteo* means "bones"; *poros* means "porous"); also known as **adult bone loss.**

CALCIUM

DRI RECOMMENDED INTAKES:
Adults: 1,000 mg/day (19–50 yr)
1,200 mg/day (>51 yr)

TOLERABLE UPPER INTAKE LEVEL:
Adults: 2,500 mg/day

CHIEF FUNCTIONS: Mineralization of bones and teeth; muscle contraction and relaxation, nerve functioning, blood clotting

DEFICIENCY: Stunted growth and weak bones in children; bone loss (osteoporosis) in adults

TOXICITY: Constipation; interference with absorption of other minerals; increased risk of kidney stone formation

*These foods provide 10 percent or more of the calcium Daily Value in a serving. For a 2,000-calorie diet, the DV is 1,000 mg/day.

ᵃAlthough broccoli, kale, and some other cooked green leafy vegetables fall short of supplying 10 percent of calcium in a serving, these foods are important sources of bioavailable calcium. Almonds also supply calcium. Other greens, such as spinach and chard, contain calcium in an unabsorbable form. Note that the amounts for green vegetables exceed ½ c serving of the USDA Food Guide. Some calcium-rich mineral waters may also be good sources.

GOOD SOURCES*

SARDINES (with bones)
3 oz = 324 mg

MILK
1 c = 300 mg

TOFU (calcium set)
½ c = 275 mg

BLACK-EYED PEAS (cooked)
½ c = 105 mg

CHEDDAR CHEESE
1½ oz = 306 mg

TURNIP GREENS (cooked)
1 c = 197 mg

WAFFLE (whole grain)
1 waffle = 196 mg

BROCCOLIᵃ (cooked)
1½ c = 93 mg

▶ **KEY POINT** Calcium makes up bone and tooth structure and plays roles in nerve transmission, muscle contraction, and blood clotting. Calcium absorption rises when there is a dietary deficiency or an increased need such as during growth.

Phosphorus

■ The mineral is *phosphorus*. The adjective form is spelled with an *-ous* (as in phospho-rous salts).

Phosphorus is the second most abundant mineral in the body, but its concentration in the blood is less than half that of calcium. About 85 percent of the body's phosphorus is found combined with calcium in the crystals of the bones and teeth. The rest is everywhere else:

- Phosphorous salts are critical buffers, helping to maintain the acid-base balance of cellular fluids.
- Phosphorus is part of the DNA and RNA of every cell and thus is essential for growth and renewal of tissues.
- Phosphorous compounds carry, store, and release energy in the metabolism of energy nutrients.
- Phosphorous compounds assist many enzymes and vitamins in extracting the energy from nutrients.
- Phosphorus forms part of the molecules of the phospholipids that are principal components of cell membranes (discussed in Chapter 5).
- Phosphorus is present in some proteins.

PHOSPHORUS

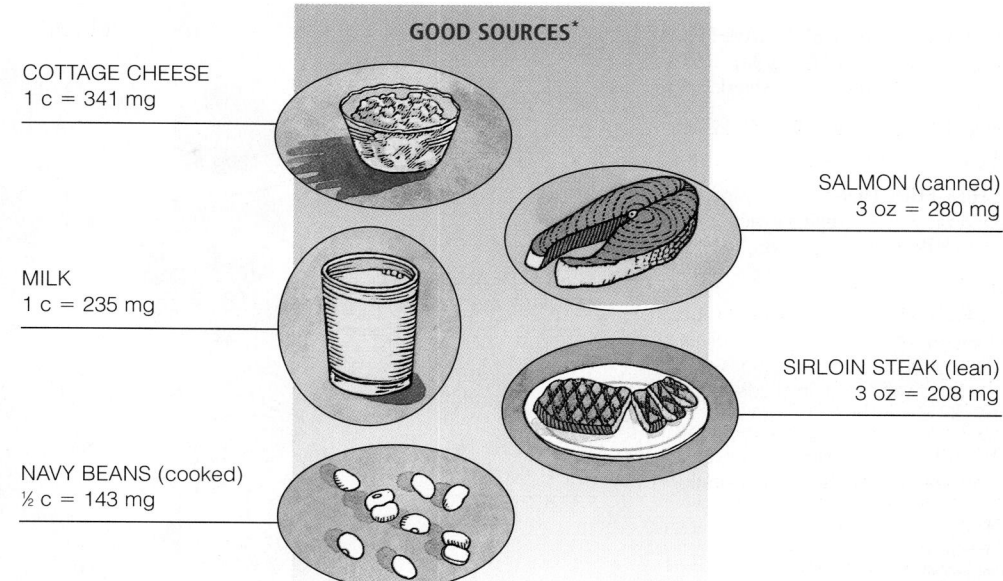

DRI RECOMMENDED INTAKE:
Adults: 700 mg/day

TOLERABLE UPPER INTAKE LEVEL:
Adults (19–70 yr): 4,000 mg/day

CHIEF FUNCTIONS: Mineralization of bones and teeth; part of phospholipids, important in genetic material, energy metabolism, and buffering systems

DEFICIENCY: Muscular weakness, bone pain[a]

TOXICITY: Calcification of soft tissues, particularly the kidneys

*These foods provide 10 percent or more of the phosphorus Daily Value in a serving. For a 2,000-calorie diet, the DV is 1,000 mg/day.

[a]Dietary deficienccy rarely occurs, but some drugs can bind with phosphorus making it unavailable.

GOOD SOURCES*

COTTAGE CHEESE
1 c = 341 mg

SALMON (canned)
3 oz = 280 mg

MILK
1 c = 235 mg

SIRLOIN STEAK (lean)
3 oz = 208 mg

NAVY BEANS (cooked)
½ c = 143 mg

Despite all of these critical roles, the body's need for phosphorus are easily met by almost any diet, and deficiencies are unknown. As Snapshot 8-2 shows, animal protein is the best source of phosphorus (because phosphorus is abundant in the cells of animals).

▶ **KEY POINT** Most of the phosphorus in the body is in the bones and teeth. Phosphorus helps maintain acid-base balance, is part of the genetic material in cells, assists in energy metabolism, and forms part of cell membranes. Under normal circumstances, deficiencies of phosphorus are unknown.

Magnesium

Magnesium barely qualifies as a major mineral: only about 1 ounce is present in the body of a 130-pound person, over half of it in the bones. Most of the rest is in the muscles, heart, liver, and other soft tissues, with only 1 percent in the body fluids. The supply of magnesium in the bones can be tapped to maintain a constant blood level whenever dietary intake falls too low. The kidneys can also act to conserve magnesium.

Like phosphorus, magnesium is critical to many cell functions. It assists in the operation of more than 300 enzymes, is needed for the release and use of energy from the energy-yielding nutrients, and directly affects the metabolism of potassium, calcium, and vitamin D. Magnesium acts in the cells of all the soft tissues, where it forms part of the protein-making machinery and is necessary for the release of energy. Magnesium and calcium work together for proper functioning of the muscles: calcium promotes contraction, and magnesium helps the muscles relax afterward. In the teeth, magnesium promotes resistance to tooth decay by holding calcium in tooth enamel.

Most Americans receive only about three-quarters of the recommended magnesium from their diets.[21] Among African Americans, whose magnesium intakes are much lower than those of other groups, chronic diseases occur in high numbers, leading to the suspicion that low intakes of magnesium-rich foods may contribute to

MAGNESIUM

DRI RECOMMENDED INTAKES:
Men (19–30 yr): 400 mg/day
Women (19–30 yr): 310 mg/day

TOLERABLE UPPER INTAKE LEVEL:
Adults: 350 mg/day[a]

CHIEF FUNCTIONS: Bone mineralization, protein synthesis, enzyme action, muscle contraction, nerve function, tooth maintenance, and immune function

DEFICIENCY: Weakness, confusion; if extreme, convulsions, uncontrollable muscle contractions, hallucinations, and difficulty in swallowing; in children, growth failure

TOXICITY: From nonfood sources only; diarrhea, pH imbalance, dehydration

[*]These foods provide 10 percent or more of the magnesium Daily Value in a serving. For a 2,000-calorie diet, the DV is 400 mg/day.

[a]From nonfood sources, in addition to the magnesium provided by food.

[b]Wheat bran provides magnesium, but refined grain products are low in magnesium.

[c]Magnesium in oysters varies.

GOOD SOURCES[*]

SPINACH (cooked)
½ c = 78 mg

BLACK BEANS (cooked)
½ c = 60 mg

SOY MILK
1 c = 46 mg

BLACK-EYED PEAS (cooked)
½ c = 44 mg

BRAN CEREAL[b]
(ready-to-eat)
1 c = 60 mg

OYSTERS[c] (steamed)
3 oz = 81 mg

YOGURT (plain)
1 c = 43 mg

disease development.[22] Snapshot 8-3 shows magnesium-rich foods. Magnesium is easily washed and peeled away from foods during processing, so slightly processed or unprocessed foods are the best sources. In some parts of the country, water contributes significantly to magnesium intakes, so people living in those regions need less from food.

A magnesium deficiency may occur as a result of inadequate intake, vomiting, diarrhea, alcoholism, or protein malnutrition. It may also occur in hospital clients who have been fed magnesium-poor fluids through a vein for too long or in people who are using diuretics. People whose drinking water provides adequate magnesium experience a lower incidence of sudden death from heart failure than other people. It seems likely that magnesium deficiency makes the heart unable to stop spasms once they start. Magnesium deficiency may also be related to cardiovascular disease, heart attack, and high blood pressure. A deficiency also causes hallucinations that can be mistaken for mental illness or drunkenness. Although intakes are often below those recommended, overt deficiency symptoms are rare in normal, healthy people.

Magnesium toxicity is rare, but it can be fatal. Toxicity occurs only with high intakes from nonfood sources such as supplements or magnesium salts. Accidental poisonings may occur in children with access to medicine chests and in older people who abuse magnesium-containing laxatives, antacids, and other medications. The consequences can be severe diarrhea, acid-base imbalance, and dehydration. For safety, use magnesium-containing medications with discretion.

> ▶ KEY POINT   Most of the body's magnesium is in the bones and can be drawn out for all the cells
> to use in building protein and using energy. Most people in the United States fail to obtain enough
> magnesium from their food.

Sodium

Salt has been known and valued throughout recorded history. "You are the salt of the earth" means that you are valuable. If "you are not worth your salt," you are worthless. Even our word *salary* comes from the Latin word for *salt*. Chemically, sodium is the positive ion in the compound sodium chloride (table salt) and makes up 40 percent of its weight: a gram of salt contains 400 milligrams of sodium.

Sodium is a major part of the body's fluid and electrolyte balance system because it is the chief ion used to maintain the volume of fluid outside cells. Sodium also helps maintain acid-base balance and is essential to muscle contraction and nerve transmission. Scientists think that 30 to 40 percent of the body's sodium is stored on the surface of the bone crystals, where the body can easily draw on it to replenish the blood concentration.

A deficiency of sodium would be harmful, but no known human diets lack sodium.[23] Most foods include more salt than is needed, and the body absorbs it freely. The kidneys filter the surplus out of the blood into the urine. They can also sensitively conserve sodium. In the rare event of a deficiency, they can return to the bloodstream the exact amount needed. Small sodium losses occur in sweat, but the amount of sodium excreted in a day equals the amount ingested that day. But, if sodium is so well controlled by the body, why do authorities urge people to limit their intakes? To understand why, you must first understand how sodium interacts with body fluids.

How Are Salt and "Water Weight" Related?   If blood sodium rises, as it will after a person eats salted foods, thirst ensures that the person will drink water until the sodium-to-water ratio is restored. Then the kidneys excrete the extra water along with the extra sodium.

Dieters sometimes think that eating too much salt or drinking too much water will make them gain weight, but they do not gain fat, of course. They gain water, but a healthy body excretes this excess water immediately. Excess salt is excreted as soon as enough water is drunk to carry the salt out of the body. From this perspective, then, the way to keep body salt (and "water weight") under control is to control salt intake and drink more, not less, water.

If blood sodium drops, body water is lost, and both water and sodium must be replenished to avert an emergency. Overly strict use of low-sodium diets in the treatment of hypertension, kidney disease, or heart disease can deplete the body of needed sodium; so can vomiting, diarrhea, or extremely heavy sweating.

Sodium Intakes   The DRI intake recommendation for sodium has been set at 1,500 milligrams for healthy active young adults, at 1,300 for people ages 51 through 70, and at 1,200 for the elderly.[24] These amounts are sufficient to ensure an overall diet that provides adequate amounts of other needed nutrients. Because U.S. intakes are much higher than these amounts, and because blood pressure rises with sodium intakes, the DRI committee set a Tolerable Upper Intake Level for sodium of 2,300 milligrams per day. The *Dietary Guidelines for Americans 2005* urge people to consume little sodium and salt and to stay within the DRI sodium guidelines (see Table 8-5).

Adults in the United States consume an average of 3,300 milligrams of sodium per day, which translates to more than 8 grams of salt (see Figure 8-7, p. 280). This amount exceeds the Tolerable Upper Intake Level of 2,300 milligrams by more than a third.[25] Asian peoples, whose staple sauces and flavorings are based on soy sauce and monosodium glutamate (MSG or Accent), may consume the equivalent of 30 to 40 grams of salt per day.

▧ To the chemist, a salt results from the neutralization of an acid and a base. Sodium chloride, table salt, results from the reaction between hydrochloric acid and the base sodium hydroxide. The positive sodium ion unites with the negative chloride ion to form the salt. The positive hydrogen ion unites with the negative hydroxide ion to form water.
Base + acid = salt + water.
Sodium hydroxide + hydrochloric acid = sodium chloride + water.

▧ Too little sodium can pose a danger to endurance athletes performing in hot, humid conditions. See Chapter 10.

▧ For a brief summary of the kidneys' actions, see Chapter 3.

TABLE 8·5   SODIUM AND SALT INTAKE GUIDELINES

DRI Recommendations
- Recommended intakes for sodium:
  Adults: (19–50 years): 1,500 mg per day.
  Adults: (51–70 years): 1,300 mg per day.
  Adults: (71 years and older): 1,200 mg per day.
- Tolerable Upper Intake Level for sodium and salt:
  Adults (19 years and older): 2,300 mg sodium, or 5.6 g salt (sodium chloride) per day.

The *Dietary Guidelines for Americans 2005* Key Recommendations:
- Consume less than 2,300 mg of sodium (approximately 1 tsp salt) per day.
- Choose and prepare foods with little salt.
- *People with hypertension, blacks, and middle-aged and older adults:* Aim to consume no more than 1,500 mg of sodium per day, and meet the potassium recommendation of 4,700 mg/day from food sources.

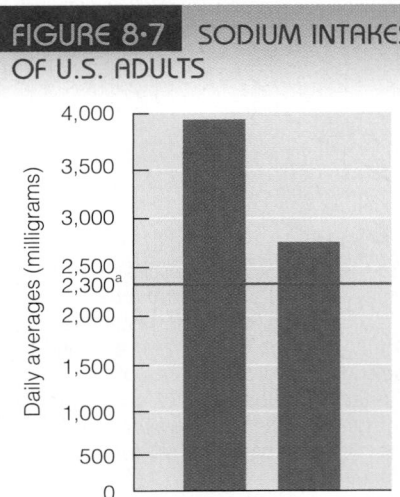

FIGURE 8·7 SODIUM INTAKES OF U.S. ADULTS

[Bar chart showing Daily averages (milligrams) on the y-axis from 0 to 4,000, with a horizontal line at 2,300ᵃ. Men bar reaches approximately 3,950; Women bar reaches approximately 2,750.]

ᵃ DRI Tolerable Upper Intake Level (UL).

SOURCE: Standing Committee on the Scientific Evaluation of *Dietary Reference Intakes, Food and Nutrition Board, Institute of Medicine, Dietary Reference Intakes for Water, Potassium, Sodium, Chloride, and Sulfate* (Washington, D.C.: National Academies Press, 2004), p. 6–46.

■ A diet designed with the DASH eating plan (Table 1-3 of Appendix I) meets the *Dietary Guidelines for Americans 2005,* including the guidelines for sodium and potassium.

The *Dietary Guidelines* also encourage people to:

• Consumer a sufficient amount of fruits and vegetables while staying within energy needs. Two cups of fruit and 2½ cups of vegetables per day are recommended in a 2,000-diet.

• Choose a variety of fruits and vegetables each day; in particular, select from all five vegetable subgroups (dark green, orange, legumes, starchy vegetables, and other vegetables) several time each week.

hypertension high blood pressure.

Sodium and Blood Pressure   Around the world, communities with high intakes of salt experience high rates of **hypertension,** cardiovascular disease, and cerebral hemorrhage, a hypertension-related stroke.[26] As sodium intakes increase, average blood pressure rises with them in a stepwise fashion.[27] As blood pressure rises, the risk of death from cardiovascular disease climbs steadily.[28]

The relationship between salt intakes and blood pressure is direct—the more salt a person eats, the higher the blood pressure goes.[29] Even small increases in sodium, even when salt intakes are low to begin with, have this effect—any amount of salt seems to raise the blood pressure. This effect occurs more strongly among more salt-sensitive people, including those with diabetes, hypertension, or kidney disease; those of African descent; those whose parents had high blood pressure; and anyone over age 50 because the blood pressure responds to salt more dramatically in older age.

Not surprisingly, genetics plays a role in the body's handling of dietary sodium and may be at the bottom of the variations in salt sensitivity observed among different groups of people. Certain genetic variations are known to cause high blood pressure, and researchers suspect that, while these relationships are complex, the genes affecting blood pressure do so by altering the kidneys' handling of sodium.[30]

Luckily, people can modify their salt response somewhat by consuming more potassium in the context of a nutritious diet. One proven dietary approach can help people to reduce their sodium and increase potassium intakes, thereby reducing their disease risks. The DASH (Dietary Approaches to Stop Hypertension) diet often achieves a lower blood pressure than restriction of sodium intake alone.[31] The DASH approach calls for greatly increased intakes of fruits and vegetables, with adequate amounts of nuts, fish, whole grains, and low-fat dairy products. At the same time, red meat, butter, and other high-fat foods and sweets are held to occasional small portions. Salt and sodium are greatly reduced.

When people consume the DASH diet with progressively lower sodium, their blood pressures fall responsively.[32] When the diet is modified to also provide foods with abundant magnesium, potassium, and calcium, as well as adequate protein and fiber, the average blood pressure drops even lower at each level of sodium intake. Low potassium intake on its own raises blood pressure, whereas high potassium intake appears to both help prevent and correct hypertension.[33] These beneficial nutrients come from consuming generous amounts of whole grains, fruits, vegetables, low-fat milk products, seeds, nuts, and legumes each day, while limiting meats and other animal products.[34] For controlling hypertension, then, actions to reduce sodium and implement the dietary changes characterizing the DASH diet seem wise (also see Chapter 11).

Many Americans have much to gain in terms of cardiovascular health and nothing to lose from cutting back on salt as part of an overall lifestyle strategy to reduce blood pressure. Physical activity should also be part of that lifestyle because regular moderate exercise reliably lowers the blood pressure.

Other valid reasons exist for most people to hold their salt intakes at or below the recommended maximum. For example, older people without clinical hypertension often die of stroke, and reducing dietary sodium may lower their blood pressure enough to reduce their stroke risk. Excess sodium in the diet also increases calcium excretion, an effect that could potentially compromise the integrity of the bones.[35] Excessive salt may also directly stress a weakened heart or aggravate kidney problems. The high salt intakes of many Asian peoples have been suggested as a possible cause for their greatly elevated rates of stomach cancer.

Controlling Salt Intake   Cutting down on salt and sodium may be easier than people believe, and Table 8-6 demonstrates how, with a few thoughtful food choices, a meal's sodium can be drastically reduced. Notice that in the meal in the left-hand column sauces, dressings, and the salt added to corn, chips, pickles, and piecrust and sprinkled on foods are the major modifiable sources of sodium. Notice, too, that even without added sauces and salt, most foods contain enough sodium to easily meet most people's need. Foods eaten without salt may seem less tasty at first, but with repetition,

TABLE 8·6   HOW TO CUT SODIUM FROM A BARBECUE LUNCH

Lunch #1 exceeds the whole day's Tolerable Upper Intake Level of 2,300 milligrams sodium. With careful substitutions, the sodium drops dramatically in the second lunch, but it still provides over 40 percent of the suggested maximum intake. In lunch #3, just two small changes—omitting the sauce and salt—cut the sodium by half again.

Lunch #1: Highest	Sodium (mg)	Lunch #2: Lower	Sodium (mg)	Lunch #3: Lowest	Sodium (mg)
• Chopped pork sandwich, with sauce	950	• Sliced pork sandwich, with 1 tbs sauce	400	• Sliced pork sandwich (no sauce)	210
• Creamed corn, ½ c	460	• Corn, 1 cob, soft margarine, salt	190	• Corn, 1 cob, soft margarine	50
• Potato chips, 2.5 oz	340	• Coleslaw, ½ c	180	• Green salad, oil and vinegar	10
• Dill pickle, ½ medium	420	• Watermelon, slice	10	• Watermelon, slice	10
• Milk, low fat, 1 c	120	• Milk, reduced fat, 1 c	120	• Milk, reduced fat, 1 c	120
• Pecan pie, slice	480	• Ice cream, low fat, ½ c	80	• Ice cream, low fat, ½ cup	80
Total 2,770		**Total 980**		**Total 480**	

tastes adjust and the natural flavor becomes the preferred taste. Also, remember that the recommendation is to consume little sodium, not eliminate salt altogether.

While an obvious step is to control the saltshaker, this source may contribute as little as 15 percent of the total salt consumed. As Figure 8-8 (p. 282) indicates, a more productive step is to cut down on processed and fast foods, the source of almost 75 percent of salt in the U.S. diet. Often, the least processed foods in each food group are not only lowest in sodium but also highest in potassium. As mentioned earlier, low potassium intakes are thought to play an important role in the development of hypertension. Many people are unaware that foods high in sodium do not always taste salty. Who could guess by taste alone that half a cup of instant chocolate pudding provides almost one-fifth of the daily upper limit for sodium? Moral: Read the labels.

▶ **KEY POINT**   Sodium is the main positively charged ion outside the body's cells. Sodium attracts water. Thus, too much sodium (or salt) raises blood pressure and aggravates hypertension. Diets rarely lack sodium.

Potassium

Potassium is the principal positively charged ion *inside* the body's cells. It plays a major role in maintaining fluid and electrolyte balance and cell integrity, and it is critical to maintaining the heartbeat. The sudden deaths that occur during fasting or severe diarrhea and in children with kwashiorkor or people with eating disorders are thought to be due to heart failure caused by potassium loss.

Dehydration leads to a loss of potassium from inside cells. This condition is dangerous because when the cells of the brain lose potassium, the person loses the ability to notice the need for water. Adults are warned not to take diuretics (water pills) that cause potassium loss or to give them to children, except under a physician's supervision. Physicians prescribing diuretics will tell clients to eat potassium-rich foods to compensate for the losses. Depending on the diuretic, physicians may also advise a lower sodium intake. When taking diuretics, a person should alert all other health-care providers.

In healthy people, almost any reasonable diet provides enough potassium to prevent the dangerously low blood potassium that indicates a severe deficiency. A typical U.S. diet, however, with its low intakes of fruits and vegetables, provides only about half of the daily 4,700 milligrams of potassium recommended by the DRI committee.[36] Though blood potassium may remain normal on such a diet, chronic diseases are more likely to occur. With low potassium intake, hypertension becomes worse, glucose

▣ Kwashiorkor is described in Chapter 6.

▣ Unlike sodium, potassium may exert a positive effect against hypertension and related ills.

FIGURE 8·8 SOURCES OF SODIUM IN THE U.S. DIET

Unprocessed Foods

Those that are low in sodium contribute less than 10 percent of the total sodium in the U.S. diet.

Fresh foods higher in sodium
 Milk, 120 mg per 1 c
 Scallops, 260 mg per 3 oz
Fresh meats, about 30 to 70 mg per 3 oz
 Chicken, beef, fish, lamb, pork
Fresh vegetables, about 30 to 50 mg per 1/2 c
 Celery, Chinese cabbage, sweet potatoes
Fresh vegetables, about 10 to 20 mg per 1/2 c
 Broccoli, brussels sprouts, carrots, corn, green beans, legumes, potatoes, salad greens
Grains (cooked without salt), about 0 to 10 mg per 1/2 c
 Barley, oatmeal, pasta, rice

Salt

Salt added at home, in cooking or at the table, contributes 15 percent of the total sodium in the U.S. diet. Many seasonings and sauces also contribute salt and sodium.

Salts, about 2,000 mg per teaspoon
 Salt, sea salt, seasoned salt, onion salt, garlic salt[a]
Soy sauce, about 300 mg per teaspoon
Condiments and sauces, about 100 to 200 mg per tablespoon
 Barbecue sauce, ketchup, mustard, salad dressings, sweet pickle relish, taco sauce, Worcestershire sauce

[a]Note that herb seasoning blends may or may not contain substantial sodium; read the labels.

Processed Foods

These contribute 75 percent of the sodium in the U.S. diet.

Dry soup mixes (prepared), about 1,000 to 2,000 mg per 1 c
 Bouillon cube or canned, noodle soups, onion soup, ramen
Smoked and cured meats, about 700 to 2,000 mg per 2 oz
 Canned ham products, corned or chipped beef, ham, lunchmeats
Fast foods and TV dinners, about 700 to 1,500 mg per serving
 Breakfast biscuit (cheese, egg, and ham), cheeseburger, chicken wings (10 spicy wings), frozen TV dinners, pizza (2 slices), taco, vegetarian soy burger (on bun)
Canned soups (prepared), about 700 to 1,500 mg per 1 c
 Bean soup, beef or chicken soups, "hearty" soups, tomato soup, vegetable soup
Canned pasta, about 800 to 1,000 mg per serving
 Beefaroni, macaroni and cheese, ravioli
Hot dogs, about 500 to 700 mg per 2 oz
 Hot dogs, smoked sausages
Foods prepared in brine, about 300 to 800 mg per serving
 Anchovies (2 fillets), dill pickles (1), olives (5), sauerkraut 1/2 c
Cheeses, processed, about 550 mg per 1 1/2 oz
 American, cheddar, Swiss
Pudding, instant, about 420 mg per 1/2 c
 All flavors
Canned vegetables, about 200 to 450 mg per 1/2 c
 Carrots, corn, green beans, legumes, peas, potatoes
Cereals, dry ready-to-eat, about 180 to 260 mg per 1 oz
 Cheerios, cornflakes, corn bran, Cocoa Puffs, Total, others

© Matthew Farruggio (all)

tolerance is impaired, metabolic acidity increases, calcium losses from bones accelerate, and kidney stone formation becomes more likely. The amount of fruits and vegetables recommended in both the USDA Food Guide and the DASH diet are sufficient to provide all the needed potassium. Potassium chloride, a salt substitute for people with hypertension who must strictly limit salt, provides potassium but does not reverse many of the conditions associated with diets that lack potassium-rich foods.[37]

POTASSIUM

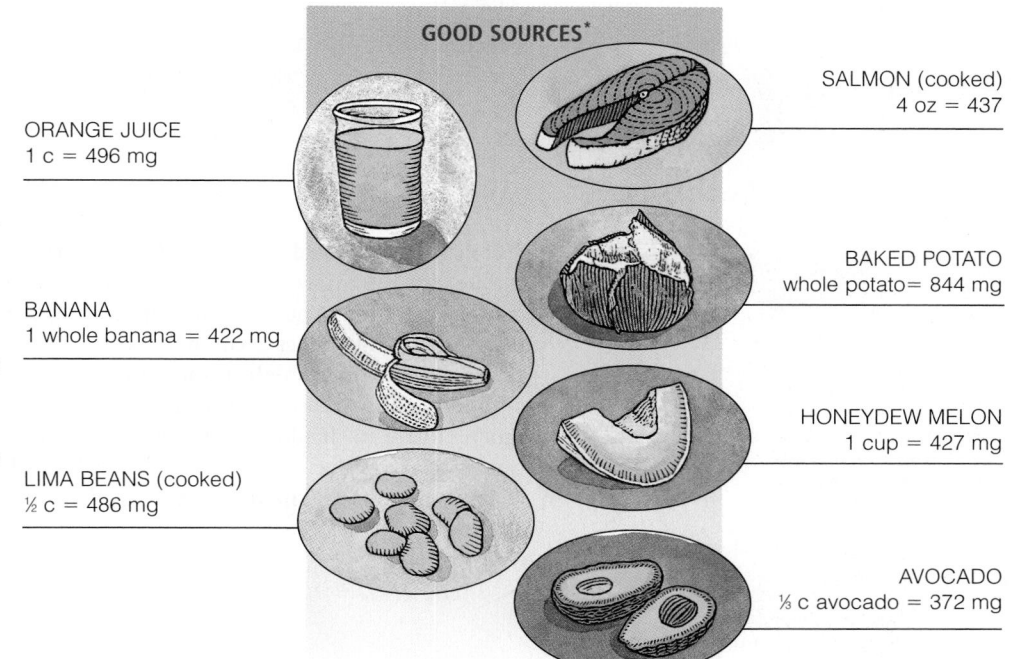

DRI RECOMMENDED INTAKE:
Adults: 4,700 mg/day

CHIEF FUNCTIONS: Maintains normal fluid and electrolyte balance; facilitates chemical reactions; suppports cell integrity; assists in nerve functioning and muscle contractions

DEFICIENCY:[a] Muscle weakness, paralysis, confusion

TOXICITY: Muscle weakness; vomiting; for an infant given supplements, or when injected into a vein in an adult, potassium can stop the heart

*These foods provide 10 percent or more of the potassium Daily Value in a serving. For a 2,000-calorie diet, the DV is 3,500 mg/day.

[a]Deficiency accompanies dehydration.

GOOD SOURCES*

ORANGE JUICE
1 c = 496 mg

BANANA
1 whole banana = 422 mg

LIMA BEANS (cooked)
½ c = 486 mg

SALMON (cooked)
4 oz = 437

BAKED POTATO
whole potato= 844 mg

HONEYDEW MELON
1 cup = 427 mg

AVOCADO
⅓ c avocado = 372 mg

Because potassium is found inside all living cells and because cells remain intact unless foods are processed, the richest sources of potassium are fresh, whole foods (see Snapshot 8-4). Most vegetables and fruits are outstanding. Bananas, despite their fame as the richest potassium source, are only one of many rich sources, which also include spinach, cantaloupe, and almonds. Nevertheless, bananas are readily available, are easy to chew, and have a sweet taste that almost everyone likes, so health-care professionals often recommend them.

Potassium from foods is safe, but potassium injected into a vein can stop the heart. Potassium chloride pills are available over-the-counter and are sold in health food stores without a warning label, but they should *not* be used except on a physician's advice. Potassium overdoses normally are not life-threatening as long as they are taken by mouth because the presence of excess potassium in the stomach triggers a vomiting reflex that expels the unwanted substance. A person with a weak heart, however, should not go through this trauma, and a baby may not be able to withstand it. Several infants have died when well-meaning parents overdosed them with potassium supplements.

▶ **KEY POINT**   Potassium, the major positive ion inside cells, is important in many metabolic functions. Fresh whole foods are the best sources of potassium. Diuretics can deplete the body's potassium and so can be dangerous; potassium excess can also be dangerous.

Chloride

In its elemental form, chlorine forms a deadly green gas. In the body, the chloride ion plays important roles as the major negative ion. In the fluids outside the cells, it accompanies sodium and so helps to maintain the crucial fluid balances (acid-base

■ The DRI committee set recommended intake values for chloride; see inside front cover.

and electrolyte balances). The chloride ion also plays a special role as part of hydrochloric acid, which maintains the strong acidity of the stomach necessary to digest protein. The principal food source of chloride is salt, both added and naturally occurring in foods, and no known diet lacks chloride.

▶ **KEY POINT** Chloride is the body's major negative ion; it is responsible for stomach acidity and assists in maintaining proper body chemistry. No known diet lacks chloride.

Sulfate

Sulfate is the oxidized form of sulfur as it exists in food and water. The body requires sulfate for synthesis of many important sulfur-containing compounds. Sulfur-containing amino acids play an important role in helping strands of protein assume their functional shapes. Skin, hair, and nails contain some of the body's more rigid proteins, which have high sulfur contents.

There is no recommended intake for sulfate, and deficiencies are unknown. Too much sulfate in drinking water, either naturally occurring or from contamination, causes diarrhea and may damage the colon. The summary table at the end of this chapter presents the main facts about sulfate and the other major minerals.

▶ **KEY POINT** Sulfate is a necessary nutrient used to synthesize sulfur-containing body compounds.

The Trace Minerals

An obstacle to determining the precise roles of the trace elements in humans has been the difficulty of providing an experimental diet lacking in the one element under study. Thus, research in this area is limited mostly to the study of laboratory animals, which can be fed highly refined, purified diets in environments free of all contamination. New laboratory techniques have enabled scientists to detect minerals in smaller and smaller quantities in living cells, and research is now rapidly expanding our knowledge about them. Intake recommendations for human beings have been established for nine trace minerals—see Table 8-7. Others are recognized as essential nutrients for some animals, but have not been proved to be required for human beings.

Iodine

The body needs only an infinitesimally small quantity of iodine, but obtaining this amount is critical. Iodine is a part of thyroxine, the hormone made by the thyroid gland that is responsible for regulating the basal metabolic rate. Iodine must be available for thyroxine to be synthesized.

In iodine deficiency, the cells of the thyroid gland enlarge in an attempt to trap as many particles of iodine as possible. Sometimes the gland enlarges until it makes a visible lump in the neck, a **goiter.** People with iodine deficiency this severe become sluggish and gain weight. Severe iodine deficiency during pregnancy causes extreme and irreversible mental and physical retardation in the infant, known as **cretinism.** Much of the mental retardation can be averted if the woman's deficiency is detected and treated within the first six months of pregnancy, but if treatment comes too late or not at all, the child may have an IQ as low as 20 (100 is average). Iodine deficiency is one of the world's most common and most preventable causes of mental retardation.

goiter (GOY-ter) enlargement of the thyroid gland due to iodine deficiency is *simple goiter;* enlargement due to an iodine excess is *toxic goiter.*

cretinism (CREE-tin-ism) severe mental and physical retardation of an infant caused by the mother's iodine deficiency during pregnancy.

TABLE 8-7	TRACE MINERALS	
Human Intake Recommendations Established	Known Essential for Animals; Human Requirements under Study	Known Essential for Some Animals; No Evidence That Intake by Humans Is Ever Limiting
Iodine	Arsenic	Cobalt
Iron	Boron	
Zinc	Nickel	
Selenium	Silicon	
Fluoride	Vanadium	
Chromium		
Copper		
Manganese		
Molybdenum		

NOTE: The evidence for requirements and essentiality is weak for the trace minerals cadmium, lead, lithium, and tin.

In iodine deficiency, the thyroid gland enlarges—a condition known as goiter.

Researchers hope to reverse the high rates of cretinism and goiter reported in developing nations by adding iodine to community food or water supplies.

The iodine in food varies because it reflects the soil in which plants are grown or on which animals graze. Iodine is plentiful in the ocean, so seafood is a dependable source. In the central parts of the United States that were never under the ocean, the soil is poor in iodine. In those areas, the use of iodized salt and the consumption of foods shipped in from iodine-rich areas have wiped out the iodine deficiency that once was widespread. Surprisingly, sea salt delivers little iodine because iodine becomes a gas and flies off into the air during the salt-drying process. In the United States, salt labels state whether the salt is iodized; in Canada, all table salt is iodized.

Excessive intakes of iodine can enlarge the thyroid gland just as a deficiency can. U.S. intakes are above the recommended intake of 150 micrograms, but still below the Tolerable Upper Intake Level of 1,100 micrograms per day for an adult.[38] Like chlorine and fluorine, iodine is a deadly poison in large amounts.

Much of the iodine in U.S. diets today comes from fast-food and other restaurant establishments, which use iodized salt with a liberal hand, and from bakery products and milk. The baking industry uses iodine-containing dough conditioners, and most dairies use iodine to disinfect milking equipment. One cup of milk supplies nearly half of one day's recommended intake of iodine, and less than a half-teaspoon of iodized salt meets the entire recommendation.

An iodine-containing medication, **potassium iodide,** effectively blocks damage to the thyroid gland caused by radioactive iodine released during nuclear radiation emergencies.* When given in the correct dosage within a certain timeframe relative to radiation exposure, potassium iodide can greatly reduce the likelihood of thyroid cancer development.[39] When given in the wrong dosage or with faulty timing, potassium iodide is useless or toxic. For this reason, concerned people who live near nuclear power plants are urged to rely on health professionals for guidance and to ignore quacks selling "anti-radiation pills" on the Internet and elsewhere.[40]

▶ **KEY POINT** Iodine is part of the hormone thyroxine, which influences energy metabolism. The deficiency diseases are goiter and cretinism. Iodine occurs naturally in seafood and in foods grown on land that was once covered by oceans; it is an additive in milk and bakery products. Large amounts are poisonous. Potassium iodide, appropriately administered, blocks some radiation damage to the thyroid during radiation emergencies.

potassium iodide a medication approved by the FDA as safe and effective for the prevention of thyroid cancer caused by radioactive iodine that may be released during radiation emergencies.

*Potassium iodide works only to prevent thyroid cancer after exposure to radioactive iodine; it is not a general radioprotective agent. Read more at the National Academies Press website, www.nap.edu.

Iron

Every living cell, whether plant or animal, contains iron. Most of the iron in the body is a component of two proteins: **hemoglobin** in red blood cells and **myoglobin** in muscle cells. Hemoglobin in the red blood cells carries oxygen from the lungs to tissues throughout the body. Myoglobin carries and stores oxygen for the muscles. Iron helps these proteins to hold and carry oxygen and then release it.

All the body's cells need oxygen to combine with the carbon and hydrogen atoms the cells release as they break down energy nutrients. The oxygen combines with these atoms to form the waste products carbon dioxide and water; thus, the body constantly needs fresh oxygen to keep the cells going. As cells of tissues use up and excrete their oxygen (as carbon dioxide and water), red blood cells shuttle in fresh oxygen supplies from the lungs. In addition to this major task, iron helps many enzymes to use oxygen, and iron is needed to make new cells, amino acids, hormones, and neurotransmitters.

Iron is clearly the body's gold, a precious mineral to be hoarded. The liver packs iron sent from the bone marrow into new red blood cells and ships them out to the bloodstream. Red blood cells live for about three to four months. When they die, the spleen and liver break them down, salvage their iron for recycling, and send it back to the bone marrow to be kept until it is reused. The body does lose iron in nail clippings, hair cuttings, and shed skin cells, but only in tiny amounts. Bleeding can cause significant iron loss from the body, however.

The body has special provisions for obtaining iron. Only about 10 to 15 percent of dietary iron is absorbed; but if the body's supply of iron is diminished or if the need increases (say, during pregnancy), absorption can increase several-fold.[41] Once inside the body, iron is difficult to excrete, so absorption of iron is carefully controlled.[42] Just as absorption increases when iron is deficient, absorption declines when iron is abundant.[43]

Special measures are also needed to contain iron in the body. Left free, iron is a powerful oxidant that can start free-radical reactions that could damage cellular structures. Consequently, the body guards against iron's renegade nature. Protein carriers guard the body's iron molecules and keep them away from vulnerable body compounds, thereby preventing damaging reactions.[44] Iron's actions are thus tightly controlled.

The Genes and Iron Absorption   Control of iron absorption depends partly on the activities of genes in the absorptive cells that line the intestine. Researchers believe that messages arising when the body lacks iron trigger these genes to produce more of the proteins that work to absorb iron.[45] With more iron-absorbing proteins, the intestinal cells can absorb more iron from each meal.[46] As the body's iron supply builds up in response, the genes receive the opposite message—to produce fewer of these proteins to prevent absorption of too much iron. By this method, the genes of the intestinal cells help to balance the absorption of iron from the diet with the body's need for iron.

What Happens to a Person Who Lacks Iron?   If absorption cannot compensate for losses or low dietary intakes, then iron stores are used up and iron deficiency sets in. **Iron deficiency** and **iron-deficiency anemia** are not one and the same, though they often occur together. The distinction between iron deficiency and its anemia is a matter of degree. People may be iron deficient, meaning that they have depleted iron stores, without being anemic, or they may be iron deficient *and* anemic. With regard to iron, the term **anemia** refers to severe depletion of iron stores resulting in low blood hemoglobin.

A body severely deprived of iron becomes unable to make enough hemoglobin to fill new blood cells, and anemia results. A sample of iron-deficient blood examined under the microscope shows cells that are smaller and lighter red than normal (see Figure 8-9). The undersized cells contain too little hemoglobin and thus deliver too little oxygen to the tissues. The diminished supply of oxygen limits the cells' energy metabolism and causes tiredness, apathy, and a tendency to feel cold.

■ Chapter 7 described free radicals and their control.

■ Chapter 6 described the influence of iron stores on the production of the protein hemoglobin.

hemoglobin (HEEM-oh-globe-in) the oxygen-carrying protein of the blood; found in the red blood cells (*hemo* means "blood"; *globin* means "spherical protein").

myoglobin (MYE-oh-globe-in) the oxygen-holding protein of the muscles (*myo* means "muscle").

iron deficiency the condition of having depleted iron stores, which, at the extreme, causes iron-deficiency anemia.

iron-deficiency anemia a form of anemia caused by a lack of iron and characterized by red blood cell shrinkage and color loss. Accompanying symptoms are weakness, apathy, headaches, pallor, intolerance to cold, and inability to pay attention. (For other anemias, see the index.)

anemia the condition of inadequate or impaired red blood cells; a reduced number or volume of red blood cells along with too little hemoglobin in the blood. The red blood cells may be immature and, therefore, too large or too small to function properly. Anemia can result from blood loss, excessive red blood cell destruction, defective red blood cell formation, and many nutrient deficiencies. Anemia is not a disease, but a symptom of another problem; its name literally means "too little blood."

FIGURE 8·9 NORMAL AND ANEMIC BLOOD CELLS

© L. V. Bergman & Associates/Corbis (both)

Normal red blood cells. Both size and color are normal.

Blood cells in iron-deficiency anemia. These cells are small and pale because they contain less hemoglobin.

TABLE 8·8 THE MENTAL SYMPTOMS OF ANEMIA

Apathy, listlessness
Behavior disturbances
Clumsiness
Hyperactivity
Irritability
Lack of appetite
Learning disorders (vocabulary, perception)
Low scores on latency and associative reactions
Lowered IQ
Reduced physical work capacity
Repetitive hand and foot movements
Shortened attention span

NOTE: These symptoms are not caused by anemia itself but by iron deficiency in the brain. Children with much more severe anemias from other causes, such as sickle-cell anemia and thalassemia, show no reduction in IQ when compared with children without anemia.

Even slightly lowered iron levels cause fatigue and impair physical work capacity and productivity.[47] Many of the symptoms associated with iron deficiency are easily mistaken for behavioral or motivational problems (see Table 8-8). With reduced energy, people work less, play less, and think or learn less eagerly. (Lack of energy does not always mean an iron deficiency—see the Think Fitness feature.) Children deprived of iron become restless, irritable, unwilling to work or play, and unable to pay attention, and they may fall behind their peers academically. Some symptoms, such as irritability, disappear when iron intake improves but others, such as academic failure, may linger after iron repletion.[48]

THINK FITNESS   Exercise-Deficiency Fatigue

ON HEARING ABOUT symptoms of iron deficiency, tired people may jump to the conclusion that they need to take iron supplements to restore their pep. More likely, they can obtain help by simply getting to bed on time and getting enough exercise. Few realize that too little exercise over weeks and months is as exhausting as too much—the less you do, the less you're able to do, and the more fatigued you feel. The condition even has a name: "sedentary inertia."

Feeling fatigued, weak, and apathetic is a sign that something is wrong. It is not a sign that you necessarily need iron or other supplements. Three actions are called for:

• First, get your diet in order.
• Second, get some exercise.
• Third, if symptoms persist for more than a week or two, consult a physician for a diagnosis.

A curious symptom seen in some people with iron deficiency is an appetite for ice, clay, paste, soil, or other nonnutritious substances. This consumption of nonfood substances, or **pica,** is most often observed in poverty-stricken women and children, in the mentally ill, and in people with kidney failure who must have their blood cleansed by machine.[49] In some cases, pica clears up within days after iron is given, even before the red blood cells have had a chance to respond. Other times, pica is unresponsive to iron.

Because pica sometimes occurs with iron deficiency and some soils contain iron, folklore has it that pica develops because the body craves what it needs—iron. Iron-rich clays and soils often contain substances that interfere with iron absorption, however, so eating clay and soil is unlikely to benefit iron status.[50] Instead, nutrition and health suffer when nonfood items displace nutritious foods from the diet and when toxic contaminants such as lead and dangerous blood-depleting parasites gain entrance to the body by way of contaminated clay and soil.[51]

■ Iron deficiency makes children more susceptible to lead poisoning. See Chapter 14.

pica (PIE-ka) a craving for nonfood substances. Also known as *geophagia* (gee-oh-FAY-gee-uh) when referring to clay eating and *pagophagia* (pag-oh-FAY-gee-uh) when referring to ice craving (*geo* means "earth"; *pago* means "frost"; *phagia* means "to eat").

Causes of Iron Deficiency and Anemia Iron deficiency is usually caused by malnutrition, that is, inadequate iron intake, either from sheer lack of food or from high consumption of the wrong foods. In the developed countries, overconsuming foods rich in sugar and fat and poor in nutrients is often responsible for low iron intakes. Snapshot 8-5 shows some foods that are good or excellent sources of iron.

Among nonnutritional causes of anemia, blood loss is number one. Because 80 percent of the iron in the body is in the blood, losing blood means losing iron. Because of menstrual losses, women need one and a half times as much iron as men do. Women are especially vulnerable to iron deficiency because they not only need more iron than men but they also, on average, eat less food. Infants over six months of age, young children, adolescents, menstruating women, and pregnant women all have increased need for iron to support the growth of new body tissues or replace losses or, in the case of adolescent girls, both.

Worldwide, iron deficiency is the most common nutrient deficiency, affecting more than 1.2 billion people.[52] In developing countries, parasitic infections of the digestive tract cause people to lose blood daily. For their entire lives, they may feel fatigued and listless but never know why. Digestive tract problems such as ulcers, sores, and even inflammation can also cause blood loss severe enough to cause anemia.[53] Almost half of the preschool children and pregnant women in these countries suffer from iron-deficiency anemia.[54]

In the United States, iron deficiency is less prevalent but still affects 10 percent of toddlers, adolescent girls, and women of childbearing age; preventing and correcting iron deficiency are high national priorities.[55] Happily, the iron status of U.S. infants and young children has improved over the last two decades, thanks to more widespread

SNAPSHOT 8·5

IRON

DRI RECOMMENDED INTAKES:
Men: 8 mg/day
Women (19–50 yr): 18 mg/day
Women (51+): 8 mg/day

TOLERABLE UPPER INTAKE LEVEL:
Adults: 45 mg/day

CHIEF FUNCTIONS: Carries oxygen as part of hemoglobin in blood or myoglobin in muscles; required for cellular energy metabolism

DEFICIENCY: Anemia: weakness, fatigue, headaches; impaired mental and physical work performance; impaired immunity; pale skin, nailbeds, and mucous membranes; concave nails; chills; pica

TOXICITY: GI distress; with chronic iron overload, infections, fatigue, joint pain, skin pigmentation, organ damage

*These foods provide 10 percent or more of the iron Daily Value in a serving. For a 2,000-calorie diet, the DV is 18 mg/day.

NOTE: Dried figs contain 0.6 mg per ¼ cup; raisins contain 0.8 mg per ¼ cup.

ªSome clams may contain less, but most types are iron-rich foods.

ᵇLegumes contain phytates that reduce iron absorption. Soaking in water before cooking reduces phytates, and consuming legumes with vitamin C or meats increases iron absorption.

ᶜEnriched cereals vary widely in iron content.

GOOD SOURCES*

CLAMSª (steamed)
3 oz = 23.8 mg

ENRICHED CEREALᶜ
(ready-to-eat)
1 c = 8.1 mg

BEEF STEAK (lean)
3 oz = 2.6 mg

SPINACH (cooked)
½ c = 3.2 mg

NAVY BEANSᵇ (cooked)
½ c = 2.3 mg

SWISS CHARD (cooked)
½ c = 3.2 mg

BLACK BEANS (cooked)
½ c = 1.8 mg

BEEF LIVER (cooked)
3 oz = 5.24 mg

breastfeeding, which promotes iron absorption, and greater use of iron-fortified infant formula and cereals. For low-income families, the Special Supplemental Food Program for Women, Infants, and Children (WIC) provides coupons redeemable for foods high in iron, giving another boost to the iron status of many U.S. children.

Can a Person Take in Too Much Iron?   Iron is toxic in large amounts, and once absorbed inside the body, it is difficult to excrete. The healthy body defends against **iron overload** by controlling its entry: the intestinal cells trap some of the iron and hold it within their boundaries. When they are shed, these cells carry out of the intestinal tract the excess iron that they collected during their brief lives.[56] In healthy people, when iron stores fill up, less iron is absorbed, protecting them against iron overload.[57]

Iron overload, which has a strong genetic component, may occur more often than previously suspected, especially among Caucasian people.[58] In hereditary iron overload, the intestine continues to absorb iron at a high rate despite the excess iron building up in the body tissues.[59] Early symptoms are general and vague, such as fatigue, mental depression, and abdominal pains. Later, tissue damage occurs, with liver failure, abnormal heartbeats, and diabetes.[60] Infections are also likely because bacteria thrive on iron-rich blood. The effects are most severe in alcohol abusers because alcohol damages the intestine, impairing its defense against absorbing too much iron.

The danger of iron overload is an argument against high-level iron fortification of foods. Susceptible people would have trouble following a low-iron diet if most foods were doused with iron. A susceptible man choosing a single ounce of fortified cereal for breakfast, an ordinary ham sandwich at lunch, and a cup of chili with meat for dinner would consume almost double his recommended iron for the day in only about 800 calories of food. Worsening the picture is the U.S. population's love of vitamin C supplements because vitamin C greatly enhances iron absorption.[61] For healthy people, however, even fortified foods pose virtually no risk for iron toxicity.[62]

■ Table 8-11 (p. 297) summarizes the effects of iron toxicity.

As for iron supplements, they can reverse iron-deficiency anemia from dietary causes in short order. However, they may create oxidative reactions in the digestive tract that may damage its linings, particularly in those suffering with inflammation of these tissues.[63] Iron supplements are a leading cause of fatal accidental poisonings among U.S. children under six years old.[64] High-dose iron pills may soon come packaged in individually sealed units to help prevent such poisonings. For now, keep iron supplements out of children's reach.

Iron Recommendations and Sources   Men need 8 milligrams of iron each day, and so do women past age 51. For women of childbearing age, the recommendation is higher—18 milligrams—to replace menstrual losses. During pregnancy, a woman needs significantly more—27 milligrams. Adult men rarely experience iron-deficiency anemia. If a man has a low hemoglobin concentration, his health-care provider should examine him for a blood-loss site. Vegetarians, because the iron in their foods is not well absorbed and because their diets lack factors from meat that enhance iron absorption, are advised to obtain 1.8 times the normal requirement (see the margin). Table 8-9 (p. 290) sums up iron recommendations.

To calculate the DRI recommendation for vegetarians, multiply by 1.8:
- Vegetarian men:
  8 mg × 1.8 = 14 mg/day.
- Vegetarian women (19 to 50 yr):
  18 mg × 1.8 = 32 mg/day.

To meet your iron needs, it is best to rely on foods because the iron from supplements is much less well absorbed than that from food. The usual Western mixed diet provides only about 5 to 6 milligrams of iron in each 1,000 calories. An adult male who eats 2,500 calories or more a day has no trouble obtaining his needed 8 milligrams or more, but a woman who eats fewer calories and needs more iron cannot obtain her needed 18 milligrams unless she selects high-iron, low-calorie foods from each food group. Pregnant women need an iron supplement.[65] No one should take iron supplements without a physician's recommendation, however.

Absorbing Iron   Iron occurs in two forms in foods. Some is bound into **heme,** the iron-containing part of hemoglobin and myoglobin in meat, poultry, and fish (look back at Figure 6-4 in Chapter 6, p. 183). Some is nonheme iron, found in foods from plants and in the nonheme iron in meats. The form affects absorption. Heme iron is much more

iron overload the state of having more iron in the body than it needs or can handle, usually arising from a hereditary defect. Also called *hemochromatosis.*

heme (HEEM) the iron-containing portion of the hemoglobin and myoglobin molecules.

TABLE 8·9 IRON INTAKE
GUIDELINES

DRI Recommendations
- Recommended intakes:
  Men: 8 mg per day.
  Women:
  Childbearing years: 18 mg per day.
  51 years and older: 8 mg per day.
  Pregnancy: 27 mg per day.
- Tolerable Upper Intake Levels:
  Infants and children: 40 mg per day.
  Adolescents and adults: 45 mg per day.

This chili provides iron and MFP factor from meat, iron from legumes, and vitamin C from tomatoes. The combination of heme iron, nonheme iron, MFP factor, and vitamin C helps to achieve maximum iron absorption.

Dietary factors that increase iron absorption:
- Vitamin C.
- MFP factor.

Factors that hinder iron absorption:
- Tea.
- Coffee.
- Calcium and phosphorus.
- Phytates, tannins, and fiber.

MFP factor a factor present in meat, fish, and poultry that enhances the absorption of nonheme iron present in the same foods or in other foods eaten at the same time.

tannins compounds in tea (especially black tea) and coffee that bind iron. Tannins also denature proteins.

phytates (FYE-tates) compounds present in plant foods (particularly whole grains) that bind iron and may prevent its absorption.

reliably absorbed than nonheme iron. Healthy people with adequate iron stores absorb heme iron at a rate of about 23 percent over a wide range of meat intakes. People absorb nonheme iron at rates of 2 to 20 percent, depending on dietary factors and iron stores.

Meat, fish, and poultry contain a factor (**MFP factor**) that promotes the absorption of nonheme iron from other foods eaten at the same time. Vitamin C can triple nonheme iron absorption from foods eaten in the same meal.[66] A system of calculating the amount of iron absorbed from a meal, based on these factors, is presented in Table 8-10.

Some substances impair iron absorption. They include the **tannins** of tea and coffee, the calcium and phosphorus in milk, and the **phytates** that accompany fiber in lightly processed legumes and whole-grain cereals.[67] Ordinary black tea is exceptional in its efficiency at reducing absorption of iron—clinical dietitians advise people with iron overload to drink it with their meals. For those who need more iron, the opposite advice applies—drink tea between meals, not with food.[68]

Thus, the amount of iron *absorbed* from a meal depends partly on the interaction between promoters and inhibitors of iron absorption. When you eat meat with legumes (for example, ham and beans or chili with beans and meat), the iron from the meat is well absorbed, and MFP factor enhances iron absorption from the beans. The vitamin C from a slice of tomato and a leaf of lettuce in a sandwich will enhance iron absorption from the bread. The bit of vitamin C in dried fruit, strawberries, or watermelon helps absorb the nonheme iron in these foods. The meat and tomato in spaghetti sauce help absorb the iron from the spaghetti.

TABLE 8·10 CALCULATION OF IRON ABSORBED FROM MEALS

You need to know three factors to calculate the amount of iron absorbed from a meal:
1. How much of the iron in the meal was heme iron, and how much was nonheme iron?
2. How much vitamin C was in the meal?
3. How much total meat, fish, and poultry (MFP factor) was consumed?
(It is assumed that your iron stores are moderate; otherwise, you'd have to take this into consideration, too.) Write down the foods you eat at a typical meal, look up their iron content in the Table of Food Composition, Appendix A, and then answer these questions:
1. How much iron was from animal tissues (MFP)? _____ mg
2. 40% of (1), on the average, is heme iron:
   (1) _____ mg × 0.40 = _____ mg heme iron
3. How much iron was from other sources? _____ mg
4. Add (3) and 60% of (1) to get nonheme iron:
   (3) _____ mg + 0.60 × (1) _____ mg = _____ mg nonheme iron
5. How much vitamin C was in the meal? _____ mg   Less than 25 mg is low; 25 to 75 mg is medium; more than 75 mg is high: _____
6. How much MFP factor was in the meal? _____ oz   Less than 1 oz lean MFP is low; 1 to 3 oz is medium; more than 3 oz is high: _____
7. Now you can calculate the amount of each type of iron that was absorbed. Start with the heme iron. You absorbed 25% of the heme iron:
   (2) _____ mg × 0.25 = _____ mg heme iron absorbed
8. Consider your scores from (5) and (6). If either vitamin C or MFP factor was high or if both were medium, the availability of your nonheme iron was **high.** If neither was high, but one was medium, the availability of your nonheme iron was **medium.** If both were low, your nonheme iron had **poor** availability. You absorbed:
   - High availability: 10% of the nonheme iron
   - Medium availability: 8% of the nonheme iron
   - Poor availability: 5% of the nonheme iron
9. Now you can calculate the nonheme iron absorbed. You absorbed _____ % of the nonheme iron, or:
   (4) _____ mg × (8) _____ % = _____ mg nonheme iron absorbed
10. Add the heme and nonheme iron from (7) and (9) together:
    (7) _____ mg heme iron absorbed
    + (9) _____ mg nonheme iron absorbed
    Total = _____ mg iron absorbed.

According to the DRI committee, if you are a man age 19 or older or a woman age 51 or older, you need to absorb about 1 mg per day. If you are a woman 11 to 50 years old, you need to absorb 1.5 mg per day on average. If you have higher menstrual losses than the average woman, you may need still more.

Cooking the sauce in an iron pan also adds more iron. Foods cooked in iron pans contain iron salts somewhat like those in supplements. The iron content of 100 grams of spaghetti sauce simmered in a glass dish is 3 milligrams, but it increases to 87 milligrams when the sauce is cooked in a black iron skillet. In the short time it takes to scramble eggs, a cook can triple the eggs' iron content by scrambling them in an iron pan. This iron salt is not as well absorbed as iron from meat, but some does get into the body, especially if the meal also contains MFP factor or vitamin C.

▶ KEY POINT   Most iron in the body is contained in hemoglobin and myoglobin or occurs as part of enzymes in the energy-yielding pathways. Iron-deficiency anemia is a problem worldwide; too much iron is toxic. Iron is lost through menstruation and other bleeding; reduced absorption and the shedding of intestinal cells protect against overload. For maximum iron absorption, use meat, other iron sources, and vitamin C together.

Zinc

Zinc occurs in a very small quantity in the human body, but it works with proteins in every organ, helping nearly 100 enzymes to:

- Make parts of the cells' genetic material.
- Make heme in hemoglobin.
- Assist the pancreas with its digestive functions.
- Help metabolize carbohydrate, protein, and fat.
- Liberate vitamin A from storage in the liver.

Besides helping enzymes function, zinc helps to regulate gene expression in protein synthesis. Zinc also affects behavior and learning; assists in immune function; and is essential to wound healing, sperm production, taste perception, fetal development, and growth and development in children. Zinc is needed to produce the active form of vitamin A in visual pigments. A protective role for zinc in oxidative damage is also under investigation.[69] When zinc deficiency occurs, it packs a wallop to the body, impairing all these functions. Even a mild zinc deficiency can result in impaired immunity, abnormal taste, and abnormal vision in the dark.

Problem: Too Little Zinc   Zinc deficiency in human beings was first reported in the 1960s from studies with growing children and adolescent boys in the Middle East. Their native diets were typically low in animal protein and high in whole grains and beans; consequently, the diets were high in fiber and phytates, which bind zinc as well as iron. Furthermore, the bread was not **leavened;** in leavened bread, yeast breaks down phytates as the bread rises.

Since the first reports, zinc deficiency has been recognized elsewhere, and it affects much more than growth. It alters digestive function profoundly and causes diarrhea, which worsens the malnutrition already present, with respect not only to zinc but to all nutrients. It drastically impairs the immune response, making infections likely.[70] Infections of the intestinal tract worsen malnutrition, including zinc malnutrition. Even mild zinc deficiency, brought on after one month of consuming a low-zinc diet, causes imbalances in the body's immune system that can increase susceptibility to infections. In developing countries, zinc treatment for children with infectious diseases reduces diarrhea and death.[71]

Normal vitamin metabolism depends on zinc, so zinc-deficiency symptoms often include vitamin-deficiency symptoms. Zinc deficiency also disturbs thyroid function and slows the body's energy metabolism, causing loss of appetite and slowing wound healing. In laboratory animals, a mild deficiency may reduce physical activity, memory, and attention span.[72] The symptoms are so pervasive that when faced with zinc deficiency, physicians are more likely to diagnose it as general malnutrition and sickness than as zinc deficiency.

Courtesy of Ray Stanyard

The old-fashioned iron skillet adds supplemental iron to foods.

leavened (LEV-end) literally, "lightened" by yeast cells, which digest some carbohydrate components of the dough and leave behind bubbles of gas that make the bread rise.

H. Sanstead, University of Texas-Galveston

How old does the boy in the picture appear to be? He is 17 years old but is only 4 feet tall, the height of a seven-year-old in the United States. His genitalia are like those of a six-year-old. The retardation is rightly ascribed to zinc deficiency because it is partially reversible when zinc is restored to the diet. The photo was taken in Egypt.

Although severe zinc deficiencies are not widespread in developed countries, they occur among some groups, including pregnant women, young children, the elderly, and the poor. When pediatricians or other health workers evaluating children's health note poor growth accompanied by poor appetite, they should think zinc.

Problem: Too Much Zinc   Zinc is toxic in large quantities, and zinc supplements can cause serious illness or even death in high enough doses. Regular doses of zinc only a few milligrams above the recommended intake, taken over time, block copper absorption and lower the body's copper content. In animals, this effect leads to degeneration of the heart muscle. In high doses, zinc may reduce the concentration of beneficial high-density lipoproteins (HDL) in the blood.

High doses of zinc can also inhibit iron absorption from the digestive tract. A protein in the blood that carries iron from the digestive tract to tissues also carries some zinc. If this protein is burdened with excess zinc, little or no room is left for iron to be picked up from the intestine. The opposite is also true: too much iron leaves little room for zinc to be picked up, thus impairing zinc absorption. Zinc and iron are often found together in foods, but food sources are safe and never cause imbalances in the body. Supplements, in contrast, can easily do so. Zinc from lozenges and sprays sold for treatment of the common cold may or may not provide the intended relief, but their use contributes supplemental zinc to the body.[73]

Unlike excess iron, excess zinc has a normal escape route from the body. The pancreas secretes zinc-rich juices into the digestive tract, and some of these are excreted. Still, large overdoses from zinc supplements can overwhelm the escape route and cause toxicity.

Food Sources of Zinc   Meats, shellfish, and poultry and milk and milk products are among the top providers of zinc in the U.S. diet (see Snapshot 8-6).[74] Among plant sources, some legumes and whole grains are rich in zinc, but the zinc is not as well absorbed as from meat. Most people meet the recommended 11 milligrams per day for men and 8 milligrams per day for women. Vegetarians are advised to eat varied diets that include zinc-enriched cereals or whole-grain breads well leavened with yeast, which helps make zinc available for absorption.

▶ KEY POINT   Zinc assists enzymes in all cells. Deficiencies in children cause growth retardation with sexual immaturity. Zinc supplements can reach toxic doses, but zinc in foods is nontoxic. Animal foods are the best sources.

Selenium

Selenium has attracted the attention of the world's scientists for its role in protecting vulnerable body chemicals against oxidative destruction. Selenium assists a group of enzymes that, in concert with vitamin E, work to prevent the formation of free radicals and prevent oxidative harm to cells and tissues.[75] For example, cells of the immune system generate oxidizing compounds when they destroy foreign invaders, and the selenium-dependent enzymes reduce these compounds to harmless by-products that can be safely metabolized by body tissues. Selenium also plays roles in activating thyroid hormone, the hormone that regulates the body's rate of metabolism. Low blood selenium correlates with the development of some forms of cancers, especially prostate cancer in men.

Prostate cancer is the fourth leading cancer in men worldwide, and black men in the United States suffer the highest rate of all. Men who have adequate selenium in their bloodstreams contract prostate cancer less often than men whose blood measures are low.[76] Supplement supporters tout results like these as proof that men should take selenium supplements to ward off cancer, but no benefit is observed from taking extra selenium. Furthermore, supplements may slightly raise the risk of a form of skin cancer.[77] Studies reveal that only men whose blood is low in selenium in the first place benefit from supplements that correct their deficiency.[78] Clearly, adequate selenium is important

■ Chapter 3 described how immune cells attack invaders with an "oxidative burst."

ZINC

DRI RECOMMENDED INTAKES:
Men: 11 mg/day
Women: 8 mg/day

TOLERABLE UPPER INTAKE LEVEL:
Adults: 40 mg/day

CHIEF FUNCTIONS: Activates many enzymes; associated with hormones; synthesis of genetic material and proteins, transport of vitamin A, taste perception, wound healing, reproduction

DEFICIENCY:[a] Growth retardation, delayed sexual maturation, impaired immune function, hair loss, eye and skin lesions, loss of appetite

TOXICITY: Loss of appetite, impaired immunity, reduced copper and iron absorption, low HDL cholesterol (a risk factor for heart disease)

*These foods provide 10 percent or more of the zinc Daily Value in a serving. For a 2,000-calorie diet, the DV is 15 mg/day.

[a]A rare inherited form of zinc malabsorption causes additional and more severe symptoms.

[b]Some oysters contain more or less than this amount, but all types are zinc-rich foods.

[c]Enriched cereals vary widely in zinc content.

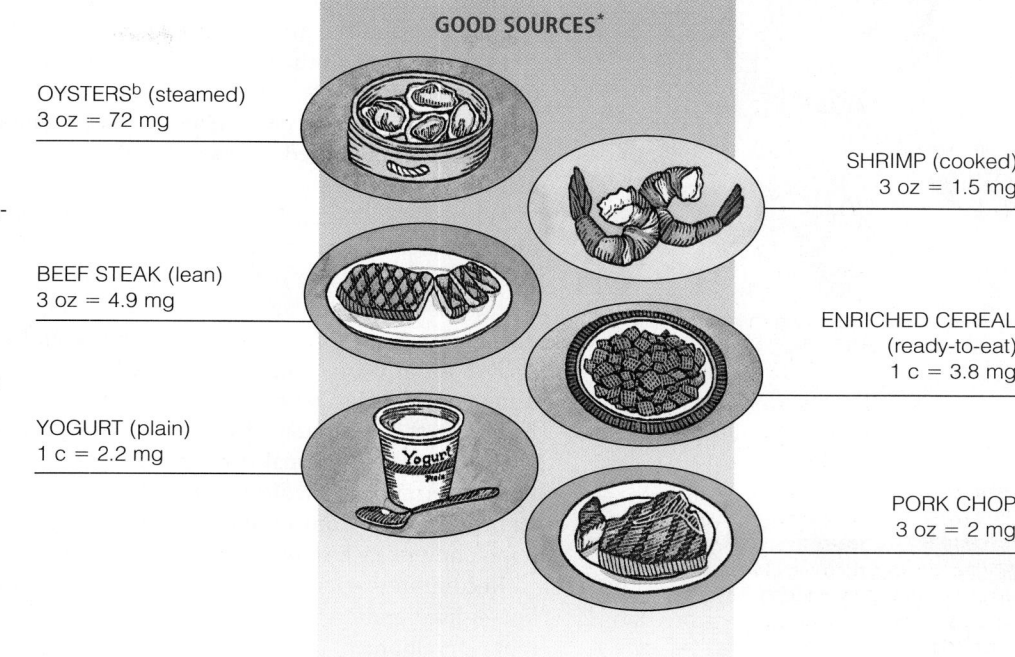

GOOD SOURCES*

OYSTERS[b] (steamed)
3 oz = 72 mg

SHRIMP (cooked)
3 oz = 1.5 mg

BEEF STEAK (lean)
3 oz = 4.9 mg

ENRICHED CEREAL
(ready-to-eat)
1 c = 3.8 mg

YOGURT (plain)
1 c = 2.2 mg

PORK CHOP
3 oz = 2 mg

but research is lacking to support taking selenium supplements in excess of the DRI recommended intake.

A deficiency of selenium can open the way for a specific type of heart disease (unrelated to the heart disease discussed in Chapters 5 and 11). The condition, first identified in China among people from areas with selenium-deficient soils, prompted researchers to place this mineral among the essential nutrients (see the inside front cover for selenium's intake recommendation).

If you eat a normal diet composed of mostly unprocessed foods, you need not worry about selenium. It is widely distributed in foods such as meats and shellfish and in vegetables and grains grown on selenium-rich soil. Soils in the United States and Canada vary in selenium, but foods from many locations across these nations mingle together on supermarket shelves, ensuring that consumers are well supplied with selenium.

Toxicity is possible when people take selenium supplements over a long period. Selenium toxicity brings on symptoms such as hair loss, diarrhea, and nerve abnormalities. The Tolerable Upper Intake Level for selenium is set at 400 micrograms per day.

▶ **KEY POINT** Selenium works with an enzyme system to protect body compounds from oxidation. A deficiency induces a disease of the heart. Deficiencies are rare in developed countries, but toxicities can occur from overuse of supplements.

Fluoride

Fluoride is not essential to life, but it is beneficial in the diet because of its ability to inhibit the development of dental caries in both children and adults. Only a trace of fluoride occurs in the human body, but the crystalline deposits in bones and teeth are

Fluoride helps prevent caries in three ways:

In developing teeth:
• Forms decay-resistant crystals.

In erupted teeth:
• Promotes remineralization.
• Reduces acidity of plaque.

To prevent fluorosis, young children should not swallow toothpaste.

Caroline Fleischer

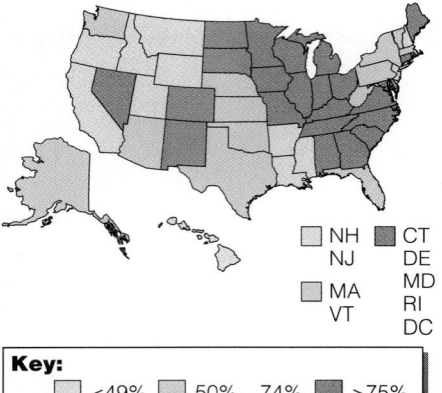

FIGURE 8·10 PERCENTAGE OF STATE POPULATIONS WITH ACCESS TO FLUORIDATED WATER THROUGH PUBLIC WATER SYSTEMS

NH   CT
NJ   DE
     MD
MA   RI
VT   DC

Key:
<49%   50% – 74%   >75%

SOURCE: Centers for Disease Control and Prevention, Recommendations for using fluoride to prevent and control dental caries in the United States, *Morbidity and Mortality Weekly Report* (supplement), August 17, 2001, p. 10.

fluorosis (floor-OH-sis) discoloration of the teeth due to ingestion of too much fluoride during tooth development.

larger and more perfectly formed because this fluoride replaces the hydroxy portion of hydroxyapatite, forming the more decay-resistant fluorapatite in developing teeth. Once teeth have erupted through the gums, fluoride helps prevent dental caries by promoting the remineralization of early lesions of the enamel that might otherwise progress to form caries. Fluoride also acts directly on the bacteria of plaque, suppressing their metabolism and reducing the amount of acid they produce.

Drinking water is the usual source of fluoride. Over 65 percent of the U.S. population has access to water with an optimal fluoride concentration, which typically delivers about 1 milligram per person per day, or about 1 part per million.[79] Fluoride is rarely present in bottled waters unless it was added at the source, as in bottled municipal tap water.

Where fluoride is lacking, the incidence of dental decay is very high, and fluoridation of water is recommended for public dental health.[80] Fluoridation is a practical, safe, and cost-effective way to help prevent dental caries in the young, and its widespread use has been a major factor in reducing U.S. dental caries.[81] Sufficient fluoride during the tooth-forming years of infancy and childhood gives lifetime protection against tooth decay. Some uninformed fluoride opponents claim that communities using fluoridated water have an increased cancer rate, but studies show no connection. Based on the accumulated evidence of its beneficial effects, fluoridation has been endorsed by the National Institute of Dental Health, the American Dietetic Association, the American Medical Association, the National Cancer Institute, and the Centers for Disease Control and Prevention. Figure 8-10 shows the percentage of the population in each state with access to fluoridated water.

In communities where the water contains too much fluoride—2 to 8 parts per million—discoloration of the teeth, or **fluorosis,** may occur. Fluorosis occurs only during tooth development, never after the teeth have formed—and it is irreversible. Widespread availability of fluoridated toothpaste and mouthwash, foods made with fluoridated water, and fluoride-containing supplements has led to an increase in the mildest form of fluorosis. In this condition, characteristic white spots form in the tooth enamel; a more severe form is shown in Figure 8-11. To prevent fluorosis, people in areas with fluoridated water should limit other sources, such as fluoride-enriched formula for infants and fluoride supplements for infants or children unless prescribed by a physician. Children younger than six years should use only a pea-sized squeeze of toothpaste and should be told not to swallow their toothpaste when brushing their teeth. The Tolerable Upper Intake Level for fluoride for all people older than 8 years is 10 mg per day.

▶ **KEY POINT** Fluoride stabilizes bones and makes teeth resistant to decay. Excess fluoride discolors teeth; large doses are toxic.

Chromium

Chromium works closely with the hormone insulin to regulate and release energy from glucose. When chromium is lacking, insulin action is impaired, resulting in a diabetes-like condition of high blood glucose that resolves with chromium supplementation. Supplements of chromium cannot treat or cure the common forms of diabetes and most people with diabetes are not chromium deficient.[82] Therefore, supplements are not recommended for management of type 2 diabetes. Diets high in simple sugars and low in whole, nutrient-dense foods deplete the body's supply of chromium.

Chromium-containing supplements will *not* build extra muscle tissue or melt off body fat or ward off its regain as popular magazines may profess. Chromium supplements have been reported to slightly increase lean body mass in laboratory animals and sometimes in human beings tested under laboratory conditions. These results led to exaggerated claims for chromium's ability to bring about weight loss and muscle gain, but follow-up studies have shown no effect of chromium on body fat and lean tissue.

Likewise, chromium supplementation does not lower blood cholesterol in people, an effect that has been implied in the "information" provided by supplement marketers.

Chromium compounds used in various industrial processes are known carcinogens and are responsible for many cases of cancer in exposed workers. The form of chromium in foods and supplements is nontoxic by comparison, and amounts of 200 micrograms per day seem to be safe. Supplements may cause skin eruptions, however, and taking large doses is ill-advised.[83]

Chromium is widely distributed in the food supply, especially in unrefined foods and whole grains. It exists in foods in complexes with other compounds that make it easily controlled and used by the body. Researchers use the terms *biologically active chromium* or *glucose tolerance factor* to describe these chromium-containing compounds. Chromium is lost during food processing, and chromium deficiencies become more likely as people depend more heavily on refined foods. The best chromium food sources are liver, whole grains, nuts, and cheeses. It is estimated that 90 percent of U.S. adults consume less than the recommended minimum intake of 50 micrograms a day.

▶ KEY POINT   Chromium works with the hormone insulin to control blood glucose concentrations. Chromium is present in a variety of unrefined foods.

FIGURE 8·11   FLUOROSIS

The brown mottled stains on these teeth indicate exposure to high concentrations of fluoride during development.

Copper

One of copper's most vital roles is to help form hemoglobin and collagen. In addition, many enzymes depend on copper for its oxygen-handling ability. Copper plays roles in the body's handling of iron and, like iron, assists in reactions leading to the release of energy.[84] One copper-dependent enzyme helps to control damage from free-radical activity in the tissues.* Researchers are investigating the possibility that a low-copper diet may contribute to heart disease by suppressing the activity of this enzyme.

Copper deficiency is rare but not unknown: it has been seen in severely malnourished infants fed a copper-poor milk formula.[85] Deficiency can severely disturb growth and metabolism, and in adults, it can impair immunity and blood flow through the arteries. Excess zinc interferes with copper absorption and can cause deficiency.

Copper toxicity from foods is unlikely, but supplements can cause it. The Tolerable Upper Intake Level for adults is set at 10,000 micrograms (10 milligrams) per day. The best food sources of copper include organ meats, seafood, nuts, and seeds. Water may also supply copper, especially where copper plumbing pipes are used. In the United States, copper intakes are thought to be adequate.[86]

▶ KEY POINT   Copper is needed to form hemoglobin and collagen and assists in many other body processes. Copper deficiency is rare.

Other Trace Minerals and Some Candidates

DRI intake recommendations have been established for two other trace minerals, molybdenum and manganese. Molybdenum functions as part of several metal-containing enzymes, some of which are giant proteins. Manganese works with dozens of different enzymes that facilitate body processes.

Several other trace minerals are now recognized as important to health. Boron influences the activity of many enzymes, and research suggests that a low intake of boron may enhance susceptibility to osteoporosis by way of its effects on calcium

---

*The enzyme is superoxide dismutase.

metabolism.[87] The richest food sources of boron are noncitrus fruits, leafy vegetables, nuts, and legumes. Cobalt is the mineral in the vitamin B_{12} molecule; the alternative name for vitamin B_{12}, *cobalamin,* reflects cobalt's presence. Nickel is important for the health of many body tissues; deficiencies harm the liver and other organs. Silicon is known to be involved in bone calcification in animals. Future research may reveal key roles played by other trace minerals including barium, cadmium, lead, lithium, mercury, silver, tin, and vanadium. Even arsenic, a known poison and carcinogen, may turn out to be essential in tiny quantities.

All trace minerals are toxic in excess, and Tolerable Upper Intake Levels exist for boron, nickel, and vanadium (see the inside front cover, p. C). Overdoses are most likely to occur in people who take multiple nutrient supplements. The way to obtain the trace minerals is from food, which is not hard to do—just eat a variety of whole foods in the amounts recommended in Chapter 2. Some claim that organically grown foods contain more trace minerals than those grown with chemical fertilizers. Organic fertilizers do contain more trace minerals than do refined chemical fertilizers, and plants do take up some of the minerals they are given. Controversy 12 considers the merits and demerits of foods grown organically.

Research on the trace minerals is uncovering many interactions among them: an excess of one may cause a deficiency of another. A slight manganese overload, for example, may aggravate an iron deficiency. A deficiency of one mineral may open the way for another to cause a toxic reaction. Iron deficiency, for example, makes the body much more susceptible to lead poisoning. Good food sources of one are poor food sources of another, and factors that cooperate with some trace elements oppose others. The continuous outpouring of new information about the trace minerals is a sign that we have much more to learn. Table 8-11 sums up what this chapter has said about the minerals and fills in some additional information.

▶ KEY POINT   Many different trace elements play important roles in the body. All of the trace minerals are toxic in excess.

TABLE 8·11 THE MINERALS—A SUMMARY

Mineral and Chief Functions in the Body	Deficiency Symptoms	Toxicity Symptoms	Significant Sources
Major Minerals			
Calcium			
The principal mineral of bones and teeth. Also acts in normal muscle contraction and relaxation, nerve functioning, regulation of cell activities, blood clotting, blood pressure, and immune defenses.	Stunted growth in children; adult bone loss (osteoporosis).	Constipation; urinary tract stone formation; kidney dysfunction; interference with absorption of other minerals.	Milk and milk products, oysters, small fish (with bones), calcium-set tofu (bean curd), certain leafy greens, broccoli, legumes.
Phosphorus			
Mineralizataion of bones and teeth; important in cells' genetic material, in cell membranes as phospholipids, in energy transfer, and in buffering systems.	Appetite loss, bone pain, muscle weakness, impaired growth, and rickets in infants.[a]	Calcification of nonskeletal tissues, particularly the kidney.	Foods from animal sources. some lelgumes
Magnesium			
A factor involved in bone mineralization, the building of protein, enzyme action, normal muscular function, transmission of nerve impulses, proper immune function and maintenance of teeth.	Weakness; muscle twitches; appetite loss; confusion; if extreme, convulsions, bizarre movements (especially of eyes and face), hallucinations, and difficulty in swallowing. In children, growth failure.[b]	Excess magnesium from abuse of laxatives (Epsom salts) causes diarrhea with fluid and electrolyte and pH imbalances.	Nuts, legumes, whole grains, dark green vegetables, seafoods, chocolate, cocoa.
Sodium			
Sodium, chloride, and potassium (electrolytes) maintain normal fluid balance and acid-base balance in the body. Sodium is critical to nerve impulse transmission.	Muscle cramps, mental apathy, loss of appetite.	Hypertension.	Salt, soy sauce, seasoning mixes, processed foods, condiments, fast foods.
Potassium			
Potassium facilitates reactions, including the making of protein; the maintenance of fluid and electrolyte balance; the support of cell integrity; the transmission of nerve impulses; and the contraction of muscles, including the heart.	Deficiency accompanies dehydration; causes muscular weakness, paralysis, and confusion; can cause death.	Causes muscular weakness; triggers vomiting; if given into a vein, can stop the heart.	All whole foods: meats, milk, fruits, vegetables, grains, legumes.
Chloride			
Chloride is part of the hydrochloric acid found in the stomach, necessary for proper digestion. Helps maintain normal fluid and electrolyte balance.	Growth failure in children; muscle cramps, mental apathy, loss of appetite; can cause death (uncommon).	Normally harmless (the gas chlorine is a poison but evaporates from water); can cause vomiting.	Salt, soy sauce; moderate quantities in whole, unprocessed foods, large amounts in processed foods.

[a]Seen only rarely in infants fed phosphorus-free formula or in adults taking medications that interact with phosphorus.
[b]A still more severe deficiency causes tetany, an extreme, prolonged contraction of the muscles similar to that caused by low blood calcium.

Continued on next page.

TABLE 8·11   THE MINERALS—A SUMMARY—continued

Mineral and Chief Functions in the Body	Deficiency Symptoms	Toxicity Symptoms	Significant Sources
Major Minerals			
Sulfate A contributor of sulfur to many important compounds, such as certain amino acids, antioxidants, and the vitamins biotin and thiamin; stabilizes protein shape by forming sulfur-sulfur bridges (see Figure 6-10 in Chapter 6, p. 189).	None known; protein deficiency would occur first.	Would occur only if sulfur amino acids were eaten in excess; this (in animals) depresses growth.	All protein-containing foods.
Trace Minerals			
Iodine A component of the thyroid hormone thyroxine, which helps to regulate growth, development, and metabolic rate.	Goiter, cretinism.	Depressed thyroid activity; goiter-like thyroid enlargement.	Iodized salt; seafood; bread; plants grown in most parts of the country and animals fed those plants.
Iron Part of the protein hemoglobin, which carries oxygen in the blood; part of the protein myoglobin in muscles, which makes oxygen available for muscle contraction; necessary for the use of energy.	Anemia: weakness, fatigue, pale skin and mucus membranes, pale concave nails, headaches, inability to concentrate, impaired cognitive function (children), lowered cold tolerance.	Iron overload: fatigue, abdominal pain, infections, liver injury, joint pain, skin pigmentation, growth retardation in children, bloody stools, shock.	Red meats, fish, poultry, shellfish, eggs, legumes, green leafy vegetables, dried fruits.
Zinc Associated with hormones; needed for many enzymes; involved in making genetic material and proteins, immune cell activation, transport of vitamin A, taste perception, wound healing, the making of sperm, and normal fetal development.	Growth failure in children, dermatitis, sexual retardation, loss of taste, poor wound healing.	Nausea, vomiting, diarrhea, loss of appetite, headache, immune suppression, decreased HDL, reduced iron and copper status.	Protein-containing foods: meats, fish, shellfish, poultry, grains, yogurt.
Selenium Assists a group of enzymes that defend against oxidation.	Predisposition to a form of heart disease characterized by fibrous cardiac tissue (uncommon).	Nausea; abdominal pain; nail and hair changes; nerve, liver, and muscle damage.	Seafoods, organ meats; other meats, whole grains, and vegetables depending on soil content.
Fluoride Helps form bones and teeth; confers decay resistance on teeth.	Susceptibility to tooth decay.	Fluorosis (discoloration) of teeth, nausea, vomiting, diarrhea, chest pain, itching.	Drinking water if fluoride containing or fluoridated; tea; seafood.
Chromium Associated with insulin; needed for energy release from glucose.	Abnormal glucose metabolism.	Possibly skin eruptions.	Meat, unrefined grains, vegetable oils.
Copper Helps form hemoglobin; part of several enzymes.	Anemia; bone abnormalities.	Vomiting, diarrhea; liver damage.	Organ meats, seafood, nuts, seeds, whole grains, drinking water.

Meeting the Need for Calcium

THE AVERAGE WOMAN consumes just a third of her recommended amount of calcium; men do somewhat better, with calcium intakes close to three-fourths of the recommendation. Low calcium intakes are associated with all sorts of major illnesses, including adult bone loss (see the following Controversy), high blood pressure and colon cancer (see Chapter 11), kidney stones, and even lead poisoning.

Consumption of one of the best sources of calcium—milk—has decreased in recent years while consumption of other beverages, such as soft drinks, has increased dramatically. One national dairy group calls the situation a "national calcium crisis," and representatives of many national health and nutrition organizations agree. This Food Feature focuses on sources of calcium in the diet and provides guidance about how to meet the need for calcium.

Milk, Yogurt, and Cheese Group

Milk and milk products are traditional sources of calcium for people who can tolerate them (see Figure 8-12). Table 8-12 shows the current milk recommendations that help to meet the calcium needs of various age groups. People who do not use milk because of lactose intolerance, dislike, or allergy can obtain calcium from other sources, but care is needed; *wise* substitutions must be made.[88] This is especially true for children. Children who don't drink milk often have lower calcium intakes and poorer bone health than those who drink milk regularly, and they may also be smaller in stature.[89] Most of milk's many relatives are recommended choices: yogurt, **kefir,** buttermilk, cheese (especially the low-fat or fat-free varieties), and, for people who can afford the calories, ice milk. Cottage cheese and frozen yogurt desserts contain about half the calcium of milk: 2 cups are needed to provide the amount of calcium in 1 cup of milk. Butter, cream, and cream cheese are almost pure fat and contain negligible calcium.

Tinker with milk products to make them more appealing. Add cocoa to milk and fruit to yogurt, make your own fruit smoothies from milk or yogurt, or add fat-free milk powder to any dish. The cocoa powder added to make chocolate milk does contain a small amount of oxalic acid, which binds with some of milk's calcium and inhibits its absorption, but the effect is insignificant. Sugar lends both sweetness and calories to chocolate milk, so mix your chocolate milk at home where you control the amount of sugary chocolate added to the milk.

Vegetables

Among vegetables, rutabaga, broccoli, beet greens, turnip greens, mustard greens, bok choy (a Chinese cabbage), and kale are good sources of available calcium. So are collard greens, green cabbage, kohlrabi, watercress, parsley, and probably some seaweeds, such as the **nori** popular in Japanese cookery. Certain other foods, including spinach, Swiss chard, and rhubarb, appear equal to milk in calcium content but provide very little or no calcium to the body because they contain binders that prevent calcium's absorption (see Figure 8-13, p. 300). The presence of calcium binders does not make spinach an inferior food. Spinach is also rich in iron, beta-carotene, and dozens of other essential nutrients and potentially helpful phytochemicals. Just don't rely on it for calcium. Dark greens of all kinds are superb sources of riboflavin and indispensable for the vegan or anyone else who does not drink milk.

Calcium in Other Foods

For the many people who cannot use milk and milk products, small fish such as canned sardines and other canned fishes prepared with their bones are rich sources of calcium. One-third cup of almonds

TABLE 8·12	SUGGESTED MINIMUM DAILY FLUID MILK INTAKES
Young children	2 cups
Teenagers	3 cups
Adults	3 cups
Pregnant or lactating women	3 cups
Women past menopause	3 cups

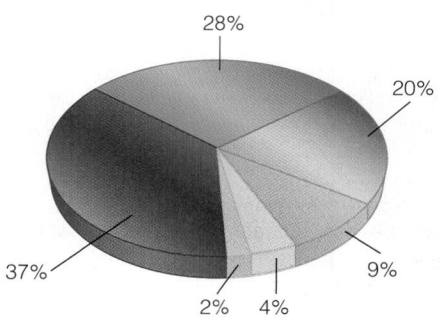

FIGURE 8·12 FOOD SOURCES OF CALCIUM IN THE U.S. DIET

Milk and milk products contribute over half of the calcium in a typical U.S. diet.

28%
20%
37%
2%   4%
9%

Key:
- Milk
- Cheese
- Yeast bread
- Ice cream, sherbet, frozen yogurt
- Cakes, cookies, quick breads, doughnuts
- Other sources[a]

[a] Other sources include foods contributing at least 1% in descending order: yogurt, ready-to-eat cereal, soft drinks, tortillas, eggs, dried beans and lentils, canned tomatoes, meal replacements and protein supplements, corn bread and corn muffins, hot breakfast cereal, and coffee.

SOURCE: Data from P. A. Cotton and coauthors, Dietary sources of nutrients among US Adults, 1994–1996, *Journal of the American Dietetic Association* 104 (2004): 921–930, supplemental Table 21 from www.eatright.org

kefir a yogurt-based beverage.

nori a type of seaweed popular in Asian, particularly Japanese, cooking.

FIGURE 8·13 CALCIUM ABSORPTION FROM FOOD SOURCES

≥ 50% absorbed	cauliflower, watercress, Chinese cabbage, head cabbage, brussels sprouts, rutabaga, kolhrabi, kale, mustard greens, bok choy, broccoli, turnip greens
≈ 30% absorbed	milk, yogurt, cheese, calcium-fortified soy milk, calcium-set tofu, calcium-fortified juices and drinks
≈ 20% absorbed	almonds, sesame seeds, beans (pinto, red, and white)
≤ 5% absorbed	spinach, rhubarb, Swiss chard

stone-ground flour flour made by grinding kernels of grain between heavy wheels made of limestone, a kind of rock derived from the shells and bones of marine animals. As the stones scrape together, bits of the limestone mix with the flour, enriching it with calcium.

supplies about 100 milligrams of calcium along with almost 300 calories—a high-energy calcium source. Stocks or extracts made from bones can be another rich source. The Vietnamese tradition of making fish stock from the bones helps account for their adequate calcium intake without the use of milk. Calcium-rich mineral water may also be a useful calcium source. Recent evidence seems to indicate that the calcium from mineral water, including hard tap water, may be as absorbable as the calcium from milk.[90] Many other foods contribute smaller, but still significant, amounts of calcium to the diet.

Calcium-Fortified Foods

Next in order of preference among non-milk sources of calcium are foods that contain large amounts of calcium salts by an accident of processing or by intentional fortification. In the processed category are soybean curd, or tofu (calcium salt is often used to coagulate it, so check the label); canned tomatoes (firming agents donate 63 milligrams per cup of tomatoes); **stone-ground flour** and self-rising flour; stone-ground whole and self-rising cornmeal; and blackstrap molasses.

Some food products available to U.S. consumers are fortified to add calcium to people's diets. The richest in calcium is high-calcium milk, that is, milk with extra calcium added; it provides more calcium per cup than any natural milk, 500 milligrams per 8 ounces. Then comes calcium-fortified orange juice, with 300 milligrams per 8 ounces, a good choice because the bioavailability of its calcium is comparable to that of milk. Calcium-fortified soy milk can also be prepared so that it contains more calcium than whole cow's milk. Soy-based infant formula is fortified with calcium, and no law prevents adults from using it in cooking for themselves, although it may be somewhat higher in fat than some other calcium-rich choices.

Finally, calcium supplements are available, sold mostly to people hoping to ward off osteoporosis. Controversy 8 points out, however, that, while often useful, supplements are not magic bullets against bone loss.

Making Meals Rich in Calcium

For those who tolerate milk, many cooks slip extra calcium into meals by sprinkling a tablespoon or two of fat-free dry milk into almost everything. The added calorie value

is small, changes to the taste and texture of the dish are practically nil, but each 2 tablespoons adds about 100 extra milligrams of calcium and moves people closer to meeting the recommendation to obtain three cups of milk each day (see Figure 8-14). Here are some more tips for including calcium-rich foods in your meals:

At Breakfast

- Choose calcium-fortified orange or vegetable juice.
- Serve tea or coffee, hot or iced with milk.
- Choose cereals, hot or cold, with milk.
- Cook hot cereals with milk instead of water; then mix in 2 tablespoons of fat-free dry milk.
- Make muffins or quick breads with milk and extra fat-free powdered milk.
- Add milk to scrambled eggs.
- Moisten cereals with flavored yogurt.

At Lunch

- Add low-fat cheeses to sandwiches, burgers, or salads.
- Use a variety of green vegetables, such as watercress or kale, in salads and on sandwiches.
- Drink fat-free milk or calcium-fortified soy milk as a beverage or in a smoothie.
- Drink calcium-rich mineral water as a beverage (studies suggest significant calcium absorption).
- Marinate cabbage shreds or broccoli spears in low-fat Italian dressing for an interesting salad that provides calcium.
- Choose coleslaw over potato and macaroni salads.
- Mix the mashed bones of canned salmon into salmon salad or patties.
- Eat sardines with their bones.
- Stuff potatoes with broccoli and low-fat cheese.
- Try pasta such as ravioli stuffed with low-fat ricotta cheese instead of meat.
- Sprinkle parmesan cheese on pasta salads.

At Supper

- Toss a handful of thinly sliced green vegetables, such as kale or young turnip greens, with hot pasta; the greens wilt pleasingly in the steam of the freshly cooked pasta.
- Serve a green vegetable every night and try new ones—how about kohlrabi? It tastes delicious when cooked like broccoli.
- Learn to stir-fry Chinese cabbage and other Asian foods.

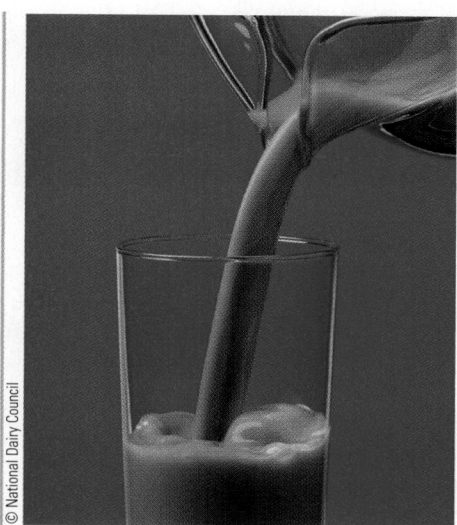

Chocolate milk is an excellent source of calcium for those who can afford the extra calories.

- Try tofu (the calcium-set kind); this versatile food has inspired whole cookbooks devoted to creative uses.
- Add fat-free powdered milk to almost anything—meat loaf, sauces, gravies, soups, stuffings, casseroles, blended beverages, puddings, quick breads, cookies, brownies. Be creative.
- Choose frozen yogurt, ice milk, or custards for dessert.

Here is a shortcut for tracking the amount of calcium in a day's meals. To start, memorize these two facts:

1. A cup of milk provides about 300 milligrams of calcium.

2. Adults need 1,000 to 1,200 milligrams each day. Broken down in terms of "cups of milk," the need is 3⅓ to 4 cups each day.

To estimate calcium from an entire day's foods, not just milk, assign "cups of milk" points to various calcium sources. The goal is to achieve 3½ to 4 points per day:

- 1 point = 1 cup milk, yogurt, or calcium-fortified beverage, or 1½ ounces cheese.
- 1 point = 4 ounces canned fish with bones.
- ½ point = 1 cup ice cream, cottage cheese, or calcium-rich vegetables (see the text).

Also, because bits of calcium are present in many foods (a bagel has about 50 milligrams, for example):

- 1 point = a well-balanced, adequate, and varied diet.

Example: Say a day's calcium-rich foods include cereal and a cup of milk, a ham and cheese sandwich, and a broccoli and pasta salad.

1 point (cup of milk) + 1 point (cheese) + ½ point (broccoli) = 2½ points.

Add 1 point for the other foods eaten that day:

1 point + 2½ points = 3½ points.

This day's foods provided a calcium intake approximately equal to the lower of the DRI committee's recommendation. The tips in this section offer many ways to aim higher.

FIGURE 8·14 MILK, YOGURT, AND CHEESE GROUP AVERAGE INTAKES[a]

On average, people in the United States fall far short of meeting the recommendation to obtain three cups of milk, yogurt, or cheese each day. The picture is worse for the dark green vegetables that supply calcium—only 3 percent of the vegetables consumed each day meet this description.

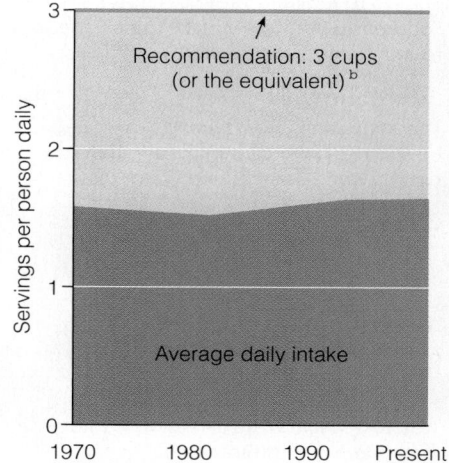

[a]Includes all forms of milk, yogurt, cheese, and frozen dairy desserts.
[b]Recommended amount for adults from the 2005 USDA Food Guide. Details in Figure 2-4 of Chapter 2.
SOURCE: Intake data from U.S. Department of Agriculture, Economic Research Service.

Nutrition Connections

The Nutrition Connections CD provides access to these additional resources.

Do It!
The Do It! exercise for this chapter, *Find the Minerals in Snack Foods,* asks you to evaluate the calories and minerals in foods found at convenience stores.

Interaction
The animation, *Maintaining Water Balance in the Body,* shows how water balance is maintained in the body through equalization of water sources and water losses.

Chapter Quiz
Take the quiz for this chapter to assess your understanding of chapter concepts.

Nutrition on the Net
Annotated links to reputable nutrition-related websites.

Online Glossary
Test your knowledge of key vocabulary through this comprehensive, interactive glossary.

Internet Connections

A variety of helpful study tools for this chapter is available at our website
http://nutrition.wadsworth.com/ncc10

Practice Test
Take the practice test for this chapter and gauge your knowledge of key concepts; answers are provided.

Student Notebook
A lecture notebook is available to help organize note taking while reading or attending lecture.

ABC Video Clip
"Mineral Water" (4:51) explores commonly consumed types of bottled waters.

Study Tools
Study resources for students include Chapter Summaries, Crossword puzzles, Flashcards, Glossary, Internet Exercises, and Learning Objectives.

SELF CHECK

Answers to these Self Check questions are in Appendix G.

1. Water balance is governed by the:
   a. liver
   b. kidneys
   c. brain
   d. (b) and (c)

2. Which two minerals are the major constituents of bone?
   a. calcium and zinc
   b. phosphorus and calcium
   c. sodium and magnesium
   d. magnesium and calcium

3. All of the following are correct concerning zinc *except:*
   a. in high doses it may reduce the HDL concentration in the blood
   b. in high doses it may inhibit iron absorption from the digestive tract
   c. fruits and vegetables are the best sources for zinc
   d. deficiencies in children retard growth

4. A deficiency of which mineral is a leading cause of mental retardation worldwide?
   a. iron
   b. iodine
   c. zinc
   d. chromium

5. Which mineral supplement is a leading cause of fatal accidental poisonings of U.S. children under six years old?
   a. iron
   b. sodium
   c. chloride
   d. potassium

6. After about 40 years of age, bones begin to lose density.   T F

7. The best way to control salt intake is to cut down on processed and fast foods.   T F

8. The most abundant mineral in the body is iron.   T F

9. Dairy foods such as butter, cream, and cream cheese are good sources of calcium whereas vegetables such as broccoli are poor sources.   T F

10. Bottled water must meet higher standards for purity and sanitation than U.S. tap water.   T F

CONTROVERSY 8

OSTEOPOROSIS: CAN LIFESTYLE CHOICES REDUCE THE RISKS?

AN ESTIMATED 44 MILLION people in the United States—the majority of them women—are suffering from or developing osteoporosis.[*1] Each year, a million and a half people—30 percent of them men—break a hip, leg, arm, hand, ankle, or other bone as a result of having osteoporosis. Of these, hip fractures prove most serious. The break is rarely clean; the bone explodes into fragments that cannot be reassembled. Just removing them is a struggle, and replacing them with an artificial joint requires major surgery. Many elderly people with hip fracture never walk or live independently again. About a fifth die from related complications within a year. By the year 2020, the number of hip fractures could double or even triple in the United States.[2] Both men and women are urged to do whatever they can to prevent fractures related to osteoporosis.

Fractures from osteoporosis occur during the later years, but osteoporosis itself develops silently much earlier. Few young adults are aware that osteoporosis is sapping the strength of their bones; then suddenly, 40 years later, the hip gives way. People say, "She fell and broke her hip," but in fact the hip may have been so fragile that it broke *before* she fell.

The causes of osteoporosis are tangled. Insufficient dietary calcium and vitamin D certainly play roles, but physical activity, genetics, and other factors are also major potential players. This Controversy addresses several questions about osteoporosis: What is it? Who gets it? What can people do to reduce their risks? And where do dietary calcium and calcium supplements fit into the picture?

TABLE C8·1 OSTEOPOROSIS TERMS

- **bone density** a measure of bone strength; the degree of mineralization of the bone matrix.
- **cortical bone** the ivorylike outer bone layer that forms a shell surrounding trabecular bone and that comprises the shaft of a long bone.
- **trabecular** (tra-BECK-you-lar) **bone** the weblike structure composed of calcium-containing crystals inside a bone's solid outer shell. It provides strength and acts as a calcium storage bank.

The Development of Osteoporosis

To understand how the skeleton loses minerals in later years, you must first know a few things about bones. Table C8-1 offers definitions of relevant terms. The photograph on this page shows a human leg bone sliced lengthwise, exposing the lattice of calcium-containing crystals (the **trabecular bone**) inside that are part of the body's calcium bank. Invested as savings during the milk-drinking years of youth, these deposits provide a nearly inexhaustible fund of calcium. **Cortical bone** is the dense, ivorylike bone that forms the exterior shell of a bone and the shaft of a long bone (look closely at the photograph). Both types of bone are crucial to overall bone strength. Cortical bone forms a sturdy outer wall, and trabecular bone provides strength along the lines of stress.

The two types of bone handle calcium in different ways. The lacy crystals of the trabecular bone are tapped to raise blood calcium when the supply from the day's diet runs short; the calcium crystals are rede-posited in bone when dietary calcium is plentiful. Trabecular bone, generously supplied with blood vessels, is more metabolically active than cortical bone, is more sensitive to hormones that govern calcium deposits and withdrawals from day to day, and readily gives up its minerals at the necessary rate whenever blood calcium needs replenishing. Losses of trabecular bone begin to be significant for men and women in their mid-20s. Cortical bone's calcium can also be withdrawn, but slowly. Cortical bone loss begins at about age 40, and bone tissue dwindles steadily thereafter.

As bone loss continues and osteoporosis progresses (Figure C8-1, p. 304), **bone density** declines, and bones become so fragile that the body's weight can overburden the spine; vertebrae may suddenly disintegrate and crush down, painfully pinching major nerves. Or they may compress into wedges, forming what is insensitively called "dowager's hump," the bent posture of many men and women as they "grow shorter" (see Figure C8-2, p. 304). Wrists may break as trabecula-rich bone ends weaken, and teeth may loosen or fall out as the trabecular bone of the jaw recedes. As the cortical bone shell weakens as well, breaks often occur in the hip.

Toward Prevention— Understanding the Causes of Osteoporosis

Scientists are searching for ways to prevent osteoporosis, but they must first establish its causes. In addition to the

Photograph of sectioned bone.

*Reference notes are found in Appendix F.

FIGURE C8·1 LOSSES OF TRABECULAR BONE

© Fred Hossler/Visuals Unlimited

Electron micrograph of healthy trabecular bone.

© Alan Boyde/Visuals Unlimited

Electron micrograph of trabecular bone affected by osteoporosis.

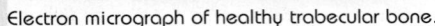

obvious factors of gender and advanced age, a person's chances of developing osteoporosis also depend both on genetic inheritance and on the environment. Environmental factors under study for their roles in lowering bone density include:

- Poor nutrition involving calcium and vitamin D.
- Estrogen deficiency in women.
- Lack of physical activity.
- Being underweight.
- Use of tobacco and alcohol.
- Possibly, excess protein, sodium, caffeine, and soft drinks and inadequate vitamin K.

Bone Density and the Genes

Science may soon reveal a genetic basis for variations in bone density and development of osteoporosis.[3] Over 30 genes are under investigation in this regard, and each may interact with others and with environmental factors, such as vitamin D and calcium nutrition. Genes influence the activities of bone-forming cells and bone-dismantling cells, the cellular mechanisms that make collagen (the structural protein of bone formation), the mechanisms for absorbing

and employing vitamin D, and many other contributors to bone metabolism.[4] Researchers hope that, once unsnarled, this tangle of genetic leads will help to

explain many of the remaining questions surrounding osteoporosis.

A classic approach to genetics includes studies of identical twins. Because such siblings share identical genes, a disease arising from genetic inheritance reliably appears in both twins. However, a study of elderly Finnish twins suggests that the fracture rates vary between twin siblings, underscoring the importance of environmental factors in bone breaks.[5] By one estimate, genetic factors may account for about 35 percent of the variation in rates of serious fractures among the elderly, with environmental factors accounting for the rest.[6]

Genetic inheritance appears to most strongly influence the maximum bone mass attainable during growth.[7] It also influences the extent of a woman's bone loss during menopause, the time when women's estrogen production declines and menstruation ceases.

Risks of osteoporosis differ by race and ethnicity. People of African descent have denser bones than do people of northern European extraction, and African American women may lose bone at just half the rate of white

FIGURE C8·2 LOSS OF HEIGHT IN A WOMAN CAUSED BY OSTEOPOROSIS

The woman on the left is about 50 years old. On the right, she is 80 years old. Her legs have not grown shorter; only her back has lost length, due to collapse of her spinal bones (vertebrae). When collapsed vertebrae cannot protect the spinal nerves, the pressure of bones pinching the nerves causes excruciating pain.

6 inches lost

50 years old   80 years old

women throughtout aging.[8] Mexican Americans' bone density falls somewhere in between. These differences in bone density hold true for both sexes of all ages: hip fractures are three times more likely in 80-year-old white women than in black women of the same age.

Some ethnic groups have lower bone densities than do northern Europeans. Asians from China and Japan, Hispanics from Central and South America, and Inuits from St. Lawrence Island all have lower bone density than do northern Europeans. Do lower bone densities forecast a higher rate of hip fractures in these groups? Not always. Chinese people living in Singapore have low bone density, but their hip fracture rates are among the lowest in the world. It may be that Chinese people have particularly strong hip bones or some other advantage that prevents fractures.

The message in all of these findings: though your genes may provide an inherited tendency for strong or weak bones, this genetic potential is tempered by individual life experiences. For example, those who attend to nutrition and physical activity attain their maximum bone density during growth, whereas those who overuse alcohol or use tobacco accelerate their bone losses, regardless of their genes.

Calcium and Vitamin D

An environmental factor that affects bone deposition and withdrawal is calcium and vitamin D nutrition during childhood, adolescence, and early adult life. Preteen children who consume extra calcium together with adequate vitamin D lay more calcium into the structure of their bones than children with less adequate intakes.

When people reach the bone-losing years of middle age, those who formed dense bones during youth have more bone tissue to lose before suffering ill effects—see Figure C8-3. Therefore, whatever factors help build strong bones in youth, including calcium nutrition, also protect against osteoporosis much later.

Women who seldom drank milk as children or teenagers have lower bone density and greater risk of fractures

than those who drank milk regularly.[9] Even in childhood, those who avoid drinking milk may be more prone to fractures than their milk-drinking peers.[10] Simply put, growing children who do not get enough calcium do not develop strong bones.[11] Once compromised, bone strength may remain in jeopardy years later during adulthood. Milk is not the only food rich in calcium, but milk and milk products supply the most calcium to the U.S. diet by far. In addition, new research suggests that a protein in milk may increase bone formation by supporting the activities of the bone-building cells.[*][12]

Although dietary calcium and vitamin D in later life cannot make up for earlier deficiencies, they can help to slow the rate of bone loss. Unfortunately, older people take in less calcium and vitamin D than others. After about age 65, they absorb less calcium, too, probably because aging skin is less efficient at making vitamin D. Also, many older people fail to go outdoors and so are deprived of the sunlight necessary to form vitamin D. Some of the hormones that regulate bone maintenance and calcium metabolism also change with age and accelerate bone mineral withdrawal.

Gender and Hormones

Gender is a powerful predictor of osteoporosis: men have greater bone density than women at maturity, and women have great losses during menopause. Women thus account for more than two out of three cases of osteoporosis. These facts may lull some men into believing that osteoporosis is a "woman's disease," but men suffering fractures each year from osteoporosis number in the millions.[13]

In women, bone dwindles rapidly when the hormone estrogen diminishes

---

[*] The protein is lactoferrin.

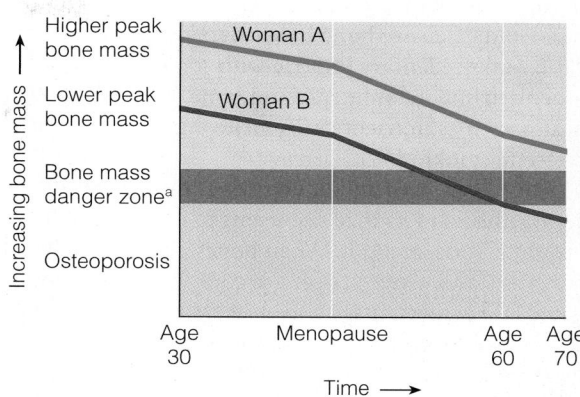

Woman A entered adulthood with enough calcium in her bones to last a lifetime. Woman B had less bone mass starting out and so suffered ill effects from bone loss later on.

[a] People with a moderate degree of bone mass reduction are said to have osteopenia and are at increased risk of fractures.

SOURCE: Data from Standing Committee on the Scientific Evaluation of Dietary Reference Intakes, Food and Nutrition Board, Institute of Medicine, *Dietary Reference Intakes for Calcium, Phosphorus, Magnesium, Vitamin D, and Fluoride* (Washington, D.C.: National Academy Press, 1997), pp. 4–10.

as menstruation ceases. Accelerated losses continue for six to eight years following menopause and then taper off, so women again lose bone at the same rate as their male counterparts. Losses of bone minerals continue throughout the remainder of a woman's lifetime, but not at the free-fall pace of the menopause years (see Figure C8-3).

Young women who cease menstruating (amenorrhea), lose bone rapidly. In some cases, diseased ovaries are to blame and must be removed; in others, the ovaries fail to produce sufficient estrogen because the women suffer from anorexia nervosa (see Controversy 9). The bone loss of anorexia nervosa remains long after recovery from the eating disorder.

If estrogen deficiency is a major cause of osteoporosis in women, what is the cause of bone loss in men? Men produce only a little estrogen, yet they resist osteoporosis better than women. It may be that the male sex hormone testosterone plays a role—men suffer more fractures after removal of diseased testes or when their testes lose function with aging. Thus, both male and female sex hormones appear to play roles in the development of osteoporosis.

Physical Activity

When people lie idle—for example, when they are confined to bed—the bones lose strength just as the muscles do. Astronauts who live without gravity for days or weeks at a time experience rapid and extensive bone losses. Table C8-2 ranks sedentary lifestyle with other detrimental influences on bone health such as nutrient deficiencies and cigarette smoking.

Muscle strength and bone strength go together, and muscle use seems to promote bone strength. When bones of sedentary and active people are compared, the active bones are denser by far.[14] The hormones that promote synthesis of new muscle tissue also favor the building of bone; flexibility and muscle strength gained through physical activity also improve balance and help to prevent falls from occurring.

To keep the bones healthy and to prevent falls, include weight-bearing exercises such as calisthenics, dancing, jogging, vigorous walking, or weight training every day.[15] Even sports or gardening, performed with vigor, can build bone strength and improve balance. In adolescence, regular physical activity as well as an adequate calcium intake is critical to maximize bone density.[15] Later on in adulthood, bone density benefits from a regular program of

These young people are putting bone in the bank.

weight training.[16] Even past menopause when most women are losing bone, a strong, fit body may help maintain the bones.[17]

Body Weight

After age and gender, the next risk factor for osteoporosis is being underweight or losing weight.[18] Women who are thin throughout life, and especially those who lose 10 percent or more of their body weight after the age of menopause, face a hip fracture rate twice as high as that of most other women. Heavier body weights and higher body fatness stress the bones and promote their maintenance.[19] Also, fat tissue serves as a storage depot for hormones, and abundant body fat may mean greater hormone stores. An appetite-controlling hormone, leptin, produced by fat cells may also have an effect on bone biology (more about leptin in Chapter 9).[20]

Tobacco Smoke and Alcohol

Smoking is hard on the bones. Bones of smokers are less dense than those of nonsmokers—even after controlling for differences in age, body weight, and physical activity habits.[21] Blood levels of vitamin D and bone-related hormones in smokers favor decreased calcium absorption and increased bone destruction.[22] Fortunately, quitting can reverse these damaging effects. Blood indicators shift beneficially after just six weeks of smoking cessation.[23] In time, bone density is similar for former smokers and nonsmokers.

People who are addicted to alcohol also experience more frequent fractures, and drinking can contribute to accidents and falls. Because alcohol (a diuretic) causes fluid excretion, it may induce excessive calcium losses through the urine. Heavy drinking may also upset the hormonal balance required

TABLE C8-2 RISK AND PROTECTIVE FACTORS THAT CORRELATE WITH OSTEOPOROSIS

Risk Factors	Protective Factors
High Correlation	
Advanced age	Black race
Alcoholism, heavy drinking	Estrogens, long-term use
Chronic steroid use	
Female gender	
Rheumatoid arthritis	
Surgical removal of ovaries or testes	
Thinness or weight loss	
White race	
Moderate Correlation	
Chronic thyroid hormone use	Having given birth
Cigarette smoking	High body weight
Diabetes (insulin-dependent, type 1)	High-calcium diet
Early menopause	Regular physical activity
Excessive antacid use	
Family history of osteoporosis	
Low-calcium diet	
Sedentary lifestyle	
Vitamin D deficiency	
Probably Important but Not Yet Proved	
Alcohol taken in moderation	Adequate vitamin K intake
Caffeine intake	Low-sodium diet (later years)
High-fiber diet	
High blood homocysteine	
High-protein diet	
Lactose intolerance	

for healthy bones or may be directly toxic to the bone-building cells, preventing their reproduction and diminishing their numbers.

Protein

Excess dietary protein causes the body to excrete calcium in the urine.[24] This finding suggests that a lifetime of consuming excess dietary protein could accelerate bone loss, especially when calcium intake is low.

Does a high-protein diet cause bone loss? Research is not entirely clear on this point, but some findings support the idea that it may, especially among well-nourished people eating a diet rich in animal protein but low in vegetables.[25] In experiments, people fed increasing amounts of purified animal protein developed an increasingly negative calcium balance (they lost more calcium in the urine than they consumed from food). Figure C8-4 shows that in women around the world, hip fracture incidence drops when the ratio

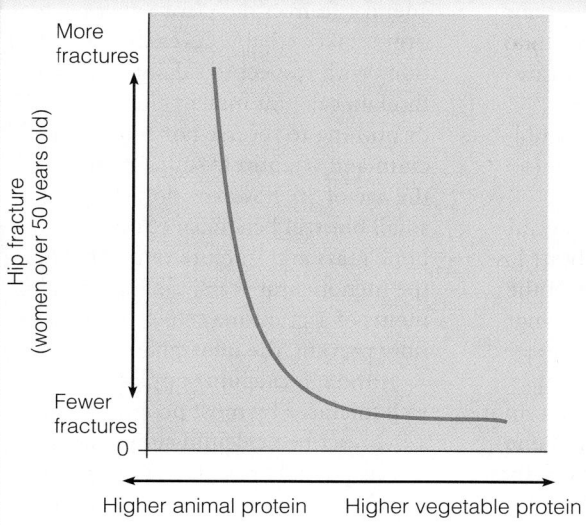

FIGURE C8·4 RATIO OF VEGETABLE TO ANIMAL PROTEIN IN THE DIET AND HIP FRACTURE INCIDENCE WORLDWIDE

SOURCE: Adapted from L. A. Frassetto and coauthors, Worldwide incidence of hip fractures in elderly women: Relation to consumption of animal to vegetable foods, *Journals of Gerontology Series A: Biological Sciences and Medical Sciences* 55 (2000): M585–M592.

of vegetable to animal protein in the diet increases.[26]

Soy foods in particular, by virtue of their protein and phytochemicals, may help stem the rapid bone losses of the menopause years.[27] Recent research suggests that soy may indeed offer some protection, but foods, and not supplements, are the best choice, as Controversy 2 made clear.[28]

Vegetables themselves, and not their protein, may be the protective factor, however, because of their role in providing alkaline substances to the body. Before the body can use or store energy from the sulfur-containing amino acids in proteins, it must first cleave off the sulfur, which then forms sulfuric acid, one of the strongest acids known.[29] The body must act quickly to neutralize this excess acidity before it can upset the body's delicate acid-base balance. One way it does so is by withdrawing alkaline calcium salts from the bones and using the noncalcium part to buffer the acid. The calcium part is detected as "excess calcium" by the kidneys, which toss it out in the urine. Vegetables are rich sources of alkaline constituents that help to neutralize acid, reducing the need for bone calcium withdrawal.

On the other side of the spectrum, obtaining sufficient protein is essential to bone health because protein deprivation also stimulates calcium losses and weakens bones.[30] Remember from the chapter that the mineral crystals of bone form on a protein matrix—collagen, mentioned earlier. Because the body stores no significant quantities of protein, it must be supplied by the diet. Especially in malnourished elderly people, restoring protein from animal or vegetable sources can often improve bone status and reduce the incidence of hip fractures. Also, some protein-rich foods from animals do not seem to produce a negative calcium balance. Milk is a good example. A milk-rich diet provides both protein and calcium and may help to *oppose* withdrawal of calcium from the skeleton. With protein, as with other nutrients, then, it seems wise to follow a now familiar nutrition principle—a vegetable-rich diet based on adequacy, moderation, and variety is best.[31]

Sodium, Caffeine, Soft Drinks, Vitamin K, Homocysteine, and Others

A high sodium intake is also associated with urinary calcium excretion, and lowering sodium intakes seems to lessen calcium losses.[32] Study subjects eating the DASH diet (the high-vegetable, high-dairy diet described in Chapter 8) at three sodium levels were compared with controls consuming a more "typical" American diet.[33] Calcium losses were greatest in those consuming the highest levels of sodium and smallest in the lower sodium groups. In addition, the DASH diet is higher in calcium than most diets and had a beneficial effect on bone metabolism—blood indicators suggest that bone loss was slowing while bone accretion was increasing.[34] Whether these effects were due to the abundant calcium or fruits and vegetables in the DASH diet is unknown.

Long-term evidence showing that a reduced sodium intake prevents osteoporosis is lacking, but no harm can come from recommending that people reduce sodium intakes.[35] Also, increasing potassium may help—some research shows that potassium may counteract the effects of sodium on calcium excretion.[36] The person who wishes to lower sodium and increase potassium should choose a diet rich in unprocessed or lightly processed foods such as fruits and vegetables while restricting heavily processed, convenience, or fast foods.

Heavy users of caffeinated beverages, such as coffee, tea, and colas, should be aware that some evidence links caffeine use and osteoporosis, although other findings tend to absolve caffeine use from posing a risk. It may be that ordinary caffeine intakes, say the amount in two cups of coffee, increase calcium losses only when calcium intakes are low. Such losses may be so small that the calcium in just one or two *tablespoons* of milk is enough to replace them.[37]

Cola beverages and other soft drinks may also have adverse effects on calcium, although the reasons why are unclear. One idea is that the high concentration of fructose sweetener in the drinks may cause calcium loss.[38] Soft drinks also displace milk from the diet, especially in children and adolescents.

Rats given colas instead of water increase their fluid intake by three times, lose calcium in their urine, and dramatically reduce their intake of solid food.[39] Despite reduced food intake, the rats gain more weight than water-drinking controls, and their bones are significantly less dense. In children, an unexplained tendency to develop bone problems such as stress fractures has been reported to accompany high intakes of fruit juices or cola beverages, but not other soft drinks.

Vitamin K plays important roles in the production of at least one bone protein that participates in bone maintenance.* People with hip fractures often

---

*The vitamin K–dependent bone protein is osteocalcin.

have low intakes of vitamin K.[40] In a study of female athletes in whom strenuous exercise had lowered estrogen production, an increased vitamin K intake reduced markers of bone loss and increased markers of bone formation.

Elevated blood concentration of the amino acid homocysteine indicates a dietary deficit of folate, vitamin B_{12}, and vitamin B_6. Recently, a study from the Netherlands revealed that people with the highest homocysteine levels experienced twice as many hip and wrist fractures as people with lower levels.[41] Likewise, in the United States the rate of hip fractures quadrupled in men with the highest homocysteine levels and doubled in women over those with lower homocysteine levels.[42] No one can say yet whether elevated blood homocysteine contributes to osteoporosis, results from the disease process, or exists as an innocent bystander in those with osteoporosis.[43]

Magnesium also helps to maintain bone mineral density.[44] Vitamin A is needed in the bone-remodeling process, but too much may be associated with osteoporosis.[45] Clearly, a well-balanced diet that supplies abundant fruits and vegetables and a full array of nutrients is central to bone health.[46] In contrast, diets containing too much salt, candy, or colas and possibly caffeine are associated with bone losses.[47]

Review Table C8-2 (p. 306) and the risk factors covered here. The more risk factors that apply to you, the greater your chances of developing osteoporosis in the future and the more seriously you should take the advice offered in this Controversy.

Diagnosis and Medical Treatment

Diagnosis of osteoporosis includes measuring bone density using an advanced form of X-ray (DEXA, described in Chapter 9) or ultrasound.[†] A thorough examination also includes evaluating risk factors for fractures, such as race, family history, and physical inactivity.[48] Men with a family history of osteoporosis and all women should have a bone density test after they reach age 50.

Estrogen therapy can help nonmenstruating women prevent further bone loss and reduce the incidence of fractures.[49] Estrogen therapy may also increase the risks for heart disease and breast cancer, however, so women and their health-care providers must carefully weigh any potential benefits against the potential risks.[50] A combination of drugs may be an option for some women.[51]

Several drugs are proving powerful allies in the struggle to reverse bone loss. Such drugs inhibit the activities of the bone-dismantling cells, thus allowing the bone-building cells to slowly shore up bone tissue with new calcium deposits.[‡] The drugs have worked minor miracles in reversing even severe bone loss in some people, but for others, they are not tolerated or not effective.

Calcium Recommendations

As described earlier, bone strength later in life depends most on how well the bones were developed and maintained during youth. Adequate calcium nutrition during the growing years is essential to achieving optimal peak bone mass.[52] Yet only 10 percent of girls and 25 percent of boys meet the recommendation for calcium during their bone-forming years.[53] The DRI committee recommends 1,300 milligrams of calcium, the amount in about 4 cups of milk, each day for everyone 9 through 18 years of age, 1,000 milligrams through age 50, and 1,200 milligrams thereafter.

How should you obtain this calcium? Nutritionists strongly recommend foods and beverages as your source of calcium and that you take supplements only when advised to do so by a physician. People can best support their bones' health by following the lifetime recommendations for healthy bones in Table C8-3. Calcium supplements cannot equal any of the actions listed in the table. By the way, some miraculous powers for preventing diseases have been attributed to calcium derived from marine coral, but this is just a marketing gimmick. "Coral calcium" may contain toxic metals or trigger allergic reactions in people with allergy to shellfish. The advertiser was fined and

banned from advertising further false health claims about coral calcium.[54]

Bone loss is not a calcium-deficiency disease comparable to iron-deficiency anemia. In iron-deficiency anemia, high iron intakes reliably reverse the condition. With respect to calcium balance, though, calcium intakes alone do little or nothing to reverse bone loss. Calcium and vitamin D supplements after the age of 50, however, do produce small but still beneficial effects on the bone mass and fracture rates. During the menopausal years, calcium supplements of 1 gram may slow, but cannot fully prevent, the inevitable bone loss.

Although calcium supplements are well tolerated by most people, taking self-prescribed calcium supplements entails possible risks (see Table C8-4, p. 310) and cannot take the place of sound food choices and other healthy habits. The next section provides some details about the variety of calcium supplements and the benefits and risks of taking them.

Calcium Supplements

Calcium supplements are available in three chemical forms. Simplest are the purified **calcium compounds,** such as calcium carbonate, citrate, gluconate, hydroxide, lactate, malate, or phosphate, and compounds of calcium with amino acids (called **amino acid chelates**). Second are mixtures of calcium with other compounds, such as calcium carbonate with magnesium carbonate, with aluminum salts (as in some **antacids**), or with vitamin D. Third are powdered, calcium-rich materials such as **bone meal, powdered bone, oyster shell,** or **dolomite** (limestone). See Table C8-5, (p. 310) for supplement terms.

Question 1: How well does the body absorb and use the calcium from various supplements? Based on research to date, many people seem to absorb calcium reasonably well—and about as well as from milk—from amino acid chelates and from calcium compounds such as calcium acetate, calcium carbonate, calcium citrate, calcium gluconate, calcium lactate, and calcium phosphate dibasic. People absorb calcium less well from a mixture of calcium and magnesium carbonates, from oyster

---

[†]DEXA stands for *dual X-ray absorptiometry.*

[‡]The drugs are bisphosphonates.

TABLE C8·3 A LIFETIME PLAN FOR HEALTHY BONES

Childhood

Ages	Goal	Guidelines
2 through 12 or 13 years (sexual maturity)	Grow strong bones.	• Use milk as the primary beverage to meet the need for calcium within a balanced diet that provides all nutrients. • Play actively in sports or other activities. • Limit television and other sedentary entertainment. • Do not start smoking or drinking alcohol. • Drink fluoridated water.

Adolescence through Young Adulthood

Ages	Goal	Guidelines
13 or 14 through 30 years	Achieve peak bone mass.	• Choose milk as the primary beverage, or if milk causes distress, include other calcium sources. • Commit to a lifelong program of physical activity. • Do not smoke or drink alcohol—if you have started, quit. • Drink fluoridated water.

Mature Adult

Ages	Goal	Guidelines
31 through 50 years	Maximize bone retention.	• Continue as for 13- to 30-year-olds. • Adopt bone-strengthening exercises. • Obtain the recommended amount of calcium from food. • Take calcium supplements only if calcium needs cannot be met through foods.

Mature Adult

Ages	Goal	Guidelines
51 years and above	Minimize bone loss.	• Continue as for 13- to 30-year-olds. • Continue striving to meet the calcium need from diet. • Continue bone-strengthening exercises. • Obtain a bone density test; follow physician's advice concerning bone-restoring medications and supplements.

NOTE: The exact ages of cessation of bone accretion and onset of loss vary among people, but in general, data indicate that the skeleton continues to accrete mass for approximately 10 years after adult height is achieved and begins to lose bone around age 35.

SOURCE: Based on data from Committee on Dietary Reference Intakes, Food and Nutrition Board, Institute of Medicine, *Dietary Reference Intakes for Calcium, Phosphorus, Magnesium, Vitamin D, and Fluoride* (Washington D.C.: National Academy Press, 1997), pp. 91–112.

shell calcium fortified with inorganic magnesium, from a chelated calcium-magnesium combination, and from calcium carbonate fortified with vitamins and iron. Some people absorb calcium better from milk and milk products than from even the most absorbable supplements. Calcium from mineral water may also be highly absorbable.[55] Table C8-6 (p. 310) lists some sources of supplemental calcium, the amount of calcium delivered, and the form of the calcium along with the number of calories in a dose or serving.

Question 2: How much calcium does the supplement provide? The Tolerable Upper Intake Level for calcium has been set at 2,500 milligrams. To be safe, sup-plements should provide less than this, as foods also provide calcium. Read the label to find out how much a dose supplies. Calcium carbonate is 40 percent elemental calcium, whereas calcium gluconate is only 9 percent. A healthy young adult (with the exception of pregnant or lactating women in whom absorption increases) absorbs about 25 percent of the available calcium.

Question 3: Will the supplement be digested and the calcium be available for absorption? Manufacturers com-press large quantities of calcium into small pills, and stomach acid often has difficulty penetrating the pill. To test whether a supplement will dissolve, drop it into a 6-ounce cup of vinegar and stir occasionally. A high-quality for-mulation will dissolve within half an hour. The chewable kind, because they are chewed into bits before swallowing, and calcium-fortified foods and bever-ages are not prone to this problem.

One last pitch: Think one more time before you decide to take supplements for calcium. The Consensus Conference on Osteoporosis recommends milk. The American Society for Bone and Mineral Research recommends foods as the source of calcium in preference to sup-plements. The National Institutes of Health concludes that foods are best and recommends supplements only when intake from food is insufficient. The authors of this book are so impressed

TABLE C8·4 CALCIUM SUPPLEMENT RISKS

People who take calcium supplements risk:

- *Impaired iron status.* Calcium inhibits iron absorption.
- *Accelerated calcium loss.* Calcium-containing antacids that also contain aluminum and magnesium hydroxide cause a net calcium loss.
- *Urinary tract stones or kidney damage in susceptible individuals.* People who have a history of kidney stones should be monitored by a physician and choose calcium citrate if they must take supplements.
- *Exposure to contaminants.* Some preparations of bone meal and dolomites are contaminated with hazardous amounts of arsenic, cadmium, mercury, and lead.
- *Vitamin D toxicity.* Vitamin D, which is present in many calcium supplements, can be toxic. Users must eliminate other concentrated vitamin D sources.
- *Excess blood calcium.* This complication is seen only with doses of calcium fourfold or more greater than customarily prescribed.
- *Milk alkali syndrome.* This condition is rare, but not absent. It is characterized by high blood calcium, metabolic alkalosis, and renal failure. Early symptoms include irritability, headaches, and apathy.
- *Other nutrient interactions.* Calcium inhibits absorption of magnesium, phosphorus, and zinc.
- *Drug interactions.* Calcium and tetracycline form an insoluble complex that impairs both mineral and drug absorption.
- *GI distress.* Constipation, intestinal bloating, and excess gas are common.

with the importance of using abundant, calcium-rich foods that we have worked out ways to do so at every meal. Seldom do nutritionists agree so unanimously.

TABLE C8·5 CALCIUM SUPPLEMENT TERMS

- **amino acid chelates** (KEY-lates) compounds of minerals (such as calcium) combined with amino acids in a form that favors their absorption. A *chelating agent* is a molecule that surrounds another molecule and can then either promote or prevent its movement from place to place (*chele* means "claw").
- **antacids** acid-buffering agents used to counter excess acidity in the stomach. Calcium-containing preparations (such as Tums) contain available calcium. Antacids with aluminum or magnesium hydroxides (such as Rolaids) can accelerate calcium losses.
- **bone meal** or **powdered bone** crushed or ground bone preparations intended to supply calcium to the diet. Calcium from bone is not well absorbed and is often contaminated with toxic materials such as arsenic, mercury, lead, and cadmium.
- **calcium compounds** the simplest forms of purified calcium. They include calcium carbonate, citrate, gluconate, hydroxide, lactate, malate, and phosphate. These supplements vary in the amount of calcium they contain, so read the labels carefully. A 500-milligram tablet of calcium gluconate may provide only 45 milligrams of calcium, for example.
- **dolomite** a compound of minerals (calcium magnesium carbonate) found in limestone and marble. Dolomite is powdered and is sold as a calcium-magnesium supplement, but may be contaminated with toxic minerals, is not well absorbed, and interacts adversely with absorption of other essential minerals.
- **oyster shell** a product made from the powdered shells of oysters that is sold as a calcium supplement, but is not well absorbed by the digestive system.

TABLE C8·6 A SAMPLING OF SUPPLEMENTAL CALCIUM SOURCES

Calcium Source	Typical Amount per Serving	Calories
Antacid medication, regular strength ("Tums" type)	500 mg per 2 tablets	10
Breads with "more" calcium[a]	80 mg per slice	70
Calcium-fortified candies or chewable candy supplements	500 mg per doze	20
Calcium-fortified or "100% nutrient" cereals	1,000 mg per serving	110
Calcium-fortified fat-free milk and milk products	500 mg per 8 oz serving	100
Calcium-fortified orange juice or other fruit beverages	350 mg per 8 oz serving	110
Calcium pills	A wide variety of pills provide varying doses. Read the label.	Negligible
Meal replacer: cereal bars "with milk"	250 mg	160
Meal replacer: "complete nutrition" drinks	200–350 mg per 8 oz drink	360
Meal replacer: "energy" bars	300 per one bar	230

[a]Bread, though not rich in calcium, is heavily consumed and may contribute significantly to many people's intakes.

Energy Balance and Healthy Body Weight

© Pierre August Renoir (1841–1919 French), *Luncheon of the Boating Party*, 1881. Phillips Collection, Washington D.C./SuperStock, Inc.

CONTENTS

Do You Ever . . .

Wish you could control your body weight, once and for all?

Feel tempted by a favorite treat when you don't feel hungry?

Wonder how extra calories from food become fat on your body?

Try fad diets to lose weight?

KEEP READING.

In 1991, only four states had an obesity rate (BMI ≥30) of greater than 15 percent. By 2002, all states except Colorado reported an obesity rate of at least 15 percent, many reported a rate of greater than 20 percent, and the trend is steadily upward.

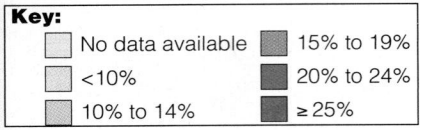

Key:

☐ No data available	☐ 15% to 19%
☐ <10%	☐ 20% to 24%
☐ 10% to 14%	☐ ≥ 25%

1991

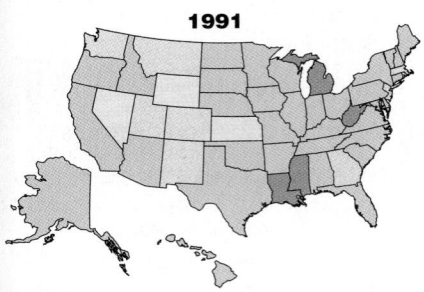

In 1991, only 4 states had obesity rates >15 percent.

2002

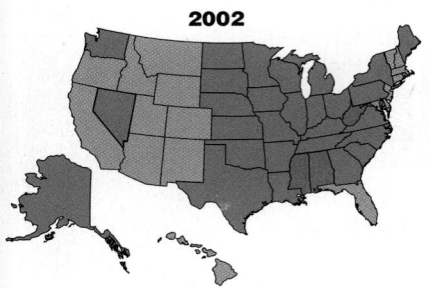

In 2002, 29 states had obesity rates of 20 to 24 percent and rates of 3 more topped 25 percent.

SOURCE: U.S. Obesity Trends 1985 to 2002, Centers for Disease Control and Prevention, available at www.cdc.gov/nccdphp/dnpa/obesity/trend/maps/slide/001.htm.

overweight overfatness of a moderate degree; defined as a body mass index (BMI) of 25.0 through 29.9. BMI is defined later.

underweight too little body fat for health; defined as having a body mass index of less than 18.5.

body composition the proportions of muscle, bone, fat, and other tissue that make up a person's total body weight.

ARE YOU PLEASED WITH your body weight? If you answered yes, you are a rare individual. Nearly all people in our society think they should weigh more or less (mostly less) than they do. Their primary concern is usually appearance, but they often correctly perceive that physical health is somehow related to weight. At the extremes, both **overweight** and **underweight** present risks to health.*[1]

People also think of their weight as something they should control, once and for all. Three misconceptions in that sentence frustrate their efforts, however—the focus on weight, the focus on *controlling* weight, and the focus on a short-term endeavor. Simply put, it isn't your weight you need to control; it's the fat in your body in proportion to the lean—your **body composition.** And controlling body composition directly isn't possible—you can only control your *behavior.* Sporadic bursts of activity, such as "dieting," are not effective; the behaviors that achieve and maintain a healthy body weight take a lifetime of commitment.

This chapter starts out by presenting the problems associated with deficient and excessive body fatness and then examines how the body manages its energy budget. The following sections show how to judge body weight on the sound basis of health and describe theories about the causes of obesity. Finally, the chapter reveals how the body gains and loses weight and suggests lifestyle strategies for achieving and maintaining a healthy body weight.

The Problems of Too Little or Too Much Body Fat

Both deficient and excessive body fat present risks to health. In the United States, too little body fat is not a widespread problem. **Obesity,** in contrast, is an escalating epidemic—see Figure 9-1.[2] In the year 2000, 64 percent of U.S. adults were overweight, and about 30 percent met the definition of obesity (see Figure 9-2).[3] Additionally, one of every seven children and teenagers in the United States is overweight.[4] The *Dietary Guidelines 2005*, and the *Healthy People 2010* objectives include mandates to reduce obesity and overweight, and the Food and Drug Administration (FDA) has taken steps to confront this nationwide crisis.[5]

While obesity and overweight have taken the spotlight, the problem of *underweight* also poses threats to those who drop below a healthy body weight. Risks to life increase for those at either extreme of body weight (Figure 9-3, p. 313).

What Are the Risks from Underweight?

Thin people die first during a siege or in a famine. Overly thin people are also at a disadvantage in the hospital, where their nutrient status can easily deteriorate if they have to go without food for days at a time while undergoing tests or surgery. Underweight also increases the risk for any person fighting a **wasting** disease. People with cancer often die, not from the cancer itself, but from starvation. Thus, excessively underweight people are urged to gain body fat as an energy reserve and to acquire protective amounts of all the nutrients that can be stored.

▶ **KEY POINT** Deficient body fatness threatens survival during a famine or in wasting diseases.

*Reference notes are found in Appendix F.

What Are the Risks from Overweight?

People who are obese more commonly suffer from serious diseases, such as hypertension, diabetes, and heart disease, than do people who stay lean. Even those just moderately overweight may increase their risks, assuming that the excess weight is fat, not muscle.[6] Body weight is not the only factor in illness causation, however; some obese people remain healthy and live long despite their excess body fatness.[7] Researchers suspect that genetic inheritance, abstinence from smoking, and cardiovascular fitness help determine who among the obese stays well and who falls ill.[8] The trouble is, no one can tell to which group he or she belongs until serious diseases set in.

To underestimate the threat from obesity is to invite calamity. An estimated 300,000 people in the United States die each year from obesity-related diseases, and the risk of dying increases proportionally with increasing weight.[9] Over 70 percent of obese people suffer from at least one other major health problem. Excess weight causes up to half of all cases of hypertension, thereby increasing the risk of heart attack and stroke.[10] Cardiovascular disease rates rise with central obesity, as described next. Currently, only tobacco contributes to more preventable illnesses and premature deaths.

The health risks of overfatness are so many that obesity has been declared a chronic disease. In addition to hypertension and cardiovascular disease, obesity triples a person's risk of developing diabetes and all of its associated ills. Chapter 4 presented maps (page 121) depicting increasing U.S. rates of diabetes over the last decade, which bear a striking resemblance to the maps of Figure 9-1 presented here. Even modest weight gain raises diabetes risk, and the risk appears greater for people of European descent than for African Americans.[11] If hypertension, cardiovascular disease, or diabetes runs in your family, you urgently need to attend to controlling body fatness.

Obese adults are also threatened by other risks. Among them are abdominal hernias, arthritis, complications in pregnancy and surgery, flat feet, gallbladder disease, gout, high blood lipids, kidney stones, liver malfunction, respiratory problems (including Pickwickian syndrome, a breathing blockage linked to sudden death), sleep apnea (dangerous abnormal breathing during sleep), some cancers, varicose veins, and even a high accident rate.[12] Moreover, after the effects of diagnosed diseases are taken into account, the risk of death from other causes remains almost twice as high for people with lifelong obesity as for others. Even modest weight loss often improves these maladies. An older obese man, for example, often gains relief from arthritis pain and reduces his blood pressure with the loss of 10 percent of his body weight.

▶ **KEY POINT** Most obese people suffer illnesses, and obesity is considered a chronic disease.

Central Obesity Even more than total fatness, fat that collects deep within the central abdominal area of the body, called **visceral fat,** may be especially dangerous with regard to risks of diabetes, stroke, hypertension, and coronary artery disease (Figure 9-4, page 314).[13] The risk of death from *all* causes may be higher in those with **central obesity** than in those whose fat accumulates elsewhere in the body. The health risks of obesity seem to run on a continuum: normal weight brings no extra risk, central obesity carries severe risks, and other forms of obesity fall somewhere in between.

Why should fat in the abdomen bring extra risk to the heart? Some researchers suspect that differences in fat mobility are part of the explanation. Visceral fat, which is readily released into the bloodstream, may make a significant contribution to the blood's daily burden of cholesterol-carrying lipoproteins, LDL, thereby increasing heart disease risk.[14] Fat layers lying just beneath the skin (**subcutaneous fat**) of the abdomen, thighs, hips, and legs also release fat, but sluggishly and so, theoretically, may contribute less to blood lipids.[15]

Men of all ages and women who are past menopause are more prone to develop the "apple" profile of central obesity, whereas women in their reproductive years develop more of a "pear" profile (fat around the hips and thighs). Some women change profile

FIGURE 9·2 BODY WEIGHTS AMONG U.S. ADULTS

BMI stands for body mass Index, an index of a person's weight in relation to height, associated with degree of health risk. BMI is defined on p. 314.

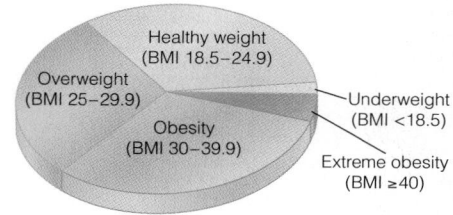

FIGURE 9·3 UNDERWEIGHT, OVERWEIGHT, AND MORTALITY

This J-shaped curve describes the relationship between body mass index (BMI), and mortality. It shows that both underweight and overweight present risks of a premature death.

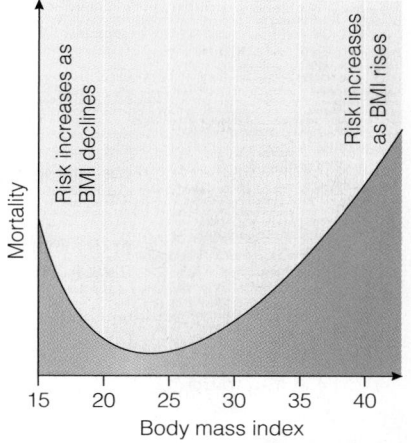

obesity overfatness with adverse health effects, as determined by reliable measures and interpreted with good medical judgment. Obesity is officially defined as a body mass index of 30 or higher.

wasting the progressive, relentless loss of the body's tissues that accompanies certain diseases and shortens survival time.

visceral fat fat stored within the abdominal cavity in association with the internal abdominal organs; also called *intra-abdominal fat.*

central obesity excess fat in the abdomen and around the trunk.

subcutaneous fat fat stored directly under the skin (*sub* means "beneath"; *cutaneous* refers to the skin).

Factors affecting body fat distribution:

- Menopause in women.
- Smoking.
- Alcohol intake.
- Physical activity.

at menopause, and lifelong "pears" may suddenly face increased risks of diseases that accompany excess visceral fat. Smokers, too, may carry more of their body fat centrally. Although a smoker may weigh less than the average nonsmoker, the smoker's waist measurement may be greater, leading to the theory that smoking directly affects body fat distribution. Two other factors also affect body fat distribution. Moderate-to-high intakes of alcohol have a positive association with central obesity, and higher levels of physical activity have a negative association.[16] A later section explains how to judge whether a person carries too much fat around the middle.

▶ KEY POINT   Central obesity may be more hazardous to health than other forms of obesity.

How Fat Is Too Fat for Health?   People want to know exactly how fat is too fat for health, and although the answer is not the same for everyone, scientists have developed guidelines. Obesity experts commonly evaluate risks to health from obesity using three indicators, discussed in more detail in a later section. The first of these is a person's **BMI,** or **body mass index.** The BMI, which defines average relative weight for height in people older than 20 years, often correlates with degree of body fatness and disease risks.[17]

The second indicator is waist circumference, reflecting the degree of visceral fatness in proportion to body fatness (see Table 9-1).* The third indicator is the person's disease risk profile, which takes into account whether the person has hypertension, type 2 diabetes, or elevated blood cholesterol; whether the person smokes; and so forth (see Table 9-2). The more risk factors and the greater the obesity, the greater the urgency to control body fatness.

Fitness of the heart and lungs gained through regular physical activity improves the health of the heart and longevity, independent of BMI.[18] Normal-weight fit people have the lowest risk of mortality from chronic diseases, but among those who are overweight, fitness may be a greater determinant of the risk of death than overfatness.[19] A healthy body weight supports good health, but fitness also clearly contributes.

▶ KEY POINT   Experts estimate health risks from obesity using BMI values, waist circumference, and a disease risk profile. Fit people are healthier than others, regardless of body fatness.

*The National Heart, Lung, and Blood Institute has replaced waist-hip ratio with waist circumference for assessment of obesity health risks.

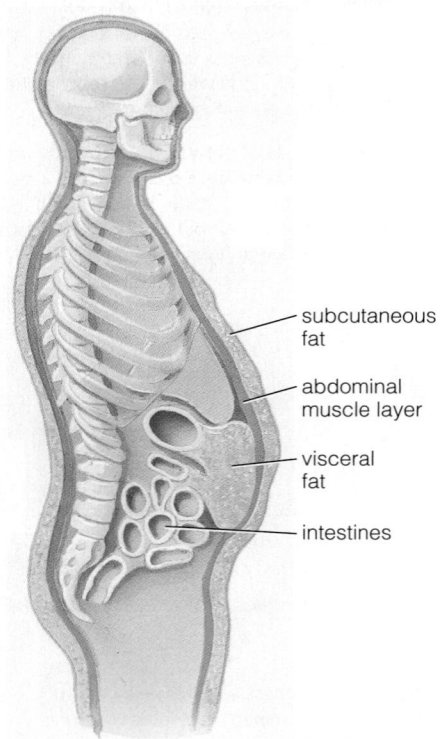

FIGURE 9·4   VISCERAL FAT AND SUBCUTANEOUS FAT
The fat deep within the body's abdominal cavity may pose an especially high risk to health.

- subcutaneous fat
- abdominal muscle layer
- visceral fat
- intestines

body mass index (BMI) an indicator of obesity or underweight, calculated by dividing the weight of a person by the square of the person's height.

TABLE 9·1   DISEASE RISKS ACCORDING TO BMI VALUES AND WAIST CIRCUMFERENCE^a

The degree of risk is heightened by the presence of specific diseases, other risk factors (such as elevated blood LDL cholesterol, see Figure 11-6, p. 402), or smoking (see the margin).

BMI		Waist ≤40 in (Men) or ≤35 in (Women)	Waist ≥40 in (Men) or ≥35 in (Women)
18.5 or less	Underweight	Low	—
18.5–24.9	Normal	Low	—
25.0–29.9	Overweight	Increased	High
30.0–34.9	Obese, class I	High	Very high
35.0–39.9	Obese, class II	Very high	Very high
40 or greater	Extremely obese, class III	Extremely high	Extremely high

^aRisk for type 2 diabetes, hypertension, and cardiovascular disease.

SOURCE: National Heart, Lung, and Blood Institute, National Institutes of Health, *The Practical Guide: Identification, Evaluation, and Treatment of Overweight and Obesity in Adults,* NIH publication no. 00-4084 (Washington, D.C.: Government Printing Office, 2000).

TABLE 9·2 INDICATORS OF AN URGENT NEED FOR WEIGHT LOSS

The National Heart, Lung, and Blood Institute states that aggressive treatment may be needed for critically obese people who also have any of the following:
- Established cardiovascular disease (CVD).
- Established type 2 diabetes, or impaired glucose tolerance.
- Sleep apnea, a disturbance of breathing in sleep, including temporary stopping of breathing.

The same urgency for treatment exists for an obese person with any *three* of the following:
- Hypertension.
- High LDL.
- Smoking.
- Low HDL cholesterol.
- Sedentary lifestyle.
- Age older than 45 years (men) or 55 years (women).
- Heart disease of an immediate family member before age 55 (male) or 65 (female).

SOURCE: National Heart, Lung, and Blood Institute, National Institutes of Health, *The Practical Guide: Identification, Evaluation, and Treatment of Overweight and Obesity in Adults*, NIH publication no. 00-4084 (Washington, D.C.: Government Printing Office, 2000).

Being active—even if overweight—is healthier than being sedentary. With a BMI of 36, aerobics instructor Jennifer Portnick is considered obese, but her daily workout routine helps to reduce the risks to her health.

Social and Economic Costs of Obesity Although some overfat people escape health problems, no one who is fat in our society quite escapes the social and economic handicaps. Fat people are more likely to be judged on their appearance than on their character.[20] Our society places enormous value on thinness, especially for women, and fat people are less sought after for romance, less often hired, and less often admitted to college. They pay higher insurance premiums, and they pay more for clothing.

An estimated 30 to 40 percent of all U.S. women (and 20 to 25 percent of U.S. men) are trying to lose weight at any given time, spending up to $40 billion each year to do so. The assumption is that every overweight person can and should achieve slenderness. Yet most overweight people cannot—for whatever reason—become slender. By one reckoning, only 5 percent of all people who successfully lose weight maintain their losses; other estimates are more optimistic but still predict that the majority of dieters' lost weight will creep back over time.[21]

Prejudice defines obese people by their appearance rather than by their character, often stereotyping them as lazy, stupid, and self-indulgent. Obese people suffer emotional pain when others treat them with hostility and contempt. Subtle blaming for an apparent lack of the discipline to resolve their weight problem often becomes internalized as guilt and self-deprecation. Even health-care professionals, including dietitians, can be among the chief offenders. To free our society of its obsession with body weight and prejudice against obesity, activists are speaking out for acceptance of body weight and respect for individuals.

▶ KEY POINT Overfatness presents social and economic handicaps as well as physical ills. Judging people by their body weight is a form of prejudice in our society.

The Body's Energy Balance

What happens inside the body when you eat too much or too little food? The body ends up with an unbalanced energy budget—you have taken in more or less food energy than you spent. The mechanisms by which the body handles its energy underlie changes that occur in body composition.

When more food energy is consumed than is needed, excess fat accumulates in the fat cells in the body's **adipose tissue** where it is stored. When energy supplies run low, stored fat is withdrawn. The daily energy balance can therefore be stated like this:

adipose tissue the body's fat tissue. Adipose tissue performs several functions, including the synthesis and secretion of the hormone leptin involved in appetite regulation.

Change in energy stores equals food energy taken in minus energy spent on metabolism and muscle activities.

More simply:

$$\text{Change in energy stores} = \text{energy in} - \text{energy out}.$$

Too much or too little fat on the body today does not necessarily reflect today's energy budget. Small imbalances in the energy budget compound over time.

Energy In and Energy Out

The energy in foods and beverages is the only contributor to the "energy in" side of the energy balance equation. Before you can decide how much food will supply the energy you need in a day, you must first become familiar with the amounts of energy in foods and beverages. One way to do so is to look up calorie amounts associated with foods and beverages in the Table of Food Composition (Appendix A). Alternatively, computer programs can provide this information in the blink of an eye for those with computer access. Such numbers are always fascinating to people concerned with managing body weight.

For example, an apple gives you 70 calories from carbohydrate; a regular-size candy bar gives you about 250 calories mostly from fat and carbohydrate. You may already know that for each 3,500 calories you eat in excess of expenditures, you store approximately 1 pound of body fat.

On the "energy out" side of the equation, no easy method exists for determining the energy an individual spends and therefore needs. Energy expenditures vary so widely among individuals that estimating an individual person's need requires knowing something about the person's lifestyle and metabolism.

> **KEY POINT** The "energy in" side of the body's energy budget is measured in calories taken in each day in the form of foods and beverages. The number of calories in foods and beverages can be obtained from published tables or computer diet analysis programs. No easy method exists for determining the "energy out" side of a person's energy balance equation.

Estimated Energy Requirements

A person wishing to know how much energy he or she needs in a day might look up his or her **Estimated Energy Requirement (EER)** value listed on the inside front cover of this book. The list of numbers there seems to imply that for the given age and gender groups, the number of calories needed to meet the daily requirement can be stated as precisely as, say, the recommended intake for vitamin A. These values, however, are appropriate only for those exactly matching the characteristics of the "reference man and woman" (see the margin). People who deviate in any way from these characteristics must use other methods for determining energy requirements, and almost everyone deviates.

Taller people need proportionately more energy than shorter people to balance their energy budgets because their greater surface area allows more energy to escape as heat. Older people generally need less than younger people due to slowed metabolism and reduced muscle mass, which occur in part because of reduced physical activity. (As Chapter 14 points out, these losses may not be inevitable for people who stay active.) On average, energy need diminishes by 5 percent per decade beyond the age of 30 years.

In reality, no one is average. In any group of 20 similar people with similar activity levels, one may expend twice as much energy per day as another. A 60-year-old person who bikes, swims, or walks briskly each day may need as many calories as a sedentary person of 30. Clearly, with such a wide range of variation, a necessary step in determining any person's energy need is to study that person.

■ 1 lb body fat = 3,500 cal.

■ Pure fat is worth 9 calories per gram. A pound of it (450 g), then, would store 4,050 calories. A pound of *body* fat is not pure fat, though; it contains water, protein, and other materials of living tissue—hence the lower calorie value.

The DRI committee sets EER for a reference man and woman.

- Reference man: "Active" physical activity level, 22.5 BMI, 5 ft 10 in tall, weighing 154 lb.
- Reference woman: "Active" physical activity level, 21.5 BMI, 5 ft 4 in tall, weighing 126 lb.

■ Instructions for determining whether you are sedentary, lightly active, active, or very active are provided in Appendix H. About 80% of people in the United States and Canada fall into the sedentary or lightly active categories.

Estimated Energy Requirement (EER) the DRI recommendation for energy intake, accounting for age, gender, weight, height, and physical activity. Also defined in Chapter 2.

▶ **KEY POINT** The DRI committee sets Estimated Energy Requirements for a reference man and woman. People's energy needs vary greatly.

How Many Calories Do I Need Each Day?

Simply put, you need to take in enough calories to cover your energy expenditure each day—your energy budget must balance. One way to estimate your energy need is to monitor your food intake and body weight over a period of time in which your activities are typical and are sufficient to maintain your health. If you keep an accurate record of all the foods and beverages you consume and if your weight is in a healthy range and has not changed during the past few months, you can conclude that your energy budget is balanced. Your average daily calorie intake is sufficient to meet your daily output—your need, therefore, is the same as your current intake.[22] At least three, and preferably seven, days or more of honest record keeping are necessary because intakes and activities fluctuate from day to day.

An alternative method of determining energy need is based on energy output. The two major ways in which the body spends energy are (1) to fuel its **basal metabolism** and (2) to fuel its **voluntary activities.** Basal metabolism requires energy to support the body's work that goes on all the time without our conscious awareness. A third energy component, the body's metabolic response to food or the **thermic effect of food,** uses up about 5 to 10 percent of a meal's energy value in stepped-up metabolism in the five or so hours after the meal.

Basal metabolism consumes a surprisingly large amount of fuel, and the **basal metabolic rate (BMR)** varies from person to person (see Figure 9-5). Depending on activity level, a person whose total energy needs are 2,000 calories a day may spend as many as 1,000 to 1,600 of them to support basal metabolism. The hormone thyroxine directly controls basal metabolism—the less secreted, the lower the energy requirements for basal functions. The rate is lowest during sleep.[*] Many other factors also affect the BMR (see Table 9-3).

People often wonder whether they can speed up their metabolism to spend more daily energy. You cannot speed up your BMR very much *today.* You can, however, amplify

---

[*]A measure of energy output taken while the person is awake but relaxed yields a slightly higher number called the *resting metabolic rate,* sometimes used in research.

FIGURE 9·5 COMPONENTS OF ENERGY EXPENDITURE

In most people, basal metabolism represents the person's largest expenditure of energy, followed by physical activity and the thermic effect of food.

25–50% physical activity[a]

5–10% thermic effect of food

50–65% BMR

[a]For a sedentary person, physical activities may account for less than half as much energy as basal metabolism, whereas a very active person's activities may equal the energy cost of basal metabolism.

SOURCE: Data from the Standing Committee on the Scientific Evaluation of Dietary Reference Intakes, Food and Nutrition Board, Institute of Medicine, *Dietary Reference Intakes for Energy, Carbohydrate, Fiber, Fat, Fatty Acids, Cholesterol, Protein, and Amino Acids* (Washington, D.C.: National Academies Press, 2002), p. 5-7.

■ Chapter 10 is devoted to presenting details of how the body spends its energy during activity.

TABLE 9·3 FACTORS THAT AFFECT THE BMR

Factor	Effect on BMR
Age	The BMR is higher in youth; as lean body mass declines with age, the BMR slows. Continued physical activity may prevent some of this decline.
Height	Tall people have a larger surface area, so their BMRs are higher.
Growth	Children and pregnant women have higher BMRs.
Body composition	The more lean tissue, the higher the BMR. A typical man has greater lean body mass than a typical woman, making his BMR higher.
Fever	Fever raises the BMR.
Stress	Stress hormones raise the BMR.
Environmental temperature	Adjusting to either heat or cold raises the BMR.
Fasting/starvation	Fasting/starvation hormones lower the BMR.
Malnutrition	Malnutrition lowers the BMR.
Thyroxine	The thyroid hormone thyroxine is a key BMR regulator; the more thyroxine produced, the higher the BMR.

basal metabolism the sum total of all the involuntary activities that are necessary to sustain life, including circulation, respiration, temperature maintenance, hormone secretion, nerve activity, and new tissue synthesis, but excluding digestion and voluntary activities. Basal metabolism is the largest component of the average person's daily energy expenditure.

voluntary activities intentional activities (such as walking, sitting, or running) conducted by voluntary muscles.

thermic effect of food (TEF) the body's speeded-up metabolism in response to having eaten a meal; also called *diet-induced thermogenesis.*

basal metabolic rate (BMR) the rate at which the body uses energy to support its basal metabolism.

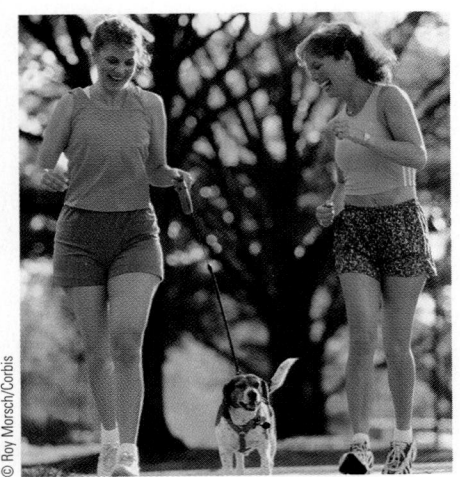
Balancing food energy intake with physical activity can add to life's enjoyment.

the second component of your energy expenditure—your voluntary activities. If you do, you will spend more calories today, and if you keep doing so day after day, your BMR will also increase.[23] Lean tissue is more metabolically active than fat tissue, so a way to speed up your BMR to the maximum possible rate is to make endurance and strength-building activities a daily habit so that your body composition moves toward the lean. A warning: some ads for weight-loss diets claim that certain substances, such as grapefruit or herbs, can elevate the BMR and thus promote weight loss. This claim is false. Any meal temporarily steps up energy expenditure due to the thermic effect of food; in the context of a mixed diet, the differences among foods are not large enough to be worth notice.

In voluntary activities, energy spent depends somewhat on your personal style. In general, the heavier the weight of the body parts you move in your activity and the longer the time you invest, the more calories you expend.

▶ **KEY POINT** Two major components of the "energy out" side of the body's energy budget are basal metabolism and voluntary activities. A third component of energy expenditure is the thermic effect of food. Many factors influence the basal metabolic rate.

The DRI Method of Estimating Energy Requirements

The DRI committee sets EER values by taking into account the ways in which energy is spent, and by whom:

- *Gender.* Women generally have less lean body mass than men; in addition, women's menstrual hormones influence the BMR, raising it just prior to menstruation.
- *Age.* Because the BMR declines by an average of 5 percent per decade, age is a determining factor when calculating EER values.
- *Physical activity.* To estimate the energy spent on physical activity, the DRI committee clusters various activities according to the typical intensity of a day's efforts.
- *Body size and weight.* The higher BMR of taller and heavier people calls for height and weight to be factored in when estimating a person's EER.

Figure 9-6 explains how to determine your own EER, sufficient to maintain your present weight. It requires a factor for physical activity that you can calculate by following the instructions given in Appendix H. Alternatively, the margin offers a quick and easy but more approximate way to determine your energy need.

▶ **KEY POINT** The DRI committee has established a method for determining an individual's approximate energy requirement.

Body Weight versus Body Fatness

For most people, weighing on a scale provides a convenient and accessible way to measure body fatness, but researchers and health-care providers require more accurate assessments. This section describes some of the preferred methods of assessing overweight and underweight.

Body Mass Index (BMI)

BMI values correlate significantly with body fatness, and experts use them to help evaluate a person's health risks associated with underweight or overweight. The inside back cover of this book provides tables of values in which to find and evaluate BMI values of both adults and adolescents.

A quick and easy estimate of energy need:

- Men: kg body weight × 24 = cal/day.
- Women: kg body weight × 22 = cal/day.

To convert pounds to kilograms (kg), divide by 2.2.

■ BMI was defined earlier on page 314.

No one can tell you exactly how much you should weigh; but with health as a value, you have a starting framework in the BMI table. Your weight should fall within the range that best supports your health. As a general guideline, overweight for adults is defined as BMI of 25.0 through 29.9 and obesity as BMI equal to or greater than 30. Racial differences are apparent when health risks are evaluated by way of BMI values—

FIGURE 9·6	HOW TO CALCULATE YOUR ESTIMATED ENERGY REQUIREMENT (EER)

1. Find the equation for your gender below.
2. Insert your age in years, your weight (wt) in kilograms, and your height (ht) in meters. To convert pounds to kilograms, divide by 2.2 (see the example below). Convert inches to meters by dividing by 39.37 (see example).
3. The final number needed to complete the equation is your physical activity factor (PA) as listed in the table at the end of this figure. To determine whether you are typically sedentary, low active, active, or very active, turn to Appendix H and follow the instructions to obtain your PA factor. The DRI committee provided these instructions because people tend to greatly overestimate their activity levels.
4. With your PA factor in place, proceed with your calculations.

Equations:
Women age 19 years and older:

$$EER = 662 - 6.91 \times age + PA \times [(9.36 \times wt) + (726 \times ht)].$$

Men age 19 years and older:

$$EER = 662 - 9.53 \times age + PA \times [(15.91 \times wt) + (539.6 \times ht)].$$

Example:
An active, 30-year-old man, 5 feet 11 inches (or 71 inches) tall, weighing 178 pounds.

Convert pounds (lb) to kilograms and inches (in) to meters, and insert them, along with his age, into the equation:

$$178 \text{ lb} \div 2.2 = 80.9 \text{ kg.}$$
$$71 \text{ in} \div 39.37 = 1.8 \text{ m.}$$

$$EER = 662 - 9.53 \times age + PA \times [(15.91 \times wt) + (539.6 \times ht)].$$
$$EER = 662 - 9.53 \times 30 + PA \times [(15.91 \times 80.9) + (539.6 \times 1.8)].$$

Determine his PA factor by following the instructions in Appendix H. This man's daily activities place him in the "Active" category on the table below. Insert his PA factor of 1.25 (from the second column of the table into the equation.

$$EER = 662 - 9.53 \times 30 + 1.25 \times [(15.91 \times 80.9) + (539.6 \times 1.8)].$$

With all of the appropriate numbers in place, it is now possible to calculate as follows: (A reminder: perform functions within brackets and parentheses first, and multiplication before addition and subtraction.)

$$EER = 662 - 9.53 \times 30 + 1.25 \times [(15.91 \times 80.9) + (539.6 \times 1.8)].$$
$$EER = 662 - 9.53 \times 30 + 1.25 \times (1,287 + 971).$$
$$EER = 662 - 9.53 \times 30 + 1.25 \times 2,258.$$
$$EER = 662 - 286 + 2,823.$$
$$EER = 3,199.$$

The EER for this man is about 3,200 calories per day. His actual requirement probably falls within a range of perhaps 10 percent above and below this value.

Physical Activity Levels and PA Factors[a]

Description	Men's PA Factors	Women's PA Factors
Sedentary	1.0	1.0
Low active	1.11	1.12
Active	1.25	1.27
Very active	1.48	1.45

[a] Appendix H provides instructions for determining you physical activity level.

At 6 feet 3 inches tall and 245 pounds, Mike O'Hearn would be judged to be obese by BMI standards alone. Further measures reveal that his body contains only 8 percent of its weight as fat, less than the average fat percentage for men, and that his waist circumference is within a healthy range.

For a fair indication of whether you develop fat centrally, measure your waist as shown in Figure 9-9 (p. 322); then compare your measurement with these cutoff points:

• Men: 102 cm (40 in).
• Women: 88 cm (35 in).

Anyone with a waist measurement larger than these standards may carry an increased risk of disease along with the extra girth.

fatfold test measurement of the thickness of a fold of skin on the back of the arm (over the triceps muscle), below the shoulder blade (subscapular), or in other places, using a caliper (depicted in Figure 9-8); also called *skinfold test*.

underwater weighing a measure of density and volume used to determine body fat content.

bioelectrical impedance (im-PEE-dense) a technique for measuring body fatness by measuring the body's electrical conductivity.

dual energy X-ray absorptiometry (ab-sorp-tee-OM-eh-tree) a noninvasive method of determining total body fat, fat distribution, and bone density by passing two low-dose X-ray beams through the body. Also used in evaluation of osteoporosis. Abbreviated *DEXA*.

the risks associated with a high BMI appear to be greater for whites than for blacks.[24] In fact, the health risks associated with obesity may not become apparent in some black women until a BMI of 37.[25]

The BMI values are most accurate in assessing degrees of obesity and are less useful for evaluating nonobese people's body fatness. The BMI values have two major drawbacks: they fail to indicate how much of a person's weight is fat and where that fat is located. These drawbacks limit the value of the BMI for use with:

• Athletes (because their highly developed musculature falsely increases their BMI values).
• Pregnant and lactating women (because their increased weight is normal during childbearing).
• Adults over 65 (because BMI values are based on data collected from younger people and because people "grow shorter" with age).

The bodybuilder in the margin proves this point: with a BMI over 30, he would be classified as obese by BMI standards alone. Yet a clinician would find that his percentage of body fat is well below average and his waist circumference is within a healthy range. A diagnosis of obesity or overweight requires a BMI value *plus* some measure of body composition and fat distribution. There is no easy way to look inside a living person to measure bones and muscles, but several indirect measures can provide an approximation.

▶ **KEY POINT** The BMI values mathematically correlate heights and weights with risks to health. They are especially useful for evaluating health risks of obesity but fail to measure body composition or fat distribution.

Measures of Body Composition and Fat Distribution

A person who stands about 5 feet 10 inches tall and weighs 150 pounds carries about 30 of those pounds as fat. The rest is mostly water and lean tissues: muscles; organs such as the heart, brain, and liver; and the bones of the skeleton (Figure 9-7). This lean tissue is vital to health. The person who seeks to lose weight wants to lose fat, not this precious lean tissue. And for someone who wants to gain weight, it is desirable to gain lean and fat in proportion, not just fat.

Techniques for estimating body fatness include these:

• *Anthropometry.* Direct body measurements include the **fatfold test** and waist circumference. Fatfold measurements taken by a trained technician with standard calipers provide an accurate estimate of total body fat and a fair assessment of the fat's location (Figure 9-8).[26] Waist circumference indicates visceral fatness (Figure 9-9, p. 322); above a certain girth, disease risks rise—even when BMI values are normal.[27]
• *Density* (the measurement of body weight compared with volume). Lean tissue is denser than fat tissue, so the denser a person's body is, the more lean tissue it must contain. Density can be determined by **underwater weighing** or air displacement methods.
• *Conductivity.* Only lean tissue and water conduct electrical current; **bioelectrical impedance** measures how well a tiny harmless electrical charge is conducted through the lean tissue of the body (Figure 9-8).
• *Radiographic techniques.* New technology yields images of body tissues and an assessment of body composition. For example, **dual energy X-ray absorptiometry (DEXA)** measures two beams of X-ray energy as they pass harmlessly through body tissues, giving high-quality assessments of total body fatness, fat distribution, and bone density (Figure 9-8).[28]

Each technique has strengths and weaknesses, but in all cases, the accuracy of the results depends on the skill of the clinician employing the technique and interpreting the results.

FIGURE 9·7 · AVERAGE BODY COMPOSITION OF MEN AND WOMEN

The substantially greater fat tissue of women is normal and necessary for reproduction.

36% muscle

24% organs

27% fat

13% bone

45% muscle

25% organs

15% fat

15% bone

SOURCE: Data from R. E. C. Wildman and D. M. Medeiros, *Advanced Human Nutrition* (Boca Raton, Fla.: CRC Press, 2000), pp. 321–323.

▶ **KEY POINT** A clinician can determine the percentage of fat in a person's body by measuring fatfolds, body density, or other parameters. Distribution of fat can be estimated by radiographic techniques, and central adiposity can be assessed by measuring waist circumference.

FIGURE 9·8 · THREE METHODS OF ASSESSING BODY FATNESS[a]

Fatfold measures can yield accurate results when a trained technician measures body fat by using a caliper to gauge the thickness of a fold of skin. Measurements are taken on the back of the arm (over the triceps), below the shoulder blade (subscapular), and in other places (including lower body sites) and are then compared with standards.

Dual energy X-ray absorptiometry (DEXA) employs two low-dose X-rays that differentiate among fat-free soft tissue (lean body mass), fat tissue, and bone tissue, providing a precise measurement of total fat and its distribution in all but extremely obese subjects.

Bioelectrical impedance is simple, painless, and accurate when properly administered; the method determines body fatness by measuring conductivity. Lean tissue conducts a mild electric current; fat tissue does not.

[a]Other methods inlcude underwater weighing (hydrodensitometry), computed tomography, and magnetic resonance imaging.

FIGURE 9·9 MEASURING WAIST CIRCUMFERENCE

Using a nonstretching tape measure, measure around the body near the belly button. (The skeleton shows the tape position relative to the hip bone.) Exhale normally while taking the measurement. A healthy waist circumference for men is no larger than 102 centimeters (40 inches); for women, no larger than 88 centimeters (35 inches).

Research has linked obesity with:

- Birth order, number of brothers.
- Divorced/single parents, nonprofessional or unemployed parents.
- Early menstruation.
- Ethnicity.
- Exposure to a variety of foods, fast-food consumption.
- Fat intake, protein intake, carbohydrate intake.
- Increased wealth (in developing nations).
- Less leisure time, international travel, geographic location.
- Lower education level, lower social class.
- Maternal famine or obesity during gestation.
- Meal skipping, meals eaten away from home.
- Napping habits, sleep deprivation.
- Less frequent alcohol intake, high alcohol intake.
- Sedentary behavior, television watching.
- Substandard housing.
- Many more.

SOURCE: Adapted from L. Grivetti, Psychology and cultural aspects of energy, *Nutrition Reviews* 59 (2001): S5–S12.

How Much Body Fat Is Ideal?

After you have a body fatness estimate, the question arises: What is the "ideal" amount of fat for a body to have? This prompts another question: Ideal for what? If the answer is "society's approval," be aware that fashion is fickle and many popular body shapes are not achievable by most people. Overweight children in particular suffer when they learn to dislike the bodies they have (see this chapter's Controversy).[29]

If the answer is "health," then the ideal depends partly on your gender and your age. A man of normal weight will have between 12 and 20 percent of body weight as fat, and a woman will have between 20 and 30 percent. Researchers draw the line when body fat exceeds 22 percent in young men, 25 percent in older men, 32 percent in younger women, and 35 percent in older women; age 40 is the dividing line between younger and older.

Besides gender and age, standards differ because of lifestyle and stage of life. For example, competitive endurance athletes need just enough body fat to provide fuel, insulate the body, and permit normal fat-soluble hormone activity, but not so much fat as to weigh them down. An Alaskan fisherman, in contrast, needs a blanket of extra fat to insulate against the cold. For a woman starting pregnancy, the outcome is compromised if she begins with too much or too little body fat. Below a threshold for body fat content set by heredity, some individuals become infertile, develop depression or abnormal hunger regulation, or become unable to keep warm. These thresholds are not the same for each function or in all individuals, and much remains to be learned about them.

▶ **KEY POINT** No single body composition or weight suits everyone; needs vary by gender, lifestyle, and stage of life.

The Mystery of Obesity

Why do some people get fat? Why do some stay thin? Is weight controlled by hereditary metabolic factors or by environmental influences? Is it a matter of eating behaviors—and if so, what directs these behaviors, internal controls or a person's free will? Many factors, some of them conflicting, *correlate* with obesity (the margin lists some of them for interest), but *cause* is elusive. This section sorts through the pieces of the obesity puzzle, but no law says that only one cause must prevail. In all likelihood, internal and external factors operate together and in different combinations, beginning with the appetite and its controls.

Why Did I Eat That?

To repeat, food intake represents the "energy in" side of the body's energy budget. Scientists hope that by discovering how food intake is regulated in the body, they can devise effective strategies against obesity.[30] Eating behavior seems to be regulated by a series of signals that fall into two broad functional categories: "go" mechanisms that stimulate eating and "stop" mechanisms that signal the body to cease or refrain from eating. One view of the process of food intake regulation is summarized in Figure 9-10.

"Go" Signals—Hunger and Appetite Most people recognize **hunger** as a strong, unpleasant sensation that signals a need for food, prompting them to search out food and eat. Hunger makes itself known roughly four to six hours after eating, after the food has left the stomach and much of the nutrient mixture has been absorbed by the intestine. Hunger is triggered by a contracting empty stomach, an empty small intestine, the stomach hormone **ghrelin** produced between meals, and

FIGURE 9·10 HUNGER, APPETITE, SATIATION, AND SATIETY

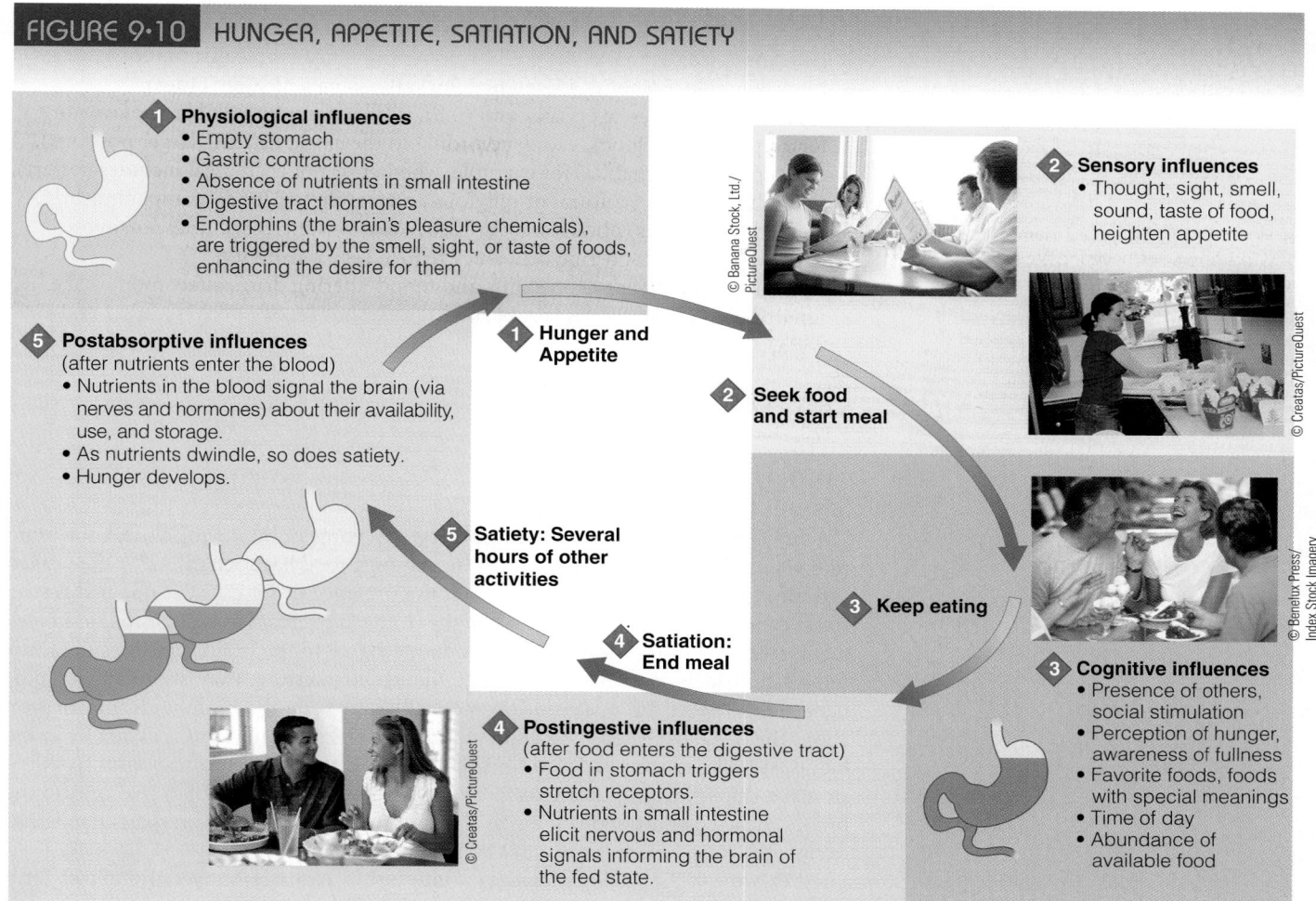

1 Physiological influences
- Empty stomach
- Gastric contractions
- Absence of nutrients in small intestine
- Digestive tract hormones
- Endorphins (the brain's pleasure chemicals), are triggered by the smell, sight, or taste of foods, enhancing the desire for them

5 Postabsorptive influences
(after nutrients enter the blood)
- Nutrients in the blood signal the brain (via nerves and hormones) about their availability, use, and storage.
- As nutrients dwindle, so does satiety.
- Hunger develops.

1 Hunger and Appetite

2 Seek food and start meal

5 Satiety: Several hours of other activities

4 Satiation: End meal

3 Keep eating

4 Postingestive influences
(after food enters the digestive tract)
- Food in stomach triggers stretch receptors.
- Nutrients in small intestine elicit nervous and hormonal signals informing the brain of the fed state.

2 Sensory influences
- Thought, sight, smell, sound, taste of food, heighten appetite

3 Cognitive influences
- Presence of others, social stimulation
- Perception of hunger, awareness of fullness
- Favorite foods, foods with special meanings
- Time of day
- Abundance of available food

chemical and nervous signals in the brain. Other factors influencing hunger may include the nutrients present in the bloodstream, the size and composition of the preceding meal, the weather (heat reduces food intake; cold increases it), exercise, sex hormones, physical and mental illnesses, and others.

The body's hunger response adapts quickly to changes in food intake. A person who restricts the amount of food consumed at each meal may feel extra hungry for a few days, but then hunger may diminish for a time. During this period, a large meal may make the person feel uncomfortably full, partly because the stomach's capacity has adapted to a smaller quantity of food. At this time a dieter may report "My stomach has shrunk," but the stomach organ itself doesn't shrink except in prolonged starvation.

Stomach adaptation may seem to be good news for dieters, but at some point in food deprivation, hunger returns with a vengeance and can lead to bouts of overeating that more than make up for the calories lost during the deprivation period. And just as the stomach's capacity can adapt to small meals, it quickly adapts to larger and larger quantities of food, until a meal of normal size no longer satisfies. This observation may partly explain why obesity is on the rise: popular demand has led to larger and larger servings of food, and stomachs have adapted to accommodate them.

Hunger is just one signal determining whether a person will eat. **Appetite** also initiates eating and, sometimes, overeating. A person can experience appetite without hunger. For example, seeing and smelling a freshly baked apple pie after finishing a big meal can stimulate release of the brain's **endorphins,** pleasure molecules that create an appetite for the pie despite an already full stomach. In contrast, a person who is ill or under stress may physically need food but have no appetite.

■ Chapter 3 described the brain's hypothalamus.

hunger the physiological need to eat, experienced as a drive for obtaining food; an unpleasant sensation that demands relief.

ghrelin (GREH-lin) a hormone released by the stomach that signals the hypothalamus of the brain to stimulate eating.

appetite the psychological desire to eat; a learned motivation and a positive sensation that accompanies the sight, smell, or thought of appealing foods.

endorphins (en-DOOR-fins), **endogenous opiates** compounds of the brain whose actions mimic those of opiate drugs (morphine, heroin) in reducing pain and producing pleasure. In appetite control, endorphins are released on seeing, smelling, or tasting delicious food and are believed to enhance the drive to eat or continue eating.

Other factors affecting appetite:

- Hormones (for example, the sex hormones influence food consumption).
- Inborn appetites (inborn preferences for fatty, salty, and sweet tastes).[31]
- Learned preferences, aversions, and timings (for example, cravings for favorite foods, fear of new foods, eating according to the clock, desiring larger portions).
- Customary eating habits (for example, whether or not each meal includes dessert).
- Social interactions (cultural or religious acceptability of foods, companionship).
- Some disease states (obesity may be associated with increased taste sensitivity, whereas colds, flu, and zinc deficiency reduce taste sensitivity).
- Appetite stimulants or depressants and mood-altering drugs affect food intake.
- Environmental conditions (people often prefer hot foods in cold weather and vice versa).

■ Figure 2-8 in Chapter 2 demonstrated how portion sizes have increased over recent decades.

▶ KEY POINT   Hunger is stimulated by an absence of food in the digestive tract. Appetite can occur with or without hunger.

"Stop" Signals—Satiation and Satiety   At some point during a meal, the brain receives messages from several sources that enough food has been eaten.[32] Often called **satiation,** this condition originates from the presence of food in the upper digestive tract (consult Figure 9-10 again). When the stomach stretches to accommodate a meal, nerve receptors in the stomach fire, sending a signal to the brain that the stomach is full. As nutrients from the meal enter the small intestine, they stimulate receptor nerves and trigger the release of hormones that provide the brain with information about the meal just eaten. The brain also detects absorbed nutrients passing by in the blood and releases neurotransmitters that suppress food intake in response. Together, stomach distention and nutrients in the small intestine trigger signals that inform the brain's hypothalamus about the size and nature of the meal. The response: satiation occurs; the eater feels full and stops eating.

After a meal, the feeling of **satiety** continues to suppress hunger and allows for a period of some hours in which the person is free to dance, study, converse, wonder, fall in love, and concentrate on endeavors other than eating. At some later point, signals from the digestive tract once again sound the alert that more food is needed.

Hunger and satiety are not equal in strength. Hunger, a life-or-death drive for survival, strongly stimulates eating behavior, whereas satiation and satiety exert weaker control over food intake and often can be ignored. The regulation of the human appetite is thus unbalanced, tipping in favor of food consumption.[33] People attempting to reduce their food intakes often must fight against this imbalance. After hours of annoying hunger pains, they may overeat when mealtime arrives but regret it later, vowing to "do better" by starving again. A destructive cycle of starving and binge eating ensues, with no weight loss to show for the effort. Researchers are exploring both the body's satiety mechanisms and the satiety value of foods in hopes of devising ways of assisting people who must reduce their food intakes for health.

▶ KEY POINT   Satiation occurs when the digestive organs signal the brain that enough food has been consumed. Satiety is the feeling of fullness that lasts until the next meal. Hunger outweighs satiety in the appetite control system.

■ A *peptide* is a molecule made of two or more amino acids linked together (see Chapter 6).

satiation (SAY-she-AY-shun) the perception of fullness that builds throughout a meal, eventually reaching the degree of fullness and satisfaction that halts eating. Satiation generally determines how much food is consumed at one sitting.

satiety (sah-TIE-eh-tee) the perception of fullness that lingers in the hours after a meal and inhibits eating until the next mealtime. Satiety generally determines the length of time between meals.

leptin an appetite-suppressing hormone produced in the fat cells that conveys information about body fatness to the brain; believed to be involved in the maintenance of body composition (*leptos* means "slender").

Leptin: A Satiety Hormone   **Leptin** is a peptide hormone primarily produced by the adipose tissue. Leptin travels to the brain via the bloodstream and is directly linked with both appetite control and body fatness. Some leptin is also produced in the stomach where it may contribute to satiety.[34] As part of the brain's appetite-regulating chemistry, leptin suppresses appetite and food intake between meals.[35] Leptin operates on a feedback mechanism—a gain in body fatness stimulates leptin production, which in turn reduces food consumption, resulting in fat loss. Fat loss brings the opposite effect—suppression of leptin and increased appetite.[36] Thus, the fat tissue that produces leptin is ultimately controlled by it.[37]

The discovery of leptin dramatically changed the scientific view of adipose tissue. Once viewed as a metabolically sluggish storage depot for lipids, fat tissue has gained respect as a hormonally active regulatory tissue with widespread effects on the body.[38] In addition to its appetite function, leptin signals the female reproductive system about body fat reserves, may assist in the sexual maturation process, stimulates growth of new blood vessels in the cornea of the eye and elsewhere, has a constricting effect on the arteries, assists in the formation of blood cells in bone marrow, and helps support a normal immune response.[39] Scientists hope to one day harness the power of leptin for treating obesity, but so far, clinical trials have been disappointing.[40]

The mouse on the right is genetically obese—it lacks the gene for producing leptin. The mouse on the left is also genetically obese but remains lean because it receives leptin.

▶ KEY POINT The adipose tissue hormone leptin suppresses the appetite in response to a gain in body fat.

Energy Nutrients and Satiety A popular idea today is that foods high in fat, protein, or slowly digested carbohydrates may sustain satiety longer than foods made from refined grains—that is, a food's satiety value may be influenced by its glycemic index.[41] Thus far, research results are mixed: a recent review of the literature describes at least 15 studies in which a diet ranking low on the glycemic index reduced or delayed hunger and another 16 studies that found the opposite effect or no effect.[42] In the negative studies, a diet based on refined grain foods such as soft breads, crackers and chips, certain cereals, and potatoes was just as satisfying as the diets based on low–glycemic index foods, and weight loss was practically identical on both diets.[43]

■ The glycemic index was defined in Chapter 4 (page 119).

Of the three energy-yielding nutrients, protein is perhaps the most satiating.[44] A meal providing protein may quickly lend a feeling of fullness, an effect that could account for some of the popularity of high-protein weight-loss diets. (The Consumer Corner later in the chapter addresses these diets.) Fat in food is also well known for its satiety effects. Fat or protein in a meal triggers the release of a hormone produced by the intestine that slows stomach emptying and prolongs the feeling of fullness after a meal.[45] Researchers have also reported increased satiety from foods high in fiber or water and even from foods that have been puffed up with air.[46] As dieters await news of foods that might be useful against hunger, researchers have not yet identified any one food, nutrient, or attribute that is especially effective.

▶ KEY POINT Some foods may confer greater satiety than others.

Inside-the-Body Causes of Obesity

Although interesting and important, findings about appetite regulation do not fully explain why some people gain too much body fatness while others stay lean. On the opposite side of the energy budget, the "energy out" side, many theories have emerged to explain obesity in terms of metabolic function. Whenever discussions turn to metabolism, the topic of genetics follows closely behind.

Selected Metabolic Theories of Obesity Metabolic theories attempt to explain variations in the ease with which people gain or lose weight when eating more

TABLE 9·4 SELECTED THEORIES OF METABOLIC CAUSES OF OBESITY

Theory	Mechanism of Action
Enzyme theory	Excess fat storage may stem from elevated concentrations of an enzyme, **lipoprotein lipase (LPL),** that enables fat cells to store triglycerides. The more LPL, the more easily fat cells store lipid, and the more likely the body will remain obese. The fat cells of obese people contain more LPL than the fat cells of lean people and therefore reach a large size quickly.
Fat cell number theory	Body fatness is determined by both the number and the size of fat cells. Fat cells increase in number during the growing years, tapering off in adulthood. Fat cell number may increase more rapidly in obese children than in lean children, leading to a lifelong tendency toward obesity.
Set-point theory	The body may "choose" a weight it wants to be and defend that weight by regulating behaviors and metabolic activities. Just as a thermostat setting triggers a heater to run when air temperature falls and turn off when warmth is restored, whenever weight is lost or gained, the set-point mechanism changes metabolic energy expenditure to restore the "chosen" body weight. The theories of thermogenesis, below, explain possible mechanisms by which the body defends its set point.
Thermogenesis I: Energy-wasting proteins and brown fat theory	Proteins control the body's heat production, or thermogenesis. A type of adipose tissue, brown fat, has abundant energy-wasting proteins that specialize in converting energy to heat. Whereas regular white fat cells have a sluggish metabolism and conserve and store fat energy, brown fat cells actively metabolize fat, releasing its stored energy as heat. Brown fat is more abundant in lean animals than in fat ones, and this theory states that a person with more brown fat and therefore more energy-wasting proteins may stay leaner. Human infants have abundant brown fat, but the amount dwindles with age.
Thermogenesis II: Adaptive thermogenesis theory	Many tissues, such as muscle, spleen, and bone marrow, convert stored energy into heat in response to cold temperature, physical conditioning, overfeeding, starvation, trauma, and other stress. Heat is also produced to "waste" fuel without useful work when energy supplies are too high; conversely, with low energy supplies, energy is conserved. Genetic inheritance is thought to determine the efficiency of this system. Dieters' efforts are often thwarted when, on reducing food intake, metabolism slows and heat production diminishes.
Thermogenesis III: Diet-induced thermogenesis theory	The thermic effect of food varies between obese and nonobese people. In lean people who have just eaten a meal, energy use speeds up for a while, but in many obese people, no change in energy use occurs after eating. In theory, this small difference in energy expenditure may account for an accumulation of body fat, but no studies have shown this conclusively. Overweight people often spend more energy each day than lean people do because their heavier bodies require more energy to move and maintain.

or less food energy than they expend. When given an extra 1,000 calories of food a day for 100 days, some people gain 30 pounds, but others gain less than 10 pounds. Those who gain more use calories efficiently: they have a "thrifty" metabolism. Similarly, some people lose more weight faster than others on comparable exercise regimens. The **set-point theory,** which states that the body somehow strives to achieve and maintain a particular body weight, offers an explanation for this phenomenon (see Table 9-4).

One area of difference in energy expenditure involves production of heat, or **thermogenesis.** In metabolic processes, enzymes "waste" a small percentage of energy, radiated away as heat. Some processes, however, expend copious energy in thermogenesis, producing abundant heat without performing any useful work. In radiating more energy away as heat, the body can spend more, rather than store more, excess energy. It may seem logical, then, to strive to manipulate metabolic enzymes to step up thermogenesis, but caution is required. At a level of activity not far beyond that of normal functioning, energy-wasting activity kills the cells. [47]

▶ KEY POINT   Metabolic theories attempt to explain obesity on the basis of molecular functioning.

lipoprotein lipase (LPL) an enzyme mounted on the surface of fat cells that splits triglycerides in the blood into fatty acids and glycerol to be absorbed into the cells for reassembly and storage.

set-point theory the theory that the body tends to maintain a certain weight by means of its own internal controls.

thermogenesis the generation and release of body heat associated with the breakdown of body fuels. *Adaptive thermogenesis* describes adjustments in energy expenditure related to changes in environment such as cold and to physiological events such as underfeeding or trauma.

Genetics and Obesity
Genetic makeup greatly influences how efficiently the body uses up or stores energy. A few relatively rare types of obesity arise primarily from genetic causes. That is, a few known genetic differences change metabolic processes in ways that reliably cause excessive gains of body fat. [48] In the great majority of obesity cases, however, an inherited *tendency* toward developing obesity is evident but is not the only determinant.

The role of genetics in obesity is demonstrated in children, whose body fatness often closely resembles that of their biological parents.[49] For someone with at least one obese parent, the chance of becoming obese is estimated to fall between 40 and 70 percent. Adopted children tend to be similar in weight to their biological parents, not to their adoptive parents. Studies of twins bear this out: identical twins (assumed to have identical genes) are twice as likely as fraternal twins to weigh the same when reared apart in similar households. When identical twins are reared apart in household environments that *differ* in terms of smoking, eating behaviors, nutrient intakes, and physical activity, however, their weights are more often related to adoptive parents' lifestyle habits.[50] Such findings demonstrate that although genetic inheritance strongly influences a person's tendency to become obese, lifestyle choices determine whether that tendency is realized.[51]

If certain genes carry the instructions for making proteins involved in energy metabolism, then these genes might reasonably be expected to vary between fat and lean people. So far, however, the search for new obesity treatments through genetic research has been disappointing. For example, great scientific interest followed the discovery of a gene whose protein product seemed promising for influencing thermogenesis or fat metabolism in ways that oppose obesity.[52] In animals, evidence of this protein is widespread in the body tissues of the leanest animals. In some children, a slight variation of the gene that produces the protein is associated with overfatness.[53] Further study, however, dashed early hopes for a genetic switch to control obesity. The protein's function now appears more related to immunity than to energy regulation or fat metabolism.[54]

■ To review protein synthesis, see Figure 6-6 of Chapter 6.

Genes clearly influence a person's tendency to gain weight or stay lean, but does this mean that obesity is inevitable in those whose genes predispose them to develop it? This fatalistic view is put to rest by a simple observation: obesity rates have greatly increased in recent decades wherever people enjoy conditions of prosperity and abundance. During the same time, the human genome has remained constant. This means that although an individual's genetic inheritance may make obesity likely, the disease of obesity cannot develop unless the environment—factors outside the body—provides the means of doing so.

▶ KEY POINT   A person's genetic inheritance greatly influences the likelihood of obesity. Genetic researchers are identifying potential genetic links with obesity development.

Outside-the-Body Causes of Obesity

Food is a source of pleasure. Being creatures of free will, people can override signals of satiety and hunger and eat whenever they wish, especially when tempted with delicious options and large servings. People also seek ease, and our increasing dependence on labor-saving inventions, such as automobiles and elevators, greatly decreases the energy we spend in activities required for daily living.

External Cues to Overeating   Almost everyone has had the experience of walking into a food store, not feeling particularly hungry, and, after viewing the vast array of goodies, walking out snacking on a favorite treat. A classic experiment showed that *variety* influences animals to eat even if they are not hungry. Rats, known to maintain body weight with precision when fed standard rat chow, rapidly became obese when fed "cafeteria style" on a variety of rich, palatable foods. Many people, too, are prone to overconsume when presented with a wide variety of delectable foods, such as sweets, snacks, condiments, and main dishes, often without being aware of doing so.[55] Conversely, consumption of a wide variety of *vegetables* but few treats correlates with lower body fatness.[56]

Overeating also occurs in response to complex human sensations such as loneliness, yearning, craving, addiction, or compulsion. Some people experience food cravings when feeling down or depressed. Favorite foods pick them up for a while. Other people

■ Controversy 11 explores some issues regarding roles of society in the current obesity epidemic and discusses potential actions to reduce the nation's fatness.

respond to other external stimuli such as the time of day ("I'm not hungry, but it's time for lunch").

Any kind of stress can cause overeating ("What do I do when I'm grieving? Eat. What do I do when I'm concentrating? Eat!"). The familiar foods that stressed people favor even have a calming name—*comfort* foods.[57] Although food availability, cravings, and stress can lead to the overconsumption of food energy, they cannot fully explain obesity development because even many thin people are susceptible to them and other people cannot eat at all when under stress.

Food Pricing, Availability, and Advertising   High-calorie fast foods are relatively inexpensive, widely available, heavily advertised, and wonderfully delicious, but a steady diet of them correlates with obesity.[58] By one estimate, 40 percent of the recent jump in U.S. body weights may be due to low food prices alone.[59] Controversy 11 revisits these issues, but consider the following: If price, availability, and advertising attract consumers of fast foods, can they do the same for nutritious, low-calorie choices such as fruit, vegetables, or yogurt? At least one experiment seems to indicate that they can. When researchers dropped the price of more nutritious options by half and made them readily available in workplace vending machines and school cafeterias, adults and students alike quadrupled their purchases of fresh fruit and doubled their purchases of baby carrot sticks.[60]

Enjoying an occasional calorie-rich treat or meal does not destine the eater for obesity, but such foods are widely advertised, available, and inexpensive, making overconsumption of calories likely. Moderation remains a bedrock of nutrition common sense and holds true with regard to all kinds of foods.

Physical Inactivity   Some people may be obese, not because they eat too much, but because they move too little. Diet histories from obese people often report energy intakes that are similar to, or even less than, those of others. (Diet histories may not be accurate records of actual intakes, though; people commonly underreport intakes of high-calorie foods.)[61] Some obese people are so inactive, however, that even when they eat less than lean people, they still have an energy surplus.

Such inactivity is a recent phenomenon. One hundred years ago, 30 percent of the energy used in farm and factory work came from human muscle power; today, only 1 percent does. The same trend follows at home, at work, at school, at play, and in transportation.[62]

Conversely, people who are sufficiently active can eat enough food to obtain the nutrients they need without the threat of weight gain. This activity must be the active kind, not passive motion such as being jiggled by a machine at a health spa or being massaged. It takes active moving of muscles to affect energy balance.

Despite benefits to body composition and health from regular physical activity, the United States seems locked in an epidemic of inactivity. For many people, television watching, video games, and computer entertainment have all but replaced outdoor work and play as the major leisure time activity. The Think Fitness feature underscores the importance of physical activity in weight management, and Table 9-5 lists the energy costs of some activities.

End of Story?   Anyone involved in a good mystery wants to know how it ends. In the case of the causes of obesity, no one yet knows which of the suspects is the real culprit, and until evidence proves otherwise, any or all may be guilty as charged. In real life, the best way for most people to attain a healthy body weight boils down to control in three areas: diet, physical activity, and behavior modification. Later sections focus on these three areas. The next section delves into the details of how, exactly, the body loses and gains weight.

▶ **KEY POINT**   Studies of human behavior identify stimuli that lead to overeating. Food pricing, availability, and advertising influence food choices. Physical inactivity is clearly linked with overfatness.

TABLE 9·5   ENERGY SPENT IN ACTIVITIES

To determine the calorie cost of an activity, multiply the number listed by your weight in pounds. Then multiply by the number of minutes spent performing the activity. Example: Jessica (125 pounds) rode a horse at a trot for 25 minutes: .052 × 125 = 6.5; 6.5 × 25 = 163.

Activity	Cal/lb body weight/min[a]
Aerobic dance (vigorous)	.062
Basketball (vigorous, full court)	.097
Bicycling	
13 mph	.045
15 mph	.049
17 mph	.057
19 mph	.076
21 mph	.090
23 mph	.109
25 mph	.139
Canoeing (flat water, moderate pace)	.045
Cross-country skiing	
8 mph	.104
Golf (carrying clubs)	.045
Handball	.078
Horseback riding (trot)	.052
Rowing (vigorous)	.097
Running	
5 mph	.061
6 mph	.074
7.5 mph	.094
9 mph	.103
10 mph	.114
11 mph	.131
Soccer (vigorous)	.097
Studying	.011
Swimming	
20 yd/min	.032
45 yd/min	.058
50 yd/min	.070
Table tennis (skilled)	.045
Tennis (beginner)	.032
Walking (brisk pace)	
3.5 mph	.035
4.5 mph	.048
Weight lifting	
light-to-moderate effort	.024
vigorous effort	.048
Wheelchair basketball	.084
Wheeling self in wheelchair	.030

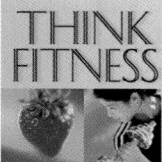

THINK FITNESS

Activity for a Healthy Body Weight

ACCORDING TO THE National Institutes of Health, exercise using about 150 calories of energy per day, or about 1,000 calories per week, constitutes moderate activity that can help build and maintain lean tissue while reducing body fatness and helping to protect against cardiovascular disease.

Some people believe that physical activity must be long and arduous to achieve fat loss. Not so. A brisk, 30-minute walk each day can help significantly.[63] To achieve the DRI committee's "active lifestyle" category requires walking for an hour a day.[64] A useful strategy is to incorporate bits of physical activity into your daily schedule in many simple, small-scale ways. Work in the garden; work your abdominal muscles while you stand in line; stand up straight; walk up stairs; fidget while sitting down; tighten your buttocks each time you get up from your chair. Small energy expenditures add up to significant contributions. The margin provides some tips for staying active, and many more details are found in Chapter 10.

Physical activity for weight loss or maintenance:

1. Choose moderate activities.
2. Move large muscle groups.
3. Invest longer times in physical activity.
4. Adopt informal strategies to be more active.

Physical activity for building lean body mass:

1. Choose strength-building exercises.
2. Use a balanced exercise routine.
3. Perform exercises with increasing intensity.
4. Adopt informal strategies to be more active.

How the Body Loses and Gains Weight

The causes of obesity may be complex, but the body's energy balance is straightforward. The balance between the energy you take in and the energy you spend determines whether you will gain, lose, or maintain body *fat*. A change in body *weight* of a pound or two may not indicate a change in body fat—it can reflect shifts in body fluid content, in bone minerals, in lean tissues such as muscles, or in the contents of the bladder or digestive tract. A change often correlates with the time of day: people generally weigh the least before breakfast. One of the most important things for people concerned with weight control to realize is that quick, large changes in weight are usually not changes in fat alone—or even at all.

The type of tissue lost or gained depends on how you go about losing or gaining it. To lose fluid, for example, you can take a "water pill" (diuretic), which will cause the kidneys to siphon extra water from the blood into the urine. Or you can engage in intense exercise while wearing heavy clothing in hot weather and lose abundant fluid in sweat. (Both practices are dangerous and are not being recommended here.) To gain water weight, you can overconsume salt and water; for a few hours, your body will retain water until it manages to excrete the salt. (This, too, is not recommended.) Most quick weight-change schemes promote large changes in body fluids that register dramatic, but temporary, changes on the scale and accomplish little weight change in the long run.

One other practice is not recommended: smoking. Each year, many adolescents, especially girls, take up smoking to control weight.[65] Nicotine blunts feelings of hunger, so when hunger strikes, a smoker can reach for a cigarette instead of food. Fear of weight gain sometimes deters people from quitting smoking. Smokers do tend to weigh less than nonsmokers, and many gain weight when they stop smoking. The best advice to smokers wanting to quit seems to be to adjust diet and exercise habits to maintain weight during and after cessation. The best advice to a person flirting with the idea of taking up smoking for weight control is don't do it—many thousands of people who became addicted as teenagers die from tobacco-related illnesses each year.

■ Chapter 8 gave details about the body's water balance.

Smoking may keep some people's weight down, but at what cost?

- Cancer.
- Chronic lung diseases.
- Heart disease.
- Low-birthweight babies.
- Miscarriage.
- Osteoporosis.
- Shortened life span.
- Sudden infant death.
- Many others.

Moderate Weight Loss versus Rapid Weight Loss

Being able to eat periodically, store fuel, and then use up that fuel between meals is a great advantage. The between-meal interval is normally about 4 to 6 waking hours—about the length of time the body takes to use up most of the readily available fuel—or 12 to 18 hours at night, when body systems slow down and the need is less.

When you eat less food energy than you need, your body draws on its stored fuel to keep going. If a person exercises appropriately, moderately restricts calories, and consumes an otherwise balanced diet that meets protein and carbohydrate needs, the body is forced to use up its stored fat for energy. Gradual weight loss will occur. This is preferred to rapid weight loss because lean body mass is spared and fat is lost.

The Body's Response to Fasting   If a person doesn't eat for, say, three whole days, then the body makes one adjustment after another. Less than a day into the fast, the liver's glycogen is essentially exhausted.[66] Where, then, can the body obtain *glucose* to keep its nervous system going? Not from the muscles' glycogen because that is reserved for the muscles' own use. Not from the abundant fat stores most people carry because these are of no use to the nervous system. Fat cannot be converted to glucose—the body lacks enzymes for this conversion.* The muscles, heart, and other organs use fat as fuel, but at this stage the nervous system needs glucose. The body does, however, possess enzymes that can convert protein to glucose. Therefore, the underfed body sacrifices the proteins in its lean tissue to supply raw materials from which to make glucose.

If the body were to continue to consume its lean tissue unchecked, death would ensue within about ten days. After all, in addition to skeletal muscle, the blood proteins, liver, digestive tract linings, heart muscle, and lung tissue—all vital tissues—are being burned as fuel. (Fasting or starving people remain alive only until their stores of fat are gone or until half their lean tissue is gone, whichever comes first.) To prevent this, the body plays its last ace: it converts fat into compounds that the nervous system can adapt for use and so forestalls the end. This process is ketosis, first mentioned in Chapter 4 as an adaptation to prolonged fasting or carbohydrate deprivation.

In ketosis, instead of breaking down fat molecules all the way to carbon dioxide and water, the body takes partially broken-down fat fragments and combines them to make **ketone bodies,** compounds that are normally kept to low levels in the blood. It converts some amino acids—those that cannot be used to make glucose—to ketone bodies, too. These ketone bodies circulate in the bloodstream and help to feed the brain; about half of the brain's cells can make the enzymes needed to use ketone bodies for energy. After about ten days of fasting, the brain and nervous system can meet most of their energy needs using ketone bodies.

Thus, indirectly, the nervous system begins to feed on the body's fat stores. Ketosis reduces the nervous system's need for glucose, spares the muscle and other lean tissue from being devoured quickly, and prolongs the starving person's life. Thanks to ketosis, a healthy person starting with average body fat content can live totally deprived of food for as long as six to eight weeks. Figure 9-11 reviews how energy is used during both feasting and fasting.

Respected, wise people in many cultures have practiced fasting as a periodic discipline. The body tolerates short-term fasting, and at least in animal studies, short-term fasting seems to benefit the body in some ways (see Chapter 14) although there is no evidence that the body becomes internally "cleansed," as some believe.[67] Fasting may harm the body, however, when ketosis upsets the acid-base balance of the blood or when fasting promotes excessive mineral losses in the urine. In as little as 24 hours of fasting, the intestinal lining begins to lose its integrity.[68] Food deprivation also leads to a tendency to overeat or even binge when food becomes available. The effect seems to last beyond the point when weight is restored to normal; people with eating disorders often report that a fast or a severely restricted diet heralded the beginning of their loss of control over eating. This indictment applies to extreme dieting and fasting, but not to the moderate weight-management strategies described later in this chapter.[69]

If you want to lose weight, fasting is not the best way. While the body's lean tissues continue to be degraded, tissues are deprived of nutrients they need to assemble new

In early food deprivation:
- The nervous system cannot use fat as fuel; it can only use glucose.
- Body fat cannot be converted to glucose.
- Body protein can be converted to glucose.

In later food deprivation:
- Ketone bodies help feed the nervous system and so help spare tissue protein.

ketone bodies acidic compounds derived from fat and certain amino acids. Normally rare in the blood, they help to feed the brain during times when too little carbohydrate is available. Also defined in Chapter 4.

---

*Glycerol, which makes up 5 percent of fat, can yield glucose but is a negligible source.

FIGURE 9·11 FEASTING AND FASTING

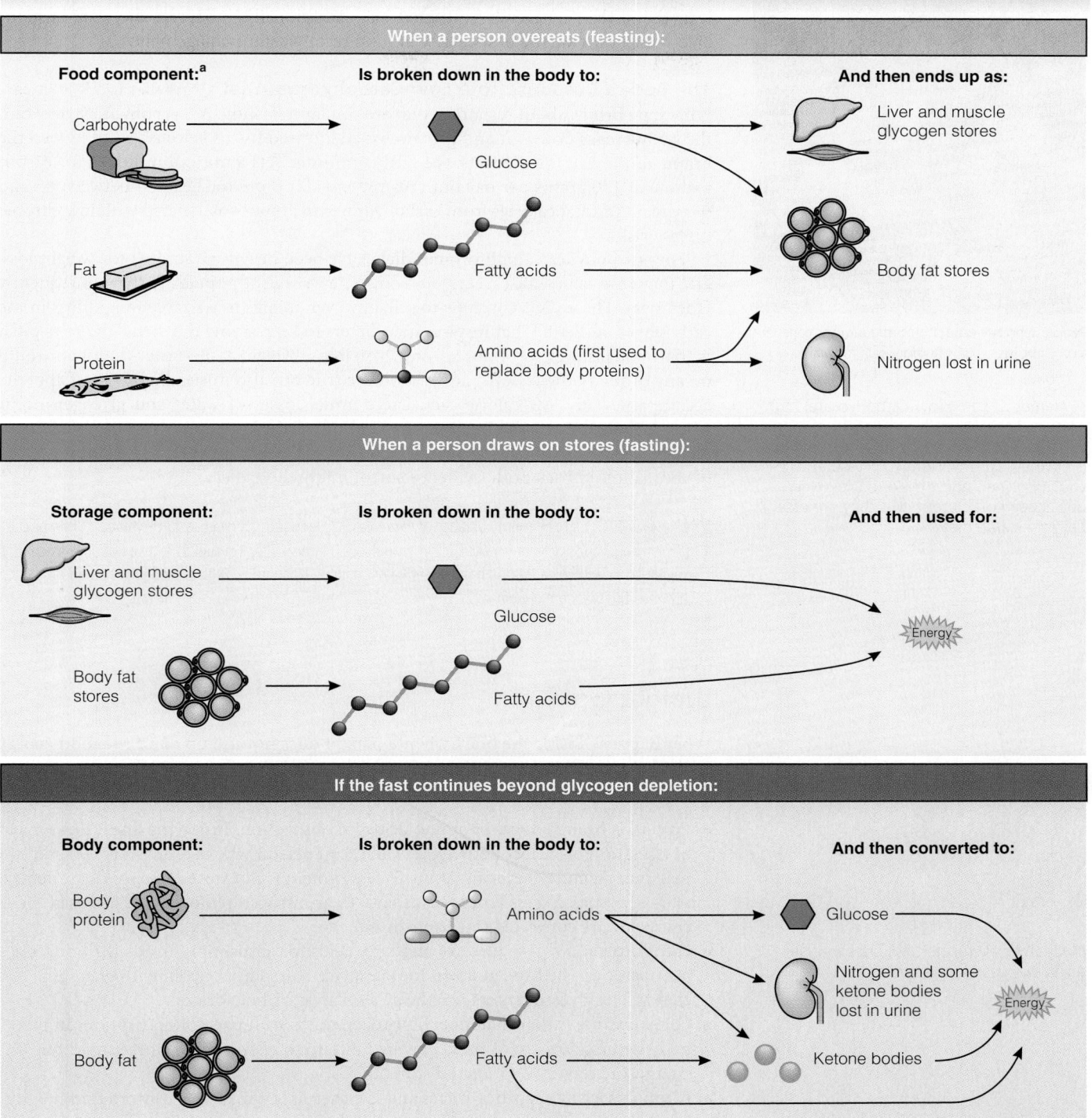

When a person overeats (feasting):

Food component:[a]   Is broken down in the body to: And then ends up as:

Carbohydrate → Glucose → Liver and muscle glycogen stores

Fat → Fatty acids → Body fat stores

Protein → Amino acids (first used to replace body proteins) → Nitrogen lost in urine

When a person draws on stores (fasting):

Storage component: Is broken down in the body to: And then used for:

Liver and muscle glycogen stores → Glucose → Energy

Body fat stores → Fatty acids → Energy

If the fast continues beyond glycogen depletion:

Body component: Is broken down in the body to: And then converted to:

Body protein → Amino acids → Glucose → Energy
Amino acids → Nitrogen and some ketone bodies lost in urine

Body fat → Fatty acids → Ketone bodies → Energy

[a]Alcohol is not included because it is a toxin and not a nutrient, but it does contribute energy to the body. After detoxifying the alcohol, the body uses the remaining two-carbon fragments to build fatty acids and stores them as fat.

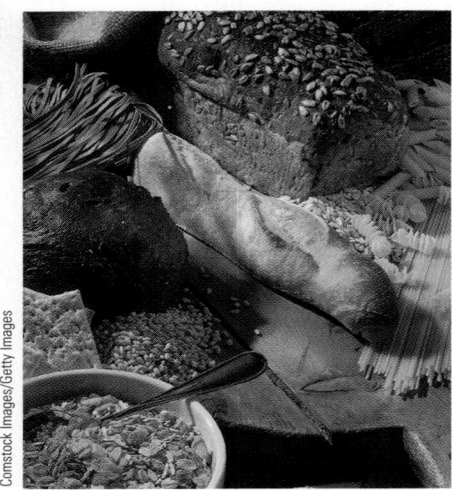

People who have healthy body weight consume more, not less, carbohydrate-rich foods.

■ Names of some low-carbohydrate, high-protein diets: Atkins New Diet Revolution, Calories Don't Count Diet, Protein-Power Diet, the Carbohydrate Addict's Diet, the Lo-Carbo Diet, and the Zone Diet. New ones keep coming out, but they are essentially the same diet.

enzymes, red and white blood cells, and other vital components. The body also slows its metabolism to conserve energy. A diet only moderately restricted in calories promotes a greater rate of *weight* loss, a faster rate of *fat* loss, and the retention of more lean tissue than a severely restricted fast. Additionally, moderate diets are sustainable—they provide the best chance at long-term weight management.

The Body's Response to a Low-Carbohydrate Diet Any diet too low in carbohydrate brings about responses that are similar to fasting. As carbohydrate runs out, the body breaks down fat and protein for energy and forms ketone bodies to feed the brain. To prevent these effects, the DRI committee sets a minimum intake for carbohydrate at 130 grams per day but recommends far more for health—between 45 and 65 percent of total calories from carbohydrate, an amount associated with low chronic disease risks.

Low-carbohydrate, high-protein diets have been heavily promoted for weight loss, and they take many guises, each enjoying a new surge of popularity that subsequently fades away. These diets continue to sell, however, thanks to a sizable weight loss in the early stages of dieting that inspires new dieters to keep trying the diets. The sales pitch is that "you'll never feel hungry" and "you'll lose weight fast—faster than you would on any ordinary diet." Both these claims are true but also misleading. Loss of appetite accompanies any low-calorie diet. Large initial losses of water and glycogen occur when carbohydrate is lacking, but this kind of weight loss rapidly reverses when people begin eating normally again. The Consumer Corner on page 334 examines some of the claims and scientific evidence surrounding these diets.[70]

▶ **KEY POINT** When energy balance is negative, glycogen returns glucose to the blood. When glycogen runs out, body protein is called upon for glucose. Fat also supplies fuel as fatty acids. If glucose runs out, fat supplies fuel as ketone bodies, but ketosis can be dangerous. Both prolonged fasts and low-carbohydrate diets are ill-advised.

Weight Gain

What happens inside the body when a person does not use up all of the food energy taken in? Previous chapters have already provided the answer—the energy-yielding nutrients contribute the excess to body stores as follows:

■ Chapter 10 further discusses muscle gains in response to exercise.

● Protein is broken down to amino acids for absorption. Inside the body, these may be used to replace lost body *protein* and, in a person who is exercising, to build new muscle and other lean tissue. Excess protein is not stored in specific protein storage tissues. Excess amino acids have their nitrogen removed and are used for energy or are converted to *glucose* or *fat*.
● Fat is broken down to glycerol and fatty acids for absorption. Inside the body, the fatty acids can be broken down for energy or stored as body *fat* with great efficiency. The glycerol enters a pathway similar to carbohydrate.
● Carbohydrate (other than fiber) is broken down to sugars for absorption. In the body tissues, excesses of these may be built up to *glycogen* and stored, used for energy, or converted to *fat* and stored.
● Alcohol is easily absorbed intact and is either used for fuel or converted into body fat for storage.[71]

Although three kinds of energy-yielding nutrients and alcohol may enter the body, they become only two kinds of energy stores: glycogen and fat. Glycogen stores amount to about three-fourths of a pound; fat stores can amount to many pounds. Remember from Chapter 6 that when excess protein is converted to fat, it cannot be recovered later as protein because the nitrogen is stripped from the amino acids and excreted in the urine. Thus, if you eat enough of any food, whether it's steak, brownies, or baked beans, any excess will be turned to fat within hours. Weight gain comes from spending less food energy than is taken in. Weight can be gained as body fat or as lean tissue, depending mostly upon whether the eater is also exercising.

Ethanol, the alcohol of alcoholic beverages, has been shown to slow down the body's use of fat for fuel by as much as a third, causing more fat to be stored. This storage is primarily in the visceral fat tissue of the "beer drinker's belly" and also on the thighs, legs, or anywhere the person tends to store surplus fat. Alcohol therefore is fattening, both through the calories it provides and through its effects on fat metabolism.* The obvious conclusion is that weight control and abundant alcohol intake cannot easily coexist.

These points are worth repeating:

- Any food can make you fat if you eat enough of it. A net excess of energy is almost all stored in the body as fat in fat tissue.
- Fat from food is particularly easy for the body to store as fat tissue.
- Protein is not stored in the body except in response to exercise; it is present only as working tissue. Protein is converted to glucose to help feed the brain when carbohydrate is lacking; excess protein can be converted to fat.
- Alcohol both delivers calories and encourages storage of body fat.
- Too little physical activity encourages body fat accumulation.

> **KEY POINT**   When energy balance is positive, carbohydrate is converted to glycogen or fat, protein is converted to fat, and food fat is stored as fat. Alcohol delivers calories and encourages fat storage.

© Corbis

Each gram of alcohol presents 7 calories of energy to the body—energy that is easily stored as body fat.

Achieving and Maintaining a Healthy Body Weight

Before setting out to change your body weight, think about your motivation for doing so. Many people in our society are dissatisfied with their body weight, not because of potential health risks or compromised daily living, but because their weight fails to meet society's ideals of attractiveness. Unfortunately, this kind of thinking sets people up for disappointment because they set unrealistic goals. The human body is not infinitely malleable—few overweight people will ever become rail-thin, even with the right diet, exercise habits, and behaviors. Likewise, most underweight people will remain on the slim side even after putting on some heft.

Research clarifies the disparity between lofty expectations and reasonable goals.[72] At the start of a year on a weight-loss program, obese women named four potential outcomes: their "dream," "happy," "acceptable," and "disappointing" weights (see Figure 9-12). All of these, even the "disappointing" weights, were far lower than reasonable goals recommended by experts. By the end of the year, most women had lost an average of 35 pounds, or 16 percent of their starting weight. Despite experiencing remarkable physical, social, and psychological benefits, the women felt discouraged because they had not met even their "disappointing" weight. In truth, their discouragement reflected only their unrealistic expectations.

The truth is that overweight takes years to accumulate and cannot be changed overnight. Achieving a healthy weight is possible, but it takes time, patience, and perseverance. People willing to take one step at a time, even if it feels like just a baby step, are on the path toward meeting their individual goals.

Modest weight loss, even for the person who is still overweight, can lead to rapid improvements in control over diabetes, blood pressure, and blood lipids. As fitness builds, stair climbing, walking, and other tasks of daily living become noticeably easier. Adopting health or fitness as the ideal rather than some ill-conceived image of beauty

FIGURE 9·12   REASONABLE GOALS VERSUS UNREASONABLE EXPECTATIONS

Obese women achieved remarkable success during a year's weight-loss program, but they were disappointed because they had set unrealistic expectations at the outset.

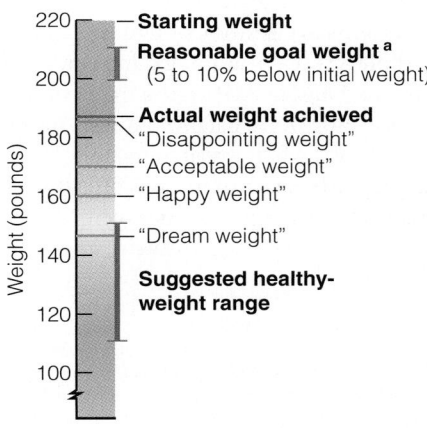

[a]Reasonable goal weights reflect pounds lost over one year's time. Given more time, reasonable goals may eventually fall within the suggested healthy-weight range.
SOURCE: Adapted from G. D. Foster and coauthors, What is a reasonable weight loss? Patients' expectations and evaluations of obesity treatment outcomes, *Journal of Consulting and Clinical Psychology* 65 (1997): 79–85.

---

*People addicted to alcohol are often overly thin because of diseased organs, depressed appetite, and subsequent malnutrition.

CONSUMER CORNER

Popular High-Protein, Low-Carbohydrate Diets

MANY POPULAR DIET books have hooked millions of overweight, meat-loving readers with this tempting advice: eat unlimited calories of protein and fat, and you'll lose weight and be healthy provided that you avoid carbohydrates.* Do these book authors really present "revolutionary" new evidence, as they claim?

An assumption of the low-carbohydrate diets seems to be that if carbohydrate intakes are low, then calories in other energy-yielding nutrients will somehow fail to increase body fatness (see Table 9-6). Will eating more protein-rich and fat-rich foods lead to weight loss? According to the *Dietary Guidelines for Americans 2005*, for weight loss, calories count, not the proportion of carbohydrate, fat, and protein in the diet.

Weight Loss

Scientific evidence is mixed on whether a high-protein, low-carbohydrate diet produces leanness. Some results follow.

Studies Reporting Negative Findings

In a population study of almost 28,000 people, those consuming a high-protein diet had *higher* BMI values than people consuming more balanced diets.[1] In the laboratory, when researchers hold energy intakes constant, no difference is observed in weight loss between subjects eating a high-protein, low-

carbohydrate diet and those eating a diet lower in protein and higher in carbohydrate.[2]

Studies Reporting Positive Findings

The picture changes in clinical studies where energy intakes vary—that is, when people living outside the laboratory freely choose how much food (and how many calories) to consume. Severely obese people choosing from a high-protein, low-carbohydrate diet lost more weight on average during a six-month study than did subjects on a low-fat adequate-carbohydrate diet.[3] In a 12-month study of this type, weight loss was achieved on both diets, but the average losses were greater on the low-carbohydrate diet, especially during the first three months.[4] At six months, the weight-loss gap between the groups had narrowed, and by the end of a year, it had all but vanished. At 12 months, both groups were regaining their lost weight, but the rate of regain differed significantly: the low-carbohydrate group regained rapidly, while the low-fat group remained more stable.

The Science Behind the Findings

On further analysis, several other points can be made about the studies just described: the weight-loss difference in these studies was small; one study was of short duration; and in both, a large number of subjects dropped out before data collection was complete. Those dropping out are likely to have not been losing weight, and omitting these negative data from the analysis may have slanted the results.[5]

How do scientists explain the greater initial weight loss observed with high-protein diets? Two partial explanations

have been put forward. The first concerns water and glycogen: when glycogen stores are lost, as in low-carbohydrate diets, substantial water weight loss follows. Water can account for much, but not all, of the extra weight lost and rapidly regained by people on a low-carbohydrate weight-loss diet. Unfortunately, the studies did not investigate whether the lost weight was mainly water, lean tissue, or fat.

A second partial explanation is that diet records indicate that people on low-carbohydrate diets generally consume fewer calories.[6] This may reflect the higher satiety value of protein and fat—those who feel fuller after a meal due to the hormonal response of the digestive tract to protein or fat may be less tempted by between-meal snacks, or they may eat less food at subsequent meals. Simple arithmetic also comes into play: people omitting calories of carbohydrate-rich foods do not replace them with additional calories of high-protein foods.[7] When they strip a burrito of its carbohydrate-rich beans, tortilla wrapper, and chopped vegetables, they are left with a tiny pile of ground beef and a sprinkle of cheese for their meal. Likewise, when ordering a steak dinner, they consume a lone piece of meat, without the calories (and nutrients) of the milk, potato, and whole-grain roll to accompany it.

Thus, when weight loss occurs on high-protein, low-carbohydrate diets, it is because people lose water and eat fewer calories by eating less food, and not because of some metabolic hocus pocus. According to a diet evaluation by scientists from the USDA, most people who recently lost substantial weight and kept it off were eating diets high in car-

---

*For scientific evaluations of popular diets, see B. Liebman, Weigh the diet books, *Nutrition Action Healthletter*, January/February 2004, pp. 1–8; S. T. St. Jeor and coauthors, Dietary protein and weight reduction: A statement for healthcare professionals from the nutrition committee of the council on nutrition, physical activity, and metabolism of the American Heart Association, *Circulation* 104 (2001): 1869–1874.

bohydrates and low in fat.[8] In addition to producing weight loss, those diets received high scores for nutrition quality.*

Health Concerns

Diet books recommend replacing carbohydrate-rich foods in the diet with those rich in protein. To discuss the health effects of doing so requires addressing the effects of both high intakes of protein and low intakes of carbohydrate.

High Protein and Animal Fat Intakes

The world's nutrition authorities agree that a steady diet of foods high in protein often presents too much saturated fat and cholesterol in the form of excessive bacon, eggs, hamburger, sausages, and cheeses—staple foods of high-protein, low-carbohydrate diets. Such a diet clearly raises the risk for heart and artery disease.[9]

In addition, saturated fats from animal sources are implicated in worsening a woman's risk of breast cancer.[10] Although early results were mixed, the link seems more certain following adjustments in research methods.[11] For example, in a study of over 90,000 nurses aged 26 to 46, breast cancer risk was one-third higher in those with the highest intakes of animal (but not vegetable) fats.[12] The DRI committee recognizes links between high-protein diets and increased risks of heart disease, osteoporosis, kidney stones and worsening of kidney disease, cancer, and obesity.[13]

Low Carbohydrate Intakes

The low-carbohydrate diet books suggest that by lowering carbohydrate intake, the dieter can lose weight, lower blood cholesterol, lower blood pressure, and so lower the risk of diseases.

*Based on the Healthy Eating Index, a measure of how well the diet meets U.S. dietary recommendations (see Chapter 2).

TABLE 9·6 ASSUMPTIONS AND SCIENCE CONCERNING HIGH-PROTEIN DIETS

Assumption: *Restricting carbohydrates alone will shift metabolism and cause weight loss.*
Science: Weight loss follows restricted food energy intake or increased energy output; shifting the proportion of energy nutrients in meals without reducing calories does not produce weight loss.

Assumption: *Eating more protein makes people lean.*
Science: In population studies, the higher the protein intake, the higher the BMI.

Assumption: *Insulin release causes obesity and disease.*
Science: Insulin plays roles in the transport of glucose into cells and in the storage of excess nutrients, including fat. Insulin cannot cause fat storage and weight gain in a person whose energy budget is balanced.

Assumption: *High-protein foods cause greater energy expenditure.*
Science: The thermic effect of food is slightly higher for protein than for carbohydrate or fat, but the increase is so slight as to be insignificant to weight-loss efforts.

SOURCES: A. Trichopoulou and coauthors, Lipid, protein and carbohydrate intake in relation to body mass index, *European Journal of Clinical Nutrition* 56 (2002): 37–43. J. C. Brüning and coauthors, Role of brain insulin receptor in control of body weight and reproduction, *Science* 289 (2000): 2122–2125.

Researchers are investigating these health claims, but so far they report mixed results or results that are hard to interpret with regard to heart disease. What is absolutely certain, however, is that consumption of more whole grains, fruits, vegetables, and low-fat milk products correlates with lean body composition and robust good health—evidence from many studies confirms this. Fad diet authors often promote and profit from supplements or meal replacers that they claim will provide the elements missing from their diets, but such concoctions cannot replace the benefits lost when people eliminate whole classes of nutritious foods.[14]

Further, a diet low enough in carbohydrate to produce chronic ketosis presents problems of its own, such as deficiencies of vitamins and minerals, impaired mood, and inadequate glycogen stores to feed the brain or support vigorous physical activity, as discussed in Chapter 4.[15] And diet book claims about damage to healthy individuals from the body's normal insulin response to starchy or sweet foods are incorrect. In healthy people, insulin protects the body from the buildup of glucose; only in those with abnormalities such as insulin resistance does elevated blood insulin signify a potential threat. Healthful diets that include starchy or sweet foods do not cause insulin resistance, but a diet enriched with saturated fats may worsen it.[16] Table 9-6 summarizes some of the high-protein diet claims and presents the science that disputes those claims.

Conclusion

Claims that weight and health can be improved by eliminating or greatly reducing intakes of whole grains, vegetables, milk, and fruit are baseless. To omit such foods is to lose out on nutrients, fibers, and phytochemicals with proven health benefits. High-fiber whole grains may support weight-loss efforts better than highly refined starches and sweet treats, but this is probably due to the satiety value of fiber.[17] Paradoxically, the fastest path to lasting weight management turns out to be the slowest one—planning a diet based on nutrient adequacy and other sound diet principles, making time to exercise, and changing behaviors permanently.[18]

TABLE 9·7   TIPS FOR ACCEPTING A HEALTHY BODY WEIGHT

- Value yourself and others for human attributes other than body weight. Realize that prejudging people by weight is as harmful as prejudging them by race, religion, or gender.
- Use only positive, nonjudgmental descriptions of your body; never use degrading negative descriptions.
- Accept positive comments from others.
- Avoid checking your weight or appearance frequently; focus on your whole self including your intelligence, social grace, and professional and scholastic accomplishments.
- Accept that no magic diet exists.
- Stop dieting to lose weight. Adopt a healthy eating and exercise life-style permanently.

- Follow the USDA Food Guide (Chapter 2). Never restrict food intake below the minimum levels that meet nutrient needs.
- Become physically active, not because it will help you get thin but because it will enhance your health.
- Seek support from loved ones. Tell them of your plan for a healthy life in the body you have been given.
- Seek professional counseling, *not* from a weight-loss counselor, but from someone who can help you make gains in self-esteem without weight as a factor.
- Join with others to fight weight discrimination and fashion stereotypes. (Search your local paper, or see the Nutrition Resources appendix for names of groups.)

can avert much misery, and Table 9-7 offers some tips to that end. The rest of this chapter stresses health and fitness as goals and uses weight only as a convenient gauge for progress. As mentioned, changes in three realms produce results: diet and physical activity (next sections), and behavior modification (see the Food Feature, p. 348).

What Diet Strategies Are Best for Weight Loss?

Dietary Guidelines for Americans 2005 Key Recommendations for weight management:

- To maintain body weight in a healthy range, balance calories from foods and beverages with calories expended.
- To prevent gradual weight gain over time, make small decreases in food and beverage calories and increase physical activity.

This section reveals diet-related changes that most often lead to successful weight loss and maintenance. Setting appropriate goals is step number one. The key recommendations for weight management from the *Dietary Guidelines for Americans 2005* are listed in the margin.[73] The FDA has also proposed a nationwide plan that may help dieters achieve their goals (see the margin).

Setting Goals   For the overweight person, a reasonable first goal might be to prevent further weight gain. With this accomplishment, a subsequent goal might be to reduce body weight by about 5 to 10 percent over a year's time. New goals can be built on prior achievements, and a lifetime goal may be to maintain the new healthier body weight. Weight maintenance often proves the most difficult.

Once you have identified your overall target, set smaller step goals for the dietary, physical activity, and behavioral changes necessary to achieve the desired result. These changes do not produce a dramatic weight loss overnight, but if you faithfully employ them, you can lose a pound or two of body fat each week, safely and effectively. Losses greater or faster than these are not recommended because they are almost invariably followed by rapid regain. Also, rapid weight loss through excessive restriction can cause gallbladder stones or dangerous electrolyte imbalances. It's better to take your time and achieve a lasting change.

The FDA's plan to confront the obesity epidemic:

- Require food labels to display calorie counts prominently and to use meaningful serving sizes.
- Launch a "Calories Count" consumer education campaign.
- Encourage restaurants to provide nutrition information to customers.
- Ensure greater accuracy of food label nutrient data.
- Accelerate development of obesity drugs and step up research activities.

Keeping Records   Keeping records is critical. Recording your food intake and exercise can help you to spot trends and identify areas needing improvement. Changes in body weight can provide a rough estimate of changes in body fatness. In addition to weight, measure your waist circumference to track changes in central adiposity.

It's Your Diet, So You'd Better Plan It   Contrary to the claims of faddists, no particular food plan is magical, and no particular food must be either included or excluded. You are the one who will have to live with the plan, so you had better be the one to design it. Remember, you are adopting a healthy eating plan for the long run, so it must consist of satisfying foods that you like, that are readily available, and that you can afford.

As for fad diets, only those that reduce calorie intake produce weight loss.[74] Table 9-8 exposes some of the untrue claims made by the sellers of fad diets, and Table 9-9 (p. 338) provides a way to judge weight-loss plans according to standard nutrition principles.

SOURCE: *Counting Calories: Report of the Working Group on Obesity,* March 12, 2004, available at www.cfsan.fda.gov/~dms/owg-rpt.html; U.S. Department of Health and Human Services, *Healthy People 2010,* 2nd ed., available at www.health.gov/healthypeople or call (800) 367-4725.

TABLE 9-8 LIES AND TRUTHS OF FAD DIETS

Lie: *You'll lose weight fast without counting calories or exercising because the diet or product alters metabolism.*
Truth: No known trick of metabolism produces significant weight loss without diet or exercise.

Lie: *On this diet, you can eat all you want and still lose weight.*
Truth: Unless the diet is composed entirely of celery or lettuce, basic laws governing energy disprove this claim—energy consumed must be used to fuel the body or stored as fat.

Lie: *You'll never regain the weight, even after you stop using the diet or product.*
Truth: Maintenance of a new lower weight requires lifelong changes in diet and exercise.

Lie: *Eat all the fat, sugar, or other foods you want and lose pounds every day because the product blocks absorption of energy nutrients.*
Truth: Losing 2 pounds of body fat per week through malabsorption would require passing about 1,000 calories of undigested fat or carbohydrate out of the digestive tract each day, an implausible outcome and losing 2 pounds *a day* in this manner is physically impossible.

Lie: *Lose more than 3 pounds per week without medical supervision.*
Truth: Weight loss in this range carries substantial risks to health and even to life, making medical supervision prudent.

Lie: *This product is 100 percent successful in producing weight loss.*
Truth: The causes of obesity are multiple, and even prescription medications and stomach-shrinking surgeries are not 100 percent effective.

Lie: *You'll lose weight just by wearing the product or rubbing it on the skin.*
Truth: No over-the-counter patch, cream, wrap, ring, bracelet, other jewelry, shoe inserts, or other gimmick is known to cause loss of weight or fat.

Lie: *Reset your genetic code to be thin.*
Truth: You inherited your genes and no diet can alter them.

Lie: *Stress hormones make you fat.*
Truth: Supplements sold to block stress hormones and produce weight loss do neither.

Lie: *High-protein diets are so popular because they are the best way to lose weight.*
Truth: See this chapter's Consumer Corner.

Lie: *High-protein diets energize the brain.*
Truth: The brain depends on carbohydrate for energy.

Lie: *Dietitians know nothing about "modern" nutrition.*
Truth: Dietitians are, by training and experience, nutrition experts who rely on scientific approaches and cannot be swayed by the claims of quacks.

SOURCES: Federal Trade Commission, *Deception in Weight-Loss Advertising Workshop: Seizing Opportunities and Building Partnerships to Stop Weight-Loss Fraud,* a Federal Trade Commission Staff Report, December 2003, available at www.ftc.gov; Food and Drug Administration, *Counting Calories: Report of the Working Group on Obesity,* March 12, 2004, available at www.cfsan.fda.gov/~dms/owg-rpt.html; S. Barrett, Impossible weight-loss claims: Summary of a Federal Trade Commission report, December 16, 2003, available at www.quackwatch.org.

Choosing Realistic Calorie Intakes   Guidelines for a weight-loss diet are outlined in Table 9-10 (p. 339). For those with a BMI greater than 35, reducing intake by 500 to 1,000 calories per day will produce the desired loss. This should amount to an intake of about 1,000 to 1,200 calories per day for most women and 1,200 to 1,600 calories per day for most men.[75] Such an intake will allow most people to lose weight while still meeting nutrient needs, as demonstrated in Table 2-3 in Chapter 2 (page 39). Diets providing energy intakes lower than about 800 calories—the so-called very-low-calorie diets—are notoriously unsuccessful at achieving lasting weight loss, lack necessary nutrients, and may set in motion the unhealthy behaviors of eating disorders (see the Controversy), and so are not recommended. For those with a BMI ranging from 27 to 35, reducing intake by 300 to 500 calories per day will result in a loss of ½ to 1 pound per week. If you plan resolutely to include the amount of food from each food group that you need each day, you will find that you will have little appetite left for high-calorie foods.

TABLE 9·9 RATING SOUND AND UNSOUND WEIGHT-LOSS SCHEMES

Each diet or program starts with 160 points and is rated on 12 factors. Whenever a plan falls short of ideals, subtract points in the third column as instructed. A plan that loses more than 20 points might still be of value, but deserves careful scrutiny.

Factor	Does the Diet or Program:	Start:
		160 points
Calories	Provide a reasonable number of calories (not fewer than 1,200 calories for an average-size person)? If not, give it a minus 10.	_____
Protein	Provide enough, but not too much, protein (at least the recommended intake, but not more than twice that much)? If no, minus 10.	_____
Fat	Provide enough fat for balance but not so much fat as to go against current recommendations (about 30% of calories from fat)? If no, minus 10.	_____
Carbohydrate	Provide enough carbohydrate to spare protein and prevent ketosis (100 grams of carbohydrate for the average-size person)? Is it mostly complex carbohydrate (not more than 10% of the calories as concentrated sugar)? If no to either or both, minus 10.	_____
Vitamins and minerals	Offer a balanced assortment of vitamins and minerals by including foods from all food groups and subgroups of Figure 2-4 in Chapter 2? If it omits a food group (for example, meats), does it provide a suitable food (not supplement) substitute? For each food group omitted and not adequately substituted for, subtract 10 points.	_____
Variety	Offer variety, in the sense that different foods can be selected each day? If you'd classify it as boring or monotonous, give it a minus 10.	_____
Ordinary foods	Consist of ordinary foods that are available locally (for example, in the main grocery stores) at the prices people normally pay? Or does the dieter have to buy special, expensive, or unusual foods to adhere to the diet? If you would class it as "bizarre" or "requiring special foods," minus 10.	_____
False promises	Promise dramatic, rapid weight loss (substantially more than 1% of total body weight per week)? If yes, minus 10.	_____
Lifestyle changes	Encourage permanent, realistic lifestyle changes, including regular exercise and the behavioral changes needed for weight maintenance? If not, minus 10.	_____
Reasonable costs	Misrepresent salespeople as "counselors" supposedly qualified to give guidance in nutrition and/or general health without a profit motive, or collect large sums of money at the start, or require that clients sign contracts for expensive, long-term programs? If so, minus 10.	_____
Warnings of risks	Fail to inform clients about the risks associated with weight loss in general or the specific program being promoted? If so, minus 10.	_____
No gimmicks or mandatory supplements	Promote unproven or spurious weight-loss aids such as human chorionic gonadotrophin hormone (hormones can be dangerous; reject such a plan immediately), starch blockers, diuretics, sauna belts, body wraps, passive exercise, ear stapling, any type of injections, acupuncture, electric muscle stimulating devices, spirulina, amino acid supplements (e.g., arginine, ornithine), glucomannan, appetite suppressants, "unique" ingredients, and so forth? If so, minus 10.	_____
	Total points:	_____

The DRI recommends:
- 45 to 65% calories from carbohydrate.
- 20 to 35% calories from fat.
- 10 to 35% calories from protein.

Balancing Carbohydrates, Fats, and Protein Healthy diets based on abundant fresh fruits and vegetables, low-fat milk products, legumes, lean meats, fish, poultry, and whole grains are high in carbohydrates, adequate in protein and fiber, and low in the kinds of fats associated with diseases. They are also best for managing weight. Earlier chapters described the importance of each of the energy-yielding nutrients to health. Therefore, diets for weight management should provide all three within the DRI recommended intake ranges (see the margin).

Crunchy, wholesome, high-fiber, unprocessed or lightly processed foods offer bulk and satiety for far fewer calories than smooth, quickly consumed, refined foods. Thus, choosing whole grains and starchy vegetables in place of most refined grains and added fats and sugars benefits both weight and nutrition. Choose fats sensibly by avoiding sources of saturated and *trans* fats and including enough of the health-supporting fats (details in Controversy 5) to provide satiety but not so much as to oversupply calories. Lean meats or other low-fat protein sources also play an important role: an ounce of lean ham contains about the same number of calories as an ounce of bread, but the ham produces greater satiety. Limit these foods but don't elim-

TABLE 9·10	RECOMMENDATIONS FOR A WEIGHT-LOSS DIET
Nutrient	**Recommended Intake**
Calories	
For people with BMI ≥35	Approximately 500 to 1,000 calories per day reduction from usual intake
For people with BMI between 27 and 35	Approximately 300 to 500 calories per day reduction from usual intake
Total fat	35% or less of total calories
Saturated fatty acids[a]	8 to 10% of total calories
Monounsaturated fatty acids	Up to 15% of total calories
Polyunsaturated fatty acids	Up to 10% of total calories
Cholesterol[a]	<300 mg per day
Protein[b]	Approximately 15% of total calories
Carbohydrate[c]	55% or more of total calories
Sodium/Sodium chloride[d]	No more than 2,300 mg of sodium or approximately 6 g of sodium chloride (salt) per day
Calcium	1,000 to 1,500 mg per day
Fiber[c]	20 to 30 g per day

[a]People with high blood cholesterol should aim for less than 7 percent calories from saturated fat and 200 milligrams cholesterol per day.
[b]Protein should be derived from plant sources and lean sources of animal protein.
[c]Carbohydrates and fiber should be derived from vegetables, fruits, and whole grains.
[d]Value from the *Dietary Guidelines for Americans 2005;* the Tolerable Upper Intake Level for salt is 5.8 grams.

inate them. Do strictly limit alcohol, which provides abundant calories but no nutrients. Furthermore, alcohol reduces inhibitions and can sabotage even the most committed dieter's plans, at least temporarily.

Of Critical Importance: Portion Sizes Remember to pay careful attention to portion sizes—the monstrous helpings served by restaurants and sold in packages are the enemy of the person striving to control weight. Most people choose portions that are familiar, not scientific, so almost every dieter needs to retrain by using measuring cups for a while to learn to judge portion sizes.[76] Losing track of amounts of fat is especially to be avoided—fat grams add up quickly and contribute more calories than do grams of carbohydrate or protein.

Most people will eat more at a sitting when presented with larger portions than when provided smaller portions.[77] Yet, at the end of the meal, they report about the same sensation of fullness whether the portions were large or small. The trick seems to be to eat just until satisfied, and then to stop because satisfaction diminishes with continued eating while unneeded calories mount up.

Keep in mind that eating large portions of reduced-calorie foods can defeat their purpose—stick with a reduced-calorie cookie or two, not half the bag. Stay focused on calories and portion sizes—don't be distracted by product claims for reduction of a particular nutrient, be it fat or carbohydrate. Read labels and compare calories per serving.

Using the Concept of Energy Density To lower calorie intakes, most people must learn how to reduce the calories in their diet. One way to do this is to reduce the **energy density** of the diet.[78] As Figure 9-13 (p. 340) demonstrates, foods containing substantial water or fiber and those low in fat help to lower a meal's energy density, providing more food and greater satiety for fewer calories. For example, selecting grapes with their high water content instead of the same weight of their dehydrated counterparts (raisins) decreases the energy density. Likewise, a cupful of fiber-rich, water-rich broccoli delivers about a quarter of the energy (calories) in a cupful of starch-rich potatoes.

■ Chapter 1 described the similar concept of nutrient density—the nutrient contribution of a food per calorie.

energy density a measure of the energy provided by a food relative to its weight (calories per gram).

FIGURE 9·13 EXAMPLES OF ENERGY DENSITY

The larger meals on the right weigh more, provide more fiber, and take far more time to enjoy than the meals on the left, yet the meals in both columns provide equal calories. Much of the additional weight of the meals on the right is made up of water. Note that the hamburger is a modest serving of very lean beef, not the huge fatty burger typical of fast-food restaurants. Keep in mind that even foods of lower energy density can be overconsumed, so watch total calories as well as energy density of foods.

HIGHER ENERGY DENSITY

OR

LOWER ENERGY DENSITY

Caesar salad; croutons (fast-food size).

$$\frac{500 \text{ calories}}{280 \text{ g total weight}} = 1.78 \text{ energy density}$$

Mixed salad with 3 oz chicken breast, mixed vegetables, almonds, cranberries, and 3 tbs low-calorie dressing; whole-wheat dinner roll with 1.5 oz lean ham.

$$\frac{500 \text{ calories}}{570 \text{ g total weight}} = 0.88 \text{ energy density}$$

OR

Large hot dog on bun; ½ c potato salad.

$$\frac{650 \text{ calories}}{265 \text{ g total weight}} = 2.45 \text{ energy density}$$

Homemade very lean (8% fat) 3 oz hamburger on whole-wheat bun; grilled vegetables with 1 tsp margarine; ½ c pork and beans; slice watermelon.

$$\frac{650 \text{ calories}}{810 \text{ g total weight}} = 0.8 \text{ energy density}$$

© Matthew Farruggio (all)

■ In general, foods high in fat or low in water, such as cookies or chips, rank high in energy density; foods high in water and fiber, such as fruits and vegetables, rank lower.

A way to evaluate foods for their energy density is to compare them mathematically according to their calories per gram. To demonstrate, compare carrot sticks with french fries. To calculate the energy density of the carrot sticks delivering 31 calories and weighing 72 grams, divide calories (31 cal) by weight in grams (72 g):

$$\frac{31 \text{ cal}}{72 \text{ g}} = 0.43 \text{ cal/g}.$$

Do the same for the fries, contributing 167 calories and weighing 50 grams:

$$\frac{167 \text{ cal}}{50 \text{ g}} = 3.34 \text{ cal/g}.$$

The higher the result, the more calories per gram and the greater the energy density. Importantly, the *energy* density of foods does not reflect their *nutrient* density (nutrients per calorie). This distinction is demonstrated in beverages, which because of their high water content all weigh about the same amount. The energy density of low-fat milk, for example, almost equals that of sugary soft drinks but these beverages rank far apart in measures of nutrient density, and therefore in their contributions toward a nutritious diet.

Consider Milk and Milk Products   In a recent study, calcium intake in the form of low-fat milk and milk products correlated with a healthy body weight; in several other studies, higher calcium intake correlates with lower body fatness.[79] Whether or not milk

FIGURE 9·14 CALORIES IN TWO SETS OF MEALS

About 3,400 cal
2% milk, 1 c, 121 cal
Orange juice, 1 c 112 cal
Whole grain waffles, 2 each, 402 cal
   Soft margarine, 2 tsp, 68 cal
   Syrup, 4 tbs, 210 cal
Banana slices, ½ c, 69 cal
                Breakfast total: 982

About 2,300 cal
Fat-free milk, 1 c, 83 cal
Orange juice, 1 c, 112 cal
Whole grain waffle, 1 each, 201 cal
   Soft margarine, 1 tsp, 34 cal
   Syrup, 2 tbs, 105 cal
Banana slices, ½ c, 69 cal
                Breakfast total: 604

2% milk, 1 c, 121 cal
Hamburger, quarter pound, 430 cal
French fries, large (about 50), 540 cal
Ketchup, 2 tbs, 32 cal
Apple pie, 1 each, 225 cal
                Lunch total: 1,348

Fat-free milk, 1 c, 83 cal
Cheeseburger, small, 330 cal
Green salad, 1 c, with light dressing, 1 tbs,
   67 cal; croutons, ½ c, 50 cal
French fries, regular (about 30), 210 cal
Ketchup, 1 tbs, 16 cal
Gelatin dessert, sugar-free, 20 cal
                Lunch total: 776

Italian bread, 2 slices, 162 cal
   Soft margarine, 2 tsp, 68 cal
Stewed skinless chicken breast, 4 oz, 202 cal
Tomato sauce, ½ c, 40 cal
Brown rice, 1 c, 216 cal
Mixed vegetables, ½ c, 59 cal
Regular cheese sauce, ¼ c, 121 cal
Brownie, 1 each, 224 cal
           Supper total: 1,092
           Day's total: 3,422

Italian bread, 1 slice, 81 cal
   Soft margarine, 1 tsp, 34 cal
Stewed skinless chicken breast, 4 oz, 202 cal
Tomato sauce, ½ c, 40 cal
Brown rice, 1 c, 216 cal
Mixed vegetables, ½ c, 59 cal
Low-fat cheese sauce, ¼ c, 85 cal
Brownie, 1 each, 224 cal
           Supper total: 941
           Day's total: 2,321

NOTE: Data from ESHA Research, The Food Processor Nutrition and Fitness Software version 8.3, 2004.

products or calcium turn out to affect fat metabolism directly, low-fat yogurt, fat-free milk, and fat-free cheeses remain nutritious lower-calorie foods for the weight-loss dieter. When fat-free milk replaces sugary soft drinks and punches, the benefits may be greater still.[81] Among 50,000 nurses, those who increased their consumption of soda or fruit punch by as little as 12 ounces daily gained much more weight on average than those who consumed less than one soda per month.[82]

Demonstration Diet The meals shown in Figure 9-14 demonstrate how meals look, before and after trimming 1,100 calories. The full-calorie meals, shown on the left side of the figure, have been modified in both portion sizes and energy density to produce the meals on the right. Some 350 calories were trimmed from the 3,400-calorie meals by reducing added sugars—reducing syrup served at breakfast and swapping apple pie for gelatin at lunch. (The diet planner kept the brownie at supper, however—pleasure matters, too.) These changes alone, repeated each day for one month, produce a calorie reduction more than sufficient to make a 3-pound difference in the person's body weight.

Meal Spacing Three meals a day is standard in our society, but no law says you can't have four or five—be sure they are smaller, of course. People who eat small, frequent meals are reported to be successful at weight loss and maintenance. Make sure that mild hunger, not appetite, is prompting you to eat. Also, eat regularly, before you become extremely hungry. When you do decide to eat, eat the entire meal you have planned for yourself. Then don't eat again until the next meal or snack. Save calorie-free or favorite foods or beverages for a planned snack at the end of the day if you need insurance against late-evening hunger.

One meal you should strive to include is breakfast. People who skip breakfast are more often overweight than breakfast eaters.[83] Those who consume the majority of their daily calories after six o'clock in the evening also often find it harder to lose weight than people who eat earlier in the day.[84] Much evidence supports the health effects of breakfast, and eating breakfast may reduce food intake all day long.

▶ KEY POINT   To achieve and maintain a healthy body weight, set realistic goals, keep records, and expect to progress slowly. Watch energy density, make the diet adequate and balanced, limit calories, reduce alcohol, and eat regularly, especially at breakfast.

Physical Activity for Weight Loss

To prevent weight gain and support loss, a person typically must spend about 30 to 60 minutes in moderate physical activity each day in addition to the activities of daily life.[85] People who combine diet and exercise typically lose more fat, retain more muscle, and regain less weight than dieters who do not exercise. Often, regular exercise seems to help people follow their diet plans more closely—and maintain weight losses more effectively.[86] Physical activity also reduces abdominal obesity, and this change improves blood pressure, insulin resistance, and fitness of the heart and lungs, even without weight loss.[87]

Increasing Metabolism and Reducing Appetite   In addition to burning off calories directly, activity also contributes to energy expenditure in an indirect way—by speeding up metabolism. This speeding up occurs both immediately and over the long term. For several hours following intense and prolonged exercise, metabolism remains slightly elevated, thus raising the calorie cost of performing the activity by about 15 percent. Over the long term, as more metabolically active lean tissue gradually increases, so does energy spent on basal metabolism. More beneficial changes in body composition invariably follow: body fat decreases while lean body mass increases, with concurrent gains in health.[88]

Physical activity may also help control appetite. Many people fear that exercising will increase their hunger, but this is not entirely true. Active people do have healthy appetites, but immediately after a workout, appetite is suppressed by the circulating lipids and glucose released from storage to fuel exercise—a state that mimics the fed state from the body's point of view. Activity also helps reduce stress, even stress caused by dieting. And stress often leads to inappropriate eating.

Choosing Activities   The best activities are those you enjoy and are safe to perform. What schedule of physical activity is best? It doesn't seem to matter much from a weight-loss point of view; benefits come from many kinds of routines—and any activity is better than being sedentary.[89] An expenditure of at least 2,000 calories per week in some sort of physical activity is especially helpful for weight management.[90] Most important: perform at a comfortable pace within your current abilities. Those rushing to improve are practically guaranteed an injury.

Health-care professionals frequently advise people to engage in activities of low-to-moderate intensity for a long duration, such as an hour-long brisk walk. The reasoning behind such advice is that people exercising at low-to-moderate intensity are more likely to stick with their activity for longer times and are less likely to injure themselves. For those at higher fitness levels, higher-intensity activities may be performed for shorter periods to gain similar benefits.[91] A 175-pound person who replaces a 30-minute television program with a 2-mile daily walk can spend enough energy to lose (or at least not gain) 18 pounds in a year.

■ Chapter 10 presents the many health benefits of physical activity.

Benefits of physical activity in a weight-management program:
- Short-term increase in energy expenditure (from exercise and from a temporary rise in metabolism).
- Long-term increase in BMR (from an increase in lean tissue).
- Improved body composition.
- Appetite control.
- Stress reduction and control of stress eating.
- Physical, and therefore psychological, well-being.

■ For an active life, limit sedentary activities, engage in strength and flexibility activities often, engage in vigorous activities regularly, and be as active as possible every day.

■ Estimated energy expended when walking at a moderate pace = 1 cal per mile per kilogram body weight.

In addition to planned exercise, fitness can be gained through hundreds of energy-spending activities required for daily living: taking the stairs instead of the elevator, biking instead of driving, raking leaves instead of using a blower, and many, many others. One suggested goal is to take 10,000 steps a day. Wearing a pedometer makes tracking a day's activities easy. However you do it, be active. Walk. Swim. Dance. Cycle. Skip. Enjoy moving—and move often.

Spot Reducing   People sometimes ask about "spot reducing." Unfortunately, muscles do not "own" the fat that surrounds them. Fat cells all over the body release fat in response to the demand of physical activity for use by whatever muscles are active. Exercise cannot remove the fat from any one particular area.

Exercise can help with trouble spots in two other ways, though. During aerobic exercise, abdominal fat readily releases its stores, providing fuel to the physically active body and reducing fat in the abdomen. Another way exercise can help is by improving the strength and tone of muscles in a trouble area, improving the overall appearance of the area. Strength and flexibility will help cure associated posture problems. A combination of aerobic, strength, and flexibility workouts is best for improving health, fitness, and, as a side benefit, body contours.

▶ **KEY POINT**   Physical activity greatly augments diet in weight-loss efforts. Improvements in health and body composition follow an active lifestyle.

What Strategies Are Best for Weight Gain?

Should a thin person try to gain weight? Not necessarily. If you are healthy at your present weight, stay there. If your physician has advised you to gain, if you are excessively tired, if you are unable to keep warm, if you fall into the "underweight" category of the BMI table (see inside back cover), or if, for women, you have missed at least three consecutive menstrual periods, you may be in danger from a too-low body weight.

Physical Activity to Gain Muscle and Fat   A healthful weight gain is best achieved through physical activity, particularly strength training (see Chapter 10 for details), combined with a high-calorie diet. Diet alone can bring about weight gain, but the gain will be mostly fat. For someone facing a wasting disease, the gain of fat tissue may be a welcome sign of improvement. For an athlete, however, such a gain can impair performance. For most people, physical activity is an essential component of a sound weight-gain plan. Many an underweight person has simply been too busy (for months) to eat or to exercise enough to gain or to maintain weight.

As important to weight gain as exercise are the calories to support that activity—otherwise you will lose weight. If you eat just enough to fuel the activity, you will build muscle, but at the expense of body fat; that is, fat will be burned to support the muscle building. If you eat more, you will gain both muscle and fat. To gain a pound of muscle and fat requires taking in about 3,000 extra calories.* Conventional advice on diet to the person building muscle is to eat about 700 to 1,000 calories a day above normal energy needs; this range supports both the added activity and the formation of new muscle.

Choose Foods with High Energy Density   The weight gainer needs *nutritious* energy-dense foods. No matter how many sticks of celery you consume, you won't gain weight because celery simply doesn't offer enough calories per bite. Energy-dense foods

---

*Theoretically, it takes an excess of 2,000 to 2,500 calories to gain a pound of only lean tissue and about 3,500 calories to gain a pound of fat.

(the very ones the weight-loss dieter is trying to avoid) are often high in fat, but their energy is spent in building new tissue and if the fat is mostly unsaturated, such foods will not contribute to heart disease. Be sure your choices are nutritious; a steady diet of chips and candy may add pounds but also threaten nutrient deficiencies.

Choose peanut butter instead of lean meat, avocado instead of cucumber, olives instead of pickles, whole-wheat muffins instead of whole-wheat bread, and flavored milk drinks instead of milk. When you do eat celery, stuff it with tuna salad (use oil-packed tuna); choose flavored coffee drinks over plain coffee; use olive oil or canola oil dressings on salads, whipped toppings on fruit, and soft or liquid margarine on potatoes. Because fat contains more than twice as many calories per teaspoon as sugar, it adds calories without adding much bulk, and its energy is in a form that is easy for the body to store.

Portion Sizes and Meal Spacing   Increasing portion sizes increases calorie intakes. Choose extra slices of meats and cheeses on sandwiches; use larger plates, bowls, and glasses to disguise the appearance of the larger portions. Expect to feel full. Most underweight individuals are accustomed to small quantities of food. When they begin eating significantly more food, they complain of uncomfortable fullness. This feeling is normal, and it passes as the stomach gradually adapts to the extra food.

Eat frequently. Make three sandwiches in the morning and eat them between classes in addition to the day's three regular meals. Spend time making foods appealing—the more varied and palatable, the better. If you fill up fast during a meal, start with the main course or a meat- or cheese-filled appetizer, not carrot sticks. Drink between meals, not with them, to save space for higher-calorie foods. Make milkshakes of milk, a frozen banana, a tablespoon of vegetable oil, and flavorings to drink between meals. Always finish with dessert.

Weight-Gain Supplements   Most "weight-gain" supplements advertised to add body weight are useless without physical activity and confer no special benefits beyond adding calories and a few nutrients. Ordinary items like instant breakfast powders or milk flavoring powders mixed with milk can do the same thing for a fraction of the supplement cost.

Of the weight gained in a day, only half an ounce to an ounce is protein tissue, so no special protein supplements can help speed gain of lean tissue. Ordinary food in abundance along with exercise to work the nutrients into place supports efforts to gain muscle and weight.

■ People's hormone status also affects their ability to gain weight, but taking steroids and other drugs to enhance weight gain is a bad idea—see the Controversy section of Chapter 10.

Avoid Tobacco   Smoking tobacco depresses the appetite and makes taste buds and olfactory (smelling) organs less sensitive. A person who smokes should quit before trying to gain weight. Quitters find that appetite picks up, food tastes and smells better, and the body reaps numerous benefits.

▶ **KEY POINT**   Weight gain requires a diet of calorie-dense foods, eaten frequently throughout the day. Physical activity builds lean tissue, and no special supplements can speed the process.

Drugs and Surgery to Treat Obesity

To someone fatigued from years of battling overweight, the idea of taking pills or undergoing surgery might seem attractive. These approaches can save the lives of obese people at critical risk, but they also carry serious risks of their own.

Each year, a million and a half U.S. citizens take prescription weight-loss medications.[92] Of these, a quarter are not overweight. In pursuit of a particular body form, physicians and patients alike seem willing to take the considerable risks that drug side effects present. Table 9-11 presents some of the known side effects and other details about weight-loss medications currently on the market.

TABLE 9·11 PHARMACEUTICAL TREATMENTS OF OBESITY[a]

Names	Actions	Known Side Effects	Comments
Prescription Drugs			
Sibutramine Trade name: Meridia	Suppresses appetite.	Dry mouth, headache, constipation, insomnia, and high blood pressure[b]	The FDA advises those with high blood pressure against its use; others should monitor their blood pressure.
Orlistat Trade name: Xenical	Inhibits pancreatic lipase activity, thus blocking dietary fat absorption by about 30%.	Gas, frequent bowel movements, and reduced absorption of fat-soluble vitamins	Most effective with a nutritionally balanced, reduced-calorie, low-fat diet
Over-the-Counter Drugs and Products			
Benzocaine Trade names: Diet Ayds (candy) or Slim Mint (gum)	Anesthetizes the tongue, reducing taste sensations	None known	Only over-the-counter weight-loss medication with FDA approval
Phenylpropanolamine (PPA) (also called norephedrine)	Appetite suppressant; nasal and sinus decongestant	Dry mouth, rapid pulse, nervousness, sleeplessness, hypertension, irregular heartbeat, kidney failure, liver damage, liver failure, seizures, and hemorrhagic strokes (bleeding in the brain)	The FDA has removed PPA from drug products and has warned consumers not to consume products containing it (check labels).
Ephedrine, ephedra, or ma huang	Enhances effects of the stress hormone norepinephrine, including reduced appetite.	Nervousness, headache, insomnia, dizziness, palpitations, skin flush, serious heart problems; almost 1,400 reported adverse events, including death	Prohibited by the FDA but available via the Internet. Consumers owning products should stop taking them (check labels).
Bitter orange extract Trade names: Xenadrine EFX, Metabolife Ultra, NOW Diet Support	Often a replacement for ephedrine, this stimulant mimics ephedra in chemical composition and function.	High blood pressure; increased risk of heart arrhythmias, heart attack, stroke	The FDA has currently taken no action against bitter orange extract.
"Carb blockers," "fat blockers" or "binders," chromium picolinate, chitosan, many others	None known	Not studied	The FDA and the Federal Trade Commission have begun taking action against products falsely claimed to produce weight or fat loss.

[a]For answers to drug-related questions, call FDA Information toll free: (888) 463-6332 or visit www.fda.gov.
[b]Sibutramine may also be associated with memory impairment.

People with a BMI of 30 or above and those with elevated disease risks may benefit from prescription medication, along with diet, exercise, and behavior therapy, to bring their weight down. A person with **extreme obesity,** that is, someone whose BMI is 40 or above (35 with coexisting disease), urgently needs to reduce body fatness, and surgery may be an option for those healthy enough to withstand it.[93] Surgical procedures effectively limit food intake by reducing the size of the stomach and delaying the passage of food from the stomach into the intestine (see Figure 9-15, p. 346), leading to significant weight loss in most patients.[94] Weight loss, in turn, often brings rapid improvements to such threats as diabetes, high blood cholesterol, hypertension, and sleep apnea.[95] The surgery is not a sure cure for obesity, however. A few people do not lose the expected pounds, and some who lose initially regain all the lost weight in a few years' time.

The long-term safety and effectiveness of gastric surgery depend, in large part, on compliance with dietary instructions. Complications immediately following surgery often include infections, nausea, vomiting, and dehydration; in the long term, vitamin and mineral deficiencies and psychological problems may develop. Lifelong medical supervision is necessary for those who choose the surgical route, but in many suitable candidates, the benefits of weight loss may prove worth the substantial risks.

extreme obesity clinically severe overweight, presenting very high risks to health; the condition of having a BMI of 40 or above; also called *morbid obesity.*

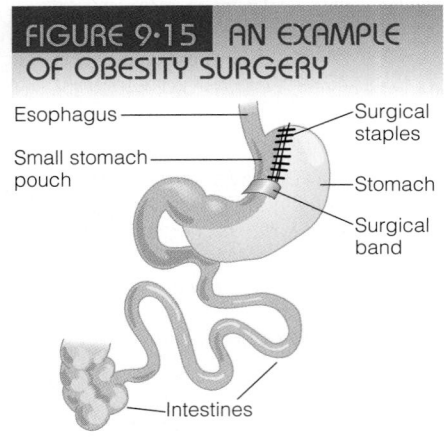

FIGURE 9·15 AN EXAMPLE OF OBESITY SURGERY

Esophagus

Small stomach pouch

Surgical staples

Stomach

Surgical band

Intestines

In one common surgery, the surgeon constructs a small stomach pouch and restricts the outlet from the stomach to the intestine.

Another surgical procedure is used not primarily to treat obesity but to remove external body fat. Plastic surgeons can extract some fat deposits by lipectomy, or "liposuction." People consenting to this cosmetic procedure expect an improved body shape, but an unusually large number of lawsuits against surgeons performing it indicate that many people are disappointed with the outcome.[96] If the fat is gained back after liposuction, as often happens, it can form a lumpy, dimpled layer that looks worse than the original fat. Lipectomy is popular in part because it seems safe, but there can be serious complications, even death, from the surgery.[97]

Herbal Products   Herbal weight-loss products are wildly popular, but their effectiveness and safety have not been proved. People may falsely believe that "natural" herbs are never harmful to the body, but many herbs contain poisonous toxins. Belladonna and hemlock are infamous examples, but many lesser-known herbs, such as sassafras, contain toxins as well. Furthermore, because weight-loss herbs are marketed as "dietary supplements," manufacturers need not present scientific evidence of their safety or effectiveness to the FDA before marketing them. Evidence about safety is gathered only through reports of consumers who sicken or die after purchasing and using herbal remedies.

A now familiar example is ephedra (also called *ma huang*), an herb that showed promise as a weight-loss drug in preliminary studies. Immediately, ephedra-containing pills and preparations for dieters and athletes flooded the market. Nearly 1,400 consumers of these products reported ill effects, including cardiac arrest, abnormal heartbeats, hypertension, strokes, and seizures; 81 people died. Canada banned ephedra, and after many years of gathering required documentation, the FDA has also banned sales of dietary supplements containing ephedra and its active constituent, ephedrine.[98] The World Health Organization has called for worldwide controls on ephedra sales.[99] Another diet aid to avoid is known as TRIAC, a powerful hormone that interferes with normal thyroid functioning and has caused heart attack and stroke (see the Controversy of Chapter 10).* Many other "herbal" additives have turned out to be kidney or liver toxins or cancer-causing agents.

Herbal laxatives containing senna, aloe, rhubarb root, cascara, castor oil, or buckthorn are sold as "dieter's tea" because they can cause a temporary water loss of a pound or two. Users commonly report nausea, vomiting, diarrhea, cramping, and fainting. Such "teas" are suspected of contributing to the deaths of four women who used them and also drastically reduced their food intakes. Some herbs may be useful for other purposes, but herbs and supplements do not produce weight loss and they clearly fail the safety test.[100] Read labels and don't take products containing these substances or any others not proved safe in laboratory studies. The risks of doing so are too high.

Other Gimmicks   Steam baths and saunas do not melt the fat off the body, although they may dehydrate you so that you lose water weight. Brushes, sponges, wraps, creams, and massages intended to move, burn, or break up **"cellulite"** are useless for fat loss. Cellulite—the rumpled, dimpled fat tissue on the thighs and buttocks—is simply fat, awaiting the body's call for energy. Recently, the FDA sent letters warning supplement distributors to stop claiming that their products:

- Block starch, fat, or sugar absorption.
- "Neutralize" starch, fat, or sugar in a meal.
- Promote weight loss with no effort.[101]

Such misleading claims distract people from the serious business of planning helpful weight-management strategies.

▶ **KEY POINT**   For people whose obesity threatens their health, medical science offers drugs and surgery. The effectiveness of herbal products and other gimmicks has not been demonstrated, and they may prove hazardous.

cellulite a term popularly used to describe dimpled fat tissue on the thighs and buttocks; not recognized in science.

*The product name was Triax Metabolic Accelerator.

Once I've Changed My Weight, How Can I Stay Changed?

One reason gimmicks fail at weight control is that they fail to produce lasting change. Millions have experienced the frustration of achieving a desired change in weight only to see their hard work visibly slipping away: "I have lost 200 pounds, but I was never more than 20 pounds overweight." Disappointment, frustration, and self-condemnation are common in dieters who find they have slipped back to their original weight or even higher.[102] What makes the difference between a successful, long-term weight-control program and one that doesn't stick? How can you maintain a healthy body weight?

A key to weight maintenance is accepting it as a lifelong endeavor, and not a goal to be achieved and then forgotten. Acceptance helps prepare the mind for making permanent changes. People who maintain a loss continue to employ the behaviors that reduce calorie intakes and increase expenditures through exercise. They cultivate the habits of people who maintain a healthy weight, such as eating diets higher in carbohydrate, lower in fat, and higher in fruits and vegetables than average. Those who maintain healthy weight also generally:

- Are more physically active than most people, for example, walking about 4 miles each day.[103]
- Monitor fat grams, calorie intake, and body weight.
- Do not vary much in food intake from weekdays to weekends.[104]
- Believe they have the ability to control their weight, an attribute known as **self-efficacy,** even in the face of previous failures.
- Develop social support systems.
- Eat controlled portions at planned times, and eat them at a leisurely pace.
- Eat high-fiber foods, particularly whole grains, and consume sufficient water each day.[105]
- Cultivate and honor realistic expectations regarding body size and shape.

The importance of exercise cannot be overstated. Those who endeavor to lose weight without exercise often become trapped in endless repeating rounds of weight loss and regain—"yo-yo" dieting. Weight-loss efforts often precede large weight gains, and a history of such **weight cycling** can predict a person's future success (or lack thereof) in maintaining weight.[106] Weight cycling may even damage the dieter's health by weakening the immune response, making diseases more likely to occur.[107]

Self-acceptance and self-efficacy also predict success, while self-hate and feelings of inadequacy predict failure. A paradox of behavior change is that it takes an attitude of self-acceptance (loving the overweight self) to lay the foundation for changing that self. Thinking habits turn out to be as important to achieving a healthy body weight as eating habits are, and thinking habits can be changed.* "Positive self-talk" is a concept worth cultivating—many people succeed because their mental dialogue supports, rather than degrades, their efforts. The Food Feature explores how a person can modify daily behaviors into healthy lifelong habits.

▶ **KEY POINT**   People who succeed at maintaining lost weight keep to their eating routines, keep exercising, and keep track of calorie and fat intakes and body weight. The more traits related to positive self-image and self-efficacy a person possesses or cultivates, the more likely that person will succeed.

Dietary Guidelines for Americans 2005
Key Recommendations for weight management:

- To sustain weight loss in adulthood: Participate in at least 60 to 90 minutes of daily moderate-intensity physical activity while not exceeding caloric intake requirements. Some people may need to consult a healthcare provider before participating in this level of activity.

self-efficacy a person's belief in his or her ability to succeed in an undertaking.

weight cycling repeated rounds of weight loss and subsequent regain, with reduced ability to lose weight with each attempt; also called *yo-yo dieting.*

Behavior Modification for Weight Control

Activities and rewards to substitute for eating:

- Attending sporting events.
- Enjoying leisure activities.
- Exercising or playing sports.
- Gardening.
- Getting praise from others.
- Going to a movie or play.
- Listening to music.
- Napping.
- Praising yourself.
- Reading.
- Receiving token rewards (stickers, stars).
- Redecorating.
- Relaxing.
- Saving money for future treats.
- Shopping.
- Taking a bubble bath.
- Telephoning.
- Tidying your room or house.
- Vacationing.
- Working on hobbies or crafts.

behavior modification alteration of behavior using methods based on the theory that actions can be controlled by manipulating the environmental factors that cue, or trigger, the actions.

SUPPORTING BOTH diet and exercise is the technique of **behavior modification,** which cements into place all the behaviors that lead to and perpetuate the desired body composition. Behavior modification is based on the knowledge that habits drive behaviors.

How Does Behavior Modification Work?

Suppose a friend tells you about a shortcut to class. To take it, you must make a left-hand turn at a corner where you now turn right. You decide to try the shortcut the next day, but when you arrive at the familiar corner, you turn right as always. Not until you arrive at class do you realize that you failed to turn left, as you had planned. You can learn to turn left, of course, but at first you will have to make an effort to remember to do so. After a while, the new behavior will become as automatic as the old one was.

For those striving to lose weight, learning to say "No, thank you" might be among the first habits to establish. Learning not to "clean your plate" might be another. Once you identify the behaviors you need to change, do not attempt to modify all of them at once. No one who attempts too many changes at one time is successful. Set your priorities and begin with a behavior you can handle—then practice it until it becomes habitual and automatic. Then select another.

Applying Behavior Modification

Behavior researchers have identified six elements of behavior modification to use to replace old eating habits with new ones:

1. Eliminate inappropriate eating cues.
2. Suppress the cues you cannot eliminate.
3. Strengthen cues to appropriate eating and exercise.
4. Repeat the desired eating and exercise behaviors.
5. Arrange or emphasize negative consequences for inappropriate eating.
6. Arrange or emphasize positive consequences for appropriate eating and exercise behaviors.

Table 9-12 provides specific examples of putting these six elements into action. Before doing so, however, you must establish a baseline, a record of your present eating behaviors against which to measure future progress. Keep a diary so that you can learn what particular eating stimuli, or cues, affect you.

To begin, set about eliminating or suppressing the cues that prompt you to eat inappropriately. An overeater's life may include many such cues: watching television, talking on the telephone, entering a convenience store, studying late at night. Resolve that you will no longer respond to such cues by eating. Respond only to one set of cues designed by you, in one particular place in one particular room. If some cues to inappropriate eating behavior cannot be eliminated, suppress them, as described in Table 9-12; then strengthen the appropriate cues, and reward yourself for doing so. The list in the margin suggests some activities and rewards to substitute for eating.

In addition, be aware that the food marketing industry spends huge sums each year to modify consumers' behaviors in the opposite direction—toward increasing their food consumption (details in Controversy 11). Watch out particularly for foods sold in large packages or large portions on platter-size plates. The larger the package or portion, the greater the amount of food the average person consumes at a sitting.[108]

As you progress in your new behaviors of physical activity and sensible eating, enjoy your new, emerging fit and healthy self.

TABLE 9·12 APPLYING BEHAVIOR MODIFICATION TO CONTROL BODY FATNESS

1. Eliminate inappropriate eating cues:
   - Don't buy problem foods.
   - Eat only in one room at the designated time.
   - Shop when not hungry.
   - Avoid vending machines, fast-food restaurants, and convenience stores.
   - Turn off the television, video games, and computer.

2. Suppress the cues you cannot eliminate:
   - Serve individual plates; don't serve "family style."
   - Measure your portions; avoid large servings or packages of food.
   - Make small portions look large by spreading them over the plate.
   - Create obstacles to consuming problem foods—wrap them and freeze them, making them less quickly accessible.
   - Control deprivation; plan and eat regular meals.
   - Likewise, plan to spend only one hour in sedentary activities, such as watching television or using a computer.

3. Strengthen cues to appropriate eating and exercise:
   - Share appropriate foods with others.
   - Store appropriate foods in convenient spots in the refrigerator.
   - Learn appropriate portion sizes.
   - Plan appropriate snacks.
   - Keep sports and play equipment by the door.

4. Repeat the desired eating and exercise behaviors:
   - Slow down eating—put down utensils between bites.
   - Always use utensils.
   - Leave some food on your plate.
   - Move more—shake a leg, pace, stretch often.
   - Join groups of active people and participate.

5. Arrange or emphasize negative consequences for inappropriate eating:
   - Ask that others respond neutrally to your deviations (make no comments—even negative attention is a reward).
   - If you slip, don't punish yourself.

6. Arrange or emphasize positive consequences for appropriate eating and exercise behaviors:
   - Buy tickets to sports events, movies, concerts, or other nonfood amusement.
   - Indulge in a new small purchase.
   - Get a massage; buy some flowers.
   - Take a hot bath; read a good book.
   - Treat yourself to a lesson in a new active pursuit such as horseback riding, handball, or tennis.
   - Praise yourself; visit friends.
   - Nap; relax.

Nutrition Connections

The Nutrition Connections CD provides access to these additional resources.

Do It!
The Do It! exercise for this chapter, *Control the Calories in a Day's Meals,* gives you practice in modifying meals to reduce caloric intake.

Interaction
The animation, *Energy Balance and Body Composition,* examines the effect of too much or too little energy on body composition.

Chapter Quiz
Take the quiz for this chapter to assess your understanding of chapter concepts.

Nutrition on the Net
Annotated links to reputable nutrition-related websites.

Online Glossary
Test your knowledge of key vocabulary through this comprehensive, interactive glossary.

Internet Connections

A variety of helpful study tools for this chapter are available at our website
http://nutrition.wadsworth.com/ncc10

Practice Test
Take the practice test for this chapter and gauge your knowledge of key concepts; answers are provided.

Student Notebook
A lecture notebook is available to help organize note taking while reading or attending lecture.

ABC Video Clips
"Diet Revolution" (11:16) discusses the pros and cons of diets for weight loss, such as low-fat and high-protein diets. "Anorexia in the Whole Family" (5:09) looks at the possible role of genetics in anorexia nervosa.

Study Tools
Study resources for students include Chapter Summaries, Crossword Puzzles, Flashcards, Glossary, Internet Exercises, and Learning Objectives.

SELF CHECK

Answers to these Self Check questions are in Appendix G.

1. All of the following are health risks associated with excessive body fat except:
   a. respiratory problems
   b. sleep apnea
   c. gallbladder disease
   d. low blood lipids

2. Which of the following statements about basal metabolic rate (BMR) is correct?
   a. The greater a person's age, the higher the BMR.
   b. The more thyroxine produced, the higher the BMR.
   c. Fever lowers the BMR.
   d. Pregnant women have lower BMRs.

3. Body density (the measurement of body weight compared with volume) is determined by which technique?
   a. fatfold test
   b. bioelectrical impedance
   c. underwater weighing
   d. all of the above

4. The obesity theory stating that the body chooses to be at a specific desired weight is the:
   a. set-point theory
   b. enzyme theory
   c. fat cell number theory
   d. external cue theory

5. Which of the following explains the great initial weight loss with a high-protein diet?
   a. increased basal metabolic rate
   b. increased thermic effect of protein
   c. lost glycogen and water
   d. urinary ketone loss

6. Which of the following is a possible physical consequence of fasting?
   a. loss of lean body tissues
   b. lasting weight loss
   c. body cleansing
   d. all of the above

7. The thermic effect of food plays a major role in energy expenditure.   T   F

8. The nervous system cannot use fat as fuel; it can only use glucose.   T   F

9. The BMI standard is an excellent tool for evaluating obesity in athletes and the elderly.   T   F

10. A diet too low in carbohydrate brings about responses that are similar to fasting.   T   F

CONTROVERSY 9

THE PERILS OF EATING DISORDERS

AN ESTIMATED 5 million people in the United States, primarily girls and women, suffer from the **eating disorders** of **anorexia nervosa** and **bulimia nervosa.** Many more suffer from **binge eating disorder** or other related conditions that imperil the sufferer's well-being. White women are most likely to have an eating disorder, although men and ethnic women are by no means immune.*[1]

An estimated 85 percent of eating disorders start during adolescence.[2] Among U.S. adolescents, some markers of disordered eating such as restrained eating, binge eating, purging, fear of fatness, and distorted body image are extraordinarily common. In a national survey of over 6,700 adolescents in grades 5 through 12, almost half of the girls and a fifth of the boys reported having dieted to lose weight.[3] Disordered eating occurred among 13 percent of the girls and 7 percent of the boys. A survey at a major university

*Reference notes are found in Appendix F.

Anorexia nervosa.

found that only 8 percent of the students were overweight by objective measure (BMI), yet more than half of the students reported themselves to be overweight. Half of the students found to be *underweight* according to the BMI charts made that claim. These behaviors and attitudes are much less prevalent in most other societies, although the incidence often increases when body image becomes central to self-worth.[4]

Why do so many people in our society suffer from eating disorders? Excessive pressure to be thin is at least partly to blame. When low body weight becomes an important goal, people begin to view normal, healthy body weight as too fat, and some take unhealthy actions to lose weight. Severe restriction of food intake can create intense stress and extreme hunger that lead to binges.[5] Painful emotions such as anger, jealousy, or disappointment may be turned inward by youngsters who express dissatisfaction with body weight or say they "feel fat." As weight loss and diet restraint become more and more a focus, psychological problems worsen, and the likelihood of developing full-blown eating disorders intensi-

fies.[6] Table C9-1 defines eating disorder terms.

Eating Disorders in Athletes

Athletes and dancers are at special risk for eating disorders. In females, three associated medical problems form the **female athlete triad:** disordered eating, amenorrhea (cessation of menstruation), and osteoporosis.[7] In males, disordered eating brings on many of the same physical problems affecting female counterparts.

The Female Athlete Triad

At age 14, Suzanne was a top contender for a spot on the state gymnastics team. Each day her coach reminded team members that they would not qualify for competition if they weighed more than a few ounces above their assigned weights. The coach chastised gymnasts who gained weight. Suzanne weighed herself several times a day to make sure that she had not exceeded her 80-pound limit. Suzanne dieted and exercised to an extreme, and unlike many of her friends, she never began to menstruate. A few months before her fifteenth birthday, Suzanne's coach dropped her back to the second-level team. Suzanne blamed her poor performance on a slow-healing stress fracture. Mentally stressed and physically exhausted, she quit gymnastics and began overeating between periods of self-starvation.

TABLE C9·1   EATING DISORDER TERMS

- **anorexia nervosa** an eating disorder characterized by a refusal to maintain a minimally normal body weight, self-starvation to the extreme, and a disturbed perception of body weight and shape; seen (usually) in teenage girls and young women (*anorexia* means "without appetite"; *nervos* means "of nervous origin").
- **binge eating disorder** an eating disorder whose criteria are similar to those of bulimia nervosa, excluding purging or other compensatory behaviors.
- **bulimia** (byoo-LEEM-ee-uh) **nervosa** recurring episodes of binge eating combined with a morbid fear of becoming fat; usually followed by self-induced vomiting or purging.
- **cathartic** a strong laxative.
- **cognitive therapy** psychological therapy aimed at changing undesirable behaviors by changing underlying thought processes contributing to these behaviors; in anorexia, a goal is to replace false beliefs about body weight, eating, and self-worth with health-promoting beliefs.
- **eating disorder** a disturbance in eating behavior that jeopardizes a person's physical or psychological health.
- **emetic** (em-ETT-ic) an agent that causes vomiting.
- **female athlete triad** a potentially fatal triad of medical problems seen in female athletes: disordered eating, amenorrhea, and osteoporosis.

Suzanne had developed the dangerous combination of problems that characterize the female athlete triad—disordered eating, amenorrhea, and weakening of the bones.[8]

Female athletes, in keeping with their coaches' recommendations, often compare themselves to unsuitable weight standards. An ultraslim appearance has long been considered desirable in activities such as dancing, gymnastics, and figure skating. This puts the athlete in a bind since an athlete's body must be heavier for a given height than a nonathlete's body because it contains more healthy muscle and dense bone tissue and less fat. Most weight standards appropriate for the general population fall far from the mark with regard to athletes' bodies—the photo in the preceding chapter of an athlete with a high BMI demonstrated this. For athletes, body composition measures such as fatfold measures yield more useful information.

The prevalence of amenorrhea among premenopausal women in the United States is about 2 to 5 percent overall, but it may be as high as 66 percent among female athletes. Amenorrhea is *not* a normal adaptation to strenuous physical training but a symptom of something going wrong, and it is particularly hazardous to the bones.[9] For most people, weight-bearing exercise helps to protect bones against the calcium losses of osteo-

porosis. For young women with anorexia nervosa, however, strenuous activity can imperil their bones, even after recovery from the eating disorder.[10] Vigorous training along with low food energy intakes and other stressors may cause bone loss even without obvious menstrual irregularities.[11] Such bone losses may increase the risks of stress fractures today and of osteoporosis in later life (see Figure C9-1).

Male Athletes and Eating Disorders

Male athletes and dancers who face pressure to achieve a certain body weight often develop eating disorders, although they may deny having them in the mistaken belief that the disorders strike only women.[12] On average, male teenagers carry about 15 percent of body weight as fat, but some high school wrestlers, gymnasts, and figure skaters strive for only 5 percent body fat.

Wrestlers, for example, are required to "make weight" to compete in the lowest possible weight class to face the smallest possible opponents. To that end, wrestlers starve themselves, don rubber suits, sweat in steam rooms, and take diuretics to shed water weight before weighing in for competition. These practices were responsible for the deaths of three college athletes in recent years and have caused untold misery and harm to many others. Athletes engaging in these practices actually compromise their athletic abilities. The diminished anaerobic strength, reduced endurance, decreased oxygen capacity, and general weakness caused by food deprivation and dehydration can hobble performance, an effect lasting days after food and water are replenished.

Male athletes are also susceptible to weight-*gain* problems, in which athletes with well-muscled bodies see themselves as underweight and weak. Such a distorted body image leads to frequent weighing, excessive exercise, overuse

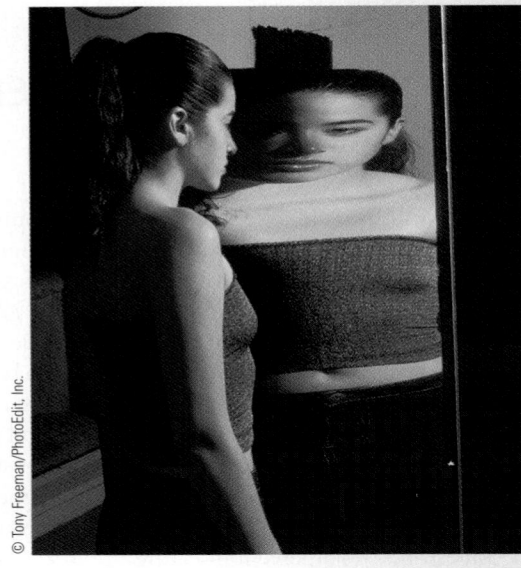

Women with anorexia nervosa see themselves as fat, even when they are dangerously underweight.

of special diets or protein supplements, or even the abuse of steroid drugs in the attempt to bulk up their muscles.

Young athletes and their coaches must be educated about links between inappropriate body-weight ideals, improper weight-loss techniques, eating disorder development, effective sports nutrition, and safe weight-control methods. For all young people's activities, idealistic artistic standards based on slim appearance or low body weight should be replaced with performance-based standards. Table C9-2 provides some suggestions to help athletes and dancers protect themselves against developing eating disorders. The next sections describe eating disorders that anyone, athlete or nonathlete, may experience.

Anorexia Nervosa

Julie is 18 years old and is a superachiever in school. She watches her diet with great care, and she exercises daily, maintaining a heroic schedule of self-discipline. She is thin, but she is determined to lose more weight. She is 5 feet 6 inches tall and weighs 85 pounds. She has anorexia nervosa.

Characteristics of Anorexia Nervosa

Julie is unaware that she is undernourished, and she sees no need to obtain

---

FIGURE C9·1 THE FEMALE ATHLETE TRIAD

In the female athlete triad, extreme weight loss causes both cessation of menstruation (amenorrhea) and excessive loss of calcium from the bones. The hormone disturbances associated with amenorrhea also contribute to osteoporosis, making the female athlete triad extraordinarily harmful to the bones.

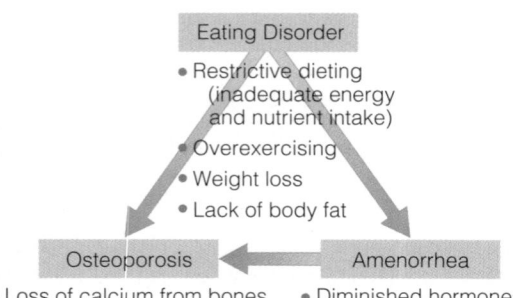

Eating Disorder
- Restrictive dieting (inadequate energy and nutrient intake)
- Overexercising
- Weight loss
- Lack of body fat

Osteoporosis
- Loss of calcium from bones

Amenorrhea
- Diminished hormones

TABLE C9•2 TIPS FOR COMBATING EATING DISORDERS

General Guidelines

- Never restrict food intakes to below the amounts suggested for adequacy by the USDA Food Guide (Chapter 2).
- Eat frequently. People often do not eat frequent meals because of time constraints, but eating can be incorporated into other activities, such as snacking while studying or commuting. The person who eats frequently never gets so hungry as to allow hunger to dictate food choices.
- If not at a healthy weight, establish a reasonable weight goal based on a healthy body composition. (Chapter 9 provides help in doing so.)
- Allow a reasonable time to achieve the goal. A reasonable rate for losing excess fat is about 1% of body weight per week.
- Establish a weight-maintenance support group with people who share interests.

Specific Guidelines for Athletes and Dancers

- Replace weight-based goals with performance-based goals.
- Remember that eating disorders impair physical performance. Seek confidential help in obtaining treatment if needed.
- Restrict weight-loss activities to the off-season.
- Focus on proper nutrition as an important facet of your training, as important as proper technique.

treatment. She stopped menstruating several months ago and is moody and chronically depressed. She insists that she is too fat, although her eyes are sunk in deep hollows in her face. Although she is close to physical exhaustion, she no longer sleeps easily. Her family is concerned, and although reluctant to push her, they have finally insisted that she see a psychiatrist. Julie's psychiatrist has prescribed group therapy as a start, but warns that if Julie does not begin to gain weight soon, she will need to be hospitalized.

Most anorexia nervosa victims come from middle- or upper-class families. Men account for only 1 or 2 in 20 cases in the general population, although the incidence among male athletes and dancers may be much higher.

No one knows for certain what causes anorexia nervosa. Central to its diagnosis is a distorted body image that overestimates body fatness. When Julie looks at herself in the mirror, she sees her 85-pound body as fat. The more Julie overestimates her body size, the more resistant she is to treatment, and the more unwilling to examine her faulty values and misconceptions. Malnutrition is known to affect brain functioning and judgment in this way. People with anorexia nervosa cannot

recognize it in themselves; only professionals can diagnose it. Table C9-3 shows the criteria that experts use.

The Role of the Family

Certain family attitudes, and especially parental attitudes, stand accused of contributing to eating disorders. Families of persons with anorexia nervosa are likely to be critical and to overvalue outward appearances while undervaluing inner self-worth. Parents may oppose one another's authority and vacillate between defending the anorexic child's behavior and condemning it, confusing the child and disrupting normal parental control. In the extreme, parents may even be sexually abusive or abusive in other ways.

Julie is a perfectionist, just as her parents are: She identifies so strongly with her parents' ideals and goals that she cannot get in touch with her own identity. She is respectful of authority but sometimes feels like a robot, and she may act that way, too: polite but controlled, rigid, and unspontaneous.[13] For Julie, rejecting food is a way of gaining control.

Self-Starvation

How can a person as thin as Julie continue to starve herself? Julie uses tremendous discipline to strictly limit her portions of low-calorie foods. She will deny her hunger, and having become accustomed to so little food, she feels full after eating only a half-dozen carrot sticks. She can recite the calorie contents of dozens of foods and the calorie costs of as many exercises. If she feels that she has gained an ounce of

TABLE C9•3 CRITERIA FOR DIAGNOSIS OF ANOREXIA NERVOSA

A person with anorexia nervosa demonstrates the following:

A. Refusal to maintain body weight at or above a minimal normal weight for age and height, e.g., weight loss leading to maintenance of body weight less than 85% of that expected; or failure to make expected weight gain during period of growth, leading to body weight less than 85% of that expected.
B. Intense fear of gaining weight or becoming fat, even though underweight.
C. Disturbance in the way in which one's body weight or shape is experienced; undue influence of body weight or shape on self-evaluation, or denial of the seriousness of the current low body weight.
D. In females past puberty, amenorrhea, i.e., the absence of at least three consecutive menstrual cycles. (A woman is considered to have amenorrhea if her periods occur only following hormone, e.g., estrogen, administration.)

Two types of anerexia nervosa include:

- Restricting type: during the episode of anorexia nervosa, the person does not regularly engage in binge eating or purging behavior (i.e., self-induced vomiting or the misuse of laxatives, diuretics, or enemas).
- Binge eating/purging type: during the episode of anorexia nervosa, the person regularly engages in binge eating or purging behavior (i.e., self-induced vomiting or the misuse of laxatives, diuretics, or enemas).

SOURCE: Reprinted with permission from the Diagnostic and Statistical Manual of Mental Disorders, Fourth Edition, Text Revision. Copyright 2000 American Psychiatric Association.

weight, she runs or jumps rope until she is sure she has exercised it off. She drinks water incessantly to fill her stomach, risking dangerous mineral imbalances. If she fears that the food energy she has eaten exceeds the exercise she has done, she takes laxatives to hasten the passage of food from her system, not knowing that laxatives reduce water absorption, but not food energy absorption. Her other methods of staying thin are so effective that she is unaware that laxatives have no effect on body fat. She is desperately hungry. In fact, she is starving, but she doesn't eat because her need for self-control dominates other needs.

Physical Perils

From the body's point of view, anorexia nervosa is starvation and thus brings the same damage as classic protein-energy malnutrition (described in Chapter 6). The person with anorexia depletes the body tissues of needed fat and protein.[14] In young people, growth ceases and normal development falters. They lose so much lean tissue that basal metabolic rate slows. In athletes, the loss of lean tissue handicaps physical performance. The heart pumps inefficiently and irregularly, the heart muscle becomes weak and thin, the heart chambers diminish in size, and the blood pressure falls.[15] Electrolytes that help to regulate the heartbeat go out of balance. Many deaths in people with anorexia are due to heart failure.

Starvation brings neurological, digestive, and circulation consequences as well. The brain loses significant amounts of tissue, nerves function abnormally, the electrical activity of the brain becomes abnormal, and insomnia is common. Digestive functioning becomes sluggish, the stomach empties slowly, and the lining of the intestinal tract shrinks. The ailing digestive tract fails to digest food adequately, even if the victim does eat. The pancreas slows its production of digestive enzymes. Diarrhea sets in, further worsening malnutrition.

Changes in the blood include anemia, impaired immune response, altered blood lipids, high concentrations of vitamins A and E, and low

blood proteins. Dry skin, low body temperature, and the development of fine body hair (the body's attempt to keep warm) also occur. In adulthood, both women and men lose their sex drives. Mothers with anorexia nervosa may severely underfeed their children who then fail to grow and suffer the other harms typical of starvation.

Treatment of Anorexia Nervosa

Treatment of anorexia nervosa requires a multidisciplinary approach that addresses two issues and behaviors: those relating to food and weight and those involving relationships with oneself and others.[16] Teams of physicians, nurses, psychiatrists, family therapists, and dietitians work together to treat people with anorexia nervosa. Appropriate diet is crucial for normalizing body weight and must be crafted individually.[17] Clients are seldom willing to eat for themselves, but if they are, chances are they can recover without other interventions.

Professionals classify clients based on the risks posed by the degree of malnutrition present.* Clients with low risks may benefit from family counseling, **cognitive therapy,** behavior modification, and nutrition guidance; those with greater risks may also need other forms of psychotherapy and supplemental formulas to provide extra nutrients and energy. High-risk clients may require involuntary hospitalization and may need to be force-fed by tube at first to forestall death. This step causes psychological trauma. Drugs are commonly prescribed, but to date, their usefulness is limited.

Stopping weight loss is a first goal of treatment; establishing regular eating patterns is next. At first, progress is slow partly owing to a speeded-up metabolic rate and an increased thermic response to food that occur upon refeeding. As small gains of body fat occur, blood concentration of the appetite-suppressing hormone leptin begins creeping up, too, causing researchers to speculate that leptin may contribute to difficulties in weight restoration.

---
*Indicators of malnutrition include a low percentage of body fat, low blood proteins, and impaired immune response.

Few people with anorexia nervosa seek treatment on their own, and denial makes treatment difficult. Almost half of the women who are treated can maintain their body weight within 15 percent of a healthy weight; at that weight, many of them begin menstruating again.[18] The other half have poor or fair outcomes of treatment, and two-thirds of those treated continue a mental battle with recurring morbid thoughts about food and body weight. Many, fearing the growing pads of fat on hips and abdomen, relapse into abnormal eating behaviors.[19] About 1,000 women die of anorexia nervosa each year, mostly from heart abnormalities brought on by malnutrition or from suicide.[20] Anorexia nervosa has one of the highest mortality rates among psychiatric disorders.[21]

Before drawing conclusions about someone who is extremely thin, remember that diagnosis of anorexia nervosa requires professional assessment. People seeking help for anorexia nervosa for themselves or for others should visit the National Eating Disorders Association website or call to get started. Appendix E lists the needed resources.

Bulimia Nervosa

Sophia is a 20-year-old flight attendant, and although her body weight is healthy, she thinks constantly about food. She alternately starves herself and then secretly binges; when she has eaten too much, she vomits. Few people would fail to recognize that these symptoms signify bulimia nervosa.

Characteristics of Bulimia Nervosa

Bulimia nervosa is distinct from anorexia nervosa and is much more prevalent, although the true incidence is difficult to establish. People with bulimia nervosa often suffer in secret and, when asked, may deny the existence of a problem. More men suffer from bulimia nervosa than from anorexia nervosa, but bulimia nervosa is still most common in women. Based on a questionnaire, one study estimates that 19 percent of female college students experience bulimic symptoms. A true diagnosis of bulimia nervosa is based on the criteria listed in Table C9-4.

TABLE C9·4 CRITERIA FOR DIAGNOSIS OF BULIMIA NERVOSA

A person with bulimia nervosa demonstrates the following:

A. Recurrent episodes of binge eating. An episode of binge eating is characterized by both of the following:
  1. Eating, in a discrete period of time (e.g., within any two-hour period), an amount of food that is definitely larger than most people would eat during a similar period of time and under similar circumstances, and,
  2. A sense of lack of control over eating during the episode (e.g., a feeling that one cannot stop eating or control what or how much one is eating).
B. Recurrent inappropriate compensatory behavior in order to prevent weight gain, such as self-induced vomiting; misuse of laxatives, diuretics, enemas, or other medications; fasting; or excessive exercise.
C. Binge eating and inappropriate compensatory behaviors that both occur, on average, at least twice a week for three months.
D. Self-evaluation unduly influenced by body shape and weight.
E. The disturbance does not occur exclusively during episodes of anorexia nervosa.

Two types:

• Purging type: the person regularly engages in self-induced vomiting or the misuse of laxatives, diuretics, or enemas.
• Nonpurging type: the person uses other inappropriate compensatory behaviors, such as fasting or excessive exercise, but does not regularly engage in self-induced vomiting or the misuse of laxatives, diuretics, or enemas.

SOURCE: Reprinted with permission from the Diagnostic and Statistical Manual of Mental Disorders, Fourth Edition, Text Revision. Copyright 2000 American Psychiatric Association.

FIGURE C9·2 THE CYCLE OF BINGEING, PURGING, AND NEGATIVE SELF-PERCEPTION

Each of these factors helps to perpetuate disordered eating.

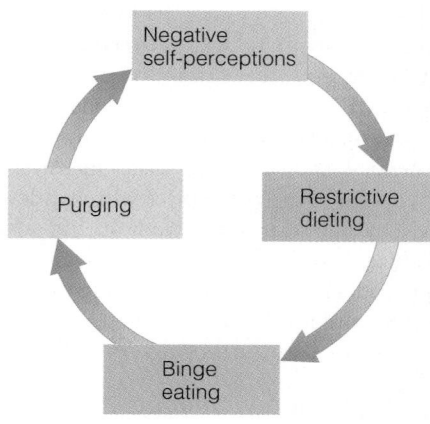

Like the typical person with bulimia nervosa, Sophia is single, female, and white. She is well educated and close to her ideal body weight, although her weight fluctuates over a range of 10 pounds or so every few weeks.

Sophia seldom lets her bulimia nervosa interfere with her work or other activities. From early childhood she has been a high achiever but emotionally dependent on her parents. As a young teen, Sophia cycled on and off crash diets. She feels anxious at social events and cannot easily establish close relationships. She is usually depressed, is often impulsive, and has low self-esteem.[22] When crisis hits, Sophia responds by replaying events, worrying excessively, seeking solace in alcohol and tobacco, and blaming herself but never asking for help—behaviors that are barriers to effective coping.[23]

The Role of the Family

Families of bulimic people are observed to be externally controlling, but emotionally uninvolved with their children, resulting in a stifling negative self-image believed to perpetuate bulimia (see Figure C9-2). Dieting, arguments, criticism of body shape or weight, minimal affection and caring, and other weaknesses are common in the families of people with bulimia.[24] Typically, the family has "secrets" that are hidden from outsiders. Bulimic women who report having been abused sexually or physically by family members or friends may continually suffer a sense of being unable to gain control.

Should the member with bulimia nervosa begin making the needed changes toward recovery, others in the family may feel threatened. Family cooperation is important, however, because making changes within a family requires effort from everyone. Such effort is well spent, for changing destructive family interactions can greatly benefit the person who has begun to fight against bulimia nervosa.

Binge Eating and Purging

A bulimic binge is unlike normal eating, and the food is not consumed for its nutritional value. During a binge, Sophia's eating is accelerated by her hunger from previous calorie restriction (Figure C9-2). She regularly takes in extra food approaching 1,000 calories at each binge, and she may have several binges in a day.[25] Typical binge foods are easy-to-eat, low-fiber, smooth-textured, high-fat, and high-carbohydrate foods, such as cookies, cakes, and ice cream; and she eats the entire bag of cookies, the whole cake, and every spoonful in a carton of ice cream. By the end of the binge, she has vastly overcorrected for her attempts at calorie restriction at other times.

A typical binge consists of easy-to-eat, low-fiber, smooth-textured, high-calorie foods.

The binge is a compulsion and usually occurs in several stages: "anticipation and planning, anxiety, urgency to begin, rapid and uncontrollable consumption of food, relief and relaxation, disappointment, and finally shame or disgust." Then, to purge the food from her body, she may use a **cathartic**—a strong laxative that can injure the lower intestinal tract. Or she may induce vomiting, using an **emetic**—a drug intended as first aid for poisoning. After the binge she pays the price with hands scraped raw against the teeth during induced vomiting, swollen neck glands and reddened eyes from straining to vomit, and the bloating, fatigue, headache, nausea, and pain that follow.

Physical and Psychological Perils

Purging may seem to offer a quick and easy solution to the problems of unwanted calories and body weight, but bingeing and purging have serious physical consequences. Fluid and electrolyte imbalances caused by vomiting or diarrhea can lead to abnormal heart rhythms and injury to the kidneys. Urinary tract infections can lead to kidney failure. Vomiting causes irritation and infection of the pharynx, esophagus, and salivary glands; erosion of the teeth; and dental caries. The esophagus or stomach may rupture or tear. Overuse of emetics can lead to death by heart failure.

Unlike Julie, Sophia is aware that her behavior is abnormal, and she is deeply ashamed of it. She wants to recover, and this makes recovery more likely for her than for Julie, who clings to denial.

Treatment of Bulimia Nervosa

To gain control over food and establish regular eating patterns requires adherence to a structured eating plan. The person may also benefit from a regular exercise program.[26] Restrictive dieting is forbidden, for it almost always precedes and may even trigger binges. Steady maintenance of weight and prevention of relapse into cyclic gains and losses are the goals. Many a former bulimia nervosa sufferer has taken a major step toward recovery by learning to consistently eat enough food to satisfy hunger needs (at least 1,600 calories a day).

TABLE C9·5 DIET STRATEGIES FOR COMBATING BULIMIA NERVOSA

Planning principles:
- Plan meals and snacks; record plans in a food diary prior to eating.
- Plan meals and snacks that require eating at the table and using utensils.
- Refrain from finger foods.
- Refrain from "dieting" or skipping meals.

Nutrition principles:
- Eat a well-balanced diet and regularly timed meals consisting of a variety of foods.
- Include raw vegetables, salad, or raw fruit at meals to prolong eating times.
- Choose whole-grain, high-fiber breads, pasta, rice, and cereals to increase bulk.
- Consume adequate fluid, particularly water.

Other tips:
- Choose meals that provide protein and fat for satiety, and bulky, fiber-rich carbohydrates for immediate feelings of fullness.
- Try including soups and other water-rich foods (at the top of the list on page 268 in Chapter 8) for satiety.
- Consume the amounts of food specified in the USDA Food Guide (Chapter 2 and Appendix I).
- For convenience (and to reduce temptation) select foods that naturally divide into portions. Select one potato, rather than rice or pasta that can be overloaded onto the plate; purchase yogurt and cottage cheese in individual containers; look for small packages of precut steak or chicken; choose frozen dinners with metered portions.
- Include 30 minutes or more of physical activity on most days—exercise may be an important tool in controlling bulimia.

Table C9-5 offers some ways to begin correcting the eating problems of bulimia nervosa. About half of women receiving a diagnosis of bulimia nervosa may recover completely after five to ten years, with or without treatment, but treatment probably speeds the recovery process. If Sophia's depression deepens, she may benefit from antidepressant medication.

Binge Eating Disorder

Anorexia nervosa and bulimia nervosa are distinct eating disorders, yet they sometimes overlap. People with both conditions share an overconcern with body weight and the tendency to drastically undereat. Both may purge. The two disorders can also appear in the same person, or one can lead to the other. Other people have eating disorders that fall short of anorexia nervosa or bulimia nervosa, but share some of their features, such as fear of body fatness. One such condition is binge eating disorder (defined earlier in Table C9-1).

Up to half of all people who restrict eating to lose weight periodically binge without purging, including about one-third of obese people who regularly engage in binge eating. Obesity itself, however, does not constitute an eating disorder. Table C9-6 lists the official diagnostic criteria for binge eating disorder.

Clinicians note differences between people with bulimia nervosa and those with binge eating disorder.[27] People with binge eating disorder consume less during a binge, rarely purge, and exert less restraint during times of dieting. Similarities also exist, including feeling out of control, disgusted, depressed, embarrassed, guilty, or distressed because of their self-perceived gluttony.[28] Binge eating behavior responds more readily to treatment than other eating disorders, and resolving such behaviors can be a first step to authentic weight control. Successful treatment also improves physical health, mental health, and the chances of breaking the cycle of rapid weight losses and gains.

Eating Disorders in Society

Most experts agree that eating disorders have many causes: sociocultural, psychological, hereditary, and probably also neurochemical. Proof that society plays a role in eating disorders is their demographic distribution: They are known only in developed nations, and they become more prevalent as wealth increases and food becomes plentiful.

TABLE C9•6	CRITERIA FOR DIAGNOSIS OF BINGE EATING DISORDER

A person with a binge eating disorder demonstrates the following:

A. Recurrent episodes of binge eating. An episode of binge eating is characterized by both of the following:
   1. Eating, in a discrete period of time (e.g., within any two-hour period) an amount of food that is definitely larger than most people would eat in a similar period of time under similar circumstances.
   2. A sense of lack of control over eating during the episode (e.g., a feeling that one cannot stop eating or control what or how much one is eating).

B. Binge eating episodes are associated with at least three of the following:
   1. Eating much more rapidly than normal.
   2. Eating until feeling uncomfortably full.
   3. Eating large amounts of food when not feeling physically hungry.
   4. Eating alone because of being embarrassed by how much one is eating.
   5. Feeling disgusted with oneself, depressed, or very guilty after overeating.

C. The binge eating causes marked distress.

D. The binge eating occurs, on average, at least twice a week for six months.

E. The binge eating is not associated with the regular use of inappropriate compensatory behaviors (e.g., purging, fasting, excessive exercise) and does not occur exclusively during the course of anorexia nervosa or bulimia nervosa.

SOURCE: Reprinted with permission from the Diagnostic and Statistical Manual of Mental Disorders, Fourth Edition, Text Revision. Copyright 2000 American Psychiatric Association.

No doubt our society sets unrealistic ideals for body weight, especially for women, and devalues those who do not conform to them. The Miss America beauty pageant, for example, puts forth a role model of female desirability. No winner has ever been overweight, and thinner and thinner women have worn the Miss America crown over the years.[29] Magazines and other media convey the message that to be thin is to be happy; eating disorders are not a form of rebellion against these unrealistic ideals, but rather an exaggerated acceptance of them. Body dissatisfaction is a primary factor in the development of eating disorders.[30]

Even professionals, including physicians and dietitians, tend to praise people for losing weight and to suggest weight loss to people who do not need it. As a result, normal-weight girls as young as five years old fear that they are too fat and are placed "on diets."[31] In addition, an increasing number of girls younger than 18 are poorly nourished.[32] Some eat so little food that normal growth ceases; thus, they miss out on their adolescent growth spurts and may never catch up.

Perhaps a young person's best defense against these disorders is to learn about normal, expected growth patterns, especially the characteristic weight gain of adolescence (see Chapter 14), and to learn respect for the inherent wisdom of the body. When people discover and honor the body's real needs for nutrition and exercise, they become unwilling to sacrifice health for conformity.

Nutrients, Physical Activity, and the Body's Responses

© Jean Metzinger (1883–1956, French). *Le Coureur Cycliste (The Racing Cyclist)* ca. 1914. © 2005 Artists Rights Society (ARS), New York/ADAGP, Paris. SuperStock, Inc.

CONTENTS

Do You Ever . . .

Wonder if physical activity can help you live longer?

Wish for foods or beverages to help you feel stronger and go longer when competing?

Take vitamin pills right before a race or a game to improve your performance?

Drink sports drinks instead of water, but wonder why?

KEEP READING.

IN THE BODY, nutrition and physical activity are interactive—each influences the other. The working body demands all three energy-yielding nutrients—carbohydrate, lipids, and protein—to fuel activity. The body also needs protein and a host of supporting nutrients to build lean tissue. Physical activity, in turn, benefits the body's nutrition by helping to regulate the use of fuels, by pushing the body composition toward the lean, and by increasing the daily calorie allowance. With more calories come more nutrients and other beneficial constituents of foods.

For those just beginning to increase fitness, be assured that improvement is not only possible but an inevitable result of becoming more active. As you improve your physical fitness, you not only *feel* better and stronger, but you *look* better, too. Physically fit people walk with confidence and purpose because posture and self-image improve along with physical fitness.

If you are already physically fit, the following description applies: You move with ease and balance. You have endurance, and your energy lasts for hours. You are strong and meet daily physical challenges without strain. What's more, you are prepared to meet mental and emotional challenges, too, because physical fitness also supports mental and emotional energy and resilience. Sounds good, doesn't it?

This chapter is written for athletes and for active people who train like athletes. Casual athletes (those who compete only with their own goals) and competitive athletes (those who compete with others) are cut from the same cloth with regard to their food and fluid needs. The chapter refers to "you" to make the connection between academic thinking and personal choices. To understand the interactions between physical activity and nutrition, you must first know a few things about fitness, its benefits, and **training** to develop fitness.

Fitness

Fitness depends on a certain minimum amount of **physical activity** or **exercise.** Physical activity and exercise both involve bodily movement, muscle contraction, and enhanced energy expenditure, but by definition, a distinction is made between the two terms. Exercise is often considered to be a vigorous, structured, and planned type of physical activity. Because this chapter focuses on the active body's use of energy nutrients—whether that body is pedaling a bike across campus or pedaling a stationary bike in a gym—for our purposes, the terms *physical activity* and *exercise* will be used interchangeably.

Benefits of Fitness

People who regularly engage in just moderate physical activity live longer on average than those who are physically inactive.*[1] Yet, despite an increasing awareness of the health benefits that physical activity confers, 25 percent of adults in the United States are completely inactive.[2] A sedentary lifestyle ranks with smoking and obesity as a powerful risk factor for developing the major killer diseases of our time—cardiovascular disease, some forms of cancer, stroke, diabetes, and hypertension.[3] Every year an estimated $24 billion is spent on health-care costs attributed to physical inactivity in the United States.[4]

As a person becomes physically active, the health of the entire body improves. Compared with unfit people, physically fit people enjoy:

- *More restful sleep.* Rest and sleep occur naturally after periods of physical activity. During rest, the body repairs injuries, disposes of wastes generated during activity, and builds new physical structures.
- *Better nutritional health.* Physical activity expends energy and thus allows people to eat more food. If they choose wisely, active people will consume more nutrients and be less likely to develop nutrient deficiencies.

training regular practice of an activity, which leads to physical adaptations of the body with improvement in flexibility, strength, or endurance.

physical activity bodily movement produced by muscle contractions that substantially increase energy expenditure.

exercise planned, structured, and repetitive bodily movement that promotes or maintains physical fitness.

*Reference notes are found in Appendix F.

- *Improved body composition.* A balanced program of physical activity limits body fat and increases or maintains lean tissue. Thus, physically active people have relatively less body fat than sedentary people at the same body weight.[5]
- *Improved bone density.* Weight-bearing physical activity builds bone strength and protects against osteoporosis.[6]
- *Enhanced resistance to colds and other infectious diseases.* Fitness enhances immunity.*[7]
- *Lower risks of some types of cancers.* Lifelong physical activity may help to protect against colon cancer, breast cancer, and some other cancers.[8]
- *Stronger circulation and lung function.* Physical activity that challenges the heart and lungs strengthens both the circulatory and the respiratory system.
- *Lower risks of cardiovascular disease.* Physical activity lowers blood pressure, slows resting pulse rate, lowers total blood cholesterol, and raises HDL cholesterol, thus reducing the risks of heart attacks and strokes.[9] Some research suggests that physical activity may reduce the risk of cardiovascular disease in another way as well—by reducing intra-abdominal fat stores.[10]
- *Lower risks of type 2 diabetes.* Physical activity normalizes glucose tolerance.[11] Regular physical activity reduces the risk of developing type 2 diabetes and benefits those who already have the condition.
- *Reduced risk of gallbladder disease (women).* Regular physical activity reduces women's risk of gallbladder disease—perhaps by facilitating weight control and lowering blood lipid levels.[12]
- *Lower incidence and severity of anxiety and depression.* Physical activity may improve mood and enhance the quality of life by reducing depression and anxiety.[13]
- *Stronger self-image.* The sense of achievement that comes from meeting physical challenges promotes self-confidence.
- *Longer life and higher quality of life in the later years.* Active people have a lower mortality rate than sedentary people.[14] Even a two-mile walk daily can add years to a person's life. In addition to extending longevity, physical activity supports independence and mobility in later life by reducing the risk of falls and minimizing the risk of injury should a fall occur.[15]

You don't have to run marathons to reap the health rewards of physical activity.[16] For health's sake, the American College of Sports Medicine (ACSM), the *Dietary Guidelines for Americans 2005,* and the DRI committee specify that people need to spend an accumulated minimum of 30 minutes in some sort of physical activity on most days of each week.[17] All three authorities, however, advises that 30 minutes of physical activity each day is not enough for adults to maintain a healthy body weight (BMI of 18.5 to 24.9) and recommend at least 60 minutes of moderately intense activity such as walking or jogging each day.[18] The hour or more of activity can be split into shorter sessions throughout the day—two 30-minute sessions or four 15-minute sessions, for example.[19]

For many people, the health benefits of regular, moderate physical activity are reward enough. Others, however, seek the kinds and amount of physical activity that will not only benefit health but improve their physical fitness or their performance in sports. In 1998, the ACSM issued recommendations for the quantity and quality of physical activity needed to develop and maintain fitness in healthy adults (see Table 10-1, p. 362). The kinds and amounts of physical activity that improve physical fitness also provide still greater health benefits (further reduction of cardiovascular disease risk and improved body composition, for example).[20]

Dietary Guidelines for Americans 2005 key recommendations for physical activity:

- Engage in regular physical activity and reduce sedentary activities to promote health, psychological well-being, and a healthy body weight.
- Engage in at least 30 minutes of moderate-intensity physical activity, above usual activity, at work or home on most days of the week.
- For most people, greater health benefits can be obtained by engaging in physical activity of more vigorous intensity or longer duration.
- Achieve physical fitness by including cardiovascular conditioning, stretching exercises for flexibility, and resistance exercises for muscle strength and endurance.

▶ KEY POINT   Physical activity and fitness benefit people's physical and psychological well-being and improve their resistance to disease. Physical activity to improve physical fitness offers additional personal benefits.

---

*Moderate physical activity can stimulate immune function. Intense, vigorous, prolonged activity such as marathon running, however, may compromise immune function.

TABLE 10·1 GUIDELINES FOR PHYSICAL FITNESS

	Cardiorespiratory	Strength	Flexibility
Type of Activity	Aerobic activity that uses large-muscle groups and can be maintained continuously	Resistance activity that is performed at a controlled speed and through a full range of motion	Stretching activity that uses the major muscle groups
Frequency	3 to 5 days per week	2 to 3 days per week	2 to 3 days per week
Intensity	55 to 90% of maximum heart rate	Enough to enhance muscle strength and improve body composition	Enough to develop and maintain a full range of motion
Duration	20 to 60 minutes	8 to 12 repetitions of 8 to 10 different exercises (minimum)	4 repetitions of 10 to 30 seconds per muscle group (minimum)
Examples	Running, cycling, swimming, in-line skating, rowing, power walking, cross-country skiing, kickboxing, jumping rope; sports activities such as basketball, soccer, racquetball, tennis, volleyball	Pull-ups, push-ups, weight lifting, pilates	Yoga

flexibility the capacity of the joints to move through a full range of motion; the ability to bend and recover without injury.

muscle strength the ability of muscles to work against resistance.

muscle endurance the ability of a muscle to contract repeatedly within a given time without becoming exhausted.

cardiorespiratory endurance the ability to perform large-muscle dynamic exercise of moderate-to-high intensity for prolonged periods.

overload an extra physical demand placed on the body; an increase in the frequency, duration, or intensity of an activity. A principle of training is that for a body system to improve, it must be worked at frequencies, durations, or intensities that increase by increments.

hypertrophy (high-PURR-tro-fee) an increase in size (for example, of a muscle) in response to use.

atrophy (AT-tro-fee) a decrease in size (for example, of a muscle) because of disuse.

aerobic (air-ROH-bic) requiring oxygen. Aerobic activity strengthens the heart and lungs by requiring them to work harder than normal to deliver oxygen to the tissues.

The Essentials of Fitness

To be physically fit, you need to develop enough **flexibility, muscle strength, muscle endurance,** and **cardiorespiratory endurance** to allow you to meet the everyday demands of life with some to spare, and you need to achieve a reasonable body composition. A person who practices a physical activity *adapts* by becoming better able to perform it after each session—with more flexibility, more strength, and more endurance.

How Do My Muscles Become Physically Fit? People shape their bodies by what they choose to do and not do. Muscle cells and tissues respond to an **overload** of physical activity by gaining strength and size, a response called **hypertrophy.** The opposite is also true: if not called on to perform, muscles dwindle and weaken, a response called **atrophy.** Thus, cyclists often have well-developed legs but less arm or chest strength; a tennis player may have one superbly strong arm, while the other is just average. A variety of physical activities produces the best overall fitness, and to this end, people need to work different muscle groups from day to day. For balanced fitness, stretching enhances flexibility, weight training develops muscle strength and endurance, and **aerobic** activity improves cardiorespiratory endurance. It makes sense to give muscles a rest, too, because it takes a day or two to replenish muscle fuel supplies and to repair wear and tear incurred through physical activity.

Periodic rest also gives muscles time to adapt to an activity. During rest, muscles build more of the equipment required to perform the activity that preceded the rest. The muscle cells of a superbly trained weight lifter, for example, store extra granules of glycogen, build up strong connective tissues, and add bulk to the special proteins that contract the muscles, thereby increasing the muscles' ability to perform.* In the same way, the muscle cells of a distance swimmer develop huge stocks of **myoglobin,** the muscles' oxygen-handling protein, and other equipment needed to burn fat and to sustain prolonged exertion. Therefore, if you wish to become a better jogger, swimmer, or biker, you should train mostly by jogging, swimming, or biking. Your performance will improve as your muscles develop the specific equipment they need to do the activity.

Bodies are shaped by the activities they perform.

▶ **KEY POINT** The components of fitness are flexibility, muscle strength, muscle endurance, and cardiorespiratory endurance. To build fitness, a person must engage in physical activity. Muscles adapt to activities they are called upon to perform repeatedly.

How Does Weight Training Benefit Health and Fitness?
Weight training has long been recognized as a method to build lean body mass and develop and maintain muscle strength and endurance. Additional benefits of weight training have recently emerged: progressive weight training also helps prevent and manage several chronic diseases, including cardiovascular disease, and enhances psychological well-being.[21]

By promoting strong muscles in the back and abdomen, weight training can improve posture and reduce the risk of back injury. Weight training can also help prevent the decline in physical mobility that often accompanies aging. Older adults, even those in their eighties, who participate in weight training programs not only gain muscle strength but also improve their muscle endurance, which enables them to walk significantly longer before exhaustion. Leg strength and walking endurance are powerful indicators of an older adult's physical abilities.

Yet another benefit is that weight training can help to maximize and maintain bone mass.[22] Research shows that even in women past menopause (when most women are losing bone), a one-year program of weight training improves bone density; the more weight lifted, the greater the improvement.[23]

Weight training can emphasize either muscle strength or muscle endurance. To emphasize muscle strength, combine high resistance (heavy weight) with a low number of repetitions. To emphasize muscle endurance, combine less resistance (lighter weight) with more repetitions. Weight training enhances performance in other sports, too. Swimmers can develop a more efficient stroke, and tennis players a more powerful serve, when they train with weights.[24]

▶ **KEY POINT** Weight training offers health and fitness benefits to adults. Weight training reduces the risk of cardiovascular disease, improves older adults' physical mobility, and helps maximize and maintain bone mass.

How Does Cardiorespiratory Training Benefit the Heart?
Although weight training provides some cardiovascular benefits, the kind of exercise most famous for improving the health of the heart is cardiorespiratory endurance training. You have felt your heartbeat pick up its pace during physical activity. Cardiorespiratory endurance determines how long you can remain active with an elevated heart rate—it

■ Extremes in physical activity, together with severely restricted energy intakes, may be detrimental to bone health in some young women. Such women risk developing the "female athlete triad," discussed in Controversy 9.

myoglobin the muscles' iron-containing protein that stores and releases oxygen in response to the muscles' energy needs.

weight training the use of free weights or weight machines to provide resistance for developing muscle strength and endurance. A person's own body weight may also be used to provide resistance as when a person does push-ups, pull-ups, or sit-ups. Also called *resistance training.*

---

*All muscles contain a variety of muscle fibers, but there are two main types—slow-twitch (also called *red fibers*) and fast-twitch (also called *white fibers*). Slow-twitch fibers contain extra metabolic equipment to perform fat-burning aerobic work; the fast-twitch type store extra glycogen for anaerobic work. Muscle fibers of one type take on some of the characteristics of the other as an adaptation to exercise.

Cardiorespiratory endurance is characterized by:

- Increased cardiac output and oxygen delivery.
- Increased heart strength and stroke volume.
- Slowed resting pulse.
- Increased breathing efficiency.
- Improved circulation.
- Reduced blood pressure.

■ The importance of HDL to heart health is a topic of the next chapter.

To take your resting pulse:

- Sit down and relax for five minutes before you begin. Using a watch or clock with a second hand, place your hand over your heart or your finger firmly over an artery at the underside of the wrist or side of the throat under the jawbone. Start counting your pulse at a convenient second, and continue counting for ten seconds. If a heartbeat occurs exactly on the tenth second, count it as one-half beat. Multiply by 6 to obtain the beats per minute. To ensure a true count:
- Use only fingers, not your thumb, on the pulse point (the thumb has a pulse of its own).
- Press just firmly enough to feel the pulse. Too much pressure can interfere with the pulse rhythm.

■ Epinephrine, discussed and defined in Chapter 3, is the major hormone that elicits the body's stress response, mobilizing fuels and readying the body for action.

cardiac output the volume of blood discharged by the heart each minute.

stroke volume the amount of oxygenated blood ejected from the heart toward body tissues at each beat.

is the ability of the heart and lungs to sustain a given physical demand. Working muscles need abundant oxygen to produce energy, and the heart and lungs work together to provide that oxygen. Thus, cardiorespiratory endurance training is aerobic.

The body's adaptation to the demands of regular aerobic activity involves a complex sequence of heart-healthy events. As cardiorespiratory endurance improves, the body delivers oxygen more efficiently. With cardiorespiratory endurance, the total blood volume and the number of red blood cells increase, so the blood can carry more oxygen. The heart muscle becomes stronger and larger, and its **cardiac output** increases. Each beat empties the heart's chambers more completely, so the heart pumps more blood per beat—its **stroke volume** increases. This makes fewer beats necessary, so the pulse rate falls. The muscles that inflate and deflate the lungs gain strength and endurance, so breathing becomes more efficient. Blood moves easily through the blood vessels because the muscles of the heart contract powerfully, and contraction of the skeletal muscles pushes the blood through the veins. Such improvements keep resting blood pressure normal. Figure 10-1 shows the major relationships among the heart, lungs, and muscles. The improvements that come with cardiorespiratory endurance also raise blood HDL, the lipoprotein associated with lower heart disease risk.

Which activities produce these beneficial changes? Effective activities elevate the heart rate, are sustained for longer than 20 minutes, and use most of the large-muscle groups of the body (legs, buttocks, and abdomen). Examples are swimming, cross-country skiing, rowing, fast walking, jogging, fast bicycling, soccer, hockey, basketball, in-line skating, lacrosse, and rugby. The ACSM guidelines for developing and maintaining cardiorespiratory fitness were given in Table 10-1 on page 362.

An informal pulse check can give you some indication of how conditioned your heart is. The average resting pulse rate for adults is around 70 beats per minute. Active people can have resting pulse rates of 50 or even lower. To take your pulse, follow the directions in the margin.

> ▶ **KEY POINT** Cardiorespiratory endurance training enhances the ability of the heart and lungs to deliver oxygen to the muscles. With cardiorespiratory endurance training, the heart becomes stronger, breathing becomes more efficient, and the health of the entire body improves.

The rest of this chapter describes the interactions between nutrients and physical activity. Nutrition alone cannot endow you with fitness or athletic ability, but along with the right mental attitude, it complements your effort to obtain them. Conversely, unwise food selections can stand in your way.

The Active Body's Use of Fuels

The fuels that support physical activity are glucose (from carbohydrate), fatty acids (from fat), and, to a small extent, amino acids (from protein). The body uses different mixtures of fuels depending on the intensity and duration of its activities and depending on its own prior training.

During rest, the body derives a little more than half of its energy from fatty acids, most of the rest from glucose, and a little from amino acids. During physical activity, the body adjusts its fuel mix to use the stored glucose of muscle glycogen. In the early minutes of an activity, muscle glycogen provides the majority of energy the muscles use to go into action. As activity continues, messenger molecules, including the hormone epinephrine, flow into the bloodstream to signal the liver and fat cells to liberate their stored energy nutrients, primarily glucose and fatty acids. Thus, hormones set the table for the muscles' energy feast, and the muscles help themselves to the fuels passing by in the blood.

FIGURE 10·1 DELIVERY OF OXYGEN BY THE HEART AND LUNGS TO THE MUSCLES

The more fit a muscle is, the more oxygen it draws from the blood. This oxygen comes from the lungs, so the person with more fit muscles extracts oxygen from inhaled air more efficiently than a person with less fit muscles. The cardiovascular system responds to increased demand for oxygen by building up its capacity to deliver oxygen. Researchers can measure cardiovascular fitness by measuring the amount of oxygen a person consumes per minute while working out. This measure of fitness, which indicates the person's maximum rate of oxygen consumption, is called $VO_{2\,max}$.

Air (O_2, CO_2), other gases

1. The respiratory system delivers oxygen to the blood.

2. The circulatory system carries oxygenated blood throughout the body.

4. The blood carries the carbon dioxide back to the lungs.

3. The muscles and other tissues obtain oxygen from the blood and release carbon dioxide into it.

Glucose Use and Storage

Both the liver and muscles store glucose as glycogen; the liver can also make glucose from fragments of other nutrients. Muscles hoard their glycogen stores—they do not release their glucose into the bloodstream to share with other body tissues, as the liver does. This hoarding is fortunate because a muscle that conserves its glycogen is prepared to act in emergencies, say, when running from danger, because muscle glucose fuels quick action. As activity continues, glucose from the liver's stored glycogen and dietary glucose absorbed from the digestive tract also become important sources of fuel for muscle activity.

$VO_{2\,max}$ the maximum rate of oxygen consumption by an individual (measured at sea level).

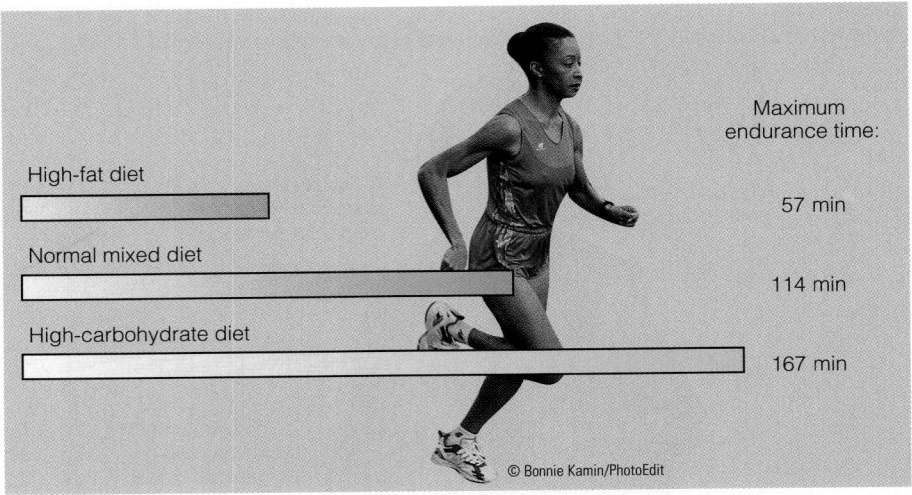

FIGURE 10·2 THE EFFECT OF DIET ON PHYSICAL ENDURANCE

A high-carbohydrate diet can increase an athlete's endurance. In this study, the high-fat diet provided 94 percent of calories from fat and 6 percent from protein; the normal mixed diet provided 55 percent of calories from carbohydrate; and the high-carbohydrate diet provided 83 percent of calories from carbohydrate.

High-fat diet

Normal mixed diet

High-carbohydrate diet

Maximum endurance time:

57 min

114 min

167 min

© Bonnie Kamin/PhotoEdit

The body constantly uses and replenishes its glycogen. The more carbohydrate a person eats, the more glycogen muscles store (up to a limit), and the longer the stores will last to support physical activity.

A classic report compared fuel use during physical activity by three groups of runners, each on a different diet. For several days before testing, one of the groups ate a normal mixed diet (55 percent of calories from carbohydrate); a second group ate a high-carbohydrate diet (83 percent of calories from carbohydrate); and the third group ate a high-fat diet (94 percent of calories from fat). As Figure 10-2 shows, the high-carbohydrate diet enabled the athletes to work longer before exhaustion. This study and many others established that a high-carbohydrate diet enhances an athlete's endurance by ensuring ample glycogen stores.

▶ KEY POINT   Glucose is supplied by dietary carbohydrate or made by the liver. It is stored in both liver and muscle tissue as glycogen. Total glycogen stores affect an athlete's endurance.

Activity Intensity, Glucose Use, and Glycogen Stores

The body's glycogen stores are much more limited than its fat stores. Glycogen supplies can easily support everyday activities but are limited to less than 2,000 calories of energy.[25] Fat stores, however, can usually provide more than 70,000 calories and fuel hours of activity without running out. How long a person's glycogen will last while exercising depends not only on diet but also on the intensity of the activity.

Anaerobic Use of Glucose   Intense activity—the kind that makes it difficult "to catch your breath," such as a quarter-mile race—uses glycogen quickly. Muscles must begin to rely more heavily on glucose, which can be partially broken down by **anaerobic** metabolism. Thus, the muscles begin drawing more heavily on their limited glycogen supply.

As the upper portion of Figure 10-3 shows, glucose can yield energy quickly in anaerobic metabolism. Anaerobic breakdown of glycogen yields energy to muscle tis-

anaerobic (AN-air-ROH-bic) not requiring oxygen. Anaerobic activity may require strength but does not work the heart and lungs very hard for a sustained period.

FIGURE 10·3 GLUCOSE AND FATTY ACIDS IN THEIR ENERGY-RELEASING PATHWAYS

Blood glucose or glycogen stores

Glucose enters here

Glucose fragments

Oxygen required here

O_2

Fatty acid fragments enter here

Some quick energy

Anaerobic part of metabolism

The first few steps of glucose breakdown yield a small amount of energy.

Aerobic part of metabolism

Glucose fragments and fatty acid fragments are broken down completely to yield a large amount of energy.

CO_2
H_2O

Many more steps yield more energy.

sue when energy demands outstrip the body's ability to provide energy aerobically, but it does so by lavishly spending the muscles' glycogen reserves.

Aerobic Use of Glucose   In contrast, *moderate* physical activity, such as easy jogging, uses glycogen slowly. The individual breathes easily, and the heart beats at a faster pace than at rest but steadily—the activity is aerobic. As the bottom half of Figure 10-3 shows, during aerobic metabolism muscles extract their energy from both glucose and fatty acids. By depending partly on fatty acids, moderate aerobic activity conserves glycogen stores. Joggers still use glycogen, however, and eventually they can run out of it.

Lactic Acid   During intense activity, anaerobic breakdown of glucose produces **lactic acid.** When production of lactic acid exceeds the ability of the muscles to use it, they release it, and it travels in the blood to the liver. There, liver enzymes convert the lactic acid back into glucose. Glucose can then return to the muscles to fuel additional activity.

At low intensities, lactic acid is readily cleared from the blood by the liver, but at higher intensities, lactic acid accumulates. When the rate of lactic acid production exceeds the rate of clearance, intense activity can be maintained for only one to three minutes (as in a 400- or 800-meter race or a boxing match). Lactic acid was long

lactic acid a compound produced during the breakdown of glucose in anaerobic metabolism.

blamed for a type of muscle fatigue, but recent thought disputes this idea. Muscles produce lactic acid during a type of fatigue, but the lactic acid does not cause the fatigue.[26]

▶ KEY POINT   The more intense an activity, the more glucose it demands. During anaerobic metabolism, the body spends glucose rapidly and accumulates lactic acid.

Activity Duration Affects Glucose Use

Glucose use during physical activity depends on the *duration* of the activity as well as its *intensity*. In the first 10 minutes or so of an activity, the active muscles rely almost entirely on their own stores of glycogen. Within the first 20 minutes or so of moderate activity, a person uses up about one-fifth of the available glycogen. As the muscles devour their own glycogen, they become ravenous for more glucose and increase their uptake of blood glucose dramatically. During moderate activity, blood glucose declines slightly, reflecting its use by the muscles.

A person who exercises moderately for longer than 20 minutes begins to use less glucose and more fat for fuel. Still, glucose use continues, and if the activity goes on long enough and at a high enough intensity, muscle and liver glycogen stores will run out almost completely. Glycogen depletion generally occurs after about two hours of vigorous exercise.* Physical activity can continue for a short time thereafter only because the liver scrambles to produce some glucose from available lactic acid and certain amino acids. This minimum amount of glucose may briefly forestall exhaustion, but when hypoglycemia accompanies glycogen depletion, it brings nervous system function almost to a halt, making activity impossible. Marathon runners call this "hitting the wall."

Maintaining Blood Glucose for Activity   To postpone exhaustion, endurance athletes must maintain their blood glucose concentrations for as long as they can. Three dietary strategies and one training strategy can help maintain glucose concentrations. One diet strategy is to eat a high-carbohydrate diet on a daily basis (see this chapter's Food Feature). Another is to take in some glucose during the activity, usually in fluid (see the next section). The third is to eat carbohydrate-rich foods (approximately 60 grams of carbohydrate) within a couple of hours after activity to boost the storage of glycogen. The training strategy involves training the muscles to store as much glycogen as they can, while supplying enough dietary glucose to enable them to do so (called *carbohydrate loading*, described in a later section).

Glucose during Activity   Glucose ingested before and during exhausting endurance activities (lasting more than 45 minutes) makes its way from the digestive tract to the working muscles, augmenting dwindling internal glucose supplies from the muscle and liver glycogen stores.[27] Especially during games such as soccer or hockey, which last for hours and demand repeated bursts of intense activity, athletes benefit from carbohydrate-containing drinks taken during the activity.[28]

Before concluding that sugar might be good for your own performance, consider first whether you engage in *endurance* activity. Do you run, swim, bike, or ski nonstop at a rapid pace for more than 45 minutes at a time, or do you compete in games lasting for hours? If not, the sugar picture changes. For an everyday jog or swim lasting less than 60 minutes, sugar probably won't help (or harm) performance. Even in athletes, extra carbohydrate does not benefit those who engage in sports in which fatigue is unrelated to blood glucose, such as 100-meter sprinting, baseball, casual basketball, and weight lifting.

■ Hypoglycemia, which is abnormally low blood glucose, was discussed in Chapter 4.

■ Four strategies can help to maintain blood glucose to support sports performance (for endurance athletes only):

1. Eat a high-carbohydrate diet regularly.
2. Take glucose (usually in sports drinks) periodically during endurance activities.
3. Eat carbohydrate-rich foods after performance.
4. Train the muscles to maximize glycogen stores.

For perspective, snack ideas providing 60 g of carbohydrate:

- 16 oz sports drink and a small bagel.
- 16 oz milk and four oatmeal-raisin cookies.
- 8 oz pineapple juice and a granola bar.

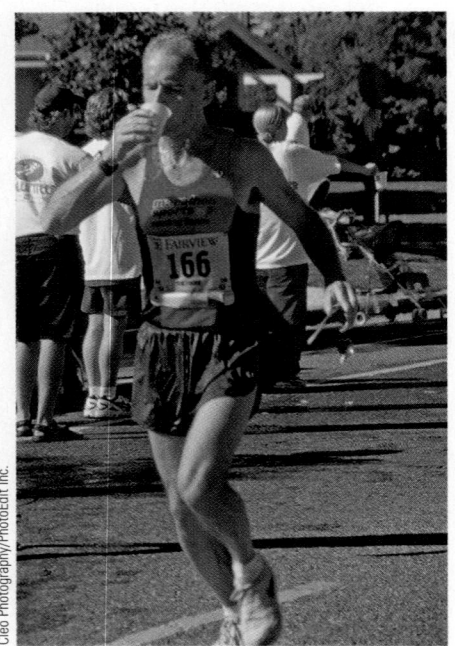

Those who compete in endurance activities require fluid and carbohydrate fuel.

---

*Here "vigorous exercise" means exercise at 75 percent of $VO_{2\,max}$.

TABLE 10-2 CARBOHYDRATE LOADING

Before the Event	Training Intensity	Training Duration	Dietary Carbohydrate
6 days	Moderate[a]	90 min	Normal (5 g/kg body weight)
5 days 4 days	Moderate[a]	40 min	Normal (5 g/kg body weight)
3 days 2 days	Moderate[a]	20 min	High-carbohydrate (10 g/kg body weight)
1 day	Rest	—	High-carbohydrate (10 g/kg body weight)

[a]Moderate intensity equals 70 percent $VO_{2\ max}$.

Carbohydrate Loading   Athletes whose sports routinely exhaust their glycogen stores sometimes use a technique called **carbohydrate loading** to trick their muscles into storing extra glycogen before competition. Carbohydrate loading can nearly double muscle glycogen concentrations. In general, the athlete tapers training during the week before the competition and then eats a high-carbohydrate diet during the three days just prior to the event.[29] See the carbohydrate-loading plan in Table 10-2.

In the carbohydrate loading plan in Table 10-2, glycogen storage occurs slowly, and athletes must alter their training in the days just before the event. In contrast, a group of researchers designed a quick method of carbohydrate loading that has produced promising preliminary results. The researchers found that athletes could attain above-normal concentrations of muscle glycogen by eating a high-carbohydrate diet (10 grams per kilogram of body weight) after a short (three minutes), but very intense bout of exercise.[30] More studies are needed to confirm these findings and to answer such questions as whether an exercise session of lesser intensity and shorter duration would accomplish the same result.

Extra glycogen gained through carbohydrate loading can benefit an athlete who must keep going for 90 minutes or longer. Those who exercise for shorter times simply need a regular high-carbohydrate diet. In a hot climate, extra glycogen confers an additional advantage: as glycogen breaks down, it releases water, which helps to meet the athlete's fluid needs.

Glucose after Activity   Eating high-carbohydrate foods *after* physical activity also enlarges glycogen stores. Train normally; then, within two hours after physical activity, consume a high-carbohydrate meal, such as a glass of orange juice and some graham crackers, toast, or cereal. This method accelerates the rate of glycogen storage by 300 percent for a while. This is especially important to athletes who train hard more than once a day. Timing is important—eating the meal after two hours have passed reduces the glycogen synthesis rate by almost half. For athletes who don't feel like eating right after exercise, **high-carbohydrate energy drinks** are available. These fruit-flavored drinks are higher in calories and carbohydrate than the regular sports drinks discussed in the Consumer Corner.

Chapter 4 introduced the glycemic effect and discussed some possible health benefits of eating a diet ranking low on the glycemic index. For athletes wishing to maximize muscle glycogen synthesis after strenuous training, however, eating foods with a high glycemic index (see the margin) may restore glycogen most rapidly.[31]

▶ **KEY POINT**   Physical activity of long duration places demands on the body's glycogen stores. Carbohydrate ingested before and during long-duration activity may help to forestall hypoglycemia and fatigue. Carbohydrate loading is a regimen of physical activity and diet that enables an athlete's muscles to store larger-than-normal amounts of glycogen to extend endurance. After strenuous training, eating foods with a high glycemic index may help restore glycogen most rapidly.

■ To make glycogen, muscles need carbohydrate, but they also need rest. Vary daily activity routines to work different muscles on different days.

Foods with a high glycemic index:
- Cornflakes.
- Mashed potatoes.
- Short-grain rice.
- Waffles.
- Watermelon.
- White bread.

carbohydrate loading a regimen of moderate exercise, followed by eating a high-carbohydrate diet, that enables muscles to temporarily store glycogen beyond their normal capacity; also called *glycogen loading* or *glycogen supercompensation*.

high-carbohydrate energy drinks fruit-flavored commercial beverages used to restore muscle glycogen after exercise or as a pregame beverage.

Degree of Training Affects Glycogen Use

Factors that affect glucose use during physical activity:

- Carbohydrate intake.
- Intensity and duration of the activity.
- Degree of training.

■ Chapter 4 described the action of insulin on blood sugar.

Training affects glycogen use during activity in at least two ways. First, muscles that deplete their glycogen stores through work adapt to store greater amounts of glycogen to support that work. Second, trained muscles burn more fat, and at higher intensities, than untrained muscles, so they require less glucose to perform the same amount of work.[32] A person attempting an activity for the first time uses up much more glucose per minute than an athlete trained to perform it. A trained person can work at high intensities for longer periods than an untrained person while using the same amount of glycogen.

People with diabetes should know how the moderating effect of physical training can influence their glucose metabolism. Those who must take insulin or insulin-eliciting drugs sometimes find that as their muscles adapt to physical activity, they can reduce their daily drug doses. Physical activity may also improve type 2 diabetes by helping the body lose excess fat.

▶ **KEY POINT** Highly trained muscles use less glucose and more fat than do untrained muscles to perform the same work, so their glycogen lasts longer.

To Burn More Fat during Activity, Should Athletes Eat More Fat?

As Figure 10-2 showed, the importance of a high-carbohydrate diet for endurance has long been recognized. When endurance athletes "fat load" by consuming high-fat, low-carbohydrate diets for one to three days, their performance is impaired because their small glycogen stores are depleted quickly.[33] Endurance athletes who adhere to a high-fat, low-carbohydrate diet for more than a week, however, adapt by relying more on fat to fuel activity. Even with fat adaptation, however, performance benefits have not been consistently shown.[34] In some cases, athletes on high-fat diets experience greater fatigue and perceive the activity to be more strenuous than athletes on high-carbohydrate diets.[35]

High-fat diets carry risks of heart disease, too. Physical activity offers some protection against cardiovascular disease, but even athletes can suffer heart attacks and strokes. Most nutrition experts agree that the potential for adverse health effects from prolonged high-fat diets makes them an unwise choice for athletes.

A diet that overly restricts fat is not recommended either. Athletes who restrict fat below 20 percent of total energy intake may fail to consume adequate energy and nutrients. Sports nutrition experts recommend that endurance athletes consume 20 to 30 percent of their energy from fat.[36] One expert says the message is "not that high-fat diets improve performance, but rather that very low-fat diets inhibit performance."*

As fuel for activity, body fat stores are more important than fat in the diet. Unlike the body's limited glycogen stores, fat stores can fuel hours of activity without running out; body fat is (theoretically) an unlimited source of energy. Even the lean bodies of elite runners carry enough fat to fuel several marathon runs.

Early in activity, muscles begin to draw on fatty acids from two sources—fats stored within the working muscles and fats from fat deposits such as the fat under the skin. Areas with the most fat to spare donate the greatest amounts of fatty acids to the blood (although they may not be the areas that one might choose to lose fat from). This is why "spot reducing" doesn't work: muscles do not own the fat that surrounds them. Fat cells release fatty acids into the blood for all the muscles to share. Proof is found in a tennis player's arms: the fatfolds measure the same in both arms, even though one arm has better-developed muscles than the other.

---

*The quotation is attributed to David R. Pendergast, in Cutting fat may crimp performance in endurance athletes, *Nutrition and the M.D.*, December 2000, pp. 3–4.

Intensity and Duration Affect Fat Use The *intensity* of physical activity also affects the percentage of energy contributed by fat because fat can be broken down for energy only by aerobic metabolism. When the intensity of activity becomes so great that energy demands surpass the ability to provide energy aerobically, the body cannot burn more fat. Instead, it burns more glucose.

The *duration* of activity also matters to fat use. At the start of activity, the blood fatty acid concentration falls, but a few minutes into an activity, the neurotransmitter norepinephrine signals the fat cells to break apart their stored triglycerides and to liberate fatty acids into the blood. After about 20 minutes of activity, the blood fatty acid concentration rises above the normal resting concentration. Only after the first 20 minutes, during this phase of sustained, submaximal activity, do the fat cells begin to shrink in size as they empty out their fat stores.

Factors that affect fat use during physical activity:
- Fat intake.
- Intensity and duration of the activity.
- Degree of training.

Degree of Training Affects Fat Use Training—repeated aerobic activity—stimulates the muscles to develop more fat-burning enzymes. Aerobically trained muscles burn fat more readily than untrained muscles. With aerobic training, the heart and lungs also become stronger and better able to deliver oxygen to the muscles during high-intensity activities. This improved oxygen supply, in turn, enables the muscles to burn more fat. Intense, prolonged activity may also increase your basal metabolic rate (BMR), as the Think Fitness feature explains.

▶ KEY POINT   Athletes who eat high-fat diets may burn more fat during endurance activity, but the risks to health outweigh any possible performance benefits. The intensity and duration of activity, as well as the degree of training, affect fat use.

THINK FITNESS

Can Physical Training Speed Up an Athlete's Metabolism?

ATHLETES IN TRAINING, whether endurance athletes or power athletes, expend huge amounts of energy each day while practicing. Common sense tells us that the harder an athlete works, the more energy the athlete spends. But what about *after* the work is done? Does the athlete continue to spend more energy at rest than a sedentary person or a casual exerciser? Research suggests the answer may be yes, for a limited time after intense, prolonged activity.[37] For example, intense endurance activity (at greater than 70 percent of $VO_{2\,max}$) seems to increase basal metabolic rate for anywhere from minutes to hours depending on the intensity and duration of the activity. The greater the intensity and the longer the duration of the activity, the longer the metabolic rate remains elevated.

Using Protein and Amino Acids to Build Muscles and Fuel Activity

Athletes use protein to build and maintain muscle and other lean tissue structures and, to a small extent, to fuel activity. The body handles protein differently during activity than during rest.

Protein for Building Muscle Tissue In the hours of rest that follow physical activity, muscles speed up their rate of protein synthesis—they build more of the proteins they need to perform the activity. Research shows that just as eating high-carbohydrate foods immediately after exercise accelerates muscle glycogen storage, eating protein, together with carbohydrate, enhances muscle protein synthesis.[38] And whenever the body rebuilds a part of itself, it must tear down the old structures to

Creatas/PictureQuest

Physical activity itself triggers the building of muscle proteins.

make way for the new ones. Physical activity, with just a slight overload, calls into action both the protein-dismantling and the protein-synthesizing equipment of individual muscle cells that work together to remodel muscles.

Dietary protein provides the needed amino acids for synthesis of new muscle proteins. As Chapter 6 pointed out, however, the true director of synthesis of muscle protein is physical activity itself. Repeated activity signals the muscle cells' genetic material to begin producing more of the proteins needed to perform the work at hand.

The genetic protein-making equipment inside the nuclei of muscle cells seems to "know" when proteins are needed. Furthermore, it knows *which* proteins are needed to support each type of physical activity. Apparently, the intensity and pattern of muscle contractions initiate signals that direct the muscles' genetic material to make particular proteins. For example, a weight lifter's workout sends the information that muscle fibers need added bulk for strength and more enzymes for making and using glycogen. A jogger's workout stimulates production of proteins needed for aerobic oxidation of fat and glucose. Muscle cells are exquisitely responsive to the need for proteins, and they build them conservatively only as needed.

Finally, after muscle cells have made all the decisions about which proteins to build and when, protein nutrition comes into play. During active muscle-building phases of training, a weight lifter might add between ¼ ounce and 1 ounce (between 7 and 28 grams) of protein to existing muscle mass each day. This extra protein comes from ordinary food.

Protein for Fuel　Not only do athletes retain more protein, but they also use a little more protein as fuel. Studies of nitrogen balance show that the body speeds up its use of amino acids for energy during physical activity, just as it speeds up its use of glucose and fatty acids. Protein contributes about 10 percent of the total fuel used, both during activity and during rest.

Diet Affects Protein Use during Activity　The factors that regulate how much protein is used during activity seem to be the same ones that regulate the use of glucose and fat. One factor is diet—a carbohydrate-rich diet spares protein from being used as fuel. Some amino acids can be converted into glucose when needed. If your diet is low in carbohydrate, much more protein will be used in place of glucose.

Intensity and Duration Affect Protein Use　The intensity and duration of the activity also affect protein use. Endurance athletes who train for over an hour a day, engaging in aerobic activity of moderate intensity and long duration, may deplete their glycogen stores by the end of their training and become more dependent on body protein for energy. In contrast, anaerobic strength training does not use more protein for energy but does demand more protein to build muscle. Thus, the protein needs of both endurance and strength athletes are higher than those of sedentary people, but not as high as the protein intakes many athletes consume.

Factors that affect protein use during physical activity:

- Carbohydrate intake.
- Intensity and duration of the activity.
- Degree of training.

Degree of Training Affects Protein Use　Finally, the extent of training also affects the use of protein. Particularly in strength athletes such as bodybuilders, the higher the degree of training, the less protein a person uses during activity at a given intensity.

▶ KEY POINT　Physical activity stimulates muscle cells to break down and synthesize protein, resulting in muscle adaptation to activity. Athletes use protein both for building muscle tissue and for energy. Diet, intensity and duration of activity, and training affect protein use during activity.

How Much Protein Should an Athlete Consume?

Although most athletes need somewhat more protein than do sedentary people, average protein intakes in the United States are high enough to cover those needs. Therefore, athletes in training should attend to protein needs but should back up the protein

TABLE 10·3 RECOMMENDED PROTEIN INTAKES FOR ATHLETES

	Recommendations (g/kg/day)	Protein Intakes (g/day)	
		Males	Females
DRI intake recommendation for adults	0.8	56	44
Recommended intake for power (strength or speed) athletes	1.6–1.7	112–119	88–94
Recommended intake for endurance athletes	1.2–1.6	84–112	66–88
U.S. average intake		95	65

NOTE: Daily protein intakes are based on a 70-kilogram (154-pound) man and a 55-kilogram (121-pound) woman.
SOURCES: Standing Committee on the Scientific Evaluation of Dietary Reference Intakes, Food and Nutrition Board, Institute of Medicine, *Dietary Reference Intakes for Energy, Carbohydrate, Fiber, Fat, Fatty Acids, Cholesterol, Protein, and Amino Acids* (Washington, D.C.: National Academies Press, 2002), pp. 10-52–10-53; Position of the American Dietetic Association, Dietitians of Canada, and the American College of Sports Medicine: Nutrition and athletic performance, *Journal of the American Dietetic Association* 100 (2000): 1543–1556.

with ample carbohydrate. Otherwise, they will burn off as fuel the very protein they wish to retain in muscle.

The DRI committee does not recommend greater-than-normal protein intakes for athletes, but other authorities do.[39] A joint position paper from the American Dietetic Association (ADA) and the Dietitians of Canada (DC) recommends protein intakes somewhat higher than the 0.8 gram of protein per kilogram of body weight recommended for sedentary people.[40] Table 10-3 compares these recommendations with the DRI recommended intake and average U.S. intakes and translates them into daily intakes for athletes.

After considering these recommendations, you may wonder whether your diet provides the protein you need. This chapter's Food Feature answers questions about choosing a performance diet. Meanwhile, relax. Athletes who eat a balanced, high-carbohydrate diet that provides enough total energy also consume enough protein—they do not need special foods, protein shakes, or supplements.

▶ KEY POINT   Although athletes need more protein than sedentary people, a balanced, high-carbohydrate diet provides sufficient protein to cover an athlete's needs.

Vitamins and Minerals— Keys to Performance

Many vitamins and minerals assist in releasing energy from fuels and transporting oxygen. In addition, vitamin C is needed for the formation of the protein collagen, the foundation material of bones and the cartilage that forms the linings of the joints and other connective tissues. Folate and vitamin B_{12} help build the red blood cells that carry oxygen to working muscles. Calcium and magnesium help make muscles contract, and so on. Do active people need extra nutrients to support their work? Do they need supplements?

Do Nutrient Supplements Benefit Athletic Performance?

Many athletes take supplements in the hope of improving their performance. A meta-analysis of more than 10,000 athletes involved in 15 sports at all levels found that about half of the athletes use vitamin-mineral supplements.[41] Elite athletes use supplements to a greater extent than college athletes, who, in turn, use supplements more than high school athletes. Supplement use by women exceeds that of men, and use by athletes exceeds that of the general population. One of the most common reasons athletes give for supplement use is "to improve performance."

Nutrient supplements do not enhance the performance of well-nourished athletes or active people. Deficiencies of vitamins and minerals, however, do impede performance. Regular, strenuous physical activity increases the demand for energy, but athletes and active people who eat enough nutrient-dense foods to meet energy needs also meet their vitamin and mineral needs. Active people eat more food; it stands to reason that with the right choices, they'll get more nutrients.

Some athletes believe that taking vitamin or mineral supplements just before competition will enhance performance. These beliefs are contrary to scientific reality. Most vitamins and minerals function as small parts of larger working units. After entering the blood, they have to wait for the cells to combine them with their appropriate other parts so that they can do their work. This takes time—hours or days. Vitamins or minerals taken right before an event do not improve performance, even if the person is actually suffering deficiencies of those nutrients.

■ Stringent weight requirements pose a risk of developing eating disorders. See Controversy 9.

Athletes who lose weight to meet low body-weight requirements, however, may consume so little food that they fail to obtain all the nutrients they need.[42] The practice of "making weight" is opposed by many health and fitness organizations, but for athletes who choose this course of action, a single daily multivitamin-mineral tablet that provides no more than the DRI recommendations for nutrients can be beneficial. In addition, some athletes do not eat enough food to maintain body weight during times of intense training or competition. For these athletes, too, a daily multivitamin-mineral supplement can be helpful.

Nutrients of Special Concern

In general, then, active people who eat well-balanced meals do not need vitamin or mineral supplements. Two nutrients, vitamin E and iron, do merit special attention, however, for different reasons. Vitamin E is addressed because so many athletes take supplements of it. Iron is discussed because some female athletes may be unaware that they need supplements.

■ The Tolerable Upper Intake Level (UL) for vitamin E is 1,000 mg per day.

Vitamin E   During prolonged, high-intensity physical activity, the muscles' consumption of oxygen increases tenfold or more, enhancing the production of damaging free radicals in the body.[43] Vitamin E is a potent fat-soluble antioxidant that vigorously defends cell membranes against oxidative damage. Some athletes take megadoses of vitamin E in hopes of preventing such oxidative damage to muscles. Supplementation with vitamin E does seem to protect against exercise-induced oxidative stress, but there is little evidence that vitamin E supplements can improve performance.[44]

■ Chapter 7 specified the risks of toxicity from vitamin pills and supplements.

Some evidence of possible benefit of vitamin E comes from a study of sled dogs competing in an extreme endurance race in Alaska.[45] In this race, the winning dog team averages 120 miles per day for about ten days. Researchers compared prerace blood vitamin E concentrations and performance in the dogs. Dogs with the highest vitamin E concentrations before the race were more likely to finish the race, and had less risk of being withdrawn for every mile run, than dogs with lower vitamin E concentrations. The vitamin E status of the dogs had no effect on speed, however. The authors speculated that vitamin E may have reduced fatigue in the dogs but stated that more research is needed to confirm their speculation. Clearly, more research is needed before drawing conclusions about vitamin E supplements for human athletes.

Iron and Performance Physically active young women, especially those who engage in endurance activities such as distance running, are prone to iron deficiency.[46] Physical activity can affect iron status in any of several ways. For one, iron losses in sweat may contribute to deficiency. For another, physical activity may cause small blood losses through the digestive tract, at least in some athletes. Perhaps more significant than these losses are the muscles' high demands for iron to make the iron-containing molecules of aerobic metabolism. Habitually low intakes of iron-rich foods, high iron losses through menstruation, as well as through the other routes mentioned, and extra demands can contribute to iron deficiency in young female athletes.

In contrast, a type of red blood cell destruction associated with physical activity does not impair iron status. Blood cells are squashed when body tissues (such as the soles of the feet) make high-impact contact with an unyielding surface (such as the ground). The iron released from the bursting, or hemolysis, of these red blood cells is recycled, however, and is not lost in the urine. Thus, exertional hemolysis, as it is called, rarely, if ever, contributes to anemia in athletes.[47]

Vegetarian female athletes are particularly vulnerable to iron insufficiency.[48] The bioavailability of iron is often poor in plant-based diets because such diets are high in fiber and phytic acid and because the nonheme iron in plant foods is not absorbed as well as the heme iron in animal-derived foods. Vegetarian diets are usually rich in vitamin C, however, which enhances iron absorption. To protect against iron deficiency, vegetarian athletes need to pay close attention to their intake of good dietary sources of iron (fortified cereals, legumes, nuts, and seeds) and include vitamin C–rich foods with each meal.[49] As long as vegetarian athletes, like all athletes, consume enough nutrient-dense foods, they can perform as well as anyone.

Iron deficiency impairs performance because iron helps deliver the muscles' oxygen. Insufficient oxygen delivery reduces aerobic work capacity, so the person tires easily. Whether marginal deficiency without clinical signs of anemia hinders physical performance is less clear.[50]

Early in training, athletes may develop low blood hemoglobin. This condition, sometimes called "sports anemia," is not a true iron-deficiency condition. Strenuous training promotes destruction of the more fragile, older red blood cells, and the resulting cleanup work reduces the blood's iron content temporarily. Strenuous activity also promotes increases in the fluid of the blood; with more fluid, the red blood cell count in a unit of blood drops. Most researchers view sports anemia as an *adaptive*, temporary response to endurance training. True iron-deficiency anemia requires treatment with prescribed iron supplements, but sports anemia goes away by itself, even with continued training.

The best strategy concerning iron is to determine individual needs. Many menstruating women border on iron deficiency even without the additional iron demand and losses incurred by physical activity. Teens of both genders, because they are growing, have high iron needs, too. For women and teens, then, prescribed supplements may be needed to correct a deficiency of iron that is confirmed by tests. (Medical testing is needed to eliminate nondietary causes of anemia, such as internal bleeding or cancer.)

> ▶ **KEY POINT**   Vitamins are essential for releasing the energy trapped in energy-yielding nutrients and for other functions that support physical activity. Active people can meet their vitamin needs if they eat enough nutrient-dense foods to meet their energy needs. Vitamin E may protect against oxidative stress, but there is little evidence that it can improve performance. Iron-deficiency anemia impairs physical performance because iron is the blood's oxygen handler. Sports anemia is a harmless temporary adaptation to physical activity.

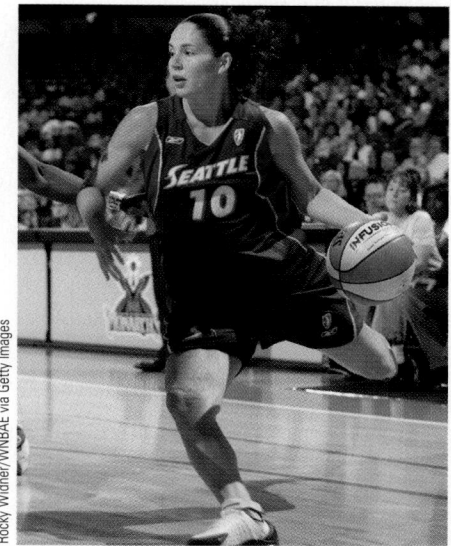

Female athletes may be at special risk of iron deficiency.

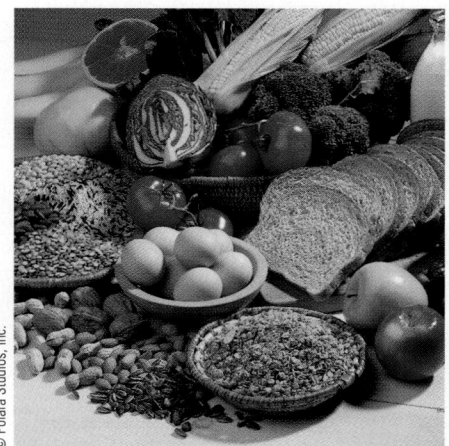

Foods like these are packed with the nutrients that active people need.

Fluids and Temperature Regulation in Physical Activity

The body's need for water far surpasses its need for any other nutrient. If the body loses too much water, its life-supporting chemistry is compromised.

The exercising body loses water primarily via sweat; second to that, breathing costs water, exhaled as vapor. During physical activity, both routes can be significant, and dehydration is a real threat. The first symptom of dehydration is fatigue. A water loss of even 1 to 2 percent of body weight can reduce a person's capacity to do muscular work.[51] A person with a water loss of about 7 percent is likely to collapse. The athlete who arrives at an event even slightly dehydrated starts out at a competitive disadvantage.

Temperature Regulation

As Chapter 8 pointed out, sweat cools the body. The conversion of water to vapor uses up a great deal of heat, so as sweat evaporates, it cools the skin's surface and the blood flowing beneath it.

In hot, humid weather, sweat may fail to evaporate because the surrounding air is already laden with water. Little cooling takes place and body heat builds up. In such conditions, athletes must take precautions to avoid **heat stroke.** Heat stroke is an especially dangerous accumulation of body heat with accompanying loss of body fluid. Three measures to prevent heat stroke are to drink enough fluid before and during the activity, rest in the shade when tired, and wear lightweight clothing that encourages evaporation.[52] The rubber or heavy suits sold with promises of weight loss during physical activity are dangerous because they promote profuse sweating, prevent sweat evaporation, and invite heat stroke. If you experience any of the symptoms of heat stroke listed in the margin, stop your activity, sip cold fluids, seek shade, and ask for help. The condition demands medical attention—it can kill.

In cold weather, **hypothermia,** or loss of body heat, can pose as serious a threat as heat stroke does in hot weather. Inexperienced runners participating in long races on cold or wet, chilly days are especially vulnerable to hypothermia. Slow runners can produce too little heat to keep warm, especially if their clothing is inadequate. Early symptoms of hypothermia include shivering and euphoria. As body temperature continues to fall, shivering stops, and weakness, disorientation, and apathy set in. People with these symptoms soon become helpless to protect themselves from further body heat losses. Even in cold weather, the body still sweats and needs fluids, but the fluids should be warm or at room temperature to help prevent hypothermia.

> ▶ KEY POINT   Evaporation of sweat cools the body. Heat stroke can be a threat to physically active people in hot, humid weather. Hypothermia threatens those who exercise in the cold.

Fluid Needs during Physical Activity

Endurance athletes can lose 2 or more quarts of fluid in every hour of activity, but the digestive system can absorb only about a quart or so an hour. Hence the athlete must hydrate before and rehydrate during and after activity to replace all the lost fluid. In hot weather, the digestive tract may not be able to absorb enough water fast enough to keep up with an athlete's sweat losses, and some degree of dehydration becomes inevitable. Athletes who are preparing for competition are often advised to drink extra fluids in the last few days of training before the event. The extra fluid is not stored in the body, but drinking extra ensures maximum tissue hydration at the start of the event. Any coach or athlete who withholds fluids during practice for any reason takes a great risk and is subject to sanctions by the American College of Sports Medicine.

Symptoms of heat stroke:
- Clumsiness.
- Confusion or loss of consciousness.
- Dizziness.
- Headache.
- Internal (rectal) temperature above 104° Fahrenheit.
- Nausea.
- Stumbling.
- Sudden cessation of sweating (hot, dry skin).

heat stroke an acute and life-threatening reaction to heat buildup in the body.

hypothermia a below-normal body temperature.

TABLE 10·4	HYDRATION SCHEDULE FOR PHYSICAL ACTIVITY	
When to Drink	Approximate Amount of Fluid
2 hr before activity	2 to 3 c
15 min before activity	1 to 2 c
Every 15 min during activity	1 to 1½ c
After activity	2 to 3 c for each pound of body weight lost

SOURCES: R. Murray, Fluid and electrolytes, in C. A. Rosenbloom, ed., *Sports Nutrition: A Guide for the Professional Working with Active People,* 3rd ed. (Chicago: The American Dietetic Association, 2000), pp. 95–106; National Athletic Trainer's Association, Position statement: Fluid replacement for athletes, *Journal of Athletic Training* 35 (2000): 212–224.

Active people need extra fluid, even in cold weather.

Athletes who rely on thirst to govern fluid intake can easily become dehydrated. During activity thirst becomes detectable only *after* fluid stores are depleted. Don't wait to feel thirsty before drinking. Table 10-4 presents one schedule of hydration for physical activity. To find out how much water you need to replenish losses, weigh yourself before and after the activity. The difference is all water. Two cups (16 ounces) of fluid weigh about a pound.

▶ KEY POINT   Physically active people lose fluids and must replace them to avoid dehydration. Thirst indicates that water loss has already occurred.

Water

What is the best fluid to support physical activity? The best drink for most active bodies is just plain cool water, for two reasons: (1) water rapidly leaves the digestive tract to enter the tissues, and (2) it cools the body from the inside out. Endurance athletes are an exception: they need more from their fluids than water alone. The first priority for endurance athletes should always be replacement of fluids to prevent life-threatening heat stroke. But endurance athletes also need carbohydrate to supplement their limited glycogen stores, so glucose is important, too. This chapter's Consumer Corner compares water and sports drinks as fluid sources for endurance athletes.

▶ KEY POINT   Water is the best drink for most physically active people, but endurance athletes need drinks that supply glucose as well as fluids.

Electrolyte Losses and Replacement

During physical activity, the body loses electrolytes—the minerals sodium, potassium, and chloride—in sweat. Beginners lose these electrolytes to a much greater extent than do trained athletes. The body's adaptation to physical activity includes better conservation of these electrolytes. To replenish lost electrolytes, a person ordinarily needs only to eat a regular diet that meets energy and nutrient needs. In events lasting more than 45 minutes, sports drinks may be needed to replace fluids and electrolytes. Salt tablets can worsen dehydration and impair performance; they increase potassium losses, irritate the stomach, and cause vomiting. Athletes should avoid them.

▶ KEY POINT   The body adapts to compensate for sweat losses of electrolytes. Athletes are advised to use foods, not supplements, to make up for these losses.

Fluid replacement tips:

- To ensure adequate fluid intake without being distracted during an event, try this technique. Before the event, fill a 32 oz (4 c) water bottle and place two colored rubber bands to mark the bottle into thirds. Finish off the first segment of the bottle in the first 30 minutes of activity; finish the next segment in the next 30 minutes, and the remainder in the next. Have someone refill the bottle if activity lasts longer than 90 minutes.

- The urine of a person who is adequately hydrated is the color of pale lemonade. Urine the color of apple juice indicates slight dehydration.

SOURCE: Ideas from J. Berning, nutrition professor and sports nutrition consultant, personal communication, 1999.

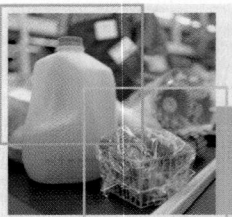

CONSUMER CORNER

What Do Sports Drinks Have to Offer?

MORE THAN 20 **sports drinks** or **fluid replacers** (see Table 10-5) compete for their share of the $1 billion market. What do sports drinks offer? First, and most important, sports drinks offer fluids to help you offset the loss of fluids during physical activity, but plain water can do this, too.

Second, sports drinks supply glucose. A beverage that supplies glucose in some form can be useful during endurance activity lasting 45 minutes or more, during intense activity, or during prolonged competitive games that demand repeated intermittent activity.[1] Not just any sweet beverage can meet this need, however, because a carbohydrate concentration greater than 8 percent can delay fluid emptying from the stomach and thereby slow down the delivery of water to the tissues. Most sports drinks contain an appropriate amount to ensure water absorption— about 7 percent glucose (about half the sugar of ordinary soft drinks, or about 5 teaspoons in each 12 ounces).

Third, sports drinks offer sodium and other electrolytes to help replace those lost during physical activity. Sodium in sports drinks also helps to improve palatability and fluid retention and maintains the osmotic drive for drinking fluid. This makes sense physiologically because the sensation of thirst is a function of changes in blood sodium concentration (and plasma osmolality).[2] Most sports drinks are relatively low in sodium (55 to 110 milligrams per serving), however, so healthy people who choose to use these beverages run little risk of excessive intake. Most athletes do not need to replace the other minerals lost in sweat immediately; a meal eaten within hours of competition replaces these minerals soon enough.

In addition, most sports drinks taste good. Manufacturers reason that if a drink tastes good, people will drink more, thereby ensuring adequate hydration. Fluids that are flavored, sweetened, and cool stimulate fluid intake. Finally, sports drinks can also provide a psychological edge to people who associate them with success in sports.

Thus, for athletes who exercise intensely or for 45 minutes or more, sports drinks offer an advantage over water. **Fitness water** does not provide glucose and electrolytes as sports drinks do, but some active people may prefer its light flavor to plain water.

TABLE 10·5 TERMS RELATED TO SPORT DRINKS

- **fitness water** water that is lightly flavored to enhance taste; often contains small amounts of vitamins.
- **sports drinks (fluid replacers)** beverages specifically developed for athletes to replace fluids and electrolytes and to provide glucose before, during, and after physical activity, especially endurance activity.

Sodium Depletion

When athletes compete in endurance sports lasting longer than three hours, replenishing electrolytes becomes crucial. If athletes sweat profusely over a long period of time and do not replace lost sodium, a dangerous condition of sodium depletion, known as **hyponatremia**, may result. The symptoms of hyponatremia are similar to, but not the same as, those of dehydration (see the margin, p. 379). Recent research shows that some athletes who sweat profusely may also lose more sodium in their sweat than others—and are prone to debilitating heat cramps.[53] These athletes lose twice as much sodium in sweat as athletes who don't cramp. Depending on individual variation, exercise intensity, and changes in ambient temperature and humidity, sweat rates can exceed 3 quarts per hour.[54]

Hyponatremia may also occur when endurance athletes drink such large amounts of water over the course of a long event that they overhydrate, diluting the body's fluids to such an extent that the sodium concentration becomes extremely low.[55] During long competitions, when athletes lose sodium through heavy sweating *and* consume excessive amounts of liquids, especially water, hyponatremia becomes likely. Water intoxication, introduced in Chapter 8, can result in life-threatening hyponatremia.

Some athletes may be vulnerable to hyponatremia even when they drink sports drinks during an event.[56] Sports drinks do contain sodium, but as the Consumer Cor-

■ Table 8-1 on page 267 lists the symptoms of dehydration.

hyponatremia (HIGH-poh-na-TREE-mee-ah) a decreased concentration of sodium in the blood (*hypo* means "below"; *natrium* means "sodium [Na]"; *emia* means "blood").

ner points out, their sodium content is low. In some cases, it is too low to replace sweat losses. Still, sports drinks do offer more sodium than plain water.

To prevent hyponatremia, endurance athletes need to replace sodium during prolonged events. They should favor sports drinks over water and eat pretzels in the last half of a long race.[57] Some may need beverages with higher sodium concentrations than commercial sports drinks. In the days before the event, especially an event in the heat, athletes should not restrict salt in their diets.

▶ KEY POINT During events lasting longer than three hours, athletes need to pay special attention to replacing sodium losses to prevent hyponatremia.

Other Beverages

Some drinks, such as iced tea, deliver caffeine along with fluid. Moderate doses of caffeine (about the amount in 2 cups of coffee) one hour prior to activity sometimes seem to assist athletic performance and other times seem to have no effect. Athletic competitions limit the amount of caffeine that can be consumed within two hours of an event. (More about caffeine's effects on performance can be found in this chapter's Controversy, and the amounts of caffeine in foods and beverages are listed in Controversy 14.)

Carbonated beverages are not a good choice for meeting an athlete's fluid needs. Although they are composed largely of water, the air bubbles from the carbonation make a person feel full quickly and so may limit fluid intake.

Athletes sometimes drink beverages that contain alcohol, but these beverages are inappropriate as fluid replacements. Alcohol is a diuretic. It promotes the excretion of water; of vitamins such as thiamin, riboflavin, and folate; and of minerals such as calcium, magnesium, and potassium—exactly the wrong effects for fluid balance and nutrition. It is hard to overstate alcohol's detrimental effects on physical activity. It impairs temperature regulation, making hypothermia or heat stroke much more likely. It alters perceptions and slows reaction time. It depletes strength and endurance and deprives people of their judgment, thereby compromising their safety in sports. Many sports-related fatalities and injuries each year involve alcohol.

▶ KEY POINT Caffeine-containing drinks within limits may not impair performance, but water and fruit juice are preferred. Alcohol use can impair performance in many ways and is not recommended.

Symptoms of hyponatremia:

- Bloating, puffiness from water retention (shoes tight, rings tight).
- Confusion.
- Seizure.
- Severe headache.
- Vomiting.

Beer facts:

- Beer is not carbohydrate-rich. Beer is calorie-rich, but only one-third of its calories are from carbohydrates. The other two-thirds are from alcohol.
- Beer is mineral-poor. Beer contains a few minerals, but to replace those lost in sweat, athletes need good sources such as fruit juices.
- Beer is vitamin-poor. Beer contains tiny traces of some B vitamins, but it cannot compete with rich food sources.
- Beer causes fluid losses. Beer is a fluid, but alcohol is a diuretic and causes the body to lose more fluid in urine than is provided by the beer.

■ Read about alcohol's effects on the brain in Controversy 3.

Choosing a Performance Diet

MANY DIFFERENT DIETS can support an athlete's performance. Food choices must obey the rules for diet planning, however.

Nutrient Density

First, athletes need a diet composed mostly of nutrient-dense foods, the kind that supply a maximum of vitamins and minerals for the energy they provide. When athletes eat mostly refined, processed foods that have suffered nutrient losses and contain too much added sugar and solid fat, their nutrition status suffers. Even if foods are fortified or enriched, manufacturers cannot replace the whole range of nutrients and nonnutrients lost in refining. For example, manufacturers mill out much of a food's original magnesium and chromium but do not replace them. This doesn't mean that athletes can never choose a white bread, bologna, and mayonnaise sandwich, but only that later they should eat a large salad or big portions of vegetables and whole grains and drink a glass of milk to compensate. The nutrient-dense foods will provide the magnesium and chromium; the bologna sandwich provides extra energy, mostly from fats.

Balance

Athletes must eat for energy, and their energy needs can be immense. Athletes need full glycogen stores, and they need to strive to prevent heart disease and cancer by limiting fat, especially saturated fat. To serve these special needs, a diet that is high in carbohydrate (60 to 70 percent of total calories), moderate in fat (20 to 30 percent), and adequate in protein (10 to 20 percent) works best. Even if you do not compete in glycogen-depleting events, such a diet provides adequate fiber while supplying abundant nutrients and energy.

With these principles in mind, compare the two 500-calorie sandwich meals in the margin. The trick to getting enough carbohydrate energy is easy, at least in theory: just reduce the amount of fat and meat in a meal, and let carbohydrate-rich foods fill in for them.

Adding carbohydrate-rich foods is a sound and reasonable option for increasing energy intake, up to a point. It becomes unreasonable when the person cannot eat enough food to met energy needs. At that point, the person can add more food energy into the diet by adding refined sugars, oils, or liquid meals. Still, these energy-rich additions must be superimposed on nutrient-rich choices; energy alone is not enough.

Some athletes use commercial high-carbohydrate liquid supplements to obtain the carbohydrate and energy needed for heavy training and top performance. Most of these products contain **glucose polymers** and about 18 to 24 percent carbohydrate. These supplements do not *replace* regular food; they are meant to be used in *addition* to it. Unlike the sports beverages discussed in the Consumer Corner, these high-carbohydrate supplements are too concentrated in carbohydrate to be used for fluid replacement.

Protein

In addition to carbohydrate, athletes need protein. Meats and milk products head the list of protein-rich foods, but suggesting that athletes eat more than the recommended servings of meat would be shortsighted advice. Athletes must protect themselves from heart disease, and even lean meats contain saturated fat. Besides, the extra servings of carbohydrate-rich foods such as legumes, grains, and vegetables that an athlete needs to meet energy requirements also boost protein intakes.

Earlier in this chapter, Table 10-3 (p. 373) showed recommended protein intakes for a 55-kilogram female athlete and a 70-kilogram male athlete. An athlete weighing 70 kilograms who engages in vigorous physical activity on a daily basis could require 3,000 to 5,000 calories per day. As a general rule, endurance athletes should aim for an average intake of 50 calories per kilogram (2.2 pounds) of body weight (23 calories per pound of body weight). Others may need more. To meet such an energy requirement, an athlete should select from a variety of nutrient-dense foods. Figure 10-4 provides an example of how foods that provide the extra nutrients athletes need can be added to a lower calorie eating pattern to attain a 3,300-calorie diet. These meals sup-

■ Small daily choices, when made consistently, enhance an athlete's nutritional health.

Compare and decide which best meets your needs:

- 1 sandwich of 2 slices bologna, 2 slices white bread, 2 tbs mayonnaise (525 cal, 9% protein, 23% carbohydrate, 68% fat).

or

- 2 sandwiches of 2 slices lean ham, 4 slices whole-wheat bread, 2 tsp mayonnaise (503 cal, 20% protein, 51% carbohydrate, 29% fat).

■ Looking for an amino acid supplement that rates a perfect score of 100 for protein quality? Try 1 oz of chicken breast—it provides almost 10,000 mg of amino acids in perfect complement for use by the human body.

glucose polymers compounds that supply glucose, not as single molecules, but linked in chains somewhat like starch. The objective is to attract less water from the body into the digestive tract.

FIGURE 10·4   HIGH-CARBOHYDRATE MEALS FOR ATHLETES

2600 Calories

Breakfast:
1 c shredded wheat.
1 c 1% low-fat milk.
1 small banana.
1 c orange juice.

Modifications   **3300 Calories**

The regular breakfast *plus*:
2 pieces whole-wheat toast.
1/2 c orange juice.
4 tsp jelly.

Lunch:
1 turkey sandwich on
   whole-wheat bread.
1 c 1% low-fat milk

The regular lunch *plus*:
1 turkey sandwich.
1/2 c 1% low-fat milk.
Large bunch of grapes.

Snack:
2 c plain popcorn.
A smoothie made from:
   1 1/2 c apple juice.
   1 1/2 frozen banana.

The regular snack *plus*:
1 c popcorn.

Dinner:
Salad:
   1 c spinach, carrots, and
      mushrooms.
   1/2 c garbanzo beans.
   1 tbs sunflower seeds.
   1 tbs ranch dressing.
1 c spaghetti with meat sauce.
1 c green beans.
1 slice Italian bread.
2 tsp soft margarine.
1 1/4 c strawberries.
1 c 1% low-fat milk.

© Polara Studios, Inc. (all)

The regular dinner *plus*:
1 corn on the cob.
1 slice Italian bread.
2 tsp soft margarine.
1 piece angel food cake.
1 tbs whipping cream.

Total cal: 2,600
62% cal from carbohydrate
23% cal from fat
15% cal from protein

Total cal: 3,300
63% cal from carbohydrate
22% cal from fat
15% cal from protein

All vitamin and mineral intakes exceed the
recommendations for both men and women.

ply about 125 grams of protein, equivalent to the highest recommended intake for an athlete weighing 160 pounds. For those with reasonable diets, protein is rarely a problem.

The meals in Figure 10-4 provide 63 percent of their calories from carbohy-drate. Athletes who train exhaustively for endurance events may want to aim for somewhat higher carbohydrate levels— from 65 to 75 percent. Notice that break-fast, though low in saturated fat, is filling and hearty. Current thinking supports the idea that athletes benefit from such a morning start. If you train early in the morning, try splitting breakfast into two parts. An hour or so before training, eat some toast, juice, and fruit. Later, after your workout, come back for the cereal and milk.

Planning an Athlete's Meals

Table 10-6 shows some sample eating patterns for athletes at various high-energy and high-carbohydrate intakes. These plans are effective only if the user chooses foods to provide nutrients as well as energy: extra milk for calcium and riboflavin; many servings of fruit for folate and vitamin C; energy-rich vegetables such as sweet potatoes, peas, and legumes; modest portions of lean meat for iron and other vitamins and minerals; and whole grains for B vitamins, magnesium, zinc, and chromium. In addition, these foods provide plenty of electrolytes.

A trick used by professional sports nutritionists to maximize athletes' intakes of energy and carbohydrates is to make sure that vegetable and fruit choices are as dense as possible in both nutrients and energy. A whole cupful of iceberg lettuce supplies few calories or nutrients, but a half-cup portion of cooked sweet potatoes is a powerhouse of vitamins, minerals, and carbohydrate energy. Similarly, it takes a whole cup of cubed melon to equal the calories and carbohydrate in a half-cup of canned fruit. Small choices like these, made consistently, can contribute significantly to nutrient, energy, and carbohydrate intakes.

Before competition, athletes may eat particular foods or practice rituals that convey psychological advantages. One eats steak the night before; another spoons up honey at the start of the event. As long as these foods or rituals remain harmless, they should be respected. Still, science has recommendations for the **pregame meal.** The foods should be carbohydrate-rich and the meal light (300 to 800 calories). It should be easy to digest and should contain fluids. Breads, potatoes, pasta, and fruit juices—carbohydrate-rich foods low in fat, protein, and fiber—form the basis of the pregame meal (see Figure 10-5 for some examples). Bulky, fiber-rich foods such as raw vegetables and high-fiber cereals, although usually desirable, are best avoided just before competition. Such foods can cause stomach discomfort during performance. The competitor should finish eating three to four hours before competition to allow time for the stomach to empty before exertion.

What about drinks or candylike sport bars claiming to provide "complete" nutrition? These mixtures of carbohydrate, protein (usually amino acids), fat, some fiber, and certain vitamins and minerals may taste good and provide additional food energy for a game or for weight gain. They fall short of providing "complete" nutrition,

pregame meal a meal eaten three to four hours before athletic competition.

TABLE 10·6 HIGH-CARBOHYDRATE EATING PATTERNS FOR ATHLETES

	Number of Servings for a Daily Energy Intake of:					
Food Group	1,500 cal	2,000 cal	2,500 cal	3,000 cal	3,500 cal	4,000[a] cal
Milk (c)	3	3	4	4	4	4
Fruit (c)	2½	3	3½	4½	5	6
Vegetable (c)	1½	2½	1½	2½	3	3½
Grain (oz)	7	11	16	18	20	24
Oils (tsp)[b]	2	3	5	6	8	10
Meat (oz)	5	5	5	5	6	6
Percent carbohydrate:	58%	58%	63%	64%	60%	62%

[a]A way to add more energy to the diet without adding much bulk is to snack on milkshakes or "complete meal" liquid supplements (see the text).
[b]Soft margarine, oil, or the equivalent.

FIGURE 10·5 EXAMPLES OF HIGH-CARBOHYDRATE PREGAME MEALS

Pregame meals should be eaten three to four hours before the event and provide 300 to 800 calories, primarily from carbohydrate-rich foods. Each of these sample meals provides at least 65 percent of total calories from carbohydrate.

300-calorie meal
1 large apple
4 saltine crackers
1½ tbs reduced-fat
   peanut butter

500-calorie meal
1 large whole-wheat bagel
2 tbs jelly
1½ c low-fat milk

750-calorie meal
1 large baked potato
2 tsp soft margarine
1 c steamed broccoli
1 c mixed carrots and green
   peas
5 vanilla wafers
1½ c apple or pineapple juice

however, because they lack many of real food's nutrients and the nonnutrients that benefit health. These products may provide one single advantage for active people—they are easy to eat in the hours before competition. They are expensive, however.

As for "complete" drinks, Table 10-7 demonstrates that there is no point in paying high prices for fancy brand-name drinks. Homemade shakes are inexpensive and easy to prepare, and they perform every bit as well as commercial products. Don't drop a raw egg in the blender, though, because raw eggs often carry bacteria that cause food poisoning.

If you want to excel physically, apply the most accurate nutrition knowledge along with dedication to rigorous training. A diet that provides ample fluid and consists of a variety of nutrient-dense foods in quantities to meet energy needs will enhance not only athletic performance but overall health as well. Training and genetics being equal, who would win a competition—the person who habitually consumes less than the amounts of nutrients needed or the one who arrives at the event with a long history of full nutrient stores and well-met metabolic needs?

TABLE 10·7 COMMERCIAL AND HOMEMADE MEAL REPLACERS COMPARED

	Cost (U.S.)	Energy (cal)	Protein (g)	Carbohydrate (g)	Fat (g)
12-ounce commercial liquid meal replacer[a]	about $2 per serving	360	15 (17% of calories)	55 (61%)	9 (22%)
12-ounce homemade milkshake[b]	about 50¢ per serving	330	15 (18% of calories)	53 (63%)	7 (19%)

[a]Average values for three commercial formulas.
[b]Home recipe: 8 oz fat-free milk, 4 oz ice milk, 3 heaping tsp malted milk powder. For even higher carbohydrate and calorie values, blend in ½ mashed banana or ½ c other fruit. For athletes with lactose intolerance, use lactose-reduced milk or soy milk and chocolate or other flavored syrup, with mashed banana or other fruit blended in.

MEDIA MENU

@ 🏠 ◀ ▶

Nutrition Connections

The Nutrition Connections CD provides access to these additional resources.

Do It!
The Do It! exercise for this chapter, *Choose What to Drink during Physical Activity,* gives you practice in evaluating the nutritional content of sports drinks.

Interaction
The animation, *Exercise Intensity and Fuel Use,* explores the types of energy used to fuel the body during exercise and the way fuel sources vary depending upon the intensity of the exercise.

Chapter Quiz
Take the quiz for this chapter to assess your understanding of chapter concepts.

Nutrition on the Net
Annotated links to reputable nutrition-related websites.

Online Glossary
Test your knowledge of key vocabulary through this comprehensive, interactive glossary.

Internet Connections

A variety of helpful study tools for this chapter are available at our website
http://nutrition.wadsworth.com/ncc10

Practice Test
Take the practice test for this chapter and gauge your knowledge of key concepts; answers are provided.

Student Notebook
A lecture notebook is available to help organize note taking while reading or attending lecture.

Study Tools
Study resources for students include Chapter Summaries, Crossword Puzzles, Flashcards, Glossary, Internet Exercises, and Learning Objectives.

SELF CHECK

Answers to these Self Check questions are in Appendix G.

1. Which of the following provides most of the energy the muscles use in the early minutes of activity?
   a. fat
   b. protein
   c. glycogen
   d. b and c

2. Which diet has been shown to increase an athlete's endurance?
   a. high-fat diet
   b. normal mixed diet
   c. high-carbohydrate diet
   d. Diet has not been shown to have any effect.

3. Which of the following stimulates synthesis of muscle cell protein?
   a. physical activity
   b. a high-carbohydrate diet
   c. a high-protein diet
   d. amino acid supplementation

4. All of the following statements concerning beer are correct *except:*
   a. beer is poor in minerals
   b. beer is poor in vitamins
   c. beer causes fluid losses
   d. beer gets most of its calories from carbohydrates

5. A person who exercises moderately longer than 20 minutes begins to:
   a. use less glucose and more fat for fuel
   b. use less fat and more protein for fuel
   c. use less fat and more glucose for fuel
   d. use less protein and more glucose for fuel

6. Weight training to improve muscle strength and endurance has no effect on maintaining bone mass.   T   F

7. The average resting pulse rate for adults is around 70 beats per minute, but the rate is higher in active people.   T   F

8. An athlete should drink extra fluids in the last few days of training before an event in order to ensure proper hydration.   T   F

9. Research does not support the idea that athletes need supplements of vitamins to enhance their performance.   T   F

10. Aerobically trained muscles burn fat more readily than untrained muscles.   T   F

ERGOGENIC AIDS: BREAKTHROUGHS, GIMMICKS, OR DANGERS?

ATHLETES CAN BE sitting ducks for quacks. Many are willing to try almost anything that is sold with promises of producing a winning edge, so long as they perceive it to be safe.*[1] Store shelves and the Internet abound with heavily advertised **ergogenic aids,** each striving to appeal to performance-conscious people: protein powders, amino acid supplements, caffeine pills, steroid replacers, "muscle-builders," vitamins, and more. Some athletes spend huge sums of money on these products, often heeding advice from a "voice of experience" such as a trusted coach or mentor. Table C10-1 (pp. 386–387) defines the terms in boldface type in this section and lists many more substances promoted as ergogenic aids. Do these products work as advertised? And most importantly, are they safe?

This Controversy focuses on the scientific evidence for and against a few of the most common dietary supplements for athletes and exercisers. In light of the evidence, this section concludes with what most people already know: consistent training and sound nutrition serve an athlete better than any pill, powder, or other supplement.

Paige and DJ

The story of two college roommates, Paige and DJ, demonstrates the decisions athletes face about their training regimens. After enjoying a freshman year when the first things on their agendas were tailgate parties and the last thing—the very last thing—was exercise, Paige and DJ have taken up running to shed the "freshman 15" pounds that have crept up on them. Their friendship, once defined by bonding over extra cheese pizzas and fried chicken wing snacks, now focuses on five-kilometer races. Both young athletes now compete to win.

Paige and DJ take their nutrition regimens and prerace preparations seriously, but they are as opposite as the sun and moon: DJ takes a traditional approach, sticking to the tried-and-true advice of her older brother, an all-state track and field star. He tells her to train hard, eat a nutritious diet, get enough sleep, drink plenty of fluid on race day, and warm up lightly for ten minutes before the starting gun. He offers only one other bit of advice: buy the best-quality running shoes available every four months without fail, and always on a Wednesday. Many an athlete admits laughingly to such superstitions as wearing "lucky socks" for the mental boost of a good luck charm.

Paige finds DJ's routine boring and woefully out-of-date. Paige surfs the Internet for the latest supplements and ergogenic aids advertised in her fitness magazines. She mixes carnitine and protein powders into her complete meal replacement drinks for the promised bonus muscle tissue to help at the weight bench, and she takes a handful of "ergogenic" supplements to get "pumped up" for a race. Her counter is cluttered with bottles of amino acids, caffeine pills, chromium picolinate, and even herbal steroid replacers. Sure, it takes money (a *lot* of money) to purchase the products and time to mix the potions and return the occasional wrong shipment—often cutting into her training time. And the high cost leaves little room in her budget for extras like new running shoes. Still, Paige feels smugly smart in her modern approach. Surely, she will win the most races.

Ergogenic Aids

Is Paige right in thinking that she can gain an athletic edge from supplements? Is she safe in taking them? For the large majority of ergogenic aids, research findings do not support the claims

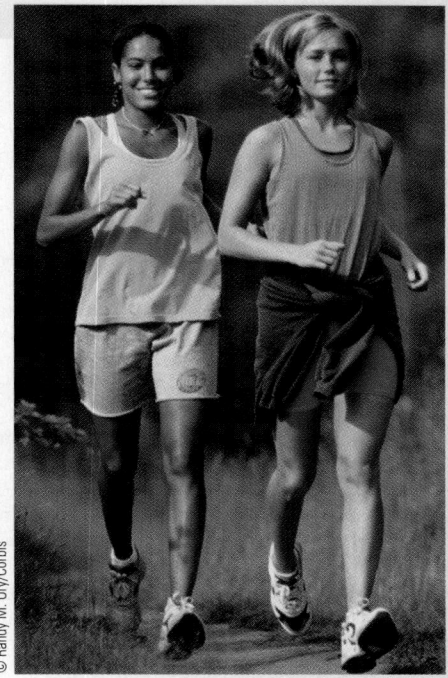
Training serves an athlete better than any pills or powders.

made for them. Athletes who hear that a product is ergogenic should ask who is making the claim and who will profit from the sale. Sometimes, even with careful deliberations, savvy shoppers find it difficult to distinguish valid claims from bogus ones. It's easy to see why Paige is misled by advertisements in fitness magazines—they often appear to be informative articles, and they present a mixture of valid and invalid ideas that is hard to sort out. Colorful anatomical figures, graphs, and tables appear scientific. Some ads even include "reviews of literature" citing such venerable sources as the *American Journal of Clinical Nutrition* and the *Journal of the American Medical Association.* Such ads create the illusion of credibility to gain readers' trust. Keep in mind, however, that ads are created, not to teach, but to *sell* (the Controversy section of Chapter 1 addresses the sales tactics of quackery).

Also keep in mind many substances are sold under the label of "dietary supplement" with scant oversight by regulating authorities (see Controversy 7 for details). This loophole in the law means that, with regard to supplements, athletes are on their own to evaluate them.

*Reference notes are found in Appendix F.

TABLE C10-1 PRODUCTS PROMOTED AS ERGOGENIC AIDS

- **anabolic steroid hormones** chemical messengers related to the male sex hormone testosterone that stimulate building up of body tissues (*anabolic* means "promoting growth"; *sterol* refers to compounds chemically related to cholesterol).
- **androstenedione** (AN-droh-STEEN-dee-own) a precursor of testosterone that elevates both testosterone and estrogen in the blood of both males and females. Often called *andro*, it is sold with claims of producing increased muscle strength, but controlled studies disprove such claims.
- **arginine** a nonessential amino acid falsely promoted as enhancing the secretion of human growth hormone, the breakdown of fat, and the development of muscle.
- **bee pollen** a product consisting of bee saliva, plant nectar, and pollen that confers no benefit on athletes and may cause an allergic reaction in individuals sensitive to it.
- **boron** a nonessential mineral that is promoted as a "natural" steroid replacement.
- **branched-chain amino acids (BCAA)** the amino acids leucine, isoleucine, and valine, which are present in large amounts in skeletal muscle tissue; supplements are falsely promoted as necessary for exercising muscles.
- **brewer's yeast** a preparation of yeast cells, containing a concentrated amount of B vitamins and some minerals; falsely promoted as an energy booster.
- **caffeine** a stimulant that may produce alertness and reduced reaction time in small doses, but creates fluid losses with a larger dose. Overdoses cause headaches, trembling, an abnormally fast heart rate, and other undesirable effects. More about caffeine appears in Controversy 11.
- **carnitine** a nitrogen-containing compound, formed in the body from lysine and methionine, that helps transport fatty acids across the mitochondrial membrane. Carnitine is claimed to "burn" fat and spare glycogen during endurance events, but it does neither.
- **cell salts** a mineral preparation supposedly prepared from living cells. No scientific evidence supports benefits from such preparations.
- **chaparral** an herb, promoted as an antioxidant (see also Table 11-7 in Chapter 11).
- **chromium picolinate** a trace element supplement; falsely promoted to increase lean body mass, enhance energy, and burn fat.
- **coenzyme Q10** a cell constituent important to energy metabolism and shown to improve exercise performance in heart disease patients, but not effective in improving performance of healthy athletes.
- **creatine** a nitrogen-containing compound that combines with phosphate to burn a high-energy compound stored in muscle. Claims that creatine safely enhances energy and stimulates muscle growth are unconfirmed but often reported by users; digestive side effects are common.
- **desiccated liver** dehydrated liver powder that supposedly contains all the nutrients found in liver in concentrated form; possibly not dangerous, but has no particular nutritional merit and is considerably more expensive than fresh liver.
- **DHEA (dehydroepiandrosterone)** a hormone made in the adrenal glands that serves as a precursor to the male hormone testosterone; recently banned by the FDA because it poses the risk of life-threatening diseases, including cancer. Falsely promoted to burn fat, build muscle, and slow aging.
- **DNA** and **RNA (deoxyribonucleic acid** and **ribonucleic acid)** the genetic materials of cells necessary in protein synthesis; falsely promoted as ergogenic aids.
- **"energy drinks"** sugar-sweetened beverages with supposedly "ergogenic" ingredients, such as vitamins, amino acids, caffeine, guarana, carnitine, ginseng, and others. The drinks are not regulated by the FDA, are often high in caffeine, and may still contain the dangerous banned stimulant ephedrine.
- **ephedra (ephedrine)** a dangerous and sometimes lethal "herbal" supplement previously sold for weight loss, muscle building, athletic performance, and other purposes. Now banned by the FDA.
- **epoetin** a drug derived from the human hormone erythropoietin and marketed under the trade name Epogen; illegally used to increase oxygen capacity.
- **ergogenic** (ER-go-JEN-ic) **aids** products that supposedly enhance performance, although none actually do so; the term *ergogenic* implies "energy giving" (*ergo* means "work"; *genic* means "give rise to").
- **gelatin** a soluble form of the protein collagen, used to thicken foods; sometimes falsely promoted as a strength enhancer.
- **ginseng** a root purported to increase work capacity through a number of mechanisms, none demonstrated by research. The bioavailability of the active constituent has been called into question and products vary widely in composition, often substituting other less costly stimulants for ginseng. (See also Table 11-7 in Chapter 11.)
- **glandular products** extracts or preparations of raw animal glands and organs; sold with the false claim of boosting athletic performance, but may present disease hazards if collected from infected animals.

Amino Acid Supplements

Some athletes—particularly body-builders and weight lifters—believe that consuming large doses of amino acids will help build muscles. Amino acid supplements are unnecessary. Healthy athletes eating a well-balanced diet never need them, and in a few unfortunate cases, these supplements have proved dangerous (see the Consumer Corner of Chapter 6). Taking amino acid supplements puts the body in a too-much-too-little bind. Amino acids compete for carriers, and an overdose of one can limit the availability of some other needed amino acids. Supplements can also lead to digestive disturbances and excess water accumulation in the digestive tract.[2]

Specifically, **branched-chain amino acids (BCAA)** are advertised as a source of fuel for the exercising body. What the ads leave out is that compared to glucose and fatty acids, BCAA provide very little fuel to working muscles, and when they *are* needed, well-fed trained muscles have plenty on hand. No consistent research indicates a performance benefit from supplemental BCAA.[3] What is known, though, is that a diet too low in carbohydrates or energy triggers activity of an enzyme that breaks down BCAA for energy. Conversely, the athlete who consumes adequate carbohydrates and calories conserves BCAA in the tissues. Perhaps more importantly, large doses of BCAA can raise plasma ammonia concentrations, causing fatigue—not an effect valued by endurance athletes.[4]

- **glycine** a nonessential amino acid, promoted as an ergogenic aid because it is a precursor of creatine.
- **growth hormone releasers** herbs or pills that supposedly regulate hormones; falsely promoted as enhancing athletic performance.
- **guarana** a reddish berry found in Brazil's Amazon basin that contains seven times as much caffeine as its relative, the coffee bean. It is used as an ingredient in carbonated sodas, and taken in powder or tablet form, it supposedly enhances speed and endurance and serves as an aphrodisiac, a "cardiac tonic," an "intestinal disinfectant," and a "smart drug" touted to improve mental functions. High doses may stress the heart and can cause panic attacks.
- **herbal steroids** or **plant sterols** mixtures of compounds from herbs that supposedly enhance human hormone activity. Products marketed as herbal steroids include astragalus, damiana, dong quai, fo ti teng, ginseng root, licorice root, palmetto berries, sarsaparilla, schizardra, unicorn root, yohimbe bark, and yucca.
- **HMB (beta-hydroxy-beta-methylbutyrate)** a metabolite of the branched-chain amino acid leucine. Claims that HMB increases muscle mass and strength stem from "evidence" from the company that developed HMB as a supplement.
- **human growth hormone (HGH)** a hormone produced by the brain's pituitary gland that regulates normal growth and development (see text discussion); also called *somatotropin*.
- **inosine** an organic chemical that is falsely said to "activate cells, produce energy, and facilitate exercise." Studies have shown that it actually reduces the endurance of runners.
- **ma huang** an herbal preparation sold with promises of weight loss and increased energy, but contains ephedrine, a banned cardiac stimulant with serious adverse effects (see Table C7-2 in Controversy 7).
- **niacin** a B vitamin that when taken in excess rushes blood to the skin, producing vascularity and a red tint—physical attributes bodybuilders strive to attain prior to performance. These attributes do not enhance performance, and excess niacin can cause headaches and nausea.
- **octacosanol** an alcohol extracted from wheat germ, often falsely promoted as enhancing athletic performance.
- **ornithine** a nonessential amino acid falsely promoted as enhancing the secretion of human growth hormone, the breakdown of fat, and the development of muscle.
- **oryzanol** a plant sterol that supposedly provides the same physical responses as anabolic steroids without the adverse side effects; also known as *ferulic acid, ferulate,* or *FRAC*.
- **pangamic acid** also called vitamin B_{15} (but not a vitamin, nor even a specific compound—it can be anything with that label); falsely claimed to speed oxygen delivery.
- **phosphate salt** a product demonstrated to increase the levels of a metabolically important phosphate compound (diphosphoglycerate) in red blood cells and the potential of the cells to deliver oxygen to the body's muscle cells. However, it does not extend endurance or increase efficiency of aerobic metabolism, and it may cause calcium losses from the bones if taken in excess.
- **plant sterols** lipid extracts of plants, called ferulic acid, oryzanol, phytosterols, or "adaptogens," marketed with false claims that they contain hormones or enhance hormonal activity.
- **pyruvate** a 3-carbon compound derived during the metabolism of glucose, certain amino acids, and glycerol; falsely promoted as burning fat and enhancing endurance. Common side effects include intestinal gas and diarrhea and possibly reduced physical performance.
- **royal jelly** a substance produced by worker bees and fed to the queen bee; often falsely promoted as enhancing athletic performance.
- **sodium bicarbonate** baking soda; an alkaline salt believed to neutralize blood lactic acid and thereby reduce pain and enhance possible workload. "Soda loading" may cause intestinal bloating and diarrhea.
- **spirulina** a kind of alga ("blue-green manna") that supposedly contains large amounts of protein and vitamin B_{12}, suppresses appetite, and improves athletic performance. It does none of these things and is potentially toxic.
- **succinate** a compound synthesized in the body and involved in the TCA cycle; falsely promoted as a metabolic enhancer.
- **superoxide dismutase (SOD)** an enzyme that protects cells from oxidation. When it is taken orally, the body digests and inactivates this protein; it is useless to athletes.
- **THG** an unapproved drug, once sold as an ergogenic aid, now banned by the FDA.
- **wheat germ oil** the oil from the wheat kernel; often falsely promoted as an energy aid.
- **whey protein** a by-product of cheese production; falsely promoted as increasing muscle mass. As for whey, it is the liquid left when most solids are removed from milk.

Paige's heavy use of amino acid supplements is unlikely to be helpful.[5] The supplements have not been scientifically demonstrated to be effective and may not even be safe. Her effort would be better spent on eating a nutritious diet adequate in carbohydrate, protein, and energy instead.

Caffeine

Many athletes find that, just as **caffeine** provides mental stimulation during late-night study sessions, the drug seems to provide a physical boost during endurance sports. Although some research supports this idea, the effectiveness of caffeine seems to depend on the activity.[6] For example, caffeine (3 to 6 milligrams per kilogram of body weight) has been demonstrated to improve endurance activities, such as cycling and rowing.[7] In contrast, sprinters, weight lifters, and other athletes performing high-intensity, short-duration activities derive little or no performance edge from caffeine.[8]

No one has yet shown exactly how caffeine benefits performance. One line of thinking, that caffeine could beneficially alter fuel use to spare glycogen, has not been supported by research.[9] Researchers conclude that while athletes may enjoy a "wake-up" effect from caffeine, it does not alter energy fuel use.[10]

Potential benefits from caffeine must be weighed against its known adverse effects—stomach upset, nervousness, irritability, headaches, dehydration, and diarrhea. High doses of caffeine also constrict the arteries and raise blood pressure above normal, making the heart work harder to pump blood to the working muscles, an effect potentially detrimental to sports performance.

Competitors should be aware that college, national, and international athletic competitions prohibit the use of caffeine in amounts greater than the equivalent of 5 or 6 cups of coffee consumed in a two-hour period prior to competition. Athletes are disqualified if urine tests detect more than this amount. Controversy 14 lists caffeine doses in common foods, beverages, and pills.

Instead of taking caffeine pills to obtain a possible performance advantage, Paige might be better off engaging in some light activity before an event, as DJ does. Activity stimulates the release of fatty acids, and a little pregame exercise warms up the muscles and connective tissues, making them flexible and resistant to injury. Caffeine does not offer these benefits. And remember that caffeine is a diuretic. DJ enjoys a cup or two of coffee before her races, but she isn't likely to suffer dehydration from this small amount taken in beverages.

Carnitine

Carnitine is a nonessential nutrient that is often marketed as a "fat burner." In the body, carnitine does help to transfer fatty acids across the membrane that encases the cell's mitochondria. (Recall from Figure 3-1 of Chapter 3 that the mitochondria are structures in cells that release energy from fatty acids and other nutrients.) So carnitine marketers use this logic: "the more carnitine, the more fat burned, the more energy produced" —but the argument is not valid. In scientific studies, carnitine supplementation for 7 to 14 days neither raised muscle carnitine concentrations nor influenced fat or carbohydrate oxidation. (Paige found out the hard way that carnitine often produces diarrhea in those taking it, just the wrong effect for sports performance.) Nor do carnitine supplements enhance exercise performance.

For those concerned about obtaining adequate carnitine, milk and meat products are good sources, but more importantly, carnitine is a *nonessential*

nutrient. This means that the body makes plenty for itself when needed.

Chromium Picolinate

Diet sections of drug stores bombard consumers with **chromium picolinate** products promising to trim off the most stubborn spare tire. Photos of impossibly fit people, supposedly the "after" shots of those taking chromium picolinate supplements, tempt people despite their knowledge that fitness transformations never result from taking a pill.

Chromium is an essential trace mineral involved in carbohydrate and lipid metabolism. One or two initial studies reported that the supplements reduced body fatness and increased lean body mass in men who trained with weights.[11] A flurry of studies of chromium picolinate followed, but the great majority show no effects of chromium picolinate on body fatness, lean body mass, strength, or, for that matter, fatigue.[12]

The safety record of chromium picolinate is not unblemished. One athlete who ingested 1,200 micrograms of chromium picolinate over two days' time developed a dangerous condition of muscle degeneration, with the supplement strongly suspected as the cause. Chromium-sensitive people may have allergic reactions to chromium picolinate supplements.[13] Also, the release of chromium from chromium picolinate creates molecular free radicals that can, theoretically, contribute to potentially harmful levels of oxidative stress in body tissues.[14]

Creatine

Interest in—and use of—**creatine** supplements to enhance performance during intense activity has grown dramatically in the last few years. Power athletes such as weight lifters use creatine supplements in the belief that they enhance stores of the high-energy compound creatine phosphate (or phosphocreatine) in muscles. Theoretically, the more creatine phosphate in muscles, the higher the intensity at which an athlete can train.

The outcomes of some studies suggest that creatine supplementation may enhance performance of high-intensity

strength activity such as weight lifting or repeated sprinting.[15] Other studies have found no effect of creatine supplements on strength performance, however, and the potential underlying mechanisms for such an effect remain obscure. Researchers tested creatine in U.S. Navy combat swimmers, SEALS, who need both strength and endurance to perform demanding physical tasks.* In a timed four-station obstacle course, no benefit from creatine over the placebo was evident.

More investigation of the effectiveness and safety of creatine supplements is required; in particular, appropriate long-term studies on creatine safety are lacking.[16] Immediate side effects such as cramping and gastrointestinal distress seem to occur with about the same dosages reported to benefit performance.[17] Even short-term (five to seven days) creatine supplementation may pose risks to athletes with kidney disease or other conditions. Medical and fitness experts voice concern that creatine is being taken in huge doses (5 to 30 grams per day) and that children as young as nine years old are taking it with unknown consequences. Creatine levels from foods, even diets high in creatine-rich foods like red meat, do not approach the amount athletes take in supplement form.

Despite the uncertainties, creatine supplements are not illegal in international competition. The smart competitor, however, will pass up creatine until all the important questions about its safety and effectiveness are answered.

Protein Powders

Like many other athletes, Paige is a big consumer of protein powders, especially **whey protein**.[18] Whey is one of nature's protein sources, and like lean meat, milk, and legumes, it can supply amino acids to the body, but it offers no special benefits beyond those provided by ordinary milk or yogurt.

This being the case, what do athletes hope to gain from added protein? Paige believes that because the body builds muscle protein from amino acids, eating extra protein will stimulate her

*SEAL stands for Sea, Air, and Land combat teams of the U.S. Navy.

muscles to grow, but this idea is false. She has been taken in by advertisements implying that "more is better." Muscle growth is stimulated by physically demanding activity, not by excess protein. Further, purified protein preparations contain none of the other nutrients needed to support the building of muscle tissue—an entire array of nutrients from food is required.

The body of an athlete who eats adequate food does not use the extra protein from supplements as such. Dutifully, the body dismantles the extra protein, removes the nitrogen from the amino acids, uses what it can for energy, and converts the rest to body fat for storage. The processing required to handle excess amino acids places an extra burden on the kidneys to excrete unused nitrogen.

Complete Meal Replacers

Specialty drinks and candy bars, packed with vitamins, minerals, and other healthy-sounding goodies, appeal to athletes by claiming to provide "complete" meals in convenient "to-go" packages. Although these bars and drinks usually taste good and provide extra food energy, largely as added fats and sugars, they fall far short of providing "complete" nutrition.

What are they good for? A nutritionally "complete" drink may help a nervous athlete who cannot tolerate solid food on the day of an event. In that case, a liquid meal two or three hours before competition can supply some of the fluid and carbohydrate needed in a pregame meal. A shake of fat-free milk or juice (such as apple or papaya) and ice milk or frozen fruit (such as strawberries or bananas), however, can do the same thing at a fraction of the cost (see Table 10-7, p. 383). The bottom line is that this form of nutrition supplement can be useful as a pregame meal or a between-meal snack but is inferior to nutritious foods for meeting the high nutrient needs of athletes.

Recently, DJ, who never bothers with such products, placed ahead of Paige in seven of their ten shared competitions. In one of these races, Paige pulled out because of light-headedness—perhaps a consequence of one or a combination of her ergogenic aids? Still, Paige remains convinced that to win, she must have chemical help, and she is venturing over the danger line by considering hormone-related products. What she doesn't know is very likely to hurt her.

Hormone Preparations

The dietary supplements discussed so far are controversial in the sense that they may or may not enhance athletic performance. Although it is always wise to err on the side of caution when it comes to issues of health, most such supplements—in the doses commonly taken by healthy adults—probably pose little serious threat except to the pocketbook. The next group of substances, however, is clearly damaging to the body. Don't consider using these products—just steer clear.

Anabolic Steroid Hormones

Among the most dangerous and illegal ergogenic practices is the taking of **anabolic steroid hormones.** Often, athletes take these drugs and related products without any medical supervision or testing, simply taking someone's word for their safety.[19]

The testes and adrenal glands in men and the adrenal glands in women make anabolic steroid hormones naturally. Synthetic versions of these natural hormones combine the masculinizing effects of male hormones and the adrenal steroid growth stimulation of female hormones. In the body, these steroids produce accelerated muscle bulking in response to physical activity in both men and women. Injections of these "fake" hormones produce muscle size and strength far beyond that attainable by training alone, but at the price of great risks to health.

The list of adverse reactions to steroids is long and continues to grow amid only a slight decline in use of the drugs. Figure C10-1 (p. 390) lists the side effects of steroids. The American Academy of Pediatrics and the American College of Sports Medicine condemn athletes' use of anabolic steroids, and the International Olympic Committee bans their use. Besides citing the known toxic side effects, these authorities maintain that taking these drugs is a form of cheating. Nevertheless, in professional circles where monetary rewards for excellence are high, steroid use is common. Athletes who lack superstar genetic material and who would normally never break into the ranks of the elite can, with the help of steroids, suddenly compete with true champions.

Steroid use has an unfortunate, domino effect on the entire athletic community. Other athletes are put in the difficult position of either conceding an unfair advantage to competitors who use steroids or taking the drugs and accepting the risk of harmful side effects or discovery and banning from their sport. Young athletes should not be forced to make such a choice.

If swollen appearance, heart disease, or liver tumors are not frightening enough possibilities, add to the mix the urge to hurt oneself or someone else. Steroids produce changes in the brain that, in some people, bring on frightening exhibitions of overly aggressive behavior, aptly nicknamed *"steroid rage."*[20] Abusers of anabolic steroids with no previous history of mental illnesses are especially likely to die a violent death, when their impulsive, aggressive behavior evokes an attack from others, or to die of suicide because of severe depression.[21] A number of bodybuilders, including a former Mr. Universe, are behind bars for the murders of their girlfriends, fiancées, and spouses, committed while under the influence of steroids. Upon quitting steroids, one world-class bodybuilder reported "feeling suicidal and having the sensation of melting away," as the body readjusted to normal by dissolving pounds of muscle.[22] Physicians know to wean people slowly from steroid medications to prevent complications.

In sum, steroids are not simple drugs that build bigger muscles, but complex chemicals to which the body and mind react in many ways, particularly when bodybuilders and other athletes take them in large amounts. The safest, most effective way to build muscle has always been through consistent training and a sound diet, and—despite naïve misconceptions—it still is.

Steroid Alternative Supplements

A substantial percentage of athletes, and particularly school-age athletes, have tried steroid "alternative" supplements such as

FIGURE C10·1 PHYSICAL RISKS OF TAKING STEROID HORMONE DRUGS

Mind
- Extreme aggression with hostility ("steroid rage"); mood swings; migraine headaches; anxiety; dizziness; drowsiness; unpredictability; insomnia; psychotic depression; personality changes; suicidal thoughts; epilepsy

Face and Hair
- Swollen appearance; greasy skin; severe, scarring acne; mouth and tongue soreness; yellowing of whites of eyes (jaundice)
- In females, male-pattern baldness and increased growth of facial and body hair; in males, baldness

Voice
- In females, irreversible deepening of voice

Chest
- In males, breathing difficulty; breast enlargement and development
- In females, breast atrophy; loss of female body contour

Heart
- Heart disease; elevated or reduced heart rate; heart attack; stroke; hypertension

Abdominal Organs
- Nausea; vomiting; bloody diarrhea; pain; edema; liver tumors (possibly cancerous); liver damage, disease, or rupture leading to fatal liver failure; kidney stones and damage; gallstones; frequent urination; possible rupture of aneurysm or hemorrhage

Blood
- Increased red blood cells; blood clots; increased LDL cholesterol; reduced HDL cholesterol; increased triglycerides; high risk of blood poisoning; those who share needles risk contracting diseases; septic shock (from injections); glucose intolerance

Reproductive System
- In males, permanent shrinkage of testes; early puberty in adolescents; prostate enlargement with increased risk of cancer; sexual dysfunction; loss of fertility; excessive and painful erections
- In females, loss of menstruation and fertility; increased libido; early puberty in adolescents; permanent enlargement of external genitalia; thickening of uterine lining; fetal damage, if pregnant

Muscles, Bones, and Connective Tissues
- Weight gain; altered body composition; increased susceptibility to injury with delayed recovery times; cramps; tremors; seizurelike movements; injury at injection site
- In adolescents, failure to grow to normal height

Other
- Fatigue; edema; increased risk of liver and uterine cancer; sleep, breathing disorders

© Cleve Bryant/PhotoEdit

© Thinkstock/PictureQuest

recently banned "andro" (**androstene-dione**) or **DHEA** or a number of herbal preparations. Claims made for these substances—that they reduce fat, build muscle, slow aging, and others miracles—are unsupported by research.[23] In truth, these substances are converted to active hormones by body tissues, and users incur many of the same serious risks as for steroid drugs (listed in Figure C10-1).[24]

People respond unpredictably to these substances.[25] Generally, in males, estrogens increase linearly with increasing doses while testosterone, the desired muscle-building hormone, elevates only a little.[26] Females may experience a greater proportional surge in testosterone along with increased estrogens.

Some rapidly occurring effects of steroid alternatives include oily skin, acne, body hair growth, liver enlargement, irreversible masculine changes in females, irreversible feminine changes in males, and aggressive behavior. Long-term effects, such as serious liver damage, may occur after years of use. Particularly vulnerable to these effects are children and adolescent athletes, who may take steroid alternatives on the advice of misinformed parents or coaches.

Recently, the Food and Drug Administration (FDA) sent letters to producers of dietary supplements warning them that products containing androstenedione are considered to be adulterated and therefore are illegal to sell. The FDA warned that firms that continue to distribute the products could be subject to criminal penalties. The National Collegiate Athletic Association, the National Football League, and the International Olympic Committee have banned the use of androstenedione and DHEA in competition. The American Academy of Pediatrics and many other medical professional groups have spoken out against the use of these and other "hormone replacement" substances.

As for the herbal or even insect sterols hawked as "natural" substitutes for steroid drugs, these are useless in sports. The body cannot convert them into human steroids, nor do they stimulate the body's own steroid production. These products may contain toxins, however. Remember: "natural" doesn't mean "harmless."

Human Growth Hormone

Though not a steroid, **human growth hormone (HGH)** can induce huge body size and is less readily detected in drug tests than steroids. Short or average-size athletes who are still growing sometimes use this hormone to build lean tissue and increase their height. Athletes in power sports such as weight lifting and judo are most likely to experiment with HGH, believing the injectable hormone will provide the benefits of anabolic steroids without the dangerous side effects. Alternatively, they may take growth hormone "stimulators," such as the amino acids **ornithine** and **arginine.**

Use of this hormone and related substances is a lose-lose proposition. The amino acids ornithine and arginine are useless in the form sold to athletes and do not stimulate growth hormone release. As for HGH itself, it causes the disease acromegaly, characterized by a widened jawline, widened nose, protruding brow, and buck teeth. The body of someone with acromegaly becomes huge, and the organs and bones enlarge abnormally. Other effects include diabetes, thyroid disorder, heart disease, menstrual irregularities, diminished sexual desire, and an increased likelihood of death before age 50.

Athletes who have paid the price of hormone abuse—even some for whom the drugs made careers in sports possible—have come forward to warn young athletes away from growth hormones. They say that even the rewards of sports success are not worth the side effects of the drugs. The U.S. Olympic Committee bans HGH use and maintains that it is a form of cheating, undermining the quest for physical excellence and seducing other athletes into joining the abuse.

A safe way to maximize the body's natural growth hormone production does exist: rest. Growth hormone is released during sleep, especially after physical activity, so make sure to get enough rest between periods of adequate training.

Drugs Posing as Supplements

Some ergogenic aids sold as dietary supplements turn out to be powerful drugs. A potent thyroid hormone known as TRIAC has been recalled by the FDA.[27] TRIAC interferes with normal thyroid functioning and has caused heart attack and stroke.* Another is **THG,** a potentially harmful synthetic steroid.[28] The FDA has reclassified both TRIAC and THG as drugs, but products containing them are still making their way into the hands of athletes. Such synthetic steroid derivatives, or "designer steroids," are designed to sneak steroid use past detection tests in athletic events. Although the FDA has banned these two, others are likely to crop up to take their place because the demand is strong and profits high.[29]

Conclusion

The general scientific response to ergogenic claims is "let the buyer beware." In a survey of advertisements in a dozen popular health and bodybuilding magazines, researchers identified over 300 products containing 235 different ingredients advertised as beneficial, mostly for muscle growth. None had been scientifically shown to be effective. What *has* weathered the test of time in priming athletes for success—from Little League players to Olympic-level gymnasts—is the basic combination of consistent training and sound nutrition.

Athletes like Paige who fall for the promises of better performance through supplements are taking a gamble with their money, their health, or both. They move from product to product, abandoning one after another when the promised performance miracles do not materialize. DJ, who takes the scientific approach reflected in this Controversy, faces a problem: How does she inform Paige of the hoaxes and still preserve their friendship?

Explaining to someone that a long-held belief is not true involves a risk: the person often becomes angry with the one delivering the truth, rather than with the source of the misinformation. To avoid this painful outcome, DJ decides to mention only the supplements in Paige's routine that are most likely to cause harm—the chromium picolinate, the overdoses of caffeine, and the hormone replacers. As for the meal replacers, protein powders, and

---

*Two of the products containing TRIAC (triiodothyroacetic acid, or tiratricol) have the trade names BioPharm *T-Cuts* and *Triax Metabolic Accelerator.*

other supplements that are probably just a waste of money, DJ decides to keep her own counsel. Perhaps they may serve as harmless superstitions.

As for those occasions when Paige believes her performance was boosted by a new concoction, DJ understands that chances are the effect came from the power of the mind over the body. Don't discount that power—it is formidable. You don't have to rely on useless supplements for an extra edge because you already have a real one—your mind. And you can use the extra money you save to buy a great pair of running shoes—perhaps on a Wednesday.

Diet and Health

© Isy Ochoa (b. 1961, French), *En Cuisine I*, 2002. © Isy Ochoa/SuperStock, Inc.

CONTENTS

Do You Ever . . .

Wish that your diet could strengthen your immune system?

Wonder whether your food choices are damaging your heart?

Use herbs or alternative medicine to improve your health?

Choose "natural" foods without additives to avoid developing cancer?

KEEP READING.

CAN YOUR DIET affect your risk of developing a disease? It depends on the disease. Two main kinds of diseases afflict people around the world: **infectious diseases** and **degenerative diseases.*** Infectious diseases such as tuberculosis, smallpox, influenza, and polio have been major killers of humankind since before the dawn of history. In any society not well defended against them, infectious diseases can cut life so short that the average person dies at 20, 30, or 40 years of age.

With the advent of vaccines and antibiotics, people in developed countries had become complacent about infectious diseases—until recently. Scientists now warn of growing infectious threats: **bioterrorism;** the emergence and rapid global spread of new diseases such as sudden acute respiratory syndrome (SARS); and a rising death toll from once-conquered diseases such as tuberculosis and foodborne infections that have become resistant to antibiotic drugs.[†1] While scientists work to develop new controls for these perils, government and health agencies hasten to strengthen existing mechanisms for emergency preparedness and to protect our food and water supplies.

Individuals can take steps to protect themselves, too. Each of us encounters millions of microbes each day, and some of these can cause diseases. Although nutrition cannot directly prevent or cure infectious diseases, it can strengthen or weaken your body's defenses against them.[2] One warning: many "immune strengthening" foods, dietary supplements, and herbs are hoaxes. For healthy, well-fed people, supplements cannot trigger extra immune power to fend off dangerous infections. The best help you can provide to your immune system is to nourish it properly to ensure that it works to its capacity every day.

In developed nations such as the United States and Canada, degenerative diseases far outrank infections as the leading causes of death and illness. Examples are heart disease, cancer, and diabetes, as Figure 11-1 shows.[3] The longer a person dodges life's other perils, the more likely that these diseases will take their toll.

Degenerative diseases do not arise from a straightforward cause such as infection but from a mixture of three factors: genetic predisposition, personal medical history, and lifestyle choices. The first two, inherited susceptibility and prior disease, people cannot control. Daily life choices that people control directly, however, can delay or prevent the onset of some diseases. Young people choose whether to nourish their bodies well, to smoke, to exercise, or to abuse alcohol. As people age, their bodies accumulate the effects of these choices, and in the later years these impacts can make the difference between a life of health or one of chronic disability.[4] Thus, degenerative diseases are often called *chronic* diseases. After a discussion of nutrition's impact on the immune system, the rest of this chapter is devoted to the diet-related factors that affect the degenerative diseases that develop over a lifetime.

Nutrition and Immunity

Without your awareness, your immune system continuously stands guard against thousands of attacks mounted against you by microorganisms and cancer cells. If your immune system falters, you become vulnerable to disease-causing agents, and disease invariably follows.

These facts underscore nutrition's importance to immunity:

- Deficient intakes of many vitamins and minerals are associated with impaired disease resistance, as are some excessive intakes.[5]
- Immune tissues are among the first to be impaired in the course of a nutrient deficiency or toxicity.
- Some deficiencies are more immediately harmful to immunity than others; the speed of the impact is affected by whether another nutrient can perform some of

---

■ Chapter 12 comes back to the topic of the safety of the U.S. food supply.

■ Diabetes was a topic of Chapter 4.

■ If there is any deficiency in food or exercise the body will fall sick.
—Hippocrates, a Greek physician, c. 400 B.C.

infectious diseases diseases that are caused by bacteria, viruses, parasites, and other microbes and can be transmitted from one person to another through air, water, or food; by contact; or through vector organisms such as mosquitoes and fleas.

degenerative diseases chronic, irreversible diseases characterized by degeneration of body organs due in part to such personal lifestyle elements as poor food choices, smoking, alcohol use, and lack of physical activity. Also called *lifestyle diseases, chronic diseases,* or the *diseases of old age.*

bioterrorism the intentional spreading of disease-causing organisms or agricultural pests as a political weapon to produce fear and intimidate others.

---

*The term *disease* is also used to refer to conditions such as birth defects, alcoholism, obesity, and mental disorders.
†Reference notes are found in Appendix F.

FIGURE 11·1 LEADING CAUSES OF DEATH—UNITED STATES, 2001ᵃ

The causes identified with the red bars are related to nutrition; those with green bars are alcohol related.

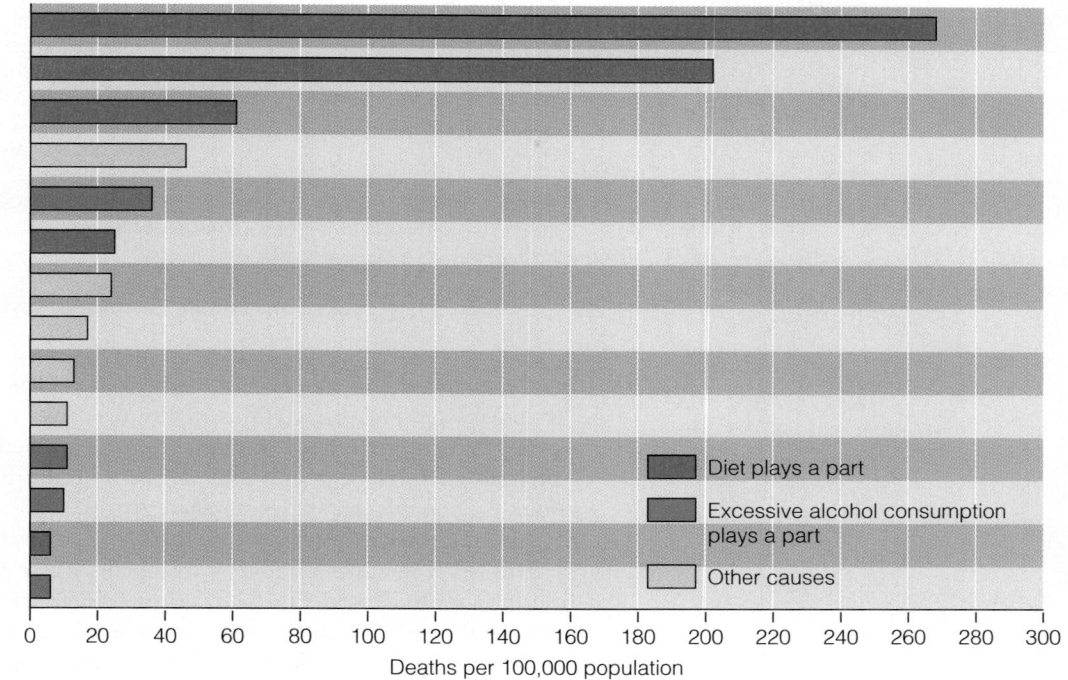

Heart disease
Cancers
Strokes
Chronic lower respiratory diseases
Motor vehicle and other accidentsᵇ
Diabetes mellitus
Pneumonia and influenza
Alzheimer's disease
Kidney diseases
Infections of the blood
Suicide
Liver disease and cirrhosis
Hypertension and hypertensive kidney disease
Homicide

Diet plays a part
Excessive alcohol consumption plays a part
Other causes

0 20 40 60 80 100 120 140 160 180 200 220 240 260 280 300

Deaths per 100,000 population

ᵃRates are age adjusted to allow relative comparisons of mortality among groups and over time.

ᵇAccidents are the leading cause of death among people aged 15–24, followed by homicide, suicide, cancer, heart disease, birth defects, lung disease, pneumonia and influenza, and stroke. Alcohol contributes to about half of all accident fatalities; together, alcohol-related deaths amount to the third leading cause of mortality. See Controversy 3.

SOURCE: Data from National Center for Health Statistics, 2003.

the metabolic tasks of the missing nutrient, how severe the deficiency is, whether an infection has already taken hold, and the person's age.

Once a person becomes malnourished, malnutrition often worsens disease, which, in turn, worsens malnutrition. The cycle often begins when impaired immunity opens the way for disease; then disease impairs food assimilation, and nutrition status suffers further. Drugs become necessary, and many of them impair nutrition status (see Chapter 14's Controversy). Other treatments, such as surgery, take a further toll.[6] Thus, disease and poor nutrition together form a downward spiral that must be broken for recovery to occur (see Figure 11-2, p. 396).

Certain groups of people are more likely than others to be caught in the downward spiral of malnutrition and weakened immunity. Among them are people who restrict their food intakes, whether because of lack of appetite, eating disorders, desire for weight loss, or any other reason. Also susceptible are those who are one or more of the following: very young or old, poor, hospitalized, or malnourished. Rates of sickness and death increase dramatically when medical tests of a malnourished person indicate weakened immunity.

In protein-energy malnutrition (PEM), indispensable tissues and cells of the immune system dwindle in size and number, making the whole body vulnerable to infection. Table 11-1 (p. 396) shows PEM's effects on body defenses. The skin and body linings, the first line of defense against infections, become thinner because their connective tissue is broken down, allowing agents of disease easy access to body tissues. For example, the digestive system normally musters a formidable defense—its linings are heavily laced with active immune tissues that work locally at the absorptive site and also form cells and antibodies that travel to protect other organs, such as the liver,

■ The term *immunonutrition* is used to describe the influence of nutrients on the functioning of the immune system, especially in medical nutrition therapies.

FIGURE 11·2 MALNUTRITION AND DISEASE

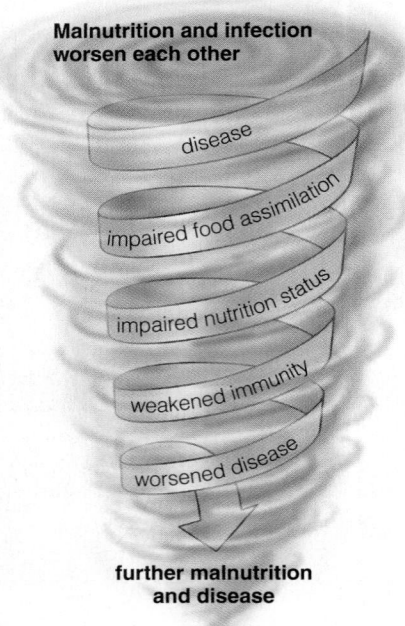

Malnutrition and infection worsen each other

disease

impaired food assimilation

impaired nutrition status

weakened immunity

worsened disease

further malnutrition and disease

Deficiencies (↓) and toxicities (↑) known to impair immunity:

- Protein (↓).
- Energy (↓).
- Vitamin A (↓).
- Vitamin E (↓).
- Vitamin D (↓).
- B vitamins (↓).
- Folate (↓).
- Vitamin C (↓).
- Iron (↓↑).
- Zinc (↓↑).
- Copper (↓).
- Magnesium (↓).
- Selenium (↓).

AIDS acquired immune deficiency syndrome; caused by infection with human immunodeficiency virus (HIV), which is transmitted primarily by sexual contact, contact with infected blood, needles shared among drug users, or fluids transferred from an infected mother to her fetus or infant.

risk factors factors known to be related to (or correlated with) diseases but not proved to be causal.

TABLE 11·1 EFFECTS OF PROTEIN-ENERGY MALNUTRITION (PEM) ON THE BODY'S DEFENSE SYSTEMS

System Component	Effects of PEM
Skin	Skin becomes thinner, with less connective tissue to serve as a barrier for protection of underlying tissues; skin sensitivity reaction to antigens is delayed.
Digestive tract membrane and other body linings	Antibody secretions and immune cell numbers are reduced.
Lymph tissues	Immune system organs[a] are reduced in size; cells of immune defense are depleted.
General response	Invader kill time is prolonged; circulating immune cells are reduced; immune response is impaired.

[a]Thymus gland, lymph nodes, and spleen.

pancreas, mammary glands, and uterus. When PEM sets in, the number of cells and antibodies present in secretions of the digestive tract diminishes. Thus, infectious agents that enter the body through linings weakened by PEM encounter diminished internal defensive responses and can easily cause illnesses of greater severity than would be possible in a well-nourished person.

Malnutrition can result not only from a lack of available food but also from diseases, such as **AIDS** and cancer, and their treatments. These alter the appetite and metabolism, causing a wasting away of the body's tissues similar to that seen in the last stages of starvation—the body uses its fat and protein reserves for survival. In people with AIDS, wasting or nutrient deficiencies can shorten survival, making medical nutrition therapy a critical need.[7] Nutrients cannot cure AIDS, of course, but an adequate diet may improve responses to drugs, shorten hospital stays, promote independence, and improve the quality of life. And exercise that strengthens muscles may hold wasting to a minimum.[8] In addition, food safety is paramount because common food bacteria and viruses can easily overwhelm a compromised immune system. Food safety rules are found in Chapter 12.

A deficiency or a toxicity of just a single nutrient can seriously weaken even a healthy person's immune defenses. For example, in vitamin A deficiency, the body's skin and membranous linings become unhealthy and unable to ward off infectious organisms. A vitamin C deficiency robs white blood cells of their killing power. Too little vitamin E may impair several aspects of immunity, especially among the aged.[9] Both deficient and excessive zinc impair immunity by reducing the number of effective white blood cells in the first case and impairing the immune response in the second.[10] The obvious conclusion is that a well-balanced diet is the cornerstone in building the best possible immune system defense.

As mentioned earlier, a *diet* of foods that supplies adequate nutrients ensures the proper functioning of the immune system, but extra daily doses of nutrients, herbs, or other substances do not enhance it. Furthermore, toxic doses clearly diminish it.

▶ KEY POINT Adequate nutrition is a key component in maintaining a healthy immune system to defend against infectious diseases. Medical nutrition therapy can improve the course of wasting diseases. Both deficient and excessive nutrients can harm the immune system.

Lifestyle Choices and Risks of Degenerative Disease

In contrast to the infectious diseases, each of which has a distinct microbial cause such as a bacterium or virus, the degenerative diseases of adulthood have suspected contributors known as **risk factors.** Risk factors show a correlation with a disease, and

FIGURE 11·3 DIET/LIFESTYLES RISK FACTORS AND DEGENERATIVE DISEASES

This chart shows that the same risk factor can affect many degenerative diseases. Notice, for example, how many diseases have been linked to a sedentary lifestyle. The chart also shows that a particular disease, such as atherosclerosis, can have several risk factors.

This flowchart shows that many of these conditions are themselves risk factors for other degenerative diseases. For example, a person with diabetes is likely to develop atherosclerosis and hypertension. These two conditions, in turn, worsen each other. Notice how all of these degenerative diseases are linked to obesity.

although they are candidates for causes, they have not yet been voted in or out. We can say with confidence that a virus causes influenza, but we cannot name the cause of heart disease with such confidence.

An analogy may help clarify the concept of risk factors. A risk factor is like a person who is often seen lurking around the scene of a particular type of crime, say, arson. The police may suspect that person of setting fires, but it may very well be that another, sneakier individual who goes unnoticed is actually pouring the fuel and lighting the match. The evidence against the known suspect is only circumstantial. The police can be sure of guilt only when they observe the criminal in the act. Risk factors have not yet been caught in the act of causing diseases (the mechanisms are still largely unknown). The presence of risk factors often predicts the occurrence of diseases, however, and researchers are working on theories of how risk factors are related to disease causation.

Among the risk factors are environmental, behavioral, social, and genetic factors that tend to occur in clusters and interact with each other. For instance, food behaviors underlie many risk factors.[11] Choosing to eat a diet too high in saturated fat, salt, and calories, for example, is choosing to risk becoming obese and contracting atherosclerosis, diabetes, diverticulosis, cancer, **hypertension,** or other diseases.[12] Figure 11-3 shows connections among some of the risk factors associated with today's major degenerative diseases and highlights the diet-related behaviors that contribute to them. Figure 11-3 also shows that in many cases, one disease or condition intensifies the risk of another.

The exact contribution diet makes to each disease is hard to estimate. Many experts believe that diet accounts for about a third of all cases of coronary heart disease. The links between diet and cancer incidence are harder to pin down because cancer's different forms associate with different dietary factors. General trends, however, support

hypertension high blood pressure.

These conditions in parents, grand-parents, or siblings, especially occurring early in life, may raise a warning flag for you:

- Alcoholism.
- Cancer.
- Diabetes.
- Heart and artery diseases.
- Hypertension.
- Liver disease (cirrhosis).
- Osteoporosis.

Minutes and seconds count when heart attack strikes. Should any of these signs occur, call for emergency medical help; in most areas, dial 911.

In men and women:

- Chest discomfort. Discomfort in the center of the chest that lasts more than a few minutes or that goes away and comes back. An uncomfortable pressure, squeezing, fullness, or pain in the chest.
- Discomfort in other areas of the upper body. Pain or discomfort in one or both arms, the back, neck, jaw, or stomach.
- Shortness of breath. Accompanying chest discomfort or preceding it.
- Other signs. Cold sweat, nausea, or lightheadedness.

Women may instead experience:

- Breathlessness.
- Cold sweat.
- Dizziness.
- Nausea.
- Neck, shoulder, or abdominal pain.
- Unusual fatigue.
- Vomiting.
- Weakness.

Also call 911 immediately for these sudden warning signs of stroke:

- Sudden numbness or weakness of the face, arm, or leg, especially on one side of the body.
- Sudden confusion, trouble speaking, or trouble understanding.
- Sudden trouble seeing in one or both eyes.
- Sudden trouble walking, dizziness, or loss of balance or coordination.
- Sudden severe headache with no known cause.

atherosclerosis (ath-er-oh-scler-OH-sis) the most common form of cardiovascular disease; characterized by plaques along the inner walls of the arteries (*scleros* means "hard"; *osis* means "too much"). The term *arteriosclerosis* refers to all forms of hardening of the arteries and includes some rare diseases.

many links between diet and cancer, and in some cases the evidence is convincing.[13] Many other suspected causes of cancer, such as exposure to radiation or environmental contamination, are often beyond direct control by individuals, but people can control their own food choices. If a dietary change can't hurt and might help, why not make it?

Some choices, such as avoiding tobacco, are important to almost everyone's health. Other choices, such as those relating to diet, are more important for people who are genetically predisposed to certain diseases. To pinpoint your own areas of concern, search your family's medical history for diseases common to your forebears. Any condition that shows up in several close blood relatives may be a special concern for you.* Also, after your next physical examination, find out which test results are out of line. Family history and laboratory test results together are powerful predictors of disease.

Accepting that everyone has certain unchangeable "givens," an effective strategy is to look to the risk factors that can be changed and choose the most influential among them. For example, a person whose parents, grandparents, or other close blood relatives suffered from diabetes and heart disease is urgently advised to avoid becoming obese and not to smoke. The guidelines presented in later sections of this chapter can benefit most people.

▶ **KEY POINT** The same diet and lifestyle risk factors may contribute to several degenerative diseases. A person's family history and laboratory test results can reveal strategies for disease prevention.

Cardiovascular Diseases

In the United States today, more than 61 million men and women suffer some form of disease of the heart and blood vessels (cardiovascular disease, abbreviated CVD), and many die from these causes.[†14] Worldwide, CVD accounts for more deaths among people of developed nations each year than any other single cause, mostly by way of heart attacks and strokes, with about a million yearly deaths from these causes in the United States alone.[15] A long-standing myth is that CVD is a man's disease; in fact, more women than men die of CVD in all its forms each year. Men have more heart attacks than women do, but women still have them in large numbers.

The margin lists the symptoms of the major CVD killers, heart attack and stroke. Note that women may or may not experience classic symptoms such as chest discomfort. Learning to recognize the symptoms can be lifesaving because prompt medical attention is most effective.

How can you minimize your risks of heart attack and stroke? Or, more positively, what actions can you take to help maintain your cardiovascular health and vigor throughout life? Many people have changed their lifestyle to lower their risk: they have quit smoking or refrained from starting; they have also changed their diets, consuming less saturated fat, less sodium, more fruits and vegetables, and more fiber. Many people are still reluctant to exercise, however.[16] And record numbers of people are obese (see Chapter 9).

At the root of most forms of CVD is **atherosclerosis.** Atherosclerosis is the common form of hardening of the arteries.

How Atherosclerosis Develops

No one is free of atherosclerosis. The question is not whether you have it but how far advanced it is and what you can do to retard or reverse it. Atherosclerosis usually begins with the accumulation of soft, fatty streaks along the inner walls of the arteries,

---

*A free tool, "My Family Health Portrait," can help organize family health information; available at www.hhs.gov/familyhistory/order.html.
†Deaths from CVD in the United States include 54 percent from coronary heart disease, 18 percent from stroke, 5 percent from congestive heart failure, 5 percent from hypertension, 4 percent from diseases of the arteries, and the remainder from rheumatic fever/heart disease, congenital cardiovascular defects, and other causes.

FIGURE 11·4   THE FORMATION OF PLAQUES IN ATHEROSCLEROSIS

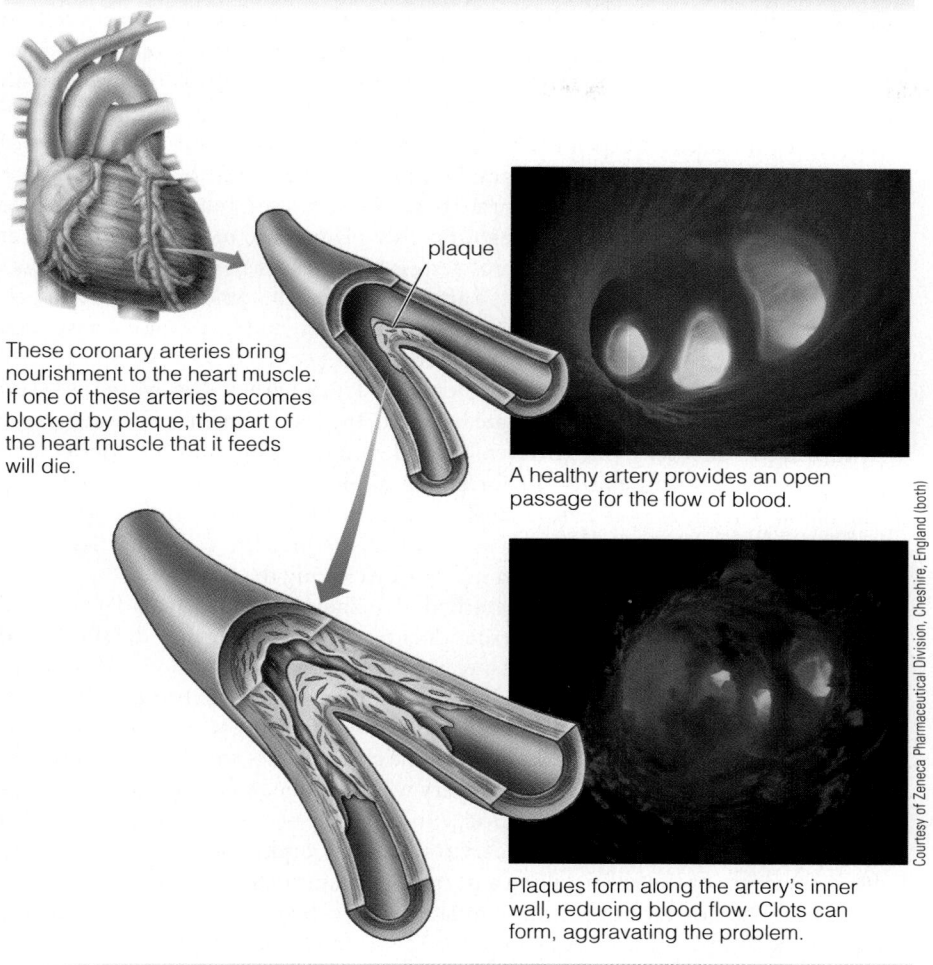

These coronary arteries bring nourishment to the heart muscle. If one of these arteries becomes blocked by plaque, the part of the heart muscle that it feeds will die.

plaque

A healthy artery provides an open passage for the flow of blood.

Plaques form along the artery's inner wall, reducing blood flow. Clots can form, aggravating the problem.

Courtesy of Zeneca Pharmaceutical Division, Cheshire, England (both)

especially at branch points. These gradually enlarge and become hardened fibrous **plaques** that damage artery walls, making them inelastic and narrowing the passage through them (see Figure 11-4). Most people have well-developed plaques by the time they reach age 30.[17]

What causes the plaques to form? A diet high in saturated fat is a major contributor to the development of plaques and the progression of atherosclerosis.[18] But atherosclerosis is much more than the simple accumulation of lipids within the artery wall—it is a complex inflammatory response to tissue damage. Protein markers of inflammation and a host of other previously obscure compounds in the blood may one day help detect heart disease before a life-threatening event occurs.*

The damage may begin from a number of factors interacting with the cells that line the arteries: high LDL cholesterol, hypertension, toxins from cigarette smoking, elevated blood concentrations of the amino acid homocysteine, or some viral and bacterial infections.[19] Such damage produces an inflammatory response that provokes the immune system to send white blood cells to the site to try to repair the damage. Soon, particles of LDL cholesterol become trapped in the blood vessel walls and become oxidized by free radicals produced during inflammatory responses.[20] The white blood cells—**macrophages**—flood the scene to remove the oxidized LDL,

plaques (PLACKS) mounds of lipid material mixed with smooth muscle cells and calcium that develop in the artery walls in atherosclerosis (*placken* means "patch"). The same word is also used to describe the accumulation of a different kind of deposits on teeth, which promote dental caries.

macrophages (MACK-roh-fah-jes) large scavenger cells of the immune system that engulf debris and remove it (*macro* means "large"; *phagein* means "to eat"). Also defined in Chapter 3.

*Examples are c-reactive protein, adiponectin, interleukin-6, and tumor necrosis factor-alpha.

■ Homocysteine was defined in Chapter 7.

A blood clot in an artery, such as this fatal heart embolism, blocks the blood flow to tissues fed by that artery.

aneurysm (AN-you-rism) the ballooning out of an artery wall at a point that is weakened by deterioration.

aorta (ay-OR-tuh) the large, primary artery that conducts blood from the heart to the body's smaller arteries.

platelets tiny cell-like fragments in the blood, important in blood clot formation (*platelet* means "little plate").

thrombus a stationary blood clot.

thrombosis a thrombus that has grown enough to close off a blood vessel. A *coronary thrombosis* closes off a vessel that feeds the heart muscle. A *cerebral thrombosis* closes off a vessel that feeds the brain (*coronary* means "crowning" [the heart]; *thrombo* means "clot"; the cerebrum is part of the brain).

embolus (EM-boh-luss) a thrombus that breaks loose and travels through the blood vessels (*embol* means "to insert").

embolism an embolus that causes sudden closure of a blood vessel.

heart attack the event in which the vessels that feed the heart muscle become closed off by an embolism, thrombus, or other cause with resulting sudden tissue death. A heart attack is also called a *myocardial infarction* (*myo* means "muscle"; *cardial* means "of the heart"; *infarct* means "tissue death").

stroke the sudden shutting off of the blood flow to the brain by a thrombus, embolism, or the bursting of a vessel (hemorrhage).

but to no avail. As the macrophages become engorged with oxidized LDL, they become known as foam cells, which themselves become sources of oxidation that attract more immune scavengers to the scene. Muscle cells of the arterial wall proliferate in an attempt to heal the oxidative damage, but they mix with the foam cells to form hardened areas of plaque. Mineralization increases hardening of the plaques. The process is repeated until many inner artery walls become virtually covered with disfiguring plaques.[21]

Normally, the arteries expand with each heartbeat to accommodate the pulses of blood that flow through them. Arteries hardened and narrowed by plaques cannot expand, however, so the blood pressure rises. The increased pressure damages the artery walls further and strains the heart. Because plaques are more likely to form at damage sites, the development of atherosclerosis becomes a self-accelerating process.

As pressure builds up in an artery, the arterial wall may become weakened and balloon out, forming an **aneurysm.** An aneurysm can burst, and in a major artery such as the **aorta,** this leads to massive bleeding and death.

Abnormal blood clotting can also threaten life. Clots form and dissolve in the blood all the time, and when these processes are balanced, the clots do no harm. That balance is disturbed in atherosclerosis, however. Arterial damage, plaques in the arteries, and the inflammatory response favor the formation of blood clots. Small, cell-like bodies in the blood, known as **platelets,** normally cause clots to form when they encounter injuries in blood vessels. In atherosclerosis, when the platelets encounter hardened plaques, they respond to them as to an injury—by clotting the blood.

A clot, once formed, may remain attached to a plaque in an artery and grow until it shuts off the blood supply to the surrounding tissue. That tissue may die slowly and be replaced by nonfunctional scar tissue. The stationary clot is called a **thrombus.** When it has grown large enough to close off a blood vessel, it is a **thrombosis.** A clot can also break loose, becoming an **embolus,** and travel along the system until it reaches an artery too small to allow its passage. There the clot becomes stuck and is referred to as an **embolism.** The tissues fed by this artery will be robbed of oxygen and nutrients and will die suddenly. Such a clot can lodge in an artery of the heart, causing sudden death of part of the heart muscle, a **heart attack** (see margin photo). A clot may also lodge in an artery of the brain, killing a portion of brain tissue, a **stroke.**

Opposing the clot-forming actions of platelets are active products of omega-3 fatty acids, described in Chapter 5. A diet lacking the seafoods that contain these essential fatty acids may contribute to clot formation.[22]

On many occasions, heart attacks and strokes occur with no apparent blockage. An artery may go into spasms, restricting or cutting off the blood supply to a portion of the heart muscle or brain. Much research today is devoted to finding out what causes plaques to form, what causes arteries to go into spasms, what governs the activities of platelets, and why the body allows clots to form unopposed by clot-dissolving cleanup activity.

Hypertension and atherosclerosis are twin demons that worsen CVD, and each worsens the other. Hypertension worsens atherosclerosis because a stiffened artery, already strained by each pulse of blood surging through it, is stressed further by high internal pressure. Injuries multiply, more plaques grow, and more weakened vessels become likely to burst and bleed.

Atherosclerosis also worsens hypertension. Since hardened arteries cannot expand, the heart's beats raise the blood pressure. Hardened arteries also fail to let blood flow freely through the kidneys, which control blood pressure. The kidneys sense the reduced flow of blood and respond as if the blood pressure were too low; they take steps to raise it further (see the discussion of hypertension later in the chapter).

▶ **KEY POINT** Plaques of atherosclerosis induce hypertension and trigger abnormal blood clotting, leading to heart attacks or strokes. Abnormal vessel spasms can also cause heart attacks and strokes.

Risk Factors for CVD

Efforts to fight atherosclerosis and resulting diseases have led to discoveries about their prevention. An expert panel of the National Cholesterol Education Program defined major heart disease risk factors already listed briefly in Chapter 5; they are presented in full in Table 11-2.[23] Many of the same factors also predict the occurrence of stroke. All people reaching middle age exhibit at least one of these factors (middle age is a risk factor), and many people have several factors, silently increasing their risk.[24] It befits a nutrition book to focus on dietary strategies to reduce these risks, but Table 11-2 shows that diet is not the only, and perhaps not even the most important, factor in the development of heart disease or stroke. Age, gender, cigarette smoking, certain diseases, and physical inactivity predict their development as well. The next few sections address these factors; discussions of diet and physical activity for CVD prevention follow.

Age, Gender, and Family History Three of the major risk factors for CVD cannot be modified by lifestyle choices: age, gender, and family history. The increasing risk associated with growing older reflects the steady progression of atherosclerosis in most people as they age.[25]

In men, aging becomes a significant risk factor for heart disease at age 45 years or older; in women, the risk increases after age 55. Up to age 45, a woman's risk of developing heart disease is lower than a man's of the same age. But for women past menopause, rates of heart disease increase two to three times over premenopause rates.[26] Thus, CVD is not a man's disease—men die earlier of heart attacks than women, but women also die of them, especially in later life. In all its forms, CVD kills more U.S. women than any other disease.

As for family history, early heart disease in immediate family members (siblings or parents) is a major risk factor for developing it.[27] The more family members affected and the earlier the age at which they became ill, the greater the risk to the individual.[28] While these findings suggest a genetic component to CVD, scientists are encountering great complexities in trying to clarify these associations.[29]

High LDL and Low HDL Cholesterol Low-density lipoprotein (LDL) cholesterol in the blood is strongly linked to a person's risk of developing atherosclerosis and heart disease—the higher the LDL, the greater the risk (see Figure 11-5). LDL carry cholesterol to the cells, including the cells that line the arteries, where it can build up as part

FIGURE 11·5 LDL, HDL, AND RISK OF HEART DISEASE

Low HDL relative to LDL increases risk

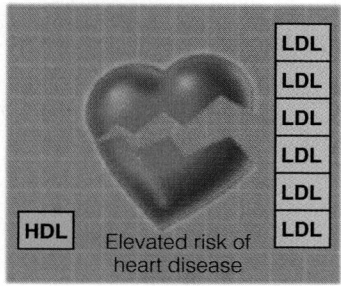

Elevated risk of heart disease

High HDL relative to LDL decreases risk

Reduced risk of heart disease

TABLE 11·2 MAJOR RISK FACTORS FOR HEART DISEASE

See Figure 11-6 (p. 402) for standards by which to judge blood lipids, obesity, and blood pressure.

Risk factors that cannot be modified:
- Increasing age.
- Male gender.
- Family history of premature heart disease.

Risk factors that can be modified:
- High blood LDL cholesterol.
- Low blood HDL cholesterol.
- High blood pressure (hypertension).
- Diabetes.
- Obesity (especially central obesity).
- Physical inactivity.
- Cigarette smoking.
- An "atherogenic" diet (high in saturated fats including *trans* fats and low in vegetables, fruits, and whole grains).

NOTE: Risk factors highlighted in color have relationships with diet.
SOURCE: Expert Panel on Detection, Evaluation, and Treatment of High Blood Cholesterol in Adults (Adult Treatment Panel III), *Third Report of the National Cholesterol Education Program (NCEP)*, NIH publication no. 02-5215 (Bethesda, Md.: National Heart, Lung, and Blood Institute, 2002), pp. II-15–II-20.

Cholesterol is carried in several lipoproteins, chief among them LDL and HDL (see Chapter 5 for details). Remember them this way:

- LDL = **L**ow-density lipoproteins = **L**ess healthy.
- HDL = **H**igh-density lipoproteins = **H**ealthy.

of the plaques of atherosclerosis described earlier. In clinical trials, lowering LDL greatly reduces the incidence of heart disease. By one estimate, for every percentage point drop in LDL cholesterol, the risk of heart disease falls proportionately (Figure C5-1 on page 176 demonstrated this effect). For this reason, the National Heart, Lung, and Blood Institute significantly lowered both the LDL cholesterol goal for those at high risk for heart disease and the threshold LDL level at which to begin LDL-lowering medication.[30] Stroke is less consistently associated with elevated LDL cholesterol.[31] Figure 11-6 sums up the values considered normal and those that exceed a healthy level for blood lipids, body mass index (BMI), and blood pressure.

High-density lipoproteins (HDL) also carry cholesterol, but HDL carry cholesterol away from the cells to the liver for other uses or disposal. Elevated HDL indicates a *reduced* risk of atherosclerosis and heart attack, with growing evidence that it may prevent stroke as well.[32] High LDL and low HDL correlate *directly* with heart disease; low LDL and high HDL correlate *inversely* with risk. For this reason, heart disease risk assessment inventories, such as the one in Figure 11-7, give extra credit for having a high HDL value.

Exactly how elevated LDL increases the risk of heart disease remains unclear, but one mechanism may be through reducing plaque stability. As LDL cholesterol is incorporated into plaques, the plaques weaken and become unstable. When this happens, they can rupture, causing a heart attack. In advanced atherosclerosis, a goal of treatment is to lower LDL cholesterol to stabilize existing plaques while slowing the development of new ones.

Hypertension Chronic hypertension worsens CVD. The higher the blood pressure above normal, the greater the risk of heart attack or stroke. The relationship between hypertension and disease risk holds for men and women, young and old. Hypertension injures the artery walls and accelerates plaque formation, thus initiating or worsening the progression of atherosclerosis. A later section gives details about hypertension because it constitutes a major threat to health on its own.

Diabetes Diabetes constitutes a major independent risk factor for all forms of cardiovascular disease, substantially increasing the risk of death from these causes.[33] In diabetes, atherosclerosis progresses rapidly, blocking blood vessels and diminishing

FIGURE 11·6 ADULT STANDARDS FOR BLOOD LIPIDS, BODY MASS INDEX (BMI), AND BLOOD PRESSURE

	Total blood cholesterol (mg/dL)	LDL cholesterol (mg/dL)	HDL cholesterol (mg/dL)	Triglycerides, fasting (mg/dL)	Body mass index (BMI)[f]	Blood pressure systolic / diastolic (mm Hg)
Unhealthy	≥240	160–189[a]	<40	200–499[b]	≥30	≥140 / ≥90
Borderline	200–239	130–159[c]	59–40	150–199	25–29.9	120/80–139/89[d]
Healthy	<200	<100[e]	≥60	<150	18.5–24.9	<120 / <80

[a] >190 mg/dL LDL indicates a very high risk.
[b] >500 mg/dL triglycerides indicates a very high risk.
[c] LDL cholesterol–lowering medication may be needed at 130 mg/dL, depending on other risks.
[d] These values indicate prehypertension.
[e] 100–129 mg/dL LDL indicates a near or above optimal level.
[f] Body Mass Index (BMI) was defined in Chapter 9; BMI standards are found on the inside back cover.

FIGURE 11·7 HOW TO ASSESS YOUR RISK OF HEART DISEASE

Do you know your heart disease risk score? This assessment estimates your ten-year risk for heart disease using charts from the Framingham Heart Study.[*] Be aware that a high score does not mean that you *will* develop heart disease, but it should warn you of the possibility and prompt you to consult a physician about your health. You will need to know your blood cholesterol (ideally, the average of at least two recent measurements) and blood pressure (ideally, the average of several recent measurements). With this information in hand, find yourself in the charts below and add the points for each risk factor.

Age (years)

	Points	
	Men	Women
20–34	−9	−7
35–39	−4	−3
40–44	0	0
45–49	3	3
50–54	6	6
55–59	8	8
60–64	10	10
65–69	11	12
70–74	12	14
75–79	13	16

HDL (mg/dL)

	Points	
	Men	Women
≥60	−1	−1
50–59	0	0
40–49	1	1
<40	2	2

Systolic Blood Pressure (mm Hg)

	Points			
	Untreated		Treated	
	Men	Women	Men	Women
<120	0	0	0	0
120–129	0	1	1	3
130–139	1	2	2	4
140–159	1	3	2	5
≥160	2	4	3	6

Total Cholesterol (mg/dL)

	Points									
	Age 20–39		Age 40–49		Age 50–59		Age 60–69		Age 70–79	
	Men	Women	Men	Women	Men	Women	Men	Women	Men	Women
<160	0	0	0	0	0	0	0	0	0	0
160–199	4	4	3	3	2	2	1	1	0	1
200–239	7	8	5	6	3	4	1	2	0	1
240–279	9	11	6	8	4	5	2	3	1	2
≥280	11	13	8	10	5	7	3	4	1	2

Smoking (any cigarette smoking in the past month)

Smoker	8	9	5	7	3	4	1	2	1	1
Nonsmoker	0	0	0	0	0	0	0	0	0	0

Scoring Your Heart Disease Risk

Add up your total points: _____ . Now find your total in the first column for your gender in the chart at the right and then look to the next column for your approximate risk of developing heart disease within the next ten years. Depending on your risk category, the following strategies can help reduce your risk:

- *>20% = High risk.* Try to lower LDL using all lifestyle changes and, most likely, lipid-lowering medications as well.

- *10–20% = Moderate risk.* Try to lower LDL using all lifestyle changes and, possibly, lipid-lowering medications.
- *<10% = Low risk.* Maintain or initiate lifestyle choices that help prevent elevation of LDL to prevent future heart disease.

Men		Women	
Total Points	**Risk**	**Total Points**	**Risk**
<0	<1%	<9	<1%
0–4	1%	9–12	1%
5–6	2%	13–14	2%
7	3%	15	3%
8	4%	16	4%
9	5%	17	5%
10	6%	18	6%
11	8%	19	8%
12	10%	20	11%
13	12%	21	14%
14	16%	22	17%
15	20%	23	22%
16	25%	24	27%
≥17	≥30%	≥25	≥30%

[*]An electronic version of this assessment is available on the ATP III page of the National Heart, Lung, and Blood Institute's website (www.nhlbi.nih.gov/guidelines/cholesterol). Another risk inventory is available from the American Heart Association (www.americanheart.org).

SOURCE: Adapted from Expert Panel on Detection, Evaluation, and Treatment of High Blood Cholesterol in Adults (Adult Treatment Panel III), Third Report of the National Cholesterol Education Program (NCEP), NIH publication no. 02-5215 (Bethesda, Md.: National Heart, Lung, and Blood Institute, 2002), section III.

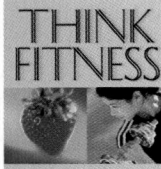
Ways to Include Physical Activity in a Day

THE BENEFITS OF physical activity are compelling, so why not tie up your athletic shoes, head out the door, and get going? Here are some ideas to get you started:

- Coach a sport.
- Garden.
- Hike, bike, or walk to nearby stores or to classes.
- Mow, trim, and rake by hand.
- Park a block from your destination and walk.
- Play a sport.
- Play with children.
- Take classes for credit in dancing, sports, conditioning, or swimming.
- Take the stairs, not the elevator.
- Walk a dog.
- Walk 10,000 steps per day. This amounts to about 5 miles, enough to meet the daily activity level that defines an active person. An inexpensive pedometer can record your steps.
- Wash your car with extra vigor, or bend and stretch to wash your toes in the bath.
- Work out at a fitness club.
- Work out with friends to help one another stay fit.

Also, try these:
- Give two laborsaving devices to someone who wants them.
- Lift small hand weights while talking on the phone or watching TV.
- Stretch often during the day.
- If you have access to the Internet, try the President's Challenge for improving fitness: www.presidentschallenge.org.

circulation. For many people with diabetes, the risk of a future heart attack is roughly equal to that of a person with a confirmed diagnosis of heart disease—two to four times as high as that of a person without diabetes.[34] When heart disease occurs in conjunction with diabetes, the condition is likely to be severe.

Few people with diabetes recognize that, left uncontrolled, diabetes holds a grave threat of all forms of CVD.[35] Even insulin resistance without diabetes may elevate the risk. A symptomless type of insulin resistance may greatly increase the likelihood of a heart attack—*without* abnormal blood glucose or lipid values that might otherwise warn the person. Chapter 4 provided some details about minimizing the damage from diabetes and insulin resistance.

Obesity and Physical Inactivity Obesity, especially central obesity, and physical inactivity amplify a person's risk of CVD by elevating LDL cholesterol while lowering HDL cholesterol and worsening hypertension and diabetes.[36] Conversely, weight loss and physical activity lower LDL, raise HDL, improve insulin sensitivity, and lower blood pressure.[37]

Routine physical activity strengthens muscles, including the muscles of the heart and arteries, and improves the heart's response to everyday demands.[38] Physical activity expands the volume of blood the heart can pump to the tissues at each beat, thereby reducing the heart's workload. It also stimulates development of new arteries to nourish the heart muscle, which may be a factor in the excellent recovery seen in some heart attack victims who exercise. In addition, physical activity favors lean over fat tissue for a healthy body composition. If pursued daily, even 30 minutes of light, balanced exercise, performed at intervals throughout the day, can improve the odds against heart disease considerably, and an hour a day, as recommended by the DRI committee, benefits the heart even more. The Think Fitness feature offers suggestions for incorporating physical activity into your daily routine.

■ Insulin resistance was defined in Chapter 4.

■ Read more about central obesity and weight-loss diets in Chapter 9; the benefits of fitness are a topic of Chapter 10.

Smoking   Cigarette smoking is a powerful risk factor for heart disease and other forms of CVD. The more a person smokes, the higher the CVD risk—a relationship that holds for both men and women. Smoking damages the heart directly with toxins and burdens it by raising the blood pressure. Body tissues starved for oxygen by smoke demand more heartbeats to deliver blood, thereby increasing the heart's workload. At the same time, smoking deprives the heart muscle itself of the oxygen it needs to maintain a steady beat. Smoking also damages platelets, making blood clots likely. Toxins in cigarette smoke directly damage the linings of the blood vessels, making atherosclerosis likely. When people quit smoking, their risk of heart disease begins to drop within a few months; a year later, their risk has dropped by half, and after 15 years of staying smoke-free, their risks equal those of lifetime nonsmokers.[39]

Atherogenic Diet   Diet influences the risk of CVD. An "atherogenic diet"—high in saturated fats, *trans* fats, and cholesterol—increases LDL cholesterol. Fortunately, a well-chosen diet often lowers the risk of CVD, and does so to a greater degree than might be expected from its effects on blood lipids alone. Any of a number of beneficial factors in such diets may take the credit, among them the vitamins, minerals, antioxidant phytochemicals, and omega-3 fatty acids. Strategies for choosing such a diet follow this section.

Metabolic Syndrome   People suffering from a characteristic cluster of CVD risk factors, including central obesity, high fasting blood glucose or insulin resistance, low blood HDL cholesterol, and hypertension, plus one other factor—elevated blood triglycerides—have an especially high risk of developing cardiovascular disease.[40] Most of these factors elevate CVD risk independently, and when they occur together, they synergistically elevate the risk.[41] (On its own, an elevated triglycerides level is not a risk factor for CVD.) This deadly cluster, called **metabolic syndrome**, approaches the power of high LDL cholesterol in raising the risk of CVD.[42] About 47 million people in the United States face the threat of metabolic syndrome, but many remain unaware of it and thus do not seek treatment.[43]

Features of metabolic syndrome:
- Central obesity
- Abnormal blood lipids: low HDL and high triglycerides.
- Elevated blood pressure.
- Elevated fasting blood glucose or insulin resistance.

Other Risk Factors   Factors other than the major ones listed earlier in Table 11-2 may also influence a person's risk. These factors, known as emerging risk factors, can help round out the assessment of an individual's risk of developing CVD. For example, research suggests emerging links among the B vitamins, homocysteine, and atherosclerosis, but authorities are not yet ready to declare homocysteine a risk factor for CVD, and B vitamins have yet to prove themselves effective in heart disease prevention.[44]

> ▶ **KEY POINT**   Major risk factors for CVD put forth by the National Cholesterol Education Program include age, gender, family history, high LDL cholesterol and low HDL cholesterol, hypertension, diabetes, obesity, physical inactivity, smoking, and an atherogenic diet. Other potential risk factors are under investigation.

Diet to Reduce CVD Risk

Now that you know which factors increase risks of developing CVD, what role can diet play in minimizing your risks? The answer focuses primarily on how *diet* relates to high blood cholesterol. Diet's effects are felt in two opposing ways: first, a diet high in saturated fat and *trans* fatty acids contributes to high blood LDL cholesterol; second, reducing those fats in the diet lowers blood LDL cholesterol and may reduce the risk of CVD.[45] Table 11-3, p. 406 demonstrates the power of diet-related factors to reduce LDL cholesterol.

Wherever in the world diets are high in saturated fat and low in fish, fruits, and vegetables, blood cholesterol is high, and heart disease takes a great toll on health and life.[46] Conversely, wherever dietary fat consists mostly of unsaturated fats with abundant fish, fruits, and vegetables, blood cholesterol and the rate of death from heart disease are low.

metabolic syndrome a combination of characteristic factors—high fasting blood glucose or insulin resistance, central obesity, hypertension, low blood HDL cholesterol, and elevated blood triglycerides—that greatly increase a person's risk of developing CVD. Also called *insulin resistance syndrome* or *syndrome X.*

TABLE 11·3 HOW MUCH DOES CHANGING THE DIET CHANGE LDL CHOLESTEROL?[a]

Diet-Related Component	Modification	Possible LDL Reduction
Saturated fat	<7% of calories	8–10%
Dietary cholesterol	<200 mg/day	3–5%
Weight reduction (if overweight)	Lose 10 lb	5–8%
Soluble, viscous fiber	5–10 g/day	3–5%

[a]See Table 11-4, p. 408 for other dietary changes believed to influence risk of CVD.

SOURCE: Expert Panel on Detection, Evaluation, and Treatment of High Blood Cholesterol in Adults (Adult Treatment Panel III), *Third Report of the National Cholesterol Education Program (NCEP)*, NIH publication no. 02-5215 (Bethesda, Md.: National Heart, Lung, and Blood Institute, 2002), p. V-21.

◼ Food sources of saturated fats and *trans* fatty acids were listed in Chapter 5; Controversy 5 distinguishes between fats that harm the heart and those thought to be safer.

◼ People with elevated triglycerides should limit simple sugars and highly refined starchy foods, which often cause triglycerides to rise.

◼ Chapter 5, pages 154–156, explained mechanisms by which fish oils may affect heart health.

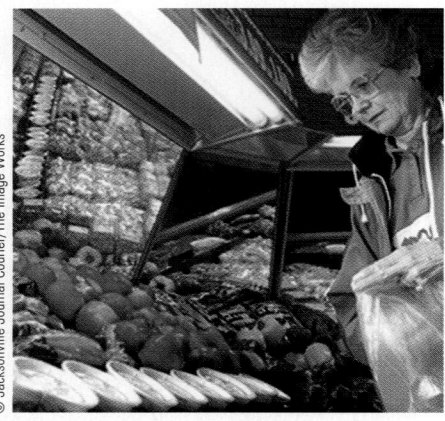

When diets are rich in vegetables and fruits, life expectancies are long.

Controlling Dietary Lipids Nutrition authorities agree on this point: lowering intakes of saturated fat and *trans* fat lowers blood LDL cholesterol and reduces heart disease risks.[47] For healthy people living in the United States and Canada, saturated fat, including *trans* fatty acids, in the diet should account for no more than 10 percent of calories. According to the *Dietary Guidelines for Americans 2005*, the diet should also contain no more than 35 percent of calories from total fat; other authorities recommend lower intakes of no more than 30 percent of calories from fat.[48]

The *Dietary Guidelines* committee, the World Health Organization, and many other experts also recommend that healthy people limit cholesterol intake to 300 milligrams a day in the context of a diet low in saturated fat and *trans* fatty acids. Although cholesterol plays a lesser role than saturated fat and *trans* fatty acids, it still elevates blood cholesterol. People with heart disease often consume diets rich in all three.[49]

Strictly limiting all kinds of dietary fats is probably not in the best interest of the health of the heart. The *Dietary Guidelines* state that healthy people choosing a diet moderate in fat—up to 35 percent of total calories—do not necessarily invite heart disease, as long as the diet is low in saturated fat and *trans* fat, controlled in energy (calories), and ample in fish, fruit, vegetables, and milk products.[50] Most people in the United States, however, eat diets rich in meats and hydrogenated fats, so if they reduce their *total* fat intakes, their *saturated* fat and *trans* fatty acids will follow suit.

It matters, however, what people choose to eat instead of saturated fats. Simple carbohydrates (sugars) seem to elevate blood triglycerides and reduce HDL cholesterol—effects opposite of those desired. Choosing polyunsaturated or monounsaturated fats or complex carbohydrates from whole grains or vegetables improves the lipoprotein profile.[51]

Fish oils, rich in omega-3 polyunsaturated fatty acids, lower triglycerides, prevent blood clots, and may reduce the risk of sudden death associated with CVD.[52] For these reasons, the American Heart Association recommends two meals of fish per week.[53] Plant sources of essential fatty acids, which include flaxseed and flaxseed oil, canola oil, soybean oil, and nuts, also confer benefits on the heart.[54] Evidence leans against the use of fish oil supplements, however, which carry their own risks (see the Consumer Corner of Chapter 5, page 162).

Effects of Fiber, Nutrients, and Phytochemicals In addition to limiting fats, the heart-healthy diet encourages consumption of foods rich in complex carbohydrates, and especially whole grains and other whole foods.[55] The viscous (soluble) fiber richly supplied by oats, barley, legumes, and pectin-rich fruits and vegetables helps to improve blood lipids.[56] These fibers bind cholesterol and bile in the intestine, reducing their absorption. An extra 5 to 10 grams of viscous fiber daily lowers LDL cholesterol levels by about 5 percent.[57] Additionally, a high-fiber diet may reduce a blood marker of inflammation thought to indicate an elevated risk of having a heart attack.[*][58] Foods rich in viscous fiber also provide minerals to help control blood pres-

*The marker is c-reactive protein.

sure (described later), antioxidants to help protect against LDL oxidation, and an array of vitamins and minerals, making them extraordinarily beneficial to health.

Diets rich in legumes, fruits, vegetables, and whole grains with their abundant fibers, nutrients, and phytochemicals correlate with low CVD rates. Some food constituents may help lower LDL cholesterol, protect it from oxidation, or lower blood pressure, but these effects are observed consistently in the body only when whole foods supply the needed compounds. Supplements of nutrients or phytochemicals are ineffective for these purposes, as earlier chapters made clear. The Food Feature of this chapter provides details needed to help in choosing a protective diet.

Alcohol People want to know whether moderate consumption of alcohol will reduce their CVD risk. Research on middle-aged and older people who drink one or two drinks a day with no binge drinking supports the idea.[59] For this group, moderate alcohol consumption has been reported to raise HDL cholesterol concentrations and reduce the risk of blood clots, thereby reducing the likelihood of heart attack.[60] The effect is small, however, and cannot reverse the effects of other risk factors such as imprudent diet or physical inactivity.

Heavy alcohol use (more than three drinks a day) is known to elevate blood pressure, to damage the heart muscle, to elevate the risk of stroke, and to have many other deleterious effects on the body's organs.[61] Heart attacks among apparently healthy young people have been associated with alcohol intoxication from heavy weekend drinking. The bottom line for young people is that the risks from alcohol greatly outweigh any benefit to the heart.

Other Dietary Factors Other dietary factors, such as soy products, also improve heart disease risk in various ways. Table 11-4 (p. 408) recounts them. A potentially helpful innovation is the sterol and stanol esters (introduced in Controversy 2) that have been added to foods such as margarine and orange juice. These compounds block absorption of cholesterol from the intestine, dropping blood cholesterol by about 7 to 10 percent. When these compounds accompany a diet low in saturated fats and high in ordinary plant foods, the effect may be as powerful as some medications in lowering blood LDL cholesterol.[62] Some foods, such as corn oil, naturally contain compounds related to sterol or stanol esters, but the amount present in a typical serving is too small to affect the health of the arteries. Sterol and stanol esters also reduce absorption of some potentially beneficial phytochemicals, so their use may be best reserved for those who fail to lower their elevated cholesterol by other means.

More Strategies against CVD Periodically, the media repopularize the idea that the vitamin niacin can lower blood cholesterol. Experimentally, pharmaceutical doses of a specific form of niacin act like a drug in lowering blood cholesterol and prolonging life, but other drugs effective for this purpose probably have fewer side effects. Some of the most common of these drugs, the "statin" drugs, efficiently lower the blood LDL values of users. They work best in association with a cholesterol-lowering diet and exercise.[63] Ordinary niacin supplements are useless in lowering blood cholesterol.

Authors of some diet books claim that low-carbohydrate diets, with their high intakes of meats and saturated fats, lower blood cholesterol. Shifts in blood lipid values have been reported, especially reduced blood triglycerides and reduced HDL, but this profile is not known to benefit the heart.[64] When the diets are low enough in calories to produce weight loss, blood cholesterol diminishes along with body fatness, but this occurs with weight loss achieved on any kind of diet. Standard cholesterol-lowering diets that are low in saturated fats improve or maintain the beneficial HDL, while lowering LDL cholesterol, and may produce beneficial weight loss if calories are kept low.

Although diet and exercise are not the easy route to heart health that everyone hopes for, they form a powerful and safe combination for improving health. Needed weight loss

■ Table 4-2 on page 106 lists fibers, characteristics, actions in the body, and health benefits. Controversy 2 gave details about phytochemicals.

■ Controversy 3 gave details about the health effects of alcohol.

On an average day:
- 4 college students die from alcohol-related accidents.
- 192 college students are raped by their dates or sexually assaulted after drinking alcohol.
- 1,370 college students suffer drinking-related injuries.

Over 80% of college students say they drink.

SOURCE: National Institute of Alcohol Abuse and Alcoholism, *A Call to Action: Changing the Culture of Drinking at U.S. Colleges*, 2002, available at www.collegedrinkingprevention.gov/Reports/.

■ Controversy 6 discussed potential benefits of soybeans and soy products to heart health.

TABLE 11·4 OTHER DIETARY FACTORS THAT MAY PROTECT AGAINST CVD

In addition to the dietary interventions mentioned in Table 11-3, these effects of dietary factors are supported by some research.

Dietary Factor	Protection against CVD
Omega-3 fatty acids (fish oils from 2 meals of fatty fish per week)	• Limit clot formation • Prevent irregular heartbeats • Decrease blood triglycerides; slightly increase HDL cholesterol • Decrease inflammation • Relax the arteries, reducing blood pressure
Folate, vitamin B$_6$, vitamin B$_{12}$ (DRI recommended intakes)	• Reduce homocysteine
Vitamin E (DRI recommended intake)	• Reduces arterial injuries and inflammation • Limits LDL oxidation
Soy foods including soy protein, soy "milk" products, and soybeans	• Contain antioxidants • Lower LDL cholesterol
Sterol esters, stanol esters	• Block cholesterol absorption and lower blood cholesterol
Alcohol (for older adults, in moderation)	• Raises HDL cholesterol • Limits clot formation

often reduces blood pressure. So does eating a diet low in fat, restricted in cholesterol, and high in complex carbohydrates, whole grains, fruits, and vegetables. In a recent study of 15,000 male physicians, those consuming at least one and a half cups of dark green, deep yellow, or red vegetables daily had a 23 percent lower risk of heart disease than men who ate less than half a cup each day.[65] And even if such a diet does not lower cholesterol or blood pressure, it will help by normalizing blood glucose (diabetes). Remember, diabetes is a major risk factor for CVD. A meal or two of fish each week can help by favoring the right fatty acid balance so that clot formation is unlikely. The pattern of protection from the recommended diet and exercise regimen becomes clear—the effects of each small choice add to the beneficial whole. While you are at it, don't smoke. Relax. Meditate or pray. Control stress. Play. Relaxed, happy people have lower rates of heart attack.[66]

▶ KEY POINT   High blood cholesterol indicates a risk of heart disease, and diet can contribute to lowering blood cholesterol and thus the risk of heart disease. Dietary measures to lower LDL cholesterol include reducing intakes of saturated fat, *trans* fat, and cholesterol, along with obtaining the fiber, nutrients, and phytochemicals of fruits, vegetables, legumes, and whole grains. Even for people on prescription medication, diet and exercise are key.

Nutrition and Hypertension

People with healthy low blood pressure generally enjoy a long life and suffer less often from heart disease. Chronic high blood pressure, or hypertension, remains one of the most prevalent forms of CVD, affecting almost one-third of the entire U.S. adult population.[67] For people age 65 or older, the lifetime risk of developing it approaches 90 percent. Hypertension contributes to half a million strokes and over a million heart attacks each year, and its rate has been rising steadily.[68] The higher above normal the blood pressure, the greater the risk of heart disease. Hypertension is especially threatening when paired with atherosclerosis, as it often is.

You cannot tell if you have high blood pressure—it presents no symptoms you can feel. The most effective single step you can take to protect yourself from hypertension is to find out whether you have it. During a checkup, a health-care professional can take an accurate resting blood pressure reading. Self-test machines in drugstores and other public places are often inaccurate. If your resting blood pressure is above normal, the reading should be repeated before confirming the diagnosis of hypertension. Thereafter blood pressure should be checked at regular intervals.

When blood pressure is measured, two numbers are important: the pressure during contraction of the heart's ventricles (large pumping chambers) and the pressure during their relaxation. The numbers are given as a fraction, with the first number representing the **systolic pressure** (ventricular contraction) and the second number the **diastolic pressure** (relaxation). Return to Figure 11-6 (p. 402) to see how to interpret your resting blood pressure.

Ideal resting blood pressure is 120 over 80 or lower, but a reading of 130 over 85 can be considered borderline normal, especially in the absence of other risk factors for CVD, such as diabetes.[69] Above this level, though, the risks of heart attacks and strokes rise in direct proportion to increasing blood pressure.

▶ KEY POINT   Hypertension is silent, progressively worsens atherosclerosis, and makes heart attacks and strokes likely. All adults should know their blood pressure.

The most effective single step you can take against hypertension is to learn your own blood pressure.

How Does Blood Pressure Work in the Body, and What Makes It Too High?

Blood pressure is vital to life. It pushes the blood through the major arteries into smaller arteries and finally into tiny capillaries whose thin walls permit exchange of fluids between the blood and the tissues (see Figure 11-8, p. 410). When the pressure is right, the cells receive a constant supply of nutrients and oxygen and can release their wastes.

The Role of the Kidneys   For the kidneys to filter waste materials out of the blood and into the urine, blood pressure has to be high enough to force the blood's fluid out of the capillaries into the kidneys' filtering networks. If the blood pressure is too low, the kidneys act to increase it—they send hormones to constrict blood vessels and to retain water and salt in the body. Dehydration sets these actions in motion, and in this case they are beneficial because when the blood volume is low, higher blood pressure is needed to deliver substances to the tissues. By constricting the blood vessels and conserving water and sodium, the kidneys ensure that normal blood pressure is maintained until the dehydrated person can drink water.

Atherosclerosis also sets this process in motion, however, and this is not beneficial. By obstructing blood vessels, atherosclerosis fools the kidneys, which react as if there were a water deficiency. The kidneys raise the blood pressure high enough to get the blood they need, but in the process they may make the pressure too high for the arteries and heart to withstand. Hypertension also aggravates atherosclerosis by mechanically injuring the artery linings, increasing the likelihood that plaques will form; plaques restrict blood flow to the kidneys, which may then raise the blood pressure still further; and the problem snowballs.

The Roles of Risk Factors   Primary among the risk factors that precipitate or aggravate hypertension are atherosclerosis, obesity (particularly central obesity), and insulin resistance (which leads to type 2 diabetes). Excess fat means miles of extra capillaries through which the blood must be pumped.

Epidemiological studies have identified several other risk factors that predict hypertension. One is age: most people who develop hypertension do so in their 50s and 60s. Another is inherited genes, and researchers are working to identify the genetic determinants of hypertension.[70] A family history of hypertension and heart disease raises the risk of developing hypertension two to five times, and African Americans are likely to develop more severe hypertension earlier in life than people of European or Asian descent. Hypertension also bears some relation to insulin resistance, probably through a common genetic link, and some measures to prevent diabetes also protect against hypertension.

systolic (sis-TOL-ik) **pressure** the first figure in a blood pressure reading (the "dupp" of the heartbeat is heard), which reflects arterial pressure caused by the contraction of the heart's left ventricle.

diastolic (dye-as-TOL-ik) **pressure** the second figure in a blood pressure reading (the "lubb" of the heartbeat is heard), which reflects the arterial pressure when the heart is between beats.

FIGURE 11·8 THE BLOOD PRESSURE

Three major factors contribute to the pressure inside an artery. First, the heart pushes blood into the artery. Second, the small-diameter arteries and capillaries at the other end resist the blood's flow (peripheral resistance). Third, the volume of fluid in the circulatory system, which depends on the number of dissolved particles in that fluid, adds pressure.

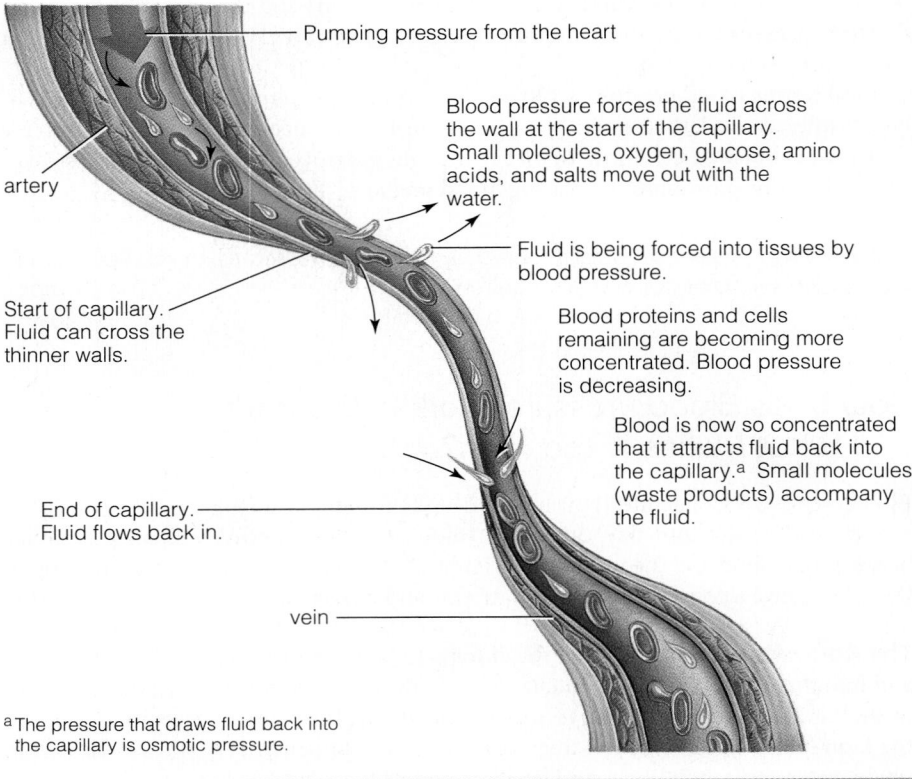

Pumping pressure from the heart

Blood pressure forces the fluid across the wall at the start of the capillary. Small molecules, oxygen, glucose, amino acids, and salts move out with the water.

artery

Fluid is being forced into tissues by blood pressure.

Start of capillary. Fluid can cross the thinner walls.

Blood proteins and cells remaining are becoming more concentrated. Blood pressure is decreasing.

Blood is now so concentrated that it attracts fluid back into the capillary.[a] Small molecules (waste products) accompany the fluid.

End of capillary. Fluid flows back in.

vein

[a] The pressure that draws fluid back into the capillary is osmotic pressure.

Environmental factors in the United States may also favor the development of hypertension. Africans living in Africa have a much lower rate of hypertension than African Americans living in the United States.

▶ **KEY POINT** Atherosclerosis, obesity, insulin resistance, age, family background, and race contribute to hypertension risks. Prevention and treatment both deserve high-priority efforts.

How Does Nutrition Affect Hypertension?

Even mild hypertension can be dangerous, but individuals who adhere to treatment are less likely to suffer illness or early death. Some people need medications to bring their blood pressure down, but diet and exercise alone can bring improvements for many and prevent hypertension for many others. Important nutrition factors are lowering salt intake; losing weight for those who are overweight; using moderation with regard to alcohol consumption; increasing intakes of fruit, vegetables, fish, and low-fat dairy products; and reducing intakes of fat. Calcium, potassium, magnesium, and other nutrients seem to play roles as does physical activity.

A blanket recommendation for prevention of hypertension, then, would center on controlling weight, consuming a nutritious diet, exercising regularly, controlling intakes of alcohol, and holding sodium intakes to prescribed levels. A classic example of such a diet, known as DASH (Dietary Approaches to Stop Hypertension) recommends significant increases in fruit and vegetable intakes, provides 30 percent of its calories from fat, emphasizes legumes over red meats, restricts sodium, and meets other recommendations of the *Dietary Guidelines for Americans 2005* besides (see Table 11-5 in the margin). The DASH diet and other similar diets consistently improve blood pressure in both study subjects whose diets are provided by researchers and those freely choosing and preparing their own foods in accordance with instructions.[71]

Weight Control and Physical Activity   For people who are overweight and hypertensive, a weight loss of as little as 10 pounds can significantly lower blood pressure.[72] Those who are taking medication to control their blood pressure can often cut down their doses if they lose weight.

Moderate physical activity helps in weight loss and also helps to reduce hypertension directly.[73] The right kind of regular physical activity can lower blood pressure in almost everyone, even in those without hypertension. Even a single session of exercise reduces the blood pressure, an effect that lasts 12 hours or more and intensifies as training improves physical condition.[74] The "right kind" of activity can be walking, jogging, cycling, or others that also increase blood HDL and lower LDL.[75] The exercise need not be strenuous—even walking at any pace (about 10,000 steps) per day seems to be effective in bringing down high blood pressure.[76] The activity may even be broken up into manageable 10-minute segments throughout the day and still provide benefits.[77] Aerobic activity for cardiovascular fitness was described in Chapter 10.

Physical activity also changes the hormonal climate in which the body does its work. By reducing stress, physical activity reduces the secretion of stress hormones, and this lowers blood pressure. Physical activity also redistributes body water and eases transit of the blood through the peripheral arteries.

Salt, Sodium, and Blood Pressure   High dietary intakes of salt and sodium are associated with hypertension. As salt intakes decrease, blood pressure drops in a stepwise fashion.[78] This direct relationship is reported at all levels of intake, from very low to much higher than average. Although the benefit from reducing salt intake to prevent and treat hypertension is beyond question, some people respond to salt more sensitively, including African Americans, people with a family history of hypertension, people with kidney problems or diabetes, or older people. People with hypertension often find that by lowering their salt (or sodium) intakes sufficiently, they can reduce or even eliminate blood-pressure-reducing medication.[79] Also increasing dietary potassium may help to heighten salt sensitivity in those who do not respond as vigorously to salt restriction alone.[80]

The World Health Organization estimates that a significant reduction in sodium intake could reduce by half the number of people requiring therapy for hypertension and greatly reduce deaths from CVD. The recommendation of professionals and agencies, based on available research, is that everyone should moderately restrict salt and sodium intake to a level not exceeding the DRI committee's Tolerable Upper Intake Level—that is, no more than 2,300 milligrams of sodium per day—and follow other standard dietary advice to help avoid hypertension.[81]

Alcohol   In moderate doses, alcohol initially relaxes the peripheral arteries and so reduces blood pressure, but high doses definitely raise blood pressure.[82] Hypertension is common among people with alcoholism and is apparently caused directly by the alcohol. Hypertension caused by alcohol leads to CVD, the same as hypertension caused by any other factor. Furthermore, alcohol may cause strokes—even *without* hypertension. The *Dietary Guidelines for Americans 2005* urge a sensible, moderate approach for those who drink alcohol. *Moderation* means no more than one drink a day for women or two drinks a day for men, an amount that seems safe relative to blood pressure.

Calcium, Potassium, Magnesium, and Vitamin C   Other dietary factors may help to regulate blood pressure. A diet providing enough calcium may be one such factor—increasing calcium often reduces blood pressure in both healthy people and those with

Food Group	Recommended Number of Daily Servings[a]
Grains	7–8
Vegetables	4–5
Fruits	4–5
Milk (fat-free/ low-fat)	2–3
Meat (lean)[b]	2 or less
Calories	2,000

TABLE 11·5 THE DASH EATING PLAN

NOTE: The DASH eating plan recommends that fats, oils, and sweets be used sparingly.
[a]For details on serving sizes turn to Table I-3 of Appendix I.
[b]The DASH eating plan also includes recommended servings for nuts, seeds, and dry beans (4 to 5 per week).

■ Chapter 8 and Appendix I also address the DASH diet.

■ Details concerning sodium, hypertension, and food sources of sodium were presented in Chapter 8.

hypertension. By one estimate, increasing the milk intake of the U.S. population to the recommended three to four cups per day could reduce hypertension by 40 percent, cut health-care costs by $14 billion per year, and save many human lives in the process.[83] If you are concerned about your blood pressure, include more calcium-rich foods in your diet.

Adequate potassium and magnesium also appear to help prevent and treat hypertension in certain populations. Diets low in potassium-rich fruits and vegetables are often associated with hypertension, whereas diets adequate in such foods appear to both prevent and correct hypertension. Magnesium deficiency causes the walls of the arteries and capillaries to constrict and so may raise the blood pressure. Similarly, consuming a diet adequate in vitamin C seems to help normalize blood pressure, while vitamin C deficiency may tend to raise it.[84]

Other dietary factors may also affect blood pressure; the roles of cadmium, selenium, lead, caffeine, protein, and fat are being studied. Vitamin supplements have not shown promise for lowering blood pressure in the long term.[85]

How can people be sure of getting all of the nutrients needed to keep blood pressure low? The best answer is to consume a low-fat diet with abundant fruits, vegetables, and low-fat dairy products that provide the needed nutrients while holding sodium intake in bounds. In addition to reducing blood pressure, a diet such as the DASH diet may also lower blood cholesterol values, providing a two-way benefit to the heart.[86]

The Food Feature provides more details on dietary measures that help support normal blood pressure. Should diet and exercise fail to reduce blood pressure, drugs such as diuretics and other antihypertensive agents may be prescribed. Some of these work by increasing fluid loss in the urine, so they may also cause potassium losses. Chapter 8 specified foods that are rich in potassium, and people taking these drugs should make it a point to consume such foods daily.

People searching for an extra boost to health often combine nutrient supplements with herbs and other alternatives to mainstream medicine. The Consumer Corner near here provides a look at some of these practices.

> **KEY POINT** For most people, a healthy body weight, regular physical activity, consuming little salt and sodium, moderation for those who use alcohol, and a diet high in fruits, vegetables, fish, and low-fat dairy products and low in fat work together to keep blood pressure normal.

Nutrition and Cancer

Cancer ranks second only to heart disease as a leading cause of death and disability in the United States. Recent advances in early detection and treatment have transformed some potentially deadly cancers into curable diseases or treatable chronic illnesses. The term *cancer survivor* has now all but replaced *cancer victim* when referring to a person with a diagnosis of cancer.[87]

Though the potential for cure is exciting, prevention of cancer remains far and away preferable. Can an individual's chosen behaviors affect the risk of contracting cancer? Probably. Inherited tendencies exert only a modest effect on most people's risk of cancer development.[88] Just a very few rare cancers are known to be caused by genetic inheritance alone and will appear in members of an affected family regardless of lifestyle choices. A few more are linked with viral infections, but far more often lifestyle and environmental factors come into play. For example, if everyone in the United States quit smoking right now and stayed quit, future total cancers would probably drop by a third. Another 15 percent may be preventable by preventing overweight and obesity.[89] Lack of physical activity almost certainly plays a role in the development of colon and breast cancer and probably contributes to others.[90]

Another 20 to 50 percent of total cancers are influenced by diet, and these relationships are the focus of this section. Dietary fat, meat, alcohol, excess calories, and low

■ Environmental tobacco smoke (passive or secondhand smoke), overexposure to sun, infections, and exposure to water and air pollution or other toxic chemicals (possibly including pesticides that mimic estrogen in the body) are also responsible for a percentage of cancers.

cancer a disease in which cells multiply out of control and disrupt normal functioning of one or more organs.

intakes of fruits and vegetables have been the targets of much research with regard to the occurrence of cancer. Such constituents of the diet relate to cancer in several ways:

- Foods or their components may cause cancer.
- Foods or their components may promote cancer.
- Foods or their components may protect against cancer.

Also, for the person who has cancer, diet can make a crucial difference in recovery.

Research on diet and cancer remains challenging, and today's research methods are under scrutiny for refinement.[91] Some dietary and environmental factors currently believed to be important in cancer causation are listed in Table 11-6.

TABLE 11·6	FACTORS ASSOCIATED WITH CANCERS AT SPECIFIC SITES	
Cancer Sites	**Associated with:**	**Possible Protective Effect from:**
Bladder cancer	Cigarette smoking and alcohol; weak association with coffee and chlorinated drinking water	Fruits and vegetables (especially fruits); adequate fluid intake
Breast cancer	High intakes of food energy, alcohol intake; low vitamin A intake; obesity; sedentary lifestyle; probably high saturated fat intake; possibly high sucrose intake	Monounsaturated fats; physical activity
Cervical cancer	Folate deficiency; viral infection; possibly cigarette smoking	Adequate folate intake; possibly, fruits and vegetables
Colorectal cancer	High intakes of fat (particularly saturated fat), red meat, alcohol, and supplemental iron; low intakes of fiber, folate, vitamin D, calcium, and vegetables; obesity; inactivity; cigarette smoking	Vegetables, especially cruciferous (cabbage-type) vegetables; calcium, vitamin D, and dairy intake; possibly, whole wheat and wheat bran; high levels of physical activity
Kidney cancer	Possibly, high intakes of red meat (especially fried, sautéed, charred, burned, or cooked well-done); cigarette smoking; obesity	Fruit and vegetables, especially orange-colored and dark green ones
Mouth, throat, and esophagus cancers	Heavy use of alcohol, tobacco, and especially combined use; heavy use of preserved foods (such as pickles); low intakes of vitamins and minerals; obesity (esophageal)	Fruits and vegetables
Liver cancer	Infection with hepatitis virus; high intakes of alcohol; iron overload; toxins of a mold (aflatoxin) or other toxicity	Vegetables, especially yellow and green
Lung cancer	Smoking; low vitamin A; supplements of beta-carotene (in smokers); air pollution	Fruits and vegetables
Ovarian cancer	Possibly, high lactose intake from milk products; inversely correlated with oral contraceptive use	Vegetables, especially green leafy ones
Pancreatic cancer	Possibly, high intakes of red meat; correlated with cigarette smoking	Fruits and vegetables, especially green and yellow ones
Prostate cancer	High intakes of fats, especially saturated fats from red meats and possibly milk products	Possibly, cooked tomatoes, soybeans, soy products, and flaxseed; adequate selenium intake
Stomach cancer	High intakes of smoke- or salt-preserved foods (such as dried, salted fish); cigarette smoking; possibly, refined flour or starch; infection with ulcer-causing bacteria	Fresh fruits and vegetables, especially tomatoes

SOURCES: I. Romieu and coauthors, Carbohydrates and risk of breast cancer among Mexican women, *Cancer Epidemiology, Biomarkers, and Prevention* 13 (2004): 1283–1289; National Cancer Policy Board, Institute of Medicine, *Fulfilling the Potential of Cancer Prevention and Early Detection*, ed. S. J. Curry, T. Byers, and M. Hewitt (Washington, D.C.: National Academies Press, 2003), pp. 66–86; S. E. McCann and coauthors, Risk of human ovarian cancer is related to dietary intake of selected nutrients, phytochemicals and food groups, *Journal of Nutrition*, 133 (2003): 1937–1942; S. A. Lamprecht and M. Lipkin, Chemoprevention of colon cancer by calcium, vitamin D and folate: Molecular mechanisms, *Nature Reviews Cancer* 3 (2003): 601–614; K. Z. Guyton and coauthors, Vitamin D and vitamin D analogs as cancer chemopreventive agents, *Nutrition Reviews* 61 (2003): 227–238; S. A. Smith-Warner and coauthors, Intake of fruits and vegetables and risk of breast cancer, *Journal of the American Medical Association* 285 (2001): 769–776; R. L. Nelson, Iron and colorectal risk: Human studies, *Nutrition Reviews* 59 (2001): 140–148; M. C. Jansen and coauthors, Dietary fiber and plant foods in relation to colorectal cancer mortality: The Seven Countries Study, *International Journal of Cancer* 81 (1999): 174–179; B. S. Reddy, Role of dietary fiber in colon cancer: An overview, *American Journal of Medicine* 106 (1999): S16–S19.

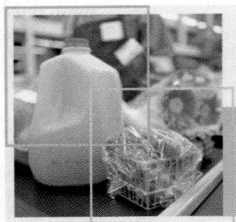

CONSUMER CORNER

Alternative Therapies and Herbal Medicine

WHERE DO YOU turn for help when illness strikes? Do you see a physician who offers treatment methods sanctioned by the established medical community? Or do you seek out an herbalist, an acupuncturist, or another practitioner of **alternative therapy?** As more people turn to alternative medicine, U.S. consumers are spending upward of $21 billion each year on such treatments.[1] Terms are defined in Table 11-7.

Some health-care professionals embrace certain helpful alternative therapies in their practices.[2] This open-minded approach, termed *complementary medicine*, often combines some humane elements of alternative therapies that fill emotional needs along with the best of proven scientific medicine, delivered by skilled physicians.

Unlike conventional therapies, alternative therapies often have been used for centuries but have not been scientifically evaluated for safety or effectiveness.[3] Most medical schools in the United States do not teach them, nor do most health insurance policies pay for them. Further, anyone can claim to be an expert in a "new" or "natural" therapy, and many practitioners act knowledgeable but are either misinformed or frauds (see Controversy 1).

Testimonials for mysterious "cures" by alternative therapies abound. The listener may think that unless the speaker is lying, the therapies really do cure diseases. But a third option also exists: remember from Chapter 1 that an ill person's belief in a treatment, even a placebo, often leads to physical healing.[4]

This is not to say that all alternative therapies are shams; several examples of effective therapies are described in this section. Nevertheless, the common belief that many of today's alternative therapies will become tomorrow's mainstream medicine is unfounded.

TABLE 11·7 ALTERNATIVE THERAPY TERMS

- **acupuncture** (ak-you-punk-chur) a technique that involves piercing the skin with long thin needles at specific anatomical points to relieve pain or illness. Acupuncture sometimes uses heat, pressure, friction, suction, or electromagnetic energy to stimulate the points.
- **alternative therapy** systems of theory and practice that have emerged independently of conventional biomedical approaches. Examples include acupuncture, biofeedback, chiropractic, faith healing, and many others. Also called *alternative medical systems.*
- **herbal medicine** use of herbs and other natural substances with the intention of preventing or curing diseases or relieving symptoms.

To tease apart potentially useful and safe alternative therapies from the worthless or harmful, the prestigious National Institutes of Health established its National Center for Complementary and Alternative Medicine (NCCAM). So far, NCCAM has found **acupuncture** useful for quelling nausea from surgery, cancer chemotherapy, and pregnancy and for relieving pain during dental procedures. Other potential uses include treating chronic headaches and migraines.[5] Acupuncture, when performed by skilled practitioners under sterile conditions, presents far less risk to the user than drugs used for these conditions.[6]

Dozens of **herbal medicines** contain effective natural drugs. For example, the resin called myrrh contains an analgesic (pain-killing) compound; willow bark contains aspirin; the herb valerian contains a tranquilizing oil; senno leaves produce a powerful laxative. The World Health Organization currently recommends an herbal Chinese drug to developing nations of the tropics as an

effective, important tool for combating malaria.*

Herbal medicines have several serious drawbacks, however. When analyzed, a majority of herb pills and supplements do not contain the species or the active ingredients stated on their labels.[7] Recently, researchers tested almost 900 samples of herbs and found that more often than not, the ingredients, potency, or variety differed significantly from the information on the labels.[8]† Another example is *Aristolochia fangchi*, an herb that causes severe kidney damage and cancer. It was mistakenly included in Chinese herbal "weight reduction pills."[9] Some preparations may even include *Aristolochia* intentionally because no safety tests are required before herbs can be marketed.

Contamination can also be a problem. The heavy metal lead found in one herbal remedy reached toxic doses.[10] Mercury and arsenic detected in traditional Chinese herb balls for treating fever, rheumatism, and cataracts have exceeded the maximum allowable levels by 20,000 and 1,000 times, respectively. Interactions of herbs with medications are also common (see Chapter 14's Controversy).

Many people undergoing cancer treatments ask about the macrobiotic diet (see Controversy 6), a restrictive low-fat diet of grains, soy beans, and certain vegetables. The diet is promoted for curing diseases but has not been proved scientifically to be beneficial.[11] Indeed, research seems to weigh against its use, partly because this bulky, low-calorie diet fails to provide the energy needed to battle cancer.[12]

*The herb is artemisinin (ar-TEM-is-in-in), derived from sweet wormwood.
†The herbs tested included echinacea, St. John's wort, ginkgo biloba, garlic, saw palmetto, ginseng, goldenseal, aloe, Siberian ginseng, and valerian. For more information, see the web site of the National Council Against Health Fraud at www.quackwatch.org.

TABLE 11·8 SELECTED HERBS: THEIR EFFECTS AND HAZARDS[a]

Hazardous

- **belladonna** any part of the deadly nightshade plant; a fatal poison.
- **hemlock** any part of the hemlock plant, which causes severe pain, convulsions, and death within 15 minutes.
- **pennyroyal** relatives of the mint family brewed as tea or extracted as oil; used as mosquito repellent, claimed to treat various conditions. Tea produced multiple organ failure in infants; ½ teaspoon of oil caused convulsions and coma; 2 tablespoons caused the death of an 18-year-old expectant mother within 2 hours, despite hospitalization.

Probably Hazardous

- **chaparral** an herbal product made from ground leaves of the creosote bush and sold in tea or capsule form; supposedly, this herb has antioxidant effects, delays aging, "cleanses" the bloodstream, and treats skin conditions—all unproven claims. Chaparral has been found to cause acute toxic hepatitis, a severe liver illness. Deaths reported.
- **comfrey** leaves and roots of the comfrey plant; believed, but not proved, to promote cell proliferation. Toxic to the liver in doses ordinarily used. Deaths reported.
- **foxglove** a plant that contains a substance used in the heart medicine digoxin.
- **germander** an evergreen bush used in small quantities as a flavoring for alcoholic beverages. Recommended for gout and other ills, it causes often-irreversible liver damage and abnormalities. Deaths reported.
- **lobelia** (low-BEE-lee-uh) dried leaves and tops of lobelia ("Indian tobacco") plant used to induce vomiting or treat a cough; abused for a mild euphoria. Causes breathing difficulty, rapid pulse, low blood pressure, diarrhea, dizziness, and tremors. Possible deaths reported.
- **sassafras** root bark from the sassafras tree; once used in beverages but now banned as an ingredient in foods or beverages because it contains cancer-causing chemicals.

Could Be Hazardous

- **echinacea** (EK-eh-NAY-see-ah) an herb popular before the advent of antibiotics for its "anti-infectious" properties and as an all-purpose remedy, especially for colds and allergy and for healing of wounds. Research is mixed on these claims. An insecticidal property opens questions about safety. Also called *cone-flower*.
- **ginkgo biloba** an extract of a tree of the same name, claimed to enhance mental alertness, but not proved to be effective or safe.
- **ginseng** (JIN-seng) a plant root containing chemicals that have stimulant drug effects. *Ginseng abuse syndrome* is a group of

symptoms associated with the overuse of ginseng, including high blood pressure, insomnia, nervousness, confusion, and depression.

- **kombucha** (KOM-boo-sha) a product of fermentation of sugar-sweetened tea by various yeasts and bacteria. Proclaimed as a treatment for everything from AIDS to cancer but lacking scientific evidence. Microorganisms in home-brewed kombucha have caused serious illnesses in people with weakened immunity. Also known as *Manchurian tea, mushroom tea,* or *Kargasok tea*.
- **skullcap** a native herb with no known medical uses but found in remedies. Other species may be harvested and sold as skullcap, so it has not been determined whether several deaths from liver toxicity reportedly from skullcap were in fact from another herb.

Hazard Undefined

- **aloe** a tropical plant with widely claimed value as a topical treatment for minor skin injury. Some scientific evidence supports this claim; evidence against its use in severe wounds also exists.
- **cat's claw** an herb from the rain forests of Brazil and Peru; claimed, but not proved, to be an "all-purpose" remedy.
- **chamomile** flowers that may provide some limited medical value in soothing menstrual, intestinal, and stomach discomforts.
- **feverfew** an herb sold as a migraine headache preventive. Some evidence exists to support this claim.
- **kava** the root of a tropical pepper plant, often brewed as a tea consumed for its calming effects. Limited scientific research supports use for anxiety relief. Adverse effects include skin rash, liver abnormalities, lethargy, and mental disorientation.
- **kudzu** a weedy vine, whose roots are harvested and used by Chinese herbalists as a treatment for alcoholism. Kudzu reportedly reduces alcohol absorption by up to 50% in rats.
- **saw palmetto** the ripe fruit or extracts of the saw palmetto plant. Claimed to relieve symptoms associated with enlarged prostate but reported as ineffective in research.
- **St. John's wort** an herb containing psychoactive substances that has been used for centuries to treat depression, insomnia, bed-wetting, and "nervous conditions." Some scientific reports find St. John's wort equal in effectiveness to standard antidepressant medication for relief of depression. Long-term safety, however, has not been established.
- **valerian** a preparation of the root of an herb used as a sedative and sleep agent. Safety and effectiveness of valerian have not been scientifically established.
- **witch hazel** leaves or bark of a witch hazel tree; not proved to have healing powers.

[a]See also Table C7-2 of Controversy 7 and Table 9-11 of Chapter 9.
SOURCES: S. Foster and V. E. Tyler, *Tyler's Honest Herbal: A Sensible Guide to the Use of Herbs and Related Remedies* (New York: Hawthorn Press, 2000); Twelve supplements you should avoid, *Consumer Reports.org,* May 2004, available at www.consumerreports.org.

To self-diagnose illnesses, and then choose herbs for treatment, is to invite problems. Few herbalists selling the pills base herbal therapy on botany, pharmacology, or human physiology, but instead rely on hearsay and folklore to plan treatments. By delaying effective medical help, this dangerous choice may allow serious conditions time to worsen. Also, perilous mistakes with herbs are extraordinarily likely. For example, most mint is safe when brewed as tea, but some may contain highly toxic pennyroyal oil. Folk medicine urges parents to soothe a colicky baby with mint tea, but one concoction laden with pennyroyal was blamed for liver and neurological injuries to at least two infants, one of whom died. Table 11-8 lists some additional herbs and their potential actions.[‡]

Intelligent, clear-minded people can fall for the allure of magical thinking when standard medical therapies fail. Many such consumers, loving life and desperate, fall prey to the worst kind of quackery on the feeblest promise of a cure.

[‡]A reliable source of information about herbs is V. E. Tyler, *The Honest Herbal* (New York: Pharmaceutical Products Press). Look for the latest edition.

How Does Cancer Develop?

Cancer arises as a disease of the genes.[92] It often begins when a cell's genetic material (DNA) sustains damage from a **carcinogen** such as a free-radical compound, radiation, or other influences. Such damage occurs every day, but most of the damage is quickly repaired. Sometimes DNA collects bits of damage here and there over time. Usually, if the damage cannot be repaired and the cell becomes unable to faithfully replicate its genome, the cell self-destructs, committing a sort of cellular suicide to avoid passing on the faulty genetic material to its progeny. Occasionally, a damaged cell loses its ability to self-destruct and also loses the ability to stop reproducing. It replicates uncontrollably, and the result is a mass of abnormal tissue—a tumor. Life-threatening cancer results when the tumor tissue overtakes the healthy organ in which it developed or disseminates its cells through the bloodstream to other parts of the body.

Simplified, cancer develops through the following steps (illustrated in Figure 11-9):

1. Exposure to a carcinogen.
2. Entry of the carcinogen into a cell.
3. **Initiation** of cancer as the carcinogen damages or changes the cell's genetic material (**carcinogenesis**).
4. Acceleration by other carcinogens, called **promoters,** so that the cell begins to multiply out of control—tumor formation.
5. Often, spreading of cancer cells via blood and lymph (**metastasis**).
6. Disruption of normal body functions.

Researchers think that the first four steps, which culminate with tumor formation, are key to cancer prevention. On hearing this, many people mistakenly believe that they should avoid eating all foods that contain carcinogens. Doing so would be impossible, however, because most carcinogens occur naturally among thousands of other chemicals and nutrients the body needs. The body is well equipped to deal with the minute amounts of carcinogens occurring naturally in foods, such as those listed in the margin.

For those who suspect food additives of being carcinogenic, be assured that additives are held to strict standards, and no additive scientifically shown to cause cancer is approved for use in the United States. Contaminants that enter foods by accident or toxins that arise naturally, for example when food becomes moldy, may indeed be powerful carcinogens, or they may be converted to carcinogens during the body's attempts to metabolize them. Most such constituents are monitored in the U.S. food supply and are generally present, if at all, in amounts well below those that may pose significant cancer risks to consumers.

> ▶ KEY POINT Cancer arises from genetic damage and develops in steps including initiation and promotion, which are thought to be influenced by diet. The body is equipped to handle tiny doses of carcinogens that occur naturally in foods.

Which Dietary Factors Most Influence a Person's Risk of Developing Cancer?

Almost certainly, diet substantially influences cancer development.[93] The cancer risk imposed by a food depends partly on the eater's genetic inheritance and the relationships between the genes, foods, and cancer have yet to be clarified.[94] Studies of populations suggest that low rates of many, but not all, kinds of cancer correlate with intakes of fiber-rich fruits, vegetables, and whole grains, particularly whole wheat. Under study for possible effects regarding cancer are excess calories, diets high in certain fats, daily use of alcohol, vitamins and minerals in foods and supplements, and diets in which meat plays a dominant role. The following sections explore current scientific thought on the effects of dietary constituents on cancer development.

Here are some selected chemicals and carcinogens occurring naturally in breakfast foods:

- Coffee: acetaldehyde, acetic acid, acetone, atractylosides, butanol, cafestol palmitate, chlorogenic acid, dimethyl sulfide, ethanol, furan, furfural, guaiacol, hydrogen sulfide, isoprene, methanol, methyl butanol, methyl formate, methyl glyoxal, propionaldehyde, pyridine, 1,3,7,-trimethylxanthine.
- Toast and coffee cake: acetic acid, acetone, butyric acid, caprionic acid, ethyl acetate, ethyl ketone, ethyl lactate, methyl ethyl ketone, propionic acid, valeric acid.

Of course, consuming coffee, toast, and coffee cake does not elevate a person's risk of developing cancer because the body detoxifies small doses found in foods.

carcinogen (car-SIN-oh-jen) a cancer-causing substance (*carcin* means "cancer"; *gen* means "gives rise to").

initiation an event, probably occurring in a cell's genetic material, caused by radiation or by a chemical carcinogen that can give rise to cancer.

carcinogenesis the origination or beginning of cancer.

promoters factors that do not initiate cancer but speed up its development once initiation has taken place.

metastasis (meh-TASS-ta-sis) movement of cancer cells from one body part to another, usually by way of the body fluids.

FIGURE 11·9 CANCER DEVELOPMENT

Malignant cells   Normal cells

Exposure to and entry of carcinogen (invisible) into normal cells.

Initiation by carcinogen that damages a cell's DNA; the cell multiplies, passing on the damage.

Promoters enhance the development of tumors which may be noncancerous (benign) or cancerous (malignant).

Cancerous tumor releases cells into the bloodstream (metastasis)

Energy Balance Restricting energy intakes inhibits cancer formation. When calorie intakes are reduced, cancer rates fall. In animal experiments, this **caloric effect** proves to be one of the most effective dietary interventions for cancer prevention. When researchers establish a cancer-causing condition and then restrict the energy in laboratory animals' feed, the onset of cancer in the restricted animals is delayed beyond the time when animals on normal feed have died. At the moment, no experimental evidence exists showing this effect in human subjects, but some population observations seem to imply that the effect seen in animals may hold true for human beings as well. This effect occurs only in cancer prevention; once started, cancer continues advancing even in a person who is starving.

It is also true that when calorie intakes rise, cancer rates rise: excess calories from carbohydrate, fat, and protein all raise cancer rates. The processes by which excess calories may stimulate cancer development remain obscure. Obesity itself is clearly a risk factor for certain types of cancer (such as colon, breast in postmenopausal women, endometrial, kidney, and esophageal) and possibly other cancers (such as ovarian and prostate) as well.[95] The risk of cancer rises with BMI. In one study, cancer deaths among people with a BMI of 40 or above were more than 50 percent higher than among people with a BMI in the normal range.[96]

Physical activity to balance energy intake may lower the risk of developing some cancers.[97] People whose lifestyles include regular, vigorous physical activity have the lowest risk of colon cancer.[98] Physical activity may also protect against breast cancer by reducing body weight and by other mechanisms.[99]

Fat and Fatty Acids Laboratory studies using animals support the idea that high dietary fat intakes correlate with development of cancer. Simply feeding fat to experimental animals is not enough to get tumors started, however; an experimenter must also expose the animals to a known carcinogen. After that exposure, animals fed the high-fat diet develop more cancers faster than animals fed low-fat diets. Thus, fat appears to be a cancer promoter in animals.

In studies of human beings, however, evidence so far is mixed as to whether a diet high in fat promotes cancer.[100] Comparisons among world populations reveal that high-fat diets often, but not always, correlate with high cancer rates.[101] For example, early studies implicated a high-fat diet as a risk factor for breast cancer, but later studies seemed to reverse this idea. Today, scientific opinion is shifting back to probable

caloric effect the drop in cancer incidence seen whenever intake of food energy (calories) is restricted.

culpability, particularly for saturated fat from red meats and high-fat dairy foods, in breast cancer causation.[102] The shift in opinion has come on the heels of a revelation that previous studies may have employed methods that failed to expose an existing link between saturated fats and breast cancer.[103]

Studying the effects of high fat intakes is complicated because an attribute of dietary fat is energy density—fat is extremely calorie dense. Because diets high in *calories* do seem to promote cancer, especially in laboratory settings, researchers must untangle the effects of fat alone from those of the energy content of the diet.

Dietary fat also tends to oxidize when exposed to high cooking temperatures. When these oxidized fat compounds enter the body, they may set up a condition of oxidative stress that may trigger cancerous changes in the tissues of the colon and rectum. Finally, the type of fat in the diet may be important. Some laboratory evidence implicates omega-6 polyunsaturated fatty acids in cancer promotion, while suggesting that omega-3 fatty acids from fish may protect against some cancers and may support recovery during treatment for cancer.[104] Moderation in fat intake and inclusion of several fish meals a week remain sound principles, if not for cancer protection, then for the prevention of obesity and the protection of the heart.

Alcohol Cancers of the head and neck correlate strongly with the combination of alcohol and tobacco use and with low intakes of green and yellow fruits and vegetables. Alcohol intake alone is associated with cancers of the mouth, throat, and breast, and alcoholism often damages the liver and precedes the development of liver cancer.

Smoked, Grilled, and Well-Done Meats Evidence from population studies spanning the globe over a period of more than 20 years links diets high in red meat with a moderately elevated risk of developing cancer of the rectum and colon.[105] In particular, meats cooked to the crispy well-done stage may be at fault, especially in genetically susceptible men who smoke.[106]

Chemical reactions during browning of meat at high temperatures, which occurs in pan-frying, broiling, or grilling, form known carcinogens in the food. Meat drippings that fall and burn on hot coals also generate carcinogens that rise in smoke and stick to the food.* Eating smoked, grilled, charbroiled, or well-browned, well-cooked meats introduces carcinogens into the digestive system and may cause problems there. Once the chemicals are absorbed and enter the body, however, they are quickly captured and detoxified by the liver's competent detoxifying system.

Remember that correlation, even consistently evident in research, is not cause. Although certain foods may appear at the scene of the colon cancer crime, no one yet knows whether eating such foods actually causes cancer or whether some other feature of a meat-rich diet is at fault. Still, health-savvy diners replace most servings of red meats with poultry, fish, or legumes and choose well-done, grilled, fried, highly browned, and smoked meats only occasionally.

Acrylamide Potatoes and other carbohydrate-rich foods develop a by-product, **acrylamide,** during high-temperature cooking, such as frying or conventional baking. Acrylamide results from a chemical reaction between the naturally occurring sugars and an amino acid in the food. High doses cause cancer in laboratory animals and are toxic to the human nervous system, but no one yet knows if acrylamide in amounts common in foods affects human beings. The Food and Drug Administration is currently investigating exposure of U.S. consumers to acrylamide, while the World Health Organization urges consumers to take a cautious approach and limit their intakes of carbohydrate-rich foods cooked at excessively high temperatures, such as french fries and fried snack chips, and baked goods that are overly browned.[107]

Fiber and Fluid Epidemiological studies often report links between eating plenty of fruits and vegetables and a low incidence of many cancers, but the reason for this is not

■ Controversy 3 addressed the topic of alcoholic beverages and cancer risks.

acrylamide (ah-KRILL-ah-mide) a chemical produced in carbohydrate-rich foods, such as potatoes and grains, when cooked at high temperatures. A known animal carcinogen, acrylamide is also toxic to the nervous system of both animals and humans. Also used in manufacturing and construction.

---

*The carcinogens of greatest concern are some of those called *heterocyclic amines*.

clear. One prominent theory is that a fiber-rich diet may protect against some cancers by increasing stool weight, diluting the colon's contents, stimulating bacterial fermentation, and speeding up the transit time of materials through the colon, thus minimizing exposure of the colon walls to cancer-causing substances in the feces. Bacterial fermentation also produces fatty acids that nourish the cells that line the colon, improving their resistance to cancerous changes. Some well-designed studies have reported no effect of high-fiber diets against colon cancer, but more recent evidence from Europe vindicates fiber in this regard.[108] Over half a million people in European countries were studied for over six years. The researchers concluded that populations with lower-than-average fiber intakes were at a greatly increased risk of cancers of the colon and rectum and that they could reduce that risk by 40 percent by doubling their fiber intakes from fruits, vegetables, grain foods, and legumes. A respected medical group concurs that the majority of the evidence supports a role for fiber-rich diets as protective against colorectal cancer.[109]

Evidence weighs in favor of eating high-fiber, low-fat foods such as whole grains for many reasons. Such foods may not only reduce the risk of some cancers but may also protect health in these ways:

- The breakdown of fibers of whole grains by digestive tract bacteria yields short-chain fatty acids thought to offer some protection against heart disease as well as cancers.
- High-fiber diets help to regulate blood glucose and blood insulin, factors that may also play roles in cancer development.
- Whole grains are rich in antioxidants and other phytochemicals that reduce oxidative stress and promote overall health.[110]

If a meat-rich, calorie-dense diet is implicated in causation of certain cancers and if a vegetable-rich, whole-grain-rich diet is associated with prevention, then shouldn't vegetarians have a lower incidence of those cancers? They do, as the many studies cited in Controversy 6 have shown.

One type of cancer, bladder cancer, may be related to intake of fluids. Men who drink about 10 cups of fluid a day have been reported to develop substantially less bladder cancer than those drinking only about half this amount.[111] The most probable explanation involves carcinogens that form naturally in urine. A greater fluid intake dilutes these carcinogens and causes more frequent urination, thus reducing the likelihood that carcinogens will interact with the tissues of the bladder. Plain water seems most beneficial in this regard, but almost any kind of fluid, save one kind, will do. The exception is alcoholic beverages, which increase the risks of many cancers.

Folate and Other Vitamins   Folate deficiency seems to make cancers of the cervix and colon more likely, and ample folate may ward off other cancers in certain populations.[112] Up to 10 percent of the U.S. population, and a much larger percentage of people with low incomes, consume a diet low enough in folate to cause breaks in DNA that make cancer likely to develop.[113] This reason alone is enough to warrant everyone attending to folate intake.

Vitamin A regulates aspects of cell division and communication that go awry in cancer, and it helps to maintain the immune system. Immune system cells can often identify cancerous cells and destroy them before cancer can develop. Vitamin D and exposure to sunshine have long been suggested as protective against cancers other than skin cancer, but the relationship is not clearly defined.[114] Cancer-opposing roles have also been suggested for vitamin B_6, vitamin B_{12}, and pantothenic acid.

Vitamin E, vitamin C, and beta-carotene received attention in Controversy 7. Suffice it to say here that taking supplements has not been proved to prevent or cure cancer. In fact, once cancer is established, antioxidants may do more harm than good according to a current line of investigation. Immune system cells normally release oxidative free radicals in an oxidative burst to kill off potentially dangerous microorganisms and cancer cells. Research suggests that some cancer cells may selectively stockpile antioxidants to defend against oxidative assaults from the immune system.[115] Researchers found that when

■ Chapters 7 and 13 give other compelling reasons to attend to folate needs.

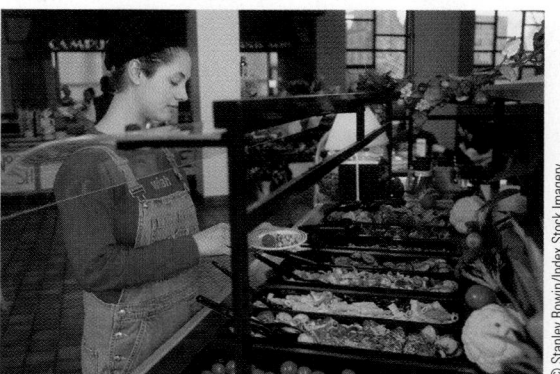

Often, whole foods like these, not individual chemicals, lower people's cancer rates.

FIGURE 11·10 PERCENTAGE OF U.S. ADULTS CONSUMING RECOMMENDED AMOUNTS OF FRUITS AND VEGETABLES A DAY, 2002

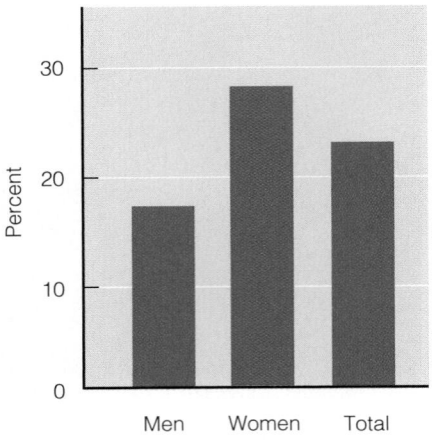

SOURCE: Centers for Disease Control and Prevention, *Physical Activity and Good Nutrition: Essential Elements to Prevent Chronic Diseaes and Obesity* (2004), At a Glance, available at www.cdc.gov/nccdphp/aag/aag_dnpa.htm.

■ Cruciferous vegetables belong to the cabbage family: bok choy, broccoli, broccoli sprouts, brussels sprouts, cabbages (all sorts), cauliflower, greens (collard, mustard, turnip), kale, kohlrabi, rutabaga, and turnip root.

anticarcinogens compounds in foods that act in any of several ways to oppose the formation of cancer.

cruciferous vegetables vegetables with cross-shaped blossoms—the cabbage family. Their intake is associated with low cancer rates in human populations. Examples include broccoli, brussels sprouts, cabbage, cauliflower, rutabagas, and turnips.

mice with brain tumors were fed antioxidant-depleted chow, the tumor growth slowed and cancer cells died off at a rapid rate.[116] Whether this intriguing laboratory finding translates into useful information for fighting cancer in people remains to be seen.

Calcium and Other Minerals Laboratory evidence suggests that a high-calcium diet may help to prevent colon cancer. Supplemental calcium seems to suppress changes in the lining of the colon associated with the onset of cancerous changes. In laboratory experiments, human colon cancer cells replicate rapidly when deprived of calcium but slow their replication when calcium is restored.[117] The findings from epidemiological studies are mixed, but a recent review of the literature indicated that, overall, people who develop colon cancer consume less calcium than people who do not develop the cancer.[118] Together, these studies have not yet proved that one can avoid colon cancer by increasing calcium intake, but with all the other points in calcium's favor, prudence dictates that everyone should arrange to meet calcium needs every day.

For years, iron has been the subject of research with regard to colon cancer. A recent review revealed that the majority of studies since 1990 confirm an association between colon cancer and both increased dietary iron intake and high body iron stores.[119] Whether iron increases a person's risk for colon cancer may depend on whether the person has inherited a tendency to store too much iron (a relatively common genetic trait), the person's gender (being female may be protective), and the degree of iron supplementation (higher intake poses greater risk).

How iron may facilitate cancer remains unanswered. Iron is a powerful oxidizing substance, and oxidation may damage DNA in ways that initiate cancer. Alternatively, iron supplements are constipating, and constipation also raises a person's risk of colon cancer. Also, meat is a generous supplier of iron in the diet, and high-meat diets often correlate with colon cancer. These findings have scientists questioning the wisdom of widespread enrichment of breads, cereals, and other grains with iron from a public health point of view. If the iron and cancer link holds up to further scrutiny, the benefits of enrichment to iron-deficient populations must be weighed against the harm to those prone to colorectal cancer.[120] Other minerals in foods, including zinc, copper, and selenium, are thought to play roles in cancer prevention, perhaps by helping antioxidant enzymes defend against its initiation.

Foods and Phytochemicals In the end, whole foods and whole diets composed of them, not single nutrients, may be most influential on cancer development. For example, the phytochemicals of some fruits and vegetables are thought to be **anticarcinogens.** Some of these may protect against cancer by acting as mild toxins that force the body to build up its arsenal of carcinogen-destroying enzymes. Then, when a potent carcinogen arrives, the prepared body deals with it swiftly. Almost without exception, population studies find that diets lacking green and yellow fruits and vegetables and citrus fruits correlate with cancers of many types. Some evidence suggests that a number of antioxidant nutrients and phytochemicals from a diet rich in fruits, vegetables, and whole grains may be necessary to minimize DNA damage from certain types of radiation, at least in the laboratory.[121] A low intake of fruits and vegetables may also leave cells unprotected against DNA damage in the initiation of bladder and other cancers.[122] Infrequent use of **cruciferous vegetables**—broccoli, brussels sprouts, cabbage, cauliflower, turnips, and the like—is common in colon cancer victims (see the margin).

One review of the literature found an almost unheard-of perfect association between reduced incidence of lung cancer and diets high in fruits and vegetables—every study included in the review reported a protective effect. Incidence of stomach cancer, too, correlates with too few vegetables in the diet: in one study, with vegetables in general; in another, with fresh vegetables; and in others, with lettuce and other fresh greens or vegetables containing vitamin C. Unfortunately, adults in the United States are slow to get the message (see Figure 11-10).

Foods that contain phytochemicals are believed to promote health and fight diseases. Such foods, often called *functional foods,* have been recognized as potentially beneficial by the National Academy of Sciences. The academy defines functional foods as *"potentially healthful products [that include] any modified food or food ingredient that may provide a health benefit beyond the traditional nutrients it contains."* Controversy 2 explored the state of the science concerning functional foods and their phytochemical constituents. The Food Feature lists dietary guidelines for cancer prevention.

> ▶ KEY POINT   Diets high in certain fats and red meats are associated with cancer development. Foods containing ample fiber, folate, calcium, many other vitamins and minerals, and phytochemicals, along with an ample intake of fluid, are thought to be protective.

Conclusion

Nutrition is often associated with promoting health, and medicine with fighting disease, but no clear line separates nutrition and medicine. Every major agency involved with health recommends a healthful diet as part of a lifestyle that provides the best possible chance for a long and healthy life.

This chapter has summarized the major forms of disease and their links with nutrition. You may have noticed a philosophical shift from previous chapters. There, we could say "a deficiency of nutrient X causes disease Y." Here we could only cite theories and discuss research that illuminates current thinking. We can say with certainty, for example, that "a diet lacking vitamin C causes scurvy," but to say that a low-fiber diet that lacks vegetables causes cancer would be inaccurate. We can, however, recommend behaviors that are prudent and increase your likelihood of a lifetime of good health. The Food Feature presents these recommendations.

FOOD FEATURE

Diet as Preventive Medicine

A REMARK BY a former surgeon general is worth repeating: If you do not smoke or drink excessively, "your choice of diet can influence your long-term health prospects more than any other action you might take." Indeed, healthy young adults today are privileged to be the first generation in history with enough knowledge now to lay a foundation for healthy years ahead through a lifetime of proper nutrition. Figure 11-11 (p. 422) illustrates this point.

An early chapter of this book presented dietary guidelines for the prevention of diseases. Chapters that followed focused on the "whys" and "hows" of those guidelines. This Food Feature comes full circle to revisit the guidelines with a broader and deeper understanding to aid the reader in applying them meaningfully, as they were intended.

Dietary Guidelines
for Disease Prevention

The more detailed our knowledge about nutrition science, it seems, the simpler the

truth becomes: people who consume the moderate, adequate, balanced, calorie-controlled diet recommended by the *Dietary Guidelines for Americans 2005* (see Table 11-9, p. 423) enjoy a longer, healthier life than those who do not.[123] The National Heart, Lung, and Blood Institute and the American Cancer Society offer suggestions specifically for disease prevention. Table 11-9 presents guidelines from these sources and shows how similar they are to the *Dietary Guidelines,* clinching the argument that it's time to get busy putting the recommendations into practice. The following paragraphs review the specifics.

Reduce Saturated Fat
and *Trans* Fat Intake

Primary among the recommendations is to choose unsaturated fats in place of saturated fat and *trans* fat. To meet this goal requires people to limit such foods as high-fat milk and dairy products, stick margarine, high-fat baked goods and convenience foods, foods with sauces, commercially

The U.S. "5 a Day" program:

- Is a shared effort among public and private organizations, such as the National Cancer Institute, the American Cancer Society, the Centers for Disease Control and Prevention, and the U.S. Department of Agriculture.

- Promotes the consumption of fruits and vegetables through supermarkets, local health agencies, schools, federal feeding programs, and businesses.

- Tracks fruit and vegetable consumption in U.S. communities and collects data on the effectiveness of the program in increasing intakes.

Visit the 5-A-Day website for more information: www.cdc.gov/nccdphp/dnpa/5Aday/index.htm

Eat 5 to 9 a Day

Tips for consuming more fruits and vegetables:

- Vow to try a new fruit or vegetable once each month. Read some cookbooks for ideas.

- Eat a rainbow of fruits and vegetables: the more reds, oranges, yellows, greens, blues, and purples the better.

- Use the salad bar to buy ready-to-eat vegetables if you are in a hurry.

- Keep a fruit bowl in plain view, filled with fresh fruit, small raisin or dried cranberry packs, and shelf-stable fruit cups.

- Add dried fruit bits to salads, cereal, or yogurt.

- Place carrot and celery sticks in a glass of water like a bouquet, and keep them in the refrigerator for crisp, healthy snacks.

- Try a grilled mushroom sandwich or "veggie burger" instead of a beef hamburger sandwich.

- Try using soy products in many ways. Soy drinks, ground and patty meat replacers, tofu, and soy snacks count as a vegetable and may offer unique benefits from soy protein, fiber, and phytochemicals.

- Drink 100 percent fruit or vegetable juices. Choose 100 percent juice bars for a frozen treat.

- Vegetable sauces count: ½ cup tomato spaghetti sauce or salsa counts as ½ cup "other vegetables" (see Chapter 2).

- Blend smoothies from bananas, fruit juice, and berries with ice or yogurt.

- Choose larger portions of lower-calorie vegetables, such as cooked leafy greens or carrots.

- Add beans to salads, stews, and meat dishes. Beans soak up the flavor of meat and stretch the servings per pound.

FIGURE 11·11 PROPER NUTRITION SHIELDS AGAINST DISEASES

A well-chosesn diet can protect your health.

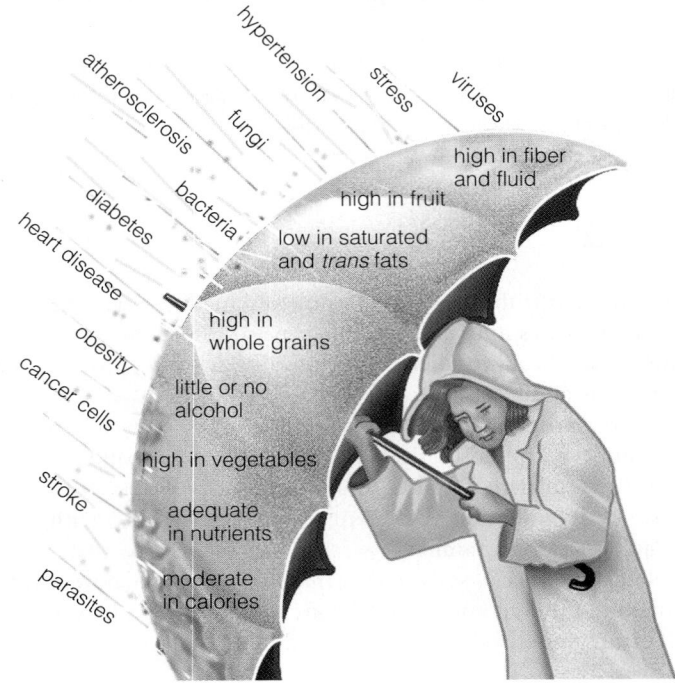

fried foods, fat-marbled meat cuts, sausages, and fatty ground beef.

Conversely, foods such as nuts, olive oil, canola oil, and liquid margarines are high in polyunsaturated fats that supply the essential fatty acids and vitamin E. Because they are high in calories, use them only within your daily energy budget. Regular meals of fish, particularly fatty fish such as salmon, help to balance intakes of omega-6 and omega-3 fatty acids. Consult the Table of Food Composition in Appendix A for further details about the fatty acid contents of your favorite foods.

Include Fruits, Vegetables, and Whole Grains

Every legitimate source of dietary advice urges people to include a variety of fruits and vegetables in the diet, not just for nutrients but also for the phytochemicals that combine synergistically to promote health.[124] The U.S. "Eat 5 to 9 a Day" and the Canadian "Reach for It" programs encourage consumers to eat enough fruits and vegetables to support health.

The U.S. "Eat 5 to 9 a Day" program reflects the new recommendations of the USDA Food Guide as it urges consumers to

eat 5 to 9 servings of fruits and vegetables each day. Research indicates that the higher number may be more supportive of health, especially for men, and for African American men in particular because of their high risk for many life-threatening chronic diseases. Sadly, few achieve this intake, but some tips for achieving it appear in the margin.

Who knows? Some of the foods still waiting for you on the produce shelves may become your favorites. An adventurous spirit is a plus in this regard.

As for whole grains, most guidelines call for their inclusion, but the authors of this book would go a step farther and ask that whole grains make up the majority of your grain choices. That way, the grain foods you consume daily will supply nutrients, fiber, and nonnutrients associated with good health.

Go for Variety

Eat foods high in potassium (fruits and vegetables), high in calcium and magnesium (milk products and appropriate substitutes), low in fat, high in fiber (whole grains, legumes, vegetables, and fruits), and ample in fluids. If you are prone to

TABLE 11·9 DIETARY GUIDELINES FOR LOWERING DISEASE RISKS

Dietary Guidelines for Americans 2005 Key Recommendations[a]	National Heart, Lung, and Blood Institute's Therapeutic Life Changes (TLC) Diet to Lower LDL Cholesterol[a]	American Cancer Society Recommendations for Nutrition and Physical Activity for Cancer Prevention
Adequate nutrients within energy needs • Consume a variety of nutrient-dense foods and beverages within and among the basic food groups; limit intakes of saturated and *trans* fats, cholesterol, added sugars, salt, and alcohol. • Plan a balanced diet with the USDA Food Guide or the DASH eating plan (Appendix I). **Weight management** • To maintain body weight in a healthy range through life, balance calories from foods and beverages with calories expended. **Physical activity** • Engage in regular physical activity and reduce sedentary activities to promote health, psychological well-being, and a healthy body weight. **Food groups to encourage** • Consume sufficient amounts of a variety of fruits and vegetables, milk and milk products, and whole grains while staying within energy needs. **Fats** • Keep saturated fat and *trans* fat to less than 10 percent of calories, and cholesterol to less than 300 miligrams. • Keep total fat intake to between 20 to 35 percent of calories from mostly unsaturated fats. **Carbohydrates** • Choose fiber-rich fruits, vegetables, and whole grains often. • Choose and prepare foods and beverages with little added sugars. **Sodium and potassium** • Choose and prepare foods with little salt (less than 2300 milligrams sodium or approximately 1 teaspoon salt). • At the same time, consume potassium-rich foods. **Alcoholic beverages** • Those who drink alcohol should do so sensibly and in moderation. Some people should abstain from alcohol. **Food safety** • Keep foods safe from microbes that cause illness.	**Carbohydrates** • 50–60% calories from carbohydrates (primarily complex carbohydrates from whole-food sources). • 20–30 g dietary fiber per day. **Lipids** • 25–35% of calories from total fats. • Up to 10% calories from polyunsaturated fats. • Up to 20% calories from monounsaturated fats. • 200 mg or less of cholesterol per day. • Keep *trans* fatty acids low. **Protein** • Approximately 15% calories from protein. These additional dietary changes can further reduce LDL cholesterol in those with higher heart disease risk:[b] • Increase intake of viscous (soluble) fiber. • Replace some animal protein sources with soy protein foods.	Eat a variety of healthful foods, with an emphasis on plant sources. • Eat five or more servings of a variety of vegetables and fruits each day. • Choose whole grains in preference to processed (refined) grains and sugars. • Limit consumption of red meats, especially those high in fat and processed meats. • Choose foods that maintain a healthful weight. Adopt a physically active lifestyle. • Adults: engage in at least moderate activity for 30 minutes or more on 5 or more days of the week; 45 minutes or more of moderate-to-vigorous activity on 5 or more days per week may further enhance reductions in the risk of breast and colon cancer. • Children and adolescents: engage in at least 60 minutes per day of moderate-to-vigorous physical activity at least 5 days per week. Maintain a healthful weight throughout life. • Balance caloric intake with physical activity. • Lose weight if currently overweight or obese. If you drink alcoholic beverages, limit consumption.

[a]Abbreviated. A more complete set can be found in Chapter 1 of this text, or online at www.health.gov/dietaryguidelines/dga2005/document/html/executivesummary.htm
[b]Also recommended for people with elevated LDL; 2 g per day of sterol or stanol esters.

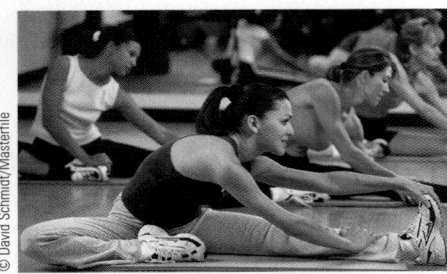
Exercise regularly, all your life.

hypertension, experts advise that you eat fewer high-sodium foods and less salt.

Advice on varying your diet is based on an important concept in the prevention of cancer initiation—dilution. Whenever you switch from food to food, you are diluting whatever is in one food with what is in the others. It is safe to eat *some* salty foods or smoked or grilled meats, but don't eat them all the time.

Be Physically Active

Expend energy in daily physical activity that you enjoy. If the threat of CVD doesn't motivate you, then exercise to improve your self-image, to improve your morale, to control your body weight, to improve your complexion, or to make friends—but do exercise.

In the end, people's choices are their own. Whoever you are, we encourage you to take the time to work out ways of making your diet meet the guidelines you know will support your health. If you are healthy and of normal weight, if you are physically active, and if your diet on most days follows the guidelines just provided, then you can indulge occasionally in a cheesy pizza, marbled steak, banana split, or even a greasy fast-food burger and fries without inflicting much damage on your health. (Once a week may be harmless, but the less frequently the better.) Especially, take time to enjoy your meals: the sights, smells, and tastes of good foods are among life's greatest pleasures. Joy, even the simple joy of eating, contributes to a healthy life.

MEDIA MENU

@ 🏠 ◀ ▶

Nutrition Connections

The Nutrition Connections CD provides access to these additional resources.

Do It!
The Do It! exercise for this chapter, *Practice Eyeballing a Meal,* provides practice in judging meals according to nutrient ideals, such as DRI intake recommendations.

Interaction
The animation, *The Development of Atherosclerosis and Hypertension,* demonstrates how these diseases develop and how each makes the other worse.

Chapter Quiz
Take the quiz for this chapter to assess your understanding of chapter concepts.

Nutrition on the Net
Annotated links to reputable nutrition-related websites.

Online Glossary
Test your knowledge of key vocabulary through this comprehensive, interactive glossary.

Internet Connections

A variety of helpful study tools for this chapter are available at our website
http://nutrition.wadsworth.com/ncc10

Practice Test
Take the practice test for this chapter and gauge your knowledge of key concepts; answers are provided.

Student Notebook
A lecture notebook is available to help organize note taking while reading or attending lecture.

ABC Video Clip
"Colon Cancer and Diet" (2:55) examines possible links between diet and colon cancer. "Obesity and Genetics" (5:58) looks at the connections between obesity and genes.

Study Tools
Study resources for students include Chapter Summaries, Crossword Puzzles, Flashcards, Glossary, Internet Exercises, and Learning Objectives.

SELF CHECK

Answers to these Self Check questions are in Appendix G.

1. By what age do most people have well-developed plaques in their arteries?
   a. 20 years
   b. 30 years
   c. 40 years
   d. 50 years

2. Which of the following is a risk factor for cardiovascular disease?
   a. high blood HDL cholesterol
   b. low blood pressure
   c. low LDL cholesterol
   d. diabetes

3. An "atherogenic diet" is high in all of the following *except:*
   a. fiber
   b. cholesterol
   c. saturated fats
   d. *trans* fats

4. Which of the following dietary factors may help to regulate blood pressure?
   a. calcium
   b. magnesium
   c. potassium
   d. all of the above

5. Which of the following have been associated with an increase in cancer?
   a. alcohol
   b. diets high in red meat
   c. high intakes of smoke-preserved foods
   d. all of the above

6. When calorie intakes rise, cancer rates also increase.   T   F

7. Laboratory evidence suggests that a high-calcium diet may increase the risk of colon cancer.   T   F

8. The DASH diet is designed for athletes who compete in sprinting events.
   T   F

9. Hypertension is more severe and occurs earlier in life among people of European or Asian descent than among African Americans.   T   F

10. Alternative therapies, such as herbal medicine, have not been well established by scientific experimentation to be safe and effective.   T   F

CONTROVERSY 11

THE OBESITY EPIDEMIC— HOW CAN WE GAIN CONTROL?

WHAT IF TOMORROW'S headlines read, "Obesity vanquished—population freed of overweight"? On that day, an estimated 97 million people would be freed from the threat of serious illness associated with obesity—heart disease, diabetes, arthritis, cancer, and others.*[1] In that year, an estimated 300,000 lives would be saved from

*Reference notes are found in Appendix F.

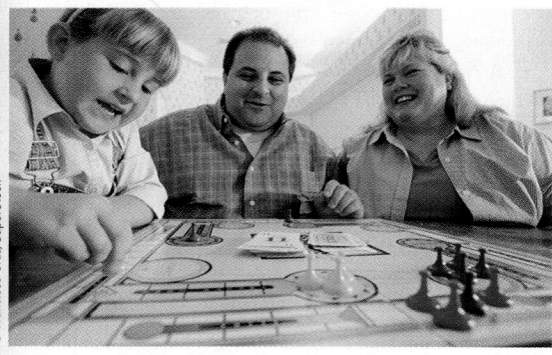

early obesity-related deaths.[2] In addition, national spending on obesity-related health care would plummet by $60 billion, freeing these immense resources for other uses.

Sadly, real headlines tell of rising rates of obesity (described in Chapter 9) that cost individuals and society dearly. How did we get into this predicament? And what can we do to get out of it?

Why Has Our Society Become Obesity-Prone?

A number of societal changes over the last 50 years have been suggested as factors driving the obesity epidemic in the United States.[3] Together, these trends seem to add up to what has been called an "obesity-promoting environment" (see Figure C11-1).

Today, many people regularly overeat on large portions of inexpensive high-calorie foods, consuming about 500 calories per day more than in 1970.[4] At the same time, they have grown accustomed to leisure time in front of DVDs, videos, computers, or television sets; laborsaving devices such as automobiles and elevators; and jobs requiring more sitting than moving. Most do not exercise enough to maintain health and fitness (see Chapter 10), and a full quarter of them get little or no physical activity.[5] Then, as they gain weight and lose fitness, stress and sleep disturbances creep into their lives, and they feel increasingly exhausted and less motivated to move.

The problem can worsen if a chronically tired person seeks relief by self-medicating with caffeine, medicines, weight-loss pills, high-calorie "energy" foods and drinks, other inappropriate supplements, alcohol, tobacco, or other drugs. Though some of these drugs may provide some relief in the short term, none solves the problems of too little exercise and too many calories from food.

As Chapter 9 made clear, some people's genetic inheritance lays the foun-

This environment promotes a sedentary lifestyle and overconsumption of high-energy foods.

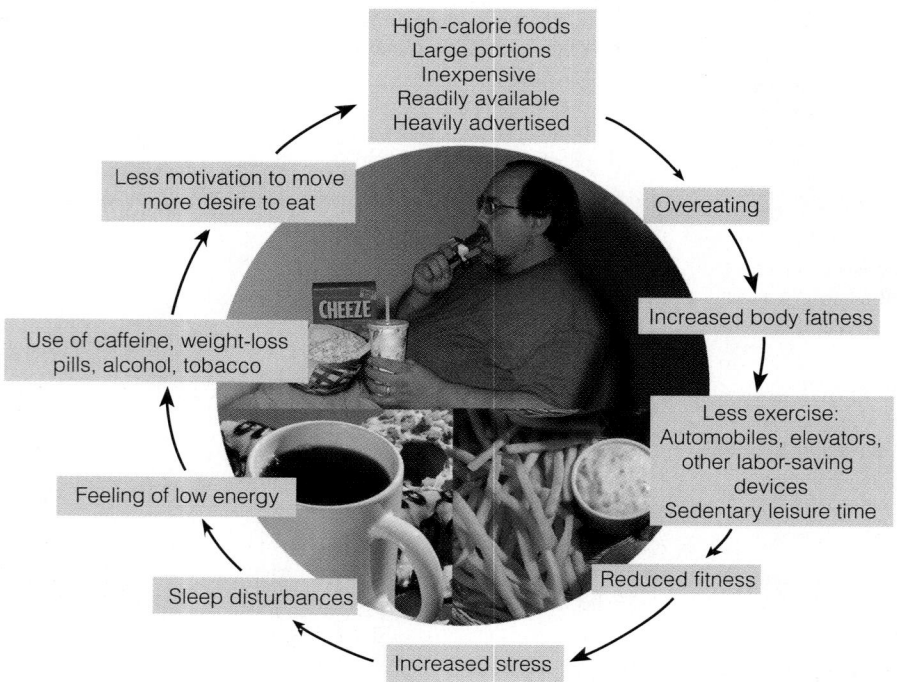

High-calorie foods
Large portions
Inexpensive
Readily available
Heavily advertised

Overeating

Increased body fatness

Less exercise:
Automobiles, elevators,
other labor-saving
devices
Sedentary leisure time

Reduced fitness

Increased stress

Sleep disturbances

Feeling of low energy

Use of caffeine, weight-loss
pills, alcohol, tobacco

Less motivation to move
more desire to eat

dation for obesity, but such tendencies remain silent unless conditions are right to support weight gain.[6] Further, the obesity-prone individual makes free-will choices each day. Personal choices leading to overweight, however, may be so heavily influenced by societal conditions that such choices become less free than most people assume.

Changing Physical Activity

Throughout history, human beings have aspired to make physical work easier and increase leisure time. Over the years, technological advances have relieved people of much of the labor of planting and harvesting food, walking or running long distances, hauling water, and, in fact, performing most of life's basic physical tasks. One famous obesity researcher had this to say about our progress in making life easier: "The war on muscular work has been a remarkable success."[7] Some researchers assign the lion's share of blame for increasing national body weights on decreasing energy expenditure.[8]

Today, most people take laborsaving inventions for granted. Builders design automobile-dependent communities that feature efficient roadways but no sidewalks. Urban sprawl locates shopping malls and workplaces within commuting distances, but not within walking or biking distances, from residences. Relatives and friends may live at great distances. Buildings have conveniently located elevators but often hide stairwells from view. Consumers who choose to walk to work or use the stairs may find that these are inconvenient and unpleasant options. Roadways are often smelly, unattractive, and hazardous for pedestrians; stairwells are often stuffy and rarely cleaned. While laborsaving advances have no doubt liberated many human hours for intellectual and leisure pursuits, their costs in terms of human health are only now being counted in full.

Changing Home Life

Changes in the family structure have influenced people's food intakes over the past four decades. By one estimate, the diet has changed more over the past 40 years than in any other comparable period of human existence.[9] As more women have entered the workforce and as the hours worked by all wage earners have increased, families have less time for the traditional activities of home-making, including meal planning, grocery shopping, and the labor-intensive preparation of meals of fresh foods. Consequently, more and more people rely on prepared or partially prepared foods to save time and effort at home. The enormous popularity of prepared foods speaks of the need for them.

Today's meals are often unplanned and are influenced more by which foods are familiar, readily available, moderately priced, and easy to prepare than by the nutrient needs of the purchasers. Chapter 1 pointed out that the farm-fresh foods consumed most often in the middle of the last century have largely been replaced by foods processed to some extent for convenience. Along with convenience, however, manufacturers often include flavorful ingredients such as salt, sugar, and fat and remove constituents such as fiber.

Thus, in exchange for convenience, the purchaser of prepared foods relinquishes to the food industry responsibility for the calorie and nutrient content of the food. Whether the industry can be trusted to act in the consumer's best interest is a question worth considering.[10]

The Changing Food Supply

Consumers now spend 45 percent of their food budget on foods purchased and eaten outside the home, an amount expected to rise more rapidly than any other food expense category (see Figure C11-2).[11] With more and more of the American diet originating with the food industry, nutritionists have no doubt that this trend powerfully influences the health of the nation.

Overall, the easiest, most accessible, most inviting, and oftentimes least expensive U.S. consumer options encourage unhealthy behaviors, especially where food choices are concerned.[12] Most visibly, fast foods are now readily available in most areas of the nation and account for the largest segment of the growing "away from home" food category for U.S. consumers.[13] Fast-food fried chicken, hamburgers, and pizza are convenient, speedy, tasty, and affordable. Although a smaller influence than fast-food outlets, more traditional sit-down "neighborhood" restaurants also offer foods appealing primarily to taste and economy, not nutrition. On average, U.S. consumers now eat three hamburger meals each week, consumption of french fries far outweighs green vegetable intakes, and soda sales have more than doubled in the past 40 years.[14]

Food Advertising

Food advertising has become an enormous business of its own. In an effort to increase sales, the food industry spends more than $33 billion each year on advertising, trade shows, special product promotions, fees to supermarkets to ensure prominent product placement, attractive packaging, games and prizes, and other efforts to attract consumer dollars.[15] Compare this $33 billion to the $1 million spent by the National Cancer Institute to promote fruit and vegetable intake. Is it any wonder that food industry promotions win influence over the American appetite?

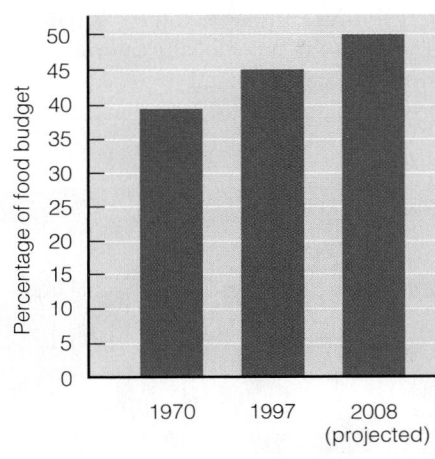

FIGURE C11-2 PERCENTAGE OF FOOD BUDGET SPENT ON MEALS AND SNACKS CONSUMED AWAY FROM HOME, 1970–2008

Consumers who purchase meals prepared away from home gain convenience but lose a degree of control over nutrient and calorie intakes.

(Percentage of food budget: 1970 ≈ 39, 1997 = 45, 2008 (projected) = 50)

The "Buy More" Strategy

Everyone already eats some kind of food every day. To increase food sales, therefore, marketers must convince us to buy their brand over others or, if we already choose it, to buy more of it.[16] The fight for brand loyalty is fierce, and much of it is aimed at children, as a later section shows. A proven strategy for prodding consumers to buy more is to appeal to their economic sense.

Thanks largely to technological advances in agriculture, a restaurant often spends less on the food it serves than on most other expenses, such as personnel, facility, or monthly utilities. The strategy of offering more food for just a little more money at each selling opportunity therefore pays off in greater profits. To a consumer making choices at a food counter, a deal often seems sweeter if significantly more food can be obtained for only a few cents more—"for just an extra 49 cents, get the bigger burger, larger fries, and super cola." A thirsty shopper buying a big 20-ounce fountain drink (twice the normal serving size, and 300 calories) for a dollar may find that the bigger 32-ounce drink (470 calories) costs only 10 cents more and that the best "bargain" is a whopping 48-ounce drink (700 calories) at only 19 cents more.

Portion sizes and obesity rates have increased proportionately over the years.[17] If you doubt that small daily decisions such as choosing a larger serving can make a difference to your body weight, try this: Turn back to page 44 in Chapter 2, and look at the foods depicted in Figure 2-9. Add up the calories in a 1970s hamburger, cola, and french fries (similar to today's "small" sizes). Do the same for the calories in the "colossal" size hamburger, cola, and french fries typical of today's meals.

SIZE DOES MATTER.
NEW! Big Chicken Fillet

A scientifically accurate food advertisement.

© Fran Webb

Now find the calorie difference between the two meals (subtract the smaller sum from the larger) and multiply the difference by 52 (for weeks in a year).

$$
\begin{aligned}
\text{Today's calorie total} &= 2{,}020 \\
-\text{1970s calorie total} &= 925 \\
\text{Calorie difference} &= 1{,}095
\end{aligned}
$$

The difference between the two meals is over a thousand calories. Multiplied by 52 weeks in a year, this adds up to an extra 56,940 calories a year. If a pound of body fat can be gained with each excess 3,500 calories, then a person who chooses a larger-sized fast-food meal *just once per week* stands to gain over 16 pounds of body fat in a year's time. Many little daily decisions such as this add up over time.

Along with restaurant food, prepared food has also gotten larger, and most common servings exceed recommended portion sizes, sometimes by huge margins.[18] For example, a store-bought chocolate chip cookie was reported to exceed the USDA standard for a cookie by 700 percent. Pasta, muffins, steaks, bagels, beer, and pizza all follow suit. The single food reportedly still in compliance with its standard is a slice of white bread.

One source of this trend may be some recent initiatives by marketers to stimulate consumers to buy more food, thus increasing food company profits. An article with the headline "Can Package Size Accelerate Usage Volume?" outlines the strategy of increasing package sizes of pasta to increase sales.[19] A study reports that larger pasta packages prompt consumers to cook more pasta at a meal—a good result for those selling pasta. The probable fate of the extra cooked pasta was explained in Chapter 9: when presented with greater quantities of food, people eat more at a sitting.[20] Sales of candy and popcorn work the same way: the bigger the container, the larger the portion consumers will buy and eat and the higher the profits to the seller.[21]

Overall, food supersizing strategies seem to be a tremendous financial success wherever they are employed. And

the effect is not all one sided: as consumers gradually adapt to larger portions, they put pressure on the food industry to supply more and more food per serving to remain competitive. All the while, obesity is skyrocketing.

Marketing to Children

The food industry spends almost $12 billion annually promoting high-calorie fast foods, fried snack foods, high-sugar cereals, candies, and soft drinks during children's television viewing hours and in other media.[22] The food industry vigorously pursues the attention of children, spurred by the influence of children on a huge portion of the nation's spending—up to $6 billion of their own pocket money each year and $132 billion more in annual family spending.[23]

A spokesperson for a commercial watchdog group speaking to the World Health Organization delineated the techniques used to sell foods and other products to youth.[24] These include manipulating children's emotional and physical needs, such as their needs for peer acceptance, love, safety, and security and their desire to be independent, to act older, and to develop an identity. Advertising agencies hire educated, talented, creative college graduates who draw on child psychology to develop effective ads. The speaker added this: "Advertisers are so successful at marketing to youth that they sometimes discuss it in terms of the battle over what they chillingly call 'mind share.'"

Some children are a captive audience for commercials brought into their classrooms by a media service that provides "free" programming to schools, paid for by selling children's classroom time to advertisers.[25] Such companies market fried snack food, fast food, and high-calorie soft drinks to students, 8 million of whom are forced to watch two minutes of such ads each school day.[26]

Do the ads work? Yes, they do. Otherwise, the billions of dollars spent annually on advertising to children would be spent elsewhere.

The Food Industry Speaks Out

The food industry has much to fear from accusations that its high-calorie products and selling tactics are fueling the obesity epidemic. Public sentiment that the industry is harming consumers for the sake of profits could direct consumer dollars elsewhere. Also, many food companies fear lawsuits that could force them to pay enormous settlements if, like tobacco firms, they are found culpable in causing harm to consumers. Finally, there is the matter of pride: many long-established food companies take pride in their traditions and products—providing the best pies and cookies, the finest steaks, or the most delicious doughnuts to consumers who are eager for them. Such companies hold that their primary goal is to meet the changing desires and demands of their customers.

Does the Food Industry Cause Obesity?

Is the food industry akin to tobacco companies in harming people's health? Products of both industries stand accused of causing illness. Both industries have marketed their wares to children. There is a difference, however. The evidence against tobacco smoke as a cause of such diseases as lung cancer and emphysema is ironclad, even when the products are used as intended. Individual foods, however, are unlikely to be found to cause obesity because many people eat them in moderation and remain lean. Tobacco smoke is a toxic, addictive, and unnecessary hazardous substance. Food is necessary for life, and even fast food provides some nutrients and seems tolerable by the healthy body when taken in moderation. Importantly, the food industry provides a service by preparing foods for today's families who cannot or choose not to prepare foods at home.

The arguments in defense of the food industry were summed up this way by the president of the national association representing restaurant owners: "Restaurants have a wide variety of choices on their menus, and people make the choice to eat what they want and when they want every day. This is all about personal responsibility and moderation."[27] Other spokespeople argue that the cause of obesity is Americans' lack of exercise, rather than the availability of certain food products or the prevalence of advertising.

New Menu of Choices

Many food companies have recently added lower-fat, lower-calorie products to their lines, while a few others have eliminated the largest of the overlarge portion options, hoping to change the view that restaurant and convenience foods damage health. Fried chicken outlets now include baked chicken on the menu. Hamburger outlets now sport vegetarian "burgers" and salad options that cost no more than the standard chicken nuggets (made mostly of ground skin and breading) and fries. The food giant Kraft (owned in part by a company known for its Philip Morris tobacco products) announced plans to offer more heart-healthy options in its food line. Still consumer choices determine whether the new offerings translate into healthier meals. At McDonald's, a new chicken sandwich on flat bread provides 470 calories. A big or fancy hamburger can provide from 500 to 800 calories, mostly from fat with much of it saturated fat.* Plain, small, fast-food salads (no dressing) can be remarkably low in calories, in the range of 100 to 300 calories. A chef's salad without dressing provides 320 calories. But served with toppings, such as bacon, cheese, chips, croutons, fried noodles, and sour cream, or drenched with the full one-third cup of oily dressing that comes with them, salads weigh in at 500 to 800 calories—with enough fat and saturated fat to rival even the greasiest burgers.

Reasonable choices other than salads are available. One sandwich chain, Subway, has built its reputation on making available some low-calorie sandwiches that can be used to support weight-loss efforts.[28] Other establishments, when asked, will omit fatty toppings from sandwiches and salads to produce a reasonably nutritious meal. The food industry states that such choices have always been available, but most people entering a fast-food restaurant associ-

*Calorie and fat values from a quick online comparison service, Fast Food Quest, available at www.cyberdiet.com/reg/ffq/show_results.cgi.

ate these establishments with a chance to indulge. Only a few have the willpower to order slimmed-down offerings when familiar full-calorie treats are within view and are the path of least resistance.

Can Government Actions Reverse Obesity Trends?

On learning of the obesity epidemic, people often respond by demanding that the government "do something" about it. Perhaps the old USDA Food Guide Pyramid was at fault, they say, or perhaps the *Dietary Guidelines* need a different focus. In truth, these documents and teaching tools are updated regularly and provide sound advice.

Many government resources and much private money have been spent over the years on developing initiatives to reverse obesity in the United States.[29] Since 1950, at least 35 consumer education reports have specified sound weight-control practices as essential to maintenance of good health. Not one of these public health efforts has succeeded in slowing obesity rates in this country, however, so it seems unlikely that more reports and documents would be effective.[30] Currently, the Food and Drug Administration is planning enhancements of food labels and launching a "Calories Count" education campaign.[31]

Previous initiatives that focused on individual responsibility failed to acknowledge the societal factors that erect barriers in the path of those who try to follow health-promoting advice. Individuals are ultimately responsible for their own behavior, but in the current social environment, few people are prepared to choose behaviors that support health when doing so is difficult (review Figure C11-1 in an earlier section).

Nutrition Education

Part of the blame for unheeded dietary advice is placed on inadequate nutrition education in today's schools. A law passed in 1977 mandates that federal funds be supplied to states to provide nutrition training for teachers and school foodservice workers and to teach basic nutrition science to the nation's schoolchildren. Why then do the vast majority of high school graduates lack even a basic understanding of important nutrition concepts? The answer may be a lack of funding. Although the law still exists, money to deliver nutrition education is inadequate.[32]

The *Healthy People 2010* report sets a goal of increasing the proportion of schools that provide a minimum of three years of nutrition education for the nation's children, including such skills as balancing food intake with physical activity. If more tax dollars were spent today to achieve this goal, perhaps fewer could be spent in the future treating preventable illnesses arising from obesity.

New Ideas for Change

When healthy foods cost less, people choose them more often.[33] Thus, policymakers and scientists have discussed how to make the healthiest choices also the most affordable and available.[34] One suggestion is to use monetary incentives to encourage both food producers and consumers to make healthier food and exercise choices.[35] This plan would raise taxes on high-calorie, high-fat, high-salt foods, sugary sodas, candies, and fried snack foods potentially generating 40 to 100 million dollars in extra tax revenue per year.[36] These tax revenues could then subsidize purchases of health-promoting salads, low-fat chili, whole-grain breads and buns, broiled chicken and fish, low-fat milk, and fresh fruit and cooked vegetables. As consumers bought more and more of the less costly items, manufacturers would respond by making them available in greater varieties and by marketing them more heavily.

Why stop at foods? According to this construct, taxing automobiles and other laborsaving devices that rob people of the exercise they need could yield needed price incentives for purchasing and using bicycles, exercise equipment, and walking shoes. This could lead to a healthy restructuring of our whole society (see Figure C11-3) into one that promotes fitness rather than obesity. In such a society, physical activity would be a natural part of everyone's day, and stress would be relieved as people interacted in small, centralized communities. They'd safely walk or bicycle to school, work, restaurants, libraries, and shops on pleasant paths in their own neighborhoods with far less traffic and pollution.[37] In this society, buildings would be designed to invite most people to use attractive stairways, while making elevators available for those with disabilities. Neighbors would meet neighbors, develop friendships, and look out for one another, making the community not only healthier but safer.

Conclusion

On hearing of economic plans such as the one just outlined, many people react with skepticism and resistance. They believe that in a free-market society, citizens are capable of making decisions in their own best interest and do not need the government to step in and restructure their lives. Consumers can choose to *live* right now, today, as though such incentives were already in place. We can:

- Turn off the television and take up a physically active recreation.

FIGURE C11·3   FITNESS-PROMOTING ENVIRONMENT

This environment promotes sound nutrition and physical activity choices that lead to fitness.

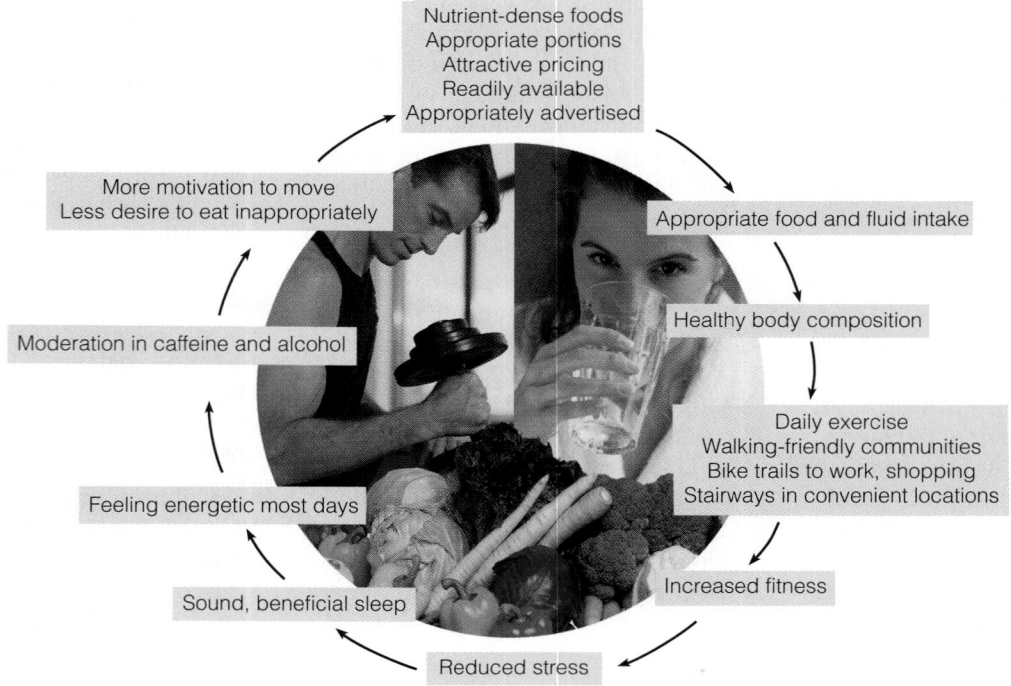

Nutrient-dense foods
Appropriate portions
Attractive pricing
Readily available
Appropriately advertised

More motivation to move
Less desire to eat inappropriately

Appropriate food and fluid intake

Moderation in caffeine and alcohol

Healthy body composition

Feeling energetic most days

Daily exercise
Walking-friendly communities
Bike trails to work, shopping
Stairways in convenient locations

Sound, beneficial sleep

Increased fitness

Reduced stress

- Choose to live close enough to shopping, work, and school to allow walking as an option.
- Park farther away than usual and walk through the parking lot.
- Lobby for laws that require landscaping, trees, and walkways in parking lots to make them safe and inviting.
- Seek out and use stairways instead of elevators.
- Take walks and get to the know the neighbors.
- Write letters to building owners, asking them to improve stairwells and access to them.
- Obtain an inexpensive pedometer and vow to walk 10,000 steps each day.
- Choose to consistently follow the advice of the experts offered in Chapters 1 and 2 of this text, even if it means paying more for food and making time to prepare meals at home.
- Make a point to remember standard portion sizes (review Figure 2-4 of Chapter 2) and daily food amounts to decide how much to eat at a sitting.
- Take some of the food home for another meal when restaurant portions are too large.
- Save enticing fatty and sugary fast foods and snacks for special occasions, while choosing healthier choices as staple, everyday foods.
- Teach the children in our lives to think differently about food.
- Lobby lawmakers to fund nutrition education classes, physical education classes, and eliminate from the classroom all commercials for high-fat, high-calorie foods.
- Pressure schools to replace candy and soda with fruit and low-fat milk options and to reverse the trend toward hiring fast-food companies to provide school lunches for children.

People who consistently make personal choices that enhance their health often find that their body weights fall into line with recommendations almost automatically. A good place to start is to take the suggestions in all of the previous chapters to heart and to put in place a plan to eat wisely and exercise every day.

Food Safety and Food Technology

© Pablo Picasso (1881–1973), *Still Life with Pitcher, Bowl and Fruit*, 1931. © 2005 Estate of Pablo Picasso/Artists Rights Society (ARS), NY. Giraudon/Art Resource, NY

CONTENTS

Do You Ever . . .

*Come down with digestive tract symptoms and maybe a headache
that you attribute to a case of "stomach flu"?*

*Assume that foods from grocery stores and restaurants are free of
harmful germs?*

Refrigerate leftover party foods such as meatballs after *the guests have
gone home?*

*Eat raw sushi or raw cookie dough but avoid foods with additives to
protect your health?*

KEEP READING.

433

CONSUMERS IN THE United States and Canada enjoy food supplies ranking among the safest, most pleasing, and most abundant in the world. With this benefit, though, comes the consumer's responsibility to distinguish between paths leading to food **safety** and those that pose a **hazard.** The Food and Drug Administration (FDA) is the major agency charged with monitoring the U.S. food supply, but other agencies are involved as well (see Table 12-1). The following list indicates the FDA's ongoing areas of concern regarding our food supply:

1. *Microbial foodborne illness* affects many millions of people every year.
2. *Natural toxins in foods* constitute a hazard whenever people consume single foods either by choice (fad diets) or by necessity (poverty).
3. *Residues in food.*
   a. *Environmental contaminants* (other than pesticides) such as household and industrial chemicals are increasing yearly in number and concentration, and their impacts are hard to foresee and to forestall.
   b. *Pesticides* are a subclass of environmental contaminants but are listed separately because they are applied intentionally to foods and, in theory, can be controlled.
   c. *Animal drugs* include hormones and antibiotics that increase growth or milk production in food animals and arsenic compounds that enhance the growth of chickens.*[1]
4. *Nutrients in foods* require close attention as more and more artificially constituted foods appear on the market.
5. *Intentional food additives* are of little concern because so much is known about them that they pose virtually no hazard to consumers and because their use is well regulated.
6. *Genetic modification of foods* is listed last because such foods undergo rigorous scrutiny before going to market.[2]

Microbial **foodborne illness,** commonly called *food poisoning,* is first on the list because episodes of food poisoning far outnumber any other kind of food contamination. The last items, food additives and genetic modification of foods, are of least concern. The others fall somewhere in between.

After the events of September 11, 2001, the threat of deliberate microbial contamination of the U.S. food supply became a pressing issue.[3] To tighten security around the nation's food supply, the FDA has developed a focused food-security program and established guidelines for firms that produce, process, transport, or otherwise handle food.†[4] The U.S. Department of Agriculture (USDA) has also created a Food Biosecu-

---

*Reference notes are found in Appendix F.

†FDA guidance documents on biosecurity are available at www.cfsan.fda.gov//dms/guidance.html.

On average, *each day:*
- Over 200,000 people in the United States fall ill with a foodborne illness.
- Of those, 14 die.

Especially vulnerable to foodborne illnesses are:
- Pregnant women.
- Newborns.
- Older adults.
- People with immune systems weakened by cancer treatments, AIDS, diabetes, and other causes.

▓ For more information on the FDA's strategies for counterterrorism: see M. Meadows, The FDA and the fight against terrorism, *FDA Consumer,* January/February 2004, pp. 20–27, available at www.fda.gov.

safety the practical certainty that injury will not result from the use of a substance.

hazard a state of danger; used to refer to any circumstance in which harm is possible under normal conditions of use.

foodborne illness illness transmitted to human beings through food and water; caused by an infectious agent *(foodborne infection)* or a poisonous substance *(food intoxication).* Also called *food poisoning.*

| TABLE 12·1 | AGENCIES THAT MONITOR THE U.S. FOOD SUPPLY |

Websites and phone numbers are listed in Appendix E.

- **CDC (Centers for Disease Control and Prevention)** a branch of the Department of Health and Human Services that is responsible for monitoring foodborne diseases.
- **EPA (Environmental Protection Agency)** the federal agency that is responsible for regulating pesticides and establishing water quality standards.
- **FDA (Food and Drug Administration)** the part of the Department of Health and Human Services' Public Health Service that is responsible for ensuring the safety and wholesomeness of all foods sold in interstate commerce except meat, poultry, and eggs (which are under the jurisdiction of the USDA); inspecting food plants and imported foods; and setting standards for food consumption.
- **USDA (U.S. Department of Agriculture)** the federal agency that is responsible for enforcing standards for the wholesomeness and quality of meat, poultry, and eggs produced in the United States; conducting nutrition research; and educating the public about nutrition.
- **WHO (World Health Organization)** an international agency that develops standards to regulate pesticide use. A related organization is the FAO (Food and Agricultural Organization).

rity Action Team to protect agriculture and other aspects of the food supply. Many agencies are taking ongoing action to protect the food supply, and the FDA has developed broad strategies for countering bioterrorism.[5]

While the bioterrorism threat is serious, the immediate assault on health and life from foodborne illnesses is a current national problem that often yields to simple preventive tactics. This chapter focuses on actions that individuals can take to promote food safety, as it addresses the first five of the FDA's areas of concern. The chapter Controversy takes up the topic of genetic modification of foods.

With the privilege of abundance comes the responsibility to choose and handle foods wisely.

Microbes and Food Safety

Some people brush off the threat from foodborne illnesses caused by **microbes** as less likely and less serious than the threat of flu, but they are misinformed—foodborne illnesses can be life threatening and are increasingly unresponsive to standard antibiotics.[6] Even normally mild foodborne illnesses can be lethal for people who are ill or malnourished; have a compromised immune system; are in an institution; have liver or stomach illnesses; or are pregnant, very old, or very young.[7] Each year in the United States, an estimated 76 million people become ill from foodborne diseases, and about 5,000 of them die.[8] Within these numbers lies some good news—because of improved safety procedures adopted by food producers, disease from a handful of problem organisms is on the decline.[9]

If digestive tract disturbances are the major or only symptoms of your next bout of "stomach flu," chances are excellent that what you really have is a foodborne illness. By learning something about these illnesses and taking a few preventive steps, you can minimize your chances of contracting a foodborne illness.

How Do Microbes in Food Cause Illness in the Body?

Microorganisms can cause foodborne illness either by infection or by intoxication. Infectious agents such as *Salmonella* bacteria or the virus that causes hepatitis A multiply and infect the tissues of the human body.[10] Other microorganisms in foods produce **enterotoxins** or **neurotoxins,** poisonous chemicals released as bacteria multiply; once absorbed into the tissues, the poisons cause various kinds of harm ranging from mild stomach pain and headache to paralysis and death. The toxins may arise in food during improper preparation or storage or within the digestive tract after a person eats contaminated food. The sources and symptoms of foodborne illnesses are listed in Table 12-2, pp. 436–438.

Although the most common source of food toxicity is the *Staphylococcus aureus* bacterium, the most infamous is undoubtedly *Clostridium botulinum*, an organism that produces a toxin so deadly that an amount as tiny as a single grain of salt can kill several people within an hour. To reproduce and release the toxin, *Clostridium botulinum* requires anaerobic conditions such as those found in improperly canned (especially home-canned) foods and homemade garlic or herb-flavored oils stored at room temperature. **Botulism** quickly paralyzes muscles making seeing, speaking, swallowing, and breathing difficult.[11] Because death can occur as soon as 24 hours later, botulism demands immediate medical attention (see the margin for warning signs). Even then, survivors may suffer the effects for months or years.

The botulinum toxin is destroyed by heat, so canned foods that contain the toxin can be rendered harmless by boiling them for ten minutes. Food can be canned safely at home if proper canning techniques are followed to the letter.[*]

microbes a shortened name for *microorganisms;* minute organisms too small to observe without a microscope, including bacteria, viruses, and others.

enterotoxins poisons that act upon mucous membranes, such as those of the digestive tract.

neurotoxins poisons that act upon the cells of the nervous system.

botulism an often-fatal food poisoning caused by botulinum toxin, a toxin produced by the *Clostridium botulinum* bacterium that grows without oxygen in nonacidic canned foods.

---

‡Complete, up-to-date, home canning instructions are available in the USDA's *Complete Guide to Home Canning*, available from the Superintendent of Documents, Government Printing Office, Washington, DC 20402.

TABLE 12·2 FOODBORNE ILLNESSES

Disease and Organism That Causes It	Yearly Occurrence and Deaths[a]	Most Frequent Food Sources	Onset and General Symptoms	Prevention Methods
Campylobacteriosis *Campylobacter jejuni* bacterium	2 million; 100 deaths	Raw poultry, beef, lamb, unpasteurized milk (foods of animal origin eaten raw or undercooked or recontaminated after cooking).	Onset: 2 to 5 days. Diarrhea, nausea, vomiting, abdominal cramps, fever; sometimes bloody stools; lasts 2 to 10 days; rarely, nervous system paralysis (Guillain-Barré syndrome).	Cook foods (especially poultry) thoroughly; use pasteurized milk; use sanitary food-handling methods.
Cryptosporidiosis *Cryptosporidium parvum* microscopic parasite	30,000; 7 deaths. (In the 1990s, a Wisconsin waterborne outbreak affected more than 400,000 people.)	Commonly, contaminated swimming or drinking water, even from treated sources. Highly chlorine-resistant. Contaminated raw produce and unpasteurized juices and ciders.	Onset: 2 to 10 days. Diarrhea, loose or watery stools, stomach cramps, upset stomach, slight fever. Symptoms may come and go for weeks or months.	Wash all raw vegetables and fruits before peeling; choose pasteurized milk and juice. Do not swallow drops of water while using pools, hot tubs, ponds, lakes, rivers, or streams for recreation.
Cyclosporiasis *Cyclospora cayetanensis* single-cell parasite	15,000; 8 deaths	Contaminated water; contaminated fresh produce.	Onset: average, 7 days. Watery diarrhea, loss of appetite, weight loss, stomach cramps, nausea, vomiting, muscle aches, fatigue. Symptoms may come and go for weeks or months.	In areas of uncertain sanitation, drink only treated or boiled water, and eat only cooked hot foods or fruits you peel yourself.
Hemolytic-uremic syndrome *Escherichia coli (E. coli)* O157:H7 bacterium	62,500; 50 deaths	Undercooked ground beef, unpasteurized milk and milk products, unpasteurized juices or cider, raw produce (especially sprouts), contaminated water, and person-to-person contact.	Onset: 1 to 8 days. Severe bloody diarrhea, abdominal pain, vomiting, acute kidney failure; death. Survivors may face kidney problems, hypertension, blindness, paralysis, and colon problems.	Cook ground beef thoroughly; avoid unpasteurized milk and juice products; use sanitary food-handling methods; use treated, boiled, or bottled water. Vulnerable people should avoid raw sprouts.
Hepatitis Hepatitis A virus	4,200; 4 deaths	Undercooked or raw shellfish; raw or lightly cooked produce; contaminated water; baked goods or other ready-to-eat foods contaminated by infected food handlers.	Onset: 15 to 50 days (28 days average). Inflammation of the liver; fatigue; dark urine; headache; nausea, vomiting, or indigestion; jaundice (yellowed skin and eyes from buildup of wastes); muscle pain.	Cook foods thoroughly; test food handlers.
Listeriosis *Listeria monocytogenes* bacterium	2,500; 500 deaths	Raw meat and seafood, raw milk, and soft cheeses.	Onset: 7 to 30 days. Mimics flu; blood poisoning; meningitis (stiff neck, severe headache, and fever); miscarriage of pregnancy; severe illness or death of newborns; blood or brain infection in elderly or other vulnerable people.	Use sanitary food-handling methods; cook foods thoroughly; use pasteurized milk.

[a]Estimated data from Centers for Disease Control and Prevention, Diseases and pathogens under surveillance, available at www.cdc.gov/foodnet/pus.htm; P. S. Mead and coauthors, Food-related illness and death in the United States, *Emerging Infectious Diseases* 5 (2000), available at www.cdc.gov/ncidod/eid/vol5no5/mead.htm. Other data from Centers for Disease Control and Prevention, Diagnosis and management of foodborne illnesses, *Morbidity and Mortality Weekly Report,* supplement, 53 (2004): 7–12.

TABLE 12-2 FOODBORNE ILLNESSES—continued

Disease and Organism That Causes It	Yearly Occurrence and Deaths[a]	Most Frequent Food Sources	Onset and General Symptoms	Prevention Methods
Salmonellosis *Salmonella* bacteria	1–34 million; 600 deaths	Raw or undercooked eggs, meats, poultry, unpasteurized milk and other dairy products, shrimp, pasta, raw produce, unpasteurized juices, contaminated water.	Onset: 1 to 3 days. Nausea, fever, chills, vomiting, abdominal cramps, diarrhea, headache; severe cases can be fatal.	Use sanitary food-handling methods; use pasteurized milk and juice; cook foods thoroughly; refrigerate foods promptly and properly.
Shigellosis *Shigella* bacteria varieties	90,000; 14 deaths	Contaminated food (may look and smell normal); raw produce and other foods contaminated by poor sanitation practices of infected farm workers or food handlers, human sewage fertilizer in growing fields, or exposure to flies. Contaminated drinking or swimming water.	Onset: 1 to 2 days. Diarrhea, fever, stomach cramps. The diarrhea is often bloody. In young children, high fever, seizures. Symptomless sufferers can spread the bacteria to others.	Frequent and careful hand washing with soap. In areas of uncertain sanitation, drink only treated or boiled water, and eat only hot cooked foods or fruits you peel yourself. Do not swallow drops of water when swimming.
"Stomach flu"[b] (mistakenly called) Noroviruses	9.2 million; 124 deaths	Foods, such as cookies, sandwiches, salads, and ice contaminated by infected food handlers. Raw produce. Shellfish from waters contaminated with human sewage, such as boat bilge.	Onset: 12 to 48 hr. Acute digestive illness, with pain, vomiting, possibly diarrhea, headache, and low-grade fever. Diarrhea more typical in adults; vomiting in children.	Choose restaurants that pass health department inspections and enforce worker sanitation. If uncertain, order cooked foods served steaming hot. Avoid raw shellfish.
"Stomach flu"[b] (mistakenly called) *Vibrio parahaemolyticus* and other *Vibrio* bacteria	5,000; 31 deaths	Raw or undercooked shellfish, often oysters. Less commonly, skin infection when an open wound is exposed to warm seawater.	Onset: 24 hr. Watery diarrhea, abdominal cramping, nausea, vomiting, fever and chills.	Cook shellfish well, especially oysters. Purchase shellfish from reputable dealer. Avoid exposing wounds to warm seawater.
Traveler's diarrhea A variety of microorganisms including *Giardia* and other protozoa.	10 million international travelers affected	Contaminated water, undercooked ground beef, raw foods, imported unpasteurized soft cheeses.	Onset: 2 days to several weeks. Loose and watery stools, nausea, vomiting bloating, abdominal cramps.	Cook foods thoroughly; use safe, treated water and pasteurized milk; wash and peel raw fruits and vegetables or avoid them in areas of uncertain sanitation.
Trichinosis *Trichinella spiralis* parasite	50 or fewer; no deaths	Raw or undercooked pork or wild game (bear or moose). Worms produce larvae that travel via the bloodstream to muscle tissue, producing cysts.	Onset: 1–2 days. Abdominal pain, nausea, vomiting, diarrhea, fatigue, and fever. Two to 8 weeks later, muscle pain, pain on breathing, edema (swelling), skin eruptions, loss of appetite, and weight loss.	Cook foods thoroughly.

[b]Though popularly called "stomach flu," the digestive disturbances caused by noroviruses and *Vibrio* organisms are unrelated to influenza.

Continued

TABLE 12·2 FOODBORNE ILLNESSES—continued

Disease and Organism That Causes It	Yearly Occurrence and Deaths[a]	Most Frequent Food Sources	Onset and General Symptoms	Prevention Methods
Botulism Botulinum toxin (produced by the *Clostridium botulinum* bacterium)	60; 4 deaths	Anaerobic environment of low acidity (canned corn, peppers, green beans, and other low-acid vegetables; canned meats, fish, and chicken; bottled garlic; packaged lunch meats, ham, and sausage; herb-flavored oils; preserved fat or fish; baked potatoes in foil and other cooked foods held at warm temperature for more than 2 hours). In infants, honey may be a source.	Onset: 12 to 72 hr. Nervous system symptoms including blurrred or double vision, inability to swallow, speech difficulty, and progressive paralysis of the respiratory system; often fatal; leaves prolonged symptoms in survivors.	Use proper canning methods for low-acid foods; avoid commercially prepared foods with leaky seals or with bent, bulging, or broken cans.
Staphylococcal food poisoning Staphylococcal toxin (produced by the *Staphylococcus aureus* bacterium)	185,100; 2 deaths	Toxin produced in meats, poultry, egg products, tuna, potato, and macaroni salads, and cream-filled pastries.	Onset: ½ to 8 hr. Diarrhea, nausea, vomiting, abdominal cramps, fatigue; mimics flu; lasts 24 to 48 hr; rarely fatal.	Use sanitary food-handling methods; cook food thoroughly; refrigerate foods promptly and properly.

▶ **KEY POINT**   Each year in the United States, many millions of people suffer mild to life-threatening symptoms caused by foodborne illness. Though relatively rare, botulism is often deadly when it strikes.

Food Safety from Farm to Table

Figure 12-1 shows that careful food handling is required to prevent microbes from becoming a problem—on the farm, in processing plants, during transportation, and at supermarkets and restaurants. The rapid globalization of the U.S. food supply has widened safety concerns in farming, processing, and transportation, especially because seafood and fresh produce lead the list of imported foods and both are common vectors of illness.[12] Equally critical to the chain of food safety, however, is the final handling of food by people who purchase it and consume it at home.

The overwhelming majority of food-poisoning cases result from errors consumers make in handling foods *after* purchase. Commercially prepared food is usually safe, but rare accidents do occur and often affect many people at once. Dairy farmers, for example, rely on **pasteurization,** a process of heating milk to kill many disease-causing organisms and make the milk safe for consumption. When, as happened in Chicago in the mid-1980s or more recently in England, a major dairy develops flaws in its pasteurization system, many thousands of cases of foodborne illness result. Other types of farming require other safeguards. Growing food usually involves soil, and soil contains abundant bacterial colonies, making contamination of food likely. Additionally, farm workers who are ill can easily transfer disease-causing organisms to foods such as produce.

Attention on *E. coli*   In the mid-1990s, a fast-food restaurant chain in the Northwest served undercooked hamburgers from meat that had been contaminated with the dangerous bacterium *E. coli* O157:H7. Four people died, and hundreds of other

pasteurization the treatment of milk with heat sufficient to kill certain pathogens (disease-causing microbes) commonly transmitted through milk; not a sterilization process. Pasteurized milk retains bacteria that cause milk spoilage. Raw milk, even if labeled "certified," transmits many foodborne diseases to people each year and should be avoided.

FIGURE 12·1· FLOW OF FOOD SAFETY: FROM FARM TO TABLE

FARMS
Farmers and workers must use safe methods of growing, harvesting, sorting, packing, and storing food to minimize contamination hazards.

PROCESSING
Processors must follow USDA and FDA guidelines concerning contamination, cleanliness, and education and training of workers. Industries must monitor for safety at critical control points (use HACCP, see text).

TRANSPORTATION
Containers and vehicles transporting food must be clean. Cold food must be kept cold at all times.

RETAIL: GROCERY AND RESTAURANT SALES
Owners and employees must follow the FDA's food code on how to prevent foodborne illnesses. Establishments must pass local health inspections and train staff in sanitation.

CONSUMER RESPONSIBILITY
Consumers must learn and use sound principles of food safety as taught in this chapter and stay mindful that foodborne illness is a real possibility.

patrons were stricken with serious illness. News coverage focused the national spotlight on two important food-safety issues: raw meats routinely contain live, disease-causing organisms of many types, and thorough cooking is necessary to make animal-derived foods safe. Revelations about *E. coli* O157:H7 have led to a much needed overhaul of the country's mechanisms for ensuring meat safety.

Infections from *E. coli* O157:H7 cause bloody diarrhea, severe intestinal cramps, and dehydration that set in a few days after eating tainted meat, raw milk, or even fresh berries or organic produce that has been contaminated. In the worst cases, *hemolytic-uremic syndrome* leads to abnormal blood clotting with kidney failure, damage to the central nervous system and other organs, and even death, especially among children.[13] Antibiotics and self-prescribed antidiarrhea medicines increase the absorption and retention of the toxin, worsening the condition.[14] Medical treatment consists of rest, fluids, and hospitalization should the kidneys fail or serious dehydration occur.

Obtain medical help for these symptoms:
- Bloody diarrhea.
- Severe intestinal cramps.
- Dehydration.

Industry Controls—Government Inspections and HACCP Inspections of U.S. meat-processing plants, performed every day by USDA inspectors, ensure that they meet government standards. Seafood, egg, produce, or processed food facilities are inspected less often by far, in some cases only once in every five years. However, such facilities are required, as meat processors are, to employ a **Hazard Analysis Critical Control Point (HACCP)** plan to help prevent food-borne illnesses at their source.[15] Each slaughterhouse, producer, packer, distributor, and transporter of susceptible foods must identify "critical control points" in its procedures where the risk of food contamination is high, and then devise and implement ways of minimizing contamination.

Since the development of the HACCP system, *Salmonella* contamination of U.S. poultry, ground beef, and pork has decreased by almost 50 percent, 40 percent, and 25 percent, respectively.[16] Illnesses from contamination in these industries are also declining, indicating that the HACCP requirement is working. Facilities providing food to people stand to benefit in terms of food safety when HACCP plans are followed. Soon, dietitians in assisted living facilities and hospitals across the nation may be called to assist in their implementation.[17]

Consumer Protection The safety of canned and packaged foods sold in grocery stores is controlled through sound food technology practices, but rare accidents do happen. Batch numbering enables the recall of contaminated foods through public

Hazard Analysis Critical Control Point (HACCP) a systematic plan to identify and correct potential microbial hazards in the manufacturing, distribution, and commercial use of food products. *HACCP* may be pronounced "HASS-ip."

Myths that often make consumers sick:

- "If it tastes okay, it's safe to eat."
- "We have always handled our food this way, and nothing has ever happened."
- "I sampled it a couple of hours ago and didn't get sick, so it's safe to eat."

announcements via newspapers, television, and radio, and the FDA monitors large food suppliers and importers. You can help protect yourself, too. Carefully inspect the seals and wrappers of packages. Reject open, leaking, or bulging cans, jars, and packages. Many jars have safety "buttons" on the lid, designed to pop up once the jar is opened; make sure that they have not "popped." If a package on the shelf looks ragged, soiled, or punctured, do not buy the product; turn it in to the store manager. A badly dented can or a mangled package is useless in protecting food from microorganisms, insects, spoilage, or even vandals. Frozen foods should be solidly frozen, and those in a chest-type freezer case should be stored below the frost line. See the margin for some food safety myths that often make consumers sick.

▶ **KEY POINT** Industry employs sound practices to safeguard the commercial food supply from microbial threats. Still, incidents of commercial foodborne illness have caused widespread harm to health.

Food Safety in the Kitchen

Large-scale commercial incidents make up only a fraction of the nation's total food-poisoning cases each year, although they are widely publicized. The vast majority of cases arise from one person's error in a small setting and affect just a few victims. Some people have come to accept a yearly bout or two of intestinal illness as inevitable, but these illnesses can and should be prevented. Just for fun, take the quiz in Table 12-3 to see how well you follow food-safety rules. Then read on to learn how you can do your part to make meals at home as safe as they can be.

Food can provide ideal conditions for bacteria to thrive or to produce toxins. Disease-causing bacteria require warmth (40°F to 140°F; 4°C to 60°C), moisture, and nutrients. To defeat bacteria, deprive them of one of these conditions. Remember these four "keepers": keep hot food hot, keep cold food cold, keep raw foods separate, and keep your hands and the kitchen clean (see Figure 12-2).

■ Requirements of disease-causing bacteria: warmth, moisture, and nutrients.

Remember these four "keepers":

- Keep hot food hot.
- Keep cold food cold.
- Keep raw foods separate.
- Keep your hands and the kitchen clean.

Keep Hot Food Hot Keeping hot food hot includes cooking foods long enough to reach an internal temperature that will kill microbes. To alert consumers to this fact, the USDA invented "Thermy," the cartoon character shown on p. 442 who urges the use of a thermometer to test the temperatures of cooked foods. Figure 12-3, p. 442, illustrates the safe internal temperatures of cooked foods and various types of thermometers. Table 12-4, on the same page is a glossary of thermometer terms.

After cooking, foods must be held at 140°F or higher until served because cooking does not destroy all bacterial toxins. Food registering 140°F on a thermometer feels hot, not just warm. Even hot cooked foods, if handled improperly prior to serving, can cause illness. Delicious looking meatballs on a buffet may harbor bacteria unless they have been kept steaming hot. After the meal, cooked foods should be refrigerated immediately or within two hours at the maximum (one hour if room temperature approaches 90°F, or 32°C). If food has been left out longer than this, toss it out.

Keep Cold Food Cold Keeping cold food cold starts when you leave the grocery store. If you are running errands, shop last so that the groceries do not stay in the car too long. (If ice cream begins to melt, it has been too long.) Upon arrival home, load foods into the refrigerator or freezer immediately. Keeping foods cold applies to defrosting foods, too. Thaw meats or poultry in the refrigerator, not at room temperature, and marinate meats in the refrigerator, too. Table 12-5, p. 444 lists some safe keeping times for foods stored at or below 40°F.

Any food with an "off" appearance or odor should not be used or even tasted. You cannot rely on your senses of smell and sight alone to warn you, however, because most hazards are not detectable by odor, taste, or appearance. As the old saying goes, "when in doubt, throw it out."

FIGURE 12·2   FIGHT BAC!

Four ways to keep food safe. The Fight Bac! website is at www.fightbac.org.

TABLE 12·3   CAN YOU PASS THE KITCHEN FOOD-SAFETY QUIZ?

How food-safety savvy are you? Give yourself 2 points for each correct answer.

1. The temperature of the refrigerator in my home is:
   A. 50°F (10°Celsius).
   B. 40°F (4°C).
   C. I don't know; I don't own a refrigerator thermometer.
2. The last time we had leftover cooked stew or other meaty food, the food was:
   A. cooled to room temperature, then put in the refrigerator.
   B. put in the refrigerator immediately after the food was served.
   C. left at room temperature overnight or longer.
3. If I use a cutting board to cut raw meat, poultry, or fish and it will be used to chop another food, the board is:
   A. reused as is.
   B. wiped with a damp cloth or sponge.
   C. washed with soap and water.
   D. washed with soap and hot water and then sanitized.
4. The last time I had a hamburger, I ate it:
   A. rare.
   B. medium.
   C. well-done.
5. The last time there was cookie dough where I live, the dough was:
   A. made with raw eggs, and I sampled some of it.
   B. store-bought, and I sampled some of it.
   C. not sampled until baked.
6. I clean my kitchen counters and food preparation areas with:
   A. a damp sponge that I rinse and reuse.
   B. a clean sponge or cloth and water.
   C. a clean cloth with hot water and soap.
   D. the same as above, then a bleach solution or other sanitizer.
7. When dishes are washed in my home, they are:
   A. cleaned by an automatic dishwasher and then air-dried.
   B. left to soak in the sink for several hours and then washed with soap in the same water.
   C. washed right away with hot water and soap in the sink and then air-dried.
   D. washed right away with hot water and soap in the sink and immediately towel-dried.
8. The last time I handled raw meat, poultry, or fish, I cleaned my hands afterward by:
   A. wiping them on a towel.
   B. rinsing them under warm tap water.
   C. washing with soap and water.
9. Meat, poultry, and fish products are defrosted in my home by:
   A. setting them on the counter.
   B. placing them in the refrigerator.
   C. microwaving and cooking promptly when thawed.
   D. soaking them in warm water.
10. I realize that eating raw seafood poses special problems for people with:
   A. diabetes.
   B. HIV infection.
   C. cancer.
   D. liver disease.

Answers

1. Refrigerators should stay at 40°F or less, so if you chose answer B, give yourself two points; zero for other answers.
2. Answer B is the best practice; give yourself two points if you picked it; zero for other answers.
3. If answer D best describes your household's practice, give yourself two points; if C, one point.
4. Give yourself two points if you picked answer C; zero for other answers.
5. If you answered A, you may be putting yourself at risk for infection from bacteria in raw shell eggs. Answer C—eating the baked product—will earn you two points and so will answer B. Commercial products are made with pasteurized eggs.
6. Answers C or D will earn you two points each; answer B, one point; answer A, zero.
7. Answers A and C are worth two points each; other answers, zero.
8. The only correct practice is answer C. Give yourself two points if you picked it; zero for others.
9. Give yourself two points if you picked B or C; zero for others.
10. This is a trick question: all of the answers apply. Give yourself two points for knowing one or more of the risky conditions.

Rating Your Home's Food Practices

20 points: Feel confident about the safety of foods served in your home.
12 to 19 points: Reexamine food-safety practices in your home. Some key rules are being violated.
11 points or below: Take steps immediately to correct food-handling, storage, and cooking techniques used in your home. Current practices are putting you and other members of your household in danger of foodborne illness.

source: Adapted from U.S. Food and Drug Administration, Can your kitchen pass the food safety test? *FDA Consumer,* October 1998.

Keep Raw Food Separate   Keeping raw food separate means preventing **cross-contamination** of foods. Raw foods, especially meats, eggs, and seafood, are likely to contain bacteria. To prevent bacteria from spreading, keep the raw foods and their juices away from ready-to-eat foods. For example, if you take burgers out to the grill on a plate, wash that plate in hot, soapy water before using it to hold the cooked burgers. If you use a cutting board to cut raw meat, wash the board, the knife, and your hands thoroughly with soap before handling other foods, and especially before making a salad or other foods that are eaten raw. Many cooks keep a separate cutting board just for raw meats.

cross-contamination the contamination of a food through exposure to utensils, hands, or other surfaces that were previously in contact with a contaminated food.

TABLE 12·4 GLOSSARY OF THERMOMETER TERMS

- **appliance thermometer** a thermometer that verifies the temperature of an appliance. An *oven thermometer* verifies that the oven is heating properly; a *refrigerator/freezer thermometer* tests for proper refrigerator (<40°F, or <4°C) or freezer temperature (0°F, or −17°C).
- **fork thermometer** a utensil combining a meat fork and an instant-read food thermometer.
- **instant-read thermometer** a thermometer that, when inserted into food, measures its temperature within seconds; designed to test temperature of food at intervals, and not to be left in food during cooking.
- **oven-safe thermometer** a thermometer designed to remain in the food to give constant readings during cooking.
- **pop-up thermometer** a disposable timing device commonly used in turkeys. The center of the device contains a stainless steel spring that "pops up" when food reaches the right temperature.
- **single-use temperature indicator** a type of instant-read thermometer that changes color to indicate that the food has reached the desired temperature. Discarded after one use, they are often used in retail food markets to eliminate cross-contamination.

FIGURE 12·3 FOOD-SAFETY TEMPERATURES (FAHRENHEIT) AND HOUSEHOLD THERMOMETERS

Different thermometers do different jobs. To choose the right one, pay attention to its temperature range: some have high temperature ranges intended to test the doneness of meats and other hot foods. Others have lower ranges for testing temperatures of refrigerators and freezers.

Keep Your Hands and the Kitchen Clean

Keeping your hands and the kitchen clean requires using freshly washed utensils and laundered towels and washing your hands properly, not just rinsing them, before and after handling raw food (see Figure 12-4). Normal, healthy skin is covered with bacteria, some of which may cause foodborne illness when deposited on moist, nutrient-rich food and allowed to multiply.[18] Remember to use a nailbrush to clean under fingernails when washing hands and tend to routine nail care—artificial nails, long nails, and chipped polish harbor more bacteria than do natural, clean, short nails.[19]

For most purposes, washing hands with ordinary soap and warm water is effective, but using an alcohol-based hand sanitizing gel can provide additional killing power against many remaining bacteria and most viruses.[20] Using a hand sanitizer provides additional protection against spreading illnesses when someone in the house is ill or when preparing food for an infant, an elderly person, or someone with a compromised immune systems.[*][21] If you are ill or have open sores, stay away from food preparation.

In the kitchen, microbes love to nestle down in small, damp spaces such as the inner cells of sponges or the pores between the fibers of wooden cutting boards. Antibacte-

The USDA's "Thermy" character was developed as a light-hearted appeal to consumers encouraging them to use food thermometers.

---

*Effective hand sanitizers contain between 60 and 70 percent ethyl alcohol.

FIGURE 12·4 PROPER HAND WASHING PREVENTS ILLNESS

Many people do not wash their hands at critical times or wash them without soap or for too short a time. You can avoid many illnesses by following these hand washing procedures before, during, and after food preparation; before eating; after using the bathroom; after handling animals or their waste; or when hands are dirty. Wash hands more frequently when someone in the house is sick.

Step 1:
Wet hands and apply liquid or clean bar soap. Place bar soap on a rack to drain between uses.

Step 2:
Dislodge germs by scrubbing hands together for about 15 seconds—about the time it takes to recite the alphabet. Scrub fingers, tops of hands, and palms, use a nailbrush to clean under fingernails.

Step 3:
Rinse hands in clean water and dry with a freshly laundered towel or paper towel.

© PhotoDisc/Getty Images

SOURCE: Centers for Disease Control and Prevention, An ounce of prevention keeps the germs away, 2002, available at www.cdc.gov/nicdod/op/handwashing.htm.

Courtesy of Food, Hands and Bacteria. Cooperative Extension Service of the University of Georgia (both).

This person's clean-looking but unwashed hand is touching a sterile nutrient-rich gel.

After 24 hours, these large colonies provide visible evidence of the microorganisms that were transferred from the hand to the gel.

Courtesy of A. Estes Reynolds, George A. Schuler, James A. Christian, and William C. Hurst. Food, Hands and Bacteria. Cooperative Extension Service of the University of Georgia, Athens, GA. 2000. Extension Publication No. 693.

rial cleaners, sponges, cloths, boards, and utensils possess a chemical additive intended to deter bacterial growth, but little science supports their use.[22] Disposable cutting sheets designed for cutting raw meats soak up raw juices and bacteria for disposal, but they are expensive and add to the environmental burden of trash. You can ensure the safety of regular cutting boards and reduce the microbes in sponges by washing them in a dishwasher or by treating them as suggested below. Alternatively, save the sponges for car washing and other heavy cleaning chores, and clean the kitchen with washable dishcloths that can be laundered often.

To eliminate microbes in your kitchen, you have four choices, each with benefits and drawbacks. One is to poison the microbes on cutting boards, sponges, and other equipment with toxic chemicals such as bleach (one teaspoon per quart of water). The benefit is that chlorine can kill even the hardiest organism. The drawback is that chlorine is toxic to handle, can ruin clothing, and washes down household drains into the water supply and forms chemicals that can harm waterways and fish.

A second option is to treat kitchen equipment with heat. Soapy water heated to 140°F kills most harmful organisms and washes most others away. This method takes effort, though, since the water must be truly scalding hot, well beyond the temperature of the tap. Third, an automatic dishwasher can combine both methods: it washes in water hotter than hands can tolerate, and most dishwasher detergents contain chlorine. Fourth, for sponges, place the wet sponge in a microwave oven and heat it until steaming hot (times vary). Pick one of these strategies to keep implements and surfaces safe for food preparation.

▶ **KEY POINT** To prevent foodborne illness, always remember that it can happen. Keep hot foods hot, keep cold foods cold, keep raw foods separate, and keep your hands and the kitchen clean.

TABLE 12·5 SAFE FOOD STORAGE TIMES: REFRIGERATOR (≤40°F)

1 to 2 Days

Raw ground meats, breakfast or other raw sausages, variety meats; raw fish or poultry; gravies

3 to 5 Days

Raw steaks, roasts, or chops; cooked meats, vegetables, and mixed dishes; ham slices; mayonnaise salads (chicken, egg, pasta, tuna)

1 Week

Hard-cooked eggs, bacon or hot dogs (opened packages); whole or half hams; smoked sausages

2 to 4 Weeks

Raw eggs (in shells); bacon or hot dogs (packages unopened); dry sausages (pepperoni, hard salami); most aged and processed cheeses (Swiss, brick)

2 Months

Mayonnaise (opened jar); most dry cheeses (parmesan, romano)

TABLE 12·6 KEEPING FOOD SAFE WHEN THE POWER GOES OUT

Keeping foods below 40°F during power outages can be challenging.
- Keep the refrigerator and freezer doors closed as much as possible to keep the cold in.
- Consume foods from the refrigerator first. Undisturbed, a full freezer will hold its temperature for about 48 hours; a refrigerator's temperature begins to rise in only 4 hours, even with the door closed.
- If available, place a block of ice or dry ice in the refrigerator; 50 pounds of dry ice can maintain the refrigerator temperature for about 2 days. Find out now where ice can be purchased in your neighborhood. If the weather is cold, fill small containers with water and put them out to freeze; use them in refrigerators, freezers, or coolers.
- Keep a week's supply of shelf-stable items, such as canned goods, bottled water, dry cereals, dry or shelf-stable milk, and ready-to-serve infant or pet foods, as needed. Use and replace these items several times a year, and keep a hand-operated can opener in the cupboard.
- Place thermometers in the freezer and refrigerator now. Then, when the power is out, the thermometer will indicate when the temperature is no longer safe and perishable food must be discarded.
- If food partially thaws (still has ice crystals) during an outage, and the temperature in the freezer never rose above 40°F, the food can safely be refrozen for later use.

Which Foods Are Most Likely to Make People Sick?

Some foods are more hospitable to microbial growth than others. Foods that are high in moisture and nutrients and those that are chopped or ground are especially favorable hosts. These foods are likely to spoil quickly without proper refrigeration. Table 12-6 provides special tips for preserving food safely in the event of a power outage.

Meats and Poultry Raw meats and poultry require special handling, and packages bear labels to instruct consumers on meat safety (see Figure 12-5).* Meats in the grocery cooler often contain all sorts of bacteria, and they provide a moist, nutritious environment that is just right for microbial growth. Ground meat or poultry is handled more than meats left whole, and grinding exposes much more surface area for bacteria to land on, so experts advise cooking these foods to well-done. Use a thermometer to test the internal temperature of poultry and meats, even hamburgers, before declaring them done. Burgers often turn brown and appear cooked before their internal temperature is high enough to kill harmful bacteria.[23]

Bovine Spongiform Encephalopathy (BSE) Though unrelated to sanitation, animal diseases such as "mad cow disease," or, more properly, **bovine spongiform encephalopathy (BSE),** pose a worry for meat eaters. A disease of cattle, BSE is linked with a rare but invariably fatal human brain disorder observed in people who consume products from infected animals.† A similar condition is suspected to be linked to consuming wild game.[24] To date, fewer than 150 people, most of them in Great Britain, have been diagnosed with or are suspected of having the disease.[25]

Scientists believe BSE to be among a handful of diseases caused by an oddly shaped protein, known as a **prion.**[26] Ingested prions seem to lie dormant in the body for many years before their deadly symptoms arise. Unlike bacteria or viruses, prions are not living things and so cannot be killed or controlled through cooking or disinfecting. Also, the body's immune system fails to mount defensive actions against prions. So little is known about how prions cause diseases or how to treat them when they arise that prevention remains the only form of control.

The prion responsible for BSE concentrates in the nervous and digestive tissues of infected animals. People living in or traveling to areas reporting BSE should avoid such foods as sausages, ground meats, and chopped meat products that may contain these heavily infected parts. In cattle, the disease probably spreads by way of cattle feed

---

■ *Anthrax* is a bacterial disease associated with large-animal farming that has been altered to become a terrorist weapon but is not considered to be a foodborne threat.

Additional USDA safeguards against BSE:
- Hold all products from cattle tested for BSE until test results are known.
- Prohibit high-risk cattle organs and tissues, such as intestines, from human use.
- Change meat-extracting procedures to exclude nervous system materials.
- Change slaughter procedures to ensure that nervous system materials do not contaminate the flesh.

bovine spongiform encephalopathy (BOW-vine SPON-jih-form en-SEH-fell-AH-path-ee) **(BSE)** an often-fatal illness of cattle affecting the nerves and brain. Also called *mad cow disease.*

prion (PREE-on) an infective agent consisting of an unusually folded protein that disrupts normal cell functioning, causing disease.

---

*The USDA's meat and poultry hotline answers questions about meat and poultry safety: 1-800-535-4555.
†The human disease is variant Creutzfeldt-Jakob disease (vCJD).

FIGURE 12·5 SAFE HANDLING INSTRUCTIONS FOR MEAT AND POULTRY

Never allow frozen meat to defrost at room temperature or in a bath of warm water. In both cases, meat thaws from outside in, and the outside meat layer can easily warm up to temperatures that permit bacterial growth before the core defrosts.

Safe Handling Instructions

THIS PRODUCT WAS PREPARED FROM INSPECTED AND PASSED MEAT AND/OR POULTRY. SOME FOOD PRODUCTS MAY CONTAIN BACTERIA THAT CAN CAUSE ILLNESS IF THE PRODUCT IS MISHANDLED OR COOKED IMPROPERLY. FOR YOUR PROTECTION, FOLLOW THESE SAFE HANDLING INSTRUCTIONS.

 KEEP REFRIGERATED OR FROZEN. THAW IN REFRIGERATOR OR MICROWAVE.

 KEEP RAW MEAT AND POULTRY SEPARATE FROM OTHER FOODS. WASH WORKING SURFACES (INCLUDING CUTTING BOARDS), UTENSILS, AND HANDS AFTER TOUCHING RAW MEAT OR POULTRY.

 COOK THOROUGHLY. KEEP HOT FOODS HOT. REFRIGERATE LEFTOVERS IMMEDIATELY OR DISCARD.

Microwave cooking of meats requires special care. Large, thick, dense foods such as roasts or meat loaves may register "cooked" on an internal meat thermometer, but may harbor cool spots in which dangerous microorganisms, such as the *Trichinella spiralis* parasite, sometimes present in pork, can survive. Such foods are best cooked by another method or divided into thin individual portions to be microwaved.

Properly cooked food hot from the oven or stove is relatively free of bacteria, but as soon as it is taken out to serve, it is reinoculated. Kitchen utensils recontaminate the food, or bacteria from the air land on its surface. Promptly after serving, even while the food is still hot, refrigerate leftovers in shallow containers for quick, even chilling. Large amounts of food refrigerated in deep containers may take hours to cool through, allowing bacteria time to multiply in the warm internal portions.

Take care when preparing meats along with foods intended to be served raw, such as chopped salads or lettuce and tomato toppers for hamburgers. A grave error is to prepare raw foods on the same board or with the same utensils as were used to prepare raw meats for cooking.

---

enriched with nutrients derived from cattle brains, blood, or other tissues. Such feed is now banned from use in most countries.

Since BSE emerged in Great Britain in the mid-1980s, tens of thousands of cattle have tested positive and four million animals were destroyed.[27] In 2003, a single U.S. dairy cow in the state of Washington, unable to walk at slaughter, was determined to have BSE. On hearing this news, many countries of the world instituted immediate bans on imports of U.S. beef and cattle. The cow, purchased from Canada, may have been fed old-style calf feed tainted with cattle parts. Since then, the USDA has enacted protective measures, such as prohibiting the use of "downer" cattle (those unable to walk) for human consumption and increasing BSE testing of cattle at slaughter.[28] These and other safeguards (see the margin) minimize the risk to consumers of U.S. beef. One substantial risk remains, however: imported supplements made from glands of animals, often sold as hormone preparations, may be made in countries with lax BSE regulations.

Eggs Raw, unpasteurized eggs are likely to be contaminated by *Salmonella* bacteria. *Salmonella* of a most virulent type have been detected in blood from food-poisoning victims and samples of illness-causing foods that contained raw eggs. To reduce illnesses from *Salmonella*, the FDA requires egg cartons to carry instruction labels urging consumers to keep eggs refrigerated, cook eggs until yolks are firm, and cook foods containing eggs thoroughly.[29]

Egg users may miss seeing these instructions, however, because they often appear in tiny type on the side of the egg carton. Healthy people can still safely enjoy favorite

A safe hamburger is cooked well-done, has juices that run clear, and has reached an internal temperature of 160°F (71°C). Place it on a clean plate when it's done.

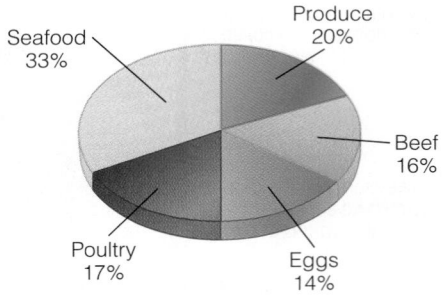

Seafood
33%

Produce
20%

Beef
16%

Eggs
14%

Poultry
17%

SOURCE: Center for Science in the Public Interest.

■ Unpasteurized or untreated juice must bear the following warning on its label:

WARNING: This product has not been pasteurized and therefore may contain harmful bacteria that can cause serious illness in children, the elderly, and persons with weakened immune systems.

For safety, when making flavored oils, wash and dry the herbs before adding them to the oil and keep the oil refrigerated.

foods that call for raw or undercooked eggs, such as rich homemade ice cream, hollandaise sauce, or raw cookie dough, by preparing them with pasteurized egg substitute.[30] For Caesar or other salad dressings, hard-cooked eggs processed in a blender or food processor work well. Even raw pasteurized egg substitutes may contain a few live bacteria that escape the pasteurization process, however, so they may not be safe for pregnant women, the elderly, the very young, or those suffering from immune dysfunction.

Raw Produce   This book champions a diet with abundant fresh fruits and vegetables, yet these foods can present a microbial threat unless they are thoroughly rinsed in cold running water to remove microbes before peeling, chopping, or eating. Just ten years ago, meats, eggs, and seafood posed the greatest foodborne illness threat by far, but today produce poses a similar threat (see Figure 12-6).[32]

Especially troublesome are foods consumed raw, such as lettuce, tomatoes, scallions, and spinach. These foods grow close to the ground, making bacterial contamination from the soil and organic fertilizers likely. Also, much produce is imported from countries where farmers may not adhere to sanitary growing and harvesting procedures and where contagious diseases are widespread. Fields may be irrigated with contaminated water, crops may be fertilized with untreated animal or human manure, or produce may be picked by infected farm workers with poor hygiene practices. Recently, more than 500 restaurant diners sickened and three died from hepatitis A transmitted by raw scallions that were harvested by infected farm workers in Mexico.[33]

Rough skins of melons such as cantaloupes provide crevices that harbor bacteria and so should be scrubbed with a brush under running water before peeling or cutting. Raspberries, other berries, and, in fact, all produce should be rinsed thoroughly under running water for at least 10 seconds (see Table 12-7). Unpasteurized or raw juices and ciders are not safe because microbes on the original fruit may multiply in the product. The label is required to carry the warning shown in the margin.

Raw sprouts (including alfalfa, clover, and radish), often eaten in salads, wraps, and other sandwiches, may not be safe.[34] Sprout seeds may harbor the dangerous *E. coli* O157:H7 bacteria that cannot be washed away, making even homegrown raw sprouts a risky food. Some experts are calling for all consumers to avoid raw sprouts for safety's sake. Also, as mentioned, oils flavored at home with fresh garlic or herbs may pose the threat of botulism unless the oils are refrigerated and used promptly.

Seafood   A variety of microbial dangers may lurk in even normal-appearing seafood: viral hepatitis; worms, flukes, and other parasites; viruses that can cause severe intestinal disorders; and naturally occurring toxins.* Hepatitis infection causes prolonged illness that persists for months or years, severely damages the liver, greatly increases the risk of developing liver cancer, and, once in the body, is transmissible to others. Many types of worms depend on the blood of their host for food and reproduction; they attack digestive membranes, sometimes causing life-threatening perforations. Flukes attack and damage the liver.

The dangers posed by seafood have grown in recent years. As population density increases along the seashores, the offshore waters are becoming more polluted, contaminating the seafood living there. Viruses that cause human diseases have been detected in some 90 percent of the waters off the U.S. coast.[31] Watchdog agencies monitor commercial fishing areas to keep harvesters out of unsafe waters, but unwholesome food can still reach the market. In a season, black-market dealers may sell millions of dollars worth of clams and oysters taken illegally from closed harvesting areas.

People who have enjoyed raw oysters and other raw seafood for years may be tempted to ignore these threats because they have never experienced serious illness. Some have heard that alcoholic beverages taken with raw seafood eliminate risks or that hot sauce kills the bacteria, but these assertions are not true. One study, long ago, did find a correlation between taking one drink of whiskey or wine and a reduced risk

---

*To speak with an expert on seafood safety, call FDA's seafood hotline: 1-800-FDA-4010.

TABLE 12·7 PRODUCE SAFETY

Cleaning Fresh Fruits and Vegetables

1. Remove and discard the outer leaves from vegetables such as lettuce and cabbage before washing.
2. Wash all fruits and vegetables (including organically grown and homegrown, regardless of place of purchase) just before cooking or eating.
3. Wash under clean running water and scrub with clean scrub brush, or with your hands. Do not use soap, detergents, or bleach solutions; commercial vegetable washing products are safe to use.
4. Dry fruits and vegetables before cutting or eating.
5. Cut away damaged or bruised areas that may contain bacteria. Toss out moldy fruit or vegetables.

Juice Safety

1. Choose chilled pasteurized juices or shelf-stable juices (canned or boxed) that have been treated with high temperature to kill microbes and check their seals to be sure no microbes have entered after processing.
2. Especially infants, children, the elderly, and people with weakened immune systems should never be given raw or unpasteurized juice products.

of disease after eating contaminated seafood, but this evidence is no guarantee of protection. Hot sauce is useless against the infectious agents that contaminate oysters. Experts unanimously agree that the risks of eating raw or lightly cooked seafood today are unacceptably high due to environmental contamination.

Consequently, adults and children alike should refrain from eating raw **sushi** or lightly steamed seafood, even when the food is prepared by an expert master chef. Not all varieties of sushi are made from raw fish; sushi made with cooked crabmeat and vegetables, avocado, and other delicacies are safe. The rumor that freezing fish will make it safe to eat raw is only partly true. Freezing fish will kill mature parasitic worms, but only cooking can kill all worm eggs and other microorganisms that can cause illness.

Honey Honey can contain dormant spores of *Clostridium botulinum* that can awaken (germinate) in the human body to produce the deadly botulinum toxin. Mature healthy adults are usually protected against this threat, but infants under one year of age should never be fed honey, which can also be contaminated with environmental pollutants picked up by the bees. Honey has been implicated in several cases of sudden infant death.

Picnics, Lunch Bags, and Take-Out Foods For safe picnics and safe packed lunches, keep these precautions in mind. Choose foods that remain safe to eat without refrigeration, such as fresh uncut fruits and vegetables, breads and crackers, shelf-stable foods, and canned spreads and cheeses that you can open and use on the spot. Aged cheeses, such as cheddar and Swiss, do well at environmental temperatures for an hour or two, but for longer periods, carry them in a cooler or thermal lunch bag. Mayonnaise is somewhat resistant to spoilage because of its acid content, but when it is mixed with chopped ingredients in pasta, meat, or vegetable salads, the mixtures spoil easily. The chopped ingredients have extensive surface areas for bacteria to invade, and foods that have been in contact with cutting boards, hands, and kitchen utensils have picked up at least a few bacteria. Start with chilled ingredients, then chill chopped salads in shallow containers before, during, and after eating. Keep meat, egg, cheese, or seafood sandwiches cold until eaten.

To keep lunch bag foods chilled, choose a thermal lunch bag and freeze beverages to pack in with the foods. As the beverages thaw in the hours before lunch, they keep the foods cold.

As for prepackaged single servings of cheese, cold cuts, and crackers promoted as lunch foods, these items keep well, but they can be extraordinarily high in saturated fat

■ In many states, containers of raw oysters must bear this warning: "There is a risk associated with consuming raw oysters or any raw animal protein. If you have chronic illness of the liver, stomach or blood or have immune disorders, you are at greater risk of serious illness from raw oysters and should eat oysters fully cooked. If unsure of your risk, consult a physician."

sushi a Japanese dish that consists of vinegar-flavored rice, seafood, and colorful vegetables, typically wrapped in seaweed. Some sushi is wrapped in raw fish; other sushi contains only cooked ingredients.

and sodium. They also often cost double or triple the price of the foods purchased separately and are excessively packaged, adding to a growing waste disposal problem.

Many people rely on take-out foods—rotisserie chicken, pizza, Chinese dishes, and the like—for parties, picnics, or days too hectic to cook. Be certain that the food is safe when you buy it: hot foods should be steaming hot, and cold foods should be thoroughly chilled. Leftover portions of restaurant meals can make a convenient lunch or dinner later on. To be sure that these foods are also safe, reheat them thoroughly (to 140°F) before eating, and discard any portion held at room temperature for longer than two hours from the time it first appears at the restaurant table until you place it in your refrigerator at home.

> ▶ **KEY POINT** Some foods pose special microbial threats and so require special handling. Raw seafood is especially likely to be contaminated. Honey is unsafe for infants. Almost all types of food poisoning can be prevented by safe food preparation, storage, and cleanliness.

Today, consumers bear much of the responsibility for staying safe from foodborne illnesses. They must cook meats and eggs to the well-done stage to kill dangerous microorganisms lurking in the raw products. They must scrub vegetables and fruits to remove bacteria, viruses, and pesticide residues. They must avoid foods like raw sprouts and raw seafood that may harbor disease-causing microorganisms. One proposal to help reduce the consumer's burden of responsibility is to expose foods to ionizing radiation in doses that kill microorganisms. The Consumer Corner on page 450 provides some details.

How Can I Avoid Illness When Traveling?

People who travel to places where cleanliness standards are lacking have a 50–50 chance of contracting a foodborne illness—commonly known as traveler's diarrhea (see Table 12-2, pages 436–438).[35] A bout of this illness can ruin a trip. To avoid foodborne illness while traveling:

- Before you travel, ask your physician which medicines to take with you in case you get sick.
- Wash your hands often with soap and water, especially before handling food or eating.
- Eat only cooked and canned foods. Eat raw fruits or vegetables only if you have washed them with your own clean hands in boiled water and peeled them yourself. Skip salads.
- Be aware that water, ice, and beverages made from water may be unsafe. Take along disinfecting tablets or an element that boils water in a cup. Drink only treated, boiled, canned, or bottled beverages, and drink them without ice, even if they are not chilled to your liking.
- Avoid using the local water, even if you are just brushing your teeth, unless you boil or disinfect it first.
- Be aware that mad cow disease poses an extremely small risk (1 in 10 billion servings) to travelers to countries where BSE has been a problem.[36] To err on the safe side, avoid eating beef altogether or select solid pieces of muscle meat that are less likely to be contaminated. Avoid variety meats such as sweetbreads, brains, and tripe, as well as sausages and ground meats that may contain these cuts.

In general, remember these rules: boil it, cook it, peel it, or forget it. If you follow these recommendations, chances are excellent that you will remain well.

> ▶ **KEY POINT** Some special food-safety concerns arise when traveling. To avoid foodborne illnesses, remember to boil it, cook it, peel it, or forget it.

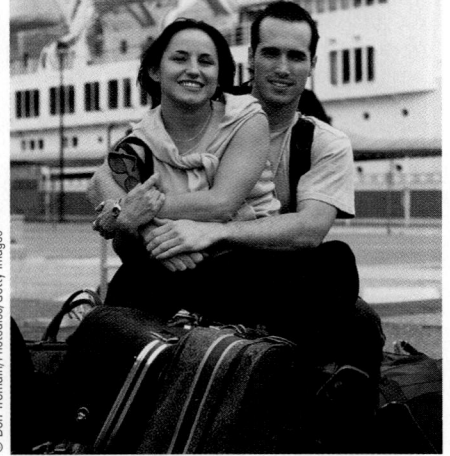

Don't let foodborne illness ruin your trip. Find information about risks in your destination at www.cdc.gov/travel/destinat.htm.

Natural Toxins in Foods

Some people think they can eliminate all poisons from their diets by eating only "natural" foods. On the contrary, nature has provided many plants with natural poisons to fend off diseases, insects, and other predators. Humans rarely suffer actual harm from such poisons, but the *potential* for harm does exist.

The herbs belladonna and hemlock have reputations as deadly poisons, but few people know that the herb sassafras contains a cancer-causing agent and is banned from use in commercially produced foods and beverages. Cabbage, turnips, mustard greens, and radishes all contain small quantities of harmful goitrogens, compounds that can enlarge the thyroid gland and aggravate thyroid problems. These effects show up only under extreme conditions when people have little but cabbage to eat. Ordinarily, cabbages and their relatives are celebrated for their nutrients and phytochemicals associated with low cancer rates.

Other natural poisons in *raw* lima beans, in the tropical root vegetable cassava, and in fruit seeds such as apricot pits are members of a group called cyanogens, which are precursors to the deadly poison cyanide. Many countries restrict commercial growers of lima beans to those varieties with the lowest cyanogen contents. Most cassava, a root vegetable eaten by hundreds of millions of people worldwide, contains just traces of cyanogens; the amount in bitter varieties, however, can be exceedingly high and poses a threat to hungry people who consume them in large quantities to escape starvation.[37] Fruit seeds are seldom deliberately eaten; an occasional swallowed seed or two presents no danger, but a couple of dozen seeds could be fatal to a small child. An infamous cyanogen is laetrile, a compound erroneously represented as a cancer cure. True, the poison laetrile kills cancer cells, but only at doses that kill the person, too. Research over the past 100 years has proved that laetrile is an ineffective cancer treatment and dangerous to the taker.

Potatoes contain many natural poisons, including solanine, a powerful, bitter, narcotic-like substance. The small amounts of solanine normally found in potatoes are harmless, but solanine can build up to toxic levels when potatoes are exposed to light during storage. Cooking does not destroy solanine, but because most of a potato's solanine is in a green layer that develops just beneath the skin, it can be peeled off, making the potato safe to eat. If the potato tastes bitter, however, throw it out.

At certain times of the year, seafood may become contaminated with the so-called red tide toxin that occurs during algae blooms. Eating seafood contaminated with red tide causes a form of food poisoning that paralyzes the eater. The FDA monitors fishing waters and closes them to fishing when red tide algae appear.

These examples of naturally occurring toxins serve as a reminder of three principles. First, any substance can be toxic when consumed in excess. Practice moderation in the use of all foods. Second, poisons are poisons, whether made by people or by nature. It is not the source of a chemical that makes it hazardous, but its chemical structure. Third, by including a variety of foods in the diet, consumers ensure that toxins in foods are diluted by the volume of the other foods eaten.

> ▶ KEY POINT   Natural foods contain natural toxins that can be hazardous if consumed in excess. To avoid poisoning by toxins, eat all foods in moderation, treat chemicals from all sources with respect, and choose a variety of foods.

Residues and Contaminants in Foods

Nutrition-conscious consumers often wonder if our nation's foods are made unsafe by contamination. The next few sections address some of their concerns.

■ Table 11-7 in Chapter 11 provides more information on potentially harmful herbs.

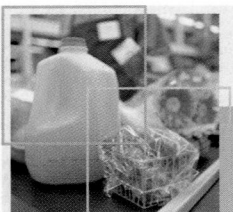

CONSUMER CORNER

Irradiation and Food Safety

SCIENTISTS WORLDWIDE support the use of food **irradiation** (see Table 12-8). (Terms are defined in Table 12-9.) According to the American Dietetic Association: "Food irradiation enhances the safety and quality of the food supply and helps protect consumers from foodborne illness."[1] The American Academy of Pediatrics states: "The science of food irradiation is mature and the scientific consensus on its efficacy and safety is strong."[2] Rarely do so many scientists agree on a single issue.

The FDA has approved irradiation for controlling microbial contamination of many foods (see Table 12-10), but so far the process is little used.[3] Can irradiation solve our food-safety and supply problems, and if so, why not use it?

Potential Benefits

Food irradiation's greatest potential benefit is its ability to kill almost all disease-producing microorganisms present in food, a characteristic that has earned it the alternative name *cold pasteurization*. It has no effect on most toxins, prions, and microbial spores, however, and thus does not solve every contamination problem.[*4]

Each year, the many millions of children, elderly, and susceptible people worldwide who sicken or die from foodborne illnesses could be spared by the power of irradiation. Raw poultry, for example, emerges from irradiation treatment 99.9 percent free of disease-causing microorganisms. Thus, chicken is safer when purchased, and the threat of cross-contamination in home or industrial kitchens is reduced. Undercooking becomes less of a health threat, too.

Growers and marketers of fresh produce also stand to benefit. Irradiation

*The spores of *Clostridium botulinum* are resistant to radiation.

eliminates the need to quarantine and spray fresh produce with pesticides before shipping to prevent the spread of plant diseases or harmful insects and their eggs. Irradiation kills mold spores, replacing the fungicides now sprayed on harvested foods. Irradiation also slows decay in fruits and vegetables, making them last longer and appear fresher.

In addition to reducing the incidence of foodborne illness, irradiation greatly reduces the destruction of food by pests and decay. Up to half of the world's food bounty is lost each year to these problems. Irradiation facilities are expensive to build, however, placing them out of reach for the world's

impoverished people who most need them—see Chapter 15 for more information on world hunger.

The Irradiation Process

Irradiation works by exposing foods to controlled doses of gamma rays from the radioactive compound cobalt 60. As radiation passes through living cells, it

---

TABLE 12·8 SCIENTIFIC GROUPS IN SUPPORT OF FOOD IRRADIATION

These groups and many others agree that irradiation, when used in conjunction with other safe food practices, can improve food safety in the United States and the rest of the world:

- American Academy of Pediatrics
- American Council on Science and Health
- American Dietetic Association
- American Medical Association
- American Veterinary Medical Association
- Centers for Disease Control and Prevention
- Council for Agricultural Science and Technology
- International Atomic Energy Agency
- Institute of Food Technologists
- Scientific Committee of the European Union
- United Nations Food and Agricultural Organization
- World Health Organization

TABLE 12·9 IRRADIATION TERMS

- **irradiation** the application of ionizing radiation to foods to reduce insect infestation or microbial contamination or to slow the ripening or sprouting process. Also called *cold pasteurization*.
- **radiolytic products** chemicals formed in foods during the irradiation process; deemed harmless by experts.

TABLE 12·10 FOODS APPROVED FOR IRRADIATION BY THE FDA

- Citrus fruits
- Eggs
- Flour
- Fresh and frozen red meats, such as lamb, beef, and pork
- Mushrooms
- Onions
- Potatoes
- Poultry
- Spices
- Strawberries
- Tomatoes
- Tropical fruits
- Wheat

SOURCE: Meat irradiation can boost food safety, *FDA Consumer*, March/April 2000, p. 4.

disrupts their internal structures and kills or deactivates the cells. Low doses can kill the growth cells in the "eyes" of potatoes and ends of onions, preventing them from sprouting. Low doses also delay ripening of bananas, avocados, and other fruits.[5] High doses can penetrate tough insect exoskeletons and mold or bacterial cell walls to destroy their life-maintaining DNA, proteins, and other molecules. Irradiation can kill most microbes even while food is in a frozen state, making irradiation uniquely useful in protecting foods such as whole turkeys that are ordinarily marketed frozen.

For perspective, compare the doses of radiation used on foods with the lethal human dose. The lowest doses of radiation needed to delay ripening and sprouting of fragile fruits and vegetables are 10 to 20 times higher than the doses that would kill human beings. The dose required to sterilize foods is many times higher still, but most foods are not completely sterilized because doses that high would destroy the food. Dried herbs and spices are notable exceptions—they can withstand sterilizing doses and are commonly irradiated before purchased by consumers in the United States.

Labeling of Irradiated Foods

The FDA requires that irradiated foods bear a label, in letters as large as those of the ingredients list, stating that the foods have been treated with radiation and display the irradiation symbol, also called the radura logo, shown here. No label is required for foods containing irradiated ingredients, including spices that are mixed with processed foods, and for irradiated foods served in restaurants.

This radura logo is the symbol for foods treated with radiation.

If Irradiation Is So Great, Why Aren't More Foods Irradiated?

The answers to this question center on cost, consumer acceptance, and fears about safety. Irradiated foods cost more to produce and package than conventionally processed foods, so the price to consumers is also higher. Irradiation also changes the flavor of foods somewhat, but most consumers have not tasted enough irradiated foods to know whether they like or dislike them.

People commonly give several reasons for fearing irradiation:

- Fear that the foods will become radioactive.
- Fear that the foods will lose substantial nutrients during irradiation.
- Fear that irradiated foods are not safe to eat.
- Fear of harmful chemicals formed during irradiation.
- Fear that the radioactive substances used to irradiate foods will endanger plant workers, the general population, and the environment.

The first concern can immediately be put to rest—properly irradiated food does not become radioactive any more than teeth become radioactive after dental X-ray procedures. The use of radioactivity demands great care, though. Correct doses are of paramount importance to improve food safety while preserving nutrients.

Irradiation's Effects on Nutrients

Most nutrients, such as proteins, fats, carbohydrates, and minerals, survive irradiation intact or sustain only insignificant losses. Nutrients sensitive to heat treatment, such as the B vitamins and ascorbic acid, are sensitive to irradiation. Even so, substantial losses occur only with high doses of radiation well beyond those permitted in the treatment of food. In general, the nutrient losses sustained during properly administered irradiation are similar to those caused by canning or cooking. The FDA deems nutrient losses of less than 2 percent of the total

as insignificant, and most losses incurred through irradiation fall within this limit.

Irradiation Safety

More than 40 years of research on animals have revealed no toxic effects from eating irradiated foods.[6] Decades ago studies of human volunteers who ate a diet composed entirely of irradiated foods found no ill effects.[†] As for chemicals formed in foods during irradiation, this issue has been resolved to the scientific community's satisfaction. These **radiolytic products** have been identified as chemicals commonly formed in foods during many forms of processing. Among them are minuscule amounts of highly reactive free radicals that could, theoretically, be damaging to health. These tiny amounts, however, are unlikely to pose a hazard to the body, whose antioxidant systems are equipped to neutralize free radicals from many sources. After much scientific evaluation, the FDA has declared irradiation safe and effective for improving the microbial safety of foods.[7]

The irradiation process itself also arouses concerns. The process necessitates transporting radioactive materials, training workers to handle them safely, and then disposing of the spent wastes, which remain radioactive for many years. Opponents of irradiation worry that workers could be exposed to nonlethal low doses of radiation that could conceivably cause ill effects. Their concerns are echoed by the food industry, which strives to safeguard both workers and consumers through strict operating standards and compliance with regulations.

Finally, some consumers worry that food manufacturers might use the technology unethically. Current law prohibits the sale of old or tainted food found to have high bacterial counts,

---

[†]The USDA posts approximately 11,000 pages of research reports supported by the Office of the Surgeon General, Department of the Army, on the topic of wholesomeness of irradiated foods on its website: www.nal.usda.gov/fnic/.Foodirad-v2/food2.html.

usually from insect or rodent droppings. The FDA condemns such foods and orders their destruction. Unscrupulous manufacturers, concerned about lost profits, could irradiate old or tainted food, killing the telltale bacteria, and then sell the food as wholesome to unsuspecting consumers. This objection raises an important point: irradiation is intended to complement other traditional food-safety methods, not to replace them. Even irradiation cannot entirely protect people from faulty food-safety practices and contamination due to poor sanitation on the farm, in the marketplace, or at home.

Consumers: The Final Authority

Only a few irradiated products are now available for use.[8] Irradiated ground beef products are approved for U.S. school lunch programs, but the higher cost of these products discourages many schools from using them. Schools that do use them are encouraged to inform students and parents about the products.[9]

Whether more irradiated foods will appear in markets depends largely upon whether consumers choose to buy them. According to a national survey of supermarket shoppers, the proportion of those who are willing to purchase irradiated foods declined during the past decade from 70 percent to about 50 percent.[10] Cost is a factor: people must be willing to pay a premium for irradiated meats, poultry, and seafood to cover the substantial expense of producing them.

In the end, acceptance may depend on taste. Flavor is high on most consumers' food priorities, and irradiation changes the taste of food slightly. Whether future refinements will positively influence consumer purchases of irradiated foods is for the future to tell.

Pesticides

Pesticides:
- Accumulate in the food chain.
- Kill pests' natural predators.
- Pollute the water, soil, and air.

Each year in the United States, 4.5 billion pounds of pesticides are applied:
- To kill pests in and around homes.
- To control pests in gardens.
- To reduce loss of farm crops to insects.
- To preserve wood products.
- To cure lice, scabies, worms, and other parasites in people and their pets.
- To repel mosquitoes and other biting insects.
- For many other uses.

pesticides chemicals used to control insects, diseases, weeds, fungi, and other pests on crops and around animals. Used broadly, the term includes *herbicides* (to kill weeds), *insecticides* (to kill insects), and *fungicides* (to kill fungi).

residues whatever remains. In the case of pesticides, those amounts that remain on or in foods when people buy and use them.

The use of **pesticides** helps to ensure the survival of food crops, but the damage pesticides do to the environment is considerable and increasing. Moreover, there is some question about whether the widespread use of pesticides has really improved the overall yield of food. Even with extensive pesticide use, the world's farmers lose large quantities of their crops to pests every year.

Do Pesticides on Foods Pose a Hazard to Consumers? Many pesticides are broad-spectrum poisons that damage all living cells, not just those of pests. Their use poses hazards to the plants and animals in natural systems, and especially to workers involved with pesticide production, transport, and application. High doses of pesticides applied to laboratory animals cause birth defects, sterility, tumors, organ damage, and central nervous system impairment. Equivalent doses are extremely unlikely to occur in human beings, however, except through accidental spills. As Figure 12-7 demonstrates, pesticide **residues** on agricultural products can survive processing and may be present in and on foods served to people. Government agencies in the United States and Canada monitor pesticide safety and residues on foods to ensure minimal risk to the consumer.

Infants and children, however, may be more susceptible than adults to adverse effects from pesticides for several reasons.[38] The immature brain cannot exclude pesticides and other chemicals to the same extent as the adult brain. Many pesticides work by interfering with normal nerve and brain chemistry, and the effects of chronic, low-dose exposure to pesticides on the developing human brain are largely unknown. Children may be exposed to pesticides through normal behavior such as playing outdoors on treated soil or lawns; handling sticks, rocks, and other potentially contaminated objects; crawling on treated carpets, furniture, and floors; placing fingers and other objects in the mouth; seldom washing their hands before eating; and using fingers instead of utensils to grasp foods. Further, children eat proportionally more food per pound of body weight than do adults, and traces of pesticides present on foods can build up quickly in their small bodies. Fortunately, compared with other sources of pesticides, the traces found on foods rarely exceed safe limits, and most of those present can be removed by washing produce thoroughly and following the other guidelines in Table 12-11, p. 454.[*]

---

[*]For answers to questions about pesticides, call the Environmental Protection Agency's 24-hour national pesticide information center: 1-800-858-PEST.

FIGURE 12·7 HOW PESTICIDE RESIDUES COULD END UP IN A FAST-FOOD MEAL

Pesticide residues may be present on these agricultural crops.

The red dots in the figure represent pesticide residues left on foods from field spraying or postharvest application. Notice that most pesticides follow fats in foods and that some processing methods, such as washing and peeling vegetables, reduce pesticide concentrations while others tend to concentrate them.

Processing affects the residues:

concentrates | reduces by washing/peeling off | reduces by washing/diluting | concentrates | reduces by washing/peeling off | no effect

fresh

extracted

milled

dried

pickled/canned

frozen

flour

FEED CORN

Hamburger Dills

Frozen French Fries

meats (especially fats)

fast-food restaurant

fries

TABLE 12·11 WAYS TO REDUCE PESTICIDE RESIDUE INTAKE

- Trim fat from meat and remove skin from poultry and fish; discard fats and oils in broths and pan drippings. (Pesticide residues concentrate in the animal's fat.)
- Vary meat, poultry, and fish choices from day to day and do not take fish oil capsules.
- Wash and trim fresh produce in water as instructed in Table 12-7, page 447.
- Use a knife to peel an orange or grapefruit; do not bite into the peel.
- Peel waxed fruit and vegetables. (Waxes don't wash off and can seal in pesticide residues.)
- Peel vegetables such as carrots and fruits such as apples when appropriate. (Peeling removes pesticides that remain in or on the peel but also removes some fibers, vitamins, and minerals.)

Regulation of Pesticides   The regulation of pesticides focuses on the legal **tolerance limit** set for each pesticide. The tolerance limits for pesticide residues in foods are low, generally 1/100 to 1/1,000 of the level found to cause no effect in laboratory animals. Over 10,000 tolerance regulations state maximum levels for the more than 300 pesticide chemicals allowed for use on various specific crops in the United States. If a pesticide is misused, growers risk fines, lawsuits, and destruction of their crops. In 25 years of testing, the FDA has seldom found crop residues above tolerance levels, so it appears that pesticides are generally used according to regulations. This makes sense because growers are not anxious to spend extra capital on unneeded chemicals.

A loophole in federal regulations allows companies in the United States to make banned pesticides and export them to other countries. The banned pesticides can occasionally slip back into the United States on imported foods. The FDA is stepping up surveillance of imported foods and can deny entry to any imports found to contain illegal residues. The overwhelming majority of both imported and domestic foods tested by the FDA are found to contain either no residues or residues within federally permitted limits.[39] A problem, however, is that few foods are tested. Budget constraints limit the FDA's testing capacity. The FDA cannot (nor can it be expected to) guarantee 100 percent safety in the food supply. Instead, it sets conditions so that substances do not become a hazard and acts promptly when problems or suspicions arise.

tolerance limit the maximum amount of a residue permitted in a food when a pesticide is used according to label directions.

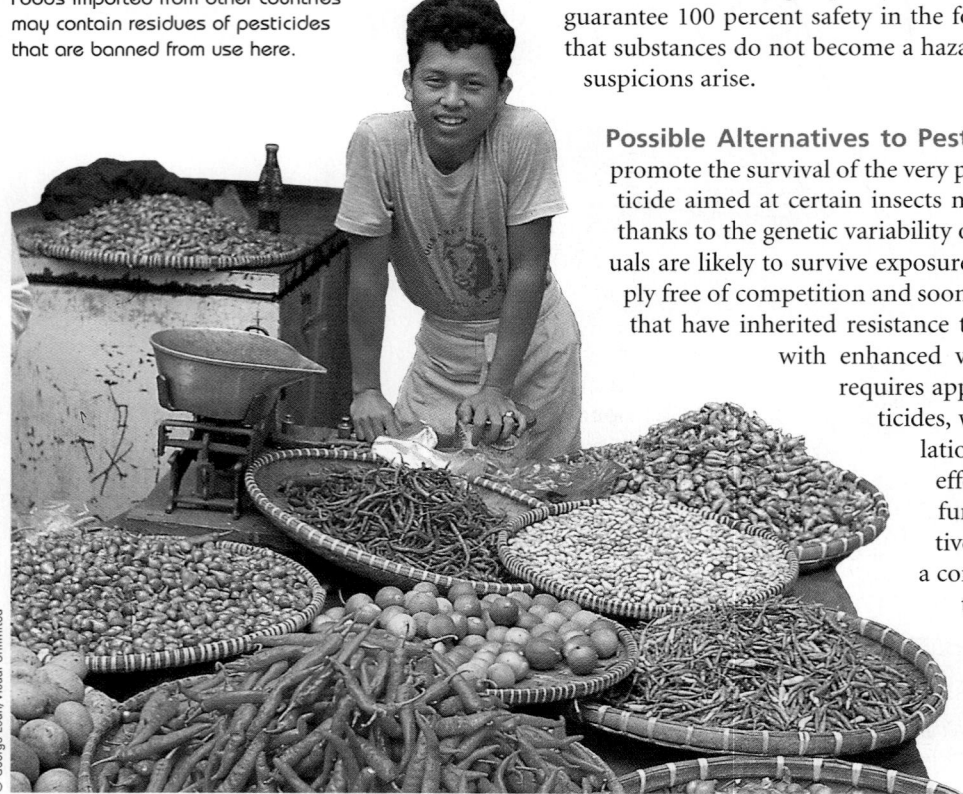

Foods imported from other countries may contain residues of pesticides that are banned from use here.

© George Loun/Visual Unlimited

Possible Alternatives to Pesticides   Ironically, some pesticides also promote the survival of the very pests they are intended to wipe out. A pesticide aimed at certain insects may kill *almost* 100 percent of them, but thanks to the genetic variability of large populations, a few hardy individuals are likely to survive exposure. These resistant insects can then multiply free of competition and soon will produce many offspring—offspring that have inherited resistance to the pesticide and can attack the crop with enhanced vigor. To control these resistant insects requires application of new and more powerful pesticides, which leads to the emergence of a population of still more resistant insects. The same effects arise from use of herbicides and fungicides. One alternative to this destructive series of events is to manage pests using a combination of natural and biological controls, as discussed in Controversy 15.

Pesticides are not produced only in laboratories; they also occur in nature. The nicotine in tobacco and psoralens in celery are examples. A bacterium from soil yields a pesticide often used in organic gardening; the genetic blueprint for producing this bacterial pesticide has been transferred to vegetables, resulting in plants that grow their own pesticides in the field. Natural pesticides are less damaging to other living things and leave less persistent residues in the environment than most human-made ones. An ideal pesticide would destroy pests in the field but vanish long before consumers ate the food. Chemical companies are working to develop such safer pesticides, and advances in **biotechnology** have reduced the need for pesticide sprays on many crops. Another possibility for consumers who want fewer pesticides in their produce and meat is to choose **organic foods;** their pros and cons are also discussed in the Controversy section.

> ▶ KEY POINT   Pesticides can be part of a safe food production process but can also be hazardous if mishandled. The FDA tests for pesticide residues in both domestic and imported foods. Consumers can take steps to minimize their ingestion of pesticide residues in foods.

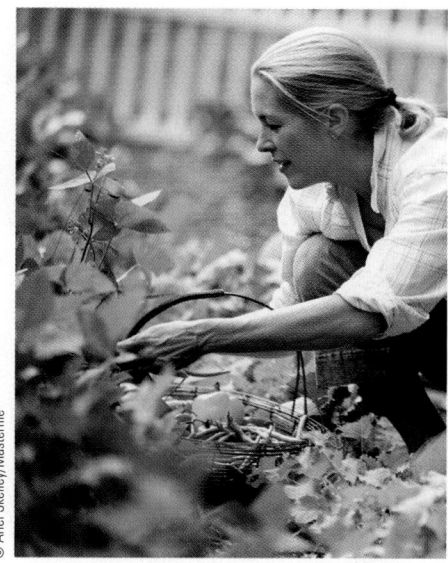

In some small gardens, handwork can take the place of pesticides.

Animal Drugs

Consumer groups express concern about drugs administered to livestock that produce food. Of particular concern to some consumers are hormones, antibiotics, and drugs that contain arsenic compounds.

Growth Hormone in Meat and Milk   Cattle producers in the United States commonly inject their herds with a form of **growth hormone—bovine somatotropin (bST)**—to promote lean tissue growth and milk production. The hormone, produced by genetically altered bacteria, was engineered to be identical to growth hormone made naturally in the pituitary gland of the animal's brain. The FDA deems the practice safe and does not require testing of food products for traces of the drugs.

Ranchers advocate the use of bST because injected animals develop more meat and less fat. The hormone may also increase milk production in dairy cows by up to 25 percent, while reducing feed requirements. Thus, the farmers can earn higher profits without incurring the high costs of more cattle, more farmhands, and more equipment. The environment may profit as well. Smaller herds can live on smaller plots of cleared land, and less feed means the use of less resources to produce and transport it (Controversy 15 gives details). Consumer groups counter that while ranchers may benefit from the use of bST, consumers do not and so its use is not justified.

The European Union and Canada ban the use of bST for milk cows on the grounds that bST stimulates the release of another bovine hormone, insulin-like growth factor I (IGF-I), and some questions have been raised about its effects on human health.[40] IGF-I is produced naturally in people, with levels declining as they age. After a thorough review of the literature, the FDA concluded that IGF-I levels in milk from bST-treated cows are within the normal range of variation seen in milk from untreated cows and therefore that IGF-I from this source presents no additional risk.[41] Because all cows make bST, the FDA prohibits milk processors from stating "hormone free" on milk labels.[42] The World Health Organization's expert committee on food additives concurs.

Antibiotics in Livestock   Ranchers and farmers often dose livestock with antibiotic drugs as part of a daily feeding regimen.[43] The drugs ward off infections that commonly afflict animals living in crowded conditions and help to promote rapid growth. Thus, farms can produce more food at lower cost.

The USDA prohibits livestock antibiotics from entering the food supply—ranchers and farmers must adhere to a drug-free waiting period before slaughter during which the drugs break down. Thus, consumers face little threat of ingesting antibiotic drugs in meats, milk, and eggs. Nevertheless, they run a substantial and increasing risk of illness from antibiotic-resistant bacteria wherever animals are treated with daily antibiotics.[44]

biotechnology the science of manipulating biological systems or organisms to modify their products or components or create new products; more properly called *genetic engineering* or *rDNA technology* (see the chapter Controversy).

organic foods products grown and processed without the use of synthetic chemicals such as pesticides, herbicides, fertilizers, and preservatives and without genetic engineering or irradiation.

growth hormone a hormone (somatotropin) that promotes growth and is produced naturally in the pituitary gland of the brain.

bovine somatotropin (bST) (so-MAT-ah-TROW-pin) growth hormone of cattle, which can be produced for agricultural use by genetic engineering. Also called *bovine growth hormone (bGH)*.

A limited number of antibiotic drugs exist—the same or related drugs used as growth promoters in livestock also treat illnesses in both animals and people. When the bacteria in an animal's intestinal tract encounter low daily doses of antibiotics, the bacteria adapt, losing their sensitivity to the drugs over time. Subsequently, such bacteria can infect consumers through food or by direct contact with ill persons. The result is a severe infectious disease that does not yield to standard antibiotic drug therapy, often ending in fatality. To safeguard the health of the world's people, the Centers for Disease Control and Prevention and the World Health Organization, along with other groups, support actions to stop the use of antibiotic drugs as animal growth promoters.

Arsenic in Food Animals   Another animal drug under surveillance by the USDA is **arsenic,** a naturally occurring element and infamous poison. Traces of arsenic are thought to be essential for normal growth in several animal species. In large amounts, arsenic leads to swelling of the brain, damage to the liver, and other deadly effects. Chronic human exposure to small amounts is associated with cancers, heart disease, diabetes, birth defects, and miscarriages.[45]

Chicken and poultry farmers often feed arsenic in tiny amounts to young flocks to control parasites that would otherwise inhibit their growth. The USDA approves this use, but it has found that current arsenic levels in young chickens exceed expected levels by three times.[46] Arsenic is also present in other foods such as fish, eggs, milk products, and other meats to a lesser extent; drinking water and environmental exposure may be the chief sources in some areas.[47] Although the arsenic in chickens is higher than expected, an adult would still have to eat more than 21 ounces of chicken each day to receive less than half of the maximum daily arsenic level deemed safe. Thus, healthy consumers run very little if any risk from ordinary portions (2 to 3 ounces) of chicken in the context of a varied diet.

> **KEY POINT**   Bovine somatotropin causes cattle to produce more meat and milk on less feed than untreated cattle. The FDA has deemed the practice safe. Antibiotic overuse fosters antibiotic resistance in bacteria, threatening human health. Arsenic drugs are used to promote growth in chickens and other livestock.

Environmental Contaminants

As populations increase worldwide and nations become more industrialized, concerns grow about the environmental contamination of foods.[48] A food **contaminant** is anything that does not belong there.

Harmfulness of Contaminants   The potential harmfulness of a contaminant depends in part on the extent to which it lingers in the environment or in the human body—that is, on how **persistent** it is. Some contaminants are short-lived because microorganisms or agents such as sunlight or oxygen can break them down. Some contaminants linger in the body for only a short time because the body can rapidly excrete them or metabolize them to harmless compounds. These contaminants present little cause for concern. Some contaminants resist breakdown, however, and interact with the body's systems without being metabolized or excreted. These contaminants can pass from one species to the next and accumulate at higher concentrations in each level of the food chain, a process called **bioaccumulation**—see Figure 12-8.

How much of a threat do environmental contaminants pose to the food supply? It depends on the contaminant. In general, the threat remains small because the FDA monitors the presence of contaminants in foods and issues warnings when contaminated foods appear in the market. In the event of an industrial spill or a natural occurrence, such as a volcano's eruption, however, the hazard can suddenly become great.

Other contaminants build in the food supply more insidiously. For example, increasing levels of the **heavy metal** mercury expelled from industrial sites have been

---

Chemical contaminants of concern in foods:

- Heavy metals:
  - Arsenic.
  - Cadmium.
  - Lead.
  - Mercury.
  - Selenium.
- Halogens and organic halogens:
  - Chlorine.
  - Ethylene dichloride.
  - Iodine.
  - Polybrominated biphenyl (PBB).
  - Polychlorinated biphenyls (PCBs).
  - Trichloroethylene (TCE).
  - Vinyl chloride.
- Others:
  - Acrylamide.
  - Antibiotics (in animal feed).
  - Asbestos.
  - Diethylstilbestrol (DES).
  - Dioxins.
  - Heat-induced mutagens.
  - Lysinoalanine.

arsenic a poisonous metallic element. In trace amounts, arsenic is believed to be an essential nutrient in some animal species. Arsenic is often added to insecticides and weed killers and, in tiny amounts, to certain animal drugs.

contaminant any substance occurring in food by accident; any food constituent that is not normally present.

persistent of a stubborn or enduring nature; with respect to food contaminants, the quality of remaining unaltered and unexcreted in plant foods or in the bodies of animals and human beings.

bioaccumulation the accumulation of a contaminant in the tissues of living things at higher and higher concentrations along the food chain.

heavy metal any of a number of mineral ions such as mercury and lead; so called because they are of relatively high atomic weight. Many heavy metals are poisonous.

FIGURE 12·8 BIOACCUMULATION OF TOXINS IN THE FOOD CHAIN

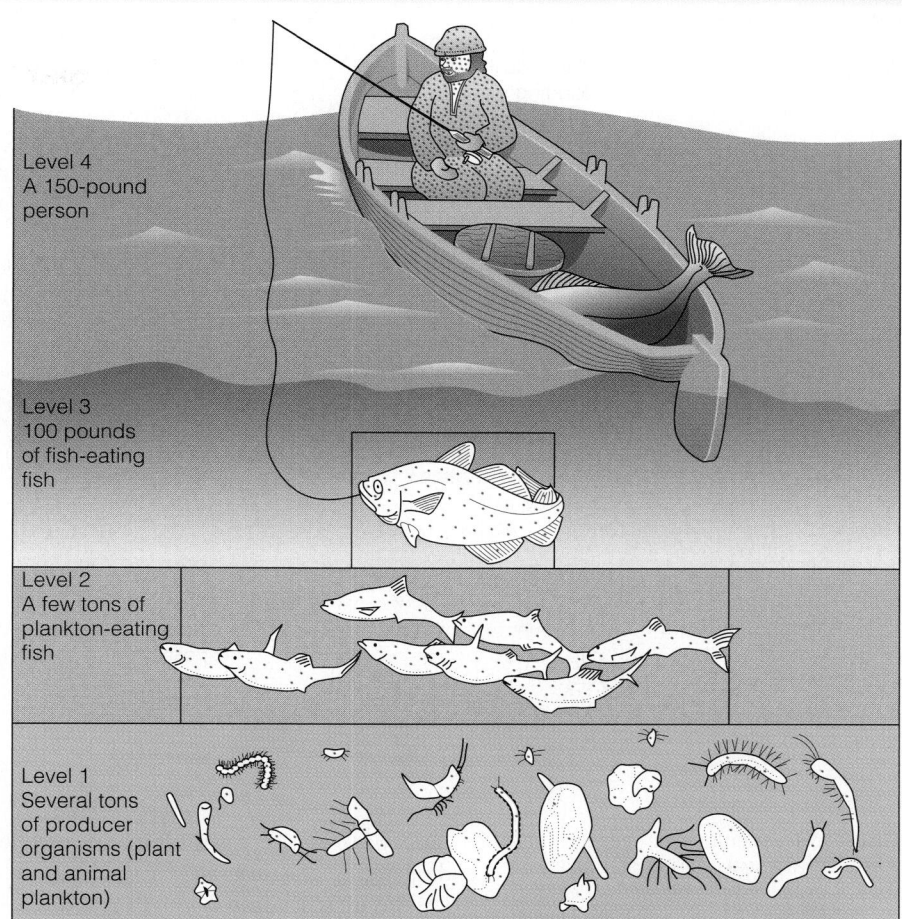

If none of the chemicals are lost along the way, one person ultimately receives all of the toxic chemicals that were present in the original several tons of producer organisms.

❹ A person whose principal animal-protein source is fish may consume about 100 pounds of fish in a year.

❸ Larger fish consume a few tons of plankton-eating fish in the course of their lifetimes—and the toxic chemicals from the small fish become more concentrated in the flesh of the larger species.

❷ The toxic chemicals become more concentrated in the plankton-eating fish that consume several tons of producer organisms in their lifetimes.

❶ Producer organisms may become contaminated with toxic chemicals.

Level 4
A 150-pound person

Level 3
100 pounds of fish-eating fish

Level 2
A few tons of plankton-eating fish

Level 1
Several tons of producer organisms (plant and animal plankton)

⌑ Toxic chemicals are represented by dots.

detected in U.S. lakes, rivers, and ocean fisheries. Virtually all fish have at least trace amounts of mercury (on average, 0.12 parts per million). Mercury, **PCBs,** chlordane, dioxins, and DDT are the toxins most responsible for fish contamination, but mercury leads the list by threefold.[49] Table 12-12 on p. 458 selects a few contaminants of great concern in foods to show how pervasively a contaminant can affect the body.

When an environmental contaminant is detected in a person's blood or urine, this does not automatically mean that the chemical will cause disease.[50] The toxicity of a chemical depends upon its dose or concentration. Small amounts may be tolerable and of no consequence to health, while larger amounts may be dangerous. For some chemicals, such as mercury, the risks of increasing blood levels are well known; for other chemicals, such as certain pesticides, the health effects of varying blood levels are yet to be established.

Mercury in Seafood Scientists learned of mercury's full potential for harm through tragedy. In 1953, a number of people in Minamata, Japan, became ill with a disease no one had seen before. By 1960, 121 cases had been reported, including 23 in infants. Mortality was high; 46 died, and the survivors suffered progressive, irreversible blindness, deafness, loss of coordination, and severely impaired mental function.* The

PCBs stable oily synthetic chemicals used in hundreds of industrial and commercial operations that persist as pollution in the environment. PCBs cause cancer in animals and a number of other serious health effects. The Environmental Protection Agency monitors their levels.

Minamata disease was named for the location of the disaster.

TABLE 12·12 EXAMPLES OF CONTAMINANTS IN FOODS

Name and Description	Sources	Toxic Effects	Typical Route to Food Chain
Cadmium (heavy metal)	Used in industrial processes including electroplating, plastics, batteries, alloys, pigments, smelters, and burning fuels. Present in cigarette smoke and in smoke and ash from volcanic eruptions.	No immediately detectable symptoms; slowly and irreversibly damages kidneys and liver.	Enters air in smokestack emissions, settles on ground, absorbed into food plants, consumed by farm animals, and eaten in vegetables and meat by people. Sewage sludge and fertilizers leave large amounts in soil; runoff contaminates shellfish.
Lead[a] (heavy metal)	Lead crystal decanters and glassware, painted china, old house paint, batteries, pesticides, old plumbing, and some food-processing chemicals.	Displaces calcium, iron, zinc, and other minerals from their sites of action in the nervous system, bone marrow, kidneys, and liver, causing failure of function.	Originates from industrial plants and pollutes air, water, and soil. Still present in soil from many years of leaded gasoline use.
Mercury (heavy metal)	Widely dispersed in gases from earth's crust; local high concentrations from industry, electrical equipment, paints, and agriculture.	Poisons the nervous system, especially in fetuses.	Inorganic mercury released into waterways by industry and acid rain is converted to methylmercury by bacteria and ingested by food species of fish (tuna, swordfish, and others).
Polychlorinated biphenyls (PCBs) (organic compounds)	No natural source; produced for use in electrical equipment (transformers, capacitors).	Long-lasting skin eruptions, eye irritations, growth retardation in children of exposed mothers, anorexia, fatigue, others.	Discarded electrical equipment; accidental industrial leakage, or reuse of PCB containers for food.

[a]For answers to questions concerning lead, call the National Lead Information Center at (800) 424-LEAD.

■ Chapter 13 presents details about mercury and other threats during pregnancy. Chapter 5 explained the benefits of eating fish and the hazards of fish oil supplements.

■ The fish highest in mercury are king mackerel, swordfish, shark, and tile fish.

methylmercury any toxic compound of mercury to which a characteristic chemical structure, a methyl group, has been added, usually by bacteria in aquatic sediments. Methylmercury is easily absorbed from the intestine and causes nerve damage in people.

cause of this misery was ultimately revealed: manufacturing plants in the region were discharging mercury into the waters of the bay, and bacteria in the water were converting the mercury into a more toxic form, **methylmercury.**[51] The fish in the bay were accumulating this poison in their bodies. Some of the poisoned people had been eating fish from the bay every day. The infants who contracted the disease had not eaten any fish, but their mothers had. The mothers were spared damage during their pregnancies because the poison had been concentrating in the tissues of their unborn babies.

Today, in the United States, the FDA and the Environmental Protection Agency (EPA) warn of unacceptably high methylmercury levels in fish and other seafood. They advise all pregnant women, women who may become pregnant, nursing mothers, and young children against eating fish species known to be high in methylmercury (Chapter 13 provides details of this warning).[52] In addition, the FDA and EPA suggest checking local advisories concerning fish caught by family and friends in lakes, rivers, streams, and coastal areas; if no advice is available, one meal of locally caught fish per week is unlikely to be harmful, but no other fish should be eaten that week. Freshwater fish often contain PCBs and other industrial contaminants.[53]

No one expects the tragic results of 1953 to occur again, but lower doses of methylmercury cause headaches, fatigue, memory loss, impaired ability to concentrate, and muscle or joint pain in adults.[54] In children, the threats may be greater and long-lasting. Methylmercury is persistent in the environment, so efforts begun today to clean up U.S. waters will take years to diminish this threat.

In an effort to limit exposure to pollutants, some consumers choose farm-raised fish. The "farms," however, are really giant ocean cages, exposed to whatever contaminants are in the water. Farm-raised salmon, especially from Europe, recently tested high in certain industrial pollutants and pesticides that are banned from use in the United States.[55] Farm-raised fish do tend to collect less methylmercury in their flesh than wild fish do, and the levels of other harmful chemicals were far below the maximums set by the FDA. Nevertheless, they demonstrate that monitoring is essential. They also serve as a reminder that our health is inextricably linked with the health of our planet (see Chapter 15).

▶ KEY POINT   Persistent environmental contaminants pose a significant, but generally small, threat to the safety of food. An accidental spill can create an extreme hazard. Mercury and other contaminants are of greatest concern during pregnancy, lactation, and childhood.

Effects of Food Processing on the Nutrients in Foods

Many consumers rely on packaged and processed foods for convenience and speed and so lose some control over exactly what their foods contain. What does processing do to foods and to their nutritional value? Food processing involves trade-offs. It makes food safer, or it gives food a longer usable lifetime, or it cuts preparation time—but at the cost of some vitamin and mineral losses. A process such as pasteurization, which makes milk safe to drink, is clearly worth that cost. Boxes of milk that can be kept at room temperature have been treated with a process called **ultrahigh temperature (UHT).** The milk is exposed to temperatures above those of pasteurization for just long enough to sterilize it. Irradiation, already discussed in this chapter's Consumer Corner, is a similar case: The price in terms of nutrient loss for gaining a safer food supply is probably trivial. Sometimes processed foods even gain a nutritional edge over their unprocessed counterparts, such as when fat is removed from milk or other foods.

Many forms of processing aim to extend the usable life of a food—that is, to preserve it. To preserve food, a process must prevent three kinds of events: (1) microbial growth, (2) oxidative changes, and (3) enzymatic destruction of food molecules. The first two have already been discussed—microbial growth earlier in this chapter and oxidative damage in Chapter 5. Enzymatic destruction occurs as active enzymes in food cells break down their internal molecular structures and cell membranes and walls. Processes involving heat denature the enzymes, and those applying cold slow enzyme activity. The next sections describe some of the most important preservation or processing techniques—modified atmosphere packaging (MAP), canning, freezing, drying, and extrusion—and their effects on nutrients.

▶ KEY POINT   Some nutrients are lost in food processing. Processing aims to protect food from microbial, oxidative, and enzymatic spoilage.

Minamata disease. The effects of mercury contamination can be severe.

Extended Shelf Life

Today, in most grocery stores, shoppers can choose bags of washed, trimmed, fresh, chilled salads and chopped vegetables. Likewise, they can pick up fully prepared meats, chickens, and other main dishes whose shelf life has been extended through innovative packaging. The secret to these foods' long shelf life is vacuum packaging or **modified atmosphere packaging (MAP).** These methods also preserve freshness and nutrients in soft pasta noodles, baked goods, prepared foods, fresh and cured meats, seafoods, dry beans and other dry products, ground and whole-bean coffee, and other foods. These convenient products are more expensive than comparable choices, but, unopened, they last much longer and so save on waste.

Food manufacturers using these techniques first package foods in plastic film or other wraps that oxygen cannot penetrate. Then, they either remove the air inside the package, creating a vacuum, or replace the air with a mixture of oxygen-free gases, such as carbon dioxide and nitrogen. Excluding oxygen:

ultrahigh temperature (UHT) a process of sterilizing food by exposing it for a short time to temperatures above those normally used in processing.

modified atmosphere packaging (MAP) a preservation technique in which a perishable food is packaged in a gas-impermeable container from which air has been removed or to which another gas mixture has been added.

TABLE 12·13 GLOSSARY OF FRESHNESS DATES ON LABELS

Food manufacturers voluntarily print the following kinds of dates on labels to inform both sellers and consumers of the products' freshness.[a]

- **Sell by:** Specifies the shelf life of the food. After this date, the food may still be safe for consumption if it has been handled and stored properly (check Table 12-5, p. 444, for safe storage times). Also called *pull date*.
- **Best if used by:** Specifies the last date the food will be of the highest quality. After this date, quality is expected to diminish, although the food may still be safe for consumption if it has been handled and stored properly (check Table 12-5 for safe storage times). Also called *freshness date* or *quality assurance date*.
- **Expiration date:** The last day the food should be consumed. All foods except eggs should be discarded after this date. For eggs, the expiration date refers to the last day the eggs may be sold as

"fresh eggs." For safety, purchase eggs before the expiration date, keep them in their original carton in the refrigerator, and use them within 30 days.
- **Open dating:** A general term referring to label dates that are stated in ordinary language that consumers can understand, as opposed to *closed dating*, which refers to dates printed in codes decipherable only by manufacturers. Open dating is used primarily on perishable foods, and closed dating on shelf-stable products such as canned goods.
- **Pack date:** The day the food was packaged or processed. When used on packages of fresh meats, pack dates can provide a general guide to freshness.

[a]Infant formula and some baby foods are required to bear a date.

SOURCE: U.S. Department of Agriculture, *FOCUS ON: Food product dating*, 2001, available at www.fsis.usda.gov/OA/pubs/dating.

- Slows ripening of fruits and vegetables.
- Reduces spoilage by mold and bacterial growth.
- Prevents discoloration of cut vegetables and fruits.
- Prevents spoilage of fats by rancidity.
- Slows development of "off" flavors from accelerated enzyme action that breaks down flavor and aroma molecules.
- Slows enzymatic breakdown of vitamins.

All foods packaged this way must be refrigerated to keep them fresh-tasting and safe.

For safety, consumers must read labels to distinguish between shelf-stable foods and those that must be refrigerated. As with other foods, improper storage may allow growth of illness-causing bacteria. Consumers must also be able to identify foods too old to use; Table 12-13 provides a list of freshness dates printed on shelf-stable and other food packages.

▶ **KEY POINT** Modified atmosphere packaging makes many fresh packaged foods available to consumers. MAP foods compare well to fresh foods in terms of nutrient quality. MAP foods may pose a threat of foodborne illness if not properly stored. Consumers can judge the freshness of packaged foods by reading the dates on their labels.

Do Canned Foods Lose Nutrients?

canning a method of preserving food by killing all microorganisms present in the food and then sealing out air. The food, container, and lid are heated until sterile; as the food cools, the lid makes an airtight seal, preventing contamination.

high-temperature–short-time (HTST) principle the rule that every 18°F (10°C) rise in processing temperature brings about an approximately tenfold increase in microbial destruction, while only doubling nutrient losses.

Canning is one of the more effective methods of protecting food against the growth of microbes (bacteria, fungi, and yeasts) that might otherwise spoil it, but canned foods do have fewer nutrients. Like other heat treatments, the canning process is based on time and temperature. Each small increase in temperature has a major killing effect on microbes and only a minor effect on nutrients. In contrast, long heating times are costly in terms of nutrient losses. Therefore, food processors employ the **high-temperature–short-time (HTST) principle** for canning.

Which nutrients does canning affect, and how? The fat-soluble vitamins and most minerals are relatively stable and are not affected much by canning. Food scientists have thus paid particular attention to three vulnerable water-soluble vitamins: thiamin, riboflavin, and vitamin C.

Acid stabilizes thiamin, but heat rapidly destroys it; therefore, the foods that lose the most thiamin during canning are the low-acid foods such as lima beans, corn, and meat. Up to half, or even more, of the thiamin in these foods can be lost during canning. Unlike thiamin, riboflavin is stable to heat, but it is sensitive to light and so is most likely to be lost from glass-packed, not canned, foods. Vitamin C's special enemy is an enzyme (ascorbic acid oxidase) present in fruits and vegetables as well as in microorganisms. By destroying this enzyme, HTST processes such as canning actually help to preserve at least some of the product's vitamin C.

Minerals are unaffected by heat, so they cannot be destroyed as vitamins can be. Both minerals and water-soluble vitamins can be lost, however, when they leach into canning or cooking water that the consumer then throws away. Losses are closely related to the extent to which a food's tissues have been broken, cut, or chopped and to the length of time the food is in the water.

Some minerals are added when foods are canned. Important in this respect is sodium chloride, table salt, which is added for flavoring. Many food companies have begun making low-salt versions of their products, which may cost more because fewer low-salt batches are made.

▶ KEY POINT   Some water-soluble vitamins are destroyed by canning, but many more diffuse into the canning liquid. Fat-soluble vitamins and minerals are not affected by canning, but minerals also leach into canning liquid.

■ The Food Feature later in this chapter gives tips on preserving nutrients during cooking.

Freezing

Freezing preserves foods because it stops bacterial reproduction and dramatically slows enzymatic reactions. The nutrient contents of frozen foods are similar to those of fresh foods—losses are minimal. The freezing process itself does not destroy any nutrients, but some losses can occur during the steps before freezing, such as the quick dunking into boiling water (blanching), washing, trimming, or grinding. Vitamin C losses are especially likely because they occur whenever tissues are broken and exposed to air (oxygen destroys vitamin C). Uncut fruits, especially if they are acidic, do not lose their vitamin C; strawberries, for example, can be kept frozen for over a year without losing any vitamin C. Mineral contents of frozen foods are much the same as for fresh.

Frozen foods may even have a nutrient advantage over fresh. Fresh foods are often shipped long distances, and to ensure that they make the trip without bruising or spoiling, they are often harvested unripe. Frozen foods are shipped frozen, so produce is allowed to ripen in the field and to develop nutrients to their fullest potential. Foods frozen and stored under proper conditions will often contain more nutrients when served at the table than fresh fruits and vegetables that have stayed in the produce department of the grocery store for even a day.

Frozen foods have to be kept solidly frozen at 0°F (-17°C), if they are to be safe and retain their nutrients. Vitamin C converts to its inactive forms rapidly at warmer temperatures. Food may seem frozen even at 36°F (2°C), but much of it is actually unfrozen, and enzyme-mediated changes can occur fast enough to completely destroy the vitamin C in only two months. If foods defrost slightly but ice crystals remain, it is probably safe to refreeze the food for later use but at a cost of substantial loss of nutrients, texture, and flavor. In general, if a food is safe to eat, it is safe to refreeze, but it will lose both nutrients and appeal.

▶ KEY POINT   Foods frozen promptly and kept frozen lose few nutrients.

freezing a method of preserving food by lowering the food's temperature to a point that halts life processes. Microorganisms do not die but remain dormant until the food is thawed.

drying a method of preserving food by removing sufficient water from the food to inhibit microbial growth.

Drying

Dried or dehydrated foods offer several advantages. **Drying** eliminates microbial spoilage (because microbes need water to grow), and it greatly reduces the weight and volume of foods (because foods are mostly water). Commercial drying does not cause major nutrient losses. Foods dried in heated ovens at home, however, may sustain dramatic nutrient losses. Vacuum puff drying and freeze drying, which take place at cold temperatures, conserve nutrients especially well.

Sulfite additives are added during the drying of fruits such as peaches, grapes (raisins), and plums (prunes) to prevent browning. Some people suffer allergic reactions when they consume sulfites. Sulfur dioxide helps to preserve vitamin C as well, but it destroys thiamin. This is of small concern, however, because most dehydrated products with added sulfur dioxide were not major sources of thiamin before processing.

▶ KEY POINT   Commercially dried foods retain most of their nutrients, but home-dried foods often sustain dramatic losses.

Extrusion

Some food products, particularly cereals and snack foods, have undergone a process known as **extrusion.** In this process, the food is heated, ground, and pushed through various kinds of screens to yield different shapes, such as breakfast "puffs," potato "crisps," the bacon-like "bits" you sprinkle on salad, and so-called food novelties. Considerable nutrient losses occur during extrusion, and nutrients are usually added to compensate. But foods this far removed from the original fresh state are still lacking significant nutrients (notably, vitamin E) and fiber, and consumers should not rely on them as staple foods. Enjoy them as occasional snacks and as additions to enhance the appearance, taste, and variety of meals.

▶ KEY POINT   Extrusion involves heat and destroys nutrients.

Food Additives

What are **additives,** why are they there, and are they dangerous in any way? In the FDA's list of concerns at the start of this chapter, food additives were not a high priority. Compared with unregulated and untested "dietary supplements" sold directly to consumers, the 3,000 food additives used in the United States are strictly controlled and pose little cause for concern.[56]

Manufacturers use food additives to give foods desirable characteristics: color, flavor, texture, stability, enhanced nutrient composition, or resistance to spoilage. Additives, classed by their functions, are listed with their definitions in Table 12-14, and some are discussed in the following sections.

Regulations Governing Additives

The FDA has the responsibility for deciding what additives shall be in foods. To obtain permission to use a new additive in food products, a manufacturer must test the additive and satisfy the FDA that:

- It is effective (it does what it is supposed to do).
- It can be detected and measured in the final food product.

Then the manufacturer must study the effects of the additive when fed in large doses to animals under strictly controlled conditions to prove that:

- It is safe for consumption (it causes no birth defects or other injury).

Finally, the manufacturer must submit all test results to the FDA. The whole process may take many years.

extrusion a process by which the form of a food is changed, such as changing corn to corn chips; not a preservation measure.

additives substances that are added to foods but are normally not consumed by themselves as foods.

TABLE 12·14 FOOD ADDITIVES BY FUNCTION

- **antimicrobial agents** preservatives that prevent spoilage by mold or bacterial growth. Familiar examples are acetic acid (vinegar) and sodium chloride (salt). Others are benzoic, propionic, and sorbic acids; nitrites and nitrates; and sulfur dioxide.
- **antioxidants** preservatives that prevent rancidity of fats in foods and other damage to food caused by oxygen. Examples are vitamins E and C, BHA, BHT, propyl gallate, and sulfites.
- **artificial colors** certified food colors, added to enhance appearance. (*Certified* means approved by the FDA.) Vegetable dyes are extracted from vegetables such as beta-carotene from carrots. Food colors are a mix of vegetable dyes and synthetic dyes approved by the FDA for use in food.
- **artificial flavors, flavor enhancers** chemicals that mimic natural flavors and those that enhance flavor.
- **bleaching agents** substances used to whiten foods such as flour and cheese. Peroxides are examples.
- **chelating agents** defined in Chapter 4 as molecules that bind other molecules. As additives, they prevent discoloration, flavor changes, and rancidity that might occur because of processing. Examples are citric acid, malic acid, and tartaric acid (cream of tartar).
- **nutrient additives** vitamins and minerals added to improve nutritive value.
- **preservatives** antimicrobial agents, antioxidants, chelating agents, radiation, and other additives that retard spoilage or preserve desired qualities, such as softness in baked goods.
- **thickening and stabilizing agents** ingredients that maintain emulsions, foams, or suspensions or lend a desirable thick consistency to foods. Dextrins (short chains of glucose formed as a breakdown product of starch), starch, and pectin are examples. (Gums such as carrageenan, guar, locust bean, agar, and gum arabic are others.)

© Polara Studios, Inc.

Without additives, bread would quickly mold and salad dressing would go rancid.

The FDA then schedules a public hearing and invites consumers to participate; at the hearing, experts present testimony for and against granting permission to use the additive. The FDA's approval of an additive does not give manufacturers free license to add it to any foods. The FDA regulation states the amounts, purposes, and foods for which the additive may be used. No additives are permanently approved; all are periodically reviewed.

The GRAS List Many substances were exempted from complying with this procedure when it was first instituted because they had been used for a long time and their use entailed no known hazards. Some 700 substances in all were put on the **generally recognized as safe (GRAS) list.** When substantial scientific evidence or public outcry has questioned the safety of a GRAS list additive, however, its safety has been reevaluated. All substances about which any legitimate question was raised have been removed or reclassified.

The Margin of Safety Decisions about an additive's safety are governed by the important distinction between **toxicity** and hazard associated with substances. Toxicity is a general property of all substances; hazard is the capacity of a substance to produce injury *under conditions of its use.** All substances can be toxic at some level of consumption, but they are called hazardous only if they are toxic in the amounts ordinarily consumed.

A food additive is supposed to have a wide **margin of safety.** Most additives that involve risk are allowed in foods only at levels 100 times below those at which the risk is still known to be zero. Experiments to determine the extent of risk involve feeding test animals the substance at different concentrations throughout their lifetimes. The additive is then permitted in foods at 1/100 the level that causes no harmful effect whatever in the animals. In many foods, *naturally* occurring toxins appear at levels that bring their margins of safety close to 1/10. Even nutrients, as you have seen, involve risks at high dosage levels. The margin of safety for vitamins A and D is 1/25 to 1/40; it may be less than 1/10 in infants. For some trace elements, it is about 1/5. People

generally recognized as safe (GRAS) list a list, established by the FDA, of food additives long in use and believed to be safe.

toxicity the ability of a substance to harm living organisms. All substances are toxic if the concentration is high enough.

margin of safety in reference to food additives, a zone between the concentration normally used and that at which a hazard exists. For common table salt, for example, the margin of safety is 1/5 (five times the concentration normally used would be hazardous).

*The Delaney Clause, a legal requirement of zero cancer risk for additives, was eliminated in 1996.

consume common table salt daily in amounts only three to five times less than those that cause serious toxicity.

Most additives used in foods offer benefits that outweigh their risks or that make the risks worth taking. In the case of color additives that only enhance the appearance of foods without improving their health value or safety, no amount of risk may be deemed worth taking. Only 10 of an original 80 synthetic color additives are still approved by the FDA for use in foods, and screening of these substances continues.

Manufacturers must comply with other regulations as well. Additives must not be used:

- In quantities larger than those necessary to achieve the needed effects.
- To disguise faulty or inferior products.
- To deceive the consumer.
- Where they significantly destroy nutrients.
- Where their effects can be achieved by economical, sound manufacturing processes.

The following sections focus on the food additives that receive the most publicity because people ask questions about them most often.

▶ KEY POINT   The FDA regulates the use of intentional additives. Additives must be safe, effective, and measurable in the final product. Additives on the GRAS list are assumed to be safe because they have long been used. Additives used must have wide margins of safety.

Antimicrobial Agents

Preservatives known as *antimicrobial agents* protect food from the growth of microbes that can spoil the food and cause foodborne illnesses. Three of these preservatives—salt, sugar, and nitrites—are commonly used.

Salt and Sugar   The best-known, most widely used antimicrobial agents are two common substances—salt and sugar. Salt has been used since before recorded history to preserve meat and fish; sugar serves the same purpose in jams, jellies, and canned and frozen fruits. (Any jam or jelly that toots its "no preservatives" horn is exaggerating. There is no need to add extra preservatives, so most makers do not.) Both salt and sugar work by withdrawing water from the food; microbes cannot grow without water. Today, other additives such as potassium sorbate and sodium propionate are also used to extend the shelf life of baked goods, cheese, beverages, mayonnaise, margarine, and many other products.

Nitrites   The *nitrites* are added to meats and meat products for three main purposes: to preserve their color (especially the pink color of hot dogs and other cured meats); to enhance their flavor by inhibiting rancidity (in cured meats); and to protect against bacterial growth. In particular, in amounts much smaller than needed to confer color, nitrites prevent the growth of the bacterium that produces the deadly botulinum toxin.

Nitrites perform important jobs, but they have been the object of controversy because they can be converted in the human stomach to nitrosamines, which cause cancer in animals. Some cured meats are available without nitrites. Reducing nitrites consumed in meats, however, would hardly make a difference in a person's overall exposure to nitrosamine-related compounds. For example, an average cigarette smoker inhales 100 times the nitrosamines that the average bacon eater ingests. Likewise, a beer drinker imbibes up to roughly five times the amount that the bacon eater receives. Even the air inside automobiles delivers measurable nitrites.

Examples of common antimicrobial additives:

- Salt.
- Sugar.
- Nitrites.

■ For more information on cancer risks from nitrites and other food constituents, search the website of the American Cancer Society at www.cancer.org/docroot/home/index.asp or call 1-800-ACS-2345.

Two long-used preservatives.

How Do Antioxidants Protect Food?

Food can also go bad when it undergoes changes in color and flavor caused by exposure to oxygen in the air (oxidation). Often these changes involve little hazard to health, but they damage the food's appearance, taste, and nutritional quality. Antioxidants are often added to vulnerable foods to prevent the damage caused by oxidation. Familiar examples of oxidative changes are sliced apples or potatoes turning brown and oil going rancid. Antioxidant preservatives protect food from this kind of spoilage. Some 27 antioxidants, including vitamin C (ascorbate) and vitamin E (tocopherol), are approved for use in food.

Sulfites   The sulfites are another group of antioxidants used to prevent oxidation in many processed foods, in alcoholic beverages (especially wine), and in drugs. Sulfites were used to keep the raw fruits and vegetables in salad bars looking fresh, but this practice was banned after a few people experienced dangerous allergic reactions to the sulfites. The FDA now prohibits sulfite use on food meant to be eaten raw, with the exception of grapes, and it requires foods and drugs to list on their labels any sulfites that are present. For most people, sulfites do not pose a hazard in the amounts used in products, but they have one other drawback. Because sulfites can destroy a lot of thiamin in foods, you can't count on a food that contains sulfites to contribute to your daily thiamin intake.

The ban on sulfites has stimulated a search for alternatives. Some producers now use honey to clarify browned apple juice. Agriculturists have created a hybrid apple that does not turn brown. A combination of four GRAS additives can also substitute for sulfites.*

BHA and BHT   Two other antioxidants in wide use are BHA and BHT, which prevent rancidity in baked goods and snack foods. BHT provides a refreshing change from the many tales of woe and cancer scares associated with other additives. Among the many tests performed on BHT were several showing that animals fed large amounts of this substance developed *less* cancer when exposed to carcinogens and lived longer than controls. BHT apparently protects against cancer through an antioxidant effect similar to that of vitamin E. To obtain this effect, though, a much larger amount of BHT must be present in the diet than the U.S. average. A caution: used experimentally at very high levels of intake, the substance has *produced* cancer.

This discussion provides the opportunity to mention an important point about additives. No two additives are alike, so generalizations about them are meaningless. No single valid statement can apply to all of the 3,000-odd substances commonly added to foods. Questions about which additives are safe and under what conditions of use must be asked and answered item by item.

Artificial Colors

As mentioned, only about ten artificial colors are still on the GRAS list; this select group has survived considerable screening. Among the most intensively investigated of all additives, artificial colors are much better known than the *natural* pigments of plants,

Raw grapes may be treated with sulfites. Wash them thoroughly before eating them.

Examples of common antioxidant additives:
- Vitamin C.
- Vitamin E (tocopherol).
- Sulfites.
- BHA and BHT.

■ Chapter 7 describes how antioxidants break the destructive chain reactions of oxidation.

---

*The four GRAS additives are citric acid, ascorbic acid, sodium acid pyrophosphate, and calcium chloride.

Color additives not only make foods attractive but identify flavors as well. Everyone agrees that yellow jellybeans should taste lemony and black ones like licorice.

■ The safety of another group of flavor additives, the artificial sweeteners, was discussed in Controversy 4.

and the limits on the safety of their use can be stated with greater certainty. Examples of natural pigments in common use are the caramel that tints cola beverages and baked goods and the carotenoids that color margarine, cheeses, and pastas. Harmless food colorants added to the feed of farm-raised salmon give their flesh an orange-red color that consumers find appealing.[57] Nevertheless, the food colors have been criticized more than almost any other group of additives. Simply stated, they only make foods pretty, whereas other additives, such as preservatives, make foods safe. Hence, with food colors we can afford to require that their use entail no risk. With other food additives, we must weigh the risks of using them against the risks of *not* using them.

▶ KEY POINT   The addition of artificial colors is tightly controlled.

Artificial Flavors and MSG

Although only a few artificial colors are currently permitted in foods, close to 2,000 artificial flavors and flavor enhancers are approved, making them the largest single group of food additives. The safety evaluation of flavoring agents is somewhat problematic because so many flavoring agents are already in use, the flavors are strong and so are used in tiny amounts unlikely to impose risks, and they occur naturally in a wide variety of foods.

A well-known flavor enhancer is monosodium glutamate, or MSG (trade name Accent), the sodium salt of the common amino acid glutamic acid. MSG is used widely in restaurants, especially Asian restaurants. In addition to enhancing other flavors, MSG itself possesses a basic taste (termed *umami*) independent of the well-known sweet, salty, bitter, and sour tastes.

In a few sensitive individuals, MSG produces adverse reactions known as the **MSG symptom complex**.[58] Symptoms may include burning sensations, chest and facial flushing or pain, and throbbing headaches. Plain broth with MSG seems most likely to bring on symptoms in sensitive people, while carbohydrate-rich meals seem to protect against them. When dining on Asian-style foods, potentially sensitive people should try ordering soups that contain noodles and eat plenty of plain rice, as do Asians themselves.

MSG has been investigated extensively enough to be deemed safe for adults to use, but it is kept out of foods for infants because very large doses have been shown to destroy brain cells in developing mice. Infants have not yet developed the capacity to fully exclude such substances from their brains. For other foods, the FDA requires that the label ingredients list itemize each additive by its full name, including MSG as *monosodium glutamate*.[59]

▶ KEY POINT   Among flavorings added to foods, the flavor enhancer MSG causes reactions in people with sensitivities to it.

Incidental Food Additives

Indirect or **incidental additives** are called *additives*, but they are really contaminants from some phase of production, processing, storage, packaging, or consumer preparation. Examples of incidental additives include tiny bits of plastic, glass, paper, tin, and the like from packages and chemicals from processing, such as the solvent used to decaffeinate some coffees.

Some microwave products are sold in "active packaging" that participates in cooking the food. Pizza, for example, may rest on a cardboard pan coated with a thin film of metal that absorbs microwave energy and may heat up to 500°F (260°C). During the intense heat, some particles of the packaging components migrate into the food. Regular plastic packages heat up less, but particles still migrate and may not be entirely safe for consumption. Avoid reusing disposable containers, such as margarine tubs or single-

MSG symptom complex the acute, temporary, and self-limiting reactions experienced by sensitive people upon ingesting a large dose of MSG. The name *MSG symptom complex*, given by the FDA, replaces the former *Chinese restaurant syndrome*.

incidental additives substances that can get into food not through intentional introduction but as a result of contact with the food during growing, processing, packaging, storing, or some other stage before the food is consumed. Also called *accidental* or *indirect additives*.

use trays from frozen microwavable meals, or ordinary plastic wraps for microwaving. Microwave-safe plastic wraps, waxed paper, cooking bags, parchment paper, and white microwave-safe paper towels are probably safe to use. The safest choice is to use only glass or ceramic containers or those plastic containers labeled as safe for microwaving.

Coffee filters, paper milk cartons, paper plates, and frozen food boxes can all be made of bleached paper and so can contaminate foods with trace amounts of compounds known as dioxins. Dioxins form during the chlorination step in making bleached paper. Dioxins can migrate into foods that come in contact with bleached paper, but the amounts entering food are infinitesimally small—one part per trillion, or the equivalent of one second in 32,000 years. Such amounts do not appear to present a health risk to people, and drinking milk from bleached cartons appears to be safe. Dioxins are persistent, however, and they leach into the environment by way of both paper mill effluent and discarded paper products in landfills, becoming more concentrated in land, water, and animals until they build up to hazardous levels.

Incidental additives sometimes find their way into foods, but adverse effects are rare. These additives are well regulated—all food packagers are required to test whether materials from packages are migrating into foods. If they are, their safety must be confirmed by strict procedures like those governing intentional additives.

▶ KEY POINT   Incidental additives are substances that get into food during processing. They are well regulated, and most present no hazard.

Nutrient Additives

Nutrients added to improve or to maintain the nutritional value of foods make up another class of additives. Among them are the enrichment nutrients added to refined grains, the iodine added to salt, vitamins A and D added to dairy products, and the nutrients used to fortify breakfast cereals, "energy" bars, liquid meal replacers, and the like. When nutrients are added to a nutrient-poor food, it may appear from its label to be nutrient-rich. It is, but only in those nutrients chosen for addition. Nutrients are sometimes also added for other purposes. Vitamins C and E used as antioxidants and beta-carotene as a colorant are examples already mentioned.

Examples of common nutrient additives:
- Thiamin, niacin, riboflavin, folate, and iron in grain products.
- Iodine in salt.
- Vitamins A and D in milk.
- Vitamin C in fruit drinks.
- Beta-carotene in cheeses.

▶ KEY POINT   Nutrients are added to foods to enrich or to fortify them. These additives do not necessarily make the foods nutritious, only rich in the vitamins and minerals that have been added.

To sum up the messages of this chapter, the U.S. food supply is safe and hazards are rare. Precautions against foodborne microbial illnesses are the most urgent measures for people to take to avoid food-related diseases. The Food Feature that follows offers pointers on the selection, storage, and cooking of foods to preserve their nutrients.

Preserving Nutrients in Foods

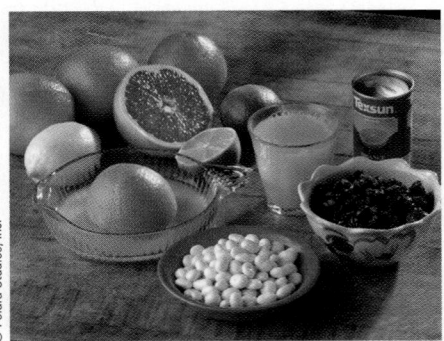

In terms of nutrient density, canned juice is almost as nutritious as fresh, but yogurt-covered raisins are not as nutritious as plain raisins.

IN GENERAL, THE more heavily processed foods are, the less nutritious they become. Does that mean that you should avoid all processed food? The answer is not simple: in each case, it depends on the food and on the process. Consider the case of orange juice and vitamin C.

The Choice of Orange Juice

Orange juice is available in several forms, each processed a different way. Fresh juice is squeezed from the orange, a process that extracts the fluid juice from the fibrous structures that contain it. Each 100 calories of the fresh-squeezed juice contains 98 milligrams of vitamin C. When this juice is condensed by heat, frozen, and then reconstituted, as is the juice from the freezer case of the grocery store, 100 calories of the reconstituted juice contain just 85 milligrams of vitamin C because vitamin C is destroyed in the condensing process. Canning is even harder on vitamin C: 100 calories of canned orange juice have 82 milligrams of vitamin C.

These figures seem to indicate that fresh juice is the superior food, but consider this: most people's recommended intake of vitamin C (75 milligrams for women or 90 milligrams for men) is fully or nearly met by 100 calories of any of the above choices. Thus, for vitamin C, the losses due to processing are not a problem. Besides, processing confers enormous convenience and distribution advantages. Fresh orange juice spoils. Shipping fresh juice to distant places in refrigerated trucks costs much more than shipping frozen juice (which takes up less space) or canned juice (which requires no refrigeration). The fresh product still contains active enzymes that continue to degrade its compounds (including vitamin C) and so cannot be stored indefinitely without compromising nutrient quality. The savings gained from shipping and storing canned and frozen juices are passed on to consumers. Without canned or frozen juice, people with limited incomes or those with no access to fresh juice would be deprived of this excellent food. Vitamin C is readily destroyed by oxygen, so whatever the processing method, orange juice and other vitamin C–rich foods and juices should be stored properly and consumed within a week of opening.[60]

Processing Mischief

Some processing stories are not so rosy. In Chapter 8, for instance, you saw how processed foods are often loaded with sodium as their potassium is leached away, exactly the wrong effect for people with hypertension. A related mischief of processing is the addition of sugar and fat—palatable, high-calorie additives that reduce nutrient density. For example, nuts and raisins covered with "natural yogurt" may sound like one healthy food being added to another, but the ingredients list shows that generous amounts of sugar and fat accompany the yogurt. About 75 percent of the weight of the product is sugar and fat; only 8 percent is yogurt. To pick one nutrient for an example, look at what happens to the iron density of the raisins: 100 calories of raisins have 0.71 milligram of iron; 100 calories of "yogurt" raisins have 0.26 milligram of iron. These foods taste so good that wishful thinking can take hold, but the reality is that sugar- and fat-coated food is candy. The word *yogurt* on the label means only that one of the ingredients of the candy coating is some small amount of yogurt.

Best Nutrient Buys

Here are two good general rules for making food choices:

- Choose whole foods to the greatest extent possible.
- Seek out among processed foods the ones that processing has improved nutritionally. For example, processing that removes saturated fat, as in fat-free milk, or pasteurization to improve the safety of juices is a benefit to the consumer.

Commercially prepared whole-grain breads, frozen cuts of meats, bags of frozen vegetables, and canned or frozen fruit juices do little disservice to nutrition and enable the consumer to eat a wide variety of foods at great savings in time and human energy. The nutrient density of processed foods exists on a continuum:

Whole-grain bread > refined white bread > sugared doughnuts.

Milk > fruit-flavored yogurt > canned chocolate pudding.

Corn on the cob > canned creamed corn > caramel popcorn.

Oranges > orange juice > orange-flavored drink.

Baked ham > deviled ham > fried bacon.

The nutrient continuum is paralleled by another continuum—the nutrition status of the consumer. The closer to the farm the foods you eat, the better nourished you are, but that doesn't mean you have to live in the fields.

Conserving Nutrients at Home

Wise food choices are half the story of smart nutrition self-care; skillful food preparation is the other half. In modern commercial processing, losses of vitamins seldom exceed 25 percent. In contrast, losses in the 60 to 75 percent range during food preparation at home are not unusual, and they can be close to 100 percent. The kinds of foods you buy make a difference, but what you do with them in your kitchen can make an even greater difference.

Preventing Enzymatic Destruction

Vitamins are organic compounds synthesized and broken down by enzymes found in the foods that contain them. Like all enzymes, the enzymes that break down nutrients in fruits and vegetables have a temperature optimum. They work best at the temperatures at which the plants grow, normally about 70°F (21°C), which is also the room temperature in most homes. Chilling fresh produce slows down enzymatic destruction of nutrients. To protect the vitamin content, most fruits and vegetables should be vine ripened (if possible), chilled immediately after picking, and kept cold until use.

Protecting from Light and Air

Besides being vulnerable to enzyme-mediated spoilage, the vitamin riboflavin is light sensitive. It can be destroyed by the ultraviolet rays of the sun or by fluorescent light. For this reason, milk is not sold (and should not be stored) in transparent glass containers. Cardboard or opaque plastic containers screen out light, protecting the riboflavin. Since grain products such as macaroni and rice are also important

sources of riboflavin, cooks who store them in glass jars should stow the jars in closed cupboards.

Some vitamins are acids or antioxidants and so are most stable in an acid solution away from air. Citrus fruits, tomatoes, and many juices are acid—as long as the skin is uncut or the can unopened, their vitamins are protected from air. If you store a cut vegetable or fruit, cover it with an airtight wrapper; close an opened carton of juice tightly and store it in the refrigerator.

Refreezing

Labels on frozen foods tell you "Do not refreeze." As food freezes, the cellular water expands into long, spiky ice crystals that puncture cell membranes and disrupt tissue structures, changing the texture of the food. There is usually no danger in eating a twice-frozen food although some nutrients are lost upon thawing and refreezing. Provided that it hasn't spoiled while it was thawed or wasn't thawed at warm temperature, the main problems with a twice-frozen food may be its reduced nutrient value and loss of taste and texture.

Preventing Nutrient Losses in Water

Minerals and water-soluble vitamins in fresh-cut vegetables readily dissolve into the water in which they are washed, boiled, or canned. If the water is discarded, as much as half of the vitamins and minerals in foods go down the drain with it. A bit of southern folk wisdom is to serve the cooking liquid with the vegetables rather than throwing it away; this liquid is known as the "pot liquor" and may be used to moisten cornbread or to make gravies or soups.

Wash the intact food vigorously and briefly, don't soak it. Cut vegetables after washing except for those such as broccoli that you have to cut to wash adequately. For peeled vegetables, such as potatoes, add them to water that is vigorously boiling, not to cold water, to minimize the length of time the vegetables are exposed to nutrient-leaching water.

Other ways to minimize cooking losses: steam vegetables over water rather than boiling them in it. For example, boiled broccoli has been demonstrated to lose more than half its original folate value, while steamed broccoli retains almost all of it even after 15 minutes of steaming.[61]

Purchase mostly whole foods or those that processing has benefited nutritionally.

Steam vegetables or cook them in a microwave oven.

Wrap foods tightly and refrigerate them. Space foods to allow chilled air to circulate around them.

■ Take care when cooking in a microwave oven. Food can become extraordinarily hot or build up steam that can scald unprotected hands or face.

■ Before cooking eggs, sausages, potatoes, or any food encased in a membrane, pierce the membrane to prevent explosion of the food.

■ Never warm baby formula or baby food in a microwave oven because hot spots can form that can scald the baby.

Likewise, stir-frying vegetables in small amounts of oil conserves nutrients. Microwave ovens are also excellent for nutrient retention. They cook fast without requiring fats or excess liquid. Some special microwaving concerns appear in the margin.

During other types of cooking, minimize the destruction of vitamins by avoiding high temperatures and long cooking times. Iron destroys vitamin C by catalyzing its oxidation, but perhaps the benefit of increasing the iron content of foods by cooking in iron utensils outweighs this disadvantage. Each of these tactics is small by itself, but saving a small percentage of the vitamins in foods each day can mean saving significant amounts in a year's time.

Meanwhile, however, a law of diminishing returns operates. Most vitamin losses under reasonable conditions are not catastrophic. You need not fret over small vitamin losses that occur in your kitchen; you may waste energy or time that is valuable to you in other ways. If you start with fresh, whole foods containing ample amounts of vitamins and are reasonably careful in their preparation, you will receive a bounty of the nutrients that they contain.

MEDIA MENU

Nutrition Connections

The Nutrition Connections CD provides access to these additional resources.

Do It!
The Do It! exercise for this chapter, *Develop a Personal HACCP (Hazard Analysis and Critical Control Point) Plan for Food Safety,* asks you to evaluate your food-handling practices and make improvements.

Interaction
The animation, *Consumer Concerns about Food Safety,* looks at the safety of commonly eaten foods.

Chapter Quiz
Take the quiz for this chapter to assess your understanding of chapter concepts.

Nutrition on the Net
Annotated links to reputable nutrition-related websites.

Online Glossary
Test your knowledge of key vocabulary through this comprehensive, interactive glossary.

Internet Connections

A variety of helpful study tools for this chapter are available at our website
http://nutrition.wadsworth.com/ncc10

Practice Test
Take the practice test for this chapter and gauge your knowledge of key concepts; answers are provided.

Student Notebook
A lecture notebook is available to help organize note taking while reading or attending lecture.

Study Tools
Study resources for students include Chapter Summaries, Crossword Puzzles, Flashcards, Glossary, Internet Exercises, and Learning Objectives.

SELF CHECK

Answers to these Self Check questions are in Appendix G.

1. Which of the following food hazards has the FDA identified as its number-one concern?
   a. pesticides in food
   b. microbial foodborne illnesses
   c. intentional food additives
   d. environmental contaminants

2. To prevent foodborne illnesses, the refrigerator's temperature should be less than:
   a. 70°F
   b. 65°F
   c. 40°F
   d. 30°F

3. Which of the following may be contracted from normal-appearing raw seafood?
   a. hepatitis
   b. worms and flukes
   c. viral intestinal disorders
   d. all of the above

4. Which of the following is correct concerning fruits that have been irradiated?
   a. They decay and ripen more slowly.
   b. They lose substantial nutrients.
   c. They are not safe to eat.
   d. They become radioactive.

5. Which of the following organisms can cause *hemolytic-uremic syndrome*?
   a. *Listeria monocytogenes*
   b. *Campylobacter jejuni*
   c. *Escherichia coli* O157:H7
   d. *Salmonella*

6. It is possible to eliminate all poisons from your diet by eating only "natural" foods.   T   F

7. Pregnant women are advised not to eat certain species of fish because the FDA and the EPA have detected unacceptably high lead levels in them.   T   F

8. The canning industry chooses treatments that employ the low-temperature–long-time (LTLT) principle for canning.   T   F

9. Artificial flavors and flavor enhancers are the largest single group of food additives.   T   F

10. Infants under one year of age should never be fed honey because it can contain spores of *Clostridium botulinum*.   T   F

CONTROVERSY 12

ORGANIC FOODS AND GENETICALLY MODIFIED FOODS: WHAT ARE THE PROS AND CONS?

Scientists and shoppers have strong opinions about groceries these days. Sales of all kinds of **certified organic foods** are skyrocketing, making organic foods one of the fastest growing segments of the U.S. food industry.[*][1] At the same time, many U.S. farms have shifted toward growing foods altered through **genetic engineering (GE).** Today, 80 percent of U.S. soybeans, 40 percent of corn, and 70 percent of cotton are GE crops.[2]

The preceding chapter addressed safety concerns surrounding conventional foods. This Controversy takes a scientific look at issues surrounding organic and **GE foods** to provide an understanding of the differences and similarities between these foods that will help in deciding whether to choose either or both of them. Some terms relating to organic and GE foods are

*Reference notes are found in Appendix F.

defined in Table C12-1. For a review of the workings of the genes, refer to Figure 3-2 in Chapter 3 and Figure 6-6 of Chapter 6.

Issues Surrounding Organic Foods

A shopper picks up two fragrant, orange-yellow mangoes, one from a bin marked "organic" and another from a regular bin; both bear stickers identifying them as the "Hayden" variety. Both may be sweet and succulent. Both may have been kept in storage or shipped from faraway destinations. In fact, both may have been harvested from the same grove, separated by only a thin strip of land. The only obvious differences are the organic label and the cost: the organic mango costs more. The not-so-obvious differences are the methods used to produce them.

A farmer wishing to grow and market organic foods must receive certifica-

tion by USDA inspectors who check everything from the seed sown in the ground, through the methods of making compost for fertilizer, to the manufacturing of the final product. To be sold or labeled as organic, or to bear the USDA organic seal (see Figure C12-1), a food must be produced according to the procedures outlined in Table C12-2. Organic foods must be free of ingredients produced by way of specified technologies, such as irradiation and genetic engineering. In contrast, foods bearing "natural" or "free-range" or other wholesome-sounding labels are not required to meet these standards.

Nutrient Composition

Supporters claim that foods produced the organic way contain more vitamins and minerals than conventional foods because organic compost fertilizes plants with more vitamins and trace minerals than those provided by synthetic chemical fertilizers. Biology tells us that plants do not take up vitamins from either compost or other fertilizers; they absorb only simple elements and water through their roots and use these substances to make complex organic molecules like vitamins that they need in their tissues. Trace minerals, on the

---

TABLE C12·1 ORGANIC FOOD AND FOOD TECHNOLOGY TERMS

- **certified organic foods** foods meeting strict USDA production regulations, including prohibition of most synthetic pesticides, herbicides, fertilizers, drugs, and preservatives, as well as genetic engineering and irradiation.
- **clone** an individual created asexually from a single ancestor, such as a plant grown from a single stem cell; a group of genetically identical individuals descended from a single common ancestor, such as a colony of bacteria arising from a single bacterial cell; in genetics, a replica of a segment of DNA, such as a gene, produced by genetic engineering.
- **GE foods** genetically engineered foods; food plants and animals altered by way of rDNA technology.
- **genetic engineering (GE)** the direct, intentional manipulation of the genetic material of living things in order to obtain some desirable trait not present in the original organism. Also called *recombinant DNA technology* and *biotechnology.*
- **genetic modification** intentional changes to the genetic material of living things brought about through a range of methods, including rDNA technology, natural cross-breeding, and agricultural selective breeding.
- **outcrossing** the unintended breeding of a domestic crop with a related wild species.
- **plant pesticides** substances produced within plant tissues that kill or repel attacking organisms.
- **recombinant DNA (rDNA) technology** a technique of genetic modification whereby scientists directly manipulate the genes of living things; includes methods of removing genes, doubling genes, introducing foreign genes, and changing gene positions to influence the growth and development of organisms.
- **selective breeding** a technique of genetic modification whereby organisms are chosen for reproduction based on their desirability for human purposes, such as high growth rate, high food yield, or disease resistance, with the intention of retaining or enhancing these characteristics in their offspring.
- **stem cell** an undifferentiated cell that can mature into any of a number of specialized cell types. A stem cell of bone marrow may mature into one of many kinds of blood cells, for example.
- **transgenic organism** an organism resulting from the growth of an embryonic, stem, or germ cell into which a new gene has been inserted.

FIGURE C12·1   USDA SEAL AND ORGANIC FOOD LABELS

The USDA organic seal on a food means that it was produced in accordance with the USDA standards for organic foods and that it is 100 percent organic.[a]

Other organic labeling regulations include:
- Labels on foods containing *only* organically produced ingredients may state that the food is "100 percent organic," or simply "organic," and they may bear the USDA organic seal.
- Labels on foods containing *70 percent or more* organic ingredients may claim "made with organic ingredients" and may bear a different mark or seal of the certifying agent, but *not* the USDA organic seal.
- Labels of foods containing *less than 70 percent* organic ingredients may not make any organic claims but may identify organic ingredients on the ingredient list of the information panel.

[a]Organic labeling information is available at www.ams.usda.gov/nop/facts/labeling.htm.

---

other hand, are absorbed into plants, and some may be more abundant in organic compost than in synthetic fertilizers, which contain only the minerals that are added to the mix.

As yet, little research exists to support or refute whether organic foods are nutritionally superior to conventional foods.[3] The nutrient differences that have been reported are so small as to be explained by the seasonal nutrient variations that normally occur in crops.[4] For consumers, the nutrients in any packaged food are easily evaluated by reading the Nutrition Facts panel. Organic candy bars, frozen soy desserts, and fried snacks are no more nutritious than ordinary treats.

Pesticide Residues

Can consumers who pay more for organic foods trust that such foods are free of pesticide contamination? Tests have revealed pesticide residues in about three-quarters of conventionally grown foods and in about a quarter of organic food samples. Pesticides may persist in soils from previous applications, drift onto organic fields from nearby sprayed fields, or contaminate organic foods during shipping and marketing.

Federal standards for pesticide residues are set far below the threshold of any known threat to human health. Further, the human body is well

equipped to handle tiny amounts of even poisonous substances without apparent harm—so long as the doses are small enough. Even if organic foods were completely pesticide-free, it is unlikely that this would constitute a health advantage for healthy adults; pesticides are more of a concern for children, however, as the previous chapter made clear.

Regardless of how foods are grown, food-safety concerns dictate that consumers vigorously wash all fruits and vegetables for 10 seconds under running water. Doing so removes many harmful microbes as well as pesticide residues that may be present on conventional or organic foods.

Environmental Benefits

Farming the organic way, promoters say, will produce food without environmental harm indefinitely into the future. In general, organic foods are grown by using the techniques of *sustainable* agriculture (see Chapter 15). Vegetables and fruits are fertilized with composted animal manure or vegetable matter with no synthetic fertilizers that can run off into waterways and pollute them. No synthetic pesticides or disease-fighting agents are applied, so accidental spills and overuse are not a threat. Pests and diseases are battled by rotating crops each season, by introducing predatory insects to kill off pests, or by picking off large insects or diseased plant parts by hand. Only pesticides derived from natural sources, such as a pesticidal peptide toxin extracted from a bacterium that lives in

---

TABLE C12·2   USDA CRITERIA FOR ORGANIC FOODS

All organic foods must be:
1. Processed or preserved with only agricultural substances or inorganic substances approved by the USDA for use in organic foods.
2. Produced without genetic modification that would not be possible under natural conditions, including rDNA technology.
3. Processed without ionizing radiation.

Food crops and feed for food-producing animals must be:
4. Produced without the use of most synthetic chemicals such as pesticides, herbicides, or fertilizers; a few low-toxicity exceptions are specified.
5. Fertilized without sewage sludge.
6. Managed in ways that preserve the integrity of farmland and surrounding environments, including reducing erosion, maintaining soil quality, and protecting drinking water sources from contamination.

Food-producing animals must be:
7. Grown without medications such as antibiotics or hormones, except appropriate vaccines to prevent diseases in healthy animals.[a]
8. Provided with living conditions similar to the animals' natural habitat, such as access to the outdoors, and maintained in conditions that reduce animal stress.
9. Grown in control-size herds, managed in ways that reduce the impact of manure and other aspects of animal ranching on the environment.
10. Fed organic feed.

[a]Under the requirements, sick animals may be treated with medications such as antibiotics, but products derived from treated animals, such as eggs, milk, or meat, may not be sold as organic.

---

soil, are allowed for use on organic fruits and vegetables.

Crop rotation and natural fertilizers not only minimize chemical impact on wildlife and human beings but are beneficial to the soil. Such techniques curtail erosion and prevent contamination of drinking water from manure runoff, which is common on commercial farms.

To produce organic eggs, dairy products, and meats, farmers and ranchers raise food-producing animals in spacious, low-stress surroundings natural to their species with access to the outdoors. Animals raised this way can grow large and stay healthy without growth hormones, daily antibiotics, and other drugs that are required when animals are stressed in overcrowded pens that make diseases likely. Without overcrowding, the threat to the nation's waterways from waste runoff is also greatly reduced.

Potential Health Risks

Although proponents of organic foods believe that these foods are inherently safer for consumers than conventional foods, this may not be the case. For example, the application of improperly composted animal manure fertilizer may expose consumers to dangerous microbial diseases, such as *E. coli* O157:H7. Chapter 12 pointed out that organic sprouts and greens have caused many cases of serious food poisoning in recent years. Unpasteurized organic juices, milk, and cheeses may also constitute microbial hazards because pasteurization is required to kill disease-causing microorganisms in these foods. Furthermore, organic foods contain no preservatives and tend to spoil faster than other foods. Those purchasing organic foods are urged to buy only amounts that can be consumed within a few days, to store and cook the food properly, to wash raw produce vigorously, and to buy only pasteurized organic dairy products and juices.

Costs

A final factor to consider with regard to organically grown foods stems from the additional measures required to produce them—they cost more than conventional foods. Nevertheless, many people are willing to pay extra for foods

they believe to be superior in nutrition, benevolent to the environment, produced with respect for animals, and low in pesticides. Some people also claim that organic fruits and vegetables taste better, and if perception of better taste encourages people to eat more fruits and vegetables, this is all to the good.

Like organic food production, another painstaking method of producing food, genetic engineering, carries its own potential set of risks and rewards. Unlike the tried and true organic foods, however, GE food varieties are new to the food supply.

Genetic Engineering

A shopper picks up two red, ripe tomatoes from two adjacent bins. Both may be delicious and nutrient-rich. Both may have been shipped a long distance. There are no obvious differences. Yet one may be a genetically altered tomato. The consumer may not be able to distinguish one from the other because genetically modified foods bear no special labels.

Food producers across the nation have welcomed **genetic modification** techniques, particularly **recombinant DNA (rDNA) technology,** to solve some age-old agricultural problems while boosting yields and profits. Many world governments and consumers, however, recoil from new genetic changes in basic foods without iron-clad assurances of safety. Further, misinformation and urban legends abound on the Internet and other media and are often indistinguishable from fact to consumers unfamiliar with genetic sciences. Although genetic engineering technologies have only recently emerged from laboratories, their roots lie in genetic events taking place in nature and centuries-old selective breeding techniques that modify the genes of living things.

Natural Cross-Breeding and Selective Breeding

Aeons ago, before being domesticated by humans, wild grains and other plants cross-pollinated randomly. Most of the resulting offspring failed to thrive, but occasionally a new plant formed with a biological advantage that ensured its survival. An example is the familiar

These colorful carrots resulted from intensive selective breeding, not rDNA technology. Researchers bred carrots with high levels of colorful phytochemicals at each generation.

wheat plant, which is the result of random crossing of wild grasses. Additionally, certain wild bacteria and viruses implant genetic material in host organism cells where it may blend into the host genome and be reproduced as native genes over generations.

Since the dawn of agriculture, season after season, farmers have been changing the genetic makeup of their crop plants and farm animals through **selective breeding** to enhance traits desirable to human beings. Today's lush, hefty, healthy agricultural crops and animals, from cabbage and squash to pigs and cattle, all demonstrate the results of those efforts.

Take corn, for example. Today's large, full, sweet cobs and high yields bear little resemblance to the original wild, native corn with its sparse two or three kernels to a stalk. Modern-day breeders use improved methods to achieve dramatic results, such as those depicted above. Selective breeding works to genetically modify living things, but slowly and imprecisely.

Genetic Engineering Basics

In contrast to the lengthy process of cross-breeding in nature and selective breeding in agriculture, genetic engineering can change the genetic makeup of an organism in a year or two of work. A desirable gene from one organism can be inserted into another organism's DNA with great economy and precision. Genetic engineering can thus change one or more characteristics of a food. Figure C12-2 compares the genetic results of selective breeding and rDNA technology. Table C12-3 presents an overview of food-related genetic engineering research.

Improving Crops for Desirable Traits

Plant cells make likely candidates for genetic engineering because a single plant cell can often be coaxed into producing an entire new plant. Each cell of the resulting plant contains an exact replica of the genetic information contained in the original cell. If scientists introduce any DNA fragments into that first single cell, those fragments will be faithfully reproduced in all of the cell's offspring. All of the resulting cells are **clone** cells—exact genetic replicas of the original.

For example, by using rDNA technology, scientists can alter the DNA of an immature cell, known as a **stem cell,** from the "eye" of a potato plant. Into that stem cell, they can insert a gene snipped from the DNA of a virus that attacks potato plants (enzymes do the snipping). This gene codes for a harmless viral protein, not the infective part. The newly created stem cell is then stimulated to grow into a **transgenic organism,** in this case, a potato plant that makes a piece of viral protein in each of its cells. The presence of the viral protein stimulates the potato plant to develop resistance and ready its defenses against an attack from the real virus in the growing field.

In addition to transferring genes among organisms, scientists may also block or suppress a gene's activity to reduce or eliminate production of unwanted proteins. An example is provided by the first GE food to be approved: a long-lasting tomato. Tomatoes produce an enzyme that softens them after picking. Scientists introduced into a tomato plant an *antisense gene,* a mirror image of the native gene that coded for the "softening" enzyme. The antisense gene blocks production of the softener and produces otherwise normal tomatoes (see Figure C12-3, p. 476). Whereas ordinary tomatoes must be harvested at the hard, green stage to withstand handling, a tomato with the antisense gene can be harvested and marketed at the most flavorful, nutritious red-ripe stage and still last longer after purchase. Although such tomatoes turned out to be too expensive or troublesome to produce commercially, the technology that produced them is being applied to alter other foods in other ways.

Animals, too, may be modified for the benefit of human beings.[5] Thanks to rDNA technology, the salmon in the photo on p. 476

grows at an astonishing rate and can be ready for market in far less time than it takes to grow an ordinary salmon.

The Promises of Genetic Engineering

Supporters hail genetic engineering as nothing short of a revolutionary means of overcoming many of the planet's pressing problems, such as food shortages, nutrient deficiencies, medicine shortages, dwindling farmland, and environmental degradation.[6] For example, animals and plants may one day join microorganisms as manufacturers of needed pharmaceutical products as well as extra nutrients and basic foods. Researchers have induced bananas and potatoes to make hepatitis vaccines and tobacco leaves to make AIDS drugs. Herds of animals that secrete vaccines into their milk could provide both nourishment and immunization to whole villages of people now suffering from the lack of both food and medical help. Many such projects hold promise for the future.

Rice, the staple food for half the world's population, may become a source of supplemental nutrients in areas where deficiencies are common. Some GE rice produces beta-carotene to fight the world's most common cause of childhood blindness: vitamin A deficiency. Other altered rice varieties, some offering 80 percent more iron and zinc than ordinary rice, could relieve much suffering from iron-deficiency anemia and zinc-deficiency diseases around the world.[7]

Other new crops can now resist insects without sprays, survive drought,

FIGURE C12·2 COMPARING TRADITIONAL BREEDING AND GENETIC ENGINEERING

Traditional Breeding—DNA is a strand of genes, much like a strand of pearls. Traditional selective breeding combines many genes from two individuals of the same species.

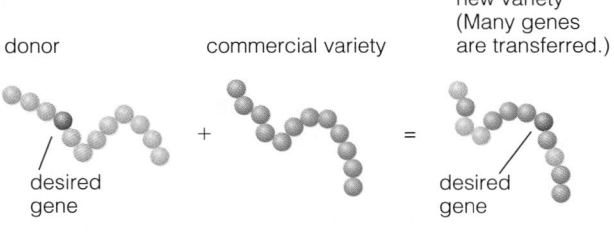

donor   commercial variety   new variety (Many genes are transferred.)

desired gene   +   =   desired gene

rDNA Technology—Through rDNA technology, a single gene or several may be transferred to the receiving DNA from the same species or others.

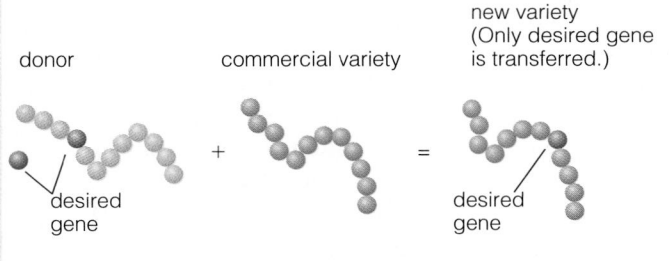

donor   commercial variety   new variety (Only desired gene is transferred.)

desired gene   +   =   desired gene

FIGURE C12·3 HOW AN ANTISENSE GENE CAN BLOCK FORMATION OF A PROTEIN

Normal tomato:

RNA

DNA

native tomato-rotting gene

An RNA strand is made from the DNA template.

This RNA guides the formation of a tomato-rotting enzyme that softens tissue and initiates rotting.

The normal tomato rots quickly.

Genetically engineered tomato:

RNA

DNA

antisense gene

Scientists insert a mirror image of the rotting gene (an antisense gene) into tomato DNA.

The antisense gene blocks the activity of the native gene, preventing production of the rotting enzyme.

The transgenic tomato lasts longer.

---

or provide more complete sources of protein for people without access to animal products or adequate vegetable proteins. Still others may one day be safer for people with food allergies because scientists have successfully suppressed certain proteins that commonly cause allergy. Bananas may soon last many days longer than ordinary bananas. Vegetables may soon grow the fish oils EPA and DHA that most people's diets lack.[8]

These salmon are all of the same age and type. The two largest fish received a growth-enhancing gene, greatly accelerating their growth rate.

In areas where people cannot afford to lose a single morsel of food, plant disease can claim up to 80 percent of a season's sweet potatoes; GE sweet potatoes can resist disease and save the crop.[9]

Crops That Withstand Herbicides and Insects

Despite the almost unlimited potential, today's first generation of genetically engineered crops fall into two main categories: herbicide resistance and insect resistance.[10] Farmers planting fields of crops modified to withstand potent herbicides can more easily control weeds by spraying entire fields with weed killers. The weeds die, leaving only the desired chemical-resistant crop in the field.

Insect pests are also less troublesome to farmers planting crops that make what the Environmental Protection Agency (EPA) calls **plant pesticides**—pesticides made within the tissues of the plants. For example, a type of GE corn used for animal feed produces a pesticide that greatly reduces crop losses caused by a common corn-destroying worm. The technology isn't a panacea, however; farmers must still spray for attacking pests *not* killed by the plant

pesticide. Also, the high cost of transgenic plants and the potential that insects will become resistant to the rDNA pesticide pose problems for researchers to solve.[11]

Modified Microorganisms

Genetically modified microorganisms are also currently at work. One bacterium, for example, was given the ability to make the enzyme rennin, a necessary enzyme in cheese production. Historically, rennin was harvested from the stomachs of calves, an expensive process. After a calf gene was spliced into the DNA of a bacterium, the resulting transgenic bacterial colony became a factory mass-producing rennin. In the same way, another transgenic bacterial factory produces human growth hormone, so more children with growth hormone deficiency can grow normally. Still another manufactures the hormone insulin for use in treatment of people with diabetes.

Other Applications

The breadth of genetic engineering reaches far beyond applications in food production. Its techniques wield the awesome power to change the most basic patterns of life in ways never before possible. Its potential progeny are under development in fields as diverse as medicine, forestry, fuel production, environmental protection, and paper and plastics manufacturing. Its applications even extend to weapons of bioterrorism, such as treatment-resistant **anthrax** and other disease-causing organisms, and biodefenses against them. Such uses and abuses lie beyond the scope of a nutrition text and must be left to ethical scientific minds to ponder and conquer.

The Future of Animal Cloning

Like plants, animals have stem cells and can therefore be cloned, as described earlier. The first and most famous barnyard clone, Dolly the sheep, developed from a single stem cell harvested from a ewe's udder. The path to useful cloning has not been smooth, however; many cloned animals were born with serious physical imperfections and often died early. Rats useful in medical experiments have now been successfully cloned after many years of failures.[12]

Some cloned animals may prove useful as food, and the FDA is currently assessing the potential risks associated with consuming meats from this source. Most likely, edible products from cloned cattle and pigs will be found to be as safe to eat as ordinary products for these reasons:

- Food animals, unlike many plants, do not produce toxins, so there is no threat of overproduction of toxins.
- Cloning does not involve gene transfer from other species, so cloned animals produce no foreign proteins.
- Malformed or otherwise imperfect animals would be destroyed and not used as food for people.

Nevertheless, until more certainty exists, the FDA requests that food producers keep cloned animals out of the food supply.[13]

Issues Surrounding GE Foods

Consumers often question whether GE foods differ substantially from other foods in their nutrient contents or safety. The next sections address these questions.

Nutrient Composition

The *potential* ranges for nutrients in GE foods are virtually without limit. Except for the intentional variations, nutrients in GE foods are identical to those of comparable traditional foods. Therefore, eating the beta-carotene–enriched GE rice mentioned earlier would be the same from the body's point of view as eating plain rice and taking a beta-carotene supplement.

The current manufacturing practice of adding enrichment nutrients to wheat, rice, other grains, and milk products may one day become obsolete if rDNA plants and animals can make the extra nutrients in their tissues, thereby eliminating the possibility of human error. Consumers may also see favorite foods such as potatoes turned into "functional foods" that sport heavy doses of disease-fighting phytochemicals made by genes borrowed from less familiar foods such as flaxseed or ginger. Thus, there is little need to worry about reduced nutrient contents of GE foods. Instead, consumers should stay alert to the possibility of nutrient or phytochemical overdoses.

Pesticide Residues

Industry scientists contend that rDNA technology could virtually eliminate pesticide residues on foods.[14] This is because both the nature of the pesticide and the amount present in the food are predetermined by the genetics of the plant, leaving no room for error or misuse by farmers.

Because plant pesticides produced by GE fruits and vegetables exist in the tissues of the food, consumers cannot wash or peel them off before eating the food. Nevertheless, it is unlikely that plant pesticides pose a danger to the body. The reason is that unlike synthetic pesticides, GE plant pesticides are made of protein. As Chapter 6 made clear, DNA governs the synthesis of protein. Plant pesticides are products of DNA activity—that is, they are peptide molecules. In the human body, peptide molecules are denatured and digested by digestive enzymes and so are rendered harmless. In addition, all pesticides are regulated as food additives by the FDA and so must be proved safe for consumption before being sold to consumers, regardless of whether they are sprayed on the plants or made by them.

As mentioned, one of the few pesticides approved for use on organic foods is derived from a bacterium in soil. The extracted bacterial pesticide is a peptide chain that is lethal to insects that attack corn crops. The GE feed corn mentioned earlier received the bacterium's gene for producing this peptide. The corn expresses the gene and produces the plant pesticide internally. It's worth noting that the plant pesticide produced by GE corn is identical to the "natural" pesticide organic farmers spray on their crops to kill the same invading insects.

Unintended Health Effects

The possibility that genetic engineering of plants or animals may have unintended and therefore unpredictable effects on human consumers is a pressing concern for the nation's genetic scientists.[15] Such effects may arise with any sort of genetic modification, from the relatively rare cross-breeding in nature, to selective breeding in agriculture, to genetic engineering technology. Further, the results may be harmful, neutral, or even beneficial to consumers. Safety evaluations most appropriately focus on each new food itself, rather than on the process used to create it.

Common celery provides an example of an unintended negative effect from genetic modification by selective breeding. Over the years, celery growers chose the most attractive celery plants for reproduction because consumers demand good-looking celery. Unbeknownst to the growers, however, the celery stayed beautiful because it repelled pests with a natural plant pesticide whose concentrations increased with successive generations. This plant pesticide causes severe skin rashes susceptible in grocery and farm workers who handle the celery and are then exposed to sunlight.

An example of a positive, but still unintended, health effect of genetic engineering involves prevention of a cancer-causing fungus that contaminates corn.[*] Scientists conferred on a corn variety the ability to produce a plant pesticide that repels attacks by common worms. After several growing seasons, the scientists noted not only less destruction by worms, but far fewer attacks by the fungus than normal. No one knew previously that the worms repelled by the GE pesticide also infected the corn with the fungus as they burrowed.

By definition, intended changes to genetically modified foods are predictable, but *unintended* consequences are not. For this reason, the National Academy of Sciences has established a framework for evaluating unintended effects of genetically modified foods before they go to market.[16] Whether created by agricultural breeding or genetic engineering, a new food's composition can be chemically analyzed and studied for any variation from the original food. Except for the genetic manipulation, the two foods should be identical with regard to nutrients, phytochemicals, toxins, nutrient antagonists, and other constituents. If unintentional changes are detected, each variation must be evaluated for biological significance.

---

[*]The fungus is of the mycoptoxin variety, including aflatoxin.

Unfortunately, even the most sophisticated chemical analysis cannot provide information about all possible health outcomes. Foods have complex compositions, and human interactions with food make safety assessments more difficult.

Factors such as the amount of the food commonly consumed and the nature of the alteration determine the degree of risk. For example, if the food is a GE spice ordinarily consumed in tiny quantities and the spice lacks a typical nutrient, little threat from this change would be predicted. People simply do not rely on spices to provide significant nutrients to the diet. If, however, the spice is heavily consumed by certain groups of people, the potential effect on the nutrient intakes of the heaviest consumers must be evaluated.

If, instead of a spice, the new food is a potato likely to be consumed in large quantities by most people, and the potato contains a constituent not typical of potatoes, the modified potato and the new constituent would require many more studies to determine their safety before marketing. Finally, post-marketing tracking systems for adverse events confirm research findings.

Environmental Effects

Advances in rDNA technology promise bumper crops of food produced on far fewer acres of land, with less loss of water and topsoil and less use of pesticides and herbicides to end up in foods and drinking water. By one estimate, genetic engineering has already led to an 80 percent reduction in insecticide use among U.S. transgenic cotton crops.[17] Other environmental worries have arisen, however.

One concern is the likelihood of **outcrossing,** the accidental cross-pollination of plant pesticide crops with related wild weeds. The failure to confine engineered genes may be irreversible—no cleanup strategy for "genetic pollution" has been developed.[18] If a weed inherits a pest-resistant trait from a neighboring field of rDNA crops, it could gain an enormous survival advantage over other wild species and crowd them out.[19]

A solution to outcrossing is available, but it has spawned problems of its own.

Scientists can ensure the destruction of all of a GE plant's offspring by using so-called *terminator technology* to alter the plant's genetic material. Terminator technology keeps transgenic plants from passing their genes to wild weeds but makes it impossible for the world's farmers to save fertile seeds from their own harvests from year to year. Thus, even the poorest subsistence farmers would be forced to buy expensive new seeds each year. A possible alternative is to modify only the genetics of structures whose DNA passes through female lines because only male genes are carried by pollen.*[20]

It is also possible that the new crops may directly damage wildlife. In the laboratory, monarch butterfly larvae die when they feed on pollen from pesticide-producing corn. In real life, wild butterflies do not seem to consume enough toxic corn pollen to be harmed, at least in the short term.[21] If butterflies remain safe in the long run, then the new technology may even protect monarchs and other harmless or beneficial insects that die when they dine on conventionally sprayed fields.[22] Still, long-term effects on butterflies and other species are unknown and require scientific vigilance.

Opposition to Biotechnology

Some fear that by tampering with the basic blueprint of life, genetic engineering will sooner or later unleash mayhem on an unsuspecting world. Opponents view biotechnology firms as naïve and profit-driven, lacking the necessary moral judgment or national laws to guide them.[23] They point out that while rDNA technology indisputably benefits biotechnology companies and giant industrial farms, it has produced no real benefits for consumers and the risks are not defined. Others object to rDNA technology on religious grounds, holding that genetic decisions are best left to nature or a higher power. Table C12-4 summarizes some of these issues.

Some people envision a biotechnology run amok and used for frivolous, greedy purposes such as cloning animals for amusement. Others fear accidental doses of GE pharmaceuticals

*The structures are chloroplasts.

ending up in their children's cereal bowls at the breakfast table. Others are alarmed at the possibility of human beings cloned for certain traits and genetic "improvements." In the United States, government funds are withheld from any laboratory involved in human cloning experiments, but independent laboratories have stated their intent to move ahead with such research.

Proponents of genetic engineering respond that most of the world's people cannot afford the luxury of rejecting the potential benefits of biotechnology and accuse protesters of living in an elitist world of fertile lands with abundant food.[24] The world's poorest citizens suffer, they say, when protesters cause delays by destroying test crops and disrupting scientific meetings. Opponents of biotechnology counter that the scope of world hunger far exceeds simple solutions such as increasing food supplies and that the potential for harm to those most in need far outweighs any potential benefits to them.[25] Chapter 15 explores the tragedy of hunger in the United States and the world.

The FDA's Position on GE Foods

The FDA has taken the position that GE foods require no special safety testing or labeling unless they are substantially different from foods already in use.[26] The FDA holds the developers of the new foods responsible for testing those that differ significantly from traditional foods. The FDA assumes that any product with an antisense gene, including the tomato described earlier, is safe because the antisense gene merely prevents the synthesis of a protein and adds nothing but a tiny fragment of genetic material. The FDA does require that any extra substances produced in the food, such as new enzymes, hormones, or resistance traits, must meet the same safety standards applied to all additives.

The case for requiring rigorous safety testing of novel constituents in GE foods is strong. Concerns surrounding the potential for diseases, allergic reactions, and unknown hazards from unique proteins must be satisfied if GE products are to gain acceptance worldwide. For example, if genetic material from a disease-producing microorganism is

TABLE C12•4 GENETIC ENGINEERING: POINT, COUNTERPOINT

Arguments in Opposition to Genetic Engineering	Arguments in Support of Genetic Engineering
1. **Ethical and moral issues.** It's immoral to "play God" by mixing genes from organisms unable to do so naturally. Religious and vegetarian groups object to genes from prohibited species occurring in their allowable foods.	1. **Ethical and moral issues.** Scientists throughout history have been persecuted and even put to death by fearful people who accuse them of playing God. Yet, today many of the world's citizens enjoy a long and healthy life of comfort and convenience due to once-feared scientific advances put to practical use.
2. **Imperfect technology.** The technology is young and imperfect; genes rarely function in just one way, their placement is often imprecise, and potential effects are impossible to predict. Toxins are as likely to be produced as the desired trait.	2. **Advanced technology.** Recombinant DNA technology is precise and reliable. Many of the most exciting recent advances in medicine, agriculture, and technology were made possible by the application of this technology.
3. **Environmental concerns.** Environmental side effects are unknown. The power of a genetically modified organism to change the world's environments is unknown until such changes actually occur—then the "genie is out of the bottle." Once out, the genie cannot be put back in the bottle because insects, birds, and the wind distribute genetically altered seed and pollen to points unknown.	3. **Environmental protection.** Genetic engineering may be the only hope of saving rain forest and other habitats from destruction by impoverished people desperate for arable land. Through genetic engineering, farmers can make use of previously unproductive lands such as salt-rich soils and arid areas.
4. **"Genetic pollution."** Other kinds of pollution can often be cleaned up with money, time, and effort. Once genes are spliced into living things, those genes forever bear the imprint of human tampering.	4. **Genetic improvements.** Genetic side effects are more likely to benefit the environment than to harm it.
5. **Crop vulnerability.** Pests and disease can quickly adapt to overtake genetically identical plants or animals around the world. Diversity is key to defense.	5. **Improved crop resistance.** Pests and diseases can be specifically fought on a case-by-case basis. Biotechnology is the key to defense.
6. **Loss of gene pool.** Loss of genetic diversity threatens to deplete valuable gene banks from which scientists can develop new agricultural crops.	6. **Gene pool preserved.** Thanks to advances in genetics, laboratories around the world are able to stockpile the genetic material of millions of species that, without such advances, would have been lost forever.
7. **Profit motive.** Genetic engineering will profit industry more than the world's poor and hungry.	7. **Everyone profits.** Industries benefit from genetic engineering, and a thriving food industry benefits the nation and its people, as witnessed by countries lacking such industries. Genetic engineering promises to provide adequate nutritious food for millions who lack such food today. Developed nations gain cheaper, more attractive, more delicious foods with greater variety and availability year round.
8. **Unproven safety for people.** Human safety testing of genetically altered products is generally lacking. The population is an unwitting experimental group in a nationwide laboratory study for the benefit of industry.	8. **Safe for people.** Human safety testing of genetically altered products is unnecessary because the products are essentially the same as the original foodstuffs.
9. **Increased allergens.** Allergens can unwittingly be transferred into foods.	9. **Control of allergens.** A few allergens can be transferred into foods, but these are known, and foods likely to contain them are clearly labeled to warn consumers.
10. **Decreased nutrients.** A fresh-looking tomato or other produce held for several weeks may have lost substantial nutrients.	10. **Increased nutrients.** Genetic modifications can easily enhance the nutrients in foods.
11. **No product tracking.** Without labeling, the food industry cannot track problems to the source.	11. **Excellent product tracking.** The identity and location of genetically altered foodstuffs are known, and they can be tracked should problems arise.
12. **Overuse of herbicides.** Farmers, knowing that their crops resist herbicide effects, will use them liberally.	12. **Conservative use of herbicides.** Farmers will not waste expensive herbicides in second or third applications when the prescribed amount gets the job done the first time.
13. **Increased consumption of pesticides.** When a pesticide is produced by the flesh of produce, consumers cannot wash it off the skin of the produce with running water as they can with most ordinary sprays.	13. **Reduced pesticides on foods.** Pesticides produced by produce in tiny amounts known to be safe for consumption are more predictable than applications by agricultural workers who make mistakes. Because other genetic manipulations will eliminate the need for postharvest spraying, fewer pesticides will reach the dinner table.
14. **Lack of oversight.** Government oversight is run by industry people for the benefit of industry—no one is watching out for the consumer.	14. **Sufficient regulation, oversight, and rapid response.** The National Academy of Sciences has established protocol for safety testing of GE foods. Government agencies are efficient in identifying and correcting problems as they occur in the industry.

inserted into a food, scientists must prove that no dangerous characteristic from the microorganism exists in the food. If the inserted genetic material comes from a food to which people are commonly allergic, such as nuts, the new food must be absolutely allergen-free or bear a label to alert those with allergies. Furthermore, newly altered genetic material may create unique proteins never before encountered by the human body. Their effects should be studied and their presence regulated to ensure that people can eat them safely.

In agreement with the FDA's position is the American Dietetic Association, which represents current scientific thinking in nutrition.[27] Many other scientific organizations also agree, contending that rDNA technology can deliver on promises for an improved

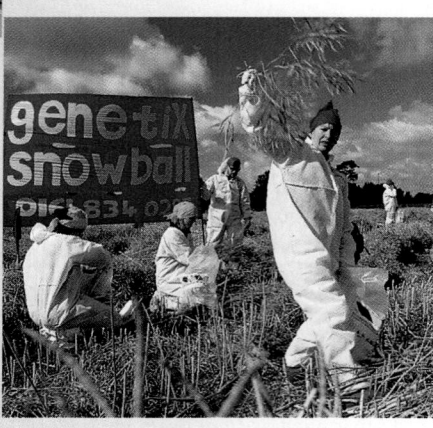
© Adrian Arbib/Still Pictures

CONTROVERSY 12

Protesters uprooting genetically altered plants.

food supply if we give it a fair chance to do so. More information is available on the Internet (see Table C12-5).

The Final Word

For those who would worry themselves into a diet of crackers and water, rest assured that eating more fruits and vegetables of any kind brings health advantages that far outweigh any slight risks. Table C12-6 sums up some pros and cons of all the major food production methods. As presented in Controversy 6 and Chapter 11 of this book, the evidence is overwhelming: those who eat the recommended amounts of fruits and vegetables each day remain healthier than those who do not, regardless of the source of those foods. Furthermore, research has repeatedly shown that the great majority of all fruits and vegetables, both domestic and imported, test well within safety margins set for pesticide contamination. Consumers can best serve their health needs, then, by washing all produce thoroughly and by handling it properly before consumption but, most of all, by eating it in abundance every day.

TABLE C12·5 GENETIC ENGINEERING INTERNET SITES

Much more information (and misinformation) is available on the Internet. Here are some sites to explore.

1. For policies and applications, search for "biotechnology" at the USDA Biotechnology Information Center: www.agnic.org
2. Get a "pro" perspective from the Council for Biotechnology Information: www.whybiotech.com
3. A scientific view is available by searching for "biotechnology" or "genetic engineering" at the International Food Information Council: http://ific.org
4. Consider the opposing perspective at the Genetic Engineering section of Greenpeace, USA: www.greenpeaceusa.org
5. Another opposition view is available at the Union of Concerned Scientists: www.ucsusa.org

TABLE C12·6 FOOD PRODUCTION METHODS COMPARED: ORGANIC, CONVENTIONAL, AND GENETIC ENGINEERING

Soil Condition and Environment

- *Organic:* Improves soil condition through crop rotation and the addition of complex fertilizers such as manure; controls erosion; highly protective of waterways and wildlife. Uses sustainable agriculture techniques.
- *Conventional:* Depletes soil; adds synthetic chemical fertilizers containing only a few key elements; can create soil erosion problems. Runoff pollutes waterways, and sprays poison wildlife such as birds and beneficial insect predators.
- *Genetic engineering:* No direct effect on soil or erosion; may require fewer pesticide sprays, thus protecting waterways and wildlife, but may harm wildlife by exposing wild species to altered genes or plant pesticides; may soon make use of salty, dry, or other currently unusable lands. May produce "genetic pollution."

Nutrients in Foods

- *Organic:* Suggestive evidence of slightly increased content of trace minerals, vitamin C, and improved amino acid balance in produce over conventionally farmed produce.
- *Conventional:* Standards for nutrient composition of foods are set by analysis of conventionally produced foods.
- *Genetic engineering:* Potential for increasing nutrient and phytochemical content, at the will of the producer.

Benefits to Consumers

- *Organic:* Reduced exposure to pesticides and other sprays and animal medications and hormones. New standards define organic techniques, with regulatory oversight. Long history of safety for human consumption of food varieties. Ethical comfort of knowing that food-producing animals are well treated.
- *Conventional:* General safety and pesticide residues monitored regularly; many varieties of foods available at low cost.
- *Genetic engineering:* Greater food production at low cost, keeping consumer prices low and availability high. Particular products may meet particular consumer demands, such as better flavor, increased vitamin or phytochemical content, or improved freshness of foods. Potential exists for helping to ease world hunger. Crops may produce medicines needed in impoverished areas of the world.

Consumer Safety Issues

- *Organic:* Consumer must wash produce well to remove possible dangerous microbial contamination and pesticides that may have "drifted" onto produce.
- *Conventional:* Consumer must wash produce well to remove possible dangerous microbial contamination and pesticides that are applied to produce.
- *Genetic engineering:* Consumer must wash produce well to remove possible dangerous microbial contamination and pesticides (especially herbicides) that are applied to produce. Internally produced plant pesticides do not wash off. Other dangers include introduction of allergens from other species and unproven safety of consuming rDNA products over a lifetime. Unknown dangers may also exist.

Life Cycle Nutrition: Mother and Infant

© Maurice Marinot (1882–1960). © Florence Marinot. *Woman with a child.* Musée d'Art Modern, Troyes, France. Réunion des Musées Nationaux/Art Resource, NY

CONTENTS

Do You Ever . . .

Think that men's lifestyle habits are of no consequence to a future pregnancy?

Wonder how much alcohol it takes to harm a developing unborn child?

Consider breastfeeding to be about the same as formula feeding for babies?

Wonder how young infants can thrive for months on only breast milk or formula without other foods?

KEEP READING.

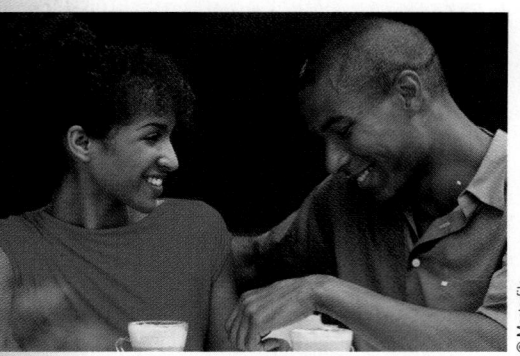

Both parents can prepare in advance for a healthy pregnancy.

ALL PEOPLE need the same nutrients, but the amounts we need change as we move through life. This chapter is the first of two on life's changing nutrient needs. It focuses on the two life stages that might be the most important to an individual's lifelong health—pregnancy and infancy.

Pregnancy: The Impact of Nutrition on the Future

We normally think of our nutrition as personal, affecting only our own lives. The woman who is pregnant, or who soon will be, must understand that her nutrition today is critical to the health of her future child throughout life. The nutrient demands of pregnancy are extraordinary.

Preparing for Pregnancy

Before she becomes pregnant, a woman must establish eating habits that will optimally nourish both the growing **fetus** and herself. She must be well nourished at the outset because early in pregnancy the **embryo** undergoes rapid and significant developmental changes that depend on good nutrition.

Fathers-to-be are also wise to examine their eating and drinking habits. For example, a sedentary lifestyle and consuming too few fruits and vegetables may affect men's **fertility** (and the fertility of their children), and men who drink too much alcohol or encounter other toxins in the weeks before conception can sustain damage to their sperm's genetic material.*[1] When both partners adopt healthy habits, they will be better prepared to meet the demands of parenting that lie ahead.

Prepregnancy Weight   Before pregnancy, all women, but underweight women in particular, should strive for appropriate body weights. A woman who starts out underweight and who fails to gain sufficiently during pregnancy is very likely to bear a baby with a dangerously low birthweight.[2] A later section comes back to the needed gains in pregnancy. Infant birthweight is the most potent single indicator of an infant's future health. A **low-birthweight** baby, defined as one who weighs less than 5½ pounds (2,500 grams), is nearly 40 times more likely to die in the first year of life than a normal-weight baby. To prevent low birthweight, underweight women are advised to gain weight before becoming pregnant and to strive to gain adequately thereafter.

When nutrient supplies during pregnancy fail to meet demands, the fetus may adapt to the sparse conditions in ways that may make obesity or chronic diseases more likely in later life (see Controversy 13).[3] Low birthweight is also associated with lower adult IQ and other brain impairments, short stature, and educational disadvantages.[4] Nutrient deficiency coupled with low birthweight is the underlying cause of more than half of all the deaths worldwide of children under five years of age.

In the United States, the infant mortality rate in 2002 was 7.0 deaths per 1,000 live births.[5] This rate, though higher than that of some other developed countries, has seen a significant decline over the last two decades and is a tribute to public health efforts aimed at reducing infant deaths (see Figure 13-1).

Not all cases of low birthweight reflect poor nutrition. Heredity, disease conditions, smoking, and drug (including alcohol) use during pregnancy all contribute.[6] Even with optimal nutrition and health during pregnancy, some women give birth to small infants for unknown reasons. But poor nutrition is the major factor in low birthweight—and an avoidable one, as later sections make clear.[7]

■ Underweight is defined as BMI <19.8. Obese is defined as BMI >29 (see Table 13-4 on page 491).

fetus (FEET-us) the stage of human gestation from eight weeks after conception until the birth of an infant.

embryo (EM-bree-oh) the stage of human gestation from the third to the eighth week after conception.

fertility the capacity of a woman to produce a normal ovum periodically and of a man to produce normal sperm; the ability to reproduce.

low birthweight a birthweight of less than 5½ pounds (2,500 grams); used as a predictor of probable health problems in the newborn and as a probable indicator of poor nutrition status of the mother before and/or during pregnancy. Low-birthweight infants are of two different types. Some are *premature infants;* they are born early and are the right size for their gestational age. Other low-birthweight infants have suffered growth failure in the uterus; they are small for gestational age (small for date) and may or may not be premature.

uterus (YOO-ter-us) the womb, the muscular organ within which the infant develops before birth.

placenta (pla-SEN-tuh) the organ of pregnancy in which maternal and fetal blood circulate in close proximity and exchange nutrients and oxygen (flowing into the fetus) and wastes (picked up by the mother's blood).

gestation the period of about 40 weeks (three trimesters) from conception to birth; the term of a pregnancy.

amniotic (AM-nee-OTT-ic) **sac** the "bag of waters" in the uterus in which the fetus floats.

*Reference notes are found in Appendix F.

Obese women are also urged to strive for healthy weights before pregnancy. The infant of an obese mother may be larger than normal and may be large even if born prematurely. The large early baby may not be recognized as premature and thus may not receive the special medical care required. The baby of an obese mother may be twice as likely to be born with a neural tube defect than others, but the reasons why are not known.[8] Obese women are more likely to require drugs to induce labor or to require surgical intervention for the birth, and they suffer gestational diabetes, hypertension, and infections after the birth more often than do women of healthy weight.[9] In addition, both overweight and obese women have a greater risk of giving birth to infants with heart defects and other abnormalities.[10] An appropriate goal for the obese woman who wishes to become pregnant is to strive to attain a healthy prepregnancy body weight in order to minimize her medical risks and those of her future child.

A Healthy Placenta and Other Organs A major reason the mother's nutrition before pregnancy is so crucial is that it determines whether her **uterus** will be able to support the growth of a healthy **placenta** during the first month of **gestation.** The placenta is both a supply depot and a waste-removal system for the fetus. If the placenta works perfectly, the fetus wants for nothing; if it doesn't, no alternative source of sustenance is available, and the fetus will fail to thrive. Figure 13-2 shows the placenta, a mass of tissue in which maternal and fetal blood vessels intertwine and exchange materials. The two bloods never mix, but the barrier between them is notably thin. To grasp how thin, picture your hands encased in skintight surgical gloves and immersed in water. Your hands represent the fetal blood vessels, the water is the pool of maternal blood, and the gloves are the tissue-thin placenta separating them. Across this thin barrier, nutrients and oxygen move from the mother's blood into the fetus's blood, and wastes move out of the fetal blood, to be excreted by the mother. The umbilical cord is the pipeline from the placenta to the fetus. The **amniotic sac** surrounds and cradles the fetus, cushioning it with fluids.

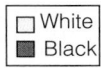

FIGURE 13·1 INFANT MORTALITY DECLINE OVER TIME

The graph shows infant deaths per 1,000 live births.

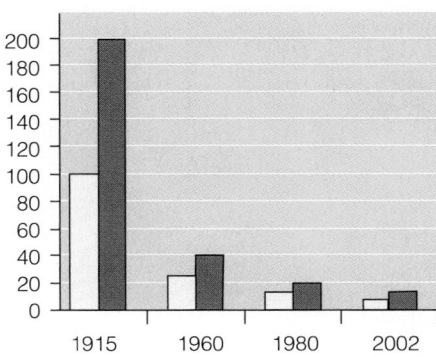

SOURCE: Data from E. Arias and coauthors, Annual summary of vital statistics—2002, *Pediatrics* 112 (2003): 1215–1230.

FIGURE 13·2 THE PLACENTA

The placenta is composed of spongy tissue in which fetal blood and maternal blood flow side by side, each in its own vessels. The maternal blood transfers oxygen and nutrients to the fetus's blood and picks up fetal wastes to be excreted by the mother. Thus, the placenta performs the nutritive, respiratory, and excretory functions that the fetus's digestive system, lungs, and kidneys will provide after birth.

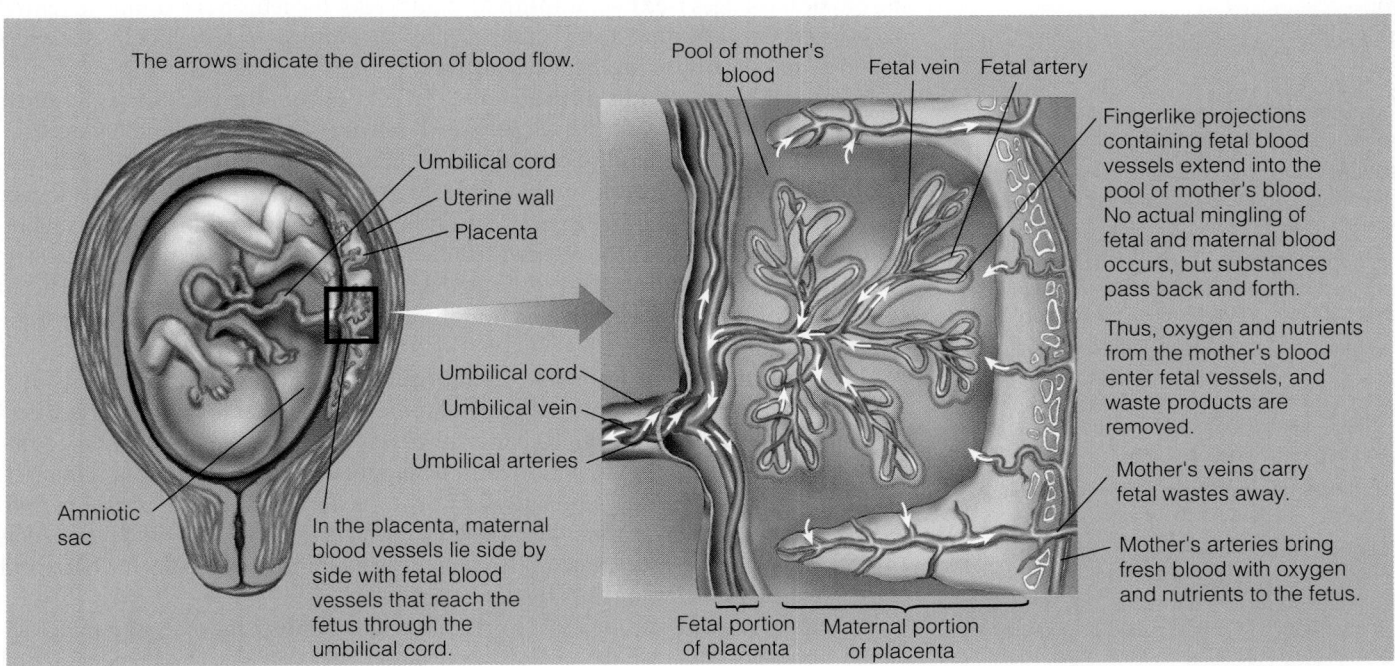

The placenta is an active metabolic organ with many responsibilities of its own. It actively gathers up hormones, nutrients, and protein molecules such as antibodies and transfers them into the fetal bloodstream. The placenta also produces a broad range of hormones that act in many ways to maintain pregnancy and prepare the mother's breasts for **lactation.**[11]

If the mother's nutrient stores are inadequate during the period when her body is developing the placenta, then the placenta will never form and function properly. As a consequence, no matter how well the mother eats later, her fetus will not receive optimal nourishment, and a low-birthweight baby with all of the associated risks is likely. After getting such a poor start on life, the child may be ill equipped, even as an adult, to store sufficient nutrients, and a girl may later be unable to grow an adequate placenta or bear healthy full-term infants. Thus, a woman's poor nutrition during her early pregnancy could affect her grandchild as well as her child.[12]

▶ **KEY POINT** Adequate nutrition before pregnancy establishes physical readiness and nutrient stores to support fetal growth. Both underweight and overweight women should strive for appropriate body weights before pregnancy. Newborns who weigh less than 5½ pounds face greater health risks than normal-weight babies. The healthy development of the placenta depends on adequate nutrition before pregnancy.

The Events of Pregnancy

The newly fertilized **ovum,** called a **zygote,** begins as a single cell and divides into many cells during the days after fertilization. Within two weeks, the zygote embeds itself in the uterine wall in a process known as **implantation,** and the placenta begins to grow inside the uterus. Minimal growth in size takes place at this time, but it is a crucial period in development. Adverse influences such as smoking, drug abuse, and malnutrition at this time lead to failure to implant or to abnormalities such as neural tube defects that can cause loss of the zygote, possibly before the woman knows she is pregnant.

The Embryo and Fetus During the next six weeks, the embryo registers astonishing physical changes (see Figure 13-3). At eight weeks, the fetus has a complete central nervous system, a beating heart, a fully formed digestive system, well-defined fingers and toes, and the beginnings of facial features.

In the last seven months of pregnancy, the fetal period, the fetus grows 50 times heavier and 20 times longer. Critical periods of cell division and development occur in organ after organ. The amniotic sac fills with fluid, and the mother's body changes. The uterus and its supporting muscles increase in size, the breasts may become tender and full, the nipples may darken in preparation for lactation, and the mother's blood volume increases by half to accommodate the added load of materials it must carry. Gestation lasts approximately 40 weeks and ends with the birth of the infant. The 40 or so weeks of pregnancy are divided into thirds, each of which is called a **trimester.**

A Note about Critical Periods Each organ and tissue type grows with its own characteristic pattern and timing. The development of each takes place only at a certain time—the **critical period.** Whatever nutrients and other environmental conditions are necessary during this period must be supplied on time if the organ is to reach its full potential. If the development of an organ is limited during a critical period, recovery is impossible. For example, the fetus's heart and brain are well developed at 14 weeks; the lungs, 10 weeks later. Therefore, early malnutrition impairs the heart and brain; later malnutrition impairs the lungs.

The effects of malnutrition during critical periods of pregnancy are seen in defects of the nervous system of the embryo (explained later), in the child's poor dental health, and in the adolescent's and adult's vulnerability to infections and possibly

lactation production and secretion of breast milk for the purpose of nourishing an infant.

ovum the egg, produced by the mother, that unites with a sperm from the father to produce a new individual.

zygote (ZYE-goat) the term that describes the product of the union of ovum and sperm during the first two weeks after fertilization.

implantation the stage of development, during the first two weeks after conception, in which the fertilized egg (fertilized ovum or zygote) embeds itself in the wall of the uterus and begins to develop.

trimester a period representing gestation. A trimester is about 13 to 14 weeks.

critical period a finite period during development in which certain events may occur that will have irreversible effects on later developmental stages. A critical period is usually a period of cell division in a body organ.

FIGURE 13·3 STAGES OF EMBRYONIC AND FETAL DEVELOPMENT

© (Photos 1, 2, 3) Petit Format/Nestle/Photo Researchers, © (Photo 4) Randy Lincks/Masterfile

(1) A newly fertilized ovum is about the size of the period at the end of this sentence. This zygote at less than one week after fertilization is not much bigger and is ready for implantation.

(3) A fetus after 11 weeks of development is just over an inch long. Notice the umbilical cord and blood vessels connecting the fetus with the placenta.

(2) After implantation, the placenta develops and begins to provide nourishment to the developing embryo. An embryo five weeks after fertilization is about ½ inch long.

(4) A newborn infant after nine months of development measures close to 20 inches in length. The average birthweight is about 7½ pounds. From eight weeks to term, this infant grew 20 times longer and 50 times heavier.

higher risks of diabetes, hypertension, stroke, or heart disease.[13] The effects of malnutrition during critical periods are irreversible: Abundant and nourishing food, fed after the critical time, cannot remedy harm already done.

Table 13-1 provides a list of factors that make nutrient deficiencies and complications likely during pregnancy. Notice that young age heads the list; a later section explains why pregnant adolescents are especially prone to malnutrition.

▶ KEY POINT Implantation, fetal development, and early critical periods depend on maternal nutrition before and during pregnancy.

Increased Need for Nutrients

During pregnancy, a woman's nutrient needs increase more for certain nutrients than for others. Figure 13-4 (p. 486) shows the percentage increase in nutrient intakes recommended for pregnant women compared to nonpregnant women. To meet the high nutrient demands of pregnancy, a woman must make careful food choices, but her body will also do its part by maximizing nutrient absorption and minimizing losses.[14]

Energy, Carbohydrate, Protein, and Fat   Energy needs vary with the progression of pregnancy. In the first trimester, the pregnant woman needs no additional energy, but her energy needs rise as pregnancy progresses. She requires an additional 340 daily calories during the second trimester and an extra 450 calories each day during the third trimester.[15] Well-nourished pregnant women meet these demands for more energy in several ways: some eat more food, some reduce their activity, and some store less of their food energy as fat.[16] A woman can easily meet the need for extra calories by selecting more nutrient dense foods from the five food groups. Table 2-3

TABLE 13·1 FACTORS PLACING PREGNANT WOMEN AT NUTRITIONAL RISK

Women likely to develop nutrient deficiencies and pregnancy complications include those who:

- Are young (adolescents).
- Have had many previous pregnancies (3 or more to mothers under age 20; 4 or more to mothers age 20 or older).
- Have short intervals between pregnancies (<18 months).
- Have a history of poor pregnancy outcomes.
- Lack nutrition knowledge, have too little money to purchase adequate food, or have too little family support.
- Consume an inadequate diet due to food faddism, preferences, weight-loss "dieting," uninformed vegetarianism, or eating disorders.
- Smoke cigarettes or use alcohol or illicit drugs.
- Are lactose intolerant or suffer chronic health conditions requiring special diets.
- Are underweight or overweight at conception.
- Are carrying twins or triplets.
- Gain insufficient or excessive weight during pregnancy.
- Have a low level of education.

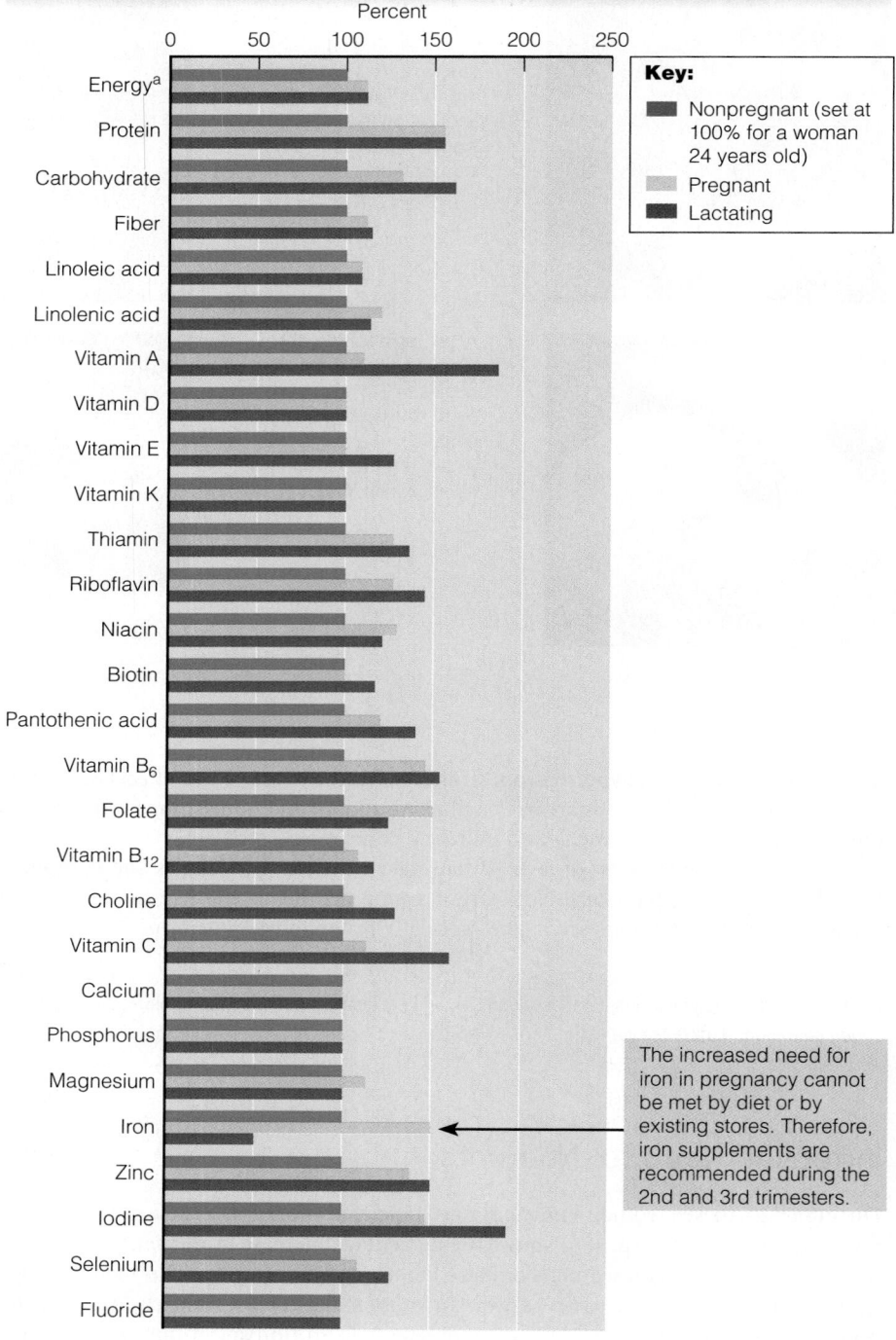

Key:
■ Nonpregnant (set at 100% for a woman 24 years old)
■ Pregnant
■ Lactating

The increased need for iron in pregnancy cannot be met by diet or by existing stores. Therefore, iron supplements are recommended during the 2nd and 3rd trimesters.

aEnergy allowance during pregnancy is for 2nd trimester; energy allowance during the 3rd trimester is slightly higher; no additional allowance is provided during the 1st trimester. Energy allowance during lactation is for the first 6 months; energy allowance during the second 6 months is slightly higher.

TABLE 13·2 FOOD CHOICES FOR PREGNANT AND LACTATING WOMEN

SAMPLE MENU

Breakfast
1 whole wheat English muffin
2 tbs peanut butter
1 c low-fat vanilla yogurt
½ c fresh strawberries
1 c orange juice

Midmorning snack
½ c cranberry juice
1 oz pretzels

Lunch
Sandwich (tuna salad on whole-wheat bread)
½ carrot (sticks)
1 c low-fat milk

Dinner
Chicken cacciatore
    3 oz chicken
    ½ c stewed tomatoes
1 c rice
½ c summer squash
1½ c salad (spinach, mushrooms, carrots)
1 tbs salad dressing
1 slice Italian bread
2 tsp soft margarine
1 c low-fat milk

NOTE: This sample meal plan provides about 2,500 calories (55 percent from carbohydrate, 20 percent from protein, and 25 percent from fat) and meets most of the vitamin and mineral needs of pregnant and lactating women.

(on p. 39) provides suggested eating patterns for several calorie levels and Table 13-2 offers a sample menu for pregnant and lactating women.

If a woman chooses less nutritious options such as sugary soft drinks or fatty snack foods to meet energy needs, she will undoubtedly come up short on nutrients. The increase in the need for nutrients is even greater than for energy, so the mother-to-be should choose nutrient-dense foods such as whole-grain breads and cereals, legumes, dark green vegetables, citrus fruits, low-fat milk and milk products, and lean meats, fish, poultry, and eggs.

Ample carbohydrate (ideally, 175 grams or more per day and certainly no less than 135 grams) is necessary to fuel the fetal brain and spare the protein needed for fetal growth. Fiber in carbohydrate-rich foods such as whole grains, vegetables, and fruit can help alleviate the constipation that many pregnant women experience.

The protein DRI recommendation for pregnancy is higher than for nonpregnant women by 25 grams per day. Most women in the United States, however, need not add protein-rich foods to their diets because they already exceed the recommended protein intake for pregnancy. Excess protein may also have adverse effects, as Chapter 6 explained.

Some vegetarian women limit or omit protein-rich meats, eggs, and dairy products from their diets. For them, meeting the recommendation for food energy each day and including several generous servings of plant-protein foods such as legumes, tofu, whole grains, nuts, and seeds are imperative. Protein supplements during pregnancy can be harmful and their use is discouraged.

The high nutrient requirements of pregnancy leave little room in the diet for excess energy from solid fats such as butter. The essential fatty acids, however, are particularly important to the growth and development of the fetus.[17] The brain is composed mainly of lipid material and depends heavily on long-chain omega-3 and omega-6 fatty acids

■ DRI nutrient and energy intake recommendations for pregnant women are listed on the inside front cover.

A pregnancy affected by a neural tube defect can occur in any woman, but these factors make it more likely:

- A previous pregnancy affected by a neural tube defect.
- Exposure to high temperatures early in pregnancy (prolonged fever or hot tub use).
- Inadequate folate intake.
- Low socioeconomic status.
- Maternal diabetes (type 1).
- Maternal obesity.
- Maternal use of antiseizure medications.
- Race/ethnicity (neural tube defects are more common among whites and Hispanics than among others).

neural tube the embryonic tissue that later forms the brain and spinal cord.

neural tube defect (NTD) a group of nervous system abnormalities caused by interruption of the normal early development of the neural tube.

anencephaly (an-en-SEFF-ah-lee) an uncommon and always fatal neural tube defect in which the brain fails to form.

spina bifida (SPY-na BIFF-ih-duh) one of the most common types of neural tube defects in which gaps occur in the bones of the spine. Often the spinal cord bulges and protrudes through the gaps, resulting in a number of motor and other impairments.

for its growth, function, and structure. (See Table 5-5 on page 157 for a list of good food sources of the essential fatty acids.)

> ▶ **KEY POINT** Pregnancy brings physiological adjustments that demand increased intakes of energy and nutrients. A balanced diet that includes more nutrient dense foods from the five food groups can help to meet these needs.

Of Special Interest: Folate and Vitamin B₁₂ The vitamins famous for their roles in cell reproduction—folate and vitamin B_{12}—are needed in large amounts during pregnancy. New cells are laid down at a tremendous pace as the fetus grows and develops. At the same time, the number of the mother's red blood cells must rise because her blood volume increases, a function requiring more cell division and therefore more vitamins. To accommodate these needs, the recommendation for folate during pregnancy increases from 400 to 600 micrograms a day.

As described in Chapter 7, folate plays an important role in preventing neural tube defects. To review, the early weeks of pregnancy are a critical period for the formation and closure of the **neural tube** that will later develop to form the brain and spinal cord. By the time a woman suspects she is pregnant, usually around the sixth week of pregnancy, the embryo's neural tube normally has closed. A **neural tube defect (NTD)** occurs when the tube fails to close properly. In the United States, about 30 of every 100,000 newborns are born with an NTD.[18] When the neural tube fails to close properly and brain development fails, a rare but lethal defect known as **anencephaly** occurs. All infants with anencephaly die shortly after birth.

In a more common NTD, the spinal cord and backbone do not develop normally—and the result is **spinal bifida** (see Figure 13-5). The membranes covering the spinal cord often protrude from the spine as a sac, and sometimes a portion of the spinal cord is contained in the sac. Spina bifida is often accompanied by varying degrees of paralysis, depending on the extent of spinal cord damage. Mild cases may not be noticed. Moderate cases may involve curvature of the spine, muscle weakness, mental handicaps, and other ills; severe cases can lead to death.

FIGURE 13·5 SPINA BIFIDA—A NEURAL TUBE DEFECT

Spina bifida

© Biophoto Associates/Photo Researchers, Inc.

Normal vertebra (top view)   Spina bifida vertebra (top view)

central chamber

supportive stacking bones of the vertebrae

Normally, the bony central chamber closes fully to encase the spinal cord and its surrounding membranes and fluid. In spina bifida, the two halves of the slender bones that should complete the casement of the cord fail to join.

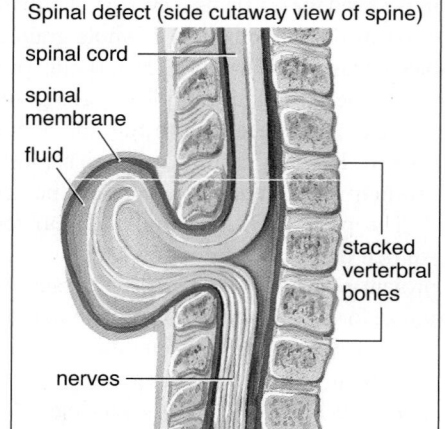

Spinal defect (side cutaway view of spine)

spinal cord

spinal membrane

fluid

stacked verterbral bones

nerves

In the serious form shown here, membranes and fluid have bulged through the gap and nerves are exposed, invariably leading to some degree of paralysis and often to mental retardation.

TABLE 13·3 RICH FOLATE SOURCES[a]

Natural Folate Sources	Fortified Folate Sources
Liver (3 oz) 221 μg	Multi-Grain Cheerios Plus cereal (1 c) 400 μg[b]
Lentils (½ c) 179 μg	Product 19 cereal (1 c) 400 μg[b]
Chickpeas or pinto beans (½ c) 145 μg	Total cereal (1 c) 400 μg[b]
Asparagus (½ c) 131 μg	Pasta, cooked (1 c) 110 μg
Spinach (1 c raw) 131 μg	Rice, cooked (1 c) 134 μg
Avocado (½ c) 45 μg	Bagel (1 small whole) 75 μg
Orange juice (1 c) 74 μg	Waffles, frozen (2) 36 μg
Beets (½ c) 68 μg	Bread, white (1 slice) 28 μg

[a]Folate amounts for these and 2,000 other foods are listed in the Table of Food Composition in Appendix A.
[b]Folate in cereals varies; read the Nutrition Facts panel of the label.

To reduce the risk of neural tube defects, women who are capable of becoming pregnant should obtain 400 micrograms of folic acid daily from supplements, fortified foods, or both, *in addition* to eating folate-rich foods (see Table 13-3).[19] The DRI committee recommends intake of synthetic folate, called folic acid, in supplements and fortified foods because it is absorbed better than the folate naturally present in foods.[20] Foods that naturally contain folate are still important, however, because they contribute to folate intakes while providing other needed vitamins, minerals, fiber, and phytochemicals.

All enriched grain products (cereal, grits, pasta, rice, bread, and the like) sold commercially in the United States are fortified with folic acid. This measure has improved folate status in women of childbearing age and lowered the number of neural tube defects that occur each year.[21] Folate fortification does raise one safety concern, however. The pregnant woman needs a greater amount of vitamin B_{12} to assist folate in the manufacture of new cells. High intakes of folate complicate the diagnosis of a vitamin B_{12} deficiency. For this reason, folate intakes should not exceed 1 milligram per day.[22]

People who eat meat, eggs, or dairy products receive all the vitamin B_{12} they need, even for pregnancy. Those who exclude all animal products from the diet need vitamin B_{12}–fortified foods or supplements.

■ Chapter 7 describes how excessive folate intakes can mask symptoms of vitamin B_{12} deficiency.

▶ KEY POINT   Due to their key roles in cell reproduction, folate and vitamin B_{12} are needed in large amounts during pregnancy. Folate plays an important role in preventing neural tube defects.

Calcium, Magnesium, Iron, and Zinc   Among the minerals, calcium, phosphorus, and magnesium are in great demand during pregnancy because they are necessary for normal development of the bones and teeth. Intestinal absorption of calcium doubles early in pregnancy, and the mineral is stored in the mother's bones. Later, when fetal bones begin to calcify, the mother's bone calcium stores are mobilized, and there is a dramatic shift of calcium across the placenta. In the final weeks of pregnancy, more than 300 milligrams of calcium a day are transferred to the fetus. Efforts to ensure an adequate calcium intake during pregnancy are aimed at conserving the mother's bone mass while supplying fetal needs.

Most women do not meet the DRI recommendations for calcium and should increase their intakes.[23] In particular, pregnant women under age 25, whose own bones are still actively depositing minerals, should strive to meet the DRI recommendation for calcium by increasing their intakes of milk, cheese, yogurt, and other calcium-rich foods. Less preferred, but still acceptable, is a daily supplement of 600 milligrams of calcium. The DRI recommendation for calcium intake is the same for nonpregnant and pregnant women in the same age group. The mineral magnesium is also essential for bone and tissue growth, and pregnancy slightly increases the need for magnesium in the diet.

FIGURE 13·6 EXAMPLE OF A PRENATAL SUPPLEMENT LABEL

Notice that vitamin A is reduced to guard against birth defects, while extra amounts of folate, iron, and other nutrients are provided to meet the specific needs of pregnant women.

Prenatal Vitamins

Supplement Facts
Serving Size 1 Tablet

Amount Per Tablet	% Daily Value for Pregnant/ Lactating Women
Vitamin A 4000 IU	50%
Vitamin C 100 mg	167%
Vitamin D 400 IU	100%
Vitamin E 11 IU	37%
Thiamin 1.84 mg	108%
Riboflavin 1.7 mg	85%
Niacin 18 mg	90%
Vitamin B6 2.6 mg	104%
Folate 800 mcg	100%
Vitamin B12 4 mcg	50%
Calcium 200 mg	15%
Iron 27 mg	150%
Zinc 25 mg	167%

INGREDIENTS: calcium carbonate, microcrystalline cellulose, dicalcium phosphate, ascorbic acid, ferrous fumarate, zinc oxide, acacia, sucrose ester, niacinamide, modified cellulose gum, di-alpha tocopheryl acetate, hydroxypropyl methylcellulose, hydroxypropyl cellulose, artificial colors (FD&C blue no. 1 lake, FD&C red no. 40 lake, FD&C yellow no. 6 lake, titanium dioxide), polyethylene glycol, starch, pyridoxine hydrochloride, vitamin A acetate, riboflavin, thiamin mononitrate, folic acid, beta carotene, cholecalciferol, maltodextrin, gluten, cyanocobalamin, sodium bisulfite.

cesarean (see-ZAIR-ee-un) **section** surgical childbirth, in which the infant is taken through an incision in the woman's abdomen.

prenatal (pree-NAY-tal) before birth.

Special Supplemental Food Program for Women, Infants, and Children (WIC) a USDA program offering low-income pregnant women and those with infants or preschool children coupons redeemable for specific foods that supply the nutrients deemed most necessary for growth and development. For more information, visit www.usda.gov/FoodandNutrition.

During pregnancy, the body avidly conserves iron—menstruation ceases and absorption of iron increases up to threefold. Despite these conservation measures, iron stores dwindle because the developing fetus draws heavily on its mother's iron to store up a supply sufficient to carry it through the first three to six months of life. Even women with inadequate iron stores transfer significant amounts of iron to the fetus, suggesting that the iron needs of the fetus have priority over those of the mother.[24] Maternal blood losses are also inevitable at birth, especially during a delivery by **cesarean section,** further draining the mother's iron supply. Few women enter pregnancy with adequate stores to meet pregnancy demands, so a daily iron supplement containing 30 milligrams is recommended during the second and third trimesters for all pregnant women. When a low hemoglobin or hematocrit is confirmed by a repeat test, more than 30 milligrams of iron may be prescribed. To enhance iron absorption, the supplement should be taken between meals and with liquids other than milk, coffee, or tea, which inhibit iron absorption.

Zinc, required for protein synthesis and cell development, is vital during pregnancy. Severe zinc deficiency during pregnancy predicts low birthweight.[25] Zinc is provided abundantly by protein-rich foods such as shellfish, meat, and nuts, but its absorption may be hindered by other trace elements or fiber in foods. For example, iron interferes with the body's absorption and use of zinc, so women taking iron supplements in excess of 30 milligrams per day may also need zinc supplements to prevent zinc deficiency.[26] Most supplements for pregnancy provide about 30 to 60 milligrams of iron a day.

▶ **KEY POINT** All pregnant women, but especially those who are less than 25 years of age, need to pay special attention to ensure adequate calcium intakes. A daily iron supplement is recommended for all pregnant women during the second and third trimesters.

Prenatal Supplements Physicians often recommend daily multivitamin-mineral supplements for pregnant women. These **prenatal** supplements typically provide more folate, iron, and calcium than regular supplements (see Figure 13-6). Prenatal supplements are especially beneficial for women who do not eat adequately and for those in high-risk groups: women carrying twins or triplets, cigarette smokers, and alcohol and drug abusers.[27] For these women, prenatal supplements may be of some help in reducing the risks of preterm delivery, low birthweights, and birth defects.[28] Supplements cannot prevent the vast majority of destruction from tobacco, alcohol, and drugs, however, as later sections explain.

▶ **KEY POINT** Women most likely to benefit from multivitamin-mineral supplements during pregnancy include those who do not eat adequately, those carrying twins or triplets, and those who smoke cigarettes or are alcohol or drug abusers.

Food Assistance Programs

Women of limited financial means often eat diets too low in calcium, iron, vitamins A and C, and protein. Often, they and their children need help in obtaining food and benefit from nutrition counseling. At the federal level, the **Special Supplemental Food Program for Women, Infants, and Children (WIC)** provides vouchers or debit cards that can be used to purchase nutritious foods, along with nutrition education, to low-income pregnant and breastfeeding women and their children.[29] The foods offered are milk and cheese, iron-fortified cereals, fruit or vegetable juices, carrots, eggs, dried beans, tuna fish, and peanut butter. These foods provide nutrients often lacking in diets of low-income women and children. For infants given infant formula, WIC also provides iron-fortified formula.

Participation in the WIC program benefits both the nutrient status and the growth and development of infants and children. WIC serves almost half of all infants in the United States, and about one-fourth of children ages one to four.[30] WIC participation during pregnancy can effectively reduce infant mortality, low birthweight, and maternal and newborn medical costs.

Federal food stamps can also help to stretch the low-income pregnant woman's grocery dollars. Many communities provide educational services and materials, including nutrition, food budgeting, and shopping information through the local agricultural extension service. Organizations such as the American Dietetic Association, the American Diabetes Association, and local hospitals also provide nutrition information.

▶ KEY POINT   Food assistance programs such as WIC can provide nutritious food for pregnant women of limited financial means.

How Much Weight Should a Woman Gain during Pregnancy?

Women must gain weight during pregnancy—fetal and maternal well-being depend on it. Ideally, a woman will have begun her pregnancy at a healthy weight, but even more importantly, she will gain within the recommended weight range based on her prepregnancy body mass index (BMI). Table 13-4 presents recommended weight gains for pregnancy. For the normal-weight woman, the ideal pattern is about 3½ pounds total during the first trimester and a pound per week thereafter. Pregnancy weight gains within the recommended ranges are associated with fewer surgical births, a greater number of healthy birthweights, and other positive outcomes for both mothers and infants, but many women do not gain within these ranges.[31]

Dieting during pregnancy is not recommended. Even an obese woman should gain about 15 pounds for the best chance of delivering a healthy infant.[32] Weight gain for a pregnant teenager must be adequate enough to accomodate her own growth and that of her fetus. Women who are carrying twins must gain more still. A sudden, large weight gain is a danger signal, however, because it may indicate the onset of preeclampsia (see the section entitled "Troubleshooting").

The weight the pregnant woman puts on is nearly all lean tissue: placenta, uterus, blood, milk-producing glands, and the fetus itself (see Figure 13-7). The fat she gains

TABLE 13·4 RECOMMENDED WEIGHT GAINS FOR PREGNANT WOMEN[a]

- Underweight (BMI <19.8):[b] 28 to 40 lb
- Healthy weight (BMI 19.8 to 26): 25 to 35 lb
- Overweight (BMI 26 to 29): 15 to 25 lb
- Obese (BMI >29): 15 lb minimum

[a]The BMI cutoff points in this table were established by the Subcommittee on Nutritional Status and Weight Gain during Pregnancy. The cutoff points defining underweight, normal weight, overweight, and obesity differ slightly from those established by the National Heart, Lung, and Blood Institute Expert Panel on the Identification, Evaluation, and Treatment of Overweight and Obesity in Adults.
[b]BMI tables are on the inside back cover.
SOURCE: Committee on Nutritional Status during Pregnancy and Lactation, Food and Nutrition Board, *Nutrition during Pregnancy* (Washington, D.C.: National Academy Press, 1990), pp. 10, 12.

FIGURE 13·7 COMPONENTS OF WEIGHT GAIN DURING PREGNANCY

	Weight gain (lb)
Increase in breast size	2
Increase in mother's fluid volume	4
Placenta	1½
Increase in blood supply to the placenta	4
Amniotic fluid	2
Infant at birth	7½
Increase in size of uterus and supporting muscles	2
Mother's necessary fat stores	7
	30

1st trimester   2nd trimester   3rd trimester

is needed later for lactation. Physical activity can help a pregnant woman cope with the extra weight, as the next section explains. Some weight is lost at delivery, but many women retain a few pounds with each pregnancy.

▶ KEY POINT   Weight gain is essential for a healthy pregnancy. A woman's prepregnancy BMI, her own nutrient needs, and the number of fetuses she is carrying help to determine appropriate weight gain.

Should Pregnant Women Be Physically Active?

An active, physically fit woman experiencing a normal pregnancy can and should continue to exercise throughout pregnancy, adjusting the intensity and duration as the pregnancy progresses. Staying active during the course of a normal, healthy pregnancy improves the fitness of the mother-to-be, facilitates labor, helps to prevent or manage gestational diabetes, and reduces psychological stress.[33] Women who remain active during pregnancy report fewer discomforts throughout their pregnancies and retain habits that help in losing excess weight and regaining fitness after the birth.

Pregnant women should take care in choosing their physical activities, however, participating in "low-impact" activities and avoiding sports in which they might fall or be hit by other people or objects (for some suggestions, see the Think Fitness on page 493). Pregnant women with medical conditions or pregnancy complications should undergo a thorough evaluation by their health-care professional before engaging in physical activity. A few guidelines are offered in Figure 13–8.[34] Several of the guidelines are aimed at preventing excessively high internal body temperature and dehydration, both of which can harm fetal development. To this end, the pregnant woman should also stay out of saunas, steam rooms, and hot whirlpools.

▶ KEY POINT   Physically fit women can continue to be physically active throughout pregnancy. Pregnant women should be cautious in their choice of activities.

FIGURE 13·8 GUIDELINES FOR PHYSICAL ACTIVITY DURING PREGNANCY

DO

Do exercise regularly (at least three times a week).

Do warm up with 5 to 10 minutes of light activity.

Do exercise for 20 to 30 minutes at your target heart rate.

Do cool down with 5 to 10 minutes of slow activity and gentle stretching.

Do drink water before, after, and during exercise.

Do eat enough to support the additional needs of pregnancy plus exercise.

Pregnant women can enjoy the benefits of physical activity.

DON'T

Don't exercise vigorously after long periods of inactivity.

Don't exercise in hot, humid weather.

Don't exercise when sick with fever.

Don't exercise while lying on your back after the first trimester of pregnancy or stand motionless for prolonged periods.

Don't exercise if you experience any pain or discomfort.

Don't participate in activities that may harm the abdomen or involve jerky, bouncy movements.

Don't scuba dive.

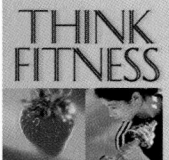

THINK FITNESS — Physical Activities for the Pregnant Woman

IS THERE AN ideal physical activity for the pregnant woman? There might be. Swimming and water aerobics offer advantages over other activities during pregnancy. Water cools and supports the body, provides a natural resistance, and lessens the impact of the body's movement, especially in the later months.[35] Research shows that water aerobics can reduce the intensity of back pain during pregnancy.[36] Other activities considered safe and comfortable for pregnant women include walking, light strength training, rowing, and climbing stairs.

Teen Pregnancy

Each year in the United States, over 800,000 adolescent girls become pregnant.[37] Of these, about half choose to continue their pregnancies. A pregnant adolescent presents a special case of intense nutrient needs. Young teenage girls have a hard enough time meeting nutrient needs for their own rapid growth and development, let alone those of pregnancy. Many teens enter pregnancy deficient in vitamins A and D, folate, iron, calcium, and zinc—deficiencies that place both mother and fetus at risk. Smoking also presents risks, and teens are more likely to smoke while pregnant than older women. Pregnant teenagers have more miscarriages, premature births, stillbirths, and low-birthweight infants than do pregnant adult women.[38] The greatest risk, though, is death of the infant: mothers under age 16 bear more infants who die within the first year than do women in any other age group. These factors combine to make teenage pregnancy a major public health problem.

Adequate nutrition is an indispensable component of prenatal care for teenagers and can substantially improve the outlook for both mother and infant. To support the needs of both mother and fetus, a pregnant teenager with a BMI in the normal range is encouraged to gain about 35 pounds to reduce the likelihood of a low-birthweight infant.[39] Pregnant and lactating teenagers would do well to follow the eating pattern presented in Table 2-3, making sure to choose a calorie level high enough to support weight gain.

▶ **KEY POINT** Of all the population groups, pregnant teenage girls have the highest nutrient needs and an increased likelihood of having problem pregnancies.

Why Do Some Women Crave Pickles and Ice Cream While Others Can't Keep Anything Down?

Does pregnancy give a woman the right to demand pickles and ice cream at 2 A.M.? Perhaps so, but not for nutrition's sake. Food cravings and aversions during pregnancy are common but do not seem to reflect real physiological needs. In other words, a woman who craves pickles is not in need of salt. Food cravings and aversions are due to changes in taste and smell sensitivities, and they quickly disappear after the birth.

Sometimes strange cravings may occur in women with nutrient-poor diets. A pregnant woman who is deficient in iron, zinc, or other nutrients may crave and eat soil, clay, ice, cornstarch, and other nonnutritious substances (pica, first mentioned in Chapter 8). Such cravings are not adaptive; the substances the woman craves do not deliver the nutrients she needs. In fact, clay and other substances can cling to the intestinal wall and form a barrier that interferes with normal nutrient absorption. Furthermore, if the soil or clay contains environmental contaminants such as lead or parasites, health and nutrition suffer.

TABLE 13·5 TIPS FOR RELIEVING COMMON DISCOMFORTS OF PREGNANCY

To alleviate the nausea of pregnancy:

- On waking, get up slowly.
- Eat dry toast or crackers.
- Chew gum or suck hard candies.
- Eat small, frequent meals whenever hunger strikes.
- Avoid foods with offensive odors.
- When nauseated, do not drink citrus juice, water, milk, coffee, or tea.

To prevent or alleviate constipation:

- Eat foods high in fiber.
- Exercise daily.
- Drink at least 8 glasses of liquids a day.
- Respond promptly to the urge to defecate.
- Use laxatives only as prescribed by a physician; avoid mineral oil—it carries needed fat-soluble vitamins out of the body.

To prevent or relieve heartburn:

- Relax and eat slowly.
- Eat small, frequent meals.
- Drink liquids between meals.
- Avoid spicy or greasy foods.
- Sit up while eating.
- Wait an hour after eating before lying down.
- Wait 2 hours after eating before exercising.

The nausea of "morning" (actually, anytime) sickness seems unavoidable and may even be a welcome sign of a healthy pregnancy because it arises from the hormonal changes of early pregnancy. Many women complain that smells, especially cooking smells, make them sick. Thus, minimizing odors can alleviate morning sickness. Sipping carbonated drinks and nibbling soda crackers or other salty snack foods before getting out of bed can sometimes prevent nausea. Some women do best by simply eating what they desire whenever they feel hungry. Table 13-5 offers some other suggestions, but morning sickness can be persistent. If morning sickness interferes with normal eating for more than a week or two, the woman should seek medical advice to prevent nutrient deficiencies.

As the hormones of pregnancy alter her muscle tone and the thriving fetus crowds her intestinal organs, an expectant mother may complain of heartburn or constipation. Raising the head of the bed with two or three pillows can help to relieve night-time heartburn. A high-fiber diet, physical activity, and a plentiful water intake will help relieve constipation. The pregnant woman should use laxatives or heartburn medications only if her physician prescribes them.

> ► KEY POINT   Food cravings usually do not reflect physiological needs, and some may interfere with nutrition. Nausea arises from normal hormonal changes of pregnancy.

Some Cautions for the Pregnant Woman

Some choices that pregnant women make or substances they encounter can harm the fetus, sometimes severely. Among these threats, smoking, medications, herbal supplements, illegal drugs, environmental contaminants, foodborne illness, vitamin-mineral megadoses, dieting, sugar substitutes, and caffeine deserve consideration. Alcohol constitutes a major threat to fetal health and is given a section of its own.

Cigarette Smoking   A surgeon general's warning states that parental smoking can kill an otherwise healthy fetus or newborn. Constituents of cigarette smoke, such as nicotine and cyanide, are toxic to a fetus. Smoking restricts the blood supply to the growing fetus and so limits the delivery of oxygen and nutrients and the removal of wastes. It slows growth, thus retarding physical development of the fetus, and it may cause behavioral or intellectual problems later. Cigarette (and cigar) smoking also impair fetal nutrition and development by adversely affecting the pregnant woman's nutrition status. Smokers tend to have lower intakes of dietary fiber, vitamin A, beta-carotene, folate, and vitamin C.

A mother who smokes is more likely to have a complicated birth, and her infant is more likely to be of low birthweight.[40] The more a mother smokes, the smaller her baby will be. Of all preventable causes of low birthweight in the United States, smoking has the greatest impact. Sudden infant death syndrome (SIDS), the unexplained deaths that sometimes occur in otherwise healthy infants, has been linked to the mother's cigarette smoking during pregnancy.[41] Research suggests that even in women who do not smoke, exposure to **environmental tobacco smoke** (**ETS,** or secondhand smoke) during pregnancy increases the risk of low birthweight and the likelihood of SIDS.[42] Unfortunately, an estimated one out of eight pregnant women smokes, and rates are even higher for unmarried women and those who have not graduated from high school.[43]

Medicinal Drugs And Herbal Supplements Medicinal drugs taken during pregnancy can cause serious birth defects. Pregnant women should not take over-the-counter drugs or any medications not prescribed by a physician. Drug labels warn: "As with any drug, if you are pregnant or nursing a baby, seek the advice of a health professional before using this product." For aspirin and ibuprofen, there is an additional warning: "It is especially important not to use aspirin (or ibuprofen) during the last three months of pregnancy unless specifically directed to do so by a doctor because it may cause problems in the unborn child or excessive bleeding during delivery." Such warnings should be taken seriously.

Some pregnant women mistakenly consider herbal supplements to be safe alternatives to medicinal drugs and take them to relieve nausea, promote water loss, alleviate depression, aid sleep, or for other reasons. Some herbal products may be safe, but almost none have been tested for safety or effectiveness during pregnancy. Pregnant women should stay away from herbal supplements, teas, or other products unless their safety during pregnancy has been ascertained.[44] The American Dietetic Association website lists more than 100 herbal supplements that may not be safe to use during pregnancy.* Chapter 11's Consumer Corner offers more information about herbal supplements and other alternative therapies.

Drugs of Abuse Research shows that women who abuse drugs such as marijuana and cocaine during pregnancy inflict serious health consequences, including nervous system disorders, on their fetuses.[45] Drugs of abuse such as cocaine easily cross the placenta and impair fetal growth and development.[46] Infants born to mothers who abuse crack and other forms of cocaine face low birthweight, heartbeat abnormalities, the pain of withdrawal, or even death as they first experience life outside the womb. Some effects of other drugs of abuse on the fetus are listed in the margin.

Environmental Contaminants Infants and young children whose mothers were exposed to environmental contaminants such as lead and mercury during pregnancy show signs of impaired cognitive development. During pregnancy, lead and mercury readily move across the placenta, inflicting severe damage on the developing fetal nervous system.[47]

Unacceptably high concentrations of mercury in fish have prompted the Food and Drug Administration (FDA) and the Environmental Protection Agency (EPA) to warn all pregnant women, women who may become pregnant, lactating mothers, and young children against eating the following large ocean fish: king mackerel, swordfish, shark, and tilefish.[48] The FDA and EPA further advise the same groups of people to:

Fetal effects of abused drugs:
- Amphetamines: Suspected nervous system damage; behavioral abnormalities.
- Barbiturates: Drug withdrawal symptoms in the newborn, lasting up to six months.
- Cocaine: Uncontrolled jerking motions; paralysis; permanent mental and physical damage.
- Marijuana: Short-term irritability at birth.
- Opiates (including heroin): Drug withdrawal symptoms in the newborn; permanent learning disability (attention deficit hyperactivity disorder).

environmental tobacco smoke (ETS) the combination of exhaled smoke (mainstream smoke) and smoke from lighted cigarettes, pipes, or cigars (sidestream smoke) that enters the air and may be inhaled by other people.

---

*Go to www.eatright.org and click on position papers, pregnancy/breastfeeding.

- Eat up to 12 ounces a week of a variety of safer fish and shellfish such as canned light tuna, salmon, pollock, catfish, and shrimp; children should be given smaller portions. Albacore "white" tuna has more mercury than canned light tuna and so should be limited to 6 ounces or less per week.
- Check local advisories about the safety of fish caught by family and friends in lakes, rivers, and coastal areas. If no advice is available, eat up to 6 ounces per week of fish from local waters, but don't eat any other fish during that week.

Chapter 12 offers more details on contaminants in foods.

Foodborne Illness   The vomiting and diarrhea caused by many foodborne illnesses (see Chapter 12) can leave a pregnant woman exhausted and dangerously dehydrated. Particularly threatening, however, is **listeriosis,** which can cause miscarriage, stillbirth, or severe brain or other infections to fetuses and newborns. According to the Centers for Disease Control and Prevention, pregnant women are "about 20 times more likely than other healthy adults to get listeriosis."[49] A woman with listeriosis may develop symptoms such as fever, vomiting, and diarrhea about 12 hours after eating a contaminated food; serious symptoms may develop a week to six weeks later. A blood test can reliably detect listeriosis, and antibiotics given promptly to the pregnant sufferer can often prevent infection of the fetus or newborn. The margin lists preventive measures pregnant women can take to avoid contracting listeriosis.

Vitamin-Mineral Megadoses   Many vitamins are toxic when taken in excess, and minerals are even more so. A single massive dose of preformed vitamin A (100 times the recommended intake) has caused birth defects. Chronic use of lower doses of vitamin A supplements (three to four times the recommended intake) may also cause birth defects. Intakes before the seventh week of pregnancy appear to be the most damaging. For this reason, additional vitamin A is not recommended during pregnancy, and the vitamin is prescribed in the first trimester of pregnancy only upon evidence of deficiency, which is rare.

Dieting   Dieting, even for short periods, is hazardous during pregnancy. Low-carbohydrate diets or fasts that cause ketosis deprive the growing fetal brain of needed glucose and may impair its development. Such diets are also likely to be deficient in other nutrients vital to fetal growth. Energy restriction during pregnancy is dangerous, regardless of the woman's prepregnancy weight or the amount of weight gained in the previous month.

Sugar Substitutes   Artificial sweeteners have been studied extensively and found to be acceptable during pregnancy if used within the FDA's guidelines (see Controversy 4).[50] Still, pregnant women would be prudent to use sweeteners in moderation and within an otherwise nutritious and well-balanced diet. Women with phenylketonuria should not use aspartame, as Controversy 4 explains.

Caffeine   Caffeine crosses the placenta, and the fetus has only a limited ability to metabolize it. No firm limit for caffeine intake is yet available, but pregnant women who drink more than three cups of coffee a day may increase their risk of spontaneous abortion.[51] In light of this evidence, the most sensible course is to limit caffeine consumption to the equivalent of one cup of coffee or two 12-ounce cola beverages a day. Caffeine amounts in food and beverages are listed in Controversy 14, on page 556.

To protect their fetuses and newborns from listeriosis, pregnant women should:
- Avoid the following Mexican soft cheeses: queso blanco, queso fresco, queso de hoja, queso de crema, and asadero. Also avoid feta cheese, brie, Camembert, and blue-veined cheeses like Roquefort.
- Use only pasteurized dairy products.
- Eat only thoroughly cooked meat, poultry, and seafood.
- Before eating hot dogs and luncheon or deli meats, including cured meats like salami, thoroughly reheat them until steaming hot.
- Wash all fruits and vegetables.
- Do not eat refrigerated smoked seafood, such as salmon or trout, or any fish labeled "nova-style," "lox," or "kippered," unless it is an ingredient in a cooked dish.
- Do not eat refrigerated pâté or meat spreads. Canned or shelf-stable pâté and meat spreads are safer.

listeriosis a serious foodborne infection that can cause severe brain infection or death in a fetus or a newborn; caused by the bacterium *Listeria monocytogenes,* which is found in soil and water.

▶ **KEY POINT**   Abstaining from smoking and other drugs, limiting intake of foods known to contain unsafe levels of contaminants such as mercury, taking precautions against foodborne illness, avoiding large doses of nutrients, refraining from dieting, using artificial sweeteners in moderation, and limiting caffeine use are recommended during pregnancy.

Drinking during Pregnancy

Alcohol is arguably the most hazardous drug to future generations because it is legally available, heavily promoted, and widely abused. Society sends mixed messages concerning alcohol. Beverage companies promote an image of drinkers as healthy and active. Opposing this image, health authorities warn that alcohol can be injurious to health, especially during pregnancy (see Figure 13-9). Every container of beer, wine, or liquor for sale in the United States is required to warn pregnant women of the dangers of drinking during pregnancy.

Alcohol's Effects

Women of childbearing age need to know about alcohol's harmful effects on a fetus. Alcohol crosses the placenta freely and is directly toxic:[52]

- A sudden dose of alcohol can halt the delivery of oxygen through the umbilical cord. Oxygen is indispensable on a minute-to-minute basis to the development of the fetus's central nervous system.
- Alcohol slows cell division, reducing the number of cells produced and inflicting abnormalities on those that are produced and all of their progeny.
- During the first month of pregnancy, the fetal brain is growing at the rate of 100,000 new brain cells a minute. Even a few minutes of alcohol exposure during this critical period can exert a major detrimental effect.
- Alcohol interferes with placental transport of nutrients to the fetus and can cause malnutrition in the mother; then, all of malnutrition's harmful effects compound the effects of the alcohol.
- Before fertilization, alcohol can damage the ovum or sperm in the mother- or father-to-be, leading to abnormalities in the child.

▶ KEY POINT Alcohol limits oxygen delivery to the fetus, slows cell division, and reduces the number of cells organs produce. Alcoholic beverages must bear warnings to pregnant women.

Fetal Alcohol Syndrome

Drinking alcohol during pregnancy threatens the fetus with irreversible brain damage, growth retardation, mental retardation, facial abnormalities, vision abnormalities, a low **Apgar score,** and more than 40 identifiable health problems—a cluster of symptoms known as **fetal alcohol syndrome** or **FAS.** The fetal brain is extremely vulnerable to a glucose or oxygen deficit, and alcohol causes both by disrupting placental functioning. The lifelong mental retardation and other tragedies of FAS can be prevented by abstaining from drinking alcohol during pregnancy. Once the damage is done, however, the child remains impaired.

Figure 13-10 (p. 498) shows the facial abnormalities of FAS, which are easy to depict. A visual picture of the internal harm is impossible, but that damage seals the fate of the child. An estimated 2 to 15 of every 10,000 children are victims of this preventable damage, making FAS one of the leading known preventable causes of mental retardation in the world.[53]

Even when a child does not develop full FAS, prenatal exposure to alcohol can lead to less severe, but nonetheless serious, mental and physical problems. The cluster of mental problems associated with prenatal alcohol exposure is known as **alcohol-related neurodevelopmental disorder** (ARND) and the physical malformations are referred to as **alcohol-related birth defects** (ARBD). Some ARND and ARBD children show no outward sign of impairment, but others are short in stature or display subtle facial abnormalities. Most perform poorly in school and in social interactions and suffer a subtle form of brain damage. Mood disorders and problem behaviors, such as aggression, are common.[54]

FIGURE 13-9 MIXED MESSAGES IN ALCOHOL ADVERTISEMENTS

Labels on alcoholic beverages often display "healthy" images, but their warnings tell the truth.

Controversy 3 defined "a drink" as:

- 5 oz wine (12% alcohol).
- 10 oz wine cooler.
- 12 oz beer.
- 1½ oz hard liquor (80 proof).

Apgar score a system of scoring an infant's physical condition right after birth. Heart rate, respiration, muscle tone, response to stimuli, and color are ranked 0, 1, or 2. A low score indicates that medical attention is required to facilitate survival.

fetal alcohol syndrome (FAS) the cluster of symptoms inlcuding brain damage, growth retardation, mental retardation, and facial abnormalities seen in an infant or child whose mother consumed alcohol during her pregnancy.

alcohol-related neurodevelopmental disorder (ARND) behavioral, cognitive, or central nervous system abnormalities associated with prenatal alcohol exposure.

alcohol-related birth defects (ARBD) malformations in the skeletal and organ systems (heart, kidneys, eyes, ears) associated with prenatal alcohol exposure.

© James W. Hanson, M.D./NICHD

A child with FAS.

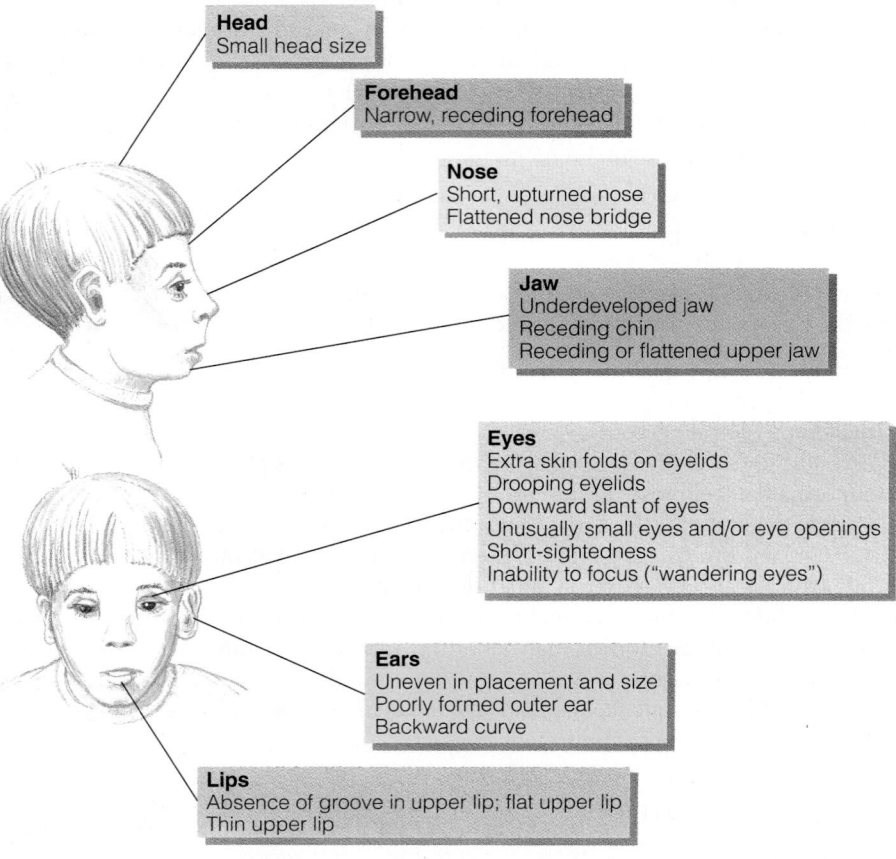

TYPICAL FACIAL CHARACTERISTICS OF FAS

The severe facial abnormalities shown here are just outward signs of severe mental impairments and internal organ damage. These defects, though hidden, may create major health problems later.

Head
Small head size

Forehead
Narrow, receding forehead

Nose
Short, upturned nose
Flattened nose bridge

Jaw
Underdeveloped jaw
Receding chin
Receding or flattened upper jaw

Eyes
Extra skin folds on eyelids
Drooping eyelids
Downward slant of eyes
Unusually small eyes and/or eye openings
Short-sightedness
Inability to focus ("wandering eyes")

Ears
Uneven in placement and size
Poorly formed outer ear
Backward curve

Lips
Absence of groove in upper lip; flat upper lip
Thin upper lip

For every child diagnosed with FAS, three or four with ARND or ARBD go undiagnosed until problems develop in the preschool years. Upon reaching adulthood, such children are ill equipped for employment, relationships, and the other facets of life most adults take for granted. Anyone exposed to alcohol before birth may always respond differently to it, and also to certain drugs, than if no exposure had occurred, making addictions likely.

▶ KEY POINT   The birth defects of fetal alcohol syndrome arise from severe damage to the fetus caused by alcohol. Lesser conditions, ARND and ARBD, may be harder to diagnose but also rob the child of a normal life.

Experts' Advice

Despite alcohol's potential for harm, 12 percent of pregnant women drink alcohol sometime during pregnancy; more than 3 percent report "frequent" drinking (seven or more drinks per week); and more than 2 percent admit to binge drinking (five or more drinks on one occasion).[55] Almost half of all pregnancies are unintended, and many are conceived during a binge drinking episode.[56]

For women who know they are pregnant and wonder how much alcohol is too much, some research findings may be of interest. Research using animals shows that one-fifth of the amount of alcohol needed to produce major visible defects will produce learning impairments or other defects in the offspring.[57] Compared to women who drink less than one drink *per week,* a sizable and significant increase in stillbirths occurs in women who drink five or more drinks per week.[58] Low birthweight is reported among infants born to women who have two drinks per day during pregnancy, and FAS is also known to occur with as few as two drinks a day. Birth defects have been reliably observed among the children of women who have four drinks of alcohol daily during pregnancy. The most severe impact is likely to occur in the first two months, when the woman may not even be aware that she is pregnant.

Given such evidence, the American Academy of Pediatrics takes the position that women should stop drinking as soon as they *plan* to become pregnant.[59] This step is important for fathers-to-be as well. Researchers have looked for a "safe" alcohol intake limit during pregnancy and have found none. Their conclusion: Abstinence from alcohol is the best policy for pregnant women. The authors of this book recommend this choice, too. After the birth of a healthy baby, celebrate with one glass of champagne, if you choose to drink.

For a pregnant woman who has already been drinking alcohol, the advice is "stop now." A woman who has drunk heavily during the first two-thirds of her pregnancy can still prevent some organ damage by stopping heavy drinking during the third trimester.

▶ **KEY POINT**   Abstinence from alcohol is critical to prevent irreversible damage to the fetus.

Troubleshooting

Disease during pregnancy can endanger the health of the mother and the health and growth of the fetus. If discovered early, many diseases can be controlled—another reason early prenatal care is recommended.

Gestational Diabetes   Some women are prone to develop a pregnancy-related form of diabetes, **gestational diabetes.** Gestational diabetes usually resolves after the infant is born, but some women go on to develop diabetes (usually type 2) later in life, especially if they are overweight.[60] Gestational diabetes can lead to fetal or infant sickness or death. If it is identified early and managed properly, however, the most serious risks fall dramatically.[61] More commonly, gestational diabetes leads to surgical birth and high infant birthweight.[62] The American Diabetes Association recommends that all women be assessed for risk of gestational diabetes at their first prenatal examination. Those with elevated risks should undergo further testing and treatment.[63]

Preeclampsia   A certain degree of **edema** is to be expected in late pregnancy, and some women also develop hypertension during that time. If a rise in blood pressure is mild, it may subside after childbirth and cause no harm. In some cases, however, hypertension may signal the onset of **preeclampsia,** a condition characterized not only by high blood pressure but by protein in the urine and fluid retention (edema). The edema of preeclampsia is a severe, whole-body edema, distinct from the localized fluid retention women normally experience late in pregnancy. The normal edema of pregnancy is a response to gravity; fluid from blood pools in the ankles. The edema of preeclampsia causes swelling of the face and hands as well as of the feet and ankles.

Preeclampsia afflicts 1 in 20 pregnancies, usually first pregnancies, and most often during the final trimester.[64] Both men and women who were born of pregnancies complicated by preeclampsia are more likely to have a child born of a pregnancy complicated by preeclampsia, suggesting a genetic predisposition.[65] Black women have a much greater risk of preeclampsia than white women.

■ The *Dietary Guidelines for Americans 2005* lists pregnant women and women who may become pregnant among the groups who should not drink alcohol at all. See Controversy 3, page 91 for others.

Risk factors for gestational diabetes:
- Glucose in the urine.
- Obesity.
- Personal history of gestational diabetes.
- Strong family history of diabetes.

These racial and ethnic groups are more prone to gestational diabetes:
- African American.
- Asian American.
- Hispanic American.
- Native American.
- Pacific Islander.

Warning signs of preeclampsia:
- Blurred vision.
- Dizziness.
- Headaches.
- Persistent abdominal pain.
- Sudden weight gain.
- Swelling, especially facial swelling.

gestational diabetes abnormal glucose tolerance appearing during pregnancy.

edema (eh-DEE-mah) accumulation of fluid in the tissues (also defined in Chapter 6).

preeclampsia (PRE-ee-CLAMP-see-uh) a potentially dangerous condition during pregnancy characterized by edema, hypertension, and protein in the urine.

Preeclampsia affects almost all of the mother's organs—the circulatory system, liver, kidneys, and brain. If the condition progresses and she experiences convulsions, the condition is called eclampsia. Maternal mortality during pregnancy is rare in developed countries, but eclampsia is the most common cause. Preeclampsia demands prompt medical attention. Dietary factors have been studied over the years, but so far none convincingly prevent preeclampsia.[66] Calcium supplementation may reduce the incidence of preeclampsia in some women, however.[67]

▶ KEY POINT   Gestational diabetes and preeclampsia are common medical problems associated with pregnancy. These should be managed to minimize associated risks.

Lactation

As the time of childbirth nears, a woman must decide whether she will feed her baby breast milk, infant formula, or both. These options are the only foods recommended for an infant during the first four to six months of life. A woman who plans to breastfeed her baby should begin to prepare toward the end of her pregnancy. No elaborate or expensive preparations are needed, but the expectant mother can read one of the many handbooks available on breastfeeding or consult a **certified lactation consultant,** employed at many hospitals.* Part of the preparation involves learning what dietary changes are needed because adequate nutrition is essential to successful lactation.

In rare cases, women produce too little milk to nourish their infants adequately. Severe consequences, including infant dehydration, malnutrition, and brain damage, can occur should the condition go undetected for long. Early warning signs of insufficient milk are dry diapers (a well-fed infant wets about six diapers a day) and infrequent bowel movements.

Nutrition during Lactation

A nursing mother produces about 25 ounces of milk a day, with considerable variation from woman to woman and in the same woman from time to time, depending primarily on the infant's demand for milk. Producing this milk costs a woman almost 500 calories per day above her regular need during the first six months of lactation. To meet this energy need, the woman is advised to eat an extra 330 calories of food each day. The other 170 calories can be drawn from the fat stores she accumulated during pregnancy. The food energy consumed by the nursing mother should carry with it abundant nutrients. Look back at Figure 13-4 (page 486) for a lactating woman's nutrient recommendations and at Table 13-2 on page 487 for a sample menu.

The volume of breast milk produced depends on how much milk the baby demands, not on how much fluid the mother drinks. The nursing mother is nevertheless advised to drink plenty of liquids each day (about 13 cups) to protect herself from dehydration. To help themselves remember to drink enough liquid, many women make a habit of drinking a glass of milk, juice, or water each time the baby nurses as well as at mealtimes.

A common question is whether a mother's milk may lack a nutrient if she fails to get enough in her diet. The answer differs from one nutrient to the next, but in general, the effect of nutritional deprivation of the mother is to reduce the *quantity,* not the *quality,* of her milk. Women can produce milk with adequate protein, carbohydrate, fat, folate, and most minerals, even when their own supplies are limited. For these nutrients, milk quality is maintained at the expense of maternal stores. This is

■ The DRI recommendation for *total* water intake during lactation is 3.8 L/day. This includes 3.1 L or about 13 c as total beverages, including drinking water.

certified lactation consultant a health-care provider, often a registered nurse or a registered dietitian, with specialized training and certification in breast and infant anatomy and physiology who teaches the mechanics of breastfeeding to new mothers.

*La Leche League is an international organization that helps women with breastfeeding concerns. See Appendix E for its address and website.

most evident in the case of calcium: dietary calcium has no effect on the calcium concentration of breast milk, but maternal bones lose some of their density during lactation. Such losses are generally made up quickly when lactation ends, and breastfeeding has no long-term harmful effects on women's bones.[68]

Any excess water-soluble vitamins the mother takes in are excreted in the urine; the body does not release them into the milk. The amounts of some vitamins in human milk, however, are affected by the mother's excessive or deficient intakes. For example, large doses of vitamin A raise the concentration of this vitamin in breast milk.

Some infants may be sensitive to foods such as cow's milk, onions, or garlic in the mother's diet and become uncomfortable when she eats them. Nursing mothers are advised to eat whatever nutritious foods they choose. If a particular food seems to cause an infant discomfort, the mother can eliminate that food from her diet for a few days and see if the problem goes away.

Another common question is whether breastfeeding promotes a more rapid loss of the extra body fat accumulated during pregnancy. Studies on this question have not provided a definitive answer. When breastfeeding continues for three months or longer, lactation does seem to accelerate a woman's weight loss, but factors such as percentage of body fat and weight gain during pregnancy also play a role.[69] This does not mean that a breastfeeding woman can eat unlimited food and return to prepregnancy weight. Breastfeeding costs energy, true, but diet and physical activity are still the cornerstones of weight control. Physical activity in particular helps to reduce body fatness and improve fitness while having little effect on a woman's milk production or her infant's weight gain. A gradual weight loss (1 pound per week) is safe and does not reduce milk output.[70] Too large an energy deficit, especially soon after birth, will inhibit lactation.

▶ KEY POINT   The lactating woman needs extra fluid and enough energy and nutrients to make sufficient milk each day. Malnutrition most often diminishes the quantity of the milk produced without altering quality. Lactation may facilitate loss of the extra fat gained during pregnancy.

When Should a Woman Not Breastfeed?

Some substances impair maternal milk production or enter breast milk and interfere with infant development, making breastfeeding an unwise choice. Some medical conditions also prohibit breastfeeding.

Alcohol, Nicotine, and Other Drugs   Alcohol enters breast milk and can adversely affect production, volume, composition, and ejection of breast milk as well as overwhelm an infant's immature alcohol-degrading system. Alcohol concentration peaks within one hour after ingestion of even moderate amounts (equivalent to a can of beer). This amount may alter the taste of the milk to the disapproval of the nursing infant, who may, in protest, drink less milk than normal. Drug addicts, including alcohol abusers, can take such high doses that their infants become addicts by way of breast milk. In these cases, breastfeeding is contraindicated.

As for cigarette smoking, research shows that lactating women who smoke produce less milk, and milk with a lower fat content, than mothers who do not smoke. Thus, their infants gain less weight than infants of nonsmokers. A lactating woman who smokes not only transfers nicotine and other chemicals to her infant via her breast milk but also exposes the infant to secondhand smoke. Babies who are "smoked over" experience a wide array of health problems—poor growth, hearing impairment, vomiting, breathing difficulties, and even unexplained death.

Excess caffeine can make an infant jittery and wakeful. Caffeine consumption should be moderate when breastfeeding.

Many medicines pose no danger during breastfeeding, but others cannot be used because they suppress lactation or are secreted into breast milk and can harm the infant.[71] If a nursing mother must take medication that is secreted in breast milk and is known to affect the infant, then breastfeeding must be put off for the duration of treatment. Meanwhile, the flow of milk can be sustained by pumping the breasts and discarding the milk. A nursing mother should consult with her physician before taking medicines or herbal supplements.

Many women wonder about using oral contraceptives during lactation. One type that combines the hormones estrogen and progestin seems to suppress milk output, lower the nitrogen content of the milk, and shorten the duration of breastfeeding. In contrast, progestin-only pills have no effect on breast milk or breastfeeding and are considered appropriate for lactating women.

Environmental Contaminants   A woman sometimes hesitates to breastfeed because she has heard warnings that contaminants in fish, water, and other foods (see page 495) may enter breast milk and harm her infant. Although some contaminants do enter breast milk, others may be filtered out. Because formula is made with water, formula-fed infants consume any contaminants that may be in the water supply. Any woman who is concerned about breastfeeding on this basis can consult with a physician or dietitian familiar with the local circumstances. With the exception of rare, massive exposure to a contaminant, the many benefits of breastfeeding far outweigh the risk associated with environmental hazards in the United States.[72]

Maternal Illness   If a woman has an ordinary cold, she can continue nursing without worry. The infant will probably catch it from her anyway, and thanks to immunological protection, a breastfed baby may be less susceptible than a formula-fed baby. With appropriate treatment, a woman who has an infectious disease such as hepatitis or tuberculosis can breastfeed; transmission is rare.[73]

The human immunodeficiency virus (HIV), responsible for causing AIDS, can be passed from an infected mother to her infant during pregnancy, at birth, or through breast milk, especially during the early months of breastfeeding.[74] The American Academy of Pediatrics and the Centers for Disease Control advise that only the complete avoidance of breastfeeding by HIV-infected women can absolutely prevent the transmission of HIV via breastfeeding.[75] Where safe alternatives are available, women who have tested positive for HIV should not breastfeed their infants.

In developing countries, where feeding inappropriate or contaminated formulas causes 1.5 million infant deaths each year, breastfeeding can be critical to infant survival. This advantage, however, must be weighed against the 200,000 to 300,000 infants who become infected with HIV each year by way of breastfeeding.[76] Whether HIV-infected women in developing countries should breastfeed comes down to a delicate decision between risks and benefits. For HIV-positive women in developing countries who are literate, have access to safe water, and an uninterrupted supply of infant formula, replacement feeding may reduce the risk of infant illness and death by AIDS. For those without those assets, interventions to decrease the risk of breastfeeding transmission of HIV are urgently needed.

▶ **KEY POINT**   Breastfeeding is not advised if the mother's milk is contaminated with alcohol, drugs, or environmental pollutants. Most ordinary infections such as colds have no effect on breastfeeding. Where safe alternatives are available, HIV-infected women should not breastfeed their infants.

Feeding the Infant

Early nutrition affects later development, and early feedings establish eating habits that influence nutrition throughout life. Trends change and experts may argue the fine points, but nourishing a baby is relatively simple. Common sense and a nurturing, relaxed environment go far to promote the infant's well-being.

Nutrient Needs

A baby grows faster during the first year of life than ever again, as Figure 13-11 shows. Pediatricians carefully monitor the growth of infants and children because growth directly reflects their nutrition status. An infant's birthweight doubles by about five months of age and triples by the age of one year. (If a 150-pound adult were to grow like this, the person would weigh 450 pounds after a single year.) The infant's length changes more slowly than weight, increasing about 10 inches from birth to one year. By the end of the first year, the growth rate slows considerably; an infant typically gains less than 10 pounds during the second year and grows about 5 inches in height.

Not only do infants grow rapidly, but their basal metabolic rate is remarkably high—about twice that of an adult's, based on body weight. The rapid growth and metabolism of the infant demand an ample supply of all the nutrients. Of special importance during infancy are the energy nutrients and the vitamins and minerals critical to the growth process, such as vitamin A, vitamin D, and calcium.

Because they are small, babies need smaller *total* amounts of these nutrients than adults do, but as a percentage of body weight, babies need more than twice as much of most nutrients. Infants require about 100 calories per kilogram of body weight per day; most adults require fewer than 40 (see Table 13-6). Figure 13-12 (p. 504) compares a five-month-old baby's needs (per unit of body weight) with those of an adult man. You can see that differences in vitamin D and iodine, for instance, are extraordinary. Around six months of age, energy needs begin to increase less rapidly as the growth rate begins to slow down, but some of the energy saved by slower growth is spent in increased activity. When their growth slows, infants spontaneously reduce their energy intakes. Parents should expect their babies to adjust their food intakes downward when appropriate and should not force or coax them to eat more.

Vitamin K nutrition for newborns presents a unique case. A newborn's digestive tract is sterile, and vitamin K–producing bacteria take weeks to establish themselves in the baby's intestines. To prevent bleeding in the newborn, the American Academy of Pediatrics (AAP) recommends that a single dose of vitamin K be given at birth.[77]

One of the most important nutrients for infants, as for everyone, is water. The younger a child is, the more of its body weight is water. Breast milk or infant formula normally provides enough water to replace fluid losses in a healthy infant. Even in hot, dry climates, neither breastfed nor bottle-fed infants need supplemental water.[78] Because proportionately more of an infant's body water than an adult's is between the cells and in the vascular space, this water is easy to lose. Conditions that cause rapid fluid loss, such as vomiting or diarrhea, require an electrolyte solution designed for infants.

> ▶ **KEY POINT** Infants' rapid growth and development depend on adequate nutrient supplies, including water from breast milk or formula.

Why Is Breast Milk So Good for Babies?

Both the AAP and the Canadian Pediatric Society stand behind this statement: "Breastfeeding is strongly recommended for full term infants, except in the few instances where specific contraindications exist." The American Dietetic Association

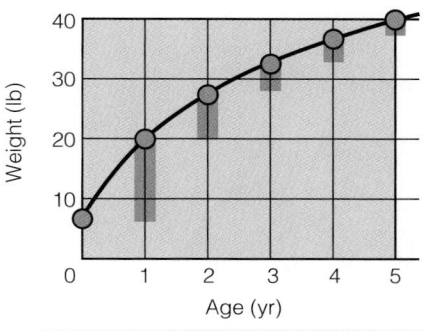

FIGURE 13·11 WEIGHT GAIN OF HUMAN INFANTS AND CHILDREN IN THE FIRST FIVE YEARS OF LIFE

The colored vertical bars show how the yearly increase in weight gain slows its pace over the years.

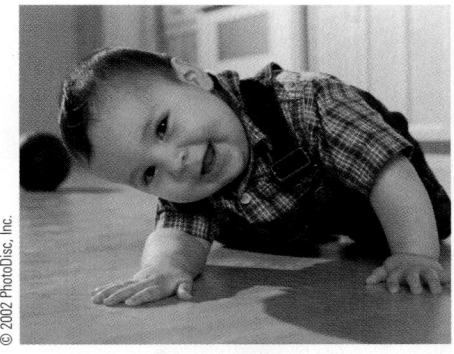

After six months of age, the energy saved by slower growth is spent on increased activity.

TABLE 13·6 INFANT AND ADULT HEART RATE, RESPIRATION RATE, AND ENERGY NEEDS COMPARED		
	Infants	Adults
Heat rate (beats/minutes)	120 to 140	70 to 80
Respiration rate (breaths/minute)	20 to 40	15 to 20
Energy needs (cal/body weight)	45/lb (100/kg)	<18/lb (<40/kg)

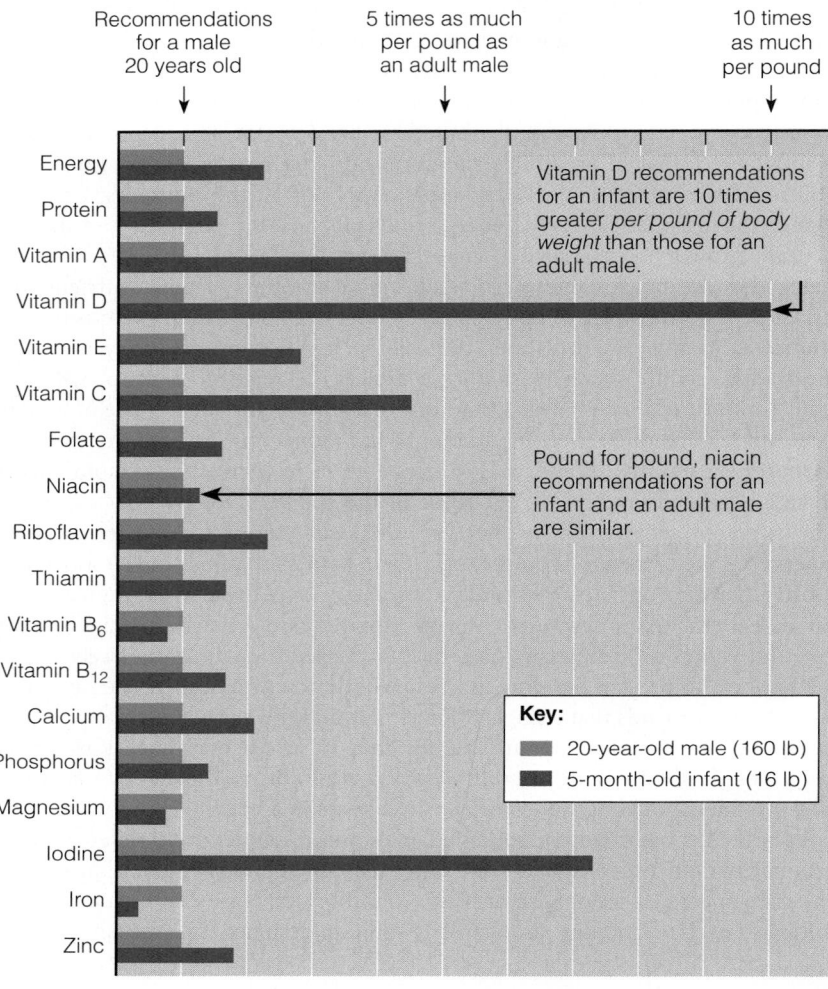

FIGURE 13·12 NUTRIENT RECOMMENDATIONS FOR A FIVE-MONTH-OLD INFANT AND AN ADULT MALE COMPARED ON THE BASIS OF BODY WEIGHT

Infants may be relatively small and inactive, but they use large amounts of energy and nutrients in proportion to their body size to keep all their metabolic processes going.

Recommendations for a male 20 years old

5 times as much per pound as an adult male

10 times as much per pound

Energy
Protein
Vitamin A
Vitamin D
Vitamin E
Vitamin C
Folate
Niacin
Riboflavin
Thiamin
Vitamin B$_6$
Vitamin B$_{12}$
Calcium
Phosphorus
Magnesium
Iodine
Iron
Zinc

Vitamin D recommendations for an infant are 10 times greater *per pound of body weight* than those for an adult male.

Pound for pound, niacin recommendations for an infant and an adult male are similar.

Key:
- 20-year-old male (160 lb)
- 5-month-old infant (16 lb)

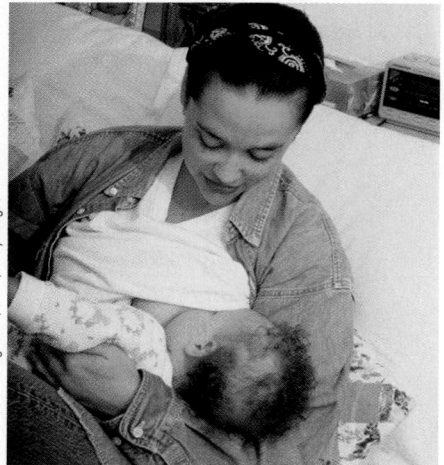

Breastfeeding is a natural extension of pregnancy—the mother's body continues to nourish the infant.

(ADA) advocates breastfeeding for the nutritional health it confers on the infant as well as for the physiological, social, economic, and other benefits it gives to the mother (see Table 13-7).[79] The AAP and the ADA recognize exclusive breastfeeding for 6 months, and breastfeeding with complementary foods for at least 12 months, as an optimal feeding pattern for infants.[80] All legitimate nutrition authorities share this view, but some makers of baby formula try to convince women otherwise—see the Consumer Corner on page 508.

Breast milk excels as a source of nutrients for the young infant. With the exception of vitamin D (discussed later), breast milk provides all the nutrients a healthy infant needs for the first six months of life.[81] Breast milk also conveys immune factors, which both protect an infant against infection and inform its body about the outside environment.

Energy Nutrients in Breast Milk The energy-nutrient balance of breast milk differs dramatically from that recommended for adults (see Figure 13-13). Yet, for infants, breast milk is the most nearly perfect food, affirming that people at different stages of life have different nutrient needs.

TABLE 13·7 BENEFITS OF BREASTFEEDING

For infants:
- Provides the appropriate composition and balance of nutrients with high bioavailability.
- Provides hormones that promote physiological development.
- Improves cognitive development.
- Protects against a variety of infections.
- May protect against some chronic diseases, such as diabetes (type 1) and hypertension, later in life.
- Protects against food allergies.

For mothers:
- Contracts the uterus.
- Delays the return of regular ovulation, thus lengthening birth intervals. (It is not, however, a dependable method of contraception.)
- Conserves iron stores (by prolonging amenorrhea).
- May protect against breast and ovarian cancer.

Other:
- Provides cost savings from not needing medical treatment for childhood illnesses or time off work to care for sick children.
- Provides cost savings from not needing to purchase formula (even after adjusting for added foods in the diet of a lactating mother).[a]
- Provides environmental savings to society from not needing to manufacture, package, and ship formula or dispose of packaging.

[a]A nursing mother produces over 35 gallons of milk during the first six months, saving her roughly $450 in formula costs.

FIGURE 13·13 PERCENTAGES OF ENERGY-YIELDING NUTRIENTS IN BREAST MILK AND IN RECOMMENDED ADULT DIETS

The proportions of energy-yielding nutrients in human breast milk differ from those recommended for adults.[a]

	Breast milk	Recommended adult diets
Protein	6%	12%
Fat	55%	30%
Carbohydrate	39%	58%

[a]The values listed for adults represent the midpoints of the acceptable ranges for protein (10 to 35 percent), fat (20 to 35 percent), and carbohydrate (45 to 65 percent).

The carbohydrate in breast milk (and standard infant formula) is lactose. Besides being easily digested, lactose enhances calcium absorption.

The lipids in breast milk—and infant formula—provide the main source of energy in the infant's diet. Breast milk contains a generous proportion of the essential fatty acids linoleic acid and linolenic acid, as well as their longer-chain derivatives, arachidonic acid and DHA (defined in Chapter 5). Formula makers have recently begun adding arachidonic acid and DHA to formulas.[82] Infants can produce some arachidonic acid and DHA from linoleic and linolenic acid, but some infants may need more than they can make.

Arachidonic acid and DHA are found abundantly in the brain and the retina of the eye.[83] Breastfed infants generally score higher on tests of mental development than formula-fed infants do, and researchers are investigating whether this difference can be attributed to DHA and arachidonic acid in breast milk, and whether adding these lipids to infant formula will close the gap.[84]

The protein in breast milk is largely **alpha-lactalbumin,** a protein the human infant can easily digest. Another breast milk protein, **lactoferrin,** is an iron-gathering compound that helps absorb iron into the infant's bloodstream, keeps intestinal bacteria from getting enough iron to grow out of control, and kills certain bacteria.[85]

Vitamins and Minerals in Breast Milk With the exception of vitamin D, the vitamin content of the breast milk of a well-nourished mother is ample. Even vitamin C, for which cow's milk is a poor source, is supplied generously. The concentration of vitamin D in breast milk is low, however, and vitamin D deficiency impairs bone mineralization. Vitamin D deficiency is most likely in infants who are not exposed to sunlight daily, have darkly pigmented skin, and receive breast milk without vitamin D supplementation.[86] Reports of infants in the United States developing the vitamin D–deficiency disease rickets and recommendations by the AAP to keep infants under six months of age out of direct sunlight have prompted new vitamin D guidelines. The AAP now recommends a vitamin D supplement for all infants who are breastfed exclusively and for those who do not receive at least 500 milliliters (15 ounces) per day of vitamin D–fortified formula.[87]

As for minerals, the calcium content of breast milk is ideal for infant bone growth, and the calcium is well absorbed. Breast milk is also low in sodium. The limited

alpha-lactalbumin (lact-AL-byoo-min) the chief protein in human breast milk. The chief protein in cow's milk is *casein* (CAY-seen).

lactoferrin (lack-toe-FERR-in) a factor in breast milk that binds iron and keeps it from supporting the growth of the infant's intestinal bacteria.

amount of iron in breast milk is highly absorbable, and its zinc, too, is absorbed better than from cow's milk, thanks to the presence of a zinc-binding protein.

After four to six months, an exclusively breastfed baby needs iron. Before four months, supplemental iron is unnecessary. Most babies are born with enough iron in their livers to last about half a year, and iron deficiency is rarely seen in very young infants. As lactation progresses, the iron in breast milk dwindles, making iron a concern for the four- to six-month-old. By six months, feeding the breastfed infant iron-fortified cereals is desirable. If the water supply is severely deficient in fluoride, both breastfed and formula-fed infants require fluoride supplementation after six months of age.

Immune Factors in Breast Milk   Breast milk offers an infant unsurpassed protection against infection by providing antiviral agents, antibacterial agents, and infection inhibitors.[88] For example, immune factors in breast milk interfere with the growth of bacteria that could otherwise attack the infant's vulnerable digestive tract linings. Breastfed babies are less prone to develop stomach and intestinal disorders during the first few months of life and so experience less vomiting and diarrhea than formula-fed babies.[89] Breast milk contains antibodies and other factors against the most common cause of diarrhea in infants and young children (rotavirus).[*] Breastfeeding reduces the severity and duration of symptoms associated with this infection.

Breastfeeding also protects against other common illnesses of infancy such as middle ear infection and respiratory illness.[90] Breast milk may also offer protection against the development of cardiovascular disease. Compared with formula-fed infants, breastfed infants have lower blood cholesterol as adults.[91]

During the first two or three days of lactation, the breasts produce **colostrum,** a premilk substance containing antibodies and white cells from the mother's blood. Because it contains immunity factors, colostrum helps protect the newborn infant from those infections against which the mother has developed immunity—precisely those in the environment likely to infect the infant. Maternal antibodies from colostrum inactivate harmful bacteria within the infant's digestive tract. Later, breast milk also delivers antibodies, although not as many as colostrum.

In addition to their protective features, colostrum and breast milk contain hormones and other factors that stimulate the development of the infant's digestive tract. Clearly, breast milk is a very special substance.

Other Potential Benefits   Researchers are investigating whether breastfeeding may also help protect against obesity in childhood and later years. This promising research is described in the Controversy section at the end of the chapter.

As suggested earlier, breastfeeding may also have a positive effect on later intelligence.[92] In one study, young adults who had been breastfed as long as nine months scored higher on two different intelligence tests than those who had been breastfed less than one month. Many other studies suggest a beneficial effect of breastfeeding on intelligence, but when subjected to strict standards of methodology (for example, large sample size and appropriate intelligence testing), the evidence is less convincing.[93] Nevertheless, the possibility that breastfeeding may positively affect later intelligence is intriguing. It may be that some specific component of breast milk, such as DHA, contributes to brain development or that certain factors associated with the feeding process itself promote intellect.

▶ KEY POINT   Breast milk is the ideal food for infants because it provides the needed nutrients in the right proportions and protective factors as well.

colostrum (co-LAHS-trum) a milklike secretion from the breasts during the first day or so after delivery before milk appears; rich in protective factors.

---
[*]More children are hospitalized for rotavirus infection than for any other single cause.

TABLE 13·8 HUMAN MILK COMPARED WITH INFANT FORMULA FOR SELECTED NUTRIENTS		
Content	Mature Human Milk	Fortified Infant Formula
Energy (cal/L)	680	680
Protein (% of cal)	6	8
Fat (% of cal)	50	50
Carbohydrate (% of cal)	42	43
Iron (mg/L)	0.5	4–12
Vitamin A (µg/L)	675	660
Niacin (mg/L)	1.5	7.1
Vitamin D (µg/L)	0.5	10
Inositol (mg/L)	149	32

SOURCE: Committee on Nutrition, American Academy of Pediatrics, *Pediatric Nutrition Handbook*, 5th ed., ed. R. E. Kleinman (Elk Grove, Ill.: American Academy of Pediatrics, 2004), Appendix E.

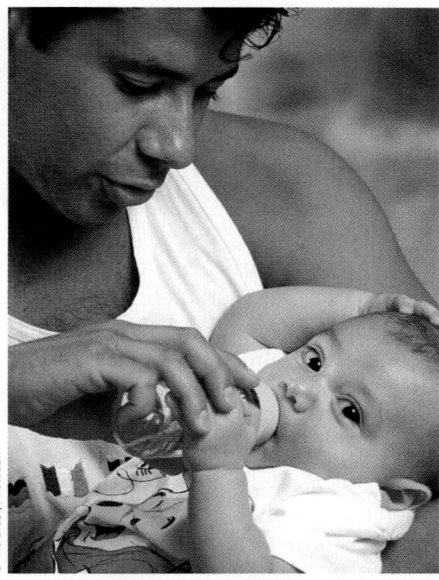

The infant thrives on formula offered with affection.

Formula Feeding

The substitution of formula feeding for breastfeeding involves striving to copy nature as closely as possible. Human milk and cow's milk differ; cow's milk is significantly higher in protein, calcium, and phosphorus, for example, to support the calf's faster growth rate. Thus, to prepare a formula from cow's milk, the formula makers must first dilute the milk and then add carbohydrate and nutrients to make the proportions comparable to those of human milk (see Table 13-8 for a comparison of human milk and standard formulas).

Standard formulas are inappropriate for some infants. For example, premature babies require special formulas, and infants allergic to milk protein can drink special **hypoallergenic formulas** or formulas based on soy protein.[94] Soy formulas are lactose-free and so can be used for infants with lactose intolerance; they are also useful as an alternative to milk-based formulas for vegetarian families. For infants with other special needs, many other variations are available.

The AAP recommends iron-fortified formulas for all formula-fed infants.[95] Low-iron formulas have no role in infant feeding. Use of iron-fortified formulas has risen in recent decades and is credited with the decline of iron-deficiency anemia in U.S. infants.[96]

Formula feeding offers an acceptable alternative to breastfeeding. Nourishment for an infant from formula is adequate, and parents can choose this course with confidence. One advantage is that parents can see how much milk the infant drinks during feedings. Another is that other family members can participate in feeding sessions, giving them a chance to develop the special closeness that feeding fosters. Mothers who return to work early after giving birth may choose formula for their infants, but they have another option. Breast milk can be pumped into bottles and given to the baby in day care. At home, mothers may breastfeed as usual. Many mothers use both methods—they breastfeed at first but wean to formula later on.

For as long as breast milk or formula is the baby's major food (until the first birthday), unmodified cow's milk is an inappropriate replacement because milk provides little iron and vitamin C. If an infant's digestive tract is sensitive to the protein content of cow's milk, it may bleed and worsen iron deficiency. Thus, plain cow's milk both causes iron loss and fails to replace iron. Also, the infant's immature kidneys are stressed by plain cow's milk. Once the baby is obtaining at least two-thirds of total daily food energy from a balanced mixture of cereals, vegetables, fruits, and other foods (usually after 12 months of age), whole cow's milk, fortified with vitamins A and D, is an acceptable accompanying beverage. Reduced-fat milk is not recommended before the age of two years.

Formula options:

- Liquid concentrate (inexpensive, relatively easy)—mix with equal part water.
- Powdered formula (cheapest, lightest for travel)—follow label directions.
- Ready-to-feed (easiest, most expensive)—pour directly into clean bottles.

Never an option—whole cow's milk before 12 months of age.

With the first birthday comes the possibility of tasting whole, unmodified cow's milk for the first time.

hypoallergenic formulas clinically tested infant formulas that do not provoke reactions in 90% of infants or children with confirmed cow's milk allergy.

▶ **KEY POINT** Infant formulas are designed to resemble breast milk and must meet an AAP standard for nutrient composition. Special formulas are available for premature infants, allergic infants, and others. Formula should be replaced with milk only after the baby's first birthday.

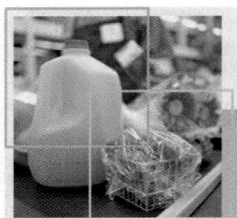

CONSUMER CORNER

Formula's Advertising Advantage

MOST WOMEN are free to choose whatever feeding method best suits their needs. For only a few is breastfeeding either prohibited for medical reasons or medically indicated for special needs of the infant. With the strong scientific consensus that breastfeeding is preferable for most infants, why do women who could breastfeed their infants choose formula? Some women find the time and logistics of breastfeeding burdensome. For many women, though, the decision to forgo breastfeeding is influenced by aggressive advertising of formulas.

Advertisers of infant formulas often strive to create the illusion that formula is identical to human milk. No formula can match the nutrients, agents of immunity, and environmental information conveyed to infants through human milk, but the ads are convincing: "Like mother's milk, our formula provides complete nutrition" or "Our brand is scientifically formulated to meet your baby's needs." Recently, the FDA warned consumers about an infant supplement misleadingly labeled with the words "Better than Formula." The product is *not* an approved infant formula and certainly is *not* better than approved formulas on the market. The label lists several ingredients that have not been evaluated for safety.[1] That this supplement with its misleading label reached market shelves and the hands of some consumers underscores the need for parents to scrutinize the labels of infant formulas and avoid any that include unusual ingredients.

To increase market share, formula manufacturers give coupons and samples of free formula to pregnant women. After childbirth, women in the hospital may receive "goody bags" with more coupons to tempt them to receive their "formula gifts." More coupons arrive by mail a couple of months later, at a time when many women give up breastfeeding, even though nutrition authorities urge continued breastfeeding for several more months. Aggressive marketing tactics can undermine a woman's confidence concerning her breastfeeding choice, and lack of confidence has a significant influence on early discontinuation of breastfeeding.[2]

National efforts to promote breastfeeding seem to be working to some extent.[3] Many hospitals employ certified lactation consultants who specialize in helping new mothers establish a healthy breastfeeding relationship with their newborns. Table 13–9 lists ten steps hospitals and birth centers can take to promote successful long-term breastfeeding. An encouraging trend of breastfeeding initiation is emerging, with 65 percent of women initiating breastfeeding today, up from 50 percent in 1990. Despite this trend toward increasing breastfeeding, the percentage of women breastfeeding their infants still falls short of the goal of *Healthy People 2010*. Lagging even further behind national goals are the number of infants breastfed beyond about two months of age.[4]

Formula-fed infants in developed nations are healthy and grow normally, but they miss out on the breastfeeding advantages described in the text. In developing nations, however, the consequence of choosing not to breastfeed can be tragic. Feeding formula is often fatal to the infant in nations where poverty limits access to formula mixes, clean water is unavailable for safe formula preparation, and medical help is limited. The World Health Organization (WHO) strongly supports breastfeeding for the world's infants in its "baby-friendly" initiative and opposes the marketing of infant formulas to new mothers.

Women are free to choose between breast and bottle, but the decision should be made by weighing valid factual information and not be influenced by sophisticated advertising ploys.

TABLE 13·9 TEN STEPS TO SUCCESSFUL BREASTFEEDING

To promote breastfeeding, every maternity facility should:

- Develop a written breastfeeding policy that is routinely communicated to all health-care staff.
- Train all health-care staff in the skills necessary to implement the breastfeeding policy.
- Inform all pregnant women about the benefits and management of breastfeeding.
- Help mothers initiate breastfeeding within ½ hour of birth.
- Show mothers how to breastfeed and how to maintain lactation, even if they need to be separated from their infants.
- Give newborn infants no food or drink other than breast milk, unless medically indicated.
- Practice rooming-in, allowing mothers and infants to remain together 24 hours a day.
- Encourage breastfeeding on demand.
- Give no artificial nipples or pacifiers to breastfeeding infants.[a]
- Foster the establishment of breastfeeding support groups and refer mothers to them at discharge from the facility.

[a]Compared with nonusers, infants who use pacifiers breastfeed less frequently and stop breastfeeding at a younger age.
SOURCE: United Nations Children's Fund, the World Health Organization, the Breastfeeding Hospital Initiative Feasibility Study Expert Work Group, and Baby Friendly U.S.A.

An Infant's First Foods

Foods can be introduced into the diet as the infant becomes physically ready to handle them. This readiness develops in stages. A newborn can swallow only liquids that are well back in the throat. Later (at four months or so), the tongue can move against the palate to swallow semisolid food such as cooked cereal. The stomach and intestines are immature at first; they can digest milk sugar (lactose) but not starch. At about four months, most infants can begin to digest starchy foods. Still later, the first teeth erupt, but not until sometime during the second year can a baby begin to handle chewy food.

When to Introduce Solid Food   Infants who are ready for solid foods thrive on receiving them and develop new skills through handling the foods. Indications of readiness for solid foods include:

- The infant can sit with support and can control its head movements.
- The infant is about six months old.

Infants develop according to their own schedules, and although Table 13-10 presents a suggested sequence, individuality is important. Three considerations are relevant: the baby's nutrient needs, the baby's physical readiness to handle different forms of foods, and the need to detect and control allergic reactions. With respect to nutrient needs, the nutrient needed most is iron, then vitamin C.

TABLE 13·10   INFANT DEVELOPMENT AND RECOMMENDED FOODS

Note: Because each stage of development builds on the previous stage, the foods from an earlier stage continue to be included in all later stages.

Age (mo)	Feeding Skill	Foods Introduced into the Diet
0–4	Turns head toward any object that brushes cheek. Initially swallows using back of tongue; gradually begins to swallow using front of tongue as well. Strong reflex (extrusion) to push food out during first 2 to 3 months.	Feed breast milk or infant formula.
4–6	Extrusion reflex diminishes, and the ability to swallow nonliquid foods develops. Indicates desire for food by opening mouth and leaning forward. Indicates satiety or disinterest by turning away and leaning back. Sits erect with support at 6 months. Begins chewing action. Brings hand to mouth. Grasps objects with palm of hand.	Begin iron-fortified cereal mixed with breast milk, formula, or water. Begin pureed vegetables and fruits.
6–8	Able to feed self with fingers. Develops pincher (finger to thumb) grasp. Begins to drink from cup.	Begin mashed vegetables and fruits. Begin plain baby food meats. Begin plain, unsweetened fruit juices from cup.
8–10	Begins to hold own bottle. Reaches for and grabs food and spoon. Sits unsupported.	Begin breads and cereals from table. Begin yogurt. Begin pieces of soft, cooked vegetables and fruit from table. Gradually begin finely cut meats, fish, casseroles, cheese, eggs, and legumes.
10–12	Begins to master spoon, but still spills some.	Include breads and cereals from table, in addition to infant cereal; soft or cooked fruits and vegetables; and finely chopped or ground meat, fish, or poultry; eggs; and mashed legumes.[a]

[a]Portions of foods for infants and young children are smaller than those for an adult. For example, a grain serving might be ½ slice of bread instead of 1 slice, or ¼ cup rice instead of ½ cup.
SOURCE: Adapted in part from Committee on Nutrition, American Academy of Pediatrics, *Pediatric Nutrition Handbook,* 5th ed., ed. R. E. Kleinman (Elk Grove Village, Ill.: American Academy of Pediatrics, 2004), pp. 103–115.

Foods such as iron-fortified cereals and formulas, mashed legumes, and strained meats provide iron.

To prevent choking, do not give infants or young children:

- Carrots.
- Cherries.
- Gum.
- Hard or gel-type candies.
- Hot dog slices.
- Marshmallows.
- Nuts.
- Peanut butter.
- Popcorn.
- Raw celery.
- Whole beans.
- Whole grapes.

Keep these nonfood items out of their reach:

- Balloons.
- Coins.
- Pen tops.
- Small balls.
- Other items of similar size.

■ Chapter 14 offers more information on allergies.

■ Appendix A includes the nutrient composition of commercial baby foods.

milk anemia iron-deficiency anemia caused by drinking so much milk that iron-rich foods are displaced from the diet.

Foods to Provide Iron and Vitamin C   Iron ranks highest on the list of nutrients most needing attention in infant nutrition. An infant's stored iron supply from before birth runs out after the birthweight doubles, long before the end of the first year. Iron deficiency is prevalent in children between the ages of six months and three years due to their rapid growth rate and the significant place that milk has in their diets. Excessive milk consumption (more than 3½ cups a day) can displace iron-rich foods and lead to iron-deficiency anemia, popularly called **milk anemia.**

To prevent iron deficiency, breast milk or iron-fortified formula, then iron-fortified cereals, and then meat or meat alternates such as legumes are recommended. Once infants are eating iron-fortified cereals, parents or caregivers should begin selecting vitamin C–rich foods to go with meals to enhance absorption. The best sources of vitamin C are fruits and vegetables.

Many fruit juices are naturally rich in vitamin C, and others may be fortified. A danger lies in overfeeding juice, however: some babies and young children may fail to grow and thrive when juice displaces other nutrient- and energy-dense foods from their diets. AAP recommendations set upper limits on juice consumption for infants and children at 4 to 6 ounces per day.[97] Fruit juices should be served in a cup, not a bottle, and not before the infant is six months of age.

Physical Readiness for Solid Foods   Foods introduced at the right times contribute to an infant's physical development. When the baby can sit up, can handle finger foods, and is teething, hard crackers and other finger foods may be introduced under the watchful eye of an adult. These foods promote the development of manual dexterity and control of the jaw muscles, but the caregiver must be careful that the infant does not choke on them. Babies and young children cannot safely chew and swallow any of the foods listed in the margin; they can easily choke on these foods, a risk not worth taking. Nonfood items of small size should always be kept out of the infant's reach to prevent choking.[98]

Some parents want to feed solids as early as possible on the theory that "stuffing the baby" at bedtime will promote sleeping through the night. There is no proof for this theory. Babies start to sleep through the night when they are ready, no matter when solid foods are introduced.

Food Allergies   New foods should be introduced one at a time so that allergies or other sensitivities can be detected. For example, when fortified baby cereals are introduced, try rice cereal first for several days; it causes allergy least often. Try wheat-containing cereal last; it is a common offender. Introduce egg whites, soy products, cow's milk, and citrus fruits still later for the same reason. If a food causes an allergic reaction (irritability due to skin rash, digestive upset, or respiratory discomfort), discontinue its use before going on to the next food. About nine times out of ten, the allergy won't be evident immediately but will manifest itself in vague symptoms occurring up to five days after the offending food is eaten. Wait a month or two to try the food again; many sensitivities disappear with maturity. If your family history indicates allergies, apply extra caution in introducing new foods. Parents or caregivers who detect allergies early in an infant's life can spare the whole family much grief.

Choice of Infant Foods   Commercial baby foods in the United States and Canada are safe, and except for mixed dinners with added starch fillers and heavily sweetened desserts, they have high nutrient density. Brands vary in their use of starch and sugar—check the ingredients lists. Parents or caregivers should not feed directly from the jar—remove portions to a dish for feeding in order not to contaminate the leftovers that will be stored in the jar.

An alternative to commercial baby food is to process a small portion of the family's table food in a blender, food processor, or baby food grinder. This necessitates cooking without salt or sugar, though, as the best baby food manufacturers do. Adults can season their own food after taking out the baby's portion. Pureed food can be frozen in an ice cube tray to yield a dozen or so servings that can be quickly thawed, heated, and served on a busy day.

Foods to Omit   Sweets of any kind (including baby food "desserts") have no place in a baby's diet. The added food energy can promote obesity, and they convey few or no nutrients to support growth. Products containing sugar alcohols such as sorbitol should also be limited, as these may cause diarrhea. Canned vegetables are inappropriate for babies because they often contain too much salt. Awareness of foodborne illness and precautions against it are imperative. Honey should never be fed to infants because of the risk of botulism.

■ Chapter 12 provides details about botulism.

Foods at One Year   For the infant weaned to whole milk after one year of age, whole milk can supply most of the needed nutrients: 2 to 3½ cups a day meet those needs. A variety of other foods—meat and meat alternates, iron-fortified cereal, enriched or whole-grain bread, fruits, and vegetables—should be supplied in amounts sufficient to round out total energy needs. Ideally, the one-year-old sits at the table, eats many of the same foods everyone else eats, and drinks liquids from a cup, not a bottle. A meal plan that meets the requirements for a one-year-old is shown in Table 13-11.

▶ KEY POINT   Solid food additions to an infant's diet should begin at about six months and should be governed by the infant's nutrient needs and readiness to eat. By one year, the baby should be receiving foods from all food groups.

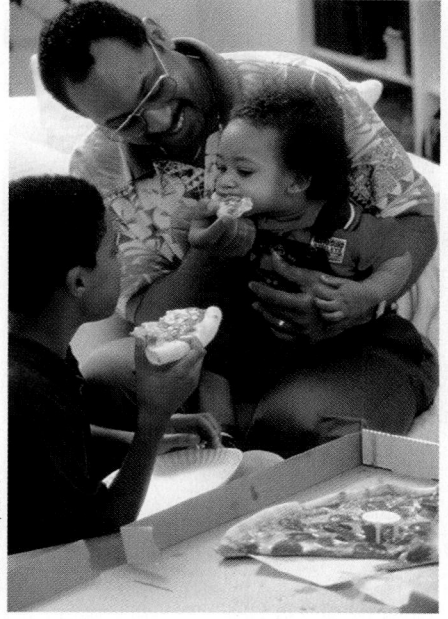

© Jose L. Pelaez/Corbis

Children love to eat what their families eat.

Looking Ahead

The first year of life is the time to lay the foundation for future health. From the nutrition standpoint, the problems most common in later years are obesity and dental disease. Prevention of obesity may also help prevent the obesity-related diseases: atherosclerosis, diabetes, and cancer.

The most important single measure to undertake during the first year is to encourage eating habits that will support continued normal weight as the child grows. This means introducing a variety of nutritious foods in an inviting way, not forcing the baby to finish the bottle or baby food jar, avoiding concentrated sweets and empty-calorie foods, and encouraging physical activity. Parents should not teach babies to seek food as a reward, to expect food as comfort for unhappiness, or to associate food deprivation with punishment. If they cry for thirst, give them water, not milk or juice. If they cry for companionship, pick them up—don't feed them. If they are hungry, by all means, feed them appropriately. More pointers are offered in this chapter's Food Feature.

TABLE 13·11 MEAL PLAN FOR A ONE-YEAR-OLD	
Breakfast	**Afternoon snack**
½ c whole milk	½ c whole milk
½ c iron-fortified cereal	½ slice toast
½ c orange juice	1 tbs apple butter
Morning snack	**Dinner**
½ c yogurt	1 c whole milk
¼ c fruit[a]	2 oz chopped meat or well-cooked mashed legumes
Lunch	¼ c potato, rice, or pasta
½ c whole milk	½ c vegetables[b]
½ c vegetables[b]	¼ c fruit[a]
1 egg or ¼ c tofu	
½ c noodles	

[a]Include citrus fruits, melons, and berries.
[b]Include dark green, leafy, and deep yellow vegetables.

Nursing bottle syndrome in an early stage.

Nursing bottle syndrome, an extreme example. The upper teeth have decayed all the way to the gum line.

An irrational fear of obesity leads some parents to underfeed their infants, depriving them of the energy and nutrients they need to grow. Others wonder if they should feed their infants a low-fat diet to reduce heart disease risk, but the AAP recommends fat intakes of 40 to 50 percent of total calories for infants. A diet too low in fat hinders growth and development even when energy from carbohydrate and protein is ample. With rare exceptions, to be identified by physicians, babies from age one to two years need the food energy and fat of whole milk. They also need frequent servings of food containing the essential fatty acids.

The same strategies promote normal dental development: supply nutritious foods, avoid sweets, and discourage the association of food with reward or comfort. Dentists strongly discourage the practice of giving a baby a bottle as a pacifier. Sucking for long periods of time pushes the normal jawline out of shape and causes a bucktoothed profile: protruding upper and receding lower teeth. Prolonged sucking on a bottle of milk or juice also bathes the upper teeth in a carbohydrate-rich fluid that favors the growth of bacteria that produce acid that dissolves tooth material. Babies regularly put to bed with a bottle sometimes have teeth decayed all the way to the gum line, a condition known as nursing bottle syndrome, shown in the margin photos.

▶ KEY POINT   The early feeding of the infant lays the foundation for lifelong eating habits. It is desirable to foster preferences that will support normal development and health throughout life.

FOOD FEATURE

Mealtimes with Infants

THE WISE PARENT or caregiver of a one-year-old offers nutrition and affection together. "Feeding with love" produces better growth in both weight and height than feeding the same food in an emotionally negative climate.

Foster a Sense of Autonomy

The person feeding a one-year-old has to be aware that the child's exploring and experimenting are normal and desirable behaviors. The child is developing a sense of autonomy that, if allowed to develop, will provide the foundation for later assertiveness in choosing when and how much to eat and when to stop eating. The child's self-direction, if consistently overridden, can later turn into shame and self-doubt.

Some Feeding Guidelines

In light of the developmental and nutrient needs of one-year-olds and in the face of their often contrary and willful behavior, a few feeding guidelines may be helpful:

- Discourage unacceptable behavior (such as standing at the table or throwing food) by removing the child from the table to wait until later to eat. Be consistent and firm, not punitive. The child will soon learn to sit and eat.
- Let the child explore and enjoy food. This may mean the child eats with fingers for a while. Use of the spoon will come in time.
- Don't force food on children. Provide children with nutritious foods, and let them choose which ones and how much they will eat. Gradually, they will acquire a taste for different foods. If children refuse milk, provide cheese, cream soups, and yogurt.
- Limit sweets strictly. Infants have little room in their 1,000-calorie daily energy allowance for empty-calorie sweets.

These recommendations reflect a spirit of tolerance that best serves the emotional and physical interests of the infant. This attitude, carried throughout childhood, helps the child to develop a healthy relationship with food. The next chapter finishes the story of growth and nutrition.

Nutrition Connections

The Nutrition Connections CD provides
access to these additional resources.

Interaction
The animation, *Nutrient Needs of Women and Infants,* explores
how pregnancy and lactation alter the nutrient needs of women,
and examine the nutrient needs of infants.

Chapter Quiz
Take the quiz for this chapter to assess your understanding of
chapter concepts.

Nutrition on the Net
Annotated links to reputable nutrition-related websites.

Online Glossary
Test your knowledge of key vocabulary through this comprehensive, interactive glossary.

Internet Connections

A variety of helpful study tools for this chapter
are available at our website
http://nutrition.wadsworth.com/ncc10

Practice Test
Take the practice test for this chapter and gauge your knowledge of key concepts; answers are provided.

Student Notebook
A lecture notebook is available to help organize note taking
while reading or attending lecture.

Study Tools
Study resources for students include Chapter Summaries, Crossword puzzles, Flashcards, Glossary, Internet Exercises, and Learning Objectives.

SELF CHECK

*Answers to these Self Check questions are in
Appendix G.*

1. A pregnant woman needs an extra 450
   calories above the allowance for nonpregnant women during which trimester(s)?
   a. first
   b. second
   c. third
   d. first, second, and third

2. A deficiency of which nutrient appears to
   be related to an increased risk of neural
   tube defects in the newborn?
   a. vitamin B$_6$
   b. folate
   c. calcium
   d. niacin

3. Which of the following preventative
   measures should a pregnant woman
   take to avoid contracting listeriosis?
   a. avoid feta cheese
   b. avoid pasteurized milk
   c. thoroughly heat hot dogs
   d. (a) and (c)

4. Breastfed infants may need
   supplements of:
   a. fluoride, iron, and vitamin D
   b. zinc, iron, and vitamin C
   c. vitamin E, calcium, and fluoride
   d. vitamin K, magnesium. and potassium

5. Which of the following foods poses a
   choking hazard to infants and small
   children?
   a. pudding
   b. marshmallows
   c. hot dog slices
   d. (b) and (c)

6. Sweets of any kind (including baby food
   "desserts") have no place in a baby's diet.
   T F

7. A major reason why a woman's nutrition
   before pregnancy is crucial is that it determines whether her uterus will support the
   growth of a normal placenta.
   T F

8. Fetal alcohol syndrome (FAS) is the leading known cause of mental retardation in
   the world.   T F

9. In general, the effect of nutritional deprivation on a breastfeeding mother is to
   reduce the quality of her milk.   T F

10. A sure way to get a baby to sleep through
    the night is to feed solid foods as soon as
    the baby can swallow them.   T F

CONTROVERSY 13

CHILDHOOD OBESITY AND EARLY DEVELOPMENT OF CHRONIC DISEASES

WHEN MOST PEOPLE think of health problems in children and adolescents, they most often think of measles and acne, not type 2 diabetes and hypertension. Today, however, unprecedented numbers of U.S. children are being diagnosed with obesity and the serious "adult diseases" such as type 2 diabetes that accompany overweight.[*1] U.S. children are not alone—galloping rates of obesity are injuring the health of an alarming number of children around the globe.[2]

This trend bodes ill for some 60 million children who, without immediate intervention, are destined to suffer type 2 diabetes and hypertension in childhood followed by cardiovascular disease (CVD) in young adulthood. Most overweight children become overweight adolescents and adults, and face increased risks of many chronic diseases.[3] Without preventive measures, many will die young of their disease.

For the sake of today's children and of future societies, prevention or treatment of childhood obesity is of critical importance. Yet, despite numerous advances in the prevention and cure of other childhood diseases, such as measles and even some forms of leukemia, reversing obesity remains an unanswered challenge.[4]

The Childhood Obesity Problem

Obesity, high blood cholesterol, and hypertension stand with diabetes at the top of the list of factors associated with the development of CVD (recall from Chapter 11). When these conditions appear in childhood, CVD may set in soon afterward—and much sooner than most people expect. Education is urgently needed in this regard. Most overweight children and their parents discount these deadly threats, focusing instead on appearance and the social costs of obesity.[5]

*Reference notes are found in Appendix F.

Characteristics of Childhood Obesity

Children are heavier today than they were 30 or so years ago. Since the late 1970s, the prevalence of overweight has doubled for children and shows no sign of slowing (see Figure C13-1). Typically, obese children:

- Are female.
- Have a family history of type 2 diabetes.
- Are of non-European descent.
- Were born to mothers who had diabetes while pregnant with them.
- Have metabolic syndrome, including hypertension, impaired glucose tolerance, elevated blood triglycerides, and reduced HDL cholesterol.[6]
- Have a low family income.
- Are more sedentary than their peers.[7]

Additionally, parental obesity predicts excessive weight gain during childhood and more than doubles the chance that a child will become an obese adult.

Influence of Genetics

For obesity, as well as for heart disease, hypertension, and type 2 diabetes, genetics does not appear to play a *determining* role; that is, most people are not simply destined at birth to develop them. Instead, genetics appears to play a largely *permissive role*—the potential is inherited and will then develop if given a push by factors in the environment such as poor diet, sedentary lifestyle, or cigarette smoking.

Genes help to govern the complex system of appetite control (described in Chapter 9).[8] If the tendency to consume lesser or greater amounts of food is inherited through the genes, then such genes may influence who among the pop-

ulation stays lean and who becomes obese. Other genes may influence whether an individual spends more or less energy in metabolism. Enzymes (proteins, produced by the genes) determine the body's ability to conserve its energy and store body fat. Such metabolic tendencies are probably in place from birth, but some researchers believe that genes alone are insufficient to explain such tendencies and are exploring whether events before and shortly after birth may affect their expression.

Fetal Theories of Obesity Development

A theory called *fetal programming* or *fetal origins of disease* states that maternal malnutrition or other harmful conditions at a critical period of fetal development could have lifelong effects on an individual's pattern of genetic expression and therefore on the tendency to develop obesity and certain diseases.[9]

An infant born with a high birthweight is likely to remain heavy into adulthood.[10] According to the fetal programming theory, *overnutrition* during fetal life may be to blame. A fetus with a rich supply of energy nutrients may adapt to its energy-rich environment by producing more of the enzyme systems active in forming and storing body fat. Later, when the fetus emerges as an independent individual, this overabun-

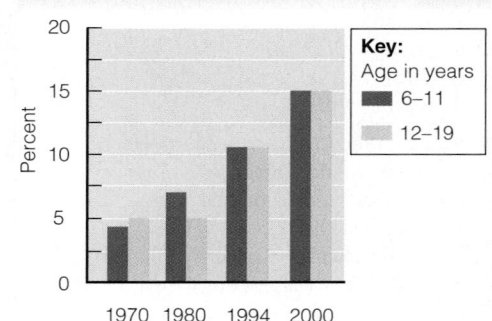

FIGURE C13·1 PREVALENCE OF OVERWEIGHT AMONG CHILDREN AND ADOLESCENTS

Key:
Age in years
6–11
12–19

SOURCE: National Center for Health Statistics, *NCHS Health E Stats—Prevalence of Overweight among Children and Adolescents: United States, 1999–2000,* available at www.cdc.gov/nchs/products/pubs/pubd/hestats/over99fig1.htm.

dance of fat-storing enzymes makes weight gain likely. This theory may also explain why infants born to women with uncontrolled diabetes during pregnancy generally have high birthweights—diabetes causes a buildup of energy fuels in the mother's bloodstream, thereby providing an energy-rich environment for the fetus.

A twist on the fetal obesity theory states that *underfed* fetuses may also adapt to their environment by becoming adept at conserving energy, but for different reasons from the overfed fetus.[11] An underfed fetus must maximize every molecule of energy nutrient it receives in order to grow and develop, so it becomes predominantly "thrifty" in its energy metabolism. Such a thrifty fetus would have an advantage if born into a famine environment, but when the newborn instead emerges into a food-rich world, the child faces a greatly increased risk for obesity and the many diseases it foreshadows.[12] When a female child who was starved in the womb then grows up to bear children of her own, they may also carry the thrifty traits, making the consequences of nutrition deprivation during pregnancy known across several generations.[13]

A related hypothesis also attributes chronic diseases to fetal origins and adverse circumstances shortly after birth.[†] Investigations into English birth records from the early twentieth century revealed that, as a group, people who were born small for date or who failed to thrive as infants suffered in later life from higher blood pressure, lower bone density, and higher rates of type 2 diabetes than other people. They were also more likely to have altered blood lipids, thicker left heart walls indicating circulatory difficulties, less elastic arteries, and even greater stress responses.[14] More studies are needed to confirm or refute these ideas.

Breastfeeding

Researchers are also exploring whether infant feeding methods influence later body fatness, but the findings are not yet clear on this point. In a well-controlled survey of over 15,000 adolescent children and their mothers, infants who were mostly breastfed for the first six months of life were less likely to be overweight than infants fed mostly formula.[15] An explanation may be that most breastfed infants spontaneously take in just enough energy to meet their needs and thus learn self-regulation of food intake at the earliest stage of life.[16] Bottle-fed infants may be encouraged to overfeed by anxious parents. In a study of much younger children (three to five years of age), however, researchers found no clear effect of the duration of exclusive breastfeeding on body weight.[17] It may be that the effect becomes evident only in later childhood, or perhaps some other factors are at work.

Energy Intake and Sedentary Behavior

Children learn food behaviors largely from their families, and entire families may be eating too much, dieting inappropriately, and exercising too little, a pattern often noted among mothers and daughters.[18] When researchers ask, "Are today's children consuming significantly more calories than those of 30 years ago?" the answer comes back, "Yes." Some researchers report an increase of 100 to 200 calories a day for all age groups, enough to account for significant weight gain.[19] The answer to another popular question, "Are children eating too many carbohydrates?" is a qualified "no"—children with the highest carbohydrate intakes, with the exception of added sugars, are leaner than other children.[20]

Research links added sugars, and especially high-fructose corn syrup—the easily consumed, energy-dense liquid sugar added to soft drinks—with excess body fatness in children. Controversy 4 pointed out that just over two cans of sugary soft drinks—an amount consumed every day by many adolescents—provide 400 calories.[21] According to one estimate, the risk of obesity increases by 60 percent with each daily sugary drink consumed by overweight children.[22]

Although the tremendous increase in soft drink consumption may play some role in the obesity epidemic, much of the epidemic can be explained by lack of exercise. Children have grown more sedentary, and sedentary children are more often overweight. A child who spends more than an hour or two in front of a television, computer monitor, or other media can become obese and develop unhealthy blood lipids even while eating fewer calories than a more active child (Figure C13-2, page 516).[23] Physically active children have higher HDL, lower LDL, and lower blood pressure than sedentary children, and these positive findings often persist into adulthood.

Early Development of Type 2 Diabetes

An estimated 85 percent of the children diagnosed with type 2 diabetes are obese.[24] Most are diagnosed during puberty, but as children become more obese and less active, the trend is shifting toward younger and younger children. Type 2 diabetes is most likely to attack those who are obese and sedentary and have a family history of diabetes.

Rates of type 2 diabetes vary widely among U.S. ethnic groups. For example, only 8 percent of white children are diagnosed with type 2 diabetes, but 45 percent of Pima Indian children of Arizona have the condition. African American, Asian American, and Hispanic children in the United States also frequently develop type 2 diabetes.

In type 2 diabetes, the cells become insulin-resistant—that is, insulin can no longer escort glucose from the blood into the cells. The combination of obesity and insulin resistance produces a cluster of symptoms, including high blood lipids and high blood pressure,

More and more children with obesity are being diagnosed with type 2 diabetes.

[†]This theory is named the *Barker hypothesis* after its originator.

FIGURE C13·2 PREVALENCE OF OBESITY BY HOURS OF TV PER DAY, CHILDREN AGES 10–15 YEARS

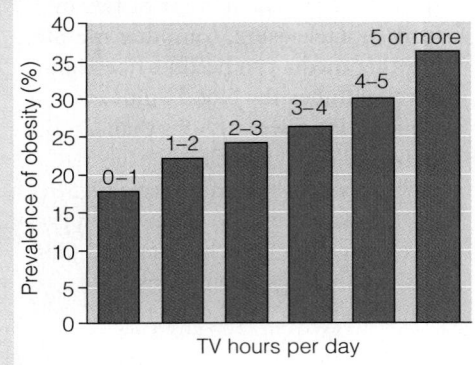

SOURCE: Centers for Disease Control and Prevention, Youth Risk Behavior Survey, 1999, available at www.cdc.gov.

which, in turn, promotes the development of atherosclerosis and the early development of heart disease.[25] Other common problems evident by early adulthood include kidney disease, blindness, and miscarriages. Details about diabetes are in Chapter 4.

Determining exactly how many children suffer from type 2 diabetes is tricky. For one thing, the symptoms of type 2 and type 1 diabetes differ only subtly in children. The child with type 2 diabetes may lack classic telltale diabetes symptoms, such as glucose in the urine, ketones in the blood, weight loss, and excessive thirst and urination, so the condition often advances undetected. Weight loss is particularly difficult to detect because childhood type 2 diabetes is associated with overweight and the child's obesity may mask any weight loss caused by the diabetes. Also, children's physicians expect to find type 1 diabetes (once called *juvenile-onset* diabetes), and not type 2 in children, making misdiagnosis likely when signs of type 2 diabetes present themselves. Undiagnosed diabetes means that children suffering with the condition are left undefended against its ravages.

Prevention and treatment of type 2 diabetes depend on weight management, which can be particularly difficult in a youngster's world of food advertising, video games, and pocket money for candy bars. The activity and dietary sug-

gestions to help defend against heart disease later in this Controversy apply to type 2 diabetes as well.

Early Development of Heart Disease

By adolescence, most children have formed fatty streaks in their coronary arteries. By early adulthood, the fibrous plaques may begin to calcify and become raised lesions causing arterial damage and blockage that make heart attacks and strokes likely. In short, the consequences of atherosclerosis, which become apparent only in adulthood, have their beginnings in the first decades of life.[26]

Children with the highest risks of developing heart disease are sedentary and obese, with diabetes, high blood pressure, and high blood cholesterol.[27] In contrast, children with the lowest risks of heart disease are physically active and of normal weight, with low blood pressure and favorable lipid profiles.

Blood Cholesterol

As blood cholesterol rises, atherosclerosis increases. Cholesterol values at birth are similar in all populations, but differences emerge in early childhood. Standard values for cholesterol in children and adolescents (ages 2 to 18 years) are listed in Table C13-1.

In general, blood cholesterol tends to rise as dietary saturated fat intakes increase. In recent years, for example, Japanese children have adopted a diet more like that of the United States—higher in saturated fat—and their blood cholesterol has increased.[28]

Blood cholesterol also correlates with childhood obesity, especially central obesity.[29] LDL cholesterol rises with obesity, and HDL declines. These relationships are apparent in childhood and increase in magnitude with age.

Children who are both overweight and have high blood cholesterol are likely to have parents who develop heart disease early.[30] For this reason, screening is recommended for children and adolescents whose parents (or grandparents) have heart disease; whose parents have elevated blood cholesterol; and whose family history is unavailable,

TABLE C13·1 CHOLESTEROL VALUES FOR CHILDREN AND ADOLESCENTS

Disease Risk	Total Cholesterol (mg/dL)	LDL Cholesterol (mg/dL)
Acceptable	<170	<110
Borderline	170–199	110–129
High	≥200	≥130

NOTE: Adult values appear in Chapter 11.

especially if other risk factors are evident.[31] Because blood cholesterol in children is a good predictor of their adult cholesterol, some experts recommend universal cholesterol screening for all children, and particularly for those who are overweight, smoke, are sedentary, or consume diets high in saturated fat.[32] Early—but not advanced—artery lesions are reversible, making screening and education a high priority.

Blood Pressure

Pediatricians routinely monitor blood pressure in children and adolescents. High blood pressure may signal an underlying disease or the early onset of hypertension. Hypertension may develop in the first decades of life, especially among obese children, and worsen with time.[33] Hypertension accelerates the development of atherosclerosis.

Breastfeeding and Hypertension

Every authority making recommendations for infants promotes breast milk as the ideal food for normal, healthy infants. A recent report ran counter to this wisdom, however, by reporting increased stiffness of the arteries in 20-year-old subjects who as infants had been exclusively breastfed for more than four months.[34] Less flexible arteries can lead to hypertension, elevating heart disease risk (see Chapter 11). Another report linked elevated blood cholesterol with breastfeeding for more than a year and also with bottle feeding alone.[35] No practical meaning can be derived from these findings, and parents should be assured that the benefits of breastfeeding far outweigh any suggestion of risk.

Controlling Hypertension

Children with hypertension can often make dramatic improvements by participating in regular aerobic activity and by losing weight or maintaining their weight as they grow taller. Evidence is needed to clarify whether restricting sodium in children's and adolescents' diets lowers blood pressure. It may be worth watching caffeine intakes, however. In one recent study, teens, and particularly African Americans, who drank two to four cans of caffeinated soft drinks each day had higher blood pressure readings than those drinking mostly other beverages.[36] No one yet knows if caffeine, soft drinks, or other factors in a lifestyle that includes such choices bear responsibility for this effect.

Prevention of hypertension may be of special urgency for adolescents because their response to blood pressure–lowering medication may be less consistently favorable than in adults.[37] Standard values for hypertension in children and adolescents are given in Table C13-2.

Lifestyle Recommendations for Children

Because their health is imperiled, the most critical efforts to modify lifestyles focus on children who are already overweight.[38] Children with high blood cholesterol should first be treated with diet and physical activity. If blood cholesterol remains high in children ten years and older after 6 to 12 months of lifestyle intervention, then drugs can lower blood cholesterol without threatening growth or development.[39] First, though, an honest evaluation of the lifestyle choices affecting the child can be of help in planning necessary changes.

Diet Moderation, Not Deprivation

All children should eat a variety of foods and maintain a desirable weight, that much is clear. There is less agreement as to whether it is wise to restrict fat in a healthy child's diet, but most experts recommend that children over age two receive at least 25 percent and no more than 35 percent of total energy from fat.[40] Restricting saturated fat and *trans* fat may be especially beneficial. Such measures appear to improve blood

TABLE C13-2 BLOOD PRESSURE STANDARDS FOR CHILDREN AND ADOLESCENTS

	Systolic over Diastolic Pressure (mm Hg)			
	6 to 9 yr	10 to 12 yr	13 to 15 yr	16 to 18 yr
Mild hypertension	111–121 over 70–77	117–125 over 75–81	124–135 over 77–85	127–141 over 80–91
Moderate hypertension	122–129 over 70–85	126–133 over 82–89	136–143 over 86–91	142–149 over 92–97
Severe hypertension	>129 over >85	>133 over >89	>143 over >91	>149 over >97

lipids without compromising nutrient adequacy, physical growth, or neurological development.[41] Pediatricians warn parents and caregivers to avoid extremes, though; they caution that while intentions may be good, excessive fat restriction may create nutrient deficiencies, impair growth, and spark unnecessary battles about food.

Healthy children over age two can continue eating age-appropriate portion sizes, with low-fat choices and more fruits and vegetables replacing some high-fat foods.[42] Healthy meals should occasionally include moderate portions of a child's favorite foods, even if they are high-fat treats such as french fries or ice cream. A steady diet of offerings from some "children's menus" in restaurants, such as commercially fried chicken nuggets, hot dogs, and french fries, easily exceeds a prudent intake of saturated fat, *trans* fat, and calories, however, and invites both nutrient shortages and gains of body fat.[43] Most restaurant chains are changing children's menus to include steamed vegetables and broiled or grilled poultry—additions welcomed by busy parents who often dine out or purchase take-out foods.[44]

Other fatty foods, such as nuts, vegetable oils, and safer varieties of fish, such as light canned tuna or salmon, are important for their essential fatty acids. Low-fat milk products deserve special attention in a child's diet for the needed calcium and other nutrients they sup-

ply; in addition, some evidence supports a favorable effect of milk products on body weights of children.[45]

In Britain, educators taught schoolchildren to reduce their soft drink consumption, resulting in lower rates of overweight at several school locations.[46] In Hawaii, researchers determined that adolescent girls who consumed more milk had lower measures of body fatness than girls drinking more sugary soft drinks and punches.[47] The best beverage choice for dense bones, adequate nutrients, and perhaps even lean body composition is low-fat milk. Sugary soft drinks and punches are best limited to occasional treats.[48]

Physical Activity

Active children have a better lipid profile and lower blood pressure than more sedentary children. Additionally, the effects of combining a nutritious calorie-controlled diet with exercise can be seen in observable improvements in children's outer measures of health, such as reduced waist circumference and increased muscle strength, along with the inner benefits of greatly improved condition of the heart and arteries.[49]

Just as blood cholesterol and obesity track over the years, so does a youngster's level of physical activity. Inactive children are likely to grow up to be inactive adults. Similarly, those who are physically active now tend to remain so. Compared with inactive teens, those who are physically active weigh less, smoke less, eat diets lower in saturated fats, and have better blood lipid profiles. Conversely, teens who pass their spare time watching television or playing video games have higher rates of obesity and more atherogenic blood cholesterol. The message is clear: physical activity offers numerous health benefits, and children who are active today are most likely to be active for years to come.

Smoking

Smoking poses so serious a threat to children and adults that it deserves attention, even in a nutrition text. Each day 3,000 children light up for the first time—typically in *grade* school. Almost two out of three high school students have tried smoking, and one in seven smokes regularly.[50] Approximately 80 percent of all adult smokers began smoking before the age of 18.[51]

Of those teenagers who continue smoking, half will eventually die of smoking-related causes. Efforts to teach children about the dangers of smoking seem most effective when they focus on immediate health consequences, such as shortness of breath when playing sports, or other consequences, such as ruined clothing, spent pocket money, and bad breath. Whatever the context, the message to all children and teens should be clear: "Don't start smoking. If you've already started, quit now."

Moving Toward Solutions

Treatment of established obesity is notoriously unsuccessful, making prevention of childhood obesity and the diseases it engenders a high national priority.[52] Exactly how to proceed in this regard remains uncertain, however.[53] Education is clearly needed. Classroom lessons that are reinforced by lunchroom offerings and other school policies often help children change their eating habits for the better.[54] When school nutrition policy conflicts with classroom teachings and adults set poor

---

TABLE C13·3   U.S. SURGEON GENERAL'S SCHOOL POLICIES FOR NUTRITION AND PHYSICAL ACTIVITY

A report of the U.S. Surgeon General recommends that school administrators employ these actions to achieve a unified approach for promoting a healthy body weight in schoolchildren.

- Provide age-appropriate, culturally sensitive instruction in health education with an emphasis on helping students develop lifelong healthy eating habits.
- Ensure that school breakfasts and lunches meet nutrient intake standards.
- Ensure that all food and beverages available on school grounds and at school events are consistent with the *Dietary Guidelines for Americans*.
- Provide food options that are low in fat, calories, and added sugars, such as fruits, vegetables, whole grains, and low-fat or fat-free milk products.
- Provide healthful snacks and foods in vending machines, school stores, and other venues within the school's control.
- Prohibit student access to vending machines, school stores, and other venues that compete with healthful school meals in elementary schools; restrict access in higher grades.
- Provide adequate time for breakfast and lunch at reasonable times of day.

In addition, the Surgeon General recommends these policies to promote physical activity:

- Provide quality physical education to all children grades kindergarten through grade 12 to develop lifelong physical activity.
- Provide daily recess periods in elementary schools.
- Provide extracurricular physical activity programs, especially intramural programs and physical activity clubs.
- Encourage use of school facilities for organized physical activity programs outside school hours.

SOURCE: U.S. Surgeon General's Call to Action to Prevent and Decrease Overweight and Obesity, 2001, available at www.surgeongeneral.gov.

---

examples, however, children are likely to follow their taste buds, not nutrition teachings. Table C13-3 offers some actions schools can take for an effective, unified approach to nutrition.

Children are an appropriate focus of national obesity concerns, but to help them, help must also reach the adults in their lives who are growing fatter and sicker themselves. Adults not only feed this and future generations of children but also set examples that children almost invariably follow.[55]

To tackle the problem in earnest may require looking honestly at global trends, such as greater accessibility to large portions of inexpensive, highly palatable, high-calorie foods (especially in schools); greater use of automobiles; and diminishing incentives to be physically active (Controversy 11 shed light on these issues). Recently, the federal government moved to make more of

the nation's fruit and vegetable bounty available to schoolchildren at lunch time, a small but positive step. Schools would also do well to provide sound nutrition and physical education for every able child in their care.[56]

It is tempting to look for simple answers to complex problems, but in truth, if everyone today would take up the advice of every legitimate health agency worldwide, many of these problems would disappear. Today, the best advice remains the easiest to give, and perhaps the most difficult to follow: don't smoke, choose a diet in accord with the *Dietary Guidelines for Americans 2005*, follow the USDA Food Guide (Chapter 2), and make it a habit to be physically active each day. In the end, parents and other significant adults can help to mold children's behaviors by the examples they set.[57]

Child, Teen, and Older Adult

Paul Gauguin (1848–1903 French), *The Meal (The Bananas)* 1891, Musée d'Orsay, Paris, France. SuperStock, Inc.

CONTENTS

Do You Ever . . .

Or will you ever provide nourishment to children?

Suspect that symptoms you feel may be from a food allergy?

Think that teenagers are old enough to decide for themselves what to eat?

Wonder whether nutrition can help you live longer?

KEEP READING.

TO GROW AND to function well in the adult world, children need a firm background of sound eating habits, which begin during babyhood with the introduction of solid foods. At that point the person's nutrition story has just begun; the plot thickens. Nutrient needs change throughout life into old age, depending on the rate of growth, gender, activities, and many other factors. Nutrient needs also vary from individual to individual, but generalizations are possible and useful.

A great majority of children in the United States (81 percent) consume diets ranked "poor" or "needs improvement," as measured by the Healthy Eating Index (see Figure 14-1). The consequences of such diets may not be evident to the casual observer, but nutritionists know that nutrient deficiencies during growth often have far-reaching effects on physical and mental development.[*1] Anyone who cares about the children in their lives, now or in the future, would profit from knowing how to provide the nutrients children require to reach their potential.

■ The Healthy Eating Index is a research tool designed to assess diets as described in Chapter 2.

Early and Middle Childhood

Imagine growing 10 inches taller in just one year, as the average healthy infant does during the first dramatic year of life. At age one, infants have just learned to stand and toddle, and growth has slowed by half; by two years, they can take long strides with solid confidence and are learning to run, jump, and climb. These new accomplishments reflect the accumulation of a larger mass, greater density of bone and muscle tissue, and refinement of nervous system coordination. These same growth trends, a lengthening of the long bones and an increase in musculature, continue until adolescence, but unevenly and more slowly.

Mentally, too, the child is making rapid advances, and proper nutrition is critical to normal brain development. The child malnourished at age 3 often demonstrates

---

*Reference notes are found in Appendix F.

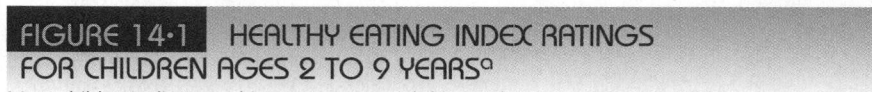

FIGURE 14·1   HEALTHY EATING INDEX RATINGS FOR CHILDREN AGES 2 TO 9 YEARS[a]

Most children's diets need improvement, and the need grows greater as children age.

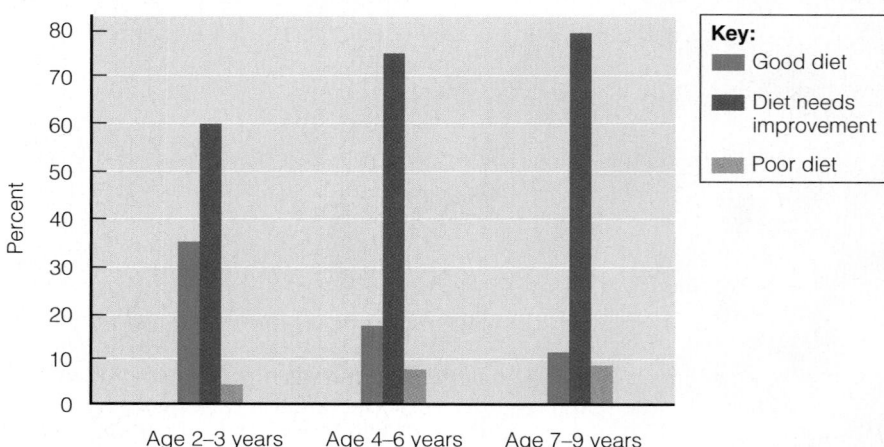

[a]The Healthy Eating Index assesses a diet's adherence to the principles of the Food Guide Pyramid, the *Dietary Guidelines for Americans*, and variety; see Chapter 2.

SOURCE: Data from USDA Center for Nutrition Policy and Promotion, Report card on the diet quality of children ages 2 to 9, *Nutrition Insights* 25 (September 2001), available at www.usda.gov/cnpp/Insights/Insight25.pdf.

FIGURE 14·2 COMPOSITION OF WEIGHT GAIN, INFANTS AND TODDLERS

These graphs demonstrate that a young infant deposits much more fat than lean tissue, but a toddler deposits more lean than fat. Water follows lean tissue, demonstrated by the water gains in the toddler. You can see that the body shape of a one year-old (left photo) changes dramatically by age two (right photo). The two-year-old has lost much baby fat; the muscles (especially in the back, buttocks, and legs) have firmed and strengthened; and the leg bones have lengthened.

Key:
  Protein
  Fat
  Water
  Other[a]

Weight gain of infants (age 0 to 6 months)
4% 11%
34%
51%

Weight gain of toddlers (age 12 to 24 months)
3% 19%
71% 7%

© Anthony Vannelli (both)

[a]"Other" consists of carbohydrate and minerals.

SOURCE: Data from K. L. McConahy and M. F. Picciano, How to grow a healthy toddler—12 to 24 months, *Nutrition Today* 38 (2003): 156–163.

diminished mental capacities at age 11, even when other life circumstances are comparable to those of peers.[2]

Feeding a Healthy Young Child

At no time in life does the human diet change faster than during the second year.[3] From 12 to 24 months, a child's diet changes from infant foods consisting of mostly formula or breast milk to mostly modified adult foods. This doesn't mean, of course, that milk loses its importance in the toddler's diet—it remains a central source of calcium, protein, and other nutrients. Nevertheless, the rapid growth and changing body composition (see Figure 14-2) during this remarkable period demand more nutrients than can be provided by milk alone. Further, the toddling years are marked by bustling activity made possible by new muscle tissue and refined neuromuscular coordination. To support both their activity and growth, toddlers need nutrients and plenty of them.

Energy and Protein An infant's appetite decreases markedly near the first birthday and fluctuates thereafter. At times children seem to be insatiable, and at other times they seem to live on air and water. Parents and other caregivers need not worry: given a selection of nutritious foods and limited treats, internal appetite regulation in children of normal weight guarantees that their overall energy intakes will be right for each stage of growth. This ideal situation depends upon the restriction of treats, however. Today's children too often consume foods high in added sugars, saturated fat, and calories.[4] Faced with a tempting array of such foods, children may disregard internal satiety signals and overconsume calories, inviting obesity. The *Dietary Guidelines for Americans 2005* are safe and appropriate goals for the diets of healthy children two years of age and older and do not compromise nutrient intakes or growth.[5]

■ The *Dietary Guidelines for Americans 2005* appear in Chapter 1. Details about appetite are found in Chapter 9.

Individual children's energy needs vary widely, depending on their growth and physical activity. A one-year-old child needs about 800 calories a day; at age six, the child needs about 800 calories more. By age ten, about 2,000 calories a day support normal growth and activity without causing excess storage of body fat. As children age, the total number of calories needed increases, but per pound of body weight, the need declines from the extraordinarily high demand of infancy.

As for protein, total needs increase slightly as a child grows larger. On a pound-for-pound basis, however, the older child's need for protein is actually slightly lower than the younger child's (see the DRI values, inside front cover). Children's protein needs are well covered by a typical U.S. diet.

Carbohydrate and Fiber Carbohydrate recommendations are based on glucose use by the brain. A one-year old's brain is large for the size of the body, so the 130 grams of glucose demanded by the one-year-old falls in the adult range (see inside front cover).[6] Fiber recommendations derive from adult intakes and should be adjusted downward for children who are picky eaters and take in little energy.[7] In general, children's fiber intakes should equal their "age plus 5 grams."[8]

Fat and Fatty Acids Keeping fat intake within bounds helps to control saturated fat and so may help protect children from developing early signs of adult diseases. Taken to extremes, however, a low-fat diet can lack essential nutrients and energy needed for growth. Children's small stomachs can hold only so much food, and fat provides a concentrated source of food energy needed for growth. In addition, essential fatty acids are critical to proper development of nerve, eye, and other tissues. The DRI recommended range for fat in a child's diet assumes that energy is sufficient.[9] Specific DRI recommendations are on the inside front cover.

Vitamins and Minerals As a child grows larger, so does the demand for vitamins and minerals. On a pound-for-pound basis, a five-year-old's need for, say, vitamin A is about double the need of an adult man (see the margin). A balanced diet of nutritious foods can meet children's needs for these nutrients, with the notable exception of iron. Iron deficiency is a major problem worldwide; it occurs in about 7 percent of U.S. toddlers one to two years of age.[10] During the second year of life, toddlers progress from a diet of iron-rich infant foods such as breast milk, iron-fortified formula, and iron-fortified infant cereal to a diet of adult foods and iron-poor cow's milk. At the same time, the stores of iron from birth diminish. Compounding the problem is the variability in toddlers' appetites; some two-year-olds are finicky while others prefer milk and juice and reject solid foods for a time.[11] All of these factors—switching to whole milk and unfortified foods, diminished iron stores, and unreliable food consumption—make iron deficiency likely at a time when iron is critically needed for normal brain growth and development. A later section comes back to iron deficiency and its consequences

To prevent iron deficiency, children's foods must deliver from 7 to 10 milligrams of iron per day. To achieve this goal, snacks and meals should include iron-rich foods. Milk intake, though critical for the calcium needed for dense healthy bones, should not exceed daily recommendations to avoid displacing lean meats, fish, poultry, eggs, legumes, and whole-grain or enriched grain products from the diet. Table 14-1 in the margin lists some iron-rich foods that many children like to eat.

The Food Guide Pyramid for Young Children The Food Guide Pyramid for Young Children, shown in Figure 14-3, displays one means of providing the needed nutrients to children. This plan designates a set number of servings from each food group rather than amounts in cups and ounces as for adults (see Chapter 2). Children two to six years old need at least the specified number of servings, but the serving sizes should vary according to age to keep calorie intakes moderate.[12] Children two to three years old should receive servings that are about two-thirds the size of an adult serving. Children in the four- to six-year-old group can generally eat the serving sizes recom-

---

The DRI range for total fat intakes for children:

- 30 to 40% of energy for children 1 to 3 years of age.
- 25 to 35% of energy for children 4 to 18 years of age.

■ A 174-pound adult male needs 900 mg of vitamin A, or 5.2 mg per pound. A 44-pound five-year-old needs 500 mg of vitamin A, or 11.4 mg per pound.

TABLE 14·1 IRON-RICH FOODS KIDS LIKE[a]

Breads, Cereals, and Grains
Canned macaroni (½ c)
Canned spaghetti (½ c)
Cream of wheat (¼ c)
Fortified dry cereals (1 oz)[b]
Noodles, rice, or barley (½ c)
Tortillas (1 flour or whole wheat, 2 corn)
Whole-wheat, enriched, or fortified bread (1 slice)

Vegetables
Baked flavored potato skins (½ skin)
Cooked mung bean sprouts or snow peas (½ c)[c]
Cooked mushrooms (½ c)
Green peas (½ c)
Mixed vegetable juice (1 c)

Fruits
Canned plums (3 plums)
Cooked dried apricots (¼ c)
Dried peaches (4 halves)
Raisins (1 tbs)

Meats and Legumes
Bean dip (¼ c)
Canned pork and beans (⅓ c)
Lean chopped roast beef or cooked ground beef (1 oz)
Liverwurst on crackers (½ oz)
Meat casseroles (½ c)
Mild chili or other bean/meat dishes (¼ c)
Peanut butter and jelly sandwich (½ sandwich)
Sloppy joes (½ sandwich)

[a]Each serving provides at least 1 milligram iron, or one-tenth of a child's iron recommendation. Vitamin C–rich foods included with these snacks increase iron absorption.
[b]Some fortified breakfast cereals contain more than 10 milligrams iron per half-cup serving (read the labels).
[c]Raw sprouts may pose a bacterial hazard to young children.

FIGURE 14·3 FOOD GUIDE PYRAMID FOR YOUNG CHILDREN

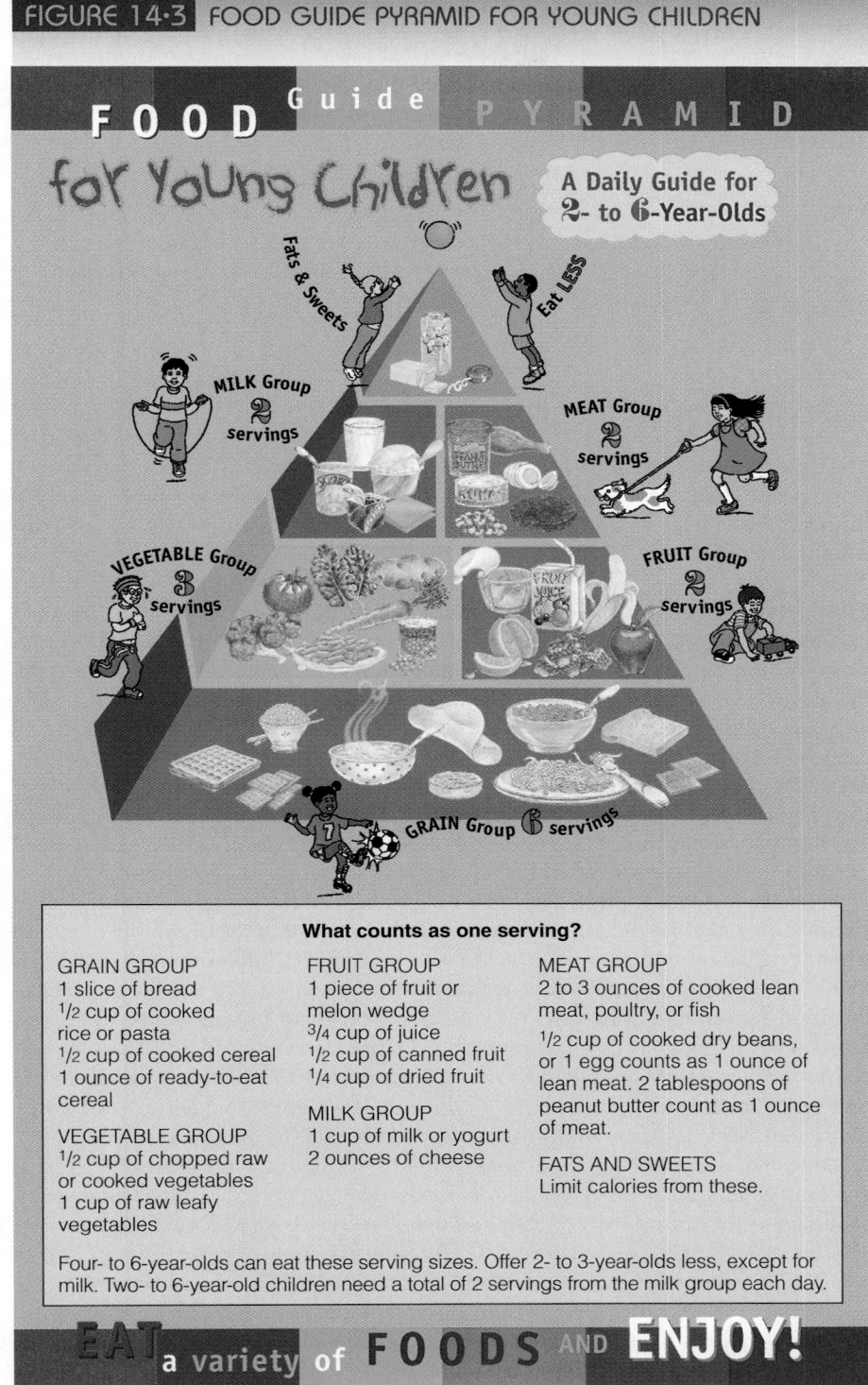

What counts as one serving?

GRAIN GROUP
1 slice of bread
1/2 cup of cooked
rice or pasta
1/2 cup of cooked cereal
1 ounce of ready-to-eat
cereal

VEGETABLE GROUP
1/2 cup of chopped raw
or cooked vegetables
1 cup of raw leafy
vegetables

FRUIT GROUP
1 piece of fruit or
melon wedge
3/4 cup of juice
1/2 cup of canned fruit
1/4 cup of dried fruit

MILK GROUP
1 cup of milk or yogurt
2 ounces of cheese

MEAT GROUP
2 to 3 ounces of cooked lean
meat, poultry, or fish

1/2 cup of cooked dry beans,
or 1 egg counts as 1 ounce of
lean meat. 2 tablespoons of
peanut butter count as 1 ounce
of meat.

FATS AND SWEETS
Limit calories from these.

Four- to 6-year-olds can eat these serving sizes. Offer 2- to 3-year-olds less, except for milk. Two- to 6-year-old children need a total of 2 servings from the milk group each day.

EAT a variety of **FOODS** AND **ENJOY!**

■ The Food Guide Pyramid for Young Children will be revised according to the 2005 USDA Food Guide.

SOURCE: USDA Center for Nutrition and Policy Promotion, March 1999, Program AID 1649.

mended for adults. Older children and adolescents need additional servings. Few children's diets follow this nutritious pattern, although some improvements in terms of slightly greater intakes of vegetables, fruit, and certain grains have been noted.[13] Still, intakes of nutrients such as calcium and zinc often fall far below recommendations.[14]

Active, normal-weight children may enjoy occasional treats of high-calorie but nutritious foods. From the milk group, ice cream or pudding is good now and then; from the grains group, whole-grain or enriched cakes, cookies, or doughnuts are an acceptable occasional addition to a balanced diet. These foods encourage a child to learn that pleasure in eating is important. Too many of these treats, however, can lead only to nutrient deficiencies, obesity, or both.

▶ **KEY POINT** Children's nutrient needs reflect their stage of growth. For a healthy child, use the DRI recommended intakes, the *Dietary Guidelines for Americans 2005,* and the *Food Guide Pyramid for Young Children* as guides to establish food patterns that provide adequate nourishment for growth while defending against obesity and chronic diseases.

Mealtimes and Snacking

The childhood years are the parents' last chance to influence their child's food choices. Appropriate eating habits and attitudes toward food, set in place by parents, can help future adults emerge with healthy eating habits that reduce risks of degenerative diseases in later life.

Children's Preferences   Children naturally like nutritious foods in all the food groups, with one exception—vegetables, which some young children refuse. Here presentation and variety may be the key. The more nutritious choices presented to a child, the more likely the child will choose adequately.

Many children prefer vegetables that are mild flavored, slightly undercooked and crunchy, bright colored, and easy to eat. Cooked foods should be served warm, not hot, because a child's mouth is much more sensitive than an adult's. The mild flavors of carrots, peas, and corn are often preferred over sharper tasting broccoli or turnips because a child has more taste buds. Smooth foods such as grits, oatmeal, mashed potatoes, and pea soup are often well received.

Fear of new foods is practically universal among children. Suggesting, rather than commanding, that a child try small amounts of new foods at the beginning of a meal when the child is hungry seems to work best. Offering the child samples of new foods that adults are enjoying can stimulate the child's natural curiosity and often produces the desired result: the child tastes the new food. Forcing or bribing a child to eat certain foods, for example, by allowing extra television time as a reward for eating vegetables, produces the opposite of the desired effect: the child will likely not develop a preference for those foods.[15] Likewise, when children are forbidden to eat favorite foods, they yearn for them more—the reverse of the well-meaning caretaker's goal.[16]

Little children prefer small portions of food served at little tables. If offered large portions, children may fill up on favorite foods, ignoring others. Toddlers often go on food jags—consecutive days of eating only one or two favored foods. For food jags lasting a week or so, make no response because two-year-olds regard any form of attention as a reward. After two weeks of serving the favored foods, try serving tiny portions of many foods, including the favored items. Invite the child's friends to occasional meals, and make other foods as attractive as possible.

Just as parents are entitled to their likes and dislikes, a child who genuinely and consistently rejects a food should be allowed the same privilege. Children should be believed when they say they are full: the "clean-your-plate" dictum should be stamped out for all time. Children who are forced to override their own satiety signals are in training for obesity. Honoring preferences does not mean that children should dictate the diet, however, because many children will choose an abundance of heavily advertised snack chips, crackers, sugary cereals, fast foods, and sugary drinks if allowed to do so.[17]

Little children like to eat small portions of food at little tables.

© Myrleen Ferguson Cate/PhotoEdit

Today's rushed parents may be tempted to abdicate food decisions to a child's taste buds, but a hungry child offered between-meal sweets and fatty treats will reliably fill up on them, rather than on nutritious foods served at meal times. The parent is responsible for *what* the child is offered to eat, but the child should decide *how much* and even *whether* to eat.

Choking A child who is choking may make no sound, so an adult should keep an eye on children when they are eating. A child who is coughing most often dislodges the food and recovers without help. To prevent choking, encourage the child to sit when eating—choking is more likely when children are running or reclining. Round foods such as grapes, nuts, hard candies, and pieces of hot dog can become lodged in a child's small windpipe. Other potentially dangerous foods include tough meat, popcorn, chips, and peanut butter eaten by the spoonful.

■ Chapter 3 describes actions to prevent choking and those to take should choking occur.

Snacking and Other Healthy Habits Parents today often find that their children snack so much that they are not hungry at mealtimes. This is not a problem if children are taught how to snack—nutritious snacks are just as health promoting as small meals. Keep snack foods simple and available: milk, cheese, crackers, fruit, vegetable sticks, yogurt, peanut butter sandwiches, and whole-grain cereal.

A bright, unhurried atmosphere free of conflict is conducive to good appetite and provides a climate in which a child can learn to enjoy eating. Parents who beg, cajole, and demand that their children eat make power struggles inevitable. A child may find mealtimes unbearable if they are accompanied by a barrage of accusations—"Susie, your hands are filthy . . . your report card . . . and clean your plate!" The child's stomach recoils as both body and mind react to stress of this kind.

Children love to be included in meal preparation, and they like to eat foods they helped to prepare (see Table 14-2 in the margin). A positive experience is most likely when tasks match developmental abilities and are undertaken in a spirit of enthusiasm and enjoyment, not criticism or drudgery. Praise for a job well done (or at least well attempted) expands a child's sense of pride and helps to develop skills and positive feelings toward healthy foods.

Many parents overlook perhaps the single most important influence on their child's food habits—their own habits.[18] Parents who don't prepare, serve, and eat carrots shouldn't be surprised when their child refuses to eat carrots.

▶ **KEY POINT** Healthy eating habits and positive relationships with food are learned in childhood. Parents teach children best by example. Choking can often be avoided by supervision during meals and avoiding hazardous foods.

Can Nutrient Deficiencies Impair a Child's Thinking or Cause Misbehavior?

A child who suffers from nutrient deficiencies exhibits physical and behavioral symptoms: the child feels sick and out of sorts. Diet-behavior connections are of keen interest to caretakers who both feed children and live with them.

Deficiencies of protein, energy, vitamin A, iron, and zinc plague children the world over. In developing nations, such deficiencies cause or contribute to nearly half the deaths of children under four and inflict blindness, stunted growth, and vulnerability to infections on millions more.

In developed countries such as the United States and Canada, most deficiencies have subtle, even unnoticeable, effects. A study of seemingly healthy British children revealed that about 40 percent of them had intakes of less than half the recommended amounts of folate, vitamin D, calcium, iron, magnesium, selenium, zinc, and other minerals. The researchers gave multinutrient supplements to some of these children and later

TABLE 14·2 FOOD SKILLS OF PRESCHOOLERS[a]

Age 1–2 years, when large muscles develop, the child:
- uses short-shanked spoon.
- helps feed self.
- lifts and drinks from cup.
- helps scrub, tear, break, or dip foods.

Age 3 years, when medium hand muscles develop, the child:
- spears food with fork.
- feeds self independently.
- helps wrap, pour, mix, shake, or spread foods.
- helps crack nuts with supervision.

Age 4 years, when small finger muscles develop, the child:
- uses all utensils and napkin.
- helps roll, juice, mash, or peel foods.
- cracks egg shells.

Age 5 years, when fine coordination of fingers and hands develops, the child:
- helps measure, grind, grate, and cut (soft foods with dull knife).
- uses hand-cranked egg beater with supervision.

[a]These ages are approximate. Healthy, normal children develop at their own pace.
SOURCES: Adapted from M. Sigman-Grant, Feeding preschoolers: Balancing nutrition and developmental needs, *Nutrition Today,* July/August 1992, pp. 13–17; A. A. Hertzler, Preschoolers' food handling skills—Motor development, *Journal of Nutrition Education* 21 (1989): 100B–100C.

administered intelligence tests to all of them. Those who had received the supplements scored significantly higher on the tests than the others did. The researchers interpreted the findings to mean that brain function may be sensitive to borderline deficiencies of some nutrients, even in children who are well nourished with protein and some vitamins. This conclusion has been supported by other findings.[19]

Iron deficiency remains common among children and adolescents despite iron fortification of foods and other programs to combat this deficiency. Besides carrying oxygen in the blood, iron works as part of large molecules to release energy within cells and plays key roles in many molecules of the brain and nervous system. A lack of iron not only causes an energy crisis but also affects behavior, mood, attention span, and learning ability.

Iron deficiency is diagnosed by a deficit of iron in the *blood,* after anemia has developed. A child's *brain,* however, is sensitive to slightly lowered iron concentrations long before the blood effects appear. Distinguishing the effects of iron deficiency from those of other factors in children's lives is difficult, but studies have found connections between iron deficiency and behavior. Iron deficiency may lessen the motivation to persist at intellectually challenging tasks, shorten the attention span, and reduce overall intellectual performance. Iron-deficient children may be irritable, aggressive, and disagreeable or sad and withdrawn. They may be labeled "hyperactive," "depressed," or "unlikable." Furthermore, a child who had iron-deficiency anemia *as an infant* may continue to perform poorly as he or she grows older, even with improvement of iron status.[20] No one knows whether the poverty and poor health often associated with early iron deficiency or some lingering effect of the deficiency itself is to blame for the later cognition and behavior problems.[21]

The diet of a disruptive or apathetic child should be examined by a registered dietitian who can identify problems and suggest actions to correct them. Only a health-care provider should make the decision to give iron supplements, and supplements should be kept out of children's reach. Iron toxicity is a leading cause of poisoning each year in toddlers and other children who accidentally ingest iron pills.[22]

■ Earlier, Table 14-1 listed some iron-rich foods that children will often accept.

▶ **KEY POINT**   The detrimental effects of nutrient deficiencies in children in developed nations can be subtle. Iron deficiency is the most widespread nutrition problem of children and causes abnormalities in both physical health and behavior. Iron toxicity is a major form of poisoning in children.

The Problem of Lead

Another form of metal poisoning arises from ingestion of lead. Lead poisoning often occurs because babies love to explore and put everything into their mouths, including things that may harm them, such as chips of old paint, pieces of metal, and other unlikely substances. Lead may also leach into a home's water supply from old lead pipes. Recently, the water flowing into older homes in some Washington, D.C., neighborhoods was discovered to exceed the maximum for lead allowed by the Environmental Protection Agency.[23] An analysis of almost 85,000 blood tests revealed somewhat higher blood lead levels among people living in the homes with old lead water pipes than in those in homes with nonlead pipes.

Lead is an indestructible metal element; the body cannot alter it. Lead can build up so silently in a child's body that caretakers will not notice unusual symptoms until much later, after toxicity has set in. Tragically, once symptoms set in, even today's most effective medical treatments may not reverse all of the functional damage.[24] Impaired thinking, reasoning, perception, and other academic skills, as well as hearing impairments and decreased growth, are associated with even very low levels of lead toxicity.[25]

As lead toxicity slowly injures the kidneys, nerves, brain, bone marrow, and other organs, the child may slip into coma, have convulsions, and possibly even die if an accurate diagnosis is not made in time to prevent it.[26] Older children with high blood

Old paint is the main source of lead in most children's lives.

lead may be suffering physical consequences but be mislabeled as delinquent, aggressive, or learning disabled.

Infants and young children absorb five to ten times as much lead as adults do. Malnutrition makes lead poisoning more likely because children absorb more lead from empty stomachs or if they lack calcium, zinc, vitamin C, vitamin D, or iron.[27] A child with iron-deficiency anemia is three times as likely to have elevated blood lead as a child with normal iron status. The chemical properties of lead are similar to those of nutrient minerals like iron, calcium, and zinc, and lead displaces these minerals from their sites of action in body cells but cannot perform their biological functions.

Bans on leaded gasoline, leaded house paint, and lead-soldered food cans have dramatically reduced the amount of lead in the U.S. environment in recent years and have produced a steady decline in children's average blood lead concentrations (see Figure 14-4).[28] A nationwide lead-monitoring system is now in place, and aggressive community programs test and treat children for lead poisoning. Lead still threatens children, however, especially in low-income families who live in older houses that still contain lead-based paint.[29] Even chips of such paint mixed into neighborhood soil can be a problem. By one estimate, more than 400,000 children—most of them under age six—currently have enough lead in their blood to cause mental, behavioral, and other health problems.[30] Some tips for avoiding lead toxicity are offered in Table 14-3.

▶ **KEY POINT**   Lead poisoning has declined dramatically over the past two decades, but when it occurs it can inflict severe, irreparable damage on growing children. Higher awareness of the remaining sources of lead poisoning can help to reduce its present rate of occurrence.

Food Allergy, Intolerance, and Aversion

Food **allergy** is frequently blamed for physical and behavioral abnormalities in children, but just 3 to 5 percent of children are diagnosed with true food allergies.[31] Children sometimes "grow out" of their food allergies (notably, allergy to peanuts may fade with time) until in adulthood food allergies affect only about 1 or 2 percent of the population.[32]

A true food allergy occurs when a food protein or other large molecule enters body tissues. Recall that most proteins from food are dismantled to smaller fragments in the digestive tract before absorption. Some, however, enter the bloodstream before being

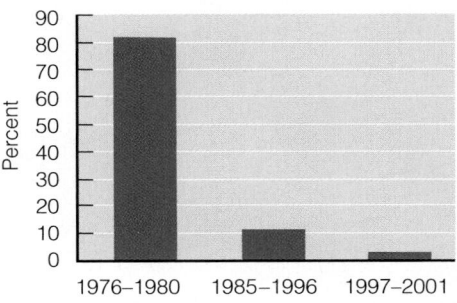

FIGURE 14·4   BLOOD LEAD IN CHILDREN, 1976–2001
The percentage of children aged one through five years with ≥ 10 µg/dL blood lead has declined substantially since 1976.

SOURCES: Data from the Centers for Disease Control and Prevention, National Center for Health Statistics, *National Health and Nutrition Examination Survey II and III;* Surveillance for elevated blood lead levels among children—United States, 1997–2001, *Morbidity and Mortality Weekly Report* 52 (2003): 1–21.

■ The Environmental Protection Agency (EPA) provides this toll-free hotline for lead information: 1-800-LEAD-FYI (1-800-532-3394).

TABLE 14·3   STEPS TO PREVENT LEAD POISONING

To protect children:

- If your home was built before 1978, wash floors, windowsills, and other surfaces weekly with warm water and detergent to remove dust released by old lead paint; clean up flaking paint chips immediately.
- Feed children balanced, timely meals with ample iron and calcium.
- Prevent children from chewing on old painted surfaces.
- Wash children's hands, bottles, and toys often.
- Wipe soil off shoes before entering the home.
- Ask a pediatrician whether your child should be tested for lead.

To safeguard yourself:

- Avoid daily use of handmade, imported, or old ceramic mugs or pitchers for hot or acidic beverages, such as juices, coffee, or tea. Commercially made U.S. ceramic, porcelain, and glass dishes or cups are safe. If ceramic dishes or cups become chalky, use them for decorative purposes only.
- Do not use lead crystal decanters for storing alcoholic or other beverages.
- If your home is old and may have lead pipes, run the water for a minute before using, especially before the first use in the morning.
- Remove lead foil from wine bottles, and wipe the mouth of the bottle before pouring.

allergy an immune reaction to a foreign substance, such as a component of food. Also called *hypersensitivity* by researchers.

fully digested. Once inside the body of an allergic person, the foreign molecules trigger a reaction from the immune system similar to the defense it launches against any other **antigen**: it releases **antibodies, histamine,** or other defensive agents to attack the invaders. In some people, the result is the life-threatening food allergy reaction of **anaphylactic shock.** Peanuts, tree nuts, milk, eggs, wheat, soybeans, fish, and shellfish are the substances most likely to trigger this extreme reaction.[33]

Managing Food Allergies If a child reacts to allergens with a life-threatening response, the family and school must guard against any exposure to the allergen.[34] Parents must teach the child which foods to avoid, but avoiding allergens can be tricky because they often sneak into foods in unexpected ways. For example, a pork chop (an innocent food) may be breaded (wheat allergy) and dipped in egg (egg allergy) before being fried in peanut oil (peanut allergy); a chocolate cookie may contain hydrogenated soybean oil shortening (soybean allergy); marshmallow candies contain eggs; lunchmeats may contain milk protein binders; and so forth.

Most caretakers of allergic children must pack safe lunches and snacks at home and ask school officials to strictly enforce a "no swapping" policy in the lunchroom. To prevent nutrient deficiencies, caretakers must also provide adequate substitutes that supply the essential nutrients in the omitted foods.[35] For example, a child allergic to milk must be supplied with calcium from fortified foods, such as calcium-rich orange juice.

The allergic child must learn to recognize the symptoms of impending anaphylactic shock, such as a tingling of the tongue, throat, or skin or difficulty breathing (see the margin). Finally, the child and the school staff should be prepared to administer injections of **epinephrine,** which prevents anaphylaxis after exposure to the allergen.[36] Too many preventable deaths occur each year among people with food allergies who accidentally ingest the allergen but have no access to epinephrine.[37]

As of 2006, food labels must announce the presence of common allergens in plain language, using the names of the eight most common allergy-causing foods.[38] For example, a food containing "textured vegetable protein" must say "soy" on its label. Similarly, "casein" must be identified as "milk protein." Food producers must also prevent cross-contamination during production and clearly label the foods in which it is likely to occur.[39] Equipment used for making peanut butter must be scrupulously clean before being used to pulverize cashew nuts for cashew butter to protect unsuspecting cashew butter consumers from peanut allergens.

Technology may soon offer new solutions. Drugs under development may interfere with the immune response that causes allergic reactions.[40] Through genetic engineering, scientists may one day banish allergens from peanuts, soybeans, and other foods to make them safer.

Detecting Food Allergy Allergies have one or two components. They *always* involve antibodies; they *sometimes* involve symptoms. Therefore, allergies cannot be diagnosed from symptoms alone. A starting point for diagnosis may entail eliminating suspected foods from the diet for a week or two and reintroducing them one at a time. If the symptoms disappear and then recur with reintroduction of a food, medical tests can then confirm the allergy. (Anyone who has suffered anaphylaxis or other severe symptoms should not reintroduce suspected foods but should go instead for medical testing.) A positive result from a blood test for antibodies can confirm a diagnosis.[41] So can a skin prick test in which a clinician applies droplets of food extracts to the skin and then lightly pricks or scratches the skin. When a food allergy exists, raised, red, itchy bumps appear within minutes.

Identifying a food that causes an immediate allergic reaction is easy because symptoms correlate with the time of eating the food. Sometimes the reaction is delayed up to 24 hours, however, making identification of the offending food more difficult because other foods will have been eaten by the time the symptoms appear.

Quick and easy scientific-sounding allergy quackery abounds on the market and may deceive people seeking relief from everything from itchy skin to mental depres-

---

These symptoms can occur in minutes or hours after ingesting an allergen:

- Tingling sensation in the mouth.
- Swelling of the tongue and throat.
- Irritated, reddened eyes.
- Difficulty breathing, wheezing, asthma.
- Hives, swelling, rashes.
- Vomiting, abdominal cramps, diarrhea.
- Drop in blood pressure.
- Loss of consciousness.
- Death.

antigen a substance foreign to the body that elicits the formation of antibodies or an inflammation reaction from immune system cells. Food antigens are usually large proteins. Also defined in Chapter 3.

antibodies large protein molecules that are produced in response to the presence of antigens and then help to inactivate the antigens. Also defined in Chapters 3 and 6.

histamine a substance that participates in causing inflammation; produced by cells of the immune system as part of a local immune reaction to an antigen.

anaphylactic (an-ah-feh-LACK-tick) **shock** a life-threatening whole-body allergic reaction to an offending substance.

epinephrine (epp-eh-NEFF-rin) a hormone of the adrenal gland that counteracts anaphylactic shock by opening the airways and maintaining heartbeat and blood pressure.

sion. Beware of "food sensitivity testing." It involves fake blood or other body fluid tests that supposedly determine appropriate dietary changes and indicate which expensive supplements the "patient" should buy from the test giver to relieve the "allergy."[42]

Food Intolerance and Food Aversion   A **food intolerance** is characterized by unpleasant symptoms that reliably occur after consumption of certain foods—lactose intolerance is an example. Unlike allergy, a food intolerance does not involve an immune response. A **food aversion** is an intense dislike of a food that may be a biological response to a food that once caused trouble. Parents are advised to watch for signs of food aversion and to take them seriously. Such a dislike may turn out to be a whim or fancy, but it may turn out to be an allergy or other valid reason to avoid a certain food. Don't prejudge. Test. Then, if an important staple food must be excluded from the diet, find other foods to provide the omitted nutrients.

These eight normally wholesome foods—milk, shellfish, fish, peanuts, tree nuts, eggs, wheat, and soybeans (and soy products)—may cause life-threatening symptoms in people with allergies.

Foods are often unjustly blamed when behavior problems arise, but children who are sick from any cause are likely to be cranky. The next section singles out one such type of misbehavior.

▶ **KEY POINT**   Food allergies can cause serious illness. Diagnosis is based on the presence of antibodies, and tests are imperative to determine whether allergy exists. Food aversions can be related to food allergies or to adverse reactions to food.

Does Diet Affect Hyperactivity?

Hyperactivity, or attention-deficit/hyperactivity disorder (ADHD), is a **learning disability** that occurs in 3 to 5 percent of young, school-aged children—at least 1 in every classroom of 30 children. ADHD is characterized by the chronic inability to pay attention, along with overly active behavior and poor impulse control (see the margin). It can delay growth, lead to academic failure, and cause major behavioral problems. Although some children improve with age, many reach the college years or adulthood before they receive a diagnosis and with it the possibility of treatment.

Food allergies have been blamed for ADHD, but research to date has shown no connection. Research has also all but dismissed the idea that sugar makes children hyperactive (see Controversy 4 for details). Over a decade ago, one study did find an association between doses of the food colorant tartrazine and increased irritability, restlessness, and sleep disturbances in a small percentage of hyperactive children.[43] Parents who wish to avoid tartrazine can find it listed with the ingredients on food labels.

Parents hope that a new diet or some other simple solution may improve children's behavior. Unfounded dietary "treatments" may seem to help for a while due to the placebo effect, but they fail to provide lasting cures.

Common sense says that all children get unruly and "hyper" at times. A child who often fills up on caffeine-containing colas and chocolate, misses lunch, becomes too cranky to nap, misses out on outdoor play, and spends hours in front of a television suffers stresses that can trigger chronic patterns of crankiness. This cycle of tension and fatigue resolves itself when the caretakers begin insisting on regular hours of sleep, regular mealtimes, a nutritious diet and regular outdoor exercise.

Living with chronic hunger is a cause of misbehavior in some children; simply lifting the children from poverty often significantly improves their behavior.[44] An estimated 12 million U.S. children are hungry at least some of the time. Such children often lack iron,

■ Chapter 4 provides details about lactose intolerance.

A child with ADHD often:

- Has a short attention span, even while playing.
- Has trouble with tasks that require sustained mental effort.
- Has trouble learning and earns failing grades in school.
- Has poor impulse control and acts physically or verbally before thinking.
- Angers friends by not taking turns or playing by the rules.
- Runs instead of walks, climbs instead of sitting still, talks excessively.

These symptoms are clues to ADHD, but they are not enough to make a diagnosis. Tests by a knowledgeable clinician are needed for diagnosis and treatment.

food intolerance an adverse reaction to a food or food additive not involving an immune response.

food aversion an intense dislike of a food, biological or psychological in nature, resulting from an illness or other negative experience associated with that food.

hyperactivity (in children) a syndrome characterized by inattention, impulsiveness, and excessive motor activity; usually diagnosed before age seven, lasts six months or more, and usually does not entail mental illness or mental retardation. Properly called *attention-deficit/hyperactivity disorder (ADHD)* and may be associated with minimal brain damage.

learning disability a condition resulting in an altered ability to learn basic cognitive skills such as reading, writing, and mathematics.

Cranky or rambunctious children may:

- Be chronically hungry.
- Be overstimulated.
- Consume too much caffeine from colas or chocolate.
- Desire attention.
- Lack exercise.
- Lack sleep.

■ The Controversy section following this chapter presents a table of the caffeine in common foods and beverages.

Family lifestyle choices to help the over-weight child:

- Do not use foods to reward or punish behaviors.
- Have fun and play vigorously outdoors every day, as a family or with friends.
- Involve children in shopping for and preparing family meals.
- Learn and use appropriate food portions.
- Limit high-sugar, high-fat foods, including sugar-sweetened soft drinks.
- Provide an appropriate nutritious break-fast every day.
- Provide a wide variety of nutritious snacks that are low in fat and sugar.
- Provide fruit juices in the amounts recommended in guidelines, such as the USDA Food Guide but no more than this amount.
- Serve family meals that control both the calorie density of the foods and the portions offered.
- Set a good example, and demonstrate positive behaviors for children to imitate.
- Slow down eating, and pause to enjoy table companions; stop eating when full.

Community actions to help correct and prevent obesity:

- Provide safe walking and biking routes to school.
- Make fruits and vegetables available and affordable.
- Provide recreational facilities with planned physical activities.
- Increase opportunities for physical activity.
- Advertise to promote energy balance and health body weight.
- Change governing policies to promote energy balance.

and a link between iron deficiency and ADHD may be emerging.[45] Hunger issues are many; see Chapter 15. Once the obvious causes of misbehavior are eliminated, a physician can recommend other strategies such as special educational programs or psychological counseling and, in many cases, prescription medication.[46]

▶ **KEY POINT** Hyperactivity, properly named attention-deficit/hyperactivity disorder (ADHD), is not caused by food allergies; temporary "hyper" behavior may reflect excess caffeine consumption or inconsistent care. A wise parent will limit children's caffeine intakes and meet their needs for structure. Poverty may cause behavior problems.

Time Use, Television, and Children's Nutrition Problems

The average child in the United States watches many hours of television each week. In addition, video and computer games and other media add significantly to children's passive viewing time, while, active free play time decreases substantially.[47] Television exerts four major adverse impacts on children's nutrition.

First, television viewing requires no energy. It seems to reduce the metabolic rate below the resting level, requiring even less energy than daydreaming. Second, it consumes time that could be spent in energetic play. An inactive child can become obese even while eating less food than an active child.[48] Third, more television watching correlates with more between-meal snacking and with eating the high-calorie fatty and sugary foods most heavily advertised on children's programs, increasing the risk of dental caries. Fourth, children who watch more than four hours of television a day, or watch during meals, are least likely to eat fruits and vegetables and most likely to be obese.[49] The average U.S. child sees an estimated 30,000 TV commercials a year, many for high-sugar, high-fat foods such as sugar-coated cereals, candies, fried snack chips, fast foods, and sugar-sweetened soft drinks. Children's intakes of sugary foods and soft drinks and fatty treats have soared in recent years while intakes of other more nutritious foods have fallen off.[50] To oppose the effects of marketing on children's appetites, it often helps to talk with children to help them to recognize the motivations and marketing techniques of advertisements.

Some blame a sedentary lifestyle for the upswing in obesity in children and adults. Fewer and fewer children actively participate in physical education classes.[51] At home, limiting television watching and other sedentary activities is probably a good idea.[52] Healthy children should never have television sets in their bedrooms—the many who do (up to 65 percent of 8- to 18-year-olds) spend even more time watching television and are very likely to be overweight.[53]

The Problem of Childhood Obesity Obesity is increasing among the world's children and poses hazards to the health of children both now and in the future (Controversy 13 provides details). Meanwhile, medical science has little to offer in the way of a cure.[54] A few obese children may remain untouched by the adverse effects of over-weight, but the majority must cope with health problems and altered physical and psychological development. Early in adolescence, an obese child's high blood pressure, insulin resistance, and blood lipid profile may warn of impending cardiovascular disease (Chapter 11 provides details about these connections). More often than thinner children, obese children develop diseases such as diabetes and asthma.[55]

Obese children often mature and develop earlier than their peers and develop greater bone and muscle mass needed to carry their extra weight. Consequently, they tend to remain "stocky" even after losing their excess body fatness.

Obese children also suffer psychologically.[56] Adults may discriminate against them, and peers may make thoughtless comments or reject them. An obese child may develop a poor self-image, a sense of failure, and a passive approach to life. Television shows, a major influence on children, stereotype the fat person as a misfit. Children have few defenses against these unfair portrayals and may internalize an unfavorable, negative attitude toward bulky body sizes.

Prevention and Treatment of Childhood Obesity An integrated approach involving diet, physical activity, psychological support, and behavioral changes can help the overweight child.[57] Consulting a professional, such as a registered dietitian who specializes in childhood nutrition and weight management, can be useful. Overweight in children must be sensitively addressed, however. Children are impressionable and may come to believe that their worth or lovability is somehow tied to their weight.

The best solutions usually involve the whole family and in fact, the whole community.* Parents are among the most influential forces shaping the self-concept, weight concerns, and dieting practices of children.[58] Thus, they play key roles in preventing or managing obesity and in helping the overweight child make appropriate lifestyle changes. Beyond this, parents can be powerful tools in helping to instill healthy habits and a strong sense of self-worth that can benefit a child through a lifetime. In addition, community efforts can support the lessons learned at home.

An initial goal is to slow the obese child's rate of gain—that is, to hold weight steady while the child grows taller. Weight loss ordinarily is not recommended because diet restriction can easily interfere with normal growth.[59] By feeding the whole family balanced meals, offering appropriate portion sizes, restricting treats, and boosting physical activity, the goal is often accomplished, and the child does not feel singled out.[60] The margin of the previous page lists some other strategies that often prove effective in producing change.

Dental Caries Sticky, high-carbohydrate snack foods cling to the teeth and provide an ideal environment for the growth of mouth bacteria that cause caries. Parents can help prevent tooth damage by helping children to:

- Limit between-meal snacking.
- Brush and floss daily, and brush or rinse after eating meals and snacks.
- Choose foods that don't stick to teeth and are swallowed quickly.
- Snack on crisp or fibrous foods to stimulate the release and rinsing action of saliva.

Table 14-4 lists foods that promote dental health and those that require speedy removal from the teeth.

▶ **KEY POINT** The nation's children are growing fatter and face advancing risks of diseases. Childhood obesity demands careful family-centered management. Television viewing can contribute to obesity through lack of exercise and by promoting overconsumption of calorie-dense snacks. The sugary snacks advertised on children's TV programs also contribute to dental caries.

Is Breakfast Really the Most Important Meal of the Day for Children?

Elders have long held that breakfast is the most important meal of the day, and for children, this bit of wisdom is now backed by science. A nutritious breakfast is a central feature of a child's diet that supports healthy growth and development. When a child consistently skips breakfast or is allowed to choose sugary foods (candy or marshmallows) in place of nourishing ones (whole-grain cereals), the child will fail to get enough of several nutrients. Nutrients missed from a skipped breakfast won't be "made up" at lunch and dinner but will be left out completely that day.[61]

Children who eat no breakfast are more likely to be overweight, perform poorly in tasks requiring concentration, have shorter attention spans, achieve lower test scores, and be tardy or absent more often than their well-fed peers.[62] Common sense tells us that it is unreasonable to expect anyone to study and learn when no fuel has been provided. Even children who have eaten breakfast suffer from distracting hunger by late morning. Chronically underfed children suffer more intensely.

■ Concern is growing for the future health of children worldwide. See Controversy 13.

TABLE 14·4 THE CARIES POTENTIAL OF FOODS
Low Caries Potential

These foods are less damaging to teeth:

- Eggs, legumes
- Fresh fruit, fruits packed in water
- Lean meats, fish, poultry
- Milk, cheese, plain yogurt
- Most cooked and raw vegetables
- Pizza
- Popcorn, pretzels
- Sugarless gum and candy,[a] diet soft drinks
- Toast, hard rolls, bagels

High Caries Potential

Brush teeth after eating these foods:

- Cakes, muffins, doughnuts, pies
- Candied sweet potatoes
- Chocolate milk
- Cookies, granola or "energy" bars, crackers
- Dried fruits (raisins, figs, dates)
- Frozen or flavored yogurt
- Fruit juices or drinks
- Fruits in syrup
- Ice cream or ice milk
- Jams, jellies, preserves
- Lunch meats with added sugar
- Meats or vegetables with sugary glazes
- Oatmeal, oat cereals, oatmeal baked goods[b]
- Peanut butter with added sugar
- Potato and other snack chips
- Ready-to-eat sugared cereals
- Sugared gum, soft drinks, candies, honey, sugar, molasses, syrups
- Toaster pastries

[a]Cariogenic bacteria cannot efficiently metabolize the sugar alcohols in these products, so they do not contribute to dental caries.
[b]The soluble fiber in oats makes this grain particularly sticky and therefore cariogenic.

---

*An informative reference is Food and Nutrition Board, Institute of Medicine, *Preventing Childhood Obesity: Health in the Balance* (Washington, D.C.: National Academics Press, 2005).

Breakfast ideas for rushed mornings:

- Make ahead and freeze sandwiches to thaw and serve with juice. Fillings may include peanut butter, low-fat cream cheese, other cheeses, jams, fruit slices, or meats. Or use flour tortillas with cheese; roll up, wrap, and freeze for later heating in a toaster oven or microwave oven.

- Teach school-aged children to help themselves to dry cereals, milk, and juice. Keep unbreakable bowls and cups in low cupboards, and keep milk and juice in small unbreakable pitchers on a low refrigerator shelf.

- Keep a bowl of fresh fruit and small containers of shelled nuts, trail mix (the kind without candy), or roasted peanuts for grabbing. Granola or other grain cereal poured into an 8-ounce yogurt tub is easy to eat on the run. So are plain toasted whole-grain frozen waffles—no syrup needed.

- Nontraditional choices are often acceptable. Purchase or make ahead enough carrot sticks to divide among several containers; serve with yogurt or bean dip. Leftover casseroles, stews, or pasta dishes are nutritious choices that children can eat hot or cold.

■ The nutritive values of selected fast foods are presented in the Table of Food Composition, Appendix A.

■ Controversy 13 lists the U.S. surgeon general's recommendations for school policies to promote healthy body weights in children.

The U.S. government funds several programs to provide nutritious, high-quality meals, including breakfast, to children at school. Children who eat school breakfast consume less fat and more magnesium during the day and are more replete with vitamin C and folate.[63] Schools that begin to participate in the federal school meal programs observe improved academic, behavioral, emotional, and social functioning of students along with lower tardiness and absentee rates.[64]

▶ **KEY POINT** Breakfast is critical to school performance. Not all children start the day with an adequate breakfast, but school breakfast programs help to fill the need.

How Nourishing Are the Lunches Served at School?

For over 50 years, school lunches have been meeting the midday nutrient needs of the nation's children. The nationally prescribed school lunches provide servings of milk, protein-rich foods (meat, poultry, fish, cheese, eggs, legumes, or peanut butter), vegetables, fruits, and breads or other grain foods each day.[65] The lunches are designed to meet at least a third of the recommended intake for specified nutrients and are often more nutritious than typical lunches brought from home.[66] Students who regularly eat school lunches have higher intakes of many nutrients and fiber than students who do not.[67] Even children's ability to learn seems to benefit from participating in national school food programs.[68] Table 14-5 shows school lunch patterns for different ages.

Children don't always like what they are served, so school lunch programs must strike a balance between what children want to eat and what will nourish them and guard their health. Additionally, short lunch periods and long waiting lines prevent some students from eating a school lunch and leave others with too little time to complete their meal.[69]

To help reduce cardiovascular risk, government-funded meals served at schools must follow the *Dietary Guidelines for Americans*. Often, however, private vendors offer unregulated meals, even heavily advertised fast foods, that compete side-by-side with the nutritious school lunches.[70] U.S. children develop a taste for such foods early in life and may reject nutritious school meals when offered a choice of meals higher in fat, sugar, and salt.[71] Those who do so consume fewer fruits and vegetables not only at lunchtime but later on at supper as well.[72] Children receive a mixed message when they are left on their own to choose between the health-supporting school lunch and the less optimal foods that their taste buds may prefer.

Students in some high schools also face the additional temptations of soft drinks, frozen confections, candies, and other low-nutrient treats from school snack bars, vending machines, or school stores. Health professionals and administrators of the school lunch program have proposed limiting or banishing such treat sales from school grounds, but they face opposition from the food industries and individual school districts that profit from students' pocket money.[73] Nutrition professionals advocate that all foods available on school grounds or at school functions, even those from vending machines or school stores, be chosen to help meet the *Dietary Guidelines* standards.[74] Reducing the prices of nutritious foods greatly increases the likelihood that students will purchase them.[75]

A word about juice: authorities recommend limiting fruit juices in children's diets to about half a cup a day because juices are high in calories and natural sugars and so could contribute to obesity or dental caries.[76] Of course, consuming too much of any high-calorie food could lead to weight gain, but evidence is mixed on whether juice intakes are linked with obesity.[77] Taken in moderation, juice is a nutritious food and may even crowd sugary soft drinks out of the diet—and soft drinks *are* associated with childhood obesity.[78] Frequent exposure of the teeth to the sugars in fruit juice increases the likelihood of dental caries, however, so children should be taught to brush their teeth after consuming juice.

TABLE 14·5 SCHOOL LUNCH PATTERNS FOR DIFFERENT AGES

Food Group	Preschool (age in years)		Grade School through High School[a]		
	Age 1 to 2	Age 3 to 4	Grade K to 3	Grade 4 to 6	Grade 7 to 12
Milk					
1 serving of fluid milk[b]	¾ c	¾ c	1 c	1 c	1 c
Meat or Meat Alternate					
1 serving:					
Lean meat, poultry, or fish	1 oz	1½ oz	1½ oz	2 oz	3 oz
Cheese	1 oz	1½ oz	1½ oz	2 oz	3 oz
Large egg(s)	½	¾	¾	1	1½
Cooked dry beans or peas	¼ c	⅜ c	⅜ c	½ c	¾ c
Peanut butter	2 tbs	3 tbs	3 tbs	4 tbs	6 tbs
Peanuts, soynuts, tree nuts, or seeds[c]	½ oz	¾ oz	¾ oz	1 oz	1½ oz
Vegetable and/or Fruit					
2 or more servings	½ c	½ c	½ c	¾ c (plus ½ c extra over a week)	¾ c
Bread or Bread Alternate Servings[d]	5 per week (minimum ½ per day)	8 per week (minimum 1 per day)	8 per week (minimum 1 per day)	8 per week (minimum 1 per day)	10 per week (minimum 1 per day)

[a]These patterns may be used so long as the meals served meet the *Dietary Guidelines for Americans* and provide one-third of the child's recommendations for nutrients.
[b]Whole milk and unflavored low-fat milk must be offered; flavored milks or fat-free milk may also be offered.
[c]These foods may meet no more than one-half serving of meat and must be accompanied by other meat or alternate in the meal.
[d]A serving is 1 slice of whole-grain or enriched bread; a whole-grain or enriched biscuit, roll, muffin, or the like; or ½ cup cooked rice, pasta, or other grain.
SOURCE: U.S. Department of Agriculture.

Budget cutbacks in recent years have all but eliminated nutrition education from the nation's classrooms. Meeting the nutrition and education needs of children is critical to supporting their healthy growth and development.[79] The most effective education plans involve whole families and focus on changing behaviors to improve food choices at home as well as at school.[80]

▶ **KEY POINT** School lunches are designed to provide at least a third of the nutrients needed daily by growing children and to stay within limits set by the *Dietary Guidelines for Americans*. Soda and snack vending machines, fast-food and snack bars, and school stores tempt students with foods high in fats and sugars. Fruit juice is a healthy food but may cause dental problems if used to excess.

The Teen Years

Teenagers are not fed; they eat. Nutrient needs are high during adolescence, and choices made during the teen years profoundly affect health, both now and in the future. Teens need reliable nutrition information to enable them to make healthy food choices. In the face of many demands on their time, including afterschool jobs, social activities, and home responsibilities, they easily fall into irregular eating habits, relying on quick snacks or fast foods for meals.[81] Only about a third of adolescents take evening meals at home with their families, but almost 80 percent say that family meals are important to them.[82] The adolescent who does eat at home with family members consumes more nutritious fruits, vegetables, grains, and calcium-rich foods and fewer soft drinks than others.[83]

The teen years bring a search for identity, acquired largely through trial and error. Among influences shaping a teenager's self-concept, the most important are parents, peers, and the media.[84] Teens face tremendous pressures regarding body image, and many adopt fads and scams offering promises of slenderness, good-looking muscles,

The tragedy of drug or alcohol abuse often begins in adolescence and causes multiple nutrition problems:

- Money is spent on drugs or alcohol instead of food.
- Substance-induced euphoria depresses the appetite.
- A drug- or alcohol-focused lifestyle does not include nutrition.
- Infections from sharing needles raise nutrient needs.
- Medical treatment for drug abuse may alter nutrition needs.

See Controversy 3 for details concerning alcohol and nutrition.

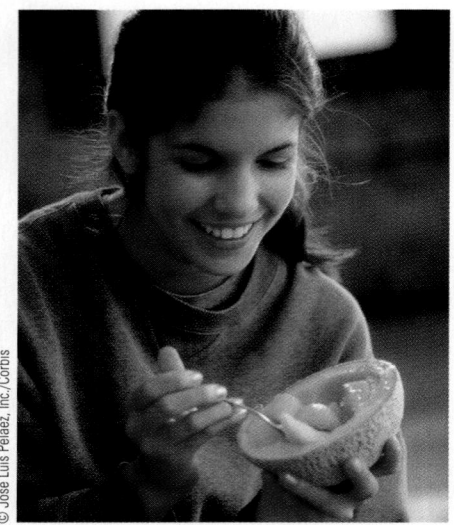
Nutritious snacks play an important role in an active teen's diet.

■ Food sources of iron are listed in Chapter 8 (p. 288).

Iron DRI intake goals for adolescent boys:

- 9–13 years: 8 mg/day.
  If in growth spurt, 10.9 mg/day.
- 14–18 years: 11 mg/day.
  If in growth spurt, 13.9 mg/day.

Iron DRI intake goals for adolescent girls:

- 9–13 years: 8 mg/day.
  If menstruating, 10.5 mg/day.
  If menstruating and in growth spurt, 11.6 mg/day.
- 14–18 years: 15 mg/day.
  If in growth spurt, 16.1 mg/day.

■ Eating disorders may arise during the teen years and often prevent the adolescent from obtaining not only energy but also sufficient calcium and a number of other needed nutrients. Controversy 9 presents details.

growth spurt the marked rapid gain in physical size usually evident around the onset of adolescence.

epiphyseal (eh-PIFF-ih-seal) **plate** a thick, cartilage-like layer that forms new cells that are eventually calcified, lengthening the bone (*epiphysis* means "growing" in Greek).

freedom from acne, or control over symptoms that may accompany menstruation. A negative self-assessment may open the door to taking risks such as using alcohol, tobacco, and drugs of abuse such as marijuana.

Growth and Nutrient Needs of Teenagers

Needs for vitamins, minerals, the energy-yielding nutrients, and, in fact, all nutrients are greater during adolescence than at any other time of life except pregnancy and lactation (see the DRI table on the inside front cover). The need for iron is particularly high as all teenagers gain lean body mass and girls begin menstruation. Calcium needs are high to support the development of peak bone mass.

The Special Case of Iron   The increase in need for iron during adolescence occurs across the genders, but for different reasons. A boy needs more iron at this time to develop extra lean body mass, whereas a girl needs extra not only to gain lean body mass but also to support menstruation. Because menstruation continues throughout a woman's childbearing years, her need stays high until older age. As boys become men, their iron needs drop back to the preadolescent value during early adulthood.

An interesting detail about adolescent iron requirements is that the need increases during the **growth spurt,** regardless of the age of the adolescent.[85] This shifting requirement makes pinpointing an adolescent's need tricky, as the margin list demonstrates.

Iron intakes often fail to keep pace with increasing needs, especially for girls, who typically consume less iron-rich foods such as meat and fewer total calories than boys. Not surprisingly, iron deficiency is most prevalent among adolescent girls, and their performance on standardized tests often suffers.[86]

▶ **KEY POINT**   The need for iron increases during adolescence for males and females. Women's iron needs remain higher as long as they continue to menstruate.

Adolescence and the Bones   Adolescence is a crucial time for bone development. The bones are growing longer at a rapid rate (see Figure 14-5), thanks to a special bone structure, the **epiphyseal plate,** that disappears as a teenager reaches adult height. At the same time, the bones are gaining density, laying down the calcium needed later in life. Low calcium intakes have reached crisis proportions: 85 percent of girls and 64 percent of boys ages 12 to 19 years have too-low intakes. Paired with lack of physical activity, low calcium intakes can compromise the development of peak bone mass, greatly increasing the risk of osteoporosis and other bone diseases later on.[87]

Teens often choose soft drinks as their primary beverage (see Figure 14-6), a choice that displaces calcium-rich milk from the diet.[88] Conversely, increasing milk products to meet calcium needs increases bone density.[89] In teenage girls especially, soft drink intake soars while milk intake—and therefore calcium—declines sharply just when calcium needs are greatest (see Figure 14-7, p. 536).[90] Regular soft drink consumption is also linked with overweight in adolescents.[91]

In addition to needing dietary calcium, bones grow stronger with physical activity, but few high schools require students to attend physical activity classes (see Figure 14-8, p. 536), so most teenagers must make a point to be physically active during leisure hours. Attainment of maximal bone mass during youth and adolescence is the best protection against age-related bone loss and fractures in later life.

▶ **KEY POINT**   Sufficient calcium intake is crucial during adolescence to support normal bone growth and density. When teens choose soft drinks and abandon milk, they increase their chances of bone disease later on in life.

FIGURE 14·5 GROWTH OF LONG BONES

Bones grow longer as new cartilage cells accumulate at the top portion of the epiphyseal plate and older cartilage cells at the bottom of the plate are calcified.

Cartilage
Bone
Epiphyseal plate
Marrow cavity

Bone grows longer

New cell layer forms here

Calcification takes place here

© D. M. Phillips/Visuals Unlimited

The Body Changes of Adolescence The adolescent growth spurt brings rapid growth and hormonal changes that affect every organ of the body, including the brain. An average girl's growth spurt begins at 10 or 11 years of age and peaks at about 12 years. Boys' growth spurts begin at 12 or 13 years and peak at about 14 years, slowing down at about 19. Two boys of the same age may vary in height by a foot, but if growing steadily, then each is fulfilling his genetic destiny according to an inborn schedule of events. Weight standards meant for adults are useless for adolescents. Parents should watch only for smooth progress and guard against comparisons that can diminish the child's self-image.

The energy needs of adolescents vary tremendously depending on growth rate, gender, body composition, and physical activity.[92] Energy balance is often difficult to regulate in this society—an estimated 15 percent of U.S. children and adolescents 6 to 19 years of age are overweight.[93] An active, growing boy of 15 may need 3,500 calories or more a day just to maintain his weight, but an inactive girl of the same age whose growth has slowed may need fewer than 1,700 calories to keep from becoming obese.

Girls normally develop a somewhat higher percentage of body fat than boys do, a fact that causes much needless worry about becoming overweight. Healthy, normal-weight teenagers are often "on diets" and make all sorts of unhealthy weight-loss attempts—even taking up smoking.[94] Some teens may benefit from lower-calorie diets that increase fruit, vegetables, fat-free milk and other nutritious foods while limiting cookies, cakes, soft drinks, fried snacks, and other less healthy choices. A few teens without diagnosable eating disorders have been reported to "diet" so severely that they stunted their own growth. Most weight-loss dieting undertaken by adolescents, particularly girls, is self-prescribed and generally unhealthful and can easily lead to nutrient deficiencies.[95] Such dieting may even promote *gain* of excess weight in the long run.[96]

Girls face a major change with the onset of menstruation. The hormones that regulate the menstrual cycle affect not just the uterus and the ovaries but the metabolic rate, glucose tolerance, appetite, food intake, mood, and behavior as well. Most women live easily with the cyclic rhythm of the menstrual cycle, but some are afflicted with physical and emotional pain prior to menstruation, a condition called **premenstrual syndrome,** or **PMS** (see the Consumer Corner, p. 537). Many teens struggle with outbreaks of acne.

FIGURE 14·6 INCREASE IN SOFT DRINK CONSUMPTION BY U.S. ADOLESCENTS

Average soft drink consumption by adolescents more than doubled between 1978 and the present.

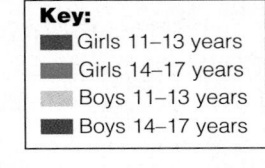

Key:
■ Girls 11–13 years
■ Girls 14–17 years
■ Boys 11–13 years
■ Boys 14–17 years

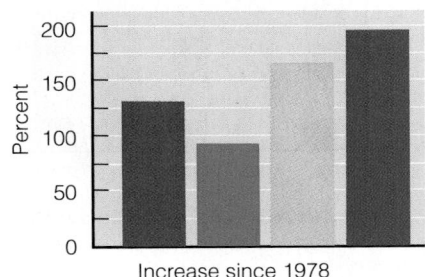

Percent

Increase since 1978

SOURCE: S. A. French, B.-H. Lin, and J. F. Guthrie, National trends in soft drink consumption among children and adolescents age 6–17 years: Prevalence, amounts, and sources, 1977/1978 to 1994/1998, *Journal of the American Dietetic Association* 103 (2003): 1326–1331.

premenstrual syndrome (PMS) a cluster of symptoms that some women experience prior to and during menstruation. They include, among others, abdominal cramps, back pain, swelling, headache, painful breasts, and mood changes.

FIGURE 14·7 MILK CONSUMPTION AMONG ADOLESCENT GIRLS

Teenage girls report drinking less milk over the years and rarely meet their need for calcium.

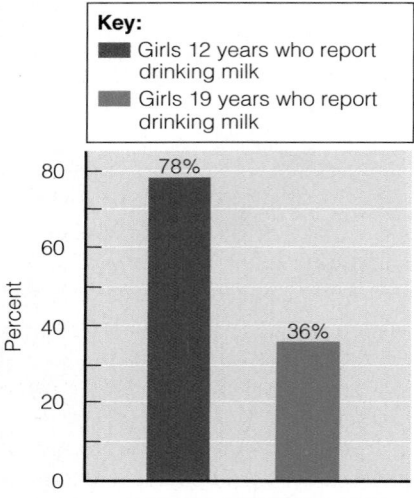

SOURCE: Data from S. A. Bowman, Beverage choices of young females: Changes and impact on nutrient intakes, *Journal of the American Dietetic Association* 102 (2002): 1234–1239.

FIGURE 14·8 U.S. HIGH SCHOOL STUDENTS ATTENDING DAILY PHYSICAL EDUCATION CLASSES, 1991–2001

SOURCE: Centers for Disease Control, National Youth Risk Behavior Surveillance—United States, 2001, available at www.cdc.gov.

acne chronic inflammation of the skin's follicles and oil-producing glands, which leads to an accumulation of oils inside the ducts that surround hairs; usually associated with the maturation of young adults.

gatekeeper with respect to nutrition, a key person who controls other people's access to foods and thereby affects their nutrition profoundly. Examples are the spouse who buys and cooks the food, the parent who feeds the children, and the caretaker in a day-care center.

▶ KEY POINT   The adolescent growth spurt increases the need for energy and nutrients. The normal gain of body fat during adolescence may be mistaken for obesity, particularly in girls. Some self-prescribed diets are detrimental to health and growth.

Acne   No one knows why some people get **acne** while others do not, but heredity plays a role—acne runs in families. The hormones of adolescence also play a role by stimulating the oil glands in the skin. The skin's natural oil is made in deep glands and is supposed to flow out through tiny ducts to the skin's surface. In acne, the ducts become clogged, and oily secretions build up in the ducts causing irritation, inflammation, and breakouts of acne.

Various foods are often charged with aggravating acne, but research has demonstrated that two often-accused foods, chocolate and sugar, do *not* worsen acne. Evidence is less conclusive for cola beverages, fatty or greasy foods, milk, nuts, and foods or salt containing iodine, but research has not shown that any of these is an aggravating factor. Psychological stress, though, clearly worsens acne. Vacations from school often bring acne relief. Sun and swimming also help, perhaps because they are relaxing, the sun's rays kill bacteria, and water cleanses the skin. Prescribed antibiotic pills and ointments work for some.

The oral prescription medicine Accutane, made from vitamin A, is effective against deep lesions from a severe form of acne. Accutane is highly toxic, though, and causes serious birth defects if taken during pregnancy. Although medicines made from vitamin A are successful in treating acne, vitamin A itself has no effect, and supplements of the vitamin can be toxic. Quacks, undaunted by these facts, market potentially toxic vitamin A and related compounds as supplements to young people who hope to cure acne.

One remedy always works: time. While waiting, attend to basic needs. Petal-smooth, healthy skin reflects a tended, cared-for body whose owner provides it with nutrients and fluids to sustain it, exercise to stimulate it, and rest to restore its cells.

▶ KEY POINT   Although no foods have been proved to aggravate acne, stress can worsen it. Supplements are useless against acne, but sunlight, proven medications, and relief from stress can help.

Eating Patterns and Nutrient Intakes

During adolescence food habits change for the worse, and teenagers often miss out on nutrients they need. Teens may begin to skip breakfast; choose less milk, fruits, juices, and vegetables; and consume more soft drinks each day.[97] Ideally, the adult becomes a **gatekeeper**, controlling the type and availability of food in the teenager's environment. Teenage sons and daughters and their friends should find plenty of nutritious, easy-to-grab food in the refrigerator (meats for sandwiches, raw vegetables, milk, fruit and fruit juices) and more in the cupboards (breads, peanut butter, nuts, popcorn, cereals). In reality, in many households today, all the adults work outside the home, and teens perform some of the gatekeeper's roles, such as shopping for groceries or choosing fast foods or prepared foods.

On average, about a fourth of a teenager's total daily energy intake comes from snacks, which, if chosen carefully, can contribute some of the needed nutrients. Often, however, teens choose foods that are too high in saturated fat and sodium and too low in fiber to support the future health of their arteries. Their calcium intakes often fall short unless they snack on dairy products, and they often fail to obtain enough iron and vitamin A. For iron and other nutrients, a teen could snack on iron-containing meat sandwiches, low-fat bran muffins, or tortillas with spicy bean spread along with a glass of orange juice to help maximize the iron's absorption.

Teenagers love fast food, and fortunately, some fast-food establishments are offering more nutritious choices than the standard hamburger meal. The gatekeeper can help the teenager choose wisely by delivering nutrition information in a way that is meaningful to the individual teen. Those who are prone to gain weight will often open their ears to news about calories in fast foods. Others attend best to information about

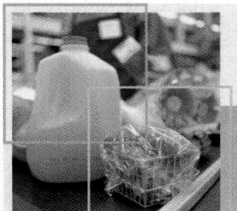

CONSUMER CORNER

Nutrition and PMS

PMS AFFECTS UP to 40 percent of menstruating women and adolescent girls.[1] The symptoms can include any or all of these: cramps and aches in the abdomen, back pain, headache, acne, swelling of the face and limbs associated with water retention, food cravings (especially for chocolate and other sweets), abnormal thirst, pain and lumps in the breasts, diarrhea, and mood changes, including both nervousness and depression. In a small number of severe cases, diagnosed as *premenstrual dysphoric disorder,* uncontrollable irritability, anxiety, and even panic can cause misery.[2]

An official diagnosis of ordinary PMS requires that such symptoms impair some facet of a woman's life. This is true for about 5 to 10 percent of menstruating women.[3] The timing and pattern of symptoms are also important, and women who seek help must keep daily records.

The cause of PMS is unknown. One proposed mechanism is an altered response to the two major regulatory hormones of the menstrual cycle: estrogen and progesterone.[4] In particular, the hormone estrogen can affect the brain's neurotransmitters and thereby alter mood. Adequate serotonin in the brain buoys a person's mood, but a serotonin deficiency creates a depressed mood and is related to major depressive disorders. In PMS, the natural rise and fall of estrogen levels in the blood may affect the activities of serotonin in the brain

during the last half of each menstrual cycle. Taking oral contraceptives, which supply estrogen, often improves mood by eliminating hormonal peaks and valleys. Taking antidepressant drugs that magnify the effects of serotonin may help in some cases, too.[5]

The major connections between PMS and nutrition concern energy metabolism and intakes of vitamin B_6 and calcium. Scientists believe that during the two weeks prior to menstruation, two events affect a woman's metabolism:

- The basal metabolic rate during sleep speeds up.
- Appetite and calorie intakes increase.[6]

Most studies indicate that women take in an average of 300 calories a day more during the ten days prior to menstruation than during the ten days after it. As a consequence, a woman who wishes to control her weight may find it easier to restrict calories during the two weeks following menstruation. Limiting calories may be harder during the two weeks before the next menstruation, however, because she is fighting a natural, hormone-governed increase in appetite.

A link with calcium is intriguing: in a few studies, calcium intakes of 1,000 to 1,300 milligrams per day significantly improved the irritability, cramping, and other symptoms of PMS.[7] Another review concluded that high-dose supplements of vitamin B_6 (100 milligrams per day) may alleviate some PMS symptoms,

but this amount is also the DRI Tolerable Upper Intake Level, indicating that over the long term such supplements may harm health. The following have not proved consistently useful in research: taking multivitamins, magnesium, or manganese supplements; cutting down on alcohol or sodium; or taking diuretics to relieve water retention.[8] If women retain sodium and water just before menstruation, that effect may be normal and desirable; diuretic drugs eliminate sodium and water but also cause a loss of potassium, potentially worsening PMS.

Consumption of black tea has been positively linked with PMS. Which component of tea—the caffeine, pigments, or other substances—may worsen PMS is not known, but evidence indicates a potential role for caffeine. Data from questionnaires administered to more than 800 women correlated caffeine intakes with PMS in a linear fashion: the more caffeine-containing beverages the women reported drinking, up to 10 cups per day, the more symptoms of PMS they reported suffering. Thus, a woman who finds menstrual symptoms troublesome might try a caffeine-free lifestyle for a while and see if her symptoms improve.

The woman with PMS should examine her total lifestyle, of which diet is only a part. Adequate sleep and physical activity help, and controlling stress may be important. She should also be on guard against bogus, possibly hazardous, PMS "cures" that abound in the marketplace.

the negative effects of an ill-chosen diet on sport performance. Still others are fascinated to learn of the skin's need for vitamins. Rather than dictating a list of dos and don'ts, the wise gatekeeper does more listening than talking.[98] When asked, teens often identify for themselves the factors blocking healthy behaviors, and acknowledging such factors is the first step in eliminating them.

The gatekeeper can set a good example, provide an environment with plenty of nutritious foods, keep lines of communication open, and stand by with reliable nutrition information and advice, but the rest is up to the teens themselves. Ultimately, they make the choices.

How will you age?

- In what ways do you expect your appearance to change as you age?
- What physical activities do you see yourself enjoying at age 70?
- What will be your financial status? Will you be independent?
- What will your sex life be like? Will others see you as sexy?
- How many friends will you have? What will you do together?
- Will you be happy? Cheerful? Curious? Depressed? Uninterested in life or new things?

Tips for productive aging:

1. Simplify your life; identify priorities and trim the superfluous.
2. Pay attention to yourself—body, mind, and spirit.
3. Continue to teach and learn; take up leisure activities (painting, woodwork, nature).
4. Be flexible; learn to navigate change.
5. Be charitable; make it a practice to give (wisdom, experience, money, time, yourself).
6. Be financially astute; invest early for retirement.
7. Find and participate in activities that interest you; you'll live better in retirement if you do.
8. Commit to good nutrition and exercise, no matter what.
9. Think about your past and future; deal with your mortality.
10. Be involved; be positive; link with others.

SOURCE: Adapted with permission from H. Kerschner and J. M. Pegues, Productive aging: A quality of life agenda, *Journal of the American Dietetic Association* 98 (1998): 1445–1448.

life expectancy the average number of years lived by people in a given society.

life span the maximum number of years of life attainable by a member of a species.

longevity long duration of life.

▶ KEY POINT   With planning, the gatekeeper can encourage teens to meet nutrient requirements by providing nutritious snacks.

The Later Years

The title may imply a section about older people, but it is relevant even if you are only 20 years old. How you live and think at 20 years of age affects the quality of your life at 60 or 80 years. According to an old saying, "as the twig is bent, so grows the tree." Unlike a tree, however, you can bend your own twig.

Before you will adopt nutrition behaviors to enhance your health in old age, you must accept on a personal level that you yourself are aging. To learn what negative and positive views you hold about aging, try answering the questions in the margin. Your answers reveal not only what you think of older people now but also what will probably become of you. When older adults were asked to give tips to younger people on how to live life fully in the later years, they offered ten suggestions, also listed in the margin.

The "graying" of America is evident. Since 1950, the population over age 65 has doubled, and people over 85 years old are the fastest-growing age group.[99] People reaching and exceeding age 100 have doubled in number in the last decade, and not just in the United States. Many of the world's populations have followed similar trends.

In the United States, the **life expectancy** is 80 years for white women and 75 years for black women; for white men, it is 75 years and for black men, 68 years—all record highs, and much higher than the life expectancy of 47 years in 1900.[100] Once a person survives the perils of youth and middle age to reach age 80, women can expect to survive an additional nine years, on average; men, an additional seven. One expert opinion estimates that 70 to 80 percent of the average person's life expectancy may depend on individual health-related behaviors, with genes determining the remaining 20 to 30 percent.[101]

The biological schedule that we call aging cuts off life at a genetically fixed point in time. The **life span** (the maximum length of life possible for a species) of human beings is believed to be 130 years. Even this limit may one day be challenged with advances in medical and genetic technologies. One caution: to date, scientists who study the aging process have found no specific diet or nutrient supplement that will increase **longevity,** although there are hundreds of dubious claims to the contrary.

▶ KEY POINT   Life expectancy for U.S. adults increased in the twentieth century. Life choices can greatly affect how long a person lives and the quality of life in the later years. No diet or supplement can extend life.

Nutrition in the Later Years

Nutrient needs become more individual with age, depending on genetics and individual medical history. For example, one person's stomach acid secretion, which helps in iron absorption, may decline, so that person may need more iron. Another person may excrete more folate due to past liver disease and thus need a higher dose. Table 14-6 lists some changes that can affect nutrition.

Energy and Activity

Energy needs often decrease with advancing age. One reason is that the number of active cells in each organ decreases, reducing the body's overall metabolic rate, although much of this loss may not be inevitable. Another reason is that older people often reduce their physical activity, and their lean tissue diminishes.

TABLE 14·6	PHYSICAL CHANGES OF AGING THAT AFFECT NUTRITION
Digestive Tract	Intestines lose muscle strength resulting in sluggish motility that leads to constipation. Stomach inflammation, abnormal bacterial growth, and greatly reduced acid output impair digestion and absorption. Pain and fear of choking may cause food avoidance or reduced intake.
Hormones	For example, the pancreas secretes less insulin and cells become less responsive, causing abnormal glucose metabolism.
Mouth	Tooth loss, gum disease, and reduced salivary output impede chewing and swallowing. Choking may become likely; pain may cause avoidance of hard-to-chew foods.
Sensory Organs	Diminished sight can make food shopping and preparation difficult; diminished senses of smell and taste may reduce appetite, although research is needed to clarify this effect.
Body Composition	Weight loss and decline in lean body mass lead to lowered energy requirements. May be preventable or reversible through physical activity.

After about the age of 50, the intake recommendation for energy assumes about a 5 percent reduction in energy output per decade. As in other groups, obesity is increasingly a problem. For those who must limit energy intake, there is little leeway in the diet for foods of low nutrient density such as sugars, fats, and alcohol.

Current thinking refutes the idea that declining energy needs are unavoidable, however. Physical activity and a diet adequate in nutrients and rich in phytochemicals are key not only to maintaining energy needs but also to upholding other functions, such as a healthy immune response and mental functioning.[102] Physical activity and diet also oppose a destructive spiral of sedentary behavior and mental and physical losses in the elderly, sometimes called the "dwindles."[103] The "dwindles" refers to a complex of interacting failures in the elderly including:

- Decreased physical ability to function.
- Diminished mental function.
- Malnutrition.
- Social withdrawal.
- Weight loss.

Involuntary weight loss deserves immediate attention in the older person. It could be the result of some easily treatable condition, such as ulcers.[104] For people older than 70 years, the best health and lowest risk of death have been observed in those who maintain a body mass index (BMI) between 25 and 32, which is higher than the optimal BMI for younger people (18.5 to 25); see the inside back cover.

The Think Fitness feature (page 540) emphasizes the importance of physical activity to maintaining body tissue integrity throughout life.[105] An expert in the nutrition of aging phrases it this way:

> We now know that physically active elders can build and rebuild muscle mass. Even the frail elderly can improve function by a remarkable 200 percent on a short, focused exercise regimen. No single feature of aging can more dramatically affect basal metabolism, insulin sensitivity, calorie intake, appetite, breathing, ambulation, mobility, and independence than muscle mass.[106]

Some people in their *nineties* have gained muscle mass, regained or improved their balance, and added pep to their walking steps and regained some precious independence after just eight weeks of weight training. People spending energy in physical activity can also eat more food, gaining nutrients. Sadly, over 90 percent of older adults fail to meet national exercise objectives and miss the opportunity for more robust health and fitness in their later years.[107]

The photos on page 540 dramatize this point: they compare cross sections of the thigh of a young woman and of an older woman to demonstrate the muscle loss typical of sedentary aging, which brings with it destructive weakness, poor balance, and deterioration of health and vigor.[108] Strength training helps to prevent at least some of this muscle loss.

▓ The DRI nutrient intake standards provide separate recommendations for those 51 to 70 years and for those 70 and older. See the inside front cover.

▓ Body mass index was discussed in Chapter 9.

▓ *Healthy People 2010* objective: Increase to 30% the proportion of adults who perform ≥2 days per week the physical activities that enhance and maintain muscular strength and endurance.

Courtesy of Dr. William Evans

Cross sections of two thighs. These two women's thighs may appear to be about the same size from the outside, but the 20-year-old woman's thigh (left) is dense with muscle tissue (dark areas). The 64-year-old woman's thigh (right) has lost muscle and gained fat, changes that may be largely preventable with strength-building physical activities.

Some unscrupulous practitioners try to sell elderly people a shortcut to muscle and bone tissue retention in the form of growth hormone (GH) "therapy." Although secretion of GH does decline with age, science does not yet support its usefulness or safety in reversing the tissue loss of aging.

▶ **KEY POINT** Energy needs decrease with age, but exercise burns off excess fuel, maintains lean tissue, and brings health benefits.

© Hamilton County General Health District

These chair volleyball players are working their upper bodies and having fun, while remaining safely seated.

THINK FITNESS Benefits of Physical Activity for the Older Adult

PHYSICALLY ACTIVE older adults have more flexibility and endurance, more lean body mass, a better sense of balance, greater blood flow to the brain, and stronger immune systems; they suffer fewer falls and broken bones, experience fewer symptoms of arthritis, better retain cognitive abilities, enjoy better overall health, and even live longer than their less-fit peers. Older people should feel free to exercise in their own way, at their own pace, and should attend to fluid and nutrient needs to support activity. Any exercise, even a ten-minute walk a day or strength and flexibility training while seated, provides progressive benefits. Great achievements are possible and improvements are inevitable. In fact, older runners, and especially women, have logged better personal time improvements than younger runners in the New York City Marathon over the past two decades.[109] An aging person unavoidably loses some capacity to perform exercise, however, so routines should be tailored to abilities.

Protein Needs

Protein needs remain about the same for older people as for young adults. Adequate daily protein is needed to replace losses, but too much protein can be hard on ailing kidneys of older adults because of the extra burden of excreting its nitrogen. Which protein-rich foods elders choose to eat takes on extra importance, too. For older people who have lost their teeth, chewing tough foods is next to impossible, and they need soft cooked beans or meats or chopped foods. Individuals with chronic constipation, heart disease, or diabetes may benefit from fiber-rich low-fat legumes and grains as sources of protein. Should a flagging appetite reduce food intake, liquid nutritional formulas between meals can supply needed energy, protein, and other nutrients.[110]

> ▶ KEY POINT   Protein needs remain about the same through adult life, but choosing low-fat fiber-rich protein foods may help control other health problems.

Carbohydrates and Fiber

The recommendation to obtain ample amounts of mostly whole-grain breads, cereals, rice, and pasta holds true for older people. As for younger people, a steady supply of carbohydrate is essential for optimal brain functioning.[111] With age, fiber takes on extra importance for its role against constipation, a common complaint among older adults and among nursing home residents in particular. Fruits and vegetables supply viscous (soluble) fibers thought to help ward off diseases of aging, but factors such as transportation problems, limited cooking facilities, and chewing problems limit some elderly people's intakes of fresh fruits and vegetables.[112] Most older adults do not obtain the recommended daily 25 or more grams of fiber (14 grams per 1,000 calories).[113] When low fiber intakes are combined with low fluid intakes, inadequate exercise, and constipating medications, constipation becomes inevitable.

> ▶ KEY POINT   Generous carbohydrate intakes are recommended for older adults. Including fiber in the diet is important to avoid constipation.

Fats and Arthritis

Older adults must attend to fat intakes for several reasons. Consuming enough of the essential fatty acids supports continued good health, and limiting intakes of saturated and *trans* fats is a priority to minimize the risk of heart disease. Many of the foods lowest in saturated fat are richest in vitamins, minerals, and phytochemicals. In addition, certain fats may affect one type of **arthritis,** a painful deterioration and swelling of the joints. High-fat diets also correlate with obesity, an arthritis risk factor.

Two kinds of arthritis attack the joints: osteoarthritis and rheumatoid arthritis. The more common type, osteoarthritis, affects upward of 40 percent of adults; it results from being overweight or from unknown causes as people age.[114] During movement, the ends of healthy bones are protected by small sacs of fluid that act as lubricants. With arthritis, the sacs erode, cartilage and bone ends disintegrate, and joints become malformed and painful to move. Nutrition does not seem to play a role in the causation of osteoarthritis. High intakes of vitamins E and C, in particular, do not slow progression or ease pain.[115] Low intake of vitamin D may speed its progression, however. Loss of body weight often brings relief, particularly in the knees.

Rheumatoid arthritis can strike at any age. It probably arises from a malfunction of the immune system—the immune system mistakenly attacks the bone coverings as if they were foreign tissue. One positive nutrition link centers around Mediterranean-style diets rich in antioxidants and the omega-3 fatty acid, EPA, found in fish oil.[116] EPA may interfere with activities of hormonelike chemicals involved in inflammation. The antioxidants in vegetables, fruits, nuts, and olive oil may also interfere with inflammation by reducing oxidative stress in the joints.[117] Supplemental doses of vitamin C may worsen the condition, if results of animal studies also prove true for people.[118] The same diet recommended for heart health—one low in saturated fats and high in fruit, vegetables, whole grains, and oils from fish—may help prevent or reduce the inflammation in the joints that makes arthritis so painful.[119]

No one universally effective diet for arthritis relief is known. Weight loss helps those who are overweight; strength training may help others.[120] Many *ineffective* or

Bogus or unproven arthritis treatments:

- Alfalfa tea.
- Aloe vera liquid.
- Any of the amino acids.
- Burdock root.
- Calcium.
- Celery juice.
- Copper or copper complexes.
- Dimethyl sulfoxide (DMSO).
- Fasting.
- Fresh fruit.
- Honey.
- Inositol.
- Kelp.
- Lecithin.
- Melatonin.
- Para-aminobenzoic acid (PABA).
- Raw liver.
- Selenium.
- Superoxide dismutase (SOD).
- Vitamin or mineral supplements.
- Watercress.
- Yeast.
- Zinc.
- 100 other substances.

arthritis a usually painful inflammation of joints caused by many conditions, including infections, metabolic disturbances, or injury; usually results in altered joint structure and loss of function.

■ Medical drugs used to relieve arthritis can impose nutrition risks. Controversy 14 explains.

unproven "cures" are sold, however, as the margin list on the previous page shows. Traditional medical interventions for arthritis include medication and surgery. Two popular supplements—glucosamine and chondroitin—may indeed relieve pain and improve mobility, and they may even slow the progression of osteoarthritis if substantial additional research bears out preliminary findings.[121]

> **KEY POINT** A diet high in fruits and vegetables and low in fats of meats and dairy products may improve some symptoms of arthritis. Omega-3 fatty acids may also have a positive effect.

Vitamin Needs

Vitamin A stands alone among the vitamins in that its absorption appears to increase with aging. For this reason some researchers have proposed lowering the vitamin A requirement for aged populations. Others resist this proposal because foods containing vitamin A and its precursor beta-carotene are under study for preventing oxidative damage to body tissues, an effect described in Chapter 7.

As people age, vitamin D synthesis declines fourfold, setting the stage for deficiency. Many older adults drink little or no vitamin D–fortified milk and get little or no exposure to sunlight. Thus, the recommendation for vitamin D intake has been doubled to 10 micrograms daily for people aged 51 to 70 and tripled to 15 micrograms for people 71 and over. Every elderly person should obtain this amount of vitamin D and get outside more often (or sit by a sunny open window for a while).[122]

The DRI committee has recommended that adults aged 51 years and older obtain 2.4 micrograms of vitamin B_{12} daily *and* that vitamin B_{12}–fortified foods (such as fortified cereals) or supplements be used to meet much of this intake.[123] The committee's recommendation reflects the finding that many people older than 50 years lose the ability to produce enough stomach acid to make the protein-bound form of vitamin B_{12} available for absorption. Vitamin B_{12} may be a problem for 10 to 15 percent of those over 60 who may suffer a serious deficiency that goes unrecognized and so untreated.[124] Synthetic vitamin B_{12} is reliably absorbed, however, and much misery can be averted by preventing deficiencies of vitamin B_{12} in elderly people.[125] In addition to other functions, a sufficiency of vitamin B_{12} along with two other B vitamins, folate and vitamin B_6, may prevent some loss of mental ability that commonly occurs among older people.[126] Other nutrients, including antioxidants such as vitamin E, may also play roles in conserving immunity, mental functions, and eyesight in the aged.[127]

Several links have emerged between the nutrients and phytochemicals in foods and age-related changes in the eyes. One theory concerns the leading cause of permanent blindness in people over age 60—macular degeneration, described in Chapter 7.[128] People with lifelong high intakes of vegetables, particularly the carotenoid-rich dark green leafy vegetables such as spinach and collard greens, rarely suffer from macular degeneration. These vegetables are rich in certain carotenoid phytochemicals that may protect the eyes from this destructive disease.[*129] Diets high in fat may also pose a risk for macular degeneration, but the fatty acids of fish oils may be protective.[130]

■ The macula of the eye was described in Chapter 7.

Another problem facing older people is **cataracts.** A cataract is a clouding of the lens that impairs vision and leads to blindness. Only 5 percent of people younger than 50 years have cataracts; by age 65, the percentage jumps to over 50 percent.

The lens of the eye is easily oxidized. People who shun fruits and green vegetables obtain too few antioxidants that may protect against both macular degeneration and cataracts. Some studies suggest that a diet providing ample carotenoids, vitamin C, and vitamin E may be especially important for preventing early onset of cataracts.[131] Also, people who follow the *Dietary Guidelines for Americans* are reported to have fewer cataracts, but cataracts can occur even in well-nourished individuals due to injury, sun exposure, or other trauma.[132] Most cataracts are vaguely called senile cataracts, meaning "caused by aging."

cataracts (CAT-uh-racts) clouding of the lens of the eye that can lead to blindness. Cataracts can be caused by injury, viral infection, toxic substances, genetic disorders, and, possibly, some nutrient deficiencies or imbalances.

---

*The carotenoids are lutein and zeaxanthin, which help to form pigments of the macula of the eye.

Water and the Minerals

Dehydration is a major risk for older adults: The thirst mechanism may become imprecise, and older people may go for long periods without drinking fluids. The kidneys also become less efficient in recapturing water before it is lost as urine. This water loss causes some problems and worsens others, such as constipation and bladder problems. In a person with asthma, dehydration thickens mucus in the lungs, which may then block airways. To prevent dehydration, older adults need to drink at least 6 cups of fluids each day to provide the needed water.[133]

A person we know uses this trick to ensure getting enough water: he keeps six inexpensive 8-ounce cups in the cupboard. Through the day he uses each to drink one cup of fluid, including juices and other beverages, collecting the used cups in the dish drain. (For clear soups and other hot beverages, he simply moves one of his cups for each 8 ounces consumed from his bowl or mug.) In the afternoon he checks the cupboard and drinks from any remaining cups. For him, drinking enough fluid has become a habit, and seldom are any cups left in the cupboard after supper.

Iron   Iron status generally improves in later life, especially in women after menstruation ceases and in those who take iron supplements, eat red meat regularly, and include vitamin C–rich fruits in their daily diet.[134] When iron-deficiency anemia does occur, diminished appetite with low food intake is often the cause. Aside from diet, other factors make iron deficiency likely in older people:

- Chronic blood loss from ulcers or hemorrhoids.
- Poor iron absorption due to reduced stomach acid secretion.
- Antacid use, which interferes with iron absorption.
- Use of medicines that cause blood loss, including anticoagulants, aspirin, and arthritis medicines.

Older people take more medicines than others, and drug and nutrient interactions are common.

Zinc   Zinc deficiencies are also common in older people. Zinc deficiency can depress the appetite and blunt the sense of taste, thereby leading to low food intakes and worsening of zinc status. Many medications interfere with the body's absorption or use of zinc, and older adults' medicine load can worsen zinc deficiency.

Research on zinc supplements demonstrates that nutrient supplements taken by the elderly can bring unexpected results. Researchers studying the immune response of elderly people sometimes observe a reduced immune response, sometimes an enhanced response, and sometimes no effect in those given supplements of zinc.[135]

Calcium   With aging, calcium absorption declines;; at the same time an estimated 75 percent of adult women in the United States fail to obtain enough calcium and many men do, too.[136] If fresh milk causes stomach discomfort, as the majority of older people report, then lactose-reduced milk or other calcium-rich foods should take its place.

Elderly people often benefit from a balanced low-dose vitamin and mineral supplement.[137] Older people taking such supplements suffer fewer sicknesses caused by infection. Vitamin A has been seen to depress the immunity of elders, while vitamin

Adults of all ages need 6 to 8 glasses of water each day.

Water recommendation for adults:
- 1 oz/kg actual body weight, with a minimum of 50 oz (6¼ c) daily.

◼ These foods provide iron and zinc together: meat, poultry, liver, oysters, whole grains, fortified breakfast cereals,* and legumes.

*Cereals fortified with iron and zinc may not be available in Canada.

◼ Calcium-rich foods are listed in Chapter 8, and Controversy 8 discusses the threat to the bones from osteoporosis.

TABLE 14•7 SUMMARY OF NUTRIENT CONCERNS IN AGING

Nutrient	Effects of Aging	Comments
Energy	Need decreases.	Physical activity moderates the decline.
Fiber	Low intakes make constipation likely.	Inadequate water intakes and physical activity, along with some medications, compound constipation.
Protein	Needs stay the same.	Low-fat, high-fiber legumes and grains meet both protein and other needs.
Vitamin A	Absorption increases.	Supplements normally not needed.
Vitamin D	Increased likelihood of inadequate intake; skin synthesis declines.	Daily moderate exposure to sunlight may be of benefit.
Vitamin B_{12}	Malabsorption of some forms.	Foods fortified with synthetic vitamin B_{12} or a low-dose supplement may be of benefit in addition to a balanced diet.
Water	Lack of thirst and increased urine output make dehydration likely.	Mild dehydration is a common cause of confusion.
Iron	In women, status improves after menopause; deficiencies linked to chronic blood losses and low stomach acid output.	Stomach acid required for absorption; antacid or other medicine use may aggravate iron deficiency; vitamin C and meat enhance absorption.
Zinc	Intakes are often inadequate and absorption may be poor, but needs may also increase.	Medications interfere with absorption; deficiency may depress appetite and sense of taste.
Calcium	Intakes may be low; osteoporosis becomes common.	Lactose intolerance commonly prevents milk intake; substitutes are needed.

TABLE 14•8 CHANGES WITH AGE YOU PROBABLY MUST ACCEPT

These changes are probably beyond your control:

✔ Graying of hair
✔ Balding
✔ Some drying and wrinkling of skin
✔ Impairment of near vision
✔ Some loss of hearing
✔ Reduced taste and smell sensitivity
✔ Reduced touch sensitivity
✔ Slowed reactions (reflexes)
✔ Slowed mental function
✔ Diminished visual memory
✔ Menopause (women)
✔ Loss of fertility (men)
✔ Loss of joint elasticity

Differences in maximum life span between animals eating normally and those that are energy restricted:

• Rats:
  Normal diet, 33 months.
  Restricted diet, 47 months.
• Spiders:
  Normal diet, 100 days.
  Restricted diet, 139 days.
• Single-celled animals (protozoans):
  Normal diet, 13 days.
  Restricted diet, 25 days.

E may enhance it. A summary of the effects of aging on nutrient needs appears in Table 14-7 above.

Can Nutrition Help People to Live Longer?

The evidence concerning nutrition and longevity is intriguing. In a classic study, researchers in California observed nearly 7,000 adults and noticed that some were young for their ages, while others were old for their ages.[138] To uncover what made the difference, the researchers focused on health habits and identified six factors that affect physiological age. Three of the six factors were related to nutrition:

• Abstinence from, or moderation in, alcohol use.
• Regular nutritious meals.
• Weight control.

The other three were regular adequate sleep, abstinence from smoking, and regular physical activity. The physical health of those who engaged in all six positive health practices was comparable to that of people 30 years younger who engaged in few or none. Numerous studies have confirmed the benefits of such lifestyle factors that people can control.[139] These findings suggest that even though people cannot alter the year of their birth, they can alter the probable length and quality of their lives. Table 14-8 in the margin and Table 14-9 on the facing page list some changes of aging that are beyond control and some that may yield to lifestyle influences.

Evidence that diet might influence life span emerged more than a half-century ago from experiments on rats. Researchers fed a group of young rats diets extremely low in energy, while control rats ate normally. The starved rats stopped growing while the control rats grew normally; when the researchers increased food energy, growth resumed. Many of the starved group died young from malnutrition. The few survivors, though permanently deformed from their ordeal, remained alive far beyond the normal life span for such animals and developed diseases of aging much later than normal.

Present-day studies have repeated these findings with more moderate diets that mildly restrict energy and provide adequate nutrients and do not inflict physical malformations. Restricted animals retain youthfulness longer and develop fewer of the factors associated with chronic diseases such as high blood pressure, changes in immune functions, and glucose intolerance.[140] News from an ongoing study of energy-restricted monkeys reports that as the animals age, the restricted group has retained healthy blood glucose and insulin responses and has healthier blood lipid profiles than the freely fed controls.[141] Even occasional fasting or energy restriction without lifetime restriction may provide benefits, at least in laboratory mice.[142] Evidence from other species (see the margin of the facing page) suggests that this effect spans many biological systems.

As for whether such animal findings can be applied to human beings, the answer may be yes. Human beings share a significant portion of their genome with other species, making human beings more or less metabolically similar to them (see the margin). Many of the physiological responses seen in rats during energy restriction are also observed in human beings who moderately restrict energy intakes. Researchers recently measured blood pressure, blood LDL concentration, and other atherosclerosis risk factors in a group of 18 people who have for years voluntarily reduced their caloric intakes by 1,100 to 1,900 calories a day. The researchers concluded that "calorie restriction results in profound and sustained beneficial effects on the major atherosclerosis risk factors."[143] Much more evidence must be collected, however, before such findings can be applied to the general population.

Investigators have proposed several mechanisms to explain how energy restriction prolongs life in rats, but none has been proved. A delay in the onset of age-related diseases has already been mentioned. Genetics may play a role: genes that modulate the aging process seem to become less active when food energy is scarce.[144] Current research is focused on the genetic response of both young and old rats to energy restriction in hopes of identifying genes that play key roles in aging. Some predict that future drugs and treatments may one day mimic the effects of energy restriction at the genetic level and so confer its life-extending effects without the drawbacks of calorie restriction.[145]

A free-radical hypothesis blames damage from oxidative stress for the physical deterioration associated with aging. The body's internal antioxidant enzymes diminish with age. Many "age-related" degenerative diseases are linked to free-radical damage while the antioxidant-rich Mediterranean diet is linked with longevity.[146] However, the storm of "life-extending" pills, supplements, and treatments on the market are worthless. Research proves that it's better to spend money on fresh fruit and green and yellow vegetables, which are natural rich sources of antioxidants and are linked with many health benefits (see Controversy 2 and Controversy 5).

> ▶ KEY POINT   Lifestyle factors can make a difference in aging. In rats and other species, food energy deprivation may lengthen the lives of individuals. Claims for life extension through antioxidants or other supplements are common hoaxes.

Can Foods or Supplements Affect the Course of Alzheimer's Disease?

Today, Alzheimer's disease affects 4.5 million people in the United States, and that number is expected to almost triple by the year 2050.[147] In Alzheimer's disease, the most prevalent form of **senile dementia,** abnormal deterioration occurs in the areas of the brain that coordinate memory and cognition. In Alzheimer's disease, the brain is littered with clumps of abnormal protein fragments that clog the brain and damage or kill certain nerve cells.* The defining symptom of Alzheimer's is impairment of memory and reasoning powers, but it may be accompanied by loss of the ability to communicate, loss of physical capabilities, anxiety, delusions, depression, inappropriate behavior, irritability, sleep disturbance, and eventually loss of life itself.[148] Once the destruction begins, the outlook for its reversal is most often bleak.[149] More research is needed, and quickly, to perfect a drug or vaccine to block the destructive progression of the disease.[150]

Research has revealed only weak links between nutrition and Alzheimer's disease (see Table 14-10, page 546). For example, a causal connection with a buildup of the mineral

---
*The protein fragments are called beta-amyloid.

According to the National Institutes of Health, the human genome is:

- 30% similar to yeast.
- 40% similar to a worm.
- 50% similar to a banana.
- 60% similar to a housefly.
- 90% similar to a mouse.
- 98.4% similar to a monkey.
- 99.9% similar to other human beings (except for identical twins who share an identical genome).

TABLE 14·9   CHANGES WITH AGE YOU PROBABLY CAN SLOW OR PREVENT

By exercising, eating an adequate diet, reducing stress, and planning ahead, you may be able to slow or prevent:

- ✔ Wrinkling of skin due to sun damage
- ✔ Some forms of mental confusion
- ✔ Elevated blood pressure
- ✔ Accelerated resting heart rate
- ✔ Reduced lung capacity and oxygen uptake
- ✔ Increased body fatness
- ✔ Elevated blood cholesterol
- ✔ Slowed energy metabolism
- ✔ Decreased maximum work rate
- ✔ Loss of sexual functioning
- ✔ Loss of joint flexibility
- ✔ Diminished oral health: loss of teeth, gum disease
- ✔ Bone loss
- ✔ Digestive problems, constipation

■ Some degree of memory loss is often simply a function of aging and is termed *benign* (meaning "harmless") *senescent* (meaning "of aging") *forgetfulness.* Occasional forgetful moments generally do not forecast the development of Alzheimer's disease in an older person.

senile dementia the loss of brain function beyond the normal loss of physical adeptness and memory that occurs with aging.

TABLE 14•10 POSSIBLE LINKS BETWEEN NUTRITION AND ALZHEIMER'S DISEASE

Researchers are exploring the following links between nutrition and Alzheimer's disease.

- *Aluminum in the brain tissues.* Brain aluminum exceeds normal brain aluminum by some 10 to 30 times, but blood and hair aluminum remains normal, indicating that the accumulation is caused by something in the brain itself, not by high aluminum in the diet.
- *Elevated levels of copper, iron, and zinc in the brain tissues.* Such elements may accelerate the progression of the disease, possibly by increasing the formation of free radicals that produce oxidative stress.[a] Their accumulation, however, may simply be a result of the disease process.
- *Low antioxidant nutrient intake.* Free radicals may attack brain tissue, damaging DNA, cell membranes, and proteins. This effect could plausibly be worsened by a diet low in fruits and vegetables, antioxidant nutrients, or soy foods.[b]
- *Low fish oil and fish intakes.* A number of observations suggest that low fish intake is often linked with Alzheimer's disease incidence. A recent study of over 800 elderly people concludes that intake of omega-3 fatty acids and weekly meals of fish may reduce the risk.
- *Obesity and overweight.* In one study, the effect of body weight in the elderly on development of Alzheimer's disease, particularly among women, was dramatic—disease risk increased 36% for each 1.0-point increase in body mass index (see inside back cover).[c]
- *Fats and saturated fats.* No clear consensus exists on whether a high intake of saturated or *trans* fat may increase risk or whether unsaturated and unhydrogenated fats may reduce it.[d]
- *Gene and diet interactions.* Several genes responsible for rare, inherited Alzheimer's disease account for about 5% of cases, but investigations into the roles of other genes suggest they may increase susceptibility. Dietary factors, such as a high-calorie diet or folate deficiency, may foster the expression of the disease in those with a genetic predisposition to develop it.[e]

[a]M. J. Engelhart and coauthors, Dietary intake of antioxidants and risk of Alzheimer disease, *Journal of the American Medical Association* 287 (2002): 3223–3229; M. C. Morris, Dietary intake of antioxidant nutrients and the risk of incident Alzheimer disease in a biracial community study, *Journal of the American Medical Association* 287 (2002): 3230–3237; Y. Christen, Oxidative stress and Alzheimer disease, *American Journal of Clinical Nutrition* 71 (2000): 621S–629S.

[b]R. H. Liu, Health benefits of fruit and vegetables are from additive and synergistic combinations of phytochemicals, *American Journal of Clinical Nutrition* 78 (2003): 517S–520S; A. Martin, Antioxidant vitamins E and C and risk of Alzheimer's disease, *Nutrition Reviews* 61 (2003): 69–73; J. A. Joseph and coauthors, Blueberry supplementation enhances signaling and prevents behavioral deficits in an Alzheimer disease model, *Nutritional Neuroscience* 6 (2003): 153–162; S. E. File, New studies: Soy appears to benefit cognitive function, *The Soy Connection,* July 2003, pp. 1–2; L. R. White and coauthors, Brain aging and midlife tofu consumption, *Journal of the American College of Nutrition* 19 (2000): 207–209; M. Grundman, Vitamin E and Alzheimer disease: The basis for additional clinical trials, *American Journal of Clinical Nutrition* 71 (2000): 630S–636S.

[c]D. Gustafson and coauthors, An 18-year follow-up of overweight and risk of Alzheimer disease, *Archives of Internal Medicine* 163 (2003): 1524–1528.

[d]M. C. Morris and coauthors, Dietary fats and the risk of incident Alzheimer disease, *Archives of Neurology* 60 (2003): 194–200; M. J. Engelhart and coauthors, Diet and risk of dementia: Does fat matter? The Rotterdam Study, *Neurology* 59 (2002): 1915–1921.

[e]A. Rocchi and coauthors, Causative and susceptibility genes for Alzheimer's disease: A review, *Brain Research Bulletin* 61 (2003): 1–24; M. P. Mattson, Gene-diet interactions in brain aging and neurodegenerative disorders, *Annals of Internal Medicine* 139 (2003): 441–444.

aluminum seems unlikely. There is conflicting evidence as to whether supplements of zinc or other trace minerals worsen Alzheimer's disease, so to err on the safe side, food sources, not concentrated supplements, of trace minerals are advisable for people with the disease.[151] Also, supplements of choline and lecithin (first mentioned in Chapter 5), once investigated as potential treatments, have had no consistent effect on memory, mental functioning, or the progression of Alzheimer's. Fish oils may hold promise, however—learning and memory depend on DHA, an omega-3 fatty acid. Diets deficient in omega-3 fatty acids produce degeneration of certain brain proteins in mice, a finding researchers hope will yield treatments to avert some Alzheimer's damage in people.[152]

The results of one small study seemed to indicate a modest benefit from the medicinal herb ginkgo biloba in cognitive and social functioning in Alzheimer's patients. More recent results from well-controlled studies do not support this result, however.[153] To date, no proven benefits are available from herbs or other remedies, but claims from quacks are all too commonplace.

■ "Smart" drugs, drinks, and supplements, sold with promises of brainpower enhancement, are generally useless except for the caffeine or other stimulant "wake-up" effects.

Preventing weight loss is an important nutrition concern for the person suffering with Alzheimer's disease.[154] Depression and forgetfulness can lead to skipped meals and poor food choices. Caregivers can help by providing well-liked, well-balanced, and well-tolerated meals and snacks served in a cheerful, peaceful atmosphere.

▶ KEY POINT   Alzheimer's disease causes some degree of brain deterioration in many people past age 65. Current treatment helps only marginally; dietary aluminum is probably unrelated. The importance of nutrition care increases as the disease progresses.

Food Choices of Older Adults

Most older people are independent, socially sophisticated, mentally lucid, fully participating members of society who report themselves to be happy and healthy. The quality of life among the 85 and older group has improved, and their chronic disabilities have declined dramatically in recent years.[155] Results of national surveys indicate that many older people have heard and heeded nutrition messages: they have cut down on saturated fats in dairy foods and meats and are eating slightly more vegetables and whole-grain breads. Those who choose a wide variety of foods are found to have better nutrient status than those who eat monotonous diets.[156] Grocers assist the elderly by prominently displaying good-tasting, low-fat, nutritious foods in easy-to-open, single-serving packages with labels that are easy to read.

Many nutrient and other supplements are marketed for older adults. Whether to take a supplement is a personal choice, but evidence supports the idea that a single low-dose multivitamin-mineral tablet a day can improve resistance to disease in older people. The best choice would be a supplement low in vitamin A, with ample amounts of all of the other vitamins and minerals for which DRI values are set (see the inside front cover).

Shared meals can be the high point of the day.

Obstacles to Adequacy   Many factors affect the food choices and eating habits of older people, including whether they live alone or with others, at home or in an institution. Men living alone, for example, are likely to consume poorer-quality diets than those living with spouses. Older people who have difficulty chewing because of tooth loss or loss of taste sensitivity may no longer seek a wide variety of foods. Medical conditions and functional losses can also adversely affect food choices and nutrition.[157] Many older people become weak when unintentional reductions in food intake result in weight loss and loss of muscle tissue, events often followed by illness or death.[158] It may be that some of these outcomes could have been prevented or delayed if the person had consumed an adequate diet.[159]

Two other factors seem to make older people vulnerable to malnutrition: use of multiple medications and abuse of alcohol. People over age 65 take about a fourth of all the medications, both prescription and over-the-counter, sold in the United States. Although these medications enable people with health problems to live longer and more comfortably, they also pose a threat to nutrition status because they may interact with nutrients, depress the appetite, or alter the perception of taste (see Controversy 14).

The incidence of alcoholism, alcohol abuse, or problem drinking among the elderly in the United States is estimated at between 2 and 10 percent. Evidence is mounting that loneliness, isolation, and depression in the elderly accompany overuse of alcohol. It isn't possible to say whether the depression or the alcohol abuse comes first, for each worsens the other, and both detract from nutrient intakes. Table 14-11 (p. 548) and the margin list provide means of identifying those who might be at risk for malnutrition.

The DETERMINE predictors of malnutrition in the elderly:

- **D**isease.
- **E**ating poorly.
- **T**ooth loss or oral pain.
- **E**conomic hardship.
- **R**educed social contact.
- **M**ultiple medications.
- **I**nvoluntary weight loss or gain.
- **N**eed of assistance with self-care.
- **E**lderly person older than 80 years.

Programs That Help   Federal programs can provide help for older people.[160] Social Security provides income to retired people over age 62 who paid into the system during their working years. The Food Stamp Program assists the very poor by supplementing

Circle the number to the right if the statement applies to you.

Statement	Yes
I have an illness or condition that makes me eat different kinds and/or amounts of food.	2
I eat fewer than 2 meals per day.	3
I eat few fruits or vegetables and use few milk products.	2
I have 3 or more drinks of beer, liquor, or wine almost every day.	2
I have tooth or mouth problems that make it hard for me to eat.	2
I don't always have enough money to buy the food I need.	4
I eat alone most of the time.	1
I take 3 or more different prescribed or over-the-counter drugs a day.	1
Without wanting to, I have lost or gained 10 pounds in the last 6 months.	2
I am not always physically able to shop, cook, and/or feed myself.	2
	Total ___

Score:
0–2: Good. Recheck your score in 6 months.
3–5: Moderate nutritional risk. Visit your local office on aging, senior nutrition program, senior citizens center, or health department for tips on improving eating habits.
6 or more: High nutritional risk. See your doctor, dietitian, or other health-care professional for help in improving your nutrition status.

NOTE: The Nutrition Screening Initiative is part of a national effort to identify and treat nutrition problems in older Americans.

■ The Senior Nutrition Program is part of the Child and Adult Care Food Program (CACFP), which is designed to help public and private nonresidential child and adult day-care programs provide nutritious meals to those younger than age 12 or older than 65, or people with disabilities.

Federal Sources of support for the elderly:

• Social Security.
• Food Stamps.
• Senior Nutrition Program.
• Meals on Wheels.

their monthly food budgets with a card similar to a credit card, encrypted with benefits and redeemable for food. The Senior Nutrition Program provides nutritious meals in a social congregate setting, education and shopping assistance, counseling and referral to other needed services, and transportation to necessary appointments. An estimated 25 percent of the nation's elderly poor benefit from meals provided by the program. For the homebound, Meals on Wheels volunteers deliver meals to the door, a benefit even though the recipients miss out on the social atmosphere of the congregate meals. Nutritionists are wise not to focus solely on nutrient and food intakes of the elderly because enjoyment and social interactions may be as important as food itself.[161]

Many older people, even able-bodied ones with financial resources, find themselves unable to perform cooking, cleaning, and shopping tasks. For anyone living alone and particularly for those of advanced age, it is important to work through the problems that food preparation presents. This chapter's Food Feature presents some ideas.

▶ KEY POINT   Food choices of the elderly are affected by aging, altered health status, and changed life circumstances. Assistance programs can help by providing nutritious meals, offering opportunities for social interactions, and easing financial problems.

FOOD FEATURE

Single Survival and Nutrition on the Run

WHEN IT COMES to feeding themselves wisely, singles of all ages face problems ranging from selection of restaurant foods to the purchasing, storing, and preparing of food from the grocery store. Whether the single person is a busy student in a college dormitory, an elderly person in a retirement apartment, or a professional in an efficiency suite, the problems of preparing nourishing meals are often the same. People who live in places without kitchens and freezers find storing foods problematic. Following is a collection of ideas gathered from single people who have devised answers to some of these problems.

Is Eating in Restaurants the Answer?

For the single person as for others, restaurants mean convenience. On average, 45 percent of a U.S. household's food budget is spent on foods prepared and eaten away from home. And in any given month, up to 70 percent of households consume food from a carry-out restaurant. Restaurant foods may be the quickest, easiest, and least taxing way to satisfy hunger at mealtime, but can they meet your body's nutrient needs or support health as well as homemade foods? The answer is "perhaps," if the diner makes the effort to meet nutritional needs.[162]

A few chefs and restaurant owners are concerned with the nutritional health of their patrons, but more often chefs strive to please the palate and leave nutrition-conscious diners on their own. Restaurant foods are often overly endowed with calories, fat, saturated fat, and salt, yet not overly generous with needed constituents such as fiber, iron, or calcium.[163] Meat portions may exceed a whole day's allowance in the USDA Food Guide. Nevertheless, restaurants can provide both convenience and nutrition if you follow these suggestions: restrict your portions to sizes that do not exceed your energy needs, ask that excess portions be placed in take-out containers, and make judicious choices of foods that stay within intake guidelines for fat and salt. The Food Feature of Chapter 5 offered specific suggestions for ordering fast food and other foods with an eye to keeping fat intakes within bounds, and Chapter 8 presented a list of high-sodium foods.

Grocery Store Take-Out Choices

Take-out delicatessen-style foods offer convenience—they can be purchased while shopping for other items—and a modicum of control over their nutrient contents. They also often cost substantially less than similar foods from restaurants. Another bonus is control: you can specify the amount you need and portion it onto your plate at home.

Grocery store take-out can be an excellent bargain in terms of nutrient density, too. Choose from among roast chicken, smoked seafood, pasta with tomato sauce, steamed vegetables, precut salads and fruit without dressings, cooked beans, and plain baked potatoes for convenient nutrient bargains. Table 14-12 (p. 550) offers additional tips for using grocery store foods to best advantage. Be aware that stuffing, macaroni and cheese, meat loaf and gravy, vegetables with creamy sauces, mayonnaise-dressed mixed salads, and fried chicken and fish can be as laden with saturated fat and calories as any traditional fast food.

Keep in mind that ready-to-eat take-out foods can spoil. Transport them home quickly and refrigerate them or reheat them within two hours of purchase (within an hour if the temperature is above 90° Fahrenheit).[164] Reheat the foods to an internal temperature of 165°F, and bring soups, gravies, and sauces to a boil to prevent foodborne illness.

More Grocery Store Know-How

Singles often face the quantity problem in the grocery store. Large packages of meat and vegetables, whether fresh or frozen, are suitable for a family of four or more, and even a head of lettuce can spoil before one person can use it all.

Buy only what you will use. Don't be timid about asking the grocer to break open a family-size package of wrapped meat or fresh vegetables. Look for bags of prepared salad greens to take the place of lettuce in both salads and sandwiches. Purchasing prepared salads and other small containers of food may be expensive, but it is also expensive to let the unused portion of a large container spoil. Buy only three pieces of each kind of fresh fruit: a ripe one, a medium-ripe one, and a green one. Eat the first right away and the second soon, and let the last one ripen to eat days later.

Think up ways to use a vegetable that you must buy in large quantity. For example, you can divide a head of cauliflower into thirds. Cook one-third and eat it as a hot vegetable. Toss another third into a salad dressing marinade for use as an appetizer. Blend up the rest, cooked, in a creamy soup.

Buy fresh milk in the size you can best use. If your grocer doesn't carry pints or quarts of milk, try a convenience store. If you eat lunch in a cafeteria, buy two pints of milk—one to drink and one to take home and store. Buy a loaf of bread and immediately store half, well wrapped, in the freezer (not the refrigerator, which will make it stale).

Food Preparation Hints

A wise person once said, "An hour spent organizing can save three hours later on."* This holds true in food preparation.

---

*That wise person was the late Eva May Hamilton, one of the original authors of this textbook.

■ Chapter 12 has more about preventing foodborne illnesses.

For shelf-stable items, prepare a space for rows of glass jars (jars from spaghetti sauce, applesauce, or other foods work well). Use the jars to store pasta, rice, lentils, other dry beans, flour, cornbread or biscuit mix, dry fat-free milk, and cereal. Light destroys riboflavin, so use opaque jars for enriched pasta and dry milk. Cut the directions-for-use label from the package of each item and store it near the jar. Place each jar, tightly sealed, in the freezer for a few days to kill any eggs or organisms before storing it on the shelf. Then the jars will keep bugs out of the foods indefinitely. The jars are also pretty to look at and will remind you of possibilities for variety in your menus.

Experiment with stir-fried foods. A large fry pan works well to stir-fry a variety of vegetables and meats. Inexpensive vegetables such as cabbage and celery are delicious when crisp cooked in a little oil with soy sauce or lemon juice. Interesting frozen vegetable mixtures are available, or cooked leftover vegetables can be dropped into a stir-fry at the last minute. A bonus of a stir-fried meal is that you have only one pan to wash.

Make mixtures using what you have on hand. A thick stew prepared from any leftover vegetables and bits of meat, with some added frozen onions, peppers, celery, and potatoes, makes a complete and balanced meal, except for milk. If you like creamed gravy, add some fat-free dry milk to your stew.

If you can afford a microwave oven, buy one. Cooking times are quick, and you'll use fewer pots and pans. Be sure to use containers designed for the microwaving, however. Margarine tubs, plastic bowls, and storage bags and containers can release potentially harmful chemicals into food when they are heated in the microwave oven. Use glass, or buy plastic containers that are labeled as safe for microwaving.

Depending on your freezer space, make a regular-size recipe of a dish that takes time to prepare: a casserole, vegetable pie, or meat loaf. Freeze individual portions in containers that can be heated later. Date these so you will use the oldest first.

Dealing with Loneliness

For nutrition's sake, it is important to attend to loneliness at mealtimes. The person who is living alone must learn to connect food with socializing. Invite guests, and make enough food so that you can enjoy the leftovers later on. If you know an older person who eats alone, you can bet that person would love to join you for a meal now and then.

TABLE 14·12 TIME-SAVING TIPS TO TURN CONVENIENCE FOODS INTO NUTRITIOUS MEALS

- Add extra nutrients and a fresh flavor to canned stews and soups by tossing in some frozen ready-to-use mixed vegetables. Choose vegetables frozen without salty, fatty sauces—prepared foods generally contain enough salt to season the whole dish including added vegetables.
- Buy frozen vegetables in a bag, toss in a variety of herbs, and use as needed. Vary your choices to prevent boredom.
- When grilling burgers, wrap a mixture of frozen broccoli, onion, and carrots in a foil packet with a tablespoon of Italian dressing and grill alongside the meat for seasoned grilled vegetables.
- Use canned fruits in their own juices as desserts. Toss in some frozen berries or peach slices and top with flavored yogurt for an instant fruit salad.
- Prepared rice or noodle dishes are convenient, but those claiming to contain broccoli, spinach, or other vegetables really contain just a trifle—not nearly enough to contribute meaningfully to the day's need for vegetables. Pump up the nutrient value by adding a half-cup of your frozen vegetables per serving of pasta or rice just before cooking.
- Purchase frozen onion, mushroom, and pepper mixtures to embellish jarred spaghetti sauce or small frozen pizzas. Top with parmesan cheese.
- Use frozen shredded potatoes, sold for hash browns, in soups or stews, or mix with a handful of shredded reduced-fat cheese or a can of fat-free "cream of anything" soup and bake for a quick and hearty casserole.

MEDIA MENU @ 🏠 ◀ ▶

Nutrition Connections

The Nutrition Connections CD provides access to these additional resources.

Interaction
The animation, *Nutrient Needs in Later Years,* looks at nutrition issues unique to the older adult.

Chapter Quiz
Take the quiz for this chapter to assess your understanding of chapter concepts.

Nutrition on the Net
Annotated links to reputable nutrition-related websites.

Online Glossary
Test your knowledge of key vocabulary through this comprehensive, interactive glossary.

Internet Connections

A variety of helpful study tools for this chapter are available at our website
http://nutrition.wadsworth.com/ncc10

Practice Test
Take the practice test for this chapter and gauge your knowledge of key concepts; answers are provided.

Student Notebook
A lecture notebook is available to help organize note taking while reading or attending lecture.

ABC Video Clip
"Lunches for Kids" (6:03) examines the nutritional quality of lunches served at schools. "Overweight Kids" (5:55) examines childhood obesity.

Study Tools
Study resources for students include Chapter Summaries, Crossword Puzzles, Flashcards, Glossary, Internet Exercises, and Learning Objectives.

SELF CHECK

Answers to these Self Check questions are in Appendix G.

1. Children naturally like nutritious foods in all the food groups, *except:*
   a. dairy
   b. meats
   c. vegetables
   d. fruits

2. Which of the following can contribute to choking in children?
   a. peanut butter eaten by the spoonful
   b. hot dogs and tough meat
   c. grapes and hard candy
   d. all of the above

3. Which of the following is most commonly deficient in children and adolescents?
   a. folate
   b. zinc
   c. iron
   d. vitamin D

4. Which of the following may worsen symptoms of PMS?
   a. adequate vitamin B_6
   b. physical activity
   c. caffeine
   d. calcium

5. Which of the following have been shown to improve acne?
   a. avoiding chocolate and fatty foods
   b. Accutane
   c. vitamin A supplements
   d. stress

6. Physical changes of aging that can affect nutrition include:
   a. reduced stomach acid
   b. increased saliva output
   c. tooth loss and gum disease
   d. a and c

7. Research to date supports the idea that food allergies or intolerances are common causes of hyperactivity in children.   T F

8. Nutrition does not seem to play a role in the causation of osteoarthritis.   T F

9. Vitamin A absorption decreases with age.   T F

10. Herbal supplements have been shown to slow down the progression of Alzheimer's disease.   T F

CONTROVERSY 14

NUTRIENT-DRUG INTERACTIONS: WHO SHOULD BE CONCERNED?

A 45-YEAR-OLD Chicago business executive attempts to give up smoking with the help of nicotine gum. She replaces smoking breaks with beverage breaks, drinking frequent servings of tomato juice, coffee, and colas. She is discouraged when her stomach becomes upset and her craving for tobacco continues unabated despite the nicotine gum. Problem: nutrient-drug interaction.

A 14-year-old girl develops frequent and prolonged respiratory infections. Over the past six months, she has suffered constant fatigue despite adequate sleep, has had trouble completing school assignments, and has given up playing volleyball because she runs out of energy on the court. During the same six months, she has been taking huge doses of antacid pills each day because she heard this was a sure way to lose weight. Her pediatrician has diagnosed iron-deficiency anemia. Problem: nutrient-drug interaction.

A 30-year-old schoolteacher who benefits from antidepressant medication attends a faculty wine and cheese party. After sampling the cheese with a glass or two of red wine, his face becomes

flushed. His behavior prompts others to drive him home. In the early morning hours, he awakens with severe dizziness, a migraine headache, vomiting, and trembling. An ambulance delivers him to an emergency room where a physician takes swift action to save his life. Problem: nutrient-drug interaction.

Medicines and Nutrition

People sometimes think that medical drugs do only good, not harm. As the opening stories illustrate, however, both prescription and over-the-counter (OTC) medicines can have unintended consequences, causing harm when they interact with the body's normal use of nutrients.[*1] As Figure C14-1 shows, drugs can interact with nutrients in the following ways:

- Foods or nutrients can enhance, delay, or prevent drug absorption.

---

*Reference notes are found in Appendix F.

- Drugs can enhance, delay, or prevent nutrient absorption.
- Nutrients can alter the distribution of a drug among body tissues or interfere with its metabolism, transport, or elimination from the body.
- Drugs can alter the distribution of a nutrient among body tissues or interfere with its metabolism, transport, or excretion.[2]
- Drugs also often modify taste, appetite, or food intake.
- Herbs can also modify drug effects.

These interactions do not occur every time a person takes a drug. Some people are more vulnerable than others to nutrient-drug interactions. The potential for undesirable nutrient-drug interactions is greatest for those who:

- Take drugs (or medicines) for long times.
- Take two or more drugs at the same time.
- Are poorly nourished to begin with or are not eating well.

Alcohol is also infamous for its interactions with nutrients (see Controversy 3).

Many people seeking the healing power of medicines also turn to herbs, but few inform their physicians. Some known herb-drug interactions are found

© Michael Newman/PhotoEdit

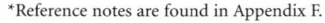

FIGURE C14·1 FOOD, DRUG, AND HERB INTERACTIONS

Foods, nutrients, and herbs

Drugs, including prescription, over-the-counter, tobacco, caffeine, and others

Enhance/delay/prevent absorption

Nutrients increase/decrease drug action/metabolism/excretion

Drugs increase/decrease nutrient action or excretion

Drugs modify appetite and taste

Herbs modify the actions of drugs

TABLE C14·1 HERB AND DRUG INTERACTIONS

Herb	Drug	Interaction
Bilberry, dong Quai, feverfew, garlic, ginger, ginkgo biloba, ginseng, meadowsweet, St. John's wort, turmeric, and willow	Warfarin, coumarin (anticlotting drugs, "blood thinners"); aspirin, ibuprofen, and other nonsteroidal anti-inflammatory drugs	Prolonged bleeding time; danger of hemorrhage
Black tea, St. John's wort, saw palmetto	Iron supplement; antianxiety drug	Tannins in herbs inhibit iron absorption; St. John's wort speeds antianxiety drug clearance.
Borage, evening primrose oil	Anticonvulsants	Seizures
Chinese herbs (xaio chai hu tang)	Prednisone (steroid drug)	Decreased blood concentrations of the drug
Echinacea (possible immunostimulant)	Cyclosporine and corticosteroids (immunosuppressants)	May reduce drug effectiveness
Feverfew	Aspirin, ibuprofen, and other nonsteroidal anti-inflammatory drugs	Drugs negate the effect of the herb for headaches.
Garlic supplements	Protease inhibitors (HIV-AIDS[a] drugs)	Decreased blood concentrations of the drug
Ginseng	Estrogens, corticosteroids	Enhanced hormonal response
Ginseng, hawthorn, kyushin, licorice, plantain, St. John's wort, uzara root	Digoxin (cardiac antiarrhythmic drug derived from the herb foxglove)	Herbs interfere with drug action and monitoring.
Ginseng, karela	Blood glucose regulators	Herbs affect blood glucose levels.
Kelp (iodine source)	Synthroid or other thyroid hormone replacers	Herb may interfere with drug action.
Licorice	Corticosteroids (oral and topical ointments)	Overreaction to drug (potentiation)
Panax ginseng	Antidepressants	Over excitability, mania.
St. John's wort	Increased enzymatic destruction of many drugs; Cyclosporin (immunosuppressant); anti-retroviral drugs (HIV[a] drugs), warfarin (anticoagulant, used to reduce blood clotting)	Decreased drug effectiveness; increased organ transplant rejection; reduced effectiveness of drugs to treat AIDS[a], reduced anticoagulant effect.
	MAOIs (used to treat depression, tuberculosis, or high blood pressure)[b]	Potentiation, with serotonin syndrome (mild): sweating, chills, blood pressure spike, nausea, abnormal heartbeat, muscle tremors, seizures
Valerian	Barbiturates (sedatives)	Enhanced sedation

[a] Acquired Immune Deficiency Syndrome, caused by HIV infection (human immunodeficiency virus).
[b] MAOI stands for monoamine oxidase inhibitors.
NOTE: A valuable free resource for reliable online information about herbs is offered by the Memorial Sloan-Kettering Cancer Center at www.mskcc.org/aboutherbs.

in Table C14-1. For example, many herbs, including ginkgo biloba, feverfew, and willow, have the potential to increase the anti-blood clotting effect of OTC pain and fever reducers, especially aspirin and its relatives.[3]

Absorption of Drugs and Nutrients

The business executive described earlier felt the effects of the first type of interaction in the list above. Acid from the tomato juice, coffee, and colas she drank before chewing the nicotine gum kept the nicotine from being absorbed into the bloodstream through the lining of her mouth. With absorption blocked, the nicotine could not quell her craving but instead traveled to her stomach and caused nausea (see Table C14-2 for other interactions with nicotine gum).

An interaction that can have serious consequences occurs when dairy products or calcium-fortified juices interfere with the absorption of certain antibiotics.[4] Without absorption of the proper dose, the antibiotics fail to do their jobs, and dangerous infectious diseases can worsen. Even the stomach acid normally secreted in response to eating can destroy some antibiotics, thereby reducing the dose. Drug labels include instructions for avoiding most such interactions, such as "Take on an empty stomach" or "Do not combine with dairy products."

Drugs can also interfere with the small intestine's absorption of nutrients, particularly minerals. This interaction explains the experience of the tired 14-year-old. Her overuse of antacids eliminated the stomach's normal acidity, on which iron absorption depends. The medicine bound tightly to the iron molecules, forming an insoluble, unabsorbable complex. Her iron stores already bordered on deficiency, as iron stores for young girls typically do, so her misuse of antacids pushed her over the edge into outright deficiency.

Chronic laxative use can also lead to malnutrition. Laxatives can carry nutrients through the intestines so rapidly that many vitamins have no time to be

TABLE C14·2 FOODS AND BEVERAGES THAT LIMIT THE EFFECTIVENESS OF NICOTINE GUM

- Apple juice
- Beer
- Coffee
- Colas
- Grape juice
- Ketchup
- Lemon-lime soda
- Mustard
- Orange juice
- Pineapple juice
- Soy sauce
- Tomato juice

absorbed. Mineral oil, a laxative the body cannot absorb, can rob a person of fat-soluble vitamins. Vitamin D deficiencies can occur this way; calcium can also be excreted with the oil, potentially accelerating adult bone loss.

Metabolic Interactions and Nutrient Excretion

The teacher who landed in the emergency room was taking an antidepressant medicine, one of the monoamine oxidase inhibitors (MAOI). At the party, he suffered a dangerous chemical interaction between the medicine and the compound tyramine in his cheese and wine. Tyramine is produced during the fermenting process in cheese and wine manufacturing.

The MAOI medication works by depressing the activity of enzymes that destroy the brain neurotransmitter dopamine. With less enzyme activity, more dopamine is left, and depression lifts. At the same time, the drug also depresses enzymes in the liver that destroy tyramine. Ordinarily, the man's liver would have quickly destroyed the tyramine from the cheese and wine. But due to the MAOI medication, tyramine built up too high in the man's body and caused the potentially fatal reaction.

Other culprits that affect the metabolism of medication include grapefruit juice, soy milk, and one of the most popular herbal supplements in the United States, ginkgo biloba. A chemical constituent of grapefruit juice suppresses an enzyme responsible for breaking down more than 20 kinds of medical drugs. With less drug breakdown, doses build up in the blood to levels that can have undesirable effects on the body. For example, in a drinker of grapefruit juice, a normal dosage of the blood-thinning drug coumarin can lead to dangerously prolonged bleeding and delayed clotting of blood. Soy milk seems to have the opposite effect: it reduces blood levels of a related blood-thinning drug.[5] As for ginkgo, people take it as a supplement in hopes of improving memory, but this effect is unproved. Takers may not know that it has been found to stimulate the activity of liver enzymes responsible for metabolizing many medications, and so may diminish their effects.

Drugs often cause nutrient losses, too. Many people take large quantities of aspirin (10 to 12 tablets each day) to relieve the pain of arthritis, backaches, and headaches. This much aspirin can speed up blood loss from the stomach by as much as ten times, enough to cause iron-deficiency anemia in some people. People who take aspirin regularly should eat iron-rich foods regularly as well. Table C14-3 lists some examples of other possible nutrient-drug interactions, including both prescription and OTC medications. Details on some common interactions follow.

Oral Contraceptives

Millions of women use oral contraceptives, daily doses of hormones that prevent pregnancy. Oral contraceptive interactions with nutrients illustrate the complexity of nutrient-drug interactions.

Each nutrient responds differently to oral contraceptive use (see Table C14-3). The vitamin B_{12} status of oral contraceptive users may be slightly lower than in others. Beta-carotene values may also be reduced, leading researchers to wonder whether the lower levels of this antioxidant might influence some disease risks. Vitamin D levels, on the other hand, may be higher in oral contraceptive users, with unknown effects. At first glance these findings seem to indicate that women using oral contraceptives are on their way to suffering deficiencies of some nutrients and have somehow enlarged their body stores of others. Research has yielded conflicting results, however, so any such assumptions are premature.

Significantly, oral contraceptives alter blood lipids, possibly increasing the risk of cardiovascular disease for menstruating women.[6] Especially in women older than about 35 years, most oral contraceptives raise total cholesterol and triglyceride concentrations and lower HDL, amplifying the risk of stroke and heart disease. A few women using oral contraceptives also experience mild hypertension.

Some women lose weight when taking oral contraceptives, but others may gain as much as 20 pounds or more from fat deposited in the hips, thighs, and breasts or from retained fluid. Some lean tissue is also deposited in response to an androgenic (steroid) effect of the pills. Sometimes a switch to another form of pill can normalize body weight.

As with oral contraceptives, women's responses to estrogen replacement drugs must be assessed individually. Some women suffer edema because estrogen promotes sodium conservation by the kidneys; dietary sodium restriction can correct this condition. Others develop low blood folate or vitamin B_6, indicating a need to include more vitamin-rich, nutrient-dense foods in the diet. All women taking estrogen either in oral contraceptives or in hormone replacement therapy should be aware that vitamin C doses of a gram or more may elevate serum estrogen and falsely suggest that a lower dose is needed.

If a woman taking any form of estrogen thinks she may have a nutrient deficiency, she should refrain from taking individual supplements and seek testing and a diagnosis from a healthcare professional to rule out other causes of her symptoms. For most women, a nutritious diet is all that is needed. If a woman feels compelled to take a supplement, however, a standard multivitamin-mineral supplement is probably harmless, as long as it accompanies a well-balanced diet.

Caffeine

The well-known "wake-up" effect of caffeine is the primary reason people in every society use it in some form. Com-

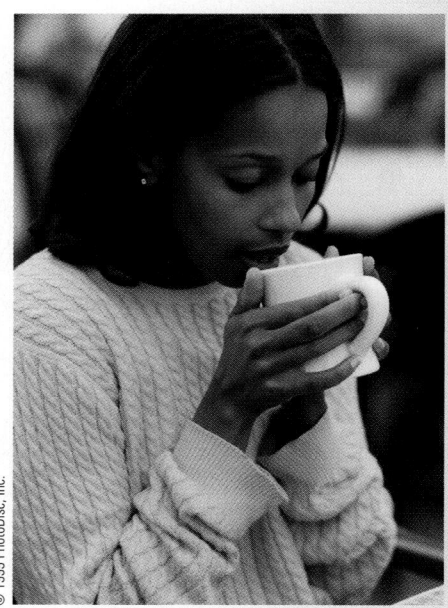

© 1999 PhotoDisc, Inc.

TABLE C14·3 NUTRITION EFFECTS OF A FEW COMMONLY USED MEDICAL DRUGS

Medicines and Caffeine	Effects on Absorption	Effects on Excretion	Effects on Metabolism
Antacids (aluminum containing)	Reduce iron absorption	Increase calcium and phosphorus excretion	May accelerate destruction of thiamin
Antibiotics (long-term usage)	Reduce absorption of fats, amino acids, folate, fat-soluble vitamins, vitamin B_{12}, calcium, copper, iron, magnesium, potassium, phosphate, zinc	Increase excretion of folate, niacin, potassium, riboflavin, vitamin C	Destroy vitamin K–producing bacteria and reduce vitamin K production
Antidepressants (monoamine oxidase inhibitors, MAOI)			Slows breakdown of tyramine, with dangerous blood pressure spike and other symptoms on consuming tyramine-rich foods or drinks: *alcoholic beverages* (sherry, vermouth, red wines, some beers); *cheeses* (aged and processed); *some meats* (caviar, pickled herring, liver, smoked and cured sausages, lunch meats); *fermented products* (soy sauce, miso, sauerkraut); *others* (brewer's yeast, yeast supplements, yeast paste; baked goods made with baker's yeast are safe; foods past their expiration date)
Aspirin (large doses, long-term usage)	Lowers blood concentration of folate	Increases excretion of thiamin, vitamin C, vitamin K; causes iron and potassium losses through gastric blood loss	
Caffeine		Increases secretion of small amounts of calcium and magnesium	Stimulates release of fatty acids into the blood
Cholesterol-lowering "statin" drugs (Zocor, Lipitor)			Grapefruit juice slows drug metabolism, causing buildup of high drug levels; potentially life-threatening muscle toxicity can result.
Diuretics		Raise blood calcium and zinc; lower blood folate, chloride, magnesium, phosphorus, potassium, vitamin B_{12}; increase excretion of calcium, sodium, thiamin, potassium, chloride, magnesium	Interfere with storage of zinc
Estrogen replacement therapy	May reduce absorption of folate	Causes sodium retention	May raise blood glucose, triglycerides, vitamin A, vitamin E, copper, and iron; may lower blood vitamin C, folate, vitamin B_6, riboflavin, calcium, magnesium, and zinc
Laxatives (effects vary with type)	Reduce absorption of glucose, fat, carotene, vitamin D, other fat-soluble vitamins, calcium, phosphate, potassium	Increase excretion of all unabsorbed nutrients	
Oral contraceptives	Reduce absorption of folate, may improve absorption of calcium	Cause sodium retention	Raise blood vitamin A, vitamin D, copper, iron; may lower blood beta-carotene, riboflavin, vitamin B_6, vitamin B_{12}, vitamin C; may elevate requirements for riboflavin and vitamin B_6; alter blood lipids elevating risk of heart disease in smokers and older women

TABLE C14·4 CAFFEINE CONTENT OF BEVERAGES AND FOODS

Most labels of caffeinated foods or beverages fail to list caffeine among their ingredients because it occurs naturally in the food or beverage.

Drinks and Foods	Average (mg)	Range (mg)
Coffee (5 oz cup)		
Brewed, drip method	130	110–150
Brewed, percolator	94	64–124
Instant	74	40–108
Instant "lite"	30	no data
Decaffeinated, brewed or instant	3	1–5
Tea (5 oz cup)		
Brewed, major U.S. brands	40	20–90
Brewed, imported brands	60	25–110
Instant	30	25–50
Iced (12 oz glass)	70	67–76
Herb teas (caffeine-free)	0	0
Soft drinks (12 oz can)		
Dr. Pepper		40
Colas and cherry colas:		
Regular		30–46
Diet		50 (average)
Clear and caffeine-free		0–trace
Extra caffeine (Jolt)		75–100
Mountain Dew, Mello Yello		52
Big Red		38
Fresca, 7-Up, Sprite, Squirt, Sunkist Orange, seltzers, root beers		0
Cocoa beverage (5 oz cup)	4	2–20
Chocolate milk beverage (8 oz)	5	2–7
Milk chocolate candy (1 oz)	6	1–15
Dark chocolate, semisweet (1 oz)	20	5–35
Baker's chocolate (1 oz)	26	26
Chocolate-flavored syrup (1 oz)	4	4
Carob	0	0

NOTE: Many over-the-counter medications such as pain relievers and cold medicines also contain caffeine. Their labels must list the milligram amounts of caffeine per dose of medicine. Read medicine labels carefully.

pared with the drugs discussed so far, though, caffeine's interactions with foods and nutrients are subtle. Yet caffeine's relationship to nutrition is important because caffeine is so widespread that people may be unaware that they are consuming it—see Table C14-4 for the caffeine contents of many beverages and foods. Many OTC cold and headache remedies contain caffeine because, in addition to being a mild pain reliever in its own right, caffeine remedies the headache caused by caffeine withdrawal that no other pain reliever can touch. Caffeine is present in chocolate bars, colas, and other foods children favor, and children are more sensitive to caffeine's effects because

they are small and, at first, not adapted to its use.

Caffeine is the most popular and widely consumed drug in the United States. One in three U.S. citizens consumes about 200 milligrams of caffeine per day (the amount in 10 ounces of coffee, less than the smallest serving from popular coffee chains), but many others consume much more.[7] Many people's intake patterns fulfill some of the accepted criteria for a diagnosis of drug dependence.

Many studies indicate that a 200-milligram dose of caffeine significantly improves the ability to pay attention, especially if subjects are sleepy, but more caffeine is probably not better.[8] A

single dose of 500 milligrams has been shown to worsen thinking abilities in almost everyone, and more than this may present some risk to health through its actions as a stimulant.

An individual's reaction to caffeine may depend in part on daily caffeine habits. In a regular user, deprivation of caffeine often causes headache and fatigue, and worsens mental performance and mood; restoration of the drug greatly improves these measures.[9] When deprived regular users were restored to ordinary intakes and then given an additional dose, their performance and mood does not benefit.[10]

Caffeine is a true stimulant drug. Like all stimulants, it increases the respiratory rate, heart rate, blood pressure, and secretion of stress and other hormones. A moderate dose of caffeine may speed up metabolic energy expenditures for several hours, and it stimulates the digestive tract, promoting efficient elimination. Because caffeine is a diuretic, it promotes some water loss from the body as well (see Chapter 8 for details).

Despite caffeine's tremendous popularity, many people today are consuming less because they fear possible harm to their health. Research in the last decade has yielded sporadic reports linking caffeine to health problems such as cancer, birth defects, and hypertension. Much other research, however, refutes any links between caffeine and cancer or birth defects and finds a weak link between caffeine and elevated blood pressure, an effect that may be significant for those with diagnosed hypertension.[11] Coffee and tea contain phytochemicals other than caffeine, some of which may reduce heart disease risk while others may increase it.[12]

Caffeine seems relatively harmless when used in moderation (2 cups of coffee a day). One study reports that men consuming 2 to 3 cups of coffee a day run a 40 percent *lower* risk of developing active gallstone disease than men who avoid coffee.[13] *Cola* consumption, however, may increase the risk.[14] In higher doses, caffeine can cause symptoms associated with anxiety: sweating, tenseness, and inability to concentrate. High doses may also accelerate bone loss in women past mid-life, and caffeine may contribute to painful but benign fibrocystic breast disease. More controlled studies are needed to determine whether eliminating caffeine can help to reverse breast disease.[15]

If you like caffeine-containing foods or beverages, the most reasonable approach is to limit your intake to the equivalent of 10 ounces of coffee per day. For most people, this is enough to reduce drowsiness and sharpen awareness without paying too high a price. Pregnant women should exercise moderation in using caffeine, and parents should monitor and control their children's intakes.

Tobacco

Cigarette and other tobacco use causes thousands of people to suffer from cancer and other diseases of the cardiovascular, digestive, and respiratory systems. These effects are beyond nutrition's scope, but smoking does depress hunger and body fatness and change nutrient status, and the nutrition effects are also linked to lung cancer. Chapter 9 provided details on smoking and body fatness.

Nutrient intakes of smokers and nonsmokers differ. Smokers have lower intakes of dietary fiber, vitamins, and minerals, even when their energy intakes are quite similar to those of nonsmokers. The association between smoking and low vitamin intake may be important because studies have shown that smoking alters the metabolism of vitamin C. (Research has just begun on other nutrients.)

Research shows that the vitamin C requirement of smokers exceeds that of nonsmokers. Smokers break down vitamin C faster and so must take in more vitamin C–containing foods to achieve steady body pools comparable to those of nonsmokers. The effect is apparently related to an increase in oxidative stress produced by smoking and is not related to the tobacco drug nicotine.[16] The evidence is so strong that the vitamin C

recommendation is set higher for smokers—smokers require an extra 35 milligrams of vitamin C per day.

Illicit Drugs

People know that illicit drugs are harmful, but many choose to abuse them anyway in spite of the risks. Like OTC and prescription drugs, illegal drugs modify body functions. Unlike medicines, however, no watchdog agency such as the Food and Drug Administration monitors them for safety, effectiveness, or even purity.

Smoking a marijuana cigarette affects several senses including the sense of taste. It produces an enhanced enjoyment of eating, especially of sweets. Why or how this effect occurs is not known. Despite higher food intakes, marijuana abusers often consume fewer nutrients than do nonabusers because the extra foods they choose tend to be high-calorie, low-nutrient snack foods. Besides the nutrition effects, regular marijuana users face the same risk of lung cancer as people who smoke a pack of cigarettes a day.

Many other drugs of abuse elicit effects such as intense euphoria, restlessness, heightened self-confidence, irritability, insomnia, and loss of appetite. Weight loss is a common side effect, and unlike marijuana, most other drugs of abuse cause serious malnutrition. The stronger the craving for the drug, the less a drug abuser wants

nutritious food. Rats given unlimited access to cocaine will choose the drug over food until they die of starvation. The effects of addictive drugs vary somewhat, but many are similar to the effects of cocaine, listed in Table C14-5. Drug abusers face multiple nutrition problems, and an important aspect of addiction recovery is the identification and correction of nutrition problems.

Personal Strategy

In conclusion, when you need to take a medicine, do so wisely. Ask your physician, pharmacist, or other health-care provider for specific instructions about the doses, times, and how to take the medication—for example, with meals or on an empty stomach. If you notice new symptoms or if a drug seems not to be working well, consult your physician. The only instruction people need about illicit drugs is to avoid them altogether for countless reasons. As for smoking and chewing tobacco, the same advice applies: don't take these habits up, or if you already have, take steps to quit. For drugs with lesser consequences to health, such as caffeine, use moderation.

Try to live life in a way that requires less chemical assistance. If you are sleepy, try a 15-minute nap or stretching exercises instead of a 15-minute coffee break. The coffee will stimulate your nerves for an hour, but the alternatives will refresh your attitude for the rest of the day. If you suffer constipation, try

TABLE C14·5	NUTRITION EFFECTS OF FOUR NONMEDICAL DRUGS
Drug of Abuse	Possible Effects on Nutrition Status
Cocaine	Reduces intakes of nutritious foods; increases intakes of alcohol, coffee, and fat; may induce or aggravate eating disorders
Heroin	Heightens and delays insulin response to glucose; reduces intakes of nutritious foods
Marijuana	Increases intakes of foods, especially sweets; may cause weight gain
Nicotine[a]	Reduces intake of sweet foods and water; increases intakes of fat; reduces fetal weight; lowers blood concentration of beta-carotene.

[a]Other effects of smoking include increased vitamin C requirements.

SOURCES: Data from M. E. Mohs, R. R. Watson, and T. Leonard-Green, Nutritional effects of marijuana, heroin, cocaine, and nicotine, *Journal of the American Dietetic Association* 90 (1990): 1261–1267; G. van Poppel, S. Spanhaak, and T. Ockhuizen, Effects of beta carotene on immunological indexes in healthy male smokers, *American Journal of Clinical Nutrition* 57 (1993): 402–407.

getting enough exercise, fiber, and water for a few days. Chances are that a laxative will be unnecessary. The strategy being suggested here is to take control of your body, allowing your reliable, self-healing nature to make fine adjustments that you need not force with chemicals. Bodies have few requests: adequate nutrition, rest, exercise, and hygiene. Give your body what it asks for, and let it function naturally, day-to-day, without interference from drugs.

Hunger and the Global Environment

Victor Vasquez Temó. *Harvesting and Maintaining Onions.* Tz'utuhil Maya artist, Guatemala. www.artemaya.com

Do You Ever . . .

Read about starvation and strife in remote locations and feel well insulated here at home?

Feel overwhelmed when you think about the problems facing our world?

Purchase a meal, considering only its monetary price?

Wish that somebody would do something about the contaminants in air, water, and food that make headlines occasionally?

KEEP READING.

One person in every five worldwide experiences persistent **hunger**—not the healthy appetite triggered by anticipation of a hearty meal, but the painful condition caused by a chronic lack of food. Tens of thousands die of starvation each day—one every two seconds. Table 15-1 presents the current best estimate of the number of hungry people worldwide. For perspective, the table also includes world incidences of vitamin and mineral deficiencies and overnutrition (obesity). The presence of any one of these conditions—hunger, nutrient deficiencies, or overnutrition—does not exclude the coexistence of any of the others within the same country, the same area, or even the same family.[*1] Up to 80 percent of hungry children live in countries that produce surplus food but the decisions of policymakers in those areas largely determine who among the population has access to the bounty.

In the United States, where most people enjoy a life of relative abundance, millions of children are hungry at least some of the time.[2] Under the broad definition of **food insecurity,** over 13 million U.S. children do not know where their next meal is coming from or when it will come. Terms relating to hunger are defined in Table 15-2.

The world's chronically hungry suffer from undernutrition, a condition of energy and nutrient deficiency that causes general weakness and fatigue. The greater the caloric deficit, the greater the susceptibility to nutrition-related health risks. Undernutrition stymies mental and physical development in children and makes people susceptible to potentially fatal infections such as dysentery, whooping cough, and tuberculosis.[3] Consequences of chronic hunger in children include infant mortality, stunted growth, iron-deficiency anemia, poor learning, extreme weakness, clinical signs of protein-energy malnutrition (PEM), increased susceptibility to disease, loss of the ability to stand or walk, and premature death.[4]

The tragedy described on these pages may seem at first to be beyond the influence of the ordinary person. What possible difference can one person make? Can one person's choice to recycle a bottle or to serve a meal to the homeless or to join a hunger-relief organization make a difference? In truth, such choices produce several benefits. For one, a single person's awareness and example, shared with others, can influence many people over time. For another, an action repeated becomes a beneficial habit. For still another, making choices with awareness of their impacts lends a sense of control over those impacts. That sense of personal control, in turn, helps people to take effective action in many areas.

Students can play a powerful role in bringing about change. Students everywhere are helping to change governments, human predicaments, and environmental problems for the better. Student movements persuaded 127 universities and many institutions, corporations, and government agencies to put pressure on South Africa and succeeded in ending apartheid. Student pressure opened the way for the first deaf president at a university for the deaf. Students offer major services to communities through soup kitchens, home repair programs, and child education. The young people of today are the world's single best hope for a better tomorrow.

---

*Reference notes are found in Appendix F.

TABLE 15·1 GLOBAL UNDERNUTRITION AND OVERNUTRITION, 2003

Condition	Global Incidence
Hunger, protein-energy malnutrition	≥800 million people
Vitamin and mineral deficiencies[a]	2.0–3.5 billion people (about half the world's population)
Overnutrition, obesity	≥1.1 billion people

[a]Vitamin and mineral deficiencies occur in both underfed and overfed populations.
SOURCES: *The State of Food Insecurity in the World 2003*, Food and Agriculture Organization, available at www.fao.org; G. Gardner and B. Halweil, *Underfed and Overfed: The Global Epidemic of Malnutrition*, World Watch Paper 150 (Washington, D.C.: Worldwatch Institute, 2000), p. 7.

TABLE 15·2 HUNGER TERMS

- **famine** widespread and extreme scarcity of food in an area that causes starvation and death in a large portion of the population.
- **food insecurity** the condition of limited or uncertain access to food of sufficient quality or quantity to sustain a healthy, active life.
- **food poverty** hunger occurring when enough food exists in an area but some of the people cannot obtain it because they lack money, are being deprived for political reasons, live in a country at war, or suffer from other problems such as lack of transportation.
- **food security** reliable access to enough nutritious food at all times to sustain a healthy, active life.
- **food shortage** hunger occurring when an area of the world lacks enough total food to feed its people.
- **hunger** lack or shortage of basic foods needed to provide the energy and nutrients that support health.

The Challenge to Change

Banishing food insecurity for all of the world's citizens poses two major challenges. The first is to provide enough food to meet the needs of the earth's expanding population, without destroying natural resources needed to continue producing food. The second challenge is to ensure **food security**—that is, to make sure all people have access to enough food to live active, healthy lives. By all accounts, today's total **world food supply** can abundantly feed the entire current population, but this does not guarantee that people in need are able to get adequate food. If people do not have enough money to buy food or to buy the land, seeds, and tools to grow food or if natural or human made disasters such as drought or war prevent them from getting food, hunger remains a problem.

Concerns for the future also exist. Many forces compound to threaten world food production and distribution in the next decade, and we may face more global food insecurity because of them:

- *Hunger, poverty, and population growth.* Millions of the world's people are starving. Fifteen children die of malnutrition every 30 seconds, but 125 children are born in that same 30 seconds. Every day, the earth gains another 220,000 new residents to feed (see Figure 15-1 p. 562), most of them born in impoverished areas.[5]
- *Loss of food-producing land.* Food-producing land is becoming saltier, eroding, and being paved over. Each year, the world's farmers try to feed some 85 million additional people with 24 billion fewer tons of topsoil. This loss threatens overall food security.
- *Accelerating fossil fuel use.* Fossil fuel use is growing rapidly, with attendant pollution of air, soil, and water; ozone depletion; and global climate changes.
- *Increasing air pollution.* As populations and human activities increase, air quality diminishes in many areas around the globe.*
- *Atmosphere and climate changes, droughts, and floods.* Climbing atmospheric levels of heat-trapping carbon dioxide are a concern. The concentration of carbon dioxide is now 26 percent higher than 200 years ago. Researchers suspect that a worldwide warming trend caused the record-breaking heat in the summer of 2003 that cost many thousands of European lives.[6] Climate changes also cause droughts and floods that destroy crops and people's homelands.[7] Diseases such as malaria that are now confined to tropical areas are expected to spread to other climate zones as they warm.

© Syracuse Newspapers/Stephen D. Cannerelli/The Image Works

Each person's efforts can help to bring about needed changes.

world food supply the quantity of food, including stores from previous harvests, available to the world's people at a given time.

---

*Many older references have been removed to save space, but they are available in older editions of this book.

FIGURE 15·1   WORLD POPULATION GROWTH

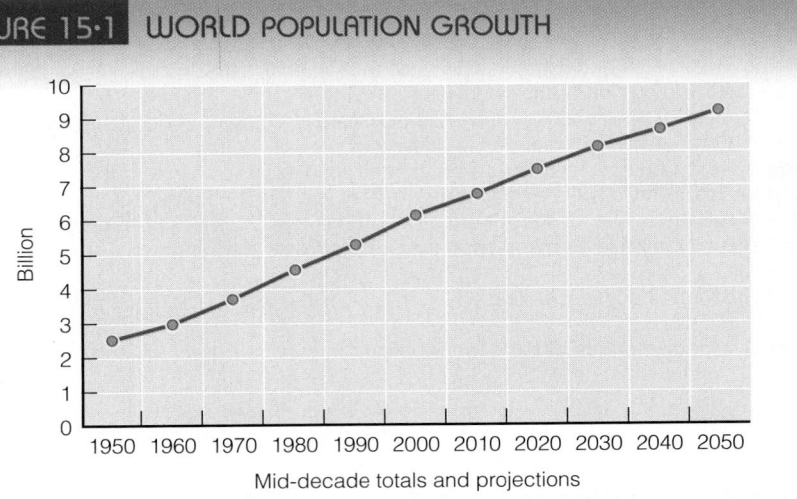

Mid-decade totals and projections

- *Ozone loss from the outer atmosphere.* The outer atmosphere's protective ozone layer is growing thinner, permitting harmful radiation from the sun to penetrate. This radiation increases the likelihood of skin cancers and cataracts in people and animals and potentially may damage crops and ecosystems.
- *Water shortages.* The world's supplies of fresh water are dwindling and becoming polluted.[8]
- *Deforestation and desertification.* Forests are shrinking and deserts are growing.
- *Ocean pollution.* Ocean pollution is killing fish in large "dead zones" along the world's coasts; overfishing is depleting the fish that remain.[9] The U.S. Commission on Ocean Policy has deemed the U.S. systems of managing oceans and coasts fragmented and insufficient for the task.[10] Many other nations have weaker policies still or no policies at all.
- *Extinctions of species.* More than 140 species of animals and plants are becoming extinct each day. Another 20 percent of all species are expected to die out in the next ten years. Thousands of animals and plants, including many whales, birds, giant mammals, and colorful butterflies, will never again be seen in the universe.

These global problems are all related and, often, so are their solutions. To think positively, this means that any initiative a person takes to help solve one problem will help solve many others. In particular, control of the earth's population is urgent, as a later section spells out. This chapter's Controversy shows how U.S. consumers and agricultural practices affect the world's resources.

▶ **KEY POINT**   The world's chronically hungry people suffer the effects of undernutrition, and many in the United States live with food insecurity. Many forces combine to threaten the world's future food supply and its distribution.

Hunger

The hunger of concern is a chronic, painful hunger people feel when no food or too little food is available. Severe deficiencies of vitamins and minerals accompany this hunger, afflicting more than 40 percent of the world's people to some degree. An estimated 2 billion people, mostly women and children, suffer the effects of iron-deficiency anemia, and many more suffer milder degrees of insufficiency. Iodine deficiency remains the single greatest cause of preventable brain damage and mental retardation; 750 million adults suffer from goiter. Deficiency of vitamin A takes a terrible toll on the world's children—it stands out as the world's leading cause of blindness in young children and robs many millions of the ability to fight off infections. Over half of the deaths among the world's children are attributable to malnutrition.[11] Worldwide, three-fourths of those who die each year from starvation and related illnesses are children.[12]

In developed countries, the primary cause of hunger is **food poverty.** People are hungry not because there is no food nearby to purchase, but because they lack sufficient money with which to buy nutritious food and pay for other necessities, such as clothing, housing, medicines, and utilities. More than 12 percent of the population of the United States lives in a general state of poverty.[13] The likelihood of *food poverty* increases with problems such as abuse of alcohol and other drugs, mental or physical illness, depression, lack of awareness of or access to available food programs, and the reluctance of people to accept what some perceive as "government handouts" or charity.[14]

Ideally, all people would always have enough nutritious food to support an active, healthy life; in other words, they would experience food security. Yet almost 34 million people in the United States, including 13 million children, live in poverty and cannot afford to buy enough food to maintain good health (see Figure 15-2).[15] Food poverty reaches into many segments of society, affecting not only the chronic poor (migrant workers, the unskilled and unemployed, the homeless, and some elderly) but also the so-called working poor. Some of these people are displaced farm families who work at minimum-wage jobs. Some are former blue-collar and white-collar workers forced out of their trades and professions. These people outnumber the chronic poor, and they are not on welfare—they have jobs, but the pay is too low to meet basic needs. The pay scale for low-skill jobs has increased only slightly in recent years, but living expenses have increased significantly, effectively reducing the real wages earned by workers.[16] Such families cannot afford to buy sufficient amounts of nourishing foods, even if they are skilled in food shopping.

To stretch meager food supplies, adults may skip meals or cut their portions. When desperate, they may be forced to break social rules—begging from strangers, stealing from markets, harvesting dead animals from roadsides, or scavenging through garbage cans, for example.[17] Such foods may be spoiled or contaminated and eating them may easily inflict dangerous foodborne illnesses on people already bordering on malnutrition.

Children in such families sometimes go hungry for an entire day until the adults can obtain food. Significant numbers of poor U.S. children consume enough calories but from a steady diet of inexpensive foods, such as white bread, fats, and crackers, with few of the fruits, vegetables, milk servings, and other nutritous foods children need to be healthy.[18] The more severe their circumstances, the more likely children are to be in poor or just fair health, and the greater the likelihood that they will be hospitalized.[19] Well-documented adverse effects suffered by children in poverty include cognitive deficits, behavioral problems, slow wound healing, and impaired immunity. These children often misbehave because of malnutrition or in rebellion against their circumstances, and relieving their poverty often improves their behavior.[20]

Hunger is not always easy to recognize. Table 15-3 (p. 564) shows how national surveys identify it in the United States. Responses to these questions provide crude but necessary data to estimate the degree of hunger in this country.[21] The American Dietetic Association (ADA), a professional organization of dietitians and other nutrition professionals, calls for aggressive action to bring an end to domestic hunger and to achieve food and nutrition security for all residents of the United States.[22] The ADA also holds that access to adequate amounts of safe, nutritious and culturally appropriate food at all times is a fundamental human right. Eradicating hunger is in everyone's interest because the hunger of individual families affects the nation as a whole.

▶ **KEY POINT** Chronic hunger causes many deaths worldwide, especially among children. Intermittent hunger is frequently seen in U.S. children. The immediate cause of hunger is poverty.

During 2003,

- The USDA spent $41.7 billion on domestic food assistance programs.
- 21 million people received credits with which to buy food from the Food Stamp Program.
- 16.4 million schoolchildren ate free or reduced-price school lunches and breakfasts.
- 4.3 million U.S. households received food from a food pantry.
- 1.1 million people per month ate a meal at an emergency kitchen (2001 data).

SOURCE: L. Tiehen, Emergency providers help poor households put food on the table, *Amber Waves*, June 2004, available at www.ers.usda.gov/Amberwaves/June04/Features/EmergencyProv.htm.

■ An estimated one out of six children in the United States lives in poverty.

In the United States, about **34 million people** go hungry for some part of every month.

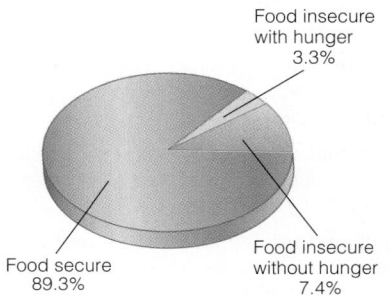

FIGURE 15·2 U.S. FOOD SECURITY AND HUNGER, 2001

Food insecure with hunger
3.3%

Food secure
89.3%

Food insecure without hunger
7.4%

SOURCE: Economic Research Service, U.S. Department of Agriculture, www.ers.usda.gov/publications/fanrr29, posted October 2002 and visited on June 3, 2003.

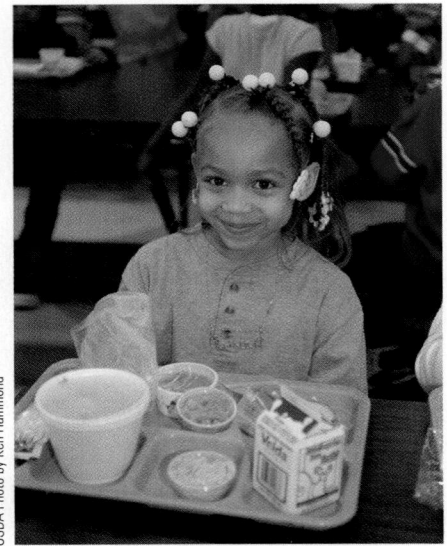

School breakfasts and lunches provide low-income children with nourishment at little or no cost.

TABLE 15·3 HOW TO IDENTIFY FOOD INSECURITY IN A U.S. HOUSEHOLD

TABLE 15·3 HOW TO IDENTIFY FOOD INSECURITY IN A U.S. HOUSEHOLD

Questions like these are asked on surveys to determine the extent of food insecurity in a household. The more questions that receive a "Yes" answer, the more intense the hunger the household is experiencing.

- Do you often go hungry?
- Do you often have too little food to eat because you have no money, transportation, or kitchen appliances that work?
- Do you ever rely on nutritionally inferior foods to feed yourself or your children because you lack any of these resources?
- Do you ever eat less than you feel you should because you lack any of these resources?
- Do you ever skip meals or cut the size of meals because you lack any of these resources?
- Do you ever rely on neighbors, friends, relatives, or schools to feed any of your children because there is not enough food in the house?
- Do your children ever say they are hungry because there is not enough food in the house?
- Do you or any of your children ever go to bed hungry because there is not enough food in the house?

SOURCES: Adapted from C. A. Wehler, R. I. Scott, and J. J. Anderson, The Community Childhood Hunger Identification Project: A model of domestic hunger—Demonstration project in Seattle, Washington, *Journal of Nutrition Education* (1 supplement), January/February 1992, pp. 29S–35S; and R. R. Briefel and C. E. Woteki, Development of food sufficiency questions for the Third National Health and Nutrition Examination Survey, *Journal of Nutrition Education* (1 supplement), January/February 1992, pp. S24–S28.

U.S. federal and state food programs include:

- Special Supplemental Food Program for Women, Infants, and Children (WIC, see Chapter 13).
- WIC Farmers' Market Nutrition Program.
- Food Stamp Program.
- National School Lunch and Breakfast Programs (see Chapter 14).
- Emergency Food Assistance Program.
- Commodity Supplemental Food Program.
- Commodity Distribution to Charitable Institutions.
- Senior Nutrition Program (see Chapter 14).
- Food Distribution Program on Indian Reservations.
- Nutrition Assistance to Puerto Rico.
- Temporary Assistance to Needy Families (often called "welfare").

food recovery collecting wholesome surplus food for distribution to low-income people who are hungry.

food banks facilities that collect and distribute food donations to authorized organizations feeding the hungry.

food pantries community food collection programs that provide groceries to be prepared and eaten at home.

emergency kitchens programs that provide prepared meals to be eaten on-site; often called *soup kitchens*.

What U.S. Food Programs Are Directed at Stopping Domestic Hunger?

An extensive network of food assistance programs delivers life-giving food daily to millions of U.S. citizens living in poverty. One of every six Americans receives food assistance of some kind, at a total cost of almost $33 billion per year.[23] Even so, the programs are not fully successful at preventing hunger, even among those who receive their benefits.

Programs described in earlier chapters include children's school lunch and school breakfast programs, child-care and elder-care food programs, programs to supply low-income pregnant women and mothers with nourishing food (WIC), and food assistance programs for older adults such as congregate meals and Meals on Wheels. In particular, the WIC program is effective at improving the health of mothers and their infants. Participation during pregnancy is associated with increased weight and longer gestation, which result in healthier infants in a low-income population. Children of WIC families have higher intakes of iron, folate, and vitamin B_6 than children in non-participating but eligible families.[24]

The centerpiece of U.S. food programs for low-income people is the Food Stamp Program, administered by the U.S. Department of Agriculture (USDA). It provides assistance to more than 20 million people at a cost of over $20 billion per year; over half of the recipients are children.[25] Eligible households receive coupons or debit cards similar to credit cards through state social services or welfare agencies. Recipients can use the coupons or cards like cash to purchase food and food-bearing plants and seeds, but not to buy tobacco, cleaning items, alcohol, or other nonfood items. Although food stamps help millions, many millions more are thought to be eligible to receive them.[26] Of the homeless people in the United States who are eligible for food assistance, only 15 percent of single adults and 50 percent of families receive food stamps.

To assist where government programs fall short, concerned citizens in many communities work through local agencies and churches to help deliver food to hungry people. National **food recovery** programs have made a dramatic difference. The largest program, Second Harvest, coordinates the efforts of more than 250 **food banks, food pantries, emergency kitchens,** and homeless shelters in providing more than 1 billion pounds of food to 45,000 local agencies that feed over 23 million people a year.

1. Establish a community-based emergency food-delivery network.
2. Assess food insecurity problems and evaluate community services. Create strategies for responding to unmet needs.
3. Establish a group of individuals, including low-income participants, to develop and to implement policies and programs to combat food insecurity; monitor responsiveness of existing services; and address underlying causes of hunger.
4. Participate in federally assisted nutrition programs that are easily accessible to targeted populations.
5. Integrate public and private resources, including local businesses, to relieve food insecurity.
6. Establish an education program that addresses the food needs of the community and the need for increased local citizen participation in activities to alleviate food insecurity.
7. Provide information and referral services for accessing both public and private programs and services.
8. Support programs to provide transportation and assistance in food shopping, where needed.
9. Identify high-risk populations, and target services to meet their needs.
10. Provide adequate transportation and distribution of food from all resources.
11. Coordinate food services with parks and recreation programs and other community-based outlets to which area residents have easy access.
12. Improve public transportation to human services agencies and food resources.
13. Establish nutrition education programs for low-income citizens to enhance their food purchasing and preparation skills and to make them aware of the connections between diet and health.
14. Establish a program for collecting and distributing nutritious foods, either agricultural commodities in farmers' fields or prepared foods that would have been wasted.

SOURCE: House Select Committee on Hunger, legislation introduced by Tony P. Hall, excerpted in *Seeds*, Sprouts edition, January 1992, p. 3 with permission (SEEDS Magazine; P.O. Box 6170; Waco, TX 76706). For more on developing a hunger-free community, write Hunger Free, House Select Committee on Hunger, 505 Ford House Office Building, Washington, DC 20515.

Four common methods of food recovery are:

- *Field gleaning:* Collecting crops from fields that either have already been harvested or are not profitable to harvest.
- *Perishable food rescue or salvage:* Collecting perishable produce from wholesalers and markets.
- *Prepared food rescue:* Collecting prepared foods from commercial kitchens.
- *Nonperishable food collection:* Collecting processed foods from wholesalers and markets.

■ For more information about gleaning, call the USDA's hunger hotline: (800) GLEAN-IT.

Each year, an estimated one-fifth of our food supply is wasted in fields, commercial kitchens, grocery stores, and restaurants—enough food to feed 49 million people. Food recovery programs collect and distribute good food that would otherwise go to waste; those donating the food often qualify for tax deductions for their donations. Table 15-4 presents a 14-step program for developing a hunger-free community. Although these efforts provide emergency relief to hungry people, they leave unsolved the greater problems of low wages and poverty among people who lack higher education or training.

▶ KEY POINT Poverty and hunger coexist with affluence and bounty in the United States; working people as well as the unemployed experience poverty and hunger. Government programs to relieve poverty and hunger are tremendously helpful, if not fully successful.

What Is the State of World Hunger?

In the developing world, hunger and poverty are even more intense, and the causes more diverse. The primary form of hunger is still food poverty, but the poverty is more extreme. Grasping the severity of poverty in the developing world can be difficult, but some statistics may help. One-fifth of the world's 6 billion people have no land and no possessions *at all.* The "poorest poor" survive on less than one dollar a day each, they lack water that is safe to drink, and they cannot read or write. Many spend about 80 percent of all they earn on food, but still they are hungry and malnourished. The average U.S. housecat eats twice as much protein every day as one of these people, and the yearly cost of keeping that cat is greater than that person's annual income.

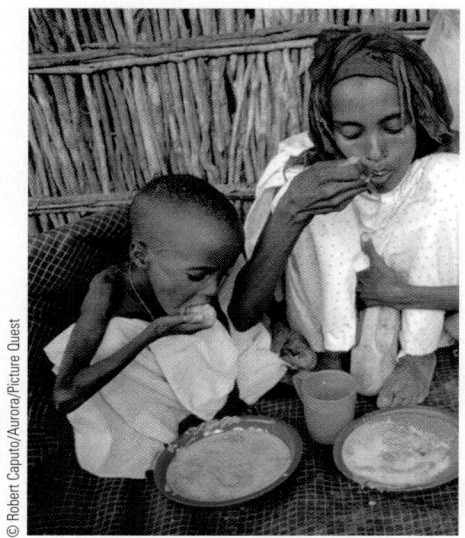

The primary cause of hunger is poverty.

Poverty causes hunger, but hunger also worsens poverty by robbing a person of the good health and the physical and mental energy needed to be active and productive. The less active and productive people become, the lower their pay. Hungry people simply cannot work hard enough to get themselves out of poverty, and most have no borrowing power to obtain credit. Economists calculate that reducing world hunger and malnutrition by half would generate more than $120 billion in productivity.[27] Figure 15-3 highlights causes and outcomes of hunger in the world.

The majority of people currently living with food emergencies are concentrated in the nations of Africa and Asia. Recent advances in agricultural technology, economic devel-

---

FIGURE 15·3 CAUSES AND OUTCOMES OF HUNGER

Regional quantity, quality, and availability of resources
- Is enough high-quality food available for all people in the region?
- Is drought or environmental degradation preventing production of adequate food?

Economic, organizational, and group discrimination factors that influence resource distribution
- Do unemployment and lack of borrowing power prevent escape from poverty?
- Does greed, government corruption, racial discrimination, ethnic or religious hatred, or a class system prevent equitable distribution of resources?
- Does the economy support necessary infrastructure, such as passable roads and fuel for food delivery?

An individual household's access to available resources
- Does the household have sufficient resources with which to obtain enough food?
- Is the household barred in some way from access to the food available?

Access to clean water and health service
- Does the family have access to enough clean water for its needs?
- Are immunizations and other basic health care available to prevent disease and disability?

Individual child-care practices and knowledge within the family
- Are some children fed, while others go hungry?
- Do mothers fail to breastfeed their newborns?
- Do parents make inappropriate substitutions for formula or mix formula with unclean water?
- Do parents understand the basic nutrition needs of their children and pregnant and lactating women?

Inadequate food or nutrient intakes lead to malnutrition, weakness, disease, and depressed appetite
- Are children and adults too weakened to compete for food?
- Are their appetites depressed as an effect of malnutrition?

Disease conditions worsen malnutrition
- When the resources the body uses to fight disease or obtain food are exhausted, death from disease or starvation results.

SOURCE: Based on ideas in Position of the American Dietetic Association: Addressing world hunger, malnutrition, and food insecurity, *Journal of the American Dietetic Association* 103 (2003): 1046–1057.

opment, and commitment to eradicating hunger have begun to improve some problems in some areas. In other areas, however, the number of hungry people continues to increase, and hunger remains an enormous challenge. In these areas, hunger and poverty, population growth, political strife, armed conflicts, natural disasters, and environmental degradation are all linked, and they tend to worsen each other (see Figure 15-4, p. 568).

Food Shortage and Armed Conflict The most visible form of hunger is **famine,** a true **food shortage** in an area that causes multitudes of people to starve and die (for definitions, see Table 15-2, p. 561). The natural causes of famine—drought, flood, and pests—have, in recent years, taken second place behind the political and social causes. In parts of Africa, killer famines recur whenever human conflict converges with drought on a country such as Sudan that has little food in reserve even in a peaceful year. Today, violence in the region has brought the population of western Sudan to the brink of mass starvation.[28] Additionally, racial discrimination and ethnic and religious hatred often underlie the food deprivation of whole groups of people.[29] For people of marginal existence, a sudden increase in food prices, a drop in workers' incomes, or a change in government policy can quickly leave millions hungry.

Since the 1990s, the violence of armed conflict has been a dominant cause of all the famines reported worldwide. Farmers become warriors and agricultural fields become battlegrounds while citizens go hungry. Warring factions often repel famine relief efforts in hopes of starving their opponents before they succumb to starvation themselves.

During natural disasters without war, food aid from other countries has provided a safety net for countries whose crops fail.[30] But food aid now does more than just off-set poor harvests; it also delivers food relief to countries, such as Ethiopia, that are chronically short of food and without resources to buy it. Some people are concerned that as many nations cut their foreign aid, the supplies of needed food may become insufficient.

Chronic Hunger Though we usually associate world hunger with famine, the numbers affected by famine are relatively small compared with those suffering from less severe but chronic hunger. Over 800 million people, mostly women and children, are chronically malnourished from a lack of food.[31] Many more have food but lack iron, iodine, and vitamin A.[32] The prevalence and consequences of these deficiencies are enormous. The ravages to the body of nutrient deficiencies were spelled out in earlier chapters of this book.

Tens of thousands die of malnutrition every day. Most children who die of malnutrition do not starve to death—they die because their health has been compromised by dehydration from infections that cause diarrhea.[33] Currently, **oral rehydration therapy (ORT)** is saving an estimated one million lives each year by helping to stop the destructive spiral in which infection worsens diarrhea and diarrhea causes dehydration. The ORT solution increases a body's ability to absorb fluids 25-fold. Clean or boiled drinking water is essential for ORT, however, because contaminated water will reinfect the child.

Focus on Women Malnourished women in poverty bear sickly infants who cannot fend off the diseases of poverty, and many succumb within the first years of life. One child in six in the world is born underweight, and 10 million die by age five, half from malnutrition-related causes.[34] Breastfeeding helps prolong an infant's life, but eventually the child must be weaned to thin gruels of scant quantity made with unclean water. All too often, children sicken and die soon after weaning. Almost 80 million young children (under age five) suffer from symptoms of vitamin A deficiency—blindness, growth retardation, and poor resistance to common childhood infections such as measles. Because of poverty, infection, and malnutrition, the life expectancy in some African countries averages 50 years; in Uganda it is only 42 years, little more than half of the U.S. life expectancy.

When crops fail or violence erupts in an already impoverished area, women are first to suffer. Seven out of ten of the world's hungry people are women and girls, yet they receive only about half of the available food aid and must use it to feed their children

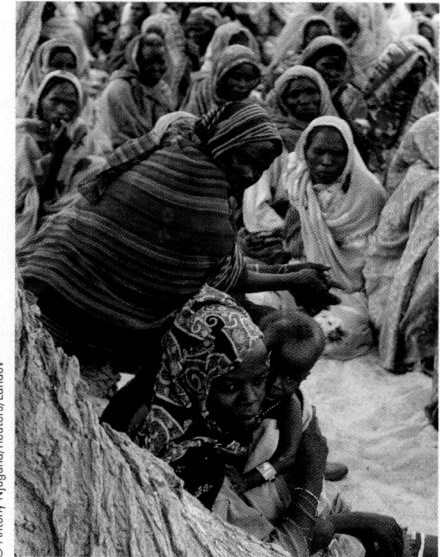

The hunger of conflict—people desperate for food in western Sudan.

■ Symptoms of malnutrition vary according to the nutrients lacking and the individual's stage of life. See Chapter 6 for effects of protein and energy deficiency; Chapters 7 and 8 for vitamin and mineral deficiencies; Chapter 11 for effects on immunity; Chapter 13 for effects on newborns and pregnant women; and Chapter 14 for effects on children, teens, and the elderly.

To prevent death from diarrheal disease, provide:

• Adequate sanitation.

• Safe water.

• Oral rehydration therapy (ORT). A simple recipe for ORT calls for 1 c boiled water, 2 tsp sugar, and a pinch of salt.

oral rehydration therapy (ORT) oral fluid replacement for children with severe diarrhea caused by infectious disease. ORT enables parents to mix a simple solution for their child from substances that they have at home.

FIGURE 15·4 HUNGER HOTSPOTS

Key:
- No food emergencies
- Economic problems
- Natural disasters (mostly drought)
- Past conflict
- Conflict

NOTE: Areas with stripes of color have multiple causes of hunger.

as well as themselves. These facts are offered by the World Food Programme (WFP) to justify targeting women as direct recipients of food relief:

- Even when women are starving, they are likely to give what food is available to their hungry children. In contrast, when food is delivered to government agencies, much of it may be diverted from its intended recipients.
- In Asia and Africa, 60 to 80 percent of women are engaged in farming.
- In a third of households worldwide, women are the sole breadwinners.
- Given low-interest loans, basic equipment, and access to land ownership along with supports such as child care and education, women engage in sustainable activities that improve conditions in the community.

Universal education for girls and women is also a high priority.

World Food Supply Most disturbingly, misery and starvation exist side by side with adequate world food supplies. The world produces sufficient food to feed all of its hungry people; what is lacking is the political will to do so.[35]

In the decades after 1960, world food production grew faster than the world's population. These gains were largely the result of agricultural advances known as the green revolution along with efforts to increase and diversify crop yields in agriculturally less advanced regions of the world. Today, the world's supply of grain, an index of the sufficiency of the world food supply, can still feed the world for several months. Wheat and corn, for example, the staple foods of many nations, are abundant and now cost less than half as much as 40 years ago.

The future may not be so bright, however. At its present rate of growth, the world's population will soon outstrip the current rate of food production.[36] The green revolution has passed, and old technology will not generate the greater crop yields needed to keep pace with the increasing numbers of people being born. A 2001 United Nations document states: "Only by doubling food production, improving distribution, and protecting the environment can we ensure food security for the 8 billion people that will inhabit the planet in 2025. Research suggests that the world's farmers will have to produce 40 percent more grain by 2020 to meet rising demand." Environmental degradation and dwindling water supplies may ultimately prevent further growth in the world's food output in many agricultural areas.

No part of the world is safely insulated against future food shortages. Developed countries may be the last to feel the effects, but they will ultimately go as the world goes.

▶ **KEY POINT**   Natural causes such as drought, flood, and pests and political and social causes such as armed conflicts and overpopulation all contribute to hunger and poverty in developing countries. To meet future demands for food, technology must continue to improve food production, food must be fairly distributed, and birthrates need to decline. The world's women and girls are major allies in the effort to fight hunger.

Environmental Degradation and Hunger

Hunger and poverty interact with a third force—environmental degradation. Poor people often destroy the very resources they need for survival. Desperate to obtain money for food, they sell everything they own—even the seeds that would have produced next year's crops. They cut the available trees for firewood or timber to sell, then lose the soil to erosion. Without these resources, they become still poorer. Thus, poverty causes environmental ruin, and the ruin leads to hunger.

Soil Erosion   Soil erosion affects agriculture in every nation. Deforestation of the world's rain forests dramatically adds to land loss. Without the forest covering to hold the soil in place, it washes off the rocks beneath, drastically reducing the land's productivity.

Around the world, irrigation and fertilizer can no longer compensate for these losses by improving crop yields because all the land that can benefit from these measures is already receiving them. Compounding the problem, continuous irrigation leaves deposits of salt in the soil, and rising salt concentrations are lowering yields on close to a quarter of the world's irrigated cropland.

The U.S. government offers monetary incentives to farmers and ranchers who conserve wetlands and employ soil conserving techniques on highly erodible croplands. In recent years, welcome evidence of slowing of soil erosion has been attributed partly to conservation incentive policies and even more to agricultural innovations, as described in this chapter's Controversy.[37]

Grazing Lands and Fisheries   Meat and fish outputs are also endangered. Grasslands for growing beef are already being fully used or overused on every continent. Despite persistent expansion of the world's fishing industry, the yield of fish from the oceans has been declining in recent years due to overfishing and pollution.[38] Researchers report that populations of big fish, such as tuna, swordfish, cod, halibut, and shark, have declined by 90 percent over the last 50 years.[39] According to the Food and Agriculture Organization (FAO), an agency of the United Nations that monitors the world's food supplies, about 47 to 50 percent of major marine fish stocks are currently fully exploited, with no room for further expansion of fishing. Another 28 percent are overexploited or depleted and in danger of extinction unless given relief from overfishing.[40] The solution is simple: stop overfishing.[41] The problem is how to do so.

Not only are ocean fish being overfished but, as previous chapters have pointed out, they are so contaminated from industrial pollution that pregnant women and young

▦ More about overgrazing appears in Controversy 15.

To protect overfished species, consumers can choose these fish options:
- Alaskan halibut.
- Alaskan salmon.
- Farm-raised tilapia and other farm-raised fish.
- Sardines.
- White sea bass.

children must strictly limit their intakes, a fact that makes some dietitians hesitant to recommend fish for health. In California, when the blood of fish-eating people was tested, blood mercury exceeded established upper safe levels, sometimes enormously so, in 89 percent of participants.[42] Children were especially affected: average blood mercury levels of children in the study were more than *40 times* higher than the national average. On reducing fish intake, the subjects' mercury levels dropped quickly. The problem spans the globe: almost every major bay in Japan suffers serious pollution sufficient to interrupt the normal breeding cycles of food species in the region.[43]

Inland fisheries have also suffered tremendous drops in yield as a result of environmental damage. In the early 1990s, 14,000 Canadian lakes were declared biologically dead as a result of acid rain. The Food and Drug Administration warns that eating too much fish from the Great Lakes threatens health because their flesh contains high levels of the industrial contaminant, PCBs, once widely used in electrical equipment and as a lubricant.

Some ideas are now forthcoming to save the world's fish stocks. Seasonal quotas limit the number and type of fish that may be removed from a region to allow species to begin to recover their numbers; this measure has already proved successful for Alaskan halibut. Another essential step is to mark and enforce "no fishing zones" in large areas of the oceans, creating safe sanctuaries for fish breeding and development. Cracking down on illegal harvesting by commercial fisheries and reducing the size of fishing fleets would also reduce the pressure. Developing commercial fish aquaculture ("fish farms") and altering wild species through recombinant DNA technology may also help meet human demand. The FAO predicts that, unless tactics like these are put into place quickly and firmly, the world demand for fish will very soon outstrip the supply.

Climate, Air, and Fresh Water Both air pollution and the resulting climate change also reduce food outputs. According to the United Nations Intergovernmental Panel on Climate Change, a major international collaboration involving more than 2,500 scientists from around the world, changes in climate are expected to result from a buildup of so-called greenhouse gases, such as carbon dioxide, methane, and nitrous oxide, and airborne particles.[44] These pollutants are produced by human industry, agriculture, and transportation activities. A rise of only a degree or so in average global temperature may reduce soil moisture, impair pollination of major food crops such as rice and corn, slow growth, weaken crops' resistance to disease, and disrupt many other factors affecting crop yields.

Dwindling supplies of fresh water are now limiting the numbers of people who can survive in some areas. Poor water management causes many of the world's water problems.[45] Each day, people dump 2 million tons of waste into the world's rivers, lakes, and streams. As such pollution grows along with the population, vast quantities of the earth's fresh water will be unusable by the year 2050.[46] By 2025, if present patterns continue, two of every three persons on earth will live in water-stressed conditions.[47]

Overpopulation The world's population reached 6 billion in 1999, but the rate of growth has begun to taper off somewhat.[48] Still, at the present rate of increase, the human population will exceed the earth's estimated **carrying capacity** by 2033. Many authorities in many fields—and more every year—are calling for a reduction in the rate at which the world's population increases. Overpopulation may well be the most serious threat that humankind faces today.

The sheer magnitude of our annual population increase is difficult to comprehend. Each month the world adds the equivalent of another New York City. During six months of the terrible 1992 famine in Somalia, an estimated 300,000 people starved to death. Yet it took the world only 29 *hours* to replace their numbers.

Population stabilization is one of the most pressing needs of our time because it appears to be the only way to enable the world's food output to keep up with demands. Without population stabilization, the world can neither support the lives of people already born nor halt environmental deterioration around the globe. And before the

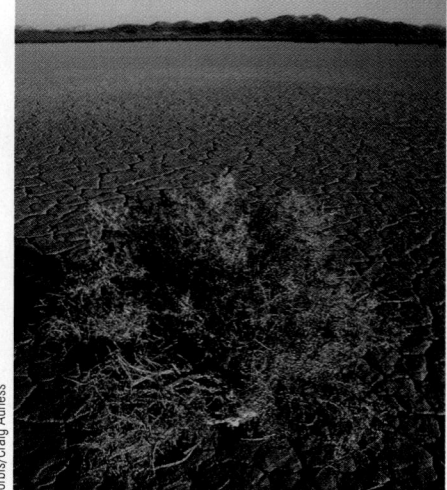

As groundwater is used up, deserts spread.

◼ Worldwide 1.2 billion people live without access to clean, safe water.

◼ Years needed for the world's population to reach . . .

Its 1st billion	2,000,000 years
2nd billion	105 years
3rd billion	30 years
4th billion	15 years
5th billion	12 years
6th billion	11 years

Is it any wonder that food and fresh water supplies may fall behind?

carrying capacity the total number of living organisms that a given environment can support without deteriorating in quality.

FIGURE 15·5 INCOME AND BIRTHRATE

Greater wealth means lower rates of birth. Each dot represents a country.

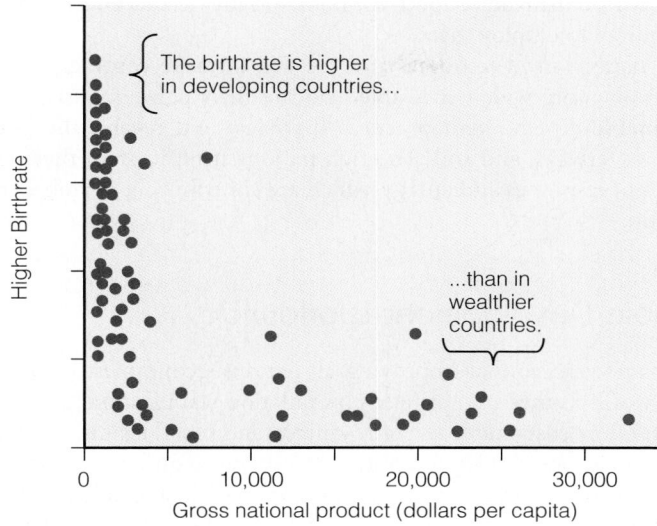

The birthrate is higher in developing countries...

...than in wealthier countries.

Higher Birthrate

Gross national product (dollars per capita)

population problem can be resolved, it may be necessary to remedy the poverty problem. Of the many millions added to the population each year, 98 percent are born in the most poverty-stricken areas of the world.

Poverty and hunger exert an ironic effect on people, driving them to bear more children. Figure 15-5 shows the high correlation of low income and high birthrate. Poverty and hunger are also correlated with lack of education, which includes lack of knowledge about controlling family size. A family in poverty also depends on its children to farm the land, haul water, and care for the adults in their old age. If a family faces ongoing poverty, and its young children are among the most likely to die from disease and other causes, the parents will choose to have many children as a form of "insurance" that some will survive to adulthood.

Relieving poverty and hunger, then, may be a necessary first step in curbing population growth. When people attain better access to health care, education, and family planning, the death rate falls. After a time, the birthrate follows suit. Thus, improvements in living standards help stabilize the population. Wealth distribution matters, too. In countries where economic growth has benefited only the rich, population growth has remained high. Examples include Brazil, Mexico, the Philippines, and Thailand, where large families continue to be a major economic asset for the poor.

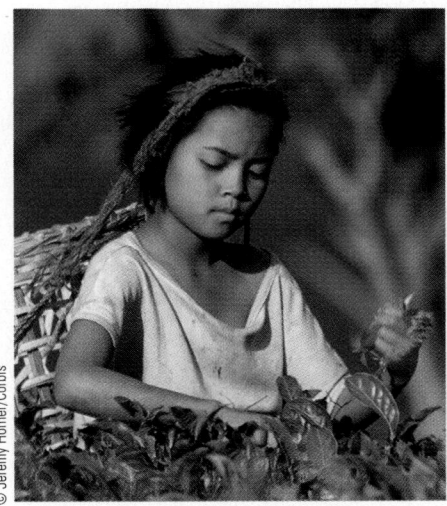

© Jeremy Horner/Corbis

In some countries, every pair of little hands is needed to help feed the family.

▶ KEY POINT Environmental degradation caused by the impacts of growing numbers of people is threatening the world's future ability to feed all of its citizens. Improvements in agriculture can no longer keep up with people's growing numbers. Human population growth is an urgent concern. Controlling population growth requires improving people's economic status and providing them with health care, education, and family planning.

Moving toward Solutions

Slowly but surely, improvements are becoming evident in developing nations. For example, most nations have seen a rise in their gross domestic product, a key measure of economic well-being. Adult literacy rates have increased by more than 50 percent in some areas since 1970, and the proportion of children being sent to school has risen, while the proportion of chronically undernourished people has declined. Today, optimism abounds, and keys to solving the world's environmental, poverty, and hunger problems are within the reach of both the poor and the rich nations—if they will make the effort required to employ them.

The poor nations need resources and the will to make contraceptive technology and information more widely available, educate their citizens, assist the poor, and adopt **sustainable** development practices that slow and reverse the destruction of their forests, waterways, and soil. The rich nations need to stem their wasteful and polluting uses of resources and energy, which are contributing to global environmental degradation.

Sustainable Development Worldwide

Many nations now agree that improving all nations' economies is a prerequisite to meeting the world's other urgent needs: population stabilization, arrest of environmental degradation, sustainable use of resources, and relief of hunger. At a summit of over a hundred nations, the United Nations Conference on Environment and Development, many nations agreed to a set of principles of sustainable development. The conferees defined sustainable development as development that would equitably meet both the economic and the environmental needs of present and future generations. To rephrase a well-known adage: If you give a man a fish, he will eat for a day. If you teach him to fish so that he can buy and maintain his own gear and bait, he will eat for a lifetime and help to feed you.

How Can People Engage in Activism and Simpler Lifestyles at Home?

Every segment of our society can play a role in the fight against poverty, hunger, and environmental degradation. The federal government, the states, local communities, big business and small companies, educators, and all individuals, including dietitians and foodservice managers, have many opportunities to forward the effort.[49]

Government Action   Government policies can change to promote sustainability. For example, the government can devote tax dollars and other resources to encouraging energy conservation and crop protection and to national and international education on sustainable development techniques.

Private and Community Enterprises   Businesses can take initiatives to help; some already have—AT&T, Prudential, and Kraft General Foods are major supporters of antihunger programs. Restaurants and other food facilities can participate in the nation's gleaning effort by giving their fresh leftover foods to community distribution centers. Food producers are more often choosing to produce their goods sustainably to meet a growing demand for products produced with integrity.[50]

Educators and Students   Educators, including nutrition educators, have a crucial role to play. They can teach others about the underlying social and political causes of poverty, the root cause of hunger. At the college level, they can teach the relationships between hunger and birthrate, hunger and environmental degradation, hunger and the status of women, and hunger and global economics. Students can share the knowledge they gain with families, friends, and communities and take action at local and national levels.

sustainable able to continue indefinitely. In this context, the use of resources in ways that maintain both natural resources and human life; the use of natural resources at a pace that allows the earth to replace them. Examples: cutting trees no faster than new ones grow and producing pollutants at a rate with which the environment and human cleanup efforts can keep pace. In a sustainable economy, resources do not become depleted, and pollution does not accumulate.

Food and Nutrition Professionals   Dietitians and foodservice managers can promote sustainable production of food and the saving of resources through reuse, recycling (including composting), energy conservation, and water conservation in both their professional and their personal lives.[51] In addition, the ADA urges its members to work for policy changes in private and government food assistance programs, to intensify education about hunger, and to be advocates on the local, state, and national levels to help end hunger in the United States.

Individuals All individuals can become involved in these large trends. Many small decisions each day add up to large impacts on the environment. The Consumer Corner that follows sums up some of these decisions and actions.

> ▶ **KEY POINT** Government, business, educators, and all individuals have many opportunities to promote sustainability worldwide and wise resource use at home.

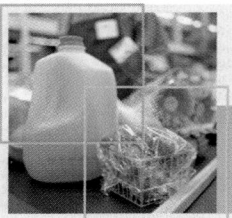

CONSUMER CORNER

Saving Money and Protecting the Environment

CONSUMERS CAN "tread lightly on the earth" through their daily choices. Consider this list:

- Shop "carless" and plan to make fewer shopping trips. Motor vehicles constitute the single largest source of air pollution, causing lung problems, reduced crop yields, and acid rain that damages forests.
- Ride a bike to work or classes.
- Choose foods that are low on the food chain more often (see the chapter Controversy).
- Limit use of imported canned beef products, including stews, chili, corned beef, and pet foods. Many of these foods come at the expense of cleared rain forest land: 200 square feet of rain forest are lost *permanently* for every pound of beef produced.
- Choose small fish more often. Small fish eat tiny aquatic animals and plants—that is, they eat low on the food chain.
- Choose chicken from local farms.
- Shop at farmers' markets and roadside stands for local foods grown close to home.[1] Locally grown foods require less transportation, packaging, and refrigeration than shipped foods. Try picking your own from local farms—it's fun and saves money, too.[2]
- Avoid overly packaged items; buy bulk items with minimal packaging or reusable or recyclable packaging. Each can, foam tray, waxed or clay-coated cardboard container, plastic bottle, or glass jar requires land and many other resources to produce, and its disposal pollutes and costs more land.
- Use reusable pans and dishes, rather than disposable items that are used

once and thrown away. Use pumps instead of spray cans, which are hard to recycle because they are made of many materials.

- Carry reusable string or cloth grocery sacks or bring plastic sacks back to the store and refill them. Production of paper and plastic grocery bags represents a huge drain on resources. Paper factories use chemicals such as toxic forms of chlorine bleach, which are released into waterways in quantities so large that the chemicals can destroy whole bays and fisheries.
- Use fast cooking methods. Stir-frying, pressure cooking, and microwaving all use less energy than conventional stovetop or oven cooking methods.
- Reduce use of aluminum foil, paper towels, plastic wraps, plastic storage bags, and other disposable items. Find permanent reusable replacements for each, such as reusable storage containers and washable cloths.
- Use fewer electric gadgets. Mix batters, chop vegetables, and open cans by hand.
- Purchase the most efficient large appliances possible—look for the Energy Star logo (see Figure 15-6). Products that rank highest in their category for energy efficiency earn this logo from the Environmental Protection Agency. By purchasing Energy Star products, consumers can save many energy dollars each year.
- Insulate the home.
- Consider using solar power, especially to heat water.
- Reduce, reuse, recycle.

The personal rewards of all these behaviors are many, from saving money

to the satisfaction of knowing that you are enjoying and preserving the earth (see Figure 15-7 for other suggestions). But do they really help? They do, if enough people join in. To make the greatest impact, people can also support organizations that lobby for changes in economic policies toward developing countries. Another way to help solve these problems is to join with others to work for international hunger-relief organizations. Table 15-5 (page 576) lists some of the major ones. Local food pantries welcome volunteers, as well.

■ "We do not inherit the earth from our ancestors, we borrow it from our children." Ascribed to Chief Seattle, a nineteenth-century Native American leader.

FIGURE 15·6 ENERGY STAR

Products bearing the U.S. government's Energy Star logo rank highest for energy efficiency. For example, a ten-year-old refrigerator uses as much energy as two refrigerators with the Energy Star label. Energy Star products range from large appliances to light bulbs and building materials.

By choosing products with the Energy Star logo when replacing old equipment, the typical household would save almost $400 per year in energy costs; if everyone chose nothing but Energy Star products over the next ten years, the national energy bill would be reduced by about $100 billion. The reduction in greenhouse gas emissions (carbon dioxide) would be equivalent to taking 17 million cars off the road for each of those ten years.

Courtesy of NASA

Money Isn't All You're Saving

FIGURE 15·7 INDIVIDUAL RESPONSIBILITY AND RESPECT FOR THE ENVIRONMENT

Reduce resource use, reduce fuel use, and reduce pollution with these individual actions.

Eat lower on the food chain—more grains, local chicken, and small fish, and less meat from large animals.

Eat local foods—visit your farmers' market.

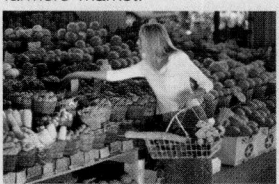

© 2002 PhotoDisc/Getty Images

Buy bulk items to save packaging.

© 2002 PhotoDisc/Getty Images

Recycle glass, cans, paper, and plastic.

Buy fruits of differing ripeness. Eat most perishable foods first.

Shop carless. It can be a pleasure and a great source of exercise.

© Corbis Images/PictureQuest

Buy recycled goods to close the loop.

Use reusable items . . .

Use reusable bags instead of throwaway bags.

© Burke/Triolo/Brand X Pictures/PictureQuest

Use items that don't use energy . . .

. . . instead of nonreusable items.

. . . instead of those that do use energy (even small appliances).

Use appliances that take less energy.

Run your refrigerator efficiently.

Even today's efficient refrigerators use substantial energy because they run day and night. Consumers can take several steps to minimize the energy a refrigerator uses:

- Set it at 37° to 40°F; set the freezer at 0°F.
- Clean the coils and the insulating gaskets around the doors regularly.
- Keep it in good repair.

Courtesy, Bradford White Corporation

The water heater can also waste a lot of energy. Keep the water heater set at 120° to 130°F (no hotter) to save energy. For safe household dishes, sterilization is not necessary. Water of 120° to 130°F enhances the action of dishwashing detergents, making microorganisms slippery and removing them from the dishes. These measures will keep food fresh and clean while keeping energy use low.

TABLE 15·5 SOME HUNGER-RELIEF ORGANIZATIONS PEOPLE CAN JOIN

Action without Borders
350 Fifth Ave., Suite 6614
New York, NY 10118
(212) 843-3973
www.idealist.org

America's Second Harvest
35 E. Wacker Dr. #2000
Chicago, IL 60601
(800) 771-2303
www.secondharvest.org

Bread for the World
50 F St. NW, Suite 500
Washington, DC 20010
(800) 82-BREAD or
(800) 822-7323
(202) 639-9400;
  fax (202) 639-9401
www.bread.org

Children's Hunger
Relief Fund
182 Farmer's Lane
Suite 200
Santa Rosa, CA 95405
(888) 781-1585
www.childrenshungerrelief.org

Congressional Hunger Center
2291/2 Pennsylvania Ave.
Washington, DC 20003
(202) 547-7022
www.hungercenter.org

Food Research and Action Center
1875 Connecticut Ave.
Suite 540
Washington, D.C. 20009
www.frac.org

Foodchain
912 Baltimore, #300
Kansas City, MO 64105
(800) 845-3008
www.foodchain.org

OXFAM America
26 West St.
Boston, MA 02111-1206
(800) 77-OXFAM or
(800) 776-9326
www.oxfam.america.org

Pan American Health
Organization
525 23rd St. NW
Washington, DC 20037
(202) 974-3000
www.paho.org

Society of St. Andrew
3383 Sweet Hollow Rd.
Big Island, VA 24526
(800) 333-4597
www.endhunger.org

United Nations Food and
Agriculture Organization (FAO)
1001 22nd St. NW, Suite 300
Washington, DC 20437
(202) 653-2400
www.fao.org

United Nations International
Children's Emergency Fund
(UNICEF)
3 United Nations Plaza
New York, NY 10017-4414
(212) 326-7035
www.unicef.org

United Nations World Food
Program
Via Cesare Giulio
Viola, 68
Parco dé Medici
Rome, Italy 00148
www.wfp.org

World Health Organization
(WHO)
525 23rd St. NW
Washington, DC 20037
(202) 861-3200
www.who.org

World Hunger Program
Brown University
Box 1831
Providence, RI 02912
(401) 863-2700
www.brown.edu/Departments/
World_Hunger_Program/
hungerweb/WHP/overview.html

World Hunger Year
505 Eighth Ave., 21st Floor
New York, NY 10018-6582
(800) GleanIt
www.worldhungeryear.org

MEDIA MENU

@ 🏠 ◀ ▶

Nutrition Connections

The Nutrition Connections CD provides access to these additional resources.

Chapter Quiz
Take the quiz for this chapter to assess your understanding of chapter concepts.

Nutrition on the Net
Annotated links to reputable nutrition-related websites.

Online Glossary
Test your knowledge of key vocabulary through this comprehensive, interactive glossary.

Internet Connections

A variety of helpful study tools for this chapter are available at our website
http://nutrition.wadsworth.com/ncc10

Practice Test
Take the practice test for this chapter and gauge your knowledge of key concepts; answers are provided.

Student Notebook
A lecture notebook is available to help organize note taking while reading or attending lecture.

Study Tools
Study resources for students include Chapter Summaries, Crossword Puzzles, Flashcards, Glossary, Internet Exercises, and Learning Objectives.

SELF CHECK

Answers to these Self Check questions are in Appendix G.

1. Which of the following is a symptom of food insecurity?
   a. You worry about gaining weight.
   b. You rely on neighbors to feed your children because there is not enough food in the house.
   c. You shop daily to get the best prices.
   d. You buy fresh rather than frozen foods.

2. Which of the items can be purchased with Food Stamps?
   a. bread
   b. cigarettes
   c. dishwashing liquid
   d. alcohol

3. Today, famine is most often a result of?
   a. poverty
   b. drought
   c. social causes such as war
   d. flood

4. Which of the following is an example of environmental degradation?
   a. soil erosion
   b. diminished grazing lands
   c. air pollution
   d. all of the above

5. Which of the following activities are recommended due to the small impact they have on the environment?
   a. Use the oven whenever possible.
   b. Line pans with aluminum foil to reduce cleanup effort.
   c. Use a pressure cooker or microwave to cook foods.
   d. Carry groceries home in paper bags rather than plastic.

6. Poverty and hunger drive people to bear more children.   T   F

7. Most children who die of malnutrition starve to death.   T   F

8. Grains require the least energy to produce.
    T   F

9. The higher a nation's economic status, the faster its population grows over the long run.   T   F

10. Deficiency of vitamin A is the world's leading cause of blindness in young children.
    T   F

CONTROVERSY 15

AGRIBUSINESS AND FOOD PRODUCTION: HOW TO GO FORWARD?

WHILE SOME individuals are making their own personal lifestyles more environmentally benign, as suggested in the chapter, others are seeking ways to improve whole sectors of human enterprise, such as agriculture. To date, large agricultural enterprises have been among the world's biggest polluters and resource users. Is it possible for agriculture to become sustainable? Do our new technologies hold promise for advancing sustainability? How are small farmers faring? This Controversy addresses these questions.

Costs of Producing Food

The environmental and social costs of agriculture and the food industry take many forms. Among them are resource waste and pollution, energy overuse, and tolls on human workers in farm communities. Table C15-1 offers some terms important to these concepts.

Impacts on Land and Water

Producing food has always cost the earth dearly. To grow food, we clear land—prairie, wetland, or forest—causing losses of native ecosystems and wildlife. Then we plant crops or graze animals on the land. The soil loses

---

TABLE C15·1 AGRICULTURAL AND ENVIRONMENTAL TERMS

- **agribusiness** agriculture practiced on a massive scale by large corporations owning vast acreages and employing intensive technological, fuel, and chemical inputs.
- **alternative (low-input,** or **sustainable) agriculture** agriculture practiced on a small scale using individualized approaches that vary with local conditions so as to minimize technological, fuel, and chemical inputs.
- **integrated pest management (IPM)** management of pests using a combination of natural and biological controls and minimal or no application of pesticides.

---

nutrients as each crop is taken from it, so fertilizer is applied. Some fertilizer runs off and pollutes the waterways causing overgrowth of algae and other imbalances. Some plowed soil runs off, clouds the water, and interferes with the growth of aquatic plants and animals.

Then, to protect crops against weeds and pests, herbicides and pesticides are applied. In addition to killing weeds and pests, most herbicides and pesticides kill native plants, native insects, and animals that eat those plants and insects. Widespread use of pesticides and herbicides also causes resistant pests and weeds to evolve. Pesticides pose hazards for farm workers who handle and apply them, and pesticide residues can become a problem for people who consume them along with foods.

Agricultural pesticides and herbicides, if not used conservatively, also pollute rivers, lakes, and groundwater. Pollution from "point sources," such as sewage plants or factories, is relatively easy to control, but runoff from fields and pastures enters waterways across broad

regions and is nearly impossible to control.

Finally, we irrigate, a practice that adds salts to the soil in many areas. The water evaporates, but the salts do not. As soils become salty, plant growth fails. Irrigation can also deplete the water supply over time because water is pulled from surface waters or from underground and then evaporates or runs off. This process, carried to an extreme, can dry up whole rivers and lakes and lower the water table of entire regions. The lower the water table, the more farmers must irrigate; and the more they irrigate, the more groundwater they use up.

Soil Depletion and Losses of Species

The soil can also be depleted by other agricultural practices, particularly indiscriminate land clearing (deforestation) and overuse by cattle (overgrazing). In just the past 40 years, human agricultural activities have ruined more than 10 percent of the earth's fertile land, an area the size of China and India combined. Over 20 million acres have been so damaged that they may be impossible to reclaim. If soil erosion proceeds unchecked, by 2025 the world may be struggling to feed many more people with 40 percent less food-producing land per person.

Unsustainable agriculture has already destroyed many once-fertile regions, where high civilizations formerly flourished. The dry, salty deserts of North Africa were once plowed and irrigated wheat fields, the breadbasket of the Roman Empire. Today's mistreatment of soil and water is causing destruction on a scale never known before.

Agriculture is also weakening its own underpinnings by failing to conserve species diversity. By the year 2050, some 40,000 plant species, existing today, may go extinct. The United Nations Food and Agriculture Organization attributes many of the losses, which are occurring daily, to modern farming practices, as well as to population growth. The growing uniformity of global eating habits also contributes. As people everywhere eat the same limited array of foods, demand for local, genetically diverse, native plants is insufficient to make

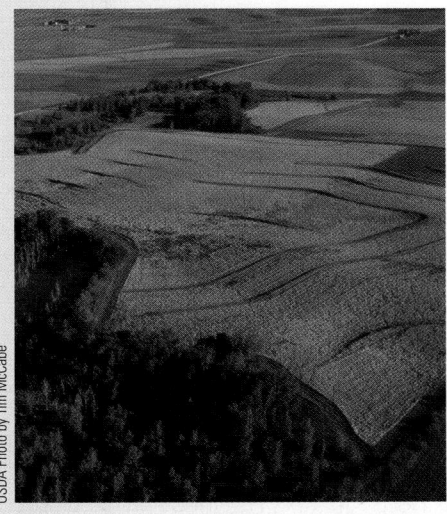
Vast areas under plow are exposed to erosion, and those that must be irrigated can, over time, become salty and unusable.

them financially worth preserving. Yet, in the future, as the climate warms, those very plants may be needed as food sources. A wild species of corn that grows in a dry climate, for example, might contain just the genetic information necessary to help make the domestic corn crop resistant to drought.

Energy

Massive fossil fuel use is threatening our planet by causing ozone depletion, water pollution, ocean pollution, and other ills and by making global warming likely. In the United States, the food industry consumes about 20 percent of all the energy the nation uses. Each year we spend 1,500 liters (over 350 gallons) of energy from oil per person to produce, process, distribute, and prepare our food. Energy is used to run farm machinery and to produce fertilizers and pesticides. Energy is also used to prepare, package, transport, refrigerate, and otherwise store, cook, and wash our foods.

The Problems of Livestock and Fishing

Raising livestock also takes a toll. Like plant crops, herds of livestock occupy land that once maintained itself in a natural state. The land suffers losses of native plants and animals, soil erosion, water depletion, and desert formation. If animals are raised in concentrated areas such as cattle feedlots or giant hog

"farms" instead, huge masses of animal wastes produced in these overcrowded, factory-style farms leach into local soils and water supplies, polluting them. In an effort to control this source of pollution, the Environmental Protection Agency (EPA) offers incentives to livestock farmers who agree to clean up their wastes and allow their operations to be monitored for pollution.[*1] The EPA may soon require more sustainable handling practices for animal wastes, such as diluting them for use as fertilizer on fields.[2] In addition to the waste problem, animals in such feedlots still have to be fed; grain is grown for them on other land (Figure C15-1 compares the grain required to produce various foods). That grain may require fertilizers, herbicides, pesticides, and irrigation, too. In the United States, one-fifth of all cropland is used to produce grain for livestock—more land than is used to produce grain for people.

Other environmental costs attend fishing. Fishing easily becomes overfishing and depletes stocks of the very fish that people need to eat. Most nets also collect many nonfood species that are killed during harvest but returned to the sea instead of being put to use. Other aquatic animals are also vulnera-

---

[*] Reference notes are found in Appendix F.

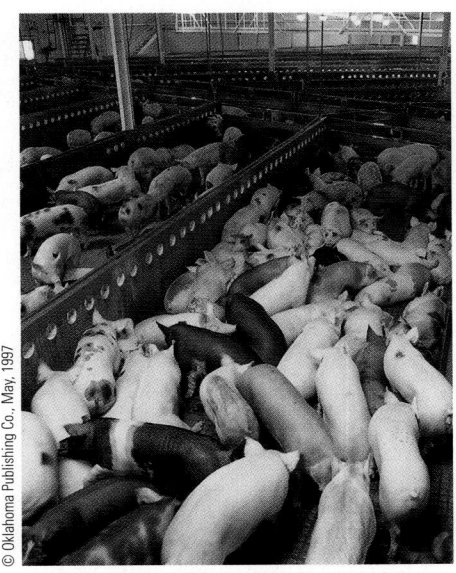
Industrial farms generate huge masses of wastes that can contaminate local soil and water.

© Oklahoma Publishing Co., May, 1997

FIGURE C15·1 POUNDS OF GRAIN NEEDED TO PRODUCE ONE POUND OF BREAD AND ONE POUND OF ANIMAL WEIGHT GAIN

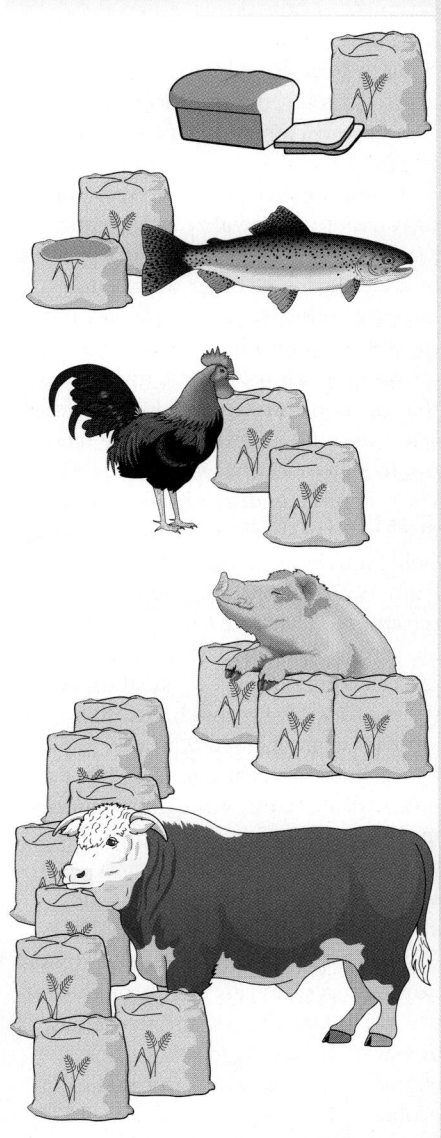

ble to injury and death from fishing activities, and populations of ecologically important nonfood animals, such as dolphins, are diminished. In short, our ways of producing foods are, for the most part, not sustainable.

Agribusiness

Farmers in the United States face serious challenges: not only are their costs for energy and other inputs increasing, but markets for U.S. farm products abroad are declining as a result of increased agricultural production and exports from other countries and economic problems that prevent grain-importing nations from purchasing U.S. products.[3] Many farms, especially medium-size family farms, are struggling to be profitable because of competition from foreign producers and a trend toward large food-producing operations.[4]

Faring somewhat better economically are huge centralized farms and ranches, many of which are being operated in Mexico or other developing nations; collectively, they are part of the massive food-producing enterprise called **agribusiness.** These operations in the United States tend to use little local labor, and the profits they make tend not to stay in local communities. Their workers often endure unsafe conditions: nearly 500 pesticide poisonings among California farm workers were reported from 1997 to 2000, and many more unreported cases are believed to have occurred.[5] Agribusinesses located in foreign countries hire local laborers willing to work for much less than laborers in the United States.

Agribusinesses also tend to place a higher priority on producing abundant, inexpensive food than on protecting soil, water, and local biodiversity. As a result, no country on earth has more affordable abundant food than the United States.[6] When these large operations accomplish this feat through overuse of land, fertilizers, and pesticides, however, the resulting soil erosion, wasted irrigation water, and pollution problems can be enormous. Small U.S. farmers may be driven to adopt similar unsustainable practices or use their creativity to stay competitive.

Because of economies of scale, agribusinesses can price their products more attractively than can smaller, local farms. Thus, local U.S. grocers offer broccoli from Mexico, carrots from California, pineapples from Hawaii, and bananas from Central America at prices no local farmers can match, even if they could grow those products.

If food prices had to include a "tax" to pay for pollution cleanup, water pro-

tection, and land restoration, the prices of foods produced unsustainably would be much higher. If they included a living wage, education, and benefits for the migrant farm workers, they would be higher still.

The Future Is Now

For each of the problems just described, solutions are being devised, and some are already being put into practice. To fully exploit these new sustainable agriculture techniques across the country will require some new learning and change. Sustainable agriculture is not one system but a set of practices that can be matched to particular needs in local areas. The crop yields from farms that employ these practices often compare favorably with those from farms using less sustainable methods.[7] The first of these ideas, sustainable, **alternative,** or **low-input,** agriculture, emphasizes careful use of natural processes wherever possible, rather than chemically intensive methods.

Low-Input Agriculture

One form of low-input agriculture is **integrated pest management (IPM).** Farmers using this system employ many techniques, such as crop rotation and natural predators, to control pests rather than depending on heavy use of pesticides alone. Not all crops can grow reliably without pesticides, but many can. Table C15-2 contrasts low-input agriculture methods with unsustainable methods. Many sustainable techniques are not really new—they would be familiar to our great-grandparents. Many farmers today are rediscovering the benefits of old techniques as they adapt and experiment with them in the search for sustainable methods.

Low-input agriculture has some apparent disadvantages, but advantages offset them. For example, as chemical use falls, yields per acre also fall somewhat, but costs per acre also fall, so the return per acre may be the same as or greater than before. More money goes to farmers and less to the fuels, fertilizers, pesticides, and irrigation. The end result of such farming is to make both farmers and consumers better off financially and environmentally.

Pure rivers represent irreplaceable water resources.

Low-input agriculture works. As the world's population grows, and its land and water dwindle, the need to adopt sustainable agriculture and development around the globe grows urgent. More than 30,000 U.S. farmers are successfully using sustainable techniques such as those described in Table C15-2. They see it as a food production system that can indefinitely sustain a healthy food supply, restore soil and water resources, and revitalize farming communities, while reducing reliance on fossil fuels.

Precision Agriculture

An exciting development in agriculture is the application of powerful new computer technologies to food production. Through techniques collectively known as *precision agriculture*, farmers can adjust soil and crop management to meet the precise needs of various areas of the farm. For example, a farmer growing crops in a field with hills, which tend to stay drier, and with low-lying areas, which tend to stay wetter, can adjust irrigation water to meet the specific needs of each part of the field. Similarly, if one section of a field needs nitrogen fertilizer while another needs a different mix, the farmer can preprogram a computer to apply fertilizer of just the right type and amount for each

TABLE C15·2 HIGH-INPUT AND LOW-INPUT AGRICULTURAL TECHNIQUES COMPARED

Unsustainable Practice	Sustainable Practice
• Growing the same crop repeatedly on the same patch of land. This takes more and more nutrients out of the soil, makes fertilizer use necessary; favors soil erosion; and invites weeds and pests to become established, making pesticide use necessary.	• Rotate crops. This increases nitrogen in the soil so there is less need to buy fertilizers. If used with appropriate plowing methods, crop rotation reduces soil erosion. Crop rotation also reduces weeds and pests.
• Using fertilizers generously. Excess fertilizer pollutes ground and surface water and costs both farmers' household money and consumers' tax money.	• Reduce the use of fertilizers and use livestock manure more effectively. Store manure during the nongrowing season and apply it during the growing season.
	• Alternate nutrient-devouring crops with nutrient-restoring crops, such as legumes.
	• Compost on a large scale, including all plant residues not harvested. Plow the compost into the soil to improve its water-holding capacity.
• Feeding livestock in feedlots where their manure produces major water and soil pollutant problems. Piled in heaps, manure also releases methane, a global-warming gas.	• Feed livestock or buffalo on the open range where their manure will fertilize the ground on which plants grow and will release no methane. Alternatively, at least collect feedlot animals' manure and use it as fertilizer, or, at the very least, treat it before release.
• Spraying herbicides and pesticides over large areas to wipe out weeds and pests.	• Apply technology in weed and pest control. Use precision agriculture techniques if affordable or use rotary hoes twice instead of herbicides once. Spot treat weeds by hand.
	• Rotate crops to foil pests that lay their eggs in the soil where last year's crop was grown.
	• Use genetically resistant crops.
	• Use biological controls such as predators that destroy the pests.
• Plowing the same way everywhere, allowing unsustainable water runoff and erosion.	• Plow in ways tailored to different areas. Conserve both soil and water by using cover crops, crop rotation, no-till planting, and contour plowing.
• Injecting animals with antibiotics to prevent disease in livestock.	• Maintain animals' health so that they can resist disease.
• Irrigating on a large scale.	• Irrigate only during dry spells and only where needed.

area.[8] Likewise, pesticide application can be programmed to prescribed patterns, avoiding areas too close to streams or other water sources. The preprogrammed system turns off the pesticide flow when it comes to a designated safety zone.

The *global positioning satellite (GPS)* system is at the heart of precision farming. In the GPS system, satellites beam accurate information about land positions and elevations of an area, such as a field, to receivers placed on farm equipment here on earth. The GPS system delivers a grid map, pinpointing locations on a farm. Farmers can use the GPS information grid to target, within a meter's accuracy, areas that need treatments. They can then program computerized farm equipment to apply chemicals or other treatments accordingly. Farmers can also use the information to adjust the depths to which they till the soil. The goal is to till deeply enough to prepare seedbeds properly and control weeds, but to avoid excessive tilling that wastes fuel

and worsens erosion. Finally, at harvest, a GPS system produces an accurate accounting of crop yield, acre by acre, so that spot adjustments can be made in the next planting season.

The future of precision agriculture seems bright, and the potential savings to farmers in terms of water, fertilizers, and pesticides are enormous. The accompanying reduction in polluting chemicals introduced into the environment means that everyone benefits. The high initial costs of the equipment pose a potential barrier to some farmers, but incentive programs can often offset some of the costs.

Soil Conservation and Incentive Programs

The U.S. Conservation Reserve Program provides federal assistance to farmers and ranchers who wish to improve their conservation of soil, water, and related natural resources on environmentally sensitive lands. It encourages farmers to plant native grasses, wildlife habitat, or trees instead of cash crops on highly

erodible cropland, wetlands, or other environmentally sensitive acreage. It also encourages conservation techniques such as shallow tilling and planting grassy strips to move water off fields. In exchange, farmers receive annual rental payments and other assistance over a 10- to 15-year contract period. The goals of the program are to:

- Reduce soil erosion.
- Protect production of food and fiber.
- Reduce sedimentation in streams and lakes.
- Improve water quality.
- Establish wildlife habitat.
- Enhance forest and wetland resources.

About 25 percent of all U.S. land is considered highly erodible and thus qualifies for conservation incentives.

Other conservation programs offer incentives for improving air quality, water quality, or animal habitat or for instituting measures to achieve goals similar to those listed above. Others may purchase sensitive lands for conservation. Private foundations or other groups may fund such programs with help from local, state, or federal agencies.

Measuring the success of any conservation program presents challenges because it entails measuring the nature and extent of the changes in the practices of farmers and ranchers that occur in response to program incentives.[9] In addition, a program's success may be affected by weather conditions at the sites, changing local regulations, technology, and dozens of other variable factors. Based on soil loss alone, though, conservation programs do seem to be working to preserve U.S. soils. The amount of cropland lost annually to erosion fell by almost 40 percent between 1982 and 1997.[10] Every year, 295 million tons of the 331 million tons of soil conserved on lands in conservation programs is believed attributable to conservation incentives.

Direct Sales from Farm to Consumer

Today's small farmer must cut costs and find creative means of boosting profits. One avenue to improved balance sheets is direct marketing to consumers. Farmers selling their broccoli, carrots, and apples at farmers' markets and roadside stands often net a higher profit, especially when consumers perceive an intrinsic benefit from making the extra effort to obtain such foods. For example, local food may offer a measure of protection against foodborne illnesses.[11] A recent outbreak of hepatitis from imported green onions was traced to infected farm workers in Mexico for example. Local food has passed through fewer hands.

Another way small farmers can boost profits is to create extra value by packaging or processing their products in ways that benefit consumers. For example, when a grape grower establishes a winery that produces local wines or an apple producer begins making candy apples and apple pies, the farmers use their own produce to create a new "added value" product line to increase profits.[12] Consumers appreciate knowing how their food and other products are grown and handled. Urban consumers often relish a trip to "the country" to pick up fresh foods grown locally.

Equally promising is the growing trend toward organic farming. Consumers pay substantially higher prices for certified organic products, so small farmers reap greater profits and the environment benefits, too.

Agricultural Biotechnology

Although not every farmer worldwide may be in a position to reap benefits from the technologies of precision agriculture, the advances of biotechnology may prove to be an essential part of a worldwide move toward sustainable agriculture. If health and safety issues are addressed and resolved (see Controversy 12), rDNA technology promises economic, environmental, and agricultural benefits by shrinking the acreage needed for crops, reducing soil losses, minimizing use of chemical insecticides, and bettering crop protection.

Genetically modified microbes can offer benefits to sustainable agriculture if research and testing can ensure that rDNA microorganisms released into the environment will not turn out to be more harmful than the products they are intended to replace. Bioengineered microbes could contribute to the continuous renewal of the soil's fertility by fixing nitrogen and releasing other nutrients into the soil, thereby lessening the need for chemical fertilizers and easing the environmental burden. Scientific laboratories are also working to engineer microbes that can recycle agricultural, industrial, and household wastes into fertilizers, an obvious boon to the environment. Bacterial and fungal herbicides, fungicides, and insecti-

cides are in advanced experimental stages. Such technologies promise to augment other integrated pest management systems of low-input agriculture such as crop rotation.

Energy Efficiency

Some 6,560 calories of fuel are used to produce a can of corn (including the can and transportation), and 7,980 calories are needed to produce a package of frozen corn (including packaging, freezing, and transportation). Much of this energy input could be reduced, as Table C15-3 shows. The last item in the table suggests that consumers should center their diets on foods that require low energy inputs, a choice that is described next.

Eating Lower on the Food Chain

Studies of energy use in the U.S. food system have revealed which foods require the most and least energy to produce. The least energy is needed for grain: about one-third calorie of fuel is burned to grow each calorie of grain. Fruits and vegetables are intermediate, and most animal protein requires from 10 to 90 calories of fossil energy per calorie of usable food. An exception is livestock raised on the open range; these animals eat grass and require low energy inputs as do most plant foods.[13] So much of our beef is grain fed, rather than range fed, however, that the average energy requirement for beef production is high.

To support our meat intake, we maintain several billion livestock, about four times our own weight in animals.

TABLE C15-3 SUSTAINABLE ENERGY-SAVING AGRICULTURAL TECHNIQUES

- Use machinery scaled to the job at hand and operate it at efficient speeds.
- Combine operations. Harrow, plant, and fertilize in the same operation.
- Use diesel fuel. Use solar and wind energy on farms. Use methane from manure. Be open-minded to alternative energy sources.
- Use new disease- and pest-resistant plant varieties developed through genetic engineering.
- Save on technological and chemical inputs and spend some of the savings paying people to do manual jobs. Increasing labor inputs has been considered inefficient. Reverse this thinking: creating more jobs is preferable to using more machinery and fuel.
- Partially return to the techniques of using animal manure and crop rotation. This would save energy because chemical fertilizers require large energy inputs to produce.
- Choose crops that require low energy inputs (fertilizer, pesticides, irrigation).
- Educate people to cook food efficiently and to eat low on the food chain.

FIGURE C15·2 ECOLOGICAL FOOTPRINTS

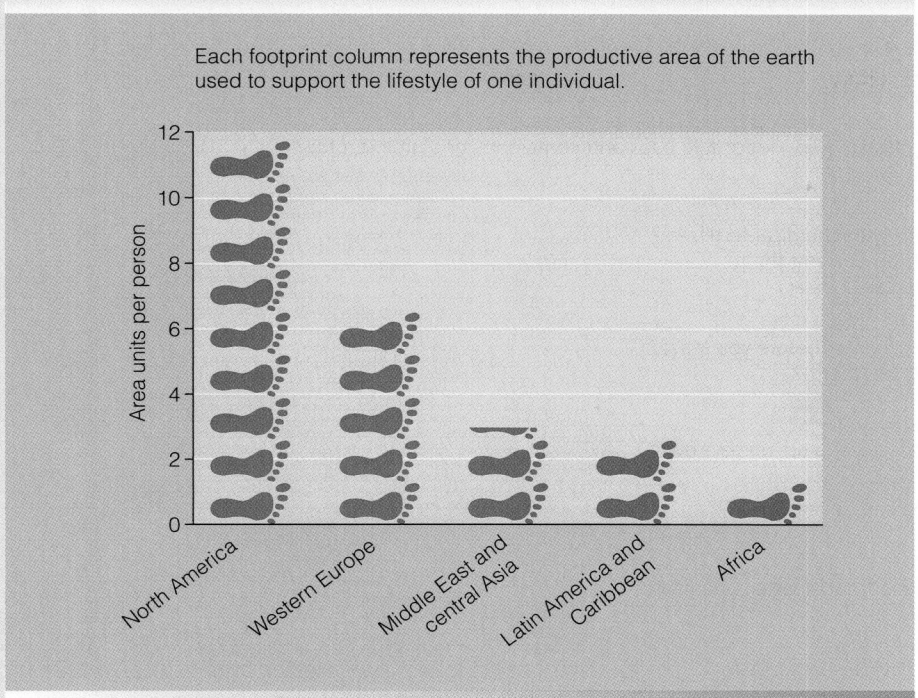

Each footprint column represents the productive area of the earth used to support the lifestyle of one individual.

Livestock consume ten times as much grain each day as we do. We could use much of that grain to make grain products for ourselves and share them. Such a shift in consumption could free up enough grain to feed 400 million people, would necessitate burning less fuel and using less water, and could also free up much more land. According to the United Nations, the "ecological footprint"—the productive land area needed to support a person's lifestyle—of each individual is four times larger in an industrialized country than in a developing one (see Figure C15-2). To help size up your own ecological footprint, take the quiz in Table C15-4 (p. 584).

Some meat eaters are choosing to cut down on their meat portions or to eat range-fed beef or buffalo only. "Range-burger" buffalo also offers nutrition advantages over grain-fed beef because it is lower in fat, and the fat has more polyunsaturated fatty acids, including the omega-3 type. Some people are switching to nonmeat, and even pure vegan, diets. The fish farming industry shows promise of being able to feed large numbers of people in the future and could help greatly to provide nutritious meat at a price people and the environment could afford.

Conclusion

Although many problems are global in scope, the actions of individual people lie at the heart of their solutions. Do what you can to tread lightly on the earth. Beware of a perfectionist attitude, however, because believing that you "should" do more than a realistic amount can lead to defeat. Small improvements add up to large accomplishments, so any amount of progress is well worth celebrating. Celebrate the changes that are possible today by making them a permanent part of your life; do the same with changes that become possible tomorrow and every day thereafter. The results may add up to more than you dared to hope for.

TABLE C15·4 HOW BIG IS YOUR ECOLOGICAL FOOTPRINT?

This quiz can help you evaluate your impact on the earth. The higher your score, the smaller your "footprint."

At home, do you:
1. Recycle everything you can: newspapers, cans, glass bottles and jars, scrap metal, used oil, etc.?
2. Use cold water in the washer whenever possible?
3. Avoid using appliances (such as electric can openers) to do things you can do by hand?
4. Reuse grocery bags to line your wastebasket? Reuse or recycle bread bags, butter tubs, etc.?
5. Store food in reusable containers rather than plastic wrap, disposable bags and containers, or aluminum foil?

In the yard, do you:
6. Pull weeds instead of using herbicides?
7. Fertilize with manure and compost, rather than with chemical fertilizers?
8. Compost your leaves and yard debris, rather than burning them?
9. Take extra plastic and rubber pots back to the plant nursery?

On vacation, do you:
10. Turn down the heat and turn off the hot water heater before you leave?
11. Carry reusable cups, dishes, and flatware (and use them)?
12. Dispose of trash appropriately (never litter)?
13. Buy no souvenirs made from wild or endangered animals?
14. Stay on roads and trails, and not trample dunes and fragile undergrowth?

About your car, do you:
15. Keep your car tuned up for maximum fuel efficiency?
16. Use public transit whenever possible?
17. Ride your bike or walk whenever possible?
18. Plan to replace your car with a more fuel-efficient model when you can?
19. Recycle your engine oil?

At school or work, do you:
20. Recycle paper whenever possible?
21. Use scrap paper for notes to yourself and others?
22. Print or copy on both sides of the paper?
23. Reuse large envelopes and file folders?
24. Use the stairs instead of the elevator whenever you can?

When buying, do you:
25. Buy as little plastic and foam packaging as possible?
26. Buy permanent, rather than disposable, products?
27. Buy paper rather than plastic, if you must buy disposable products?
28. Buy fresh produce grown locally?
29. Buy in bulk to avoid unnecessary packaging?

In other areas, do you:
30. Volunteer your time to conservation projects?
31. Encourage your family, friends, and neighbors to save resources, too?
32. Write letters to support conservation issues?

Scoring:
First, give yourself 4 points for answering this quiz: ___
Then, give yourself 1 point each for all the habits you know people should adopt. This is to give you credit for your awareness, even if you haven't acted on it, yet (total possible points = 32): ___
Finally, give yourself 2 more points for each habit you have adopted—or honestly would if you could (total possible points = 64): ___
Total score:
1 to 25: You are a beginner in stewardship of the earth. Try to improve.
26 to 50: You are on your way and doing better than many consumers.
51 to 75: Good. Pat yourself on the back, and keep on improving.
76 to 100: Excellent. You are a shining example for others to follow.

SOURCE: Adapted from *Conservation Action Checklist,* produced by the Washington Park Zoo, Portland, Oregon, and available from Conservation International, 1015 18th St. NW, Suite 1000, Washington, D.C. 20036: 1-800-406-2306. (website: www.conservation.org). Call or write for copies of the original or for more information.

Appendixes

Table of Food Composition

This edition of the table of food composition has been updated to reflect current nutrient data for foods, to remove outdated foods, and to add foods that are new to the marketplace.* The nutrient database for this appendix is compiled from a variety of sources, including the USDA Standard Reference database (Release 17) and manufacturers' data. The USDA database provides data for a wider variety of foods and nutrients than other sources. Because laboratory analysis for each nutrient can be quite costly, manufacturers tend to provide data only for those nutrients mandated on food labels. Consequently, data for their foods are often incomplete; any missing information on this table is designated as a dash. Keep in mind that a dash means only that the information is unknown and should not be interpreted as a zero. A zero means that the nutrient is not present in the food.

Whenever using nutrient data, remember that many factors influence the nutrient contents of foods. These factors include the mineral content of the soil, the diet fed to the animal or the fertilizer used on the plant, the season of harvest, the method of processing, the length and method of storage, the method of cooking, the method of analysis, and the moisture content of the sample analyzed. With so many influencing factors, users should view nutrient data as a close approximation of the actual amount.

For updates, corrections, and a list of 8,000 foods and codes found in the diet analysis software that accompanies this text, visit www.wadsworth.com/nutrition and click on *Diet Analysis*.

- *Fats* Total fats, as well as the breakdown of total fats to saturated, monounsaturated, and polyunsaturated fats, are listed in the table. The fatty acids seldom add up to the total in part due to rounding but also because values are derived from a variety of laboratories and other fatty acid components, such as *trans*-fatty acids and glycerol, are not included in these basic categories. *Trans*-fatty acids can comprise a large share of the total fat in margarine and shortening (hydrogenated oils) and in any foods that include them as ingredients.
- *Vitamin A and Vitamin E* In keeping with the 2001 RDA for vitamin A, this appendix presents data for vitamin A in micrograms (µg) RAE. Similarly, because the 2000 RDA for vitamin E is based only on the alpha-tocopherol form of vitamin E, this appendix reports vitamin E data in milligrams (mg) alpha-tocopherol, listed on the table as Vit E (mg).
- *Bioavailability* Keep in mind that the availability of nutrients from foods depends not only on the quantity provided by a food, but also on the amount absorbed and used by the body—the bioavailability. The bioavailability of folate from fortified foods, for example, is greater than from naturally occurring sources. Similarly, the body can make niacin from the amino acid tryptophan, but niacin values in this table (and most databases) report preformed niacin only.
- *Using the Table* The foods and beverages in this table have been organized into several categories, which are listed at the head of each right-hand page. Page numbers have been provided, and each group has been color-coded to make it easier to find individual foods.

---

*This food composition table has been prepared by Wadsworth Publishing Company. The nutritional data are supplied by First Data Bank.

A

DA + Code	Food Description	Quantity	Measure	Wt (g)	H₂O (g)	Ener (cal)	Prot (g)	Carb (g)	Dietary Fiber (g)	Fat (g)	Fat Breakdown (g) Sat	Mono	Poly
	BREADS, BAKED GOODS, CAKES, COOKIES, CRACKERS, CHIPS, PIES												
	Bagels												
8534	Cinnamon & raisin	1	item(s)	71	23	195	7	39	2	1	0.19	0.12	0.48
4910	Enriched, all varieties	1	item(s)	71	23	195	7	38	2	1	0.16	0.09	0.49
4911	Plain, enriched, toasted	1	item(s)	66	18	195	7	38	2	1	0.16	0.09	0.49
8538	Oat bran	1	item(s)	71	23	181	8	38	3	1	0.14	0.18	0.35
12079	Whole grain	1	item(s)	85	—	170	9	35	6	2.5	0	—	—-
	Biscuits												
25008	Biscuits	1	item(s)	41	16	121	3	16	1	5	1.40	1.41	1.82
16729	Scone	1	item(s)	42	11	149	4	19	1	6	2.01	2.55	1.26
25166	Wheat biscuits	1	item(s)	55	21	162	4	22	1	7	1.90	1.92	2.51
	Bread												
325	Boston brown, canned	1	slice(s)	45	21	88	2	19	2	1	0.13	0.09	0.25
8716	Bread sticks, plain	4	item(s)	24	1	99	3	16	1	2	0.34	0.86	0.87
25176	Cornbread	1	piece(s)	55	26	141	5	18	1	5	2.09	1.44	1.50
327	Cracked wheat	1	slice(s)	25	9	65	2	12	1	1	0.23	0.48	0.17
9079	Croutons, plain	¼	cup(s)	8	<1	31	1	6	<1	<1	0.11	0.23	0.10
8582	Egg	1	slice(s)	40	14	115	4	19	1	2	0.64	0.92	0.44
8585	Egg, toasted	1	slice(s)	37	10	117	4	19	1	2	0.60	1.11	0.43
329	French	1	slice(s)	25	9	69	2	13	1	1	0.16	0.30	0.17
8591	French, toasted	1	slice(s)	23	7	69	2	13	1	1	0.16	0.30	0.17
8597	Indian fry	1	item(s)	90	24	296	6	48	2	9	2.08	3.59	2.33
332	Italian	1	slice(s)	30	11	81	3	15	1	1	0.26	0.24	0.42
1393	Mixed grain	1	slice(s)	26	10	65	3	12	2	1	0.21	0.40	0.24
8604	Mixed grain, toasted	1	slice(s)	24	8	65	3	12	2	1	0.21	0.40	0.24
8605	Oat bran	1	slice(s)	30	13	71	3	12	1	1	0.21	0.48	0.51
8608	Oat bran, toasted	1	slice(s)	27	10	70	3	12	1	1	0.21	0.47	0.50
8609	Oatmeal	1	slice(s)	27	10	73	2	13	1	1	0.19	0.43	0.46
8613	Oatmeal, toasted	1	slice(s)	25	8	73	2	13	1	1	0.19	0.43	0.46
1409	Pita	1	item(s)	60	19	165	5	33	1	1	0.10	0.06	0.32
7905	Pita, whole wheat	1	item(s)	64	20	170	6	35	5	2	0.26	0.22	0.68
338	Pumpernickel	1	slice(s)	32	12	80	3	15	2	1	0.14	0.30	0.40
334	Raisin, enriched	1	slice(s)	26	9	71	2	14	1	1	0.28	0.60	0.18
8625	Raisin, toasted	1	slice(s)	24	7	71	2	14	1	1	0.28	0.60	0.18
10168	Rice, white	1	slice(s)	42	—	140	1	21	1	6	0.50	—	—
8653	Rye	1	slice(s)	32	12	83	3	15	2	1	0.20	0.42	0.26
8654	Rye, toasted	1	slice(s)	29	9	82	3	15	2	1	0.20	0.42	0.25
336	Rye, light	1	slice(s)	25	9	65	2	12	2	1	0.20	0.30	0.30
8588	Sourdough	1	slice(s)	25	9	69	2	13	1	1	0.16	0.30	0.17
8592	Sourdough, toasted	1	slice(s)	23	7	69	2	13	1	1	0.16	0.30	0.17
491	Submarine or hoagie roll	1	item(s)	135	41	400	11	72	4	8	1.80	3.00	2.20
8596	Vienna, toasted	1	slice(s)	23	7	69	2	13	1	1	0.16	0.30	0.17
8670	Wheat	1	slice(s)	25	9	65	2	12	1	1	0.22	0.43	0.23
8671	Wheat, toasted	1	slice(s)	23	7	65	2	12	1	1	0.22	0.43	0.23
340	White	1	slice(s)	25	9	67	2	13	1	1	0.18	0.17	0.34
1395	Whole wheat	1	slice(s)	46	15	128	4	24	3	2	0.37	0.53	1.35
	Cakes												
386	Angel food, from mix	1	slice(s)	50	16	129	3	29	<1	<1	0.02	0.01	0.06
8772	Butter pound, ready to eat, commercially prepared	1	slice(s)	75	18	291	4	37	<1	15	8.67	4.43	0.80
8737	Carrot, cream cheese frosting, from mix	1	slice(s)	111	23	484	5	52	1	29	5.43	7.24	15.10
4931	Chocolate, chocolate icing, commercially prepared	1	slice(s)	64	15	235	3	35	2	10	3.05	5.61	1.18
8756	Chocolate, from mix	1	slice(s)	95	23	340	5	51	2	14	5.16	5.74	2.62
393	Devil's food cupcake, chocolate frosting	1	item(s)	35	8	120	2	20	1	4	1.80	1.60	0.60
8757	Fruitcake, ready to eat, commercially prepared	1	piece(s)	43	11	139	1	26	2	4	0.45	1.81	1.43
1397	Pineapple upside down, from mix	1	slice(s)	115	37	367	4	58	1	14	3.35	5.97	3.77
411	Sponge, from mix	1	slice(s)	63	19	187	5	36	<1	3	0.82	0.99	0.41
8817	White, coconut frosting, from mix	1	slice(s)	112	23	399	5	71	1	12	4.36	4.14	2.42
8819	Yellow, chocolate frosting, ready to eat, commercially prepared	1	slice(s)	64	14	243	2	35	1	11	2.98	6.14	1.35

PAGE KEY: A–4 = Breads/Baked Goods   A–8 = Cereal/Rice/Pasta   A–12 = Fruit   A–18 = Vegetables/Legumes   A–28 = Nuts/Seeds   A–30 = Vegetarian
A–32 = Dairy   A–38 = Eggs   A–40 = Seafood   A–42 = Meats   A–44 = Poultry   A–46 = Processed meats   A–48 = Beverages   A–52 = Fats/Oils
A–54 = Sweets   A–54 = Sauces/Condiments/Spices   A–58 = Mixed Foods/Soups/Sandwiches   A–64 = Fast food   A–82 = Convenience   A–84 = Baby foods

A

Chol (mg)	Calc (mg)	Iron (mg)	Magn (mg)	Pota (mg)	Sodi (mg)	Zinc (mg)	Vit A (RAE) (µg)	Thia (mg)	Vit E (mg)	Ribo (mg)	Niac (mg)	Vit B₆ (mg)	Fola (µg)	Vit C (mg)	Vit B₁₂ (µg)	Sele (µg)
0	13	2.70	20	105	229	0.80	15	0.27	0.11	0.20	2.19	0.04	79	<1	0	22
0	53	2.53	21	72	379	0.62	0	0.38	0.03	0.22	3.24	0.04	75	0	0	23
0	53	2.52	20	72	379	0.62	0	0.31	0.02	0.20	2.91	0.03	64	0	0	23
0	9	2.19	22	82	360	0.64	1	0.24	0.10	0.24	2.10	0.03	70	<1	0	24
0	200	1.08	120	0	200	4.5	0	0.44	0	0.5	8	0.6	—	0	1.79	0
<1	33	1.01	6	37	205	0.27	9	0.13	0.03	0.12	1.08	0.01	26	0	<.1	7
49	80	1.31	7	48	288	0.29	—	0.15	0.72	0.16	1.20	0.03	8	<.1	<1	—
<1	57	1.22	16	81	321	0.42	12	0.16	0.11	0.13	1.49	0.03	29	<.1	<.1	12
<1	32	0.95	28	143	284	0.23	11	0.01	0.25	0.05	0.50	0.04	5	0	<.1	10
0	5	1.03	8	30	158	0.21	0	0.14	0.36	0.13	1.27	0.02	39	0	0	9
21	88	1.01	10	59	209	0.57	38	0.13	0.58	0.16	0.98	0.04	36	2	<1	6
0	11	0.70	13	44	135	0.31	0	0.09	0.15	0.06	0.92	0.08	15	0	<.1	6
0	6	0.31	2	9	52	0.07	0	0.05	—	0.02	0.41	0.00	10	0	0	3
20	37	1.22	8	46	197	0.32	25	0.18	0.24	0.17	1.94	0.03	42	0	<.1	12
21	38	1.24	8	47	200	0.32	26	0.14	0.31	0.16	1.77	0.02	36	0	<.1	12
0	19	0.63	7	28	152	0.22	0	0.13	0.07	0.08	1.19	0.01	37	0	0	8
0	19	0.63	7	28	152	0.22	0	0.10	0.06	0.07	1.07	0.01	22	0	0	8
0	210	3.24	14	67	626	0.45	0	0.39	0.69	0.27	3.27	0.02	67	0	0	21
0	23	0.88	8	33	175	0.26	0	0.14	0.11	0.09	1.31	0.01	57	0	0	8
0	24	0.90	14	53	127	0.33	0	0.11	0.17	0.09	1.13	0.09	31	<.1	<.1	8
0	24	0.90	14	53	127	0.33	0	0.08	0.16	0.08	1.02	0.08	28	<.1	<.1	8
0	20	0.94	11	44	122	0.27	1	0.15	0.19	0.10	1.45	0.01	24	0	0	9
0	19	0.93	9	33	121	0.28	1	0.12	0.12	0.09	1.29	0.01	19	0	0	9
0	18	0.73	10	38	162	0.28	1	0.11	0.16	0.06	0.85	0.02	17	0	<.1	7
0	18	0.74	10	39	163	0.28	1	0.09	0.09	0.06	0.77	0.02	13	<.1	<.1	7
0	52	1.57	16	72	322	0.50	0	0.36	0.02	0.20	2.78	0.02	64	0	0	16
0	10	1.96	44	109	340	0.97	0	0.22	0.58	0.05	1.82	0.17	22	0	0	28
0	22	0.92	17	67	215	0.47	0	0.10	0.14	0.10	0.99	0.04	30	0	0	8
0	17	0.75	7	59	101	0.19	0	0.09	0.13	0.10	0.90	0.02	28	<.1	0	5
0	17	0.76	7	59	102	0.19	0	0.07	0.20	0.09	0.81	0.02	24	<.1	0	5
0	40	1.08	—	45	160	—	0	0.23	—	0.14	1.20	—	40	0	—	—
0	23	0.91	13	53	211	0.36	0	0.14	0.12	0.11	1.22	0.02	35	<1	0	10
0	23	0.90	12	53	210	0.36	0	0.11	0.18	0.10	1.09	0.02	30	<.1	0	10
0	20	0.70	4	51	175	0.18	0	0.10	0.14	0.08	0.80	0.01	5	0	<.1	8
0	19	0.63	7	28	152	0.22	0	0.13	0.07	0.08	1.19	0.01	37	0	0	8
0	19	0.63	7	28	152	0.22	0	0.10	0.06	0.07	1.07	0.01	22	0	0	8
0	100	3.80	—	128	683	—	0	0.54	0.24	0.33	4.50	0.05	—	0	—	42
0	19	0.63	7	28	152	0.22	0	0.10	0.06	0.07	1.07	0.01	22	0	0	8
0	26	0.83	12	50	133	0.26	0	0.10	0.14	0.07	1.03	0.02	23	0	0	8
0	26	0.83	12	50	132	0.26	0	0.08	0.14	0.06	0.93	0.02	19	0	0	8
0	38	0.94	6	25	170	0.19	0	0.11	0.10	0.08	1.10	0.02	28	0	0	4
0	15	1.43	37	144	159	0.69	0	0.14	0.61	0.10	1.83	0.09	30	0	0	18
0	42	0.12	4	68	255	0.07	0	0.05	0.00	0.00	0.09	0.00	10	0	<.1	8
166	26	1.04	8	89	299	0.35	112	0.10	0.17	0.98	0.03	31	0	<1	7	
60	28	1.39	20	124	273	0.54	—	0.15	4.69	0.17	1.13	0.08	13	1	<1	—
27	28	1.41	22	128	214	0.44	—	0.02	1.42	0.09	0.37	0.03	11	<.1	<.1	2
55	57	1.53	30	133	299	0.66	38	0.13	1.51	0.20	1.08	0.04	26	<1	<1	11
19	21	0.70	—	46	92	—	—	0.04	0.63	0.05	0.30	—	2	0	—	2
2	14	0.89	7	66	116	0.12	3	0.02	0.71	0.04	0.34	0.02	9	<1	<.1	1
25	138	1.70	15	129	367	0.36	71	0.18	1.54	0.18	1.37	0.04	30	1	<.1	11
107	26	1.00	6	89	144	0.37	49	0.10	—	0.19	0.76	0.04	25	0	<1	12
1	101	1.30	13	111	318	0.37	13	0.14	0.80	0.21	1.19	0.03	35	<1	<.1	12
35	24	1.33	19	114	216	0.40	21	0.08	1.45	0.10	0.80	0.02	14	0	<1	2

Food Composition

(Computer code number is for Wadsworth Diet Analysis program)　　　(For purposes of calculations, use "0" for t, <1, <.1, <.01, etc.)

DA + Code	Food Description	Quantity	Measure	Wt (g)	H₂O (g)	Ener (cal)	Prot (g)	Carb (g)	Dietary Fiber (g)	Fat (g)	Fat Breakdown (g)		
											Sat	Mono	Poly
	BREADS, BAKED GOODS, CAKES, COOKIES, CRACKERS, CHIPS, PIES—Continued												
8822	Yellow, vanilla frosting, ready to eat, commercially prepared	1	slice(s)	64	14	239	2	38	<1	9	1.52	3.91	3.30
	Snack cakes												
8791	Chocolate snack cake, creme filled, w/frosting	1	item(s)	50	10	188	2	30	<1	7	1.43	2.85	2.62
25010	Cinnamon coffee cake	1	piece(s)	72	23	231	4	36	1	8	2.19	2.65	2.99
16777	Funnel cake	1	item(s)	90	37	278	7	29	1	14	2.77	4.46	6.33
8794	Sponge snack cake, creme filled	1	item(s)	43	9	155	1	27	<1	5	1.09	1.73	1.40
	Snacks, chips, pretzels												
29428	Bagel chips, plain	3	item(s)	29	—	130	3	19	1	5	0.50	—	—
29429	Bagel chips, toasted onion	3	item(s)	29	—	130	4	20	1	5	0.50	—	—
38192	Chex traditional snack mix	1	cup(s)	46	—	198	3	33	2	6	0.76	—	—
654	Potato chips, salted	20	item(s)	28	1	152	2	15	1	10	3.11	2.79	3.46
8816	Potato chips, unsalted	20	item(s)	28	1	152	2	15	1	10	3.11	2.79	3.46
4641	Tortilla chips, plain	6	item(s)	28	1	142	2	18	2	7	1.43	4.39	1.03
5096	Pretzels, plain, hard, twists	5	item(s)	30	1	114	3	24	1	1	0.23	0.41	0.37
4632	Pretzels, whole wheat	1	ounce(s)	28	1	103	3	23	2	1	0.16	0.29	0.24
	Cookies												
8859	Animal crackers	12	piece(s)	30	0	134	2	22	<1	4	1.03	2.29	0.56
8876	Brownie, prepared from mix	1	item(s)	24	3	112	1	12	1	7	1.76	2.60	2.26
25207	Chocolate chip cookies	1	item(s)	30	4	140	2	16	1	8	2.09	3.26	2.09
8915	Chocolate sandwich cookie, extra creme filling	1	item(s)	13	<1	65	<1	9	<1	3	0.50	1.39	1.22
14145	Fig Newtons	1	item(s)	16	—	55	1	10	1	1	0.50	0.50	0.00
8920	Fortune cookie	1	item(s)	8	1	30	<1	7	<1	<1	0.05	0.11	0.04
25208	Oatmeal cookies	1	item(s)	69	12	234	6	45	3	4	0.70	1.28	1.85
25213	Peanut butter cookies	1	item(s)	35	4	163	4	17	1	9	1.65	4.72	2.43
33095	Sugar cookies	1	item(s)	16	4	61	1	7	<1	3	0.63	1.27	0.87
9002	Vanilla sandwich cookie, creme filling	1	item(s)	10	<1	48	<1	7	<1	2	0.30	0.84	0.76
	Crackers												
9008	Cheese crackers (mini)	30	item(s)	30	1	151	3	17	1	8	2.81	3.63	0.74
9010	Cheese crackers (mini), low salt	30	item(s)	30	1	151	3	17	1	8	2.82	2.70	1.44
9012	Cheese cracker sandwich w/peanut butter	4	item(s)	28	1	139	3	16	1	7	1.23	3.64	1.43
8928	Honey graham crackers	4	item(s)	28	1	118	2	22	1	3	0.43	1.14	1.07
9016	Matzo crackers, plain	1	item(s)	28	1	112	3	24	1	<1	0.06	0.04	0.17
9024	Melba toast	3	item(s)	15	1	59	2	11	1	<1	0.07	0.12	0.19
14189	Ritz crackers	5	item(s)	16	<1	80	1	10	1	4	0.50	1.50	0.00
9014	Rye crispbread crackers	1	item(s)	10	1	37	1	8	2	<1	0.01	0.02	0.06
9028	Rye melba toast	3	item(s)	15	1	58	2	12	1	1	0.07	0.14	0.20
9040	Rye wafer	1	item(s)	11	1	37	1	9	3	<.1	0.01	0.02	0.04
432	Saltine crackers	5	item(s)	15	1	65	1	11	<1	2	0.44	0.96	0.25
9046	Saltine crackers, low salt	5	item(s)	15	1	65	1	11	<1	2	0.44	0.96	0.25
9048	Snack crackers, round	10	item(s)	30	1	151	2	18	<1	8	1.13	3.19	2.86
9050	Snack crackers, round, low salt	10	item(s)	30	1	151	2	18	<1	8	1.13	3.19	2.86
9052	Snack cracker sandwich, cheese filling	4	item(s)	28	1	134	3	17	1	6	1.72	3.15	0.72
9054	Snack cracker sandwich, peanut butter filling	4	item(s)	28	1	138	3	16	1	7	1.38	3.86	1.30
9044	Soda crackers	5	item(s)	15	1	65	1	11	<1	2	0.44	0.96	0.25
9055	Wheat crackers	10	item(s)	30	1	142	3	19	1	6	1.55	3.43	0.84
9057	Wheat crackers, low salt	10	item(s)	30	1	142	3	19	1	6	1.55	3.43	0.84
9059	Wheat cracker sandwich, cheese filling	4	item(s)	28	1	139	3	16	1	7	1.16	2.90	2.57
9061	Wheat cracker sandwich, peanut butter filling	4	item(s)	28	1	139	4	15	1	7	1.29	3.29	2.48
9022	Whole wheat crackers	7	item(s)	28	1	124	2	19	3	5	0.95	1.65	1.85
	Pastry												
16754	Apple fritter	1	item(s)	17	6	62	1	6	<1	4	0.87	1.69	1.13
5118	Cinnamon sweet roll w/icing, from refrigerator dough	1	item(s)	30	7	109	2	17	1	4	1.00	2.23	0.52
4945	Croissant, butter	1	item(s)	57	13	231	5	26	1	12	6.59	3.15	0.62
9096	Danish pastry, nut	1	item(s)	65	13	280	5	30	1	16	3.78	8.90	2.78
4947	Doughnut, cake	1	item(s)	47	10	198	2	23	1	11	1.70	4.37	3.70

PAGE KEY: A–4 = Breads/Baked Goods   A–8 = Cereal/Rice/Pasta   A–12 = Fruit   A–18 = Vegetables/Legumes   A–28 = Nuts/Seeds   A–30 = Vegetarian   A–32 = Dairy   A–38 = Eggs   A–40 = Seafood   A–42 = Meats   A–44 = Poultry   A–46 = Processed meats   A–48 = Beverages   A–52 = Fats/Oils   A–54 = Sweets   A–54 = Sauces/Condiments/Spices   A–58 = Mixed Foods/Soups/Sandwiches   A–64 = Fast food   A–82 = Convenience   A–84 = Baby foods

A

Chol (mg)	Calc (mg)	Iron (mg)	Magn (mg)	Pota (mg)	Sodi (mg)	Zinc (mg)	Vit A (RAE) (μg)	Thia (mg)	Vit E (mg)	Ribo (mg)	Niac (mg)	Vit B$_6$ (mg)	Fola (μg)	Vit C (mg)	Vit B$_{12}$ (μg)	Sele (μg)
35	40	0.68	4	34	220	0.16	12	0.06	—	0.04	0.32	0.02	17	0	<.1	4
9	37	1.68	21	61	213	0.26	3	0.11	1.68	0.15	1.21	0.01	20	0	<.1	1
26	50	1.46	10	81	277	0.38	35	0.14	1.06	0.16	1.17	0.02	30	<.1	<1	10
63	128	1.86	18	154	273	0.64	—	0.24	2.54	0.32	1.86	0.05	14	<1	<1	—
7	19	0.55	3	37	155	0.12	2	0.07	0.86	0.06	0.52	0.01	17	<.1	<.1	1
0	0	0.72	—	45	70	—	0	—	—	—	—	—	—	0	0	—
0	0	0.72	—	50	300	—	0	—	—	—	—	—	—	0	0	—
0	0	0.55	0	76	623	0.00	0	0.09	0.00	0.05	1.22	0.00	12	0	—	—
0	7	0.46	19	362	169	0.31	0	0.05	1.39	0.06	1.09	0.19	13	9	0	2
0	7	0.46	19	362	2	0.31	0	0.05	1.39	0.06	1.09	0.19	13	9	0	2
0	44	0.43	25	56	150	0.43	1	0.02	0.39	0.05	0.36	0.08	3	0	0	2
0	11	1.30	11	44	515	0.26	0	0.14	0.06	0.19	1.58	0.03	51	0	0	2
0	8	0.76	9	122	58	0.18	0	0.12	—	0.08	1.86	0.08	15	<1	0	—
0	13	0.82	5	30	1118	0.19	—	0.10	0.44	0.09	1.04	0.00	50	0	<.1	—
18	14	0.44	13	42	82	0.23	42	0.03	—	0.05	0.24	0.02	7	<.1	<.1	3
13	11	0.70	12	62	109	0.24	27	0.07	1.75	0.06	0.82	0.02	16	<.1	<.1	4
0	3	0.37	4	16	64	0.08	0	0.01	0.58	0.02	0.20	0.00	6	0	<.1	<1
0	5	0.36	—	40	60	—	4	0.03	—	0.04	0.22	—	—	<1	—	—
<1	1	0.12	1	3	22	0.01	<.1	0.01	0.03	0.01	0.15	0.00	5	0	<.1	<1
<.1	26	1.94	49	177	311	1.43	48	0.23	0.97	0.12	1.24	0.09	30	<1	<.1	17
13	28	0.67	22	104	157	0.46	51	0.08	3.35	0.09	1.81	0.05	21	<.1	<.1	5
18	5	0.32	2	13	50	0.08	31	0.04	0.45	0.04	0.28	0.01	8	<.1	<.1	3
0	3	0.22	1	9	35	0.04	0	0.03	0.36	0.02	0.27	0.00	5	0	0	<1
4	45	1.43	11	44	299	0.34	9	0.17	0.79	0.13	1.40	0.17	46	0	<1	3
4	45	1.44	11	32	137	0.33	—	0.18	0.30	0.12	1.41	0.18	8	0	<1	
0	14	0.76	16	61	199	0.29	0	0.15	1.05	0.08	1.63	0.04	26	0	<.1	2
0	7	1.04	8	38	169	0.23	0	0.06	0.57	0.09	1.15	0.02	13	0	0	3
0	4	0.90	7	32	1	0.19	0	0.11	0.02	0.08	1.11	0.03	5	0	0	10
0	14	0.56	9	30	124	0.30	0	0.06	0.01	0.04	0.62	0.01	19	0	0	5
0	20	0.72	3	10	135	0.23	—	0.07	—	0.04	0.45	0.01	10	1	0	—
0	3	0.24	8	32	26	0.24	0	0.02	0.13	0.01	0.10	0.02	5	0	0	4
0	12	0.55	6	29	135	0.20	0	0.07	0.10	0.04	0.71	0.01	13	0	0	6
0	4	0.65	13	54	87	0.31	0	0.05	0.16	0.03	0.17	0.03	5	<.1	0	3
0	18	0.81	4	19	195	0.12	0	0.08	0.23	0.07	0.79	0.01	19	0	0	2
0	18	0.81	4	109	95	0.12	0	0.08	0.23	0.07	0.79	0.01	19	0	0	3
0	36	1.08	8	40	254	0.20	0	0.12	1.36	0.10	1.21	0.02	27	0	0	2
0	36	1.08	8	107	112	0.20	0	0.12	1.36	0.10	1.21	0.02	27	0	0	2
1	72	0.67	10	120	392	0.17	5	0.12	0.72	0.19	1.05	0.01	28	<.1	<.1	6
0	23	0.78	15	60	201	0.32	0	0.14	1.05	0.08	1.71	0.04	24	0	<.1	3
0	18	0.81	4	19	195	0.12	0	0.08	0.23	0.07	0.79	0.01	19	0	0	2
0	15	1.32	19	55	239	0.48	0	0.15	0.97	0.10	1.49	0.04	35	0	0	2
0	15	1.32	19	61	85	0.48	0	0.15	0.97	0.10	1.49	0.04	15	0	0	10
2	57	0.73	15	86	256	0.24	5	0.10	—	0.12	0.89	0.07	18	<1	<.1	7
0	48	0.75	11	83	226	0.23	0	0.11	—	0.08	1.65	0.04	20	0	0	6
0	14	0.86	28	83	185	0.60	0	0.06	0.31	0.03	1.27	0.05	8	0	0	4
14	9	0.25	2	24	7	0.09	—	0.03	0.52	0.04	0.23	0.01	2	<1	<.1	—
0	10	0.80	4	19	250	0.10	—	0.12	0.29	0.07	1.09	0.01	14	<.1	<.1	—
38	21	1.16	9	67	424	0.43	101	0.22	0.25	0.14	1.25	0.03	35	<1	<.1	13
30	61	1.17	21	62	236	0.57	6	0.14	2.37	0.16	1.50	0.07	54	1	<1	9
17	21	0.92	9	60	257	0.26	—	0.10	1.80	0.11	0.87	0.03	22	<.1	<1	0

A

DA + Code	Food Description	Quantity	Measure	Wt (g)	H₂O (g)	Ener (cal)	Prot (g)	Carb (g)	Dietary Fiber (g)	Fat (g)	Fat Breakdown (g)		
											Sat	Mono	Poly
	BREADS, BAKED GOODS, CAKES, COOKIES, CRACKERS, CHIPS, PIES—Continued												
9105	Doughnut, cake, chocolate glazed	1	item(s)	42	7	175	2	24	1	8	2.16	4.74	1.04
9115	Doughnut, creme filling	1	item(s)	85	32	307	5	26	1	21	4.62	10.27	2.62
437	Doughnut, glazed	1	item(s)	60	15	242	4	27	1	14	3.49	7.72	1.74
9117	Doughnut, jelly filling	1	item(s)	85	30	289	5	33	1	16	4.12	8.69	2.02
10617	Toaster pastry, brown sugar cinnamon	1	item(s)	50	5	210	3	35	1	6	1.00	4.00	1.00
30928	Toaster pastry, cream cheese	1	item(s)	54	—	200	3	23	1	11	3.50	—	—
	Muffins												
25015	Blueberry	1	item(s)	63	30	160	3	23	1	6	0.87	1.48	3.25
4997	Bran, from mix	1	item(s)	50	18	138	3	23	2	5	1.18	2.34	0.72
9189	Corn, ready to eat	1	item(s)	57	19	174	3	29	2	5	0.77	1.20	1.83
9121	English muffin, plain, enriched	1	item(s)	57	24	134	4	26	2	1	0.15	0.17	0.51
29582	English, toasted	1	item(s)	50	19	128	4	25	1	1	0.14	0.16	0.48
9145	English, wheat	1	item(s)	57	24	127	5	26	3	1	0.16	0.16	0.48
	Granola bars												
38161	Kudos milk chocolate w/fruit & nuts	1	item(s)	28	—	90	2	15	1	3	1.00	—	—
38196	Nature Valley banana nut crunchy	1	item(s)	21	—	95	2	14	1	4	0.50	—	—
38187	Nature Valley fruit n nut trail mix	1	item(s)	35	—	140	3	25	2	4	0.50	—	—
1383	Plain, hard	1	item(s)	25	1	115	2	16	1	5	0.58	1.07	2.95
4606	Plain, soft	1	item(s)	28	2	126	2	19	1	5	2.06	1.08	1.51
	Pies												
454	Apple pie, from home recipe	1	slice(s)	155	73	411	4	58	2	19	4.73	8.36	5.17
470	Pecan pie, from home recipe	1	slice(s)	122	24	503	6	64	0	27	4.87	13.64	6.97
472	Pumpkin pie, from home recipe	1	slice(s)	155	91	316	7	41	0	14	4.92	5.73	2.81
9007	Pie crust, frozen, ready to bake, enriched, baked	1	slice(s)	16	2	82	1	8	<1	5	1.69	2.51	0.65
5052	Pie crust, prepared w/water, baked	1	slice(s)	20	2	100	1	10	<1	6	1.54	3.46	0.77
	Rolls												
8555	Crescent dinner roll	1	item(s)	28	10	80	2	14	1	1	0.34	0.70	0.25
489	Hamburger roll or bun, plain	1	item(s)	43	15	120	4	21	1	2	0.47	0.48	0.85
490	Hard roll	1	item(s)	57	18	167	6	30	1	2	0.35	0.65	0.98
5127	Kaiser roll	1	item(s)	57	18	167	6	30	1	2	0.35	0.65	0.98
5130	Whole wheat roll or bun	1	item(s)	28	9	76	2	15	2	1	0.24	0.34	0.62
	Sport bars												
37026	Balance original chocolate	1	item(s)	50	—	200	14	22	1	6	3.50	—	—
37024	Balance original peanut butter	1	item(s)	50	—	200	14	22	1	6	2.50	—	—
36580	Clif Bar chocolate brownie energy bar	1	item(s)	68	—	240	10	41	6	4	1.00	—	—
36583	Clif Bar crunchy peanut butter energy bar	1	item(s)	68	—	240	12	39	5	5	0.50	—	—
36584	Clif Luna tropical crisp energy bar	1	item(s)	48	—	180	10	24	2	5	3.50	0.00	0.00
12005	Powerbar apple cinnamon	1	item(s)	65	—	230	10	45	3	3	0.50	1.50	0.50
16078	Powerbar banana	1	item(s)	65	—	230	9	45	3	2	0.50	1.00	0.50
16080	Powerbar chocolate	1	item(s)	65	—	230	10	45	3	2	0.50	0.50	1.00
16079	Powerbar mocha	1	item(s)	65	—	230	10	45	3	3	1.00	1.00	0.50
	Tortillas												
1391	Corn tortillas, soft	1	item(s)	26	11	58	1	12	1	1	0.09	0.17	0.29
1669	Flour tortilla	1	item(s)	32	9	104	3	18	1	2	0.56	1.21	0.34
1390	Taco shells, hard	1	item(s)	13	1	62	1	8	1	3	0.43	1.19	1.13
	Pancakes, waffles												
8926	Pancakes, blueberry, from recipe	3	item(s)	114	61	253	7	33	1	10	2.26	2.64	4.74
5037	Pancakes, from mix w/egg & milk	3	item(s)	114	60	249	9	33	2	9	2.33	2.36	3.33
9219	Waffle, plain, frozen, toasted	2	item(s)	66	28	174	4	27	2	5	0.95	2.12	1.84
500	Waffle, plain, from recipe	1	item(s)	75	<.1	218	6	25	2	11	2.14	2.64	5.08
30311	Waffle, 100% whole grain	1	item(s)	75	32	201	7	25	2	8	2.35	3.38	2.06
	CEREAL, FLOUR, GRAIN, PASTA, NOODLES, POPCORN												
	Grain												
2861	Amaranth, dry	½	cup(s)	98	10	365	14	65	15	6	1.62	1.40	2.82
1953	Barley, pearled, cooked	½	cup(s)	79	54	97	2	22	3	<1	0.07	0.04	0.17
1956	Buckwheat groats, cooked, roasted	½	cup(s)	84	64	77	3	17	2	1	0.11	0.16	0.16
1957	Bulgur, cooked	½	cup(s)	91	71	76	3	17	4	<1	0.04	0.03	0.09
1963	Couscous, cooked	½	cup(s)	79	57	88	3	18	1	<1	0.02	0.02	0.05
1967	Millet, cooked	½	cup(s)	120	86	143	4	28	2	1	0.21	0.22	0.61

PAGE KEY: A–4 = Breads/Baked Goods   A–8 = Cereal/Rice/Pasta   A–12 = Fruit   A–18 = Vegetables/Legumes   A–28 = Nuts/Seeds   A–30 = Vegetarian   A–32 = Dairy   A–38 = Eggs   A–40 = Seafood   A–42 = Meats   A–44 = Poultry   A–46 = Processed meats   A–48 = Beverages   A–52 = Fats/Oils   A–54 = Sweets   A–54 = Sauces/Condiments/Spices   A–58 = Mixed Foods/Soups/Sandwiches   A–64 = Fast food   A–82 = Convenience   A–84 = Baby foods

A

Chol (mg)	Calc (mg)	Iron (mg)	Magn (mg)	Pota (mg)	Sodi (mg)	Zinc (mg)	Vit A (RAE) (µg)	Thia (mg)	Vit E (mg)	Ribo (mg)	Niac (mg)	Vit B₆ (mg)	Fola (µg)	Vit C (mg)	Vit B₁₂ (µg)	Sele (µg)
24	89	0.95	14	45	143	0.24	5	0.02	1.13	0.03	0.20	0.01	19	<.1	<.1	2
20	21	1.56	17	68	263	0.68	9	0.29	2.35	0.13	1.91	0.06	60	0	<1	9
4	26	0.36	13	65	205	0.46	2	0.53	1.83	0.04	0.39	0.03	13	<.1	<.1	5
22	21	1.50	17	67	249	0.64	14	0.27	2.09	0.12	1.82	0.09	58	0	<1	11
0	0	1.80	—	70	190	—	—	0.15	0.00	0.17	2.00	0.20	40	0	0	—
15	0	1.08	—	—	230	—	—	—	—	—	—	—	—	0	—	—
20	50	1.15	7	56	288	0.39	20	0.14	1.28	0.15	1.14	0.03	29	<1	<1	9
34	16	1.27	29	74	234	0.57	—	0.10		0.12	1.44	0.09	33	0	<.1	—
15	42	1.60	18	39	297	0.31	30	0.16	1.05	0.19	1.16	0.05	46	0	<.1	9
0	30	1.43	12	75	264	0.40	0	0.25	0.10	0.16	2.21	0.02	42	0	<.1	—
0	95	1.36	11	72	252	0.38	0	0.19	0.06	0.14	1.90	0.02	15	<.1	<.1	—
0	101	1.64	21	106	218	0.61	0	0.25	0.28	0.17	1.91	0.05	36	0	0	17
0	200	0.36	—	—	60	—	0	—	—	—	—	—	—	0	0	—
0	10	0.54	—	60	80	—	0	—	—	—	—	—	—	0	—	—
0	0	0.00	—	—	95	—	0	—	—	—	—	—	—	0	—	—
0	15	0.72	24	82	72	0.50	2	0.06	0.00	0.03	0.39	0.02	6	<1	0	4
<1	30	0.73	21	92	79	0.43	0	0.08	—	0.05	0.15	0.03	7	0	<1	5
0	11	1.74	11	122	327	0.29	17	0.23	—	0.17	1.91	0.05	37	3	0	12
106	39	1.81	32	162	320	1.24	100	0.23	—	0.22	1.03	0.07	32	<1	<1	15
65	146	1.97	29	288	349	0.71	660	0.14	—	0.31	1.21	0.07	33	3	<1	11
0	3	0.36	3	18	104	0.05	0	0.04	0.82	0.06	0.39	0.01	9	0	<.1	<1
0	12	0.43	3	12	146	0.08	0	0.06	—	0.04	0.47	0.01	20	0	0	—
0	39	0.89	6	39	157	0.17	0	0.14	—	0.09	1.10	0.01		0	<.1	—
0	59	1.43	9	40	206	0.28	0	0.17	0.67	0.14	1.79	0.03	48	0	<.1	8
0	54	1.87	15	62	310	0.54	0	0.27	0.19	0.19	2.42	0.02	54	0	0	22
0	54	1.87	15	62	310	0.54	0	0.27	0.19	0.19	2.42	0.02	54	0	0	22
0	30	0.69	24	78	136	0.57	0	0.07	0.39	0.04	1.05	0.06	9	0	0	14
3	100	4.50	40	160	180	3.75	—	0.38	20.13	0.43	5.00	0.50	100	60	2	18
3	100	4.50	40	130	230	3.75	—	0.38	20.13	0.43	5.00	0.50	100	60	2	18
0	250	5.40	120	260	150	3.75	—	0.38	20.13	0.26	4.00	0.40	80	60	1	18
0	250	5.40	120	300	290	3.75	—	0.38	20.13	0.34	6.00	0.40	100	60	1	14
0	350	6.30	140	120	135	5.25	—	1.50	20.13	1.70	20.00	2.00	400	60	6	25
0	300	6.30	140	110	90	5.25	0	1.50	20.13	1.70	20.00	2.00	400	60	6	—
0	300	6.30	140	200	90	5.25	0	1.50	20.13	1.70	20.00	2.00	400	60	6	—
0	300	6.30	140	150	90	5.25	0	1.50	20.13	1.70	20.00	2.00	400	60	6	—
0	46	0.36	17	40	42	0.24	0	0.03	0.04	0.02	0.39	0.06	26	0	0	1
0	40	1.06	8	42	153	0.23	0	0.17	0.29	0.09	1.14	0.02	33	0	0	7
0	21	0.33	14	24	49	0.19	0	0.03	0.48	0.01	0.18	0.04	17	0	0	2
64	235	1.96	18	157	470	0.62	57	0.22	—	0.31	1.74	0.06	41	3	<1	16
81	245	1.48	25	227	576	0.86	82	0.23	1.29	0.36	1.40	0.12	105	1	<1	—
16	153	2.95	15	84	519	0.38	253	0.25	0.55	0.31	2.93	0.59	36	0	2	11
52	191	1.73	14	119	383	0.50	49	0.19	—	0.26	1.55	0.04	51	<1	<1	35
71	196	1.56	30	173	374	0.85	—	0.15	1.06	0.25	1.47	0.09	14	<1	<1	—
0	149	7.40	259	357	20	3.10	0	0.08	1.00	0.20	1.25	0.22	48	4	0	—
0	9	1.04	17	73	2	0.64	0	0.07	0.04	0.05	1.62	0.09	13	0	0	7
0	6	0.67	43	74	3	0.51	0	0.03	0.20	0.03	0.79	0.06	12	0	0	2
0	9	0.87	29	62	5	0.52	0	0.05	0.03	0.03	0.91	0.08	16	0	0	1
0	6	0.30	6	46	4	0.20	0	0.05	0.01	0.02	0.77	0.04	12	0	0	22
0	4	0.76	53	74	2	1.09	0	0.13	0.07	0.10	1.60	0.13	23	0	0	1

Table A–1

A | **Food Composition** | (Computer code number is for Wadsworth Diet Analysis program) | (For purposes of calculations, use "0" for t, <1, <.1, <.01, etc.)

DA + Code	Food Description	Quantity	Measure	Wt (g)	H₂O (g)	Ener (cal)	Prot (g)	Carb (g)	Dietary Fiber (g)	Fat (g)	Fat Breakdown (g)		
											Sat	Mono	Poly
	CEREAL, FLOUR, GRAIN, PASTA, NOODLES, POPCORN—Continued												
1969	Oat bran, dry	½	cup(s)	47	3	116	8	31	7	3	0.62	1.12	1.30
1972	Quinoa, dry	½	cup(s)	85	8	318	11	59	5	5	0.50	1.30	1.99
	Rice												
129	Brown, long grain, cooked	½	cup(s)	98	71	108	3	22	2	1	0.18	0.32	0.31
2863	Brown, medium grain, cooked	½	cup(s)	97.5	0.07	109.19	2.26	22.92	1.75	0.8	0.16	0.29	0.28
37488	Jasmine, saffroned, cooked	½	cup(s)	280	—	340	8	78	0	0	0.00	—	—
30280	Pilaf, cooked	½	cup(s)	103	74	129	2	22	1	3	0.67	1.61	0.95
28066	Spanish, cooked	½	cup(s)	120	3	25	2	1	<1	<1	0.33	0.07	18.31
2867	White glutinous, cooked	½	cup(s)	87	67	84	2	18	1	<1	0.03	0.06	0.06
482	White, instant long grain, enriched, boiled	½	cup(s)	83	63	81	2	18	<1	<1	0.04	0.04	0.04
484	White, long grain, boiled	½	cup(s)	79	54	103	2	22	<1	<1	0.06	0.07	0.06
486	White, long grain, enriched, parboiled, cooked	½	cup(s)	88	63	100	2	22	<1	<1	0.06	0.07	0.06
1194	Wild brown, cooked	½	cup(s)	82	0.06	82.81	3.27	17.49	1.47	0.27	0.04	0.04	0.17
	Flour & grain fractions												
505	All purpose flour, self rising, enriched	½		63	7	221	6	46	2	1	0.10	0.05	0.26
503	All purpose flour, white, bleached, enriched	½	cup(s)	63	7	228	6	48	2	1	0.10	0.05	0.26
1643	Barley flour	½	cup(s)	56	6	198	4	45	2	1	0.16	0.10	0.38
383	Buckwheat flour, whole groat	½	cup(s)	60	7	201	8	42	6	2	0.41	0.57	0.57
504	Cake wheat flour, enriched	½	cup(s)	55	7	197	4	43	1	<1	0.07	0.04	0.21
426	Cornmeal, degermed, enriched	½	cup(s)	69	8	253	6	54	5	1	0.16	0.28	0.49
424	Cornmeal, yellow whole grain	½	cup(s)	61	6	221	5	47	4	2	0.31	0.58	1.00
1644	Masa corn flour, enriched	½	cup(s)	57	5	208	5	43	5	2	0.30	0.57	0.98
1976	Rice flour, brown	½	cup(s)	79	9	287	6	60	4	2	0.44	0.80	0.79
1645	Rice flour, white	½	cup(s)	79	9	289	5	63	2	1	0.30	0.35	0.30
1978	Rye flour, dark	½	cup(s)	64	7	207	9	44	14	2	0.20	0.21	0.77
1980	Semolina, enriched	½	cup(s)	84	11	301	11	61	3	1	0.13	0.10	0.36
2827	Soy flour, raw	½	cup(s)	43	2	186	15	15	4	9	1.27	1.94	4.96
1990	Wheat germ, crude	2	tablespoon(s)	14	2	52	3	7	2	1	0.24	0.20	0.86
506	Whole wheat flour	½	cup(s)	60	6	203	8	44	7	1	0.19	0.14	0.47
	Breakfast bars												
39230	Atkins Morning Start apple crisp	1	item(s)	37	—	170	11	12	6	9	4.00	—	—
10574	Health Valley fat free apple	1	item(s)	38	—	110	2	26	3	0	0.00	0.00	0.00
10647	Nutri-Grain blueberry cereal bar	1	item(s)	37	5	140	2	27	1	3	0.50	2.00	0.50
10648	Nutri-Grain raspberry cereal bar	1	item(s)	37	5	140	2	27	1	3	0.50	2.00	0.50
10649	Nutri-Grain strawberry cereal bar	1	item(s)	37	5	140	2	27	1	3	0.50	2.00	0.50
	Breakfast cereals, hot												
363	Corn grits, white, regular & quick, enriched, cooked w/water & salt	½	cup(s)	121	103	71	2	16	<1	<1	0.03	0.06	0.10
8636	Corn grits, yellow, regular & quick, enriched, cooked w/salt	½	cup(s)	121	103	71	2	16	<1	<1	0.03	0.06	0.10
1260	Cream of Wheat, instant, prepared	½	cup(s)	121	106	61	2	13	<1	<.1	0.01	0.01	0.04
365	Farina, enriched, cooked w/water & salt	½	cup(s)	117	102	56	2	12	<1	<.1	0.01	0.01	0.03
8657	Oatmeal, cooked w/water	½	cup(s)	117	100	74	3	13	2	1	0.19	0.37	0.44
5500	Oatmeal, maple & brown sugar, instant, prepared	1	item(s)	198	150	200	5	40	2	2	0.42	0.74	0.85
5510	Oatmeal, ready to serve, packet	1	item(s)	186	158	112	4	20	3	2	0.38	0.66	0.76
	Breakfast cereals, ready to eat												
1197	All-Bran	1	cup(s)	62	2	160	8	46	20	2	0.00	0.00	1.00
1200	All-Bran Buds	1	cup(s)	91	3	212	6	73	42	3	—	—	—
1199	Apple Jacks	1	cup(s)	33	1	130	1	30	1	1	—	—	—
13633	Bran Flakes, Post	1	cup(s)	40	1	133	4	32	7	1	0.00	0.00	0.71
1204	Cap'n Crunch	1	cup(s)	36	1	144	2	30	1	2	0.53	0.39	0.27
1205	Cap'n Crunch Crunchberries w/wildberry colors	1	cup(s)	35	1	139	2	29	1	2	0.49	0.39	0.28
1206	Cheerios	1	cup(s)	30	1	110	3	22	3	2	0.00	0.50	0.50
3415	Cocoa Puffs	1	cup(s)	30	1	120	1	26	0	1	—	—	—
1207	Cocoa Rice Krispies	1	cup(s)	41	1	160	1	36	1	1	0.67	0.00	0.00
5522	Complete wheat bran flakes	1	cup(s)	39	1	120	4	31	7	1	—	—	—
1211	Corn Flakes	1	cup(s)	28	1	100	2	24	1	0	0.00	0.00	0.00

PAGE KEY: A–4 = Breads/Baked Goods   A–8 = Cereal/Rice/Pasta   A–12 = Fruit   A–18 = Vegetables/Legumes   A–28 = Nuts/Seeds   A–30 = Vegetarian
A–32 = Dairy   A–38 = Eggs   A–40 = Seafood   A–42 = Meats   A–44 = Poultry   A–46 = Processed meats   A–48 = Beverages   A–52 = Fats/Oils
A–54 = Sweets   A–54 = Sauces/Condiments/Spices   A–58 = Mixed Foods/Soups/Sandwiches   A–64 = Fast food   A–82 = Convenience   A–84 = Baby foods

A

Chol (mg)	Calc (mg)	Iron (mg)	Magn (mg)	Pota (mg)	Sodi (mg)	Zinc (mg)	Vit A (RAE) (µg)	Thia (mg)	Vit E (mg)	Ribo (mg)	Niac (mg)	Vit B$_6$ (mg)	Fola (µg)	Vit C (mg)	Vit B$_{12}$ (µg)	Sele (µg)
0	27	2.54	110	266	2	1.46	0	0.55	0.80	0.10	0.44	0.08	24	0	0	21
0	51	7.86	179	629	18	2.81	0	0.17	—	0.34	2.49	0.19	42	0	0	—
0	10	0.41	42	42	5	0.61	0	0.09	0.21	0.02	1.49	0.14	4	0	0	10
0	9.75	0.51	42.9	77.02	0.97	0.6	0	0.09	0.04	0.01	1.29	0.14	3.9	0	0	38
0	—	2.16	—	—	780	—	—	—	—	—	—	—	—	—	—	—
0	13	1.16	9	55	403	0.38	—	0.13	0.53	0.02	1.24	0.06	4	<1	<.1	—
1	47	167.40	48	1	13	0.13	<1	0.03	1.62	0.19	8.71	10.30	<.1	7	<.1	9
0	2	0.12	4	9	4	0.36	0	0.02	0.03	0.01	0.25	0.02	1	0	0	5
0	7	0.52	4	3	2	0.20	0	0.06	0.04	0.04	0.73	0.01	58	0	0	3
0	8	0.95	9	28	1	0.39	0	0.13	0.04	0.01	1.17	0.07	46	0	0	6
0	17	0.99	11	32	3	0.27	0	0.22	0.04	0.02	1.23	0.02	67	0	0	7
0	2.46	0.49	26.23	82.81	2.46	1.09	0	0.04	0.15	0.07	1.05	0.11	21.31	0	0	0.65
0	211	2.92	12	78	794	0.39	0	0.42	0.07	0.26	3.65	0.03	123	0	0	22
0	9	2.90	14	67	1	0.44	0	0.49	0.08	0.31	3.69	0.03	114	0	0	21
0	16	0.71	45	186	4	1.05	0	0.07	0.07	0.03	2.57	0.16	13	0	0	2
0	25	2.44	151	346	7	1.87	0	0.25	0.62	0.11	3.69	0.35	32	0	0	3
0	8	3.99	9	57	1	0.34	0	0.49	0.02	0.23	3.70	0.02	101	0	0	3
0	3	2.85	28	112	2	0.50	8	0.49	0.23	0.28	3.47	0.18	161	0	0	5
0	4	2.10	77	175	21	1.11	7	0.23	0.41	0.12	2.22	0.19	15	0	0	9
0	80	4.11	63	170	3	1.01	0	0.81	0.14	0.43	5.61	0.21	133	0	0	9
0	9	1.56	88	228	6	1.94	0	0.35	0.57	0.06	5.01	0.58	13	0	0	—
0	8	0.28	28	60	0	0.63	0	0.11	0.10	0.02	2.05	0.34	3	0	0	12
0	36	4.13	159	467	1	3.60	1	0.20	1.65	0.16	2.73	0.28	38	0	0	23
0	14	3.64	39	155	1	0.88	0	0.68	0.22	0.48	5.00	0.09	153	0	0	75
0	88	2.71	183	1070	6	1.67	3	0.25	0.83	0.49	1.84	0.20	147	0	0	3
0	6	0.90	34	128	2	1.77	0	0.27	—	0.07	0.98	0.19	40	0	0	11
0	20	2.33	83	243	3	1.76	0	0.27	0.74	0.13	3.82	0.20	26	0	0	42
0	200	—	—	90	70	—	—	0.23	3.02	0.26	3.00	—	—	9	—	—
0	0	0.72	—	160	25	—	—	0.09	—	0.03	0.40	—	—	1	—	—
0	200	1.80	8	75	110	1.50	—	0.38	0.00	0.43	5.00	0.50	40	0	0	—
0	200	1.80	8	70	110	1.50	—	0.38	0.00	0.43	5.00	0.50	40	0	0	—
0	200	1.80	8	55	110	1.50	—	0.38	0.00	0.43	5.00	0.50	40	0	0	—
0	4	0.73	6	25	270	0.08	0	0.10	0.02	0.07	0.87	0.03	40	0	0	4
0	4	0.73	6	25	270	0.08	2	0.10	0.02	0.07	0.87	0.03	40	0	0	3
0	27	8.60	2	17	1	0.10	0	0.07	0.02	0.04	0.60	0.01	357	0	0	—
0	5	0.58	2	15	383	0.09	0	0.07	—	0.05	0.57	0.01	40	0	0	11
0	9	0.80	28	66	1	0.57	0	0.13	0.12	0.02	0.15	0.02	5	0	0	9
0	26	6.84	50	126	404	1.04	0	1.02	0.42	0.05	1.57	0.31	30	0	0	11
0	21	3.96	45	112	241	0.93	0	0.60	0.22	0.05	0.78	0.19	19	0	0	4
0	300	9.00	200	700	160	3.00	300	0.75	—	0.85	10.00	4.00	800	12	12	6
0	0	13.64	182	909	606	4.55	455	1.14	—	1.29	15.15	6.06	1212	18	18	26
0	0	4.50	8	35	150	1.50	150	0.38	—	0.43	5.00	0.50	100	15	2	2
0	0	10.77	80	253	293	2.00	—	0.50	—	0.57	6.65	0.67	133	0	2	—
0	5	6.00	20	72	269	4.99	3	0.51	0.20	0.57	6.66	0.67	133	0	0	7
<.1	7	6.14	19	71	242	5.12	2	0.51	0.23	0.57	6.66	0.67	133	<.1	0	7
0	100	8.10	40	95	280	3.75	150	0.38	0.21	0.43	5.00	0.50	200	6	2	11
0	100	4.50	8	50	170	3.75	0	0.38	0.14	0.43	5.00	0.50	100	6	2	2
0	53	5.99	11	67	253	2.00	200	0.50	—	0.57	6.65	0.67	133	20	2	6
0	0	23.94	53	226	279	19.95	299	2.00	26.78	2.26	26.60	2.66	532	80	8	4
0	0	8.10	3	25	200	0.17	150	0.38	0.04	0.43	5.00	0.50	100	6	2	1

Table A-1

Food Composition   (Computer code number is for Wadsworth Diet Analysis program)   (For purposes of calculations, use "0" for t, <1, <.1, <.01, etc.)

DA + Code	Food Description	Quantity	Measure	Wt (g)	H₂O (g)	Ener (cal)	Prot (g)	Carb (g)	Dietary Fiber (g)	Fat (g)	Sat	Mono	Poly
	CEREAL, FLOUR, GRAIN, PASTA, NOODLES, POPCORN—Continued												
1247	Corn Pops	1	cup(s)	31	1	120	1	28	0	0	0.00	0.00	0.00
1937	Cracklin' Oat Bran	1	cup(s)	65	0	266	5	47	7	9	2.70	4.70	1.33
1220	Froot Loops	1	cup(s)	32	1	120	1	28	1	1	0.50	0.00	0.00
38214	Frosted Cheerios	1	cup(s)	30	—	120	2	25	1	1	0.00	0.00	0.00
372	Frosted Flakes	1	cup(s)	41	1	160	1	37	1	0	0.00	0.00	0.00
38215	Frosted Mini Chex	1	cup(s)	40	—	146	1	36	0	0	0.00	0.00	0.00
10268	Frosted Mini-Wheats	5	item(s)	51	3	180	5	41	5	1	0.00	0.00	0.50
38216	Frosted Wheaties	1	cup(s)	40	—	146	1	36	<1	0	0.00	0.00	0.00
1223	Granola, prepared	½	cup(s)	61	0	299	9	32	5	15	2.76	4.7	6.53
13334	Granola, Quaker 100% natural, oats & honey	½	cup(s)	48	0	219	5	31	3	9	3.83	4.0	1.19
13335	Granola, Quaker 100% natural, oats, honey & raisins	½	cup(s)	51	0	225	5	34	3	9	3.57	3.80	1.10
2415	Honey Bunches of Oats honey roasted	1	cup(s)	40	1	160	3	33	1	2	0.67	1.20	0.13
1227	Honey Nut Cheerios	1	cup(s)	30	1	120	3	24	2	2	0.00	0.50	0.00
2424	Honeycomb	1	cup(s)	22	<1	83	2	20	<1	<1	0.00	—	—
10286	Kashi puffed	1	cup(s)	25	—	70	3	13	2	1	0.00	—	—
1231	Kix	1	cup(s)	23	<1	90	2	20	1	<1	0.00	0.00	0.00
30569	Life	1	cup(s)	43	2	160	4	33	3	2	0.35	0.64	0.61
1233	Lucky Charms	1	cup(s)	30	1	120	2	25	1	1	0.00	0.00	0.00
1201	Multi-Bran Chex	1	cup(s)	58	1	200	4	49	7	2	0.00	0.00	0.00
38220	Multi Grain Cheerios	1	cup(s)	30	—	110	3	24	3	1	0.00	0.00	0.00
1238	Nutri-Grain golden wheat	1	cup(s)	40	—	133	4	31	5	1	0.00	0.00	0.67
1241	Product 19	1	cup(s)	30	1	100	2	25	1	0	0.00	0.00	0.00
32432	Puffed rice, fortified	1	cup(s)	14	<1	56	1	13	<1	<.1	0.02	—	—
32433	Puffed wheat, fortified	1	cup(s)	12	0	43.68	1.76	9.55	0.52	0.14	0.02	—	—
2420	Raisin Bran	1	cup(s)	59	5	190	4	47	8	1	0.00	0.10	0.36
1244	Rice Chex	1	cup(s)	25	1	96	2	22	<1	0	0.00	0.00	0.00
1245	Rice Krispies	1	cup(s)	26	1	96	2	23	0	0	0.00	0.00	0.00
5593	Shredded Wheat	1	cup(s)	25	1	88	3	20	3	1	0.04	0.01	0.10
1248	Smacks	1	cup(s)	36	1	133	3	32	1	1	0.00	0.00	0.00
1246	Special K	1	cup(s)	31	1	110	7	22	1	0	0.00	0.00	0.00
3428	Total, corn flakes	1	cup(s)	23	1	83	2	18	1	0	0.00	0.00	0.00
1253	Total whole grain	1	cup(s)	40	1	146	3	31	4	1	0.00	0.00	0.00
1254	Trix	1	cup(s)	30	1	120	1	27	1	1	0.00	0.00	0.00
382	Wheat germ, toasted	2	tablespoon(s)	14	0	53.95	4.11	7	2.13	1.51	0.25	0.21	0.93
1257	Wheaties	1	cup(s)	30	1	110	3	24	3	1	0.00	0.00	0.00
	Pasta, noodles												
449	Chinese chow mein noodles, cooked	½	cup(s)	23	<1	119	2	13	1	7	0.99	1.73	3.90
1995	Corn pasta, cooked	½	cup(s)	70	48	88	2	20	3	1	0.07	0.13	0.23
448	Egg noodles, enriched, cooked	½	cup(s)	80	55	106	4	20	1	1	0.25	0.34	0.33
440	Macaroni, enriched, cooked	½	cup(s)	70	46	99	3	20	1	<1	0.07	0.06	0.19
1996	Pasta, plain, fresh-refrigerated, cooked	½	cup(s)	64	44	84	3	16	0	1	0.10	0.08	0.27
1725	Ramen noodles, cooked	½	cup(s)	114	95	104	3	15	1	4	0.19	0.22	0.21
2878	Soba noodles, cooked	½	cup(s)	95	69	94	5	20	0	<.1	0.02	0.02	0.03
2879	Somen noodles, cooked	½	cup(s)	88	60	115	4	24	0	<1	0.02	0.02	0.06
493	Spaghetti, al dente, cooked	½	cup(s)	65	42	95	4	20	1	1	0.05	0.05	0.15
2884	Spaghetti, whole wheat, cooked	½	cup(s)	70	47	87	4	19	3	<1	0.07	0.05	0.15
1563	Spinach egg noodles, enriched, cooked	½	cup(s)	80	55	105	4	19	2	1	0.29	0.39	0.28
2000	Tricolor vegetable macaroni, enriched, cooked	½	cup(s)	67	46	86	3	18	3	<.1	0.01	0.01	0.03
	Popcorn												
476	Air popped	1	cup(s)	8	<1	31	1	6	1	<1	0.05	0.09	0.15
4619	Caramel	1	cup(s)	35	1	152	1	28	2	5	1.27	1.01	1.58
4620	Cheese flavored	1	cup(s)	37	1	196	3	19	4	12	2.38	3.61	5.72
477	Popped in oil	1	cup(s)	33	1	165	3	19	3	9	1.61	2.70	4.43
	FRUIT AND FRUIT JUICES												
	Apples												
223	Raw medium, w/peel	1	item(s)	138	118	72	<1	19	3	<1	0.04	0.01	0.07
224	Slices	½	cup(s)	55	47	29	<1	8	1	<.1	0.02	0.00	0.03

Chol (mg)	Calc (mg)	Iron (mg)	Magn (mg)	Pota (mg)	Sodi (mg)	Zinc (mg)	Vit A (RAE) (µg)	Thia (mg)	Vit E (mg)	Ribo (mg)	Niac (mg)	Vit B_6 (mg)	Fola (µg)	Vit C (mg)	Vit B_{12} (µg)	Sele (µg)
0	0	1.80	2	25	120	1.50	150	0.38		0.43	5.00	0.50	100	6	2	2
0	27	2.38	80	293	186	2.00	299	0.49	0.33	0.56	6.65	0.67	218	20	2	14
0	0	4.50	8	35	150	1.50	150	0.38	0.12	0.43	5.00	0.50	100	15	2	2
0	100	4.50	16	55	210	3.75	—	0.38	—	0.43	5.00	0.50	100	6	2	—
0	0	5.99	4	27	200	0.21	200	0.50	0.05	0.57	6.65	0.67	133	8	2	2
0	133	11.97	—	33	266	3.99	—	0.50	—	0.57	6.65	0.67	266	8	2	
0	0	15.30	60	170	5	1.50	0	0.38	0.46	0.43	5.00	0.50	100	0	2	2
0	133	10.77	0	47	266	9.98	—	1.00	—	1.13	13.30	1.33	532	8	4	
0	48	2.59	107	328	13	2.5	2	0.44	6.28	0.17	1.29	0.18	51	1	0	16.95
1	61	1.21	51	225	20	1.04	1	0.12	0.4	0.11	0.81	0.07	17	<1	0.1	8.3
1	59	1.24	49	250	19	0.99	<1	0.12	0.37	0.11	0.8	0.07	16	<1	0.1	8.82
0	0	3.59	21	67	253	0.40	—	0.50	0.28	0.57	6.65	0.67	133	0	2	—
0	100	4.50	24	95	270	3.75	—	0.38	0.31	0.43	5.00	0.50	200	6	2	7
0	0	2.03	6	26	165	1.13	—	0.28	0.09	0.32	3.74	0.37	75	0	1	—
0	0	0.72	—	35	0	—	0	0.03	—	0.03	0.80	0.00	—	0	—	—
0	113	6.08	6	26	203	2.81	113	0.28	0.06	0.32	3.75	0.38	150	5	1	5
0	124	11.92	41	121	218	5.32	1	0.53	0.23	0.60	7.10	0.70	142	0	0	11
0	100	4.50	16	60	210	3.75	—	0.38	0.13	0.43	5.00	0.50	200	6	2	6
0	100	16.20	60	220	390	3.75	158	0.38	0.67	0.03	5.00	0.50	100	6	2	5
0	100	18.00	24	85	200	15.00	—	1.50	20.13	1.70	20.00	2.00	400	15	6	—
0	0	1.46	32	146	279	4.99	0	0.50	7.14	0.57	6.65	0.67	133	20	2	9
0	0	18.00	16	50	210	15.00	225	1.50	20.13	1.70	20.00	2.00	400	60	6	4
0	1	4.44	4	16	<1	0.14	0	0.36	—	0.25	4.94	0.01	3	0	0	1
0	3.35	3.8	17.39	41.75	0.47	0.28	0	0.31	—	0.21	4.23	0.02	3.83	0	0	14.77
0	20	10.80	80	340	300	2.25	—	0.53	1.37	0.60	7.00	0.70	140	0	2	—
0	80	7.20	7	28	232	3.00	—	0.30	0.03	0.34	4.00	0.40	160	5	1	1
0	0	1.44	13	32	256	0.48	120	0.30	0.03	0.34	4.80	0.40	80	5	1	4
0	10	1.08	31	92	2	0.70	0	0.07	0.13	0.06	1.77	0.10	12	0	0	1
0	0	0.48	11	53	67	0.40	200	0.50	0.18	0.57	6.65	0.67	133	8	2	17
0	0	8.70	16	60	220	0.90	225	0.53	7.05	0.60	7.00	2.00	400	15	6	7
0	750	13.50	0	23	158	11.25	113	1.13	15.00	1.28	15.00	1.50	300	45	5	1
0	1330	23.94	32	120	253	19.95	200	2.00	26.60	2.26	26.60	2.66	532	80	8	2
0	100	4.50	0	15	190	3.75	150	0.38	0.60	0.43	5.00	0.50	100	6	2	6
0	6.35	1.28	45.2	133.76	0.56	2.35	0	0.23	2.04	0.11	0.78	0.13	49.72	0.84	0	9.18
0	0	8.10	32	110	220	7.50	150	0.75	0.37	0.85	10.00	1.00	200	6	3	1
0	5	1.06	12	27	99	0.32	0	0.13	0.04	0.09	1.34	0.02	20	0	0	10
0	1	0.18	25	22	0	0.44	2	0.04	0.23	0.02	0.39	0.04	4	0	0	2
26	10	1.27	15	22	6	0.49	5	0.15	0.04	0.07	1.19	0.03	51	0	<.1	17
0	5	0.98	13	22	1	0.37	0	0.14	0.09	0.07	1.17	0.02	54	0	0	15
21	4	0.73	12	15	4	0.36	4	0.13	—	0.10	0.63	0.02	41	0	<.1	—
18	9	0.89	9	34	415	0.31	—	0.08	0.11	0.05	0.71	0.03	4	<.1	<.1	—
0	4	0.45	9	33	57	0.11	0	0.09	—	0.02	0.48	0.04	7	0	0	—
0	7	0.46	2	25	141	0.19	0	0.02	—	0.03	0.09	0.01	2	0	0	—
0	7	1.00	12	52	1	0.35	0	0.12	0.04	0.07	0.90	0.04	8	0	0	40
0	11	0.74	21	31	2	0.57	0	0.08	0.04	0.03	0.49	0.06	4	0	0	18
26	15	0.87	19	30	10	0.50	4	0.20	0.04	0.10	1.18	0.09	51	0	<1	17
0	7	0.33	13	21	4	0.29	3	0.08	0.03	0.04	0.72	0.02	44	0	0	13
0	1	0.22	11	24	<1	0.28	1	0.02	0.01	0.02	0.16	0.02	2	0	0	1
2	15	0.61	12	38	73	0.20	1	0.02	0.42	0.02	0.77	0.01	2	0	<.1	1
4	42	0.83	34	97	331	0.75	14	0.05	0.04	0.09	0.54	0.09	4	<1	<.1	4
0	3	0.92	36	74	292	0.87	3	0.04	0.04	0.04	0.51	0.07	6	<.1	0	2
0	8	0.17	7	148	1	0.06	4	0.02	0.44	0.04	0.13	0.06	4	6	0	0
0	3	0.07	3	59	1	0.02	2	0.01	0.18	0.01	0.05	0.02	2	3	0	0

Table A–1

Food Composition   (Computer code number is for Wadsworth Diet Analysis program)   (For purposes of calculations, use "0" for t, <1, <.1, <.01, etc.)

DA + Code	Food Description	Quantity	Measure	Wt (g)	H₂O (g)	Ener (cal)	Prot (g)	Carb (g)	Dietary Fiber (g)	Fat (g)	Fat Breakdown (g)		
											Sat	Mono	Poly
	FRUIT AND FRUIT JUICES—Continued												
946	Slices w/o skin, boiled	½	cup(s)	85	73	45	<1	12	2	<1	0.05	0.01	0.09
948	Dried, sulfured	½	cup(s)	22	7	52	<1	14	2	<.1	0.01	0.00	0.02
952	Juice, from frozen concentrate	½	cup(s)	120	105	56	<1	14	<1	<1	0.02	0.00	0.04
225	Juice, unsweetened, canned	½	cup(s)	124	109	58	<.1	14	<1	<1	0.02	0.01	0.04
226	Applesauce, sweetened, canned	½	cup(s)	128	101	97	<1	25	2	<1	0.04	0.01	0.07
227	Applesauce, unsweetened, canned	½	cup(s)	122	108	52	<1	14	1	<.1	0.01	0.00	0.02
38492	Crabapples	1	item(s)	35	28	27	<1	7	1	<1	0.02	0.00	0.03
	Apricot												
228	Fresh w/o pits	4	item(s)	140	121	67	2	16	3	1	0.04	0.24	0.11
230	Halves, dried, sulfured	¼	cup(s)	33	10	79	1	21	2	<1	0.01	0.02	0.02
229	Halves w/skin, canned in heavy syrup	½	cup(s)	129	100	107	1	28	2	<1	0.01	0.04	0.02
	Avocado												
233	California, whole, w/o skin or pit	1	item(s)	170	<1	284	3	15	12	26	3.59	16.61	3.42
234	Florida, whole, w/o skin or pit	1	item(s)	304	<1	365	7	24	17	31	5.90	16.70	5.00
2998	Pureed	⅛	cup(s)	29	21	46	1	2	2	4	0.61	2.82	0.52
	Banana												
235	Fresh whole, w/o peel	1	item(s)	118	88	105	1	27	3	<1	0.13	0.04	0.09
4580	Dried chips	¼	cup(s)	55	2	287	1	32	4	19	16.00	1.08	0.35
	Blackberries												
237	Raw	½	cup(s)	72	63	31	1	7	4	<1	0.01	0.03	0.20
958	Unsweetened, frozen	½	cup(s)	76	62	48	1	12	4	<1	0.01	0.03	0.18
	Blueberries												
238	Raw	½	cup(s)	72	61	41	1	10	2	<1	0.02	0.03	0.11
959	Canned in heavy syrup	½	cup(s)	128	98	113	1	28	2	<1	0.03	0.06	0.18
960	Unsweetened, frozen	½	cup(s)	78	67	40	1	10	2	1	0.04	0.07	0.22
	Boysenberries												
961	Canned in heavy syrup	½	cup(s)	128	98	113	1	29	3	<1	0.01	0.02	0.09
962	Unsweetened, frozen	½	cup(s)	66	57	33	1	8	3	<1	0.01	0.02	0.10
35576	**Breadfruit**	1	item(s)	384	271	396	4	104	17	1	0.00	0.00	0.00
	Cherries												
3000	Sour red, raw	½	cup(s)	78	67	39	1	9	1	<1	0.05	0.06	0.07
967	Sour red, canned in water	½	cup(s)	122	110	44	1	11	1	<1	0.03	0.03	0.04
240	Sweet, raw	½	cup(s)	73	60	46	1	12	2	<1	0.03	0.03	0.04
3004	Sweet, canned in heavy syrup	½	cup(s)	127	98	105	1	27	2	<1	0.04	0.05	0.06
969	Sweet, canned in water	½	cup(s)	124	108	57	1	15	2	<1	0.03	0.04	0.05
	Cranberries												
3007	Chopped, raw	½	cup(s)	55	48	25	<1	7	3	<.1	0.01	0.01	0.03
1638	Cranberry juice cocktail	½	cup(s)	127	108	72	0	18	<1	<1	0.01	0.02	0.06
241	Cranberry juice cocktail, low calorie, w/saccharin	½	cup(s)	127	120	24	<.1	6	0	<.1	0.00	0.00	0.00
1717	Cranberry apple juice drink	½	cup(s)	123	100	87	<.1	22	<1	<.1	0.00	0.00	0.00
242	Cranberry sauce, sweetened, canned	¼	cup(s)	69	42	105	<1	27	1	<1	0.01	0.01	0.05
	Dates												
244	Domestic, chopped	¼	cup(s)	44.5	0	126	1	33	4	<1	0.01	0.01	0
243	Domestic, whole	¼	cup(s)	44.5	0	126	1	33	4	<1	0.01	0.01	0
	Figs												
973	Raw, medium	2	item(s)	101	80	74	1	19	3	<1	0.06	0.07	0.14
975	Canned in heavy syrup	½	cup(s)	130	99	114	<1	30	3	<1	0.03	0.03	0.06
974	Canned in water	½	cup(s)	124	106	66	<1	17	3	<1	0.02	0.03	0.06
	Fruit cocktail & salad												
245	Fruit cocktail, canned in heavy syrup	½	cup(s)	124	100	91	<1	23	1	<.1	0.01	0.02	0.04
978	Fruit cocktail, canned in juice	½	cup(s)	119	104	55	1	14	1	<.1	0.00	0.00	0.00
977	Fruit cocktail, canned in water	½	cup(s)	119	108	38	<1	10	1	<.1	0.01	0.00	0.02
979	Fruit salad, canned in water	½	cup(s)	123	112	37	<1	10	1	<.1	0.01	0.02	0.03
	Gooseberries												
981	Raw	½	cup(s)	75	66	33	1	8	3	<1	0.03	0.04	0.24
982	Canned in light syrup	½	cup(s)	126	101	92	1	24	3	<1	0.02	0.02	0.14
	Grapefruit												
3022	Raw, pink or red	½	cup(s)	115	<.1	48	1	12	2	<1	0.02	0.02	0.04
247	Raw, white	½	item(s)	118	107	39	1	10	1	<1	0.02	0.02	0.03
251	Juice, pink, sweetened, canned	½	cup(s)	125	109	58	1	14	<1	<1	0.02	0.02	0.03

PAGE KEY: A–4 = Breads/Baked Goods   A–8 = Cereal/Rice/Pasta   A–12 = Fruit   A–18 = Vegetables/Legumes   A–28 = Nuts/Seeds   A–30 = Vegetarian
A–32 = Dairy   A–38 = Eggs   A–40 = Seafood   A–42 = Meats   A–44 = Poultry   A–46 = Processed meats   A–48 = Beverages   A–52 = Fats/Oils
A–54 = Sweets   A–54 = Sauces/Condiments/Spices   A–58 = Mixed Foods/Soups/Sandwiches   A–64 = Fast food   A–82 = Convenience   A–84 = Baby foods

A

Chol (mg)	Calc (mg)	Iron (mg)	Magn (mg)	Pota (mg)	Sodi (mg)	Zinc (mg)	Vit A (RAE) (µg)	Thia (mg)	Vit E (mg)	Ribo (mg)	Niac (mg)	Vit B₆ (mg)	Fola (µg)	Vit C (mg)	Vit B₁₂ (µg)	Sele (µg)
0	4	0.16	3	75	1	0.03	2	0.01	0.01	0.01	0.08	0.04	1	<1	0	<1
0	3	0.30	3	97	19	0.04	0	0.00	0.12	0.03	0.20	0.03	0	1	0	<1
0	7	0.31	6	151	8	0.05	0	0.00	0.01	0.02	0.05	0.04	0	1	0	<1
0	9	0.46	4	148	4	0.04	0	0.03	0.01	0.02	0.12	0.04	0	1	0	<1
0	5	0.45	4	78	4	0.05	1	0.02	0.01	0.04	0.24	0.03	1	2	0	<1
0	4	0.15	4	92	2	0.04	1	0.02	0.01	0.03	0.23	0.03	1	1	0	<1
0	6	0.13	2	68	<1	—	0	0.01	0.21	0.01	0.04	—	2	3	0	—
0	18	0.55	14	363	1	0.28	134	0.04	1.25	0.06	0.84	0.08	13	14	0	<1
0	18	0.88	11	383	3	0.13	59	0.00	1.44	0.02	0.85	0.05	3	<1	0	1
0	12	0.39	9	181	5	0.14	80	0.03	1.15	0.03	0.49	0.07	3	4	0	<1
0	22	1.00	49	861	14	1.12	104	0.12	1.77	0.24	3.24	0.47	105	15	0	1
0	30	0.50	73	1067	6	1.20	185	0.00	—	0.10	2.00	0.20	106	53	0	0
0	3	0.16	8	139	2	0.18	2	0.02	0.39	0.04	0.50	0.07	17	3	0	<1
0	6	0.31	32	422	1	0.18	4	0.04	0.32	0.09	0.78	0.43	24	10	0	1
0	10	0.69	42	296	3	0.41	2	0.05	2.98	0.01	0.39	0.14	8	3	0	1
0	21	0.45	14	117	1	0.38	8	0.01	0.51	0.02	0.47	0.02	18	15	0	<1
0	22	0.60	17	106	1	0.19	5	0.02	0.54	0.03	0.91	0.05	26	2	0	<1
0	4	0.20	4	55	1	0.12	2	0.03	0.72	0.03	0.30	0.04	4	7	0	<.1
0	6	0.42	5	51	4	0.09	3	0.04	1.28	0.07	0.14	0.05	3	1	0	<1
0	6	0.14	4	42	1	0.06	2	0.03	0.78	0.03	0.41	0.05	6	2	0	0
0	23	0.55	14	115	4	0.24	3	0.03	0.91	0.04	0.29	0.05	44	8	0	1
0	18	0.56	11	92	1	0.15	2	0.03	0.30	0.02	0.51	0.04	42	2	0	<1
0	65	2.07	96	1882	8	0.46	8	0.42	4.30	0.12	3.46	0.00	54	111	0	2
0	12	0.25	7	134	2	0.08	50	0.02	0.10	0.03	0.31	0.05	6	8	0	0
0	13	1.67	7	120	9	0.09	46	0.02	0.16	0.05	0.22	0.05	10	3	0	0
0	9	0.26	8	161	0	0.05	2	0.02	0.09	0.02	0.11	0.04	3	5	0	0
0	11	0.44	11	183	4	0.13	10	0.03	0.08	0.05	0.50	0.04	5	5	0	0
0	14	0.45	11	162	1	0.10	10	0.03	0.16	0.05	0.51	0.04	5	3	0	0
0	4	0.14	3	47	1	0.06	2	0.01	0.06	0.01	0.06	0.03	1	7	0	<.1
0	4	0.19	3	23	3	0.09	0	0.01	0.00	0.01	0.04	0.02	0	45	0	0
0	11	0.05	3	32	4	0.03	0	0.00	0.00	0.00	0.01	0.00	0	41	0	0
0	6	0.15	2	34	9	0.22	0	0.01	0.00	0.02	0.07	0.03	0	39	0	0
0	3	0.15	2	18	20	0.03	1	0.01	0.07	0.01	0.07	0.01	1	1	0	<1
0	17	0.45	19	292	1	0.12	1	0.02	0.03	0.02	0.56	0.07	9	<1	0	1
0	17	0.45	19	292	1	0.12	1	0.02	0.03	0.02	0.56	0.07	9	<1	0	1
0	35	0.37	17	233	1	0.15	7	0.06	0.89	0.05	0.40	0.11	6	2	0	<1
0	35	0.36	13	128	1	0.14	3	0.03	1.15	0.05	0.55	0.09	3	1	0	<1
0	35	0.36	12	128	1	0.15	2	0.03	1.10	0.05	0.55	0.09	2	1	0	<1
0	7	0.36	6	109	7	0.10	12	0.02	0.36	0.02	0.46	0.06	4	2	0	1
0	9	0.25	8	113	5	0.11	18	0.01	0.24	0.02	0.48	0.06	4	3	0	1
0	6	0.30	8	111	5	0.11	15	0.02	0.34	0.01	0.43	0.06	4	2	0	1
0	9	0.37	6	96	4	0.10	27	0.02	—	0.03	0.46	0.04	4	2	0	1
0	19	0.23	8	149	1	0.09	11	0.03	0.28	0.02	0.23	0.06	5	21	0	<1
0	20	0.42	8	97	3	0.14	9	0.03	0.47	0.07	0.19	0.02	4	13	0	1
0	25	0.09	10	155	0	0.08	30	0.05	—	0.03	0.23	0.06	15	36	0	<1
0	14	0.07	11	175	0	0.08	2	0.04	0.30	0.02	0.32	0.05	12	39	0	2
0	10	0.45	13	203	3	0.08	0	0.05	0.06	0.03	0.40	0.03	13	34	0	<1

Food Composition

(Computer code number is for Wadsworth Diet Analysis program)   (For purposes of calculations, use "0" for t, <1, <.1, <.01, etc.)

DA + Code	Food Description	Quantity	Measure	Wt (g)	H₂O (g)	Ener (cal)	Prot (g)	Carb (g)	Dietary Fiber (g)	Fat (g)	Sat	Mono	Poly
	FRUIT AND FRUIT JUICES—Continued												
249	Juice, white	½	cup(s)	124	111	48	1	11	<1	<1	0.02	0.02	0.03
248	Sections, canned in light syrup	½	cup(s)	127	106	76	1	20	1	<1	0.02	0.02	0.03
983	Sections, canned in water	½	cup(s)	122	<.1	44	1	11	<1	<1	0.02	0.02	0.03
	Grapes												
255	American, slip skin	½	cup(s)	46	37	31	<1	8	<1	<1	0.05	0.01	0.05
256	European, red or green, adherent skin	½	cup(s)	80	<.1	55	1	14	1	<1	0.04	0.01	0.04
259	Juice, sweetened, added vitamin C, from frozen concentrate	½	cup(s)	125	109	64	<1	16	<1	<1	0.04	0.01	0.03
3159	Juice drink, canned	½	cup(s)	125	109	63	<1	16	0	0	0.00	0.00	0.00
3060	Raisins, seeded, packed	¼	cup(s)	41	7	122	1	32	3	<1	0.07	0.01	0.07
987	**Guava, raw**	1	item(s)	90	77	46	1	11	5	1	0.15	0.05	0.23
35593	**Guava, strawberry**	1	item(s)	6	5	4	<.1	1	<1	<.1	0.01	0.00	0.02
3027	**Jackfruit**	½	cup(s)	83	61	78	1	20	1	<1	0.05	0.04	0.07
8458	**Kiwi fruit**	1	item(s)	77	63	53	1	11	3	1	0.02	0.03	0.19
	Lemon												
992	Raw	1	item(s)	108	94	22	1	12	5	<1	0.04	0.01	0.10
262	Juice	1	tablespoon(s)	15	14	4	<.1	1	<.1	0	0.00	0.00	0.00
993	Peel	1	teaspoon(s)	2	2	1	<.1	<1	<1	<.1	0.00	0.00	0.00
	Lime												
994	Raw	1	item(s)	67	61	15	<1	6	2	<.1	0.01	0.01	0.02
269	Juice	1	tablespoon(s)	15	14	4	<.1	1	<.1	<.1	0.00	0.00	0.00
995	**Loganberries, frozen**	½	cup(s)	74	62	40	1	10	4	<1	0.01	0.02	0.13
	Mandarin orange												
1038	Canned in juice	½	cup(s)	125	111	46	1	12	1	<.1	0.00	0.01	0.01
1039	Canned in light syrup	½	cup(s)	126	105	77	1	20	1	<1	0.02	0.02	0.03
999	**Mango**	½	item(s)	104	85	67	1	18	2	<1	0.07	0.10	0.05
1005	**Nectarine, raw, sliced**	½	cup(s)	69	60	30	1	7	1	<1	0.02	0.06	0.08
	Melons												
271	Cantaloupe	½	cup(s)	80	72	27	1	7	1	<1	0.04	0.00	0.07
1000	Casaba melon	½	cup(s)	85	78	24	1	6	1	<.1	0.02	0.00	0.03
272	Honeydew	½	cup(s)	89	80	32	<1	8	1	<1	0.03	0.00	0.05
318	Watermelon	½	cup(s)	77	71	23	<1	6	<1	<1	0.01	0.03	0.04
	Orange												
273	Raw	1	item(s)	131	114	62	1	15	3	<1	0.02	0.03	0.03
3040	Peel	1	teaspoon(s)	2	1	2	<.1	1	<1	<.1	0.00	0.00	0.00
274	Sections	½	cup(s)	90	78	43	1	11	2	<1	0.01	0.02	0.02
275	Juice	½	cup(s)	124	109	56	1	13	<1	<1	0.03	0.04	0.05
29630	Juice, fresh squeezed	½	cup(s)	124	109	56	1	13	<1	<1	0.03	0.04	0.05
14414	Juice w/calcium & extra vitamin C	½	cup(s)	125	55	1	13	<1	0	0.00	0.00	0.00	0
278	Juice, unsweetened, from frozen concentrate	½	cup(s)	125	110	56	1	13	<1	<.1	0.01	0.01	0.01
	Papaya												
282	Raw	½	cup(s)	70	62	27	<1	7	1	<.1	0.03	0.03	0.02
16830	Dried, strips	2	item(s)	46	12	119	2	30	5	<1	0.13	0.12	0.09
35640	**Passion fruit, purple**	1	item(s)	18	13	17	<1	4	3	<1	0.00	0.00	0.00
	Peach												
283	Raw, medium	1	item(s)	98	87	38	1	9	1	<1	0.02	0.07	0.08
285	Halves, canned in heavy syrup	½	cup(s)	131	104	97	1	26	2	<1	0.01	0.05	0.06
286	Halves, canned in water	½	cup(s)	122	114	29	1	7	2	<.1	0.01	0.03	0.03
290	Slices, sweetened, frozen	½	cup(s)	125	93	118	1	30	2	<1	0.02	0.06	0.08
	Pear												
291	Raw	1	item(s)	166	139	96	1	26	5	<1	0.01	0.04	0.05
8672	Asian	1	item(s)	122	108	51	1	13	4	<1	0.01	0.06	0.07
293	Danjou	1	item(s)	200	168	120	1	30	5	1	0.00	0.20	0.20
294	Halves, canned in heavy syrup	½	cup(s)	133	107	98	<1	25	2	<1	0.01	0.04	0.04
1012	Halves, canned in juice	½	cup(s)	124	107	62	<1	16	2	<.1	0.00	0.02	0.02
1017	**Persimmon**	1	item(s)	25	16	32	<1	8	0	<1	0.01	0.02	0.02
	Pineapple												
295	Raw, diced	½	cup(s)	78	67	37	<1	10	1	<.1	0.01	0.01	0.03
3053	Canned in extra heavy syrup	½	cup(s)	130	101	108	<1	28	1	<1	0.01	0.02	0.05
1019	Canned in juice	½	cup(s)	125	104	75	1	20	1	<.1	0.01	0.01	0.04

PAGE KEY: A–4 = Breads/Baked Goods   A–8 = Cereal/Rice/Pasta   A–12 = Fruit   A–18 = Vegetables/Legumes   A–28 = Nuts/Seeds   A–30 = Vegetarian
A–32 = Dairy   A–38 = Eggs   A–40 = Seafood   A–42 = Meats   A–44 = Poultry   A–46 = Processed meats   A–48 = Beverages   A–52 = Fats/Oils
A–54 = Sweets   A–54 = Sauces/Condiments/Spices   A–58 = Mixed Foods/Soups/Sandwiches   A–64 = Fast food   A–82 = Convenience   A–84 = Baby foods

A

Chol (mg)	Calc (mg)	Iron (mg)	Magn (mg)	Pota (mg)	Sodi (mg)	Zinc (mg)	Vit A (RAE) (µg)	Thia (mg)	Vit E (mg)	Ribo (mg)	Niac (mg)	Vit B_6 (mg)	Fola (µg)	Vit C (mg)	Vit B_{12} (µg)	Sele (µg)
0	11	0.25	15	200	1	0.06	2	0.05	0.06	0.02	0.25	0.05	12	47	0	<1
0	18	0.51	13	164	3	0.10	0	0.05	0.32	0.03	0.31	0.03	11	27	0	1
0	18	0.50	12	161	2	0.11	0	0.05	0.31	0.03	0.30	0.02	11	27	0	1
0	6	0.13	2	88	1	0.02	2	0.04	0.16	0.03	0.14	0.05	2	2	0	<.1
0	8	0.29	6	153	2	0.06	6	0.06	0.45	0.06	0.15	0.07	2	9	0	<.1
0	5	0.13	5	26	3	0.05	0	0.02	0.06	0.03	0.16	0.05	1	30	0	<1
0	4	0.13	4	41	1	0.03	0	0.01	0.00	0.02	0.09	0.02	1	20	0	<1
0	12	1.07	12	340	12	0.07	0	0.05	0.29	0.08	0.46	0.08	1	2	0	<1
0	18	0.28	9	256	3	0.21	28	0.05	1.01	0.05	1.08	0.13	13	165	0	1
0	1	0.01	1	18	2	—	—	0.00	—	0.00	0.04	0.00	—	2	0	—
0	28	0.50	31	251	2	0.35	12	0.02	0.12	0.09	0.33	0.09	12	6	0	<1
0	30	0.38	14	251	2	0.10	4	—	1.17	0.02	0.25	0.05	<.1	74	0	—
0	66	0.76	13	157	3	0.11	2	0.05	—	0.04	0.22	0.12	—	83	0	1
0	1	0.00	1	19	<1	0.01	<1	0.00	0.01	0.00	0.02	0.01	2	7	0	<.1
0	3	0.02	<1	3	<1	0.01	<.1	0.00	0.00	0.00	0.01	0.00	<1	3	0	<.1
0	9	0.06	5	78	1	0.05	1	0.02	0.16	0.01	0.10	0.03	7	20	0	<.1
0	1	0.00	1	17	<1	0.01	<1	0.00	0.01	0.00	0.02	0.01	1	5	0	<.1
0	19	0.47	15	107	1	0.25	1	0.04	1.62	0.02	0.62	0.05	19	11	0	<1
0	14	0.34	14	166	6	0.63	54	0.10	0.62	0.04	0.55	0.05	6	43	0	<1
0	9	0.47	10	98	8	0.30	53	0.07	0.43	0.06	0.56	0.05	6	25	0	1
0	10	0.13	9	161	2	0.04	39	0.06	1.16	0.06	0.60	0.14	14	29	0	1
0	4	0.19	6	139	0	0.12	12	0.02	0.61	0.02	0.78	0.02	3	4	0	0
0	7	0.17	10	215	13	0.14	136	0.03	0.12	0.02	0.59	0.06	17	30	0	<1
0	9	0.29	9	155	8	0.06	0	0.01	0.13	0.03	0.20	0.14	7	19	0	<1
0	5	0.15	9	203	16	0.08	3	0.03	0.13	0.01	0.37	0.08	17	16	0	1
0	5	0.19	8	86	1	0.08	22	0.03	0.12	0.02	0.14	0.03	2	6	0	<1
0	52	0.13	13	237	0	0.09	14	0.11	0.31	0.05	0.37	0.08	39	70	0	1
0	3	0.02	<1	4	<.1	0.01	<1	0.00	0.00	0.00	0.02	0.00	1	3	0	<.1
0	36	0.09	9	164	0	0.06	10	0.08	0.22	0.04	0.26	0.05	27	48	0	<1
0	14	0.25	14	248	1	0.06	12	0.11	0.11	0.04	0.50	0.05	37	62	0	<1
0	14	0.25	14	248	1	0.06	—	0.11	0.11	0.04	0.50	0.05	38	62	0	—
—	175	—	—	225	0	—	5	0.08	—	—	0.40	0.06	30	54	0	—
0	11	0.12	12	237	1	0.06	6	0.10	0.24	0.02	0.25	0.05	55	48	0	<1
0	17	0.07	7	180	2	0.05	39	0.02	0.78	0.02	0.24	0.01	27	43	0	<1
0	73	0.30	30	783	9	0.21	—	0.06	3.41	0.09	0.93	0.05	58	38	0	—
0	2	0.29	5	63	5	—	—	0.00	0.20	0.02	0.27	—	3	5	0	<1
0	6	0.25	9	186	0	0.17	16	0.02	0.69	0.03	0.79	0.02	4	6	0	<.1
0	4	0.35	7	121	8	0.12	22	0.01	1.17	0.03	0.80	0.02	4	4	0	<1
0	2	0.39	6	121	4	0.11	33	0.01	1.09	0.02	0.64	0.02	4	4	0	<1
0	4	0.46	6	163	8	0.06	18	0.02	1.11	0.04	0.82	0.02	4	118	0	1
0	15	0.28	12	198	2	0.17	2	0.02	0.83	0.04	0.26	0.05	12	7	0	<1
0	5	0.00	10	148	0	0.02	0	0.01	0.61	0.01	0.27	0.03	10	5	0	<1
0	22	0.50	12	250	0	0.24	—	0.04	—	0.08	0.20	0.04	15	8	0	1
0	7	0.29	5	86	7	0.11	0	0.01	0.67	0.03	0.32	0.02	1	1	0	0
0	11	0.36	9	119	5	0.11	0	0.01	0.62	0.01	0.25	0.02	1	2	0	0
0	7	0.63	—	78	<1	—	—	—	—	—	—	—	—	17	0	0
0	10	0.22	9	89	1	0.08	2	0.06	0.08	0.02	0.38	0.09	12	28	0	<.1
0	18	0.49	20	133	1	0.14	1	0.12	—	0.03	0.37	0.10	7	9	0	—
0	17	0.35	17	152	1	0.12	2	0.12	0.12	0.02	0.35	0.09	6	12	0	<1

Table A–1

Food Composition   (Computer code number is for Wadsworth Diet Analysis program)   (For purposes of calculations, use "0" for t, <1, <.1, <.01, etc.)

DA + Code	Food Description	Quantity	Measure	Wt (g)	H₂O (g)	Ener (cal)	Prot (g)	Carb (g)	Dietary Fiber (g)	Fat (g)	Sat	Mono	Poly
	FRUIT AND FRUIT JUICES—Continued												
296	Canned in light syrup	½	cup(s)	126	108	66	<1	17	1	<1	0.01	0.02	0.05
1018	Canned in water	½	cup(s)	123	112	39	1	10	1	<1	0.01	0.01	0.04
299	Juice, unsweetened, canned	½	cup(s)	125	107	70	<1	17	<1	<1	0.01	0.01	0.04
1024	**Plantain, cooked**	½	cup(s)	77	52	89	1	24	2	<1	0.05	0.01	0.03
300	**Plum, raw, large**	1	item(s)	83	72	38	1	9	1	<1	0.01	0.11	0.04
1027	**Pomegranate**	1	item(s)	154	125	105	1	26	1	<1	0.06	0.07	0.10
	Prunes												
5644	Dried	2	item(s)	17	5	40	<1	11	1	<.1	0.01	0.06	0.02
305	Dried, stewed	½	cup(s)	119	<.1	128	1	33	4	<1	0.00	0.15	0.04
306	Juice, canned	1	cup(s)	256	208	182	2	45	3	<.1	0.01	0.05	0.02
	Raisins, *see* grapes												
	Raspberries												
309	Raw	½	cup(s)	62	53	32	1	7	4	<1	0.01	0.04	0.23
310	Red, sweetened, frozen	½	cup(s)	125	91	129	1	33	6	<1	0.01	0.02	0.11
311	**Rhubarb, cooked with sugar**	½	cup(s)	120	82	140	1	38	3	<.1	0.00	0.00	0.05
	Strawberries												
313	Raw	½	cup(s)	72	65	23	<1	6	1	<1	0.01	0.03	0.11
315	Sweetened, frozen, thawed	½	cup(s)	128	100	99	1	27	2	<1	0.01	0.02	0.09
16828	**Tangelo**	1	item(s)	95	82	45	1	11	2	<1	0.01	0.02	0.02
	Tangerine												
316	Raw	1	item(s)	84	74	37	1	9	2	<1	0.02	0.03	0.03
1040	Juice	½	cup(s)	124	110	53	1	12	<1	<1	0.03	0.04	0.05
	VEGETABLES, LEGUMES												
	Amaranth												
1042	Leaves, raw	1	cup(s)	28	26	6	1	1	0	<.1	0.03	0.02	0.04
1043	Leaves, boiled, drained	½	cup(s)	66	60	14	1	3	0	<1	0.03	0.03	0.05
8683	**Arugula leaves, raw**	1	cup(s)	20	18	5	1	1	<1	<1	0.02	0.01	0.06
	Artichoke												
1044	Boiled, drained	1	item(s)	120	101	60	4	13	6	<1	0.04	0.01	0.08
2885	Hearts, boiled, drained	½	cup(s)	84	71	42	3	9	5	<1	0.03	0.00	0.06
	Asparagus												
566	Boiled, drained	½	cup(s)	90	0.08	20	2	4	2	0.19	0.06	0	0.12
568	Canned, drained	½	cup(s)	121	114	23	3	3	2	1	0.18	0.03	0.34
565	Tips, frozen, boiled, drained	½	cup(s)	90	82	25	3	4	1	<1	0.09	0.01	0.17
	Bamboo shoots												
1048	Boiled, drained	½	cup(s)	60	58	7	1	1	1	<1	0.03	0.00	0.06
1049	Canned, drained	½	cup(s)	65	62	12	1	2	1	<1	0.06	0.01	0.12
	Beans												
1801	Adzuki beans, boiled	½	cup(s)	115	76	147	9	28	8	<1	0.04	—	—
511	Baked beans w/franks, canned	½	cup(s)	129	89	182	9	20	9	8	3.02	3.64	1.07
512	Baked beans w/pork in tomato sauce, canned	½	cup(s)	127	92	124	7	25	6	1	0.50	0.56	0.17
513	Baked beans w/pork in sweet sauce, canned	½	cup(s)	127	89	140	7	27	7	2	0.71	0.80	0.24
1805	Black beans, boiled	½	cup(s)	86	57	114	8	20	7	<1	0.12	0.04	0.20
14597	Chickpeas, garbanzo beans, or bengal gram, boiled	½	cup(s)	82	49	134	7	22	6	2	0.22	0.48	0.95
569	Fordhook lima beans, frozen, boiled, drained	½	cup(s)	85	62	88	5	16	5	<1	0.07	0.02	0.14
1806	French beans, boiled	½	cup(s)	89	59	114	6	21	8	1	0.07	0.05	0.40
2773	Great northern beans, boiled	½	cup(s)	89	61	104	7	19	6	<1	0.12	0.02	0.17
2736	Hyacinth beans, boiled, drained	½	cup(s)	44	38	22	1	4	0	<1	0.05	0.06	0.00
515	Lima beans, boiled, drained	½	cup(s)	85	57	105	6	20	5	<1	0.06	0.02	0.13
570	Lima beans, baby, frozen, boiled, drained	½	cup(s)	90	65	95	6	18	5	<1	0.06	0.02	0.13
579	Mung beans, sprouted, boiled, drained	½	cup(s)	62	<.1	13	1	3	<1	<.1	0.02	0.00	0.02
510	Navy beans, boiled	½	cup(s)	91	57	129	8	24	6	1	0.13	0.05	0.22
32816	Pinto beans, boiled, drained, no salt added	½	cup(s)	114	106	25	2	5	0	<1	0.04	0.03	0.21
1052	Pinto beans, frozen, boiled, drained	½	cup(s)	47	27	76	4	15	4	<1	0.03	0.02	0.13
514	Red kidney beans, canned	½	cup(s)	128	99	109	7	20	8	<1	0.06	0.03	0.24

PAGE KEY: A–4 = Breads/Baked Goods A–8 = Cereal/Rice/Pasta A–12 = Fruit A–18 = Vegetables/Legumes A–28 = Nuts/Seeds A–30 = Vegetarian A–32 = Dairy A–38 = Eggs A–40 = Seafood A–42 = Meats A–44 = Poultry A–46 = Processed meats A–48 = Beverages A–52 = Fats/Oils A–54 = Sweets A–54 = Sauces/Condiments/Spices A–58 = Mixed Foods/Soups/Sandwiches A–64 = Fast food A–82 = Convenience A–84 = Baby foods

Chol (mg)	Calc (mg)	Iron (mg)	Magn (mg)	Pota (mg)	Sodi (mg)	Zinc (mg)	Vit A (RAE) (µg)	Thia (mg)	Vit E (mg)	Ribo (mg)	Niac (mg)	Vit B$_6$ (mg)	Fola (µg)	Vit C (mg)	Vit B$_{12}$ (µg)	Sele (µg)
0	18	0.49	20	132	1	0.15	3	0.11	0.13	0.03	0.37	0.09	6	9	0	1
0	18	0.49	22	156	1	0.15	2	0.11	0.12	0.03	0.37	0.09	6	9	0	<1
0	21	0.33	16	168	1	0.14	0	0.07	0.03	0.03	0.32	0.12	29	13	0	<1
0	2	0.45	25	358	4	0.10	35	0.04	0.11	0.04	0.58	0.18	20	8	0	1
0	5	0.14	6	130	0	0.08	14	0.02	0.50	0.02	0.34	0.02	4	8	0	0
0	5	0.46	5	399	5	0.18	8	0.05	0.85	0.05	0.46	0.16	9	9	0	1
0	9	0.42	8	125	1	0.09	17	0.01	0.00	0.03	0.33	0.04	1	1	0	<1
0	23	0.46	21	383	1	0.19	37	0.00	0.00	0.12	0.85	0.23	0	3	0	<1
0	31	3.02	36	707	10	0.54	0	0.04	0.03	0.18	2.01	0.56	0	10	0	2
0	15	0.42	14	93	1	0.26	1	0.02	0.28	0.02	0.37	0.03	13	16	0	<1
0	19	0.81	16	143	1	0.23	4	0.02	0.56	0.06	0.29	0.04	33	21	0	<1
0	174	0.25	16	115	1	—	—	0.02	—	0.03	0.25	—	—	4	0	—
0	12	0.30	9	110	1	0.10	1	0.02	0.10	0.02	0.28	0.03	17	42	0	<1
0	14	0.60	8	125	1	0.06	1	0.02	0.34	0.10	0.37	0.04	5	50	0	1
0	38	0.10	10	172	0	0.07	—	0.08	0.23	0.04	0.27	0.06	29	51	0	—
0	12	0.08	10	132	1	0.20	29	0.09	0.20	0.02	0.13	0.06	17	26	0	<1
0	22	0.25	10	220	1	0.04	16	0.07	0.11	0.02	0.12	0.05	6	38	0	<1
0	60	0.65	15	171	6	0.25	0	0.01	0.00	0.04	0.18	0.05	24	12	0	<1
0	138	1.49	36	423	14	0.58	92	0.01	—	0.09	0.37	0.12	38	27	0	1
0	32	0.29	9	74	5	0.09	24	0.01	0.09	0.02	0.06	0.01	19	3	0	<.1
0	54	1.55	72	425	114	0.59	11	0.08	0.23	0.08	1.20	0.13	61	12	0	<1
0	38	1.08	50	297	80	0.41	8	0.05	0.16	0.06	0.84	0.09	43	8	0	<1
0	20.7	0.81	12.6	201.6	12.6	0.54	48.59	0.14	0.27	0.12	0.97	0.07	134.1	6.92	0	5.48
0	19	0.73	12	208	347	0.48	50	0.07	0.52	0.12	1.15	0.13	116	22	0	2
0	21	0.58	12	196	4	0.50	—	0.06	1.13	0.09	0.93	0.02	121	22	0	4
0	7	0.14	2	320	2	0.28	0	0.01	—	0.03	0.18	0.06	1	0	0	<1
0	5	0.21	3	52	5	0.43	1	0.02	0.25	0.02	0.09	0.09	2	1	0	<1
0	32	2.30	60	612	9	2.04	0	0.13	—	0.07	0.82	0.11	139	0	0	1
8	62	2.22	36	302	553	2.40	5	0.07	0.60	0.07	1.16	0.06	39	3	0	8
9	71	4.15	44	380	557	7.41	5	0.07	0.68	0.06	0.63	0.09	29	4	0	6
9	77	2.10	43	336	425	1.90	1	0.06	0.68	0.08	0.44	0.11	47	4	0	6
0	23	1.81	60	305	1	0.96	0	0.21	—	0.05	0.43	0.06	128	0	0	1
0	40	2.37	39	239	6	1.25	1	0.10	0.29	0.05	0.43	0.11	141	1	0	3
0	26	1.55	36	258	59	0.63	9	0.06	0.25	0.05	0.91	0.10	18	11	0	1
0	56	0.96	50	327	5	0.57	0	0.12	—	0.05	0.48	0.09	66	1	0	1
0	60	1.89	44	346	2	0.78	0	0.14	—	0.05	0.60	0.10	90	1	0	4
0	18	0.33	18	114	1	0.17	3	0.02	—	0.04	0.21	0.01	20	2	0	1
0	27	2.08	63	485	14	0.67	16	0.12	0.12	0.08	0.88	0.16	22	9	0	2
0	25	1.76	50	370	26	0.50	7	0.06	0.58	0.05	0.69	0.10	14	5	0	2
0	7	0.40	9	63	6	0.29	1	0.03	0.00	0.06	0.50	0.03	18	7	0	<1
0	64	2.26	54	335	1	0.96	0	0.18	—	0.06	0.48	0.15	127	1	0	5
0	17	0.75	20	111	58	0.19	0	0.08	—	0.07	0.82	0.06	146	7	0	1
0	24	1.27	25	304	39	0.32	0	0.13	—	0.05	0.30	0.09	16	<1	0	1
0	31	1.61	36	329	436	0.70	0	0.13	0.06	0.11	0.58	0.03	65	1	0	2

DA + Code	Food Description	Quantity	Measure	Wt (g)	H₂O (g)	Ener (cal)	Prot (g)	Carb (g)	Dietary Fiber (g)	Fat (g)	Fat Breakdown (g) Sat	Mono	Poly
	VEGETABLES, LEGUMES—Continued												
1810	Refried beans, canned	½	cup(s)	127	96	119	7	20	7	2	0.60	0.71	0.19
1053	Shell beans, canned	½	cup(s)	123	111	37	2	8	4	<1	0.03	0.02	0.13
1670	Soybeans, boiled	½	cup(s)	86	54	149	14	9	5	8	1.12	1.70	4.36
1108	Soybeans, green, boiled, drained	½	cup(s)	90	62	127	11	10	4	6	0.67	1.09	2.71
1807	White beans, small, boiled	½	cup(s)	90	57	127	8	23	9	1	0.15	0.05	0.25
575	Yellow snap, string or wax beans, boiled, drained	½	cup(s)	62	<.1	22	1	5	2	<1	0.04	0.00	0.09
576	Yellow snap, string or wax beans, frozen, boiled, drained	½	cup(s)	68	<.1	19	1	4	2	<1	0.02	0.00	0.05
	Beets												
580	Whole, boiled, drained	2	item(s)	100	87	44	2	10	2	<1	0.03	0.04	0.06
581	Sliced, boiled, drained	½	cup(s)	85	74	37	1	8	2	<1	0.02	0.03	0.05
583	Sliced, canned, drained	½	cup(s)	85	77	26	1	6	1	<1	0.02	0.02	0.04
2730	Pickled, canned with liquid	½	cup(s)	114	93	74	1	18	3	<.1	0.01	0.02	0.03
584	Beet greens, boiled, drained	½	cup(s)	72	64	19	2	4	2	<1	0.02	0.03	0.05
585	**Cowpeas or black-eyed peas, boiled, drained**	½	cup(s)	83	0.06	80	2.61	16.76	4.12	0.31	0.07	0.02	0.13
	Broccoli												
587	Raw, chopped	½	cup(s)	44	39	15	1	3	1	<1	0.02	0.00	0.02
588	Chopped, boiled, drained	½	cup(s)	78	70	27	2	6	3	<1	0.06	0.03	0.13
590	Frozen, chopped, boiled, drained	½	cup(s)	92	83	26	3	5	3	<1	0.02	0.01	0.05
16848	**Broccoflower, raw, chopped**	½	cup(s)	32	29	10	1	2	1	<.1	0.01	0.01	0.04
	Brussels sprouts												
591	Boiled, drained	½	cup(s)	78	69	28	2	6	2	<1	0.08	0.03	0.20
592	Frozen, boiled, drained	½	cup(s)	78	67	33	3	6	3	<1	0.06	0.00	0.16
	Cabbage												
594	Raw, shredded	1	cup(s)	70	65	17	1	4	2	<.1	0.01	0.01	0.04
595	Boiled, drained, no salt added	1	cup(s)	150	140	33	2	7	3	1	0.08	0.05	0.29
35611	Chinese (pak choi or bok choy), boiled w/salt, drained	1	cup(s)	170	162	20	3	3	2	<1	0.04	0.02	0.13
16869	Kim chee	1	cup(s)	150	138	31	2	6	2	<1	0.04	0.02	0.15
596	Red, shredded, raw	1	cup(s)	70	63	22	1	5	1	<1	0.02	0.01	0.09
597	Savoy, shredded, raw	1	cup(s)	70	64	19	1	4	2	<.1	0.01	0.00	0.03
11710	**Capers**	1	teaspoon(s)	5	—	0	0	0	0	0	0.00	0.00	0.00
	Carrots												
600	Raw	½	cup(s)	61	54	25	1	6	2	<1	0.02	0.01	0.06
8691	Raw, baby	8	item(s)	80	72	28	1	7	1	<1	0.02	0.01	0.05
601	Grated	½	cup(s)	55	49	23	1	5	2	<1	0.02	0.01	0.06
602	Sliced, boiled, drained	½	cup(s)	78	0.07	27.29	0.59	6.41	2.33	0.14	0.02	0	0.08
1055	Juice, canned	½	cup(s)	123	109	49	1	11	1	<1	0.03	0.01	0.09
32725	**Cassava or manioc**	½	cup(s)	103	61	165	1	39	2	<1	0.08	0.08	0.05
	Cauliflower												
605	Raw, chopped,	½	cup(s)	50	46	13	1	3	1	<1	0.02	0.01	0.05
606	Boiled, drained	½	cup(s)	62	58	14	1	3	2	<1	0.04	0.02	0.13
607	Frozen, boiled, drained	½	cup(s)	90	85	17	1	3	2	<1	0.03	0.01	0.09
	Celery												
609	Diced	½	cup(s)	60	58	8	<1	2	1	<1	0.03	0.02	0.05
608	Stalk	2	item(s)	80	76	11	1	2	1	<1	0.03	0.03	0.06
	Chard												
1056	Swiss chard, raw	1	cup(s)	36	33	7	1	1	1	<.1	0.01	0.01	0.03
1057	Swiss chard, boiled, drained	½	cup(s)	88	81	18	2	4	2	<.1	0.01	0.01	0.02
	Collard greens												
610	Boiled, drained	½	cup(s)	95	87	25	2	5	3	<1	0.04	0.02	0.16
611	Frozen, chopped, boiled, drained	½	cup(s)	85	75	31	3	6	2	<1	0.05	0.02	0.18
	Corn												
29614	Yellow corn, fresh, cooked	1	item(s)	100	0.06	107.37	3.3	24.96	2.78	1.27	0.19	0.37	0.59
612	Yellow sweet corn, boiled, drained	½	cup(s)	82	57	89	3	21	2	1	0.16	0.31	0.49
614	Yellow sweet corn, frozen, boiled, drained	½	cup(s)	82	63	66	2	16	2	1	0.08	0.16	0.26
615	Yellow creamed sweet corn, canned	½	cup(s)	128	101	92	2	23	2	1	0.08	0.16	0.25
618	**Cucumber**	¼	item(s)	75	72	11	<1	3	<1	<.1	0.03	0.00	0.04

PAGE KEY: A–4 = Breads/Baked Goods A–8 = Cereal/Rice/Pasta A–12 = Fruit A–18 = Vegetables/Legumes A–28 = Nuts/Seeds A–30 = Vegetarian A–32 = Dairy A–38 = Eggs A–40 = Seafood A–42 = Meats A–44 = Poultry A–46 = Processed meats A–48 = Beverages A–52 = Fats/Oils A–54 = Sweets A–54 = Sauces/Condiments/Spices A–58 = Mixed Foods/Soups/Sandwiches A–64 = Fast food A–82 = Convenience A–84 = Baby foods

A

Chol (mg)	Calc (mg)	Iron (mg)	Magn (mg)	Pota (mg)	Sodi (mg)	Zinc (mg)	Vit A (RAE) (µg)	Thia (mg)	Vit E (mg)	Ribo (mg)	Niac (mg)	Vit B_6 (mg)	Fola (µg)	Vit C (mg)	Vit B_{12} (µg)	Sele (µg)
10	44	2.10	42	338	378	1.48	0	0.03	0.00	0.02	0.40	0.18	14	8	0	2
0	36	1.21	18	134	409	0.33	13	0.04	0.04	0.07	0.25	0.06	22	4	0	1
0	88	4.42	74	443	1	0.99	0	0.13	1.68	0.25	0.34	0.20	46	1	0	6
0	131	2.25	54	485	13	0.82	7	0.23	0.01	0.14	1.13	0.05	100	15	0	1
0	65	2.54	61	414	2	0.98	0	0.21	—	0.05	0.24	0.11	123	0	0	1
0	29	0.80	16	187	2	0.22	5	0.05	0.14	0.06	0.38	0.03	21	6	0	<1
0	33	0.59	16	85	6	0.32	7	0.02	0.08	0.06	0.26	0.04	16	3	0	<1
0	16	0.79	23	305	77	0.35	2	0.03	0.30	0.04	0.33	0.07	80	4	0	1
0	14	0.67	20	259	65	0.30	2	0.02	0.26	0.03	0.28	0.06	68	3	0	1
0	13	1.55	14	126	165	0.18	1	0.01	0.26	0.03	0.13	0.05	26	3	0	<1
0	12	0.47	17	168	300	0.30	1	0.01	—	0.05	0.28	0.06	31	3	0	1
0	82	1.37	49	654	174	0.36	276	0.08	0.22	0.21	0.36	0.10	10	18	0	1
0	105.59	0.92	42.9	344.85	3.29	0.84	65.17	0.08	0.14	0.12	1.15	0.05	104.77	1.81	0	2.06
0	21	0.32	9	139	15	0.18	15	0.03	0.73	0.05	0.28	0.08	28	39	0	1
0	31	0.52	16	229	32	0.35	76	0.05	1.32	0.10	0.43	0.16	84	51	0	1
0	30	0.56	12	131	10	0.26	52	0.05	1.52	0.07	0.42	0.12	52	37	0	1
0	11	0.23	6	96	7	0.20	0	0.03	0.01	0.03	0.23	0.07	18	28	0	—
0	28	0.94	16	247	16	0.26	30	0.08	0.66	0.06	0.47	0.14	47	48	0	1
0	20	0.37	14	225	12	0.19	36	0.08	0.45	0.09	0.42	0.22	78	35	0	<1
0	33	0.41	11	172	13	0.13	6	0.04	0.07	0.03	0.21	0.07	30	23	0	1
0	47	0.26	12	146	12	0.14	11	0.09	0.16	0.08	0.42	0.17	30	30	0	1
0	158	1.77	19	631	459	0.29	360	0.05	—	0.11	0.73	0.28	70	44	0	1
0	145	1.28	27	375	995	0.36	—	0.07	0.24	0.10	0.75	0.34	88	80	0	—
0	32	0.56	11	170	19	0.15	39	0.04	0.07	0.05	0.29	0.15	13	40	0	<1
0	25	0.28	20	161	20	0.19	35	0.05	0.07	0.02	0.21	0.13	56	22	0	1
0	—	—	—	—	105	—	—	—	—	—	—	—	—	—	0	—
0	20	0.18	7	195	42	0.15	367	0.04	0.28	0.04	0.60	0.08	12	4	0	<.1
0	26	0.71	8	190	62	0.14	552	0.02	—	0.03	0.44	0.08	26	7	0	1
0	18	0.17	7	177	38	0.13	333	0.04	0.25	0.03	0.54	0.08	11	3	0	<.1
0	23.39	0.26	7.8	183.3	45.24	0.15	1914.9	0.05	0.26	0.03	0.5	0.11	10.92	2.8	0	0.54
0	30	0.57	17	359	36	0.22	1176	0.11	0.01	0.07	0.47	0.27	5	10	0	1
0	16	0.28	22	279	14	0.35	1	0.09	0.20	0.05	0.88	0.09	28	21	0	1
0	11	0.22	8	152	15	0.14	1	0.03	0.02	0.03	0.26	0.11	29	23	0	<1
0	10	0.20	6	88	9	0.11	1	0.03	0.02	0.03	0.25	0.11	27	27	0	<1
0	15	0.37	8	125	16	0.12	0	0.03	0.04	0.05	0.28	0.08	37	28	0	1
0	24	0.12	7	157	48	0.08	13	0.01	0.22	0.03	0.19	0.04	22	2	0	<1
0	32	0.16	9	208	64	0.10	18	0.02	0.29	0.05	0.26	0.06	29	2	0	<1
0	18	0.65	29	136	77	0.13	110	0.01	0.68	0.03	0.14	0.04	5	11	0	<1
0	51	1.98	75	480	157	0.29	268	0.03	1.65	0.08	0.32	0.07	8	16	0	1
0	133	1.10	19	110	15	0.22	386	0.04	0.84	0.10	0.55	0.12	88	17	0	<1
0	179	0.95	26	213	43	0.23	489	0.04	0.43	0.10	0.54	0.10	65	22	0	1
0	2.12	0.6	31.81	247.59	242.45	0.47	21.87	0.21	0.07	0.07	1.6	0.05	—	6.16	0	—
0	2	0.50	26	204	14	0.39	11	0.18	0.07	0.06	1.32	0.05	38	5	0	<1
0	2	0.39	23	191	1	0.52	8	0.02	0.07	0.05	1.08	0.08	29	3	0	1
0	4	0.49	22	172	365	0.68	5	0.03	0.12	0.07	1.23	0.08	58	6	0	1
0	12	0.21	10	111	2	0.15	4	0.02	0.06	0.02	0.07	0.03	5	2	0	<1

A

Table A–1
Food Composition   (Computer code number is for Wadsworth Diet Analysis program)   (For purposes of calculations, use "0" for t, <1, <.1, <.01, etc.)

DA + Code	Food Description	Quantity	Measure	Wt (g)	H₂O (g)	Ener (cal)	Prot (g)	Carb (g)	Dietary Fiber (g)	Fat (g)	Fat Breakdown (g) Sat	Mono	Poly
	VEGETABLES, LEGUMES—Continued												
16870	**Cucumber, kim chee**	½	cup(s)	75	68	16	1	4	1	<.1	0.02	0.00	0.03
	Dandelion greens												
2734	Raw	1	cup(s)	55	47	25	1	5	2	<1	0.09	0.01	0.17
620	Chopped, boiled, drained	½	cup(s)	53	47	17	1	3	2	<1	0.08	0.01	0.14
1066	**Eggplant, boiled, drained**	½	cup(s)	48	43	17	<1	4	1	<1	0.02	0.01	0.04
621	**Endive or escarole, chopped, raw**	1	cup(s)	53	49	9	1	2	2	<1	0.03	0.00	0.05
8784	**Jicama or yambean**	½	cup(s)	65	59	25	<1	6	3	<.1	0.01	0.00	0.03
	Kale												
29313	Raw	1	cup(s)	67	57	34	2	7	1	<1	0.06	0.03	0.23
623	Frozen, chopped, boiled, drained	½	cup(s)	65	59	20	2	3	1	<1	0.04	0.02	0.15
	Kohlrabi												
1071	Raw	1	cup(s)	135	123	36	2	8	5	<1	0.02	0.01	0.06
1072	Boiled, drained	½	cup(s)	83	74	24	1	6	1	<.1	0.01	0.01	0.04
	Leeks												
1073	Raw	1	cup(s)	89	74	54	1	13	2	<1	0.04	0.00	0.15
1074	Boiled, drained	½	cup(s)	52	47	16	<1	4	1	<1	0.01	0.00	0.06
	Lentils												
522	Boiled	½	cup(s)	99	69	115	9	20	8	<1	0.05	0.06	0.17
1075	Sprouted	1	cup(s)	77	52	82	7	17	0	<1	0.04	0.08	0.17
	Lettuce												
624	Butterhead, boston, or bibb	1	cup(s)	55	53	7	1	1	1	<1	0.02	0.00	0.06
625	Butterhead leaves	11	piece(s)	83	79	11	1	2	1	<1	0.02	0.01	0.10
626	Iceberg	1	cup(s)	55	53	6	<1	1	1	<.1	0.01	0.00	0.03
628	Iceberg, chopped	1	cup(s)	55	53	6	<1	1	1	<.1	0.01	0.00	0.03
629	Looseleaf	1	cup(s)	56	54	8	1	2	1	<.1	0.01	0.00	0.05
1665	Romaine, shredded	1	cup(s)	56	53	10	1	2	1	<1	0.02	0.01	0.09
	Mushrooms												
15585	Crimini (about 6)	3	ounce(s)	85	28	4	3	2	0	0.00	0.00	0.00	0
8700	Enoki	30	item(s)	90	80	31	2	6	2	<1	0.04	0.01	0.14
630	Mushrooms, raw	½	cup(s)	35	32	8	1	1	<1	<1	0.02	0.00	0.05
1079	Mushrooms, boiled, drained	½	cup(s)	78	71	22	2	4	2	<1	0.05	0.01	0.14
1080	Mushrooms, canned, drained	½	cup(s)	78	71	20	1	4	2	<1	0.03	0.00	0.09
15587	Portobello, raw	1	item(s)	85	30	3	4	3	0	0.00	0.00	0.00	0
2743	Shiitake, cooked	½	cup(s)	73	61	40	1	10	2	<1	0.04	0.05	0.02
	Mustard greens												
29319	Raw	1	cup(s)	56	51	15	2	3	2	<1	0.01	0.05	0.02
2744	Frozen, boiled, drained	½	cup(s)	75	70	14	2	2	2	<1	0.01	0.08	0.04
	Okra												
632	Sliced, boiled, drained	½	cup(s)	80	74	18	1	4	2	<1	0.04	0.02	0.04
32742	Frozen, boiled, drained, no salt added	½	cup(s)	92	84	26	2	5	3	<1	0.07	0.05	0.07
16866	Batter coated, fried	11	piece(s)	83	55	160	2	13	2	11	1.50	2.80	6.37
	Onions												
633	Raw, chopped	½	cup(s)	80	71	34	1	8	1	<.1	0.02	0.02	0.05
635	Chopped, boiled, drained	½	cup(s)	106	93	47	1	11	1	<1	0.03	0.03	0.08
2748	Frozen, boiled, drained	½	cup(s)	106	98	30	1	7	2	<1	0.02	0.01	0.04
16850	Red onions, sliced, raw	½	cup(s)	58	52	22	1	5	1	<.1	0.02	0.01	0.04
636	Scallions, green or spring onions	2	item(s)	30	27	10	1	2	1	<.1	0.01	0.01	0.02
1081	Onion rings, breaded & pan fried, frozen, heated	11	item(s)	78	22	318	4	30	1	21	6.70	8.49	3.99
16860	**Palm hearts, cooked**	½	cup(s)	73	51	75	2	19	1	<1	0.03	0.00	0.07
637	**Parsley, chopped**	1	tablespoon(s)	4	3	1	<1	<1	<1	<.1	0.01	0.01	0.00
638	**Parsnips, sliced, boiled, drained**	½	cup(s)	78	63	55	1	13	3	<1	0.04	0.09	0.04
	Peas												
639	Green peas, canned, drained	½	cup(s)	85	69	59	4	11	3	<1	0.05	0.03	0.14
641	Green peas, frozen, boiled, drained	½	cup(s)	80	64	62	4	11	4	<1	0.04	0.02	0.10
35694	Pea pods, boiled w/salt, drained	½	cup(s)	80	71	34	3	6	2	<1	0.04	0.02	0.08
1082	Peas & carrots, canned w/liquid	½	cup(s)	128	112	48	3	11	3	<1	0.06	0.03	0.16
1083	Peas & carrots, frozen, boiled, drained	½	cup(s)	80	69	38	2	8	2	<1	0.06	0.03	0.16
640	Snow or sugar peas, raw	½	cup(s)	32	28	13	1	2	1	<.1	0.01	0.01	0.03
2750	Snow or sugar peas, frozen, boiled, drained	½	cup(s)	80	69	42	3	7	2	<1	0.06	0.03	0.13

PAGE KEY: A–4 = Breads/Baked Goods   A–8 = Cereal/Rice/Pasta   A–12 = Fruit   A–18 = Vegetables/Legumes   A–28 = Nuts/Seeds   A–30 = Vegetarian
A–32 = Dairy   A–38 = Eggs   A–40 = Seafood   A–42 = Meats   A–44 = Poultry   A–46 = Processed meats   A–48 = Beverages   A–52 = Fats/Oils
A–54 = Sweets   A–54 = Sauces/Condiments/Spices   A–58 = Mixed Foods/Soups/Sandwiches   A–64 = Fast food   A–82 = Convenience   A–84 = Baby foods

A

Chol (mg)	Calc (mg)	Iron (mg)	Magn (mg)	Pota (mg)	Sodi (mg)	Zinc (mg)	Vit A (RAE) (µg)	Thia (mg)	Vit E (mg)	Ribo (mg)	Niac (mg)	Vit B_6 (mg)	Fola (µg)	Vit C (mg)	Vit B_{12} (µg)	Sele (µg)
0	7	3.62	6	88	766	0.38	—	0.02	0.12	0.02	0.35	0.08	17	3	0	—
0	103	1.71	20	219	42	0.23	137	0.11	1.38	0.14	0.45	0.14	15	19	0	<1
0	74	0.95	13	122	23	0.15	260	0.07	1.31	0.09	0.27	0.08	7	9	0	<1
0	3	0.12	5	59	<1	0.06	1	0.04	0.01	0.01	0.29	0.04	7	1	0	<.1
0	27	0.44	8	165	12	0.41	57	0.04	0.23	0.04	0.21	0.01	75	3	0	<1
0	8	0.39	8	98	3	0.10	1	0.01	0.30	0.02	0.13	0.03	8	13	0	<1
0	90	1.14	23	299	29	0.29	515	0.07	0.54	0.09	0.67	0.18	19	80	0	1
0	90	0.61	12	209	10	0.12	478	0.03	0.12	0.07	0.44	0.06	9	16	0	1
0	32	0.54	26	473	27	0.04	3	0.07	0.65	0.03	0.54	0.20	22	84	0	1
0	21	0.33	16	281	17	0.26	2	0.03	1.38	0.02	0.32	0.13	10	45	0	1
0	53	1.87	25	160	18	0.11	74	0.05	0.82	0.03	0.36	0.21	57	11	0	1
0	16	0.57	7	45	5	0.03	1	0.01	0.00	0.01	0.10	0.06	13	2	0	<1
0	19	3.30	36	365	2	1.26	0	0.17	0.11	0.07	1.05	0.18	179	1	0	3
0	19	2.47	28	248	8	1.16	2	0.18	—	0.10	0.87	0.15	77	13	0	<1
0	19	0.69	7	132	3	0.11	92	0.03	0.24	0.03	0.20	0.05	40	2	0	<1
0	29	1.02	11	196	4	0.17	137	0.05	0.36	0.05	0.29	0.07	60	3	0	<1
0	11	0.19	4	84	5	0.09	9	0.02	0.15	0.01	0.07	0.03	31	2	0	<1
0	11	0.19	4	84	5	0.09	9	0.02	0.15	0.01	0.07	0.03	31	2	0	<1
0	20	0.48	7	109	16	0.10	208	0.04	0.25	0.05	0.21	0.05	21	10	0	<1
0	19	0.55	8	139	5	0.13	163	0.04	0.25	0.04	0.18	0.04	77	14	0	<1
0	0.67	—	—	33	—	0	—	—	—	—	—	—	—	0	0	—
0	1	0.80	14	343	3	0.51	0	0.08	—	0.09	3.28	0.04	27	11	0	14
0	1	0.18	3	110	1	0.18	0	0.03	0.04	0.15	1.35	0.04	6	1	<.1	3
0	5	1.36	9	278	2	0.68	0	0.06	0.09	0.23	3.48	0.07	14	3	0	9
0	9	0.62	12	101	332	0.56	0	0.07	0.09	0.02	1.24	0.05	9	0	0	3
40	0.36	—	—	10	—	0	—	—	—	—	—	—	—	0	0	—
0	2	0.32	10	85	3	0.96	0	0.03	0.09	0.12	1.09	0.12	15	<1	0	18
0	58	0.82	18	199	14	0.11	295	0.05	1.13	0.06	0.45	0.10	105	39	0	1
0	76	0.84	10	104	19	0.15	266	0.03	1.31	0.04	0.19	0.08	53	10	0	<1
0	62	0.22	29	108	5	0.34	11	0.11	0.55	0.04	0.70	0.15	37	13	0	<1
0	88	0.62	47	215	3	0.57	16	0.09	0.63	0.11	0.72	0.04	134	11	0	1
2	54	1.13	32	170	110	0.44	—	0.16	2.72	0.13	1.29	0.11	34	9	<.1	—
0	18	0.15	8	115	2	0.13	0	0.04	0.25	0.02	0.07	0.12	15	5	0	<1
0	23	0.26	12	177	3	0.22	0	0.04	0.14	0.02	0.18	0.14	16	6	0	1
0	17	0.32	6	115	13	0.07	0	0.02	0.20	0.03	0.15	0.07	14	3	0	<1
0	11	0.13	6	90	2	0.11	0	0.02	0.07	0.01	0.09	0.07	11	4	0	—
0	22	0.44	6	83	5	0.12	15	0.02	0.04	0.02	0.16	0.02	19	6	0	<1
0	24	1.32	15	101	293	0.33	9	0.22	—	0.11	2.82	0.06	52	1	0	3
0	13	1.23	7	1318	10	2.72	—	0.03	0.37	0.13	0.62	0.53	15	5	0	—
0	5	0.24	2	21	2	0.04	16	0.00	0.07	0.00	0.05	0.00	6	5	0	<.1
0	29	0.45	23	286	8	0.20	0	0.06	0.78	0.04	0.56	0.07	45	10	0	1
0	17	0.81	14	147	214	0.60	23	0.10	0.32	0.07	0.62	0.05	37	8	0	1
0	19	1.22	18	88	58	0.54	84	0.23	0.14	0.08	1.18	0.09	47	8	0	1
0	34	1.58	21	192	192	0.30	43	0.10	—	0.06	0.43	0.12	23	38	0	1
0	29	0.96	18	128	332	0.74	368	0.09	—	0.07	0.74	0.11	23	8	0	1
0	18	0.75	13	126	54	0.36	374	0.18	0.26	0.05	0.92	0.07	21	6	0	1
0	14	0.66	8	63	1	0.09	17	0.05	0.12	0.03	0.19	0.05	13	19	0	<1
0	47	1.92	22	174	4	0.39	53	0.05	0.17	0.10	0.45	0.14	28	18	0	1

A

Table A–1

Food Composition   (Computer code number is for Wadsworth Diet Analysis program)   (For purposes of calculations, use "0" for t, <1, <.1, <.01, etc.)

DA + Code	Food Description	Quantity	Measure	Wt (g)	H₂O (g)	Ener (cal)	Prot (g)	Carb (g)	Dietary Fiber (g)	Fat (g)	Fat Breakdown (g)		
											Sat	Mono	Poly
	VEGETABLES, LEGUMES—Continued												
29324	Split peas, sprouted	½	cup(s)	60	37	77	5	17	0	<1	0.07	0.04	0.20
	Peppers												
643	Green bell or sweet, raw	½	cup(s)	75	70	15	1	3	1	<1	0.04	0.01	0.05
644	Green bell or sweet, boiled, drained	½	cup(s)	68	62	19	1	5	1	<1	0.02	0.01	0.07
1664	Green hot chili	1	item(s)	45	39	18	1	4	1	<.1	0.01	0.00	0.05
1663	Green hot chili, canned w/liquid	½	cup(s)	68	63	14	1	3	1	<.1	0.01	0.00	0.04
1086	Jalapeno, canned w/liquid	½	cup(s)	68	60	18	1	3	2	1	0.07	0.04	0.35
8703	Yellow bell or sweet	1	item(s)	186	171	50	2	12	2	<1	0.06	0.03	0.21
1087	**Poi**	½	cup(s)	122	87	136	<1	33	<1	<1	0.04	0.01	0.07
	Potatoes												
5791	Baked, flesh & skin	1	item(s)	202	144	220	5	51	4	<1	0.05	0.00	0.09
645	Baked, flesh only	½	cup(s)	61	46	57	1	13	1	<.1	0.02	0.00	0.03
1088	Baked, skin only	1	item(s)	58	27	115	2	27	5	<.1	0.02	0.00	0.06
5794	Boiled, drained, skin & flesh	1	item(s)	150	116	129	3	30	2	<1	0.04	0.00	0.06
647	Boiled, flesh only	½	cup(s)	78	60	67	1	16	1	<.1	0.02	0.00	0.03
5795	Boiled in skin, drained, flesh only	1	item(s)	136	105	118	3	27	2	<1	0.04	0.00	0.06
2759	Microwaved	1	item(s)	202	146	212	5	49	5	<1	0.05	0.00	0.09
5804	Microwaved, skin only	1	item(s)	58	37	77	3	17	4	<.1	0.02	0.00	0.02
2760	Microwaved in skin, flesh only	½	cup(s)	78	57	78	2	18	1	<.1	0.02	0.00	0.03
1089	Au gratin, prepared w/butter	½	cup(s)	123	91	162	6	14	2	9	5.80	2.63	0.34
1090	Au gratin mix, prepared w/water, whole milk, & butter	½	cup(s)	114	90	106	3	15	1	5	2.94	1.34	0.15
648	French fried, deep fried, prepared from raw	14	item(s)	70	32	190	3	24	2	10	1.93	4.21	2.97
649	French fried, frozen, heated	14	item(s)	70	40	140	2	22	2	5	0.88	3.33	0.55
1091	Hashed brown	½	cup(s)	78	37	207	2	27	2	10	1.11	3.13	2.78
653	Mashed, from dehydrated granules w/milk, water, & margarine	½	cup(s)	105	80	122	2	17	1	5	1.27	2.05	1.41
652	Mashed, w/margarine & whole milk	½	cup(s)	105	79	119	2	18	2	4	1.05	1.83	1.27
1097	Potato puffs, frozen, heated	½	cup(s)	64	34	142	2	20	2	7	3.26	2.79	0.51
1093	Scalloped, prepared w/butter	½	cup(s)	123	99	105	4	13	2	5	2.76	1.27	0.20
1094	Scalloped mix, prepared w/water, whole milk, & butter	½	cup(s)	114	90	106	2	15	1	5	2.99	1.38	0.22
	Pumpkin												
1773	Boiled, drained	½	cup(s)	123	115	25	1	6	1	<.1	0.05	0.01	0.00
656	Canned	½	cup(s)	123	110	42	1	10	4	<1	0.18	0.05	0.02
	Radicchio										<.1			
2498	Raw	1	cup(s)	40	37	9	1	2	<1	<1	0.02	0.00	0.04
8731	Raw, leaves	10	item(s)	80	75	18	1	4	1	<1	0.05	0.01	0.09
657	**Radishes**	6	item(s)	27	26	4	<1	1	<1	<.1	0.01	0.00	0.01
1099	**Rutabaga, boiled, drained**	½	cup(s)	85	76	33	1	7	2	<1	0.02	0.02	0.08
658	**Sauerkraut, canned**	½	cup(s)	114	105	22	1	5	3	<1	0.04	0.01	0.07
	Seaweed												
1102	Kelp	½	cup(s)	41	33	17	1	4	1	<1	0.10	0.04	0.02
1104	Spirulina, dried	½	cup(s)	8	<1	22	4	2	<1	1	0.20	0.05	0.16
1106	**Shallots**	3	tablespoon(s)	30	24	22	1	5	0	<.1	0.01	0.00	0.01
	Soybeans												
1670	Boiled	½	cup(s)	86	0.05	148.77	14.31	8.53	5.15	7.71	1.11	1.7	4.35
2825	Dry roasted	½	cup(s)	86	1	388	34	28	7	19	2.69	4.11	10.50
2824	Roasted, salted	½	cup(s)	86	2	405	30	29	15	22	3.16	4.82	12.33
30282	Soup (miso)	1	cup(s)	240	218	85	6	8	2	3	0.59	1.05	1.47
8739	Sprouted, stir fried	3	ounce(s)	85	57	106	11	8	1	6	0.84	1.37	3.41
	Soy products												
1813	Soy milk	1	cup(s)	240	214	118	9	11	3	5	0.51	0.78	2.00
2838	Tofu, dried, frozen (koyadofu)	3	ounce(s)	85	5	408	41	12	6	26	3.73	5.70	14.57
13844	Tofu, extra firm	3	ounce(s)	79	—	80	8	2	1	4	0.50	0.87	2.60
13843	Tofu, firm	3	ounce(s)	79	—	80	8	2	1	4	0.50	0.87	2.17
1816	Tofu, firm, w/calcium sulfate & magnesium chloride (nigari)	3	ounce(s)	85	0.07	65.48	6.83	2.52	0.34	3.79	0.54	0.83	2.14
1817	Tofu, fried	3	ounce(s)	85	43	230	15	9	3	17	2.48	3.79	9.69
13841	Tofu, silken	3	ounce(s)	91	—	30	6	0	1	1	0.50	0.51	1.52
13842	Tofu, soft	3	ounce(s)	91	—	30	6	1	1	1	0.50	1.00	2.00

PAGE KEY: A–4 = Breads/Baked Goods   A–8 = Cereal/Rice/Pasta   A–12 = Fruit   A–18 = Vegetables/Legumes   A–28 = Nuts/Seeds   A–30 = Vegetarian   A–32 = Dairy   A–38 = Eggs   A–40 = Seafood   A–42 = Meats   A–44 = Poultry   A–46 = Processed meats   A–48 = Beverages   A–52 = Fats/Oils   A–54 = Sweets   A–54 = Sauces/Condiments/Spices   A–58 = Mixed Foods/Soups/Sandwiches   A–64 = Fast food   A–82 = Convenience   A–84 = Baby foods

A

Chol (mg)	Calc (mg)	Iron (mg)	Magn (mg)	Pota (mg)	Sodi (mg)	Zinc (mg)	Vit A (RAE) (µg)	Thia (mg)	Vit E (mg)	Ribo (mg)	Niac (mg)	Vit B_6 (mg)	Fola (µg)	Vit C (mg)	Vit B_{12} (µg)	Sele (µg)
0	22	1.36	34	229	12	0.63	5	0.14	—	0.09	1.85	0.16	86	6	0	<1
0	7	0.25	7	130	2	0.10	13	0.04	0.51	0.02	0.36	0.17	8	60	0	0
0	6	0.31	7	113	1	0.08	10	0.04	0.47	0.02	0.32	0.16	11	51	0	<1
0	8	0.54	11	153	3	0.14	27	0.04	0.31	0.04	0.43	0.13	10	109	0	<1
0	5	0.34	10	127	798	0.12	24	0.01	0.47	0.03	0.54	0.10	7	46	0	<1
0	16	1.28	10	131	1136	0.23	58	0.03	0.47	0.03	0.27	0.13	10	7	0	<1
0	20	0.86	22	394	4	0.32	19	0.05	—	0.05	1.66	0.31	48	341	0	1
0	19	1.07	29	223	15	0.27	4	0.16	0.22	0.05	1.34	0.33	26	5	0	1
0	20	2.75	55	844	16	0.65	0	0.22	0.10	0.07	3.32	0.70	22	26	0	2
0	3	0.21	15	239	3	0.18	0	0.06	0.02	0.01	0.85	0.18	5	8	0	<1
0	20	4.08	25	332	12	0.28	1	0.07	0.02	0.06	1.78	0.36	13	8	0	<1
0	13	1.27	34	572	7	0.47	0	0.15	0.08	0.03	2.13	0.44	15	18	0	—
0	6	0.24	16	256	4	0.21	0	0.08	0.04	0.01	1.02	0.21	7	6	0	<1
0	7	0.42	30	515	5	0.41	0	0.14	0.07	0.03	1.96	0.41	14	18	0	<1
0	22	2.50	55	903	16	0.73	0	0.24	—	0.06	3.46	0.69	24	31	0	1
0	27	3.45	21	377	9	0.30	0	0.04	—	0.04	1.29	0.29	10	9	0	<1
0	4	0.32	20	321	5	0.26	0	0.10	—	0.02	1.27	0.25	9	12	0	<1
28	146	0.78	25	485	530	0.85	78	0.08	—	0.14	1.22	0.21	13	12	0	3
17	94	0.36	17	249	499	0.27	59	0.02	—	0.09	1.07	0.05	8	4	0	3
0	9	1.02	28	731	8	0.53	0	0.10	—	0.05	1.90	0.33	13	21	0	—
0	6	0.87	15	293	21	0.28	0	0.08	0.13	0.02	1.46	0.22	8	7	0	<1
0	11	0.43	27	449	267	0.37	0	0.13	0.15	0.03	1.80	0.37	12	10	0	<1
2	34	0.22	21	163	181	0.25	49	0.09	—	0.09	0.91	0.17	8	7	<1	6
1	21	0.27	20	342	350	0.32	43	0.10	0.32	0.05	1.23	0.26	9	11	<.1	1
0	19	1.00	12	243	477	0.19	0	0.13	0.03	0.05	1.38	0.15	11	4	0	<1
15	70	0.70	23	463	410	0.49	39	0.08	—	0.11	1.29	0.22	13	13	0	2
13	41	0.43	16	231	388	0.28	40	0.02	0.17	0.06	1.17	0.05	11	4	0	2
0	18	0.70	11	282	1	0.28	306	0.04	1.30	0.10	0.51	0.05	11	6	0	<1
0	32	1.70	28	252	6	0.21	953	0.03	1.30	0.07	0.45	0.07	15	5	0	<1
0	8	0.23	5	121	9	0.25	<1	0.01	0.90	0.01	0.10	0.02	24	3	0	<1
0	15	0.46	10	242	18	0.50	1	0.01	1.81	0.02	0.20	0.05	48	6	0	1
0	7	0.09	3	63	11	0.08	0	0.00	0.00	0.01	0.07	0.02	7	4	0	<1
0	41	0.45	20	277	17	0.30	0	0.07	0.13	0.03	0.61	0.09	13	16	0	1
0	34	1.67	15	193	751	0.22	1	0.02	0.11	0.02	0.16	0.15	27	17	0	1
0	68	1.16	49	36	94	0.50	2	0.02	0.35	0.06	0.19	0.00	73	1	0	<1
0	9	2.14	15	102	79	0.15	2	0.18	0.38	0.28	0.96	0.03	7	1	0	1
0	11	0.36	6	100	4	0.12	18	0.02	—	0.01	0.06	0.10	10	2	0	<1
0	87.72	4.42	73.95	442.89	0.86	0.98	0.86	0.13	1.34	0.24	0.34	0.2	46.43	1.46	0	6.27
0	120	3.40	196	1173	2	4.10	0	0.37	—	0.65	0.91	0.19	176	4	0	17
0	119	3.35	125	1264	140	2.70	9	0.09	1.68	0.12	1.21	0.18	181	2	0	16
0	64	1.89	37	361	988	0.87	—	0.06	0.69	0.16	2.61	0.17	57	4	<1	—
0	70	0.34	82	482	12	1.79	1	0.36	—	0.16	0.94	0.14	108	10	0	1
0	10	1.39	46	338	29	0.55	5	0.39	0.02	0.17	0.35	0.10	5	0	0	3
0	310	8.28	50	17	5	4.17	22	0.42	—	0.27	1.01	0.24	78	1	0	46
0	60	1.08	78	—	0	—	0	—	0.00	—	—	—	—	0	0	—
0	60	1.08	52	—	0	—	0	—	0.00	—	—	—	—	0	0	—
0	137.78	1.23	39.12	149.68	6.8	0.85	0.85	0.07	—	0.08	0	0.05	28.06	0.17	0	7.99
0	316	4.14	51	124	14	1.69	1	0.14	0.03	0.04	0.09	0.08	23	0	0	24
0	300	0.73	35	—	65	—	0	—	6.04	—	—	—	—	0	2	—
0	300	0.72	33	—	65	—	0	—	6.04	—	—	—	—	0	2	—

A

Table A–1
Food Composition   (Computer code number is for Wadsworth Diet Analysis program)   (For purposes of calculations, use "0" for t, <1, <.1, <.01, etc.)

DA + Code	Food Description	Quantity	Measure	Wt (g)	H₂O (g)	Ener (cal)	Prot (g)	Carb (g)	Dietary Fiber (g)	Fat (g)	Fat Breakdown (g)		
											Sat	Mono	Poly
	VEGETABLES, LEGUMES—Continued												
1671	Tofu, soft, w/calcium sulfate & magnesium chloride (nigari)	3	ounce(s)	85	0.07	51.88	5.57	1.53	0.17	3.13	0.45	0.69	1.76
	Spinach												
659	Raw, chopped	1	cup(s)	30	27	7	1	1	1	<1	0.02	0.00	0.05
663	Canned, drained	½	cup(s)	108	100	25	3	4	3	1	0.09	0.02	0.23
660	Chopped, boiled, drained	½	cup(s)	90	82	21	3	3	2	<1	0.04	0.01	0.10
661	Chopped, frozen, boiled, drained	½	cup(s)	95	84	30	4	5	4	<1	0.09	0.00	0.20
662	Leaf, frozen, boiled, drained	½	cup(s)	95	84	30	4	5	4	<1	0.09	0.00	0.20
8470	Trimmed leaves	1	cup(s)	32	27	3	1	<.1	3	<.1	—	—	—
	Squash												
1662	Acorn, baked	½	cup(s)	103	85	57	1	15	5	<1	0.03	0.01	0.06
29702	Acorn, boiled, mashed	½	cup(s)	123	110	42	1	11	3	<.1	0.02	0.01	0.04
1661	Butternut, baked	½	cup(s)	103	90	41	1	11	3	<.1	0.02	0.01	0.04
29451	Butternut, frozen, boiled	½	cup(s)	132	116	51	2	13	2	<.1	0.02	0.01	0.04
32773	Butternut, frozen, boiled, mashed, no salt added	½	cup(s)	122	<1	47	1	12	0	<.1	0.02	0.00	0.03
29700	Crookneck & straightneck, boiled, drained	½	cup(s)	90	0.08	18	0.81	3.87	1.25	0.27	0.05	0.02	0.11
29703	Hubbard, baked	v	cup(s)	103	87	51	3	11	0	1	0.13	0.05	0.27
1660	Hubbard, boiled, mashed	½	cup(s)	118	107	35	2	8	3	<1	0.09	0.03	0.18
29704	Spaghetti, boiled, drained, or baked	½	cup(s)	78	72	21	1	5	1	<1	0.05	0.02	0.10
664	Summer, all varieties, sliced, boiled, drained	½	cup(s)	90	84	18	1	4	1	<1	0.06	0.02	0.12
665	Winter, all varieties, baked, mashed	½	cup(s)	103	91	38	1	9	3	<1	0.13	0.05	0.27
1112	Zucchini, boiled, drained	½	cup(s)	90	85	14	1	4	1	<.1	0.01	0.00	0.02
1113	Zucchini, frozen, boiled, drained	½	cup(s)	113	107	19	1	4	1	<1	0.03	0.01	0.06
	Sweet potatoes												
666	Baked, peeled	½	cup(s)	100	76	90	2	21	3	<1	0.03	0.00	0.06
667	Boiled, mashed	½	cup(s)	166	133	126	2	29	4	<1	0.05	0.00	0.10
668	Candied, home recipe	½	cup(s)	84	56	115	1	23	2	3	1.13	0.53	0.12
670	Canned, vacuum pack	½	cup(s)	100	76	91	2	21	2	<1	0.04	0.01	0.09
2765	Frozen, baked	½	cup(s)	88	65	88	2	21	2	<1	0.02	0.00	0.05
1136	Yams, baked or boiled, drained	½	cup(s)	68	48	79	1	19	3	<.1	0.02	0.00	0.04
32785	**Taro shoots, cooked, no salt added**	½	cup(s)	70	67	10	1	2	0	<.1	0.01	0.00	0.02
	Tomatillo												
8774	Raw	2	item(s)	68	62	22	1	4	1	1	0.09	0.11	0.28
8777	Raw, chopped	½	cup(s)	66	60	21	1	4	1	1	0.09	0.10	0.28
	Tomato												
671	Fresh, ripe, red	1	item(s)	123	0.11	22.13	1.08	4.82	1.47	0.24	0.05	0.06	0.16
16846	Fresh, cherry	5	item(s)	85	0.07	17.85	0.72	3.94	0.93	0.28	0.03	0.04	0.11
3952	Diced, red	½	cup(s)	90	85	16	1	4	1	<1	0.04	0.05	0.12
1118	Boiled, red	½	cup(s)	120	113	22	1	5	1	<1	0.02	0.02	0.05
675	Juice, canned	½	cup(s)	122	115	21	1	5	<1	<.1	0.01	0.01	0.03
75	Juice, no salt added	½	cup(s)	122	115	21	1	5	<1	<.1	0.01	0.01	0.03
1699	Paste, canned	2	tablespoon(s)	33	24	27	1	6	1	<1	0.04	0.03	0.07
1700	Puree, canned	¼	cup(s)	63	55	24	1	6	1	<1	0.02	0.02	0.05
1125	Sauce, canned	¼	cup(s)	61	55	20	1	5	1	<1	0.02	0.02	0.06
1120	Stewed, canned, red	½	cup(s)	128	117	33	1	8	1	<1	0.03	0.04	0.10
8778	Sun dried	½	cup(s)	27	4	70	4	15	3	1	0.12	0.13	0.30
8783	Sun dried in oil, drained	¼	cup(s)	28	15	59	1	6	2	4	0.52	2.38	0.57
	Turnips												
677	Turnips, cubed, boiled, drained	½	cup(s)	78	73	17	1	4	2	<.1	0.01	0.00	0.03
678	Turnip greens, chopped, boiled, drained	½	cup(s)	72	67	14	1	3	3	<1	0.04	0.01	0.07
679	Turnip greens, frozen, chopped, boiled, drained	½	cup(s)	82	74	24	3	4	3	<1	0.08	0.02	0.14
	Vegetables, mixed												
1132	Canned, drained	½	cup(s)	82	71	40	2	8	2	<1	0.04	0.01	0.10
680	Frozen, boiled, drained	½	cup(s)	91	76	59	3	12	4	<1	0.03	0.01	0.07
7489	Vegetable juice, V8 100%	½	cup(s)	120	113	25	1	5	1	0	0.00	0.00	0.00
7490	Vegetable juice, V8 low sodium	½	cup(s)	120	113	25	0	7	1	0	0.00	0.00	0.00
7491	Vegetable juice, V8 spicy hot	½	cup(s)	120	113	25	1	5	1	0	0.00	0.00	0.00

PAGE KEY: A–4 = Breads/Baked Goods   A–8 = Cereal/Rice/Pasta   A–12 = Fruit   A–18 = Vegetables/Legumes   A–28 = Nuts/Seeds   A–30 = Vegetarian   A–32 = Dairy   A–38 = Eggs   A–40 = Seafood   A–42 = Meats   A–44 = Poultry   A–46 = Processed meats   A–48 = Beverages   A–52 = Fats/Oils   A–54 = Sweets   A–54 = Sauces/Condiments/Spices   A–58 = Mixed Foods/Soups/Sandwiches   A–64 = Fast food   A–82 = Convenience   A–84 = Baby foods

A

Chol (mg)	Calc (mg)	Iron (mg)	Magn (mg)	Pota (mg)	Sodi (mg)	Zinc (mg)	Vit A (RAE) (µg)	Thia (mg)	Vit E (mg)	Ribo (mg)	Niac (mg)	Vit B6 (mg)	Fola (µg)	Vit C (mg)	Vit B12 (µg)	Sele (µg)
0	94.4	0.94	22.96	102.05	6.8	0.54	0.85	0.03	0	0.03	0.45	0.04	37.42	0.17	0	7.56
0	30	0.81	24	167	24	0.16	141	0.02	0.57	0.06	0.22	0.06	58	8	0	<1
0	138	2.49	82	375	29	0.50	531	0.02	1.41	0.15	0.42	0.11	106	16	0	2
0	122	3.21	78	419	63	0.68	472	0.09	0.86	0.21	0.44	0.22	131	9	0	1
0	145	1.86	78	287	92	0.47	573	0.07	0.91	0.17	0.42	0.13	115	2	0	5
0	145	1.86	78	287	92	0.47	573	0.07	0.91	0.17	0.42	0.13	115	2	0	5
0	25	2.13	25	134	38	0.18	—	0.03		0.06	0.18	0.07	<.1	8	0	—
0	45	0.95	44	448	4	0.17	22	0.17	—	0.01	0.90	0.20	19	11	0	1
0	32	0.69	32	322	4	0.13	50	0.12	—	0.01	0.65	0.14	13	8	0	<1
0	42	0.62	30	291	4	0.13	572	0.07	—	0.02	0.99	0.13	19	15	0	1
0	25	0.77	12	176	3	0.16	—	0.07	—	0.05	0.61	0.09	22	5	0	1
0	23	0.70	11	162	2	0.14	406	0.05		0.05	0.56	0.08	19	4	0	1
0	18.2	0.41	18.2	183.73	1.73	0.25	29.46	0.04	—	0.03	0.39	0.09	19.93	7.28	0	0.17
0	17	0.48	23	367	8	0.15	310	0.08	—	0.05	0.57	0.18	16	10	0	1
0	12	0.33	15	253	6	0.12	236	0.05	0.14	0.03	0.39	0.12	12	8	0	<1
0	16	0.26	9	91	14	0.16	5	0.03	0.09	0.02	0.63	0.08	6	3	0	<1
0	24	0.32	22	173	1	0.35	10	0.04	0.11	0.04	0.46	0.06	18	5	0	<1
0	23	0.45	13	448	1	0.23	268	0.02	0.12	0.07	0.51	0.17	21	10	0	<1
0	12	0.32	20	228	3	0.16	50	0.04	0.11	0.04	0.39	0.07	15	4	0	<1
0	19	0.54	15	219	2	0.23	11	0.05	0.34	0.05	0.44	0.05	9	4	0	<1
0	38	0.69	27	475	36	0.32	961	1.45	0.28	0.11	1.49	0.29	6	20	0	<1
0	45	1.20	30	382	45	0.33	1310	0.09	0.47	0.08	0.89	0.27	10	21	0	<1
7	22	0.95	9	159	59	0.13	176	0.02	0.00	0.04	0.33	0.03	9	6	0	1
0	22	0.89	22	312	53	0.18	399	0.04	0.25	0.06	0.74	0.19	17	26	0	1
0	31	0.48	18	332	7	0.26	722	0.06	0.24	0.05	0.49	0.16	19	8	0	1
0	10	0.36	12	458	5	0.14	4	0.06	0.11	0.02	0.38	0.16	11	8	0	<1
0	10	0.29	6	241	1	0.38	2	0.03	—	0.04	0.57	0.08	2	13	0	1
0	5	0.42	14	182	1	0.15	4	0.03	0.26	0.02	1.26	0.04	5	8	0	<1
0	5	0.41	13	177	1	0.15	4	0.03	0.25	0.02	1.22	0.04	5	8	0	<1
0	12.3	0.33	13.52	291.51	6.15	0.2	76.26	0.04	0.37	0.02	0.73	0.09	18.45	15.62	0	0
0	4.25	0.37	9.35	188.69	7.65	0.07	52.7	0.05	0.25	0.03	0.53	0.07	—	16.23	0	—
0	9	0.24	10	213	5	0.15	38	0.03	0.34	0.02	0.53	0.07	14	11	0	0
0	13	0.82	11	262	13	0.17	29	0.04	0.46	0.03	0.64	0.09	16	27	0	1
0	12	0.52	13	279	328	0.18	28	0.06	1.11	0.04	0.82	0.14	24	22	0	<1
0	12	0.52	13	279	12	0.18	28	0.06	1.11	0.04	0.82	0.14	24	22	0	<1
0	12	0.98	14	333	259	0.21	25	0.02	1.41	0.05	1.01	0.07	4	7	0	2
0	11	1.11	14	274	249	0.23	16	0.02	1.58	0.05	0.92	0.08	7	7	0	3
0	8	0.62	10	203	321	0.12	10	0.01	0.86	0.04	0.60	0.06	6	4	0	<1
0	43	1.70	15	264	282	0.22	11	0.06	0.48	0.05	0.91	0.02	6	10	0	1
0	30	2.45	52	925	566	0.54	12	0.14	0.00	0.13	2.44	0.09	18	11	0	1
0	13	0.74	22	430	73	0.21	18	0.05	—	0.11	1.00	0.09	6	28	0	1
0	26	0.14	7	138	12	0.09	0	0.02	0.02	0.02	0.23	0.05	7	9	0	<1
0	99	0.58	16	146	21	0.10	274	0.03	1.24	0.05	0.30	0.13	85	20	0	1
0	125	1.59	21	184	12	0.34	441	0.04	2.39	0.06	0.38	0.05	32	18	0	1
0	22	0.86	13	237	121	0.33	474	0.04	0.49	0.04	0.47	0.06	20	4	0	<1
0	23	0.75	20	154	32	0.45	195	0.06	0.33	0.11	0.77	0.07	17	3	0	<1
0	20	0.54	13	270	310	0.24	50	0.05	0.38	0.03	0.87	0.17	—	30	0	—
0	20	0.36	—	420	70	—	63	0.02	—	0.02	0.75	—	—	30	0	—
0	20	0.36	13	255	370	0.24	50	0.05	0.38	0.03	0.88	0.17	—	18	0	—

A

DA + Code	Food Description	Quantity	Measure	Wt (g)	H₂O (g)	Ener (cal)	Prot (g)	Carb (g)	Dietary Fiber (g)	Fat (g)	Fat Breakdown (g)		
											Sat	Mono	Poly
	VEGETABLES, LEGUMES—Continued												
	Water chestnuts												
31073	Sliced, drained	½	cup(s)	75	70	20	<1	5	1	0	0.00	0.00	0.00
31087	Whole	½	cup(s)	75	70	20	<1	5	1	0	0.00	0.00	0.00
1135	**Watercress**	1	cup(s)	34	32	4	1	<1	<1	<.1	0.01	0.00	0.01
	NUTS, SEEDS, AND PRODUCTS												
	Almonds												
32886	Blanched	¼	cup(s)	36	2	211	8	7	4	18	1.41	11.70	4.37
32887	Dry roasted, no salt added	¼	cup(s)	35	1	206	8	7	4	18	1.40	11.61	4.36
29724	Dry roasted, salted	¼	cup(s)	35	1	206	8	7	4	18	1.40	11.61	4.36
29725	Oil roasted, salted	¼	cup(s)	39	1	238	8	7	4	22	1.65	13.66	5.31
508	Slivered	¼	cup(s)	34	2	195	7	7	4	17	1.31	10.85	4.12
1137	Almond butter, no salt added	1	tablespoon(s)	16	<1	101	2	3	1	9	0.90	6.14	1.98
32940	Almond butter, salt added	1	tablespoon(s)	16	<1	101	2	3	1	9	0.90	6.14	1.98
1138	**Beechnuts, dried**	¼	cup(s)	57	4	327	4	19	5	28	3.25	12.43	11.41
517	**Brazil nuts, unblanched, dried**	¼	cup(s)	35	1	230	5	4	3	23	5.30	8.59	7.20
1166	**Breadfruit seeds, roasted**	¼	cup(s)	57	28	118	4	23	3	2	0.41	0.20	0.82
1139	**Butternuts, dried**	¼	cup(s)	30	1	184	7	4	1	17	0.39	3.13	12.82
	Cashews												
1140	Dry roasted	¼	cup(s)	34	1	197	5	11	1	16	3.14	9.36	2.68
518	Oil roasted	¼	cup(s)	33	1	189	5	10	1	16	2.76	8.42	2.78
32889	Cashew butter, no salt added	1	tablespoon(s)	16	<1	94	3	4	<1	8	1.56	4.66	1.34
32931	Cashew butter, salt added	1	tablespoon(s)	16	<1	94	3	4	<1	8	1.56	4.66	1.34
	Coconut												
32896	Dried, not sweetened	¼	cup(s)	60	2	393	4	14	10	38	34.06	1.63	0.42
1153	Dried, shredded, sweetened	¼	cup(s)	24	3	122	1	12	1	9	7.68	0.37	0.09
520	Shredded	¼	cup(s)	21	10	75	1	3	2	7	6.27	0.30	0.08
	Chestnuts												
1152	Chinese, roasted	¼	cup(s)	57	23	136	3	30	0	1	0.10	0.35	0.17
32895	European, boiled & steamed	¼	cup(s)	57	39	74	1	16	0	1	0.15	0.27	0.31
32911	European, roasted	¼	cup(s)	57	23	139	2	30	3	1	0.23	0.43	0.49
32922	Japanese, boiled & steamed	¼	cup(s)	57	49	32	<1	7	0	<1	0.02	0.06	0.03
32923	Japanese, roasted	¼	cup(s)	57	28	114	2	26	0	<1	0.07	0.24	0.12
4958	**Flaxseeds or linseeds**	¼	cup(s)	57	5	276	11	19	16	19	1.79	3.85	12.54
32904	**Ginkgo nuts, dried**	¼	cup(s)	57	7	197	6	41	0	1	0.22	0.42	0.42
	Hazelnuts or filberts												
32901	Blanched	¼	cup(s)	57	3	357	8	10	6	35	2.65	27.32	3.15
32902	Dry roasted, no salt added	¼	cup(s)	57	1	366	9	10	5	35	2.56	26.43	4.80
1156	**Hickorynuts, dried**	¼	cup(s)	30	1	197	4	5	2	19	2.11	9.78	6.57
	Macadamias												
1157	Raw	¼	cup(s)	34	<1	241	3	5	3	25	4.04	19.72	0.50
32905	Dry roasted, no salt added	¼	cup(s)	34	1	241	3	4	3	25	4.00	19.86	0.50
32932	Dry roasted, salt added	¼	cup(s)	34	1	240	3	4	3	25	4.00	19.86	0.50
	Mixed nuts												
1159	With peanuts, dry roasted	¼	cup(s)	34	1	203	6	9	3	18	2.36	10.75	3.69
32933	With peanuts, dry roasted, salt added	¼	cup(s)	34	1	203	6	9	3	18	2.36	10.75	3.69
32906	Without peanuts, oil roasted, no salt added	¼	cup(s)	36	1	221	6	8	2	20	3.27	11.93	4.12
	Peanuts												
2807	Dry roasted	¼	cup(s)	37	0	214	9	8	3	18	2.51	8.99	5.72
2806	Dry roasted, salted	¼	cup(s)	37	0	214	9	8	3	18	2.51	8.99	5.72
1763	Oil roasted, salted	¼	cup(s)	36	0	216	10	5	3	19	3.12	9.33	5.49
2804	Raw	¼	cup(s)	37	2	207	9	6	3	18	2.49	8.92	5.68
1884	Peanut butter, chunky	1	tablespoon(s)	16	<1	94	4	3	1	8	1.53	3.77	2.27
30303	Peanut butter, low sodium	1	tablespoon(s)	16	<1	95	4	3	1	8	1.66	3.88	2.21
30305	Peanut butter, reduced fat	1	tablespoon(s)	18	<1	94	5	6	1	6	1.33	2.91	1.85
524	Peanut butter, smooth	1	tablespoon(s)	16	<1	96	4	3	1	8	1.60	3.96	2.38
	Pecans												
32907	Dry roasted, no salt added	¼	cup(s)	57	1	403	5	8	5	42	3.56	24.92	11.66
32936	Dry roasted, salt added	¼	cup(s)	57	1	403	5	8	5	42	3.56	24.92	11.66
1162	Halves, oil roasted	¼	cup(s)	28	<1	197	3	4	3	21	1.99	11.27	6.49

PAGE KEY: A–4 = Breads/Baked Goods   A–8 = Cereal/Rice/Pasta   A–12 = Fruit   A–18 = Vegetables/Legumes   A–28 = Nuts/Seeds   A–30 = Vegetarian   A–32 = Dairy   A–38 = Eggs   A–40 = Seafood   A–42 = Meats   A–44 = Poultry   A–46 = Processed meats   A–48 = Beverages   A–52 = Fats/Oils   A–54 = Sweets   A–54 = Sauces/Condiments/Spices   A–58 = Mixed Foods/Soups/Sandwiches   A–64 = Fast food   A–82 = Convenience   A–84 = Baby foods

A

Chol (mg)	Calc (mg)	Iron (mg)	Magn (mg)	Pota (mg)	Sodi (mg)	Zinc (mg)	Vit A (RAE) (µg)	Thia (mg)	Vit E (mg)	Ribo (mg)	Niac (mg)	Vit B_6 (mg)	Fola (µg)	Vit C (mg)	Vit B_{12} (µg)	Sele (µg)
0	7	0.23	—	—	6	—	0	—	—	—	—	—	—	2	—	—
0	7	0.23	—	—	6	—	0	—	—	—	—	—	—	2	—	—
0	41	0.07	7	112	14	0.04	80	0.03	0.34	0.04	0.07	0.04	3	15	0	<1
0	78	1.35	100	249	10	1.13	0	0.07	9.06	0.20	1.33	0.04	11	0	0	1
0	92	1.56	99	257	<1	1.22	0	0.03	9.07	0.30	1.33	0.04	11	0	0	1
0	92	1.56	99	257	117	1.22	0	0.03	9.07	0.30	1.33	0.04	11	0	0	1
0	114	1.44	108	274	133	1.20	0	0.04	10.31	0.31	1.44	0.05	11	0	0	1
0	84	1.45	93	246	<1	1.13	0	0.08	8.84	0.27	1.32	0.04	10	0	0	1
0	43	0.59	48	121	2	0.49	0	0.02	3.25	0.10	0.46	0.01	10	<1	0	—
0	43	0.59	48	121	72	0.49	0	0.02	3.24	0.10	0.46	0.01	10	<1	0	1
0	1	1.40	0	578	22	0.20	0	0.17	—	0.21	0.50	0.39	64	9	0	4
0	56	0.85	132	231	1	1.42	0	0.22	2.66	0.01	0.10	0.04	8	<1	0	671
0	49	0.51	35	615	16	0.59	9	0.23	—	0.14	4.20	0.24	34	4	0	8
0	16	1.21	71	126	<1	0.94	2	0.11	1.05	0.04	0.31	0.17	20	1	0	5
0	15	2.06	89	194	5	1.92	0	0.07	0.20	0.07	0.48	0.09	24	0	0	4
0	14	1.97	89	205	4	1.74	0	0.12	0.48	0.07	0.56	0.10	8	<.1	0	7
0	7	0.80	41	87	2	0.83	0	0.05	0.25	0.03	0.26	0.04	11	0	0	2
0	7	0.80	41	87	98	0.83	0	0.05	0.25	0.03	0.26	0.04	11	0	0	2
0	15	1.98	54	323	22	1.20	0	0.04	0.80	0.06	0.36	0.18	5	1	0	11
0	4	0.47	12	82	64	0.44	0	0.01	0.33	0.00	0.12	0.07	2	<1	0	4
0	3	0.51	7	75	4	0.23	0	0.01	0.15	0.00	0.11	0.01	5	1	0	2
0	11	0.85	51	271	2	0.53	0	0.09	—	0.05	0.85	0.25	41	22	0	4
0	26	0.98	31	405	15	0.14	1	0.08	0.00	0.06	0.41	0.13	22	15	0	—
0	16	0.52	19	336	1	0.32	1	0.14	0.68	0.10	0.76	0.28	40	15	0	1
0	6	0.30	10	67	3	0.23	1	0.07	—	0.03	0.31	0.06	10	5	0	—
0	20	1.19	36	242	11	0.81	2	0.26	—	—	0.40	0.24	33	16	0	—
0	111	3.48	203	381	19	2.34	0	0.10	2.80	0.09	0.78	0.52	156	1	0	3
0	11	0.91	30	566	7	0.38	31	0.24	—	0.10	6.65	0.36	60	17	0	0
0	84	1.87	91	373	0	1.25	1	0.27	10.15	0.06	0.88	0.33	44	1	0	2
0	70	2.48	98	428	0	1.42	2	0.19	8.75	0.07	1.16	0.35	50	2	0	2
0	18	0.64	52	131	<1	1.29	2	0.26	1.56	0.04	0.27	0.06	12	1	0	2
0	28	1.24	44	123	2	0.44	0	0.40	0.18	0.05	0.83	0.09	4	<1	0	1
0	23	0.89	40	122	1	0.43	0	0.24	0.19	0.03	0.76	0.12	3	<1	0	1
0	23	0.89	40	122	89	0.43	0	0.24	0.19	0.03	0.76	0.12	3	<1	0	4
0	24	1.27	77	204	4	1.30	<1	0.07	2.06	0.07	1.61	0.10	17	<1	0	1
0	24	1.27	77	204	229	1.30	0	0.07	2.06	0.07	1.61	0.10	17	<1	0	3
0	38	0.93	90	196	4	1.68	<1	0.18	2.16	0.17	0.71	0.06	20	<1	0	—
0	20	0.82	64	240	2	1.20	0	0.15	2.27	0.03	4.93	0.09	53	0	0	3
0	20	0.82	64	240	297	1.20	0	0.15	2.16	0.03	4.93	0.09	53	0	0	3
0	22	0.54	63	261	115	1.18	0	0.03	2.13	0.03	4.97	0.16	43	<1	0	1
0	34	1.67	61	257	7	1.19	0	0.23	3.33	0.05	4.40	0.13	88	0	0	3
0	8	0.33	31	101	75	0.52	0	0.02	—	0.02	2.19	0.07	15	0	0	1
0	6	0.29	25	107	3	0.47	0	0.01	1.60	0.02	2.14	0.07	12	0	0	—
0	6	0.34	31	120	97	0.50	0	0.05	1.20	0.01	2.63	0.06	11	0	0	—
0	8	0.30	28	88	80	0.47	0	0.01	1.60	0.02	2.14	0.07	12	0	0	1
0	41	1.59	75	240	1	2.87	4	0.26	2.13	0.06	0.66	0.11	9	<1	0	2
0	41	1.59	75	240	217	2.87	4	0.26	2.13	0.06	0.66	0.11	9	<1	0	2
0	18	0.68	33	108	<1	1.23	1	0.13	1.42	0.03	0.33	0.05	4	<1	0	2

Table A–1

Food Composition   (Computer code number is for Wadsworth Diet Analysis program)   (For purposes of calculations, use "0" for t, <1, <.1, <.01, etc.)

DA + Code	Food Description	Quantity	Measure	Wt (g)	H₂O (g)	Ener (cal)	Prot (g)	Carb (g)	Dietary Fiber (g)	Fat (g)	Fat Breakdown (g) Sat	Mono	Poly
	NUTS, SEEDS, AND PRODUCTS—Continued												
526	Raw	¼	cup(s)	27	1	187	2	4	3	19	1.67	11.02	5.84
12973	**Pine nuts or pignolia, dried**	1	tablespoon(s)	9	<1	58	1	1	<1	6	0.42	1.61	2.93
	Pistachios												
1164	Dry roasted	¼	cup(s)	32	1	183	7	9	3	15	1.78	7.75	4.45
32938	Dry roasted, salt added	¼	cup(s)	32	1	182	7	9	3	15	1.78	7.75	4.45
1167	**Pumpkin or squash seeds, roasted**	¼	cup(s)	57	4	296	19	8	2	24	4.52	7.43	10.90
	Sesame												
1169	Sesame seeds, whole, roasted, toasted	3	teaspoon(s)	9	<1	51	2	2	1	4	0.60	1.63	1.89
32912	Sesame butter paste	1	tablespoon(s)	16	<1	95	3	4	1	8	1.14	3.07	3.57
32941	Tahini or sesame butter	1	tablespoon(s)	15	<1	89	3	3	1	8	1.11	3.00	3.48
	Soy nuts												
34173	Deep sea salted	¼	cup(s)	56	—	240	24	18	10	8	2.00	—	—
34174	Unsalted	¼	cup(s)	56	—	240	24	18	10	8	2.00	—	—
	Sunflower seeds												
528	Kernels, dried	¼	cup(s)	36	2	205	8	7	4	18	1.87	3.41	11.78
29721	Kernels, dry roasted, salted	¼	cup(s)	32	<1	186	6	8	3	16	1.67	3.04	10.52
29723	Kernels, toasted, salted	¼	cup(s)	34	<1	207	6	7	4	19	1.99	3.63	12.56
32928	Sunflower seed butter, salt added	1	tablespoon(s)	16	<1	93	3	4	0	8	0.80	1.46	5.04
	Trail mix												
4646	Trail mix	¼	cup(s)	38	3	173	5	17	2	11	2.08	4.70	3.62
4647	Trail mix with chocolate chips	¼	cup(s)	38	2	182	5	17	0	12	2.29	5.08	4.23
4648	Tropical trail mix	¼	cup(s)	35	3	142	2	23	0	6	2.97	0.87	1.81
	Walnuts												
529	Dried black, chopped	¼	cup(s)	31	1	193	8	3	2	18	1.05	4.69	10.96
531	English or persian	¼	cup(s)	30	1	196	5	4	2	20	1.84	2.68	14.15
	VEGETARIAN FOODS												
	Prepared												
34222	Brown rice & tofu stir-fry (vegan)	8	ounce(s)	227	183	228	12	13	3	16	1.25	4.03	9.54
34368	Cheese enchilada casserole (lacto)	8	ounce(s)	227	86	410	18	41	4	19	10.06	6.54	1.24
34247	Five bean casserole (vegan)	8	ounce(s)	228	178	178	6	26	5	6	1.11	2.49	1.96
34261	Lentil stew (vegan)	8	ounce(s)	228	152	125	8	24	7	<1	0.08	0.07	0.21
34397	Macaroni & cheese (lacto)	8	ounce(s)	226	163	181	8	17	<1	9	4.37	2.88	0.89
34238	Steamed rice & vegetables (vegan)	8	ounce(s)	228	100	265	5	40	3	10	1.84	3.91	4.07
34308	Tofu rice burgers (ovo-lacto)	1	piece(s)	218	78	435	22	68	6	8	1.69	2.39	3.52
34276	Vegan spinach enchiladas (vegan)	1	piece(s)	82	59	93	5	15	2	2	0.34	0.55	1.27
34243	Vegetable chow mein (vegan)	8	ounce(s)	227	163	166	6	22	2	6	0.65	2.66	2.47
34454	Vegetable lasagna (lacto)	8	ounce(s)	225	154	177	12	25	2	4	1.92	0.93	0.34
34339	Vegetable marinara (vegan)	8	ounce(s)	229	182	94	3	15	1	3	0.36	1.32	0.92
34356	Vegetable rice casserole (lacto)	8	ounce(s)	227	172	230	9	24	4	12	4.67	3.48	2.96
34311	Vegetable strudel (ovo-lacto)	8	ounce(s)	227	100	756	19	51	4	54	18.24	26.38	6.17
34371	Vegetable taco (lacto)	1	item(s)	227	147	365	13	43	9	17	6.45	5.81	4.02
34282	Vegetarian chili (vegan)	8	ounce(s)	227	196	116	6	21	7	2	0.24	0.29	0.74
34367	Vegetarian vegetable soup (vegan)	8	ounce(s)	226	204	92	3	14	2	4	0.77	1.67	1.30
	Boca burger												
32067	All American flamed grilled patty	1	item(s)	71	—	110	14	6	4	4	1.00	—	—
32070	Bigger chef max's favorite	1	item(s)	99	—	130	18	11	5	4	1.00	1.00	1.50
32069	Bigger vegan	1	item(s)	99	—	120	18	11	6	0	0.00	0.00	0.00
32074	Boca chik'n nuggets	4	item(s)	87	—	190	16	16	2	7	2.00	—	—
32075	Boca meatless ground burger	½	cup(s)	57	—	70	11	7	4	1	0.00	—	—
32073	Boca tenders	1	item(s)	85	—	140	20	9	3	3	0.00	2.00	1.00
32072	Breakfast links	2	item(s)	45	—	100	10	6	5	4	0.00	—	—
32071	Breakfast patties	1	item(s)	38	—	80	8	5	3	4	0.00	—	—
32068	Roasted garlic patty	1	item(s)	71	—	100	14	7	5	2	0.50	—	—
32066	Vegan original patty	1	item(s)	71	—	90	13	4	0	1	0.00	—	—
	Gardenburger												
37810	Bbq chik'n with sauce	1	item(s)	142	—	250	14	30	5	8	1.00	—	—
39661	Black bean burger	1	item(s)	71	—	80	8	11	4	2	0.00	—	—
39666	Buffalo chick'n wing	3	item(s)	95	—	180	9	8	5	12	1.50	—	—
37808	Chik'n grill	1	item(s)	71	—	100	13	5	3	3	0.00	—	—

PAGE KEY: A–4 = Breads/Baked Goods   A–8 = Cereal/Rice/Pasta   A–12 = Fruit   A–18 = Vegetables/Legumes   A–28 = Nuts/Seeds   A–30 = Vegetarian
A–32 = Dairy   A–38 = Eggs   A–40 = Seafood   A–42 = Meats   A–44 = Poultry   A–46 = Processed meats   A–48 = Beverages   A–52 = Fats/Oils
A–54 = Sweets   A–54 = Sauces/Condiments/Spices   A–58 = Mixed Foods/Soups/Sandwiches   A–64 = Fast food   A–82 = Convenience   A–84 = Baby foods

A

Chol (mg)	Calc (mg)	Iron (mg)	Magn (mg)	Pota (mg)	Sodi (mg)	Zinc (mg)	Vit A (RAE) (µg)	Thia (mg)	Vit E (mg)	Ribo (mg)	Niac (mg)	Vit B_6 (mg)	Fola (µg)	Vit C (mg)	Vit B_{12} (µg)	Sele (µg)
0	19	0.68	33	111	0	1.22	1	0.18	1.09	0.04	0.32	0.06	6	<1	0	1
0	1	0.48	22	51	<1	0.55	<.1	0.03	0.30	0.02	0.38	0.01	6	<.1	0	<.1
					<1											
0	35	1.34	38	333	<1	0.74	4	0.27	1.36	0.05	0.46	0.41	16	1	0	3
0	35	1.34	38	333	<1	0.74	4	0.27	1.36	0.05	0.46	0.41	16	1	0	3
0	24	8.48	303	457	10	4.22	11	0.12	0.57	0.18	0.99	0.05	32	1	0	3
0	89	1.33	32	43	1	0.64	0	0.07	—	0.02	0.41	0.07	9	0	0	1
0	154	3.07	58	93	2	1.17	<1	0.04	0.36	0.03	1.07	0.13	16	0	0	1
0	21	0.66	14	69	5	0.69	<1	0.24	0.34	0.02	0.85	0.02	15	1	0	<1
0	120	2.16	—	—	300	—	0	—	—	—	—	—	—	0	—	—
0	120	2.16	—	—	20	—	0	—	—	—	—	—	—	0	—	—
0	42	2.44	127	248	1	1.82	1	0.82	18.10	0.09	1.62	0.28	82	1	0	21
0	22	1.22	41	272	250	1.69	<1	0.03	16.09	0.08	2.25	0.26	76	<1	0	25
0	19	2.28	43	164	205	1.78	0	0.11	—	0.10	1.41	0.27	80	<1	0	21
0	20	0.76	59	12	83	0.85	<1	0.05	—	0.05	0.85	0.13	38	<1	0	—
0	29	1.14	59	257	86	1.21	<1	0.17	—	0.07	1.77	0.11	27	1	0	—
2	41	1.27	60	243	45	1.18	1	0.15	—	0.08	1.65	0.10	24	<1	0	—
0	20	0.92	34	248	4	0.41	1	0.16	—	0.04	0.52	0.11	15	3	0	—
0	19	0.98	63	163	1	1.05	1	0.02	1.47	0.04	0.15	0.18	10	1	0	5
0	29	0.87	47	132	1	0.93	<1	0.10	0.87	0.05	0.34	0.16	29	<1	0	1
0	266	4.73	88	375	112	1.51	121	0.14	1.29	0.12	1.08	0.28	32	18	0	11
42	468	2.58	37	204	1219	1.96	107	0.33	0.96	0.38	2.38	0.11	77	22	<1	22
0	48	1.71	42	367	618	0.60	54	0.09	1.69	0.08	0.93	0.11	33	8	<.1	4
0	23	2.35	31	380	289	0.87	18	0.14	1.08	0.10	1.50	0.16	61	13	0	9
22	187	0.77	20	120	768	1.11	82	0.15	0.46	0.24	1.02	0.04	39	<.1	<1	16
0	41	1.43	68	358	1403	0.91	86	0.16	2.23	0.12	2.76	0.30	28	13	<.1	8
51	468	4.78	90	455	2454	2.07	82	0.27	0.32	0.27	3.43	0.30	99	2	<1	43
0	117	1.13	40	168	134	0.68	26	0.07	0.26	0.07	0.54	0.11	46	1	0	5
0	190	3.65	28	302	371	0.74	8	0.13	1.00	0.12	1.43	0.15	47	7	0	6
10	144	1.91	33	393	637	1.06	31	0.20	1.13	0.27	2.07	0.21	64	15	<1	19
0	15	0.85	17	180	378	0.35	18	0.13	1.31	0.08	1.25	0.11	41	20	0	10
16	176	1.72	28	395	609	1.19	121	0.16	2.18	0.29	1.93	0.18	92	54	<1	6
46	318	3.36	39	299	813	1.98	288	0.45	6.23	0.50	4.52	0.16	123	27	<1	31
21	231	2.58	83	550	893	1.80	81	0.23	2.00	0.18	1.48	0.25	132	12	<1	10
<1	68	2.42	41	532	383	0.78	46	0.13	1.48	0.13	1.26	0.18	58	16	0	5
0	37	1.32	28	443	503	0.44	109	0.11	1.69	0.08	1.54	0.22	38	24	<.1	1
3	150	1.80	—	—	370	—	0	—	—	—	—	—	—	0	—	—
5	150	2.70	—	—	400	—	—	—	—	—	—	—	—	0	—	—
0	60	1.80	—	—	380	—	0	—	—	—	—	—	—	2	—	—
0	80	1.80	—	220	570	—	0	—	—	—	—	—	—	0	—	—
0	80	1.44	—	—	220	—	0	—	—	—	—	—	—	0	—	—
0	80	1.08	—	—	440	—	0	—	—	—	—	—	—	0	—	—
0	60	1.44	—	—	330	—	0	—	—	—	—	—	—	0	—	—
0	60	1.44	—	—	260	—	0	—	—	—	—	—	—	0	—	—
3	100	1.80	—	—	400	—	0	—	—	—	—	—	—	1	—	—
0	80	1.80	—	—	350	—	0	—	—	—	—	—	—	1	—	—
0	150	1.08	—	—	890	—	0	—	—	—	—	—	—	0	—	—
0	40	1.44	—	—	330	—	0	—	—	—	—	—	—	0	—	—
0	40	0.72	—	—	1000	—	0	—	—	—	—	—	—	0	—	—
0	60	3.60	—	—	360	—	0	—	—	—	—	—	—	0	—	—

A

Table A–1

Food Composition   (Computer code number is for Wadsworth Diet Analysis program)   (For purposes of calculations, use "0" for t, <1, <.1, <.01, etc.)

DA + Code	Food Description	Quantity	Measure	Wt (g)	H₂O (g)	Ener (cal)	Prot (g)	Carb (g)	Dietary Fiber (g)	Fat (g)	Fat Breakdown (g)		
											Sat	Mono	Poly
	VEGETARIAN FOODS—Continued												
39665	Country fried chicken w/creamy pepper gravy	1	item(s)	142	—	190	9	16	2	9	1.00	—	—
37805	Crispy nuggets	6	item(s)	82	—	180	4	22	3	9	1.50	—	—
39663	Homestyle classic burger	1	item(s)	71	—	110	12	6	4	5	0.50	—	—
37807	Meatless breakfast sausage	1	item(s)	43	—	50	5	2	2	4	0.00	—	—
37809	Meatless meatballs	6	item(s)	85	—	110	12	8	4	5	1.00	—	—
37806	Meatless riblets w/sauce	1	item(s)	142	—	210	17	11	4	5	0.00	—	—
29913	Original	3	ounce(s)	85	—	132	7	19	4	4	1.80	1.80	0.60
31707	Santa Fe	3	ounce(s)	85	—	156	—	24	5	3	1.20	—	—
29915	Veggie medley	3	ounce(s)	85	—	108	6	22	4	0	0.00	0.00	0.00
	Loma Linda												
9311	Big franks	1	item(s)	51	30	110	10	2	2	7	1.00	2.00	4.00
9315	Chik'n nuggets	5	item(s)	85	40	240	14	13	4	15	2.00	4.50	8.00
9317	Corn dogs	1	item(s)	71	31	150	7	22	3	4	0.50	1.00	2.50
9323	Fried chik'n with gravy	2	piece(s)	80	46	150	12	5	2	10	1.50	2.50	5.00
9326	Linketts, canned	1	item(s)	35	21	70	7	1	1	5	0.50	1.00	2.50
9336	Redi-Burger patties, canned	1	slice(s)	85	50	120	18	7	4	3	0.50	0.50	1.50
9354	Tender Rounds meatball substitute, canned in gravy	6	piece(s)	80	54	120	13	6	1	5	0.50	1.00	2.50
	Morningstar Farms												
33707	America's Original Veggie Dog links	1	item(s)	57	—	80	11	6	1	1	0.00	0.00	0.00
9362	Better n Eggs egg substitute	¼	cup(s)	57	50	20	5	0	0	0	0.00	0.00	0.00
9368	Breakfast links	2	item(s)	45	27	80	9	3	2	3	0.50	0.50	2.00
9371	Breakfast strips	2	item(s)	16	7	60	2	2	1	5	0.50	1.00	3.00
33705	Chik Nuggets	4	piece(s)	86	—	180	13	17	5	6	0.50	1.50	4.00
11587	Chik Patties	1	item(s)	71	36	150	9	16	2	6	1.00	1.50	2.50
2531	Garden veggie patties	1	item(s)	67	40	100	10	9	4	3	0.50	0.50	1.50
9412	Natural Touch low fat vegetarian chili, canned	1	cup(s)	230	173	170	18	21	11	1	—	—	—
33702	Spicy black bean veggie burger	1	item(s)	78	47	150	11	16	5	5	0.50	1.50	2.50
	Worthington												
9422	Chik Stiks	1	item(s)	47	27	110	10	4	2	6	1.00	1.00	3.00
9424	Chili, canned	1	cup(s)	230	167	290	19	21	9	15	2.50	3.50	9.00
9432	Crispychik patties	1	item(s)	71	37	150	9	16	2	6	1.00	1.50	3.50
9440	Dinner roast, frozen	1	slice(s)	85	53	180	12	5	3	12	1.50	5.00	5.00
9442	Fillets, frozen	2	piece(s)	85	48	180	16	8	4	9	1.00	3.50	4.50
9478	Meatless smoked beef, sliced	6	slice(s)	57	—	130	11	7	1	7	1.00	2.00	4.00
9480	Meatless smoked turkey, sliced	3	slice(s)	57	—	140	10	5	0	9	1.00	2.50	5.00
9462	Prosage links	2	item(s)	45	27	80	9	3	2	3	0.50	0.50	2.00
9486	Stripples bacon substitute	2	item(s)	16	7	60	2	2	1	5	0.50	1.00	2.50
9496	Vegetable Skallops	½	cup(s)	85	65	90	15	3	3	2	0.50	0.50	0.00
9434	Vegetarian cutlets	1	slice(s)	61	43	70	11	3	2	1	—	—	—
	DAIRY												
	Butter: *see* Fats & Oils												
	Cheese												
1433	Blue, crumbled	1	ounce(s)	28	12	100	6	1	0	8	5.29	2.21	0.23
884	Brick	1	ounce(s)	28	12	104	7	1	0	8	5.25	2.41	0.22
885	Brie	1	ounce(s)	28	14	94	6	<1	0	8	4.87	2.24	0.23
34821	Camembert	1	ounce(s)	29	15	87	6	<1	0	7	4.43	2.04	0.21
888	Cheddar or colby	1	ounce(s)	28	11	110	7	1	0	9	5.66	2.60	0.27
32096	Cheddar or colby, low fat	1	ounce(s)	28	18	49	7	1	0	2	1.23	0.59	0.06
5	Cheddar, shredded	¼	cup(s)	28	10	114	7	<1	0	9	5.96	2.65	0.27
889	Edam	1	ounce(s)	28	12	100	7	<1	0	8	4.92	2.28	0.19
890	Feta	1	ounce(s)	28	15	74	4	1	0	6	4.18	1.29	0.17
891	Fontina	1	ounce(s)	28	11	109	7	<1	0	9	5.37	2.43	0.46
8527	Goat, soft	1	ounce(s)	28	17	76	5	<1	0	6	4.14	1.37	0.14
893	Gouda	1	ounce(s)	28	12	100	7	1	0	8	4.93	2.17	0.18
894	Gruyere	1	ounce(s)	28	9	116	8	<1	0	9	5.30	2.81	0.49
895	Limburger	1	ounce(s)	28	14	92	6	<1	0	8	4.69	2.41	0.14
896	Monterey jack	1	ounce(s)	28	11	104	7	<1	0	8	5.34	2.45	0.25

PAGE KEY: A–4 = Breads/Baked Goods   A–8 = Cereal/Rice/Pasta   A–12 = Fruit   A–18 = Vegetables/Legumes   A–28 = Nuts/Seeds   A–30 = Vegetarian
A–32 = Dairy   A–38 = Eggs   A–40 = Seafood   A–42 = Meats   A–44 = Poultry   A–46 = Processed meats   A–48 = Beverages   A–52 = Fats/Oils
A–54 = Sweets   A–54 = Sauces/Condiments/Spices   A–58 = Mixed Foods/Soups/Sandwiches   A–64 = Fast food   A–82 = Convenience   A–84 = Baby foods

A

Chol (mg)	Calc (mg)	Iron (mg)	Magn (mg)	Pota (mg)	Sodi (mg)	Zinc (mg)	Vit A (RAE) (µg)	Thia (mg)	Vit E (mg)	Ribo (mg)	Niac (mg)	Vit B₆ (mg)	Fola (µg)	Vit C (mg)	Vit B₁₂ (µg)	Sele (µg)
5	40	1.44	—	—	550	—	—	—	—	—	—	—	—	0	—	—
5	60	0.72	—	—	570	—	—	—	—	—	—	—	—	5	—	—
0	80	1.44	—	—	380	—	—	—	—	—	—	—	—	0	—	—
0	20	0.72	—	—	120	—	—	—	—	—	—	—	—	0	—	—
0	60	1.80	—	—	400	—	—	—	—	—	—	—	—	0	—	—
0	60	1.80	—	—	720	—	—	—	—	—	—	—	—	4	—	—
24	72	0.00	37	232	672	1.07	0	0.12	0.24	0.18	1.30	0.10	12	0	<1	8
24	96	0.00	—	—	336	—	0	—	—	—	—	—	—	0	—	0
0	48	0.00	32	218	336	0.55	—	0.08	0.31	0.10	1.08	0.11	13	0	<.1	5
0	0	0.77	—	50	240	0.89	0	0.23	—	0.43	1.60	0.04	—	0	1	—
0	20	1.44	—	210	410	0.43	0	0.75	—	0.51	6.00	0.90	—	0	3	—
0	0	1.08	—	60	500	0.43	0	0.72	—	0.61	1.47	0.87	—	0	2	—
0	20	1.80	—	70	430	0.34	0	1.05	—	0.34	4.00	0.30	—	0	2	—
0	0	0.36	—	15	160	0.46	0	0.12	—	0.20	0.40	0.20	—	0	1	—
0	0	1.06	—	140	450	1.11	0	0.23	—	0.34	6.00	0.40	—	0	2	—
0	20	1.08	—	80	340	0.66	0	0.75	—	0.17	2.00	0.16	—	0	1	—
0	0	0.72	—	60	580	—	0	—	—	—	—	—	—	0	—	—
0	20	0.63	—	75	90	0.60	75	0.03	0.81	0.34	0.00	0.08	24	0	1	—
0	0	1.44	—	50	320	0.36	0	1.80	—	0.17	2.00	0.30	—	0	3	—
0	0	0.27	—	15	220	0.05	0	0.75	—	0.04	0.40	0.07	—	0	<1	—
0	40	3.60	—	330	590	—	0	1.20	—	0.26	5.00	0.40	—	0	3	—
0	0	1.80	—	210	540	0.31	0	1.80	—	0.17	2.00	0.20	—	0	1	—
0	40	0.72	—	180	350	0.58	—	6.47	—	0.10	0.00	0.00	—	0	0	—
0	40	1.80	—	480	870	1.36	—	0.60	—	0.21	0.00	0.30	—	0	0	—
0	40	1.80	44	320	470	0.93	0	—	—	0.14	0.00	0.21	—	0	<.1	—
0	20	1.80	—	100	300	0.31	0	0.60	—	0.17	6.00	0.40	—	0	2	—
0	40	3.60	—	420	1130	1.24	0	0.06	—	0.07	2.00	0.70	—	0	2	—
0	0	1.80	—	170	440	0.33	0	1.80	—	0.17	2.00	0.20	—	0	1	—
3	40	0.36	—	55	580	0.64	0	1.80	—	0.26	6.00	0.60	—	0	2	—
0	0	1.80	—	130	750	0.92	0	0.68	—	0.14	0.80	0.40	—	0	3	—
0	20	1.80	—	180	510	0.14	0	1.80	—	0.17	6.00	0.40	—	0	2	—
0	100	2.70	—	60	490	0.23	0	1.80	—	0.17	6.00	0.40	—	0	3	—
0	0	1.44	—	50	320	0.36	0	1.80	—	0.17	2.00	0.30	—	0	3	—
0	0	0.36	—	15	220	0.05	0	0.75	—	0.03	0.40	0.08	—	0	<1	—
0	0	0.72	—	10	410	0.67	0	0.03	—	0.03	0.00	0.01	—	0	0	—
0	0	0.00	—	30	340	0.43	0	0.03	—	0.03	0.00	0.04	—	0	0	—
21	150	0.09	7	73	395	0.75	56	0.01	0.18	0.11	0.29	0.05	10	0	<1	4
26	189	0.12	7	38	157	0.73	82	0.00	0.14	0.10	0.03	0.02	6	0	<1	4
28	52	0.14	6	43	176	0.67	49	0.02	0.18	0.15	0.11	0.07	18	0	<1	4
21	112	0.10	6	54	244	0.69	—	0.01	0.19	0.14	0.18	0.07	18	0	<1	4
27	192	0.21	7	36	169	0.86	74	0.00	0.10	0.11	0.03	0.02	5	0	<1	4
6	118	0.12	5	19	174	0.52	17	0.00	0.02	0.06	0.01	0.01	3	0	<1	4
30	204	0.19	8	28	175	0.88	75	0.01	0.10	0.11	0.02	0.02	5	0	<1	4
25	205	0.12	8	53	270	1.05	68	0.01	0.21	0.11	0.02	0.02	4	0	<1	4
25	138	0.18	5	17	312	0.81	35	0.04	0.01	0.24	0.28	0.12	9	0	<1	4
32	154	0.06	4	18	224	0.98	73	0.01	0.10	0.06	0.04	0.02	2	0	<1	4
13	40	0.54	5	7	105	0.26	82	0.02	0.13	0.11	0.12	0.07	3	0	<.1	1
32	196	0.07	8	34	229	1.09	46	0.01	0.10	0.09	0.02	0.02	6	0	<1	4
31	283	0.05	10	23	94	1.09	76	0.02	0.10	0.08	0.03	0.02	3	0	<1	4
25	139	0.04	6	36	224	0.59	95	0.02	0.18	0.14	0.04	0.02	16	0	<1	4
25	209	0.20	8	23	150	0.84	55	0.00	0.10	0.11	0.03	0.02	5	0	<1	4

A

Table A–1

Food Composition   (Computer code number is for Wadsworth Diet Analysis program)   (For purposes of calculations, use "0" for t, <1, <.1, <.01, etc.)

DA + Code	Food Description	Quantity	Measure	Wt (g)	H₂O (g)	Ener (cal)	Prot (g)	Carb (g)	Dietary Fiber (g)	Fat (g)	Sat	Mono	Poly
	DAIRY—Continued												
13	Mozzarella, part skim milk	1	ounce(s)	28	15	71	7	1	0	4	2.83	1.26	0.13
12	Mozzarella, whole milk	1	ounce(s)	28	14	84	6	1	0	6	3.68	1.84	0.21
897	Muenster	1	ounce(s)	28	12	103	7	<1	0	8	5.35	2.44	0.19
898	Neufchatel	1	ounce(s)	28	17	73	3	1	0	7	4.14	1.90	0.18
14	Parmesan, grated	1	tablespoon(s)	5	1	22	2	<1	0	1	0.87	0.42	0.06
17	Provolone	1	ounce(s)	28	11	98	7	1	0	7	4.78	2.07	0.22
19	Ricotta, part skim milk	¼	cup(s)	62	46	85	7	3	0	5	3.03	1.42	0.16
18	Ricotta, whole milk	¼	cup(s)	62	44	107	7	2	0	8	5.10	2.23	0.24
20	Romano	1	tablespoon(s)	5	2	19	2	<1	0	1	0.86	0.39	0.03
900	Roquefort	1	ounce(s)	28	11	103	6	1	0	9	5.39	2.37	0.37
21	Swiss	1	ounce(s)	28	10	106	8	2	0	8	4.98	2.04	0.27
	Imitation cheese												
7998	Shredded imitation cheddar	¼	cup(s)	28	—	90	5	2	0	7	1.50	—	—
8028	Shredded imitation mozzarella	¼	cup(s)	28	—	80	6	1	0	6	1.00	—	—
	Cottage Cheese												
9	Low fat, 1% fat	½	cup(s)	113	93	81	14	3	0	1	0.73	0.33	0.04
8	Low fat, 2% fat	½	cup(s)	113	90	102	16	4	0	2	1.38	0.62	0.07
	Cream cheese												
11	Cream cheese	2	tablespoon(s)	29	16	101	2	1	0	10	6.37	2.85	0.37
17366	Fat free cream cheese	2	tablespoon(s)	30	23	29	4	2	0	<1	0.27	0.10	0.02
10438	Tofutti Better Than Cream Cheese	2	tablespoon(s)	30	—	80	1	1	0	8	2.00	—	6.00
	Processed cheese												
22	American cheese, processed	1	ounce(s)	28	11	106	6	<1	0	9	5.58	2.54	0.28
24	American cheese food, processed	1	ounce(s)	28	12	94	5	2	0	7	4.23	2.05	0.31
25	American cheese spread, processed	1	ounce(s)	28	14	82	5	2	0	6	3.78	1.77	0.18
9110	Kraft deluxe singles pasteurized process American cheese	1	ounce(s)	28	—	110	5	1	0	9	6.00	—	—
23	Swiss cheese, processed	1	ounce(s)	28	12	95	7	1	0	7	4.55	2.00	0.18
	Soy cheese												
10430	Nu Tofu cheddar flavored cheese alternative	1	ounce(s)	28	—	70	6	1	0	4	0.50	2.50	1.00
10435	Nu Tofu mozzarella flavored cheese alternative	1	ounce(s)	28	—	70	6	2	0	4	0.50	2.50	1.00
	Cream												
26	Half & half	1	tablespoon(s)	15	12	20	<1	1	0	2	1.07	0.50	0.06
28	Light coffee or table, liquid	1	tablespoon(s)	15	11	29	<1	1	0	3	1.80	0.84	0.11
30	Light whipping cream, liquid	1	tablespoon(s)	15	10	44	<1	<1	0	5	2.90	1.36	0.13
32	Heavy whipping cream, liquid	1	tablespoon(s)	15	9	52	<1	<1	0	6	3.45	1.60	0.21
34	Whipped cream topping, pressurized	1	tablespoon(s)	4	2	10	<1	<1	0	1	0.52	0.24	0.03
	Sour cream												
36	Sour cream	2	tablespoon(s)	24	17	51	1	1	0	5	3.13	1.45	0.19
30556	Fat free sour cream	2	tablespoon(s)	32	26	24	1	5	0	0	0.00	0.00	0.00
	Imitation cream												
3659	Coffeemate nondairy creamer, liquid	1	tablespoon(s)	16	—	20	0	2	0	1	0.00	0.50	0.00
40	Cream substitute, powder	1	teaspoon(s)	2	<.1	11	<.1	1	0	1	0.65	0.02	0.00
35972	Nondairy coffee whitener, liquid, frozen	1	tablespoon(s)	16	12	22	<1	2	0	2	0.31	1.20	0.00
35975	Nondairy dessert topping, pressurized	1	tablespoon(s)	5	3	12	<.1	1	0	1	0.88	0.09	0.01
35976	Nondairy dessert topping, frozen	1	tablespoon(s)	5	3	16	<.1	1	0	1	1.09	0.08	0.03
904	Imitation sour cream	2	tablespoon(s)	24	17	50	1	2	0	5	4.27	0.14	0.01
	Fluid milk												
57	Fat free, nonfat, or skim	1	cup(s)	245	223	83	8	12	0	<1	0.29	0.12	0.02
58	Fat free, nonfat, or skim, w/nonfat milk solids	1	cup(s)	245	221	91	9	12	0	1	0.40	0.16	0.02
54	Low fat, 1%	1	cup(s)	244	219	102	8	12	0	2	1.54	0.68	0.09
55	Low fat, 1%, w/nonfat milk solids	1	cup(s)	245	220	105	9	12	0	2	1.48	0.69	0.09
60	Low fat buttermilk	1	cup(s)	245	221	98	8	12	0	2	1.34	0.62	0.08
51	Reduced fat, 2%	1	cup(s)	244	218	122	8	11	0	5	2.35	2.04	0.17
52	Reduced fat, 2%, w/nonfat milk solids	1	cup(s)	245	218	125	9	12	0	5	2.93	1.36	0.17
50	Whole, 3.3%	1	cup(s)	244	216	146	8	11	0	8	4.55	1.98	0.48
	Canned												
61	Whole evaporated	2	tablespoon(s)	32	23	42	2	3	0	2	1.45	0.74	0.08

PAGE KEY: A–4 = Breads/Baked Goods A–8 = Cereal/Rice/Pasta A–12 = Fruit A–18 = Vegetables/Legumes A–28 = Nuts/Seeds A–30 = Vegetarian
A–32 = Dairy A–38 = Eggs A–40 = Seafood A–42 = Meats A–44 = Poultry A–46 = Processed meats A–48 = Beverages A–52 = Fats/Oils
A–54 = Sweets A–54 = Sauces/Condiments/Spices A–58 = Mixed Foods/Soups/Sandwiches A–64 = Fast food A–82 = Convenience A–84 = Baby foods

Chol (mg)	Calc (mg)	Iron (mg)	Magn (mg)	Pota (mg)	Sodi (mg)	Zinc (mg)	Vit A (RAE) (µg)	Thia (mg)	Vit E (mg)	Ribo (mg)	Niac (mg)	Vit B$_6$ (mg)	Fola (µg)	Vit C (mg)	Vit B$_{12}$ (µg)	Sele (µg)
18	219	0.06	6	24	173	0.77	36	0.01	0.12	0.08	0.03	0.02	3	0	<1	4
22	141	0.12	6	21	176	0.82	50	0.01	0.10	0.08	0.03	0.01	2	0	1	5
27	201	0.11	8	38	176	0.79	83	0.00	0.13	0.09	0.03	0.02	3	0	<1	4
21	21	0.08	2	32	112	0.15	83	0.00	—	0.05	0.04	0.01	3	0	<.1	1
4	55	0.05	2	6	76	0.19	6	0.00	0.04	0.02	0.01	0.00	1	0	<1	1
19	212	0.15	8	39	245	0.90	66	0.01	0.10	0.09	0.04	0.02	3	0	<1	4
19	167	0.27	9	77	77	0.82	66	0.01	0.13	0.11	0.05	0.01	8	0	<1	10
31	127	0.23	7	65	52	0.71	74	0.01	0.22	0.12	0.06	0.03	7	0	<1	9
5	53	0.04	2	4	60	0.13	5	0.00	0.04	0.02	0.00	0.00	<1	0	<.1	1
25	185	0.16	8	25	507	0.58	82	0.01	0.18	0.16	0.21	0.03	14	0	<1	4
26	221	0.06	11	22	54	1.22	62	0.02	0.14	0.08	0.03	0.02	2	0	1	5
0	150	0.00	—	—	420	—	—	—	—	—	—	—	—	0	—	—
0	150	0.00	8	—	320	1.20	—	0.00	—	0.26	0.00	0.00	40	0	<1	—
5	69	0.16	6	97	459	0.43	12	0.02	0.12	0.19	0.14	0.08	14	0	1	10
9	78	0.18	7	108	459	0.47	24	0.03	0.06	0.21	0.16	0.09	15	0	1	12
32	23	0.35	2	35	86	0.16	106	0.00	0.27	0.06	0.03	0.01	4	0	<1	1
2	56	0.05	4	49	164	0.26	84	0.02	0.01	0.05	0.05	0.02	11	0	<1	1
0	0	0.00	—	—	135	—	0							0	—	—
27	156	0.05	8	48	422	0.81	72	0.01	0.13	0.10	0.02	0.02	2	0	<1	4
23	162	0.16	9	83	359	0.91	57	0.02	0.20	0.15	0.05	0.02	2	0	<1	5
16	160	0.09	8	69	382	0.74	49	0.01	0.20	0.12	0.04	0.03	2	0	<1	3
25	150	0.00	0	25	450	0.90	84	—	—	0.10	—	—	—	0	<1	—
24	219	0.17	8	61	388	1.02	56	0.00	0.19	0.08	0.01	0.01	2	0	<1	5
0	200	0.36	—	—	190	—	—	—	—	—	—	—	—	0	—	—
0	150	0.36	—	—	190	—	—	—	—	—	—	—	—	0	—	—
6	16	0.01	2	20	6	0.08	15	0.01	0.02	0.02	0.01	0.01	<1	<1	<.1	<1
10	14	0.01	1	18	6	0.04	27	0.00	0.02	0.02	0.01	0.00	<1	<1	<.1	<1
17	10	0.00	1	15	5	0.04	42	0.00	0.09	0.02	0.00	0.00	1	<.1	<.1	<.1
21	10	0.00	1	11	6	0.03	62	0.00	0.09	0.02	0.01	0.00	1	<.1	<.1	<.1
3	4	0.00	<1	6	5	0.01	7	0.00	0.02	0.00	0.00	0.00	<1	0	<.1	<.1
11	28	0.01	3	35	13	0.06	42	0.01	0.14	0.04	0.02	0.00	3	<1	<.1	1
3	40	0.00	3	41	45	0.16	—	0.01	0.00	0.05	0.02	0.01	4	0	<.1	—
0	0	0.00	—	30	0	—	0	0.02	—	0.02	0.20	—	—	0	—	—
0	<1	0.02	<.1	16	4	0.01	<.1	0.00	0.01	0.00	0.00	0.00	0	0	0	<.1
0	1	0.00	<.1	30	13	0.00	—	0.00	0.26	0.00	0.00	0.00	0	0	0	<1
0	<1	0.00	<.1	1	3	0.00	—	0.00	0.01	0.00	0.00	0.00	0	0	0	<.1
0	<1	0.01	<.1	1	1	0.00	—	0.00	0.01	0.00	0.00	0.00	0	0	0	<1
0	1	0.09	1	39	24	0.28	0	0.00	0.04	0.00	0.00	0.00	0	0	0	1
5	223	1.23	22	238	108	2.08	149	0.11	0.10	0.45	0.23	0.09	12	0	1	8
5	316	0.12	37	419	130	1.00	149	0.10	0.10	0.43	0.22	0.11	12	2	1	5
12	264	0.85	27	290	122	2.12	142	0.05	0.10	0.45	0.23	0.09	12	0	1	8
10	314	0.12	34	397	127	0.98	145	0.10	0.10	0.42	0.22	0.11	12	2	1	6
10	284	0.12	27	370	257	1.03	17	0.08	0.15	0.38	0.14	0.08	12	2	1	5
20	271	0.24	27	342	115	1.17	134	0.10	0.17	0.45	0.22	0.09	12	<1	1	6
20	314	0.12	34	397	127	0.98	137	0.10	0.17	0.42	0.22	0.11	12	2	1	6
24	246	0.07	24	325	105	0.93	68	0.11	0.24	0.45	0.26	0.09	12	0	1	9
9	82	0.06	8	95	33	0.24	20	0.01	0.06	0.10	0.06	0.02	3	1	<.1	1

A

Table A–1

Food Composition   (Computer code number is for Wadsworth Diet Analysis program)   (For purposes of calculations, use "0" for t, <1, <.1, <.01, etc.)

DA + Code	Food Description	Quantity	Measure	Wt (g)	H₂O (g)	Ener (cal)	Prot (g)	Carb (g)	Dietary Fiber (g)	Fat (g)	Fat Breakdown (g)		
											Sat	Mono	Poly
	DAIRY—Continued												
62	Fat free, nonfat, or skim evaporated	2	tablespoon(s)	32	25	25	2	4	0	<.1	0.04	0.02	0.00
63	Sweetened condensed	2	tablespoon(s)	38	10	123	3	21	0	3	2.10	0.93	0.13
	Dried Milk												
64	Dried buttermilk	¼	cup(s)	30	1	118	10	15	0	2	1.09	0.51	0.07
65	Instant nonfat dry milk w/added vitamin A	¼	cup(s)	17	1	63	6	9	0	<1	0.08	0.03	0.00
5234	Skim milk powder	¼	cup(s)	18	1	64	6	9	0	<1	0.08	0.03	0.01
907	Whole dry milk	¼	cup(s)	32	1	161	9	12	0	9	5.43	2.57	0.22
909	Goat milk	1	cup(s)	244	212	168	9	11	0	10	6.51	2.71	0.36
	Chocolate milk												
69	Low fat	1	cup(s)	250	211	158	8	26	1	3	1.54	0.75	0.09
68	Reduced fat	1	cup(s)	250	209	180	8	26	1	5	3.10	1.47	0.18
67	Whole milk	1	cup(s)	250	206	208	8	26	2	8	5.26	2.48	0.31
33156	Chocolate syrup, fortified, prepared w/milk	1	cup(s)	263	220	197	8	24	<1	8	5.22	2.44	0.31
908	Cocoa, hot, prepared w/milk	1	cup(s)	250	206	193	9	27	3	6	3.58	1.69	0.09
33184	Cocoa mix with aspartame, added sodium & vitamin A, no added calcium or phosphorus, prepared with water	1	cup(s)	192	177	56	2	10	1	<1	0.00	0.15	0.01
70	**Eggnog**	1	cup(s)	254	189	343	10	34	0	19	11.29	5.67	0.86
	Breakfast drinks												
10093	Carnation Instant Breakfast classic chocolate malt, prepared w/skim milk, no sugar added	1	cup(s)	243	—	142	11	21	<1	1	0.89	—	—
10091	Carnation Instant Breakfast strawberry creme, prepared w/skim milk	1	cup(s)	273	—	220	13	39	0	<1	0.40	—	—
10094	Carnation Instant Breakfast strawberry creme, prepared w/skim milk, no sugar added	1	cup(s)	243	—	134	12	21	0	<1	0.45	—	—
10092	Carnation Instant Breakfast vanilla creme, prepared w/skim milk, no sugar added	1	cup(s)	273	—	220	13	39	0	<1	0.40	—	—
1417	Ovaltine rich chocolate flavor, prepared w/skim milk	1	cup(s)	243	—	134	12	21	0	<1	0.45	—	—
8539	**Malted milk, chocolate mix, fortified, prepared w/milk**	1	cup(s)	265	216	223	9	29	1	9	4.95	2.17	0.54
	Milkshakes												
73	Chocolate	1	cup(s)	227	164	270	7	48	1	6	3.81	1.77	0.23
74	Vanilla	1	cup(s)	227	169	254	9	40	0	7	4.28	1.98	0.26
	Ice cream												
4776	Chocolate	½	cup(s)	66	37	143	3	19	1	7	4.49	2.12	0.27
16514	Chocolate, soft serve	½	cup(s)	87	50	177	3	24	1	8	5.17	2.43	0.31
12137	Chocolate fudge, fat free no sugar added	½	cup(s)	71	—	100	4	22	0	0	0.00	0.00	0.00
82	Light vanilla	½	cup(s)	66	42	109	4	18	<1	3	1.71	0.57	0.10
78	Light vanilla, soft serve	½	cup(s)	86	60	108	4	19	0	2	1.40	0.65	0.09
16523	Sherbet, all flavors	½	cup(s)	97	64	133	1	29	<1	2	1.12	0.51	0.08
4778	Strawberry	½	cup(s)	66	40	127	2	18	1	6	3.43	—	—
76	Vanilla	½	cup(s)	66	40	133	2	16	<1	7	4.48	1.96	0.30
12146	Vanilla chocolate swirl, fat free, no sugar added	½	cup(s)	71	—	100	4	20	0	0	0.00	0.00	0.00
	Soy desserts												
10694	Tofutti low fat vanilla fudge nondairy frozen dessert	½	cup(s)	70	—	120	2	24	0	2	1.00	—	—
15721	Tofutti premium chocolate supreme nondairy frozen dessert	½	cup(s)	60	—	180	3	18	0	11	2.00	—	—
15720	Tofutti premium vanilla nondairy frozen dessert	½	cup(s)	60	—	190	2	20	0	11	2.00	—	—
	Ice milk												
16516	Flavored, not chocolate	½	cup(s)	66	45	91	2	15	0	3	1.72	0.81	0.11
16517	Chocolate	½	cup(s)	66	43	95	3	17	<1	2	1.29	0.61	0.08
	Pudding												
25032	Chocolate	½	cup(s)	144	110	154	5	23	1	5	2.78	1.94	0.23
1923	Chocolate, sugar free, prepared w/2% milk	½	cup(s)	133	—	100	5	14	<1	3	1.50	—	—

PAGE KEY: A–4 = Breads/Baked Goods   A–8 = Cereal/Rice/Pasta   A–12 = Fruit   A–18 = Vegetables/Legumes   A–28 = Nuts/Seeds   A–30 = Vegetarian
A–32 = Dairy   A–38 = Eggs   A–40 = Seafood   A–42 = Meats   A–44 = Poultry   A–46 = Processed meats   A–48 = Beverages   A–52 = Fats/Oils
A–54 = Sweets   A–54 = Sauces/Condiments/Spices   A–58 = Mixed Foods/Soups/Sandwiches   A–64 = Fast food   A–82 = Convenience   A–84 = Baby foods

A

Chol (mg)	Calc (mg)	Iron (mg)	Magn (mg)	Pota (mg)	Sodi (mg)	Zinc (mg)	Vit A (RAE) (µg)	Thia (mg)	Vit E (mg)	Ribo (mg)	Niac (mg)	Vit B_6 (mg)	Fola (µg)	Vit C (mg)	Vit B_{12} (µg)	Sele (µg)
1	93	0.09	9	106	37	0.29	38	0.01	0.00	0.10	0.06	0.02	3	<1	<.1	1
13	109	0.07	10	142	49	0.36	28	0.03	0.08	0.16	0.08	0.02	4	1	<1	6
21	360	0.09	33	484	157	1.22	15	0.12	0.12	0.48	0.27	0.10	14	2	1	6
3	215	0.05	20	298	96	0.77	124	0.07	0.00	0.30	0.16	0.06	9	1	1	5
3	222	0.06	21	307	99	0.79	0	0.07	0.00	0.31	0.16	0.06	9	1	1	5
31	296	0.15	28	431	120	1.08	83	0.09	0.35	0.39	0.21	0.10	12	3	1	5
27	327	0.12	34	498	122	0.73	139	0.12	0.22	0.34	0.68	0.11	2	3	<1	3
8	288	0.60	33	425	153	1.03	145	0.10	0.07	0.42	0.32	0.10	13	2	1	5
18	285	0.60	33	423	150	1.03	138	0.09	0.13	0.41	0.32	0.10	13	2	1	5
30	280	0.60	33	418	150	1.03	65	0.09	0.23	0.41	0.31	0.10	13	2	1	5
34	292	2.68	32	460	147	0.92	—	0.09	—	0.55	6.53	0.11	13	2	1	5
20	263	1.20	58	493	110	1.58	128	0.10	0.26	0.46	0.33	0.10	13	1	1	7
<1	90	0.75	33	405	171	0.52	27	0.04	0.06	0.21	0.16	0.05	2	<1	<1	2
150	330	0.51	48	419	137	1.17	114	0.09	0.58	0.48	0.27	0.13	3	4	1	11
9	445	4.01	89	632	196	3.38	—	0.35	4.48	0.45	4.45	0.45	4	27	1	8
9	500	4.47	100	638	360	3.75	—	0.38	5.03	0.51	5.08	0.48	100	30	1	9
9	445	4.01	89	570	187	3.38	—	0.33	4.48	0.45	4.45	0.45	89	27	1	8
9	500	4.50	100	630	240	3.75	—	0.38	5.03	0.51	5.00	0.50	100	30	2	9
9	445	4.01	89	570	187	3.38	—	0.33	4.48	0.45	4.45	0.45	89	27	1	8
27	339	3.76	45	578	231	1.17	904	0.76	0.33	1.32	11.08	1.01	19	32	1	12
25	299	0.70	36	508	252	1.09	41	0.11	0.23	0.50	0.28	0.06	11	0	1	4
27	331	0.23	27	415	215	0.88	57	0.07	0.23	0.44	0.33	0.10	16	0	1	5
22	72	0.61	19	164	50	0.38	78	0.03	0.22	0.13	0.15	0.04	11	<1	<1	2
22	103	0.33	19	192	44	0.48	—	0.04	0.23	0.13	0.11	0.03	5	1	<1	—
0	80	0.36	—	—	60	—	—	—	—	—	—	—	—	0	—	—
17	77	0.05	9	137	49	0.48	91	0.02	0.00	0.11	0.06	0.02	3	<1	<1	1
10	135	0.05	12	190	60	0.46	25	0.04	0.00	0.17	0.10	0.04	5	1	<1	3
5	52	0.14	8	93	44	0.46	—	0.02	0.06	0.07	0.09	0.03	4	4	<1	—
19	79	0.14	9	124	40	0.22	63	0.03		0.17	0.11	0.03	8	5	<1	1
29	84	0.06	9	131	53	0.46	78	0.03	0.14	0.16	0.08	0.03	3	<1	<1	1
0	80	0.00	50	0												
0	0	0.00	—	8	90	—	0	—	—	—	—	—	—	0	—	—
0	0	0.00	—	7	180	—	0	—	—	—	—	—	—	0	—	—
0	0	0.00	—	2	210	—	0	—	—	—	—	—	—	0	—	—
9	91	0.07	10	138	56	0.29	—	0.04	0.00	0.17	0.06	0.04	4	1	<1	—
6	94	0.17	13	155	41	0.38	—	0.03	0.06	0.12	0.09	0.03	4	<1	<1	—
35	138	1.04	29	211	135	1.07	73	0.04	0.20	0.25	0.18	0.03	7	<1	<1	5
10	150	0.72	—	330	310	—	—	0.06	—	0.26	—	—	—	0	—	—

Table A–1

A

Food Composition   (Computer code number is for Wadsworth Diet Analysis program)   (For purposes of calculations, use "0" for t, <1, <.1, <.01, etc.)

DA + Code	Food Description	Quantity	Measure	Wt (g)	H₂O (g)	Ener (cal)	Prot (g)	Carb (g)	Dietary Fiber (g)	Fat (g)	Fat Breakdown (g) Sat	Mono	Poly
	DAIRY—Continued												
1722	Rice	½	cup(s)	113	73	175	6	26	1	6	1.99	2.14	0.88
4747	Tapioca, ready to eat	1	item(s)	142	105	169	3	28	<1	5	0.85	2.24	1.93
25031	Vanilla	½	cup(s)	136	110	116	5	17	<.1	3	1.31	1.21	0.16
1924	Vanilla, sugar free, prepared w/2% milk	½	cup(s)	133	90	4	12	<1	2	1.50	10	150	0.00
	Frozen yogurt												
4785	Chocolate, soft serve	½	cup(s)	72	46	115	3	18	2	4	2.61	1.26	0.16
1747	Fruit varieties	½	cup(s)	113	80	144	3	24	0	4	2.63	1.11	0.11
4786	Vanilla, soft serve	½	cup(s)	72	47	117	3	17	0	4	2.46	1.14	0.15
	Milk substitutes												
	Lactose free												
16081	Fat free calcium fortified milk	1	cup(s)	240	—	90	9	13	0	0	0.00	—	—
36486	Low fat milk	1	cup(s)	240	—	110	8	13	0	3	1.50	—	—
36487	Reduced fat milk	1	cup(s)	240	—	130	8	13	0	5	3.00	—	—
36488	Whole milk	1	cup(s)	240	—	160	8	12	0	9	5.00	—	—
	Rice												
10083	Rice Dream carob rice beverage	1	cup(s)	240	—	150	1	32	0	3	0.00	—	—
10087	Rice Dream vanilla enriched rice beverage	1	cup(s)	240	—	130	1	28	0	2	0.00	—	—
17089	Rice Dream original rice beverage, enriched	1	cup(s)	240	—	120	1	25	0	2	0.00	—	—
	Soy												
34750	Soy Dream chocolate enriched soy beverage	1	cup(s)	240	—	210	7	37	1	4	0.50	—	—
34749	Soy Dream vanilla enriched soy beverage	1	cup(s)	240	—	150	7	22	0	4	0.50	—	—
13840	Vitasoy light chocolate soymilk	1	cup(s)	237	—	100	4	17	0	2	0.50	0.50	1.00
13839	Vitasoy light vanilla soymilk	1	cup(s)	237	—	70	4	10	0	2	0.50	0.50	1.00
13836	Vitasoy rich chocolate soymilk	1	cup(s)	237	—	160	7	24	1	4	0.50	1.00	2.50
13835	Vitasoy vanilla delite soymilk	1	cup(s)	237	—	120	8	13	1	4	0.50	1.00	2.50
	Yogurt												
3615	Custard style, fruit flavors	6	ounce(s)	170	127	190	7	32	0	4	2.00	—	—
3617	Custard style, vanilla	6	ounce(s)	170	134	190	7	32	0	4	2.00	0.94	0.10
32101	Fruit, low fat	1	cup(s)	245	184	243	10	46	0	3	1.82	0.77	0.08
29638	Fruit, nonfat, sweetened w/low calorie sweetener	1	cup(s)	241	208	122	11	19	1	<1	0.21	0.10	0.04
93	Plain, low fat	1	cup(s)	245	208	154	13	17	0	4	2.45	1.04	0.11
94	Plain, nonfat	1	cup(s)	245	209	137	14	19	0	<1	0.28	0.12	0.01
32100	Vanilla, low fat	1	cup(s)	245	194	208	12	34	0	3	1.97	0.84	0.09
5242	Yogurt beverage	1	cup(s)	245	200	172	6	33	0	2	1.39	0.59	0.06
38202	Yogurt smoothie, nonfat, all flavors	1	item(s)	325	—	290	10	60	6	0	0.00	0.00	0.00
	Soy yogurt												
10453	White Wave plain silk cultured	8	ounce(s)	227	—	120	5	22	1	3	0.00	—	—
34616	Stonyfield Farm Osoy chocolate-vanilla pack organic cultured	1	serving(s)	113	—	90	4	15	3	2	0.00	—	—
34617	Stonyfield Farm Osoy strawberry-peach pack organic cultured	1	serving(s)	113	—	90	4	15	3	2	0.00	—	—
	EGGS												
96	Raw, whole	1	item(s)	50	38	74	6	<1	0	5	1.55	1.91	0.68
97	Raw, white	1	item(s)	33	29	17	4	<1	0	<.1	0.00	0.00	0.00
98	Raw, yolk	1	item(s)	17	9	53	3	1	0	4	1.59	1.95	0.70
99	Fried	1	item(s)	46	32	92	6	<1	0	7	1.98	2.92	1.22
100	Hard boiled	1	item(s)	50	37	78	6	1	0	5	1.63	2.04	0.71
101	Poached	1	item(s)	50	38	74	6	<1	0	5	1.54	1.90	0.68
102	Scrambled, prepared w/milk & butter	2	item(s)	122	89	203	14	3	0	15	4.49	5.82	2.62
	Egg Substitute												
920	Frozen	¼	cup(s)	60	44	96	7	2	0	7	1.16	1.46	3.74
918	Liquid	¼	cup(s)	63	52	53	8	<1	0	2	0.41	0.56	1.01
4028	Egg Beaters	¼	cup(s)	61	—	30	6	1	0	0	0.00	0.00	0.00

PAGE KEY: A–4 = Breads/Baked Goods   A–8 = Cereal/Rice/Pasta   A–12 = Fruit   A–18 = Vegetables/Legumes   A–28 = Nuts/Seeds   A–30 = Vegetarian
A–32 = Dairy   A–38 = Eggs   A–40 = Seafood   A–42 = Meats   A–44 = Poultry   A–46 = Processed meats   A–48 = Beverages   A–52 = Fats/Oils
A–54 = Sweets   A–54 = Sauces/Condiments/Spices   A–58 = Mixed Foods/Soups/Sandwiches   A–64 = Fast food   A–82 = Convenience   A–84 = Baby foods

A

Chol (mg)	Calc (mg)	Iron (mg)	Magn (mg)	Pota (mg)	Sodi (mg)	Zinc (mg)	Vit A (RAE) (µg)	Thia (mg)	Vit E (mg)	Ribo (mg)	Niac (mg)	Vit B_6 (mg)	Fola (µg)	Vit C (mg)	Vit B_{12} (µg)	Sele (µg)
71	130	1.21	21	250	253	0.61	—	0.10	0.60	0.26	0.73	0.08	14	1	<1	—
1	119	0.33	11	136	226	0.38	0	0.03	0.13	0.14	0.44	0.03	4	1	<1	2
35	133	0.25	14	173	134	0.63	73	0.03	0.14	0.24	0.11	0.03	6	<1	<1	5
—	190	380	—	—	0.03	—	—	0.17	—	—	—	0	—	—	—	—
4	106	0.90	19	188	71	0.35	32	0.03	0.10	0.15	0.22	0.05	8	<1	<1	2
15	113	0.52	11	176	71	0.32	—	0.05	0.09	0.20	0.08	0.05	5	1	<.1	—
1	103	0.22	10	152	63	0.30	42	0.03	0.04	0.16	0.21	0.06	4	1	<1	2
3	500	0.00	—	—	130	—	100	—	0.10	—	—	—	—	0	—	—
15	300	0.00	—	—	125	—	100	—	—	—	—	—	—	0	—	—
20	300	0.00	—	—	125	—	98	—	—	—	—	—	—	0	—	—
35	300	0.00	—	—	125	—	58	—	—	—	—	—	—	0	—	—
0	20	0.72	—	—	100	—	—	—	0.81	—	—	—	—	1	—	—
0	300	0.00	—	—	90	—	—	—	0.81	—	—	—	—	0	2	—
0	300	0.00	13	60	90	0.24	—	0.07	0.81	0.00	0.84	0.08	—	0	2	—
0	300	1.80	60	350	160	0.60	33	0.15	5.03	0.07	0.80	0.12	60	0	3	—
0	300	1.80	40	260	140	0.60	33	0.15	5.03	0.07	0.80	0.12	60	0	3	—
0	300	0.72	24	200	140	0.90	0	0.09	—	0.34	—	—	24	0	1	—
0	300	0.72	24	200	110	0.90	0	0.09	—	0.34	—	—	24	0	1	—
0	300	1.08	40	320	150	0.90	0	0.15	—	0.34	—	—	60	0	1	—
0	40	0.72	—	320	115	—	0	—	—	—	—	—	—	0	—	—
15	200	0.00	16	310	90	—	0	—	0.05	0.26	—	—	—	0	—	—
15	200	0.00	16	300	90	—	0	—	0.08	0.26	—	—	—	0	—	—
12	338	0.15	32	434	130	1.64	27	0.08	0.08	0.40	0.21	0.09	22	1	1	7
3	370	0.62	41	550	139	1.83	—	0.10	0.17	0.45	0.50	0.11	32	26	1	—
15	448	0.20	42	573	172	2.18	34	0.11	0.10	0.52	0.28	0.12	27	2	1	8
5	488	0.22	47	625	189	2.38	5	0.12	0.01	0.57	0.30	0.13	29	2	1	9
12	419	0.17	39	537	162	2.03	29	0.10	0.08	0.49	0.26	0.11	27	2	1	12
13	260	0.22	39	399	98	1.10	—	0.11	0.01	0.51	0.30	0.15	29	2	2	—
5	300	2.70	100	580	290	2.25	—	0.38	5.03	0.43	5.00	0.50	100	15	2	—
0	700	0.90	—	—	30	—	—	—	—	—	—	—	—	0	0	—
0	100	0.72	—	—	20	—	0	—	—	—	—	—	—	0	—	—
0	100	0.72	—	—	20	—	0	—	—	—	—	—	14	0	—	—
212	27	0.92	6	67	70	0.56	70	0.03	0.53	0.24	0.04	0.07	24	0	1	16
0	2	0.03	4	54	55	0.01	0	0.00	0.00	0.15	0.04	0.00	1	0	<.1	7
205	21	0.45	1	18	8	0.38	63	0.03	0.52	0.09	0.00	0.06	24	0	<1	9
210	27	0.91	6	68	94	0.55	91	0.03	0.75	0.24	0.04	0.07	23	0	1	16
212	25	0.60	5	63	62	0.53	85	0.03	0.53	0.26	0.03	0.06	22	0	1	15
211	27	0.92	6	67	147	0.55	70	0.03	0.53	0.24	0.04	0.07	24	0	1	16
429	87	1.46	15	168	342	1.22	174	0.06	1.60	0.53	0.10	0.14	37	<1	1	27
1	44	1.19	9	128	119	0.59	7	0.07	1.27	0.23	0.08	0.08	10	<1	<1	25
1	33	1.32	6	207	111	0.82	11	0.07	0.30	0.19	0.07	0.00	9	0	<1	16
0	20	1.08	4	85	115	0.60	113	0.15	0.81	0.85	0.20	0.08	60	0	1	—

Food Composition (Computer code number is for Wadsworth Diet Analysis program) (For purposes of calculations, use "0" for t, <1, <.1, <.01, etc.)

DA + Code	Food Description	Quantity	Measure	Wt (g)	H₂O (g)	Ener (cal)	Prot (g)	Carb (g)	Dietary Fiber (g)	Fat (g)	Fat Breakdown (g)		
											Sat	Mono	Poly
	SEAFOOD												
	Fish												
	Cod												
6040	Atlantic cod or scrod, baked or broiled	3	ounce(s)	44	34	46	10	0	0	<1	0.07	0.05	0.13
1573	Atlantic cod, cooked, dry heat	3	ounce(s)	85	65	89	19	0	0	1	0.14	0.11	0.25
2905	**Eel, raw**	3	ounce(s)	85	58	156	16	0	0	10	2.01	6.12	0.81
	Fish fillets												
25079	Baked	3	ounce(s)	84	80	99	22	0	0	1	0.08	0.07	0.26
8615	Batter coated or breaded, fried	3	ounce(s)	85	0.04	197.19	12.46	14.42	0.42	10.44	2.39	2.19	5.32
25082	Broiled fish steaks	3	ounce(s)	86	69	129	24	0	0	3	0.37	0.87	0.84
25083	Poached fish steaks	3	ounce(s)	86	68	112	21	0	0	2	0.33	0.76	0.74
25084	Steamed fish fillets	3	ounce(s)	86	73	80	17	0	0	1	0.12	0.08	0.22
25089	**Flounder, baked**	3	ounce(s)	85	65	114	15	<1	<.1	6	1.15	2.17	1.44
1825	**Grouper, cooked, dry heat**	3	ounce(s)	85	62	100	21	0	0	1	0.25	0.23	0.34
	Haddock												
6049	Baked or broiled	3	ounce(s)	44	33	50	11	0	0	<1	0.07	0.07	0.14
1578	Cooked, dry heat	3	ounce(s)	85	63	95	21	0	0	1	0.14	0.13	0.26
1886	**Halibut, Atlantic & Pacific, cooked, dry heat**	3	ounce(s)	85	61	119	23	0	0	2	0.35	0.82	0.80
1582	**Herring, Atlantic, pickled**	4	piece(s)	60	33	157	9	6	0	11	1.43	7.17	1.01
1587	**Jack mackerel, solids, canned, drained**	2	ounce(s)	57	39	88	13	0	0	4	1.05	1.26	0.94
8580	**Octopus, common, cooked, moist heat**	3	ounce(s)	85	51	139	25	4	0	2	0.39	0.28	0.41
1831	**Perch, mixed species, cooked, dry heat**	3	ounce(s)	85	62	99	21	0	0	1	0.20	0.17	0.40
1592	**Pacific rockfish, cooked, dry heat**	3	ounce(s)	85	62	103	20	0	0	2	0.40	0.38	0.50
	Salmon												
29727	Smoked chinook (lox)	2	ounce(s)	57	<.1	66	10	0	0	2	0.52	1.14	0.56
1594	Broiled or baked w/butter	3	ounce(s)	85	54	155	23	0	0	6	1.16	2.29	2.33
2938	Coho, farmed, raw	3	ounce(s)	85	60	136	18	0	0	7	1.54	2.83	1.58
154	**Sardines, Atlantic, with bones, canned in oil**	2	item(s)	24	<.1	50	6	0	0	3	0.36	0.92	1.23
	Scallops												
155	Mixed species, breaded, fried	3	item(s)	47	<.1	100	8	5	0	5	1.24	2.09	1.32
1599	Steamed	3	ounce(s)	85	65	90	14	2	0	3	—	—	—
1839	**Snapper, mixed species, cooked, dry heat**	3	ounce(s)	85	60	109	22	0	0	1	0.31	0.27	0.50
	Squid												
1868	Mixed species, fried	3	ounce(s)	85	55	149	15	7	0	6	1.60	2.34	1.82
16617	Steamed or boiled	3	ounce(s)	85	63	90	15	3	0	1	0.35	0.11	0.51
1570	**Striped bass, cooked, dry heat**	3	ounce(s)	85	62	105	19	0	0	3	0.55	0.72	0.85
1601	**Sturgeon, steamed**	3	ounce(s)	85	59	111	17	0	0	4	0.97	2.04	0.73
1840	**Surimi, formed**	3	ounce(s)	85	65	84	13	6	0	1	0.16	0.13	0.38
1842	**Swordfish, cooked, dry heat**	3	ounce(s)	85	58	132	22	0	0	4	1.20	1.68	1.00
1846	**Tuna, yellowfin or ahi, raw**	3	ounce(s)	85	60	92	20	0	0	1	0.20	0.13	0.24
	Tuna, canned												
159	Light, canned in oil, drained	2	ounce(s)	57	34	113	17	0	0	5	0.87	1.68	1.64
355	Light, canned in water, drained	2	ounce(s)	57	42	66	14	0	0	<1	0.13	0.09	0.19
33211	Light, no salt, canned in oil, drained	2	ounce(s)	57	34	112	17	0	0	5	0.87	1.67	1.64
33212	Light, no salt, canned in water, drained	2	ounce(s)	57	43	66	14	0	0	<1	0.13	0.09	0.19
2961	White, canned in oil, drained	2	ounce(s)	57	36	105	15	0	0	5	0.73	1.85	1.69
351	White, canned in water, drained	2	ounce(s)	57	41	73	13	0	0	2	0.45	0.44	0.63
33213	White, no salt, canned in oil, drained	2	ounce(s)	57	36	105	15	0	0	5	0.94	1.41	1.92
33214	White, no salt, canned in water, drained	2	ounce(s)	57	42	73	13	0	0	2	0.45	0.44	0.63
	Yellowtail												
2970	Mixed species, raw	2	ounce(s)	57	42	83	13	0	0	3	0.73	1.13	0.81
8548	Mixed species, cooked, dry heat	3	ounce(s)	85	0.05	158.94	25.21	0	0	5.71	1.44	2.21	1.52
	Shellfish, meat only												
1857	Abalone, mixed species, fried	3	ounce(s)	85	51	161	17	9	0	6	1.40	2.33	1.42
16618	Abalone, steamed or poached	3	ounce(s)	85	41	177	29	10	0	1	0.25	0.18	0.18
	Crab												
1851	Blue crab, canned	2	ounce(s)	57	43	56	12	0	0	1	0.14	0.12	0.25
1852	Blue crab, cooked, moist heat	3	ounce(s)	85	66	87	17	0	0	2	0.19	0.24	0.58
8562	Dungeness crab, cooked, moist heat	3	ounce(s)	85	62	94	19	1	0	1	0.14	0.18	0.35

PAGE KEY: A–4 = Breads/Baked Goods   A–8 = Cereal/Rice/Pasta   A–12 = Fruit   A–18 = Vegetables/Legumes   A–28 = Nuts/Seeds   A–30 = Vegetarian
A–32 = Dairy   A–38 = Eggs   A–40 = Seafood   A–42 = Meats   A–44 = Poultry   A–46 = Processed meats   A–48 = Beverages   A–52 = Fats/Oils
A–54 = Sweets   A–54 = Sauces/Condiments/Spices   A–58 = Mixed Foods/Soups/Sandwiches   A–64 = Fast food   A–82 = Convenience   A–84 = Baby foods

A

Chol (mg)	Calc (mg)	Iron (mg)	Magn (mg)	Pota (mg)	Sodi (mg)	Zinc (mg)	Vit A (RAE) (µg)	Thia (mg)	Vit E (mg)	Ribo (mg)	Niac (mg)	Vit B$_6$ (mg)	Fola (µg)	Vit C (mg)	Vit B$_{12}$ (µg)	Sele (µg)
24	6	0.22	19	108	35	0.26	—	0.04	0.13	0.03	1.11	0.13	5	<1	<1	17
47	12	0.42	36	207	66	0.49	12	0.07	0.26	0.07	2.14	0.24	7	1	1	32
107	17	0.43	17	231	43	1.38	887	0.13	3.40	0.03	2.98	0.06	13	2	3	6
44	8	0.32	29	489	86	0.49	10	0.03	0.28	0.05	2.48	0.46	8	3	1	44
28.89	15.3	1.79	20.39	272	452.2	0.37	10.19	0.09	—	0.09	1.78	0.08	17	0	0.94	7.73
37	55	0.99	98	529	64	0.49	55	0.06	1.00	0.08	6.88	0.36	13	0	1	43
33	48	0.86	85	460	55	0.43	48	0.06	0.87	0.08	5.97	0.33	12	0	1	37
42	13	0.30	25	323	42	0.35	12	0.07	0.22	0.06	1.92	0.22	6	1	1	32
44	19	0.35	47	225	281	0.21	39	0.06	1.84	0.08	2.03	0.19	7	3	2	34
40	18	0.97	31	404	45	0.43	43	0.07	—	0.01	0.32	0.30	9	0	1	40
33	19	0.60	22	177	39	0.21	—	0.02	0.41	0.02	2.05	0.15	4	0	1	18
63	36	1.15	43	339	74	0.41	16	0.03	—	0.04	3.94	0.29	11	0	1	34
35	51	0.91	91	490	59	0.45	46	0.06	0.93	0.08	6.05	0.34	12	0	1	40
8	46	0.73	5	41	522	0.32	155	0.02	0.60	0.08	1.98	0.10	1	0	3	35
45	137	1.16	21	110	215	0.58	74	0.02	0.79	0.12	3.50	0.12	3	1	4	21
82	90	8.11	51	536	391	2.86	77	0.05	1.02	0.06	3.21	0.55	20	7	31	76
98	87	0.99	32	292	67	1.22	9	0.07	—	0.10	1.62	0.12	5	1	2	14
37	10	0.45	29	442	65	0.45	60	0.04	1.06	0.07	3.33	0.23	9	0	1	40
13	6	0.48	10	99	1134	0.17	15	0.01	—	0.05	2.67	0.15	1	0	2	22
40	15	1.02	27	377	99	0.56	—	0.14	1.54	0.05	8.33	0.19	4	2	2	41
43	10	0.29	26	383	40	0.37	48	0.08	—	0.09	5.79	0.56	11	1	2	11
34	108	0.70	9	95	121	0.31	16	0.01	0.05	0.05	1.25	0.04	3	0	2	13
28	20	0.38	27	155	216	0.49	10	0.01	—	0.05	0.69	0.06	23	1	1	13
27	21	0.22	—	238	366	—	—	—	—	—	—	—	—	2	—	—
40	34	0.20	31	444	48	0.37	30	0.05	—	0.00	0.29	0.39	5	1	3	42
221	33	0.86	32	237	260	1.48	9	0.05	—	0.39	2.21	0.05	12	4	1	44
227	31	0.63	29	192	356	1.49	—	0.02	1.17	0.32	1.70	0.04	4	3	1	—
88	16	0.92	43	279	75	0.43	26	0.10	—	0.03	2.17	0.29	9	0	4	40
63	11	0.59	30	239	389	0.36	—	0.07	—	0.07	8.31	0.19	14	0	2	—
26	8	0.22	37	95	122	0.28	17	0.02	—	0.02	0.19	0.03	2	0	1	24
43	5	0.88	29	314	98	1.25	35	0.04	—	0.10	10.02	0.32	2	1	2	52
38	14	0.62	43	378	31	0.44	15	0.37	0.43	0.04	8.33	0.77	2	1	<1	31
10	7	0.79	18	118	202	0.51	13	0.02	0.68	0.07	7.06	0.06	3	0	1	43
17	6	0.87	15	134	192	0.44	10	0.02	0.30	0.04	7.53	0.20	2	0	2	46
10	7	0.79	18	117	28	0.51	13	0.02	—	0.07	7.03	0.06	3	0	1	43
17	6	0.87	15	134	28	0.44	10	0.02	0.30	0.04	7.53	0.20	2	0	2	46
18	2	0.37	19	189	225	0.27	3	0.01	—	0.04	6.63	0.24	3	0	1	34
24	8	0.55	19	134	214	0.27	3	0.00	0.90	0.02	3.29	0.12	1	0	1	37
18	2	0.37	19	189	28	0.27	14	0.01	—	0.04	6.63	0.24	3	0	1	34
24	8	0.55	19	134	28	0.27	3	0.00	0.90	0.02	3.29	0.12	1	0	1	37
31	13	0.28	17	238	22	0.29	16	0.08	—	0.02	3.86	0.09	2	2	1	21
60.34	24.64	0.53	32.29	457.29	42.5	0.56	26.35	0.14	—	0.04	7.41	0.15	3.4	2.46	1.06	39.77
80	31	3.23	48	241	502	0.81	2	0.19	—	0.11	1.62	0.13	12	2	1	44
143	50	4.85	69	295	980	1.38	—	0.29	6.74	0.13	1.90	0.22	6	3	1	—
50	57	0.48	22	212	189	2.28	1	0.05	0.57	0.05	0.78	0.09	24	2	<1	18
85	88	0.77	28	275	237	3.59	2	0.09	0.85	0.04	2.81	0.15	43	3	6	34
65	50	0.37	49	347	321	4.65	26	0.05	—	0.17	3.08	0.15	36	3	9	40

Table A–1

Food Composition   (Computer code number is for Wadsworth Diet Analysis program)   (For purposes of calculations, use "0" for t, <1, <.1, <.01, etc.)

DA + Code	Food Description	Quantity	Measure	Wt (g)	H₂O (g)	Ener (cal)	Prot (g)	Carb (g)	Dietary Fiber (g)	Fat (g)	Fat Breakdown (g)		
											Sat	Mono	Poly
	SEAFOOD—Continued												
1860	**Clams, cooked, moist heat**	3	ounce(s)	85	54	126	22	4	0	2	0.16	0.15	0.47
1853	**Crayfish, farmed, cooked, moist heat**	3	ounce(s)	85	69	74	15	0	0	1	0.18	0.21	0.35
	Oysters												
8720	Baked or broiled	3	ounce(s)	85	69	90	6	3	0	6	1.38	2.18	1.88
152	Eastern, farmed, raw	3	ounce(s)	85	73	50	4	5	0	1	0.38	0.13	0.50
8715	Eastern, wild, cooked, moist heat	3	ounce(s)	85	60	116	12	7	0	4	1.31	0.53	1.65
8584	Pacific, cooked, moist heat	3	ounce(s)	85	55	139	16	8	0	4	0.87	0.66	1.52
1865	Pacific, raw	3	ounce(s)	85	70	69	8	4	0	2	0.43	0.30	0.76
1854	**Lobster, northern, cooked, moist heat**	3	ounce(s)	85	65	83	17	1	0	1	0.09	0.14	0.08
1862	**Mussels, blue, cooked, moist heat**	3	ounce(s)	85	52	146	20	6	0	4	0.72	0.86	1.03
	Shrimp												
1855	Mixed species, cooked, moist heat	3	ounce(s)	85	66	84	18	0	0	1	0.25	0.17	0.37
158	Mixed species, breaded, fried	3	ounce(s)	85	0.04	205.69	18.18	9.74	0.34	10.43	1.77	3.24	4.32
	BEEF, LAMB, PORK												
	Beef												
4450	Breakfast strips, cooked	2	slice(s)	23	0	101.47	7.07	0.31	0	7.77	3.24	3.8	0.35
174	Corned, canned	3	ounce(s)	85	49	213	23	0	0	13	5.25	5.07	0.54
33147	Cured, thin sliced	2	ounce(s)	57	31	87	18	2	0	1	0.54	0.48	0.04
4581	Jerky	1	ounce(s)	28	0	116.44	9.42	3.12	0.51	7.27	3.08	3.21	0.28
	Ground												
4411	Extra lean, broiled, well	3	ounce(s)	85	46	225	24	0	0	13	5.28	5.88	0.50
4417	Lean, broiled, medium	3	ounce(s)	85	47	231	21	0	0	16	6.16	6.87	0.59
4418	Lean, broiled, well	3	ounce(s)	85	45	238	24	0	0	15	5.89	6.56	0.56
4423	Regular, broiled, medium	3	ounce(s)	85	46	246	20	0	0	18	6.91	7.70	0.65
	Rib												
4183	Rib, whole, lean & fat, ¼" fat, roasted	3	ounce(s)	85	39	320	19	0	0	27	10.71	11.42	0.94
	Roast												
4264	Bottom round, lean & fat, ¼" fat, braised	3	ounce(s)	85	44	241	24	0	0	15	5.71	6.63	0.58
169	Bottom round, separable lean, ¼" fat, roasted	3	ounce(s)	85	0.05	160.64	24.45	0	0	6.26	2.13	2.83	0.24
4147	Chuck, arm pot roast, lean & fat, ¼" fat, braised	3	ounce(s)	85	41	282	23	0	0	20	7.97	8.68	0.77
4161	Chuck, blade roast, lean & fat, ¼" fat, braised	3	ounce(s)	85	40	293	23	0	0	22	8.70	9.44	0.78
5853	Chuck, blade roast, separable lean, ¼" trim, pot roasted	3	ounce(s)	85	0.04	209.1	27.45	0	0	10.15	3.94	4.37	0.33
4295	Eye of round, lean, ¼" fat, roasted	3	ounce(s)	85	55	149	25	0	0	5	1.76	2.06	0.15
4285	Eye of round, lean & fat, ¼" fat, roasted	3	ounce(s)	85	51	195	23	0	0	11	4.23	4.66	0.39
	Steak												
1757	Rib, small end, lean, ¼" fat, broiled	3	ounce(s)	85	49	188	24	0	0	10	3.84	4.01	0.27
4349	Short loin, T-bone steak, lean, ¼" fat, broiled	3	ounce(s)	85	52	174	23	0	0	9	3.05	4.23	0.26
4348	Short loin, T-bone steak, lean & fat, ¼" fat, broiled	3	ounce(s)	85	43	274	19	0	0	21	8.29	9.58	0.75
4360	Top loin, prime, lean & fat, ¼" fat, broiled	3	ounce(s)	85	43	275	22	0	0	20	8.16	8.61	0.73
	Variety												
188	Liver, pan fried	3	ounce(s)	85	53	149	23	4	0	4	1.27	0.56	0.49
4447	Tongue, simmered	3	ounce(s)	85	49	236	16	0	0	19	6.91	8.59	0.56
	Lamb												
	Chop												
3275	Loin, domestic, lean & fat, ¼" fat, broiled	3	ounce(s)	85	44	269	21	0	0	20	8.36	8.25	1.43
3287	Shoulder, arm, domestic, lean & fat, ¼" fat, braised	3	ounce(s)	85	38	294	26	0	0	20	8.39	8.65	1.45
3290	Shoulder, arm, domestic, lean, ¼" fat, braised	3	ounce(s)	85	42	237	30	0	0	12	4.28	5.24	0.78
	Leg												
3264	Domestic, lean & fat, ¼" fat, cooked	3	ounce(s)	85	46	250	21	0	0	18	7.51	7.50	1.28

PAGE KEY: A–4 = Breads/Baked Goods A–8 = Cereal/Rice/Pasta A–12 = Fruit A–18 = Vegetables/Legumes A–28 = Nuts/Seeds A–30 = Vegetarian A–32 = Dairy A–38 = Eggs A–40 = Seafood A–42 = Meats A–44 = Poultry A–46 = Processed meats A–48 = Beverages A–52 = Fats/Oils A–54 = Sweets A–54 = Sauces/Condiments/Spices A–58 = Mixed Foods/Soups/Sandwiches A–64 = Fast food A–82 = Convenience A–84 = Baby foods

A

Chol (mg)	Calc (mg)	Iron (mg)	Magn (mg)	Pota (mg)	Sodi (mg)	Zinc (mg)	Vit A (RAE) (µg)	Thia (mg)	Vit E (mg)	Ribo (mg)	Niac (mg)	Vit B$_6$ (mg)	Fola (µg)	Vit C (mg)	Vit B$_{12}$ (µg)	Sele (µg)
57	78	23.77	15	534	95	2.32	145	0.13	—	0.36	2.85	0.09	25	19	84	54
116	43	0.94	28	202	82	1.26	13	0.04	—	0.07	1.42	0.11	9	<1	3	29
42	37	5.30	38	126	418	72.22	60	0.07	1.27	0.06	1.04	0.05	8	3	15	—
21	37	4.91	28	105	151	32.23	7	0.09	—	0.06	1.08	0.05	15	4	14	54
89	77	10.19	81	239	359	154.37	46	0.16	—	0.15	2.11	0.10	12	5	30	61
85	14	7.82	37	257	180	28.25	124	0.11	—	0.38	3.08	0.08	13	11	24	131
43	7	4.35	19	143	90	14.14	69	0.06	0.72	0.20	1.71	0.04	9	7	14	65
61	52	0.33	30	299	323	2.48	22	0.01	0.85	0.06	0.91	0.07	9	0	3	36
48	28	5.71	31	228	314	2.27	77	0.26	—	0.36	2.55	0.09	65	12	20	76
166	33	2.63	29	155	190	1.33	58	0.03	0.43	0.03	2.20	0.11	3	2	1	34
150.44	56.95	1.07	34	191.25	292.39	1.17	47.59	0.1	—	0.11	2.6	0.08	20.39	1.27	1.58	35.44
26.89	2.03	0.7	6.1	93.11	509.17	1.43	0	0.02	0.03	0.05	1.46	0.07	1.8	0	0.77	6.05
73	10	1.77	12	116	855	3.03	0	0.02	0.13	0.12	2.07	0.11	8	0	1	36
45	3	1.58	11	140	1582	2.49	0	0.03	0.08	0.12	1.85	0.16	5	0	1	13
13.63	5.67	1.53	14.48	169.54	628.49	2.3	0	0.04	0.11	0.04	0.49	0.05	38.05	0	0.28	3.03
84	8	2.35	21	314	70	5.47	0	0.06	0.15	0.27	4.97	0.27	9	0	2	19
74	9	1.79	18	256	65	4.56	0	0.04	0.17	0.18	4.39	0.22	8	0	2	25
86	10	2.08	20	297	76	5.27	0	0.05	0.17	0.20	5.07	0.26	9	0	2	22
77	9	2.07	17	248	71	4.40	0	0.03	0.20	0.16	4.90	0.23	8	0	2	16
72	9	1.96	16	252	54	4.45	0	0.06	0.20	0.14	2.86	0.20	6	0	2	19
82	5	2.65	19	240	43	4.17	0	0.06	0.15	0.20	3.17	0.28	9	0	2	27
66.3	4.25	2.66	23.79	332.35	56.09	3.92	0	0.06	—	0.2	3.45	0.31	10.19	0	2.29	23.29
84	9	2.64	16	209	51	5.81	0	0.06	0.19	0.20	2.70	0.24	8	0	3	21
88	11	2.64	16	196	54	7.07	0	0.06	0.17	0.20	2.06	0.22	4	0	2	21
73.94	11.05	3.12	19.54	223.55	60.34	8.72	0	0.06	0.09	0.23	0	0.24	—	0	2.09	22.69
59	4	1.66	23	336	53	4.03	0	0.08	—	0.14	3.19	0.32	6	0	2	23
61	5	1.56	20	308	50	3.69	0	0.07	0.15	0.14	2.97	0.30	6	0	2	22
68	11	2.18	23	335	59	5.94	0	0.09	0.12	0.19	4.08	0.34	7	0	3	19
50	5	3.11	22	278	65	4.34	0	0.09	0.12	0.21	3.94	0.33	7	0	2	9
58	7	2.56	18	234	58	3.56	0	0.08	0.19	0.18	3.29	0.28	6	0	2	10
67	8	1.89	20	294	54	3.85	0	0.07	—	0.15	3.96	0.31	6	0	2	19
324	5	5.24	19	298	65	4.45	6582	0.15	0.54	2.91	14.85	0.87	221	1	71	28
112	4	2.22	13	156	55	34.77	0	0.02	0.30	0.25	2.97	0.13	6	1	3	11
85	17	1.54	20	278	65	2.96	0	0.09	0.11	0.21	6.04	0.11	15	0	2	23
102	21	2.03	22	260	61	5.17	0	0.06	0.13	0.21	5.66	0.09	15	0	2	32
103	22	2.30	25	287	65	6.21	0	0.06	0.15	0.23	5.38	0.11	19	0	2	32
82	14	1.60	20	264	61	3.79	0	0.09	0.12	0.21	5.66	0.11	15	0	2	22

Food Composition   (Computer code number is for Wadsworth Diet Analysis program)   (For purposes of calculations, use "0" for t, <1, <.1, <.01, etc.)

DA + Code	Food Description	Quantity	Measure	Wt (g)	H₂O (g)	Ener (cal)	Prot (g)	Carb (g)	Dietary Fiber (g)	Fat (g)	Fat Breakdown (g)		
											Sat	Mono	Poly
	BEEF, LAMB, PORK—Continued												
	Rib												
183	Domestic, lean, ¼" fat, broiled	3	ounce(s)	85	50	200	24	0	0	11	3.95	4.43	1.00
182	Domestic, lean & fat, ¼" fat, broiled	3	ounce(s)	85	40	307	19	0	0	25	10.80	10.30	2.01
	Shoulder												
187	Arm & blade, domestic, choice, lean, ¼" fat, roasted	3	ounce(s)	85	54	173	21	0	0	9	3.47	3.71	0.81
186	Arm & blade, domestic, choice, lean & fat, ¼" fat, roasted	3	ounce(s)	85	48	235	19	0	0	17	7.17	6.94	1.38
	Variety												
3375	Brain, pan fried	3	ounce(s)	85	52	232	14	0	0	19	4.82	3.42	1.94
3406	Tongue, braised	3	ounce(s)	85	49	234	18	0	0	17	6.66	8.50	1.06
	Pork												
	Cured												
161	Bacon, cured, broiled, pan fried or roasted	2	slice(s)	13	2	68	5	<1	0	5	1.73	2.33	0.57
29229	Bacon, Canadian style, cured	2	ounce(s)	57	38	89	12	1	0	4	1.26	1.79	0.36
35422	Breakfast strips, cured, cooked	3	slice(s)	34	9	156	10	<1	0	12	4.34	5.58	1.92
16561	Ham, smoked or cured, lean, cooked	1	slice(s)	42	28	66	11	0	0	2	0.77	1.06	0.27
189	Ham, cured, boneless, 11% fat, roasted	3	ounce(s)	85	55	151	19	0	0	8	2.65	3.77	1.20
1316	Ham, cured, extra lean, 5% fat, roasted	3	ounce(s)	85	58	123	18	1	0	5	1.54	2.23	0.46
29215	Ham, cured, extra lean, 4% fat, canned	2	ounce(s)	57	42	68	10	0	0	3	0.86	1.25	0.22
	Chop												
32671	Loin, blade, lean & fat, pan fried	3	ounce(s)	85	42	291	18	0	0	24	8.65	9.97	2.64
32672	Loin, center cut, lean & fat, pan fried	3	ounce(s)	85	45	236	25	0	0	14	5.11	6.00	1.62
32682	Loin, center rib, boneless, lean & fat, braised	3	ounce(s)	85	49	217	22	0	0	13	5.21	6.13	1.12
32603	Loin, center rib, lean, broiled	3	ounce(s)	85	48	186	26	0	0	8	2.94	3.78	0.53
32481	Loin, whole, lean, braised	3	ounce(s)	85	52	174	24	0	0	8	2.87	3.54	0.60
32478	Loin, whole, lean & fat, braised	3	ounce(s)	85	50	203	23	0	0	12	4.35	5.15	1.00
	Leg or ham												
32471	Rump portion, lean & fat, roasted	3	ounce(s)	85	48	214	25	0	0	12	4.47	5.42	1.17
32468	Whole, lean & fat, roasted	3	ounce(s)	85	47	232	23	0	0	15	5.50	6.70	1.43
	Ribs												
32696	Loin, country style, lean, roasted	3	ounce(s)	85	49	210	23	0	0	13	4.52	5.49	0.94
32693	Loin, country style, lean & fat, roasted	3	ounce(s)	85	43	279	20	0	0	22	7.83	9.36	1.71
	Shoulder												
32629	Arm picnic, lean, roasted	3	ounce(s)	85	51	194	23	0	0	11	3.66	5.09	1.02
32626	Arm picnic, lean & fat, roasted	3	ounce(s)	85	44	270	20	0	0	20	7.47	9.12	2.00
	Rabbit												
3366	Domesticated, roasted	3	ounce(s)	85	52	167	25	0	0	7	2.04	1.84	1.33
3367	Domesticated, stewed	3	ounce(s)	85	50	175	26	0	0	7	2.13	1.93	1.39
	Veal												
3391	Liver, braised	3	ounce(s)	85	51	163	24	3	0	5	1.69	0.97	0.88
3319	Rib, lean only, roasted	3	ounce(s)	85	55	150	22	0	0	6	1.77	2.26	0.57
1732	**Deer or venison, roasted**	3	ounce(s)	85	55	134	26	0	0	3	1.06	0.75	0.53
	POULTRY												
	Chicken												
29562	Flaked, canned	2	ounce(s)	57	0.03	97.47	10.37	0.05	0	5.87	1.62	2.32	1.29
	Fried												
29632	Breast, meat only, breaded, baked or fried	3	ounce(s)	85	44	193	25	7	<1	7	1.62	2.66	1.73
35327	Broiler breast, meat only, fried	3	ounce(s)	85	51	159	28	<1	0	4	1.10	1.46	0.91
36413	Broiler breast, meat & skin, flour coated, fried	3	ounce(s)	85	48	189	27	1	<.1	8	2.08	2.98	1.67
35389	Broiler drumstick, meat only, fried	3	ounce(s)	85	53	166	24	0	0	7	1.81	2.50	1.68
36414	Broiler drumstick, meat & skin, flour coated, fried	3	ounce(s)	85	48	208	23	1	<.1	12	3.11	4.61	2.75
35406	Broiler leg, meat only, fried	3	ounce(s)	85	52	177	24	1	0	8	2.12	2.92	1.89
35484	Broiler wing, meat only, fried	3	ounce(s)	85	51	179	26	0	0	8	2.13	2.62	1.76
29580	Patty, fillet, or tenders, breaded, cooked	3	ounce(s)	85	42	241	14	13	<1	15	4.62	7.25	1.87

PAGE KEY: A–4 = Breads/Baked Goods   A–8 = Cereal/Rice/Pasta   A–12 = Fruit   A–18 = Vegetables/Legumes   A–28 = Nuts/Seeds   A–30 = Vegetarian
A–32 = Dairy   A–38 = Eggs   A–40 = Seafood   A–42 = Meats   A–44 = Poultry   A–46 = Processed meats   A–48 = Beverages   A–52 = Fats/Oils
A–54 = Sweets   A–54 = Sauces/Condiments/Spices   A–58 = Mixed Foods/Soups/Sandwiches   A–64 = Fast food   A–82 = Convenience   A–84 = Baby foods

A

Chol (mg)	Calc (mg)	Iron (mg)	Magn (mg)	Pota (mg)	Sodi (mg)	Zinc (mg)	Vit A (RAE) (µg)	Thia (mg)	Vit E (mg)	Ribo (mg)	Niac (mg)	Vit B$_6$ (mg)	Fola (µg)	Vit C (mg)	Vit B$_{12}$ (µg)	Sele (µg)
77	14	1.88	25	266	72	4.48	0	0.09	0.15	0.21	5.57	0.13	18	0	2	26
84	16	1.60	20	230	65	3.40	0	0.08	0.10	0.19	5.95	0.09	12	0	2	20
74	16	1.81	21	225	58	5.13	0	0.08	0.15	0.22	4.90	0.13	21	0	2	24
78	17	1.67	20	213	56	4.45	0	0.08	0.12	0.20	5.23	0.11	18	0	2	22
2128	18	1.73	19	304	133	1.70	0	0.14	—	0.31	3.87	0.20	6	20	20	10
161	9	2.24	14	134	57	2.54	0	0.07	—	0.36	3.14	0.14	3	6	5	24
14	1	0.18	4	71	291	0.44	1	0.05	0.07	0.03	1.40	0.04	<1	0	<1	8
28	5	0.39	10	195	799	0.79	0	0.43	0.15	0.10	3.53	0.22	2	0	<1	14
36	5	0.67	9	158	714	1.25	0	0.25	0.10	0.13	2.58	0.12	1	0	1	8
23	3	0.40	9	133	557	1.08	0	0.29	0.11	0.11	2.11	0.20	2	0	<1	—
50	7	1.14	19	348	1275	2.10	0	0.62	0.22	0.28	5.23	0.26	3	0	1	17
45	7	1.26	12	244	1023	2.45	0	0.64	0.22	0.17	3.42	0.34	3	0	1	17
22	3	0.53	10	206	712	1.09	0	0.47	0.15	0.13	3.01	0.26	3	0	<1	8
72	26	0.75	18	282	57	2.71	3	0.53	0.22	0.25	3.36	0.29	3	1	1	30
78	23	0.77	25	361	68	1.96	2	0.97	0.22	0.26	4.76	0.40	5	1	1	33
62	4	0.78	14	329	34	1.76	2	0.45	—	0.21	3.67	0.26	3	<1	<1	28
69	26	0.70	24	357	55	2.02	2	0.95	0.22	0.28	5.25	0.40	3	<1	1	40
67	15	0.96	17	329	43	2.11	2	0.56	0.22	0.23	3.90	0.33	3	1	<1	41
68	18	0.91	16	318	41	2.02	2	0.54	0.22	0.22	3.76	0.31	3	1	<1	39
82	10	0.89	23	318	53	2.40	3	0.64	0.22	0.28	3.96	0.27	3	<1	1	40
80	12	0.86	19	299	51	2.52	3	0.54	0.22	0.27	3.89	0.34	9	<1	1	39
79	25	1.10	20	297	25	3.24	2	0.49	—	0.29	3.97	0.37	4	<1	1	36
78	21	0.90	20	293	44	2.01	3	0.76	—	0.29	3.67	0.38	4	<1	1	32
81	8	1.21	17	299	68	3.46	2	0.49	—	0.30	3.67	0.35	4	<1	1	33
80	16	1.00	14	276	60	2.93	2	0.44	—	0.26	3.33	0.30	3	<1	1	29
70	16	1.93	18	326	40	1.93	0	0.08	—	0.18	7.17	0.40	9	0	7	33
73	17	2.01	17	255	31	2.01	0	0.05	0.67	0.14	6.09	0.29	8	0	6	33
434	5	4.34	17	280	66	9.55	17973	0.15	0.29	2.43	11.18	0.78	281	1	72	16
98	10	0.82	20	264	82	3.82	0	0.05	0.31	0.25	6.38	0.23	12	0	1	9
95	6	3.80	20	285	46	2.34	0	0.15	—	0.51	5.70	—	—	0	—	11
35.34	7.98	0.9	6.84	148.19	410.39	0.8	19.37	0	—	0.07	3.6	0.19	—	0	0.16	—
67	19	1.05	25	223	450	0.84	—	0.08	0.72	0.10	10.98	0.47	4	0	<1	—
77	14	0.97	26	235	67	0.92	—	0.07	0.36	0.11	12.57	0.54	3	0	<1	22
76	14	1.01	26	220	65	0.94	—	0.07	0.39	0.11	11.69	0.49	5	0	<1	20
80	10	1.12	20	212	82	2.74	—	0.07	0.42	0.20	5.23	0.33	8	0	<1	17
77	10	1.14	20	195	76	2.46	—	0.07	0.66	0.19	5.13	0.30	9	0	<1	16
84	11	1.19	21	216	82	2.53	—	0.07	0.38	0.21	5.69	0.33	8	0	<1	16
71	13	0.97	18	177	77	1.80	—	0.04	0.41	0.11	6.16	0.50	3	0	<1	22
51	14	1.06	17	209	452	0.88	—	0.08	1.66	0.12	5.71	0.26	9	<1	<1	—

Table A–1

Food Composition   (Computer code number is for Wadsworth Diet Analysis program)   (For purposes of calculations, use "0" for t, <1, <.1, <.01, etc.)

DA + Code	Food Description	Quantity	Measure	Wt (g)	H₂O (g)	Ener (cal)	Prot (g)	Carb (g)	Dietary Fiber (g)	Fat (g)	Fat Breakdown (g) Sat	Mono	Poly
	POULTRY—Continued												
	Roasted, meat only												
35409	Broiler chicken leg	3	ounce(s)	85	55	162	23	0	0	7	1.95	2.59	1.68
35486	Broiler chicken wing	3	ounce(s)	85	53	173	26	0	0	7	1.92	2.22	1.51
35138	Roasting chicken, dark meat	3	ounce(s)	85	57	151	20	0	0	7	2.07	2.82	1.70
35136	Roasting chicken, light meat	3	ounce(s)	85	58	130	23	0	0	3	0.92	1.29	0.79
35132	Roasting chicken	3	ounce(s)	85	57	142	21	0	0	6	1.54	2.13	1.28
	Stewed												
3174	Meat only, stewed	3	ounce(s)	85	0.05	150.44	23.19	0	0	5.7	1.56	2.03	1.3
1268	Gizzard, simmered	3	ounce(s)	85	58	124	26	0	0	2	0.57	0.45	0.30
1270	Liver, simmered	3	ounce(s)	85	57	142	21	1	0	6	1.75	1.20	1.08
	Duck												
1286	Domesticated, meat & skin, roasted	3	ounce(s)	85	44	286	16	0	0	24	8.22	10.97	3.10
1287	Domesticated, meat only, roasted	3	ounce(s)	85	55	171	20	0	0	10	3.54	3.15	1.22
	Goose												
35507	Domesticated, meat & skin, roasted	3	ounce(s)	85	44	259	21	0	0	19	5.84	8.72	2.14
35524	Domesticated, meat only, roasted	3	ounce(s)	85	49	202	25	0	0	11	3.88	3.69	1.31
1297	Liver pâté, smoked, canned	4	tablespoon(s)	52	19	240	6	2	0	23	7.51	13.32	0.44
	Turkey												
3256	Ground turkey, cooked	3	ounce(s)	85	51	200	23	0	0	11	2.88	4.16	2.75
222	Roasted, fryer roaster breast, meat only	3	ounce(s)	85	58	115	26	0	0	1	0.20	0.11	0.17
219	Roasted, dark meat, meat only	3	ounce(s)	85	54	159	24	0	0	6	2.06	1.39	1.84
220	Roasted, light meat, meat only	3	ounce(s)	85	56	133	25	0	0	3	0.88	0.48	0.73
3263	Patty, batter coated, breaded, fried	1	item(s)	94	47	266	13	15	<1	17	4.41	7.02	4.43
1302	Turkey roll, light meat	2	slice(s)	57	41	83	11	<1	0	4	1.15	1.42	0.99
1303	Turkey roll, light & dark meat	2	slice(s)	57	40	84	10	1	0	4	1.16	1.30	1.01
	PROCESSED MEATS												
	Beef												
1331	Corned beef loaf, jellied, sliced	2	slice(s)	57	39	87	13	0	0	3	1.47	1.52	0.18
	Bologna												
13458	Made w/chicken, pork, & beef	1	slice(s)	28	15	90	3	1	0	8	3.00	4.05	1.10
13461	Light, made w/pork, chicken, & beef	1	slice(s)	28	18	60	3	2	0	4	1.50	2.04	0.43
13459	Beef	1	slice(s)	28	15	90	3	1	0	8	3.50	4.26	0.31
13565	Turkey	1	slice(s)	28	19	50	3	1	0	4	1.00	1.09	0.98
	Chicken												
13562	Oven roasted white chicken	1	slice(s)	28	20	40	4	1	0	3	0.50	—	—
	Ham												
13581	Honey glazed, traditional carved	2	slice(s)	45	—	50	8	1	0	2	0.50	0.68	0.18
13777	Deli sliced cooked	1	slice(s)	28	—	30	5	1	0	1	0.50	0.39	0.11
13778	Deli sliced honey	1	slice(s)	28	—	35	5	1	0	1	0.50	0.39	0.11
8614	**Pork & beef mortadella, sliced**	2	slice(s)	46	24	143	8	1	0	12	4.37	5.23	1.44
1323	Pork olive loaf	2	slice(s)	57	33	133	7	5	0	9	3.32	4.47	1.10
1324	**Pork pickle & pimento loaf**	2	slice(s)	57	32	149	7	3	0	12	4.45	5.45	1.47
	Sausages & frankfurters												
37296	Beerwurst beef beer salami (bierwurst)	1	slice(s)	29	17	74	4	1	0	6	2.50	2.69	0.21
37257	Beerwurst pork beer salami	1	slice(s)	21	13	50	3	<1	0	4	1.32	1.89	0.50
35338	Berliner, pork & beef	1	ounce(s)	28	17	65	4	1	0	5	1.72	2.27	0.45
37299	Braunschweiger pork liver sausage	1	slice(s)	15	0	51.34	1.97	0.34	0	4.48	1.52	2.08	0.52
37298	Bratwurst pork, cooked	1	piece(s)	74	42	181	10	2	0	14	5.15	6.73	1.51
1329	Cheesefurter or cheese smokie, beef & pork	1	item(s)	43	23	141	6	1	0	12	4.52	5.89	1.30
1330	Chorizo, beef & pork	2	ounce(s)	57	18	258	14	1	0	22	8.15	10.43	1.96
8600	Frankfurter, beef	1	item(s)	45	23	149	5	2	0	13	5.26	6.44	0.53
202	Frankfurter, beef & pork	1	item(s)	57	32	174	7	1	1	16	6.14	7.79	1.56
1293	Frankfurter, chicken	1	item(s)	45	26	116	6	3	0	9	2.49	3.82	1.82
3261	Frankfurter, turkey	1	item(s)	45	28	102	6	1	0	8	2.65	2.51	2.25
37275	Italian sausage, pork, cooked	1	item(s)	68	34	220	14	1	0	17	6.14	8.13	2.23
37307	Kielbasa, kolbassa, pork & beef	2⅛	ounce(s)	61	37	135	10	2	0	9	3.40	4.44	1.06
1333	Knockwurst or knackwurst, beef & pork	2	ounce(s)	57	31	174	6	2	0	16	5.79	7.26	1.66
37285	Pepperoni, beef & pork	1	slice(s)	11	3	55	2	<1	0	5	1.77	2.32	0.48

PAGE KEY: A–4 = Breads/Baked Goods   A–8 = Cereal/Rice/Pasta   A–12 = Fruit   A–18 = Vegetables/Legumes   A–28 = Nuts/Seeds   A–30 = Vegetarian   A–32 = Dairy   A–38 = Eggs   A–40 = Seafood   A–42 = Meats   A–44 = Poultry   A–46 = Processed meats   A–48 = Beverages   A–52 = Fats/Oils   A–54 = Sweets   A–54 = Sauces/Condiments/Spices   A–58 = Mixed Foods/Soups/Sandwiches   A–64 = Fast food   A–82 = Convenience   A–84 = Baby foods

A

Chol (mg)	Calc (mg)	Iron (mg)	Magn (mg)	Pota (mg)	Sodi (mg)	Zinc (mg)	Vit A (RAE) (µg)	Thia (mg)	Vit E (mg)	Ribo (mg)	Niac (mg)	Vit B$_6$ (mg)	Fola (µg)	Vit C (mg)	Vit B$_{12}$ (µg)	Sele (µg)
80	10	1.11	20	206	77	2.43	—	0.06	0.23	0.20	5.37	0.32	7	0	<1	19
72	14	0.99	18	179	78	1.82	—	0.04	0.23	0.11	6.22	0.50	3	0	<1	21
64	9	1.13	17	191	81	1.81	14	0.05	—	0.16	4.88	0.26	6	0	<1	17
64	11	0.92	20	201	43	0.66	7	0.05	0.23	0.08	8.90	0.46	3	0	<1	22
64	10	1.03	18	195	64	1.29	10	0.05	—	0.13	6.70	0.35	4	0	<1	21
70.55	11.89	0.99	17.85	153	59.5	1.69	12.75	0.04	0.18	0.13	5.19	0.22	5.09	0	0.18	17.76
315	14	2.71	3	152	48	3.76	0	0.02	1.01	0.18	2.65	0.06	4	0	1	35
479	9	9.89	21	224	65	3.38	3384	0.25	1.22	1.69	9.39	0.64	491	24	14	70
71	9	2.30	14	173	50	1.58	54	0.15	0.60	0.23	4.10	0.15	5	0	<1	17
76	10	2.30	17	214	55	2.21	20	0.22	0.60	0.40	4.34	0.21	9	0	<1	19
77	11	2.41	19	280	60	2.23	18	0.07	1.48	0.28	3.55	0.32	2	0	<1	19
82	12	2.44	21	330	65	2.70	10	0.08	—	0.33	3.47	0.40	10	0	<1	22
78	36	2.86	7	72	362	0.48	521	0.05	—	0.16	1.31	0.03	31	0	5	23
87	21	1.64	20	230	91	2.43	0	0.05	0.29	0.14	4.10	0.33	6	0	<1	32
71	10	1.30	25	248	44	1.48	0	0.04	0.08	0.11	6.37	0.48	5	0	<1	27
72	27	1.98	20	247	67	3.79	0	0.05	0.54	0.21	3.10	0.31	8	0	<1	35
59	16	1.15	24	259	54	1.73	0	0.05	0.08	0.11	5.81	0.46	5	0	<1	27
58	13	2.07	14	259	752	1.35	10	0.09	2.25	0.18	2.16	0.19	26	0	<1	19
24	23	0.73	9	142	277	0.88	0	0.05	0.08	0.13	3.97	0.18	2	0	<1	13
31	18	0.77	10	153	332	1.13	0	0.05	0.19	0.16	2.72	0.15	3	0	<1	17
27	6	1.16	6	57	540	2.32	0	0.00	0.11	0.06	1.00	0.07	5	0	1	10
30	0	0.36	6	43	290	0.40	0	—	—	—	—	—	—	0	—	—
15	0	0.36	6	46	310	0.45	0	—	—	—	—	—	—	0	—	—
20	0	0.36	4	47	310	0.57	0	0.01	—	0.03	0.68	0.05	4	0	<1	—
20	40	0.36	6	43	270	0.52	0	—	—	—	—	—	—	0	—	—
15	0	0.36	7	85	350	0.32	0	—	—	—	—	—	—	0	—	—
25	0	0.72	—	—	560	—	0	—	—	—	—	—	—	0	—	—
15	0	0.00	—	—	240	—	0	—	—	—	—	—	—	0	—	—
15	0	0.00	—	—	240	—	0	—	—	—	—	—	—	0	—	—
26	8	0.64	5	75	573	0.97	0	0.05	0.10	0.07	1.23	0.06	1	0	1	10
22	62	0.31	11	169	843	0.78	34	0.17	0.14	0.15	1.04	0.13	1	0	1	9
21	54	0.58	10	193	789	0.80	12	0.17	0.14	0.14	1.17	0.11	3	0	1	8
18	3	0.44	4	67	265	0.71	0	0.02	0.06	0.04	0.99	0.05	1	0	1	5
12	2	0.16	3	53	261	0.36	0	0.12	—	0.04	0.69	0.07	1	0	<1	—
13	3	0.33	4	80	368	0.70	0	0.11	0.06	0.06	0.88	0.06	1	0	1	4
23.69	1.36	1.42	1.67	27.49	131.54	0.42	641.01	0.03	0.04	0.23	1.27	0.05	—	0	3.05	8.81
44	33	0.96	11	157	412	1.70	0	0.37	0.19	0.14	2.37	0.16	1	1	1	16
29	25	0.46	6	89	465	0.97	20	0.11	0.14	0.07	1.25	0.06	1	0	1	7
50	5	0.90	10	226	700	1.93	0	0.36	0.12	0.17	2.91	0.30	1	0	1	12
24	6	0.68	6	70	513	1.11	0	0.02	0.09	0.07	1.07	0.04	2	0	1	4
29	6	0.66	6	95	638	1.05	10	0.11	0.14	0.07	1.50	0.07	2	0	1	8
45	43	0.90	5	38	617	0.47	18	0.03	0.10	0.05	1.39	0.14	2	0	<1	8
48	48	0.83	6	81	642	1.40	0	0.02	0.28	0.08	1.86	0.10	4	0	<1	7
53	16	1.02	12	207	627	1.62	0	0.42	0.17	0.16	2.83	0.22	3	1	1	15
41	27	0.88	10	169	566	1.23	0	0.14	0.13	0.13	1.75	0.11	3	0	1	11
34	6	0.37	6	113	527	0.94	0	0.19	0.32	0.08	1.55	0.10	1	0	1	8
9	1	0.15	2	38	224	0.28	0	0.04	0.02	0.03	0.55	0.03	<1	0	<1	—

A
Food Composition   (Computer code number is for Wadsworth Diet Analysis program)   (For purposes of calculations, use "0" for t, <1, <.1, <.01, etc.)

DA + Code	Food Description	Quantity	Measure	Wt (g)	H₂O (g)	Ener (cal)	Prot (g)	Carb (g)	Dietary Fiber (g)	Fat (g)	Fat Breakdown (g)		
											Sat	Mono	Poly
	PROCESSED MEATS—Continued												
37313	Polish sausage, pork	2	slice(s)	57	31	163	8	2	—	14	4.91	6.42	1.46
206	Salami, beef, cooked, sliced	2	slice(s)	46	28	119	6	1	0	10	4.54	4.90	0.48
37272	Salami, pork, dry or hard	1	slice(s)	13	5	52	3	<1	0	4	1.52	2.05	0.48
3262	Salami, turkey	2	slice(s)	57	31	125	8	11	<.1	5	1.98	1.80	1.43
7162	Sausage, breakfast, turkey	2½	ounce(s)	100	67	190	17	<1	0	13	3.90	6.23	3.33
8620	Smoked sausage, beef & pork	2	ounce(s)	57	31	181	7	1	0	16	5.54	6.94	2.23
8619	Smoked, sausage, pork	2	ounce(s)	57	22	221	13	1	0	18	6.42	8.30	2.13
37273	Smoked, sausage, pork link	1	piece(s)	76	30	295	17	2	—	24	8.58	11.09	2.85
1336	Summer sausage, thuringer, or cervelat, beef & pork	2	ounce(s)	57	29	190	9	<1	0	17	6.82	7.35	0.68
37294	Vienna sausage, cocktail, beef & pork, canned	1	piece(s)	16	10	45	2	<1	0	4	1.49	2.01	0.27
	Spreads												
32419	Pork & beef sandwich spread	4	tablespoon(s)	60	36	141	5	7	<1	10	3.59	4.57	1.54
1318	Ham salad spread	¼	cup(s)	60	38	130	5	6	0	9	3.04	4.32	1.62
	Turkey												
16049	Breast, hickory smoked, slices	1	slice(s)	56	—	50	11	1	0	0	0.00	0.00	0.00
13606	Breast, hickory smoked fat free	1	slice(s)	28	—	25	4	1	0	0	0.00	0.00	0.00
16047	Breast, honey roasted, slices	1	slice(s)	56	—	60	11	2	0	0	0.00	0.00	0.00
16048	Breast, oven roasted, slices	1	slice(s)	56	—	50	11	1	0	0	0.00	0.00	0.00
13583	Breast, traditional carved	2	slice(s)	45	—	40	9	0	0	1	0.00	0.07	0.14
13604	Breast, oven roasted, fat free	1	slice(s)	28	—	25	4	1	0	0	0.00	0.00	0.00
13567	Turkey ham, 10% water added	1	slice(s)	28	20	35	5	0	0	1	0.00	0.22	0.31
13596	Turkey pastrami	2	ounce(s)	56	—	70	11	1	0	2	1.00	—	—
13597	Turkey salami	2	ounce(s)	56	—	120	8	1	0	9	2.50	2.92	2.30
	BEVERAGES												
	Alcoholic												
	Beer												
866	Ale, mild	12	fluid ounce(s)	360	332	148	1	13	1	0	0.00	0.00	0.00
686	Beer	12	fluid ounce(s)	356	336	118	1	6	<1	<1	0.00	0.00	0.00
869	Beer, light	12	fluid ounce(s)	354	337	99	1	5	0	0	0.00	0.00	0.00
16886	Beer, nonalcoholic	12	fluid ounce(s)	360	353	32	1	5	0	0	0.00	0.00	0.00
31608	Budweiser beer	12	fluid ounce(s)	355	328	143	1	11	0	0	0.00	0.00	0.00
31609	Bud Light beer	12	fluid ounce(s)	355	335	110	1	7	0	0	0.00	0.00	0.00
31613	Michelob Beer	12	fluid ounce(s)	355	323	155	1	13	0	0	0.00	0.00	0.00
31614	Michelob Light beer	12	fluid ounce(s)	355	330	134	1	12	0	0	0.00	0.00	0.00
	Gin, rum, vodka, whiskey												
687	Distilled alcohol, 80 proof	1	fluid ounce(s)	28	19	64	0	0	0	0	0.00	0.00	0.00
688	Distilled alcohol, 86 proof	1	fluid ounce(s)	28	18	70	0	<.1	0	0	0.00	0.00	0.00
689	Distilled alcohol, 90 proof	1	fluid ounce(s)	28	17	73	0	0	0	0	0.00	0.00	0.00
856	Distilled alcohol, 94 proof	1	fluid ounce(s)	28	17	76	0	0	0	0	0.00	0.00	0.00
857	Distilled alcohol, 100 proof	1	fluid measure(s)	28	16	82	0	0	0	0	0.00	0.00	0.00
	Liqueurs												
3142	Coffee liqueur, 63 proof	1	fluid ounce(s)	35	14	107	<.1	11	0	<1	0.04	0.01	0.04
33187	Coffee liqueur, 53 proof	1	fluid ounce(s)	35	11	117	<.1	16	0	<1	0.04	0.01	0.04
736	Cordials, 54 proof	1	fluid ounce(s)	30	9	106	<.1	13	0	<.1	0.02	0.01	0.04
	Wine												
858	Champagne, domestic	5	fluid ounce(s)	150	—	105	<1	4	0	0	0.00	0.00	0.00
861	Red wine, California	5	fluid ounce(s)	150	133	125	<1	4	0	0	0.00	0.00	0.00
690	Sweet dessert wine	5	fluid ounce(s)	150	106	240	<1	21	0	0	0.00	0.00	0.00
1481	White wine	5	fluid ounce(s)	148	132	100	<1	1	0	0	0.00	0.00	0.00
1811	Wine cooler	10	fluid ounce(s)	300	270	150	<1	18	<.1	<.1	0.01	0.00	0.02
	Carbonated												
692	Club soda	12	fluid ounce(s)	355	355	0	0	0	0	0	0.00	0.00	0.00
12010	Coca-Cola Classic cola soda	12	fluid ounce(s)	360	—	146	0	41	0	0	0.00	0.00	0.00
12031	Coke diet cola soda	12	fluid ounce(s)	360	—	2	0	<1	0	0	0.00	0.00	0.00
693	Cola	12	fluid ounce(s)	426	380	179	<1	46	0	0	0.00	0.00	0.00
9522	Cola soda, decaffeinated	12	fluid ounce(s)	372	331	156	<1	40	0	0	0.00	0.00	0.00
1415	Cola, low calorie w/aspartame	12	fluid ounce(s)	355	354	4	<1	<1	0	0	0.00	0.00	0.00

PAGE KEY: A–4 = Breads/Baked Goods   A–8 = Cereal/Rice/Pasta   A–12 = Fruit   A–18 = Vegetables/Legumes   A–28 = Nuts/Seeds   A–30 = Vegetarian   A–32 = Dairy   A–38 = Eggs   A–40 = Seafood   A–42 = Meats   A–44 = Poultry   A–46 = Processed meats • A–48 = Beverages   A–52 = Fats/Oils   A–54 = Sweets   A–54 = Sauces/Condiments/Spices   A–58 = Mixed Foods/Soups/Sandwiches   A–64 = Fast food   A–82 = Convenience   A–84 = Baby foods

A

Chol (mg)	Calc (mg)	Iron (mg)	Magn (mg)	Pota (mg)	Sodi (mg)	Zinc (mg)	Vit A (RAE) (µg)	Thia (mg)	Vit E (mg)	Ribo (mg)	Niac (mg)	Vit B_6 (mg)	Fola (µg)	Vit C (mg)	Vit B_{12} (µg)	Sele (µg)
40	7	0.82	8	102	546	1.10	0	0.29	—	0.08	1.96	0.11	1	1	1	10
33	3	1.01	6	86	524	0.81	0	0.05	0.09	0.09	1.49	0.08	1	0	1	7
10	2	0.17	3	48	289	0.54	0	0.12	0.04	0.04	0.72	0.07	<1	0	<1	3
45	42	0.87	15	225	616	1.76	1	0.24	0.32	0.17	2.26	0.24	6	12	1	11
92	57	2.20	18	188	665	2.07	0	0.04	0.24	0.12	3.55	0.29	5	1	<1	—
33	7	0.43	7	101	517	0.71	7	0.11	0.12	0.06	1.67	0.09	1	0	<1	0
39	17	0.66	11	191	851	1.60	0	0.40	0.14	0.15	2.57	0.20	3	1	1	12
52	23	0.88	14	255	1137	2.14	0	0.53	0.19	0.20	3.43	0.27	4	0	1	16
43	7	1.44	8	154	704	1.45	0	0.09	0.12	0.19	2.44	0.15	1	0	3	12
8	2	0.14	1	16	152	0.26	0	0.01	0.04	0.02	0.26	0.02	1	0	<1	3
23	7	0.47	5	66	608	0.61	16	0.10	1.04	0.08	1.04	0.07	1	0	1	6
22	5	0.35	6	90	547	0.66	0	0.26	1.04	0.07	1.26	0.09	1	0	<1	11
25	0	0.72	—	—	730	—	0	—	—	—	—	—	—	0	—	—
10	0	0.00	—	—	300	—	0	—	—	—	—	—	—	0	—	—
20	0	0.72	—	—	640	—	0	—	—	—	—	—	—	0	—	—
20	0	0.72	—	—	620	—	0	—	—	—	—	—	—	0	—	—
20	0	0.72	—	—	540	—	0	—	—	—	—	—	—	0	—	—
10	0	0.00	—	—	330	—	0	—	—	—	—	—	—	0	—	—
20	0	0.36	6	81	310	0.73	0	—	—	—	—	—	—	0	—	—
40	0	0.72	—	—	590	—	0	—	—	—	—	—	—	0	—	—
50	40	0.72	—	—	500	—	0	—	—	—	—	—	—	0	—	—
0	18	0.11	—	—	18	—	0	0.02	—	0.10	1.63	—	—	0	<.1	—
0	18	0.07	21	89	14	0.04	0	0.02	0.00	0.09	1.61	0.18	21	0	<.1	2
0	18	0.14	18	64	11	0.11	0	0.03	0.00	0.11	1.39	0.12	14	0	<.1	2
0	25	0.04	32	90	18	0.04	—	0.02	0.00	0.10	1.63	0.18	22	0	<.1	—
0	18	0.11	21	89	9	0.07	0	0.02	0.00	0.09	1.61	0.18	21	0	<.1	4
0	18	0.14	18	64	9	0.11	0	0.03	0.00	0.11	1.39	0.12	15	0	<.1	4
0	18	0.11	21	89	9	0.07	0	0.02	0.00	0.09	1.61	0.18	21	0	<.1	4
0	18	0.14	18	64	9	0.11	0	0.03	0.00	0.11	1.39	0.12	15	0	<.1	4
0	0	0.01	0	1	<1	0.01	0	0.00	0.00	0.00	0.00	0.00	0	0	0	0
0	0	0.01	0	1	<1	0.01	0	0.00	0.00	0.00	0.00	0.00	0	0	0	0
0	0	0.01	0	1	<1	0.01	0	0.00	0.00	0.00	0.00	0.00	0	0	0	0
0	0	0.01	0	1	<1	0.01	0	0.00	—	0.00	0.00	0.00	0	0	0	0
0	0	0.01	0	1	<1	0.01	0	0.00	—	0.00	0.00	0.00	0	0	0	0
0	<1	0.02	1	10	3	0.01	0	0.00	—	0.00	0.05	0.00	0	0	0	<1
0	<1	0.02	1	10	3	0.01	0	0.00	0.00	0.00	0.05	0.00	0	0	0	<1
0	<1	0.02	<1	5	2	0.01	0	0.00	—	0.00	0.02	0.00	0	0	0	—
0	—	—	—	—	—	—	—	—	—	—	0.00	—	—	—	0	—
0	12	1.43	16	171	15	0.15	0	0.02	0.00	0.04	0.12	0.05	1	0	<.1	—
0	12	0.36	14	138	14	0.11	0	0.03	0.00	0.03	0.32	0.00	0	0	0	1
0	13	0.47	15	118	7	0.10	0	0.01		0.01	0.10	0.02	0	0	0	<1
0	17	0.81	16	135	25	0.17	0	0.01	0.02	0.02	0.13	0.04	4	5	<.1	—
0	18	0.04	4	7	75	0.36	0	0.00	0.00	0.00	0.00	0.00	0	0	0	0
0	—	—	—	0	50	—	0	—	0.00	—	—	—	—	0	—	—
0	—	—	—	18	42	—	0	—	0.00	—	—	—	—	0	—	—
0	13	0.09	4	4	17	0.04	0	0.00	0.00	0.00	0.00	0.00	0	0	0	<1
0	11	0.07	4	4	15	0.04	0	0.00	0.00	0.00	0.00	0.00	0	0	0	<1
0	11	0.11	4	21	18	0.00	0	0.02	0.00	0.08	0.00	0.00	0	0	0	0

A | **Food Composition** (Computer code number is for Wadsworth Diet Analysis program) (For purposes of calculations, use "0" for t, <1, <.1, <.01, etc.)

DA + Code	Food Description	Quantity	Measure	Wt (g)	H₂O (g)	Ener (cal)	Prot (g)	Carb (g)	Dietary Fiber (g)	Fat (g)	Sat	Mono	Poly
	BEVERAGES—Continued												
9524	Cola, decaffeinated, low calorie w/aspartame	12	fluid ounce(s)	355	354	4	<1	<1	0	0	0.00	0.00	0.00
1412	Cream soda	12	fluid ounce(s)	371	321	189	0	49	0	0	0.00	0.00	0.00
31899	Diet 7 Up	12	fluid ounce(s)	360	—	0	0	0	0	0	0.00	0.00	0.00
695	Ginger ale	12	fluid ounce(s)	366	334	124	0	32	0	0	0.00	0.00	0.00
694	Grape soda	12	fluid ounce(s)	372	330	160	0	42	0	0	0.00	0.00	0.00
1876	Lemon lime soda	12	fluid ounce(s)	368	330	147	0	38	0	0	0.00	0.00	0.00
29392	Mountain Dew diet soda	12	fluid ounce(s)	360	—	0	0	0	0	0	0.00	0.00	0.00
29391	Mountain Dew soda	12	fluid ounce(s)	360	—	170	0	46	0	0	0.00	0.00	0.00
3145	Orange soda	12	fluid ounce(s)	372	326	179	0	46	0	0	0.00	0.00	0.00
1414	Pepper-type soda	12	fluid ounce(s)	368	329	151	0	38	0	<1	0.26	0.00	0.00
2391	Pepper-type or cola soda, low calorie w/saccharin	12	fluid ounce(s)	355	354	0	0	<1	0	0	0.00	0.00	0.00
29389	Pepsi diet cola soda	12	fluid ounce(s)	360	—	0	0	0	0	0	0.00	0.00	0.00
29388	Pepsi regular cola soda	12	fluid ounce(s)	360	—	150	0	41	0	0	0.00	0.00	0.00
696	Root beer	12	fluid ounce(s)	370	330	152	0	39	0	0	0.00	0.00	0.00
31898	7 Up	12	fluid ounce(s)	360	—	240	0	59	0	0	0.00	0.00	0.00
12034	Sprite diet soda	12	fluid ounce(s)	360	—	4	0	0	0	0	0.00	0.00	0.00
12044	Sprite soda	12	fluid ounce(s)	360	—	144	0	39	0	0	0.00	0.00	0.00
	Coffee												
731	Brewed	8	fluid ounce(s)	237	236	9	<1	0	0	2	0.00	0.00	0.00
9520	Brewed, decaffeinated	8	fluid ounce(s)	237	235	5	<1	1	0	0	0.00	0.00	0.00
16882	Cappuccino	8	fluid ounce(s)	240	224	78	4	6	<1	4	2.53	1.18	0.15
16883	Cappuccino, decaffeinated	8	fluid ounce(s)	240	224	78	4	6	<1	4	2.53	1.18	0.15
16880	Espresso	8	fluid ounce(s)	237	235	5	<1	1	0	0	0.00	0.00	0.00
16881	Espresso, decaffeinated	8	fluid ounce(s)	237	235	5	<1	1	0	0	0.00	0.00	0.00
732	Instant, prepared	8	fluid ounce(s)	239	237	5	<1	1	0	0	0.00	0.00	0.00
	Fruit drinks												
29357	Crystal Light low calorie lemonade drink	8	fluid ounce(s)	240	—	5	0	0	0	0	0.00	0.00	0.00
6012	Fruit punch drink w/added vitamin C, canned	8	fluid ounce(s)	276	242	129	0	33	<1	<.1	0.01	0.01	0.01
260	Grape drink, canned	8	fluid ounce(s)	250	221	113	<.1	29	0	0	0.00	0.00	0.00
266	Lemonade, from frozen concentrate	8	fluid ounce(s)	248	213	131	<1	34	<1	<1	0.02	0.00	0.04
268	Limeade, from frozen concentrate	8	fluid ounce(s)	247	220	104	<.1	26	0	<.1	0.00	0.00	0.00
31143	Gatorade Thirst Quencher, all flavors	8	fluid ounce(s)	240	—	50	0	14	0	0	0.00	0.00	0.00
17372	Kool-Aid (lemonade/punch/fruit drink)	8	fluid ounce(s)	248	220	108	<1	28	<1	<.1	0.01	0.01	0.02
17225	Kool-Aid sugar free, low calorie tropical punch mix, prepared	8	fluid ounce(s)	240	—	5	0	0	0	0	0.00	0.00	0.00
14266	Odwalla strawberry 'c' monster fruit drink	8	fluid ounce(s)	240	—	150	2	34	1	1	0.00	—	—
10080	Odwalla strawberry lemonade quencher	8	fluid ounce(s)	240	—	120	1	28	1	0	0.00	0.00	0.00
10099	Snapple fruit punch	8	fluid ounce(s)	240	—	110	0	29	0	0	0.00	0.00	0.00
10096	Snapple kiwi strawberry	8	fluid ounce(s)	240	211	110	0	28	0	0	0.00	0.00	0.00
	Slim Fast ready to drink shake												
16056	Dark chocolate fudge	11	fluid ounce(s)	325	—	220	10	42	5	3	1.00	1.50	0.50
16054	French vanilla	11	fluid ounce(s)	325	—	220	10	40	5	3	0.50	1.50	0.50
16055	Strawberries n cream	11	fluid ounce(s)	325	—	220	10	40	5	3	0.50	1.50	0.50
	Tea												
733	Tea, prepared	8	fluid ounce(s)	237	236	2	0	1	0	0	0.00	0.00	0.01
33179	Decaffeinated, prepared	8	fluid ounce(s)	237	236	2	0	1	0	0	0.00	0.00	0.01
1877	Herbal, prepared	8	fluid ounce(s)	237	236	2	0	<1	0	0	0.00	0.00	0.01
734	Instant tea mix, unsweetened, prepared	8	fluid ounce(s)	237	236	2	<.1	<1	0	0	0.00	0.00	0.00
735	Instant lemon flavored tea mix w/sugar, prepared	8	fluid ounce(s)	259	236	88	<1	22	0	<.1	0.01	0.00	0.02
	Water												
1413	Mineral water, carbonated	8	fluid ounce(s)	237	237	0	0	0	0	0	0.00	0.00	0.00
33183	Poland spring water, bottled	8	fluid ounce(s)	237	237	0	0	0	0	0	0.00	0.00	0.00
1	Tap water	8	fluid ounce(s)	237	237	0	0	0	0	0	0.00	0.00	0.00
1879	Tonic water	8	fluid ounce(s)	244	222	83	0	21	0	0	0.00	0.00	0.00

PAGE KEY: A–4 = Breads/Baked Goods   A–8 = Cereal/Rice/Pasta   A–12 = Fruit   A–18 = Vegetables/Legumes   A–28 = Nuts/Seeds   A–30 = Vegetarian   A–32 = Dairy   A–38 = Eggs   A–40 = Seafood   A–42 = Meats   A–44 = Poultry   A–46 = Processed meats   A–48 = Beverages   A–52 = Fats/Oils   A–54 = Sweets   A–54 = Sauces/Condiments/Spices   A–58 = Mixed Foods/Soups/Sandwiches   A–64 = Fast food   A–82 = Convenience   A–84 = Baby foods

A

Chol (mg)	Calc (mg)	Iron (mg)	Magn (mg)	Pota (mg)	Sodi (mg)	Zinc (mg)	Vit A (RAE) (µg)	Thia (mg)	Vit E (mg)	Ribo (mg)	Niac (mg)	Vit B$_6$ (mg)	Fola (µg)	Vit C (mg)	Vit B$_{12}$ (µg)	Sele (µg)
0	14	0.11	4	0	21	0.28	0	0.02	0.00	0.08	0.00	0.00	0	0	0	<1
0	19	0.19	4	4	44	0.26	0	0.00	0.00	0.00	0.00	0.00	0	0	0	0
0	—	—	—	116	53	—	—	—	—	—	—	—	—	—	—	—
0	11	0.66	4	4	26	0.18	0	0.00	0.00	0.00	0.00	0.00	0	0	0	<1
0	11	0.30	4	4	56	0.26	0	0.00	0.00	0.00	0.00	0.00	0	0	0	0
0	7	0.26	4	4	41	0.18	0	0.00	0.00	0.00	0.06	0.00	0	0	0	0
0	—	—	—	70	35	—	—	—	—	—	—	—	—	—	—	—
0	—	—	—	0	70	—	—	—	—	—	—	—	—	—	—	—
0	19	0.22	4	7	45	0.37	0	0.00	0.00	0.00	0.00	0.00	0	0	0	0
0	11	0.15	0	4	37	0.15	0	0.00	0.00	0.00	0.00		0	0	0	<1
0	14	0.07	4	14	57	0.11	0	0.00	0.00	0.00	0.00	0.00	0	0	0	<1
0	—	—	—	30	35	—	—	—	—	—	—	—	—	—	—	—
0	—	—	—	0	35	—	—	—	—	—	—	—	—	—	—	—
0	18	0.18	4	4	48	0.26	0	0.00	0.00	0.00	0.00	0.00	0	0	0	<1
0	—	—	—	0	113	—	—	—	—	—	—	—	—	—	—	—
0	—	—	—	110	36	—	0	—	—	0.00	—	—	—	0	—	—
0	—	—	—	0	71	—	0	—	—	0.00	—	—	—	0	—	—
0	2	0.02	5	114	2	0.02	0	0.00	0.00	0.12	0.00	0.00	5	0	0	0
0	5	0.12	12	128	5	0.05	0	0.00	0.00	0.00	0.53	0.00	<1	0	0	0
17	152	0.26	22	250	62	0.50	—	0.04	0.12	0.20	0.37	0.05	5	1	<1	—
17	152	0.26	22	250	62	0.50	—	0.04	0.12	0.20	0.37	0.05	5	1	<1	—
0	5	0.12	12	128	5	0.05	0	0.00	0.00	0.00	0.53	0.00	<1	0	0	—
0	5	0.12	12	128	5	0.05	0	0.00	0.00	0.00	0.53	0.00	<1	0	0	—
0	10	0.10	7	72	5	0.02	0	0.00	0.00	0.00	0.56	0.00	0	0	0	<1
0	0	0.00	—	160	20	—	0	—	—	—	—	—	—	0	—	—
0	22	0.58	6	69	61	0.33	—	0.06	0.00	0.06	0.06	0.00	4	99	0	0
0	5	0.45	3	30	15	0.30	0	0.00	0.00	0.01	0.03	0.01	0	85	0	<1
0	10	0.52	5	50	7	0.07	0	0.02	—	0.07	0.05	0.02	2	13	0	<1
0	7	0.02	2	22	5	0.02	0	0.00	—	0.01	0.02	0.01	2	6	0	<1
0	10	0.18	—	30	110	—	—	—	—	—	—	—	—	1	—	—
0	14	0.46	5	50	31	0.20	—	0.04	—	0.05	0.05	0.01	4	42	0	1
0	0	0.00	—	10	10	—	0	—	—	—	—	—	—	6	—	—
0	20	1.44	—	330	40	—	—	—	—	—	—	—	—	600	0	—
0	20	0.00	—	70	30	—	0	—	—	—	—	—	—	60	—	—
0	0	0.00	—	20	10	—	0	—	—	—	—	—	—	0	0	—
0	0	0.00	—	40	10	—	0	—	—	—	—	—	—	0	0	—
5	400	2.70	140	600	220	2.25	—	0.53	20.13	0.60	7.00	0.70	120	60	2	18
5	400	2.70	140	600	220	2.25	—	0.53	20.13	0.60	7.00	0.70	120	60	2	18
5	400	2.70	140	600	220	2.25	—	0.53	20.13	0.60	7.00	0.70	120	60	2	18
0	0	0.05	7	88	7	0.05	0	0.00	0.00	0.03	0.00	0.00	12	0	0	0
0	0	0.05	7	88	7	0.05	0	0.00	0.00	0.03	0.00	0.00	12	0	0	0
0	5	0.19	2	21	2	0.09	0	0.02	0.00	0.01	0.00	0.00	2	0	0	0
0	7	0.05	5	47	7	0.02	0	0.00	0.00	0.00	0.09	0.00	0	0	0	0
0	5	0.05	5	49	8	0.03	0	0.00	—	0.04	0.09	0.01	0	<1	0	<1
0	33	0.00	0	0	2	0.00	0	0.00	—	0.00	0.00	0.00	0	0	0	0
0	2	0.02	2	0	2	0.00	0	0.00	—	0.00	0.00	0.00	0	0	0	0
0	4.74	0.00	2.37	0	4.74	0	0	0	0	0	0	0	0	0	0	0
0	2	0.02	0	0	10	0.24	0	0.00	0.00	0.00	0.00	0.00	0	0	0	0

Table A–1

Food Composition   (Computer code number is for Wadsworth Diet Analysis program)   (For purposes of calculations, use "0" for t, <1, <.1, <.01, etc.)

DA + Code	Food Description	Quantity	Measure	Wt (g)	H₂O (g)	Ener (cal)	Prot (g)	Carb (g)	Dietary Fiber (g)	Fat (g)	Fat Breakdown (g)		
											Sat	Mono	Poly
	FATS AND OILS												
	Butter												
104	Butter	1	tablespoon(s)	15	2	108	<1	<.1	0	12	6.13	5.00	0.43
921	Unsalted	1	tablespoon(s)	15	3	108	<1	<.1	0	12	7.71	3.15	0.46
107	Whipped	1	tablespoon(s)	11	2	82	<.1	<.1	0	9	5.76	2.67	0.34
944	Whipped, unsalted	1	tablespoon(s)	11	2	82	<.1	<.1	0	9	5.76	2.67	0.34
2522	Butter Buds, dry butter substitute	1	teaspoon(s)	2	—	8	0	2	0	0	0.00	0.00	0.00
	Fats, cooking												
2671	Beef tallow, semisolid	1	tablespoon(s)	13	0	115	0	0	0	13	6.37	5.35	0.51
922	Chicken fat	1	tablespoon(s)	13	<.1	115	0	0	0	13	3.81	5.72	2.68
5454	Household shortening w/vegetable oil	1	tablespoon(s)	13	0	115	0	0	0	13	3.39	5.56	2.75
111	Lard	1	tablespoon(s)	13	0	114	0	0	0	13	4.94	5.68	1.41
	Margarine												
114	Margarine	1	tablespoon(s)	14	2	101	<1	<1	0	11	2.23	5.05	3.58
116	Soft	1	tablespoon(s)	14	2	101	<1	<.1	0	11	1.95	4.02	4.88
117	Soft, unsalted	1	tablespoon(s)	14	3	101	<1	<1	0	11	1.95	5.26	3.62
928	Unsalted	1	tablespoon(s)	14	3	101	<.1	<.1	0	11	2.12	5.17	3.53
119	Whipped	1	tablespoon(s)	9	1	64	<.1	<.1	0	7	1.17	3.25	2.51
	Spreads												
16164	I Can't Believe It's Not Butter! whipped spread	1	tablespoon(s)	14	4	60	0	0	0	7	1.50	1.50	2.50
16157	Promise vegetable oil spread, stick	1	tablespoon(s)	14	4	90	0	0	0	10	2.50	2.00	4.00
	Oils												
2681	Canola	1	tablespoon(s)	14	0	120	0	0	0	14	0.97	8.01	4.03
120	Corn	1	tablespoon(s)	14	0	120	0	0	0	14	1.73	3.29	7.98
122	Olive	1	tablespoon(s)	14	0	119	0	0	0	14	1.82	9.98	1.35
124	Peanut	1	tablespoon(s)	14	0	119	0	0	0	14	2.28	6.24	4.32
2693	Safflower	1	tablespoon(s)	14	0	120	0	0	0	14	0.84	10.15	1.95
923	Sesame	1	tablespoon(s)	14	0	120	0	0	0	14	1.93	5.40	5.67
130	Soybean w/cottonseed oil	1	tablespoon(s)	14	0	120	0	0	0	14	2.45	4.01	6.54
128	Soybean, hydrogenated	1	tablespoon(s)	14	0	120	0	0	0	14	2.03	5.85	5.11
2700	Sunflower	1	tablespoon(s)	14	0	120	0	0	0	14	1.77	6.28	4.95
357	**Pam original no stick cooking spray**	1	serving(s)	0	—	0	0	0	0	0	0.00	0.00	0.00
	Salad dressing												
132	Blue cheese	2	tablespoon(s)	31	10	154	1	2	0	16	3.03	3.76	8.51
133	Blue cheese, low calorie	2	tablespoon(s)	32	25	32	2	1	0	2	0.82	0.57	0.78
1764	Caesar	2	tablespoon(s)	30	10	158	<1	1	<.1	17	2.64	4.05	9.86
29654	Creamy, reduced calorie, fat free, cholesterol free, sour cream and/or buttermilk & oil	2	tablespoon(s)	32	24	34	<1	6	0	1	0.16	0.21	0.46
29617	Creamy, reduced calorie, sour cream and/or buttermilk & oil	2	tablespoon(s)	30	22	48	<1	2	0	4	0.63	0.98	2.40
134	French	2	tablespoon(s)	31	11	143	<1	5	0	14	1.76	2.63	6.56
135	French, low fat	2	tablespoon(s)	33	18	76	<1	10	<1	4	0.36	1.92	1.64
136	Italian	2	tablespoon(s)	29	17	86	<1	3	0	8	1.32	1.86	3.80
137	Italian, diet	2	tablespoon(s)	30	25	23	<1	1	0	2	0.14	0.66	0.51
139	Mayonnaise type	2	tablespoon(s)	29	12	115	<1	7	0	10	1.44	2.65	5.29
942	Oil & vinegar	2	tablespoon(s)	31	15	140	0	1	0	16	2.84	4.62	7.52
1765	Ranch	2	tablespoon(s)	30	12	146	<1	2	<.1	16	2.32	3.85	8.92
3666	Ranch, reduced calorie	2	tablespoon(s)	30	21	62	<1	2	<.1	6	1.13	1.79	2.89
940	Russian	2	tablespoon(s)	31	11	151	<1	3	0	16	2.23	3.61	9.00
939	Russian, low calorie	2	tablespoon(s)	33	21	46	<1	9	<.1	1	0.20	0.29	0.75
941	Sesame seed	2	tablespoon(s)	31	12	136	1	3	<1	14	1.90	3.64	7.68
142	Thousand island	2	tablespoon(s)	31	15	115	<1	5	<1	11	1.59	2.46	5.68
143	Thousand island, low calorie	2	tablespoon(s)	31	19	62	<1	7	<1	4	0.23	1.98	0.82
	Sandwich spreads												
138	Mayonnaise w/soybean oil	1	tablespoon(s)	14	2	99	<1	1	0	11	1.64	2.70	5.89
2708	Mayonnaise w/soybean & safflower oils	1	tablespoon(s)	14	0	98.94	0.15	0.37	0	10.95	1.18	1.79	7.59
140	Mayonnaise, low calorie	1	tablespoon(s)	16	10	37	<.1	3	0	3	0.53	0.72	1.70
141	Tartar sauce	2	tablespoon(s)	28	9	144	<1	4	<.1	14	2.14	4.13	7.57

PAGE KEY: A–4 = Breads/Baked Goods   A–8 = Cereal/Rice/Pasta   A–12 = Fruit   A–18 = Vegetables/Legumes   A–28 = Nuts/Seeds   A–30 = Vegetarian
A–32 = Dairy   A–38 = Eggs   A–40 = Seafood   A–42 = Meats   A–44 = Poultry   A–46 = Processed meats   A–48 = Beverages   A–52 = Fats/Oils
A–54 = Sweets   A–54 = Sauces/Condiments/Spices   A–58 = Mixed Foods/Soups/Sandwiches   A–64 = Fast food   A–82 = Convenience   A–84 = Baby foods

A

Chol (mg)	Calc (mg)	Iron (mg)	Magn (mg)	Pota (mg)	Sodi (mg)	Zinc (mg)	Vit A (RAE) (µg)	Thia (mg)	Vit E (mg)	Ribo (mg)	Niac (mg)	Vit B$_6$ (mg)	Fola (µg)	Vit C (mg)	Vit B$_{12}$ (µg)	Sele (µg)
32	4	0.00	<1	4	86	0.01	103	0.00	0.24	0.01	0.01	0.00	<1	0	<.1	<1
32	4	0.00	<1	4	2	0.01	103	0.00	0.24	0.01	0.01	0.00	<1	0	<.1	<1
25	3	0.02	<1	3	94	0.01	78	0.00	0.18	0.00	0.00	0.00	<1	0	<.1	<1
25	3	0.02	<1	3	1	0.01	—	0.00	0.18	0.00	0.01	0.00	<1	0	<.1	—
0	0	0.00	0	2	70	0.00	0	0.00	0.00	0.00	0.00	0.00	<1	0	0	—
14	0	0.00	0	0	0	0.00	0	0.00	0.35	0.00	0.00	0.00	0	0	0	<.1
11	0	0.00	0	0	0	0.00	0	0.00	0.35	0.00	0.00	0.00	0	0	0	<.1
0	0	0.00	0	0	0	0.00	0	0.00	2.10	0.00	0.00	0.00	0	0	0	—
12	0	0.00	0	0	0	0.01	0	0.00	0.15	0.00	0.00	0.00	0	0	0	<.1
0	4	0.01	<1	6	133	0.00	115	0.00	1.80	0.01	0.00	0.00	<1	<.1	<.1	0
0	4	0.00	<1	5	152	0.00	103	0.00	1.69	0.00	0.00	0.00	<1	<.1	<.1	0
0	4	0.00	<1	5	4	0.00	103	0.00	1.23	0.00	0.00	0.00	<1	<.1	<.1	0
0	2	0.00	<1	4	<1	0.00	115	0.00	1.80	0.00	0.00	0.00	<1	<.1	<.1	0
0	2	0.00	<1	3	97	0.00	—	0.00	1.08	0.00	0.00	0.00	<.1	<.1	<.1	—
0	10	0.18	—	4	70	—	—	1.65	0.00	0.00	0.00	—	—	1	—	—
0	10	0.18	—	9	90	—	—	0.00	3.02	0.00	0.00	—	—	1	—	—
0	0	0.00	0	0	0	0.00	0	0.00	2.85	0.00	0.00	0.00	0	0	0	0
0	0	0.00	0	0	0	0.00	0	0.00	2.87	0.00	0.00	0.00	0	0	0	0
0	<1	0.09	0	<1	<1	0.00	0	0.00	1.67	0.00	0.00	0.00	0	0	0	0
0	0	0.00	0	0	0	0.00	0	0.00	1.74	0.00	0.00	0.00	0	0	0	0
0	0	0.00	0	0	0	0.00	0	0.00	4.68	0.00	0.00	0.00	0	0	0	0
0	0	0.00	0	0	0	0.00	0	0.00	0.56	0.00	0.00	0.00	0	0	0	0
0	0	0.00	0	0	0	0.00	0	0.00	3.84	0.00	0.00	0.00	0	0	0	0
0	0	0.00	0	0	0	0.00	0	0.00	2.47	0.00	0.00	0.00	0	0	0	0
0	0	0.00	0	0	0	0.00	0	0.00	6.94	0.00	0.00	0.00	0	0	0	0
0	0	0.00	—	0	0	—	0	—	—	—	—	—	—	0	0	—
5	25	0.06	0	11	335	0.08	21	0.00	2.85	0.03	0.03	0.01	9	1	<.1	<1
<1	28	0.16	2	2	384	0.08	—	0.01	0.29	0.03	0.02	0.01	1	<.1	<.1	—
1	7	0.05	1	9	323	0.03	—	0.00	1.56	0.00	0.01	0.00	1	0	<.1	—
0	12	0.08	2	43	320	0.06	0	0.00	0.21	0.02	0.01	0.01	1	0	0	—
0	2	0.04	1	11	307	0.01	—	0.00	0.57	0.00	0.01	0.01	4	<1	<.1	—
0	7	0.25	2	21	261	0.09	7	0.01	2.63	0.02	0.06	0.00	0	0	<.1	0
0	4	0.28	3	35	262	0.07	9	0.01	0.39	0.02	0.15	0.02	1	0	0	1
0	2	0.19	1	14	486	0.04	1	0.00	3.05	0.01	0.00	0.02	0	0	0	1
2	3	0.20	1	26	410	0.06	<1	0.00	0.45	0.00	0.00	0.02	0	0	0	2
8	4	0.06	1	3	209	0.05	19	0.00	1.18	0.01	0.00	0.00	2	0	<.1	<1
0	0	0.00	0	2	<1	0.00	0	0.00	2.75	0.00	0.00	0.00	0	0	0	0
1	4	0.03	1	8	354	0.01	—	0.00	3.22	0.00	0.00	0.00	<1	<.1	<.1	—
<1	5	0.01	1	8	414	0.02	—	0.00	0.28	0.01	0.01	0.00	<1	<1	<.1	—
6	6	0.18	1	48	266	0.13	5	0.02	3.12	0.02	0.18	0.01	3	2	<.1	<1
2	6	0.20	0	51	283	0.03	1	0.00	0.25	0.00	0.00	0.00	1	2	<.1	1
0	6	0.18	0	48	306	0.03	1	0.00	1.53	0.00	0.00	0.00	0	0	0	<1
8	5	0.37	2	33	269	0.08	3	0.45	0.36	0.02	0.13	0.00	0	0	0	<1
<1	5	0.28	2	62	254	0.06	5	0.01	0.36	0.01	0.13	0.00	0	0	0	0
5	2	0.07	<1	5	78	0.02	12	0.00	0.32	0.00	0.00	0.08	1	0	<.1	<1
8.14	2.48	0.06	0.13	4.69	78.38	0.01	11.59	0	2.42	0	0	0.07	1.1	0	0.03	0.22
4	<.1	0.00	<.1	2	80	0.02	0	0.00	1.03	0.00	0.00	0.00	0	0	0	—
11	6	0.21	1	10	200	0.05	—	0.00	2.04	0.00	0.01	0.08	2	<1	<.1	—

Food Composition   (Computer code number is for Wadsworth Diet Analysis program)   (For purposes of calculations, use "0" for t, <1, <.1, <.01, etc.)

DA + Code	Food Description	Quantity	Measure	Wt (g)	H₂O (g)	Ener (cal)	Prot (g)	Carb (g)	Dietary Fiber (g)	Fat (g)	Sat	Mono	Poly
	SWEETS												
4799	**Butterscotch or caramel topping**	2	tablespoon(s)	41	13	103	1	27	<1	<.1	0.05	0.01	0.00
	Candy												
1786	Almond Joy candy bar	1	item(s)	49	5	240	2	29	2	13	9.00	3.63	0.74
1785	Bit-o-Honey candy	6	item(s)	40	2	170	1	34	0	3	2.00	0	20
33375	Butterscotch candy	2	piece(s)	12	1	47	<.1	11	0	<1	0.25	0.10	0.01
1701	Chewing gum, stick	1	item(s)	3	<.1	7	0	2	<.1	<.1	0.00	0.00	0.00
33378	Chocolate fudge w/nuts, prepared	2	piece(s)	38	3	175	2	26	1	7	2.29	1.41	2.81
1787	Jelly beans	15	item(s)	43	3	159	0	40	<.1	<.1	0.00	0.00	0.00
1784	Kit Kat wafer bar	1	item(s)	42	1	220	3	27	1	11	7.00	3.53	0.34
4674	Krackel candy bar	1	item(s)	41	1	220	3	26	1	11	6.00	3.94	0.37
4934	Licorice	4	piece(s)	44	7	147	1	34	1	1	0.18	0.07	0.00
1780	Life Savers candy	1	item(s)	2	—	8	0	2	0	<.1	0.00	—	—
1790	Lollipop	1	item(s)	28	—	108	0	28	0	0	0.00	0.00	0.00
4679	M & Ms peanut chocolate candy, small bag	1	item(s)	49	1	250	5	30	2	13	5.00	5.42	2.07
1781	M & Ms plain chocolate candy, small bag	1	item(s)	48	1	240	2	34	1	10	6.00	3.30	0.30
4673	Milk chocolate bar	1	item(s)	91	1	483	8	53	2	28	16.69	7.20	0.63
1783	Milky Way bar	1	item(s)	58	4	270	2	41	1	10	5.00	3.50	0.35
1788	Peanut brittle	1½	ounce(s)	43	<1	206	3	30	1	8	1.76	3.43	1.94
1789	Reese's peanut butter cups	2	piece(s)	45	1	250	5	25	1	14	5.00	6.17	2.34
4689	Reese's pieces candy, small bag	1	item(s)	46	1	230	6	26	1	11	7.00	0.97	0.46
33399	Semisweet chocolate candy, made w/butter	½	ounce(s)	14	<.1	68	1	9	1	4	2.49	1.41	0.13
1782	Snickers bar	1	item(s)	59	3	280	4	35	1	14	5.00	6.13	2.89
4694	Special Dark chocolate bar	1	item(s)	41	<1	220	2	24	3	13	8.00	4.59	0.41
4695	Starburst fruit chews, original fruits	1	package	59	4	240	0	48	0	5	1.00	2.10	1.83
4698	Taffy	3	piece(s)	45	2	169	<.1	41	0	1	0.92	0.43	0.05
4699	Three Musketeers bar	1	item(s)	60	4	260	2	46	1	8	4.50	2.59	0.27
4702	Twix caramel cookie bars	2	item(s)	58	2	280	3	37	1	14	5.00	7.75	0.49
4705	York peppermint pattie	1	item(s)	42	4	170	1	34	1	3	2.00	1.32	0.12
	Frosting, icing												
4760	Chocolate frosting, ready to eat	2	tablespoon(s)	28	5	112	<1	18	<1	5	1.55	2.54	0.60
4771	Creamy vanilla frosting, ready to eat	2	tablespoon(s)	28	4	118	0	19	<.1	5	0.84	1.37	2.24
17291	Dec-a-Cake variety pack candy decoration	1	teaspoon(s)	4	—	15	0	3	0	1	0.00	—	—
536	White icing	2	tablespoon(s)	40	3	163	<1	32	0	4	0.86	2.07	1.19
	Gelatin												
13697	Gelatin snack, all flavors	1	item(s)	99	97	70	1	17	0	0	0.00	0.00	0.00
2616	Mixed fruit gelatin mix, sugar free, low calorie, prepared	½	cup(s)	121	—	10	1	0	0	0	0.00	0.00	0.00
548	**Honey**	1	tablespoon(s)	21	4	64	<.1	17	<.1	0	0.00	0.00	0.00
545	**Marshmallows**	4	item(s)	29	5	92	1	23	<.1	<.1	0.02	0.02	0.01
4800	**Marshmallow cream topping**	2	tablespoon(s)	28	6	91	<1	22	<.1	<.1	0.02	0.02	0.01
555	**Molasses**	1	tablespoon(s)	20	4	58	0	15	0	<.1	0.00	0.01	0.01
4780	**Popsicle or ice pop**	1	item(s)	59	47	42	0	11	0	0	0.00	0.00	0.00
	Sugar												
559	Brown, packed	1	teaspoon(s)	5	<.1	17	0	4	0	0	0.00	0.00	0.00
563	Powdered, sifted	⅓	cup(s)	33	<.1	130	0	33	0	<.1	0.01	0.01	0.02
561	White granulated	1	teaspoon(s)	4	<.1	15	0	4	0	0	0.00	0.00	0.00
	Sugar Substitute												
1760	Equal sweetener, packet	1	item(s)	1	<.1	4	<.1	1	0	0	0.00	0.00	0.00
13029	Splenda granular no calorie sweetener	1	teaspoon(s)	1	—	2	0	1	0	0	0.00	0.00	0.00
1759	Sweet n Low sugar substitute, packet	1	item(s)	1	<.1	4	0	1	0	0	0.00	0.00	0.00
	Syrup												
3148	Chocolate	2	tablespoon(s)	38	12	105	1	24	1	<1	0.19	0.11	0.01
29676	Maple	¼	cup(s)	80	26	209	0	54	0	<1	0.03	0.05	0.08
4795	Pancake	¼	cup(s)	80	30	187	0	49	1	0	0.00	0.00	0.00
	SAUCES, SPICES, CONDIMENTS												
	Spices												
807	Allspice, ground	1	teaspoon(s)	2	<1	5	<1	1	<1	<1	0.05	0.01	0.04

PAGE KEY: A–4 = Breads/Baked Goods A–8 = Cereal/Rice/Pasta A–12 = Fruit A–18 = Vegetables/Legumes A–28 = Nuts/Seeds A–30 = Vegetarian A–32 = Dairy A–38 = Eggs A–40 = Seafood A–42 = Meats A–44 = Poultry A–46 = Processed meats A–48 = Beverages A–52 = Fats/Oils A–54 = Sweets A–54 = Sauces/Condiments/Spices A–58 = Mixed Foods/Soups/Sandwiches A–64 = Fast food A–82 = Convenience A–84 = Baby foods

A

Chol (mg)	Calc (mg)	Iron (mg)	Magn (mg)	Pota (mg)	Sodi (mg)	Zinc (mg)	Vit A (RAE) (µg)	Thia (mg)	Vit E (mg)	Ribo (mg)	Niac (mg)	Vit B$_6$ (mg)	Fola (µg)	Vit C (mg)	Vit B$_{12}$ (µg)	Sele (µg)
<1	22	0.08	3	34	143	0.08	11	0.00	—	0.04	0.02	0.01	1	<1	<.1	0
3	20	0.36	33	138	70	0.40	0	0.02	—	0.08	0.24	—	—	0	—	—
0.00	—	—	85	—	0	—	—	—	—	—	—	0	—	0	—	—
1	<1	0.00	<1	<1	47	0.00	3	0.00	0.01	0.00	0.00	0.00	0	0	0	<.1
0	0	0.00	0	<.1	<.1	0.00	0	0.00	0.00	0.00	0.00	0.00	0	0	0	<.1
5	21	0.75	21	68	16	0.54	14	0.03	0.25	0.04	0.12	0.03	6	<.1	<.1	1
0	1	0.06	1	16	21	0.02	0	0.00	0.00	0.00	0.00	0.00	0	0	0	<1
3	40	0.36	16	126	25	0.52	8	0.07	—	0.23	1.07	0.05	60	0	<.1	2
3	60	0.37	—	169	80	—	0	—	—	—	—	—	—	0	—	—
0	3	0.13	3	28	109	0.07	0	0.01	0.08	0.02	0.04	0.00	0	0	0	—
0	<1	0.04	—	0	1	—	0	0.00	—	0.00	0.00	—	0	0	—	0
0	0	0.00	—	—	11	—	0	0.00	—	0.00	0.00	—	0	0	—	1
5	40	0.36	36	171	25	1.13	15	0.03	1.06	0.07	1.60	0.04	17	1	<.1	2
5	40	0.36	20	127	30	0.46	15	0.03	0.41	0.07	0.11	0.01	3	1	<1	1
22	228	0.83	61	399	92	1.00	20	0.06	0.19	0.26	0.15	0.10	11	2	<1	—
5	60	0.18	20	140	95	0.41	15	0.02	0.40	0.07	0.20	0.03	6	1	<1	3
5	11	0.52	18	71	189	0.37	17	0.06	1.17	0.02	1.13	0.03	20	0	<.1	1
3	20	0.36	40	233	140	0.82	7	0.11	—	0.08	2.08	0.07	25	0	<.1	2
0	40	0.00	20	182	90	0.35	25	0.04	—	0.07	1.31	0.03	13	0	<.1	1
3	5	0.44	16	52	2	0.23	<1	0.01	—	0.01	0.06	0.01	<1	0	0	<1
5	40	0.36	42	—	140	1.38	15	0.03	1.53	0.07	1.60	0.05	23	1	<.1	3
3	0	0.72	46	136	0	0.60	0	0.01	—	0.03	0.16	0.01	1	0	0	1
0	10	0.18	1	1	0	0.00	—	0.00	0.89	0.00	0.00	0.00	0	30	0	<1
4	1	0.03	<1	2	40	0.02	—	0.00	—	0.01	0.01	0.00	0	0	<.1	—
5	20	0.36	18	80	110	0.33	14	0.02	0.28	0.03	0.20	0.01	0	1	<1	2
5	40	0.36	18	117	115	0.45	15	0.09	0.40	0.13	0.69	0.02	14	1	<1	1
0	0	0.36	25	71	10	0.31	0	0.01	—	0.04	0.34	0.01	2	0	<.1	—
0	2	0.40	6	55	51	0.08	0	0.00	0.67	0.00	0.03	0.00	<1	0	0	<1
0	1	0.04	<1	10	52	0.02	0	0.00	1.33	0.08	0.06	0.00	2	0	0	<.1
0	0	0.00	—	—	15	—	0	—	—	—	—	—	—	0	—	—
<1	5	0.02	—	7	92	—	—	0.00	—	0.01	0.00	—	—	<.1	—	—
0	0	0.00	—	0	40	—	0	—	—	—	—	—	—	0	—	—
0	0	0.00	0	0	50	0.00	0	0.00	0.00	0.00	0.00	0.00	0	0	0	—
0	1	0.09	<1	11	1	0.05	0	0.00	0.00	0.01	0.03	0.01	<1	<1	0	<1
0	1	0.07	1	1	23	0.01	0	0.00	0.00	0.00	0.02	0.00	<1	0	0	<1
0	1	0.06	1	1	23	0.01	0	0.00	0.00	0.00	0.02	0.00	<1	0	0	1
0	41	0.94	48	293	7	0.06	0	0.01	0.00	0.00	0.19	0.13	0	0	0	4
0	0	0.00	1	2	7	0.01	0	0.00	0.00	0.00	0.00	0.00	0	0	0	0
0	4	0.09	1	16	2	0.01	0	0.00	0.00	0.00	0.00	0.00	<.1	0	0	<.1
0	<1	0.01	0	1	<1	0.00	0	0.00	0.00	0.00	0.01	0.00	0	0	0	<1
0	<.1	0.00	0	<.1	0	0.00	0	0.00	0.00	0.00	0.00	0.00	0	0	0	<.1
0	0	0.00	0	0	0	0.00	0	0.00	0.00	0.00	0.00	0.00	0	0	0	0
0	10	0.18	—	—	<1	—	—	0.02	—	0.02	0.20	—	—	1	0	—
0	0	0.00	0	—	0	0.00	0	0.00	0.00	0.00	0.00	0.00	0	0	0	0
0	5	0.79	24	84	27	0.27	0	0.00	0.01	0.02	0.12	0.00	1	<.1	0	1
0	54	0.96	11	163	7	3.33	0	0.00	0.00	0.01	0.02	0.00	0	0	0	<1
0	2	0.02	2	12	66	0.06	0	0.00	0.00	0.01	0.01	0.00	0	0	0	0
0	13	0.13	3	20	1	0.02	1	0.00	0.02	0.00	0.05	0.00	1	1	0	<.1

A

Table A–1

Food Composition　　(Computer code number is for Wadsworth Diet Analysis program)　　(For purposes of calculations, use "0" for t, <1, <.1, <.01, etc.)

DA + Code	Food Description	Quantity	Measure	Wt (g)	H₂O (g)	Ener (cal)	Prot (g)	Carb (g)	Dietary Fiber (g)	Fat (g)	Fat Breakdown (g)		
											Sat	Mono	Poly
	SAUCES, SPICES, CONDIMENTS—Continued												
1171	Anise seeds	1	teaspoon(s)	2	<1	7	<1	1	<1	<1	0.01	0.21	0.07
729	Baker's yeast active	1	teaspoon(s)	4	<1	12	2	2	1	<1	0.02	0.10	0.00
683	Baking powder, double acting, w/phosphate	1	teaspoon(s)	5	<1	2	<.1	1	<.1	0	0.00	0.00	0.00
1611	Baking soda	1	teaspoon(s)	5	<.1	0	0	0	0	0	0.00	0.00	0.00
8552	Basil	1	teaspoon(s)	1	1	<1	<.1	<.1	<.1	<.1	0.00	0.00	0.00
34959	Basil, fresh	1	piece(s)	1	<1	<1	<.1	<.1	<.1	<.1	0.059	0.00	0.00
808	Basil, ground	1	teaspoon(s)	1	<.1	4	<1	1	1	<.1	0.00	0.01	0.03
809	Bay leaf	1	teaspoon(s)	1	<.1	2	<.1	<1	<1	<.1	0.01	0.01	0.01
11720	Betel leaves	1	ounce(s)	28	—	17	2	2	0	<.1	—	—	—
818	Black pepper	1	teaspoon(s)	2	<1	5	<1	1	1	<.1	0.02	0.02	0.02
730	Brewer's yeast	3	<1	8	1	1	1	0	0.00	0.00	0.00			
35417	Capers	1	teaspoon(s)	4	—	2	0	0	0	0	0.00	0.00	0.00
1172	Caraway seeds	1	teaspoon(s)	2	<1	7	<1	1	1	<1	0.01	0.15	0.07
819	Cayenne pepper	1	teaspoon(s)	2	<1	6	<1	1	<1	<1	0.06	0.05	0.15
1173	Celery seeds	1	teaspoon(s)	2	<1	8	<1	1	<1	1	0.04	0.32	0.07
1174	Chervil, dried	1	teaspoon(s)	1	<.1	1	<1	<1	<.1	<.1	0.00	0.01	0.01
810	Chili powder	1	teaspoon(s)	3	<1	8	<1	1	1	<1	0.08	0.09	0.19
8553	Chives, chopped	1	teaspoon(s)	1	1	<1	<.1	<.1	<.1	<.1	0.00	0.00	0.00
8556	Cilantro	1	teaspoon(s)	2	1	<1	<.1	<.1	<.1	<.1	0.00	0.00	0.00
811	Cinnamon, ground	1	teaspoon(s)	2	<1	6	<.1	2	1	<.1	0.01	0.01	0.01
812	Cloves, ground	1	teaspoon(s)	2	<1	7	<1	1	1	<1	0.11	0.03	0.15
1175	Coriander leaf, dried	1	teaspoon(s)	1	<.1	2	<1	<1	<.1	<.1	0.00	0.01	0.00
1176	Coriander seeds	1	teaspoon(s)	2	<1	5	<1	1	1	<1	0.02	0.24	0.03
1706	Cornstarch	1	tablespoon(s)	8	1	30	<.1	7	<.1	<.1	0.00	0.00	0.00
11729	Cumin, ground	1	teaspoon(s)	5	—	11	<1	1	1	<1	—	—	—
1177	Cumin seeds	1	teaspoon(s)	2	<1	8	<1	1	<1	<1	0.03	0.29	0.07
1178	Curry powder	1	teaspoon(s)	2	<1	7	<1	1	1	<1	0.04	0.11	0.05
1179	Dill seeds	1	teaspoon(s)	2	<1	6	<1	1	<1	<1	0.02	0.20	0.02
1180	Dill weed, dried	1	teaspoon(s)	1	<.1	3	<1	1	<1	<.1	0.00	0.01	0.00
34949	Dill weed, fresh	5	piece(s)	1	1	<1	<.1	<.1	<.1	<.1	0.00	0.01	0.00
4949	Fennel leaves, fresh	1	teaspoon(s)	1	1	<1	<.1	<.1	0	<.1	0.00	0.00	0.00
1181	Fennel seeds	1	teaspoon(s)	2	<1	7	<1	1	1	<1	0.01	0.20	0.03
1182	Fenugreek seeds	1	teaspoon(s)	4	<1	12	1	2	1	<1	0.05	—	—
11733	Garam masala, powder	1	ounce(s)	28	—	107	4	13	0	4	—	—	—
1067	Garlic clove	1	item(s)	3	2	4	<1	1	<.1	<.1	0.00	0.00	0.01
813	Garlic powder	1	teaspoon(s)	3	<1	9	<1	2	<1	<.1	0.00	0.00	0.01
1183	Ginger, ground	1	teaspoon(s)	2	<1	6	<1	1	<1	<1	0.03	0.02	0.02
1068	Ginger root	2	teaspoon(s)	4	3	3	<.1	1	<.1	<.1	0.01	0.01	0.01
35497	Leeks, bulb & lower leaf, freeze-dried	¼	cup(s)	1	<.1	3	<1	1	<.1	<.1	0.00	0.00	0.01
1184	Mace, ground	1	teaspoon(s)	2	<1	8	<1	1	<1	1	0.16	0.19	0.07
1185	Marjoram, dried	1	teaspoon(s)	1	<.1	2	<.1	<1	<1	<.1	0.00	0.01	0.03
1186	Mustard seeds, yellow	1	teaspoon(s)	3	<1	15	1	1	<1	1	0.05	0.65	0.18
814	Nutmeg, ground	1	teaspoon(s)	2	<1	12	<1	1	<1	1	0.57	0.07	0.01
2747	Onion flakes, dehydrated	1	teaspoon(s)	2	<.1	6	<1	1	<1	<.1	0.00	0.00	0.00
1187	Onion powder	1	teaspoon(s)	2	<1	7	<1	2	<1	<.1	0.00	0.00	0.01
815	Oregano, ground	1	teaspoon(s)	2	<1	5	<1	1	1	<1	0.04	0.01	0.08
816	Paprika	1	teaspoon(s)	2	<1	6	<1	1	1	<1	0.04	0.03	0.17
817	Parsley, dried	1	teaspoon(s)	0	<.1	1	<.1	<1	<.1	<.1	0.00	0.01	0.00
1189	Poppy seeds	1	teaspoon(s)	3	<1	15	1	1	<1	1	0.14	0.18	0.86
1190	Poultry seasoning	1	teaspoon(s)	2	<1	5	<1	1	<1	<1	0.05	0.02	0.03
1191	Pumpkin pie spice, powder	1	teaspoon(s)	2	<1	6	<.1	1	<1	<1	0.11	0.02	0.01
1192	Rosemary, dried	1	teaspoon(s)	1	<1	4	<.1	1	1	<1	0.09	0.04	0.03
11723	Rosemary, fresh	1	teaspoon(s)	1	<1	1	<.1	<1	<.1	<.1	0.02	0.01	0.01
2722	Saffron powder	1	teaspoon(s)	1	<.1	2	<.1	<1	<.1	<.1	0.01	0.00	0.01
11724	Sage	1	ounce(s)	28	—	34	1	4	0	1	—	—	—
1193	Sage, ground	1	teaspoon(s)	1	<.1	2	<.1	<1	<1	<.1	0.05	0.01	0.01
822	Salt, table	¼	teaspoon(s)	2	<.1	0	0	0	0	0	0.00	0.00	0.00
30189	Salt substitute	¼	teaspoon(s)	1	—	<.1	0	<.1	0	0	0.00	0.00	0.00
30190	Salt substitute, seasoned	¼	teaspoon(s)	1	—	1	<.1	<1	0	<.1	0.00	—	—
1194	Savory, ground	1	teaspoon(s)	1	<1	4	<.1	1	1	<.1	0.05	—	—

PAGE KEY: A–4 = Breads/Baked Goods   A–8 = Cereal/Rice/Pasta   A–12 = Fruit   A–18 = Vegetables/Legumes   A–28 = Nuts/Seeds   A–30 = Vegetarian
A–32 = Dairy   A–38 = Eggs   A–40 = Seafood   A–42 = Meats   A–44 = Poultry   A–46 = Processed meats   A–48 = Beverages   A–52 = Fats/Oils
A–54 = Sweets   A–54 = Sauces/Condiments/Spices   A–58 = Mixed Foods/Soups/Sandwiches   A–64 = Fast food   A–82 = Convenience   A–84 = Baby foods

A

Chol (mg)	Calc (mg)	Iron (mg)	Magn (mg)	Pota (mg)	Sodi (mg)	Zinc (mg)	Vit A (RAE) (µg)	Thia (mg)	Vit E (mg)	Ribo (mg)	Niac (mg)	Vit B$_6$ (mg)	Fola (µg)	Vit C (mg)	Vit B$_{12}$ (µg)	Sele (µg)
0	14	0.78	4	30	<1	0.11	<1	0.01	0.02	0.01	0.06	0.01	<1	<1	0	<1
0	3	0.66	4	80	2	0.26	0	0.09	0.00	0.22	1.59	0.06	94	<.1	<.1	1
0	339	0.52	2	<1	363	0.00	0	0.00	0.00	0.00	0.00	0.00	0	0	0	<.1
0	0	0.00	0	0	1259	0.00	0	0.00	0.00	0.00	0.00	0.00	0	0	0	<.1
0	1	0.03	1	4	<.1	0.01	2	0.00	0.00	0.00	0.01	0.00	1	<1	0	<.1
0	1	—	<1	2	<.1	0.00	—	0.00	0.00	0.00	0.01	0.00	<1	—	0	<.1
0	30	0.59	6	48	<1	0.08	7	0.00	0.02	0.00	0.10	0.03	4	1	0	<.1
0	5	0.26	1	3	<1	0.02	2	0.00	0.01	0.00	0.01	0.01	1	<1	0	<.1
0	110	2.29	—	156	2	—	—	0.04	—	0.07	0.20	—	—	1	0	—
0	9	0.61	4	26	1	0.03	<1	0.00	0.02	0.01	0.02	0.01	<1	<1	0	<.1
0	6	0.47	6	51	3	0.21	0	0.42	0.00	0.11	1.00	0.07	104	0	0	0
0	0	0.00	—	—	140	—	0	—	—	—	—	—	—	0	—	—
0	14	0.34	5	28	<1	0.12	<1	0.01	0.05	0.01	0.08	0.01	<1	<1	0	<1
0	3	0.14	3	36	1	0.04	37	0.01	0.09	0.02	0.16	0.04	2	1	0	<1
0	35	0.90	9	28	3	0.14	<.1	0.01	0.02	0.01	0.06	0.02	<1	<1	0	<1
0	8	0.19	1	28	<1	0.05	2	0.00	0.01	0.00	0.03	0.01	2	<1	0	<1
0	7	0.37	4	50	26	0.07	39	0.01	0.03	0.02	0.21	0.10	3	2	0	<1
0	1	0.02	<1	3	<.1	0.01	2	0.00	0.00	0.01	0.00	1	1	0	<.1
0	1	0.03	<1	8	1	0.00	—	0.00	0.03	0.00	0.02	0.00	1	1	0	<.1
0	28	0.88	1	12	1	0.05	<1	0.00	0.00	0.00	0.03	0.01	1	1	0	<.1
0	14	0.18	6	23	5	0.02	1	0.00	0.04	0.01	0.03	0.01	2	2	0	<1
0	7	0.25	4	27	1	0.03	2	0.01	0.01	0.01	0.06	0.00	2	3	0	<1
0	13	0.29	6	23	1	0.08	0	0.00	—	0.01	0.04	—	0	<1	0	<1
0	<1	0.04	<1	<1	1	0.00	0	0.00	0.00	0.00	0.00	0.00	0	0	0	<1
0	20	—	—	44	5	—	—	—	—	—	—	—	—	—	0	—
0	20	1.39	8	38	4	0.10	1	0.01	0.02	0.01	0.10	0.01	<1	<1	0	<1
0	10	0.59	5	31	1	0.08	1	0.01	0.01	0.01	0.07	0.02	3	<1	0	<1
0	32	0.34	5	25	<1	0.11	<.1	0.01	0.02	0.01	0.06	0.01	<1	<1	0	<1
0	18	0.49	5	33	2	0.03	3	0.00	—	0.00	0.03	0.02	2	1	0	0
0	2	—	1	7	1	0.01	—	0.00	—	0.00	0.02	0.00	2	—	0	—
0	1	0.03	—	4	<.1	—	—	0.00	—	0.00	0.01	0.00	—	<1	0	—
0	24	0.37	8	34	2	0.07	<1	0.01	—	0.01	0.12	0.01	—	<1	0	0
0	7	1.24	7	28	2	0.09	<1	0.01	—	0.01	0.06	0.02	2	<1	0	<1
0	215	9.25	94	411	28	1.07	—	0.10	—	0.09	0.71	—	0	0	0	—
0	5	0.05	1	12	1	0.03	0	0.01	0.00	0.00	0.02	0.04	<.1	1	0	<1
0	2	0.08	2	31	1	0.07	0	0.00	0.00	0.00	0.02	0.08	<.1	1	0	1
0	2	0.21	3	24	1	0.08	<1	0.00	0.01	0.00	0.09	0.02	1	<1	0	1
0	1	0.02	2	17	1	0.01	0	0.00	0.01	0.00	0.03	0.01	<1	<1	0	<.1
0	3	0.06	1	19	<1	0.01	<1	0.01	—	0.00	0.03	0.01	3	1	0	<.1
0	4	0.24	3	8	1	0.04	1	0.01	0.04	0.01	0.02	0.00	1	<1	0	<.1
0	12	0.50	2	9	<1	0.02	2	0.00	0.01	0.00	0.02	0.01	2	<1	0	<.1
0	17	0.33	10	23	<1	0.19	<.1	0.02	0.08	0.01	0.26	0.01	3	<.1	0	4
0	4	0.07	4	8	<1	0.05	<1	0.01	0.06	0.00	0.03	0.00	2	<.1	0	<.1
0	4	0.03	2	27	<1	0.03	<.1	0.01	0.02	0.00	0.02	0.03	3	1	0	<.1
0	8	0.05	3	20	1	0.05	0	0.01	0.00	0.00	0.01	0.03	3	<1	0	<.1
0	24	0.66	4	25	<1	0.07	5	0.01	0.03	0.00	0.09	0.02	4	1	0	<.1
0	4	0.50	4	49	1	0.09	55	0.01	0.01	0.04	0.32	0.08	2	1	0	<.1
0	4	0.29	1	11	1	0.01	2	0.00	0.01	0.00	0.02	0.00	1	<1	0	<.1
0	41	0.26	9	20	1	0.29	0	0.02	0.08	0.00	0.03	0.01	2	<.1	0	<1
0	15	0.53	3	10	<1	0.05	2	0.00	0.02	0.00	0.04	0.02	2	<1	0	<1
0	12	0.34	2	11	1	0.04	<1	0.00	0.02	0.00	0.04	0.02	1	<1	0	<1
0	15	0.35	3	11	1	0.04	2	0.01	0.02	0.01	0.01	0.02	4	1	0	<.1
0	2	0.05	1	5	<1	0.01	1	0.00	0.00	0.00	0.01	0.00	1	<1	0	—
0	1	0.08	2	12	1	0.01	<1	0.00	0.01	0.01	0.01	0.01	1	1	0	<.1
0	170	—	45	110	1	0.48	—	0.03	—	—	—	—	—	—	0	—
0	12	0.20	3	7	<.1	0.03	2	0.01	0.01	0.00	0.04	0.02	2	<1	0	<.1
0	<1	0.00	<.1	<1	581	0.00	0	0.00	0.00	0.00	0.00	0.00	0	0	0	<.1
0	7	0.00	<.1	604	<.1	—	0	—	—	—	—	—	—	0	—	—
0	0	0	476	<1	—	0	—	—	—	—	—	—	0	—	—	—
0	30	0.53	5	15	<1	0.06	4	0.01	—	—	0.06	0.03	—	1	0	<.1

Table A–1

Food Composition

(Computer code number is for Wadsworth Diet Analysis program)   (For purposes of calculations, use "0" for t, <1, <.1, <.01, etc.)

DA + Code	Food Description	Quantity	Measure	Wt (g)	H₂O (g)	Ener (cal)	Prot (g)	Carb (g)	Dietary Fiber (g)	Fat (g)	Sat	Mono	Poly
	SAUCES, SPICES, CONDIMENTS—Continued												
820	Sesame seed kernels, toasted	1	teaspoon(s)	3	<1	15	<1	1	<1	1	0.18	0.49	0.57
11725	Sorrel	1	tablespoon(s)	9	—	2	<1	<1	<.1	<.1	0.00	—	—
11721	Spearmint	1	teaspoon(s)	2	2	1	<.1	<1	<1	<.1	0.00	0.00	0.01
35498	Sweet green peppers, freeze-dried	¼	cup(s)	2	<.1	5	<1	1	<1	<.1	0.01	0.00	0.03
11726	Tamarind leaves	1	ounce(s)	28	—	33	2	5	0	1	—	—	—
11727	Tarragon	1	ounce(s)	28	—	14	1	2	0	<1	—	—	—
1195	Tarragon, ground	1	teaspoon(s)	2	<1	5	<1	1	<1	<1	0.03	0.01	0.06
11728	Thyme, fresh	1	teaspoon(s)	1	1	1	<.1	<1	<1	<.1	0.00	0.00	0.00
821	Thyme, ground	1	teaspoon(s)	1	<1	4	<1	1	1	<1	0.04	0.01	0.02
1196	Turmeric, ground	1	teaspoon(s)	2	<1	8	<1	1	<1	<1	0.07	0.04	0.05
11995	Wasabi	1	tablespoon(s)	14	11	11	1	2	<1	<.1	—	—	—
1188	White pepper	1	teaspoon(s)	2	<1	7	<1	2	1	<.1	0.02	0.02	0.01
	Condiments												
674	Catsup or ketchup	1	tablespoon(s)	15	11	14	<1	4	<1	<.1	0.01	0.01	0.04
703	Dill pickle	1	ounce(s)	28	26	5	<1	1	<1	<.1	0.01	0.00	0.02
1641	Horseradish sauce, prepared	1	teaspoon(s)	5	3	10	<1	<1	<.1	1	0.59	0.28	0.04
140	Mayonnaise, low calorie	1	tablespoon(s)	16	10	37	<.1	3	0	3	0.53	0.72	1.70
138	Mayonnaise w/soybean oil	1	tablespoon(s)	14	2	99	<1	1	0	11	1.64	2.70	5.89
1682	Mustard, brown	1	teaspoon(s)	5	4	5	<1	<1	<.1	<1	—	—	—
700	Mustard, yellow	1	teaspoon(s)	5	4	3	<1	<1	<1	<1	0.01	0.11	0.03
706	Sweet pickle relish	1	tablespoon(s)	15	9	20	<.1	5	<1	<.1	0.01	0.03	0.02
141	Tartar sauce	2	tablespoon(s)	28	9	144	<1	4	<.1	14	2.14	4.13	7.57
	Sauces												
685	Barbecue sauce	2	tablespoon(s)	31	25	23	1	4	<1	1	0.08	0.24	0.21
834	Cheese sauce	¼	cup(s)	70	49	121	5	5	<1	9	4.19	2.67	1.81
32123	Chili enchilada sauce, green	2	tablespoon(s)	57	53	15	1	3	1	<1	0.04	0.04	0.13
32122	Chili enchilada sauce, red	2	tablespoon(s)	32	24	27	1	5	2	1	0.08	0.05	0.43
29688	Hoisin sauce	1	tablespoon(s)	16	7	35	1	7	<1	1	0.09	0.15	0.27
16670	Mole poblano sauce	½	cup(s)	133	103	155	5	11	2	11	2.67	5.15	2.91
29689	Oyster sauce	1	tablespoon(s)	16	13	8	<1	2	<.1	<.1	0.01	0.01	0.01
1655	Pepper sauce or tabasco	1	teaspoon(s)	5	5	1	<.1	<.1	<.1	<.1	0.01	0.00	0.02
347	Salsa	2	tablespoon(s)	16	14	4	<1	1	<1	<.1	0.00	0.00	0.02
841	Soy sauce	1	tablespoon(s)	18	13	10	1	2	0	<.1	0.00	0.00	0.01
839	Sweet & sour sauce	2	tablespoon(s)	39	30	37	<.1	9	<.1	<.1	0.00	0.00	0.00
1613	Teriyaki sauce	1	tablespoon(s)	18	12	15	1	3	<.1	0	0.00	0.00	0.00
25294	Tomato sauce	½	cup(s)	112	100	46	2	8	2	1	0.18	0.29	0.72
728	White sauce, medium	¼	cup(s)	63	47	92	2	6	<1	7	1.78	2.78	1.79
1654	Worcestershire sauce	1	teaspoon(s)	6	4	4	0	1	0	0	0.00	0.00	0.00
	Vinegar												
30853	Balsamic	1	tablespoon(s)	15	—	10	0	2	0	0	0.00	0.00	0.00
727	Cider	1	tablespoon(s)	15	14	2	0	1	0	0	0.00	0.00	0.00
1673	Distilled	1	tablespoon(s)	15	14	2	0	1	0	0	0.00	0.00	0.00
15439	Tarragon	1	tablespoon(s)	16	—	0	0	0	0	0	0.00	0.00	0.00
	MIXED FOODS, SOUPS, SANDWICHES												
	Mixed Dishes												
16652	Almond chicken	1	cup(s)	242	186	280	22	16	3	15	1.91	6.07	5.62
25224	Barbecued chicken	2	piece(s)	177	100	325	27	15	<1	17	4.63	6.78	3.71
25227	Bean burrito	1	item(s)	149	82	327	17	33	6	15	8.30	4.73	0.85
9516	Beef & vegetable fajita	1	item(s)	223	144	397	23	35	3	18	5.50	7.53	3.45
16796	Beef or pork egg roll	2	item(s)	128	85	227	10	19	1	12	2.88	5.96	2.64
177	Beef stew w/vegetables, prepared	1	cup(s)	245	201	220	16	15	3	11	4.40	4.50	0.50
30233	Beef stroganoff w/noodles	1	cup(s)	256	190	343	20	23	2	19	7.37	5.62	4.47
16651	Cashew chicken	1	cup(s)	242	131	644	43	17	3	46	7.75	20.83	14.47
475	Cheese pizza	2	slice(s)	126	60	281	15	41	0	6	3.08	1.98	0.98
30330	Cheese quesadilla	1	item(s)	54	19	183	6	18	1	10	3.49	3.42	2.16
215	Chicken & noodles, prepared	1	cup(s)	240	170	365	22	26	1	18	5.10	7.10	3.90
30239	Chicken & vegetables w/broccoli, onion, bamboo shoots in soy based sauce	1	cup(s)	162	112	287	22	6	1	19	5.13	7.65	4.68
25093	Chicken cacciatore	1	cup(s)	230	166	266	28	5	1	14	3.98	5.78	3.11

PAGE KEY: A–4 = Breads/Baked Goods   A–8 = Cereal/Rice/Pasta   A–12 = Fruit   A–18 = Vegetables/Legumes   A–28 = Nuts/Seeds   A–30 = Vegetarian   A–32 = Dairy   A–38 = Eggs   A–40 = Seafood   A–42 = Meats   A–44 = Poultry   A–46 = Processed meats   A–48 = Beverages   A–52 = Fats/Oils   A–54 = Sweets   A–54 = Sauces/Condiments/Spices   A–58 = Mixed Foods/Soups/Sandwiches   A–64 = Fast food   A–82 = Convenience   A–84 = Baby foods

A

Chol (mg)	Calc (mg)	Iron (mg)	Magn (mg)	Pota (mg)	Sodi (mg)	Zinc (mg)	Vit A (RAE) (µg)	Thia (mg)	Vit E (mg)	Ribo (mg)	Niac (mg)	Vit B$_6$ (mg)	Fola (µg)	Vit C (mg)	Vit B$_{12}$ (µg)	Sele (µg)
0	4	0.21	9	11	1	0.28	<.1	0.03	0.06	0.01	0.15	0.00	3	0	0	<.1
0	—	—	—	—	<1	—	—	—	—	—	—	—	—	—	—	—
0	4	0.23	1	9	1	0.02	4	0.00	0.01	0.00	0.02	0.00	2	<1	0	—
0	2	0.17	3	51	3	0.04	3	0.02	0.06	0.02	0.12	0.04	4	30	0	<.1
0	85	1.48	20	—	—	—	—	0.07	—	0.03	1.16	—	—	1	0	—
0	48	—	14	128	3	0.17	—	0.04	—	—	—	—	—	1	0	—
0	18	0.52	6	48	1	0.06	3	0.00	0.03	0.02	0.14	0.04	4	1	0	<.1
0	3	0.14	1	5	<.1	0.01	2	0.00	—	0.00	0.01	0.00	<1	1	0	—
0	26	1.73	3	11	1	0.09	3	0.01	0.02	0.01	0.07	0.01	4	1	0	<.1
0	4	0.91	4	56	1	0.10	0	0.00	0.00	0.01	0.11	0.04	1	1	0	<.1
0	13	0.11	—	—	—	—	—	0.02	—	0.01	0.07	—	—	11	0	—
0	6	0.34	2	2	<1	0.03	0	0.00	0.06	0.00	0.01	0.00	<1	1	0	<.1
0	3	0.08	3	57	167	0.04	7	0.00	0.22	0.07	0.23	0.02	2	2	0	<.1
0	3	0.15	3	33	363	0.04	3	0.00	0.05	0.01	0.02	0.00	<1	1	0	0
2	5	0.00	1	7	15	0.01	—	0.00	0.03	0.01	0.00	0.00	1	<.1	<.1	—
4	<.1	0.00	<.1	2	80	0.02	0	0.00	1.03	0.00	0.00	0.00	0	0	0	—
5	2	0.07	<1	5	78	0.02	12	0.00	0.32	0.00	0.00	0.08	1	0	<.1	<1
0	6	0.09	1	7	68	0.02	0	0.00	0.22	0.00	0.01	0.00	<1	<.1	0	—
0	4	0.09	2	8	56	0.03	<1	0.00	0.09	0.01	0.02	0.00	<1	<1	0	2
0	<1	0.13	1	4	122	0.02	1	0.00	0.01	0.00	0.03	0.00	<1	<1	0	0
11	6	0.21	1	10	200	0.05	—	0.00	2.04	0.00	0.01	0.08	2	<1	<.1	—
0	6	0.28	6	54	255	0.06	<1	0.01	0.35	0.01	0.28	0.02	1	2	0	<1
20	128	0.15	6	21	578	0.68	56	0.00	0.22	0.08	0.02	0.01	3	<1	<.1	2
0	5	0.36	9	126	62	0.11	—	0.03	0.21	0.02	0.63	0.06	6	44	0	0
0	7	1.05	11	231	114	0.15	—	0.02	0.42	0.22	0.61	0.34	7	<1	0	<1
<1	5	0.16	4	19	258	0.05	0	0.00	0.00	0.03	0.19	0.01	4	<.1	0	<1
1	37	1.51	57	283	305	0.95	—	0.07	1.77	0.09	1.82	0.09	14	5	<.1	—
0	5	0.03	1	9	437	0.01	0	0.00	0.01	0.02	0.24	0.00	2	<.1	<.1	1
0	1	0.06	1	6	32	0.01	4	0.00	0.69	0.00	0.01	0.01	<.1	<1	0	—
0	5	0.16	2	34	69	0.04	5	0.01	0.10	0.01	0.13	0.02	3	2	0	<.1
0	3	0.36	6	32	1029	0.07	0	0.01	0.00	0.02	0.61	0.03	3	0	0	—
0	5	0.20	1	8	98	0.01	0	0.00	—	0.01	0.12	0.04	<1	0	0	—
0	5	0.31	11	41	690	0.02	0	0.01	0.00	0.01	0.23	0.02	4	0	0	<1
0	21	1.08	19	431	199	0.30	48	0.05	1.55	0.05	1.18	0.13	15	15	0	1
4	74	0.21	9	98	221	0.26	—	0.04	0.85	0.12	0.25	0.03	3	1	<1	—
0	6	0.30	1	45	56	0.01	—	0.00	0.00	0.01	0.04	0.00	0	1	0	—
0	0	0.00	—	—	0	—	0	—	—	—	—	—	—	0	—	—
0	1	0.09	3	15	<1	0.00	0	0.00	0.00	0.00	0.00	0.00	0	0	0	<.1
0	1	0.09	0	2	<1	0.00	0	0.00	0.00	0.00	0.00	0.00	0	0	0	5
0	0	0.00	—	0	0	—	0	—	—	—	—	—	—	0	0	—
40	69	1.97	60	549	526	1.62	—	0.09	3.80	0.20	9.48	0.44	26	7	<1	—
120	26	1.64	31	387	477	2.69	69	0.07	1.12	0.37	6.92	0.39	15	5	<1	19
38	331	2.95	45	384	514	1.92	119	0.24	1.52	0.29	1.82	0.15	115	4	<1	18
45	84	3.74	37	476	757	3.51	—	0.39	1.73	0.30	5.37	0.38	23	27	2	—
74	30	1.66	20	248	547	0.91	—	0.32	1.66	0.25	2.55	0.19	20	4	<1	—
71	29	2.90	—	613	292	—	—	0.15	1.35	0.17	4.70	—	—	17	<.1	15
74	70	3.26	37	393	818	3.66	—	0.21	1.64	0.31	3.80	0.21	17	1	2	—
96	74	2.92	94	640	1355	2.24	—	0.23	5.83	0.22	19.76	0.88	64	11	<1	—
19	233	1.16	32	219	672	1.63	147	0.37	—	0.33	4.96	0.09	69	3	1	27
13	132	1.21	13	77	230	0.64	—	0.13	1.05	0.14	1.09	0.04	6	15	<.1	—
103	26	2.20	—	149	600	—	—	0.05	0.43	0.17	4.30	—	—	0	—	29
84	22	1.38	29	344	962	1.70	—	0.08	0.94	0.17	7.90	0.32	13	8	<1	—
103	45	2.21	37	444	451	2.01	53	0.10	1.18	0.21	9.20	0.54	15	8	<1	22

A

Table A–1

Food Composition   (Computer code number is for Wadsworth Diet Analysis program)   (For purposes of calculations, use "0" for t, <1, <.1, <.01, etc.)

DA + Code	Food Description	Quantity	Measure	Wt (g)	H₂O (g)	Ener (cal)	Prot (g)	Carb (g)	Dietary Fiber (g)	Fat (g)	Sat	Mono	Poly
	MIXED FOODS, SOUPS, SANDWICHES—Continued												
28020	Chicken fried turkey steak	3	ounce(s)	85	48	122	13	12	1	2	0.59	0.37	0.78
218	Chicken pot pie	1	cup(s)	252	154	542	23	42	3	31	9.79	12.52	7.03
30240	Chicken teriyaki	1	cup(s)	244	163	339	51	13	1	7	1.78	2.03	1.71
25119	Chicken waldorf salad	½	cup(s)	100	68	178	14	6	1	11	1.76	3.18	5.05
25099	Chili con carne	¾	cup(s)	215	175	197	14	21	7	7	2.55	2.83	0.54
1062	Coleslaw	¾	cup(s)	90	73	62	1	11	1	2	0.35	0.64	1.22
1896	Combination pizza, w/meat & vegetables	2	slice(s)	158	75	368	26	43	5	11	3.07	5.09	1.83
1574	Crab cakes, from blue crab	1	item(s)	60	43	93	12	<1	0	5	0.89	1.69	1.36
32144	Enchiladas w/green chili sauce (enchiladas verdes)	1	item(s)	144	104	207	9	18	3	12	6.35	3.65	0.96
2793	Falafel patty	3	item(s)	51	18	170	7	16	0	9	1.22	5.19	2.12
28546	Fettuccine alfredo	1	cup(s)	222	81	247	11	42	1	3	1.61	0.79	0.43
32146	Flautas	3	item(s)	162	78	438	25	36	4	22	8.22	8.80	2.29
29629	Fried rice w/meat or poultry	1	cup(s)	198	129	329	12	41	1	12	2.27	3.53	5.69
16649	General tso chicken	1	cup(s)	146	91	293	19	16	1	17	3.98	6.27	5.27
1826	Green salad	¾	cup(s)	104	99	17	1	3	2	<.1	0.01	0.00	0.04
1814	Hummus	½	cup(s)	123	80	218	6	25	5	11	1.38	6.04	2.56
16650	Kung pao chicken	1	cup(s)	162	88	431	29	11	2	31	5.19	13.95	9.69
16622	Lamb curry	1	cup(s)	236	188	256	28	3	1	14	3.93	4.92	3.35
25253	Lasagna w/ground beef	1	cup(s)	237	157	288	18	22	2	15	7.47	4.84	0.84
442	Macaroni & cheese	1	cup(s)	200	122	393	15	40	1	19	8.18	6.72	2.66
25105	Meat loaf	1	slice(s)	115	85	244	17	7	<1	16	6.15	6.89	0.83
16646	Moo shi pork	1	cup(s)	151	77	512	19	5	1	46	6.84	14.80	22.07
16788	Nachos w/beef, beans, cheese, tomatoes, & onions	7	item(s)	551	284	1496	40	119	19	99	22.34	40.19	30.69
1668	Pepperoni pizza	2	slice(s)	142	66	362	20	40	1	14	4.47	6.28	2.33
655	Potato salad	½	cup(s)	125	95	179	3	14	2	10	1.79	3.10	4.67
29637	Ravioli, meat filled, w/tomato or meat sauce, canned	1	cup(s)	251	196	220	9	38	2	4	1.58	1.49	0.41
25109	Salisbury steaks w/mushroom sauce	1	serving(s)	135	102	251	17	9	1	15	5.98	6.67	0.76
16637	Shrimp creole w/rice	1	cup(s)	243	176	311	27	28	1	9	1.83	3.79	2.88
497	Spaghetti & meat balls w/tomato sauce, prepared	1	cup(s)	248	174	330	19	39	3	12	3.90	4.40	2.20
28585	Spicy thai noodles (pad thai)	8	ounce(s)	231	74	222	9	36	3	6	0.83	3.33	1.83
33073	Stir fried pork & vegetables w/rice	1	cup(s)	235	173	349	15	34	2	16	5.55	6.87	2.62
28588	Stuffed shells	2½	item(s)	299	189	292	18	33	3	10	3.81	3.57	1.62
16821	Sushi w/egg in seaweed	6	piece(s)	156	117	190	9	20	<1	8	2.09	3.02	1.55
16819	Sushi w/vegetables & fish	6	piece(s)	156	102	217	8	44	2	1	0.16	0.14	0.20
16820	Sushi w/vegetables in seaweed	6	piece(s)	156	110	182	3	41	1	<1	0.10	0.11	0.11
25266	Sweet & sour pork	¾	cup(s)	249	206	264	29	17	1	8	2.59	3.51	1.48
16824	Tabouli, tabbouleh, or tabuli	1	cup(s)	160	124	199	3	16	4	15	2.04	10.83	1.37
25276	Three bean salad	½	cup(s)	99	82	95	2	10	3	6	0.76	1.41	3.48
160	Tuna salad	½	cup(s)	103	65	192	16	10	0	9	1.58	2.96	4.23
25241	Turkey & noodles	1	cup(s)	319	228	271	24	21	1	9	2.39	3.48	2.27
16794	Vegetable egg roll	2	item(s)	128	90	202	5	20	2	12	2.46	5.71	2.65
16818	Vegetable sushi, no fish	6	piece(s)	156	99	225	5	50	2	<1	0.11	0.10	0.14
	Sandwiches												
1744	Bacon, lettuce & tomato w/mayonnaise	1	item(s)	164	97	349	11	34	2	19	4.54	7.22	6.07
30287	Bologna & cheese w/margarine	1	item(s)	111	46	350	13	28	1	20	8.55	8.40	2.28
30286	Bologna w/margarine	1	item(s)	83	34	256	7	26	1	13	4.08	6.31	2.07
16546	Cheese	1	item(s)	83	31	262	10	27	1	13	5.59	4.77	1.67
8789	Cheeseburger, large, plain	1	item(s)	185	72	609	30	47	0	33	14.84	12.74	2.44
8624	Cheeseburger, large, w/bacon, vegetables, & condiments	1	item(s)	195	85	608	32	37	2	37	16.24	14.49	2.71
1745	Club w/bacon, chicken, tomato, lettuce, & mayonnaise	1	item(s)	246	137	555	31	48	3	26	5.94	—	—
1908	Cold cut submarine w/cheese & vegetables	1	item(s)	228	132	456	22	51	2	19	6.81	8.23	2.28
30247	Corned beef	1	item(s)	130	75	268	19	25	2	10	3.75	3.96	0.80
25283	Egg salad	1	item(s)	126	72	278	10	29	1	13	2.96	3.97	4.79
16686	Fried egg	1	item(s)	96	50	226	10	26	1	9	2.29	3.51	1.64

PAGE KEY: A–4 = Breads/Baked Goods   A–8 = Cereal/Rice/Pasta   A–12 = Fruit   A–18 = Vegetables/Legumes   A–28 = Nuts/Seeds   A–30 = Vegetarian   A–32 = Dairy   A–38 = Eggs   A–40 = Seafood   A–42 = Meats   A–44 = Poultry   A–46 = Processed meats   A–48 = Beverages   A–52 = Fats/Oils   A–54 = Sweets   A–54 = Sauces/Condiments/Spices   A–58 = Mixed Foods/Soups/Sandwiches   A–64 = Fast food   A–82 = Convenience   A–84 = Baby foods

A

Chol (mg)	Calc (mg)	Iron (mg)	Magn (mg)	Pota (mg)	Sodi (mg)	Zinc (mg)	Vit A (RAE) (µg)	Thia (mg)	Vit E (mg)	Ribo (mg)	Niac (mg)	Vit B₆ (mg)	Fola (µg)	Vit C (mg)	Vit B₁₂ (µg)	Sele (µg)
27	69	1.34	19	197	139	1.08	5	0.15	0.08	0.18	3.46	0.22	21	<1	<1	16
69	64	3.38	38	393	651	1.93	607	0.40	2.73	0.40	7.24	0.24	31	11	<1	—
157	52	3.27	67	589	3209	3.75	—	0.15	0.67	0.37	16.69	0.89	23	6	1	—
42	20	0.78	24	197	246	1.13	21	0.04	1.04	0.10	4.05	0.25	15	2	<1	11
27	43	3.16	50	646	865	2.44	25	0.13	1.65	0.23	3.01	0.18	56	10	1	10
7	41	0.53	9	163	21	0.18	48	0.06	—	0.06	0.24	0.11	24	29	0	1
41	202	3.07	36	357	765	2.23	117	0.43	—	0.35	3.92	0.19	65	3	1	22
90	63	0.65	20	194	198	2.45	34	0.05	—	0.05	1.74	0.10	32	2	4	24
27	266	1.08	38	251	276	1.27	—	0.07	0.46	0.16	1.28	0.18	45	59	<1	6
0	28	1.74	42	298	150	0.77	1	0.07	—	0.08	0.53	0.06	47	1	0	1
9	153	1.88	32	123	386	1.48	51	0.35	0.17	0.34	2.60	0.06	103	1	<1	35
73	146	2.66	61	223	886	3.44	0	0.10	0.33	0.17	3.00	0.27	96	0	1	37
102	36	2.66	31	182	821	1.42	—	0.30	2.19	0.19	3.51	0.24	24	3	<1	—
65	27	1.49	24	250	906	1.40	—	0.10	1.78	0.19	6.28	0.28	17	12	<1	—
0	13	0.65	11	178	27	0.22	59	0.03	—	0.05	0.57	0.08	38	24	0	<1
0	60	1.93	36	213	298	1.34	0	0.11	1.23	0.06	0.49	0.49	73	10	0	3
64	49	1.96	63	428	907	1.50	—	0.15	3.90	0.15	13.23	0.59	43	8	<1	—
89	36	2.97	40	495	495	6.62	—	0.09	1.20	0.28	8.05	0.20	27	1	3	—
68	222	2.33	40	437	493	2.81	108	0.19	1.15	0.29	3.02	0.20	50	10	1	22
30	323	2.26	42	263	800	1.95	327	0.25	1.45	0.40	2.18	0.10	12	<1	<1	—
85	54	2.09	21	278	423	3.55	27	0.08	0.28	0.29	3.77	0.13	20	<1	2	17
172	30	1.45	26	330	1078	1.83	—	0.50	8.51	0.38	2.90	0.31	22	8	1	—
82	699	6.71	205	1067	1611	7.55	—	0.31	14.23	0.50	5.62	0.85	59	14	1	—
28	129	1.87	17	305	534	1.04	105	0.27	—	0.47	6.09	0.11	74	3	<1	26
85	24	0.81	19	318	661	0.39	40	0.10	—	0.08	1.11	0.18	9	13	0	5
17	28	2.04	23	337	1354	1.19	—	0.22	0.41	0.20	2.88	0.14	17	22	<1	—
60	64	2.21	23	282	370	3.66	27	0.11	0.22	0.30	4.00	0.13	22	<1	2	17
181	101	4.44	64	439	381	1.73	—	0.29	3.99	0.10	4.77	0.22	12	18	1	—
89	124	3.70	—	665	1009	—	82	0.25	1.94	0.30	4.00	—	—	22	—	22
37	32	1.58	50	187	598	1.08	38	0.18	1.31	0.13	1.88	0.17	44	22	<.1	3
46	39	2.65	32	394	574	2.07	80	0.51	1.19	0.20	5.07	0.30	102	18	<1	23
35	241	3.18	63	462	543	1.68	280	0.32	1.30	0.36	4.64	0.30	109	15	<1	36
217	42	1.63	18	128	527	0.98	—	0.12	0.89	0.29	1.33	0.13	29	2	<1	—
11	24	2.18	25	204	340	0.79	—	0.26	0.58	0.07	2.77	0.15	14	4	<1	—
0	20	1.54	20	99	153	0.70	—	0.20	0.13	0.04	1.86	0.14	10	2	0	—
74	41	1.78	35	622	624	2.53	64	0.80	0.98	0.37	6.69	0.65	14	10	1	50
0	29	1.25	36	246	799	0.48	—	0.08	2.10	0.05	1.14	0.11	31	29	0	—
0	26	0.96	15	144	224	0.31	12	0.04	1.41	0.06	0.26	0.06	31	9	0	3
13	17	1.03	19	182	412	0.57	25	0.03	—	0.07	6.87	0.08	8	2	1	42
77	60	2.69	33	379	576	2.64	108	0.23	1.30	0.32	6.40	0.30	60	1	1	34
60	29	1.61	18	193	548	0.51	—	0.16	1.77	0.21	1.59	0.10	27	6	<1	—
0	23	2.40	23	158	369	0.84	—	0.28	0.19	0.06	2.44	0.13	15	4	0	—
20	76	2.54	27	328	837	0.98	—	0.39	1.87	0.27	3.81	0.20	31	15	<1	—
35	221	2.18	24	185	940	1.68	—	0.30	0.95	0.33	2.77	0.12	21	<.1	1	—
16	60	1.96	15	112	598	0.85	—	0.29	0.76	0.21	2.73	0.08	19	<.1	<1	—
19	216	1.75	20	135	655	1.14	—	0.25	0.92	0.29	2.04	0.07	19	<.1	<1	—
96	91	5.46	39	644	1589	5.55	185	0.48	—	0.57	11.17	0.28	74	0	3	39
111	162	4.74	45	332	1043	6.83	82	0.31	—	0.41	6.63	0.31	86	2	2	33
72	116	4.05	47	463	855	1.65	—	0.61	2.59	0.44	11.92	0.59	48	9	1	—
36	189	2.51	68	394	1651	2.58	71	1.00	—	0.80	5.49	0.14	87	12	1	31
46	67	2.67	20	187	1177	2.24	—	0.24	0.26	0.25	3.23	0.10	22	2	1	—
217	107	2.60	18	147	494	0.94	94	0.26	0.95	0.43	2.27	0.16	82	1	1	24
207	80	2.25	17	120	433	0.85	—	0.27	0.89	0.41	2.06	0.10	34	0	<1	—

Table A–1

Food Composition   (Computer code number is for Wadsworth Diet Analysis program)   (For purposes of calculations, use "0" for t, <1, <.1, <.01, etc.)

DA + Code	Food Description	Quantity	Measure	Wt (g)	H₂O (g)	Ener (cal)	Prot (g)	Carb (g)	Dietary Fiber (g)	Fat (g)	Fat Breakdown (g)		
											Sat	Mono	Poly
	MIXED FOODS, SOUPS, SANDWICHES—Continued												
16547	Grilled cheese	1	item(s)	83	27	292	10	27	1	16	6.22	6.29	2.54
16659	Gyro w/onion & tomato	1	item(s)	105	67	170	12	21	1	4	1.53	1.41	0.43
1906	Ham & cheese	1	item(s)	146	74	352	21	33	2	15	6.44	6.74	1.38
31890	Ham w/mayonnaise	1	item(s)	112	55	282	14	27	1	13	3.06	5.04	3.79
756	Hamburger, double patty, large, w/condiments & vegetables	1	item(s)	226	121	540	34	40	0	27	10.52	10.33	2.80
8793	Hamburger, large, plain	1	item(s)	137	58	426	23	32	2	23	8.38	9.88	2.14
8795	Hamburger, large, w/vegetables & condiments	1	item(s)	218	121	512	26	40	3	27	10.42	11.42	2.20
25134	Hot chicken salad	1	item(s)	98	49	239	16	23	1	9	2.83	2.61	2.76
1411	Hot dog w/bun, plain	1	item(s)	98	53	242	10	18	2	15	5.11	6.85	1.71
25133	Hot turkey salad	1	item(s)	98	50	221	16	23	1	7	2.23	1.76	2.28
30249	Pastrami	1	item(s)	134	71	331	14	27	2	18	6.18	8.74	1.02
16701	Peanut butter	1	item(s)	93	24	344	13	37	3	17	3.55	8.16	4.58
30306	Peanut butter & jelly	1	item(s)	93	24	330	11	42	3	15	3.00	6.87	3.82
1910	Roast beef, plain	1	item(s)	139	68	346	22	33	1	14	3.61	6.80	1.71
1909	Roast beef submarine w/mayonnaise & vegetables	1	item(s)	216	127	410	29	44	—	13	7.09	1.84	2.61
1907	Steak w/mayonnaise & vegetables	1	item(s)	204	104	459	30	52	2	14	3.81	5.34	3.35
25288	Tuna salad	1	item(s)	179	102	414	24	29	2	22	3.61	5.46	11.43
31891	Turkey w/mayonnaise	1	item(s)	143	75	330	29	26	1	11	2.61	3.25	4.40
30283	Turkey submarine w/cheese, lettuce, tomato, & mayonnaise	1	item(s)	277	156	583	37	51	3	25	7.15	8.03	7.81
	Soups												
25296	Bean	1	cup(s)	301	253	191	14	29	6	2	0.67	0.83	0.53
711	Bean with pork, condensed, prepared w/water	1	cup(s)	265	223	180	8	24	9	6	1.59	2.28	1.91
713	Beef noodle, condensed, prepared w/water	1	cup(s)	244	224	83	5	9	1	3	1.15	1.24	0.49
825	Cheese, condensed, prepared w/milk	1	cup(s)	251	207	231	9	16	1	15	9.11	4.09	0.45
826	Chicken broth, condensed, prepared w/water	1	cup(s)	244	234	39	5	1	0	1	0.39	0.59	0.27
25297	Chicken noodle	1	cup(s)	286	258	117	11	11	1	3	0.78	1.10	0.66
827	Chicken noodle, condensed, prepared w/water	1	cup(s)	241	222	75	4	9	1	2	0.65	1.11	0.55
724	Chicken noodle, dehydrated, prepared w/water	1	cup(s)	252	237	58	2	9	<1	1	0.31	0.52	0.39
823	Cream of asparagus, condensed, prepared w/milk	1	cup(s)	248	213	161	6	16	1	8	3.32	2.08	2.23
824	Cream of celery, condensed, prepared w/milk	1	cup(s)	248	214	164	6	15	1	10	3.94	2.46	2.65
708	Cream of chicken, condensed, prepared w/milk	1	cup(s)	248	210	191	7	15	<1	11	4.64	4.46	1.64
715	Cream of chicken, condensed, prepared w/water	1	cup(s)	244	221	117	3	9	<1	7	2.07	3.27	1.49
709	Cream of mushroom, condensed, prepared w/milk	1	cup(s)	248	210	203	6	15	<1	14	5.13	2.98	4.61
716	Cream of mushroom, condensed, prepared w/water	1	cup(s)	244	220	129	2	9	<1	9	2.44	1.71	4.22
25298	Cream of vegetable	1	cup(s)	285	251	165	7	15	2	9	1.56	4.62	1.92
16689	Egg drop	1	cup(s)	244	229	73	8	1	0	4	1.15	1.52	0.59
25138	Golden squash	1	cup(s)	258	224	144	8	21	2	4	0.84	2.18	0.88
16663	Hot & sour	1	cup(s)	244	210	161	15	5	1	8	2.72	3.40	1.20
28054	Lentil chowder	1	cup(s)	229	188	150	11	27	12	<1	0.09	0.08	0.22
28560	Macaroni & bean	1	cup(s)	229	129	136	6	21	5	3	0.48	2.06	0.59
714	Manhattan clam chowder, condensed, prepared w/water	1	cup(s)	244	224	78	2	12	1	2	0.38	0.38	1.29
28561	Minestrone	1	cup(s)	230	177	99	4	16	5	2	0.32	1.30	0.43
717	Minestrone, condensed, prepared w/water	1	cup(s)	241	220	82	4	11	1	3	0.55	0.70	1.11
28038	Mushroom & wild rice	1	cup(s)	230	188	81	4	12	2	<1	0.05	0.02	0.15
828	New England clam chowder, condensed, prepared w/milk	1	cup(s)	248	211	164	9	17	1	7	2.95	2.26	1.09
28036	New England style clam chowder	1	cup(s)	229	207	83	3	15	2	<1	0.08	0.03	0.05

PAGE KEY: A–4 = Breads/Baked Goods   A–8 = Cereal/Rice/Pasta   A–12 = Fruit   A–18 = Vegetables/Legumes   A–28 = Nuts/Seeds   A–30 = Vegetarian   A–32 = Dairy   A–38 = Eggs   A–40 = Seafood   A–42 = Meats   A–44 = Poultry   A–46 = Processed meats   A–48 = Beverages   A–52 = Fats/Oils   A–54 = Sweets   A–54 = Sauces/Condiments/Spices   A–58 = Mixed Foods/Soups/Sandwiches   A–64 = Fast food   A–82 = Convenience   A–84 = Baby foods

A

Chol (mg)	Calc (mg)	Iron (mg)	Magn (mg)	Pota (mg)	Sodi (mg)	Zinc (mg)	Vit A (RAE) (µg)	Thia (mg)	Vit E (mg)	Ribo (mg)	Niac (mg)	Vit B$_6$ (mg)	Fola (µg)	Vit C (mg)	Vit B$_{12}$ (µg)	Sele (µg)
19	219	1.76	21	137	696	1.15	—	0.19	1.38	0.28	1.86	0.06	13	<.1	<1	—
34	46	1.85	21	209	272	2.30	—	0.24	0.29	0.21	3.14	0.13	18	4	1	—
58	130	3.24	16	291	771	1.37	96	0.31	0.29	0.48	2.69	0.20	76	3	1	23
36	59	2.10	23	245	1033	1.50	—	0.71	1.09	0.31	4.89	0.26	19	0	<1	—
122	102	5.85	50	570	791	5.67	5	0.36	—	0.38	7.57	0.54	77	1	4	26
71	74	3.58	27	267	474	4.11	0	0.29	—	0.29	6.25	0.23	60	0	2	27
87	96	4.93	44	480	824	4.88	24	0.41	—	0.37	7.28	0.33	83	3	2	34
39	114	1.93	20	150	470	1.22	28	0.20	1.09	0.23	4.93	0.20	54	<1	<1	17
44	24	2.31	13	143	670	1.98	0	0.24	—	0.27	3.65	0.05	48	<.1	1	26
37	113	2.04	22	167	459	1.09	23	0.19	1.04	0.21	4.36	0.23	54	<1	<1	20
51	68	2.64	23	243	1335	2.69	—	0.29	0.33	0.27	4.77	0.13	21	2	1	—
1	80	2.47	62	272	479	1.25	0	0.33	3.12	0.25	6.46	0.17	43	0	<.1	—
1	68	2.11	53	239	409	1.06	—	0.27	2.63	0.21	5.45	0.15	37	<1	<.1	—
51	54	4.23	31	316	792	3.39	11	0.38	—	0.31	5.87	0.26	57	2	1	29
73	41	2.81	67	330	845	4.38	30	0.41	—	0.41	5.96	0.32	71	6	2	26
73	92	5.16	49	524	798	4.53	20	0.41	—	0.37	7.30	0.37	90	6	2	42
53	100	3.29	35	302	795	1.08	46	0.26	1.27	0.26	12.29	0.48	70	1	2	71
69	78	3.10	34	315	490	2.94	—	0.30	1.23	0.33	6.64	0.46	24	0	<1	—
70	324	3.88	51	552	2408	2.66	—	0.53	2.67	0.49	12.50	0.54	46	5	2	—
5	80	3.08	61	590	690	1.41	26	0.27	0.28	0.15	3.61	0.23	139	3	<1	8
3	85	2.15	48	421	996	1.09	48	0.09	0.08	0.03	0.59	0.04	34	2	<.1	8
5	15	1.10	5	100	952	1.54	7	0.07	0.00	0.06	1.07	0.04	20	<1	<1	7
48	289	0.80	20	341	1019	0.68	359	0.06	0.25	0.33	0.50	0.08	10	1	<1	7
0	10	0.51	2	210	776	0.24	0	0.01	0.04	0.07	3.35	0.02	5	0	<1	0
24	26	1.34	16	335	776	0.77	49	0.15	0.20	0.16	5.57	0.13	40	1	<1	10
7	17	0.77	5	55	1106	0.39	36	0.05	0.07	0.06	1.39	0.03	22	<1	<1	6
10	5	0.50	8	33	577	0.20	3	0.20	0.10	0.08	1.09	0.03	18	0	<.1	10
22	174	0.87	20	360	1042	0.92	62	0.10	0.84	0.28	0.88	0.06	30	4	<1	8
32	186	0.69	22	310	1009	0.20	114	0.07	0.97	0.25	0.44	0.06	7	1	<1	5
27	181	0.67	17	273	1047	0.67	179	0.07	0.24	0.26	0.92	0.07	7	1	1	8
10	34	0.61	2	88	986	0.63	163	0.03	0.20	0.06	0.82	0.02	2	<1	<.1	7
20	179	0.60	20	270	918	0.64	35	0.08	1.34	0.28	0.91	0.06	10	2	<1	4
2	46	0.51	5	100	881	0.59	15	0.05	1.24	0.09	0.72	0.01	5	1	<.1	1
1	68	1.38	17	312	784	0.74	100	0.12	5.13	0.20	3.27	0.12	37	10	<1	5
103	21	0.75	5	220	729	0.48	—	0.02	0.29	0.19	3.03	0.05	15	0	<1	—
4	203	1.63	39	412	500	1.72	454	0.17	2.46	0.38	1.15	0.15	32	10	1	8
34	29	1.89	29	382	1561	1.51	—	0.27	0.15	0.25	4.97	0.20	13	1	<1	—
<1	47	4.07	55	590	26	1.44	163	0.21	0.54	0.12	1.69	0.30	164	13	0	3
<1	64	1.86	32	254	489	0.46	174	0.15	0.73	0.13	1.36	0.09	59	7	0	9
2	27	1.63	12	188	578	0.98	56	0.03	0.73	0.04	0.82	0.10	10	4	4	9
0	68	1.76	31	273	423	0.38	138	0.10	0.57	0.10	0.69	0.07	47	12	0	4
2	34	0.92	7	313	911	0.75	118	0.05	0.07	0.04	0.94	0.10	36	1	0	8
0	27	1.08	26	332	267	0.87	4	0.06	0.22	0.21	2.97	0.14	18	4	<.1	4
22	186	1.49	22	300	992	0.79	57	0.07	0.15	0.24	1.03	0.13	10	3	10	13
2	69	1.29	26	430	236	0.66	34	0.07	0.61	0.12	1.02	0.20	17	12	3	4

A

Table A–1

Food Composition   (Computer code number is for Wadsworth Diet Analysis program)   (For purposes of calculations, use "0" for t, <1, <.1, <.01, etc.)

DA + Code	Food Description	Quantity	Measure	Wt (g)	H₂O (g)	Ener (cal)	Prot (g)	Carb (g)	Dietary Fiber (g)	Fat (g)	Fat Breakdown (g)		
											Sat	Mono	Poly
	MIXED FOODS, SOUPS, SANDWICHES—Continued												
28566	Old country pasta	1	cup(s)	228	164	135	6	20	3	3	1.17	1.60	0.63
725	Onion, dehydrated, prepared w/water	1	cup(s)	246	237	27	1	5	1	1	0.12	0.32	0.07
16667	Shrimp gumbo	1	cup(s)	244	206	171	10	19	3	7	1.34	3.02	2.05
28037	Southwestern corn chowder	1	cup(s)	229	202	102	5	18	2	<1	0.12	0.12	0.20
25140	Split pea	1	cup(s)	165	117	85	4	19	2	<1	0.07	0.03	0.18
718	Split pea with ham, condensed, prepared w/water	1	cup(s)	253	207	190	10	28	2	4	1.77	1.80	0.63
710	Tomato, condensed, prepared w/milk	1	cup(s)	248	210	161	6	22	3	6	2.90	1.61	1.12
719	Tomato, condensed, prepared w/water	1	cup(s)	244	220	85	2	17	<1	2	0.37	0.44	0.95
726	Tomato vegetable, dehydrated, prepared w/water	1	cup(s)	253	237	56	2	10	1	1	0.38	0.30	0.08
28595	Turkey noodle	1	cup(s)	228	203	106	8	14	2	2	0.27	1.06	0.67
28051	Turkey vegetable	1	cup(s)	227	203	98	11	8	2	1	0.32	0.17	0.30
720	Vegetable beef, condensed, prepared w/water	1	cup(s)	244	224	78	6	10	<1	2	0.85	0.81	0.12
28598	Vegetable gumbo	1	cup(s)	229	168	153	4	26	3	4	0.61	2.93	0.56
25141	Vegetable	1	cup(s)	252	225	96	5	20	4	—	0.06	0.04	0.16
721	Vegetarian vegetable, condensed, prepared w/water	1	cup(s)	241	223	72	2	12	—	2	0.29	0.82	0.72
	FAST FOOD												
	Arby's												
36094	Au jus sauce	1	serving(s)	85	—	5	<1	1	<.1	<.1	0.02	—	—
751	Beef 'n cheddar sandwich	1	item(s)	198	—	480	23	43	2	24	8.00	—	—
9279	Cheddar curly fries	1	serving(s)	170	—	460	6	54	4	24	6.00	—	—
36131	Chocolate shake	1	serving(s)	397	—	480	10	84	0	16	8.00	—	—
36045	Curly fries, large	1	serving(s)	198	—	620	8	78	7	30	7.00	—	—
36044	Curly fries, medium	1	serving(s)	128	—	400	5	50	4	20	5.00	—	—
9265	Fish fillet sandwich	1	item(s)	220	—	529	23	50	2	27	7.00	9.20	10.60
752	Ham 'n cheese sandwich	1	item(s)	170	—	340	23	35	1	13	4.50	—	—
36048	Homestyle fries, large	1	serving(s)	213	—	560	6	79	6	24	6.00	—	—
36047	Homestyle fries, medium	1	serving(s)	142	—	370	4	53	4	16	4.00	—	—
33465	Homestyle fries, small	1	serving(s)	113	—	300	3	42	3	13	3.50	—	—
9267	Italian sub sandwich	1	item(s)	312	—	780	29	49	3	53	15.00	—	—
36041	Market Fresh grilled chicken caesar salad w/o dressing	1	serving(s)	338	—	230	33	8	3	8	3.50	—	—
9291	Roast beef deluxe sandwich, light	1	item(s)	182	—	296	18	33	6	10	3.00	5.00	2.00
9251	Roast beef sandwich, giant	1	item(s)	228	—	480	32	41	3	23	10.00	—	—
9249	Roast beef sandwich, junior	1	item(s)	129	—	310	16	34	2	13	4.50	—	—
750	Roast beef sandwich, regular	1	item(s)	157	—	350	21	34	2	16	6.00	—	—
2009	Roast beef sandwich, super	1	item(s)	245	—	470	22	47	3	23	7.00	—	—
9269	Roast beef sub sandwich	1	item(s)	334	—	760	35	47	3	48	16.00	—	—
9295	Roast chicken deluxe sandwich, light	1	item(s)	194	—	260	23	33	3	5	1.00	—	—
9293	Roast turkey deluxe sandwich, light	1	item(s)	194	—	260	23	33	3	5	0.50	—	—
36132	Strawberry shake	1	serving(s)	397	—	500	11	87	0	13	8.00	—	—
9273	Turkey sub sandwich	1	item(s)	306	—	630	26	51	2	37	9.00	—	—
36130	Vanilla shake	1	serving(s)	397	—	470	10	83	0	15	7.00	—	—
	Auntie Anne's												
35371	Cheese dipping sauce	1	serving(s)	35	—	100	3	4	0	8	4.00	—	—
35353	Cinnamon sugar soft pretzel	1	item(s)	120	—	350	9	74	2	2	0.00	—	—
35354	Cinnamon sugar soft pretzel w/butter	1	item(s)	120	—	450	8	83	3	9	5.00	—	—
35372	Marinara dipping sauce	1	serving(s)	35	—	10	0	4	0	0	0.00	0.00	0.00
35357	Original soft pretzel	1	item(s)	120	—	340	10	72	3	1	0.00	—	—
35358	Original soft pretzel w/butter	1	item(s)	120	—	370	10	72	3	4	2.00	—	—
35359	Parmesan herb soft pretzel	1	item(s)	120	—	390	11	74	4	5	2.50	—	—
35360	Parmesan herb soft pretzel w/butter	1	item(s)	120	—	440	10	72	9	13	7.00	—	—
35361	Sesame soft pretzel	1	item(s)	120	—	350	11	63	3	6	1.00	—	—
35362	Sesame soft pretzel w/butter	1	item(s)	120	—	410	12	64	7	12	4.00	—	—
35364	Sour cream & onion soft pretzel	1	item(s)	120	—	310	9	66	2	1	0.00	—	—
35366	Sour cream & onion soft pretzel w/butter	1	item(s)	120	—	340	9	66	2	5	3.00	—	—
35373	Sweet mustard dipping sauce	1	serving(s)	35	—	60	1	8	0	2	1.00	—	—

PAGE KEY: A–4 = Breads/Baked Goods   A–8 = Cereal/Rice/Pasta   A–12 = Fruit   A–18 = Vegetables/Legumes   A–28 = Nuts/Seeds   A–30 = Vegetarian   A–32 = Dairy   A–38 = Eggs   A–40 = Seafood   A–42 = Meats   A–44 = Poultry   A–46 = Processed meats   A–48 = Beverages   A–52 = Fats/Oils   A–54 = Sweets   A–54 = Sauces/Condiments/Spices   A–58 = Mixed Foods/Soups/Sandwiches   A–64 = Fast food   A–82 = Convenience   A–84 = Baby foods

A

Chol (mg)	Calc (mg)	Iron (mg)	Magn (mg)	Pota (mg)	Sodi (mg)	Zinc (mg)	Vit A (RAE) (µg)	Thia (mg)	Vit E (mg)	Ribo (mg)	Niac (mg)	Vit B$_6$ (mg)	Fola (µg)	Vit C (mg)	Vit B$_{12}$ (µg)	Sele (µg)
6	51	2.32	47	434	319	0.69	114	0.20	0.85	0.15	2.42	0.23	65	17	<.1	9
0	12	0.15	5	64	849	0.05	0	0.03	0.10	0.06	0.48	0.00	2	<1	0	2
51	99	2.34	51	515	515	0.93	—	0.19	2.38	0.10	2.54	0.19	59	26	<1	—
1	65	1.10	24	374	200	0.73	46	0.08	0.20	0.14	1.65	0.22	27	37	<1	2
0	30	1.25	33	352	608	0.57	112	0.12	0.13	0.09	1.67	0.21	61	9	0	<1
8	23	2.28	48	400	1007	1.32	23	0.15	—	0.08	1.47	0.07	3	2	<1	8
17	159	1.81	22	449	744	0.30	64	0.13	2.60	0.25	1.52	0.16	17	68	<1	2
0	12	1.76	7	264	695	0.24	29	0.09	2.49	0.09	1.42	0.11	15	66	0	<1
0	8	0.63	20	104	1146	0.18	10	0.06	0.81	0.05	0.79	0.05	10	6	0	5
24	27	1.40	22	200	372	0.67	81	0.20	0.54	0.11	2.68	0.15	45	5	<1	13
20	36	1.30	22	383	328	0.90	110	0.08	0.19	0.09	3.33	0.27	21	10	<1	9
5	17	1.12	5	173	791	1.54	95	0.04	0.32	0.05	1.03	0.08	10	2	<1	4
0	52	1.90	35	313	471	0.56	15	0.17	1.13	0.07	1.59	0.16	51	18	0	4
0	41	2.45	38	688	674	0.78	118	0.12	2.32	0.13	2.37	0.27	33	23	0	5
0	22	1.08	7	210	822	0.46	116	0.05	0.80	0.05	0.92	0.06	10	1	0	4
0	0	0.00	—	—	386	—	0	—	—	—	—	—	—	0	—	—
90	100	3.60	—	—	1240	—	0	—	0.41	—	—	—	—	1	—	—
5	60	1.80	—	—	1290	—	0	—	—	—	—	—	—	15	—	—
45	500	0.72	—	—	370	—	38	—	—	—	—	—	—	2	—	—
0	0	2.70	—	—	1540	—	0	—	—	—	—	—	—	21	—	—
0	0	1.80	—	—	990	—	0	—	—	—	—	—	—	15	—	—
43	90	3.78	—	450	864	—	10	0.35	—	0.31	5.60	—	—	1	—	—
90	150	2.70	—	—	1450	—	20	—	—	—	—	—	—	1	—	—
0	0	1.80	—	—	1070	—	0	—	—	—	—	—	—	30	—	—
0	0	1.08	—	—	710	—	0	—	—	—	—	—	—	21	—	—
0	0	0.72	—	—	570	—	0	—	—	—	—	—	—	15	—	—
120	250	2.70	—	—	2440	—	—	—	—	—	—	—	—	2	—	—
80	200	1.80	—	—	920	—	—	—	—	—	—	—	—	42	—	—
42	130	4.50	—	392	826	—	40	0.27	—	0.49	8.40	—	—	8	—	—
110	60	5.40	—	—	1440	—	0	—	—	—	—	—	—	0	—	—
70	60	2.70	—	—	740	—	0	—	—	—	—	—	—	0	—	—
85	60	3.60	—	—	950	—	0	—	—	—	—	—	—	0	—	—
85	80	3.60	—	—	1130	—	40	—	—	—	—	—	—	1	—	—
130	300	4.50	—	—	2230	—	40	—	—	—	—	—	—	4	—	—
40	100	2.70	—	—	1010	—	—	—	—	—	—	—	—	2	—	—
40	80	1.80	—	—	980	—	—	—	—	—	—	—	—	1	—	—
15	350	0.36	—	—	340	—	36	—	—	—	—	—	—	1	—	—
100	200	0.36	—	—	2170	—	—	—	—	—	—	—	—	2	—	—
45	500	1.08	—	—	360	—	39	—	—	—	—	—	—	2	—	—
10	100	0.00	—	—	510	—	—	—	—	—	—	—	—	0	—	—
0	20	1.98	—	—	410	—	0	—	—	—	—	—	—	0	—	—
25	30	2.34	—	—	430	—	—	—	—	—	—	—	—	0	—	—
0	0	0.00	—	—	180	—	0	—	—	—	—	—	—	0	—	—
0	30	2.34	—	—	900	—	0	—	—	—	—	—	—	0	—	—
10	30	2.16	—	—	930	—	0	—	—	—	—	—	—	0	—	—
10	80	1.80	—	—	780	—	—	—	—	—	—	—	—	1	—	—
30	60	1.80	—	—	660	—	—	—	—	—	—	—	—	1	—	—
0	20	2.88	—	—	840	—	0	—	—	—	—	—	—	0	—	—
15	20	2.70	—	—	860	—	—	—	—	—	—	—	—	0	—	—
0	30	1.98	—	—	920	—	—	—	—	—	—	—	—	0	—	—
10	40	2.16	—	—	930	—	—	—	—	—	—	—	—	0	—	—
40	0	0.00	—	—	120	—	0	—	—	—	—	—	—	0	—	—

A

Table A–1

Food Composition   (Computer code number is for Wadsworth Diet Analysis program)   (For purposes of calculations, use "0" for t, <1, <.1, <.01, etc.)

DA + Code	Food Description	Quantity	Measure	Wt (g)	H₂O (g)	Ener (cal)	Prot (g)	Carb (g)	Dietary Fiber (g)	Fat (g)	Sat	Mono	Poly
	FAST FOOD—Continued												
35367	Whole wheat soft pretzel	1	item(s)	120	—	350	11	72	7	2	0.00	—	—
35368	Whole wheat soft pretzel w/butter	1	item(s)	120	—	370	11	72	7	5	1.50	—	—
	Boston Market												
34975	Bbq baked beans	¾	cup(s)	201	—	270	8	48	12	5	2.00	—	—
34976	Black beans & rice	1	cup(s)	227	—	300	8	45	5	10	1.50	—	—
34978	Butternut squash	¾	cup(s)	193	—	150	2	25	6	6	4.00	—	—
35006	Caesar side salad	1	serving(s)	119	—	300	5	13	1	26	4.50	—	—
34979	Chicken gravy	1	ounce(s)	28	—	15	0	2	0	1	0.00	—	—
34973	Chicken pot pie	1	item(s)	425	—	750	26	57	2	46	14.00	—	—
35007	Cole slaw	¾	cup(s)	184	—	300	2	30	3	19	3.00	—	—
35057	Cornbread	1	item(s)	68	—	200	3	33	1	6	1.50	—	—
35008	Cranberry walnut relish	¾	cup(s)	210	—	350	3	75	3	5	0.00	—	—
34980	Creamed spinach	¾	cup(s)	181	—	260	9	11	2	20	13.00	—	—
34981	Glazed carrots	¾	cup(s)	153	—	280	1	35	4	15	3.00	—	—
34983	Green bean casserole	¾	cup(s)	170	—	80	1	9	2	5	1.50	—	—
34982	Green beans	¾	cup(s)	85	—	70	1	6	2	4	0.50	—	—
34967	Half chicken, w/skin	1	item(s)	277	—	590	70	4	0	33	10.00	—	—
34984	Homestyle mashed potatoes	¾	cup(s)	173	—	210	4	30	2	9	5.00	—	—
34985	Homestyle mashed potatoes & gravy	1	cup(s)	201	—	230	4	32	3	9	5.00	—	—
34969	Honey glazed ham	5	ounce(s)	142	—	210	24	10	0	8	3.00	—	—
34988	Hot cinnamon apples	¾	cup(s)	181	—	250	0	56	3	5	0.50	—	—
34989	Macaroni & cheese	¾	cup(s)	192	—	280	13	33	1	11	6.00	—	—
34970	Meatloaf	5	ounce(s)	142	—	282	20	15	1	17	7.28	—	—
35012	Old-fashioned potato salad	¾	cup(s)	150	—	200	3	22	2	12	2.00	—	—
34965	Quarter chicken, dark meat, no skin	1	item(s)	95	—	190	22	1	0	10	3.00	—	—
34966	Quarter chicken, dark meat, w/skin	1	item(s)	125	—	320	30	2	0	21	6.00	—	—
34963	Quarter chicken, white meat, no skin or wing	1	item(s)	140	—	170	33	2	0	4	1.00	—	—
34964	Quarter chicken, white meat, w/skin & wing	1	item(s)	152	—	280	40	2	0	12	3.50	—	—
34993	Rice pilaf	1	cup(s)	137	—	140	2	24	1	4	0.50	—	—
34968	Rotisserie turkey breast, skinless	5	ounce(s)	142	—	170	36	3	0	1	0.00	—	—
34998	Savory stuffing	1	cup(s)	132	—	190	4	27	2	8	1.50	—	—
34999	Squash casserole	¾	cup(s)	187	—	330	7	20	3	24	13.00	—	—
35003	Steamed vegetables	1	cup(s)	102	—	30	2	6	2	0	0.00	—	—
35004	Sweet potato casserole	¾	cup(s)	181	—	280	3	39	2	13	4.50	—	—
35005	Whole kernel corn	¾	cup(s)	146	—	180	5	30	2	4	0.50	—	—
	Burger King												
29731	Biscuit with sausage, egg, & cheese	1	item(s)	189	—	650	20	38	1	46	14.00	—	—
3739	BK Broiler chicken sandwich	1	item(s)	258	—	550	30	52	3	25	5.00	—	—
14249	Cheeseburger	1	item(s)	133	—	360	19	31	2	17	8.00	—	—
14251	Chicken sandwich	1	item(s)	224	—	660	25	53	3	39	8.00	—	—
3808	Chicken Tenders, 8 pieces	1	serving(s)	123	—	340	22	20	1	19	5.00	—	—
14259	Chocolate shake, small	1	item(s)	333	—	620	12	72	2	32	21.00	—	—
29732	Croissanwich w/sausage & cheese	1	item(s)	107	—	420	14	23	1	31	11.00	—	—
14261	Croissanwich w/sausage, egg, & cheese	1	item(s)	157	—	520	19	24	1	39	14.00	—	—
3809	Double cheeseburger	1	item(s)	189	—	540	32	32	2	31	15.00	—	—
14244	Double Whopper	1	item(s)	374	—	980	52	52	4	62	22.00	—	—
14245	Double Whopper w/cheese	1	item(s)	399	—	1070	57	53	4	70	27.00	—	—
14250	Fish Fillet sandwich	1	item(s)	185	—	520	18	44	2	30	8.00	—	—
14255	French fries, medium, salted	1	item(s)	117	—	360	4	46	4	18	5.00	—	—
14262	French toast sticks	1	serving(s)	112	—	390	6	46	2	20	4.50	—	—
14248	Hamburger	1	item(s)	121	—	310	17	31	2	13	5.00	—	—
14263	Hash brown rounds, small	1	serving(s)	75	—	230	2	23	2	15	4.00	—	—
14256	Onion rings, medium	1	serving(s)	91	—	320	4	40	3	16	4.00	—	—
39000	Tendercrisp chicken sandwich	1	item(s)	310	—	810	28	72	6	47	8.00	—	—
14258	Vanilla shake, small	1	item(s)	305	—	560	11	56	1	32	21.00	—	—
1736	Whopper	1	item(s)	291	—	710	31	52	4	43	13.00	—	—
14243	Whopper w/cheese	1	item(s)	316	—	800	36	53	4	50	18.00	—	—
	Carl's Jr												
10801	Carl's Catch fish sandwich	1	item(s)	201	—	530	18	55	2	28	7.00	—	1.89

PAGE KEY: A–4 = Breads/Baked Goods   A–8 = Cereal/Rice/Pasta   A–12 = Fruit   A–18 = Vegetables/Legumes   A–28 = Nuts/Seeds   A–30 = Vegetarian   A–32 = Dairy   A–38 = Eggs   A–40 = Seafood   A–42 = Meats   A–44 = Poultry   A–46 = Processed meats   A–48 = Beverages   A–52 = Fats/Oils   A–54 = Sweets   A–54 = Sauces/Condiments/Spices   A–58 = Mixed Foods/Soups/Sandwiches   A–64 = Fast food   A–82 = Convenience   A–84 = Baby foods

A

Chol (mg)	Calc (mg)	Iron (mg)	Magn (mg)	Pota (mg)	Sodi (mg)	Zinc (mg)	Vit A (RAE) (µg)	Thia (mg)	Vit E (mg)	Ribo (mg)	Niac (mg)	Vit B$_6$ (mg)	Fola (µg)	Vit C (mg)	Vit B$_{12}$ (µg)	Sele (µg)
0	30	1.98	—	—	1100	—	0	—	—	—	—	—	—	0	—	—
10	30	2.34	—	—	1120	—	—	—	—	—	—	—	—	0	—	—
0	100	3.60	—	—	540	—	42	—	—	—	—	—	—	6	—	—
0	40	1.80	—	—	1050	—	0	—	—	—	—	—	—	4	—	—
20	80	1.08	—	—	560	—	1150	—	—	—	—	—	—	30	—	—
15	100	0.72	—	—	690	—	—	—	—	—	—	—	—	9	—	—
0	0	0.00	—	—	180	—	0	—	—	—	—	—	—	0	—	—
110	40	4.50	—	—	1530	—	—	—	—	—	—	—	—	1	—	—
20	60	0.72	—	—	540	—	108	—	—	—	—	—	—	36	—	—
25	0	1.08	—	—	390	—	0	—	—	—	—	—	—	0	—	—
0	0	5.40	—	—	0	—	0	—	—	—	—	—	—	0	—	—
55	250	2.70	—	—	740	—	—	—	—	—	—	—	—	9	—	—
0	40	1.08	—	—	80	—	1000	—	—	—	—	—	—	1	—	—
5	20	0.72	—	—	670	—	—	—	—	—	—	—	—	2	—	—
0	40	0.36	—	—	250	—	30	—	—	—	—	—	—	5	—	—
290	0	2.70	—	—	1010	—	0	—	—	—	—	—	—	0	—	—
25	40	0.36	—	—	590	—	53	—	—	—	—	—	—	15	—	—
25	60	0.36	—	—	780	—	—	—	—	—	—	—	—	15	—	—
75	0	1.08	—	—	1460	—	0	—	—	—	—	—	—	0	—	—
0	20	0.36	—	—	45	—	—	—	—	—	—	—	—	0	—	—
30	300	1.44	—	—	890	—	—	—	—	—	—	—	—	0	—	—
68	91	2.46	—	—	592	—	—	—	—	—	—	—	—	1	—	—
15	60	1.08	—	—	450	—	0	—	—	—	—	—	—	6	—	—
115	0	1.08	—	—	440	—	0	—	—	—	—	—	—	0	—	—
155	0	1.80	—	—	500	—	0	—	—	—	—	—	—	0	—	—
85	0	0.72	—	—	480	—	0	—	—	—	—	—	—	0	—	—
135	0	1.08	—	—	510	—	0	—	—	—	—	—	—	0	—	—
0	20	1.08	—	—	520	—	—	—	—	—	—	—	—	4	—	—
100	20	1.80	—	—	850	—	0	—	—	—	—	—	—	0	—	—
5	40	1.44	—	—	620	—	—	—	—	—	—	—	—	2	—	—
70	200	0.72	—	—	1110	—	—	—	—	—	—	—	—	5	—	—
0	40	0.35	—	—	135	—	389	—	—	—	—	—	—	18	—	—
10	40	1.08	—	—	190	—	—	—	—	—	—	—	—	9	—	—
0	0	0.36	—	—	170	—	20	—	—	—	—	—	—	5	—	—
190	150	2.70	—	—	1600	—	90	—	—	—	—	—	—	0	—	—
105	60	3.60	—	—	1110	—	—	0.46	3.81	0.23	10.50	—	—	6	—	—
50	150	3.60	—	—	790	—	63	0.25	—	0.32	4.18	—	—	1	—	—
70	80	2.70	—	—	1330	—	—	0.47	—	0.30	9.59	—	—	0	—	—
50	20	0.72	—	—	840	—	—	0.14	0.41	0.12	10.93	—	—	0	—	—
95	350	1.08	—	—	310	—	42	0.11	—	0.56	0.24	—	—	0	—	—
45	100	3.60	—	—	840	—	—	—	—	—	—	—	—	0	—	—
210	300	4.50	—	—	1090	—	140	0.36	—	0.42	4.35	—	—	0	—	—
100	250	4.50	—	—	1050	—	100	0.26	1.89	0.45	6.37	—	—	1	—	—
160	150	9.00	—	—	1070	—	—	0.40	—	0.60	11.08	—	—	9	—	—
185	300	9.00	—	—	1500	—	—	0.40	—	0.67	11.07	—	—	9	—	—
55	150	2.70	—	—	840	—	14	—	—	—	—	—	—	1	—	—
0	20	0.72	—	—	640	—	0	0.16	—	0.48	2.32	—	—	9	—	—
0	60	1.80	—	—	440	—	0	0.19	—	0.22	2.86	—	—	0	—	—
40	76	3.60	—	—	580	—	9	0.25	—	0.29	4.26	—	—	1	—	—
0	0	0.36	—	—	450	—	0	0.11	—	0.07	2.11	—	—	1	—	—
0	97	0.00	—	—	460	—	0	0.14	—	0.09	2.33	—	—	0	—	—
60	80	4.50	—	—	1800	—	—	—	—	—	—	—	—	9	—	—
95	300	0.36	—	—	220	—	39	0.11	—	0.64	0.22	—	—	0	—	—
85	150	6.30	—	—	980	—	52	0.39	4.25	0.44	7.33	—	—	9	—	—
110	250	6.30	—	—	1420	—	157	0.39	—	0.51	7.31	—	—	9	—	—
80	150	1.80	—	—	1030	—	60	—	—	—	—	—	—	2	—	—

Table A–1

Food Composition   (Computer code number is for Wadsworth Diet Analysis program)   (For purposes of calculations, use "0" for t, <1, <.1, <.01, etc.)

DA + Code	Food Description	Quantity	Measure	Wt (g)	H₂O (g)	Ener (cal)	Prot (g)	Carb (g)	Dietary Fiber (g)	Fat (g)	Fat Breakdown (g) Sat	Mono	Poly
	FAST FOOD—Continued												
10862	Carl's Famous Star hamburger	1	item(s)	254	—	590	24	50	3	32	9.00	—	—
10866	Charboiled chicken salad-to-go	1	item(s)	350	—	200	25	12	4	7	3.00	—	1.02
10855	Charboiled Sante Fe chicken sandwich	1	item(s)	220	—	540	28	37	2	31	8.00	—	—
10790	Chicken stars (6 pieces)	6	item(s)	90	—	260	13	14	1	16	4.50	—	1.71
34864	Chocolate shake, small	1	item(s)	595	—	530	14	96	0	10	7.00	—	—
10797	Crisscut fries	1	serving(s)	139	—	410	5	43	4	24	5.00	—	—
10799	Double western bacon cheeseburger	1	item(s)	308	—	920	51	65	3	50	21.00	—	6.55
34855	Famous bacon cheeseburger	1	item(s)	279	—	700	31	51	3	41	13.00	—	—
14238	French fries, small	1	serving(s)	92	—	290	5	37	3	14	3.00	—	—
10798	French toast dips w/o syrup	1	serving(s)	105	—	370	6	42	1	20	2.50	—	1.35
34856	Hamburger	1	item(s)	119	—	280	14	36	1	9	3.50	—	—
10802	Onion rings	1	serving(s)	127	—	430	7	53	3	22	5.00	—	0.84
38925	Six Dollar burger	1	item(s)	539	—	1000	39	72	6	82	25.00	—	—
34858	Spicy chicken sandwich	1	item(s)	198	—	480	14	47	2	26	5.00	—	—
34867	Strawberry shake, small	1	item(s)	595	—	510	14	91	0	10	7.00	—	—
10865	Super Star hamburger	1	item(s)	345	—	790	41	51	3	47	15.00	—	—
10818	Vanilla shake, small	1	item(s)	595	—	470	15	78	0	11	7.00	—	—
10770	Western bacon cheeseburger	1	item(s)	225	—	660	31	64	3	30	12.00	—	4.85
	Chick Fil-A												
38746	Biscuit w/bacon, egg, & cheese	1	item(s)	155	—	430	16	38	1	24	9.00	—	—
38747	Biscuit w/egg	1	item(s)	135	—	340	11	38	1	16	4.50	—	—
38748	Biscuit w/egg & cheese	1	item(s)	148	—	390	13	38	1	21	7.00	—	—
38753	Biscuit w/gravy	1	item(s)	191	—	310	5	44	1	13	3.50	—	—
38752	Biscuit w/sausage, egg, & cheese	1	item(s)	189	—	540	18	43	1	33	13.00	—	—
38741	Biscuit, plain	1	item(s)	78	—	260	4	38	1	11	2.50	—	—
38771	Carrot & raisin salad	1	item(s)	91	—	130	1	22	2	5	1.00	—	—
38761	Chargrilled chicken cool wrap	1	item(s)	245	—	380	29	54	3	6	3.00	—	—
38766	Chargrilled chicken garden salad	1	item(s)	275	—	180	22	9	3	6	3.00	—	—
38758	Chargrilled chicken sandwich	1	item(s)	157	—	280	26	30	1	7	1.50	—	—
38759	Chargrilled deluxe chicken sandwich	1	item(s)	195	—	290	27	31	2	7	1.50	—	—
38742	Chicken biscuit	1	item(s)	137	—	400	16	43	2	18	4.50	—	—
38743	Chicken biscuit w/cheese	1	item(s)	151	—	450	19	43	2	23	7.00	—	—
38762	Chicken caesar wrap	1	item(s)	227	—	460	36	52	2	10	6.00	—	—
38757	Chicken deluxe sandwich	1	item(s)	208	—	420	28	39	2	16	3.50	—	—
38764	Chicken salad sandwich	1	item(s)	153	—	350	20	32	5	15	3.00	—	—
38756	Chicken sandwich	1	item(s)	170	—	410	28	38	1	15	3.50	—	—
38768	Chick-n-Strip salad	1	item(s)	331	—	390	34	22	4	18	5.00	—	—
38763	Chick-n-Strips	4	item(s)	127	—	290	29	14	1	13	2.50	—	—
38770	Coleslaw	1	item(s)	105	—	210	1	14	2	17	2.50	—	—
38755	Hash browns	1	serving(s)	84	—	170	2	20	2	9	4.50	—	—
38765	Hearty breast of soup	1	cup(s)	241	—	140	8	18	1	4	1.00	—	—
38778	Icedream, small cone	1	item(s)	135	—	160	4	28	0	4	2.00	—	—
38774	Icedream, small cup	1	serving(s)	213	—	230	5	38	0	6	3.50	—	—
38775	Lemonade	1	cup(s)	255	—	170	0	41	0	1	0.00	—	—
38776	Lemonade, diet	1	cup(s)	255	—	25	0	5	0	0	0.00	0.00	0.00
38777	Nuggets	8	item(s)	113	—	260	26	12	1	12	2.50	—	—
38769	Side salad	1	item(s)	108	—	60	3	4	2	3	1.50	—	—
38767	Southwest chargrilled salad	1	item(s)	303	—	240	22	17	5	8	3.50	—	—
38772	Waffle potato fries, small, salted	1	serving(s)	85	—	280	3	37	5	14	5.00	—	—
	Cinnabon												
39569	Caramel Pecanbon	1	item(s)	272	—	1100	16	141	8	56	10.00	—	—
39572	Caramellata Chill w/whipped cream	16	fluid ounce(s)	480	—	406	10	61	0	14	8.00	—	—
39571	Cinnapoppers	1	serving(s)	74	—	368	4	41	2	21	11.00	—	—
39567	Classic roll	1	item(s)	221	—	813	15	117	4	32	8.00	—	—
39568	Minibon	1	item(s)	92	—	339	6	49	2	13	3.00	—	—
39573	Mochalatta chill w/whipped cream	16	fluid ounce(s)	480	—	362	9	55	0	13	8.00	—	—
39570	Stix	5	item(s)	85	—	379	6	41	1	21	6.00	—	—
	Dairy Queen												
1466	Banana split	1	item(s)	369	—	510	8	96	3	12	8.00	3.00	0.50
38552	Brownie Earthquake	1	serving(s)	304	—	740	10	112	0	27	16.00	—	—

PAGE KEY: A–4 = Breads/Baked Goods   A–8 = Cereal/Rice/Pasta   A–12 = Fruit   A–18 = Vegetables/Legumes   A–28 = Nuts/Seeds   A–30 = Vegetarian   A–32 = Dairy   A–38 = Eggs   A–40 = Seafood   A–42 = Meats   A–44 = Poultry   A–46 = Processed meats   A–48 = Beverages   A–52 = Fats/Oils   A–54 = Sweets   A–54 = Sauces/Condiments/Spices   A–58 = Mixed Foods/Soups/Sandwiches   A–64 = Fast food   A–82 = Convenience   A–84 = Baby foods

Chol (mg)	Calc (mg)	Iron (mg)	Magn (mg)	Pota (mg)	Sodi (mg)	Zinc (mg)	Vit A (RAE) (µg)	Thia (mg)	Vit E (mg)	Ribo (mg)	Niac (mg)	Vit B_6 (mg)	Fola (µg)	Vit C (mg)	Vit B_{12} (µg)	Sele (µg)
70	100	4.50	—	—	910	—	—	—	—	—	—	—	—	6	—	—
75	150	1.80	—	—	440	—	—	—	—	—	—	—	—	5	—	—
95	200	2.70	—	—	1210	—	—	—	—	—	—	—	—	6	—	—
40	20	1.08	—	—	480	—	0	—	—	—	—	—	—	0	—	—
45	600	1.08	—	—	350	—	0	—	—	—	—	—	—	0	—	—
0	20	1.80	—	—	950	—	0	—	—	—	—	—	—	12	—	—
155	300	7.20	—	—	1770	—	—	—	—	—	—	—	—	1	—	—
95	200	5.40	—	—	1310	—	102	—	—	—	—	—	—	6	—	—
0	0	1.08	—	—	180	—	0	—	—	—	—	—	—	21	—	—
0	40	1.08	—	—	430	—	0	0.26	—	0.24	2.00	—	—	0	—	—
35	80	2.70	—	—	480	—	0	—	—	—	—	—	—	1	—	—
0	20	0.72	—	—	700	—	0	—	—	—	—	—	—	4	—	—
135	350	5.40	—	—	1690	—	—	—	—	—	—	—	—	21	—	—
40	100	2.70	—	—	1220	—	—	—	—	—	—	—	—	6	—	—
45	600	0.00	—	—	330	—	0	—	—	—	—	—	—	0	—	—
130	100	7.20	—	—	980	—	—	—	—	—	—	—	—	9	—	—
50	600	0.00	—	—	350	—	0	—	—	—	—	—	—	0	—	—
85	200	5.40	—	—	1410	—	40	—	—	—	—	—	—	1	—	—
265	150	3.60	—	—	1070	—	—	—	—	—	—	—	—	0	—	—
245	80	2.70	—	—	740	—	—	—	—	—	—	—	—	0	—	—
260	150	2.70	—	—	960	—	—	—	—	—	—	—	—	0	—	—
5	60	1.80	—	—	930	—	0	—	—	—	—	—	—	0	—	—
280	150	3.60	—	—	1030	—	—	—	—	—	—	—	—	0	—	—
0	60	1.80	—	—	670	—	0	—	—	—	—	—	—	0	—	—
0	20	0.36	—	—	90	—	—	—	—	—	—	—	—	4	—	—
70	200	2.70	—	—	1060	—	—	—	—	—	—	—	—	6	—	—
70	150	0.72	—	—	660	—	—	—	—	—	—	—	—	30	—	—
70	80	1.80	—	—	980	—	0	—	—	—	—	—	—	2	—	—
70	80	1.80	—	—	990	—	—	—	—	—	—	—	—	5	—	—
30	60	2.70	—	—	1200	—	0	—	—	—	—	—	—	0	—	—
45	150	2.70	—	—	1430	—	—	—	—	—	—	—	—	0	—	—
80	500	2.70	—	—	1390	—	—	—	—	—	—	—	—	1	—	—
60	100	2.70	—	—	1300	—	—	—	—	—	—	—	—	2	—	—
65	150	1.80	—	—	880	—	—	—	—	—	—	—	—	0	—	—
60	100	2.70	—	—	1300	—	—	—	—	—	—	—	—	0	—	—
80	200	0.36	—	—	860	—	—	—	—	—	—	—	—	30	—	—
65	20	0.36	—	—	730	—	—	—	—	—	—	—	—	1	—	—
20	40	0.36	—	—	180	—	—	—	—	—	—	—	—	27	—	—
10	0	0.72	—	—	350	—	—	—	—	—	—	—	—	0	—	—
25	40	1.08	—	—	900	—	—	—	—	—	—	—	—	0	—	—
15	100	0.36	—	—	80	—	—	—	—	—	—	—	—	0	—	—
25	150	0.00	—	—	100	—	—	—	—	—	—	—	—	0	—	—
0	0	0.36	—	—	10	—	0	—	—	—	—	—	—	15	—	—
0	0	0.36	—	—	5	—	0	—	—	—	—	—	—	15	—	—
70	40	1.08	—	—	1090	—	0	—	—	—	—	—	—	0	—	—
10	100	0.00	—	—	75	—	—	—	—	—	—	—	—	15	—	—
60	200	1.08	—	—	770	—	—	—	—	—	—	—	—	24	—	—
15	20	0.00	—	—	105	—	0	—	—	—	—	—	—	21	—	—
63	—	—	—	—	600	—	—	—	—	—	—	—	—	—	—	—
46	—	—	—	—	187	—	—	—	—	—	—	—	—	—	—	—
62	—	—	—	—	104	—	—	—	—	—	—	—	—	—	—	—
67	—	—	—	—	801	—	—	—	—	—	—	—	—	—	—	—
27	—	—	—	—	337	—	—	—	—	—	—	—	—	—	—	—
46	100	0.00	—	—	252	—	—	—	—	—	—	—	—	0	—	—
16	—	—	—	—	413	—	—	—	—	—	—	—	—	—	—	—
30	250	1.80	—	860	180	—	—	0.15	—	0.60	0.20	—	—	15	—	—
50	250	1.80	—	—	350	—	—	—	—	—	—	—	—	0	—	—

A

DA + Code	Food Description	Quantity	Measure	Wt (g)	H₂O (g)	Ener (cal)	Prot (g)	Carb (g)	Dietary Fiber (g)	Fat (g)	Fat Breakdown (g) Sat	Mono	Poly
	FAST FOOD—Continued												
38561	Chocolate chip cookie dough blizzard, small	1	item(s)	319	—	720	12	105	0	28	14.00	—	—
1464	Chocolate malt, small	1	item(s)	418	—	650	15	111	0	16	10.00	—	—
38541	Chocolate shake, small	1	item(s)	397	—	560	13	93	1	15	10.00	—	—
17257	Chocolate soft serve	½	cup(s)	94	—	150	4	22	0	5	3.50	—	—
1463	Chocolate sundae, small	1	item(s)	163	—	280	5	49	0	7	4.50	1.00	1.00
1462	Dipped cone, small	1	item(s)	156	—	340	6	42	1	17	9.00	4.00	3.00
38555	Oreo cookies blizzard, small	1	item(s)	283	—	570	11	83	1	21	10.00	—	—
38547	Royal Treats Peanut Buster parfait	1	item(s)	305	—	730	16	99	2	31	17.00	—	—
17256	Vanilla soft serve	½	cup(s)	94	—	140	3	22	0	5	3.00	—	—
	Domino's												
31606	Barbeque wings	1	item(s)	25	—	50	6	2	<1	2	0.65	—	—
31604	Breadsticks	1	item(s)	37	—	116	3	18	1	4	0.79	—	—
37551	Buffalo chicken kickers	1	item(s)	24	14	47	4	3	<1	2	0.39	—	—
37548	Cinnastix	1	item(s)	32	8	122	2	15	1	6	1.15	—	—
	Classic hand tossed pizza												
31573	America's favorite feast, 12"	2	slice(s)	205	99	508	22	57	4	22	9.20	—	—
31574	America's favorite feast, 14"	2	slice(s)	283	138	697	30	79	5	30	12.70	—	—
37543	Bacon cheeseburger feast, 12"	2	slice(s)	198	60	549	25	55	3	26	11.62	—	—
37545	Bacon cheeseburger feast, 14"	2	slice(s)	275	121	762	35	75	4	36	16.10	—	—
37546	Barbeque feast, 12"	2	slice(s)	192	85	506	22	62	3	20	9.08	—	—
37547	Barbeque feast, 14"	2	slice(s)	262	115	691	30	85	4	27	12.24	—	—
31569	Cheese, 12"	2	slice(s)	159	—	375	15	55	3	11	4.81	—	—
31570	Cheese, 14"	2	slice(s)	219	—	516	21	75	4	15	6.72	—	—
37538	Deluxe feast, 12"	2	slice(s)	201	102	465	20	57	3	18	7.66	—	—
37540	Deluxe feast, 14"	2	slice(s)	273	138	627	26	78	5	24	10.20	—	—
31685	Deluxe, 12"	2	slice(s)	213	—	465	20	57	3	18	7.65	—	—
31694	Deluxe, 14"	2	slice(s)	273	—	627	26	78	5	24	10.20	—	—
31686	Extravaganzza, 12"	2	slice(s)	245	127	576	27	59	4	27	11.56	—	—
31695	Extravaganzza, 14"	2	slice(s)	329	171	773	36	88	5	36	15.42	—	—
31575	Hawaiian feast, 12"	2	slice(s)	204	105	450	21	58	3	16	7.20	—	—
31576	Hawaiian feast, 14"	2	slice(s)	283	147	623	29	80	5	22	10.09	—	—
31687	Meatzza, 12"	2	slice(s)	213	—	560	26	57	3	26	11.40	—	—
31696	Meatzza, 14"	2	slice(s)	293	139	753	35	78	5	34	15.24	—	—
31571	Pepperoni feast, extra pepperoni & cheese, 12"	2	slice(s)	196	87	534	24	56	3	25	10.92	—	—
31572	Pepperoni feast, extra pepperoni & cheese, 14"	2	slice(s)	270	121	732	33	77	4	34	15.00	—	—
31577	Vegi feast, 12"	2	slice(s)	203	107	439	19	57	4	16	7.09	—	—
31578	Vegi feast, 14"	2	slice(s)	278	147	304	27	78	5	22	9.89	—	—
37549	Dot cinnamon	1	item(s)	28	8	99	2	15	1	4	0.68	—	—
31605	Double cheesy bread	1	item(s)	35	11	123	4	13	1	6	2.06	—	—
31607	Hot wings	1	item(s)	25	—	45	5	1	<1	2	0.65	—	—
	Thin crust pizza												
31583	America's favorite, 12"	¼	item(s)	159	—	408	19	34	2	23	9.77	—	—
31584	America's favorite, 14"	¼	item(s)	202	—	557	26	47	3	31	13.19	—	—
31579	Cheese, 12"	¼	item(s)	106	—	273	12	31	2	12	9.37	—	—
31580	Cheese, 14"	¼	item(s)	148	—	382	17	43	2	17	6.72	—	—
31688	Deluxe, 12"	¼	item(s)	159	—	363	16	34	2	19	7.64	—	—
31697	Deluxe, 14"	¼	item(s)	202	—	494	22	47	3	25	10.20	—	—
31689	Extravaganzza, 12"	¼	item(s)	159	—	425	20	34	3	24	9.41	—	—
31698	Extravaganzza, 14"	¼	item(s)	202	—	571	27	48	4	31	12.44	—	—
31585	Hawaiian, 12"	¼	item(s)	159	—	349	18	35	2	16	7.20	—	—
31586	Hawaiian, 14"	¼	item(s)	202	—	489	25	48	3	23	10.09	—	—
31690	Meatzza, 12"	¼	item(s)	159	—	458	23	33	2	27	11.39	—	—
31699	Meatzza, 14"	¼	item(s)	202	—	619	31	46	3	36	15.24	—	—
31581	Pepperoni, extra pepperoni & cheese 12"	¼	item(s)	159	—	420	20	32	2	24	10.46	—	—
31582	Pepperoni, extra pepperoni & cheese 14"	¼	item(s)	202	—	586	28	45	3	34	14.55	—	—
31587	Vegi, 12"	¼	item(s)	159	—	338	16	34	3	17	7.08	—	—
31588	Vegi, 14"	¼	item(s)	202	—	471	22	47	3	23	9.89	—	—

PAGE KEY: A–4 = Breads/Baked Goods A–8 = Cereal/Rice/Pasta A–12 = Fruit A–18 = Vegetables/Legumes A–28 = Nuts/Seeds A–30 = Vegetarian A–32 = Dairy A–38 = Eggs A–40 = Seafood A–42 = Meats A–44 = Poultry A–46 = Processed meats A–48 = Beverages A–52 = Fats/Oils A–54 = Sweets A–54 = Sauces/Condiments/Spices A–58 = Mixed Foods/Soups/Sandwiches A–64 = Fast food A–82 = Convenience A–84 = Baby foods

A

Chol (mg)	Calc (mg)	Iron (mg)	Magn (mg)	Pota (mg)	Sodi (mg)	Zinc (mg)	Vit A (RAE) (µg)	Thia (mg)	Vit E (mg)	Ribo (mg)	Niac (mg)	Vit B6 (mg)	Fola (µg)	Vit C (mg)	Vit B12 (µg)	Sele (µg)
50	350	2.70	—	—	370	—	—	—	—	—	—	—	—	1	—	—
55	450	1.80	—	—	370	—	—	—	—	—	—	—	—	2	—	—
50	450	1.44	—	—	280	—	—	0.12	—	—	—	—	—	2	—	—
15	100	0.72	—	—	75	—	—	—	—	—	—	—	—	0	—	—
20	200	1.08	—	278	140	—	—	0.06	—	0.24	0.20	—	—	0	—	—
20	200	1.08	—	290	130	—	—	0.06	—	0.26	0.20	—	—	1	—	—
40	350	2.70	—	—	430	—	—	—	—	—	—	—	—	1	—	—
35	300	1.80	—	—	400	—	—	—	—	—	—	—	—	1	—	—
15	150	0.72	—	—	70	—	150	—	—	—	—	—	—	0	—	—
26	6	0.32	—	—	175	—	—	—	—	—	—	—	—	<.1	—	—
0	<.1	0.87	—	—	152	—	—	—	—	—	—	—	—	6	—	—
9	3	0.00	—	—	163	—	—	—	—	—	—	—	—	0	—	—
0	6	0.70	—	—	110	—	—	—	—	—	—	—	—	<.1	—	—
49	202	3.70	—	—	1221	—	—	—	—	—	—	—	—	1	—	—
68	281	5.10	—	—	1685	—	—	—	—	—	—	—	—	1	—	—
60	293	3.56	—	—	1274	—	—	—	—	—	—	—	—	0	—	—
84	395	4.96	—	—	1809	—	—	—	—	—	—	—	—	0	—	—
46	—	—	—	—	1206	—	—	—	—	—	—	—	—	—	—	—
63	393	4.42	—	—	1672	—	—	—	—	—	—	—	—	2	—	—
23	187	2.99	—	—	776	—	131	—	—	—	—	—	—	0	—	—
32	261	4.13	—	—	1080	—	184	—	—	—	—	—	—	0	—	—
40	199	3.56	—	—	1063	—	—	—	—	—	—	—	—	1	—	—
53	276	4.84	—	—	1432	—	—	—	—	—	—	—	—	2	—	—
40	199	3.56	—	—	1063	—	—	—	—	—	—	—	—	1	—	—
53	276	4.85	—	—	1432	—	—	—	—	—	—	—	—	2	—	—
60	290	4.08	—	—	1348	—	—	—	—	—	—	—	—	1	—	—
89	403	5.48	—	—	1780	—	—	—	—	—	—	—	—	2	—	—
41	274	3.30	—	—	1102	—	—	—	—	—	—	—	—	2	—	—
57	384	4.57	—	—	1544	—	—	—	—	—	—	—	—	3	—	—
344	282	3.71	—	—	1463	—	—	—	—	—	—	—	—	<1	—	—
85	393	5.04	—	—	1947	—	—	—	—	—	—	—	—	<1	—	—
57	279	3.36	—	—	1349	—	155	—	—	—	—	—	—	<1	—	—
78	390	4.66	—	—	1855	—	233	—	—	—	—	—	—	<1	—	—
34	279	3.44	—	—	987	—	—	—	—	—	—	—	—	1	—	—
47	389	4.71	—	—	1369	—	—	—	—	—	—	—	—	2	—	—
0	6	0.59	—	—	86	—	—	—	—	—	—	—	—	<.1	—	—
6	47	0.66	—	—	164	—	—	—	—	—	—	—	—	<1	—	—
26	5	0.30	—	—	354	—	—	—	—	—	—	—	—	1	—	—
51	318	1.52	—	—	1285	—	—	—	—	—	—	—	—	<1	—	—
69	444	2.07	—	—	1751	—	—	—	—	—	—	—	—	1	—	—
23	225	0.97	—	—	835	—	125	—	—	—	—	—	—	0	—	—
32	315	1.36	—	—	1172	—	175	—	—	—	—	—	—	0	—	—
40	237	1.54	—	—	1123	—	—	—	—	—	—	—	—	1	—	—
53	330	2.08	—	—	1523	—	—	—	—	—	—	—	—	2	—	—
53	245	1.95	—	—	1408	—	—	—	—	—	—	—	—	1	—	—
69	340	2.59	—	—	1871	—	—	—	—	—	—	—	—	2	—	—
41	312	1.28	—	—	1162	—	—	—	—	—	—	—	—	2	—	—
57	437	1.80	—	—	1635	—	—	—	—	—	—	—	—	3	—	—
64	320	1.69	—	—	1523	—	—	—	—	—	—	—	—	<1	—	—
454	446	2.27	—	—	2039	—	—	—	—	—	—	—	—	<1	—	—
54	316	1.34	—	—	1362	—	162	—	—	—	—	—	—	<1	—	—
76	442	1.87	—	—	1900	—	227	—	—	—	—	—	—	<1	—	—
34	317	1.42	—	—	1047	—	—	—	—	—	—	—	—	1	—	—
47	442	1.94	—	—	1460												

Table of Food Composition • **A-71**

Table A–1

Food Composition   (Computer code number is for Wadsworth Diet Analysis program)   (For purposes of calculations, use "0" for t, <1, <.1, <.01, etc.)

DA + Code	Food Description	Quantity	Measure	Wt (g)	H₂O (g)	Ener (cal)	Prot (g)	Carb (g)	Dietary Fiber (g)	Fat (g)	Fat Breakdown (g) Sat	Mono	Poly
	FAST FOOD—Continued												
	Ultimate deep dish pizza												
31596	America's favorite, 12"	2	slice(s)	235	—	617	26	59	4	33	12.88	—	—
31702	America's favorite, 14"	2	slice(s)	311	—	851	36	84	5	44	17.35	—	—
31590	Cheese, 12"	2	slice(s)	181	—	482	19	56	3	22	7.91	—	—
31591	Cheese, 14"	2	slice(s)	257	—	677	26	80	5	30	10.88	—	—
31589	Cheese, 6"	1	item(s)	215	—	598	23	68	4	28	9.94	—	—
31691	Deluxe, 12"	2	slice(s)	235	—	527	23	59	4	29	10.75	—	—
31700	Deluxe, 14"	2	slice(s)	311	—	788	31	84	5	38	14.36	—	—
31692	Extravaganzza, 12"	2	slice(s)	235	—	635	27	59	4	34	12.52	—	—
31701	Extravaganzza, 14"	2	slice(s)	311	—	866	36	85	6	45	16.60	—	—
31599	Hawaiian, 12"	2	slice(s)	235	—	558	24	60	4	26	10.31	—	—
31600	Hawaiian, 14"	2	slice(s)	311	—	784	35	85	5	36	14.25	—	—
31693	Meatzza, 12"	2	slice(s)	235	—	667	30	58	4	37	14.50	—	—
31703	Meatzza, 14"	2	slice(s)	311	—	914	40	83	5	49	19.40	—	—
31593	Pepperoni, extra pepperoni & cheese 12"	2	slice(s)	235	—	629	26	57	4	34	13.57	—	—
31594	Pepperoni, extra pepperoni & cheese 14"	2	slice(s)	311	—	880	37	82	5	47	18.71	—	—
31602	Vegi, 12"	2	slice(s)	235	—	547	22	59	4	26	10.19	—	—
31603	Vegi, 14"	2	slice(s)	311	—	765	32	84	6	36	14.05	—	—
31598	With ham & pineapple tidbits, 6"	1	item(s)	430	—	619	25	70	4	28	10.19	—	—
31595	With Italian sausage, 6"	1	item(s)	430	—	642	25	70	4	31	11.33	—	—
31592	With pepperoni, 6"	1	item(s)	430	—	647	25	69	4	32	11.70	—	—
31601	With vegetables, 6"	1	item(s)	430	—	619	23	71	5	29	10.11	—	—
	In-n-Out Burger												
34374	Cheeseburger	1	item(s)	268	—	480	22	39	3	27	10.00	—	—
34391	Cheesburger w/mustard & ketchup	1	item(s)	268	—	400	22	41	3	18	9.00	—	—
34390	Cheeseburger, lettuce leaves instead of buns	1	item(s)	300	—	330	18	11	2	25	9.00	—	—
34377	Chocolate shake	1	item(s)	425	—	690	9	83	0	36	24.00	—	—
34375	Double-Double cheeseburger	1	item(s)	328	—	670	37	40	3	41	18.00	—	—
34393	Double-Double cheeseburger w/mustard & ketchup	1	item(s)	328	—	590	37	42	3	32	17.00	—	—
34392	Double-Double cheeseburger, lettuce leaves instead of buns	1	item(s)	361	—	520	33	11	2	39	17.00	—	—
34376	French fries	1	item(s)	125	—	400	7	54	2	18	5.00	—	—
34373	Hamburger	1	item(s)	243	—	390	16	39	3	19	5.00	—	—
34389	Hamburger w/mustard & ketchup	1	item(s)	243	—	310	16	41	3	10	4.00	—	—
34388	Hamburger, lettuce leaves instead of buns	1	item(s)	275	—	240	12	10	2	17	4.50	—	—
34379	Strawberry shake	1	item(s)	425	—	690	8	91	2	33	22.00	—	—
34378	Vanilla shake	1	item(s)	425	—	680	9	78	2	37	25.00	—	—
	Jack in the Box												
30392	Bacon ultimate cheeseburger	1	item(s)	353	—	1120	52	59	2	55	28.00	—	—
1740	Breakfast Jack	1	item(s)	133	—	310	14	34	1	14	5.00	—	—
14074	Cheeseburger	1	item(s)	116	—	300	14	31	2	13	6.00	—	—
14106	Chicken breast pieces	5	piece(s)	150	—	360	27	24	1	17	3.00	—	—
37241	Chicken club salad	1	item(s)	535	—	310	28	15	5	16	6.00	—	—
14111	Chocolate ice cream shake	1	item(s)	315	—	660	11	89	1	29	18.00	—	—
14075	Double cheeseburger	1	item(s)	155	—	410	20	32	1	22	11.00	—	—
14098	French fries, jumbo	1	serving(s)	142	—	410	4	55	4	20	4.50	—	—
14099	French fries, super scoop	1	serving(s)	198	—	580	6	77	6	28	6.00	—	—
14073	Hamburger	1	item(s)	104	—	250	12	30	2	9	3.50	—	—
14090	Hash browns	1	serving(s)	57	—	150	1	13	2	10	2.50	—	—
14072	Jack's Spicy Chicken sandwich	1	item(s)	253	—	580	24	53	3	31	6.00	—	—
1468	Jumbo Jack hamburger	1	item(s)	269	—	600	22	58	3	31	11.00	—	—
1469	Jumbo Jack hamburger w/cheese	1	item(s)	294	—	690	26	60	3	38	16.00	—	—
1470	Onion rings	1	serving(s)	119	—	500	6	51	3	30	5.00	—	—
33141	Sausage, egg, & cheese biscuit	1	item(s)	223	—	760	25	33	2	60	20.00	—	—
14095	Seasoned curly fries	1	serving(s)	125	—	400	6	45	5	23	5.00	—	—
14077	Sourdough Jack	1	item(s)	244	—	700	30	36	3	49	16.00	—	—
37249	Southwest chicken salad	1	serving(s)	598	—	340	28	31	9	13	6.00	—	—

PAGE KEY: A–4 = Breads/Baked Goods   A–8 = Cereal/Rice/Pasta   A–12 = Fruit   A–18 = Vegetables/Legumes   A–28 = Nuts/Seeds   A–30 = Vegetarian
A–32 = Dairy   A–38 = Eggs   A–40 = Seafood   A–42 = Meats   A–44 = Poultry   A–46 = Processed meats   A–48 = Beverages   A–52 = Fats/Oils
A–54 = Sweets   A–54 = Sauces/Condiments/Spices   A–58 = Mixed Foods/Soups/Sandwiches   A–64 = Fast food   A–82 = Convenience   A–84 = Baby foods

Chol (mg)	Calc (mg)	Iron (mg)	Magn (mg)	Pota (mg)	Sodi (mg)	Zinc (mg)	Vit A (RAE) (µg)	Thia (mg)	Vit E (mg)	Ribo (mg)	Niac (mg)	Vit B$_6$ (mg)	Fola (µg)	Vit C (mg)	Vit B$_{12}$ (µg)	Sele (µg)
58	334	4.43	—	—	1573	—	—	—	—	—	—	—	—	1	—	—
78	464	6.24	—	—	2155	—	—	—	—	—	—	—	—	1	—	—
30	241	3.88	—	—	1123	—	151	—	—	—	—	—	—	<1	—	—
41	335	5.53	—	—	1575	—	210	—	—	—	—	—	—	1	—	—
36	295	4.67	—	—	1341	—	174	—	—	—	—	—	—	1	—	—
47	253	4.45	—	—	1410	—	—	—	—	—	—	—	—	2	—	—
62	349	6.25	—	—	1927	—	—	—	—	—	—	—	—	2	—	—
60	261	4.86	—	—	1696	—	—	—	—	—	—	—	—	2	—	—
78	359	6.76	—	—	2275	—	—	—	—	—	—	—	—	2	—	—
48	328	4.19	—	—	1449	—	—	—	—	—	—	—	—	2	—	—
67	457	5.97	—	—	2039	—	—	—	—	—	—	—	—	3	—	—
379	336	4.60	—	—	1810	—	—	—	—	—	—	—	—	1	—	—
501	466	6.44	—	—	2443	—	—	—	—	—	—	—	—	1	—	—
61	332	4.25	—	—	1650	—	187	—	—	—	—	—	—	1	—	—
85	462	6.04	—	—	2304	—	260	—	—	—	—	—	—	1	—	—
41	333	4.33	—	—	1334	—	—	—	—	—	—	—	—	2	—	—
57	462	6.11	—	—	1864	—	—	—	—	—	—	—	—	2	—	—
43	298	4.84	—	—	1498	—	—	—	—	—	—	—	—	1	—	—
45	302	4.89	—	—	1478	—	—	—	—	—	—	—	—	1	—	—
47	299	4.81	—	—	1524	—	168	—	—	—	—	—	—	1	—	—
36	307	5.10	—	—	1472	—	—	—	—	—	—	—	—	5	—	—
60	200	3.60	—	—	1000	—	188	—	—	—	—	—	—	15	—	—
55	200	3.60	—	—	1080	—	182	—	—	—	—	—	—	15	—	—
60	200	1.08	—	—	720	—	—	—	—	—	—	—	—	18	—	—
95	300	0.72	—	—	350	—	143	—	—	—	—	—	—	0	—	—
120	350	5.40	—	—	1430	—	184	—	—	—	—	—	—	15	—	—
115	350	5.40	—	—	1510	—	229	—	—	—	—	—	—	15	—	—
120	350	1.08	—	—	1160	—	275	—	—	—	—	—	—	18	—	—
0	20	1.80	—	—	245	—	0	—	—	—	—	—	—	0	—	—
40	40	3.60	—	—	640	—	50	—	—	—	—	—	—	15	—	—
35	40	3.60	—	—	720	—	75	—	—	—	—	—	—	15	—	—
40	40	1.08	—	—	370	—	—	—	—	—	—	—	—	18	—	—
85	250	0.00	—	—	280	—	134	—	—	—	—	—	—	0	—	—
90	300	0.00	—	—	390	—	145	—	—	—	—	—	—	0	—	—
160	300	7.20	—	600	2260	—	—	—	—	—	—	—	—	1	—	—
210	150	3.60	—	210	770	—	—	—	—	—	—	—	—	4	—	—
40	150	3.60	—	180	840	—	40	—	—	—	—	—	—	0	—	—
80	20	1.80	—	430	970	—	—	—	—	—	—	—	—	1	—	—
65	300	3.60	—	1010	890	—	—	—	—	—	—	—	—	54	—	—
110	350	0.36	—	720	270	—	215	—	—	—	—	—	—	0	—	—
70	250	4.50	—	280	920	—	—	—	—	—	—	—	—	1	—	—
0	20	1.08	—	550	690	—	0	—	—	—	—	—	—	6	—	—
0	20	1.44	—	770	960	—	0	—	—	—	—	—	—	9	—	—
30	100	3.60	—	155	610	—	0	—	—	—	—	—	—	0	—	—
0	10	0.18	—	190	230	—	0	—	—	—	—	—	—	0	—	—
60	150	1.80	—	470	950	—	—	—	—	—	—	—	—	9	—	—
45	164	4.92	—	390	980	—	—	—	—	—	—	—	—	10	—	—
75	250	4.50	—	420	1360	—	—	—	—	—	—	—	—	9	—	—
0	40	2.70	—	140	420	—	40	—	—	—	—	—	—	18	—	—
280	100	2.70	—	240	1390	—	—	—	—	—	—	—	—	0	—	—
0	40	1.80	—	580	890	—	—	—	—	—	—	—	—	0	—	—
80	200	4.50	—	450	1220	—	—	—	—	—	—	—	—	9	—	—
60	300	4.50	—	1020	920	—	—	—	—	—	—	—	—	48	—	—

A
Table A–1

Food Composition   (Computer code number is for Wadsworth Diet Analysis program)   (For purposes of calculations, use "0" for t, <1, <.1, <.01, etc.)

DA + Code	Food Description	Quantity	Measure	Wt (g)	H₂O (g)	Ener (cal)	Prot (g)	Carb (g)	Dietary Fiber (g)	Fat (g)	Fat Breakdown (g)		
											Sat	Mono	Poly
	FAST FOOD—Continued												
14112	Strawberry ice cream shake	1	item(s)	313	—	640	10	84	0	28	18.00	—	—
14078	Ultimate cheeseburger	1	item(s)	328	—	990	41	59	2	66	28.00	—	—
14110	Vanilla ice cream shake	1	item(s)	285	—	570	12	65	0	29	18.00	—	—
	Jamba Juice												
31646	Banana berry smoothie	24	fluid ounce(s)	719	—	470	5	112	5	2	0.50	—	—
31647	Caribbean passion smoothie	24	fluid ounce(s)	730	—	440	4	102	4	2	1.00	—	—
38422	Carrot juice	16	fluid ounce(s)	472	—	100	3	23	0	1	0.00	—	—
31648	Chocolate mood smoothie	24	fluid ounce(s)	612	—	690	16	142	2	8	4.50	—	—
31649	Citrus squeeze smoothie	24	fluid ounce(s)	729	—	450	4	105	5	2	1.00	—	—
31650	Coffee mood smoothie	24	fluid ounce(s)	560	—	596	13	121	1	6	4.00	—	—
31651	Coldbuster smoothie	24	fluid ounce(s)	724	—	430	5	100	5	3	1.00	—	—
31652	Cranberry craze smoothie	24	fluid ounce(s)	731	—	420	6	97	4	2	1.00	—	—
31654	Jamba powerboost smoothie	24	fluid ounce(s)	730	—	440	6	103	7	2	0.00	—	—
38423	Lemonade	16	fluid ounce(s)	483	—	300	1	75	0	0	0.00	0.00	0.00
31656	Lime sublime smoothie	24	fluid ounce(s)	721	—	450	3	104	6	2	1.00	—	—
31657	Mango-a-go-go smoothie	24	fluid ounce(s)	739	—	500	4	117	4	2	1.00	—	—
38424	Orange juice, freshly squeezed	16	fluid ounce(s)	496	—	220	3	52	1	1	0.00	—	—
38426	Orange/carrot juice	16	fluid ounce(s)	484	—	160	3	37	0	1	0.00	—	—
31660	Orange-a-peel smoothie	24	fluid ounce(s)	726	—	440	9	102	5	1	0.00	—	—
31665	Protein berry pizzaz smoothie	24	fluid ounce(s)	710	—	440	20	92	6	2	0.00	—	—
31667	Raspberry refresher smoothie	24	fluid ounce(s)	636	—	442	3	101	8	3	0.90	—	—
31668	Razzmatazz smoothie	24	fluid ounce(s)	730	—	480	3	112	4	2	1.00	—	—
31669	Strawberries wild smoothie	24	fluid ounce(s)	725	—	450	6	105	4	0	0.00	—	—
38421	Strawberry tsunami smoothie	24	fluid ounce(s)	740	—	530	4	128	4	2	1.00	—	—
38427	Vibrant C juice	16	fluid ounce(s)	448	—	210	2	50	1	0	0.00	0.00	0.00
38428	Wheatgrass juice, freshly squeezed	1	ounce(s)	32	—	5	1	1	0	0	0.00	0.00	0.00
	Kentucky Fried Chicken (KFC)												
31850	BBQ baked beans	1	serving(s)	156	—	190	6	33	6	3	1.00	—	—
31853	Biscuit	1	item(s)	56	—	180	4	20	1	10	2.50	—	—
31851	Coleslaw	1	serving(s)	142	—	232	2	26	3	14	2.00	—	—
31842	Colonel's Crispy Strips	3	item(s)	150	—	340	28	20	0	16	4.50	—	—
31849	Corn on the cob	1	item(s)	162	—	150	5	35	2	2	0.00	—	—
3761	Extra Crispy chicken, breast	1	item(s)	162	—	470	34	19	0	28	8.00	—	—
3762	Extra Crispy chicken, drumstick	1	item(s)	60	—	160	12	5	0	10	2.50	—	—
3763	Extra Crispy chicken, thigh	1	item(s)	114	—	370	21	12	0	26	7.00	—	—
3764	Extra Crispy chicken, whole wing	1	item(s)	52	—	190	10	10	0	12	3.50	—	—
31833	Honey BBQ wing pieces	6	item(s)	189	—	607	33	33	1	38	10.00	—	—
10810	Hot & spicy chicken, breast	1	item(s)	179	—	450	33	20	0	27	8.00	—	—
10813	Hot & spicy chicken, drumstick	1	item(s)	60	—	140	13	4	0	9	2.50	—	—
10811	Hot & spicy chicken, thigh	1	item(s)	128	—	390	22	14	0	28	8.00	—	—
10812	Hot & spicy chicken, whole wing	1	item(s)	55	—	180	11	9	0	11	3.00	—	—
10859	Hot wings pieces	6	piece(s)	135	—	471	27	18	2	33	8.00	—	—
31848	Macaroni & cheese	1	serving(s)	153	—	180	7	21	2	8	3.00	—	—
31847	Mashed potatoes with gravy	1	serving(s)	136	—	120	1	17	2	6	1.00	—	—
10825	Original Recipe chicken, breast	1	item(s)	161	—	370	40	11	0	19	6.00	—	—
10826	Original Recipe chicken, drumstick	1	item(s)	59	—	140	14	4	0	8	2.00	—	—
10827	Original Recipe chicken, thigh	1	item(s)	126	—	360	22	12	0	25	7.00	—	—
10828	Original Recipe chicken, whole wing	1	item(s)	47	—	145	11	5	0	9	2.50	—	—
3760	Original Recipe chicken sandwich w/sauce	1	item(s)	200	—	450	29	33	2	22	5.00	—	—
31834	Original Recipe chicken sandwich w/o sauce	1	item(s)	187	—	360	29	21	1	13	3.50	—	—
31852	Potato salad	1	serving(s)	160	—	230	4	23	3	14	2.00	—	—
10845	Potato wedges	1	serving(s)	156	—	376	6	53	5	15	4.20	—	—
10853	Rotisserie Gold chicken, breast & wing w/skin	4	ounce(s)	114	—	218	26	1	0	12	3.51	—	—
10851	Rotisserie Gold chicken, thigh & leg w/skin	4	ounce(s)	114	—	260	23	1	0	18	5.15	—	—
10852	Rotisserie Gold chicken, thigh & leg w/o skin	4	ounce(s)	117	—	217	27	0	0	12	3.50	—	—
31843	Spicy Crispy Strips	3	item(s)	115	—	335	25	23	1	15	4.00	—	—
10854	Tender Roast chicken, breast w/o skin	1	item(s)	118	—	169	31	1	0	4	1.20	—	—

PAGE KEY: A–4 = Breads/Baked Goods   A–8 = Cereal/Rice/Pasta   A–12 = Fruit   A–18 = Vegetables/Legumes   A–28 = Nuts/Seeds   A–30 = Vegetarian   A–32 = Dairy   A–38 = Eggs   A–40 = Seafood   A–42 = Meats   A–44 = Poultry   A–46 = Processed meats   A–48 = Beverages   A–52 = Fats/Oils   A–54 = Sweets   A–54 = Sauces/Condiments/Spices   A–58 = Mixed Foods/Soups/Sandwiches   A–64 = Fast food   A–82 = Convenience   A–84 = Baby foods

A

Chol (mg)	Calc (mg)	Iron (mg)	Magn (mg)	Pota (mg)	Sodi (mg)	Zinc (mg)	Vit A (RAE) (µg)	Thia (mg)	Vit E (mg)	Ribo (mg)	Niac (mg)	Vit B$_6$ (mg)	Fola (µg)	Vit C (mg)	Vit B$_{12}$ (µg)	Sele (µg)
110	350	0.00	—	610	220	—	202	—	—	—	—	—	—	0	—	—
130	300	7.20	—	480	1670	—	—	—	—	—	—	—	—	1	—	—
115	400	0.00	—	630	220	—	218	—	—	—	—	—	—	0	—	—
5	200	1.08	32	1000	85	0.30	—	0.06	0.40	0.26	1.20	0.40	33	15	0	0
5	100	1.80	24	810	60	0.30	—	0.09	0.81	0.26	5.00	0.50	100	78	0	1
0	150	2.70	80	1030	250	0.90	0	0.53	1.61	0.26	5.00	0.70	80	18	0	6
25	500	1.08	32	760	280	0.60	0	0.09	0.00	0.85	0.40	0.08	9	6	1	4
5	150	1.80	60	1150	50	0.30	—	0.30	0.50	0.26	1.90	0.40	100	168	0	1
28	455	0.30	49	634	429	1.50	—	0.10	0.20	0.60	0.30	0.10	18	7	1	3
5	100	1.08	60	1240	35	15.00	—	0.38	22.15	0.34	3.00	0.40	122	1302	0	1
5	250	1.44	16	500	90	0.30	—	0.03	0.80	0.26	5.00	0.50	100	54	0	1
0	1100	1.44	480	1110	40	15.00	—	5.25	22.15	5.78	66.00	6.80	640	294	10	70
0	20	0.00	8	200	10	0.00	0	0.03	0.00	0.17	14.00	1.80	320	36	0	0
5	150	1.80	32	660	75	0.60	—	0.12	0.40	0.26	7.00	0.80	160	66	<1	1
5	100	1.08	24	800	60	0.30	—	0.15	2.01	0.26	5.00	0.70	120	72	0	1
0	60	1.08	60	990	0	0.30	0	0.45	0.40	0.14	2.00	0.20	160	246	0	0
0	100	1.80	60	1010	125	0.60	0	0.45	0.81	0.26	3.00	0.50	120	132	0	3
0	250	1.80	60	1350	100	0.30	—	0.38	0.81	0.43	3.00	0.40	140	240	0	1
0	1100	2.62	39	650	240	0.58	0	0.09	0.39	0.10	1.55	0.40	58	60	0	4
3	104	2.20	56	806	47	0.80	—	0.10	0.50	0.30	1.60	0.40	43	35	<1	1
5	150	1.80	32	790	70	0.60	—	0.09	0.40	0.26	6.00	0.90	160	60	0	1
0	250	1.80	32	1020	115	0.30	—	0.03	0.40	0.34	1.20	0.20	32	60	0	1
5	100	1.08	24	480	10	0.30	0	0.06	0.81	0.34	14.00	1.80	320	90	0	1
0	20	1.08	40	720	0	0.30	0	0.30	0.40	0.10	1.60	0.40	80	678	0	0
0	0	1.80	8	80	0	0.00	0	0.03	0.00	0.03	0.40	0.04	16	4	0	3
5	80	1.80	—	—	760	—	—	—	—	—	—	—	—	1	—	—
0	20	1.08	—	—	560	—	—	—	—	—	—	—	—	1	—	—
8	30	0.18	—	—	284	—	65	—	—	—	—	—	—	34	—	—
70	10	0.72	—	—	1140	—	—	—	—	—	—	—	—	1	—	—
0	10	0.18	—	—	20	—	10	—	—	—	—	—	—	4	—	—
135	19	1.44	—	—	1230	—	—	—	0.91	—	—	—	—	1	—	—
70	9	0.65	—	—	415	—	—	—	0.46	—	—	—	—	1	—	—
120	19	1.04	—	—	710	—	—	—	0.53	—	—	—	—	1	—	—
55	9	0.34	—	—	390	—	—	—	0.48	—	—	—	—	1	—	—
193	40	1.44	—	—	1145	—	—	—	—	—	—	—	—	5	—	—
130	10	1.07	—	—	1450	—	—	—	—	—	—	—	—	1	—	—
65	20	0.68	—	—	380	—	—	—	—	—	—	—	—	1	—	—
125	10	1.44	—	—	1240	—	—	—	—	—	—	—	—	1	—	—
60	10	0.72	—	—	420	—	—	—	—	—	—	—	—	1	—	—
150	40	1.44	—	—	1230	—	—	—	—	—	—	—	—	1	—	—
10	150	0.18	—	—	860	—	350	—	—	—	—	—	—	1	—	—
1	10	0.36	—	—	440	—	—	—	—	—	—	—	—	1	—	—
145	20	1.14	—	—	1145	—	—	—	—	—	—	—	—	1	—	—
75	10	0.70	—	—	440	—	—	—	—	—	—	—	—	1	—	—
165	10	1.00	—	—	1060	—	—	—	—	—	—	—	—	1	—	—
60	10	0.36	—	—	370	—	—	—	—	—	—	—	—	1	—	—
70	40	1.80	—	—	940	—	—	—	2.81	—	—	—	—	1	—	—
60	40	1.80	—	—	890	—	—	—	—	—	—	—	—	1	—	—
15	20	2.70	—	—	540	—	100	—	—	—	—	—	—	1	—	—
4	36	1.55	—	—	1323	—	—	—	—	—	—	—	—	8	—	—
102	7	0.12	—	—	718	—	—	—	—	—	—	—	—	1	—	—
127	8	0.14	—	—	764	—	—	—	—	—	—	—	—	1	—	—
128	10	0.18	—	—	772	—	—	—	—	—	—	—	—	1	—	—
70	20	0.90	—	—	1140	—	—	—	—	—	—	—	—	1	—	—
112	10	0.18	—	—	797	—	—	—	—	—	—	—	—	1	—	—

Table A–1

A **Food Composition** (Computer code number is for Wadsworth Diet Analysis program) (For purposes of calculations, use "0" for t, <1, <.1, <.01, etc.)

DA + Code	Food Description	Quantity	Measure	Wt (g)	H₂O (g)	Ener (cal)	Prot (g)	Carb (g)	Dietary Fiber (g)	Fat (g)	Fat Breakdown (g)		
											Sat	Mono	Poly
	FAST FOOD—Continued												
	Long John Silver												
39392	Baked cod	1	serving(s)	101	—	120	22	1	0	5	1.00	—	—
3777	Batter dipped fish sandwich	1	item(s)	177	—	440	17	48	3	20	5.00	—	—
37568	Battered fish	1	item(s)	92	—	230	11	16	0	13	4.00	—	—
37569	Breaded clams	1	serving(s)	85	—	240	8	22	1	13	2.00	—	—
39404	Clam chowder	1	item(s)	227	—	220	9	23	0	10	4.00	—	—
39398	Cocktail sauce	1	ounce(s)	28	—	25	0	6	0	0	0.00	0.00	0.00
3770	Coleslaw	1	serving(s)	113	—	200	1	15	4	15	2.50	1.76	4.10
39394	Crunchy shrimp basket	21	item(s)	114	—	340	12	32	2	19	5.00	—	—
39400	French fries, large	1	item(s)	142	—	390	4	56	5	17	4.00	—	—
3774	Fries regular	1	serving(s)	85	—	230	3	34	3	10	2.50	7.40	5.10
3779	Hushpuppy	1	piece(s)	23	—	60	1	9	1	3	0.50	—	—
3781	Shrimp batter-dipped	1	piece(s)	14	—	45	2	3	0	3	1.00	—	—
39399	Tartar sauce	1	ounce(s)	28	—	100	0	4	0	9	1.50	—	—
39395	Ultimate fish sandwich	1	item(s)	199	—	500	20	48	3	25	8.00	—	—
	McDonald's												
2247	Barbecue sauce	1	serving(s)	28	—	45	0	10	0	0	0.00	0.00	0.00
737	Big Mac hamburger	1	item(s)	216	—	590	24	47	3	34	11.00	—	—
738	Cheeseburger	1	item(s)	121	—	330	15	36	2	14	6.00	—	—
29775	Chicken McGrill sandwich	1	item(s)	213	—	400	25	37	2	17	3.00	—	—
3792	Chicken McNuggets	4	item(s)	72	—	210	10	12	1	13	2.50	—	—
1873	Chicken McNuggest	6	item(s)	108	—	310	15	18	2	20	4.00	—	—
73	Chocolate milkshake	8	fluid ounce(s)	227	164	270	7	48	1	6	3.81	1.77	0.23
29774	Crispy chicken sandwich	1	item(s)	219	—	500	22	46	2	26	4.50	—	—
743	Egg McMuffin	1	item(s)	138	—	300	18	29	2	12	4.50	—	—
742	Filet-o-fish sandwich	1	item(s)	156	—	470	15	45	1	26	5.00	—	—
2257	French fries, large	1	serving(s)	176	—	540	8	68	6	26	4.50	—	—
1872	French fries, small	1	serving(s)	68	—	210	3	26	2	10	1.50	—	—
2244	French fries, super size	1	serving(s)	198	—	610	9	77	7	29	5.00	—	—
33822	Fruit n' yogurt parfait	1	item(s)	338	—	380	10	76	2	5	2.00	—	—
2251	Garden salad	1	item(s)	177	—	35	2	7	3	0	0.00	0.00	0.00
739	Hamburger	1	item(s)	107	—	280	12	35	2	10	4.00	—	—
2003	Hash browns	1	item(s)	53	—	130	1	14	1	8	1.50	—	—
2249	Honey sauce	1	item(s)	14	—	45	0	12	0	0	0.00	0.00	0.00
33816	McSalad Shaker chef salad	1	item(s)	206	—	150	17	5	2	8	3.50	—	—
33817	McSalad Shaker garden salad	1	item(s)	149	—	100	7	4	2	6	3.00	—	—
33818	McSalad Shaker grilled chicken caesar salad	1	item(s)	163	—	100	17	3	2	3	1.50	—	—
38396	Newman's Own cobb salad dressing	1	item(s)	59	—	120	2	9	0	9	1.50	—	—
38397	Newman's Own creamy caesar salad dressing	1	item(s)	59	—	190	2	4	0	18	3.50	—	—
38398	Newman's Own low fat balsamic vinaigrette salad dressing	1	item(s)	44	—	40	0	4	0	3	0.00	—	—
38399	Newman's Own ranch salad dressing	1	item(s)	59	—	290	1	4	0	30	4.50	—	—
1874	Plain hotcakes w/syrup & margarine	3	item(s)	228	—	600	9	104	0	17	3.00	—	—
740	Quarter Pounder hamburger	1	item(s)	172	—	430	23	37	2	21	8.00	—	—
741	Quarter Pounder hamburger w/cheese	1	item(s)	200	—	530	28	38	2	30	13.00	—	—
2005	Sausage McMuffin w/egg	1	item(s)	164	—	450	20	29	2	28	10.00	—	—
3163	Strawberry milkshake	8	fluid ounce(s)	226	168	256	8	43	1	6	3.93	—	—
74	Vanilla milkshake	8	fluid ounce(s)	227	169	254	9	40	0	7	4.28	1.98	0.26
	Pizza Hut												
39009	Hot chicken wings	2	item(s)	57	—	110	11	1	0	6	2.00	—	—
14025	Meat Lovers hand tossed pizza	1	slice(s)	125	—	320	16	30	2	15	7.00	—	—
14026	Meat Lovers pan pizza	1	slice(s)	130	—	360	16	29	2	20	7.00	—	—
31009	Meat Lovers stuffed crust pizza	1	slice(s)	188	—	500	25	44	3	25	11.00	—	—
14024	Meat Lovers thin 'n crispy pizza	1	slice(s)	112	—	310	15	22	2	18	8.00	—	—
14031	Pepperoni Lovers hand tossed pizza	1	slice(s)	114	—	300	15	30	2	14	7.00	—	—
14032	Pepperoni Lovers pan pizza	1	slice(s)	119	—	350	15	29	2	19	8.00	—	—
31011	Pepperoni Lovers stuffed crust pizza	1	slice(s)	171	—	480	23	44	3	24	11.00	—	—
14030	Pepperoni Lovers thin 'n crispy pizza	1	slice(s)	94	—	270	13	22	2	14	7.00	—	—
10834	Personal Pan pepperoni pizza	1	slice(s)	59	—	150	7	18	—	6	2.50	—	—

PAGE KEY: A–4 = Breads/Baked Goods A–8 = Cereal/Rice/Pasta A–12 = Fruit A–18 = Vegetables/Legumes A–28 = Nuts/Seeds A–30 = Vegetarian A–32 = Dairy A–38 = Eggs A–40 = Seafood A–42 = Meats A–44 = Poultry A–46 = Processed meats A–48 = Beverages A–52 = Fats/Oils A–54 = Sweets A–54 = Sauces/Condiments/Spices A–58 = Mixed Foods/Soups/Sandwiches A–64 = Fast food A–82 = Convenience A–84 = Baby foods

A

Chol (mg)	Calc (mg)	Iron (mg)	Magn (mg)	Pota (mg)	Sodi (mg)	Zinc (mg)	Vit A (RAE) (µg)	Thia (mg)	Vit E (mg)	Ribo (mg)	Niac (mg)	Vit B_6 (mg)	Fola (µg)	Vit C (mg)	Vit B_{12} (µg)	Sele (µg)
90	20	0.72	—	—	240	—	—	—	—	—	—	—	—	0	—	—
35	60	3.60	—	—	1120	—	—	—	—	—	—	—	—	9	—	—
30	20	1.80	—	—	700	—	—	—	—	—	—	—	—	5	—	—
10	20	1.08	—	—	1110	—	—	—	—	—	—	—	—	0	—	—
25	150	0.72	—	—	810	—	—	—	—	—	—	—	—	0	—	—
0	0	0.00	—	—	250	—	—	—	—	—	—	—	—	0	—	—
20	40	0.36	—	223	340	0.70	34	0.07	—	0.08	2.35	—	—	18	—	—
105	500	1.80	—	—	720	—	—	—	—	—	—	—	—	1	—	—
0	0	0.00	—	—	580	—	—	—	—	—	—	—	—	24	—	—
0	0	0.00	—	370	350	0.30	—	0.09	—	0.02	1.60	—	—	15	—	—
0	20	0.36	—	—	200	—	—	—	—	—	—	—	—	0	—	—
15	0	0.00	—	—	125	—	—	—	—	—	—	—	—	1	—	—
15	0	0.00	—	—	250	—	—	—	—	—	—	—	—	0	—	—
50	150	3.60	—	—	1310	—	—	—	—	—	—	—	—	9	—	—
0	10	0.18	—	45	250	—	3	—	1.53	—	—	—	—	4	—	—
85	300	4.50	—	430	1090	—	60	—	—	—	—	—	—	4	—	—
45	250	2.70	—	250	830	—	60	—	0.57	—	—	—	—	2	—	—
60	200	2.70	—	440	890	—	—	—	—	—	—	—	—	6	—	—
35	20	0.72	—	180	460	—	—	—	—	—	—	—	—	1	—	—
50	20	0.72	—	260	680	—	—	—	—	—	—	—	—	1	—	—
25	299	0.70	36	508	252	1.09	41	0.11	0.23	0.50	0.28	0.06	11	0	1	4
50	200	2.70	—	400	1100	—	—	—	—	—	—	—	—	6	—	—
235	300	2.70	—	210	830	—	—	—	0.56	—	—	—	—	1	—	—
50	200	1.80	—	280	890	—	40	—	—	—	—	—	—	1	—	—
0	20	1.44	—	1210	350	—	—	—	0.30	—	—	—	—	21	—	—
0	10	0.36	—	470	135	—	—	—	—	—	—	—	—	9	—	—
0	20	1.44	—	1370	390	—	—	—	—	—	—	—	—	24	—	—
15	300	1.80	—	550	240	—	—	—	—	—	—	—	—	24	—	—
0	40	1.09	—	410	20	—	—	—	0.61	—	—	—	—	24	—	—
30	200	2.70	—	230	590	—	5	—	0.43	—	—	—	—	2	—	—
0	10	0.36	—	210	330	—	—	—	0.10	—	—	—	—	2	—	—
0	10	0.18	—	7	0	—	—	—	—	—	—	—	—	1	—	—
95	150	1.44	—	360	740	—	323	—	—	—	—	—	—	15	—	—
75	150	1.08	—	290	120	—	273	—	—	—	—	—	—	15	—	—
40	100	1.08	—	420	240	—	—	—	—	—	—	—	—	12	—	—
10	40	0.18	—	13	440	—	—	—	—	—	—	—	—	1	—	—
20	60	0.18	—	16	500	—	—	—	—	—	—	—	—	1	—	—
0	10	0.18	—	9	730	—	—	—	—	—	—	—	—	2	—	—
20	40	0.18	—	64	530	—	—	—	—	—	—	—	—	1	—	—
20	100	4.50	—	280	770	—	—	—	—	—	—	—	—	1	—	—
70	200	4.50	—	370	840	—	10	—	—	—	—	—	—	2	—	—
95	350	4.50	—	420	1310	—	100	—	—	—	—	—	—	2	—	—
255	300	2.70	—	260	930	—	115	—	2.30	—	—	—	—	1	—	—
25	256	0.25	29	412	188	0.82	59	0.10	—	0.44	0.40	0.10	7	2	1	5
27	331	0.23	27	415	215	0.88	57	0.07	0.23	0.44	0.33	0.10	16	0	1	5
70	0	0.36	—	—	450	—	—	—	—	—	—	—	—	0	—	—
40	150	1.80	—	—	830	—	—	—	—	—	—	—	—	6	—	—
40	150	2.70	—	—	810	—	—	—	—	—	—	—	—	6	—	—
65	250	2.70	—	—	1450	—	—	—	—	—	—	—	—	9	—	—
45	150	1.80	—	—	880	—	—	—	—	—	—	—	—	9	—	—
40	200	1.80	—	—	730	—	58	—	—	—	—	—	—	2	—	—
40	200	2.70	—	—	710	—	58	—	—	—	—	—	—	2	—	—
65	300	2.70	—	—	1300	—	—	—	—	—	—	—	—	4	—	—
40	200	1.44	—	—	700	—	58	—	—	—	—	—	—	2	—	—
15	80	1.44	—	—	340	—	38	—	—	—	—	—	—	1	—	—

DA + Code	Food Description	Quantity	Measure	Wt (g)	H₂O (g)	Ener (cal)	Prot (g)	Carb (g)	Dietary Fiber (g)	Fat (g)	Fat Breakdown (g)		
											Sat	Mono	Poly
	FAST FOOD—Continued												
10842	Personal Pan supreme pizza	1	slice(s)	73	—	170	8	19	1	7	3.00	—	—
39013	Personal Pan Veggie Lovers pizza	1	slice(s)	69	—	150	6	19	1	6	2.00	—	—
14028	Veggie Lovers hand tossed pizza	1	slice(s)	120	—	220	10	31	2	6	3.00	—	—
14029	Veggie Lovers pan pizza	1	slice(s)	125	—	260	10	31	2	12	4.00	—	—
31010	Veggie Lovers stuffed crust pizza	1	slice(s)	181	—	370	17	45	3	14	7.00	—	—
14027	Veggie Lovers thin 'n crispy pizza	1	slice(s)	110	—	190	8	23	2	7	3.00	—	—
39012	Wing blue cheese dipping sauce	1	item(s)	43	—	230	2	2	0	24	5.00	—	—
39011	Wing ranch dipping sauce	1	item(s)	43	—	210	1	4	0	22	3.50	—	—
	Starbucks												
38042	Apple cider, tall steamed	12	fluid ounce(s)	360	—	180	0	45	0	0	0.00	0.00	0.00
38052	Cappuccino, tall	12	fluid ounce(s)	360	—	120	7	10	0	6	4.00	—	—
38053	Cappuccino, tall nonfat	12	fluid ounce(s)	360	—	80	7	11	0	0	0.00	0.00	0.00
38054	Cappuccino, tall soy milk	12	fluid ounce(s)	360	—	100	5	13	1	3	0.00		
38059	Cinnamon spice mocha, tall nonfat w/o whipped cream	12	fluid ounce(s)	360	—	170	11	32	0	0	0.50	0.00	0.00
38057	Cinnamon spice mocha, tall w/whipped cream	12	fluid ounce(s)	360	—	320	10	31	0	17	11.00	—	—
38051	Espresso, single shot	1	fluid ounce(s)	30	—	5	0	1	0	0	0.00	0.00	0.00
38088	Flavored syrup, 1 pump	1	serving(s)	10	—	20	0	5	0	0	0.00	0.00	0.00
32562	Frappuccino coffee drink, lite mocha	9½	fluid ounce(s)	281	—	100	7	12	3	3	2.00	—	—
38079	Frappuccino, grande chocolate malt	16	fluid ounce(s)	480	—	470	15	87	2	10	3.50	—	—
38075	Frappuccino, grande mocha malt	12	fluid ounce(s)	360	—	430	14	91	1	7	4.00	—	—
32561	Frappuccino low fat coffee drink, all flavors	9½	fluid ounce(s)	281	—	190	6	39	0	3	2.00	—	—
38067	Frappuccino, tall caramel	12	fluid ounce(s)	360	—	210	4	43	0	3	1.50	—	—
38078	Frappuccino, tall chocolate	12	fluid ounce(s)	360	—	290	13	52	1	5	1.00	—	—
38069	Frappuccino, tall chocolate brownie	12	fluid ounce(s)	360	—	270	5	51	1	7	4.50	—	—
38070	Frappuccino, tall coffee	12	fluid ounce(s)	360	—	190	4	38	0	3	1.50	—	—
38071	Frappuccino, tall espresso	12	fluid ounce(s)	360	—	160	4	33	0	2	1.50	—	—
38073	Frappuccino, mocha	12	fluid ounce(s)	360	—	220	5	44	0	3	1.50	—	—
38072	Frappuccino, tall mocha coconut	12	fluid ounce(s)	360	—	300	5	58	2	7	5.00	—	—
38080	Frappuccino, tall vanilla	12	fluid ounce(s)	360	—	260	11	47	0	4	1.00	—	—
38074	Frappuccino, tall white chocolate	12	fluid ounce(s)	360	—	240	5	48	0	4	2.50	—	—
33111	Latte, tall w/nonfat milk	12	fluid ounce(s)	360	335	123	12	17	0	1	0.40	0.16	0.02
33112	Latte, tall w/whole milk	12	fluid ounce(s)	360	325	212	11	17	0	11	6.90	3.24	0.42
33109	Macchiato, tall caramel w/nonfat milk	12	fluid ounce(s)	360	—	140	7	27	0	1	0.40	—	—
33110	Macchiato, tall caramel w/whole milk	12	fluid ounce(s)	360	—	190	6	27	0	7	4.00	—	—
33107	Mocha coffee drink, tall nonfat, w/o whipped cream	12	fluid ounce(s)	360	—	180	12	33	1	2	1.50	0.68	0.08
38089	Mocha syrup	1	serving(s)	17	—	25	1	6	0	1	0.00	—	—
33108	Mocha, tall w/whole milk	12	fluid ounce(s)	360	—	340	12	33	1	20	12.00	3.48	0.44
38084	Tazo chai black tea, tall	12	fluid ounce(s)	360	—	210	6	36	0	5	3.50	—	—
38083	Tazo chai black tea, tall nonfat	12	fluid ounce(s)	360	—	170	6	37	0	0	0.00	0.00	0.00
38087	Tazo chai black tea, tall soy milk	12	fluid ounce(s)	360	—	190	4	39	1	2	0.00	—	—
38063	Tazo chai creme frappuccino, tall	12	fluid ounce(s)	360	—	280	11	51	0	4	1.00	—	—
38076	Tazo iced tea, tall	12	fluid ounce(s)	360	—	60	0	16	0	0	0.00	0.00	0.00
38077	Tazo tea, grande lemonade	16	fluid ounce(s)	480	—	120	0	31	0	0	0.00	0.00	0.00
38065	Tazoberry creme frappuccino, tall	12	fluid ounce(s)	360	—	240	4	54	1	1	0.00	—	—
38066	Tazoberry frappuccino, tall	12	fluid ounce(s)	360	—	140	1	36	1	0	0.00	0.00	0.00
38045	Vanilla creme steamed nonfat milk, tall w/whipped cream	12	fluid ounce(s)	360	—	180	12	32	0	0	0.00	0.00	0.00
38046	Vanilla creme steamed soy milk, tall w/whipped cream	12	fluid ounce(s)	360	—	300	8	37	1	12	6.00	—	—
38044	Vanilla creme steamed whole milk, tall w/whipped cream	12	fluid ounce(s)	360	—	340	10	31	0	18	12.00	—	—
38090	Whipped cream	1	serving(s)	27	—	100	0	2	0	9	6.00	—	—
38062	White chocolate mocha, tall nonfat w/o whipped cream	12	fluid ounce(s)	360	—	260	12	45	0	4	3.00	—	—
38061	White chocolate mocha, tall w/whipped cream	12	fluid ounce(s)	360	—	410	11	44	0	20	13.00	—	—
38048	White hot chocolate, tall w/o whipped cream	12	fluid ounce(s)	360	—	300	15	51	0	5	3.50	—	—

PAGE KEY: A–4 = Breads/Baked Goods   A–8 = Cereal/Rice/Pasta   A–12 = Fruit   A–18 = Vegetables/Legumes   A–28 = Nuts/Seeds   A–30 = Vegetarian   A–32 = Dairy   A–38 = Eggs   A–40 = Seafood   A–42 = Meats   A–44 = Poultry   A–46 = Processed meats   A–48 = Beverages   A–52 = Fats/Oils   A–54 = Sweets   A–54 = Sauces/Condiments/Spices   A–58 = Mixed Foods/Soups/Sandwiches   A–64 = Fast food   A–82 = Convenience   A–84 = Baby foods

A

Chol (mg)	Calc (mg)	Iron (mg)	Magn (mg)	Pota (mg)	Sodi (mg)	Zinc (mg)	Vit A (RAE) (µg)	Thia (mg)	Vit E (mg)	Ribo (mg)	Niac (mg)	Vit B$_6$ (mg)	Fola (µg)	Vit C (mg)	Vit B$_{12}$ (µg)	Sele (µg)
15	80	1.86	—	—	400	—	—	—	—	—	—	—	—	4	—	—
10	80	1.80	—	—	280	—	—	—	—	—	—	—	—	4	—	—
15	150	1.80	—	—	490	—	—	—	—	—	—	—	—	9	—	—
15	150	2.70	—	—	470	—	—	—	—	—	—	—	—	9	—	—
35	250	2.70	—	—	980	—	—	—	—	—	—	—	—	12	—	—
15	150	1.44	—	—	480	—	—	—	—	—	—	—	—	12	—	—
25	20	0.00	—	—	550	—	0	—	—	—	—	—	—	0	—	—
10	0	0.00	—	—	340	—	0	—	—	—	—	—	—	0	—	—
0	0	1.08	—	—	15	—	0	—	—	—	—	—	—	0	0	—
25	250	0.00	—	—	95	—	0	—	—	—	—	—	—	1	0	—
3	200	0.00	—	—	100	—	0	—	—	—	—	—	—	0	0	—
0	250	0.72	—	—	75	—	0	—	—	—	—	—	—	0	0	—
5	300	0.72	—	—	150	—	0	—	—	—	—	—	—	0	0	—
70	350	1.08	—	—	140	—	0	—	—	—	—	—	—	2	0	—
0	0	0.00	—	—	0	—	0	—	—	—	—	—	—	0	0	—
0	0	0.00	—	—	0	—	0	—	—	—	—	—	—	0	0	—
13	200	1.08	—	—	80	—	—	—	—	—	—	—	—	0	—	—
15	250	2.70	—	—	420	—	0	—	—	—	—	—	—	12	0	—
20	250	1.08	—	—	390	—	0	—	—	—	—	—	—	0	0	—
12	220	0.00	—	—	110	—	—	—	—	—	—	—	—	0	—	—
10	150	0.00	—	—	180	—	0	—	—	—	—	—	—	0	0	—
3	400	1.80	—	—	300	—	0	—	—	—	—	—	—	5	0	—
10	150	1.44	—	—	220	—	0	—	—	—	—	—	—	0	0	—
10	150	0.00	—	—	180	—	0	—	—	—	—	—	—	0	0	—
10	100	0.00	—	—	160	—	0	—	—	—	—	—	—	0	0	—
10	150	0.72	—	—	180	—	0	—	—	—	—	—	—	0	0	—
10	150	1.08	—	—	220	—	0	—	—	—	—	—	—	0	0	—
3	400	0.00	—	—	280	—	0	—	—	—	—	—	—	4	0	—
10	150	0.00	—	—	210	—	0	—	—	—	—	—	—	0	0	—
6	420	0.18	40	—	174	1.35	—	0.12	0.20	0.47	0.36	0.14	18	4	1	—
46	400	0.18	47	254	165	1.28	—	0.13	0.40	0.54	0.35	0.14	17	3	1	—
25	250	0.36	—	—	110	—	—	—	—	—	—	—	—	2	—	—
25	200	0.36	—	—	105	—	—	—	—	—	—	—	—	1	—	—
5	350	2.70	—	—	150	—	—	—	—	—	—	—	—	2	—	—
0	0	0.72	—	—	0	—	0	—	—	—	—	—	—	0	0	—
47	300	0.18	—	—	169	—	—	—	—	—	—	—	—	2	—	—
20	200	0.36	—	—	85	—	0	—	—	—	—	—	—	1	0	—
5	200	0.36	—	—	95	—	0	—	—	—	—	—	—	0	0	—
0	200	0.72	—	—	70	—	0	—	—	—	—	—	—	0	0	—
3	400	0.00	—	—	280	—	0	—	—	—	—	—	—	4	0	—
0	0	0.00	—	—	0	—	0	—	—	—	—	—	—	0	0	—
0	0	0.00	—	—	15	—	0	—	—	—	—	—	—	5	0	—
0	150	0.00	—	—	125	—	0	—	—	—	—	—	—	1	0	—
0	0	0.00	—	—	30	—	0	—	—	—	—	—	—	0	0	—
5	350	0.00	—	—	170	—	0	—	—	—	—	—	—	0	0	—
30	400	1.44	—	—	130	—	0	—	—	—	—	—	—	0	0	—
75	40	0.00	—	—	160	—	0	—	—	—	—	—	—	2	0	—
40	0	0.00	—	—	10	—	0	—	—	—	—	—	—	0	0	—
5	400	0.00	—	—	210	—	0	—	—	—	—	—	—	0	0	—
70	400	0.00	—	—	210	—	0	—	—	—	—	—	—	2	0	—
10	450	0.00	—	—	250	—	0	—	—	—	—	—	—	0	0	—

Table A–1

Food Composition   (Computer code number is for Wadsworth Diet Analysis program)   (For purposes of calculations, use "0" for t, <1, <.1, <.01, etc.)

DA + Code	Food Description	Quantity	Measure	Wt (g)	H₂O (g)	Ener (cal)	Prot (g)	Carb (g)	Dietary Fiber (g)	Fat (g)	Fat Breakdown (g)		
											Sat	Mono	Poly
	FAST FOOD—Continued												
38047	White hot chocolate, tall w/whipped cream	12	fluid ounce(s)	360	—	460	13	50	0	22	15.00	—	—
38050	White hot chocolate soy milk, tall w/whipped cream	12	fluid ounce(s)	360	—	420	11	56	1	16	9.00	—	—
	Subway												
34023	Asiago caesar chicken wrap	1	item(s)	244	—	413	22	47	2	15	3.00	—	—
38622	Atkins-friendly chicken bacon ranch wrap	1	item(s)	213	—	480	40	19	11	27	9.00	—	—
38623	Atkins-friendly turkey bacon melt wrap	1	item(s)	199	—	430	32	22	12	25	9.00	—	—
34029	Bacon & egg breakfast sandwich	1	item(s)	127	—	302	14	29	1	15	4.00	—	—
32045	Chocolate chip cookie	1	item(s)	48	—	209	3	29	1	10	3.50	—	—
32048	Chocolate chip M&M cookie	1	item(s)	48	—	210	2	29	1	10	3.00	—	—
32049	Chocolate chunk cookie	1	item(s)	48	—	210	2	30	1	10	3.00	—	—
4024	Classic Italian B.M.T. sandwich, 6", white bread	1	item(s)	250	—	453	21	40	3	24	8.00	—	—
16397	Club salad	1	item(s)	323	—	145	17	12	3	4	1.00	—	—
3422	Club sandwich, 6", white bread	1	item(s)	253	—	294	22	40	3	5	1.50	—	—
4030	Cold cut trio sandwich, 6", white bread	1	item(s)	254	—	415	19	40	3	20	7.00	—	—
34030	Ham & egg breakfast sandwich	1	item(s)	147	—	291	15	30	1	12	3.00	—	—
3885	Ham sandwich, 6", white bread	1	item(s)	219	—	261	17	39	3	5	1.50	—	—
34026	Honey mustard melt sandwich, 6", Italian bread	1	item(s)	258	—	373	23	47	3	11	5.00	—	—
34027	Horseradish roast beef sandwich, 6", Italian bread	1	item(s)	230	—	401	18	42	3	17	3.00	—	—
4651	Meatball sandwich, 6", white bread	1	item(s)	284	—	501	23	46	4	25	10.00	—	—
15839	Melt sandwich, 6", white bread	1	item(s)	256	—	380	23	41	3	15	5.00	—	—
32046	Oatmeal raisin cookie	1	item(s)	48	—	197	3	29	1	8	2.00	—	—
32047	Peanut butter cookie	1	item(s)	48	—	220	3	26	1	12	3.00	—	—
3957	Roast beef sandwich, 6", white bread	1	item(s)	220	—	264	18	39	3	5	1.00	—	—
16403	Roasted chicken breast salad	1	item(s)	304	—	137	16	12	3	3	0.50	—	—
16378	Roasted chicken breast sandwich, 6", white bread	1	item(s)	234	—	311	25	40	3	6	1.50	—	—
34028	Southwest steak & cheese sandwich, 6", Italian bread	1	item(s)	255	—	412	23	42	4	18	6.00	—	—
4032	Spicy italian sandwich, 6", white bread	1	item(s)	213	—	458	19	42	2	24	9.00	—	—
4031	Steak & cheese sandwich, 6", white bread	1	item(s)	253	—	362	23	41	4	13	4.50	—	—
34024	Steak & cheese wrap	1	item(s)	245	—	353	22	46	3	9	4.00	—	—
32050	Sugar cookie	1	item(s)	48	—	222	2	28	1	12	3.00	—	—
16402	Tuna salad	1	item(s)	314	—	238	13	11	3	16	4.00	—	—
15844	Tuna sandwich, 6", white bread	1	item(s)	252	—	419	18	39	3	21	5.00	—	—
15834	Turkey breast & ham sandwich, 6", white bread	1	item(s)	229	—	267	18	40	3	5	1.00	—	—
34025	Turkey breast & bacon wrap	1	item(s)	228	—	318	19	45	2	7	2.50	—	—
16376	Turkey breast sandwich, 6", white bread	1	item(s)	220	—	254	16	39	3	4	1.00	—	—
16375	Veggie delite, 6", white bread	1	item(s)	163	—	200	7	37	3	3	0.50	—	—
32051	White macadamia nut cookie	1	item(s)	48	—	221	2	27	1	12	3.00	—	—
	Taco Bell												
29906	7-layer burrito	1	item(s)	283	—	530	18	67	10	22	8.00	—	—
744	Bean burrito	1	item(s)	198	—	370	14	55	8	10	3.50	—	—
749	Beef burrito supreme	1	item(s)	248	—	440	18	51	7	18	8.00	—	—
33417	Beef chalupa supreme	1	item(s)	153	—	390	14	31	3	24	10.00	—	—
29910	Beef gordita supreme	1	item(s)	153	—	310	14	30	3	16	7.00	—	—
2014	Beef soft taco	1	item(s)	99	—	210	10	21	2	10	4.50	—	—
10860	Beef soft taco supreme	1	item(s)	134	—	260	11	22	3	14	7.00	—	—
2018	Big beef burrito supreme	1	item(s)	291	—	510	23	52	11	23	9.00	6.55	1.61
14467	Big chicken burrito supreme	1	item(s)	255	—	460	27	50	3	17	6.00	—	—
34472	Chicken burrito supreme	1	item(s)	248	—	410	21	50	5	14	6.00	—	—
33418	Chicken chalupa supreme	1	item(s)	153	—	370	17	30	1	20	8.00	—	—
29900	Chicken fajita wrap supreme	1	item(s)	255	—	510	20	53	3	24	7.76	—	—
29895	Choco taco ice cream dessert	1	item(s)	113	—	310	3	37	1	17	10.00	—	—
10794	Cinnamon twists	1	serving(s)	35	—	160	0	28	0	5	1.00	—	—

PAGE KEY: A–4 = Breads/Baked Goods A–8 = Cereal/Rice/Pasta A–12 = Fruit A–18 = Vegetables/Legumes A–28 = Nuts/Seeds A–30 = Vegetarian
A–32 = Dairy A–38 = Eggs A–40 = Seafood A–42 = Meats A–44 = Poultry A–46 = Processed meats A–48 = Beverages A–52 = Fats/Oils
A–54 = Sweets A–54 = Sauces/Condiments/Spices A–58 = Mixed Foods/Soups/Sandwiches A–64 = Fast food A–82 = Convenience A–84 = Baby foods

A

Chol (mg)	Calc (mg)	Iron (mg)	Magn (mg)	Pota (mg)	Sodi (mg)	Zinc (mg)	Vit A (RAE) (µg)	Thia (mg)	Vit E (mg)	Ribo (mg)	Niac (mg)	Vit B$_6$ (mg)	Fola (µg)	Vit C (mg)	Vit B$_{12}$ (µg)	Sele (µg)
75	500	0.00	—	—	250	—	0	—	—	—	—	—	—	4	0	—
35	500	1.44	—	—	210	—	0	—	—	—	—	—	—	0	0	—
46	40	2.70	—	—	1320	—	—	—	—	—	—	—	—	15	—	—
90	350	2.70	—	—	1340	—	—	—	—	—	—	—	—	7	—	—
65	300	2.70	—	—	1650	—	—	—	—	—	—	—	—	5	—	—
185	60	1.80	—	—	480	—	—	—	—	—	—	—	—	15	—	—
12	0	1.00	—	—	135	—	0	—	—	—	—	—	—	0	—	—
13	0	1.00	—	—	135	—	0	—	—	—	—	—	—	0	—	—
12	0	1.00	—	—	150	—	0	—	—	—	—	—	—	0	—	—
56	100	2.70	—	—	1740	—	—	—	—	—	—	—	—	24	—	—
30	40	1.80	—	—	1070	—	—	—	—	—	—	—	—	30	—	—
30	40	3.60	—	—	1250	—	60	—	—	—	—	—	—	24	—	—
57	150	3.60	—	—	1670	—	100	—	—	—	—	—	—	24	—	—
189	60	2.70	—	—	700	—	67	—	—	—	—	—	—	15	—	—
25	40	2.70	—	—	1260	—	—	—	—	—	—	—	—	24	—	—
41	100	2.70	—	—	1570	—	—	—	—	—	—	—	—	24	—	—
27	40	3.60	—	—	880	—	—	—	—	—	—	—	—	24	—	—
56	100	3.60	—	—	1350	—	—	—	—	—	—	—	—	24	—	—
41	100	2.70	—	—	1690	—	—	—	—	—	—	—	—	24	—	—
14	0	1.00	—	—	180	—	0	—	—	—	—	—	—	0	—	—
0	0	1.00	—	—	200	—	0	—	—	—	—	—	—	0	—	—
20	40	3.60	—	—	840	—	60	—	—	—	—	—	—	24	—	—
36	40	1.08	—	—	730	—	—	—	—	—	—	—	—	30	—	—
48	60	3.60	—	—	880	—	—	—	—	—	—	—	—	24	—	—
44	100	6.30	—	—	1120	—	—	—	—	—	—	—	—	24	—	—
57	30	3.00	—	—	1498	—	—	—	—	—	—	—	—	13	—	—
37	100	6.30	—	—	1200	—	—	—	—	—	—	—	—	24	—	—
37	150	7.20	—	—	1400	—	—	—	—	—	—	—	—	15	—	—
18	0	1.00	—	—	170	—	0	—	—	—	—	—	—	0	—	—
42	100	1.08	—	—	880	—	177	—	—	—	—	—	—	30	—	—
42	100	2.70	—	—	1180	—	100	—	—	—	—	—	—	24	—	—
23	40	2.70	—	—	1210	—	—	—	—	—	—	—	—	24	—	—
24	60	2.70	—	—	1490	—	—	—	—	—	—	—	—	15	—	—
15	40	2.70	—	—	1000	—	—	—	—	—	—	—	—	24	—	—
0	40	1.80	—	—	500	—	—	—	—	—	—	—	—	24	—	—
13	0	1.00	—	—	140	—	0	—	—	—	—	—	—	0	—	—
25	300	3.59	—	—	1360	—	—	—	—	—	—	—	—	5	—	—
10	200	2.69	—	—	1200	—	53	—	—	—	—	—	—	5	—	—
40	200	2.70	—	—	1330	—	351	—	—	—	—	—	—	9	—	—
40	150	1.80	—	—	600	—	—	—	—	—	—	—	—	5	—	—
35	150	2.70	—	—	590	—	—	—	—	—	—	—	—	5	—	—
25	100	1.80	—	—	620	—	44	—	—	—	—	—	—	2	—	—
40	150	1.80	—	—	630	—	73	—	—	—	—	—	—	5	—	—
60	150	2.70	—	493	1500	—	877	—	0.79	0.07	—	—	—	5	—	—
70	101	1.46	—	—	1200	—	—	—	—	—	—	—	—	2	—	—
45	200	2.70	—	—	1270	—	—	—	—	—	—	—	—	9	—	—
45	100	1.08	—	—	530	—	—	—	—	—	—	—	—	5	—	—
57	165	1.52	—	—	1182	—	—	—	—	—	—	—	—	7	—	—
20	60	0.72	—	—	100	—	—	—	—	—	—	—	—	0	—	—
0	0	0.37	—	—	150	—	0	—	—	—	—	—	—	0	—	—

Food Composition   (Computer code number is for Wadsworth Diet Analysis program)   (For purposes of calculations, use "0" for t, <1, <.1, <.01, etc.)

DA + Code	Food Description	Quantity	Measure	Wt (g)	H₂O (g)	Ener (cal)	Prot (g)	Carb (g)	Dietary Fiber (g)	Fat (g)	Fat Breakdown (g) Sat	Mono	Poly
	FAST FOOD—Continued												
14465	Grilled chicken burrito	1	item(s)	198	—	390	19	49	3	13	4.00	—	—
29911	Grilled chicken gordita supreme	1	item(s)	153	—	290	17	28	2	12	5.00	—	—
14463	Grilled chicken soft taco	1	item(s)	99	—	190	14	19	0	6	2.50	—	—
29912	Grilled steak gordita supreme	1	item(s)	153	—	290	16	28	2	13	6.00	—	—
29904	Grilled steak soft taco	1	item(s)	127	—	280	12	21	1	17	4.50	—	—
29905	Grilled steak soft taco supreme	1	item(s)	135	—	240	15	20	2	11	5.00	—	—
2021	Mexican pizza	1	serving(s)	216	—	550	21	46	7	31	11.00	—	—
2011	Nachos	1	serving(s)	99	—	320	5	33	2	19	4.50	—	—
2012	Nachos bellgrande	1	serving(s)	308	—	780	20	80	12	43	13.00	—	—
34473	Steak burrito supreme	1	item(s)	248	—	420	19	50	6	16	7.00	—	—
33419	Steak chalupa supreme	1	item(s)	153	—	370	15	29	2	22	8.00	—	—
29899	Steak fajita wrap supreme	1	item(s)	255	—	510	21	52	3	25	8.00	—	—
747	Taco	1	item(s)	78	—	170	8	13	3	10	4.00	—	—
2015	Taco salad w/salsa, with shell	1	serving(s)	533	—	790	31	73	13	42	15.00	—	—
14459	Taco supreme	1	item(s)	113	—	220	9	14	3	14	7.00	—	—
748	Tostada	1	item(s)	170	—	250	11	29	7	10	4.00	—	—
29901	Veggie fajita wrap supreme	1	item(s)	255	—	470	11	55	3	22	7.00	—	—
	CONVENIENCE MEALS												
	Banquet												
29961	Barbeque chicken meal	1	item(s)	281	—	330	16	37	2	13	3.00	—	—
14788	Boneless white fried chicken meal	1	item(s)	234	—	490	14	49	2	27	7.00	—	—
29960	Fish sticks meal	1	item(s)	187	—	270	13	31	3	10	3.00	—	—
29957	Lasagna with meat sauce meal	1	item(s)	312	—	320	15	46	7	9	4.00	—	—
14777	Macaroni & cheese meal	1	item(s)	340	—	420	15	57	5	14	8.00	—	—
1741	Meatloaf meal	1	item(s)	269	—	240	14	20	4	11	4.00	—	—
39418	Pepperoni pizza meal	1	item(s)	191	—	480	11	56	5	23	8.00	—	—
33759	Roasted white turkey meal	1	item(s)	255	—	230	14	30	5	6	2.00	—	—
1743	Salisbury steak meal	1	item(s)	269	197	380	12	28	3	24	12.00	—	—
	Budget Gourmet												
1914	Cheese manicotti w/meat sauce	1	item(s)	284	194	420	18	38	4	22	11.00	6.00	1.34
1915	Chicken w/fettucini	1	item(s)	284	—	380	20	33	3	19	10.00	—	—
3986	Light beef stroganoff	1	item(s)	248	177	290	20	32	3	7	4.00	—	—
3996	Light sirloin of beef in herb sauce	1	item(s)	269	214	260	19	30	5	7	4.00	2.30	0.31
3987	Light vegetable lasagna	1	item(s)	298	227	290	15	36	5	9	1.79	0.89	0.60
	Healthy Choice												
36979	Bowls chicken teriyaki with rice	1	item(s)	298	—	330	19	50	5	6	2.00	2.00	2.00
9425	Cheese French bread pizza	1	item(s)	170	—	360	20	57	5	5	1.50	—	—
9306	Chicken enchilada suprema meal	1	item(s)	320	252	360	13	59	8	7	3.00	2.00	2.00
9316	Lemon pepper fish meal	1	item(s)	303	—	280	11	49	5	5	2.00	1.00	2.00
9322	Traditional salisbury steak meal	1	item(s)	354	250	360	23	45	5	9	3.50	4.00	1.00
9359	Traditional turkey breasts meal	1	item(s)	298	—	330	21	50	4	5	2.00	1.50	1.50
9451	Zucchini lasagna	1	item(s)	383	—	280	13	47	5	4	2.50	—	—
	Stouffers												
2363	Cheese enchiladas with mexican rice	1	serving(s)	276	—	370	12	48	5	14	5.00	—	—
2313	Cheese French bread pizza	1	serving(s)	294	—	370	14	43	3	16	6.00	—	—
11138	Cheese manicotti w/tomato sauce	1	item(s)	255	—	330	17	35	3	13	8.00	—	—
2366	Chicken pot pie	1	item(s)	284	—	740	23	56	4	47	18.00	12.41	10.48
11116	Homestyle baked chicken breast w/mashed potatoes & gravy	1	item(s)	252	—	260	19	21	1	11	3.00	—	—
11146	Homestyle beef pot roast & potatoes	1	item(s)	252	—	270	16	25	3	12	4.50	—	—
11152	Homestyle roast turkey breast w/stuffing & mashed potatoes	1	item(s)	273	—	300	16	34	2	11	3.00	—	—
11043	Lean Cuisine Cafe Classics baked chicken & whipped potatoes w/stuffing	1	item(s)	227	—	240	17	33	3	5	1.50	1.50	1.00
11046	Lean Cuisine Cafe Classics honey mustard chicken	1	item(s)	213	—	260	18	37	1	4	1.50	1.00	1.00
360	Lean Cuisine Everyday Favorites chicken chow mein w/rice	1	item(s)	255	—	210	12	33	2	3	1.00	1.00	0.50
9467	Lean Cuisine Everyday Favorites fettucini alfredo	1	item(s)	262	—	280	13	40	2	7	3.50	2.00	1.00

PAGE KEY: A–4 = Breads/Baked Goods   A–8 = Cereal/Rice/Pasta   A–12 = Fruit   A–18 = Vegetables/Legumes   A–28 = Nuts/Seeds   A–30 = Vegetarian
A–32 = Dairy   A–38 = Eggs   A–40 = Seafood   A–42 = Meats   A–44 = Poultry   A–46 = Processed meats   A–48 = Beverages   A–52 = Fats/Oils
A–54 = Sweets   A–54 = Sauces/Condiments/Spices   A–58 = Mixed Foods/Soups/Sandwiches   A–64 = Fast food   A–82 = Convenience   A–84 = Baby foods

A

Chol (mg)	Calc (mg)	Iron (mg)	Magn (mg)	Pota (mg)	Sodi (mg)	Zinc (mg)	Vit A (RAE) (µg)	Thia (mg)	Vit E (mg)	Ribo (mg)	Niac (mg)	Vit B$_6$ (mg)	Fola (µg)	Vit C (mg)	Vit B$_{12}$ (µg)	Sele (µg)
40	151	1.44	—	—	1240	—	—	—	—	—	—	—	—	2	—	—
45	100	1.80	—	—	530	—	—	—	—	—	—	—	—	5	—	—
30	100	1.08	—	—	550	—	15	—	—	—	—	—	—	1	—	—
35	100	2.70	—	—	520	—	—	—	—	—	—	—	—	4	—	—
30	100	1.44	—	—	650	—	29	—	—	—	—	—	—	4	—	—
35	100	1.08	—	—	510	—	29	—	—	—	—	—	—	4	—	—
45	350	3.60	—	—	1030	—	—	—	—	—	—	—	—	6	—	—
4	80	0.72	—	—	530	—	0	—	—	—	—	—	—	0	—	—
35	200	2.70	—	—	1300	—	162	—	—	—	—	—	—	6	—	—
35	200	2.70	—	—	1260	—	789	—	—	—	—	—	—	9	—	—
35	100	1.44	—	—	520	—	—	—	—	—	—	—	—	4	—	—
50	150	1.80	—	—	1200	—	—	—	—	—	—	—	—	6	—	—
25	60	1.08	—	—	350	—	44	—	—	—	—	—	—	2	—	—
65	400	6.23	—	—	1670	—	—	—	—	—	—	—	—	21	—	—
40	80	1.44	—	—	360	—	73	—	—	—	—	—	—	5	—	—
15	150	1.44	—	—	710	—	281	—	—	—	—	—	—	5	—	—
30	150	1.44	—	—	990	—	—	—	—	—	—	—	—	6	—	—
50	40	1.08	—	—	1210	—	0	—	—	—	—	—	—	5	—	—
65	60	1.08	—	—	1150	—	—	—	—	—	—	—	—	0	—	—
30	60	1.44	—	—	690	—	—	—	—	—	—	—	—	2	—	—
20	100	2.70	—	—	1170	—	—	—	—	—	—	—	—	0	—	—
20	150	1.44	—	—	1330	—	0	—	—	—	—	—	—	0	—	—
30	0	1.80	—	—	1040	—	0	—	—	—	—	—	—	0	—	—
35	150	1.80	—	—	870	—	0	—	—	—	—	—	—	0	—	—
25	60	1.80	—	—	1070	—	—	—	—	—	—	—	—	4	—	—
60	40	1.44	—	—	1140	—	0	—	—	—	—	—	—	0	—	—
85	300	2.70	45	484	810	2.29	—	0.45	1.95	0.51	4.00	0.23	31	0	1	—
85	100	2.70	—	—	810	—	—	0.15	—	0.43	6.00	—	—	0	—	—
35	40	1.80	39	280	580	4.71	—	0.17	2.16	0.37	4.28	0.27	19	2	3	—
30	40	1.80	58	540	850	4.81	—	0.16	0.62	0.29	5.53	0.37	38	6	2	—
15	283	3.03	79	420	780	1.39	—	0.22	2.42	0.45	3.13	0.32	75	59	<1	—
40	20	0.72	—	—	600	—	—	—	—	—	—	—	—	15	—	—
10	350	3.60	—	—	600	—	—	—	—	—	—	—	—	12	—	—
30	40	1.44	—	—	580	—	—	—	—	—	—	—	—	4	—	—
30	40	0.36	—	—	580	—	—	—	—	—	—	—	—	30	—	—
45	80	2.70	—	—	580	—	—	—	—	—	—	—	—	21	—	—
35	40	1.44	—	—	600	—	—	—	—	—	—	—	—	0	—	—
10	200	1.80	—	—	310	—	—	—	—	—	—	—	—	0	—	—
25	200	1.44	—	360	890	—	—	—	—	—	—	—	—	12	—	—
15	200	1.80	—	240	880	—	—	—	—	—	—	—	—	0	—	—
40	350	1.08	—	430	810	—	—	—	—	—	—	—	—	1	—	—
65	150	2.70	—	—	1170	—	—	—	—	—	—	—	—	2	—	—
50	20	0.72	—	500	760	—	0	—	—	—	—	—	—	0	—	—
35	20	1.80	—	790	820	—	—	—	—	—	—	—	—	6	—	—
35	40	0.72	—	450	1190	—	0	—	—	—	—	—	—	0	—	—
30	80	0.72	—	480	690	—	—	—	—	—	—	—	—	0	—	—
35	60	0.36	—	370	640	—	—	—	—	—	—	—	—	0	—	—
30	20	0.36	—	310	620	—	—	—	—	—	—	—	—	0	—	—
20	200	0.36	—	260	670	—	0	—	—	—	—	—	—	0	—	—

Table A–1

Food Composition   (Computer code number is for Wadsworth Diet Analysis program)   (For purposes of calculations, use "0" for t, <1, <.1, <.01, etc.)

DA + Code	Food Description	Quantity	Measure	Wt (g)	H₂O (g)	Ener (cal)	Prot (g)	Carb (g)	Dietary Fiber (g)	Fat (g)	Sat	Mono	Poly
	CONVENIENCE MEALS—Continued												
11055	Lean Cuisine Everyday Favorites lasagna w/meat sauce	1	item(s)	291	—	300	19	41	3	8	4.00	2.00	0.50
9479	Lean Cuisine French bread deluxe pizza	1	item(s)	174	—	330	18	44	3	9	3.50	1.50	1.00
	Weight Watchers												
11164	Smart Ones chicken enchiladas suiza entree	1	serving(s)	255	—	270	15	33	2	9	3.50	—	—
11155	Smart Ones garden lasagna entree	1	item(s)	312	—	270	14	36	5	7	3.50	—	—
11187	Smart Ones pepperoni pizza	1	item(s)	158	—	390	23	46	4	12	4.00	—	—
31514	Smart Ones spicy penne pasta & ricotta	1	item(s)	289	—	280	11	45	4	6	2.00	—	—
31512	Smart Ones spicy szechuan style vegetables & chicken	1	item(s)	255	—	220	11	39	3	2	0.50	—	—
	BABY FOODS												
787	Apple juice	4	fluid ounce(s)	127	112	60	0	15	<1	<1	0.02	0.00	0.04
778	Applesauce, strained	4	tablespoon(s)	64	55	31	<1	8	1	<1	0.02	0.01	0.04
779	Bananas w/tapioca, strained	4	tablespoon(s)	60	50	34	<1	9	1	<.1	0.02	0.01	0.01
604	Carrots, strained	4	tablespoon(s)	56	52	15	<1	3	1	<.1	0.01	0.00	0.03
770	Chicken noodle dinner, strained	4	tablespoon(s)	64	55	42	2	6	1	1	0.38	0.55	0.30
801	Green beans, strained	4	tablespoon(s)	60	0.05	15	0.77	3.53	1.13	0.05	0.01	0	0.03
910	Human milk, mature	2	fluid ounce(s)	62	54	43	1	4	0	3	1.24	1.02	0.31
760	Mixed cereal, prepared w/whole milk	4	ounce(s)	114	85	128	5	18	1	4	2.19	1.25	0.43
772	Mixed vegetable dinner, strained	2	ounce(s)	57	50	23	1	5	1	<.1	0.00	0.00	0.06
762	Rice cereal, prepared w/whole milk	4	ounce(s)	114	85	131	4	19	<1	4	2.64	1.02	0.16
758	Teething biscuits	1	item(s)	11	1	43	1	8	<1	<1	0.17	0.16	0.09

PAGE KEY: A–4 = Breads/Baked Goods   A–8 = Cereal/Rice/Pasta   A–12 = Fruit   A–18 = Vegetables/Legumes   A–28 = Nuts/Seeds   A–30 = Vegetarian   A–32 = Dairy   A–38 = Eggs   A–40 = Seafood   A–42 = Meats   A–44 = Poultry   A–46 = Processed meats   A–48 = Beverages   A–52 = Fats/Oils   A–54 = Sweets   A–54 = Sauces/Condiments/Spices   A–58 = Mixed Foods/Soups/Sandwiches   A–64 = Fast food   A–82 = Convenience   A–84 = Baby foods

A

Chol (mg)	Calc (mg)	Iron (mg)	Magn (mg)	Pota (mg)	Sodi (mg)	Zinc (mg)	Vit A (RAE) (µg)	Thia (mg)	Vit E (mg)	Ribo (mg)	Niac (mg)	Vit B$_6$ (mg)	Fola (µg)	Vit C (mg)	Vit B$_{12}$ (µg)	Sele (µg)
30	200	1.08	—	590	650	—	—	—	—	—	—	—	—	5	—	—
20	100	1.80	—	390	630	—	—	—	—	—	—	—	—	9	—	—
50	250	1.08	—	—	660	—	—	—	—	—	—	—	—	4	—	—
30	350	1.80	—	—	610	—	—	—	—	—	—	—	—	6	—	—
45	450	1.80	—	320	650	—	55	—	—	—	—	—	—	5	—	—
5	150	2.70	—	250	400	—	—	—	—	—	—	—	—	6	—	—
10	150	1.80	—	—	730	—	—	—	—	—	—	—	—	2	—	—
0	5	0.72	4	115	4	0.04	1	0.01	0.76	0.02	0.11	0.04	0	73	0	<1
0	3	0.14	2	45	1	0.01	1	0.01	0.38	0.02	0.04	0.02	1	25	0	<1
0	3	0.12	6	53	5	0.04	1	0.01	0.36	0.02	0.11	0.07	4	10	0	<1
0	12	0.21	5	110	21	0.08	321	0.01	0.29	0.02	0.26	0.04	8	3	0	<1
10	17	0.41	9	89	15	0.35	70	0.03	0.14	0.04	0.46	0.04	7	<.1	<.1	2
0	23.39	0.44	14.39	94.8	1.2	0.12	27	0.01	0.24	0.05	0.2	0.02	21	3.11	0	0.18
9	20	0.02	2	31	10	0.10	38	0.01	0.55	0.02	0.11	0.01	3	3	<.1	1
12	250	11.85	31	226	53	0.81	28	0.49	—	0.66	6.56	0.07	12	1	<.1	—
0	12	0.19	6	69	5	0.09	77	0.01	—	0.02	0.29	0.04	5	2	0	<1
12	272	13.85	51	216	52	0.73	25	0.53	—	0.57	5.91	0.13	9	1	<1	4
0	29	0.39	4	36	40	0.10	3	0.03	0.05	0.06	0.48	0.01	5	1	<.1	3

B Canadiana: Recommendations, Choice System, and Labels

This appendix presents the *Nutrition Recommendations for Canadians* and Canada's Food Guide to Healthy Eating. It also offers the Choice System for Meal Planning, a food exchange system. Last, Canada's food label is included with a brief explanation of its meaning. Visit the Internet website of Health Canada at www.hc-sc.gc.ca for more details. Appendix E includes addresses of Canadian governmental agencies and professional organizations that may provide additional information.

Canadian Guidelines

The *Nutrition Recommendations for Canadians* (see Table B-1) are intended to assist Canadians in choosing foods wisely for good health. The recommendations undergo periodic updates; obtain the latest edition from the Health Canada website, www.hc-sc.gc.ca/english/new.html. In addition, Canada's Food Guide to Healthy Eating urges Canadians to obtain nutrients and other health-promoting food constituents from servings of foods in the grouops show in Figure B-1.

TABLE B-1 NUTRITION RECOMMENDATIONS FOR CANADIANS

- The Canadian diet should provide energy consistent with the maintenance of *body weight* within the recommended range.
- The Canadian diet should include *essential nutrients* in amounts specified in the Recommended Nutrient Intakes.
- The Canadian diet should include no more than 30% of energy as *fat* (33 grams/1,000 calories or 39 grams/5,000 kilojoules) and no more than 10% as saturated fat (11 grams/1,000 calories or 13 grams/5,000 kilojoules).
- The Canadian diet should provide 55% of energy as *carbohydrate* (138 grams/1,000 calories or 165 grams/5,000 kilojoules) from a variety of sources.
- The *sodium* content of the Canadian diet should be reduced.
- The Canadian diet should include no more than 5% of total energy as *alcohol,* or two drinks daily, whichever is less.
- The Canadian diet should contain no more *caffeine* than the equivalent of four regular cups of coffee per day.
- Community water supplies containing less than 1 milligram per liter should be *fluoridated* to that level.

NOTE: Italics added to highlight areas of concern.

SOURCE: Health and Welfare Canada, *Nutrition Recommendations: The Report of the Scientific Review Committee* (Ottawa: Canadian Government Publishing Centre, updated 2003).

Health and Welfare   Santé et Bien-être social
Canada   Canada

CANADA'S
Food Guide
TO HEALTHY EATING

Enjoy a variety
of foods from each
group every day.

Choose lower-
fat foods
more often.

Grain Products
Choose whole grain
and enriched
products more
often.

Vegetables & Fruit
Choose dark green and
orange vegetables and
orange fruit more often.

Milk Products
Choose lower-fat
milk products more
often.

Meat & Alternatives
Choose leaner meats,
poultry and fish, as well
as dried peas, beans and
lentils more often.

Different People Need Different Amounts of Food

The amount of food you need every day from the 4 food groups and other foods depends on your age, body size, activity level, whether you are male or female and if you are pregnant or breast-feeding. That's why the Food Guide gives a lower and higher number of servings for each food group. For example, young children can choose the lower number of servings, while male teenagers can go to the higher number. Most other people can choose servings somewhere in between.

Grain Products
5-12 SERVINGS PER DAY

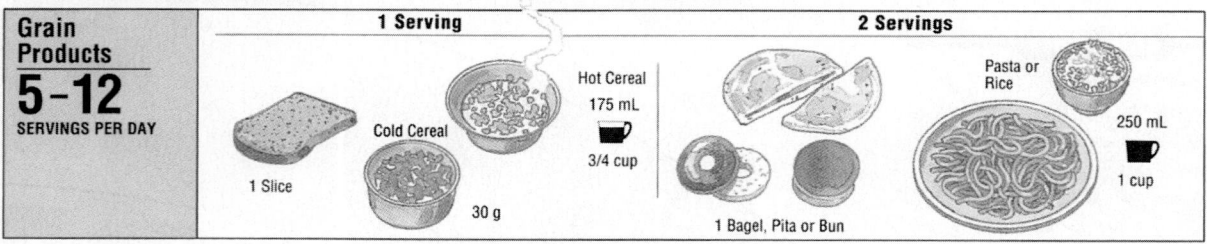

1 Serving — 1 Slice; Cold Cereal 30 g; Hot Cereal 175 mL 3/4 cup
2 Servings — 1 Bagel, Pita or Bun; Pasta or Rice 250 mL 1 cup

Vegetables & Fruit
5-10 SERVINGS PER DAY

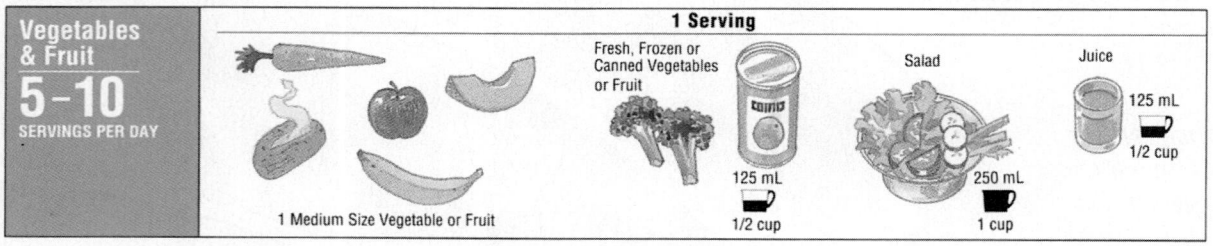

1 Serving — 1 Medium Size Vegetable or Fruit; Fresh, Frozen or Canned Vegetables or Fruit 125 mL 1/2 cup; Salad 250 mL 1 cup; Juice 125 mL 1/2 cup

Milk Products
SERVINGS PER DAY
Children 4–9 years: 2–3
Youth 10–16 years: 3–4
Adults: 2–4
Pregnant & Breast-feeding
Women: 3–4

1 Serving — Milk 250 mL 1 cup; Cheese 3"x1"x1" 50 g; 2 Slices 50 g; Yogourt 175 g 3/4 cup

Other Foods

Taste and enjoyment can also come from other foods and beverages that are not part of the 4 food groups. Some of these foods are higher in fat or Calories, so use these foods in moderation.

Meat & Alternatives
2-3 SERVINGS PER DAY

1 Serving — Meat, Poultry or Fish 50-100 g; Fish 1/3–2/3 Can 50–100 g; 1-2 Eggs; Beans 125-250 mL; Tofu 100 g 1/3 cup; Peanut Butter 30 mL 2 tbsp

Enjoy eating well, being active and feeling good about yourself. That's VITALIT

© Minister of Supply and Services Canada 1992 Cat. No. H39-252/1992E No changes permitted. Reprint permission not required.

Choice System for Meal Planning

The *Good Health Eating Guide* is the Canadian choice system of meal planning.[*] It is similar to the U.S. exchange system (Appendix D) in the following ways:

- Foods are divided into lists according to carbohydrate, protein, and fat content.
- Foods are interchangeable within a group.
- Most foods are eaten in measured amounts.
- An energy value is given for each food group.

Tables B-2 through B-9 present the Canadian choice system.

---

[*]The tables for the Canadian choice system are adapted from the *Good Health Eating Guide Resource*, copyright 1994, with permission of the Canadian Diabetes Association.

TABLE B·2 CANADIAN CHOICE SYSTEM: STARCH FOODS

1 starch choice = 15 g carbohydrate (starch), 2 g protein, 290 kJ (68 kcal)

Food	Measure	Mass (Weight)
Breads		
Bagels	½	30 g
Bread crumbs	50 mL (¼ c)	30 g
Bread cubes	250 mL (1 c)	30 g
Bread sticks	2	20 g
Brewis, cooked	50 mL (¼ c)	45 g
Chapati	1	20 g
Cookies, plain	2	20 g
English muffins, crumpets	½	30 g
Flour	40 mL (2½ tbs)	20 g
Hamburger buns	½	30 g
Hot dog buns	½	30 g
Kaiser rolls	½	30 g
Matzo, 15 cm	1	20 g
Melba toast, rectangular	4	15 g
Melba toast, rounds	7	15 g
Pita, 20 cm (8″) diameter	¼	30 g
Pita, 15 cm (6″) diameter	½	30 g
Plain rolls	1 small	30 g
Pretzels	7	20 g
Raisin bread	1 slice	30 g
Rice cakes	2	30 g
Roti	1	20 g
Rusks	2	20 g
Rye, coarse or pumpernickel	½ slice	30 g
Soda crackers	6	20 g
Tortilla, corn (taco shell)	1	10 g
Tortilla, flour (9″ diameter)	1	30 g
White (French and Italian)	1 slice	25 g
Whole-wheat, cracked-wheat, rye, white enriched	1 slice	30 g
Cereals		
Bran flakes, 100% bran	125 mL (½ c)	30 g
Cooked cereals, cooked	125 mL (½ c)	125 g
Dry	30 mL (2 tbs)	20 g
Cornmeal, cooked	125 mL (½ c)	125 g
Dry	30 mL (2 tbs)	20 g
Ready-to-eat unsweetened cereals	125 mL (½ c)	20 g
Shredded wheat biscuits, rectangular or round	1	20 g
Shredded wheat, bite size	125 mL (½ c)	20 g
Wheat germ	75 mL (⅓ c)	30 g
Cornflakes	300 mL (1¼ c)	30 g
Rice Krispies	250 mL (1 c)	30 g

(continued on the next page)

1 starch choice = 15 g carbohydrate (starch), 2 g protein, 290 kJ (68 kcal)

Food	Measure	Mass (Weight)
Cereals—continued		
Cheerios	200 mL (¾ c)	20 g
Muffets	1	20 g
Puffed rice	300 mL (1¼ c)	15 g
Puffed wheat	425 mL (1⅔ c)	20 g
Grains		
Barley, cooked	125 mL (½ c)	120 g
Dry	30 mL (2 tbs)	20 g
Bulgur, kasha, cooked, moist	125 mL (½ c)	70 g
Cooked, crumbly	75 mL (⅓ c)	40 g
Dry	30 mL (2 tbs)	20 g
Rice, cooked, brown and white (short and long grain)	75 mL (⅓ c)	60 g
Rice, cooked, wild	75 mL (⅓ c)	60 g
Tapioca, pearl and granulated, quick cooking, dry	30 mL (2 tbs)	15 g
Couscous, cooked moist	125 mL (½ c)	70 g
Dry	30 mL (2 tbs)	20 g
Quinoa, cooked moist	125 mL (½ c)	70 g
Dry	30 mL (2 tbs)	20 g
Pastas		
Macaroni, cooked	125 mL (½ c)	70 g
Noodles, cooked	125 mL (½ c)	80 g
Spaghetti, cooked	125 mL (½ c)	70 g
Starchy Vegetables		
Beans and peas, dried, cooked	125 mL (½ c)	98 g
Breadfruit	1 slice	75 g
Corn, canned, whole kernel	125 mL (½ c)	85 g
Corn on the cob	½ medium cob	140 g
Cornstarch	30 mL (2 tbs)	15 g
Plantains	⅓ small	50 g
Popcorn, air-popped, unbuttered	750 mL (3 c)	20 g
Potatoes, whole (with or without skin)	½ medium	95 g
Yams, sweet potatoes, (with or without skin)	½	75 g

Food	Choices per Serving	Measure	Mass (Weight)
NOTE: Food items found in this category provide more than 1 starch choice:			
Bran flakes	1 starch + ½ sugar	150 mL (⅔ c)	24 g
Croissant, small	1 starch + 1½ fats	1 small	35 g
Large	1 starch + 1½ fats	½ large	30 g
Corn, canned creamed	1 starch + ½ sugar	12 mL (½ c)	113 g
Potato chips	1 starch + 2 fats	15 chips	30 g
Tortilla chips (nachos)	1 starch + 1½ fats	13 chips	30 g
Corn chips	1 starch + 2 fats	30 chips	30 g
Cheese twists	1 starch + 1½ fats	30 chips	30 g
Cheese puffs	1 starch + 2 fats	27 chips	30 g
Tea biscuit	1 starch + 2 fats	1	30 g
Pancakes, homemade using 50 mL (¼ c) batter (6" diameter)	1½ starches + 1 fat	1 medium	50 g
Potatoes, french fried (homemade or frozen)	1 starch + 1 fat	10 regular size	35 g
Soup, canned* (prepared with equal volume of water)	1 starch	250 mL (1 c)	260 g
Waffles, packaged	1 starch + 1 fat	1	35 g

*Soup can vary according to brand and type. Check the label for Food Choice Values and Symbols or the core nutrient listing.

TABLE B·3 CANADIAN CHOICE SYSTEM: FRUITS AND VEGETABLES

1 fruits and vegetables choice = 10 g carbohydrate, 1 g protein, 190 kJ (44 kcal)

Food	Measure	Mass (Weight)
Fruits (fresh, frozen, without sugar, canned in water)		
Apples, raw (with or without skin)	½ medium	75 g
Sauce unsweetened	125 mL (½ c)	120 g
Sweetened (see Combined Food Choices, Table B-9)		
Apple butter	20 mL (4 tsp)	20 g
Apricots, raw	2 medium	115 g
Canned, in water	4 halves, plus 30 mL (2 tbs) liquid	110 g
Bake-apples (cloudberries), raw	125 mL (½ c)	120 g
Bananas, with peel	½ small	75 g
Peeled	½ small	50 g
Berries (blackberries, boysenberries, raspberries)	250 mL (1 c)	150 g
(blueberries, loganberries, huckleberries)	125 mL (½ c)	70 g
Raw	125 mL (½ c)	70 g
Cantaloupe, wedge with rind	¼	160 g
Cubed or diced	175 mL (⅔ c)	135 g
Cherries, raw, with pits	10	75 g
Raw, without pits	10	70 g
Canned, in water, with pits	75 mL (⅓ c), plus 30 mL (2 tbs) liquid	90 g
Canned, in water, without pits	75 mL (⅓ c), plus 30 mL (2 tbs) liquid	85 g
Crabapples, raw	1 small	55 g
Cranberries, raw	250 mL (1 c)	100 g
Figs, raw	1 medium	50 g
Canned, in water	3 medium, plus 30 mL (2 tbs) liquid	100 g
Foxberries, raw	250 mL (1 c)	100 g
Fruit cocktail, canned, in water	125 mL (½ c), plus 30 mL (2 tbs) liquid	120 g
Fruit, mixed, cut-up	125 mL (½ c)	120 g
Gooseberries, raw	250 mL (1 c)	150 g
Canned, in water	250 mL (1 c), plus 30 mL (2 tbs) liquid	230 g
Grapefruit, raw, with rind	½ small	185 g
Raw, sectioned	125 mL (½ c)	100 g
Canned, in water	125 mL (½ c), plus 30 mL (2 tbs) liquid	120 g
Grapes, raw, slip skin	125 mL (½ c)	75 g
Raw, seedless	125 mL (½ c)	75 g
Canned, in water	75 mL (⅓ c), plus 30 mL (2 tbs) liquid	115 g
Guavas, raw	½	50 g
Honeydew melon, raw, with rind	⅟₁₀ wedge	130 g
Cubed or diced	175 mL (⅔ c)	115 g
Kiwis, raw, with skin	1 medium	76 g
Kumquats, raw	3	60 g
Loquats, raw	8	130 g
Lychee fruit, raw	8	120 g
Mandarin oranges, raw, with rind	1	135 g
Raw, sectioned	125 mL (½ c)	100 g
Canned, in water	125 mL (½ c), plus 30 mL (2 tbs) liquid	100 g
Mangoes, raw, without skin and seed,		
Diced	75 mL (⅓ c)	65 g
Nectarines	½ medium	75 g
Oranges, raw, with rind	1 small	130 g
Raw, sectioned	125 mL (½ c)	95 g
Papayas, raw, with skin and seeds	¼ medium	150 g
Raw, without skin and seeds	¼ medium	100 g
Cubed or diced	125 mL (½ c)	100 g
Peaches, raw, with seed and skin	1 large	100 g
Raw, sliced or diced	125 mL (½ c)	85 g
Canned in water, halves or slices	125 mL (½ c), plus 30 mL (2 tbs) liquid	120 g

(continued on the next page)

1 fruits and vegetables choice = 10 g carbohydrate, 1 g protein, 190 kJ (44 kcal)

Food	Measure	Mass (Weight)
Pears, raw, with skin and core	½	90 g
Raw, without skin and core	½	85 g
Canned, in water, halves	1 half plus 30 mL (2 tbs) liquid	90 g
Persimmons, raw, native	1	30 g
Raw, Japanese	¼	50 g
Pineapple, raw	1 slice	75 g
Raw, diced	125 mL (½ c)	75 g
Canned, in juice, diced	75 mL (⅓ c), plus 15 mL (1 tbs) liquid	55 g
Canned, in juice, sliced	1 slice, plus 15 mL (1 tbs) liquid	55 g
Canned, in water, diced	125 mL (½ c), plus 30 mL (2 tbs) liquid	100 g
Canned, in water, sliced	2 slices, plus 15 mL (1 tbs) liquid	100 g
Plums, raw	2 small	60 g
Damson	6	65 g
Japanese	1	70 g
Canned, in apple juice	2, plus 30 mL (2 tbs) liquid	70 g
Canned, in water	3, plus 30 mL (2 tbs) liquid	100 g
Pomegranates, raw	½	140 g
Rhubarb	250 mL (1 c)	150 g
Strawberries, raw	250 mL (1 c)	150 g
Frozen/canned, in water	250 mL (1 c), plus 30 mL (2 tbs) liquid	240 g
Tangelos, raw	1	205 g
Tangerines, raw	1 medium	115 g
Raw, sectioned	125 mL (½ c)	100 g
Watermelon, raw, with rind	1 wedge	310 g
Cubed or diced	250 mL (1 c)	160 g
Dried Fruit		
Apples	5 pieces	15 g
Apricots	4 halves	15 g
Banana flakes	30 mL (2 tbs)	15 g
Currants	30 mL (2 tbs)	15 g
Dates, without pits	2	15 g
Peaches	½	15 g
Pears	½	15 g
Prunes, raw, with pits	2	15 g
Raw, without pits	2	10 g
Stewed, no liquid	2	20 g
Stewed, with liquid	2, plus 15 mL (1 tbs) liquid	35 g
Raisins	30 mL (2 tbs)	15 g
Juices (no sugar added or unsweetened)		
Apricot, grape, guava, mango, prune	50 mL (¼ c)	55 g
Apple, carrot, papaya, pear, pineapple, pomegranate	75 mL (⅓ c)	80 g
Clamato (see Sugars, Table B-5)		
Cranberry (see Sugars, Table B-5)		
Grapefruit, loganberry, orange, raspberry, tangelo, tangerine	125 mL (½ c)	130 g
Tomato, tomato-based mixed vegetables	250 mL (1 c)	255 g
Vegetables (fresh, frozen, or canned)		
Artichokes, French, globe	2 small	50 g
Beets, diced or sliced	125 mL (½ c)	85 g
Carrots, diced, cooked or uncooked	125 mL (½ c)	75 g
Chestnuts, fresh	5	20 g
Parsnips, mashed	125 mL (½ c)	80 g
Peas, fresh or frozen	125 mL (½ c)	80 g
Canned	75 mL (⅓ c)	55 g
Pumpkin, mashed	125 mL (½ c)	45 g
Rutabagas, mashed	125 mL (½ c)	85 g
Sauerkraut	250 mL (1 c)	235 g
Snow peas	250 mL (1 c)	135 g
Squash, yellow or winter, mashed	125 mL (½ c)	115 g

(continued on the next page)

TABLE B·3   CANADIAN CHOICE SYSTEM: FRUITS AND VEGETABLES—continued

1 fruits and vegetables choice = 10 g carbohydrate, 1 g protein, 190 kJ (44 kcal)

Food	Measure	Mass (Weight)
Succotash	75 mL (⅓ c)	55 g
Tomatoes, canned	250 mL (1 c)	240 g
Tomato paste	50 mL (¼ c)	55 g
Tomato sauce*	75 mL (⅓ c)	100 g
Turnips, mashed	125 mL (½ c)	115 g
Vegetables, mixed	125 mL (½ c)	90 g
Water chestnuts	8 medium	50 g

*Tomato sauce varies according to brand name. Check the label or discuss with your dietitian.

TABLE B·4   CANADIAN CHOICE SYSTEM: MILK

Type of Milk	Carbohydrate (g)	Protein (g)	Fat (g)	Energy
Fat-free (0%)	6	4	0	170 kJ (40 kcal)
1%	6	4	1	206 kJ (49 kcal)
2%	6	4	2	244 kJ (58 kcal)
Whole (4%)	6	4	4	319 kJ (76 kcal)

Food	Measure	Mass (Weight)
Buttermilk (higher in salt)	125 mL (½ c)	125 g
Evaporated milk	50 mL (¼ c)	50 g
Milk	125 mL (½ c)	125 g
Powdered milk, regular	30 mL (2 tbs)	15 g
Instant	50 mL (¼ c)	15 g
Plain yogurt	125 mL (½ c)	125 g

Food	Choices per Serving	Measure	Mass (Weight)
NOTE: Food items found in this category provide more than 1 milk choice:			
Milkshake	1 milk + 3 sugars + ½ protein	250 mL (1 c)	300 g
Chocolate milk, 2%	2 milks 2% + 1 sugar	250 mL (1 c)	300 g
Frozen yogurt	1 milk + 1 sugar	125 mL (½ c)	125 g

B | TABLE B·5 CANADIAN CHOICE SYSTEM: SUGARS

1 sugar choice = 10 g carbohydrate (sugar), 167 kJ (40 kcal)

Food	Measure	Mass (Weight)
Beverages		
Condensed milk	15 mL (1 tbs)	
Flavoured fruit crystals*	75 mL (⅓ c)	
Iced tea mixes*	75 mL (⅓ c)	
Regular soft drinks	125 mL (½ c)	
Sweet drink mixes*	75 mL (⅓ c)	
Tonic water	125 mL (½ c)	
Miscellaneous		
Bubble gum (large square)	1 piece	5 g
Cranberry cocktail	75 mL (⅓ c)	80 g
Cranberry cocktail, light	350 mL (1⅓ c)	260 g
Cranberry sauce	30 mL (2 tbs)	
Hard candy mints	2	5 g
Honey, molasses, corn and cane syrup	10 mL (2 tsp)	15 g
Jelly beans	4	10 g
Licorice	1 short stick	10 g
Marshmallows	2 large	15 g
Popsicle	1 stick (½ popsicle)	
Powdered gelatin mix (Jello®) (reconstituted)	50 mL (¼ c)	
Regular jam, jelly, marmalade	15 mL (1 tbs)	
Sugar, white, brown, icing, maple	10 mL (2 tsp)	10 g
Sweet pickles	2 small	100 g
Sweet relish	30 mL (2 tbs)	

Food	Choices per Serving	Measures	Mass (Weight)
NOTE: The following food items provide more than 1 sugar choice:			
Brownie	1 sugar + 1 fat	1	20 g
Clamato juice	1½ sugars	175 mL (⅔ c)	
Fruit salad, light syrup	1 sugar + 1 fruits & vegetables	125 mL (½ c)	130 g
Aero® bar	2½ sugars + 2½ fats	1 bar	43 g
Smarties®	4½ sugars + 2 fats	1 box	60 g
Sherbet	3 sugars + ½ fat	125 mL (½ c)	95 g

*These beverages have been made with water.

1 protein choice = 7 g protein, 3 g fat, 230 kJ (55 kcal)

Food	Measure	Mass (Weight)
Cheese		
Low-fat cheese, about 7% milkfat	1 slice	30 g
Cottage cheese, 2% milkfat or less	50 mL (¼ c)	55 g
Ricotta, about 7% milkfat	50 mL (¼ c)	60 g
Fish		
Anchovies (see Extras, Table B-8)		
Canned, drained (e.g., mackerel, salmon, tuna packed in water)	50 mL (¼ c)	30 g
Cod tongues, cheeks	75 mL (⅓ c)	50 g
Fillet or steak (e.g., Boston blue, cod, flounder, haddock, halibut, mackerel, orange roughy, perch, pickerel, pike, salmon, shad, snapper, sole, swordfish, trout, tuna, whitefish)	1 piece	30 g
Herring	⅓ fish	30 g
Sardines, smelts	2 medium or 3 small	30 g
Squid, octopus	50 mL (¼ c)	40 g
Shellfish		
Clams, mussels, oysters, scallops, snails	3 medium	30 g
Crab, lobster, flaked	50 mL (¼ c)	30 g
Shrimp, fresh	5 large	30 g
Frozen	10 medium	30 g
Canned	18 small	30 g
Dry pack	50 mL (¼ c)	30 g
Meat and Poultry (e.g., beef, chicken, goat, ham, lamb, pork, turkey, veal, wild game)		
Back, peameal bacon	3 slices, thin	30 g
Chop	½ chop, with bone	40 g
Minced or ground, lean or extra-lean	30 mL (2 tbs)	30 g
Sliced, lean	1 slice	30 g
Steak, lean	1 piece	30 g
Organ Meats		
Hearts, liver	1 slice	30 g
Kidneys, sweetbreads, chopped	50 mL (¼ c)	30 g
Tongue	1 slice	30 g
Tripe	5 pieces	60 g
Soyabean		
Bean curd or tofu	½ block	70 g
Eggs		
In shell, raw or cooked	1 medium	50 g
Without shell, cooked or poached in water	1 medium	45 g
Scrambled	50 mL (¼ c)	55 g

Food	Choices per Serving	Measures	Mass (Weight)
NOTE: The following choices provide more than 1 protein exchange:			
Cheese			
Cheeses	1 protein + 1 fat	1 piece	25 g
Cheese, coarsely grated (e.g., cheddar)	1 protein + 1 fat	50 mL (¼ c)	25 g
Cheese, dry, finely grated (e.g., parmesan)	1 protein + 1 fat	45 mL	15 g
Cheese, ricotta, high fat	1 protein +1 fat	50 mL (¼ c)	55 g
Fish			
Eel	1 protein +1 fat	1 slice	50 g
Meat			
Bologna	1 protein + 1 fat	1 slice	20 g
Canned lunch meats	1 protein + 1 fat	1 slice	20 g
Corned beef, canned	1 protein + 1 fat	1 slice	25 g
Corned beef, fresh	1 protein + 1 fat	1 slice	25 g
Ground beef, medium-fat	1 protein + 1 fat	30 mL (2 tbs)	25 g

(continued on the next page)

TABLE B·6   CANADIAN CHOICE SYSTEM: PROTEIN FOODS—continued

1 protein choice = 7 g protein, 3 g fat, 230 kJ (55 kcal)

Food	Choices per Serving	Measures	Mass (Weight)
Meat—continued			
Meat spreads, canned	1 protein + 1 fat	45 mL	35 g
Mutton chop	1 protein + 1 fat	½ chop, with bone	35 g
Pâté (see Fats and Oils, Table B-7)			
Sausages, garlic, Polish or knockwurst	1 protein + 1 fat	1 slice	50 g
Sausages, pork, links	1 protein + 1 fat	1 link	25 g
Spareribs or shortribs, with bone	1 protein + 1 fat	1 large	65 g
Stewing beef	1 protein + 1 fat	1 cube	25 g
Summer sausage or salami	1 protein + 1 fat	1 slice	40 g
Weiners, hot dog	1 protein + 1 fat	½ medium	25 g
Miscellaneous			
Blood pudding	1 protein + 1 fat	1 slice	25 g
Peanut butter	1 protein + 1 fat	15 mL (1 tbs)	15 g

1 fat choice = 5 g fat, 190 kJ (45 kcal)

Food	Measure	Mass (Weight)	Food	Measure	Mass (Weight)
Avocado*	⅛	30 g	Nuts (continued):		
Bacon, side, crisp*	1 slice	5 g	Sesame seeds	15 mL (1 tbs)	10 g
Butter*	5 mL (1 tsp)	5 g	Sunflower seeds		
Cheese spread	15 mL (1 tbs)	15 g	Shelled	15 mL (1 tbs)	10 g
Coconut, fresh*	45 mL (3 tbs)	15 g	In shell	45 mL (3 tbs)	15 g
Coconut, dried*	15 mL (1 tbs)	10 g	Walnuts	4 halves	10 g
Cream, Half and half			Oil, cooking and salad	5 mL (1 tsp)	5 g
(cereal), 10%*	30 mL (2 tbs)	30 g	Olives, green	10	45 g
Light (coffee), 20%*	15 mL (1 tbs)	15 g	Ripe black	7	57 g
Whipping, 32 to 37%*	15 mL (1 tbs)	15 g	Pâté, liverwurst,	15 mL (1 tbs)	15 g
Cream cheese*	15 mL (1 tbs)	15 g	meat spreads		
Gravy*	30 mL (2 tbs)	30 g	Salad dressing: blue,	10 mL (2 tsp)	10 g
Lard*	5 mL (1 tsp)	5 g	French, Italian,		
Margarine	5 mL (1 tsp)	5 g	mayonnaise,		
Nuts, shelled:			Thousand Island	5 mL (1 tsp)	5 g
Almonds	8	5 g	Salad dressing,	30 mL (2 tbs)	30 g
Brazil nuts	2	10 g	low-calorie		
Cashews	5	10 g	Salt pork, raw	5 mL (1 tsp)	5 g
Filberts, hazelnuts	5	10 g	or cooked*		
Macadamia	3	5 g	Sesame oil	5 mL (1 tsp)	5 g
Peanuts	10	10 g	Sour cream		
Pecans	5 halves	5 g	12% milkfat	30 mL (2 tbs)	30 g
Pignolias, pine nuts	25 mL (5 tsp)	10 g	7% milkfat	60 mL (4 tbs)	60 g
Pistachios, shelled	20	10 g	Shortening*	5 mL (1 tsp)	
Pistachios, in shell	20	20 g			
Pumpkin and	20 mL (4 tsp)	10 g			
squash seeds					

*These items contain higher amounts of saturated fat.

B

TABLE B·8 CANADIAN CHOICE SYSTEM: EXTRAS

Extras have no more than 2.5 g carbohydrate, 60 kJ (14 kcal)

Vegetables 125 mL (½ c)
Artichokes
Asparagus
Bamboo shoots
Bean sprouts, mung or soya
Beans, string, green, or yellow
Bitter melon (balsam pear)
Bok choy
Broccoli
Brussels sprouts
Cabbage
Cauliflower
Celery
Chard
Cucumbers
Eggplant
Endive
Fiddleheads
Greens: beet, collard, dandelion, mustard, turnip, etc.
Kale
Kohlrabi
Leeks
Lettuce
Mushrooms
Okra
Onions, green or mature
Parsley
Peppers, green, yellow or red
Radishes
Rapini
Rhubarb
Sauerkraut
Shallots
Spinach
Sprouts: alfalfa, radish, etc.
Tomato wedges
Watercress
Zucchini

Free Foods (may be used without measuring)
Artificial sweetener, such as
   cyclamate or aspartame
Baking powder, baking soda
Bouillon from cube,
   powder, or liquid
Bouillon or clear broth
Chowchow, unsweetened
Coffee, clear
Consommé
Dulse
Flavorings and extracts
Garlic
Gelatin, unsweetened
Ginger root
Herbal teas, unsweetened
Horseradish, uncreamed
Lemon juice or lemon wedges

Lime juice or lime wedges
Marjoram, cinnamon, etc.
Mineral water
Mustard
Parsley
Pimentos
Salt, pepper, thyme
Soda water, club soda
Soya sauce
Sugar-free Crystal Drink
Sugar-free Jelly Powder
Sugar-free soft drinks
Tea, clear
Vinegar
Water
Worcestershire sauce

Condiments	Measure
Anchovies	2 fillets
Barbecue sauce	15 mL (1 tbs)
Bran, natural	30 mL (2 tbs)
Brewer's yeast	5 mL (1 tsp)
Carob powder	5 mL (1 tsp)
Catsup	5 mL (1 tsp)
Chili sauce	5 mL (1 tsp)
Cocoa powder	5 mL (1 tsp)
Cranberry sauce, unsweetened	15 mL (1 tbs)
Dietetic fruit spreads	5 mL (1 tsp)
Maraschino cherries	1
Nondairy coffee whitener	5 mL (1 tsp)
Nuts, chopped pieces	5 mL (1 tsp)
Pickles	
Unsweetened dill	2
Sour mixed	11
Sugar substitutes, granular	5 mL (1 tsp)
Whipped toppings	15 mL (1 tbs)

Food	Choices per Serving	Measure	Mass (Weight)
Angel food cake	½ starch + 2½ sugars	½ cake	50 g
Apple crisp	½ starch + 1½ fruits & vegetables + 1 sugar + 1–2 fats	125 mL (½ c)	
Applesauce, sweetened	1 fruits & vegetables + 1 sugar	125 mL (½ c)	
Beans and pork in tomato sauce	1 starch + ½ fruits & vegetables + ½ sugar + 1 protein	125 mL (½ c)	135 g
Beef burrito	2 starches + 3 proteins + 3 fats		110 g
Brownie	1 sugar + 1 fat	1	20 g
Cabbage rolls*	1 starch + 2 proteins	3	310 g
Caesar salad	2–4 fats	20 mL dressing (4 tsp)	
Cheesecake	½ starch + 2 sugars + ½ protein + 5 fats	1 piece	80 g
Chicken fingers	1 starch + 2 proteins + 2 fats	6 small	100 g
Chicken and snow pea Oriental	2 starches + ½ fruits & vegetables + 3 proteins + 1 fat	500 mL (2 c)	
Chili	1½ starches + ½ fruits & vegetables + 3½ protein	300 mL (1¼ c)	325 g
Chips			
Potato chips	1 starch + 2 fats	15 chips	30 g
Corn chips	1 starch + 2 fats	30 chips	30 g
Tortilla chips	1 starch + 1½ fats	13 chips	
Cheese twist	1 starch + 1½ fats	30 chips	30 g
Chocolate bar			
Aero®	2½ sugars + 2½ fats	bar	43 g
Smarties®	4½ sugars + 2 fats	package	60 g
Chocolate cake (without icing)	1 starch + 2 sugars + 3 fats	⅒ of a 8″ pan	
Chocolate devil's food cake (without icing)	2 starches + 2 sugars + 3 fats	¹⁄₁₂ of a 9″ pan	
Chocolate milk	2 milks 2% + 1 sugar	250 mL (1 c)	300 g
Clubhouse (triple-decker) sandwich	3 starches + 3 proteins + 4 fats		
Cookies			
Chocolate chip	½ starch + ½ sugar + 1½ fats	2	22 g
Oatmeal	1 starch + 1 sugar + 1 fat	2	40 g
Donut (chocolate glazed)	1 starch + 1½ sugars + 2 fats	1	65 g
Egg roll	1 starch + ½ protein + 1 fat		75 g
Four bean salad	1 starch + ½ protein + 1 fat	125 mL (½ c)	
French toast	1 starch + ½ protein + 2 fats	1 slice	65 g
Fruit in heavy syrup	1 fruits & vegetables + 1½ sugars	125 mL (½ c)	
Granola bar	½ starch + 1 sugar + 1–2 fats		30 g
Granola cereal	1 starch + 1 sugar + 2 fats	125 mL (½ c)	45 g
Hamburger	2 starches + 3 proteins + 2 fats	junior size	
Ice cream and cone, plain flavour			
Ice cream	½ milk + 2–3 sugars + 1–2 fats		100 g
Cone	½ sugar		4 g
Lasagna			
Regular cheese	1 starch + 1 fruits & vegetables + 3 proteins + 2 fats	3″ × 4″ piece	
Low-fat cheese	1 starch + 1 fruits & vegetables + 3 proteins	3″ × 4″ piece	
Legumes			
Dried beans (kidney, navy, pinto, fava, chick peas)	2 starches + 2 protein	250 mL (1 c)	175 g
Dried peas	2 starches + 2 protein	250 mL (1 c)	196 g
Lentils	2 starches + 2 protein	250 mL (1 c)	196 g
Macaroni and cheese	2 starches + 2 proteins + 2 fats	250 mL (1 c)	210 g
Minestrone soup	1½ starches + ½ fruits & vegetables + ½ fat	250 mL (1 c)	
Muffin	1 starch + ½ sugar + 1 fat	1 small	45 g

*If eaten with sauce, add ½ fruits & vegetables exchange.

(continued on the next page)

Food	Choices per Serving	Measure	Mass (Weight)
Nuts (dry or roasted without any oil added).			
Almonds, dried sliced	½ protein + 2 fats	50 mL (¼ c)	22 g
Brazil nuts, dried unblanched	½ protein + 2½ fats	5 large	23 g
Cashew nuts, dry roasted	½ starch + ½ protein + 2 fats	50 mL (¼ c)	28 g
Filbert hazelnut, dry	½ protein + 3½ fats	50 mL (¼ c)	30 g
Macadamia nuts, dried	½ protein + 4 fats	50 mL (¼ c)	28 g
Peanuts, raw	1 protein + 2 fats	50 mL (¼ c)	30 g
Pecans, dry roasted	½ fruits & vegetables + 3 fats	50 mL (¼ c)	22 g
Pine nuts, pignolia dried	1 protein + 3 fats	50 mL (¼ c)	34 g
Pistachio nuts, dried	½ fruits & vegetables + ½ protein + 2½ fats	50 mL (¼ c)	27 g
Pumpkin seeds, roasted	2 proteins + 2½ fats	50 mL (¼ c)	47 g
Sesame seeds, whole dried	½ fruits & vegetables + ½ protein + 2½ fats	50 mL (¼ c)	30 g
Sunflower kernel, dried	½ protein + 1½ fats	50 mL (¼ c)	17 g
Walnuts, dried chopped	½ protein + 3 fats	50 mL (¼ c)	26 g
Perogies	2 starches + 1 protein + 1 fat	3	
Pie, fruit	1 starch + 1 fruits & vegetables + 2 sugars + 3 fats	1 piece	120 g
Pizza, cheese	1 starch + 1 protein + 1 fat	1 slice (⅛ of a 12")	50 g
Pork stir-fry	½ to 1 fruits & vegetables + 3 proteins	200 mL (¾ c)	
Potato salad	1 starch + 1 fat	125 mL (½ c)	130 g
Potatoes, scalloped	2 starches + 1 milk + 1–2 fats	200 mL (¾ c)	210 g
Pudding, bread or rice	1 starch + 1 sugar + 1 fat	125 mL (½ c)	
Pudding, vanilla	1 milk + 2 sugars	125 mL (½ c)	
Raisin bran cereal	1 starch + ½ fruits & vegetables + ½ sugar	175 mL (⅔ c)	40 g
Rice krispie squares	½ starch + 1½ sugars + ½ fat	1 square	30 g
Shepherd's pie	2 starches + 1 fruits & vegetables + 3 proteins	325 mL (1⅓ c)	
Sherbet, orange	3 sugars 1 ½ fat	125 mL (½ c)	
Spaghetti and meat sauce	2 starches + 1 fruits & vegetables + 2 proteins + 3 fats	250 mL (1 c)	
Stew	2 starches + 2 fruits & vegetables + 3 proteins + ½ fat	200 mL (¾ c)	
Sundae	4 sugars + 3 fats	125 mL (½ c)	
Tuna casserole	1 starch + 2 proteins + ½ fat	125 mL (½ c)	
Yogurt, fruit bottom	1 fruits & vegetables + 1 milk + 1 sugar	125 mL (½ c)	125 g
Yogurt, frozen	1 milk + 1 sugar	125 mL (½ c)	125 g

Food Labels

Consumers can gather a lot of information from a nutrition label. Figure B-2 points out changes to the Canadian food label, and Table B-10 (below) and Table B-11 (next page) define terms.

FIGURE B·2   EXAMPLES OF CANADA'S FOOD LABELS

English Label

Nutrition Facts
Per 1 cup (264g)

Amount	% Daily Value
Calories 260	
Fat 13g	20%
Saturated Fat 3g + Trans Fat 2g	25%
Cholesterol 30mg	
Sodium 660mg	28%
Carbohydrate 31g	10%
Fibre 0g	0%
Sugars 5g	
Protein 5g	
Vitamin A 4% • Vitamin C 2%	
Calcium 15% • Iron 4%	

Bilingual Label

Nutrition Facts
Valeur nutritive
Per 1 cup (264g)
pour 1 tasse (264g)

Amount Quantité□	% Daily Value % valeur quotidienne
Calories / Calories 260	
Fat / Lipides 13g	20%
Saturated / saturés 3g + Trans / trans 2g	25%
Cholesterol / Cholesté rol 30mg	
Sodium / Sodium 660mg	28%
Carbohydrate / Glucides 31g	10%
Fibre / Fibres 0g	0%
Sugars / Sucres 5g	
Protein / Proté ines 5g	
Vitamin A / Vitamine A	4%
Vitamin C / Vitamine C	2%
Calcium / Calcium	15%
Iron / Fer	4%

French Label

Valeur nutritive
pour 1 tasse (264g)

Quantité□	% valeur quotidienne
Calories 260	
Lipides 13g	20%
Saturés 3g + Trans 2g	25%
Cholestéol 30mg	
Sodium 660mg	28%
Glucides 31g	10%
Fibres 0g	0%
Sucres 5g	
Protênes 5g	
Vitamine A 4% • Vitamine C 2%	
Calcium 15% • Fer 4%	

The nutrition label includes:

- A title: *Nutrition Facts.*
- Consistent serving sizes on which nutrient information is based.
- A list of nutrients.
- A standardized format that is bold, clear, and easy to read.

- Consistent appearance from product to product.
- Clearly identified nutrient information.
- The Daily Value, which gives a context to the nutrient values.

For more information, visit the website:

www.hc-sc.gc.ca/hppb/nutrition/labels/e_before.html

TABLE B·10   TERMS ON FOOD LABELS

Energy

- **calorie reduced** 50% or fewer calories than the regular version.
- **light** term may be used to describe anything (for example, light in colour, texture, flavour, taste, or calories); read the label to find out what is "light" about the product.
- **low calorie** calorie-reduced and no more than 15 calories per serving.

Fat and Cholesterol

- **low cholesterol** no more than 3 mg of cholesterol per 100 g of the food and low in saturated fat; does not always mean low in total fat.
- **low fat** no more than 3 g of fat per serving; does not always mean low in calories.
- **lower fat** at least 25% less fat than the comparison food; be aware that 80% fat-free still means the food is 20% fat.

Carbohydrates: Fibre and Sugar

- **carbohydrate reduced** not more than 50% of the carbohydrate found in the regular version; does not always mean the product is lower in calories because other ingredients such as fat may have increased.
- **source of dietary fibre** a product that provides 2–4 g of fibre.
- **high source of dietary fibre** a product that provides 4–6 g of fibre.
- **very high source of fibre** a product that provides 6 g (or more) of fibre.
- **sugar free** low in carbohydrates and calories; can be used as an extra food in the exchange system.
- **unsweetened or no sugar added** no sugar was added to the product; sugar may be found naturally in the food (for example, fruit canned in its own juice).

B | TABLE B·11 NUTRIENT CONTENT CLAIMS

Certain nutrient content claims are allowable on labels if the food meets certain criteria compared with a reference amount and serving of stated size. For example, a food claiming to be "high fibre" would have to contain 4 grams or more of fibre per reference amount and serving of stated size. For prepackaged meals or entrées for which no reference amount exists, the criteria for most claims would be based per 100 grams food for most claims.

Nutrient Content Claim	Compositional Criteria
	(per reference amount and serving of stated size)[a]
Calories	
Calorie-free	less than 5 calories
Low in calories	40 calories or less
Reduced or lower in calories	at least 25% less energy[b]
Source of calories	at least 100 calories
Protein	
Source of protein	protein rating of 20 or more[c]
Excellent source of protein	protein rating of 40 or more[c]
More protein	protein rating of 20 or more[c], at least 25% or more protein[c], and 7 g or more protein
Fat	
Fat-free	less than 0.5 g fat
Low in fat	3 g or less fat
Reduced or lower in fat	at least 25% less fat[b]
100% fat-free	less than 0.5 g fat per 100 g, no added fat, and "free of fat"
(naming the %) fat-free	"low in fat"
Saturated Fatty Acids	
Saturated fatty acid–free	less than 0.2 g saturated fatty acids and less than 0.2 g trans fatty acids
Low in saturated fatty acids	2 g or less saturated fatty acids and *trans* fatty acids combined and 15% or less energy from saturated fatty acids plus *trans* fatty acids
Reduced or lower in saturated fatty acids	at least 25% less saturated fatty acids and *trans* fatty acids not increased[b]
***Trans* Fatty Acids**	
Free of *trans* fatty acids	less than 0.2 g *trans* fatty acids and "low in saturated fatty acids"
Reduced or lower in *trans* fatty acids	at least 25% less *trans* fatty acids and saturated fatty acids not increased[b]
Polyunsaturated Fatty Acids	
Source of omega-3 polyunsaturated fatty acids	0.3 g or more omega-3 polyunsaturated fatty acids
Source of omega-6 polyunsaturated fatty acids	2 g or more omega-6 polyunsaturated fatty acids
Cholesterol	
Cholesterol-free	less than 2 mg cholesterol and "low in saturated fatty acids"
Low in cholesterol	20 mg or less cholesterol and "low in saturated fatty acids"
Reduced or lower in cholesterol	at least 25% less cholesterol[b] and "low in saturated fatty acids"
Sodium	
Sodium-free or salt-free	less than 5 mg sodium or salt
Low in sodium or salt	140 mg or less sodium or salt
Reduced or lower in sodium or salt	at least 25% less sodium or salt[b]
No added sodium or salt	no salt or other sodium salts added during processing
Lightly salted	at least 50% less added sodium or salt[b]
Sugars	
Sugar-free	less than 0.5 g sugars and (except chewing gum) "free of energy"
Reduced or lower in sugar	at least 25% less sugars[b]
No added sugar	no sugars added in processing or packaging, including ingredients that contain added sugars or ingredients that functionally substitute for added sugars (e.g., concentrated fruit juice), and sugars not increased through some other means
Fibre	
Source of fibre	2 g or more fibre or of each identified fibre
High source of fibre	4 g or more fibre or of each identified fibre
Very high source of fibre	6 g or more fibre or of each identified fibre
More fibre	at least 25% more fibre[b] and at least 2 g fibre
Light	
Light	"reduced in energy" or "reduced in fat"

[a]Criteria for "reduced" or "lower in" claims are based on the reference amount of the food, and for protein claims, criteria are given per reasonable daily intake of food (Schedule K of the Regulations) or, in the case of breakfast cereals, per 30 g of breakfast cereal combined with 125 mL of milk.
[b]As compared with the reference amount of a reference food. Reference food must not be "low" in subject nutrient.
[c]As determined by official method FO-1, Determination of Protein Rating, Oct. 15, 1981.

Aids to Calculation

C

Mathematical problems have been worked out for you as examples at appropriate places in the text. This appendix aims to help with the use of the metric system and with those problems not fully explained elsewhere.

Conversion Factors

Conversion factors are useful mathematical tools in everyday calculations, like the ones encountered in the study of nutrition. A conversion factor is a fraction in which the numerator (top) and the denominator (bottom) express the same quantity in different units. For example, 2.2 pounds (lb) and 1 kilogram (kg) are equivalent; they express the same weight. The conversion factor used to change pounds to kilograms or vice versa is:

$$\frac{2.2\ \text{lb}}{1\ \text{kg}} \quad \text{or} \quad \frac{1\ \text{kg}}{2.2\ \text{lb}}$$

Because both factors equal 1, measurements can be multiplied by the factor without changing the value of the measurement. Thus, the units can be changed.

The correct factor to use in a problem is the one with the unit you are seeking in the numerator (top) of the fraction. Following are some examples of problems commonly encountered in nutrition study; they illustrate the usefulness of conversion factors.

Example 1

Convert the weight of 130 pounds to kilograms.

1. Choose the conversion factor in which the unit you are seeking is on top:

$$\frac{1\ \text{kg}}{2.2\ \text{lb}}$$

2. Multiply 130 pounds by the factor:

$$130\ \text{lb} \times \frac{1\ \text{kg}}{2.2\ \text{lb}} = \frac{130\ \text{kg}}{2.2}$$

$$= 59\ \text{kg (rounded off to the nearest whole number)}$$

Example 2

How many grams (g) of saturated fat are contained in a 3-ounce (oz) hamburger?

1. Appendix A shows that a 4-ounce hamburger contains 7 grams of saturated fat. You are seeking grams of saturated fat; therefore, the conversion factor is:

$$\frac{7\ \text{g saturated fat}}{4\ \text{oz hamburger}}$$

2. Multiply 3 ounces of hamburger by the conversion factor:

$$3\ \text{oz hamburger} \times \frac{7\ \text{g saturated fat}}{4\ \text{oz hamburger}} = \frac{3 \times 7}{4} = \frac{21}{4}$$

$$= 5\ \text{g saturated fat (rounded off to the nearest whole number)}$$

Energy Units

1 calorie* (cal) = 4.2 kilojoules
1 millijoule (MJ) = 240 cal
1 kilojoule (kJ) = 0.24 cal
1 gram (g) carbohydrate = 4 cal = 17 kJ
1 g fat = 9 cal = 37 kJ
1 g protein = 4 cal = 17 kJ
1 g alcohol = 7 cal = 29 kJ

Nutrient Unit Conversions

Sodium

To convert milligrams of sodium to grams of salt:

$$\text{mg sodium} \div 400 = \text{g of salt}$$

The reverse is also true:

$$\text{g salt} \times 400 = \text{mg sodium}$$

Folate

To convert micrograms (µg) of synthetic folate in supplements and enriched foods to Dietary Folate Equivalents (µg DFE):

$$\text{µg synthetic folate} \times 1.7 = \text{µg DFE}$$

For naturally occurring folate, assign each microgram folate a value of 1 µg DFE:

$$\text{µg folate} = \text{µg DFE}$$

Example 3

Consider a pregnant woman who takes a supplement and eats a bowl of fortified cornflakes, 2 slices of fortified bread, and a cup of fortified pasta.

1. From the supplement and fortified foods, she obtains synthetic folate:

Supplement	100 µg folate
Fortified cornflakes	100 µg folate
Fortified bread	40 µg folate
Fortified pasta	60 µg folate
	300 µg folate

2. To calculate the DFE, multiply the amount of synthetic folate by 1.7:

$$300\ \text{µg} \times 1.7 = 510\ \text{µg DFE}$$

3. Now add the naturally occurring folate from the other foods in her diet—in this example, another 90 µg of folate.

$$510\ \text{µg DFE} + 90\ \text{µg} = 600\ \text{µg DFE}$$

Notice that if we had not converted synthetic folate from supplements and fortified foods to DFE, then this woman's intake would appear to fall short of the 600 µg recommendation for pregnancy (300

---

*Throughout this book and in the appendixes, the term *calorie* is used to mean kilocalorie. Thus, when converting calories to kilojoules, do not enlarge the calorie values—they are kilocalorie values.

μg + 90 μg = 390 μg). But as this example shows, her intake does meet the recommendation. At this time, supplement and fortified food labels list folate in μg only, not μg DFE, making such calculations necessary.

Vitamin A
Equivalencies for vitamin A:

1 μg RAE = 1 μg retinol
= 12 μg beta-carotene
= 24 μg other vitamin A carotenoids

1 international unit (IU) = 0.3 μg retinol
= 3.6 μg beta-carotene
= 7.2 μg other vitamin A carotenoids

To convert older RE values to micrograms RAE:
1 μg RE retinol = 1 μg RAE retinol
6 μg RE beta-carotene = 12 μg RAE beta-carotene
12 μg RE other vitamin A carotenoids = 24 μg RAE other vitamin A carotenoids

International Units (IU)
To convert IU to:
- μg vitamin D: divide by 40 or multiply by 0.025.
- 1 IU natural vitamin E = .67 mg alpha-tocopherol.
- 1 IU synthetic vitamin E = .45 mg alpha-tocopherol.
- vitamin A, see above.

Percentages

A percentage is a comparison between a number of items (perhaps your intake of energy) and a standard number (perhaps the number of calories recommended for your age and gender—your energy DRI). The standard number is the number you divide by. The answer you get after the division must be multiplied by 100 to be stated as a percentage (*percent* means "per 100").

Example 4
What percentage of the DRI recommendation for energy is your energy intake?

1. Find your energy DRI value on the inside front cover. We'll use 2,368 calories to demonstrate.
2. Total your energy intake for a day—for example, 1,200 calories.
3. Divide your calorie intake by the DRI value:

$$1{,}200 \text{ cal (your intake)} \div 2{,}368 \text{ cal (DRI)} = 0.507$$

4. Multiply your answer by 100 to state it as a percentage:

$$0.507 \times 100 = 50.7 = 51\% \text{ (rounded off to the nearest whole number)}$$

In some problems in nutrition, the percentage may be more than 100. For example, suppose your daily intake of vitamin A is 3,200 and your DRI is 900 μg. Your intake as a percentage of the DRI is more than 100 percent (that is, you consume more than 100 percent of your recommendation for vitamin A). The following calculations show your vitamin A intake as a percentage of the DRI value:

$$3{,}200 \div 900 = 3.6 \text{ (rounded)}$$
$$3.6 \times 100 = 360\% \text{ of DRI}$$

Example 5
Food labels express nutrients and energy contents of foods as percentages of the Daily Values. If a serving of a food contains 200 milligrams of calcium, for example, what percentage of the calcium Daily Value does the food provide?

1. Find the calcium Daily Value on the inside front cover, page c.

2. Divide the milligrams of calcium in the food by the Daily Value standard:

$$\frac{200}{1{,}000} = 0.2$$

3. Multiply by 100:

$$0.2 \times 100 = 20\% \text{ of the Daily Value}$$

Example 6
This example demonstrates how to calculate the percentage of fat in a day's meals.

1. Recall the general formula for finding percentages of calories from a nutrient:

$$\text{(one nutrient's calories} \div \text{total calories)} \times 100 = \text{the percentage of calories from that nutrient}$$

2. Say a day's meals provide 1,754 calories and 54 grams of fat. First, convert fat grams to fat calories:

$$54 \text{ g} \times 9 \text{ cal per g} = 486 \text{ cal from fat}$$

3. Then apply the general formula for finding percentage of calories from fat:

$$\text{(fat calories} \div \text{total calories)} \times 100 = \text{percentage of calories from fat}$$

$$(486 \div 1{,}754) \times 100 = 27.7 \text{ (28\%, rounded)}$$

Weights and Measures

Length
1 inch (in) = 2.54 centimeters (cm)
1 foot (ft) = 30.48 cm
1 meter (m) = 39.37 in

Temperature

	Steam	100°C	212°F	Steam	
	Body temperature	37°C	98.6°F	Body temperature	
	Ice	0°C	32°F	Ice	
		Celsius‡		Fahrenheit	

- To find degrees Fahrenheit (°F) when you know degrees Celsius (°C), multiply by 9/5 and then add 32.
- To find degrees Celsius (°C) when you know degrees Fahrenheit (°F), subtract 32 and then multiply by 5/9.

Volume
Used to measure fluids or pourable dry substances such as cereal.
1 milliliter (ml) = ⅕ teaspoon or 0.034 fluid ounce or ¹⁄₁₀₀₀ liter
1 deciliter (dl) = ¹⁄₁₀ liter
1 teaspoon (tsp or t) = 5 ml or about 5 grams (weight) salt
1 tablespoon (tbs or T) = 3 tsp or 15 ml
1 ounce, fluid (fl oz) = 2 tbs or 30 ml
1 cup (c) = 8 fl oz or 16 tbs or 250 ml
1 quart (qt) = 32 fl oz or 4 c or 0.95 liter
1 liter (L) = 1.06 qt or 1,000 ml
1 gallon (gal) = 16 c or 4 qt or 128 fl oz or 3.79 L

Weight
1 microgram (μg or mcg) = ¹⁄₁₀₀₀ milligram
1 milligram (mg) = 1,000 mg or ¹⁄₁₀₀₀ gram
1 gram (g) = 1,000 mg or ¹⁄₁₀₀₀ kilogram
1 ounce, weight (oz) = about 28 g or ¹⁄₁₆ pound
1 pound (lb) = 16 oz (wt) or about 454 g
1 kilogram (kg) = 1,000 g or 2.2 lb

‡Also known as centigrade.

U.S. Food Exchange System

D

The U.S. food exchange system is intended to help people with diabetes control the levels of glucose and lipids in the blood by controlling the grams of carbohydrate and fat they consume. Other diet planners have found the system invaluable for achieving calorie control and moderation.

Planning a Diet

The exchange system sorts foods into three main groups by their proportions of carbohydrate, fat, and protein. These three groups—the carbohydrate group, the fat group, and the meat and meat substitute group (protein)—are each subdivided into several exchange lists of foods (Table D-1).

Portion Sizes

All of the food portions in a given list provide approximately the same amounts of energy nutrients (carbohydrate, fat, and protein) and the same number of calories. Portion sizes are strictly defined so that every item on a given list provides roughly the same amount of energy. Any food on a list can then be exchanged, or traded, for any other food on that same list without affecting a plan's balance or total calories.

To apply the system successfully, users must become familiar with portion sizes. A convenient way to remember the portion sizes and energy values is to keep in mind a typical item from each list. Table D-1 includes some representative portion sizes.

The Foods on the Lists

Foods are not always on the exchange list where you might first expect them to be because they are grouped according to their energy-nutrient contents rather than by their source (such as milks), their outward appearance, or their vitamin and mineral contents. For example, cheeses are grouped with meats in the exchange system because, like meats, cheeses contribute energy from protein and fat but provide negligible carbohydrate. (In the food group plans presented earlier, cheeses are classed with milk because they are milk products with a similar calcium content.)

For similar reasons, starchy vegetables such as corn, green peas, and potatoes are listed on the starch list in the exchange system, rather than with the vegetables. Likewise, olives are not classed as a "fruit" as a botanist would claim; they are classified as a "fat" because their fat content makes them more similar to butter than to berries. Bacon is also on the fat list to remind users of its high fat content. These groupings permit you to see the characteristics of foods that are significant to energy intake.

Users of the exchange lists learn to view mixtures of foods, such as casseroles and soups, as combinations of foods from different exchange lists. They also learn to interpret food labels with the exchange system in mind.

TABLE D·1 THE EXCHANGE GROUPS AND LISTS

Group/Lists	Typical Item/Portion Size	Carbohydrate (g)	Protein (g)	Fat (g)	Energy[a] (cal)
Carbohydrate Group					
Starch[b]	1 slice bread	15	3	0–1	80
Fruit	1 small apple	15	—	—	60
Milk					
Fat-free, low-fat	1 c fat-free milk	12	8	0–3	90
Reduced-fat	1 c reduced-fat milk	12	8	5	120
Whole	1 c whole milk	12	8	8	150
Other carbohydrates[c]	2 small cookies	15	varies	varies	varies
Vegetable (nonstarchy)	½ c cooked carrots	5	2	—	25
Meat and Meat Substitute Group[d]					
Meat					
Very lean	1 oz chicken (white meat, no skin)	—	7	0–1	35
Lean	1 oz lean beef	—	7	3	55
Medium-fat	1 oz ground beef	—	7	5	75
High-fat	1 oz pork sausage	—	7	8	100
Fat Group					
Fat	1 tsp butter	—	—	5	45

[a]The energy value for each exchange list represents an approximate average for the group and does not reflect the precise number of grams of carbohydrate, protein, and fat. For example, a slice of bread contains 15 grams of carbohydrate (60 calories), 3 grams protein (12 calories), and a little fat—rounded to 80 calories for ease in calculating. A half-cup of vegetables (not including starchy vegetables) contains 5 grams carbohydrate (20 calories) and 2 grams protein (8 more), which has been rounded down to 25 calories.

[b]The starch list includes cereals, grains, breads, crackers, snacks, starchy vegetables (such as corn, peas, and potatoes), and legumes (dried beans, peas, and lentils).

[c]The other carbohydrates list includes foods that contain added sugars and fats such as cakes, cookies, doughnuts, ice cream, potato chips, pudding, syrup, and frozen yogurt.

[d]The meat and meat substitutes list includes legumes, cheeses, and peanut butter.

Knowing that foods on the starch list provide 15 grams of carbohydrate and those on the vegetable list provide 5, you can interpret the label of a lasagna dinner that lists 37 grams of carbohydrate as "2 starches" (mostly noodles) and "1 vegetable" (the sauce).

Controlling Energy, Fat, and Sodium

The exchange system helps people control their energy intakes by paying close attention to portion sizes. A portion of any food on a given list provides roughly the same amount of energy nutrients and total calories. The portion sizes have been adjusted so that all portions have the same energy value. For example, 17 grapes count as one fruit portion, as does ½ grapefruit. A whole grapefruit counts as two portions.

A *portion* in the exchange system is not the same as a serving equivalent presented in Chapter 2, especially when it comes to meats. The exchange system lists meats and most cheeses in single ounces; that is, 1 *portion* (or *exchange*) of meat is 1 ounce, whereas one *serving* is 2 to 3 ounces. Calculating meat by the ounce encourages the planner to keep close track of the exact amounts eaten. This, in turn, helps control energy and fat intakes.

By allocating items like bacon and avocados to the fat list, the exchange system alerts consumers to foods that are unexpectedly high in fat. Even the starch list specifies which grain products contain added fat (such as biscuits, muffins, and waffles). In addition, the exchange system encourages users to think of fat-free milk as milk and of whole milk as milk with added fat, and to think of very lean meats as meats and of lean, medium-fat, and high-fat meats as meats with added fat. To that end, foods on the milk and meat lists are separated into categories based on their fat contents.

Control of food energy and fat intake can be highly successful with the exchange system. Exchange plans do not, however, guarantee adequate intakes of vitamins and minerals. Food group plans work better from that standpoint because the food groupings are based on similarities in vitamin-mineral content. In the exchange system, for example, meats are grouped with cheeses, yet the meats are iron-rich and calcium-poor, whereas the cheeses are iron-poor and calcium-rich. To take advantage of the strengths of both food group plans and exchange patterns, and to compensate for their weaknesses, diet planners often combine these two diet-planning tools, as the following section shows.

People wishing to control the sodium in their diets can begin by eliminating any foods bearing this symbol [✏] found on any exchange list. The symbol identifies each food that, in one exchange, provides 400 milligrams or more of sodium. Other foods may also contribute substantially to sodium, however (consult Chapter 8 for details).

Combining Food Group Plans and Exchange Lists

A diet planner may find that using a food group plan together with the exchange lists eases the task of choosing foods that will provide all the nutrients. The food group plan ensures that all classes of nutritious foods are included, thus promoting adequacy, balance, and variety. The exchange system classifies the food selections by their energy-yielding nutrients, thus controlling energy and fat intakes.

Table D-2 shows how to use the USDA Food Guide together with the exchange lists to plan a diet. The USDA Food Guide ensures that a certain amount of food is chosen from each of the food groups (see the first column of the table). The second column translates the amounts of food into exchanges. With the addition of a small amount of oil and added sugars, this sample diet plan provides about 1,750 calories. Many people can meet their needs for all the nutrients within this reasonable energy allowance. (Table D-3 shows patterns for other energy intakes.) The next step in diet planning is to assign the exchanges to meals and snacks. The final plan might look like the one in Table D-4. To aid you in the development of your own diet plan, Tables D-5 through D-13 present the U.S. exchange system in detail.

Next, a person could begin to fill in the plan with real foods to create a menu. For example, the breakfast plan calls for 2 starches, 1 fruit, and 1 fat-free milk. A person might select a bowl of shredded wheat with banana slices and milk (1 cup shredded wheat = 2 starches, 1 small banana = 1 fruit, and 1 cup fat-free milk = 1 milk); or a bagel and a bowl of cantaloupe pieces topped with yogurt (1 bagel = 2 starches, ⅛ cantaloupe melon = 1 fruit, and ¾ cup fat-free plain yogurt = 1 milk). A person who wanted butter on the bagel could move a fat exchange or two from dinner to breakfast. If willing to use two fat exchanges at breakfast, the person could have pancakes with strawberries and milk (4 small pancakes = 2 starches plus 2 fats, 1¼ cup strawberries = 1 fruit, and a cup of fat-free milk = 1 milk). Then the person could move on to complete the menu for lunch, dinner and snacks.

U.S. Exchange Lists for Meal Planning[a]

[a]SOURCE: The Exchange Lists are the basis of a meal planning system designed by a committee of the American Dietetic Association. © 2003 American Diabetes Association, Inc.

TABLE D·2 DIET PLANNING WITH THE EXCHANGE SYSTEM USING THE USDA FOOD GUIDE PATTERN	
Pattern from USDA Food Guide (see Chapter 2)	Selections Made Using the Exchange System
Grains (breads and cereals)—4½ ounces	Starch list—select 9 exchanges
Vegetables—2 cups	Vegetable list—select 4 exchanges
Fruits—1½ cups	Fruit list—select 3 exchanges
Meat and beans[a]—6 ounces	Meat list—select 6 lean exchanges
Milk—2 cups	Milk list—select 2 nonfat exchanges
Oils—5 teaspoons	Fat list—select 5 monounsaturated or polyunsaturated exchanges

[a]Or the equivalent of seeds, eggs, or nuts.

TABLE D·3 DIET PATTERNS FOR DIFFERENT ENERGY INTAKES

Exchange	Energy Level (cal)						
	1,200	1,500	1,800	2,000	2,200	2,600	3,000
Starch	6	7	8	9	11	13	15
Meat (lean)	4	5	6	6	6	7	8
Vegetable	3	4	5	5	5	6	6
Fruit	2	3	4	4	4	5	6
Milk (fat-free)	2	2	2	3	3	3	3
Fat	3	5	6	7	8	10	12

NOTE: These patterns follow the Daily Food Guide plan and supply less than 30 percent of calories as fat.

TABLE D·4 A SAMPLE DIET PLAN

Exchange	Breakfast	Lunch	Snack	Dinner	Evening Snack
9 starch	2	2	1	3	1
4 vegetable				4	
3 fruit	1	1	1		
6 lean meat		2		4	
2 fat-free milk	1	1			1
5 fat		1		4	

NOTE: This diet plan is one of many possibilities. It meets dietary recommendations to provide 45 to 65 percent of its calories from carbohydrate, 10 to 35 percent from protein, and less than 35 percent from fat.

D | TABLE D·5 U.S. EXCHANGE SYSTEM: STARCH LIST

1 starch exchange = 15 g carbohydrate, 3 g protein, 0–1 g fat, and 80 cal

NOTE: In general, one starch exchange is ½ c cooked cereal, grain, or starchy vegetable; ⅓ c cooked rice or pasta; 1 oz of bread; ¾ to 1 oz snack food.

Serving Size	Food
Bread	
¼ (1 oz)	Bagel, 4 oz
2 slices (1½ oz)	Bread, reduced-calorie
1 slice (1 oz)	Bread, white (including French and Italian), whole-wheat, pumpernickel, rye
4 (⅔ oz)	Bread sticks, crisp, 4″ × ½″
½	English muffin
½ (1 oz)	Hot dog or hamburger bun
¼	Naan, 8″ × 2″
1	Pancake, 4″ across, ¼″ thick
½	Pita, 6″ across
1 (1 oz)	Plain roll, small
1 slice (1 oz)	Raisin bread, unfrosted
1	Tortilla, corn, 6″ across
1	Tortilla, flour, 6″ across
⅓	Tortilla, flour, 10″ across
1	Waffle, 4″ square or across, reduced-fat
Cereals and Grains	
½ c	Bran cereals
½ c	Bulgur, cooked
½ c	Cereals, cooked
¾ c	Cereals, unsweetened, ready-to-eat
3 tbs	Cornmeal (dry)
⅓ c	Couscous
3 tbs	Flour (dry)
¼ c	Granola, low-fat
¼ c	Grape nuts
½ c	Grits, cooked
½ c	Kasha
⅓ c	Millet
¼ c	Muesli
½ c	Oats
⅓ c	Pasta, cooked
1½ c	Puffed cereals
⅓ c	Rice, white or brown, cooked
½ c	Shredded wheat
½ c	Sugar-frosted cereal
3 tbs	Wheat germ
Starchy Vegetables	
⅓ c	Baked beans
½ c	Corn
½ cob (5 oz)	Corn on cob, large
1 c	Mixed vegetables with corn, peas, or pasta

Serving Size	Food
½ c	Peas, green
½ c	Plantains
½ medium (3 oz) or ½ c	Potato, boiled
¼ large (3 oz)	Potato, baked with skin
½ c	Potatoes, mashed
1 c	Squash, winter (acorn, butternut, pumpkin)
½ c	Yams, sweet potatoes, plain
Crackers and Snacks	
8	Animal crackers
3	Graham crackers, 2½″ square
¾ oz	Matzoh
4 slices	Melba toast
24	Oyster crackers
3 c	Popcorn (popped, no fat added or low-fat microwave)
¾ oz	Pretzels
2	Rice cakes, 4″ across
6	Saltine-type crackers
15–20 (¾ oz)	Snack chips, fat-free or baked (tortilla, potato)
2–5 (¾ oz)	Whole-wheat crackers, no fat added
Beans, Peas, and Lentils (count as 1 starch + 1 very lean meat)	
½ c	Beans and peas, cooked (garbanzo, lentils, pinto, kidney, white, split, black-eyed)
⅔ c	Lima beans
3 tbs	Miso
Starchy Foods Prepared with Fat (count as 1 starch + 1 fat)	
1	Biscuit, 2½″ across
½ c	Chow mein noodles
1 (2 oz)	Cornbread, 2″ cube
6	Crackers, round butter type
1 c	Croutons
1 c (2 oz)	French-fried potatoes (oven baked)
¼ c	Granola
⅓ c	Hummus
⅙ (1 oz)	Muffin, 5 oz
3 c	Popcorn, microwave
3	Sandwich crackers, cheese or peanut butter filling
9–13 (¾ oz)	Snack chips (potato, tortilla)
⅓ c	Stuffing, bread (prepared)
2	Taco shells, 6″ across
1	Waffle, 4½″ square or across
4–6 (1 oz)	Whole-wheat crackers, fat added

 = 400 mg or more of sodium per serving.

D

1 fruit exchange = 15 g carbohydrate and 60 cal

NOTE: In general, one fruit exchange is 1 small fresh fruit; ½ c canned or fresh fruit or unsweetened fruit juice; ¼ c dried fruit.

Serving Size	Food	Serving Size	Food
1 (4 oz)	Apple, unpeeled, small	½ (8 oz) or 1 c cubes	Papaya
½ c	Applesauce, unsweetened	1 (4 oz)	Peach, medium, fresh
4 rings	Apples, dried	½ c	Peaches, canned
4 whole (5½ oz)	Apricots, fresh	½ (4 oz)	Pear, large, fresh
8 halves	Apricots, dried	½ c	Pears, canned
½ c	Apricots, canned	¾ c	Pineapple, fresh
1 (4 oz)	Banana, small	½ c	Pineapple, canned
¾ c	Blackberries	2 (5 oz)	Plums, small
¾ c	Blueberries	½ c	Plums, canned
⅓ melon (11 oz) or 1 c cubes	Cantaloupe, small	3	Plums, dried (prunes)
12 (3 oz)	Cherries, sweet, fresh	2 tbs	Raisins
½ c	Cherries, sweet, canned	1 c	Raspberries
3	Dates	1¼ c whole berries	Strawberries
1½ large or 2 medium (3½ oz)	Figs, fresh	2 (8 oz)	Tangerines, small
1½	Figs, dried	1 slice (13½ oz) or 1¼ c cubes	Watermelon
½ c	Fruit cocktail		
½ (11 oz)	Grapefruit, large	**Fruit Juice, unsweetened**	
¾ c	Grapefruit sections, canned	½ c	Apple juice/cider
17 (3 oz)	Grapes, small	⅓ c	Cranberry juice cocktail
1 slice (10 oz) or 1 c cubes	Honeydew melon	1 c	Cranberry juice cocktail, reduced-calorie
1 (3½ oz)	Kiwi	⅓ c	Fruit juice blends, 100% juice
¾ c	Mandarin oranges, canned	⅓ c	Grape juice
½ (5½ oz) or ½ c	Mango, small	½ c	Grapefruit juice
1 (5 oz)	Nectarine, small	½ c	Orange juice
1 (6½ oz)	Orange, small	½ c	Pineapple juice
		⅓ c	Prune juice

TABLE D·7 U.S. EXCHANGE SYSTEM: MILK LIST

NOTE: In general, one milk exchange is 1 c milk or yogurt.

Serving Size	Food
Fat-Free and Low-Fat Milk	
1 fat-free/low-fat milk exchange = 12 g carbohydrate, 8 g protein, 0–3 g fat, 90 cal	
1 c	Fat-free milk
1 c	½% milk
1 c	1% milk
1 c	Fat-free or low-fat buttermilk
½ c	Evaporated fat-free milk
⅓ c dry	Fat-free dry milk
1 c	Soy milk, low-fat or fat-free
⅔ c (6 oz)	Yogurt, fat-free or low-fat, flavored, sweetened with nonnutritive sweetener and fructose
⅔ c (6 oz)	Yogurt, plain fat-free

Serving Size	Food
Reduced-Fat Milk	
1 reduced-fat milk exchange = 12 g carbohydrate, 8 g protein, 5 g fat, 120 cal	
1 c	2% milk
1 c	Soy milk
1 c	Sweet acidophilus milk
¾ c	Yogurt, plain low-fat
Whole Milk	
1 whole milk exchange = 12 g carbohydrate, 8 g protein, 8 g fat, 150 cal	
1 c	Whole milk
½ c	Evaporated whole milk
1 c	Goat's milk
1 c	Kefir
¾ c	Yogurt, plain (made from whole milk)

1 other carbohydrate exchange = 15 g carbohydrate, or 1 starch, or 1 fruit, or 1 milk exchange

Food	Serving Size	Exchanges per Serving
Angel food cake, unfrosted	¹⁄₁₂ cake (2 oz)	2 carbohydrates
Brownies, small, unfrosted	2″ square (1 oz)	1 carbohydrate, 1 fat
Cake, unfrosted	2″ square (1 oz)	1 carbohydrate, 1 fat
Cake, frosted	2″ square (2 oz)	2 carbohydrates, 1 fat
Cookies or sandwich cookies with creme filling	2 small (⅔ oz)	1 carbohydrate, 1 fat
Cookies, sugar-free	3 small or 1 large (¾–1 oz)	1 carbohydrate, 1–2 fats
Cranberry sauce, jellied	¼ c	1½ carbohydrates
Cupcake, frosted	1 small (2 oz)	2 carbohydrates, 1 fat
Doughnut, plain cake	1 medium (1½ oz)	1½ carbohydrates, 2 fats
Doughnut, glazed	3¾″ across (2 oz)	2 carbohydrates, 2 fats
Energy, sport, or breakfast bar	1 bar (1⅓ oz)	1½ carbohydrates, 0–1 fat
Energy, sport, or breakfast bar	1 bar (2 oz)	2 carbohydrates, 1 fat
Fruit cobbler	½ c (3½ oz)	3 carbohydrates, 1 fat
Fruit juice bar, frozen, 100% juice	1 bar (3 oz)	1 carbohydrate
Fruit snacks, chewy (pureed fruit concentrate)	1 roll (¾ oz)	1 carbohydrate
Fruit spreads, 100% fruit	1½ tbs	1 carbohydrate
Gelatin, regular	½ c	1 carbohydrate
Gingersnaps	3	1 carbohydrate
Granola or snack bar, regular or low-fat	1 bar (1 oz)	1½ carbohydrates
Honey	1 tbs	1 carbohydrate
Ice cream	½ c	1 carbohydrate, 2 fats
Ice cream, light	½ c	1 carbohydrate, 1 fat
Ice cream, low-fat	½ c	1½ carbohydrates
Ice cream, fat-free, no sugar added	½ c	1 carbohydrate
Jam or jelly, regular	1 tbs	1 carbohydrate
Milk, chocolate, whole	1 c	2 carbohydrates, 1 fat
Pie, fruit, 2 crusts	⅙ of 8″ commercially prepared pie	3 carbohydrates, 2 fats
Pie, pumpkin or custard	⅛ of 8″ commercially prepared pie	2 carbohydrates, 2 fats
Pudding, regular (made with reduced-fat milk)	½ c	2 carbohydrates
Pudding, sugar-free (made with fat-free milk)	½ c	1 carbohydrate
Reduced-calorie meal replacement (shake)	1 can (10–11 oz)	1½ carbohydrates, 0–1 fats
Rice milk, low-fat or fat-free, plain	1 c	1 carbohydrate
Rice milk, low-fat, flavored	1 c	1½ carbohydrates
Salad dressing, fat-free	¼ c	1 carbohydrate
Sherbet, sorbet	½ c	2 carbohydrates
Spaghetti or pasta sauce, canned	½ c	1 carbohydrate, 1 fat
Sports drinks	8 oz (1 c)	1 carbohydrate
Sugar	1 tbs	1 carbohydrate
Sweet roll or danish	1 (2½ oz)	2½ carbohydrates, 2 fats
Syrup, light	2 tbs	1 carbohydrate
Syrup, regular	1 tbs	1 carbohydrate
Syrup, regular	¼ c	4 carbohydrates
Vanilla wafers	5	1 carbohydrate, 1 fat
Yogurt, frozen	½ c	1 carbohydrate, 0–1 fat
Yogurt, frozen, fat-free	⅓ c	1 carbohydrate
Yogurt, low-fat with fruit	1 c	3 carbohydrates, 0–1 fat

= 400 mg or more sodium per exchange.

TABLE D·9 U.S. EXCHANGE SYSTEM: NONSTARCHY VEGETABLE LIST

1 vegetable exchange = 5 g carbohydrate, 2 g protein, 0 g fat, and 25 cal
NOTE: In general, one vegetable exchange is ½ c cooked vegetables or vegetable juice; 1 c raw vegetables. Starchy vegetables such as corn, peas, and potatoes are on the starch list (Table D-5).

Artichokes
Artichoke hearts
Asparagus
Beans (green, wax, Italian)
Bean sprouts
Beets
Broccoli
Brussels sprouts
Cabbage
Carrots
Cauliflower
Celery
Cucumbers
Eggplant
Green onions or scallions
Greens (collard, kale, mustard, turnip)
Kohlrabi
Leeks
Mixed vegetables (without corn, peas, or pasta)

Mushrooms
Okra
Onions
Pea pods
Peppers (all varieties)
Radishes
Salad greens (endive, escarole, lettuce, romaine, spinach)
Sauerkraut 🖋
Spinach
Summer squash (crookneck)
Tomatoes
Tomatoes, canned
Tomato sauce 🖋
Tomato/vegetable juice 🖋
Turnips
Water chestnuts
Watercress
Zucchini

🖋 = 400 mg or more sodium per exchange.

NOTE: In general, a meat exchange is 1 oz meat, poultry, or cheese; ½ c dried beans (weigh meat and poultry and measure beans after cooking).

Serving Size	Food

Very Lean Meat and Substitutes

1 very lean meat exchange = 7 g protein, 0–1 g fat, 35 cal

1 oz	Poultry: Chicken or turkey (white meat, no skin), Cornish hen (no skin)
1 oz	Fish: Fresh or frozen cod, flounder, haddock, halibut, trout, lox (smoked salmon) ; tuna, fresh or canned in water
1 oz	Shellfish: Clams, crab, lobster, scallops, shrimp, imitation shellfish
1 oz	Game: Duck or pheasant (no skin), venison, buffalo, ostrich
	Cheese with ≤1g fat/oz:
¼ c	Fat-free or low-fat cottage cheese
1 oz	Fat-free cheese
1 oz	Processed sandwich meats with ≤1 g fat/oz (such as deli thin, shaved meats, chipped beef , turkey ham)
2	Egg whites
¼ c	Egg substitutes, plain
1 oz	Hot dogs with ≤1 g fat/oz
1 oz	Kidney (high in cholesterol)
1 oz	Sausage with ≤1 g fat/oz

Count as 1 very lean meat + 1 starch exchange:

½ c	Beans, peas, lentils (cooked)

Lean Meat and Substitutes

1 lean meat exchange = 7 g protein, 3 g fat, 55 cal

1 oz	Beef: USDA Select or Choice grades of lean beef trimmed of fat (round, sirloin, and flank steak); tenderloin; roast (rib, chuck, rump); steak (T-bone, porterhouse, cubed), ground round
1 oz	Pork: Lean pork (fresh ham); canned, cured, or boiled ham; Canadian bacon ; tenderloin, center loin chop
1 oz	Lamb: Roast, chop, leg
1 oz	Veal: Lean chop, roast
1 oz	Poultry: Chicken, turkey (dark meat, no skin), chicken (white meat, with skin), domestic duck or goose (well drained of fat, no skin)
	Fish:
1 oz	Herring (uncreamed or smoked)
6 medium	Oysters
1 oz	Salmon (fresh or canned), catfish
2 medium	Sardines (canned)
1 oz	Tuna (canned in oil, drained)
1 oz	Game: Goose (no skin), rabbit
	Cheese:
¼ c	4.5%-fat cottage cheese

Serving Size	Food
2 tbs	Grated parmesan
1 oz	Cheeses with ≤3 g fat/oz
1½ oz	Hot dogs with ≤3 g fat/oz
1 oz	Processed sandwich meat with ≤3 g fat/oz (turkey pastrami or kielbasa)
1 oz	Liver, heart (high in cholesterol)

Medium-Fat Meat and Substitutes

1 medium-fat meat exchange = 7 g protein, 5 g fat, and 75 cal

1 oz	Beef: Most beef products (ground beef, meatloaf, corned beef, short ribs, Prime grades of meat trimmed of fat, such as prime rib)
1 oz	Pork: Top loin, chop, Boston butt, cutlet
1 oz	Lamb: Rib roast, ground
1 oz	Veal: Cutlet (ground or cubed, unbreaded)
1 oz	Poultry: Chicken (dark meat, with skin), ground turkey or ground chicken, fried chicken (with skin)
1 oz	Fish: Any fried fish product
	Cheese with ≤5 g fat/oz:
1 oz	Feta
1 oz	Mozzarella
¼ c (2 oz)	Ricotta
1	Egg (high in cholesterol, limit to 3/week)
1 oz	Sausage with ≤5 g fat/oz
1 c	Soy milk
¼ c	Tempeh
4 oz or ½ c	Tofu

High-Fat Meat and Substitutes

1 high-fat meat exchange = 7 g protein, 8 g fat, 100 cal

1 oz	Pork: Spareribs, ground pork, pork sausage
1 oz	Cheese: All regular cheeses (American , cheddar, Monterey Jack, swiss)
1 oz	Processed sandwich meats with ≤8 g fat/oz (bologna, pimento loaf, salami)
1 oz	Sausage (bratwurst, Italian, knockwurst, Polish, smoked)
1 (10/lb)	Hot dog (turkey or chicken)
3 slices (20 slices/lb)	Bacon
1 tbs	Peanut butter (contains unsaturated fat)

Count as 1 high-fat meat + 1 fat exchange:

1 (10/lb)	Hot dog (beef, pork, or combination)

= 400 mg or more of sodium per serving.

D

TABLE D·11   U.S. EXCHANGE SYSTEM: FAT LIST

1 fat exchange = 5 g fat and 45 cal
NOTE: In general, one fat exchange is 1 tsp regular butter, margarine, or vegetable oil; 1 tbs regular salad dressing. Many fat-free and reduced-fat foods are on the Free Foods List (Table D-12).

Serving Size	Food
Monounsaturated Fats	
2 tbs (1 oz)	Avocado
1 tsp	Oil (canola, olive, peanut)
8 large	Olives, ripe (black)
10 large	Olives, green, stuffed 🖊
6 nuts	Almonds, cashews
6 nuts	Mixed nuts (50% peanuts)
10 nuts	Peanuts
4 halves	Pecans
½ tbs	Peanut butter, smooth or crunchy
1 tbs	Sesame seeds
2 tsp	Tahini or sesame paste
Polyunsaturated Fats	
4 halves	English walnuts
1 tsp	Margarine, stick, tub, or squeeze
1 tbs	Margarine, lower-fat spread (30% to 50% vegetable oil)
1 tsp	Mayonnaise, regular
1 tbs	Mayonnaise, reduced-fat
1 tsp	Oil (corn, safflower, soybean)
1 tbs	Salad dressing, regular 🖊
2 tbs	Salad dressing, reduced-fat
2 tsp	Mayonnaise type salad dressing, regular
1 tbs	Mayonnaise type salad dressing, reduced-fat
1 tbs	Seeds (pumpkin, sunflower)
Saturated Fats*	
1 slice (20 slices/lb)	Bacon, cooked
1 tsp	Bacon, grease
1 tsp	Butter, stick
2 tsp	Butter, whipped
1 tbs	Butter, reduced-fat
2 tbs (½ oz)	Chitterlings, boiled
2 tbs	Coconut, sweetened, shredded
1 tbs	Coconut milk
2 tbs	Cream, half and half
1 tbs (½ oz)	Cream cheese, regular
1½ tbs (¾ oz)	Cream cheese, reduced-fat
	Fatback or salt pork† 🖊
1 tsp	Shortening or lard
2 tbs	Sour cream, regular
3 tbs	Sour cream, reduced-fat

🖊 = 400 mg or more sodium per exchange
*Saturated fats can raise blood cholesterol levels.
†Use a piece 1″ × 1″ × ¼″ if you plan to eat the fatback cooked with vegetables. Use a piece 2″ × 1″ × ½″ when eating only the vegetables with the fatback removed.

NOTE: A serving of free food contains less than 20 calories or no more than 5 grams of carbohydrate; those with serving sizes should be limited to 3 servings a day whereas those without serving sizes can be eaten freely.

Serving Size	Food
Fat-Free or Reduced-Fat Foods	
1 tbs (½ oz)	Cream cheese, fat-free
1 tbs	Creamers, nondairy, liquid
2 tsp	Creamers, nondairy, powdered
4 tbs	Margarine spread, fat-free
1 tsp	Margarine spread, reduced-fat
1 tbs	Mayonnaise, fat-free
1 tsp	Mayonnaise, reduced-fat
1 tbs	Mayonnaise type salad dressing, fat-free
1 tsp	Mayonnaise type salad dressing, reduced-fat
	Nonstick cooking spray
1 tbs	Salad dressing, fat-free or low-fat
2 tbs	Salad dressing, fat-free, Italian
1 tbs	Sour cream, fat-free, reduced-fat
1 tbs	Whipped topping, regular
2 tbs	Whipped topping, light or fat-free
Sugar-Free Foods	
1 piece	Candy, hard, sugar-free
	Gelatin dessert, sugar-free
	Gelatin, unflavored
	Gum, sugar-free
2 tsp	Jam or jelly, light
	Sugar substitutes
2 tbs	Syrup, sugar-free
Drinks	
	Bouillon, broth, consommé
	Bouillon or broth, low-sodium
	Carbonated or mineral water
	Club soda
1 tbs	Cocoa powder, unsweetened

Serving Size	Food
	Coffee
	Diet soft drinks, sugar-free
	Drink mixes, sugar-free
	Tea
	Tonic water, sugar-free
Condiments	
1 tbs	Horseradish
	Ketchup
	Lemon juice
	Lime juice
	Mustard
1 tbs	Pickle relish
1½ medium	Pickles, dill
2 slices	Pickles, sweet (bread and butter)
¾ oz	Pickles, sweet (gherkin)
¼ c	Salsa
1 tbs	Soy sauce, regular or light
1 tbs	Taco sauce
	Vinegar
2 tbs	Yogurt
Seasonings	
	Flavoring extracts
	Garlic
	Herbs, fresh or dried
	Hot pepper sauces
	Pimento
	Spices
	Wine, used in cooking
	Worcestershire sauce

= 400 mg or more of sodium per serving.

TABLE D·13 U.S. EXCHANGE SYSTEM: COMBINATION FOODS LIST

Food	Serving Size	Exchanges per Serving
Entrées		
Tuna noodle casserole, lasagna, spaghetti with meatballs, chili with beans, macaroni and cheese ✐	1 c (8 oz)	2 carbohydrates, 2 medium-fat meats
Chow mein (without noodles or rice)	2 c (16 oz)	1 carbohydrate, 2 lean meats
Tuna or chicken salad	½ c (3½ oz)	½ carbohydrate, 2 lean meats, 1 fat
Frozen Entrées and Meals		
Dinner-type meal ✐	Generally 14–17 oz	3 carbohydrates, 3 medium-fat meats, 3 fats
Entrée or meal with <340 cal ✐	About 8–11 oz	2–3 carbohydrates, 1–2 lean meats
Meatless burger, soy based	3 oz	½ carbohydrate, 2 lean meats
Meatless burger, vegetable and starch based	3 oz	1 carbohydrate, 1 lean meat
Pizza, cheese, thin crust ✐	¼ of 120 (6 oz)	2 carbohydrates, 2 medium-fat meats, 1 fat
Pizza, meat topping, thin crust ✐	¼ of 120 (6 oz)	2 carbohydrates, 2 medium-fat meats, 2 fats
Pot pie ✐	1 (7 oz)	2½ carbohydrates, 1 medium-fat meat, 3 fats
Soups		
Bean ✐	1 c	1 carbohydrate, 1 very lean meat
Cream (made with water) ✐	1 c (8 oz)	1 carbohydrate, 1 fat
Instant ✐	6 oz prepared	1 carbohydrate
Instant with beans/lentils ✐	8 oz prepared	2½ carbohydrates, 1 very lean meat
Split pea (made with water) ✐	½ c (4 oz)	1 carbohydrate
Tomato (made with water)	1 c (8 oz)	1 carbohydrate
Vegetable beef, chicken noodle, or other broth-type ✐	1 c (8 oz)	1 carbohydrate
Fast Foods ✐		
Burrito with beef ✐	1 (5–7 oz)	3 carbohydrates, 1 medium-fat meat, 1 fat
Chicken nuggets ✐	6	1 carbohydrate, 2 medium-fat meats, 1 fat
Chicken breast and wing, breaded and fried ✐	1 each	1 carbohydrate, 4 medium-fat meats, 2 fats
Chicken sandwich, grilled ✐	1	2 carbohydrates, 3 very lean meats
Chicken wings, hot	6 (5 oz)	1 carbohydrate, 3 medium-fat meats, 4 fats
Fish sandwich/tartar sauce ✐	1	3 carbohydrates, 1 medium-fat meat, 3 fats
French fries ✐	1 medium serving (5 oz)	4 carbohydrates, 4 fats
Hamburger, regular	1	2 carbohydrates, 2 medium-fat meats
Hamburger, large ✐	1	2 carbohydrates, 3 medium-fat meats, 1 fat
Hot dog with bun ✐	1	1 carbohydrate, 1 high-fat meat, 1 fat
Individual pan pizza ✐	1	5 carbohydrates, 3 medium-fat meats, 3 fats
Pizza, cheese, thin crust ✐	¼ of 12″ (about 6 oz)	2½ carbohydrates, 2 medium-fat meats
Pizza, meat, thin crust ✐	¼ of 12″ (about 6 oz)	2½ carbohydrates, 2 medium-fat meats, 1 fat
Soft serve cone	1 small (5 oz)	2½ carbohydrates, 1 fat
Submarine sandwich ✐	1 sub (6″)	3 carbohydrates, 1 vegetable, 2 medium-fat meats, 1 fat
Submarine sandwich (<6 g fat) ✐	1 sub (6″)	2½ carbohydrates, 2 lean meats
Taco, hard or soft shell	1 (3–3½ oz)	1 carbohydrate, 1 medium-fat meat, 1 fat

✐ = 400 mg or more sodium per exchange.

Nutrition Resources

CONTENTS

Books

Journals

Addresses

People interested in nutrition often want to know where they can find reliable nutrition information. Wherever you live, there are several sources you can turn to:

- The Department of Health may have a nutrition expert.
- The local extension agent is often an expert.
- The food editor of your local paper may be well informed.
- The dietitian at the local hospital had to fulfill a set of qualifications before becoming an RD.
- There may be knowledgeable professors of nutrition or biochemistry at a nearby college or university.

In addition, you may be interested in building a nutrition library of your own. Books you can buy, journals you can subscribe to, and addresses you can contact for general information are given below.

Books

For students seeking to establish a personal library of nutrition references, the authors of this text recommend the following books:

- B. A. Bowman and R. M. Russel, eds., *Present Knowledge in Nutrition,* 8th ed. (Washington, D.C.: International Life Sciences Institute—Nutrition Foundation, 2001).

This 760-page paperback has a chapter on each of 65 topics, including energy, obesity, each of the nutrients, several diseases, malnutrition, growth and its assessment, immunity, alcohol, fiber, exercise, drugs, and toxins. Watch for an update; new editions come out every few years.

- M. E. Shils and coeditors, *Modern Nutrition in Health and Disease,* 9th ed. (Baltimore: Williams & Wilkins, 1999).

This reference book contains encyclopedic articles on the nutrients, foods, diet, metabolism, malnutrition, age-related needs, and nutrition in disease.

- Committee on Dietary Reference Intakes, *Dietary Reference Intakes for Calcium, Phosphorus, Magnesium, Vitamin D, and Fluoride* (Washington, D.C.: National Academy Press, 1997).
- Committee on Dietary Reference Intakes, *Dietary Reference Intakes for Thiamin,*

Riboflavin, Niacin, Vitamin B$_6$, Folate, Vitamin B$_{12}$, Pantothenic Acid, Biotin, and Choline (Washington, D.C.: National Academy Press, 1998).

- Committee on Dietary Reference Intakes, *Dietary Reference Intakes for Vitamin C, Vitamin E, Selenium, and Carotenoids* (Washington, D.C.: National Academy Press, 2000).
- Committee on Dietary Reference Intakes, *Dietary Reference Intakes for Vitamin A, Vitamin K, Arsenic, Boron, Chromium, Copper, Iodine, Iron, Manganese, Molybdenum, Nickel, Silicon, Vanadium, and Zinc* (Washington, D.C.: National Academy Press, 2001).
- Committee on Dietary Reference Intakes, *Dietary Reference Intakes for Energy, Carbohydrate, Fiber, Fat, Fatty Acids, Cholesterol, Protein, and Amino Acids* (Washington, D.C.: National Academies Press, 2002).
- Committee on Dietary Reference Intakes, *Dietary Reference Intakes for Water, Potassium, Sodium, Chloride, and Sulfate* (Washington, D.C.: National Academies Press, 2004).

These reports review the function of each nutrient, dietary sources, and deficiency and toxicity symptoms as well as provide recommendations for intakes. Watch for additional reports on the Dietary Reference Intakes for the remaining nutrients and other food components.

- Report of a Joint WHO/FAO Expert Consultation, *Diet, Nutrition and the Prevention of Chronic Diseases: Report of a WHO Study Group,* WHO Technical Report Series no. 916 (Geneva, Switzerland: World Health Organization, 2003), available free of charge on the Internet at wwwwh.int/hpr/NPH/docs/who_fao_expert_report.pdf.

This 160-page report presents a world view of the relationship between diet and chronic disease prevention. It provides research evidence supporting the links between dietary patterns and disease development worldwide.

- 2005 Report of the Dietary Guidelines Advisory Committee, available at www.health.gov/dietaryguidelines/dga2005/report.

This lengthy report presents the science base for forming the 2005 *Dietary Guidelines for*

Americans and summarizes current research on many nutrition issues in the United States.

- S. S. Gropper, *The Biochemistry of Human Nutrition: A Desk Reference,* 2nd ed. (Belmont, Calif.: Wadsworth/ Thomson Learning, 2000).

This 263-page paperback presents the biochemical concepts necessary for an understanding of nutrition. It is a handy reference book for those who are learning biochemistry for the first time.

Journals

Nutrition Today is an excellent magazine for the interested layperson. It makes a point of raising controversial issues and providing a forum for conflicting opinions. Six issues per year are published. Order from Lippincott Williams & Wilkins, P.O. Box 1600, Hagerstown, MD 21741.

The *Journal of the American Dietetic Association,* the official publication of the ADA, contains articles of interest to dietitians and nutritionists, news of legislative action on food and nutrition, and a very useful section of abstracts of articles from many other journals of nutrition and related areas. There are 12 issues per year, available from the American Dietetic Association (see "Addresses," later).

Nutrition Reviews does much of the work for the library researcher, compiling recent evidence on current topics and presenting extensive bibliographies. Twelve issues per year are available from International Life Sciences Institute, Subscription Office, P.O. Box 830430, Birmingham, AL 35283.

Nutrition and the M.D. is a monthly newsletter that provides up-to-date, easy-to-read, practical information on nutrition for health-care providers. It is available from Lippincott Williams & Wilkins, P.O. Box 1600, Hagerstown, MD 21741.

Other journals that deserve mention here are *Food Technology, Journal of Nutrition, American Journal of Clinical Nutrition, Nutrition Research,* and *Journal of Nutrition Education. FDA Consumer,* a government publication with many articles of interest to the consumer, is available from the Food and Drug Administration (see "Addresses," below). Many other journals of value are referred to throughout the book.

Nutrition Resources • **E-1**

E Addresses

Many of the organizations listed below will provide publication lists free on request. Government and international agencies and professional nutrition organizations are listed first, followed by organizations in the following areas: aging, alcohol and drug abuse, consumer organizations, fitness, food safety, health and disease, infancy and childhood, pregnancy and lactation, trade and industry organizations, weight control and eating disorders, and world hunger.

U.S. Government

Federal Trade Commission (FTC)
Public Reference Branch
600 Pennsylvania Avenue NW
Washington, DC 20580
(202) 326-2222
www.ftc.gov

Food and Drug Administration (FDA)
Office of Consumer Affairs, HFE 1
Room 16-85
5600 Fishers Lane
Rockville, MD 20857
(301) 443-1726
www.fda.gov

FDA Consumer Information Line
(888) INFO-FDA or (888) 463-6332

FDA Center for Food Safety & Applied Nutrition
5100 Paint Branch Parkway
College Park, MD 20740
(888) SAF FOOD or (888) 723-3366
www.cfsan.fda.gov

Food and Nutrition Information Center, National Agricultural Library
10301 Baltimore Avenue
Beltsville, MD 20705-2351
(301) 504-5719; fax (301) 504-6409
www.nal.usda.gov/fnic

Superintendent of Documents
U.S. Government Printing Office
732 North Capitol Street NW
Washington, DC 20401
(888) 293-6498
www.gpo.access.gov

U.S. Department of Agriculture (USDA)
14th Street and Independence Avenue SW
Washington, DC 20250
(202) 720-5964
www.fns.usda.gov/fncs

USDA Center for Nutrition Policy and Promotion (Dietary Guidelines and Food Guide Pyramid)
3101 Park Center Drive, Room 1034
Alexandria, VA 22302-1594
(703) 305-7600 or (800) 687-2258
fax (703) 305-3400
www.cnpp.usda.gov

USDA Food Safety and Inspection Service
Food Safety Education Office

1400 Independence Avenue SW,
Room 2932-S
Washington, DC 20250
(800) 535-4555
www.usda.gov/fsis

U.S. Department of Education (DOE)
Accreditation and State Liaison
Accrediting Agency Evaluation
400 Maryland Avenue SW
Washington, DC 20202
(800) USA-LEARN or (800) 872-5327
fax (202) 401-0689
www.ed. gov/offices/OPE/accreditation

U.S. Department of Health and Human Services
200 Independence Avenue SW
Washington, DC 20201
(877) 696-6775
www.os.dhhs.gov

U.S. Environmental Protection Agency (EPA)
1200 Pennsylvania Avenue NW
Washington, DC 20460
(202) 272-0167
www.epa.gov

U.S. National Library of Medicine
8600 Rockville Pike
Bethesda, MD 20894
www.nlm.nih.gov

Assistant Secretary of Health
Office of Public Health and Science
Department of Health and Human Sciences
200 Independence Avenue SW, Room 725-H
Washington, DC 20201
(202) 690-7694

Canadian Government
Federal

Bureau of Nutritional Sciences Food Directorate
Health Protection Branch
3-West
Sir Frederick Banting Research Centre, AL0904A
Tunney's Pasture
Ottawa, ON K1A 0K9
(613) 957-2991; fax (613) 941-5366
www.hc-sc.gc.ca

Canadian Food Inspection Agency
Agriculture and Agri-Food Canada
930 Carling Avenue
Ottawa, ON K1A 0C5
(613) 225-CFIA or (613) 225-2342
fax (613) 228-6653
www.agr.ca

Office of Nutrition Policy and Promotion
Health Products and Food Branch
AL 3303D
Tower A, Qualicum Towers
2936 Baseline Road, 3rd Floor
Ottawa, ON K1A 0K9

(613) 957-8329
www.hc-sc.gc.ca/hpfb-dgpsa/onpp-bppn/index_e.html

Provincial and Territorial

Health and Wellness
323 Legislature Building 10800
97 Avenue
Edmonton, AB T5K 2B6
(780) 427-3665 ; fax (780) 415-0961

Prevention and Health Promotion Strategies,
Ministry of Health, B-1
1520 Blanshard Street
Victoria, BC V8W 3C8
(800) 465-4911

Chief Medical Officer of Health
Health Programs
4th Floor, 300 Carlton Street
Winnipeg, MB R3B 3M9
(204) 788-6666; fax (204) 948-2204

Project Manager
Public Health Management Services
Health and Community Services
Carlton Place, Floor 2
P.O. Box 5100
Fredericton, NB E3B 5G8

Centre for Health Information
Newfoundland and Labrador
1st Floor, Crosbie Building
1 Crosbie Place
St. John's, NL A1B 3Y8
(709) 757-2400; fax (709) 757-2411

Consultant, Nutrition
Health & Wellness Promotion
Population Health
Department of Health and Social Services
Government of the Northwest Territories
Centre Square Tower, 6th Floor
P.O. Box 1320
Yellowknife, NT X1A 2L9

Office of Health Promotion
10th Floor, Joe Howe Building
1690 Hollis Street
P.O. Box 488
Halifax, NS B3J 2R8
(902) 424-4807; fax (902) 424-4716

Ministry of Health and Long-Term Care
Suite M1-57, MacDonald Block
900 Bay Street
Toronto, ON M7A 1N3

Responsables de la santé cardiovasculaire et de la nutrition
Ministère de la Santé et des Services sociaux, Service de la Prévention en Santé
5e étage
1075, chemin Sainte-Foy
Québec (Québec) G1S 2M1
(418) 266-7005; fax (418) 266-7024

Health Promotion Unit
Population Health Branch
Saskatchewan Health

T.C. Douglas Building
3475 Albert Street
Regina, SK S4S 6X6
(306) 787-8847; fax (306) 787-3112

Health Promotion
Department of Health and Social Services
Government of Yukon
P.O. Box 2703
Whitehorse, Yukon Y1A 2C6
(867) 667-8394; fax (867) 667-8338

International Agencies

Food and Agriculture Organization of
the United Nations (FAO)
Liaison Office for North America
2175 K Street, Suite 300
Washington, DC 20437
(202) 653-2400; fax (202) 653-5760
www.fao.org

International Food Information Council
Foundation
1100 Connecticut Avenue NW, Suite 430
Washington, DC 20036
(202) 296-6540
www.ific.org

UNICEF
3 United Nations Plaza
New York, NY 10017
(212) 326-7000
www.unicef.org

World Health Organization (WHO)
Regional Office
525 23rd Street NW
Washington, DC 20037
(202) 974-3000
www.who.org

Professional Organizations

American Dietetic Association (ADA)
120 South Riverside Plaza, Suite 2000
Chicago, IL 60606-6995
(800) 877-1600; (312) 899-0040
www.eatright.org

ADA, Consumer Nutrition Hotline
(800) 366-1655

American Society for Clinical Nutrition
9650 Rockville Pike
Bethesda, MD 20814-3998
(301) 530-7110; fax (301) 571-1863
www.faseb.org/ascn

American Society of Nutritional Sciences
9650 Rockville Pike
Bethesda, MD 20814
(301) 634-7050; fax (301) 634-7892
www.asns.org

Dietitians of Canada
480 University Avenue, Suite 604
Toronto, ON M5G 1V2
(416) 596-08571 fax (416) 596-0603
www.dietitians.ca

International Life Sciences Institute
One Thomas Circle, 9th Floor

Washington, DC 20005
(202) 659-0074; fax (202) 659-3859
www.ilsi.org

National Academy of Sciences/National
Research Council (NAS/NRC)
500 Fifth Street NW
Washington, DC 20001
(202) 334-2000
www.nas.edu

National Institute of Nutrition
408 Queen Street, 3rd Floor
Ottawa, ON K1R 5A7
(613) 235-3355; fax (613) 235-7032
www.nin.ca

Society for Nutrition Education
9202 N. Meridian, Suite 200
Indianapolis, IN 46260
(800) 235-6690; fax (317) 571-5603
www.sne.org

Aging

Administration on Aging
330 Independence Avenue SW
Washington, DC 20201
(202) 619-0724
www.aoa.dhhs.gov

American Association of Retired Persons
(AARP)
601 E Street NW
Washington, DC 20049
(800) 424-3410
www.aarp.org

Center for Communication and
Consumer Services
1 Massachussetts Avenue NW
Washington, DC 20001
(202) 619-7501; fax (202) 401-7620
www.aoa.dhhs.gov/naic

National Institute on Aging
Public Information Office
31 Center Drive, MSC 2292
Bethesda, MD 20892
(800) 222-2225
www.nih.gov/nia

Alcohol and Drug Abuse

Al-Anon Family Groups, Inc.
1600 Corporate Landing Parkway
Virginia Beach, VA 23454-5617
(888) 4AL-ANON or (888) 425-2666
(757) 563-1600; fax (757) 563-1655
www.al-anon.alateen.org

Alcohol & Drug Abuse Information Line
Adcare Hospital
(800) 252-6465

Alcoholics Anonymous (AA)
Grand Central Station
P.O. Box 459
New York, NY 10163
(212) 870-3400
www.alcoholics-anonymous.org

Narcotics Anonymous (NA)
P.O. Box 9999

Van Nuys, CA 91409
(818) 773-9999; fax (818) 700-0700
www.wsoinc.com

National Clearinghouse for Alcohol
and Drug Information (NCADI)
P.O. Box 2345
Rockville, MD 20847-2345
(800) 729-6686
www.health.org

National Council on Alcoholism and
Drug Dependence (NCADD)
20 Exchange Place, Suite 2902
New York, NY 10005
(800) NCA-CALL or (800) 622-2255
(212) 269-7797; fax (212) 269-7510
www.ncadd.org

U.S. Center for Substance Abuse
Prevention
1010 Wayne Avenue, Suite 850
Silver Spring, MD 20910
(301) 459-1591 ext. 244
fax (301) 495-2919
www.covesoft.com/csap.html

Consumer Organizations

Center for Science in the Public Interest
(CSPI)
1875 Connecticut Avenue NW, Suite 300
Washington, DC 20009-5728
(202) 332-9110; fax (202) 265-4954
www.cspinet.org

Federal Citizen Information Center
Pueblo, CO 81009
(888) 8-PUEBLO or (888) 878-3256
(800) FED-INFO or (800) 333-4636
www.pueblo.gsa.gov

Consumers Union of US, Inc.
101 Truman Avenue
Yonkers, NY 10703-1057
(914) 378-2000
www.consumersunion.org

National Council Against Health Fraud,
Inc. (NCAHF)
119 Foster Street
Peabody, MA 01960
(978) 532-9383
www.ncahf.org

Partnership for Caring
1620 Eye Street NW, Suite 202
Washington, DC 20006
(800) 989-WILL or (800) 989-9455
(202) 296-8071; fax (202) 296-8352
www.partnershipforcaring.org

Fitness

American College of Sports Medicine
401 West Michigan Street
Indianapolis, IN 46206-3233
(317) 637-9200; fax (317) 634-7817
www.acsm.org

American Council on Exercise (ACE)
4851 Paramount Drive
San Diego, CA 92123

(800) 825-3636; fax (858) 279-8064
www.acefitness.org

Shape Up America!
15009 Native Dancer Road
North Potomac, MD 20878
(240) 631-6533; fax (240) 632-1075
www.shapeup.org

Food Safety

Alliance for Food & Fiber
Food Safety Hotline
10866 Wilshire Blvd., Suite 550
Los Angeles, CA 90024
(310) 446-1827 or (800) 262-4189
fax (310) 446-1896

National Lead Information Center
(800) LEAD-FYI or (800) 532-3394
(800) 424-LEAD or (800) 424-5323

National Pesticide Telecommunications
Network (NPTN)
Oregon State University
333 Weniger Hall
Corvallis, OR 97331-6502
(800) 858-7378
www.ace.orst.edu/info/nptn

USDA Meat and Poultry Hotline
(800) 535-4555

U.S. EPA Safe Drinking Water Hotline
(800) 426-4791

Health and Disease

Alzheimer's Disease Education and
Referral Center
P.O. Box 8250
Silver Spring, MD 20907-8250
(800) 438-4380
www.alzheimers.org

Alzheimer's Disease Information and
Referral Service
225 North Michigan Avenue, Suite 1700
Chicago, IL 60611-7633
(800) 272-3900
www.alz.org

American Academy of Allergy, Asthma,
and Immunology
611 East Wells Street
Milwaukee, WI 53202
(414) 272-6071; fax (414) 272-6070
www.aaaai.org

American Cancer Society
National Home Office
1599 Clifton Road NE
Atlanta, GA 30329-4251
(800) ACS-2345 or (800) 227-2345
www.cancer.org

American Council on Science and Health
1995 Broadway, 2nd Floor
New York, NY 10023-5860
(212) 362-7044; fax (212) 362-4919
www.acsh.org

American Dental Association
211 East Chicago Avenue

Chicago, IL 60611
(312) 440-2500; fax (312) 440-2800
www.ada.org

American Diabetes Association
1701 North Beauregard Street
Alexandria, VA 22311
(800) 232-3472 or (703) 549-1500
(800) DIABETES or (800) 342-2383
www.diabetes.org

American Heart Association
Box BHG, National Center
7272 Greenville Avenue
Dallas, TX 75231
(800) 242-8721
www.americanheart.org

American Institute for Cancer Research
1759 R Street NW
Washington, DC 20009
(800) 843-8114 or (202) 328-7744
fax (202) 328-7226
www.aicr.org

American Medical Association
515 North State Street
Chicago, IL 60610
(312) 464-5000
www.ama-assn.org

American Public Health Association
(APHA)
801 I Street NW
Washington, DC 20001-3710
(202) 777-2742; fax (202) 777-2534
www.apha.org

American Red Cross
2025 East Street NW
Washington, DC 20006
(202) 303-4498
www.redcross.org

Arthritis Foundation
P.O. Box 7669
Atlanta, GA 30357
(800) 283-7800
www.arthritis.org

Canadian Diabetes Association
15 Toronto Street, Suite 800
Toronto, ON M5C 2E3
(800) BANTING or (800) 226-8464
(416) 363-3393
www.diabetes.ca

Canadian Public Health Association
1565 Carling Avenue, Suite 400
Ottawa, ON K1Z 8R1
(613) 725-3769; fax (613) 725-9826
www.cpha.ca

Centers for Disease Control and
Prevention(CDC)
1600 Clifton Road NE
Atlanta, GA 30333
(404) 639-3311
www.cdc.gov

The Food Allergy Network
11781 Lee Jackson Highway, Suite 160
Fairfax, VA 22030-3309

(800) 929-4040; fax (703) 691-2713
www.foodallergy.org

Internet Health Resources
www.ihr.com

Mayo Clinic Health Oasis
www.mayohealth.org

National AIDS Hotline (CDC)
(800) 342-AIDS (English)
(800) 344-SIDA (Spanish)
(800) 2437-TTY (Deaf)
(900) 820-2437

National Cancer Institute
Suite 3036A
6116 Executive Boulevard, MSC 8322
Bethesda, MD 20892
(800) 4-CANCER or (800) 422-6237
www.nci.nih.gov

National Health Information Center
(NHIC)
P.O. Box 1133
Washington, DC 20013
(800) 336-4797
www.health.gov/nhic

National Heart, Lung, and Blood
Institute Information Center
P.O. Box 30105
Bethesda, MD 20824-0105
(301) 592-8573
www.nhlbi.nih.gov

National Institute of Allergy and
Infectious Diseases
Office of Communications
31 Center Drive, MSC 2520
Bethesda, MD 20892-2520
(301) 496-5717
www.niaid.nih.gov

National Institute of Dental Research
(NIDR)
45 Center Drive, MSC 6400
Bethesda, MD 20892-2190
(301) 496-4261
www.nidr.nih.gov

National Institute of Diabetes and
Digestive and Kidney Diseases (NIDDK)
31 Center Drive, MSC 2560
Bethesda, MD 20892-2560
(301) 654-3810
www.niddk.nih.gov

National Institutes of Health (NIH)
9000 Rockville Pike
Bethesda, MD 20892
(301) 496-2433
www.nih.gov

National Library of Medicine Health
Hotline List
healthhotlines.nlm.nih.gov/index.html

National Osteoporosis Foundation
1232 22 Street NW
Washington, DC 20037
(202) 223-2226
www.nof.org

Office of Disease Prevention and
Health Promotion

200 Independence Avenue SW,
Room 738G
Washington, DC 20201
(202) 205-8611; fax (202) 205-9478
odphp.osophs.dhhs.gov

Osteoporosis and Related Bone Diseases
2 AMS Circle
Bethesda, MD 20892
(800) 624-BONE or (800) 624-2663
www.osteo.org

Infancy and Childhood

American Academy of Pediatrics
141 Northwest Point Boulevard
Elk Grove Village, IL 60007-1098
(847) 434-4000; fax (847) 434-8000
www.aap.org

Birth Defect Research for Children
930 Woodcock Road, Suite 225
Orlando, FL 32803
(407) 895-0802; fax (407) 895-0824
www.birthdefects.org

Canadian Paediatric Society
100-2204 Walkley Road
Ottawa, ON K1G 4G8
(613) 526-9397; fax (613) 526-3332
www.cps.ca

National Center for Education in Maternal & Child Health
Georgetown University
P.O. Box 571272
Washington, DC 20057
(202) 784-9770
www.ncemch.org

Pregnancy and Lactation

American College of Obstetricians and
Gynecologists Resource Center
409 12th Street SW
Washington, DC 20090
(202) 638-5577
www.acog.org

La Leche International, Inc.
1400 N. Meacham Road
Schaumburg, IL 60173
(847) 519-7730
www.lalecheleague.org

March of Dimes Birth Defects Foundation
1275 Manaroneck Avenue
White Plains, NY 10605
(888) MoDimes or (888) 663-4637
www.modimes.org

Trade Organizations

Beech-Nut Nutrition Corporation
100 South 4th Street
St. Louis, MO 63102
(800) BEECHNUT or (800) 233-2468
www.beechnut.com

Campbell Soup Company
Consumer Response Center
Campbell Place, Box 26B

Camden, NJ 08103-1701
(800) 257-8443
www.campbellssoup.com

Equal
P.O. Box 1280
South Bend, IN 46624
(800) 323-5316
www.equal.com

General Mills, Inc.
1 General Mills Boulevard
Minneapolis, MN 55440
(800) 328-6787
www.generalmills.com

Kellogg Company
P.O. Box CAMB
Battle Creek, MI 49016-3599
(800) 962-1413
www.kelloggs.com

Kraft Foods
Consumer Response and
Information Center
One Kraft Court
Glenview, IL 60025
(800) 323-0768
www.kraftfoods.com

Mead Johnson Nutritionals
2400 West Lloyd Expressway
Evansville, IN 47721
(800) 247-7893
www.meadjohnson.com

National Dairy Council
10255 West Higgins Road, Suite 900
Rosemond, IL 60018-5616
(847) 803-2000
www.dairyinfo.com

Pillsbury Company
1 General Mills Boulevard
Minneapolis, MN 55440
(800) 775-4777
www.pillsbury.com

Procter and Gamble Company
213 N Miami Street
Trenton, OH 45067
(513) 988-6229
www.pg.com

Ross Products, Abbott Laboratories
625 Cleveland Avenue
Columbus, OH 43215-1724
(800) 227-5767
www.ross.com or www.abbott.com

Sunkist Growers
Consumer Affairs
Fresh Fruit Division
14130 Riverside Drive
Sherman Oaks, CA 91423
(818) 986-4800
www.sunkist.com

United Fresh Fruit and Vegetable
Association
1901 Pennsylvania Avenue NW,
Suite 1100
Washington, DC 20006

(202) 303-3400; fax (202) 303-3433
www.uffva.org

USA Rice Federation
4301 North Fairfax Drive, Suite 305
Arlington, VA 22203
(703) 351-8161
www.usarice.com

Weight Control and Eating Disorders

American Obesity Association
1250 24 Street NW, Suite 300
Washington, DC 20037
(800) 98-OBESE or (800) 986-2373
www.obesity.org

National Association of Anorexia
Nervosa and Associated Disorders, Inc.
(ANAD)
P.O. Box 7
Highland Park, IL 60035
(847) 831-3438; fax (847) 433-4632
www.anad.org

National Eating Disorders Association
603 Stewart Street, Suite 803
Seattle, WA 98101
(206) 382-3587
www.nationaleatingdisorders.org

National Eating Disorder
Information Centre
Toronto General Hospital
200 Elizabeth Street, CW 1-211
Toronto, ON M5G 2C4
(416) 340-4156 or (866) 633-4220
fax (416) 340-4736
www.nedic.ca

National Institute of Diabetes and
Digestive Diseases Weight-Control
Information Network (WIN)
31 Center Drive, MSC 2560
Bethesda, MD 20892
(800) WIN-8098
www.niddk.nih.gov/health/nutrit/win.htm

Overeaters Anonymous (OA)
World Service Center
P.O. Box 44020
Rio Rancho, NM 87174
(505) 891-2664;
fax (505) 891-4320
www.overeatersanonymous.org

TOPS (Take Off Pounds Sensibly)
4575 South Fifth Street
P.O. Box 070360
Milwaukee, WI 53207-0360
(800) 932-8677 or (414) 482-4620
www.tops.org

Weight Watchers International, Inc.
Consumer Affairs Department/IN
175 Crossways Park West
Woodbury, NY 11797
(800) 651-6000;
fax (516) 390-1632
www.weightwatchers.com

E World Hunger

Bread for the World
50 F Street NW, Suite 500
Washington, DC 20001
(800) 82-BREAD or (800) 822-7323
(202) 639-9400;
fax (202) 639-9401
www.bread.org

Freedom from Hunger
P.O. Box 2000

1644 DaVinci Court
Davis, CA 95616
(800) 708-2555 or (530) 758-6200
fax (530) 758-6241
www.freefromhunger.org

Oxfam America
26 West Street
Boston, MA 02111-1206
(800) 77-OXFAM or (800) 776-9326
www.oxfamamerica.org

Worldwatch Institute
1776 Massachusetts Avenue NW
Washington, DC 20036
(202) 452-1999
www.worldwatch.org

Notes

F

Chapter 1

1. M. J. Stampfer and coauthors, Primary prevention of coronary heart disease in women through diet and lifestyle, *New England Journal of Medicine* 343 (2000): 16–22; A. K. Kant and coauthors, A prospective study of diet quality and mortality in women, *Journal of the American Medical Association* 283 (2000): 2109–2115.

2. M. Merti, Bridging genomics and genetics, *BioMedNet Conference Reporter,* American Society of Human Genetics 2000, available at http://news.bmn.com/conferences.

3. D. Shattuck, Nutritional genomics, *Journal of the American Dietetic Association* 103 (2003): 16, 18; P. Trayhurn, Nutritional genomics—"Nutrigenomics," *British Journal of Nutrition* 89 (2003): 1–2; F. P. Guengerich, Functional genomics and proteomics applied to the study of nutritional metabolism, *Nutrition Reviews* 59 (1999): 259–263.

4. B. van Ommen and R. Stierum, Nutrigenomics: Exploiting systems biology in the nutrition and health arena, *Current Opinion in Biotechnology* 13 (2002): 517–521.

5. N. Fogg-Johnson and J. Kaput, Nutrigenomics: An emerging scientific discipline, *Food Technology* 57 (2003): 60–67.

6. FDA Public Health Advisory, Reports of blue discoloration and death in patients receiving enteral feedings tinted with the dye, FD&C blue no. 1, September 29, 2003, available at www.cfsan.fda.gov.

7. Position of the American Dietetic Association: Food fortification and dietary supplements, *Journal of the American Dietetic Association* 101 (2001): 115–125.

8. C. D. Johnson and coauthors, Route of nutrition influences generation of antibody-forming cells and initial defense to an active viral infection in the upper respiratory tract, *Annals of Surgery* 237 (2003): 565–573; W. L. Biffl, E. E. Moore, and J. B. Haenel, Nutrition support of the trauma patient, *Nutrition* 18 (2002): 960–965.

9. F. B. Hu, Dietary pattern analysis: A new direction in nutritional epidemiology, *Current Opinion in Lipidology* 13 (2002): 3–9.

10. Standing Committee on the Scientific Evaluation of Dietary Reference Intakes, Food and Nutrition Board, Institute of Medicine, *Dietary Reference Intakes for Energy, Carbohydrate, Fiber, Fat, Fatty Acids, Cholesterol, Protein, and Amino Acids* (Washington, D.C.: National Academy Press, 2002), pp. 11-1–11-88.

11. J. A. Foote and coauthors, Dietary variety increases the probability of nutrient adequacy among adults, *Journal of Nutrition* 134 (2004): 1779–1785.

12. U.S. Department of Agriculture and U.S. Department of Health and Human Services, *Dietary Guidelines for Americans, 6th edition, 2005,* available at www.usda.gov/cnpp.

13. J. E. Tillotson, Our ready-prepared, ready-to-eat nation, *Nutrition Today* 37 (2002): 36-38; F. Katz, "How nutritious?" meets "How convenient?" *Food Technology* 53 (1999): 44–50.

14. A. E. Sloan, What, when, and where Americans eat: 2003, *Food Technology* 57 (2003): 48–66.

15. L. L. Birch, Development of food preferences, *Annual Review of Nutrition* 19 (1999): 41–62; M. B. M. van den Bree, L. J. Eaves, and J. T. Dwyer, Genetic and environmental influences on eating patterns of twins aged ≥ 50 y, *American Journal of Clinical Nutrition* 70 (1999): 456–465.

16. S. S. Smith, NCHS launches latest National Health and Nutrition Examination Survey, *Public Health Reports* 114 (1999): 190–192.

17. N. S. Wellman and coauthors, Do we facilitate the scientific process and the development of dietary guidance when findings from single studies are publicized? An American Society for Nutritional Sciences Controversy Session Report, *American Journal of Clinical Nutrition* 70 (1999): 802–805.

18. U.S. Department of Agriculture and U.S. Department of Health and Human Services, *Dietary Guidelines for Americans 2005*, 6th ed., Home and Garden Bulletin no. 232 (Washington, D.C.: 2005), available online at www.usda.gov/cnpp or call (888) 878-3256.

19. U.S. Department of Health and Human Services, *Healthy People 2010,* 2nd ed. (Washington, D.C.: Government Printing Office, 2000), available online at www.health.gov/healthypeople or call (800) 367-4725.

20. D. S. Satcher, U.S. Surgeon General, as cited in C. Marwick, Healthy People 2010 initiative launched, *Journal of the American Medical Association* 283 (2000): 989–990.

21. D.S. Satcher, Healthy People at 2000, *Public Health Reports* 114 (1999): 563–564.

22. U.S. Department of Agriculture and U.S. Department of Health and Human Services, Nutrition and Your Health: Dietary Guidelines for Americans, 6th edition, 2005, available online at www.usda.gov/cnpp or call (888) 878-3256.

23. J. D. Anding, R. R. Suminski, and L. Boss, Dietary intake, body mass index, exercise, and alcohol: Are college women following the *Dietary Guidelines for Americans*? *Journal of American College Health* 49 (2001): 167–171.

Consumer Corner 1

1. E. M. Hackman and G. L. Moe, Evaluation of newspaper reports of nutrition-related research, *Journal of the American Dietetic Association* 99 (1999): 1564–1566.

Controversy 1

1. Position of the American Dietetic Association: Food and nutrition misinformation, *Journal of the American Dietetic Association* 102 (2002): 260–266.

2. Quackery-related definitions are available from the National Council Against Health Fraud, at www.ncahf.org/pp/definitions.html.

3. E. Cunningham and W. Marcason, Internet hoaxes: How to spot them and how to debunk them, *Journal of the American Dietetic Association* 101 (2001): 460.

4. J. A. Schulman, Nutrition education in medical schools: Trends and implications for health educators, *Med Ed Online,* www.med-ed-Online.org/ f0000015.htm.

Chapter 2

1. Standing Committee on the Scientific Evaluation of Dietary Reference Intakes, Food and Nutrition Board, Institute of Medicine, *Dietary Reference Intakes: Applications in Dietary Assessment* (Washington, D.C.: National Academy Press, 2000), pp. 5–7.

2. S. Murphy, S. I. Barr, and M. I. Poos, Using the new Dietary Reference Intakes to assess diets: A map to the maze, *Nutrition Reviews* 60 (2002): 267–275.

3. Committee on Use of Dietary Reference Intakes in Nutrition Labeling, *Dietary Reference Intakes: Guiding Principles for Nutrition Labeling and Fortification* (Washington, D.C.: National Academies Press, 2003), pp. ES1–ES3; Standing Committee on the Scientific Evaluation of Dietary Reference Intakes, Food and Nutrition Board, Institute of Medicine, *Dietary Reference Intakes: Applications in Dietary Planning* (Washington, D.C.: National Academies Press, 2003), pp. 51–52.

4. U.S. Department of Agriculture and U.S. Department of Health and Human Services, *Dietary Guidelines for Americans 2005*, 6th ed., available online at www.healthierus.gov.

5. S. J. Weinstein and coauthors, Healthy Eating Index scores are associated with blood nutrient concentrations in the Third National Health and Nutrition Examination Survey, *Journal of the American Dietetic Association* 104 (2004): 576–584.

6. P. P. Basiotis and coauthors, Healthy Eating Index: 1999–2000, 2002, U.S. Department of Agriculture, Center for Nutrition Policy and Promotion, available at www.usda.gov/cnpp/Pubs/HEI/HEI99-00report.pdf.

Consumer Corner 2

1. M. Mitka, Food fight over product label claims: Critics say proposed changes will confuse consumers, *Journal of the American Medical Association* 290 (2003): 871–875.

2. Food and Drug Administration, Food labeling: Health claims; dietary guidance, *Federal Register,* Docket no. 2003N-0496, CFR part 101 (2003), pp. 66040–66048; N. Hellmich, FDA to allow qualified health claims on foods, *USA Today,* July 11, 2003, available at www.USATODAY.com.

Controversy 2

1. L. Hilakivi-Clarke and coauthors, Maternal and prepubertal diet, mammary development and breast cancer risk, *Journal of Nutrition* 131 (2001): 154S–157S.

2. A. H. Wu and coauthors, Green tea and risk of breast cancer in Asian Americans, *International Journal of Cancer* 106 (2003): 574–579; L. Le Marchand and coauthors, Intake of flavonoids and lung cancer, *Journal of the National Cancer Institute* 92 (2000): 154–160; P. Knekt and coauthors, Quercetin intake and the incidence of cerebrovascular disease, *European Journal of Clinical Nutrition* 54 (2000): 415–417.

3. J. A. Ross and C. M. Kasum, Dietary flavonoids: Bioavailability, metabolic effects, and safety, *Annual Review of Nutrition* 22 (2002): 19–34.

4. J. H. M. de Vries and coauthors, Red wine is a poor source of bioavailable flavonols in men, *Journal of Nutrition* 131 (2001): 745–748; Y. Scheider and coauthors, Anti-proliferative effect of resveratrol, a natural component of grapes and wine, on human colonic cancer cells, *Cancer Letter* 158 (2000): 85–91.

5. K. T. Howitz and coauthors, Small molecule activators of sirtuins extend *Saccharomyces cerevisiae lifespan*, *Nature* 425 (2003): 191–196.

6. A. Drewnowski and C. Gomez-Carneros, Bitter taste, phytonutrients, and the consumer: A review, *American Journal of Clinical Nutrition* 72 (2000): 1424–1435.

7. U. Wenzel and coauthors, Dietary flavone is a potent apoptosis inducer in human colon carcinoma cells, *Cancer Research* 60 (2000): 3823–3831.

8. D. Rein and coauthors, Epicatechin in human plasma: In vivo determination and effect of chocolate consumption on plasma oxidation status, *Journal of Nutrition* 130 (2000): 2109S–2114S.

9. J. F. Wang and coauthors, A dose-response effect from chocolate consumption on plasma epicatechin and oxidative damage, *Journal of Nutrition* 130 (2000): 2115S–2119S.

10. F. M. Steinberg, M. M. Bearden, and C. L. Keen, Cocoa and chocolate flavonoids: Implications for cardiovascular health, *Journal of the American Dietetic Association* 103 (2003): 215–223; D. Rein and coauthors, Cocoa and wine polyphenols modulate platelet activation and function, *Journal of Nutrition* 130 (2000): 2120S–2126S.

11. Steinberg, Bearden, and Keen, 2003.

12. T. L. Dillinger and coauthors, Food of the gods: Cure for humanity? A cultural history of the medicinal and ritual use of chocolate, *Journal of Nutrition* 130 (2000): 2057S–2072S.

13. K. E. Wangen and coauthors, Soy isoflavones improve plasma lipids in normo-cholesterolemic and mildly hypercholesterolemic postmenopausal women, *American Journal of Clinical Nutrition* 73 (2001): 225–231.

14. L. Keinan-Boker and coauthors, Dietary phytoestrogens and breast cancer risk, *American Journal of Clinical Nutrition* 79 (2004): 282–288; A. Cassidy, Potential risks and benefits of phytoestrogen-rich diets, *International Journal of Vitamin Nutrition Research* 73 (2003): 120–126.

15. C. Nagata and coauthors, Soy product intake and hot flashes in Japanese women: Results from a community-based prospective study, *American Journal of Epidemiology* 153 (2001): 790–793.

16. A. Cotter and K. D. Cashman, Genistein appears to prevent early post-menopausal bone loss as effectively as hormone replacement therapy, *Nutrition Reviews* 61 (2003): 346–351.

17. S. Kreijkamp-Kaspers and coauthors, Effect of soy protein containing isoflavones on cognitive function, bone mineral density, and plasma lipids in postmenopausal women, *Journal of the American Medical Association* 292 (2004): 65–74; J. A. Tice and coauthors, Phytoestrogen supplements for the treatment of hot flashes: The Isoflavone Clover Extract Study, *Journal of the American Medical Association* 290 (2003): 207–214; I. Demonty, B. Lamarche, and P. J. H. Jones, Role of isoflavones in the hypo-cholesterolemic effect of soy, *Nutrition Reviews* 61 (2003): 189–203; J. Rymer, R. Wilson, and K. Ballard, Making decisions about hormone replacement therapy, *British Journal of Medicine* 326 (2003): 322–326; J. J. Anderson and coauthors, Soy isoflavones: No effects on bone mineral content and bone mineral density in healthy, menstruating young women after one year, *Journal of the American College of Nutrition* 21 (2002): 388–393; H. D. Nelson and coauthors, Postmenopausal hormone replacement therapy, *Journal of the American Medical Association* 288 (2002): 872–881; A. St. Germain and coauthors, Isoflavone-rich or isoflavone-poor soy protein does not reduce menopausal symptoms during 24 weeks of treatment, *Menopause* 8 (2001): 17–26.

18. C. D. Allred and coauthors, Soy diets containing varying amounts of genistein stimulate growth of estrogen-dependent (MCF-7) tumors in a dose-dependent manner, *Cancer Research* 61 (2001): 5045–5050.

19. R. R. Newbold and coauthors, Uterine adenocarcinoma in mice treated neonatally with genistein, *Cancer Research* 61 (2001): 4325–4328.

20. M. S. Kurzer, Photyestrogen supplement use by women, *Journal of Nutrition* 133 (2003): 1983S–1986S.

21. Cassidy, 2003.

22. C. Munro and coauthors, Soy isoflavones: A safety review, *Nutrition Reviews* 61 (2003): 1–33.

23. M. R. Ogborn and coauthors, Dietary flax oil reduces renal injury, oxidized LDL content, and tissue n-6/n-3 FA ratio in experimental polycystic kidney disease, *Lipids* 37 (2002): 1059–1065.

24. P. E. Bown, Evaluating the health claim of flaxseed and cancer prevention, *Nutrition Today* 36 (2001): 144–158.

25. E. Giovannucci and coauthors, A prospective study of tomato products, lycopene, and prostate cancer risk, *Journal of the National Cancer Institute* 94 (2002): 391–398; L. Chen and coauthors, Oxidative DNA damage in prostate cancer patients consuming tomato sauce-based entrees as a whole-food intervention, *Journal of the National Cancer Institute* 93 (2001): 1872–1879.

26. P. Prakash, R. M. Russell, and N. I. Krinsky, In vitro inhibition of proliferation of estrogen-dependent and estrogen-independent human breast cancer cells treated with carotenoids or retinoids, *Journal of Nutrition* 131 (2001): 1574–1580.

27. P. Toniolo and coauthors, Serum carotenoids and breast cancer, *American Journal of Epidemiology* 153 (2001): 1142–1147.

28. T. H. Rissanen and coauthors, Serum lycopene concentrations and carotid atherosclerosis: The Kuopio Ischaemic Heart Disease Risk Factor Study, *American Journal of Clinical Nutrition* 77 (2003): 133–138; J. H. Dwyer and coauthors, Oxygenated carotenoid lutein and progression of early atherosclerosis: The Los Angeles Atherosclerosis Study, *Circulation* 103 (2001): 2922–2927; L. Arab and S. Steck, Lycopene and cardiovascular disease, *American Journal of Clinical Nutrition* 71 (2000): 1691S–1695S.

29. R. L. Sedjo and coauthors, Vitamin A, carotenoids, and risk of persistent oncogenic human papillomavirus infection, *Cancer Epidemiology, Biomarkers and Prevention* 11 (2002): 876–884.

30. M. A. Schiff and coauthors, Serum carotenoids and risk of cervical intraepithelial neoplasia in southwestern American Indian women, *Cancer Epidemiology, Biomarkers and Prevention* 10 (2001): 1219–1222.

31. C. Borek, Antioxidant health effects of aged garlic extract, *Journal of Nutrition* 131 (2001): 1010S–1015S.

32. H. Amagase and coauthors, Intake of garlic and its bioactive components, *Journal of Nutrition* 131 (2001): 955S–962S.

33. J. A. Milner, A historical perspective on garlic and cancer, *Journal of Nutrition* 131 (2001): 1027S–1031S.

34. D. L. Lamm and D. R. Riggs, Enhanced immunocompetence by garlic: Role in bladder cancer and other malignancies, *Journal of Nutrition* 131 (2001): 1067S–1070S.

35. E. Kyo and coauthors, Immunomodulatory effects of aged garlic extract, *Journal of Nutrition* 131 (2001): 1075S–1079S; G. P. Sivam, Protection against *Helicobacter pylori* and other bacterial infections by garlic, *Journal of Nutrition* 131 (2001): 1106S–1108S.

36. H. Matsuura, Saponins in garlic as modifiers of the risk of cardiovascular disease, *Journal of Nutrition* 131 (2001): 1000S–1005S.

37. K. Rahman, Historical perspective on garlic and cardiovascular disease, *Journal of Nutrition* 131 (2001): 977S–979S.

38. Borek, 2001.

39. S. K. Banerjee, P. K. Mukherjee, and S. K. Maulik, Garlic as an antioxidant: The good, the bad, and the ugly, *Phytotherapy Research* 17 (2003): 97–106.

40. C. H. Halsted, Dietary supplements and functional foods: 2 sides of a coin? *American Journal of Clinical Nutrition* 77 (2003): 1001S–1007S.

41. Position of the American Dietetic Association: Functional foods, *Journal of the American Dietetic Association* 104 (2004): 814–826.

42. R. E. Ostlund, Phytosterols in human nutrition, *Annual Review of Nutrition* 22 (2002): 533–549; N. B. Cater, Plant stanol ester foods: New tools in the dietary management of cholesterol, *Nutrition and the MD*, November 1999, pp. 1–4.

43. T. Sudhop, B. M. Gottwald, and K. von Bergmann, Serum plant sterols as potential risk factor for coronary heart disease, *Metabolism: Clinical and Experimental* 51 (2002): 1519–1521; Ostlund, 2002; H. Gylling and coauthors, Retinol, vitamin D, carotenes and alpha-tocopherol in serum of a moderately hypercholesterolemic population consuming sitostanol ester margarine, *Atherosclerosis* 145 (1999): 279–285.

44. M. E. Sanders, Probiotics: Considerations for human health, *Nutrition Reviews* 61 (2003): 91–99; J. E. Teitelbaum and W. A. Walker, Nutritional impact of pre- and probiotics as protective gastrointestinal organisms, *Annual Review of Nutrition* 22 (2002): 107–138; J. M. Saavedra, Clinical applications of probiotic agents, *American Journal of Clinical Nutrition* 73 (2001): 1147S–1151S; J. Schrezenmeir and M. de Vrese, Probiotics, prebiotics, and synbiotics—approaching a definition, *American Journal of Clinical Nutrition* 73 (2001): 361S–364S.

45. L. Sherwood and M. D. Gorbach, Probiotics and gastrointestinal health, *American Journal of Gastroenterology* 95 (2000): S2–S4J.

46. O. Adolfsson, S. N. Meydani, and R. M. Russell, Yogurt and gut function, *American Journal of Clinical Nutrition* 80 (2004): 245–256.

47. J. M. Saavedra and A. Tschernia, Human studies with probiotics and prebiotics: Clinical implications, *British Journal of Nutrition* 87 (2002): S241–S246; P. Marteau and M. C. Boutron-Ruault, Nutritional advantages of probiotics and prebiotics, *British Journal of Nutrition* 87 (2002): S153–S157; G. T. Macfarlane and J. H. Cummings, Probiotics, infection and immunity, *Current Opinion in Infectious Diseases* 15 (2002): 501–506; L. Kopp-Hoolihan, Prophylactic and therapeutic uses of probiotics: A review, *Journal of the American Dietetic Association* 101 (2001): 229–238; M. de Vrese and coauthors, Probiotics—compensation for lactase insufficiency, *American Journal of Clinical Nutrition* 73 (2001): 421S–429S; M. B. Roberfroid, Prebiotics and probiotics: Are they functional foods? *American Journal of Clinical Nutrition* 71 (2000): 1682S–1687S.

48. J. W. Lampe and S. Peterson, Brassica, biotransformation and cancer risk: Genetic polymorphisms alter the preventive effects of cruciferous vegetables, *Journal of Nutrition* 132 (2002): 2991–2994.

49. A. B. Howell and B. Foxman, Cranberry juice and adhesion of antibiotic resistant uropathogens, *Journal of the American Medical Association* 287 (2002): 3082–3083; T. Kontiokari and coauthors, Randomised trial of cranberry-lingonberry juice and *Lactobacillus* GG drink for the prevention of urinary tract infections in women, *British Medical Journal* 322 (2001): 1571, available at http://bmj.com/cgi/content/full/322/7302/1571?view=full&pmid=11431298.

50. P. F. Jacques and coauthors, Long-term nutrient intake and early age-related nuclear lens opacities, *Archives of Ophthalmology* 119 (2001): 1009–1019; J. T. Landrum and R. A. Bone, Lutein, zeaxanthin, and the macular pigment, *Archives of Biochemistry and Biophysics* 385 (2001): 28–40.

51. B. A. Corl and coauthors, Cis-9, trans-11 CLA derived endogenously from *trans*-11 18:1 reduces cancer risk in rats, *Journal of Nutrition* 133 (2003): 2893–2900; M. A. Belury, Dietary conjugated linoleic acid in health: Physiological effects and mechanisms of action, *Annual Review of Nutrition* 22 (2002): 505–531; L. D. Whigham, M. E. Cook, and R. L. Atkinson, Conjugated linoleic acid: Implications for human health, *Pharmacological Research* 42 (2000): 503–510.

52. J. Sher and coauthors, Dietary conjugated linoleic acid lowers plasma cholesterol during cholesterol supplementation, but accentuates the atherogenic lipid profile during the acute phase response in hamsters, *Journal of Nutrition* 133 (2003): 456–460; M. A. Belury, A. Mahon, and S. Banni, The conjugated linoleic acid (CLA) isomer, t10c12-CLA, is inversely associated with changes in body weight and serum leptin in subjects with type 2 diabetes mellitus, *Journal of Nutrition* 133 (2003): 257S–260S.

53. Steinberg, Bearden, and Keen, 2003.

54. M. K. Terris, M. M. Issa, and J. R. Tacker, Dietary supplementation with cranberry concentrate tablets may increase the risk of nephrolithiasis, *Urology* 57 (2001): 26–29.

55. J. W. Erdman and coauthors, Not all soy products are created equal: Caution needed in interpretation of research results, *Journal of Nutrition* 134 (2004): 1229S–1233S; C. Atkinson and coauthors, Urinary equol excretion in relation to 2-hydroxyestrone and 16alpha-hydroxyestrone concentrations: An observational study of young to middle-aged women, *Journal of Steroid Biochemistry and Molecular Biology* 86 (2003): 71–77.

56. K. D. Setchell and coauthors, Bioavailability, disposition, and dose-response effects of soy isoflavones when consumed by healthy women at physiologically typical dietary intakes, *Journal of Nutrition* 133 (2003): 1027–1035.

57. C. Hasler and coauthors, How to evaluate the safety, efficacy, and quality of functional foods and their ingredients, *Journal of the American Dietetic Association* 101 (2002): 733–736; B. Brophy and D. Schardt, Functional foods, *Nutrition Action Healthletter*, April 1999, pp. 3–7.

Chapter 3

1. L. L. Birch, Development of food preferences, *Annual Reviews of Nutrition* 19 (1999): 41–62.

2. R. J. Laheij and coauthors, Risk of community-acquired pneumonia and use of gastric acid-suppressing drugs, *Journal of the American Medical Association* 292 (2004): 1955-1960.

3. J. L. Smith, The role of gastric acid in preventing foodborne disease and how bacteria overcome acid conditions, *Journal of Food Protection* 66 (2003): 1310–1325.

4. FDA approves an implant for gastroesophageal reflux disease, *FDA Talk Paper*, 2003, available at www.fda.gov/bbs/topics/ANSWERS/2003/ANS01217.html.

5. N. Uemura and coauthors, *Helicobacter pylori* infection and the development of gastric cancer, *New England Journal of Medicine* 345 (2001): 784–789; J. G. Fox and T. C. Wang, *Helicobacter pylori*—Not a good bug after all, *New England Journal of Medicine* 345 (2001): 829–832.

6. S. T. Mayne and S. A. Navarro, Diet, obesity, and reflux in the etiology of adenocarcinomas of the esophagus and gastric cardia in humans, *Journal of Nutrition* 132 (2002): 3467S–3470S; T. Gislason and coauthors, Respiratory symptoms and nocturnal gastroesophageal reflux: A population-based study of young adults in three European countries, *Chest* 121 (2002): 158–163.

7. M. C. Roberts and coauthors, Constipation, laxative use, and colon cancer in a North Carolina population, *American Journal of Gastroenterology* 98 (2003): 857–864.

8. M. Watford and A. G. Goodridge, Regulation of fuel utilization, in M. A. Stipanuk, ed., *Biochemical and Physiological Aspects of Human Nutrition* (Philadelphia: W. B. Saunders, 2000), p. 387.

Controversy 3

1. F. W. Goodhart and coauthors, *Binge* drinking: Not the word of choice, *Journal of American College Health* 52 (2003): 44–46.

2. M. W. Westerterp-Plantenga and C. R. Verwegen, The appetizing effect of an aperitif in overweight and normal-weight humans, *American Journal of Clinical Nutrition* 69 (1999): 205–212.

3. H. Wechsler and coauthors, College alcohol use: A full or empty glass? *Journal of American College Health* 47 (1999): 247–252.

4. Standing Committee on the Scientific Evaluation of Dietary Reference Intakes, Food and Nutrition Board, Institute of Medicine, *Dietary Reference Intakes for Energy, Carbohydrate, Fiber, Fat, Fatty Acids, Cholesterol, Protein, and Amino Acids* (Washington, D.C.: National Academy Press, 2002), pp. 5–3; Y. Schutz, Role of substrate utilization and thermogenesis on body-weight control with particular reference to alcohol, *Proceedings of the Nutrition Society* 59 (2000): 511–517.

5. C. S. Leiber, Alcohol: Its metabolism and interaction with nutrients, *Annual Review of Nutrition* 20 (2000): 395–430.

6. X.-D. Wang, Mechanisms of cancer chemoprevention: Retinoids and alcohol-related carcinogenesis, *Journal of Nutrition* 133 (2003): 287S–290S.

7. K. W. Singletary and S. M. Gapstur, Alcohol and breast cancer: Review of epidemiologic and experimental evidence and potential mechanisms, *Journal of the American Medical Association* 286 (2001): 2143–2151.

8. X. Wang, Chronic alcohol intake interferes with retinoid metabolism and signaling, *Nutrition Reviews* 57 (1999): 51–59.

9. M. L. Ambrose, S. C. Bowden, and G. Whelan, Thiamin treatment and working memory function of alcohol-dependent people: Preliminary findings, *Alcoholism: Clinical and Experimental Research* 25 (2001): 112–116.

10. J. W. Miller, Does lowering plasma homocysteine reduce vascular disease risk? *Nutrition Reviews* 59 (2001): 242–244; N. M. van der Put and coauthors, Folate, homocysteine and neural tube defects: An overview, *Experimental Biology and Medicine* 226 (2001): 243–270.

11. Centers for Disease Control and Prevention, Alcohol-attributable deaths and years of potential life lost—United States, 2001, *Morbidity and Mortality Weekly Reports* 53 (2004): 866–870.

12. U.S. Department of Agriculture and U.S. Department of Health and Human Services, *Dietary Guidelines for Americans 2005, 6th edition*, available online at www.usda.gov/cnpp.

13. H. D. Sesso and coauthors, Seven-year changes in alcohol consumption and subsequent risk of cardiovascular disease in men, *Archives of Internal Medicine* 160 (2000): 2605–2612.

14. M. Bobak, A. Skodova, and M. Marmot, Effect of beer drinking on risk of myocardial infarction: Population based case-control study, *British Medical Journal* 320 (2000): 1378–1379.

15. C. L. Hart and coauthors, Alcohol consumption and mortality from all causes, coronary heart disease, and stroke: Results from a prospective cohort study of Scottish men with 21 years of follow up, *British Medical Journal* 318 (1999): 1725–1729.

16. S. Stranges and coauthors, Relationship of alcohol drinking pattern to risk of hypertension: A population-based study, *Hypertension* 44 (2004): 813–819; K. Reynolds and coauthors, Alcohol consumption and risk of stroke: A meta-analysis, *Journal of the American Medical Association* 289 (2003): 579–588.

17. A. A. Bertelli and coauthors, Oxidative stress and inflammatory reaction modulation by white wine, *Annals of the New York Academy of Science* 957 (2002): 295–301.

18. J. H. M. de Vries and coauthors, Red wine is a poor source of bioavailable flavonols in men, *Journal of Nutrition* 131 (2001): 745–748.

19. E. Waddington, I. B. Puddey, and K. D. Croft, Red wine polyphenolic compounds inhibit atherosclerosis in apolipoprotein E-deficient mice indepen-dently of effects on lipid peroxidation, *American Journal of Clinical Nutrition* 79 (2004): 54–61.

Chapter 4

1. B. V. McCleary, Dietary fibre analysis, *Proceedings of the Nutrition Society* 62 (2003): 3–9.

2. T. Maeder, Sweet medicines, *Scientific American* 287 (2002): 40–47; J. Travis, The true sweet science—Researchers develop a taste for the study of sugars, *Science News* 161 (2002): 232–233; multiple articles in Carbohydrates and glycobiology—Searching for medicine's sweet spot, *Science* 291 (2001): 2338–2378.

3. R. O. Bonow and R. H. Eckel, Diet, obesity and cardiovascular risk, *New England Journal of Medicine* 348 (2003): 2057–2058; D. M. Bravata and coauthors, Efficacy and safety of low-carbohydrate diets, *Journal of the American Medical Association* 289 (2003): 1837–1850.

4. U.S. Food and Drug Administration, *Counting Calories: Report of the Working Group on Obesity*, March 12, 2004, available at www.cfsan.fda.gov/~dms/owg-rpt.html; Standing Committee on the Scientific Evaluation of Dietary Reference Intakes, Food and Nutrition Board, Institute of Medicine, *Dietary Reference Intakes for Energy, Carbohydrate, Fiber, Fat, Fatty Acids, Cholesterol, Protein, and Amino Acids* (Washington, D.C.: National Academy Press, 2002): pp. 13–17; A. Trichopoulou and coauthors, Lipid, protein and carbohydrate intake in relation to body mass index, *European Journal of Clinical Nutrition* 56 (2002): 37–43.

5. S. K. Fried and S. P. Rao, Sugars, hypertriglyceridemia, and cardiovascular disease, *American Journal of Clinical Nutrition* 78 (2003): 873S–880S; D. R. Lineback and J. M. Jones, Sugars and health workshop: Summary and conclusions, *American Journal of Clinical Nutrition* 78 (2003): 893S–897S; M. K. Hellerstein, Carbohydrate-induced hypertriglyceridemia: Modifying factors and implications for cardiovascular risk, *Current Opinion in Lipidology* 13 (2002): 33–40; E. J. Parks, Changes in fat synthesis influenced by dietary macronutrient content, *Proceedings of the Nutrition Society* 61 (2002): 281–286.

6. Joint WHO/FAO Expert Consultation, *Diet, Nutrition and the Prevention of Chronic Diseases* (Geneva, Switzerland: World Health Organization, 2003), pp. 57–58; S. H. F. Vermunt and coauthors, Effects of sugar intake on body weight: A review, *Obesity Reviews* 4 (2003): 91–99; National Heart, Lung, and Blood Institute, Facts about the Dash Eating Plan, NIH publication no. 03-4082, 2003; Standing Committee on the Scientific Evaluation of Dietary Reference Intakes, 2002, pp. 13–17; Third Report of the National Cholesterol Education Program (NCEP) Expert Panel on Detection, Evaluation, and Treatment of High Blood Cholesterol in Adults (Adult Treatment Panel III), 2002, NIH publication no. 02-5215, pp. V-1–V-4; Nutrition Committee of the American Heart Association, AHA dietary guidelines revision 2000: A statement for healthcare professionals from the nutrition committee of the American Heart Association, *Circulation* 102 (2000): 2284–2299.

7. Standing Committee on the Scientific Evaluation of Dietary Reference Intakes, 2002, pp. 7–9.

8. S. Liu and coauthors, Relation between changes in intakes of dietary fiber and grain products and changes in weight and development of obesity among middle-aged women, *American Journal of Clinical Nutrition* 78 (2003): 920–927.

9. F. B. Hu and W. C. Willett, Optimal diets for prevention of coronary heart disease, *Journal of the American Medical Association* 288 (2002): 2569–2578; N. M. McKeown and coauthors, Whole-grain intake is favorably associated with metabolic risk factors for type 2 diabetes and cardiovascular disease in the Framingham Offspring Study, *American Journal of Clinical Nutrition* 76 (2002): 390–398; S. Liu and coauthors, Whole-grain consumption and risk of coronary heart disease: Results from the Nurses' Health Study, *American Journal of Clinical Nutrition* 70 (1999): 412–419.

10. Position of the American Dietetic Association, Health implications of dietary fiber, *Journal of the American Dietetic Association* 102 (2002): 993–999; L. Brown and coauthors, Cholesterol-lowering effects of dietary fiber: A meta-analysis, *American Journal of Clinical Nutrition* 69 (1999): 30–42.

11. M. L. Fernandez, Soluble fiber and nondigestible carbohydrate effects on plasma lipids and cardiovascular risk, *Current Opinion in Lipidology* 12 (2001): 35–40; Brown and coauthors, 1999.

12. B. M. Davy and coauthors, High-fiber oat cereal compared with wheat cereal consumption favorably alters LDL-cholesterol subclass and particle numbers in middle-aged and older men, *American Journal of Clinical Nutrition* 76 (2002): 351–358; D. J. A. Jenkins and coauthors, Soluble fiber intake at a dose approved by the US Food and Drug Administration for a claim of health benefits: Serum lipid risk factors for cardiovascular disease assessed in a randomized controlled crossover trial, *American Journal of Clinical Nutrition* 75 (2002): 834–839.

13. T. T. Fund and coauthors, Whole-grain intake and the risk of type 2 diabetes: A prospective study in men, *American Journal of Clinical Nutrition* 76 (2002): 535–540.

14. W. Aldoori and M. Ryan-Harshman, Preventing diverticular disease: Review of recent evidence on high-fibre diets, *Canadian Family Physician* 48 (2002): 1632–1637.

15. U. Peters and coauthors, Dietary fibre and colorectal adenoma in a colorectal cancer early detection programme, *Lancet* 361 (2003): 1491–1495; S. A. Bingham and coauthors, Dietary fibre in food and protection against colorectal cancer in the European Prospective Investigation into Cancer and Nutrition (EPIC): An observational study, *Lancet* 361 (2003): 1496–1501; T. Honda, I. Kai, and G. Ohi, Fat and dietary fiber intake and colon cancer mortality: A chronological comparison between Japan and the United States, *Nutrition and Cancer* 33 (1999): 95–99; B. S. Reddy, Role of dietary fiber in colon cancer: An overview, *American Journal of Medicine* 106 (1999): S16–S19.

16. A. Schatzkin and coauthors, Lack of effect of a low-fat, high-fiber diet on the recurrence of colorectal adenomas, *New England Journal of Medicine* 342 (2000): 1149–1155; D. S. Alberts and coauthors, Lack of effect of a high-fiber cereal supplement on the recurrence of colorectal adenomas, *New England Journal of Medicine* 342 (2000): 1156–1162; F. Macrae, Wheat bran fiber and development of adenomatous polyps: Evidence from randomized, controlled clinical trials, *American Journal of Medicine* 106 (1999): 38S–42S; D. Kritchevsky, Protective role of wheat bran fiber: Preclinical data, *American Journal of Medicine* 106 (1999): 28S–31S; C. S. Fuchs and coauthors, Dietary fiber and risk of colorectal cancer and adenoma in women, *New England Journal of Medicine* 340 (1999): 169–176.

17. Bingham and coauthors, 2003.

18. Standing Committee on the Scientific Evaluation of Dietary Reference Intakes, 2002, p. 7-8.

19. A. Andoh, T. Tsujikawa, and Y. Fujiyama, Role of dietary fiber and short-chain fatty acids in the colon, *Current Pharmaceutical Design* 9 (2003): 347–358; L. McMillan and coauthors, Opposing effects of butyrate and bile acids on apoptosis of human colon adenoma cells: Differential activation of PKC and MAP kinases, *British Journal of Cancer* 88 (2003): 748–753; M. E. Rodriguez-Cabezas and coauthors, Dietary fiber down-regulates colonic tumor necrosis factor alpha and nitric oxide production in trinitrobenzenesulfonic acid-induced colitic rats, *Journal of Nutrition* 132 (2002): 3263–3271; J. L. Slavin, Mechanisms for the impact of whole grain foods on cancer risk, *Journal of the American College of Nutrition* 19 (2000): 300S–307S.

20. American Gastroenterological Association medical position statement: Impact of dietary fiber on colon cancer occurrence, *Gastroenterology* 118 (2000): 1233–1234.

21. N. C. Howarth, E. Saltzman, and S. B. Roberts, Dietary fiber and weight regulation, *Nutrition Reviews* 59 (2001): 129–139; A. Sparti and coauthors, Effect of diets high or low in unavailable and slowly digestible carbohydrates on the pattern of 24-h substrate oxidation and feelings of hunger in humans, *American Journal of Clinical Nutrition* 72 (2000): 1461–1468.

22. Standing Committee on the Scientific Evaluation of Dietary Reference Intakes, 2002, p. 7-4.

23. Position of the American Dietetic Association, 2002.

24. Standing Committee on the Scientific Evaluation of Dietary Reference Intakes, 2002, p. 7-4.

25. K. N. Englyst and coauthors, Rapidly available glucose in foods: An in vitro measurement that reflects the glycemic response, *American Journal of Clinical Nutrition* 69 (1999): 448–454.

26. J. G. Muir and coauthors, Combining wheat bran with resistant starch has more beneficial effects on fecal indexes than does wheat bran alone, *American Journal of Clinical Nutrition* 79 (2004): 1020–1028.

27. B. V. McCleary, Dietary fiber analysis, *Proceedings of the Nutrition Society* 62 (2003): 3–9. S. Kimura, Glycemic carbohydrate and health: Background and synopsis of the symposium, *Nutrition Reviews* 61 (2003): S1–S4.

28. D. L. Swagerty, A. D. Walling, and R. M. Klein, Lactose intolerance, *American Family Physician* 65 (2002): 1845–1850.

29. D. Savaiano, Lactose intolerance: A self-fulfilling prophecy leading to osteoporosis? *Nutrition Reviews* 61 (2003): 221–223.

30. H. J. Kalkwarf, J. C. Khoury, and B. P. Lanphear, Milk intake during childhood and adolescence, adult bone density, and osteoporotic fractures in US women, *American Journal of Clinical Nutrition* 77 (2003): 257–265.

31. P. Marteau and M. C. Boutron-Ruault, Nutritional advantages of probiotics and prebiotics, *British Journal of Nutrition* 87 (2002): S153–S157; M. de Vrese and coauthors, Probiotics—Compensation for lactase insufficiency, *American Journal of Clinical Nutrition* 73 (2001): 421S–429S; L. Kopp-Hoolihan, Prophylactic and therapeutic uses of probiotics: A review, *Journal of the American Dietetic Association* 101 (2001): 229–238, 241; J. M. Saavedra, Clinical applications of probiotic agents, *American Journal of Clinical Nutrition* 73 (2001): 1147S–1151S.

32. M. Levitt, as quoted in Who ya gonna call? Gas busters, *Nutrition Action Healthletter*, May 2003, pp. 3–5.

33. T. B. VanItallie and T. H. Nufert, Ketones: Metabolism's ugly duckling, *Nutrition Reviews* 61 (2003): 327–341.

34. J. Achten and coauthors, Higher dietary carbohydrate content during intensified running training results in better maintenance of performance and mood state, *Journal of Applied Physiology* 96 (2004): 1331–1340; Standing Committee on the Scientific Evaluation of Dietary Reference Intakes, 2002, p. 6-1; S. T. Reddy and coauthors, Effect of low-carbohydrate high-protein diet on acid-base balance, stone-forming propensity, and calcium metabolism, *American Journal of Kidney Disease* 40 (2002): 265–274; H. R. Lieberman, C. M. Falco, and S. S. Slade, Carbohydrate administration during a day of sustained aerobic activity improves vigilance, as assessed by a novel ambulatory monitoring device, and mood, *American Journal of Clinical Nutrition* 76 (2002): 120–127.

35. J. L. Dorman, Pediatric ketogenic diets for intractable seizures, *Today's Dietitian* 5 (2003): 16–20.

36. Standing Committee on the Scientific Evaluation of Dietary Reference Intakes, 2002, p. 6-9.

37. R. Gruetter, Glycogen: The forgotten cerebral energy store, *Journal of Neuroscience Research* 74 (2003): 179–183; I. Y. Choi, E. R. Seaquist, and R. Gruetter, Effect of hypoglycemia on brain glycogen metabolism in vivo, *Journal of Neuroscience Research* 72 (2003): 25–32.

38. G. Nantel, Glycemic carbohydrate: An international perspective, *Nutrition Reviews* 61 (2003): S34–S39; T. M. S. Wolever, Carbohydrate and the regulation of blood glucose and metabolism, *Nutrition Reviews* 61 (2003): S40–S48; D. J. A. Jenkins and coauthors, Glycemic index: Overview of implications in health and disease, *American Journal of Clinical Nutrition* 76 (2002): 266S–273S.

39. J. C. Brand-Miller, Glycemic load and chronic disease, *Nutrition Reviews* 61 (2003): S49–S55.

40. J. C. Brand-Miller, Physiological validation of the concept of glycemic load in lean young adults, *Journal of Nutrition* 133 (2003): 2728–2732; Jenkins and coauthors, 2002; S. Liu and coauthors, Dietary glycemic load assessed by food-frequency questionnaire in relation to plasma high-density lipoprotein cholesterol and fasting plasma triacylglycerols in postmenopausal women, *American Journal of Clinical Nutrition* 73 (2001): 560–566; S. Liu and coauthors, A prospective study of dietary glycemic load, carbohydrate intake, and risk of coronary heart disease in US women, *American Journal of Clinical Nutrition* 71 (2000): 1455–1461.

41. Brand-Miller, 2003; D. S. Ludwig, The glycemic index: Physiological mechanisms relating to obesity, diabetes, and cardiovascular disease, *Journal of the American Medical Association* 287 (2002): 2414–2423; L. E. Spieth and coauthors, A low-glycemic index diet in the treatment of pediatric obesity, *Archives of Pediatrics & Adolescent Medicine* 154 (2000): 947–951; D. S. Ludwig and coauthors, Dietary fiber, weight gain, and cardiovascular disease risk factors in young adults, *Journal of the American Medical Association* 282 (1999): 1539–1546.

42. S. Liu and coauthors, Relation between a diet with a high glycemic load and plasma concentrations of high-sensitivity C-reactive protein in middle-aged women, *American Journal of Clinical Nutrition* 75 (2002): 492–498; D. S. Ludwig and coauthors, High glycemic index foods, overeating, and obesity, *Pediatrics* 103 (1999): e26 (www.pediatrics.org).

43. S. Schenk and coauthors, Different glycemic indexes of breakfast cereals are not due to glucose entry into blood but to glucose removal by tissue, *American Journal of Clinical Nutrition* 77 (2003): 742–748.

44. Ludwig, 2002; F. X. Pi-Sunyer, Glycemic index and disease, *American Journal of Clinical Nutrition* 76 (2002): 290S–298S; A. E. Buyken and coauthors, Glycemic index in the diet of European outpatients with type 1 diabetes: Relations to glycated hemoglobin and serum lipids, *American Journal of Clinical Nutrition* 73 (2001): 574–581.

45. Wolever, 2003.

46. W. Willett, J. Manson, and S. Liu, Glycemic index, glycemic load, and risk of type 2 diabetes, *American Journal of Clinical Nutrition* 76 (2002): 274S–280S.

47. International Food Information Council, Glycemic index: The ups and downs of indexing blood sugar, *Food Insight*, May/June 2003, available at www.ific.org/foodinsight 2003; K. L. Johnson, M. N. Clifford, and L. M. Morgan, Coffee acutely modifies gastrointestinal hormone secretion and glucose tolerance in humans: Glycemic effects of chlorogenic acid and caffeine, *American Journal of Clinical Nutrition* 78 (2003): 728–733.

48. C. Beebe, Diets with a low glycemic index: Not ready for practice yet! *Nutrition Today* 34 (1999): 82–86.

49. O. Lammert and coauthors, Effects of isoenergetic overfeeding of either carbohydrate or fat in young men, *British Journal of Nutrition* 88 (2002): 331–332.

50. W. H. M. Saris, Sugars, energy metabolism, and body weight control, *American Journal of Clinical Nutrition* 78 (2003): 850S–857S.

51. J. M. Schwartz and coauthors, Hepatic de novo lipogenesis in normoinsulinemic and hyperinsulinemic subjects consuming high-fat, low-carbohydrate and low-fat, high carbohydrate isoenergetic diets, *American Journal of Clinical Nutrition* 77 (2003): 43–50.

52. Standing Committee on the Scientific Evaluation of Dietary Reference Intakes, 2002, p. 11-1.

53. A. H. Mokdad and coauthors, Diabetes trends in the U.S.: 1990–1998, *Diabetes Care* 23 (2000): 1278–1283; K. M. V. Narayan and coauthors, Diabetes—A common, growing, serious, costly, and potentially preventable public health problem, *Diabetes Research and Clinical Practice* 50 (2000): S77–S84.

54. National Center for Chronic Disease Prevention and Health Promotion, Diabetes: Disabling, deadly, and on the rise, 2003, available at www.cdc.gov/ nccdphp/aag/ aag_ddt.htm.

55. National Center for Chronic Disease Prevention and Health Promotion, 2003.

56. Canadian Diabetes Association Clinical Practice Guidelines Expert Committee, Canadian Diabetes Association 2003 Clinical Practice Guidelines for the Prevention and Management of Diabetes in Canada, *Canadian Journal of Diabetes* 27 (2003), available at www.diabetes.ca/cpg2003/chapters.aspx.

57. National Centers for Chronic Disease Prevention and Health Promotion, Diabetes: Disabling, Deadly, and on the Rise, *At A Glance*, 2003, available at www.cdc.gov/nccdphp; E. A. Gale, The rise of childhood type 1 diabetes in the 20th century, *Diabetes* 51 (2003): 3353–3361.

58. K. Sadeharju and coauthors, Enterovirus infections as a risk factor for type 1 diabetes: Virus analyses in a dietary intervention trial, *Clinical and Experimental Immunology* 132 (2003): 271–277; C. D. Berdanier, Mitochondrial gene expression in diabetes mellitus: Effect of nutrition, *Nutrition Reviews* 59 (2001): 61–70; F. A. Darlsson and coauthors, Beta-cell activity and destruction in type 1 diabetes, *Upsala Journal of Medical Sciences* 105 (2000): 85–95.

59. J. Hopkins, Treating diabetes with aerosolized insulin, *Chest* 120 (2001): 99S–106S.

60. D. B. Kaufman and W. L. Lowe, Clinical islet transplantation, *Current Diabetes Reports* 3 (2003): 344–350; M. Atkinson and G. Eisenbarth, Type 1 diabetes: New perspectives on disease pathogenesis and treatment, *Lancet* 358 (2001): 221–229.

61. E. Roche and coauthors, Bio-engineering insulin-secreting cells from embryonic stem cells: A review of progress, *Medical and Biological Engineering and Computing* 41 (2003): 384–391.

62. National Center for Chronic Disease Prevention and Health Promotion, 2003.

63. S. Klein and coauthors, Weight management through lifestyle modification for the prevention and management of type 2 diabetes: rationale and strategies. A statement of the American Diabetes Association, the North American Association for the Study of Obesity, and the American Society for Clinical Nutrition, *American Journal of Clinical Nutrition* 80 (2004): 257–263; K. Silander and coauthors, A large set of Finnish affected sibling pair families with type 2 diabetes suggests susceptibility loci on chromosomes 6, 11, and 14, *Diabetes* 53 (2004): 821–829; C. N. Rotimi and coauthors, A genome-wide search for type 2 diabetes susceptibility genes in West Africans: The Africa America Diabetes Mellitus (AADM) Study, *Diabetes* 53 (2004): 838–841.

64. K. L. Cox and coauthors, Independent and additive effects of energy restriction and exercise on glucose and insulin concentrations in sedentary overweight men, *American Journal of Clinical Nutrition* 80 (2004): 308–316; F. Collins, as quoted by E. Lane, Researchers find potential gene link to diabetes, *Newsday.Com*, March 18, 2004, available at www.newsday.com.

65. National Center for Chronic Disease Prevention and Health Promotion, 2003; R. M. van Dam and coauthors, Dietary patterns and risk for type 2 diabetes mellitus in U.S. men, *Annals of Internal Medicine* 136 (2002): 201–209; F. B. Hu and coauthors, Diet, lifestyle, and the risk of type 2 diabetes mellitus in women, *New England Journal of Medicine* 345 (2001): 790–797.

66. D. H. Amschler, The alarming increase of type 2 diabetes in children, *Journal of School Health* 72 (2002): 39–41; A. Fagot-Campagna, K. M. Narayan, and G. Imperatore, Type 2 diabetes in children exemplifies the growing problem of chronic diseases, *British Medical Journal* 322 (2001): 377–378; American Diabetes Association Consensus Statement, Type 2 diabetes in children and adolescents, *Diabetes Care* 23 (2000): 381–389, available at www.diabetes.org/ada/ Consensus/pg381.htm; American Diabetes Association, Type 2 diabetes in children and adolescents, *Pediatrics* 105 (2000): 671–680.

67. U. B. Pajvani and P. E. Scherer, Adiponectin: Systemic contributor to insulin sensitivity, *Current Diabetes Reports* 3 (2003): 207–213; M. Li and coauthors, Small molecule insulin receptor activators potentiate insulin action in insulin-resistant cells, *Diabetes* 50 (2001): 2323–2328; I. D. Goldfine, Unraveling the riddle of insulin resistance (editorial), *Journal of Laboratory and Clinical Medicine* 134 (1999): 100–102.

68. J. P. Girod and D. J. Brotman, The metabolic syndrome as a vicious cycle: Does obesity beget obesity? *Medical Hypotheses* 60 (2003): 584–589.

69. R. Kikkawa, Chronic complications in diabetes mellitus, *British Journal of Nutrition* 84 (2000): S183–S185.

70. J. L. Evans and coauthors, Are oxidative stress-activated signaling pathways mediators of insulin resistance and beta-cell dysfunction? *Diabetes* 52 (2003): 1–8; I. G. Obrosova, How does glucose generate oxidative stress in peripheral nerve? *International Review of Neurobiology* 50 (2002): 3–35.

71. P. Gaude and coauthors, Multifactorial interventions and cardiovascular disease in patients with type 2 diabetes, *New England Journal of Medicine* 348 (2003): 383–393.

72. S. Tesfaye and coauthors, Vascular risk factors and diabetic neuropathy, *New England Journal of Medicine* 352 (2005): 341–350; J. G. Pastors, How effective is medical nutrition therapy in diabetes care? *Journal of the American Dietetic Association* 103 (2003): 827–831.

73. Position of the American Dietetic Association: Total diet approach to communicating food and nutrition information, *Journal of the American Dietetic Association* 102 (2002): 100–108.

74. American Diabetes Association, Evidence-based nutrition principles and recommendations for the treatment and prevention and related complications, *Diabetes Care* 26 (2003): S51–S61.

75. M. C. Gannon and coauthors, An increase in dietary protein improves the blood glucose response in persons with type 2 diabetes, *American Journal of Clinical Nutrition* 78 (2003): 734–741; M. J. Franz and M. L. Wheeler, Nutrition therapy for diabetic nephropathy, *Current Diabetes Reports* 3 (2003): 412–417; E. M. Wrone and coauthors, Association of dietary protein intake with microalbuminuria in healthy adults: Third National Health and Nutrition Examination Survey, *American Journal of Kidney Diseases* 41 (2003): 580–587.

76. H. T. Pigman, D. X. Gan, and M. A. Krousel-Wood, Role of exercise for type 2 diabetic patient management, *Southern Medical Journal* 95 (2002): 72–77.

77. N. F. Sheard, Moderate changes in weight and physical activity can prevent or delay the development of type 2 diabetes mellitus in susceptible individuals, *Nutrition Reviews* 61 (2003): 76–79; A. M. Swartz and coauthors, Increasing daily walking improves glucose tolerance in overweight women, *Preventive Medicine* 37 (2003): 356–362.

78. L. Dye, A. Lluch, and J. E. Blundell, Macronutrients and mental performance, *Nutrition* 16 (2000): 1021–1034.

79. J. F. Brun, C. Fedou, and J. Mercier, Postprandial reactive hypoglycemia, *Diabetes and Metabolism* 26 (2000): 337–351.

80. G. Pourmotabbed and A. E. Kitabchi, Hypoglycemia, *Obstetrics and Gynecology Clinics of North America* 28 (2001): 383–400.

81. M. V. Apte and J. S. Wilson, Alcohol-induced pancreatic injury, *Best Practice and Research: Clinical Gastroenterology* 17 (2003): 593–612.

82. U.S. Department of Agriculture and U.S. Department of Health and Human Services, *Dietary Guidelines for Americans 2005,* available online at www.healthierus.gov.

Consumer Corner 4

1. S. Liu, Intake of refined carbohydrates and whole grain foods in relation to risk of type 2 diabetes mellitus and coronary heart disease, *Journal of the American College of Nutrition* 21 (2002): 298–306.

2. Committee on Use of Dietary Reference Intakes in Nutrition Labeling, *Dietary Reference Intakes: Guiding Principles for Nutrition Labeling and Fortification* (Washington, D.C.: National Academies Press, 2003), pp. ES1–ES5.

Controversy 4

1. A. S. Levine, C. M. Kotz, and B. A. Gosnell, Sugars and fats: The neurobiology of preference, *Journal of Nutrition* 133 (2003): 831S–834S.

2. Economic Research Service, Estimating consumption of caloric sweeteners, *Amber Waves,* April 2003, available at www.ers.usda.gov/AmberWaves/April03/Indicators/behinddata.htm.

3. B. M. Popkin and S. J. Nielsen, The sweetening of the world's diet, *Obesity Research* 11 (2003): 1325–1332.

4. U.S. Department of Agriculture and U.S. Department of Health and Human Services, *2005 Dietary Guidelines Committee Report, 2005,* available online at www.health.gov/dietaryguidelines/dga2005report/

5. U.S. Department of Agriculture and U.S. Department of Health and Human Services, *Dietary Guidelines for Americans, 6th edition, 2005,* available online at www.healthierus.gov or call (888) 878-3256.

6. M. Nestle, as quoted by O. Dyer, U.S. government rejects WHO's attempts to improve diet, *British Medical Journal* 328 (2004): 185.

7. Joint WHO/FAO Expert Consultation, 2003, pp. 57–58; S. H. F. Vermunt and coauthors, Effects of sugar intake on body weight: A review, *Obesity Reviews* 4 (2003): 91–99.

8. A. K. Kant, Consumption of energy-dense, nutrient-poor foods by adult Americans: Nutrition and health implications, The Third National Health and Nutrition Examination Survey, 1988–1994, *American Journal of Clinical Nutrition* 72 (2000): 929–936.

9. L. Harnack, J. Stang, and M. Story, Soft drink consumption among U.S. children and adolescents: Nutritional consequences, *Journal of the American Dietetic Association* 99 (1999): 436–441.

10. A. M. Coulston and R. K. Johnson, Sugar and sugars: Myths and realities, *Journal of the American Dietetic Association* 102 (2002): 351–353; J. F. Guthrie and J. F. Morton, Food sources of added sweeteners in the diets of Americans, *Journal of the American Dietetic Association* 100 (2000): 43–48.

11. D. S. Ludwig, K. E. Peterson, and L. S. Gortmaker, Relation between consumption of sugar-sweetened drinks and childhood obesity: A prospective, observational analysis, *Lancet* 357 (2001): 505–508; R. P. Troiano and coauthors, Energy and fat intakes of children and adolescents in the United States: Data from the National Health and Nutrition Examination Surveys, *American Journal of Clinical Nutrition* 72 (2000): 1343S–1353S.

12. D. P. DiMeglio and R. D. Mattes, Liquid versus solid carbohydrate: Effects of food intake and body weight, *International Journal of Obesity* 24 (2000): 794–800.

13. G. A. Bray, S. J. Nielsen, and B. M. Popkin, Consumption of high-fructose corn syrup in beverages may play a role in the epidemic of obesity, *American Journal of Clinical Nutrition* 79 (2004): 537–543. S. J. Nielsen and B. M. Popkin, Changes in beverage intake between 1977 and 2001, *American Journal of Preventive Medicine* 27 (2004): 205–210.

14. Bray, Nielsen, and Popkin, 2004; S. S. Elliott and coauthors, Fructose, weight gain, and the insulin resistance syndrome, *American Journal of Clinical Nutrition* 76 (2002): 911–922.

15. Standing Committee on the Scientific Evaluation of Dietary Reference Intakes, 2002, p. 6-35.

16. L. Lissner, B. L. Heitmann, and C. Bengtsson, Population studies of diet and obesity, *British Journal of Nutrition* 83 (2000): S21–S24.

17. Popkin and Nielsen, 2003.

18. A. S. Levine, C. M. Kotz, and B. A. Gosnell, Sugars: Hedonic aspects, neuroregulation, and energy balance, *American Journal of Clinical Nutrition* 78 (2003): 834S–842S;

M. Kanazawa and coauthors, Effects of a high-sucrose diet on body weight, plasma triglycerides, and stress tolerance, *Nutrition Reviews* 61 (2003): S27–S33.

19. Levine, Kotz, and Gosnell, Sugars: Hedonic aspects, 2003.

20. Joint WHO/FOA Expert Consultation, 2003, pp. 72–80.

21. K. A. Meyer and coauthors, Carbohydrates, dietary fiber, and incident type 2 diabetes in older women, *American Journal of Clinical Nutrition* 71 (2000): 921–930.

22. M. B. Schulze and coauthors, Sugar-sweetened beverages, weight gain, and incidence of type 2 diabetes in young and middle-aged women, *Journal of the American Medical Association* 292 (2004): 927–934; Centers for Disease Control and Prevention, Prevalence of overweight and obesity among adults with diagnosed diabetes—United States, 1988–1994 and 1999–2002, *Morbidity and Mortality Weekly Report* 53 (2004): 1066–1068.

23. M. K. Hellerstein, Carbohydrate-induced hypertriglyceridemia: Modifying factors and implications for cardiovascular risk, *Current Opinion in Lipidology* 13 (2002): 33–40; E. J. Parks, Changes in fat synthesis influenced by dietary macronutrient content, *Proceedings of the Nutrition Society* 61 (2002): 281–286.

24. Kanazawa and coauthors, 2003; A. Raben and A. Astrup, Ad libitum intake of low-fat diets rich in either starchy foods or sucrose: Effects on blood lipids, factor VII coagulant activity, and fibrinogen, *Metabolism: Clinical and Experimental* 49 (2000): 731–735; J. P. Bantle and coauthors, Effects of dietary fructose on plasma lipids in healthy subjects, *American Journal of Clinical Nutrition* 72 (2000): 1128–1134; F. Abbasi and coauthors, High carbohydrate diets, triglyceride-rich lipoproteins, and coronary heart disease risk, *American Journal of Cardiology* 85 (2000): 45–48.

25. E. J. Parks and M. K. Hellerstein, Carbohydrate-induced hypertriacylglycerolemia: Historical perspective and review of biological mechanisms, *American Journal of Clinical Nutrition* 71 (2000): 412–433.

26. Standing Committee on the Scientific Evaluation of Dietary Reference Intakes, 2002, p. 6-29.

27. J. M. Schwarz and coauthors, Hepatic de novo lipogenesis in normoinsulinemic and hyperinsulinemic subjects consuming high-fat, low-carbohydrate and low-fat, high-carbohydrate isoenergetic diets, *American Journal of Clinical Nutrition* 77 (2003): 43–50; A. T. Erkkilä and coauthors, APOE polymorphism and the hypertriglyceridemic effect of dietary sucrose, *American Journal of Clinical Nutrition* 73 (2001): 746–752.

28. S. Dosreis and coauthors, Parental perceptions and satisfaction with stimulant medication for attention-deficit hyperactivity disorder, *Journal of Developmental and Behavioral Pediatrics* 24 (2003): 155–162.

29. Dosreis and coauthors, 2003.

30. Standing Committee on the Scientific Evaluation of Dietary Reference Intakes, 2002, p. 6-24.

31. D. Benton and S. Nabb, Carbohydrate, memory, and mood, *Nutrition Reviews* 61 (2003): S61–S67; C. E. Greenwood, Dietary carbohydrate, glucose regulation, and cognitive performance in elderly persons, *Nutrition Reviews* 61 (2003): S68–S74.

32. S. Gibson and S. Williams, Dental caries in pre-school children: Associations with social class, toothbrushing habit and consumption of sugars and sugar-containing foods. Further analysis of data from the National Diet and Nutrition Survey of children aged 1.5–4.5 years, *Caries Research* 33 (1999): 101–113.

33. S. Kashket and D. P. DePaola, Cheese consumption and the development and progression of dental caries, *Nutrition Reviews* 60 (2002): 97–103; Department of Health and Human Services, *Oral Health in America: A Report of the Surgeon General* (Rockville, Md.: National Institutes of Health, 2000), pp. 250–251.

34. Joint WHO/FAO Expert Consultation, 2003, p. 119.

35. Position of the American Dietetic Association: Oral health and nutrition, *Journal of the American Dietetic Association* 103 (2003): 615–625.

36. Joint WHO/FAO Expert Consultation, 2003, p. 119.

37. K. McNutt, What clients need to know about sugar replacers, *Journal of the American Dietetic Association* 100 (2000): 466–469.

38. Fact Sheet: The report on carcinogens, 9th ed., National Institutes of Health News Release, available at www.nih.gov/news/pr/may2000/niehs-15.htm.

39. National Cancer Institute, Artificial sweeteners, *Cancer Facts,* 2003, available at http://cis.nci.nih.gov/fact/3_19.htm.

40. S. M. Cohen and coauthors, Calcium phosphate–containing precipitate and the carcinogenicity of sodium salt in rats, *Carcinogenesis* 21 (2000): 783–792.

41. A bibliography of 167 research articles on aspartame can be found in J. Van de Kamp, Adverse effects of aspartame, *Current Bibliographies in Medicine* (Washington, D.C.: Government Printing Office, 1991).

42. Food and Drug Administration, Agency response letter GRAS notice no. GRN 000078, October 25, 2001, available at www.cfsan.fda.gov.

43. E. Koyama and coauthors, Absorption and metabolism of glycosidic sweeteners of stevia mixture and their aglycone, steviol in rats and humans, *Food and Chemical Toxicology* 41 (2003): 875–883.

44. P. Chan and coauthors, A double-blind placebo-controlled study of the effectiveness and tolerability of oral stevioside in human hypertension, *British Journal of Clinical Pharmacology* 50 (2000): 215–220.

45. A. Raben and coauthors, Sucrose compared with artificial sweeteners: Different effects on ad libitum food intake and body weight after 10 wk of supplementation in overweight subjects, *American Journal of Clinical Nutrition* 76 (2002): 721–729.

46. Raben and coauthors, 2002.

47. M.-P. St-Onge and S. B. Heymsfield, Usefulness of artificial sweeteners for body weight control, *Nutrition Reviews* 61 (2003): 219–220.

48. G. H. Anderson and D. Woodend, Effect of glycemic carbohydrates on short-term satiety and food intake, *Nutrition Reviews* 61 (2003): S17–S26.

Chapter 5

1. E. Faloia and coauthors, Adipose tissue as an endocrine organ? A review of some recent data, *Eating and Weight Disorders* 5 (2000): 116–123.

2. A. Baylin and coauthors, Adipose tissue biomarkers of fatty acid intake, *American Journal of Clinical Nutrition* 76 (2002): 750–757.

3. J. Zhang and H. Kesteloot, Differences in all-cause, cardiovascular and cancer mortality between Hong Kong and Singapore: Role of nutrition, *European Journal of Epidemiology* 17 (2001): 469–477.

4. F. J. Sanchez-Muniz and coauthors, Dietary fat saturation affects apolipoprotein AII levels and HDL composition in postmenopausal women, *Journal of Nutrition* 132 (2002): 50–54; D. O. Edem, Palm oil: Biochemical, physiological, nutritional, hematological, and toxicological aspects: A review, *Plant Foods for Human Nutrition* 57 (2002): 319–341.

5. A. H. Stark and Z. Madar, Olive oil as a functional food: Epidemiology and nutritional approaches, *Nutrition Reviews* 60 (2002): 170–176.

6. C. T. Phan and P. Tso, Intestinal lipid absorption and transport, *Frontiers in Bioscience* 6 (2001): d299–d319.

7. U.S. Department of Agriculture and U.S. Department of Health and Human Services, *Dietary Guidelines for Americans 2005*, available online at www.healthierus.gov.

8. S. A. Bingham and coauthors, Are imprecise methods obscuring a relation between fat and breast cancer? *Lancet* 362 (2003): 212–214; C. E. Spiegelman and coauthors, Premenopausal fat intake and risk of breast cancer, *Journal of the National Cancer Institute* 95 (2003): 1079–1085; P. L. Zock, Dietary fats and cancer, *Current Opinion in Lipidology* 12 (2001): 5–10.

9. A. L. Lichtenstein and P. J. H. Jones, Lipids: Absorption and transport, in *Present Knowledge in Nutrition*, ed. B. A. Bowman and R. M. Russell (Washington, D.C.: ILSI Press, 2001), p. 99; C. F. L. Semenkovich, Nutrient and genetic regulation of lipoprotein metabolism, in *Modern Nutrition in Health and Disease*, ed. M. E. Shils and coeditors (Baltimore: Williams & Wilkins, 1999), p. 1196.

10. Executive Summary of the Third Report of the National Cholesterol Education Program (NCEP) Expert Panel on Detection, Evaluation, and Treatment of High Blood Cholesterol in Adults (Adult Treatment Panel III), *Journal of the American Medical Association* 285 (2001): 2486–2497.

11. U.S. Department of Agriculture and U.S. Department of Health and Human Services, *Dietary Guidelines for Americans 2005*, 6th ed., available online at www.usda.gov/cnpp.

12. K. L. Herron and coauthors, Men classified as hypo- or hyperresponders to dietary cholesterol feeding exhibit differences in lipoprotein metabolism, *Journal of Nutrition* 133 (2003): 1036–1042; K. L. Herron and coauthors, Pre-menopausal women, classified as hypo- or hyperresponding, do not alter their LDL/HDL ratio following a high dietary cholesterol challenge, *Journal of the American College of Nutrition* 21 (2002): 250–258.

13. J. F. Mauger and coauthors, Effect of different forms of dietary hydrogenated fats on LDL particle size, *American Journal of Clinical Nutrition* 78 (2003): 370–375; J. T. Judd and coauthors, Dietary cis and trans monounsaturated and saturated FA and plasma lipids and lipoproteins in men, *Lipids* 37 (2002): 123–131; Standing Committee on the Scientific Evaluation of Dietary Reference Intakes, Food and Nutrition Board, Institute of Medicine, *Dietary Reference Intakes for Energy, Carbohydrate, Fiber, Fat, Fatty Acids, Cholesterol, Protein, and Amino Acids* (Washington, D.C.: National Academy Press, 2002), pp. 9-1–9-32.

14. M. Rantala and coauthors, Apolipoprotein B gene polymorphisms and serum lipids: Meta-analysis of the role of genetic variation in responsiveness to diet, *American Journal of Clinical Nutrition* 71 (2000): 713–724.

15. Executive Summary of the Third Report of the National Cholesterol Education Program (NCEP) Expert Panel on Detection, Evaluation, and Treatment of High Blood Cholesterol in Adults (Adult Treatment Panel III), 2001.

16. Joint WHO/FAO Expert Consultation, *Diet, Nutrition and the Prevention of Chronic Diseases* (Geneva, Switzerland: World Health Organization, 2003), p. 56.

17. D. Hwang, Fatty acids and immune responses—A new perspective in searching for clues to mechanism, *Annual Review of Nutrition* 20 (2000): 431–456.

18. H. Tapiero and coauthors, Polyunsaturated fatty acids (PUFA) and eicosanoids in human health and pathologies, *Biomedicine and Pharmacotherapy* 56 (2002): 215–222.

19. J. P. Middaugh, Cardiovascular deaths among Alaskan Natives, 1980–1986, *American Journal of Public Health* 80 (1990): 282–285; J. Dyerberg, Linolenate-derived polyunsaturated fatty acids and prevention of atherosclerosis, *Nutrition Reviews* 44 (1986): 125–134.

20. J. N. Din, D. E. Newby, and A. D. Flapan, Omega 3 fatty acids and cardiovascular disease—Fishing for a natural treatment, *British Medical Journal* 328 (2004): 30–35; E. Dewailly and coauthors, n-3 fatty acids and cardiovascular disease risk factors among the Inuit of Nunavik, *American Journal of Clinical Nutrition* 74 (2001): 464–473.

21. E. Dewailly and coauthors, 2001.

22. T. A. Mori and L. J. Beilin, Long-chain omega 3 fatty acids, blood lipids and cardiovascular risk reduction, *Current Opinion in Lipidology* 12 (2001): 11–17.

23. P. Marckmann, Fishing for heart protection, *American Journal of Clinical Nutrition* 78 (2003): 781–782.

24. P. M. Kris-Etherton, W. S. Harris, and L. J. Appel, Fish consumption, fish oil, omega-3 fatty acids, and cardiovascular disease, *Circulation* 106 (2002): 2747–2757.

25. V. Wijendran and K. C. Hayes, Dietary n-6 and n-3 fatty acid balance and cardiovascular health, *Annual Review of Nutrition* 24 (2004): 597–615; G. Zhao and coauthors, Dietary α-linolenic acid reduces inflammatory and lipid cardiovascular risk factors in hypercholesterolemic men and women, *Journal of Nutrition* 134 (2004): 2991–2997; S. C. Larsson and coauthors, Dietary long-chain n-3 fatty acids for the prevention of cancer: A review of potential mechanisms, *American Journal of Clinical Nutrition* 79 (2004): 935–945; S. Kew and coauthors, Lack of effect of foods enriched with plant- or marine-derived n-3 fatty acids on human immune function, *American Journal of Clinical Nutrition* 77 (2003): 1287–1295.

26. Fish oil, American Heart Association Recommendation, 2000, available at www.americanheart.org; S. L. Connor and W. E. Connor, Are fish oils beneficial in the prevention and treatment of coronary artery disease? *American Journal of Clinical Nutrition* 66 (1997): S1031–S1020.

27. Standing Committee on the Scientific Evaluation of Dietary Reference Intakes, Food and Nutrition Board, Institute of Medicine, 2002, p. 8-26.

28. F. Thies and coauthors, Dietary supplementation with eicosapentaenoic acid, but not with other long-chain n-3 or n-6 polyunsaturated fatty acids, decreases natural killer cell activity in healthy subjects aged >55 y, *American Journal of Clinical Nutrition* 73 (2001): 539–548; V. M. Montori and coauthors, Fish oil supplementation in type 2 diabetes: A quantitative systematic review, *Diabetes Care* 23 (2000): 1407–1415.

29. M. T. Nakamura and coauthors, Metabolism and functions of highly unsaturated fatty acids: An update, *Lipids* 36 (2001): 961–964.

30. Backgrounder for the 2004 FDA/EPA Consumer Advisory: What you need to know about mercury in fish and shellfish, 2004, available at www.fda.gov/oc/opacom/hottopics/mercury/backgrounder.html; S. L. Schantz and coauthors, Impairment of memory and learning in older adults exposed to polychlorinated biphenyls via consumption of Great Lakes fish, *Environmental Health Perspectives* 110 (2002): A70–A72.

31. H. M. Chan and G. M. Egeland, Fish consumption, mercury exposure, and heart diseases, *Nutrition Reviews* 62 (2004): 68–72.

32. B. M. Popkin and coauthors, Where's the fat? Trends in U.S. diets 1965–1996, *Preventive Medicine* 32 (2001): 245–254.

33. S. L. Booth and coauthors, Effects of a hydrogenated form of vitamin K on bone formation and resorption, *American Journal of Clinical Nutrition* 74 (2001): 783–790.

34. D. J. Baer and coauthors, Dietary fatty acids affect plasma markers of inflammation in healthy men fed controlled diets: A randomized crossover study, *American Journal of Clinical Nutrition* 79 (2004): 969–973; M. B. Katan, *Trans* fatty acids and plasma lipoproteins, *Nutrition Reviews* 58 (2000): 188–191.

35. Bingham and coauthors, 2003.

36. Standing Committee on the Scientific Evaluation of Dietary Reference Intakes, Food and Nutrition Board, Institute of Medicine, 2002, p. 8-46; D. B. Allison and coauthors, Estimated intakes of *trans* fatty and other fatty acids in the US population, *Journal of the American Dietetic Association* 99 (1999): 166–174.

37. Revealing *trans* fats, *FDA Consumer*, September/October 2003, pp. 20–26.

38. A. H. Lichtenstein and coauthors, Effects of different forms of dietary hydrogenated fats on serum lipoprotein cholesterol levels, *New England Journal of Medicine* 340 (1999): 1933–1940.

39. E. A. Bell and B. J. Rolls, Energy density of foods affects energy intake across multiple levels of fat content in lean and obese women, *American Journal of Clinical Nutrition* 73 (2001): 1010–1018; M. Raeini-Sarjaz and coauthors, Comparison of the effect of dietary fat restriction with that of energy on human lipid metabolism, *American Journal of Clinical Nutrition* 73 (2001): 262–267.

40. Popkin and coauthors, 2001.

41. P. A. Cotton and coauthors, Dietary sources of nutrients among U.S. adults, 1994–1996, *Journal of the American Dietetic Association* 104 (2004): 921–930.

42. Committee on the Scientific Evaluation of Dietary Reference Intakes, Food and Nutrition Board, Institute of Medicine, 2002, p. 11-46.

43. T. Zind, Out to lunch, *Prepared Foods*, May 2003, available at www.preparedfoods.com.

F

Consumer Corner 5

1. Position of the American Dietetic Association: Fat replacers, *Journal of the American Dietetic Association* 105 (2005): 266–275; G. Inglett, Food ingredients from cereals with nutraceutical properties, ARS online publications, October 1, 2002, from the USDA Agricultural Research Service research project: *Heart Healthy Foods: Enzymatic and Physical Modifications of Carbohydrates*, available at www.ars.usda.gov/research/publications/publications.htm?SEQ_NO_115=140228.

2. R. S. Sandler and coauthors, Gastrointestinal symptoms in 3181 volunteers ingesting snack foods containing olestra or triglycerides: A 6-week randomized, placebo-controlled trial, *Annals of Internal Medicine* 130 (1999): 253–261.

3. G. S. Allgood and coauthors, Postmarketing surveillance of new food ingredients: Results from the program with the fat replacer olestra, *Regulatory Toxicology and Pharmacology* 33 (2001): 224–233.

4. FDA changes labeling requirement for olestra, *FDA Talk Paper* #T03-59, August 2003.

5. E. Kennedy and S. Bowman, Assessment of the effect of fat-modified foods on diet quality in adults, 19 to 50 years, using data from the Continuing Survey of Food Intake by Individuals, *Journal of the American Dietetic Association* 101 (2001): 455–460.

6. H. J. Roy and coauthors, Effect on body weight of replacing dietary fat with olestra for two or ten weeks in healthy men and women, *Journal of the American College of Nutrition* 21 (2002): 259–267.

7. R. E. Patterson and coauthors, Changes in diet, weight, and serum lipid levels associated with olestra consumption, *Archives of Internal Medicine* 160 (2000): 2600–2604.

Controversy 5

1. M. de Lorgeril and coauthors, Mediterranean diet, traditional risk factors, and the rate of cardiovascular complications after myocardial infarction: Final report of the Lyon Diet Heart Study, *Circulation* 99 (1999): 779–785.

2. F. B. Hu, J. E. Manson, and W. C. Willett, Types of dietary fat and risk of coronary heart disease: A critical review, *Journal of the American College of Nutrition* 20 (2001): 5–19.

3. M. Kanazawa and coauthors, Effects of a high-sucrose diet on body weight, plasma triglycerides, and stress tolerance, *Nutrition Reviews* 61 (2003): S27–S33; E. J. Parks, Changes in fat synthesis influenced by dietary macronutrient content, *Proceedings of the Nutrition Society* 61 (2002): 281–286.

4. *Third Report of the National Cholesterol Education Program (NCEP) Expert Panel on Detection, Evaluation, and Treatment of High Blood Cholesterol in Adults (Adult Treatment Panel III)*, 2002, National Institutes of Health publication no. 02-1205.

5. A. Keys, *Seven Countries: A Multivariate Analysis of Death and Coronary Heart Disease* (Cambridge, Mass.: Harvard University Press, 1980).

6. *Third Report of the National Cholesterol Education Program (NCEP) Expert Panel on Detection, Evaluation, and Treatment of High Blood Cholesterol in Adults (Adult Treatment Panel III)*, 2002; Committee on the Scientific Evaluation of Dietary Reference Intakes, Food and Nutrition Board, Institute of Medicine, *Dietary Reference Intakes for Energy, Carbohydrate, Fiber, Fat, Fatty Acids, Cholesterol, Protein, and Amino Acids* (Washington, D.C.: National Academy Press, 2002).

7. Committee on the Scientific Evaluation of Dietary Reference Intakes, Food and Nutrition Board, Institute of Medicine, 2002, p. 11-3.

8. A. H. Stark and Z. Madar, Olive oil as a functional food: Epidemiology and nutritional approaches, *Nutrition Reviews* 60 (2002): 170–176.

9. P. M. Kris-Etherton and coauthors, High-monounsaturated fatty acid diets lower both plasma cholesterol and triacyglycerol concentrations, *American Journal of Clinical Nutrition* 70 (1999): 1009–1015.

10. R. L. Hargrove and coauthors, Low fat and high monounsaturated fat diets decrease human low density lipoprotein oxidative susceptibility in vitro, *Journal of Nutrition* 131 (2001): 1758–1763.

11. C. M. Williams, Beneficial nutritional properties of olive oil: Implications for postprandial lipoproteins and factor VII, *Nutrition, Metabolism, and Cardiovascular Diseases* 11 (2001): 51–56; J. P. De La Cruz and coauthors, Antithrombotic potential of olive oil administration in rabbits with elevated cholesterol, *Thrombosis Research* 100 (2000): 305–315.

12. F. Visioli and coauthors, Virgin Olive Oil Study (VOLOS): Vasoprotective potential of extra virgin olive oil in mildly dyslipidemic patients, *European Journal of Nutrition* 44 (2005): 121–117.

13. L. A. Ferrara and coauthors, Olive oil and reduced need for antihypertensive medications, *Archives of Internal Medicine* 160 (2000): 837–842.

14. L. Van Horn and N. Ernst, A summary of the science supporting the new National Cholesterol Education program dietary recommendations: What dietitians should know, *Journal of the American Dietetic Association* 101 (2001): 1148–1154.

15. A. Trichopoulou and coauthors, Adherence to a Mediterranean diet and survival in a Greek population, *New England Journal of Medicine* 348 (2003): 2599–2608; A. T. Lada and L. L. Rudel, Dietary monounsaturated versus polyunsaturated fatty acids: Which is really better for protection from coronary heart disease? *Current Opinion in Lipidology* 14 (2003): 41–46.

16. D. B. Panageotakos, as quoted in Greek diet reduces inflammatory protein, *Science News* 164 (2003): 164.

17. D. R. Jacobs Jr. and L. M. Steffen, Nutrients, foods, and dietary patterns as exposures in research: A framework for food synergy, *American Journal of Clinical Nutrition* 78 (2003): 508S–513S.

18. P. M. Kris-Etherton and coauthors, The effects of nuts on coronary heart disease risk, *Nutrition Reviews* 59 (2001): 103–111.

19. F. B. Hu and M. J. Stampfer, Nut consumption and risk of coronary heart disease: A review of epidemiologic evidence, *Current Atherosclerosis Reports* 1 (1999): 204–209.

20. E. B. Feldman, The scientific evidence for a beneficial health relationship between walnuts and coronary heart disease, *Journal of Nutrition* 132 (2002): 1062S–1101S; D. Zambón and coauthors, Substituting walnuts for monounsaturated fat improves the serum lipid profile of hypercholesterolemic men and women: A randomized crossover trial, *Annals of Internal Medicine* 132 (2000): 538–546.

21. D. J. Jenkins and coauthors, Dose response of almonds on coronary heart disease risk factors: blood lipids, oxidized low-density lipoproteins, lipoprotein (a), homocysteine, and pulmonary nitric oxide: a randomized, controlled, crossover trial, *Circulation* 106 (2002): 1327–1332.

22. G. Zhao and coauthors, Dietary α-linolenic acid reduces inflammatory and lipid cardiovascular risk factors in hypercholeseterolemic men and women, *Journal of Nutrition* 134 (2004): 2991–2997.

23. C.-J. Tsai and coauthors, Frequent nut consumption and decreased risk of cholecystectomy in women, *American Journal of Clinical Nutrition* 80 (2004): 76–81.

24. M. A. Hallikainen and M. I. Uusitupa, Effects of 2 low-fat stanol ester-containing margarines on serum cholesterol concentrations as part of a low-fat diet in hypercholesterolemic subjects, *American Journal of Clinical Nutrition* 69 (1999): 403–410.

25. F. B. Hu and coauthors, Fish and omega-3 fatty acid intake and risk of coronary heart disease in women, *Journal of the American Medical Association* 287 (2002): 1815–1821; C. M. Albert and coauthors, Blood levels of long-chain n-3 fatty acids and the risk of sudden death, *New England Journal of Medicine* 346 (2002): 1113–1118.

26. H. Iso and coauthors, Intake of fish and omega-3 acids and risk of stroke in women, *Journal of the American Medical Association* 285 (2001): 304–312.

27. *Third Report of the National Cholesterol Education Program (NCEP) Expert Panel on Detection, Evaluation, and Treatment of High Blood Cholesterol in Adults (Adult Treatment Panel III)*, 2002, p.V-8.

28. Committee on the Scientific Evaluation of Dietary Reference Intakes, Food and Nutrition Board, Institute of Medicine, 2002, p. 11-22.

29. Committee on the Scientific Evaluation of Dietary Reference Intakes, Food and Nutrition Board, Institute of Medicine, 2002, pp. 11–46 and G-1.

30. Committee on the Scientific Evaluation of Dietary Reference Intakes, Food and Nutrition Board, Institute of Medicine, 2002, p. 11-19.

31. K. T. B. Knoops and coauthors, Mediterranean diet, lifestyle factors, and 10-year mortality in elderly European men and women, *Journal of the American Medical Association* 292 (2004): 1433–1439.

Chapter 6

1. Standing Committee on the Scientific Evaluation of Dietary Reference Intakes, Food and Nutrition Board, Institute of Medicine, *Dietary Reference Intakes for Energy, Carbohydrate, Fiber, Fat, Fatty Acids, Cholesterol, Protein, and Amino Acids* (Washington, D.C.: National Academy Press, 2002), pp. 10-2–10-3.

2. Standing Committee on the Scientific Evaluation of Dietary Reference Intakes, Food and Nutrition Board, Institute of Medicine, 2002, pp. 10-5–10-6; M. A. McNurlan and P. J. Garlick, Protein synthesis and degradation, in M. H. Stipanuk, ed., *Biochemical and Physiological Aspects of Human Nutrition* (Philadelphia: Saunders, 2000), pp. 211–232.

3. A. Tomer and coauthors, Thrombogenesis in sickle cell disease, *Journal of Laboratory and Clinical Medicine* 137 (2001): 398–407; S. T. Miller and coauthors, Prediction of adverse outcomes in children with sickle cell disease, *New England Journal of Medicine* 342 (2000): 83–89.

4. R. N. Kulkarni and C. R. Kahn, HNFs—linking the liver and pancreatic islets in diabetes, *Science* 303 (2004): 1311–1312; N. Moustaid-Moussa and C. D. Berdanier, *Nutrient-Gene Interactions in Health and Disease* (Boca Raton, Fla.: CRC Press, 2001).

5. M. J. Koury and P. Ponka, New insights into erythropoiesis: The roles of folate, vitamin B_{12}, and iron, *Annual Review of Nutrition* 24 (2004): 105–131.

6. B. R. Stevens, Digestion and absorption of protein, in M. H. Stipanuk, ed., *Biochemical and Physiological Aspect of Human Nutrition* (Philadelphia: Saunders, 2000), pp. 121–122.

7. T. R. Ziegler and coauthors, Trophic and cytoprotective nutrition for intestinal adaptation, mucosal repair, and barrier function, *Annual Review of Nutrition* 23 (2003): 229–261.

8. Standing Committee on the Scientific Evaluation of Dietary Reference Intakes, Food and Nutrition Board, Institute of Medicine 2002, p. 10-13.

9. Standing Committee on the Scientific Evaluation of Dietary Reference Intakes, Food and Nutrition Board, Institute of Medicine 2002, p. 10-14; M. A. McNurlan and P. J. Garlick, Protein synthesis and degradation, in M. H. Stipanuk, ed., *Biochemical and Physiological Aspects of Human Nutrition* (Philadelphia: Saunders, 2000), p. 306.

10. Position of the American Dietetic Association and Dietitians of Canada: Vegetarian diets, *Journal of the American Dietetic Association* 103 (2003): 748–765.

11. W. M. Rand, P. L. Pellett, and V. R. Young, Meta-analysis of nitrogen balance studies for estimating protein requirements in healthy adults, *American Journal of Clinical Nutrition* 77 (2003): 109–127.

12. L. D. Plank and G. L. Hill, Energy balance in critical illness, *Proceedings of the Nutrition Society* 62 (2003): 545–552.

13. J. L. McCrory and coauthors, Locomotion in simulated microgravity: Gravity replacement loads, *Aviation, Space, and Environmental Medicine* 73 (2002): 625–631.

14. B. Woodward, Protein, calories, and immune defenses, *Nutrition Reviews* 56 (1998): S84–S92.

15. L. Combaret, D. Taillandier, and D. Attaix, Nutritional and hormonal control of protein breakdown, *American Journal of Kidney Diseases* 37 (2001): S108–S111.

16. S. M. Day and D. H. DeHeer, Reversal of the detrimental effects of chronic protein malnutrition on long bone fracture healing, *Journal of Orthopaedic Trauma* 15 (2001): 47–53.

17. M. Reid and coauthors, The acute-phase protein response to infection in edematous and nonedematous protein-energy malnutrition, *American Journal Clinical Nutrition* 76 (2002): 1409–1415.

18. V. Scherbaum and P. Furst, New concepts on nutritional management of severe malnutrition: The role of protein, *Current Opinion in Clinical Nutrition and Metabolic Care* 3 (2000): 31–38.

19. S. M. Grantham-McGregor, S. P. Walker, and S. Chang, Nutritional deficiencies and later behavioral development, *Proceedings of the Nutrition Society* 59 (2000): 47–54.

20. T. Liu and coauthors, Kwashiorkor in the United States: Fad diets, perceived and true milk allergy, and nutritional ignorance, *Archives of Dermatology* 137 (2001): 630–636; N. F. Carvalho and coauthors, Severe nutritional deficiencies in toddlers resulting from health food milk alternatives (electronic article), *Pediatrics* 107 (2001): e46.

21. Position of the American Dietetic Association and Dietitians of Canada: Nutrition intervention in the care of persons with human immunodeficiency virus infection, *Journal of the American Dietetic Association* 100 (2000): 708–717.

22. Federal Interagency Forum on Child and Family Statistics, America's Children: Key National Indicators of Well-Being, 2001, available at www.childstats.gov.

23. B. A. Underwood and S. Smitasiri, Micronutrient malnutrition: Policies and programs for control and their implications, *Annual Review of Nutrition* 19 (1999): 303–324; C. G. Victora and coauthors, Potential interventions for the prevention of childhood pneumonia in developing countries: Improving nutrition, *American Journal of Clinical Nutrition* 70 (1999): 309–320.

24. A. Trichopoulou and coauthors, Lipid, protein and carbohydrate intake in relation to body mass index, *European Journal of Clinical Nutrition* 56 (2002): 37–43.

25. S. Tonstad, K. Smerud, and L. Hoie, A comparison of the effects of 2 doses of soy protein or casein on serum lipids, serum lipoproteins, and plasma total homocysteine in hypercholesterolemic subjects, *American Journal of Clinical Nutrition* 76 (2002): 78–84.

26. E. L. Knight and coauthors, The impact of protein intake on renal function decline in women with normal or mild renal insufficiency, *Annals of Internal Medicine* 138 (2003): 460–467.

27. B. Dawson-Hughes and S. S. Harris, Calcium intake influences the association of protein intake with rates of bone loss in elderly men and women, *American Journal of Clinical Nutrition* 75 (2002): 773–779.

28. S. T. St. Jeor and coauthors, Dietary protein and weight reduction, *Circulation* 104 (2001): 1869–1874.

Consumer Corner 6

1. H. Moriwaki and coauthors, Branched-chain amino acids as a protein- and energy-source in liver cirrhosis, *Biochemical and Biophysical Research Communications* 313 (2004): 405–409; M. A. Choudhry and coauthors, Enteral nutritional supplementation prevents mesenteric lymph node T-cell suppression in burn injury, *Critical Care Medicine* 31 (2003): 1764–1770; G. S. Sacks, L. Genton, and K. A. Kudsk, Controversy of immunonutrition for surgical critical-illness patients, *Current Opinion in Critical Care* 9 (2003): 300–305.

2. U. Suchner, D. K. Heyland, and K. Peter, Immune-modulatory actions of arginine in the critically ill, *British Journal of Nutrition* 87 (2002): S121–S132.

3. J. Hernández-Rodriguez and G. Manjarrez-Guitiérrez, Macronutrients and neurotransmitter formation during brain development, *Nutrition Reviews* 59 (2001): S49–S59.

4. Standing Committee on the Scientific Evaluation of Dietary Reference Intakes, Food and Nutrition Board, Institute of Medicine, *Dietary Reference Intakes for Energy, Carbohydrate, Fiber, Fat, Fatty Acids, Cholesterol, Protein, and Amino Acids* (Washington, D.C.: National Academy Press, 2002), pp. 10-105–10-106.

5. Impurities confirmed in dietary supplement 5-hydroxy-L-tryptophan, FDA Talk Paper, August 1998, available from Food and Drug Administration, U.S. Department of Health and Human Services, Public Health Service, 5600 Fishers Lane, Rockville, MD 20857.

6. T. L. Schwenk and C. D. Costley, When food becomes a drug: Nonanabolic nutritional supplement use in athletes, *American Journal of Sports Medicine* 30 (2002): 907–916.

7. P. Garlick, Assessment of the safety of glutamine and other amino acids, *Journal of Nutrition* 131 (2001): S2556–S2561.

Controversy 6

1. P. N. Singh, J. Sabate, and G. E. Fraser, Does low meat consumption increase life expectancy in humans? *American Journal of Clinical Nutrition* 78 (2003): 526S–532S.

2. S. I. Barr and G. E. Chapman, Perceptions and practices of self-defined current vegetarian, former vegetarian, and nonvegetarian women, *Journal of the American Dietetic Association* 102 (2002): 354–360; R. Weinsier, Use of the term vegetarian, *American Journal of Clinical Nutrition* 71 (2000): 1211–1212.

3. Position of the American Dietetic Association and Dietitians of Canada: Vegetarian diets, *Journal of the American Dietetic Association* 103 (2003): 748–765; U.S. Department of Agriculture and U.S. Department of Health and Human Services, *Nutrition and Your Health: Dietary Guidelines for Americans,* 5th ed., Home and Garden Bulletin no. 232 (Washington, D.C.: 2000), available online at www.usda.gov/cnpp or call (888) 878-3256.

4. E. H. Haddad and J. S. Tanzman, What do vegetarians in the United States eat? *American Journal of Clinical Nutrition* 78 (2003): 626S–632S; N. Brathwaite and coauthors, Obesity, diabetes, hypertension, and vegetarian status among Seventh-Day Adventists in Barbados: Preliminary results, *Ethnicity and Disease* 13 (2003): 34–39; J. Sabante, The contribution of vegetarian diets to health and disease: A paradigm shift? *American Journal of Clinical Nutrition* 78 (2003): 502S–507S.

5. G. E. Fraser, Associations between diet and cancer, ischemic heart disease, and all-cause mortality in non-Hispanic white California Seventh-Day Adventists, *American Journal of Clinical Nutrition* 70 (1999): 532S–538S; P. N. Appleby and coauthors, The Oxford Vegetarian Study: An overview, *American Journal of Clinical Nutrition* 70 (1999): 525S–531S.

6. F. B. Hu, Plant-based foods and prevention of cardiovascular disease: An overview, *American Journal of Clinical Nutrition* 78 (2003): 544S–551S.

7. Expert Panel on Detection, Evaluation, and Treatment of High Blood Cholesterol in Adults (Adult Treatment Panel III), Third Report of the National Cholesterol Education Program (NCEP) NIH Publication no. 02-5215 (Bethesda, Md.: National Heart, Lung, and Blood Institute, 2002).

8. Appleby and coauthors, 1999.

9. C. D. Gardner and coauthors, The effect of soy protein with or without isoflavones relative to milk protein on plasma lipids in hypercholesterolemic postmenopausal women, *American Journal of Clinical Nutrition* 73 (2001): 667–668.

10. E. L. Ashton, F. S. Dalais, and M. J. Ball, Effect of meat replacement by tofu on CHD risk factors including copper induced LDL oxidation, *Journal of the American College of Nutrition* 19 (2000): 761–767.

11. L. J. Appel, The effects of protein intake on blood pressure and cardiovascular disease, *Current Opinion in Lipidology* 14 (2003): 55–59; T. B. Clarkson, Soy, soy phyto-estrogens, and cardiovascular disease, *Journal of Nutrition* 132 (2002): 566S–569S; S. Tonstad, K. Smerud, and L. Hoie, A comparison of the effects of 2 doses of soy protein or casein on serum lipids, serum lipoproteins, and plasma total homocysteine in hypercholesterolemic subjects, *American Journal of Clinical Nutrition* 76 (2002): 78–84.

12. M. R. Adams and coauthors, The inhibitory effect of soy protein isolate on atherosclerosis in mice does not require the presence of LDL receptors or alteration of plasma lipoproteins, *Journal of Nutrition* 132 (2002): 43–49.

13. Clarkson, 2002.

14. M. Messina, C. Gardner, and S. Barnes, Gaining insight into the health effects of soy but a long way still to go: Commentary on the Fourth International Symposium on the Role of Soy in Preventing and Treating Chronic Disease, *Journal of Nutrition* 132 (2002): 547S–551S; S. R. Teixeira and coauthors, Effects of feeding 4 levels of soy protein for 3 and 6 wk on blood lipids and apoliproproteins in moderately hypercholesterolemic men, *American Journal of Clinical Nutrition* 71 (2000): 1077–1084.

15. Clarkson, 2002.

J. Jenkins and coauthors, High-protein diets in hyperlipidemia: Effect of wheat gluten on serum lipids, uric acid, and renal function, *American Journal of Clinical Nutrition* 74 (2001): 57–63.

17. Gardner and coauthors, 2001; M. Pfeuffer and J. Schrezenmeir, Bioactive substances in milk with properties decreasing risk of cardiovascular diseases, *British Journal of Nutrition* 84 (2000): S155–S159.

18. D. A. Lieberman and coauthors, Risk factors for advanced colonic neoplasia and hyperplastic polyps in asymptomatic individuals, *Journal of the American Medical Association* 290 (2003): 2959–2967.

19. E. Giovannucci, Modifiable risk factors for colon cancer, *Gastroenterology Clinics of North America* 31 (2002): 925–943.

20. C. Van Der Meer-Van Kraaij and coauthors, Mucosal pentraxin (Mptx), a novel rat gene 10-fold down-regulated in colon by dietary heme, *The Federation of American Societies for Experimental Biology* 17 (2003): 1277–1285; A. R. Eynard and C. B. Lopez, Conjugated linoleic acid (CLA) versus saturated fats/cholesterol: Their proportion in fatty and lean meats may affect the risk of developing colon cancer, *Lipids in Health and Disease* 2 (2003): 6; F. Pierre and coauthors, Meat and cancer: Haemoglobin and haemin in a low calcium diet promote colorectal carcinogenesis at the aberrant crypt state in rats, *Carcinogenesis* 24 (2003): 1683–1690.

21. H. Chen and coauthors, Dietary patterns and adenocarcinoma of the esophagus and distal stomach, *American Journal of Clinical Nutrition* 75 (2002): 137–144.

22. Lieberman and coauthors, 2003.

23. Lieberman and coauthors, 2003.

24. E. Gonzalez de Mejia, T. Bradford, and C. Hasler, The anticarcinogenic potential of soybean lectin and lunasin, *Nutrition Reviews* 61 (2003): 239–246; W. H. Xu and coauthors, Soya food intake and risk of endometrial cancer among Chinese women in Shanghai: population based case-control study, *British Medical Journal* 328 (2004): 1285–1288; E. Riboli and T. Norat, Epidemiologic evidence of the protective effect of fruit and vegetables on cancer risk, *American Journal of Clinical Nutrition* 78 (2003): 559S–569S; C. C. Yeh and coauthors, Risk factors for colorectal cancer in Taiwan: A hospital-based case control study, *Journal of the Formosan Medical Association* 102 (2003): 305–312; Giovannucci, 2002.

25. S. P. Stabler and R. H. Allen, Vitamin B12 deficiency as a worldwide problem, *Annual Review of Nutrition* 24 (2004): 299–326.

26. A. C. Antony, Vegetarianism and vitamin B-12 (cobalamin) deficiency, *American Journal of Clinical Nutrition* 78 (2003): 3–6; W. Herrmann and coauthors, Vitamin B-12 status, particularly holotranscobalamin II and methylmalonic acid concentrations, and hyperhomocysteinemia in vegetarians, *American Journal of Clinical Nutrition* 78 (2003): 131–136.

27. Neurologic impairment in children associated with maternal dietary deficiency of cobalamin—Georgia, 2001, *Morbidity and Mortality Weekly Reports* 52 (2003): 61–64.

28. M. Hebbelinck, P. Clarys, and A. DeMalsche, Growth, development, and physical fitness of Flemish vegetarian children, adolescents, young adults, *American Journal of Clinical Nutrition* 70 (1999): 579S–585S.

29. Position of the American Dietetic Association and Dietitians of Canada, 2003.

30. Position of the American Dietetic Association and Dietitians of Canada, 2003.

31. M. van Dusseldorp and coauthors, Risk of persistent cobalamin deficiency in adolescents fed a macrobiotic diet in early life, *American Journal of Clinical Nutrition* 69 (1999): 664–671.

32. A. R. Mangels and V. Messina, Considerations in planning vegan diets: Children, *Journal of the American Dietetic Association* 101 (2001): 661–669.

33. C. L. Perry and coauthors, Adolescent vegetarians: How well do their dietary patterns meet the Healthy People 2010 objectives? *Archives of Pediatrics and Adolescent Medicine* 156 (2002): 431–437.

34. C. A. Venti and C. S. Johnston, Modified food guide pyramid for lactovegetarians and vegans, *Journal of Nutrition* 132 (2002): 1050–1054; Messina and Mangels, 2001; T. J. Parsons and coauthors, Are levels of bone turnover related to lower bone mass of adolescents previously fed a macrobiotic diet? *Experimental and Clinical Endocrinology and Diabetes* 109 (2001): 288–293.

35. M. W. Louwman and coauthors, Signs of impaired cognitive function in adolescents with marginal cobalamin status, *American Journal of Clinical Nutrition* 72 (2000): 762–769.

36. D. Milea, N. Cassoux, and P. LeHoang, Blindness in a strict vegan, *New England Journal of Medicine* 342 (2000): 897–898.

37. C. L. Perry and coauthors, Characteristics of vegetarian adolescents in a multiethnic urban population, *Journal of Adolescent Health* 29 (2001): 406–416; Y. Martins, P. Pliner, and R. O'Connor, Restrained eating among vegetarians: Does a vegetarian eating style mask concerns about weight? *Appetite* 32 (1999): 145–154.

38. S. A. Klopp, C. J. Heiss, and H. S. Smith, Self-reported vegetarianism may be a marker for college women at risk for disordered eating, *Journal of the American Dietetic Association* 103 (2003): 745–747.

39. Venti and Johnston, 2002; Messina and Mangels, 2001.

40. Position of the American Dietetic Association and Dietetians of Canada, 2003.

41. J. R. Hunt, Moving toward a plant-based diet: Are iron and zinc at risk? *Nutrition Reviews* 60 (2002): 127–134.

42. J. R. Hunt and Z. K. Roughead, Nonheme-iron absorption, fecal ferritin excretion, and blood indexes of iron status in women consuming controlled lactoovovegetarian diets for 8 wk, *American Journal of Clinical Nutrition* 69 (1999): 944–952.

43. J. Kandiah, Impact of tofu or tofu + orange juice on hematological indices of lacto-ovo vegetarian females, *Plant Foods for Human Nutrition* 57 (2002): 197–204.

44. C. L. Larsson and G. K. Johansson, Dietary intake and nutritional status of young vegans and omnivores in Sweden, *American Journal of Clinical Nutrition* 76 (2002): 100–106; M. J. Ball and M. A. Bartlett, Dietary intake and iron status of Australian vegetarian women, *American Journal of Clinical Nutrition* 70 (1999): 353–358.

45. Hunt, 2002.

46. Milea, Cassoux, and LeHoang, 2000.

47. T. A. Outila and coauthors, Dietary intake of vitamin D in premenopausal, healthy vegans was insufficient to maintain concentrations of serum 25-hydroxyvitamin D and intact parathyroid hormone within normal ranges during the winter in Finland, *Journal of the American Dietetic Association* 100 (2000): 434–441.

48. C. M. Oomen and coauthors, α-Linolenic acid intake is not beneficially associated with 10-y risk of coronary artery disease incidence: The Zutphen Elderly Study, *American Journal of Clinical Nutrition* 74 (2001): 457–463.

49. W. E. Connor, n-3 Fatty acids from fish and fish oil: Panacea or nostrum? (editorial), *American Journal of Clinical Nutrition* 74 (2001): 415–416.

50. V. Messina, V. Melina, and A. R. Mangels, A new food guide for North American vegetarians, *Journal of the American Dietetic Association* 103 (2003): 771–775; Venti and Johnston, 2002; E. H. Haddad, J. Sabaté, and C. G. Whitten, Vegetarian food guide pyramid: A conceptual framework, *American Journal of Clinical Nutrition* 70 (1999): 615S–619S.

51. Venti and Johnston, 2002.

Chapter 7

1. R. Rodriguez-Melendez, J. B. Griffin, and J. Zempleni, Biotin supplementation increases expression of the cytochrome P_{450} 1B1 in Jurkat cells, increasing the occurrence of single-stranded DNA breaks. *Journal of Nutrition* 134 (2004): 2222–2228; B. N. Ames and P. Wakimoto, Are vitamin and mineral deficiencies a major cancer risk? *Nature Reviews: Cancer* 2 (2002): 694–704.

2. S. Kato, Molecular mechanism of transcriptional control by nuclear vitamin receptors, *British Journal of Nutrition* 84 (2000): S229–S233.

3. S. Perrotta and coauthors, Vitamin A and infancy: Biochemical, functional, and clinical aspects, *Vitamins and Hormones* 66 (2003): 457–591; M. Clagett-Dame and H. F. DeLuca, The role of vitamin A in mammalian reproduction and embryonic development, *Annual Review of Nutrition* 22 (2002): 347–381.

4. J. Bastien and C. Rochette-Egly, Nuclear retinoid receptors and the transcription of retinoid-target genes, *Gene* 328 (2004): 1–16; L. N. Wei, Retinoid receptors and their coregulators, *Annual Review of Pharmacology and Toxicology* 43 (2003): 47–72; R. Zolfaghari and A. C. Ross, Recent advances in molecular cloning of fatty acid desaturase genes and the regulation of their expression by dietary vitamin A and retinoic acid, *Prostaglandins, Leukotrienes and Essential Fatty Acids* 68 (2003): 171–179; J. E. Balmer and R. Blomhoff, Gene expression regulation by retinoic acid, *Journal of Lipid Research* 43 (2002): 1773–1808.

5. Balmer and Blomhoff, 2002.

6. G. L. Johanning and C. J. Piyathilake, Retinoids and epigenetic silencing in cancer, *Nutrition Reviews* 61 (2003): 284–289.

7. R. Reifen, Vitamin A as an anti-inflammatory agent, *Proceedings of the Nutrition Society* 61 (2002): 397–400.

8. M. Maden and M. Hind, Retinoic acid, a regeneration-inducing molecule, *Developmental Dynamics* 226 (2003): 237–244; L. Bonet and coauthors, Vitamin A and the regulation of fat reserves, *Cellular and Molecular Life Sciences* 60 (2003): 1311–1321; Perrotta and coauthors, 2003; Clagett-Dame and DeLuca, 2002.

9. A. C. Ross, Advances in retinoid research: Mechanisms of cancer chemoprevention, symposium introduction, *Journal of Nutrition* 133 (2003): 271S–272S.

10. C. Ballew and coauthors, Serum retinol distributions in residents of the United States: Third National Health and Nutrition Examination Survey, 1988–1994, *American Journal of Clinical Nutrition* 73 (2001): 586–593.

11. Standing Committee on the Scientific Evaluation of Dietary Reference Intakes, Food and Nutrition Board, Institute of Medicine, *Dietary Reference Intakes for Vitamin A, Vitamin K, Arsenic, Boron, Chromium, Copper, Iodine, Iron, Manganese, Molybdenum, Nickel, Silicon, Vanadium, and Zinc* (Washington, D.C.: National Academy Press, 2001), pp. 4-9–4-10.

12. M. A. Dijkhuizen and coauthors, Concurrent micronutrient deficiencies in lactating mothers and their infants in Indonesia, *American Journal of Clinical Nutrition* 73 (2001): 786–791.

13. E. Villamor and coauthors, Vitamin A supplements ameliorate the adverse effect of HIV-1, malaria, and diarrheal infections on child growth, *Pediatrics* 109 (2002): available at www.pediatrics.org/cgi/content/full/109/1/e6; C. Duggan and W. Fawzi, Micronutrients and child health: Studies in international nutrition and HIV infection, *Nutrition Reviews* 59 (2001): 358–369.

14. P. S. Genaro and L. A. Martini, Vitamin A supplementation and risk of skeletal fracture, *Nutrition Reviews* 62 (2004): 65–67; K. Michaelsson and coauthors, Serum retinol levels and the risk of fractures, *New England Journal of Medicine* 348 (2003): 287–294; S. Johnasson and coauthors, Subclinical hypervitaminosis A causes fragile bones in rats, *Bone* 31 (2002): 685–689; D. Feskanich and coauthors, Vitamin A intake and hip fractures among postmenopausal women, *Journal of the American Medical Association* 287 (2002): 47–54; N. Binkley and D. Krueger, Hypervitaminosis A and bone, *Nutrition Reviews* 58 (2000): 138–144.

15. Genaro and Martini, 2004; K. L. Penniston and S. A. Tanumihardjo, Vitamin A in dietary supplements and fortified foods: Too much of a good thing? *Journal of the American Dietetic Association* 103 (2003): 1185–1187.

16. L. H. Allen and M. Haskell, Estimating the potential for vitamin A toxicity in women and young children, *Journal of Nutrition* 132 (2003): 2907S–2919S.

17. S. J. Hickenbottom and coauthors, Variability in conversion of ß-carotene to vitamin A in men as measured by using a double-tracer study design, *American Journal of Clinical Nutrition* 75 (2002): 900–907; K. J. Yeum and R. M. Russell, Carotenoid bioavailability and bioconversion, *Annual Review of Nutrition* 22 (2002): 483–504.

18. A. Mazzone and A. dal Canton, Images in clinical medicine—Hypercarotenemia, *New England Journal of Medicine* 346 (2002): 821.

19. P. F. Jacques and coauthors, Long-term nutrient intake and early age-related nuclear lens opacities, *Archives of Ophthalmology* 119 (2001): 1009–1019; J. T. Landrum and R. A. Bone, Lutein, zeaxanthin, and the macular pigment, *Archives of Biochemistry and Biophysics* 385 (2001): 28–40.

20. B. R. Hammond Jr., B. R. Wooten, and J. Curran-Celentano, Carotenoids in the retina and lens: Possible acute and chronic effects on human visual performance, *Archives of Biochemistry and Biophysics* 385 (2001): 41–46.

21. C. Moore and coauthors, Vitamin D intake in the United States, *Journal of the American Dietetic Association* 104 (2004): 980–983; M. S. Calvo and S. J. Whiting, Prevalence of vitamin D insufficiency in Canada and the United States: Importance to health status and efficacy of current food fortification and dietary supplement use, *Nutrition Reviews* 61 (2003): 107–113; S. Nesby-O'Dell and coauthors, Hypovitaminosis D prevalence and determinants among African American and white women of reproductive age: Third National Health and Nutrition Examination Survey, 1988–1994, *American Journal of Clinical Nutrition* 76 (2002): 187–192.

22. S. Kato, Molecular mechanism of transcriptional control by nuclear vitamin receptors, *British Journal of Nutrition* 84 (2000): S229–S233.

23. M. T. Cantorna and coauthors, Vitamin D status, 1, 25-dihydroxyvitamin D_3, and the immune system, *American Journal of Clinical Nutrition* 80 (2004): 1717S–1720S; M. F. Holick, Vitamin D: Importance in the prevention of cancers, type 1 diabetes, heart disease, and osteoporosis, *American Journal of Clinical Nutrition* 79 (2004): 362–371; J. B. Zella and H. F. DeLuca, Vitamin D and autoimmune diabetes, *Journal of Cell Biochemistry* 88 (2003): 216–222; A. Zitterman, Vitamin D in preventive medicine: Are we ignoring the evidence? *British Journal of Nutrition* 89 (2003): 552–572; I. A. van der Mei and coauthors, Past exposure to sun, skin phenotype, and risk of multiple sclerosis: Case-control study, *British Medical Journal* 327 (2003): 316–332; G. Wolf, Intestinal bile acids can bind to and activate the vitamin D receptor, *Nutrition Reviews* 60 (2002): 281–288.

24. P. Weisberg and coauthors, Nutritional rickets among children in the United States: Review of cases reported between 1986 adn 2003, *Ameircan Journal of Clinical Nutrition* 80 (2004): 1697S–1705S; S. R. Kreiter and coauthors, Nutritional rickets in African American breast-fed infants, *Journal of Pediatrics* 137 (2000): 143–145.

25. M. K. M. Lehtonen-Veromaa and coauthors, Vitamin D and attainment of peak bone mass among peripubertal Finnish girls: A 3-y prospective study, *American Journal of Clinical Nutrition* 76 (2002): 1446–1453; T. A. Outila, M. U. M. Kärkkäinen, and C. J. E. Lamberg-Allardt, Vitamin D status affects serum parathyroid hormone concentrations during winter in female adolescents: Associations with forearm bone mineral density, *American Journal of Clinical Nutrition* 74 (2001): 206–210.

26. B. Liebman, Soaking up the D's: An interview with Michael F. Holick, *Nutrition Action Healthletter*, December 2003, pp. 1, 3–6.

27. Calvo and Whiting, 2003.

28. Calvo and Whiting, 2003; V. Tangpricha and coauthors, Vitamin D insufficiency among free-living adults, *American Journal of Medicine* 112 (2002): 659–662.

29. M. F. Holic, Sunlight and vitamin D for bone health and prevention of autoimmune diseases, cancers, and cardiovascular disease, *American Journal of Clinical Nutrition* 80 (2004): 1678S–1688S; R. P. Heaney and coauthors, Human serum 25-hydroxycholecalciferol response to extended oral dosing with cholecalciferol, *American Journal of Clinical Nutrition* 77 (2003): 204–210.

30. E. R. Miller and coauthors, Meta-Analysis: High-dosage vitamin E supplementation may increase all-cause mortality, *Annals of Internal Medicine* 142 (2005), available from www.annals.org; P. M. Kris-Etherton and coauthors, Antioxidant vitamin supplements and cardiovascular disease, *Circulation* 110 (2004): 637–641; J. E. Manson, S. S. Bassuk, and M. J. Stampfer, Does vitamin E supplementation prevent cardiovascular events? *Journal of Women's Health* 12 (2003): 123–136; L. Kritharides and R. Stocker, The use of antioxidant supplements in coronary heart disease, *Atherosclerosis* 164 (2002): 211–219.

31. U.S. Preventive Services Task Force, Routine vitamin supplementation to prevent cancer and cardiovascular disease: Recommendations and rationale, *Annals of Internal Medicine* 139 (2003): 51–55; C. D. Morris and S. Carson, Routine vitamin supplementation to prevent cardiovascular disease: A summary of the evidence for the U.S. Preventive Services Task Force, *Annals of Internal Medicine* 139 (2003): 56–70.

32. M. A. Beck, Nutritionally induced oxidative stress: Effect on viral disease, *American Journal of Clinical Nutrition* 71 (2000): 1676S–1679S.

33. E. R. Miller and coauthors, Meta-Analysis: High-dosage vitamin E supplementation may increase all-cause mortality, *Annals of Internal Medicine* 142 (2005), e-pub available from www.annals.org.

34. C. T. Taylor and coauthors, Vitamin K to reverse excessive anticoagulation: A review of the literature, *Pharmacotherapy* 19 (1999): 1415–1425.

35. S. L. Booth and coauthors, Vitamin K intake and bone mineral density in women and men, *American Journal of Clinical Nutrition* 77 (2003): 512–516.

36. N. C. Binkley and coauthors, A high phylloquinone intake is required to achieve maximal osteocalcin γ-carboxylation, *American Journal of Clinical Nutrition* 76 (2002): 1055–1060.

37. S. J. Padayatty and coauthors, Vitamin C as an antioxidant: Evaluation of its role in disease prevention, *Journal of the American College of Nutrition* 22 (2003): 18–35.

38. U.S. Preventive Services Task Force, 2003; Padayatty, 2003.

39. Standing Committee on the Scientific Evaluation of Dietary Reference Intakes, Food and Nutrition Board, Institute of Medicine, 2000, p. 101.

40. A. Alberg, The influence of cigarette smoking on circulating concentrations of antioxidant micronutrients, *Toxicology* 180 (2002): 121–137.

41. A. M. Preston and coauthors, Influence of environmental tobacco smoke on vitamin C status in children, *American Journal of Clinical Nutrition* 77 (2003): 167–172; R. S. Strauss, Environmental tobacco smoke and serum vitamin C levels in children, *Pediatrics* 107 (2001): 540–542.

42. C. S. Johnston and D. L. Bowling, Stability of ascorbic acid in commercially available orange juices, *Journal of the American Dietetic Association* 102 (2002): 525–529.

43. K. J. Carpenter, Acute versus marginal deficiencies of nutrients, *Nutrition Reviews* 60 (2002): 277–280.

44. H. J. Powers, Riboflavin (vitamin B-2) and health, *American Journal of Clinical Nutrition* 77 (2003): 1352–1360.

45. P. A. Cotton and coauthors, Dietary sources of nutrients among US Adults, 1994–1996, *Journal of the American Dietetic Association* 104 (2004): 921–930.

46. J. L. Parra and K. R. Reddy, Hepatotoxicity of hypolipidemic drugs, *Clinics in Liver Disease* 7 (2003): 415–433.

47. P. J. Stover, Physiology of folate and vitamin B_{12} in health and disease, *Nutrition Reviews* 62 (2004): S3–S4.

48. L. B. Bailey, G. C. Rampersaud, and G. P. A. Kauwell, Folic acid supplements and fortification affect the risk for neural tube defects, vascular disease and cancer: Evolving science, *Journal of Nutrition* 133 (2003): 1961S–1968S; G. C. Rampersaud, L. B. Bailey, and G. P. A. Kauwell, Relationship of folate to colorectal and cervical cancer: Review and recommendations for practitioners, *Journal of the American Dietetic Association* 102 (2002): 1273–1282; S. W. Choi and J. B. Mason, Folate and carcinogenesis: An integrated scheme, *Journal of Nutrition* 130 (2000): 129–132; Y. I. Kim, Methylenetetrahydrofolate reductase polymorphisms, folate, and cancer risk: A paradigm of gene-nutrient interactions in carcinogenesis, *Nutrition Reviews* 58 (2000): 205–209.

49. A. Fleming, The role of folate in the prevention of neural tube defects: Human and animal studies, *Nutrition Reviews* 59 (2001): S13–S23.

50. Centers fort Disease Control and Prevention, Spina bifida and anencephaly before and after folic acid mandate—United States, 1995–1996 and 1999–2000, *Morbidity and Mortality Weekly Report* 53 (2004): 362–365; E. P. Quinlivan and J. F. Gregory III, Effect of food fortification on folic acid intake in the United States, *American Journal of Clinical Nutrition* 77 (2003): 221–225.

51. I. A. Brouwer and coauthors, Low-dose folic acid supplementation decreases plasma homocysteine concentrations: A randomized trial, *American Journal of Clinical Nutrition* 69 (1999): 99–104.

52. S. P. Rothenberg, Increasing the dietary intake of folate: Pros and cons, *Seminars in Hematology* 36 (1999): 65–74.

53. J. L. Mills and coauthors, Low vitamin B-12 concentrations in patients without anemia: The effect of folic acid fortification of grain, *American Journal of Clinical Nutrition* 77 (2003): 1474–1477.

54. S. P. Stabler and R. H. Allen, Vitamin B$_{12}$ deficiency as a worldwide problem, *Annual Review of Nutrition* 24 (2004): 299–326.

55. W. Hermann and J. Geisel, Vegetarian lifestyle and monitoring of vitamin B-12 status, *Clinica Chimica Acta; International Journal of Clinical Chemistry* 326 (2002): 47–59.

56. R. Muhammad, Neurologic impairment in children associated with maternal dietary deficiency of cobalamin—Georgia, 2001, *Morbidity and Mortality Weekly Report* 52 (2003): 61–64.

57. K. M. Fairfield and R. H. Fletcher, Vitamins for chronic disease prevention in adults, *Journal of the American Medical Association* 287 (2002): 3116–3126.

58. E. Aufiero and coauthors, Pyridoxine hydrochloride treatment of carpal tunnel syndrome: A review, *Nutrition Reviews* 62 (2004): 96–104; A. A. Gerritsen and coauthors, Conservative treatment options for carpal tunnel syndrome: A systematic review of randomized controlled trials, *Journal of Neurology* 249 (2002): 272–280; R. Luboshitzky and coauthors, The effect of pyridoxine administration on melatonin secretion in normal men, *Neuroendocrinology Letters* 23 (2002): 213–217.

59. Fairfield and Fletcher, 2002; J. W. Miller, Does lowering plasma homocysteine reduce vascular disease risk? *Nutrition Reviews* 59 (2001): 242–244.

60. F. V. van Oort and coauthors, Folic acid and reduction of plasma homocysteine concentrations in older adults: A dose-response study, *American Journal of Clinical Nutrition* 77 (2003): 1318–1323; D. S. Wald, M. Law, and J. K. Morris, Homocysteine and cardiovascular disease: Evidence on causality from a meta-analysis, *British Medical Journal* 325 (2002): 1202–1208; B. J. Venn and coauthors, Dietary counseling to increase natural folate intake: A randomized placebo-controlled trial in free-living subjects to assess effects on serum folate and plasma total homocysteine, *American Journal of Clinical Nutrition* 76 (2002): 758–765.

61. S. Friso and coauthors, Low plasma vitamin B-6 concentrations and modulation of coronary artery disease risk, *American Journal of Clinical Nutrition* 79 (2004): 992–998; Bailey, Rampersaud, and Kauwell, 2003.

62. N. M. van der Put and coauthors, Folate, homocysteine and neural tube defects: An overview, *Experimental Biology and Medicine* 226 (2001): 243–270; P. F. Jacques and coauthors, The effect of folic acid fortification on plasma folate and total homocysteine concentrations, *New England Journal of Medicine* 340 (1999): 1449–1454.

63. Lengthy jail sentence for vendor of laetrile—a quack medication to treat cancer patients, *FDA News*, June 22, 2004, available at www.fda.gov/bbs/topics/news/2004/NEW01080.html.

Consumer Corner 7

1. L. Pauling, *Vitamin C and the Common Cold* (San Francisco: Freeman, 1970).

2. B. Takkouche and coauthors, Intake of vitamin C and zinc and risk of common cold: A cohort study, *Epidemiology* 13 (2002): 38–44; H. Hemilä and coauthors, Vitamin C, vitamin E, and beta-carotene in relation to common cold incidence in male smokers, *Epidemiology* 13 (2002): 32–37.

3. R. M. Douglas, E. B. Chalker, and B. Tracey, Vitamin C for preventing and treating the common cold (Cochrane Review), *The Cochrane Library* 2 (2001): available at www.update-software.com/abstracts/ab000980.htm; M. Van Straten and P. Josling, Preventing the common cold with a vitamin C supplement: A double-blind, placebo-controlled survey, *Journal of Manipulative and Physiological Therapeutics* 22 (1999): 530–533.

4. H. Hemilä, Vitamin C supplementation and common cold symptoms: Factors affecting the magnitude of the benefit, *Medical Hypotheses* 52 (1999): 171–178.

5. C. J. Field, I. R. Johnson, and P. D. Schley, Nutrients and their role in host resistance to infection, *Journal of Leukocyte Biology* 71 (2002): 16–32; V. M. Victor, N. Guayerbas, and F. M. De, Changes in the antioxidant content of mononuclear leukocytes from mice with endotoxin-induced oxidative stress, *Molecular Cell Biochemistry* 229 (2002): 107–111.

6. M. Farva and coauthors, The problem of the placebo response in clinical trials for psychiatric disorders: Culprits, possible remedies, and a novel study design approach, *Psychotherapy and Psychosomatics* 72 (2003): 115–127.

7. Standing Committee on the Scientific Evaluation of Dietary Reference Intakes, Food and Nutrition Board, Institute of Medicine, *Dietary Reference Intakes for Vitamin C, Vitamin E, Selenium, and Carotenoids* (Washington, D.C.: National Academy Press, 2000).

Controversy 7

1. R. H. Fletcher and K. M. Fairfield, Vitamins for chronic disease prevention in adults: Clinical applications, *Journal of the American Medical Association* 287 (2002): 3127–3129; W. C. Willet and M. J. Stampfer, What vitamins should I be taking, doctor? *New England Journal of Medicine* 345 (2001): 1819–1824.

2. K. Michaelsson and coauthors, Serum retinol levels and the risk of fracture, *New England Journal of Medicine* 348 (2003): 287–294; D. A. Bender, Daily doses of multivitamin tablets: Regular consumption will probably do you no good, with a few exceptions, *British Medical Journal* 325 (2002): 173–174.

3. Position of the American Dietetic Association: Food fortification and dietary supplements, *Journal of the American Dietetic Association* 101 (2001): 115–125.

4. U.S. Preventive Services Task Force, Routine vitamin supplementation to prevent cancer and cardiovascular disease, AHRQ Publication no. APPIP03-0012, June 2003, available at www.ahrq.gov.

5. Michaelsson and coauthors, 2003.

6. Committee on the Framework for Evaluating the Safety of Dietary Supplements, Institute of Medicine and National Research Council, *Dietary Supplements: A Framework for Evaluating Safety* (Washington, D.C.: National Academy Press, 2004), pp. ES1–ES14.

7. C. L. Taylor, Regulatory frameworks for functional foods and dietary supplements, *Nutrition Reviews* 62 (2004): 55–59.

8. Acting FDA Commissioner Dr. Lester M. Crawford outlines science-based plan for dietary supplement enforcement, *FDA News*, April 19, 2004, available at www.fda.gov/bbs/topics/news/2004/NFW01055.html.

9. M. Meadows, Ensuring the safety of dietary supplements, *FDA Consumer*, July/August 2004, available at www.fda.gov/fdac/features/2004/404_supp.html.

10. U.S. Food and Drug Administration, Center for Food Safety and Applied Nutrition, Claims that can be made for conventional foods and dietary supplements, September 2003, available at www.cfsan.fda.gov.

11. B. N. Ames, The metabolic tune-up: Metabolic harmony and disease prevention, *Journal of Nutrition* 133 (2003): 1544S–1548S.

12. R. J. Hillstrom, A. K. Yacapin-Ammons, and S. M. Lynch, Vitamin C inhibits lipid oxidation in human HDL, *Journal of Nutrition* 133 (2003): 3047–3051.

13. C. D. Morris and S. Carson, Routine vitamin supplementation to prevent cardiovascular disease: A summary of the evidence for the U.S. Preventive Services Task Force, *Annals of Internal Medicine* 139 (2003): 56–70.

14. Ames, 2003.

15. Ames, 2003.

16. Fletcher and Fairfield, 2002.

17. P. M. Kris-Etherton and coauthors, Antioxidant vitamin supplements and cardiovascular disease: AHA science advisory, *Circulation* 110 (2004): 637–641.

18. U. S. Preventive Services Task Force, 2003.

19. L. T. Chylack and coauthors, The Roche European American Cataract Trial (REACT): A randomized clinical trial to investigate the efficacy of an oral antioxidant micronutrient mixture to slow progression of age-related cataract, *Ophthalmic Epidemiology* 9 (2002): 49–80.

20. A. Martin and coauthors, Roles of vitamins E and C on neurodegenerative diseases and cognitive performance, *Nutrition Reviews* 60 (2002): 308–326; H. Chen and coauthors, Dietary patterns and adenocarcinoma of the esophagus and distal stomach, *American Journal of Clinical Nutrition* 75 (2002): 137–144.

21. E. R. Berton and coauthors, A population-based case-control study of carotenoid and vitamin A intake and ovarian cancer (United States), *Cancer Causes and Control* 12 (2001): 83–90; Standing Committee on the Scientific Evaluation of Dietary Reference Intakes, Food and Nutrition Board, Institute of Medicine, *Dietary Reference Intakes for Vitamin C, Vitamin E, Selenium, and Carotenoids* (Washington, D.C.: National Academy Press, 2000), p. 346; D. S. Michaud and coauthors, Intake of specific carotenoids and risk of lung cancer in 2 prospective US cohorts, *American Journal of Clinical Nutrition* 72 (2000): 990–997; M. L. Slattery and coauthors, Carotenoids and colon cancer, *American Journal of Clinical Nutrition* 71 (2000): 575–582; N. McKeown and coauthors, Antioxidants and breast cancer, *Nutrition Reviews* 57 (1999): 321–324; E. Giovannucci, Tomatoes, tomato-based products, lycopene, and cancer: Review of the epidemiologic literature, *Journal of the National Cancer Institute* 91 (1999): 317–331.

22. I. Min Lee and coauthors, ß-carotene supplementation and incidence of cancer and cardiovascular disease: The Women's Health Study, *Journal of the National Cancer Institute* 91 (1999): 2102–2106.

23. R. A. Jacob, Evidence that diet modification reduces in vivo oxidant damage, *Nutrition Reviews* 57 (1999): 255–258.

24. A. Iannuzzi and coauthors, Dietary and circulating antioxidant vitamins in relation to carotid plaques in middle-aged women, *American Journal of Clinical Nutrition* 76 (2002): 582–587; M. Meydani, Effect of functional food ingredients: Vitamin E modulation of cardiovascular diseases and immune status in the elderly, *American Journal of Clinical Nutrition* 71 (2000): 1665S–1668S; L. A. Yochum, A. R. Folsom, and L. H. Kushi, Intake of antioxidant vitamins and risk of death from stroke in postmenopausal women, *American Journal of Clinical Nutrition* 72 (2000): 476–483.

25. S. Devaraj, A. Harris, and I. Jialal, Modulation of monocyte-macrophage function with α-tocopherol: Implications for atherosclerosis, *Nutrition Reviews* 60 (2002): 8–14; L. J. van Tits and coauthors, α-Tocopherol supplementation decreases production of superoxide and cytokines by leukocytes ex vivo in both normolipidemic and hypertriglyceridemic individuals, *American Journal of Clinical Nutrition* 71 (2000): 458–464; M. Meydani, Vitamin E and prevention of heart disease in high-risk patients, *Nutrition Reviews* 58 (2000): 278–281.

26. P. Knekt and coauthors, Antioxidant vitamins and coronary heart disease risk: A pooled analysis of 9 cohorts, *American Journal of Clinical Nutrition* 80 (2004): 1508–1520; D. D. Waters and coauthors, Effects of hormone replacement therapy and antioxidant vitamin supplements on coronary atherosclerosis in postmenopausal women: A randomized controlled trial, *Journal of the American Medical Association* 288 (2002): 2432–2440; Collaborative Group of the Primary Prevention Proj-ect (PPP), Low-dose aspirin and vitamin E in people at cardiovascular risk: A randomized trial in general practice, *Lancet* 357 (2001): 89–95.

27. U.S. Preventive Services Task Force, 2003; L. Liu and coauthors, Vitamin C preserves endothelial function in patients with coronary heart disease after a high-fat meal, *Clinical Cardiology* 25 (2002): 219–224.

28. E. R. Miller and coauthors, Meta-analysis: highdosage vitamin E supplementation may increase all-cause mortality, *Annals of Internal Medicine* 142 (2005), e-pub available at www.annalas.org; S. L. Booth and coauthors, Effect of vitamin E supplementation on vitamin K status in adults with normal coagulation status, *American Journal of Clinical Nutrition* 80 (2004): 143–148.

29. D. Lee and coauthors, Does supplemental vitamin C increase cardiovascular disease risk in women with diabetes? *American Journal of Clinical Nutrition* 80 (2004): 1194–1200.

30. R. Rodriguez-Melendez, J. B. Griffin, and J. Zempleni, Biotin supplementation increases expression of the cytochrome P_{450} 1B1 in Jurkat cells, increasing the occurrence of single-stranded DNA breaks, *Journal of Nutrition* 134 (2004): 2222–2228.

31. B. Liebman, Spin the bottle: How to pick a multivitamin, *Nutrition Action Health Letter*, January/February 2003, pp. 3–9.

32. L. Troppmann, K. Gray-Donald, and T. Johns, Supplement use: Is there any nutritional benefit? *Journal of the American Dietetic Association* 102 (2002): 818–825.

Chapter 8

1. Standing Committee on the Scientific Evaluation of Dietary Reference Intakes, Food and Nutrition Board, Institute of Medicine, *Dietary Reference Intakes: Water, Potassium, Sodium, Chloride, and Sulfate* (Washington, D.C.: National Academy Press, 2004), p. 4-7.

2. Standing Committee on the Scientific Evaluation of Dietary Reference Intakes, Food and Nutrition Board, Institute of Medicine, 2004, p. 4-26.

3. Standing Committee on the Scientific Evaluation of Dietary Reference Intakes, Food and Nutrition Board, Institute of Medicine, 2004, p. 4-29.

4. J. W. Gardner, Death by water intoxication, *Military Medicine* 167 (2002): 432–434.

5. Standing Committee on the Scientific Evaluation of Dietary Reference Intakes, Food and Nutrition Board, Institute of Medicine, 2004, p. 4-49.

6. A. C. Grandjean and coauthors, The effect of caffeinated, non-caffeinated, caloric and non-caloric beverages on hydration, *Journal of the American College of Nutrition* 19 (2000): 591–600.

7. Standing Committee on the Scientific Evaluation of Dietary Reference Intakes, Food and Nutrition Board, Institute of Medicine, 2004, p. 4-55.

8. M. P. Sauvant and D. Pepin, Geographic variation of the mortality from cardiovascular disease and drinking water in a French small area (Puy de Dome), *Environmental Research* 84 (2000): 219–227.

9. H. Bohmer, H. Muller, and K. L. Resch, Calcium supplementation with calcium-rich mineral waters: A systematic review and meta-analysis of its bioavailability, *Osteoporosis International* 11 (2000): 938–943; R. Maheswaran and coauthors, Magnesium in drinking water supplies and mortality from acute myocardial infarction in north west England, *Heart* 82 (1999): 455–460.

10. S. Satarug and coauthors, A global perspective on cadmium pollution and toxicity in non-occupationally exposed population, *Toxicology Letters* 137 (2003): 65–83.

11. R. S. Barwick and coauthors, Surveillance for waterborne-disease outbreaks—United States, 1997–1998, *Morbidity and Mortality Weekly Report* 49 (2000): 1–35.

12. N. W. Woodruff and coauthors, Human cell mutagenicity of chlorinated and unchlorinated water and the disinfection byproduct 3-chloro-4-(dichloromethyl)-5-hydroxy-2(5H)-furanone (MX), *Mutation Research* 495 (2001): 157–168; S. F. Thai and coauthors, Detection of early gene expression changes by differential display in the livers of mice exposed to dichloroacetic acid, *Carcinogenesis* 22 (2001): 1317–1322.

13. K. P. Cantor and coauthors, Drinking water source and chlorination byproducts in Iowa. III. Risk of brain cancer, *American Journal of Epidemiology* 150 (1999): 552–560.

14. D. S. Michaud and coauthors, Fluid intake and the risk of bladder cancer in men, *New England Journal of Medicine* 340 (1999): 1390–1397.

15. A. Flood and coauthors, Calcium from diet and supplements is associated with reduced risk of colorectal cancer in a prospective cohort of women, *Cancer Epidemiology, Biomarkers and Prevention* 14 (2005): 126–132; V. Chia and P. A. Newcomb, Calcium and colorectal cancer: Some questions remain, *Nutrition Reviews* 62 (2004): 115–120.

16. M. B. Zemel and S. L. Miller, Dietary calcium and dairy modulation of adiposity and obesity risk, *Nutrition Reviews* 62 (2004): 125–131; S. J. Parikh and J. A. Yanovski, Calcium and adiposity, *American Journal of Clinical Nutrition* 77 (2003): 281–287; D. Teegarden, Calcium intake and reduction in weight or fat mass, *Journal of Nutrition* 133 (2003): 249S–251S; R. P. Heaney, K. M. Davies, and M. J. Barger-Lux, Calcium and weight: Clinical studies, *Journal of the American College of Nutrition* 21 (2002): 152–155.

17. M. E. Wastney and coauthors, Changes in calcium kinetics in adolescent girls induced by high calcium intake, *Journal of Clinical Endocrinology and Metabolism* 85 (2000): 4470–4475.

18. Standing Committee on the Scientific Evaluation of Dietary Reference Intakes, Food and Nutrition Board, Institute of Medicine, *Dietary Reference Intakes for Calcium, Phosphorus, Magnesium, Vitamin D, and Fluoride* (Washington, D.C.: National Academy Press, 1997), p. 72.

19. Standing Committee on the Scientific Evaluation of Dietary Reference Intakes, Food and Nutrition Board, Institute of Medicine, 1997, pp. 106–107.

20. Standing Committee on the Scientific Evaluation of Dietary Reference Intakes, Food and Nutrition Board, Institute of Medicine, 1997, pp. 101–106.

21. E. S. Ford and A. H. Mokkad, Dietary magnesium intake in a national sample of U.S. adults, *Journal of Nutrition* 133 (2003): 2879–2882.

22. Ford and Mokkad, 2003.

23. L. Cordain and coauthors, Origins and evolution of the Western diet: Health implications for the 21st century, *American Journal of Clinical Nutrition* 81 (2005): 341–354.

24. Standing Committee on the Scientific Evaluation of Dietary Reference Intakes, Food and Nutrition Board, Institute of Medicine, 2004, pp. 6-1–6-2.

25. F. M. Sacks and coauthors, Effects on blood pressure of reduced dietary sodium and the Dietary Approaches to Stop Hypertension (DASH) diet, *New England Journal of Medicine* 344 (2001): 3–10.

26. J. Tuomilehto and coauthors, Urinary sodium excretion and cardiovascular mortality in Finland: A prospective study, *Lancet* 357 (2001): 848–851; L. Liu and coauthors, Comparative studies of diet-related factors and blood pressure among Chinese and Japanese: Results from the China-Japan Cooperative Research of the WHO-CARDIAC Study. Cardiovascular Disease and Alimentary Comparison, *Hypertension Research* 23 (2000): 413–420.

27. Standing Committee on the Scientific Evaluation of Dietary Reference Intakes, Food and Nutrition Board, Institute of Medicine, 2004, p. 6-2.

28. J. Stamler, The INTERSALT Study: Background, methods, findings, and implications, *American Journal of Clinical Nutrition* 65 (1997): 626S–642S.

29. U.S. Department of Agriculture and U.S. Department of Health and Human Services, *2005 Dietary Guidelines Committee Report*, 2005, available online at www.health.gov/dietaryguidelines/dga2005report/

30. F. C. Luft, Present status of genetic mechanisms in hypertension, *Medical Clinics of North America* 88 (2004): 1–18; S. T. Turner and E. Boerwinkle, Genetics of blood pressure, hypertensive complications, and antihypertensive drug responses, *Pharmacogenomics* 4 (2003): 53–65.

31. Engler and coauthors, 2003; N. M. Karanja and coauthors, Descriptive characteristics of the dietary patterns used in the Dietary Approaches to Stop Hypertension Trial, *Journal of the American Dietetic Association* 99 (1999): S19–S27; L. P. Svetkey and coauthors, Effects of dietary patterns on blood pressure: Subgroup analysis of the Dietary Approaches to Stop Hypertension (DASH) randomized clinical trial, *Archives of Internal Medicine* 159 (1999): 285–293.

32. Sacks and coauthors, 2001.

33. F. J. He and G. A. MacGregor, Beneficial effects of potassium, *British Medical Journal* 323 (2001): 497–501.

34. P.-H. Lin and coauthors, Food group sources of nutrients in the dietary patterns of the DASH-Sodium trial, *Journal of the American Dietetic Association* 103 (2003): 488–496.

35. P.-H. Lin and coauthors, The DASH diet and sodium reduction improve markers of bone turnover and calcium metabolism in adults, *Journal of Nutrition* 133 (2003): 3130–3136.

36. Standing Committee on the Scientific Evaluation of Dietary Reference Intakes, Food and Nutrition Board, Institute of Medicine, 2004, p. 5-6.

37. Standing Committee on the Scientific Evaluation of Dietary Reference Intakes, Food and Nutrition Board, Institute of Medicine, 2004, p. 5-35.

38. Standing Committee on the Scientific Evaluation of Dietary Reference Intakes, Food and Nutrition Board, Institute of Medicine, *Dietary Reference Intakes for Vitamin A, Vitamin K, Arsenic, Boron, Chromium, Copper, Iodine, Iron, Manganese, Molybdenum, Nickel, Silicon, Vanadium, and Zinc* (Washington, D.C.: National Academy Press, 2001), p. 258.

39. Committee to Assess the Distribution and Administration of Potassium Iodide in the Event of a Nuclear Incident, *Distribution and Administration of Potassium Iodide in the Event of a Nuclear Incident* (Washington, D.C.: National Academies Press, 2003), pp. 1–5.

40. FDA attacks "anti-radiation pills," *Consumer Health Digest,* September 2003.

41. Standing Committee on the Scientific Evaluation of Dietary Reference Intakes, Food and Nutrition Board, Institute of Medicine, 2001, p. 9-14.

42. N. C. Andrews, Disorders of iron metabolism, *New England Journal of Medicine* 341 (1999): 1986–1995.

43. S. Miret, R. J. Simpson, and A. T. McKie, Physiology and molecular biology of dietary iron absorption, *Annual Review of Nutrition* 23 (2003): 283–301.

44. M. B. Reddy and L. Clark, Iron, oxidative stress, and disease risk, *Nutrition Reviews* 62 (2004):120–124.

45. R. D. Baynes and M. H. Stipanuk, Iron, in *Biochemical and Physiological Aspects of Human Nutrition* (Philadelphia: Saunders, 2000), pp. 711–740.

46. Chung and Wessling-Ressnick, Lessons learned from genetic and nutrition iron deficiencies, *Nutrition Reviews* 62 (2004): 212–220.

47. F. Verdon and coauthors, Iron supplementation for unexplained fatigue in non-anaemic women: Double blind randomised placebo controlled trial, *British Medical Journal* 326 (2003): available at www.bmj.com; J. Beard, Iron deficiency alters brain development and functioning, *Journal of Nutrition* 133 (2003): 1468S–1472S; E. M. Ross, Evaluation and treatment of iron deficiency in adults, *Nutrition in Clinical Care* 5 (2002): 220–224; J. D. Haas and T. Brownlie IV, Iron deficiency and reduced work capacity: A critical review of the research to determine a causal relationship, *Journal of Nutrition* 131 (2001): 676S–690S; S. Horton and C. Levin, Commentary on "evidence that iron deficiency anemia causes reduced work capacity," *Journal of Nutrition* 131 (2001): 691S–696S.

48. D. Pinero, B. Jones, and J. Beard, Variations in dietary iron alter behavior in developing rats, *Journal of Nutrition* 131 (2001): 311–318; Standing Committee on the Scientific Evaluation of Dietary Reference Intakes, Food and Nutrition Board, Institute of Medicine, 2001, pp. 9-5–9-6.

49. C. I. Obialo and coauthors, Clay pica has no hematologic or metabolic correlate in chronic hemodialysis patients, *Journal of Renal Nutrition* 11 (2001): 32–36.

50. P. W. Harvey, P. B. Dexter, and I. Darnton-Hill, The impact of consuming iron from non-food sources on iron status in developing countries, *Public Health Nutrition* 3 (2000): 375–383.

51. M. Shannon, Severe lead poisoning in pregnancy, *Ambulatory Pediatrics* 3 (2003): 37–39; E. A. Rose, J. H. Porcerelli, and A. V. Neale, Pica: Common but commonly missed, *Journal of the American Board of Family Practice* 13 (2000): 353–358; J. F. Magnaval and coauthors, Highlights of human toxocariasis, *Korean Journal of Parasitology* 39 (2001): 1–11.

52. J. L. Beard and J. R. Connor, Iron status and neural functioning, *Annual Review of Nutrition* 23 (2003): 41–58.

53. B. Annibale and coauthors, Reversal of iron deficiency anemia after *Helicobacter pylori* eradication in patients with asymptomatic gastritis, *Annals of Internal Medicine* 131 (1999): 668–672.

54. World Health Organization, www.who.int/nut/ida.htm.

55. Iron deficiency—United States, 1999–2000, *Morbidity and Mortality Weekly Report* 51 (2002): 897–899.

56. M. Wessling-Resnick, Iron transport, *Annual Review of Nutrition* 20 (2000): 129–151.

57. A. L. M. Heath and S. J. Fairweather-Tait, Health implications of iron overload: The role of diet and genotype, *Nutrition Reviews* 61 (2003): 45–62; J. R. Hunt and Z. K. Roughead, Adaptation of iron absorption in men consuming diets with high or low iron bioavailability, *American Journal of Clinical Nutrition* 71 (2000): 94–102.

58. L. M. Neff, Current directions in hemochromatosis research: Towards an understanding of the role of iron overload and the *HFE* gene mutations in the development of clinical disease, *Nutrition Reviews* 61 (2003): 38–42; M. R. Borgaonkar, Hemochromatosis: More common than you think, *Canadian Family Physician* 49 (2003): 36–43.

59. M. J. Nowicki and B. R. Bacon, Hereditary hemochromatosis in siblings: Diagnosis by genotyping, *Pediatrics* 105 (2000): 426–429; A. S. Tavill, Clinical implications of the hemochromatosis gene, *New England Journal of Medicine* 341 (1999): 755–757.

60. Neff, 2003.

61. J. D. Cook and M. B. Reddy, Effect of ascorbic acid intake on nonheme-iron absorption from a complete diet, *American Journal of Clinical Nutrition* 73 (2001): 93–98.

62. Heath and Fairweather-Tait, 2003.

63. A. E. O. Fisher and D. P. Naughton, Iron supplements: The quick fix with long-term consequences, *Nutrition Journal* 3 (2004), available at www.nutritionj.com/ content/3/1/2.

64. M. Shannon, Ingestion of toxic substances by children, *New England Journal of Medicine* 342 (2000): 186–191; C. C. Morris, Pediatric iron poisonings in the United States, *Southern Medical Journal* 93 (2000): 352–358.

65. Standing Committee on the Scientific Evaluation of Dietary Reference Intakes, Food and Nutrition Board, Institute of Medicine, 2001, p. 347.

66. M. B. Reddy, R. F. Hurrell, and J. D. Cook, Estimation of nonheme-iron bioavailability from meal composition, *American Journal of Clinical Nutrition* 71 (2000): 937–943.

67 . A. S. Sandberg, Bioavailability of minerals in legumes, *British Journal of Nutrition* 88 (2002): S281–S285.

68. I. M. Zijp, O. Korver, and L. B. Tijburg, Effect of tea and other dietary factors on iron absorption, *Critical Reviews in Food Science and Nutrition* 40 (2000): 371–398.

69. R. A. DiSivestro, Zinc in relation to diabetes and oxidative disease, *Journal of Nutrition* 130 (2000): 1509S–1511S.

70. C. F. Walker and R. E. Black, Zinc and the risk for infectious disease, *Annual Review of Nutrition* 24 (2004): 255–275.

71. T. A. Strand and coauthors, Effectiveness and efficacy of zinc for the treatment of acute diarrhea in young children, *Pediatrics* 109 (2002): 898–903; N. Bhandari and coauthors, Substantial reduction in severe diarrheal morbidity by daily zinc supplementation in young North Indian children, *Pediatrics* 109 (2002): e86; C. Duggan and W. Fawzi, Micronutrients and child health: Studies in international nutrition and HIV infection, *Nutrition Reviews* 59 (2001): 358–369; The Zinc Investigators' Collaborative Group, Therapeutic effects of oral zinc in acute and persistent diarrhea in children in developing countries: Pooled analysis of randomized controlled trials, *American Journal of Clinical Nutrition* 72 (2000): 1516–1522.

72. K. A. Keller, A. Grider, and J. A. Coffield, Age-dependent influence of dietary zinc restriction on short-term memory in male rats, *Physiology and Behavior* 72 (2001): 339–348.

73. B. H. McElroy and S. P. Miller, Effectiveness of zinc gluconate glycine lozenges (Cold-Eeze) against the common cold in school-aged subjects: A retrospective chart review, *American Journal of Therapeutics* 9 (2002): 472–475; I. Marshall, Zinc for the common cold, *Cochrane Database of Systematic Reviews* 2 (2000): CD001364; J. L. Jackson, E. Lesho, and C. Peterson, Zinc and the common cold: A meta-analysis revisited, *Journal of Nutrition* 130 (2000): 1512S–1515S; A. S. Prasad and coauthors, Duration of symptoms and plasma cytokine levels in patients with the common cold treated with zinc acetate: A randomized, double-blind, placebo-controlled trial, *Annals of Internal Medicine* 133 (2000): 245–252; R. B. Turner and W. E. Cetnarowski, Effect of treatment with zinc gluconate or zinc acetate on experimental and natural colds, *Clinical Infectious Diseases* 31 (2000): 1202–1208.

74. P. A. Cotton and coauthors, Dietary sources of nutrients among US Adults, 1994–1996, *Journal of the American Dietetic Association* 104 (2004): 921–930.

75. D. H. Holben and A. M. Smith, The diverse role of selenium within selenoproteins: A review, *Journal of the American Dietetic Association* 99 (1999): 836–843.

76. T. M. Vogt and coauthors, Serum selenium and risk of prostate cancer in U.S. blacks and whites, *International Journal of Cancer* 103 (2003): 664–670.

77. A. J. Duffield-Lillico and coauthors, Selenium supplementation and secondary prevention of nonmelanoma skin cancer in a randomized trial, *Journal of the National Cancer Institute* 95 (2003): 1477–1481.

78. M. A. Moyad, Selenium and vitamin E supplements for prostate cancer: Evidence or embellishment? *Urology* 59 (2002): 9–19.

79. Populations receiving optimally fluoridated public drinking water—United States, 2000, *Morbidity and Mortality Weekly Report* 51 (2002): 144–147.

80. Achievements in public health, 1900–1999: Fluoridation of drinking water to prevent dental caries, *Morbidity and Mortality Weekly Report* 48 (1999): 933–940.

81. Centers for Disease Control and Prevention, Recommendations for using fluoride to prevent and control dental caries in the United States, *Morbidity and Mortality Weekly Report* 50 (2001): 1–42; Position of the American Dietetic Association: The impact of fluoride on health, *Journal of the American Dietetic Association* 100 (2000): 1208–1213.

82. G. Y. Yeh and coauthors, Systematic review of herbs and dietary supplements for glycemic control in diabetes, *Diabetes Care* 26 (2003): 1277–1294; M. D. Althuis and coauthors, Glucose and insulin reponses to dietary chromium supplement: a meta-analysis, *American Journal of Clinical Nutrition* 76 (2002): 148–55.

83. J. F. Fowler, Jr., Systemic contact dermatitis caused by oral chromium picolinate, *Cutis* 65 (2000): 116.

84. P. L. Fox, The copper-iron chronicles: The story of an intimate relationship, *Biometals* 16 (2003): 9–40; N. E. Hellman and J. D. Gitlin, Ceruloplasmin metabolism and function, *Annual Review of Nutrition* 22 (2002): 439–458.

85. Standing Committee on the Scientific Evaluation of Dietary Reference Intakes, Food and Nutrition Board, Institute of Medicine, 2001, p. 227.

86. Standing Committee on the Scientific Evaluation of Dietary Reference Intakes, Food and Nutrition Board, Institute of Medicine, 2001, p. 245.

87. T. A. Devirian and S. L. Volpe, The physiological effects of dietary boron, *Critical Reviews in Food Science and Nutrition* 43 (2003): 219–231.

88. U.S. Department of Agriculture and U.S. Department of Health and Human Services, *Dietary Guidelines for Americans 2005*, available online at www.healthierus.gov.

89. R. E. Black and coauthors, Children who avoid drinking cow milk have low dietary calcium intakes and poor bone health, *American Journal of Clinical Nutrition* 76 (2002): 675–680.

90. A. Azoulay, P. Garzon, and M. J. Eisenberg, Comparison of the mineral content of tap water and bottled waters, *Journal of General Internal Medicine* 16 (2001): 168–175; Bohmer, Muller, and Resch, 2000.

Consumer Corner 8

1. Water sold state to state safe to drink, agency says, *FDA Consumer,* July/August 1999, p. 6.

2. Natural Resources Defense Council, *Bottled Water: Pure Drink or Pure Hype?* 1999, available at www.nrdc.org/water/drinking/nbw.asp.

3. National Sanitation Foundation, Consumer guide to bottled water, *Information for Consumers,* March 1998.

4. P. Garzon and M. J. Eisenberg, Variation in the mineral content of commercially available bottled waters: Implications for health and disease, *American Journal of Medicine* 105 (1998): 125–130.

5. M. Sabatier and coauthors, Meal effect on magnesium bioavailability from mineral water in healthy women, *American Journal of Clinical Nutrition* 75 (2002): 65–71.

Controversy 8

1. U.S. Department of Health and Human Services, *Bone Health and Osteoporosis: A Report of the Surgeon General* (Rockville, MD: Government Printing Office, 2004), available at www.surgeongeneral.gov/library

2. U.S. Department of Health and Human Servies, 2004.

3. Y.-Z. Liu and coauthors, Molecular studies of identification of genes for osteoporosis: The 2002 update, *Journal of Endocrinology* 177 (2003): 147–196.

4. S. L. Teitelbaum and F. P. Ross, Genetic regulation of osteoclast development and function, *Nature Reviews: Genetics* 4 (2003): 638–649.

5. P. Kannus and coauthors, Genetic factors and osteoporotic fractures in elderly people: Prospective 25 year follow up of a nationwide cohort of elderly Finnish twins, *British Medical Journal* 319 (1999): 1334–1337.

6. A. J. MacGregor, H. Snieder, and T. D. Spector, Genetic factors and osteoporotic fractures in elderly people, letter, *British Medical Journal* 320 (2000): 1669.

7. S. M. Runyan and coauthors, Familial resemblance of bone mineralization, calcium intake, and physical activity in early-adolescent daughters, their mothers, and maternal grandmothers, *Journal of the American Dietetic Association* 103 (2003): 1320–1325.

8. J. A. Cauley and coauthors, Longitudinal study of changes in hip bone mineral density in Caucasian and African American women, *Journal of the American Geriatric Society* 53 (2005): 183–189.

9. H. J. Kalkwarf, J. C. Khoury, and B. P. Lanphear, Milk intake during childhood and adolescence, adult bone density, and osteoporotic fractures in US women, *American Journal of Clinical Nutrition* 77 (2003): 257–265.

10. A. Goulding and coauthors, Children who avoid drinking cow's milk are at increased risk for prepubertal bone fractures, *Journal of the American Dietetic Association* 104 (2004): 250–253.

11. J. S. Volek and coauthors, Increasing fluid milk favorably affects bone mineral density responses to resistance training in adolescent boys, *Journal of the American Dietetic Association* 103 (2003): 1353–1356; R. E. Black and coauthors, Children who avoid drinking cow milk have low dietary calcium intakes and poor bone health, *American Journal of Clinical Nutrition* 76 (2002): 675–680.

12. J. Cornish and coauthors, Lactoferrin is a potent regulator of bone cell activities and increases bone formation in vivo, *Endocrinology* 145 (2004): 4366–4374.

13. J. M. Campion and M. J. Maricic, Osteoporosis in men, *American Family Physician* 67 (2003): 1551–1526.

14. Runyan and coauthors, 2003.

15. W. Kemmler and coauthors, The Erlangen Fitness Osteoporosis Prevention Study: A controlled exercise trial in early postmenopausal women with low bone density—first-year results, *Archives of Physical Medicine and Rehabilitation* 84 (2003): 673–682; M. C. Wang and coauthors, Diet in midpuberty and sedentary activity in prepuberty predict peak bone mass, *American Journal of Clinical Nutrition* 77 (2003): 495–503; S. J. Stear and coauthors, Effect of a calcium and exercise intervention on the bone mineral status of 16–18-y-old adolescent girls, *American Journal of Clinical Nutrition* 77 (2003): 985–992.

16. J. E. Layne and M. E. Nelson, The effects of progressive resistance training on bone density: A review, *Medicine and Science in Sports and Exercise* 31 (1999): 25–30.

17. E. C. Cussler and coauthors, Weight lifted in strength training predicts bone change in postmenopausal women, *Medicine and Science in Sports and Exercise* 35 (2003): 10–17.

18. L. M. Salamone and coauthors, Effect of a lifestyle intervention on bone mineral density in premenopausal women: A randomized trial, *American Journal of Clinical Nutrition* 70 (1999): 97–103.

19. S. Gillette-Guyonnet and coauthors, Body composition and osteoporosis in elderly women, *Gerontology* 46 (2000): 189–193.

20. J. C. Fleet, Leptin and bone: Does the brain control bone biology? *Nutrition Reviews* 58 (2000): 209–211.

21. P. Gerdhem and K. J. Obrant, Effects of cigarette-smoking on bone mass as assessed by dual-energy X-ray absorptiometry and ultrasound, *Osteoporosis International* 13 (2002): 932–936; K. D. Ward and R. C. Klesges, A meta-analysis of the effects of cigarette smoking on bone mineral density, *Calcified Tissue International* 68 (2001): 259–270.

22. P. B. Rapuri and coauthors, Smoking and bone metabolism in elderly women, *Bone* 27 (2000): 429–436; A. P. Hermann and coauthors, Premenopausal smoking and bone density in 2015 perimenopausal women, *Journal of Bone and Mineral Research* 15 (2000): 780–787.

23. C. Oncken and coauthors, Effects of smoking cessation or reduction on hormone profiles and bone turnover in postmenopausal women, *Nicotine and Tobacco Research* 4 (2002): 451–458.

24. B. Dawson-Hughes, Interaction of dietary calcium and protein in bone health in humans, *Journal of Nutrition* 133 (2003): 852S–854S.

25. D. E. Sellmyer and coauthors, A high ratio of dietary animal to vegetable protein increases the rate of bone loss and the risk of fracture in postmenopausal women, *American Journal of Clinical Nutrition* 73 (2001): 118–122.

26. L. A. Frassetto and coauthors, Worldwide incidence of hip fracture in elderly women: Relation to consumption of animal to vegetable foods, *Journals of Gerontology Series A: Biological Sciences and Medical Sciences* 55 (2000): M585–M592.

27. R. Brynin, Soy and its isoflavones: A review of their effects on bone density, *Alternative Medicine Review* 7 (2002): 317–327.

28. B. H. Arjmandi and coauthors, Soy protein has a greater effect on bone in postmenopausal women not on hormone replacement therapy, as evidenced by reducing bone resorption and urinary calcium excretion, *Journal of Clinical Endocrinology and Metabolism* 88 (2003): 1048–1054; T. Uesugi, Y. Fukui, and Y. Yamori, Beneficial effects of soybean isoflavone supplementation on bone metabolism and serum lipids in postmenopausal Japanese women: A four-week study, *Journal of the American College of Nutrition* 21 (2002): 97–102.

29. L. K. Massey, Dietary animal and plant protein and human bone health: A whole foods approach, *Journal of Nutrition* 133 (2003): 862S–865S.

30. J. Bell, Elderly women need dietary protein to maintain bone mass, *Nutrition Reviews* 60 (2002): 337–341; B. Dawson-Hughes and S. S. Harris, Calcium intake influences the association of protein intake with rates of bone loss in elderly men and women, *American Journal of Clinical Nutrition* 75 (2002): 773–779; J. H. E. Promislow and coauthors, Protein consumption and bone mineral density in the elderly: The Rancho Bernardo Study, *American Journal of Epidemiology* 155 (2002): 636–644.

31. J. Raloff, Do meat and dairy harm aging bones? *Science News* 159 (2001): 20.

32. M. Harrington and K. D. Cashman, High salt intake appears to increase bone resorption in postmenopausal women but high potassium intake ameliorates this adverse effect, *Nutrition Reviews* 61 (2003): 179–183.

33. P.-H. Lin and coauthors, The DASH diet and sodium reduction improve markers of bone turnover and calcium metabolism in adults, *Journal of Nutrition* 133 (2003): 3130–3136.

34. L. Doyle and K. D. Cashman, The DASH diet may have beneficial effects on bone health, *Nutrition Reviews* 62 (2004): 215–220.

35. Standing Committee on the Scientific Evaluation of Dietary Reference Intakes, Food and Nutrition Board, Institute of Medicine, *Dietary Reference Intakes for Water, Potassium, Sodium, Chloride, and Sulfate* (Washington, D.C.: National Academies Press, 2004), pp. 6-89–6-92; F. P. Cappuccio and coauthors, Unraveling the links between calcium excretion, salt intake, hypertension, kidney stones and bone metabolism, *Journal of Nephrology* 13 (2000): 169–177.

36. D. E. Sellmyer, M. Schloetter, and A. Sebastin, Potassium citrate prevents increased urine calcium excretion and bone resorption induced by a high sodium chloride diet, *Journal of Clinical Endocrinology and Metabolism* 87 (2002): 2008–2012.

37. R. P. Heaney, Effects of caffeine on bone and the calcium economy, *Food and Chemical Toxicology* 40 (2002): 1263–1270.

38. D. B. Milne and F. H. Nielsen, The interaction between dietary fructose and magnesium adversely affects macromineral homeostasis in men, *Journal of the American College of Nutrition* 19 (2000): 31–37.

39. F. Garcia-Contreras and coauthors, Cola beverage consumption induces bone mineralization reduction in ovariectomized rats, *Archives of Medical Research* 31 (2000): 360–365.

40. N. C. Binkley and coauthors, A high phylloquinone intake is required to achieve maximal osteocalcin γ-carboxylation, *American Journal of Clinical Nutrition* 76 (2002): 1055–1060; S. L. Booth and coauthors, Dietary vitamin K intakes are associated with hip fracture but not with bone mineral density in elderly men and women, *American Journal of Clinical Nutrition* 71 (2000): 1201–1208.

41. J. B. J. van Meurs and coauthors, Homocysteine levels and the risk of osteoporotic fracture, *New England Journal of Medicine* 350 (2004): 2033–2041.

42. R. R. McLean and coauthors, Homocysteine as a predictive factor for hip fracture in older persons, *New England Journal of Medicine* 350 (2004): 2042–2049.

43. L. G. Raisz, Homocysteine and osteoporotic fractures: Culprit or bystander? *New England Journal of Medicine* 350 (2004): 2089–2090.

44. K. L. Tucker and coauthors, Potassium, magnesium, and fruit and vegetable intakes are associated with greater bone mineral density in elderly men and women, *American Journal of Clinical Nutrition* 69 (1999): 727–736.

45. K. Michaelsson and coauthors, Serum retinol levels and the risk of fractures, *New England Journal of Medicine* 348 (2003): 287–294; D. Feskanich and coauthors, Vitamin A intake and hip fractures among postmenopausal women, *Journal of the American Medical Association* 287 (2002): 47–54; S. Johnasson and coauthors, Subclinical hypervitaminosis A causes fragile bones in rats, *Bone* 31 (2002): 685–689; N. Binkley and D. Krueger, Hypervitaminosis A and bone, *Nutrition Reviews* 58 (2000): 138–144.

46. K. L. Tucker and coauthors, Bone mineral density and dietary patterns in older adults: The Framingham Osteoporosis Study, *American Journal of Clinical Nutrition* 76 (2002): 245–252; S. A. New and coauthors, Dietary influences on bone mass and bone metabolism: Further evidence of a positive link between fruit and vegetable consumption and bone health? *American Journal of Clinical Nutrition* 71 (2000): 142–151.

47. Harrington and Cashman, 2003; Tucker and coauthors, 2002; P. B. Rapuri and coauthors, Caffeine intake increases the rate of bone loss in elderly women and interacts with vitamin D receptor genotypes, *American Journal of Clinical Nutrition* 74 (2001): 694–700.

48. D. J. van der Voort and coauthors, Screening for osteoporosis using easily obtainable biometrical data: Diagnostic accuracy of measured, self-reported and recalled BMI, and related costs of bone mineral density measurements, *Osteoporosis International* 11 (2000): 233–239.

49. H. J. Kloosterboer and A. G. Ederveen, Pros and cons of existing treatment modalities in osteoporosis: A comparison between tibolone, SERMs and estrogen (+/− progestogen) treatments, *Journal of Steroid Biochemistry and Molecular Biology* 83 (2002): 157–165; R. A. Sayegh and P. G. Stubblefield, Bone metabolism and the perimenopause overview, risk factors, screening, and osteoporosis preventive measures, *Obstetrics and Gynecology Clinics of North America* 29 (2002): 495–510.

50. R. T. Chlebowski and coauthors, Influence of estrogen plus progestin on breast cancer and mammography in healthy postmenopausal women: The Women's Health Initiative Randomized Trial, *Journal of the American Medical Association* 289 (2003): 3243–3253; C. G. Solomon and R. G. Dluhy, Rethinking postmenopausal hormone therapy, *New England Journal of Medicine* 348 (2003): 579–580; Writing Group for the Women's Health Initiative Investigators, Risks and benefits of estrogen plus progestin in healthy postmenopausal women: Principal results from the Women's Health Initiative Randomized Controlled Trial, *Journal of the American Medical Association* 288 (2002): 321–333; O. Ylikorkala and M. Metsa-heikkila, Hormone replacement therapy in women with a history of breast cancer, *Gynecological Endocrinology* 16 (2002): 469–478.

51. S. L. Greenspan, N. M. Resnick, and R. A. Parker, Combination therapy with hormone replacement and alendronate for prevention of bone loss in elderly women: A randomized controlled trial, *Journal of the American Medical Association* 289 (2003): 2525–2533.

52. D. Teegarden and coauthors, Previous milk consumption is associated with greater bone density in young women, *American Journal of Clinical Nutrition* 69 (1999): 1014–1017.

53. NIH Consensus Development Panel, Osteoporosis prevention, diagnosis, and therapy, *Journal of the American Medical Association* 285 (2001): 785–795.

54. Federal Trade Commission, Kevin Trudeau banned from infomercials, September 7, 2004, available at www.ftc.gov/opa/2004/09/trudeaucoral.htm

55. J. Guillemant and coauthors, Mineral water as a source of dietary calcium: Acute effects on parathyroid function and bone resorption in young men, *American Journal of Clinical Nutrition* 71 (2000): 999–1002.

Chapter 9

1. A. Thorogood and coauthors, Relation between body mass index and mortality in an unusually slim cohort, *Journal of Epidemiology and Community Health* 57 (2003) 130–133; P. T. Katzmarzyk, C. L. Craig, and C. Bouchard, Underweight, overweight and obesity: Relationships with mortality in the 13-year follow-up of the Canada Fitness Study, *Journal of Clinical Epidemiology* 54 (2001): 916–920.

2. T. E. Kottke, L. A. Wu, and R. S. Hoffman, Economic and psychological implications of the obesity epidemic, *Mayo Clinic Proceedings* 78 (2003): 92–94; A. H. Mokdad and coauthors, Centers for Disease Control and Prevention, The continuing epidemic of obesity in the United States (letter), *Journal of the American Medical Association* 284 (2000): 1650–1651.

3. Food and Drug Administration, *Counting Calories: Report of the Working Group on Obesity,* March 12, 2004, available at www.cfsan.fda.gov/~dms/owg-rpt.html.

4. Prevalence of Overweight among Children and Adolescents: United States, 1999–2000, available at www.cdc.gov/nchs/products/pubs/pubd/hestats/overwght99.htm.

5. U.S. Department of Agriculture and U.S. Department of Health and Human Services, *2005 Dietary Guidelines Committee Report, 2005,* available online at www.health.gov/dietaryguidelines/dga2005report/; Food and Drug Administration, 2004; U.S. Department of Health and Human Services, *Health People 2010,* 2nd ed. (Washington, D.C.: Government Printing Office, 2000), available at www.health.gov/healthypeople or call (800) 367–4725.

6. J. K. Alexander, Obesity and coronary heart disease, *American Journal of the Medical Sciences* 321 (2001): 215–224.

7. A. H. Mokdad and coauthors, Prevalence of obesity, diabetes, and obesity-related health risk factors, 2001, *Journal of the American Medical Association* 289 (2003): 76–79; K. R. Fontaine and coauthors, Years of life lost due to obesity, *Journal of the American Medical Association* 289 (2003): 187–193.

8. J. C. Peters, The challenge of managing body weight in the modern world, *Asia Pacific Journal of Clinical Nutrition* 11 (2002): S714–S717.

9. M.-P. St-Onge and S. B. Heymsfield, Overweight and obesity status are linked to lower life expectancy, *Nutrition Reviews* 61 (2003): 313–316; D. B. Allison and coauthors, Annual deaths attributable to obesity in the United States, *Journal of the American Medical Association* 282 (1999): 1530–1538.

10. U.S. Preventive Services Task Force, *Screening for Obesity in Adults,* update of *Guide to Clinical Preventive Service,* 2nd ed. (2003), available at www.ahrq.gov/clinic/uspstf/uspsobes.htm#summary.

11. Centers for Disease Control and Prevention, Prevalence of overweight and obesity among adults with diagnosed diabetes—United States, 1988–1994 and 1999–2002, *Morbidity and Mortality Weekly Reports* 53 (2004): 1066–1068.

12. G. R. Dagenais and coauthors, Prognosstic impact of body weight and abdominal obesity in women and men with cardiovascular disease, *American Heart Journal* 149 (2005): 54–60; E. N. Taylor, M. J. Stampfer, and G. C. Curhan, Obesity, weight gain, and the risk of kidney stones, *Journal of the American Medical Association* 293 (2005): 455–462; E. E. Calle and coauthors, Overweight, obesity, and mortality from cancer in a prospectively studied cohort of U.S. adults, *New England Journal of Medicine* 348 (2003): 1625–1638.

13. T. B. Nguyen-Duy and coauthors, Visceral fat and liver fat are independent predictors of metabolic risk factors in men, *American Journal of Physiology. Endocrinology and Metabolism* 284 (2003): 558–561; G. Davì and coauthors, Platelet activation in obese women—Role of inflammation and oxidant stress, *Journal of the American Medical Association* 288 (2002): 2008–2014; J. M. Oppert and coauthors, Anthropometric estimates of muscle and fat mass in relation to cardiac and cancer mortality in men: The Paris Prospective Study, *American Journal of Clinical Nutrition* 75 (2002): 1107–1113.

14. F. X. Pi-Sunyer, The epidemiology of central fat distribution in relation to disease, *Nutrition Reviews* 62 (2004): S120–S126.

15. P. R. Jones and D. A. Edwards, Areas of fat loss in overweight young females following an 8-week period of energy intake reduction, *Annals of Human Biology* 26 (1999): 151–162.

16. R. A. Breslow and B. A. Smothers, Drinking patterns and body mass index in never smokers, *American Journal of Epidemiology* 161 (2005): 368–376; C. A. Holcomb, D. L. Heim, and T. M. Loughin, Physical activity minimizes the association of body fatness with abdominal obesity in white, premenopausal women: results from the Third National Health and Nutrition Examination Survey, *Journal of the American Dietetic Association* 104 (2004): 1859–1862; J. M. Dorn and coauthors, Alcohol drinking patterns differentially affect central adiposity as measured by abdominal height in women and men, *Journal of Nutrition* 133 (2003): 2655–2662.

17. J. Stevens and coauthors, Evaluation of WHO and NHANES II standards for overweight using mortality rates, *Journal of the American Dietetic Association* 100 (2000): 825–827.

18. T. R. Wessel and coauthors, Relationship of physical fitness vs. body mass index with coronary artery disease and cardiovascular events in women, *Journal of the American Medical Association* 292 (2004): 1179–1187; S. W. Farrell and coauthors, The relation of body mass index, cardiorespiratory fitness, and all-cause mortality in women, *Obesity Research* 10 (2002): 417–423; C. D. Lee and S. N. Blair, Cardiorespiratory fitness and smoking-related total cancer mortality in men, *Medicine and Science in Sports and Exercise* 34 (2002): 735–739; C. D. Lee and S. N. Blair, Cardiorespiratory fitness and stroke mortality in men, *Medicine and Science in Sports and Exercise* 34 (2002): 592–595.

19. T. S. Church and coauthors, Exercise capacity and body composition as predictors of mortality among men with diabetes, *Diabetes Care* 27 (2004): 83–88.

20. N. S. Wellman and B. Friedberg, Causes and consequences of adult obesity: Health, social and economic impacts in the United States, *Asia Pacific Journal of Clinical Nutrition* 11 (2002): S705–S709.

21. M. R. Lowe, K. Miller-Kovach, and S. Phelan, Weight-loss maintenance in overweight individuals one to five years following successful completion of a commercial weight-loss program, *International Journal of Obesity and Related Metabolic Disorders* 25 (2001): 325–331.

22. Standing Committee on the Scientific Evaluation of Dietary Reference Intakes, Food and Nutrition Board, Institute of Medicine, *Dietary Reference Intakes for Energy, Carbohydrate, Fiber, Fat, Fatty Acids, Cholesterol, Protein, and Amino Acids* (Washington, D.C.: National Academies Press, 2002), p. 13-12.

23. M. Gilliat-Wimberely and coauthors, Effects of habitual physical activity on the resting metabolic rates and body compositions of women aged 35 to 50 years, *Journal of the American Dietetic Association* 101 (2001): 1181–1188.

24. J. Stevens and coauthors, The effect of decision rules on the choice of a body mass index cutoff for obesity: Examples from African American and white women, *American Journal of Clinical Nutrition* 75 (2002): 986–992; J. Stevens, Obesity and mortality in African-Americans, *Nutrition Reviews* 58 (2000): 346–353.

25. J. E. Manson and S. S. Bassuk, Obesity in the United States: A fresh look at its high toll, *Journal of the American Medical Association* 289 (2003): 229–230.

26. I. Janssen and coauthors, Body mass index and waist circumference independently contribute to the prediction of nonabdominal, abdominal subcutaneous, and visceral fat, *American Journal of Clinical Nutrition* 75 (2002): 683–688.

27. I. Lofgren and coauthors, Waist circumference is a better predictor than body mass index of coronary heart disease risk in overweight premenopausal women, *Journal of Nutrition* 134 (2004): 1071–1076.

28. G. Panotopoulos and coauthors, Dual X-ray absorptiometry, bioelectrical impedance, and near infrared interactance in obese women, *Medicine and Science in Sports and Exercise* 33 (2001): 665–670.

29. J. Wardle, J. Waller, and E. Fox, Age of onset and body dissatisfaction in obesity, *Addictive Behaviors* 27 (2002): 561–573.

30. L. J. Harnack, R. W. Jeffrey, and K. N. Boutelle, Temporal trends in energy intake in the United States: An ecologic perspective, *American Journal of Clinical Nutrition* 71 (2000): 1478–1484.

31. A. S. Levine, C. M. Kotz, and B. A. Gosnell, Sugars and fats: The neurobiology of preference, *Journal of Nutrition* 133 (2003): 831S–834S.

32. D. E. Gerstein and coauthors, Clarifying concepts about macronutrients' effects on satiation and satiety, *Journal of the American Dietetic Association* 104 (2004): 1151–1153; S. C. Woods, Gastrointestinal satiety signals I. An overview of gastrointestinal signals that influence food intake, *American Journal of Physiology. Gastrointestinal and Liver Physiology* 286 (2004): G7–G13.

33. A. Prentice and S. Jebb, Energy intake/physical activity interactions in the homeostasis of body weight regulation, *Nutrition Reviews* 62 (2004): S98–S104.

34. C. Pico and coauthors, Gastric leptin: A putative role in the short-term regulation of food intake, *British Journal of Nutrition* 90 (2003): 735–741; M. J. Lewin and A. Bado, Gastric leptin, *Microscopy Research and Technique* 53 (2001): 372–376.

35. R. B. Ceddia, W. N. William, Jr., and R. Curi, The response of skeletal muscle to leptin, *Frontiers in Bioscience* 6 (2001): D90–D97.

36. J. Williams and S. Mobarhan, A critical interaction: Leptin and ghrelin, *Nutrition Reviews* 61 (2003): 391–393; D. E. Cummings and M. W. Schwartz, Genetics and pathophysiology of human obesity, *Annual Reviews of Medicine* 54 (2003): 453–471.

37. E. Faloia and coauthors, Adipose tissue as an endocrine organ? A review of some recent data, *Eating and Weight Disorders* 5 (2000): 116–123.

38. M. W. Rajala and P. E. Scherer, Minireview: The adipocyte—at the crossroads of energy homeostasis, inflammation, and atherosclerosis, *Endocrinology* 144 (2003): 3765–3773; I. Sadaf Farooqi and coauthors, Beneficial effects of leptin on obesity, T cell hyporesponsiveness, and neuroendocrine/metabolic dysfunction of human congenital leptin deficiency, *Journal of Clinical Investigation* 110 (2002): 1093–1103; R. L. Bradley, K. A. Cleveland, and B. Cheatham, The adipocyte as a secretory organ: Mechanisms of vesicle transport and secretory pathways, *Recent Progress in Hormone Research* 56 (2001): 329–358.

39. C. K. Welt and coauthors, Recombinant human leptin in women with hypothalamic amenorrhea, *New England Journal of Medicine* 351 (2004): 987–997; S. Shalitin and M. Phillip, Role of obesity and leptin in the pubertal process and pubertal growth—a review, *International Journal of Obesity and Related Metabolic Disorders* 27 (2003): 869–874.

40. Ceddia, William, Jr., and Curi, 2001; M. S. Westerterp-Plantenga and coauthors, Effects of weekly administration of pegylated recombinant human OB protein on appetite profile and energy metabolism in obese men, *American Journal of Clinical Nutrition* 74 (2001): 426–434.

41. M. Yao and S. B. Roberts, Dietary energy and weight regulation, *Nutrition Reviews* 59 (2001): 247–258.

42. W. J. Pasman and coauthors, Effect of two breakfasts, different in carbohydrate composition, on hunger and satiety and mood in healthy men, *International Journal of Obesity and Related Metabolic Disorders* 27 (2003): 663–668; D. B. Pawlak, C. B. Ebbeling, and D. S. Ludwig, Should obese patients be counselled to follow a low-glycaemic index diet? Yes, *Obesity Reviews* 3 (2002): 235–243; J. C. Brand-Miller and coauthors, Glycemic index and obesity, *American Journal of Clinical Nutrition* 76 (2002): 281S–285S.

43. A. Raben, Should obese patients be counselled to follow a low-glycaemic index diet? No, *Obesity Reviews* 3 (2002): 245–256.

44. A. R. Skov and coauthors, Randomized trial on protein vs carbohydrate in ad libitum fat reduced diet for the treatment of obesity, *International Journal of Obesity and Related Metabolic Disorders* 23 (1999): 528–536.

45. T. H. Moran and K. P. Kinzig, Gastrointestinal satiety signals II. Cholecystokinin, *American Journal of Physiology. Gastrointestinal and Liver Physiology* 286 (2004): G183–G188.

46. E. M. Kovacs and coauthors, The effect of guar gum addition to a semisolid meal on appetite related to blood glucose, in dieting men, *European Journal of Clinical Nutrition* 56 (2002): 771–778; N. C. Howarth, E. Saltzman, and S. B. Roberts, Dietary fiber and weight regulation, *Nutrition Reviews* 59 (2001): 129–139; B. J. Rolls, E. A. Bell, and B. A. Waugh, Increasing the volume of food by incorporating air increases satiety in men, *American Journal of Clinical Nutrition* 72 (2000): 361–368; B. J. Rolls, E. A. Bell, and M. L. Thorwart, Water incorporated into a food but not served with a food decreases energy intake in lean women, *American Journal of Clinical Nutrition* 70 (1999): 448–455.

47. L. P. Kozak and M. E. Harper, Mitochondrial uncoupling proteins in energy expenditure, *Annual Review of Nutrition* 20 (2000): 339–363.

48. S. O'Rahilly and coauthors, Minireview: Human obesity—lessons from monogenic disorders, *Endocrinology* 144 (2003): 3757–3764.

49. O'Rahilly and coauthors, 2003; M. S. Treuth and coauthors, Familial resemblance of body composition in prepubertal girls and their biological parents, *American Journal of Clinical Nutrition* 74 (2001): 529–533; L. Pérusse and C. Bouchard, Genotype-environment interaction in human obesity, *Nutrition Reviews* 57 (1999): S31–S38.

50. P. Hakala and coauthors, Environmental factors in the development of obesity in identical twins, *International Journal of Obesity and Related Metabolic Disorders* 23 (1999): 746–753.

51. B. E. Levin and A. A. Dunn-Meynell, Defense of body weight depends on dietary composition and palatability in rats with diet-induced obesity, *American Journal of Physiology: Regulatory, Integrative and Comparative Physiology* 282 (2002): R46–R54; L. Pérusse and C. Bouchard, Gene-diet interactions in obesity, *American Journal of Clinical Nutrition* 72 (2000): 1285S–1290S.

52. D. E. Cummings and M. W. Schwartz, Genetics and pathophysiology of human obesity, *Annual Review of Medicine* 54 (2003): 453–471; J. Altman, Weight in the balance, *Neuroendocrinology* 76 (2002): 131–136; G. Wolf, The uncoupling proteins UCP2 and UCP3 in skeletal muscle, *Nutrition Reviews* 59 (2001): 56–57.

53. J. A. Yanovski and coauthors, Associations between uncoupling protein 2, body composition, and resting energy expenditure in lean and obese African American, white, and Asian children, *American Journal of Clinical Nutrition* 71 (2000): 1405–1412.

54. J. Nedergaard and B. Cannon, The "novel" "uncoupling" proteins UCP2 and UCP3: What do they really do? Pros and cons for suggested functions, *Experimental Physiology* 88 (2003): 65–84; G. Argyropoulos and M. Harper, Molecular biology of thermoregulation, invited review: Uncoupling proteins and thermoregulation, *Journal of Applied Physiology* 92 (2001): 2187–2198.

55. B. Wansink, as quoted in Food illusions: Why we eat more than we think, *Nutrition Action Healthletter*, March 2004, pp. 3–6; M. A. McCrory and coauthors, Dietary variety within food groups: Association with energy intake and body fatness in men and women, *American Journal of Clinical Nutrition* 69 (1999): 440–447.

56. E. Kennedy, Diet diversity, diet quality, and body weight regulation, *Nutrition Reviews* 62 (2004): S78–S79.

57. M. F. Dallman and coauthors, Chronic stress and obesity: A new view of "comfort food," *Proceedings of the National Academy of Sciences of the United States of America* 100 (2003): 11696–11701.

58. J. K. Binkley, J. Eales, and M. Jekanowski, The relation between dietary change and rising U.S. obesity, *International Journal of Obesity and Related Metabolic Disorders* 24 (2000): 1032–1039.

59. T. Philipson and coauthors, The economics of obesity: A report on the workshop held at the USDA's Economic Research Service, May 2004, available at www.ers.usda.gov/publications/efan04004/.

60. J. N. Variyam, The price is right: Economics and the rise in obesity, *Amber Waves*, February 2005, available at www.ers.usda.gov/AmberWaves; S. A. French, Pricing effects on food choices, *Journal of Nutrition* 133 (2003): 841S–843S.

61. A. H. C. Goris, M. S. Westerterp-Plantenga, and K. R. Westerterp, Undereating and underrecording of habitual food intake in obese men: Selective underreporting of fat intake, *American Journal of Clinical Nutrition* 71 (2000): 130–134.

62. M. Wei and coauthors, The association between cardiorespiratory fitness and impaired glucose and type 2 diabetes mellitus in men, *Annals of Internal Medicine* 130 (1999): 89–96.

63. Even moderate amounts of exercise can prevent weight gain, *FDA Consumer,* March/April, 2004, p. 5.

64. C. A. Slentz and coauthors, Effects of the amount of exercise on body weight, body composition, and measures of central obesity: STRRIDE—a randomized controlled study, *Archives of Internal Medicine* 164 (2004): 31–39; Committee on the Scientific Evaluation of Dietary Reference Intakes, Food and Nutrition Board, Institute of Medicine, 2002, p. 12-1.

65. J. Cawley, S. Markowitz, and J. Tauras, Lighting up and slimming down: The effects of body weight and cigarette prices on adolescent smoking initiation, *Journal of Health Economics* 23 (2004): 293–311; S. E. Saarni and coauthors, Intentional weight loss and smoking in young adults, *International Journal of Obesity and Related Metabolic Disorders* 28 (2004): 796–802.

66. M. M. McGrane, Carbohydrate metabolism—Synthesis and oxidation, in M. H. Stipanuk, ed., *Biochemical and Physiological Aspects of Human Nutrition* (Philadelphia: W. B. Saunders, 2000), p. 192.

67. R. M. Anson and coauthors, Intermittent fasting dissociates beneficial effects of dietary restriction on glucose metabolism and neuronal resistance to injury from calorie intake, *Proceedings of the National Academy of Sciences* 100 (2003): 6216–6220; R. Wan, S. Camandola, and M. P. Mattson, Intermittent fasting and dietary supplementation with 2-deoxy-D-glucose improve functional and metabolic cardiovascular risk factors in rats, *The Journal of FASEB* 17 (2003): 1133–1134.

68. R. P. Ferraris and H. V. Carey, Intestinal transport during fasting and malnutrition, *Annual Review of Nutrition* 20 (2000): 195–219.

69. National Task Force on the Prevention and Treatment of Obesity, Dieting and the development of eating disorders in overweight and obese adults, *Archives of Internal Medicine* 260 (2000): 2581–2589.

70. J. Eisenstein and coauthors, High-protein weight-loss diets: Are they safe and do they work? A review of the experimental and epidemiologic data, *Nutrition Reviews* 60 (2002): 189–200.

71. A. Raben and coauthors, Meals with similar energy densities but rich in protein, fat, carbohydrate, or alcohol have different effects on energy expenditure and substrate metabolism but not on appetite and energy intake, *American Journal of Clinical Nutrition* 77 (2003): 91–100.

72. G. D. Foster and coauthors, Obese patients' perceptions of treatment outcomes and the factors that influence them, *Archives of Internal Medicine* 161 (2001): 2133–2139.

73. Report of the Dietary Guidelines Advisory Committee, August 2005, Section E, available at www.health.gov/dietaryguidelines/dga2005/report.

74. U.S. Department of Agriculture, Coordinated nutrition research program on health and nutrition effects of popular weight-loss diets, 2001.

75. National Heart, Lung, and Blood Institute, National Institutes of Health, *The Practical Guide: Identification, Evaluation, and Treatment of Overweight and Obesity in Adults,* NIH publication no. 00-4084 (Washington, D.C.: Government Printing Office, 2000), p. 26.

76. T. V. Kral, L. S. Roe, and B. J. Rolls, Does nutrition information about the energy density of meals affect food intake in normal-weight women? *Appetite* 39 (2002): 137–145; B. J. Rolls and E. A. Bell, Dietary approaches to the treatment of obesity, *Medical Clinics of North America* 84 (2000): 401–418, vi.

77. E. A. Bell and B. J. Rolls, Energy density of foods affects energy intake across multiple levels of fat content in lean and obese women, *American Journal of Clinical Nutrition* 73 (2001): 1010–1018.

78. A. Drewnowski and coauthors, Dietary energy density and body weight: is there a relationship? *Nutrition Reviews* 62 (2004): 403–413; T. V. E. Kral, L. S. Roe, and B. J. Rolls, Combined effects of energy density and portion size on energy intake in women, *American Journal of Clinical Nutrition* 79 (2004): 962–968.

79. M. B. Zemel and S. L. Miller, Dietary calcium and dairy modulation of adiposity and obesity risk, *Nutrition Reviews* 62 (2004): 125–131; S. J. Parikh and J. A. Yanovski, Calcium and adiposity, *American Journal of Clinical Nutrition* 77 (2003): 281–287; D. Teegarden, Calcium intake and reduction in weight or fat mass, *Journal of Nutrition* 133 (2003): 249S–251S; R. P. Heaney, K. M. Davies, and M. J. Barger-Lux, Calcium and weight: Clinical studies, *Journal of the American College of Nutrition* 21 (2002): 152–155.

80. M. P. St-Onge, Dietary fats, teas, dairy, and nuts: Potential functional foods for weight control? *American Journal of Clinical Nutrition* 81 (2005): 7–15; M. B. Zemel, Mechanisms of dairy modulation of adiposity, *Journal of Nutrition* 133 (2003): 252S–256S.

81. G. A. Bray, S. J. Nielsen, and B. M. Popkin, Consumption of high-fructose corn syrup in beverages may play a role in the epidemic of obesity, *American Journal of Clinical Nutrition* 79 (2004): 537–543; R. Novotny and coauthors, Dairy intake is associated with lower body fat and soda intake with greater weight in adolescent girls, *Journal of Nutrition* 134 (2004): 1905–1909.

82. M. B. Schulze and coauthors, Sugar-sweetened beverages, weight gain, and incidence of type 2 diabetes in young and middle-aged women, *Journal of the American Medical Association* 292 (2004): 927–934.

83. M. Y. Bertone and coauthors, Association between eating patterns and obesity in a free-living US adult population, *American Journal of Epidemiology* 158 (2003): 85–92.

84. L. A. Pawlow, P. M. O'Neil, and R. J. Malcolm, Night eating syndrome: Effects of brief relaxation training on stress, mood, hunger, and eating patterns, *International Journal of Obesity and Related Metabolic Disorders* 27 (2003): 970–978; M. E. Gluck, A. Geliebter, and T. Satov, Night eating syndrome is associated with depression, low self-esteem, reduced daytime hunger, and less weight loss in obese outpatients, *Obesity Research* 9 (2001): 264–267.

85. U.S. Department of Agriculture and U.S. Department of Health and Human Services, *2005 Dietary Guidelines Committee Report, 2005,* available online at www.health.gov/dietaryguidelines/dga2005report/; S. N. Blair, M. J. LaMonte, and M. Z. Nichaman, The evolution of physical activity recommendations: How much is enough? *American Journal of Clinical Nutrition* 79 (2004): 913S–920S; Committee on the Scientific Evaluation of Dietary Reference Intakes, Food and Nutrition Board, Institute of Medicine, 2002.

86. J. W. Anderson and coauthors, Long-term weight-loss maintenance: A meta-analysis of US studies, *American Journal of Clinical Nutrition* 74 (2001): 579–584.

87. J. F. Carroll and C. K. Kyser, Exercise training in obesity lowers blood pressure independent of weight change, *Medicine and Science in Sports and Exercise* 34 (2002): 596–601; B. Gutin and coauthors, Effects of exercise intensity on cardiovascular fitness, total body composition, and visceral adiposity of obese adolescents, *American Journal of Clinical Nutrition* 75 (2002): 818–826; R. Ross and coauthors, Reduction in obesity and related comorbid conditions after diet-induced weight loss or exercise induced weight loss in men: A randomized, controlled trial, *Annals of Internal Medicine* 133 (2000): 92–103.

88. Gutin and coauthors, 2002; Ross and coauthors, 2000; G. Benedetti and coauthors, Body composition and energy expenditure after weight loss following bariatric surgery, *Journal of the American College of Nutrition* 19 (2000): 270–274.

89. W. D. Schmidt, C. J. Biwer, and L. K. Kalscheuer, Effects of long *versus* short bout exercise on fitness and weight loss in overweight females, *Journal of the American College of Nutrition* 20 (2001): 494–501.

90. American College of Sports Medicine, Position stand: Appropriate intervention strategies for weight loss and prevention of weight regain for adults, *Medicine and Science in Sports and Exercise* 33 (2001): 2145–2156.

91. National Heart, Lung, and Blood Institute, National Institutes of Health, 2000, p. 29.

92. L. K. Khan and coauthors, Use of prescription weight loss pills among U.S. adults in 1996–1998, *Annals of Internal Medicine* 134 (2001): 282–286.

93. E. C. Mun, G. L. Blackburn, and J. B. Matthews, Current status of medical and surgical therapy for obesity, *Gastroenterology* 120 (2001): 669–681.

94. FDA approves implanted stomach band to treat severe obesity, *FDA Talk Paper,* June 5, 2001, available at www.fda.gov.

95. H. Buchwald and coauthors, Bariatric surgery: a systematic review and meta-analysis, *Journal of the American Medical Association* 292 (2004): 1724–1737.

96. J. G. Bruner and R. H. de Jong, Lipoplasty claims experience of U.S. insurance companies, *Plastic and Reconstructive Surgery* 107 (2001): 1285–1291.

97. R. B. Rao, S. F. Ely, and R. S. Hoffman, Deaths related to liposuction, *New England Journal of Medicine* 340 (1999): 1471–1475.

98. Food and Drug Administration, Consumer alert: FDA plans regulation prohibiting sale of ephedra-containing dietary supplements and advises consumers to stop using these products, December 2003, available at www.fda.gov/oc/initiatives/ephedra/december2003/advisory.html.

99. S. Barrett, ed., California restricts ephedra "supplements," *Consumer Health Digest,* October 21, 2003, available at www.ncahf.org; W. K. Jones, Safety of dietary supplements containing ephedrine alkaloids, FDA Public Meeting Summary, August 2000, available at http://vm.cfsan.fda.gov/~dms/.

100. M. H. Pittler and E. Ernst, Dietary supplements for body-weight reduction: A systematic review, *American Journal of Clinical Nutrition* 79 (2004): 529–536.

101. Food and Drug Administration, FDA warns distributors of dietary supplements promoted online for weight loss, *FDA News,* April 1, 2004, available at www.fda.gov/bbs/topics/news/2004/NEW01045.html.

102. J. P. Ikeda and coauthors, Self-reported dieting experiences of women with body mass indexes of 30 or more, *Journal of the American Dietetic Association* 104 (2004): 972–974.

103. 1,800 calories, 4 miles keep the weight off, *In Diabetes Today,* November 17, 2004, available at http://www.diabetes.org/indiabetestoday.jsp.

104. A Groin and coauthors, Promoting long-term weight control: does dieting consistency matter? *International Journal of Obesity and Related Metabolic Disorders* 28 (2004): 278–281.

105. P. Koh-Banerjee and coauthors, Changes in whole-grain, bran, and cereal fiber consumption in relation to 8-y weight gain among men, *American Journal of Clinical Nutrition* 80 (2004): 1237–1245; Howarth, Saltzman, and Roberts, 2001.

106. A. E. Field and coauthors, Relation between dieting and weight change among preadolescents and adolescents, *Pediatrics* 112 (2003): 900–906; A. Kroke and coauthors, Recent weight changes and weight cycling as predictors of subsequent two year weight change in a middle-aged cohort, *International Journal of Obesity and Related Metabolic Disorders* 26 (2002): 403–409.

107. E. D. Shade and coauthors, Frequent intentional weight loss is associated with lower natural killer cell cytotoxicity in postmenopausal women: Possible long-term immune effects, *Journal of the American Dietetic Association* 104 (2004): 903–912.

108. B. Wansink, Environmental factors that increase the food intake and consumption volume of unknowing consumers, *Annual Review of Nutrition* 24 (2004): 455–479.

Consumer Corner 9

1. A. Trichopoulou and coauthors, Lipid, protein and carbohydrate intake in relation to body mass index, *European Journal of Clinical Nutrition* 56 (2002): 37–43.

2. J. Eisenstein and coauthors, High-protein weight-loss diets: Are they safe and do they work? A review of the experimental and epidemiologic data, *Nutrition Reviews* 60 (2002): 189–200.

3. F. F. Samaha and coauthors, A low-carbohydrate diet as compared with a low-fat diet in severe obesity, *New England Journal of Medicine* 348 (2003): 2074–2081.

4. G. D. Foster and coauthors, A randomized trial of a low-carbohydrate diet for obesity, *New England Journal of Medicine* 348 (2003): 2082–2090.

5. J. H. Ware, Interpreting incomplete data in studies of diet and weight loss, *New England Journal of Medicine* 348 (2003): 2136–2137.

6. A. C. Buchholz and D. A. Schoeller, Is a calorie a calorie? *American Journal of Clinical Nutrition* 79 (2004): 899S–906S.

7. D. K. Layman and coauthors, A reduced ratio of dietary carbohydrate to protein improves body composition and blood lipid profiles during weight loss in adult women, *Journal of Nutrition* 133 (2003): 411–417; D. M. Bravata and coauthors, Efficacy and safety of low-carbohydrate diets, *Journal of the American Medical Association* 289 (2003): 1837–1850; M. R. Freedman, J. King, and E. Kennedy, Popular diets: A scientific review, *Obesity Research* 9 (2001): 1S–5S; A. Golay and coauthors, Similar weight loss with low-energy food combining or balanced diets, *International Journal of Obesity and Related Metabolic Disorders* 24 (2000): 492–496.

8. E. T. Kennedy and coauthors, Popular diets: Correlation to health, nutrition, and obesity, *Journal of the American Dietetic Association* 101 (2001): 411–420.

9. L. E. Kelemen and coauthors, Associations of dietary protein with disease and mortality in a prospective study of postmenopausal women, *American Journal of Epidemiology* 161 (2005): 239–249; Joint WHO/FAO Expert Consultation, *Diet, Nutrition and the Prevention of Chronic Diseases,* (Geneva: World Health Organization, 2003); National Heart, Lung, and Blood Institute, Facts about the Dash Eating Plan, NIH publication no. 03-4082, revised 2003; Executive Summary of the Third Report of the National Cholesterol Education Program (NCEP) Expert Panel on Detection, Evaluation, and Treatment of High Blood Cholesterol in Adults (Adult Treatment Panel III), 2002, NIH publication no. 02-5215; Nutrition Committee of the American Heart Association, AHA dietary guidelines revision 2000: A statement for healthcare professionals from the nutrition committee of the American Heart Association, *Circulation* 102 (2000): 2284–2299.

10. E. Cho and coauthors, Premenopausal fat intake and risk of breast cancer, *Journal of the National Cancer Institute* 95 (2003): 1079–1085.

11. S. A. Bingham and coauthors, Are imprecise methods obscuring a relation between fat and breast cancer? *Lancet* 362 (2003): 212–214.

12. Cho and coauthors, 2003.

13. Standing Committee on the Scientific Evaluation of Dietary Reference Intakes, Food and Nutrition Board, Institute of Medicine, *Dietary Reference Intakes for Energy, Carbohydrate, Fiber, Fat, Fatty Acids, Cholesterol, Protein, and Amino Acids* (Washington, D.C.: National Academy Press, 2002).

14. S. A. Bowman and J. T. Spence, A comparison of low-carbohydrate vs. high-carbohydrate diets: Energy restriction, nutrient quality and correlation to body mass index, *Journal of the American College of Nutrition* 21 (2002): 268–274.

15. B. D. Butki, J. Baumstark, and S. Driver, Effects of a carbohydrate-restricted diet on affective responses to acute exercise among physically active participants, *Perceptual and Motor Skills* 96 (2003): 607–615; J. Achten and coauthors, Higher dietary carbohydrate content during intensified running training results in better maintenance of performance and mood state, *Journal of Applied Physiology* 96 (2004): 1331–1340; R. Gruetter, Glycogen: The forgotten cerebral energy store, *Journal of Neuroscience Research* 74 (2003): 179–183; I. Y. Choi, E. R. Seaquist, and R. Gruetter, Effect of hypoglycemia on brain glycogen metabolism in vivo, *Journal of Neuroscience Research* 72 (2003): 25–32; Standing Committee on the Scientific Evaluation of Dietary Reference Intakes, Food and Nutrition Board, Institute of Medicine, 2002, p. 6-1; S. T. Reddy and coauthors, Effect of low-carbohydrate high-protein diet on acid-base balance, stone-forming propensity, and calcium metabolism, *American Journal of Kidney Disease* 40 (2002): 265–274; H. R. Lieberman, C. M. Falco, and S. S. Slade, Carbohydrate administration during a day of sustained aerobic activity improves vigilance, as assessed by a novel ambulatory monitoring device, and mood, *American Journal of Clinical Nutrition* 76 (2002): 120–127.

16. C. Lara-Castro and W. T. Garvey, Diet, insulin resistance, and obesity: zoning in on data for Atkins dieters living in South Beach, *Journal of Clinical Endocrinology and Metabolism* 89 (2004): 4197–4205.

17. S. Liu and coauthors, Relation between changes in intakes of dietary fiber and grain products and changes in weight and development of obesity among middle-aged women, *American Journal of Clinical Nutrition* 78 (2003): 920–927.

18. FDA Working Group on Obesity, *Counting Calories: Report of the Working Group on Obesity,* March 2004, available at www.cfsanfda.gov/~dms/owg-rpt.html; W. M. Mueller-Cunningham, R. Quintana, and S. E. Kasim-Karakas, An ad libitum, very low-fat diet results in weight loss and changes in nutrient intakes in postmenopausal women, *Journal of the American Dietetic Association* 103 (2003): 1600–1606.

Controversy 9

1. C. A. Arriaza and T. Mann, Ethnic differences in eating disorder symptoms among college students: The confounding role of body mass index, *Journal of American College Health* 49 (2001): 309–315.

2. Position of the American Dietetic Association: Nutrition intervention in the treatment of anorexia nervosa, bulimia nervosa, and eating disorders not otherwise specified (EDNOS), *Journal of the American Dietetic Association* 101 (2001): 810–819.

3. A. E. Field and coauthors, Relation between dieting and weight change among preadolescents and adolescents, *Pediatrics* 112 (2003): 900–906; D. Neumark-Sztainer and P. J. Hannan, Weight-related behaviors among adolescent girls and boys: Results from a national survey, *Archives of Pediatrics and Adolescent Medicine* 154 (2000): 569–577.

4. G. Tsai, Eating disorders in the Far East, *Eating and Weight Disorders* 5 (2000): 183–197.

5. J. A. McLean, S. I. Barr, and J. C. Prior, Cognitive dietary restraint is associated with higher urinary cortisol excretion in healthy premenopausal women, *American Journal of Clinical Nutrition* 73 (2001): 7–12.

6. E. Cooley and T. Toray, Disordered eating in college freshman women: A prospective study, *Journal of American College Health* 49 (2001): 229–235.

7. K. Kazis and E. Iglesias, The female athlete triad, *Adolescent Medicine* 14 (2003): 87–95.

8. Kazis and Iglesias 2003.

9. N. H. Golden, A review of the female athlete triad (amenorrhea, osteoporosis and disordered eating), *International Journal of Adolescent Medicine and Health* 14 (2002): 9–17.

10. D. Hartman and coauthors, Bone density of women who have recovered from anorexia nervosa, *International Journal of Eating Disorders* 28 (2000): 107–112.

11. K. L. Cobb and coauthors, Disordered eating, menstrual irregularities, and bone mineral density in female runners, *Medicine and Science in Sports and Exercise* 35 (2003): 711–719.

12. Are you finding and treating males with eating disorders at your school? *School Health Professional,* August 2000, pp. 1–2.

13. V. Vidovic, N. Henigsberg, and V. Juresa, Anxiety and defense styles in eating disorders, *Collegium antropologicum* 27 (2003): 125–134.

14. K. P. Kerruish and coauthors, Body composition in adolescents with anorexia nervosa, *American Journal of Clinical Nutrition* 75 (2002): 31–37.

15. C. Romano and coauthors, Reduced hemodynamic load and cardiac hypotrophy in patients with anorexia nervosa, *American Journal of Clinical Nutrition* 77 (2003): 308–312; C. Panagiotopoulos and coauthors, Electrocardiographic findings in adolescents with eating disorders, *Pediatrics* 105 (2000): 1100–1105.

16. Committee on Adolescence, Identifying and treating eating disorders, *Pediatrics* 111 (2003): 204–211; Position of the American Dietetic Association, 2001.

17. A. E. Becker and coauthors, Eating disorders, *New England Journal of Medicine* 340 (1999): 1092–1098.

18. H. C. Steinhausen, The outcome of anorexia nervosa in the 20th century, *American Journal of Psychiatry* 159 (2002): 1284–1293.

19. M. Misra and coauthors, Regional body composition in adolescents with anorexia nervosa and changes with weight recovery, *American Journal of Clinical Nutrition* 77 (2003): 1361–1367; L. Mayer, Body composition and anorexia nervosa: Does physiology explain psychology? *American Journal of Clinical Nutrition* 73 (2001): 851–852; S. Grinspoon and coauthors, Changes in regional fat distribution and the effects of estrogen during spontaneous weight gain in women with anorexia nervosa, *American Journal of Clinical Nutrition* 73 (2001): 865–869.

20. M. B. Tamburrino and R. A. McGinnis, Anorexia nervosa: A review, *Panminerva Medica* 44 (2002): 301–311.

21. P. K. Keel and coauthors, Predictors of mortality in eating disorders, *Archives of General Psychiatry* 60 (2003): 179–183.

22. J. Gutzwiller, J. M. Oliver, and B. M. Katz, Eating dysfunctions in college women: The roles of depression and attachment to fathers, *Journal of American College Health* 52 (2003): 27–32.

23. M. L. Granner, D. A. Abood, and D. R. Black, Racial differences in eating disorder attitudes, cigarette, and alcohol use, *American Journal of Health Behavior* 25 (2001): 83–99.

24. D. Neumark-Sztainer and coauthors, Disordered eating among adolescents: Associations with sexual/physical abuse and other familial/psychosocial factors, *International Journal of Eating Disorders* 28 (2000): 249–258.

25. J. L. Guss and coauthors, Binge size increases with body mass index in women with binge-eating disorder, *Obesity Research* 10 (2002): 1021–1029.

26. J. Sundgot-Borgen and coauthors, The effect of exercise, cognitive therapy, and nutritional counseling in treating bulimia nervosa, *Medicine and Science in Sports and Exercise* 34 (2002): 190–195.

27. A. E. Dingemans, M. J. Bruna, and E. F. van Furth, Binge eating disorder: A review, *International Journal of Obesity and Related Metabolic Disorders* 26 (2002): 299–307.

28. D. M. Ackard and coauthors, Overeating among adolescents: Prevalence and associations with weight-related characteristics and psychological health, *Pediatrics* 111 (2003): 67–74.

29. S. Rubinstein and B. Caballero, Is Miss America an undernourished role model? *Journal of the American Medical Association* 283 (2000): 1569.

30. J. Polivy and C. P. Herman, Causes of eating disorders, *Annual Review of Psychology* 53 (2002): 187–213.

31. J. A. Shunk and L. L. Birch, Girls at risk for overweight at age 5 are at risk for dietary restraint, disinhibited overeating, weight concerns, and greater weight gain from 5 to 9 years, *Journal of the American Dietetic Association* 104 (2004): 1120–1126.

32. Federal Interagency Forum on Child and Family Statistics, *America's Children: Key National Indicators of Well-Being,* 2003, a report from the National Institutes of Health, available from National Maternal and Child Health Clearinghouse, 2070 Chain Bridge Road, Suite 450, Vienna, VA 22182 or on the Internet at http://childstats.gov.

Chapter 10

1. Y. A. Kesaniemi and coauthors, Dose-response issues concerning physical activity and health: An evidence-based symposium, *Medicine and Science in Sports and Exercise* 33 (2001): 351S–358S; I. M. Lee and R. S. Paffenbarger, Associations of light, moderate, and vigorous intensity physical activity with longevity: The Harvard Alumni Study, *American Journal of Epidemiology* 151 (2000): 293–299.

2. P. M. Barnes and C. A. Schoenborn, *Physical Activity among Adults: United States, 2000,* Advance Data from Vital and Health Statistics document no. 333 (2003), available from www.cdc.gov/nchs/Default.htm

3. K. R. Evenson and coauthors, The effect of cardiorespiratory fitness and obesity on cancer mortality in women and men, *Medicine and Science in Sports and Exercise* 35 (2003): 270–277; J. Dorn and coauthors, Lifetime physical activity and breast cancer risk in pre- and postmenopausal women, *Medicine and Science in Sports and Exercise* 35 (2003): 278–285; C. D. Lee and S. N. Blair, Cardiorespiratory fitness and stroke mortality in men, *Medicine and Science in Sports and Exercise* 34 (2002): 592–595; K. Moreau and coauthors, Increasing daily walking lowers blood pressure in postmenopausal women, *Medicine and Science in Sports and Exercise* 33 (2001): 1825–1831.

4. G. A. Colditz, Economic costs of obesity and inactivity, *Medicine and Science in Sports and Exercise* 31 (1999): 663S–667S.

5. U. G. Kyle and coauthors, Physical activity and fat-free and fat mass by bioelectrical impedance in 3853 adults, *Medicine and Science in Sports and Exercise* 33 (2001): 576–584.

6. L. Metcalfe and coauthors, Postmenopausal women and exercise for prevention of osteoporosis: The Bone, Estrogen, Strength Training (BEST) Study, *ACSM's Health & Fitness Journal,* May/June 2001, pp. 6–14.

7. C. E. Matthews and coauthors, Moderate to vigorous physical activity and risk of upper-respiratory tract infection, *Medicine and Science in Sports and Exercise* 34 (2002): 1242–1248.

8. C. M. Friedenreich, Physical activity and cancer: Lessons learned from nutritional epidemiology, *Nutrition Reviews* 59 (2001): 349–357.

9. American College of Sports Medicine, Position stand, Exercise and hypertension, *Medicine and Science in Sports and Exercise* 36 (2004): 533–553; M. R. Carnethon and coauthors, Cardiorespiratory fitness in young adulthood and the development of cardiovascular disease risk factors, *Journal of the American Medical Association* 290 (2003): 3092–3100.

10. S. L. Wong and coauthors, Cardiorespiratory fitness is associated with lower abdominal fat independent of body mass index, *Medicine and Science in Sports and*
Exercise 36 (2004): 286–291; A. Trichopoulou and coauthors, Physical activity and energy intake selectively predict the waist-to-hip ratio in men but not in women, *American Journal of Clinical Nutrition* 74 (2001): 574–578.

11. R. M. van Dam and coauthors, Physical activity and glucose tolerance in elderly men: The Zutphen Elderly Study, *Medicine and Science in Sports and Exercise* 34 (2002): 1132–1136; K. J. Stewart, Exercise training and the cardiovascular consequences of type 2 diabetes and hypertension: Plausible mechanisms for improving cardiovascular health, *Journal of the American Medical Association* 288 (2002): 1622–1631.

12. G. Misciagna and coauthors, Diet, physical activity, and gallstones—a population-based, case-control study in southern Italy, *American Journal of Clinical Nutrition* 69 (1999): 120–126; M. F. Leitzmann and coauthors, Recreational physical activity and the risk of cholecystectomy in women, *New En-gland Journal of Medicine* 341 (1999): 777–784.

13. W. J. Strawbridge and coauthors, Physical activity reduces the risk of subsequent depression for older adults, *American Journal of Epidemiology* 156 (2002): 328–334.

14. J. Myers and coauthors, Exercise capacity and mortality among men referred for exercise testing, *New England Journal of Medicine* 346 (2002): 793–801; Lee and Paffenbarger, 2000.

15. T. Rantanen and coauthors, Midlife hand grip strength as a predictor of old age disability, *Journal of the American Medical Association* 281 (1999): 558–560.

16. Myers and coauthors, 2002; A. L. Dunn and coauthors, Comparison of lifestyle and structured interventions to increase physical activity and cardiorespiratory fitness, *Journal of the American Medical Association* 281 (1999): 327–334.

17. U.S. Department of Agriculture and U.S. Department of Health and Human Services, *Dietary Guidelines for Americans 2005,* available online at www.usda.gov/cnpp. American College of Sports Medicine, *ACSM's Guidelines for Exercise Testing and Prescription,* 6th ed. (Philadelphia: Williams & Wilkins, 2000), pp. 137–164; U.S. Department of Agriculture and U.S. Department of Health and Human Services, *Nutrition and Your Health: Dietary Guidelines for Americans,* 5th ed., Home and Garden Bulletin no. 232 (Washington, D.C.: 2000).

18. U.S. Department of Agriculture and U.S. Department of Health and Human Services, 2005; American College of Sports Medicine, Position Stand: Appropriate intervention strategies for weight loss and prevention of weight regain for adults, *Medicine and Science in Sports and Exercise* 33 (2001): 2145–2156; Standing Committee on the Scientific Evaluation of Dietary Reference Intakes, Food and Nutrition Board, Institute of Medicine, *Dietary Reference Intakes for Energy, Carbohydrate, Fiber, Fat, Fatty Acids, Cholesterol, Protein, and Amino Acids* (Washington, D.C.: National Academies Press, 2002), pp. 12-1–12-39.

19. M. Murphy and coauthors, Accumulating brisk walking for fitness, cardiovascular risk, and psychological health, *Medicine and Science in Sports and Exercise* 34 (2002): 1468–1474; W. D. Schmidt, C. J. Biwer, and L. K. Kalscheuer, Effects of long *versus* short bout exercise on fitness and weight loss in overweight females, *Journal of the American College of Nutrition* 20 (2001): 494–501.

20. Lee and Blair, 2002.

21. American College of Sports Medicine, Position stand: Progression models in resistance training for healthy adults, *Medicine and Science in Sports and Exercise* 34 (2002): 364–380; M. L. Pollock and coauthors, AHA Science Advisory: Resistance exercise in individuals with and without cardiovascular disease: Benefits, rationale safety, and prescription, *Circulation* 101 (2000): 828–833.

22. J. E. Layne and M. E. Nelson, The effects of progressive resistance training on bone density: A review, *Medicine and Science in Sports and Exercise* 31 (1999): 25–30.

23. E. C. Cussler and coauthors, Weight lifted in strength training predicts bone change in postmenopausal women, *Medicine and Science in Sports and Exercise* 35 (2003): 10–17.

24. W. J. Kraemer and coauthors, Physiological changes with periodized resistance training in women tennis players, *Medicine and Science in Sports and Exercise* 35 (2003): 157–168.

25. J. H. Wilmore and D. L. Costill, Physical energy: Fuel metabolism, *Nutrition Reviews* 59 (2001): S13–S16.

26. T. H. Pedersen and coauthors, Intracellular acidosis enhances the excitability of working muscle, *Science* 305 (2004): 1144–1147; D. Allen and H. Westerblad, Enhanced: Lactic acid—the latest performance-enhancing drug, *Science* 305 (2004): 1112–1113; R. A. Robergs, F. Ghiasvand, and D. Parker, Biochemistry of exercise-induced metabolic acidosis, *American Journal of Physiology—Regulatory, Integrative and Comparative Physiology* 287 (2004): R502–R516.

27. A. C. Utter and coauthors, Carbohydrate supplementation and perceived exertion during prolonged running, *Medicine and Science in Sports and Exercise* 36 (2004): 1036–1041.

28. R. S. Welsh and coauthors, Carbohydrates and physical/mental performance during intermittent exercise to fatigue, *Medicine and Science in Sports and Exercise* 34 (2002): 723–731.

29. E. Coleman, Carbohydrate and exercise, in C. A. Rosenbloom, ed., *Sports Nutrition: A Guide for the Professional Working with Active People,* 3rd ed. (Chicago: The American Dietetic Association, 2000), pp. 13–31.

30. T. J. Fairchild and coauthors, Rapid carbohydrate loading after a short bout of near maximal-intensity exercise, *Medicine and Science in Sports and Exercise* 34 (2002): 980–986.

31. Coleman, 2000.

32. J. Manetta and coauthors, Fuel oxidation during exercise in middle-aged men: Role of training and glucose disposal, *Medicine and Science in Sports and Exercise* 34 (2002): 423–429.

33. L. M. Burke and J. A. Hawley, Effects of short-term fat adaptation on metabolism and performance of prolonged exercise, *Medicine and Science in Sports and Exercise* 34 (2002): 1492–1498.

34. Burke and Hawley, 2002; L. M. Burke and coauthors, Adaptations to short-term high-fat diet persist during exercise despite high carbohydrate availability, *Medicine and Science in Sports and Exercise* 34 (2002): 83–91; A. L. Staudacher and coauthors, Effects of fat adaptation and carbohydrate restoration on prolonged endurance exercise, *Journal of Applied Physiology* 91 (2001): 115–122.

35. J. W. Helge, Long-term fat diet adaptation, effects on performance, training capacity, and fat utilization, *Medicine and Science in Sports and Exercise* 34 (2002): 1499–1504; N. D. Stepto and coauthors, Effect of short-term fat adaptation on high-intensity training, *Medicine and Science in Sports and Exercise* 34 (2002): 449–455.

36. E. Coleman, Does a low-fat diet impair nutrition and performance? *Sports Medicine Digest* 22 (2000): 41; Position of the American Dietetic Association, Dietitians of Canada, and the American College of Sports Medicine: Nutrition and athletic performance, *Journal of the American Dietetic Association* 100 (2000): 1543–1556.

37. E. L. Melanson and coauthors, Resistance and aerobic exercise have similar effects on 24-hour nutrient oxidation, *Medicine and Science in Sports and Exercise* 34 (2002): 1793–1800; M. K. Thornton and J. A. Potteiger, Effects of resis-tance exercise bouts of different intensities but equal work on EPOC, *Medicine and Science in Sports and Exercise* 34 (2002): 715–722.

38. J. S. Volek, Influence of nutrition on response to resistance training, *Medicine and Science in Sports and Exercise* 36 (2004): 689–696; M. Suzuki, Glycemic carbohydrates consumed with amino acids or protein right after exercise enhance muscle formation, *Nutrition Reviews* 61 (2003): S88–S94; D. K. Levenhagen and coauthors, Postexercise protein intake enhances body and leg accretion in humans, *Medicine and Science in Sports and Exercise* 34 (2002): 828–837.

39. Standing Committee on the Scientific Evaluation of Dietary Reference Intakes, Food and Nutrition Board, Institute of Medicine, 2002, pp. 10-52–10-53.

40. Position of the American Dietetic Association, Dietitians of Canada, and the American College of Sports Medicine, 2000.

41. T. L. Schwenk and C. D. Costley, When food becomes a drug: Nonanabolic nutritional supplement use in athletes, *American Journal of Sports Medicine* 30 (2002): 907–916.

42. Position of the American Dietetic Association, Dietitians of Canada, and the American College of Sports Medicine, 2000.

43. S. K. Powers, L. L. Ji, and C. Leeuwenburgh, Exercise training-induced alterations in skeletal muscle antioxidant capacity: A brief review, *Medicine and Science in Sports and Exercise* 31 (1999): 987–997.

44. M. L. Urso and P. M. Clarkson, Oxidative stress, exercise, and antioxidant supplementation, *Toxicology* 189 (2003): 41–54; H. Itoh and coauthors, Vitamin E supplementation attenuates leakage of enzymes following 6 successive days of running training, *International Journal of Sports Medicine* 21 (2000): 369–374; P. M. Clarkson and H. S. Thompson, Antioxidants: What role do they play in physical activity and health? *American Journal of Clinical Nutrition* 72 (2000): 637S–646S.

45. R. J. Piercy and coauthors, Association between vitamin E and enhanced athletic performance in sled dogs, *Medicine and Science in Sports and Exercise* 33 (2001): 826–833.

46. J. Beard and B. Tobin, Iron status and exercise, *American Journal of Clinical Nutrition* 72 (2000): 594S–597S.

47. E. R. Eichner, Non-anemias in athletes—sports anemia and footstrike hemolysis: Friends not foes? *Sports Medicine Digest* 23 (2001): 53.

48. Position of the American Dietetic Association, Dietitians of Canada, and the American College of Sports Medicine, 2000; Beard and Tobin, 2000.

49. Position of the American Dietetic Association and Dietitians of Canada: Vegetarian diets, *Journal of the American Dietetic Association* 103 (2003): 748–765; D. C. Nieman, Physical fitness and vegetarian diets: Is there a relation? *American Journal of Clinical Nutrition* 70 (1999): 570S–575S.

50. T. Brownlie and coauthors, Marginal iron deficiency without anemia impairs aerobic adaptation among previously untrained women, *American Journal of Clinical Nutrition* 75 (2002): 734–742; E. R. Eichner, Anemia in female athletes, *Sports Medicine Digest* 22 (2000): 42–43.

51. Standing Committee on the Scientific Evaluation of Dietary Reference Intakes, Food and Nutrition Board, Institute of Medicine, *Dietary Reference Intakes for Water, Potassium, Sodium, Chloride, and Sulfate* (Washington, D.C.: National Academies Press, 2004), pp. 4-31–4-36.

52. American College of Sports Medicine, Position stand: Heat and cold illness during distance running, *Medicine and Science in Sports and Exercise* 28 (1996): i–x.

53. J. R. Stofan and coauthors, Sweat and sodium losses in NCAA Division 1 football players with a history of whole-body muscle cramping, paper presented at the annual meeting of the American College of Sports Medicine, 2003.

54. N. J. Rehrer, Fluid and exectrolyte balance in ultra-endurance sport, *Sports Medicine* 31 (2001): 701–715.

55. J. W. Gardner, Death by water intoxication, *Military Medicine* 168 (2003): 432–434.

56. M. Hsieh and coauthors, Hyponatremia in runners requiring on-site medical treatment at a single marathon, *Medicine and Science in Sports and Exercise* 34 (2002): 185–189.

57. E. R. Eichner, Exertional hyponatremia: Why so many women? *Sports Medicine Digest* 24 (2002): 54, 56.

Consumer Corner 10

1. E. Coleman, Fluid replacement for athletes, *Sports Medicine Digest* 25 (2003): 76–77; Inter-Association Task Force on Exertional Heat Illness, Consensus statement, *NATA NEWS,* June 2003.

2. R. Murray, Fluid and electrolytes, in C. A. Rosenbloom, ed., *Sports Nutrition: A Guide for the Professional Working with Active People,* 3rd ed. (Chicago: The American Dietetic Association, 2000), pp. 95–106.

Controversy 10

1. T. L. Schwenk and C. D. Costley, When food becomes a drug: Nonanabolic nutritional supplement use in athletes, *American Journal of Sports Medicine* 30 (2002): 907–916.

2. Schwenk and Costley, 2002.

3. E. Blomstrand, Amino acids and central fatigue, *Amino Acids 20* (2001): 25–34.

4. M. J. Gibala, Regulation of skeletal muscle amino acid metabolism during exercise, *International Journal of Sports Nutrition and Exercise Metabolism* 11 (2001): 87–108.

5. M. J. Rennie and K. D. Tipton, Protein and amino acid metabolism during and after exercise and the effects of nutrition, *Annual Review of Nutrition* 20 (2000): 457–483; R. R. Wolfe, Protein supplements and exercise, *American Journal of Clinical Nutrition* 72 (2000): 551S–557S.

6. C. D. Paton, W. G. Hopkins, and L. Vollebregt, Little effect of caffeine ingestion on repeated sprints in team-sport athletes, *Medicine and Science in Sports and Exercise* 33 (2001): 822–825; C. J. Sinclair and J. D. Geiger, Caffeine use in sports: A pharmacological review, *Journal of Sports Medicine and Physical Fitness* 40 (2000): 71–79.

7. D. G. Bell and T. M. McLellan, Effect of repeated caffeine ingestion on repeated exhaustive exercise endurance, *Medicine and Science in Sports and Exercise* 35 (2003): 1348–1354; T. E. Graham, Caffeine and exercise: Metabolism, endurance and performance, *Sports Medicine* 31 (2001): 785–807; C. R. Bruce and coauthors, Enhancement of 2000-m rowing performance after caffeine ingestion, *Medicine and Science in Sports and Exercise* 32 (2000): 1958–1963; M. E. Anderson and coauthors, Improved 2000-meter rowing performance in competitive oarswomen after caffeine ingestion, *International Journal of Sports Nutrition and Exercise Metabolism* 10 (2000): 464–475.

8. I. Jacobs and coauthors, Effects of ephedrine, caffeine, and their combination on muscular endurance, *Medicine and Science in Sports and Exercise* 35 (2003): 987–997; Paton, Hopkins, and Vollebregt, 2001.

9. T. E. Graham and coauthors, Caffeine ingestion does not alter carbohydrate or fat metabolism in human skeletal muscle during exercise, *Journal of Physiology* 529 (2000): 837–847.

10. D. Laurent and coauthors, Effects of caffeine on muscle glycogen utilization and the neuroendocrine axis during exercise, *Journal of Clinical Endocrinology and Metabolism* 85 (2000): 2170–2175.

11. G. W. Evans, The effect of chromium picolinate on insulin-controlled parameters in humans, *Journal of Biosocial Medicine Research* 11 (1989): 163–180.

12. J. M. Davis, R. S. Welsh, and N. A. Alerson, Effects of carbohydrate and chromium ingestion during intermittent high-intensity exercise to fatigue, *International Journal of Sports Nutrition and Exercise Metabolism* 10 (2000): 476–485.

13. J. F. Fowler, Systemic contact dermatitis caused by oral chromium picolinate, *Cutis* 65 (2000): 116.

14. J. B. Vincent, The biochemistry of chromium, *Journal of Nutrition* 130 (2000): 715–718.

15. D. Preen and coauthors, Effect of creatine loading on long-term sprint exercise performance and metabolism, *Medicine and Science in Sports and Exercise* 33 (2001): 814–821.

F

16. E. B. Feldman, Creatine: A dietary supplement and ergogenic aid, *Nutrition Reviews* 57 (1999): 45–50.

17. M. Greenwood and coauthors, Creatine supplementation patterns and perceived effects in select division I collegiate athletes, *Clinical Journal of Sports Medicine* 10 (2000): 191–194.

18. E. Ha and M. B. Zemel, Functional properties of whey, whey components, and essential amino acids: Mechanisms underlying health benefits for active people (review), *Journal of Nutritional Biochemistry* 14 (2003): 251–258.

19. M. Freeman, Scientist suspects many athletes are using undetected steroids, *New York Times Online*, October 21, 2003.

20. R. C. Daly and coauthors, Cerebrospinal fluid and behavioral changes after methyltestosterone administration, *Archives of General Psychiatry* 58 (2001): 172–177.

21. I. Thiblin, O. Lindquist, and J. Rajs, Cause and manner of death among users of anabolic androgenic steroids, *Journal of Forensic Science* 45 (2000): 16–23.

22. Andrew Taber, Roid rage, *Salon.com,* November 18, 1999, www.salon.com/health/feature/1999/11/18/steroids/index.html.

23. B. B. Rasmussen and coauthors, Androstenedione does not stimulate muscle protein anabolism in young healthy men, *Journal of Clinical Endocrinology and Metabolism* 85 (2000): 55–59.

24. R. Skinner, E. Coleman, and C. A. Rosenbloom, Ergogenic aids, in C. A. Rosenbloom, ed., *Sports Nutrition: A Guide for the Professional Working with Active People,* 3rd ed. (Chicago: The American Dietetic Association, 2000), pp. 107–146; E. Coleman, DHEA—An anabolic aid? *Sports Medicine Digest* 18 (1996): 140–141.

25. A. T. Kieman and coauthors, Effect of androstenedione ingestion on plasma testosterone in young women: A dietary supplement with potential health risks, *Clinical Chemistry* 49 (2002): 167–169; B. Z. Leder and coauthors, Oral androstenedione administration and serum testosterone concentrations in young men, *Journal of the American Medical Association* 283 (2000): 779–782; D. S. King and coauthors, Effect of oral androstenedione on serum testosterone and adaptations to resistance training in young men: A randomized controlled trial, *Journal of the American Medical Association* 281 (1999): 2020–2028.

26. Food and Drug Administration, Health effects of androstenedione, *FDA White Paper,* March 11, 2004, available at www.fda.gov/oc/whitepapers/ andro.html.

27. FDA warns against consuming dietary supplements containing tiratricol, *FDA Talk Paper,* November 2001, available from http://vm.cfsan.fda.gov.

28. FDA statement on THG, October 28, 2003, available at www.fda.gov or by phone at (888) INFO-FDA.

29. FDA statement on THG, 2003.

Chapter 11

1. Centers for Disease Control and Prevention, World TB Day—March 24, 2002, *Morbidity and Mortality Weekly Report* 51 (2002): 229; S. H. Gillespie, Antibiotic resistance in the absence of selective pressure, *International Journal of Antimicrobial Agents* 17 (2001): 171–176; D. K. Warren and V. J. Fraser, Infection control measures to limit antimicrobial resistance, *Critical Care Medicine* 29 (2001): N128–N134.

2. G. T. Keusch, The history of nutrition: Malnutrition, infection and immunity, *Journal of Nutrition* 133 (2003): 336S–340S; P. Bhaskaram, Micronutrient malnutrition, infection, and immunity: An overview, *Nutrition Reviews* 60 (2002): S40–S45.

3. A. M. Miniño and coauthors, Deaths: Final data for 2000, *National Vital Statistics Report,* September 16, 2002.

4. M. J. Stampfer and coauthors, Primary prevention of coronary heart disease in women through diet and lifestyle, *New England Journal of Medicine* 343 (2000): 16–22; A. K. Kant and coauthors, A prospective study of diet quality and mortality in women, *Journal of the American Medical Association* 283 (2000): 2109–2115.

5. C. J. Field, I. R. Johnson, and P. D. Schley, Nutrients and their role in host resistance to infection, *Journal of Leukocyte Biology* 71 (2002): 16–32; Bhaskaram, 2002.

6. P. C. Calder, Immunonutrition may have beneficial effects in surgical patients, editorial, *British Medical Journal* 327 (2003): 117–118.

7. W. W. Fawzi and coauthors, A randomized trial of multivitamin supplements and HIV disease progression and mortality, *New England Journal of Medicine* 351 (2004): 23–32.

8. Position of the American Dietetic Association and Dietitians of Canada: Nutrition intervention in the care of persons with human immunodeficiency virus infection, *Journal of the American Dietetic Association* 100 (2000): 708–717.

9. M. Serafini, Dietary vitamin E and T cell-mediated function in the elderly: Effectiveness and mechanism of action, *International Journal of Developmental Neuroscience* 18 (2000): 401–410; C. Y. Lee and J. Man-Fan Wan, Vitamin E supplementation improves cell-mediated immunity and oxidative stress of Asian men and women, *Journal of Nutrition* 130 (2000): 2932–2937.

10. Standing Committee on the Scientific Evaluation of Dietary Reference Intakes, Food and Nutrition Board, Institute of Medicine, *Dietary Reference Intakes for Vitamin A, Vitamin K, Arsenic, Boron, Chromium, Copper, Iodine, Iron, Manganese, Molybdenum, Nickel, Silicon, Vanadium, and Zinc* (Washington, D.C.: National Academy Press, 2001), pp. 12–30; J. C. Fleet, Zinc, copper, and manganese, in M. H. Stipanuk, ed., *Biochemical and Physiological Aspects of Human Nutrition* (Philadelphia: Saunders, 2000), p. 753.

11. Standing Committee on the Scientific Evaluation of Dietary Reference Intakes, Food and Nutrition Board, Institute of Medicine, *Dietary Reference Intakes for Energy, Carbohydrate, Fiber, Fat, Fatty Acids, Cholesterol, Protein, and Amino Acids* (Washington, D.C.: National Academies Press, 2002), p. 3-1.

12. A. K. Kant, Dietary patterns and health outcomes, *Journal of the American Dietetic Association* 104 (2004): 615–635.

13. A. J. Blumenfeld and coauthors, Nutritional aspects of prostate cancer: A review, *Canadian Journal of Urology* 7 (2000): 927–935.

14. American Heart Association, *Heart and Stroke Statistics—2003 Update* (Dallas: American Heart Association, 2002), printable version available at www.americanheart.org.

15. American Heart Association, 2002.

16. Centers for Disease Control and Prevention, State-specific prevalence of selected chronic disease-related characteristics—behavioral risk factor surveillance system, 2001, *Morbidity and Mortality Weekly Report* 52 (2003): 4.

17. H. C. McGill and coauthors, Origin of atherosclerosis in childhood and adolescence, *American Journal of Clinical Nutrition* 72 (2000): 1307S–1325S.

18. F. B. Hu and W. C. Willett, Optimal diets for prevention of coronary heart disease, *Journal of the American Medical Association* 288 (2002): 2569–2578; R. P. Lauber and N. F. Sheard, The American Heart Association Dietary Guidelines for 2000: A summary report, *Nutrition Reviews* 59 (2001): 298–306.

19. F. Pellegatta and coauthors, Different short- and long-term effects of resveratrol on nuclear factor-kB phosphorylation and nuclear appearance in human endothelial cells, *American Journal of Clinical Nutrition* 77 (2003): 1220–1228; R. Ross, Atherosclerosis—an inflammatory disease, *New England Journal of Medicine* 340 (1999): 115–126.

20. G. A. A. Ferns and D. J. Lamb, What does the lipoprotein oxidation phenomenon mean? *Biochemical Society Transactions* 32 (2004): 160–163.

21. McGill and coauthors, 2000.

22. Standing Committee on the Scientific Evaluation of Dietary Reference Intakes, Food and Nutrition Board, Institute of Medicine, 2002, pp. 11-40–11-43; H. Iso and coauthors, Intake of fish and omega-3 fatty acids and risk of stroke in women, *Journal of the American Medical Association* 285 (2001): 304–312.

23. Third report of the National Cholesterol Education Program (NCEP) Expert Panel on Detection, Evaluation, and Treatment of High Blood Cholesterol in Adults (Adult Treatment Panel III), *Journal of the American Medical Association* 285 (2001): 2486–2497.

24. E. S. Ford, W. H. Giles, and W. H. Dietz, Prevalence of the metabolic syndrome among US adults: Findings from the Third National Health and Nutrition Examination Survey, *Journal of the American Medical Association* 287 (2002): 356–359.

25. Expert Panel on Detection, Evaluation, and Treatment of High Blood Cholesterol in Adults (Adult Treatment Panel III), *Third Report of the National Cholesterol Education Program (NCEP),* NIH publication no. 02-5215 (Bethesda, Md.: National Heart, Lung, and Blood Institute, 2002), p. II-18.

26. American Heart Association, 2002, p. 11.

27. D. M. Lloyd-Jones and coauthors, Parental cardiovascular disease as a risk factor for cardiovascular disease in middle-aged adults, *Journal of the American Medical Association* 291 (2004): 2204–2211.

28. Expert Panel on Detection, Evaluation, and Treatment of High Blood Cholesterol in Adults (Adult Treatment Panel III), 2002, p. II-19.

29. J. M. Ordovas, Cardiovascular disease genetics: A long and winding road, *Current Opinion in Lipidology* 14 (2003): 47–54.

30. S. M. Grundy and coauthors, NCEP Report: Implications of recent clinical trials for the National Cholesterol Education Program Adult Treatment Panel III Guidelines, *Circulation* 110 (2004): 227–239.

31. K. He and coauthors, Dietary fat intake and risk of stroke in male U.S. healthcare professionals: 14 year prospective cohort study, *British Medical Journal* 327 (2003): 777–783.

32. A. W. Weverling-Rijnsburger and coauthors, High-density vs low-density lipoprotein cholesterol as the risk factor for coronary artery disease and stroke in old age, *Archives of Internal Medicine* 163 (2003): 1549–1554; R. L. Sacco and coauthors, High-density lipoprotein cholesterol and ischemic stroke in the el-derly: The Northern Manhattan Stroke Study, *Journal of the American Medical Association* 285 (2001): 2729–2735.

33. J. A. Beckman, M. A. Creager, and P. Libby, Diabetes and atherosclerosis: Epidemiology, pathophysiology, and management, *Journal of the American Medical Association* 287 (2002): 2570–2581.

34. Expert Panel on Detection, Evaluation, and Treatment of High Blood Cholesterol in Adults (Adult Treatment Panel III), 2002, pp. II-16, 11-50–11-53.

35. American Heart Association, 2002, p. 34.

36. S. Bioletto and coauthors, Acute hyperinsulinemia and very-low-density and low-density lipoprotein subfractions in obese subjects, *American Journal of Clinical Nutrition* 71 (2000): 443–449.

37. P. T. Katzmaryzyk and coauthors, Targeting the metabolic syndrome with exercise: Evidence from the HERITAGE Family Study, *Medicine and Science in Sports and Exercise* 35 (2003): 1703–1709.

38. M. L. Pollock and coauthors, Resistance exercise in individuals with and without cardiovascular disease, *Circulation* 101 (2000): 828–833.

39. American Heart Association, 2002, p. 26; Expert Panel on Detection, Evaluation, and Treatment of High Blood Cholesterol in Adults (Adult Treatment Panel III), 2002, p. II-16.

40. G. A. Bray and C. M. Champagne, Obesity and the metabolic syndrome: Implications for dietetics practitioners, *Journal of the American Dietetic Association* 104 (2004): 86–89.

41. H. M. Lakka and coauthors, The metabolic syndrome and total and cardiovascular disease mortality in middle-aged men, *Journal of the American Medical Association* 288 (2002): 2709–2716; D. L. Sprecher and G. L. Pearce, How deadly is the "deadly quartet"? A post-CABG evaluation, *Journal of the American College of Cardiology* 36 (2000): 1159–1165.

42. National Institutes of Health, National Heart, Lung, and Blood Institute, NCEP issues major new cholesterol guidelines (press release), 2001, available at www.nhlbi.nih.gov.

43. Ford, Giles, and Dietz, 2002.

44. J. W. Miller, Does lowering plasma homocysteine reduce vascular disease risk? *Nutrition Reviews* 59 (2001): 242–244.

45. A. Ascherio and coauthors, *Trans*-fatty acids and coronary heart disease, *New England Journal of Medicine* 340 (1999): 1994–1998.

46. S. Yusuf and coauthors, Effect of potentially modifiable risk factors associated with myocardial infarction in 52 countries (the INTERHEART study): Case-control study, *Lancet* 364 (2004): 937–952.

47. Standing Committee on the Scientific Evaluation of Dietary Reference Intakes, Food and Nutrition Board, Institute of Medicine, 2002, pp. 11–46.

48. Joint WHO/FAO Expert Consultation, *Diet, Nutrition and the Prevention of Chronic Diseases* (Geneva, Switzerland: World Health Organization, 2003).

49. K. L. Herron and coauthors, Men classified as hypo- or hyperresponders to dietary cholesterol feeding exhibit differences in lipoprotein metabolism, *Journal of Nutrition* 133 (2003): 1036–1042.

50. Standing Committee on the Scientific Evaluation of Dietary Reference Intakes, Food and Nutrition Board, Institute of Medicine, 2002, p. 11-3.

51. P. M. Kris-Etherton, K. D. Hecker, and A. E. Binkoski, Polyunsatured fatty acids and cardiovascular health, *Nutrition Reviews* 62 (2004): 414–426; M. Kanazawa and coauthors, Effects of a high-sucrose diet on body weight, plasma triglycerides, and stress tolerance, *Nutrition Reviews* 61 (2003): S27–S33; A. Raben and A. Astrup, Ad libitum intake of low-fat diets rich in either starchy foods or sucrose: Effects on blood lipids, factor VII coagulant activity, and fi-brinogen, *Metabolism: Clinical and Experimental* 49 (2000): 731–735; J. P. Bantle and coauthors, Effects of dietary fructose on plasma lipids in healthy subjects, *American Journal of Clinical Nutrition* 72 (2000): 1128–1134.

52. R. N. Lemaitre and coauthors, n-3 polyunsaturated fatty acids, fatal ischemic heart disease, and nonfatal myocardial infarction in older adults: The Cardiovascular Health Study, *American Journal of Clinical Nutrition* 77 (2003): 319–325; C. M. Albert and coauthors, Blood levels of long chain n-3 polyunsaturated fatty acids and the risk of sudden death, *New England Journal of Medicine* 346 (2002): 1113–1118; P. Nestel and coauthors, The n-3 fatty acids eicosapentaenoic acid and docosahexaenoic acid increase systemic arterial compliance in humans, *American Journal of Clinical Nutrition* 76 (2002): 326–330; W. S. Harris, Cardioprotective effects of ω-3 fatty acids, *Nutrition in Clinical Practice* 16 (2001): 6–12; C. R. Harper and T. A. Jacobson, The role of omega-3 fatty acids in the prevention of coronary heart disease, *Archives of Internal Medicine* 161 (2001): 2185–2192.

53. Fish oil: AHA recommendation, *Heart and Stroke A to Z Guide*, 2000, available at www.americanheart.org.

54. P. M. Kris-Etherton, K. D. Hecker, and A. E. Binkoski, 2004; L. Djousse and coauthors, Dietary linolenic acid and carotid atherosclerosis: The National Heart, Lung, and Blood Institute Family Heart Study, *American Journal of Clinical Nutrition* 77 (2003): 819–825; P. M. Kris-Etherton, W. S. Harris, and L. J. Appel, Fish consumption, fish oil, omega-3 fatty acids, and cardiovascular disease, *Circulation* 106 (2002): 2747–2757.

55. L. M. Steffen and coauthors, Associations of whole-grain, refined-grain, and fruit and vegetable consumption with risks of all-cause mortality and incident coronary artery disease and ischemic stroke: The Atherosclerosis Risk in Communities (ARIC) Study, *American Journal of Clinical Nutrition* 78 (2003): 383–390; S. Liu, Intake of refined carbohydrates and whole grain foods in relation to risk of type 2 diabetes mellitus and coronary heart disease, *Journal of the American College of Nutrition* 21 (2002): 298–306.

56. S. Liu and coauthors, Is intake of breakfast cereals related to total and cause-specific mortality in men? *American Journal of Clinical Nutrition* 77 (2003): 594–599; L. A. Bazzano and coauthors, Fruit and vegetable intake and risk of cardiovascular disease in US adults: The first National Health and Nutrition Examination Survey Epidemiologic Follow-up Study, *American Journal of Clinical Nutrition* 76 (2002): 93–99.

57. Expert Panel on Detection, Evaluation, and Treatment of High Blood Cholesterol in Adults (Adult Treatment Panel III), 2002, p. V-13.

58. U. A. Ajani, E. S. Ford, and A. H. Mokdad, Dietary fiber and c-reactive protein: Findings from National Health and Nutrition Examination Survey data, *Journal of Nutrition* 134 (2004): 1181–1185.

59. C. T. Valmadrid and coauthors, Alcohol intake and the risk of coronary heart disease mortality in persons with older-onset diabetes mellitus, *Journal of the American Medical Association* 282 (1999): 239–246.

60. B.-H. Chung and coauthors, Alcohol-mediated enhancement of postprandial lipemia: A contributing factor to an increase in plasma HDL and a decrease in risk of cardiovascular disease, *American Journal of Clinical Nutrition* 78 (2003): 391–399; D. J. Baer and coauthors, Moderate alcohol consumption lowers risk factors for cardiovascular disease in postmenopausal women fed a controlled diet, *American Journal of Clinical Nutrition* 75 (2002): 593–599.

61. K. Reynolds and coauthors, Alcohol consumption and risk of stroke: A meta-analysis, *Journal of the American Medical Association* 289 (2003): 579–588.

62. D. J. A. Jenkins and coauthors, A dietary portfolio approach to cholesterol reduction: Combined effects of plant sterols, vegetable proteins, and viscous fibers in hypercholesterolemia, *Metabolism: Clinical and Experimental* 51 (2002): 1596–1604.

63. Expert Panel on Detection, Evaluation, and Treatment of High Blood Cholesterol in Adults (Adult Treatment Panel III), 2002.

64. R. O. Bonow and R. H. Eckel, Diet, obesity and cardiovascular risk, *New England Journal of Medicine* 348 (2003): 2057–2058; D. M. Bravata and coauthors, Efficacy and safety of low-carbohydrate diets, *Journal of the American Medical Association* 289 (2003): 1837–1850.

65. S. Liu and coauthors, Intake of vegetables rich in carotenoids and risk of coronary heart disease in men: The Physicians' Health Study, *International Journal of Epidemiology* 30 (2001): 130–135.

66. A. Rosengren and coauthors, Association of psychosocial risk factors with risk of acute myocardial infarction in 11,119 cases and 13,648 controls from 52 countries (the INTERHEART study): Case-control study, *Lancet* 364 (2004): 953–962; L. A. Chaput and coauthors, Hostility predicts recurrent events among postmenopausal women with coronary heart disease, *American Journal of Epidemiology* 15 (2002): 1092–1099.

67. L. E. Fields and coauthors, The burden of adult hypertension in the United States 1999 to 2000: A rising tide, *Hypertension* 44 (2004):398–404.

68. I. Hajjar and T. A. Kotchen, Trends in prevalence, awareness, treatment, and control of hypertension in the United States, 1988–2000, *Journal of the American Medical Association* 290 (2003): 199–206.

69. Expert Panel on Detection, Evaluation, and Treatment of High Blood Cholesterol in Adults (Adult Treatment Panel III), 2002.

70. J. M. Lalouel and A. Rohrwasser, Development of genetic hypotheses in essential hypertension, *Journal of Human Genetics* 46 (2001): 299–306; K. W. Sellers and coauthors, Gene therapy to control hypertension: Current studies and future perspectives, *American Journal of the Medical Sciences* 322 (2001): 1–6.

71. C. A. Nowson and coauthors, Blood pressure response to dietary modifications in free-living individuals, *Journal of Nutrition* 134 (2004): 2322–2329.

72. V. J. Stevens and coauthors, Long-term weight loss and changes in blood pressure: Results of the Trials of Hypertension Prevention, phase II, *Annals of Internal Medicine* 134 (2001): 1–11.

73. L. S. Pescatello and coauthors, American College of Sports Medicine position stand: Exercise and hypertension, *Medicine and Science in Sports and Exercise* 36 (2004): 533–553.

74. P. D. Thompson and coauthors, The acute versus the chronic response to exercise, *Medicine and Science in Sports and Exercise* 33 (2001): S452–S453.

75. Thompson and coauthors, 2001.

76. M. Iwane and coauthors, Walking 10,000 steps/day or more reduces blood pressure and sympathetic nerve activity in mild essential hypertension, *Hypertension Research* 23 (2000): 573–580.

77. T. S. Altena and coauthors, Single sessions of intermittent and continuous exercise and postprandial lipemia, *Medicine and Science in Sports and Exercise* 36 (2004): 1364–1371.

78. U.S. Department of Agriculture and U.S. Department of Health and Human Services, *2005 Dietary Guidelines Committee Report, 2005*, available online at www.health.gov/dietaryguidelines/dga2005report/; Joint WHO/FAO Expert Consultation, 2003, pp. 85–86.

79. L. Hooper and coauthors, Reduced dietary salt for prevention of cardiovascular disease, *Cochrane Database System Review* 3 (2003): CD003656.

80. U.S. Department of Agriculture and U.S. Department of Health and Human Services, 2005; Joint WHO/FAO Expert Consultation, 2003, pp. 85–86.

81. Standing Committee on the Scientific Evaluation of Dietary Reference Intakes, Food and Nutrition Board, Institute of Medicine, *Dietary Reference Intakes for Water, Potassium, Sodium, Chloride, and Sulfate* (Washington, D.C.: National Academies Press, 2002): p. 6-2; F. M. Sacks and coauthors, Effects on blood pressure of reduced dietary sodium and the Dietary Approaches to Stop Hypertension (DASH) diet, *New England Journal of Medicine* 344 (2001): 3–10.

82. F. D. Fuchs and coauthors, Alcohol consumption and the incidence of hypertension: The Atherosclerosis Risk in Communities Study, *Hypertension* 37 (2001): 1242–1250.

83. D. A. McCarron and R. P. Heaney, Estimated healthcare savings associated with adequate dairy food intake, *American Journal of Hypertension* 17 (2004): 88–97.

84. G. Block, Ascorbic acid, blood pressure, and the American diet, *Annals of the New York Academy of Sciences* 959 (2002): 180–187.

85. I. M. Hajjar and coauthors, A randomized, double-blind, controlled trial of vitamin C in the management of hypertension and lipids, *American Journal of Therapeutics* 10 (2003): 289–293; M. K. Kim and coauthors, Lack of long-term effect of vitamin C supplementation on blood pressure, *Hypertension* 40 (2002): 797–791.

86. E. Obarzanek and coauthors, Effects on blood lipids of a blood pressure-lowering diet: The Dietary Approaches to Stop Hypertension (DASH) trial, *American Journal of Clinical Nutrition* 74 (2001): 80–89.

87. Centers for Disease Control and Prevention, Cancer survivorship—United States, 1971–2001, *Morbidity and Mortality Weekly Report* 53 (2004): 526–529.

88. P. Lichtenstein and coauthors, Environmental and heritable factors in the causation of cancer: Analyses of cohorts of twins from Sweden, Denmark, and Finland, *New England Journal of Medicine* 343 (2000): 78–85.

89. E. E. Calle and coauthors, Overweight, obesity and mortality from cancer in a prospectively studied cohort of U.S. adults, *New England Journal of Medicine* 348 (2003): 1625–1638.

90. National Cancer Policy Board, Institute of Medicine, *Fulfilling the Potential of Cancer Prevention and Early Detection,* ed. S. J. Curry, T. Byers, and M. Hewitt (Washington, D.C.: National Academies Press, 2003), pp. 58–61.

91. M. R. Forman and coauthors, Nutrition and cancer prevention: A multidisciplinary perspective on human trials, *Annual Review of Nutrition* 24 (2004): 223–254.

92. Cell-cycle regulation and the genetics of cancer, in L. Hartwell and coauthors, *Genetics: From Genes to Genomes* (Boston: McGraw-Hill, 2000), pp. 590–622.

93. Position of the American Dietetic Association and Dietitians of Canada: Nutrition and women's health, *Journal of the American Dietetic Association* 104 (2004): 984–1001.

94. S. A. Nowell, J. Ahn, and C. B. Ambrosone, Gene-nutrient interactions in cancer etiology, *Nutrition Reviews* 62 (2004): 427–438.

95. Calle and coauthors, 2003; National Cancer Policy Board, Institute of Medicine, 2003, pp. 61–66.

96. Calle and coauthors, 2003.

97. Y. Mao and coauthors, Physical inactivity, energy intake, obesity and the risk of rectal cancer in Canada, *International Journal of Cancer* 105 (2003): 831–837; A. S. Furberg and I. Thune, Metabolic abnormalities (hypertension, hyperglycemia and overweight) lifestyle (high energy intake and physical inactivity) and endometrial cancer risk in a Norwegian cohort, *International Journal of Cancer* 104 (2003): 669–676; E. Giovannucci, Diet, body weight, and colorectal cancer: A summary of the epidemiologic evidence, *Journal of Women's Health* 12 (2003): 173–182; H. Vainio, R. Kaaks, and F. Bianchini, Weight control and physical activity in cancer prevention: International evaluation of the evidence, *European Journal of Cancer Prevention* 2 (2002): S94–S100.

98. National Cancer Policy Board, Institute of Medicine, 2003, pp. 58–61; M. L. Slattery and coauthors, Lifestyle and colon cancer: An assessment of factors associated with risk, *American Journal of Epidemiology* 150 (1999): 869–877.

99. National Cancer Policy Board, Institute of Medicine, 2003, pp. 59–60; J. B. Barnett, The relationship between obesity and breast cancer risk and mortality, *Nutrition Reviews* 61 (2003): 73–76.

100. P. L. Zock, Dietary fats and cancer, *Current Opinions in Lipidology* 12 (2001): 5–10.

101. Zock, 2001.

102. E. Cho and coauthors, Premenopausal fat intake and risk of breast cancer, *Journal of the National Cancer Institute* 95 (2003): 1079–1085.

103. S. A. Bingham and coauthors, Are imprecise methods obscuring a relation between fat and breast cancer? *Lancet* 362 (2003): 212–214.

104. S. C. Larsson and coauthors, Dietary long chain n-3 fatty acids for the prevention of cancer: A review of potential mechanisms, *American Journal of Clinical Nutrition* 79 (2004): 935–945; A. Nkondjock and coauthors, Specific fatty acids and human colorectal cancer: An overview, *Cancer Detection and Prevention* 27 (2003): 55–66; National Cancer Policy Board, Institute of Medicine, 2003, p. 77; P. D. Terry, T. E. Rohan, and A. Wolk, Intakes of fish and marine fatty acids and the risks of cancers of the breast and prostate and of other hormone-related cancers: A review of the epidemiologic evidence, *American Journal of Clinical Nutrition* 77 (2003): 532–543.

105. S. A. Nowell, J. Ahn, and C. B. Ambrosone, 2004; M. A. Murtaugh and coauthors, Meat consumption patterns and preparation, genetic variants of metabolic enzymes, and their association with rectal cancer in men and women, *Journal of Nutrition* 134 (2004): 776–784; L. M. Butler and coauthors, Heterocyclic amines, meat intake, and association with colon cancer in a population-based study, *American Journal of Epidemiology* 157 (2003): 434–445.

106. L. Le Marchand and coauthors, Well-done red meat, metabolic phenotypes and colorectal cancer in Hawaii, *Mutation Research* 506–507 (2002): 205–214.

107. U.S. Food and Drug Administration, FDA action plan for acrylamide in food, March 2004, available at www.cfsan.fda.gov.

108. S. A. Bingham and coauthors, Dietary fiber in food and protection against colorectal cancer in the European Prospective Investigation into Cancer and Nutrition (EPIC): An observational study, *Lancet* 361 (2003): 1496–1501; I. A. Schatzkin and coauthors, Lack of effect of a low-fat, high-fiber diet on the recurrence of colorectal adenomas, *New England Journal of Medicine* 342 (2000): 1149–1155.

109. American Gastroenterological Association, Medical position statement: Impact of dietary fiber on colon cancer occurrence, *Gastroenterology* 118 (2000): 1233–1234.

110. J. L. Slavin, Mechanisms for the impact of whole grain foods on cancer risk, *Journal of the American College of Nutrition* 19 (2000): S300–S307.

111. D. S. Michaud and coauthors, Fluid intake and the risk of bladder cancer in men, *New England Journal of Medicine* 340 (1999): 1390–1397.

112. S. W. Choi and J. B. Mason, Folate and carcinogenesis: An integrated scheme, *Journal of Nutrition* 130 (2000): 129–132; K. Young-In, Folate and cancer prevention: A new medical application for folate beyond hyperhomocysteinemia and neural tube defects, *Nutrition Reviews* 57 (1999): 314–324.

113. B. N. Ames, Micronutrient deficiencies: A major cause of DNA damage, *Annals of the New York Academy of Sciences* 889 (1999): 87–106.

114. K. Z. Guyton, T. W. Kensler, and G. H. Posner, Vitamin D and vitamin D analogs as cancer chemopreventive agents, *Nutrition Reviews* 61 (2003): 227–238.

115. A. S. Vrablic and coauthors, Altered mitochondrial function and overgeneration of reactive oxygen species precede the induction of apoptosis by 1-O-octadecyl-2-methyl-rac-glycero-3-phophocholine in p53-defective hepatocytes, *FASEB Journal* 15 (2001): 1739–1744.

116. R. I. Salganik and coauthors, Dietary antioxidant depletion: Enhancement of tumor apoptosis and inhibition of brain tumor growth in transgenic mice, *Carcinogenesis* 21 (2000): 909–914.

117. E. Kallay and coauthors, Dietary calcium and growth modulation of human colon cancer cells: Role of the extracellular calcium-sensing receptor, *Cancer Detection and Prevention* 24 (2000): 127–136.

118. E. Kampman and coauthors, Calcium, vitamin D, sunshine exposure, dairy products and colon cancer risk (United States), *Cancer Causes and Control* 11 (2000): 459–466.

119. A.-L. M. Heath and S. J. Fairweather-Tait, Health implications of iron overloads: The role of diet and genotype, *Nutrition Reviews* 61 (2003): 45–62; R. L. Nelson, Iron and colorectal cancer risk: Human studies, *Nutrition Reviews* 59 (2001): 140–148.

120. Norat and Riboli, 2001; Nelson, 2001.

121. J. F. Weiss and M. R. Landauer, Protection against ionizing radiation by antioxidant nutrients and phytochemicals, *Toxicology* 189 (2003): 1–20.

122. M. Peluso and coauthors, White blood cell DNA adducts and fruit and vegetable consumption in bladder cancer, *Carcinogenesis* 21 (2000): 183–187.

123. Joint WHO/FAO Expert Consultation, 2003, pp. 30–53; Kant and coauthors, 2000.

124. R. H. Liu, Health benefits of fruit and vegetables are from additive and synergistic combinations of phytochemicals, *American Journal of Clinical Nutrition* 78 (2003): 517S–520S.

Consumer Corner 11

1. R. J. Lamarine, Alternative medicine: More than a harmless option, *Journal of School Health* 71 (2001): 114–116.

2. M. A. Frenkel and J. M. Borkan, An approach for integrating complementary-alternative medicine into primary care, *Family Practice* 20 (2003): 324–332.

3. National Center for complementary and Alternative Medicine, nccam.nih.gov/news/2001, February 2001.

4. R. del la Fuente-Fernandez, M. Schultzer, and A. J. Stoessl, The placebo effect in neurological disorders, *Lancet Neurology* 1 (2002): 85–91.

5. A. J. Vickers and coauthors, Acupuncture for chronic headache in primary care: Large, pragmatic, randomised trial, *British Medical Journal* 328 (2004): 744–750;

D. Wonderling and coauthors, Cost effectiveness analysis of a randomised trial of acupuncture for chronic headache in primary care, *British Medical Journal* 328 (2004): 747–752.

6. A. White and coauthors, Adverse events following acupuncture: Prospective survey of 32,000 consultations with doctors and physiotherapists, *British Medical Journal* 323 (2001): 485–486; H. MacPherson and coauthors, The York acupuncture safety study: Prospective survey of 34,000 treatments by traditional acupuncturists, *British Medical Journal* 323 (2001): 486–487.

7. C. M. Gilroy and coauthors, Echinacea and truth in labeling, *Archives of Internal Medicine* 163 (2003): 699–704; A. H. Feifer, N. E. Fleshner, and L. Klotz, Analytical accuracy and reliability of commonly used nutritional supplements in prostate disease, *Journal of Urology* 168 (2002): 150–154.

8. J. Garrard, S. Harms, L. E. Eberly, and A. Matiak, Varations in produce choices of frequently purchased herbs: caveat emptor,, *Archives of Internal Medicine* 163 (2003): 2290–2295.

9. U.S. Food and Drug Administration, Letter to health professionals regarding safety concerns related to the use of botanical products containing aristolochic acid, April 2001, available at www.cfsan.fda.gov.

10. Centers for Disease Control and Prevention, Lead poisoning associated with ayurvedic medications—five states, 2000–2003, *Morbidity and Mortality Weekly Report* 53 (2004): 582–584; G. Trochet, Warning! High levels of lead (a poisonous metal) have been found in an Ayurvedic pill taken to increase fertility, *Sacramento County Department of Heath and Human Services,* available from Sacramento County Department of Health and Human Services, Childhood Illness and Injury Prevention Program at (916) 875–5869.

11. L. H. Kushi and coauthors, The macrobiotic diet in cancer, *Journal of Nutrition* 131 (2001): 3056S–3064S.

12. A. Nkondjock and coauthors, Specific fatty acids and human colorectal cancer: An overview, *Cancer Detection and Prevention* 27 (2003): 55–66; Macrobiotic proponent dies of cancer, *Consumer Health Digest,* July 9, 2001, available at www.ncahf.org.

Controversy 11

1. National Cancer Policy Board, Institute of Medicine, *Fulfilling the Potential of Cancer Prevention and Early Detection,* ed. S. J. Curry, T. Byers, and M. Hewitt (Washington, D.C.: National Academies Press, 2003), p. 41; Expert Panel on Detection, Evaluation, and Treatment of High Blood Cholesterol in Adults (Adult Treatment Panel III), *Third Report of the National Cholesterol Education Program (NCEP),* NIH publication no. 02-5215 (Bethesda, Md.: National Heart, Lung, and Blood Institute, 2002), p. II-16–II-17.

2. A. H. Mokdad and coauthors, The continuing epidemics of obesity and diabetes in the United States, *Journal of the American Medical Association* 286 (2001): 1195–1200.

3. J. Foreyt and G. K. Goodrick, Dieting and weight loss: The energy perspective, *Nutrition Reviews* 59 (2001): S25–S26.

4. J. J. Putnam and J. E. Allshouse, *Food Consumption, Prices, and Expenditures,* USDA Economic Research Service, 1999, Statistical Bulletin no. 965, available at www.USDA.gov.

5. P. M. Barnes and C. A. Schoenborn, *Physical Activity among Adults: United States, 2000,* Advance Data from Vital and Health Statistics document no. 333, 2003, available at www.cdc.gov/nchs/Default.htm.

6. L. Pérusse and C. Bouchard, Gene-diet interactions in obesity, *American Journal of Clinical Nutrition* 72 (2000): 1285S–1290S; D. B. Allison and M. S. Faith, Genetic and environmental influences on human body weight: Implications for the behavior therapist, *Nutrition Today* 35 (2000): 18–21.

7. C. Bouchard, *Physical Activity and Obesity* (Champaign, Ill.: Human Kinetics Publishers, 2000), p. 14.

8. D. Lakdawalla and T. Philipson, The Growth of Obesity and Technological Change: A Theoretical and Empirical Examination, a paper from the National Bureau of Economic Research, October 8, 2001, available at http://papers.nber.org/papers/w8946.

9. E. Schlosser, *Fast Food Nation* (Boston: Houghton Mifflin, 2001), p. 7.

10. M. Nestle, *Food Politics: How the Food Industry Influences Nutrition and Health* (Berkeley: University of California Press, 2002).

11. A. Clauson, Share of food spending for eating out reaches 47 percent, *Food Review* 22 (1999): 20–22, as cited by L. R. Young and Marion Nestle, The contribution of expanding portion sizes to the US obesity epidemic, *American Journal of Public Health* 92 (2002): 246–249; USDA Baseline Projections, February 1999, available from www.usda.gov.

12. S. L. Booth and coauthors, Environmental and societal factors affect food choice and physical activity: Rationale, influences, and leverage points, *Nutrition Reviews* 59 (2001): S21–S39.

13. L. Harnack and S. French, Fattening up on fast food, *Journal of the American Dietetic Association* 103 (2003): 1296–1297.

14. M. A. McCrory, V. M. M. Suen, and S. B. Roberts, Biobehavioral influences on energy intake and adult weight gain, *Journal of Nutrition* 132 (2002): 3830S–3834S; USDA Economic Research Service, *Food Consumption: Food Supply and Use,* updated October 2001, available at http://ers.usda.gov/breifing/consumption/supply.htm; M. Nestle and M. F. Jacobson, Halting the obesity epidemic: A public health policy approach, *Public Health Reports* 115 (2000): 12–24; L. J. Harnack, R. W. Jeffery, and K. N. Boutelle, Temporal trends in energy intake in the United States: An ecologic perspective, *American Journal of Clinical Nutrition* 71 (2000): 1478–1484.

15. Nestle and Jacobson, 2000.

16. Nestle, 2002.

17. B. J. Rolls, The supersizing of America: Portion size and the obesity epidemic, *Nutrition Today,* March/April, 2003, pp. 42–53.

18. Young and Nestle, 2002.

19. B. Wansink, Environmental factors that increase the food intake and consumption volume of unknowing consumers, *Annual Review of Nutrition* 24 (2004): 455–479.

20. B. J. Rolls, E. L. Morris, and L. S. Roe, Portion size of food affects energy intake in normal-weight and overweight men and women, *American Journal of Clinical Nutrition* 76 (2002): 1207–1213.

21. Wansink, 2004.

22. Gary Ruskin, Executive Director of Commercial Alert to the World Health Organization Conference on Health Marketing and Youth, Treviso, Italy, April, 17, 2002.

23. Nestle, 2002.

24. Ruskin, 2002.

25. Ralph Nader, Testimony on "Channel One" before the U.S. Senate Committee on Health, Education, Labor, and Pensions, 1999, available at www.commercialalert.org.

26. Ruskin, 2002.

27. Steven Anderson, President of the National Restaurant Association, as quoted in Tobacco suits light way to fast-food litigation, *Foodservice News and Editorial,* 2002, available at www.foodserve.com.

28. C. H. Powers and M. A. Hess, A message to the restaurant industry: It's time to "step up to the plate," *Journal of the American Dietetic Association* 103 (2003): 1136–1138.

29. J. G. Dausch, The obesity epidemic: What's being done? *Journal of the American Dietetic Association* 102 (2002): 638–639.

30. Nestle and Jacobson, 2000.

31. FDA Working Group on Obesity, *Counting Calories: Report of the Working Group on Obesity,* March 2004, available at www.cfsan.fda.gov/~dms/owg-rpt.html.

32. USDA, Nutrition Education and Training Program, available at www.fns.usda.gov and www.ncseonline.org/NLE/CRSreports/legislative/leg-25.cfm.

33. S. A. French, Pricing effects on food choices, *Journal of Nutrition* 133 (2003): 841S–843S.

34. F. Kuchler, A. Tegene, and J. M. Harris, Taxing snack foods: what to expect for diet and tax revenues, *Current Issues in Economics of Food Markets,* Agriculture Information Bulletin No. 747-08, August 2004, available at www.ers.usda.gov. N. J. Temple and A. L. Balay-Karperien, Nutrition in cancer prevention: An integrated approach, *Journal of the American College of Nutrition* 21 (2002): 79–83.

35. Nestle and Jacobson, 2000.

36. C. E. Staunton, D. Hubsmith, and W. Kallins, Promoting safe walking and biking to school: The Marin County success story, *American Journal of Public Health* 93 (2003): 1431–1434; K. E. Powell, I. M. Martin, and P. P. Chowdhury, Places to walk: Convenience and regular physical activity, *American Journal of Public Health* 93 (2003): 1519–1521.

Chapter 12

1. B. Harder, Chicken little? Study cites arsenic in poultry, *Science News* 164 (2003): 259–260.

2. National Research Council and Institute of Medicine, *Safety of Genetically Engineered Foods: Approaches to Assessing Unintended Health Effects* (Washington, D.C.: National Academies Press, 2004), pp. 103–125.

3. B. Bruemmer, Food biosecurity, *Journal of the American Dietetic Association* 103 (2003): 687–691; T. Peregrin, Bioterrorism and food safety: What nutrition professionals need to know to educate the American public, *Journal of the American Dietetic Association* 102 (2002): 14, 16; Food and Drug Administration, Food security guidance: Availability, *Federal Register* 67 (2002): 1224–1225; J. Sobel, A. S. Khan, and D. L. Swerdlow, Threat of a biological terrorist attack on the US food supply: The CDC perspective, *Lancet* 359 (2002): 874–880.

4. HHS, FDA implement food security research program, *FDA Consumer,* September/October 2003, p. 9.

5. M. Meadows, The FDA and the fight against terrorism, *FDA Consumer,* January/February 2004, pp. 20–27; Protecting the food supply: FDA actions on new bioterrorism legislation, October 2003, available at www.cfsan.fda.gov.

6. Centers for Disease Control and Prevention, Preliminary FoodNet data on the incidence of foodborne illnesses—selected sites, United States, *Morbidity and Mortality Weekly Report* 50 (2001): 241–246; R. G. Villar and coauthors, Investigation of multidrug-resistant *Salmonella* serotype *typhimurium* DT104 infections linked to raw-milk cheese in Washington state, *Journal of the American Medical Association* 281 (1999): 1811–1816.

7. Position of the American Dietetic Association: Food and water safety, *Journal of the American Dietetic Association* 103 (2003): 1203–1218.

8. Centers for Disease Control and Prevention, Diagnosis and management of foodborne illnesses, *Morbidity and Mortality Weekly Report,* supplement, 53 (2004).

9. Centers for Disease Control and Prevention, Preliminary FoodNet data on the incidence of infection with pathogens transmitted commonly through food—selected sites, United States, 2003, *Morbidity and Mortality Weekly Report* 53 (2004): 338–343; A. Harrison and S. Cohen, Tests show salmonella in meat and poultry products declines 66 percent, USDA News Release, November 24, 2003, available at www.usda.gov/news/releases/2003/11/0396.htm.

10. Consumers advised that recent hepatitis A outbreaks have been associated with green onions, *FDA Talk Paper,* November 15, 2003, available at www.fda.gov/bbs/topics/ANSWERS/2003/ANS01262.html.

11. E. A. Coleman and M. E. Yergler, Botulism, *American Journal of Nursing* 102 (2002): 44–47.

12. Position of the American Dietetic Association, 2003.

13. W. L. Chandler and coauthors, Prothrombotic coagulation abnormalities preceding the hemolytic-uremic syndrome, *New England Journal of Medicine* 346 (2002): 23–32; E. F. Grabowski, The hemolytic-uremic syndrome: Toxin, thrombin, and thrombosis, *New England Journal of Medicine* 346 (2002): 58–61.

14. L. B. Zimmerhackl, *E. coli,* antibiotics, and the hemolytic-uremic syndrome, *New England Journal of Medicine* 342 (2000): 1990–1991.

15. C. S. DeWaal and K. Barlow, *Outbreak Alert! Closing the Gaps in Our Federal Food-Safety Net,* a publication of the Center for Science in the Public Interest, March 2004, available on the Internet at www.cspinet.org, Committee on the Review of the Use of Scientific Criteria and Performance Standards for Safe Food, Food and Nutrition Board, *Scientific Criteria to Ensure Safe Food* (Washington, D.C.: National Academies Press, 2003).

16. USDA, Second progress report on *Salmonella* testing for raw meat and poultry products, *FSIS Backgrounder,* January 21, 1999.

17. J. Sneed, C. Strohbehn, and S. H. Gilmore, Food safety practices and readiness to implement HACCP programs in assisted-living facilities in Iowa, *Journal of the American Dietetic Association* 104 (2004): 1678–1683; C. Strohbehn, S. H. Gilmore, and J. Sneed, Food safety practices and HACCP implementation: perceptions of registered dietitians and dietary managers, *Journal of the American Dietetic Association* 104 (2004): 1693–1699.

18. J. M. Boyd and D. Pittet, Centers for Disease Control and Prevention, Guideline for hand hygiene in health-care settings, *Morbidity and Mortality Weekly Report* 51 (2002): 1–44.

19. Boyd and Pittet, 2002.

20. B. Hammond and coauthors, Effect of hand sanitizer use on elementary school absenteeism, *American Journal of Infection Control* 28 (2000): 340–346.

21. W. Picheansathian, A systematic review on the effectiveness of alcohol-based solutions for hand hygiene, *International Journal of Nursing Practice* 10 (2004): 3–9; C. M. Lin and coauthors, A comparison of hand washing techniques to remove *Escherichia coli* and caliciviruses under natural or artificial fingernails, *Journal of Food Protection* 66 (2003): 2296–2301; D. J. Weber and coauthors, Efficacy of selected hand hygiene agents used to remove *Bacillus atrophaeus* (a surrogate of *Bacillus anthracis*) from contaminated hands, *Journal of the American Medical Association* 289 (2003): 1274–1277; J. M. Boyce, Using alcohol for hand antisepsis: Dispelling old myths, *Infection Control and Hospital Epidemiology* 21 (2000): 438–442.

22. E. Larson and coauthors, Effect of antibacterial home cleaning and handwashing products on infectious disease symptoms, *Annals of Internal Medicine* 140 (2004): 321–329.

23. J. Henkel, "Thermy" promotes thorough food cooking, *FDA Consumer,* September/October 2000, p. 35.

24. Centers for Disease Control and Prevention, Fatal degenerative neurologic illnesses in men who participated in wild game feasts—Wisconsin, 2002, *Morbidity and Mortality Weekly Report* 52 (2003): 125–127.

25. G. Legname and coauthors, Synthetic mammalian prions, *Science* 305 (2004): 673–676; Centers for Disease Control and Prevention, Probable variant Creutzfeldt-Jakob disease in a U.S. resident—Florida, 2002, *Morbidity and Mortality Weekly Report* 51 (2002): 927–929.

26. Executive summary, *Advancing Prion Research: Guidance for the National Prion Research Program* (Washington, D.C.: National Academies Press, 2004).

27. C. A. Donnelly, Bovine spongiform encephalopathy in the United States—an epidemiologist's view, *New England Journal of Medicine* 350 (2004): 539–542.

28. L. Bren, Agencies work to corral mad cow disease, *FDA Consumer,* May/June 2004, pp. 29–35.

29. FDA Statement on "Scientific Criteria to Ensure Safe Food," a Report by the National Academy of Sciences, *FDA Talk Paper,* 2003, available at www.fda.gov/bbs/topics/ANSWERS/2003/ANS01217.html.

30. L. Bren, Homemade ice cream: A safe summertime treat? *FDA Consumer,* July/August 2004, available at www.fda.gov/fdac/features/2004/404_summer.html.

31. Viruses—just a flush away? *Science News* 155 (1999): 107.

32. U.S. Food and Drug Administration, Center for Food Safety and Applied Nutrition, *Produce Safety from Production to Consumption: 2004 Action Plan to Minimize Foodborne Illness Associated with Fresh Produce Consumption,* October 2004, available from www.cfsan.fda.gov. FDA issues alert on foodborne illness associated with certain basil and mesculin/spring mix salad products, *FDA News,* May 21, 2004, available at www.fda.gov/bbs/topics/news/2004/NEW01071.html.

33. FDA update on recent hepatitis A outbreaks associated with green onions from Mexico, *FDA Statement,* December 9, 2003, available at www.fda.gov/bbs/topics/NEWS/2003/NEW00993.html; Consumers advised that recent hepatitis A outbreaks have been associated with green onions, *FDA Talk Paper,* November 15, 2003, available at www.fda.gov/bbs/topics/ANSWERS/2003/ ANS01262.html.

34. Consumers in Oregon area advised of risks associated with raw sprouts, *FDA Statement,* November 26, 2003, available at www.cfsan.fda.gov/~lrd/ fpsprout.html.

35. E. T. Ryan, M. E. Wilson, and K. C. Kain, Illness after international travel, *New England Journal of Medicine* 347 (2002): 505–516.

36. CDC National Center for Infectious Diseases, Travels' health: Bovine spongiform encephalopathy (mad cow disease) and new variant Creutzfeldt-Jakob disease (nvCJD), January 29, 2001, available at www. cdc.gov/travel/madcow.htm.

37. M. Sheffer, ed., *Hydrogen Cyanide and Cyanides: Human Health Aspects* (Geneva, Switzerland: World Health Organization, 2004), pp. 4–9.

38. B. Weiss, S. Amler, and R. W. Amler, Pesticides, *Pediatrics* 113 (2004): 1030–1036.

39. FDA, *Food and Drug Administration Pesticide Program—Residue Monitoring,* a report obtainable from the FDA, HFF-420, 200 C Street S.W., Washington, DC 20204.

40. J. Ma and coauthors, Milk intake, circulating levels of insulin-like growth factor-I, and risk of colorectal cancer in men, *Journal of the National Cancer Institute* 93 (2001): 1330–1336.

41. FDA Center for Veterinary Medicine, Report on the Food and Drug Administration's review of the safety of recombinant bovine somatotropin, available at www.fda.gov/cvm/index/bst/RBRPTFNL.htm.

42. Milk processors warned to remove 'hormone free' claims from the labeling of dairy products, *FDA News,* September 12, 2003, available at www.fda.gov/ bbs/topics/NEWS/2003/NEW00943.html.

43. K. H. Mathews, Jr. and coauthors, USDA Economic Research Service, Livestock drugs: More questions than answers? *Agricultural Outlook,* September 2001, pp. 18–21.

44. No more antibiotics says WHO, *Nutrition Today* 38 (2003): 163.

45. T. Lasky and coauthors, Mean total arsenic concentrations in chicken 1989–2000 and estimated exposures for consumers of chicken, *Environmental Health Perspectives* 112 (2004): 18–21.

46. Lasky and coauthors, 2004.

47. Standing Committee on the Scientific Evaluation of Dietary Reference Intakes, Food and Nutrition Board, Institute of Medicine, 2001.

48. C. P. Dougherty and coauthors, Dietary exposures to food contaminants across the United States, *Environmental Research* 84 (2000): 170–185.

49. EPA Fact Sheet, Update: National listing of fish and wildlife advisories, May 2002, available at www.epa.gov/ost/fish.

50. Centers for Disease Control and Prevention, *Third National Report on Human Exposure to Environmental Chemicals,* January 2005, available at www.cdc.gov/exposurereport.

51. K. M. Smith and N. R. Sahyoun, Fish consumption: Recommendations versus advisories, can they be reconciled? *Nutrition Reviews* 63 (2005): 39–46.

52. Joint Federal Advisory for Mercury in Fish, Backgrounder for the 2004 FDA/EPA Consumer Advisory: What you need to know about mercury in fish and shellfish, available at http://epa.gov/waterscience/fishadvice/factsheet.html.

53. EPA national advice on mercury in fish caught by family and friends for women who are or may become pregnant, nursing mothers, and young children, *EPA National Advisory,* 2001, available from www.epa.gov/ost/fish.

54. J. M. Hightower and D. Moore, Mercury levels in high-end consumers of fish, *Environmental Health Perspectives* 111 (2003): 604–608.

55. R. A. Hites and coauthors, Global assessment of organic contaminants in farmed salmon, *Science* 303 (2004): 226–229.

56. International Food Information Council, What's this doing in my food? A guide to food ingredients, *Food Insight,* November/December 2000, available at www.ific.org.

57. Color: Salmon of the Americas, 2003, a fact sheet available from www.salmonoftheamericas.com.

58. M. Meadows, MSG: A common flavor enhancer, *FDA Consumer* 37 (2003): 35.

59. International Food Information Council, Glutamate and monosodium glutamate: Examining the myths, *IFIC Review*, November 2001, available at www.ific.org/relatives/17660.pdf.

60. C. S. Johnston and D. L. Bowling, Stability of ascorbic acid in commercially available orange juices, *Journal of the American Dietetic Association* 102 (2002): 525–529.

61. D. J. McKillop and coauthors, The effect of different cooking methods on folate retention in various foods that are amongst the major contributors to folate intake in the UK diet, *British Journal of Nutrition* 88 (2002): 681–688.

Consumer Corner 12

1. Position of the American Dietetic Association: Food irradiation, *Journal of the American Dietetic Association* 100 (2000): 246–253.

2. K. M. Shea, Technical report: Irradiation of food, *Pediatrics* 106 (2000): 1505–1510.

3. Meat irradiation can boost food safety, *FDA Consumer*, March/April 2000, p. 4.

4. M. T. Osterholm and A. P. Norgan, The role of irradiation in food safety, *New England Journal of Medicine* 350 (2004): 1898–1901.

5. P. Loaharanu, *Irradiated Foods* (New York: American Council on Science and Health, 2003).

6. Position of the American Dietetic Association: Food and water safety, *Journal of the American Dietetic Association* 103 (2003): 1203–1218.

7. U.S. Food and Drug Administration, Food irradiation: A safe measure, January 2000, available at www.fda.gov/opacom/catalog/irradbro.html.

8. A. L. Young, Food irradiation: After 35 years, have we made progress? A government perspective, *Environmental Science and Pollution Research International* 10 (2003): 82–88.

9. J. Daniels and M. Abrams, USDA releases specifications for the purchase of irradiated ground beef in the national school lunch program, USDA News Release, May 29, 2003.

10. P. D. Frenzen and coauthors, Consumer acceptance of irradiated meat and poultry in the United States, *Journal of Food Protection* 64 (2001): 2020–2026.

Controversy 12

1. J. Scheel, The health of organic foods, *Prepared Foods*, May 2003, available at www.preparedfoods.com.

2. L. Bren, The future of foods? *FDA Consumer*, November/December 2003, pp. 28–34; M. Mansour, A perspective from the food industry, *Nutrition Reviews* 61 (2003): S117–S123.

3. C. M. Williams, Nutritional quality of organic food: Shades of grey or shades of green? *Proceedings of the Nutrition Society* 61 (2002): 19–34.

4. Organic foods: More nutritious? *Nutrition and the MD*, August 2002, pp. 6–7.

5. C. Lewis, A new kind of fish story: The coming of biotech animals, *FDA Consumer*, January/February 2001, pp. 15–20.

6. P. W. Phillips, Biotechnology in the global agri-food system, *Trends in Biotechnology* 20 (2002): 376–381.

7. G. Khush, Productivity improvements in rice, *Nutrition Reviews* 61 (2003): S114–S116.

8. B. Qi and coauthors, Production of very long chain polyunsaturated omega-3 and omega-6 fatty acids in plants, *Nature Biotechnology* 10 (May 16, 2004), available at www.nature.com.

9. F. M. Wambugu, Development and transfer of genetically modified virus-resistant sweet potato for subsistence farmers in Kenya, *Nutrition Reviews* 61 (2003): S110–S113.

10. A. Pollack, Narrow path for new biotech food crops, *New York Times*, May 20, 2004, available at www.nytimes.com/2004/05/20/business/20crop.html?pagewanted=print&position=.

11. H. C. Sharma and coauthors, Prospects for using transgenic resistance to insects in crop improvement, *EJB Electronic Journal of Biotechnology* 3, no. 2 (2000), available at www.ejb.org/contrent/vol3/issue2/full/3/index/html.

12. J. Travis, Rats join the roster of clones, *Science News* 164 (2003): 237.

13. Safety of food animal clones: Updates, *FDA Consumer*, January/February 2004, pp. 6–7; L. Rudenko, Animal cloning: A risk assessment, conclusions for food safety, Fall 2003, an FDA slide presentation available at www.fda.gov/cvm/ index/vmac/VMAC-Fall03Pres/7Rudenko.ppt.

14. J. Huang, C. Pray, and S. Rozelle, Enhancing the crops to feed the poor, *Nature* 418 (2002): 678–684.

15. National Research Council and Institute of Medicine, *Safety of Genetically Engineered Foods: Approaches to Assessing Unintended Health Effects* (Washington, D.C.: National Academies Press, 2004), pp. 1–15.

16. National Research Council and Institute of Medicine, 2004, pp. 175–177.

17. Amy Ridenour, director of the National Center for Public Policy Research in Washington, D.C.

18. Committee on Biological Confinement of Genetically Engineered Organisms of the National Research Council, *Biological Confinement of Genetically Engineered Organisms* (Washington, D.C.: National Academies Press, 2004), pp. 10–13.

19. S. Milius, When genes escape: Does it matter to crops and weeds? *Science News* 164 (2003): 232–233.

20. Sharma and coauthors, 2000.

21. M. K. Sears and coauthors, Impact of BT corn pollen on monarch butterfly populations: A risk assessment, *Proceedings of the National Academy of Sciences* 98 (2001): 11937–11942.

22. D. E. Stanley-Horn and coauthors, Assessing the impact of Cry1Ab-expressing corn pollen on monarch butterfly larvae in field studies, *Proceedings of the National Academy of Sciences* 98 (2001): 11931–11936.

23. J. Walsh, Brave new farm, *Time*, January 11, 1999, pp. 86–88.

24. H. Adamu, Nigeria's minister of agriculture and rural development, "Hungry Africans want biotech crops, so get out of the way," *International Herald Tribune*, September 12, 2000, p. 8.

25. A. Bakshi, Potential adverse health effects of genetically modified crops, *Journal of Toxicology and Environmental Health. Part B, Critical Reviews* 6 (2003): 211–215; B. C. Babcock and C. A. Francis, Solving global nutrition challenges requires more than new biotechnologies, *Journal of the American Dietetic Association* 100 (2000): 1308–1311.

26. R. Formanek Jr., Proposed rule issued for bioengineered foods, *FDA Consumer*, March/April 2001, pp. 9–11.

27. Position of the American Dietetic Association: Biotechnology and the future of food, *Journal of the American Dietetic Association* 95 (1995): 1429–1432.

Chapter 13

1. W. Y. Wong and coauthors, New evidence of the influence of exogenous and endogenous factors on sperm count in man, *European Journal of Obstetrics, Gynecology, and Reproductive Biology* 110 (2003): 49–54; R. M. Sharpe and S. Franks, Enviromental, lifestyle, and infertility—an inter-generational issue, *Nature Cell Biology* 4 (2002): S33–S40.

2. M. S. Kramer, The epidemiology of adverse pregnancy outcomes: An overview, *Journal of Nutrition* 133 (2003): 1592S–1596S.

3. C. M. Law and coauthors, Fetal, infant, and childhood growth and adult blood pressure: A longitudinal study from birth to 22 years of age, *Circulation* 105 (2002): 1088–1092; K. M. Rasmussen, The "fetal origins" hypothesis: Challenges and opportunities for maternal and child nutrition, *Annual Review of Nutrition* 21 (2001): 73–95.

4. M. Hack and coauthors, Outcomes in young adulthood for very-low-birth-weight infants, *New England Journal of Medicine* 346 (2002): 149–157.

5. K. D. Kochanek and B. L. Smith, Deaths: Preliminary data for 2002, *National Vital Statistics Reports* 52 (2004): 1–30.

6. Kramer, 2003.

7. U. Ramakrishnan, Nutrition and low birth weight: From research to practice, *American Journal of Clinical Nutrition* 79 (2004): 17–21.

8. M. L. Watkins and coauthors, Maternal obesity and risk for birth defects, *Pediatrics* 111 (2003): 1152–1158; F. Galtier-Dereure, C. Boegner, and J. Bringer, Obesity and pregnancy: Complications and cost, *American Journal of Clinical Nutrition* 71 (2000): 1242S–1248S.

9. Position of the American Dietetic Association: Nutrition and lifestyle for a healthy pregnancy outcome, *Journal of the American Dietetic Association* 102 (2002): 1479–1490; J. M. Baeten, E. A. Bukusi, and M. Lambe, Pregnancy complications and outcomes among overweight and obese nulliparous women, *American Journal of Public Health* 91 (2001): 436–440.

10. M. C. Lacroix and coauthors, Placental growth hormones, *Endocrine* 1 (2002): 73–79.

11. A. J. Drake and B. R. Walker, The intergenerational effects of fetal programming: Non-genomic mechanisms for the inheritance of low birth weight and cardiovascular risk, *Journal of Endocrinology* 180 (2004): 1–16.

12. A. Singhal and coauthors, Programming of lean body mass: A link between birth weight, obesity, and cardiovascular disease? *American Journal of Clinical Nutrition* 77 (2003): 726–730; B. E. Birgisdottir and coauthors, Size at birth and glucose intolerance in a relatively genetically homogeneous, high-birth weight population, *American Journal of Clinical Nutrition* 76 (2002): 399–403.

13. M. F. Picciano, Pregnancy and lactation: Physiological adjustments, nutritional requirements and the role of dietary supplements, *Journal of Nutrition* 133 (2003): 1997S–2002S; J. C. King, Physiology of pregnancy and nutrient metabolism, *American Journal of Clinical Nutrition* 71 (2000): 1218S–1225S.

14. Standing Committee on the Scientific Evaluation of Dietary Reference Intakes, Food and Nutrition Board, Institute of Medicine, *Dietary Reference Intakes for*

F

Energy, Carbohydrate, Fiber, Fat, Fatty Acids, Cholesterol, Protein, and Amino Acids (Washington, D.C.: National Academies Press, 2002), pp. 5-59–5-64.

16. R. M. Pitkin, Energy in pregnancy, *American Journal of Clinical Nutrition* 69 (1999): 583; L. E. Kopp-Hoolihan and coauthors, Longitudinal assessment of energy balance in well-nourished, pregnant women, *American Journal of Clinical Nutrition* 69 (1999): 697–704.

17. S. J. Otto and coauthors, Changes in the maternal essential fatty acid profile during early pregnancy and the relation of the profile to diet, *American Journal of Clinical Nutrition* 73 (2001): 302–307; G. Hornstra, Essential fatty acids in mothers and their neonates, *American Journal of Clinical Nutrition* 71 (2000): 1262S–1269S.

18. T. J. Mathews, Trends in spina bifida and anencephalus in the United States, 1991–2001, *National Center for Health Statistics Health and Stats* (2003), available at www.cdc.gov/nchs/products/pubs/pubd/hestats/spine-anen.htm.

19. Standing Committee on the Scientific Evaluation of Dietary Reference Intakes, Food and Nutrition Board, Institute of Medicine, *Dietary Reference Intakes for Thiamin, Riboflavin, Niacin, Vitamin B_6, Folate, Vitamin B_{12}, Pantothenic Acid, Biotin, and Choline* (Washington, D.C.: National Academy Press, 1998), pp. 196–305.

20. H. McNulty, G. J. Cuskelly, and M. Ward, Response of red blood cell folate to intervention: Implications for folate recommendations for the prevention of neural tube defects, *American Journal of Clinical Nutrition* 71 (2000): 1308S–1311S.

21. J. Erickson, Folic acid and prevention of spina bifida and anencephaly, *Morbidity and Mortality Weekly Report* 51 (2002): 1–3; M. A. Honein and coauthors, Impact of folic acid fortification of the US food supply on the occurrence of neural tube defects, *Journal of the American Medical Association* 285 (2001): 2981–2986.

22. Standing Committee on the Scientific Evaluation of Dietary Reference Intakes, Food and Nutrition Board, Institute of Medicine, 1998.

23. A. Prentice, Maternal calcium metabolism and bone mineral status, *American Journal of Clinical Nutrition* 71 (2000): 1312S–1316S.

24. K. O. O'Brien and coauthors, Maternal iron status influences iron transfer to the fetus during the third trimester of pregnancy, *American Journal of Clinical Nutrition* 77 (2003): 924–930.

25. J. C. King, Determinants of maternal zinc status during pregnancy, *American Journal of Clinical Nutrition* 71 (2000): 1334S–1343S.

26. C. S. Chung and coauthors, A single 60-mg iron dose decreases zinc absorption in lactating women, *Journal of Nutrition* 132 (2002): 1903–1905; King, 2000.

27. Position of the American Dietetic Association, 2002.

28. M. M. Werler and coauthors, Multivitamin supplementation and risk of birth defects, *American Journal of Epidemiology* 150 (1999): 675–682.

29. American Academy of Pediatrics, WIC program, *Pediatrics* 108 (2001): 1216–1217.

30. V. Oliveira and M. Prell, Sharing the economic burden: Who pays for WIC's infant formula? *Amber Waves*, available at www.ers.usda.gov/AmberWaves/September04/features/infantformula.htm.

31. B. Abrams, S. L. Altman, and K. E. Pickett, Pregnancy weight gain: Still controversial, *American Journal of Clinical Nutrition* 71 (2000): 1233S–1241S.

32. Position of the American Dietetic Association, 2002.

33. R. Artal and M. O'Toole, Guidelines of the American College of Obstetricians and Gynecologists for exercise during pregnancy and the postpartum period, *British Journal of Sports Medicine* 37 (2003): 6–12; American College of Obstetricians and Gynecologists, Exercise during pregnancy and postpartum period, ACOG Committee Opinion no. 267, *Obstetrics and Gynecology* 99 (2002): 171–173.

34. American College of Obstetricians and Gynecologists, 2002.

35. K. Kullick and L. Dugan, Exercise and pregnancy, in C. A. Rosenbloom, ed., *Sports Nutrition: A Guide for the Professional Working with Active People,* 3rd ed., (Chicago: The American Dietetic Association, 2000), pp. 463–476.

36. M. Kihlstrand and coauthors, Water-gymnastics reduced the intensity of back/low back pain in pregnant women, *Acta Obstetrica Gynecologica Scandinavica* 78 (1999): 180–185.

37. National Center for Health Statistics, Centers for Disease Control (2003), available at www.cdc.gov/nchs, National and state-specific pregnancy rates among adolescents—United States, 1995–1997, *Morbidity and Mortality Weekly Report* 49 (2000): 605–611.

38. D. S. Elfenbein and M. E. Felice, Adolescent pregnancy, *Pediatric Clinics of North America* 50 (2003): 781–800, viii; R. J. Trissler, The child within: A guide to nutrition counseling for pregnant teens, *Journal of the American Dietetic Association* 99 (1999): 916–917.

39. D. J. Hunt and coauthors, Effects of nutrition education programs on anthropometric measurements and pregnancy outcomes of adolescents, *Journal of the American Dietetic Association* 102 (2002): S100–S102.

40. J. M. Lightwood, C. S. Phibbs, and S. A. Glantz, Short-term health and economic benefits of smoking cessation: Low birth weight, *Pediatrics* 104 (1999): 1312–1320.

41. American Academy of Pediatrics, Task Force on Infant Sleep Position and Sudden Infant Death Syndrome, Changing concepts of sudden infant death syndrome: Implication for infant sleeping environment and sleep position, *Pediatrics* 105 (2000): 650–656.

42. J. R. DiFranza, C. A. Aligne, and M. Weitzman, Prenatal and postnatal enviromental tobacco smoke exposure and children's health, *Pediatrics* 113 (2004): 1007–1015; K. I. McMartin and coauthors, Lung tissue concentrations of nicotine in sudden infant death syndrome (SIDS), *Journal of Pediatrics* 140 (2002): 205–209.

43. S. J. Ventura and coauthors, Trends and variations in smoking during pregnancy and low birth weight: Evidence from the birth certificate, 1990–2000, *Pediatrics* 111 (2003): 1176–1180.

44. Position of the American Dietetic Association, 2002.

45. M. S. Scher, G. A. Richardson, and N. L. Day, Effects of prenatal cocaine/crack and other drug exposure on electroencephalographic sleep studies at birth and one year, *Pediatrics* 105 (2000): 39–48.

46. L. T. Singer and coauthors, Cognitive and motor outcomes of cocaine-exposed infants, *Journal of the American Medical Association* 287 (2002): 1952–1960; E. S. Bandstra and coauthors, Intrauterine growth of full-term infants: Impact of prenatal cocaine exposure, *Pediatrics* 108 (2001): 1309–1319.

47. S. E. Schober and coauthors, Blood mercury levels in US children and women of childbearing age, 1999–2000, *Journal of the American Medical Association* 289 (2003): 1667–1674; A. Gomaa and coauthors, Maternal bone lead as an independent risk factor for fetal neurotoxicity: A prospective study, *Pediatrics* 110 (2002): 110–118.

48. FDA and EPA announce the revised consumer advisory on methylmercury in fish, *News*, 2004, available from www.cfsan.fda.gov.

49. Center for the Evaluation of Risks to Human Reproduction (2003), available at http://cerhr.niehs.nih.gov/genpub/topics/listeria-ccae.html.

50. Position of the American Dietetic Association: Use of nutritive and nonnutritive sweeteners, *Journal of the American Dietetic Association* 104 (2004): 255–275.

51. S. Cnattingius and coauthors, Caffeine intake and the risk of first-trimester spontaneous abortion, *New England Journal of Medicine* 343 (2000): 1839–1845; M. A. Klebanoff and coauthors, Maternal serum paraxanthine, a caffeine metabolite, and the risk of spontaneous abortion, *New England Journal of Medicine* 341 (1999): 1639–1644.

52. J. W. Olney and coauthors, The enigma of fetal alcohol neurotoxicity, *Annals of Medicine* 34 (2002): 109–119; C. Ikonomidou and coauthors, Ethanol-induced apoptotic neurodegeneration and fetal alcohol syndrome, *Science* 287 (5455) (2000): 947–948.

53. Centers for Disease Control, Fetal alcohol syndrome (2003), available at www.cdc.gov/ncbddd/fas/fasask.htm.

54. M. J. O'Connor and coauthors, Psychiatric illness in a clinical sample of children with prenatal alcohol exposure, *American Journal of Drug and Alcohol Abuse* 28 (2002): 743–754; B. Sooc and coauthors, Prenatal alcohol exposure and childhood behavior at age 6 to 7 years: Dose-response effect, *Pediatrics* 108 (2001): e34.

55. Alcohol use among women of childbearing age—United States 1991–1999, *Morbidity and Mortality Weekly Report* 51 (2002): 273–276.

56. T. S. Naimi and coauthors, Binge drinking in the preconception period and the risk of unintended pregnancy: Implications for women and their children, *Pediatrics* 111 (2003): 1136–1141.

57. Committee on Substance Abuse and Committee on Children with Disabilities, American Academy of Pediatrics, Fetal alcohol syndrome and alcohol-related neurodevelopmental disorders, *Pediatrics* 106 (2000): 358–361.

58. U. Kesmodel and coauthors, Moderate alcohol intake during pregnancy and the risk of stillbirth and death in the first year of life, *American Journal of Epidemiology* 155 (2002): 305–312.

59. Committee on Substance Abuse and Committee on Children with Disabilities, 2000.

60. Position statement from the American Diabetes Association: Gestational diabetes mellitus, *Diabetes Care* 26 (2003): S103–S105.

61. Report of the Expert Committee on the Diagnosis and Classification of Diabetes Mellitus, *Diabetes Care* 26 (2003): S5–S20.

62. Position statement from the American Diabetes Association, 2003.

63. Report of the Expert Committee on the Diagnosis and Classification of Diabetes Mellitus, 2003.

64. B. M. Sibai, Diagnosis and management of gestational hypertension and preeclampsia, *Obstetrics and Gynecology* 102 (2003): 181–192; R. Skjaerven, A. J. Wilcox, and R. T. Lie, The interval between pregnancies and the risk of preeclampsia, *New England Journal of Medicine* 346 (2002): 33–38.

65. M. S. Esplin and coauthors, Paternal and maternal components of the predisposition to preeclampsia, *New England Journal of Medicine* 344 (2001): 867–872.

66. D. Maine, Role of nutrition in the prevention of toxemia, *American Journal of Clinical Nutrition* 72 (2000): 298S–300S.

67. J. Villar and coauthors, Nutritional interventions during pregnancy for the prevention or treatment of maternal morbidity and preterm delivery: An overview of randomized controlled trials, *Journal of Nutrition* 133 (2003): 1606S–1625S.

F

68. L. M. Paton and coauthors, Pregnancy and lactation have no long-term deleterious effect on measures of bone mineral in healthy women: A twin study, *American Journal of Clinical Nutrition* 77 (2003): 707–714.

69. G. Kac and coauthors, Breast feeding and postpartum weight retention in a cohort of Brazilian women, *American Journal of Clinical Nutrition* 79 (2004): 487–493; L. N. Haiek and coauthors, Postpartum weight loss and infant feeding, *Journal of the American Board of Family Practice* 14 (2001): 85–94.

70. M. A. McCrory, Does dieting during lactation put infant growth at risk? *Nutrition Reviews* 59 (2001): 18–27; C. A. Lovelady and coauthors, The effect of weight loss in overweight, lactating women on the growth of their infants, *New England Journal of Medicine* 342 (2000): 449–453.

71. S. Ito and A. Lee, Drug excretion into breast milk—overview, *Advanced Drug Delivery Reviews* 55 (2003): 617–627; Committee on Drugs, American Academy of Pediatrics, The transfer of drugs and other chemicals into human milk, *Pediatrics* 108 (2001): 776–789.

72. R. M. Lawrence and R. A. Lawrence, Given the benefit of breastfeeding, what contraindications exist? *Pediatric Clinics of North America* 48 (2001): 235–251.

73. J. S. Wang, Q. R. Zhu, and X. H. Wang, Breastfeeding does not pose any additional risk of immunoprophylaxis failure on infants of HBV carrier mothers, *International Journal of Clinical Practice* 57 (2003): 100–102; J. B. Hill and coauthors, Risk of hepatitis B transmission in breast-fed infants of chronic hepatitis B carriers, *Obstetrics and Gynecology* 99 (2002): 1049–1052; P. Ormrod, Tuberculosis in pregnancy and the puerperium, *Thorax* 56 (2001): 494–499.

74. R. Nduati and coauthors, Effect of breastfeeding and formula feeding on transmission of HIV-1: A randomized clinical trial, *Journal of the American Medical Association* 283 (2000): 1167–1174.

75. J. S. Read and the Committee on Pediatric AIDS, American Academy of Pediatrics Technical Report, Human milk, breastfeeding, and transmission of human immunodeficiency virus type 1 in the United States, *Pediatrics* 112 (2003): 1196–1205.

76. J. Humphrey and P. Iliff, Is breast not best? Feeding babies born to HIV-positive mothers: Bringing balance to a complex issue, *Nutrition Reviews* 59 (2001): 119–127.

77. American Academy of Pediatrics, Policy statement, Controversies concerning vitamin K and the newborn, *Pediatrics* 112 (2003): 191–192.

78. Committee on Nutrition, American Academy of Pediatrics, *Pediatric Nutrition Handbook*, 5th ed., ed. R. E. Kleinman (Elk Grove, Ill.: American Academy of Pediatrics, 2004), pp. 110–111.

79. Position of the American Dietetic Association: Breaking the barriers to breastfeeding, *Journal of the American Dietetic Association* 101 (2001): 1213–1220.

80. Position of the American Dietetic Association, 2001; Committee on Nutrition, American Academy of Pediatrics, Hypoallergenic infant formulas, *Pediatrics* 106 (2000): 346–349.

81. M. S. Kramer and R. Kakuma, Optimal duration of exclusive breastfeeding, *Cochrane Database of Systematic Reviews* 1 (2002): CD003517.

82. J. D. Carver, Advances in nutritional modifications of infant formulas, *American Journal of Clinical Nutrition* 77 (2003): 1550S–1554S.

83. I. B. Helland and coauthors, Maternal supplementation with very-long-chain n-3 fatty acids during pregnancy and lactation augments children's IQ at 4 years of age, *Pediatrics* 111 (2003): e39; N. Auestad and coauthors, Growth and development in term infants fed long-chain polyunsaturated fatty acids: A double-masked, randomized, parallel, prospective, multivariate study, *Pediatrics* 108 (2001): 372–381; J. W. Anderson, B. M. Johnstone, and D. T. Remley, Breast-feeding and cognitive development: A meta-analysis, *American Journal of Clinical Nutrition* 70 (1999): 525–535.

84. W. W. Koo, Efficacy and safety of docosahexaenoic acid and arachidonic acid addition to infant formulas: Can one buy better vision and intelligence? *Journal of the American College of Nutrition* 22 (2003): 101–107; E. E. Birch and coauthors, A randomized controlled trial of long-chain polyunsaturated fatty acid supplementation of formula in term infants after weaning at 6 wk of age, *American Journal of Clinical Nutrition* 75 (2002): 570–580.

85. B. Lonnerdal, Nutritional and physiologic significance of human milk proteins, *American Journal of Clinical Nutrition* 77 (2003): 1537S–1543S.

86. L. M. Gartner, F. R. Greer, and the Section on Breastfeeding and Committee on Nutrition, Prevention of rickets and vitamin D deficiency: New guidelines for vitamin D intake, *Pediatrics* 111 (2003): 908–910; S. Fitzpatrick and coauthors, Vitamin D–deficient rickets: A multifactorial disease, *Nutrition Reviews* 58 (2000): 218–222.

87. Gartner, Greer, and the Section on Breastfeeding and Committee on Nutrition, 2003.

88. Position of the American Dietetic Association, 2001.

89. J. Raisler and coauthors, Breast-feeding and infant illness: A dose-response relationship? *American Journal of Public Health* 89 (1999): 25–30.

90. M. Gdalevich, D. Mimouni, and M. Mimouni, Breastfeeding and the risk of bronchial asthma in childhood: A systematic review with meta-analysis of prospective studies, *Journal of Pediatrics* 139 (2001): 261–266; S. Arifeen and coauthors, Exclusive breastfeeding reduces acute respiratory infection and diarrhea deaths among infants in Dhaka slums, *Pediatrics* 108 (2001): 309–320.

91. C. G. Owen and coauthors, Infant feeding and blood cholesterol: A study in adolescents and a systematic review, *Pediatrics* 110 (2002): 597–608.

92. E. L. Mortensen and coauthors, The association between duration of breastfeeding and adult intelligence, *Journal of the American Medical Association* 287 (2002): 2365–2371.

93. A. Jain, J. Concato, and J. M. Leventhal, How good is the evidence linking breastfeeding and intelligence? *Pediatrics* 109 (2002): 1044–1053.

94. Committee on Nutrition, American Academy of Pediatrics, 2000.

95. Committee on Nutrition, American Academy of Pediatrics, Iron fortification of infant formulas, *Pediatrics* 104 (1999): 119–123.

96. Committee on Nutrition, American Academy of Pediatrics, 1999.

97. J. D. Skinner and B. R. Carruth, A longitudinal study of children's juice intake and growth: The juice controversy revisited, *Journal of the American Dietetic Association* 101 (2001): 432–437; Committee on Nutrition, American Academy of Pediatrics, The use and misuse of fruit juice in pediatrics, *Pediatrics* 107 (2001): 1210–1213.

98. Centers for Disease Control and Prevention, Nonfatal choking—related episodes among children—United States, 2001, *Morbidity and Mortality Weekly Report* 51 (2002): 945–948.

Consumer Corner 13

1. FDA warns against feeding dietary supplement product to infants, *FDA Consumer,* May/June, 2004, p. 2.

2. I. Ozturk, N. Votto, and J. M. Leventhal, The timing and predictors of the early termination of breastfeeding, *Pediatrics* 107 (2001): 543–548.

3. B. L. Phillip and coauthors, Baby-friendly hospital initiative improves breastfeeding initiation rates in a U.S. hospital setting, *Pediatrics* 108 (2001): 677–681.

4. R. Li and coauthors, Prevalence of breastfeeding in the United States: The 2001 National Immunization Survey, *Pediatrics* 111 (2003): 1198–1201.

Controversy 13

1. T. Lobstein, L. Baur, and R. Uauy, Obesity in children and young people: A crisis in public health, *Obesity Reviews* 5 (2004): 4–85; T. Aye and L. L. Levitsky, Type 2 diabetes: An epidemic disease in childhood, *Current Opinion in Pediatrics* 15 (2003): 411–415; G. D. Ball and L. J. McCargar, Childhood obesity in Canada: A review of prevalence estimates and risk factors for cardiovascular diseases and type 2 diabetes, *Canadian Journal of Applied Physiology* 28 (2003): 117–140; M. L. Story and coauthors, Demographic and lifestyle factors associated with body mass index among children and adolescents, *International Journal of Food Sciences and Nutrition* 54 (2003): 491–503.

2. M. Kohn and M. Booth, The worldwide epidemic of obesity in adolescents, *Adolescent Medicine* 14 (2003): 1–9; L. S. Lieberman, Dietary, evolutionary, and modernizing influences on the prevalence of type 2 diabetes, *Annual Review of Nutrition* 23 (2003): 345–377.

3. A. Must, Does overweight in childhood have an impact on adult health? *Nutrition Reviews* 61 (2003): 139–142; D. S. Freedman, Clustering of coronary heart disease risk factors among obese children, *Journal of Pediatric Endocrinology and Metabolism* 15 (2002): 1099–1108.

4. J. P. Kaplan, C. T. Liverman, and V. I. Kraak, eds., *Preventing Childhood Obesity: Health in the Balance* (Washington, D.C.: National Academies Press, 2005), available at www.nap.edu; M. I. Goran, Metabolic precursors and effects of obesity in children: A decade of progress, 1990–1999, *American Journal of Clinical Nutrition* 73 (2001): 158–171.

5. A. N. Jeffery and coauthors, Parents' awareness of overweight in themselves and their children: Cross-sectional study within a cohort (EarlyBird 21), *British Medical Journal* 330 (2005): 23–24; S. T. Borra and coauthors, Developing health messages: Qualitative studies with children, parents, and teachers help identify communications opportunities for healthful lifestyles and the prevention of obesity, *Journal of the American Dietetic Association* 103 (2003): 721–728.

6. R. Weiss and coauthors, Obesity and the metabolic syndrome in children and adolescents, *New England Journal of Medicine* 350 (2004): 2362–2374.

7. Storey and coauthors, 2003.

8. J. M. de Castro, Genes, the environment, and the control of food intake, *British Journal of Nutrition* 92 (2004): S59–S62.

9. G. Wu and coauthors, Maternal nutrition and fetal development, *Journal of Nutrition* 134 (2004): 2169–2172.

10. R. Martorell, A. D. Stein, and D. G. Schroeder, Early nutrition and later adiposity, *Journal of Nutrition* 131 (2001): 874S–880S.

11. A. J. Drake and B. R. Walker, The intergenerational effects of fetal programming: Non-genomic mechanisms for the inheritance of low birth weight and cardiovascular

F

risk, *Journal of Endocrinology* 189 (2004): 1–16; Committee on Nutrition, American Academy of Pediatrics, 2003.

12. M. Hanson and coauthors, Report on the 2nd World Congress on Fetal Origins of Adult Disease, Brighton, U.K., *Pediatric Research* 55 (2004): 894–897; J. V. Neel, The "thrifty genotype" in 1998, *Nutrition Reviews* 57 (1999): S2–S9.

13. Drake and Walker, 2004.

14. D. Barker, The midwife, the coincidence, and the hypothesis, *British Medical Journal* 327 (2003): 1428–1430.

15. M. W. Gillman and coauthors, Risk of overweight among adolescents who were breastfed as infants, *Journal of the American Medical Association* 285 (2001): 2461–2467.

16. K. G. Dewey, Is breastfeeding protective against childhood obesity? *Journal of Human Lactation* 19 (2004): 9–18.

17. M. L. Hediger and coauthors, Association between infant breastfeeding and overweight in young children, *Journal of the American Medical Association* 285 (2001): 2453–2460.

18. T. M. Cutting and coauthors, Like mother, like daughter: Familiar patterns of overweight are mediated by mothers' dietary disinhibition, *American Journal of Clinical Nutrition* 69 (1999): 608–613.

19. S. Kranz, A. M. Siega-Riz, and A. H. Herring, Changes in diet quality of American preschoolers between 1977 and 1998, *American Journal of Public Health* 94 (2004): 1525–1530; S. J. Nielsen, A. M. Siega-Riz, and B. M. Popkin, Trends in energy intake in U.S. between 1977 and 1996: Similar shifts seen across age groups, *Obesity Research* 10 (2002): 370–378.

20. Story and coauthors, 2003.

21. L. Harnack, J. Stang, and M. Story, Soft drink consumption among U.S. children and adolescents: Nutritional consequences, *Journal of the American Dietetic Association* 99 (1999): 436–441.

22. D. S. Ludwig, K. E. Peterson, and L. S. Gortmaker, Relation between consumption of sugar-sweetened drinks and childhood obesity: A prospective, observational analysis, *Lancet* 357 (2001): 505–508; R. P. Troiano and coauthors, Energy and fat intakes of children and adolescents in the United States: Data from the National Health and Nutrition Examination Surveys, *American Journal of Clinical Nutrition* 72 (2000): 1343S–1353S.

23. M. H. Proctor and coauthors, Television viewing and change in body fat from preschool to early adolescence: The Framingham Children's Study, *International Journal of Obesity and Related Metabolic Disorders* 27 (2003): 827–833.

24. D. S. Ludwig and C. B. Ebbeling, Type 2 diabetes mellitus in children: Primary care and public health considerations, *Journal of the American Medical Association* 286 (2001): 1427–1430; American Diabetes Association, Type 2 diabetes in children and adolescents, *Pediatrics* 105 (2000): 671–680.

25. R. Kohen-Avramoglu, A. Theriault, and K. Adeli, Emergence of the metabolic syndrome in childhood: An epidemiological overview and mechanistic link to dyslipidemia, *Clinical Biochemistry* 36 (2003): 413–420.

26. S. Li and coauthors, Childhood cardiovascular risk factors and carotid vascular changes in adulthood: The Bogalusa Heart Study, *Journal of the American Medical Association* 290 (2003): 2271–2276; K. B. Keller and L. Lemberg, Obesity and the metabolic syndrome, *American Journal of Clinical Care* 12 (2003): 167–170; H. C. McGill Jr. and coauthors, Origin of atherosclerosis in childhood and adolescence, *American Journal of Clinical Nutrition* 72 (2000): 1307S–1315S.

27. V. N. Muratova and coauthors, The relation of obesity to cardiovascular risk factors among children: The CARDIAC project, *West Virginia Medical Journal* 98 (2002): 263–267.

28. S. C. Couch and coauthors, Rapid westernization of children's blood cholesterol in 3 countries: Evidence for nutrient-gene interactions? *American Journal of Clinical Nutrition* 72 (2000): 1266S–1274S.

29. O. Fiedland and coauthors, Obesity and lipid profiles in children and adolescents, *Journal of Pediatric Endocrinology and Metabolism* 15 (2002): 1011–1016; T. Dwyer and coauthors, Syndrome X in 8-y-old Australian children: Stronger associations with current body fatness than with infant size or growth, *International Journal of Obesity and Related Metabolic Disorders* 26 (2002): 1301–1309.

30. B. Glowinska, M. Urban, and A. Koput, Cardiovascular risk factors in children with obesity, hypertension and diabetes: Lipoprotein (a) levels and body mass index correlate with family history of cardiovascular disease, *European Journal of Pediatrics* 161 (2002): 511–518.

31. A. Wiegman and coauthors, Family history and cardiovascular risk in familial hypercholesterolemia: Data in more than 1000 children, *Circulation* 107 (2003): 1473–1478.

32. Committee on Nutrition, American Academy of Pediatrics, 1998.

33. Dwyer and coauthors, 2002.

34. C. P. M. Leeson and coauthors, Duration of breast feeding and arterial distensibility in early adult life: Population based study, *British Medical Journal* 322 (2001): 643–647.

35. C. H. Fall and coauthors, Relation of infant feeding to adult serum cholesterol concentration and death from ischaemic heart disease, *British Medical Journal* 304 (1992): 427–431, as cited by Y. S. Chong, Human milk is still best, letter, *British Medical Journal* 327 (2003): 683.

36. M. R. Savoca and coauthors, The association of caffeinated beverages with blood pressure in adolescents, *Archives of Pediatrics and Adolescent Medicine* 158 (2004): 473–477.

37. U. S. Sieber and coauthors, How good is blood pressure control among treated hypertensive children and adolescents? *Journal of Hypertension* 21 (2003): 633–637.

38. Committee on Nutrition, American Academy of Pediatrics, 2003.

39. S. de Jongh and coauthors, Efficacy and safety of statin therapy in children with familial hypercholesterolemia: A randomized, double-blind, placebo-controlled trial with simvastatin, *Circulation* 106 (2002): 2231–2237.

40. R. E. Olson, Is it wise to restrict fat in the diets of children? *Journal of the American Dietetic Association* 100 (2000): 28–32; E. Satter, A moderate view on fat restriction, *Journal of the American Dietetic Association* 100 (2000): 32–36; L. A. Lytle, In defense of a low-fat diet for healthy children, *Journal of the American Dietetic Association* 100 (2000): 39–41.

41. E. Obarzanek and coauthors, Long-term safety and efficacy of a cholesterol-lowering diet in children with elevated low-density lipoprotein cholesterol: Seven-year results of the Dietary Intervention Study in Children (DISC), *Pediatrics* 107 (2001): 256–264; L. Rask-Nissilä and coauthors, Neurological development of 5-year-old children receiving a low-saturated fat, low-cholesterol diet since infancy: A randomized controlled study, *Journal of the American Medical Association* 284 (2000): 993–1000; R. M. Lauer and coauthors, Efficacy and safety of lowering dietary intake of total fat, saturated fat, and cholesterol in children with elevated LDL cholesterol: The Dietary Intervention Study in Children, *American Journal of Clinical Nutrition* 72 (2000): 1332S–1342S.

42. J. Orlet Fisher, B. J. Rolls, and L. L. Birch, Children's bite size and intake of an entrée are far greater with large portions than with age-appropriate or self-selected portions, *American Journal of Clinical Nutrition* 77 (2003): 1164–1170.

43. J. Hurley and B. Liebman, Kids' cuisine: "What would you like with your fries?" *Nutrition Action Healthletter* 31 (2004): 12–15.

44. N. Hellmich, Children's menus hold the fries, dish up broccoli, *USA Today,* May 26, 2004, available at www.usatoday.com/news/health/2004-05-26-kids-menus_x.htm.

45. S. M. Phillips and coauthors, Dairy food consumption and body weight and fatness studied longitudinally over the adolescent period, *International Journal of Obesity and Related Metabolic Disorders* 27 (2003): 1106–1113; C. M. Weaver and C. J. Boushey, Milk—good for bones, good for reducing childhood obesity? *Journal of the American Dietetic Association* 103 (2003): 1598–1599; J. D. Skinner and coauthors, Longitudinal calcium intake is negatively related to children's body fat indexes, *Journal of the American Dietetic Association* 103 (2003): 1626–1631.

46. J. A. Welsh and coauthors, Overweight among low-income preschool children associated with the consumption of sweet drinks: Missouri, 1999–2002, *Pediatrics* 115 (2005): 223–229; J. James and coauthors, Preventing childhood obesity by reducing consumption of carbonated drinks: Cluster randomised controlled trial, *British Medical Journal* 328 (2004): 1236.

47. R. Novotny and coauthors, Dairy intake is associated with lower body fat and soda intake with greater weight in adolescent girls, *Journal of Nutrition* 134 (2004): 1905–1909.

48. G. A. Bray, S. J. Nielsen, and B. M. Popkin, Consumption of high-fructose corn syrup in beverages may play a role in the epidemic of obesity, *American Journal of Clinical Nutrition* 79 (2004): 537–543.

49. K. S. Woo and coauthors, Effects of diet and exercise on obesity-related vascular dysfunction in children, *Circulation* 109 (2004): 1981–1986.

50. Trends in cigarette smoking among high school students—United States, 1991–2001, *Morbidity and Mortality Weekly Report* 51 (2002): 409–412.

51. Youth tobacco surveillance—United States, 2000, *Morbidity and Mortality Weekly Report* 50 (2001): entire supplement.

52. Committee on Nutrition, American Academy of Pediatrics, 2003.

53. P. Wilson and coauthors, The prevention and treatment of childhood obesity, *Quality and Safety in Health Care* 12 (2003): 65–74.

54. S. M. Gross and B. Cinelli, Coordinated school health program and dietetics professionals: Partners in promoting healthful eating, *Journal of the American Dietetic Association* 104 (2004): 793–798.

55. J. O. Fisher and coauthors, Parental influences on young girls' fruit and vegetable, micronutrient, and fat intakes, *Journal of the American Dietetic Association* 102 (2002): 58–64.

56. R. Lowry and coauthors, Recent trends in participation in physical education among U.S. high school students, *Journal of School Health* 71 (2001): 145–152; Committee on Sports Medicine and Fitness and Committee on School Health, Physical fitness and activity in schools, *Pediatrics* 105 (2000): 1156–1157.

57. A. Moag-Stahlberg, A. Miles, and M. Marcello, What kids say they do and what parents think kids are doing: The ADAF/Knowledge Networks 2003 Family Nutrition and Physical Activity Study, *Journal of the American Dietetic Association* 103 (2003): 1541–1546.

Chapter 14

1. F. E. Viteri and H. Gonzalez, Adverse outcomes of poor micronutrient status in childhood and adolescence, *Nutrition Reviews* 60 (2002): S77–S83.

2. J. Liu and coauthors, Malnutrition at age 3 years and lower cognitive ability at age 11 years: Independence from psychosocial adversity, *Archives of Pediatrics and Adolescent Medicine* 157 (2003): 593–600.

3. K. L. McConahy and M. F. Picciano, How to grow a healthy toddler—12 to 24 months, *Nutrition Today* 38 (2003): 156–163.

4. S. Kranz, A. M. Siega-Riz, and A. H. Herring, Changes in diet quality of American preschoolers between 1977 and 1998, *American Journal of Public Health* 94 (2004): 1525–1530; S. J. Nielsen, A. M. Siega-Riz, and B. M. Popkin, Trends in energy intake in U.S. between 1977 and 1996: Similar shifts seen across age groups, *Obesity Research* 10 (2002): 370–378.

5. U.S. Department of Agriculture and U.S. Department of Health and Human Services, *Dietary Guidelines for Americans 2005, 6th edition,* available online at www.usda.gov/cnpp or call (888) 878-3256; Position of the American Dietetic Association: Dietary guidance for healthy children ages 2 to 11 years, *Journal of the American Dietetic Association* 104 (2004): 660–677.

6. Committee on the Scientific Evaluation of Dietary Reference Intakes, Food and Nutrition Board, Institute of Medicine, *Dietary Reference Intakes for Energy, Carbohydrate, Fiber, Fat, Fatty Acids, Cholesterol, Protein, and Amino Acids* (Washington, D.C.: National Academies Press, 2002), Chapter 6.

7. Committee on the Scientific Evaluation of Dietary Reference Intakes, Food and Nutrition Board, Institute of Medicine, 2002, Chapter 7.

8. Position of the American Dietetic Association: Health implications of dietary fiber, *Journal of the American Dietetic Association* 102 (2002): 993–1000.

9. Position of the American Dietetic Association, 2004; Committee on the Scientific Evaluation of Dietary Reference Intakes, Food and Nutrition Board, Institute of Medicine, 2002, Chapter 8.

10. Centers for Disease Control and Prevention, Iron deficiency—United States, 1999–2000, *Morbidity and Mortality Weekly Report* 51 (2002): 897–899; A. N. Eden, Iron deficiency and the toddler: An ongoing dilemma, *Nutrition & the M.D.* 27 (2001): 1–3.

11. S. L. Johnson, Children's food acceptance patterns: The interface of ontogeny and nutrition needs, *Nutrition Reviews* 60 (2002): S91–S94; Eden, 2001.

12. K. L. McConahy and coauthors, Portion size of common foods predicts energy intake among preschool-aged children, *Journal of the American Dietetic Association* (2004): 975–979.

13. Kranz, Siega-Riz, and Herring, 2004.

14. S. B. Roberts and M. B. Heyman, Micronutrient shortfalls in young children's diets: Common, and owing to inadequate intakes both at home and at child care centers, *Nutrition Reviews* 58 (2000): 27–29.

15. L. L. Birch, Development of food preferences, *Annual Review of Nutrition* 19 (1999): 41–62.

16. J. O. Fisher and L. L. Birch, Restricting access to palatable foods affects children's behavioral response, food selection, and intake, *American Journal of Clinical Nutrition* 69 (1999): 1264–1272.

17. J. D. Skinner and coauthors, Meal and snack patterns of infants and toddlers, *Journal of the American Dietetic Association* 104 (2004): S65–S70.

18. J. O. Fisher and coauthors, Parental influences on young girls' fruit and vegetable, micronutrient, and fat intakes, *Journal of the American Dietetic Association* 102 (2002): 58–64.

19. J. Bryan and coauthors, Nutrients for cognitive development in school-aged children, *Nutrition Reviews* 62 (2004): 295–306; M. Singh, Role of micronutrients for physical growth and mental development, *Indian Journal of Pediatrics* 71 (2004): 59–62.

20. E. K. Hurtado, A. H. Claussen, and K. G. Scott, Early childhood anemia and mild or moderate mental retardation, *American Journal of Clinical Nutrition* 69 (1999): 115–119.

21. J. Liu and coauthors, Malnutrition at age 3 years and externalizing behavior problems at ages 8, 11, and 17 years, *American Journal of Psychiatry* 161 (2004): 2005–2013; D. Pinero, B. Jones, and J. Beard, Variations in dietary iron alter behavior in developing rats, *Journal of Nutrition* 131 (2001): 311–318; S. M. Grantham-McGregor, S. P. Walker, and S. Chang, Nutritional deficiencies and later behavioral development, *Proceedings of the Nutrition Society* 59 (2000): 47–54.

22. B. D. Anderson and coauthors, Retrospective analysis of ingestions of iron containing products in the United States: Are there differences between chewable vitamins and adult preparations? *Journal of Emergency Medicine* 19 (2000): 255–258;

C. C. Morris, Pediatric iron poisonings in the United States, *Southern Medical Journal* 93 (2000): 352–358.

23. L. Stokes and coauthors, Blood lead levels in residents of homes with elevated lead and tap water—District of Columbia, 2004, *Morbidity and Mortality Weekly Report* 53 (2004): 268–270.

24. K. Kalia and S. J. Flora, Strategies for safe and effective therapeutic measures for chronic arsenic and lead poisoning, *Journal of Occupational Health* 47 (2005): 1–21; X. Liu and coauthors, Do children with falling blood lead levels have improved cognition? *Pediatrics* 110 (2002): 787–791.

25. W. J. Rogan and coauthors, The effect of chelation therapy with succimer on neuropsychological development in children exposed to lead, *New England Journal of Medicine* 344 (2001): 1421–1426; J. F. Rosen and P. Mushak, Primary prevention of childhood lead poisoning—the only solution, *New England Journal of Medicine* 344 (2001): 1470–1471.

26. Fatal pediatric lead poisoning—New Hampshire, 2000, *Morbidity and Mortality Weekly Report* 50 (2001): 457–459.

27. T. D. Matte, Reducing blood lead levels: Benefits and strategies, *Journal of the American Medical Association* 281 (1999): 2340–2342; J. A. Simon and E. S. Hudes, Relationship of ascorbic acid to blood lead levels, *Journal of the American Medical Association* 281 (1999): 2289–2293.

28. P. A. Meyer and coauthors, Surveillance for elevated blood lead levels among children—United States, 1997–2001, *Morbidity and Mortality Weekly Report* 52 (2003): 1–21; M. J. Brown and coauthors, The effectiveness of housing policies in reducing children's lead exposure, *American Journal of Public Health* 91 (2001): 621–624.

29. Advisory Committee on Childhood Lead Poisoning Prevention, Recommendations for blood lead screening of young children enrolled in Medicaid: Targeting a high-risk group, *Morbidity and Mortality Weekly Report* 49 (2000): 1–13.

30. Centers for Disease Control and Prevention, Childhood lead poisoning, available at www.cdc.gov/nceh/lead/factsheets/childhoodlead.htm.

31. Food Allergy and Intolerances, National Institutes of Health Fact Sheet (2003), available at www.niaid.nih.gov/factsheets/food.htm; R. Formanek Jr., Food allergies: When food becomes the enemy, *FDA Consumer,* July/August 2001, pp. 10–16.

32. H. S. Skolnick and coauthors, The natural history of peanut allergy, *Journal of Allergy and Clinical Immunology* 107 (2001): 367–374.

33. K. J. Falci, K. L. Gombas, and E. L. Elliot, Food allergen awareness: An FDA priority, *Food Safety Magazine,* February/March, 2001, available at www.cfsan.fda.gov/~dms/.

34. B. Wuthrich, Lethal or life-threatening allergic reactions to food, *Journal of Investigational Allergology and Clinical Immunology* 10 (2000): 59–65.

35. L. Christie and coauthors, Food allergies in children affect nutrient intake and growth, *Journal of the American Dietetic Association* 102 (2002): 1648–1651.

36. G. S. Rhim and M. S. McMorris, School readiness for children with food allergies, *Annals of Allergy, Asthma and Immunololgy* 86 (2001): 172–176.

37. S. A. Bock, S. Munoz-Furlong, and H. A. Sampson, Fatalities due to anaphylactic reactions to foods, *Journal of Allergy and Clinical Immunology* 107 (2001): 191–193.

38. Food Allergen Labeling and Consumer Protection Act of 2004, available at http://thomas.loc.gov/cgi-bin/query/F?c108:6:./temp/~c108Dz8zuL:e48634.

39. Formanek Jr., 2001.

40. B. Merz, Studying peanut anaphylaxis, *New England Journal of Medicine* 348 (2003): 975–976; H. Metzger, Two approaches to peanut allergy, *New England Journal of Medicine* 348 (2003): 1046–1048; X. M. Li and coauthors, Persistent protective effect of heat-killed *Escherichia coli* producing "engineered," recombinant peanut proteins in a murine model of peanut allergy, *Journal of Allergy and Clinical Immunology* 112 (2003): 159–167.

41. K. Beyer, Characterization of allergenic food proteins for improved diagnostic methods, *Current Opinion in Allergy and Clinical Immunology* 3 (2003): 189–197.

42. S. Barrett, Allergies: Dubious diagnosis and treatment, August 23, 2003, available at www.quackwatch.org.

43. K. S. Rowe and K. J. Rowe, Synthetic food coloring and behavior: A dose response effect in a double-blind, placebo-controlled, repeated measure study, *Journal of Pediatrics* 125 (1994): 691–698.

44. E. J. Costello and coauthors, Relationships between poverty and psychopathology: A natural experiment, *Journal of the American Medical Association* 290 (2003): 2023–2029.

45. E. Konofal and coauthors, Iron deficiency in children with attention-deficit/hyperactivity disorder, *Archives of Pediatric and Adolescent Medicine* 158 (2004): 1113–1115; Federal Interagency Forum on Child and Family Statistics, *America's Children: Key National Indicators of Well-Being 2002,* available at www.childstats.gov.

46. S. Parmet, C. Lynm, and R. M. Glass, Attention-deficit/hyperactivity disorder, *Journal of the American Medical Association* 288 (2002): 1804.

47. R. Sturm, Childhood obesity—what we can learn from existing data on societal trends, part 1, *Preventing Chronic Disease: Public Health Research, Practice, and Policy,* January 2005, available at www.cdc.gov/pcd/issues/ 2005/jan/04_0038.htm;

Committee on Public Education, American Academy of Pediatrics, Policy statement: Children, adolescents, and television, *Pediatrics* 107 (2001): 423–426.

48. R. Chatrath and coauthors, Physical fitness of urban American children, *Pediatric Cardiology* 23 (2002): 608–612; A. D. Salbe and coauthors, Assessing risk factors for obesity between childhood and adolescence: II. Energy metabolism and physical activity, *Pediatrics* 110 (2002): 307–314.

49. R. Boynton-Jarrett and coauthors, Impact of television viewing patterns on fruit and vegetable consumption among adolescents, *Pediatrics* 112 (2003): 1321–1326; K. A. Coon and coauthors, Relationships between use of television during meals and children's food consumption patterns, *Pediatrics* 107 (2001): 167; C. J. Crespo and coauthors, Television watching, energy intake, and obesity in U.S. children: Results from the third National Health and Nutrition Examination Survey, 1988–1994, *Archives of Pediatrics and Adolescent Medicine* 155 (2001): 360–365.

50. C. W. Enns, S. J. Mickle, and J. D. Goldman, Trends in food and nutrient intakes by children in the United States, *Family Economics and Nutrition Review* 14 (2002): 56–68.

51. R. Lowry and coauthors, Recent trends in participation in physical education among U.S. high school students, *Journal of School Health* 71 (2001): 145–152.

52. T. N. Robinson, Reducing children's television viewing to prevent obesity: A randomized controlled trial, *Journal of the American Medical Association* 282 (1999): 1561–1567.

53. Committee on Public Education, American Academy of Pediatrics, 2001; B. A. Dennison, T. A. Erb, and P. L. Jenkins, Television viewing and television in bedroom associated with overweight risk among low-income preschool children, *Pediatrics* 109 (2002): 1028–1035.

54. M. I. Goran, Metabolic precursors and effects of obesity in children: A decade of progress, 1990–1999, *American Journal of Clinical Nutrition* 73 (2001): 158–171.

55. A. Must and S. E. Anderson, Effects of obesity on morbidity in children and adolescents, *Nutrition in Clinical Care* 6 (2003): 4–12; R. Sinha and coauthors, Prevalence of impaired glucose tolerance among children and adolescents with marked obesity, *New England Journal of Medicine* 346 (2002): 802–810.

56. J. B. Schwimmer, T. M. Burwinkle, and J. W. Varni, Health-related quality of life of severely obese children and adolescents, *Journal of the American Medical Association* 289 (2003): 1813–1819.

57. American Academy of Pediatrics, Policy statement, Prevention of pediatric overweight and obesity, *Pediatrics* 112 (2003): 424–430.

58. D. Spruijt-Metz and coauthors, Relation between mothers' child-feeding practices and children's adiposity, *American Journal of Clinical Nutrition* 75 (2002): 581–586; A. E. Field and coauthors, Peer, parent, and media influences on the development of weight concerns and frequent dieting among preadolescent and adolescent girls and boys, *Pediatrics* 107 (2001): 54–60; K. K. Davison and L. L. Birch, Weight status, parent reaction, and self-concept in five-year-old girls, *Pediatrics* 107 (2001): 46–53.

59. National Institute of Diabetes and Digestive and Kidney Diseases, National Institutes of Health, *Helping Your Overweight Child,* available at www.niddk.nih.gov.

60. B. J. Rolls, D. Engell, and L. L. Birch, Serving portion size influences 5-year-old but not 3-year-old children's food intakes, *Journal of the American Dietetic Association* 100 (2000): 232–234.

61. T. A. Nicklas and coauthors, Breakfast consumption with and without vitamin-mineral supplement use favorably impacts daily nutrient intake of ninth-grade students, *Journal of Adolescent Health* 27 (2000): 314–321.

62. C. S. Berkey and coauthors, Longitudinal study of skipping breakfast and weight change in adolescents, *International Journal of Obesity and Related Metabolic Disorders* 27 (2003): 1258–1266.

63. J. Bhattacharya, J. Currie, and S. J. Haider, Evaluating the impact of school nutrition programs: Final Report, a report of the Economic Research Service, July 2004, available at www.ers.usda.gov.

64. S. M. Gross and B. Cinelli, Coordinated school health programs and dietetics professionals: Partners in promoting healthful eating, *Journal of the American Dietetic Association* 104 (2004): 793–798.

65. Position of the American Dietetic Association, 2004.

66. A. J. Rainville, Nutritional quality of reimbursable school lunches compared with lunches brought from home in elementary schools in two southeastern Michigan districts, *Journal of Child Nutrition and Management* 1 (2000): 13–18.

67. P. M. Gleason and C. W. Suitor, Eating at school: How the National School Lunch Program affects children's diets, *American Journal of Agricultural Economics* 85 (2003): 1047–1051.

68. Position of the American Dietetic Association, Society of Nutrition Education, and American School Food Service Association, Nutrition services: An essential component of comprehensive school health programs, *Journal of the American Dietetic Association* 103 (2003): 505–514.

69. E. A. Bergman and coauthors, Time spent by schoolchildren to eat lunch, *Journal of the American Dietetic Association* 100 (2000): 696–698.

70. Position of the American Dietetic Association: Local support for nutrition integrity in schools, *Journal of the American Dietetic Association* 100 (2000): 108–111; K. W. Cullen and coauthors, Effect of a la carte and snack bar foods at school on children's lunchtime intake of fruits and vegetables, *Journal of the American Dietetic Association* 100 (2000): 1482–1486.

71. M. B. Wildey and coauthors, Fat and sugar levels are high in snacks purchased from student stores in middle schools, *Journal of the American Dietetic Association* 100 (2000): 319–322; L. Harnack and coauthors, Availability of a la carte food items in junior and senior high schools: A needs assessment, *Journal of the American Dietetic Association* 100 (2000): 701–703.

72. K. W. Cullen and I. Zakeri, Fruits, vegetables, milk, and sweetened beverages consumption and access to a la carte/snack bar meals at school, *American Journal of Public Health* 94 (2004): 463–467; M. Y. Kubik and coauthors, The association of the school food environment with dietary behaviors of young adolescents, *American Journal of Public Health* 93 (2003): 1168–1173.

73. Committee on School Health, American Academy of Pediatrics, Soft drinks in schools, *Pediatrics* 113 (2004): 152–154; J. L. Kramer-Atwood and coauthors, Fostering healthy food consumption in schools: Focusing on the challenges of competitive foods, *Journal of the American Dietetic Association* 102 (2002): 1228–1233.

74. Position of the American Dietetic Association, Society for Nutrition Education, and American School Food Service Association, 2003.

75. B. P. Roberts, A. S. Blinkhorn, and J. T. Duxbury, The power of children over adults when obtaining sweet snacks, *International Journal of Pediatric Dentistry* 13 (2003): 76–84; S. A. French, Pricing effects on food choices, *Journal of Nutrition* 133 (2003): 841S–843S.

76. U.S. Department of Agriculture and U.S. Department of Health and Human Services, 2005; Committee on Nutrition, American Academy of Pediatrics, The use and misuse of fruit juice in pediatrics, *Pediatrics* 107 (2001): 1210–1213.

77. H. R. Melgar-Quiñonez and L. L. Kaiser, Relationship of child-feeding practices to overweight in low-income Mexican-American preschool-aged children, *Journal of the American Dietetic Association* 104 (2004): 1110–1119; P. K. Newby and coauthors, Beverage consumption is not associated with changes in weight and body mass index among low-income preschool children in North Dakota, *Journal of the American Dietetic Association* 104 (2004): 1086–1094; J. D. Skinner and B. R. Carruth, A longitudinal study of children's juice intake and growth: The juice controversy revisited, *Journal of the American Dietetic Association* 101 (2001): 432–437.

78. D. S. Ludwig, K. E. Peterson, and S. L. Gortmaker, Relation between consumption of sugar-sweetened drinks and childhood obesity: A prospective, observational analysis, *Lancet* 357 (2001): 505–508.

79. Position of the American Dietetic Association, Society of Nutrition Education, and American School Food Service Association, 2003.

80. Position of the American Dietetic Association, Society for Nutrition Education, and American School Food Service Association, 2003.

81. M. Story, D. Neumark-Sztainer, and S. French, Individual and environmental influences on adolescent eating behaviors, *Journal of the American Dietetic Association* 102 (2002): S40–S51.

82. Story, Neumark-Sztainer, and French, 2002.

83. D. Neumark-Sztainer and coauthors, Family meal patterns: Associations with sociodemographic characteristics and improved dietary intake among adolescents, *Journal of the American Dietetic Association* 103 (2003): 317–322.

84. Field and coauthors, 2001; Davison and Birch, 2001.

85. Committee on the Scientific Evaluation of Dietary Reference Intakes, Food and Nutrition Board, Institute of Medicine, *Dietary Reference Intakes for Vitamin A, Vitamin K, Arsenic, Boron, Chromium, Copper, Iodine, Iron, Manganese, Molybdenum, Nickel, Silicon, Vanadium, and Zinc* (Washington, D.C.: National Academy Press, 2001), pp. 290–393.

86. J. S. Halterman and coauthors, Iron deficiency and cognitive achievement among school-aged children and adolescents in the United States, *Pediatrics* 107 (2001): 1381–1386.

87. Federal Update, Milk matters, *Journal of the American Dietetic Association* 102 (2002): 469.

88. G. Mrdjenovic and D. A. Levitsky, Nutritional and energetic consequences of sweetened drink consumption in 6- to 13-year-old children, *Journal of Pediatrics* 142 (2003): 604–610.

89. H. J. Kalkwarf, J. C. Khoury, and B. P. Lanphear, Milk intake during childhood and adolescence, adult bone density, and osteoporotic fractures in US women, *American Journal of Clinical Nutrition* 77 (2003): 257–265.

90. S. A. Bowman, Beverage choices of young females: Changes and impact on nutrient intakes, *Journal of the American Dietetic Association* 102 (2002): 1234–1239.

91. J. James and coauthors, Preventing childhood obesity by reducing consumption of carbonated drinks: Cluster randomised controlled trial, *British Medical Journal* 328 (2004): 1237; Ludwig, Peterson, and Gortmaker, 2001.

92. Committee on the Scientific Evaluation of Dietary Reference Intakes, Food and Nutrition Board, Institute of Medicine, 2002, Chapter 5.

93. Food and Nutrition Board, Institute of Medicine, *Preventing Childhood Obesity: Health in the Balance* (Washington, D.C.: National Academics Press, 2005), p. 56; C. L. Ogden and coauthors, Prevalence and trends in overweight among US children and adolescents, 1999–2000, *Journal of the American Medical Association* 288 (2002): 1728–1732.

94. D. Neumark-Sztainer and coauthors, Weight-control behaviors among adolescent girls and boys: Implications for dietary intake, *Journal of the American Dietetic Association* 104 (2004): 913–920.

95. Neumark-Sztainer and coauthors, 2004.

96. A. E. Field and coauthors, Relation between dieting and weight change among preadolescents and adolescents, *Pediatrics* 112 (2003): 900–906.

97. L. A. Lytle and coauthors, How do children's eating patterns and food choices change over time? Results from a cohort study, *American Journal of Health Promotion* 14 (2000): 222–228.

98. M. Sigman-Grant, Strategies for counseling adolescents, *Journal of the American Dietetic Association* 102 (2002): S32–S39.

99. Federal Interagency Forum on Aging-Related Statistics, *Older Americans 2004: Key Indicators of Well-Being*, November 2004, available at www.agingstats.gov: Trends in aging—United States and worldwide, *Morbidity and Mortality Weekly Report* 52 (2003): 101–106.

100. National vital statistics report (2003), available at www.cdc.gov/nchs.

101. T. Perls, Genetic and environmental influences on exceptional longevity and the AGE nomogram, *Annals of the New York Academy of Sciences* 959 (2002): 1–13.

102. C. E. Greenwood, Dietary carbohydrate, glucose regulation, and cognitive performance in elderly persons, *Nutrition Reviews* 61 (2003): S68–S74.

103. E. W. Gregg and coauthors, Relationship of changes in physical activity and mortality among older women, *Journal of the American Medical Association* 289 (2003): 2379–2386; W. J. Strawbridge and coauthors, Physical activity reduces the risk of subsequent depression for older adults, *American Journal of Epidemiology* 156 (2002): 328–334.

104. K. M. Tarpenning and coauthors, Endurance training delays age of decline in leg strength and muscle morphology, *Medicine and Science in Sports and Exercise* 36 (2004): 74–78; J. Haberer and coauthors, Clinical problem solving: A gut feeling, *New England Journal of Medicine* 349 (2003): 73–78; P. A. Ades and coauthors, Resistance training on physical performance in disabled older female cardiac patients, *Medicine and Science in Sports and Exercise* 35 (2003): 1265–1270.

105. R. D. Hansen and B. J. Allen, Habitual physical activity, anabolic hormones, and potassium content of fat-free mass in postmenopausal women, *American Journal of Clinical Nutrition* 75 (2002): 314–320.

106. A. S. Nicolas and coauthors, Successful aging and nutrition, *Nutrition Reviews* 59 (2001): S88–S92.

107. J. Kruger and coauthors, Strength training among adults aged ≥65 years—United States, 2001, *Morbidity and Mortality Weekly Report* 53 (2004): 25–28.

108. H. K. Kamel, Sarcopenia and aging, *Nutrition Reviews* 61 (2003): 157–167; C. W. Bales and C. S. Ritchie, Sarcopenia, weight loss, and nutritional frailty in the elderly, *Annual Review of Nutrition* 22 (2002): 309–323.

109. P. Jokl, P. M. Sethi, and A. J. Cooper, Master's performance in the New York City Marathon 1983–1999, *British Journal of Sports Medicine* 38 (2004): 408–412.

110. M. M. G. Wilson, R. Purushothaman, and J. E. Morley, Effect of liquid dietary supplements on energy intake in the elderly, *American Journal of Clinical Nutrition* 75 (2002): 944–947.

111. D. Benton and S. Nabb, Carbohydrate, memory, and mood, *Nutrition Reviews* 61 (2003): S61–S67.

112. N. R. Sahyoun and E. Krall, Lower dietary quality among older adults with self-perceived ill-fitting dentures, *Journal of the American Dietetic Association* 103 (2003): 1494–1499.

113. Committee on the Scientific Evaluation of Dietary Reference Intakes, Food and Nutrition Board, Institute of Medicine, 2002.

114. Centers for Disease Control and Prevention, Prevalence of doctor-diagnosed arthritis and possible arthritis—30 states, 2002, *Morbidity and Mortality Weekly Report* 53 (2004): 383–386.

115. A. E. Wluka and coauthors, Supplementary vitamin E does not affect loss of cartilage volume in knee osteoarthritis: A 2 year double blind randomized placebo controlled study, *Journal of Rheumatology* 29 (2002): 2585–2591.

116. L. Hagfors and coauthors, Antioxidant intake, plasma antioxidants and oxidative stress in a randomized, controlled, parallel, Mediterranean dietary intervention study on patients with rheumatoid arthritis, *Nutrition Journal* 2 (2003), available at www.nutritionj.com/content/2/1/5; O. Adam, Dietary fatty acids and immune reactions in synovial tissue, *European Journal of Medical Research* 8 (2003): 381–387; L. Cleland, M. James, and S. Proudman, The role of fish oils in the treatment of rheumatoid arthritis, *Drugs* 63 (2003): 845–853.

117. L. Skoldstam, L. Hagfors, and G. Johansson, An experimental study of a Mediterranean diet intervention for patients with rheumatoid arthritis, *Annals of the Rheumatic Diseases* 62 (2003): 208–214.

118. V. B. Kraus and coauthors, Ascorbic acid increases the severity of spontaneous knee osteoarthritis in a guinea pig model, *Arthritis and Rheumatism* 50 (2004): 1822–1831.

119. D. J. Pattison, D. P. Symmons, and A. Young, Does diet have a role in the aetiology of rheumatoid arthritis? *Proceedings of the Nutrition Society* 63 (2004): 137–143; J. M. Kremer, n-3 Fatty acid supplements in rheumatoid arthritis, *American Journal of Clinical Nutrition* 71 (2000): 349S–351S.

120. S. P. Messier and coauthors, Exercise and dietary weight loss in overweight and obese older adults with knee osteoarthritis: The Arthritis, Diet, and Activity Promotion Trial, *Arthritis and Rheumatism* 50 (2004): 1501–1510; K. R. Baker and coauthors, The efficacy of home based progressive strength training in older adults with knee osteoarthritis: A randomized controlled trial, *Journal of Rheumatology* 28 (2001): 1655–1665.

121. J. Y. Reginster and coauthors, Naturocetic (glucosamine and chondroitin sulfate) compounds as structure-modifying drugs in the treatment of osteoarthritis, *Current Opinion in Rheumatology* 15 (2003): 651–655; National Institute of Arthritis and Musculoskeletal and Skin Diseases, National Institutes of Health, Glucosamine/Chondroitin Arthritis Intervention Trial (GAIT) begins patient recruitment, Press releases, 2000, available at www.niams.nih.gov/ ne/press/2000/12_11a.htm; T. E. McAlindon and coauthors, Glucosamine and chondroitin for treatment of osteoarthritis: A systematic quality assessment and meta-analysis, *Journal of the American Medical Association* 283 (2000): 1469–1475.

122. Standing Committee on the Scientific Evaluation of Dietary Reference Intakes, Food and Nutrition Board, Institute of Medicine, *Dietary Reference Intakes for Calcium, Phosphorus, Magnesium, Vitamin D, and Fluoride* (Washington, D.C.: National Academy Press, 1997).

123. C. Ho and coauthors, Practitioners' guide to meeting the vitamin B_{12} Recommended Dietary Allowance for people aged 51 years and older, *Journal of the American Dietetic Association* 99 (1999): 725–727.

124. T. S. Dharmarajan, G. U. Adiga, and E. P. Norkus, Vitamin B12 deficiency: Recognizing subtle symptoms in older adults, *Geriatrics* 58 (2003): 30–34, 37–38; H. W. Baik and R. M. Russell, Vitamin B_{12} deficiency in the elderly, *Annual Review of Nutrition* 19 (1999): 357–377.

125. M. A. Johnson and coauthors, Hyperhomocysteinemia and vitamin B-12 deficiency in elderly using Title IIIc nutrition services, *American Journal of Clinical Nutrition* 77 (2003): 211–220; Baik and Russell, 1999.

126. I. H. Rosenberg, B vitamins, homocysteine , and neurocognitive function, *Nutrition Reviews* 59 (2001): S69–S74.

127. M. Meydani, Antioxidants and cognitive function, *Nutrition Reviews* 59 (2001): S75–S82.

128. J. L. Gottlieb, Age-related macular degeneration, *Journal of the American Medical Association* 288 (2002): 2233–2236.

129. N. I. Krinsky, J. T. Landrum, and R. A. Bone, Biologic mechanisms of the protective role of lutein and zeaxanthin in the eye, *Annual Review of Nutrition* 23 (2003): 171–201.

130. E. Cho and coauthors, Prospective study of dietary fat and the risk of age-related macular degeneration, *American Journal of Clinical Nutrition* 73 (2000): 209–218.

131. C. Chitchumroonchokchai and coauthors, Xanthophylls and α-tocopherol decrease UVB-induced lipid peroxidation and stress signaling in human lens epithelial cells, *Journal of Nutrition* 134 (2004): 3225–3232; A. Taylor and coauthors, Long-term intake of vitamins and carotenoids and odds of early age-related cortical and posterior subcapsular lens opacities, *American Journal of Clinical Nutrition* 75 (2002): 540–549; J. A. Simon and E. S. Hudes, Serum ascorbic acid and other correlates of self-reported cataract among older Americans, *Journal of Clinical Epidemiology* 52 (1999): 1207–1211.

132. S. M. Moeller and coauthors, Overall adherence to the Dietary Guidelines for Americans is associated with reduced prevalence of early age-related nuclear lens opacities in women, *Journal of Nutrition* 134 (2004): 1812–1819.

133. D. H. Holben and coauthors, Fluid intake compared with established standards and symptoms of dehydration among elderly residents of a long-term-care facility, *Journal of the American Dietetic Association* 99 (1999): 1447–1450.

134. D. J. Fleming and coauthors, Dietary factors associated with the risk of high iron stores in the elderly Framingham Heart Study cohort, *American Journal of Clinical Nutrition* 76 (2002): 1375–1384.

135. F. Girodon and coauthors, Impact of trace elements and vitamin supplementation on immunity and infections in institutionalized elderly patients: A randomized controlled trial, *Archives of Internal Medicine* 159 (1999): 748–754.

136. B. E. C. Nordin and coauthors, Effect of age on calcium absorption in postmenopausal women, *American Journal of Clinical Nutrition* 80 (2004): 998–1002;

F

Position of the American Dietetic Association and Dietitians of Canada: Nutrition and women's health, *Journal of the American Dietetic Association* 104 (2004): 984–1001.

137. R. H. Fletcher and K. M. Fairfield, Vitamins for chronic disease prevention in adults, *Journal of the American Medical Association* 287 (2002): 3127–3129; W. C. Willett and M. J. Stampfer, What vitamins should I be taking, doctor? *New England Journal of Medicine* 345 (2001): 1819–1824.

138. N. B. Belloc and L. Breslow, Relationship of physical health status and health practices, *Preventive Medicine* 1 (1972): 409–421.

139. G. E. Vaillant and K. Mukamal, Successful aging, *American Journal of Psychiatry* 158 (2001): 839–847.

140. C. A. Jolly, Dietary restriction and immune function, *Journal of Nutrition* 134 (2004): 1853–1856; B. T. Larson and coauthors, Improved glucose tolerance with lifetime diet restriction favorably affects disease and survival in dogs, *Journal of Nutrition* 133 (2003): 2887–2892.

141. T. A. Gresl and coauthors, Dietary restriction and glucose regulation in aging rhesus monkeys: A follow-up report at 8.5 yr, *American Journal of Physiology: Endocrinology and Metabolism* 281 (2001): E757–E765; J. J. Ramsey and coauthors, Dietary restriction and aging in rhesus monkeys: The University of Wisconsin study, *Experimental Gerontology* 35 (2000): 1131–1149.

142. O. Lamas, J. A. Martinez, and A. Marti, Energy restriction restores the impaired immune response in overweight (cafeteria) rats, *Journal of Nutritional Biochemistry* 15 (2004): 418–425; R. M. Anson and coauthors, Intermittent fasting dissociates beneficial effects of dietary restriction on glucose metabolism and neuronal resistance to injury from calorie intake, *Proceedings of the National Academy of Sciences* 100 (2003): 6216–6220.

143. L. Fontana and coauthors, Long-term calorie restriction is highly effective in reducing the risk for atherosclerosis in humans, *Proceedings of the National Academy of Sciences* 101 (2004): 6659–6663.

144. W. S. Browner and coauthors, The genetics of human longevity, *American Journal of Medicine* 117 (2004): 882–883; J. M. Dhahbi and coauthors, Caloric restriction alters the feeding response of key metabolic enzyme genes, *Mechanics of Aging and Development* 122 (2001): 1033–1048.

145. S. X. Cao and coauthors, Genomic profiling of short- and long-term caloric restriction effects in the liver of aging mice, *Procedings of the National Academy of Sciences* 98 (2001): 10630–10635.

146. K. T. B. Knoops and coauthors, Mediterranean diet, lifestyle factors, and 10-year mortality in elderly European men and women, *Journal of the American Medical Association* 292 (2004): 1433–1439; G. Barja, Endogenous oxidative stress: Relationship to aging, longevity and caloric restriction, *Ageing Research and Reviews* 1 (2002): 397–411; B. J. Merry, Molecular mechanisms linking calorie restriction and longevity, *International Journal of Biochemistry and Cell Biology* 34 (2002): 1340–1354.

147. L. E. Hebert and coauthors, Alzheimer disease in the US population: Prevalence estimates using the 2000 census, *Archives of Neurology* 60 (2003): 1119–1122.

148. G. T. Grossberg, Diagnosis and treatment of Alzheimer's disease, *Journal of Clinical Psychiatry* 64 (2003): 3–6; R. C. Green and coauthors, Depression as a risk factor for Alzheimer's disease: The MIRAGE study, *Archives of Neurology* 60 (2003): 753–759; J. L. Cummings and G. Cole, Alzheimer disease, *Journal of the American Medical Association* 287 (2002): 2335–2338.

149. A. M. Clarfield, The decreasing prevalence of reversible dementias: An updated meta-analysis, *Archives of Internal Medicine* 163 (2003): 2219–2229.

150. C. Janus, Vaccines for Alzheimer's disease: How close are we? *CNS Drugs* 17 (2003): 457–474.

151. M. A. Lovely, C. Xie, and W. R. Markesbery, Protections against amyloid beta peptide toxicity by zinc, *Brain Research* 823 (1999): 88–95.

152. F. Calon and coauthors, Docosahexaenoic acid protects from dendritic pathology in an Alzheimer's disease mouse model, *Neuron* 43 (2004): 633–645.

153. M. von Dongen and coauthors, Ginkgo for elderly people with dementia and age-associated memory impairment: A randomized clinical trial, *Journal of Clinical Epidemiology* 56 (2003): 367–376.

154. S. Gillette-Guyonnet and coauthors, Weight loss in Alzheimer disease, *American Journal of Clinical Nutrition* 71 (2000): 637S–642S; E. T. Poehlman and R. V. Dvorak, Energy expenditure, energy intake, and weight loss in Alzheimer disease, *American Journal of Clinical Nutrition* 71 (2000): 650S–655S.

155. V. A. Freedman, L. G. Martin, and R. F. Schoeni, Recent trends in disability and functioning among older adults in the United States: A systematic review, *Journal of the American Medical Association* 288 (2002): 3137–3146.

156. M. A. Bernstein and coauthors, Higher dietary variety is associated with better nutritional status in frail elderly people, *Journal of the American Dietetic Association* 102 (2002): 1096–1104.

157. B. Bartali and coauthors, Age and disability affect dietary intake, *Journal of Nutrition* 133 (2003): 2868–2873.

158. B. A. Horwitz, C. A. Blanton, and R. B. McDonald, Physiologic determinants of the anorexia of aging: Insights from animal studies, *Annual Review of Nutrition* 22 (2002): 417–438; C. W. Bales and C. S. Ritchie, Sarcopenia, weight loss, and nutritional frailty in the elderly, I *Annual Review of Nutrition* 22 (2002): 309–323.

159. S. B. Roberts, Energy regulation and aging: Recent findings and their implications, *Nutrition Reviews* 58 (2000): 91–97; F. Landi and coauthors, Body mass index and mortality among older people living in the community, *Journal of the American Geriatrics Society* 47 (1999): 1072–1076.

160. Position of the American Dietetic Association: Nutrition, aging, and the continuum of care, *Journal of the American Dietetic Association* 100 (2000): 580–595.

161. Position of the American Dietetic Association: Liberalized diets for older adults in long-term care, *Journal of the American Dietetic Association* 102 (2002): 1316–1323.

162. Economic Research Service of the U.S. Department of Agriculture, USDA report encourages Americans to remember nutritional needs when eating out, USDA News Release no. 0060.99 (1999), available at www.econ.ag.gov/whatsnew/ news/diet.htm.

163. L. H. Clemens, D. L. Slawson, and R. C. Klesges, The effect of eating out on quality of diet in premenopausal women, *Journal of the American Dietetic Association* 99 (1999): 442–444.

164. Safe handling of take-out foods, a pamphlet from the Food and Safety Inspection Service, U.S. Department of Agriculture, September 2003.

Consumer Corner 14

1. American Dietetic Association position paper, 2004.

2. M. Landen and E. Eriksson, How does premenstrual dysphoric disorder relate to depression and anxiety disorders? *Depression and Anxiety* 17 (2003): 122–129.

3. ACOG News Release, ACOG issues guidelines on diagnosis and treatment of PMS, March 31, 2001, available at www.acog.org.

4. U. Halbreich, The etiology, biology, and evolving pathology of premenstrual syndromes, *Psychoneuroendocrinology* 28 (2003): 55–99.

5. C. A. Roca and coauthors, Differential menstrual cycle regulation of hypothalamic-pituitary-adrenal axis in women with premenstrual syndrome and controls, *Journal of Clinical Endocrinology and Metabolism* 88 (2003): 3057–3063.

6. G. B. Cross and coauthors, Changes in nutrient intake during the menstrual cycle of overweight women with premenstrual syndrome, *British Journal of Nutrition* 85 (2001): 475–482.

7. American Dietetic Association position paper, 2004.

8. A. Bendick, The potential for dietary supplements to reduce premenstrual symptoms, *Journal of the American College of Nutrition* 19 (2002): 3–12.

Controversy 14

1. J. M. Sorenson, Herb-drug, food-drug, nutrient-drug, and drug-drug interactions: Mechanisms involved and their medical implications, *Journal of Alternative and Complementary Medicine* 8 (2002): 293–308.

2. L-N. Chan, Drug-nutrient interaction clinical nutrition, *Current Opinion in Clinical Nutrition and Metabolic Care* 5 (2002): 327–332.

3. W. Abebe, Herbal medication: Potential for adverse interactions with analgesic drugs, *Journal of Clinical Pharmacy and Therapeutics* 27 (2002): 391–401.

4. A. W. Wallace, J. M. Victory, and G. W. Amsden, Lack of bioequivlaence of gatifloxacin when coadministered with calcium fortified orange juice in health volunteers, *Journal of Clinical Pharmacology* 43 (2003): 92–96; L. E. Schmidt and K. Dalhoff, Food-drug interactions, *Drugs* 62 (2002): 1481–1502.

5. J. A. Cambria-Kiely, Effect of soy milk on warfarin efficacy, *Annals of Pharmacotherapy* 36 (2002): 1893–1896.

6. S. G. Stoney and coauthors, Oral contraceptive use is associated with increased cardiovascular reactivity in nonsmokers, *Annals of Behavioral Medicine* 23 (2001): 149–157.

7. International Food Information Council, Questions and answers about caffeine and health, January 2003, available at www.ific.org/publications/qa/caffqa.cfm.

8. C. Papadelis and coauthors, Effects of mental workload and caffeine on catecholamines and blood pressure compared to performance variations, *Brain and Cognition* 51 (2003): 143-154; E. De Valck, E. De Groot, and R. Cluydts, Effects of slow-release caffeine and a nap on driving simulator performance after partial sleep deprivation, *Perceptual and Motor Skills* 96 (2003): 67–78; B. J. Fine and coauthors, Effects of caffeine or diphenhydramine on visual vigilance, *Psychopharmacology (Berl)* 114 (1994): 233–238, as cited by H. R. Lieberman, The effects of ginseng, ephedrine, and caffeine on cognitive performance, mood, and energy, *Nutrition Reviews* 59 (2001): 91–102.

9. L. M. Juliano and R. R. Griffiths, A critical review of caffeine withdrawal: empirical validation of symptoms and signs, incidence, severity, and associated features, *Psychopharmacology (Berlin)* 176 (2004), e-pub, available at www.springerlink.com; M. R. Yeomans and coauthors, Effects of caffeine on performance and mood depend on the level of caffeine abstinence, *Psychopharmacology (Berl)* 164 (2002): 241–249.

10. Yeomans and coauthors, 2002.

11. D. S. Michaud and coauthors, Coffee and alcohol consumption and the risk of pancreatic cancer in two prospective United States cohorts, *Cancer Epidemiology, Biomarkers, and Prevention* 10 (2001): 429–437; T. R. Hartley and coauthors, Hypertension risk status and effect of caffeine on blood pressure, *Hypertension* 36 (2000): 137–141.

12. M. R. Olthof and coauthors, Consumption of high doses of chlorogenic acid, present in coffee, or of black tea increases plasma total homocysteine concentrations in humans, *American Journal of Clinical Nutrition* 73 (2001): 532–538.

13. M. F. Leitzmann and coauthors, A prospective study of coffee consumption and the risk of symptomatic gallstone disease in men, *Journal of the American Medical Association* 281(1999): 2106–2112.

14. A. Rodgers, Effect of cola consumption on urinary biochemical and physiochemical risk factors associated with calcium oxalate urolithianis, *Urological Research* 27 (1999): 77–81.

15. K. Neilann and coauthors, Potential mechanisms of diet therapy for fibrocystic breast conditions show inadequate evidence of effectiveness, *Journal of the American Dietetic Association* 100 (2000): 1368–1380.

16. Standing Committee on the Scientific Evaluation of Dietary Reference Intakes, Food and Nutrition Board, Institute of Medicine, *Dietary Reference Intakes for Vitamin C, Vitamin E, Selenium, and Carotenoids* (Washington, D.C.: National Academy Press, 2000), pp. 152–153.

Chapter 15

1. G. Gardner and B. Halweil, *Underfed and Overfed: The Global Epidemic of Malnutrition,* World Watch Paper 150 (Washington, D.C.: Worldwatch Institute, 2000).

2. Federal Interagency Forum on Child and Family Statistics, *America's Children: Key National Indicators of Well-Being,* 2003, available at www.childstats.gov.

3. G. Mudur, India's burden of waterborne diseases is underestimated, *British Medical Journal* 326 (2003): 1284; J. Liu and coauthors, Malnutrition at age 3 years and lower cognitive ability at age 11 years: Independence from psychosocial adversity, *Archives of Pediatrics and Adolescent Medicine* 157 (2003): 593–600.

4. S. Collins and M. Myatt, Short-term prognosis in severe adult and adolescent malnutrition during famine, *Journal of the American Medical Association* 284 (2000): 621–626.

5. World Health Organization, Executive summary, World Health Report 1998: Life in the 21st century—A vision for all, available at www.ch/whr/1998/ exsum98e.htm.

6. S. Perkins, Dead heat: The health consequences of global warming could be many, *Science News* 166 (2004): 10–12; R. F. Keeling and H. E. Garcia, The change in oceanic O_2 inventory associated with recent global warming, *Proceedings of the National Academy of Sciences* 99 (2002): 7848–7853.

7. S. Peng and coauthors, Rice yields decline with higher night temperature from global warming, *Proceedings of the National Academy of Sciences* 101 (2004): 9971–9975.

8. R. MacDonald, Providing the world with clean water, *British Medical Journal* 327 (2003): 1416–1418.

9. J. Raloff, Dead waters: Massive oxygen-starved zones are developing along the world's coasts, *Science News* 165 (2004): 360–362; J. Raloff, Limiting dead zones: How to curb river pollution and save the Gulf of Mexico, *Science News* 165 (2004): 378–380; Food and Agriculture Organization, *The State of Food and Agriculture 2000,* available at www.fao.org.

10. B. Harder, Sea change: Ocean report urges new policies, *Science News* 165 (2004): 259.

11. L. E. Caulfield and coauthors, Undernutrition as an underlying cause of child deaths associated with diarrhea, pneumonia, malaria, and measles, *American Journal of Clinical Nutrition* 80 (2004): 193–198.

12. Food and Agriculture Organization, *The State of Food Insecurity in the World 2003,* available at www.fao.org.

13. U.S. Census Bureau, *America's Community Survey 2003,* available at www.census.gov/acs.

14. S. Bartlett and N. Burstein, Food Stamp Program access study: Eligible nonparticipants, a report of the Economic Research Service, May 2004, available at www.ers.usda.gov.

15. U.S. Department of Agriculture, *Household Food Security in the United States, 2001,* ERS Food Assistance and Nutrition Research Report no. FANRR-29, October 2002, available at www.ers.usda.gov/publications/fanrr29.

16. K. Hanson and K. S. Hamrick, *Moving Public Assistance Recipients into the Labor Force, 1996–2000,* Report 40, electronic report, available at www.ers.usda.gov/publications/FANRR40/.

17. K. M. Kempson and coauthors, Food management practices used by people with limited resources to maintain food sufficiency as reported by nutrition educators, *Journal of the American Dietetic Association* 102 (2002): 1795–1799.

18. Economic Research Service, U.S. Department of Agriculture, Low-income households' expenditures on fruits and vegetables, May 28, 2004, available at www.ers.usda.gov/publications/AER833/.

19. J. T. Cook and coauthors, Food insecurity is associated with adverse health outcomes among human infants and toddlers, *Journal of Nutrition* 134 (2004): 1432–1438.

20. E. J. Costello and coauthors, Relationships between poverty and psychopathology: A natural experiment, *Journal of the American Medical Association* 290 (2003): 2023–2029.

21. J. S. Hampl and R. Hill, Dietetic approaches to US hunger and food insecurity, *Journal of the American Dietetic Association* 102 (2002): 919–923.

22. Position of the American Dietetic Association: Addressing world hunger, malnutrition, and food insecurity, *Journal of the American Dietetic Association* 103 (2003): 1047–1057; Position of the American Dietetic Association: Domestic food and nutrition security, *Journal of the American Dietetic Association* 102 (2002): 1840–1847.

23. Economic Research Service, U.S. Department of Agriculture, Food and assistance programs, Briefing Room 2000, available at www.ers.usda.gov/briefing/foodnutritionassistance/.

24. V. Oliveira and C. Gundersen, WIC increases the nutrient intakes of children, *Food Review,* January/April 2001, pp. 27–30.

25. USDA Food and Nutrition Service, 2003, available at www.fns.usda.gov.

26. USDA Office of Analysis, The decline in food stamp participation: A report to Congress, 2001, Nutrition and Evaluation, Nutrition Assistance Research Report Series no. FSP-01-WEL, available from USDA Food and Nutrition Service.

27. Food and Agriculture Organization, *State of Food Insecurity 2003,* available at www.fao.org.

28. P. Moszynski, Darfur teetering "on the verge of mass starvation," *British Medical Journal* 328 (2004): 1275.

29. Position of the American Dietetic Association, 2003.

30. S. Shapouri and S. Rosen, Fifty years of U.S. food aid and its role in reducing world hunger, *Amber Waves,* September 2004, available at www.ers.usda.gov/AmberWaves/September04/Features/usfoodaid.htm.

31. U. Ramakrishnan, Prevalence of micronutrient malnutrition worldwide, *Nutrition Reviews* 60 (2002): S46–S52; Food and Agriculture Organization, *The State of Food Insecurity in the World 2003,* available at www.fao.org.

32. U. Kapil and A. Bhavna, Adverse effects of poor micronutrient status during childhood and adolescence, *Nutrition Reviews* 60 (2002): S84–S90.

33. M. Peña and J. Bacallao, Malnutrition and poverty, *Annual Review of Nutrition* 22 (2002): 241–253.

34. World Health Organization, Meeting of Interested Parties, Nutrition risk factors, June 2001, available at www.who.int//mipfiles/2233/NHDbrochurecentrefold.pdf.

35. Food and Agriculture Organization, *The State of Food Insecurity in the World 2003,* available at www.fao.org.

36. United Nations Population Fund, Food and water: Can we meet increasing demand? *The State of World Population,* 2001, available at www.unfpa.org.

37. R. Claassen, Have conservation compliance incentives reduced soil erosion? *Amber Waves,* available at www.ers.usda.gov/AmberWaves/June04/Features/HaveConservation.htm.

38. Food and Agriculture Organization, Elements for a plan of action on fishing capacity, Preparatory meeting for the FAO consultation on the management of fishing capacity, shark fisheries, and incidental catch of seabirds in longline fisheries, July 22–24, 1998, available at www.nmfs.noaa.gov.

39. R. A. Myers and B. Worm, Rapid worldwide depletion of predatory fish communities, *Nature* 423 (2003): 280–283.

40. New international plan of action targets illegal, unregulated and unreported fishing, Press release, 2001, available at www.fao.org/WAICENT/OIS/PRESS_NE/PRESSENG/2001/pren0111.htm; Food and Agriculture Organization, Understanding the cultures of fishing communities: A key to fisheries management and food security, *The State of World Fisheries and Aquaculture,* 2000, available at www.fao.org/docrep/003/x8002e/x8002e00.htm.

41. L. R. Brown, as quoted in B. Harder, Catch zero: What can be done as marine ecosystems face a deepening crisis, *Science News* 164 (2003): 59–61.

42. J. M. Hightower and D. Moore, Mercury levels in high-end consumers of fish, *Environmental Health Perspectives* 111 (2003): 604–608.

43. B. M. Jenssen, Marine pollution: The future challenge is to link human and wildlife studies, editorial, *Environmental Health Perspectives* 111 (2003): A198; T. Suzuki, Oxygen-deficient waters along the Japanese coast and their effects upon the estuarine ecosystem, *Journal of Environmental Quality* 30 (2001): 291–302.

44. Intergovernmental Panel on Climate Change 2001: IPCC Third Assessment Report, available at www.ipcc.ch/pub/spm2201.pdf.

45. The Role of Science in Solving the World's Emerging Water Problems, Arthur M. Sackler Colloquia of the National Academy of Sciences held in Irvine, California, October 8–10, 2004.

F

46. United Nations, *Final Report: 3rd World Water Forum* (Tokyo: Nexxus Communications, 2003), available at www.world.waterforum3.com.

47. United Nations Environmental Programme, Major global trends: The state of the environment, *Global Environmental Outlook 2000,* 2000, overview available at www.unep.org.

48. United Nations Environmental Programme, 2000.

49. Position of the American Dietetic Association: Dietetic professionals can implement practices to conserve natural resources and protect the environment, *Journal of the American Dietetic Association* 101 (2001): 1221–1227.

50. A. Maitland, Shoppers pay up for ethical food, *Financial Times,* May 13, 2004, available at www.ft.com; T. Peterson, Fresh food for thought, *Business Week Online,* June 8, 2004, available at www.businessweek.com.

51. R. Robinson and C. Smith, Integrating issues of sustainably produced foods into nutrition practice: A survey of Minnesota Dietetic Association members, *Journal of the American Dietetic Association* 103 (2003): 608–611.

Consumer Corner 15

1. C. Leitzmann, Nutrition ecology: The contribution of vegetarian diets, *American Journal of Clinical Nutrition* 78 (2003): 657S–659S.

2. What can you do to support sustainable agriculture? UC Sustainable Agriculture Research and Education Program, available at www.sarep.ucdavis.edu.

Controversy 15

1. EPA and hog farmers agree, *Progressive Farmer,* January 1999, pp. 6–7.

2. Using manure as fertilizer will help dairy farmers prepare for regulatory changes, University of California Sustainable Agriculture Research and Education Program, January 2003, available at www.sarep.ucdavis.edu.

3. USDA Interagency Agricultural Projections Committee, *USDA Agricultural Baseline Projections to 2008,* 1999, available at usda.mannlib.cornell.edu.

4. USDA National Agricultural Statistics Service, *Farms and Land in Farms, 1998* (Washington, D.C.: USDA Interagency Agricultural Projections Committee, 1999).

5. M. Reeves and K. S. Schafer, Greater risks, fewer rights: U.S. farmworkers and pesticides, *International Journal of Occupational and Environmental Health* 9 (2003): 30–39.

6. Groceries cost less in the U.S., News and Notes, *Progressive Farmer,* March 2001, p. 10.

7. BIFS program overview, University of California Sustainable Agriculture Research and Education Program, available at www.sarep.ucdavis.edu.

8. J. Leidner, Sprayer mixes on the go, *Progressive Farmer,* February 2001, p. 26.

9. K. Smith and M. Weinberg, Measuring the success of conservation programs, *Amber Waves,* September 2004, available at www.ers.usda.gov.AmberWaves/September04/Features/measuring success.htm.

10. R. Claassen, Have conservation compliance incentives reduced soil erosion? *Amber Waves,* September 2004, available at www.ers.usda.gov/AmberWaves/June04/Features/HaveConservation.htm.

11. Local food systems may protect communities from food-borne illness, *News Releases,* 2003, University of California Sustainable Agriculture Research and Education Program, available at www.sarep.ucdavis.edu.

12. J. L. Ohmart, Direct marketing with value-added products or: "give me the biggest on of those berry tarts!", 2003, a study with the University of California Sustainable Agriculture Research and Education Program, available at www.sarep.ucdavis.edu; D. Miller, Cabernet among the corn, *Progressive Farmer,* Mid-March 2001, pp. 34–36.

13. J. Robinson, Grass fed basics: Key differences between conventional and pasture animal production, 2002–2003, available at www.eatwild.com/basics.html.

Answers to Self Check Questions

Chapter 1

1. a
2. a
3. c
4. b
5. d
6. b
7. False. Heart disease and cancer are influenced by many factors with genetics and diet among them.
8. True
9. False. The term "natural" has no legal definition.
10. True

Chapter 2

1. b
2. d
3. c
4. d
5. a
6. True
7. False. The DRI are estimates of the needs of healthy persons only. Medical problems alter nutrient needs.
8. False. People who choose to eat no meats or products taken from animals can still use the USDA Food Guide to make their diets adequate.
9. False. By law, food labels must state as a percentage of the Daily Values the amounts of vitamins A, C, calcium, and iron present in a food.
10. True

Chapter 3

1. a
2. d
3. c
4. c
5. d
6. False. Phagocytes are white blood cells that can ingest and destroy antigens in a process known as phagocytosis.
7. False. Hydrochloric acid initiates protein digestion and activates a protein-digesting enzyme in the stomach.
8. False. The digestive tract works efficiently to digest all foods simultaneously, regardless of composition.
9. True
10. False. Absorption of the majority of nutrients takes place across the specialized cells of the small intestine.

Chapter 4

1. b
2. a
3. c
4. b
5. a
6. True

7. False. Type I diabetes is most often controlled with insulin injections.
8. True
9. False. Whole-grain bread remains more nutritious despite the enrichment of white flour.
10. True

Chapter 5

1. c
2. a
3. c
4. b
5. d
6. True
7. False. Taking fish oil supplements is not recommended and may raise LDL cholesterol and have other harmful effects.
8. False. Consuming large amounts of *trans*-fatty acids elevates serum LDL cholesterol and thus raises the risk of heart disease and heart attack.
9. False. When olestra is present in the digestive tract, fat-soluble vitamins, including vitamin E, become unavailable for absorption.
10. True

Chapter 6

1. b
2. b
3. d
4. a
5. d
6. True
7. False. Excess protein in the diet may have adverse effects such as obesity, enlarged liver or kidneys, worsened kidney disease, and accelerated bone loss.
8. False. Impoverished people living on U.S. Indian reservations, in inner cities, and in rural areas of the United States, as well as some elderly, homeless, and ill people in hospitals, have been diagnosed with PEM.
9. True
10. True

Chapter 7

1. b
2. c
3. d
4. a
5. c
6. d
7. False. No study to date has conclusively demonstrated that vitamin C can prevent colds or reduce their severity.
8. True
9. True
10. False. Vitamin A itself has no effect on acne.

G Chapter 8

1. d
2. b
3. c
4. b
5. a
6. True
7. True
8. False (Calcium is the most abundant mineral in the body.)
9. False (Butter, cream, and cream cheese contain negligible calcium, being almost pure fat. Many vegetables, such as broccoli are good sources of available calcium.)
10. False (The standards for bottled water are substantially less rigorous than those applied to U.S. tap water.)

Chapter 9

1. d
2. b
3. c
4. a
5. c
6. a
7. False. The thermic effect of food is believed to have negligible effects on total energy expenditure.
8. True
9. False. The BMI are unsuitable for use with athletes and adults over age 65.
10. True

Chapter 10

1. c
2. c
3. a
4. d
5. a
6. False. Weight training to improve muscle strength and endurance also helps maximize and maintain bone mass.
7. False. The average resting pulse for adults is around 70 beats per minute, but the rate is lower for active people.
8. True
9. True
10. True

Chapter 11

1. b
2. d
3. a
4. d
5. d
6. True
7. False. Laboratory evidence suggests that a high-calcium diet may help to prevent colon cancer.
8. False. The DASH diet is designed for helping people with hypertension to control the disease.
9. False. Hypertension is more severe and occurs earlier in life among African Americans than those of European or Asian descent.
10. True

Chapter 12

1. b
2. c
3. d
4. a
5. c
6. False. Nature has provided many plants used for food with natural poisons to fend off diseases, insects and other predators.
7. True
8. False. The canning industry chooses treatments that employ the high-temperature-short-time (HTST) principle for canning.
9. True
10. True

Chapter 13

1. c
2. b
3. d
4. a
5. d
6. True
7. True
8. True
9. False. In general, the effect of nutritional deprivation of the mother is to reduce the quantity, not the quality, of her milk.
10. False. There is no proof for the theory that "stuffing the baby" at bedtime will promote sleeping through the night.

Chapter 14

1. c
2. d
3. c
4. c
5. b
6. d
7. False. Research to date does not support the idea that food allergies or intolerances cause hyperactivity in children, but studies continue.
8. True
9. False. Vitamin A absorption appears to increase with aging.
10. False. To date, no proven benefits are available from herbs or other remedies.

Chapter 15

1. b
2. a
3. c
4. d
5. c
6. True
7. False. Most children who die of malnutrition do not starve to death—they die because their health has been compromised by dehydration from infections that cause diarrhea.
8. True
9. False. The link between improved economic status and slowed population growth has been demonstrated in country after country.
10. True

Physical Activity and Energy Requirements

Chapter 9 described how to calculate your estimated energy requirements by using an equation that accounts for your gender, age, weight, height, and physical activity level. This appendix first helps you determine the correct physical activity factor to use in the equation, either by calculating your physical activity level or by guesstimating it. Then the appendix presents tables that provide a shortcut to estimating total energy expenditure.*

Calculating Your Physical Activity Level

To calculate your physical activity level, record all of your activities for a typical 24-hour day, noting the type of activity, the level of intensity, and the duration. Then, using a copy of Table H-1, find your activity in the first column (or an activity that is reasonably similar) and multiply the number of minutes spent on that activity by the factor in the third column. Write your answer in the last column and total the accumulated values for the day. Now add the subtotal of the last column to 1.1 (to account for basal energy and the thermic effect of food) as shown. This is your Physical Activity Level Score, and you are now close to having the number you need—the Physical Activity (PA) factor—for the equation of Chapter 9. Proceed, by using Table H-2, to find the PA factor for your gender that correlates with your Physical Activity Level Score and use it in the energy equation on page 319 of Chapter 9.

Guesstimating Your Physical Activity Level

As an alternative to recording your activities for a day, you can use the first two columns of Table H-3 to decide if your daily activity is sedentary, low active, active, or very active. Find the PA factor for your gender that correlates with your typical physical activity level and use it in the energy equation on page 319. Warning: People generally overestimate their activity, guessing "active" most of the time. In reality, most people in the United States and Canada fall below the active level.

Using a Shortcut to Estimate Total Energy Expenditure

The DRI committee has developed estimates of total energy expenditures (and, therefore, energy requirements) based on the equations presented in Chapter 9. The DRI estimated numbers are provided in Table H-4 for women and Table H-5 for men. To use these tables to determine the energy (calories) needed to maintain your current body weight:

1. Find the table appropriate for your gender.
2. Find your height in meters (or inches) in the left-hand column.
3. Follow the row across to find your weight in kilograms (or pounds). (If you can't find your exact height and weight, choose a value between the two closest ones.)
4. Look down the column to find the number of calories that corresponds to your activity level.

Importantly, the values given in the tables are for 30-year-old people. Women 19 to 29 should add 7 calories per day for each year below age 30; older women should subtract 7 calories per day for each year above age 30. Similarly, men 19 to 29 should add 10 calories per day for each year below age 30; older men should subtract 10 calories per day for each year above age 30.

---

*This appendix, including the tables, is adapted from Committee on Dietary Reference Intakes, *Dietary Reference Intakes for Energy, Carbohydrate, Fiber, Fat, Fatty Acids, Cholesterol, Protein, and Amino Acids* (Washington, D.C.: National Academies Press, 2002).

TABLE H·1 PHYSICAL ACTIVITIES AND THEIR SCORES

If your activity was equivalent to this...	Then list the number of minutes here and ...	Multiply by this factor ...	Add this column to get your Physical Activity Level Score:
Activities of Daily Living			
Gardening (no lifting)		0.0032	
Household tasks (moderate effort)		0.0024	
Lifting items continuously		0.0029	
Loading/unloading car		0.0019	
Lying quietly		0.0000	
Mopping		0.0024	
Mowing lawn (power mower)		0.0033	
Raking lawn		0.0029	
Riding in a vehicle		0.0000	
Sitting (idle)		0.0000	
Sitting (doing light activity)		0.0005	
Taking out trash		0.0019	
Vacuuming		0.0024	
Walking the dog		0.0019	
Walking from house to car or bus		0.0014	
Watering plants		0.0014	
Additional Activities			
Billiards		0.0013	
Calisthenics (no weight)		0.0029	
Canoeing (leisurely)		0.0014	
Chopping wood		0.0037	
Climbing hills (carrying 11 lb load)		0.0061	
Climbing hills (no load)		0.0056	
Cycling (leisurely)		0.0024	
Cycling (moderately)		0.0045	
Dancing (aerobic or ballet)		0.0048	
Dancing (ballroom, leisurely)		0.0018	
Dancing (fast ballroom or square)		0.0043	
Golf (with cart)		0.0014	
Golf (without cart)		0.0032	
Horseback riding (walking)		0.0012	
Horseback riding (trotting)		0.0053	
Jogging (6 mph)		0.0088	
Music (playing accordion)		0.0008	
Music (playing cello)		0.0012	
Music (playing flute)		0.0010	
Music (playing piano)		0.0012	
Music (playing violin)		0.0014	
Rope skipping		0.0105	
Skating (ice)		0.0043	
Skating (roller)		0.0052	
Skiing (water or downhill)		0.0055	
Squash		0.0106	
Surfing		0.0048	
Swimming (slow)		0.0033	
Swimming (fast)		0.0057	
Tennis (doubles)		0.0038	
Tennis (singles)		0.0057	
Volleyball (noncompetitive)		0.0018	
Walking (2 mph)		0.0014	
Walking (3 mph)		0.0022	
Walking (4 mph)		0.0033	
Walking (5 mph)		0.0067	
Subtotal			
Factor for basal energy and the thermic effect of food			1.1
Your Physical Activity Level Score			

TABLE H·2 PHYSICAL ACTIVITY LEVEL SCORES AND THEIR PA FACTORS

Physical Activity Level Score	Description	Men: PA Factor	Women: PA Factor
1.0 to 1.39	Sedentary	1.0	1.0
1.4 to 1.59	Low active	1.11	1.12
1.6 to 1.89	Active	1.25	1.27
1.9 and above	Very active	1.48	1.45

TABLE H·3 PHYSICAL ACTIVITY EQUIVALENTS AND THEIR PA FACTORS

Description	Physical Activity Equivalents	Men: PA Factor	Women: PA Factor
Sedentary	Only those physical activities required for normal independent living	1.0	1.0
	Activities equivalent to walking at a pace of 2–4 mph for the following distances:		
Low active	1.5 to 3.0 miles/day	1.11	1.12
Active	3 to 10 miles/day	1.25	1.27
Very active	10 or more miles/day	1.48	1.45

TABLE H·4 TOTAL ENERGY EXPENDITURE (TEE IN CALORIES PER DAY) FOR WOMEN 30 YEARS OF AGE[a] AT VARIOUS LEVELS OF ACTIVITY AND VARIOUS HEIGHTS AND WEIGHTS

Heights m (in)	Physical Activity Level	Weight[b] kg (lb)					
1.45 (57)		38.9 (86)	45.2 (100)	52.6 (116)	63.1 (139)	73.6 (162)	84.1 (185)
				Calories			
	Sedentary	1564	1623	1698	1813	1927	2042
	Low active	1734	1800	1912	2043	2174	2304
	Active	1946	2021	2112	2257	2403	2548
	Very active	2201	2287	2387	2553	2719	2886
1.50 (59)		41.6 (92)	48.4 (107)	56.3 (124)	67.5 (149)	78.8 (174)	90.0 (198)
				Calories			
	Sedentary	1625	1689	1771	1894	2017	2139
	Low active	1803	1874	1996	2136	2276	2415
	Active	2025	2105	2205	2360	2516	2672
	Very active	2291	2382	2493	2671	2849	3027
1.55 (61)		44.4 (98)	51.7 (114)	60.1 (132)	72.1 (159)	84.1 (185)	96.1 (212)
				Calories			
	Sedentary	1688	1756	1846	1977	2108	2239
	Low active	1873	1949	2081	2230	2380	2529
	Active	2104	2190	2299	2466	2632	2798
	Very active	2382	2480	2601	2791	2981	3171
1.60 (63)		47.4 (104)	55.0 (121)	64.0 (141)	76.8 (169)	89.6 (197)	102.4 (226)
				Calories			
	Sedentary	1752	1824	1922	2061	2201	2340
	Low active	1944	2025	2168	2327	2486	2645
	Active	2185	2276	2396	2573	2750	2927
	Very active	2474	2578	2712	2914	3116	3318

continued

[a]For each year below 30, add 7 calories/day to TEE. For each year above 30, subtract 7 calories/day from TEE.
[b]These columns represent a BMI of 18.5, 22.5, 25, 30, 35, and 40, respectively.

TABLE H·4 TOTAL ENERGY EXPENDITURE (CALORIES PER DAY) FOR WOMEN 30 YEARS OF AGE[a] AT VARIOUS LEVELS OF ACTIVITY AND VARIOUS HEIGHTS AND WEIGHTS—continued

Heights m (in)	Physical Activity Level	Weight[b] kg (lb)					
1.65 (65)		50.4 (111)	58.5 (129)	68.1 (150)	81.7 (180)	95.3 (210)	108.9 (240)
		Calories					
	Sedentary	1816	1893	1999	2148	2296	2444
	Low active	2016	2102	2556	2425	2594	2763
	Active	2267	2364	2494	2682	2871	3059
	Very active	2567	2678	2824	3039	3254	3469
1.70 (67)		53.5 (118)	62.1 (137)	72.3 (159)	86.7 (191)	101.2 (223)	115.6 (255)
		Calories					
	Sedentary	1881	1963	2078	2235	2393	2550
	Low active	2090	2180	2345	2525	2705	2884
	Active	2350	2453	2594	2794	2994	3194
	Very active	2662	2780	2938	3166	3395	3623
1.75 (69)		56.7 (125)	65.8 (145)	76.6 (169)	91.9 (202)	107.2 (236)	122.5 (270)
		Calories					
	Sedentary	1948	2034	2158	2325	2492	2659
	Low active	2164	2260	2437	2627	2817	3007
	Active	2434	2543	2695	2907	3119	3331
	Very active	2758	2883	3054	3296	3538	3780
1.80 (71)		59.9 (132)	69.7 (154)	81.0 (178)	97.2 (214)	113.4 (250)	129.6 (285)
		Calories					
	Sedentary	2015	2106	2239	2416	2593	2769
	Low active	2239	2341	2529	2731	2932	3133
	Active	2519	2634	2799	3023	3247	3472
	Very active	2855	2987	3172	3428	3684	3940
1.85 (73)		63.3 (139)	73.6 (162)	85.6 (189)	102.7 (226)	119.8 (264)	136.9 (302)
		Calories					
	Sedentary	2083	2179	2322	2509	2695	2882
	Low active	2315	2422	2624	2836	3049	3262
	Active	2605	2727	2904	3141	3378	3615
	Very active	2954	3093	3292	3562	3833	4103
1.90 (75)		66.8 (147)	77.6 (171)	90.3 (199)	108.3 (239)	126.4 (278)	144.4 (318)
		Calories					
	Sedentary	2151	2253	2406	2603	2800	2996
	Low active	2392	2505	2720	2944	3168	3393
	Active	2693	2821	3011	3261	3511	3760
	Very active	3053	3200	3414	3699	3984	4270
1.95 (77)		70.3 (155)	81.8 (180)	95.1 (209)	114.1 (251)	133.1 (293)	152.1 (335)
		Calories					
	Sedentary	2221	2328	2492	2699	2906	3113
	Low active	2470	2589	2817	3053	3290	3526
	Active	2781	2917	3119	3383	3646	3909
	Very active	3154	3309	3538	3838	4139	4439

[a]For each year below 30, add 7 calories/day to TEE. For each year above 30, subtract 7 calories/day from TEE.
[b]These columns represent a BMI of 18.5, 22.5, 25, 30, 35, and 40, respectively.

TABLE H·5 TOTAL ENERGY EXPENDITURE (CALORIES PER DAY) FOR MEN 30 YEARS OF AGE[a] AT VARIOUS LEVELS OF ACTIVITY AND VARIOUS HEIGHTS AND WEIGHTS

H

Heights m (in)	Physical Activity Level	Weight[b] kg (lb)					
1.45 (57)		38.9 (86)	47.3 (100)	52.6 (116)	63.1 (139)	73.6 (163)	84.1 (185)
				Calories			
	Sedentary	1777	1911	2048	2198	2347	2496
	Low active	1931	2080	2225	2393	2560	2727
	Active	2127	2295	2447	2636	2826	3015
	Very active	2450	2648	2845	3075	3305	3535
1.50 (59)		41.6 (92)	50.6 (107)	56.3 (124)	67.5 (149)	78.8 (174)	90.0 (198)
				Calories			
	Sedentary	1848	1991	2126	2286	2445	2605
	Low active	2009	2168	2312	2491	2670	2849
	Active	2215	2394	2545	2748	2951	3154
	Very active	2554	2766	2965	3211	3457	3703
1.55 (61)		44.4 (98)	54.1 (114)	60.1 (132)	72.1 (159)	84.1 (185)	96.1 (212)
				Calories			
	Sedentary	1919	2072	2205	2376	2546	2717
	Low active	2089	2259	2401	2592	2783	2974
	Active	2305	2496	2646	2862	3079	3296
	Very active	2660	2887	3087	3349	3612	3875
1.60 (63)		47.4 (104)	57.6 (121)	64.0 (141)	76.8 (169)	89.6 (197)	102.4 (226)
				Calories			
	Sedentary	1993	2156	2286	2468	2650	2831
	Low active	2171	2351	2492	2695	2899	3102
	Active	2397	2601	2749	2980	3210	3441
	Very active	2769	3010	3211	3491	3771	4051
1.65 (65)		50.4 (111)	61.3 (129)	68.1 (150)	81.7 (180)	95.3 (210)	108.9 (240)
				Calories			
	Sedentary	2068	2241	2369	2562	2756	2949
	Low active	2254	2446	2585	2801	3017	3234
	Active	2490	2707	2854	3099	3345	3590
	Very active	2880	3136	3339	3637	3934	4232
1.70 (67)		53.5 (118)	65.0 (137)	72.3 (159)	86.7 (191)	101.2 (223)	115.6 (255)
				Calories			
	Sedentary	2144	2328	2454	2659	2864	3069
	Low active	2338	2542	2679	2909	3139	3369
	Active	2586	2816	2961	3222	3483	3743
	Very active	2992	3265	3469	3785	4101	4417
1.75 (69)		56.7 (125)	68.9 (145)	76.6 (169)	91.9 (202)	107.2 (236)	122.5 (270)
				Calories			
	Sedentary	2222	2416	2540	2757	2975	3192
	Low active	2425	2641	2776	3020	3263	3507
	Active	2683	2927	3071	3347	3623	3900
	Very active	3108	3396	3602	3937	4272	4607
1.80 (71)		59.9 (132)	72.9 (154)	81.0 (178)	97.2 (214)	113.4 (250)	129.6 (285)
				Calories			
	Sedentary	2301	2507	2628	2858	3088	3318
	Low active	2513	2741	2875	3132	3390	3648
	Active	2782	3040	3183	3475	3767	4060
	Very active	3225	3530	3738	4092	4447	4801

continued

[a]For each year below 30, add 10 calories/day to TEE. For each year above 30, subtract 10 calories/day from TEE.
[b]These columns represent a BMI of 18.5, 22.5, 25, 30, 35, and 40, respectively.

TABLE H·5 TOTAL ENERGY EXPENDITURE (CALORIES PER DAY) FOR MEN 30 YEARS OF AGE[a] AT VARIOUS LEVELS OF ACTIVITY AND VARIOUS HEIGHTS AND WEIGHTS—continued

Heights m (in)	Physical Activity Level	Weight[b] kg (lb)					
1.85 (73)		63.3 (139)	77.0 (162)	85.6 (189)	102.7 (226)	119.8 (264)	136.9 (302)
		Calories					
	Sedentary	2382	2599	2718	2961	3204	3447
	Low active	2602	2844	2976	3248	3520	3792
	Active	2883	3155	3297	3606	3915	4223
	Very active	3344	3667	3877	4251	4625	4999
1.90 (75)		66.8 (147)	81.2 (171)	90.3 (199)	108.3 (239)	126.4 (278)	144.4 (318)
		Calories					
	Sedentary	2464	2693	2810	3066	3322	3579
	Low active	2693	2948	3078	3365	3652	3939
	Active	2986	3273	3414	3739	4065	4390
	Very active	3466	3806	4018	4413	4807	5202
1.95 (77)		70.3 (155)	85.6 (180)	95.1 (209)	114.1 (251)	133.1 (293)	152.1 (335)
		Calories					
	Sedentary	2547	2789	2903	3173	3443	3713
	Low active	2786	3055	3183	3485	3788	4090
	Active	3090	3393	3533	3875	4218	4561
	Very active	3590	3948	4162	4578	4993	5409

[a]For each year below 30, add 10 calories/day to TEE. For each year above 30, subtract 10 calories/day from TEE.
[b]These columns represent a BMI of 18.5, 22.5, 25, 30, 35, and 40, respectively.

Food Patterns to Meet the *Dietary Guidelines for Americans 2005*

Two examples of eating patterns that exemplify the *Dietary Guidelines for Americans 2005* are the USDA Food Guide of Chapter 2 (presented here as Tables I-1 and I-2) and the DASH Eating Plan shown in Table I-3.

TABLE I·1 USDA FOOD GUIDE

This table specifies the suggested amounts of food to consume each day from the basic food groups, subgroups, and oils to meet recommended nutrient intakes at 12 different calorie levels. Note that amounts given for vegetable subgroups are weekly amounts. Nutrient and energy contributions from each group are calculated according to the nutrient-dense forms of foods in each group (e.g., lean meats and fat-free milk). The table also shows the discretionary calorie allowance that can be accommodated within each calorie level, in addition to the suggested amounts of nutrient-dense forms of foods in each group.

Amount of Food to Choose Each Day From Each Group (vegetable subgroup amounts are per week)

Calorie Level	1,000	1,200	1,400	1,600	1,800	2,000	2,200	2,400	2,600	2,800	3,000	3,200
Food Group	Food group amounts shown in cup (c) or ounce-equivalents (oz-eq), with number of servings (srv) in parentheses when it differs from the other units. See note for quantity equivalents for foods in each group.[2] Oils are shown in grams (g).											
Fruits	1 c (2 srv)	1 c (2 srv)	1.5 c (3 srv)	1.5 c (3 srv)	1.5 c (3 srv)	2 c (4 srv)	2 c (4 srv)	2 c (4 srv)	2 c (5 srv)	2.5 c (5 srv)	2.5 c (5 srv)	2.5 c (5 srv)
Vegetables	1 c (2 srv)	1.5 c (3 srv)	1.5 c (3 srv)	2 c (4 srv)	2.5 c (5 srv)	2.5 c (5 srv)	3 c (6 srv)	3 c (6 srv)	3.5 c (7 srv)	3.5 c (7 srv)	4 c (8 srv)	4 c (8 srv)
Dark green veg.	1 c/wk	1.5 c/wk	1.5 c/wk	2 c/wk	3 c/wk	3 c/wk	3 c/wk	3 c/wk	3 c/wk	3 c/wk	3 c/wk	3 c/wk
Orange veg.	.5 c/wk	1 c/wk	1 c/wk	1.5 c/wk	2 c/wk	2 c/wk	2 c/wk	2 c/wk	2.5 c/wk	2.5 c/wk	2.5 c/wk	2.5 c/wk
Legumes	.5 c/wk	1 c/wk	1 c/wk	2.5 c/wk	3 c/wk	3 c/wk	3 c/wk	3 c/wk	3.5 c/wk	3.5 c/wk	3.5 c/wk	3.5 c/wk
Starchy veg.	1.5 c/wk	2.5 c/wk	2.5 c/wk	2.5 c/wk	3 c/wk	3 c/wk	6 c/wk	6 c/wk	7 c/wk	7 c/wk	9 c/wk	9 c/wk
Other veg.	4 c/wk	4.5 c/wk	4.5 c/wk	5.5 c/wk	6.5 c/wk	6.5 c/wk	7 c/wk	7 c/wk	8.5 c/wk	8.5 c/wk	10 c/wk	10 c/wk
Grains	3 oz-eq	4 oz-eq	5 oz-eq	5 oz-eq	6 oz-eq	6 oz-eq	7 oz-eq	8 oz-eq	9 oz-eq	10 oz-eq	10 oz-eq	10 oz-eq
Whole grains	1.5	2	2.5	3	3	3	3.5	4	4.5	5	5	5
Other grains	1.5	2	2.5	2	3	3	3.5	4	4.5	5	5	5
Lean meat and beans	2 oz-eq	3 oz-eq	4 oz-eq	5 oz-eq	5 oz-eq	5.5 oz-eq	6 oz-eq	6.5 oz-eq	6.5 oz-eq	7 oz-eq	7 oz-eq	7 oz-eq
Milk	2 c	2 c	2 c	3 c	3 c	3 c	3 c	3 c	3 c	3 c	3 c	3 c
Oils[1]	15 g	17 g	17 g	22 g	24 g	27 g	29 g	31 g	34 g	36 g	44 g	51 g
Discretionary calorie allowance[2]	165	171	171	132	195	267	290	362	410	426	512	648

[1] Explanataion of oils: Oils (including soft margarine with zero *trans* fat) shown in this table represent the amounts that are added to foods during processing, cooking, or at the table. Oils and soft margarines include vegetable oils and soft vegetable oil table spreads that have no *trans* fats. The amounts of oil listed in this table are not considered to be part of discretionary calories because they are a major source of the vitamin E and polyunsaturated fatty acids, including the essential fatty acids, in the food pattern. In contrast, solid fats are listed separately in Table (I-2) because, compared with oils, they are higher in saturated fatty acids and lower in vitamin E and polyunsaturated and monounsaturated fatty acids, including essential fatty acids. The amounts of each type of fat in the food intake pattern were based on 60% oils and/or soft margarines with no *trans* fats and 40% solid fat. To find teaspoons, divide grams by 5.

[2] Explanation of discretionary calorie allowance: The discretionary calorie allowance is the remaining amount of calories in each food pattern after selecting the specified number of nutrient-dense forms of foods in each food group. The number of discretionary calories assumes that food items in each food group are selected in nutrient-dense forms (that is, forms that are fat-free or low-fat and that contain no added sugars). Solid fat and sugar calories always need to be counted as discretionary calories, as in the following examples:
• The fat in low-fat, reduced fat, or whole milk or milk products or cheese and the sugar and fat in chocolate milk, ice cream, pudding, etc.
• The fat in higher fat meats (e.g., ground beef with more than 5% fat by weight, poultry with skin, higher fat luncheon meats, sausages)
• The sugars added to fruits and fruit juices with added sugars or fruits canned in syrup
• The added fat and/or sugars in vegetables prepared with added fat or sugars
• The added fats and/or sugars in grain products containing higher levels of fats and/or sugars (e.g., sweetened cereals, higher fat crackers, pies and other pastries, cakes, cookies)

Food Patterns to Meet the *Dietary Guidelines for Americans 2005* • I-1

TABLE I·2 DISCRETIONARY CALORIE ALLOWANCE IN THE USDA FOOD GUIDE

This table shows the number of discretionary calories remaining at each calorie level if nutrient-dense foods provide the base of the diet. The table shows an example of how these calories may be divided between solid fats and added sugars. Those trying to lose weight may choose to omit discretionary calories.

Discretionary calories that remain at each level

Food Guide calorie level	1,000	1,200	1,400	1,600	1,800	2,000	2,200	2,400	2,600	2,800	3,000	3,200
Discretionary calories[1]	165	171	171	132	195	267	290	362	410	426	512	648

Example of division of discretionary calories: Solid fats are shown in grams (g); added sugars in grams (g) and teaspoons (tsp).

	1,000	1,200	1,400	1,600	1,800	2,000	2,200	2,400	2,600	2,800	3,000	3,200
Solid fats[2]	11 g	14 g	14 g	11 g	15 g	18 g	19 g	22 g	24 g	24 g	29 g	34 g
Added sugars[3]	20 g	16 g	16 g	12 g	20 g	32 g	36 g	48 g	56 g	60 g	72 g	96 g
	(5 tsp)	(4 tsp)	(4 tsp)	(3 tsp)	(5 tsp)	(8 tsp)	(9 tsp)	(12 tsp)	(14 tsp)	(15 tsp)	(18 tsp)	(24 tsp)

[1] The USDA Food Guide assumes foods will be nutrient dense (that is, fat-free or low-fat and contain no added sugars). Solid fat and sugar calories must always be counted as discretionary calories. See Chapter 2 for more details.

[2] Solid fats: Amounts of solid fats listed in the table represent about 7 to 8% of calories from saturated fat. Solid fats shown in this table represent the amounts of fats that may be added in cooking or at the table, and fats consumed when higher fat items are selected from the food groups. Most oils are not considered to be part of the discretionary calorie allowance because they are a major source of the essential fatty acids and vitamin E in the food pattern.

[3] The amounts of added sugars suggested in the example are NOT specific recommendations for amounts of added sugars to consume, but rather represent the amounts that can be included in each calorie level without over-consuming calories. The suggested amounts of added sugars may be helpful as part of the Food Guide to allow for some sweetened foods or beverages, without exceeding energy needs. This use of added sugars as a calorie balance requires two assumptions: (1) that selections are made from all food groups in accordance with the suggested amounts and (2) that additional fats are used in the amounts shown, which, together with the fats in the core food groups, represent about 27–30% of calories from fat.

The number of daily servings to choose from each food group depends on a person's energy requirement (see Chapter 9).

Food Groups	1,600 Calories	2,000 Calories	2,600 Calories	3,100 Calories	Serving Sizes	Examples and Notes	Significance of Each Food Group to the DASH Eating Plan
Grains[b]	6 servings	7–8 servings	10–11 servings	12–13 servings	1 slice bread, 1 oz dry cereal,[c] ½ cup cooked rice, pasta, or cereal	Whole wheat bread, English muffin, pita bread, bagel, cereals, grits, oatmeal, crackers, unsalted pretzels, and popcorn	Major sources of energy and fiber
Vegetables	3–4 servings	4–5 servings	5–6 servings	6 servings	1 cup raw leafy vegetable, ½ cup cooked vegetable, 6 oz vegetable juice	Tomatoes, potatoes, carrots, green peas, squash, broccoli, turnip greens, collards, kale, spinach, artichokes green beans, lima beans, sweet potatoes	Rich sources of potassium, magnesium, and fiber
Fruits	4 servings	4–5 servings	5–6 servings	6 servings	6 oz fruit juice 1 medium fruit ¼ cup dried fruit ½ cup fresh, frozen, or canned fruit	Apricots, bananas, dates, grapes, oranges, orange juice, grapefruit, grapefruit juice, mangoes, melons, peaches, pineapples, prunes, raisins, strawberries, tangerines	Important sources of potassium, magnesium, and fiber
Low-fat or fat-free dairy foods	2–3 servings	2–3 servings	3 servings	3–4 servings	8 oz milk 1 cup yogurt 1½ oz cheese	Fat-free or low-fat milk, fat-free or low-fat buttermilk, fat-free or low-fat regular or frozen yogurt, low-fat and fat-free cheese	Major sources of calcium and protein
Meat, poultry, fish	1–2 servings	2 or less servings	2 servings	2–3 servings	3 oz cooked meats, poultry, or fish	Select only lean; trim away visible fats; broil, roast, or boil instead of frying; remove skin from poultry	Rich sources of protein and magnesium
Nuts, seeds, legumes	3–4 servings/ week	4–5 servings/ week	1 serving	1 serving	⅓ cup or 1½ oz nuts, 2 Tbsp or ½ oz seeds ½ cup cooked dry beans or peas	Almonds, filberts, mixed nuts, peanuts, walnuts, sunflower seeds, kidney beans, lentils	Rich sources of energy, magnesium, potassium, protein, and fiber
Fat and oils[d]	2 servings	2–3 servings	3 servings	4 servings	1 tsp soft margarine 1 Tbsp low-fat mayonnaise 2 Tbsp light salad dressing, 1 tsp vegetable oil	Soft margarine, low-fat mayonnaise, light salad dressing, vegetable oil (such as olive, corn, canola, or safflower)	DASH has 27 percent of calories as fat (low in saturated fat), including fat in or added to foods
Sweets	0 servings	5 servings/ week	2 servings	2 servings	1 Tbsp sugar 1 Tbsp jelly or jam, ½ oz jelly beans, 8 oz lemonade	Maple syrup, sugar, jelly, jam, fruit-flavored gelatin, jelly beans, hard candy, fruit punch sorbet, ices	Sweets should be low in fat

[a]NIH publication No. 03–4082; Karanja NM et al. *JADA* 8:S19–27, 1999.

[b]Whole grains are recommended for most servings to meet fiber recommendations.

[c]Equals ½–1¼ cups, depending on cereal type. Check the product's Nutrition Facts Label.

[d]Fat content changes serving counts for fats and oils: For example, 1 Tbsp of regular salad dressing equals 1 serving; 1 Tbsp of a low-fat dressing equals ½ serving; 1 Tbsp of a fat-free dressing equals 0 servings.

SOURCE: U.S. Department of Agriculture and U.S. Department of Health and Human Services, *Dietary Guidelines for Americans 2005*, 6th edition, available online at www.healthierus.gov or call (888) 878-3256.

Glossary

A

absorb to take in, as nutrients are taken into the intestinal cells after digestion; the main function of the digestive tract with respect to nutrients.

acceptable daily intake (ADI) the estimated amount of sweetener that can be consumed daily over a person's lifetime without any adverse effects.

accredited approved; in the case of medical centers or universities, certified by an agency recognized by the U.S. Department of Education.

acesulfame (AY-sul-fame) **potassium**, also called **acesulfame-K** a zero-calorie sweetener approved by the FDA and Health Canada.

acetaldehyde (ass-et-AL-deh-hide) a substance to which ethanol is metabolized on its way to becoming harmless waste products that can be excreted.

acid reducers prescription and over-the-counter drugs that reduce the acid output of the stomach; effective for treating severe, persistent forms of heartburn but not for neutralizing acid already present. Side effects are frequent and include diarrhea, other gastrointestinal complaints, and reduction of the stomach's capacity to destroy alcohol, thereby producing higher-than-expected blood alcohol levels from each drink. Also called *acid controllers.*

acid-base balance equilibrium between acid and base concentrations in the body fluids.

acidosis (acid-DOH-sis) the condition of excess acid in the blood, indicated by a below-normal pH (osis means "too much in the blood").

acids compounds that release hydrogens in a watery solution.

acne chronic inflammation of the skin's follicles and oil-producing glands, which leads to an accumulation of oils inside the ducts that surround hairs; usually associated with the maturation of young adults.

acrylamide (ah-KRILL-ah-mide) a chemical produced in carbohydrate-rich foods, such as potatoes and grains, when cooked at high temperatures. A known animal carcinogen, acrylamide is also toxic to the nervous system of both animals and humans. Also used in manufacturing and construction.

acupuncture (ak-you-punk-chur) a technique that involves piercing the skin with long thin needles at specific anatomical points to relieve pain or illness. Acupuncture sometimes uses heat, pressure, friction, suction, or electromagnetic energy to stimulate the points.

added sugars sugars and syrups added to a food for any purpose, such as to add sweetness or bulk or to aid in browning (baked goods). Also called *carbohydrate sweeteners,* they include glucose, fructose, corn syrup, concentrated fruit juice, and other sweet carbohydrates.

additives substances that are added to foods but are normally not consumed by themselves as foods.

adequacy the dietary characteristic of providing all of the essential nutrients, fiber, and energy in amounts sufficient to maintain health and body weight.

adipose tissue the body's fat tissue, consisting of masses of fat-storing cells and blood vessels to nourish them.

advertorials lengthy advertisements in newspapers and magazines that read like feature articles but are written for the purpose of touting the virtues of products and may or may not be accurate.

aerobic (air-ROH-bic) requiring oxygen. Aerobic activity strengthens the heart and lungs by requiring them to work harder than normal to deliver oxygen to the tissues.

agribusiness agriculture practiced on a massive scale by large corporations owning vast acreages and employing intensive technological, fuel, and chemical inputs.

AIDS acquired immune deficiency syndrome; caused by infection with human immunodeficiency virus (HIV), which is transmitted primarily by sexual contact, contact with infected blood, needles shared among drug users, or fluids transferred from an infected mother to her fetus or infant.

alcohol dehydrogenase (dee-high-DRAH-gen-ace) **(ADH)** an enzyme system that breaks down alcohol. The antidiuretic hormone listed below is also abbreviated ADH.

alcoholism a dependency on alcohol marked by compulsive uncontrollable drinking with negative effects on physical health, family relationships, and social health.

alcohol-related birth defects (ARBD) malformations in the skeletal and organ systems (heart, kidneys, eyes, ears) associated with prenatal alcohol exposure.

alcohol-related neurodevelopmental disorder (ARND) behavioral, cognitive, or central nervous system abnormalities associated with prenatal alcohol exposure.

alitame a noncaloric sweetener formed from the amino acids L-aspartic acid and L-alanine. In the United States, the FDA is considering its approval.

alkalosis (al-kah-LOH-sis) the condition of excess base in the blood, indicated by an above-normal blood pH (alkalinity—*alka* means "base"; *osis* means "too much in the blood").

allergy an immune reaction to a foreign substance, such as a component of food. Also called *hypersensitivity* by researchers.

aloe a tropical plant with widely claimed value as a topical treatment for minor skin injury. Some scientific evidence supports this claim; evidence against its use in severe wounds also exists.

alpha-lactalbumin (lact-AL-byoo-min) the chief protein in human breast milk. The chief protein in cow's milk is *casein* (CAY-seen).

alternative (low-input, or sustainable) agriculture agriculture practiced on a small scale that uses individualized approaches that vary with local conditions so as to minimize technological, fuel, and chemical inputs.

alternative therapy systems of theory and practice that have emerged independently of conventional biomedical approaches. Examples include acupuncture, biofeedback, chiropractic, faith healing, and many others. Also called *alternative medical systems*

American Dietetic Association (ADA) the professional organization of dietitians in the United States. The Canadian equivalent is the Dietitians of Canada (DC), which operates similarly.

amine (a-MEEN) **group** the nitrogen-containing portion of an amino acid.

amino (a-MEEN-o) **acids** the building blocks of protein. Each has an amine group at one end, an acid group at the other, and a distinctive side chain.

amino acid chelates (KEY-lates) compounds of minerals (such as calcium) combined with amino acids in a form that favors their absorption. A *chelating agent* is a molecule that surrounds another molecule and can then either promote or prevent its movement from place to place (*chele* means "claw").

amino acid pools amino acids dissolved in the body's fluids that provide cells with ready raw materials from which to build new proteins or other molecules.

amniotic (AM-nee-OTT-ic) **sac** the "bag of waters" in the uterus in which the fetus floats.

anabolic steroid hormones chemical messengers related to the male sex hormone testosterone that stimulate building up of body tissues (*anabolic* means "promoting growth"; *sterol* refers to compounds chemically related to cholesterol).

anaerobic (AN-air-ROH-bic) not requiring oxygen. Anaerobic activity may require strength but does not work the heart and lungs very hard for a sustained period.

anaphylactic (an-ah-feh-LACK-tick) **shock** a life-threatening whole-body allergic reaction to an offending substance.

androstenedione (AN-droh-STEEN-dee-own) a precursor of testosterone that elevates both testosterone and estrogen in the blood of both males and females. Often called *andro*, it is sold with claims of producing increased muscle strength, but controlled studies disprove such claims.

anecdotal evidence information based on interesting and entertaining, but not scientific, personal accounts of events.

anemia the condition of inadequate or impaired red blood cells; a reduced number or volume of red blood cells along with too little hemoglobin in the blood. The red blood cells may be immature and, therefore, too large or too small to function properly. Anemia can result from blood loss, excessive red blood cell destruction, defective red blood cell formation, and many nutrient deficiencies. Anemia is not a disease, but a symptom of another problem; its name literally means "too little blood."

anencephaly (an-en-SEFF-ah-lee) an uncommon and always fatal neural tube defect in which the brain fails to form.

aneurysm (AN-you-rism) the ballooning out of an artery wall at a point that is weakened by deterioration.

anorexia nervosa an eating disorder characterized by a refusal to maintain a minimally normal body weight, self-starvation to the extreme, and a disturbed perception of body weight and shape; seen (usually) in teenage girls and young women (*anorexia* means "without appetite"; *nervos* means "of nervous origin").

antacids medications that react directly and immediately with the acid of the stomach, neutralizing it. Antacids are most suitable for treating occasional heartburn.

antibodies (AN-te-bod-ees) large proteins of the blood, produced by the immune system in response to an invasion of the body by foreign substances (antigens). Antibodies combine with and inactivate the antigens.

anticarcinogens compounds in foods that act in any of several ways to oppose the formation of cancer.

antidiuretic (AN-tee-dye-you-RET-ick) **hormone (ADH)** a hormone produced by the pituitary gland in response to dehydration (or a high sodium concentration in the blood). It stimulates the kidneys to reabsorb more water and so to excrete less. (This hormone should not be confused with the enzyme alcohol dehydrogenase, which is also abbreviated ADH.)

antigen a substance foreign to the body that elicits the formation of antibodies or an inflammation reaction from immune system cells. Food antigens are usually large proteins.

antimicrobial agents preservatives that prevent spoilage by mold or bacterial growth. Familiar examples are acetic acid (vinegar) and sodium chloride (salt). Others are benzoic, propionic, and sorbic acids; nitrites and nitrates; and sulfur dioxide.

antioxidant nutrients vitamins and minerals that oppose the effects of oxidants on human physical functions. The antioxidant vitamins are vitamin E, vitamin C, and beta-carotene. The mineral selenium also participates in antioxidant activities.

antioxidants (anti-OX-ih-dants) compounds that protect other compounds from damaging reactions involving oxygen by themselves reacting with oxygen (*anti* means "against"; *oxy* means "oxygen"). *Oxidation* is a potentially damaging effect of normal cell chemistry involving oxygen.

aorta (ay-OR-tuh) the large, primary artery that conducts blood from the heart to the body's smaller arteries.

Apgar score a system of scoring an infant's physical condition right after birth. Heart rate, respiration, muscle tone, response to stimuli, and color are ranked 0, 1, or 2. A low score indicates that medical attention is required to facilitate survival.

appendicitis inflammation and/or infection of the appendix, a sac protruding from the intestine.

appetite the psychological desire to eat; a learned motivation and a positive sensation that accompanies the sight, smell, or thought of appealing foods.

appliance thermometer a thermometer that verifies the temperature of an appliance. An *oven thermometer* verifies that the oven is heating properly; a *refrigerator/freezer thermometer* tests for proper refrigerator (40°F, or 4°C) or freezer temperature (0°F, or 217°C).

aquifers underground rock formations containing water that can be drawn to the surface for use.

arachidonic (ah-RACK-ih-DON-ik) **acid** an omega-6 fatty acid derived from linoleic acid.

arginine a nonessential amino acid falsely promoted as enhancing the secretion of human growth hormone, the breakdown of fat, and the development of muscle.

aristolochic acid a Chinese herb ingredient known to attack the kidneys and to cause cancer; U.S. consumers have required kidney transplants and must take lifelong anti-rejection medication after use. Banned by the FDA but available in supplements sold on the Internet.

arsenic a poisonous metallic element. In trace amounts, arsenic is believed to be an essential nutrient in some animal species. Arsenic is often added to insecticides and weed killers and, in tiny amounts, to certain animal drugs.

arteries blood vessels that carry blood containing fresh oxygen supplies from the heart to the tissues.

artesian water water drawn from a well that taps a confined aquifer in which the water is under pressure.

arthritis a usually painful inflammation of joints caused by many conditions, including infections, metabolic disturbances, or injury; usually results in altered joint structure and loss of function.

artificial colors certified food colors, added to enhance appearance. (*Certified* means approved by the FDA.) Vegetable dyes are extracted from vegetables such as beta-carotene from carrots. Food colors are a mix of vegetable dyes and synthetic dyes approved by the FDA for use in food.

artificial fats zero-energy fat replacers that are chemically synthesized to mimic the sensory and cooking qualities of naturally occurring fats, but are totally or partially resistant to digestion. Also called *fat analogues*.

artificial flavors, flavor enhancers chemicals that mimic natural flavors and those that enhance flavor.

ascorbic acid one of the active forms of vitamin C (the other is *dehydroascorbic* acid); an antioxidant nutrient.

aspartame a compound of phenylalanine and aspartic acid that tastes like the sugar sucrose but is much sweeter. It is used in both the United States and Canada.

atherosclerosis (ath-er-oh-scler-OH-sis) the most common form of cardiovascular disease; characterized by plaques along the inner walls of the arteries (*scleros* means "hard"; *osis* means "too much"). The term *arteriosclerosis* refers to all forms of hardening of the arteries and includes some rare diseases.

atrophy (AT-tro-fee) a decrease in size (for example, of a muscle) because of disuse.

B

baby water ordinary bottled water treated with ozone to make it safe but not sterile.

balance the dietary characteristic of providing foods of a number of types in proportion to each other, such that foods rich in some nutrients do not crowd out the diet foods that are rich in other nutrients. Also called *proportionality*.

balance study a laboratory study in which a person is fed a controlled diet and the intake and excretion of a nutrient are measured. Balance studies are valid only for nutrients like calcium (chemical elements) that do not change while they are in the body.

basal metabolic rate (BMR) the rate at which the body uses energy to support its basal metabolism.

basal metabolism the sum total of all the involuntary activities that are necessary to sustain life, including circulation, respiration, temperature maintenance, hormone secretion, nerve activity, and new tissue synthesis, but excluding digestion and voluntary activities. Basal metabolism is the largest component of the average person's daily energy expenditure.

bases compounds that accept hydrogens from solutions.

basic foods milk and milk products; meats and similar foods such as fish and poultry; vegetables, including dried beans and peas; fruits; and grains. These foods are generally considered to form the basis of a nutritious diet. Also called *whole foods*.

B-cells lymphocytes that produce antibodies. *B* stands for bursa, an organ in the chicken where B-cells were first identified.

bee pollen a product consisting of bee saliva, plant nectar, and pollen that confers no benefit on athletes and may cause an allergic reaction in individuals sensitive to it.

beer belly central-body fatness associated with alcohol consumption.

behavior modification alteration of behavior using methods based on the theory that actions can be controlled by manipulating the environmental factors that cue, or trigger, the actions.

belladonna any part of the deadly nightshade plant; a fatal poison.

beriberi (berry-berry) the thiamin-deficiency disease; characterized by loss of sensation in the hands and feet, muscular weakness, advancing paralysis, and abnormal heart action.

best if used by: Specifies the last date the food will be of the highest quality. After this date, quality is expected to diminish, although the food may still be safe for consumption if it has been handled and stored properly. Also called *freshness date* or *quality assurance date*.

beta-carotene an orange pigment with antioxidant activity; a vitamin A precursor made by plants and stored in human fat tissue.

bicarbonate a common alkaline chemical; a secretion of the pancreas; also, the active ingredient of baking soda.

bile a cholesterol-containing digestive fluid made by the liver, stored in the gallbladder, and released into the small intestine when needed. It emulsifies fats and oils to ready them for enzymatic digestion.

binge drinkers people who drink four or more drinks in a short period.

binge eating disorder an eating disorder whose criteria are similar to those of bulimia nervosa, excluding purging or other compensatory behaviors.

bioaccumulation the accumulation of a contaminant in the tissues of living things at higher and higher concentrations along the food chain.

bioelectrical impedance (im-PEE-dense) a technique for measuring body fatness by measuring the body's electrical conductivity.

biotechnology the science of manipulating biological systems or organisms to modify their products or components or create new products; more properly called *genetic engineering* or *rDNA technology.*

bioterrorism the intentional spreading of disease-causing organisms or agricultural pests as a political weapon to produce fear and intimidate others.

biotin (BY-o-tin) a B vitamin; a coenzyme necessary for fat synthesis and other metabolic reactions.

bladder the sac that holds urine until time for elimination.

bleaching agents substances used to whiten foods such as flour and cheese. Peroxides are examples.

blind experiment an experiment in which the subjects do not know whether they are members of the experimental group or the control group. In a *double-blind experiment*, neither the subjects nor the researchers know to which group the members belong until the end of the experiment.

blood the fluid of the cardiovascular system; composed of water, red and white blood cells, other formed particles, nutrients, oxygen, and other constituents.

body composition the proportions of muscle, bone, fat, and other tissue that make up a person's total body weight.

body mass index (BMI) an indicator of obesity or underweight, calculated by dividing the weight of a person by the square of the person's height.

body system a group of related organs that work together to perform a function. Examples are the circulatory system, respiratory system, and nervous system.

bone density a measure of bone strength; the degree of mineralization of the bone matrix.

bone meal or **powdered bone** crushed or ground bone preparations intended to supply calcium to the diet. Calcium from bone is not well absorbed and is often contaminated with toxic materials such as arsenic, mercury, lead, and cadmium.

boron a nonessential mineral that is promoted as a "natural" steroid replacement.

bottled water drinking water sold in bottles.

botulism an often-fatal food poisoning caused by botulinum toxin, a toxin produced by the *Clostridium botulinum* bacterium that grows without oxygen in nonacidic canned foods.

bovine somatotropin (bST) (so-MAT-ah-TROW-pin) growth hormone of cattle, which can be produced for agricultural use by genetic engineering. Also called *bovine growth hormone (bGH).*

bovine spongiform encephalopathy (BOW-vine SPON-jih-form en-SEH-fell-AH-path-ee) **(BSE)** an often-fatal illness of cattle affecting the nerves and brain. Also called *mad cow disease.*

bran the protective fibrous coating around a grain; the chief fiber donator of a grain.

branched-chain amino acids (BCAA) the amino acids leucine, isoleucine, and valine, which are present in large amounts in skeletal muscle tissue; supplements are falsely promoted as necessary for exercising muscles.

brewer's yeast a preparation of yeast cells, containing a concentrated amount of B vitamins and some minerals; falsely promoted as an energy booster.

broccoli sprouts the sprouted seed of *Brassica italica,* or the common broccoli plant; believed to be a functional food by virtue of its high phytochemical content.

brown bread bread containing ingredients such as molasses that lend a brown color; may be made with any kind of flour, including white flour.

brown sugar white sugar with molasses added, 95% pure sucrose.

buffers molecules that can help to keep the pH of a solution from changing by gathering or releasing H ions.

bulimia (byoo-LEEM-ee-uh) **nervosa** recurring episodes of binge eating combined with a morbid fear of becoming fat; usually followed by self-induced vomiting or purging.

butyrate (BYOO-tier-ate) a small fat fragment produced by the fermenting action of bacteria on viscous, soluble fibers; the preferred energy source for the colon cells.

C

caffeine water bottled water with caffeine added.

caffeine a stimulant that may produce alertness and reduced reaction time in small doses, but creates fluid losses with a larger dose. Overdoses cause headaches, trembling, an abnormally fast heart rate, and other undesirable effects.

CAGE questions a set of four questions often used internationally for initial screening for alcoholism.

calcium compounds the simplest forms of purified calcium. They include calcium carbonate, citrate, gluconate, hydroxide, lactate, malate, and phosphate. These supplements vary in the amount of calcium they contain, so read the labels carefully. A 500-milligram tablet of calcium gluconate may provide only 45 milligrams of calcium, for example.

caloric effect the drop in cancer incidence seen whenever intake of food energy (calories) is restricted.

calorie control control of energy intake; a feature of a sound diet plan.

calorie free fewer than 5 calories per serving.

calories units of energy. Strictly speaking, the unit used to measure the energy in foods is a kilocalorie (*kcalorie* or *Calorie*): it is the amount of heat energy necessary to raise the temperature of a kilogram (a liter) of water 1 degree Celsius. This book follows the common practice of using the lowercase term *calorie* (abbreviated *cal*) to mean the same thing.

cancer a disease in which cells multiply out of control and disrupt normal functioning of one or more organs.

canning a method of preserving food by killing all microorganisms present in the food and then sealing out air. The food, container, and lid are heated until sterile; as the food cools, the lid makes an airtight seal, preventing contamination.

capillaries minute, weblike blood vessels that connect arteries to veins and permit transfer of materials between blood and tissues.

carbohydrate loading a regimen of moderate exercise, followed by eating a high-carbohydrate diet, that enables muscles to temporarily store glycogen beyond their normal capacity; also called *glycogen loading* or *glycogen super-compensation.*

carbohydrates compounds composed of single or multiple sugars. The name means "carbon and water," and a chemical shorthand for carbohydrate is CHO, signifying carbon (C), hydrogen (H), and oxygen (O).

carbonated water water that contains carbon dioxide gas, either naturally occurring or added, that causes bubbles to form in it; also called *bubbling* or *sparkling water.* Seltzer, soda, and tonic waters are legally soft drinks and are not regulated as water.

carcinogen (car-SIN-oh-jen) a cancer-causing substance (*carcin* means "cancer"; *gen* means "gives rise to").

carcinogenesis the origination or beginning of cancer.

cardiac output the volume of blood discharged by the heart each minute.

cardiorespiratory endurance the ability to perform large-muscle dynamic exercise of moderate-to-high intensity for prolonged periods.

cardiovascular disease (CVD) disease of the heart and blood vessels; disease of the arteries of the heart is called *coronary heart disease (CHD).*

carnitine a nonessential nutrient that functions in cellular activities. Carnitine is claimed to "burn" fat and spare glycogen during athletic events, but it does neither.

carotenoid (CARE-oh-ten-oyd) a member of a group of pigments in foods that range in color from light yellow to reddish orange and are chemical relatives of beta-carotene. Many with a degree of vitamin A activity in the body.

carpal tunnel syndrome a pinched nerve at the wrist, causing pain or numbness in the hand. It is often caused by repetitive motion of the wrist.

carrying capacity the total number of living organisms that a given environment can support without deteriorating in quality.

case studies studies of individuals. In clinical settings, researchers can observe treatments and their apparent effects. To prove that a treatment has produced an effect requires simultaneous observation of an untreated similar subject (a *case control*).

cat's claw an herb from the rain forests of Brazil and Peru; claimed, but not proved, to be an "all-purpose" remedy.

cataracts (CAT-uh-racts) clouding of the lens of the eye that can lead to blindness. Cataracts can be caused by injury, viral infection, toxic substances, genetic disorders, and, possibly, some nutrient deficiencies or imbalances.

cathartic a strong laxative.

CDC (Centers for Disease Control and Prevention) a branch of the Department of Health and Human Services that is responsible for monitoring food-borne diseases.

cell differentiation (dih-fer-en-she-AY-shun) the process by which immature cells are stimulated to mature and gain the ability to perform functions characteristic of their cell type.

cell salts a mineral preparation supposedly prepared from living cells. No scientific evidence supports benefits from such preparations.

cells the smallest units in which independent life can exist. All living things are single cells or organisms made of cells.

cellulite a term popularly used to describe dimpled fat tissue on the thighs and buttocks; not recognized in science.

central obesity excess fat in the abdomen and around the trunk.

certified lactation consultant a health-care provider, often a registered nurse or a registered dietitian, with specialized training and certification in breast and infant anatomy and physiology who teaches the mechanics of breastfeeding to new mothers.

certified organic foods foods meeting strict USDA production regulations, including prohibition of most synthetic pesticides, herbicides, fertilizers, drugs, and preservatives, as well as genetic engineering and irradiation.

cesarean (see-ZAIR-ee-un) **section** surgical childbirth, in which the infant is taken through an incision in the woman's abdomen.

chamomile flowers that may provide some limited medical value in soothing menstrual, intestinal, and stomach discomforts.

chaparral an herbal product made from ground leaves of the creosote bush and sold in tea or capsule form; supposedly, this herb has antioxidant effects, delays aging, "cleanses" the bloodstream, and treats skin conditions—all unproven claims. Chaparral has been found to cause acute toxic hepatitis, a severe liver illness. Deaths reported.

chelating (KEE-late-ing) **agents** molecules that attract or bind with other molecules and are therefore useful in either preventing or promoting movement of substances from place to place.

chlorophyll the green pigment of plants that captures energy from sunlight for use in photosynthesis.

cholesterol (koh-LESS-ter-all) a member of the group of lipids known as sterols; a soft, waxy substance made in the body for a variety of purposes and also found in animal-derived foods.

cholesterol free less than 2 mg of cholesterol *and* 2 g or less saturated fat and *trans* fat combined per serving.

choline (KOH-leen) a nonessential nutrient used to make the phospholipid lecithin and other molecules.

chromium picolinate a trace element supplement; falsely promoted to increase lean body mass, enhance energy, and burn fat.

chronic diseases long-duration degenerative diseases characterized by deterioration of the body organs. Examples include heart disease, cancer, and diabetes.

chylomicrons (KYE-low-MY-krons) clusters formed when lipids from a meal are combined with carrier proteins in the cells of the intestinal lining. Chylomicrons transport food fats through the watery body fluids to the liver and other tissues.

chyme (KIME) the fluid resulting from the actions of the stomach upon a meal.

cirrhosis (seer-OH-sis) advanced liver disease, often associated with alcoholism, in which liver cells have died, hardened, turned an orange color, and permanently lost their function.

clone an individual created asexually from a single ancestor, such as a plant grown from a single stem cell; a group of genetically identical individuals descended from a single common ancestor, such as a colony of bacteria arising from a single bacterial cell; in genetics, a replica of a segment of DNA, such as a gene, produced by genetic engineering.

coenzyme (co-EN-zime) a small molecule that works with an enzyme to promote the enzyme's activity. Many coenzymes have B vitamins as part of their structure (*co* means "with").

coenzyme Q-10 an enzyme made by cells and important for its role in energy metabolism. With diminished coenzyme Q-10 function, oxidative stress increases, as may occur in aging. Preliminary research suggests that it may be of value for treating certain conditions; toxicity in animals appears to be low. No safe intake levels for human beings have been established.

cognitive therapy psychological therapy aimed at changing undesirable behaviors by changing underlying thought processes contributing to these behaviors; in anorexia, a goal is to replace false beliefs about body weight, eating, and self-worth with health-promoting beliefs.

collagen (COLL-a-jen) the chief protein of most connective tissues, including scars, ligaments, and tendons, and the underlying matrix on which bones and teeth are built.

colon the large intestine.

colostrum (co-LAHS-trum) a milklike secretion from the breasts during the first day or so after delivery before milk appears; rich in protective factors.

comfrey leaves and roots of the comfrey plant; believed, but not proved, to promote cell proliferation. Toxic to the liver in doses ordinarily used. Deaths reported.

complementary proteins two or more proteins whose amino acid assortments complement each other in such a way that the essential amino acids missing from one are supplied by the other.

complex carbohydrates long chains of sugar units arranged to form starch or fiber; also called *polysaccharides*.

concentrated fruit juice sweetener a concentrated sugar syrup made from dehydrated, deflavored fruit juice, commonly grape juice; used to sweeten products that can then claim to be "all fruit."

conditionally essential amino acid an amino acid that is normally nonessential, but must be supplied by the diet in special circumstances when the need for it exceeds the body's ability to produce it.

confectioner's sugar finely powdered sucrose, 99.9% pure.

congeners (CON-jen-ers) chemical substances other than alcohol that account for some of the physiological effects of alcoholic beverages, such as appetite, taste, and aftereffects.

conjugated linoleic acid (CLA) a type of fat in butter, milk, and other dairy products believed by some to have biological activity in the body. Not a phytochemical, but a biologically active chemical produced by animals.

constipation difficult, incomplete, or infrequent bowel movements, associated with discomfort in passing dry, hardened feces from the body.

contaminant any substance occurring in food by accident; any food constituent that is not normally present.

control group a group of individuals who are similar in all possible respects to the group being treated in an experiment but who receive a sham treatment instead of the real one. Also called *control subjects*. See also *experimental group* and *intervention studies*.

corn sweeteners corn syrup and sugar solutions derived from corn.

corn syrup a syrup, mostly glucose, partly maltose, produced by the action of enzymes on cornstarch. *High-fructose corn syrup (HFCS)* is a mixture of fructose, glucose (dextrose), and maltose.

cornea (KOR-nee-uh) the hard, transparent membrane covering the outside of the eye.

correlation the simultaneous change of two factors, such as the increase of weight with increasing height (a *direct* or *positive* correlation) or the decrease of cancer incidence with increasing fiber intake (an *inverse* or *negative* correlation). A correlation between two factors suggests that one may cause the other, but does not rule out the possibility that both may be caused by chance or by a third factor.

correspondence school a school that offers courses and degrees by mail. Some correspondence schools are accredited; others are *diploma mills*.

cortex the outermost layer of something. The brain's cortex is the part of the brain where conscious thought takes place.

cortical bone the ivorylike outer bone layer that forms a shell surrounding trabecular bone and that comprises the shaft of a long bone.

creatine a nitrogen-containing compound that combines with phosphate to burn a high-energy compound stored in muscle. Claims that creatine safely enhances energy and stimulates muscle growth are unconfirmed but often reported by users; digestive side effects are common.

cretinism (CREE-tin-ism) severe mental and physical retardation of an infant caused by the mother's iodine deficiency during pregnancy.

critical period a finite period during development in which certain events may occur that will have irreversible effects on later developmental stages. A critical period is usually a period of cell division in a body organ.

cross-contamination the contamination of a food through exposure to utensils, hands, or other surfaces that were previously in contact with a contaminated food.

cruciferous vegetables vegetables with cross-shaped blossoms—the cabbage family. Their intake is associated with low cancer rates in human populations. Examples include broccoli, brussels sprouts, cabbage, cauliflower, rutabagas, and turnips.

cuisines styles of cooking.

cyclamate a zero-calorie sweetener under consideration for use in the United States and used with restrictions in Canada.

D

Daily Values nutrient standards that are printed on food labels. Based on nutrient and energy recommendations for a general 2,000-calorie diet, they allow consumers to compare the nutrient and energy contents of packaged foods.

degenerative diseases chronic, irreversible diseases characterized by degeneration of body organs due in part to such personal lifestyle elements as poor food choices, smoking, alcohol use, and lack of physical activity. Also called *lifestyle diseases, chronic diseases,* or the *diseases of old age.*

dehydration loss of water. The symptoms progress rapidly, from thirst to weakness to exhaustion and delirium, and end in death.

denaturation the irreversible change in a protein's shape brought about by heat, acids, bases, alcohol, salts of heavy metals, or other agents.

dental caries decay of the teeth (*caries* means "rottenness").

desiccated liver dehydrated liver powder that supposedly contains all the nutrients found in liver in concentrated form; possibly not dangerous, but has no particular nutritional merit and is considerably more expensive than fresh liver.

dextrose an older name for glucose.

DHEA (dehydroepiandrosterone) a hormone made in the adrenal glands that serves as a precursor to the male hormone testosterone; recently banned by the FDA because it poses the risk of life-threatening diseases, including cancer. Falsely promoted to burn fat, build muscle, and slow aging.

diabetes (dye-uh-BEET-eez) a disease (technically termed *diabetes mellitus*) characterized by elevated blood glucose and inadequate or ineffective insulin, which impairs a person's ability to regulate blood glucose normally.

dialysis (die-AL-ih-sis) in kidney disease, treatment of the blood to remove toxic substances or metabolic wastes; more properly, *hemodialysis*, meaning "dialysis of the blood."

diarrhea frequent, watery bowel movements usually caused by diet, stress, or irritation of the colon. Severe, prolonged diarrhea robs the body of fluid and certain minerals, causing dehydration and imbalances that can be dangerous if left untreated.

diastolic (dye-as-TOL-ik) **pressure** the second figure in a blood pressure reading (the "lubb" of the heartbeat is heard), which reflects the arterial pressure when the heart is between beats.

diet the foods (including beverages) a person usually eats and drinks.

dietary antioxidant (anti-OX-ih-dant) a substance in food that significantly decreases the damaging effects of reactive compounds, such as reactive forms of oxygen and nitrogen on tissue functioning (*anti* means "against"; *oxy* means "oxygen").

dietary folate equivalent (DFE) a unit of measure expressing the amount of folate available to the body from naturally occurring sources. The measure mathematically equalizes the difference in absorption between less absorbable food folate and highly absorbable synthetic folate added to enriched foods and found in supplements.

Dietary Reference Intakes (DRI) a set of four lists of values for measuring the nutrient intakes of healthy people in the United States and Canada. The four lists are Estimated Average Requirements (EAR), Recommended Dietary Allowances (RDA), Adequate Intakes (AI), and Tolerable Upper Intake Levels (UL).

dietary supplement a product, other than tobacco, that is added to the diet and contains one of the following ingredients: a vitamin, mineral, herb, botanical (plant extract), amino acid, metabolite, constituent, or extract, or a combination of any of these ingredients.

dietetic technician a person who has completed a two-year acadmic degree from an accredited college or university and an approved dietetic technician program. A **dietetic technician, registered** (DTR) has also passed a national examination and maintains registration through continuing professional education.

dietitian a person trained in nutrition, food science, and diet planning. See also *registered dietitian.*

digest to break molecules into smaller molecules; a main function of the digestive tract with respect to food.

digestive system the body system composed of organs that break down complex food particles into smaller, absorbable products. The *digestive tract* and *alimentary canal* are names for the tubular organs that extend from the mouth to the anus. The whole system, including the pancreas, liver, and gallbladder, is sometimes called the *gastrointestinal*, or *GI*, system.

dipeptides (dye-PEP-tides) protein fragments that are two amino acids long (*di* means "two").

diploma mill an organization that awards meaningless degrees without requiring its students to meet educational standards.

disaccharides pairs of single sugars linked together (*di* means "two").

discretionary calorie allowance the balance of calories remaining in a person's energy allowance after accounting for the number of calories needed to meet recommended nutrient intakes through consumption of nutrient-dense foods.

distilled water water that has been vaporized and recondensed, leaving it free of dissolved minerals.

diuretic (dye-you-RET-ic) a compound, usually a medication, causing increased urinary water excretion; a "water pill."

diverticula (dye-ver-TIC-you-la) sacs or pouches that balloon out of the intestinal wall, caused by weakening of the muscle layers that encase the intestine. The painful inflammation of one or more of these diverticula is known as *diverticulitis.*

DNA and **RNA (deoxyribonucleic acid** and **ribonucleic acid)** the genetic materials of cells necessary in protein synthesis.

dolomite a compound of minerals (calcium magnesium carbonate) found in limestone and marble. Dolomite is powdered and is sold as a calcium-magnesium supplement, but may be contaminated with toxic minerals, is not well absorbed, and interacts adversely with absorption of other essential minerals.

drink a dose of any alcoholic beverage that delivers 1/2 ounce of pure ethanol.

drug any substance that when taken into a living organism may modify one or more of its functions.

drying a method of preserving food by removing sufficient water from the food to inhibit microbial growth.

dual energy X-ray absorptiometry (ab-sorp-tee-OM-eh-tree) a noninvasive method of determining total body fat, fat distribution, and bone density by passing two low-dose X-ray beams through the body. Also used in evaluation of osteoporosis. Abbreviated *DEXA.*

dysentery (DISS-en-terry) an infection of the digestive tract that causes diarrhea.

E

eating disorder a disturbance in eating behavior that jeopardizes a person's physical or psychological health.

echinacea (EK-eh-NAY-see-ah) an herb popular before the advent of antibiotics for its "anti-infectious" properties and as an all-purpose remedy, especially for colds and allergy and for healing of wounds. Research is mixed on these claims. An insecticidal property opens questions about safety. Also called *cone-flower.*

edema (eh-DEEM-uh) swelling of body tissue caused by leakage of fluid from the blood vessels; seen in protein deficiency (among other conditions).

eicosanoids (eye-COSS-ah-noyds) biologically active compounds that regulate body functions.

electrolytes compounds that partly dissociate in water to form ions, such as the potassium ion (K^+) and the chloride ion (Cl^-).

electrons parts of an atom; negatively charged particles. Stable atoms (and molecules, which are made of atoms) have even numbers of electrons in pairs. An atom or molecule with an unpaired electron is an unstable *free radical.*

elemental diets diets composed of purified ingredients of known chemical composition; intended to supply all essential nutrients to people who cannot eat foods.

embolism an embolus that causes sudden closure of a blood vessel.

embolus (EM-boh-luss) a thrombus that breaks loose and travels through the blood vessels (*embol* means "to insert").

embryo (EM-bree-oh) the stage of human gestation from the third to the eighth week after conception.

emergency kitchens programs that provide prepared meals to be eaten on-site; often called *soup kitchens.*

emetic (em-ETT-ic) an agent that causes vomiting.

emulsification the process of mixing lipid with water by adding an emulsifier.

emulsifier (ee-MULL-sih-fire) a compound with both water-soluble and fat-soluble portions that can attract fats and oils into water, combining them.

endorphins (en-DOOR-fins), **endogenous opiates** compounds of the brain whose actions mimic those of opiate drugs (morphine, heroin) in reducing pain and producing pleasure. In appetite control, endorphins are released on seeing, smelling, or tasting delicious food and are believed to enhance the drive to eat or continue eating.

endosperm the bulk of the edible part of a grain, the starchy part.

energy the capacity to do work. The energy in food is chemical energy; it can be converted to mechanical, electrical, heat, or other forms of energy in the body. Food energy is measured in calories, defined on page 5.

energy density a measure of the energy provided by a food relative to its weight (calories per gram).

"energy drinks" sugar-sweetened beverages with supposedly "ergogenic" ingredients, such as vitamins, amino acids, caffeine, guarana, carnitine, ginseng, and others. The drinks are not regulated by the FDA, are often high in caffeine, and may still contain the dangerous banned stimulant ephedrine.

energy-yielding nutrients the nutrients the body can use for energy. They may also supply building blocks for body structures.

enriched foods and **fortified foods** foods to which nutrients have been added. If the starting material is a whole, basic food such as milk or whole grain, the result may be highly nutritious. If the starting material is a concentrated form of sugar or fat, the result may be less nutritious.

enriched, fortified refers to the addition of nutrients to a refined food product. As defined by U.S. law, these terms mean that specified levels of thiamin, riboflavin, niacin, folate, and iron have been added to refined grains and grain products. The terms *enriched* and *fortified* can refer to the addition of more nutrients than just these five; read the label.

enterotoxins poisons that act upon mucous membranes, such as those of the digestive tract.

environmental tobacco smoke (ETS) the combination of exhaled smoke (mainstream smoke) and smoke from lighted cigarettes, pipes, or cigars (sidestream smoke) that enters the air and may be inhaled by other people.

enzymes (EN-zimes) protein catalysts. A catalyst is a compound that facilitates a chemical reaction without itself being altered in the process.

EPA (Environmental Protection Agency) the federal agency that is responsible for regulating pesticides and establishing water quality standards.

EPA, DHA eicosapentaenoic (EYE-cossa-PENTA-ee-NO-ick) acid, docosahexaenoic (DOE-cossa-HEXA-ee-NO-ick) acid; omega-3 fatty acids made from linolenic acid in the tissues of fish.

ephedra (ephedrine) a dangerous and sometimes lethal "herbal" supplement previously sold for weight loss, muscle building, athletic performance, and other purposes. Now banned by the FDA.

epidemiological studies studies of populations; often used in nutrition to search for correlations between dietary habits and disease incidence; a first step in seeking nutrition-related causes of diseases.

epinephrine (EP-ih-NEFF-rin) the major hormone that elicits the stress response.

epiphyseal (eh-PIFF-ih-seal) **plate** a thick, cartilage-like layer that forms new cells that are eventually calcified, lengthening the bone (*epiphysis* means "growing" in Greek).

epithelial (ep-ith-THEE-lee-ull) **tissue** the layers of the body that serve as selective barriers to environmental factors. Examples are the cornea, the skin, the respiratory tract lining, and the lining of the digestive tract.

epoetin a drug derived from the human hormone erythropoietin and marketed under the trade name Epogen; illegally used to increase oxygen capacity.

ergogenic (ER-go-JEN-ic) **aids** products that supposedly enhance performance, although none actually do so; the term *ergogenic* implies "energy giving" (*ergo* means "work"; *genic* means "give rise to").

erythrocyte (eh-REETH-ro-sight) **hemolysis** (HE-moh-LIE-sis, he-MOLL-ih-sis) rupture of the red blood cells, caused by vitamin E deficiency (*erythro* means "red"; *cyte* means "cell"; *hemo* means "blood"; *lysis* means "breaking").

essential amino acids amino acids that either cannot be synthesized at all by the body or cannot be synthesized in amounts sufficient to meet physiological need. Also called *indispensable amino acids*.

essential fatty acids fatty acids that the body needs but cannot make in amounts sufficient to meet physiological needs.

essential nutrients the nutrients the body cannot make for itself (or cannot make fast enough) from other raw materials; nutrients that must be obtained from food to prevent deficiencies.

Estimated Energy Requirement (EER) the average dietary energy intake predicted to maintain energy balance in a healthy adult of a certain age, gender, weight, height, and level of physical activity consistent with good health.

ethanol the alcohol of alcoholic beverages, produced by the action of microorganisms on the carbohydrates of grape juice or other carbohydrate-containing fluids.

ethnic foods foods associated with particular cultural subgroups within a population.

euphoria (you-FOR-ee-uh) an inflated sense of well-being and pleasure brought on by a moderate dose of alcohol and some other drugs.

evaporated cane juice raw sugar from which impurities have been removed.

exchange system a diet-planning tool that organizes foods with respect to their nutrient contents and calorie amounts. Foods on any single exchange list can be used interchangeably. See the U.S. Exchange System, Appendix D, for details.

exercise planned, structured, and repetitive bodily movement that promotes or maintains physical fitness.

experimental group the people or animals participating in an experiment who receive the treatment under investigation. Also called *experimental subjects*. See also *control group* and *intervention studies*.

expiration date: The last day the food should be consumed. All foods except eggs should be discarded after this date. For eggs, the expiration date refers to the last day the eggs may be sold as "fresh eggs." For safety, purchase eggs before the expiration date, keep them in their original carton in the refrigerator, and use them within 30 days.

extra lean less than 5 g of fat *and* less than 2 g of saturated fat and *trans* fat combined, *and* less than 95 mg of cholesterol per serving.

extracellular fluid fluid residing outside the cells that transports materials to and from the cells.

extreme obesity clinically severe overweight, presenting very high risks to health; the condition of having a BMI of 40 or above; also called *morbid obesity*.

extrusion a process by which the form of a food is changed, such as changing corn to corn chips; not a preservation measure.

F

famine widespread and extreme scarcity of food in an area that causes starvation and death in a large portion of the population.

fast foods restaurant foods that are available within minutes after customers order them—traditionally, hamburgers, french fries, and milkshakes; more recently, salads and other vegetable dishes as well. These foods may or may not meet people's nutrient needs, depending on the selections made and on the energy allowances and nutrient needs of the eaters.

fasting hypoglycemia hypoglycemia that occurs after 8 to 14 hours of fasting.

fat cells cells that specialize in the storage of fat and form the fat tissue. Fat cells also produce enzymes that metabolize fat and hormones involved in appetite and energy balance.

fat free less than 0.5 g of fat per serving.

fat replacers ingredients that replace some or all of the functions of fat and may or may not provide energy. Often used interchangeably with *fat substitutes*, but the latter technically applies only to ingredients that replace all of the functions of fat and provide no energy.

fatfold test measurement of the thickness of a fold of skin on the back of the arm (over the triceps muscle), below the shoulder blade (subscapular), or in other places, using a caliper; also called *skinfold test*.

fats lipids that are solid at room temperature (70°F, or 25°C).

fatty acids organic acids composed of carbon chains of various lengths. Each fatty acid has an acid end and hydrogens attached to all of the carbon atoms of the chain.

fatty liver an early stage of liver deterioration seen in several diseases, including kwashiorkor and alcoholic liver disease, in which fat accumulates in the liver cells.

FDA (Food and Drug Administration) the part of the Department of Health and Human Services' Public Health Service that is responsible for ensuring the safety and wholesomeness of all foods sold in interstate commerce except meat, poultry, and eggs (which are under the jurisdiction of the USDA); inspecting food plants and imported foods; and setting standards for food consumption.

feces waste material remaining after digestion and absorption are complete; eventually discharged from the body.

female athlete triad a potentially fatal triad of medical problems seen in female athletes: disordered eating, amenorrhea, and osteoporosis.

fertility the capacity of a woman to produce a normal ovum periodically and of a man to produce normal sperm; the ability to reproduce.

fetal alcohol syndrome (FAS) the cluster of symptoms including brain damage, growth retardation, mental retardation, and facial abnormalities seen in an infant or child whose mother consumed alcohol during her pregnancy.

fetus (FEET-us) the stage of human gestation from eight weeks after conception until the birth of an infant.

feverfew an herb sold as a migraine headache preventive. Some evidence exists to support this claim.

fibers the indigestible parts of plant foods, largely nonstarch polysaccharides that are not digested by human digestive enzymes, although some are digested by resident bacteria of the colon. Fibers include cellulose, hemicelluloses, pectins, gums, and mucilages and the nonpolysaccharide lignin.

fibrosis (fye-BROH-sis) an intermediate stage of alcoholic liver deterioration. Liver cells lose their function and assume the characteristics of connective tissue cells (fibers).

fight-or-flight reaction the body's instinctive hormone- and nerve-mediated reaction to danger. Also known as the *stress response*.

filtered water water treated by filtration, usually through *activated carbon filters* that reduce the lead in tap water, or by *reverse osmosis* units that force pressurized water across a membrane removing lead, arsenic, and some microorganisms from tap water.

fitness water lightly flavored bottled water enhanced with vitamins, supposedly to enhance athletic performance.

flavonoids (FLAY-von-oyds) yellow pigments in foods; phytochemicals that may exert physiological effects on the body. *Flavus* means "yellow."

flaxseed the small brown seed of the flax plant; used in baking, cereals, or other foods and valued by industry as a source of linseed oil and fiber.

flexibility the capacity of the joints to move through a full range of motion; the ability to bend and recover without injury.

fluid and electrolyte balance the distribution of fluid and dissolved particles among body compartments.

fluid and electrolyte imbalance failure to maintain the proper amounts and kinds of fluids and minerals in every body compartment; a medical emergency.

fluorapatite (floor-APP-uh-tight) a crystal of bones and teeth, formed when fluoride displaces the "hydroxy" portion of hydroxyapatite. Fluorapatite resists being dissolved back into body fluid.

fluorosis (floor-OH-sis) discoloration of the teeth due to ingestion of too much fluoride during tooth development.

folate (FOH-late) a B vitamin that acts as part of a coenzyme important in the manufacture of new cells. The form added to foods and supplements is *folic acid.*

food medically, any substance that the body can take in and assimilate that will enable it to stay alive and to grow; the carrier of nourishment; socially, a more limited number of such substances defined as acceptable by each culture.

food aversion an intense dislike of a food, biological or psychological in nature, resulting from an illness or other negative experience associated with that food.

food banks facilities that collect and distribute food donations to authorized organizations feeding the hungry.

food group plan a diet-planning tool that sorts foods into groups based on their nutrient content and then specifies that people should eat certain minimum numbers of servings of foods from each group.

food insecurity the condition of limited or uncertain access to food of sufficient quality or quantity to sustain a healthy, active life.

food intolerance an adverse reaction to a food or food additive not involving an immune response.

food pantries community food collection programs that provide groceries to be prepared and eaten at home.

food poverty hunger occurring when enough food exists in an area but some of the people cannot obtain it because they lack money, are being deprived for political reasons, live in a country at war, or suffer from other problems such as lack of transportation.

food recovery collecting wholesome surplus food for distribution to low-income people who are hungry.

food security reliable access to enough nutritious food at all times to sustain a healthy, active life.

food shortage hunger occurring when an area of the world lacks enough total food to feed its people.

foodborne illness illness transmitted to human beings through food and water; caused by an infectious agent (*foodborne infection*) or a poisonous substance (*food intoxication*). Also called *food poisoning.*

foodways the sum of a culture's habits, customs, beliefs, and preferences concerning food.

fork thermometer a utensil combining a meat fork and an instant-read food thermometer.

formaldehyde a substance to which methanol is metabolized on the way to being converted to harmless waste products that can be excreted.

foxglove a plant that contains a substance used in the heart medicine digoxin.

fraud or **quackery** the promotion, for financial gain, of devices, treatments, services, plans, or products (including diets and supplements) that alter or claim to alter a human condition without proof of safety or effectiveness. (The word *quackery* comes from the term *quacksalver,* meaning a person who quacks loudly about a miracle product—a lotion or a salve.)

free radicals atoms or molecules with one or more unpaired electrons that make the atom or molecule unstable and highly reactive.

free, without, no, zero none or a trivial amount. *Calorie free* means containing fewer than 5 calories per serving; *sugar free* or *fat free* means containing less than half a gram per serving.

freezing a method of preserving food by lowering the food's temperature to a point that halts life processes. Microorganisms do not die but remain dormant until the food is thawed.

fresh raw, unprocessed, or minimally processed with no added preservatives.

fructose (FROOK-tose) a monosaccharide; sometimes known as fruit sugar (*fruct* means "fruit"; *ose* means "sugar").

fructose, galactose, glucose the monosaccharides.

fruitarian includes only raw or dried fruits, seeds, and nuts in the diet.

functional foods a term that reflects an attempt to define as a group the foods known to possess nutrients or nonnutrients that might lend protection against diseases. However, all nutritious foods can support health in some ways.

G

galactose (ga-LACK-tose) a monosaccharide; part of the disaccharide lactose (milk sugar).

garlic oil an extract of garlic; may or may not contain the chemicals associated with garlic; claims for health benefits unproved.

gastric juice the digestive secretion of the stomach.

gastroesophageal reflux disease (GERD) (GAS-tro-eh-SOFF-ah-jee-al) a severe and chronic splashing of stomach acid and enzymes into the esophagus, throat, mouth, or airway that causes inflammation and injury to those organs. Untreated GERD may increase the risk of esophageal cancer; treatment may require surgery or management with medication.

gatekeeper with respect to nutrition, a key person who controls other people's access to foods and thereby affects their nutrition profoundly. Examples are the spouse who buys and cooks the food, the parent who feeds the children, and the caretaker in a day-care center.

GE foods genetically engineered foods; food plants and animals altered by way of rDNA technology.

gelatin a soluble form of the protein collagen, used to thicken foods; sometimes falsely promoted as a strength enhancer.

generally recognized as safe (GRAS) list a list, established by the FDA, of food additives long in use and believed to be safe.

genes units of a cell's inheritance, made of the chemical DNA (deoxyribonucleic acid). Each gene directs the making of one or more proteins, which perform important tasks in the body.

genetic engineering (GE) the direct, intentional manipulation of the genetic material of living things in order to obtain some desirable trait not present in the original organism. Also called *recombinant DNA technology* and *biotechnology.*

genetic modification intentional changes to the genetic material of living things brought about through a range of methods, including rDNA technology, natural cross-breeding, and agricultural selective breeding.

genistein (GEN-ih-steen) a phytoestrogen found primarily in soybeans that both mimics and blocks the action of estrogen in the body.

genome (GEE-nome) the full complement of genetic material in the chromosomes of a cell. The study of genomes is *genomics.*

germ the nutrient-rich inner part of a grain.

germander an evergreen bush used in small quantities as a flavoring for alcoholic beverages. Recommended for gout and other ills, it causes often-irreversible liver damage and abnormalities. Deaths reported.

gestation the period of about 40 weeks (three trimesters) from conception to birth; the term of a pregnancy.

gestational diabetes abnormal glucose tolerance appearing during pregnancy.

ghrelin (GREH-lin) a hormone released by the stomach that signals the hypothalamus of the brain to stimulate eating.

ginkgo biloba an extract of a tree of the same name, claimed to enhance mental alertness, but not proved to be effective or safe.

ginseng (JIN-seng) a plant root containing chemicals that have stimulant drug effects. *Ginseng abuse syndrome* is a group of symptoms associated with the overuse of ginseng, including high blood pressure, insomnia, nervousness, confusion, and depression.

glandular products extracts or preparations of raw animal glands and organs; sold with the false claim of boosting athletic performance, but may present disease hazards if collected from infected animals.

glucagon (GLOO-cah-gon) a hormone secreted by the pancreas that stimulates the liver to release glucose into the blood when blood glucose concentration dips.

glucose (GLOO-cose) a single sugar used in both plant and animal tissues for energy; sometimes known as blood sugar or *dextrose.*

glucose polymers compounds that supply glucose, not as single molecules, but linked in chains somewhat like starch. The objective is to attract less water from the body into the digestive tract.

glycemic index (GI) a ranking of foods according to their potential for raising blood glucose relative to a standard such as glucose or white bread.

glycemic load a mathematical expression of both the glycemic index and the carbohydrate content of a food, meal, or diet (glycemic index multiplied by grams carbohydrate).

glycerol (GLISS-er-all) an organic compound, three carbons long, of interest here because it serves as the backbone for triglycerides.

glycine a nonessential amino acid, promoted as an ergogenic aid because it is a precursor of creatine.

glycogen (GLY-co-gen) a highly branched polysaccharide composed of glucose that is made and stored by liver and muscle tissues of human beings and animals as a storage form of glucose. Glycogen is not a significant food source of carbohydrate and is not counted as one of the complex carbohydrates in foods.

goiter (GOY-ter) enlargement of the thyroid gland due to iodine deficiency is *simple goiter;* enlargement due to an iodine excess is *toxic goiter.*

good source 10 to 19% of the Daily Value per serving.

good source of fiber 2.5 g to 4.9 g per serving.

gout (GOWT) a painful form of arthritis caused by the abnormal buildup of the waste product uric acid in the blood, with uric acid salt deposited as crystals in the joints.

grams units of weight. A gram (g) is the weight of a cubic centimeter (cc) or milliliter (ml) of water under defined conditions of temperature and pressure. About 28 grams equal an ounce.

granulated sugar common table sugar, crystalline sucrose, 99.9% pure.

granules small grains. Starch granules are packages of starch molecules. Various plant species make starch granules of varying shapes.

green pills, fruit pills pills containing dehydrated, crushed vegetable or fruit matter. An advertisement may claim that each pill equals a *pound* of fresh produce, but in reality a pill may equal one small forkful—minus nutrient losses incurred in processing.

ground water water that comes from underground aquifers.

growth hormone a hormone (somatotropin) that promotes growth and is produced naturally in the pituitary gland of the brain.

growth hormone releasers herbs or pills that supposedly regulate hormones; falsely promoted as enhancing athletic performance.

growth spurt the marked rapid gain in physical size usually evident around the onset of adolescence.

guarana a reddish berry found in Brazil's Amazon basin that contains seven times as much caffeine as its relative, the coffee bean. It is used as an ingredient in carbonated sodas, and taken in powder or tablet form, it supposedly enhances speed and endurance and serves as an aphrodisiac, a "cardiac tonic," an "intestinal disinfectant," and a "smart drug" touted to improve mental functions. High doses may stress the heart and can cause panic attacks.

H

hard water water with high calcium and magnesium concentrations.

Hazard Analysis Critical Control Point (HACCP) a systematic plan to identify and correct potential microbial hazards in the manufacturing, distribution, and commercial use of food products. *HACCP* may be pronounced "HASS-ip."

hazard a state of danger; used to refer to any circumstance in which harm is possible under normal conditions of use.

health claims claims linking food constituents with disease states; allowable on labels within the criteria established by the Food and Drug Administration.

healthy low in fat, saturated fat, *trans* fat, cholesterol, and sodium and containing at least 10% of the Daily Value for vitamin A, vitamin C, iron, calcium, and fiber.

Healthy Eating Index (HEI) a dietary assessment tool that evaluates a diet's adherence to the principles of the USDA Food Guide and the *Dietary Guidelines for Americans,* as well as the variety of foods the diet contains.

heart attack the event in which the vessels that feed the heart muscle become closed off by an embolism, thrombus, or other cause with resulting sudden tissue death. A heart attack is also called a *myocardial infarction* (*myo* means "muscle"; *cardial* means "of the heart"; *infarct* means "tissue death").

heartburn a burning sensation in the chest (in the area of the heart) area caused by backflow of stomach acid into the esophagus.

heat stroke an acute and life-threatening reaction to heat buildup in the body.

heavy metal any of a number of mineral ions such as mercury and lead; so called because they are of relatively high atomic weight. Many heavy metals are poisonous.

heme (HEEM) the iron-containing portion of the hemoglobin and myoglobin molecules.

hemlock any part of the hemlock plant, which causes severe pain, convulsions, and death within 15 minutes.

hemoglobin (HEEM-oh-globe-in) the oxygen-carrying protein of the blood; found in the red blood cells (*hemo* means "blood"; *globin* means "spherical protein").

hemorrhoids (HEM-or-oids) swollen, hardened (varicose) veins in the rectum, usually caused by the pressure resulting from constipation.

herbal medicine use of herbs and other natural substances with the intention of preventing or curing diseases or relieving symptoms.

herbal steroids or **plant sterols** mixtures of compounds from herbs that supposedly enhance human hormone activity. Products marketed as herbal steroids include astragalus, damiana, dong quai, fo ti teng, ginseng root, licorice root, palmetto berries, sarsaparilla, schizardra, unicorn root, yohimbe bark, and yucca.

hernia a protrusion of an organ or part of an organ through the wall of the body chamber that normally contains the organ. An example is a *hiatal* (high-AY-tal) *hernia,* in which part of the stomach protrudes up through the diaphragm into the chest cavity, which contains the esophagus, heart, and lungs.

hiccups spasms of both the vocal cords and the diaphragm, causing periodic, audible, short, inhaled coughs. Can be caused by irritation of the diaphragm, indigestion, or other causes. Hiccups usually resolve in a few minutes, but can have serious effects if prolonged. Breathing into a paper bag (inhaling carbon dioxide) or dissolving a teaspoon of sugar in the mouth may stop them.

high fiber 5 g or more per serving. (Foods making high-fiber claims must fit the definition of low fat, or the level of total fat must appear next to the high-fiber claim.)

high in 20% or more of the Daily Value for a given nutrient per serving; synonyms include "rich in" or "excellent source."

high-carbohydrate energy drinks fruit-flavored commercial beverages used to restore muscle glycogen after exercise or as a pregame beverage.

high-density lipoproteins (HDL) lipoproteins that return cholesterol from the tissues to the liver for dismantling and disposal; contain a large proportion of protein.

high-quality proteins dietary proteins containing all the essential amino acids in relatively the same amounts that human beings require. They may also contain nonessential amino acids.

high-temperature–short-time (HTST) principle the rule that every 18°F (10°C) rise in processing temperature brings about an approximately tenfold increase in microbial destruction, while only doubling nutrient losses.

histamine a substance that participates in causing inflammation; produced by cells of the immune system as part of a local immune reaction to an antigen.

HMB (beta-hydroxy-beta-methylbutyrate) a metabolite of the branched-chain amino acid leucine. Claims that HMB increases muscle mass and strength stem from "evidence" from the company that developed HMB as a supplement.

homocysteine (hoe-moe-SIS-teen) an amino acid produced as an intermediate compound during amino acid metabolism. A buildup of homocysteine in the blood is associated with deficiencies of folate and other B vitamins and may increase the risk of diseases.

honey a concentrated solution primarily composed of glucose and fructose, produced by enzymatic digestion of the sucrose in nectar by bees.

hormones chemicals that are secreted by glands into the blood in response to conditions in the body that require regulation. These chemicals serve as messengers, acting on other organs to maintain constant conditions.

human growth hormone (HGH) a hormone produced by the brain's pituitary gland that regulates normal growth and development (see text discussion); also called *somatotropin.*

hunger the physiological need to eat, experienced as a drive for obtaining food; an unpleasant sensation that demands relief.

hunger lack or shortage of basic foods needed to provide the energy and nutrients that support health.

husk the outer, inedible part of a grain.

hydrogenation (high-dro-gen-AY-shun) the process of adding hydrogen to unsaturated fatty acids to make fat more solid and resistant to the chemical change of oxidation.

hydroxyapatite (hi-DROX-ee-APP-uh-tight) the chief crystal of bone, formed from calcium and phosphorus.

hyperactivity (in children) a syndrome characterized by inattention, impulsiveness, and excessive motor activity; usually diagnosed before age seven, lasts six months or more, and usually does not entail mental illness or mental retardation. Properly called *attention-deficit/hyperactivity disorder (ADHD)* and may be associated with minimal brain damage.

hypertension high blood pressure.

hypertrophy (high-PURR-tro-fee) an increase in size (for example, of a muscle) in response to use.

hypoallergenic formulas clinically tested infant formulas that do not provoke reactions in 90% of infants or children with confirmed cow's milk allergy.

hypoglycemia (HIGH-poh-gly-SEE-mee-uh) a blood glucose concentration below normal, a symptom that may indicate any of several diseases, including impending diabetes.

hyponatremia (HIGH-poh-na-TREE-mee-ah) a decreased concentration of sodium in the blood (*hypo* means "below"; *natrium* means "sodium [Na]"; *emia* means "blood").

hypothalamus (high-poh-THAL-uh-mus) a part of the brain that senses a variety of conditions in the blood, such as temperature, glucose content, salt content, and others. It signals other parts of the brain or body to adjust those conditions when necessary.

hypothermia a below-normal body temperature.

I

immune system a system of tissues and organs that defend the body against antigens, foreign materials that have penetrated the skin or body linings.

immunity protection from or resistance to a disease or infection by development of antibodies and by the actions of cells and tissues in response to a threat.

impaired glucose tolerance blood glucose levels higher than normal but not high enough to be diagnosed as diabetes; sometimes called *prediabetes*.

implantation the stage of development, during the first two weeks after conception, in which the fertilized egg (fertilized ovum or zygote) embeds itself in the wall of the uterus and begins to develop.

incidental additives substances that can get into food not through intentional introduction but as a result of contact with the food during growing, processing, packaging, storing, or some other stage before the food is consumed. Also called *accidental* or *indirect additives*.

infectious diseases diseases that are caused by bacteria, viruses, parasites, and other microbes and can be transmitted from one person to another through air, water, or food; by contact; or through vector organisms such as mosquitoes and fleas.

infomercials feature-length television commercials that follow the format of regular programs but are intended to convince viewers to buy products, and not to educate or entertain them. The statements made may or may not be accurate.

initiation an event, probably occurring in a cell's genetic material, caused by radiation or by a chemical carcinogen that can give rise to cancer.

inosine an organic chemical that is falsely said to "activate cells, produce energy, and facilitate exercise." Studies have shown that it actually reduces the endurance of runners.

inositol (in-OSS-ih-tall) a nonessential nutrient found in cell membranes.

insoluble fibers the tough, fibrous structures of fruits, vegetables, and grains; indigestible food components that do not dissolve in water.

instant-read thermometer a thermometer that, when inserted into food, measures its temperature within seconds; designed to test temperature of food at intervals, and not to be left in food during cooking.

insulin a hormone secreted by the pancreas in response to a high blood glucose concentration. It assists cells in drawing glucose from the blood.

insulin resistance a condition in which a normal or high level of insulin produces a less-than-normal response by the tissues; thought to be a metabolic consequence of obesity.

integrated pest management (IPM) management of pests using a combination of natural and biological controls and minimal or no application of pesticides.

intervention studies studies of populations in which observation is accompanied by experimental manipulation of some population members—for example, a study in which half of the subjects (the *experimental subjects*) follow diet advice to reduce fat intakes while the other half (the *control subjects*) do not, and both groups' heart health is monitored.

intestine the body's long, tubular organ of digestion and the site of nutrient absorption.

intracellular fluid fluid residing inside the cells that provides the medium for cellular reactions.

intrinsic factor a factor found inside a system. The intrinsic factor necessary to prevent pernicious anemia is now known to be a compound that helps in the absorption of vitamin B_{12}.

invert sugar a mixture of glucose and fructose formed by the splitting of sucrose in an industrial process. Sold only in liquid form and sweeter than sucrose, invert sugar forms during certain cooking procedures and works to prevent crystallization of sucrose in soft candies and sweets.

ions (EYE-ons) electrically charged particles, such as sodium (positively charged) or chloride (negatively charged).

iron deficiency the condition of having depleted iron stores, which, at the extreme, causes iron-deficiency anemia.

iron overload the state of having more iron in the body than it needs or can handle, usually arising from a hereditary defect. Also called *hemochromatosis*.

iron-deficiency anemia a form of anemia caused by a lack of iron and characterized by red blood cell shrinkage and color loss. Accompanying symptoms are weakness, apathy, headaches, pallor, intolerance to cold, and inability to pay attention. (For other anemias, see the index.)

irradiation the application of ionizing radiation to foods to reduce insect infestation or microbial contamination or to slow the ripening or sprouting process. Also called *cold pasteurization*.

irritable bowel syndrome intermittent disturbance of bowel function, especially diarrhea or alternating diarrhea and constipation; associated with diet, lack of physical activity, or psychological stress.

isomalt, lactitol, maltitol, mannitol, sorbitol, xylitol sugar alcohols that can be derived from fruits or commercially produced from a sugar; absorbed more slowly and metabolized differently than other sugars in the human body and not readily used by ordinary mouth bacteria.

IU (international unit) a measure of fat-soluble vitamin activity sometimes used on supplement labels.

J

jaundice (JAWN-dis) yellowing of the skin due to spillover of the bile pigment bilirubin (bill-ee-ROO-bin) from the liver into the general circulation.

K

kava the root of a tropical pepper plant, often brewed as a tea consumed for its calming effects. Limited scientific research supports use for anxiety relief. Adverse effects include skin rash, liver abnormalities, lethargy, and mental disorientation.

kefir (KEE-fur) a liquid form of yogurt, based on milk, probiotic microorganisms, and flavorings.

kelp tablets tablets made from dehydrated kelp, a kind of seaweed used by the Japanese as a foodstuff.

keratin (KERR-uh-tin) the normal protein of hair and nails.

keratinization accumulation of keratin in a tissue; a sign of vitamin A deficiency.

ketone bodies acidic compounds derived from fat and certain amino acids. Normally rare in the blood, they help to feed the brain during times when too little carbohydrate is available.

ketosis (kee-TOE-sis) an undesirable high concentration of ketone bodies, such as acetone, in the blood or urine.

kidneys a pair of organs that filter wastes from the blood, make urine, and release it to the bladder for excretion from the body.

kombucha (KOM-boo-sha) a product of fermentation of sugar-sweetened tea by various yeasts and bacteria. Proclaimed as a treatment for everything from AIDS to cancer but lacking scientific evidence. Microorganisms in home-brewed kombucha have caused serious illnesses in people with weakened immunity. Also known as *Manchurian tea, mushroom tea,* or *Kargasok tea*.

kudzu a weedy vine, whose roots are harvested and used by Chinese herbalists as a treatment for alcoholism. Kudzu reportedly reduces alcohol absorption by up to 50% in rats.

kwashiorkor (kwash-ee-OR-core, kwashee-or-CORE) a disease related to protein malnutrition, with a set of recognizable symptoms, such as edema.

L

laboratory studies studies that are performed under tightly controlled conditions and are designed to pinpoint causes and effects. Such studies often use animals as subjects.

lactase the intestinal enzyme that splits the disaccharide lactose to monosaccharides during digestion.

lactation production and secretion of breast milk for the purpose of nourishing an infant.

lactic acid a compound produced during the breakdown of glucose in anaerobic metabolism.

lactoferrin (lack-toe-FERR-in) a factor in breast milk that binds iron and keeps it from supporting the growth of the infant's intestinal bacteria.

lacto-ovo vegetarian includes dairy products, eggs, vegetables, grains, legumes, fruits, and nuts; excludes flesh and seafood.

lactose intolerance impaired ability to digest lactose due to reduced amounts of the enzyme lactase.

lactose a disaccharide composed of glucose and galactose; sometimes known as milk sugar (*lact* means "milk"; *ose* means "sugar").

lactose, maltose, sucrose the disaccharides.

lacto-vegetarian includes dairy products, vegetables, grains, legumes, fruits, and nuts; excludes flesh, seafood, and eggs.

large intestine the portion of the intestine that completes the absorption process.

lean on a food label, less than 10 g of fat *and* less than 4 g of saturated fat and *trans* fat combined, *and* less than 95 mg of cholesterol per serving.

learning disability a condition resulting in an altered ability to learn basic cognitive skills such as reading, writing, and mathematics.

leavened (LEV-end) literally, "lightened" by yeast cells, which digest some carbohydrate components of the dough and leave behind bubbles of gas that make the bread rise.

lecithin (LESS-ih-thin) a phospholipid manufactured by the liver and also found in many foods; a major constituent of cell membranes.

legumes (leg-GOOMS, LEG-yooms) beans, peas, and lentils, valued as inexpensive sources of protein, vitamins, minerals, and fiber that contribute little fat to the diet.

leptin an appetite-suppressing hormone produced in the fat cells that conveys information about body fatness to the brain; believed to be involved in the maintenance of body composition (*leptos* means "slender").

less saturated fat 25% or less saturated fat and *trans* fat combined than the comparison food.

less, fewer, reduced containing at least 25% less of a nutrient or calories than a reference food. This may occur naturally or as a result of altering the food. For example, pretzels, which are usually low in fat, can claim to provide less fat than potato chips, a comparable food.

levulose an older name for fructose.

license to practice permission under state or federal law, granted on meeting specified criteria, to use a certain title (such as *dietitian*) and to offer certain services. Licensed dietitians may use the initials LD after their names.

life expectancy the average number of years lived by people in a given society.

life span the maximum number of years of life attainable by a member of a species.

light this descriptor has three meanings on labels: (1) A serving provides one-third fewer calories or half the fat of the regular product. (2) A serving of a low-calorie, low-fat food provides half the sodium normally present. (3) The product is light in color and texture, so long as the label makes this intent clear, as in "light brown sugar."

lignans phytochemicals present in flaxseed, but not in flax oil, that are converted to phytosterols by intestinal bacteria and are under study as possible anticancer agents.

limiting amino acid an essential amino acid that is present in dietary protein in an insufficient amount, thereby limiting the body's ability to build protein.

linoleic (lin-oh-LAY-ic) **acid** and **linolenic** (lin-oh-LEN-ic) **acid** polyunsaturated fatty acids that are essential nutrients for human beings. The full name of linolenic acid is *alpha-linolenic acid*.

lipid (LIP-id) a family of organic (carbon-containing) compounds soluble in organic solvents but not in water. Lipids include triglycerides (fats and oils), phospholipids, and sterols.

lipoic (lip-OH-ic) **acid** a nonessential nutrient.

lipoprotein lipase (LPL) an enzyme mounted on the surface of fat cells that splits triglycerides in the blood into fatty acids and glycerol to be absorbed into the cells for reassembly and storage.

lipoproteins (LYE-poh-PRO-teens, LIH-poh-PRO-teens) clusters of lipids associated with protein, which serve as transport vehicles for lipids in blood and lymph. Major lipoprotein classes are the chylomicrons, the VLDL, the LDL, and the HDL.

listeriosis a serious foodborne infection that can cause severe brain infection or death in a fetus or a newborn; caused by the bacterium *Listeria monocytogenes,* which is found in soil and water.

liver a large, lobed organ that lies just under the ribs. It filters the blood, removes and processes nutrients, manufactures materials for export to other parts of the body, and destroys toxins or stores them to keep them out of the circulation.

lobelia (low-BEE-lee-uh) dried leaves and tops of lobelia ("Indian tobacco") plant used to induce vomiting or treat a cough; abused for a mild euphoria. Causes breathing difficulty, rapid pulse, low blood pressure, diarrhea, dizziness, and tremors. Possible deaths reported.

longevity long duration of life.

low birthweight a birthweight of less than 5½ pounds (2,500 grams); used as a predictor of probable health problems in the newborn and as a probable indicator of poor nutrition status of the mother before and/or during pregnancy. Low-birthweight infants are of two different types. Some are *premature infants;* they are born early and are the right size for their gestational age. Other low-birthweight infants have suffered growth failure in the uterus; they are small for gestational age (small for date) and may or may not be premature.

low calorie 40 calories or fewer per serving.

low cholesterol 20 mg or less of cholesterol *and* 2 g or less saturated fat per serving.

low fat 3 g or less fat per serving.

low saturated fat 1 g or less saturated fat and less than 0.5 g of *trans* fat per serving.

low sodium 140 mg or less sodium per serving.

low-density lipoproteins (LDL) lipoproteins that transport lipids from the liver to other tissues such as muscle and fat; contain a large proportion of cholesterol.

lungs the body's organs of gas exchange. Blood circulating through the lungs releases its carbon dioxide and picks up fresh oxygen to carry to the tissues.

lutein (LOO-teen) a plant pigment of yellow hue; a phytochemical believed to play roles in eye functioning and health.

lycopene (LYE-koh-peen) a pigment responsible for the red color of tomatoes and other red-hued vegetables; a phytochemical that may act as an antioxidant in the body.

lymph (LIMF) the fluid that moves from the bloodstream into tissue spaces and then travels in its own vessels, which eventually drain back into the bloodstream.

lymphocytes (LIM-foh-sites) white blood cells that participate in the immune response; B-cells and T-cells.

M

ma huang an herbal preparation sold with promises of weight loss and increased energy, but contains ephedrine, a banned cardiac stimulant with serious adverse effects.

macrobiotic diet a vegan diet composed mostly of whole grains, beans, and certain vegetables; taken to extremes, macrobiotic diets have resulted in malnutrition and even death.

macrophages (MACK-roh-fah-jes) large scavenger cells of the immune system that engulf debris and remove it (*macro* means "large"; *phagein* means "to eat").

macular degeneration a common, progressive loss of function of the part of the retina that is most crucial to focused vision (the macula is shown on page 216). This degeneration often leads to blindness.

major minerals essential mineral nutrients found in the human body in amounts larger than 5 grams.

malnutrition any condition caused by excess or deficient food energy or nutrient intake or by an imbalance of nutrients. Nutrient or energy deficiencies are classed as forms of undernutrition; nutrient or energy excesses are classed as forms of overnutrition.

maltose a disaccharide composed of two glucose units; sometimes known as malt sugar.

maple sugar a concentrated solution of sucrose derived from the sap of the sugar maple tree, mostly sucrose. This sugar was once common but is now usually replaced by sucrose and artificial maple flavoring.

marasmus (ma-RAZ-mus) the calorie-deficiency disease; starvation.

margin of safety in reference to food additives, a zone between the concentration normally used and that at which a hazard exists. For common table salt, for example, the margin of safety is 1/5 (five times the concentration normally used would be hazardous).

medical nutrition therapy nutrition services used in the treatment of injury, illness, or other conditions; includes assessment of nutrition status and dietary intake, and corrective applications of diet, counseling, and other nutrition services.

melatonin a hormone of the pineal gland believed to help regulate the body's daily rhythms, to reverse the effects of jet lag, and to promote sleep. Claims for life extension or enhancement of sexual prowess are without merit.

MEOS (microsomal ethanol oxidizing system) a system of enzymes in the liver that oxidize not only alcohol but also several classes of drugs.

metabolic syndrome a combination of characteristic factors—high fasting blood glucose or insulin resistance, central obesity, hypertension, low blood HDL cholesterol, and elevated blood triglycerides—that greatly increase a person's risk of developing CVD. Also called *insulin resistance syndrome* or *syndrome X*.

metabolism the sum of all physical and chemical changes taking place in living cells; includes all reactions by which the body obtains and spends the energy from food.

metastasis (meh-TASS-ta-sis) movement of cancer cells from one body part to another, usually by way of the body fluids.

methanol an alcohol produced in the body continually by all cells.

methylmercury any toxic compound of mercury to which a characteristic chemical structure, a methyl group, has been added, usually by bacteria in aquatic sediments. Methylmercury is easily absorbed from the intestine and causes nerve damage in people.

MFP factor a factor present in meat, fish, and poultry that enhances the absorption of nonheme iron present in the same foods or in other foods eaten at the same time.

microbes a shortened name for *microorganisms;* minute organisms too small to observe without a microscope, including bacteria, viruses, and others.

microbes bacteria, viruses, or other organisms invisible to the naked eye, some of which cause diseases. Also called *microorganisms.*

microvilli (MY-croh-VILL-ee, MY-croh-VILL-eye) tiny, hairlike projections on each cell of every villus that greatly expand the surface area available to trap nutrient particles and absorb them into the cells (singular: *microvillus*).

milk anemia iron-deficiency anemia caused by drinking so much milk that iron-rich foods are displaced from the diet.

mineral water water from a spring or well that typically contains 250 to 500 parts per million (ppm) of minerals. Minerals give water a distinctive flavor. Many mineral waters are high in sodium.

minerals naturally occurring, inorganic, homogeneous substances; chemical elements.

miso fermented soybean paste used in Japanese cooking. Soy products are considered to be functional foods.

moderate drinkers people who do not drink excessively and do not behave inappropriately because of alcohol. A moderate drinker's health is not harmed by alcohol over the long term.

moderation the dietary characteristic of providing constituents within set limits, not to excess.

modified atmosphere packaging (MAP) a preservation technique in which a perishable food is packaged in a gas-impermeable container from which air has been removed or to which another gas mixture has been added.

molasses a syrup left over from the refining of sucrose from sugarcane; a thick, brown syrup. The major nutrient in molasses is iron, a contaminant from the machinery used in processing it.

monoglycerides (mon-oh-GLISS-er-ides) products of the digestion of lipids; consist of glycerol molecules with one fatty acid attached (*mono* means "one"; *glyceride* means "a compound of glycerol").

monosaccharides (mon-oh-SACK-ah-rides) single sugar units (*mono* means "one"; *saccharide* means "sugar unit").

monounsaturated fats triglycerides in which most of the fatty acids have one point of unsaturation (are monounsaturated).

monounsaturated fatty acid a fatty acid containing one point of unsaturation.

more or **added fiber** at least 2.5 g more per serving than a reference food.

more, extra at least 10% more of the Daily Value than in a reference food. The nutrient may be added or may occur naturally.

MSG symptom complex the acute, temporary, and self-limiting reactions experienced by sensitive people upon ingesting a large dose of MSG. The name *MSG symptom complex,* given by the FDA, replaces the former *Chinese restaurant syndrome.*

mucus (MYOO-cus) a slippery coating of the digestive tract lining (and other body linings) that protects the cells from exposure to digestive juices (and other destructive agents). The adjective form is *mucous* (same pronunciation). The digestive tract lining is a *mucous membrane.*

muscle endurance the ability of a muscle to contract repeatedly within a given time without becoming exhausted.

muscle strength the ability of muscles to work against resistance.

mutual supplementation the strategy of combining two incomplete protein sources so that the amino acids in one food make up for those lacking in the other food. Such protein combinations are sometimes called *complementary proteins.*

myoglobin (MYE-oh-globe-in) the oxygen-holding protein of the muscles (*myo* means "muscle").

N

natural foods a term that has no legal definition, but is often used to imply wholesomeness.

natural water water obtained from a spring or well that is certified to be safe and sanitary. The mineral content may not be changed, but the water may be treated in other ways such as with ozone or by filtration.

naturally occurring sugars sugars that are not added to a food but are present as its original constituents, such as the sugars of fruit or milk.

neotame (NEE-oh-tame) an artificial sweetener composed of two amino acids (phenylalanine and aspartic acid) linked in such a way as to make them indigestible by human enzymes.

nephrons (NEFF-rons) the working units in the kidneys, consisting of intermeshed blood vessels and tubules.

neural tube defects abnormalities of the brain and spinal cord apparent at birth and believed to be related to a woman's folate intake before and during pregnancy.

neural tube the embryonic tissue that later forms the brain and spinal cord.

neurotoxins poisons that act upon the cells of the nervous system.

neurotransmitters chemicals that are released at the end of a nerve cell when a nerve impulse arrives there. They diffuse across the gap to the next cell and alter the membrane of that second cell to either inhibit or excite it.

niacin a B vitamin needed in energy metabolism. Niacin can be eaten preformed or can be made in the body from tryptophan, one of the amino acids. Other forms of niacin are *nicotinic acid, niacinamide,* and *nicotinamide.*

niacin equivalents (NE) the amount of niacin present in food, including the niacin that can theoretically be made from its precursor tryptophan that is present in the food.

night blindness slow recovery of vision after exposure to flashes of bright light at night; an early symptom of vitamin A deficiency.

nitrogen balance the amount of nitrogen consumed compared with the amount excreted in a given time period.

nonalcoholic a term used on labels of beverages, such as wine or beer, indicating that the product contains less than 0.5% alcohol. The terms *dealcoholized* and *alcohol removed* mean the same thing. *Alcohol free* means that the product contains no detectable alcohol.

nonnutrients a term used in this book to mean compounds other than the six nutrients that are present in foods and have biological activity in the body.

norepinephrine (NOR-EP-ih-NEFF-rin) a compound related to epinephrine that helps to elicit the stress response.

nori a type of seaweed popular in Asian, particularly Japanese, cooking.

nutraceutical a term that has no legal or scientific meaning, but is sometimes used to refer to foods, nutrients, or dietary supplements believed to have medicinal effects. Often used to sell unnecessary or unproved supplements.

nutrient additives vitamins and minerals added to improve nutritive value.

nutrient claims claims using approved wording to describe the nutrient values of foods, such as a claim that a food is "high" in a desirable constituent or "low" in an undesirable one.

nutrient density a measure of nutrients provided per calorie of food.

nutrients components of food that are indispensable to the body's functioning. They provide energy, serve as building material, help maintain or repair body parts, and support growth. The nutrients include water, carbohydrate, fat, protein, vitamins, and minerals.

Nutrition Facts on a food label, the panel of nutrition information required to appear on almost every packaged food. Grocers may also provide the information for fresh produce, meats, poultry, and seafoods.

nutrition the study of the nutrients in foods and in the body; sometimes also the study of human behaviors related to food.

nutritional genomics the science of how nutrients affect the activities of genes and how genes affect the activities of nutrients. Also called *molecular nutrition* or *nutrigenomics*.

nutritional yeast a preparation of yeast cells, often praised for its high nutrient content. Yeast is a source of B vitamins as are many other foods. Also called *brewer's yeast;* not the yeast used in baking.

nutritionist someone who engages in the study of nutrition. Some nutritionists are RDs, whereas others are self-described experts whose training is questionable and who are not qualified to give advice. In states with responsible legislation, the term applies only to people who have masters of science (MS) or doctor of philosophy (PhD) degrees from properly accredited institutions.

O

obesity overfatness with adverse health effects, as determined by reliable measures and interpreted with good medical judgment. Obesity is officially defined as a body mass index of 30 or higher.

octacosanol an alcohol extracted from wheat germ, often falsely promoted as enhancing athletic performance.

oils lipids that are liquid at room temperature (70°F, or 25°C).

olestra a noncaloric artificial fat made from sucrose and fatty acids; formerly called *sucrose polyester*.

omega-3 fatty acid a polyunsaturated fatty acid with its endmost double bond three carbons from the end of the carbon chain. Linolenic acid is an example.

omega-6 fatty acid a polyunsaturated fatty acid with its endmost double bond six carbons from the end of the carbon chain. Linoleic acid is an example.

omnivores people who eat foods of both plant and animal origin, including animal flesh.

open dating: A general term referring to label dates that are stated in ordinary language that consumers can understand, as opposed to *closed dating,* which refers to dates printed in codes decipherable only by manufacturers. Open dating is used primarily on perishable foods, and closed dating on shelf-stable products such as canned goods.

oral rehydration therapy (ORT) oral fluid replacement for children with severe diarrhea caused by infectious disease. ORT enables parents to mix a simple solution for their child from substances that they have at home.

organ and glandular extracts dried or extracted material from brain, adrenal, pituitary, or other glands or tissues providing few nutrients but posing a theoretical risk of "mad cow disease."

organic foods products grown and processed without the use of synthetic chemicals such as pesticides, herbicides, fertilizers, and preservatives and without genetic engineering or irradiation.

organic carbon containing. Four of the six classes of nutrients are organic: carbohydrate, fat, protein, and vitamins. Strictly speaking, organic compounds include only those made by living things and do not include carbon dioxide and a few carbon salts.

organosulfur compounds a large group of phytochemicals containing the mineral sulfur. Organosulfur phytochemicals are responsible for the pungent flavors and aromas of foods belonging to the onion, leek, chive, shallot, and garlic family and are thought to stimulate cancer defenses in the body.

organs discrete structural units made of tissues that perform specific jobs. Examples are the heart, liver, and brain.

ornithine a nonessential amino acid falsely promoted as enhancing the secretion of human growth hormone, the breakdown of fat, and the development of muscle.

oryzanol a plant sterol that supposedly provides the same physical responses as anabolic steroids without the adverse side effects; also known as *ferulic acid, ferulate,* or *FRAC*.

osteomalacia (OS-tee-o-mal-AY-shuh) the adult expression of vitamin D–deficiency disease, characterized by an overabundance of unmineralized bone protein (*osteo* means "bone"; *mal* means "bad"). Symptoms include bending of the spine and bowing of the legs.

osteoporosis (OSS-tee-oh-pore-OH-sis) a reduction of the bone mass of older persons in which the bones become porous and fragile (*osteo* means "bones"; *poros* means "porous"); also known as **adult bone loss.**

outcrossing the unintended breeding of a domestic crop with a related wild species.

oven-safe thermometer a thermometer designed to remain in the food to give constant readings during cooking.

overload an extra physical demand placed on the body; an increase in the frequency, duration, or intensity of an activity. A principle of training is that for a body system to improve, it must be worked at frequencies, durations, or intensities that increase by increments.

overweight overfatness of a moderate degree; defined as a body mass index (BMI) of 25.0 through 29.9. BMI is defined later.

ovo-vegetarian includes eggs, vegetables, grains, legumes, fruits, and nuts; excludes flesh, seafood, and milk products.

ovum the egg, produced by the mother, that unites with a sperm from the father to produce a new individual.

oxidants compounds (such as oxygen itself) that oxidize other compounds. Compounds that prevent oxidation are called *anti*oxidants, whereas those that promote it are called *pro*oxidants (*anti* means "against"; *pro* means "for").

oxidation interaction of a compound with oxygen; in this case, a damaging effect by a chemically reactive form of oxygen.

oxidative stress damage inflicted on living systems by free radicals.

oyster shell a product made from the powdered shells of oysters that is sold as a calcium supplement, but is not well absorbed by the digestive system.

P

pack date: The day the food was packaged or processed. When used on packages of fresh meats, pack dates can provide a general guide to freshness.

pancreas an organ with two main functions. One is an endocrine function—the making of hormones such as insulin, which it releases directly into the blood (*endo* means "into" the blood). The other is an exocrine function—the making of digestive enzymes, which it releases through a duct into the small intestine to assist in digestion (*exo* means "out" into a body cavity or onto the skin surface).

pancreatic juice fluid secreted by the pancreas that contains both enzymes to digest carbohydrate, fat, and protein and sodium bicarbonate, a neutralizing agent.

pangamic acid also called vitamin B_{15} (but not a vitamin, nor even a specific compound—it can be anything with that label); falsely claimed to speed oxygen delivery.

pantothenic (PAN-to-THEN-ic) **acid** a B vitamin.

partial vegetarian a term sometimes used to mean an eating style that includes seafood, poultry, eggs, dairy products, vegetables, grains, legumes, fruits, and nuts; excludes or strictly limits certain meats, such as red meats.

partitioned foods foods composed of parts of whole foods, such as butter (from milk), sugar (from beets or cane), or corn oil (from corn). Partitioned foods are generally overused and provide few nutrients with many calories.

pasteurization the treatment of milk with heat sufficient to kill certain pathogens (disease-causing microbes) commonly transmitted through milk; not a sterilization process. Pasteurized milk retains bacteria that cause milk spoilage. Raw milk, even if labeled "certified," transmits many foodborne diseases to people each year and should be avoided.

PCBs stable oily synthetic chemicals used in hundreds of industrial and commercial operations that persist as pollution in the environment. PCBs cause cancer in animals and a number of other serious health effects. The Environmental Protection Agency monitors their levels.

peak bone mass the highest attainable bone density for an individual; developed during the first three decades of life.

pellagra (pell-AY-gra) the niacin-deficiency disease (*pellis* means "skin"; *agra* means "rough"). Symptoms include the "4 Ds": diarrhea, dermatitis, dementia, and death.

pennyroyal relatives of the mint family brewed as tea or extracted as oil; used as mosquito repellent, claimed to treat various conditions. Tea produced multiple organ failure in infants; 1/2 teaspoon of oil caused convulsions and coma; 2 tablespoons caused the death of an 18-year-old expectant mother within 2 hours, despite hospitalization.

peptide bond a bond that connects one amino acid with another, forming a link in a protein chain.

percent fat free may be used only if the product meets the definition of *low fat* or *fat free*. Requires disclosure of grams of fat per 100 g food.

peristalsis (perri-STALL-sis) the wavelike muscular squeezing of the esophagus, stomach, and small intestine that pushes their contents along.

pernicious (per-NISH-us) **anemia** a vitamin B_{12}–deficiency disease, caused by lack of intrinsic factor and characterized by large, immature red blood cells and damage to the nervous system (*pernicious* means "highly injurious or destructive").

persistent of a stubborn or enduring nature; with respect to food contaminants, the quality of remaining unaltered and unexcreted in plant foods or in the bodies of animals and human beings.

pesco-vegetarian same as partial vegetarian, but eliminates poultry.

pesticides chemicals used to control insects, diseases, weeds, fungi, and other pests on crops and around animals. Used broadly, the term includes *herbicides* (to kill weeds), *insecticides* (to kill insects), and *fungicides* (to kill fungi).

pH a measure of acidity on a point scale. A solution with a pH of 1 is a strong acid; a solution with a pH of 7 is neutral; a solution with a pH of 14 is a strong base.

phagocytes (FAG-oh-sites) white blood cells that can ingest and destroy antigens. The process by which phagocytes engulf materials is called *phagocytosis.* The Greek word *phagein* means "to eat."

phosphate salt a product demonstrated to increase the levels of a metabolically important phosphate compound (diphosphoglycerate) in red blood cells and the potential of the cells to deliver oxygen to the body's muscle cells. However, it does not extend endurance or increase efficiency of aerobic metabolism, and it may cause calcium losses from the bones if taken in excess.

phospholipids (FOSS-foh-LIP-ids) one of the three main classes of dietary lipids. These lipids are similar to triglycerides, but each has a phosphorus-containing acid in place of one of the fatty acids. Phospholipids are present in all cell membranes.

photosynthesis the process by which green plants make carbohydrates from carbon dioxide and water using the green pigment chlorophyll to capture the sun's energy (*photo* means "light"; *synthesis* means "making").

physical activity bodily movement produced by muscle contractions that substantially increase energy expenditure.

phytates (FYE-tates) compounds present in plant foods (particularly whole grains) that bind iron and may prevent its absorption.

phytochemicals nonnutrient compounds in plant-derived foods that have biological activity in the body (*phyto* means "plant").

phytoestrogens (FIGH-toe-ESS-troh-gens) phytochemicals structurally similar to mammalian hormones, such as the female sex hormone estrogen. Phytoestrogens weakly mimic hormone activity in the human body.

pica (PIE-ka) a craving for nonfood substances. Also known as *geophagia* (gee-oh-FAY-gee-uh) when referring to clay eating and *pagophagia* (pag-oh-FAY-gee-uh) when referring to ice craving (*geo* means "earth"; *pago* means "frost"; *phagia* means "to eat").

placebo a sham treatment often used in scientific studies; an inert harmless medication. The *placebo effect* is the healing effect that the act of treatment, rather than the treatment itself, often has.

placenta (pla-SEN-tuh) the organ of pregnancy in which maternal and fetal blood circulate in close proximity and exchange nutrients and oxygen (flowing into the fetus) and wastes (picked up by the mother's blood).

plant pesticides substances produced within plant tissues that kill or repel attacking organisms.

plant sterols lipid extracts of plants, called ferulic acid, oryzanol, phytosterols, or "adaptogens," marketed with false claims that they contain hormones or enhance hormonal activity.

plaques (PLACKS) mounds of lipid material mixed with smooth muscle cells and calcium that develop in the artery walls in atherosclerosis (*placken* means "patch"). The same word is also used to describe the accumulation of a different kind of deposits on teeth, which promote dental caries.

plasma the cell-free fluid part of blood and lymph.

platelets tiny cell-like fragments in the blood, important in blood clot formation (*platelet* means "little plate").

point of unsaturation a site in a molecule where the bonding is such that additional hydrogen atoms can easily be attached.

polypeptides (POL-ee-PEP-tides) protein fragment of many (more than ten) amino acids bonded togeteher (*poly* means "many"). A peptide is a strand of amino acids. A strand of between four and ten amino acids is called an *oligopeptide.*

polysaccharides another term for complex carbohydrates; compounds composed of long strands of glucose units linked together (*poly* means "many"). Also called *complex carbohydrates.*

polyunsaturated fats triglycerides in which most of the fatty acids have two or more points of unsaturation (are polyunsaturated).

polyunsaturated fatty acid (PUFA) a fatty acid with two or more points of unsaturation.

pop-up thermometer a disposable timing device commonly used in turkeys. The center of the device contains a stainless steel spring that "pops up" when food reaches the right temperature.

postprandial hypoglycemia an unusual drop in blood glucose that follows a meal and is accompanied by symptoms such as anxiety, rapid heartbeat, and sweating; also called *reactive hypoglycemia.*

potassium iodide a medication approved by the FDA as safe and effective for the prevention of thyroid cancer caused by radioactive iodine that may be released during radiation emergencies.

precursors, provitamins compounds that can be converted into active vitamins.

preeclampsia (PRE-ee-CLAMP-see-uh) a potentially dangerous condition during pregnancy characterized by edema, hypertension, and protein in the urine.

pregame meal a meal eaten three to four hours before athletic competition.

premenstrual syndrome (PMS) a cluster of symptoms that some women experience prior to and during menstruation. They include, among others, abdominal cramps, back pain, swelling, headache, painful breasts, and mood changes.

prenatal (pree-NAY-tal) before birth.

preservatives antimicrobial agents, antioxidants, chelating agents, radiation, and other additives that retard spoilage or preserve desired qualities, such as softness in baked goods.

prion (PREE-on) an infective agent consisting of an unusually folded protein that disrupts normal cell functioning, causing disease.

probiotics consumable products containing live microorganisms in sufficient numbers to alter the bacterial colonies of the body in ways believed to benefit health. A *prebiotic* product is a substance that may not be digestible by the host, such as fiber, but serves as food for probiotic bacteria and thus promotes their growth.

problem drinkers or **alcohol abusers** people who suffer social, emotional, family, job-related, or other problems because of alcohol. A problem drinker is on the way to alcoholism.

processed foods foods subjected to any process, such as milling, alteration of texture, addition of additives, cooking, or others. Depending on the starting material and the process, a processed food may or may not be nutritious.

promoters factors that do not initiate cancer but speed up its development once initiation has taken place.

proof a statement of the percentage of alcohol in an alcoholic beverage. Liquor that is 100 proof is 50% alcohol, 90 proof is 45%, and so forth.

prooxidant a compound that triggers reactions involving oxygen.

protein digestibility–corrected amino acid score (PDCAAS) a measuring tool used to determine protein quality. The PDCAAS reflects a protein's digestibility as well as the proportions of amino acids that it provides.

protein turnover the continuous breakdown and synthesis of body proteins involving the recycling of amino acids.

protein-energy malnutrition (PEM) the world's most widespread malnutrition problem, including both marasmus and kwashiorkor and states in which they overlap; also called *protein-calorie malnutrition (PCM).*

proteins compounds composed of carbon, hydrogen, oxygen, and nitrogen and arranged as strands of amino acids. Some amino acids also contain the element sulfur.

protein-sparing action the action of carbohydrate and fat in providing energy that allows protein to be used for purposes it alone can serve.

public health nutritionist a dietitian or other person with an advanced degree in nutrition who specializes in public health nutrition.

public water water from a municipal or county water system that has been treated and disinfected.

purified water water that has been treated by distillation or other physical or chemical processes that remove dissolved solids. Because purified water contains no minerals or contaminants, it is useful for medical and research purposes.

pyloric (pye-LORE-ick) **valve** the circular muscle of the lower stomach that regulates the flow of partly digested food into the small intestine. Also called *pyloric sphincter.*

pyruvate a 3-carbon compound derived during the metabolism of glucose, certain amino acids, and glycerol; falsely promoted as burning fat and enhancing endurance. Common side effects include intestinal gas and diarrhea and possibly reduced physical performance.

R

radiolytic products chemicals formed in foods during the irradiation process; deemed harmless by experts.

raw sugar the first crop of crystals harvested during sugar processing. Raw sugar cannot be sold in the United States because it contains too much filth (dirt, insect fragments, and the like). Sugar sold as "raw sugar" is actually evaporated cane juice.

recombinant DNA (rDNA) technology a technique of genetic modification whereby scientists directly manipulate the genes of living things; includes methods of removing genes, doubling genes, introducing foreign genes, and changing gene positions to influence the growth and development of organisms.

reduced calorie at least 25% lower in calories than a "regular," or reference, food.

reduced or **less cholesterol** at least 25% less cholesterol than a reference food *and* 2 g or less saturated fat per serving.

reduced saturated fat at least 25% less saturated fat *and* reduced by more than 1 g saturated fat per serving compared with a reference food.

reduced sodium at least 25% lower in sodium than the regular product.

refined refers to the process by which the coarse parts of food products are removed. For example, the refining of wheat into flour involves removing three of the four parts of the kernel—the chaff, the bran, and the germ—leaving only the endosperm, composed mainly of starch and a little protein.

registered dietitian (RD) a dietitian who has graduated from a university or college after completing a program of dietetics. The program must be approved or accredited by the American Dietetic Association (or Dietitians of Canada). The dietitian must serve in an approved internship, coordinated program, or preprofessional practice program to practice the necessary skills; pass the five parts of the association's *registration* examination; and maintain competency through continuing education. Many states also require licensing for practicing dietitians.

registration listing with a professional organization that requires specific course work, experience, and passing of an examination.

requirement the amount of a nutrient that will just prevent the development of specific deficiency signs; distinguished from the DRI recommended intake value, which is a generous allowance with a margin of safety.

residues whatever remains. In the case of pesticides, those amounts that remain on or in foods when people buy and use them.

resistant starch the fraction of starch in a food that is digested slowly, or not at all, by human enzymes.

retina (RET-in-uh) the layer of light-sensitive nerve cells lining the back of the inside of the eye.

retinol activity equivalents (RAE) a new measure of the vitamin A activity of beta-carotene and other vitamin A precursors that reflects the amount of retinol that the body will derive from a food containing vitamin A precursor compounds.

retinol one of the active forms of vitamin A made from beta-carotene in animal and human bodies; an antioxidant nutrient. Other active forms are *retinal* and *retinoic acid.*

rhodopsin (roh-DOP-sin) the light-sensitive pigment of the cells in the retina; it contains vitamin A (*rod* refers to the rod-shaped cells; *opsin* means "visual protein").

riboflavin (RIBE-o-flay-vin) a B vitamin active in the body's energy-releasing mechanisms.

rickets the vitamin D–deficiency disease in children; characterized by abnormal growth of bone and manifested in bowed legs or knock-knees, outward-bowed chest, and knobs on the ribs.

risk factors factors known to be related to (or correlated with) diseases but not proved to be causal.

royal jelly a substance produced by worker bees and fed to the queen bee; often falsely promoted as enhancing athletic performance.

S

saccharin a zero-calorie sweetener used freely in the United States but restricted in Canada.

safety the practical certainty that injury will not result from the use of a substance.

salts compounds composed of charged particles (ions). An example is potassium chloride (K^+Cl^-).

SAM-e an amino acid derivative that may have an antidepressant effect on the brain in some people, but is not recommended as a substitute for standard antidepressant therapy.

sassafras root bark from the sassafras tree; once used in beverages but now banned as an ingredient in foods or beverages because it contains cancer-causing chemicals.

satiation (SAY-she-AY-shun) the perception of fullness that builds throughout a meal, eventually reaching the degree of fullness and satisfaction that halts eating. Satiation generally determines how much food is consumed at one sitting.

satiety (sah-TIE-eh-tee) the perception of fullness that lingers in the hours after a meal and inhibits eating until the next mealtime. Satiety generally determines the length of time between meals.

saturated fat free less than 0.5 g of saturated fat *and* less than 0.5 g of *trans* fat.

saturated fats triglycerides in which most of the fatty acids are saturated.

saturated fatty acid a fatty acid carrying the maximum possible number of hydrogen atoms (having no points of unsaturation).

saw palmetto the ripe fruit or extracts of the saw palmetto plant. Claimed to relieve symptoms associated with enlarged prostate but reported as ineffective in research.

scurvy the vitamin C–deficiency disease.

selective breeding a technique of genetic modification whereby organisms are chosen for reproduction based on their desirability for human purposes, such as high growth rate, high food yield, or disease resistance, with the intention of retaining or enhancing these characteristics in their offspring.

self-efficacy a person's belief in his or her ability to succeed in an undertaking.

sell by: Specifies the shelf life of the food. After this date, the food may still be safe for consumption if it has been handled and stored properly. Also called *pull date.*

senile dementia the loss of brain function beyond the normal loss of physical adeptness and memory that occurs with aging.

serotonin (SER-oh-tone-in) a neurotransmitter important in sleep regulation, appetite control, and mood regulation, among other roles. Serotonin is synthesized in the body from the amino acid tryptophan with the help of vitamin B_6.

set-point theory the theory that the body tends to maintain a certain weight by means of its own internal controls.

side chain the unique chemical structure attached to the backbone of each amino acid that differentiates one amino acid from another.

simple carbohydrates sugars, including both single sugar units and linked pairs of sugar units. The basic sugar unit is a molecule containing six carbon atoms, together with oxygen and hydrogen atoms.

single-use temperature indicator a type of instant-read thermometer that changes color to indicate that the food has reached the desired temperature. Discarded after one use, they are often used in retail food markets to eliminate cross-contamination.

skullcap a native herb with no known medical uses but found in remedies. Other species may be harvested and sold as skullcap, so it has not been determined whether several deaths from liver toxicity reportedly from skullcap were in fact from another herb.

small intestine the 20-foot length of small-diameter intestine, below the stomach and above the large intestine, that is the major site of digestion of food and absorption of nutrients.

smoking point the temperature at which fat gives off an acrid blue gas.

social drinkers people who drink only on social occasions. Depending on how alcohol affects a social drinker's life, the person may be a moderate drinker or a problem drinker.

sodium bicarbonate baking soda; an alkaline salt believed to neutralize blood lactic acid and thereby reduce pain and enhance possible workload. "Soda loading" may cause intestinal bloating and diarrhea.

sodium free less than 5 mg per serving.

soft water water with a high sodium concentration.

soluble fibers food components that readily dissolve in water and often impart gummy or gel-like characteristics to foods. An example is pectin from fruit, which is used to thicken jellies. Soluble fibers are indigestible by human enzymes but may be broken down to absorbable products by bacteria in the digestive tract.

soy drink a milk-like beverage made from soybeans, claimed to be a functional food. Soy drink should be fortified with vitamin A, vitamin D, riboflavin, and calcium to approach the nutritional equivalency of milk. Also called *soy milk.*

Special Supplemental Food Program for Women, Infants, and Children (WIC) a USDA program offering low-income pregnant women and those with infants or preschool children coupons redeemable for specific foods that supply the nutrients deemed most necessary for growth and development. For more information, visit www.usda.gov/FoodandNutrition.

sphincter (SFINK-ter) a circular muscle surrounding, and able to close, a body opening.

spina bifida (SPY-na BIFF-ih-duh) one of the most common types of neural tube defects in which gaps occur in the bones of the spine. Often the spinal cord bulges and protrudes through the gaps, resulting in a number of motor and other impairments.

spirulina a kind of alga ("blue-green manna") that supposedly contains large amounts of protein and vitamin B_{12}, suppresses appetite, and improves athletic performance. It does none of these things and is potentially toxic.

spring water water originating from an underground spring or well. It may be bubbly (carbonated), or "flat" or "still," meaning not carbonated. Brand names such as "Spring Pure" do not necessarily mean that the water comes from a spring.

St. John's wort an herb containing psychoactive substances that has been used for centuries to treat depression, insomnia, bedwetting, and "nervous conditions." Some scientific reports find St. John's wort equal in effectiveness to standard antidepressant medication for relief of depression. Long-term safety, however, has not been established.

staple foods foods used frequently or daily, for example, rice (in East and Southeast Asia) or potatoes (in Ireland). If well chosen, these foods are nutritious.

starch a plant polysaccharide composed of glucose. After cooking, starch is highly digestible by human beings; raw starch often resists digestion.

stem cell an undifferentiated cell that can mature into any of a number of specialized cell types. A stem cell of bone marrow may mature into one of many kinds of blood cells, for example.

sterol esters compounds derived from vegetable oils that lower blood cholesterol in human beings by competing with cholesterol for absorption from the digestive tract. The term *sterol esters* often refers to both stanol esters and sterol esters.

sterols (STEER-alls) one of the three main classes of dietary lipids. Sterols have a structure similar to that of cholesterol.

stevia (STEEV-ee-uh) the sweet-tasting leaves of a shrub sold as a dietary supplement, but lacking FDA approval as a sweetener.

stomach a muscular, elastic, pouchlike organ of the digestive tract that grinds and churns swallowed food and mixes it with acid and enzymes, forming chyme.

stone ground refers to a milling process using limestone to grind any grain, including refined grains, into flour.

stone-ground flour flour made by grinding kernels of grain between heavy wheels made of limestone, a kind of rock derived from the shells and bones of marine animals. As the stones scrape together, bits of the limestone mix with the flour, enriching it with calcium.

stroke the sudden shutting off of the blood flow to the brain by a thrombus, embolism, or the bursting of a vessel (hemorrhage).

stroke volume the amount of oxygenated blood ejected from the heart toward body tissues at each beat.

subclinical, or **marginal, deficiency** a nutrient deficiency that has no outward clinical symptoms. The term is often used to market unneeded nutrient supplements to consumers.

subcutaneous fat fat stored directly under the skin (*sub* means "beneath"; *cutaneous* refers to the skin).

succinate a compound synthesized in the body and involved in the TCA cycle; falsely promoted as a metabolic enhancer.

sucralose a noncaloric sweetener derived from a chlorinated form of sugar that travels through the digestive tract unabsorbed. Approved for use in the United States and Canada.

sucrose (SOO-crose) a disaccharide composed of glucose and fructose; sometimes known as table, beet, or cane sugar and, often, simply *sugar*.

sucrose polyester any of a family of compounds in which fatty acids are bonded with sugars or sugar alcohols. Olestra is an example.

sugars simple carbohydrates; that is, molecules of either single sugar units or pairs of those sugar units bonded together. By common usage, *sugar* most often refers to sucrose.

superoxide dismutase (SOD) an enzyme that protects cells from oxidation. When it is taken orally, the body digests and inactivates this protein; it is useless to athletes.

surface water water that comes from lakes, rivers, and reservoirs.

sushi a Japanese dish that consists of vinegar-flavored rice, seafood, and colorful vegetables, typically wrapped in seaweed. Some sushi is wrapped in raw fish; other sushi contains only cooked ingredients.

sustainable able to continue indefinitely. In this context, the use of resources in ways that maintain both natural resources and human life; the use of natural resources at a pace that allows the earth to replace them. Examples: cutting trees no faster than new ones grow and producing pollutants at a rate with which the environment and human cleanup efforts can keep pace. In a sustainable economy, resources do not become depleted, and pollution does not accumulate.

systolic (sis-TOL-ik) **pressure** the first figure in a blood pressure reading (the "dupp" of the heartbeat is heard), which reflects arterial pressure caused by the contraction of the heart's left ventricle.

T

tagatose an incompletely absorbed monosaccharide sweetener derived from lactose with a caloric value of 1.5 calories per gram. About 80% of the ingested tagatose travels to the large intestine where bacterial colonies ferment it. Tagatose is not readily used by mouth bacteria, and so does not promote dental caries.

tannins compounds in tea (especially black tea) and coffee that bind iron. Tannins also denature proteins.

T-cells lymphocytes that attack antigens. *T* stands for the thymus gland of the neck, where the T-cells are stored and matured.

textured vegetable protein processed soybean protein used in products formulated to look and taste like meat, fish, or poultry.

thermic effect of food (TEF) the body's speeded-up metabolism in response to having eaten a meal; also called *diet-induced thermogenesis*.

thermogenesis the generation and release of body heat associated with the breakdown of body fuels. *Adaptive thermogenesis* describes adjustments in energy expenditure related to changes in environment such as cold and to physiological events such as underfeeding or trauma.

THG an unapproved drug, once sold as an ergogenic aid, now banned by the FDA.

thiamin (THIGH-uh-min) a B vitamin involved in the body's use of fuels.

thickening and stabilizing agents ingredients that maintain emulsions, foams, or suspensions or lend a desirable thick consistency to foods. Dextrins (short chains of glucose formed as a breakdown product of starch), starch, and pectin are examples. (Gums such as carrageenan, guar, locust bean, agar, and gum arabic are others.)

thrombosis a thrombus that has grown enough to close off a blood vessel. A *coronary thrombosis* closes off a vessel that feeds the heart muscle. A *cerebral thrombosis* closes off a vessel that feeds the brain (*coronary* means "crowning" [the heart]; *thrombo* means "clot"; the cerebrum is part of the brain).

thrombus a stationary blood clot.

tissues systems of cells working together to perform specialized tasks. Examples are muscles, nerves, blood, and bone.

tocopherol (tuh-KOFF-er-all) a kind of alcohol. The active form of vitamin E is alpha-tocopherol.

tofu (TOE-foo) a curd made from soybeans that is rich in protein, often rich in calcium, and variable in fat content; used in many Asian and vegetarian dishes in place of meat.

tolerance limit the maximum amount of a residue permitted in a food when a pesticide is used according to label directions.

toxicity the ability of a substance to harm living organisms. All substances are toxic if the concentration is high enough.

trabecular (tra-BECK-you-lar) **bone** the weblike structure composed of calcium-containing crystals inside a bone's solid outer shell. It provides strength and acts as a calcium storage bank.

trace minerals essential mineral nutrients found in the human body in amounts less than 5 grams.

training regular practice of an activity, which leads to physical adaptations of the body with improvement in flexibility, strength, or endurance.

trans **fat free** less than 0.5 g of *trans* fat *and* less than 0.5 g of saturated fat per serving.

trans **fats** fats that contain unusual fatty acids, *trans* fatty acids, formed during processing; a later section provides details.

trans **fatty acids** fatty acids with unusual shapes that can arise when polyunsaturated oils are hydrogenated.

transgenic organism an organism resulting from the growth of an embryonic, stem, or germ cell into which a new gene has been inserted.

triglycerides (try-GLISS-er-ides) one of the three main classes of dietary lipids and the chief form of fat in foods and in the human body. A triglyceride is made up of three units of fatty acids and one unit of glycerol (fatty acids and glycerol are defined later). Triglycerides are also called *triacylglycerols*.

trimester a period representing gestation. A trimester is about 13 to 14 weeks.

tripeptides (try-PEP-tides) protein fragments that are three amino acids long (*tri* means "three").

turbinado (ter-bih-NOD-oh) **sugar** raw sugar from which the filth has been washed; legal to sell in the United States.

type 1 diabetes the type of diabetes in which the pancreas produces no or very little insulin; often diagnosed in childhood, although some cases arise in adulthood. Formerly called *juvenile-onset* or *insulin-dependent diabetes*.

type 2 diabetes the type of diabetes in which the pancreas makes plenty of insulin, but the body's cells resist insulin's action; often diagnosed in adulthood. Formerly called *adult-onset* or *noninsulin-dependent diabetes*.

U

ulcer an erosion in the topmost, and sometimes underlying, layers of cells that form a lining. Ulcers of the digestive tract commonly form in the esophagus, stomach, or upper small intestine.

ultrahigh temperature (UHT) a process of sterilizing food by exposing it for a short time to temperatures above those normally used in processing.

unbleached flour a beige-colored endosperm flour with texture and nutritive qualities that approximate those of regular white flour.

underwater weighing a measure of density and volume used to determine body fat content.

underweight too little body fat for health; defined as having a body mass index of less than 18.5.

unsaturated fatty acid a fatty acid that lacks some hydrogen atoms and has one or more points of unsaturation. An unsaturated fat is a triglyceride that contains one or more unsaturated fatty acids.

urban legends stories, usually false, that may travel rapidly throughout the world via the Internet gaining strength of conviction solely on the basis of repetition.

urea (yoo-REE-uh) the principal nitrogen-excretion product of protein metabolism; generated mostly by removal of amine groups from unneeded amino acids or from amino acids being sacrificed to a need for energy.

urethane a carcinogenic compound that commonly forms in alcoholic beverages.

USDA (U.S. Department of Agriculture) the federal agency that is responsible for enforcing standards for the wholesomeness and quality of meat, poultry, and eggs produced in the United States; conducting nutrition research; and educating the public about nutrition.

uterus (YOO-ter-us) the womb, the muscular organ within which the infant develops before birth.

V

valerian a preparation of the root of an herb used as a sedative and sleep agent. Safety and effectiveness of valerian have not been scientifically established.

variety the dietary characteristic of providing a wide selection of foods—the opposite of monotony.

vegan includes only food from plant sources: vegetables, grains, legumes, fruits, seeds, and nuts; also called *strict vegetarian*.

vegetarians people who exclude from their diets animal flesh and possibly other animal products such as milk, cheese, and eggs.

veins blood vessels that carry blood, with the carbon dioxide it has collected, from the tissues back to the heart.

very low sodium 35 mg or less sodium per serving.

very-low-density lipoproteins (VLDL) lipoproteins that transport triglycerides and other lipids from the liver to various tissues in the body.

villi (VILL-ee, VILL-eye) fingerlike projections of the sheets of cells that line the intestinal tract. The villi make the surface area much greater than it would otherwise be (singular: *villus*).

visceral fat fat stored within the abdominal cavity in association with the internal abdominal organs; also called *intra-abdominal fat*.

viscous (VISS-cuss) having a sticky, gummy, or gel-like consistency that flows relatively slowly.

vitamin B₆ a B vitamin needed in protein metabolism. Its three active forms are *pyridoxine*, *pyridoxal*, and *pyridoxamine*.

vitamin B₁₂ a B vitamin that helps to convert folate to its active form and also helps maintain the sheath around nerve cells. Vitamin B₁₂'s scientific name, not often used, is *cyanocobalamin*.

vitamin water bottled water with a few vitamins added; does not replace vitamins from a balanced diet and may worsen overload in people receiving vitamins from enriched food, supplements, and other enriched products such as "energy" bars.

vitamins organic compounds that are vital to life and indispensable to body functions, but are needed only in minute amounts; noncaloric essential nutrients.

VO₂ max the maximum rate of oxygen consumption by an individual (measured at sea level).

voluntary activities intentional activities (such as walking, sitting, or running) conducted by voluntary muscles.

W

wasting the progressive, relentless loss of the body's tissues that accompanies certain diseases and shortens survival time.

water balance the balance between water intake and water excretion, which keeps the body's water content constant.

water intoxication a dangerous dilution of the body's fluids resulting from excessive ingestion of plain water. Symptoms are headache, muscular weakness, lack of concentration, poor memory, and loss of appetite.

weight cycling repeated rounds of weight loss and subsequent regain, with reduced ability to lose weight with each attempt; also called *yo-yo dieting*.

weight training the use of free weights or weight machines to provide resistance for developing muscle strength and endurance. A person's own body weight may also be used to provide resistance as when a person does push-ups, pull-ups, or sit-ups. Also called *resistance training*.

well water water drawn from ground water by tapping into an aquifer.

Wernicke-Korsakoff (VER-nik-ee KOR-sah-koff) **syndrome** a cluster of symptoms involving nerve damage arising from a deficiency of the vitamin thiamin in alcoholism. Characterized by mental confusion, disorientation, memory loss, jerky eye movements, and staggering gait.

wheat flour any flour made from wheat, including white flour.

wheat germ oil the oil from the wheat kernel; often falsely promoted as an energy aid.

whey protein a by-product of cheese production; falsely promoted as increasing muscle mass. As for whey, it is the liquid left when most solids are removed from milk.

white flour an endosperm flour that has been refined and bleached for maximum softness and whiteness.

white sugar pure sucrose, produced by dissolving, concentrating, and recrystallizing raw sugar.

WHO (World Health Organization) an international agency that develops standards to regulate pesticide use. A related organization is the FAO (Food and Agricultural Organization).

whole grain refers to a grain milled in its entirety (all but the husk), not refined.

whole-wheat flour flour made from whole-wheat kernels; a whole-grain flour.

witch hazel leaves or bark of a witch hazel tree; not proved to have healing powers.

world food supply the quantity of food, including stores from previous harvests, available to the world's people at a given time.

X

xerophthalmia (ZEER-ahf-THALL-me-uh) progressive hardening of the cornea of the eye in advanced vitamin A deficiency that can lead to blindness (*xero* means "dry"; *ophthalm* means "eye").

xerosis (zeer-OH-sis) drying of the cornea; a symptom of vitamin A deficiency.

Z

zygote (ZYE-goat) the term that describes the product of the union of ovum and sperm during the first two weeks after fertilization.

Index

The page letters A, B, and C that stand alone refer to the tables beginning on the inside front cover. The page letters Y and Z refer to the tables on the last two pages of the book. The page numbers preceded by A through G are appendix page numbers. The boldfaced page numbers indicate definitions. Terms are also defined in the glossary. Page numbers followed by *n* indicate footnotes. Page numbers followed by *t* indicate tables. Page numbers followed by *f* indicate figures.

A

Abdominal fat, 313–14, 314f, 320
Abortion, spontaneous, 95, 496
Absorb, defined, **75**
Absorption. *See also* Digestion; *specific nutrients*
  by active transport, 190f
  of alcohol, 79, 93–4, 93f, 95
  of amino acids, 186–7, 187f, 193
  of carbohydrates, 101, 114, 115f
  digestive system anatomy and, 83, 84f, 115f, 186–7, 187f
  of fats, 146f, 147, 148f
  medications and, 86, 552, 553–4, 555t
  olestra and vitamins, 162–2
  transport of nutrients, 83, 84f, 190f
  of vitamins, 162–3, 215, 542–3
Acceptable Daily Intake (ADI), 30, **135t**, 137
Acceptable Macronutrient Distribution Ranges (AMDR), **29t**, 30
Accredited (schools), 25, 25n, **26t**
Accutane, 536
Acesulfame potassium (acesulfame-K), **135t**, 136t, 137
Acetaldehyde, **91**, 96
Acid-base balance, **190**, **273**. *See also* PH
  fasting and, 330
  ketosis and, 117
  low-carbohydrate diets and, 117
  minerals and, 273, 307
  regulation of, 190–1, 307
Acid controllers. *See* Acid reducers
"Acid indigestion." *See* Heartburn
Acidosis, **190**
Acid reducers, **86**
Acids, **190**. *See also* PH
Acne, 218, **536**
Acromegaly, 391
Acrylamide, **418**
Active transport, 190f
Acupuncture, **414t**
ADA. *See* American Dietetic Association (ADA)
Additives, **462–7**. *See also specific additives*
  cancer risk, 416, 463n, 464, 465
  FDA regulation of, 462–4
  hyperactivity and, 529
  incidental (accidental), **466**–7
  nutrient additives, **463t**, 467
  salt as, 464
  sugar as, 127, 464. *See also* Sugars, added
  sulfites, 462, 465
  types of, by function, 463t
Adequacy (dietary), **8**, 34, 37
Adequate Intake (AI), 29, **29t**, 31n, A–B. *See also* DRI (Dietary Reference Intakes)

ADH (alcohol dehydrogenase), **91**, 94–5, 94n
ADH (antidiuretic hormone), **91**, 94
ADHD (attention-deficit/hyperactivity disorder), 529–30
Adiponectin, 399n
Adipose tissue, **88**, **315**. *See also* Fat (body); Fat cells
  central obesity, **313**–14, 314f, 320
  diabetes and, 122
  functions of, 140, 140t
  leptin and, 306, 324–5
  lipoprotein lipase, **326**, 326t
  storage capacity, 88, 120
Adolescence, 533–8. *See also* Children
  alcohol/drug use, 533
  blood lipid profiles, 516t
  BMI and, Z
  body changes, 535
  body weight, 533–4
  bone density, 309t, 534, 535f
  calcium/milk needs, 299t
  disordered eating, 351, 534
  DRI for, A–B
  eating patterns of, 536–8
  energy needs, 534, C, Y
  hypertension, 516t, 517
  milk consumption, 536f
  obesity in, 514f
  pregnancy and, 493
  smoking/tobacco use, 329, 518
  soft drink consumption, 535t
  vegetarian diets and, 210
Adrenaline, 118n. *See also* Epinephrine
Adult bone loss. *See* Osteoporosis
Adult-onset diabetes. *See* Diabetes, type 2
Adult rickets. *See* Osteomalacia
Adverse reactions to foods. *See* Allergies
Advertising. *See also* Information on nutrition, validity of
  of alcohol, 497f
  of amino acid supplements, 186
  of athletic supplements, 391
  of digestive enzymes, 186
  of elemental diets, 6
  of ergonomic aids, 388
  false information in, 22, 23f
  of foods for children, 530
  of infant formulas, 508
  manipulating emotions, 428
  of medications, 85–6
  obesity and, 328, 348, 427–9
  in schools, 428
Advertorials, **22t**
Aerobic, **362**
Aerobic activity. *See also* Physical activity
  benefits of, 362, 364
  energy metabolism, 367, 367f, 372
  fatty acids and, 367
Aflatoxin, 477n
Africa, hunger in, 567, 568f
African Americans. *See also* Race/ethnicity
  hypertension and, 409–10
  infant mortality, 483f
  lactose intolerance, 116
  osteoporosis and, 304
Age, physiological, 544–5

Agencies for health and nutrition, 15, 24t, E-1–E-6
Aging (older adults), 538–50
  alcohol and, 98
  Alzheimer's disease, 545–7, E-4
  antioxidant supplements for, 259
  body weight, 320, 539
  bone mass, 275, 303, 306, 309t. *See also* Osteoporosis
  calcium absorption, 305
  dehydration in, 543
  energy needs, 318, 538–9, C, Y
  food choices/eating habits, 547–50
  free radicals, 545
  heart disease risk and, 401, 403f
  hypertension and, 408, 409
  life expectancy, **538**, 544–5
  nutrient concerns in, 223, 305, 540–9, 544t
  physical activity and, 539, 540, 540f
  physical changes, 538–40, 539t, 542, 544t, 545, 545t
  productive aging tips, 538
  programs for, 547–8
  resources for, E-3
  vitamin D and, 223, 305
  water needs, 543
  weight training benefits, 363
Agribusiness, 578–83, **578t**, 579f, 581t
Agriculture
  alternative/low-input/sustainable, **578t**, 580, 581t, 582t
  direct marketing, 582
  diversity of species, 578–9
  eggs, 474
  fish farming, 458–9
  genetically modified microbes in, 582
  genetic engineering, 472, 474–80, 475f, 476f, 480t
  integrated pest management, **578t**, 580
  meat and poultry
    antibiotics in, 455–6
    BSE (mad cow disease), **444**–5, 448
    buffalo, 583
    cloning, 476–7
    growth hormones, 455
    organically produced, 473t, 474
    production issues, 579, 579f, 582–3
    USDA hotline, 444n
  organically produced foods, 472–4, 473t, 480t
  pesticides, 473, 477, 578, 580
  precision agriculture, 580–1
  selective breeding, **472**, 474
  small farms, 580, 582
  soil erosion/depletion, 561, 569, 578, 581–2
  terminator technology, 478
  USDA contact information, E-2
  vegetables, organically produced, 473t
  world food supply and, 568–9
AI (Adequate Intake), 29, **29t**, 31n, A–B. *See also* DRI (Dietary Reference Intakes)
AIDS, **396**, 502
Air pollution, 561, 570
Alcohol, **91–8**. *See also* Alcohol abuse; Alcoholism
  absorption of, 79, 93–4, 93f, 95
  advertising of, 497f
  amount in beverages, 93, 93f
  antioxidants in, 98
  benefits of, 97–8

protein and, 201
stress fractures, 307
vegetable *vs.* animal proteins and, 306–7, 307f
vitamin A supplements and, 255–6
Bone loss. *See* Bone density; Osteoporosis
Bone mass, **275**. *See also* Bone density
Bone meal, 308, **310t**
Bones. *See also* Osteoporosis
calcium and, 264, 273–4, 275
calcium and protein, 201, 306–7, 307f
cortical, **304t**
density of. *See* Bone density
fluoride and, 274
growth of, 534, 535f
minerals in, 273–4, 275
rickets, 221–2, **221–2**, 222f, 505
strength of, 274, 303
structure of, 274f, 303
trabecular, 304f, **304t**
vitamin A and, 217
vitamin D deficiency, 221–2, 245t
vitamin K and, 227
vitamins and, 245–9t
Book recommendations, E-1
Borage, 553t
Boron, 295–6, **386t**
Bottled water, **270**
Botulism, **435**, 438t, 446, 447, 450n
Bovine somatotropin (bST), **455**
Bovine spongiform encephalopathy (BSE), **444**–5, 448
Bowel movements. *See also* Diarrhea
constipation, **86, 108,** 494
feces defined, **78**
fiber and, 86–7
olestra problems, 162
Brain
alcohol's effects on, 94f, 95, 96
Alzheimer's disease, 545–7, E-4
anatomy of, 72f
appetite and, 323, 324
fetal development and, 484, 487–8
fetal development and alcohol, 497–9
hunger and, 72–3, 323, 323f
iron deficiency in children, 526
malnutrition and, 520–1, 526
omega-3 fatty acids and, 546
pesticides and, 452
sugar's effects on, 132
supplements and, 546
Bran, 107f, 110f, **110t**, 112
Branched-chain amino acids, 386–7, **386t**
Bread, cereal, rice and pasta food group
in Canadian Choice System, B-1–B-2, B-3–B-4
Breads, 110–12. *See also* Grains
digestion of, 113–14
enriched, **110t**
in exchange system, D-4t
fiber in, 111t, 126f
folate fortification, 489, 489t
fortified, 110–12, **110t,** 111f
labeling of, 112f
low-carb, 112f
nutrients in, 110–12, 111f, 112f
refined, **110t**
terminology, 110t
whole-grain, 111f, 112f
Breakfast, 342, 416, 531–2
Breakfast bars, 168
Breast cancer, 96, 413t, 417
Breastfeeding, 500–7
advertising's effect on, 508
alcohol use, 501
benefits of, 503–6, 505t
blood pressure and, 516

breast milk composition, 500–1, 504–5, 505f, 507t
caffeine and, 501
certified lactation consultants, **500, 508**
diabetes risk and, 499, 515
drug use, illicit, 501
energy needs during, 500
environmental contaminants, 502
fish consumption and, 458
future weight and, 515
infant sensitivity to mother's foods, 501
intelligence and, 505, 506
lactation, defined, **484**
maternal illness and, 502
maternal nutrition and, 487t, 500–1
maternal weight and, 501
medications and, 501–2
milk volume, 500
success strategies, 508t
vegan diets and, 209
websites/resources, E-5
by working mothers, 507
"Breathalyzer" test, 87
Brewer's yeast, **386t**
Broccoli sprouts, **57t**, 63
Brown bread, **110t**
Brown fat, 326
BSE (bovine spongiform encephalopathy), **444**–5, 448
bST (bovine somatotropin), **455**
Buffalo, 583
Buffers, **190, 273**
Bulimia nervosa, 351t, 354–6, 355f, 355t, E-5. *See also* Eating disorders
Butter, 127, 166, 175, 299
Butterflies, 478
Butylated hydroxyanisole (BHA), 465
Butylated hydroxytoluene (BHT), 465
Butyrate, **109**
B vitamins, 233–44. *See also specific B vitamins*
deficiencies, 233–5, 247–9t
energy metabolism and, 233
food irradiation and, 451
functions of, 233, 234f, 247–9t
homocysteine and heart disease, 243
listed, 214, 233
"non-B vitamins," 244
toxicities, 247–9t

C

Cabbage family, 57t, 63, 420, 449
Cadmium, 269, 458t
Caffeine, **386t**
athletic activity and, 379, 387–8
in beverages, 556t
bone density and, 307
for children, 133, 556
diuretic effects, 268, 556
lactation and, 501
nutrient interactions, 554, 555t, 556–7
PMS and, 537
pregnancy and, 496
Caffeine water, **271t**
CAGE questions, **91,** 92t
Calcitonin, 275n
Calcium, 273–6, 297t
absorption of, 83, 275, 299, 300f, 309, 489
acid-base balance, 307
adolescents' needs, 534
in body composition, 264, 264f
in bones. *See also* Osteoporosis
pregnancy and, 489
protein and, 201, 306–7, 307f
strength and, 308–9, 309t
structure and, 273–4, 275
cancer and, 274, 420

colon cancer and, 420
coral calcium, 308
deficiency, 8, 275, 276t, 297t
functions of, 273–5, 276f
hypertension and, 411–12
intake recommendations, 275, 276f, 276t, 308
lactose intolerance and, 116
magnesium and, 277
meal planning for, 299, 300–1
for older adults, 543, 544t
osteomalacia and, 222
osteoporosis and, 275, 305
PMS and, 537
pregnancy and, 489
sodium and, 280
sources of, 8, 204, 210–11, 211n, 276f, 276t, 297t, 299–300, 299f
supplements, 308–10, 310t
in tooth formation, 274, 276f
toxicity, 276t, 297t
vegetarian diets, 210–11
vitamin D and, 222
Calcium compounds, 308, 309, **310t**
Calculation, aids to, C-1–C-2
Caloric content
of alcohol, 96, 97t, 333
of artificial sweeteners, 134–5, 135t, 136
of carbohydrates, 5, 6t, 104, 192f
of chocolate, 60
of dairy products, 38
of fat, body, 316
of fat, dietary, 5, 6t, 142, 170t, 171f
of fat replacers, 163t
of fatty foods, 45, 153f
of nuts, 174–5
of protein, 5, 6t, 192f
of restaurant meals, 250t, 428
of sugar/sugary foods, 129t
Caloric effect, **417**
Calorie control, **8,** 9. *See also* Calories, discretionary; Weight loss; Weight management
dietary fat reduction tips, 153f, 168–9, 169t, 170t
energy density, **339**–40, 340f, 343–4
portion control tips, 44
USDA Food Guide and, 37–8, I-1–I-2
Calories, **5**. *See also* Energy; Energy metabolism; Nutrient density
for adolescents, 534, C, Y
for adults, 318, 538–9, C, Y
alcohol and, 6t
athletes' needs, 380
in body fat, 316, 343n
during breastfeeding, C, Y
calculating needs, 317–18
cancer and caloric intake, 417
for children, 515, 521–2, C, Y
daily needs, 38–9, 39t
for elderly adults, 538–9, 544–5, C, Y
"empty," 129t
energy density, **339**–40, 340f
exercise and expenditure of. *See* Physical activity and energy expenditure
fat intake in U.S., 161
on labels of foods, 48f, 49, 50t, 51t
in lean tissue, 343n
pregnancy, needs during, C, Y
weight loss plans and, 337
Calories, discretionary, **37**. *See also* Nutrient density
overview, 37–8, 37f
added sugars, 38, 127, 131, I-2
fatty foods, 161, I-1n, I-2
physical activity and, 38
USDA Food Guide recommendations, 39t, 40t, I-1–I-2
vegetables and fruits, 38
Campylobacteriosis, 436t

advertising food to, 428
allergies in, 527–9
attention-deficit/hyperactivity disorder
  (ADHD), 529
behavior of, 133, 525–6, 529, 563
BMI and, Z
body composition changes, 521f
caffeine and, 556
calcium/milk needs, 299, 299t, 305
choking in, 525
chronic diseases, early development of, 514–18
diabetes and, 121, 122, 515–16
diet planning for, 517, 520–5, 520f, 523f, 531–2,
  533t
DRI for, A–B
energy needs of, 521–2, C, Y
fats (dietary), 522
fish consumption, 157–8
food aversion, **529**
food preferences, 524–5
food programs for, 490–1, 564
food skills of, 525t
fruit juices, 510
growth, 209, 503f
guidelines for bone health, 309t
Healthy Eating Index, 520f
heart disease and, 516–17, 516t, 517t
hypertension in, 516–17
infectious diseases and, 200, 217
iron and, 522, 522t
iron-deficiency, 287
iron-deficiency anemia, 522
iron supplements, 289
lead toxicity, 526–7, 527f, 527t
malnutrition. *See* Children, hunger and malnutrition in
mealtime guidelines, 524–5
obesity in, 132, 312, 327, 517–18, 524, 530–1
pesticides and, 452
physical activity for, 515, 517–18
pollutants and, 458
protein needs, 197
smoking, 518
television and nutrition problems, 530
vegetarian diets and, 209
vitamin A deficiency in, 217
vitamin/mineral needs, 522, A–B
of women with anorexia nervosa, 354
Children, hunger and malnutrition in
behavior and, 526, 529–30, 563
causes of, 560, 562–3, 566t
consequences of, 560
death from, 198, 561, 567
death rates, 561
PEM, 198–201, 200t
poverty and birthrate, 570–1, 571f
in the U.S., 560, 563, 563f
vitamin A, 216, 217
Chinese cuisine, 42–3f
Chinese herbs, 553t
Chinese restaurant syndrome. *See* MSG symptom
  complex
Chloride
in body composition, 264f
deficiency and toxicity, 297t
functions of, 283–4, 297t
Chlorine, in water, 269–70
Chlorophyll, **101**
Chloroplasts, 478n
Chocolate, 59–60, 60f, 134, 556t
Chocolate milk, 299, 301f, 556t
Choking
Heimlich maneuver, 87f
infants/children, 510, 525
vs. normal swallowing, 86f
prevention of, 85

Cholesterol, **140**, 145–6, 150n. *See also* Blood lipid
  profiles
Cholesterol (blood). *See also* HDL; LDL
in childhood, 516, 516t, 517t
chromium and, 295
diet and, 243t, 406, 406t, 408t
vs. dietary cholesterol, 151–2
diet *vs.* medications, 407
factors affecting, 151–2
fiber and, 106–7, 108f
functional foods and, 62
garlic and, 62
genetics and, 151, 152
"good" *vs.* "bad," 150
heart disease and, 149–51, 402t, 403f
lipoprotein composition, 149–50, 151f
lowering of, 152, 407
measurement of. *See* Blood lipid profiles
medications for, 555t
phytochemicals and, 175
standards for, 402t
terminology, 150n
trans fats and, 159
Cholesterol (dietary)
vs. blood cholesterol, 151–2
heart disease and, 149–53
intake recommendations, 150t, 151–3, 406
on labels, 49, 50t
sources of, 151
Choline, **244**
Chromium, 294–5, 298t
Chromium picolinate, **386t,** 388
Chromosomes. *See* DNA
Chronic diseases, **3.** *See also* Degenerative diseases
Chylomicrons, **147**
Chyme, **78**
Circulatory systems. *See* Cardiovascular system;
  Lymphatic system
Cirrhosis, **91,** 95
Citrate malate, 201n
Clay eating. *See* Pica
Climate change, 561, 570
Clinical dietitian, responsibilities, 26t
Clones, **472t,** 475, 476–7
Cobalamin. *See* Vitamin B$_{12}$
Cobalt, 296
Cocaine, 227, 495, 557t
Coconut oil, 144, 145f, 161n
Coenzyme Q-10, **257t, 386t**
Coenzymes, **233,** 233f, 234f
Coffee
caffeine content, 556t
chemicals in, 416
gallstones and, 556
Cognitive therapy, **351t,** 354
Cola beverages, 129t, 307, 557
Colds, 231, 292
Collagen, **182, 229,** 275f
Colon/colorectal cancer
diet and, 208
exercise and, 417
fiber and, 108–9, 419
iron and, 420
meat and, 418
risk/protective factors, 413t
Colon (large intestine), **78**
diverticula, **108,** 109f
fiber and, 81t
function of, 76f, 78–9, 81t, 82
Colostrum, **506**
Combination foods
in Canadian Choice System, B-13–B-14t
in exchange system, D-1, D-12t
"Combining foods," myth of, 81–2
Comfrey, **415t**
Complementary medicine, 414–15, 414t, 415t

Complementary proteins, **195–6,** 195f
Complex carbohydrates, **100,** 104
Composition of foods, A-1t
Composition of the body. *See* Body composition
Condiments, in Canadian Choice System, B-12t
Conditionally essential amino acids, **181**
Congeners, **91,** 96
Conjugated linoleic acid, **57t,** 63
Constipation, **86, 108,** 494
Consumers
costs of healthy foods, 430
environmental issues, 574, 574f, 575f
foodborne illness safety, 438–40, 439f
food irradiation and, 452
food labels, 47–51
GM foods, 479t
organizations/resources, 24t
USDA Food Guide and, 41
website/resources, 24t, E-3, E-4
Contaminants, **456.** *See also* Environmental contaminants; Foodborne illnesses; Pesticides;
  Toxicity; *specific contaminants*
cadmium, 269, 458t
cancer and, 416
chemical, 452–5, 456–9
in fish, 157–8, 175, 446–7, 457–9, 457f, 495–6,
  569–70
in fish oil supplements, 157
in herbal supplements, 414
incidental food additives, **466–7**
lead, 269, 458t, 526–7, 527f, 527t
mercury, 141, 157, 456–8, 495–6
radiation emergencies, 285
in water, 269–70
Continuing Survey of Food Intakes by Individuals
  (CSFII), 13
Contraceptives, 502, 537, 554, 555t
Control groups, **12f, 13t**
Convenience foods. *See* Processed foods
Conversion factors, C-1–C-2
Cooking/food preparation. *See also* Processed
  foods
calcium tips, 299, 300–1
cancer and, 418, 418n
children and, 524
convenience, 10, 20
digestibility of proteins, 194
fat reduction tips, 153f, 168–9, 169f, 170t
foodborne illness prevention, 415t, 439, 440–3,
  441t, 442f, 443f, 445f
iron from cookware, 291, 470
meat/poultry, 418, 438–9, 444, 445f
microwave cooking, 466, 470
nutrient loss, preventing, 460–1, 468–70
omega-3 fatty acids, 169
sugar guidelines, 129
tips for singles, 548–50, 550t
trans-fatty acids and, 160
vitamins and, 225, 226, 228
Cooperative extension agencies, 25n
Copper
absorption and zinc, 292
in body composition, 264f, 264f
deficiency, 295, 298t
functions of, 264, 295, 298t
toxicity, 295, 298t
Coral calcium, 308
Corn, 237, 477
Cornea, **216**
Corn sweeteners, **128**
Corn syrup, **128,** 132
Coronary arteries, 399f
Correlation (research), **12f, 13t,** 418
Correspondence schools, 25, 25n, **26t**
Cortex, **72,** 72f
Cortical bone, 303, **304t**

Cottage cheese, 299
Coumadin (warfarin), 227, 553t, 554
Cranberry juice, 63
Cranberry tablets, 63
Cravings
    for nonfood substances, 287, 493
    during pregnancy, 493–4
C-reactive protein, 243, 399n, 407n
Cream, 127, 166, 299
Creatine, **386t**, 388
Credentials, fake, 25–6
Cretinism, **284**
Creutzfeldt-Jacob disease, variant, 444n
Crib death. *See* Sudden infant death syndrome
    (SIDS)
Critical periods (fetal development), **484**–5
Cross-contamination, **441**, 528
Cruciferous vegetables, **420**. *See also* Cabbage fam-
    ily
Cryptosporidium, 269, 436t
CSFII (Continuing Survey of Food Intakes by
    Individuals), 13
Cucurmin, 58t
Cuisine, defined, **10**
Culture. *See* Ethnic foods; Race/ethnicity
Cyanogens, 449
Cyclamate, **135t**, 137–8
Cyclosporiasis, 436t
Cytochromes. *See* DNA; Genetics
Cytokines, 197n

D

Daily Food Guide. *See* USDA Food Guide
Daily Values (DV), **28**, 29t, C
    carbohydrates and fiber intake, 53–4f
    fat and cholesterol intake, 53–4f
    for food labels, 33, 48f, 49
    for meal planning, 53–4f
    proteins (dietary), 53–4f
    purpose of, 33, 49
Dairy products. *See* Cheese; Milk and milk prod-
    ucts
DASH diet, 280, 410–11, 411t, I-3
Dates on food labels, 460t
Death. *See also* Longevity
    alcohol and, 3, 94, 97, 395n
    athletes "making weight," 352
    body weight and, 313
    from eating disorders, 354, 356
    ephedra/ephedrine and, 256, 346
    from foodborne illnesses, 435, 436–8t
    from heart disease, 173, 395f, 398
    from heat stroke, 376
    infant mortality, 482, 483f, 493
    from infections/malnutrition, 199–200, 217
    iron supplements, 289
    leading causes of, 3t, 395f
    liquid protein diets, 193
    from malnutrition, 198, 525, 560, 561, 567
    obesity and, 313, 426
    potassium supplements, 283
    from scurvy, 229
    vitamin D and, 222
Defecation. *See* Bowel movements; Diarrhea
Deficiencies. *See also* Malnutrition; *specific nutrients*
    alcohol abuse and, 97
    behavior and, 525–6
    cancer and, 258–9
    diagnosis of, 241
    DRI and, 30f
    essential fatty acids, 155
    grain enrichment and, 110–11
    hypertension and, 412
    immunity and, 394–6, 396f
    incidence of, 560t, 567
    intelligence and, 526

marginal, 258–9, **258t**
    of minerals, 297–8t
    during pregnancy, 482
    subclinical, 258, **258t**
Deforestation, 562
Degenerative diseases, 394–431. *See also* Cardio-
    vascular disease (CVD); Diabetes; Disease
    risk/prevention; Hypertension
    overview of causes, 3–4, 3f, 394
    cause *vs.* correlation, 396–8
    children's nutrition and, 514–18, 530
    chronic diseases, defined, **3**
    death rates from, 3t, 395f
    folate and, 238
    health claims on labels, 51t
    lifestyle choices and, 3, 4
    Mediterranean diet and, 173, 174, 177f
    nutrition and, 3, 3f
    prevention and the DRI, 30
    risk factors, **396**–8, 397f
    supplements and, 258–60
    vegetarians and, 207, 208–9
    websites/resources, E-4–E-5
Dehydration, **266**–8
    blood pressure and, 409
    constipation and, 87
    health effects of, 267t
    in older adults, 543
    oral rehydration therapy (ORT), **567**
    physical activity and, 376–7, 377t
    potassium and electrolyte balance, 281
    signs of, 267, 267t, 376, 377
Dehydroepiandrosterone (DHEA), 257n, **257t**,
    386t, 390f, 391
Delaney Clause, 463n
Denaturation, **184**
Dental caries, **134t**. *See also* Teeth
    diet/lifestyle and, 397f
    fluoride and, 294
    foods, caries potential of, 531, 531t
    infants/children, 512, 531, 531t
    sugar and, 133–4, 133f, 512
    websites/resources, E-4
Dental plaque, 134, **134t**
Deoxyribonucleic acid. *See* DNA
Depression, 537
Dermatitis of pellagra, 237f
DES (diethylstilbestrol), 61n
Desiccated liver, **386t**
Development, sustainable, **572**
DEXA (dual energy X-ray absorptiometry), 308,
    308n, **320**, 321f
Dextrose, **128**. *See also* Glucose
DHA (docosahexaenoic acid), **155**, 157t, 175, 505
DHEA (dehydroepiandrosterone), 257n, **257t**,
    386t, 390f, 391
Diabetes, type 1, **121**–2, 122t
Diabetes, type 2, **122**–4, 122t
    causes of, 515–16
    in children, 515–16
    diet/lifestyle and, 397f
    heart disease risk, 402, 404
    obesity and, 122, 514
    race/ethnicity, 515
Diabetes (diabetes mellitus), **121**–4. *See also*
    Insulin resistance
    effects of, 121, 123
    fiber and, 108
    food exchange system, D-1–D-12t
    gestational diabetes, **499**, 515
    glycemic index and, 119, 133
    heart disease risk, 402, 404
    management of, 123–4
    obesity and, 123, 123f, 132, 514
    physical activity and, 123–4, 370
    prevalence of, 121f, 514f, 515

sugar and, 132–3
    symptoms of, 516
    vitamin C and, 260
    warning signs of, 121, 122t
    websites/resources, E-4
Diadzein, 60f
Diarrhea, **86**
    cause/treatment, 87, 567
    dysentery, **199**–200
    traveler's diarrhea, 437t, 448
Diastolic pressure, **409**
Dietary antioxidants, **152, 221**. *See also* Antioxi-
    dants
Dietary balance, **8**–9, 34, 37, 380
Dietary Folate Equivalents (DFE), **239**
Dietary Guidelines for Americans 2005
    overview, 14–18, 16f
    carbohydrates and fiber intake, 105t, 125
    for disease prevention, 423t
    drinking and, 92, 92t
    exercise guidelines, 361
    fat and cholesterol intake, 150t
    food patterns to meet recommendations, I-1–I-2
    sugar intake, 131, 134
    weight loss, 336
Dietary Reference Intakes (DRI). *See* DRI (Dietary
    Reference Intakes)
Dietary Supplement Health and Education Act of
    1994 (DSHEA), 256–7
Dietary supplements, **5, 257t**. *See also* Supple-
    ments
Dietetic technician, 25, **26t**
Dietetic technician, registered (DTR), 25, **26t**
Dietitians, 24–5, **26t**
    clinical, responsibilities of, 26t
    hunger activism, 572–3
    nutritionists, **26t**
    professional organizations, 24, 24t, E-3
    websites of, E-3
Dietitians of Canada, E-3
Diet planning, 27–55. *See also* Diet therapy; DRI
    (Dietary Reference Intakes); Exchange sys-
    tems; USDA Food Guide
    for athletes, 380–3, 381f, 382t, 383f, 383t
    for calcium, 299, 300–1
    caloric needs and, I-1–I-2
    Canadian recommendations, 17t, 37t, B-0–
        B-14t
    characteristics of nutritious diets, 6–11
    for children, 299t, 517, 523t, 525t, 531–2, 533t
    Daily Values and, 53–4f, C
    for diabetics, 123, D-1–D-12t
    Dietary Guidelines for Americans, 14–18, 16f,
        I-1–I-2
    disease risk/prevention, 421–4, 422f, 423t
    energy density and, **339**–40, 340f
    with exchange system, D-1–D-12t
    excuses for poor diet, 9t
    fat intake and, 142f
    Food Guide Pyramids, 41, 211f
    glycemic index and, 119–20
    grocery shopping and, 549–50, 550t
    Healthy Eating Index, **45**–6
    for infants, 509–12, 509t, 511t
    international standards, 33
    meal spacing, 341–2, 344
    nutrient density and, **18**, 19–20, 19f, 38
    for obesity. *See* Obesity treatment
    for older adults, 547–50, 548t
    for physical activity, 120
    during pregnancy, 485–90, 487t, 496
    restaurant meals, 250t, 548–9
    sample menu, 53–4f
    for single people, 548–50, 550t
    USDA Food Guide for, 34–40, 52, 53–4f, 55t,
        I-1–I-2

variety, importance of, **8,** 9–10, 37, 251
  for vegetarians, 211–12, 211f
  for vitamins, 250–1, 250t, 252–3f
Diet(s), **2.** *See also* Obesity treatment; Weight loss
  atherogenic diet, 405
  DASH diet, 280, 410–11, 411t, I-3
  elemental diets, **6**
  ethnic. *See* Ethnic foods
  evaluating, 6–11
  excuses for poor diet, 9t
  fad diets, 337t
  "food combining," 81–2
  heart disease risk reduction, 405–8, 423t
  high-carbohydrate, 173, 366, 366f, 382t, 383f
  high-protein, 201–2
  high-protein, low-carbohydrate, 202, 332, 334–5, 335t
  lactose intolerance, 116
  liquid, 5–6, 193, 382–3, 383f, 389
  liquid protein, 193
  low-calorie, 330
  low-carbohydrate, 117–18, 261, 330, 332, 335, 407
  low-fat, 173, 370, 522
  malnutrition and, 201
  TLC diet, 423t
  vegetarian. *See* Vegans; Vegetarian diets
  for weight loss, 336–42, 338t, 339t
  for weight maintenance, 336–57
Diet therapy. *See also* Obesity treatment
  for bulimia nervosa, 356, 356t
  for diabetes, 123, D-1–D-12
  for high cholesterol, 406, 406t
  medical nutrition therapy, 26t
Digest, defined, **75**
Digestion. *See also* Absorption; Digestive system
  of carbohydrates, 79–80, 81t, 101, 113–14, 115f
  chemical aspects of, 79–82, 81t
  of fats, 79–80, 81t, 146–7, 146f, 147f
  mechanical aspects of, 77–9
  of protein, 79–80, 81t, 184, 186, 187f
  transit time, 82f
Digestive juices, 79–82, 80f, 81t. *See also* Enzymes
Digestive system, 74–87, **75.** *See also* Digestion; Small intestines
  allergies and, 527–8
  anatomy of, 75–9, 76f, 77f, 78f, 83, 84f
  diseases/disorders of, 83, 85–7. *See also* Foodborne illnesses
  excretion of waste, 87–8
  fiber and, 106, 108–9
  intervenous feeding and, 6
  malnutrition and, 396f
  in older adults, 539t
  olestra and, 162–3
  taste preferences, 74–5, 75f
  vitamins and, 245–9t
Dioxins, 467
Dipeptides, **186,** 186f, 187f
Diploma mills, 25–6, **26t**
Disaccharides, **101**
  digestion and absorption, 101, 113, 114, 115f
  structure of, 100, 101, 101f
Discretionary calorie allowance, **37,** I-2. *See also* Calories, discretionary
Disease. *See also specific diseases*
  chronic/degenerative. *See* Degenerative diseases
  defined, 394n
  foodborne. *See* Foodborne illnesses
  genetic. *See* Genetic disorders
  infectious. *See* Infectious diseases
Disease risk/prevention. *See also* Safety
  overview of nutrition's influence on, 3–4, 3f
  breast milk and, 503–6
  cancer. *See* Cancer
  cause *vs.* correlation, 396–8

diabetes. *See* Diabetes
diet as preventive medicine, 421–4, 422f, 423t
DRI and, 30
DRI and fat intake, 173
exercise overview, 361
fats, healthy and, 173–7
fiber and, 106–9, 106f
functional foods and, 62–4
heart disease. *See* Cardiovascular disease (CVD)
hypertension. *See* Hypertension
lifestyle and. *See* Lifestyle choices
obesity and, 312–15, 314f
physical fitness and, 312
phytochemicals, 58t, 59–62, 60f
smoking. *See* Smoking/tobacco use
sugar and, 131–4
supplements and, 258–60
underweight, 312
vegetarian diets and, 207–9
websites on, E-2–E-6
Diseases of old age. *See* Degenerative diseases
Distance Education and Training Council, 25n
Distilled water, **271t**
Diuretics, **268**
  alcohol as, 379
  caffeine as, 268
  nutrient-drug interactions, 555t
  PMS and, 537
  potassium and electrolyte balance, 281
Diverticula, **108,** 109f
Diverticulosis, 397f
DNA, **4.** *See also* Genetics
  antioxidants and cancer, 258–9
  as ergogenic aid, **386t**
  folate and, 238, 419
  fruit/vegetable intake and, 420
  genes, function of, 67–8, 67f
  iron and, 420
  nutrients and gene expression, 183–4
  protein synthesis and, 185f
  recombinant DNA technology, **472t,** 475–6, 475f, 476f, 479t
Docosahexaenoic acid (DHA), **155,** 157t, 175, 505
Dolomite, 308, **310t**
Dopamine, 554
DRI (Dietary Reference Intakes), **28**–33, A–B. *See also specific nutrients*
  carbohydrate/fat/protein proportions (AMDR), 30, 338
  carbohydrates and fiber intake, 30, 105t, 109
  component categories, 28–30, 29t
  energy requirements, estimating, 318
  exercise guidelines, 361
  fat and cholesterol intake, 30, 150t, 173, 406
  goals of, 28–30
  method of setting, 31–3
  protein intake, 30, 197t, 201
  purpose/uses of, 28, 30–1
  safe *vs.* toxic nutrient intakes, 29–30, 30f
Dried foods, 461–2
Drink, **91,** 93
Drug-nutrient interactions, 86, 281, 552–8, 552f, 553t, 555t
Drugs, **57t**
Drugs, therapeutic. *See* Medications
Drug use/abuse. *See also* Alcohol abuse
  effects of, 557, 557t
  lactation and, 501
  nutrition problems and, 533
  pregnancy and, 495
  resources for, E-3
Drying (of foods), **461**
Dual energy X-ray absorptiometry (DEXA), **320,** 321f
DV. *See* Daily Values (DV)
Dysentery, **199**

E

E. coli infection, 436t, 438–9
EAR. *See* Estimated Average Requirement (EAR)
Eating disorders, 351–7
  anorexia nervosa, **351t,** 352–4, 353t
  in athletes, 351–2, 354
  binge eating, **351t**
  binge eating disorder, 356, 357t
  bulimia nervosa, **351t,** 354–6, 355f, 355t
  dieting/fasting and, 330
  society's role in, 356–7
  vegetarian diets and, 210
  websites/resources, E-5
Echinacea, **415t,** 553t
Edema, **189, 499**
Education. *See also* Schools
  accreditation information, 25, 25n, 26t
  correspondence schools, 25, 25n, **26t**
  fake credentials, 25–6
  of health care professionals, 24
  on hunger issues, 572
  on nutrition in high schools, 429–30
  of nutrition professionals, 24–6, 25n
EER (Estimated Energy Requirement), **316**–17, 318, 319f
Eggs, 151, 184, 445–6, 474. *See also* Meat, poultry, fish, dry peas and beans, eggs, and nuts food group
Eicosanoids, **154**–5. *See also* EPA (eicosapentaenoic acid)
Elderly people. *See* Aging (older adults)
Electrolytes, **272.** *See also* Fluid and electrolyte balance
  athletes and, 377
  balance, 200, 273f
  potassium, 281
  in sports drinks, 378–9
  water flow and, 272–3, 272f
Electrons, **258t**
Elemental diets, **6**
Ellagic acid, 58n
Embolism, **400**
Embolus, **400**
Embryo, **482,** 485f
Emergency kitchens, **564**
Emetics, **351t**
Emotions, 428. *See also* Psychological problems
"Empty" calorie foods, 129t
EMS (eosinophilia myalgia syndrome), 193
Emulsification, **145,** 146, 146f
Emulsifiers, **79, 145**
Endogenous opiates, **323**
Endorphins, **323**
Endosperm, 110f, **110t**
Endurance, 366f, 368, 370
Energy, **4.** *See also* Caloric content; Calories; Energy metabolism; Metabolism
  amino acids and, 191, 192f
  calorie needs, calculating, 316–18
  children's needs, 521–2
  Estimated Energy Requirement (EER), **33,** 316–17, 318, 319f
  in exchange system, D-1, D-1t, D-2, D-2t, D-3t
  intake recommendations, C, Y
  on labels, 50t
  for lactating women, 500
  for older adults, 538–9, 544t
  for physical activity, 364–73, H-1, H-2–H-6t
  recommended intake, B-0
  storage of, 88
  from sun/plants, 4, 100, 100f
Energy balance, 315–18
  for athletes, 380
  calculation of, 316
  cancer and, 417

Intracellular fluid, **68**
Intrinsic factor, **240,** 241
Invert sugar, **128**
Iodine, 284–5
  in body composition, 264, 264f
  deficiency, 284–5, 285f, 298t, 562–3
  functions of, 264, 284, 298t
  sources of, 285, 298t
  toxicity, 285, 298t
Ions, **272**
Iron, 286–91, 298t
  absorption of. *See* Iron absorption
  adolescents' needs, 534
  for athletes, 375
  bioavailability of, 375
  in body composition, 264, 264f
  in bread, 111f
  children's needs, 522
  colon cancer and, 420
  functions of, 286, 288f, 298t
  infants' need for, 506, 510
  intake recommendations, 261, 288f, 289, 290t,
      534
  for older adults, 544t
  in pregnancy, 486f, 490
  sources of, 8, 8n, 288f, 289, 298t, 522t
  in sugar/sugary foods, 127, 128–9, 129t
  supplements, 8n, 289
  toxicity, 288f, 289, 298t
  vegetarian diets, 210, 375
  vitamin C and, 470
  zinc absorption and, 292
Iron absorption, 289–91
  antacids and, 552
  calculation of, 290t
  control of, 286
  cooking pans and, 291, 470
  genetics and, 286
  increasing/reducing, 289–90, 490
  meat/MPF factor, 289–90, 291
  soy products and, 204
  from supplements, 8n, 289
  vitamin C and, 229, 230, 289, 290
Iron cookware, 290–1, 470
Iron deficiency, **286**–7, 298t
  in athletes, 375
  behavior of children, 287, 526
  causes of, 288
  in infants, 506
  symptoms of, 287, 287t, 288f, 298t
Iron-deficiency anemia, **286**–7, 287f
  aspirin and, 554
  causes of, 3, 8, 288, 375
  lead levels and, 527
  in older adults, 543
  physical activity and, 375
  prevalence of, 562
  red blood cells in, 287f
  symptoms of, 8
Iron overload, 230, 260, **289,** 290
Irradiation (of foods), **450**–2, 450t
Irritable bowel syndrome, **87**
Isomalt, 134, 135t, **135t**
Isothiocyanates, 58t
IU (international units), **219**

J

Jam, 127, 128f
Jaundice, **228**
Journals, 13t, 15, E-1–E-2
Juices. *See* Fruit juices

K

Karela, 553t
Kava, 257, **415t**

Kefir, 57t, 63, **299**
Kelp, 553t
Kelp tablets, **257t**
Keratin, **216,** 217
Keratinization, **216,** 217, 245n
Ketchup, 127
Ketogenic diets. *See* Ketosis
Ketone bodies, **117, 330**
Ketosis, **117,** 148, 330, 335
Kidney cancer, 413t
Kidney dialysis machines, 265n
Kidney disease
  cranberry tablets and, 63
  diabetes and, 123
  dialysis and, **123**
  proteins (dietary) and, 201
Kidneys, **69**
  amino acids and, 389
  blood pressure regulation, 409
  fluid balance, 267, 273
  function of, 69f, 87–8
  sodium and, 279
Kidney stones, 63
Kombucha, **415t**
Kudzu, **415t**
Kwashiorkor, **199,** 199n, 200, 200t. *See also*
      Protein-energy malnutrition (PEM)

L

Labeling, 47–51, 48f
  allergy-producing foods, 528
  of bottled water, 271t
  of breads, 112f
  Canadian system, B-15f, B-16f
  cholesterol content, 49, 50t
  Daily Values, 33, 47, 48f, 79, C
  date terms, 460t
  energy (calories), 49, 50t
  fat content, 49, 50t, 160, 164, 168
  fiber, 51t
  functional foods, 63t
  health claims, **47t,** 50, 51t
  ingredient list, 48f, 49
  of irradiated foods, 451
  of meats, 165f
  nutrient amounts, 49
  nutrient claims, **47t,** 49–50, B-16t
  Nutrition Facts, **47t**
  of organic foods, 473f
  of phytochemical supplements, 62
  protein quality and, 196
  of reduced-fat foods, 164
  regulations for, 47–51
  servings, 47, 48, 48f
  sodium, 49, 51t
  structure/function claims, 257–8
  sugar content, 49
  sugar terminology, 128f, 128t
  of supplements, 256f, 257–8
  terminology, 47t, **50t,** 51t, 460t, B-15t
  trans-fatty acids, 160
  of water, 271t
  of wine, 98
Laboratory studies, **12f, 13t**
Lactase, **116**
Lactation, **484,** 500–2. *See also* Breastfeeding
Lactation consultants, certified, **500,** 508
Lactic acid, **367**–8
Lactitol, 135t, **135t**
Lactobacillus, 63
Lacto-ovo vegetarians, **207t,** 209
Lactose, **101,** 101f, 116
Lactose content of foods, 116
Lactose intolerance, **114**

calcium sources, 116, 299
  formulas for infants with, 507
  symptoms of, 116
Lacto-vegetarians, **207t**
Laetrile, 244, 449
La Leche League International, 500n, E-5
Large intestine, **78.** *See also* Colon
Laxatives
  cathartics, **351t**
  constipation and, 86–7
  as "dieter's tea," 346
  eating disorders and, 354, 356
  magnesium and, 278
  malnutrition and, 553–4
  nutrient-drug interactions, 553–4, 555t
LDL (low-density lipoprotein), **149**
  atherosclerosis and, 399–400
  fish oil supplements, 157
  free radicals and, 260
  function of, 149–50
  HDL and, 150, 151f, 401–2, 401f
  heart disease and, 149–52, 175–6, 176f, 401–2,
      401f
  lowering blood level, 152, 173
  oxidation of, 152
  standards for, 402t
  structure of, 151f
Lead
  in foods, 458t
  in herbal remedies, 414
  information center on, 458n
  iron-deficiency and, 527
  malnutrition and, 527
  protection against, 527t
  sources and effects, 458t
  toxicity in children, 526–7, 527f, 527t
  in water, 269
Learning disability, **529**–30
Leavened, **291**
Lecithin, **140,** 145
Legumes, **8, 194.** *See also* Beans
  digestibility of, 194
  fiber content of, 107f, 126f
  fixing nitrogen, 205f
  nutrients in, 202
  protein quality, 202, 203
  proteins, complementary combinations, 195f
Leptin, 306, **324**–5
Levulose, **128**
License to practice, 24, **26t**
Life expectancy, **538,** 544–5, 567
Life span, **538,** 544–5
Lifestyle choices
  aging and, 544–5
  behavior modification for weight control, 348t,
      358
  cancer risk, 259–60, 413t
  children and chronic disease, 517–18
  chronic/degenerative diseases and, 3, 4, 396–8,
      397f
  diabetes and, 123
  environmental considerations, 574, 575f, 583,
      583f
  heart disease and, 149
  obesity and, 327–8, 426–31, 426f, 430f
  obesity in children, 530–1
  vegetarians and health, 207–12
  for weight control, 347
Lifestyle diseases. *See* Degenerative diseases
"Light," **51t**
Lignans, 57t, 58t, 60f, 61
Lima beans, 449
Limiting amino acids, **194**–5
Limonene, 60f
Linoleic acid, **154.** *See also* Omega-6 fatty acids
  conjugated, 57t, 63

chemistry of, 102, 103f
digestion and absorption, 81t, 113–14, 115f
in exchange system, D-1, D-3t, D-4t
resistant, **114**
Starvation. *See* Eating disorders; Fasting; Hunger (chronic/world); Malnutrition
Statins, 407, 555t
Stem cells, **472t**, 475
Steroid hormone, anabolic, **386t**, 389, 390f
Sterol esters, **57t**, 62, 175, 407, 408t
Sterols, **140**. *See also* Cholesterol
phytosterols, **57t**
structure of, 145–6
Stevia, **135t**, 138
Stomach, **77**
acidity of, 79, 85–6, 184
anatomy of, 78f
carbohydrate digestion, 81t
fat digestion, 81t
function of, 76f, 79, 82
hunger sensation and, 72
problems with, 85–6
protein digestion, 81t
Stomach cancer, 86, 280, 413t, 420
"Stomach flu," 435, 437t, 537n
Stone ground, **110t**, 111
Stone ground flour, **300**
Storage of nutrients, 88
Strength training, 362t
Stress
exercise and, 342
hormonal response to, 73, 118–19
overeating and, 328
supplements and, 261
Stress, oxidative. *See* Oxidative stress
Stress fractures, 307
Stroke, **400**
alcohol and, 98
diet/lifestyle and, 397f
signs of, 398
sodium and, 280
vitamin E and smoking, 226
Stroke volume, **364**
Structure/function claims, 257–8
Student movements and social change, 560
Studies. *See* Research
Subclinical deficiency, 258, **258t**
Subcutaneous fat, **313**, 314f
Succinate, **387t**
Sucralose, **135t**, 136t, 137
Sucrose, **101**, 101f. *See also* Sugars
Sucrose polyester, **162t**–164
Sudden infant death syndrome (SIDS), 495
Sugar, raw, **128**
Sugar alcohols, 134–5, 135t
Sugars. *See also* Carbohydrates; Sweets
alternatives to, 129, 134–8, 135t
as antimicrobial agents, 464
behavior and, 132, 133
blood lipid profiles and, 406
in Canadian Choice System, B-8t
chemistry of, 100–1, 100f, 101f
dental caries and, 133–4, 133f
digestion and absorption of, 101, 114, 115f
disaccharides. *See* Disaccharides
as "empty-calorie food," 129t
food labels and, 49, 128f, 128t
functions of, 104
health effects of, 131–4
intake in U.S., 131, 131f, 131n
intakes guidelines, 105t, 127–9, 131, 134
monosaccharides. *See* Monosaccharides
nutrient density and, 104
obesity and, 131–2, 515
in processed foods, 127, 127f, 128f, 468
in soft drinks, 127, 129t, 131, 132

structure of, 100–1, 101f
taste preferences, 74–5, 75f
types of/terms describing, 100–1, 127–9, 128f, 128t, 129t
Sugars, added, **127**
childhood obesity and, 515
consumption statistics, 131, 131f
Dietary Guidelines, 134
on food labels, 128, 128f, 128t
health effects of, 131–4
intake recommendations, 105t, 134, I-2
overeating and, 132
terms identifying, 128t
in USDA Food Guide, 36f
Sugars, free, 131n
Sugars, naturally occurring, **127**
Sulfites, 462, 464
Sulforaphane, 60f
Sulfur/sulfate, 264f, 284
Sunette, 136t, 137
Sunlight in synthesis of vitamin D, 221, 222–3
Superoxide dismutase (SOD), 295n, **387t**
Supplements, **5, 257t**. *See also* Functional foods
Alzheimer's disease and, 545
amino acids, 186, 193, 386–7
for athletes, 193, 373–5, 385–92
beta-carotene, 61–2, 220–1
brain function and, 545
calcium, 308–10, 310t
cancer and, 62, 221, 419
carnitine, 388
chromium, 295–6, 388
contaminants in, 157
creatine, 388
definition, legal, 257n
dosages, 261–2
elemental diets, **6**
fish oil, 156–7
foods *vs.*, 57–64, 260, 292, 407
garlic, 62
growth hormone, 540
herbal. *See* Herbal supplements
high-carbohydrate liquid, 380
hormonal, 389–92, 390f, 391f, 540
for infants, 227, 503, 506, 508
iron, 289
labeling of, 256f, 262
for malnourished children, 216, 217
need for, 6, 255, 258–60t
for older adults, 259, 543, 546
for physical activity, 373–5
phytochemicals, 57–60, 61, 62
phytoestrogens, 61
potassium chloride, 282–3
during pregnancy, 487, 489, 490, 490t, 496
pros and cons of, 256–60, 260t
protein, 193, 388–9
reasons for taking, 255, 255t, 260t
regulation of, 256–8, 257f
safety of, 193, 256–7, 257t, 259–60
selection of, 260–2
selenium, 293
smokers and, 62
toxicity of, 256, 260. *See also specific nutrients*
tryptophan, 193
types of, 257t, 386–7t
for vegetarians, 255
vitamin A, 216, 217
vitamin C, 230–1
vitamin D, 222
vitamin E, 224
vitamin K, 227, 503
for weight gain, 344
for weight loss, 257, 345t, 346
zinc, 292
Surface water, **270**

Sushi, **447**
Sustainable, defined, **572**
Sustainable agriculture. *See under* Agriculture
Sustainable development, 571–2
Swallowing, 86f
Sweat
body temperature and, 265, 376
mineral loss, 279, 377, 378
water needs and, 268
Sweeteners
artificial, 131, 134–8, 135t, 136t
on labels, 128f
terminology for, 128t, 135t
Sweeteners, carbohydrate, 127, **128**
Sweet One, 136t, 137
Sweets. *See also* Sugar
children and, 532
in exchange system, D-7t
in food group plans, 35–6f, B-8t, D-7t
for infants, 511, 512
Systolic pressure, **409**

T

Tagalose, **135t**, 136t, 137
Tanning booths, 223
Tannins, 58t, **290**
Tartrazine, 529
Taste buds, 74
Taste preferences, 74–5, 75f
T-cells, **74**, 74f
Teas. *See also* Caffeine
caffeine content, 556t
"dieters' tea," 346
drug-nutrient interactions, 553t
iron absorption and, 290
phytochemicals in, 59, 60f
PMS and, 537
Technology. *See* Biotechnology; Food industry; GE foods; Genetic engineering
Teenagers. *See* Adolescence
Teeth. *See also* Dental caries
calcium and, 274, 275f
fluoride and, 274
magnesium and, 277
structure of, 133f, 274, 275f
vitamin C and, 237f, 261
Television. *See also* Advertising; Mass media
children and, 515, 530
obesity and, 515, 516f, 530
Temperature for food safety, 440, 442f, 444t
Temperature regulation
alcohol and, 379
dehydration and, 267
physical activity and, 376
sweating and, 265
Terminator technology, 478
Terrorism, 434
Testosterone, 95, 305
Textured vegetable protein, **204**
Therapy. *See* Diet therapy
Thermic effect of food, **317**, 317f, 335t, 354
Thermogenesis, **326**, 326t, 327
Thermometers, 442f, **442t**
THG, **387t**, 391
Thiamin, **235**, 236f. *See also* B vitamins
in bread, 111f
in canned foods, 460–1
deficiency, 97, 235, 235f, 236f, 247t
functions of, 234f, 235, 247t
intake recommendations, 235, 236f
in restaurant meals, 250t
sources of, 235, 236f, 247t, 251, 252f
sulfites and, 465
Thickening and stabilizing agents, **463t**
Thirst, 267, 268, 377, 378

Daily Values for Food Labels

The Daily Values are standard values developed by the Food and Drug Administration (FDA) for use on food labels. The values are based on 2000 calories a day for adults and children over 4 years old. Chapter 2 provides more details.

Nutrient	Amount	Nutrient	Amount
Protein[a]	50 g	Vitamin K	80 μg
Thiamin	1.5 mg	Calcium	1000 mg
Riboflavin	1.7 mg	Iron	18 mg
Niacin	20 mg NE	Zinc	15 mg
Biotin	300 μg	Iodine	150 μg
Pantothenic acid	10 mg	Copper	2 mg
Vitamin B_6	2 mg	Chromium	120 μg
Folate	400 mg	Selenium	70 μg
Vitamin B_{12}	6 mg	Molybdenum	75 μg
Vitamin C	60 mg	Manganese	2 mg
Vitamin A	5000 IU[b]	Chloride	3400 mg
Vitamin D	400 IU[b]	Magnesium	400 mg
Vitamin E	30 IU[b]	Phosphorus	1000 mg

[a]The Daily Values for protein vary for different groups of people: pregnant women, 60 g; nursing mothers, 65 g; infants under 1 year, 14 g; children 1 to 4 years, 16 g.
[b]Equivalent values for nutrients expressed as IU are: vitamin A, 1500 RAE (assumes a mixture of 40% retinol and 60% beta-carotene); vitamin D, 10 μg; vitamin E, 20 mg.

Food Component	Amount	Calculation Factors
Fat	65 g	30% of calories
Saturated fat	20 g	10% of calories
Cholesterol	300 mg	Same regardless of calories
Carbohydrate (total)	300 g	60% of calories
Fiber	25 g	11.5 g per 1000 calories
Protein	50 g	10% of calories
Sodium	2400 mg	Same regardless of calories
Potassium	3500 mg	Same regardless of calories

Body Mass Index (BMI)

Find your height along the left-hand column and look across the row until you find the number that is closest to your weight. The number at the top of that column identifies your BMI. Chapter 9 describes how BMI correlates with disease risks and defines obesity. The area shaded in blue represents healthy weight ranges.

Height	18	19	20	21	22	23	24	25	26	27	28	29	30	31	32	33	34	35	36	37	38	39	40
													Body Weight (pounds)										
4'10"	86	91	96	100	105	110	115	119	124	129	134	138	143	148	153	158	162	167	172	177	181	186	191
4'11"	89	94	99	104	109	114	119	124	128	133	138	143	148	153	158	163	168	173	178	183	188	193	198
5'0"	92	97	102	107	112	118	123	128	133	138	143	148	153	158	163	168	174	179	184	189	194	199	204
5'1"	95	100	106	111	116	122	127	132	137	143	148	153	158	164	169	174	180	185	190	195	201	206	211
5'2"	98	104	109	115	120	126	131	136	142	147	153	158	164	169	175	180	186	191	196	202	207	213	218
5'3"	102	107	113	118	124	130	135	141	146	152	158	163	169	175	180	186	191	197	203	208	214	220	225
5'4"	105	110	116	122	128	134	140	145	151	157	163	169	174	180	186	192	197	204	209	215	221	227	232
5'5"	108	114	120	126	132	138	144	150	156	162	168	174	180	186	192	198	204	210	216	222	228	234	240
5'6"	112	118	124	130	136	142	148	155	161	167	173	179	186	192	198	204	210	216	223	229	235	241	247
5'7"	115	121	127	134	140	146	153	159	166	172	178	185	191	198	204	211	217	223	230	236	242	249	255
5'8"	118	125	131	138	144	151	158	164	171	177	184	190	197	203	210	216	223	230	236	243	249	256	262
5'9"	122	128	135	142	149	155	162	169	176	182	189	196	203	209	216	223	230	236	243	250	257	263	270
5'10"	126	132	139	146	153	160	167	174	181	188	195	202	209	216	222	229	236	243	250	257	264	271	278
5'11"	129	136	143	150	157	165	172	179	186	193	200	208	215	222	229	236	243	250	257	265	272	279	286
6'0"	132	140	147	154	162	169	177	184	191	199	206	213	221	228	235	242	250	258	265	272	279	287	294
6'1"	136	144	151	159	166	174	182	189	197	204	212	219	227	235	242	250	257	265	272	280	288	295	302
6'2"	141	148	155	163	171	179	186	194	202	210	218	225	233	241	249	256	264	272	280	287	295	303	311
6'3"	144	152	160	168	176	184	192	200	208	216	224	232	240	248	256	264	272	279	287	295	303	311	319
6'4"	148	156	164	172	180	189	197	205	213	221	230	238	246	254	263	271	279	287	295	304	312	320	328
6'5"	151	160	168	176	185	193	202	210	218	227	235	244	252	261	269	277	286	294	303	311	319	328	336
6'6"	155	164	172	181	190	198	207	216	224	233	241	250	259	267	276	284	293	302	310	319	328	336	345

Under-weight (<18.5)	Healthy Weight (18.5–24.9)	Overweight (25–29.9)	Obese (≥30)

Body Mass Index-for-Age Percentiles: Boys and Girls, Age 2 to 20

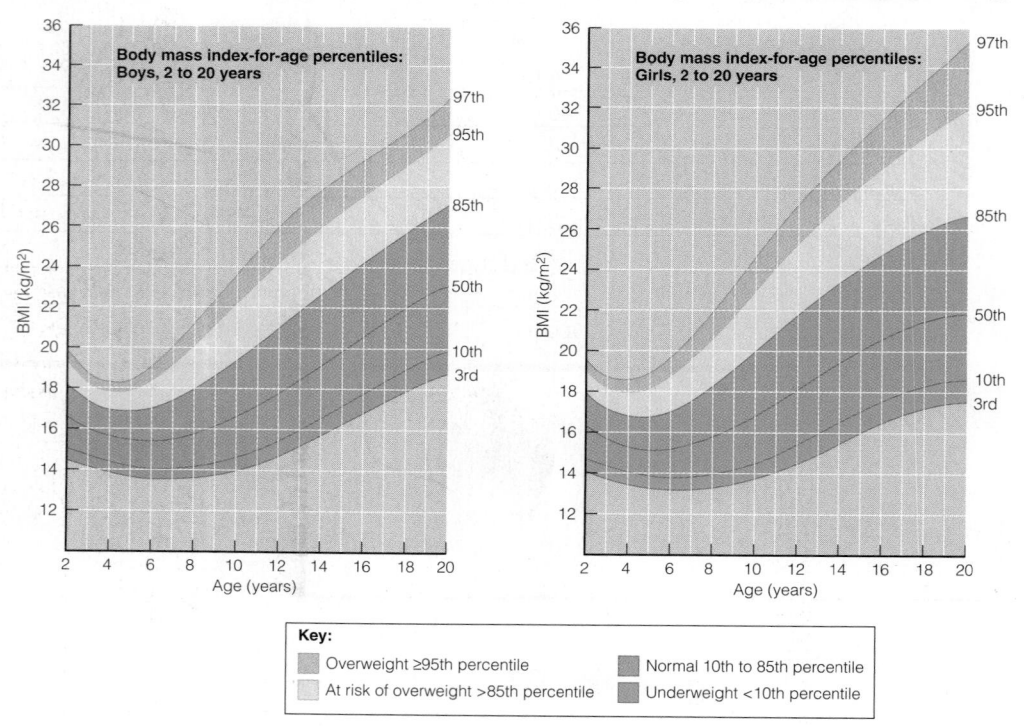

Key:
- Overweight ≥95th percentile
- At risk of overweight >85th percentile
- Normal 10th to 85th percentile
- Underweight <10th percentile